McGraw-Hill Encyclopedia of Science and Technology

7

ice-lyt

VOLUME

7

ice-lyt

McGraw-Hill **Encyclopedia of**

Science and Technology

AN INTERNATIONAL REFERENCE WORK IN FIFTEEN VOLUMES INCLUDING AN INDEX

McGRAW-HILL BOOK COMPANY

NEW YORK   ST. LOUIS   SAN FRANCISCO

AUCKLAND	NEW DELHI
BOGOTA	PANAMA
DUSSELDORF	PARIS
JOHANNESBURG	SAO PAULO
LONDON	SINGAPORE
MADRID	SYDNEY
MEXICO	TOKYO
MONTREAL	TORONTO

On preceding pages:

*Left. Computer drawing, executed with the Gerber
digital plotter, of plan view of a sculpture that can
be folded from four flat sheets of metal. (R. D.
Resch, Associate Research Professor, Computer
Science, University of Utah)*

*Right. American beech (Fagus grandifolia); surface
is uneven and also contains spikelike hairs. (From
S. B. Carpenter and N. D. Smith, The hidden world
of leaves, Amer. Forests, 81(5):28–31, 1975)*

Library of Congress Cataloging in Publication Data:

McGraw-Hill encyclopedia of science and technology.

    1. Science—Dictionaries. 2. Technology—Dictionaries.   I. Title:
Encyclopedia of science and technology.
Q121.M3   1977   503   76–44232
ISBN 0-07-079590-8

This Encyclopedia presents pertinent information in every area of modern science and technology. Although the fourth edition is considerably enlarged, by about 750 pages and 500,000 words, the organization follows the plan that was so successful in the three previous editions.

The plan calls for broad survey articles to give even the uninitiated reader the basic concepts or rudiments of a subject. Thus there is a survey or lead article for each of the disciplines or large subject areas. From the lead article the reader may proceed to other articles that are more restricted in scope. For example,

Petroleum engineering

is a survey article. Cross references in the survey article lead the reader to the more specific articles in this field of which there are, among many others,

Cable-tool drill
Geophysical exploration
Oil field model

The pattern of proceeding from the general to the specific has been employed not only in the plan of the Encyclopedia, but in each article as well. Each article begins with a definition of the subject, followed by sufficient background material to give the reader a frame of reference and permit him or her to move into the detailed body of the article. Usually, only the scientific or technological aspect of the subject is discussed. Within the text are boldface and italic subheadings which outline the article; they are intended to enhance understanding, and can guide the user who prefers to read selectively the sections of a long article.

Though each article is as complete in itself as practical, every article has been provided with cross references to related articles in the Encyclopedia. These cross references are set in small capitals for emphasis and are inserted at the points in the text where they are relevant. In all, some 50,000 cross references are given. Often these references may lead to areas which may not have occurred to the reader. For example, the article **Chlorine** has such diverse cross references as HALOGEN ELEMENTS, ANTIMICROBIAL AGENTS, BLEACHING, and HYPOCHLORITE.

The articles are arranged in alphabetical sequence in 14 volumes, and the reader can often find what is wanted by simply choosing the volume bearing the appropriate letter.

Some article titles contain parenthetical information for clarification:

Cell (biology)
Land drainage (agriculture)

Some article titles are inverted so that subject matter can be grouped:

Earth, heat flow in
Earth, interior of

Alphabetization is by word, not by letter, with a comma providing a stop, so that the two article titles just above are followed by

Earth inductor
Earth sciences
Earthmover
Earthquake

Since biology offers the most elaborate subject matter, every phylum, class, and order in the plant and animal kingdoms is allotted a separate article. Many of the common families, genera, and species are covered either within an article on the order or a separate article under the scientific or common name.

Every article has the contributor's full name, not only as a credit but because the name of the author is often an important data element in itself. Each author is further identified in an alphabetical Contributors List in vol. 15. It cites the university, laboratory, or business with which the author is affiliated and the titles of the articles written.

Most of the articles contain bibliographies citing useful sources. The bibliographies are placed at the ends of articles or sometimes at the ends of major sections in long articles. For additional bibliographies, the reader should refer to related articles (indicated by cross references).

Thus, the alphabetical arrangement of article titles, the subheadings, the cross references, and the bibliographies permit the reader to pursue a particular interest by simply taking a volume from the shelf. However, the reader can also find information in the Encyclopedia by using the Analytical Index and the Topical Index in vol. 15. The Analytical Index contains each important term, concept, and person—140,000 entries in all—mentioned throughout the 14 text volumes. It guides the researcher to the volume numbers and page numbers concerned with a specific point. The reader wishing to consult everything in the Encyclopedia on a particular aspect of a subject will find that the Analytical Index is the best approach. A broader approach may be made through the Topical Index, which groups all article titles of the Encyclopedia under nearly a hundred general headings. For example, under "Geophysics" more than 80 articles are listed, and under "Photography" more than 50. The Topical Index thus enables the reader quickly to identify all articles in the Encyclopedia in a particular subject area.

A feature of the Encyclopedia is the section "Scientific Notation in the Encyclopedia" in vol. 15. It clarifies usage of symbols, abbreviations, and nomenclature, and is especially valuable in making conversions between International System, U.S. Customary, and metric measurements.

McGraw-Hill Encyclopedia of Science and Technology

Ice cream

A frozen dessert prepared by freezing a mixture of milk products and flavoring, and made smooth by stirring during freezing. Commercial ice cream varies in composition, depending largely on state laws. However, the Federal Frozen Desserts Standards of Identity, as amended Dec. 31, 1965, to which most states subscribe, set certain minimum standards for composition of ice cream as follows: It must contain a minimum of 10% milk fat and a minimum total solids content of 20%. For ice creams flavored with bulky flavors such as chocolate, fruit, or nuts, lower fat and milk solids are specified by appropriate allowable reduction factors for the added ingredients, but in no case may the fat of these flavored ice creams be less than 8% or the total milk solids be less than 16% by weight of the finished product. The finished ice cream must contain not less than 1.6 lb of total solids per gallon and weigh not less than 4.5 lb/gal. The optional ingredients, quantity requirement and restrictions for optional ingredients, and labeling requirements for ice cream are included in the standards.

The usual fat content is 10–14%, and the nonfat milk solids average about 10.5%. Sweetness is supplied by cane or beet sugar (sucrose) to the amount of 15%. This is supplemented with corn sugar (dextrose) and corn syrup. To protect ice cream against heat shocks during marketing, 0.2–0.4% of colloidal material, such as gelatin or vegetable gum, is added, as well as 0.1–0.2% of an emulsifier, such as mono- or diglyceride.

Commercial process. Commercial ice cream is always made from pasteurized and homogenized mix (see illustration). Both batch and continuous freezers are used. The holding system of pasteurization (68.3–71.1°C for 30 min) is commonly used by small operators; however, large operators usually use high-temperature short-time treatments that make possible continuous flow of the mix. The U.S. Public Health Service has set a minimum pasteurization temperature of 79.5°C, with a minimum holding time at this temperature of 25 sec, for this method, but some manufacturers use temperatures as high as 127°C. The higher temperatures denature enough of the whey proteins to produce a heavier body in the mix and a smoother texture in the ice cream without damaging the delicate flavor of the milk products. Excessively high temperatures, however, tend to lessen the action of some stabilizers.

At first the freezing was done in batch machines, but continuous freezing has almost entirely replaced the earlier method. The continuous operation is more efficient and produces an ice cream with a smoother texture and body. Continuous freezers vary in capacity from 85 to 2500 gal/hr. Numerous flavors are added to ice cream. The most common is vanilla. Chocolate, strawberry, butter pecan, peach, coffee, and peppermint stick also are popular.

During the freezing process, air is incorporated into the ice cream mix. This is termed overrun. A 100% overrun means that the volume of the mix has been doubled or the weight per unit volume has been halved. After packaging, the ice cream is hardened by passing it through freezing tunnels or refrigerated plate hardeners or by placing it in low-temperature (−28.9°C) rooms where the air is usually circulated to facilitate heat transfer.

During World War II the shortage of milk fat led to the consideration of vegetable oils in the manufacture of frozen desserts. The oils commonly used are cottonseed, coconut, and soybean. The use of these oils for this purpose is illegal in some states, but others permit their use if the frozen product is properly labeled. Manufacturers in Texas, a leading state in the use of vegetable oils in frozen desserts, coined the name Mellorine for such products. There are certain economic advantages in the use of vegetable oils in place of milk fat.

Ice milk. In the period prior to World War II considerable interest was shown by consumers in a low-fat frozen product which came to be called ice milk. The Federal Frozen Desserts Standards of Identity provides that ice milk comply with all the provisions in the standards for ice cream, except that (1) its content of milk fat should be more than 2% but less than 7% of the weight of the finished product; (2) its total milk solids should be not less than 11%; (3) the provision for reduction in fat content and total milk-solids content by the addition of bulky flavor ingredients does not apply; (4) the quantity of food solids per gallon should not be less than 1.3 lb.

Products in the market vary from 3 to 6% fat, 11 to 14% milk solids – nonfat, 13 to 18% sugar solids, and 0.4 to 0.5% stabilizer. Ice milk is frozen, packaged, and hardened much the same as is ice cream or sold soft directly from the freezer.

[PAUL H. TRACY]

Sources of microorganisms. The cream, condensed- and dry-milk products, egg products, sugar, stabilizers, flavoring and coloring materials,

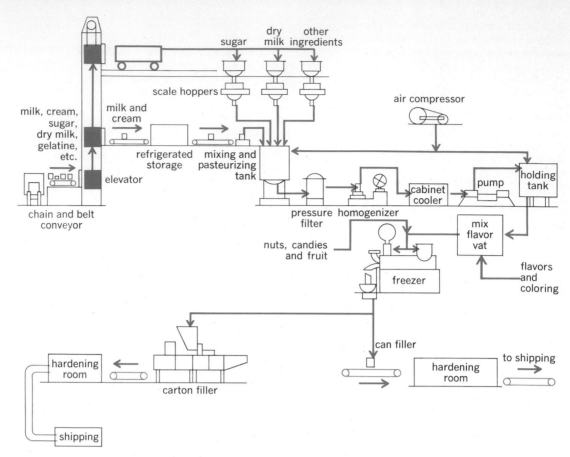

Ice cream manufacture. (*Adapted from S. C. Prescott and B. E. Proctor, Food Technology, McGraw-Hill, 1937*)

and fruits and nuts used as ingredients in ice cream may be contaminated with microorganisms. Therefore the ice cream mix must be pasteurized before freezing because a low temperature in itself is not sufficient to kill microorganisms. Other possible sources of microorganisms are equipment and utensils used in manufacture, scoops and dippers, vendors, employees, and air. Stringent regulations must be established for all manipulations of ice cream from producer to consumer in order to avoid recontamination, especially with regard to pathogens. Once incorporated, microorganisms survive a long time in ice cream. This can be seen from the abridged data in Table 1. *See* FOOD MANUFACTURING; PASTEURIZATION.

Kinds of microorganisms. A wide variety of microorganisms may be found, including streptococci, micrococci, coliform bacteria, sporeformers, yeasts, and molds. No selection of species by freezing takes place. Since no growth and only slow destruction of microorganisms in the final product occur, the flora of ice cream consists of organisms surviving pasteurization or incorporated by contamination. With well-controlled sanitation during production, plate counts may be as low as a few thousand; with careless methods they may be in the millions. Normally a slight increase of the bacterial count takes place during manufacture, especially after homogenization and freezing, due to the breaking up of clumps of bacteria (Table 2). A plate count not greater than 50,000–100,000 per gram is regarded as acceptable from a sanitary point of view in most states. *See* BACTERIA; FUNGI.

Table 1. Persistence of Salmonella typhosa (Bacterium typhosum) in ice cream*

Time after freezing, days	Typhoid bacteria, $\times 10^3$ per milliliter
Freshly frozen	51,000
5	10,000
20	2,200
104	900
165	640
260	57
342	51
544	13
730	6

*M. J. Prucha and J. M. Brannon, Viability of *Bacterium typhosum* in ice cream, *J. Bacteriol.*, 11:27–29, 1926.

Table 2. Effect of various steps in manufacture on standard plate count of ice cream*

Stage of manufacture	Unsupervised preparation (12 mixes)	Supervised preparation (11 mixes)
Before pasteurization	10,758,566	4,617,535
Before homogenization	105,748	11,850
After homogenization	200,745	18,643
Before aging	289,341	31,381
After aging	354,300	33,227
After freezing	458,325	58,136
After hardening	390,225	39,127

*N. E. Olson and A. C. Fay, The bacterial content of ice cream: A report of experiments in bacterial control in six commercial plants, *J. Dairy Sci.*, 8:415, 1925.

Hygienic measures. These include proper pasteurization of the mix and adequate sanitizing of equipment. Pasteurization must be more intensive with the mix than with milk because increased milk solids protect microorganisms to some degree. At least 68.4°C for 30 min or 79.5°C for 25 sec are applied to obtain a sufficient margin of safety. In this way all pathogens are killed and total bacterial counts of 10,000 or less per gram are usually reached. Higher counts after pasteurization are due to thermoduric organisms. The source of these organisms is inadequately cleaned farm or plant equipment. Sanitizing of equipment is achieved by cleaning and sterilizing operations with heat or chemicals. *See* DAIRY MACHINERY; STERILIZATION.

Bacteriological control. For determining the hygienic quality of ice cream, the laboratory tests used are similar to those employed for milk. Of these, the standard plate count and the test for coliform bacteria have found widespread application. *See* MICROBIOLOGICAL METHODS; MILK.

[WILLIAM C. WINDER]

Bibliography: W. S. Arbuckle, *Ice Cream*, 1966; E. M. Foster et al., *Dairy Microbiology*, 1957.

Ice field

A network of interconnected glaciers or ice streams, with common source area or areas, in contrast to ice sheets and ice caps. The German word *Eisstromnetz*, which translates literally to ice-stream net, is sometimes used for glacial systems of moderate size (such as less than 3000 mi²) and is most applicable to mountainous regions. Being generally associated with terrane of substantial relief, ice-field glaciers are mostly of the broad-basin, cirque, and mountain-valley type. Thus, different sections of an ice field are often separated by linear ranges, bedrock ridges, and nunataks.

Contrast with ice sheet. An ice sheet is a broad, cakelike glacial mass with a relatively flat surface and gentle relief. Ice sheets are not confined or controlled by valley topography and usually cover broad topographic features such as a continental plateau (for example, much of the antarctic ice sheet), or a lowland polar archipelago (such as the Greenland ice sheet). Although ice sheets are generally of very large dimension, in some regions small, rather flat ice bodies have been called ice sheets because they are thinned remnants of once large masses of this form. Small ice sheets and even ice fields are sometimes incorrectly referred to by casual or lay observers as "ice caps," even though their configurations have been well characterized.

Contrast with ice cap. Ice caps are properly defined as domelike glacial masses, usually at high elevation. They may, for example, make up the central nourishment area of an ice field at the crest of a mountain range, or they may exist in isolated positions as separate glacial units in themselves. The latter type is characterized by a distinctly convex summit dome, bordered by contiguous glacial slopes with relatively regular margins not dissected by outlet valleys or abutment ridges.

Similarities and gradations. There are all gradations between ice caps, ice fields, and ice sheets. Over a period of time, a morphogenetic gradational sequence may also develop in any one region. Major ice sheets, for example, probably originate from the thickening and expansion of ice fields and the coalescence of bordering piedmont glaciers. Conversely, ice fields can develop through the thinning and retraction of a large ice sheet overlying mountainous terrane. *See* GLACIATED TERRANE; GLACIOLOGY.

[MAYNARD M. MILLER]

Ice island

One of the massive bodies of floating ice in the Arctic Ocean, from 20 to 200 ft thick, irregular in shape, and from a few square miles to 300 mi² in area. Ice islands originate in the landfast ice along the high-latitude northern shores of the Canadian archipelago and Greenland. This distinguishes them from the smaller and more rugged icebergs which originate in the glaciers of Greenland. The unbroken appearance of the ice islands contrasts greatly with the surrounding pack ice, which normally almost completely covers the Arctic Ocean with a maximum thickness of 20 ft. It was the unbroken appearance and the elevation of ice islands above the pack ice which first attracted the attention of U.S. Air Force weather reconnaissance planes in 1946 (Fig. 1). Since that time about 100 ice islands have been observed, mostly in the numerous bays and straits of the Canadian archipelago.

Character and formation. Ice islands are tabular features without the pressure ridges found on pack ice. Long shallow drainage channels are most evident in summer when they are filled with meltwater. The presence of rock piles on the surface and of dust and dirt layers within the ice, as well as plant and animal remains, testify that the ice islands were close to land at one time.

The principal source of ice islands is the Ward Hunt Ice Shelf, a floating body of landfast ice about 60 by 10 mi in size located between McClintock and Markham bays on northern Ellesmere Island. Extensively studied, it has been found to be similar in all respects to the ice islands. The Ward Hunt Ice Shelf has decreased considerably in area since it was first visited by G. S. Nares in 1875 and R. E. Peary in 1906. A dramatic formation of ice islands occurred during the winter of 1961–1962 when the entire northern part (some 200 mi²) of the Ward Hunt Ice Shelf broke off to form five large ice islands. The largest of these, WH-5 (about 11 by 5 mi), drifted eastward, eventually turning southward and entering Robeson Channel. It continued southward through Baffin Bay and eventually disintegrated off Labrador and Newfoundland in 1964. The other four islands drifted westward and southward, skirting the northern edge of the Canadian islands.

The ice islands probably break from the shelf during years of generally warm Arctic climate. The growth of this landfast shelf ice to thicknesses of 100 ft or more takes centuries and depends on general climatic conditions. *See* ICEBERG; SEA ICE.

Scientific endeavor. In March, 1952, a landing was made on one of these islands, designated T-3 (T for radar target), also called Fletcher's Ice Island for Col. J. O. Fletcher, who was in charge of the first operations there. It was occupied from March, 1952, to May, 1954; April, 1955, to September, 1955; and from April, 1957, to

ICE ISLAND

Fig. 1. Ice Island T-3. (*Official photograph, USAF Cambridge Research Center, Mass.*)

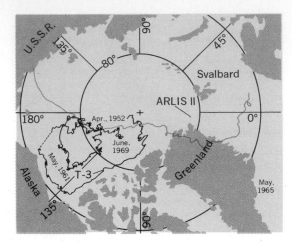

Fig. 2.   Drift of T-3 and ARLIS II during occupancy.

September, 1961, when it ran aground on the continental shelf 80 mi northwest of Point Barrow, Alaska. It was reoccupied in February, 1962, after floating free again and has been continually occupied since then (as of 1969). In 1952 a landing was made on T-1, the first island discovered and the largest of the known ice islands, but no permanent facilities were established. The Soviet Union located one of its North Pole stations, NP-6, on an ice island from 1956 until 1959, without encountering fracturing of the ice or danger to the camp, thus showing the usefulness of ice islands as research platforms; eight other Soviet stations on pack ice were reestablished a total of 57 times. After 1961 the operation of United States drifting ice stations in the Arctic Ocean shifted from the Air Force to the Navy. In 1961 ARLIS II (Arctic Research Laboratory Ice Station II) was established by the Naval Arctic Research Laboratory of Barrow, Alaska, on an ice island that measured 3.25 by 2.0 mi and 20–80 ft in thickness.

The ice islands are excellent platforms for Arctic scientific studies. They drift under the influence of winds and currents in an erratic course through the Arctic Ocean (Fig. 2). There are two patterns of drift: a clockwise drift around the western Arctic, for example, T-3, or a direct drift across the basin and out into the Greenland Sea, for example, ARLIS II. T-3 has completed two revolutions around the western Arctic Basin since 1952, averaging about 9 years for one revolution. ARLIS II drifted from the area north of Alaska across the Arctic Ocean and down the coast of Greenland. It was evacuated by icebreaker in May, 1965, in Denmark Strait between Iceland and Greenland, and its breakup was later observed in the waters off southern Greenland.

Camps on ice islands have been extensively used as scientific research stations. T-3, ARLIS II, and NP-6 have been bases for investigations in the meteorological, oceanographic, geophysical, and upper atmospheric phenomena of the north polar regions. T-3 was the only ice island being used as a research base in 1969. The research program conducted from T-3 in the Arctic Ocean is similar to that on an oceanographic research vessel in other oceans: The ice island position is determined by celestial methods and with the U.S. Navy Satellite Navigation System; echo soundings of

ocean depths are recorded continuously; the Earth's magnetic and gravity fields are recorded regularly; surface and upper-air meteorological observations are taken several times per day; the ocean is sampled with nets and water bottles lowered from a winch on a cable through a hole cut in the pack ice near the island; and sediment cores and photographs of the ocean floor are obtained regularly. *See* ARCTIC OCEAN.

[KENNETH L. HUNKINS]

Bibliography: H. Landsberg (ed.), *Advances in Geophysics*, vol. 3, 1956; J. Sater (ed.), *Arctic Drifting Stations: A Report on Activities Supported by the Office of Naval Research*, 1968; B. Staib, *On Skis Toward the North Pole*, 1965; T. Weeks and R. Mather, *Ice Island: Polar Science and the Arctic Research Laboratory*, 1965.

Ice manufacture

Commercial production of manufactured artificial ice from water, or of dry ice from the solidification of carbon dioxide, by mechanical refrigeration. Of greatest economic importance is the manufacture of water ice for use in refrigerator cars, fishing boats, fish- and meat-packing plants, and dairies. The National Association of Refrigerated Warehouses reported ice production in the United States in 1963 as 19,920,000 tons (valued at $210,-178,479), compared to 23,950,550 tons in 1957. Commercial ice production thus shows a falling-off at the rate of about 670,000 tons per year as direct mechanical refrigeration makes further inroads on ice markets.

However, there has been a sharp increase in the trend toward cubing, sizing, and packaging of ice. The domestic ice market practically vanished with development of the automatic household refrigerator. *See* DRY ICE.

Most ice is made in galvanized cans that are partially immersed in brine in an ice-making tank (Fig. 1). Brine made of sodium chloride or calcium chloride is used. The brine is cooled by ammonia as the refrigerant evaporates in pipe coils or brine coolers submerged in the ice tank. The ice cans are filled with raw water, or treated water if it initially contains large amounts of impurities. The water is usually precooled. Cold brine circulates around the ice cans to freeze the water. Commercial ice is frozen in 300- or 400-lb blocks for which the freezing time with brine at 12°F is 38–42 hr. Freezing time depends largely on the thickness of the ice cake and the brine temperature. In large plants, cans are arranged in group frames for harvesting as many as 34 cans at a time. A traveling crane picks up a frozen group, transports and drops it into a dip tank for thawing, then moves it to a can dump where the ice cakes slide out and into a storage room. The empty cans are refilled with fresh water and are returned to the ice tank by the crane.

If clear ice is required, the water in each can must be agitated with air during freezing; otherwise opaque ice is formed. Because the water must be cooled to 32°F before freezing can start and because of system losses, about 1.6 tons of refrigeration is required to make 1 ton of ice when the raw water enters at 70°F.

The manufacture of ice in slush, flake, or cube form by continuous ice-makers (Fig. 2) has taken

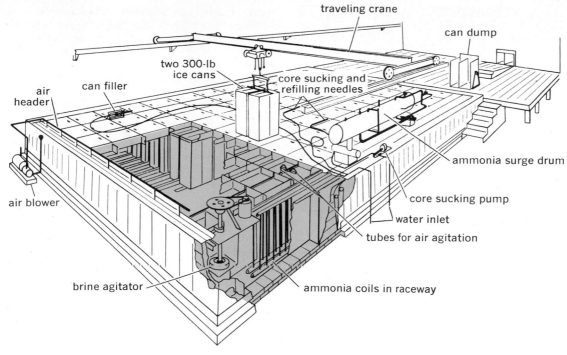

Fig. 1. Typical can ice plant. (*Worthington Pump and Machinery Corp.*)

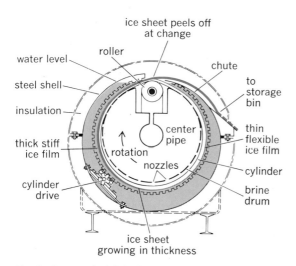

Fig. 2. Commerical flake ice machine.

over a large portion of the commercial market. Also, small, fully automatic, self-contained, cube and flake ice-makers have been developed for use in bars, restaurants, hotels, motels, clubs, and hospitals.

Ice is also frozen artificially in a flat horizontal sheet for ice-skating. Such skating rinks may be indoor or outdoor, permanent or portable. The rink floor is covered with pipe coils through which cold brine is circulated and over which water is sprayed until a frozen sheet 1/2 to 3/4 in. thick is obtained. The brine is cooled in brine coolers by a refrigerant such as ammonia or Freon-12. *See* REFRIGERATION.

[CARL F. KAYAN]

Bibliography: American Society of Heating, Refrigerating, and Air-Conditioning Engineers, *Guide and Data Book: Applications Volume*, 1968.

Ice point

The temperature at which liquid and solid water are in equilibrium under atmospheric pressure. The ice point is by far the most important "fixed point" for defining temperature scales and for calibrating thermometers. A closely related point is the triple point, where liquid, solid, and gaseous water are in equilibrium. It is 0.01° higher on the Kelvin scale than the ice point. The triple point has gained favor as the primary standard since it can be attained with great accuracy in a simple closed vessel, isolated from the atmosphere. Readings are reproducible to about 0.0001°K, but dissolved gases or other foreign matter may raise the error to 0.001° or more. *See* TEMPERATURE; TRIPLE POINT (THERMODYNAMICS).

The triple-point apparatus shown in the figure consists of a thermometer well that is filled with liquid water and jacketed by a cavity containing the three phases of water under a pressure of about 0.006 atm. The ice, initially deposited by prechilling the well, melts during heat transfer from the thermometer. [RALPH A. BURTON]

Bibliography: A. L. King, *Thermophysics*, 1962; H. C. Wolfe (ed.), *Temperature, Its Measurement and Control in Science and Industry*, vol. 2, 1955.

Iceberg

A large mass of glacial ice broken off and drifted from parent glaciers or ice shelves along polar seas. Icebergs should be distinguished from polar pack ice which is sea ice, or frozen sea water, though rafted or hummocked fragments of the later may resemble small bergs. *See* GLACIOLOGY; SEA ICE.

Characteristics and types. The continental or island icecaps of both Arctic and Antarctic regions produce icebergs where the icecaps extend to the sea in the form of glaciers or ice shelves. The

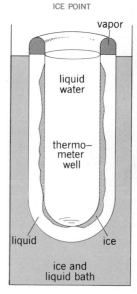

ICE POINT

Arrangement for determining triple point.

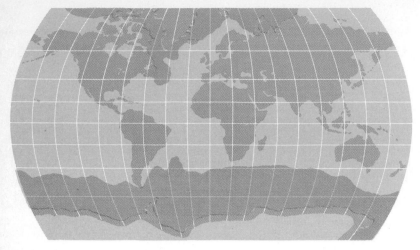

Fig. 1.   Normal extent of iceberg drift.

normal extent of iceberg drift is shown in Fig. 1.

Arctic icebergs. In the Arctic, icebergs originate chiefly from glaciers along Greenland coasts. It is estimated that a total of about 16,000 bergs are calved annually in the Northern Hemisphere, of which over 90% are of Greenland origin; but only about half of these have a size or source location to enable them to achieve any significant drift. The majority of the latter stem from some 20 glaciers along the west coast of Greenland between the 65th and 80th parallels of latitude. The most productive glacier is the Jacobshavn Glacier at latitude 68°N, calving about 1400 bergs yearly, and the largest is the Humboldt Glacier at latitude 79° with a seaward front extending 65 mi. The remainder of the Arctic berg crop comes from East Greenland and the island icecaps of Ellesmere Island, Iceland, Spitzbergen, and Novaya Zemlya, with almost no sizable bergs produced along the Eurasian or Alaskan Arctic coasts. No icebergs are discharged or drift into the North Pacific Ocean or its adjacent seas, except a few small bergs each year that calve from the piedmont glaciers along the Gulf of Alaska. These achieve no significant drift. *See* ARCTIC OCEAN; ATLANTIC OCEAN.

Ocean currents of the Arctic and adjacent seas determine the drift and ultimate distribution of icebergs, wind having little effect except on small, sail-shaped fragments. The dominant drift along the East Greenland coast is southward around the tip of Greenland and then northward along the west coast. Here the drifting bergs join the main body of West Greenland bergs and drift in a counterclockwise gyral across Davis Strait and Baffin Bay. The bergs are then swept southward along the coasts of Baffin Island, Labrador, and Newfoundland by the Labrador Current. This drift terminates along the Grand Banks of Newfoundland, where the waters of the Labrador Current mix with the warm Gulf Stream and even the largest of bergs melt within 2–3 weeks. Freak iceberg drifts have been reported where bergs or remaining fragments were sighted off Scotland, Nova Scotia, Bermuda, and even the Azores Islands. Such reports, however, are extremely rare. About 400 bergs each year are carried past Newfoundland as survivors of the estimated 3-year journey from West Greenland. The remainder become stranded along Arctic coasts and shoals and are ultimately destroyed by wave erosion and summer melting.

Icebergs in the Northern Hemisphere rarely

"calving" of a large iceberg is one of nature's greatest spectacles, considering that a Greenland berg may weigh over 1,000,000 tons and that Antarctic bergs are many times larger. An iceberg consists of glacial ice which is compressed snow having a variable specific gravity that averages about 0.89. This results in an above-water mass of from one-eighth to one-seventh of the entire mass. However, spires and peaks of an eroded or weathered berg will result in height to depth ratios of between 1–6 and 1–3. Tritium age experiments with melted Greenland berg ice indicate these bergs may be of the order of 50,000 years old. Minute air bubbles imprisoned in glacial ice impart to bergs a snow-white color and cause it to effervesce when immersed. *See* SEA WATER; TRITIUM.

Icebergs are classified by shape and size. The terms used are arched, blocky, dome, pinnacled, tabular, valley, and weathered for berg discription, and bergy-bit and growler for berg fragments ranging smaller than cottage size above water. The lifespan of an iceberg may be indefinite while the berg remains in cold polar waters, eroding only slightly during summer months. But under the influence of ocean currents, an iceberg that drifts into warmer water will disintegrate rapidly, its life being measured in weeks in sea temperatures between 40–50°F and in days in sea temperatures over 50°F. A notable feature of icebergs is their long and distant drift which may carry them into steamship tracks, where they become hazards to navigation. The

Fig. 2.   Arctic iceberg, eroded to form a valley or dry-dock type; grotesque shapes are common to the glacially produced icebergs of the North. Note the brash and small floes of sea ice surrounding the berg.

Fig. 3.   Antarctic iceberg, tabular type. Such bergs develop from great ice shelves along Antarctica and may reach over 100 mi in length. The U.S. Coast Guard icebreaker *Westwind* is in the foreground.

reach proportions larger than 2000 ft in breadth or 400 ft in height above the water (Fig. 2). However, true glacial ice islands several miles in extent are occasionally found and have even served as floating bases for scientific studies. The origin of these rare counterparts of the common Antarctic type is uncertain but is thought to be an ice shelf along northern Ellesmere Island. *See* ICE ISLAND.

Antarctic icebergs. In the Southern Ocean, bergs originate from the giant ice shelves all along the Antarctic continent. These result in huge, tabular bergs (Fig. 3) or ice islands several hundred feet high and often over a hundred miles in length, which frequent the entire waters of the Antarctic seas. The most active iceberg-producing regions are the Ross and Filchener ice shelves in the Ross and Weddell seas. The large size of these bergs and influence of the Antarctic Circumpolar Current give them an indeterminant life-span. When weathered, Antarctic icebergs attain a deep bluish hue of great beauty rarely seen in the Arctic. *See* ANTARCTIC OCEAN.   [ROBERTSON P. DINSMORE]

Ichneumon

A member of the hymenopteran family Ichneumonidae. Included are more than 10,000 species of minute to large parasitic wasps which have a worldwide distribution. One of the smallest species, measuring about one-tenth inch in length, is *Pecomachus philpotti*, a wingless form which occurs in New Zealand. In contrast, *Megarhyssa nortoni* may reach a length of about 9 in. The ichneumons serve a beneficial function in that they prey upon the immature stages of many injurious species of agricultural and forest insects such as beetles, butterflies, moths, flies, and other hymenopterans. *See* ENTOMOLOGY, ECONOMIC.

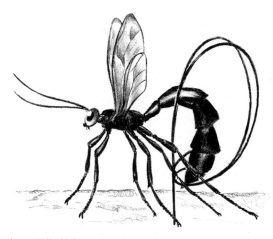

Long-tailed ichneumon (*Megarhyssa lunator*), about 6 in., of which more than 4 in. is "tail."

Ichneumons parasitize their victims by depositing eggs on or within the host. Endoparasitic forms undergo a complete metamorphosis in the host body with a number of instars or developmental stages; the adults emerge from the pupal or chrysalis stage of the host. Various species of *Megarhyssa* (see illustration), *Rhyssa*, and *Thalessa*, which parasitize members of the hymenopteran family Siricidae, must bore through wood to oviposit since the Siricidae larvae are wood-boring species. Adults of *Agriotypus* swim beneath the

water to find the appropriate larvae of caddis flies to parasitize. *See* HYMENOPTERA; INSECTA.
                                        [CHARLES B. CURTIN]

Ichthyopterygia

A subclass of extinct reptiles composed of predatory fish-finned and sea-swimming forms of the Mesozoic Era, widely divergent from land forms, although surely descended from some ancient terrestrial group as yet unrecognized. All are much alike in body form: short-necked, streamlined, swordfishlike or porpoiselike in body, yet unrelated to either fishes or porpoises; a noteworthy example of convergence in evolution. A single order is included in this subclass. *See* ICHTHYOSAURIA; REPTILIA.   [CHARLES L. CAMP]

Ichthyornithiformes

An order of ancient fossil birds from the Upper Cretaceous marine chalk beds of the Niobrara formation. Two families are included: Ichthyornithidae, with seven species known from Kansas and Texas, and Apathornithidae, with one species each from Kansas and Wyoming. These birds are segregated by some authorities into their own superorder Ichthyornithes on the basis of the presumed presence of biconcave vertebrae, unknown in all modern birds. Most likely, these vertebrae have been incorrectly associated with the material of *Ichthyornis*. Moreover, the Ichthyornithiformes were long associated with the Hesperornithiformes in the superorder Odontognathae in the belief that *Ichthyornis* possessed teeth. J. T. Gregory concluded that the toothed lower jaw assigned to *Ichthyornis* is reptilian, probably that of a small mosasaur. While this reidentification has not been universally accepted, the mandible in question appears too large to be associated with known specimens of *Ichthyornis* and does not appear to be avian. While the systematic position of these birds must still be considered tentative, many workers regard them to be relatives of the Charadriiformes. The Ichthyornithiformes possess all skeletal characteristics of modern birds (with the possible exceptions noted) and appear to be strong flying birds of inland seas, somewhat gull-like in their proportions and presumably in their habits. *See* AVES; CHARADRIIFORMES; HESPERORNITHIFORMES.   [WALTER BOCK]

Ichthyosauria

The only order of the reptilian subclass Ichthyopterygia. Included here are the extinct fish-lizards, the most highly adapted of all reptiles to life in the sea. Well-preserved fossils show the entire body form; there are impressions of tail flukes, paddles dorsal fins, even of embryonic skeletons lined up within the mother's body, ready when born to swim in the open sea in the manner of young whales. A specimen has been recovered in which birth was occurring at the time of the mother's death, which shows that the embryos were not young ichthyosaurs which she had swallowed for food.

Early ichthyosaurs had long-toothed, fish-catching jaws similar to the gar pikes and gavials. Some had blunt teeth for crushing shellfish. Late Cretaceous forms lost the teeth and had beaks like those of the swordfishes and marlins, which they greatly resembled. All were predacious, and some larger

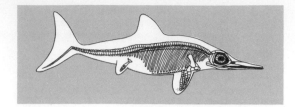

A Jurassic ichthyosaur, much reduced. (*Simplified from E. von Stromer, as used by A. S. Romer, Vertebrate Paleontology, 3d ed., University of Chicago Press, 1966*)

ones probably fed on other ichthyosaurs. The earliest had strong fluted teeth set in sockets. Later these sockets ran together into grooves and still later the teeth became small and loosely set. The eye was exceptionally large to seek prey in moderately deep water.

In early genera the elongate tail was turned down toward the tip where it was expanded into a narrow, elongate fin. Later, in the Jurassic, the tail became shorter and was turned abruptly down at the tip for the ventral support of an enlarged terminal fin (see illustration). This tail fin was the main propulsive organ, which transmitted its great force through the broad flat vertebrae of the heavy spinal column. Vertebrae, up to 200 in number, were developed much as in whales.

The paired fins or paddles, derived from the feet of land ancestors, were of two main types, broad and narrow, and were used as rudders and elevators or depressors when the creatures rose to the surface to breathe or submerged to feed. There was also a stabilizing dorsal fin without supporting skeleton, as in whales and sharks, to prevent sideslip while swimming. This structure, along with the entire outline of the body, is preserved in the remarkable fossils found abundantly in Holzmaden, Germany.

Ichthyosaurs ranged in length from about 2 to 50 ft. The giant forms of the Late Triassic and Early Jurassic, the largest animals of their day, were the last representatives of the long-tailed forms and were superseded by smaller, probably more active, short-tailed types. Toothless forms, late in the Cretaceous, rapidly decreased in numbers, and extinction took place before the close of the Cretaceous when the ichthyosaurs were evidently supplanted by the mosasaurs of the order Squamata, subclass Lepidosauria. *See* LEPIDOSAURIA; SQUAMATA.

[CHARLES L. CAMP]

Ichthyostegalia

An order of labyrinthodont amphibia found in Late Devonian deposits in East Greenland and hence the oldest known representatives of the class. The limbs and girdles are of an amphibian pattern, although small and primitive in structure, but there is a definite, if reduced, caudal fin reminiscent of

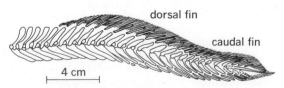

Tail of *Ichthyostega*, labyrinthodont amphibian, showing fin rays retained from fish ancestors. (*After A. Jarvik*)

piscine ancestors (see illustration). The vertebrae are comparable to those of crossopterygian fishes on the one hand and rhachitomous amphibians on the other. The skull appears to be specialized in a greater reduction of the otic notch than would be expected in a truly ancestral amphibian, and the intertemporal bone is absent. The skull is remarkably primitive in other features, particularly in the retention of a large canal for the notochord through the base of the occipital region; this construction, characteristic of crossopterygians, is absent in all later amphibians. *See* AMPHIBIA; CROSSOPTE-RYGII; LABYRINTHODONTIA.   [ALFRED S. ROMER]

Ictidosauria

An order composed of a few extremely mammal-like reptiles (subclass Synapsida). Since the included genera probably stemmed from more than a single source among the therapsid reptiles, cynodonts, and therocephalians, the order probably is phylogenetically artificial. Although very close to mammals, ictidosaurs are technically excluded by the reptilian nature of their lower jaws and middle ears. Representatives occur in the uppermost Triassic (probably also Lower Jurassic) deposits. One group, the tritylodonts, is known from North America, Asia, Europe, and South Africa. Ictidosaurs proper are known from South Africa only. All ictidosaurs were small animals, no more than 3 ft in length. The order included both carnivores and herbivores and seems to have been restricted to terrestrial life. Ictidosaurs represent one of the sources of mammals. *See* ANIMAL EVOLUTION; MAMMALIA; SYNAPSIDA; THERAPSIDA.

[EVERETT C. OLSON]

Ideal aerodynamics

The oldest branch of analytical aerodynamics, also referred to as ideal fluid dynamics. Although limited in its application, as pointed out below, it is useful for explaining some airflow problems and for providing approximate answers. It is also still studied because many other analytical methods in aerodynamics presuppose a knowledge of this earlier work.

The use of a set of five basic simplifying assumptions characterizes ideal aerodynamics. There are other assumptions made in other branches of analytical aerodynamics, but the five used in classical ideal analytical aerodynamics are:

1. The flow is two-dimensional; that is, there exists a direction in which the flow characteristics do not change.

2. The flow is steady; that is, the flow characteristics do not change with time.

3. The flow is incompressible; that is, density does not change.

4. The fluid is nonviscous; that is, the effects of viscosity are ignored except indirectly, as explained below.

5. The flow is irrotational; that is, the curl of the velocity vector is zero.

The assumption of constant density limits application to flows with a Mach number below about 0.6, depending on the accuracy desired. *See* LA-PLACE'S IRROTATIONAL MOTION; FLUID-FLOW PROPERTIES; SUBSONIC FLIGHT.

Complex potential and velocity. The velocity potential, which exists for irrotational flow, is a scalar function such that its gradient is the velocity vector, from which it follows that its partial derivative in any direction is equal to the velocity component in that direction. The stream function for two-dimensional, steady incompressible flow is a scalar function such that its partial derivative with respect to x equals minus the y component and its partial derivative with respect to y equals the x component of velocity. When all the five basic simplifying assumptions are used, each of these functions not only exists but is harmonic, that is, satisfies Laplace's equation. Let these functions be denoted respectively by $\phi(x,y)$ and $\psi(x,y)$. A complex function is defined in Eq. (1). This function is called the complex potential. It can be

$$w(z) = \phi(x,y) + \psi(x,y)i \qquad (1)$$

tion is called the complex potential. It can be shown that its derivative, called the complex velocity, is given by Eq. (2), where v_x and v_y are the ve-

$$\frac{dw}{dz} = v_x - v_y i \qquad (2)$$

locity components. Since ϕ and ψ are harmonic, $w(z)$ is an analytic function. This is an important result for ideal aerodynamics. Three examples of complex potentials are needed: uniform flow, Eq. (3); two-dimensional doublet flow, Eq. (4); and two-dimensional vortex flow, Eq. (5). Here α and β are

$$w = \alpha z + \beta \qquad (3)$$

$$w = \frac{\mu}{2\pi} \frac{1}{z} \qquad (4)$$

$$w = -i\frac{\Gamma}{2\pi} \log(z - \alpha) \qquad (5)$$

complex constants, μ is the doublet strength, and Γ is the circulation. *See* COMPLEX NUMBERS AND COMPLEX VARIABLES; DOUBLET FLOW; HYDRODYNAMICS; VORTEX.

Ideal flow past a circle. Superimpose a uniform flow parallel to the x axis, a doublet flow, centered at the origin, and a vortex, also centered at the origin. The complex potential, found by adding those in Eqs. (3), (4), and (5) and choosing constants appropriately, is given in Eq. (6), where V is the ve-

$$w = V\left(z + \frac{a^2}{z}\right) + i\frac{\Gamma}{2\pi}\log z \qquad (6)$$

locity magnitude of the uniform flow and a is a real constant. If the imaginary part of Eq. (6) is found (the stream function of the flow) and is set equal to a family of constants, the graphs of the resulting family of equations will represent the streamlines of the flow. They will be seen to represent the ideal flow past a circular cross section of radius a.

It is possible to find the stagnation points in this flow, the points where the velocity is zero, by differentiating Eq. (6) and setting the resulting expression (the complex velocity) equal to zero. Solving the resulting quadratic equation, the stagnation points are given by Eqs. (7)–(9) for the three possible cases, $\Gamma < 4\pi aV$, $\Gamma = 4\pi aV$, and $\Gamma > 4\pi aV$, depending on the strength of the circulation, where $\gamma = \arcsin \Gamma/4\pi aV$.

$$\begin{array}{ll} x = a\cos\gamma & y = -a\sin\gamma \\ x = -a\cos\gamma & y = -a\sin\gamma \end{array} \qquad (7)$$

$$x = 0 \qquad y = -a \qquad (8)$$

$$x = 0 \qquad y = \frac{-\Gamma \pm \sqrt{\Gamma^2 - 16\pi^2 a^2 V^2}}{4\pi V} \qquad (9)$$

Having found the ideal flow past a circle, by conformal transformation the circle and its streamlines can be transformed into various airfoil profiles and their streamlines. The strength of the circulation, Γ, is determined by the Joukowski hypothesis, that one stagnation point is to be placed at the trailing edge of the airfoil profile. This is also known as the Kutta condition. *See* CONFORMAL MAPPING; STREAMLINING.

Blasius theorem. Once the nature of the flow past a profile is determined, the aerodynamic force can be determined. Physically this determination in ideal aerodynamics is based on the relation between velocity magnitude and pressure as given by Bernoulli's formula (ignoring gravity effects), Eq. (10), where p and v are the local pressure and ve-

$$p + \frac{1}{2}\rho v^2 = p_0 \qquad (10)$$

locity magnitude, ρ is the mass density of the fluid, and p_0 is the stagnation pressure. *See* BERNOULLI'S THEOREM.

With the pressure determined, the force is found by multiplying pressure by area. Since the pressure varies along the surface, this means integration. Analytically, a general result is found by means of the Blasius theorem. Suppose that the complex potential for a flow past a profile is given by $w = f(z)$. Let X and Y denote the force components in the x and y directions and let M be the moment about the origin, plus counterclockwise. Let w' denote the derivative of w or f with respect to z, that is, the complex velocity. Let C denote the profile or any closed curve containing the profile with no singularity between the profile and the curve. Then the theorem can be stated by Eqs. (11) and (12), where $\mathscr{R}$ means "real part of." The inte-

$$X - Yi = \frac{1}{2}i\rho\int_C (w')^2 dz \qquad (11)$$

$$M = \mathscr{R}\left[-\frac{1}{2}\rho\int_C z(w')^2 dz\right] \qquad (12)$$

gral can ordinarily be evaluated by the method of residues.

Kutta-Joukowski theorem. By applying the Blasius theorem to a specific function $f(z)$, an interesting result is obtained. Since in incompressible flow the velocity at a great distance from the profile is the free-stream velocity V, a boundary condition can be written, Eq. (13),

$$\text{as} \quad |z| \to \infty, \quad w' \to V \qquad (13)$$

A general function satisfying Eq. (13) is given by Eq. (14). Then Eq. (15) holds.

$$w = Vz + A\log z - \frac{B}{z} - \frac{C}{2z^2} - \frac{D}{3z^3} - \cdots \qquad (14)$$

$$w' = V + \frac{A}{z} + \frac{B}{z^2} + \frac{C}{z^3} + \frac{D}{z^4} + \cdots \qquad (15)$$

Since a circulation term in the clockwise direction is needed, Eq. (5) shows that A should be taken as in Eq. (16). Use Eq. (16) in Eq. (15) and substi-

$$A = \frac{\Gamma i}{2\pi} \qquad (16)$$

tute in the Blasius theorem Eqs. (11) and (12). Let the free stream velocity make the angle of attack α with the x axis, and the results for lift, drag, and

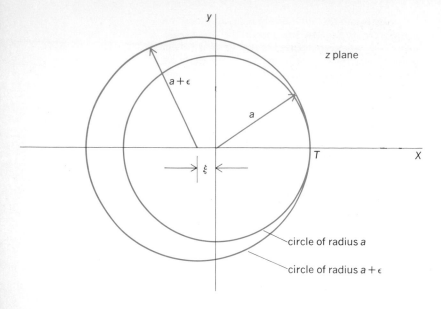

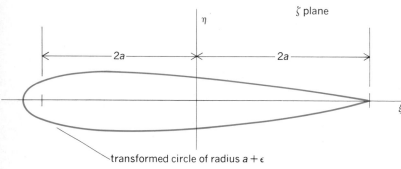

Example of Joukowski transformation.

moment are given by Eqs. (17), (18), and (19), respectively. Equation (17) is known as the Kutta-Joukowski theorem. In a given case the circulation Γ is determined by the Joukowski hypothesis.

$$L = \rho V \Gamma \qquad (17)$$
$$D = 0 \qquad (18)$$
$$M = -2\pi\rho VB \sin \alpha \qquad (19)$$

When this result is applied to a straight-line profile (flat plate airfoil) of length c, it is found that the circulation must be as in Eq. (20). The angle of

$$\Gamma = c\pi V \sin \alpha \qquad (20)$$

attack must be small to avoid stall, so it is customary to replace $\sin \alpha$ by α in radians. By use of the definition for two-dimensional lift coefficient, Eq. (21) is obtained. *See* AIRFOIL PROFILE.

$$c_L = 2\pi\alpha \qquad (21)$$

Transformations from circle. The basic transformation from a circle into other profiles, due to N. Joukowski, is given by Eq. (22), where a circle

$$w = z + \frac{a^2}{z} \qquad (22)$$

in the z plane is transformed into some profile in the w plane. The nature of the profile depends on the radius of the circle, r, and on the location of the center of the circle. It is found that the trailing edge of the profile will be sharp (mathematically, a cusp) if the circle passes through the point $x =$

$a, y = 0$. Note that the derivative of Eq. (22) is Eq. (23), the complex velocity, so that when $z = \pm a$, the

$$w' = 1 - \frac{a^2}{z^2} \qquad (23)$$

velocity is zero, thus making these points stagnation points. In what follows, the circle will always be taken to pass through $z = a$, and the point $z = -a$ will not be outside the circle. Four different kinds of locations of the center are considered. Let ϵ and η be positive quantities less than a, usually much less.

Case 1. Center at origin, so that $r = a$. The profile in the w plane is a straight line of length $4a$.

Case 2. Center at $x = 0, y = \eta$, so that $r = \sqrt{a^2 + \eta^2}$. The profile is a portion of a circle above the real axis. The length of the chord is $4a$ and the maximum distance above the real axis is 2η.

Case 3. Center at $x = -\epsilon, y = 0$, so that $r = a + \epsilon$. The profile is a symmetrical shape of the nature of an airfoil section. The chord length is given by Eq. (24) and the maximum thickness is about 5.2ϵ (see illustration).

$$c = 4\left(a + \frac{\epsilon^2}{a}\right) \qquad (24)$$

Case 4. Center at $x = -\epsilon, y = \eta$, so that $r = \sqrt{(a + \epsilon)^2 + \eta^2}$. The profile is then an unsymmetrical airfoil cross section with a circle arc as a camber line.

These Joukowski profiles have been generalized in at least two ways. The von Mises generalization adds additional terms to Eq. (22), thus allowing more constants to be determined and more variety in the resulting profile. The transformation can be written as Eq. (25). The von Mises profiles still

$$w = z + \frac{a_1}{z} + \frac{a_2}{z^2} + \cdot \cdot \cdot + \frac{a_n}{z^n} \qquad (25)$$

have cusps at the trailing edges. If a nonzero angle between the upper and lower surfaces at the trailing edge is wanted, the Kármán-Trefftz generalization is used. To explain this change, first write the Joukowski transformation, Eq. (22), in the equivalent form shown in Eq. (26). It is found that the

$$\frac{w - 2a}{w + 2a} = \left(\frac{z - a}{z + a}\right)^2 \qquad (26)$$

exponent on the right side determines the zero trailing edge angle. Generalize Eq. (26), making the exponent equal b in Eq. (27). This is the Kármán-

$$\frac{w - 2a}{w + 2a} = \left(\frac{z - a}{z + a}\right)^b \qquad (27)$$

Trefftz transformation. It can be shown that, if the trailing edge angle is denoted by τ, Eq. (28) holds.

$$\tau = (2 - b)\pi \qquad (28)$$

Vortex distribution method. A different analytic method, which ignores the profile thickness, is based on replacing the mean line by an infinite number of vortexes with infinitesimal strength. Let γ denote the vortex strength per unit length along the mean line. Consider the vortexes to be vortex lines perpendicular to the plane of the profile, then the Biot-Savart formula can be used to find the velocity on the mean line due to the vortexes. *See* BIOT-SAVART LAW.

Let v be the total velocity at the mean line, V the free stream velocity, and V_i the induced velocity due to the vortexes. Then Eq. (29) and, by the Biot-

$$v = V_i + V \qquad (29)$$

Savart formula, Eq. (30) hold where r is distance

$$dV_i = \frac{1}{2\pi} \frac{\gamma}{r} ds \qquad (30)$$

from vortex center and s is distance along the mean line. The essence of the method is to determine γ as a function of x (distance along chord line) so that v satisfies the following conditions: (1) As distance from profile approaches infinity, v approaches V. (2) The mean line is a streamline. (3) The Joukowski hypothesis is satisfied; that is, γ is zero at trailing edge.

It is found that if Eq. (31), involving a Fourier

$$\gamma(\theta) = 2V \left[A_0 \left(\frac{1 + \cos\theta}{\sin\theta} \right) + \sum_{n=1}^{\infty} A_n \sin n\theta \right] \qquad (31)$$

series, is used for γ, where the variable x has been replaced by θ, defined in Eq. (32), then the slope of

$$x = \frac{1}{2} c (1 - \cos\theta) \qquad (32)$$

the mean line is given by Eq. (33). If the slope of

$$\frac{dy}{dx} = (\alpha - A_0) + \sum_{n=1}^{\infty} A_n \cos n\theta \qquad (33)$$

the mean line is given, then the formulas for the coefficients in a Fourier series give Eqs. (34) and

$$A_0 = \alpha - \frac{1}{\pi} \int_0^{\pi} \frac{dy}{dx} d\theta \qquad (34)$$

(35). Thus the function γ in Eq. (31) is determined.

$$A_n = \frac{2}{\pi} \int_0^{\pi} \frac{dy}{dx} \cos n\theta \, d\theta \qquad (35)$$

It can also be shown that the two-dimensional lift coefficient is given by Eq. (36) and that the moment

$$c_L = 2\pi A_0 + \pi A_1 \qquad (36)$$

coefficient about the quarter-chord point is given by Eq. (37).

$$c_M = \frac{\pi}{4} (A_2 - A_1) \qquad (37)$$

[FRANK MC L. MALLETT]

Bibliography: R. V. Churchill, *Complex Variables and Applications*, 2d ed., 1960; H. Glauert, *Aerofoil and Airscrew Theory*, 1926; A. W. Kuethe and J. D. Schetzer, *Foundations of Aerodynamics*, 2d ed., 1959; M. Rauscher, *Introduction to Aeronautical Dynamics*, 1953; A. Robinson and J. A. Laurman, *Wing Theory*, 1956; V. L. Streeter, *Fluid Dynamics*, 1948.

Idocrase

A sorosilicate mineral of complex composition crystallizing in the tetragonal system; also known by the name vesuvianite. Crystals, frequently well formed, are usually prismatic with pyramidal terminations (see illustration). It commonly occurs in columnar aggregates but may be granular or massive. The luster is vitreous to resinous; the color is usually green or brown but may be yellow, blue, or

Idocrase. (*a*) Crystal, Christiansand, Norway (*specimen from Department of Geology, Bryn Mawr College*). (*b*) Crystal habits (*from C. S. Hurlbut, Jr., Dana's Manual of Mineralogy, 17th ed., Wiley, 1959*).

red. Hardness is $6\frac{1}{2}$ on Mohs scale; specific gravity is 3.35–3.45. *See* SILICATE MINERALS.

The composition of idocrase is expressed by the formula $Ca_{10}Al_4(Mg,Fe)_2Si_9O_{34}(OH)_4$. Magnesium and ferrous iron are present in varying amounts, and boron or fluorine is found in some varieties. Beryllium has been reported in small amounts.

Idocrase is found characteristically in crystalline limestones resulting from contact metamorphism. It is there associated with other contact minerals such as garnet, diopside, wollastonite, and tourmaline. Noted localities are Zermatt, Switzerland; Christiansand, Norway; River Vilui, Siberia; and Chiapas, Mexico. In the United States it is found at Sanford, Maine; Franklin, N.J.; Amity, N.Y.; and at many contact metamorphic deposits in western states. A compact green variety resembling jade is found in California and is called californite.   [CORNELIUS S. HURLBUT, JR.]

Igneous rocks

Those rocks which have congealed from a molten mass. They may be composed of crystals or glass or both, depending on the conditions of formation. The molten matter from which they come is called magma; where erupted to the surface, it is commonly known as lava. Solidification of the hot rock melt occurs in response to loss of heat. Generated at depth the magma tends to rise. It commonly breaks through the Earth's crust and spills out on the Earth's surface or ocean floor to form volcanic or extrusive rocks. At the surface where cooling is rapid, fine-grained or glassy rocks are formed.

Where unable to reach the surface, magma cools more slowly, insulated by the overlying rocks; and a coarser texture develops. The resulting igneous rocks appear intrusive relative to adjacent rocks. In general, deeply formed (plutonic) rocks display the coarsest texture. Igneous rocks formed at shallow depths (hypabyssal) display features somewhat intermediate between those of volcanic and plutonic types. *See* MAGMA; PLUTON; VOLCANO; VOLCANOLOGY.

Textures. Texture refers to the mutual relation of the rock constituents within a uniform aggregate. It is dependent upon the relative amounts of crystalline and amorphous (glassy) matter as well as the size, shape, and arrangement of the constituents.

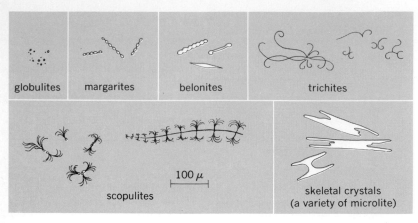

Fig. 1. Crystallites and microlites.

talline if their constituent mineral grains can be distinguished as individual entities by the naked eye. All other igneous rocks are aphanitic.

Phaneritic rocks are divided on the basis of their average grain diameter as follows: fine-grained, grains less than 1 mm; medium-grained, grains 1–5 mm; coarse-grained, grains 5–30 mm; very coarse-grained (pegmatitic), grains more than 3 cm.

Aphanitic rocks are microcrystalline if individual constituents can be distinguished only with the microscope. They are cryptocrystalline if constituents are submicroscopically crystalline. Dominantly glassy rocks are considered aphanitic. Aphanitic rocks rich in light-colored (felsic) minerals are termed felsitic. *See* FELSITE.

Grain shape. In igneous rocks grain shape is controlled by many factors. In highly glassy rocks the rate of growth is important. Crystallites (Fig. 1) are the most rudimentary forms and abound in glassy rocks in which rapid consolidation has arrested further growth. They are too small to polarize light and cannot be identified as to mineral species. These embryonic forms are perhaps most varied and beautifully displayed in the glassy rock, pitchstone. *See* OBSIDIAN.

Microlites are slightly larger, elongate crystals. They polarize light and can usually be identified specifically under the microscope. Many have grown rapidly but imperfectly to form skeletal crystals (Fig. 1).

In most rocks, grain shape is controlled largely by sequence of mineral crystallization and the nature and variety of associated minerals. A grain is said to be euhedral if bounded by its characteristic crystal faces and anhedral if crystal faces are absent. Intermediate forms are subhedral. Crystals developed early in a magma tend to be euhedral. Late crystals, however, meet interference from numerous adjacent grains and are forced to assume irregular mutual boundaries.

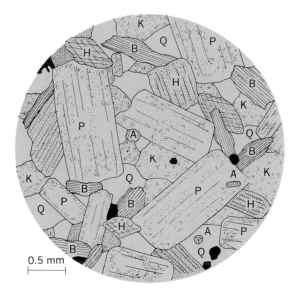

Fig. 2. Hypidiomorphic granular texture in granodiorite. Euhedral plagioclase (P), subhedral hornblende (H), biotite (B), anhedral quartz (Q), and potash feldspar (K). Accessory minerals include euhedral apatite (A) and subhedral magnetite (black).

Rock textures are highly significant; they shed light on the problem of rock genesis, and tell much about the conditions and environment under which the rock formed.

Crystallinity. This property expresses the proportion of crystalline to amorphous material in an igneous rock. Most igneous rocks, such as granite, are composed entirely of crystalline material and are called holocrystalline. Entirely glassy, or holohyaline, rocks such as obsidian are extremely rare. Many rocks such as rhyolite or vitrophyre contain both glass and crystals and are called hypocrystalline or hypohyaline.

Glass may be considered an amorphous solid with no systematic arrangement of its constituent atoms. Crystals form as the temperature of a magma falls and atoms begin to arrange themselves into orderly, repetitive groups. With rapid cooling there may be no opportunity for crystals to develop, and a magma will congeal as glass.

Granularity or grain size. In igneous rocks grain size ranges widely and depends in large part upon rate of cooling. Rocks are phaneric or phanerocrys-

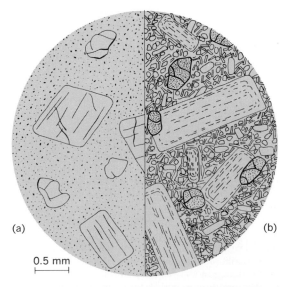

Fig. 3. Porphyritic textures. (a) Porphyritic rhyolite showing euhedral phenocrysts of sanidine and resorbed crystals of quartz in a submicroscopically crystalline matrix. (b) Porphyritic basalt showing euhedral phenocrysts of plagioclase and subhedral olivine in a matrix of granular pyroxene and feldspar microlites.

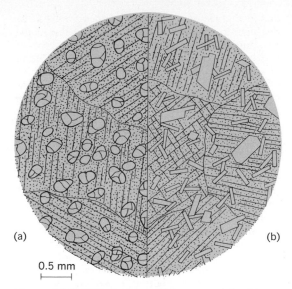

(a)   (b)

0.5 mm

Fig. 4. (a) Poikilitic texture with large crystals of hornblende enclosing abundant small grains of olivine. (b) Orphitic texture with large crystals of pyroxene enclosing small laths of plagioclase.

It is not necessary that all mutually interfering grains develop anhedral forms. Some mineral species possess a greater power of growth (a greater form energy) and are capable of maintaining their characteristic crystal form in competition with adjacent minerals.

Most igneous rocks show a grainy or granular texture in which the majority of crystals are roughly equidimensional. Rarely, grains with euhedral outline dominate and give the rock an idiomorphic granular texture. More commonly, nearly all grains are anhedral and the rock texture is allotriomorphic granular. Most rocks show an intermediate or hypidiomorphic granular texture (Fig. 2).

Porphyritic texture. The grain size of some igneous rocks is extremely uniform (equigranular texture), but that of others may be highly inequigranular. Rocks in which relatively large crystals (phenocrysts) are dispersed in a matrix or groundmass of finer-grained or glassy material are said to be porphyritic or phyric (Fig. 3). Porphyritic glasses with abundant phenocrysts are known specifically as vitrophyres.

Porphyritic rocks may form in a number of ways. (1) Phenocrysts may have grown early and slowly while the magma was deeply buried. The groundmass may have congealed later after the magma was erupted to higher levels where rapid cooling ensued. (2) Phenocrysts in many rocks (some granites) may develop late and still attain large dimensions if their growth rate is sufficiently greater than that of adjacent minerals. (3) The large crystals of some plutonic rocks are probably more properly classed as porphyroblasts. They may have formed essentially in solid rock by recrystallization aided by residual fluids from the solidifying magma. (4) Large crystals in many rocks (certain porphyries and lamprophyres) may not be indigenous. They may have been incorporated during intrusion of the magma. (5) Phenocrysts might develop by inoculation or by disturbance of supersaturated magma. *See* LAMPROPHYRE; PHENOCRYST; PORPHYRY.

Poikilitic texture. This texture involves numerous small grains of one mineral, in random orientation, enclosed by single large crystals of another (Fig. 4a). Conditions favoring development of poikilitic texture are not well understood. In some rocks this texture may have developed by direct crystallization of magma. In other rocks this texture may represent recrystallization of magmatic rocks.

Ophitic texture. This is a special type of poikilitic texture and is characteristic of the rock diabase (Fig. 4b). The texture involves lath-shaped crystals of plagioclase feldspar enclosed by large anhedral grains or plates of pyroxene (augite or pigeonite). If the length of the feldspar crystals exceeds that of the pyroxene, enclosure is only partial and the texture is called subophitic. *See* DIABASE.

Other textures, more or less related to ophitic, are characteristic of very fine-grained and glassy rocks of basaltic composition. *See* BASALT.

Implication or intergrown textures. These are formed by the mutual penetration of two or more mineral phases. The intergrowth may be so intimate that one phase appears disintegrated into smaller grains which are isolated by the other. Within small domains, however, grains of one phase show optical and crystallographic continuity.

Graphic or micrographic textures may develop between almost any mineral pair where one member, in cuneiform masses resembling runic inscriptions, is enclosed by the other (Fig. 5a). Micropegmatitic texture is essentially a micrographic texture involving only quartz and potash feldspar (Fig. 5b). If the intergrowth is more varied and involves plumose, fringing, radial, or micropegmatitic patterns, the texture is granophyric (Fig. 5c). In myrmekitic texture plagioclase (generally oligoclase) grains enclose vermicular quartz (Fig. 5d). Perthitic texture is extremely common in feldspars and takes on a wide variety of forms (Fig. 6). It usually consists of tiny masses of sodic plagioclase enclosed by potash feldspar. Various proportions of the two constituents may exist.

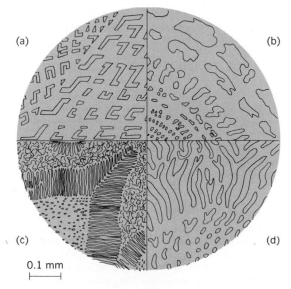

(a)   (b)

(c)   (d)

0.1 mm

Fig. 5. Implication or intergrown textures. (a) Micrographic texture. (b) Micropegmatitic texture. (c) Granophyric texture. (d) Myrmekitic texture.

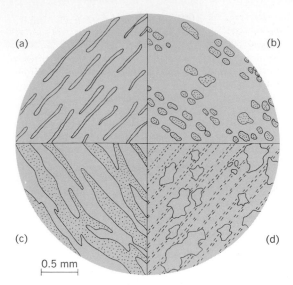

Fig. 6. Perthitic textures. (*a*) String perthite. (*b*) Patch perthite. (*c*) Vein perthite. (*d*) Antiperthite.

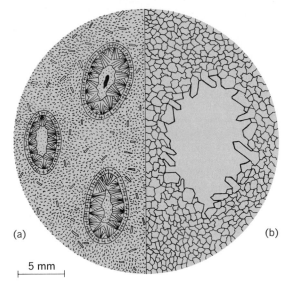

Fig. 7. (*a*) Amygdaloidal structure showing former gas cavities (bubbles) in lava filled with later minerals. (*b*) Miarolitic cavity in fine-grained granite with allotriomorphic granular texture. Euhedral outline is visible only where crystals form cavity boundary.

Where potash feldspar is more abundant and constitutes the host mineral, the material is known as perthite. Where plagioclase predominates, the material is called antiperthite.

Some implication textures may develop by simultaneous crystallization of two constituents. Others may form by exsolution in the solid state (some perthite). Still other textures may be due to the partial replacement of one mineral phase by another.

Structures. Structure as applied to igneous rocks is easily confused with texture. In general, however, structure refers to a geometrical form or architectural feature in a rock. Structure emphasizes the heterogeneous nature of a rock or mineral aggregate; texture emphasizes homogeneity. Certain large-scale structures, such as faults, folds,

IGNEOUS ROCKS

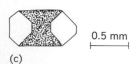

Fig. 8. (*a*) Zoned crystals of pyroxene. (*b*) Zoned plagioclass. (*c*) Hourglass structure as seen in pyroxene.

and joints, are common to most rock types. They are perhaps more properly classed as geologic structures. Like textures, the structures of igneous rocks may tell much about the history or conditions of formation of the rocks themselves.

Vesicular structures. These structures are common in many volcanic rocks. They form when magma is brought to or near the Earth's surface. Here the low pressure permits partial release and expansion of dissolved water or other volatiles and the formation of steam bubbles which may be preserved as small cavities when the magma congeals. In highly viscous lavas (rhyolitic) much gas may be trapped, but only tiny bubbles may form. Rapid cooling of this frothy liquid produces a pumiceous structure (characteristic of the rock pumice). In less viscous lavas (basaltic) integration of tiny bubbles produces a coarser, spongy, or scoriaceous structure (characteristic of the rock scoria). Vesiculation of some basaltic lavas produces well-formed, ellipsoidal cavities. These may later be filled with minerals (such as quartz, calcite, epidote, zeolite) deposited from fluids which permeated the rock. Such fillings are called amygdules in allusion to their almond shape. The structure is known as amygdaloidal (Fig. 7*a*).

Miarolitic openings. These are the most common cavernous structures found in plutonic rocks. They are irregular, range up to several inches across, and appear crusted with beautifully formed crystals (quartz and feldspar). These crystals are not truly encrusting a cavity wall; their bases constitute an integral part of the rock (Fig. 7*b*). This fact indicates that each cavity (vug) formed as a small interspace in the crystal aggregate and filled with residual magmatic fluid against which bounding grains readily developed their euhedral outline. The presence of muscovite, tourmaline, topaz, and apatite suggests that volatiles (water, fluorine, and

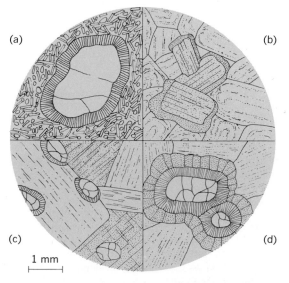

Fig. 9. Reaction rims. (*a*) Pyroxene around quartz grain in basalt. (*b*) Hornblende around three grains of orthopyroxene in norite. Most of rock is composed of plagioclase. (*c*) Rim of orthopyroxene formed between small olivine grains and plagioclase in diabase. No rim exists between olivine in norite. (*d*) Reaction rims (inner, orthopyroxene; outer, amphibole) around olivine in norite.

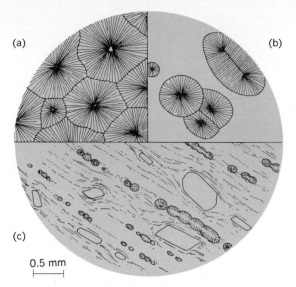

(a)

(b)

(c)

0.5 mm

Fig. 10. (a) Spherulitic structure. (b) Spherulites in volcanic glass. (c) Fluidal structure showing trains of spherulites, phenocrysts, microlites, and crystallites.

boron), which concentrate in residual magmatic fluids, have played an important role in the development of miarolitic structures. These cavities are most likely to form in rocks crystallizing at shallow depth where confining pressure is relatively low.

Zoned crystals. Crystals possessing zonal structures are common and appear to be built up of concentric shells or zones of different composition which follow the general crystal outline (Fig. 8). Though minute, these zonal structures are readily detected in thin sections under the microscope. Individual zones may be thick or thin, and zoned boundaries may be sharp or gradational. Compositional changes from the crystal's center outward may be great or slight; they may be progressive, reversed, interrupted, oscillatory, or repetitive. Zoning is characteristic of minerals belonging to solid solution series (for example, plagioclase, pyroxene, and amphibole).

Numerous theories have been proposed to explain various types of zoning in minerals. These are based on physicochemical principles relating to supersaturation of the magma and changes in composition, pressure, temperature, and volatile content of the magma. Movement of crystals from one part of the magma chamber to another may have been important. The most common type of zoning (progressive) appears because of incomplete reaction between solid and liquid phases during the crystallization of an isomorphous series.

Hourglass stucture. This structure, somewhat related to zoning, is most frequently displayed by crystals of pyroxene. Certain sections through a crystal possessing the structure have the appearance of an hourglass (Fig. 8c). This structure probably demonstrates the minute differences in energy involved at different faces of a growing crystal. It may be due to selective adsorption of ions by different faces during crystal growth.

Reaction rims. These rims or zones, in which one mineral envelopes another, are believed to have formed by reaction and are common in some

rocks (Fig. 9). They may develop by reaction between early formed crystals and surrounding magma (pyroxene rims on early formed olivine crystals). Reaction between two incompatible minerals, induced by residual fluids in the late stages of magma consolidation, may produce similar effects. Pyroxene or amphibole may form by reaction around olivine crystals where they would otherwise come in contact with plagioclase. Some petrologists refer to rims of primary origin as coronas and those of secondary origin as kelyphitic borders.

Spherulites. These are radial aggregates of needlelike crystals. They are roughly spherical and usually less than a centimeter across (Fig. 10). They abound in silica-rich lavas, particularly rhyolitic glass, and are composed principally of quartz, tridymite, and alkali feldspar.

Somewhat similar aggregates in basaltic rocks, called varioles, consist of radial plagioclase crystals with interstitial glass or granules of olivine or pyroxene.

Spherulites consisting of concentric shells with cavernous interspaces are known as lithophysae, or stone bubbles. In many, the tiny annular cavities are lined with delicate crystals of cristobalite, quartz, and feldspar.

Inclusions or enclosures. Inclusions are common in most varieties of igneous rocks. These masses of extraneous-looking material vary widely in size, shape, constitution, and origin. Inclusions demonstrated to be foreign rock fragments enclosed and trapped by congealing magma or lava may be specifically designated as xenoliths. Incorporated foreign crystals are known as xenocrysts. *See* XENOLITH.

If an earlier consolidated portion of a magma is ruptured and fragments of it become enclosed by portions which solidify later, the older rock bodies are known as autoliths. Enclosures formed by selective accretion of minerals, either during or after consolidation of a magma, are termed segregations.

Orbicular structures. These structures, found in some plutonic rocks (granite, granodiorite, and diorite), are orblike masses generally up to a few inches across. They show concentric shells of different mineral composition and thickness which may envelope xenolithic cores (Fig. 11). Most commonly dark mineral shells (rich in biotite, hornblende, or pyroxene) alternate with light shells (rich in feldspar). Individual shells may be sharply or vaguely defined, and the minerals within may be

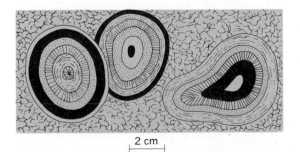

2 cm

Fig. 11. Orbicular structure.

granular or elongate and in radial or tangential arrangement.

Orbicular structures may develop by reaction (between xenoliths and magma) involving chemical reconstitution of the solid fragments or rhythmic crystallization around xenoliths. Many orbicular structures may represent products of metamorphism and metasomatism of solid rock. *See* META-MORPHIC ROCKS; METAMORPHISM; METASOMA-TISM.

Pillow or ellipsoidal structure. This type of structure is a peculiar feature of certain lava flows (basalt, spilite). Rocks exhibiting this structure appear to be composed of closely packed, pillow-shaped masses up to several feet across. Individual pillows have a very fine-grained crust or margin which carries abundant vesicles, commonly arranged concentrically with the pillow surface.

The pillows are so perfectly fitted together as to suggest that they were assembled in a plastic state. Relatively little matrix occurs; and it consists commonly of chert, limestone, or shale. The close association of pillow lavas with sedimentary rocks is in agreement with the popular belief that they are of subaqueous origin. Pillow lavas have been observed to form both on dry land and in water, but the precise conditions favoring their formation are still not well understood. *See* PRE-CAMBRIAN; SPILITE.

Flow structure. This is a nongenetic term for a number of directive features in rocks. The structure may be formed by flowage during crystallization of a magma (primary flow structure). Post-solidification flow (secondary) may develop similar features, but these are classed as metamorphic in origin.

The structure takes the form of parallel streaks or lenses of different minerals or textures; or it may result from parallel arrangement of elongate or platy minerals (mica, hornblende, or feldspar). Some flow structures consist of abundant slabby

inclusions or xenoliths in parallel orientation. Flowage may be expressed by flow lines (lineation) or flow layers (some foliation). These may be straight or contorted.

Fluidal structure and fluxion structure. These are genetic terms and specifically imply flowage of lava or magma (Fig. 10c).

Schlieren. Schlieren are irregular streaks, patches, or layers having more or less blended outlines and measuring up to many feet in length. They are generally composed of the same minerals as the enclosing rock but in different proportions. Schlieren may represent early segregation drawn out by magma flow. Some may be xenoliths more or less digested and reworked by magma. Others may represent residual magmatic liquors of different composition injected into already crystallized portions. Schlieren formed in solid rocks are more properly metamorphic or metasomatic features. *See* GRANITIZATION.

Banding. Banding is exhibited by rocks composed of alternating layers of different composition, texture, or color. The term is merely descriptive, not genetic. If flowage is to be implied, the term flow banding is used.

Classification. Schemes for classifying igneous rocks are numerous. Prior to the advent of the polarizing microscope (roughly 1870), rock classifications were based on megascopic characteristics, many of them misleading. These systems were gradually improved as chemical analyses were more commonly employed.

Today three principal methods of classification are used. (1) Megascopic schemes are based on the appearance of the rock-in-hand specimen or as seen with a magnifying glass (hand lens). Such schemes are useful in the field study of rocks. (2) Microscopic schemes (largely mineralogical) are employed in laboratory investigations where more detailed information is needed. (3) Chemical schemes are very useful but have more limited

Table 1. Classification of igneous rock families*

	Feldspar	Quartz	Quartz or feldspathoid (5%)	Nepheline	Leucite
Alkali feldspar and plagioclase (oligoclase, andesine)	Alkali feldspar 50% of the feldspar	Granite (P) Rhyolite (A)	Syenite (P) Trachyte (A)	Nepheline syenite (P) Phonolite (A)	Leucite phonolite (A)
	Alkali feldspar 5–50% of the feldspar	Granodiorite (P) Quartz latite (A)	Monzonite (P) Latite (A)		
	Alkali feldspar 5% of the feldspar	Quartz diorite (P) (tonalite) Dacite (A)	Diorite (P) Andesite (A)		
			──── Anorthosite (P) ────		
Labradorite-anorthite			Gabbro (P) Basalt (A)	Theralite (P) Tephrite (A)	Leucite tephrite (A)
			Diabase (H)	Basanite (A)	Leucite basanite (A)
			Periodotite (P)		
Plagioclase < 10%			Pyroxenite (P) Hornblendite (P)	Nephelinite (A) Nepheline basalt (A)	Leucitite (A) Leucite basalt (A)

*In this table P indicates phaneritic (plutonic) rock; A, aphanitic (volcanic) rock; and H, hypabyssal rock.

application. The mineral content and texture of a rock generally tell much more about the rock's origin than does a bulk chemical analysis. For example, granite, quartz porphyry, rhyolite, and obsidian may all have the same chemical composition; but the geologic conditions under which each forms may be very different. Granite solidifies slowly at depth and under high pressure. The porphyry may crystallize in two stages, one at depth and a later one nearer the surface. The other two rocks are of surficial origin; the obsidian solidifies most rapidly and as glass.

Igneous rocks show great variations chemically, mineralogically, texturally, and structurally with few if any natural boundaries. This accounts in large part for the great disagreement among petrologists as to how igneous rocks should be classified. The following subsections discuss plutonic, volcanic, and hypabyssal types.

Plutonic rocks. Plutonic rocks occur in large intrusive masses (batholiths, stocks, and other large plutons). They form at great depth and, therefore, are often referred to as abyssal rocks. Generated from large bodies of magma which has cooled slowly, they characteristically show a phaneritic, holocrystalline texture.

Under deep-seated conditions where confining pressure is high, volatiles dissolved in the magma are retained until the last stage of crystallization. These act as fluxes and reduce the temperature of crystallization. Consequently plutonic rocks, as compared with volcanic rocks, may carry relatively low-temperature mineral phases.

Volcanic rocks. These are formed as lava flows or as pyroclastic rocks (heterogeneous accumulations of volcanic ash and coarser fragmental matter). They have solidified rapidly to develop an aphanitic texture with more or less glass. Volatiles are readily lost as the lava reaches the Earth's surface. Therefore, crystallization tends to proceed within a relatively high-temperature range, so high-temperature minerals such as sanidine and high-temperature plagioclase are characteristic. Expanding gas bubbles formed by escaping volatiles frequently create highly porous rocks.

Volcanic rocks frequently show evidence of two stages of cooling, an early, deep-seated stage (intratelluric) and a later, effusive stage. Slow cooling in the first stage may produce a few large crystals. These become suspended in the lava and frozen into an aphanitic matrix during the effusive stage. This accounts for the porphyritic texture so commonly encountered in volcanic rocks.

Hypabyssal rocks. These rocks exhibit characteristics more or less intermediate between those of volcanic and plutonic types. They differ from volcanic rocks in that they are intrusive and generally free from glass and vesicular structures. They differ from plutonic rocks in that they occur in small bodies (dikes and sills) or in larger bodies formed at shallow depths (laccoliths) and they have textures characteristically resulting from more rapid cooling. Hypabyssal rocks cannot be sharply distinguished from volcanic rocks on the one hand and plutonic rocks on the other. The recognition of such a group, therefore, is perhaps of greater value in field studies, where mode of occurrence is known, than in the laboratory.

The rocks of each subdivision (plutonic, volcanic, and hypabyssal) may be further divided into families (groups with the same or closely allied composition and relatively limited textural variation). Families are sometimes grouped into rock clans. A clan includes all families with the same chemical composition. Thus, the gabbro family (plutonic), diabase family (hypabyssal), and basalt family (volcanic) have the same chemical composition and, therefore, belong to the same clan (gabbro clan). The clan name is derived from the plutonic family member.

The principal families of the phaneritic (plutonic) and aphanitic (volcanic) rocks are shown in Table 1. This scheme expresses something of the bulk chemical composition of the rock and the conditions of formation. For discussion of rock types, see articles under specific names.

More specific rock types may be indicated by prefixing some pertinent mineral, textural, or structural term to the appropriate family name (such as biotite granite, graphic granite, orbicular granite). *See* PEGMATITE.

Chemical composition. By averaging a large number of chemical rock analyses, one may derive a representative composition for each of the rock families. Average analyses for the more common igneous rocks are shown in Table 2. Such average values are very useful standards for comparison in spite of the fact that the variations within families may be greater than those between families.

Mineral composition. Igneous rock-forming minerals may be classed as primary or secondary. The primary minerals are those formed by direct crystallization from the magma. Secondary minerals may form at any subsequent time.

Essential primary minerals. The principal primary minerals are relatively few and may be classed as light-colored (felsic) or dark-colored (mafic) varieties. Felsic is a mnemonic term for feldspathic minerals (feldspar and feldspathoids) and silica (quartz, tridymite, and cristobalite). Mafic is mnemonic for magnesium and iron-rich minerals (biotite, amphibole, pyroxene, and olivine). Felsic minerals are composed largely of silica, alumina, and alkalies. Mafics are rich in iron, magnesium, and calcium.

Table 3 summarizes the essential primary constituents of the more common plutonic rocks. The percentage ranges are highly generalized. Individual rock specimens may depart radically from these values, but the averages are fairly representative and useful for comparison. The mineral composition of the corresponding volcanic rocks is roughly similar to the values in the table. Major departures will be encountered particularly in the glassy rocks.

Accessory minerals. Accessory minerals are those occurring in very small or trace amounts. They consist principally of magnetite, ilmenite, pyrite, hematite, apatite, zircon, rutile, and sphene. Most generally these are widely distributed as tiny grains or crystals.

Secondary minerals. Included in this group are minerals formed by addition of material subsequent to solidification of the rock or by alteration of minerals already present in the rock. The addition of fluorine and boron, which tend to concen-

Table 2. Average chemical compositions of igneous rocks (totals reduced to 100%)*

Components	Plutonic rocks							
	Granite	Grano-diorite	Quartz diorite	Syenite	Monzo-nite	Diorite	Gabbro	Nepheline syenite
SiO_2	70.18	65.01	61.59	60.19	56.12	56.77	48.24	54.63
TiO_2	0.39	0.57	0.66	0.67	1.10	0.84	0.97	0.86
Al_2O_3	14.47	15.94	16.21	16.28	16.96	16.67	17.88	19.89
Fe_2O_3	1.57	1.74	2.54	2.74	2.93	3.16	3.16	3.37
FeO	1.78	2.65	3.77	3.28	4.01	4.40	5.95	2.20
MnO	0.12	0.07	0.10	0.14	0.16	0.13	0.13	0.35
MgO	0.88	1.91	2.80	2.49	3.27	4.17	7.51	0.87
CaO	1.99	4.42	5.38	4.30	6.50	6.74	10.99	2.51
Na_2O	3.48	3.70	3.37	3.98	3.67	3.39	2.55	8.26
K_2O	4.11	2.75	2.10	4.49	3.76	2.12	0.89	5.46
H_2O	0.84	1.04	1.22	1.16	1.05	1.36	1.45	1.35
P_2O_5	0.19	0.20	0.26	0.28	0.47	0.25	0.28	0.25

Components	Aphanitic rocks							
	Rhyolite	Quartz latite	Dacite	Trachyte	Latite	Andesite	Basalt	Phonolite
SiO_2	72.80	62.43	65.68	60.68	57.65	59.59	49.06	57.45
TiO_2	0.33	0.85	0.57	0.38	1.00	0.77	1.36	0.41
Al_2O_3	13.49	16.15	16.25	17.74	16.68	17.31	15.70	20.60
Fe_2O_3	1.45	4.04	2.38	2.64	2.29	3.33	5.38	2.35
FeO	0.88	1.20	1.90	2.62	4.07	3.13	6.37	1.03
MnO	0.08	0.09	0.06	0.06	0.10	0.18	0.31	0.13
MgO	0.38	1.74	1.41	1.12	3.22	2.75	6.17	0.30
CaO	1.20	4.24	3.46	3.09	5.74	5.80	8.95	1.50
Na_2O	3.38	3.34	3.97	4.43	3.59	3.58	3.11	8.84
K_2O	4.46	3.75	2.67	5.74	4.39	2.04	1.52	5.23
H_2O	1.47	1.90	1.50	1.26	0.91	1.26	1.62	2.04
P_2O_5	0.08	0.27	0.15	0.24	0.36	0.26	0.45	0.12

*After Daly.

Table 3. Approximate mineral composition of the common plutonic rocks

Rock	Felsic minerals, %	Total felsic, %	Mafic minerals	%
Granite	Potassium feldspar, 35–45 Sodic plagioclase, 20–30 Quartz, 20–30	80–95	Biotite, hornblende	5–20
Granodiorite	Potassium feldspar, 15–25 Sodic plagioclase, 35–45 Quartz, 15–25	75–90	Biotite, hornblende	10–25
Quartz diorite	Oligoclase, andesine, 55–65 Quartz, 15–25 Potassium feldspar, 0–5	70–85	Hornblende, biotite, pyroxene	15–30
Syenite	Potassium feldspar, 60–70 Sodic plagioclase, 10–20 Quartz or nepheline, 0–5	70–90	Biotite, hornblende, pyroxene	10–30
Monzonite	Potassium feldspar, 20–30 Sodic plagioclase, 45–55 Quartz or nepheline, 0–5	65–85	Biotite, hornblende, pyroxene	15–35
Diorite	Oligoclase, andesine, 60–70 Potassium feldspar, 0–5 Quartz or nepheline, 0–5	60–80	Hornblende, biotite, pyroxene	20–40
Gabbro	Labradorite, bytownite, 45–70 Potassium feldspar, 0–5 Quartz or nepheline, 0–5	45–75	Pyroxene, olivine, hornblende, biotite	25–55

trate in the residual magmatic liquids, to already crystallized portions of the rock may form small crystals of fluorite, topaz, or tourmaline. Alteration in which certain minerals become more or less reconstituted is common and widespread. It is generally believed to occur during the last stages of solidification while hot, residual fluids (for exam-ple, water and carbon dioxide) permeate the crystal aggregate and convert water-free silicate minerals into hydrous forms. This hydrothermal or deuteric effect may be so intense that virtually all igneous characteristics of the rock are lost. The common alteration products derived from the essential primary minerals are listed as follows.

Primary mineral	Secondary mineral
Quartz	Not altered
Potash feldspar	Kaolinite, sericite
Plagioclase	Kaolinite, sericite (paragonite), epidote, zoisite, calcite
Nepheline	Cancrinite, analcite, natrolite
Leucite	Nepheline and potash feldspar
Sodalite	Analcite, cancrinite
Biotite	Chlorite, sphene, epidote, rutile, iron oxide
Hornblende	Actinolite, biotite, chlorite, epidote, calcite
Orthopyroxene	Antigorite, actinolite, talc
Clinopyroxene	Hornblende, actinolite, biotite, chlorite, epidote, antigorite
Olivine	Serpentine, magnetite, talc, magnesite

Density. Density is a significant rock property and is a function largely of mineralogical composition and porosity. Chemical composition alone is not a reliable indication of density because different minerals (with different densities) may form from a single bulk composition.

Table 4 gives the approximate average and common range of density for the more abundant plutonic rocks. Densities of volcanic equivalents

Table 4. Approximate densities of common plutonic rocks

Rock	Average	Common range
Granite	2.67	2.60–2.73
Granodiorite	2.71	2.65–2.77
Syenite	2.76	2.65–2.85
Quartz diorite	2.82	2.72–2.92
Diorite	2.85	2.75–2.95
Gabbro	2.99	2.85–3.15

are generally slightly lower due to higher porosity and greater amount of glass. Highly porous volcanic rocks (pumice and scoria) may be so vesicular as to float on water. The density of completely glassy rocks is approximately 6% less than that of the corresponding holocrystalline (entirely crystalline) type.   [CARLETON A. CHAPMAN]

Bibliography: B. Bayly, *Introduction to Petrology*, 1968; E. W. Heinrich, *Microscopic Petrography*, 1956; W. W. Moorhaus, *The Study of Rocks in Thin Sections*, 1959; H. H. Read and J. Watson, *Introduction to Geology*, 1962; L. E. Spock, *Guide to the Study of Rocks*, 1962.

Ignition system

The ignition system of an internal combustion engine initiates the chemical reaction between fuel and air in the cylinder charge. For maximum efficiency it is necessary to ignite the charge in each cylinder shortly before the piston reaches the top of its compression stroke. The best timing is determined experimentally for various engine speeds and loadings, and the ignition system is designed to provide this timing automatically. For smooth running, multicylinder engines are usually built so that the various pistons arrive at their firing top center positions in evenly spaced intervals. The sequence in which the cylinders reach their firing points depends upon the geometrical arrangement of the cylinders and crankshaft, and is

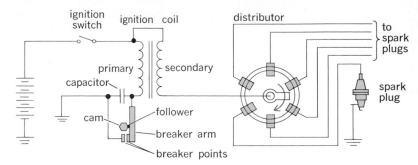

Fig. 1.   Diagram of battery ignition system. (*Modified from A. R. Rogowski, Elements of Internal-combustion Engines, McGraw-Hill, 1953*)

called the firing order of the engine. The ignition system distributes the ignition impulse to the cylinders in this order.

A large class of engines, including most automobile engines, operate on a premixed charge of fuel vapor and air. In these engines the charge is ignited by passing a high-voltage electric current between two electrodes in the combustion chamber. The electrodes are incorporated in a removable unit. When a spark of sufficient energy jumps the gap between the electrodes, a self-propagating flame is produced in the fuel-air mixture which spreads rapidly throughout the charge. *See* DISTRIBUTOR; SPARK PLUG.

Battery system. The electric energy for the spark is obtained from a storage battery in practically all spark-ignition engines. The battery, usually 6 or 12 volts, is part of a low-voltage primary circuit, which includes a switch, the primary winding of the ignition coil, a capacitor (or condenser), and breaker points (Fig. 1). When the switch and breaker points are closed, current flows from the battery through the primary winding, through the points, and back to the battery through the engine frame or other ground connection. The coil is wound about a soft iron core. The current in the primary winding induces a magnetic field in and around this core. When the breaker points open as the engine-driven cam rotates, the current, which had been passing through the points, now flows into the capacitor. As the capacitor becomes charged, the primary current falls rapidly, and the magnetic field collapses.

The secondary winding consists of many turns of fine wire wound on the same core with the primary winding. The rapid collapse of the magnetic field in the core induces a very high voltage (10,000 to 20,000 volts) in the secondary winding.

Without the capacitor the primary voltage caused by the rapid collapse of the magnetic field around the primary winding would cause an arc across the breaker points. The arc would burn the points and soon destroy them. The capacitor also assists the rapid drop in the primary current and collapse of the magnetic field, both of which are needed to produce the high secondary voltage.

The secondary voltage is led to the spark plugs in proper sequence by a rotary switch called the distributor, which is driven by the same shaft that drives the breaker point cam. From the head of the distributor, well-insulated wires carry the secondary voltage to the central electrodes of the spark plugs. The discharge which takes place between

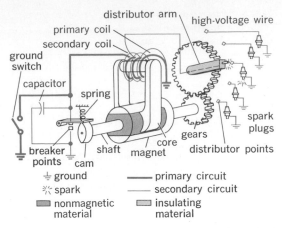

Fig. 2. Diagram of magneto ignition system. (*From C. H. Chatfield, C. F. Taylor, and S. Ober, The Airplane and Its Engine, 5th ed., McGraw-Hill, 1949*)

the central electrode and the grounded electrode inside the combustion chamber ignites the cylinder charge. The timing of the ignition spark is controlled by the opening of the breaker points. The ignition timing may be varied by rotating the plate upon which the breaker points are mounted relative to the cam. Timing may also be varied by changing the angular relationship between the breaker cam and the shaft that drives it.

Failures. Ignition system failures are usually due to breaker points which have become overheated by the primary current or to spark plugs which are fouled by carbon or other electrically conducting cylinder deposits. Ignition systems are being considered in which a solid-state switching device, called a transistor, is used in place of the breaker points. The transistor is capable of interrupting the primary current without wear or burning. Breaker points are still used but only to control the transistor, and the control current is very small. Another type of ignition system that has been found superior in certain applications where plug fouling is common consists of circuitry which charges a capacitor to several hundred volts. When an ignition spark is required, the capacitor is discharged into the primary of the ignition coil. A transistor, controlled by breaker points or other triggering devices, is used to make the connection between the capacitor and the coil. The sudden application of high voltage to the primary coil produces a much more rapid rise in secondary voltage than is obtainable with the conventional ignition system. Because of the rapid voltage rise, the spark jumps the spark plug gap before the electric charge has had time to leak away through the conducting carbon deposits on the plug. *See* TRANSISTOR.

Magneto system. The magneto system is similar to the battery system except that the voltage required to cause a flow of current in the primary winding is generated by the rotation of a set of permanent magnets, instead of being supplied by a battery. The magneto assembly normally includes, within one housing, the rotating magnets, ignition coil windings, iron core, breaker points, cam, capacitor, and distributor (Fig. 2).

Rotation of the magnet assembly completely reverses the direction of the magnetic flux in the

soft iron core about which the primary and secondary coils are wound. This rapid change in flux intensity and direction drives a current through the primary circuit. When the primary current reaches a high value and the contribution to the total flux is principally from the primary current, and not from the permanent magnets, the breaker points open with the same result as in the battery system described. The magneto is often used with small compact installations such as lawnmowers, chain saws, and small outboard engines, where a battery would represent undersirable weight and bulk.

Compression ignition. In the diesel or compression ignition engine, the inducted cylinder air is heated sufficiently by the compression stroke of the piston so that ignition takes place shortly after the fuel is sprayed into the cylinder. Ignition timing is thus controlled by the phasing of the fuel injection pump, while the firing order of the cylinders is established by the geometry of the pump. In compression-ignition engines, the fuel burns as soon after ignition as it is able to find air with which to combine, so that less attention need be paid to automatic injection timing control. When control is used, pump shaft position relative to engine shaft can be changed automatically by centrifugal weights and a mechanical linkage. *See* COMBUSTION CHAMBER.

Sometimes in small engines a premixed charge of fuel and air is ignited near the end of the compression stroke by the temperature of compression. Under these conditions the entire charge, once ignited, burns rapidly, since all parts of the charge are equally heated and prepared for combustion. The result is rough, pounding combustion due to the high reaction rate. No precise way exists to control the exact timing of ignition with this method, so that the combustion may occur too early or too late depending upon such factors as the temperature of the cylinder and the kind of fuel being used; efficiency with such ignition is likely to be poor.   [A. R. ROGOWSKI]

Bibliography: T. Baumeister (ed.), *Standard Handbook for Mechanical Engineers*, 7th ed., 1967; J. Carroll and D. Fink, *Standard Handbook for Electrical Engineers*, 10th ed., 1968; W. H. Crouse, *Automotive Mechanics*, 4th ed., 1960; M. Tepper, *Transistor Ignition Systems*, 1965; B. Ward, Jr., *Transistor Ignition Systems Handbook*, 1963.

Ignitron

A single-anode mercury-pool gas tube operating in the arc region. An ignitor is employed to initiate the arc cathode spot before each conducting period. *See* GAS TUBE.

While many tubes of this type are still in service, only limited numbers are now manufactured because they have been supplemented by semiconductor rectifiers of either the diode or silicon-controlled-rectifier (SCR and thyristor) type. *See* CONTROLLED RECTIFIER.

Principle. The ignitor is a stationary electrode of semiconducting material, such as boron carbide, which extends into the mercury pool. A current pulse through the ignitor starts the arc at a desired time in each cycle of the ac anode voltage by forming a cathode spot on the mercury surface. Either a thyratron tube or a semiconductor rectifier may be

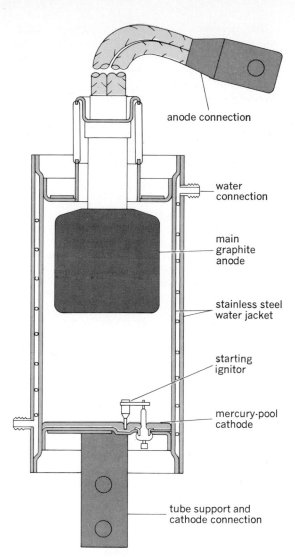

Fig. 1. Diagram of a typical sealed ignitron for welding service. (*General Electric Co.*)

tion at rated loads. The combination of a mercury-pool cathode and short anode-to-cathode spacing makes it possible for the excess vapor generated by the overload current to fill the discharge space just when needed and then condense on the walls in time for the next cycle of normal operation.

Types. The two classes of ignitrons especially designed for different kinds of service are the welder and the rectifier ignitrons.

Welder ignitrons. These tubes are used in pairs which, connected inverse-parallel, operate as ac contactors and control the primary current supplied to resistance welding transformers.

Welding ignitrons are of relatively open construction with little shielding, and are designed specifically to carry the high currents encountered in resistance welding. In welding service the anode current wave is sinusoidal, and the current decreases to zero so slowly that the residual vapor blast and ionization present at the beginning of the inverse period are low enough that they do not cause arc-back. Furthermore, only the tube that conducts last is subjected to inverse voltage immediately following conduction. These conditions permit the use of tubes without splash baffles, deionizing baffles, and grids. The construction of a typical sealed ignitron for welding service is shown in Fig. 1.

used to deliver the firing impulse to the ignitor.

The tube is nonconducting, regardless of anode potential, until a source of electrons is supplied to initiate electrical breakdown of the gas within the tube. The cathode spot formed by the ignitor supplies the needed electrons to create an electron avalanche, which terminates as an arc. Once the arc has been formed, the cathode spot is maintained by the arc current and the action of the ignitor is terminated.

The ignitron remains conducting until the anode potential is removed, and for some microseconds thereafter until all the ions have become deionized.

The repetitive starting technique employed with ignitrons makes possible a short anode-to-cathode spacing, resulting in low voltage drop and high efficiency. In addition, single-anode tubes like the ignitron and the excitron are unique among gas-filled tubes in their ability to carry high overload and short-circuit currents without vapor starvation surges. This results from the use of the excess vapor pressure generated by the overload. The normal vapor pressure is controlled, usually by water cooling, at a value sufficient for best opera-

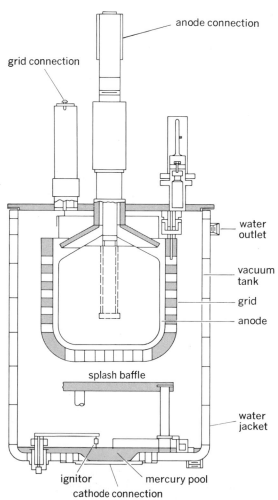

Fig. 2. Diagram of a single-grid, pumpless rectifier ignitron. (*General Electric Co.*)

IGNITRON

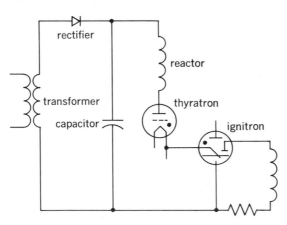

Fig. 3. Elementary anode firing circuit.

Rectifier ignitrons. These ignitrons are generally used in sets of six to convert power from ac to dc. The smaller tubes are of sealed construction. The larger units are built in tanks which are either vacuum-sealed or continuously vacuum-pumped.

Rectifier ignitrons are equipped with baffles in the space between the anode and cathode to prevent the vapor blast from throwing mercury drops against the face of the anode and to help deionize the mercury vapor. In the larger sizes, the anode is surrounded by one or more grids to increase the rate of deionization of the space adjacent to the anode and to provide additional control of the start of anode conduction. The construction of a typical high-current ignitron for rectifier service is shown in Fig. 2.

Firing circuits. Ignitors require a pulse of current of 5–50 amp in the forward direction from the ignitor to the mercury to start a cathode spot. Since the ignitor resistance varies from a few to several hundred ohms, a positive voltage of several hundred volts must be impressed on it.

The voltage and current applied to the ignitor in the reverse direction must be limited to low values to avoid damage. Special circuits have been devised to deliver the firing pulse to the ignitor and provide control of the time of firing.

Anode firing circuit. The power circuit voltage, which appears between anode and cathode just before conduction, is used to fire the ignitor in the circuit shown in Fig. 3. This circuit is widely used in ignitron contactor (resistance welding) service.

Capacitor firing circuit. A capacitor that has been charged from an auxiliary ac power source is discharged through the ignitor at the desired moment in the cycle by a thyratron tube in the circuit shown in Fig. 4.

Fig. 4. Elementary capacitator-firing circuit.

Magnetic firing circuit. Many ignitron rectifiers are equipped with magnetic firing to obtain longer life and greater reliability than is possible with thyratron firing tubes. In these circuits the capacitor is discharged through the ignitor by saturating a reactor, which is connected in series with the capacitor.

Ignitron tubes and tanks are built in sizes ranging from 2 to 20 in. in diameter. Ratings range from 25 to 1000 amp average per tube at voltages from 250 to 20,000 volts and higher.

[CARL C. HERSKIND]

Bibliography: W. G. Dow, *Fundamentals of Engineering Electronics*, 1952; A. Schure, *Gas Tubes*, 1958; J. Slepian and L. R. Ludwig, A new method of initiating the cathode of an arc, *Trans. AIEE*, 52(2):693–698, 1933.

Iguanid

A member of the reptilian family Iguanidae, in the order Squamata. The 400 species, found in the New World only, are adapted to various modes of life: terrestrial, arboreal, and fossorial (burrowing); some species are semiaquatic. They range in length from 4 in. to 7 ft. They are mainly insectivorous, although there are herbivorous and omnivorous forms. The teeth are fixed to the inner edge of the jaws; the fleshy tongue is nonretractile; the body is compressed; the tail is long but rarely prehensile; the limbs have five clawed toes; and many species have dorsal or caudal spines, lappets, and crests. *See* DENTITION (VERTEBRATE).

Many iguanids have the ability to change color and are commonly but erroneously referred to as chameleons. One such species is the green anole (*Anolis carolinensis*), which is found from North Carolina to Texas. Anoles are arboreal lizards that feed on insects, arachnids, and other similar invertebrates. The flattened digits are adhesive (by the mechanism of microscopic suction cups), and thus aid in climbing; the prehensile outer toe is opposable to the other digits. Males have a throat sac that may be inflated when the lizard is excited.

Coloration depends on brown and green pigment granules found in chromatophores distributed in the skin. Contraction of the chromatophores results in a green color, while the reverse activity results in a grayish to brown color. Color changes

Fig. 1. Common iguana (*Iguana iguana*) has characteristic dewlap under the throat and a dorsal crest of soft spines; maximum length is 6 ft.

Fig. 2. Horned toad (*Phrynosoma*), a lizard, uses its tongue to catch insects while on the run.

are dependent on various environmental conditions, such as Sun intensity and temperature: The animal is usually brown with extremes of temperature and a brilliant green in moderate temperatures. In combat, both adversaries are grayish; when combat terminates, the loser becomes dark yellow and the victor a bright green. *See* CHAMELEON; CHROMATOPHORE.

The basilisk (*Basiliscus basiliscus*) is a semiaquatic species found along riverbanks in the lowlands of Mexico and Central America. It is agile when in the branches of trees and bushes and can run over the surface of water on its hindlegs for short distances. Its food consists exclusively of insects and plants.

Iguana. There are two species of true iguanas, the common iguana (*Iguana iguana*) (Fig. 1) and *I. delicatessima*. Both are herbivorous, arboreal species found in tropical forest areas of South America. When alarmed, they readily take to water. They leave their burrows, which are hollows in the riverbanks, only to feed or to deposit their eggs in holes at tree bases.

The marine iguana (*Amblyrhynchus cristatus*) and the Galapagos land iguana (*Conolophus subcristatus*) are found only on the Galapagos Islands. The marine iguana, the only marine lizard, is about $4\frac{1}{2}$ ft long and weighs about 22 lb. It lives by the sea and is a vegetarian, feeding on seaweed and other marine plants. The Galapagos land iguana is smaller (up to 4 ft long) and aggressive. It burrows in the sand away from the sea and feeds primarily on plants and grasshoppers. Both species appear to be in danger of extinction as a result of predation of cats and dogs that have been released on the island. *See* ISLAND FAUNAS AND FLORAS.

Horned toads. There are 15 species of lizards in the desert areas of southwestern North America that belong to the genus *Phrynosoma*. All are reptiles but resemble toads except for being less bulky (Fig. 2). When at rest they bury themselves in the sand and are protected by their coloration, which resembles that of the terrain. When excited, these animals apparently have a rapid rise in blood pressure which causes the capillaries of the conjunctiva of the eye to rupture and tears of blood to be shot at an enemy over a distance of several feet. One species is adapted to low temperatures and is found in the high plateaus of Mexico up to altitudes of 10,000 ft. It is ovoviviparous, giving birth to young that have almost completed their development. *See* PROTECTIVE COLORATION; REPTILIA; SQUAMATA.   [CHARLES B. CURTIN]

Ileitis

Regional ileitis is a sharply demarcated inflammation of the lower portion of the small intestine. Actually, other portions of the small and large intestine may be involved, so regional enteritis or regional enterocolitis may be preferred terms.

The disease is marked by intestinal obstruction, pain, cramps, diarrhea, and constipation. There is often associated weight loss or anemia. Late in the disease, after a long but often vague history, the bowel appears thickened and firm and has a nodular mucosal lining which shows variable ulceration and granulomatous inflammation. Adjacent segments may show little or no change.

The cause is unknown, but faulty protein absorption, psychic factors, and abnormal lymph drainage have been implicated by various investigators. Although regional ileitis is found most often in young adults, no age group is immune.

The clinical course is typically erratic but progressive in most individuals. Neither medical nor surgical treatment offers a guaranteed cure because extension to previously unaffected bowel segments may occur.

In contrast to regional ileitis, ordinary ileitis is any inflammation of the ileum from specific causes such as trauma or infections such as tuberculosis, typhoid fever, and dysentery. *See* INTESTINE (VERTEBRATE); INTESTINE DISORDERS.

[EDWARD G. STUART/N. KARLE MOTTET]

Illite

The term illite is not a specific clay-mineral name, but is a general term for the mica-type clay minerals. It is commonly used for any nonexpanding clay mineral with a 10-A *c*-axis spacing. *See* CLAY MINERALS.

Illite clays are used for the manufacture of structural clay products, such as brick and tile. Some degraded high plastic illites are used for bonding molding sands. *See* CLAY, COMMERCIAL; CLAY PRODUCTS, ARCHITECTURAL.

Structural characteristics. By definition, all illites have the mica-type structure. The basic structural unit is composed of two silica tetrahedral sheets with a central octahedral sheet. This unit is essentially the same as that of montmorillonite, except that there is always some replacement (15%±) of silicon by aluminum in the tetrahedral sheets. This substitution results in a charge deficiency which is balanced by potassium ions between the illite unit layers. The stacking of adjacent illite layers is such that the potassium just fits and is always surrounded by a total of 12 oxygens, thus balancing the structural charge and binding the layers together without the possibility of expansion (see illustration). Illites may be either dioctahedral or trioctahedral, depending on the population of possible octahedral positions. They differ from well-crystallized micas in that they exhibit less substitution of aluminum for silicon, resulting in a higher silicon-to-aluminum molecular ratio.

Nonstructural characteristics. The size of naturally occurring illite particles is very small, yet they are larger and thicker than montmorillonite particles and have better-defined edges.

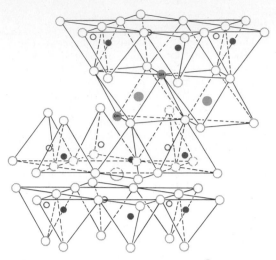

○ oxygens ● hydroxyls ● aluminum ○ potassium

○ ● silicons (one-fourth replaced by aluminums)

Diagrammatic sketch of the structure of illite. *(From R. E. Grim, Clay Mineralogy, McGraw-Hill, 1953)*

The illites have a moderately low cation-exchange capacity (20–30 meq/100 g). It is primarily due to broken bonds, but lattice substitutions may also be a cause in poorly crystallized varieties. The rate of the exchange reaction is likely to be slow, since a small part of the exchange occurs through partial replacement of the firmly held interlayer potassium ions. Dehydration curves for illites show the presence of a small amount of interlayer water which is lost below 100°C. The OH lattice water is lost between 300 and 600°C. Following the loss of structure between 850 and 950°C, spinel and mullite phases form prior to fusion at about 1400°C.

Occurrence. Illite is a common product of weathering if potash is present in the environment of alteration. It is a frequent constituent in many soil types, and may form in soils under certain conditions as a consequence of the addition of potash fertilizers. This is possible because illite has the ability to "fix" potassium. Much illite in soils loses its potassium through leaching and becomes degraded. On the addition of potassium to such material, K+ goes back into these normal positions and the illite is rebuilt. *See* SOIL CHEMISTRY.

Illite is common in recent sediments and is particularly abundant in deep-sea clays. It probably forms from montmorillonite and other minerals during marine diagenesis. Because of its stability, illite is often found in ancient sediments. It is the dominant clay mineral in many, probably most, shales that have been studied. *See* DIAGENESIS; MARINE SEDIMENTS; SHALE.

[FLOYD M. WAHL; RALPH E. GRIM]

Illuminance

A term expressing the density of luminous flux incident on a surface. This word has been proposed by the Colorimetry Committee of the Optical Society of America to replace the term illumination. The definitions are the same. The symbol of illumination is E, and the equation is $E = dF/dA$, where A is the area of the illuminated surface and F is the luminous flux. *See* ILLUMINATION; PHOTOMETRY.

[RUSSELL C. PUTNAM]

Illumination

In a general sense, the science of the application of lighting. Radiation in the range of wavelengths of 0.38–0.76 micron (μ) produces the visual effect commonly called light by the response of the average human eye for normal brilliance levels. Illumination engineering pertains to the sources of lighting and the design of lighting systems which distribute light to effect a comfortable and effective environment for seeing.

In a specific quantitative sense, illumination (sometimes called illuminance) is the combination of the spatial density of radiant power received at a surface and the effectiveness of that radiation in producing a visual effect. *See* ILLUMINANCE.

Luminosity curve of human eye. A standard response of the human eye to radiation at normal brilliance levels was established by international agreement in 1924 by La Commission Internationale de l'Eclairage through the function known as the CIE values of visibility. (The letters ICI for the English name are also used.) The ordinates of this function are shown plotted against the wavelength of radiation in Fig. 1 and are tabulated in the table. The ordinates are commonly called relative luminosity factors. *See* LUMINOSITY FACTOR.

At threshold levels of seeing the spectral response function of the eye is shifted toward lower wavelengths. This effect is known as the Purkinje effect.

Photometric definition. Illumination (or illuminance) is the density of radiant power incident upon a receiving surface evaluated throughout the spectrum in terms of the ICI values of visibility. For a line-spectrum component of irradiation, a simple multiplication of the irradiation component G_i and the value of relative luminosity factor v_i taken at the proper value of wavelength i yields the component illumination. For continuous-spectrum irradiation G_λ is a function of wavelength; therefore, the product of the two functions v_λ and G_λ must be integrated. Equation (1) states mathemati-

$$E = k \sum_1^n v_i G_i + k \int_0^\infty v_\lambda G_\lambda \, d\lambda \qquad (1)$$

cally a combination of line spectra and a continuous-spectrum source. Here G_i is the irradiation due to the ith component of the line spectra, v_i is the relative luminosity factor at a wavelength corresponding to the wavelength of the ith component irradiation, n is the number of component line-spectra radiations, G_λ is the spectral ir-

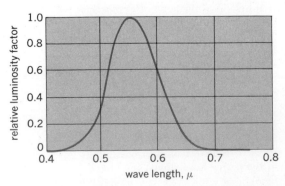

Fig. 1.   Human-eye luminosity curve. *(From W. B. Boast, Illumination Engineering, 2d ed., McGraw-Hill, 1953)*

6

ILLUMINATION 25

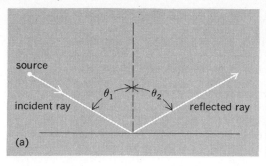

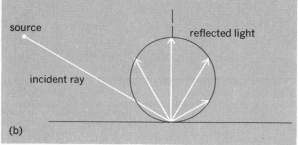

Fig. 2. Ideal reflection. (*a*) Specular, such as from a polished surface. (*b*) Diffuse, such as from a mat surface of microscopic roughness. (*From W. B. Boast, Illumination Engineering, 2d ed., McGraw-Hill, 1953*)

radiation of the continuous spectrum at λ wavelength, v_λ is the relative luminosity factor at the corresponding wavelengths, and k is a constant equal to 680 lumens/watt.

The standard upon which all quantitative units in the field of illumination are based is the brightness, or luminance, of a blackbody radiator at the freezing point of platinum (2042°K). The brightness has been standardized by international agreement at 60 candelas/cm² (or 60π lamberts). This value, together with Planck's equation for the radiation density, when evaluated in terms of the luminosity function, yields the constant k of 680 lumens/watt. *See* LAMBERT; LUMINANCE.

The lumen is the unit of luminous flux that results when radiant power is evaluated through the luminosity function. Any one of several units of area may be used to express the density of this luminous flux, or the illumination. In the English system, the unit of illumination is the lumen/ft². The term footcandle was coined in earlier evaluation techniques and probably will continue to persist in the terminology. The lumen/ft² and the footcandle are identical units. *See* FOOTCANDLE; LUMEN; LUMINOUS FLUX; LUX; PHOT.

Light sources. Nature's source of radiation, the Sun, produces radiant power on the Earth extending from wavelengths below 0.3 μ in the ultraviolet region to well over 3 μ in the infrared region of the spectrum. The Sun's spectral radiation per unit of wavelength is greatest in the region of 0.4–0.9 μ. The response of the human eye is well matched to this range of wavelengths. *See* SOLAR RADIATION; VISION.

Oil-flame and gas lighting were used before the advent of the electric light, but since about 1900 electric energy has been the source of essentially all modern lighting devices.

Incandescent lamps. These electric lamps operate by virtue of the incandescence of a filament heated by electric current. The filament usually is composed of tungsten and in the conventional lamp is contained in either a vacuum or inert-gas-filled glass bulb. More recent designs use a halogen regenerative cycle. Presently iodine or bromide vapors are used in a quartz bulb which in some designs is internal to an outer, conventional inert-gas-filled bulb. The evaporated tungsten reacts chemically to form a halogen compound, which returns the tungsten to the filament. Bulb blackening is prevented, higher temperature filaments with increased luminous efficacy result, and higher luminous output can be achieved in relatively small bulbs. Much more efficient optical designs can be achieved for projection equipment, where the source size of light should be small for efficient utilization of the luminous flux. *See* INCANDESCENT LAMP; INFRARED LAMP.

Vapor lamps. These operate by the passage of an electric current through a gas or vapor. In some lamps the light may be produced by incandescence of one or both electrodes. In others the radiation results from luminescent phenomena in the

Relative luminosity factors (value of unity at 0.554 μ wavelength)

Wavelength, μ	Factor	Wavelength, μ	Factor	Wavelength, μ	Factor
0.38	0.00004	0.51	0.503	0.64	0.175
0.39	0.00012	0.52	0.710	0.65	0.107
0.40	0.0004	0.53	0.862	0.66	0.061
0.41	0.0012	0.54	0.954	0.67	0.032
0.42	0.0040	0.55	0.995	0.68	0.017
0.43	0.0116	0.56	0.995	0.69	0.0082
0.44	0.023	0.57	0.952	0.70	0.0041
0.45	0.038	0.58	0.870	0.71	0.0021
0.46	0.060	0.59	0.757	0.72	0.00105
0.47	0.091	0.60	0.631	0.73	0.00052
0.48	0.139	0.61	0.503	0.74	0.00025
0.49	0.208	0.62	0.381	0.75	0.00012
0.50	0.323	0.63	0.265	0.76	0.00006

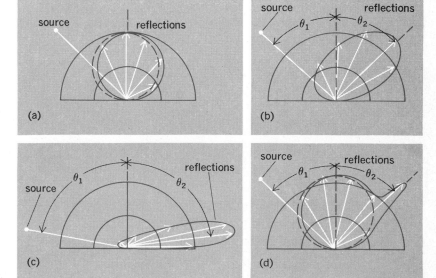

Fig. 3. Typical reflection characteristics. (*a*) Practical mat surface. (*b*) Semimat surface. (*c*) Semimat surface at large angle of incidence. (*d*) Porcelain-enameled steel. Dashed curves in (*a*) and (*d*) illustrate ideal cosine distribution. (*From W. B. Boast, Illumination Engineering, 2d ed., McGraw-Hill, 1953*)

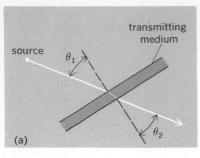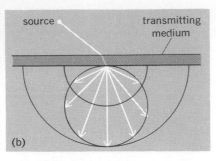

Fig. 4. Spatial characteristics of ideal transmission media. (*a*) Specular. (*b*) Diffuse. (*From W. B. Boast, Illumination Engineering, 2d ed., McGraw-Hill, 1953*)

electrically excited gas itself. *See* ARC LAMP; VAPOR LAMP.

Fluorescent lamps. These are usually in the form of a glass tube, either straight or curved, coated internally with one or more fluorescent powders called phosphors. Electrodes are located at each end of the tube. The lamp is filled to a low pressure with an inert gas and a small amount of mercury is added. The electric current passing through the gas and vapor generates ultraviolet radiation, which in turn excites the phosphors to emit light. If the emission of light continues only during the excitation, the process is called fluorescence. If the materials continue to emit light after the source of excitation energy is removed, the process is called phosphorescence. Phosphors for fluorescent lamps are chosen to accentuate the fluorescent action. *See* FLUORESCENT LAMP.

Reflection and transmission. The control of light is of primary importance in illumination engineering because light sources rarely have inherent characteristics of distribution, brightness, or color

desirable for direct application. Modification of light may be provided in a number of ways, all of which may be grounded under the general topics of reflection and transmission. Reflection from a surface and transmission through it each may be classified according to their spatial and spectral characteristics.

Spatial characteristics. Spatially a surface may exhibit reflection conditions ranging from a regular, or specular, reflection to an ideally diffuse characteristic. Similarly, transmission may range spatially from complete transparency to an idealized diffuse transmission.

Figure 2 illustrates the extremes of specular reflection such as would be obtained from polished metal or silvered glass, and of an ideal mat-finished surface possessing microscopic roughness of minute crystals or pigment particles. The specular reflector gives a direct image of the source, with the angle of reflection equal to the angle of incidence. The plot shown for the diffuse reflector is the small area intensity distribution curve; it exhibits a cosine distribution in the idealized case.

Practical surfaces possess spatial-reflection characteristics intermediate between these ideals, particularly at the diffuse end of the range. Typical distribution curves are shown in Fig. 3. In all of these illustrations the light source is small. Figure 3*a* demonstrates that even with the best practical mat surfaces, such as dull-finished metals or those painted with flat paint, the location of the source of light has some slight influence upon the intensity distribution as evidenced by an irregularity of the cosine distribution of intensity of the small area of the surface. Figure 3*b* and *c* demonstrates the pronounced influence of the location of the light source upon the intensity distribution for surfaces covered with semimat materials, such as satin-finish paint. Figure 3*d* illustrates a mixed-reflection phenomenon resulting when a diffusing surface is overlaid with a surface possessing sheen.

Spatial characteristics of transmission are illustrated in Fig. 4. Figure 4*a* shows regular or transparent transmission, such as occurs with clear glass or plastic. Figure 4*b* shows the diffuse transmission that would result with an idealized diffusing material. In Fig. 4*a* a refraction of the transparent image occurs, but the angles θ_1 and θ_2 are equal.

Practical transmission materials possess spatial transmission characteristics intermediate between these ideals and exhibit intensity distribution curves illustrated by Fig. 5. Figure 5*a* demonstrates that a perfect cosine intensity distribution is not obtained even with a solid opal- or milk-glass material. A more direct transmission is shown in Fig. 5*b* for a flashed opal-glass medium, in which the majority of the base material is clear glass. Sandblasted or frosted-glass material is much less diffusing, and the location of the light source greatly influences the intensity distribution curves of Fig. 5*c* and *d*.

Spectral characteristics. If a homogeneous radiation of wavelength λ impinges upon a surface, the reflection characteristic or the transmission characteristic will result geometrically according to the spatial considerations of the preceding section. The magnitude of total power density reflected, or transmitted, to that received is

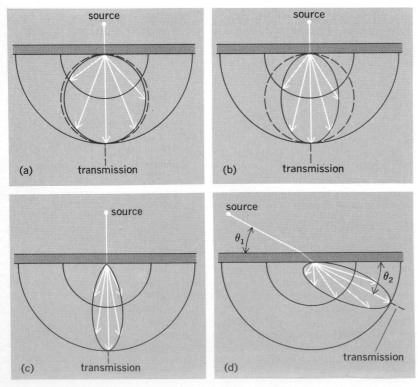

Fig. 5. Transmission characteristics. (*a*) Solid opal glass. (*b*) Flashed opal glass. (*c*) Sand-blasted glass. (*d*) Sand-blasted glass, incident light at 30° from plane at surface. (*From W. B. Boast, Illumination Engineering, 2d ed., McGraw-Hill, 1953*)

defined as the spectral reflection factor ρ_λ or the spectral transmission factor τ_λ respectively. Thus $\rho_\lambda = J_\lambda/G_\lambda$ and $\tau_\lambda = J'_\lambda/G_\lambda$, where J_λ is the reflected spectral emission at wavelength λ, J'_λ is the transmitted spectral emission at wavelength λ, and G_λ is the spectral irradiation at wavelength λ.

A device for measuring ρ_λ or τ_λ is called a spectrophotometer. Comparison of J_λ or J'_λ with G_λ (or its essential equivalence) may be accomplished by visual or photoelectric methods of photometry. *See* PHOTOMETER.

Either continuous- or line-spectra sources of irradiation may constitute a practical source of radiant-power density. The resulting secondary source of radiant-power density J_λ or J'_λ will be modified, compared with the original, depending upon the spectral reflection or transmission characteristics of the reflecting or transmitting material.

An evaluation of J_λ or J'_λ using the relative luminosity factor v_λ may be effected for the secondary sources of radiant-power density in the same manner as is done for incident radiant-power density conditions. Both line- and continuous-spectrum effects may be included as demonstrated for the evaluation of the illumination. The evaluation of the reflected or transmitted luminous flux density yields the luminance, symbolized by L for the reflecting surface and L' for the transmitting surface. The functional operations are shown graphically in the series of curves of Fig. 6 for the reflection resulting from a continuous-spectrum source. The ratio of L or L' to the illumination E results in $\rho = L/E$ and $\tau = L'/E$, where ρ is the reflection factor of the surface and τ is the transmission factor of the material. The reflection and transmission factors are dependent upon the source of illumination as well as the spectral characteristics of the surface or the transmitting medium. Such factors are frequently published for tungsten-filament light and are useful as guides and indications of magnitude if the sources of illumination do not deviate too greatly from this spectral distribution.

Light-control methods. Either a control of the character of the reflecting surface or transmitting material or a control of the primary source of luminous flux determines the degree of light control possible. If a light ray is reflected or transmitted diffusely, the shape of the reflecting or transmitting surface is of little importance, although its size may be of great importance. A larger diffuse surface reduces the brightness of the primary source. If a light ray is reflected from a specular surface, more accurate control of the light can be effected.

Parabolic reflectors. These are probably the most useful of all reflector forms. The rays emerge essentially parallel from such a reflector if the light source is placed at its focal point. Searchlights use parabolic mirror reflectors.

Elliptic reflectors. A relatively large amount of light may be made to pass through a small opening before spreading out by using an elliptic reflector. A light source is placed at one of the focal points of a nearly complete elliptical reflector. A small opening is placed at the second focal point where the light is focused. Such a design is used for the pinhole spotlight. For further discussion of reflectors *see* PROJECTOR, LIGHT.

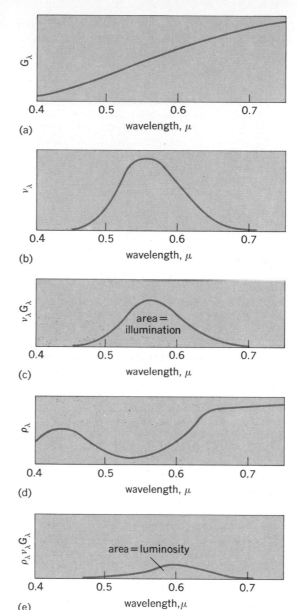

Fig. 6. The process of evaluating the luminance of a reflecting surface after evaluating the illumination. (*a*) Spectral irradiation G_λ curve. (*b*) Relative luminosity factor v_λ curve. (*c*) Curve of $v_\lambda G_\lambda$. Illumination is the area under this curve, the integration of $v_\lambda G_\lambda$. (*d*) Spectral reflection factor ρ_λ curve. (*e*) Curve of $\rho_\lambda v_\lambda G_\lambda$. Luminance is the area under this curve, the integration at $\rho_\lambda v_\lambda G_\lambda$. (*From W. B. Boast, Illumination Engineering, 2d ed., McGraw-Hill, 1953*)

Lens control. Refraction of light rays at the boundary between glass and air may be used to control transmitted light in a manner similar to reflector control. Simple thick lenses may be cut away if the cutaway piece is duplicated in the new equivalent surface. Slight irregularities in the field of the light beam occur. These may be of little consequence for searchlights or floodlights, but they would be undesirable for high-quality picture projection. *See* LENS, OPTICAL.

Polarization of light. Light rays emitted by common sources may be considered as waves vibrating at all angles in planes at right angles to the direction of the ray. As they pass through some substances or are reflected from specular surfaces, particularly at certain angles, wave components in some directions are absorbed more than those in other directions. Such light is said to be polarized.

Polarized light may be controlled by a transmitting section of a polarizing material. When the absorbing section is oriented to permit passage of the polarized light, minimum control is exercised. If the absorbing section is rotated 90°, the polarized light is essentially completely absorbed. Because sunlight reflected from water or specular surfaces is highly polarized, glare from such sources is effectively controlled by polarized sunglasses or camera filters. Polarized light is used also for detection of strains and defects in glass or plastics. The strained areas appear as color fringes in the material. *See* POLARIZED LIGHT.

Control of primary source. Control of the source of luminous flux can be achieved by controlling the voltage supplied and thus controlling the magnitude of current for some types of light sources or by controlling the time at which the current is permitted to begin on each half cycle of the electric supply for other types of light sources.

For simple incandescent lamps the control of the magnitude of alternating voltage and current can be controlled by variable transformer-type dimmers. Very small silicon controlled rectifiers (SCRs), a modern solid-state development, can accomplish the same control function with equipment which is so small that a unit can be installed in the same space as would be occupied by a conventional, residential "on-off" switch and with very small inherent heat loss caused by the control itself. SCRs are also used in larger capacities for control of large lighting installations.

The dimming control of fluorescent lamps or other sources employing a gas or vapor discharge is more involved because of the relatively high striking voltage of the vapor discharge. To achieve a wide range of dimming control the timing on each half-cycle at which current is permitted to begin can be controlled through electronic means and thus produce dimming down to low levels. The complexity and cost of such control equipment, however, is greater than the control equipment needed for incandescent lamp dimming.

Interior lighting design. Attainment of a comfortable visual environment is the ultimate goal of interior lighting design. This requires an adequate distribution of illumination and control of brightness of all visible surfaces to avoid excessive contrasts. The adequate distribution of illumination requires that a sufficient number of lighting sources be installed in a manner that produces an adequate level and a proper distribution of illumination. Brightness is controlled by adequate surface reflection or transmission areas associated with the lighting sources and with proper reflecting surfaces within the room itself. Control should ensure that the brightest surfaces of the room are not more than three times the brightness of the work and the brightness of the work not more than three times that of the darkest surfaces in the room. When such brightness-ratio limitations are satisfied, glare is absent and a satisfactory visual environment is achieved. Proper application of color in the interior enhances overall appearance.

The function of a building greatly influences the manner in which lighting is applied. A particular visual task may require a certain level of illumination, but the lighting design is also influenced by the economics, appearance, and quality of desired results. For example, application techniques are designated as industrial lighting, commercial lighting, and school lighting. Certain application experience and consumer acceptance assume importance in the methods of achieving the level of illumination and control of brightness contrasts. The *IES Lighting Handbook* treats in detail many of these specialized concepts.

Location and spacing of luminaires. Many interior lighting designs utilize general lighting to distribute luminous flux throughout the room on an imaginary working plane, which is usually 30 in. above the floor. To accomplish a reasonably uniform illumination on this plane, certain limitations in spacing between luminaires and in spacing between luminaires and walls must not be exceeded. The height of luminaires above the plane may be an important factor for some interiors. For indirect-type luminaires the distance from the ceiling to the luminaire must provide a proper distribution of light over the ceiling. *See* LUMINAIRE.

Room coefficients. The shape of a room influences the fraction of the emitted luminous flux that will be received on the working plane. Equations and empirical tabulations have been developed to classify rooms according to their shape. Large, broad rooms are more efficient than small or narrow rooms for transferring luminous flux from the luminaire to the working plane, particularly if the reflection factor of the side walls is low.

Lumen method of illumination design. This empirical method of design gives the average illumination on the working plane for a particular number of luminaires arranged in a symmetric pattern in a room. The illumination that may be expected to be maintained in the room is given in Eq. (2),

$$E = \frac{N\phi_L K_u M}{A} \qquad (2)$$

where E is illumination in footcandles (or lumens/ft²), N is number of luminaires, ϕ_L is initial luminous flux per luminaire in lumens, K_u is coefficient of utilization, M is maintenance factor, and A is area of room in square feet.

Tables of the coefficient of utilization are obtained empirically. Usually these tables pertain to a particular luminaire. The coefficient of utilization is a function of the room coefficient and the reflection factor of principally the ceiling and side walls. Also called the flux-of-light method, this was originally designed as a component method whereby the candle-power distribution curve of any luminaire could be resolved into three component curves and each calculated according to an equation similar to that shown. However, modern use of the method is principally through coefficients of utilization for each particular luminaire.

The maintenance factor M is a number less than one to account for lamp darkening and the collection of dirt within the luminaire, which reduce the illumination emitted from the luminaire.

Zonal-cavity method. The interreflection of luminous flux within a room is accounted for in the empirical data of the lumen method. An exact mathematical calculation of the distribution of luminous flux within rectangular rooms becomes extremely involved. Methods developed in the 1940s used various simplying assumptions and permitted an approximate solution for both the illumination upon the working plane and the brightness of various surfaces within the room. More recently computer-derived flux-transfer data have resulted in a design method called the zonal-cavity method, which was the method recognized by the Illuminating Engineering Society (IES) as of 1969. Calculation procedures are presented in the *IES Lighting Handbook.* Conformance with the 3:1 or 1:3 brightness-ratio criterion for a comfortable visual environment becomes a standard checking procedure in the application of the zonal-cavity method.

Luminous architectural elements. The original emphasis in the lumen method was upon discrete luminaires arranged symmetrically within the room. The development of the interreflection method of design presumed surfaces of initial brightness on ceiling, walls, and floor as the sources of light within the room. Ingenuity in the application of both methods permits them to be applied both to designs involving luminaires within the room and to designs incorporating many types of luminous architectural elements, such as louvered ceilings, luminous ceilings, and cove systems of lighting.

Natural lighting. Daylight may be an important factor in building design. For comfortable visual conditions any openings for the admission of daylight are planned in position, size, and shape as carefully as any other part of a building. A window can be treated as any other type of light source and included as a part of the illumination and brightness design according to the conventional design procedures. It is usually necessary to provide electric lighting systems when daylight is not available.

Supplementary lighting. Local lighting, such as spotlighting, is frequently used to supplement general lighting where a high level of illumination is desirable on an isolated task. Local lighting that does not supplement general lighting usually does not permit conformance with the limitations on brightness contrasts for interior lighting designs.

Exterior lighting design. The nighttime lighting of exterior areas serves many useful purposes. Such lighting extends into the fields of advertising and recreational pursuits as well as utilitarian fields, such as the lighting of industrial yards, parking areas, airports, streets, and highways. The *IES Lighting Handbook* and many textbooks treat in detail many of these specialized applications.

Automobile headlamps. Headlamps use two beams of different characteristics under the control of the driver. The aiming and intensity of one beam is designed to provide distant illumination. The other beam is aimed low enough and in such directions to prevent glare for oncoming drivers.

Sealed-beam headlights have been used since about 1940. When a filament burns out the entire lamp is replaced, thus restoring the parabolic reflector and condition of alignment of the light source better than was possible for a separate lamp in a permanent reflector.

Recommended standards. Illumination levels have been established by the IES to bring all visual tasks to the same level of visibility for normal observers. Some visual tasks are more arduous than others, and higher illuminations provide a compensating factor in bringing such tasks to a satisfactory level of visibility. Standards are established not only for interior lighting designs but also for exterior systems including recreational lighting. For tables of recommended illumination levels consult references listed in the bibliography.

Colored light. The greatest precision in color specifications is obtained by the use of the spectral curves as a function of wavelength of the radiation. However, for visual considerations it has been found that any color made up of complex spectral components may be matched visually by proportioning three properly chosen component radiations. The relative magnitudes of these three component radiations are called the trichromatic coefficients of the color. Other designations of dominant wavelength and spectral purity may be obtained also from these trichromatic coefficients. *See* COLOR.

A laser may be used to obtain a powerful, highly directional beam of coherent monochromatic light. *See* LASER.   [WARREN B. BOAST]

Bibliography: C. L. Amick, *Fluorescent Lighting Manual,* 2d ed., 1947; W. E. Barrows, *Light, Photometry, and Illuminating Engineering,* 3d ed., 1951; W. B. Boast, *Illumination Engineering,* 2d ed., 1953; J. Carroll and D. G. Fink, *Standard Handbook for Electrical Engineers,* 10th ed., 1968; Illuminating Engineering Society, *IES Lighting Handbook,* 4th ed., rev., 1968; J. O. Kraehenbuehl, *Electric Illumination,* 2d ed., 1951; P. H. Moon and D. E. Spencer, *Lighting Design,* 1948; H. Pender and W. A. Del Mar (eds.), *Electrical Engineers' Handbook,* vol. 1, 4th ed., 1949; H. M. Sharp, *Introduction to Lighting,* 1951.

Ilmenite

A rhombohedral mineral, space group $R\bar{3}$, with composition $Fe^{2+}Ti^{4+}O_3$ It is derived from the hematite crystal structure by ordering iron and titanium atoms over the octahedral sites, thus resulting in lower symmetry. The arrangement is such that the iron-centered oxygen octahedrons share edges and faces only with titanium-centered

Ilmenite crystals, from Arendal, Norway. (*Specimen from Department of Geology, Bryn Mawr College*)

oxygen octahedrons. The hardness is $5\frac{1}{2}$ on Mohs scale, specific gravity 4.72. Color is black, and there is no cleavage. Crystals are thick tabular parallel to {0001} (see illustration) but the mineral usually occurs massive or in thin plates. Mg^{2+} and Mn^{2+} often substitute for Fe^{2+} in ilmenite; two other minerals belonging to the ilmenite structure type are geikielite, $MgTiO_3$, and pyrophanite, $MnTiO_3$.

Ilmenite is the most abundant titanium mineral in igneous rocks and the most important ore of titanium. It occurs as an early differentiate in anorthosite, norite, and gabbroic rocks, sometimes in large quantities; also in metamorphic rocks such as gneisses and marbles. Ilmenite resists weathering, and alluvial accumulations may occur in sufficient quantity to constitute ore. Important occurrences include the Ilmen Mountains, Soviet Union (whence the name); Kragerø, Norway; and Allard Lake, Quebec. *See* HEMATITE; TITANIUM.

[PAUL B. MOORE]

Image, acoustical

If a point source of sound is placed on one side of an extended reflecting surface, it may be considered to have an "image" at an equivalent distance on the other side of the surface along a perpendicular projection, analogous to the familiar optical image. The effects of reflecting surfaces on acoustic waves frequently can be predicted by the use of such images. A source in front of a very reflective wall (for example, plaster, concrete, and masonry reflect 97–99% of incident sound energy) will have an image of "strength" almost equal to that of the source, vibrating in phase with the source. To obtain the total effect at any point due to the combined action of a source and such a reflecting surface, the effects due to the source itself and another source of equal strength placed at the image point are added. If there is a second reflecting surface present, this so-called first-order image will have an image, called the second-order image of the source, at an equivalent distance on the other side of the second reflecting surface. Similarly, an image of the second-order image is said to be a third-order image, and so on. Although this method is precise only when the reflecting surface is nonabsorptive, it is often of considerable help in investigations of the action of sound waves in rooms. *See* ARCHITECTURAL ACOUSTICS; IMAGE, OPTICAL.

[CYRIL M. HARRIS]

Image, optical

The image formed by the light rays from a self-luminous or an illuminated object that traverse an optical system. The image is said to be real if the light rays converge to a focus on the image side and virtual if the rays seem to come from a point within the instrument (see illustration).

The optical image of an object is given by the light distribution coming from each point of the object at the image plane of an optical system. The ideal image of a point according to geometrical optics is obtained when all rays from an object point unite in a single image point. However, diffraction theory teaches that even in this case the image is not a point but a minute disk. The diameter of this disk is about $1.22 \lambda/A$, where λ is the wavelength of the light considered and A is the numerical aperture, the sine of the largest cone angle on the image side multiplied by its refractive index (which is usually equal to unity).

Aberrations. From the standpoint of geometrical optics, if this most desirable type of image formation cannot be achieved, the next best objective is to have the image free from all but aperture errors (spherical aberration). In this case the light distribution in the image plane is still circular, resembling the point image; there is a true coordination of object point and image, although the image may be slightly unsharp. If the aperture errors are small, or if the image is viewed from a distance, such an image formation may be very satisfactory.

Asymmetry and deformation errors may be very disturbing if not held in check, because the light distribution of the image of a point in this case has a decidedly undesirable shape.

When the image of an axis point is considered, the rays through a fixed aperture circle converge to an axis point. For this type of imagery, the term half-sharp image will be used. A small object at the object point is then imaged by a circular stop at the focus of the image bundle with a magnification as given by Eq. (1) where u and u' are the angles of

$$m = (n \sin u)/(n' \sin u') \qquad (1)$$

the imaging cone in object and image space, respectively, and n and n' are the corresponding refractive indices.

If the axis point is sharply imaged, an object of finite extent is sharply imaged if, and only if, $m = m_o$ (the Gaussian magnification) for all values of u (sine condition).

In the case of aperture errors, the most desirable image formation for an axis point is attained when the different images appear under the same angle from the exit pupil. If k' is the distance of the

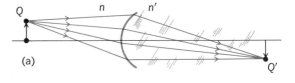

(a)

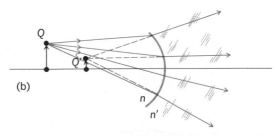

(b)

Optical images. (a) Real image. Rays leaving object point Q and passing through the refracting surface separating media n and n' are brought to a focus at the image point Q'. (b) Virtual image. Rays leaving Q and refracted by the concave surface separating n and n' appear to be coming from the virtual image point Q'. As the rays are diverging, they cannot be focused at any point. (*Modified from F. A. Jenkins and H. E. White, Fundamentals of Optics, 3d ed., McGraw-Hill, 1957*)

image point from the exit pupil, $\Delta s'$ is the aperture aberration, and if Eq. (2) gives the magnification

$$\Delta m = \frac{n \sin u}{n' \sin u'} - m_o \qquad (2)$$

error compared with the magnification m_o on the axis, the condition is given by Eq. (3). The

$$\Delta s'/k' - \Delta m/m_o = \text{constant} \qquad (3)$$

fulfillment of this condition gives equal quality for an object near the axis of a system with rotation symmetry.

Corresponding conditions can be ascertained for the image of an off-axis element if all the asymmetry errors and deformation errors are balanced. *See* ABERRATION, OPTICAL.

Resolution. Two points are resolved by an optical system if the two images lie apart. Photometric analysis of an image may indicate the existence of two object points even if their images overlap, but in such an analysis the illumination of the object, as well as the imagery, plays a role.

In interference experiments, it is found that the image of two self-luminous points (that is, two light sources that are sufficiently separated) is incoherent; that is, the intensities of the two beams simply add. If the two object points are illuminated by the same light source, however, the phase relation of the light at the two points has to be taken into consideration. This is of the greatest importance for microscopes and telescopes, which image very small or distant objects. In this case an artificial change of phase by phase plates and apodization may improve resolution. *See* DIFFRACTION; RESOLVING POWER (OPTICS).

Resolving power is not the only consideration in image formation. The eye recognizes only contrast differences, and therefore objects may not be discerned if the contrast difference is too small. Again, for the image of a point, or an object illuminated by a point light source, means can be found to change the apparent contrast, making it possible to discern biological objects, for example, having small differences of refractive index.

Image analysis. Methods have been suggested for obtaining information about optical images by sine-wave analysis. A sinusoidal test object is imaged by an optical system as a sinusoidal image, but altered in phase and amplitude. A large number of sinusoidal test objects with different frequencies (number of maxima per millimeter) are imaged, and the amplitude and phase are measured.

The curve of amplitude versus frequency gives a measure for resolving power and contrast as a function of the frequency of the test object, whereas the adjusted curve of phase versus frequency describes the lack of symmetry in the image. These amplitude-frequency curves can be measured as well as calculated from the spot diagrams, onto which the effects of diffraction can be superimposed if necessary.

[MAX HERZBERGER]

Bibliography: E. U. Condon and H. Odishaw, *Handbook of Physics*, 1967; A. E. Conrady, *Applied Optics and Optical Design*, 1929; M. Herzberger, *Modern Geometrical Optics*, 1958; I. Strong, *Concepts of Classical Optics*, 1958.

Image orthicon

A high-sensitivity television camera tube that can be used to pick up scenes of widely varying light values. It is used singly for monochrome television or in sets of three for color television. An ordinary camera lens produces an optical image of the scene being televised on a sensitive photoelectric surface at one end of the tube. Photoelectrons released from the back of this surface are electrically focused on one side of a separate storage target. The other side of the target is scanned by an electron beam produced by an electron gun at the far end of the tube. Beam electrons are reflected back from the target in proportion to the charge at each point on the target image, and are returned to an electron multiplier surrounding the electron gun. The electron multiplier amplifies the signal many thousands of times to produce the video output signal of the camera tube. Image orthicons are used in broadcast television for both indoor and outdoor work. *See* TELEVISION CAMERA TUBE.

[JOHN MARKUS]

Image processing

A term that can mean many things, from adjusting the contrast on a television set to sophisticated nonlinear enhancement of moon images or automated detection of black lung disease from chest radiographs. This article is restricted to image enhancement systems in which the input into the system is an image and the output is also an image that has been improved in some way.

An image is any two-dimensional representation of a physical quantity. Thus, it may be a photographic negative in which the darkening of the emulsion represents brightness in an object scene. Or it can be a medical thermogram in which the color at any point is determined by the temperature at a corresponding point on a patient's skin. Or it can be a nuclear medicine image consisting of a collection of dots, with the density of dots related to the distribution of a radioactive pharmaceutical in the body. Even the restriction to two-dimensional representations is arbitrary, because many of the image-processing techniques discussed here can also be profitably applied to one-dimensional data such as spectrograms.

An image-processing system may have one of two goals: it either makes the image more pleasing or more useful to a human observer, or it actually performs some of the interpretation and recognition tasks usually performed by humans. The latter goal is often referred to as image analysis or pattern recognition.

Hardware. Of the various types of hardware that can be used for image enhancement, the digital computer is the most generally useful and perhaps also the easiest to understand conceptually. Suppose that the initial image is a photographic transparency. The first step is to convert the density levels in the image into numbers that can be entered into the computer memory. This can be accomplished with a microdensitometer in which a small, rectangular aperture is used to isolate a single picture element, or pixel. The light transmitted through this element is measured with a photodetector, digitized, and stored. Then the aperture is

The processing of a blurred image. (*a*) Original object. (*b*) Blurred image of original object. (*c*) Processed image with linear deblurring filter. (*d*) Processed image using iterative nonlinear method. (*Courtesy of J. Burke and R. Hershel, University of Arizona, Optical Sciences Center*)

moved over by its width, and the density of the adjacent picture element is recorded. In this way the entire picture is converted into a matrix of numbers suitable for computer manipulation. The number of picture elements required to faithfully represent the image depends on both the overall size of the image and the size of the fine detail recorded in it. *See* DIGITAL COMPUTER.

Once the computer manipulations are complete, the processed image may be displayed on a cathode-ray tube, a mechanical printer, or any other suitable display device.

Operation types. The operations that can be performed on images can be classified into point, global and local operations.

Point operations. In these operations the image value at a point in the output image depends on the value at only a single point in the input image. The most common example of this type of operation is contrast enhancement, which is very useful if the input image is underexposed or depicts a low-contrast scene. A second example is a pseudocolor display, in which each gray level in the input is encoded as a color in the output, the theory being that the eye can distinguish colors more easily than shades of gray.

Global operation. At the opposite extreme from point operations are global operations, in which the value at one picture element in the output depends on all, or almost all, of the picture elements in the input. An example is reconstruction of an image from a hologram.

Local operations. However, the most useful operations are frequently the local operations in which a relatively small neighborhood of a point on the input contributes to the value at a point on the output. To illustrate, consider the problem of edge detection. If the image were ideal, the following sequence of numbers might be generated when the aperture of the microdensitometer is scanned across the edge:

$$3, 3, 3, 3, 3, 3, 6, 6, 6, 6, 6, 6.$$

The edge is readily apparent between the sixth and seventh picture elements. Suppose, however, that noise is present, so that there is an uncertainty in the value at each element. Then the input data might look like:

$$4, 3, 2, 5, 2, 3, 5, 7, 6, 4, 6, 8.$$

The position of the edge is now less apparent. One simple smoothing operation is to replace each number by its average with its two neighbors. Mathematically, this corresponds to convolution of the data with the sequence (1/3, 1/3, 1/3). The result of performing this operation on the noisy data above is

$$2, 3, 3, 3, 3, 3, 5, 6, 6, 5, 6, 5,$$

where all numbers have been rounded (quantized) to the nearest integer. Note that the noise has been reduced considerably, but the edge is not quite as abrupt as in the original object.

Now suppose that the problem is not noise but blur (see illustration). Here the data might look like

$$30, 30, 30, 31, 34, 41, 50, 56, 59, 60, 60, 60,$$

where now two digits are used to represent each picture element for reasons that will become ap-

parent shortly. To sharpen up this edge, the data can be convolved with $(-2, 5, -2)$. In other words, the value at each picture element is multiplied by 5, and twice the sum of its neighbors is subtracted from it. This produces the following result:

$$90, 30, 28, 31, 26, 37, 56, 62, 63, 60, 60, 180.$$

It is now clear that the edge is between the sixth and seventh elements. Note also the large values at the ends. The edge detection routine has found the edges of the data.

Suppose the original blurred data had been rounded off to one digit, so that it looked like

$$3, 3, 3, 3, 3, 4, 5, 6, 6, 6, 6, 6.$$

Now application of the deblurring filter $(-2, 5, -2)$ yields

$$9, 3, 3, 3, 1, 4, 5, 8, 6, 6, 6, 18.$$

The round-off error has been greatly magnified, and the processed image is poorer than the original. Thus, the deblurring filter $(-2, 5, -2)$ improves sharpness but increases the noise level, whereas the smoothing filter (1/3, 1/3, 1/3) has just the opposite effect.

The action of these filters can also be described in terms of spatial frequencies. Just as an electrical signal can be decomposed by Fourier analysis into a sum of sine waves of varying amplitude, phase, and frequency, so too can a two-dimensional image be broken up into a superposition of sinusoidal bar patterns. However, the spatial frequency of one of these patterns is not uniquely specified by a single number, such as the number of bars per centimeter. The orientation of the pattern must also be given; spatial frequency is a vector.

But leaving this matter of dimensionality aside, a strong formal analogy exists between spatial and temporal frequencies. The smoothing filter described above is a low-pass filter that suppresses the higher spatial frequencies in the image. Conversely, the deblurring filter emphasizes the higher frequencies.

The practical utility of the spatial frequency concept lies in the ease with which modern computers can perform Fourier analysis. Using the so-called fast Fourier transform algorithms, a 128×128 image can be decomposed into its frequency components in a matter of seconds, even on a small computer. Then each frequency component can be weighted appropriately, and the output image is easily reassembled. *See* FOURIER SERIES AND INTEGRALS.

As powerful as these frequency-domain methods are, they still do not represent the ultimate in image-processing technology. The best results to date have been obtained with iterative nonlinear techniques in which the computer systematically searches for the "best" output image that is consistent with the input data and with any other information that is known about the object. Although spectacular results can be obtained by these nonlinear methods, they frequently require large machines and exorbitant computing times.

Many practical image-processing tasks can also be performed without the use of digital computers. For example, there is a commercial instrument that uses a closed-circuit television system to sharpen radiographs. And rather striking examples of both image enhancement and pattern recognition have been produced by holographic filtering. *See* HOLOGRAPHY.

Thus, image processing is an active and fruitful area of research, with the problems ranging from basis mathematics to practical hardware engineering.   [HARRISON H. BARRETT]

Bibliography: H. C. Andrews, *Computer Techniques in Image Processing*, 1970; A. Rosenfeld, *Picture Processing by Computer*, 1969.

Image tube (astronomy)

A photoelectric device for intensifying faint astronomical images. The photographic emulsion (plate or film), which is the time-honored method for recording astronomical images, has many excellent characteristics for the purpose. It has high resolution; reasonably high signal-to-noise ratio images; good dimensional stability and permanence; and wide spectral response. However, it lacks one characteristic of greatest importance to the astronomer: high quantum efficiency, or the ability to record a high percentage of the light quanta originally in the incident image. Even the most sensitive photographic emulsions are able to record only a few light photons in every thousand that the telescope collects and focuses into an image. All the rest, and the information they bring, are lost. The photographic process has two other drawbacks for astronomical applications. The sensitivity, already low, becomes lower during the very long exposures that astronomers frequently use. Ordinarily one thinks of an emulsion as responding to a given total amount of light representing the exposure regardless of whether it comes in the form of an intense burst of short duration or as a fainter image exposed for a correspondingly longer time. But when exposures exceed a few seconds, the intensity and exposure time are no longer inversely proportional, a phenomenon called reciprocity failure. The net effect of reciprocity failure is that the emulsion speed decreases drastically as exposures become very long. The other drawback of the photographic process is its nonlinearity. There is a linear relation between exposure and resulting emulsion density only over a relatively narrow range of exposures, and beyond this range for either higher or lower exposures the response is nonlinear. *See* ASTRONOMICAL INSTRUMENTS.

Photoemissive process. The photoemissive process embodied in modern photomultiplier tubes is free from the drawbacks cited above. In these tubes the photosensitive surface, or photocathode, receives light and converts it into electrons that are emitted from the cathode into the vacuum of the tube where, by means of electric and magnetic fields, they can be collected for measurement. *See* CATHODE-RAY TUBE.

The process of converting light photons to photoelectrons can be quite efficient, in some cases yielding as much as one electron for every four incident light quanta, though the average yield over a reasonably wide region of the spectrum would be more like one electron per 10–20 photons, an amount that is still impressive compared with photographic efficiencies. The photoemissive process also responds linearly in terms of photocurrent versus light intensity over an extremely wide range, and brightness ranges of more than a billion to one can be accommodated without difficulty; and equally important, the photoemissive process does not have reciprocity failure. It is no wonder, then, that the photomultiplier has become an essential astronomical research tool, for its characteristics are best just where the photographic emulsion is most deficient. Unfortunately, the photomultiplier can look at only one small part of the image at a time, whereas the photographic emulsion may receive light from millions of tiny picture elements simultaneously. In this respect the photographic emulsion is far superior, and photography and photoelectric photometry have continued to coexist as complementary methods for obtaining astronomical data. *See* PHOTOELECTRIC DEVICES; PHOTOEMISSION; PHOTOMETRY; PHOTOTUBE, MULTIPLIER.

The photoelectric image tube represents a device which in principle combines the best features of photography and photoelectric photometry. An image tube is conceptually like a vast array of microscopic phototubes independently and simultaneously receiving information from each picture element as the photographic emulsion does, but handling it with the efficiency and linearity of the photomultiplier. The potential advantages of such devices for astronomy are tremendous; they combine the high resolution and simultaneous reception from all elements of an image that characterize the photographic emulsion with the high quantum efficiency and wide, linear dynamic range available from photoemissive devices.

Types of tubes. Basically the image tube is very simple. It consists of a photoemissive surface inside a vacuum tube on which a light image from the telescope or its associated spectrograph falls. Electrons emitted from this surface represent an electronic replica of the original light image. This electron image is electrically accelerated and focused down the tube by means of electric or combined electric and magnetic fields (Fig. 1). Up to this point all image tubes are very much the same. The different types are distinguished largely by what happens to the electron image at its focal plane down the tube.

Electronographic tube. One way to handle the electron image is to record it directly upon film or plates just as is done in the electron microscope. Direct recording of electron images is called electronography, and image tubes using this method of recording are referred to as electronographic tubes. This recording technique can be very efficient if the electrons strike the photographic emulsion with energies corresponding to acceleration through voltages of 15–40 kV, and it is possible in this way to record nearly every electron in the electron image. Not only is recording efficiency high in electronography, but very-fine-grain, high-resolution emulsions can be used. These emulsions are generally too insensitive to light to be useful for direct astronomical photography, but some of them have sufficiently good response to fast electrons to be valuable for electronographic recording. Furthermore, some of the emulsions have a linear relationship between density and exposure to electrons, thus preserving the linearity of response of the photocathode. Electronographic cameras using high-resolution plates are capable

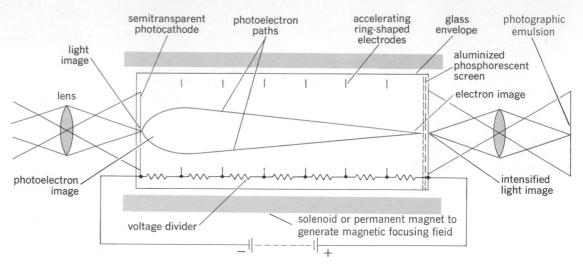

Fig. 1. A single stage, magnetically focused image tube. (*U.S. Army Engineer Development Corps*)

of producing excellent results with resolutions close to 100 line pairs per millimeter. Unfortunately, the electronographic camera has a technical drawback that has not been fully surmounted. The photographic emulsion when placed in vacuum gives off gases such as water vapor that destroy the photocathode of the image tube in a short time. Various schemes have been successfully employed to prevent gas evolved from the emulsion from getting to the photosurface. However, because of the complexity of electronographic tubes, they are used only at a very few observatories which have the necessary staff and equipment to operate them successfully. *See* ASTRONOMICAL PHOTOGRAPHY; ELECTRON EMISSION; ELECTRON TUBE.

Fiber-optically coupled multistage tube. Another type of image tube converts the electron image back into a light image by means of a phosphor screen placed at the focus of the electron image. Since the electrons are accelerated and gain energy by electric fields in the tube, the final light image can be appreciably brighter than the original, and this device functions as an image intensifier. The intensification process is usually repeated one

or more times by coupling single-stage image tubes so the output of the first is imaged on the photocathode of the second, and so on.

It is inefficient to transfer the image between stages by means of lenses that collect only about 5% of the light from the phosphor screen. To overcome this loss, the individual stages can be fiber-optically coupled. In this case the output window of the first tube is a vacuum-tight fiber-optic faceplate, and the phosphor screen of the first tube is put directly upon it. Likewise, the photocathode of the second tube is deposited on an input fiber-optic faceplate, and the tubes are operated in series with their fiber-optic windows in optical contact. The image transfer efficiency is much higher with this arrangement, with values around 50% common. The electron multiplication (number of secondary electrons emitted per initial input photoelectron) is about 50–100 at each phosphor-photocathode sandwich. The final intensified image is transferred from the output phosphor screen to a recording detector either by a fiber-optic faceplate or by a fast transfer lens. Figure 2 shows a tube of this type.

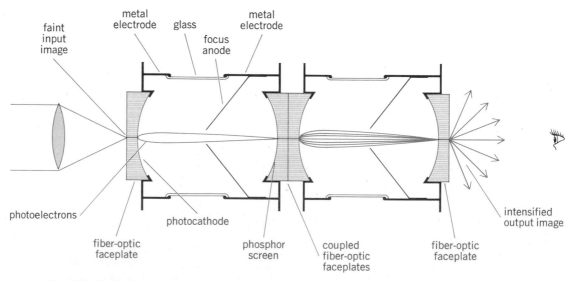

Fig. 2. Two-stage electrostatically focused image tube, fiber-optically coupled.

Cascade image tube. Alternately, the phosphor of one stage may be deposited on a very thin transparent membrane of mica, glass, aluminum oxide, or similar material, with the photocathode of the following stage deposited on the other side of the membrane which is enclosed in the vacuum tube. Close proximity between phosphor and photocathode in this thin sandwich gives very good light transfer efficiency without serious loss of resolution, provided the membrane is thin enough. The cascading of two or more stages may be accomplished in this manner to obtain high electron multiplication and high brightness gain.

TSEM-type tube. Another way to provide intensification is to focus the electrons on a succession of very thin membranes (dynodes) in the tube, each causing high-energy primary electrons to knock out several secondary electrons directly (without converting electrons into photons, and photons back into electrons). This process is known as transmission secondary emission multiplication (TSEM). Although TSEM dynodes are simpler than the phosphor-photocathode sandwiches, dynodes used in past tubes have had the disadvantage of producing, on an average, only five to seven secondary electrons for each primary electron. Thus, more stages of amplification are required to produce the same overall gain. Worse, one primary electron may produce 10 secondaries, whereas another primary electron may produce only one or two—or zero (indeed, zero is the most probable number). For this reason, high-gain multistage TSEM intensifiers show poor multiplication statistics, with photoelectron scintillations at the final output phosphor screen varying in brightness over a range of 100:1. Because of their inferior performance, TSEM tubes are no longer used in astronomy, but new dynode materials may alter this in the future.

Channel plate multiplier tube. Another means of obtaining electron multiplication within an image tube is to focus the electrons on a channel plate multiplier. The channel plate multiplier (Fig. 3) consists of an array of closely spaced tubes, each coated on the inside with a secondary-electron-emitting material, so that when a primary electron enters the channel, it collides with the side wall and releases several secondary electrons. The secondary electrons are then accelerated further down the channel by an electric field, and they, in turn, collide with the side wall and are multiplied in a similar fashion. The process is repeated several times in a given channel, so that by the time the secondary electrons exit the rear of the channel plate they have been multiplied some 1000–10,000 times. These electrons are then focused on an output phosphor screen.

The channel plate multiplier suffers from a type of poor multiplication statistics similar to that of the TSEM dynode, and for this reason the technique is not widely used in astronomical detectors. However, because of the much shorter overall length and lower voltage requirements of this device compared to the others, channel plates have been used where size or operating voltage are important considerations.

Compared with the electronographic tube, all other types of tubes described above suffer some loss of resolution, and resolutions obtained in prac-

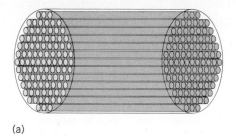

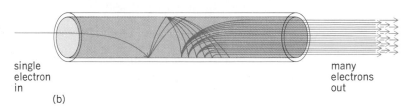

single
electron
in

(b)

many
electrons
out

Fig. 3. Channel plate multiplier. (*a*) Array of tubes. (*b*) Details of a single channel.

tice generally fall in the range of 25–45 line pairs/mm.

Recording images. Except for the electronographic tube, the image tubes described above do not incorporate a means of recording their output image; they merely intensify the incoming light. The output light must be transferred by means of a fiber-optic faceplate or a transfer lens to another detector to make a permanent record of the image. The photographic emulsion is perhaps the most straightforward and convenient detector to use for this purpose. When a photographic emulsion is optimally combined with a high-quality image tube, the result is a photographic detector that has very nearly the same high quantum efficiency as the image tube photocathode and also has a large number of resolution elements. For detecting the very faintest images, multistage intensifiers can provide sufficient gain so that each photoelectron is recorded as a discrete clump of photographic grains. Then the overall photographic system behaves like a photon counter.

Steps have been made to record image tube outputs with various electronic detectors, thus eliminating the photographic emulsion altogether. Once the output is recorded electronically, the imagery data are much more readily available for quantitative study, including computer analysis. Image tube systems have been developed that use photodiode arrays, image dissector tubes, and various types of television camera tubes, all with promising success.

Several advances have been made in image tube technology since 1960, and the field of astronomy has greatly benefited from these advances. Image intensifiers are widely used at nearly every major observatory for the detection of faint astronomical sources. Knowledge about quasistellar objects, distant galaxies, star clusters, and many other astronomical objects has been considerably enhanced through the use of image tube detectors. *See* PHOTOTUBE; VACUUM TUBE.

[RICHARD H. CROMWELL]

Bibliography: J. D. McGee et al. (eds.), *Advances in Electronics and Electron Physics*, vols. 16, 22, 28, and 33, 1962, 1966, 1969, and 1972.

Imhoff tank

A sewage treatment tank named after its developer, Karl Imhoff. Imhoff tanks differ from septic tanks in that digestion takes place in a separate compartment from that in which settlement occurs. The tank was introduced in the United States in 1907 and was widely used as a primary treatment process and also in preceding trickling filters. Developments in mechanized equipment have lessened its popularity, but it is still valued as a combination unit for settling sewage and digesting sludge. *See* SEPTIC TANK; SEWAGE.

The Imhoff tank is constructed with the flowing-through chamber on top and the digestion chamber on the bottom (see illustration). The upper chamber is designed according to the principles of a sedimentation unit. Sludge drops to the bottom of the tank and through a slot along its length into the lower chamber. As digestion takes place, scum is formed by rising sludge in which gas is trapped. The scum chamber, or gas vent, is a third section of the tank located above the lower chamber and beside the upper chamber. As gases escape, sludge from the scum chamber returns to the lower chamber. The slot is so constructed that particles cannot rise through it. A triangle or sidewall

deflector below the slot prevents vertical rising of gas-laden sludge. Sludge in the lower chamber settles to the bottom, which is in the form of one or more steep-sloped hoppers. At intervals the sludge can be withdrawn. The overall height of the tank is 30–40 ft, and sludge can be expelled under hydraulic pressure of the water in the upper tank. Large tanks are built with means for reversing flow in the upper chamber, thus making it possible to distribute the settled solids more evenly over the digestion chamber.

Design. Detention period in the upper chamber is usually about $2\frac{1}{2}$ hr. The surface settling rate is usually 600 gal/(ft²)(day). The weir overflow rate is not over 10,000 gal/ft of weir per day. Velocity of flow is held below 1 ft/sec. Tanks are dimensioned with a length-width ratio of 5:1–3:1 and with depth to slot about equal to width. Multiple units are built rather than one large tank to carry the entire flow. Two flowing-through chambers can be placed above one digester unit. The digestion chamber is normally designed at 3–5 ft³ per capita of connected sewage load. When industrial wastes include large quantities of solids, additional allowance must be made. Ordinarily sludge withdrawals are scheduled twice per year. If these are to be less frequent, an increase in capacity is desirable.

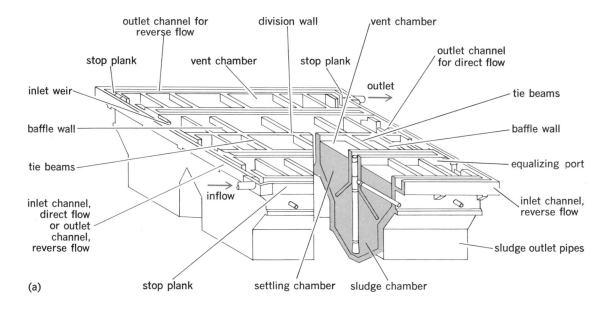

(a)

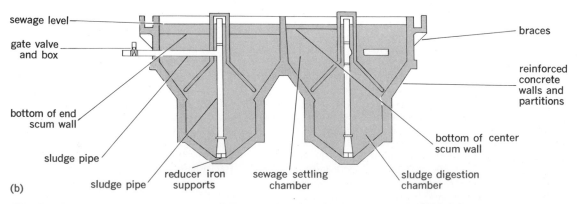

(b)

Diagram of typical large Imhoff tank for sewage treatment. (*a*) General arrangement. (*b*) Cross section. (*From H. E. Babbitt and E. R. Baumann, Sewerage and Sewage Treatment, 8th ed., Wiley, 1958*)

Some chambers have been provided with up to 6.5 ft³ per capita. The scum chamber should have a surface area 25–30% of the horizontal surface of the digestion chamber. Vents should be 24 in. wide. Top freeboard should be at least 2 ft to contain rising scum. Water under pressure must be available to combat foaming and knockdown scum.

Efficiency. The efficiency of Imhoff tanks is equivalent to that of plain sedimentation tanks. Effluents are suitable for treatment on trickling filters. The sludge is dense, and when withdrawn it may have a moisture content of 90–95%. Imhoff sludge has a characteristic tarlike odor and a black granular appearance. It dries easily and when dry is comparatively odorless. It is an excellent humus but not a fertilizer. Gas vents may occasionally give off offensive odors.   [WILLIAM T. INGRAM]

Imidazole

One of a group of organic heterocyclic compounds (also called iminazoles, glyoxalines, and 1,3-diazoles) containing a five-membered diunsaturated ring with two nonadjacent nitrogen atoms as part of the ring. Imidazole (I) is a typical member of the group. *See* AZOLE; HETEROCYCLIC COMPOUNDS.

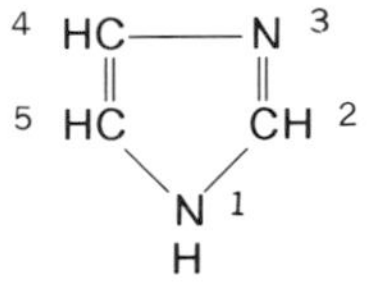

(I)

The imidazole ring system is found in a number of natural products, for example, in the α-amino acid histidine (II) and in the alkaloid pilocarpine (III). Histamine (IV) is associated with allergic response, and ostensibly is the biological target against which the antagonistic action of synthetic antihistaminics is directed. The biologically impor-

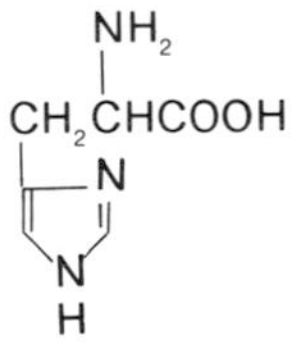

(II) 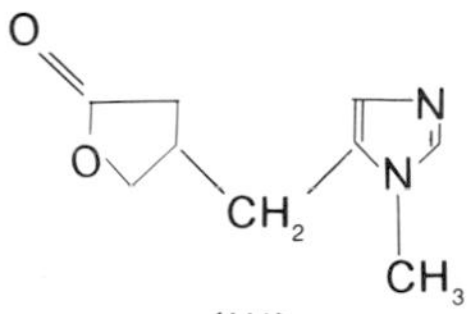   (III)

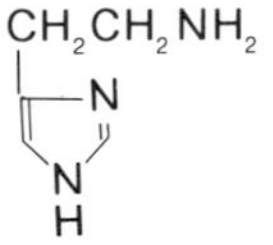

(IV)

tant purine system contains an imidazole ring fused to pyrimidine. The imidazole ring, present in enzyme proteins as the histidine side chain, is involved in enzyme-catalyzed reactions, presumably by serving as an efficient acyl-transfer agent. The same kind of imidazole ring in the blood protein, globin, holds heme and globin together by coordinating with the iron atom of the heme.

Properties and preparation. Imidazole itself is a water-soluble solid, mp 90°C, which is basic enough (pK_a 6.95 at 25°) to form stable salts with both organic and inorganic acids. Imidazole is also weakly acidic, since the hydrogen at the 1 position may be replaced by metal. The low volatility of

imidazole, bp 256°C, is indicative of considerable association by intermolecular hydrogen bonding.

Imidazole is a resonance system showing the chemical behavior of a moderately aromatic ring. The system is stable to oxidation by nitric acid, hexavalent chromium, and alkaline permanganate, and it is, in general, resistant to ring reduction. Amino groups at position 4 may be diazotized normally. Imidazole undergoes electrophilic substitutions such as bromination, nitration, or sulfonation at position 4, and azo coupling at position 2. The imidazole ring is opened by attack of peroxide and peracids. Alkylation on nitrogen is possible to give first 1-alkylimidazoles and then 1,3-dialkylimidazolium cations. Hot alkali disrupts the ring in the quarternary imidazolium compounds.

Several general methods of synthesis are known. Combination of α-halo ketones with amidines (V) gives imidazoles, a process emphasizing

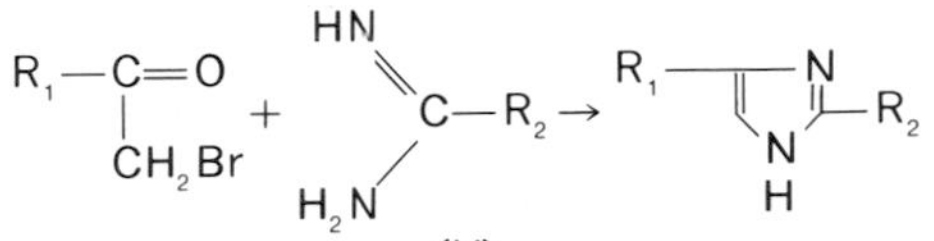

(V)

the amidine structure of imidazole. α-Dicarbonyl compounds react with ammonia and aldehydes to give imidazoles (VI). Further, α-amino ketones or

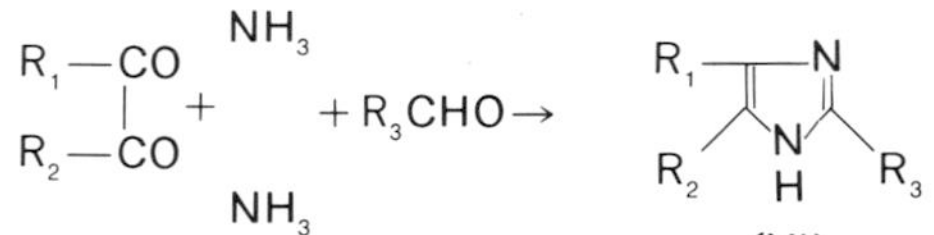

(VI)

aldehydes (VII) condense with thiocyanate to give 2-mercaptoimidazoles (VIII).

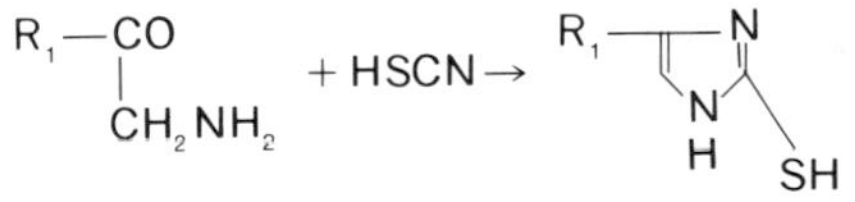

(VII)   (VIII)

Important derivatives. Imidazolines and imidazolidines are the names assigned to dihydro- and tetrahydroimidazoles, respectively. Such compounds are generally formed by ring-closure processes involving positions 1, 2, or 3. Ring carbonyl derivatives are known. Biotin (IX), for example, is a condensed 2-imidazolidone.

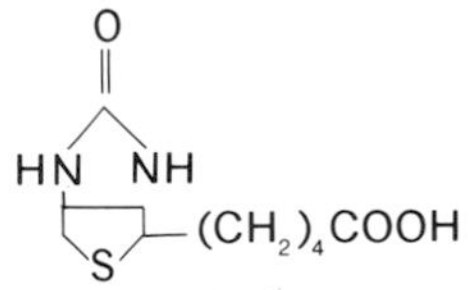

(IX)

Hydantoins or 2,4-diketoimidazolidines (X) are prepared by condensation of a ketone or aldehyde with potassium cyanide and ammonium carbonate.

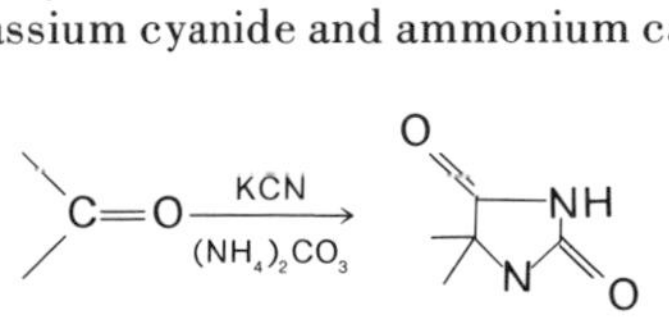

(X)

Hydantoins have been studied extensively in connection with the physiological activity of 5,5-

disubstituted derivatives and because of the possibility of converting hydantoins to α-amino acids. The 5,5-diphenyl derivative, Dilantin sodium (XI), is used as an anticonvulsant in treatment of epilepsy. The 5-ethyl-5-phenyl derivative, Nirvanol (XII),

is an effective hypnotic. Complete hydrolysis of hydantoins generates an α-amino acid (XIII, XIV).

This route constitutes one of the standard syntheses for α-amino acids. 1,3-Dibromo- and 1,3-dichloro-5,5-dimethylhydantoin are convenient sources of positive halogen.

Allantoin, or 5-ureidohydantoin (XV), is the end product of purine metabolism in most mammals

(but not including man) and is an intermediate in the purine metabolism of crustaceans and amphibia. Creatinine (XVI) is formed irreversibly from creatine and is excreted.   [WALTER J. GENSLER]

Bibliography: R. N. Acheson, *An Introduction to the Chemistry of Heterocyclic Compounds*, 1967; R. C. Elderfield (ed.), *Heterocyclic Compounds*, vol. 5, 1957; K. Hofmann, *Imidazole and Its Derivatives*, pt. 1, 1953.

Immunity

A term used in medicine to denote a state of resistance to an agent, the parasite, that normally produces an infection in one or more host species. Analogous immune responses also occur to tumors (cancers), in allergies and delayed hypersensitivity states, in autoimmune diseases, and in incompatible blood transfusions or tissue and organ transplantations. These responses are specifically lacking in immune deficiency diseases. Immunity is a complex resultant of many components, some native and hereditable, others acquired. In acquired immunity both the invading parasite and the host may alter their offensive and resistive capacities with time. Many pathogenic organisms possess chemical components—enzymes, toxins, or surface antigens—which adversely affect the suceptible host. After contact with these substances (antigens) the host may, in turn, produce neutralizing or inhibiting substances (generally, specific antibodies), which check the infection. Alternatively, specifically sensitized lymphocytes may produce an effective cellular immunity. Meanwhile, the pathogen may mutate so that its surface antigens or toxins vary chemically, and host resistance is then not effective until new antibodies are formed. Influenza constitutes a notorious example among viruses; for example, the 1968–1969 epidemic of Asian and Hong Kong flu was due to new antigenic subtypes.

Natural immunity. Domestic cats, like nearly all mammals, never contract measles; they are naturally insusceptible. In contrast, humans and monkeys, on an experimental basis, readily acquire the disease, but after a single infection quite generally resist subsequent infection. It is remarkable that animals, including humans, are generally insusceptible to even the most virulent plant pathogens, and plants are likewise insusceptible to animal pathogens.

The conditions of exposure must be taken into account in analyzing natural immunity. For example, mice are not known to contract pneumococcal pneumonia under any natural conditions, and normal mice may be kept indefinitely, without infection, in cages filled with mice dying of an experimental pneumococcal infection. The natural barriers to acquiring or transmitting the disease appear absolute here; nevertheless, if these barriers are bypassed by the intraperitoneal injection of even one virulent organism, the mouse becomes exceedingly susceptible to a fatal infection.

Immunity mechanisms. An invading pathogen may fail to win a foothold because the environment lacks one or more growth factors or physical conditions necessary for multiplication, as in the case of an adenine-requiring enteric bacterium that cannot infect the mouse unless the chemical adenine is also injected. Alternatively, the natural resistance of dogs to anthrax appears to be due to active destruction of the organism by substances uniformly present in dog tissues.

Some barriers, such as skin or the mucous membrane of the nasal and respiratory tract, may mechanically prevent entrance of microorganisms, or they may contain chemical components that prevent multiplication of bacteria or actively destroy them (lysozyme, spermine, phagocytin, and so forth). Even if they do not constitute absolute barriers, both actions may delay the process of infection until other defenses can be brought into play. Foremost among these are the various fixed and wandering cells that comprise the phagocytic systems. Their actions are complex: (1) They may engulf and destroy certain microorganisms directly; or (2) they may engulf and kill other virulent organisms only if aided by accessory humoral factors, such as antibody and complement; and (3) in other cases, while they may engulf microorganisms such as the gonococcus, the latter may then remain in the cell, secure from the accessory substances or sometimes chemotherapeutic drugs, and the disease may then progress to a chronic state. Antibody, and complement may help not only to prepare microorganisms for effective phagocytosis but also may have direct bactericidal effects by themselves. True antibodies together with sensitized lymphocytes have been found only in vertebrate species. While invertebrate animals may produce relatively nonspecific agglutinins and lysins, their primary defense rests with phagocytic cells which can ingest or encapsulate microorganisms. Sensitized lymphocytes can destroy tumor cells and certain microorganisms as well as produce delayed hypersensitivity and graft rejection reactions. *See* ANTIBODY; COMPLEMENT, SERUM; PHAGOCYTOSIS.

Although the various barriers, phagocytic systems, and humoral and cellular factors appear to constitute the active mechanisms by which a level of immunity is achieved, their qualitative and quantitative activities are governed by a variety of other host factors, among which a genetic component has been identified. While this is difficult to evaluate for man, statistical evidence suggests that man's susceptibility to rheumatic fever, poliomyelitis, tuberculosis, leprosy, and various allergies is influenced by heredity. More decisive evidence for a genetic factor in immunity has come from experiments with animals and plants. Crops with greater resistance to wheat rusts and other diseases are important results of agricultural research. Age, diet, stress, hormones, radiation, and associated intestinal flora also have significant influences on immunity.

Active immunity. Recovery from certain virus diseases, such as yellow fever, measles, and chickenpox, is usually attended by a lifelong immunity effective under the conditions of natural exposure. At the other extreme, immunity following recovery from the common cold is very shortlived or nonexistent. In many bacterial diseases, such as pneumococcus pneumonia or diphtheria, effective immunity extends only for a few years at best, unless a renewed antigenic stimulus is applied. This is often accomplished naturally through the carrier state, in which either virulent or nonvirulent but immunizing strains are harbored by the subject without evidence of clinical disease. Under favorable conditions, the carrier state may also effect a primary immunity.

As a special case, immunity to new infection (superinfection) may hold only as long as the original infection persists. This is known as infection-immunity or premunition. Also, in certain cases infection with one viral species may temporarily prevent establishment of a second infection with a related virus or a more virulent strain of the original virus, a phenomenon known as virus interference. In many but not all cases, interference is due to the production by the animal cell of one of a chemical group of substances (interferons) which restrict further virus synthesis. The therapeutic possibilities of interferons are currently under investigation.

Immunization. It is an objective of preventive medicine to produce a prophylactic immunity through vaccination without the inconveniences and often dangers of an initial attack of the disease. The oldest procedure, vaccination against smallpox, was introduced by E. Jenner in 1796. It employs a living agent, a calfpox, antigenically related to the smallpox virus. Living attenuated vaccines against poliomyelitis, tuberculosis, yellow fever, and bubonic plague have also been widely employed. Nonliving vaccines are commonly used for the prophylaxes of the bacterial diseases pertussis, typhoid, and cholera; the viral disease influenza; and the bacterial intoxications of diphtheria, tetanus, and botulinus. Repeated or booster doses of vaccine are commonly given from 1–5 years after the first immunization course, or oftener under conditions of special risk.

Passive immunization. Since protective levels of antibody are not formed in response to an antigen until some weeks or months after birth, the newborn would be at a disadvantage unless this transition period were provided for. In humans and in some animals this occurs during pregnancy through the passive transfer, across the placenta, of antibodies circulating in the maternal blood. In other animals antibody is transferred via the first milk (colostrum) taken after birth. These antibodies are only slowly eliminated.

Passive transfer of antibody may also be accomplished artificially when there is need of specific protection within the 1–4 weeks normally required for antibody synthesis after vaccination. Before the advent of the sulfonamides and antibiotics, extensive use was made of immune animal serums in the treatment of pneumococcal pneumonia, and continued use is made of diphtheria and tetanus antitoxins. A concentrated preparation of human serum gamma globulin containing antibodies against measles and mumps viruses is frequently employed in the prophylaxes of these diseases. The protection afforded is only temporary. Serum from the homologous species is effectively eliminated, but without untoward reaction, in 2–3 weeks. Serum from a heterologous species, although effective for 7–10 days, is then rapidly eliminated as a foreign antigen, while often producing the syndrome of serum sickness. On the repeated administration of a foreign serum, a sensitized individual may give an immediate anaphylactic reaction that may be severe or even fatal if the proper precautions are not observed. *See* IMMUNOGLOBULIN.

Cellular immunity to a specific material can also be passively transferred either with sensitized lymphoid cells or—to humans—with soluble extracts of these (transfer factors). Functional deficiencies (as in immune deficiency diseases) have been restored through transfers of spleen or other lymphoid cells, but care must be taken to avoid graft-versus-host rejection reactions. *See* IMMUNOLOGY, CELLULAR.

Immunity and chemotherapy. Chemotherapy in an overt infectious disease usually does not exclude the development of significant immunity. If, however, exposure to the microbial antigens is insufficient, the patient may still remain susceptible, and if the risk of reinfection is high, a more active immunity should be provided by vaccination. This is especially pertinent if chemotherapy is applied during the incubation of diseases such as scrub typhus, since only a suppressive action results and, unless antibody is present, relapses occur when the drug is discontinued.

Hypersensitivity reactions. It may be expected that the reactions mediated by host antibody produced to foreign antigens will be beneficial to the host. This is not always true; in fact, some antigen-antibody reactions result in extensive host pathology and even death. Such reactions, known collectively as hypersensitivity reactions, may be subdivided as follows: immediate systemic or cutaneous anaphylaxis, serum sickness, and

Arthus reactions, all mediated by conventional antibodies (such as γG); atopic allergies, such as poison ivy or pollen sensitivities, mediated by γE; and the various delayed reactions, such as tuberculin-type reactions and graft-rejection reactions, mediated only by cell-bound antibodies. In addition, there is a highly varied group of so-called autoimmune diseases such as thyroiditis, systemic lupus erythematosus, or rheumatoid arthritis in which antibodies are produced to some host constituent. *See* HYPERSENSITIVITY.

Immunologic suppression. In view of these often unfavorable results of antigen-antibody reactions, extensive research is being conducted on the reduction or suppression of antibody production, particularly in graft transplants. Among the agents currently used are x-irradiation, chemicals such as folic acid analogs, corticosteroid hormones, and antilymphocyte antisera. Their use is, however, attended with some danger of general immune suppression and consequent bacterial infection. A different approach makes use of the phenomenon of immune tolerance. *See* ANTIGEN-ANTIBODY REACTION; IMMUNOLOGICAL TOLERANCE, ACQUIRED; IMMUNOPATHOLOGY.

[HENRY P. TREFFERS]

Bibliography: B. D. Davis et al., *Microbiology*, 1973; K. Maramorosch and R. E. Shope, *Invertebrate Immunity*, 1975; R. Roitt, *Essential Immunology*, 1971; M. Samter (ed.), *Immunological Diseases*, 1971; S. Sell, *Immunology, Immunopathology and Immunity*, 1975; G. A. Wilson and A. A. Miles (eds.), *Topley and Wilson's Principles of Bacteriology and Immunity*, 2 vols., 1975.

Immunoglobulin

Any of a set of serum glycoproteins which have the ability to bind other molecules with a high degree of specificity. Molecules which are foreign or non-self are called antigens; and when an antigen is introduced into a vertebrate, the immunoglobulin induced is called an antibody. The critical property of this antibody is that it will combine specifically with the inducing antigen.

In the living animal, antibodies perform the two coordinated functions of binding to a foreign molecule or antigen and, ideally, eliminating it as a threat to the host. Binding to antigen is achieved through the combining site, a cleft or groove which is complementary to the size, shape, and charge characteristics of the antigen. Bound antigen is frequently inactivated by one or more of the effector functions of the antibody such as opsonization or complement fixation. *See* COMPLEMENT, SERUM; OPSONIN.

By using simple antigens under carefully controlled experimental conditions, antibodies have been induced which are homogeneous in affinity, charge, and amino acid sequence. The other form of homogeneous immunoglobulin is the myeloma paraprotein. These are immunoglobulins of antibody-producing cells which have undergone neoplastic transformation and may be secreted in such copious amounts that their serum concentration may exceed that of albumin. Human myeloma proteins were the first structurally discrete immunoglobulins known, and they have played a central role in the elucidation of the antigenic, structural, and functional features of immunoglobulins. Myeloma proteins appear to result from the malignant expansion of a normal clone of antibody-producing cells; but since they are spontaneous or nonspecifically induced, they have no predictable antibody activity.

Classes of immunoglobulins. Immunoglobulins are heterogeneous with respect to charge, size, antigenicity, and function. Although charge heterogeneity was initially used to define the "γ-globulins," charge is the least useful of these criteria for precise characterization. Antibodies and myeloma proteins exhibit gamma or slow beta electrophoretic mobility, but groupings arranged by migration do not conform closely to those determined by other criteria. The first separation of immunoglobulins by size was in 1937 when horse antipneumoccal antibody was separated into two different molecular weight groups: 900,000 daltons and 150,000 daltons. The larger antibody has been named immunoglobulin macro (IgM) and the smaller immunoglobulin gamma (IgG), reflecting its electrophoretic mobility. *See* ELECTROPHORESIS.

Antigenic analysis is the most practical approach in differentiating the classes and subclasses of immunoglobulins. The three most abundant classes in normal human serum, IgG, IgA, and IgM, are easily detected and resolved by heterologous antisera to whole serum. Two other classes of antigenically distinct immunoglobulins, IgD and IgE, were first recognized as myeloma proteins because their concentration in normal human serum was too low to detect by the then available techniques.

Some patients with multiple myeloma secrete a protein into their urine which shares some of the antigenic determinants of myeloma proteins. These Bence-Jones proteins exist as two antigenic types, kappa and lambda. All immunoglobulins share determinants with either kappa- or lambda-type Bence-Jones proteins, and a complete description of an immunoglobulin molecule requires identification of both class and type.

IgG. IgG is the most abundant immunoglobulin in serum, generally being present at around 1200 mg %. The IgG molecule has a molecular weight of 150,000 and consists of two pairs of polypeptide chains which are covalently linked by disulfide bonds.

The subunit structure of IgG was elucidated by data from two independent lines of investigation. When IgG is broken into subunits by reduction of disulfide bonds and the subunits are separated, the 150-kilodalton structure is resolved into two polypeptides, one of 50 kdal molecular weight and one of 25 kdal molecular weight. Since the polypeptides were obtained in equal yield, it was apparent that the intact molecule consisted of two of the 50-kdal heavy chains (H) and two 25-kdal light chains (L).

The other means used to determine the subunit structure was proteolytic cleavage. Pooled rabbit IgG digested with papain results in two major fragments. One of the fragments is present in twice the yield of the other, contains the light chain, and is able to bind univalently to antigen. This is called the Fab, for "fragment, antigen binding." The other fragment is crystallizable, contains no light chain, and does not bind antigen. This is called Fc, for "fragment, crystallizable." The relationship

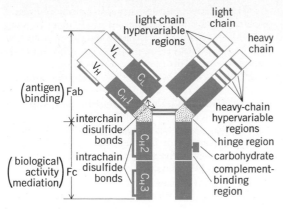

Fig. 1. Topology and functional architecture of the gamma-G molecule. V_L and V_H are variable regions; C_L and C_H are constant regions.

between the chain structure and the proteolytic fragments is well known, and the model shown in Fig. 1 is a current conceptualization of the IgG molecule. This basic subunit structure is the prototype for all immunoglobulins. IgG molecules are a monomer of this H_2L_2 four-chain configuration.

The complete amino acid sequence of an IgG myeloma has been achieved. The heavy chain is 446 amino acids long and consists of four homology regions of approximately 110 residues, each with an intrachain disulfide loop. The light chain is 224 amino acids long and is homologous to previously sequenced Bence-Jones proteins.

The knowledge of these homology units taken with information on the susceptibility of certain regions of the molecule to enzymatic attack and the functional segregation in the Fab and Fe led to the postulation of the domain hypothesis. This hypothesis proposed that each homology region within a disulfide loop evolved to serve a unique function. The hypothesis has been supported by work which showed that an IgG molecule could be cleaved into domains and that each domain carries separate discrete functions.

Apart from the antigen-binding function, antibodies possess certain biologic effector functions which are instrumental to the protection of the host. These functions include complement fixation, opsonization, skin fixation, fixation to macrophages, and membrane transport. While the molecular events responsible for these functions have not been clearly defined, biologic activities presumably result from the interaction of a portion of the IgG molecule and a membrane or enzyme receptor site.

IgA. IgA is the second most abundant immunoglobulin in human serum but is the primary immunoglobulin of secretions. The nature of IgA differs widely among various mammals and between serum and secretions.

The most common form of IgA in human serum is a four-chain monomer following the H_2L_2 prototype of IgG. IgA also exists in several polymeric forms, ranging up 18S in human serum. Polymeric IgA contains an additional polypeptide called J chain which serves to join the monomers.

In human secretions—milk, saliva, tears, and bronchial fluids—the predominant immunoglobulin is dimeric secretory IgA (Fig. 2), which is made up of two IgA monomers, a J chain, and a glyco-

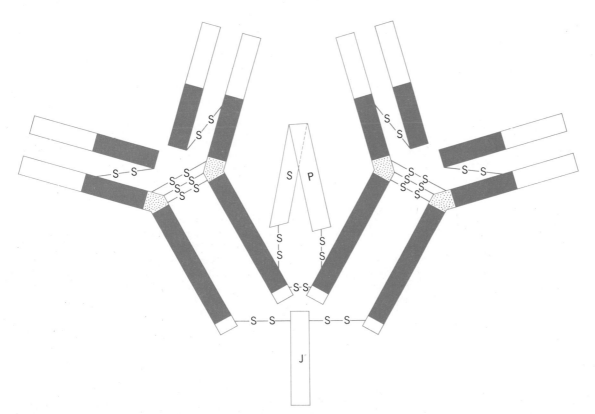

Fig. 2. Schematic representation of human secretory IgA. The model is approximately to scale and indicates the location of attachment of secretory piece (SP) and J chain (J).

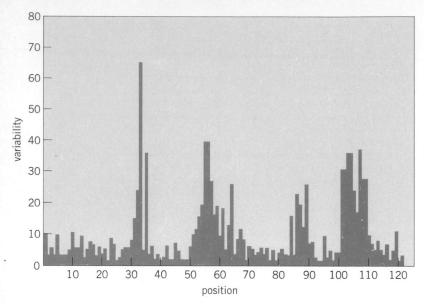

Fig. 3. Variability analysis of the amino acid sequences of human immunoglobulin heavy-chain variable regions. Four distinct regions of hypervariability are evident.

peptide called secretory component.

Increased incidence of disease in IgA-deficient individuals had for many years been the only evidence, albeit indirect, of the biologic effector activity of IgA. However, IgA has been shown to bind to neutrophils suggesting opsonic activity and to fix complement via the alternate pathway.

J chain. J chain is a polypeptide of unusual shape and charge which is present in polymeric but not monomeric immunoglobulins. Both IgA and IgM contain a single J chain per polymer which is disulfide-linked to the penultimate cysteine of the heavy chain. The J chain is necessary for polymerization and binds two monomeric units, forming the complete IgA dimer but only serving to initiate polymerization of the IgM pentamer.

Secretory component. Secretory component is synthesized by epithelial cells and becomes associated with IgA during transport through the human gastrointestinal lumen. In the human, secretory piece is covalently bound to the alpha chain. There appears to be one secretory piece per SIgA dimer. The function of secretory piece is unknown, but the possibilities include action as a transport protein or a shield from the enzymatic activity of the gut.

IgM. Immunoglobulin macro or IgM is the largest immunoglobulin, with a molecular weight of about 900 kdal. The IgM molecule is a polymer of five IgG-like H_2L_2 monomers joined by a J chain.

Electron-microscopic studies show that the monomeric units are arranged like the spokes of a wheel, with the Fc portions at the hub and the antigen-binding Fab portions directed outward.

The mu chain is approximately 110 residues longer than the gamma chain and has an additional domain. The complete amino acid sequence of a mu chain has been achieved. Comparison of the mu chain with the previously sequenced gamma chain and limited sequence data on the alpha chain shows that the mu and alpha chains are more homologous than either the mu and gamma or the gamma and alpha pairs.

IgM is the first immunoglobulin to appear during the primary immune response and is usually present only transiently. IgM also binds to Clq, the first component of complement, and thereby fixes complement via the classical pathway.

IgD and IgE. IgD and IgE are present in minute amounts in normal human serum: IgD, .03 mg/ml; IgE, 0.0001 mg/ml. Most of the structural information available on these immunoglobulins was obtained by examining human myeloma proteins. IgD consists of two light and two heavy chains. The delta heavy chain has a molecular weight of approximately 60 kdal, but 11.3% of that is carbohydrate and the weight of the peptide portion is close to that of the gamma chain. No function has as yet been attributed to IgD. However, evidence suggests it may play a prominent role as a cell surface immunoglobulin and may serve as an antigen receptor.

An immunoglobulin class carrying reaginic (immediate hypersensitivity or allergy) activity was postulated on the basis of fractionations of atopic sera. This fraction of serum with atopic activity was shown to be structurally related to a myeloma protein and distinct from the four known classes of immunoglobulin. IgE has a molecular weight of 196 kdal, and the epsilon chain has five domains like the mu chain.

IgE functions as the mediator of immediate hypersensitivity. After binding to antigen, it causes the degranulation of mast cells and the release of histamine. There is an as yet unlocalized site on the Fc portion of the molecule which binds to basophils and mast cells prior to the reaction with antigen. *See* HYPERSENSITIVITY.

Antibody combining site. Initial structural studies of IgG placed the antigen-binding portion of the molecule in the Fab region. The binding site was more specifically localized when the Fab fragment of a mouse myeloma protein with anti-DNP activity was shown to maintain its antibody activity after further cleavage by pepsin. The pepsin cleavage product consisted of the amino terminal domains of the light and heavy chain. Thus, the V regions alone possess full antibody activity.

Variable regions. The unique characteristic of immunoglobulin chains became obvious as sequence data from Bence-Jones proteins was obtained. Proteins of a given antigenic type were identical in the carboxyterminal portion but were markedly different near the amino terminus. In further studies it was determined that the region of sequence variability extends to positions 105–107 in the light chain and that a similar region extends in the heavy chain to approximately residue 124. The amino terminal domains of both light and heavy chains are characterized by this variability and are termed variable regions V_L and V_H.

Extent of variability. Despite the great variability of the variable regions, there are stretches of both light and heavy chain which are invariant. The position of the disulfide loop of human kappa chains is invariant, and the immediately adjacent residues show a very low variability. Thirteen human myeloma heavy-chain variable regions had been sequenced by 1975. About 65% of the positions in human heavy chains of the V_HIII subgroup exhibit limited variability over the amino terminal 124 residues.

Within the relatively constant portion or framework of the variable regions are several hypervariable regions. Light chains have three hypervariable regions around positions 28, 50, and 96. Human heavy chains exhibit four hypervariable regions: 31–37, 52–60, 86–91, and 101–110 (see Fig. 3).

X-ray crystallography. The ultimate description of the antigen-binding site is possibly by crystallographic analysis. A 2.8 A (0.28 nm) Fourier map of the Fab′ fragment of a human myeloma defines the structure of this fragment. The hypervariable regions of both the light and heavy chains are in close proximity at the surface of the molecule. The hypervariable regions form a shallow groove, 15×6 A (1.5×0.6 nm) deep. A similar study of a mouse myeloma with antiphosphorylcholine activity with and without bound hapten has been achieved. By difference map, the hapten is shown to be located in a crevice formed by the position of the light- and heavy-chain V regions. *See* ANTIBODY; ANTIGEN; IMMUNOLOGY; PROTEIN.

[J. DONALD CAPRA]

Bibliography: J. D. Capra and J. M. Kehoe, Variable region sequences of five human immunoglobulin heavy chains of the V$_H$III subgroup. Definitive identification of four heavy chain hypervariable regions, *Proc. Nat. Acad. Sci.*, 71:845–848, 1974; R. J. Poljak et al., The three-dimensional structure of the Fab′ fragment of a human myeloma immunoglobulin at 2.0 A resolution, *Proc. Nat. Acad. Sci.*, 71:3440–3444, 1974; C. E. Wilde and M. E. Koshland, Molecular size and shape of the J chain from polymeric immunoglobulins, *Biochemistry*, 12:3218–3224, 1973.

Immunological tolerance, acquired

Failure of immunological responsiveness, brought about by exposure to antigen, in which antigen-sensitive cells become unable to initiate synthesis of a restricted range of antibodies. Through induced tolerance, animals can be prevented from rejecting grafts of foreign cells (Figs. 1 and 2); rendered more sensitive to infection with pathogens; protected from experimentally induced autoimmune disease; and prevented from developing hypersensitivity to drugs. Tolerance derives its principal practical interest from possible future applications in surgery. It is hoped that permanent survival of transplanted organs will be obtained by inducing tolerance of donor antigen in the host, as can already be done in animal experiments. Other possible applications include treatment of allergy and autoimmune disease. *See* ANTIBODY; ANTIGEN; HYPERSENSITIVITY; IMMUNITY.

The terms immunological tolerance and immunological paralysis can be used interchangeably, although the former is applied mainly to tissue transplants and the latter only to purified antigens. Unresponsiveness to both types of antigen can be acquired in the same way, and probably involves the same cellular mechanism.

Purified protein and carbohydrate antigens induce tolerance if they can be maintained in the body for long enough and at a high enough concentration, and provided that an immune response is not provoked (low-zone tolerance). Conditions which reduce the likelihood of immunization therefore favor induction of tolerance, including treat-

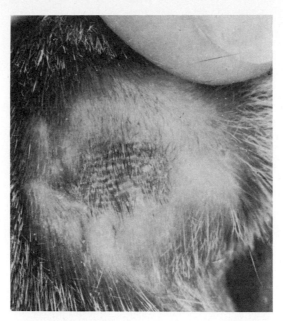

Fig. 1. Tolerated graft of A-strain skin growing on a CBA-strain mouse. Tolerance was induced by the injection of 5,000,000 A-strain lymphoid cells into the animal at birth. (*Courtesy of N. A. Mitchison*)

ment of the newborn, treatment with radiation or immunosuppressive drugs, and avoidance of nonspecific agents (adjuvants) which enhance the immune response. Some protein antigens do not immunize, but can still induce tolerance if administered simply in small enough doses. Tolerance can still be induced after immunization, but much larger quantities of antigen are then required (high-zone tolerance).

Recovery. Recovery from a state of tolerance proceeds spontaneously after antigen has been eliminated from the body. If induced by transplantation of cells, tolerance may persist indefinitely; the foreign cells survive and the animal is termed a chimera. Otherwise, recovery proceeds more rapidly in younger animals, and can be prevented by surgical removal of the thymus gland (thymectomy). Recovery is therefore thought to proceed by thymus-mediated recruitment of antigen-sensitive cells. *See* MOSAICISM AND CHIMERISM; THYMUS GLAND (VERTEBRATE).

Specificity. In the context of transplantation immunity tolerance is fully specific: If CBA-strain

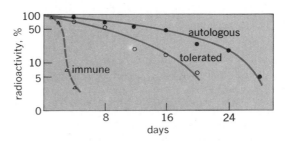

Fig. 2. Elimination of Cr^{51}-labeled erythrocytes in the chicken. Autologous: taken from the same individual. Immune: taken from a member of another strain. Tolerated: taken from a member of another strain to which the recipient had been rendered tolerant by repeated blood transfusions commencing at hatching.

cells are injected into a newborn A-strain mouse, when that mouse grows up it will be found to accept any homograft of CBA origin, although it will continue to reject homografts from third parties, for example, from mice of strains C3H or C57. In the context of tolerance of protein antigens the specificity mechanism recognizes molecules as a whole, rather than the individual antigen determinants, which antibodies recognize. Thus, for example, rabbits tolerant of human serum albumin can be immunized with bovine serum albumin, and will then produce antibodies not only against those determinants which were not present on the original tolerance inducer, but also against determinants held in common by the two antigens.

Cellular basis. That tolerance is due to a failure on the part of antigen-sensitive cells (lymphocytes) to respond has been demonstrated by cell transfers between inbred animals. Normal lymphocytes can be transferred adoptively into tolerant animals, which are thereby restored to normal immunological reactivity. Lymphocytes from tolerant donors will not act in this way, nor will they restore immunological reactivity to irradiated hosts or perform graft-versus-host reactions. Phagocytic cells from tolerant animals are normal in function.

Theories. These findings on the whole support the theory of Sir MacFarlane Burnet that tolerance is the consequence of eliminating the clones of cells competent to respond to the inducing antigen. Whether elimination or stimulation of the clone takes place is thought to depend on the mode of presentation of antigen, a point at which the intervention of macrophages may exercise a decisive influence. This theory is not universally accepted, and the problem of tolerance is the subject of lively but as yet inconclusive discussion. *See* IMMUNOLOGY; IMMUNOPATHOLOGY; TRANSPLANTATION BIOLOGY.   [N. A. MITCHISON]

Bibliography: D. W . Dresser and N. A. Mitchison, *Advan. Immunol.*, 8:129, 1968; H. W. Florey (ed.), *General Pathology*, 3d ed., 1962; J. L. Gowans and D. D. McGregor, *Progr. Allergy*, 9:1, 1965; J. H. Humphrey and R. G. White, *Immunology for Students of Medicine*, 1964; W. C. Topley and G. S. Wilson, *Principles of Bacteriology and Immunity*, 5th ed., 1964.

Immunology

The division of biological science concerned with the native or acquired resistance of higher living forms, or hosts, to infection with microorganisms. As the formal science of immunology has developed, it has remained restricted to the study of human or animal resistance, the corresponding reactions of plants being generally considered as falling within the scope of plant pathology. Immunology is an eclectic science drawing on many other branches of knowledge. The comprehensive study of infection and resistance may be divided, arbitrarily, into data relating to the pathogen, the host, and their interactions during infection. *See* ANTIGEN; DISEASE; INFECTION; PATHOGEN; PLANT DISEASE; TOXIN, BACTERIAL.

For the pathogen, data are needed on its classification and its potentialities for causing disease, as well as its mode of host entry and exit. Its survival in the host after active infection (carrier state) is of consequence for both immunity and public health. The antigens, toxins, and other specialized chemical constituents of pathogens have received considerable study.

Important information on the host includes its species (a prime variable) and the anatomical, physical, genetic, and nutritional factors that may influence infection and resistance. Host-parasite interactions include, first, some classification of disease types, because one microbial species (such as the hemolytic streptococcus, *Streptococcus pyogenes*) may cause multiple diseases (septicemia, boils, scarlet fever, pneumonia, and rheumatic fever), with diverse immunological consequences. Several types of biometric designs exist for quantitating the extent of infection. Many manifestations of pathology (for example, fever and inflammation) influence the reactivity of both host and parasite. Of prime concern to immunology are the cellular factors (phagocytes) and the humoral factors (antibodies, complement, and properdin) that are significant aids in host resistance. The protective role of allergy is much less clear. *See* ANTIBODY; BIOASSAY; BIOMETRICS; COMPLEMENT, SERUM; FEVER; PATHOLOGY.

Immunology is also heavily concerned with assaying the immune status of the host through a variety of serological procedures, and in devising methods of increasing host resistance through prophylactic vaccination. There has also been much important investigation of induced resistance and tolerance to transplants of skin and organs, including tumors. *See* AGGLUTININ; ANTIGEN-ANTIBODY REACTION; BIOLOGICALS; BLOOD GROUPS; COMPLEMENT-FIXATION TEST; HYPERSENSITIVITY; IMMUNITY; IMMUNOLOGICAL TOLERANCE, ACQUIRED; INTERFERON; ISOANTIGEN; PHAGOCYTOSIS; PRECIPITIN; QUELLUNG REACTION; SEROLOGY.   [HENRY P. TREFFERS]

Bibliography: B. D. Davis et al., *Microbiology*, 1967.

Immunology, cellular

The study of the cells of the lymphoid organs, which are the main agents of immune reactions in all the vertebrates. Immune responses to bacteria, viruses, or other pathogens (antigen, Ag) may be humoral (production of antibody, Ab) or cellular (cell-mediated immunity, CMI). Ab molecules are proteins in the circulating blood; at least five classes which differ in molecular size, carbohydrate content, and combining affinity for Ag may be formed. CMI is expressed by local mobilization of highly phagocytic inflammatory cells at sites of Ag penetration. Both Ab formation and CMI are specific in that they react only with the original Ag or a closely related form. Both show "memory," expressed by an accelerated and enhanced response on second exposure to the same Ag. Both are hedged about with elaborate regulatory mechanisms to prevent excessive reactions and, particularly, reactions against self-components ("tolerance" to auto-Ags). The study of the cells which underlie these phenomena, their origin, character, life span, and reactions, and especially their interactions (both cooperative and antagonistic) has become the fastest-growing field in biomedical science. *See* ANTIBODY; ANTIGEN.

Lymphocytes. The central cell of immunology is the lymphocyte, a small, round, featureless cell, common in the blood and lymph and often found wandering in the tissues. Two types of lymphocytes are derived from stem cells in the yolk sac, fetal liver, and bone marrow. T lymphocytes undergo maturation and proliferation in the thymus, an organ in the upper chest derived from the third and fourth pharyngeal pouches, before entering the "pool" of peripheral immunocompetent lymphocytes. B lymphocytes mature and proliferate in the marrow or fetal liver or (in birds) in a special organ, the bursa of Fabricius, similar to the thymus but derived from an outpouching of the cloaca. Experts disagree as to the importance of gastrointestinal lymphoid organs such as the appendix and Peyer's patches as a possible "bursa-equivalent" in mammals. Whether there are other types of lymphocytes, so-called null cells, also remains unclear. *See* BLOOD; THYMUS GLAND (VERTEBRATE).

The maturation of stem cells in the thymus to become T cells requires the interaction of mesenchymal and epithelial elements and the probable intervention of a thymus hormone derived from the latter. Research suggests that this hormone may act by activating adenyl cyclase in the lymphocytic stem cells, thus raising intracellular levels of cyclic AMP, which in turn activates certain specific intracellular kinases. Little is known as yet about B-cell maturation mechanisms. *See* ADENYLIC ACID, CYCLIC.

Both cells, when mature, enter the bloodstream. Some recirculate, that is, pass back and forth between the bloodstream and peripheral lymphoid organs, the spleen and lymph nodes. Others become sessile or fixed in specific areas of these organs. Some are very long-lived—T cells in humans have been identified by chromosomal analysis techniques (karyotyping) over as long as 30 years without cell division—while others are short-lived, dividing repeatedly.

The two lymphocyte populations, once they enter the peripheral pool, have entirely different roles. The B cell, when adequately stimulated becomes an Ab-producing and secreting cell. The T cell, on the other hand, becomes the immune effector cell of CMI. Some T cells also play an important role in cooperation, helping B lymphocytes to respond to antigen, and in suppression of excessive or unwanted responses. There are thus several subpopulations of T cells with different functional attributes: effector cells of CMI, helper cells, and suppressor cells. Studies suggest that T cells as they leave the thymus may already be committed to one or another of these subpopulations. Functionally distinct B-cell subpopulations also exist but are less well defined.

Lymphocytic markers. T and B lymphocytes and their subpopulations have not only distinct functional attributes but also markers which permit their identification and enumeration, for example, in different tissues or in certain disease states. Some are purely empirical: thus, human T cells have a surface membrane component which binds washed sheep red blood cells (SRBC) and can therefore be easily enumerated by mixing lymphocytes with SRBC and counting the percentage which form obvious rosettes. Other markers are distinctive Ag's, present in one type of cell only: one can immunize rabbits against human T cells and use the antibody to distinguish T cells from B cells, which do not have the same Ag's. Still others are surface components which can be identified chemically or by other means. Of the lymphocytes in normal human blood, approximately 70% are T and 30% B.

Lymphocyte receptors. B lymphocytes all have conventional Ab in their plasma membrane, usually of a single specific molecular type, reactive with one Ag. Thus these cells can be triggered only by the correct Ag; conversely, a given Ag selects and triggers only those B cells which are predetermined to make Ab reactive with it. Researchers have been concerned with the number of possible antibody molecules of differing specificities (estimates range between 10^5 and 10^6) and whether their diversity is due to a diversity of genes governing the highly variable peptide sequences found in the Ab combining sites or to somatic "mutation" events. It seems clear that a large pool of genes developed during evolution exists, but there appears to be a somatic event superimposed as well, occurring during the development of the individual B cell.

T lymphocytes have a different type of receptor governed by so-called immune response (I region) genes. While I-region genes are on a different chromosome from the genes governing Ab, the chemical evidence suggests that both Ab and T-cell receptor are derived from primitive recognition units found in the membranes of all types of cells. T cells also have a fixed Ab, immunoglobulin T (IgT), in their cell membrane whose role is considered below.

Lymphocyte transformation. When T or B lymphocytes in culture react with Ag or with other ligands which can bind some component of the cell membrane, they are stimulated (triggered) to metamorphose into large dividing cells called blasts. A wide variety of agents can cause blast transformation: Ag, Ab against surface components of the lymphocyte, certain lectins (plant extracts) which happen to bind similarly, bacterial products such as endotoxin, heavy metals, periodate, and so forth. B-cell blasts ultimately become Ab-forming plasma cells, whereas T blasts ultimately return to a small lymphocytic state. Transformation is considered a tissue culture model for events which occur when antigen triggers lymphocytes in the intact individual.

The lymphocyte has a mechanism, also seen with a variety of other, nonlymphoid cells, for clearing its surface of complexes between surface components and ligands (a sort of primitive dispose-all). The complexes may aggregate to form a polar cap (capping) and be ingested and destroyed, or they may be discarded into the medium (shedding). Neither of these events is directly related to triggering, but either or both may proceed whether transformation takes place or not.

An important aspect of transformation is that blasts, T blasts especially, produce and release a wide variety of protein mediators (lymphokines, LK) which affect the behavior of other cells. Over 30 such LK have been identified. Some of these may later be found to be identical, but many have been unequivocally shown to be distinct molecules. Among them are chemotactic molecules for several types of cells (macrophages, leukocytes, eosinophils, basophils, other lymphocytes); stimulatory molecules for macrophages, lymphocytes, and the like; toxic molecules (lymphotoxin); inhibitors of DNA synthesis and cell proliferation; and so forth. An exciting development is that these LK fall into two or more large classes. Some are actually a shed complex of antigen with specific plasma membrane receptor. Others are nonspecific. Still others may prove to be small molecules, such as the prostaglandins.

Cooperation and suppression. Certain Ag's, of large molecular size and repeating structure, can stimulate B lymphocytes directly, in tissue culture or in the intact organism, presumably by delivering multiple signals to their surface via the Ab receptor molecules. These are "thymus-independent" Ag's. Most Ag's are "thymus-dependent", that is, can stimulate only when reactive T lymphocytes are also present (cooperation). The T lymphocyte stimulated by Ag in the vicinity of the B cell which has reacted with the same Ag releases both a nonspecific mitogenic LK and a specific complex of Ag with IgT; both of these provide additional signals to the B cell and help to turn it on. In this process the T cell usually reacts with a different chemical configuration of the Ag than does the B cell—these are referred to as carrier and determinant, respectively—but the antibody formed is of course reactive only with the determinant.

The same type of cooperation takes place between two T cells in the generation of CMI, one reacting to a carrier and the other to a determinant of the Ag. There is an additional cooperation between T lymphocytes and macrophages, the phagocytic cells which first take up Ag, perhaps via mechanisms similar to T-B cooperation. Both T-B cooperation and macrophage-T cooperation require I-region genetic identity between the participating cells. This suggests that recognition of one cell by the other is required to permit close association which, in turn, permits the LK mentioned above to act.

Suppression, which is the almost exact converse of cooperation, is seen in a variety of experimental and real-life situations; among these are tolerance to self-antigens and competition of Ag's (when two or more Ags are administered at once). Mediation is again, via LK, both a nonspecific inhibitor of DNA synthesis and specific complexes of Ag with T-cell receptor (I-region gene product). The existence of multple mediators having the same effect suggests that the biologic importance of regulation requires efficient backup mechanisms. Indeed, defects in tolerance which permit immune responses against self-antigens result in serious forms of autoimmune disease. The biology of suppressor T cells is one of the most active research areas in immunology, along with the genetic control of surface receptors and of regulatory LK. See IMMUNO-PATHOLOGY.

Peripheral immune response. Ab formation and the generation of immune cells of CMI take place in the spleen and lymph nodes, and for certain special types of Ab, in loose connective tissue of mucous membranes of the respiratory, gastrointestinal, and genitourinary tracts, and the mammary glands. Ab-forming cells are blasts and plasma cells having well-developed rough endoplasmic reticulum for the production and secretion of protein. Levels in the blood of the different immunoglobulins (Ig) reflect the actual numbers of cells making each of the molecular types of Ab: IgM, IgG, IgA, IgD, and IgE. Tumors of plasma cells which produce any particular Ig are seen with a frequency which also parallels the frequency of the corresponding normal cells. The effector cells of CMI are small lymphocytes, which show certain qualitative differences from their precursors, the immunocompetent T lymphocytes. Both the T and B "memory cells" are formed in the peripheral lymphoid organs. See LYMPHATIC SYSTEM (VERTEBRATE).

Antibody synthesis is like other forms of protein synthesis. The component peptide chains of the Ig molecule are synthesized on polyribosomes, assembly takes place in the endoplasmic reticulum, and a carbohydrate moiety is added to the molecule at successive stages of secretion. Little is yet known about the biochemical events associated with LK production or with the capping and ingestion or shedding of surface receptor molecules. See PROTEIN; RIBOSOMES.

Adjuvants and immunosuppression. Agents which enhance immune responses (adjuvants) include substances which act in quite different ways. Depot materials, like the aluminum phosphate gel used in the conventional vaccine against diphtheria, pertussis, and tetanus, slow absorption of Ag and thus prolong its action. Many agents act by stimulating phagocytic macrophages and thus enhancing one component of macrophage-T-B cooperation. Endotoxic lipopolysaccharide acts as a direct stimulant to B lymphocytes and thus provides additional signals to complement the effect of Ag.

Other agents are immunosuppressive, that is, they diminish one or more types of immune responses. The most effective are agents which remove or destroy lymphocytes, such as x-ray, nitrogen mustards, antilymphocyte antiserum, and thoracic duct drainage. These are effective only if used before attempted immunization. Drugs which kill dividing cells (antimitotic agents) are effective immunosuppressants if given a day or two after Ag, when blast transformation and division of the blasts are at their height. Antimetabolites which act to inhibit RNA or protein synthesis are immunosuppressive in tissue cultures and can be used in some situations in the intact organism. Finally, lymphoid cells can be killed with asparaginase since they do not synthesize their own asparagine.

Cell-mediated immunity. The peripheral expression of CMI usually represents a cooperation between immune T lymphocytes and macrophages, mediated by such LK as chemotactic factor or macrophage activating factor. As an example, an individual immunized against tuberculosis has immune T cells in his blood specifically reactive to Ag proteins of the tubercle bacillus. If infection occurs and these bacilli enter the body, circulating immune T-cells recognize the Ag, undergo blast transformation at the site, and elaborate LK.

These attract large numbers of highly activated macrophages, which ingest and kill the tubercle bacilli and thus protect the host.

In transplantation and especially tumor immunity, a type of immune T lymphocyte is produced which kills tumor cells directly. Another type of T cell acts by producing a toxic LK, lymphotoxin, which can kill tumor cells. *See* IMMUNITY; TRANSPLANTATION BIOLOGY.

Methods and applications. Advances in knowledge of this complex field have depended on development of new materials and techniques. The most important of these has been the use of inbred lines of mice, in which cells and tissues can be freely moved from one animal to another genetically identical recipient. By transfer of different purified cell populations, one can analyse their role in immunity. Also, genetic differences between inbred lines has permitted identification of immune response genes and chromosomal mapping of other significant elements of the immune response.

The development of very sophisticated techniques of separating and purifying lymphocyte subpopulations has also been a major advance. Among these are column techniques, the use of density gradient centrifugation, and use of antisera specific for certain cell markers.

Finally, the extensive use of improved methods of cell culture has permitted analysis of lymphocytic responses under simplified conditions. Most of the points described in this article were discovered in tissue cultures, and then their significance in the intact organism was established by transfer studies with purified cells, often in immunosuppressed recipients.

By 1975, immunologists were in the middle of a period of rapid application of the new techniques and new insights to problems of human medicine. A great challenge facing researchers in this field is the use of immunologic methods to alleviate the large range of autoimmune diseases, to permit free grafting of tissues and organs as needed, and to enhance immunity to tumors. *See* IMMUNOLOGY.

[BYRON H. WAKSMAN]

Bibliography: L. Brent and J. Holborow (eds.), *Progress in Immunology II*, vols. 1–5, 1974; F. J. Dixon and H. G. Kunkel (eds.), *Advances in Immunology*, vols. 1–20, 1961–1975; R. A. Good and D. W. Fisher (eds.), *Immunobiology*, 1971; R. T. McCluskey and S. Cohen (eds.), *Mechanisms of Cell-Mediated Immunity*, 1974; I. Roitt, *Essential Immunology*, 2d ed., 1974; B. H. Waksman, *Atlas of Experimental Immunobiology and Immunopathology*, 1970.

Immunopathology

A term used to describe various human and animal (experimental) diseases (autoimmune diseases) where humoral and cellular immune factors appear to be important in causing pathological damage to cells, tissues, and the host. The same immune mechanisms which cause immunopathology are also involved in protecting the host against exogenous toxic substances or microorganisms. *See* INFLAMMATION.

Mechanisms. The concept of the mechanisms of immune reaction (or immunopathology) to antigens that was set forth in 1968 is as follows: A substance, antigen (microorganism, protein, carbohydrate, nucleic acid, or a complex of these), is introduced into the host and is recognized by genetic mechanisms as being foreign to the host. The antigen will then either be phagocytosed by macrophages, bound to a special ribonucleic acid (RNA), and transferred as an antigen-RNA complex to immunocompetent cells to induce the production of antibodies; or, in an unknown manner, it will sensitize lymphocytes (probably not via a circulating antibody) so that these cells subsequently recognize these antigens as foreign and interact with them. *See* RIBONUCLEIC ACID (RNA).

Once antibodies reach the circulation, they react with the antigen to form an immune (antigen-antibody [Ag-Ab]) complex. This Ag-Ab complex interacts with the complement (C′) system, binding first C′1 and then reacting with C′4, C′2, and then C′3. This Ag-AbC′1423 complex is active in immune adherence (making platelets, red blood cells, and white blood cells stick to the complex) and enhances the phagocytosis of the complex; that is, there is more phagocytosis of Ag-AbC′1423 than of Ag-Ab alone. In addition, this Ag-Ab complement interaction causes the release of vasodilator principle, anaphylotoxin, from C′3. The Ag-AbC′1423 complex then reacts with subsequent complement components, C′5, 6, 7, to cause the formation of a chemotactic factor which attracts granulocytes (phagocytes) to the site of the lesion (Ag-Ab complex). The Ag-Ab complex also causes vasopermeability by a number of different routes, namely histamine and SRS-A (slowly reactive substance in anaphylaxis), via the action of immunoglobulins of different classes (or subgroups). *See* IMMUNOGLOBULIN.

Human disease conditions. Certain human diseases have been thought to involve these immunological mechanisms, especially where any of the following conditions exist: (1) Serological (antibodies, complement) abnormalities appear during the course of the diseases; (2) gamma globulin and complement are found at the site of tissue injury (pathology); (3) "delayed hypersensitivity" (cellular immunity) appears; or (4) immunocompetent (lymphocytes and plasma cells) cells predominate at the site of tissue injury.

Protective mechanisms. Immune mechanisms protect against infections by microorganisms. Shortly after the invasion of the host, antibodies appear in the circulation, react with the offending organism, and form an Ag-Ab immune complex that reacts with complement promoting the phagocytosis and elimination of the organism. In addition, the complex causes vasopermeability, permitting increased amounts of enzymes to reach the site of infection (pathology) and help in cleaning it up (healing).

In some diseases, however, such as tuberculosis, antibodies develop but do not appear to be instrumental in enhancing the elimination of antigens from the host. Individuals harboring tubercle baccilli (TBC) when skin-tested with killed TBC generally develop a "delayed" hypersensitive reaction (in contrast to "immediate," antibody-mediated, anaphylactoid reactions). This delayed reaction is mediated by sensitized lymphocytes and can be transferred from one individual to another via these lymphocytes—but not by serum (antibodies). At the site of pathology, in this case a tuberculoma, lymphocytes abound around the central necrotic lesion, and are probably instrumental in the phagocytosis and elimination of the antigen. *See* HYPERSENSITIVITY; SKIN TEST.

Absence of antibody, lymphocytes, and complement. As can be readily foreseen, absence of any of the above protective immunological mechanisms will result in a situation where individuals might be very prone to develop infections. Such is the case in those disorders associated with low serum levels of gamma globulin; these individuals cannot make protective antibodies to antigens. Such disorders are congenital agammaglobulinemia (now subdivided into those groups where different classes of immunoglobulins are absent, or nearly so); and various malignancies (multiple myeloma, leukemia, and lymphoma) where there is a decreased production of normal gamma globulins. *See* AGAMMAGLOBULINEMIA.

A number of children have been described who have virtually no lymphocytes, no plasma cells, or none of both; these children are usually deficient also in gamma globulins, are especially prone to develop infections, and tend to die at early ages because of infections. Patients with malignancies of lymphocytes (leukemia and lymphomas) also are prone to infection. However, immunoglobulin A deficiency, as well as congenital deficiency of the second complement component ($C'2$) and congenital abnormalities of $C'3$, is not associated with any known disease. Therefore, immunological deficiencies are not necessarily associated with disease.

Rheumatoid arthritis. Most patients with rheumatoid arthritis (RA) have a circulating immune complex consisting of a macroglobulin (1 9S gamma M) antibody, the rheumatoid factor, and an antigen, the patient's own 7S gamma G globulin. This rheumatoid factor is therefore considered to be an autoantibody, that is, an antibody directed to a molecule of the host, to which the host, somehow deciding it is foreign, has made an antibody. The rheumatoid factor is an atypical antibody, for it will not precipitate with antigen, but will, however, combine with 5 moles of the antigen (7S gamma G globulin) to form a 22S complex. In addition, this reaction appears to fix complement weakly, which may explain the clinical observations of low levels of complement in the inflammatory joint effusions found so often in this condition.

Systemic lupus erythematosus. All patients with systemic lupus erythematosus (SLE) have antibodies to nuclear antigens (deoxyribonucleic acid, nucleohistone, and glycoproteins), and many have antibodies to cytoplasmic organelles (mitochondria, lysosomes, and ribosomes). These antibodies are recognized by the lupus erythematosus cell test, precipitation in agar, complement fixation, or the immunofluorescent technique. Many patients with SLE develop renal disease, especially those with antibodies to deoxyribonucleic acid. Pathological studies of kidneys obtained from such patients have demonstrated gamma globulin (including anti-DNA antibodies), complement, and antigen (deoxyribonucleic acid) on the renal glomerulus. Serum complement levels are also depressed during active nephritis. These observations suggest that circulating immune complexes fix complement and cause the nephritis. In addition, it is very possible that these immune complexes also cause the fibrin deposition and attract the leukocytes which are also found at the site of pathology. *See* LUPUS ERYTHEMATOSUS.

Glomerulonephritis. Patients with poststreptococcal glomerulonephritis also have depression of serum complement levels during acute disease. In addition, streptococcal antigen, gamma globulin, and complement have been found at the site of nephritis. Therefore, it appears that the etiology of the nephritis is a circulating immune complex that settles on the renal glomerulus, attracts complement, granulocytes, and fibrin, and causes nephritis, just as in SLE. *See* KIDNEY DISORDERS.

Acute rheumatic fever. Pathological studies of acute rheumatic fever have demonstrated gamma globulin and complement on the myocardium. The disease is thought to involve the host making antibodies to some streptococcal antigens during an infection with streptococci which then crossreact with a closely related myocardial antigen. Serum complement levels are elevated in this condition, probably reflecting increased production, and possibly increased consumption, of complement. *See* COMPLEMENT, SERUM.

Thyroiditis. Autoimmune thyroiditis can be induced in rabbits by removing half of the thyroid gland of a rabbit, grinding it up, mixing it with (Freund's) adjuvant, and then reinjecting it into the rabbit. Many such animals will develop thyroiditis (inflammation of the thyroid) in the remaining half of the gland. This is accompanied by the production of antibodies and lymphocytes sensitized to thyroid antigens. Similar observations have been made in human thyroiditis, where there are antithyroglobulin antibodies and sensitized lymphocytes. It is probable that a combination of both these serological and cellular factors are instrumental in causing the thyroid inflammation which is also characterized by infiltration of lymphocytes.

Studies on experimental animal counterparts have led to a better understanding of immunological mechanisms and thereby a better understanding of human disease. With better understanding of mechanism, improved treatment can be given. *See* ANTIBODY; ANTIGEN; ANTIGEN-ANTIBODY REACTION; IMMUNOLOGY.   [PETER H. SCHUR]

Bibliography: F. J. Dixon and H. G. Kunkel (eds.), *Advances in Immunology*, 8 vols., 1960–1968; P. Grabar and P. Miescher (eds.), *Immunopathology: International Symposium on Immunopathology*, 5 vols., 1958–1968; M. Samter and H. L. Alexander (eds.), *Immunological Diseases*, 1965.

Impact (impulsive force)

A force which acts only during a short time interval but which is sufficiently large to cause an appreciable change in the momentum of the system on which it acts. The momentum change produced by the impulsive force is described by the momentum-impulse relation. For a discussion of this relation *see* IMPULSE (MECHANICS).

The concept of impulsive force is most useful when the time in which the force acts is so short that the system which it acts on does not move appreciably during this time. Under these conditions the momentum of the system is changed rapidly by a finite amount. The details of the way in which the force varies with time are unimportant, since only the impulse determines the momentum change.

Ordinarily the forces occurring in collisions are impulsive forces. In the processes in which impulsive forces occur, mechanical energy can be dissipated, and attempts to apply the conservation of mechanical energy may lead to incorrect results. *See* COLLISION; CONSERVATION OF ENERGY.

A phenomenon known as impulsive loading occurs when materials are subjected to high-speed impacts or explosive charges. The study of failure of materials under impulsive loads is increasing in technological importance.   [PAUL W. SCHMIDT]

Bibliography: R. A. Becker, *Introduction to Theoretical Mechanics*, 1954; J. S. Rinehart and J. Pearson, *Behavior of Metals under Impulsive Loads*, 1954.

Impedance, acoustic

At a given surface, the complex ratio of effective sound pressure averaged over the surface to the effective flux (volume velocity or particle velocity multiplied by the surface area) through it. The unit is the newton-second/meter^5, or the mks acoustic ohm. In the cgs system the unit is the dyne-second/centimeter^5. *See* SOUND PRESSURE.

Specific acoustic impedance is the complex ratio of the effective sound pressure at a point to the effective particle velocity at a point. The unit is the newton-second/meter^3, or the mks rayl. In the cgs system the unit is the dyne-second/centimeter^3, or the rayl. The difference between specific acoustic impedance and acoustic impedance is in the specification of impedance at a point, as compared to the average over a surface. The specific acoustic impedance is generally employed in acoustical analyses, with the acoustic impedance being computed from it when required.

Characteristic acoustic impedance is the ratio of effective sound pressure at a point to the particle velocity at that point in a free, progressive wave. This ratio is equal to the product of the density of the medium ρ_0 times the speed of sound c in the medium. The characteristic impedance of a sound wave is analogous to the characteristic electrical impedance of an infinitely long, dissipationless transmission line. It is common in acoustical analyses to represent specific acoustic impedances in terms of their ratio to the characteristic impedance of air. For example, the specific acoustic impedance Z of a heavy drapery material may be written as $Z = 2\rho_0 c$, meaning that it is twice the characteristic impedance of air.

Impedance analogies. Acoustic impedance, being a complex quantity, can have real and imaginary components analogous to those in an electrical impedance. In applying this analogy, the real part of the acoustic impedance is termed acoustic resistance, and the imaginary part is termed acoustic reactance. *See* IMPEDANCE, ELECTRICAL.

The analogy between acoustic and electrical impedances is useful in the solution of many acoustical problems because it permits the analysis to be conducted by the techniques of electrical circuit theory. In these analyses sound pressure is usually taken as analogous to voltage, and volume velocity as analogous to current. Various parts of acoustical circuits can be associated directly with their electrical counterparts by employing these analogs.

Acoustic resistance is associated with the dissipative losses occurring when there is a viscous movement of a quantity of gas through a thin tube of mesh. It is analogous to electrical resistance.

Acoustic mass, associated with a mass of air accelerated by a net force which acts to displace the gas without appreciably compressing it, assumes the role of an inductance.

Acoustic compliance, associated with a volume of air that is compressed by a net force without an appreciable average displacement of the center of gravity of the air in the volume, acts as a capacitance. Both acoustic mass and acoustic compliance are reactive portions of acoustic impedance, in analogy with their electrical counterparts.

Examples of reactances. As simple examples of acoustical reactances, consider the low-frequency approximation to the impedances of an open-ended tube and that of a simple container having a given volume. If end corrections are neglected, the impedance of the tube is an acoustic mass M_a given by Eq. (1), where ρ_0 is the density of air, l is the

$$M_a = \frac{\rho_0 l}{S} \qquad (1)$$

length of the tube, and S is its cross-sectional area. The impedance of the container is an acoustic compliance C_a given by Eq. (2), where V is the vol-

$$C_a = \frac{V}{\rho_0 c^2} \qquad (2)$$

ume, ρ_0 the density of air, and c the velocity of sound.

The absorption of sound by a material is often described in terms of its acoustical impedance. For example, the absorption coefficient α of a material exposed to a normally incident plane wave of sound in air is given by Eq. (3), where Z is the

$$\alpha = 1 - \left(\frac{Z - \rho_0 c}{Z + \rho_0 c}\right)^2 \qquad (3)$$

specific acoustic impedance at the surface of the material and $\rho_0 c$ is the characteristic impedance of air. *See* ABSORPTION OF SOUND.

[WILLIAM J. GALLOWAY]

Bibliography: L. L. Beranek, *Acoustics*, 1954; C. G. Officer, *Introduction to the Theory of Sound Transmission*, 1958.

Impedance, electrical

The total opposition that a circuit presents to an alternating current. Impedance, measured in ohms, may include resistance R, inductive reactance X_L, and capacitive reactance X_C. *See* REACTANCE; RESISTANCE, ELECTRICAL.

The impedance of the series RLC circuit is given by Eq. (1).

$$Z = \sqrt{R^2 + (X_L - X_C)^2} \text{ ohms (magnitude)} \qquad (1)$$

In terms of complex quantities, this impedance is given by Eq. (2). The two components of Z are

$$Z = R + j(X_L - X_C) \qquad (2)$$

at right angles to each other in an impedance diagram. Therefore, impedance also has an associated angle, given by Eq. (3). The angle is called the

$$\theta = \arctan \frac{X_L - X_C}{R} \qquad (3)$$

phase, or power-factor, angle of the circuit. The current lags or leads the voltage by angle θ depending upon whether X_L is greater than, or less than, X_C.

Impedance may also be defined as the ratio of the rms voltage to the rms current, $Z = E/I$. This is a form of Ohm's law for ac circuits. For further discussion of impedance *see* ALTERNATING-CURRENT CIRCUIT THEORY.   [BURTIS L. ROBERTSON]

Impedance, mechanical

For a system executing simple harmonic motion, the mechanical impedance is the ratio of force to particle velocity. If the force is that which drives the system and the velocity is that of the point of application of the force, the ratio is the input or driving-point impedance. If the velocity is that at some other point, the ratio is the transfer impedance corresponding to the two points.

As in the case of electrical impedance, to which it is analogous, mechanical impedance is a complex quantity. The real part, the mechanical resistance, is independent of frequency if the dissipative forces are proportional to velocity; the imaginary part, the mechanical reactance, varies with frequency, becoming zero at the resonant and infinite at the antiresonant frequencies of the system. *See* FORCED OSCILLATION; HARMONIC MOTION; IMPEDANCE, ACOUSTIC; IMPEDANCE, ELECTRICAL.

[MARTIN GREENSPAN]

Bibliography: N. W. McLachlan, *Theory of Vibrations*, 1951; W. T. Thompson, *Mechanical Vibrations*, 1964.

Impedance matching

The use of electric circuits and devices to establish the condition in which the impedance of a load is equal to the internal impedance of the source. This condition of impedance match provides for the maximum transfer of power from the source to the load. In a radio transmitter, for example, it is desired to deliver maximum power from the power amplifier to the antenna. In an audio amplifier, the requirement is to deliver maximum power to the loudspeaker. *See* IMPEDANCE, ELECTRICAL.

The maximum power transfer theorem of electric network theory states that at any given frequency the maximum power is transferred from the source to the load when the load impedance is equal to the conjugate of the generator impedance. Thus, if the generator is a resistance, the load must be a resistance equal to the generator resistance for maximum power to be delivered from the generator to the load. When these conditions are satisfied, the power is delivered with 50% efficiency; that is, as much power is dissipated in the internal impedance of the generator as is delivered to the load.

Impedance matching network. In general, the load impedance will not be the proper value for maximum power transfer. A network composed of inductors and capacitors may be inserted between the load and the generator to present to the genera-

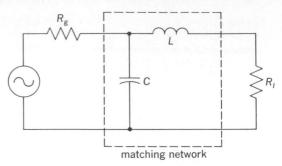

L-section impedance matching network.

tor an impedance that is the conjugate of the generator impedance. Since the matching network is composed of elements which, in the ideal case of no resistance in the inductors and perfect capacitors, do not absorb power, all of the power delivered to the matching network is delivered to the load. An example of an L-section matching network is illustrated. Matching networks of this type are used in radio-frequency circuits. The values of inductance and capacitance are chosen to satisfy the requirements of the maximum power transfer theorem. The power dissipated in the matching network is a small fraction of that delivered to the load, because the elements used are close approximations of ideal components.

Transformers. The impedance measured at the terminals of one winding of an iron-cored transformer is approximately the value of the impedance connected across the other terminals multiplied by the square of the turns ratio. Thus, if the load and generator impedances are resistances, the turns ratio can be chosen to match the load resistance to the generator resistance for maximum power transfer. If the generator and load impedances contain reactances, the transformer cannot be used for matching because it cannot change the load impedance to the conjugate of the generator impedance (the L-section matching network can). The turns ratio can be chosen, however, to deliver maximum power under the given conditions, this maximum being less than the theoretical one.

Iron-cored transformers are used for impedance matching in the audio and supersonic frequency range. The power dissipated in the core increases with frequency because of hysteresis. Above the frequency range at which iron-cored transformers can be used, the air-core transformer or transformers with powdered-iron slugs can be used effectively. However, in these cases the turns-ratio-squared impedance-transforming property is no longer true. Since the transformer is usually part of a tuned circuit, other factors influence the design of the transformer.

The impedance-transforming property of an iron-cored transformer is not always used to give maximum power transfer. For example, in the design of power-amplifier stages in audio amplifiers, the impedance presented to the transistor affects distortion. A study of a given circuit can often show that at a given output power level, usually the maximum expected, there is a value for the load resistance which will minimize a harmonic component in the harmonic distortion, such as the second or third harmonic. The transformer turns ratio is selected to present this resistance to the transistor. *See* TRANSFORMER.

Emitter follower. In electronic circuitry a signal source of large internal impedance must often be connected to a low-impedance load. If the source were connected directly to the load, attenuation of the signal would result. To reduce this attenuation, an emitter follower is connected between the source and the load. The input impedance of the emitter follower is high, more nearly matching the large source impedance, and the output impedance is low, more nearly matching the low load impedance. If the object were the delivery of maximum power to the load, it might be possible to design the emitter follower to have an output resistance equal to the load resistance, assuming that the load is a resistance. (Special audio amplifiers have been designed to use emitter followers, rather than a transformer, to connect the loudspeaker to the power amplifier.) In many cases, maximum power transfer is not the goal; the emitter follower is introduced primarily to reduce to a minimum the attenuation of the signal.

There exist a number of applications where the emitter follower is not useful as an impedance matching circuit. For example, if a very-low-impedance source must be matched to a high-impedance load, then a transistor is used in the common base configuration. *See* VOLTAGE AMPLIFIER.

[CHRISTOS C. HALKIAS]

Bibliography: E. W. Kimbark, *Electrical Transmission of Power and Signals*, 1949; J. Millman and C. C. Halkias, *Electronic Devices and Circuits*, 1967; M. E. Van Valkenburg, *Network Analysis*, 1964.

Impedance measurements, high-frequency

The electrical measurement of the complex ratio of voltage to current in a given circuit at frequencies from several hundred kilohertz (kHz) to 100,000 megahertz (MHz). This frequency range includes medium- and high-frequency bands. *See* ELECTRICAL MEASUREMENTS.

At lower frequencies impedances may be accurately measured by standard techniques for measuring resistance, capacitance, and inductance. The most precise measurements are those in which the unknown impedance is compared with a resistance, capacitance, or inductance standard of nearly equal value; at lower frequencies such standards can be very accurate. *See* CAPACITANCE MEASUREMENT; INDUCTANCE MEASUREMENT; RESISTANCE MEASUREMENT.

At higher frequencies, details of measurement-standard shape, terminal geometry, and component-interconnection wiring provide series inductance and shunt capacitance as well as stray couplings that may dominate the situation unless great care is used. These undesired parameters cause particular difficulty if they have different values each time a connection is made. Highly repeatable connections may be made at high frequencies by the use of precision connectors. These connectors are coaxial to remove external effects and make a butt joint to ensure repeatable inductance and capacitance (Fig. 1). Some resistance, inductance, and capacitance standards using modified low-frequency construction are usable up into vhf range. Distributed-parameter standards consisting of coaxial transmission lines are useful to over 10 GHz. At the highest frequencies, where mechanical dimensions become comparable with the wavelength, resonant cavities and waveguides are used because they incorporate simple boundary conditions for field calculations.

For measurements over the whole high-frequency range from below 1 MHz to over 10 GHz, the measurement methods used can be classified as voltmeter-ammeter methods, resonance methods, null methods, standing-wave methods, and reflection methods.

VOLTMETER-AMMETER METHODS

The impedance of a device is defined as the ratio of the voltage across it to the current through it, and a basic method of impedance measurement is, therefore, to measure voltage and current on appropriate meters and to calculate impedance. A modification of this method is to supply a known and constant voltage or current and measure the resulting current or voltage. These methods can give a direct indication of admittance or impedance, respectively.

Vector impedance meter. This instrument not only determines the ratio between the voltage and current to give impedance magnitude but also determines the phase difference between these quantities to give the phase angle of the component under test. This principle has been applied to several instruments at low frequencies and is used to 108 MHz in the Hewlett-Packard rf vector impedance meter. Because it is difficult to perform the necessary measurement operations at high frequencies, a synchronous sampling technique is used to convert the measured voltage and current to a 5-kHz intermediate frequency. The sampled current is used to control the level of the applied test current so that the voltage measured is a direct indication of impedance magnitude and is displayed on one meter. A phase detector drives another meter which indicates phase angle. Such methods are limited to accuracy of the circuitry performing the measurement operations and of the meter movement itself. Basic accuracy of the vector impedance meter is 4% of full scale. While other methods are more accurate, the wide range and ease of use makes such instruments popular.

Potentiometer method. Another modification of the basic method defined above is to put a known impedance in series with the unknown impedance, drive them with an appropriate source, and measure the voltage across each separately. Because the same current is passed through them, the ratio of the measured voltages is the ratio of the magnitudes of the two impedances. This is the same principle as the potentiometer method of resistance measurement, long used at direct current; the advent of low-level rf voltmeters makes this method practical to quite high frequencies. If the standard and unknown are approximately equal, the error caused by the voltmeter is small because both readings are on the same part of the scale. *See* POTENTIOMETER (VOLTAGE METER).

RESONANCE METHODS

Resonance is a typical phenomenon at radio frequencies. It is readily observed and reproduced

Fig. 1. An rf capacitance standard fitted with a GR900 precision connector. (*General Radio Co.*)

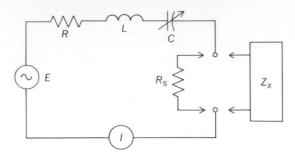

Fig. 2. Series-resonance circuit for resistance-variation and reactance-variation measurements.

and defines a circuit condition for which the inter-relationship among the component impedances is known. It is therefore an excellent indicator for measurement purposes. Either series-resonant or parallel-resonant circuits can be used. Series-resonant circuits are best suited for measuring low impedances, and conversely, parallel-resonant circuits for low admittances. Both methods determine the quadrature or reactive component of impedance from the change in resonant capacitance when the unknown impedance is inserted into the circuit. They differ in the method of measurement of the real, or resistive, component of impedance. *See* RESONANCE (ALTERNATING-CURRENT CIRCUITS).

Series-resonance methods. Two methods of obtaining the necessary data to permit solving for the unknown impedance are employed.

Resistance-variation method. A short-circuiting link is first connected across the terminals shown in Fig. 2, and the circuit is tuned to resonance as indicated by a maximum current reading. This current I_1 is then given by Eq. (1), where E is the

$$I_1 = \frac{E}{R} \qquad (1)$$

source voltage and R the total circuit resistance.

The short-circuiting line is then replaced by the unknown impedance Z_x and resonance reestablished. The new current I_2 is given by Eq. (2),

$$I_2 = \frac{E}{R + R_x} \qquad (2)$$

where R_x is the resistive component of the unknown impedance Z_x.

Finally, the unknown impedance is replaced by a known standard resistance R_s and resonance is reestablished. The current I_3 is given by Eq. (3).

$$I_3 = \frac{E}{R + R_s} \qquad (3)$$

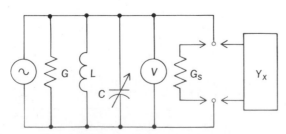

Fig. 3. Parallel-resonance circuit for conductance-variation and susceptance-variation measurements.

Combining Eqs. (1), (2), and (3), one obtains for the unknown resistance Eq. (4).

$$R_x = R_s \frac{I_3(I_1 - I_2)}{I_2(I_1 - I_3)} \qquad (4)$$

The unknown reactance X_x is given by Eq. (5),

$$X_x = \frac{1}{\omega}\left(\frac{1}{C_2} - \frac{1}{C_1}\right) \qquad (5)$$

where C_1 and C_2 are the settings of the variable capacitor for resonance with the short-circuiting link and the unknown impedance in circuit, respectively, and ω is the angular frequency.

Reactance-variation method. This method differs from the resistance-variation method only in the measurement of the unknown resistive component. The circuit resistance is deduced from the capacitance change necessary to detune the circuit by a known amount.

The circuit is first tuned to resonance and the current I noted; then the circuit is detuned and the capacitance values C' and C'' for which the current becomes $0.707I$ are determined. At these settings, one on each side of resonance, the circuit reactance equals the circuit resistance R, and it can be shown that Eq. (6) holds.

$$R = \frac{1}{2\omega}\left(\frac{1}{C'} - \frac{1}{C''}\right) \qquad (6)$$

A measurement with the short-circuiting link in place yields a resistance R_1 which is equal to the total circuit resistance R. A second measurement, with the short-circuiting link replaced by the unknown impedance Z_x yields a resistance R_2 which is equal to $R + R_x$. This substitution measurement then gives Eqs. (7) and (8).

$$R_x = R_2 - R_1 \qquad (7)$$

$$X_x = \frac{1}{\omega}\left(\frac{1}{C_2} - \frac{1}{C_1}\right) \qquad (8)$$

Parallel-resonance methods. The parallel-resonance methods are duals of the series-resonance methods. The same general techniques are used except that resonance is defined in terms of maximum voltage and the circuit losses are measured by the open-circuit conductance G instead of the short-circuit resistance R (Fig. 3).

Conductance-variation method. This method is the dual of the resistance-variation method. Measurements with the terminals open-circuited, with the unknown admittance Y_x connected, and then with the unknown admittance replaced by a known conductance standard G_s give Eqs. (9) and (10),

$$G_x = G_s \frac{V_3(V_1 - V_2)}{V_2(V_1 - V_3)} \qquad (9)$$

$$B_x = \omega(C_1 - C_2) \qquad (10)$$

where the subscripts refer to the three measurements, respectively.

Susceptance-variation method. This method is the dual of the reactance-variation method. At the capacitance settings C' and C'' for which the circuit susceptance equals the circuit conductance G, the voltage becomes $0.707V$ and Eq. (11) holds.

$$G = \frac{\omega}{2}(C'' - C') \qquad (11)$$

The same sequence of measurements as in the reactance-variation method then yields Eqs. (12) and (13).

$$G_x = G_2 - G_1 \qquad (12)$$

$$B_x = \omega(C_1 - C_2) \qquad (13)$$

The resonance methods give accurate results with fairly simple equipment but are not readily made direct-reading and are therefore not used as the basis for commercial instruments.

Resonant-rise method. A circuit that has been widely used commercially to measure the storage factor Q of coils is shown in Fig. 4. Commercially the instrument is known as a Q meter. *See* Q METER.

The resonant current I in this series-resonant circuit, as before, is given by $I = E/R$, and the voltage V across the tuning capacitor C by Eq. (14),

$$V = IX_C = IX_L \qquad (14)$$

where R is the total circuit resistance and X_C and X_L are the reactances of the capacitor C and inductor L, respectively. The voltage ratio V/E is therefore given by Eq. (15).

$$\frac{V}{E} = \frac{X_L}{R} \qquad (15)$$

If the resistance and inductance of the rest of the circuit are made negligibly small compared with the resistance and inductance of the coil L, the voltage V will be directly proportional to the storage factor Q of the coil. In many Q meters the voltmeter scale is calibrated directly in Q, and the value of the unknown inductance is determined from the calibrated capacitor setting by Eq. (16). When L is known, the effective resistance of the inductance is given by Eq. (17).

$$L = \frac{1}{\omega^2 C} \qquad (16)$$

tance of the inductance is given by Eq. (17).

$$R_L = \frac{\omega L}{Q} \qquad (17)$$

Q of tuned circuit or cavity. Several methods of measuring Q are available.

Resonance-curve method. If frequency is varied so that the current in a tuned circuit or cavity goes through resonance, and the frequencies f' and f'' for which the current is reduced to 0.707 of its maximum (half-power points) are noted, it can be shown that Eq. (18) holds. Here the subscript 0

$$Q_0 = \frac{\omega_0}{\omega'' - \omega'} = \frac{f_0}{f'' - f'} \qquad (18)$$

refers to values at resonance. For high-Q cavities at microwave frequencies this measurement is usually performed by coupling the generator and detector to the magnetic field with small pickup coils, so oriented that direct pickup from one to the other is negligible. If the couplings are too strong, the resonance curve will be broadened by losses coupled in from the generator and detector. As they are weakened, however, this broadening will disappear and the observed curve will be that of the cavity alone.

Decrement method. The storage factor Q of a resonant device represents the ratio of the maximum energy stored in the electric or magnetic field

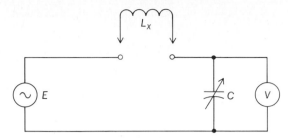

Fig. 4. Series-resonance circuit used for the resonant-rise method of measuring Q.

during a cycle to the amount of energy dissipated in that cycle. It can also be measured, therefore, by observing the decay in oscillation amplitude when the exciting signal is cut off. The current in the tuned circuit then follows the law in Eq. (19),

$$I = I_0 \epsilon^{-(R/2L)t} = I_0 \epsilon^{-(\omega_0/2Q_0)t} \qquad (19)$$

where t is time, I_0 the initial resonant current, and the other symbols carry their previous connotation. The cavity Q_0 is then related to the time interval Δt during which the current decays by a factor of $\epsilon = 2.71828 \ldots$ by Eq. (20). *See* TIME CONSTANT.

$$Q_0 = \frac{\omega_0 \Delta t}{2} \qquad (20)$$

This measurement is carried out with a pulse-modulated generator, a detector having a large bandwidth compared with that of the cavity to be measured, and a cathode-ray oscilloscope. The detected signal, which measures the cavity current, is applied to the vertical plates of the oscilloscope, and the horizontal deflection is synchronized with the modulating pulse to produce a stationary pattern. This pattern may be scaled directly off the screen for rough measurements, or in a more complex setup, the pattern may be compared with the discharge curve of an RC network excited by the modulating pulse.

R_0/Q_0 of resonant cavity. At microwave frequencies, where a resonant cavity may be difficult to analyze as an equivalent LC resonant circuit, it is often necessary to measure the quantity in Eq.

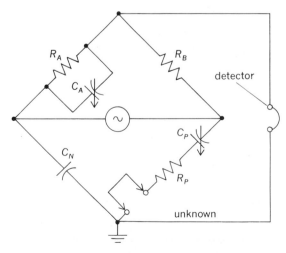

Fig. 5. Circuit schematic of General Radio radio-frequency bridge.

Fig. 6. Radio-frequency bridge. (*General Radio Co.*)

(21) to obtain, in conjunction with a measurement

$$\frac{R_0}{Q_0} = \sqrt{\frac{L}{C}} = \omega_0 L = \frac{1}{\omega_0 C} \qquad (21)$$

of Q_0, a value for R_0, the effective shunt resistance of the cavity. This is most often done by the perturbation method. If C can be varied in an LC circuit, it can be shown that Eq. (22) holds.

$$\frac{R_0}{Q_0} = -\frac{2}{\omega_0^2} \frac{d\omega}{dC} \qquad (22)$$

Analysis of the field in a microwave cavity when a perturbing object is introduced can, by analogy, be used to relate the resultant change in frequency to R_0/Q_0.

NULL METHODS

A phenomenon at least equal in importance to resonance as an indicator of prescribed circuit conditions is balance. The precision with which the difference between two alternating voltages can be reduced to zero can easily reach a few parts in a million. Null methods are almost universally used for the most precise measurements.

Radio-frequency bridges. At radio frequencies the problems arising from residual parameters make use of variable air capacitors as impedance standards in substitution methods as desirable for null devices as for resonant circuits. Special circuits particularly well adapted for these frequencies have therefore been developed. Three widely used commercial instruments are discussed.

General Radio rf bridge. The balance conditions for this bridge, shown in Figs. 5 and 6, yield the unknown resistive component R_x and the unknown reactive component X_x of impedance in terms of the known bridge components R_B, C_N, C_A, and C_P, as in Eqs. (23) and (24). As indicated, both

$$R_x = \frac{R_B}{C_N}(C_{A2} - C_{A1}) \qquad (23)$$

$$X_x = \frac{1}{\omega}\left(\frac{1}{C_{P2}} - \frac{1}{C_{P1}}\right) \qquad (24)$$

the resistive and reactive components are measured in terms of capacitance differences; the subscripts 1 and 2 refer respectively to balances with the terminals, first, short-circuited, and second, connected to the unknown impedance. The dial of

the capacitor C_A is calibrated in ohms, independent of frequency, and the dial of the capacitor C_P in ohms at a frequency of 1 MHz. At other frequencies, the reading of this reactance dial must be divided by the frequency in megahertz. The instrument covers the frequency range from 400 kHz to 60 MHz; other versions extend these limits to 50 kHz and 120 MHz.

Wayne-Kerr rf bridge. This bridge uses a center-tapped transformer with a high degree of coupling between windings to develop equal and opposite voltages in the standard and unknown arms (Fig. 7). A similar transformer with one winding reversed couples the output to the detector. As shown in Fig. 7, taps are used to modify the effectiveness of the capacitance and conductance standards, C_s and G_s, respectively. In terms of the number of turns n between the center tap and the unknown admittance (n_1), the conductance standard (n_2), and the capacitance standard (n_3), the balance conditions are given by Eqs. (25) and (26).

$$G_x = \frac{n_2}{n_1} G_s \qquad (25)$$

$$B_x = \frac{n_3}{n_1} \omega C_s \qquad (26)$$

Combinations of switched fixed-value standards and continuously adjustable standards yield scales that are calibrated directly in conductance and capacitance, independent of frequency. Similar taps on the unknown side of the center tap are used to switch admittance ranges. The instrument covers the frequency range from 15 kHz to 5 MHz; other versions extend the range from 15 kHz to 250 MHz.

Bridges of this kind are particularly well adapted to the measurement of the direct component of balanced and other three-terminal admittances because the shunt components can be thrown across the low-impedance transformer windings, by grounding of the center tap, and eliminated from the measurement. They can also be adapted to the measurement of the transfer impedance of four-terminal devices.

Hewlett-Packard RX meter. This bridge uses the same configuration of bridge arms as the General Radio bridge but measures the unknown as an admittance in the A arm rather than an impedance in the P arm. For this inversion, the balance equations become Eqs. (27) and (28). The *RX* meter

$$G_x = \frac{C_N}{R_B}\left(\frac{1}{C_{P2}} - \frac{1}{C_{P1}}\right) \qquad (27)$$

$$B_x = \omega(C_{A1} - C_{A2}) \qquad (28)$$

differs from conventional bridges in that the bridge arms are excited by out-of-phase voltages from a transformer. At balance the junction-point voltages are equal in magnitude but opposite in phase. Null voltage is obtained between the center point of a capacitive voltage divider and ground (Fig. 8).

Microwave null devices. At frequencies so high that the distributed nature of parameters must be taken into account, the principle of null comparison can still be used to effect precise adjustment. Two commercial instruments for this frequency range are discussed here.

General Radio admittance bridge. This instrument (Figs. 9 and 10) samples the magnetic field

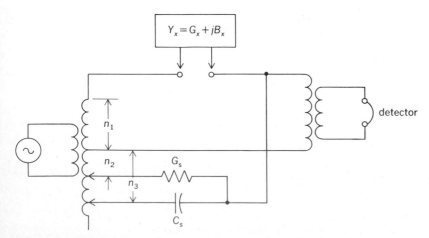

$Y_x = G_x + jB_x$

detector

n_1

n_2

n_3

G_s

C_s

Fig. 7. Simplified inductive-ratio-arm bridge.

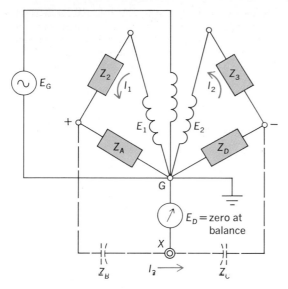

Fig. 8. Circuit schematic of Hewlett-Packard *RX* meter. Three-winding transformer couples out-of-phase voltages from generator to two halves of bridge; two capacitors Z_B and Z_C provide a center tap to feed the detector.

arising from the current in each of three coaxial transmission lines through adjustable loops (M_B, M_G, and M_x) coupled to their center conductors at their junction point in the T configuration shown in the figures. All the lines are fed at this point from a common voltage, and thus the currents bear the same relationship to each other as the respective admittances seen looking into each of the lines. One of the lines is terminated in its characteristic impedance Z_0 to present a standard conductance G_s, equal to $1/Z_0$; one is terminated in an eighth-wavelength transmission line to present a standard susceptance B_s, equal to $1/jZ_0$; and the third is terminated in the unknown admittance Y_x, equal to $G_x + jB_x$. The coupling loops can be rotated by means of shafts carrying dials, the first two being calibrated in conductance and susceptance, respectively, and the third in admittance range. The loop that couples to the susceptance line is adjustable over a 180° range to indicate either positive or negative susceptances; the eighth-wavelength line

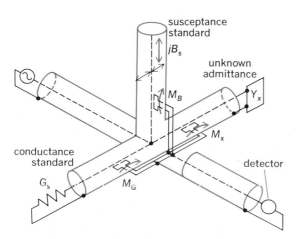

Fig. 9. General Radio admittance bridge. The three loops M_B, M_G, and M_x sample magnetic fields in susceptance, conductance, and unknown-admittance arms, respectively, and combine outputs in detector arm.

is set to the proper length for the operating frequency, so that both conductance and susceptance scales are direct-reading, independent of frequency. If the unknown is connected to the instrument by a quarter-wave line, the readings of these scales become directly proportional to impedance. The instrument is direct-reading for the frequency range from 40 to 1000 MHz.

Transfer-function and immittance bridge. A modification of the admittance bridge compares the output current of a four-terminal device with admittance standards in a T configuration similar to that described above, thereby making possible the measurement of transfer admittance by a null method. Through the use of quarter- and half-wave transmission lines in the input and output circuits, it is possible to measure the ratio of output current to input voltage, output voltage to input current, output to input voltage, and output to input current. An interchangeable head converts the instrument to an admittance meter, making possible measurement of two-terminal admittance and impedance over the frequency range from 25 to 1000 MHz.

STANDING-WAVE METHODS

Standing-wave detectors are devices used at microwave frequencies to measure impedances in terms of electromagnetic field distributions in guided-wave systems.

It is desirable to make such measurements with devices of simple geometrical configuration so that their performance can be readily analyzed in terms of field theory. At the lower microwave frequencies, where transverse dimensions are small compared to the wavelength, the performance of coaxial lines is easy to compute, and these devices can be conductively connected to circuits characterized by two-terminal connections. At the higher microwave frequencies, where all dimensions are comparable with the wavelength and the concept of lumped, or even distributed, parameters is no longer readily usable; waveguides have simple boundary conditions that facilitate computation and visualization of field distributions. They can be coupled to other waveguides or to cavities through electric or magnetic fields rather than with conductive connections. *See* TRANSMISSION LINES; WAVEGUIDE.

Impedance can be measured with these transmission lines by analysis of the behavior of electric fields propagating along their axes. In an idealized coaxial line or waveguide with lossless, unity-dielectric-constant insulation and perfectly conducting walls, a plane wave will propagate along the axis at the speed of light without alteration so long as the cross-sectional dimensions remain uniform. The relationship between the electric and magnetic fields will remain constant, and because they are respectively proportional to voltage and current, the ratio of these fields will define a characteristic impedance for the line.

When the wave reaches the end of the transmission line, a portion determined by the impedance discontinuity at that point is reflected backwards toward the source. The incident and reflected waves then add together to form a stationary interference pattern, or standing wave, that has maxima and minima occurring alternately at intervals of a quarter wavelength. The maxima occur at

Fig 10 Admittance meter with variable capacitor connected as susceptance standard. (*General Radio Co.*)

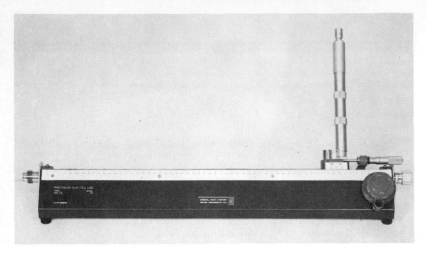

Fig. 11. A precision slotted line fitted with a precision rf connector. (*General Radio Co.*)

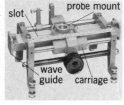

slot   probe mount

wave
guide   carriage

Fig. 12. Hewlett-Packard slotted section for microwave measurements. Different waveguide elements can be inserted in the carriage for frequency ranges between 3950 and 18,000 MHz. (*Hewlett-Packard Co.*)

Fig. 13. Network analyzer for impedance measurements from 110 MHz to 12.4 GHz. A directional coupler is used as measuring element. (*Hewlett-Packard Co.*)

those points at which the waves reinforce each other and measure the sum of the two amplitudes; the minima, conversely, measure their difference. The ratio of maximum to minimum voltage is called the voltage–standing-wave ratio (VSWR).

The amount of reflection depends upon the relation between the impedance in which the line is terminated and the characteristic impedance of the line. If the impedance is resistive and equal to the characteristic impedance, the current will flow into it just as it would into a further extension of the line itself and there is no reflection; if the transmission line is short-circuited, the reflected voltage will be equal in magnitude, but of reversed polarity, to the incident wave, because the net voltage must be zero at the termination; if the transmission line is open-circuited, the reflected voltage will again be equal in magnitude, but of like polarity, to the incident wave, so that the voltage doubles at the termination. When the terminating impedance equals the characteristic impedance, the VSWR is unity and the line is said to be matched. When the line is short-circuited or open-circuited, the VSWR would be infinite if the line were lossless. The distance from the termination to the first minimum is respectively zero and a quarter wavelength for these terminations.

It can be shown that these two quantities, the VSWR and the distance to the first minimum, uniquely define the terminating impedance. They are easy to determine experimentally, but the mathematical conversion to the conventional impedance components is somewhat involved. Graphical methods of interpretation have therefore been developed, and one known as the Smith chart is in wide use, for determining impedances from standing-wave measurements and analyzing the effects of finite lengths of connecting lines.

Slotted line. A commercial slotted line for measuring impedances at the lower microwave frequencies is shown in Fig. 11. It comprises a cylindrical coaxial line having a slot in the outer conductor into which a small capacitive probe extends to sample the electric field. As the probe slides along the line, its position can be measured to find the distance from the termination to the first voltage minimum, and the maximum and minimum voltages determined to find the VSWR. This line

is suitable for measurements from about 300 to 8500 MHz.

Slotted section. A commercial slotted section for measuring impedances at the higher microwave frequencies is shown in Fig. 12. Its function is similar to that of the coaxial slotted line, but it differs in certain practical respects. Coaxial lines cover the frequency range from direct current to the frequencies at which the higher-order modes of propagation used in waveguides occur. Waveguides, on the other hand, are restricted to the relatively narrower frequency ranges over which a single, selected dominant mode is useful. Replaceable slotted sections are therefore provided so that a wide frequency range can be covered with a single carriage mechanism. These various sections cover the frequency range 2600–40,000 MHz with two models of carriage.

REFLECTION METHODS

Reflection methods are based on a separation of the incident and reflected waves in a transmission line. The individual amplitudes are measured and the ratio of these amplitudes is equal to the magnitude of the reflection coefficient $|\Gamma|$ of the terminating impedance. The VSWR can be determined from Eq. (29).

$$\text{VSWR} = \frac{1 + |\Gamma|}{1 - |\Gamma|} \qquad (29)$$

To measure the terminating impedance itself, an additional measurement of phase shift at the point of reflection is necessary. The reflection coefficient can then be represented as a vectorial quantity $\Gamma\underline{/\theta}$ related to the terminating impedance $\mathbf{Z}_L$ and the characteristic impedance $\mathbf{Z}_0$ of the transmission line by Eq. (30), from which Eq. (31) is obtained.

$$\mathbf{\Gamma} = \Gamma\underline{/\theta} = \frac{\mathbf{Z}_L - \mathbf{Z}_0}{\mathbf{Z}_L + \mathbf{Z}_0} \qquad (30)$$

$$\mathbf{Z}_L = \frac{\mathbf{\Gamma} + 1}{\mathbf{\Gamma} - 1}\mathbf{Z}_0 \qquad (31)$$

A basic component of most reflectometers is the directional coupler, which performs the actual separation of the two waves. Two of these, coupled to the transmission line in opposite directions, yield independent measurements of the two waves from which their amplitude ratio can be computed. A convenient arrangement uses a ratio meter to display the magnitude of reflection coefficient directly. The angle θ of the reflection coefficient can be measured with an auxiliary slotted line to determine the distance from the termination to the first voltage minimum; with an auxiliary capacitance probe to indicate the vector sum of the incident and reflected waves at a known point on the line; or by comparing the phases of the outputs of the directional couplers themselves.

Similar separation of incident and reflected signals is accomplished with hybrid and hybridlike devices which utilize a "balun" (a balanced-to-unbalanced transformer).

Some commercial examples of instruments that measure impedance by a reflection method are described below.

Hewlett-Packard network analyzer. This instrument (Fig. 13) employs a directional coupler as the measuring element. The frequency of the incident and reflected signals from the coupler are converted by a phase-coherent sampling technique to a 20-MHz intermediate frequency, where the phase-and-amplitude measurement is performed. This system and the ones described below operate on a sweep-frequency basis with readout on a meter or cathode-ray tube.

General radio reflectometer. This instrument (Fig. 14) employs a hybrid bridge to cover the 20–1500-MHz band and a directional coupler to cover the 0.5–7-GHz band. It provides amplitude measurements of the reflected signals by the use of square-law detectors.

Automatic plotter. The automatic impedance and transfer-character plotter of the Alford Manufacturing Co. employs a hybridlike device (Hybridge) as the measurement element. The phase-and-amplitude measurements of impedance are made directly at the high frequency by means of a broadband phase detector. A number of such devices cover the frequency range 0.025–7000 MHz.

Impedance plotter. Waveguide impedance plotters covering waveguide bands from 5.4 to 18.0 GHz are offered by Rantec Corp. They utilize directional couplers as measuring elements and broadband rf phase detectors to generate the phase and amplitude of the impedance.

[MARK G. FOSTER]

Bibliography: G. J. Alonzo, R. H. Blackwell, and H. V. Marantz, Direct-reading, fully-automatic vector impedance meters, *Hewlett-Packard J.*, January, 1967; R. W. Anderson and O. T. Dennison, An advanced new network analyzer for sweep-measuring amplitude and phase from 0.1 to 12.4 GHz, *Hewlett-Packard J.*, February, 1967; R. Calvert, New technique in bridge measurements, *Electron. Eng.*, 20(239):28–29, 1948; E. L. Ginzton, *Microwave Measurements*, 1957; B. Hague, *Alternating Current Bridge Methods*, 5th ed., 1957; T. E. MacKenzie, Calibration standards for precision coaxial lines, *Gen. Radio Exp.*, vol. 40, no. 5, 1966; W. W. Mumford, Directional couplers, *Proc. IRE*, 35(2):160–166, 1947; R. W. Orr, Capacitance standards with precision connectors, *Gen. Radio. Exp.*, vol. 41, no. 9, 1967; D. B. Sinclair, A radio-frequency bridge for impedance measurements from 400 kilocycles to 60 megacycles, *Proc. IRE*, 28(11):497–503, 1940; D. B. Sinclair, Parallel-resonance methods for precise measurements of high impedances at radio frequencies and comparison with ordinary series-resonance methods, *Proc. IRE*, 26(12):1466–1497, 1938; P. H. Smith, Transmission-line calculator, *Electronics*, 12(1):29–31, 1939; J. Zorzy, Precision coaxial equipment, *Gen. Radio Exp.*, vol. 37, no. 11, 1963.

Impetigo

A contagious skin disease commonly seen in children but also found in adults. It is caused by either streptococcal or staphylococcal infections, or both. The disease is autoinoculable; that is, it can be quickly spread from an original site to any part of the body. No immunity is conferred by exposure. *See* STAPHYLOCOCCUS; STREPTOCOCCUS.

The lesions are superficial, thin-walled blisters which may develop into pustules. These become covered with a thick, adherent crust and the surrounding skin is red and itchy. The face, scalp, and extremities are most often involved, but lesions may appear anywhere, including the mucous membranes of the eyes, nose, and mouth.

Predisposing factors include poor hygiene, inadequate diet, and crowded living conditions. In nurseries impetigo may quickly become epidemic since the newborn are especially susceptible.

Impetigo frequently follows other disorders, particularly those which produce severe itching and subsequent scratching, such as scabies and pediculosis.

Only rarely is the disease serious insofar as mortality is concerned, except in infants. Treatment employs frequent cleansing and local anti-infective agents such as some antibiotics. *See* ANTIBIOTIC.

[EDWARD G. STUART/N. KARLE MOTTET]

Impsonite

A black, naturally occurring carbonaceous material having specific gravity 1.10–1.25 and fixed carbon 50–85%. Impsonite is infusible and is insoluble in carbon disulfide. A vein of impsonite 10 ft wide occurs in La Flore County, Okla., and there are other deposits in Arkansas, Nevada, Michigan, Peru, Argentina, Brazil, and Australia. The Peruvian impsonite may be derived from grahamite by weathering. The origin of impsonite is not well understood, but it appears to be derived from a fluid bitumen that polymerized after it filled the vein in which it is found. Impsonites are known to contain unusually high percentages of vanadium. *See* ALBERTITE; ASPHALT AND ASPHALTITE; ELATERITE; WURTZILITE.

[IRVING A. BREGER]

Impulse (mechanics)

The integral of a force over an interval of time. For a force $\mathbf{F}$, the impulse $\mathbf{J}$ over the interval from t_0 to t_1 can be written as Eq. (1). The impulse thus rep-

$$\mathbf{J} = \int_{t_0}^{t_1} \mathbf{F} \, dt \qquad (1)$$

resents the product of the time interval and the average force acting during the interval. Impulse is a vector quantity with the units of momentum.

The momentum-impulse relation states that the change in momentum over a given time interval equals the impulse of the resultant force acting during that interval. This relation can be proved by integration of Newton's second law over the time interval from t_0 to t_1. Let $\mathbf{P}$ represent the momentum at time t, with $\mathbf{P}_0$ and $\mathbf{P}_1$ being the values of $\mathbf{P}$ at times t_0 and t_1, respectively. Then Eq. (2) holds.

$$\mathbf{J} = \int_{t_0}^{t_1} \mathbf{F} \, dt = \int_{t_0}^{t_1} (d\mathbf{P}/dt) \, dt$$
$$= \int_{\mathbf{P}_0}^{\mathbf{P}_1} d\mathbf{P} = \mathbf{P}_1 - \mathbf{P}_0 \qquad (2)$$

If, as is ordinarily true, the mass m is constant, the momentum change can be expressed in terms of the velocities $\mathbf{v}_1$ and $\mathbf{v}_0$ at times t_1 and t_0, respectively, giving Eq. (3).

$$\mathbf{J} = m(\mathbf{v}_1 - \mathbf{v}_0) \qquad (3)$$

The concept of impulse is ordinarily most useful when the forces are large but act only for a short period. In most of these cases it is necessary to know only the momentum change, which is determined by the impulse. The relation between momentum and impulse thus has the advantage of

Fig. 14. Reflectometer for voltage—standing-wave ratio measurements in coaxial systems from 20 MHz to 7 GHz. (*General Radio* Co.)

eliminating the need for a detailed knowledge of how the forces, which can be very complicated, change with time. Forces which occur during collisions are of this type. *See* COLLISION.

A further simplification of this type of system is obtained if the time interval is short enough for the system to be considered essentially stationary during the time of action of the force, so that the momentum change occurs almost instantaneously. Except for the changed momentum, the motion of the system is treated as it would have been if no impulse had occurred. The effect of the impulse can thus be considered to provide a set of initial conditions for the motion. In a ballistic galvanometer, for example, there is an electric current for only a short time, during which the galvanometer coil, originally at rest, is given an impulse by the force associated with the current. The only effect of the current on the later motion of the coil, which can be considered stationary during the time there is a current, is to provide an initial velocity for the subsequent motion. *See* GALVANOMETER; IMPACT (IMPULSIVE FORCE); MOMENTUM.

[PAUL W. SCHMIDT]

Bibliography: J. W. Campbell, *An Introduction to Mechanics*, rev. ed., 1947.

Impulse generator

An electrical apparatus which produces very short surges of high-voltage, or high-current, power. High impulse voltages are used to test the strength of insulators and of power equipment against lightning and switching surges. High-current impulses are produced by the discharge of capacitors connected in parallel. Such current surges may be used to magnetize permanent magnets or to pro-

duce the rising magnetic field in circular particle accelerators. *See* PARTICLE ACCELERATOR.

High-voltage impulse generators commonly employ the principle, originally suggested by E. Marx, of charging capacitors in parallel and discharging them in series. The figure shows a four-stage Marx circuit in which the capacitors are first charged in parallel through charging resistors R, then connected in series and discharged through the test piece by the simultaneous sparkover of spark gaps G. The discharge is precipitated by placing a sufficient voltage on the middle electrode of the three-electrode gap between the first and second capacitor banks.

Although 1,000,000 to 2,000,000 volts (peak) is most common, over 7,500,000 volts to ground has been obtained. The waveform shows a rapid rise followed by a less rapid decline to zero, expressed by $v \propto (e^{-mt} - e^{-nt})$, where v is the instantaneous voltage t seconds after onset of the discharge, and m and n are the exponential decay constants of the circuit. Typical industrial laboratory waveforms are the 0.5−5, the 1−10, and the 1.5−40, in which the first number is the time in microseconds to the peak of the voltage wave, and the second is the time to one-half voltage of the tail of the wave. These discharges simulate the transient voltages induced in electrical conductors by natural lightning.

Impulses of still shorter rise times and duration are now used to produce short intense bursts of ionizing radiation for radiation damage studies. By the discharge of a single high-voltage condenser, such as the terminal of a Van de Graaff electrostatic generator insulated in compressed gas, voltage pulses which last less than 0.05 μsec have been applied to an electron acceleration tube. Transient electron pulses have been attained with energies of several million volts with currents of more than 25,000 amp. *See* RADIATION BIOLOGY.

[JOHN G. TRUMP]

Bibliography: T. E. Allibone and F. R. Perry, Standardization of impulse-voltage testing, *J. Inst. Elec. Eng. (London)*, 78:257−284, 1936; J. D. Craggs and J. M. Meek, *High Voltage Laboratory Technique*, 1954; E. Marx, Investigations in the testing of insulators with impact voltages, *Elektrotech. Z.* 45:652, 1924; J. F. McPartland, *Electrical Systems for Power and Light*, 1964.

Impulse turbine

A prime mover in which fluid (water, steam, or hot gas) under pressure enters a stationary nozzle where its pressure (potential) energy is converted to velocity (kinetic) energy. The accelerated fluid then impinges on the blades of a rotor, imparting its energy to the blades to produce rotation and overcome the connected rotor resistance. The impulse principle is basic to many turbines.

The impulse principle can be distinguished from the reaction principle by considering the flow of water from a hole near the bottom of a bucket (Fig. 1). The hole is a nozzle that serves to convert potential energy to the kinetic form ΔE; the impulse force F_i in the issuing jet is given by the expression $\Delta E = F_i v/2$, where v is the velocity of the jet.

If the jet is allowed to impinge on a series of vanes mounted on the periphery of a wheel, the impulse force can overcome the resistance con-

IMPULSE TURBINE

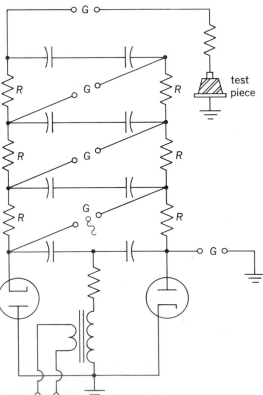

Fig. 1. Water escaping through a nozzle near the base of a bucket, free to swing, illustrating the impulse F_i and reaction F_r forces of the jet issuing with a velocity v.

Typical four-stage Marx impulse generator circuit.

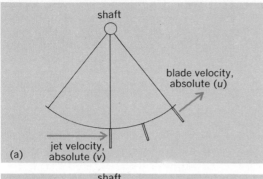

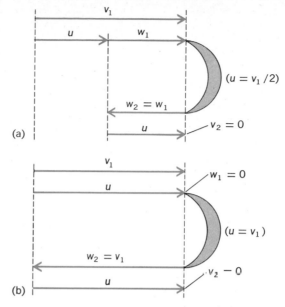

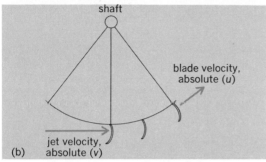

Fig. 2. Diagrams of a jet impinging on a series of (a) flat blades and (b) curved blades mounted on the periphery of a wheel.

Fig. 4. Velocity vector diagrams for idealized axial fluid flow on (a) impulse-turbine blading and (b) reaction-turbine blading; 180° jet reversal. In b the fluid is accelerated to w across the moving blades (nozzles). In both cases the leaving absolute fluid velocity v is zero, giving maximum efficiency.

nected to the shaft (Fig. 2). The efficiency of the device is ideally dependent upon the vane curvature and the absolute vane velocity. For flat blades (Fig. 2a), the efficiency cannot exceed 50%. With curved blades and complete reversal of the jet (Fig. 2b), the efficiency will be 100% when the vane velocity u is one-half the jet velocity v.

If the bucket in Fig. 1 is suspended and free to move, there will be, by Newton's third law of motion, a reaction force F_r equal and opposite to the impulse force F_i. For maximum efficiency (100%) the swinging bucket will have to move with an absolute tip velocity u equal to the jet velocity, v. This is the reaction principle and is demonstrated by the Barker's mill (Fig. 3).

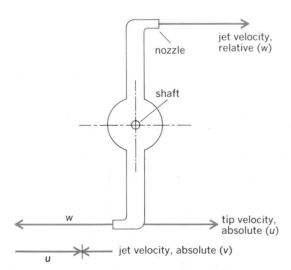

Fig. 3. Barker's mill, a reaction-type jet device, illustrating the application of moving nozzles and the resultant absolute and relative velocities.

The basic difference between the reaction principle and the impulse-turbine principle is determined by the presence or absence of moving nozzles. A nozzle is a throat section device in which there is a drop in pressure with consequent acceleration of the emerging fluid. An impulse turbine has stationary nozzles only. A reaction turbine must have moving nozzles but may have stationary nozzles also so that the fluid can reach the moving nozzle. This is the usual condition for any practical reaction turbine.

The idealized vector diagrams of Fig. 4 demonstrate distinguishing features for the construction which uses a row of blades mounted on the periphery of a wheel and for which flow is axially through the blade passages from one side of the wheel to the other. The theoretical condition further presupposes complete (180°) reversal of the jet and no friction losses. In Fig. 4, v is the absolute velocity of the fluid, u is the absolute velocity of the moving blade, and w is the relative velocity of the fluid with respect to the moving blade. Subscripts 1 and 2 apply to entrance and exit conditions, respectively. The vectors of the illustration demonstrate that maximum efficiency (zero residual absolute fluid velocity v_2) obtains (1) when the blade speed is half the jet speed for impulse turbines and (2) when the blade speed is equal to the jet speed for reaction turbines.

Figure 5 shows the situation more practically as complete reversal of the jet is not realistic. The vector diagrams of Fig. 5 show speed ratios of 0.49, 0.88, and 1.22 for reasonable degrees of jet reversal and no friction losses. By varying the angle of entrance α_1 to the moving blades, a wide range of speed ratios is available to the designer. In each case shown in Fig. 5, the stationary nozzle entrance angle β_1 is fixed at 15°; the exit vector triangle from the moving blades is identical in all

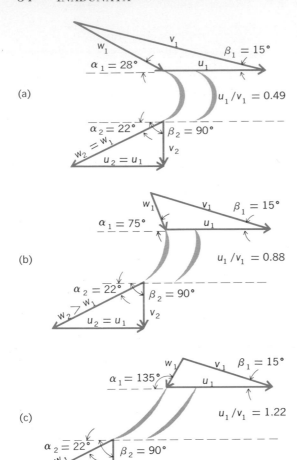

Fig. 5.   Vector diagrams for idealized (frictionless) axial fluid flow on (a) impulse-turbine blading and (b, c) reaction-turbine blading; jet reversal less than 180°. Exit vector triangles are identical with consequent equal efficiencies. Resultant speed ratios (u_1/v_1) are 0.49, 0.88, and 1.22, respectively.

cases. The data in Fig. 5c demonstrate that it is entirely possible to have speed ratios greater than unity with the reaction principle. This condition is utilized in many designs of hydraulic turbines, for example, propeller and Kaplan units.

The choice of details is at the discretion of the turbine designer. He can determine the extent to which the basic principles of impulse and reaction should be applied for a practical, reliable, economic unit in the fields of hydraulic turbines, gas turbines, and steam turbines. *See* GAS TURBINE; HYDRAULIC TURBINE; PELTON WHEEL; REACTION TURBINE; STEAM TURBINE; TURBINE PROPULSION.

[THEODORE BAUMEISTER]

Bibliography: T. Baumeister (ed.), *Standard Handbook for Mechanical Engineers*, 7th ed., 1967; G. T. Csanady, *Theory of Turbomachines*, 1964; V. L. Streeter, *Handbook of Fluid Dynamics*, 1961.

Inadunata

An extinct subclass of stalked Crinoidea comprising some 300 Paleozoic genera, ranging from the Upper Cambrian (Tremadocian) to the Permian.

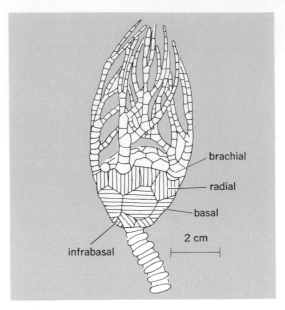

Sketch of *Cyathocrinus*, a dicyclic inadunate crinoid, showing the main plates. (*After F. Bather 1900*)

The arms (3–5 or more) were completely free and in no way incorporated into the calyx (see illustration). They were either branched or simple, and sometimes bore pinnules. The mouth and ambulacral grooves were usually roofed over by a series of small plates, and the tegmen was covered by plates. The calyx was either monocyclic or dicyclic, with conspicuous radials and basals, and all the thecal plates were sutured together into a rigid system. In some forms the anus lay at the upper end of a long plated siphon rising from the tegmen. The group contains the oldest recognizable crinoid, *Ramseyocrinus*, from the Tremadocian of Wales. *See* CRINOIDEA; ECHINODERMATA.

[HOWARD B. FELL]

Inarticulata

A class of phylum Brachiopoda, constituting about eight percent of the known genera. It is presently divided into four orders: Paterinida, Obolellida, Acrotretida, and Lingulida. *See* ACROTRETIDA; LINGULIDA; OBOLELLIDA; PATERINIDA.

All four orders are represented in the Early Cambrian; Obolellida is confined to the period, while Paterinida extends into the Ordovician. Lingulida and Acrotretida are still extant, but both orders were formerly more abundant and had their maximum diversity in the Ordovician.

Lingula and its allies are exceptional in that they are the only brachiopods known to burrow and are able to withstand considerable variations in salinity for short periods. They appear to be well-adjusted to their environment, for species of *Lingula* occur in shallow-water sediments of all ages from the Silurian to the Recent.

The two valves of the shell are typically not articulated and are held together only by soft tissue of the living animal. A few genera, notably *Linnarssonella* and *Dicellomus*, developed hinge mechanisms posteriorly; these may have been functionally comparable with the teeth-and-socket arrangement of the Articulata, but are not believed

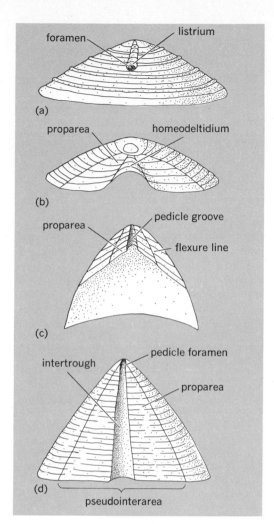

Fig. 1. Modifications of posterior of pedicle valve of some inarticulate brachiopods. (*a*) Discinacean *Orbiculoidea*. (*b*) Paterinacean *Paterina*. (*c*) Lingulacean *Lingulella*. (*d*) Acrotretacean *Prototreta*. (*R. C. Moore, ed., Treatise on Invertebrate Paleontology, pt. H, Geological Society of America, Inc., and University of Kansas Press, 1965*)

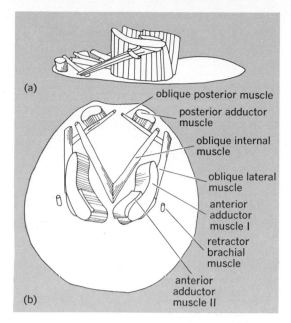

Fig. 2. Muscle system of the acrotretide *Discinisca*. (*a*) Lateral view. (*b*) Dorsal view. (*R. C. Moore, ed., Treatise on Invertebrate Paleontology, pt. H, Geological Society of America, Inc., and University of Kansas Press, 1965*)

to have been homologous because they evolved differently.

The posterior sector of one or both valves is commonly flattened, forming a pseudointerarea. The ventral pseudointerarea may be divided by a structure associated with the pedicle opening into two proparaes (Fig. 1). Internally, muscle scars and mantle canal impressions are commonly the only features present; a median ridge or elevated septum may be developed, particularly in the brachial valve. As an exception, the muscle scars are impressed on raised platforms, which are extravagantly developed in later trimerellids (Fig. 2).

Beneath the organic periostracum, the shell of the majority of inarticulates consists dominantly of calcium phosphate and chitin. These may form distinct alternating bands as in the lingulids, or be more homogeneously disposed as in the acrotretid discinaceans. A minority of genera have valves consisting principally of calcium carbonate. Such a composition seems to have evolved independently in four distinct stocks, and the Paterinida are the only order in which calcareous shelled repre-

sentatives are unknown. In all inarticulates, the fibrous secondary layer characteristic of the Articulata is undeveloped. *See* ARTICULATA (BRACHIOPODA); BRACHIOPODA.

[A. J. ROWELL]

Incandescence

The emission of visible radiation by a hot body. A theoretically perfect radiator, called a blackbody, will emit radiant energy according to Planck's radiation law at any temperature. Prediction of the visual brightness requires additional consideration of the sensitivity of the eye, and the radiation will be visible only for temperatures of the blackbody which are above some minimum. The relation between brightness and temperature is plotted in the figure. As shown, the minimum temperature for incandescence for the dark-adapted eye is about 390°C. Under these ideal observing conditions, the incandescence appears as a colorless glow. The dull red light commonly associated with incandescence of objects in a lighted room requires a temperature of about 500°C. *See* BLACKBODY; HEAT RADIATION; VISION.

Not all sources of light are incandescent. A cold gas under electrical excitation may emit light, as in the so-called neon tube or the low-pressure mercury-vapor lamp. Ultraviolet light from mercury vapor may excite visible light from a cold solid, as in the fluorescent lamp. Luminescence is the term used to refer to the emission of light due to causes other than high temperature, and includes thermoluminescence, in which emission of previously trapped energy occurs on moderate heating. *See* FLUORESCENT LAMP; MERCURY-VAPOR LAMP; THERMOLUMINESCENCE.

Flames are made luminous by incandescent particles of carbon. Gas flames can be made to produce intense light by the use of a gas mantle of

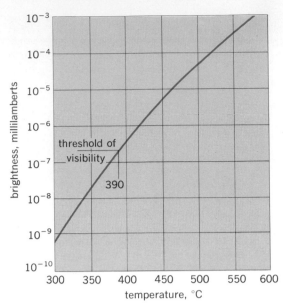

Graph showing the relation between the brightness of a blackbody and temperature.

Approximate color temperatures of common light sources

Source	Color temperature, °K
Candle	1925
Kerosine lamp	2000
Common tungsten-filament 100-watt electric light bulb	2800
Carbon arc	4000
Sun	5800

thoria containing a small amount of ceria. This mantle is a good emitter of visible light, but a poor emitter of infrared radiation. As less heat is lost in the long waves, the mantle operates at a higher temperature than a blackbody would and, hence, produces more intense visible light.

A useful criterion of an incandescent source is its color temperature, the temperature at which a blackbody has the same color, although not necessarily the same brightness. The color temperature of common light sources depends upon operating conditions. Approximate values are given in the accompanying table. See COLOR TEMPERATURE; INCANDESCENT LAMP.

[H. W. RUSSELL/GEORGE R. HARRISON]

Bibliography: M. M. Benarie, Optical pyrometry below red heat, *J. Opt. Soc. Amer.*, 47:1005–1009, 1957; R. W. Ditchburn, *Light*, 2 vols., 1963; F. A. Jenkins and H. E. White, *Fundamentals of Optics*, 3d ed., 1957.

Incandescent lamp

An electric lamp that produces light by heating a metallic filament to intense heat by passing an electric current through it. It is designed to produce light in the visible portion (at wavelengths of 380–760 mμ) of the electromagnetic spectrum. The filament is prepared of special materials and is enclosed in either an evacuated enclosure or one filled with an inert gas. In addition to radiation in the visible spectrum, infrared and ultraviolet energy are emitted, thus lowering the luminous

efficiency of the lamp. When either of these radiations is accentuated, however, the lamp may be used as a source of that radiant energy. The light-source efficacy, formerly called light-source efficiency, is expressed in lumens per watt (lu/w). Luminous efficiency is the ratio of the luminous flux to the radiant flux. *See* LIGHT.

Lamp construction. The important parts of an incandescent lamp are the lamp enclosure or bulb, the filament, and the base. The parts of a typical incandescent lamp are shown in Fig. 1.

Standard lamps have various bulb shapes, bases, and filament constructions. The bulb may be clear, colored, inside-frosted, or coated with diffusing or reflecting material. Most lamps have soft-glass bulbs; hard glass is used when the lamp will be subjected to sudden and severe temperature changes. In addition, lamps are available with a variety of bulb shapes, base types, and filament structures, as shown in Fig. 2. These vary according to the type of service planned, the need for easy replacement, and other environmental and service conditions.

The efficient design of an incandescent lamp centers about obtaining a high temperature at the filament without the loss of heat or disintegration of the filament. The early selection of carbon, which has the highest melting point of any element (6510°F) was a natural one. Carbon evaporates from its solid phase (sublimates) below this temperature, so carbon filaments must be operated at relatively low temperatures to obtain reasonable life. Two other elements, osmium (mp 4890°F) and tantalum (mp 5250°F), claimed attention because they could be operated at a high temperature with a longer life and less evaporation. With the advent of ductile tungsten a nearly perfect filament material was discovered. Ductile tungsten has a tensile strength four times that of steel, its melting point is high (6120°F), and it has relatively low evaporation. Hot tungsten is an efficient light radiator; it has a continuous spectrum closely following that of a blackbody radiator with a relatively high portion of the radiation in the visible spectrum. Because of its strength, ductility, and workability, it may be formed into coils, these coils again recoiled (for coil-coil filaments), and these again recoiled for cathodes in fluorescent lamps. If tungsten could be held at its melting point, 52 lu/w would be radiated. Because of physical limitations, however, 22 lu/w is the highest practical radiation for general-service lamps; some special lamps reach 35.8 lu/w. The higher the lumens per watt, the higher the filament temperature and the color temperature and the whiter the light.

Because the temperature of the filament controls the life of the lamp and its efficiency, it also controls the economics of lighting. For an economical installation, the factors affecting lamp life are weighed against the cost of the lamp and its installation and the cost of operation. This type of economic study, however, is rarely made of a lighting installation. The usual practice is to select a desired lamp size from the stock of the supplier for the accepted regional voltage.

Vaporization of the filament is reduced as much as possible. A small amount remains, however, and causes blackening of the bulb. The evaporated tungsten particles are carried to the upper part of

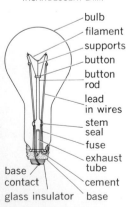

INCANDESCENT LAMP

bulb
filament
supports
button
button rod
lead in wires
stem
seal
fuse
exhaust tube
base contact
cement
glass insulator
base

Fig. 1. Parts of an incandescent lamp. (*General Electric Co.*)

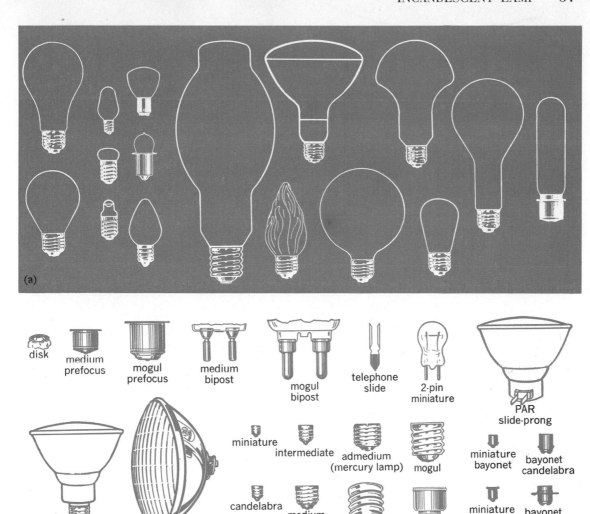

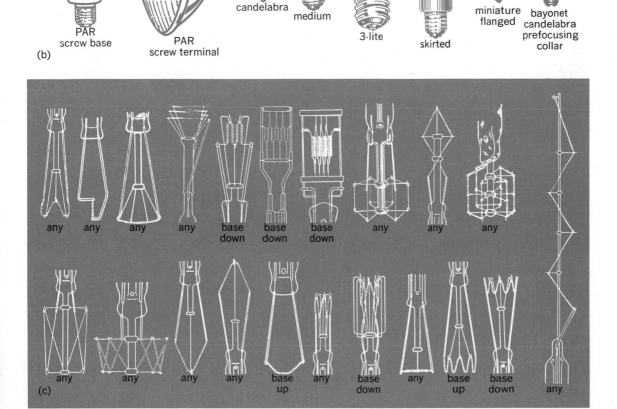

Fig. 2. Variations of incandescent lamps. (a) Bulb shapes. (b) Types of lamp bases. (c) Filament structures (notes designate burning position for each example). (*General Electric Co.*)

the bulb by convection currents. With the lamp in a base-up position the blackening is confined to the socket area, and the light output is only slightly affected. In a base-down position the blackening reduces the output a few percent. To reduce blackening, the inner atmosphere of the lamp is kept as clean as possible by use of a getter, which combines chemically with the tungsten particles. In lamps in which a getter is not adequate, tungsten powder is enclosed in the bulb and used to scour the surface by shaking. In some lamps a grid is placed to attract and hold the evaporated tungsten particles.

When cement failures are likely because of the heat, mechanical fastening is used to hold the base to the glass. When large electric currents are present, either mechanical fastening or bipost construction is used. Bipost and prefocus bases allow accurate placement of the filament with respect to the equipment for which it is designed. There are two common failures of the lamp bulb: in projectors, when the filament image is focused upon the glass and the bulb blisters; and when the hot bulb comes in contact with some low-temperature medium and thermal cracks develop. To protect the electric circuit from some lamp failures, a fuse may be placed in the lead-in wire.

Lamp ratings. Lamps are built for various voltage conditions, the most common being 115, 120, and 125 volts. High-voltage lamps are designed for ratings of 220–260 volts and low-voltage lamps for 6–64 volts. Lamps for use in 525–625 volt systems are designed to operate in groups of five in series across the line. Christmas-tree lights are designed to operate in parallel, or in series with eight lamps placed across the line. Street lamps are of the series type operating on 6.6 amp, except in series systems where individual lamp transformers are used with a lamp current of 15–20 amp. In a series street-lighting system continuity is maintained by a device that short-circuits the lamp when the filament burns out.

Lamp characteristics. Two characteristics of the incandescent lamp of particular interest are operation and color. Figure 3 shows the relative energy in the various color regions for the major types of incandescent lamp. Figure 4 shows the

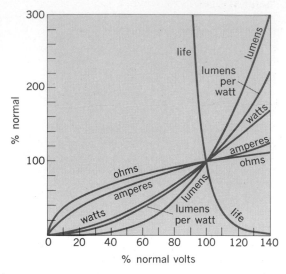

Fig. 4. Characteristic curves for incandescent lamps.

change of the various operating characteristics with a change of voltage.

At times it is necessary to determine lamp characteristics under other than normal conditions. Equations (1)–(5) with exponents given in Table 1 permit solution for actual service conditions as compared to normal operation.

$$\frac{\text{life}}{\text{LIFE}} = \left(\frac{\text{LUMENS}}{\text{lumens}}\right)^a = \left(\frac{\text{LUMENS/WATT}}{\text{lumens/watt}}\right)^b$$

$$= \left(\frac{\text{VOLTS}}{\text{volts}}\right)^d = \left(\frac{\text{AMPERES}}{\text{amperes}}\right)^u \quad (1)$$

$$\frac{\text{lumens}}{\text{LUMENS}} = \left(\frac{\text{volts}}{\text{VOLTS}}\right)^k = \left(\frac{\text{lumens/watt}}{\text{LUMENS/WATT}}\right)^h$$

$$= \left(\frac{\text{watts}}{\text{WATTS}}\right)^s = \left(\frac{\text{amperes}}{\text{AMPERES}}\right)^y$$

$$= \left(\frac{\text{ohms}}{\text{OHMS}}\right)^z \quad (2)$$

$$\frac{\text{LUMENS/WATT}}{\text{lumens/watt}} = \left(\frac{\text{LUMENS}}{\text{lumens}}\right)^f = \left(\frac{\text{VOLTS}}{\text{volts}}\right)^g$$

$$= \left(\frac{\text{AMPERES}}{\text{amperes}}\right)^j \quad (3)$$

$$\frac{\text{amperes}}{\text{AMPERES}} = \left(\frac{\text{volts}}{\text{VOLTS}}\right)^t \quad (4)$$

$$\frac{\text{watts}}{\text{WATTS}} = \left(\frac{\text{volts}}{\text{VOLTS}}\right)^n \quad (5)$$

To use Eqs. (1)–(5) it is necessary first to know the normal characteristics of the lamp, such as life, lumens, lumens per watt, volts, amperes, watts, and ohms, if all the characteristics are to be studied under abnormal conditions. By substituting into the equations the specific normal characteristics (upper case letters) and the special conditions (lower case letters) with the proper exponent from Table 1, the unknown characteristic may be determined. The exponents in Table 1 are changed at times, but for practical purposes and for normal voltage range those given will suffice.

Table 2 gives the effect, in percentage, of operat-

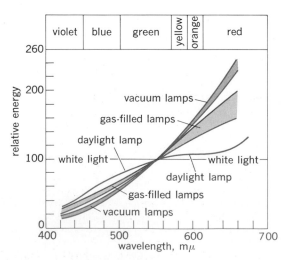

Fig. 3. Spectral energy distribution for important types of incandescent lamps.

Table 1. Exponents for Eqs. (1)–(5) in text

Exponent	Gas-filled	Vacuum
a	3.86	3.85
b	7.1	7.0
d	13.1	13.5
u	24.1	23.3
k	3.38	3.51
h	1.84	1.82
s	2.19	2.22
y	6.25	6.05
z	7.36	8.36
f	9.544	0.550
g	1.84	1.93
j	3.40	3.33
t	0.541	0.580
n	1.54	1.58

ing incandescent lamps below normal voltage. Low voltage increases lamp life, but the economic sacrifice of light advises against such practice.

Applications and special types. Incandescent lamps have been developed for many services. Most common are those used in general service and the miniature lamp. Special types have been developed for rough service applications, bake oven use, severe vibration applications, showcase lamps, multiple lights (three-way lamp), sign lamps, spotlights, floodlights, and insect-control lamps.

Series street-lighting lamps are designed with heavy filaments. The circuit operates at a constant current. The voltage is adjusted to the number of burning lamps to keep the current constant.

Reflector-type lamps are designed with built-in light control, that is, silvering placed on a portion of the lamp's outside surface. The reflecting surface is protected and is effective for the life of the lamp.

Projector-type lamps have molded-glass reflectors, silvered inside the lamp cavity, with either a clear glass cover or a molded control-lens cover. The reflector and the cover are sealed together, forming a lamp with an internal reflector. The parts are of hard glass, and the lamps may be

Table 2. Typical performance of large incandescent lamps burned below rated voltage

	Percentage of normal operation		
Voltage	Light output	Watts	Efficiency
100.0	100.0	100.0	100.0
99.2	97.3	98.8	98.2
98.3	94.4	97.4	96.9
97.5	91.8	96.1	95.4
96.7	89.2	95.0	93.8
95.8	86.4	93.6	92.3
95.0	84.0	92.4	90.8
94.2	81.5	91.2	89.4
93.3	78.0	89.8	87.7
92.5	76.6	88.7	86.2
91.7	74.1	87.5	84.8
90.0	69.5	85.0	81.7
88.3	65.0	82.5	79.4
86.7	60.8	80.3	75.8
85.0	56.6	77.9	72.8
83.3	52.0	75.5	69.7

used for outdoor service in floodlight and spotlight installations. This is also the type of lamp used in the sealed-beam headlight for automobiles, locomotives, and airplanes. A high degree of accuracy—just short of optical accuracy—is achieved by molding the contours of the reflector, thereby obtaining accurate beam control. With this sturdy structure the filament can be positioned for the best use of the lens, and the lamp has little depreciation during its life. Being constructed of hard glass, it lends itself to high-wattage use.

Miniature incandescent lamps are used in many fields and in many pieces of equipment, from the ordinary flashlight to the "grain of wheat" lamps used in surgical and dental instruments. These lamps are designed to give the highest efficiency consistent with the nature of the power source employed.

Special picture-projection lamps are designed for accurate filament location in the focal plane of the optical system, with the filament concentrated as much as possible in a single plane and in a small area. These precision lamps use a prefocus base for accurate positioning of the filament with respect to the base. Projector lamps run at high temperatures, and forced ventilation is frequently required.

A special class of lamp is designed for the photographic field, where the chief requirement is actinic quality. Frequently the most important rating is the color temperature, with little regard for economic efficiency or life. Photoflood lamps give high illumination for a short life, obtaining twice the lumens from high filament temperature, and high color temperature with three times the photographic effectiveness. The "daylight blue" photoflood lamp gives a very white light at 4800° K color temperature at 35.8 lu/w.

For other types of incandescent lamps see ARC LAMP; INFRARED LAMP.

[JOHN O. KRAEHENBUEHL]

Bibliography: Illuminating Engineering Society, *IES Lighting Handbook,* 4th ed., 1968; also see various manufacturers' publications.

Incendiary

One of a number of flammable materials and devices that are used to set fire to tactical and strategic targets, such as buildings, industrial installations, and fuel and ammunition dumps. Flame warfare extends also to antipersonnel use in the case of flamethrowers and fire bombs.

Modern incendiaries can be classified as either those which owe their effect to a self-supporting chemical reaction or those which depend on atmospheric oxygen to support combustion.

Metallic incendiaries. Thermite is a mixture of powdered iron oxide, Fe_2O_3, and powdered or granular aluminum. When it is heated, the reaction given below occurs. This reaction, some-

$$Fe_2O_3 + 2Al \rightarrow Al_2O_3 + 2Fe + 185,000 \text{ calories}$$

times called the Goldschmidt reaction, is sufficiently exothermic that maximum temperatures can exceed 4000°F. The iron formed melts and easily ignites any combustible material contacted. Consequently, the reaction is self-sustaining.

Metallic magnesium, igniting at about 1120°F, requires oxygen to support combustion. Once started, the fire is extremely difficult to put out.

Mechanized flamethrower.

Water is decomposed by burning magnesium and adds hydrogen to the combustion. Burning is stopped only by excluding oxygen or by cooling the metal below its ignition temperature. Bomb casings of magnesium are usually ignited by a thermite core.

Petroleum incendiaries. These are based on gasoline as a fuel. The gasoline may be either used alone or mixed with other petroleum fuels. It must, however, be thickened to be an effective incendiary. This thickening is necessary to confine the burning material to the target and, when used in flamethrowers, to increase the range of the ejected rod of fuel and to prevent its being consumed before reaching the target. Thickened fuel is sometimes called jellied gasoline.

One of the first practical fuel thickeners was napalm, a mixed aluminum soap in which the organic acids are derived from coconut oil (50%), naphthenic acids (25%), and oleic acid (25%). It may be used in quantities ranging from 4 to 12%, depending on the thickness desired. Gasoline thickened with napalm becomes a firm jelly when undisturbed, but in motion, such as when forced through a flamethrower nozzle, it acts as a viscous liquid (see illustration). This thixotropy is characteristic of all thickened fuels.

Flamethrowers are devices which force petroleum fuels through nozzles, igniting them as they emerge. The driving force is usually compressed air carried in a small tank which is an integral part of the device. Portable flamethrowers carried by soldiers have a range of over 50 yd under ideal conditions. Mechanized flamethrowers mounted on vehicles can throw fuel over 150 yd. Flamethrowers are primarily antipersonnel weapons.

Other incendiaries. White phosphorus, which ignites spontaneously in the presence of atmospheric oxygen to produce a dense cloud of phosphorus pentoxide, has been considered by some as an incendiary, although its principal use has been as a smoke producer. Bursting shells or grenades that scatter small drops of burning white phosphorus have been effective as antipersonnel incendiaries. *See* CHEMICAL WARFARE; FIRE EXTINGUISHER; SCREENING SMOKE.   [SEYMOUR D. SILVER]

Bibliography: H. Ellern, *Modern Pyrotechnics*, 1961; G. J. B. Fisher, *Incendiary Warfare*, 1946; U. S. Department of the Army, *Military Chemistry and Chemical Agents*, Tech. Manual TM 3–215, December, 1963.

Inclined plane

A plane surface inclined at an angle with the line of action of the force that is to be exerted or overcome. The usual application is illustrated in the diagram. A relatively small force acting parallel to the surface holds an object in place, or moves it at a constant speed.

In the free-body diagram shown here, three forces act on the object when no friction is present. The forces are its weight W, the force F_p parallel to the surface, and a force F_n normal to the surface. The summation of the forces acting in any direction on a body in static equilibrium equals zero; therefore, the summation of forces parallel to and forces normal to the surface are given by Eqs. (1) and (2). A force slightly greater than $W \sin \theta$ moves

$$F_p - W \sin \theta = 0 \qquad (1)$$

$$F_n - W \cos \theta = 0 \qquad (2)$$

the object up the incline, but the inclined plane supports the greater part of the weight of the object. The principal use of the inclined plane is as ramps for moving goods from one level to another. Wheels may be added to the object to be raised to decrease friction but the principle remains the same. The wedge and screw are closely related to the inclined plane and find wide application. *See* SIMPLE MACHINE.   [RICHARD M. PHELAN]

Inclinometer

An instrument for measuring magnetic dip, or inclination, which is defined as the angle between the magnetic field vector and the horizontal plane.

In the form of a simple dip needle, the instrument is a thin, pointed, mechanically balanced bar magnet mounted on horizontal bearings. When oriented so that its plane of rotation is in the magnetic meridian, the needle aligns itself with the direction of the magnetic field and indicates on a circular scale the angle of dip. It is used principally in mine surveying where the minerals have a high content of magnetic materials.

The dip circle also uses balanced needles, but of finer construction and provided with accurately ground and polished round shafts bearing on horizontal agate plates. The direction of magnetization of each needle is artificially reversed several times during a set of observations to reduce the residual effect of mechanical unbalance, and the positions of the ends of the needles are read with microscopes mounted on a vernier-equipped vertical circle. The dip circle, having an accuracy of no better than 3–5' of arc, has largely been replaced by other instruments in magnetic observatories and for first quality magnetic field work. *See* EARTH INDUCTOR; MAGNETOMETER.

[JAMES H. NELSON]

Inclusion blennorrhea

Inclusion conjunctivitis of the newborn; an inflammation of the conjunctiva of the baby's eye caused by an agent of the PLT-Bedsonia group. The infectious units of this conjunctivitis are the elementary bodies which form cytoplasmic inclusions in the conjunctival cells. The adult form of this disease is also called inclusion conjunctivitis. The differentiation of these diseases from trachoma rests on the dissimilarity of the clinical pictures; trachoma is more severe and more likely to cause pannus and scarring. Entropion, trichiasis, and the later visual loss in trachoma do not occur with inclusion conjunctivitis. *See* PLT-BEDSONIA GROUP.

INCLINED PLANE

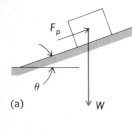

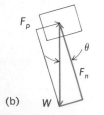

(a)

(b)

Weight resting on an inclined plane (a) with principal forces applied, and (b) their resolution into normal force.

So-called swimming-pool conjunctivitis is of the same etiology as conjunctivitis of the newborn. Infection of the newborn's eyes comes from the cervix of the mother, and since the reservoir is the genital tract, water of swimming pools is probably infected by discharges from the genital tracts of infected bathers. Chlorination of the water eliminates this source of adult conjunctivitis.

The agent of inclusion conjunctivitis may cause cervicitis in the female and nonspecific urethritis in the male. Exposure to the discharge will produce eye disease in adults.

Sulfonamides or tetracyclines given by mouth are effective against inclusion conjunctivitis; topically applied sulfadiazine or tetracycline ointment are successful in treatment and prevention in infants. *See* TRACHOMA.

[K. F. MEYER; JULIUS SCHACHTER]

Inclusion bodies, viral

Abnormal structures which appear within the cell nucleus, the cytoplasm, or both, during the course of virus multiplication. They usually have characteristic staining properties. In general, the inclusion bodies are concerned with the developmental processes of the virus. In some virus infections, such as molluscum contagiosum, inclusion bodies may be simply masses of maturing virus particles. In other infections (herpes simplex), typical inclusion bodies do not appear until after the virus has multiplied. Such inclusions may be remnants of the process of virus multiplication. *See* HERPES SIMPLEX.

The presence of inclusion bodies is often important in diagnosis. A cytoplasmic inclusion in nerve cells, the Negri body, is pathognomic for rabies. *See* RABIES; VIRUS.   [JOSEPH L. MELNICK]

Incompressible flow

Fluid motion without change in density. For practical purposes liquids are assumed to flow incompressibly. At low velocity this is nearly the case; however, even for liquids, abrupt changes in velocity produce compression or rarefaction. Usually a liquid flows under the action of gravity to occupy the lower portion of an open vessel. This property is taken as the distinguishing feature of a liquid. A gas, in contrast, flows compressibly to occupy any closed space to which it is confined, regardless of the initial volumes of gas and space. This property distinguishes a gas. As with liquids, slow flow of a gas is closely approximated by assuming it to be incompressible. In general, whenever fluid velocities are low (less than a fourth) relative to the rate of propagation of a pressure wave in the fluid, density variations during flow will be negligible and the flow will be effectively incompressible. *See* WATER HAMMER.

Analysis. Incompressible flow is quite frequently analyzed by supplementing the solution for an inviscid or "perfect" fluid with the effects of fluid viscosity. Simple flows such as uniform flow, sources, sinks, and vortexes may be represented by mathematical expressions defining flow velocity. These solutions may be superimposed to form realistic representation of complex inviscid flows, such as an airfoil moving through air or a boat hull through water. The results are mathematical expressions for velocity magnitude and direction at all points of the flow field. Pressure (p) may then be related to velocity (V) at a point in the flow by means of Bernoulli's equation, given below.

$$p + \tfrac{1}{2}\rho V^2 = \text{constant}$$

Here ρ is the constant fluid density. Forces on the boundary due to pressure may then be calculated. It remains then to determine the manner in which viscosity affects the flow field, hence the pressure distribution, and the additional force tangential to the boundary caused by fluid friction.

In this regime of incompressible flow, viscosity plays an important role, since it determines the behavior of the fluid adjacent to the boundaries of the flow (the boundary layer) and in the regions in which the fluid flow does not follow the boundaries (separated regions). Reynolds number, a nondimensional ratio of the inertia and viscous force in the fluid, provides a measure of flow characteristic which is quite useful in correlating experimental data with theory. *See* VISCOUS FLOW.

Applications. There is a vast variety of practical problems which may be evaluated through use of both inviscid and viscous incompressible flow theory and experimental data. One might first think of aircraft moving at low airspeeds, hovercraft (air-cushion vehicles), helicopters and balloons passing through the atmosphere, boats of various types passing through water (here only the flow beneath the surface is pertinent to this regime), surface vehicles such as cars and trains, and wind effects on structures causing unusual loads or oscillations. Other equally important applications of incompressible flow theory include heating and air conditioning design, conveying of solid particles and liquid droplets, and airflow in various industrial processes such as steelmaking. *See* AERODYNAMIC FORCE; COMPRESSIBLE FLOW; FLUID-FLOW PROPERTIES; GAS DYNAMICS; MACH NUMBER; REYNOLDS NUMBER.   [JAMES E. MAY]

Bibliography: J. E. Allen, *Aerodynamics*, 1963; T. von Karman, *Aerodynamics*, 1954, J. Lachnitt, *Aerodynamics*, 1957.

Indene

A colorless, liquid hydrocarbon (C_9H_8, also called benzocyclopentadiene) with the structure shown below, which boils at 181°C and freezes at −2°C. It is usually obtained from the light-oil frac-

tion (boiling point 178−182°C) produced in the carbonization of coal but is also obtained by pyrolysis of certain petroleum fractions.

Indene resembles cyclopentadiene in that one hydrogen of the methylene (CH_2) group may be replaced by sodium. Indene may be oxidized to phthalic acid or reduced to indan (C_9H_{10}). In the presence of acid, indene polymerizes. Copolymers with benzofuran have been manufactured on a small scale for use in coatings and floor coverings. *See* POLYNUCLEAR HYDROCARBON.

[CHARLES K. BRADSHER]

Index fossil

The ancient remains and traces of a plant or animal that lived during a particular span of geologic time and that geologically dates the containing rocks.

Index fossils are almost exclusively confined to sedimentary rocks which originated in such diverse environments as open oceans, tropical lagoons, coral reefs, beaches, lakes, and rivers. They represent a variety of organic forms, such as microscopic algae, plant pollen, oyster shells, shark teeth, geometric cavities in the rock that are clearly organic but otherwise unknown, or such large and obvious examples as the teeth of extinct elephants. Index fossils are a very necessary means for comparing the geologic age of sedimentary rock formations and are an everyday tool in the search for petroleum, coal, and metallic ores.

Criteria. The choice of a fossil as an index depends on several criteria. In general, the fossil represents a group that evolved rapidly. Through the mechanisms of evolution, plant and animal forms have appeared in a determinable order, persisted for awhile, and then been replaced by new forms, the old never to return. They thus have formed a general succession of floras and faunas in which individual genera and species serve as detailed time markers. The greater the rate of evolution, the shorter the period of time represented by any given index fossil and the narrower the limits of relative age assigned to the rocks containing the index. Commonly, the span of geologic time during which a fossil lived is referred to as its range, and the thickness of rocks through which a particular index fossil or selected group of fossils occurs is referred to as a faunal zone. Ranges and zones take their names from the fossils whose occurrence they represent, that is, the range of *Fusulinella* or the *Siphonodella duplicata* Zone.

An index fossil also must be present in the rocks in sufficient numbers to be found with reasonable effort, must be relatively easy to collect or identify, and must be geographically extensive so that the zone it defines is widely applicable. These requirements also imply that it must be resistant to burial pressures, to solution by groundwater percolation, or to surface weathering in order to be preserved. Fossils commonly are incompletely preserved or cannot be removed from the rock without considerable effort or possible destruction. Features that permit the fossil to be readily identified under such circumstances are of great advantage. The usefulness of a fossil is increased, however, if it is relatively easy to separate from the rock. Conodonts, spores, and pollen, for example, are resistant to acid and can thus be chemically separated from the rocks. Small size is likewise an advantage, inasmuch as many fossils must be identified in the rock chips of drilled well cuttings. Such fossils, mainly foraminifers, spores, and pollen, are widely used by the petroleum industry.

The presence or absence of the index fossil ideally should not reflect environmental conditions. Many fossil groups that qualify as indices in other respects are unsuitable in application because they were especially adapted to a restricted environment and thus are present only where that particular fossil environment is represented in the rocks. Such "facies" fossils were not under enough environmental pressure to evolve as rapidly as many other groups; their value lies in interpretation of environment rather than in age determination.

Common flora and fauna. The fossil groups most useful as index fossils are generally marine and either floaters or open ocean swimmers, such as cephalopods, or bottom dwellers that had a floating or swimming stage in their life cycles, such as the medusa stage in the brachiopods. Such characteristics are necessary for rapid dispersal of newly evolved forms. On land, such mobile forms as the horses or wind-borne pollen and spores were relatively unrestricted by environmental barriers and became widely dispersed. All of these groups have provided biochronological zones of worldwide extent.

Sequence of evolution. The recognition of evolutionary sequences of fossil species is a prime factor in the determination of precise geologic age. In the 19th century general faunal succession received considerable attention, but present-day workers emphasize the use of phylogenetic lineages wherein species complexes grade from ancestor to descendant, and fossil populations are recognized as possessing the same kinds of inter- and intraspecific variation as living populations. Thus in modern use the individual index fossil is no longer entirely adequate for detailed geologic age determinations, and close examination of multiple phylogenetic sequences is necessary for precise dating and correlation of rock strata. Variations from the normal evolutionary sequence are keys to differences in sedimentation rates, gaps in the geologic record, or recognition of special events. Modern paleontologists, using foraminifera, discoasters, and conodonts in particular, are commonly able to date the occurrence of geologic events to within a million years.

Indices for biochronology. Relatively few groups of fossil organisms embody all of the above characteristics, but several groups have proved to be practical indices and have been widely accepted as a basis for world biochronological systems. Foremost among the forms are spores and pollen, discoasters, foraminifers, graptolites, ammonoid cephalopods (see illustration), and conodonts.

During the Cambrian Period (600,000,000 years

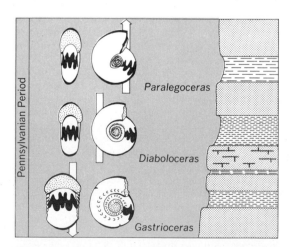

The cephalopod genera *Gastrioceras*, *Diaboloceras*, and *Paralegoceras* form an evolutional sequence. Their short ranges (indicated by bars) and distinctive suture lines (area between two sutures is shown in black) make them ideal index fossils for the early Pennsylvanian Period. (*Modified from A. K. Miller and W. M. Furnish, Middle Pennsylvanian Schistoceratidae (Ammonoidea), J. Paleontol., 32(2):254, 1958*)

before present) the oldest highly developed animals appeared; among them the trilobites provide the first important group of index fossils. They were marine bottom dwellers that evolved rapidly and are keys for both Cambrian and Ordovician time. Small plantlike floating colonial animals called graptolites have proved useful in correlating Ordovician (500,000,000 years B.P.) and Silurian (425,000,000 years B.P.) rocks in areas as widely separated as North America, Sweden, and the Soviet Union. Ammonoids, coiled cephalopods with irregular suture patterns that record the animal's history of growth as well as the species' evolutional relationships, are a classic example of the internationally useful index fossil and are important beginning in the Devonian Period (405,000,000 years B.P.) and extending to the end of the Cretaceous Period (135,000,000 years B.P.). From the Pennsylvanian Period (310,000,000 years B.P.) fusulinids, a family of Foraminiferida, and pollen and spores from the coal forests are important indices. Small phosphatic teethlike fossils known as conodonts, whose zoological affinities are unknown, have been useful for detailed zonation throughout the Paleozoic Era as well as the early part of the Mesozoic. Closer to present time, the bones and teeth of vertebrate animals serve as index fossils for the Tertiary Era while the remains of primitive man himself have been used to date the Recent past. *See* BRACHIOPODA; CEPHALOPODA; CONODONT; EVOLUTION, ORGANIC; FORAMINIFERIDA; FOSSIL; FUSULINACEA; GEOLOGICAL TIME SCALE; GRAPTOLITHINA; STRATIGRAPHIC NOMENCLATURE; STRATIGRAPHY. [CHARLES COLLINSON]

Bibliography: J. R. Beerbower, *Search for the Past*, 1968; C. O. Dunbar and J. Rodgers, *Principles of Stratigraphy*, 1958; M. Kay and E. H. Colbert, *Stratigraphy and Life History*, 1965; R. C. Moore, C. G. Lalicker, and A. G. Fischer, *Invertebrate Fossils*, 1952; J. M. Weller, *The Course of Evolution*, 1969.

Index mineral

In metamorphic petrology, a characteristic mineral which by its presence in a rock indicates the mineral facies of the rock. While an index fossil characterizes a sedimentary bed (stratum) in the geological sequence, an index mineral characterizes the mode of origin of a metamorphic rock in regard to temperature, pressure, and composition. A mineral such as quartz is not an index mineral because quartz is found in a great variety of rocks formed from low to high temperature and pressure. However, minerals with stability fields within restricted parts of the geologically important temperature-pressure range (by a given bulk composition of the containing rock) usually may be regarded as index minerals. In a crude way they are also geological thermometers and pressure gages. *See* GEOLOGIC THERMOMETRY; HIGH-PRESSURE PHENOMENA; METAMORPHIC ROCKS.

In going from an area of nonmetamorphic rocks into an area of progressively more highly metamorphic rocks, each zone of progressive metamorphism is defined by an index mineral the first appearance of which marks the outer limit of the zone in question. The sequence of index minerals in progressively metamorphosed pelitic rocks is as follows: chlorite, biotite, almandite, garnet, stau-

rolite, kyanite, and sillimanite. An index mineral for high pressure but relatively low temperature is glaucophane. Pyrope garnet and omphacite pyroxene indicate high values of both pressure and temperature. *See* ECLOGITE. [T. F. W. BARTH]

Indian Ocean

The smallest and geologically the most youthful of the three oceans. It differs from the Pacific and Atlantic oceans in two important aspects. First, it is landlocked in the north, does not extend into the cold climatic regions of the Northern Hemisphere, and consequently is asymmetrical with regard to its circulation. Second, the wind systems over its equatorial and northern portions change twice each year, causing an almost complete reversal of its circulation. During the International Indian Ocean Expedition from 1960 to 1965, the ocean was explored systematically by more than 50 ships of 18 nations.

Size and bathymetry. The eastern and western boundaries of the Indian Ocean are 147 and 20°E, respectively. In the southeastern Asian waters the boundary is usually placed across Torres Strait, and then from New Guinea along the Lesser Sunda Islands, across Sunda Strait and Singapore Strait. The surface area within these boundaries is 75,900,000 km², or about 21% of the surface of all oceans. The volume of the Indian Ocean is 293,000,000 km³, and the average depth is about 3850 m.

The ocean floor is divided into a number of basins by a system of ridges (Fig. 1). The largest is the Mid-Ocean Ridge, the greater part of which has a rather deep rift valley along its center. It lies like an inverted Y in the central portions of the ocean and ends in the Gulf of Aden. More recently discovered is the Ninety East Ridge. Most of the ocean basins separated by the ridges reach depths in excess of 5000 m. The Sunda Trench, stretching along Java and Sumatra, is the only deep-sea trench in the Indian Ocean and contains its deepest observed depth of 7455 m. The Andaman Basin, with a maximum depth of 4360 m, is separated from the open ocean by a 1400-m sill. The Red Sea has a maximum depth of 2835 m, but its entrance at the Strait of Bab-el-Mandeb is only 125 m deep. East of the Mid-Ocean Ridge, deep-sea sediments are chiefly red clay; in the western half of the ocean, globigerina ooze prevails and, near the Antarctic continent, diatom ooze. *See* MARINE GEOLOGY; MARINE SEDIMENTS.

Wind systems. Atmospheric circulation over the northern and equatorial Indian Ocean is characterized by the changing monsoons. In the southern Indian Ocean atmospheric circulation undergoes only a slight meridional shift during the year (Fig. 2). During winter in the Northern Hemisphere, from December through February, the equatorial low-pressure trough is situated at about 10°S and continues into a low-pressure system over northern Australia. A strong high-pressure system lies over Asia. This situation causes the Northeast Monsoon to blow everywhere north of the Equator. The winds cross the Equator from north to south and then usually become northwest winds before reaching the Intertropical Convergence. The subtropical high-pressure ridge of the Southern Hemisphere is situated near 35°S. North of it, the

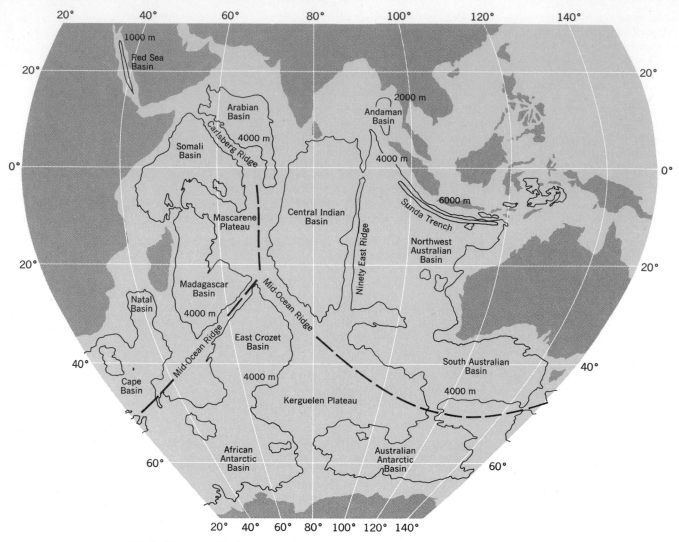

Fig. 1. Physiographic features of the Indian Ocean.

Southeast Trades blow. South of 40°S, winds from the west prevail and are associated with cyclones traveling around Antarctica.

During summer in the Northern Hemisphere, from June through August, a low-pressure system is developed over Asia with the center around Iran. The subtropical high-pressure ridge of the Southern Hemisphere has shifted slightly northward and continues into a high-pressure system over Australia. The Southeast Trades are more strongly developed during this season, cross the Equator from south to north, and become the Southwest Monsoon, bringing rainfall and the wet season to India and Burma. Atmospheric circulation during June through August is much stronger and more consistent than during February. In the Southern Hemisphere the West Wind Belt has shifted about 5° to the north, with westerly winds starting from 30°S and storms becoming stronger and more frequent during winter in that hemisphere.

Circulation. The surface circulation is caused largely by winds and changes in response to the wind systems (Fig. 3). In addition, strong boundary currents are formed, especially along the western coastline, as an effect of the Earth's rotation and of the boundaries created by the landmasses. In the

Southern Hemisphere south of 35°S, a general drift from west to east is found as a result of the prevailing west winds. Near 50°S, where west winds are strongest, the Antarctic Circumpolar Current is embedded in the general West Wind Drift. In the subtropical southern Indian Ocean, circulation is anticyclonic. It consists of the South Equatorial Current flowing west between 10 and 20°S, a flow to the south along the coast of Africa and Madagascar, parts of the West Wind Drift, and flow to the north in the eastern portions of the ocean, especially along the coast of Australia. The flow to the south between Madagascar and Africa is called the Mozambique Current. It continues along the coast of South Africa as the Agulhas Current, most of which turns east into the West Wind Drift. The circulation in the Southern Hemisphere south of 10°S changes only slightly with the seasons.

North of 10°S the changing monsoons cause a complete reversal of surface circulation twice a year. In February, during the Northeast Monsoon, flow north of the Equator is mostly to the west and the North Equatorial Current is well developed. Its water turns south along the coast of Somaliland and returns to the east as the Equatorial Countercurrent between about 2 and 10°S. In August, during the Southwest Monsoon, the South Equatorial

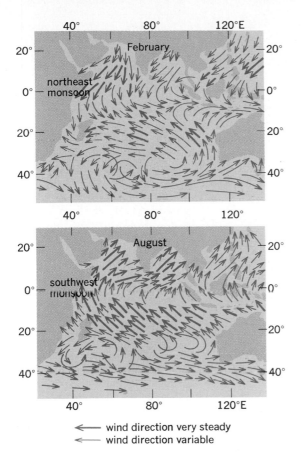

Fig. 2. Winds over the Indian Ocean.

Current extends to the north of 10°S; most of its water turns north along the coast of Somaliland, forming the strong Somali Current. North of the Equator flow is from west to east and is called the Monsoon Current. Parts of this current turn south along the coast of Sumatra and return to the South Equatorial Current.

Transports of many of these current systems have been determined. The Antarctic Circumpolar Current and the West Wind Drift carry between 100×10^6 and 140×10^6 m³/sec to the east. The South Equatorial Current transport ranges between 20×10^6 and 30×10^6 m³/sec, and that of the Agulhas Current is about 40×10^6 m³/sec. During the full development of the Southwest Monsoon, the Somali Current and the Monsoon Current transport between 30×10^6 and 40×10^6 m³/sec. Although the Antarctic Circumpolar Current reaches to great depths, probably to the bottom, most of the other currents are much shallower. The Agulhas Current reaches to approximately 1200 m, the Somali Current to 800 m, and the remainder of the circulation is limited to the upper 300 m of the ocean. Below these depths, movements are sluggish and irregular, and the spreading of water properties is a slow process accomplished chiefly by mixing. At the Equator, during the time of the Northeast Monsoon, an Equatorial Undercurrent is found flowing as a subsurface current near the 150-m depth from west to east. This current is weaker than the corresponding currents in the Pacific and Atlantic oceans. *See* OCEAN CURRENTS.

Surface temperature. The pattern of sea-surface temperatures changes considerably with the seasons (Fig. 4). During February, when the Intertropical Convergence is near 10°S, the heat equator is also in the Southern Hemisphere and most of the area between the Equator and 20°S has temperatures near 28°C. The water in the northern parts of the Bay of Bengal and of the Arabian Sea is much cooler, and temperatures below 20°C can be found in the northern portions of the Persian Gulf and the Red Sea. In the Southern Hemisphere temperatures decrease gradually from the tropics to the polar regions. Surface circulation affects the distribution of temperature, and warm water spreads south along the coast of Africa and cool water north off the west coast of Australia, causing the isotherms to be inclined from west to east.

During August high temperatures are found in the Northern Hemisphere and in the equatorial region. The Somali Current advects cool water along the coast of Africa to the north. In the Southern Hemisphere isotherms are almost 10° of latitude farther north.

Surface salinity. The distribution of surface salinity is controlled by the difference between evaporation and precipitation and by runoff from the continents (Fig. 5).

High surface salinities, which are greater than above 36 parts per thousand (%o), are found in the subtropical belt of the Southern Hemisphere, where evaporation exceeds rainfall. In contrast, the Antarctic waters are of low salinity because of heavy rainfall and melting ice. Another area of low salinities stretches from the Indonesian waters

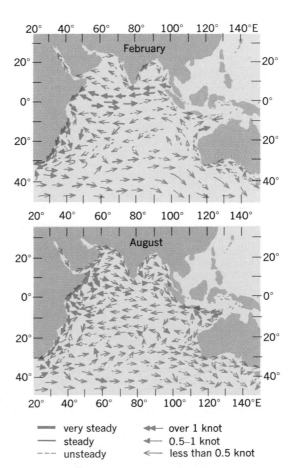

Fig. 3. Surface currents of the Indian Ocean.

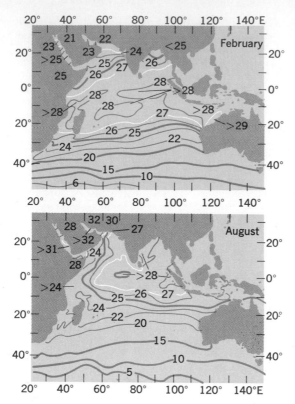

Fig. 4.   Sea-surface temperature (°C) of Indian Ocean.

along 10°S to Madagascar. It is caused by the heavy rainfall in the tropics. The Bay of Bengal has very low salinities, often less than 30°/oo, as a result of the runoff from large rivers. In contrast, because of high evaporation the Arabian Sea has salinities as high as 36.5°/oo. High salinities are also found in the Persian Gulf (>38°/oo) and in the Red Sea (41°/oo), representing the arid character of the landmasses surrounding them. The salinity distribution changes relatively little during the year; however, south of India from the Bay of Bengal to the west, a flow of low-salinity water, caused by the North Equatorial Current, can be noticed during February.

Surface water masses. The different climatic conditions over various parts of the Indian Ocean cause the formation of characteristic surface water masses. The Arabian Sea water is of high salinity, has a moderate seasonal temperature variation, and can be classified as subtropical. The water in the Bay of Bengal is of low salinity and always warm, representing tropical surface water. Another type of tropical surface water stretches from the Indonesian waters to the west and is called the Equatorial Surface Water. Subtropical Surface Water with salinities in excess of 35.5°/oo and a seasonal temperature variation of between 15 and 25°C is found in the subtropical regions of the Southern Hemisphere. Its southern boundary is the Subtropical Convergence coinciding with temperatures of about 15°C. From there, temperature and salinity decrease in the area of the transition water to about 4°C and 34°/oo at the Antarctic Polar Front. South of the Antarctic Polar Front, Antarctic Surface Water of low salinity (<34°/oo) is found; its temperature is near the freezing point (−1.9°C) in winter and is approximately 2°C in summer.

Subsurface water masses. The vertical distributions of temperature, salinity, oxygen content, and main water masses in the Indian Ocean are shown in Fig. 6, along a section from the Arabian Sea to Antarctica. Warm water of more than 15°C occupies only a very thin surface layer of less than 300 m depth. This layer contains the tropical and subtropical water masses. Subtropical water of the Southern Hemisphere spreads as a subsurface salinity maximum toward the Equator at depths between 100 and 200 m.

Water of high salinity formed in the Red Sea and in the Persian Gulf leaves these basins and spreads as a subsurface layer of high salinity throughout the Arabian Sea at depths between 600 and 1000 m. To the south it can be traced as far as Madagascar, and to the east as far as Sumatra. Water of low salinity and a temperature of approximately 4°C sinks near the Antarctic Polar Front and spreads north as a salinity minimum between 600 and 1500 m. It is called the Antarctic Intermediate Water.

The deep and bottom water of the Indian Ocean is of external origin. South of Africa, water of rather high salinity, originating south of Greenland in the North Atlantic Ocean, enters the Indian Ocean, filling the layers between 1500- and 3000-m depths. The water below the 3000-m depth is Antarctic Bottom Water originating in the Weddell Sea. Its temperature is lower than 0°C.

Because of their origin at the sea surface, both the Antarctic Intermediate Water and the Antarctic Bottom Water are of rather high oxygen content (>5 ml/liter). Since their residence time is long, the water masses of the northern Indian Ocean below the surface layer have a very low oxygen content. An oxygen minimum is associated with

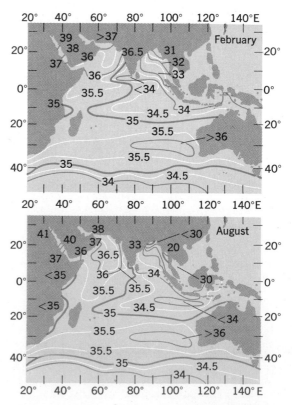

Fig. 5.   Sea-surface salinity of the Indian Ocean in parts per thousand by weight (°/oo).

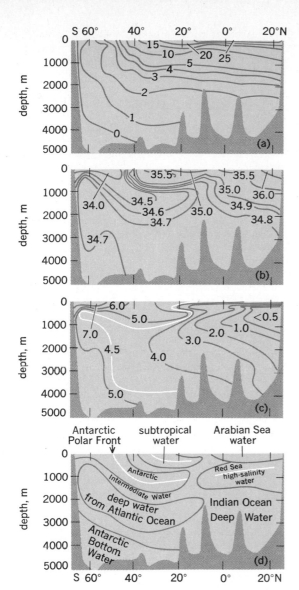

Fig. 6. The Indian Ocean from the Arabian Sea to Antarctica showing (a) temperature (°C), (b) salinity (°/oo), (c) oxygen content (ml/liter), and (d) water masses.

the Red Sea water and extends south between the Antarctic Intermediate Water and the North Atlantic Deep Water. *See* ANTARCTIC OCEAN; RED SEA; SEA WATER.

Upwelling. Several areas of upwelling are found along the shores of the Indian Ocean, but all of them are seasonal in character, in contrast to upwelling areas in the other oceans. They are especially developed during the Southwest Monsoon season from May through September, when upwelling takes place in the Banda Sea, south of Java, along the coast of Somaliland, and off Arabia. The strongest upwelling is found along the Somaliland coast, where surface temperatures during this season may be as low as 14°C. Water of high nutrient content ascends at a rate of approximately 1 m/day from as deep as 150 m and is integrated into the surface flow. These areas have a high biological production. During the Northeast Monsoon, some upwelling takes place in the Andaman Sea and off the west coast of India. *See* UPWELLING.

Tides. Both semidiurnal and diurnal tides occur in the Indian Ocean. The semidiurnal tides rotate

around three amphidromic points situated in the Arabian Sea, southeast of Madagascar, and west of Perth. The diurnal tide also has three amphidromic points: south of India, in the Mozambique Channel, and between Africa and Antarctica. It has more the character of a standing wave, oscillating between the central portions of the Indian Ocean, the Arabian Sea, and the waters between Australia and Antarctica.

Along the coast of Africa, in the Bay of Bengal, and along the northwest coast of Australia semidiurnal tides prevail, with two high waters each day. Mixed tides are found around the Arabian Sea and along the coasts of Sumatra and Java. Prevailing diurnal tides with only one high water each day occur only along southwest Australia. The ranges of spring tides are high in the Arabian Sea, 2.5 m at Aden and 5.7 m at Bombay. The Bay of Bengal has high tidal ranges, chiefly along the coast of Burma, with 7 m at Rangoon. Moderately high tides between 1 and 2 m are found along the coast of Sumatra and Java, but spring tides along the northwest coast of Australia are between 6 and 10 m. Along the coast of South Africa, tidal ranges are between 1.5 and 2 m, but in the Mozambique Channel they increase to 3–5 m. The islands in the central Indian Ocean usually have spring tidal ranges between 1 and 1.5 m. *See* TIDE.

[KLAUS WYRTKI]

Bibliography: G. Dietrich, *General Oceanography*, 1963; International Indian Ocean Expedition, *Collected Reprints*, vols. 1–4, UNESCO, 1965–1967; *Monatskarten fur den Indischen Ozean*, Ger. Hydrogr. Inst. Publ. no. 2242, 1960; G. Schott, *Geographie des Indischen und Stillen Ozeans*, 1935; H. U. Sverdrup, M. W. Johnson, and R. H. Fleming, *The Oceans: Their Physics, Chemistry, and General Biology*, 1946.

Indicator, acid-base

A substance that reveals, through characteristic color changes, the degree of acidity or basicity of solutions. Indicators are weak organic acids or bases which exist in more than one structural form (tautomers) of which at least one form is colored. Intense color is desirable so that very little indicator is needed; the indicator itself will thus not affect the acidity of the solution.

The equilibrium reaction of an indicator may be regarded typically by giving it the formula HIn. It dissociates into H^+ and In^- ions and is in equilibrium with a tautomer InH which is either a nonelectrolyte or at most ionizes very slightly. In the overall equilibrium shown as Eq. (1) the simplify-

$$InH \rightleftharpoons HIn \rightleftharpoons H^+ + In^- \qquad (1)$$
$$(I) \qquad\qquad (II)$$

ing assumption that the indicator exists only in forms (I) and (II) leads to no difficulty. The addition of acid will completely convert the indicator to form (I), which is therefore called the acidic form of the indicator although it is functioning as a base. A hydroxide base converts the indicator to form (II) with the formation of water; this is called the alkaline form. For the equilibrium between (I) and (II) the equilibrium constant is given by Eq. (2). In

$$K_{In} = \frac{[H^+][In^-]}{[InH]} \qquad (2)$$

a manner similar to the pH designation of acidity,

Common acid-base indicators

Common name	pH range	Color change (acid to base)	pK	Chemical name	Structure	Solution
Methyl violet	0–2, 5–6	Yellow to blue violet to violet		Pentamethylbenzyl-pararosaniline hydro-chloride	Base	0.25% in water
Metacresol purple	1.2–2.8, 7.3–9.0	Red to yellow to purple	1.5	m-Cresolsulfon-phthalein	Acid	0.73% in N/50 NaOH, dilute to 0.04%
Thymol blue	1.2–2.8, 8.0–9.6	Red to yellow to blue	1.7	Thymolsulfon-phthalein	Acid	0.93% in N/50 NaOH, dilute to 0.04%
Tropeoline 00 (Orange IV)	1.4–3.0	Red to yellow		Sodium p-diphenyl-aminoazobenzene-sulfonate	Base	0.1% in water
Bromphenol blue	3.0–4.6	Yellow to blue	4.1	Tetrabromophenol-sulfonphthalein	Acid	1.39% in N/50 NaOH, dilute to 0.04%
Methyl orange	2.8–4.0	Orange to yellow	3.4	Sodium p-dimethyl-aminoazobenzene-sulfonate	Base	0.1% in water
Bromcresol green	3.8–5.4	Yellow to blue	4.9	Tetrabromo-m-cresol-sulfonphthalein	Acid	0.1% in 20% alcohol
Methyl red	4.2–6.3	Red to yellow	5.0	Dimethylaminoazo-benzene-o-carboxylic acid	Base	0.57% in N/50 NaOH, dilute to 0.04%
Chlorphenol red	5.0–6.8	Yellow to red	6.2	Dichlorophenolsulfon-phthalein	Acid	0.85% in N/50 NaOH, dilute to 0.04%
Bromcresol purple	5.2–6.8	Yellow to purple	6.4	Dibromo-o-cresolsulfon-phthalein	Acid	1.08% in N/50 NaOH, dilute to 0.04%
Bromthymol blue	6.0–7.6	Yellow to blue	7.3	Dibromothymolsulfon-phthalein	Acid	1.25% in N/50 NaOH, dilute to 0.04%
Phenol red	6.8–8.4	Yellow to red	8.0	Phenolsulfonphthalein	Acid	0.71% in N/50 NaOH, dilute to 0.04%
Cresol red	2.0–3.0, 7.2–8.8	Orange to amber to red	8.3	o-Cresolsulfonphthalein	Acid	0.76% in N/50 NaOH, dilute to 0.04%
Orthocresol-phthalein	8.2–9.8	Colorless to red	—		Acid	0.04% in alcohol
Phenolphthalein	8.4–10.0	Colorless to pink	9.7	—	Acid	1% in 50% alcohol
Thymolphthalein	10.0–11.0	Colorless to red	9.9	—	Acid	0.1% in alcohol
Alizarin yellow GG	10.0–12.0	Yellow to lilac		Sodium p-nitrobenzene-azosalicylate	Acid	0.1% in warm water
Malachite green	11.4–13.0	Green to colorless		p,p'-Benzylidenebis-N,N-dimethylaniline	Base	0.1% in water

that is, $pH = -\log[H^+]$, the K_{In} is converted to pK_{In} with the result shown in Eq. (3). It is seen that the

$$pK_{In} = pH - \log\frac{[In^-]}{[InH]} \qquad (3)$$

pK of an indicator has a numerical value approximately equal to that of a specific pH level.

Use of indicators. Acid-base indicators are commonly employed to mark the end point of an acid-base titration or to measure the existing pH of a solution. For titration the indicator must be so chosen that its pK is approximately equal to the pH of the system at its equivalence point. For pH measurement, the indicator is added to the solution and also to several buffers. The pH of the solution is equal to the pH of that buffer which gives a color match. Care must be used to compare colors only within the indicator range. A color comparator may also be used, employing standard color filters instead of buffer solutions.

Indicator range. This is the pH interval of color change of the indicator. In this range there is competition between indicator and added base for the

available protons; the color change, for example, yellow to red, is gradual rather than instantaneous. Observers may, therefore, differ in selecting the precise point of change. If one assumes arbitrarily that the indicator is in one color form when at least 90% of it is in that form, there will be uncertain color in the range of 90–10% InH (that is, 10–90% In$^-$). When these arbitrary limits are substituted into the pK equation, the interval between definite colors is shown by Eqs. (4). Thus the

$$pH = pK + \log\frac{10}{90} \quad \text{to} \quad pH = pK + \log\frac{90}{10} \qquad (4)$$

pH uncertainty of the indicator is from $pK + 1$ to $pK - 1$ (approximately), and $pK \pm 1$ is called the range of the indicator. The experimentally observed ranges may differ from this prediction somewhat because of variations in color intensity.

Examples. The table lists many of the common indicators, their chemical names, pK values, ranges of pH, and directions for making solutions. Many of the weak-acid indicators are first dissolved in N/50 NaOH to the concentration shown, then further diluted with water. The weak-base indicators show some temperature variation of pH range, following approximately the temperature change of the ionization constant of water. Weak-acid indicators are more stable toward temperature change. The colors are formed by the usual chromophoric groups, for example, quinoid and azo-. One of the most common indicators is phenolphthalein, obtained by condensing phthalic anhydride with phenol. The acid form is colorless. It is converted by OH^- ion to the red quinoid form, as shown by Eq. (5).

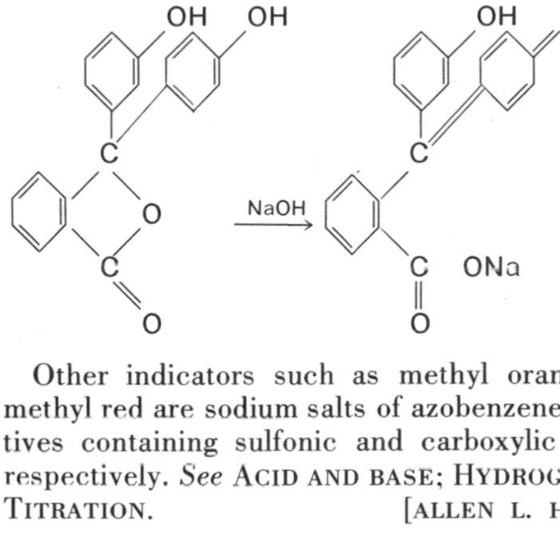

(5)

Other indicators such as methyl orange and methyl red are sodium salts of azobenzene derivatives containing sulfonic and carboxylic groups respectively. *See* ACID AND BASE; HYDROGEN ION; TITRATION.   [ALLEN L. HANSON]

Indium

A chemical element, In, atomic number 49, a member of subgroup IIIa and the fifth period of the periodic table. The valence electron notation corresponding to its ground-state term, $5s^2 5p^1$, accounts for the maximum oxidation state of III in its compounds. Compounds of oxidation state I and apparent oxidation state II are also reported. Indium has a relative atomic weight of 114.82.

Indium occurs in the Earth's crust to the extent of about $1 \times 10^{-5}\%$ and is normally found in concentrations of 0.1% or less. It is widely distributed in many ores and minerals but is largely recovered from the flue dusts and residues of zinc-processing operations.

Indium is used in soldering lead wires to germanium transistors and as a component of the intermetallic semiconductor used for germanium transistors. Indium arsenide, antimonide, and phosphide are semiconductors with unique properties. Other uses of indium are sleeve-type bearings to reduce corrosion and wear, glass-sealing alloys, and dental alloys.

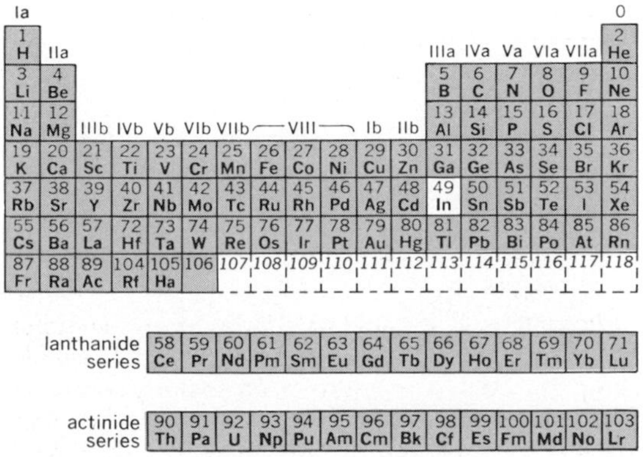

Metal. Indium is recovered from the residues of zinc processing by an acid leach followed by chemical separations from the accompanying elemental impurities such as zinc, cadmium, aluminum, arsenic, and antimony. Final purification by aqueous electrolysis at a controlled potential yields a product of 99.9% purity. Indium is a soft metal which can be easily scratched with the fingernail. It adheres to other surfaces when rubbed across them. It has a melting point of 156.4°C, a boiling point of over 2000°C, and a density of 7.31 g/cm^3 at 20°C. It crystallizes in the face-centered tetragonal structure and has a metal radius of 1.66 A.

The standard oxidation potential for the reaction, shown below, is $+0.34$, which accounts for

$$In_{(s)} \rightleftharpoons In^{3+}_{(aq)} + 3e$$

the fact that the metal dissolves in all acids to give solutions of In^{3+}. The metal reacts slowly with oxygen of the air up to its melting point, but at higher temperatures it readily yields yellow In_2O_3. Indium reacts directly with the halogens and other nonmetals such as sulfur, selenium, and phosphorus when warm.

Compounds. The trivalent chloride, bromide, and iodide are dimeric in the vapor state. The iodide is also dimeric in the solid state with approximate tetrahedral symmetry about the metal ions, whereas the bromide and iodide have a layer lattice structure in which the metal ions sit in distorted octahedra of anions. The bromide and iodide each exhibit an alternate form. The trifluoride has a structure in which the metal ion is distributed in one-third of the octahedral holes throughout three dimensions, resulting in a high-melting (1170°C) compound. The radius for the In^{3+} ion in a CN_6 environment is 0.80 A (0.08 nm). The anhydrous fluoride is only slightly soluble in water, whereas the other trihalides are very soluble and are recovered from aqueous solution as hydrates. The soluble indium trihalides hydrolyze, yielding species such as $In(OH)^{2+}$, $In_2(OH)_4^{4+}$ depending upon the metal ion concentration, and $InCl_n^{3-n}$ ($n = 1-7$) depending upon the halide ion concentration, cation, and solvent. A 0.1 M solution of InX_3 has a pH of 3; the hydrate hydroxide precipitates from such a solution at pH 3.4. Indium hydroxide dissolves neither in excess hydroxyl ion nor in the presence of ammonia. The carbonate, oxalate, and sulfide are insoluble in water.

The anydrous halides, except the fluoride, react as Lewis acids, forming 1:2 adducts with a variety of Lewis bases. Although the $MX_3 \cdot 2NMe_3$ compounds have a trigonal bipyramidal structure, $InI_3 \cdot 2DMSO$ appears to have a solid-state structure which contains $[InI_2(DMSO)_4^+][InI_4^-]$ species. On the other hand, five-coordinate complexes of InI_3 are formed with olefinic phosphines as in $InI_3 \cdot 2[P(C_6H_4)_3]$. Variable coordination with NCS^- is observed depending upon the cation present. Ph_4As^+ and $Ph_3PCH_2Ph^+$ yield five-coordinate species $[In(NCS)_5^{3-}]$, whereas Me_4N^+, Et_3NH^+, and $Bu_4^nN^+$ result in six-coordinate species $[In(NCS)_6^{3-}]$.

Indium(III) forms six-coordinate complexes with bidentate ligands such as 2,2′-dipyridyl, dicarboxylic acids, β-diketones, catechol, 8-hydroxyquinoline, and diethoxythiophosphate, while an eight-coordinate complex $[InT_4^-]$ is obtained with tropone, T⁻. The bidentate ligand, *cis*-1,2-S$_2$C$_2$-(CN)$_2^{2-}$, mnt, yields In(mnt)$_2^-$, In(mnt)$_2$X^{2-}, and In(mnt)$_3^{3-}$. Indium(I) halides form XIn[1,2,S$_2$C$_2$-(CF$_3$)$_2$].

Reduction of the anhydrous halides (except iodide) with hydrogen or a hydrogen-hydrogen halide mixture leads to compounds of composition InX$_2$. These compounds are known to be In^I-(In^{III}X$_4$). Reduction of the trihalides with indium metal results in products of composition InX. Indium monochloride has a distorted NaCl structure, while the bromide and iodide structures are unknown. Compounds of composition In$_2$Cl$_3$, In$_2$Br$_3$, and In$_7$Cl$_9$ appear to have the compositions 3In$^+$, InCl$_6^{3-}$; 2In$^+$, In$_2$Br$_6^{2-}$; and 6In$^+$, InBr$_6^{3-}$, 3Br$^-$. The reduced-state compounds are unstable in aqueous solution with respect to disproportionation to the metal and oxidation state III. InX and InX$_2$ compounds are insoluble in most organic solvents but dissolve in aniline and morpholine, from which solutions compounds of composition In(morpholine)$_2$X and In(aniline)$_4$X$_2$ are obtained. Morpholine acts as a bidentate ligand, and conductivity studies suggest that the species in solution is a 1:1 electrolyte. Aniline acts as a unidentate ligand, and in solution the species probably is [In(aniline)$_4$]$^+$[InX$_4$]$^-$.

The reaction of indium metal with mercury alkyls or aryls yields organometallic compounds of the R$_3$In class. Trimethylindium in the solid state has a unique tetrameric structure in which unsymmetrical CH$_3 \cdots$ In—CH$_3$ bridges are present. In solution the trimethylakyl is a monomer. Trimethylindium reacts with primary and secondary aliphatic phosphines to give (Me$_2$InPMe$_2$)$_n$ and methane. Trialkyl indium etherates react with triorganosilanols to give compounds of composition (X$_2$M—O—M'R$_3$)$_2$, where X = methyl or Cl; M^I = C, Si, or Ge; and R = methyl or phenyl. These compounds contain a framework of four-membered In—O rings. The trialkyls also react with some carboxylic, phosphinic, thiophosphinic, and sulfuric acids to give eight-membered rings, each containing two In—Me groups. A series of monomeric four-coordinate complexes, (C$_6$F$_5$)$_3$InL, where L = py, Ph$_3$P, Ph$_3$PO, Ph$_3$AsO, have been prepared as well as five-coordinate complexes, (C$_6$F$_5$)$_3$InL$_2$, with L = DMSO or THF. Bipyridyl, Ph$_2$P(CH$_2$)$_2$PPh$_2$, and *N,N,N',N'*-tetramethylenediamine appear to be bridging ligands in compounds of composition [(C$_6$H$_5$)$_3$In]$_2$L. Insertion into the In—C bond occurs in the reaction of (CH$_3$)$_3$In with SO$_2$ to yield In(SO$_2$Me)$_3$, while In—C bond rupture occurs when (C$_2$H$_5$)$_3$In reacts with CH$_2$X$_2$ (X = Br or I) or CCl$_4$ to produce (C$_2$H$_5$)$_2$InCH$_2$X and (C$_2$H$_5$)$_2$InCCl$_3$ respectively. Indium metal reacts with alkylhalides (RX, X = Br, I and R = Me, Et, *n*-Pr, *n*-Bu) to yield products of composition R$_3$In$_2$X$_3$, which when heated in the presence of KBr or KI give R$_2$InX. The Me$_2$In$^+$ ion is stable in aqueous solution at 0°C for several days. It appears to possess a linear structure. A series of quinolinato complexes of the type R$_2$InQ (R = Me, Et, or Bu') have been prepared. Dialkylindium compounds Et$_2$In-[C(NO$_2$)$_2$R$_2$], where R = H or Me, Me$_2$In(dtc), Et$_2$In(OX), and R$_2$InOC$_2$H$_4$NMe$_2$, where R = Me or Et, are obtained by reaction of the indium trialkyl and α-mononitroalkanes, SSCNMe$_2$(dtc), 8-hydroxyquinoline(OXH), and HOC$_2$H$_4$NMe$_2$

respectively. RInX$_2$ compounds (X = Br or I, R = Me, Et, Pr, or Bu) have been prepared by the reaction of InBr or InI with the proper alkyl halide. The iodides appear to be InMe$_2^+$InI$_4^-$, while the bromides are probably polymeric. Cyclopentadiene forms a composition C$_2$H$_5$In, which is known to be a monomer in the gaseous phase but probably is polymeric in the solid. It is not considered to be an ionic compound. Adducts of composition (Cp)In · BX$_3$ (X = F, Cl, Br, or CH$_3$) have been prepared. Spectroscopic evidence suggests that the cyclopentadienyl group is σ-bonded. Solid Cp$_3$In consists of infinite polymeric chains of σ-bonded Cp$_2$In units bridged by Cp groups. The cyclopentadienyl groups form a slightly distorted tetrahedral environment about the indium.

Numerous compounds containing indium-metal bonds have been prepared. The reaction of indium metal with Mn$_2$(CO)$_{10}$ at 140°C yields In[Mn(CO)$_5$]$_3$. In[Co(CO)$_4$]$_3$ is also known. The cobalt and manganese atoms are essentially trigonal about the indium, and the four carbon monoxides and indium atom are trigonal bipyramidal about the cobalt and manganese. The indium-metal bonds are cleaved by halogens or hydrogen halides to form X$_{3-n}$[In(Mn(CO)$_5$)]$_n$, where X = Cl or Br and n = 1 or 2. Cobalt forms similar compounds. The ions In[Mn(CO)$_5$]$_2^+$, In[Co(CO)$_4$]$_2^-$, and In[Co(CO)$_4$]$_4^-$ have been obtained. The crystal structure of Br$_3$In$_3$Co$_4$(CO)$_{15}$ consists of a six-membered ring of alternate In and Br atoms with each indium atom bonded to separate Co(CO)$_4$ groups and a central cobalt atom possessing three carbon monoxides. Indium(I) bromide reacts with Co$_2$(CO)$_8$ by insertion to give Co(CO)$_4$—InBr$_2$In-[Co(CO)$_4$], but with W(CO)$_6$ only the ion W(CO)$_5$-InBr$_3^-$ is obtained. The insertion of InX(Cl,Br) into a variety of compounds has yielded products such as XIn[Co(CO)$_4$]$_2$, XIn[CpFe(CO)$_2$]$_2$, and In-[Mo(CO$_3$)$_2$Cp]$_3$.

Analysis. Indium may be determined quantitatively with 8-hydroxyquinoline by precipitation of the compound In(C$_9$H$_6$ON)$_3$ at 70–80°C from a sodium acetate–acetic acid buffer, followed by drying at 120°C and direct weighing of the precipitate. An alternate to the direct weighing is the bromometric titration procedure after solution of the compound in warm 10–15% hydrochloric acid or the colorimetric determination at 400 mμ of a chloroform solution of the 8-hydroxyquinolate. Indium may also be determined by atomis absorption spectroscopy. *See* GALLIUM; THALLIUM.

[EDWIN M. LARSEN]

Bibliography: F. A. Cotton and G. Wilkinson, *Advanced Inorganic Chemistry*, 3d ed., 1972; R. T. Sanderson, *Inorganic Chemistry*, 1967.

Indole

One of a group of organic heterocyclic compounds in which a benzene ring is fused to a pyrrole ring. Indole (I) is a typical member of the group. *See* HETEROCYCLIC COMPOUNDS; PYRROLE.

Indolenines refer to isomeric systems, which are of interest generally only when two groups are

present at the 3 position, as in (II). 2,3-Dihydroin-
dole is called indoline (III). The dihydroisoindole
system (V) has received some study, the isoindole

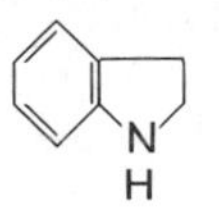

(III) 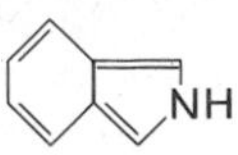 (IV) 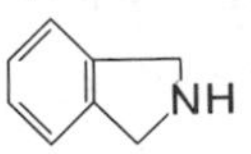 (V)

system (IV) much less. Indoles substituted with
oxygen at position 2 or 3 have special names
(VI – IX).

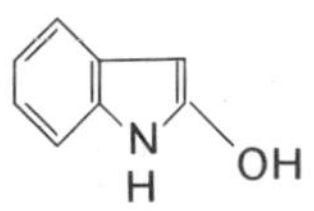

(VI) (VII)
Oxindole Indoxyl

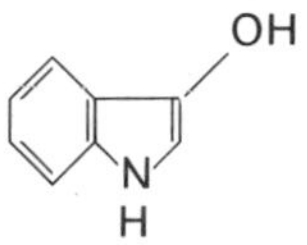

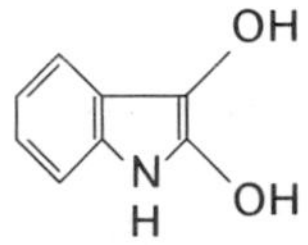

(VIII) (IX)
Dioxindole Isatin

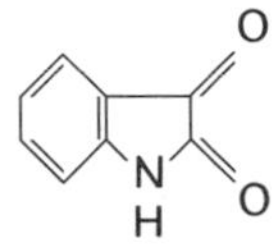

The indole skeleton occurs in many natural
products. Examples might include the blue dye
indigo (X), the plant-growth hormone heteroauxin
(XI), the amino acid tryptophan (XII), and the in-

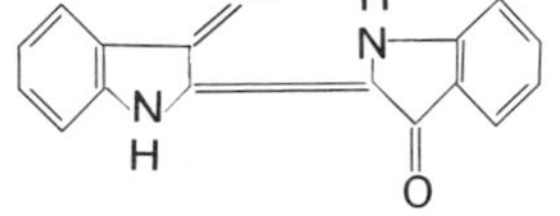

(X)

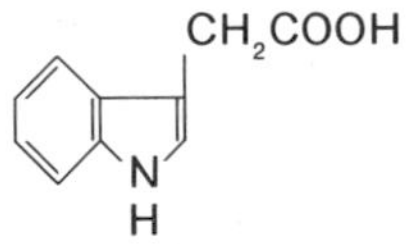

(XI) 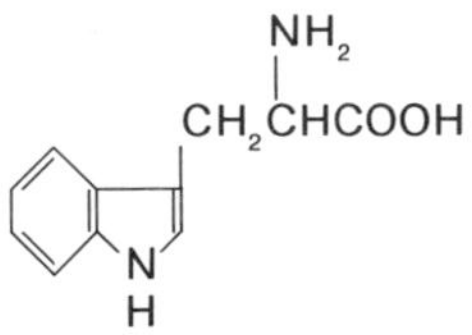 (XII)

dole alkaloids, for example, strychnine. Tryptophan not only is one of the essential amino acids
for man, but also is the biochemical precursor of
the indole plant alkaloids; it is also very likely the
progenitor of indole and 3-methylindole (skatole),
which are produced by pyrolysis or putrefaction of
protein material.

Properties and preparation. Indole (I) is a
steam-volatile, colorless solid, mp 52.5°, bp 253°C.
It is found in small amounts in coal tar, in feces,
and in flower oils. Indole is not ordinarily classed
as acidic or basic, although it is an active hydrogen
compound in which hydrogen at the 1 position is
replaceable by metal. In the absence of oxygen,
indole is stable to heat and to alkali. Indoles are
sensitive to acids and to oxidation processes; however, the more highly substituted indoles are in
general more stable. Indoles can be regarded as
relatively reactive aromatic compounds. Electrophilic substitution favors the 3 position, and to a
lesser extent, the 2 position. The experimental
resonance energy for indole is 54 kcal/mole.

Indole syntheses proceed by fusing a five-membered N-heterocycle on a benzenoid compound. Of
the many known syntheses, the Fischer indole synthesis is the most versatile. By this method, a
phenylhydrazone (XIII) of an aldehyde or ketone

on treatment with acid is converted to an indole
(XIV). For example, when the starting material is

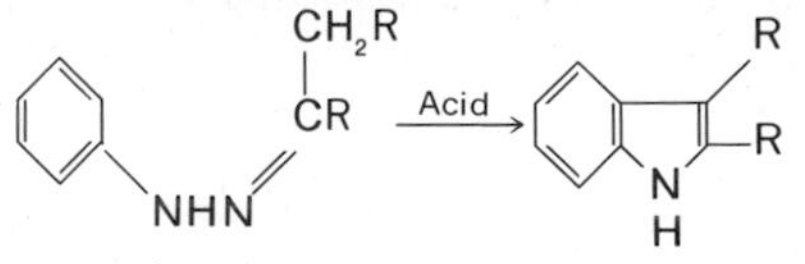

(XIII) (XIV)

the phenylhydrazone of acetone, the product is 2-
methylindole. The R groups at indole positions 2
and 3 may be varied widely. Substituted indoles
are formed when the phenylhydrazone (XIII) is
substituted on the benzene ring. Another synthesis
cyclizes o-toluide derivatives (XV) to indoles (XVI)
by heating them in the presence of strongly basic
reagents.

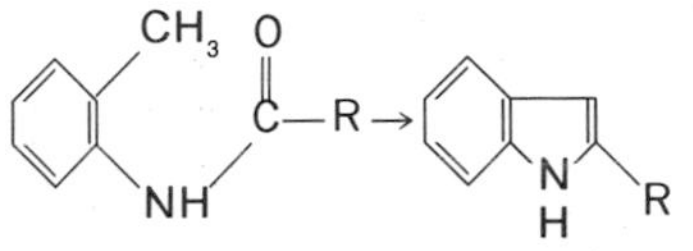

(XV) (XVI)

Indoles are attacked at the 2,3 bond by air,
ozone, peroxides, and other reagents. Catalytic
hydrogenation yields not only indoline (III), but
also, depending on conditions, octahydroindole
(XVII) and o-ethylcyclohexylamine (XVIII).

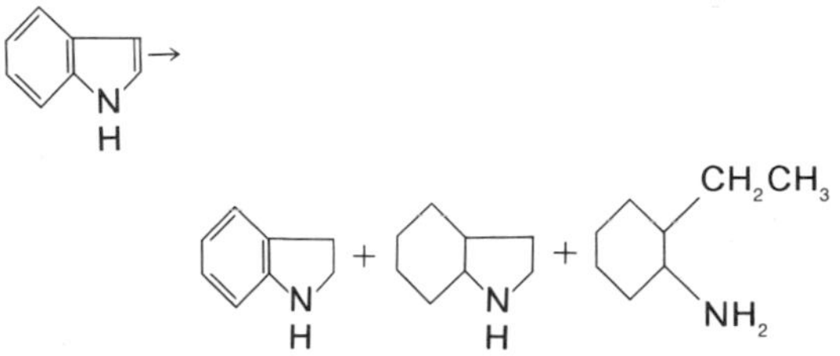

(XVII) (XVIII)

Indoles substitute at the 3 position, sometimes
at the 2 position, but only rarely in the benzene
ring. Groups inserted by substitution include
chloro, iodo, nitro, nitroso, and azo. Indole-3-
carboxaldehydes are formed in the reaction with
chloroform and alkali (Reimer-Tiemann) or with
N-methylformanilide and phosphorus oxychloride.

The 1-sodio derivative of indole reacts with
methyl iodide to give mainly 1-methylindole.
Indoles with ethylmagnesium iodide give 1-indolyl-
magnesium iodides, which combine with alkyl halides, acyl halides, carbon dioxide, ethyl chlorocar-
bonate, formaldehyde, and ethyl orthoformate to
give, respectively, 3-alkylindoles, 3-acylindoles,
indole-3-carboxylic acids, the corresponding esters, 3-hydroxymethylindoles, and indole-3-car-
boxaldehydes.

Important derivatives. Gramine, or 3-dimethyl-
aminomethylindole (XIX), is one of the simplest
indole alkaloids. Gramine is prepared by condensation of indole, dimethylamine, and formaldehyde. The dimethylamino grouping, either as
such or after quaternization, can be displaced
smoothly by nucleophilic groups. This property
has been exploited for the synthesis of various
3-substituted indoles. For example, the reaction
of gramine with sodium cyanide gives 3-cyano-
methylindole (XX), which is hydrolyzed by alkali
to give heteroauxin (XXI), or which is reduced by
hydrogen and Raney nickel catalyst to tryptamine
(XXII). Gramine with acetamidomalonic ester

gives the intermediate (XXIII), which can be converted to tryptophan (XXIV).

Serotonin or 5-hydroxytryptamine (XXV), an effective local vasoconstrictor, is distributed throughout the body. Bufotenine (XXVI), found in the skin of the common toad, is an active pressor agent. Both serotonin and bufotenine have been synthesized. 5,6-Dihydroxyindole, derived in the organism by the biochemical oxidation and decarboxylation of tyrosine, gives rise by further oxidation to the dark-skin pigment melanin. The indoline, eserine or physostigmine (XXVII), is a specific inhibitor to the action of acetylcholine esterase, and as such, is effective in blocking nerve conduction. Gliotoxin (XXVIII) is a fungus-derived antibiotic.

Indigo dye has been known for centuries. Indigo or indigotin (XXXI) and its derivatives, including the thioindigos in which S takes the place of NH, are commercially important dyes. Indican (XXIX), a glucoside of indoxyl, is found in various species of *Indigofera*. Natural indigo is obtained by enzymatic hydrolysis of indican followed by air-oxidation of the resulting indoxyl (XXX). Although many syntheses are known, the important commercial syntheses are closely related to that shown in the accompanying formulations, in which phenylgly-

cine (XXXII) is an intermediate. Substituted indigos are prepared from indigo or by total synthesis. The bromo and chloro derivatives are valuable dyes. Tyrian purple, or 6,6'-dibromoindigo, is an old dye obtained from a Mediterranean snail.

3-Hydroxyindole, or indoxyl, is a yellow compound with an unpleasant odor. It is easily oxidized with air or with ferric chloride to indigo. Reduction with sodium amalgam or with hot zinc dust affords indole. The reactive 2 position of indoxyl combines readily with aldehydes to give colored indogenides,

such as (XXXIV), and with nitrous acid to give isatin-α-monooxine (XXXIII).

Oxindole is 2-hydroxindole; however, the tautomeric amide structure (XXXVI) is probably the more appropriate formulation. Oxindole can be prepared by Friedel-Crafts cyclization of N-(chloroacetyl)aniline (XXXV), or by a two-stage reduction of isatin. The oxindole 3 position may be alkylated, nitrosated, acylated, or condensed with aldehydes.

Dioxindole (XXXVII) is produced by hydrosulfite reduction of isatin (XL) or by cyclization of o-aminomandelic acid (XXXVIII). Oxidation of dioxindole regenerates isatin; reduction gives oxindole (XXXVI).

Isatin or 2,3-dioxindoline (XL) is prepared by

oxidation of oxindole, by oxidative cleavage of indigo, or by ring synthesis. One standard isatin synthesis (Sandmeyer) proceeds by cyclization of isonitrosoacetanilide (XXXIX), which is obtained from aniline and chloral oxime. Isatin is a red, weakly acidic material. The reactive carbonyl group at the 3 position takes part in familiar carbonyl addition

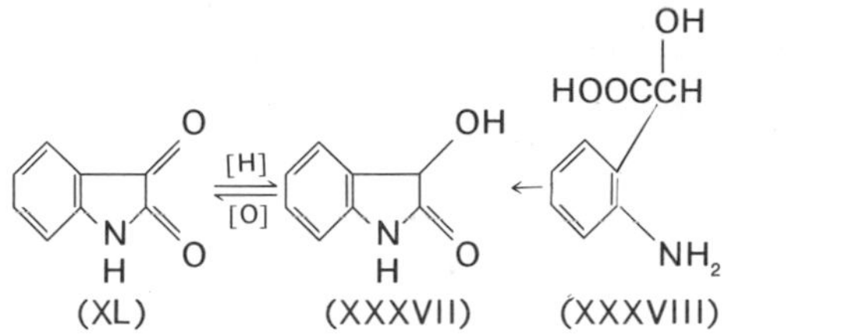

reactions. Phosphorus pentachloride converts isatin to isatin-α-chloride (XLI), which couples with indoxyl in a standard and useful indigo synthesis.

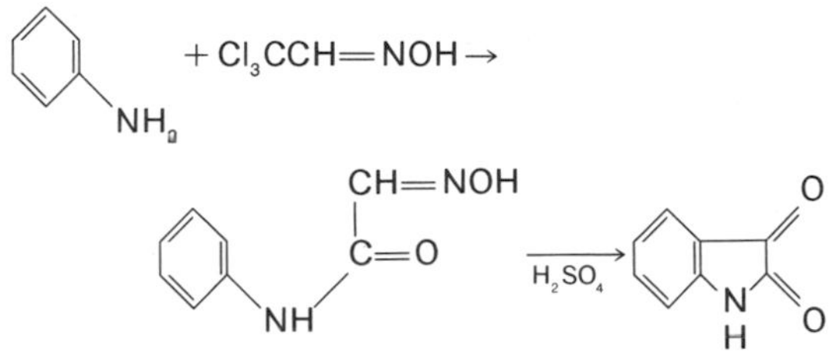

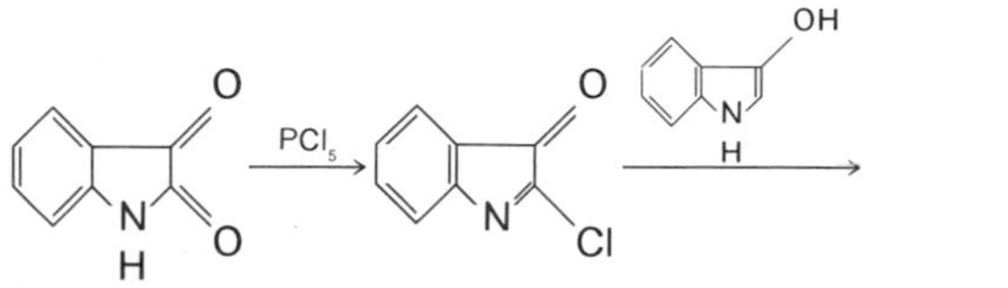

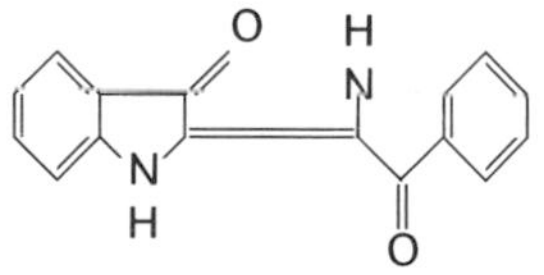

Isatin can be converted in the Pfitzinger process to quinoline derivatives.   [WALTER G. GENSLER]

Bibliography: R. C. Elderfield (ed.), *Heterocyclic Compounds*, vol. 3, 1952; R. H. Manske (ed.), *The Indole Alkaloids*, 1965; W. C. Sumpter and F. M. Miller, *Heterocyclic Compounds with Indole and Carbazole Systems*, 1954.

Inductance

That property of an electric circuit or of two neighboring circuits whereby an electromotive force is induced (by the process of electromagnetic induction) in one of the circuits by a change of current in either of them. The term inductance coil is sometimes used as a synonym for inductor, a device possessing the property of inductance. *See* ELECTROMOTIVE FORCE (EMF); INDUCTION, ELECTROMAGNETIC; INDUCTOR.

Self-inductance. For a given coil, the ratio of the electromotive force of induction to the rate of change of current in the coil is called the self-inductance L of the coil, given in Eq. (1), where e is

$$L = -\frac{e}{dI/dt} \qquad (1)$$

the electromotive force at any instant and dI/dt is

the rate of change of the current at that instant. The negative sign indicates that the induced electromotive force is opposite in direction to the current when the current is increasing (dI/dt positive) and in the same direction as the current when the current is decreasing (dI/dt negative). The self-inductance is in henrys when the electromotive force is in volts, and the rate of change of current is in amperes per second. *See* HENRY (UNIT).

An alternative definition of self-inductance is the number of flux linkages per unit current. Flux linkage is the product of the flux Φ and the number of turns in the coil N. Then Eq. (2) holds. Both

$$L = \frac{N\Phi}{I} \qquad (2)$$

sides of Eq. (2) may be multiplied by I to obtain Eq. (3), which may be differentiated with respect to t,

$$LI = N\Phi \qquad (3)$$

as in Eqs. (4). Hence the second definition is equivalent to the first.

$$L\frac{dI}{dt} = N\frac{d\Phi}{dt} = -e \qquad (4)$$

or

$$L = -\frac{e}{dI/dt}$$

Self-inductance does not affect a circuit in which the current is unchanging; however, it is of great importance when there is a changing current, since there is an induced emf during the time that the change takes place. For example, in an alternating-current circuit, the current is constantly changing and the inductance is an important factor. Also, in transient phenomena at the beginning or end of a steady unidirectional current, the self-inductance plays a part. *See* TRANSIENT, ELECTRIC.

Consider a circuit of resistance R and inductance L connected in series to a constant source of potential difference V. The current in the circuit does not reach a final steady value instantly, but rises toward the final value $I = V/R$ in a manner that depends upon R and L. At every instant after the switch is closed the applied potential difference is the sum of the iR drop in potential and the back emf $L\, di/dt$, as in Eq. (5), where i is

$$V = iR + L\frac{di}{dt} \qquad (5)$$

the instantaneous value of the current. Separating the variables i and t, one obtains Eq. (6). The solution of Eq. (6) is given in Eq. (7).

$$\frac{di}{\frac{V}{R} - i} = \frac{R}{L}\, dt \qquad (6)$$

$$i = \frac{V}{R}\left(1 - e^{-(R/L)t}\right) \qquad (7)$$

The current rises exponentially to a final steady value V/R. The rate of growth is rapid at first, then less and less rapid as the current approaches the final value.

The power p supplied to the circuit at every instant during the rise of current is given by Eq. (8).

$$p = iV = i^2R + Li\, di/dt \qquad (8)$$

The first term i^2R is the power that goes into heating the circuit. The second term $Li\,di/dt$ is the power that goes into building up the magnetic field in the inductor. The total energy W used in building up the magnetic field is given by Eq. (9). The

$$W = \int_0^t p\,dt = \int_0^t Li\frac{di}{dt}dt = \int_0^I Li\,di = \tfrac{1}{2}LI^2 \quad (9)$$

energy used in building up the magnetic field remains as energy of the magnetic field. When the switch is opened, the magnetic field collapses and the energy of the field is returned to the circuit, resulting in an induced emf. The arc that is often seen when a switch is opened is a result of this emf, and the energy to maintain the arc is supplied by the decreasing magnetic field.

Mutual inductance. The mutual inductance M of two neighboring circuits A and B is defined as the ratio of the emf induced in one circuit $\mathscr{E}$ to the rate of change of current in the other circuit, as in Eq. (10).

$$M = -\frac{\mathscr{E}_B}{(dI/dt)_A} \quad (10)$$

The mks unit of mutual inductance is the henry, the same as the unit of self-inductance. The same value is obtained for a pair of coils, regardless of which coil is taken as the starting point.

The mutual inductance of two circuits may also be expressed as the ratio of the flux linkages produced in circuit B by the current in circuit A to the current in circuit A. If Φ_{BA} is the flux threading B as a result of the current in circuit A, Eqs. (11) hold. Integration leads to Eq. (12).

$$\mathscr{E}_B = -N_B\frac{d\Phi_{BA}}{dt} = -M\frac{dI_A}{dt} \quad (11)$$

or $N_B\,d\Phi_{BA} = M\,dI_A$

$$M = \frac{N_B\Phi_{BA}}{I_A} \quad (12)$$

See INDUCTANCE MEASUREMENT.

[KENNETH V. MANNING]

Bibliography: *See* INDUCTION, ELECTROMAGNETIC.

Inductance bridge

A device for comparing inductances. The inductance bridge is a special case of an alternating-current impedance bridge. Just as the Wheatstone bridge is used to compare resistances, the impedance bridge is used to compare impedances which may contain inductance, capacitance, and resistance.

General impedance bridge. A general impedance bridge is shown in Fig. 1. Four impedances Z_a, Z_b, Z_c, and Z_d are connected into a square array. A source of ac voltage v is applied across one diagonal of the square, and a null detector D is connected across the other diagonal. The bridge is in balance when the voltage across D is zero. At balance the ac current through D is also zero, which means that Eqs. (1) hold. For the voltage

$$I_a = I_b \qquad I_c = I_d \quad (1)$$

across D to be zero, the instantaneous voltage drop across Z_b must equal that across Z_d; that is, the two instantaneous sine-wave voltages must have equal amplitudes and be in phase with each other.

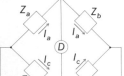

INDUCTANCE BRIDGE

Fig. 1. General impedance bridge.

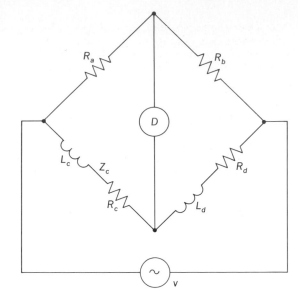

Fig. 2. General inductance bridge.

Equating the magnitudes, $Z_aI_a = Z_cI_c$, and by similar reasoning, $Z_bI_a = Z_dI_c$. Dividing the first of these equations by the second gives Eq. (2), one of the two equations of balance of the bridge.

$$Z_aZ_b = Z_c/Z_d \quad (2)$$

The voltage across Z_a leads I_a by the power-factor angle ϕ_a, and the voltage across Z_c leads I_c by the power-factor angle ϕ_c. If I_a leads I_c by the angle ϕ, then $\phi + \phi_a$ must equal ϕ_c if the two voltage drops are to be in phase. Similarly, $\phi + \phi_b = \phi_d$. By eliminating ϕ from these two equations the second equation of balance for the impedance bridge, Eq. (3), is obtained.

$$\phi_a + \phi_d = \phi_b + \phi_c \quad (3)$$

Several important properties can be recognized by considering the second equation of balance. If Z_a and Z_b are resistors, with ϕ_a and ϕ_b both equal to zero, then for balance $\phi_d = \phi_c$. This means that Z_c and Z_d must both be inductive or both capacitive for balance. If ϕ_a and ϕ_d are both equal to zero, the second equation for balance becomes $\phi_b + \phi_c = 0$. This means that Z_b is inductive and Z_c is capacitive or vice versa.

General inductance bridge. The inductance bridge of Fig. 2 has resistors R_a and R_b as ratio arms and compares an unknown Z_c to a standard consisting of R_d and L_d. If the standard L_d is variable, it and R_d are varied to reduce the detector voltage to zero. This balances the bridge, and the equations of balance become Eq. (4).

$$L_c/L_d = R_c/R_d = R_a/R_b \quad (4)$$

Sometimes a substitution method is preferred. In this case the balance is obtained, as above, with any good quality inductance for L_d. The unknown is then replaced by a standard L_s in series with R_s. The bridge is balanced a second time by varying L_s and R_s. When balance is obtained, L_s equals L_c and R_s equals R_d. For inductance standards *see* INDUCTOMETER.

If the standard is not adjustable, it becomes necessary to vary one of the ratio arms R_a or R_b or both, as well as R_d, in order to obtain balance.

Obviously the substitution method cannot be applied where only fixed standards are available.

Theoretically the condition for balance is independent of frequency. In practice, the capacitance between turns and between layers of wire in the two coils will be different, and at high frequencies the bridge can be balanced at only one frequency at a time. Under such conditions, harmonics in the supply voltage will make absolute balance impossible. When harmonics are present, the bridge should be balanced by reducing the fundamental component in the detector voltage to zero. This condition does not necessarily correspond to minimum detector voltage. A technician can sometimes achieve a balance by ear at audio frequencies. An oscilloscope provides a better means of seeing when the fundamental is reduced to zero.

Anderson bridge. A suitable variable capacitor may often be available when an appropriate inductometer is not. The Anderson bridge, shown in Fig. 3, can be used to measure the inductance L_b in terms of a standard capacitance C. The equations of balance of this bridge are Eqs. (5) and (6).

$$L_b = \left(R_d + r + \frac{rR_d}{R_c}\right) R_a C \tag{5}$$

$$R_b + r_b = \frac{R_a R_d}{R_c} \tag{6}$$

The bridge is usually balanced by varying r_b and C. The second equation indicates that for some choice of R_a, R_d, and R_c a negative r_b might be required. Consequently, if a balance does not seem possible by varying C and r_b, R_a or R_d should be increased, or R_c should be reduced. This will increase $R_a R_d / R_c$, which is needed because a balance cannot be obtained unless this quantity is at least as large as R_b.

Carey-Foster bridge. This useful bridge for determining mutual inductance is shown in Fig. 4. It is theoretically independent of frequency. When the bridge is balanced, Eqs. (7) and (8) hold.

$$M = R_a(R_M + R)C \tag{7}$$

$$L = (R_c + R_a)(R_M + R)C \tag{8}$$

If these equations are to be used directly, the resistance R_M of the coil must be measured by some other circuit arrangement. If a known large resistance R is connected in series with the coil, it may be possible to neglect R_M in comparison with R, and the equations become Eqs. (9) and (10).

$$M = R_a R C \tag{9}$$

$$L = (R_c + R_a)R C \tag{10}$$

Resistances R_a and R_c are varied to obtain balance. An obvious advantage of this bridge is that all balancing operations can be performed by varying resistances. The capacitance C may be a constant standard.

Any discussion of impedance-bridge operation should include mention of Wagner ground precautions. Each portion of a bridge has a capacitance to ground as well as to all other portions. In high-impedance circuits and at high frequencies, the effects of these capacitances are not negligible, and may make balance impossible if the detector is not at ground potential. To produce this ground, a divider, which may include reactive elements, can be placed across the source with a point near

its middle connected to ground. The divider is adjusted until the detector is brought to ground potential at the same time the bridge is balanced. This method of obtaining a ground at the detector is called the Wagner ground connection. *See* IMPEDANCE MEASUREMENTS, HIGH-FREQUENCY; INDUCTANCE MEASUREMENT.

[HARRY SOHON/EDWARD C. STEVENSON]

Bibliography: D. Bartholomew, *Electrical Measurements and Instrumentation*, 1963; F. A. Laws, *Electrical Measurement*, 2d ed., 1938; M. B. Stout, *Basic Electrical Measurements*, 1950.

Inductance measurement

The determination of an electromagnetic parameter of an electric circuit. The electric current in a circuit produces a magnetic field which is considered to consist of lines of magnetic flux that link the circuit. Whenever the magnetic field linking a circuit changes, a voltage is induced in the circuit. The faster the change in the field, the larger is the induced voltage. When there is no ferromagnetic material present, the magnetic field is proportional to the current i, and the induced voltage v is proportional to the rate of change of current, as in Eq. (1). The proportionality factor L is, by definition,

$$v = L \, di/dt \tag{1}$$

the self-inductance of the circuit. If v is measured in volts and if the rate of change of current is in amperes per second, the inductance is in henries.

The direction of the induced voltage is specified by Lenz's law, which states that the current that would be produced by the induced voltage would be in a direction that opposes the change in the magnetic field. This is a consequence of the fact that energy is stored in a magnetic field and, as it collapses, it feeds out power. To do this, the induced voltage must be in the direction of the current. Conversely, as a current builds up creating a magnetic field, the induced voltage must be opposite in direction to the current so that power will be delivered to the field. *See* INDUCTANCE; LENZ'S LAW.

If two circuits are so close together that the magnetic field of one will link the other, a changing current in one will cause a voltage proportional to the rate of change of this current to be induced in the other. The proportionality factor in this case is called the mutual inductance, and is the same whichever circuit has the changing current. *See* COUPLED CIRCUITS.

If ferromagnetic material is present, saturation and hysteresis effects may be evident for some values of current. For such cases several definitions may be introduced, or the concept may be discarded entirely. An effective inductance is defined as the ratio of the effective induced voltage to the effective rate of change of current, and is a function of a maximum current. When the current changes about a certain average value and the magnitude of the change is small compared to the average, this definition specifies an incremental inductance. This inductance is a function of the average current and the magnitude of the change.

Inductance standards. Coils constructed so that their dimensions and consequently their inductance remain constant over long periods of time are used as inductance standards. If the dimensions are known, the inductance can be com-

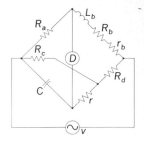

Fig. 3. Anderson bridge.

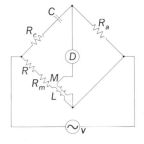

Fig. 4. Carey-Foster bridge.

puted from Eq. (2), where N is the number of turns

$$L = N^2 P \qquad (2)$$

of wire comprising the inductor and P is a proportionality coefficient. Such an inductance is called a primary standard. Standards are usually maintained at constant temperature to keep the coil size and therefore the inductance from changing in value. *See* INDUCTOMETER.

When the inductance cannot be computed precisely, it may be measured by comparing it with a primary standard, and it would then be called a secondary standard.

The effect of inductance is manifested only when the current in a circuit is changing with time. Any means of measuring inductance must employ changing current. Usually the current is an alternating current of frequency f, given by Eq. (3), with

$$i = I_m \sin 2\pi f t \qquad (3)$$

a maximum instantaneous value of I_m.

The voltage induced by the changing magnetic field is given by Eq. (4), where L is the self-inductance in henries.

$$v_x = 2\pi f L I_m \cos 2\pi f t = V_x \cos 2\pi f t \qquad (4)$$

The ratio of V_x to I_m is called the reactance of the circuit and is represented by X in Eq. (5).

$$X = 2\pi f L = V_x / I_m \qquad (5)$$

Impedance measurement. This is the determination of the total effect of a circuit element. The wire of which an inductor is wound has resistance, and there will be capacitance between turns. Thus every inductor exhibits resistance and capacitance as well as inductance. Also, every resistor exhibits some inductance and some capacitance, and every capacitor has some inductance and some resistance. Even if the capacitance of an inductor can be neglected, a measurement to determine the inductance is usually an impedance measurement giving the resistance as well as the inductance. *See* INDUCTANCE BRIDGE.

If the circuit element has only inductance L with resistance R in series, the total voltage drop across the element is given by Eq. (6), where $Z = \sqrt{R^2 + X^2}$

$$\begin{aligned} v &= R I_m \sin 2\pi f t + X I_m \cos 2\pi f t \\ &= Z I_m \sin (2\pi f t + \phi) \end{aligned} \qquad (6)$$

is the impedance of the circuit and ϕ is the phase angle or the power-factor angle of the circuit given by the relation $\tan \phi = X/R$.

It is impossible to construct an inductor with an ohmic resistance of zero; however, by careful design it is possible to minimize the resistance for a given amount of inductance. The factor of merit of any inductor is called the Q of the inductor and is defined as the ratio of the reactance to the effective resistance at any specified operating frequency, as in Eq. (7). The effective resistance R_{eff}

$$Q = \frac{2\pi f L}{R_{\text{eff}}} \qquad (7)$$

represents the combined ac resistance of the wire plus eddy-current losses in the wire at the specified operating frequency. The Q of a well-designed radio-frequency inductor will be on the order of several hundred. *See* Q METER.

If the Q of an inductor is high enough that the effective resistance may be neglected, then Eq. (8) holds.

$$L = \frac{V_x}{2\pi f I_m} \qquad (8)$$

Measurement of distributed inductance of cables, lines, and waveguides is accomplished indirectly. For very low frequencies the cable currents are distributed uniformly throughout the conductors and the inductance per unit length is a maximum. At higher frequencies the skin effect causes the current to concentrate along the conductor surfaces. This concentration does not affect the magnetic field between the conductors, but it does tend to eliminate the field within each conductor itself. This reduction in magnetic field brings about a reduction in inductance. Consequently, as the frequency is increased the inductance decreases, approaching a fixed value.

The capacitance per unit length of a cable or open-wire transmission line is independent of frequency and can be measured at low frequencies. If direct measurements of attenuation and velocity of propagation are made, the inductance per unit length of transmission line may be computed at any frequency of operation. *See* ELECTRICAL MEASUREMENTS.

[HARRY SOHON/EDWARD C. STEVENSON]

Bibliography: D. Bartholomew, *Electrical Measurements and Instrumentation*, 1963; F. A. Laws, *Electrical Measurements*, 2d ed., 1938; M. B. Stout, *Basic Electrical Measurements*, 1950.

Induction, electromagnetic

The production of an electromotive force either by motion of a conductor through a magnetic field in such a manner as to cut across the magnetic flux or by a change in the magnetic flux that threads a conductor. *See* MAGNETIC FLUX.

Motional electromotive force. A charge moving perpendicular to a magnetic field experiences a force that is perpendicular to both the direction of the field and the direction of motion of the charge. In any metallic conductor, there are free electrons, electrons that have been temporarily detached from their parent atoms.

If a conducting bar (Fig. 1) moves through a magnetic field, each free electron experiences a force due to its motion through the field. If the direction of the motion is such that a component of the force on the electrons is parallel to the conductor, the electrons will move along the conductor. The electrons will move until the forces due to the

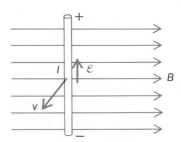

Fig. 1. Flux density B, motion v, and induced emf$_e$ when a conductor of length l moves in a uniform field. (*From R. L. Weber, M. W. White, and K. V. Manning, Physics for Science and Engineering, McGraw-Hill, 1957*)

motion of the conductor through the magnetic field are balanced by electrostatic forces that arise because electrons collect at one end of the conductor, leaving a deficit of electrons at the other. There is thus an electric field along the rod, and hence a potential difference between the ends of the rod while the motion continues. As soon as the motion stops, the electrostatic forces will cause the electrons to return to their normal distribution.

From the definition of magnetic induction (flux density) B, the force on a charge q due to the motion of the charge through a magnetic field is given by Eq. (1), where the force F is at right angles to a

$$F = Bqv \sin \theta \qquad (1)$$

plane determined by the direction of the field, and the component $v \sin \theta$ of the velocity is perpendicular to the field. When B is in webers/m², q is in coulombs, and v is in meters/sec, the force is in newtons. *See* INDUCTION, MAGNETIC.

The electric field intensity E due to this force is given in magnitude and direction by the force per unit positive charge. The electric field intensity is equal to the negative of the potential gradient along the rod. In motional electromotive force (emf), the charge being considered is negative. Thus, Eqs. (2) hold. Here l is the length of the conductor in a direction perpendicular to the field, and

$$E = \frac{F}{-q} = -Bv \sin \theta = -\frac{\mathscr{E}}{l} \qquad (2)$$

$$\mathscr{E} = Blv \sin \theta$$

$v \sin \theta$ is the component of the velocity that is perpendicular to the field. If B is in webers/m², l is in meters, and v is in meters/sec, the emf $\mathscr{E}$ is in volts.

This emf exists in the conductor as it moves through the field whether or not there is a closed circuit. A current would not be set up unless there were a closed circuit, and then only if the rest of the circuit does not move through the field in exactly the same manner as the rod. For example, if the rod slides along stationary tracks that are connected together, there will be a current in the closed circuit. However, if the two ends of the rod were connected by a wire that moved through the field with the rod, there would be an emf induced in the wire that would be equal to that in the rod and opposite in sense in the circuit. Therefore, the net emf in the circuit would be zero, and there would be no current.

Emf due to change of flux. When a coil is in a magnetic field, there will be a flux Φ threading the coil the magnitude of which will depend upon the area of the coil and its orientation in the field. The flux is given by $\Phi = BA \cos \theta$, where A is the area of the coil and θ is the angle between the normal to the plane of the coil and the magnetic field. Whenever there is a change in the flux threading the coil, there will be an induced emf in the coil while the change is taking place. The change in flux may be caused by a change in the magnetic induction of the field or by a motion of the coil. The magnitude of the induced emf, Eq. (3), depends upon the

$$\mathscr{E} = -N \frac{d\Phi}{dt} \qquad (3)$$

number of turns of the coil N and upon the rate of change of flux. The negative sign in Eq. (3) refers

to the direction of the emf in the coil; that is, it is always in such a direction as to oppose the change that causes it, as required by Lenz's law. If the change is an increase in flux, the emf would be in a direction to oppose the increase by causing a flux in a direction opposite to that of the increasing flux; if the flux is decreasing, the emf is in such a direction as to oppose the decrease, that is, to produce a flux that is in the same direction as the decreasing flux. *See* FARADAY'S LAW OF INDUCTION; LENZ'S LAW.

Consider the case of a flat coil of area A rotating with uniform angular velocity ω about an axis perpendicular to a uniform magnetic field of flux density B. For any position of the coil, the flux threading the coil is $\Phi = BA \cos \theta = BA \cos \omega t$, where the zero of time is taken when θ is zero and the normal to the plane of the coil is parallel to the field. Then the emf induced as the coil rotates is given by Eq. (4).

$$\mathscr{E} = -N \frac{d\Phi}{dt}$$

$$= -NBA \frac{d(\cos \theta)}{dt} = NBA\omega \sin \omega t \qquad (4)$$

The induced emf is sinusoidal, varying from zero when the plane of the coil is perpendicular to the field to a maximum value when the plane of the coil is parallel to the field.

Self-induction. If the flux threading a coil is produced by a current in the coil, any change in that current will cause a change in flux, and thus there will be an induced emf while the current is changing. This process is called self-induction. The emf of self-induction is proportional to the rate of change of current. The ratio of the emf of induction to the rate of change of current in the coil is called the self-inductance of the coil.

Mutual induction. The process by which an emf is induced in one circuit by a change of current in a neighboring circuit is called mutual induction. Flux produced by a current in a circuit A (Fig. 2) threads or links circuit B. When there is a change of current in circuit A, there is a change in the flux linking coil B, and an emf is induced in circuit B while the change is taking place. Transformers operate on the principle of mutual induction. *See* TRANSFORMER.

The mutual inductance of two circuits is defined

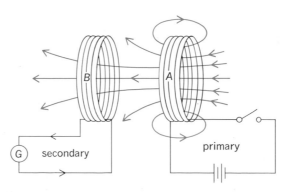

Fig. 2. Mutual induction. An emf is induced in the secondary when the current changes in the primary. (*R. L. Weber, M. W. White, K. V. Manning, Physics for Science and Engineering, McGraw-Hill, 1957*)

as the ratio of the emf induced in one circuit B to the rate of change of current in the other circuit A. For a detailed discussion of self- and mutual inductance *see* INDUCTANCE.

Coupling coefficient. This refers to the fraction of the flux of one circuit that threads the second circuit. If two coils A and B having turns N_A and N_B, respectively, are so related that all the flux of either threads both coils, the respective self-inductances are given by Eqs. (5) and the mutual inductance of the pair is given by Eq. (6). Then Eqs. (7) hold.

$$L_A = \frac{N_A \Phi_A}{I_A} \qquad L_B = \frac{N_B \Phi_B}{I_B} \qquad (5)$$

$$M = \frac{N_A \Phi_B}{I_B} = \frac{N_B \Phi_A}{I_A} \qquad (6)$$

$$M^2 = \frac{N_A N_B \Phi_A \Phi_B}{I_A I_B} = \frac{N_A \Phi_A}{I_A} \frac{N_B \Phi_B}{I_B} = L_A L_B \qquad (7)$$
$$M = \sqrt{L_A L_B}$$

In general, not all the flux from one circuit threads the second. The fraction of the flux from circuit A that threads circuit B depends upon the distance between the two circuits, their orientation with respect to each other, and the presence of a ferromagnetic material in the neighborhood, either as a core or as a shield. It follows that for the general case that Eq. (8) holds.

$$M \leqq \sqrt{L_A L_B} \qquad (8)$$

The ratio of the mutual inductance of the pair to the square root of the product of the individual self-inductances is called the coefficient of coupling K, given by Eq. (9). The coupling coefficient has a

$$K = \frac{M}{\sqrt{L_A L_B}} \qquad (9)$$

maximum value of unity if all the flux threads both circuits, zero if none of the flux from one circuit threads the other. For all other conditions, K has a value between 0 and 1.

Applications. The phenomenon of electromagnetic induction has a great many important applications in modern technology. For example, *see* COUPLED CIRCUITS; GENERATOR, ELECTRIC; INDUCTION HEATING; MICROPHONE; MOTOR, ELECTRIC; SERVOMECHANISM. [KENNETH V. MANNING]

Bibliography: S. S. Attwood, *Electric and Magnetic Fields*, 3d ed., 1949; L. Page and N. I. Adams, *Principles of Electricity and Magnetism*, 3d ed. 1958; E. M. Purcell, *Electricity and Magnetism*, vol. 2, 1965; F. W. Sears, *Principles of Physics*, vol. 2, 1951; R. P. Winch, *Electricity and Magnetism*, 1955.

Induction, electrostatic

A method whereby an electrical conductor becomes electrified when near a charged body. Its usefulness arises from the fact that for every electric charge there is somewhere an equal and opposite induced charge. For a detailed discussion *see* ELECTROSTATICS.   [RALPH P. WINCH]

Induction, magnetic

A vector quantity that is used as a quantitative measure of a magnetic field. It is defined in terms of the force on a charge moving in the field by Eq. (1), where B is the magnitude of the magnetic in-

$$B = \frac{F}{qv \sin \theta} \qquad (1)$$

duction and F is the force on the charge q, which is moving with speed v in a direction making an angle θ with the direction of the field. The direction of the vector quantity B is the direction in which the force on the moving charge is zero.

The magnetic induction may also be expressed in terms of the force F on a current element of length l and current I, as in Eq. (2). The meter-

$$B = \frac{F}{Il \sin \theta} \qquad (2)$$

kilogram-second (mks) unit of magnetic induction is derived from this equation by expressing the force in newtons, the current in amperes, and the length in meters. The unit of B is thus the newton/ampere-meter.

Magnetic flux density is the magnetic flux per unit area through a surface perpendicular to the magnetic induction. Magnetic flux density and magnetic induction are equivalent terms. *See* MAGNETIC FIELD; MAGNETIC FLUX.

The magnetic induction may be represented by lines that are drawn so that at every point in the field the tangent to the line is in the direction of the magnetic induction. To represent the magnetic induction qualitatively, as many such lines may be drawn as are necessary to portray the field. If, however, the lines are to represent the magnetic induction quantitatively, an arbitrary choice must be made for the number of lines to represent a given condition. One such choice is that in which the number of lines per square meter of a surface perpendicular to B is set equal to the value of B. These lines are called magnetic flux. One line of induction as here selected is called a flux of 1 weber. The corresponding unit of flux density is then the weber per square meter. From the manner of defining flux used here, it follows that the weber per square meter is equivalent to the newton/ampere-meter.

Another unit of flux density is defined by using centimeter-gram-second (cgs) units in both the defining equation for magnetic induction and in the area in which there is unit flux. This cgs unit of flux density is called the gauss. The relationship between the gauss and the weber is given by 1 weber/m² = 10^4 gauss. *See* GAUSS.

For discussion of an important device known as a betatron, which utilizes the principles of magnetic induction *see* PARTICLE ACCELERATOR.

[KENNETH V. MANNING]

Bibliography: D. S. Parasuis, *Magnetism*, 1961; E. R. Peck, *Electricity and Magnetism*, 1953; E. M. Purcell, *Electricity and Magnetism*, vol. 2, 1965; F. W. Sears, *Principles of Physics*, vol. 2, 1951.

Induction coil

A device for producing a high-voltage alternating current or high-voltage pulses from a low-voltage direct current. The largest modern use of the induction coil is in the ignition system of internal combustion engines, such as automobile engines. Devices of similar construction, known as vibrators, are used as rectifiers and synchronous inverters. *See* IGNITION SYSTEM; VIBRATOR.

Figure 1 shows a typical circuit diagram for an

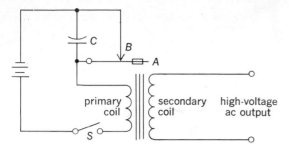

Fig. 1. Typical circuit for an induction coil.

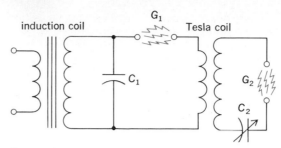

Fig. 3. Circuit diagram of Tesla coil.

induction coil. The primary coil, wound on the iron core, consists of only a few turns. The secondary coil, wound over the primary, consists of a large number of turns.

When the switch S is closed, the iron core becomes magnetized and therefore attracts the armature A. This automatically breaks the circuit to the coil through contact B and the armature. The armature is returned to its initial position by a spring and again makes contact with the contact B, restoring the circuit to the primary coil. The cycle is then repeated rapidly.

While current is flowing in the primary coil, a magnetic field is produced. When the contact between A and B is broken, the magnetic field collapses and induces a high voltage in the secondary coil, similar to transformer action. The self-inductance of the coil must be limited; therefore, the core is a straight bundle of iron wires, which minimize eddy-current losses, rather than a closed iron circuit as is used in a transformer. *See* TRANSFORMER.

The capacitor C is placed across the breaker contacts to reduce the voltage across the contacts at the moment of their opening and thus reduce sparking. Sparking is caused by the induced voltage in the primary winding resulting from the collapsing magnetic field. The capacitor allows some of the energy of the magnetic field to be converted to electrostatic energy in the capacitor, rather than into heat at the contacts.

Induction coils of a different type are used in telephone circuits to step up the voltage from the transmitter and match the impedance of the line. The direct current in the circuit varies in magnitude at speech frequencies; therefore, no interrupter contacts are necessary. The battery and primary winding are connected in series with the transmitter as in Fig. 2. The secondary winding and the receiver are connected in series with the line. This circuitry reduces the required battery voltage.

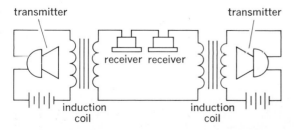

Fig. 2. Induction coils in telephone circuit.

Still another type of induction coil, called a reactor, is really a one-winding transformer designed to produce a definite voltage drop for a given current. *See* REACTOR, ELECTRIC.

In 1892 Nicola Tesla used a form of induction coil to obtain currents of very high frequencies and high voltages. The oscillatory discharge of a Leyden jar was used as the interrupter. The terminals of the secondary of an induction coil are connected, one to the inner coating and the other to the outer coating of an insulated Leyden jar C_1 (Fig. 3). The circuit is completed through the primary winding of the Tesla coil, and the primary gap G_1. The coils and the Leyden jar act as a resonant circuit in the production of the high-frequency oscillation.

The primary of the Tesla coil consists of a half-dozen turns of wires wound on a nonmagnetic core. The secondary consists of many turns. The two coils are separated by air or oil as insulation. The alternation from the Leyden jar may have a frequency of several million hertz. Hence, the current induced in the secondary is not only of high voltage but also of high frequency.

[NORMAN R. BELL]

Induction heating

The heating of a nominally electrical conducting material by currents induced by a varying electromagnetic field.

The principle of the induction heating process is similar to that of a transformer. In Fig. 1, the inductor coil can be considered the primary winding of a transformer, with the workpiece as a single-turn secondary. When an alternating current flows in the primary coil, secondary currents will be induced in the workpiece. These induced currents are called eddy currents. The current flowing in the workpiece can be considered as the summation of all of the eddy currents.

In the design of conventional electrical apparatus, the losses due to induced eddy currents are minimized because they reduce the overall efficiency. However, in induction heating, their maximum effect is desired. Therefore close spacing is used between the inductor coil and the workpiece, and high coil currents are used to obtain the maximum induced eddy currents and therefore high heating rates. *See* CORE LOSS.

Applications. Induction heating is widely employed in the metal working industry to heat metals for soldering, brazing, annealing, hardening, and for induction melting and sintering.

As compared to other conventional processes, it has these inherent advantages:

INDUCTION HEATING

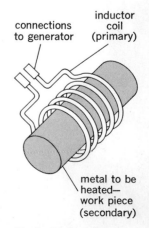

connections to generator

inductor coil (primary)

metal to be heated—work piece (secondary)

Fig. 1. Basic elements of induction heating.

1. Heating is induced directly into the material. It is therefore an extremely rapid method of heating. It is not limited by the relatively slow rate of heat diffusion in conventional processes using surface-contact or radiant heating methods.

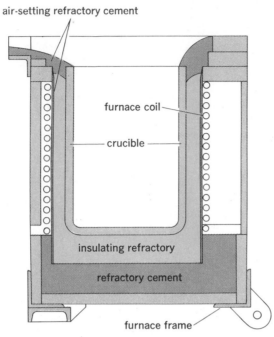

Fig. 2. Cross section of a typical induction melting furnace. (*Inductotherm Corp.*)

Fig. 3. Manually loaded induction heater for hardening of gears. Gear is loaded on spindle under coil in center. Spindle raises gear for heating and lowers it into oil quench for hardening. (*Westinghouse Electric Corp.*)

2. Because of skin effect, the heating is localized and the heated area is easily controlled by the shape and size of the inductor coil.

3. Induction heating is easily controllable, resulting in uniform high quality of the product.

4. It lends itself to automation, in-line processing, and automatic process cycle control.

5. Start-up time is short, and standby losses are low or nonexistent.

6. Working conditions are better because of the absence of noise, fumes, and radiated heat.

Heating process. The induced currents in the workpiece flow roughly parallel to the current in the inductor coil turns.

Because of the skin effect, these induced currents concentrate near the surface of the workpiece. The effective depth of current penetration is greater for lower frequencies than for higher frequencies, and is greater for high-resistivity metals than for low-resistivity metals. In magnetic materials such as steel, the depth of current penetration is less below the Curie temperature (approximately 1350°F where the steel is magnetic) than it is above the Curie temperature. *See* Curie temperature, magnetic; Skin effect (electricity).

For efficient heating, the frequency used must be high enough to make the depth of current penetration considerably less than the thickness of the workpiece, measured at right angles to the coil turns. The table shows the range of frequencies used for applications of the induction heating process. *See* Dielectric heating.

In mass heating applications the lowest frequency consistent with efficient heating is used, because as a rule the initial cost of equipment goes up with frequency. When the workpieces are small, or when it is desired to concentrate the heating near the surface, as in surface hardening, it is necessary to use higher frequencies, even though efficient power transfer could be accomplished at some lower frequency. In large production use, such as forging, two separate frequencies, with separate inductor coils, are employed in an effort to maximize production at minimum equipment cost.

Power sources. The equipment used as power sources depends on the frequency range for the application. When line frequencies (generally 60 cycles) are used, suitable transformers, power factor correction capacitors, and control equipment are required.

For higher frequencies, up to 10,000 Hz, inductor-type alternators are used. These are usually driven by induction motors and are available in ratings from $7\frac{1}{2}$ to over 300 kw.

Converters are used for the 10,000–60,000 Hz range, principally for small-scale melting. These produce the desired frequency by repeatedly charging a large capacitor from the 60-Hz line and discharging it through an output circuit tuned to the desired frequency. The output is a train of damped oscillations.

For frequencies above 200 kHz, vacuum-tube oscillators are used. These are self-excited, and are complete with high-voltage rectifier, oscillator tank circuit, controls, and instrumentation. When operating from a three-phase supply, they put out a continuous wave of rf power.

Frequencies used in induction and dielectric heating

Frequency, Hz	Source of power	Uses
60–960	Rotating generators or converters	Mass induction heating for forging, forming, extrusion, or preheating
960–10,000	Motor-generator sets	Induction heating for melting, heat-treating, and hardening
10,000–60,000	Converters	Induction heating for small-scale melting and sintering
200,000–550,000	Vacuum-tube oscillators	Surface induction heating for brazing, soldering, hardening, and strip and wire heating
2,000,000–90,000,000	Vacuum-tube oscillators	Dielectric heating

Process use. Induction heating is used for many heat processes, as shown in the table. The construction of a typical melting furnace is shown in Fig. 2. An induction heater used for hardening is shown in Fig. 3. *See* FURNACE CONSTRUCTION.

Induction heating differs from other methods of heat treating in that it heats the metals very rapidly, and that holding time at hardening temperature approaches zero. A minimum of time is therefore available for metallurgical reactions, and this has a significant influence on the selection of steel to be used. *See* HEAT TREATMENT (METALLURGY). For other electric heating methods *see* HEATING, ELECTRIC.

[CARL P. BERNHARDT]

Bibliography: D. W. Brown, *Induction Heating Practice*, 1956; G. H. Brown, C. N. Hoyler, and R. A. Bierwirth, *Theory and Application of Radio-Frequency Heating*, 1947; J. W. Cable, *Induction and Dielectric Heating*, 1954; P. G. Simpson, *Induction Heating: Coil and System Design*, 1960; C. A. Tudbury, *Basics of Induction Heating*, 1960.

Induction motor

An alternating-current motor in which the currents in the secondary winding (usually the rotor) are created solely by induction. These currents result from voltages induced in the secondary by the magnetic field of the primary winding (usually the stator). An induction motor operates slightly below synchronous speed and is sometimes called an asynchronous (meaning not synchronous) motor.

Induction motors are the most commonly used electric motors because of their simple construction, efficiency, good speed regulation, and low cost. Polyphase induction motors come in all sizes

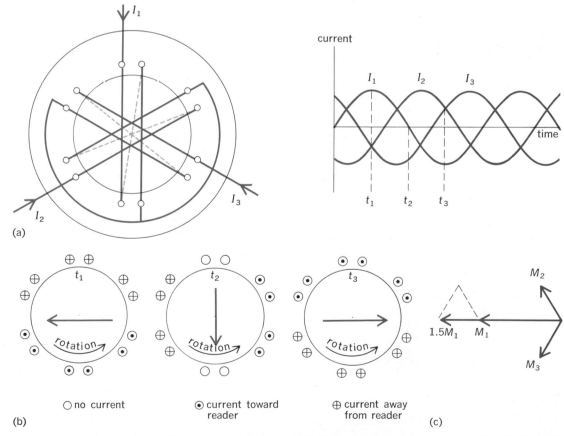

(a) no current   (•) current toward reader   (⊕) current away from reader

Fig. 1. Three-phase, two-pole, Y-connected stator of induction motor supplied with currents I_1, I_2, and I_3.

(a) Stator windings and currents. (b) Rotating field. (c) Magnetomotive forces produced by stator winding.

and find wide use where polyphase power is available. Single-phase induction motors are found mainly in fractional-horsepower sizes, and those up to 25 hp are used where only single-phase power is available.

POLYPHASE INDUCTION MOTORS

There are two principal types of polyphase induction motors: squirrel-cage and wound-rotor machines. The differences in these machines is in the construction of the rotor. The stator construction is the same and is also identical to the stator of a synchronous motor. Both squirrel-cage and wound-rotor machines can be designed for two- or three-phase current.

Stator. The stator of a polyphase induction motor produces a rotating magnetic field when supplied with balanced, polyphase voltages (equal in magnitude and 90 electrical degrees apart for two-phase motors, 120 electrical degrees apart for three-phase motors). These voltages are supplied to phase windings, which are identical in all respects. The currents resulting from these voltages produce a magnetomotive force (mmf) of constant magnitude which rotates at synchronous speed. The speed is proportional to the frequency of the supply voltage and inversely proportional to the number of poles constructed on the stator.

Figure 1 is a simplified diagram of a three-phase, two-pole, Y-connected stator supplied with currents I_1, I_2, and I_3. Each stator winding produces a pulsating mmf which varies sinusoidally with time. The resultant mmf of the three windings (Fig. 1c) is constant in magnitude and rotates at synchronous speed. Figure 1b shows the direction of the mmf in the stator for times t_1, t_2, and t_3 shown in Fig. 1a and shows how the resultant mmf rotates. The synchronous speed N_s is shown by Eq. (1), where f is the frequency in hertz and p is the

$$N_s = \frac{120f}{p} \quad \text{rpm} \tag{1}$$

number of stator poles. For any given frequency of operation, the synchronous speed is determined by the number of poles. For 60-Hz frequency, a two-pole motor has a synchronous speed of 3600 rpm; a four-pole motor, 1800 rpm; and so on. For details of stator windings *see* WINDINGS IN ELECTRIC MACHINERY.

Squirrel-cage rotor. Figure 2 shows the bars, end rings, and cooling fins of a squirrel-cage rotor. The bars are skewed or angled to prevent cogging (operating below synchronous speed) and to reduce noise. The end rings provide paths for currents that result from the voltages induced in the rotor bars by the stator flux. The number of poles on a squirrel-cage rotor is always equal to the number of poles created by the stator winding.

Figure 3 shows how the two motor elements interact. A counterclockwise rotation of the stator flux causes voltages to be induced in the top bars of the rotor in an outward direction and in the bottom bars in an inward direction. Currents will flow in these bars in the same direction. These currents interact with the stator flux and produce a force on the rotor bars in the direction of the rotation of the stator flux.

Fig. 2.   Bars, end rings, and cooling fins of a squirrel-cage rotor.

When not driving a load, the rotor approaches synchronous speed N_s. At this speed there is no motion of the flux with respect to the rotor conductors. As a result, there is no voltage induced in the rotor and no rotor current flows. As load is applied, the rotor speed decreases slightly, causing an increase in rotor voltage and rotor current and a consequent increase in torque developed by the rotor. The reduction in speed is therefore sufficient to develop a torque equal and opposite to that of the load. Light loads require only slight reductions in speed; heavy loads require greater reduction. The difference between the synchronous speed N_s and the operating speed N is the slip speed. Slip s is conveniently expressed as a percentage of synchronous speed, as in Eq. (2).

$$s = \frac{N_s - N}{N_s} \times 100\% \tag{2}$$

When the rotor is stationary, a large voltage is induced in the rotor. The frequency of this rotor

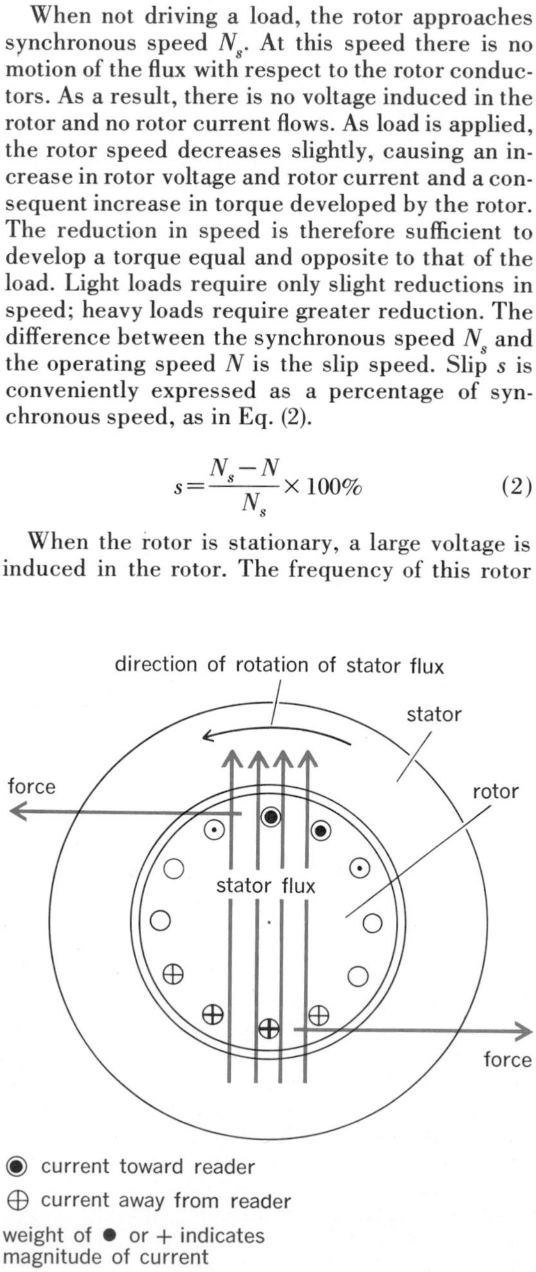

direction of rotation of stator flux

stator

force

rotor

stator flux

force

⦿ current toward reader

⊕ current away from reader

weight of ● or + indicates magnitude of current

Fig. 3. Forces on the rotor winding.

voltage is the same as that of the supply voltage. The frequency f_2 of rotor voltage at any speed is shown by Eq. (3), where f_1 is the frequency of the

$$f_2 = f_1 s \qquad (3)$$

supply voltage and s is the slip expressed as a decimal. The voltage e_2 induced in the rotor at any speed is shown by Eq. (4), where e_{2s} is the rotor

$$e_2 = (e_{2s})s \qquad (4)$$

voltage at standstill. The reactance x_2 of the rotor is a function of its standstill reactance x_{2s} and slip, as shown by Eq. (5). The impedance of the rotor at

$$x_2 = (x_{2s})s \qquad (5)$$

any speed is determined by the reactance x_2 and the rotor resistance r_2. The rotor current i_2 is shown by Eq. (6). In the equation, for small

$$i_2 = \frac{e_2}{\sqrt{r_2{}^2 + x_2{}^2}}$$
$$= \frac{(e_{2s})s}{\sqrt{r_2{}^2 + (x_{2s})^2 s^2}} = \frac{e_{2s}}{\sqrt{\left(\frac{r_2}{s}\right)^2 + (x_{2s})^2}} \qquad (6)$$

values of slip, the rotor current is small and possesses a high power factor. When slip becomes large, the r_2/s term becomes small, current increases, and the current lags the voltage by a large phase angle. Standstill (or starting) current is large and lags the voltage by 50–70°. Only in-phase, or unity-power-factor, rotor currents are in space phase with the air-gap flux and can therefore produce torque. The current i_2 contains both a unity power-factor component i_p and a reactive component i_r. The maximum value of i_p and therefore maximum torque are obtained when slip is of the correct value to make r_2/s equal to x_{2s}. If the value of r_2 is changed, the slip at which maximum torque is developed must also change. If r_2 is doubled and s is doubled, the current i_2 is not changed and the torque is unchanged.

This feature provides a means of changing the speed-torque characteristics of the motor. In Fig. 4, curve 1 shows a typical characteristic curve of an induction motor. If the resistance of the rotor bars were doubled without making any other changes in the motor, it would develop the characteristic of curve 2, which shows twice the slip of curve 1 for any given torque. Further increases in the rotor resistance could result in curve 3. When r_2 is made equal to x_{2s}, maximum torque will be developed at standstill, as in curve 4. These curves show that higher resistance rotors give higher starting torque. However, since the motor's normal operating range is on the upper portion of the curve, the curves also show that a higher-resistance rotor results in more variation in speed from no load to full load (or poorer speed regulation) than the low-resistance rotor. Higher-resistance rotors also reduce motor efficiency. Except for their characteristic low starting torque, low-resistance rotors would be desirable for most applications.

Wound rotor. A wound-rotor induction motor can provide both high starting torque and good speed regulation. This is accomplished by adding external resistance to the rotor circuit during start-

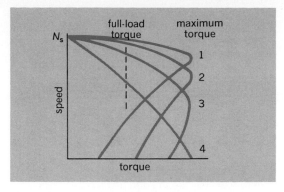

Fig. 4. Speed-torque characteristic of polyphase induction motor.

ing and removing the resistance after speed is attained.

The wound rotor has a polyphase winding similar to the stator winding and must be wound for the same number of poles. Voltages are induced in these windings just as they are in the squirrel-cage rotor bars. The windings are connected to slip rings so that connections may be made to external impedances, usually resistors, to limit starting currents, improve power factor, or control speed.

Figure 5 shows the connection of a rheostat used to bring a wound-rotor motor up to speed. The rheostat limits the starting current drawn from the supply to a value less than that required by a squirrel-cage motor. The resistance is gradually reduced to bring the motor up to speed. By leaving various portions of the starting resistances in the circuit, some degree of speed control can be obtained, as in Fig. 4. However, this method of speed control is inherently inefficient and converts the motor into a variable-speed motor, rather than an essentially constant-speed motor. For other means of controlling speed of polyphase induction motors and for other types of ac motors *see* ALTERNATING-CURRENT MOTOR.

SINGLE-PHASE INDUCTION MOTORS

Single-phase induction motors display poorer operating characteristics than polyphase machines, but are used where polyphase voltages are not available. They are most common in small siz-

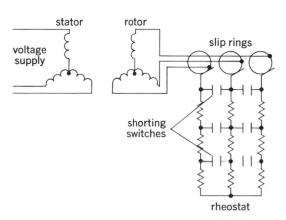

Fig. 5. Connections of wound-rotor induction motor.

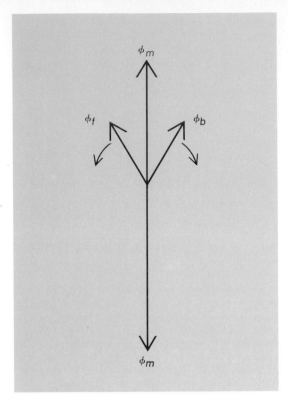

Fig. 6. Fluxes associated with the single-phase induction motor.

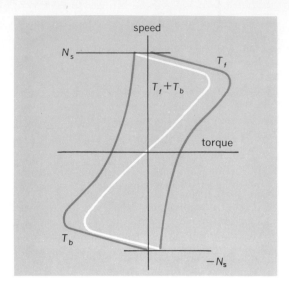

Fig. 7. Torques produced in the single-phase induction motor.

es (1/2 hp or less) in domestic and industrial applications. Their particular disadvantages are low power factor, low efficiency, and the need for special starting devices.

The rotor of a single-phase induction motor is of the squirrel-cage type. The stator has a main winding which produces a pulsating field. At standstill, the pulsating field cannot produce rotor currents that will act on the air-gap flux to produce rotor torque. However, once the rotor is turning, it produces a cross flux at right angles in both space and time with the main field and thereby produces a rotating field comparable to that produced by the stator of a two-phase motor.

An explanation of this is based on the concept that a pulsating field is the equivalent of two oppositely rotating fields of one-half the magnitude of the resultant pulsating field. In Fig. 6, ϕ_m is the maximum value of the stator flux ϕ, which is shown only by its two components ϕ_f and ϕ_b, which represent the two oppositely rotating fields of constant equal magnitudes of $\phi_m/2$. Each component ϕ_f and ϕ_b produces a torque T_f and T_b on the rotor. Figure 7 shows that the sum of these torques is zero when speed is zero. However, if started, the sum of the torques is not zero and rotation will be maintained by the resultant torque.

This machine has good performance at high speed. However, to make this motor useful, it must have some way of producing a starting torque. The method by which this starting torque is obtained designates the type of the single-phase induction motor.

Split-phase motor. This motor has two stator windings, the customary main winding and a start-ing winding located 90 electrical degrees from the main winding, as in Fig. 8a. The starting winding has fewer turns of smaller wire, to give a higher resistance-to-reactance ratio, than the main winding. Therefore their currents I_m (main winding) and I_s (starting winding) are out of time phase, as in Fig. 8c, when the windings are supplied by a common voltage V. These currents produce an elliptical field (equivalent to a uniform rotating field superimposed on a pulsating field) which causes a unidirectional torque at standstill. This torque will

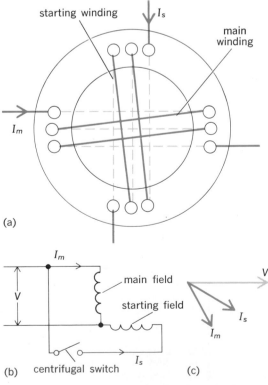

Fig. 8. Split-phase motor. (a) Windings. (b) Winding connections. (c) Vector diagram.

start the motor. When sufficient speed has been attained, the circuit of the starting winding can be opened by a centrifugal switch and the motor will operate with a characteristic illustrated by the broken-line curve of Fig. 7.

Capacitor motor. The stator windings of this motor are similar to the split-phase motor. However, the starting winding is connected to the supply through a capacitor (Fig. 9a). This results in a starting winding current which leads the applied voltage. The motor then has winding currents at standstill which are nearly 90° apart in time, as well as 90° apart in space. High starting torque and high power factor are therefore obtained. The starting winding circuit can be opened by a centrifugal switch when the motor comes up to speed. A typical characteristic is shown in Fig. 9c.

In some motors two capacitors are used. When the motor is first connected to the voltage supply, the two capacitors are used in parallel in the starting circuit. At higher speed one capacitor is removed by a centrifugal switch, leaving the other in series with the starting winding. This motor has high starting torque and good power factor.

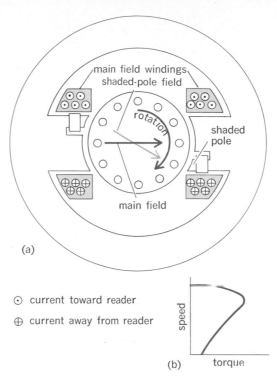

⊙ current toward reader

⊕ current away from reader

Fig. 10. Shaded-pole motor. (a) Cross-sectional view. (b) Typical characteristic.

Shaded-pole motor. This motor is used extensively where large power and large starting torque are not required, as in fans. A squirrel-cage rotor is used with a salient-pole stator excited by the ac supply. Each salient pole is slotted so that a portion of the pole face can be encircled by a short-circuited winding, or shading coil.

The main winding produces a field between the poles as in Fig. 10. The shading coils act to delay the flux passing through them, so that it lags the flux in the unshaded portions. This gives a sweeping magnetic action across the pole face, and consequently across the rotor bars opposite the pole face, and results in a torque on the rotor. This torque is much smaller than the torque of a split-phase motor, but it is adequate for many operations. A typical characteristic of the motor is shown in Fig. 10b.

For other single-phase alternating-current motors *see* REPULSION MOTOR; UNIVERSAL MOTOR. For synchronous motors built for single-phase *see* HYSTERESIS MOTOR; RELUCTANCE MOTOR.

Linear motor. Figure 11 illustrates the arrangements of the elements of the polyphase squirrel-cage induction motor. The squirrel cage (secondary) is embedded in the rotor in a manner to provide a close magnetic coupling with the stator winding (primary). This arrangement provides a small air gap between the stator and the rotor. If the squirrel cage is replaced by a conducting sheet as in Fig. 11b, motor action can be obtained. This machine, though inferior to that of Fig. 11a, will function as a motor. If the stator windings and iron are unrolled (rectangular laminations instead of circular laminations), the arrangement of the elements will take a form shown in Fig. 11c, and the field produced by polyphase excitation of the primary wind-

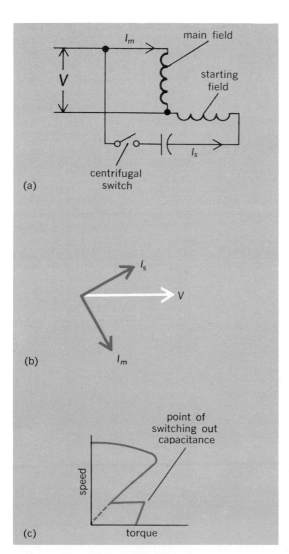

Fig. 9. Capacitor motor. (a) Winding connections. (b) Vector diagram. (c) Characteristic.

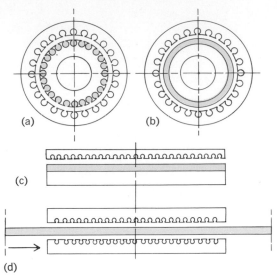

Fig. 11. Evolution of linear induction motor. (*a*) Poly-phase squirrel-cage induction motor. (*b*) Conduction sheet motor. (*c*) One-sided long-secondary linear motor with short-flux return yoke. (*d*) Double-sided long-secondary linear motor.

ing will travel in a linear direction instead of a circular direction. This field will produce a force on the conducting sheet that is in the plane of the sheet and at right angles with the stator conductors. A reversal of the phase rotation of the primary voltages will reverse the direction of motion of the air-gap flux and thereby reverse the force on the secondary sheet. No load on the motor corresponds to the condition when the secondary sheet is moving at the same speed as the field produced by the primary. For the arrangement of Fig. 11*c*, there is a magnetic attraction between the iron of the stator and the iron of the secondary sheet. For some applications, this can be a serious disadvantage of the one-sided motor of Fig. 11*c*. This disadvantage can be eliminated by use of the double-sided arrangement of Fig. 11*d*. Here the primary iron of the upper and lower sides is held together rigidly, and the forces that are normal to the plane of the sheet do not act on the sheet.

Applications. In conventional transportation systems, traction effort is dependent on the con-

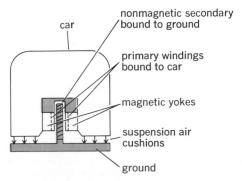

Fig. 12. Drawing of linear induction-motor configuration. Magnetic interaction supplies the forces between the car and the ground. (*Le Moteur Lineaire and Société de l'Aerotrain*)

tact of the wheels with the ground. In some cases, locomotives must be provided with heavy weights to keep the wheels from sliding when under heavy loads. This disadvantage can be eliminated by use of the linear motor. Through use of the linear motor with air cushions, friction loss can be reduced and skidding can be eliminated. Figure 12 illustrates the application of the double-sided linear motor for high-speed transportation.

Conveying systems that are operated in limited space have been driven with linear motors. Some of these, ranging from 1/2 to 1 mi (805–1609 m) in length, have worked successfully.

Because of the large effective air gap of the linear motor, its magnetizing current is larger than that of the conventional motor. Its efficiency is somewhat lower, and its cost is high. The linear motor is largely in the experimental stage.

[ALBERT G. CONRAD]

Bibliography: P. D. Agarwal and T. C. Wang, Evaluation of fixed and moving primary linear induction motor systems, *Proc. IEEE*, pp. 631–637, May 1973; P. L. Ager, *The Nature of Induction Machines*, 1964; J. H. Dannon, R. N. Day, and G. P. Kalman, A linear-induction motor propulsion system for high-speed ground vehicles, *Proc. IEEE*, pp. 621–630, May 1973; A. M. Dudley and S. F. Henderson, *Connecting Induction Motors*, 4th ed., 1960; E. R. Laithwaite, Linear electric machine: A personal view, *Proc. IEEE* pp. 220–290, February 1975; A. F. Puchstein, T. C. Lloyd, and A. G. Conrad, *Alternating-current Machines*, 3d ed., 1954; David Schieber, Principles of operation of linear induction devices, *Proc. IEEE*, pp. 647–656, May 1973.

Inductometer

A coil of wire of known inductance. The inductance may be fixed, as in the case of primary standards; or the inductance may be adjustable by means of switches, or continuously variable by means of a movable-coil construction.

Faraday's law states that the voltage v induced in a circuit is proportional to the rate of change of magnetic flux. For a flux ϕ linking N turns of a circuit, Eq. (1) holds.

$$v = N\frac{d\phi}{dt} \tag{1}$$

If there is no ferromagnetic material present, the flux θ is proportional to the magnetomotive force NI, as in Eq. (2). The proportionality factor P

$$\phi = PNI \tag{2}$$

is called the permeance of the flux path, and I is the current passing through the coil. The rate of change of flux is equal to NP times the rate of change of current, as in Eq. (3).

$$v = N^2 P\frac{dI}{dt} \tag{3}$$

Because the proportionality factor in this equation is defined as the inductance L, Eq. (4) can be written.

$$L = N^2 P \tag{4}$$

The formula, so derived, assumes that all of the flux links with all of the current. In reality there is some flux that links with only part of the current.

When the formula is corrected to include partial linkages N_p and the permeance of the path of the flux that enters into the partial linkages as well as the complete linkages N_c, Eq. (5) is obtained.

$$L = N_c^2 P_c + N_p^2 P_p \qquad (5)$$

A standard inductance is constructed by winding a coil on a form having dimensions that are stable. Some materials dry out or absorb moisture over the years and consequently change continuously in size. The material on which a standard inductance is wound must not be subject to such change. Marble and some synthetics have been found suitable. A helical groove is cut in the material, and the coil is placed in this groove.

The wire should be soft enough to conform to the groove, and annealed so that the resistivity and the current density are uniform. The coil and its form are maintained at constant temperature to reduce the tendency for rapid inductance variation. The material for the form should have a low temperature coefficient of expansion in addition to negligible long-period change in size.

If the magnetic field produced by the current in one circuit links with another circuit, the voltage–rate-of-change-of-current relation still holds. The proportionality factor in this case is called the mutual inductance between the two circuits. The mutual inductance M can be expressed as Eq. (6),

$$M = N_1 N_2 P_{12} \qquad (6)$$

where N_1 and N_2 are the numbers of turns in the respective circuits that link with the common flux, and P_{12} is the permeance of the path of the common flux. The mutual inductance is often written as Eq. (7), where L_1 and L_2 are the self-induc-

$$M = k\sqrt{L_1 L_2} \qquad (7)$$

tances of the individual circuits and k is a number, less than unity, called the coupling coefficient.

These equations, with the formula for self-inductance, give Eq. (8), where P_1 is the permeance

$$P_1 P_2 = k^2 P_{12}^2 \qquad (8)$$

of the path for flux produced by one circuit, and P_2 is the permeance of the path for the flux produced by the other circuit. If there were no leakage flux, P_1, P_2, and P_{12} would all be equal, and k would be unity.

Campbell standard mutual inductance. This has a primary winding consisting of two coils placed on a cylindrical form similar to that used for the standard self-inductance. One winding is placed in the groove near each end of the form. The resulting space between the coils is adjusted so that the magnetic field produced by one coil will exactly cancel that produced by the other on a ring concentric with the supporting cylinder. The flux density in the neighborhood of this ring is very low. A secondary winding is placed on a form that supports it centered as nearly as possible on the zero-field ring. Any inaccuracies in placement of the coil or subsequent slight displacement will have nearly negligible effect on the flux enclosed and consequently upon the mutual inductance.

Ayrton and Perry inductometer. This inductometer consists of two coils (Fig. 1). One is wound on the outside of a spherical form; the other is wound on the inside of a spherical form. The first

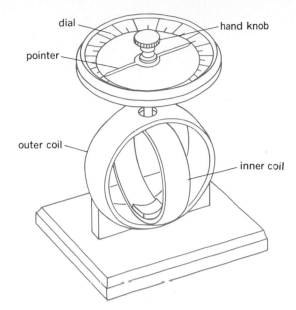

Fig. 1. Ayrton and Perry variable inductometer.

coil, on its form, is placed inside the second coil, on its form. The inner coil can be turned through 180° to change the direction of the mutual linkages and the sign of the mutual inductance. If the two coils are connected in series, the effective inductance is equal to the sum of the self-inductances plus or minus twice the magnitude of the mutual inductance, depending upon whether the magnetic fields aid or oppose each other, as in Eq. (9).

$$L = L_1 + L_2 \pm 2M \qquad (9)$$

The spherical forms make it possible to have the individual inductances, as well as the maximum mutual inductance, all nearly equal. This tends to increase the maximum effective inductance of the series combination. More important, it results in a minimum effective inductance of nearly zero. A disadvantage of this construction is that the calibration of the effective inductance L for angular displacement θ between the two coils is irregular. Interpolation is always unsatisfactory with an irregular calibration. Linear interpolation generally cannot be used. If the scale is always set on a calibrated point, to avoid interpolation, the inductometer in use is no better than one that is varied by means of switches.

Brooks variable inductometer. This inductometer provides a nearly linear scale. It consists of four fixed coils and two movable ones (Fig. 2). The two movable coils, side by side in a plane, are sandwiched between two pairs of fixed coils. With the movable coils directly between the fixed coils the mutual inductance is a maximum. By moving the coils in their plane so that the coil, previously between two particular coils, is placed between the other two coils, the mutual inductance is again a maximum but its sign has been changed. As in the Ayrton and Perry device, the series-connected inductance is expressed by Eq. (10).

$$L = L_1 + L_2 \pm 2M \qquad (10)$$

By using link-shaped coils of special design, it is possible to obtain an inductance whose calibration

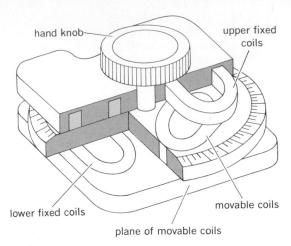

hand knob

upper fixed coils

movable coils

lower fixed coils

plane of movable coils

Fig. 2. Brooks variable inductometer.

is quite linear over most of its range, departing from linearity only slightly near the two ends of the scale.

The sandwich-type construction of the Brooks inductometer contributes to the stability of its calibration. Any tendency for axial displacement of the movable coils from two of the fixed coils will be accompanied by a similar displacement toward the other two fixed coils, making the net change in mutual inductance negligible. *See* INDUCTANCE; INDUCTANCE BRIDGE; INDUCTANCE MEASUREMENT.

[HARRY SOHON/EDWARD C. STEVENSON]

Bibliography: D. Bartholomew, *Electrical Measurements and Instrumentation*, 1963; F. A. Laws, *Electrical Measurements*, 2d ed., 1938; M. B. Stout, *Basic Electrical Measurements*, 1950.

Inductor

A device for introducing inductance into a circuit. The term covers devices with a wide range of uses, sizes, and types, including components for electric-wave filters, tuned circuits, electrical measuring circuits, and energy storage devices.

Inductors are classified as fixed, adjustable, and variable. All are made either with or without magnetic cores. Inductors without magnetic cores are called air-core coils, although the actual core material may be a ceramic, a plastic, or some other nonmagnetic material. Inductors with magnetic cores are called iron-core coils. A wide variety of magnetic materials are used, and some of these contain very little iron. Magnetic cores for inductors for low-frequency, or high-energy storage, use are most commonly made from laminations of silicon steel. Some iron-core inductors with cores of compressed powdered iron, powdered permalloy, or ferrite are more suitable for higher-frequency applications.

Fixed inductors. In fixed inductors coils are wound so that the turns remain fixed in position with respect to each other. If an iron core is used, any air gap it has is also fixed and the position of the core remains unchanged within the coil.

A toroidal coil is a fixed inductor wound uniformly around a toroidal form (see illustration). Because of the closed magnetic circuit, such an inductor has practically no leakage flux and is little affected by the presence of stray magnetic fields. High-accuracy standard inductors are commonly

INDUCTOR

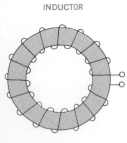

Toroidal coil.

made in this form. Powdered cores are used to increase the Q of the coil and reduce the size required for a specified inductance. Ceramic-core toroidal coils supported in cork are used as standard inductors of high stability and accuracy.

Adjustable inductors. These either have taps for changing the number of turns desired, or consist of several fixed inductors which may be switched into various series or parallel combinations.

Variable inductors. Such inductors are constructed so that the effective inductance can be changed. Means for doing this include (1) changing the permeability of a magnetic core; (2) moving the magnetic core, or part of it, with respect to the coil or the remainder of the core; and (3) moving one or more coils of the inductor with respect to one or more of the other coils, thereby changing mutual inductance. *See* ALTERNATING-CURRENT CIRCUIT THEORY; INDUCTANCE.

[BURTIS L. ROBERTSON; WILSON S. PRITCHETT]

Industrial cost control

A specific system or procedure used to keep manufacturing costs in line. It is implied that cost accounting is the basic procedure used to do this. Actually, cost accounting provides top management with an opportunity to check performance and to place responsibility. It does not, however, provide information to those responsible for costs rapidly enough for them to take corrective action before costs get seriously out of line. Therefore, to have good cost control, it is necessary to have other controls which are more current. Cost control is generally considered to cover direct labor, material, and manufacturing burden. Each of these will be discussed separately.

Direct labor control. Prices of products and their profit contributions are generally based upon an estimate of direct labor costs. This is done by determining the operations to be performed, what classes of labor perform the operations, and how long the operations should take. A company may think it does not have direct labor standards because it does not have time study standards, but as soon as it establishes a price for a product, by whatever means, it has established a direct labor standard.

To control direct labor costs, therefore, a company must so control its operations that the amount of direct labor used in setting prices is either equaled or beaten. The best way to do so is the most obvious, but is frequently overlooked. In pricing out a product it may be found, for example, that a part must be drilled on a single spindle drill press and that an operator should produce 600 pieces per hour. The drilling operation is obviously going to be performed by a drill press operator, and this job carries a rate of $2.40 per hour. The direct labor cost of the piece should then be $2.40 ÷ 600, or $0.004.

In order to control this cost, it is only necessary to know three things: the standard time, the actual time, and the labor rate of the operator who runs the part. If the actual time is equal to, or less than, the standard time and the operator who runs the job has a $2.40 rate, the direct labor cost is under control. If the time is satisfactory but a higher-rated operator is used, then the cost is out of control. If the correct operator is used but the time exceeds the standard time, the cost is out of control.

The easiest and most obvious way to control direct labor cost therefore is to compare standard time and actual time daily and to develop a daily rate variance. This can be done for each part, for each operation, and for each direct labor employee. A foreman can know the performance of his employees daily, or hourly if he so desires; a superintendent can receive departmental performances daily, and those above him can receive the same information at any time interval they feel necessary. Under this type of direct labor cost control, the cost accounting for direct labor can be quite simple and inexpensive.

Material control. The standard cost for material, like the standard cost of direct labor, is established when the product is priced. Material cost control therefore consists of first measuring the purchase price paid per unit of material against the purchase price used in establishing the selling price, and second determining whether or not the quantity of material used is the same as that used in establishing the selling price.

The following is somewhat oversimplified, but will serve to illustrate the procedure. Production control provides the operator with enough material to produce the required number of parts, with allowance for normal scrap. If the operator produces the number of pieces he is supposed to from the material provided, then the quantity usage is under control. Excessive usage can be readily and promptly determined from production counts, from scrap reports, or by not permitting an operator to get more material without a requisition signed by the foreman. The faster information is available on excess usage, the sooner corrective action can be taken. Quantity control can obviously be assessed at least daily and, if desired, more frequently.

Since the cost of material in industry is normally 40–50% of the selling price, it is extremely important to know the effectiveness of purchasing. Once material price standards have been established, it is easy to determine a daily, weekly, and monthly price variance. In some companies the president requires the purchasing agent to advise him of any price change exceeding a certain percentage. He can, then, check at his leisure to find out if it is a temporary or permanent situation. If permanent, he can determine the products which use this material, check the prices, and decide whether or not a price increase is justified. Material cost accounting, like direct labor cost accounting, can thus be simple and inexpensive.

Manufacturing burden. Manufacturing burden includes all costs involved in the manufacturing operation other than direct labor and material. It consists of such items as depreciation, heat, light and power, indirect labor, supervision, supplies, and so forth.

Since manufacturing burden is seldom less than 200% of direct labor and usually a considerably higher percentage than 200, it deserves, but seldom receives, as close or closer attention than direct labor costs.

Burden accounts are best controlled by flexible budgets which require that each account be ana-lyzed to determine whether it is fixed, variable, or semivariable in nature. The following definitions of manufacturing factors are applicable in assessing and evaluating the economics of manufacturing.

Fixed costs. Fixed costs are generally the cost of being in business. They are time costs which remain constant over an activity range of ±20%. Examples are rent, depreciation, watchmen, and supervisors' salaries.

Variable costs. Variable costs are those which vary directly with the number of units produced. Direct labor and material are examples of variable costs; their control has been previously discussed. Variable costs in manufacturing burden could be such items as supplies if, for example, a certain number of supply items were used for a certain number of items produced.

Semivariable costs. Semivariable costs are those which vary with the number of items produced, but not in direct proportion. The equation of such a cost is $y = k + ax$, where y is the semivariable cost, or the y axis; k is a constant or, in this case, the fixed portion of the expense; a is the slope of the line, which is expressed as a percentage; and x is the activity base, that is, direct labor or units of production (the x axis).

Development of burden budgets. In developing budgets for burden accounts, it is usually advisable to find out how each account has behaved in the past. This is most easily done by preparing a scatter diagram for each burden expense item. A scatter diagram is shown in the illustration.

Each dot on the scatter diagram represents 1 month that has occurred in the past. At least 8 months should be plotted, preferably 12. To illustrate, take point 1, which represents January. In January the direct labor was $1000 and the indirect labor was $2000. The dot is put at this point, and then other dots are added for February, March, April, and so on. It can be seen that a definite pattern develops, and that a line can be drawn through the plotted points, called the "line of best fit." The line represents how the expense was controlled and varied in the past. It is not necessarily the desired line. To develop the budget line, it is necessary to discuss the historical line with the individual responsible for the expense to determine why it behaved that way and where it should be. After discussion, it may be found, for example,

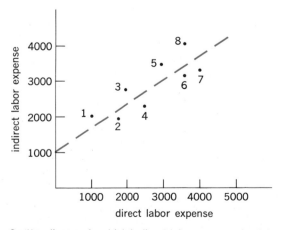

Scatter diagram in which indirect labor expense is plotted against direct labor expense.

that expenses were out of control in months 1, 3, 5, and 8 and that the budget line should be lowered to go through points 2, 4, 6 and 7. An equation for the budget line would then be developed which might, for example, be $800 per month plus 200% direct labor. The person responsible for the cost would therefore know each day how well he was controlling indirect labor by taking his actual direct labor and applying it to the equation to determine his indirect labor standard. In practice the equation is applied to standard direct labor and not actual.

Many companies make it easier for supervision by expressing their scatter charts as "things" rather than "dollars." For example, indirect labor might be expressed as the number of indirect labor people allowed as compared to the number of direct labor people; or supplies might be expressed as numbers of cleaning rags as measured against number of direct labor people. In manufacturing burden control, a budget must be established for each expense account, and it is essential that each account be "clean." By this it is meant that the account should only include what the name implies, so that if there is a variance, the thing that varies is identifiable. For example, a "people cost" should not be mixed with a "thing cost" in one expense account. To carry the example further, separate expense accounts should exist for maintenance labor and maintenance supplies.

For good cost control, it is important that one controls by responsibility. This means that the responsibility for each item of cost must be clearly spelled out and the person responsible must have a standard to measure his performance against. It is foolish to have a foreman, for instance, allocated a portion of the plant depreciation costs since he can do nothing to control such costs.

Industrial cost control can be quite simple and effective. It should be remembered that costs can only be controlled at their source, and the easier it is made for the person responsible for the cost, the better job he will do. See INDUSTRIAL ENGINEERING; INSPECTION AND TESTING; INVENTORY MANAGEMENT; MASS PRODUCTION; OPERATIONS RESEARCH; OPTIMIZATION; PERFORMANCE RATING; PERT; PROCESS ENGINEERING; PRODUCTION ENGINEERING; PRODUCTION METHODS; QUALITY CONTROL.   [ROBERT C. TRUNDLE]

Industrial engineering

As defined by the American Institute of Industrial Engineers, "Industrial engineering is concerned with the design, improvement, and installation of integrated systems of men, materials, and equipment. It draws upon the specialized knowledge and skill in the mathematical, physical, and social sciences together with the principles and methods of engineering analysis and design, to specify, predict, and evaluate the results to be obtained from such systems." Since the mid-1950s, industrial engineering has expanded its horizons and perspectives, with emphasis on the total-systems concept as applied to industry in a broader context, and now includes areas such as aerospace, hospitals, banks, transportation, utilities, and academic institutions.

While most of the industrial engineers had previously been employed in manufacturing, this is no longer true. For example, the industrial engineer may be involved in such diverse assignments as designing a space-information center for industrial development; structuring a management system for an educational institution to integrate and control the functions of planning, financing, programming students, and utilizing faculty and physical plant resources; developing the deterministic and stochastic costs for hospitals; or designing a general-purpose materials handling simulator by use of a digital computer.

The dual function of analysis and synthesis in design and decision-making demands that the industrial engineer has a foundation in engineering, mathematics and probability theory, computer science, economics, and behavioral science. A corollary to this requirement is that the industrial engineer should have a basic knowledge of disciplines which enables him to work effectively in a complex man-machine environment. See ANTHROPOLOGY, PHYSICAL; COMPUTER; ENGINEERING; PROBABILITY; PSYCHOLOGY, PHYSIOLOGICAL AND EXPERIMENTAL.

Classical approach. The term classical has been applied to the traditional industrial engineering activities which date back to the work of Frederick W. Taylor and Frank B. Gilbreth. This work includes motion-and-time study, plant layout, wage incentives, job evaluation, manufacturing processes, quality control, and tool design. These subjects have had great practical value, and for a long time have been considered the hallmarks of industrial engineering. However, these functions, as practiced in the past, do not presently comprise the forward-looking techniques of industrial engineering. While some proponents of management science have a tendency to slight such activities in modern industrial engineering, these are still important tasks. Methods, job standards, and costs are vitally important to management; in fact, the quantitative-system models need data which are often obtained by time study, synthetic time standards, or work sampling.

At the university level of education, the emphasis is generally on professional competence at the frontiers of capability, with the training of technicians proficient in methods, time study, job evaluation, and so on being done primarily by technical institutes, community colleges, and consulting firms. Universities give only minimal exposure to these traditional areas in present curricula. Industrial engineering, to be effective, must necessarily be capable of evaluating both the classical and the more sophisticated management science–operations research techniques relative to their utility in solving unstructured, real-life problems.

Systems concept. The trend in industrial engineering is from micro- to macrolevel structures, for example, from the micromotion charts in methods engineering to the total information system design of a company or the complex materials-handling function involving many interrelated departments. See SYSTEMS ENGINEERING.

Management of industrial, business, or service activities is growing increasingly complicated as the sciences and humanities become more interdependent. These organizations need not only management personnel but individuals capable of designing and installing new integrated systems of men, materials, and equipment which will function effectively in a society made more complex by the technological explosion. The industrial engineer is educated to meet this challenge: He has a solid foundation in the basic sciences, engineering, and computers, and a skill in systems analysis and design which crosses traditional disciplinary lines. The concomitant is an awareness of the demands imposed by a dynamic social system.

The logical structure of systems study is outlined below.

1. Identification of the components of a complex system. Examples include a production-scheduling system, a computer-software system, an educational scheduling system, a project-planning-and-control system for underdeveloped countries, a hospital information system, and a space-information system.

2. Development of the topological properties of the system, and the generation of mathematical models and analogs, or the adaptation of a heuristic approach when desirable. *See* TOPOLOGY.

3. Establishment of the functional relationships between the variables of the system, together with the necessary feedback required to control the operational system.

4. Selection and evaluation of optimization criteria. Examples of criteria are profit, costs, idle capacity, and information entropy.

5. Analysis and design of the nondeterministic functions to make them amenable to solution of the total system structure.

6. Integration of behavioral patterns in different environments, as required, within the physical framework of the system. Examples are the socio-political environment of a country for which the total system is designed, and the difference between behavioral attitudes of nurses in a hospital and those of production-line workers in a fabrication shop.

7. Design of simulation models which permit a rigorous critique of the parameters of a dynamic system.

8. Economic analysis of the total system and concomitant subsystems with extensions into alternative structures.

Management science and operations research. The real-world problems are complex systems configurations which require the use of more sophisticated methods than previously to present meaningful results to the decision maker. In the development of modern systems technology, whether industrial or space-oriented, the pacing factor is management. This is the function which directs, coordinates, and controls the many facets of a system. For the United States to maintain technological supremacy, it is essential that optimal use be made of all resources and that new discoveries be translated into new products in minimal time. The terms optimal and minimal are characteristic of management science and operations research. Management science and operations research are often used synonymously since the tenor of their objectives is compatible. Operations research has been referred to as a sharper kind of industrial engineering. *See* OPERATIONS RESEARCH.

While no two people define operations research

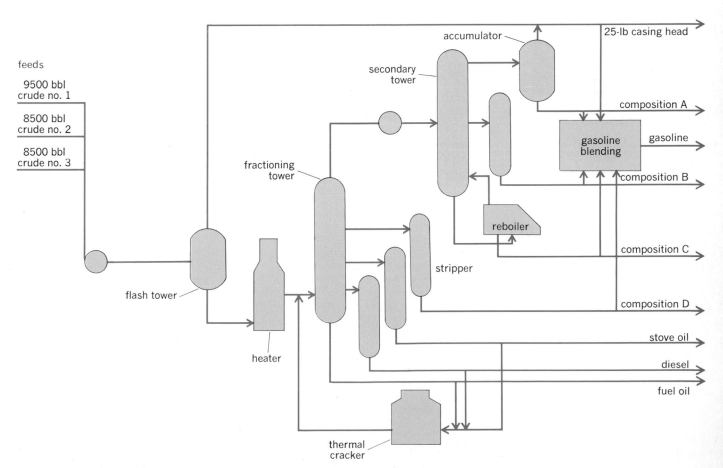

Fig. 1.   Refinery flow diagram. (*From G. B. Dantzig and R. J. Ullman, Mathematica, 1963*)

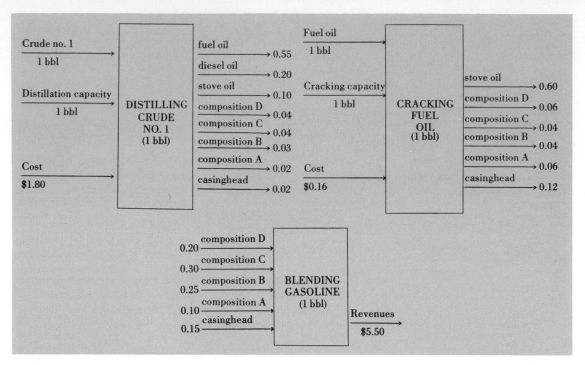

Fig. 2.   Refinery input and output quantities.

in exactly the same way, four common denominators can be abstracted from most definitions.

1. Management problems: those which involve the steering or guiding toward attainment of the company's objectives.

2. Decision-making: choices or alternative courses of action that require a decision to be made.

3. Optimization: selection of the best choice predicated on a specific criterion, such as maximizing profit, minimizing cost, or maximizing capacity.

4. Systems perspective: the inherent capability of considering the many facets resulting from interdepartmental or interplant relationships.

While the denominator optimization is a salient criterion in operations research, most solutions represent only suboptimal answers because of the intractability of large-scale problems. In fact, Thomas L. Saaty defines operations research as the art of giving bad answers to problems to which worse answers would otherwise be given. Generally, it is concerned with quantified data, something which one can measure or put a value on. Certainly it is true that there are many aspects of managerial problems which cannot be quantified. However, industrial engineers, until recently, have taken calculus as a formal requirement in the early years of their academic program, but had never utilized mathematics beyond college algebra in industrial engineering courses or practice. This has changed drastically; even at the undergraduate level, great utility has been found not only for calculus but also for differential and difference equations, matrix algebra, and probability theory. In fact, the mathematical maturity expected of industrial engineers is often greater than that of other engineering disciplines. *See* CALCULUS, DIFFERENTIAL AND INTEGRAL; MATRIX THEORY.

Operations research does not eliminate manag-

ers; it consists of a powerful set of mathematical techniques to be utilized in resolving data for certain types of decision-making problems. The existing methods and procedures used for analyzing management problems are not dissipated or eliminated but instead are augmented by advances in applied mathematics. The manager still must make decisions; however, the objective of operations research is more efficient decision-making. It has often been said that the capable executive is one who makes the right decision more than half the time. The purpose of operations research is to enable the executive to make right decisions with greater accuracy. The goal is the shrinking of the large sphere of subjective decision-making confronting today's managers.

Behavioral science. Since the industrial engineer is concerned with man in his technical environment, he focuses on what is called man-machine engineering or human factors. The industrial engineer must draw upon information from a variety of life sciences to design modern machine systems which can most effectively take advantage of man's capabilities and limitations. Typical applications include the design of information displays and controls, man as an element in servo systems, the design of machine systems for efficient maintenance, the control of the working environment, and the anthropometry of the working environment. *See* HUMAN-FACTORS ENGINEERING.

Behavioral aspects are also important in the design of the conceptual framework for the analysis of the organic functions of project management, where emphasis is placed on the execution phase of management.

Methodologies. The following techniques (although not all-inclusive) are particularly useful in a scientific approach to solving complex industrial-engineering problems. The solutions of these complex problems usually demands a large

Activities / Items	Unused available crude oil			Distillation				Cracking				Product marketing									Available, bbl/day
	Crude no. 1	Crude no. 2	Crude no. 3	Crude no. 1	Crude no. 2	Crude no. 3	Unused capacity	Fuel oil	Diesel oil	Stove oil	Unused capacity	Fuel oil	Diesel oil	Stove oil	Gasoline	Composition D	Composition C	Composition B	Composition A	Casinghead	
	X_1	X_2	X_3	X_4	X_5	X_6	X_7	X_8	X_9	X_{10}	X_{11}	X_{12}	X_{13}	X_{14}	X_{15}	X_{16}	X_{17}	X_{18}	X_{19}	X_{20}	
Crude no. 1	1			1																	= 9500
Crude no. 2		1			1																= 8500
Crude no. 3			1			1															= 8000
Distillation capacity				1	1	1	1														= 14,000
Fuel oil				−0.55	−0.61	−0.50		1				1									= 0
Diesel oil				−0.20	−0.12	−0.11			1				1								= 0
Stove oil				−0.10	−0.07	−0.14		−0.6	−0.2	1				1							= 0
Cracking capacity								1	1	1	1										= 3500
Composition D				−0.04	−0.06	−0.05		−0.06	−0.41	−0.30					0.20	1					= 0
Composition C				−0.04	−0.05	−0.08		−0.04	−0.20	−0.30					0.30		1				= 0
Composition B				−0.03	−0.04	−0.05		−0.04	−0.04	−0.04					0.25			1			= 0
Composition A				−0.02	−0.02	−0.03		−0.06	−0.12	−0.10					0.10				1		= 0
Casinghead				−0.02	−0.03	−0.04		−0.12	−0.16	−0.14					0.15					1	= 0
Profit				−1.8	−1.9	−2.0		−0.16	−0.21	−0.21		1.8	4.0	4.2	5.5	4.0	4.1	4.2	4.3	3.3	= maximum

Fig. 3. Linear programming model of a refinery.

amount of computation. A salient factor to the rapid advancement of these new, more sophisticated techniques has been the computer revolution. Stimulus to the application of operations-research techniques has been the capability of electronic digital computers to perform calculations thousands or millions of times faster than a human.

Linear programming. This technique has been applied successfully in a large number of diverse problems in which limited resources must be allocated among competing activities in an optimal manner. The constraints and the evaluation criterion are expressed in the form of a mathematical model. Typical applications include blending problems, shipping assignments, production scheduling, urban planning, and transportation patterns.

To show the linear programming approach in the development of a mathematical model for a real-life system, an example is taken from the petroleum industry. Figure 1 is a flow diagram of the refinery, showing the interrelationships among the components of the system. The input-output relationships for three of the major activities are given in Fig. 2. Figure 3 represents the mathematical model for linear programming.

Through the use of linear programming, such as the simplex method, it is possible to determine the best allocation of the limited resources to maximize profit.

Queueing theory. This involves the study of waiting lines, or queues, which is a ubiquitous problem occurring whenever the demand for service exceeds the capacity to provide that service. Since queueing structures are nondeterministic in nature, probability theory is essential in the structuring of mathematical models representing these systems. Large real-life queueing problems are often mathematically intractable and are solved by simulation. Examples of queueing theory applications include congestion in traffic, patients waiting at a hospital clinic, employees waiting for crane service, orders waiting to be serviced, and in-process items waiting for machining.

Simulation. This is the experimental component of operations research. When it is possible to express a problem in mathematical form amenable to solution, such as linear programming, the analytical approach is usually preferred to simulation. However, as stated previously, large-scale systems problems may be intractable analytically, even with a high-speed computer. While it is possible to simulate by pencil and paper, most systems are more efficiently simulated by a digital computer. The system is segregated into activities for which probability distributions can be obtained. The interrelationships between these activities are also determined. Sampling experiments are then performed by the computer on the systems model rather than on the real system. Simulation has been used successfully in the design and analysis of steel-making operations, communications systems, large-scale military operations, plant layout, and materials-handling configurations.

Network models. The basis for work in this area is graph theory, which until recently was considered as an abstract mathematical subject with some application in electrical engineering. However, network theory is now found to have application in the various management, engineering, and physical sciences. All network problems resemble each other in that their elements can be represented by a topological structure consisting of stations and links between stations. The program evaluation and review technique (PERT) and critical path method (CPM) are examples of network theory applied to the planning and control function of large projects. *See* CRITICAL PATH METHOD (CPM); PERT.

[A. G. HOLZMAN]

Bibliography: E. S. Buffa, *Production-Inventory Systems: Planning and Control*, 1967; S. E. Elmaghraby, *The Design of Production Systems*, 1966; F. S. Hillier and G. J. Lieberman, *Introduction to Operations Research*, 1967; A. G. Holzman,

H. H. Schaefer, and R. Glaser, *Matrices and Mathematical Programming*, 1962; H. B. Maynard, *Industrial Engineering Handbook*, 1963.

Industrial meteorology

The commercial application of weather information to the operational problems of business, industry, transportation, and agriculture in a manner intended to optimize the operation with respect to the weather factor. The weather information may consist of past weather records, contemporary weather data, predictions of anticipated weather conditions, or an understanding of physical processes which occur in the atmosphere. The operational problems are basically decisions in which weather exerts an influence.

Because meteorological data are not available to specify past, present, and future states of the atmosphere with absolute precision and knowledge of physical processes in the atmosphere is incomplete, the application of weather information to operational decisions is properly viewed as a specialized case of decision-making in the face of uncertainty. What is involved here is best illustrated by the following example developed by J. C. Thompson. A hypothetical construction company was confronted daily with the decision whether to pour concrete. If the concrete were poured and 0.15 in. or more of rain fell in the subsequent 36 hr, damage of $5000 would result. The cost of protecting the newly poured concrete from rain would be $400. To minimize the total expense of a series of such repetitive decisions (optimization of the operation), it follows from the principle of the calculated risk that protective measures should be taken only when $P > C/L$, where P is the probability of 0.15 in. or more of rain within 36 hr, C is the cost of protective measures ($400), and L is the contingent loss ($5000). For this particular case, under the actual weather occurring during a season's operations, the total expense (cost plus loss) would be $85,000 if no protective measures were taken, $72,800 if protective measures were taken every day regardless of anticipated weather, $32,600 if protective measures were taken only on days when there was a 50:50 chance of this amount of rain, and $24,400 if protective measures were taken only on days when the probability of the critical amount of rainfall exceeded 0.08, that is, in ratio of $400:$5000.

This approach can be generalized to include more complex decisions, and relations were developed by Thompson and G. W. Brier to measure the economic utility of weather information. In practice, less formal methods are frequently used and the fundamental principle described here is handled in a qualitative manner by close collaboration between the meteorologist and the user of weather information.

The substantial economic significance of weather in problems of business and industry has been emphasized by a survey conducted by the U.S. Weather Bureau in which an attempt was made to assign monetary values to savings or profits realized through applications of daily weather reports, forecasts, storm warnings, and past weather records. The total for the United States was on the order of $1,000,000,000 annually.

Some measure of the economic toll of severe or unusual weather conditions is provided by data assembled by the National Board of Fire Underwriters, which indicated that claims totaling $866,000,000 were paid as a result of 72 major hurricanes, tornadoes, windstorms, hailstorms, and rainstorms during the period 1949–1957, inclusive. These losses represent insurance losses, not total property damage, and include only instances in which claims within a single state exceeded $1,000,000.

The economic benefits that could be achieved in a single industry, the petroleum industry, by only a modest improvement in forecasts of anticipated average temperature conditions 1 month in advance have been analyzed with the result that potential reduction in tankage and inventory costs has been conservatively estimated at $100,000,000 per year. These savings would be realized by improvements in heating oil scheduling made possible through better estimates of expected space-heating requirements available from forecasts of anticipated temperature conditions.

As a result of the economic significance of the weather factor in a variety of industrial applications, there exists in meteorology a specialized activity on the part of professional meteorologists to serve the specific needs which lie outside the general public responsibilities of the U.S. Weather Bureau. This service is provided either by staff meteorologists employed by particular companies or by consultant meteorologists who work for several clients. Approximately 9% of the more than 7000 professional meteorologists in the United States are engaged in some facet of industrial meteorology. A substantial number of these are employed by commercial airlines where the need for specialized weather information is vital for safe and efficient operations. Gas and electric utility load estimates, highway and street maintenance, outdoor construction work, marine transportation, retail merchandising and advertising, flood control design, air pollution, building and plant design, atmospheric corrosion, agricultural planning and production scheduling, and air conditioning and heating design are but a few of the activities in which the industrial meteorologist has found a demand for his services. *See* AERONAUTICAL METEOROLOGY; AGRICULTURAL METEOROLOGY.

Annual conferences on industrial meteorology are sponsored by the Committee on Industrial Meteorology of the American Meteorological Society. That society has established a program for the certification of consulting meteorologists who meet rigorous standards of knowledge, experience, and adherence to high standards of ethical practice.

[THOMAS F. MALONE]

Bibliography: American Meteorological Society, A selective annotated bibliography on industrial meteorology, *Weatherwise*, vol. 6, no. 2, 1953; S. Petterssen, *Introduction to Meteorology*, 3d ed., 1969; J. C. Thompson, A numerical method for forecasting rainfall in the Los Angeles area, *Mon. Weather Rev.*, 78(7):113–124, 1950; J. C. Thompson and G. W. Brier, The economic utility of weather forecasts, *Mon. Weather Rev.*, 83(11): 249–254, 1955.

Industrial microbiology

The study, utilization, and manipulation of those microorganisms (in the broad sense including bacteria, yeasts and other fungi, and some algae) capable of economically producing desirable substances, or changes in substances, and controlling unwanted microorganisms.

Industrial fermentation criteria. To be industrially important, microorganisms should be easily cultivable in large quantities; grow rapidly on cheap, available substrates; and carry out the required transformations under comparatively simple and workable modifications of environmental conditions. They should maintain physiological constancy under these conditions and produce economically adequate yields. Above all, the microbiological method must be the cheapest method of producing the desired material or change. The marketing of wastes or other by-products of a process often serves to keep the process competitive. Lack of current demand for a product may be of little concern, for a market can often be created when the price of the product is low.

The potential uses of microorganisms for the good of man were not appreciated until World War I, when microorganisms were used to produce acetone, butanol, and certain enzymes. The discovery of the usefulness of penicillin during World War II caused an enormous development of the fermentation industry and a fuller realization of the value of microbiological activities. Some of the problems of space travel, pollution, and waste disposal are being solved by the use of microorganisms. *See* ACETONE-BUTANOL FERMENTATION; FERMENTATION; SPACE BIOLOGY.

Products. Since earliest times microorganisms have been used, intentionally or otherwise, as means of preserving foods or of making them more palatable, as in the making of such products as wines, cheese, butter, and sauerkraut. They have also been used for such procedures as flax retting. *See* BUTTER; CHEESE; FLAX; FOOD MANUFACTURING; FOOD PRESERVATION; WINE.

Microorganisms are used to produce chemical substances, which are then extracted from the culture mixture and purified. Included are (1) solvents such as ethanol, acetone, butanol, 2,3-butylene glycol, and other alcohols, such as glycerol, mannitol, erythritol, and polyhydric alcohols; (2) organic acids such as acetic, lactic, citric, gluconic, fumaric, propionic, itaconic, kojic, and gallic; (3) carbohydrates such as sorbose, fructose, dextrans, and other polysaccharides; (4) vitamins such as B_{12}, thiamine, and riboflavin; (5) amino acids such as lysine, arginine, and glutamic acid; (6) antibiotics such as penicillin, streptomycin, tetracycline, and chloramphenicol; (7) enzymes such as amylases, proteases, and glucose oxidases; and (8) nucleosides. *See* ACETOBACTER SUBOXYDANS FERMENTATION; ACETONE-ETHANOL FERMENTATION; ANTIBIOTIC; 2,3-BUTANEDIOL; CITRIC ACID; ENZYME, INDUSTRIAL PRODUCTION OF; ETHYL ALCOHOL; GLUCONIC ACID; LACTIC ACID; RIBOFLAVIN; THIAMINE; VINEGAR; VITAMIN B_{12}.

Microorganisms are used to produce desired changes in certain types of compounds or materials, such as in the microbiological transformations of steroids by dehydrogenation, hydrogenation, hydroxylation, oxidative cleavage, or reductive rearrangement of the steroid molecules. More ancient are the uses of microorganisms in the textile industry with cotton, wool, and flax. More modern are the applications of bacteriological processes to the generation of electric power in fuel cells and to studies of human wastes as nutrients in closed space ecologies. *See* ENZYME; STEROID; TEXTILE MICROBIOLOGY.

As with all life functions, the activities of microorganisms can be reduced to a consideration of enzymes acting on a substrate. The industrial microbiologist is concerned more with the final product or change—the sum total of the activities of a series of enzymes—than with the action of each individual enzyme as such.

Some of the uses already mentioned overlap those where the product is not extracted and purified but is used in the culture mixture. In many cases the product need not be extracted, for the whole culture mixture can be used as in fodder yeast, feed supplement production, and nitrogen-fixing organism production.

The field of animal-feed supplementation, whether with extracted chemicals, whole culture mixtures, or wastes from fermentation, is of great importance and yields large financial return. The production of growth stimulators, growth inhibitors, fungicides, and insecticides is of increasing importance. Because of the vast markets available, the production of materials for use in the agricultural industry, both animal and plant, is receiving increased attention and will probably equal or surpass production for human consumption and therapy, as well as production of industrial chemicals. *See* ANIMAL-FEED COMPOSITION.

Food technology. The leavening of bread and other bakery products; the lactic acid fermentation of pickles, olives, sauerkraut, and other products; the manufacture of cheese, butter, and certain beverages; and the manufacture of shoyu, tempeh, miso, and other Oriental foods involve the activities of one or more microorganisms. In some cases these activities provide a more palatable product, in others they provide a measure of preservation, and in still others they achieve both.

The control or eradication of unwanted microorganisms is of great importance to prevent spoilage, loss of nutritive value and flavor, and the formation of poisonous substances such as botulins and aflatoxins. *See* FOOD ENGINEERING; FOOD MICROBIOLOGY; FOOD TECHNOLOGY.

Waste and surplus disposal. It may be said that there are no pollution problems—only disposal problems. Microbiological activity may be used merely to dispose of some wastes, such as sewage; with other wastes it may be used to produce material of value. Such wastes may be from other microbiological processes, as in the fermentation of corn-steep liquor, distillers' solubles, and cheese wheys; or they may be from nonmicrobiological processes, such as sugarbeet and sugarcane molasses, waste-sulfite liquor, fish solubles, or cannery wastes. The need to dispose of various agricultural surpluses, such as potatoes and cereal grains, may make feasible their use as fermentation substrates. To facilitate disposal, biodegradable detergents have been developed. The reduction of the biological oxygen demand of waste-sulfite liquor prior to disposal, by the production of alcohol or food yeast from it, promises to lessen some pollution problems. *See* SEWAGE DISPOSAL; WATER POLLUTION.

Bioassay. Microbiological assay procedures for vitamins, amino acids, and antibiotics are in common use because of their relative simplicity, their specificity, and the speed with which they may be carried out. *See* BIOASSAY; VITAMIN.

Types of fermentations. Fermentations may be of the single organism – one product type, or of the type in which two or more organisms, grown together or in sequence, may be used. Fungi may be used aerobically to saccharify starch, the resulting sugar then being dissimilated by anaerobic bacterial activity to produce desired products. In a patented method of producing *l*-lysine, the precursor diaminopimelic acid is produced by one organism and then decarboxylated by a second organism to give excellent yields of the desired lysine. Organisms producing desired materials, when grown together in one way or another, may be induced to produce new and even more desirable substances.

While many fermentations are still done by the batch process, long-used continuous culture methods have been refined and are finding increased application, as in continuous beer fermentation and in tower fermentors. Microbiological modifications of hydrocarbons are more recent entries into the field, as are those organisms which utilize jet fuel and the many which influence the quality of petroleum products. *See* DISTILLED SPIRITS; ITACONIC ACID; ITATARTARIC ACID; KOJIC ACID; MALT BEVERAGE; PETROLEUM MICROBIOLOGY; YEAST, INDUSTRIAL.

[R. H. HASKINS]

Bibliography: C. W. Hesseltine, A millenium of fungi, food and fermentation, *Mycologia*, 57:149–197, 1965; D. J. D. Hockenhull (ed.), *Progress in Industrial Microbiology*, vol. 6, 1967; H. J. Peppler (ed.), *Microbial Technology*, 1967; S. C. Prescott and C. G. Dunn, *Industrial Microbiology*, 3d ed., 1959; Society for Industrial Microbiology, *Developments in Industrial Microbiology*, vol. 4, A.I.B.S., 1963; M. P. Starr (ed.), *Global Impacts of Applied Microbiology*, 1964; G. N. Wogan (ed.), *Mycotoxins in Foodstuffs*, 1965.

Industrial trucks

Manually propelled or powered carriers for transporting materials over level, slightly inclined, or slightly declined running surfaces. Some industrial trucks can lift and lower their loads, and others can also tier them. In any event, all such trucks maintain contact with the running surface over which they operate and, except when towed by a chain conveyor, follow variable paths of travel as distinct from conveying machines or monorails. *See* MATERIALS-HANDLING MACHINES.

Running gear. The means employed to support a truck and its load and to provide rolling-friction contact with the running surface is the running gear. Factors in the selection of running gears include load capacity, operating conditions, travel surface, kind of material to be handled, protection of load and machine, economy, and, in the case of hand trucks, ease of manipulation and the reduction of operator fatigue.

Rollers, used in dollies, are of solid or tubular steel with antifriction bearings. Rigid and swivel casters are used with the dollies and with hand and powered trucks. Swivels differ from rigid casters in that they are offset wheels that swivel about their own vertical axis, thereby easing the turning of the hand truck with its load. Steel, solid rubber, semipneumatic, and other wheels fitted with plain or antifriction bearings are designed to

meet specific requirements. Industrial wheels for heavier-duty and special automotive-type wheels in a wide selection of tire treads are used with powered trucks and tractors.

Hand trucks. The manually operated, two-wheel hand truck, regarded by some as old fashioned and inefficient, still plays a prominent part in the handling of materials in some of the most modern plants. The hand truck is inexpensive, requires a negligible amount of maintenance, and is convenient for handling light loads for short distances. Its light weight makes it ideally suited for applications where small quantities are handled, with a low frequency of movement, where floor loading restrictions preclude the use of the heavier powered trucks. The hand truck is also easily adapted for use in explosive atmospheres.

The key feature to the hand truck is the running gear. Care must be exercised in selecting the type and size of wheel to suit the carrying load and operating conditions. In many cases, depending on the loads to be carried and the surface of the operating floor, inexpensive cast-iron or steel wheels may be more suitable than higher-priced pneumatic tires or plastic-rimmed wheels.

Basic types of hand trucks and their distinctive features are shown in the table. Two-wheel hand trucks are classified broadly as eastern and western (Fig. 1). Multiwheel hand trucks are produced in many models, but platform types continue to be the most widely used in industry and distribution. Stakes, end-and-side gates, solid panels, and other superstructures add versatility to the basic machines. Low-lift types of hand trucks elevate their loads just sufficiently to clear the running surface (Fig. 2).

Powered trucks. The powered industrial truck can be defined as any self-propelled, power-driven truck or tractor used to carry, push, pull, lift, or tow loads. Relatively new in the use of industrial trucks are personnel and burden carriers for transporting messengers, watchmen, mail, and blueprints, as well as materials between buildings and in-plant areas covering considerable ground. Shop tractors are used extensively for miscellaneous towing jobs in plants and warehouses (Fig. 3). More recently an electronically controlled tractor was introduced, directed by radio, overhead wires, or wires concealed in the floor.

Whatever the type or function, these trucks depend on compact, high-capacity storage batteries or small internal combustion engines. Sources of power for internal combustion engines are gasoline, diesel fuel, and more recently liquid petroleum gas; some manufacturers offer trucks powered in this fashion as original equipment and conversion kits are available for existing equipment. Characteristics of these sources of power are briefly outlined as follows.

Storage battery. Trucks powered from storage batteries are quiet and clean in operation, have low maintenance cost and smooth acceleration, and when properly charged present no starting problems in cold weather. However, their speed is generally considered too low for long-distance travel. Battery-charging facilities are required at many points for truck travels over a large area, and battery drainage is excessive on steep slopes; additional capital costs are involved in providing batteries for multishift operations.

Handtrucks and their features

Type	Description	Capacity	Range
Pry	Lever bars with long wooden handles, short steel noses, and two wheels at the fulcrum. Used singly or in pairs for moving and spotting crates and similar heavy articles in freight cars.	Up to 5000 lb each	Very short distances
Dolly	Low platforms or specially shaped carriers mounted on rollers or combinations of fixed and swivel casters. Designated as furniture, milk can, paper roll, and so on to indicate intended uses.	From few pounds up to 80 tons for moving machinery	A few feet
Two-wheel truck	Normally constructed of wood and steel but available in aluminum and magnesium. Stevedore and warehouse models are general purpose, but many special-purpose types are available for egg crates, drums, or paper rolls.	Normally 400–500 lb, but exceptionally up to 1500 lb	Up to 150 ft
Multiwheel platform	Trucks with flat platforms mounted on various combinations of 3 (unstable), 4, or 6 rigid and swivel casters. Addition of stakes, side and end gates, solid panels, and similar accessories increases the versatility of these trucks.	Normally 1500–2000 lb for one-man operation	Weight is an important factor, but normally within a few hundred feet
Specials	A wide variety of trucks with special superstructures and running gear. Examples are box, frames for spools and sheets of metal, frames for plate glass, shelves for small parts and tools, shapes to accommodate bar stock, textile beams, and stock items.	Same as multiwheel platform	Same as multiwheel platform

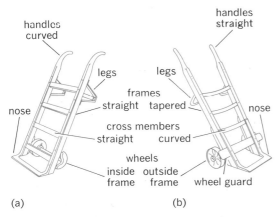

Fig. 1. Two-wheel hand trucks. (*a*) Eastern. (*b*) Western.

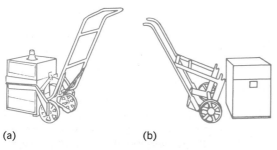

Fig. 2. Low-lift hand trucks elevate loads to clear running surface. (*a*) Carboy. (*b*) Trunnion box.

Diesel engine. Higher speed of operation is provided by diesel-powered trucks. They can withstand severe operating conditions, such as working on inclines or traveling over long distances; refueling is a simple operation; and fuel cost is low. However, operation is noisy and fumes are objectionable though nontoxic. Maintenance cost is higher than for electric-powered trucks.

Gasoline engine. Basically the characteristics are the same as the diesel engine, except that the exhaust fumes are toxic.

Liquid petroleum gas. High octane rating permits higher compression and greater efficiency; clean combustion, even at idling speed, prolongs the life of the engine; and there is less carbon monoxide in exhaust fumes than from a gasoline engine. However, gas is transported and stored in tanks under pressure requiring stringent safety precautions; the fuel is heavier than air and odorless and requires a smell additive to make a leak detectable.

Unit load principle. In industry the principle of handling materials in unit loads has developed in parallel with the increased use of powered industrial trucks, particularly the forklift type. These mobile mechanical handling aids have removed the limitations that existed when the weight and size of a load for movement and stacking depended mainly on the ability of a man to lift it manually. The unit-load principle of materials handling underlies the skid-platform and the pallet-forklift

(a)

(b)

(c)

Fig. 3. Powered nonlift platform trucks. (*a*) Personnel carrier. (*b*) Burden carrier. (*c*) Shop tractor.

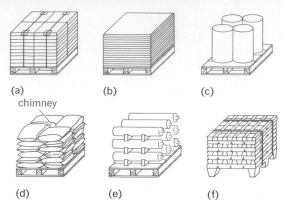

Fig. 4. Loads assembled for handling as units. (*a*) Strapped load. (*b*) Sheet or flat stack. (*c*) Barrels or drums. (*d*) Bags, pinwheel pattern. (*e*) Cylinders with dunnage. (*f*) Ingots with molded legs.

methods of operation. Both methods, especially the latter, have revolutionized handling techniques and equipment and even production equipment.

A unit load can be defined as a quantity of material (bulk or individual items) assembled and, if necessary, restrained to permit handling as a single object. When referred to in connection with industrial trucks, it is implied that a unit load will retain its shape and arrangement when deposited by the truck to allow subsequent movement, if required (Fig. 4).

In order to gain the maximum benefits from unit-load handling, the principle should be applied as far back and as far forward as possible in the manufacturing cycle. In particular, the possibility of introducing unit loads should be thoroughly investigated from raw material receipt through production operations, process storage, and warehousing. Investigation may reveal that modification of production equipment is justified. Some or all of the following benefits can be expected from the correct application of the unit load principle: (1) cheaper direct cost of unit handling because of the movement of more pieces or a larger quantity at a time; (2) maximum use of the cubic space available in storage area; (3) quicker loading and unloading and consequently a faster turn around of transport vehicles; (4) unit load lending itself to an efficient storage system and hence easier and quicker stock control; (5) well-secured unit load deterring pilfering; (6) product damage reduced by the elimination of manual handling of individual pieces; (7) reduction of packaging costs; (8) reduction of materials-handling accidents, particularly injuries, such as damaged fingers or strained backs, which result in lost time; and (9) value of correctly designed load units as a sales aid since most customers welcome receipt of goods in suitable unit loads.

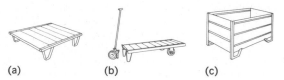

Fig. 5.   Skids for load units. (*a*) Dead skid. (*b*) Semilive skid and jack. (*c*) Corrugated metal box.

Skids are constructed in three basic types: (1) live skids, with fixed and swivel casters, which are too lively for tiering purposes; (2) semilive skids, with two fixed legs in the front and a pair of rigid casters at the rear and made mobile by means of a jack; and (3) dead skids, having either two solid runners or four metal legs (Fig. 5). All wooden dead skids and those made of hardwood platforms with metal edges and metal legs are conventional; others are made of metal such as steel or aluminum and have special superstructures designed with the most efficient configuration for the product being handled.

Pallets differ from skids in having two decks, or faces, separated by two or more lengthwise members called stringers. The National Wooden Pallet Manufacturers Association has prepared and issued specifications for lumber, fasteners, and other pallet components. Different constructions are called by descriptive names (Fig. 6); for example, those with double wings to permit the use of sling bars are the stevedore type. A newer development in pallets is the "take it or leave it" pallet (Fig. 6*h*), which permits the choice of retrieving the pallet and unit load together or of retrieving the unit load and leaving the pallet.

There are nine standard rectangular pallet sizes and three square sizes, which range from 24×32 to 88×108 in. The International Organization for Standardized Sizes recommends the following sizes: 32×40, 32×48, and 40×48 in., all of which can be carried advantageously in over-the-road trucks and trailers and in box and refrigerated railroad cars. Other sizes may be suitable for the product being handled; however, these sizes should be developed with regard to proper increments to utilize maximum space in motor or rail carriers.

The pallet was developed primarily for use with forklift trucks. If a pallet is to be used with a low-lift truck, the bottom deckboards must be so spaced that openings are left near both ends to permit the rollers in the fork arms of the truck to drop to the floor when the pallet is being elevated. Pallets so constructed are called nonreversible. Special pallets are those made of steel, magnesium, or even plastic; expendable varieties (one-way shippers) are made of fiberboard or inexpensive wooden construction and, as the name implies, are designed for extremely limited service.

Structural differences between skids (Fig. 5), with considerable clearance underneath, and pallets (Fig. 6), with stringers and bottom boards presenting obstructions, account for the use of platform trucks with the former and, of necessity, fork-equipped trucks with the latter. The double-deck feature of pallets makes them more suitable for multiple tiering than the runners or legs of skids, which may damage the supporting surface of the lower load.

Low- and high-lift trucks. Skids and pallets are handled by self-loading machines. The lift trucks pick up, transport, set down, and in the case of high-lift types tier their loads without manual handling. Powered models evolved from prototype hand machines and, because the first of the self-propelled machines were led by the operator, were called "walkies." The name persists, even though most of them are now produced as rider trucks (Fig. 7).

Low-lift trucks, hand and powered, lift their loads sufficiently to make them mobile. The elevating mechanisms of hand types may be operated by pulling the handle down one or more times. Others have a lever- or pedal-operated hydraulic mechanism or a powered hydraulic system.

Noncounterbalanced high-lift trucks evolved from hand-propelled stackers (Fig. 8). Fork-equipped types are called outrigger or straddle trucks because of the supports which straddle the pallet. All varieties come with fixed or telescoping masts. These trucks are advantageous when space is at a premium. For example, when piling vertically from aisles, these machines operate in aisles 6 ft wide, compared with the 10-ft aisles required by the counterbalanced trucks of equal capacity. Because they have no counterweight, these machines are lighter than the counterbalanced types and hence are a boon to handling operations in old multistory buildings.

Counterbalanced lift trucks are constructed by extending the wheelbase, adding counterweight, and removing the outriggers. By using a forward- and backward-tilting mast, they are used where outriggers cannot function advantageously, but the entire truck is longer than its noncounterbalanced equivalent.

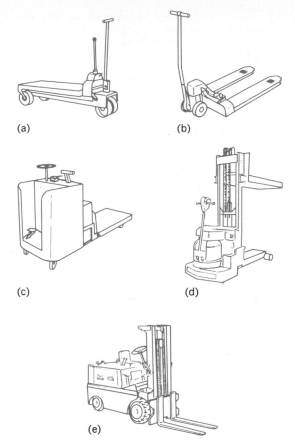

Fig. 7. Hand-operated, counterbalanced, powered, stacker-type skid and pallet trucks. (a) Hand. (b) Hand. (c) Powered low-lift skid. (d) Noncounterbalanced stacker with skid platform. (e) High-lift fork.

Forklift trucks. Conventional fork trucks are made with any desired source of power, the selection depending upon the service for which the truck is intended. Elevation of the forks, forward and backward tilt of the mast, and the power for actuating attachments are provided by the hydraulic system. The hydraulic system consists of a pump to draw oil from a reservoir and force it through the control valves, which are manipulated

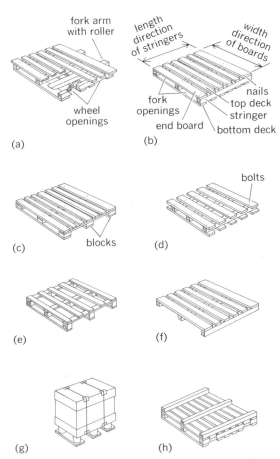

Fig. 6. Standard and special pallets. (a) Semiwing, non-reversible, two-way pallet. (b) Flush-type, reversible, two-way pallet. (c) Four-way pallet. (d) Stevedore (double-wing), reversible, two-way pallet. (e) Expendable (one-way shipper) pallet. (f) Single-faced pallet. (g) Fiberboard expendable pallet. (h) "Take it or leave it" pallet unit; load rests on stringers.

Fig. 8. Industrial high-lift truck raising a load. (*Forest Products Laboratory, Madison, Wis.*)

INDUSTRIAL TRUCKS

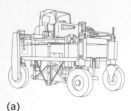

(a)

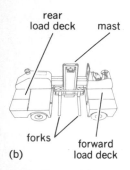

rear
load deck   mast

forks

forward
load deck

(b)

Fig. 9. Heavy-duty
industrial trucks.
(a) Straddle truck.
(b) Side-loading truck.

by hand levers selected by the machine operator, into piston actuators.

Free lift is an important factor in some operations. This is the distance through which the forks are elevated before the mast starts to rise. Where headrooms are low, as in boxcars, double-tiering is possible if sufficient free lift is provided. Attachments for use with high-lift trucks are designed to handle specific products singly or in multiple units. Many attachments eliminate the need for a pallet or other supporting carrier, a feature with direct economic advantages.

The rated capacity of a high-lift fork truck is based on the weight of the load in pounds that the machine can lift when the load center is located a specified distance from the vertical faces of the forks. The load center is the center of gravity of a uniform load. The load center is 15 in. for trucks up to 2000 lb capacity, 24 in. from 2000 to 10,000 lb, 36 in. from 10,000 to 20,000 lb, and as designated above 20,000 lb.

Trends in forklift truck design have been toward the replacement of friction clutches by torque converters in gasoline and diesel machines, more complete protection from exhaust fumes through more efficient catalytic mufflers, and changes in mast construction to provide better visibility for the operator.

Trucks and trailers. Other trucks capable of handling unit loads include end-loading (also called gantry and straddle-carries) and side-loading types (Fig. 9). The former type must straddle its load before picking up and transporting. This type unit is most suitable for handling long, unwieldy items such as pipe and lumber. These trucks, however, cannot tier their loads so that, when tiering is essential, they team up with forklift trucks. In contrast, side-loading trucks can

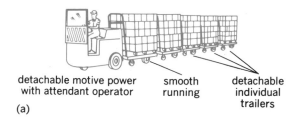

detachable motive power   smooth   detachable
with attendant operator   running   individual
                                          trailers

(a)

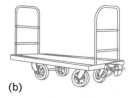

(b)

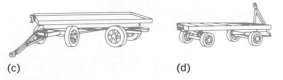

(c)   (d)

Fig. 10. Industrial trailers. (a) In-plant, tractor-trailer, trackless train. (b) Caster steer. (c) Four-wheel steer. (d) Wagon (fifth-wheel) steer.

transport and tier within the reach of their forks. Furthermore, side-loading trucks can be equipped with crane arms for handling lengths of pipe or lumber and with reel carriers for paying out cable as the truck advances.

Tractor-trailer or trackless trains are motorized versions of industrial (narrow-gage) railroads but have an advantage over the railroads in being able to follow variable paths in travel (Fig. 10). Industrial tractors, powered electrically or by internal-combustion engines, are classified as three- or four-point contact, according to the number of supporting wheels. They are used primarily for service along loading platforms and in yard operations.

Of industrial trailers, those with caster steer have good trailability (that is, they follow closely the path established by the tractor); those with four-wheel steer are rated as having excellent trailability but are difficult to maneuver manually; and those with fifth-wheel (wagon) steering are effective with heavy loads. They are offered in a wide variety of constructions, which makes it possible to select one with proper characteristics to meet given requirements.

Each type of industrial truck is used to the best advantage when carrying the particular load for which it is intended. Several types may be found working singly or in teams. For example, palletized loads may be transported on tractor-trailer trains and then distributed and tiered by forklift trucks. See TRUCK.   [ARTHUR M. PERRIN]

Bibliography: American Society of Mechanical Engineers, *Pallet Terminology and Sizes*, MHl, 1965; K. E. Booth and C. G. Chantrill, *Material Handling with Industrial Trucks*, 1962; Caster and Floor Truck Manufacturers' Association, *Engineering and Purchasing Handbook*, 1959; Industrial Truck Association, *Handbook of Powered Industrial Trucks*, 1957; National Wooden Pallet Manufacturers Association, *Technical Handbook Pallets and Palletization*, 1954.

Inert gases

The inert gases, listed in the table, constitute group 0 of the periodic table of the elements. They are now better known as the noble gases, since stable compounds of xenon have been prepared. The term rare gases is a misnomer, since argon is plentiful and helium is not rare in the United States and some other countries.

All these gases occur to some extent in the Earth's atmosphere, but the concentrations of all but argon are exceedingly low. Argon is plentiful, constituting almost 1% of the air.

All isotopes of radon are radioactive, the longest lived having a half-life of about 4 days. Each of the other inert gases has at least two stable (nonradio-

The inert gases

Name	Symbol	Atomic number	Atomic weight
Helium	He	2	4.0026
Neon	Ne	10	20.183
Argon	Ar	18	39.948
Krypton	Kr	36	83.80
Xenon	Xe	54	131.30
Radon	Rn	86	(222)

active) isotopes, in addition to one or more radio-active isotopes.

All the gases are colorless, odorless, and taste-less. They are all slightly soluble in water, the solubility increasing with increasing molecular weight. They can be liquefied at low temperatures, the boiling point being proportional to the atomic weight. All but helium can be solidified by reducing the temperature sufficiently, and helium can be solidified at temperatures of $0-1°K$ by the application of an external pressure of 25 atm or more.

The noble gases are all monatomic. The outer shell of each of the atoms (unless strongly excited by radiation, electron bombardment, or other disturbing effects) is completely filled with electrons, and the gases are generally chemically inert. However, since the outer electrons in xenon are relatively far from the nucleus, they are held rather loosely; and they are capable of interacting with the outer electrons of fluorine atoms and the atoms of some other elements to form fairly stable compounds, such as XeF_2 and XeF_4. Radon can presumably form compounds similar to those of xenon. A fluoride of krypton which is stable at $-8°C$ has been prepared. No compounds of any of the lighter noble gases have been obtained. *See* KRYPTON COMPOUNDS; XENON COMPOUNDS.

In an electric discharge, as in a mass spectrometer, very short-lived ions and molecules of all the noble-gas atoms can be formed; examples are Ar_2^+, HgHe, HgNe, HgAr, $(ArKr)^+$, and $(NeNe)^+$. In the presence of methane, $(XeCH_3)^+$ and other highly unstable ions can be formed.

All the inert gases except helium and radon are produced in concentrated form by the liquefaction and distillation of air, followed by special purification processes. Helium is obtained from certain natural gases containing 0.5% or more He. Radon is obtained by collecting the gas, called radium emanation, given off in the radioactive decay of radium. *See* ARGON; ATMOSPHERIC GASES, PRODUCTION OF; HELIUM; KRYPTON; NEON; PERIODIC TABLE; RADON; XENON.   [A. W. FRANCIS]

Bibliography: V. G. Fastovskii et al., *Inert Gases*, Moscow, 1964 (English translation available from U.S. Department of Commerce, Clearinghouse for Federal Scientific and Technical Information).

Inertia

That property of matter which manifests itself as a resistance to any change in the motion of a body. Thus when no external force is acting, a body at rest remains at rest and a body in motion continues moving in a straight line with a uniform speed (Newton's first law of motion). The mass of a body is a measure of its inertia. *See* MASS.

[LEO NEDELSKY]

Inertia of energy

The principle of inertia of energy states that the inertial properties of matter determine, and are determined by, its total energy content. If E is the total energy content, and m_0 the rest mass of a piece of matter, c being the speed of light, the mass-energy relation is $E = m_0c^2$. This formula was proposed on general grounds by H. Poincaré in 1900 and was deduced from the special theory of relativity by Albert Einstein in 1905.

If the mass of a body changes by an amount Δm, the corresponding energy change is $\Delta E = c^2 \cdot \Delta m$. The existence of disintegration processes in atomic nuclei has made it possible to test this formula with great accuracy. Energy released from nuclear reactions provides the power source in nuclear reactors, as well as the principal energy source in the Sun and other stars. The radiation emitted from the Sun is equivalent to a loss of mass of about 4,000,000 tons/sec.

The statement that the rest mass of matter is determined by its total energy content is not susceptible of a simple test since there exists no independent measure of the latter quantity. The validity of the general principle was adopted by Einstein as a cornerstone of his theory of gravitation. According to this theory, the gravitational properties of matter are determined by the distribution of energy in the universe. Matter and energy are used as interchangeable terms, all forms of energy being subject to gravitational action. Physical predictions made by Einstein on the basis of this principle are (1) light passing near a star should be deviated by the gravitational field of the star, and (2) light emitted from a massive star should lose energy in escaping from the star and consequently should appear to be slightly reddened with respect to a terrestrial source (gravitational red shift). Both these effects have been subjected repeatedly to physical test; results favor Einstein's predictions. *See* EINSTEIN SHIFT; MASS; RELATIVITY.

[E. L. HILL]

Inertial guidance system

A self-contained system which can automatically determine the position, velocity, and attitude of a moving vehicle for the purpose of directing its future course. Based on prior knowledge of time, gravitational field, initial position, initial velocity, and initial orientation relative to a known reference frame, an inertial guidance system is capable of determining its present position, velocity, and

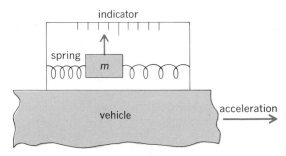

Fig. 1. Elementary accelerometer.

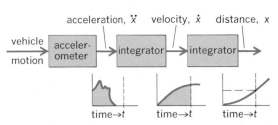

Fig. 2. Computation of distance by two integrations. Integration of acceleration gives velocity; integration of velocity gives distance.

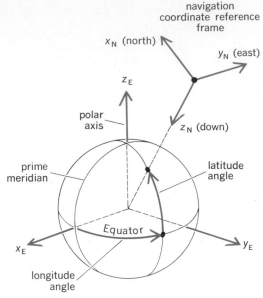

(a)

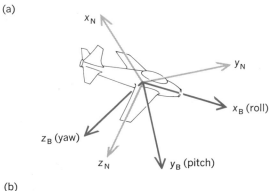

(b)

Fig. 3.   Reference frames for navigation. (a) Earth reference coordinate frame. (b) Body (vehicle) reference coordinate frame.

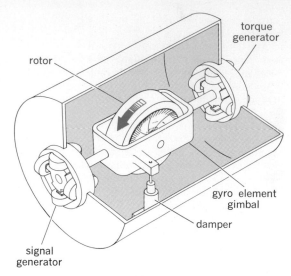

Fig. 4.   A single-degree-of-freedom gyroscope.

orientation without the aid of external information. The generated navigational data is used to determine the future course for a vehicle to follow in order to bring it to its destination. Such systems have found application in the guidance and control of submarines, ships, aircraft, missiles, and spacecraft. *See* NAVIGATION SYSTEMS, ELECTRONIC.

Principles of operation. An inertial navigator determines position by measuring the acceleration of a vehicle with accelerometers. Gyros are used to establish a reference frame in order to specify the orientation of the accelerometers. A computer performs one integration of the acceleration to determine velocity and a second integration of the velocity to determine position. Consequently, the total system consists of accelerometers, gyros, and a computer. A simple accelerometer is shown in Fig. 1. If the unit is kept level, then the displacement of the indicator is a measure of the vehicle acceleration; if the unit is not kept level, then the value of local gravity must be known and subtracted from the measurement to determine the acceleration. Figure 2 shows how distance is computed from the acceleration by two integrations. In order to determine position, the angular orientation of the accelerometers must be known relative to a reference frame, examples of which are shown in Fig. 3.

A perfect gyro maintains a fixed direction in space when no torque is applied to it and moves with a calibrated precession rate relative to space when a torque is applied. Figure 4 shows a gyro which controls angular orientation about a single axis. Three of these gyros are mounted on a platform, as shown in Fig. 5. Geometric storage of the reference frame is accomplished by initially aligning the platform with the reference frame and then commanding it to move with the desired angular velocity relative to space. The command physically results in a calibrated torque being applied to the torque motors of the gyros mounted on the platform. As a result of the torque, the gyro spin axes establish a new inertial reference with respect to space. A servo follow-up system acting through the motors M_1, M_2, and M_3 forces the platform to align with the spin axes of the gyros.

A typically used reference frame is the navigational frame defined by the local directions of north, east, and down. The navigational frame rotates with respect to space because of Earth's daily rotation and the motion of the vehicle over the Earth. The angular velocity due to these rotations depends only on the position and velocity of the vehicle relative to the Earth, both of which are computed by the system. Accelerometers are placed on the level platform with their sensitive axes in the north and east directions. Since the accelerometer axes remain perpendicular to gravity, their outputs are directly proportional to northerly and easterly accelerations. One integration produces the northerly and easterly components of velocity from which the latitude and longitude may be determined. The attitude of the vehicle is measured by the angles ϕ, θ, and ψ formed by adjacent gimbals. A photograph of an actual system is shown in Fig. 6. *See* GYROSCOPE.

Some newer inertial systems are being designed to eliminate the complex gimbal structure. In these systems the reference frame is stored mathematically by computing the angular orientation of the desired reference frame relative to the vehicle itself. This method allows the instruments to be mounted directly on the vehicle (instead of on a gimballed platform), giving easier access to individual components. This results in a larger com-

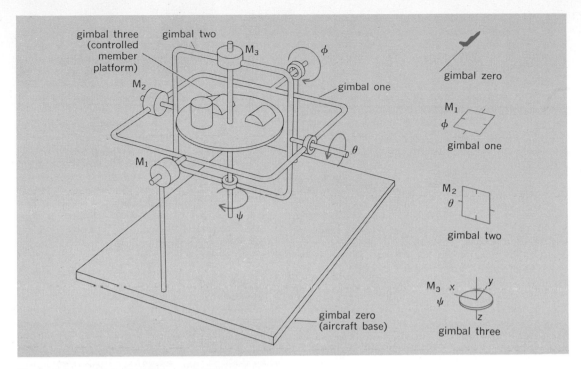

Fig. 5. Schematic of a three-gimbal platform.

Fig. 6. A typical inertial guidance system.

puter, a requirement for more rugged instruments, and some loss in accuracy.

Practical aspects. The greatest utility of inertial guidance comes from the fact that it is self-contained. This makes it useful for any kind of vehicle. It also gives continuous information, which makes it an excellent smoothing agent for other navigational data. The accuracy of an inertial system without external aids is determined primarily by the drift of the gyros, and the production of these precision instruments requires great technology.

One of the most important problems of construction and assembly of gyros is the absolute elimination of dirt. A small piece of dust weighing one-thousandth of a gram can introduce a drift rate equivalent to 1 mph in an inertial-grade gyro by shifting from one side of the instrument to the other. Consequently, gyros are assembled in "clean rooms" in which the air is filtered and hospital-operating-room cleanliness is maintained. Prior to final assembly, the gyros are inspected with a microscope for dirt. Better gyros can be built, but increased accuracy brings an increase in cost. Continuous efforts are being made to improve the performance of gyros, as well as to achieve present performance with smaller, lighter, less expensive units.   [WALTER M. HOLLISTER]

Bibliography: C. Broxmeyer, *Inertial Navigation Systems,* 1964; C. F. O'Donnell (ed.), *Inertial Navigation Analysis and Design,* 1964; G. R. Pitman, *Inertial Guidance,* 1962; J. M. Slater, *Inertial Guidance Sensors,* 1964; W. Wrigley et al., *Gyroscopic Instruments,* 1969.

Infant diarrhea

An important symptom purposely discussed as an acute specific disease in medical literature. Babies commonly exhibit diarrhea, but extensive fluid and blood chemical loss can be fatal. Treatment stresses control of these losses with replacement while the cause of the diarrhea is sought. Stools are numerous or watery, are green, and contain mucus. Blood or pus suggests specific enteritis such as bacterial or amebic dysentery. Diarrhea often denotes infection. There is evidence that a virus is also causative and that normal intestinal bacteria can become pathogenic or disease-producing under certain conditions. Other reasons include food sensitivity or allergy, and food poisoning such as by contaminated milk. Epidemic diarrhea of the newborn is the descriptive term for a serious situation, the malady which is widespread in hospital nurseries. *See* AMEBIASIS; BACILLARY DYSENTERY; BACTERIOLOGY, MEDICAL; FOOD POISONING, BACTERIAL; VIRULENCE; VIRUS.

[PAUL L. BOISVERT]

Infarction

A process of vascular obstruction resulting in a region of dead tissue (necrosis) called an infarct. The usual cause is occlusion of an artery or vein by a thrombus or embolus and sometimes by severe arteriosclerosis; the development of the infarct

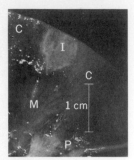

Fig. 1. Magnified gross photograph of a wedge-shaped infarct of kidney (I). Renal cortex (C), medullary pyramid (M), and renal pelvis (P) can be seen.

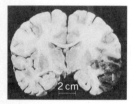

Fig. 2. Gross photograph of a brain section, showing a wedge-shaped infarct (I) that resulted from a thrombus. Secondary hemorrhage has occurred.

depends to a great extent on the collateral circulation. If the anastomotic blood supply is inadequate, or if the vessel that is obstructed by the thrombus is the sole source of blood supply to the region, an infarct results.

Following an initial period of hyperemia, the infarcted region becomes pale. Because of the branching arrangement of the vascular system, infarcts of the spleen, kidney, and lung are usually wedge-shaped, with the base of the wedge at the periphery and the apex toward the point of vascular obstruction (Fig. 1). *See* EMBOLUS; THROMBOSIS.

With occlusion of an artery the existing collateral circulation attempts to bring more blood to the region. Depending on the extent of the process, the infarcted area will be red or pale. This process varies with the organ involved. Those of the kidney and heart are pale, whereas those of the spleen and lung are usually red. With death of the tissue in the region and dilatation of the nearby vessels a secondary hemorrhage may occur (Fig. 2).

The tissue in the affected area undergoes coagulative necrosis and dies while the tissue around the infarcted region becomes inflamed (Fig. 3). Soon fibroblastic activity and organization begin and eventually the dead tissue is replaced by a shrunken depressed scar. *See* INFLAMMATION.

Infarcts commonly occur in the lungs, heart, brain, spleen, and kidneys. The common cause of infarcts of the heart is thrombosis of the coronary artery, usually secondary to arteriosclerosis. In addition, hemorrhage into an arteriosclerotic plaque can result in occlusion of the lumen.

Myocardial infarction is usually accompanied by severe prolonged substernal oppression or pain, shock, electrocardiographic changes, fever, leukocytosis, increased transaminase activity of the

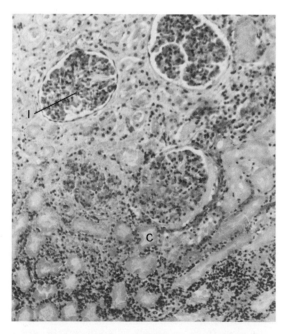

Fig. 3. Low-power view of the edge of a renal infarct. The remnant of a glomerulus and the ghosts of the dead renal tubules can still be seen in the infarcted region (I). On the left several relatively better-preserved glomeruli and tubules can also be seen. A zone of inflammatory cells (C) is seen separating the dead and relatively better-preserved tissue.

serum, and increased sedimentation rate. Myocardial infarcts usually involve the left ventricle or interventricular septum and only rarely involve the right ventricle or atria. The location of the infarct usually depends on the coronary vessel occluded. If tissue death does not occur immediately the infarct heals and is replaced by scar tissue in 5–8 weeks, depending on its size.

Complications of myocardial infarction include rupture of the necrotic portion of the heart with hemorrhage into the pericardium and rapid death; mural thrombi which form on the endocardial surface of the healing infarct and which may give rise to embolism; congestive heart failure as a result of an insufficient amount of remaining healthy muscle to carry on the work of the heart (occurring most commonly with repeated infarcts); and disturbances in heart rhythm if the pacemaker or conduction system is involved.

Infarction of a portion of bowel will result in death of the individual unless surgical intervention is forthcoming. As might be assumed because of the frequency of thrombi in the veins and right auricle, embolization to the lungs is a rather frequent occurrence. However, because of the collateral blood supply of the lungs, infarction follows only when there is some interference with the circulation, such as chronic pulmonary venous congestion. An extensive collateral circulation also exists in the liver, hence the rarity of infarcts in this organ. *See* CIRCULATION DISORDERS.

[ROMEO A. VIDONE]

Bibliography: W. A. D. Anderson (ed.), *Pathology*, 5th ed., 1966; W. Boyd, *Pathology*, 7th ed., 1961; M. Kligerman and R. A. Vidone, Intestinal infarction in the absence of occlusive mesenteric vascular disease, *Amer. J. Cardiol.*, 16:562–570, 1965; S. L. Robbins, *Pathology*, 3d ed., 1967; R. A. Vidone and A. A. Liebow, Anatomical and functional studies of the lung deprived of pulmonary arteries and veins, with an application in the therapy of transportation of the great vessels, *Amer. J. Pathol.*, 33:539–571, 1957.

Infection

A term considered by some to mean the entrance, growth, and multiplication of a microorganism (pathogen) in the body of a host, resulting in the establishment of a disease process. Others define infection as the presence of microorganism in host tissues whether or not it evolves into detectable pathologic effects. The host may be a bacterium, plant, animal, or human being, and the infecting agent may be viral, rickettsial, bacterial, fungal, or protozoan.

A differentiation is made between infection and infestation. Infestation is the invasion of a host by higher organisms such as parasitic worms. *See* BACTERIOLOGY, MEDICAL; EPIDEMIOLOGY; INFECTION, LYTIC; MYCOLOGY, MEDICAL; PARASITOLOGY, MEDICAL; PATHOGEN; VIRUS.

[DANIEL N. LAPEDES]

Infection, lytic

Infection of a bacterium by a bacteriophage with subsequent production of more phage particles and lysis, or dissolution, of the cell. The viruses reponsible are commonly called virulent phages. Lytic infection is one of the two major bacteriophagebacterium relationships, the other being lysogenic infection. *See* BACTERIOPHAGE; COLIPHAGE; LYSOGENY.

[PHILIP B. COWLES]

Infectious mononucleosis

A disease of children and young adults, characterized by fever and enlarged lymph nodes and spleen. Only in the mid-1970s did accumulated evidence permit clear linkage of this disease to EB (Epstein-Barr) herpesvirus as the causative agent, and there is still much to be learned about the manner in which infection with this virus produces the illness.

Onset of the disease is slow and nonspecific with variable fever and malaise; later, cervical lymph nodes enlarge, and in about 50% of cases the spleen also becomes enlarged. The disease lasts 4–20 days or longer. Epidemics are common in institutions where young people live. EB virus infections occurring in early childhood are usually asymptomatic. In later childhood and adolescence, the disease more often accompanies infection— although even at these ages inapparent infections are common.

Seroepidemiologic studies indicate that infection with EB virus is common in different parts of the world. In some areas, including the urban United States, about 50% of children 1 year old, 80–90% of children over age 4, and 90% of adults have antibody to EB virus. The mechanism of virus transmission is unknown, but it could take place through oropharyngeal excretion, which has been found to be very common, especially in those with overt disease, and can occur in the absence of symptoms and be prolonged.

For diagnosis, the heterophile antibody test is useful. This test is based on a nonspecific serologic reaction present at high levels in patients with infectious mononucleosis. Total white blood cell count and differential blood count are also useful in diagnosis. *See* CLINICAL PATHOLOGY; HETEROPHILE ANTIGEN.

The nature of the immune response, which terminates the disease and the shedding of virus, is not fully understood; the virus appears to persist within the host at a location inaccessible to neutralizing antibody. Unlike the short-lived heterophile antibodies, neutralizing antibodies against EB virus persist for years.

The presence of EB virus, an antigenically distinct member of the herpes group, was initially detected by means of electron microscopy in a small proportion of cells in continuous lymphoblastoid cell lines derived from Burkitt's lymphoma, a tumor indigenous to children in central Africa. EB virus has been detected in peripheral blood leukocytes of patients with infectious mononucleosis, leukocytes from normal individuals, and also in lymphoblastoid cell lines derived from nasopharyngeal carcinomas. A causative role of the virus in the cancers has not been unequivocally shown, and it is possible that the virus is merely a "passenger" that grows well in the type of cell present in the neoplasm. *See* EPSTEIN-BARR VIRUS.

[JOSEPH L. MELNICK]

Bibliography: A. S. Evans and J. C. Niederman, Infectious mononucleosis, in A. S. Evans (ed.), *Epidemiology of Viral Infections*, 1975; G. Miller, Epstein-Barr herpesvirus and infectious mononucleosis, *Progr. Med. Virol.*, 20:84–112, 1975.

Infectious-disease control

Since infection depends on factors in the host, the parasite, and the environment, there are various means by which infectious-disease control may be brought about. Immunization strengthens the defenses of the host; chemoprophylaxis is directed at the parasite; and quarantine and sanitary measures remove the parasite from the environment of susceptible people.

Immunization has been successful in controlling smallpox, diphtheria, tetanus, whooping cough, and poliomyelitis. It protects the individuals immunized, and when the number so protected has reached a high enough level, it may prevent the outbreak of an epidemic. Chemotherapy is now used successfully in the treatment of many infectious diseases and is therefore effective in controlling the secondary spread of the diseases. Chemoprophylaxis acts at an earlier stage; it may be defined as the use of drugs to prevent the development of infectious diseases. Persons exposed to tuberculosis, for instance, may take isoniazid with the objective of preventing the tubercle bacilli from becoming established. *See* CHEMOTHERAPY; ISONICOTINIC ACID HYDRAZIDE.

Quarantine may be defined as such a limitation of freedom of movement of persons who have been exposed to a communicable disease that, for a period of time equal to the incubation period of the disease, they are unable to make effective contact with persons who have not been exposed. The term may also be used for less extreme degrees of isolation. Quarantine had been effective in countries such as Australia which are distant from other centers of population, so that people who arrived by sea had the opportunity to develop, during the voyage, any infectious disease they may have been incubating. Examination at the quarantine port was then effective. Quarantine is less useful now that air travel has become common.

The basic importance of sanitary measures in the control of infectious diseases is due to the fact that certain microorganisms that infect man are excreted in the feces by patients or carriers; if sewage disposal is inadequate, contamination of the water supply is likely. Preventive measures are concentrated both on the safe disposal of sewage and on the provision of clean water. The diseases that may be controlled by these measures include typhoid fever, dysentery, hookworm, and infectious hepatitis. *See* EPIDEMIOLOGY; SEWAGE DISPOSAL.

[COLIN WHITE]

Infectious-disease transmission

The transmission of infectious disease is accomplished in several ways.

1. There may be direct contact with an infected person. The venereal diseases are almost invariably, and measles is commonly, spread in this way. *See* GONORRHEA; LYMPHOGRANULOMA VENEREUM; MEASLES; SYPHILIS; YAWS.

2. Certain diseases are spread by direct contact from animals to man, an example being undulant fever, which may be transmitted directly from cattle, goats, or swine. Animal ringworm is another

disease in this class. *See* BRUCELLOSIS; ZOONOSES.

3. Milk-borne infectious diseases include typhoid fever, scarlet fever, summer diarrhea, and bovine tuberculosis. The cow infects the milk with bovine tubercle bacilli, but in the other diseases listed the infection is from human sources. The milk responsible is almost always raw, since pasteurization, when properly carried out, can be relied upon to kill the pathogenic microorganisms. *See* MILK; PASTEURIZATION; SCARLET FEVER; TUBERCULOSIS.

4. In the past there have been many serious epidemics of water-borne diseases. They included epidemics of typhoid fever, cholera, bacillary dysentery, and amebic dysentery. As a result of chlorination these infections have been eliminated from the water supplies of large communities in many parts of the world. *See* WATER-BORNE DISEASE.

5. Gastroenteritis due to the *Salmonella* group of organisms is a food-borne disease in certain instances. Pork and poultry are particularly liable to be contaminated, and duck eggs have been incriminated on several occasions. Amebic dysentery may be transmitted by contaminated vegetables. *See* SALMONELLA.

6. A nosocomial infection is one that is spread in hospitals. An urgent problem is the spread of staphylococci, many of them antibiotic-resistant, in hospitals. *See* STAPHYLOCOCCUS.

7. Articles which become contaminated by contact with a patient are known as fomites, and these in turn may cause fomite-borne infection. Among children there is considerable "traffic in saliva" by means of toys; other contaminated objects which serve as a means of indirect contact between a case and a susceptible are soiled clothing, bedding, and dressings. Diseases transmitted in this way include streptococcal infections and diphtheria. *See* DIPHTHERIA; STREPTOCOCCUS.

8. Insect vectors are the most important means of spreading certain diseases such as yellow fever, sleeping sickness, plague, and typhus. The insects involved are, respectively, mosquitoes, tsetse flies, fleas, and lice. These insects are infected, but the housefly may cause disease by acting as a passive carrier of germs acquired from infected material such as feces. *See* PLAGUE; SLEEPING SICKNESS, AFRICAN; TYPHUS FEVER, EPIDEMIC (LOUSE-BORNE); TYPHUS FEVER, MURINE (FLEA-BORNE).

Host-parasite relationship. The transmission of disease may also be studied as a problem in host-parasite relationship. There is a dynamic balance between the efforts of the parasite to multiply and the defense of the host against infection. Environmental factors such as temperature and humidity may influence the equilibrium between the two protagonists; therefore, three sets of factors associated respectively with the parasite, the host, and the environment are considered.

Bacterial population. The bacterial population, or other microparasite population, may be fairly constant in its biological effects on the host. The virus of measles is in this category. The viruses of influenza, however, are in marked contrast, since they exhibit a large number of strains which vary in their structure and their effects on the host; a vaccine which is protective in one influenza epidemic may be only partially protective, or not

at all protective, in another. The virus of Asian influenza, for example, was unlike any influenza virus that had been encountered, though there was some evidence that older people possessed antibodies that indicated they had encountered the same virus previously. *See* INFLUENZA.

Host population. The genesis of an epidemic depends also on factors in the host population. The age distribution is important because many microorganisms are selective in the age groups they attack: meningococcal meningitis is more likely to attack younger age groups, and there have therefore been several epidemics in army camps in which numerous young recruits were living. The factor of crowding is also important in this case, and its influence may be seen again in the high rate for certain diseases in slums, overcrowded jails, and mental hospitals.

Environmental factors. Numerous environmental factors act on both the bacterial population and the host population. Provision of clean water and proper sewage disposal have been the chief factors leading to a reduction of the enteric diseases in many communities. As a result of so-called sanitary reform, typhoid fever is now virtually unknown in modern cities. *See* SEWAGE DISPOSAL.

Herd immunity. When a host is infected, he may react successfully to the parasite and in so doing produce antibodies that make him immune for some time from further attacks of the parasite. As well as considering the immunity of the individual one may also study herd immunity, that is, the immunity status of the population as a whole. If the immunity acquired to a particular parasite is lasting, the susceptibles will tend to be concentrated in the younger age group and an epidemic will mostly involve this section of the population. The common childhood infections such as measles and chicken pox illustrate this point. On the other hand, the immunity to influenza is short-lived and this disease attacks all age groups. *See* IMMUNITY.

Threshold density. If the proportion of susceptibles is below a certain critical value known as the threshold density, an infection introduced into the group will die out. This is an important factor in the cyclical behavior of epidemics. During the endemic period the density of susceptibles gradually increases and the conditions become appropriate for a further outbreak. *See* EPIDEMIC.

Nonhuman hosts. So far the host under consideration has been human, but there are hosts other than man which may, under natural conditions, harbor an infectious agent pathogenic for man. An epidemic of plague in man is always preceded by a corresponding outbreak in susceptible rodents, and when the infection rate becomes high in this host the disease spills over to man. *See* EPIDEMIOLOGY.   [COLIN WHITE]

Infective dose 50

The dose of microorganisms required to cause infection in 50% of the experimental animals; it is an important special case of the median effective dose and is also known as ID_{50}, or the median infective dose. In former times it was the practice to try to measure the least amount of infective agent that was needed to produce a definite infection, but this proved to be an extremely variable quantity. The median infective dose is a much more reproducible measure, though it is still necessary to specify in detail certain conditions to be observed during measurement. For example, the median infective dose varies widely according to the route of administration; a dose that is ineffective by mouth may cause a high rate of infection when given intravenously. *See* EFFECTIVE DOSE 50. [COLIN WHITE]

Infinity

The terms infinity and infinite have a variety of related meanings in mathematics. The adjective finite means "having an end," so infinity may be used to refer to something having no end. In order to give a precise definition, the mathematical domain of discourse must be specified.

Infinity in sets. A simple and basic example of an infinite collection is the class of natural numbers, or positive integers. A fundamental property of positive integers is that after each integer there follows a next one, so that there is no last integer. Now it is necessary in mathematics to treat the collection of all positive integers as an entity, and this entity is the simplest infinity, or infinite collection.

Suppose that two collections A and B of objects can be set into one-to-one correspondence; that is, each object in A is paired with one and only one object in B, and each object in B belongs to just one of these pairs. Then A and B are said to contain "the same number" of objects, or to have "the same cardinal number." This is equivalent to the ordinary meaning of "the same number," obtained by counting, when the collections A and B are finite, and is taken as the definition of this phrase in the general case.

Now let A consist of all the positive integers, and let B consist of the even integers. Then A and B have the same number of elements, since each element of A can be paired with its double in B; thus elements can be removed from an infinite collection without reducing the number of its elements. This property distinguishes infinite collections from finite ones, and indicates that the statement "the number of elements of a collection A is greater than the number of elements of a collection B" should be taken to mean "B can be put into one-to-one correspondence with part of A, but not with the whole of A."

Consider next the number of subsets of a given collection A. For example, let $A = \{a,b,c\}$; that is, A consists of the three letters a, b, and c. Then the sets $\{a\}$, $\{b\}$, $\{c\}$, $\{a,b\}$, $\{a,c\}$, $\{b,c\}$ are subsets of A. To these, for convenience, can be adjoined the null set, containing no elements, and the set A itself. This makes a total of $8 = 2^3$ subsets. It is readily proved that a set A having a finite number n of elements has 2^n subsets. Thus the number of subsets of a finite set is always greater than the number of elements. By a different method, the same property may be proved to hold also for infinite sets. Hence for every infinite number there is a greater number. In other words, there exist infinitely many distinct infinities. Sets having the same number of elements as the collection of positive integers are called countably infinite, and this is the smallest infinity.

Infinity in limits of functions. The term infinity appears in mathematics in a different sense in connection with limits of functions. For example, consider the function defined by $y = 1/x$. When x tends to 0, y approaches infinity, and the expression may be written as shown below.

$$\lim_{x \to 0} y = \infty$$

Precisely, this means that for an arbitrary number $a > 0$, there exists a number $b > 0$ such that when $0 < x < b$, then $y > a$, and when $-b < x < 0$, then $y < -a$. This example indicates that it is sometimes useful to distinguish $+\infty$ and $-\infty$. The points $+\infty$ and $-\infty$ are pictured at the two ends of the y axis, a line which has no ends in the proper sense of euclidean geometry.

Infinity in geometry. In geometry of two or more dimensions, it is sometimes said that two parallel lines meet at infinity. This leads to the conception of just one point at infinity on each set of parallel lines and of a line at infinity on each set of parallel planes. With such agreements, parts of euclidean geometry can be discussed in the terms of projective geometry. For example, one may speak of the asymptotes of a hyperbola as being tangent to the hyperbola at infinity. Note that the points at infinity which are adjoined to a euclidean line or plane are chosen in a manner dictated by convenience for the theory being discussed. Thus only one point at infinity is adjoined to the complex plane used for geometric representation in connection with the theory of functions of a complex variable.

Other concepts. Other types of infinities may be distinguished when properties other than the mere cardinal number of a set are being considered. For example, a line and a plane contain the same number of points, but when continuity considerations are important the line is said to be made up of a single infinity of points, whereas the plane has a double infinity, or ∞^2, points. This is because there exists a one-to-one continuous correspondence between the line and a subset of the plane, but there does not exist such a correspondence between the whole plane and the line or any part of it. As another example, an infinite set may be ordered in different ways so as to have different ordinal numbers. *See* CALCULUS, DIFFERENTIAL AND INTEGRAL. [LAWRENCE M. GRAVES]

Bibliography: L. M. Graves, *Theory of Functions of Real Variables*, 2d ed., 1956; N. Rudin, *Principles of Mathematical Analysis*, 2d ed., 1964.

Inflammation

The local response to injury, involving small blood vessels, the cells circulating within these vessels, and nearby connective tissue.

The early phases of the inflammatory response are stereotyped: A similar sequence of events occurs in a variety of tissue sites in response to a diversity of injuries. The response characteristically begins with hyperemia, edema, and margination of the circulating white blood cells. The white cells then migrate between the endothelial cells of the blood vessel into the tissue. The subsequent development of the inflammatory process is determined by factors such as type and location of injury, condition of the host, and the use of therapeutic agents.

Cardinal signs. The discomfort of inflammation may be attendant on a sunburn, a mosquito bite, a cut, an abscess, or a vaccination; so it is not surprising that historical records of inflammation date to the earliest known medical writings in the Egyptian papyri. It remained, however, for Celsus (about 30 B.C.–A.D. 38), a Roman, to enumerate the four cardinal signs of inflammation: redness, swelling, heat, and pain. A fifth cardinal sign, loss of function, was contributed by Galen. Redness is caused by hyperemia, that is, increased red blood cells (erythrocytes) in the local capillary bed. Heat results from increased flow of blood in the small blood vessels, while the edema, or swelling, represents the accumulation of extracellular fluid in the connective tissue. Stimulation of nerve endings by agents released during the inflammatory process causes pain. Loss of function is a consequence of the changes. See EDEMA.

While it is possible to observe the events of inflammation in human skin, a detailed study of the dynamic cellular events requires a more convenient site in an experimental animal. Tissues such as rat mesentery, frog foot web, or frog tongue have the virtues of accessibility, vascularity, and transparency necessary for microscopic study of the inflammatory process.

Vascular changes. Following a mild injury, there is a fleeting constriction of the smallest arterial branches (arterioles) in the viable tissue close to the injury, lasting from 5 sec to about 5 min, followed by dilation of the same arterioles. This leads to engorgement and dilation of the capillaries, a change that extends over an area larger than the constriction site. Constriction of the smallest veins (venules), a second component contributing to capillary engorgement, is more closely localized in the central area around the injury.

Along with these blood-vessel changes, there is a departure from the steady-state flow of water back and forth across the vessel walls; when the outflow of water exceeds the return, extravascular extracellular fluid collects. An excess of fluid in the extracellular compartment of the connective tissue is termed edema. The vessel walls also become more permeable to the large protein molecules that circulate in the plasma; these molecules then leak out into the tissue. Blood flow is slowed in the immediate vicinity of the injury by constriction of the venules and concentration of the blood cells. The latter phenomenon is a result of loss of water from the circulation. Actual cessation of blood flow (stasis) may occur, which may be associated with formation of a blood clot in the small vessels (thrombosis), or may be only transitory with prompt restoration of flow.

Margination of leukocytes out of the flowing bloodstream to the vessel wall commonly occurs during slowing of flow but may also occur with normal or even increased flow. The leukocytes, predominantly neutrophilic granulocytes, stick to the endothelial cells that line the capillaries and venules, and then migrate through the vessel wall into the edematous connective tissue, frequently drawn there by specific substances, chemotactic agents, released at the site of the injury.

If the tissue injury breaks small blood vessels, another mechanism comes into play. Exposure of circulating blood to collagen leads to clumping of platelets, and the resulting aggregate forms a temporary hemostatic plug preventing free bleeding from the broken vessel. The clotting mechanism is set in action somewhat more slowly and the fibrin formed thereby bolsters the plug. See BLOOD; COLLAGEN.

Cellular changes. In contrast to the vascular response, the cellular response in inflammation is varied and serves to characterize the different types of inflammation. Participating cells come from two sources, the circulating blood and the local connective tissue. The circulating leukocytes are divided into six distinguishable cell types: granulocytes (eosinophil, neutrophil, and basophil), small and large lymphocytes, and monocytes. Fibroblasts and mast cells are solely connective-tissue cells. Macrophages and giant cells arise in the inflammatory locus, probably from circulating monocytes.

Neutrophils are the most abundant leukocyte in the circulation and are the first cells to accumulate in an inflammatory site. They have the capacity to ingest and kill bacteria; the cytoplasmic granules, which identify the neutrophil, contain the enzymes responsible for killing and digesting the bacteria. When the number of circulating neutrophils is greatly decreased, as in patients after treatment with certain drugs or exposure to nuclear radiation, frequent and severe infections may occur. In spite of antibiotic therapy, such patients often succumb to infection. See PHAGOCYTOSIS.

Eosinophils are much rarer than neutrophils but are also phagocytic and may be considerably increased in patients with allergies. The basophil, least abundant of the granulocytes, is similar to the tissue mast cell; both store and release histamine and are thus responsible for hives and the symptoms of hay fever. Lymphocytes are second to neutrophils in abundance and are very important in immune responses, particularly in the rejection of grafts of foreign tissues. Monocytes, like lymphocytes, lack specific granules. This cell type when stimulated has the potential to change into a macrophage, which, like a neutrophil, can ingest and kill bacteria. Fibroblasts are responsible for synthesis of collagen and other components of the extracellular connective tissue during the healing process. See HISTAMINE.

Cause-effect relationship. An inflammatory response may be induced in a great variety of ways. Causes of inflammation includes trauma, heat, ultraviolet light, x-rays, bacteria, viruses, many chemicals such as turpentine and croton oil, and certain inherently innocuous but foreign substances (antigens) which evoke immune responses. Although many of the components of inflammation are common to a variety of inflammatory responses, particularly in the early stages, there is considerable diversity in fully developed responses; an abscess is very different from a burn, for instance. The character of the injury, its severity, and the site of injury each modifies the progress of the inflammatory response, as does therapeutic intervention. See ANTIGEN; IMMUNITY.

A local inflammatory response is usually accompanied by systemic changes: fever, malaise, and an increase in circulating leukocytes (leukocytosis). Such signals and symptoms are often helpful to the physician, first as clues to the presence of inflammation and later as an indication of its course.

Types. Inflammation is frequently described in terms of its time course. Acute inflammation develops rapidly, in a matter of hours to days, and is of relatively short duration. In acute inflammation the neutrophil is the predominant cell type, and hyperemia and edema are prominent. Chronic inflammation develops over a period of weeks to months and is characterized by an infiltrate of lymphocytes, monocytes, and plasma cells—the chief antibody-producing cells. Local differentiation and proliferation of macrophages is characteristic of chronic inflammation.

Granulomatous inflammation is a specific type of chronic inflammation in which there is a discrete nodular lesion (granuloma) formed of macrophages, lymphocytes, plasma cells, and giant cells arranged around a central mass of noncellular material. Granulomas, typical of tuberculosis and fungus infection, also occur in rheumatoid arthritis. *See* ARTHRITIS; GRANULOMA INGUINALE; TUBERCULOSIS.

Abscesses and cellulitis are specific forms of acute inflammation; the former term denotes a localized collection of pus composed of necrotic debris derived from dead tissue cells and neutrophils. Cellulitis is characterized by diffuse hyperemia and edema with an extensive neutrophil infiltrate, but much less tissue destruction. When inflammation causes erosion of an epithelial surface, the lesion is termed an ulcer. *See* ULCER.

Function. Inflammation is basically a protective mechanism. The leakage of water and protein into the injured area brings humoral factors, including antibodies, into the locale and may serve to dilute soluble toxic substances and wash them away. The margination and migration of leukocytes brings these cells to the local site to deal with the injurious agent, dead or alive. There are also instances, however, in which no causative toxic substance or infectious agent can be found to account for the inflammation. This is the case in rheumatoid arthritis and rheumatic fever. Such diseases may be examples in which an uncontrolled or misdirected inflammatory response is turned on the host. *See* IMMUNOPATHOLOGY; INFECTION.

[DAVID LAGUNOFF]

Bibliography: J. G. Adami, *Inflammation: An Introduction to the Study of Pathology*, 1909; R. J. Dubos and J. G. Hirsch (eds.), *Bacterial and Mycotic Infections in Man*, 4th ed., 1965; H. Florey (ed.), *General Pathology*, 3d ed., 1962; B. W. Zweibach, L. Grant, and R. T. McCluskey (eds.), *Inflammatory Process*, 1965.

Inflorescence

A flower cluster, or the arrangement of flowers on a plant. It is a branch system—simple if the main axis is undivided, compound if the axis divides into two or more branches. The stalk of the inflorescence is the peduncle, the axis is the rachis, and the stalks of individual flowers of the cluster are pedicels. Inflorescences may be subtended by a whorl of bracts called the involucre.

Inflorescences are classified as determinate when the terminal bud forms a flower, thereby arresting further growth of the axis, with later flowers developing in succession from the tip toward the base; or as indeterminate when the tip continues to grow, forming new flowers in succes-

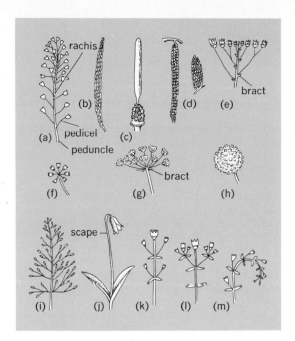

Examples of inflorescence types. (*a*) Raceme. (*b*) Spike. (*c*) Spadix (pistillate). (*d*) Catkins, male and female. (*e*) Corymb. (*f*) Simple umbel. (*g*) Compound umbel. (*h*) Head. (*i*) Panicle. (*j*) Solitary. (*k*) Simple cyme. (*l*) Compound cyme. (*m*) Helicoid cyme.

sion, the oldest at base. Indeterminate types include: raceme (illustration *a*), with pedicels of about equal length on an elongated axis; spike (illustration *b*), similar to raceme but with sessile (without pedicels) flowers; spadix (illustration *c*), a fleshy, thickened spike frequently subtended by a large bract (spathe); catkin (ament), a scaly spike, sometimes pendent (illustration *d*), of unisexual flowers; corymb (illustration *e*), with lower pedicels elongated giving a flat-topped appearance; umbel (illustration *f* and *g*), with pedicels all arising at top of peduncle and radiating like umbrella ribs; head (illustration *h*), a rounded or flattened compact cluster of sessile flowers on a very short axis or receptacle; and panicle (illustration *i*), a branched raceme.

Determinate types include: solitary (illustration *j*), a lone flower on a peduncle; cyme (illustration *k* and *l*), a loose, flat-topped cluster with central flowers opening first; and helicoid cyme (illustration *m*) often referred to as scorpioid, a coiled cluster, with flowers on only one side of axis.

Considerable confusion exists regarding the use of the terms helicoid and scorpioid. As the two types are not readily discriminated, some taxonomists prefer to use the older and commoner term, scorpioid, for all coiled inflorescences. *See* FLOWER; PLANT ORGANS; PLANT TAXONOMY.

[NELLE AMMONS]

Influenza

An acute respiratory viral infection, usually epidemic in occurrence. This disease has been one of the world's greatest killers; over 21 million deaths are attributed to the 1918–1919 pandemic.

Three immunologic types of influenza virus are known. Group A appears to be subject to continual antigenic changes, group B changes to a lesser degree, and group C seems antigenically stable. The infectious particle is about 110 nm in diameter. The innermost component is the helically coiled ribonucleoprotein, 9 nm in diameter, composed of nucleoprotein polypeptides and the viral RNA. The protein shell surrounds the nucleoprotein and forms the inner part of the virus envelope. Outside the protein is lipid, formed into a bilayer structure. The surface of the virus particle is covered with projections or spikes (approximately 10 nm long) which possess either the hemagglutinin or the neuraminidase activity of the virus (see illustration). Viral infectivity may be inactivated by heat, ultraviolet irradiation, or formaldehyde. Chick embryos are the chief laboratory hosts for study and cultivation. The viruses are capable of agglutinating red blood cells, from which they then spontaneously dissociate. Human influenza viruses, together with animal influenza viruses anf fowl plague, constitute the orthomyxovirus group, one subgroup of the myxoviruses. *See* CULTURE, EMBRYONATED EGG; MYXOVIRUS.

Pathology and diagnosis. In human infection, the virus enters the respiratory tract in airborne droplets; within 1–2 days inflammation of the respiratory tract occurs, with fever, chills, and muscular aches. The illness is usually mild, and asymptomatic infections may occur. However, a more serious course or even death may result in older persons and those with chronic debilitating diseases. Bacterial complications may produce more severe illnesses, especially pneumonia. The lethal impact of an influenza epidemic is reflected in an increase in deaths due to cardiovascular and renal disease, as well as those due to pneumonia or to influenza itself.

About 80,000,000 cases occurred during the 1957–1958 pandemic of Asian influenza (A2). The illnesses were generally mild, but the estimated number of pneumonia-influenza deaths in the United States was 60,000 above normal during the pandemic. Older age groups have the lowest incidence of influenza, but the highest case fatality

rate—especially among those who suffer from chronic debilitating diseases.

Diagnosis is by isolation of virus from throat washings inoculated into embryonated eggs, or by detection of hemagglutination-inhibiting, complement-fixing, or neutralizing antibody responses in blood. For rapid detection of influenza virus in clinical specimens, positive smears from nasal swabs and washings may be demonstrated by specific staining with fluorescein-labeled antibody. *See* COMPLEMENT-FIXATION TEST; NEUTRALIZING ANTIBODY.

Influenza occurs in successive epidemic waves and also in pandemics. The antigenic shifts of influenza viruses usually occur gradually, so that infection with strains widespread in one year confers some immunity against those prevalent in the next several years. However, sudden wide antigenic shifts do occur at 10–15-year intervals: the group A virus of the Asian influenza pandemics of 1957–1958 and 1968–1969 showed marked antigenic difference from group A strains of preceding years, and few persons had protecting antibodies. Epidemics may start with transportation of virus by travelers, accumulation of sufficient numbers of susceptibles, and mutation of the virus.

Immunization. Vaccines made up of current strains, grown in embryonated eggs and inactivated by formalin or ultraviolet light, confer immunity to the homologous strains in the vaccine, but only relatively transient resistance to other strains. Since the vaccine is prepared from infected chick embryos, persons hypersensitive to eggs or egg products may experience severe allergic reaction. Vaccination is currently recommended in the United States only for persons not allergic to eggs or egg products who have chronic debilitating conditions and are known to be at increased risk of mortality from influenza (in general, persons of any age with chronic cardiovascular-renal, bronchopulmonary, or metabolic disorders; and elderly persons with incipient chronic disease). For these groups, annual immunization is recommended. If a major epidemic is forecast, appropriately altered vaccine should be administered to the high-risk groups and their families and probably also to selected employment groups which provide essential community services.

Improvement of influenza vaccines has diminished the toxicity. Purified vaccines with less nonviral protein are available and recommended. Another type of vaccine available in the United States is made from subviral antigens; the nucleic acid has been removed, leaving a purified preparation of noninfectious virus proteins.

Attenuated live virus vaccines have been developed and used widely in the Soviet Union, and to a much lesser extent in Great Britain. One of difficulties with the live vaccine is in obtaining a sufficiently high infection rate in those vaccinated. There is also the same problem of antigenic shift that makes for obsolescence of the killed virus vaccines. It had been hoped that laboratory-manipulated influenza virus strains could be developed which would both forecast and prepare for the next step of natural antigenic drift. By growing currently prevalent virus strains in culture in the presence of homologous antiserum, the selection of new mutants might be pushed in the same direc-

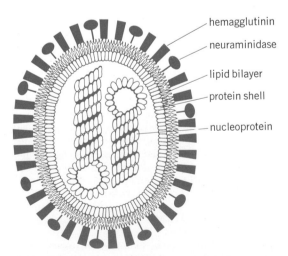

Schematic diagram of the arrangement suggested for the structural components of the influenza virion. (*From R. W. Compans and P. W. Choppin, Reproduction of Myxoviruses, in H. Fraenkel-Conrat and R. R. Wagner, eds., Comprehensive Virology, vol. 4: Reproduction of Large RNA Viruses, 1975*)

hemagglutinin

neuraminidase

lipid bilayer

protein shell

nucleoprotein

tion as future natural mutations. A vaccine has been developed in France through this approach, but it has not proven itself in the field.

Recombinant vaccines. Influenza viruses are known to recombine genetically, forming stable "hybrids" with characteristics from each parent strain. Experimental vaccines have been produced by cocultivation of strains having the properties desired (antigens of new wild variant combined with ability to grow to high yields in eggs) and by selection of the recombinant progeny in which these properties are obtained. The first "man-made" hybrid virus for human immunization, recombinant X-31, has been used in vaccine production in Denmark, Rumania, the United Kingdom, and—on a limited basis—the United States. The X-31 experimental vaccine has been tested, in parallel with a standard A2 vaccine, in volunteers who were subsequently challenged with live virus of a current wild strain and in military populations exposed to natural infections. From the serologic responses and also from the protection conferred against the wild strain, the recombinant vaccine is judged to be as effective as the standard vaccine.

Once a new wild variant is recognized, deliberate "tailoring" in the laboratory can greatly shorten the process of preparing vaccines against the new variant. It may be possible to develop a "library" of recombinant strains of all known antigentic compositions, to be at hand immediately whenever future antigenic shifts appear in the wild virus population. Recombination techniques could also be utilized to select strains for adaptation to temperature sensitivity.

Temperature-sensitive mutants. In efforts to overcome the problems encountered with previous live virus vaccines, temperature-sensitive (*ts*) mutants are being investigated. Selection of *ts* mutants unable to replicate at temperatures of the lung parenchyma (37°C) but replicating well at the lower temperatures of the nasopharyngeal mucosa (32–34°C) may permit adequate immunizing replication of a vaccine virus without severe symptoms of pulmonary pathology. The *ts* property has been associated with loss of virulence. Once a desired *ts* defect is established, recombination should permit this property to be combined with the desired antigenicity needed to combat new wild virus strains.

Antineuraminidase vaccine. The rationale for this vaccine is evidence indicating that antibody against the hemagglutinin is virus-blocking antibody which prevents infection, while antibody against the neuraminidase limits, but does not prevent, viral replication and spread of virus. By means of recombination, a hybrid virus was developed with an "irrelevant" hemagglutinin (from an equine influenza virus, not pathogenic for humans) but with a contemporary neuraminidase. The hybrid has been used to produce a vaccine. Vaccinees who develop antineuraminidase antibody should still be susceptible to infection by circulating wild influenza virus, and receive the benefits of active immunization, while being protected from serious symptoms of the infection. In 1975 this vaccine was ready for small-scale trials. During the current era of minor antigenic shifts, the neuraminidase antigen is quite stable, while the hemagglutinin mutates more often. The contemporary neuraminidase vaccine therefore should be usable for several years. *See* ANIMAL VIRUS; BIOLOGICALS.

[JOSEPH L. MELNICK]

Bibliography: A. S. Beare, Live viruses for immunization against influenza, *Progr. Med. Virol.*, 20:49–83, 1975; W. R. Dowdle et al., Natural history of influenza type A in the United States, 1957–1972, *Progr. Med. Virol.*, 17:91–135, 1974; B. A. Rubin and H. Tint, The development and use of vaccines based on studies of virus substructures, *Progr. Med. Virol.*, 21:144–157, 1975.

Information systems, hospital

The collection, evaluation or verification, storage, and retrieval of information about a patient. As in any information system, three broad areas can be distinguished in a hospital information system: input, transformation through a medium, and output. Although these three functions suffice to describe an existing system, they are insufficient to design a new system; to them must be added the action that is expected to take place on the basis of the information output from the system—thereby defining the purpose of the system—and the feedback from this action to the system input—thereby locating the system in the hospital environment. Crucial to any information system, and especially so to a hospital system, is the accuracy of the input data.

Input. The patient himself provides one set of data input, and all information sources outside the patient can be grouped as a second input. Input information is complex because the patient must be described both at present (cross-sectionally) and back to his childhood and to his parents (longitudinally). Historical information obtainable from the patient is subject to increasingly large error as he goes backward in time. Significant error in the hospital information system can arise if imperfect recall is not considered and appropriately discounted. Historical data from the patient must also be weighted according to psychologic status of the patient. Unfortunately this may easily be done only in retrospect. In addition to obtaining factual replies, a significant purpose of the physician's historical interview is to place all facts in their true light according to the subject's psychologic makeup. Data from outside sources of information are subject to the same errors, hence the value of objective over subjective data from either input source.

The longitudinal history requires substantiated observation, repeated and confirmed laboratory data, and other hard facts. The worth of data, however, is dependent on the validity of the information at the time it was acquired. Storage by itself does not increase the value of the data. It is apparent that good laboratory data obtained during a cross-sectional evaluation of the patient is the best input for longitudinal patient review and for storage for future retrieval. Thus the collection of such hard data to confirm or, better, to supplant current soft data input is to be stressed in every instance. It is important to stress data acquired directly from a subject by a mechanical system in any information system that intends to use engineering technology for processing or storage. This suggests the possibility of dedicating computers and semi-automatic data-gathering systems as basic components to a complete hospital control system. *See* ELECTRODIAGNOSIS; ELECTROENCEPHALOGRAPHY.

In today's framework much that is considered hard data is subject to error through human reinterpretation. Human variability in medical data interpretation is acknowledged to be about 20%. This implies that even the hard output of a laboratory device of known precision and accuracy is softened by the human observer. Consider a computer system reading electrocardiograms with an accuracy of 89% and with the mechanical reliability of the system at 97%. Compounding these two factors rates the performance of the total system at 86%. If the system's printout is reinterpreted by an observer with human variability, the performance of the computer-human system decreases. With 10% human variability the system could be rated at 77%, with 20% at 68%. Reinterpretation is of key importance and suggests that the appropriate role for a human in such circumstances is to concentrate on the part that cannot be computer-read. The example is a simplification but is illustrative of the relative worth of various factors in the input to a hospital information system.

It is evident that with increasing use of engineering and other technological advances, the quantity of soft data currently in use can gradually diminish. For example, one can now obtain a patient's history of a heart attack. Even such a history can be relatively soft data. Hard data would comprise the actual electrocardiogram interpretation by a computer system along with the date of processing. Availability of such minimal information markedly diminishes the requirements for preserving a large soft data file of patient-derived information.

In addition to the patient or data directly derived from him, outside sources should be used to provide the information that goes into the hospital records. Multiple sources come to mind: the family, social service, insurance records, and other hospital or clinic records. In today's system suprisingly little data are easily available from outside the immediate family, although in many instances the data may be harder in the factual sense than the patient's data. Retrieval is currently cumbersome and delayed, so that little benefit is possible to the patient as a result of acquiring it. Regrettably in patient care, a service's value is to a large extent dependent on whether the data are immediately available. Without speed, retrieval can be of little value; hence outside data are little used at present.

Transformation medium. In today's hospital information system input information can go through two different parallel routes. These routes may be mutually exclusive, although they need not be. One is administrative; the other is through the service organization. The former route includes the admitting clerk, administrative officers, pharmacists, and all those who do not have direct contact with the patient; the service route includes the physician, the nurse, and all others who contact the patient directly.

Items are handled differently and to a different degree of detail in each of these two routes. For example, the information that a patient cannot pay his bill or did not pay the previous bill might be considered by the administration in detail to set in motion a social service structure. This information might be of no significance to the nurse or physician except insofar as the psychological character of the subject is pertinent to the diagnosis or illness under consideration. It is passage of hospital information through both these routes, each to its required degree, that allows data to get to the output stage.

It is reasonable to assume that redundancy, as illustrated by parallel information routes in hospitals, can be a source of economic drain. Nonetheless, a degree of redundancy is necessary for the quality of service required. Quality in a hospital implies such features as time links, accuracy, efficiency, and comfort. Redundancy in a hospital is a fail-safe feature. It is vital for all aspects of quality. Redundancy is particularly necessary because the end doer (the physician or nurse) acts with human error or variability. Quality can be obtained either by always presenting hard information or by use of redundancy. Economics may favor redundancy over a constant supply of hard data.

Output. The output hospital information can be viewed in essence as a control system for actions by the physician or nurse. The output of the data and information, after going through the administrative and service organization structures, makes up the patient's hospital records. These are the elements of the control system. The items found in the usual hospital record include, in addition to the history and physical examination data, progress notes, nursing notes, doctor's orders, and required summaries and notations of values from physical laboratory data. *See* MEDICAL CONTROL SYSTEMS.

Generally, items found in the administrative system are not found together with the service records. In reality both represent patient information or input translated through the parallel mediums of the medical personnel and the administration. But the separation indicates the problem involved in collating various sources of information and is illustrative of the essence of the problems that face planners of good hospital information storage and retrieval.

Items of output information are interrelated to establish a diagnosis of a course of action. The feedback from that stage (often interrelations) to the patient can be called patient care.

Hospital information is intended primarily for patient care, which is indeed the primary reason for the existence of a hospital information system. As an ancillary by-product, information from the hospital record is also useful for investigation and research into disease and therapy. The administrative portion of the record is useful in planning for hospital operation to improve the quality of service. Efficiency should primarily be directed to medical areas and secondarily to administrative problems, to allow quality care for most of the population most of the time. This could imply a research by-product such as developing appropriate costing to assure best use of time, space, machines, and personnel. *See* COMMUNICATIONS, ELECTRICAL; DATA COMMUNICATIONS; DATA REDUCTION; DATA-PROCESSING SYSTEMS; MEDICINE; PATHOLOGY.   [CESAR A. CACERES]

Bibliography: E. M. Award, *Automatic Data Processing*, Data Processing Management Association, 1966; W. L. Bennett, A viable computer-based hospital information system, *Hosp. Manage.*, 103:43–47, 1967; C. P. Bourne, *Methods of Information Handling*, 1963; J. M. Buchanan, Automated hospital information systems, *Mil. Med.*, 131:1510–1512, 1968.

Information theory

A branch of communication theory devoted to problems in coding. A unique feature of information theory is its use of a numerical measure of the amount of information gained when the contents of a message are learned. Information theory relies heavily on the mathematical science of probability. For this reason the term information theory is often applied loosely to other probabilistic studies in communication theory, such as signal detection, random noise, and prediction. *See* COMMUNICATIONS, ELECTRICAL.

Information theory provides criteria for comparing different communication systems. The need for comparisons became evident during the 1940s. A large variety of systems had been invented to make better use of existing wires and radio spectrum allocations. In the 1920s the problem of comparing telegraph systems attracted H. Nyquist and R. V. L. Hartley, who provided some of the philosophy behind information theory. In 1948 C. E. Shannon published a precise general theory which formed the basis for subsequent work in information theory. *See* MODULATION; RADIO SPECTRUM ALLOCATIONS.

In information theory, communication systems are compared on the basis of signaling rate. Finding an appropriate definition of signaling rate was itself a problem, which Shannon solved by the use of his measure of information, to be explained later. Of special interest are optimal systems which, for a given set of communication facilities, attain a maximum signaling rate. Optimal systems provide communication-systems designers with useful absolute bounds on obtainable signaling rates. Although optimal systems often use complicated and expensive encoding equipment, they provide insight into the design of fast practical systems.

Communication systems. In designing a one-way communication system from the standpoint of information theory, three parts are considered beyond the control of the system designer: (1) the source, which generates messages at the transmitting end of the system, (2) the destination, which ultimately receives the messages, and (3) the channel, consisting of a transmission medium or device for conveying signals from the source to the destination. Constraints beyond the mere physical properties of the transmission medium influence the designer. For example, in designing a radio system only a given portion of the radio-frequency spectrum may be available to him. His transmitter power may also be limited. If the system is just one link in a larger system which plans to use regenerative repeaters, the designer may be restricted to pulse-transmission schemes. All such conditions are considered part of the description of the channel. The source does not usually produce messages in a form acceptable as input by the channel. The transmitting end of the system contains another device, called an encoder, which prepares the source's messages for input to the channel. Similarly the receiving end of the system will contain a decoder to convert the output of the channel into a form recognizable by the destination. The encoder and decoder are the parts to be designed. In radio systems this design is essentially the choice of a modulator and a detector.

Discrete and continuous cases. A source is called discrete if its messages are sequences of elements (letters) taken from an enumerable set of possibilities (alphabet). Thus sources producing integer data or written English are discrete. Sources which are not discrete are called continuous, for example, speech and music sources. Likewise, channels are classified as discrete or continuous according to the kinds of signals they transmit. Most transmission media (such as transmission lines and radio paths) can provide continuous channels; however, constraints (such as a restriction to use pulse techniques) on the use of these media may convert them into discrete channels.

The treatment of continuous cases is sometimes simplified by noting that a signal of finite bandwidth can be encoded into a discrete sequence of numbers. If the power spectrum of a signal $s(t)$ is confined to the band O to W hertz (cycles per second) then Eq. (1) applies. Equation (1) reconstructs

$$s(t) = \sum_{n=-\infty}^{\infty} s\left(\frac{n}{2W}\right) \frac{\sin 2\pi W\left(t - \frac{n}{2W}\right)}{2\pi W\left(t - \frac{n}{2W}\right)} \qquad (1)$$

$s(t)$ exactly from its sample values (Nyquist samples), at discrete instants $(2W)^{-1}$ sec apart. Thus, a continuous channel which transmits such signals resembles a discrete channel which transmits Nyquist samples drawn from a large finite set of signal levels and at the rate of $2W$ samples per second.

Noiseless and noisy cases. The output of a channel need not agree with its input. For example, a channel might, for secrecy purposes, contain a cryptographic device to scramble the message. Still, if the output of the channel can be computed knowing just the input message, then the channel is called noiseless. If, however, random agents make the output unpredictable even when the input is known, then the channel is called noisy. *See* PRIVACY SYSTEMS (SCRAMBLING).

Encoding and decoding. Many encoders first break the message into a sequence of elementary blocks; next they substitute for each block a representative code, or signal, suitable for input to the channel. Such encoders are called block encoders. For example, telegraph and teletype systems both use block encoders in which the blocks are individual letters. Entire words form the blocks of some commercial cablegram systems. The operation of a block encoder may be described completely by a function or table showing, for each possible block, the code that represents it.

It is generally impossible for a decoder to reconstruct with certainty a message received via a noisy channel. Suitable encoding, however, may make the noise tolerable. For illustration, consider a channel that transmits pulses of two kinds. It is customary to let binary digits 0 and 1 denote the two kinds of pulse. Suppose the source has only the four letters A, B, C, D. One might simply encode each single-letter block into a pair of binary

Three possible binary codes for four-letter alphabet

Letter	Code I	Code II	Code III
A	0 0	0 0 0	0 0 0 0 0
B	0 1	0 1 1	0 0 1 1 1
C	1 0	1 0 1	1 1 0 0 1
D	1 1	1 1 0	1 1 1 1 0

digits (code I of table). In that case the decoder would make a mistake every time noise produced an error. If code II is used, the decoder can at least recognize that a received triple of digits must contain errors if it is one of the triples 001, 010, 100, or 111 not listed in the code II column. Because an error in any one of the three pulses of code II always produces a triple that is not listed, code II provides single-error detection. Similarly code III provides double-error detection, because errors in a single pulse or pair of pulses always produce a quintuple that is not listed.

As an alternative, code III may provide single-error correction, an idea due to R. W. Hamming. In this usage, the decoder picks a letter for which code III agrees with the received quintuple in as many places as possible. If only a single digit is in error, this rule chooses the correct letter.

Even when the channel is noiseless, a variety of encoding schemes exists and there is a problem of picking a good one. Of all encodings of English letters into dots and dashes, the Continental Morse encoding is nearly the fastest possible one. It achieves its speed by associating short codes with the most common letters. A noiseless binary channel (capable of transmitting two kinds of pulse 0, 1, of the same duration) provides the following example. Suppose one had to encode English text for this channel. A simple encoding might just use 27 different five-digit codes to represent word space (denoted by #), A, B, . . . , Z; say # 00000, A 00001, B 00010, C 00011, . . . , Z 11011. The word #CAB would then be encoded into 00000000110000100010. A similar encoding is used in teletype transmission; however, it places a third kind of pulse at the beginning of each code to help the decoder stay in synchronism with the encoder. The five-digit encoding can be improved by assigning four-digit codes 0000, 0001, 0010, 0011, 0100 to the five most common letters #, E, T, A, O. There are 22 quintuples of binary digits which do not begin with any of the five four-digit codes; these may be assigned as codes to the 22 remaining letters. About half the letters of English text are #, E, T, A, or O; therefore the new encoding uses an average of only 4.5 digits per letter of message. *See* TELETYPEWRITER.

More generally, if an alphabet is encoded in single-letter blocks, using $L(i)$ digits for the ith letter, the average number of digits used per letter is shown in Eq. (2), where $p(i)$ is the probability of the

$$L = p(1)L(1) + p(2)L(2) + p(3)L(3) + \cdots \quad (2)$$

ith letter. An optimal encoding scheme will minimize L. However, the encoded messages must be decipherable, and this condition puts constraints on the $L(i)$. B. McMillan has shown that the code lengths of decipherable encodings must satisfy the relationships shown in inequality (3). The real

$$2^{-L(1)} + 2^{-L(2)} + 2^{-L(3)} + \cdots \leqq 1 \quad (3)$$

numbers $L(1)$, $L(2)$, . . . , which minimize L subject to inequality (3) are $L(i) = -\log_2 p(i)$ and the corresponding minimum L is shown in Eq. (4),

$$H = -\sum_i p(i) \log_2 p(i) \quad (4)$$

which provides a value of H equal to a number of digits per letter.

The $L(i)$ must be integers and $-\log_2 p(i)$ generally are not integers; for this reason there may be no encoding which provides $L = H$. However, Shannon showed that it is always possible to assign codes to letters in such a way that $L \leqq H + 1$. A procedure for constructing an encoding which actually minimizes L has been given by D. A. Huffman. For (27-letter) English text $H = 4.08$ digits per letter, as compared with the actual minimum 4.12 digits per letter obtained by Huffman's procedure.

By encoding in blocks of more than one letter, the average number of digits used per letter may be reduced further. If messages are constructed by picking letters independently with the probabilities $p(1)$, $p(2)$, . . . , then H is found to be the minimum of the average numbers of digits per letter used to encode these messages using longer blocks.

Information content of message. The information contained in a message unit is defined in terms of the average number of digits required to encode it. Accordingly the information associated with a single letter produced by a discrete source is defined to be the number H. Some other properties of H help to justify using it to measure information. If one of the $p(i)$ equals unity, only one letter appears in the messages. Then nothing new is learned by seeing a letter and, indeed, $H = 0$. Second, of all possible ways of assigning probabilities $p(i)$ to an N-letter alphabet, the one which maximizes H is $p(1) = p(2) = \cdots = 1/N$. This situation is the one in which the unknown letter seems most uncertain; therefore it does seem correct that learning such a letter provides the most information. The corresponding maximum value of H is $\log_2 N$. This result seems reasonable by the following argument. When two independent letters are learned, the information obtained should be $2H = 2 \log_2 N$. However, such pairs of letters may be considered to be the letters of a larger alphabet of N^2 equally likely pairs. The information associated with one of these new letters is $\log_2 N^2 = 2 \log_2 N$. Although H given by Eq. (4) is dimensionless, it is given units called bits (a contraction of binary digits). Occasionally the information is expressed in digits of other kinds (such as ternary or decimal). Then bases other than 2 are used for the logarithm in Eq. (4).

The majority of message sources do not merely pick successive letters independently. For example in English, H is the most likely letter to follow T but is otherwise not common. The source is imagined to be a random process in which the letter probabilities change, depending on what the past of the message has been. Statistical correlations between different parts of the message may be exploited by encoding longer blocks. The average number of digits per letter may thereby be reduced below the single-letter information H given by Eq.

(4). For example, by encoding English words instead of single letters, 2.1 digits/letter suffice. Encoding longer and longer blocks, the number of digits needed per letter approaches a limiting minimum value. This limit is called the entropy of the source and is interpreted as the rate, in bits per letter, at which the source generates information. If the source produces letters at some fixed average rate, n letters/sec, the entropy may also be converted into a rate in bits per second by multiplying by n. The entropy may be computed from tables giving the probabilities of blocks of N letters (N-grams). If in Eq. (4) the summation index i is extended over all N-grams, then the number H represents the information in N consecutive letters. As $N \rightarrow \infty$, H/N approaches the entropy of the source. The entropy of English has been estimated by Shannon to be about 1 bit/letter. However, an encoder might have to encode 100-grams in order to achieve a reduction to near 1 digit/letter. Comparing English with a source which produces 27 equally likely letters independently (and hence has entropy $\log_2 27 = 4.8$ bits/letter), this result is often restated: English is 80% redundant. Other common sources are also very redundant. Facsimile, for example, can be speeded by a factor of 10 by means of practical encoding techniques.

Capacity. The notion of entropy is more widely applicable than might appear from the discussion of the binary channel. Any discrete noiseless channel may be given a number C, which is called the capacity. C is defined as the maximum rate (bits per second) of all sources that may be connected directly to the channel. Shannon proved that any given source (which perhaps cannot be connected directly to the channel) of entropy H bits/letter, can be encoded for the channel and run at rates arbitrarily close to C/H letters/sec.

By using repetition, error-correcting codes, or similar techniques, the reliability of transmission over a noisy channel can be increased at the expense of slowing down the source. It might be expected that the source rate must be slowed to 0 bits/sec as the transmission is required to be increasingly error-free. On the contrary, Shannon proved that even a noisy channel has a capacity C. Suppose that errors in at most a fraction ϵ of the letters of the message can be tolerated ($\epsilon > 0$). Suppose also that a given source, of entropy H bits/letter, must be operated at the rate of at least $(C/H) - \delta$ letters/sec ($\delta > 0$). No matter how small ϵ and δ are chosen, an encoder can be found which satisfies these requirements.

For example, the symmetric binary channel has binary input and output letters; noise changes a fraction p of the 0s to 1 and a fraction p of the 1s to 0 and treats successive digits independently. The capacity of this channel is shown by Eq. (5), where

$$C = m\{1 + p \log_2 p + (1-p) \log_2 (1-p)\} \quad (5)$$

m is the number of digits per second which the channel transmits.

A famous formula is shown by Eq. (6), which

$$C = W \log_2 \left(1 + \frac{S}{N}\right) \quad (6)$$

gives the capacity C of a band-limited continuous channel. The channel consists of a frequency band W Hz wide, which contains a Gaussian noise

of power N. The noise has a flat spectrum over the band and is added to the signal by the channel. The channel also contains a restriction that the average signal power may not exceed S.

Equation (6) illustrates an exchange relationship between bandwidth W and signal-to-noise ratio S/N. By suitable encoding, a signaling system can use a smaller bandwidth, provided that the signal power is also raised enough to keep C fixed. *See* BANDWIDTH REQUIREMENTS (COMMUNICATIONS).

Typical capacity values are 20,000 bits/sec for a telephone speech circuit and 50,000,000 bits/sec for a broadcast television circuit. Speech and TV are very redundant and would use channels of much lower capacity if the necessary encodings were inexpensive. For example, the vocoder can send speech, only slightly distorted, over a 2000-bits/sec channel. Successive lines or frames in television tend to look alike. This resemblance suggests a high redundancy; however, to exploit it the encoder may have to encode in very long blocks. *See* VOCODER.

Not all of the waste in channel capacity can be attributed to source redundancies. Even with an irredundant source, such as a source producing random digits, some channel capacity will be wasted. The simplest encoding schemes provide reliable transmission only at a rate equal to the capacity of a channel with roughly 8 dB smaller signal power (the 8-dB figure is merely typical and really depends on the reliability requirements). Again, more efficient encoding to combat noise generally requires larger-sized blocks. This is to be expected. The signal is separated from the noise on the basis of differences between the signal's statistical properties and those of noise. The block size must be large enough to supply the decoder with enough data to draw statistically significant conclusions. *See* NOISE, ELECTRICAL.

Algebraic codes. Practical codes must use simple encoding and decoding equipment. Error-correcting codes for binary channels have been designed to use small digital logic circuits. These are called algebraic codes, linear codes, or group codes because they are constructed by algebraic techniques involving linear vector spaces or groups.

For example, each of the binary codes I, II, and III in the table contains four code words which may be regarded as vectors $C = (c_1, c_2, \ldots, c_n)$ of binary digits c_i. Define the sum $C + C'$ of two vectors to be the vector $(c_1 + c'_1, \ldots, c_n + c'_n)$ in which coordinates of C and C' are added modulo 2. Codes I, II, and III each have the property that the vector sum of any two code words is also a code word. Because of that, these codes are linear vector spaces and groups under vector addition. Their code words also belong to the n-dimensional space consisting of all 2^n vectors of n binary coordinates. Codes II and III, with $n = 3$ and 5, do not contain all 2^n vectors; they are only two-dimensional linear subspaces of the larger space. Consequently, in Codes II and III, the coordinates c_i must satisfy certain linear homogeneous equations. Code II satisfies $c_1 + c_2 + c_3 = 0$. Code III satisfies $c_3 + c_4 = 0$, $c_2 + c_3 + c_5 = 0$, $c_1 + c_2 = 0$, and other equations linearly dependent on these three. The sums in such equations are performed modulo 2; for this reason the equations are called parity check equations. In general, any r linearly independent parity check equations in $c_1, \ldots, c_n$ determine a linear subspace of dimension $k = n - r$. The 2^k vectors in this subspace are the code words of a linear code. *See* ALGEBRA, LINEAR; GROUP THEORY; NUMBER THEORY.

One may transform the r parity checks into a form which simplifies the encoding. This transformation consists of solving the original parity check equations for some r of the coordinates c_i as expressions in which only the remaining $n - r$ coordinates appear as independent variables. For example, the three parity check equations given for Code III are already in solved form with c_1, c_4, c_5 expressed in terms of c_2 and c_3. The $k = n - r$ independent variables are called message digits because the 2^k values of these coordinates may be used to represent the letters of the message alphabet. The r dependent coordinates, called check digits, are then easily computed by circuits which perform additions modulo 2. *See* LINEAR SYSTEMS OF EQUATIONS.

At the receiver the decoder can also do additions modulo 2 to test if the received digits still satisfy the parity check equations. The set of parity check equations that fail is called the syndrome because it contains the data that the decoder needs to diagnose the errors. The syndrome depends only on the error locations, not on which code word was sent. In general, a code can be used to correct e errors if each pair of distinct code words differ in at least $2e + 1$ of the n coordinates. For a linear code that is equivalent to requiring the smallest number d of "ones" among the coordinates of any code word (excepting the zero word $(0,0, \ldots ,0)$) to be $2e + 1$ or more. Under these conditions each pattern of $0,1, \ldots ,e-1$, or e errors produces a distinct syndrome; then the decoder can compute the error locations from the syndrome. This computation may offer some difficulty. But at least it involves only r binary variables, representing the syndrome, instead of all n coordinates.

Hamming codes. The r parity check equations may be written concisely as binary matrix equation (7). Here C^T is a column vector, the transpose

$$HC^T = 0 \qquad (7)$$

of $(c_1, \ldots ,c_n)$. H is the so-called parity check matrix, having n columns and r rows. A Hamming single-error correcting code is obtained when the columns of H are all $n = 2^r - 1$ distinct columns of r binary digits, excluding the column of all zeros. If a single error occurs, say in coordinate c_i, then the decoder uses the syndrome to identify c_i as the unique coordinate that appears in just those parity check equations that fail. *See* MATRIX THEORY.

Shift register codes. A linear shift register sequence is a periodic infinite binary sequence $\ldots$, $c_0, c_1, c_2, \ldots$ satisfying a recurrence equation expressing c_j as a modulo 2 sum of some of the b earlier digits $c_{j-b}, \ldots ,c_{j-1}$. A recurrence with two terms would be an equation $c_j = c_{j-a} + c_{j-b}$, with a equal to some integer $1,2, \ldots$, or $b - 1$. The digits of a shift register sequence can be computed, one at a time, by very simple equipment. It consists of a feedback loop, containing a shift register to store $c_{j-b}, \ldots ,c_{j-1}$ and a logic circuit performing modulo 2 additions. This equipment may be used to implement a linear code. First,

message digits $c_1, \ldots ,c_b$ are stored in the register and transmitted. Thereafter the equipment computes and transmits successively the $n - b$ check digits c_j obtained from the recurrence equation with $j = b + 1, \ldots ,n$. By choosing a suitable recurrence equation, one can make the period of the shift register sequence as large as $2^b - 1$. Then, with n equal to the period $2^b - 1$, the code consists of the zero code word $(0,0, \ldots ,0)$ and $2^b - 1$ other code words which differ from each other only by cyclic permutations of their coordinates. These latter words all contain $d = 2^{b-1}$ "ones" and so the code can correct $e = 2^{b-2} - 1$ errors. *See* SWITCHING CIRCUIT.

Intermediate codes. The Hamming codes and maximal period shift register codes are opposite extremes, correcting either one or many errors and having code words consisting either mostly of message digits or mostly of check digits. Many intermediate codes have been invented. One of them, due to R. C. Bose, D. K. Ray-Chaudhuri, and A. Hocquenghem, requires $n + 1$ to be a power of 2; say $n + 1 = 2^q$. It then uses at most qe check digits to correct e errors.

Perfect codes. Although each pattern of $0,1, \ldots ,e$ errors produces a distinct syndrome, there may be extra syndromes which occur only after more than e errors. In order to keep the number of check digits small, extra syndromes must be avoided. A code is called perfect if all 2^r syndromes can result from patterns of $0,1, \ldots ,e-1$, or e errors. Hamming codes are all perfect. M. J. E. Golay found another perfect binary code having $n = 23$, $r = 11$ check digits, and correcting $e = 13$ errors.

Orthogonal parity codes. Orthogonal parity codes are codes with especially simple decoding circuits which take a kind of majority vote. Suppose the parity check equations can be used to derive $2e + 1$ linear equations in which one digit, say c_1, is expressed with each of the remaining digits $c_2, \ldots ,c_n$ appearing in at most one equation. If at most e errors occur, then the received digits satisfy a majority of the $2e + 1$ equations if and only if c_1 was received correctly. For example, the recurrence $c_j = c_{j-2} + c_{j-3}$ generates a maximal period shift register code with $n = 7$, $r = 4$, $e = 1$. Set $j = 1, 3$, and 4 in the recurrence equation. One obtains three of the parity check equations, $c_1 = c_5 + c_6$, $c_1 = c_3 + c_7$, and $c_1 = c_2 + c_4$, after using the fact that the shift register sequence has period 7. These three equations are already in the form required for decoding c_1 by majority vote. Similar equations, obtained by permuting $c_1, \ldots ,c_7$ cyclically, apply for $c_2, \ldots ,c_7$. Then the decoder can be organized so that most of the equipment used to decode c_1 can be used again in decoding $c_2, \ldots ,c_7$. *See* COMBINATORIAL THEORY.

[EDGAR N. GILBERT]

Bibliography: R. B. Ash, *Information Theory*, 1965; E. R. Berlekamp, *Algebraic Coding Theory*, 1968; C. Cherry, *On Human Communications: A Review, a Survey, and a Criticism*, 1957; S. Lin, *An Introduction to Error-Correcting Codes*, 1970; J. R. Pierce, *Symbols, Signals, and Noise*, 1961; M. Schwartz, *Information Transmission, Modulation, and Noise*, 1959; C. E. Shannon and W. Weaver, *The Mathematical Theory of Communication*, 1949.

Information theory, biological applications of

Information is customarily analyzed with respect to meaning, origin, and value. Modern information theory adds another dimension, amount. The need for a measure of amounts of information arose in telecommunication and in the general treatment of control and communication (cybernetics). Once a precise technical definition had been developed, it was found that its domain of validity was wide and included biology. *See* BIOLOGY; CYBERNETICS.

Representation theorem. The first basic concept in information theory is that the information content of an event is not defined by what has occurred, but only with respect to what might have occurred instead. For instance, the amount of information in the phrase "four miles per hour" is defined only against a background of other possible speeds. To formalize this idea, consider the set of all actualities (things, processes) which may be relevant in a given class of situations; let these actualities be partitioned into classes, according to any principle chosen; these classes together form a pattern of information about the actualities under consideration. Amount of information is not defined with respect to the actualities themselves, but with respect to a given pattern of information about them. It is related to the task of specifying (selecting, recognizing) a particular class in a particular pattern for a particular actuality.

The information contents of two different patterns are compared by determining whether one can represent the other. If every class in a pattern *A* can be associated with at least one class in a pattern *B*, then every event as defined under pattern *A* can be represented by some event defined under pattern *B*, and *B* is said to have an information content at least as great as *A*. The comparison is merely formal; it makes no difference whether the association invoked is natural or arbitrary, and whether or not a representation of *A* by *B* is likely to occur. A common yardstick for all patterns of information is obtained by using a standard representation; it is customary to use binary symbols. In other words, the information content of a pattern *X* is determined by finding how many yes-or-no decisions are needed, on average, to specify which particular class of the pattern has occurred. As in a botanical key, the number of decisions made is not necessarily the same for all classes; it will be efficient to associate events which occur frequently with the shortest decision sequences (as in the Morse code, in which the shortest symbol groups are used for the most frequent letters). It is shown in texts on information theory that the optimum relation is achieved if each class of probability p_i is represented by a binary number of $-log_2 p_i$ digits. In general, it will be necessary to set up the representation for *N*-sized clusters of events in order to obtain a classification in which all probabilities p_i are integral powers of 1/2. Therefore, if *X* is a pattern of information comprising *r* classes with probabilities p_i (where $i = 1,2, \ldots ,r$), there exists a function $H(X)$ which is the minimum average number of binary symbols per event; its value is shown by the equation given here. $H(X)$ can be as small as 0 (when one $p_i = 1.0$) and as large as $log_2 r$ (when all *p*s are equal). It is called information content

(amount, quantity of information) of *X* and also the uncertainty of *X* (some prefer opposite signs for

$$H(X) = \sum_i p_i(-log_2 p_i)$$

these two functions). It is also called entropy (or negentropy) because it is related to physical entropy; the term "measure of specificity" can be used to avoid the association with mental processes implied by the word "information." *See* ENTROPY.

The *H* function is important in dealing with transfer of information. Any operation on information involves mapping from one pattern *A* into another pattern *B*. In many instances, the patterns are very dissimilar and the transfer operation far from perspicuous. The following theorem holds for every act of information transfer: Transfer from pattern *A* into pattern *B* is possible only if $H(B) \geqq H(A)$; and the amount of information transferred per act cannot be greater than $H(A)$. This is the content of the representation theorem, the first fundamental theorem of information theory. Its importance is that it can be used to check the possibility of an operation. It does not imply that such an operation actually occurs; it says nothing about the means of information transfer, and particularly, it does not imply that efficient binary coding is used at any stage of the process; it says nothing about causes and effects of the operation, and therefore applies just as well to conscious communication as to the transfer of genetic specification from parent to offspring.

Because informational bookkeeping can be applied to any form of interaction, the representation theorem provides a useful criterion which any proposed mechanism must satisfy. If part *A* of a system is supposed to control part *B*, then it must be assumed that part *A* can emit at least as many different control signals as there are different controlled responses in *B*. If *A* is supposed to maintain homeostasis of *B*, then the effective variety of control signals emitted by *A* must match the variety of effects which tend to move *B* away from its fixed value. Specificity of pairing of hormones with target organs, enzymes with substrates, and antigens with antibodies, in the absence of directed communication, implies that the specification of one reaction partner must be somehow "inscribed" upon the other. If the choice of 1 amino acid out of 20 is to be specified by a cluster of nucleotides taken from a set of 4, then a cluster of 2 is insufficient because it yields only 16 combinations, and a cluster of 3 yields 64, which is more than is needed; much effort has been spent, so far without success, to establish and test a system of correspondence between the two kinds of molecules. *See* AMINO ACIDS.

One might expect to find informational bookkeeping most rewarding in dealing with stimulus perception and response. However, this is true only for the peripheral parts of the processes. Information analysis is helpful in mapping stimulus properties on sense organ responses and nervous signals, and in mapping muscle action upon nerve signals. It also yields satisfactory results in analyzing the overall performance of the organism. However, it has so far been completely inadequate in dealing with the central nervous system. Incoming signals disappear into, and responses issue out of, a system that is orders of magnitude more complex in structure and function than can be accounted for in any simple fashion. It is suspected that central operations occur in a highly complex statistical fashion which would necessitate much elementary information traffic to achieve even a comparatively simple result with high reliability.

Noise and redundancy theorem. The relation between reliability and information content constitutes the second basic aspect of information theory. Consider an information transfer from X into Y; for example, let X be the input and Y the output of some system. If each transfer process is perfectly reliable, then knowing X implies knowing Y and vice versa; X and Y are said to be connected by a noiseless channel, and $H(X) = H(Y)$. If the transfer process is subject to perturbation, that is, if the channel is noisy, then some information will be lost in transit and there will be some uncertainty concerning X if Y is known. This uncertainty is called equivocation and designated $H(X/Y)$. The difference in uncertainty concerning X before and after the transfer process, that is, $H(X) - H(X/Y)$, is called the amount of information transmitted.

If control action depends upon information transmitted, then input equivocation will sometimes lead to faulty responses. The only protection against loss of information is the incorporation into the input of extra information which can be used to spot and correct errors. This is redundant information. A trivial example of redundancy is the repetition of a message. In general, the probability of an error's remaining uncorrected will be smaller the more redundant information is added, and the more efficiently it is used. It is shown in texts on information theory that it is possible to make the error probability vanish by precisely matching the amount of redundant information to the amount of information which is expected to be lost as a result of noise. This is the content of the noise and redundancy theorem, the second fundamental theorem in information theory.

Although it is always possible to eliminate errors, it is often not practical and indeed not efficient. It has been asserted that an optimum strategy for living things is to commit as many errors as is compatible with survival (Dancoff's principle). This has been used to establish relations between the probability of an organism to fail (for example, death rate) and amount of wear and tear it is subjected to in the course of ordinary aging or as a result of the administration of various deleterious agents. *See* INFORMATION THEORY.

[HENRY QUASTLER]

Bibliography: F. Attneave, *Applications of Information Theory to Psychology*, 1959; W. S. Fields and W. Abbott (eds.), *Information Storage and Neural Control*, 1963; H. Quastler, Information theory in radiobiology, *Ann. Rev. Nucl. Sci.*, 8:387–398, 1958; C. E. Shannon, *The Mathematical Theory of Communication*, 1949; H. P. Yockey (ed.), *Symposium on Information Theory in Biology*, 1958.

Infrared detector

A device for detecting electromagnetic radiation having a wavelength greater than that detectable by the human eye. *See* INFRARED RADIATION.

Infrared detectors are used in electronic equip-
ment for many purposes, such as detecting fires, detecting overheating in machinery, detecting planes, vehicles, and even people, and controlling temperature-sensitive industrial processes.

Infrared detectors may be classified by the basic mechanism of operation. One class, called thermal detectors, uses the power of the radiation to increase the temperature of the detecting element. This, in turn, causes some property of the detector, often the electrical resistance, to change. In the second class, called photodetectors, the radiation produces a direct effect on some electrical property of the detector.

The thermal class comprises the radiation thermocouple, the bolometer, and the Golay cell. The radiation thermocouple has a number of thermocouples connected in series, arranged so that the radiation falls on half of the junctions, thus causing a voltage to be generated. The bolometer functions through a resistance change in a material having a high temperature coefficient of resistance. The Golay cell utilizes the heat of the radiation to deflect a diaphragm in accordance with the amount of radiation. *See* BOLOMETER; THERMOCOUPLE.

The photodetectors can be classified by type of most efficient use as photoconductive and photovoltaic. These in turn can be classified into the compound types, such as lead sulfide, lead selenide, lead telluride, and thallium sulfide, and the elemental types, such as germanium and silicon. However, some types, such as germanium or silicon photodiodes and phototransistors, can be used as either photoconductive or photovoltaic detectors. For more detailed information *see* PHOTOCONDUCTIVE CELL; PHOTODIODE; PHOTOTRANSISTOR; PHOTOVOLTAIC CELL; RADIOMETRY.

[W. R. SITTNER]

Infrared image converter tube

A member of the class of devices for converting an invisible infrared image into a visible image. It consists of an infrared sensitive semitransparent photocathode on one end of an evacuated envelope and a phosphor screen on the other, with an electrostatic lens system between the two. A typical infrared converter tube structure is illustrated in Fig. 1. The infrared image to be viewed is focused on the photocathode which emits electrons with a density proportional to the radiation intensity. These electrons are accelerated to a high velocity

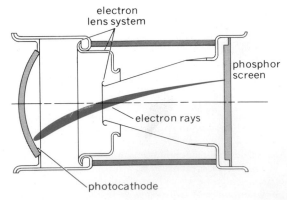

Fig. 1.   Infrared image converter tube.

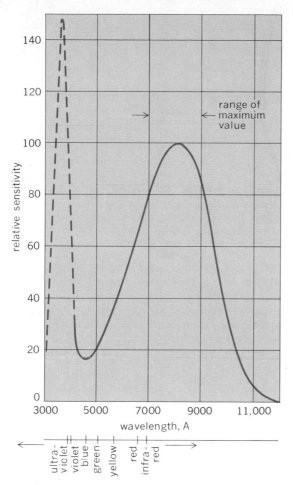

Fig. 2. Graph of spectral sensitivity of S-1 cesium-oxygen-silver photocathode.

by a potential (for example, 15 kv) applied between the photocathode and phosphor screen and focused onto the screen by the electron lens. The phosphor screen, which fluoresces in proportion to the incident electron density, displays a luminous reproduction of the image focused on the photocathode. *See* PHOTOTUBE.

The wavelength range to which the infrared converter tube is sensitive depends upon the spectral response of the photocathode. An image tube of the type in Fig. 1 employs a cesium-oxygen-silver photocathode whose relative spectral response is shown in Fig. 2. As research is continued on these devices, it can be expected that photocathodes will be developed which are sensitive to much longer wavelengths.

When combined with an objective lens for imaging the scene to be viewed onto the photocathode, and an ocular for viewing the phosphor screen, the image tube forms an infrared telescope. The long-wave limit of the photocathode does not extend far enough into the infrared to permit seeing objects close to room temperature (for example, people) by their own radiance. However, a scene illuminated with an infrared "black light" searchlight can be readily seen through the telescope, although the scene will be completely invisible to the unaided eye.

Infrared converter tubes were developed independently in a number of countries, including the United States, the Netherlands, and Germany in the mid-1930s. During World War II they were improved to a point of practical usefulness in such devices as the "Snooperscope," the "Sniperscope," and signaling telescopes. In addition to their military applications, infrared image converter tubes are useful in science for spectroscopy, viewing certain chemical processes, studying nocturnal animals, and studying emissivity and reflectivity. They have also found limited application in other areas of technology.

Other devices for making infrared images visible include the evaporagraph, the thermograph, and special television camera tubes. *See* TELEVISION CAMERA TUBE.   [G. A. MORTON]

Bibliography: P. W. Kruse et al., *Elements of Infrared Technology: Generation, Transmission, and Detection*, 1962; G. A. Morton and L. E. Flory, An infrared image tube and its military applications, *RCA Rev.*, 7:385–413, 1946; V. K. Zworykin et al., *Electron Optics and the Electron Microscope*, 1945.

Infrared lamp

An electric lamp that radiates energy at wavelengths between 0.8 and 1000 μ in the electromagnetic spectrum. Infrared lamps are essentially incandescent lamps operating at a reduced voltage with a filament temperature of 4000°F. Between 10 and 15% of the energy is converted to light. Although both carbon and tungsten filaments are used, the tungsten is more efficient. The average life is 5000 hr, and 25 different types are available. Other forms of radiant heaters generate longer wavelengths causing heat absorption by the air. The short waves generated by the infrared lamp can be concentrated and directed; the reflector-type lamp is the most satisfactory. The illustration shows the spectral distribution of energy from various incandescent filament lamps. The infrared lamps concentrate their energy in the infrared region.

Infrared lamps are used for therapeutic treatments, for heating applications in industry and homes, and for brooders in farming (heat lamps). They are effective and safe, and the heating effect is instantaneous because the air need not be heated first. The most extensive application is in industry, where they are used for baking (such as in the curing of applied finishes), for heating (such as in preparing material for processing), and for drying

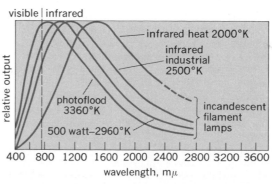

Spectral distribution of energy from various infrared sources. (*From Illuminating Engineering Society IES Lighting Handbook, 4th rev. ed., 1968*)

(or evaporating liquids). *See* INFRARED RADIATION; INFRARED RADIATION (BIOLOGY); SPECTROPHOTOMETRIC ANALYSIS.   [JOHN O. KRAEHENBUEHL]

Bibliography: Illuminating Engineering Society, *IES Lighting Handbook*, 4th rev. ed., 1968.

Infrared radiation

Electromagnetic radiation whose wavelengths lie in the range from about 0.8 μ to about 1000 μ (1 mm). The lower of these boundaries is set by the long-wavelength limit of the human eye's sensitivity to red light and the upper by the short-wavelength limit to radiation which can be generated and measured conveniently by microwave electronic devices. All solid bodies whose temperature is above absolute zero radiate some energy in the infrared, and if their effective temperature has a value up to about 3500°K, the radiation falls preponderantly in the infrared. Hence, infrared radiation is often called heat radiation or heat rays.

The infrared region was discovered around 1800 by William Herschel. He found that sunlight, dispersed into a spectrum by a glass prism, showed its greatest heating effect outside the visible part of the spectrum just beyond the red end. Herschel concluded that this effect was due to invisible radiation which was of the same nature as light except for its inability to affect the eye. In the next 50 years, experimental proof of this view slowly accumulated, and it has been generally accepted by scientists for the past 100 years.

Among the scientific, industrial, and military applications of infrared radiation are qualitative and quantitative chemical analyses by infrared spectroscopy; control of industrial processes; radiant heating; invisible signaling for burglar alarms, military messages, and the like; active and passive detection of military targets; and missile guidance.

This article discusses the spectral subdivisions, sources, detectors, and propagation of infrared radiation. For related and supplementary information *see* ELECTROMAGNETIC RADIATION; GUIDANCE SYSTEMS; GUIDED MISSILE; HEAT RADIATION; INFRARED LAMP; INFRARED RADIATION (BIOLOGY); INFRARED SPECTROSCOPY; LASER; PHOTOGRAPHY; RADIOMETRY; SPECTROPHOTOMETRIC ANALYSIS.

Infrared spectrum. The infrared region is often subdivided, but there is no general agreement about the names and boundaries of the subdivisions. The table gives a subdivision which is based on instrumental characteristics but which also corresponds in a rough way to the subdivision of natural frequencies of molecules, as shown in the last column. The boundaries of the subdivisions are rather arbitrary and approximate.

As the table implies, near-infrared radiation can be detected by photoelectric cells, and it corresponds in frequency range to the lower energy levels of electrons in molecules and semiconductors (the higher electronic levels correspond to the frequencies of visible and ultraviolet radiation). The intermediate infrared region covers the frequency range of most molecular vibrations; only the overtones of the higher vibrational frequencies fall outside it. This region is sometimes called the prism infrared because spectrometers equipped with alkali-halide prisms are commonly used here. The

Subdivisions of the infrared (IR) spectrum

Wavelength range, μ	Wave number range, cm^{-1}	Names used	Appropriate molecular motions
0.8–2.5	4000–12,500	Near IR; photoelectric IR	Low electronic levels; vibrational overtones
2.5–50	200–4000	Intermediate IR; prism IR	Molecular vibrations
50–1000	10–200	Far IR; grating IR	Molecular rotations

far-infrared, where diffraction gratings must be used in spectrometers because no suitable prism materials are known, spans the region of very low molecular vibrational frequencies and most rotational frequencies, at least of the lighter molecules. Because radiation of about 1-mm wavelength, the upper limit set for the far infrared, can also be produced and detected by microwave radar devices, there is no gap between the infrared and microwave domains, and in fact the region on either side of 1 mm is under investigation by both techniques. *See* MICROWAVE.

Sources. The usual source of infrared radiation is an incandescent solid body. Such sources emit radiant power having a continuous broad range of wavelengths. The variation of this radiant power with wavelength ordinarily has the form shown in Fig. 1, with a pronounced maximum at some wavelength λ_{max}, a rather sharp decrease on the low-wavelength side, and a gentler slope at higher wavelengths.

Infrared radiation may be emitted by systems other than hot solid bodies, for example, by gases through which an electrical discharge is passed. Although such systems are relatively inefficient emitters, they are sometimes used for special purposes. The mercury-vapor arc has been found to be one of the best sources of radiant power in the far-infrared.

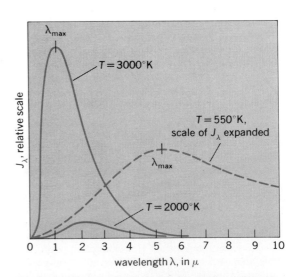

Fig. 1. Spectral distribution of J_λ, the power radiated by an incandescent blackbody (a theoretically perfect radiator) at temperature T. The ordinate J_λ is the radiant power in watts per unit area of radiating surface per unit solid angle per unit of wavelength at λ in the infinitesimal wavelength range $d\lambda$.

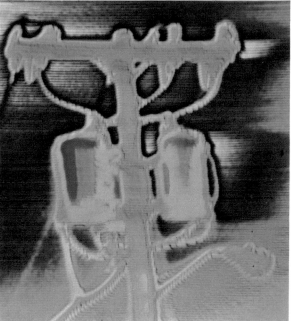

Color positives of infrared thermograms taken with Barnes Engineering Co. infrared cameras. (a) Thermal effluent entering a slow-moving stream. Flow is from left to right. (b) Ford Mercury station wagon with engine running and a person seated behind the steering wheel. (c) Two hot transformers on a telephone pole. (d) Manhattan skyline as seen from Governor's Island. Positives a, b, and d were taken with Model I-4 camera; positive c, with Model T-6. (Barnes Engineering Co.)

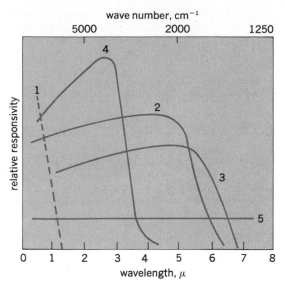

wave number, cm⁻¹

Fig. 2. Spectral sensitivity of infrared detectors. 1, infrared-sensitized photographic emulsion; 2, PbTe (lead telluride) photoconductive detector (refrigerated); 3, PbSe (lead selenide) photoconductive detector (refrigerated); 4, PbS (lead sulfide) photoconductive detector; 5, thermal detector. Responsivity scale is not the same for various detectors; approximate wavelengths at which detectors 1, 2, 3, and 4 become less responsive than a thermal detector are correctly shown.

The development of the laser has resulted in sources of infrared radiation that are sharply monochromatic and of extremely high radiant power. The most powerful continuously operating laser in the infrared is the carbon-dioxide-gas laser, which can be made to emit several hundred watts of power at a wavelength of 10.6 μ. Other infrared gas lasers are the helium-neon laser (prominent wavelengths of 0.692, 1.19 and 3.39 μ), the water-vapor laser (27.97, 47.7, 78.46 and 118.6 μ), and the hydrogen-cyanide laser (311 and 337 μ). *See* LASER.

While the gas laser is sharply monochromatic, its wavelength spread or half-width being smaller than one-millionth of its wavelength, the wavelength cannot be varied continuously over a significant range. The solid-state laser of gallium arsenide produces near- and mid-infrared radiation whose wavelength can be adjusted, or tuned, by means of hydrostatic pressure. Apart from a

tunable laser, the only way of producing monochromatic infrared radiation of any desired wavelength is to isolate it from an infrared source containing that wavelength and many others, such as a blackbody source (Fig. 1). The isolating device may be a selective optical filter or a spectrometer.

Detectors. Detectors of electromagnetic radiation may be classified broadly as quantum detectors, resonant detectors, and heat engines or thermal detectors. The first are devices which convert a quantum of the radiation in question into a proportionate signal by some process which is insensitive to quanta of less than a certain energy (for example, the mean energy of quanta emitted by a body at room temperature). Photographic emulsions, photoelectric cells, and Geiger counters are examples of quantum detectors. Resonant detectors are devices that are responsive only to radiation of the frequency to which they are tuned. Heat engines act as detectors by converting the radiation into heat and using the heat to operate a device that produces a signal which is proportionate to the amount of radiant energy received.

No resonant detectors have been constructed which can be tuned to infrared frequencies. For the photoelectric infrared (see table), quantum detectors in the form of specially sensitized photographic emulsions, photoemissive cells, and particularly photoconductive cells are usable. As can be seen from Fig. 2, the responsivity of such detectors varies considerably with frequency and drops to low values at wave numbers of about 2000–3000 cm⁻¹ (3.5–5 μ in wavelength). Photoconductive detectors are known having good responsivity below 2000 cm⁻¹ (above 5μ), but these must be operated at or near liquid helium temperatures (<10°K).

Therefore, in a large part of the infrared region, heat-engine detectors—whose responsivity is the same to all kinds of radiation provided the radiation is converted entirely to heat in the detector— are the only generally usable kind. Examples of heat-engine detectors are thermocouples and thermopiles, which produce an electromotive force when heated; bolometers, which change their electrical resistance when heated; and pneumatic radiometers, in which heat is detected by the increase in pressure of a heated gas. Because these devices are all subject to the laws of thermodynamics governing the conversion of heat into useful work (that is, into a signal), their ultimate respon-

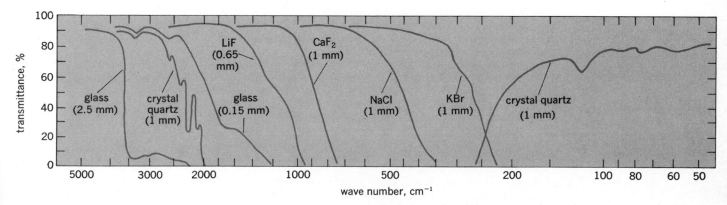

Fig. 3. Transmission curves of some widely used infrared optical materials (bottom scale is the wave number in cm⁻¹).

sivity is expected to be approximately the same, and this is found to be the case. Thermal detectors operating at room temperature have a lower limit of sensitivity of the order of 10^{-10} watt, with response times of the order of 0.1 sec. This limit can be reduced by a factor of 100 or more if the detector is capable of operation at temperatures near 1°K.

Propagation. The properties of various media for the transmission of infrared radiation are often quite different from those of light. For example, window glass is quite opaque to 5-μ radiation, whereas pure germanium crystals, which do not transmit visible radiation, are very transparent to this wavelength (apart from reflection losses which can be reduced by surface coatings). The alkali halide crystals, such as sodium chloride (common salt) and potassium bromide, are well known for their transparency in the near- and mid-infrared regions (Fig. 3).

The attenuation of infrared radiation by the atmosphere is of special interest. Nitrogen, oxygen, and the rare gases are transparent to all infrared wavelengths, but water vapor and carbon dioxide are strongly absorbing in certain regions. In the range 0.8–4.0 μ, there is irregular absorption due mainly to water vapor, with more or less open "windows" at about 1.0, 1.4, 1.6, 2.2, and 3.4–4.0 μ. From 4 to 8 μ water vapor and carbon dioxide together are strongly absorbing, but there is an extensive window from 8 to 13.5 μ. This window is of great meteorological importance because the peak of the radiation curve of the Earth falls near 10 μ. Beyond 14 out to 600 μ, there is more or less continuous absorption by atmospheric water vapor, arising from transitions between the rotational energy levels of this molecule.

Liquid water is rather generally opaque in the infrared above 2 μ in path lengths larger than 1 mm. The transmission of infrared radiation through fog is little better than that of visible light, because of a combination of scattering and absorption by the water droplets. The popular misconception that fog is transparent to infrared radiation has perhaps arisen from the well-advertised effectiveness of red-sensitive photographic film for photography through atmospheric haze of dust particles.   [RICHARD C. LORD]

Bibliography: R. S. Estey, Infra-red: New uses for an old technique, *Missiles and Rockets*, 3(7): 107–112, 1958; G. R. Harrison, R. C. Lord, and J. R. Loofbourow, *Practical Spectroscopy*, 1948; P. W. Kruse, L. D. McGlauchlin, and R. B. McQuistan, *Elements of Infrared Technology*, 1962; Ivan Simon, *Infrared Radiation*, 1966; R. A. Smith, F. E. Jones, and R. P. Chasmar, *Detection and Measurement of Infrared Radiation*, 2d ed., 1968.

Infrared radiation (biology)

Infrared radiations occupy the span between the visible spectrum and radio waves, and encompass wavelengths 7800–4,000,000 A, neither boundary being precisely delimited. All bodies above absolute zero in temperature are potential sources of infrared radiations; therefore, all organisms are continually exposed to them. About 60% of the Sun's rays are infrared. Water absorbs infrared radiations strongly, except for the band of transparency between 7800 and 14,000 A. Since protoplasm contains much water, it absorbs infrared radiations readily. A large animal absorbs infrared radiations at its surface, only the span from 7800 to 14,000 A penetrating as far as the blood vessels. *See* INFRARED RADIATION.

While many substances and even tissues selectively absorb infrared rays, and one might therefore postulate selective effects of these radiations, none have been unequivocally demonstrated except possibly in conjunction with x-rays. The reason is perhaps because quanta of infrared radiation do not excite energy states in molecules other than those excited by conducted heat. *See* X-RAYS, PHYSICAL NATURE OF.

The essential biological effect of infrared rays depends primarily upon the rise in temperature produced following their absorption, which in turn increases the rate of biological activities in proportion to the temperature change. Because of the prominence of infrared in sunlight, organisms show many adaptations to dissipate or to avoid the heat. The temperature of a large animal or plant may rise temporarily, but the heat is dissipated by transpiration in the plant and by perspiration in the animal. Submerged animals and plants are protected by the water, the temperature change of which depends upon the heat capacity of the particular body of water.

Treatment of biological materials with infrared radiations (7800–11,500 A) either before or after x-ray treatment increases the rearrangements in chromosomes induced by the x-ray in tissues of plants and animals tested. The way in which the infrared radiations do this is unknown, but a comparable amount of conducted heat does not have the same effect.

Medical practitioners make use of infrared radiations to treat sprains, strains, bursitis, peripheral vascular diseases, arthritis, muscle pain, and many other pains for which heating gives relief, probably because of vasodilation of peripheral vessels. For this purpose, glow coil radiators are generally employed, but for some purposes, the more penetrating radiations (7800–14,000 A) obtainable from incandescent sources are preferable to the glow coil to stimulate circulation deeper in the tissues. *See* BIOPHYSICS; THERMOTHERAPY.

[ARTHUR C. GIESE]

Bibliography: O. Glasser, *Medical Physics*, vol. 1, 1944; A. Hollaender (ed.), *Radiation Biology*, vol. 1, 1954; W. Summer, *Ultraviolet and Infrared Engineering*, 1962.

Infrared spectroscopy

The study of the interaction of material systems with electromagnetic radiation in the infrared region of the spectrum. The infrared region is valuable for the study of the structure of matter because the natural vibrational frequencies of atoms in molecules and crystals fall in the infrared range. Some gaseous molecules also have rotational frequencies in the far-infrared range, and certain frequencies corresponding to the energy levels of electrons in solids and in large molecules lie in the near infrared. For a detailed discussion of molecular vibration and rotation *see* MOLECULAR STRUCTURE AND SPECTRA. *See also* INFRARED RADIATION.

The infrared absorption spectrum of a molecule

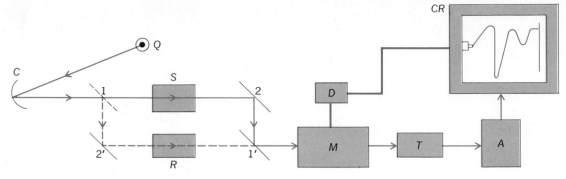

Fig. 1. Recording infrared spectrometer.

is highly characteristic, and often has been referred to as a molecular fingerprint. The spectrum can thus be used for molecular identification. Because the absorption of radiation at various infrared frequencies is quantitatively related to the number of absorbing molecules in a system, quantitative analysis is also possible.

The usefulness of an infrared absorption spectrum for identification and chemical analysis was recognized as long ago as 1890. In the early 1900s the American physicist W. W. Coblentz determined the infrared spectra of hundreds of substances and clearly demonstrated the potential value of such spectra. Unfortunately, the instrumentation of that day was cumbersome and necessarily homemade, so that few physicists and chemists were attracted by Coblentz's work. Only after the development of commercial electronic devices for amplification and recording of a continuously scanned spectrum in the 1940s was extensive use made of the technique.

Instrumentation and techniques. The usual arrangement for measurement of an infrared spectrum is shown schematically in Fig. 1. A source Q sends a beam of continuous infrared radiation to a spherical condensing mirror C, which passes the beam through S, the sample to be studied. Some of the infrared frequencies in the beam are absorbed strongly, some weakly. The reduced beam passes on and comes to a focus at the entrance slit of the monochromator M. The latter is an infrared spectrometer which disperses the radiation into a spectrum. One frequency at a time appears at the exit slit of M, from which the radiation of that frequency is passed by a suitable optical system to the detector T. The detector (a thermocouple or other device) converts the radiant energy into an electrical signal, which is amplified electronically at A and recorded by a chart recorder CR.

An infrared spectrum is a record of intensity of infrared radiation as a function of frequency or wavelength. To produce such a record, the chart recorder is driven in synchronism with the dispersing system of the monochromator M by some common driving mechanism D. In this way a given position on the chart corresponds directly to a given frequency setting of M, at which setting radiation of that frequency is emerging from M.

For many basic reasons—atmospheric absorption, variation of source intensity with frequency, changing dispersion in the spectrometer, and the like—the electrical output of the detector would not be constant even if the sample S were completely transparent. To correct for these variations it is necessary to determine two spectra, one with the sample S in the beam and one with S removed from the beam. The absorption of S as a function of frequency can then be computed from these two spectra. The individual spectra on which the computation is based are called single-beam spectra.

The computation is laborious, time consuming, and potentially unreliable because of changes in the entire system between the two determinations of spectra. These difficulties are avoided if a second optical path, shown in dotted lines in Fig. 1, is introduced. The second optical path, called the reference beam, is made as nearly like the first as possible, except for the absence of the sample. In fact, the reference beam may contain an absorption cell R which differs from S only in the absence of the sample itself. For instance, if the sample S were in solution, R would contain the same amount of solvent as S.

The operation of the double-beam spectrometer, often called a spectrophotometer, consists of a rapid switching of the beam (say 10 times per second) back and forth between S and R by alternately placing plane mirrors 1 and 1' in the optical system. The identical mirrors 2 and 2' are permanently placed. The spectrum is scanned continuously as for single-beam operation, but the beams through S and R are compared 10 times per second and the chart records the energy passing through S relative to that through R. In this way the variations mentioned cancel out.

Typical spectra. Typical mid-infrared spectra, plotted automatically as percent transmission of the sample on a linear frequency scale (wave number in cm^{-1}), are shown in Fig. 2. Samples of gases, liquids, and solids can be readily measured. Techniques for high and low temperature of sample and for small samples (down to about 1 mg or less in special cases) are in common use.

Percent transmission T, the quantity usually plotted by commercial instruments, is defined in Eq. (1). Here $I_{0,\nu}$ is the intensity of infrared radia-

$$T_\nu = 100\,I_\nu/I_{0,\nu} \qquad (1)$$

tion of frequency ν entering the sample and I_ν is the intensity of the same radiation after passing through the sample. The percent transmission T_ν at frequency ν is different in principle at different values of ν. A quantity of fundamental importance, the absorbance A_ν, is defined in Eq. (2). The ab-

$$A_\nu = \log\,(I_{0,\nu}/I_\nu) = -\log\,(T_\nu/100) \qquad (2)$$

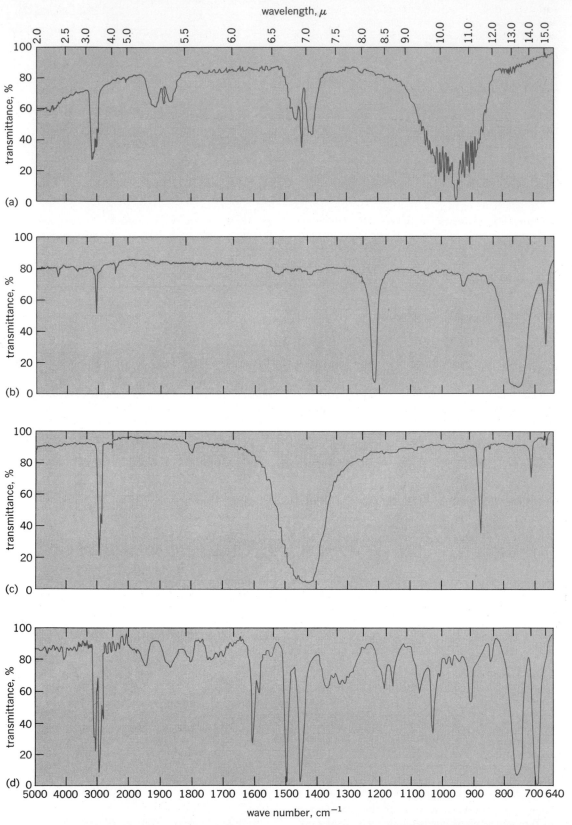

Fig. 2. Typical mid-infrared transmission spectra, recorded automatically. Note compressed scale on left
portion of abscissa. (a) Spectrum of ethylene gas. Transmission minima in range 850–1050 cm⁻¹ result from
modulation of ethylene vibrational frequency at 950
cm⁻¹ by molecular rotational frequencies. (b) Spectrum
of liquid chloroform. (c) Spectrum of powdered crystalline calcium carbonate. Powder was suspended in mineral oil to obtain spectrum. Transmission minimum at
1430 cm⁻¹ is characteristic of carbonate ion, and that at
2900 cm⁻¹ is characteristic of CH groups in the oil.
(d) Spectrum of an amorphous high polymer (polystyrene). Detail here shows why infrared spectra are sometimes called molecular fingerprints by workers in this
discipline of science.

sorbance A_ν is proportional to the number of absorbing molecules, and by evaluating the proportionality constant at frequency ν for a given kind of molecule in a particular system, the number of such molecules in other systems of the same kind may be measured quantitatively.

Interferometric methods. Infrared spectra are measured in many applications by the technique of Fourier-transform spectroscopy (FTS). In dispersive (prism or grating) spectroscopy a spectrum is recorded by continuous scanning of the spectrum at successive frequencies (Fig. 1). In FTS the entire frequency range of interest is passed simultaneously through an interferometer, which produces an output signal containing all these frequencies. The quantitative way in which this signal varies as the condition for interference within the interferometer is varied is called an interferogram (Fig. 3). The interferogram can be made to yield the spectrum as a function of frequency by the mathematical procedure known as a Fourier transform. Although this procedure is complicated, small and powerful digital computers are available so that an interferometer and a digital computer can be combined into a single unit which produces the transformed spectrum with negligible delay. *See* DIGITAL COMPUTER; FOURIER SERIES AND INTEGRALS; INTEGRAL TRANSFORM.

A block diagram of an interferometric spectrometer and auxiliary components is shown in Fig. 4. The source Q sends a beam containing a complete range of infrared frequencies into the interferometer I. The beam is first divided at the semitransparent beam splitter BS, which transmits half to the variable plane mirror M_V and reflects half to the fixed plane mirror M_F. The separate beams are returned by M_V and M_F to the beam splitter, where they are reunited after having traveled distances differing by some amount L which is continuously variable. The reunited beams interfere at BS, where they are partially reflected and sent out of the interferometer to the sample S. After passing through S, the radiation is converted to an electrical signal at the detector *det*. The output of *det* as L changes is shown in Fig. 3. This output is processed electronically at A, stored in the computer memory, and then transformed to the spectrum. The spectrum is read out in suitable form, for example, as a plot on the chart recorder CR of percent transmission versus frequency. Appropriate instructions to the computer are provided from the command post, which may be a teletypewriter. For double-beam operation the optical system at S may resemble the two-beam arrangement through S and R in Fig. 1.

The virtues of interferometric spectroscopy as compared to dispersive (that is, grating or prism) spectroscopy are:

1. Enormous superiority in the effective use of the limited radiant power in infrared sources. This superiority, which leads to much larger signal-to-noise ratios in the transformed spectra, arises from two fundamental differences between interferometers and spectrometers: first, the interferometer processes all frequencies in the input radiant power simultaneously (the multiplex advantage), whereas the dispersive spectrometer processes them one at a time. Second, the spectrometer needs narrow entrant and exit slits to do

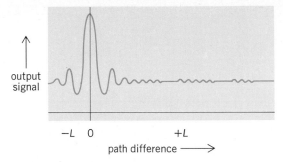

Fig. 3. A representative interferogram.

this processing, and the solid angle of radiant power accepted from the source is correspondingly restricted. The interferometer does not have this restriction and thus can accept a much larger solid angle of input radiation (the through-put advantage). The combination of these two advantages may lead to a superiority of several orders of magnitude in signal-to-noise ratio of the FTS over that of a dispersive spectrometer having the same resolution and scanning rate. Alternatively, the FTS will be able to record at the same signal-to-noise ratio and scanning rate with an increase in the resolution of more than an order of magnitude. This latter superiority is illustrated in Fig. 5, which shows the absorption spectrum of ethylene gas recorded from 940 to 960 cm^{-1} by a Digilab FTS-14 spectrometer. The effective scanning rate in cm^{-1}/s and the signal-to-noise ratio are about the same in Fig. 2a (dispersive spectrum of ethylene) and in Fig. 5, but the resolution (as measured by the reciprocal of the spectral bandpass $\Delta\nu$) is 8 times higher in the latter ($\Delta\nu \simeq 2$ cm^{-1} at 950 cm^{-1} in Fig. 2a, $\Delta\nu = 0.25$ cm^{-1} in Fig. 5). *See* RESOLVING POWER (OPTICS).

2. The interferometer does not require any filters or other order-sorting devices.

3. The wavelength or wave-number scale of the interferogram is automatically provided by the scanning parameter of the interferometer. This parameter is usually controlled to high precision by an auxiliary laser interferometer.

4. The spectral bandpass $\Delta\nu$ computed from the interferogram is constant throughout the spectrum. It is equal to, or somewhat larger than, $1/2L_{max}$ where L_{max} is the maximum excursion of the moving mirror, depending on how the data are processed.

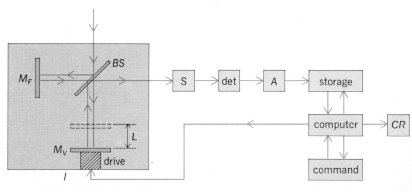

Fig. 4. Block diagram of an FTS system.

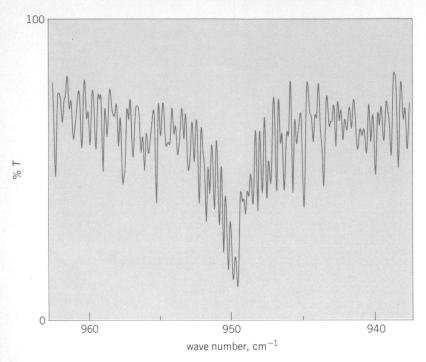

Fig. 5. FTS spectrum of ethylene gas, 940–960 cm⁻¹. Path length of sample, 10 cm; gas pressure, 30 torr (4.0×10^3 Pa); recording time, ~ 30 min; spectral bandpass $\Delta \nu$, 0.25 cm⁻¹.

5. The computer needed to calculate the Fourier transform can be used in addition for the manipulation of the spectroscopic data at any stage. It can also be programmed to control the mechanical operation of the interferometer, the electronics system, and the read-out device (for example, the recorder).

Use of tunable lasers. The sharpness of the frequency and the high power per unit solid angle and unit spectral bandpass in a laser beam make it attractive for infrared spectroscopy at ultrahigh resolution. The main problem is the tuning of the laser frequency, that is, varying some parameter in the laser system so that its frequency may be varied continuously and accurately. The table

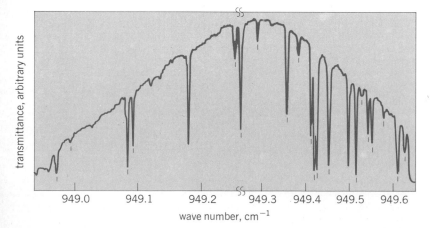

Fig. 6. Absorption spectrum of ethylene gas obtained by tuning a Pbₓ Snₗ₋ₓ Te SDL in the range 948.9–949.7 cm⁻¹. Path length, 10 cm; pressure, 0.08 torr (1.1×10^2 Pa); effective spectral bandpass, $\Delta \nu < 0.001$ cm⁻¹. (From G. P. Montgomery, Jr., and J. C. Hill, High-resolution diode-laser spectroscopy of the 942.9 cm⁻¹ band of ethylene, J. Opt. Soc. Amer., 65: 579–585, 1975)

shows the frequency ranges over which various kinds of infrared lasers can be tuned. It is apparent from the table that their extremely high resolution (small spectral bandpass) makes infrared spectroscopy with tunable lasers a quite different kind of enterprise from dispersive or Fourier transform spectroscopy. An example of this resolution is shown in Fig. 6, where the absorption spectrum of ethylene gas at 950 cm⁻¹ measured with a tunable semiconductor diode laser (SDL) is illustrated (compare Figs. 2a and 5). The triangular background with maximum at 949.3 cm⁻¹ is the slit function of the dispersive spectrometer used to eliminate unwanted SDL modes.

Tunable infrared lasers have been used mainly to measure the energy levels due to the vibration and rotation of molecules in the gas phase at low pressures. By this means, very accurate values of such molecular parameters as internuclear distances, electric dipole moments, vibrational frequencies, internal force fields, and the like may be measured. Other anticipated uses of laser infrared spectroscopy include the monitoring of the composition of the atmosphere. For example, E. D. Hinkley and associates have monitored the amount of carbon monoxide in a horizontal path of 600 m at ground level with an SDL; the sensitivity achieved was of the order of 5 parts per billion. *See* LASER.

Applications. An infrared spectrum consists of a plot of T or A as a function of ν (or of wavelength λ). The basic information provided by the spectrum is a set of ν values at which the substance is absorbing strongly, that is, at which T_ν is a minimum (Fig. 2) or A_ν is a maximum. These frequencies of maximum absorption usually correspond to the actual vibrational frequencies of the absorbing molecules or to some arithmetical combination of such vibrational frequencies. If the molecules are in the vapor phase, absorption maxima may also be observed at frequencies which are combinations of frequencies of molecular rotation and vibration. The qualitative usefulness of an infrared spectrum lies in the fact that the set of observed vibrational frequencies characterizes the absorbing molecule.

Qualitative chemical analysis. Infrared spectra can be used for the following purposes.

1. To identify pure chemical compounds by comparison of the spectrum of an unknown with previously recorded spectra of pure compounds. Catalogs of spectra are available, and in addition there are practical methods for encoding the information in the spectra and storing it in a computer memory system or on punched cards. The stored data can then be used for fast identification of an unknown spectrum.

2. To identify the constituents of mixtures. When a mixture contains only two or three constituents, it is often possible to identify them directly from the spectrum of the mixture. More commonly it is necessary to fractionate the mixture by gas chromatography or other technique and to identify the individual compounds in the mixture by their infrared spectra. For this purpose Fourier transform spectroscopy is advantageous because of its speed and superior signal-to-noise ratio; combined chromatographic and FTS equipment is commercially available.

3. To show the presence of a group of atoms, a

so-called functional group, in a molecule of un-known or doubtful structure. It has been known since the 1890s that certain groups of atoms—for example, a methyl group (CH_3), a carbonyl group (CO), a nitrate ion (NO_3^-)—have characteristic absorption frequencies that are relatively inde-pendent of the rest of the molecule or crystal in which the group occurs. Literally hundreds of such group frequencies are known. *See* QUALITATIVE CHEMICAL ANALYSIS.

Quantitative chemical analysis. There is a linear quantitative relationship between the ab-sorbance A_ν, defined in Eq. (2), and the number of absorbing molecules. Thus the quantitative anal-ysis of mixtures by infrared means is feasible. The infrared method is not particularly sensitive, the limit of detection and measurement of minor constituents being in the range 0.1–1.0%, except in favorable circumstances. An example of the latter is the detection of a component of a gaseous mixture. If the component has strong absorption in a spectral range where the rest of the mixture is transparent, much smaller quantities may be detected. In general, the higher the resolution (the narrower the spectral bandpass) of the infrared instrument, the lower the minimum concentration that may be detected, provided the width of the spectral lines of the absorbing constituent is of the same order as the spectral bandpass. The advantage of narrow spectral bandpass disappears if it is much smaller than the spectral line width. Thus the sensitivity of the SDL measurement of atmospheric carbon monoxide (5 parts per billion over a path of 600 m) would be much lower (that is, much improved) if the spectral line width of CO (~0.2 cm^{-1} at atmospheric pressure) were reduced to that of the SDL used (~0.0001 cm^{-1}).

The precision of quantitative measurement is mainly limited by the signal-to-noise ratio, and Fourier transform spectroscopy therefore offers an advantage for quantitative work. With dispersive spectroscopy, precision of measurement is seldom better than 1% of the quantity being measured and may be considerably worse. Infrared methods are especially useful in the quantitative determina-tion of isomeric substances and in measurement of constituents of a chemical equilibrium. *See* ANA-LYTICAL CHEMISTRY; QUANTITATIVE CHEMICAL ANALYSIS.

Determination of molecular structure. Struc-tures of molecules can be determined to varying degrees of refinement from infrared spectra. If only a few independent parameters (interatomic distances and bond angles) are required to specify the structure, as is the case with a small symmetri-cal molecule, these can be evaluated from the moments of inertia of the molecule, which can in turn be measured from rotational frequencies, usually observed as fine structure in a vibrational absorption. The structural parameters of carbon dioxide, methane, ethylene (Figs. 2a and 5), and ethane, for example, have been evaluated with high precision from their infrared spectra.

If the number of parameters is too large to be determined in this way, it may nevertheless be possible to draw conclusions about the molecule's shape without measuring its size. The number of vibrational frequencies which appear in the in-frared spectrum is related to the molecular sym-metry, and it is often possible to infer the symme-try from the observed spectrum. Such inferences are more reliable if they are based on combined data from both infrared and Raman spectra. *See* RAMAN EFFECT.

It is still possible to say something about the structure of large molecules of little or no symme-try from their spectra if one is content with a state-ment about the presence or absence of various functional groups. The organic chemist often finds such statements very valuable. The nature of func-tional groups in the molecules of high polymers or of natural products such as the steroids can be determined from their infrared spectra, and this permits information about their structure to be obtained.

Study of solids, catalysts, and matrices. The in-frared spectra of crystalline solids give information about modes of vibration of crystals, about hydro-gen-bond vibrations when such bonds are present in them, and about electronic energy states in semiconductors and superconductors. Fourier transform spectrometers are regularly used by solid-state physicists, particularly for their advan-tages in the spectral range 1–200 cm^{-1}. At higher frequencies Fourier transform spectrometers are useful in the study of materials adsorbed on the surfaces of catalysts. The FTS signal-to-noise ad-vantage is especially important because of the optical heterogeneity of these samples. A similar heterogeneity in samples trapped at low tempera-tures in condensed rare gases or other types of matrices likewise makes FTS the technique of choice for the investigation of trapped transient species. *See* SPECTROSCOPY.

[RICHARD C. LORD]

Bibliography: L. J. Bellamy, *Infra-red Spectra of Complex Molecules,* 3d ed., 1975; G. R. Harri-son, R. C. Lord, and J. R. Loofbourow, *Practical Spectroscopy,* 1948; G. Herzberg, *Infrared and Raman Spectra of Polyatomic Molecules,* 1945; R. T. Ku, E. D. Hinkley, and J. O. Sample, Long-path monitoring of atmospheric carbon monoxide with a tunable diode laser system, *Appl. Opt.,* 14:854–861, 1975; H. Walther (ed.), *Topics in Applied Physics: Laser Spectroscopy of Atoms and Molecules,* 1974.

Frequency ranges of tunable infrared lasers*

Type of laser	Frequency range, cm^{-1}	Minimun spectral bandpass $\Delta\nu$, cm^{-1}
Semiconductor diode laser (SDL)	300–10,000*	3×10^{-6}
Spin-flip Raman laser (SFR)	1600–2000	3×10^{-6}
Zeeman-tuned gas laser (ZTG)	700–1000	3×10^{-2}
High-pressure CO_2 laser (HPG)	1100–3300	3×10^{-3}
	900–1100	3×10^{-4}
Nonlinear devices		
Optical parametric oscillator (OPO)	900–10,000+	3×10^{-2}
Difference frequency generator (DFG)	1600–3300	5×10^{-4}
Two-photon mixer (TPM)	900–1100	3×10^{-5}
Four-photon mixer (FPM)	400–5000	1×10^{-1}

*Adapted from K. W. Nill, Tunable infrared lasers, Opt. Eng., 13:516–522, 1974.

Inhibitor (chemistry)

A substance which is capable of stopping or retarding a chemical reaction. To be technically useful, such compounds must be effective in low concentrations, usually under 1%. The type of reaction which is most easily inhibited is the free-radical chain reaction. The study of inhibitor action is often used as a diagnostic test for free-radical chain character of a reaction. Vinyl polymerization and autoxidation are two important examples of the class. Another reaction type for which inhibitors have been found is corrosion, particularly in aqueous systems. The economic importance of corrosion inhibition can scarcely be overestimated. An understanding of inhibitor action depends on an understanding of the processes which are to be interrupted.

Inhibition of vinyl polymerization. This type of inhibitor action must be considered in terms of the accepted mechanism for the polymerization process, which may be summarized as reaction sequence (1). The symbol P represents a catalyst,

$$\left.\begin{array}{l} P \rightarrow 2R\bullet \\ R\bullet + M \rightarrow \sim M\bullet \end{array}\right\} \text{initiation}$$
$$\sim M\bullet + M \rightarrow \sim M\bullet \text{ propagation} \qquad (1)$$
$$2 \sim M\bullet \rightarrow \text{termination by dimerization or} \\ \text{disproportionation}$$

often a peroxide, R• is a free radical derived from the catalyst, M is a monomer, and ∼M• is a growing polymer chain. The polymerization will be stopped or inhibited if some added substance (an inhibitor) reacts more readily than does the monomer with R• to yield a product which will not sustain the polymerization. Every reaction chain is stopped until the inhibitor is consumed. If the added substance (a retarder) is somewhat less reactive, the monomer can compete more successfully for the initiating radicals, so that the result is retardation rather than total inhibition. The difference between inhibition and retardation is one of degree rather than of kind.

Phenolic compounds and quinones. These interact with an initiating radical or a growing polymer chain either by hydrogen atom abstraction or by radical addition to an unsaturated linkage. These interactions are represented by sequence (2).

The phenoxy radicals produced are stabilized by resonance and are not sufficiently reactive to add to the vinyl linkage of another monomer molecule. The usual fate of these radicals is further hydrogen atom loss by reaction with a second polymer radi-

cal or by disproportionation with another phenoxy radical to yield a quinone and a hydroquinone, both of which may continue to act as inhibitors. All the reaction possibilities shown have been demonstrated experimentally. Although the efficiencies of phenols and quinones for the interruption of polymerizations vary with the structures of these molecules, they may be classed as inhibitors. Aromatic amines react similarly.

Nitroaromatics. As typified by trinitrobenzene, nitroaromatics function as retarders rather than as inhibitors in most polymerizations. It is necessary, however, to consider the specific reaction involved. Thus polynitroaromatics inhibit the polymerization of vinyl acetate, retard that of styrene, and have no effect on that of methyl methacrylate. No clear-cut mechanism has been established for the interaction of nitro compounds with free radicals.

Monomers. Both monomers and the radicals derived from them differ greatly among themselves in reactivity. Thus, although certain monomers may copolymerize with one another, others may actually function as inhibitors. Styrene and vinyl acetate, for example, both polymerize well when alone. Styrene, however, inhibits the polymerization of vinyl acetate. This occurs because the vinyl acetate radical and the styrene monomer are highly reactive, whereas the styrene radical and the vinyl acetate monomer are not. A small amount of styrene added to vinyl acetate will rapidly react with any vinyl acetate radicals formed when polymerization is initiated. The resulting styrenelike radical will react only very slowly with the vinyl acetate monomer. In the overall process the chain-carrying radical is converted to one which is too unreactive to carry on the chain.

Autoinhibition. This action, sometimes called allylic termination, is exhibited by monomers which contain the highly reactive allylic C—H linkage. Free radicals are capable of hydrogen atom abstraction from copresent molecules as well as of addition to an unsaturated linkage. The ease with which this abstraction reaction is carried out by a given radical is a function of the reactivity of the C—H linkage which is attacked. The reactivity of radicals containing these C—H linkages increases in the order aryl < primary alkyl < secondary alkyl < tertiary alkyl < allyl < benzyl. Because of this high reactivity, a monomer containing an allylic C—H, allyl acetate, for example, functions as its own retarder, as shown by sequence (3).

$$\sim M\bullet + CH_2 = CHCH_2OAc \qquad (3)$$
$$\sim MCH_2CHCH_2OAc \qquad \sim MH + CH_2 = CHCHOAc$$

The resonance-stabilized allylic radical will react with the monomer only very slowly. The predominant further reaction is dimerization. Not only is polymerization slowed in this case, but the molecular weight of the polymer formed is low.

Miscellaneous inhibitors. Oxygen, iodine, and nitric oxide interact rapidly with free radicals to yield stable products. Inclusion of these materials in polymerizing systems thus leads to effective inhibition. It is of particular interest that oxygen will copolymerize, under carefully controlled conditions, with certain monomers, styrene, for example, to yield high-molecular-weight polymeric peroxides. The repeating unit is shown in formula (4).

$$[-CH(C_6H_5)CH_2OO-]_n \qquad (4)$$

Iodine and nitric oxide have been used extensively to detect and in some cases to identify alkyl free radicals in nonchain as well as in chain reactions. This method has proved to be of great value in defining primary processes in photochemical decompositions of aldehydes and ketones.

Inhibition of corrosion. Metallic corrosion in conducting media is electrochemical in nature. Local electrolytic cells are set up because of the presence of impurities, crystal lattice imperfections, or strains within the metal surface. The result is dissolution of the metal from the anodic regions. Corrosion inhibitors now in use may operate at the anodes or the cathodes or provide physical protection over the entire surface.

Anodic inhibitors. These are mild oxidants which reduce the open-circuit potential difference between local anodes and cathodes and increase the polarization of the former. Sodium chromate and sodium nitrite are most commonly used. The former is used in air conditioners, refrigeration systems, automobile radiators, power plant condensers, and similar equipment. Sodium nitrite finds special use in the protection of petroleum pipelines. It is effective even on rusty, mild steel. An extension of the nitrite type is the use of nitrite salts of secondary amines as vapor-phase inhibitors. The inclusion of a salt such as dicyclohexylammonium nitrite with a packaged steel object provides effective protection against corrosion.

Cathodic inhibitors. Compounds such as calcium bicarbonate and sodium phosphate, in an aqueous medium, deposit on metal surfaces films that provide physical protection against corrosive attack.

Organic inhibitors. These are usually long-chain aliphatic acids and the soaps which are derived from them. Adsorption of these compounds on metal surfaces gives a hydrophobic film which protects the metal from corrosion by many agents. As little as 0.1% of palmitic acid, for example, is sufficient to protect mild steel from attack by nitric acid. *See* ANTIOXIDANT; CATALYSIS; CORROSION; FREE RADICAL; POLYMERIZATION.

[LEE M. MAHONEY]

Bibliography: P. G. Ashmore, *Catalysis and Chemical Inhibition of Reactions*, 1963; R. L. Le Mar, *VCI Bibliography and Abstracts*, U.S. Atomic Energy Commission, 1958.

Ink

A dispersion of a pigment or a solution of a dye in a carrier vehicle, yielding a fluid, paste, or powder to be applied to and "dried" on a substrate. Writing, marking, drawing, and printing inks are applied by several methods to paper, metal, plastic, wood, glass, fabric, or other substrates. Often inks perform communicative, decorative, and even protective functions.

As early as 2000 B.C., the Chinese and Egyptians used inks from tannins and galls in their fine arts and for communication. Much later, in the 4th century A.D., the Chinese developed a solid ink, in the form of pellets, similar to India ink. They also are credited with invention of the printing process in the 11th century, some 400 years before the invention by Johann Gutenberg in Germany.

Since printing ink of today is by far the most important and furthest-developed type of ink, it and its application methods will be discussed first.

PRINTING INK

The composition, properties, uses, and manufacture of various printing inks and the main printing processes in which they are used are described in the following sections.

According to the 1972 census of manufacturers, the ink industry consists of 407 printing ink manufacturers or establishments. Of this total, 145 are establishments with 20 or more employees. The total value of shipments in 1972 was reported to be $508,400,000, up 46% from the amount reported in the 1967 census. Industry estimates placed the value of shipments in 1974 at approximately $670,000,000. The top seven ink manufacturers reportedly account for combined sales of approximately $425,000,000, or 63% of the total industry output. The "big three" ink companies, Inmont Corporation, Sun Chemical, and Sinclair and Valentine, collectively represent approximately 45% of total industry sales.

Classification. Printing ink can be classified according to its composition and texture, application, end use, and drying manner. Composition and texture may be any of the following types: oil or paste ink, solvent or liquid (aniline) ink, or specialty ink. The applications are letterpress, lithographic, flexographic, gravure, silk screen, stencil, duplicating, electrostatic, and jet. The end uses are news, publication, commercial, folding carton, book, corrugated box, paper bag, wrapper, label, metal container, plastic container, plastic film, foil, laminating, food insert, sanitary paper, and textile. The various drying manners are oxidizing, evaporating, penetrating, precipitating, polymerizing, reactive, gelling, and cold-setting or quick-setting.

Generally, the printing is done on a single- or multicolor press in sheet- or web-fed manner at a linear speed ranging from 200 to 2000 fpm. This requires adjustment of ink rheology and strength, and careful selection of materials and dryers.

Obviously, if one were to multiply the above classes by the various color, shade, opacity, gloss, film thickness, drying speed, and functional property requirements, it would be easy to understand that some 900,000 ink formulations exist in order to meet the various needs and conditions.

To meet the ever-increasing printing-process demands, modern ink making evolved into a distinct art and science, requiring special skills and experience, in addition to the knowledge of chemistry, physics, and engineering. Current ink making is based on extensive research and development effort in the polymer, resin, colorant, and drying-system sciences, as well as engineering of dispersions. In addition, it is supported by customer service, product maintenance, formulation adjustment, and color-matching activity. Sophisticated optical, rheological, and end-use testers, as well as computers, aid in the development and control of new or repeat formulations.

Formulation. Fundamentally, inks are composed of four major material categories: (1) Colorants (which include pigments, toners, and dyes) provide the color contrast with the substrate. (2) Vehicles, or varnishes, act as carriers for the colorants during the printing operation. Upon drying, the vehicles bind the colorants to the substrate.

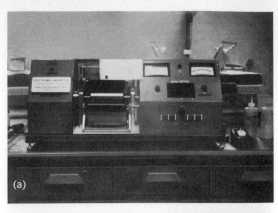

Fig. 1. Instruments for the measurement of ink tack. (a) Inkometer. (b) Tack-O-Scope.

(3) Additives influence the printability, film characteristics, drying speed, and end-use properties. (4) Solvents, besides participating in formation of the vehicles, are used to reduce ink viscosity and adjust drying ease and resin compatibility. Ingredients from these four classes are weighed, mixed, and ground (dispersed) together, or separately, according to the formulas preestablished in the laboratory.

Colorants. These are used in a dry powder, flushed paste, or liquid concentrate form. The latter two, however, are more economical, requiring a minimum of dispersing effort. Their selection influences the hue, permanency, bulk, opacity, gloss, rheology, end use, and print quality.

1. Predominant black pigments are carbon and furnace blacks.

2. The main opaque white pigments are titanium dioxide, zinc sulfide, and zinc oxide.

3. Transparent white pigments or extenders are alumina hydrate, magnesium and calcium carbonates, barium sulfate, and clay (composed of silicates).

4. Of the typical inorganic color pigments, lead chromate yellows and oranges and the molybdated orange exhibit light fastness, opacity, and high specific gravity; cadmium yellows, oranges, and reds (and mercury) are resistant to alkalinity. Iron blue (ferrocyanide complex), ultramarine blue (aluminum-sodium-silica-sulfur complex), and iron oxide reds and browns must be included in this group.

5. Important among the many organic yellow colorants are the permanent Hansa and transparent diarylide yellows and the dianisidine, diary-

lide, and Persian oranges. The reds include the para reds and toluidines, with fair permanency to light; the lithol reds, of a wide spread of shades but moderate fastness; the clean but fugitive phloxine (eosine) lakes and the clean and expensive rhodamine reds; opaque red lake C and end-use bleed-resisting Watchung or permanent 2-B reds. Organic blues include light-fast and resistant phthalocyanines, fugitive peacocks, brilliant Victoria blue, strong greenish alkali blues, and reddish methyl violet. The last three belong to the triphenylmethane family and bleed in polar solvents.

6. Daylight-fluorescent pigments have limited light fastness and tinctorial strength and predominantly find use in gravure and silk-screen processes.

7. Metallic (silver and gold) pigments are treated aluminum and bronze flakes or powders.

Pigments containing lead, mercury, cadmium, and so on as ingredients or impurities are considered less safe and desirable for food packaging and book and magazine use. *See* DYE; PIGMENT.

Vehicles. Vehicles, or varnishes, mainly consist of resin polymers dissolved in the various solvents, oils, or liquid resins. Occasionally, a single alkyd or a bodied vegetable or mineral oil constitutes a vehicle. Vehicle selection or its composition is dictated by the printing-drying process and substrate and end-use requirements. The ink vehicles are best categorized according to their drying mechanism and application. Both physical and chemical reaction phenomena take place during the solidification, or drying, step.

1. Solvent evaporation is accomplished by convective or radiant (infrared) heat and a stream, or jet, of hot air. Letterpress and lithographic heatset inks are based on calcium-and-zinc rosinates; esters of mono- and diabetic acids and pentaerythritol or glycerine; phenolic polymers; hydrocarbon polymers; and gilsonite. These are dissolved in aliphatic (12–15 carbon) hydrocarbons. Occasionally, fumarated and partly esterified rosin, dissolved in dipropylene, diethylene, or monoethylene glycol, forms a letterpress vehicle.

Flexographic and gravure inks employ singly, or in combination, polyamide, nitrocellulose, protein, acrylic, rosin ester, phenolic polymer, chlorinated rubber, and polyketone resins which are dissolved in pure, or a blend of, aliphatic (6–12 carbon) naphthas and aromatic hydrocarbons, esters, alcohols, ketones, nitroparaffins (of 2–3 carbon chain length), glycol ethers, and water. Since flexographic presses utilize rubber plates and rollers, the tolerance of stronger (aromatic, ester, and ketone) solvent is often below 10%. This fact also limits the resin selection.

2. Binder precipitation occurs when water is absorbed by a hygroscopic solvent. Steam or moisture of the air may be utilized to dilute the glycol, which ceases to be solvent for the resin. The mixture separates and penetrates into the porous substrate or evaporates. Fumarated rosin, or occasionally a maleated drying oil, which might be partly esterified or condensed with an amine, represent typical binders. This vehicle system is useful for the letterpress and dry-offset printing methods.

3. Oxidation is the reaction of atmospheric oxygen with unsaturated fatty acids present in vegetable oils (linseed, china, tall, soya) or in their alkyds. Cobalt, manganese, and lead soaps, with fatty and

aromatic acids, act as catalysts for the reaction. Letterpress and lithographic sheet-fed inks, based on pure drying oils or on vehicles consisting of phenolic and rosin ester resins dissolved in these, "dry" (solidify) by this mechanism.

4. Polymerization, or cross-linking, normally proceeds via the polyol-amine condensation path or through the Diels-Alder diene-addition mechanism. Lithographic, dry-offset, letterpress, and silk-screen inks for metal, paper, and glass most frequently utilize this. method. Alkyds with free hydroxyl groups condense with melamine resins in the presence of acid catalysts, such as para-toluene-sulfonic acid. Alkyds or oils with diene-conjugated unsaturations polymerize via the addition and oxidation mechanism simultaneously. China, dehydrated castor oil, and linseed and soya fatty acids are employed here.

5. Gelation, or fusion, takes place when the binder, composed of a nonsolvent plasticizer at room temperature and a dispersed polymer, is heated to the temperature at which solubility is achieved. Letterpress, dry-offset, and lithographic inks can be so formulated, utilizing a polyvinyl chloride resin dispersed in hercolyn or dioctyl phthalate or both.

6. Quick-setting is a two-step mechanism relying on the critical solubility of the resinous binder and the porosity of the substrate. The vehicle of borderline solubility readily releases its liquid portion to the substrate, exhibiting osmotic pressure and thereby becoming tack-free. The second drying step proceeds via the conventional oxidation process. Modern lithographic and letterpress folding-carton inks for sheet-fed presses widely utilize this drying principle. Phenolic, rosin-ester, and alkyd resins, combined with drying oils and aliphatic hydrocarbon solvents, are typical quick-set systems.

7. Absorption-setting is achieved through the penetration (sponge effect) of the total vehicle or its liquid portion into a porous substrate. Newsprint inks, based on mineral oil or on rosin, gilsonite, or melhi resin solutions, set, although never fully dry, by this method.

8. Cold-setting is based on solidification of an ink applied in a molten state. Cold-setting inks are based on a combination of mineral and synthetic waxes.

9. Radiation cure, or photopolymerization, is a newer method of drying lithographic, letterpress, flexo, and gravure printing inks, varnishes, and coatings. Irradiation with ultraviolet-rich light from medium-high-pressure mercury lamps activates the photoinitiators, which initiate free-radical polymerization among the vinyl or acrylate ester oligomers and monomers. The cross-linking results in the instantly dry, hard, and resistant films.

Polymerization induced through irradiation with "soft" (170 kV) electrons also proceeds via the free-radical mechanism. This process was in the early stage of development in the mid-1970s. Electron-beam method does not require help from photoinitiators to activate and cross-link the vinyl oligomers or monomers. Electrocurtain, developed by Energy Sciences, Inc., is a source of such soft electrons. It is safer, costs less, and is smaller than the scanning-type beam gun.

Additives. Auxiliary ink components or additives consist of metallic driers (Co, Mn, Pb); antioxi-

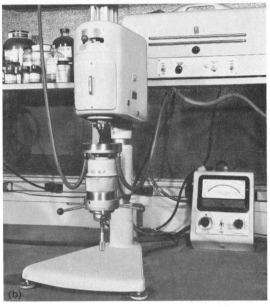

Fig. 2. Viscometers. (*a*) Laray type. (*b*) Shirley-Ferranti type.

dants (eugenol, ionol); waxes (hydrocarbon, vegetable, animal); lubricants (silicones, polyethylenes); plasticizers (phthalates); gellants (titanates and aluminates); and photoinitiators (benzophenone).

Solvents. Ink solvents are aliphatic (12–15 carbon) hydrocarbons; aromatic hydrocarbons (toluene, xylene), mineral oil; aliphatic (6–13 carbon) naphthas, with and without an aromatic admixture; glycols (ethylene, diethylene, propylene, dipropylene); normal and isoalcohols (1–4 carbon); esters (1–5 carbon alkyl acetates); and ketones (methyl 1–5 carbon alkyl ketones).

Rheology. Rheology is the science of flow which describes the behavior of liquids under stress. Printing inks are non-Newtonian liquids which behave irregularly under stress and temperature changes. In the past ink technologists were satisfied simply with measurement of ink tack on an inkometer (Fig. 1). Modern techniques, in addition to the redesigned Thwing Albert Inkometer or Tack-O-Scope, utilize viscometers made by Laray, Shirley-Ferranti, Haake, Brookfield, Inmont (formerly Interchemical), and Stormer (Fig. 2). Ink viscosity (resistance to flow), yield value (point at which ink starts to flow under stress), thixotropy (decreasing viscosity with increasing agitation), and dilatancy (opposite of thixotropy) can now be determined. Proper ink rheology is highly impor-

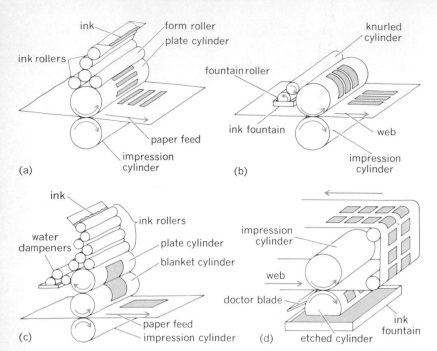

Fig. 3.   Printing processes. (a) Letterpress. (b) Flexography. (c) Offset lithography. (d) Gravure.

metal rollers (Fig. 3a). Since the total roller surface is relatively large and the ink film thin, built-in press stability against evaporation, oxidation, or moisture is required by the ink.

Linear press speed of letterpress printing ranges from 200 to 2000 fpm. Lower ink viscosity is required for higher speeds. The plates are made of metal, rubber, or plastic. Press makeready is long and costly. Oxidizing, absorptive, moisture-set, and heat-set inks are printed letterpress on newspapers, magazines, displays, labels, folding cartons, corrugated boxes, books, bags, and wrappers. *See* PRINTING; PRINTING PLATE.

Flexography. Another typographic printing method utilizing a raised plate is flexography. It differs from the letterpress by having a much shorter roller frame, consisting of two rollers and the use of a knurled cylinder for ink transfer to the rubber plate (Fig. 3 b). Flexographic inks are very volatile and fluid (20–100 cps).

Flexography offers economy, stock flexibility, and the lack of odor required in food package printing. Polyethylene bread bags, cellophane, polypropylene, and Mylar structures for snack and cheese packages are printed by flexo, as well as paper bags and wrappers, corrugated boxes, and folding cartons.

Offset lithography. This is often called a planographic printing method because the image is printed from a flat or plane surface. The system is based on a photochemically treated plate whose image areas accept ink and nonimage areas reject ink but accept acidified water (fountain solution). In offset lithography the printed image is first transferred to a rubber blanket (cylinder) from the metal plate and then to the substrate (Fig. 3c). Offset printing is done with water-repellent and acid-resistant paste inks. Oxidizing, quick-setting, heat-setting, and curable inks are used for printing of metal cans, folding cartons, labels, and rigid plastics, as well as for magazines, books, and newspapers.

Gravure. Gravure or intaglio printing is based on ink transfer from the tiny cells etched in the surface of the printing cylinder onto the substrate. Excess ink is removed from the surface of the cylinder by a doctor blade (Fig. 3d). The etched cells vary in depth and area and, thereby, the print intensity is regulated. Since no rubber rollers are in contact with the ink, the gravure process allows wide selection of strong, yet volatile, solvents and complex polymers, otherwise insoluble. Ink viscosity, drying speed, and self-resolubility are required to keep the cells active and open. Heat-set gravure inks are used in printing of folding cartons (food, soap, cigarette), labels, flexible film, newspapers, and magazines.

tant in ink transfer from roller to plate to substrate in fidelity of printing, trapping, and hold-out (gloss) characteristics. Vehicles, gellants, additives, and pigments affect the ink rheology. *See* RHEOLOGY.

Printing processes and end uses. Different printing processes and the various uses of printed products demand special ink characteristics. Five basic printing processes dominate the industry: letterpress (typographic system), flexography (typographic system), offset lithography (planographic system), gravure or intaglio (recessed engraving), and screen process (stencil system). Less important printing systems are dry offset (letterset), roll leaf stamping, decalcomania, hot transfer (therimage), electrostatic, and jet printing. The systems, inks, and products for most of these systems are discussed below.

Letterpress. This process involves the transfer of the ink paste to a type form or plate and subsequent direct transfer to the substrate. The ink transfer from the fountain to the form roller is accomplished by a series of alternating resilient and

Screen or stencil. This process involves the preparation of a screen pattern through which a liquid or paste ink is allowed to pass directly to the areas to be printed (Fig. 4). Stencil process encompasses the screen printing and a mimeograph printing. Silk-screen inks are the heat-set oxidizing and curing type. Their texture is pasty and of relatively high yield value. Silk screen is utilized for printing of posters, cards, molded plastics, glass, and some metal containers. Mimeograph inks are based on mineral oil, castor oil, and oleic acid vehicles.

Dry offset. This method is a hybrid of letterpress

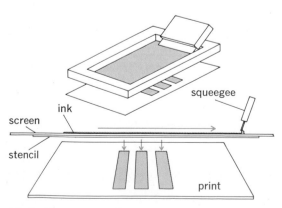

Fig. 4.   The screen or stencil method of printing.

and lithographic processes. Plastic wraparound plates are used to transfer the ink to a rubber blanket and then to the substrate. Dry offset inks are similar to the letterpress type. Folding cartons, plastic containers, and metal sheets are printed by the dry offset or letterset process.

Electrostatic. Electrostatic printing is accomplished in three ways: a screen method, a gravure-cylinder method, and a photoconductive method. The "ink," in the form of dry powder or toner suspension in a solvent, is transferred through a short distance by the electric force (Fig. 5). It thus is a contact-free printing method, which can be used for the decoration of uneven surfaces. Eventually, the powder is bound to the substrate by heat, or a solvent vapor.

In the screen and gravure methods, the charged powder is attracted through a substrate by an oppositely charged electrode. Charging of the powder can be accomplished by a metal screen, or a gravure cylinder, which also provides the image. In the photoconductive method, a latent electric image is produced by discharging of the zinc oxide paper or a selenium drum with strong source of light. The dark areas remain charged and attract the toner particles when the sheet or drum is in contact with them. The powders are compounded to retain electric charge, to respond to the attraction, and to fuse readily and adhere to the substrate. The process is being utilized in selected package printing, identification of wood planks, and in reprography (copying).

Ink-jet printing is a plateless, nonimpact, noncontact method adaptable to electronic controls and suitable for image formation from computer memory or a tape. The method offers four variations developed independently by A. B. Dick, Teletype, Mead, and Gould, Inc. It is based on ejection of tiny ink droplets from a fine nozzle and their selective deflection or attraction onto a substrate or into a collecting pan. Electronic impulses accomplish this at 10,000–20,000 droplets per second. Even though these ink spots range in size from 0.05 to 0.4 mm, quality of the jet printing is not yet considered high enough to compete with traditional methods. It is being used commercially for addressing magazines and printing teletypes or business forms. Jet printing requires a very fluid and well-screened aqueous ink, which is often colored with dyes. The ink is dried by infrared heat.

Manufacturing process. The ink manufacturing process can be related to two operations, vehicle preparation and colorant dispersion. Vehicle preparation can require complex synthesis of polymers or simple dissolving of purchased resins. Elaborate reactors, as well as simple blending kettles, are employed. Pigment dispersion is done in change-can, "dough," or jacketed mixers, in tank agitators, in ball or sand-and-pebble mills, or often in three- or two-roller mills, depending on volatility, pigment, and consistency.

SPECIAL INKS

Examples are given of inks developed for particular purposes aside from printing.

Ultraviolet (radiation) cure inks. The pollution legislation and energy crisis stimulated development of the ultraviolet (uv) ink system based on in-

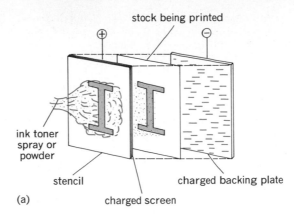

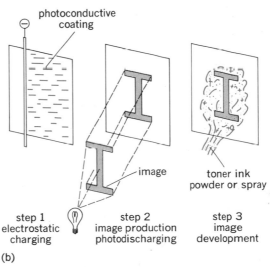

Fig. 5. Electrostatic printing methods. (*a*) Screen method. (*b*) Photoconductive method.

stant photopolymerization of monomers by a high-energy light source (Fig. 6). It completely eliminates pollution from the web and spray powder on the sheet presses, and reduces energy consumption to 20% of that required by heat-set ovens previously employed in printing on paper or metal. Its main disadvantage is 50–100% higher cost of the inks.

A uv curer (dryer) consists of a lamp (mercury,

Fig. 6. Ultraviolet curing lamp assembly on a sheet-fed press (lamp with shutter below it).

xenon, pulsating, doped, or electrodeless types), an elliptical or a parabolic reflector, a power pack, and a shielding box. The lamp irradiates uv-rich light in the 200–400-nm wavelength region with a significant peak at the 360-nm band. There were eight major suppliers of the uv curers in the United States as of 1976.

Lithographic and letterpress inks are available throughout the world in any desirable color (except fluorescent) from 12 independent producers. These are being used commercially on some 250 uv installations, of which 130 are located in the United States. The system is gaining rapid acceptance and exhibits an annual growth rate of 30%. Ultraviolet cure is finding especial application in the production of high-quality folding cartons and the decoration of metal cans, where it reduces processing cost and consumption of natural gas. Inmont Corporation, Sun Chemical, Borden, Sinclair and Valentine, and Acme are some of the main suppliers of uv inks.

Ball-point inks. Inks developed for ball-point pens are Newtonian fluids of high tinctorial strength. These must be free of particles and premature drying so as to continue the feed to the paper without clogging. Rapid penetration into the paper accomplishes "drying." They are dye solutions or pigment dispersions in oleic acid, castor oil, sulfonamide, or in aqueous solutions of gums or glues.

Stamp-pad or marking inks. These inks are impregnated into a cloth or foam rubber pad and transferred by pressure to a rubber type which is then stamped or impressed against the substrate. The inks must remain nondrying on the pad, yet rapidly penetrate into the stamped substrate. Induline dye is often dissolved in glycols, phenol, or cresol, depending upon whether paper or metal is being stamped. Marking inks are based on water, alcohol, turpentine, and wax.

[BOHDAN V. BURACHINSKY]

Bibliography: J. F. Ackerman, Mod. Lithogr. 40(3):18, 1972; E. A. Apps, Printing Ink Technology, 1964; Flexographic Technical Association, Flexography: Principles and Practices, 1970; F. A. Askew et al., Printing Ink Manual, 1969; F. J. Kamphoefner, Ink jet printing, IEEE Trans. Electr. Devices, ED-19:584, April 1972; M. W. Ranney, New Curing Techniques in the Printing, Coating and Plastic Industries, 1973.

Inland waterways transportation

The internal commercial water routes of the United States except for the Great Lakes and the commercial transportation services provided by vessels operating on these rivers, canals, bays, sounds, and lakes. Transportation on the inland waterways of the United States accounts for 10% of the nation's total movement of commerce. Barges move 7.2%; the remaining 2.8% is moved in deep-draft vessels.

The 25,380-mi network of commercially navigable inland channels is used primarily by barges which are either pushed by towboats or pulled on a hawser by tugboats. Where improvements of the channels have been made for barge transportation, a standard operating depth of 9 ft is provided. Such 9-ft depth or more is provided in 15,348 mi of channel. The rest—10,032 mi—are under 9 ft and therefore substandard. Certain sections of these inland waterways, as has been indicated, are deep enough to support oceangoing vessels. The primary routes where both barges and oceangoing vessels operate are the Hudson River to Albany, the Mississippi River to Baton Rouge, the Houston Ship Channel, and the Columbia River to Portland.

This article deals with barge transportation on the inland waterways; and for purposes of this article the Great Lakes are excluded. For a discussion of commercial transportation on the oceans and large bodies of inland water see SHIP, MERCHANT.

Barges, towboats, and tugboats. Three principal types of barges are operated: open-hopper barges with a cargo capacity range of 1000–3000 tons; covered dry-cargo barges with a capacity range of 1000–1500 tons (Fig. 1); and tank barges with a cargo capacity range of 1000–3000 tons (Fig. 2). Open-hopper barges are used to transport commodities which do not need protection from the elements. Dry cargoes which need protection are carried in covered barges. The covers may be of the lift-on, lift-off type; or they may be the telescoping type which roll fore and aft on tracks installed on the barge. Tank barges are used to carry a great variety of liquid commodities, such as petroleum and petroleum products, chemicals, and fertilizers.

Propulsion power for barges is provided by towboats and tugboats. On fully protected inland channels, the barges are made up into flotillas by lashing them together with wire rigging. Such a flotilla is lashed against the towing knees of the towboat which pushes them ahead.

Barges and towboats tied rigidly together, as they are for movement over fully protected inland channels, cannot withstand wind and wave action such as is encountered in open water. Sections of the two intracoastal waterways—the Atlantic Intracoastal Waterway and the Gulf Intracoastal Waterway—are not fully protected by surrounding land masses. On these, barges are moved by tugboats pulling them on a hawser. Obviously the number of barges which can be handled by this method of towing is less than can be handled in push-towing operations. (Barges are used for transportation in the open seas. These are also pulled on a hawser by a tug. Such barges range up to 30,000 tons capacity.)

A river towboat is a relatively flat-bottomed vessel with a square bow on which heavy, upright towing knees are fixed for the purpose of lashing the barges against the towing vessel. A tugboat has a shaped hull and bow and, unlike the towboat, is designed and constructed to provide watertight integrity.

The barge and towing industry fleet consists of 4395 towboats and tugboats having an aggregate of 3,545,821 hp; 15,380 dry-cargo barges having aggregate capacity of 16,066,302 tons; and 2781 tank barges with an aggregate capacity of 5,120,029 tons.

Cargo. Most cargoes transported by this service move in bargeload lots. Generally they are bulk commodities which are susceptible to highly efficient cargo handling techniques. Petroleum and petroleum products are the leading commodities, representing 34% of total barge commerce. These are followed by coal, which represents approximately 23.1% of barge tonnage (Fig. 3). Then

Fig. 1. Motor vessel southbound just above Baton Rouge, La., with 40 loaded grain barges (58,000 tons).

come sand and gravel, sea shells, logs, grain and grain products, industrial chemicals, iron and steel products, limestone, sulfur, cement, coal tar products, soybeans, pulpwood, fertilizer, and paper and paper products. The fastest growth in barge service is being made in the movement of industrial chemicals.

The average cost of bargeline service to the shipper is 3 mills per ton-mile, with a range from a low of 1.75 mills per ton-mile to a high of about 7 mills per ton-mile. The 1.75 mills per ton-mile rate, incidentally, applies in the case of certain petroleum movements. The average cost to the shipper for service by other modes is 15 mills per ton-mile by rail, 6.5 cents by motor truck, and 20 cents by air freight. Pipeline service costs compare very favorably with bargeline rates.

Single maximum-load movements on inland-waterways channels in the late 1940s consisted of about 10,000 tons, equivalent to the capacity of one Liberty ship or $1\frac{1}{2}$ train loads. In the late 1960s, river towboats could easily handle five times as much. The Lower Mississippi River between St. Louis and New Orleans is an open river without locks or dams and commodious enough to permit a towboat to push a string of barges carrying as much as 50,000 tons. On canalized waterways, such as the Ohio, the Upper Mississippi, and the Illinois, the payload of a towboat is smaller because the tow must pass through a system of locks and because the channels are not as commodious as the Lower Mississippi. Even on such waterways, single movements of 20,000 tons (144,000 bbl of petroleum) are not uncommon. Additional improvements, which undoubtedly will be made as the demands of commerce increase, will permit the payload of river towboats and barges to be raised on many of these waterways. A river towboat of 6000 hp, handling 50,000 tons of commerce on the Lower Mississippi, moves the equivalent of $8\frac{1}{3}$ trains of 120 cars each, with each train pulled by a 6000-hp engine; or the equivalent of 1500 of the largest motor trucks. A river towboat can haul 8 tons/hp as compared with an average of 1 ton/hp

Fig. 2. Integrated tow serving the chemical industry.

Fig. 3. Northbound tow going through Chain of Rocks Canal near Granite City, Ill., with 21,000 tons of coal.

Navigable lengths and depths of United States inland-waterway routes*

Group	Length of waterways, mi†					
	Under 6 ft	6 – 9 ft	9 – 12 ft	12 – 14 ft	14 ft and over	Total
Atlantic Coast waterways (exclusive of Atlantic Intracoastal Waterway from Norfolk, Va., to Key West, Fla.), but including New York State Barge Canal System	**1502** 1563	**1271** 1445	**593** 549	**975** 1005	**1490** 1241	**5831** 5803
Atlantic Intracoastal Waterway from Norfolk, Va., to Key West, Fla.	**–** –	**211** 158	**65** 65	**954** 1104	**–** –	**1230** 1327
Gulf Coast waterways (exclusive of Gulf Intracoastal Waterway from St. Marks River, Fla., to Mexican border)	**2048** 2142	**718** 819	**1239** 2097	**216** 305	**444** 372	**4665** 5735
Gulf Intracoastal Waterway from St. Marks River, Fla. to Mexican border (including Port Allen–Morgan City alternate route)	**–** –	**–** –	**–** –	**1177** 1177	**–** –	**1177** 1177
Mississippi River system	**2400** 4829	**684** 1491	**4449** 5058	**732** 755	**273** 268	**8538** 12,401
Pacific Coast waterways	**725** 733	**370** 515	**239** 237	**26** 27	**2182** 461	**3542** 1973
All other waterways (exclusive of Alaska)	**45** 100	**58** 148	**–** 14	**8** 8	**286** 369	**397** 639
Grand total	**6720** 9367	**3312** 4576	**6585** 8020	**4088** 4381	**4675** 2711	**25,380** 29,055

*From publications of the Corps of Engineers, United States Army.

†The mileages in bold type represent the lengths of all navigable inland channels of the United States (exclusive of the Great Lakes) including those improved by the Federal Government, other agencies, and those which have not been improved but are usable for commercial navigation. The mileages in light type represent the lengths authorized for improvement by the United States Congress in legislation known as Rivers and Harbors Acts.

hauled by a railroad locomotive and $\frac{1}{11}$ ton/hp hauled by a motor truck.

Cargoes are moved along the American waterways by three types of carriers: (1) common carriers, regulated by the Interstate Commerce Commission as to routes, rates, and commodities which they may handle; (2) nonregulated, exempt carriers, limited to carrying only bulk cargoes of an "uncountable, unpackageable, unlabelable" nature (grain, coal, and so forth); and (3) private carriers, owned and operated by the company whose freight is hauled.

The amount of cargo moved in 1967 was 500,912,733 tons, for a total of 173,300,000,000 ton-miles. Estimates for 1968 are about 3% higher.

Waterways. The system of commercially navigable inland waterways of the United States is made up of a series of Federal projects, with the exception of the New York State Barge Canal, which is a state project. Improvement, maintenance, and operation is the responsibility of the Army Corps of Engineers working with Federal funds appropriated by the Congress (see table and Fig. 4).

With the exception of the Upper Mississippi Waterway, the Missouri River, and the New York State Barge Canal, all the inland channels are open to navigation all year. Icing conditions close the three waterways named for about 4 months of the year—December through March. At times, ice forms on the Illinois Waterway, the Mississippi above St. Louis, and on the Ohio River, but it seldom impedes navigation for any length of time.

With two notable exceptions, the channels are slack water routes which have been improved for navigation by the construction of systems of locks and dams. The Mississippi is open river for 1000 mi south of St. Louis. The Missouri is open river. Yet the two present a striking difference. The Mississippi is a wide, deep, commodious river. The Missouri has a restricted 7-ft depth. Both the Atlantic and Gulf Intracoastal waterways are largely open, and both have reaches that are exposed to tidal currents and winds.

Most waterway construction projects involve work to widen or deepen channels or to modernize channels by the construction of higher dams and larger locks. See CANAL; DAM; RIVER ENGINEERING.

The redevelopment of the Ohio River, for instance, is underway. The river was canalized in 1929 by a system of 52 locks and dams. The new canalization project, which was started in 1955, will replace the present lock and dam system with 19 new high-lift locks with lock chambers 110 ft wide by 1200 ft long. The present chambers are 110 ft by 600.

Recanalization of the Warrior-Tombigbee River System in Alabama is under way. The Corps of Engineers has completed work on a study to build a second system of locks on the Illinois Waterway. Widening and deepening of the Gulf Intracoastal Waterway has been authorized and is under way. Water-resource development interests along the Upper Mississippi River are urging studies of the possibility of recanalization with higher and fewer dams and bigger lock chambers. These projects will not add any channel mileage.

The Tennessee-Tombigbee Waterway has been authorized for improvement by the Congress, but no funds have been made available. This project would connect the Tombigbee River from its headwaters in Mississippi with the Tennessee River at Pickwick Dam, thus providing a route from the Tennessee River through Mississippi and Alabama to connect with the Gulf Intracoastal Waterway at Mobile.

A proposal has been made to extend the Gulf Intracoastal Waterway from its eastern terminus at Carrabelle, Fla., along the shoreline of Florida to Tampa.

Construction is underway on the Cross-Florida Barge Canal, a 185-mi, 12-ft deep waterway with 5 single-lift locks 84 ft by 600 ft. It will provide a link from the Atlantic Intracoastal Waterway at Jacksonville, Fla., to the Gulf of Mexico through the Withlacoochee River valley, 95 mi north of Tampa on the west coast of Florida.

Construction of 19 locks and dams on the Arkansas-Verdigris River System will provide an additional 450 mi of navigable channel from the Mississippi River to Catoosa, near Tulsa, Okla.

Work is underway to complete a 9-ft channel on the Missouri River from its confluence with the Mississippi just north of St. Louis, Mo., to Sioux City, Iowa. The Missouri, from its mouth to Kansas City, now has a controlling depth of 7 ft; from Kansas City to Omaha, $6\frac{1}{2}$ ft; and from Omaha to Sioux City, 4 ft.

Construction work is underway to canalize the Alabama River from its confluence with the Mobile River just north of Mobile, Ala., to Montgomery, with prospects of extending the navigable system to Gadsden, Ala.

Fig. 4. Commercially navigable inland waterways of the United States.

The Chattahoochee River has been canalized and opened to navigation as far north as Columbus, Ga., and studies are underway on extending this navigation system to Atlanta.

[BRAXTON B. CARR]

Inoculation

The process of introducing a microorganism or suspension of microorganism into a culture medium. The medium may be (1) a solution of nutrients required by the organism or a solution of nutrients plus agar; (2) a cell suspension (tissue culture); (3) embryonated egg culture; or (4) animals, for example, rat, mouse, guinea pig, hamster, monkey, birds, or human being. When animals are used, the purpose usually is the activation of the immunological defenses against the organism. This is a form of vaccination, and quite often the two terms are used interchangeably. Both constitute a means of producing an artificial but active immunity against specific organisms, although the length of time given by such protection may vary widely with different organisms. *See* VACCINATION.

Inoculation is the natural process of acquiring

protection against disease, in that most persons are exposed to some organisms at times when no severe symptoms are displayed. The protective mechanisms of the body, especially antibody production, are stimulated by such a mild or insignificant exposure. An example of this is the discovery that the majority of adults have antibodies to poliomyelitis present, despite the absence of a history of the severe or recognizable disease form.

Inoculation may also refer to the deliberate seeding of organisms into culture media, and the introduction of fermenting bacteria, yeasts, or molds into various industrial processes that employ the chemical reactivity of these organisms. *See* IMMUNITY; MICROBIOLOGICAL METHODS.

[EDWARD G. STUART/N. KARLE MOTTET]

Inorganic chemistry

The chemical reactions and properties of all the elements and their compounds, with the exception of carbon-hydrogen compounds. Inorganic chemistry is thus defined by subtraction. The chemistry of carbon-hydrogen compounds forms that province of chemistry designated as organic chemistry. All the remaining elements in the periodic table fall in the domain of inorganic chemistry. The boundaries with other major disciplines in chemistry are not precisely defined, however, and it is often difficult to allocate a given topic to the field of inorganic chemistry or to physical chemistry. Physical chemistry may be defined as the application of quantitative and theoretical methods to chemical problems, and is a methodology rather than a specific body of knowledge. Investigations into theoretical inorganic chemistry or the study of problems in inorganic chemistry by quantitative and sophisticated physical methods may be considered to be inorganic or physical chemistry quite arbitrarily. In similar fashion, metalloorganic compounds may be considered as being either in the sphere of inorganic or organic chemistry. To an increasing extent, the inorganic chemist concerns himself with problems that once were considered the prerogative of physical or organic chemists, or even biochemists.

Because inorganic chemistry concerns itself with 100 elements in the periodic table, its scope is very broad. Nevertheless, some natural divisions exist, and it is convenient to treat the subject under the headings of synthetic inorganic chemistry; theoretical, or physical, inorganic chemistry; and applied inorganic chemistry.

SYNTHETIC INORGANIC CHEMISTRY

The reactivity of the elements of the periodic table varies enormously and over a much wider range than is encountered in organic chemistry. Consequently, the inorganic chemist must frequently employ unusual apparatus and techniques. The elements range from the rare gases, which are unreactive and form very few chemical compounds, to the extremely reactive halogens and alkali metals. Fluorine is perhaps the most reactive element known; it forms compounds with all other elements, including the rare gases. Fluorine compounds are important in the separation of the isotopes of uranium, as refrigerants, anesthetics, chemical warfare agents, high-energy rocket fuels, and for many other purposes. Because of the great reactivity of fluorine and the closely related halogen fluorides, special metal and plastic apparatus must be used in experimentation. Both fluorine and the important fluorine compound, hydrogen fluoride, attack glass, and thus the most common material of construction used by chemists cannot be used.

Another element important in synthetic inorganic chemistry since World War II is boron. The hydrides of boron were first discovered some 60 years ago, but it is only within recent years that the chemistry of these compounds has been clarified. The evolution of synthetic procedures in boron chemistry is an instructive example of the methods of the inorganic chemist. The hydrides of boron were first obtained by reaction of a metal boride with a solution of an aqueous acid. The procedure was very difficult and tedious, and it required weeks or months of labor to obtain a few cubic centimeters of the gaseous product, which turned out to be the simplest boron hydride, B_2H_6. Many years later it was discovered that yields could be increased substantially by passing boron trichloride and hydrogen through an electric discharge, but here also, the yields were very low. Under the impetus of wartime urgency, chemical syntheses for the boron hydrides were developed. Using the readily available compounds, alkali metal hydrides and boron trifluoride, as starting materials, boron hydrides were obtained in very good yield. In addition to diborane, B_2H_6, a considerable number of other boron hydrides are now known: B_4H_{10}, B_5H_9, B_5H_{11}, B_6H_{10}, B_9H_{15}, and $B_{10}H_{14}$, and many derivatives of these have also been synthesized. Compounds that contain the borohydride group, BH_4^-, have been prepared, and the metal borohydrides are now very important compounds. Aluminum borohydride, $Al(BH_4)_3$, is of interest as a possible high-energy fuel, and sodium borohydride, $NaBH_4$, is a widely used reducing agent. *See* BORANE.

The carboranes, a relatively new class of organoboron compounds, have a rich chemistry with many potential applications. Hydroboration, the addition of diborane to organic compounds containing a carbon-carbon double bond, is a preparative reaction that has many applications in organic synthesis. Many new compounds of elements such as silicon, germanium, gallium, phosphorus, and nitrogen, some of which were at one time considered too unstable to be synthesized, have yielded to the synthetic inorganic chemist. An outstanding achievement in inorganic synthesis is the noble gas fluorides. The discovery that the noble gases could form compounds with fluorine and even oxygen has resulted in a whole new area of inorganic research. *See* CARBORANE; HYDROBORATION.

Many boron compounds react violently with air and water. It is necessary, therefore, to use special equipment in studying them, and all-glass vacuum systems have been devised for carrying out experiments with these substances. Vacuum-line techniques are widely used in inorganic chemistry for the manipulation of volatile, highly reactive compounds such as the hydrides of phosphorus, silicon, and related compounds, and an array of vacuum lines often hallmarks a laboratory for synthetic inorganic chemistry. *See* VACUUM PUMP.

Organometallic compounds. Metalloorganic compounds constitute a borderline area of study lying between inorganic and organic chemistry. Until 1950, although a considerable number of metalloorganic compounds was known, all were derived from metals that belong to the principal families in the periodic table; no stable derivatives of subgroup metals or transition metals had been prepared, despite many attempts. Now a large number of metalloorganic compounds of the transition and subgroup elements with cyclopentadiene, C_5H_6, and with aromatic hydrocarbons, such as benzene, C_6H_6, have been synthesized. Because these compounds form interesting derivatives with carbon monoxide, CO, and undergo numerous other reactions, and because the existence of these compounds poses interesting theoretical problems, many investigators have concerned themselves with this field of research. Organometallic compounds of transition elements are powerful catalysts for hydrogenation, and thus find important use in synthetic organic chemistry. *See* ORGANOMETALLIC COMPOUND.

Coordination chemistry. The transition elements in the periodic table are prone to the formation of coordination or addition compounds with a great variety of electron donor molecules. Although a major portion of the research activities of inorganic chemists has been centered on these substances for many years, new discoveries continue at such a pace that it is unlikely that inorganic chemists will soon lose interest in these compounds. Cobalt and chromium coordination compounds are favorites for the study of the mechanisms of inorganic reactions involving aquation or oxidation-reduction. Many coordination compounds have color, thus making fundamental spectroscopic investigations on electronic transitions and energy levels possible. The optical activity of many coordination compounds makes them amenable to study by optical circular dichroism. Many important compounds in living organisms, such as hemoglobin, chlorophyll, and vitamin B_{12}, and various enzymes, are metal coordination compounds, and the efforts of inorganic chemists are now being directed to the study of these biologically important substances. It has been found that there are classes of coordination compounds, and organometallic compounds as well, that can interact with elemental nitrogen, opening the prospect of converting elemental nitrogen to ammonia by ordinary temperature and pressure reactions. *See* NITROGEN COMPLEXES.

Solid-state chemistry. The elements and compounds that form the subject matter of inorganic chemistry exhibit a very wide range of physical properties, and run the gamut from helium, the substance of lowest melting and boiling point, to metal oxides, carbides, and nitrides, which are among the most refractory high-melting substances known. The study of solid-state reactions and compounds has assumed great importance, and is another major area of research in inorganic chemistry. Unlike gaseous compounds and most organic compounds, which follow the law of definite proportions very closely, many solid compounds, particularly those of the transition elements, exhibit variability of composition, or as it is frequently designated, nonstoichiometry. When a solid compound deviates from simple stoichiometric relations, it contains an excess of either positively charged metal cations or negatively charged anions. Under such conditions, the solid frequently shows unusual electronic properties that are made use of in a host of solid-state devices. Solids whose electrical properties are a function of incident light belong to this group, and photoelectric devices are constructed of such substances. Transistors, thermistors, phosphors, and light-emitting diodes are steadily increasing in importance. Since these materials are solid compounds and are frequently prepared by solid-phase reactions, high temperatures may be involved in the syntheses. High-temperature chemistry has greatly expanded, not only in connection with the preparation of nonstoichiometric compounds, but also for the preparation of refractories useful at very high temperatures in rocketry and nuclear technology. Although many reactions can be forced to proceed merely by increasing the temperature sufficiently, this is not in itself adequate for all purposes. A new dimension has been added by the simultaneous use of very high pressures. With the equipment now available, it is possible to carry out solid-state reactions at temperatures of approximately 2500°C (4532°F) and 200,000 atm (3×10^6 psi). Under these conditions, ordinary carbon (graphite) can be converted into diamond, and since the pressure is actually sufficient to distort the electron orbitals, new varieties of matter can be obtained. Thus, it is possible to make a new variety of silica, which, unlike the usual ones, reacts with hydrofluoric acid at a very markedly reduced rate. High-pressure — high-temperature chemistry is among the most important areas of research in inorganic chemistry. *See* HIGH-PRESSURE CHEMISTRY; HIGH-TEMPERATURE CHEMISTRY.

For many solid-state electronic devices, it is of the greatest importance to have very pure material, or material with just the correct amount of impurity. For transistors, silicon containing the order of a few parts per billion of impurities is required. This is commonly achieved by a combination of chemical and physical procedures, including zone melting. The preparation of ultrapure inorganic compounds is also vital in the preparation of luminescent and phosphorescent materials. *See* NONSTOICHIOMETRIC COMPOUNDS; ZONE REFINING.

Geochemical aspects. Many of the synthetic procedures carried out at high temperatures and pressures have considerable interest in geochemistry. Perhaps the earliest inorganic chemistry was practiced in connection with mineralogy. Mineral syntheses, or the preparation of inorganic compounds identical with those found in nature, are increasingly occupying the attention of the inorganic chemist. Not only is this of scientific importance in explaining the sequence of chemical reactions and conditions responsible for the formation of the mineral in nature, but many minerals and gems of industrial importance can now be synthesized, and some, such as diamond, ruby, sapphire, quartz and corundum, on the large industrial scale. Either high-temperature reactions or hydrothermal reactions carried out in water at high temperature and pressure are employed. With the advent of ultrahigh-pressure equipment, it has become feasible to study chemical reactions under conditions approximating those many miles below the Earth's surface. Such studies are expected to add greatly to the understanding of geochemical phenomena. *See* HIGH-PRESSURE PHENOMENA.

Nuclear technology. The development of nuclear technology has provided a great impetus to inorganic chemistry. The efforts of the inorganic chemist have been indispensable to the successful utilization of nuclear energy. The discovery of the transuranium elements, one of the outstandingly important events in chemical science, has provided a host of problems. It has been necessary to explore in detail the chemistry of the actinide elements in order to determine how these new elements were to be incorporated into the periodic table, and for the immediate practical purpose of devising suitable methods for the isolation and purification of plutonium and other transuranic elements. The development of solvent-extraction procedures for separating the actinide elements has been applied to other inorganic ions and has had widespread repercussions in inorganic chemistry. Nuclear technology has also provided the impetus for the development of other separations procedures, for example, the separation of zirconium and hafnium, and of the rare-earth elements from each other, and has generally served to reinforce the traditional interest of the inorganic chemist in separations procedures. *See* NUCLEAR CHEMISTRY.

Applications in organic chemistry. Before leaving the discussion of these aspects of inorganic chemistry, it may be instructive to point out one of its interesting by-products. Many of the most important advances in organic chemistry that have occurred since 1900 have resulted from the introduction of inorganic substances as reagents. The introduction of magnesium metal gave rise to the vast corpus of Grignard chemistry, metal carbonyls to Reppe chemistry, and various boron compounds have found many applications in synthetic organic chemistry. Other inorganic substances that have found important use in organic chemistry are selenium, for dehydrogenation reactions; lead tetraacetate, for selective oxidations; aluminum chloride, as a catalyst for alkylation, acylation, and ring-closure reactions; anhydrous hydrogen fluoride, for diazotization, nitration, sulfonation, and isomerization reactions; and lithium aluminum hydride and various of its derivatives, for selective reduction reactions. Many inorganic compounds are of great use in organic syntheses.

THEORETICAL INORGANIC CHEMISTRY

For a long time, inorganic chemistry was essentially preparative and descriptive, but it is frequently difficult to differentiate modern inorganic chemistry from physical chemistry or chemical physics. The modern inorganic chemist has an intense interest in the structure and electronic properties of chemical compounds. From a knowledge of the interatomic distance and other geometrical data, valuable inferences may often be drawn regarding the nature of the chemical bonding involved and thus of chemical behavior. All modern methods of structure determination are employed. Electron diffraction, x-ray and neutron diffraction, infrared and ultraviolet absorption spectroscopy, magnetic susceptibility measurements, and nuclear and electron paramagnetic resonance and mass spectroscopy are all employed for structure determination. For the case of nonstoichiometric compounds, Gibbs' phase rule studies, x-ray crystallography, and various electrical measurements find particular application.

Bonding. There is an intimate relation between structural studies and detailed interpretation of chemical bonding. A knowledge of the geometry of the molecule contributes to an understanding of the forces involved in chemical bonding, and conversely, for certain classes of compounds, a knowledge of the nature of the chemical bonding helps define the geometry of the substances. In addition to the classic forms of valence, crystal field or ligand field theory has become very important in theoretical inorganic chemistry. It is essentially an extension of the electrostatic theory of chemical forces. Based on molecular orbital theory, crystal field theory considers the effect of the local electric field on the energy levels of the various orbitals involved in chemical bonding. For certain combinations of field and orbitals, splitting of the energy levels occurs, and certain orbitals gain an extra energy of stabilization. The effect of the crystal (or ligand) field may be reflected in a distortion of the molecule, so that crystal field theory is important in interpreting spectroscopic properties. *See* CHEMICAL BINDING; CRYSTAL FIELD THEORY.

One of the most important areas in which crystal field theory is being applied is in the study of coordination compounds. Crystal field theory has enjoyed considerable success in predicting the stability of complexes of different metals with different ligands, in explaining magnetic and absorption spectra of complex compounds, and in predicting rates and mechanisms of reaction of complex compounds. The stereochemistry of coordination compounds is a classic preoccupation of the inorganic chemist, and the study of the complex coordination compounds of the transition elements has been one of the most active and fruitful areas of endeavor in chemical research since 1900. The factors involved in the relative stability of the large number of coordination compounds formed from the various transition elements has occupied the attention of many inorganic chemists, and despite the great advances which have been made, particularly by the application of crystal field theory, many problems are still far from settled. Rapidly assuming great importance is the calculation of various physical properties of inorganic compounds at the onset by procedures using very accurate wave functions. *See* COMPUTER.

Reaction kinetics. The mechanism of inorganic reactions has not in the past attracted as much attention as the study of organic reactions, but this situation is rapidly changing. Organic reactions generally proceed with the skeleton of the molecule remaining intact. Inorganic gas-phase reactions are usually characterized by a complete disruption of molecular structure, followed by reorganization to form the products of the reaction. Gas-phase reactions are thus more complicated than the usual organic reactions, and despite considerable study, much remains to be done. Despite the importance of oxidation-reduction reactions and transition elements, which are characterized by a multiplicity of oxidation states, the mechanism of electron-transfer reactions in solution has only recently become the subject of serious investigation. The results so far indicate that direct electron transfer from one ion to another is an improbable event, that the transfer of electrons between cations is mediated by anions present that act as bridging groups, and that the solvent participates in the electron-transfer reaction. The mechanism of solid-state reactions has also become an important field of research, and there is every reason to believe that the study of reaction mechanisms in inorganic chemistry will be one of great importance in the future development of inorganic chemistry. An important concept in predicting the course of inorganic reactions is that of *hard* and *soft* acids and bases, where the terms hard and soft refer to the ease of polarizability of the reagents. The principle that hard acids react best with hard bases, and likewise for soft reagents, clarifies and systematizes a great deal of inorganic chemistry.

The nature of ions in solution is important in the interpretation of the properties of solutions of electrolytes and in reaction kinetics. Consequently, the study of ions in solution is vigorously pursued. Many methods are available, and these include spectrophotometric, electromotive force, and polarographic measurements, as well as other by now classic physical-chemical procedures, such as solubility measurements, ion exchange, and solvent partition studies. The object is to ascertain the extent of interaction of the metal ions with the solvent, an event which is particularly prominent when highly charged metal ions are involved and water is the solvent. The extent of interaction between cation and various anions that may be present must also be determined. Proceeding in this way, it is possible to specify the kind and amount of each ionic species present in the solution. A knowledge of the ionic species is important, particularly in the interpretation of rate studies, and the study of ions in solution is, in a sense, fundamental to further progress in the field of reaction mechanisms.

APPLIED INORGANIC CHEMISTRY

The production of inorganic chemicals is a basic aspect of the chemical industry, and the large-scale production of the heavy inorganic chemicals, sulfuric acid, ammonia, chlorine, soda ash, and phosphorus, provide indispensable starting materials for many industries. In addition to the tonnage production of inorganic chemicals, the glass, ceramic, cement, fertilizer, and metallurgical industries are essentially inorganic chemistry applied on the large scale. Many of these industries are still traditional in nature, but it is altogether likely that modern inorganic chemistry will make increasingly substantial contributions to the future evolution of inorganic technology. *See* CHEMICAL STRUCTURES; CHEMISTRY; HIGH-PRESSURE PROCESSES; MAGNETO-CHEMISTRY; ORGANIC CHEMISTRY; PERIODIC TABLE; PHYSICAL CHEMISTRY.

[JOSEPH J. KATZ]

Bibliography: F. A. Cotton and G. Wilkinson, *Advanced Inorganic Chemistry*, 2d ed., 1966.

Inositol

A crystalline water-soluble alcohol often grouped with the vitamins. Also called myoinositol, inositol has the structural formula shown in Fig. 1. Of the nine stereoisomers of hexahydroxycyclohexane, inositol is the only one which is of biological importance.

It is estimated by microbiological methods. Virtually nothing is known of its function in enzyme systems. It can be metabolized as a carbohydrate. In seeds it is found as the hexaphosphoric ester phytic acid. The phytic acid of whole wheat is of

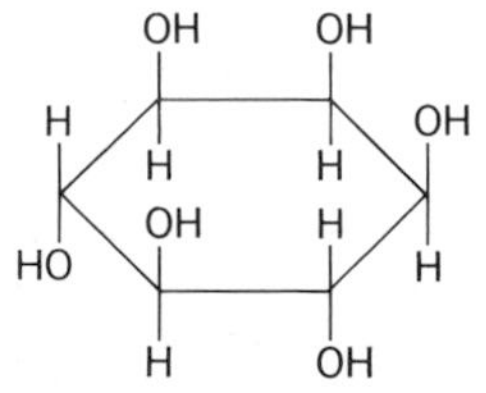

Fig. 1. Structural formula of inositol.

nutritional importance because it combines with dietary calcium, decreasing its availability. Besides being essential for the growth of microorganisms, inositol has been shown to be necessary for mice and rats under some conditions. Inositol deficiency in rodents has been characterized by poor growth and hair loss. Inositol is considered a lipotropic agent, since in a number of experimental conditions it prevents fatty livers. A considerable amount of inositol is synthesized by intestinal bacteria; yeast cells as well as mammalian tissues are capable of inositol synthesis. There is no evidence for a human dietary inositol requirement. *See* CARBOHYDRATE; LIPID; VITAMIN.

[STANLEY N. GERSHOFF]

Industrial production. Inositol is produced commercially from corn-steep water, which is a byproduct of the wet milling of corn. Inositol occurs in corn-steep water as acid salts of its hexaphosphate ester, phytic acid. Treating the steep water with lime yields a precipitate of crude calcium phytate. This is recovered and subjected to high-temperature hydrolysis under pressure, yielding inositol and calcium acid phosphate. The soluble phosphate is precipitated by neutralization with lime, and the clarified solution is concentrated until inositol crystallizes. Purification is accomplished by carbon treatment and recrystallization.

In spite of its seeming simplicity, the process is expensive, because product yield is extremely small in comparison to total material handled in the process.

[CLARENCE S. SHERMAN]

Biosynthesis in mammals. Although mammals may require dietary supplements of inositol, part of their nutritional need for the vitamin is met by their own tissues which are capable of synthesizing the compound. Many years ago organic chemists, noting the similarity in structure between inositol and its isomer glucose, succeeded in cyclizing glucose to inositol and suggested that this same transformation might occur in living tissues. Since 1964 biochemists working with plant and animal tissues have proved the accuracy of this prediction, although the intermediate steps in the living process differ markedly from those in the test tube.

Analysis by carbon labeling. To facilitate the study of a chemical reaction in a living cell, biochemists attempt to isolate the reaction from the myriad other reactions occurring simultaneously. One way to do this is to label the substance under study with an isotopic marker such as C^{14}, administer it to an animal, and then examine the tissues or excreta for compounds containing the isotope, easily detected by its radioactivity (Fig. 2). With glucose in which various of the normal C^{12} atoms were replaced synthetically by C^{14}, inositol isolated from the tissues was found by chemical analysis to contain C^{14} only in carbon atoms corresponding to those labeled in the glucose. These observations demonstrated unequivocally that inositol is derived intact from the glucose chain. Had glucose been broken down to small fragments which were then rebuilt into inositol, the label would have been distributed randomly around the inositol ring.

Enzyme isolation. A more searching means of studying a biochemical reaction is to isolate the

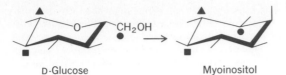

Fig. 2. Molecular conformation diagrams which illustrate the conversion of C^{14}-labeled glucose to myoinositol. The different symbols represent the corresponding label in both molecules.

enzyme which catalyzes the reaction. Since enzymes are proteins they can be separated from each other and from other tissue proteins by a variety of techniques. The tissue under study is first homogenized, a process of mechanically breaking cell walls, making the cell contents accessible to further treatment. The homogenate is then centrifuged at various speeds, separating cell particles according to their density—the heaviest particles, such as cell wall fragments and nuclei, sedimenting at lowest speed. Under the high gravitational field of the ultracentrifuge, the homogenate is separated into a transparent supernatant fluid and a gelatinous pellet of cell particles. Enzymes can be found in both fractions, those in the supernatant designated as soluble and those bound to particles as particulate. Thus far the most active animal source of the inositol-synthesizing system is the supernatant fluid prepared from tubules of rat testis. *See* CENTRIFUGATION (BIOLOGY); ENZYME.

Centrifugation is only the first step in isolating an enzyme. A common further step, successful in the isolation of the inositol enzyme, is precipitation with ammonium sulfate. Proteins fall into several categories of solubility, depending on the salt content of the solvent; by the addition of ammonium sulfate, a highly soluble and relatively innocuous salt, various proteins become insoluble at specific concentrations and can then be removed by centrifugation and redissolved for further purification. The inositol system is precipitated from a solution 30–40% saturated with ammonium sulfate.

Dialysis, a further step in the purification of the inositol enzyme system, is based on the difference in size between protein molecules and other compounds involved in chemical reactions. When a protein solution is placed in a cellophane bag, which in turn is bathed in a dilute salt solution, small molecules pass freely back and forth through the bag while the vastly larger protein molecules remain in the bag. In this way unwanted materials are removed from the protein as the outside liquid is replaced by fresh solution. Additional steps, such as chromatography or electrophoresis, may be required in enzyme purification. *See* CHROMATOGRAPHY; DIALYSIS; ELECTROPHORESIS.

Reaction sequence. Although enzymes are indispensable to biochemical reactions, they are often not the only requirement. For inositol synthesis from glucose, three other substances are essential: adenosine triphosphate (ATP), nicotinamide adenine dinucleotide (NAD), which is also known as diphosphopyridine nucleotide (DPN), and magnesium ions (Mg^{++}). Since these are all small molecules, they are lost during the purification of the enzyme. By replacing these compounds singly or

in various combinations, further insight can be gained into the detailed mechanism of the reaction. As a result of such studies, the conversion of glucose into inositol was found to be catalyzed by three separate enzymes acting sequentially: a kinase requiring ATP converts glucose into glucose-6-phosphate; the cyclase requiring NAD converts glucose-6-phosphate into inositol-1-phosphate; and a phosphatase requiring Mg^{++} converts inositol-1-phosphate into inositol. *See* ADENOSINETRIPHOSPHATE (ATP); DIPHOSPHOPYRIDINE NUCLEOTIDE (DPN).

The first reaction, phosphorylation of glucose, is common to many pathways of glucose metabolism and is the principal step by which glucose, the chief fuel, is prepared for further utilization by the organism. Glucose-6-phosphate is assayed by a specific dehydrogenase which requires nicotinamide adenine dinucleotide phosphate (NADP), also known as triphosphopyridine nucleotide (TPN), as coenzyme. In the course of dehydrogenation of glucose-6-phosphate, NADP is reduced to NADPH, which, compared to NADP, exhibits strong absorbance of ultraviolet light. Since ultraviolet light absorbance can be measured with great sensitivity, the reaction can be followed spectrophotometrically with minute amounts of material. The cyclization product of the second reaction is also measured enzymatically by its hydrolysis to inositol and inorganic phosphate, catalyzed by the purified phosphatase of the third reaction. This enzyme is highly specific so that, even in the presence of glucose-6-phosphate, inorganic phosphate is derived solely from inositol-1-phosphate and is measured with great sensitivity spectrophotometrically. *See* COENZYME; GLUCOSE; TRIPHOSPHOPYRIDINE NUCLEOTIDE (TPN).

These studies illustrate a general approach to the elucidation of a biochemical mechanism from the first superficial observation to the detailed structure arrived at after deeper probing. The earliest experiments suggested that glucose is cyclized to inositol; later it became clear that phosphorylated sugars, and not free sugars, are involved. From the NAD requirement it is clear that other intermediates are implicated and that the actual cyclizing species is probably a phosphate ester at some higher state of oxidation than glucose-6-phosphate.   [FRANK EISENBERG, JR.]

Biosynthesis in yeast. Most yeasts are able to synthesize inositol from glucose or other carbohydrates which can be converted to glucose. In this biosynthesis the entire molecule of glucose, with its six carbon atoms intact, is converted to inositol through the following sequence of enzymatic reactions.

Glucose is first phosphorylated to glucose-6-phosphate by the enzyme glucokinase in the presence of an energy source, which in this case is adenosinetriphosphate (ATP), as shown in Eq. (1).

$$\text{D-Glucose} + \text{ATP} \rightarrow \text{D-Glucose-6-P} + \text{ADP} \qquad (1)$$

The glucose-6-P then undergoes a complex reaction in which the first carbon atom becomes covalently bonded to the sixth carbon atom, giving rise to inositol-1-phosphate. This cyclization is catalyzed by the enzyme glucose-6-P cyclase and requires the participation of DPN and NH_4^+ ions. Further studies established that of the two possible stereoisomeric forms of inositol-1-P (D or L), only D-inositol-1-P is formed. Thus the cyclase catalyzes the reaction shown in Eq. (2a).

$$\text{D-Glucose-6-P} \rightarrow \text{D-Inositol-1-P} \quad (2a)$$
$$\text{Glucose-6-P} + \text{DPN} \rightarrow$$
$$\text{5-Ketoglucose-6-P} + \text{DPNH} \quad (2b)$$

During the cyclization of D-glucose-6-P to D-inositol-1-P, glucose-6-P is converted as an intact unit to D-inositol-1-P, with carbons 1 and 6 of glucose-6-P becoming carbons 6 and 1 of D-inositol-1-P, respectively. During this cyclization the phosphate group remains attached to the original carbon atom.

Little is known of the mechanism of the cyclase reaction. However, various lines of evidence have led to the proposal of the following ordered sequence of reactions for the cyclization of D-glucose-6-P to D-inositol-1-P: (1) oxidation of glucose-6-P to 5-ketoglucose-6-P with DPN as the coenzyme, as shown in Eq. (2b); (2) dissociation of a C-6-bound hydrogen and then condensation of C-1 and C-6 of glucose-6-P, leading to ring closure; (3) reduction by DPNH (formed in the oxidation of glucose-6-P) of an intermediate during or following ring closure.

Besides yeasts, the biosynthesis of inositol from glucose has been shown to occur in preparations obtained from the higher plants and neurospora. In these systems, as in yeast, D-glucose-6-P and D-inositol-1-P have been shown to be intermediates.

The next step is the dephosphorylation of D-inositol-1-P to inositol and inorganic phosphate, P_i. This reaction, shown as Eq. (3), is catalyzed by a

$$\text{D-Inositol-1-P} \rightarrow \text{Inositol} + P_i \quad (3)$$

specific phosphatase known as inositol-1-phosphatase. See YEAST.   [FRIXOS CHARALAMPOUS]

Bibliography: I. W. Chen and F. C. Charalampous, *Arch. Biochem. Biophys.*, vol. 117, 1966; I. W. Chen and F. C. Charalampous, *Biochim. Biophys. Acta*, vol. 136, 1967; I. W. Chen and F. C. Charalampous, *J. Biol. Chem.*, vol. 241, 1965; F. Eisenberg, Jr., *J. Biol. Chem.*, vol. 242, 1967; H. Kindl and O. Hoffmann-Ostenhof, *Biochem. Z.*, vol. 339, 1964; H. R. Mahler and E. H. Cordes, *Biological Chemistry*, 1966; E. Piña and E. L. Tatum, *Biochim. Biophys. Acta*, vol. 136, 1967; T. Posternak, *The Cyclitols*, 1966.

Insect control, biological

The term biological control was proposed in 1919 to apply to the use or role of natural enemies in insect population regulation. The enemies involved are termed parasites (parasitoids), predators, or pathogens. This remains preferred usage, although other biological methods of insect control have been proposed or developed, such as the release of mass-produced sterile males to mate with wild females in the field, thereby greatly reducing or suppressing the pests' production of progeny. Classical biological control is an ecological phenomenon which occurs everywhere in nature without aid from, or sometimes even understanding by, man. However, man has utilized the ecological principles involved to develop the field of applied biological control of insects, and the great majority of practical applications have been achieved with insect pests. Additionally, such diverse types of pest organisms as weeds, mites, and certain mammals have been successfully controlled by use of natural enemies.

History. Man observed the action of predacious insects early in his agricultural history, and a few crude attempts to utilize predators have been carried on for centuries. However, the necessary understanding of biological and ecological principles, especially those of population dynamics, did not begin to emerge until the 19th century. The first great applied success in biological control of an insect pest occurred 2 years after the importation into California from Australia in 1888–1889 of the predatory vedalia lady beetle. This insect feeds on the cottony-cushion scale, a notorious citrus pest that was destroying orange trees at that time, and the vedalia was credited with saving the citrus industry. This successful control firmly established the field or discipline of applied biological control as it is known today. See POPULATION DYNAMICS.

Ecological principles. Natural enemy populations have a feedback relationship to prey populations, termed density-dependence, which results in the increasing or decreasing of one group in response to changes in the density of the other group. Such reciprocal interaction prevents indefinite increase or decrease and thus results in the achievement of a typical average density or "balance." However, this balance may be either at high or low levels, depending on the inherent abilities of the natural enemies. If the natural enemy is highly density-dependent, it can regulate the prey population density at very low levels. An effective enemy achieves regulation by rapidly responding to any increase in the prey population in two different ways: killing more prey by increased feeding or parasitism, and producing more progeny for the next generation. The net effect is a more rapid increase in prey mortality as prey population tends to increase, so that first the trend is stopped and then it becomes reversed, lowering the prey population. This results in relaxing the pressure caused by the enemy so that ultimately the prey population is enabled to increase again, and the cycle is repeated over and over.

Applications of classical method. Since the vedalia beetle controlled the cottony-cushion scale some 80 years ago, many projects in applied biological control have been undertaken and a large number of successes achieved. There are about 300 recorded cases of applied biological control of insects in 70 countries. Of these recorded cases over 85 have been so completely successful that insecticides are no longer required. In the other cases the need for chemical treatment has been more or less greatly reduced. Also, there are about 47 cases of biological control of weeds by insect natural enemies. In California alone it is estimated that the agricultural industry has been saved $272,000,000 during the past 52 years because of reduction in insect pest-caused crop losses and diminished need for chemical control. A few of the many outstanding successes include the biological control of Florida red scale in Israel and subsequently in Mexico, Lebanon, South Africa, Brazil, Peru, Texas, and Florida; Oriental fruit fly in Hawaii; green vegetable bug in Australia; citrus blackfly and purple scale in Mexico; dictyospermum scale and purple scale in Greece; and olive scale in California.

All the cases of applied biological control mentioned above involved foreign exploration and importation and colonization of new exotic natural enemies of the insect pest in question. In the majority of cases the pest was an invader, being native to another country, and its natural enemies had been left behind. By searching the native habitat for effective enemies and sending them to the new home of the invader pest, the biological balance was reconstituted. This method is considered to be classical biological control, and its scientific application has been responsible for most of the outstanding results obtained. There are, however, two other major phases of biological control which are categorized under the headings of conservation and augmentation.

Conservation. This phase involves manipulation of the environment to favor survival and reproduction of natural enemies already established in the habitat, whether they be indigenous or exotic. In other words, adverse environmental factors are modified or eliminated, and requisites which are lacking such as food or nesting sites may be provided. Even though potentially effective natural enemies are present in a habitat, adverse factors may so affect them as to preclude attainment of satisfactory biological control. Insecticides commonly produce such adverse effects, causing so-called upsets in balance or pest population explosions.

Augmentation. This phase concerns direct manipulation of established enemies themselves. Potentially effective enemies may be periodically decimated by environmental extremes or other adverse factors which are not subject to man's control. For example, low winter temperatures may seriously decrease certain enemy populations each year. The major means of solving such problems has involved laboratory mass culture of enemies and their periodic colonization in the field, generally after the adverse period has passed. This practice is gaining rapidly in application. *Trichogramma* sp., common egg parasites of lepidopterous pests, have been utilized in this manner with reportedly good results in many countries, and various microorganisms likewise have been successfully used.

Advantages over chemical control. Although research and development costs are modest for biological control projects, they are high for new insecticides. Biological control application costs are minimal and nonrecurring, except where periodic colonization is utilized, whereas insecticide costs are high annually. There are no environmental pollution problems connected with biological control, whereas insecticides cause severe problems related to toxicity to man, wildlife, birds, and fish, as well as causing adverse effects in soil and water. Biological control causes no upsets in the natural balance of organisms, but these upsets are common with chemical control; biological control is permanent, chemical control is temporary, usually one to many annual applications being necessary. Additionally, pests more and more frequently are developing resistance or immunity to insecticides, but this is not a problem with pests and their enemies. Both biological and chemical control

have restrictions as far as general applicability to the control of all pest insect species is concerned, although the application of biological control remains greatly underdeveloped compared to chemical control. *See* ENTOMOLOGY, ECONOMIC; INSECTICIDE.   [PAUL DE BACH]

Bibliography: R. R. Askew, *Parasite Insects*, 1971; K. F. Baker and R. J. Cook, *Biological Control of Plant Pathogens*, 1974; C. P. Clausen, *Entomophagous Insects*, 1962; P. DeBach (ed.), *Biological Control of Insect Pests and Weeds*, 1964; P. DeBach, *Biological Control by Natural Enemies*, 1974; C. B. Huffaker (ed.), *Biological Control*, 1971; W. W. Kilgore and R. L. Doutt, *Pest Control: Biological, Physical and Selected Chemical Methods*, 1967; E. A. Steinhaus (ed.), *Insect Pathology*, 1963; L. A. Swan, *Beneficial Insects*, 1964; R. van den Bosch and P. S. Messenger, *Biological Control*, 1973.

Insect pathology

A biological discipline embracing the general principles of pathology (disease in the broadest sense) as applied to insects. It refers to man's observations and actions concerning the cause, symptomatology, gross pathology, histopathology, pathogenesis, and epizootiology of the diseases of insects; it is concerned with whatever can go wrong or become abnormal in an insect. A diseased insect may be suffering from an infectious disease caused by a microorganism or a noninfectious disease, such as a metabolic disturbance, a genetic abnormality, a nutritional deficiency, a physical or chemical injury, or injury caused by parasites or predators.

Insect pathology draws upon and contributes to the general field of microbiology and provides understanding of certain of the biological relationships existing between insects and microorganisms not pathogenic to them. Insect pathology finds applications in agriculture, medicine, and biology generally. Microbial control, the use of microorganisms in biological control, is one area of applied insect pathology. Microorganisms are introduced to control insect pests for the protection of man and other animals and agricultural crops. However, the suppression of disease in beneficial insects, such as the silkworm, honeybee, and ladybird beetle, is also of significant practical importance. *See* INSECT CONTROL, BIOLOGICAL.

Insect pathology may be viewed as a part of invertebrate pathology and a part of the branch of entomology. Rapid developments in insect pathology since 1945 have provided new impetus for research in the pathology of mollusks (clams, oysters, and snails), crustaceans (crabs and lobsters), and acarina (spiders, mites, and ticks). *See* INVERTEBRATE PATHOLOGY.

In examining the subject of diseases of insects, it should be kept in mind that society's interest in food, fiber, shelter, and health has focused attention on only a very small fraction of the invertebrate animal species, which comprise 97% of all species of animals.

History. From the time of Aristotle, the honeybee was known to suffer from disease, and maladies of the silkworm were recognized during the Middle Ages. The first recorded insect pathogen was a *Cordyceps* fungus on a lepidopterous insect, reported and illustrated by René de Reaumur in

1726. In 1826 William Kirby included a chapter on the diseases of insects in the famous treatise, *An Introduction to Entomology*, by Kirby and W. Spence. In 1835 Agostino Bassi published his great work on muscardine, a fungus disease of the silkworm. Louis Pasteur gained fame and the appreciation of his countrymen in 1870 for his monumental studies on pébrine and flacherie of the silkworm, and for devising methods of controlling the former, thereby saving the silk industry of France from ruin.

That destructive insects were also subject to disease was recognized by such early biologists and entomologists as B. D. Walsh, J. L. LeConte, H. A. Hagen, E. Metchnikoff, S. A. Forbes, F. H. Snow, L. O. Howard, and W. F. Fiske. However, prior to World War II, most information concerning diseases in insects and the nature and properties of the causative agents was provided by a small group of men, the more prominent of whom were F. d'Herelle, S. S. Metalnikov, G. F. White, A. Paillot, E. Masera, and R. W. Glaser. Their contributions did much to establish the foundations of insect pathology and to furnish basic information as to the essential character of microbial diseases in insects. The establishment of the Laboratory of Insect Pathology at the University of California in 1945 by Edward A. Steinhaus provided the first educational and research center for insect diseases and opened a new era. A research laboratory was established by the Canadian Department of Agriculture in 1946 and was followed by numerous projects and laboratories for the study of insect diseases in the United States and in many other countries of the world.

The fact that microorganisms which cause diseases in insects are harmless to humans and other animals has stimulated business interests in the United States and Europe to develop microbial insecticides. In 1960 Thuricide, a product containing spores of a bacteria pathogenic for larvae of Lepidoptera, was the first microbial insecticide that the United States Department of Agriculture and the Food and Drug Administration permitted to be used on food crops. Additional products incorporating bacterial, fungal, viral, and other pathogens may be expected to be used safely against injurious insects.

Pathogenesis and epizootiology. Many factors contribute to the condition or state described as a disease. It is important when diagnosing a disease to understand its pathogenesis, that is, how the pathology develops and how the individual host responds. Identifying the infectious microorganism or physical, chemical, or genetic factors affecting the insect is only a part of the knowledge needed for determining pathogenesis. Distinguishing one disease from another is as basically important in insect pathology as it is in the study of disease in other forms of life. One must know what is abnormal about or has killed an insect before the disease can be properly studied, controlled, or suppressed, used as a microbial control measure, its potentialities for natural spread determined, or its role in the ecology of an insect species known. *See* ECOLOGICAL INTERACTIONS.

An important facet of insect pathology is concerned with the epizootiology of diseases, which involves the study of disease in populations of insects rather than in individual animals. Any epizootic affecting an insect population is concerned with three primary natural entities: the infectious agent, the insect host, and the environment. Each of these factors has certain attributes that, when properly related to the attributes of the others, play their appropriate role in determining the initiation, rise, and decline of an epizootic. Knowledge concerning the nature of epizootics is extremely important in insect ecology generally and in an understanding of insect-microbe ecosystems. The degree to which disease-producing microorganisms are dependent upon the density of the host population, the susceptibility of the host insect to the disease, the influence of weather and other environmental conditions on epizootics, the mode of transmission of pathogens, and other factors are important aspects of the epizootiology of insect diseases. In the microbial control of an injurious insect, man can initiate and increase the rapid rate of development of an epizootic by controlling the amount, method of distribution, and time of introduction of an effective microorganism.

Infectious diseases. While the noninfectious diseases and abnormalities of insects are very important and of numerous types, most of the activity in insect pathology has been with the infectious diseases. The infectious agents responsible for diseases in insects belong to the same major groups as those that cause such diseases in other animals: the bacteria, fungi, viruses, rickettsiae, protozoans, and nematodes. In general, however, insects are not normally susceptible to those particular microorganisms that cause diseases of man, other animals, and plants. Moreover, most of the microorganisms that cause fatal diseases in insects are harmless to plants and to higher animals. Insects often can resist infection by humoral and cellular types of immunity; acquired humoral immunity is manifested in the hemolymph. However, it appears that the type of antibody immunity so characteristic of higher animals does not occur in insects.

A convenient way to consider the various infections of insects is on the basis of the etiologic agent, as will be done in the sections to follow.

BACTERIAL DISEASES

The characteristics of a bacterial disease in an insect may vary, but in general the following may be observed. As the disease develops, the insect becomes less active, has a smaller appetite, and discharges fluids from the mouth and anus. Infection may begin as a dysenteric condition with an accompanying diarrhea, but in most instances the invading bacterium eventually enters the body cavity of the insect and causes a septicemia that terminates in the death of the host. Following death, the insect's body usually darkens to brown or black. This is especially true of larvae and pupae in which the disintegration takes place rapidly.

The freshly dead insect is usually soft and flaccid. Internal tissues may disintegrate to a viscid consistency, sometimes accompanied by odor, but ordinarily they do not "melt," or liquefy, as do insects dying of certain virus infections. The cadaver usually dries and becomes shriveled, the integument remaining intact. Microscopic examinations of smears or of histological sections of an insect dead or dying of a bacterial disease usually disclose large numbers of the causative bacterium. These bacteria, if they are pathogens, must be differentiated from similar-appearing saprophytes that are normally in the insect intestine and that may flourish in the tissues of the dead insect. In some cases the bacterial pathogen may not be present in large numbers if it has killed the insect primarily through the production of toxins.

The first major study of a bacterial disease of a destructive insect was that of grasshoppers reported by d'Hirelle in 1911 from Yucatan, Mexico.

Foulbrood. The first bacterial diseases of insects to receive concentrated study were flacherie of the silkworm, which Pasteur considered at the time he was studying the protozoan disease pébrine of this same insect, and European foulbrood of the honeybee, which was studied by F. R. Cheshire and W. W. Cheyne in 1885. Later it was found that, in addition to European foulbrood, caused by a group of non-spore-forming bacteria, American foulbrood caused by the spore-forming bacterium *Bacillus larvae* also afflicted this beneficial insect. After the role of these bacteria in causing diseases of larval bees (hence the term foulbrood) was eluci-

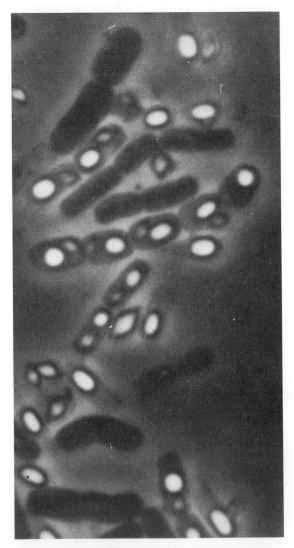

Fig. 1. *Bacillus thuringiensis* Berliner, showing spores as well as enclosed and free crystals. (*Laboratory of Insect Pathology, University of California, Berkeley*)

dated, it was possible to institute measures of control; however, in some areas these and other bee diseases still cause losses to beekeepers.

Milky diseases. In addition to the diseases of bees, two of the most important bacterial diseases of insects are the so-called milky diseases, which affect grubs (larvae) of the Japanese beetle (*Popillia japonica*) and are caused by *Bacillus popilliae* and *B. lentimorbus*; and the disease of numerous lepidopterous larvae caused by *B. thuringiensis*. All of these bacteria are sporeformers and are effective in the control of certain insect pests. Infection of Japanese beetle grubs occurs through ingestion of the bacterial spore, which germinates, penetrates the insect's intestinal wall, and develops in the body cavity, where the bacteria sporulate. The blood of the infected grub assumes the milky-white appearance that gives the disease its common name. Eventually the insect dies and decomposes, thus providing new spores which may be consumed by other grubs. In the eastern United States a dust containing milky-disease spores is used, at times in conjunction with chemicals and insect parasites, to control the Japanese beetle in lawns and turf. Distribution of this material has been carried on by the U.S. Department of Agriculture, which did much of the original work on the disease. Products containing the spores are commercially available.

Crystalliferous bacteria. *Bacillus thuringiensis* is one of the most intensively studied bacterial pathogens of insects because it has proved to be an effective microbial control agent against a number of lepidopterous pests, such as the alfalfa caterpillar and the cabbage butterfly (Fig. 1). A number of closely related species of this bacillus are known to cause diseases of beneficial insects, such as silkworms. These bacteria are characterized by the formation of a protein crystal in the sporangium at the time of spore formation. The material comprising the crystal is highly toxic for certain insects, primarily Lepidoptera, but apparently harmless for other forms of life. The crystal usually appears bipyramidal but may occur in other forms. Ordinarily each sporangium contains but one crystal; the crystal is freed from the cell wall with the spore and appears to persist indefinitely. *B. thuringiensis* and related species also produce a heat-stable, water-soluble, dialyzable substance, distinct from the crystal and from lecithinase, that is toxic for certain insects when injected, and to larval Diptera when they are exposed to it in a rearing medium. *See* BACTERIAL ENZYME.

Preparations containing the spores and crystals of *B. thuringiensis* or its varieties may be sprayed or dusted on crops to destroy lepidopterous pests. Commercial products containing the bacillus are available. One of the primary advantages of this type of control is that this and other entomophilic microorganisms are nontoxic and noninfectious for plants and higher animals.

FUNGAL DISEASES

There are a large number of species of different groups of fungi that are pathogenic for insects. Their development on and in susceptible hosts depends greatly upon the presence of adequate moisture to provide for the germination of conidia and spores. Whereas bacteria, viruses, and protozoans infect their insect hosts by way of the mouth and digestive tract, most entomophilic fungi enter their hosts directly through the integument or body wall.

Many of the fungi associated with insects are not true pathogens, or are pathogenic only under certain conditions. These saprophytic species frequently develop in or on insects that have died of other causes, since the conidia of the fungi may be closely associated with the living healthy insect. Some fungi appear to be parasitic but not lethal in their association with insects; thus the large order Laboulbeniales (Ascomycetes) is made up of species that live primarily on the external surfaces of insects. The genus *Septobasidium* consists of species associated with scale insects which, although some are parasitized, live beneath the stromata of the fungi. Some fungi, such as the ambrosia fungi of certain termites and wood-boring beetles, are definitely mutualistic in their relationship with insects. *See* ASCOMYCETES; BASIDIOMYCETES.

Fungal development. Once the germ tube from a conidium is within the body cavity of an insect, the threadlike hypha of the fungus proliferates, invades the tissues, and fills the body of the insect. In most cases the fungus then develops conidiophores to the exterior of the host body, where conidia develop to provide a means for initiating development of the fungus in new hosts. The infected insect usually assumes a dried mummylike appearance, frequently becomes covered with a conidia-bearing mycelium, or sometimes contains resting spores which enable the fungus to survive periods of adverse environmental conditions or absence of a suitable host. Toxins may be produced by the fungus during the initial period of hyphal growth.

Entomophilic fungi. Some entomophilic fungi occur in the orders Mucorales, Blastocladiales, and Chytridiales, but the most important fungi of the class Phycomycetes belong to the order Entomophthorales. In this order the single family Entomophthoraceae is made up of several genera, of which *Entomophthora* and *Massospora* are composed primarily of entomophilic species. One of the most commonly known species is *Entomophthora muscae*, which attacks the housefly. A large number of *Entomophthora* species are known. They infect many species of Diptera, Orthoptera, Lepidoptera, and other insects.

Infected houseflies are frequently found indoors attached to the walls and ceilings of buildings in a lifelike position. Close inspection of flies killed by the fungus usually reveals on the wall or windowpane a distinct halo of discharged spores (conidia) encircling the insects. This halo results from the fact that, after the conidiophores have emerged through the integument of the insect, conidia are formed which are discharged violently into the air from the terminal portion of the conidiophores, thus forming a ring of conidia about the dead insect. When the discharged conidium comes into contact with a susceptible insect in the presence of adequate moisture, it begins to germinate, sending out a conidial hypha (germ tube) that penetrates through the integument into the body cavity. *See* PHYCOMYCETES.

Blastocladiales. The entomophilic Blastocladiales are confined largely to one group, the family Coelomomycetaceae, which parasitizes mainly mosquito larvae. About 40 species are known, all

Fig. 2. Larvae of European corn borer (*Pyrausta nubilalis*) killed by *Beauveria bassiana*. Whitish coat of the fungus covers anterior parts of larvae.

in the genus *Coelomomyces*. Certain regions of the body cavity or virtually the entire hemocoel of the mosquito may become filled with the spores and mycelium of the fungus.

Hypocreales. Species of the genus *Cordyceps*, an ascomycete of the order Hypocreales, are among the first recorded entomophilic fungi. They are frequently of large size and very colorful in appearance. About 250 species are known. They are cosmopolitan in distribution and occur on representatives of several orders of insects, principally Hemiptera, Diptera, Lepidoptera, Hymenoptera, and Coleoptera. In members of the genus *Cordyceps*, the stroma arises from a sclerotium formed within the body of the infected insect. At the end of this stroma, or stem, is a fertile portion, the "head," which may be brightly colored. The life history of most *Cordyceps* is similar to that of entomophilic fungi in general.

Muscardine. One of the most important groups of entomophilic fungi is responsible for muscardine diseases, in which the fungus emerges from the body of the insect and covers the animal with a characteristic fungus mat resembling, in a way, a French bonbon or candy mint. The word muscardine was first used for white muscardine, a well-known disease of the silkworm caused by the fungus *Beauveria bassiana*. It has also been used in reference to green muscardine, a disease of the European corn borer, the wheat cockchafer, and other insects caused by *Metarrhizium anisopliae*. These fungi belong to the Fungi Imperfecti. *B. bassiana* and closely related species have been identified from a large number of other insects. In certain situations the fungus has shown some promise as a means of control of pest insects. The fungi produce toxin in the host, accelerating the death of the insect. Among their best known hosts, in addition to the silkworm, are the European corn borer (Fig. 2), codling moth, chinch bug, and the alfalfa weevil. *See* FUNGI IMPERFECTI.

VIRAL DISEASES

The insect species infected by viruses are principally in the order Lepidoptera, although a few

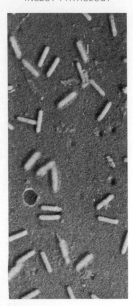

Fig. 3. Electron micrograph of virus particles from granulosis of omnivorous looper (*Sabulodes caberata*).

Hymenoptera and Diptera and one or two Coleoptera and Neuroptera are known to suffer infection with these agents. Only the immature stages, larva and pupa, are killed by virus. Adults may carry the virus, but they are not killed by it. In general, many insect viruses exhibit a fairly high degree of host specificity, but numerous instances of cross infectivity have been recorded. *See* INCLUSION BODIES, VIRAL; VIRUS.

Polyhedroses. The polyhedroses are characterized by the formation of polyhedron-shaped inclusion bodies in the infected tissues of the host. These proteinaceous inclusion bodies contain the virus particles embedded in their matrices. In the nuclear polyhedroses the virus multiplies in the nuclei of the infected cells, and in the cytoplasmic polyhedroses the virus multiplies in the cytoplasm of such cells. As far as is known, viruses causing nuclear polyhedroses are rod-shaped and contain deoxyribonucleic acid (DNA); those causing cytoplasmic polyhedroses are more or less spherical and contain ribonucleic acid (RNA).

Larvae infected with the virus of a nuclear polyhedrosis usually show few distinctive symptoms until a few hours before death. The incubation period varies in different host species between 5 and 20 days; usually it covers a period of about a week. In some species of insects, the infected larvae may cease feeding, become somewhat sluggish in movement, and become yellowish or pale in color. They may swell slightly, then become limp and flaccid. The tissues most prominently affected are the fat-body cells, epidermal cells, blood cells, and the cells of the tracheal matrix. Shortly before and after death the integument is very fragile and easily ruptured, emitting the liquefied contents, which are filled with disintegrating tissue and polyhedra. Polyhedra increase in size and number in the nucleus of the cell until the nuclear and cell membranes are disrupted. The dead larvae are usually found hanging by their prolegs from the host plant or other support. Eventually, they may burst or dry to a dark-brown or black cadaver.

The polyhedra are insoluble in water, alcohol, ether, and acetone but soluble in weak acids and alkalies. They range in size from 0.5 to 15 μ. If the polyhedra are treated properly with dilute alkali, the rod-shaped particles within them can be demonstrated. The size of the virus rods ranges from about 20 to 50 mμ in width to about 200 to 350 mμ in length. The virus rods may occur singly or in bundles of up to as many as eight rods.

The cytoplasmic polyhedroses are, in general, not so fulminating as the nuclear polyhedroses. The host integument does not become fragile, as with the nuclear polyhedroses. In the cases known, the infection is limited largely to the midgut epithelium, where the polyhedra are formed in the cytoplasm of the cells. The polyhedra are morphologically similar to those present in nuclear polyhedroses. The virus particles, on the other hand, tend to be spherical in shape, about 20 to 70 mμ in diameter, and near the surface of the polyhedron. One of the best-known cytoplasmic polyhedroses is that of the silkworm, which also suffers from a nuclear polyhedrosis. *See* DEOXYRIBONUCLEIC ACID (DNA); RIBONUCLEIC ACID (RNA).

Granuloses. The granuloses have so far been found only in larvae of Lepidoptera. They are char-acterized by the accumulation of small granular inclusion bodies, called capsules, in the infected cells. Each capsule contains a single virus rod composed of DNA (Fig. 3). Compared to the nuclear polyhedroses, the symptoms associated with the granuloses are more subdued; the diseased larvae usually become less active, somewhat flaccid, and assume a pallid or whitish translucent aspect. The integument does not become fragile. The period from infection to death ranges from 6 to 20 days. The inclusion bodies, or capsules, are about 200 by 500 mμ in size, and the virus rod ranges, depending on the virus, from 40 to 80 mμ in width by 200 to slightly over 300 mμ in length.

Other viruses. Among the insect viruses so far discovered that are not associated with an inclusion of any kind are those which infect certain Lepidoptera (army worms), Hymenoptera (honeybees), Coleoptera (beetle grubs), and Diptera (mosquitoes, crane flies). A number of the noninclusion viruses exhibit a unique physical characteristic when concentrated in large quantity. Centrifugate pellets of the virus particles appear orange or amber in color by transmitted light and iridescent or turquoise by reflected light. Another peculiarity is the fact that an exceptionally large amount of virus is produced in the host tissues. A noninclusion virus called "sigma virus" is known to occur in Drosophila. This virus renders infected flies sensitive to carbon dioxide. Among acarines, noninclusion viruses have been found in at least two species of mites.

There is considerable evidence that insect viruses frequently remain occult and cause latent infections in their hosts. Some authorities believe that a virus can survive in a host, even for several generations, without causing recognizable symptoms, but that under the influence of certain stresses or incitants, such as cold, crowding, and certain chemicals, can be triggered into causing an active or frank infection.

RICKETTSIAL DISEASES

A number of rickettsiae or rickettsialike organisms are capable of causing frank infections in insects. Three or four have been found in Coleoptera and one in Diptera. These infections appear to be slow to develop and kill their hosts in 1–4 months. The fat body of grubs assumes a characteristic bluish-green coloration, and in some host species crystalline bodies occur in association with the rickettsie. *See* RICKETTSIALES.

PROTOZOAN DISEASES

Protozoan infections in insects may be benign and cause little morbidity or mortality, but many are severe and highly fatal. Some appear in epizootic proportions, while others may be represented by only a local invasion of tissue by the protozoan. The success of reproducing populations of insects can be seriously affected by destruction of gonads without death of the host. Generally, protozoan diseases are relatively slow in developing, and frequently they become somewhat chronic in nature; but some sporozoan infections develop rapidly and kill the insect host within a short time. Infected insects may show few, if any, external signs of infection, or may be stunted in growth and development or not undergo metamorphosis, or show a change in translucency and color. Usually they become opaque and whitish because of the accumulation of spores or cysts in the internal tissues or fluids. Further, infected insects may exhibit a loss of appetite, abnormality of movement, and may remain moribund for long periods prior to death.

Only occasionally, as with *Leptomonas pyrrhocoris* in certain plant bugs, do flagellates cause actual disease in insects. A few amebas cause disease. There is an important amebic disease in the honeybee and another in grasshoppers. In both of these maladies the insect's Malpighian tubes are especially involved. Several species of Ciliata have been found infecting insects; others are frequently found adhering to the exterior of insect larvae and pupae living in aquatic environments. Perhaps the best known are ciliates associated with mosquitoes. *See* AMOEBIDA; CILIATEA; MASTIGOPHORA; TRYPANOSOMATIDAE.

Sporozoa. The most important protozoan diseases in insects are caused by Sporozoa, of which the Microsporida as a group are the most serious. Many Gregarinida are found in insects. Most of the so-called eugregarines from the alimentary canal are not pathogenic for their hosts, but the schizogregarines found in the body cavity may be highly virulent. A number of Coccidia are also pathogenic for insects. *See* SPOROZOA.

Microsporida. Of the infections caused by members of the order Microsporida, perhaps the most famous is pébrine of the silkworm, caused by *Nosema bombycis*. Another well-known infection is nosema disease of the honeybee, caused by *Nosema apis* (Fig. 4). Intensive studies of these two diseases have revealed a great deal about the Microsporida. Microsporidan infections have been found in species of at least 14 orders of Insecta. The most common insect hosts are Diptera, Lepidoptera, Coleoptera, and Ephemerida. Some of these are rather specific as to their insect hosts, while others are capable of infecting insects in different genera or even in different orders. *See* HELICOSPORIDA.

Spore stage. The most commonly seen form of any Microsporida is the spore stage, the characteristics of which are important in distinguishing the different species. The spore serves as the resistant stage of the organism and is able to tide the pathogen over periods of unfavorable environmental conditions and during the period between the change of host individuals. The average microsporidan spore is 3–6 μ long and 1–3 μ broad: some species exhibit a range of spore sizes. The shape of the spore is usually oval-spheroidal, although it may be spherical or bacilliform. The spore consists of a spore membrane or covering surrounding a sporoplasm and a polar filament coiled within the spore. The polar filament is capable of extrusion as a very fine, long thread that may be several hundred microns in length in some species.

Microsporidan development. Although the life cycles of Microsporida vary from species to species, in general, soon after the spore is ingested by a susceptible host, the polar filament is extruded. A successful infection is initiated when the tip of the polar filament deposits sporoplasm from the spore either within or between the epithelial cells of the insect's intestine. The parasite, now called a

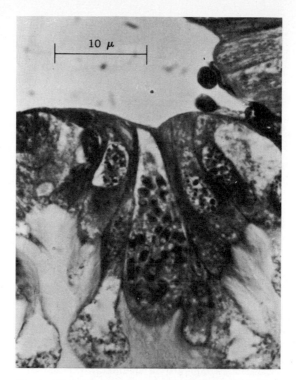

Fig. 4. Photomicrograph of epithelial cells in mid-intestine of adult honeybee. The enlarged cell in the center is filled with spores of *Nosema apis* Zander. (*Insect Pathology Laboratory, Ohio State University*)

planont, may remain in the intestinal cell or pass into the body cavity, where in either location it multiplies by binary fission. Various tissues of the body can be invaded, and the developing protozoan is now designated as a schizont or meront. In their intracellular location the spherical-to-oval meronts divide actively by fission, budding, or multiple division, filling the cytoplasm of the host cell. This period of schizogony ends in the formation of sporonts, each of which produces one sporoblast or many, depending on the genus of the microsporidian. Each sporoblast becomes a spore. This is the sporogony part of the life cycle. The entire life cycle is usually completed in about 4 days. Insects are usually infected with Microsporida through the mouth. However, some of these protozoans are transmitted from one generation to the next by way of the eggs.

NEMATODE DISEASES

Those roundworms for which insects serve as primary hosts are included in the classes Nematoda and Nematomorpha. Some members of the class Acanthocephala spend their larval stage in an insect host and the adult stage in mammals. More than 1000 species of roundworms have been reported from insects; most of these belong to Nematoda that frequently kill or seriously harm their hosts. At least 16 orders of insects are involved as hosts to nematodes (Fig. 5). About one-third of the known species have been reported from Lepidoptera. *See* ACANTHOCEPHALA; NEMATODA; NEMATOMORPHA.

All nematodes are similar in their general external appearance. Their elongate, unsegmented body has little or no variation in diameter, although

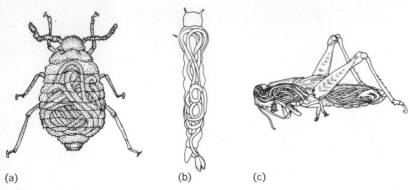

(a)   (b)   (c)

Fig. 5. Insects parasitized by nematodes. (*a*) Root aphid (*Anoecia*) infected with an undetermined nematode (*after Davis*). (*b*) Larva of *Aedes aegypti* (Linn.) infected with *Mermis* sp.; one is emerging via the anus (*after Muspratt*). (*c*) Grasshopper nymph (*Melanoplus*) containing one fully grown female *Agamermis decaudata* (*after Christie*). (*From E. A. Steinhaus, Principles of Insect Pathology, McGraw-Hill, 1949*)

it does taper toward one or both ends. There are three main stages in the developmental cycle of most nematodes: egg; juvenile, or larval, which includes four growth stages; and adult. Usually the young larvae spend a short period as free-living organisms, frequently in an aquatic environment.

The biological relations between nematodes and insects vary all the way from mere fortuitous association to obligate parasitism. For convenience these relationships may be separated into three general groups: (1) nematodes that live in the alimentary tract of the insect in a more or less commensal association; (2) nematodes that combine saprophagous and parasitic habits, that is, semiparasites; and (3) nematodes that parasitize the body cavity or tissues of their hosts, called obligate parasites. Examples of the first relationship are found in the alimentary tracts of crickets and cockroaches. These nematodes cause little, if any, harm to the insect. Of the second group, the semiparasites, two well-known members are species of *Neoaplectana*, of which *N. glasseri* infects the Japanese beetle. The second member, although a *Neoaplectana*, has not been identified definitely as to species. It is important, however, in that it has shown some promise as a biological control agent against a large number of insects, including the codling moth, corn earworm, and cabbageworm. The obligate parasites are a large group. Among the best known are the Mermithidae, which parasitize and kill grasshoppers. *See* MERMITHOIDEA.

NONINFECTIOUS DISEASES

Although most of the present activity in insect pathology appears to be concerned with the infectious diseases, increasing attention is paid to the noninfectious diseases of insects. These noninfectious diseases and abnormalities may be grouped as follows: (1) mechanical injuries such as traumata, bruises, and torn tissues; (2) injuries caused by physical agents such as burning, freezing, and drought; (3) injuries caused by poisons or chemical agents such as insecticides; (4) injuries caused by parasitization or infestation by other insects or arachnids; (5) diseases caused by a nutritional disturbance or a deficiency of proper nutrients or vitamins; (6) diseases caused by deranged physiology and metabolism; (7) inherited abnormal conditions

or genetic diseases; (8) congenital anomalies and malformations; (9) certain tumors and neoplasms; and (10) disturbances in development and in regenerative capacity of tissues. *See* INSECT PHYSIOLOGY; INSECTA; INSECTICIDE.   [JOHN D. BRIGGS]

Bibliography: L. Bailey, *Infectious Diseases of the Honey-Bee*, 1963; P. DeBach, *Biological Control of Insect Pests and Weeds*, 1964; M. Rockstein (ed.), *Physiology of the Insecta*, 1964; K. M. Smith, *Insect Virology*, 1967; E. A. Steinhaus, *Insect Pathology: An Advanced Treatise*, 1963.

Insect physiology

The study of the functional properties of insect tissues and organs. The adaptations enabling insects to live on land are often strikingly different from those found in other animal groups, such as the higher vertebrates. Accordingly, insect physiology makes a comparison of physiologic adaptations in insects which have analogous functions in other forms.

Insects, because of their diversity, abundance, and wide distribution, exert great influence on the general character of life on land. Insect physiology, therefore, contributes to the broad study of terrestrial ecology. Insects assume economic importance as crop pollinators, disease vectors, and pests. Insect physiology is also concerned with the control of certain insect species and with the mode of action of insecticides. *See* ENTOMOLOGY, ECONOMIC.

DEVELOPMENT AND GROWTH

Fertilization is internal. Unfertilized eggs normally develop into males in many ants, bees, and wasps, and into females in summer generations of aphids. In some species of walkingstick the males are either extremely rare or have never been found. Egg development is external in most cases and early cleavage in the egg is limited to the nuclei. *See* ANT; APHID; BEE.

Postembryonic growth. The period of postembryonic growth is discontinuous, being interrupted by 3–8, or more, molts. During each intermolt period the insect feeds and gains weight, but there is no increase in the linear dimensions of the hard parts. At the molt the old cuticle is shed and the new cuticle expands and hardens within an hour or less. In primitive wingless insects there are only slight changes in form at each molt. In others, such as the grasshoppers and roaches, the adults differ from the immature stages, or nymphs, by possessing wings and fully formed reproductive organs. In the higher insects, exemplified by the butterflies, bees, wasps, ants, flies, and beetles, growth takes place during a series of larval stages that are specialized for feeding. During the inactive pupal stage, the body of the adult differentiates at the expense of degenerating larval tissue and fat.

Endocrine activity. Tissues of the adult are already determined in the egg, and can be identified in the larva as relatively slow-growing imaginal disks. Proliferation and differentiation of larval or nymphal tissue precede each molt except the last. Activation of latent adult tissue takes place at the end of the last larval stage and continues during the pupal stage. These processes and the ensuing molts are stimulated by a hormone, ecdysone, periodically released from the prothoracic glands. Secretion of the prothoracic glands is initiated by a hormone produced by neurosecretory cells in the brain. The prothoracic hormone acts on larval and adult tissue alike, but during nymphal and larval stages, its action on adult tissue is largely suppressed by a juvenile hormone (neotenin) coming from the corpus allatum, a small gland near the brain. At pupation a decrease in neotenin permits full action of the prothoracic hormone on adult tissue within the larva. Thus, extra molts and abnormally large insects can be produced by transplanting corpora allata from earlier stages. Conversely, premature pupation and dwarf adults result from excision of the corpora allata from young larvae. *See* ECDYSONE; ENDOCRINE SYSTEM (INVERTEBRATE); NEUROSECRETION.

Diapause, a condition in which development stops even at normal temperatures, occurs in many insects that overwinter in immature stages. Development can be initiated by exposing diapausing pupae to a period of cold, and then rewarming them. This treatment reactivates the brain and the prothoracic gland system. In insects having several generations a year, the last fall generation may be caused to go into pupal diapause because the larvae have been exposed to progressively shorter day lengths during their later stages.

Cuticle. The epidermal cells secrete a semirigid, acellular cuticle that covers all external parts and also lines the respiratory system and the fore- and hindguts. The cuticle serves as a structural skeleton of basically cylindrical form, and all skeletal muscles pull on its inner surface. A special rubberlike and highly flexible cuticle forms the hinges of movable structures such as the wings.

The structural components of the cuticle are protein and chitin, a complex of polyacetylglucosamine. The hardening and darkening of the cuticle is due to oxidation of tyrosine and other phenols to quinones, which form cross bridges with proteins. Cuticle colors may be due to pterines, xanthines, and carotenoids of plant origin, and to iridescent effects created by fine sculpturing of the cuticle surface. *See* CHITIN.

The chitinous cuticle is traversed by fine pore canals, through which waxy or oily secretions reach its surface and spread out into a thin hydrophobic layer. On most insect cuticles, a very thin cement layer is found outside this hydrophobic layer. These layers play an important part in limiting water loss by evaporation. The effectiveness of many oil-soluble insecticides, such as DDT, depends in part on solubility in the wax layer of the cuticle. *See* INSECTICIDE.

Growth is limited by the rigidity of the cuticle. Before molting, the deeper layers of the old cuticle are dissolved by enzymes and a new pliable cuticle is laid down beneath it. Finally a molting fluid is secreted between the cuticles, and gas accumulates in the gut. This causes an increase in body bulk which splits the remains of the old cuticle along the back. Gas is retained in the gut until the new cuticle hardens.

The semirigid cuticular exoskeleton has been a major factor in determining the success (number of species, environments occupied) and limitations (body size) of the group Insecta. *See* INSECTA.

SYSTEM PHYSIOLOGY

In spite of their specialization, insects have basic physiologic mechanisms common to most forms of life. Some of these mechanisms, because of the unique manner in which they are displayed in the insects, are especially amenable to experimentation.

Respiration. Openings, or spiracles, on the sides of certain body segments connect with a system of air tubes, the tracheae, that branch to every part of the body (Fig. 1). Tracheae are often enlarged into air sacs. Internal cuticular ridges (taenidia) prevent collapse of the tracheae and their terminal branches, the tracheoles. The tracheoles end in submicroscopic branches close to or within all body cells. At certain times the tracheal system may be filled with fluid which can be partially or completely withdrawn into the tissues. This process has not been fully explained.

Exchange of oxygen and carbon dioxide may occur partly or entirely through the body cuticle particularly in small insects and eggs, or by passive diffusion through the spiracle-tracheal sytem. In large insects or during periods of muscular activity passive diffusion is augmented by active ventilation. The tracheal tubes, or air sacs, are rhythmically compressed by pumping movements of the abdomen or by contraction of the muscles during locomotion or flight. Gas flow in the main tracheal trunks may be regulated by spiracular valves sensitive to changes in oxygen or carbon dioxide pressure. The final gaseous exchange appears to be accomplished by diffusion through tracheal walls to the body cells.

Aquatic insects that remain continuously submerged have a closed gas-filled tracheal system, into which gases diffuse through gills. Diving insects return periodically to the surface to renew a film of air carried on the hairy hydrophobic cuticle or under the wing cases. Other aquatic or endoparasitic forms have a spiracular "siphon" that breaks the surface film.

Many insects tolerate low oxygen tensions for long periods. The blood has little or no respiratory function, except in the larvae of some midges living in stagnant water. The blood of these larvae contains hemoglobin in solution. Most insects are insensitive to carbon monoxide owing to the absence of hemoglobin. Except in diapausing pupae, cyanide is usually poisonous because of the presence of a cytochrome system. *See* RESPIRATORY PIGMENTS (INVERTEBRATE).

Digestion and nutrition. There is little of nutritive value that is not attacked by insects. Diets

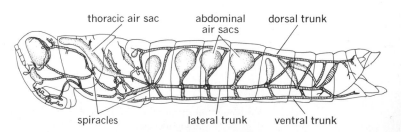

Fig. 1. Respiratory system of the grasshopper viewed from the left side. (*From T. I. Storer and R. L. Usinger, General Zoology, 4th ed., McGraw-Hill, 1965*)

range from fluids, such as plant juice and blood, through conventional animal and vegetable foods, to cellulose, wax, and wool. Mouthparts vary greatly with diet. The digestive tract is degenerate in the adults of some species in which feeding is limited to the immature stages. Many insect larvae feed as internal parasites within the bodies of other animals.

Conventional digestive enzymes are produced by the salivary glands, which secrete silk in some species, and by the midgut epithelium. Mucus is absent. In many insects the gut contents are enclosed in a thin chitinous sac, the peritrophic membrane, which is continuously secreted by the midgut. Products of digestion diffuse through the peritrophic membrane before absorption by the gut epithelium. The hindgut absorbs water from the digestive waste and plays an active part in metabolism and regulation. It is also used as a propulsive and respiratory organ in dragonfly larvae, which draw water in through the anus and then force it out again.

Keratin, the protein of wool and feathers, is digested by the larvae of some beetles and moths. The disulfide bonds of keratin are changed to sulfhydryl groups by strongly reducing conditions in the midgut. Beeswax is the normal diet of wax moth larvae; symbiotic bacteria, as well as enzymes secreted by the gut, may play some part in its digestion. Wood digestion is accomplished, in termites and some roaches, by symbiotic bacteria and protozoa always present in the gut.

In spite of their varied and often unusual diets the nutritional requirements of insects are fairly conventional. Ten amino acids, the B vitamin complex, and choline are essential for growth. Essential nutrients lacking in the diet are often synthesized by symbiotic microorganisms. Adult flies may exist for long periods on sugar alone, but require an external source of protein for the eggs to develop. Adult butterflies and moths require only sugar and water, their eggs maturing at the expense of internal sources of protein. The fat body is the main storage site for nutritive materials.

Symbiotes. A number of insects harbor microorganisms within, or among, the cells of specialized tissues, as well as within the cavity of the gut. These symbiotes include protozoa, fungi, yeasts, bacteria, and bacterialike organisms. In some cases microorganisms appear to be essential to the life of their insect host as in the termites with their intestinal protozoans. In others, they contribute to host nutrition and growth by increasing the availability of nutritional factors, such as certain vitamins. Still other insects show no obvious defects when deprived of their symbiotes.

Certain bacterialike organisms are cultured within specific cells, grouped in a large organ, the mycetome, or scattered throughout the fat body. Infection of the next host generation by these intracellular symbiotes may be accomplished by migration of certain of the host cells to the walls of the ovary, where the symbiotes are released and penetrate the developing egg cells. Intestinal symbiotes may be picked up when the host insect feeds on contaminated food. In other cases the eggs of the host become contaminated at the time of oviposition.

Circulation. The blood is propelled forward, at low pressure, by a dorsal heart. It is discharged into the body cavity and filters back to the heart through spaces between organs and tissues. Accessory hearts may promote the blood flow in elongate organs, such as antennae, legs, and wings. Since the blood has little or no respiratory function, the circulation may be stopped for hours without untoward effects. The heart rhythm is probably neurogenic, as in other arthropods. This means that the beat originates in the activity of nerve cells in the heart wall, from which nerve impulses are transmitted to the striated heart muscle. In some larvae, the direction of the heartbeat is reversed at certain times.

The blood is commonly colorless, yellowish, or green. Blood cells are variable in form and numbers, show phagocytic properties, and may aggregate during blood clotting. The osmotic pressure of blood varies widely in different species and under different conditions. The sodium-potassium ratio ranges from 10 or more in carnivorous insects (and in land vertebrates) to less than 1 in many herbivorous forms, such as the silkworm larvae. The protein content is similar to that of vertebrate blood, although the amino acid content may be 50 times higher. The uric acid content is also exceptionally high. The major carbohydrate is trehalose, a nonreducing sugar. *See* BLOOD; CIRCULATION.

Water regulation and excretion. Terrestrial insects conserve their body water by having a watertight cuticle and by reabsorbing most of the water before voiding urine and feces. Some beetles live on dry food, such as stored grain and wood, under conditions where free water is not available for drinking. They conserve, as their body fluid, the water formed as one of the end products of oxidative metabolism. Insects having a fluid diet of plant juice or blood excrete large amounts of watery urine or other fluid, like the honeydew of aphids. In fresh-water insects, the osmotic tendency to gain water is limited by low cuticular permeability and by excretion of dilute urine. Salt loss in fresh water is offset by salt intake in the food, by active salt retention, and by salt uptake through anal papillae, as seen in mosquito larvae.

Land insects are able to tolerate considerable blood loss or desiccation, blood dilution, and changes in the relative concentrations of blood sodium and potassium. Nerves and muscles depend for normal function upon a fairly constant ratio of these ions in the surrounding fluid. In insects, these tissues are protected against fluctuations in the blood composition by the Malpighian tube and rectal gland system and by a special sheath surrounding ganglia and nerves. Other regulating mechanisms are still to be discovered.

The nitrogen waste of land insects is eliminated, mainly in the form of uric acid, by the Malpighian tubes. A variable number of these long fine tubes opens into the hindgut, their closed ends lying free among other organs in the body cavity. Fluid passes from the blood into the Malpighian tubes and is concentrated in the lower parts of the tubes and in the hindgut. The urine consists mainly of a paste, or powder, or uric acid crystals. Excess salts, like potassium from plant juices or sodium from vertebrate blood, ingested during feeding, are excreted in the same manner.

Insects like the blowfly larvae, living in fluid or semifluid environments, eliminate some of their nitrogen waste as ammonia. Salts of ammonia, urates, phosphates, carbonates, and pigments derived from the food may be stored in the fat body and other tissues for part of the larval or pupal life (storage excretion). On emergence of the adult these metabolic end products may be eliminated in the first bowel movement (meconium) or they may be incorporated as pigments into the cuticle on the scales of butterfly wings. *See* EXCRETION.

RECEPTORS AND RECEPTION

Sense organs in insects respond to chemicals, water vapor, light, and mechanical stimuli and provide information for a variety of orientation mechanisms ranging from a simple turning toward or away from a stimulus to the complex light-compass orientation used by bees.

Vision. Many adult insects possess two kinds of light-sensitive organs, compound eyes and ocelli. In both types, the light-sensitive region consists of a number of cell groups (retinulae), each containing 2–8 photoreceptor cells. The photoreceptor cells of each retinula are arranged radially, with their central light-sensitive portions (rhabdomeres) either lying in close contact or fusing to form a central structure, the rhabdome. *See* EYE (INVERTEBRATE).

In compound eyes each retinula is provided with a separate optical system consisting of a convex corneal facet of cuticle and an underlying crystalline cone (Fig. 2a). These visual units (ommatidia) are partially or completely shielded from one another by light-absorbing pigments so that each rhabdome is most effectively stimulated by light entering the ommatidium with a few degrees of its long axis (Fig. 2b). The shielding pigments are often capable of active movement and may be withdrawn under conditions of low light intensity so that light entering an ommatidium may reach neighboring units as well.

Each ommatidium is a visual unit of small visual field (about 8°) and has a visual acuity, or resolving power, limited by the number of rhabdomeres (8 or less). Each compound eye may contain from a few dozen to many thousands of ommatidia, arranged radially in a convex array, covering in some cases a solid angle greater than 180°. Thus, with both compound eyes an insect can survey most of the immediate environment simultaneously, without the need for eye or head movements.

The angle between the axes of adjacent ommatidia varies in different insect species and in different regions of the same eye. Small ommatidial angles are encountered on the front of the eye in predatory insects, such as the praying mantis and dragonfly. In regions of the eye of the blowfly where this angle is about 1°, the visual field of each ommatidium is about 8°. A point within this field can be resolved by light falling on any one of the 8 rhabdomeres, but since the visual fields of neighboring ommatidia must overlap, the point may also be resolved by rhabdomeres in adjacent ommatidia. Similarly, the visual fields, covered by groups of ommatidia on the anterior aspects of the right and left compound eyes, often overlap one another.

Although the visual acuity of the compound eye may depend in some degree upon ommatidial angle, the overlapping of ommatidial fields indicates that the mechanism is certainly more complex than this. Visual acuity of insects is inferior to

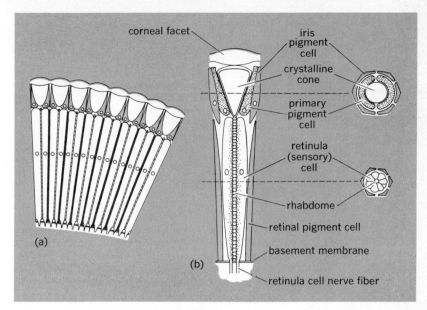

Fig. 2. Structures of a compound eye. (a) Vertical section of a portion of a compound eye. (b) Detail of a single ommatidium, the visual unit. (*From R. E. Snodgrass, Principles of Insect Morphology, McGraw-Hill, 1936*)

that of man, being about 1/100 in the honeybee and about 1/1000 in the fruitfly. Form perception, necessarily limited by visual acuity, is manifested in activities such as the learning, or recognition, of foraging territory by honeybees, wasps, and dragon flies. Though their visual acuity is poorer, insect eyes discriminate much higher frequencies of flicker than man; this may aid them as they fly rapidly at close range past flowers and leaves.

Light energy reaching the rhabdomeres is absorbed by a visual pigment. In some unknown way the photochemical reaction which attends absorption of light elicits electrical changes in the photoreceptor cell bodies and in their axons. The latter enter a complex system of optical ganglia which make up a large portion of the brain. *See* PHOTORECEPTION.

Insect eyes are more sensitive than the human eye at the ultraviolet end of the spectrum, but less sensitive at the red end. Within this broad range some insect eyes can distinguish qualitative differences among various wavelengths. Behavioral studies show that honeybees recognize four "colors," each representing a fairly wide range of wavelengths: "ultraviolet," 300–400 millimicrons (mμ); "blue," 400–480 mμ; "blue-green," 480–500 mμ; and "yellow," 500–650 mμ. A sensory basis for color vision has been demonstrated electrophysiologically but not behaviorally in compound eyes of flies and cockroaches and in dorsal ocelli of honeybees. How widely color vision is distributed among insects remains to be discovered. *See* COLOR VISION.

In some of the species two or three dorsal ocelli occur which contain many retinulae grouped beneath a single, undivided cornea. They do not form images. Their behavioral role is rather obscure, though in the cockroach they are involved in the maintenance of a diurnal activity rhythm.

Taste and smell. The senses of taste and smell are considered together since both depend on the action of specific chemicals. The common distinction between taste and smell depends upon whether the chemical reaches the sense organ in solution or in the form of a vapor.

The sense organs of taste are fine cuticular hairs that often serve in addition as touch receptors. A minute receptive area at the tip of each hair is reached by fine nerve processes from three to five sense cells at its base. Chemoreceptive hairs occur on the feet, antennae, ovipositor, and mouthparts. *See* CHEMORECEPTION.

Electrical recordings from the chemoreceptive hairs show that the sensory cells are divided into several classes. One type responds only to sugars, and the most stimulating sugars are those found in nectar and honeydew. A second type of cell responds only to salts, particularly to common table salt. A third type responds only to water. Most chemoreceptive hairs are supplied with one of each type of cell—a sugar receptor, a salt receptor, and a water receptor. A fourth cell at the base of the hair responds when the hair is bent.

The sense of taste is easily demonstrated by applying drops of solution to chemoreceptive hairs on the feet of flies, bees, or butterflies. If the insect has been deprived of water, its mouthparts will immediately extend to drink from the drop. If it has been deprived of food but has drunk water to satiation, the insect will extend its mouthparts only if the drop contains sugar above a certain concentration. If, in addition to sugar, the drop contains salt, acid, alcohol, or many other substances, the feeding response may be blocked. The taste sensitivity to sugar and many other substances has been tested by this method in insects. The sensitivity to salt, and other substances that are rejected, has also been measured by finding the lowest concentration of sugar. Taste sensitivity varies widely with the species tested and the individual's state of nutrition. In insects sensitivity range to sugar, salts, acids, and quinine covers that of human beings and may extend beyond it.

Less is known about the physiology of the olfactory mechanism. Nerve impulses can be recorded from chemosensory endings on the antennae when certain vapors are wafted over them. Among the mysteries of the olfactory sense are its sensitivity to high dilutions and its selectivity for certain chemical structures. It is estimated that only 40 molecules of the sex attractant produced by the female silkworm moth are needed to evoke a behavioral response in the male. The high potency of the female sex attractant has been used by researchers to lure the males of destructive species, such as the gypsy moth, into traps.

Knowledge of the pheromones is rapidly expanding. Pheromones are chemicals released into the air or deposited on the ground from special glands which have specific significance as a message to other animals of the same species. Sex attractants are an example. Ants have six or more of such glands, the individual secretions of which signify "enemy," "nest-mate," "food," and so on to nest-mates. *See* PHEROMONES.

A water sense, the ability to discriminate between relative humidities and to orient toward or away from water vapor, even at a distance, is well developed in insects. This sense again appears to depend upon the antennae, but nothing is known about its mechanism.

Touch, posture, and hearing. These senses depend upon mechanical stimuli. Since insects are enclosed in a semirigid cuticle, they must have special means for detecting mechanical changes in the outer world.

The sense of touch is mediated by hairs and spines of all sizes covering the cuticle of the legs, antennae, and other regions. Many of these hairs are set in hinged sockets in the cuticle. When the hair is bent, a sense cell below the cuticle at its base is stimulated, and a series of nerve impulses is transmitted inward to the segmental ganglion. Some hairs are so fine and lightly poised that gentle air currents or sound vibrations cause bending sufficient to stimulate. Others are large and stout, responding only to impacts with obstacles encountered by the insect.

A sense of the position of parts of the body relative to each other is important in posture and in a number of actions. The position of one leg joint relative to the next, or of the head relative to the body, is detected by small groups of fine hairs placed in the hinge so that they are bent in varying numbers as the joint is flexed. Part of this postural sense is determined by the force of gravity on the standing insect and by the tensions from contracting muscles. These forces cause bending, or shearing, stresses in the cuticle that are detected by groups of sense cells placed below small elliptical domes of thin cuticle in the leg joints. These campaniform organs are also found in flies at the base of the halteres (small rods surmounted by knobs) that replace the second pair of wings. During flight the halteres are vibrated some hundreds of times per second in a fixed plane. This gives the halteres the properties of small gyroscopes. Any tendency of the fly to deviate from a straight flight path causes a twisting tendency at the haltere base and stimulates the campaniform organs. The resulting nerve impulses, acting through the nervous system and flight muscles, serve to correct the flight deviation.

Other mechanical sense organs take the form of fine tissue strands attached at each end to the inner surface of the cuticle. Each strand contains a sense cell and an ending (scolops) that appears to be stimulated by vibrations or by changes in the distance between the cuticular attachments. In some insects these scolopophorous organs are especially modified for hearing. The cuticle at one attachment becomes very thin and is exposed to airborne vibrations. Specialized ears of this type are found on the legs of grasshoppers, antennae of mosquitoes, abdomen of cicadas, and abdomen or thorax of certain moths. They are sensitive to change in sound intensity but do not appear to discriminate changes in pitch. The rate at which short sound pulses are repeated appears to be an important aspect of sound to insect ears. Moths can detect sound frequencies of 100 kHz or more. Insect ears serve to warn of a predator's approach (for example, moths can hear the high-pitched cries of hunting bats) or to locate the opposite sex. The majority of insects are very sensitive to surface vibrations. These are detected by fine sensory hairs and campaniform organs as well as by scolopophorous organs.

SOUND AND LIGHT PRODUCTION

Although insects have no sound-producing organs, they can make sound by a variety of mechanisms. Many of these sounds are incidental to ac-

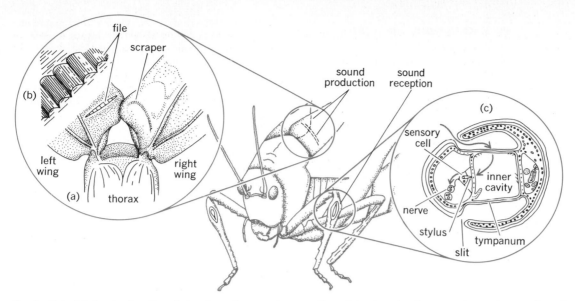

Fig. 3. Katydid showing location of structures for sound production and reception. (a) Undersurface of forewings with file and scraper. (b) Enlarged detail of file. (c) Cross section of foreleg showing sound receptor, or "ear." Arrows show path of sound waves. (*From T. I. Storer and R. L. Usinger, General Zoology, 4th ed., McGraw-Hill, 1965*)

tivities such as feeding or flight, but an increasing number are being found to have communicative significance in insect behavior. Bioluminescence, the production of light by a living organism, is characteristic of a number of species. This activity is a consequence of the chemical reactions in photogenic organs, and serves a communicative function.

Sound production. Insects make sounds by chewing on hard materials; by vibration of wings or special membranes; by rapid expulsion of gas from the tracheal system or digestive tract; by tapping on the substrate with legs, abdomen, or head; and by snapping or rubbing wings, legs, or other body parts against each other.

The songs of grasshoppers, crickets, and cicadas are familiar to all. Grasshoppers move a filelike structure on the leg against the edge of the wing; crickets and katydids draw a scraper on one forewing over a file on the other (Fig. 3); cicadas vibrate a pair of drumlike membranes on the abdomen by means of special muscles. The sounds produced are complex but consist generally of a series of pulses upon which is superimposed the higher resonant frequency of the vibrating organ.

The pulse frequency, pulse grouping, pitch, damping of vibrator, and other characteristics of these sounds are often highly species specific. Some closely related forms can be more easily recognized by their songs than by any other means. The informational content of these songs to other insects depends on differences in pulse frequency and grouping but not on differences in pitch. Male crickets have a repertoire of songs for different behavioral circumstances such as courtship or encounters with other males. In most cases insect songs serve to bring the sexes together or as part of the courting behavior.

Sounds produced by the wingbeat or by vibration of the flight muscles are common and often intense. In mosquitoes, they play a part in sex recognition. Bees produce a variety of sounds by vibrating their flight muscles. Some of these may have behavioral significance. Clicks caused by sudden movements of body segments and hissing sounds from escaping air may play some part in predator evasion.

Light production. Special photogenic organs occur in the larvae and adults of a number of beetles. They are commonly found on the ventral surface of the abdomen beneath a transparent sheet of cuticle, as in the common American fireflies, *Photuris* and *Photinus*, but they also occur on other parts of the body. The light may be yellowish green, bright green, orange, or red in color. It can take the form of a steady glow, of pulsations, or of brief flashes 0.075 sec in duration.

Photogenic organs range from loose, undifferentiated cells to an organized tissue of closely packed photogenic cells backed by a whitish reflecting layer. The latter type, characteristic of the fireflies mentioned above, has a rich tracheal supply. As they approach the photogenic cells, the tracheae narrow and branch into fine tracheoles, at which point they are encompassed by tracheal end cells.

Light is produced within the photogenic cells by the oxidation of a low-molecular-weight complex known as luciferin in the presence of an enzyme, luciferase. The production of flashes appears to be controlled by the nervous system. Nerve impulses probably act directly on an enzyme-controlled reaction within the photogenic cells, rather than indirectly by regulating the oxygen supply through valves in the tracheal end cells. *See* BIOLUMINESCENCE.

At dusk, the males of *Photinus* fly about, flashing spontaneously. The female is wingless, and flashes only in response to the flash of a nearby male. The flying male turns toward her flash and after a few further exchanges, they meet and mate. The male discriminates the female's answering flash from the flashes made by other males since it occurs exactly 2 sec after his signal. In certain parts of the tropics thousands of fireflies may gather in a tree or bush and flash in unison, lighting up the whole tree. The significance of this behavior is unknown. *See* ANIMAL COMMUNICATION.

MUSCULAR MOVEMENT AND FLIGHT

Insect muscles are generally similar to those of vertebrates, except that they exert their pull on the inner surface of the cuticular exoskeleton. Almost all insects are capable of flight, and wing movement is powered by highly specialized muscle groups.

Muscle. Each muscle is made up of multinucleate fibers, each containing numbers of parallel, cross-striated myofibrils. The fine structure of the myofibrils is similar to that found in striated muscle of other animal groups. Smooth muscle and motile cilia are lacking in insects. Tendons are absent, the muscle fibers being attached directly to the inner surface of the cuticle or to inward cuticular extensions (apodemes). A network of striated muscle fibers encloses hollow organs such as the gut, heart, and reproductive tract.

Muscle fibers of leg and most trunk muscles are tubular in form and 10–30 microns (μ) in diameter. A central core contains a number of nuclei. Around this are radial rows of myofibrils enclosed in the muscle fiber membrane. Giant mitochondria (sarcosomes) rich in oxidative enzymes are distributed among the myofibrils. Outside the muscle fiber membrane, and sometimes penetrating the fiber, is a network of tracheae and air sacs.

One to three motor nerve axons branch to all the fibers in a muscle. Terminal branches of one or all of these axons may be found in the membrane at intervals of 40–60 μ over the length of the muscle fiber. One motor axon, the "fast" fiber, supplies most or all of the muscle fibers, and one or more impulses in it generate a quick, powerful twitch. One or more "slow" motor axons supply a fraction of the fibers in a given muscle. A sequence of 15–150 impulses per second in a slow axon causes a slow, gentle, and sustained contraction. The magnitude of this gradual contraction depends upon the frequency of the impulses arriving via the slow axon. Combinations of fast and slow excitation originating in the nervous system determine movements, which may range from a quick, powerful twitch propelling the insect into the air, as in the grasshopper jumping muscle, to slow walking and the steady tensions necessary to maintain the standing posture.

Repeated stimulation of the fast system causes a continuous shortening (tetanus) that develops higher tensions than those reached during a single twitch. Individual twitches fuse into a tetanus at stimulus frequencies ranging from less than 20 to as much as 70 per second. Single twitches may occupy from 0.02 sec to several seconds. In spite of the high overall muscle tensions (20 kg/g of muscle in the locust jumping muscle), the maximum tensions developed in the individual tubular muscle fibers of insects are about the same as those encountered in frog muscle fibers. The work performance in the two groups is also comparable. *See* MUSCLE (BIOPHYSICS); MUSCLE (VERTEBRATE).

Flight. Wings had appeared in some insect groups by the Devonian. It is thought that they served first as immovable gliding surfaces. In most modern insects the wings are hinged to the meso- and metathorax in such a way that elastic deformation of the cuticle deflects the wings through a path roughly diagonal to the long axis of the body. Primitive and nymphal insects lack wings, although most adult forms have two pairs. The flies have only one pair of wings, the hindwings being replaced by halteres. Fleas, probably evolved from flylike ancestors, have lost their wings.

The flapping motion of the wings, which provides most of the power for flight, is caused by contraction of the indirect flight muscles. In the higher insects these muscles consist essentially of two fiber groups running at right angles to each other and attached to the elastic cuticular plates enclosing the thorax (Fig. 4). The thoracic deformation produced by the longitudinal horizontal muscle is reversed by contraction of the vertical muscle, and these muscles contract alternately during wing movement. The indirect flight muscles range from modified tubular leg muscles in slow-flapping primitive insects, like grasshoppers and dragonflies, to fibrillar muscle filling most of the thorax in higher insects. Fibrillar muscles are made up of large fibers, up to 1.8 mm in diameter. They are often colored yellow or pink from high concentrations of respiratory enzymes. These fibers cleave into large bundles of fibrils and contain many irregularly placed nuclei and great numbers of large mitochondria, or sarcosomes. The latter are the seat of the enzymes used in metabolism.

Superimposed on the flapping motion of the wings are changes in wing angle at the top and bottom of each wing stroke. These angular changes are essential in obtaining a resultant force of lift and forward thrust from the flapping motion. The wing movement can be likened to a helicopter with oscillating instead of rotating blades. Changes in wing angle are brought about in part by the complex wing articulation to the thorax, and in part by the pull of small muscles attached directly at the wing base. The direct flight muscles enable many insects to hover, fly backward, and make tight turns, by acting differentially on the wings.

In insects with two pairs of wings the hindwings flap in air rendered turbulent by movement of the forewings, and their aerodynamic efficiency is thereby impaired. In the butterflies, moths, bees, and wasps, the hindwings have become reduced in size, while in the flies they are lost. They may be coupled by hooks to the forewings so as to produce the effect of a single moving air foil. In beetles the forewings (elytra) serve only as covers for the hindwings, which provide the power for flight.

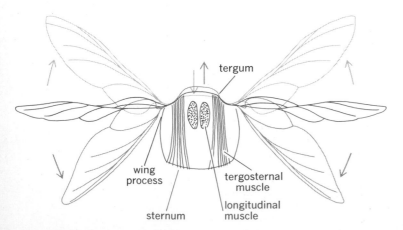

Fig. 4. Wing movements in the flight of an insect. (*From T. I. Storer and R. L. Usinger, General Zoology, 4th ed., McGraw-Hill, 1965*)

A unique characteristic of insect flight is the high frequency at which the wings may beat. This ranges from 5 beats per second in large butterflies to as high as 1000 per second in small midges. Most grasshoppers, dragonflies, and moths have wingbeat frequencies of 15–50; medium-sized flies, 100–200; bees, 200–300; and mosquitoes, 400–500 per second. Since the indirect flight muscles contract and relax once during each beat, the higher frequencies imply an unusual muscular performance.

With a few exceptions insects beating their wings less than 50 times per second have flight muscles similar in structure and physiology to the tubular muscle found in the legs and other appendages. In these each motor nerve impulse from the nervous system brings about one muscle contraction and one ensuing wingbeat. High-frequency wingbeat is associated with fibrillar muscle. One physiological peculiarity of this muscle is that it contracts a dozen or more times for each motor nerve impulse. These impulses appear to generate a critical tension at which the muscle becomes unstable, oscillates, and does work. In other words, the rhythm of fibrillar muscle is myogenic and independent of motor nerve impulse frequency. This oscillatory property of fibrillar muscle is not fully understood, but it depends upon a critical tension generated by nerve impulses, in addition to critical inertia and damping of the load, the elastic properties of the thorax, and in some cases a mechanical click mechanism in series with the muscle and its load. When the muscle tension surpasses that of the spring in the click, the latter goes abruptly from the "up" to the "down" position, and the muscle goes suddenly slack. This slackening may destroy the muscle's ability to develop tension, thus permitting the click to return to its original position. It is not entirely clear how this facilitates the oscillatory behavior of fibrillar muscle. A click seems to be involved in the flight mechanism of some flies and in the sound-producing system of cicadas. In the latter a stiff convex membrane (tymbal) is vibrated in and out 100–500 times per second by a fibrillar muscle attached to its inner surface.

The air speeds of insects rarely exceed 15–20 mph, although some dragonflies may reach speeds of about 50 mph for short periods. A tethered fruitfly is capable of a 2-hr continuous flight and 1,500,000 wingbeats before exhaustion. Tethered locusts are capable of 5–8 hr of unbroken flight. Little is known about the distances free-flying insects can travel nonstop. Favorable winds are undoubtedly responsible for reported trips of several hundred miles.

The immediate energy source for short flights is sugar. An exhausted fruitfly is capable of further flight within 30 sec of feeding on sugar solution. Fat reserves are drawn upon during the longer flights of locusts. Bees may forage several miles for nectar, which contains up to 70% sugar. Long trips are uneconomic, however, since most of the sugar is used in fueling the return flight.

The energy expenditure during flight reaches 2000 kcal/kg of muscle per hour, 10 times that of human heart muscle and twice that calculated for hummingbird flight. Sugar is oxidized within indirect flight muscle by enzymes contained in the numerous mitochondria. The oxygen consumption during flight may be 50 times that when at rest, and the temperature in the thorax may rise 10–15C° above the resting temperature.

Flight can be initiated in most insects by loss of contact between the feet and the ground. Light intensity, flicker, odors, and the relative wind passing over the insect's body all play a part in maintaining and steering flight activity. Stabilization of pitch and yaw tendencies during flight is assisted by inertial or gyroscopic torques acting on the movable head of dragonflies, the halteres of flies, and possibly on other appendages. See FLIGHT.

NERVOUS SYSTEM

Included here are the mechanisms of nerve transmission and integration lying between the sense organs and the muscles or glands. The nervous system cannot properly be considered apart from these input and output systems since they provide the pathways into it from the outer world and the means whereby its activity is exhibited as behavior. However, the nervous system does much more than merely connect sense organs and muscles. This is illustrated by what may be called its permissive property. Throughout life the nervous system is bombarded by a continuous but fluctuating barrage of sensory nerve impulses from many sense organs, yet only a small part of this input can be clearly traced to output in the form of a behavioral response. Similarly, the nervous system is connected with all skeletal muscles, yet a purposive behavioral response is made up of a highly restricted pattern and sequence of muscle contractions. Thus, the permissive quality of the nervous system can be likened to a censorship, or limitation, of the spread of nerve excitation to specific pathways.

Insects are well suited for study of behavior and other problems. Their behavior, though complex, frequently follows stereotyped and inborn patterns. Most of the nervous system, like that of other invertebrates, is spread out in a chain of ganglia lying close to the ventral side of the body. Only the brain is dorsal, lying in the head in close contact with the eyes and antennae. Each pair of ventral ganglia is a semiautonomous center for the sense organs and muscles of its body segment. Longitudinal connectives join ganglia and brain to provide pathways for the integration of their separate activities into a unified pattern of behavior (Fig. 5).

Three general approaches have provided information on the physiology of the nervous system. The surgical approach, or ablation, draws conclusions from the changes in behavior that follow the cutting of nerves or the removal of part or all of specific ganglia. The electrophysiological approach aims to isolate and to map in detail specific events in neural transmission and integration. This method depends upon precise stimulation of nerves and upon detection of the electrical signs of nerve activity. The behavioral approach leaves the animal intact and observes its reactions to measured environmental changes. See NERVOUS SYSTEM (INVERTEBRATE).

Ablation. The surgical approach goes back to the ancient Greeks, who noted the ability of a headless insect to perform many complex acts. It has revealed the following facts about the insect nervous system. A single isolated body segment,

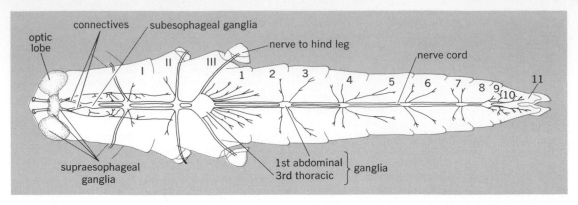

Fig. 5. Grasshopper nervous system, dorsal view. Most of the nervous system, spread in chain of ganglia, lies close to the ventral side of the body. (*From T. I. Storer and R. L. Usinger, General Zoology, 4th ed., McGraw-Hill, 1965*)

bearing a pair of legs, is capable of a variety of responses. These include grasping by the feet when they encounter an object, withdrawal of the leg from a noxious stimulus, alternate rhythmic leg movements as in walking, and respiratory movements. Some of these activities are more intense after neural isolation of a segment; others are less so. Other purposive actions that may be continuous in isolated segments or groups of segments are copulatory movements in the praying mantis, egg laying in a number of insects, and the stinging movements of bees.

In the intact insect, these activities take their proper sequence in the total pattern of behavior through the action of excitatory and inhibitory nervous influences reaching the local ganglia from the brains and other nerve centers. Thus, leg and foot movements are coordinated in the six legs so as to produce effective locomotion by sensory and nervous interactions of the three pairs of thoracic ganglia. The brain, in addition to serving as a receiving and coordinating center for the input from eyes and antennae, determines the presence and direction of locomotion by a combination of excitatory and inhibitory neural actions on the thoracic centers. The local activities of copulation, egg laying, and stinging are suppressed most of the time by higher nerve centers, and only released and steered by them under appropriate stimulus situations. The mechanisms of posture and stance, due to a steady contraction of certain muscle groups, have much the same origin in semi-independent local nerve mechanisms, subject to coordination and regulation from higher centers.

Electrophysiology. The electrophysiological approach depends to a great extent upon a knowledge of the finer internal structure of the nervous system; this is information that is much needed concerning insects. Sense cells just below the cuticle send nerve fibers (axons) into their segmental ganglia. Axons from motor nerve cells within the ganglia leave the nerve cord to terminate in muscles and glands. Internuncial nerve cells within the ganglia connect sensory and motor nerve fibers in a dense central region known as neuropile.

Nerve impulse transmission along insect axons is basically the same process as in other animals. Each impulse appears to be a digital event with a high safety factor. Although most insect nerve cells are about the same size as in larger animals, special "giant" fiber systems, with axons as much as 40 μ in diameter (largest mammalian axons are

20 μ in diameter), are concerned in rapid responses, such as escape from predators. It follows from this that the total number of nerve cells in the insect nervous system is considerably less than in vertebrates, and even large muscles, such as the jumping muscle of the grasshopper, are supplied with only two or three motor axons.

As in all animals, synapses are important as limiting mechanisms in the nervous system because of the greater instability of impulse transmission from one nerve cell to another as compared with impulse transmission along an axon. Synaptic transmission is readily modified in various ways by previous activity and by changes in the chemical environment of the nerve cells concerned. An impulse arriving at a synapse may excite or inhibit the recipient nerve cell, or it may modify previous activity or influence the response to impulses arriving over parallel synapses. For these and other reasons, synapses undoubtedly play a major part in the permissive properties of the nervous system. *See* SYNAPTIC TRANSMISSION.

There is strong evidence that recipient nerve cells are acted upon by chemical substances, mediators released at the synapses formed by impinging nerve fibers. Acetylcholine, identified at synapses in other animals, appears to be the mediator at some insect synapses. Nicotine is toxic to insects mainly because of its acetylcholinelike action on their synapses. A large dose of nicotine, acting generally on a number of synapses, serves to upset the permissive properties of the nervous system and causes general muscle spasms. An important group of modern insecticides, the organophosphorus compounds, inactivate an enzyme in nerve tissue whose normal role is to destroy acetylcholine as soon as its action at the synapse is completed. Persistent acetylcholine at the synapses causes discoordination and spasms. *See* ACETYLCHOLINE.

Transmission of nerve impulses along axons and at synapses can be detected as electrical changes. These electrical changes are due to rapid mass movements of ions, particularly sodium and potassium, along diffusion gradients across nerve cell membranes. It follows from this that critical concentrations of these ions in the fluid surrounding nerve tissues are of some importance. Since the concentrations of these ions in the blood may fluctuate with diet, the ganglia and nerves of insects are separated from the general circulation by a special sheath that appears to regulate the penetration of salts and other substances into the nervous system. *See* BIOPOTENTIALS AND ELECTROPHYSIOLOGY.

Behavior. It is impossible to make many generalizations about insect behavior since the activities of insects are as varied as their body forms. The most that can be done is to point out a few facts of insect biology that provide clues to the kind of behavior likely to be encountered.

Insects living in temperate climates are essentially seasonal in their activities. The eggs of a species are all laid at the same time of year, and one to several generations, in which all the individuals of the population are at about the same developmental stage, are completed in the course of a season. This means that in most insects there is little or no contact or interaction between immature stages and adults, and that each stage, with its own specialized way of life, may last for only a few days or weeks.

Thus, the nature of most insect life cycles leaves neither time nor opportunity for behavior patterns to be acquired through emulation of experienced individuals. This means that most of the highly complex behavior patterns associated with orientation, foraging, feeding, and reproduction are as innate and inevitable as is the unfolding of body form, which appears complete in all details as soon as there is coincidence of the appropriate developmental stage and the releasing stimulus situation. The adaptive perfection and beauty of the elaborate sequence of actions made by foraging wasps and bees tempt speculation about foresight and intelligence, but there is no evidence that these actions are determined or guided in any way other than that seen in the development of a pattern of colored scales on a butterfly's wing.

The inevitability of most insect behavior does not mean that insects cannot learn. In fact, learning and memory of a high order are parts of many behavior patterns in bees and wasps.

In many artificially designed mazes, insects learn slowly and show poor retention. However, solitary wasps have a striking capacity to learn the topography and landmarks of their foraging territory, being able to reorient themselves immediately when transported to another part of it in a darkened box. A worker honeybee returning to the hive from a rich source of nectar not only remembers and is able to retrace its path but is able to communicate by means of a dance the distance and direction of the food source to other workers within the darkened hive.

Another learning process which is of some interest is preimaginal conditioning. In many insects the site chosen for egg laying is highly specific. This may be a special food plant, or in parasitic forms, a particular host species. When there is some slight latitude in this choice, the female may be influenced by the odor of the particular food upon which she was nourished in the larval stage.

Examples of insect orientation have been given in the sections of this article on sense organs. Insects such as bees, ants, and wasps navigate with some precision over long distances by moving at a fixed angle to the Sun or its polarization pattern. For intervals of more than a few minutes between foraging trips this type of navigation is subject to an error due to displacement of the Sun from east to west. However, these and many other insects appear to have a built-in time sense that permits them to compensate for Sun displacement. This time sense is manifest in other ways. Bees fed on a rich source of nectar between 9 and 10 A.M. will reappear promptly at the same spot 24 hr later, even when the usual solar cues are lacking. If, between the first and second feeding, the hive is moved to an entirely strange area, the bees still take the original compass bearing from the hive and search at the correct distance at the proper time of day.

A time sense is also manifest in the daily activity rhythms found in most insects. For instance, cockroaches move about in search of food at night and gather in inactive groups during the day. If solar changes such as alternation of light and dark are eliminated, this 24-hr activity cycle persists for some time. Similar diurnal rhythms of even greater fixity are found in many other animals.

Most insects lead a solitary life and there is little interaction between individuals of a species except for mating. In carnivorous forms the male is likely to be attacked and eaten at this time, although special neural mechanisms or elaborate courtship behavior may increase the chances for fertilization before the male is consumed. The social insects, such as termites, ants, bees, and wasps, represent the other extreme of this situation. In them the social habit is obligatory, the whole colony being derived from, and dependent upon, the reproducing individual, or queen. Nest-mate recognition and colony coherence are determined by an odor common to all members and unique to the colony. The ubiquity of this odor, determined apparently by the food intake of the colony, is assured by frequent acts of food regurgitation and exchange between individuals. Intruders from another colony are immediately attacked, although parasites within the nest may be able to acquire this odor and thus go unmolested. *See* SOCIAL INSECTS.

[KENNETH D. ROEDER]

Bibliography: V. G. Dethier, *The Physiology of Insect Senses*, 1963; K. von Frisch, *Bees: Their Vision, Chemical Senses and Language*, 1950; J. W. S. Pringle, *Insect Flight*, Cambridge Monogr. Exp. Biol. no. 9, 1957; M. Rockstein (ed.), *The Physiology of Insecta*, 1964; K. D. Roeder (ed.), *Insect Physiology*, 1953; K. D. Roeder, *Nerve Cells and Insect Behavior*, rev. ed., 1967; V. B. Wigglesworth, *The Principles of Insect Physiology*, rev. ed., 1953.

Insecta

A class of the phylum Arthropoda sometimes called the Hexapoda. This is the largest class of animals, containing about 750,000 described species, but there are possibly as many as 5,000,000 actual species of insects. Like other arthropods, they have an external, chitinous covering. Fossil insects dating as early as the Paleozoic have been found throughout the world.

CLASSIFICATION

The class Insecta is divided into orders on the basis of the structure of the wings and the mouthparts, on the type of metamorphosis, and on various other characteristics. There are differences of opinion among entomologists as to the limits of some of the orders. Two groups treated separately here might be considered as one by other investi-

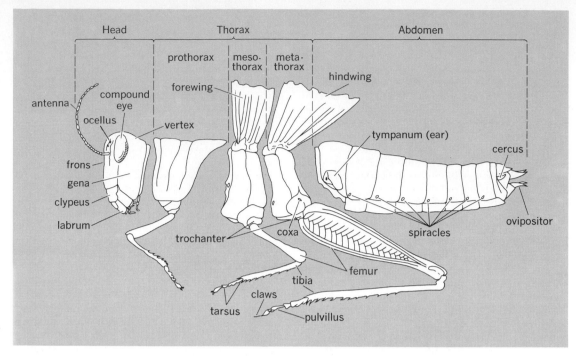

Fig. 1. Grasshopper, in lateral view, showing three main body divisions: head, thorax, and abdomen. (*From C. L. Metcalf, W. P. Flint, and R. L. Metcalf, Destructive and Useful Insects, 3d ed., McGraw-Hill, 1951*)

gators; however, a few groups treated here as a single order have been divided into two or more orders by some entomologists. The following are considered as orders of insects. Each of the items classified in the list will be found as a separate article.

Subclass Apterygota
  Order: Protura (Myrientomata) — telsontails
         Thysanura — bristletails, silverfish, firebrats
         Collembola — springtails
Subclass Pterygota
  Exopterygota — insects with simple metamorphosis
  Order: Ephemeroptera (Ephemerida, Plectoptera) — mayflies
         Odonata — dragonflies and damselflies
         Orthoptera — grasshoppers, katydids, crickets, walking sticks, cockroaches, and mantids
         Isoptera — termites
         Plecoptera — stoneflies
         Dermaptera (Euplexoptera) — earwigs
         Embioptera (Embiidina) — webspinners
         Psocoptera (Corrodentia) — psocids
         Zoraptera — zorapterans
         Mallophaga — chewing lice
         Anoplura (Siphunculata) — sucking lice
         Thysanoptera (Physopoda) — thrips
         Hemiptera (Heteroptera) — bugs
         Homoptera — cicadas, hoppers, aphids, whiteflies, scale insects
  Endopterygota — insects with complete metamorphosis
  Order: Neuroptera — dobsonflies, fishflies, snakeflies, lacewings, antlions
         Coleoptera — beetles
         Mecoptera — scorpionflies

Trichoptera — caddisflies
Lepidoptera — butterflies and moths
Diptera — true flies
Siphonaptera — fleas
Hymenoptera — sawflies, ichneumons, chalcids, wasps, ants, bees

The approximate numbers of described species of insects in the world, compiled chiefly from the USDA Yearbook for 1952, are as follows.

Insect order	World species
Protura	90
Thysanura	700
Collembola	2,000
Ephemeroptera	1,500
Odonata	4,870
Orthoptera	22,500
Isoptera	1,717
Plecoptera	1,490
Dermaptera	1,100
Embioptera	149
Psocoptera	1,100
Zoraptera	19
Mallophaga	2,675
Anoplura	250
Thysanoptera	3,170
Hemiptera	23,000
Homoptera	32,000
Neuroptera	4,670
Coleoptera	276,700
Strepsiptera	300
Mecoptera	350
Trichoptera	4,450
Lepidoptera	112,000
Diptera	85,000
Siphonaptera	1,100
Hymenoptera	103,000

MORPHOLOGY

Insects are usually elongate and cylindrical in form, and are bilaterally symmetrical. The body is segmented, and the ringlike segments are grouped into three distinct regions, the head, thorax, and abdomen (Fig. 1). The head bears the eyes, antennae, and mouthparts; the thorax bears the legs and wings, when wings are present; the abdomen usually bears no locomotor appendages but often bears some appendages at its apex. Most of the appendages of an insect are segmented.

Body wall. In vertebrate animals the skeleton is on the inside of the body and is spoken of as the endoskeleton; in an insect the skeleton is primarily on the outside of the body and is called an exoskeleton.

The body wall of an insect serves not only as a covering, but also as a supporting structure to which many important muscles are attached. The body wall of an insect is composed of three principal layers: the outer cuticula, which contains, among other chemicals, chitin; a cellular layer, the epidermis, which secretes the chitin; and a thin noncellular layer beneath the epidermis, the basement membrane. Chitin is a nitrogenous polysaccharide, $C_{32}H_{54}N_4O_{21}$, and is resistant to most weak acids and alkalies. It is secreted as a liquid and hardens. The surface of an insect's body consists of a number of hardened plates, or sclerites, separated by sutures or membranous areas, which permit bending or movement. *See* CHITIN.

Head. On the head of an adult insect are the eyes, of which there are usually two types; the mouthparts, which are variously modified according to the feeding habits of the individual, and the associated digestive glands; and the antennae, which usually consist of several segments and vary greatly in size and form.

Eyes. A pair of compound eyes usually cover a large part of the head surface. The surface of each compound eye is divided into a number of circular or hexagonal areas called facets, each of which serves as a lens for a single eye unit, called an ommatidium. The compound eye may contain from a few hundred to several thousand facets. An eye of this type is believed to have mosaic vision, forming a great number of tiny images. In addition to having compound eyes, most insects also possess two or three simple eyes, the ocelli, usually located on the upper part of the head between the compound eyes; each of these has a single lens. The ocelli apparently do not serve as organs of vision, but rather are able to detect the degree of light intensity and its source. Insect larvae do not possess compound eyes, but usually have several ocelli. Adult insects never possess more than three ocelli. *See* EYE (INVERTEBRATE).

Mouthparts. Insect mouthparts typically consist of labrum, or upper lip; a pair each of mandibles and maxillae; a labium, or lower lip; and a tonguelike structure, the hypopharynx (Fig. 2). These structures are variously modified in different insect groups and are often used in classification and identification. The type of mouthparts an insect has determines how it feeds and what type of damage it is capable of causing. A cricket is said to have chewing mouthparts because it has heavily sclerotized mandibles that move sideways and can bite off and chew particles of food. A bedbug has

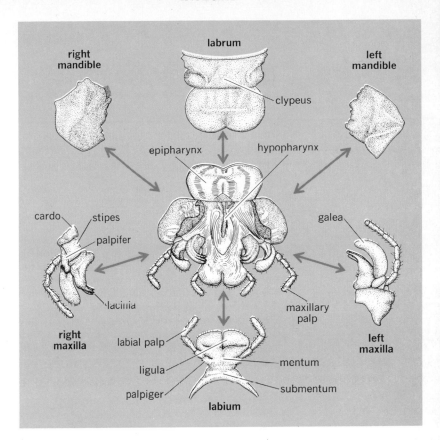

Fig. 2. Grasshopper mandibulate mouthparts. (*From C. L. Metcalf, W. P. Flint, and R. L. Metcalf, Destructive and Useful Insects, 3d ed., McGraw-Hill, 1951*)

long, slender, needlelike mandibles which can pierce the skin. Also, a pair of maxillae form a tube which is inserted through the puncture and used to suck blood.

Certain glands, such as the salivary glands, may secrete fluids which aid in digestion or which may serve as a defense mechanism against attack by other insects. These glands may also serve as a reservoir for disease organisms, such as malaria parasites. Modified salivary glands in silkworm larvae produce a fluid which, upon hardening, is known as silk.

Antennae. Several forms of antennae are recognized, to which various names are applied; they are used extensively in classification. The antennae are usually located between or below the compound eyes and are often reduced to a very small size. They are sensory in function and act as tactile organs, organs of smell, and in some cases organs of hearing.

Thorax. The thorax, or middle region, of the body bears the legs and the wings and is composed of three segments, the prothorax, mesothorax, and metathorax. Each thoracic segment usually bears a pair of legs, and the wings, when present, are borne by the mesothorax and the metathorax; if only one pair of wings is present, they are usually borne by the mesothorax. The prothorax never bears wings. In some adult insects there are no wings, and in many immature and a few adult insects there are no legs. The thorax is usually connected to the head by a membranous neck region, the cervix.

Legs. The legs of insects typically consist of five general parts: the coxa, the basal segment next to the thorax; the trochanter, a small segment, occasionally two segments; the femur, usually a long segment and often quite large; the tibia, which may also be quite long; and the tarsus, the last portion, a series of small segments. The number of tarsal segments varies from one to five in different insects. The last tarsal segment usually bears a pair of claws and often a padlike structure between the claws. The legs may be variously modified and fitted especially for running, crawling, swimming, jumping, digging or clinging to mammalian hair. The different segments of the leg may vary in size, shape, and spination. The crickets and long-horned grasshoppers have an eardrum, or tympanum, at the basal end of the front tibiae.

Wings. Insects are the only winged invertebrates, and their dominance as a group is probably due to their wings. Immature insects do not have wings, except in the mayflies, where wings develop in the subimago stage. The wings are saclike outgrowths of the body wall. In the adult they are solid structures and may be likened to the two sides of a cellophane bag that have been pressed tightly together. The form and rigidity of the wing are due to the stiff chitinous veins which support and strengthen the membranous portion. At the base are small sclerites which serve as muscle attachments and produce consequent wing movement. The wings vary in number, size, shape, texture, and venation, and in the position at which they are held at rest. Adult insects may be wingless or may have one pair of wings on the mesothorax, or, more often two pairs, one each on the mesothorax and on the metathorax. The membranous wings may bear tiny hairs or scales. In some insects the front wings are thickened, leathery, or hardened and brittle. Most insects are able to fold the wings over the abdomen when at rest, but the butterflies, mayflies, dragonflies, and damselflies cannot do this; they hold the wings outstretched or together above the body when at rest. Some insects, such as the male crickets and grasshoppers, are able to produce characteristic sounds by rubbing the two front wings together or by rubbing the front wings with the hindlegs.

There is a common basic pattern of wing venation in insects which is variously modified and in general quite specific for different large groups of insects. Much of the classification of insects depends upon these variations. A knowledge of fossil insects depends largely upon the wings, because they are among the more readily fossilized parts of the insect body.

Abdomen. The third division of the insect body is the abdomen; it has typically 11 segments. The eleventh segment is usually greatly reduced and is represented only by appendages; therefore, the maximum number of segments rarely appears to be more than 10. In many insects this number is reduced, either by a fusion of segments or by a telescoping of the terminal segments. Segments 1–7 bear appendages in various immature insects and in adult Apterygota and male Odonata. The abdominal appendages of immature insects may consist of tracheal gills (mayfly nymphs), lateral filaments (Neuroptera larvae), or prolegs (Lepidoptera larvae). These appendages are represented as styli in the Apterygota and as copulatory structures in male Odonata. When 10 complete segments are present, a pair of clasperlike or feelerlike abdominal appendages, the cerci, arise from segment 10; however, they may be absent.

The genital structures arise from segments 8 and 9. In the male these function as copulatory structures for the transfer of sperm to the female. These external genital structures are quite variable and complex and frequently serve as excellent taxonomic characters. When not in use, these pieces are usually withdrawn into the tip of the abdomen, often called the genital chamber, and frequently cannot be observed unless dissected. In the female these structures form the ovipositor or egg-deposition organ concerned with oviposition. They are often fitted for slitting open plant or animal tissues and depositing eggs in the tissue. They may be modified to form a sting in the bees and wasps. In parasitic Hymenoptera they often are quite long and are used to drill into wood, where the eggs are deposited. In some of the meadow grasshoppers the ovipositor is a bladelike structure as long as the body.

INTERNAL ANATOMY

The internal structures of insects (Fig. 3) may be considered in their major systems: respiratory, digestive, excretory, circulatory, nervous, and reproductive.

Respiratory system. The intake of oxygen, its distribution to the tissues, and the removal of carbon dioxide are accomplished by means of an intricate system of tubes called the tracheal system. The principal tubes of this system, the trachea, open externally at the spiracles. Internally they branch extensively, extend to all parts of the body, and terminate in simple cells, the tracheoles. The trachea are lined with a thin layer of cuticle, which is thickened to form spiral rings that give the trachea rigidity. The spiracles are located laterally and vary in number from 1 to 10. Usually there is a pair of spiracles on the mesothorax, a pair on the metathorax, and one pair on each of the first seven or eight abdominal segments. The oxgyen enters through the spiracles, and carbon dioxide is eliminated principally through them; probably one-fourth of the carbon dioxide is eliminated through the body wall. In some insects the oxygen is taken in the body through the thoracic spiracles and is eliminated through the abdominal spiracles. The spiracles may be partially or completely closed for extended periods by some insects. Many adaptations for carrying on respiration are known. Aquatic insects may have tracheal gills located either on the external portion of the body or in the rectal cavity. The larvae of certain insects, such as mosquitoes, have respiratory tubes at the posterior end of the body. Some aquatic insects with normal tracheal systems and active spiracles can go beneath the water and remain for extended periods of time by carrying with them a film of air on the surface of the body.

Digestive system. Insects possess an alimentary tract consisting of a tube, usually coiled, which extends from the mouth to the anus. It is differentiated into three main regions, the foregut, midgut, and hindgut. The foregut is usually composed of four parts, the pharynx, esophagus, crop, and proventriculus. The salivary glands are evaginations of the foregut. The foregut is an invagination of the ectoderm and is lined with cuticle. The

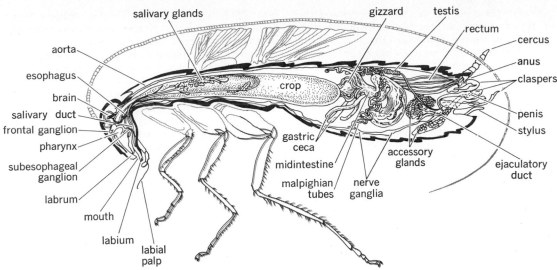

salivary glands

gizzard   testis

rectum

cercus

anus

claspers

crop

aorta

esophagus

brain

salivary duct

frontal ganglion

pharynx

subesophageal ganglion

labrum

mouth

labium

labial palp

gastric ceca

midintestine

malpighian tubes

nerve ganglia

accessory glands

penis

stylus

ejaculatory duct

Fig. 3. Sagittal section of a cockroach showing the internal anatomy of a male specimen. (*From C. L. Metcalf,* *W. P. Flint, and R. L. Metcalf, Destructive and Useful Insects, 3d ed., McGraw-Hill, 1951*)

midgut is usually undifferentiated, except for evaginations called gastric ceca, and is endodermal in origin and lined with epithelial cells. Some of the epithelial cells produce enzymes and others absorb digested food. The hindgut may be differentiated into the small intestine, or ileum; the large intestine, or colon; and the rectum. The Malpighian tubules, which are excretory in function, connect with the alimentary canal at the point of union of the hindgut and midgut. Valves between the three main divisions of the alimentary canal regulate the passage of food from one region to another.

The major functions of the digestive system are the ingestion and digestion of food. Digestion is the process of changing food chemically so that it may be taken into the blood to supply nutrient to various parts of the body. The food habits may vary in a given order. Larvae and adults usually have entirely different food habits and different types of digestive systems. Because of the great variety of insect foods, both the specific structures and the functions of the different parts vary greatly. When food is ingested, the crop may serve as a temporary storage chamber, where partial digestion may take place; before reaching the midgut, food passes through the proventriculus, which is often fitted with large or sharp needlelike teeth which may strain out or crush the larger particles of food. Digestion is completed and most of the absorption takes place in the midgut. After food is digested and for the most part absorbed, the residue passes into the hindgut, where a slight amount of absorption may take place.

Excretory system. The excretory system consists of a group of tubes with closed distal ends, the Malpighian tubules, which arise as evaginations of the anterior end of the hindgut. They vary in number from 1 to over 100, and extend into the body cavity. Various waste products are taken up from the blood by these tubules and passed out by way of the hindgut and anus.

Circulatory system. The circulatory system of an insect is an open one, as compared with a closed system composed of arteries, veins, and capillaries, as in the vertebrates. The only blood

vessel is a tube located dorsal to the alimentary tract and extending through the thorax and abdomen. The posterior portion of this tube, the heart, is divided into a series of chambers, each of which has a pair of lateral openings called ostia. The anterior part of the tube which lacks ostia, is called the dorsal aorta. Pulsations of the heart produce the circulation by pumping the blood forward, and out of the aorta in the neck region. The increased pressure in this region then causes the blood to move posteriorly through the body cavity to the ostia.

Nervous system. The nervous system consists of a brain, often called the supraesophageal ganglion, located in the head above the esophagus; a subesophageal ganglion, connected to the brain by two commissures that extend around each side of the esophagus; and a ventral nerve cord, typically double, extending posteriorly through the thorax and abdomen from the subesophageal ganglion. In the nerve cords there are enlargements, called ganglia. Typically, there is a pair to each body segment, but this number may be reduced by the fusion of ganglia.

From each ganglion of the chain, nerves extend to each adjacent segment of the body, and also extend from the brain to part of the alimentary canal. There are several minor ganglia connected directly with the brain.

Reproductive system. Reproduction in insects is nearly always sexual, and the sexes are separate. Variations from the usual reproductive pattern occur occasionally. In many social insects, such as the ants and bees, certain females, the workers, may be unable to reproduce because their sex organs are undeveloped; in some insects, individuals occasionally occur that have characters of both sexes, called gynandromorphs.

In the female the reproductive system consists primarily of a pair of ovaries, each with a number of tubes called ovarioles. The ovarioles from each ovary unite into a common tube, and the two tubes fuse to form the oviduct. This discharges through an external opening, the vulva, located on the ventral side of the eighth segment. There are special glands and other chambers that open into the oviduct. The germ cell forms at the inner end of each ovariole and passes through a series of chambers, receiving its supply of food material as it proceeds. In the last chamber it receives its chorion, or shell, which has a minute opening, the micropyle. As the completed egg proceeds into the oviduct, it passes the opening of a chamber, the spermatheca, where sperm previously received from the male is released to enter the egg and cause fertilization. The other glands may, at this point, contribute various substances used for attaching the egg to a substrate or for covering it.

In the male, there are paired testes in which the sperm are formed. From these, a pair of slender ducts, called the vasa deferentia, unite into a common ejaculatory duct, and this extends into a variously sclerotized penis. The sperm is commonly stored in swellings of the vasa deferentia to await copulation. Connected with these structures are various accessory glands.

Among the many groups of insects, various modifications exist in these basic structures and in their methods of operation. In fact, almost every method of reproduction that is known in the animal kingdom can be found among the insects. Copulation normally occurs through the vulva, but in the bedbug family Cimicidae of the order Hemiptera, it takes place by means of a special structure occurring at various points on the female body. The sperm is received in this structure and then passed through the body cavity to fertilize the egg in the ovariole. In some insects, the sperm must pass through the oviduct into the ovariole to fertilize the egg before it enters the oviduct. The fertilized immature egg develops to maturity in the ovariole. In some forms the ovarioles are not tubular but are shaped like a bunch of grapes. In a few insects the embryo is retained within the body until it is fully grown, being nourished by glandular secretions. There is even something of a placental development in some forms of insect. *See* INSECT PHYSIOLOGY.

METAMORPHOSIS

After insects hatch from an egg, they begin to grow and increase in size (growth) and will also usually change, to some degree at least, in form (metamorphosis) and often in appearance. The growth of an insect is accompanied by a series of molts, or ecdyses, in which the cuticle is shed and renewed. Prior to the actual shedding process, a new cuticular layer is secreted beneath the old layer and a molting fluid is secreted which separates the new cuticle from the old.

The molt involves not only the external layers of the body wall, the cuticula, but also the cuticular linings of the tracheae, foregut, and hindgut; the cast skins often retain the shape of the insects from which they were shed. The shedding process begins with a splitting of the old cuticle, usually along the midline on the dorsal side of the thorax. This split grows and the insect eventually wriggles out of the old cuticle. The new skin, at first wrinkled and folded, remains soft and pliable long enough for the body to expand to its fullest capacity before hardening.

Insects differ regarding the number of molts during their growing period. Many have as few as four molts; a few species have 40 or more, and

the latter continue to molt throughout life.

Insects have been grouped or classified upon the basis of the type of metamorphosis which they undergo. All entomologists, however, do not agree upon the same classification. These types of metamorphosis have been placed in two groups, under the heading of simple metamorphosis and complete metamorphosis. Some investigators include ametabolous, paurometabolous, and hemimetabolous types in simple metamorphosis, and consider complete metamorphosis to be synonymous with holometabolous. In order to cover the more detailed pattern, the following outline is used.

1. Ametabolous or primitive: No distinct external changes are evident with an increase in size.
2. Metabolous: Distinct changes, both in size and form, are evident.
   a. Paurometabolous: Direct metamorphosis that is simple and gradual; immature forms resemble the adult except in size and are referred to as nymphs.
   b. Hemimetabolous: Metamorphosis is incomplete; gills are present in aquatic larvae, or naiads.
   c. Holometabolous: Complete or indirect metamorphosis; stages in this development are the egg, larva, pupa, and adult, or imago.

In the ametabolous forms, the insects never obtain wings and have no visible external changes. In the paurometabolous forms, each insect gradually acquires wings during a period of growth and several moltings. The hemimetabolous group contains only aquatic insects which change from an aquatic, gill-breathing naiad to a spiracle-breathing adult with wings. This transformation is often spoken of as an incomplete type of metamorphosis. In the holometabolous, or complete, form of metamorphosis there is a complete change from the larva (which has only simple eyes, no wings, and frequently no legs or antennae) to the adult, usually of entirely different appearance, with wings, compound eyes, well-developed legs and antennae, and in certain groups entirely different mouthparts. Also, in this type of metamorphosis there is an extra life stage, the pupa, usually called a resting stage, in which the complete transformation takes place. The larva is the growth stage, the pupa the transformation stage, and the adult the reproductive stage. In the insects having complete metamorphosis the feeding habits of the larva and the adult of the same species are usually different.

[DWIGHT M. DE LONG]

FOSSILS

Fossil insects are the remains or traces of insects preserved in rocks. They usually consist in part of an impression in the rock matrix and in part of some altered organic substances; more rarely, as in the case of insects preserved in amber, some organic compounds such as chitin may persist without change. In general, insects become preserved as fossils by the burial of living or freshly killed specimens in soft mud, which accumulates rapidly enough to prevent disintegration of the insect. The eventual hardening of the mud forms a sedimentary rock containing the insect remains. The fossil insects are obtained by splitting the rock

Fig. 4. Fossil insects from the Miocene Epoch, found in Colorado. (a) *Prodryas persephone* Scudder, of the order Lepidoptera. (b) *Lithosmylus columbianus* (Cockerell), of the order Neuroptera.

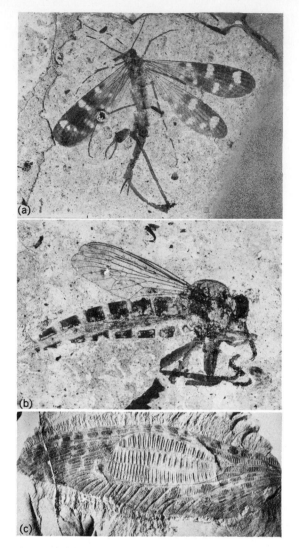

Fig. 5. Fossil insects. (a) *Holcorpa maculosa* Scudder, of the order Mecoptera, and (b) robber fly, of the order Diptera, from the Miocene of Colorado. (c) *Clatrotitan andersoni* McKeown, of the order Orthoptera, from the Triassic of New South Wales, Australia.

along planes of deposition with a hammer and chisel. *See* AMBER.

Fossil record. Fossil insects have been found in about 100 different rock formations throughout the world. One of the best known deposits is a shale bed in the town of Florissant, Colo. In the Tertiary Period, about 50,000,000 years ago, this was the site of a shallow lake, surrounded by active volcanoes. As the ash from the volcanoes fell into the water of the lake, it carried along with it whatever insects were flying over the lake, and as the ash settled on the bottom to form mud, the insects were promptly buried. The Florissant shale has yielded fully 100,000 specimens of insects (Figs. 4 and 5).

Paleozoic. The earliest known fossil insects, about 250,000,000 years old, are found in rocks of Upper Carboniferous (Pennsylvanian) age. Most of the early insects had approximately the size and appearance of the living mayflies and dragonflies, to which they were closely related. A few of the dragonflylike insects (belonging to one family of the extinct order Protodonata) were large, having wingspreads of 76 cm (30 in.). The most numerous insects of the time, at least in swampy regions, were the cockroaches, which were similar in appearance to living roaches. The early insects were less specialized than any other known insects and are usually thought to have been close to the ancestors of the winged insects.

The insects found in Permian rocks (formed about 225,000,000 years ago) represent a far more diversified fauna than that of any other period in the Earth's history, including the present. The extinct orders that existed in the Carboniferous Period survived into the Permian, and in addition several living orders were represented. Along with the huge dragonflylike insects (Protodonata) were such minute insects as the bark lice (Psocoptera), some having a wingspread of only 3.2 mm. The fossils show about an equal representation of the extinct orders of the Carboniferous and some relatively specialized living orders. Adding to this diversity were several extinct orders not found in Carboniferous strata. The living orders represented in the early Permian include such types as the dragonflies (Odonata), mayflies (Ephemerida), bark lice (Psocoptera), Hemiptera (true bugs) and scorpion flies (Mecoptera). The presence of the scorpion flies is especially interesting, since the living species have a complicated postembryonic development characteristic of all of the most highly specialized of living insects. True beetles (Coleoptera) and stoneflies (Plecoptera) appear in Late Permian deposits.

Mesozoic. At the beginning of the Mesozoic Era, the record of fossil insects changes significantly. The contrast between the archaic fauna of the Early Permian and the relatively modern one of the Triassic is as great as that between the faunas of the Triassic and the Recent (present) periods. The extinct protodonate dragonflies persisted into the Triassic but apparently died out by the end of that period. None of the other extinct orders of insects is found in deposits later than the Permian.

The insects found in Jurassic strata comprise a fauna much like that of the Triassic, except that more living families are represented. In general such families occur in those orders that were well developed in the Permian, such as the Odonata and Hemiptera. Notably absent in the Jurassic were flower insects, in particular the bees and syrphid flies. By the beginning of the next period, the Cretaceous, the flowering plants had become established and in all probability the types of insects associated with those plants promptly followed. Unfortunately, only a very few insects are known from Cretaceous rocks.

Cenozoic. Insects are more abundant in Tertiary deposits than in any other formations. The fossils show that a much higher percentage of living genera was present in Eocene and Oligocene times than in the Jurassic. Those preserved in Baltic amber, which is of early Tertiary age, are especially important in making possible the comparisons with living genera and species. Extensive studies of amber insects have shown that the amount of evolutionary change that has taken place since the early Tertiary varies greatly with the particular family of insects concerned. For example, about half of the ant genera found in the Baltic amber (totaling 43) are living genera; whereas, in the case of the bees, only a single amber genus is still living. Several species of the Baltic amber ants are indistinguishable from living species.

Significance. The study of fossil insects has made significant contributions to the understanding of the evolution of insects and of the relationships of existing orders and families. It has become apparent that the groups now living are but a small part of the total number that have existed on the Earth. A detailed study of the fossil forms provides evidence of certain progressive changes in structure and development, which, in the main, are confirmed by morphological and developmental investigations of living insects. There is, however, no fossil evidence bearing on the question of insect origin; the oldest insects known show no transition to other arthropods. Morphological and developmental studies, however, have indicated that the ancestors of the insects were terrestrial arthropods, probably related to the Symphyla. Although the time of that origin is not directly demonstrated in the rocks, the fossil record suggests that it was at least as far back as the Mississippian Period (Lower Carboniferous strata).

The fossil record, combined with morphological studies of living types, shows that insects have passed through three major evolutionary steps. The first of these was the development of wings as simple but functional outgrowths of the thoracic wall which could not be folded back over the abdomen when the insect was at rest. The dragonflies and mayflies constitute the only living orders that represent this stage in insect evolution. The second step was marked by the evolution of a complicated wing articulation, which enabled the wings to be folded back over the abdomen at rest, and by the presence of immature stages resembling the adults. The grasshoppers and their relatives (Orthoptera), the true bugs (Hemiptera), and similar living orders belong to this stage. The third step was the acquisition of a complicated postembryonic development with an immature stage (larva) very different from the adult, as in beetles and true flies. The fossils show that insects passed through these stages before the beginning of the Permian Period, some 225,000,000 years ago, when land-inhabiting vertebrates were only starting their evolutionary history.   [FRANK M. CARPENTER]

Bibliography: D. J. Borror and D. M. DeLong, *An Introduction to the Study of Insects*. 2d ed., 1963; C. T. Brues, A. L. Melander, and F. M. Carpenter, Classification of insects, *Bull. Mus. Comp. Zool.*, 108:777–827, 1954; F. M. Carpenter, *The Geological History and Evolution of Insects*, Smithson. Inst. Publ. Rep. no. 1953, 1954; J. H. Comstock, *Introduction to Entomology*, 9th rev. ed., 1940; E. O. Essig, *College Entomology*, 1942; A. D. Imms, *General Textbook of Entomology*, rev. ed., 1957; H. H. Ross, *A Textbook of Entomology*, 2d ed., 1956.

Insecticide

A material used to kill insects and related animals by disruption of vital processes through chemical action. Chemically, insecticides may be of inorganic or organic origin. The principal source is from chemical manufacturing, although a few are derived from plants. Insecticides are classified according to type of action, as stomach poisons, contact poisons, residual poisons, systemic poisons, fumigants, repellents, or attractants. Many act in more than one way. Stomach poisons are applied to plants so that they will be ingested as insects chew the leaves. Contact poisons are applied directly to insects and are used principally to control species which obtain food by piercing leaf surfaces and withdrawing liquids. Residual insecticides are applied to surfaces so that insects touching them will pick up lethal dosages. Systemic insecticides are applied to plants or animals and are absorbed and translocated to all parts of the organisms, so that insects feeding upon them will obtain lethal doses. Fumigants are applied as gases, or in a form which will vaporize to a gas, to be inhaled by insects. Repellents prevent insects from coming in contact with their hosts. Attractants induce insects to come to specific locations in preference to normal food sources.

In the United States, about 500 species of insects are of primary economic importance, and losses caused by insects range from $4,000,000,000 to $8,000,000,000 annually.

Inorganic insecticides. Prior to 1945, large volumes of lead arsenate, calcium arsenate, paris green (copper acetoarsenite), sodium fluoride, and cryolite (sodium fluoaluminate) were used. The potency of arsenicals is a direct function of the percentage of metallic arsenic contained. Lead arsenate was first used in 1892 and proved effective as a stomach poison against many chewing insects. Calcium arsenate was widely used for the control of cotton pests. Paris green was one of the first stomach poisons and had its greatest utility against the Colorado potato beetle. The amount of available water-soluble arsenic governs the utility of arsenates on growing plants, because this fraction will cause foliage burn. Lead arsenate is safest in this respect, calcium arsenate is intermediate, and paris green is the most harmful. Care must be exercised in the application of these materials to food and feed crops because they are poisonous to man and animals as well as to insects.

Sodium fluoride has been used to control chewing lice on animals and poultry, but its principal application has been for the control of household insects, especially roaches. It cannot be used on plants because of its extreme phytotoxicity. Cryolite has found some utility in the control of the Mexican bean beetle and flea beetles on vegetable crops because of its low water solubility and lack of phytotoxicity.

Organic insecticides. These began to supplant the arsenicals when DDT [2,2-bis(p-chlorophenyl)-1,1,1-trichloroethane] became available in 1945. During World War II, the insecticidal properties of γ-benzenehexachloride (γ-1,2,3,4,5,6-hexachlorocyclohexane of γ-BHC) were discovered in England and France. The two largest-volume insecticides are DDT and γ-BHC. Certain insects cannot be controlled with either, and there are situations and crops where they cannot be used. For these reasons, other chlorinated hydrocarbon insecticides have been discovered and marketed successfully. These include TDE [2,2-bis(p-chlorophenyl)-1,1-dichloroethane], methoxychlor [2,2-bis(p-methoxyphenyl)-1,1,1-trichloroethane], Dilan [mixture of 1,1-bis(p-chlorophenyl)-2-nitropropane and

1,1-bis(*p*-chlorophenyl)-2-nitrobutane], chlordane (2,3,4,5,6,7,8-8-octachloro-2,3,3*a*,4,7,7*a*-hexahydro-4,7-methanoindene), heptachlor (1,4,5,6,7,8,8-heptachloro-3*a*,4,7,7*a*-tetrahydro-4,7-methanoindene), aldrin (1,2,3,4,10,10-hexachloro-1,4,4*a*,5,8,8*a*-hexahydro-1,4-*endo*, *exo*-5,8-dimethanonaphthalene), dieldrin (1,2,3,4,10,10-hexachloro-6,7-epoxy-1,4,4*a*,5,6,7,8,8*a*-octahydro-1,4-*endo*,*exo*-5,8-dimethanonaphthalene), endrin (1,2,3,4,10,10-hexachloro-6,7-epoxy-1,4,4*a*,5,6,7,8,8*a*-octahydro-1,4-*endo*, *endo*-5,8-dimethanonaphthalene), toxaphene (camphene plus 67–69% chlorine), and endosulfan (6,7,8,9,10,10-hexachloro-1,5,5*a*,6,9,9*a*-hexahydro-6,9-methano-2,4,3-benzodioxathiepin-3-oxide). TDE is considerably less toxic than DDT but in general is also less effective. It has given outstanding results in the control of larvae of several moths, however. Methoxychlor is even less toxic than TDE and has found considerable use in the control of houseflies and also of the Mexican bean beetle which is not susceptible to DDT. Restrictions have been placed on its use on dairy cattle for the control of flies, and its consumption is declining. Dilan has found some use in the control of the Mexican bean beetle, as well as of some thrips and aphids. Chlordane was the first cyclopentadiene insecticide to reach commercial status. It has been the most effective chemical available for the control of roaches. In addition to lengthy residual properties, chlordane also possesses fumigant action. Related chemicals include heptachlor, aldrin, dieldrin, and endrin. They are effective against grasshoppers and are especially useful for the control of insects inhabiting soil. Registrations for uses of aldrin and dieldrin were reduced in 1966, following extensive investigations of their metabolism and persistence in animal tissues. Toxaphene is used principally for the control of the cotton boll weevil and other insect pests of cotton. Endosulfan, in addition to showing promise for the control of numerous insects, also shows promise of controlling a number of species of phytophagous (plant-feeding) mites that are not generally susceptible to chlorinated hydrocarbon insecticides.

Insect resistance. The resistance of insects to DDT was first observed in 1947 in the housefly. By the end of 1967, 91 species of insects had been proved to be resistant to DDT, 135 to cyclodienes, 54 to organophosphates, and 20 to other types of insecticides, including the carbamates. Among the chlorinated hydrocarbon insecticides, two types of resistance occur. One applies to DDT and its analogs, such as TDE, methoxychlor, and Dilan, and the other to the cyclodiene compounds, such as chlordane, heptachlor, aldrin, dieldrin, endrin, and also γ-BHC and toxaphene. Insecticides are not mutagenic, indicating that resistance preexists in a small part of the natural population even before exposure.

Nearly every country in the world, except mainland China, has reported the presence of resistant strains of the housefly. In many heavily populated urban areas of the United States, it is difficult to obtain control of roaches with chlordane. During 1957 and 1958, many growers of cotton in the southern states changed from toxaphene and γ-BHC to organic phosphorus chemicals because of the resistance of the cotton boll weevil to chlorinat-ed hydrocarbon insecticides. By 1967, resistant strains of the cotton bollworm and the tobacco budworm had developed and proliferated to the extent that in numerous areas the use of chlorinated hydrocarbon insecticides was of doubtful value. The onion maggot has developed widely spread strains resistant to aldrin, dieldrin, and heptachlor, especially in the northeastern United States and in Ontario, Canada. Three species of corn rootworms also are resistant to these insecticides, principally in the corn-growing regions west of the Mississippi River. The development of chlorinated hydrocarbon resistance among several species of disease-transmitting mosquitoes continues to pose a threat to world health. The control of typhus in the Far East could be at stake because strains of vector lice are resistant to DDT. Evidence for insect resistance to the plant-derived insecticides pyrethrum and rotenone has only recently been established.

Organic phosphorus insecticides. The development of this type of insecticide paralleled that of the chlorinated hydrocarbons. Since 1947, more than 50,000 organic phosphorus compounds have been synthesized in academic and industrial laboratories throughout the world for evaluation as potential insecticides. Parathion [*O*,*O*-diethyl *O*-(*p*-nitrophenyl) phosphorothioate] and methyl parathion [*O*,*O*-dimethyl *O*-(*p*-nitrophenyl) phosphorothioate] are estimated to have had a world production of 70,000,000 lb during 1966.

A great diversity of activity is found among organophosphorus insecticides. Many are extremely toxic to man and other warm-blooded animals, but a few show a very low toxicity. The more important include tetraethylpyrophosphate, dicapthon [*O*,*O*-dimethyl *O*-(3-chloro-4-nitrophenyl) phosphorothioate], malathion [*O*,*O*-dimethyl *S*-(1,2-dicarbethoxyethyl) phosphorodithioate], dichloro- [*O*,*O*-dimethyl *O*-(2,2-dichlorovinyl) phosphate], diazinon {*O*,*O*-diethyl *O*-[2-isopropyl-4-methylpyrimidyl (6)] phosphorothioate}, dioxathion [2,3-*p*-dioxanedithiol *S*,*S*-bis (*O*,*O*-diethylphosphorodithioate)], azinphosmethyl [*O*,*O*-dimethyl *S*-4-oxo-1,2,3-benzotriazin-3-(4*H*)-yl-methylphosphorodithioate], carbophenothion [*S*-(*p*-chlorophenylthiomethyl) *O*,*O*-diethylphosphorodithioate], ethion {bis[*S*-(diethoxyphosphinothioyl) mercapto] methane}, *EPN* [*O*-ethyl *O*-(*p*-nitrophenyl) phenylphosphonothionate], trichlorfon [*O*,*O*-dimethyl (2,2,2,-trichloro-1-hydroxyethyl) phosphonate], dimethoate [*O*,*O*-dimethyl *S*-(methylcarbamoylmethyl) phosphorodithioate], and fenthion {*O*,*O*-dimethyl-*O*-[4-(methylthio)-*m*-tolyl] phosphorothioate}.

Schradan [bis(dimethylamino) phosphoric anhydride] was unique among organic insecticides in that it showed systemic properties when applied to plants. By direct contact, it has a relatively low order of activity. When sprayed on plants, it is absorbed from areas receiving treatment, is translocated throughout the entire plant, and is metabolized to yield a product highly toxic to such sucking pests as aphids and phytophagous mites. It is selective in that it affects only aphids ingesting juices from treated plants and does not kill predators which destroy aphids. With schradan, it is possible to protect the growing parts of plants without resorting to frequent spraying, because the insecticide is translocated to these growing parts, whereas with most stomach poison or residual insecticides plants outgrow the protection. Several chemicals showing systemic properties have reached commercial or near commercial status. These include demeton [O,O-diethyl O (and S)-(2-ethylthio) ethylphosphorothioates], disulfoton [O,O-diethyl S-(2-ethylthio) ethylphosphorodithioate], phorate [O,O-diethyl S-(ethylthiomethyl)-phosphorodithioate], and mevinphos [2-methoxycarbonyl-1-methyl vinyl dimethyl phosphate].

The use of organic phosphorus chemicals for the systemic control of animal parasites is another facet of interest. During 1958, semicommercial application began of coumaphos [O-(3-chloro-4-methyl-2-oxo-2-H-1-benzopyran-7-yl) O,O-diethylphosphorothioate] and ronnel [O,O-dimethyl O-(2,4,5-trichlorophenyl) phosphorothioate] for the control of grubs in cattle. Coumaphos is applied externally as a spray and is absorbed and translocated to kill the cattle grubs. Ronnel is most effective when administered internally. Dimethoate, fenthion, famphur [O-p-(dimethyl-sulfamoyl)phenyl O,O-dimethylphosphorothioate], and menazon [S-(4,6-diamino-s-triazin-2-ylmethyl) O,O-dimethylphosphorodithioate] also show activity for this type of application.

Activity of organic phosphate insecticides results from the inhibition of the enzyme cholinesterase, which performs a vital function in the transmission of impulses in the nervous system. Inhibition of some phenyl esterases occurs also. Inhibition results from direct coupling of phosphate with the enzyme. Phosphorothionates are moderately active but become exceedingly potent upon oxidation to phosphates.

Other types of insecticides. Synthetic carbamate insecticides are attracting increased interest. These include dimetan (5,5-dimethyldihydroresorcinol dimethylcarbamate), Pyrolan (3-methyl-1-phenyl-5-pyrazolyl dimethylcarbamate), Isolan (1-isopropyl-3-methyl-5-pyrazolyl dimethylcarbamate), pyramat [2-n-propyl-4-methylpyrimidyl-(6)-dimethylcarbamate], carbaryl (1-naphthyl-N-methylcarbamate), Bagon, (o-isopropoxyphenyl methylcarbamate), Zectran (4-dimethylamino-3,5-xylyl methylcarbamate), TRANID [5-chloro-6-oxo-2-norbornanecarbonitrile O-(methylcarbamoyl)-oxime], dimetilan [1-(dimethylcarbamoyl)-5-methyl-3-pyrazolyl dimethylcarbamate], Furadan (2,3-dihydro-2.2-dimethyl-7-benzofuranyl methylcarbamate), and Temik [2-methyl-2-(methylthio) propionaldehyde O-(methylcarbamoyl) oxime]. They are also cholinergic.

Insecticides obtained from plants include nicotine [L-1-methyl-2-(3'-pyridyl)-pyrrolidine], rotenone, the pyrethrins, sabadilla, and ryanodine, some of which are the oldest-known insecticides. Nicotine was used as a crude extract of tobacco as early as 1763. The alkaloid is obtained from the leaves and stems of *Nicotiana tabacum* and *N. rustica*. It has been used as a contact insecticide, fumigant, and stomach poison and is especially effective against aphids and other soft-bodied insects.

Rotenone is the most active of six related alkaloids found in a number of plants, including *Derris elliptica*, *D. malaccensis*, *Lonchocarpus utilis*, and *L. urucu*. *Derris* is a native of East Asia, and *Lonchocarpus* occurs in South America. The highest concentrations are found in the roots. Rotenone is active against a number of plant-feeding pests and has found its greatest utility where toxic residues are to be avoided. Rotenone is known also as derris or cubé.

The principal sources of pyrethrum are *Chrysanthemum cinerariaefolium* and *C. coccineum*. Pyrethrins, which are purified extracts prepared from flower petals, contain four chemically different active ingredients. Allethrin is a synthetic pyrethroid. The pyrethrins find their greatest use in fly sprays, household insecticides, and grain protectants because they are the safest insecticidal materials available.

Synergists. These materials have little or no insecticidal activity but increase the activity of chemicals with which they are mixed, especially that of the pyrethrins. Piperonyl butoxide {α-[2-(2-butoxyethoxy)-ethoxy]-4,5-methylenedioxy-2-propyltoluene}, sulfoxide {1,2,methylenedioxy-4-[2-octyl-(sulfinyl),propyl] benzene}, and N-(2-ethylhexyl)-5-norbornene-2,3-dicarboximide are commercially available. Sesamex [acetaldehyde 2-(2-ethoxyethoxy)ethyl 3,4-methylenedioxyphenyl acetal] and Tropital {piperonal bis[2-(2-butoxyethoxy)-ethyl]- acetal} are active but not fully developed synergists. These synergists have their greatest utility in mixtures with the pyrethrins. Some have been shown to enhance the activity of carbamate insecticides as well.

Formulation and application. Formulation of insecticides is extremely important in obtaining satisfactory control. Common formulations include dusts, water suspensions, emulsions, and solutions. Accessory agents, including dust carriers, solvents, emulsifiers, wetting and dispersing agents, stickers, deodorants or masking agents, synergists, and antioxidants, may be required to obtain a satisfactory product. Insecticidal dusts are formulated for application as powders. Toxicant concentration is usually quite low. Water suspensions are usually prepared from wettable powders, which are formulated in a manner similar to dusts except that the insecticide is incorporated at a high concentration and wetting and dispersing agents are included. Emulsifiable concentrates are usually prepared by solution of the chemical in a satisfactory solvent to which an emulsifier is added. They are diluted with water prior to application. Granular formulations are an effective means of applying insecticides to the soil to control insects which feed on the subterranean parts of plants. Proper timing of insecticide applications is important in obtaining satisfactory control. Dusts are more easily and rapidly applied than are sprays. However, results may be more erratic and much greater attention must be paid to weather conditions than is required for sprays. Coverage of plants and insects is generally less satisfactory with dusts than with sprays. It is best to make dust applications early in the day, while the plants are covered with dew, so that greater amounts of dust will adhere. If prevailing winds are too strong, a considerable proportion of dust will be lost. Spray operations will usually require the use of heavier equipment, however. Application of insecticides should be properly correlated with the occurrence of the most susceptible stage in the life cycle of the pest involved.

During the past decade, attention has focused sharply on the impact of the highly active synthetic insecticides upon the total environment—man, domestic and wild animals and fowl, soil-inhabiting microflora and microfauna, and all forms of aquatic life. Effects of these materials upon populations of beneficial insects, particularly parasites and predators of the economic species, are being critically assessed. The study of insect control by biological means has expanded. The concepts and practices of integrated pest control and pest management are expanding rapidly. Among problems associated with insect control which must receive major emphasis during the coming years are the development of strains of insects resistant to insecticides; the assessment of the significance of small, widely distributed insecticide residues in and upon the environment; the development of better and more reliable methods for forecasting insect outbreaks; and the evolvement of control programs integrating all methods—physical, physiological, chemical, biological, and cultural—for which practicality may have been demonstrated. *See* ENTOMOLOGY, ECONOMIC; FUMIGANT; INSECT CONTROL, BIOLOGICAL; INSECT PHYSIOLOGY; INSECTA; MITICIDE; PESTICIDE.

[GEORGE F. LUDVIK]

Bibliography: A. B. Borkovec, *Insect Chemosterilants*, 1966; A. W. A. Brown, *Insect Control by Chemicals*, 1951; D. E. H. Frear, *Pesticide Handbook: Entoma*, 1966; D. E. H. Frear, *Pesticide Index*, 1963; M. Jacobson, *Insect Sex Attractants*, 1965; E. E. Kenega, Commercial and experimental organic insecticides, *Bull. Entomol. Soc. Amer.*, 12(2):161–217, 1966; W. W. Kilgore and R. L. Doutt, *Pest Control: Biological, Physical and Selected Chemical Methods*, 1967; H. Martin, *Insecticide and Fungicide Handbook*, 1965; C. L. Metcalf, W. P. Flint, and R. L. Metcalf, *Destructive and Useful Insects*, 1962; R. L. Metcalf, *Organic Insecticides*, 1955; R. D. O'Brien, *Insecticides: Action and Metabolism*, 1967; U.S. Department of Agriculture, *Agr. Handb. no. 331*, Agricultural Research Service and Forest Service, 1967.

Insectivora

An order of mammals including such familiar forms as the hedgehogs, shrews, and moles. The Insectivora are of ancient origin and are difficult to define; they are united by primitive characters that, for the most part, are primitive for all placental mammals. At various times other stocks of primitive mammals have been included in the Insec-

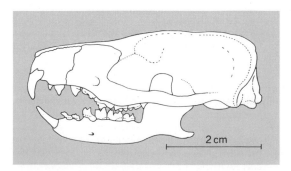

Fig. 1.   Skull and lower jaw of *Amphechinus*, a mid-Cenozoic lipotyphlan insectivore. (*After J. Viret*)

tivora, such as the elephant shrews now placed in an order of their own, the Macroscelidea. There are many fossil insectivores, some excellent examples being known from as far back as the Cretaceous, during the age of dinosaurs. Many kinds of insectivores are extinct; among these are some which were more primatelike than living insectivores. *See* MACROSCELIDEA.

Although their name implies a diet of insects, most insectivores feed, or once fed, on a variety of invertebrate and even vertebrate prey; some were even frugivorous or omnivorous. Many insectivores have spines, notably the hedgehogs and tenrecs, and generally the senses of smell and hearing are far more acute than eyesight. Echolocation is known in shrews and tenrecs, and some shrews and *Solenodon* share the distinction of being the only mammals whose bite is poisonous, at least to small prey. Almost all insectivores are nocturnal or crepuscular, some are excellent tunnelers through the soil, and a few are aquatic.

Insectivores may be divided into two groups, one clear-cut, the other less well defined. The first group is composed of insectivores in which the intestinal cecum is lacking and the stapedial artery is a major source of blood supply to the brain. These are known as the Lipotyphla, or Insectivora proper. All living insectivores (except tree shrews, Tupaiidae) and most late Cenozoic, a few early Cenozoic, and possibly Cretaceous insectivores belong to this first group. The second group, sometimes called the Proteutheria, is generally more primitive and is primarily early Cenozoic in known occurrence. Living members have an intestinal cecum, and the blood supply to the brain is still primitively complex. The stapedial artery is not as important as in the lipotyphlans. Primates and Macroscelidea apparently descended from early but still unknown representatives of this second group. *See* PRIMATES.

Lipotyphla. In this group the ancestry of the hedgehogs (Erinaceidae) (Fig. 1) has been traced back to the late Eocene or Oligocene in Eurasia without question, and still more primitive ancestors have been tentatively identified as early as the Paleocene in the United States. Closely related forms, known as early as the end of Cretaceous time, are close to the ancestral Proteutheria in known structure. The principal homeland of the hedgehogs has always been Eurasia. Most records of the family in North America represent aberrant derivatives of the more central Eurasian stock, but hedgehogs almost indistinguishable from the modern European genus lived in North America during the Miocene and Pliocene. The family never reached South America or Australia, both of which were islands during most of the Cenozoic. A side branch, the Dimylidae, flourished in the middle Cenozoic of Europe.

The living Solenodontidae and the extinct Nesophontidae are fairly closely related West Indian families of large, but nevertheless somewhat shrewlike, lipotyphlans whose remains have been found in late Quaternary deposits. Both families are believed to have become isolated in the West Indies in the earliest Cenozoic. Mainland North American Eocene and Oligocene genera (*Apternodus* and others) may have given rise to them but are now thought not to have done so.

Shrews (Soricidae) have been traced back to the late Eocene and prior to that their ancestry appears to have merged with that of hedgehogs. Fossil forms differ little from their modern counterparts in known structure except in the extinct subfamily Heterosoricinae, in which the jaw is short and the teeth somewhat more hedgehoglike than in other shrews. The fossil record of shrews is primarily in Europe and North America.

Moles (Talpidae) are known from both European and American deposits as early as the beginning of the Oligocene, but at their first known appearance the moles were already well differentiated. The characteristic digging humerus of most talpids is one of the most easily identified fossil bones in the continental Cenozoic record. Fossils of possible mole ancestors are known from the Paleocene and Eocene.

Tenrecs (Tenrecidae), whose Recent representatives are confined to Madagascar except for the aberrant subfamily Potamogalinae of West Africa, have a poorly known fossil record in the Miocene of East Africa and in the Eocene and Oligocene of North America.

Another African lineage of problematical lipotyphlans is the Chrysochloridae, or golden moles. These have a Pleistocene record in South Africa and have recently been found in the Miocene of East Africa. Their affinities probably lie with the tenrecs, though in reality little is known of the origin of these peculiar molelike animals.

Proteutheria. The second, less well-defined group of insectivores contains the living tree shrews of Southeast Asia. These are primatelike in many features and are frequently placed in the Primates by workers who regard the hedgehogs as more primitive. Supposed tupaioids have been found in the early Cenozoic of Europe and Central Asia and in the middle Paleocene of North America, but an adequate fossil record of these primatelike insectivores is still lacking.

Seven or eight primarily early Cenozoic families are tentatively referred to the Proteutheria. The Leptictidae (*Ictops*, *Diacodon*, and others) ranged from the Cretaceous to middle Oligocene in North America (Fig. 2) and from the late Paleocene to middle Eocene in Europe. The Pantolestidae (*Pantolestes*, *Aphronorus*, and others) were large, aquatic forms known from the middle Paleocene to the early Oligocene of North America and Europe and from the late Eocene of Asia. The Apatemyidae (*Apatemys*, *Heterohyus*, and others) were superficially rodentlike forms that ranged from the early Paleocene to middle Oligocene in North America and from the early Eocene to the late Eocene in Europe. The Mixodectidae are another superficially rodentlike group confined to the Paleocene of North America. The Zalambdalestidae (*Zalambdalestes* only) occur in the Late Cretaceous of Mongolia and appear to be primitive relatives of the Leptictidae, in which the anterior dentition is aberrant. The Endotheriidae (*Endotherium*) occur in the Early Cretaceous of China and may be the most primitive known proteutherians, but they are insufficiently known for meaningful appraisal. The Pappotheriidae are primitive and somewhat tenreclike but are known only from the Cretaceous of North America.

Still a third group, the Deltatheridioidea, most of

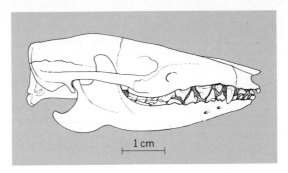

Fig. 2. Skull and lower jaw of *Ictops*, an Oligocene monotyphlan insectivore. (*After W. Scott and G. Jepsen*)

whose members possess upper teeth bearing two closely appressed main cusps, ranged from the Late Cretaceous to the late Tertiary. In the past these animals were placed in the Insectivora, but more recently they have been shown to be closely related to the hyaenodonts and oxyaenoids, often regarded as typical creodonts. Whether or not they are classed with the latter, the Deltatheridioidea are probably ancestral to them. *See* DELTATHERIDIA; EUTHERIA; HEDGEHOG; MAMMALIA; MOLE (ANIMAL); SHREW; SOLENODON; TENREC.

[D. DWIGHT DAVIS/MALCOLM C. MC KENNA]

Insectivorous plants

Plants having variously modified, highly specialized leaves which capture and digest insects. The proteins of the digested insect bodies supply nitrogen, which otherwise may be unavailable to these plants in the places where they grow. Sometimes they are also called carnivorous (flesh-eating) plants. *See* PITCHER PLANT; SARRACENIALES; SECRETORY STRUCTURES (PLANT); SUNDEW; VENUS' FLYTRAP.

[PERRY D. STRAUSBAUGH/EARL L. CORE]

Insolation

The amount of solar radiation which reaches a unit horizontal area of the Earth. It is the latitudinal variation of insolation which supplies energy for the general circulation of the atmosphere. Insolation outside the Earth's atmosphere depends on the angle of incidence of the solar beam and on the solar constant. The solar constant is the amount of energy which, in unit time, reaches a unit plane surface perpendicular to the Sun's rays outside the Earth's atmosphere when the Earth is at its mean distance from the Sun. Insolation is measured in langleys (ly) or calories per centimeter squared (ly = cal/cm^2). Outside the Earth's atmosphere a marked gradient exists, from ~900 ly/day near the Equator to zero in the dark polar areas; but in summer the insolation is quite uniform—over 1000 ly/day near the poles compared with about 800 ly/day near the Equator.

The extraterrestrial solar energy is modified by the air, the clouds, and the land and water surface of the Earth. Energy of wavelengths shorter than about 2900 A does not reach the surface but is absorbed high in the atmosphere mainly by nitrogen, oxygen, and ozone.

Air molecules also scatter energy in accordance with Rayleigh's law, and absorption by ozone, water vapor, and other gases further reduces the solar energy transmitted. *See* SCATTERING OF ELEC-TROMAGNETIC RADIATION.

Particles always present in the lower atmosphere scatter and absorb energy, too. However, clouds affect the extraterrestrial insolation more than any other atmospheric factor. The reflectivity (or albedo) of clouds varies from less than 10 to over 90% of the insolation on them. The albedo will be higher in visible light than in total solar energy, for clouds absorb mainly in the near infrared. The cloud albedo depends on the drop sizes, liquid water content, water vapor content, and thickness of the cloud; it also depends on the Sun's zenith distance Z. The smaller the drops and the greater the liquid water content, other things being the same, the greater the cloud albedo.

Clouds also absorb solar energy. Consider a cloud with a given liquid water content; if the drops are large and therefore relatively few in number, solar energy can deeply penetrate the cloud. If the cloud is also warm (high water vapor content) and occurs in deep vertical layers, it will absorb more than other cloud types.

The solar energy measured underneath clouds also depends on the reflectivity of the surface. An overcast which transmits 0.4 of the energy incident on it when the cloud is over a forest will transmit 0.7 of the incident energy when it is located over a highly reflecting snow surface. This is caused by the multiple reflections between the snow and the cloud, but the forest still absorbs more energy than the snow.

All these atmospheric effects will modify the distribution of the solar energy which reaches the surface. For example, H. G. Houghton computed the annual values given in the table, which shows that the latitudinal gradients of insolation are appreciably smaller at the Earth's surface than at the outside of the atmosphere. The high value of surface insolation near latitude 30°N is due in part to the small cloudiness in the large high-pressure areas at those latitudes. Additional longitudinal variations in insolation depend on cloudiness. Thus near latitude 40°N in summer, the average insolation exceeds 700 ly/day in desert areas and falls to 250 ly/day in the cloudy areas off Japan. *See* RADIATION, TERRESTRIAL.

The energy absorbed at the surface depends also on the surface albedo. The reflectivity of the solid Earth varies with ground cover. Forests absorb nearly all the energy incident on them. Fresh snow, on the other hand, reflects up to 90% of the incident energy. In middle latitudes, snow albedos may decrease to less than 40% with time, especially after warm spells have modified the surface. Grass, fields, and other surfaces will have intermediate albedos.

The reflectivity of the water surface of the Earth with cloudless skies depends strongly on the Sun's zenith angle Z. The albedo for the direct solar beam is only 0.02 for $Z = 0°$ and 0.13 for $Z = 70°$. It increases to 0.35 for $Z = 80°$ and to 1.00 for $Z = 90°$. For these larger values of Z, the wind may cause the albedo to decrease. An albedo value of 0.17 for $Z = 80°$ with a wind of 15 knots has been suggested. *See* ALBEDO.

When all the factors are considered, the absorption by the atmosphere and Earth's surface can be estimated, as shown in the last line of the table. It is interesting to note that the gradient of absorbed radiation, which is the driving force of the circulation, is similar to the gradient of the extraterrestrial insolation.   [SIGMUND FRITZ]

Bibliography: F. A. Berry, Jr., E. Bollay, and N. R. Beers (eds.), *Handbook of Meteorology*, 1945; S. Fritz, Solar energy on clear and cloudy days, *Sci. Mon.*, 84:55–65, 1957; N. Robinson (ed.), *Solar Radiation*, 1966.

Inspection and testing

In industrial engineering, the examination of a product to determine if it has any flaws. Commercial practice is to inspect products for outward characteristics and to test products for performance. Inspection is generally a mechanical operation; it determines the adequacy of such characteristics as appearance, dimensions, and overall operation. Testing is more penetrating; it measures performance in one form or another. It involves instrumentation to measure the functional adequacy of the product. Testing may be used to determine the quality of video and audio characteristics of a television receiver. It may be used to measure the load-carrying characteristics of a motor-generator set under varying operating conditions; information is obtained on speed-torque, load-temperature, and load-efficiency characteristics.

Inspection is the critical examination of a product to determine its conformance to applicable quality standards or specifications. The product is either accepted or rejected, depending on its degree of conformance. Inspection is a post-production operation and by itself has no effect on the quality of products; the inspector merely accepts or rejects products as submitted.

Inspection can, however, serve as the first step in a program to maintain a desired quality level. The product is inspected as soon as it is completed, the results analyzed, and appropriate action determined to adjust the manufacturing operation so as to maintain the desired quality. It is this control which assures the production of products within applicable quality restrictions and prevents the production of subquality items.

The desired level of the quality of product reaching the customer can be maintained by inspecting or testing all of, or a portion of, the final product. When inspection or testing is used for this purpose, it is referred to as a "quality assurance" activity. Thus, the data from inspection or testing can be used for either process control or quality assurance.

The purpose of testing is threefold. First, it is to evaluate a product for conformance to applicable quality standards. This is referred to as commercial testing as distinguished from a complete

Latitudinal variation of insolation

Latitude	0°N	30°N	60°N	90°N
		(annual mean, ly/day)		
Extraterrestrial	850	740	470	350
At surface				
With clear sky	570	520	320	220
With normal cloud				
cover	410	440	200	150
Total absorption (atmosphere and surface)	570	530	260	120

engineering test of the product. Second, testing is used to discover the possible existence of safety hazards. For example, electrical products are given an overpotential test to assure safety to the operator under normal operating conditions. Third, testing determines the reliability of the product. Reliability is the probability that a product will perform its intended function under environmental conditions of storage, maintenance, and operation. Life tests of the product and tests to destruction are used to determine reliability. Tests are made on a limited number of units and go beyond commercial testing.

Statistical methods are used extensively for evaluating inspection and test results, because these methods convey a tremendous amount of information regarding average quality, expected range of variation in quality, and reliability of the product. *See* GAGES; QUALITY CONTROL; STATISTICS   [DONALD S. HOLMES]

Bibliography: American Society for Quality Control, *J. Qual. Technol.*, periodically; American Society for Quality Control, *Qual. Progr.*, periodically; A. V. Feigenbaum, *Total Quality Control*, 1961; E. L. Grant, *Statistical Quality Control*, 3d ed., 1964; W. C. Holzbock, *Instruments for Measurement and Control*, 1955; C. W. Kennedy, *Inspection and Gaging*, 1951.

Instability, electrical

A persistent condition of unwanted self-oscillation in an amplifier or other electrical circuit. Instability is usually caused by excessive positive feedback from the output to the input of an active network. If, in an audio-frequency amplifier, instability is at a low audible frequency, the output will contain a putt-putt sound, from which such instability is termed motorboating. The instability may also be at a high audible frequency, or it may be at frequencies outside the audible range. Although such oscillations may not be heard directly, they produce distortion by driving the amplifier beyond its linear range of operation. *See* AMPLIFIER.

Instability can arise if the load on an amplifier has a critical phase or magnitude. Similarly, instability arises in a closed-loop control system if the damping is too light relative to the response time. Such instability may result in the system hunting about a control condition instead of remaining steadily at the condition. *See* CONTROL SYSTEMS; NEGATIVE-RESISTANCE CIRCUITS; SERVOMECHANISM.

In a power distribution system, if the mechanical load on synchronous motors exceeds the steady-state stability limit or if an abruptly changed mechanical load causes the synchronous machines to exceed the transient stability limit, the power system becomes unstable. It can also lose stability from a three-phase short circuit between generators and motors. High-speed circuit breakers and other protective devices guard against such instability. *See* CIRCUIT BREAKER.   [FRANK H. ROCKETT]

Instinctive behavior

Any species-typical pattern of responses not clearly acquired through training. The term instinctive behavior is preferred to the traditional term instinct, which denotes an innate impulse blindly impelling action appropriate to attaining certain ends. The adjective instinctive is used by some, the Freudians for example, to signify certain complex, motivated behavior, and thus leans toward the classical meaning, that of an innate drive or predisposition to given acts.

The problem. In every animal species, certain characteristic patterns of systems of adaptive action appear under certain conditions. Thus, many spiders spin webs typical of their species, most birds make species-typical nests, and beavers build dams and lodges. These activities, instinctive in the above sense, pose important questions concerning their evolutionary basis and genetics, their ontogeny and psychology, and their adaptive significance.

Criteria. Earlier theorists seeking objective approaches to this problem devised certain criteria of the instinctive in behavior, particularly, appearance shortly after birth or hatching; no essential dependence upon learning; and appearance in the individual raised in isolation. With further research, however, these criteria all met with objections. For example, the first is contradicted by evidence that species-typical behavior may appear at stages other than birth; the second, by evidence that experience often exerts its effects in ways resisting clear identification, for example, embryonic stimulation or prenatal conditioning; and the third, by the fact that isolation may not exclude extrinsic influences, that is, stimulative properties of the individual itself which are characteristic of the species. But the last two of these criteria still influence many students of comparative animal behavior who equate with "instinctive" such concepts as innate, native, and endogenous.

Focal points in research and theory. Objective study demands that the behavior patterns of representative animals be studied analytically, and that basic assumptions be tested. Research must cover environmental conditions and stimuli, the range of behavioral variation, and organic conditions underlying the development and appearance of the behavior. Both longitudinal and cross-sectional studies of behavioral ontogeny are needed, obtained by methods appropriate to the species, and consideration for genetic processes.

In theorizing, all conceptual terms must be evaluated. Skepticism about the instinct concept centers around traditional dogma which differentiates psychologically between man, conceived as the sole possessor of reason, and lower animals ruled by instinct. This idea is contradicted by much evidence in comparative psychology. The related assumption that animals possess an original, innate nature modified only secondarily (if at all) by experience is opposed by evidence on behavior development. The related idea of sharply distinguishing instinctive from intelligent behavior is in contrast to evidence that instinctive behavior is often plastic in relation to the situation. Differences in the nature of such behavior, and in its developmental basis, doubtless exist on all phyletic levels.

Other unsettled, controversial points in research and theory concern the role in such behavior of (1) neural organization, (2) nonnervous organic factors such as hormones, and (3) sensitivity and perception, together with the relation of these in ontogeny (development). Some authorities postulate the existence of innate central neural coordinations to account for the rise and control of instinctive behavior; others emphasize the role of peripheral mechanisms interwoven with neural processes in behavioral ontogeny. Classical distinctions between reflex and instinct are questioned on the ground that such functions may differ in degree rather than in kind. Theories distinguishing sharply between instinct and learning meet the objections that neither of these is sufficiently well understood, and that both may vary greatly in their forms and in their relationships to development on different phyletic levels. *See* BEHAVIOR, ONTOGENY OF.

Although all behavior is related to heredity, the function of the genes in behavior is a complex question, still unanswered. Genes in the chromosomes must exert basic influences on structural growth, and, through structure, on behavior. Examples of behavior determined or influenced by structure are readily found, as in limb structure and characteristic gait, cellular equipment in the retina of the eye and proneness for day or night activity, size of certain glands, such as the adrenal, and degree of docility or of wildness. However, it is another matter to trace such correspondences from fertilized egg through complexities of ontogeny. *See* BEHAVIOR AND HEREDITY; GENETICS.

The genetics of some of the species-predictable behavioral characteristics of animals have been worked out and are impressive. Thus, through selective breeding of parental generations, it has been possible to produce hybrids and backcrosses of certain insects, fishes, birds, and mammals differing predictably from the parents in behavioral characteristics such as level of excitement, parental behavior, and reproductive behavior.

The facts do not yet indicate how directly or in what manner the genes function in the development of behavior in any animal. Preformists postulate a direct relationship between genes and behavior; epigeneticists postulate an indirect relationship through many interlacing factors in organism and developmental situation. The traditional nature-nurture controversy has waned; instead, the emphasis is on the study of the rise of typical and atypical ranges of species behavior in evolution and in ontogeny. *See* HIBERNATION; HOARDING BEHAVIOR; HYPOTHERMIA; MIGRATORY BEHAVIOR; PSYCHOLOGY, PHYSIOLOGICAL AND EXPERIMENTAL; REPRODUCTIVE BEHAVIOR.

[T. C. SCHNEIRLA/ETHEL TOBACH]

Bibliography: R. A. Hinde, *Animal Behavior*, 1966; J. Hirsch (ed.), *Behavior-Genetic Analysis*, 1967; N. R. F. Maier and T. C. Schneirla, *Principles of Animal Psychology*, 1964; P. R. Marler and W. Hamilton, III, *Mechanisms of Animal Behavior*, 1966; W. N. Tavolga, *Principles of Animal Behavior*, 1969.

Instrument landing system (ILS)

A continuous-wave amplitude-modulated, horizontally-polarized fixed-beam system comprising the following aids to aircraft approaching a runway: (1) a vhf localizer to give azimuth guidance with reference to the runway center line, with an operating frequency in the band 108–112 MHz; (2) a uhf glide slope to give elevation guidance about an optimum approach angle, with an operating frequency in the band 328.6–335.4 MHz; and

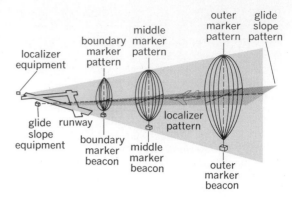

Fig. 1. Guidance from fixed-beam low-approach system.

(3) two or three marker beacons operating at 75 MHz to give range with reference to the touchdown point. Figure 1 illustrates the system.

Since 1948, performance standards and recommended practices have been agreed on and formulated by the International Civil Aviation Authority (ICAO). In 1963, ICAO laid down three categories of ILS guidance criteria to standardize operations to lower approach visibility minima:

Category I. Operation down to a height of 60 m (200 ft) and 800 m (2600 ft) visibility.

Category II. Operation down to 30 m (100 ft) and 400 m (1300 ft) visibility.

Category III. Operation down to and along the surface of the runway unrestricted by cloud base and visibility conditions.

Localizer. The antenna array is installed about 1000 ft back from the upwind end of the runway, providing guidance to touchdown and along the runway. The antenna system, generally a multielement broadside array, varies according to the nature of the airfield site and the performance category required. Signal reflections from obstructions around the runway and approach path cause interference, or bends, on the approach. Stringent ICAO limits are imposed on the amplitude of such bends. Thus it is generally necessary to restrict the main COURSE guidance signal field to a narrow azimuth sector and radiate a subsidiary signal system to provide CLEARANCE coverage over a wider azimuth, to at least ±35° for the ICAO standard. The CLEARANCE signals must provide the same guidance intelligence as the COURSE system but, by reason of a different characteristic be suppressed in the receiver when this is sampling both signals. In one method, the carrier frequencies differ by about 10 kHz; in a later method, a quadrature phase displacement between the guidance signals gives 3:1 better suppression, without requiring a second transmitter.

Two types of signals are produced in the electronic equipment. The first type is a carrier with double sidebands (CSB) of 90 and 150 Hz. The second type of signal has double sidebands only (SBO) of 90 and 150 Hz but with the phase of the 150-Hz sidebands 180° out of phase with the corresponding sidebands in the CSB. The simple array shown in Fig. 2 has been used to provide CLEARANCE coverage and a back-course, in association with a highly directional front-course system.

CSB energy is fed to the center element and SBO energy equally but in rf phase opposition to the two outer elements, the connection to one outer element being reversed. The CSB signal phase is the phase center. Figure 3 shows the resulting radiated field patterns.

For the narrow-beam highly directional multielement arrays, CSB energy is distributed in phase to all the elements and SBO in phase opposition to each half of the array, each distribution with a particular symmetrical amplitude gradient; again CSB is the phase center. The guidance field patterns of a typical dual system are shown in Fig. 4.

In the aircraft, a special superheterodyne receiver detects the localizer signals. Filters in the audio output separate the 90- and 150-Hz signals, whose relative amplitudes are compared. When these are equal, the indication on the vertical pointer of the cross-pointer instrument is that the aircraft is on course (Fig. 5). Deviation from the true course results in a difference in the depths of modulation (DDM) of the two tones, resulting in a left or right movement of the pointer. The standard sensitivity is such that a lateral displacement of ±350 ft at the runway threshold deflects the indicator to full scale. For a 10,000-ft runway, this displacement is approximately equivalent to ±2° in azimuth. Deflection is linear with angular deviation. Associated with the pointer is a warning indicator marked OFF, called the flag, which is operated by the sum of the 90 and 150-Hz signals. The flag resolves two conditions: an ON COURSE indication when no signals at all are received (flag appears); and an ON COURSE indication when all signals are present and normal (flag disappears).

Glide slope. This system utilizes the same types of signals as the localizer. The linear DDM structure is required in the vertical plane, zero DDM occurring at the desired glide-path angle, generally 2.5 or 3°. Accordingly, the antenna systems mostly employed have been vertical arrays of two or three elements, depending on signal reflection from an extensive frontal area of level ground to form lobular field patterning in the vertical plane.

The relative lobular patterning is controlled by the element heights, the combination of the different patterns determining the DDM structure. The radiation pattern is therefore affected by changes of ground characteristics and subject to interference by obstacle reflections.

The array is located offset 440–550 ft from the runway center line and a distance back from the landing threshold, such that the glide path intersects the runway center line 50 ft above the threshold. In the aircraft a separate receiver operates on the glide-path signals in a similar manner to that for the localizer. Standard DDM sensitivity gives full-scale deflection, up or down, of the horizontal pointer of the aircraft instrument, for an angular deviation equivalent to ±12 ft vertical displacement at the threshold height.

To overcome different site difficulties and provide adequately bend-free flight paths, four types of ground-dependent arrays have evolved, namely, equisignal, null-reference, sideband-reference, and capture-effect arrays. The null-reference, the preferred and simplest system, superseded the equisignal, which is asymmetric, less linear in DDM structure, and seriously affected by ground changes and has high bend potential.

The null-reference array in Fig. 6 employs two

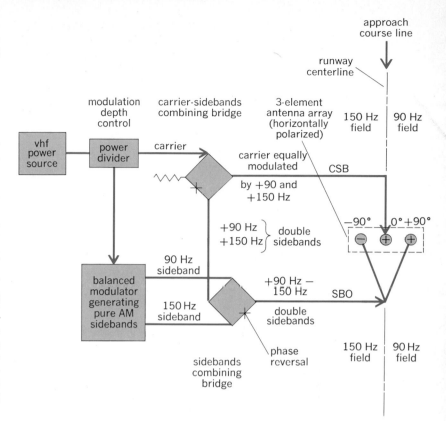

Fig. 2. Basic method of instrument landing system (ILS) signal production and simple omnidirectional localizer antenna. This is a null-reference system.

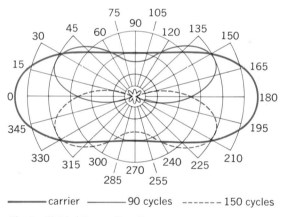

—— carrier —— 90 cycles ------ 150 cycles

Fig. 3. Field pattern of localizer.

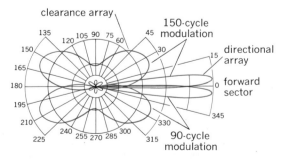

Fig. 4. Localizer field pattern of a typical dual system showing directional pattern for forward-sector approach and guidance for all other sectors.

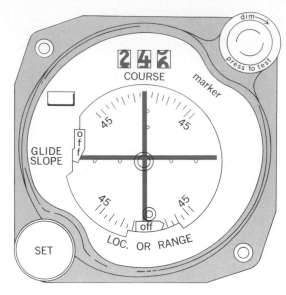

Fig. 5.   Modern cross-pointer instrument. A marker beacon light is mounted in the upper right-hand corner. In the aircraft, a special superheterodyne receiver detects the localizer signals.

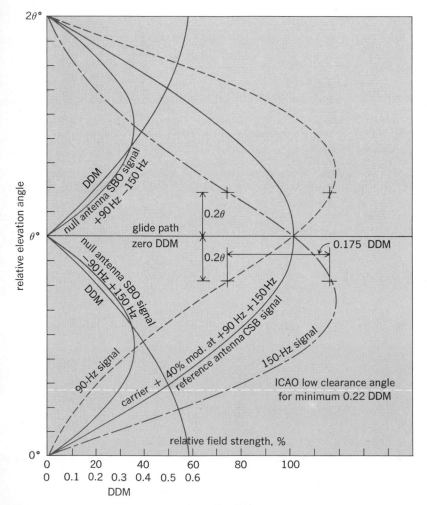

Fig. 6.   Null-reference glide-path array showing composite vertical radiation patterns. SBO amplitude is shown 2.5 times actual value, normalized to 40% modulated level represented by CSB curve, which also indicates carrier level. Conditions are for maximum International Civil Aviation Authority (ICAO) sensitivity.

antenna elements at heights h and $2h$. The CSB signal is radiated from the lower element, its first lobe peak occurring at the glide-path angle θ. The SBO signal from the upper element produces two lobes in the space of the CSB lobe. It produces a signal null at the glide-path angle with its second lobe phase reversed. The resulting DDM structure is symmetrical, very linear, and only slightly affected by ground changes.

The two-element sideband-reference and three-element capture-effect arrays, developed from the null-reference, have identical DDM characteristics. The former, of lower height, is suited to sites where a short frontal area is terminated by a valley. The latter, of greater height, has very low bend potential, which makes it suited to sites with multiple reflection hazards.

A later development, the multielement broadside array, is substantially independent of the ground. This development is a vertical array of horizontal dipoles or vertical waveguide slots, at least 70 ft high, energized like a localizer array to produce very narrow main beams of CSB and SBO signals, substantially independent of the ground. The array is tilted backwards by the glide-path angle, at which angle the CSB peak and SBO null occur.

Marker beacons. Two or three beacons indicate progress along the approach path. The outer marker is located 4.5 mi from the runway threshold and is modulated at 400 Hz. The middle marker distance is 3500 ft and modulated at 1300 Hz. The inner marker, when used for category II approaches, intercepts the glide path at about the 100-ft height to mark the overshoot decision point if the runway is still not visible. This marker is recognized by its 3000-Hz modulation.

The radiation patterns are shaped such that the signals received in the aircraft during fly-through cause lamps to be illuminated for different periods to identify which beacon is being overflown. A separate crystal-controlled receiver is used, the filtered output of which actuates the lamps and provides oral identification by the different beacon modulation tones. *See* NAVIGATION SYSTEMS, ELECTRONIC; PRECISION APPROACH RADAR (PAR).

[FRANK H. TAYLOR]

Bibliography: F. G. Fernau, Design philosophy of the STAN.37/38 category III instrument landing system, *Interavia*, December, 1966; W. L. Garfield and F. T. Norbury, Instrument low-approach system and radio altimeter for all-weather landings, *Elec. Commun.*, vol. 41, no. 2, 1966; J. S. Hall, *Radar Aids to Navigation*, 1947; F. T. Norbury, ILS equipment for cat. III all-weather landing, *World Aerosp. Systm.*, January, 1967; P. C. Sandretto, *Electronic Avigation Engineering*, 1958.

Instrument transformer

Electric transformers specifically designed for use in measurement and control circuits. The purpose of an instrument transformer is to convert primary voltages or currents to secondary values suitable for use with relays, meters, or other measuring equipment. A second purpose is to isolate the high-voltage primary circuit from the measurement circuit. Thus the construction of measuring devices and the protection of personnel using such devices

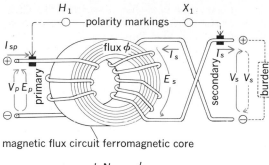

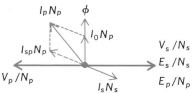

Fig. 1. A simple instrument transformer and its phasor diagram. Core-loss components and impedance drops have been omitted from diagram. (*From I. F. Kinnard, Applied Electrical Measurements, Wiley, 1956*)

are greatly simplified by the use of instrument transformers. A sketch showing the basic elements and phasor diagram for an instrument transformer is shown in Fig. 1. For general discussion of transformers *see* TRANSFORMER.

Specialized design techniques and materials are used in instrument transformers to insure a high degree of stability and reliability, because improper operation could result in economic loss in metering applications or power system damage in relaying applications.

There are two general types of instrument transformers: instrument potential transformers, which are used in voltage measurements, and instrument current transformers, which are used in current measurement.

The primary of potential transformers is connected across the line and the current that flows through this winding produces a flux in the core. Since the core links both the primary and secondary windings, a voltage is induced in the secondary circuit. The ratio of primary to secondary voltage is roughly in proportion to the number of turns in the primary and secondary windings. Usually this ratio is selected to produce 115 or 120 volts at the secondary terminals when rated voltage is applied to the primary.

The current transformer differs from the potential transformer in that the primary winding is designed for connection in series with the line. The ratio of primary to secondary current is, roughly, inversely proportional to the ratio of primary to secondary turns. Usually 5 amp of secondary current is produced by rated current in the primary.

Instrument transformers may differ considerably, since the physical construction depends upon the intended application. A potential transformer usually has a magnetic core and coil arrangement that is quite similar to that of the conventional power or distribution transformer. Since the core operates continuously at the highest flux density consistent with good magnetic design, the core is the major heat producer. Therefore, shell-type

construction usually is employed for effective heat dissipation.

In a current transformer the magnetic core density is a function of line current and, except for occasional transients caused by line faults, is usually very low. The core loss is negligible, and the windings are the major source of heat. Therefore, the core type of construction is generally employed.

Special insulations, such as arc-resistant butyl rubber, have been developed for winding insulation to ensure necessary reliability and safety.

Instrument transformers have a marked ratio, which differs from the turns ratio and more nearly represents the true ratio of output to input. However, since the true ratio changes slightly with burden, frequency, temperature changes, and other factors, the marked ratio must be multiplied by a ratio correction factor to give the true ratio under specified operating conditions.

Principles of operation. Instrument transformers approach an ideal transformer, and their principle of operation is explained by a study of the ideal. The practical differences from the ideal can then be incorporated, and equivalent circuits can be established which allow analysis of most instrument transformer problems.

In the ideal instrument transformer the output voltage or current is exactly proportional in magnitude, and exactly opposite in phase, to the corresponding input voltage or current. This means there is no core or winding loss, and no resistance or reactance voltage drops. All the flux links both primary and secondary windings.

An alternating voltage applied to the transformer produces a primary current and a resulting alternating flux ϕ in the core. The same alternating emf e is induced in each primary and secondary turn by the changing flux linkages $d\phi/dt$. For a complete winding the induced voltage equals $-N(d\phi/dt)$, where N is the number of turns.

The instantaneous flux ϕ in an ideal transformer is sinusoidal with a maximum value of ϕ_m when the applied voltage is sinusoidal, as shown by Eq. (1). The emf is shown by Eq. (2). The root-mean-square (rms) value of the voltage is shown by Eq. (3),

$$\phi = \phi_m \sin \omega t \qquad (1)$$

$$e = -\phi_m N\omega \cos \omega t \qquad (2)$$

$$E = 4.44\phi_m Nf \qquad (3)$$

where f is the frequency. This induced voltage is determined only by the applied voltage and the number of turns, since there are no other voltage drops. Since the induced voltage in the primary E_p must cancel the applied voltage V_p, its polarity is determined for all turns including the secondary, as shown by the relations given in Eq. (4). Instru-

$$E_p/E_s = -V_p/V_s = N_p/N_s \qquad (4)$$

ment transformers have polarity marks such that, when the current enters the H primary, it instantaneously leaves the X secondary, and if H has a positive potential, X is also positive.

Closing the secondary circuit of an ideal transformer results in a secondary current i_s, with an mmf $\mathscr{F}$ in the core producing a magnetizing force H. Magnetomotive force and magnetizing force

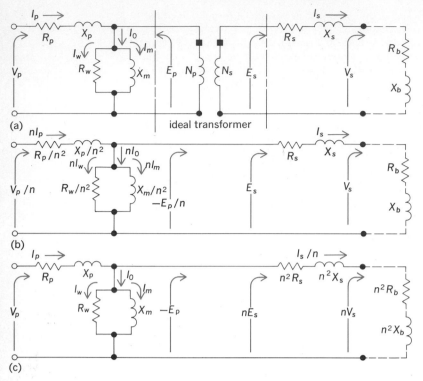

Fig. 2. Instrument transformer equivalent circuits. (a) Ideal transformer. (b) All quantities are referred to secondary. (c) All quantities are referred to primary. (*From I. F. Kinnard, Applied Electrical Measurements, Wiley, 1956*)

can be computed from Eqs. (5) and (6), respective-

$$\mathscr{F} = N_s i_s \qquad (5)$$

$$H = N_s i_s / l \qquad (6)$$

ly, where N_s is the number of secondary turns and l is the magnetic circuit length.

This magnetizing force must exactly cancel that of the primary, since no excitation flux is needed for the ideal transformer. This is shown by Eq. (7), from which Eq. (8) can be derived.

$$N_s i_s / l = -N_p i_p / l \qquad (7)$$

$$i_p / i_s = N_s / N_p \qquad (8)$$

By defining n as equal to $-N_p/N_s$, Eqs. (9) through (11) can be obtained.

$$i_s = i_p n \qquad (9)$$

$$E_s = -E_p / n \qquad (10)$$

$$V_s = V_p / n \qquad (11)$$

When the secondary circuit is closed, the secondary current is V_s/Z_b, where Z_b is the secondary burden (the load and its wiring) composed of resistance R_b and reactance X_b.

An actual instrument transformer can be simulated by the network shown in Fig. 2a. Some ampere turns are needed to produce exciting flux in the magnetic circuit. A current I_o through N_p turns produces this flux; it has a magnetizing component I_m and an eddy current and hysteresis watt loss component I_w. These are simulated in the equivalent circuit by a reactance X_m and a resistance R_w respectively. The transformer windings have resistances shown by R_p and R_s, and magnetic leakage produces small reactances shown by X_p and

X_s. Equivalent circuits eliminating the ideal transformer can be drawn by referring all values to the secondary, as in Fig. 2b, or to the primary, as in Fig. 2c.

Rules for referring quantities from the primary to the secondary are (1) multiply I_p by n, (2) divide V_p and E_p by n, and (3) divide Z_p, R_p, and X_p by n^2.

These equivalent circuits are usually satisfactory to predict transformer performance, although in the transformer the magnetization characteristic is nonlinear. Capacitive effects also exist. These should be evaluated at frequencies above 100 Hz, and it is usually necessary to change values of X_m and R_w as a function of frequency.

Accuracies. Potential transformer accuracies depend on the magnitude of primary and secondary voltage drops. With the secondary open circuited the only voltage drop is caused by the exciting current. With a burden it is possible to add drops due to the burden current in both primary and secondary to obtain the overall change in performance. Potential transformers are designed to have low exciting current and low winding impedances.

In a current transformer the difference between the secondary and primary current is the exciting current. Therefore, small errors in both ratio and phase angle exist fundamentally. These errors vary with primary current, and many compensating schemes have been used to increase accuracy. Both the magnetizing and core-loss currents are kept small, special lamination material is used, and special techniques are used to reduce losses in the transformer.

Commercial instrument transformers have extremely good accuracies, and manufacturers furnish typical curves depicting any small error. American Standard for Transformers, Regulators, and Reactors (ANSI C57) specifies the accuracy classes for specific burdens, and errors are kept within these standard accuracies for transformers of a given class. For even greater accuracies test equipment is available for accurately measuring ratios and phase angles by either direct methods or by comparison with standard transformers. *See* ELECTRICAL MEASUREMENTS. [ISAAC F. KINNARD]

Bibliography: I. F. Kinnard, *Applied Electrical Measurements*, 1956; H. M. Nordenberg, *Electronic Transformers*, 1964; A. Schure (ed.), *Transformers*, 1961.

Instrumental analysis

The use of an instrument to measure a component, to detect the completion of a quantitative reaction, or to detect a change in the properties of a system.

The presence of a measurable characteristic which is uniquely dependent upon the concentration of the substance to be determined is required. This characteristic can be absorption or emission of energy, radioactivity, oxidation or reduction behavior, nuclear or electronic properties, magnetic behavior, or thermal properties. The characteristic should be easily measured and should be simply related to the amount of material present. Examples of this application are the determinations of radium in ores by radioactivity analysis, of tungsten in steel by emission spectroscopy, of vitamin A by ultraviolet absorption, and of copper in biological materials by polarography. In a two-component system, a property of the system as a whole can often be measured and related to the desired component. For example, a water–ethyl alcohol mixture can be analyzed by density or refractive index measurements. For a sample containing n components, $n-1$ separate measurements are necessary.

An instrument can be used to detect the completion of a quantitative reaction when simpler methods, such as visual indicators, are not applicable. Usually a characteristic of the substance being determined or of the excess reagent is measured. An example is the amperometric titration of sulfate ion with lead ion where the first excess of lead beyond the end point is detected by its electrolytic reduction current.

Instrumental analysis does not require the appearance or disappearance of a characteristic. It can be only an increase or decrease. An example is the pH titration of a weak acid in water where the change in hydronium-ion concentration is measured, using the change in potential of an electrode in the solution.

Automatic instrumental analysis is applicable to any system. For example, an infrared analyzer can be built into a pipe through which a process stream flows. After the relationship between energy absorption and concentration has been established, the device reports the concentration to the control room and, in addition, makes a record on a graph, all without human supervision. Instruments can also be used to remove a measured quantity of sample, to titrate a given constituent, to clean the titration vessel, and to report the analysis with no help other than provision of standard reagent solutions and the selection of proper equivalence point conditions. *See* ANALYTICAL CHEMISTRY.

[K. G. STONE; C. RULFS]

Bibliography: H. H. Willard, L. L. Merritt, and J. A. Dean, *Instrumental Methods of Analysis*, 4th ed., 1965.

Instrumentation

Designing, manufacturing, and utilizing physical instruments or instrument systems for detection, observation, measurement, automatic control, automatic computation, communication, or data processing. Loosely, the term instrumentation is also used for the ensemble of instruments and auxiliary equipment used in an experiment, test, or process.

Instruments and instrument systems. In the broadest meaning of the word, an instrument is any device useful in accomplishing an objective. Scientific and technical instruments (to be differentiated from surgical, musical, and legal instruments) are devices used in observing, measuring, controlling, recording, computing, or communicating.

Instruments and instrument systems refine, extend, or supplement human faculties and abilities to sense, perceive, communicate, remember, calculate, or reason. The human senses, by which we feel, smell, taste, see, hear, maintain balance, estimate distance, and the like, are refined or extended by such devices as surface-roughness and contour gages, micrometers, chemical analyzers, pH meters, microscopes, telescopes, gyro-stabilized platforms, range finders, and many others. Other instruments, such as magnetometers and cosmic-ray counters, sense or measure physical quantities for which there is no physiological sense developed in human beings. Other instruments (such as cameras, correlators, simulators, and computers) perform functions of storing, transmitting or processing information signals in ways analogous to, or going beyond, human abilities to record, remember, communicate, compare, count, and systematically apply logical operations.

The simplest instruments merely provide a material standard with which the user can compare a physical quantity, as in measurement of lengths by a yardstick. Transducers are instruments which transform the quantity under observation (the measurand) into another quantity for which a standard of comparison is more readily or easily available or which may be transmitted, further transformed, recorded, or otherwise processed for utilization. A liquid-in-glass thermometer transforms temperature (the measurand) into position of the liquid in the stem with an attached scale calibrated in temperature units.

The signal—a condition, quantity or magnitude generated by the transducer as representative of the measurand—may be transformed a number of times and in a number of ways in an instrument system. For example, temperature may affect the position of a bimetallic strip, whose movement may change the electrical capacitance in a circuit, thereby changing a frequency or voltage, which may control the duration of a radiotelemetry signal (pulse-duration modulation). The telemetered signal in turn (after transmission, reception, and demodulation) may be transformed into current in a galvanometer, which changes the position of a light beam on a photographic film for recording, or it might be transformed by an analog-to-digital converter into a pattern of signals for use in a computer, or the signal might actuate a servomotor or a relay which would adjust the input of heat into the region of the bimetallic strip to control temperature.

Each of the many transformations of a signal throughout an instrument system may result in some additional lag and often in a loss in accuracy, but may be justified by convenience or necessity and even by improved over-all system accuracy or response characteristics.

The following list includes some of the more important functions of instruments or components of instrumentation systems in creating or handling signals (or information, or data): excitation, generation, modulation, detection, comparison, amplification, differentiation, integration, attenuation, conversion, switching, counting, coding, timing, programming, correlating, linearizing, correcting, displaying, recording, reducing, analyzing, computing, and controlling. Since all branches of experimental science and technology depend on instrumentation, specialized instruments, with a corresponding body of knowledge and practice, have been developed separately in many fields. Thus chemical instrumentation, aeronautical instrumentation, medical instrumentation, optical instrumentation, and many other similar terms indicate areas of specialization in various industries or professions.

Instruments are sometimes classified according

to the field of purpose or application, such as navigation instruments, surveying instruments, or oceanographic instruments; according to their functions in instrument systems, such as detection, measurement, recording, computing, controlling, signal modification, or display; or according to the physical quantity or property that is to be measured or controlled by the instrument, such as flow, temperature, pressure, force, displacement, level, viscosity, acceleration, electrical quantities (voltage, current, resistance, capacitance), or optical quantities (transmission, gloss, color, and brightness).

Another method of designating or classifying instruments, particularly those of wide applicability, is according to the operating principles. Examples of this method are x-ray (microscopes, difractometers), spectrometric of various types for various portions of the frequency spectrum or various particles (including infrared, ultraviolet, visible, microwave, x-ray, γ-ray, neutron, electron, and alpha particles), pneumatic, mechanical, electronic, electrical, magnetic, hydraulic, nuclear, acoustic, optical, and photographic.

Each of the methods of classifying instruments is useful only in specific instances. A single instrument or instrument system may utilize many combinations of different principles. A given instrument often has many applications. Any physical quantity may be measured by a number of different principles. Each of the many operations performed in handling the information flow in instrumentation systems may be accomplished by various techniques and devices. Thus no single method of classifying instruments is possible and the several methods are used where they apply.

In this encyclopedia some article headings may be found according to one of the classifications listed above, depending on common usage. Because of the many articles dealing with instruments and measurement, it is impossible to list cross references to all such articles in this encyclopedia. Most instruments that measure physical quantities or properties are covered in the article discussing that property. Other instruments used in a particular field may be found by consulting the article on that field. Specific instruments may be the subject of separate articles; if not, they can be found by consulting the index.

Scope of instrumentation. Instrumentation involves not only the design of individual instruments and components but also their integration with auxiliary or associated devices into instrument systems to perform one or many of the functions mentioned above.

In many instrument systems an auxiliary energy source, or field, is modified by the quantity to be measured, and a transducer is used to measure the modification. As an example, an x-ray beam penetrates a sheet of steel; the intensity of the emergent beam depends on, and is thus a measure of, the thickness of the sheet. An ionization chamber or a photocell transforms the ray intensity into electric current, which is further transformed into motion of the pointer on a meter. The position of the pointer is compared with a standard (the scale) which is calibrated in units of thickness. Vacuum pumps, x-ray tubes, lamps, batteries, electric signal generators (oscillators), sound or radio sources,

radioactive sources, high-voltage electron or ion accelerators, wind tunnels, and shock tubes are all examples of equipment used to provide such auxiliary fields, beams, or conditions.

The general physical and mathematical principles underlying the operation of instruments and instrument systems are drawn from all branches of physics and engineering, primarily from physics. Instrumentation as a science is concerned with the development and study of these principles and techniques, their application to design of specific instruments, and the utilization of instruments in research, engineering, testing, industrial processes, communications, defense, education, and other areas.

All instruments have inherent deficiencies. These deficiencies are broadly described by accuracy and precision characteristics of the instrument. Accuracy is the degree to which the measured value approaches the true value. Precision, which may be regarded as potential accuracy, relates to the capability of the instrument to distinguish one value of the measurand from other nearly equal values when the measurements are performed separately in space and time. No instrument responds instantaneously to a change in the measurand; the lag is dependent on the natural frequency of the instrument system and its degree of damping (internal dissipation of energy). If the damping is too great or the natural frequency too low, the response lag may be excessive and the indication or output seriously in error during an interval of several times the natural period of the system. Conversely, if the damping is too small, the dynamic response may be excessive and the output may exhibit oscillations before attaining the desired steady value. In either event a detailed analysis of the instrument characteristics is essential to make appropriate corrections.

Many instrument elements exhibit drift, a slow change in properties with time resulting in a change in output. Many elements also exhibit the phenomenon of hysteresis; that is, they exhibit a response to increasing signals which is different from their response to decreasing signals. Environmental factors, such as temperature or vibration, may also affect performance of certain elements. The instrument maker must choose materials and operating principles which minimize these effects; the user must develop measurement techniques which either avoid these effects or correct for them.

Instrument systems react upon the physical systems with which they are associated. In measurement, the presence or operation of the exciting field, sensing element, or measuring instrument may undesirably disturb or change the quantity to be measured; in automatic control, the instrument system acts positively to change the behavior of the physical system.

In automatic control systems the value of the variable to be controlled is compared with a desired value, and the difference (error) signal is utilized to initiate corrective action. Since the output measurement is used to control the input, such systems are called closed-loop or feedback control systems. For discussion of the use of instrumentation in control systems *see* CONTROL SYSTEMS; SERVOMECHANISM.

If the response characteristics of the instrument system are not appropriately matched to those of the controlled system, the parameters of the combined systems may exhibit transient or sustained oscillations. The general problem in design or analysis of automatic control systems is to determine and obtain the instrument performance which will ensure the optimum speed of response consistent with stability. For linear systems (those for which output response is proportional to input) this may be done analytically. Texts on instrument engineering, servomechanisms, or mechanics describe the several mathematical procedures commonly used for analysis of simple instruments, measuring systems, and control systems. For nonlinear systems, digital computers may provide solutions of acceptable accuracy.

Analog computers, or simulators, provide rapid nonanalytic solutions for behavior of linear (or to some extent nonlinear) systems under various conditions. Both types of computer are not only used to study systems but are increasingly included as part of systems whose behavior they predict and control. Digital computers also find major uses in rapid storage, retrieval, and processing of data of all types, and for solving scientific problems of great complexity. *See* COMPUTER.

The entire field of military technology rests largely upon modern developments in instrumentation, many of which in fact resulted from military needs. Radar, sonar, and infrared detectors are used not only for detection but also in combination with computers and control systems for automatic guidance of weapons. Homing torpedoes, self-guided missiles, and computer-controlled antimissile missiles are examples of the radical change in methods of warfare resulting from instrumentation, whereby warheads are automatically guided to, rather than merely thrown at, the enemy.

Instrumentation for measurement of nuclear radiation is the basis for all military and peaceful uses of atomic energy and radioactive isotopes.

Thus the scope of instrumentation is nearly universal; it may well be called a common denominator of all of science and technology.

Related fields. Because of the universal scope of instrumentation, there are many fields that are closely related.

Systems engineering. This term is applied to the design or analysis of process or machine systems as well as instrument systems. Instrumentation systems primarily involve the flow of signals (information), whereas the physical system or process involves primarily the flow of energy, materials, or both. Since the same general laws apply to all physical systems and since instruments are critically involved in most systems, instrument-systems engineering is a large and necessary part of systems engineering in any field, such as aircraft, missiles, oil refining, steel rolling, and communications. *See* SYSTEMS ENGINEERING.

A fundamental part of instrument-systems engineering relates to the theory of automatic control and to the practical application of automatic control equipment.

Information theory. This body of mathematical and logical knowledge, developed for analysis of the efficiency and the limitations of communication systems, is broadly applicable to all types of signals, or data, and is thus important to instrumentation. In any instrument or system the output signal varies to some extent in an unpredictable manner because of disturbing influences, some of which may be inherent in the instrument or system. The undesired signal is often referred to as noise. The ratio of the magnitude of the signal to that of the noise is important as an indication of possible error limits and detectability limits. The frequency range of signals that can be utilized by an instrument system determines the amount of useful measurement data that can be handled per unit of time. The fundamental definitions of information, frequency bandwidth, noise, channel capacity, and their interrelations are all involved in information theory; thus this field is clearly of basic importance to instrumentation, and particularly to communication instruments and systems. *See* INFORMATION THEORY.

Human engineering. This is frequently involved in instrumentation. To be operated by or to provide information to an observer, instrumentation systems should be designed to take account of his physical, physiological, and psychological characteristics. In design of dials, displays, or manipulative elements, visual acuity and body characteristics must be considered; in analysis or design of systems in which a human being provides a link between measurement and control, the reaction times the decision capabilities of the human being and his susceptibility to fatigue and to environmental factors must also be taken into account. *See* HUMAN-FACTORS ENGINEERING.

Cybernetics. Cybernetics relates to the field of man-machine communications and the application of information theory to response and automatic control in physiological systems. It is thus another field which partly includes, and is partly included in, the field of instrumentation. *See* CYBERNETICS.

Automation. The term automation has various meanings but generally implies the application of automatic controls to an operation or process. As a term applied to economic and technological trend, it implies the modernization of industrial operations to utilize mechanized equipment and instruments for automatic control, measurement, and data handling to a greater degree. *See* AUTOMATION.

Trends in instrumentation. Advances in science and technology make ever-increasing demands on instrumentation. Every new area of investigation, for example, the exploration of space, generation of power by nuclear reactors, and determination of high-temperature properties of materials, presents new instrumentation problems. In general, instrumentation research seeks to attain higher accuracies, greater sensitivities, capability of measuring extreme values, applicability under extreme conditions of use, and capability of resolving changes or effects that occur at extremely high speeds, such as nuclear phenomena. *See* PHYSICAL MEASUREMENT.

The trend to wider use of instrumentation for automatic measurement and control in industry is continually accelerating, and whole new industries are emerging which could not exist without highly sophisticated instrumentation. This trend will continue at a rate limited only by questions of cost, safety, and reliability. Reliability, particularly of electronic components, has been a serious problem, but is improving as a result of developments in solid-state circuit elements, new materials, new techniques, and more attention to overall systems engineering. *See* RELIABILITY OF EQUIPMENT.

Another significant trend is the increasing utilization of data-processing devices, recorders, computers, and similar equipment for automatic sorting, storage, and rapid selective retrieval of information and numerical data in banking, accounting, and documentation. Similar devices are being adapted for automatic reading, leading to automatic language translation.

Computer theory is being extended to the design of logic machines and self-organizing systems, which may stimulate not only the processes of remembering and reasoning, but also the process of learning. Mere extension of speed and capacity makes feasible the use of computers to compute weather predictions from analysis of meteorological data from a worldwide observation network; to analyze and forecast economic trends; to assess logistic requirements and military capabilities; and to predict the probable outcome of military ventures of political contests.

In the medical field, instrumentation for automatic testing of physiological, biochemical, and neurological parameters is emerging and is certain to become widespread. Correlation devices are being developed for diagnosis, and clinical instruments for location and identification of internal ills are growing in number and precision.

In education, audiovisual instruments are being developed intensively. Supplemented by automatic scoring devices and other instruments for learning, they promise great increases in the efficiency of the whole system.   [WILLIAM A. WILDHACK]

Bibliography: W. G. Berl (ed.), *Physical Methods in Chemical Analysis*, vol. 1, 1950; H. Chestnut and R. W. Mayer, *Servomechanisms and Regulating System Design*, 1951; D. M. Considine (ed.), *Process Instruments and Controls Handbook*, 1957; D. M. Considine and S. Ross (eds.), *Handbook of Applied Instrumentation*, 1964; B. C. Delahooke, *Industrial Control Instruments*, 1956; C. S. Draper, W. McKay, and S. Lees, *Instrument Engineering*, 3 vols., 1952; D. P. Eckman, *Industrial Instrumentation*, 1950; E. M. Grabbe, *Automation in Business and Industry*, 1957; E. M. Grabbe et al. (eds.), *Handbook of Automation Computation and Control*, 3 vols., 1958; I. A. Greenwood, J. V. Holdam, and D. MacRae, *Electronic Instruments*, vol. 21, 1948; F. K. Harris, *Electrical Measurements*, 1952; I. F. Kinnard, *Applied Electrical Measurements*, 1956; Arthur D. Little, Inc., *Electronic Data Processing Industry*, 1956; G. J. Murphy, *Basic Automatic Control Theory*, 1957; M. H. Nichols and L. L. Rauch, *Radio Telemetry*, 2d ed., 1956; H. C. Roberts, *Mechanical Measurements by Electrical Methods*, 2d ed., 1951; T. N. Whitehead, *The Design and Use of Instruments and Accurate Mechanism*, 1954.

Insulation, electric

Material of high resistivity used to confine the electrons in a conductor so that it may guide them to some useful device. For electric wiring, the insulation is limited to fairly flexible types. *See* CONDUCTOR, ELECTRIC.

INSULATION REQUIREMENTS

The applications of insulations are so varied that a choice must be made after consideration of the necessary properties and the weaknesses of the insulation and of the special circumstances of installation. No single type of insulation has all the advantages; the insulation should be chosen so that its disadvantages are not a detriment in the specific application.

Many desirable characteristics are not permanent. Therefore, the properties of the material after a period of years in the environmental conditions must also be known. New materials require time to develop their unforeseen weaknesses.

Electrical requirements. These vary with use; specific properties of interest depend upon the application. The important properties of insulation are discussed briefly. *See* DIELECTRICS.

Resistivity. Commonly measured as insulation resistance, this is based on a direct-current measurement of flow into the insulation after electrification of 1 min. The current should be small and the insulation resistance large. In itself, it is not perhaps the most important characteristic, but deterioration is often found to occur faster in low-insulation-resistance materials under weathering conditions. It is a valuable measure of consistency and uniformity in manufacture. If, in service, it should drop abruptly or fairly rapidly, a service failure is likely to follow.

Dielectric strength. This property determines the ability of the insulation to resist puncture by electric potentials. It is adversely affected by weathering or aging, by high temperature, by continued heat, and by the entrance of moisture.

Power factor. This is a measure of the power loss in the insulation and should be low. A rapid increase is a danger sign. It is important in high-voltage cables but unimportant in low-voltage cables, except those used for communication. It usually increases with a rise in temperature of the insulation.

Dielectric constant. Also known as specific inductive capacity (SIC), the dielectric constant is a measure of the charge required to bring the apparatus up to voltage, compared to that required if air were substituted for the insulation. It is principally of importance in communication circuits. Low values are preferred, except for capacitors where high dielectric constants provide high energy storage in a small volume. *See* DIELECTRIC CONSTANT.

Temperature requirements. These vary with each application and installation. A compromise is usually necessary because an insulation good for high temperatures has other shortcomings, and the same is true for low temperatures. The Electrical Code and Institute of Electrical and Electronics Engineers Standards give limiting temperatures at which specific insulations shall be used. This means the final temperature of the insulation, which is the result of the temperature of the surrounding medium, usually air, and the heat from the current in the conductor. Temperature should be maintained at a value at which the insulation will last for a long time. Most insulations are stiffer and more likely to crack on bending at lower temperatures and, conversely, to soften at higher temperatures. This applies to temperatures applied for short duration. Higher temperatures applied for long duration usually harden and possibly embrittle insulation, although some rubber compounds soften and become puttylike. In both cases the value of the insulation is impaired, and electrical failure may occur if bent, in the first case, or due to its own weight, in the latter case.

Where a nonflammable cable is required, some other desirable property must usually be sacrificed.

Mechanical requirements. Normally, electric insulation should be designed to hold electricity, not to withstand mechanical abuse. For the necessary rough treatment that may occur in installation of cable, braids, tapes, or sheaths are employed over the insulation as a protection. These can be designed to withstand the mechanical treatment likely to occur, and the insulation can be designed for optimum electrical functions. Of course, the insulation should not break at the low temperatures anticipated or soften at the high ones. Some materials will even flow at temperatures not usually considered high; these are unreliable as insulation.

Chemical requirements. These are usually determined by the specific application intended for the wire. Sometimes resistance to oil, liquids, gas fumes, or airborne powders such as those in chemical plants is required. It is usually the mechanical protection of the insulation that must be designed to withstand this environment rather than the insulation, and each case needs special consideration.

Chemicals in the soil around buried cables or in ducts and manholes around duct cables can cause deterioration. This again usually occurs on the mechanical protection of the cable but may apply to the insulation itself. About the only guide for choice is to use an insulation that has proven by many years of service its suitability for such conditions.

Water can have disastrous effects if it enters the insulating medium of a cable. This is especially true with paper insulation, either dry or oiled, or with varnished cambric. Some compounds of rubber can hold appreciable amounts of water without harm to their electrical properties, but others are seriously affected. In many cases water absorption as a criterion of serviceability of an insulation is exaggerated, and undue importance may be given to it. Its importance actually depends on its effect on the life or the electrical properties needed for the specific application of the insulating material.

Many seemingly impervious materials have been used for keeping water and oil or hydrocarbons away from insulation, with the later discovery that these liquids can pass through the barrier without adversely affecting it. It is thought that electric potentials sometimes aid in the transfer and penetration by the liquid.

FLEXIBLE INSULATION

Insulating materials are often only one ingredient in a compound finally applied to a wire as insulation. An infinite variety of mixtures can result, all based on the same primary ingredient, which often gives its name to the class. Still, the properties of the compounds may be so different as not to deserve grouping as a certain class. Such compounding may make a poor material or an outstanding one, both with the same major ingredient. In general, insulations are either thermoplastic or thermosetting. In others the effect of heat has no function. There are many insulating materials; only the more important ones are discussed here.

Thermoplastic insulation. Materials in this class can be softened by heating so they can be extruded and, when cool, return to their original condition. The general characteristics of these materials are compared in the table.

Polyethylene. This is particularly important for high-frequency applications; however, it has shortcomings. It melts at about 110°C, so that short circuits or overloads of current on a cable may cause the insulation to run off. It is very stiff and, in thick walls on wire, may prove too stiff for usefulness. Cases of cracking have occurred in service,

Some characteristics of plastic insulations

Characteristic	Poly-ethylene	Poly-styrene	Polyvinyl chloride	Mylar	Nylon	Teflon	Kel-F
Insulation resistance	High	High	Can be anything	Good	Fair	Good	Good
Dielectric strength	Good	Good	Can be anything	Good	Fair	Good	Good
Power factor	Low	Low	Rather high	Low	High	Low	Low
Dielectric constant	Low	Low	Medium	Low	Medium	Medium	Medium
Chemical resistance	Good	Good	Good to fair	Good	Good to fair	Good	Good
Water impermeability	Good	Good	Good to fair	Good	Fair	Good	Good
Resistance to sunlight	Good if black	Fair	Can be good	Fair to good	Good if black	Good	Good
Flammability	Yes	Yes	No	Yes	Yes	No	No
Tensile strength	Good	Good	Good	High	Good	Good	Good
Resistance to deformation	Fair	Good	Fair	Good	Good	Good	Good
Abrasion resistance	Good	Fair to good	Good	Good	Good	Good	Good
Stiffness	Considerable	Extreme	Anything desired		Considerable	Considerable	Considerable
Cost	Low	Low	Low	Rather high	Fair	High	High

mostly due to certain greases or hydrocarbons. Newer higher molecular weight polyethylene is designed to avoid this but may lose some of the advantages of the original polyethylene. *See* POLYOLEFIN RESINS.

Polystyrene. This also has exceptional electrical properties, but it must be used in thin films or where it will not be bent, because thick walls are stiff and brittle. Its electrical properties are less affected by change of temperature than most insulations. *See* POLYVINYL RESINS.

Polyvinyl chloride. This plastic is in wide use today. It is difficult to generalize its properties, and it can be compounded to make a single property exceptional. The table should be interpreted with that limitation in mind. It is distinctly sensitive to temperature, softening faster than polyethylene up to a point and becoming harder at low temperatures. It also is subject to flowing under pressure, even at room temperatures.

Mylar. This is not in much use as an insulation, although it can be used in thin tape form. *See* POLYESTER RESINS.

Nylon. Similar to mylar in many ways, nylon is outstanding in abrasion resistance but is not used as cable insulation. It is usually used for its physical benefits rather than as an insulation. *See* POLYAMIDE RESINS.

Teflon and Kel-F. These chlorine-fluorine compounds have limited application. They are normally stiff and so must be used in thin layers. They can withstand higher temperature (over 200°C) than any other semiflexible insulation. *See* POLYFLUOROOLEFIN RESINS.

Thermosetting materials. Materials in this class are normally soft. They can be extruded under pressure and then heat treated until they are no longer soft when cool.

Rubber compounds. For insulation, rubber compounds are changed from plastic form to elastic condition by combination with sulfur. The heat treatment is called vulcanizing. Natural rubber is used less frequently for insulation in recent years as other materials have become available.

Synthetic rubber (Buna S). Man-made and carefully controlled, synthetic rubber lacks the foreign matter of natural rubber and is more uniform. Electrical insulation can be made with either natural or synthetic rubber, but the latter should yield more uniform and reproducible results.

Rubber compounds can be made with good insulation resistance and dielectric strength, with medium power factor and SIC, and with resistance to chemicals, weathering, and moist conditions. They are more flexible and are easier to handle and install than plastics.

Usually they need protection from mechanical abuse. They burn and are usually limited to a maximum circuit voltage of 35,000 volts. The aging of such compounds usually causes them to harden, but if undisturbed they continue to function electrically.

Butyl. This synthetic rubber has not been used at the highest voltages because its dielectric strength is less than that of Buna S and of natural rubber. It will not harden under use; instead, it softens. Again, certain characteristics are accentuated in some compounds and not in others, so a generalization of characteristics cannot accurately cover every case.

Neoprene. This rubberlike material is compounded to look like a rubber compound. It does not have the electrical properties usually required and is used principally for the mechanical protection of insulation which it encloses. Such neoprene sheaths have fine tensile strength, abrasion, and crushing resistance. They resist the effect of the sun and oil. However, oil and water can penetrate them.

Silicone rubber. This is an expensive material, but it is not damaged by temperatures from 150 to 200°C. Its electrical properties are good. It usually is physically weak, although this can be improved in some varieties. *See* RUBBER; SILICONE RESINS.

Epoxy resins. These are used as insulation of joints. A liquid and a powder are mixed and heat is given off. The chemical action sets it to a solid which is not flexible but has good electrical and mechanical properties.

Heat-insensitive insulations. The effect of heat has no function in establishing the properties of the insulating materials in this group.

Paper. Paper insulation in multilayers, oil-impregnated and protected from moisture by a lead sheath, is an old standby. It is popularly used at voltages over 15,000 volts. It has good electrical characteristics, which change with age. Its weakness is the lead or similar sheath, which must be in perfect condition to keep the cable dry and operating. Lead is subject to crystallization and electrolysis, which can perforate it.

Varnished cambric. This is a series of thin layers of varnished cloth. Its electrical properties are not as good as paper-oil cable, and it requires the same moisture and mechanical protection, usually a lead sheath. It is used at lower voltages.

Nitrogen gas. This is used to a considerable extent as an insulation for high-voltage applications. It is used under pressure in a pipe containing the conductor.

Magnesium oxide. This is a novel insulation for low voltages but rather high temperatures. It is enclosed in a copper tube and carries the conductor at its center. It is good for wet and dry locations, and in special applications it may be used to 250°C.

Asbestos. Besides its heat-insulating properties, asbestos is an electric insulator to some extent, particularly if impregnated with some waxlike substance. It usually requires a water-impervious sheath and is used for relatively low voltages. It is also used in combination with plastics or varnished cambric. *See* ASBESTOS.

[ALAN S. DANA/PHILIP L. ALGER]

RIGID INSULATION

Besides their use as flexible coverings for wires and cables, insulating materials are employed in molded or built-up form as components of rigid structures. This rigid insulation must provide mechanical strength and stability of form as well as a dielectric barrier. Mica, glass, porcelain, and the thermosetting resins are the principal rigid insulating materials, but these may be used in combination with any of the flexible materials.

Mica. A mineral of finely laminated structure and easy cleavage, mica flakes are flexible, tough, and highly resistant to heat. Mica is most often employed in the form of splittings, about 1 mil thick, which may be snowed onto sheets of thin paper or glass fiber, bonded with a suitable varnish, and applied in multiple layers of tape. Finely divided mica in the form of mica paper is used with epoxy or other resins, with or without separate backing. Ground mica is used as a filler in molded insulation. *See* MICA.

Glass. An amorphous material, glass ordinarily consists of a mixture of silicates, borates, phosphates, and other materials, with silica, SiO_2, forming 50–90% of the total content. It is employed in a great variety of compositions. Blown and cast forms are used. Glass yarn or cloth made of fibers 0.2–0.3 mil in diameter is also employed. *See* GLASS AND GLASS PRODUCTS.

Porcelain. This is a hard, brittle, and impervious material made from feldspar, quartz, clay, and other minerals. The materials are finely ground, intimately mixed in a liquid state, molded into the desired shape while plastic, dipped in glaze, and fired at a high temperature. *See* PORCELAIN.

Uses of rigid insulation. Rigid electrical insulation is used for supporting conductors and spacing them apart in a gas or liquid environment where the air or other surrounding medium is relied upon to provide the needed dielectric strength, except at the supports. Here the chief requirements are high surface-creepage resistance and high strength to withstand the imposed mechanical forces, including the shocks from short-circuit currents. The insulation must be impervious to water so that it will retain its high electrical resistance when washed or exposed to the weather, and for the higher voltages it must be able to withstand arc discharges over the surface.

For the highest voltages, as for power transmission lines, the conductors are hung on strings of suspension insulators or supported on bushings. These are made of glazed porcelain with a series of skirts to lengthen the creepage paths and provide maximum resistance to flashover. For medium voltages, the bushings may be made of glass or molded plastics.

For low voltages, as for industrial control equipment and household appliances, and where many conductors need to be closely spaced, thermosetting plastic compounds are used, chosen for their structural rigidity, toughness, high surface-creepage resistance, ease of manufacture, and low cost. Besides a wide variety of synthetic compounds, such as epoxy, phenolic, melamine, and polyester resins, composite materials are used, such as phenolic resin-coated cotton or asbestos fabric and impregnated wood.

When the conductor spacing is too small for reliance on the dielectric strength of the surrounding gas or liquid, as in transformers and rotating machines, the rigid insulation must form a sealed barrier of high dielectric strength over the entire conductor surface. The requirements are especially severe for the slot-embedded conductors of rotating machines, where the insulation thickness must be held to a minimum and where part of the surface may be exposed to high-velocity cooling air, often containing fine particles of conducting material, such as carbon dust.

In general, the windings of electric machines consist of stranded conductors, arranged in series- or parallel-connected coils or both, each with one or more turns. Thus three principal kinds of insulation are required for apparatus windings, for the strands, for the turns, and for the complete coil. The strand insulation must be flexible enough to withstand bending during manufacture but only has to withstand low voltages. It is usually applied in the form of synthetic enamel or glass fiber covering, but for large machines the strands may be wrapped with glass fiber, asbestos, or mica-paper tape—or with paper in the case of oil-filled transformers. The turn insulation is formed by wrapping the conductors with tape or by interposing separators of sheet material. The insulation to ground may be formed of rigid spools on which the coils are wound, as for small transformers and field coils. In small rotating machines it may consist of slot liners made of Mylar, Kapton, or cellulose acetate film; of papers, such as dacron, Nomex, or Kraft; of mica or glass cloth; or of a combination of these. In large rotating machines the coils are wrapped with flexible sheet or tape which is thoroughly impregnated or coated with insulating resins before or after application. The coils may be molded and cured in a press, by hydraulic pressure in a hot liquid, or wound dry and subsequently impregnated and cured in place. The insulation is applied in overlapping layers to make the surface-creepage paths long and to seal interstices.

Rigid-insulation systems have an almost infinite variety of possible forms, depending on the size, type, intended use, and environment of the equipment. Any one piece of apparatus may use a variety of materials in many different forms as components of its insulation system. After selecting the most suitable and compatible materials, the designer must provide for their application and assembly in such a way that the completed structure will have adequate dielectric strength, both initially and on a long-term basis, and will withstand all the overvoltages, mechanical stresses, temperatures, and environmental exposures that may occur in service.

Voltage requirements. Usually, enclosed-conductor rigid-insulation systems are required to withstand initially a test voltage of twice the supply voltage plus 1000 volts. Additional tests, such as repeated surge impulse voltages, may be required to check the strand and turn insulation. A chief problem in meeting these requirements is to ensure that the voltage distribution along the winding, and across each thickness of insulation, does not overstress any part.

As indicated in the figure, the dielectric stress along any path made up of different materials in series is inversely proportional to their values of the dielectric constant K. K is usually 3–6 for rigid insulation and 1 for air, so that a thin layer of air between the insulation and a slot wall may have a stress three or more times that in the insulation itself. Because air breaks down at about 80 volts/mil, electric discharge, or corona, may be expected whenever the insulation stress exceeds about 30 volts/mil, unless preventive steps are taken. Therefore, the slot portions of high-voltage coils are given a semiconducting coating, which short-circuits the external air spaces; they are also made of corona-resistant materials such as mica.

Temperature requirements. Organic insulating materials undergo chemical and physical changes when exposed to high temperatures. For each 10°C increase in temperature, the rates of most chemical reactions double, such as oxidation of cellu-

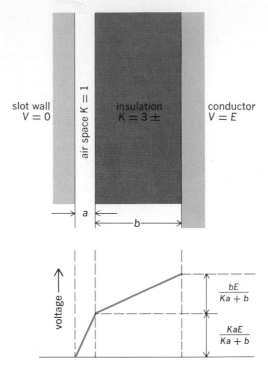

Effect of dielectric constant *K* on voltages of conductors and slot wall.

lose, and therefore the useful life of most insulation systems is roughly halved. Insulating materials are grouped in temperature classes O, A, B, F, and H, with nominal continuous temperature limits of 90, 105, 130, 155, and 180°C, respectively. Varnish-treated cotton, paper, and enamel are included in class A, whereas glass fiber and built-up mica with a small proportion of organic material have been traditionally placed in class B. Some of the best organic-film materials have been shown by test to be suitable for class B temperatures or above. Class H includes silicone compounds; silicone-treated class B materials; polymers with aromatic chains, such as aromatic polyimides, amides, and oxadiazoles; and fluorocarbons, such as Teflon. The standards permit deviations from the nominal temperature limits, depending on the size and type of equipment and the intended service. Consideration must be given to the temperature drop across the insulation thickness. For the usual insulating materials, this drop is about 200°C/watt/in.² of heat flow.

Insulation tests. Service experience is required to prove that an insulation system is fully satisfactory, but life tests on materials and on complete systems will shorten the time needed for their evaluation. The Institute of Electrical and Electronics Engineers provides a number of test procedures for this purpose; these provide for exposures of samples of insulating materials, or models, to repeated cycles of high temperature, moisture, and voltage checks until failure occurs. By plotting logarithmic life against absolute temperature on an inverse scale and extrapolating to the proposed operating temperature, the life expectancies of alternative materials and systems may be compared.

[PHILIP L. ALGER]

Bibliography: D. G. Fink and J. M. Carroll, *Standard Handbook for Electrical Engineers*, 10th ed., 1968; Institute of Electrical and Electronics Engineers, *Evaluation of Systems of Insulating Materials for Random-Wound Electric Machinery: Test Procedure*, AIEE Stand. no. 510, 1956; Institute of Electrical and Electronics Engineers, *General Principles upon Which Temperature Limits Are Based in the Rating of Electric Equipment*, AIEE Stand. no. 1, 1957; G. L. Moses, *Electrical Insulation*, 1951.

Insulation, heat

Materials whose principal purpose is to retard the flow of heat. Thermal- or heat-insulation materials may be divided into two classes, bulk insulations and reflective insulations. The class and the material within a class to be used for a given application depend upon such factors as temperature of operation, ambient conditions, mechanical strength requirements, and economics.

Examples of bulk insulation include mineral wool, vegetable fibers and organic papers, foamed plastics, calcium silicates with asbestos, expanded vermiculite, expanded perlite, cellular glass, silica aerogel, and diatomite and insulating firebrick. They retard the flow of heat, breaking up the heat-flow path by the interposition of many air spaces and in most cases by their opacity to radiant heat.

Reflective insulations are usually aluminum foil or sheets, although occasionally a coated steel sheet, an aluminumized paper, or even gold or silver surfaces are used. Refractory metals, such as tantalum, may be used at higher temperatures. Their effectiveness is due to their low emissivity (high reflectivity) of heat radiation. *See* EMISSIVITY.

Thermal insulations are regularly used at temperatures ranging from a few degrees above absolute zero, as in the storage of liquid hydrogen and helium, to above 3000°F in high-temperature furnaces. Temperatures of 4000–5000°F are encountered in the hotter portions of missiles, rockets, and aerospace vehicles. To withstand these temperatures during exposures lasting seconds or minutes, insulation systems are designed that employ radiative, ablative, or absorptive methods of heat dissipation.

Heat flow. The distinguishing property of bulk thermal insulation is low thermal conductivity. Under conditions of steady-state heat flow the empirical equation that describes the heat flow through a material is Eq. (1), where q = time rate of

$$\frac{q}{A} = -k\frac{(\theta_2 - \theta_1)}{l} \qquad (1)$$

heat flow, A = area, θ_1 = temperature of warmer side, θ_2 = temperature of colder side, l = thickness or length of heat-flow path, and k = thermal conductivity, representative values being listed in the table. For a given thickness of material exposed to a given temperature difference, the rate of heat flow per unit area is directly proportional to the thermal conductivity of the material. *See* CONDUCTION, HEAT.

In the unsteady state, or transient heat flow, the density and specific heat of a material have a strong influence upon the rate of heat flow. In such cases, thermal diffusivity $\alpha = k/\rho\, C_p$ is the important property. Here ρ = density and C_p = specific heat at constant pressure. In the simple case of

Thermal conductivities of selected solids*

Material	Density, lb/ft³		Conductivity, k† Btu/(in.) (hr) (ft²) (°F)
Asbestos cement board	120	75	4
Cotton fiber	0.8–2.0	75	0.26
Mineral wool, fibrous rock, slag, or glass	1.5–4.0	75	0.27
Insulating board, wood, or cane fiber	15	75	0.35
Foamed plastics	1.6	75	0.29
Glass			3.6–7.32
Hardwoods, typical	45	75	1.10
Softwoods, typical	32	75	0.80
Cellular glass	9	75	0.40
Fine sand (4% moisture content)	100	40	4.5
Silty clay loam (20% moisture content)	100	40	9.5
Gypsum or plaster board	50	75	1.1

*From American Society of Heating, Refrigerating, and Air Conditioning Engineers, *Heating, Ventilating and Air Conditioning Guide*, 1959.

†Typical; suitable for engineering calculations.

one-dimensional heat flow through a homogeneous material, the governing equation is Eq. (2), where

$$\frac{d\theta}{dt} = \alpha \frac{d^2\theta}{dx^2}\Big|_0^l \qquad (2)$$

$t =$ time and x is measured along the heat-flow path from 0 to l.

Thermal conductivity. In general, thermal conductivity is not a constant for the material but varies with temperature. Generally, for metals and other crystalline materials, conductivity decreases with increasing temperature; for glasses and other amorphous materials, conductivity increases with temperature. Bulk insulation materials in general behave like amorphous materials and have a positive temperature coefficient of conductivity.

Thermal conductivity of bulk insulation depends upon the nature of the gas in the pores. The conductivities of two insulations, identical except for the gases filling the pore spaces, will differ by an amount approximately proportional to the difference in the conductivities of the two gases.

Increasing the pressure of the gas in the pores of a bulk insulation has little effect on the conductivity even with pressures of several atmospheres. Decreasing the pressure has little effect until the mean free path of the gas is in the order of magnitude of the dimensions of the pores. Below this pressure the conductivity decreases rapidly until it reaches a value determined by radiation and solid conduction. A few materials have such fine pores that at atmospheric pressure their dimensions are smaller than the mean free path of air. Such insulations may have conductivities less than that of still air. *See* HEAT RADIATION; HEAT TRANSFER.

[HARRY F. REMDE]

Bibliography: H. S. Carlslaw and J. C. Jaeger, *Conduction of Heat in Solids*, 2d ed., 1959; P. E.

Glaser et al., *Investigation of Materials for Vacuum Insulators up to 4000F*, ASD TR-62-88, 1962; W. H. MacAdams, *Heat Transmission*, 3d ed., 1954; E. M. Sparrow and R. D. Cess, *Radiation Heat Transfer*, 1967; H. M. Strong, F. P. Bundy, and H. P. Bovenkerk, Flat panel vacuum thermal insulation, *J. Appl. Phys.*, 31:39, 1960; J. D. Vershoor and P. Greebler, Heat transfer by gas conduction and radiation in fibrous insulations, *Trans. ASME*, 74(6):961–968, 1952; G. B. Wilkes, *Heat Insulation*, 1950.

Insulation resistance testing

The testing of the electrical resistance of dielectric materials or insulators. Insulation resistance is measured in megohms. It consists of (1) the volume resistance resulting from the resistance to current flow through the volume of the material, and (2) the surface resistance due to the resistance to current flow over the surface of the material. These two resistances are electrically in parallel, and it is impossible to separate them in a single measurement. If basic insulating properties of a material are being studied, judicious choice of sample size and electrode configuration can emphasize one effect over the other so that reasonable evaluating data of each factor can be obtained.

If the conditions of measurement do not permit such control, as in the checking of insulation resistance of large generator windings, the testing

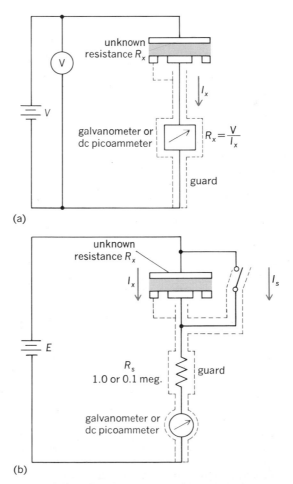

Fig. 1. Circuits for guarded deflection measurement of insulation resistance. (*a*) Voltmeter-ammeter method. (*b*) Comparison method.

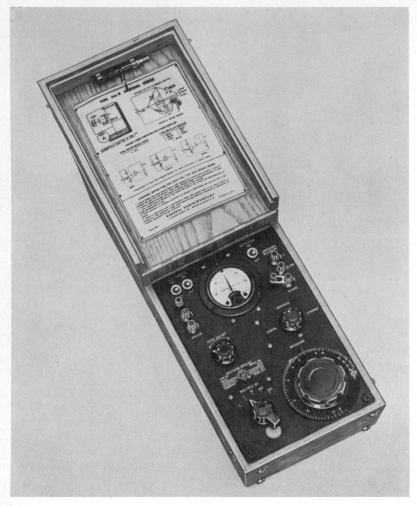

Fig. 2. Megohm bridge for insulation resistance measurement. (*General Radio Co.*)

conditions and physical arrangements must be evaluated carefully to determine which, if either, factor is dominant and whether or not the test will produce the desired evaluation of the winding insulation. Often, only the total insulation resistance is required, and individual evaluation of the surface and volume effects need not be made. Except for the simplest of measurements, such as the resistance of a capacitor terminal or the point-to-point resistance of a switch, the use of a guard circuit is essential for precise and accurate results. *See* RESISTANCE MEASUREMENT.

Usually, surface-leakage effects may be eliminated by guarding one terminal or a portion of the measuring circuit. A guard is a low-resistance conductor; it is insulated from the guarded terminal and located to maintain the guard and guarded terminal at near-equal potential and to minimize the current flow between them. The guard circuit also conducts leakage current around the measuring circuit to minimize its effect on the measuring circuit.

The determination of insulation resistance is further complicated by two other factors. First, if the capacitance between the test electrodes is high, the time constant of the circuit, RC, will be large, and the time required to charge the capacitance will be long. Second, many insulating materi-

als exhibit dielectric absorption which may continue for extended periods of time. *See* DIELECTRICS; TIME CONSTANT.

If a measurement is made without guarding and before the two conditions described above are stabilized—an interval of microseconds to minutes for the time constant effect, and seconds to days for dielectric absorption—an apparent insulation resistance will be determined. A guarded measurement made after stabilization will determine volume resistance from which resistivity may be calculated.

It is therefore necessary to completely specify all test conditions, including sample and electrode configuration and condition, test voltage, state of capacitance charge before test, time of measurement after application of test voltage, and temperature and humidity if reproducible results are to be achieved. *See* INSULATOR, ELECTRIC; RESISTANCE, ELECTRICAL; RESISTIVITY, ELECTRICAL.

Guarded deflection method. This method for measuring insulation resistance (Fig. 1*a*) uses the voltmeter-ammeter method. The voltage is measured with a voltmeter of suitable range; current is measured by a calibrated galvanometer, with an Ayrton shunt for range extension, or a vacuum-tube dc amplifier calibrated as a picoammeter with appropriate multipliers.

A modification of this method, which permits the comparison of a high resistance standard with the insulation resistance under test, is shown in Fig. 1*b*. In this method, neither the test voltage nor the currents need be known. With a constant test voltage applied, the deflection of the current measuring device is noted, first, with the sample shorted. For this condition, Eq. (1) holds,

$$E = d_1 R_s \qquad (1)$$

where $d_1 \propto I_S$. Next, with the switch open, a new deflection is noted. For this second condition, Eq. (2) holds, where $d_2 \propto I_x$, and, combined with Eq. (1), yields Eq. (3).

$$E = d_2 (R_X + R_S) \qquad (2)$$
$$R_X = R_s (d_1/d_2 - 1) \qquad (3)$$

In each of these circuits, requisite guarding of test equipment is shown by dotted lines with the guard of the sample connected to the equipment guard. It is essential that all of the equipment shown within the guard lines be mounted on the guard plate or circuit. However, the insulation requirements between guard and measuring circuit are not severe. This insulation shunts the internal resistance of the measuring circuit and therefore need only be 10–100 times this value to minimize the error from this source. If an electronic picoammeter is used, its circuits, including such components as the power supply transformer, must also be guarded.

Guarded Wheatstone bridge. This bridge requires the use of guarded ratio arms, detector, and power supply, and careful attention must be given to the details of construction, particularly with regard to component insulation. Such construction also results in a well-shielded bridge. A bridge of this type may be used for highly accurate and precise measurements even under conditions of high humidity. *See* WHEATSTONE BRIDGE.

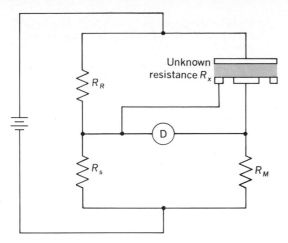

Fig. 3. An unguarded megohm bridge for measuring guarded insulation specimen.

Unguarded Wheatstone bridge. An unguarded bridge (Fig. 2) may also be used to measure guarded insulation samples if the guard connections are made as shown in Fig. 3. This connection allows the leakage current to bypass the detector. If the detector input resistance is high and the bridge resistor R_R is no larger than 100,000 ohms, little residual error will result.

Hi-pot testing. In this test, high potentials are applied to electrical materials, components, equipment or systems to determine experimentally the dielectric breakdown characteristics or to proof-test the insulation for manufacture, acceptance or service requirements at voltages higher than operating but lower than expected breakdown potential. This testing may be done with either alternating or direct current. The voltage applied and the time of application are chosen to provide the required information for specific purposes. Interpretation of hi-pot test data is complex; considerable experience is required to evaluate this data properly, particularly when solid dielectrics are involved. *See* MEGGER.

[CHARLES E. APPLEGATE]

Bibliography: American Society for Testing and Materials, *AC Loss Characteristics and Dielectric Constants of Solid Electrical Insulating Materials,* ASTM Stand. no. D-150-74, 1974, and *DC Resistance or Conductance of Insulating Materials,* ASTM Stand. no. D-257-66(72), 1966, and *Specific Resistance of Electrical Insulating Liquids,* ASTM Stand. no. D-1169-74, 1974, and *Terms Relating to Electrical Insulation,* ASTM Stand. no. D-1711-74A, 1974, and *Test for Dielectric Breakdown Voltage and Dielectric Strength of Electrical Insulating Materials at Commercial Power Frequencies,* ASTM Stand. no. D-149-64(70), 1964; W. P. Baker, *Electrical Insulation Measurements,* 1966; F. K. Harris, *Electrical Measurements,* 1958; Institute of Electrical and Electronics Engineers, *Recommended Guide for Making Dielectric Measurements in the Field,* IEEE Stand. no. 62, 1958, and *Test Code for Resistance Measurements,* IEEE Stand. no. 118, 1949, and *Recommended Practice for Testing Insulation Resistance of Rotating Machinery,* IEEE Stand. no. 41, 1974; F. A. Laws, *Electrical Measurements,* 2d ed., 1938; H. A. Sauer and W. H. Shirk, Jr., Wheatstone bridge for multi-terohm measurements, *IEEE Trans. Comm. Elec.,* 71:131–136, March, 1964; F. H. Wyeth, J. B. Higley, and W. H. Shirk, Jr., Precision, guarded resistance measuring facility, *Trans. AIEE,* pt. 1, 77(38):471–475, 1958.

Insulator, electric

Ideal insulators are substances into which static electric fields penetrate uniformly. Substances for which Ohm's law is valid are often classed as insulators if their resistivities exceed about 10^{14} ohm-cm. Insulators used primarily as media for the storage of electrostatic energy, for example, in capacitors, are called dielectrics. Important electrical properties of insulators are resistivity, dielectric constant, and breakdown voltage, the applied voltage above which a material ceases to act as an insulator. *See* DIELECTRIC CONSTANT; DIELECTRICS.

There are gaseous, liquid, and solid insulators in wide variety. A few of each type are listed here.

Gaseous
  Air
  SF_6 (sulfur hexafluoride)
  CO_2 (carbon dioxide)
Liquid
  Petroleum oils
  Halogenated aromatic hydrocarbons
  Silicone oils
Solid
  Sulfur
  Polystyrene
  Rubber
  Porcelain

Gases. At high pressures gases are excellent insulators, with higher breakdown voltages than most solids or liquids or high vacuum. Free electrons and ions in gases arise from ionizing radiation which, because of the existence of cosmic rays, is never completely absent. In an applied field, these charged particles are accelerated. The electrons colliding with atoms of the gas may produce additional electrons and ions. The ions colliding with the cathode may eject further current carriers. At sufficiently high fields these processes greatly multiply the number of carriers and breakdown occurs, the gas becoming an excellent conductor. Below the breakdown field there is a range of applied field in which the current passing through a gas between electrodes is small and independent of the applied voltage. In this region the gas is typically a good insulator, because of the small density of charge carriers.

At high pressures, the distance electrons travel between collisions with gas atoms is shortened and as a consequence they do not, on the average, acquire as much kinetic energy from the applied field as do the electrons in a gas at a lower pressure. The number of electrons capable of ionizing gas atoms is then reduced at high pressures and the breakdown field increases. Introducing electronegative gases, such as carbon tetrachloride, CCl_4, which can remove free electrons by attachment, also improves the insulating properties of gases. *See* ELECTRICAL CONDUCTION IN GASES.

Solids. In solids electric current is carried by electrons or ions. Liquids show many of the features of solids in their conducting mechanisms. An understanding of why some solids, such as metals, are excellent conductors of electricity while others, such as sulfur, are insulators is one of the most important successes of the band theory of solids. According to the quantum-mechanical theory of electron motion in solids, a consequence of the microscopic periodic structure of a solid is that certain energies are forbidden to electrons in the solid. These forbidden energy gaps separate the energy spectrum in general into bands of allowed energies, each of which may, for simple translational lattices, accommodate only $2N$ electrons, where N is the number of atoms in the solid. This restriction is a consequence of the Pauli exclusion principle. The occupation of these bands (which may overlap) by the electrons in the solid either may completely fill certain of the bands or it may leave one or more partially filled, depending on the crystal structure and the number of electrons in the solid. In the event that a band is only partially filled, there exist in the solid unoccupied electron states with energies very near those of the electrons of highest energy. A small electric field is in this case capable of accelerating these electrons and producing current flow. Such a solid is a conductor. If, on the other hand, all the electrons lie in completely filled bands, there are no such near-lying unoccupied states. Since the exclusion principle prevents multiple occupation of states, the application of a weak electric field is unable to accelerate electrons and no current flows. Such a substance is an insulator if the gap to the nearest unfilled band is large enough (a few electron volts).

A material with filled bands but with a small energy gap of the order of thermal energies (a few tenths of an electron volt) is an intrinsic semiconductor. *See* BAND THEORY OF SOLIDS; FERROELECTRICS; IONIC CRYSTALS; SEMICONDUCTOR.

[B. G. DICK]

Insulin

The hormone, produced and secreted by the beta cells of the islets (insulae) of Langerhans of the pancreas, which regulates the use and storage of foodstuffs, especially the carbohydrates. Chemically insulin is a small, simple protein. The molecular weight is close to 60,000, and the molecule consists of 51 amino acids, arranged in two linked polypeptide chains. The amino acid composition shown in the illustration depicts beef insulin. Insulins from other species differ in their composition most often at positions 8, 9, and 10 in the A chain and at position 30 in the B chain. These differences account for the fact that diabetics treated with animal insulins develop antibodies which may sometimes interfere with the action of the hormone. The structure has been verified by synthesis of insulin from pure amino acids in the laboratory. *See* IMMUNOLOGY.

Chemistry and synthesis. The three sulfur atom (S-S) linkages are of singular importance to the biological activity and shape of the hormone. Exposure to reducing agents inactivates insulin and splits it into two reduced chains. In many organs of the body an enzyme is present which splits insulin in this manner with the help of the compound reduced glutathione. Insulin, being a polypeptide, can also be broken down by many proteolytic enzymes to its constituent amino acids.

Because of these breakdown systems the turnover of insulin in the body is rapid; its "half-life" has been estimated to be 10–30 min. The liver alone is capable of destroying about 50% of the insulin passing through it on its way from the pancreas to the bodily tissues.

Insulin, being a simple protein, does not possess any unique, specific chemical grouping by means of which it can be measured quantitatively in the blood or other bodily fluids. However, animals given injections of an insulin derived from another species make specific antibodies against it. By means of such specific antisera a very sensitive technique for insulin estimation has been developed. Under normal conditions the blood serum contains about 0.001 μg of insulin per cubic centimeter before a meal; this rises to about 0.005 μg after the ingestion of a carbohydrate meal.

At one time it was assumed that insulin is made in the pancreas in two pieces; that is, the A and B chains are synthesized separately. The two chains would then join together through the reduced sulfur (SH) interchain groups. It now appears that the beta cells synthesize at first a single polypeptide chain of about 9000 mol. wt; this has been called pro-insulin. A piece of this protein of about 3000 mol. wt is then split off to give the insulin molecule (6000 mol. wt). This supports the probability that insulin synthesis is under the direction of a single gene. *See* GENE ACTION.

In spite of the comparatively small size of the insulin molecule, it took a long time to unravel its tertiary (tridimensional) structure; but this was achieved by Dorothy C. Hodgkin. Other work in-

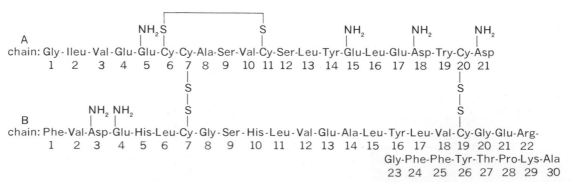

Structure of cattle insulin. The amino acids are: Gly= glycine; Ileu= Isoleucine; Val= valine; Glu= glutomine; Cy= cytosine; Ala= alanine; Ser= serine; Leu= leucine; Tyr = tyrosine; Asp = asparagine; Try = tryptophan; Phe= phenylalanine; His= histidine; Arg= argenine; Lys= lysine; and Thr= threoine.

dicates strongly that in its action on cells insulin combines first with a specific protein acceptor molecule at the cell membrane. Hodgkin's successful crystallographic work may lead to the unraveling of the active site on insulin which combines with the receptor and initiates the physiological action of the hormone. It may also be possible to determine the sites at which the hormone combines with antibodies formed against it.

Physiological activity. The role played by insulin in the body is most clearly approached by considering the abnormalities resulting from removing insulin from an organism by surgical excision of the pancreas or by the chemical destruction (for example, by alloxan) of the insulin-producing cells: A state of severe diabetes is produced. Normally the blood glucose level is about 100 mg/100 ml. A carbohydrate meal raises the blood sugar to about 150 mg and the premeal value is reached again within about $1\frac{1}{2}$ hr. The normal organism manages to dispose of food by storage and oxidation within this period because insulin is present. When food (carbohydrate and protein) reaches the upper intestine, a substance is liberated which in turn stimulates the beta cells to secrete extra insulin. Insulin acts on most tissues to speed the uptake of glucose. In the cells the glucose is burned for energy, stored as glycogen, or transformed to and stored as fat.

Insulin is a very active molecule. The human pancreas probably produces 1–2 mg of the hormone per day. This is sufficient to regulate the metabolism of more than 250 g of carbohydrate, 70 g of protein, and 75 g of fat, the usual composition of an ordinary 2000-calorie diet.

Diabetes. In diabetes the rate of glucose uptake is slowed, the level of circulating blood sugar rises, and sugar spills over into the excreted urine. Calories are wasted, more water is excreted, and there is muscular weakness and weight loss; hence urinary frequency, hunger, thirst, and fatigue. Whenever glucose metabolism is defective, stored fat is broken down to fatty acids because of the actions of adrenaline and the pituitary growth hormone. The liberated fatty acids are useful fuel materials, but whenever present in large amounts they are not only burned for energy, but form organic acids in the liver known as the beta-ketoacids. Accumulation of these in the blood reduces its alkalinity, and a state of acidosis results which depresses the brain and results in coma. In addition to an increased fat breakdown there occurs an increase in the breakdown of tissue proteins. Insulin is able to reverse all these phenomena by favoring storage and swift intake of glucose into the tissues, by decreasing the breakdown of stored fat, and by promoting protein synthesis. See DIABETES; KETOSIS.

Hypoglycemia. When insulin is secreted or given in excess, it may lower the blood sugar level much below its normal value. Hypoglycemia is dangerous because the metabolism in the brain cells depends primarily upon an adequate supply of glucose. The neurones may become irreversibly damaged when exposed to low blood sugar levels frequently or for long periods of time.

Mechanism of action. The exact biochemical mechanism of the action of insulin is still unknown. One important action site is the cell membrane, which contains a system by which the glucose molecule is carried into the cell. Insulin powerfully enhances the activity of this sugar transport system. It remains to be clarified how this action is related to the effects insulin exerts on protein and fat as well as mineral metabolism. See ENDOCRINE MECHANISMS. [RACHMIEL LEVINE]

Insulin shock

A treatment for schizophrenic psychosis introduced by Manfred Sakel in 1933. Gradually increasing doses of insulin are administered until subcoma or coma is achieved, depending upon therapeutic objectives. Subcoma is associated with sweating, hunger, drowsiness, and an altered state of consciousness, which some feel is useful in overcoming barriers to psychotherapeutic contact. Those who employ this treatment for benefits from shock alone agree that deep coma must be induced for maximum results. Coma is terminated by administration of glucose or glucagon. Approximately 50 such treatments are administered, usually at the rate of two to three a week. See INSULIN; PSYCHOSIS; SCHIZOPHRENIA.

Since this form of treatment is rarely the sole therapeutic treatment given, its effects have been difficult to evaluate. Probably it has been most effective for schizophrenics who also respond well to other treatments.

The number of centers utilizing insulin-coma therapy in the United States decreased from 214 in the 1950s to 63 in the 1960s. Pharmacotherapy, electric shock, milieu therapy, and combinations of these have replaced insulin treatment because it is time-consuming, expensive, occasionally associated with severe side effects, and not demonstrably superior to simpler methods. Some clinicians, however, continue to maintain that insulin shock has definite advantages, particularly in cases refractory to more modern treatment programs.

Sakel believed that insulin shock destroyed the sympathetic predominance associated with psychotic states and reinstituted sympathetic-parasympathetic balance in the autonomic nervous system. Despite continuing research, no hypothesis of the mechanism of action of insulin shock has won widespread acceptance. See AUTONOMIC NERVOUS SYSTEM (VERTEBRATE); PSYCHOTHERAPY.
[EDWARD C. SENAY]

Bibliography: M. Rinkel (ed.), *Biological Treatment of Mental Illness*, 1966.

Intaglio (gemology)

The name given to the type of carved gemstone in which the figure is engraved into the surface of the stone, rather than left in relief by cutting away the background, as in a cameo. Intaglios are almost as old as recorded history, for this type of carving was popular in ancient Egypt in the form of cylinders. The cylinder, as well as the more familiar form of intaglio, was popular to impress seals on sealing wax. Intaglios have been carved in a variety of gem materials, including emerald, crystalline quartz, hematite, and the various forms of chalcedony. See CAMEO.   [RICHARD T. LIDDICOAT, JR.]

Integral equation

An equation of the form typyfied by Eq. (1). The major problem is to decide when there is a function $\phi(x)$ which is a solution to the equation. Equations such as (1) arise from the analysis of ordinary differential equations. For example, it is easy to verify that the function $\phi(t)$ is a solution of differential equation (2) if and only if ϕ is a solution of integral equation (3). This equation is the same type as Eq. (1), with the function K given by Eq. (4).

$$\int_a^b K(x,y,\phi(y))dy + f(x) = a(x)\phi(x) \qquad (1)$$

$$\frac{d\phi}{dt} = F(t,\phi) \qquad \phi(0) = A \qquad (2)$$

$$\phi(t) = A + \int_0^t F(s,\phi(s))ds \qquad (3)$$

$$K(t,s,\phi) = \begin{cases} F(s,\phi) & 0 \le s \le t \\ 0 & t < s \end{cases} \qquad (4)$$

Integral equation (1) is an equation of the first kind if $a(x) \equiv 1$, and an equation of the second kind if $a(x) \equiv 0$.

The domain of integration, which in Eq. (1) is the interval $[a,b]$, can be a domain in two or more variables, a curve, or even a surface. The unknown function ϕ and the kernel K can be vector-valued functions, in which case Eq. (1) is called a system of integral equations.

Volterra equations. If $K(x,y,\phi) = 0$ for $x \le y \le b$, Eq. (1) is called a Volterra equation. Thus Eq. (3) is an example of a Volterra equation. Under mild assumptions about the smoothness of $K(x,y,\phi)$, for example, if K is continuous and if $|K(x,y,\phi) - K(x,y,\psi)| \le A|\phi - \psi|$, where A is a constant, for all x,y in the domain of integration and for all ϕ and ψ—then the Volterra equation of the second kind, Eq. (5), can be solved by the method of successive

$$\phi(x) = \int_a^x K(x,y,\phi(y))dy + f(x) \qquad (5)$$

approximations (also known as the method of Picard). The idea is to define a sequence of functions $\phi_0, \phi_1, \ldots$ according to the scheme of Eqs. (6). Under the assumed conditions, it can be shown

$$\phi_0(x) = f(x)$$
$$\phi_1(x) = \int_0^x K(x,y,\phi_0(y))dy + f(x)$$
$$\cdots \qquad (6)$$
$$\phi_{n+1}(x) = \int_0^x K(x,y,\phi_n(y))dy + f(x)$$

that this sequence converges to a function $\phi(x)$ which is the only solution to Eq. (5). The method of Picard is fundamental to the study of integral equations, but occurs also in the study of many other functional equations.

Linear integral equations. A special case of some interest is that of linear integral equations. The function $K(x,y,\phi)$ is a linear function of ϕ. The linear equations of the first and second kind are shown in Eqs. (7) and (8) respectively. The function

$$f(x) = \int_a^b K(x,y)\phi(x)dy \qquad (7)$$

$$\phi(x) = f(x) + \lambda \int_a^b K(x,y)\phi(y)dy \qquad (8)$$

$K(x,y)$ is called the kernel, and the complex number λ is called the parameter. The equation of the second kind, Eq. (8), is homogeneous if $f(x) \equiv 0$. In typical cases, the homogeneous equations will

have only the trivial solution $\phi(x) \equiv 0$. For some values of the parameter λ, however, there will be nontrivial solutions. Such a value of the parameter λ is called a characteristic value for K, and the corresponding function ϕ is called a characteristic function for K. These concepts are related to corresponding concepts in linear algebra and operator theory. The operator L_K acting on the function ϕ is defined by Eq. (9). Then L_K is an example of a lin-

$$L_K\phi(x) = \int_a^b K(x,y)\phi(y)dy \qquad (9)$$

ear operator, that is, if ϕ and ψ are functions and α and β are complex constants, then $L_K(\alpha\phi + \beta\psi) = \alpha L_K\phi + \beta L_K\psi$. If the function ϕ is a characteristic function for K with corresponding characteristic value λ, then the homogeneous equation can be rewritten as Eq. (10). In terms of linear operator

$$L_K\phi = \frac{1}{\lambda}\phi \qquad (10)$$

theory, Eq. (10) means that $1/\lambda$ is an eigenvalue (proper value) for the linear operator L_K, and ϕ is a corresponding eigenvector (proper vector), or in this case, eigenfunction. A linear Volterra equation has no characteristic values. *See* EIGENFUNCTION: MATRIX CALCULUS; MATRIX THEORY; OPERATOR THEORY.

Fredholm equations. The integral equation, Eq. (7) or (8), is called a Fredholm equation, and $K(x,y)$ is called a Fredholm kernel if $\|K\| < \infty$, where $\|K\|$, the norm of K, is defined by Eq. (11).

$$\|K\|^2 = \int_a^b \int_a^b |K(x,y)|^2 dxdy \qquad (11)$$

This is certainly true if K is continuous or even bounded. Even some infinite discontinuities are allowable: for example, if $|K(x,y)| \le M|x-y|^{-\alpha}$, where M is a constant and $\alpha < 1/2$, then K is a Fredholm kernel.

Fredholm equations of the second kind for which the parameter satisfies the inequality $|\lambda|\,\|K\| < 1$ can be shown to have unique solutions by the method of Picard. The result is Eq. (12) where the resolvent $R_K(x,y;\lambda)$ is given by the Neumann series, Eqs. (13) and (14). Equation (12) can be rewritten in terms of the operator L_K as Eq. (15), when $L_K^2 f = L_K(L_K f), \ldots, L_K^{n+1}(f) = L_K(L_K^n f)$.

$$\phi(x) = f(x) + \lambda \int R_K(x,y;\lambda)f(y)dy \qquad (12)$$

$$R_K(x,y;\lambda) = \sum_{n=1}^{\infty} \lambda^{n-1}K^{(n)}(x,y) \qquad (13)$$

$$K^{(1)}(x,y) = K(x,y)$$
$$\qquad\qquad (14)$$
$$K^{(n+1)}(x,y) = \int_a^b K(x,z)K^{(n)}(z,y)dz$$

$$\phi = \sum_{n=0}^{\infty} \lambda^n L_K^n f \qquad (15)$$

As a result there are no characteristic values for which $|\lambda| \le 1/\|K\|$. In general, for any constant $M > 0$ there are only finitely many characteristic values which satisfy $|\lambda| \le M$. Further, only finitely many linearly independent characteristic functions are associated with each characteristic value.

For the kernel K the conjugate kernel is defined

to be $K^*(x,y) = \overline{K(y,x)}$, where the bar denotes complex conjugation. It can be shown that λ is a characteristic value for the kernel K if and only if $\bar{\lambda}$ is a characteristic function for the conjugate kernel K^*. Furthermore, the number of linearly independent corresponding characteristic functions is the same for λ and K as for $\bar{\lambda}$ and K^*. Finally, inhomogeneous equation (16) has a solution if and only if f satisfies Eq. (17) for every solution ψ of conjugate homogeneous equation (18). These results are a

$$\phi(x) = f(x) + \lambda \int_a^b K(x,y)\phi(y)\,dy \qquad (16)$$

$$\int_a^b f(x)\,\overline{\psi(x)}\,dx = 0 \qquad (17)$$

$$\psi(x) = \lambda \int_a^b K^*(x,y)\psi(y)\,dy \qquad (18)$$

statement of the Fredholm alternative which may be stated more succinctly as follows: For a given value of λ, either λ is a characteristic value and the homogeneous equation has a nontrivial solution, or λ is not a characteristic value and the inhomogeneous equation has a solution for every choice of the function f. *See* DIFFERENTIAL EQUATION; INTEGRAL TRANSFORM; INTEGRATION. [JOHN POLKING]

Bibliography: J. L. Cochran, *The Analysis of Linear Integral Equations*, 1972; S. G. Mikhlin, *Integral Equations*, 1964; F. G. Tricomi, *Integral Equations*, 1957.

Integral transform

An integral relation between two classes of functions. For example, a relation such as Eq. (1) is said to define an integral transform. More generally, the integral may be a multiple integral, and the functions f, G, ϕ may depend on a larger number of variables. Equation (1) is thought of as transforming a

$$f(x) = \int_{-\infty}^{\infty} G(x,y)\phi(y)\,dy \qquad (1)$$

whole class, or space, of functions $\phi(y)$ into another class of functions $f(x)$. The function $G(x,y)$ is the kernel of the transform. One of the important uses of such a transform is based on the fact that a problem posed in one of the two spaces in question may be more easily solved in the other. For example, a differential equation to be solved for the function $\phi(y)$ may become an algebraic equation for the unknown function $f(x)$. *See* LAPLACE TRANSFORM.

The two basic problems for any integral transform are inversion and representation. In inversion the aim is to recover $\phi(y)$ from $f(x)$, the kernel $G(x,y)$ being known. That is, Eq. (1) is thought of as an integral equation (of the first kind) to be solved for the unknown function $\phi(y)$. A means of calculating $\phi(y)$ from $f(x)$ is called an inversion formula, and in its presence the transform achieves maximum utility, since explicit passage from each space to the other is thus assured. In representation the question is which functions $f(x)$ may be written or represented in the form (1). That is, one asks which functions $f(x)$ will make Eq. (1) solvable for $\phi(y)$. Usually this problem becomes more tractable when the solutions $\phi(y)$ are restricted to some subspace such as the class of positive or bounded functions.

Inversion theory. In many cases where the inversion problem has been solved, an inversion operator, such as a differential or integral operator, has been found, depending perhaps on a parameter, which accomplishes the inversion. It may be denoted by $O_t[f(x)]$. This means that for each value of the parameter t some operation O on $f(x)$ is defined. For example, t might be 3 and the operation might consist of computing the third derivative of $f(x)$. Suppose that O_t applied to Eq. (1) produces Eq. (2), where $G_t(x,y)$ is a new kernel for

$$O_t[f(x)] = \int_{-\infty}^{\infty} G_t(x,y)\phi(y)\,dy \qquad (2)$$

each t. In favorable cases the integral (2) approaches $\phi(x)$ as t approaches a suitable limit, for example, when G_t approaches the Dirac δ-function $\delta(x-y)$.

An important special case of Eq. (1) is the convolution transform, when the kernel is a function of $(x-y)$, Eq. (3). An equivalent form of Eq. (3) is Eq. (4), since the change of variable $x = e^t$, $y = e^{-u}$ car-

$$f(x) = \int_{-\infty}^{\infty} G(x-y)\phi(y)\,dy \qquad (3)$$

$$F(x) = \int_{0}^{\infty} K(xy)\Phi(y)\,dy \qquad (4)$$

ries Eq. (4) into Eq. (3) after a suitable change in notation.

Named transforms. Many of the classical transforms have been named for the original or principal investigator. A few of the more important ones can be listed with the names ordinarily attached to them.

For Eq. (1):

A.	Bilateral Laplace	$G(x,y) = e^{-xy}$	
B.	Fourier	$G(x,y) = e^{ixy}$	
C.	Mellin	$G(x,y) = y^{x-1}$	$y > 0$
		$= 0$	$y < 0$
D.	Stieltjes	$G(x,y) = (x+y)^{-1}$	$y > 0$
		$= 0$	$y < 0$

For Eq. (3):

E.	Dirichlet	$G(x) = x^{-1}\sin x$	
F.	Fejér	$G(x) = x^{-2}(\sin x)^2$	
G.	Fractional (Weyl form)	$G(x) = (-x)^{\mu-1}$	$x < 0$
		$= 0$	$x > 0$
H.	Hilbert	$G(x) = x^{-1}$	
I.	Picard	$G(x) = e^{-|x|}$	
J.	Poisson	$G(x) = (1+x^2)^{-1}$	
K.	Laplace (unilateral)	$G(x) = e^x e^{-e^x}$	
L.	Stieltjes	$G(x) = \mathrm{sech}\,(x/2)$	
M.	Weierstrass or Gauss	$G(x) = e^{-x^2}$	

For Eq. (4):

N.	Fourier cosine	$K(x) = \cos x$
O.	Fourier sine	$K(x) = \sin x$
P.	Hankel	$K(x) = \sqrt{x}\,J_\nu(x)$
Q.	Laplace (unilateral)	$K(x) = e^{-x}$
R.	Meijer	$K(x) = \sqrt{x}\,K_\nu(x)$

Here $J_\nu(x)$ is a Bessel function and $K_\nu(x)$ a modified Bessel function. Certain transforms are listed twice because of their frequent appearance in two equivalent forms. In fact, all of the above are convolution transforms except A, B, and C. It can be seen also that A and C are the same transform (set $y = e^{-t}$ in C).

Convolution inversion. The inversion theory for a large class of convolution transforms has been completed. Operational calculus is a guide to the method. If the symbol D stands for differentiation with respect to x but is nonetheless treated as a number in the familiar series represented by Eq. (5), one is led to make the definition in Eq. (6),

$$e^{tD} = \sum_{k=0}^{\infty} \frac{t^k}{k!} D^k \quad (5)$$

$$e^{tD}f(x) = \sum_{k=0}^{\infty} \frac{t^k}{k!} D^k f(x) = f(x+t) \quad (6)$$

since the series (6) is precisely the Maclaurin expansion of $f(x+t)$ if D^k is allowed to mean a kth derivative. *See* OPERATOR THEORY.

Now consider the bilateral Laplace transform of $G(t)$ in Eq. (1), denoting it by $1/E(s)$, as shown in Eq. (7). Replacing s by D and using Eq. (6), one has Eq.

$$\frac{1}{E(s)} = \int_{-\infty}^{\infty} e^{-st}G(t)\,dt \quad (7)$$

(8). If $E(D)$ were a number, one could solve Eq. (8)

$$\frac{1}{E(D)}\phi(x) = \int_{-\infty}^{\infty} e^{-tD}\phi(x)G(t)\,dt$$
$$= \int_{-\infty}^{\infty} \phi(x-t)G(t)\,dt$$
$$= \int_{-\infty}^{\infty} G(x-y)\phi(y)\,dy$$
$$= f(x) \qquad x-t=y \quad (8)$$

to obtain Eq. (9). Finally, reverting to the original meaning of D as a derivative, one has in Eq. (9) an

$$\phi(x) = E(D)f(x) \quad (9)$$

inversion of Eq. (1) by means of a differential operator $E(D)$.

This argument is meant to be exploratory only, but the result is accurate for a large class of kernels G and their corresponding inversion functions E. In summary, the inversion function is the reciprocal of the bilateral Laplace transform of the kernel. It has been shown that the result is correct if, for example, $E(s)$ is the infinite product in Eq. (10),

$$E(s) = e^{bs-cs^2} \prod_{k=1}^{\infty} \left(1 - \frac{s}{a_k}\right) e^{s/a_k} \quad (10)$$

where $c \geqq 0$ and the series of real constants $\sum_{k=1}^{\infty} a_k$ converges.

For example, if $K(x) = e^{-x}$, then Eq. (4) is the Laplace transform. Expressed as a convolution transform as in Eq. (3), it becomes Eq. (11), where G is

$$e^x F(e^x) = \int_{-\infty}^{\infty} G(x-y)\Phi(e^{-y})\,dy \quad (11)$$

given in the above list as entry K. The bilateral Laplace transform of this kernel is the familiar gamma function, Eq. (12), whose reciprocal has a

$$\Gamma(1-s) = \int_{-\infty}^{\infty} e^{-st}G(t)\,dt = \int_0^{\infty} e^{-t}t^{-s}\,dt \quad (12)$$

well-known expansion in the form of Eq. (10). In Eq. (13) γ is Euler's constant. In the present example

$$E(s) = \frac{1}{\Gamma(1-s)} = e^{-\gamma s}\prod_{k=1}^{\infty}\left(1-\frac{s}{k}\right)e^{s/k} \quad (13)$$

plc Eq. (9) becomes Eq. (14), or if $e^{-x}=t$, Eq. (15)

$$e^{-\gamma D}\prod_{k=1}^{\infty}\left(1-\frac{D}{k}\right)e^{D/k}e^x F(e^x) = \Phi(e^{-x}) \quad (14)$$

$$\lim_{k\to\infty}\frac{(-1)^k}{k!}F^{(k)}\left(\frac{k}{t}\right)\left(\frac{k}{t}\right)^{k+1} = \Phi(t) \quad (15)$$

may be written. This familiar inversion formula also serves to illustrate the operator O_t appearing in Eq. (2). In the present case the operator is a differential one, and the parameter t is an integer k which tends to ∞. *See* CONFORMAL MAPPING; INTEGRATION.

[DAVID V. WIDDER]

Bibliography: T. H. Hildebrandt, *Introduction to the Theory of Integration*, 1963; D. V. Widder and I. I. Hirschman, *The Convolution Transform*, 1955.

Integrated circuits

Miniature electronic circuits produced within and upon a single semiconductor crystal, usually silicon. Integrated circuits range in complexity from simple logic circuits and amplifiers, about $\frac{1}{20}$ in. (1.3 mm) square, to large-scale integrated circuits up to about $\frac{1}{5}$ in. (5 mm) square and containing thousands of transistors and other components. The latter provide computer memory circuits and complex logic subsystems such as microcomputer central processor units.

Since the mid-1960s, integrated circuits have become the primary components of most electronic systems. Their low cost, high reliability, and speed have been essential in furthering the wide use of digital computers. Microcomputers have spread the use of computer technology to instruments, business machines, and other equipment. Another common use of large-scale integrated circuits is in pocket calculators. For analog signal processing, integrated subsystems such as FM stereo demodulators are made. *See* CALCULATORS; DIGITAL COMPUTER.

Integrated circuits consist of the combination of active electronic devices such as transistors and diodes with passive components such as resistors and capacitors within and upon a single semiconductor crystal. The construction of these elements within the semiconductor is achieved through the introduction of electrically active impurities into well-defined regions of the semiconductor. The fabrication of integrated circuits thus involves such processes as vapor-phase deposition of semiconductors and insulators, oxidation, solid-state diffusion, and vacuum deposition.

Generally, integrated circuits are not straightforward replacements of electronic circuits assembled from discrete components. They represent an extension of the technology by which silicon planar transistors are made. Because of this, transistors or modifications of transistor structures are the primary devices of integrated circuits. Methods of fabricating good-quality resistors and capacitors have been devised, but the third major type of passive component, inductors, must be simulated with complex circuitry or added to the integrated circuit as discrete components. *See* TRANSISTOR.

Simple logic circuits were the easiest to adapt to these design changes. The first of these, such as inverters and gates, were produced in the early 1960s primarily for miniaturization of missile guidance computers and other aerospace systems. Analog circuits, called linear integrated circuits, did not become commercially practical until several years later because of their heavy dependence on passive components such as resistors and capacitors. The first good-quality operational amplifiers for analog computers and instruments were produced in 1966. *See* AMPLIFIER; ANALOG COMPUTER; LOGIC CIRCUIT.

Integrated circuits can be classified into two groups on the basis of the type of transistors which they employ: bipolar integrated circuits, in which the principal element is the bipolar junction transistor; and metal oxide semiconductor (MOS) integrated circuits, in which the principal element is the MOS transistor. Both depend upon the construction of a desired pattern of electrically active impurities within the semiconductor body, and upon the formation of an interconnection pattern of metal films on the surface of the semiconductor.

Bipolar circuits are generally used where highest logic speed is desired, and MOS for largest-scale integration or lowest power dissipation. For example, by 1975 mass-produced bipolar memory circuits stored up to 4096 bits and MOS circuits up to 16,384 bits or more in read-only memory (ROM) designs. For random-access memory (RAM), typical bipolar capacity is 256 or 1024 bits and MOS capacity 1024 or 4096 bits. A MOS development, the charge-coupled device, provides serially

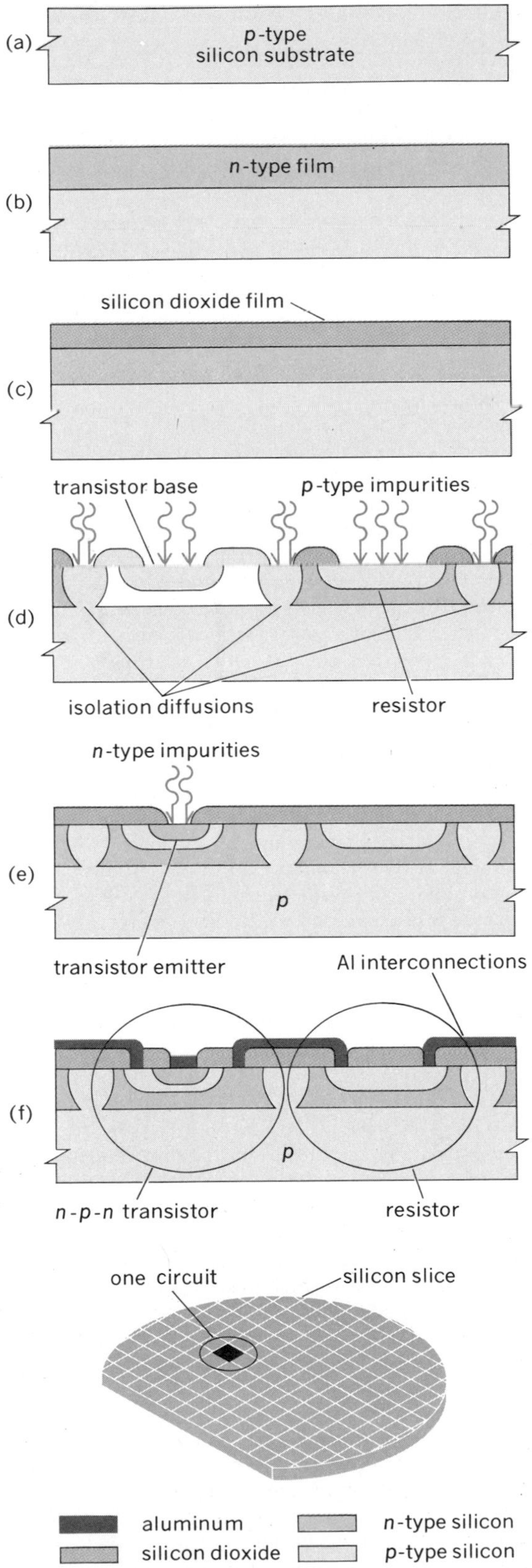

(a)
p-type
silicon substrate

(b)
n-type film

silicon dioxide film

(c)

transistor base   p-type impurities

(d)

isolation diffusions   resistor

n-type impurities

(e)
p

transistor emitter   Al interconnections

(f)
p

n-p-n transistor   resistor

one circuit   silicon slice

aluminum   n-type silicon
silicon dioxide   p-type silicon

Fig. 1. Fabrication of bipolar inverter circuit.

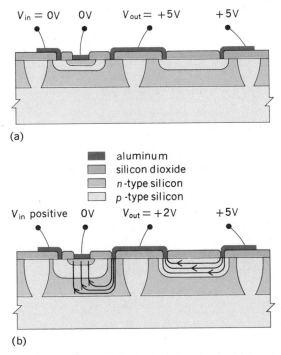

$V_{in} = 0V$ 0V $V_{out} = +5V$ +5V

(a)

aluminum
silicon dioxide
n-type silicon
p-type silicon

V_{in} positive   0V $V_{out} = +2V$ +5V

(b)

Fig. 2. Operation of bipolar inverter circuit. (*a*) Input voltage V_{in} is zero. (*b*) Positive input voltage applied. Arrows indicate direction of current flow.

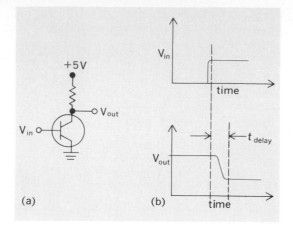

Fig. 3.   Characteristics of the inverter circuit of Fig. 2.
(a) Circuit symbol. (b) Switching waveforms.

accessible read-write memory at a typical storage capacity of 16,384 bits per circuit. Linear circuits are generally bipolar, but MOS devices are often used in them to improve precision and stability.

Bipolar integrated circuits. The principal steps involved in the fabrication of a bipolar inverter circuit, representative of the simplest logic function, are illustrated schematically in Fig. 1. An inverter requires only a transistor and resistor, shown in cross section. Complete digital integrated circuits generally contain tens to hundreds of inverters and gates interconnected as counters, arithmetic units, and other building blocks. As indicated by the inset, hundreds of such circuits may be fabricated on a single slice of silicon crystal. This feature of planar technology—simultaneous production of many circuits—is responsible for the economic advantages and wide use of integrated circuits.

The starting material is a slice of single-crystal silicon, more or less circular, up to 3 in. in diameter, and a fraction of an inch thick. Typically, this material is doped with p-type impurities. A film of semiconductor, less than 1/1000 in. thick, is then grown upon this substrate in a vapor-phase reac-

tion of a silicon-containing compound. The conditions of this reaction are such that the film maintains the single-crystal nature of the substrate. Such films are called epitaxial (Greek for "arranged upon"). By incorporating n-type impurities into the gas from which the film is grown, the resulting epitaxial film is made n-type. *See* SEMICONDUCTOR; SILICON.

Next, the silicon slice is placed into an oxygen atmosphere at high temperatures (1200°C). The silicon and oxygen react, forming a cohesive silicon dioxide film upon the surface of the slice that is relatively impervious to the electrically active impurities.

To form the particular semiconductor regions required in the fabrication of electronic devices, however, p- and n-type impurities must be introduced into certain regions of the semiconductor. In the planar technology, this is done by opening windows in the protective oxide layer by photoengraving techniques, and then exposing the slice to a gas containing the appropriate doping impurity. In the case of an integrated circuit, the isolation regions—p-type regions which, together with the p-type substrate, surround the separate pockets of the n-type film—are formed first by the diffusion of a p-type impurity. This is followed by a shorter exposure to p-type impurities during which the base region of the transistors and the resistors are formed. *See* JUNCTION TRANSISTOR.

Next, the slice is again covered with oxide, smaller windows are cut over the transistor base regions, and n-type impurities are permited to diffuse in these regions to form the emitters of the transistors. Openings are cut in the oxide layer at all places where contact to the silicon is desired; a metal, aluminum for example, is deposited over the entire slice by vacuum evaporation, and finally the undesired aluminum is removed by photoengraving, leaving behind aluminum stripes which interconnect the transistor and the resistor, as indicated in the lowest part of Fig. 1.

The preparation of the hundreds of inverter circuits on the silicon slice is now complete. The slice is cut apart, much as a pane of glass is cut, and the individual circuits are tested and packaged.

Both the transistor and the resistor are formed within a separate pocket of n-type semiconductor surrounded on all sides by p-type regions. When the inverter circuit is operated, a reverse bias develops between the n-type pocket and its surroundings. The depletion region separating the n- and p-type regions has a very high resistance; consequently, the individual transistors and resistors are electrically isolated from each other, even though both of them have been formed within the same semiconductor crystal.

The way an integrated inverter circuit operates is illustrated in Fig. 2. The input voltage V_{in} is applied to the base of the transistor. When V_{in} is zero or negative with respect to the emitter, no current flows. As a result, no voltage drop exists across the resistor, and the output voltage V_{out} will be the same as the externally applied biasing voltage, +5 volts in this example. When a positive input voltage is applied, the transistor becomes conducting. Current now flows through the transistor and hence through the resistor; as a result, the output

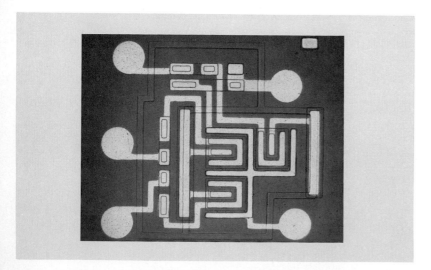

Fig. 4.   Photomicrograph of early bipolar logic gate circuit. (*Fairchild Semiconductor*)

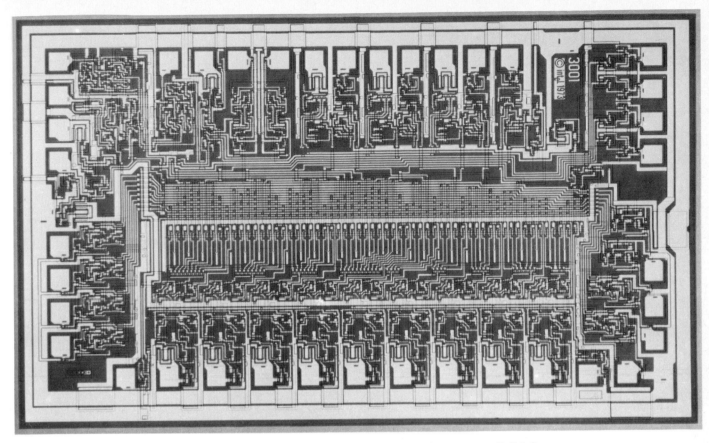

Fig. 5. Photomicrograph of bipolar logic circuit forming section of microcomputer central processor unit. (*Intel*)

voltage will become more negative. Thus, the change in input voltage appears inverted at the output.

The circuit symbol and the changes in input and output voltages during the switching process just described are illustrated in Fig. 3. Note that the change in the output voltage occurs slightly later than the change in the input voltage. This time difference, called propagation delay, is an important characteristic of all integrated circuits. Much effort has been spent on reducing it, and values less than one-billionth of a second have been achieved.

Most simple digital circuits can be fabricated much as the inverter circuit described above. As an example, a photomicrograph of an early logic gate circuit is shown in Fig. 4. This circuit is one of the earliest digital integrated circuits, introduced commercially in 1961. For comparison, a more up-to-date digital integrated circuit is shown in Fig. 5. This circuit, although hardly larger than the former, contains more than 100 times as much circuitry.

Figures 4 and 5 indicate the tendency in development of digital integrated circuits toward increasing complexity. This tendency is dictated by the economics of integrated circuit manufacturing. Because of the nature of this manufacturing process, all circuits on a slice are fabricated together. Consequently, the more circuitry accommodated on a slice, the cheaper the circuitry becomes. Because testing and packaging costs depend on the number of chips, it is desirable, in order to keep costs down, to crowd more circuitry onto a given chip rather than to increase the number of chips on a wafer.

Circuits such as the one in Fig. 4 are the building blocks of digital computers. Their development and that of large computers were closely interrelated. *See* DIGITAL COMPUTER.

When subsystem complexity is too great for production in a small area of crystal, the subsystem is "sliced" into building blocks. The regular structure of memory circuits allows memory blocks to contain several times as much logic as other circuits. For instance, a 1024-bit RAM has 1024 flip-flop circuits for data storage, plus address decoders. The random structure of circuits such as Fig. 5 requires more interconnection area, and thus contains fewer logic elements. This is a 2-bit slice of a microcomputer central processor unit (CPU). Eight such slices are simply parallel-connected to form a 16-bit CPU, for instance.

Linear circuits. Integrated circuits based on amplifiers are called linear because amplifiers usually exhibit a linearly proportional response to input signal variations. However, the category includes memory sense amplifiers, combinations of analog and digital processing functions, and other circuits with nonlinear characteristics. Some digital and analog combinations include analog-to-digital converters, timing controls, and modems (data communications modulator-demodulator units).

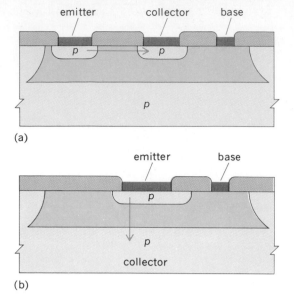

Fig. 6. Types of *pnp* transistor structures. (*a*) Lateral *pnp* transistor. (*b*) Substrate *pnp* transistor. Arrows indicate direction of useful transistor action.

Linear circuits cannot yet match the power and frequency range of discrete transistors. However, they meet most other technical requirements. One long-standing drawback was the lack of inductors for tuning and filtering. That was overcome by the use of resistor-capacitor networks and additional circuitry. One oscillator-based circuit called the phase-locked loop provides a general-purpose replacement for inductors in applications such as radio transmission demodulation. *See* PHASE-LOCKED LOOP.

At first, the development of linear circuits was slow because of the difficulty of integrating passive components and also because of undesirable interactions between the semiconductor substrate and the operating components. Thus, much greater ingenuity was required to design and use the early linear circuits.

In addition, manufacturing economics favors digital circuits. A computer can be built by repetitious use of simple inverters and gates, while analog signal processing requires a variety of specialized linear circuits.

Techniques such as the use of MOS devices to stabilize amplifiers, integration of multiple amplifiers, and development of demodulators and other complex circuits are greatly extending linear circuit use. As a result, they are coming into wide use in automobiles, consumer electronics, and other high-volume applications.

Semiconductor devices. In the continuing effort to increase the complexity and speed of digital circuits, and the performance characteristics and versatility of linear circuits, a significant role has been played by the discovery and development of new types of active and passive semiconductor devices which are suitable for use in integrated circuits. Among these devices is the *pnp* transistor which, when used in conjunction with the standard *npn* transistors described earlier, lends added flexibility to the design of integrated circuits.

Two types of *pnp* transistor structures, both

compatible with standard integrated circuit technology, are shown in Fig. 6. Figure 6*a* shows the so-called lateral *pnp*, which is a *pnp* transistor formed between two closely spaced *p*-type diffused regions. Since these can be formed using the same diffusion step by which the base of the *npn* transistor and the resistor are formed, this structure can be fabricated simultaneously with the rest of the circuit. The same is true of the substrate *pnp* shown in Fig. 6*b*, except here the *p*-type substrate is employed as the collector.

Other possibilities involve means of producing integrated circuits with resistors of high resistance values. By using the standard resistor process, as described in connection with Fig. 1, high resistances require long resistors, hence large chips are required in order to accommodate these long resistors. An alternate approach is to use the *n*-region between the diffused *p*-region and the substrate as the resistor. This semiconductor region has a significantly higher resistance than that formed by the *p*-type diffusion. However, due to the reverse bias between the *n*-region and the two *p*-regions enclosing it from top and bottom, such resistors have highly nonlinear current-voltage characteristics. An alternate scheme involves the use of a very thin film of high-resistivity metal deposited by evaporation or sputtering on top of the insulating silicon dioxide layer. Such films can display resistivities as much as a hundred times higher than that of the diffused *p*-region, and since their deposition is performed after the entire semiconductor structure is complete, they can be used in conjunction with any type of integrated circuits.

Several alternatives to *pn*-junction isolation are in use. These involve etching around the transistor regions, so that oxide or air isolates the components. Called dielectric isolation, the technique was initially used in military circuits to stop radiation-induced currents from flowing through and destroying the circuits. It is now used commercially to reduce circuit capacitance, thus speeding up operation, and to reduce the silicon area required for each transistor. It is not yet widely used, however, because it complicates the basic process.

MOS integrated circuits. The other major class of integrated circuits is called MOS because its principal device is a metal-oxide-semiconductor field-effect transistor (MOSFET). It is more suitable for large-scale integration (LSI) than bipolar circuits because MOS transistors are self-isolating and can have an average size of only a few millionths of a square inch. This has made it practical to use upward of 10,000 transistors per circuit, which is ample for many types of memory arrays or logic subsystems (Fig. 7).

Several major types of MOS device fabrication technologies have been developed since the mid-1960s. They are: metal-gate *p*-channel MOS, which uses aluminum for electrodes and interconnections (Fig. 8); silicon-gate *p*-channel MOS, employing polycrystalline silicon for gate electrodes and the first interconnection layer; *n*-channel MOS, which is usually silicon gate; and complementary MOS, which employs both *p*-channel and *n*-channel devices.

The basic principles of MOS technology can be illustrated with the metal-gate *p*-channel MOSFET in Fig. 8. It consists of two *p*-type diffused regions

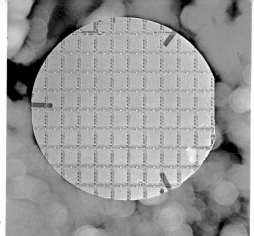

(b)

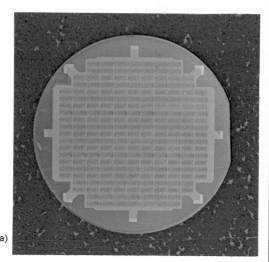

(a)

Recently the size of electronic devices has been
shrunk by a factor of 10. Now more than 1000 transistors can be connected into a circuit only one-tenth
of an inch square. The technology that produces such
high-density electronic circuits is known as large-
scale integration (LSI). (a) Beam lead circuitry;
(b) metal oxide semiconductor (MOS) slice; (c) completed LSI slice in package; (d) close-up of completed LSI slice showing three levels of metallization
(*Texas Instruments Inc.*). (e) Four-bit register chip,
a medium-scale integration (MSI) device (*Sylvania
Electric Products Inc.*).

(c)

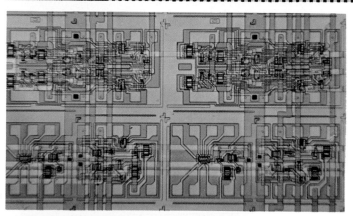

(d)

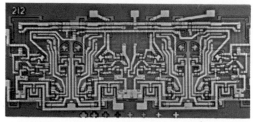

(e)

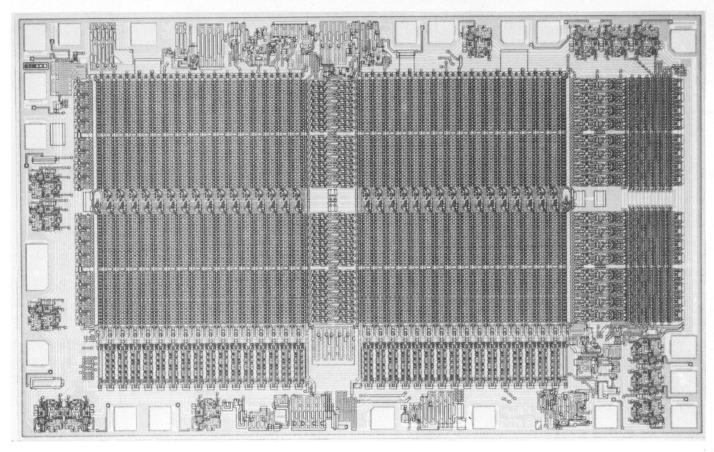

Fig. 7. Photomicrograph of MOS LSI circuit used as 4096-bit random-access memory. (*Intel*)

(such as can be obtained by the process used to form the base region of the *npn* transistors described earlier), separated by an *n*-type region which is under control of an electrode, called the gate. When there is no voltage applied to the gate, the two *p*-type regions, called source and drain, are electrically insulated from each other by the *n*-type region which surrounds them. When, however, a negative voltage is applied to the gate, the electric field induces a very thin *p*-type region at the surface of the silicon. This thin *p*-type region now connects the source and the drain, permitting the passage of current between them. The path is called a channel.

Both conceptually and structurally the MOS transistor is a much simpler device than the bipolar transistor. In fact, its principle of operation has been known since the late 1930s, and the research effort that led to the discovery of the bipolar transistor was originally aimed at developing the MOS transistor. What kept this simple device from commercial utilization until 1964 is the fact that it depends on the properties of the semiconductor surface for its operation, while the bipolar transistor depends principally on the bulk properties of the semiconductor crystal. Hence MOS transistors became practical only when understanding and control of the properties of the oxidized silicon surface had been perfected to a very great degree.

Figure 9 shows an MOS inverter circuit. One of the valuable features of MOS devices in integrated circuit applications is that with some minor modification, such as increasing the distance be-

tween the two *p*-type regions, they can also be employed as resistors with adequately high resistance values. Thus, in an MOS inverter circuit both the transistor and the resistor are realized using an MOS device of suitable geometry, as shown in Fig. 9. In this example a separate voltage, $V_{resistor}$, is applied to the gate of the resistor to set its resistance at the desired value.

The operation of an MOS inverter circuit is very similar to the operation of the bipolar inverter circuit discussed earlier. The input voltage V_{in} is applied to the gate of the transistor. When it is zero, no conduction can take place between the source and drain of the transistor, and therefore no current flows in the resistor. As a result, no voltage drop will exist across the resistor and the output

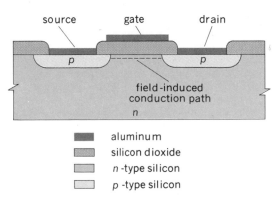

Fig. 8. Diagram of *p*-channel MOS field-effect transistor.

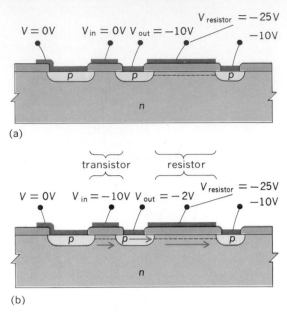

$V = 0V$ $V_{in} = 0V$ $V_{out} = -10V$ $V_{resistor} = -25V$ $-10V$

(a)

transistor   resistor

$V = 0V$ $V_{in} = -10V$ $V_{out} = -2V$ $V_{resistor} = -25V$ $-10V$

(b)

Fig. 9. Operation of a MOS inverter circuit. (a) Input voltage V_{in} is zero. (b) Input voltage applied. Arrows indicate direction of current flow.

voltage V_{out} will be the same as the externally applied biasing voltage, -10 volts in this example. When a negative input voltage is applied, a field-induced conduction path is created between source and drain. Current now flows through the transistor and hence the resistor, and as a result the output voltage will become more positive. Thus, the change in input voltage appears inverted at the output.

A comparison of the bipolar inverter circuit of Figs. 1 and 2 with its MOS counterpart in Fig. 9 reveals most of the principal advantages of the latter. Because all biased junctions are always reverse-biased, there is no need for additional means of isolation between the elements. In addition, no emitters are required in MOS integrated circuits. As a result, the fabrication process is simpler. Resistor and transistor are connected through a common p-type diffused region in the MOS case, but an added aluminum metallization stripe is needed in the bipolar case. Thus, interconnection is generally much simpler in an MOS circuit. This becomes an especially significant advantage in connection with complex circuits.

The n-channel MOS. An n-channel MOSFET has n-type source and drain diffusions in a p-type region. A positive voltage on the gate electrode converts the p-type region to a conducting n-type channel. Because n-channel properties were more difficult to control than p-channel, this form of MOS did not become commercially viable until about 1972.

The n-channel MOS is faster and more compatible with bipolar logic than p-channel MOS. It is the preferred MOS for LSI logic and memory, which are often interconnected with bipolar logic. The n-channel MOS circuits can operate at speeds approaching bipolar circuits. Moreover, switching thresholds and other operating voltages can be made the same as bipolar voltages, eliminating the special interfaces and power supplies which

are required by p-channel MOS.

Complementary MOS (CMOS). The basic CMOS logic switch is a pair of p-channel and n-channel transistors. A voltage applied to both gate electrodes makes only one of the two conductive. Thus, significant current flows through the device only during the switching operations, reducing average power dissipation to the range of microwatts per circuit. Also, like n-channel MOS, CMOS is compatible with the voltage levels used in bipolar logic circuits.

CMOS is preferred for equipment with limited power supply or cooling problems, such as battery-powered portable equipment and aerospace systems. In such equipment, its advantages often offset the added cost of fabricating the complementary transistors.

Silicon-gate MOS. The newer MOS devices generally have silicon rather than metal gates. Silicon-gate technology has greatly accelerated LSI development by simplifying the MOS process, allowing smaller MOSFETs to be made, and serving as a first layer of interconnections. The 4096-bit RAM in Fig. 7 is a silicon-gate n-channel MOS circuit.

Polycrystalline silicon is deposited on the oxidized silicon wafer prior to diffusion, then etched to form diffusion windows, electrodes, and interconnections. The result is automatic alignment of source, drain, and channel with the gate, making very small transistor geometries practical. An aluminum interconnection layer is finally deposited on the oxidized silicon-gate layer.

Charge-coupled devices. Semiconductor memory circuits have generally replaced magnetic core memories, which are randomly accessible. However, magnetic drums and discs, which are serially accessible, are generally used for bulk data storage in digital systems. The rotation of these memories is easily emulated by MOS shift registers, which store data by recirculating it through logic stages.

Since the cost per bit of storage capacity in drum and disc peripherals is relatively low, much effort has been devoted to development of shift-register circuits with very high storage capacities. The most promising is the charge-coupled device (CCD). By early 1975, a CCD storing 16,384 bits was in production, compared with 1024 bits for typical MOS LSI shift registers.

The higher storage density of CCD circuits stems from the fact that the shift-register logic stages are replaced by charge storage and transfer paths diffused into the bulk semiconductor. Tiny electrodes control recirculation, so that each electrode is effectively a replacement for a logic circuit of several transistors. While not as fast as RAMs, CCD memory circuits can be accessed at many times the speed of electromechanically rotated drums and discs. *See* COMPUTER MEMORY.

Integrated optical devices. Semiconductors have long been used as light sensors. Advances in LSI have enabled large arrays of sensors, such as solid-state television cameras, to be made commercially. MOS LSI is now preferred for sensor arrays because it permits large amounts of control logic to be fabricated in the same circuit as the sensors.

In MOS optical arrays, the sensors are MOS devices or CCD elements. The sensor portion of the circuit views the scene through a transparent window. Light variations in the scene viewed cause variations in charge, producing signals transferable in shift-register fashion to processing stages such as amplifiers. These arrays have numerous applications, from measuring and sorting objects on production lines to optical character readers, which automatically translate written information to digital computer input codes.

Microcomputers. The LSI development having the most profound effect on electronic equipment economics and design in general is the microcomputer. Introduced in 1971, microcomputer circuits such as those in Figs. 5 and 8 have thousands of applications and are made by most major producers of integrated circuits.

These programmable LSI devices are general-purpose control and data-processing subsystems. They enable equipment designers to replace tens to hundreds of electromechanical control devices or conventional circuits with a relatively few LSI logic and memory circuits.

Like those of a large computer, a microcomputer's operations are specialized by programs. But unlike most computer software, the programs are stored in read-only memories and become part of the system hardware. The equipment user is rarely involved in software preparation.

Designing a microcomputer assembly is a relatively straightforward task. However, the designer must change from traditional circuitry-oriented design to a computer design-like activity, a more radical shift than the previous changeovers from electron tube circuitry to transistors and from transistors to small-scale integrated circuits. Microcomputer-based development systems that facilitate simultaneous development of equipments and their software have further accelerated this change in design.

One effect has been the development of many highly automated electronic products. Some simple examples are cash registers that also maintain inventory records, instruments that analyze transducer signals for the doctor or other user, and computerized games and food-dispensing machines. The one major area where microcomputers are not yet used is the home. However, since some major appliances feature electronic controls, the household use of microcomputers can be anticipated.

On the other hand, personal uses of large-scale integrated circuit products have increased very rapidly. Two examples of these are the very-low-cost pocket-size personal calculator, the cost of which had declined dramatically because of progress in large-scale integrated circuit technology, and the electronic digital wristwatch, which has closely followed pocket calculators in using large-scale integration. The calculator is generally based on the use of p-channel MOS techology, the watch on CMOS technology. *See* CIRCUIT (ELECTRONICS); MICROCIRCUITRY; PRINTED CIRCUIT.

[ANDREW S. GROVE]

Bibliography: R. H. Crawford, *MOSFET in Circuit Design*, 1967; A. S. Grove, *Physics and Technology of Semiconductor Devices*, 1967; D. K. Lynn, C. S. Meyer, and D. J. Hamilton, *Analysis and Design of Integrated Circuits*, 1967; R. M. Warner, Jr., and J. N. Fordemwalt (eds.), *Integrated Circuits: Design Principles and Fabrication*, 1965.

Integration

An operation of the infinitesimal calculus which has two aspects. The roots of one go back to antiquity, for Archimedes and other Greek mathematicians used the "method of exhaustion" to compute areas and volumes. A simple example of this is the approximation to the area of a circle obtained by inscribing a regular polygon of known area, and then repeatedly doubling the number of sides. The areas of the successive polygons are computable with the help of elementary geometry. The limit of the sequence of these areas gives the area of the circle. The area of each polygon can be regarded as being made up of the sum of the areas of triangles with vertices at the center of the circle, and so the process described is a constructive definition of an integral which is the limit of a sum. Modern definitions of integrals as limits of sums are discussed in this article.

The other aspect of integration is the process of finding antiderivatives, that is, for a given function $f(x)$ to find another function $g(x)$ whose derivative is $f(x)$. This aspect is related to the first by the fundamental theorem of integral calculus, so both processes are called integration.

Sir Isaac Newton emphasized the antiderivative aspect of integration, and his work shows how much can be done in the applications of integral calculus without introducing limits of sums. However, limits of sums lead to very fruitful theoretical developments in the theory of integration, as in the notion of multiple integrals, for example, and hence lead to a wider variety of applications. Leibnitz, a 17th-century mathematician, inspired the development in this direction, but many years elapsed before the theory was given a firm logical foundation. In the early 19th century A. L. Cauchy gave a clear-cut definition of the definite integral for continuous functions and a proof of its existence. Later, G. F. B. Riemann discussed the integral for discontinuous functions and gave a necessary and sufficient condition for its existence. Thus the most generally used definition of the integral as the limit of a sum has come to be called the Riemann integral.

Riemann integral. The precise definition of the Riemann integral for a real function f of one real variable x on a finite interval $a \leq x \leq b$ may be formulated as follows. Let P be a partition of the interval $[a,b]$ into n subintervals by points t_i, where $t_{i-1} < t_i, t_0 = a, t_n = b$, and consider a sum S of the

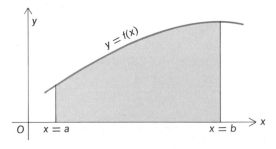

Fig. 1. Graph of $y = f(x)$.

form of Eq. (1), where $t_{i-1} \leqq x_i \leqq t_i$. The sum S

$$S = \sum_{i=1}^{n} f(x_i)(t_i - t_{i-1}) \qquad (1)$$

depends not only on the partition P but on the choice of the intermediate points x_i. It may happen that the sum S approaches a definite limit I when the maximum of the numbers $(t_i - t_{i-1})$ tends to zero, and in this case I is called the Riemann integral (or the definite integral) of f from a to b, and is denoted by the Leibnitzian symbol (2).

$$\int_a^b f(x)\, dx \qquad (2)$$

Also, f is said to be integrable on $[a,b]$. When f is a continuous function with positive values on the interval $[a,b]$, the integral has a simple geometrical interpretation as the area bounded by the x axis, the ordinates $x = a$ and $x = b$, and the graph of $y = f(x)$ (Fig. 1). It is convenient to use Eq. (3) as a definition of the symbol on the left.

$$\int_b^a f(x)\, dx = -\int_a^b f(x)\, dx \qquad (3)$$

For functions f and g which are integrable on $[a,b]$, the integral has the following properties:

(i) f is integrable on every subinterval of $[a,b]$.
(ii) For every triple of points c, d, and e in $[a,b]$,

$$\int_c^d f(x)\, dx + \int_d^e f(x)\, dx = \int_c^e f(x)\, dx$$

(iii) $f(x) + g(x)$ is integrable, and

$$\int_a^b [f(x) + g(x)]\, dx = \int_a^b f(x)\, dx + \int_a^b g(x)\, dx$$

(iv) $f(x)g(x)$ is integrable, and in particular $cf(x)$ is integrable for every real number c, and

$$\int_a^b cf(x)\, dx = c \int_a^b f(x)\, dx$$

(v) If $f(x) \leqq g(x)$ on $[a,b]$, then

$$\int_a^b f(x)\, dx \leqq \int_a^b g(x)\, dx$$

(vi) $|f(x)|$ is integrable, and

$$\left| \int_a^b f(x)\, dx \right| \leqq \int_a^b |f(x)|\, dx$$

It can be proved that a function f is integrable on $[a,b]$ if and only if the following two conditions are satisfied: f is bounded on $[a,b]$; and the set of points where f is discontinuous can be enclosed in a series (possibly infinite) of intervals, the sum of whose lengths is arbitrarily near to zero.

Antiderivatives. To develop the fundamental theorem of integral calculus let u be a variable in the interval $[a,b]$, on which f is integrable; then formula (4) defines a function of u which may be

$$\int_a^u f(x)\, dx \qquad (4)$$

noted by $I(u)$. If f is continuous on $[a,b]$, then f is integrable, and it is also true that $I(u)$ has a derivative $I'(u) = f(u)$. Now let h be any antiderivative of f, that is, $h'(u) = f(u)$ on $[a,b]$. Then $I'(u) - h'(u) = 0$, so $I(u) - h(u) = \text{constant} = -h(a)$, by the theorem of the mean for derivatives and the fact that $I(a) = 0$, so relation (5) can be written. This is the fundamental theorem of integral calculus, and it shows

that definite integrals may be calculated by the process of finding antiderivatives. For this reason antiderivatives are frequently called indefinite inte-

$$I = I(b) = h(b) - h(a) \qquad (5)$$

grals and denoted by $\int f(x)\, dx$, and special methods of finding indefinite integrals for frequently occurring functions occupy a large part of elementary calculus. The principal methods are outlined in the next section. The standard notation for an indefinite integral of $f(x)$ is formula (6).

$$\int f(x)\, dx \qquad (6)$$

Elementary methods of integration. Obviously, each formula for differentiation yields a formula for indefinite integration.

Method of substitution. From these indefinite integrals, additional formulas can be obtained by the method of substitution, which is based on the chain rule for differentiating composite functions. For example, from formula (7) one obtains Eq. (8), by the substitution $v = a^2 + x^2$, and Eq. (9), by the substitution $v = \sin x$.

$$\int x^m\, dx = \frac{x^{m+1}}{m+1} \qquad (7)$$

$$\int (a^2 + x^2)^m x\, dx = \frac{(a^2 + x^2)^{m+1}}{2(m+1)} \qquad (8)$$

$$\int \sin^m x \cos x\, dx = \frac{\sin^{m+1} x}{m+1} \qquad (9)$$

Method of partial fractions. In case $f(x)$ is a quotient of two polynomials in x, $f(x)$ may be represented as a sum of a polynomial and terms of the form, shown as notations (10) and (11), where m is a posi-

$$\frac{1}{(ax+b)^m} \qquad a \neq 0 \qquad (10)$$

$$\frac{Ax+B}{(cx^2 + dx + e)^m} \qquad c \neq 0, d^2 - 4ce < 0 \qquad (11)$$

tive integer and the coefficients are real. This is called the method of partial fractions. Its actual execution requires a knowledge of the factors of the denominator of $f(x)$. Terms of the form (10) have known indefinite integrals. Terms of the form (11) with $m = 1$ can also be integrated by elementary formulas. When $m > 1$, a term of the form (11) can be written as a sum of terms of the forms (12) and (13), where $\alpha = (2dA - 4cB)/(d^2 - 4ce)$, $\beta = -\alpha/2$, and $\gamma = (dB - 2eA)/(d^2 - 4ce)$. By the process

$$\frac{\alpha}{(\gamma x^2 + dx + e)^{m-1}} \qquad (12)$$

$$\frac{(\beta x + \gamma)(2cx + d)}{(cx^2 + dx + e)^m} \qquad (13)$$

of integration by parts, the indefinite integral of a term of the form (13) can be expressed in terms of the integral of a term such as (11), with m replaced by $m - 1$. Reduction formulas of this type are given in standard tables of integrals.

A rational function of $\sin x$ and $\cos x$ can be integrated by the method just described after the substitution $u = \tan(x/2)$ has been applied. This substitution reduces the problem to the integration of a rational function of u. Various functions involving radicals can be integrated by means of trigonometric substitutions or by other substitutions which reduce the problem to the integration of a rational function. For example, by setting $u = a + bx$, one finds Eq. (14), and when m is a nonnegative integer

$$\int x^m (a + bx)^p \, dx = \int \left(\frac{u-a}{b}\right)^m u^p \frac{du}{b} \quad (14)$$

the right side can be multiplied out and integrated by the power formula, even though p is a fraction.

Integration by parts. This is a very powerful and important method. It follows from the formula for differentiating a product, namely Eq. (15). After

$$d(fg) = f \, dg + g \, df \quad (15)$$

the terms are rearranged and integrated, Eq. (16)

$$\int f \, dg = fg - \int g \, df \quad (16)$$

or (17) may be written, where the arbitrary con-

$$\int fg' \, dx = f(x)g(x) - \int gf' \, dx \quad (17)$$

stant, as usual, has been omitted. As an example, consider Eq. (18), which is a special case of the

$$I = \int \frac{2x^2 \, dx}{(x^2 + 1)^m} \quad (18)$$

form (13). Let $f = x$, $g' = 2x/(x^2 + 1)^m$. Then $f' = 1$ and Eq. (19) can be written, and by Eq. (17), Eq. (20) follows.

$$g = \frac{1}{(1-m)(x^2+1)^{m-1}} \quad (19)$$

$$I = \frac{x}{(1-m)(x^2+1)^{m-1}}$$
$$- \frac{1}{(1-m)} \int \frac{dx}{(x^2+1)^{m-1}} \quad (20)$$

The indefinite integrals of many of the commonly occuring functions are given in tables of integrals in handbooks and textbooks.

The elementary functions are those expressible by means of a finite number of algebraic operations and trigonometric and exponential function and their inverses. The integrals of many elementary functions are known to be not elementary, so they define new functions. A number of these are sufficiently important that their values have been tabulated. Examples are Eqs. (21) and (22). $F(k,x)$

$$F(k,x) = \int_0^x \frac{dx}{\sqrt{(1-x^2)(1-k^2x^2)}} \quad (k^2 \neq 1) \quad (21)$$

$$\text{Si } x = \int_0^x \frac{\sin x \, dx}{x} \quad (22)$$

is called an elliptic integral of the first kind. The inverse of $u = F(k,x)$ is an elliptic function called the sine amplitude of u and denoted by $x = \sin u$. Some nonelementary functions have been included in tables of integrals. *See* ELLIPTIC FUNCTION AND INTEGRAL.

Improper integrals. This term is used to refer to an extension of the notion of definite integral to cases in which the integrand is unbounded or the domain of integration is unbounded. Consider first the case when $f(x)$ is integrable (in the sense defined above) and hence bounded on every interval $[a + \epsilon, b]$ for $\epsilon > 0$, but is unbounded on $[a,b]$. Then by definition, Eq. (23) is written, provided the

$$\int_a^b f(x) \, dx = \lim_{\epsilon \to 0} \int_{a+\epsilon}^b f(x) \, dx \quad (23)$$

limit on the right exists. For example, if $f(x) = x^{-2/3}$, relation (24) follows. Similarly, Eq. (25) follows.

$$\int_0^1 x^{-2/3} \, dx = \lim_{\epsilon \to 0} \int_\epsilon^1 x^{-2/3} \, dx$$
$$= \lim_{\epsilon \to 0} 3[1 - \epsilon^{1/3}] = 3 \quad (24)$$

$$\int_{-1}^0 x^{-2/3} \, dx = \lim_{\epsilon \to 0} \int_{-1}^{-\epsilon} x^{-2/3} \, dx$$
$$= \lim_{\epsilon \to 0} 3[(-\epsilon)^{1/3} + 1] = 3 \quad (25)$$

When $f(x)$ is integrable on every finite subinterval of the real axis, by definition Eqs. (26) and (27) hold, provided the limits on the right exist.

$$\int_a^{+\infty} f(x) \, dx = \lim_{b \to +\infty} \int_a^b f(x) \, dx \quad (26)$$

$$\int_{-\infty}^a f(x) \, dx = \lim_{b \to -\infty} \int_b^a f(x) \, dx \quad (27)$$

More general cases are treated by dividing the real axis into pieces, each of which satisfies one of the conditions just specified. Equation (28) is an exam-

$$\int_{-\infty}^{+\infty} x^{-5/3} \, dx = \lim_{a \to -\infty} \int_a^{-1} x^{-5/3} \, dx$$
$$+ \lim_{\delta \to 0} \int_{-1}^{-\delta} x^{-5/3} \, dx + \lim_{\epsilon \to 0} \int_\epsilon^1 x^{-5/3} \, dx$$
$$+ \lim_{b \to +\infty} \int_1^b x^{-5/3} \, dx \quad (28)$$

ple. Because the second and third of the limits on the right do not exist, integral (29) does not exist.

$$\int_{-\infty}^{+\infty} x^{-5/3} \, dx \quad (29)$$

However, it is sometimes useful to assign a value to it, called the Cauchy principal value, by replacing definition (28) by Eq. (30). In this particular case the value is zero.

$$\int_{-\infty}^{+\infty} x^{-5/3} \, dx = \lim_{e \to 0} \left[\int_{-1}^{-\epsilon} x^{-5/3} \, dx + \int_\epsilon^1 x^{-5/3} \, dx \right]$$
$$+ \lim_{b \to +\infty} \left[\int_{-b}^{-1} x^{-5/3} \, dx + \int_1^b x^{-5/3} \, dx \right] \quad (30)$$

The preceding definitions apply to cases when the integrand $f(x)$ is bounded, except in arbitrarily small neighborhoods of a finite set of points. In the closing years of the 19th century various extensions were made to more general cases. Then Henri Lebesgue produced a comparatively simple general theory for the case of absolutely convergent integrals. The integral of Lebesgue will be discussed below. The integral of Denjoy includes both nonabsolutely convergent integrals and the integral of Lebesgue.

Multiple integrals. The concept called the Riemann integral can be extended to functions of several variables. The case of a function of two variables illustrates sufficiently the additional features which arise. To begin with, let $f(x,y)$ denote a real function defined on a rectangle of the form (31).

$$R: a \leq x \leq b \quad c \leq y \leq d \quad (31)$$

Let P be a partition of R into n nonoverlapping rectangles R_i with areas A_i, and let (x_i, y_i) be a point of R_i. Define S by Eq. (32). In case the sum S tends to a definite limit I when the maximum diagonal of a rectangle R_i tends to zero, then f is said to be in-

$$S = \sum_{i=1}^{n} f(x_i, y_i) A_i \qquad (32)$$

tegrable over R, and the limit I is called the Riemann integral of f over R. It will be denoted here by the abbreviated symbol $\iint_R f$. Properties corresponding to those numbered (i) to (iv) can be proved for these double integrals, except that property (ii) should now read as follows:

(ii') If the rectangle R is the union of two rectangles S and T, then Eq. (33) can be written.

$$\iint_R f = \iint_S f + \iint_T f \qquad (33)$$

Necessary and sufficient conditions for a function $f(x,y)$ to be integrable over R can be stated as follows: f is bounded on R; and the set of points where f is discontinuous can be enclosed in a series of rectangles, the sum of whose areas is arbitrarily near to zero.

To define the integral of a function $f(x,y)$ over a more general domain D where it is defined, suppose that D is enclosed in a rectangle R, and define $F(x,y)$ by Eqs. (34). Then f is said to be inte-

$$\begin{aligned} F(x,y) &= f(x,y) \text{ in } D \\ F(x,y) &= 0 \text{ outside } D \end{aligned} \qquad (34)$$

grable over D in case F is integrable over R, and Eq. (35) holds, by definition.

$$\iint_D f = \iint_R F \qquad (35)$$

When the function $f(x,y)$ is continuous on D, and D is defined by inequalities of the form (36), where

$$a \leqq x \leqq b \qquad a(x) \leqq y \leqq \beta(x) \qquad (36)$$

the functions $\alpha(x)$ and $\beta(x)$ are continuous, the double integral of f over D always exists, and may be represented in terms of two simple integrals by Eq. (37). In many cases, this formula makes possible the evaluation of the double integral.

$$\iint_D f = \int_a^b \left[\int_{\alpha(x)}^{\beta(x)} f(x,y) \, dy \right] dx \qquad (37)$$

Improper multiple integrals have been defined in a variety of ways. There is not space to discuss these definitions and their relations here.

Line, surface, and volume integrals. A general discussion of curves and surfaces requires an extended treatise. The following outline is restricted to the simplest cases.

A curve may be defined as a continuous image of an open interval. Thus a curve C in the xy plane is given by a pair of continuous functions of one variable, Eq. (38), whereas a curve in space is given by

$$x = f(u) \quad y = g(u) \quad a < u < b \qquad (38)$$

a triple of such functions. The mapping (38) is more properly called a representation of a curve, and a curve is then defined as a suitable class of such representations. Note that a curve such as Eq. (38) is a path rather than a locus. For example, notation (39) is the path going twice around the

$$x = \cos u \quad y = \sin u \quad 0 < u < 4\pi \qquad (39)$$

unit circle, Eq. (40), in the counterclockwise direc-

tion. A curve is called smooth in case the func-

$$x^2 + y^2 = 1 \qquad (40)$$

tion f and g have continuous derivatives $f'(u)$ and $g'(u)$ which are never simultaneously zero.

The length of a smooth curve C may be defined by Eq. (41) when this integral exists (as a proper or

$$L(C) = \int_a^b \sqrt{[f'(u)]^2 + [g'(u)]^2 \, du} \qquad (41)$$

improper integral). When the integral does not exist, the length of C is said to be infinite. This definition of length in terms of the integral may be shown to coincide with the geometric definition in a much larger class than the class of smooth curves. When C has finite length, the position vector $[f(u), g(u)]$ approaches definite limits when u tends to a or to b, which are the ends of the curve C. A smooth curve has at every point a nonzero tangent vector $\mathbf{T}(u)$, with components $f'(u)$ and $g'(u)$, so that the integrand of $L(C)$ is the length $|\mathbf{T}(u)|$ of this vector. Thus one may write $ds = |\mathbf{T}(u)| \, du$, where s is the arc length measured from some convenient point on C.

Let $A(x,y)$ be a bounded continuous function defined on a set containing a plane curve C having finite length. Then A determines three functions of C called line integrals, whose symbols and definitions, Eqs. (42), follow. If $-C$ denotes the

$$\int_C A \, dx = \int_a^b A[f(u), g(u)] f'(u) \, du$$

$$\int_C A \, dy = \int_a^b A[f(u), g(u)] g'(u) \, du \qquad (42)$$

$$\int_C A \, ds = \int_a^b A[f(u), g(u)] |T(u)| \, du$$

curve C traversed in the opposite direction, as shown in Eqs. (43), a reversal of the orientation of

$$\int_{-C} A \, dx = -\int_C A \, dx$$

$$\int_{-C} A \, dy = -\int_C A \, dy \qquad (43)$$

$$\int_{-C} A \, ds = \int_C A \, ds$$

C changes the signs of the first two but not of the third of these functions.

The extension of the preceding definitions to curves in three-space is made in an obvious way. An important application of line integrals is to express the work done by a force field on a moving particle. Thus, if a force field $\mathbf{F}(x,y,z)$ has components $F_1(x,y,z)$, $F_2(x,y,z)$, and $F_3(x,y,z)$ in the directions of the x, y, and z axes, then the work done by the field on a particle moving on a curve C is given by formula (44). When $\mathbf{T}$ is the tangent vector of C, expressed in terms of a parameter u, the work W can be expressed in terms of the dot product by Eq. (45). If $\mathbf{T}_1$ denotes the tangent vector of unit

$$W = \int_C F_1 \, dx + F_2 \, dy + F_3 \, dz \qquad (44)$$

$$W = \int_a^b \mathbf{F} \cdot \mathbf{T} \, du \qquad (45)$$

length, notation (46) then can be written.

$$W = \int_C \mathbf{F} \cdot \mathbf{T}_1 \, ds \qquad (46)$$

To pass from curves to surfaces replace the open interval $a < u < b$ by a bounded connected open set D in a uv plane, which may be called the parameter plane. The domain D can be restricted to be of sufficiently simple shape so that every function which is continuous and bounded in D is integrable over D, as a multiple integral. For example, D may be the interior of a circle, or of a rectangle. Then a surface S in three-space is defined by a triple of functions continuous on D, Eqs. (47). Such

$$x = f(u,v)$$
$$y = g(u,v) \qquad (47)$$
$$z = h(u,v)$$

a surface is called smooth in case the functions f, g, and h have continuous first partial derivatives in D, and the three Jacobians listed in formulas (48),

$$J_1 = \frac{\partial(y,z)}{\partial(u,v)} \qquad J_2 = \frac{\partial(z,x)}{\partial(u,v)} \qquad J_3 = \frac{\partial(x,y)}{\partial(u,v)} \qquad (48)$$

are never simultaneously zero. A smooth surface has at every point a nonzero normal vector $\mathbf{J}(u,v)$ with components J_1, J_2, J_3, and length $|\mathbf{J}|$.

The area of a smooth surface S may be defined by the integral (49), and it is always finite when the vector $\mathbf{J}$ has bounded length.

$$\sigma(S) = \iint_D |\mathbf{J}(u,v)| \qquad (49)$$

If A (x,y,z) is a bounded continuous function defined on a set containing a surface S with finite area, four surface integrals can be defined as in notation (50). If $\mathbf{F}$ (x,y,z) is a vector field with com-

$$\iint_S A \, dy \, dz = \iint_D A \, [\, f(u,v), g(u,v), h(u,v) \,] J_1(u,v)$$

$$\iint_S A \, dz \, dx = \iint_D A \, [\, f(u,v), g(u,v), h(u,v) \,] J_2(u,v)$$

$$\iint_S A \, dx \, dy = \iint_D A \, [\, f(u,v), g(u,v), h(u,v) \,] J_3(u,v)$$

$$\iint_S A \, d\sigma = \iint_D A \, [\, f(u,v), g(u,v), h(u,v) \,] |\mathbf{J}(u,v)|$$
$$(50)$$

ponents F_1, F_2, F_3, and $\mathbf{n}$ denotes the unit vector normal to S, with components $J_1/|\mathbf{J}|, J_2/|\mathbf{J}|, J_3/|\mathbf{J}|$, then a combination, formula (51), of the first three

$$\iint_S F_1 \, dy \, dz + F_2 \, dz \, dx + F_3 \, dx \, dy \qquad (51)$$

kinds of surface integral may also be written in the form of Eq. (52), where $\mathbf{F} \cdot \mathbf{n}$ is the dot product. Ex-

$$\iint_S \mathbf{F} \cdot \mathbf{n} \, d\sigma \qquad (52)$$

pression (51) or (52) is referred to as the integral of the vector $\mathbf{F}$ over the side of the surface S to which the vector $\mathbf{n}$ points. The integral over the oposite side of S is denoted by formula (53), whose value

$$\iint_S F_1 \, dz \, dy + F_2 \, dx \, dz + F_3 \, dy \, dx \qquad (53)$$

is the negative of that in (51), by Eq. (54). The last

$$\frac{\partial(z,y)}{\partial(u,v)} = -J_1, \text{ and so on} \qquad (54)$$

integral in Eq. (50), however, is independent of a choice of side of the surface.

The important formulas of Stokes and of Gauss, Eqs. (55), are expressed in terms of line and sur-

$$\int_C \mathbf{F} \cdot \mathbf{T} \, ds = \iint_S \text{curl } \mathbf{F} \cdot \mathbf{n} \, d\sigma \quad \text{(Stokes)}$$
$$(55)$$
$$\iint_S \mathbf{F} \cdot \mathbf{n} \, d\sigma = \iiint_V \text{div } \mathbf{F} \quad \text{(Gauss)}$$

face integrals and the triple or volume integral analogous to the double integral. In the formula of Stokes, $\mathbf{F}$ is a vector field, S is a smooth surface having a piecewise smooth boundary curve C, directed so that S lies to the left of an observer proceeding along C on the side of S on which the unit normal vector $\mathbf{n}$ is chosen, and $\mathbf{T}$ is the unit tangent vector to C. In terms of the components of F, Stokes' formula may be written as Eq. (56), where

$$\int_C F_1 \, dx + F_2 \, dy + F_3 \, dz = \iint_S (F_{3y} - F_{2z}) \, dy \, dz$$
$$+ (F_{1z} - F_{3x}) \, dz \, dx + (F_{2x} - F_{1y}) \, dx \, dy \quad (56)$$

$F_{3y} = \partial F_3 / \partial y$, and so on. When the surface S lies in the xy plane, this equation reduces to Green's formula, Eq. (57). The latter is readily proved for re-

$$\int_C F_1 \, dx + F_2 \, dy = \iint_S (F_{2x} - F_{1y}) \, dx \, dy \quad (57)$$

gions S of simple shape, and from it Stokes' formula may be deduced.

In the formula of Gauss (also called the divergence theorem), V is a space domain bounded by a smooth or piecewise smooth surface S, and $\mathbf{n}$ is the exterior normal of S. In terms of the components of F, Gauss' formula may be written as Eq. (58). *See* CALCULUS OF VECTORS.

$$\iint_S F_1 \, dy \, dz + F_2 \, dz \, dx + F_3 \, dx \, dy$$

$$= \iiint_V (F_{1x} + F_{2y} + F_{3z}) \qquad (58)$$

Functions defined by integrals. Mention has already been made of elliptic functions, which may be defined as the inverse functions of elliptic integrals. Certain definite integrals may also be used to define important nonelementary functions which occur frequently in applications. For example, the gamma function may be defined by Eq. (59) for $x > 0$. The Bessel functions of integral or-

$$\Gamma(x) = \int_0^\infty e^{-t} t^{x-1} \, dt \qquad (59)$$

der may be defined by Eq. (60) for all values of x.

$$J_n(x) = \frac{1}{\pi} \int_0^\pi \cos{(nt - x \sin t)} \, dt \qquad (60)$$

The properties of these functions may be derived from these expressions with the help of various methods of advanced analysis, such as the theory of functions of a complex variable. *See* BESSEL FUNCTIONS; GAMMA FUNCTION.

In general, if $f(x,t)$ is an integrable function of t on $a \leqq t \leqq b$ for each x, then Eq. (61) is a well-

$$g(x) = \int_a^b f(x,t) \, dt \qquad (61)$$

defined function of x. When suitable conditions are satisfied by the function f, the derivative $g'(x)$ may

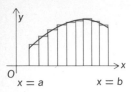

Fig. 2. Division of interval [a,b] into parts.

be calculated by formula (62). This process is called

$$g'(x) = \int_a^b \frac{\partial}{\partial x} f(x,t) \, dt \qquad (62)$$

differentiation under the integral sign. It is valid in case f and $\partial f/\partial x$ are continuous in (x,t) for $c \leqq x \leqq d$, $a \leqq t \leqq b$, where a and b are finite, and even in more general cases. However, in the simple example given in Eqs. (63), formula (62) leads to a wrong result.

$$f(x,t) = x^3 e^{-tx^2}$$
$$g(x) = \int_0^\infty f(x,t) \, dt \qquad (63)$$

Approximate and mechanical integration. The definition of the definite integral itself gives a means of calculating its value approximately. There remains the question of a suitable selection of the functional values $f(x_i)$ for use in formula (1). It is usual to divide the interval $[a,b]$ into a number n of equal parts of length h (Fig. 2). If the points x_i are taken at the midpoints of the subintervals, one obtains the midpoint formula, Eq. (64). For a curve

$$S = h \sum_{i=1}^n f[a + (i-1/2)h] \qquad (64)$$

that is concave downward, this gives too large a value.

Another formula, called the trapezoidal rule, is derived by calculating the area below a polygon inscribed in the graph of $f(x)$ (Fig. 3). This formula is Eq. (65). It gives too small a value for a curve that is concave downward.

$$T = \frac{h}{2}\left[f(a) + f(b) + 2\sum_{i=1}^{n-1} f(a+ih)\right] \quad (65)$$

The error in the approximation may sometimes be reduced without increasing the number of subintervals by use of the parabolic rule (Simpson's rule). To obtain the formula, the subarcs of the graph of $f(x)$ are replaced by arcs of parabolas rather than by line segments. Since three points determine a parabola (with vertical axis), the interval $[a,b]$ is divided into an even number n of subintervals. The area under the parabola passing through the points on the graph of $f(x)$ having abscissas given in formulas (66) is given by formula (67). Therefore, the parabolic rule gives approximation (68).

$$a + (2i-2)h \quad a + (2i-1)h \quad a + 2ih \quad (66)$$

$$\frac{h}{3}\left\{f[a+(2i-2)h]\right.$$
$$\left. + 4f[a+(2i-1)h] + f(a+2ih)\right\} \quad (67)$$

$$P = \frac{h}{3}\left\{f(a) + f(b) + 2\sum_{i=1}^{\frac{n}{2}-1} f(a+2ih)\right.$$
$$\left. + 4\sum_{i=1}^{n/2} f[a+(2i-1)h]\right\} \quad (68)$$

When a function $f(x)$ is given only by a table, its integral is most simply computed by one of these formulas or a similar formula.

When a function $f(x)$ is given by a graph, mechanical means may be used to calculate associated areas. One such means is the polar planimeter, which registers on a rotating wheel the area enclosed by a closed curve around which a tracing

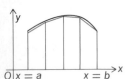

Fig. 3. Polygon inscribed in graph of f(x).

point is passed. The integraph, invented by Abdank Abakanowicz, is designed to draw the graph of an indefinite integral of $f(x)$ when a tracing point is passed over the graph of $f(x)$. These simple devices were the forerunners of more complex machines designed to solve differential equations, such as the differential analyzer of Vannevar Bush. For large-scale computations involving formulas such as (64), (65), or (68), a digital computer may be preferred.

Other methods of integration. The method of differentiation under the integral sign is often convenient for the evaluation of definite integrals, even when other methods are available. For example, Eq. (69) leads to Eq. (70). Or, if one has Eq. (71), Eq. (72) follows.

$$g(x) = \int_0^a e^{xt} \, dt = \frac{e^{xa}-1}{x} \qquad (69)$$

$$g'(x) = \int_0^a t e^{xt} \, dt = \frac{a e^{xa}}{x} - \frac{e^{xa}-1}{x^2} \qquad (70)$$

$$g(x) = \int_0^a \frac{dt}{t^2 + x^2} = \frac{1}{x} \arctan(a/x) \qquad (71)$$

$$g'(x) = -2x \int_0^a \frac{dt}{(t^2+x^2)^2}$$
$$= -\frac{1}{x^2} \arctan(a/x) - \frac{a}{x(a^2+x^2)} \qquad (72)$$

Another example in which the conditions for validity of the method still hold, though they are more difficult to verify, is given by Eq. (73), by means of which Eq. (74) can be derived, by integration by

$$g(x) = \int_0^\infty e^{-xt} \frac{\sin t}{t} \, dt \qquad 0 \leqq x < \infty \quad (73)$$

$$g'(x) = -\int_0^\infty e^{-xt} \sin t \, dt = \frac{-1}{1+x^2} \qquad (74)$$

parts twice. Hence $g(x) = -\arctan x + C$, and $C = \pi/2$, since Eq. (75) can be proved.

$$\lim_{x \to \infty} g(x) = 0 \qquad (75)$$

In particular, Eq. (76) holds.

$$g(0) = \int_0^\infty \frac{\sin t}{t} \, dt = \pi/2 \qquad (76)$$

In case an indefinite integral is not elementary but the integrand is representable by an infinite series, the method of integrating term by term gives a representation for the integral which may be used for purposes of approximation. This can be seen in Eq. (77).

$$\text{Si}(x) = \int_0^x \frac{\sin t \, dt}{t} = \sum_0^\infty \frac{(-1)^n x^{2n+1}}{(2n+1)^2 (2n)!} \quad (77)$$

Certain definite integrals may be readily evaluated by means of contour integrals, that is, integrals of analytic functions taken along curves in the complex plane.

Integral of Lebesgue. In the study of the "space" of real functions defined, for example, on the interval $a \leqq x \leqq b$, it is frequently useful to take formula (78) as the distance between the func-

$$\int_a^b |f(x) - g(x)| \, dx \qquad (78)$$

tions f and g. This distance already has a meaning when f and g are Riemann-integrable, that is, bounded and not too discontinuous, in the sense specified for the Riemann integral. There is no generally useful extension of the concept of integral to apply to all real functions on $[a,b]$, but it is desirable to extend it to apply to the functions obtained from the continuous ones by certain limiting processes. In particular, it is desirable to have correspond to each sequence (f_n) of functions satisfying the Cauchy condition for convergence in terms of the distance (78), namely Eq. (79), a func-

$$\lim_{\substack{m=\infty\\n=\infty}} \int_a^b |f_m(x)-f_n(x)|\,dx = 0 \qquad (79)$$

tion g which is integrable (in the extended sense), and for which Eq. (80) holds. An extended

$$\lim_{n=\infty} \int_a^b |f_n(x)-g(x)|\,dx = 0 \qquad (80)$$

definition of integral having this property was given by H. L. Lebesgue in his thesis. It made obsolete many of the complicated extensions of the Riemann integral which had been previously proposed. Following Lebesgue, various mathematicians have proposed other ways of defining the integral which are equivalent to that of Lebesgue.

F. Riesz devised a definition which can be stated quite simply, at least for the case of a bounded function $g(x)$. As a first definition, a point set S in the interval $[a,b]$ has measure zero in case it can be enclosed in a sequence (finite or infinite) of intervals, the sum of whose lengths is arbitrarily small. Also, a step function $f(x)$ is defined as one which is constant on each interval of a partition of $[a,b]$, as in Fig. 4. The Riemann integral of a step function is expressible as a finite sum. Then a bounded function $g(x)$ is integrable in Lebesgue's sense in case it is the limit of a uniformly bounded sequence of step functions $f_n(x)$, at each point of $[a,b]$ except those in a set S with measure zero, and by definition Eq. (81) can be written. In case the step

$$\int_a^b g(x)\,dx = \lim_n \int_a^b f_n(x)\,dx \qquad (81)$$

functions f_n in Eq. (81) are replaced by Lebesgue-integrable functions forming a uniformly bounded sequence, no new functions g are obtained.

The integral of Lebesgue is also defined for unbounded functions, but the points of infinite discontinuity do not need to be considered one by one. For each function $g(x)$ and each positive integer N let $g_N(x)$ denote the lesser of $g(x)$ and N. If each $g_N(x)$ is Lebesgue-integrable in the sense already defined, expression (82) forms a nondecreas-

ing sequence, and so if expression (82) is bounded,

$$\int_a^b g_N(x)\,dx \qquad (82)$$

it tends to a finite limit, which is taken as the value of integral (83). An arbitrary function $g(x)$ is the

$$\int_a^b g(x)\,dx \qquad (83)$$

difference of two nonnegative functions, Eqs. (84)

$$g^+(x) = \frac{|g(x)|+g(x)}{2} \qquad (84)$$

and (85), and one may write Eq. (86) whenever both

$$g^-(x) = \frac{|g(x)|-g(x)}{2} \qquad (85)$$

$$\int_a^b g(x)\,dx = \int_a^b g^+(x)\,dx - \int_a^b g^-(x)\,dx \qquad (86)$$

terms on the right have a meaning according to the definition just given. It is provable that the class of all functions for which integrals exist according to the definitions just given, does indeed have the property of completeness with respect to the Cauchy condition (79). The theory of Lebesgue yields many other useful results, including an extension of the fundamental theorem of integral calculus. One important restriction on the class of Lebesgue-integrable functions which is implicit in the definition is that when $g(x)$ is integrable so is $|g(x)|$. This restriction does not hold for the improper integrals which were described earlier.

The various methods of defining Lebesgue integrals are extensible also to functions of several variables. This extension throws light on the properties of multiple Riemann integrals and on the reduction of multiple integrals to repeated integrals.

Other definitions of integration. If $\alpha(x)$ is a fixed function defined on the interval $[a,b]$, the sum S defined by Eq. (1) may be replaced by that in Eq. (87). Then if the limit I of S exists in this case,

$$S = \sum_{i=1}^{n} f(x_i)[\alpha(t_i)-\alpha(t_{i-1})] \qquad (87)$$

it is called the Stieltjes integral of f with respect to α, and is denoted by integral (88). It has many of

$$\int_a^b f(x)\,d\alpha(x) \qquad (88)$$

the properties of the Riemann integral, especially in case the function α is nondecreasing. In addition, it may take special account of the values of f at a finite or countable set of points. For example, if $\alpha(x)=c_i$ on $u_{i-1}<x<u_i$, where $u_0<a$, $u_n>b$, $a<u_i<b$ for $i=1,\ldots,n-1$, and if $f(x)$ is continuous at each u_i, then Eq. (89) may be written. With some restrictions to ensure convergence,

$$\int_a^b f(x)\,d\alpha(x) = \sum_{i=1}^{n-1} f(u_i)[c_{i+1}-c_i] \qquad (89)$$

this may be extended to the case in which α has infinitely many discontinuities.

In the case in which α is nondecreasing, the Stieltjes integral has an extension which is similar to that of the Lebesgue integral and is called the Lebesgue-Stieltjes integral.

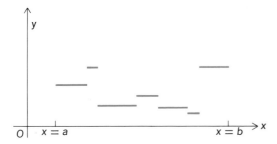

Fig. 4.   Step function $f(x)$.

Other extensions of the Lebesgue integral are the integrals of Denjoy. They apply to certain functions f for which $|f|$ is not integrable, and include the improper integrals, but not the Cauchy principal value. Numerous other types of integrals have been defined. In particular the integral of Lebesgue has been extended to cases in which the independent variable lies in a suitable space of infinitely many dimensions, or in which the functional values of f lie in such a space. *See* CALCULUS, DIFFERENTIAL AND INTEGRAL; FOURIER SERIES AND INTEGRALS; SERIES.

[LAWRENCE M. GRAVES]

Bibliography: E. J. B. Goursat, *A Course in Mathematical Analysis*, 1917; L. M. Graves, *Theory of Functions of Real Variables*, 1956; P. R. Halmos, *Measure Theory*, 1950; T. H. Hildebrandt, *Introduction to the Theory of Integration*, 1963; P. O. Peirce, *A Short Table of Integrals*, 3d ed., 1929; J. H. Williamson, *Lebesgue Integration*, 1962.

Integument

The skin, or outer covering, of the body. It is extremely diverse in its structure, varying both within species and from one species to another. It always consists of an outer epidermis and an innermost dermis (or corium), which rests on fatty connective tissue surrounding bone and muscle. The entire skin has a primary barrier function in protecting the soft internal tissues from physical trauma and bacteriological invasion, and in maintaining the internal salt and fluid levels. There are a variety of secondary functions which vary from species to species, namely, thermal insulation, sensory reception, scent secretion, respiration, attack and defense (claws, thick skins), and protective coloration. Comparative studies of unspecialized and specialized skin by many different techniques have made this organ system a popular subject for the study of patterns of cell production and differentiation, and have provided new insights into the question of how adult tissues are maintained in day-to-day existence, as well as how they respond to injury.

Gross morphology. The knowledge that all vertebrate integuments consist of an epidermis and a dermis does not help in understanding the diversity of structure of this organ system. However, vertebrate integuments can also be described as scaled or nonscaled. Since the earliest fossil vertebrates, the ostracoderms, had dermal or bony scales, it seems likely that bony scaled integuments are primitive and that nonbony scaled integuments are secondary. A "scale" in this sense is merely a fold of the epidermal-dermal boundary of the body. To clearly define integumentary structures further, it is necessary to examine the types of proteins which characterize the epidermis and the dermis, and to appreciate the fact that these two skin components have almost unlimited capacities to form appendages either separately or by complex interactions; an appendage is defined as a population of cells whose synthetic activity is clearly distinguishable from that of adjacent cells. *See* OSTRACODERM; SCALE (ZOOLOGY).

Embryonic development. The vertebrate integument consists of cells derived from three distinct regions of the embryo. The epidermis derives from the outermost ectoderm, the dermis derives from somatic mesoderm, and the pigment cells found in either the epidermis or the dermis, or both (depending on the species), derive from the neural crest.

The initial stages of integumentary differentiation are identical in all vertebrates. The elongated, flattened cells of the ectodermal epithelium become cuboidal, and a basement membrane appears beneath what can now be termed a stratum germinativum or basal layer. While the first epidermal cell divisions are giving rise to an outer covering—the periderm—the erstwhile stellate mesenchymal cells of the somatic mesoderm begin to aggregate beneath the basal layer; this is the first step in dermal differentiation. Subsequent steps in integumentary differentiation depend on the species.

In a scaled integument, the epidermal-dermal boundary of the embryo becomes thrown into a series of symmetrical folds which eventually become asymmetrical, and then the outer and inner scale surfaces become apparent; this general pattern of events is seen in most fish and reptiles, and on the legs of birds. In nonscaled integuments, the body's boundary remains relatively smooth at first, but subsequently appendages such as hairs, feathers, or glands differentiate as localized epidermal proliferations, with associated mesenchymal condensations. Such appendages may, of course, be found on scaled integuments, where they differentiate in an essentially similar fashion, although the mesenchymal condensations may be absent, or less apparent. Subsequent patterns of differentiation are mainly concerned with specific protein synthesis in the epidermis, and although dermal cells do participate in the formation of certain appendages, collagen fibers, blood vessels, and elastic tissue are universal characteristics of dermis. In general, the integuments of vertebrate embryos at birth or hatching show a nearly complete adult structure. *See* FEATHER; GLAND; HAIR (MAMMAL).

The neural crest contributes to the integument by serving as the source of the pigment cells, which are almost totally responsible for the coloration of the skin. This discussion will be confined to those cells which eventually synthesize the protein melanin. A prospective melanin-producing cell is called a melanoblast; and although such cells cannot be identified directly either while they are in the neural crest or during their migration to their ultimate integumentary locations, their presence can be demonstrated by appropriate transplantation or tissue-culture experiments. When their migrations are complete, and the cells attain their typical adult locations in either the epidermis or the dermis, they begin their patterns of specific protein synthesis.

Epidermis. During human embryonic development, the nonneural ectoderm gives rise to the epidermis of the skin, and to the buccal epithelium, as well as to a few other tissues. The fact that the human epidermis is dry and firmer than the moist buccal epithelium spotlights the bipotentiality of ectodermal cells with respect to their possible pathways of protein synthesis. This bipotentiality, which has long been known from experimental studies, is the key to understanding comparative epidermal morphology in different vertebrate groups. Other important data derive from ultrastructural analysis of epidermal cell anatomy with the electron microscope which allows resolution of units as close together as 8 A (0.8 nm) or 0.0001 μm versus a limit of 0.25 μm achievable with the best light microscope. In the study of skin, fine structural analysis can be correlated with analyses of protein structure by x-ray diffraction. *See* MICROSCOPE, ELECTRON; X-RAY DIFFRACTION.

Light-microscopic study of human epidermis (Fig. 1) shows that it is a stratified squamous epithelium in which basal cells divide to give rise to daughter cells (keratinocytes) which move outward, away from the basal layer. The keratinocytes become increasingly flattened, their nuclei less readily observed, and finally obscured, as the cells come to form part of the stratum corneum. Eventually, they are shed from the body surface (dandruff). During hair growth, mature cells adhere more closely together to form a shaft instead of a multilayered stratum corneum. This type of cell differentiation is called keratinization, referring to the process whereby epidermal cells become filled with keratinaceous proteins during their maturation. Ultrastructurally, living epidermal cells can be seen to contain all the organelles characteristic of actively synthesizing cells and, in addition, bundles of filaments approximately 60–70 A (6–7 nm) in diameter. During the maturation process in the human epidermis, the cells become increasingly filled with such bundles, and the filaments may increase in size up to 120 A (12 nm). In the mammalian epidermis, the filaments are for a time surrounded by keratohyalin granules as the cells form part of the stratum granulosum. The relationship of keratohyalin granules (and similar electron-opaque inclusions seen in other vertebrates) to the keratinization process is not understood, but the bundles of filaments are assumed to represent the fibrous elements of keratinaceous proteins. In general, similar cellular changes are seen during the maturation of "soft" epidermis, soles of feet or palms of hands, hairs (of all types), nails, and even claws. However, in spite of the diversity of such epidermal derivatives in humans, and in all mammals, x-ray diffraction analysis indicates that the same basic keratinaceous protein—the so-called alpha keratin—exists throughout.

X-ray diffraction analysis of avian feathers, and of certain parts of the epidermis of scales of reptiles and birds, reveals the presence of a second type of keratinaceous protein, typified by feathers, which for convenience will be referred to as beta keratin. Ultrastructural examination of epidermal cells which form this type of material shows that initially 70-A (7 nm) filaments of the type described above are seen; but, as differentiation proceeds, a second category of filaments, approximately 30 A (3 nm) in diameter, appear, and eventually fill the cells.

If it is assumed that there is a relationship between filaments in epidermal cells, as well as a capacity for keratinization, it is probable that this capacity is present in all vertebrates; the "horny teeth" of parasitic cyclostomes appear to contain alpha keratin, but in general, fish epidermal cells

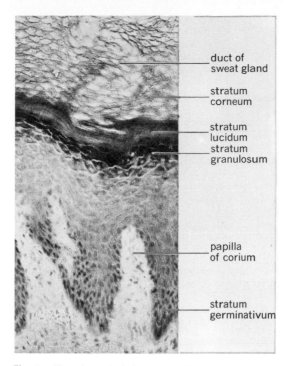

Fig. 1. The characteristic strata of thick skin of the human finger as seen in cross section at high magnification. (*From J. F. Nonidez and W. F. Windle, Textbook of Histology, McGraw-Hill, 1949*)

synthesize mucus. Modern amphibians have truly keratinized epidermal cells only under certain conditions, and it seems to be the alpha type. Only in amniotes is a truly cornified epidermis always found, probably reflecting the increased importance of this epidermis to water conservation in association with terrestrial life. The soft parts of the earliest reptiles are not available for analysis, but since all mammalian epidermal derivatives, even the baleen plates of whalebone whales, consist of alpha keratin, it must be assumed that only this fundamental synthetic capacity was retained in the cotylosaur-therapsid reptilian ancestors. In the other branch of cotylosaur evolution, which gave rise to the turtles, to all other reptiles, and ultimately to birds from the archosaurian stock, the "extra" capacity for beta keratin synthesis appeared in the epidermis. This capacity is variously expressed (Fig. 2), but no explanation of its functional significance is available. Mechanical protection cannot be the whole answer, since no one would argue that the down feathers on a bird (beta keratin) are tougher than the quills on a porcupine (alpha keratin). Tetrapod claws and nails have been inadequately studied, but they are probably of multiple evolutionary origin.

Dermis. The most conspicuous component of the dermis is collagen, an extracellular fibrous protein which has the basic molecular dimensions of 2600–3000 A (260–300 nm) in length and 15 A (1.5 nm) in width. It shows a typical "collagen period" of alternating light and dark bands which are 640 A (64 nm) in length. The molecule gives rise to complex polymers which form filaments or fibers or bundles of fibers ranging in size from units observable only with the electron microscope to those

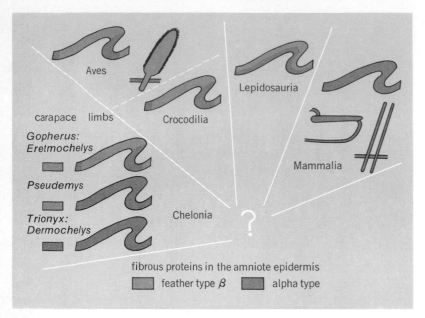

Fig. 2. The distribution of fibrous proteins in the amniote epidermis as revealed by x-ray diffraction analysis. (*From H. P. Baden and P. F. A. Maderson, Morphological and biophysical identification of fibrous protein in the amniotic epidermis, J. Exp. Zool., 174: 225–232, 1970*)

macroscopically visible units which give leather its texture. The protein is synthesized by fibroblasts scattered throughout the dermis which show organelles typical of secretory cells. The fibroblasts also synthesize elastin, which is present as extracellular filaments 120 A (12 nm) in width, and amorphous glycosaminoglycans which fill in the spaces surrounding the fibroblasts and the extracellular filamentous material. The resultant matrix not only serves as a supporting structure for blood vessels, nerves, pigment cells, and dermal ossifications but also gives the skin its "cushionlike" physical characteristics which protect the softer, deeper body tissues. *See* COLLAGEN.

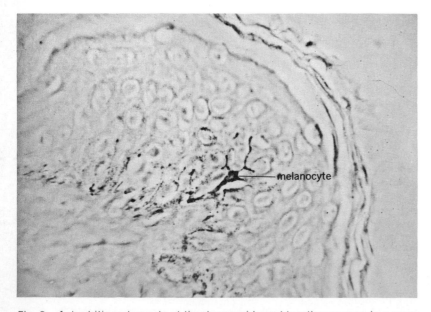

Fig. 3. A dendritic melanocyte at the dermoepidermal junction as seen in a cross section of thin human skin at high magnification (phase-contrast optics).

Melanogenic cells and epidermal coloration.

Once melanoblasts have reached their definitive adult positions, either in the dermis, or between the basal cells of the epidermis, they begin their specific processes of cytodifferentiation. In both positions, they become highly branched or dendritic (Fig. 3), and their processes extend either throughout the adjacent extracellular matrix, or between the tightly packed keratinocytes. There are fixed ratios of melanocytes-keratinocytes in different regions of the body. In the perikaryal region around the nucleus, ovoid bodies termed premelanosomes are synthesized. Eventually, tyrosinase activity can be demonstrated in these bodies, and the electron microscope reveals steadily increasing electron opacity which eventually obscures the cristae of the premelanosomes as they move away into the dendrites and melanin is deposited upon them. With the electron microscope, a fully mature melanosome ("pigment granule") is seen as a dense, electron-opaque granule with no discernible substructure. Melanin is an extremely stable protein—it has been demonstrated in the fossilized skin of ichthyosaurs 150,000,000 years old—and it does not seem to be metabolized by living cells. If melanogenic cells continued to steadily produce mature melanosomes throughout their lives, the cytoplasm would become totally filled, and eventually the cell would burst. In certain melanogenic populations, notably the pigmented retina of the eye, the pattern of organelle synthesis and differentiation which was described above occurs only for a limited period in the embryo. A similarly limited period of melanogenesis seems to characterize the dermal melanophores and helps to reveal their function in skin.

As melanosomes mature, they move away from the perikaryal region, toward the dendrites. In melanophores in the dermis, under hormonal or neural influence, or both, the melanosomes migrate back and forth between the distal dendrites and the main cell body. The differential location and density of pigment granules within the dermal cells change the light-reflecting capabilities of the dermis and, owing to an interaction between the reflected light and the other dermal pigment cells, are responsible for rapid color change in lower vertebrates.

Thus, dermal melanophores, like pigmented retinal cells, seem to synthesize only one batch of melanosomes in the embryo, and to retain them throughout life; thus the cells are described as "continent." Other cells, also called melanophores, but lacking conspicuous dendrites, may also be present in the dermis. They are probably macrophages which engulf melanosomes resulting from the death of cells occasionally involved in the color-change mechanism, but no other functional role can be ascribed to them. *See* CHROMATOPHORE.

The melanogenic cells in the basal region of the epidermis continue to synthesize new melanosomes throughout life. They are described as "incontinent" by virtue of a unique property known as cytocrine transfer. The dendrites of epidermal melanocytes appear to have a fixed distribution, running between the keratinocytes and forming an "epidermal-melanin unit" (Fig. 4). Neither the cell body nor the dendrites show the desmosomal junctions with adjacent cells that are characteristic of contact sites between keratinocytes, a fact which reflects the dynamic status of the relationship between melanocytes and keratinocytes. As new keratinocytes are produced in the basal layer, and move upward toward the body surface, melanocyte dendrites come into close apposition with their plasma membranes. Because of either phagocytic ingestion or perhaps direct injection through discontinuities in the membranes, greater or lesser numbers of melanosomes are transferred from those cells which synthesized them into another, distinctly different, type of cell. As a result, each keratinocyte acquires a complement of pigment granules. Depending on the spatial distribution throughout the cytoplasm, or the density of granules per unit of volume, or both, cornified cells, even in hairs or feathers or scales or claws, are endowed with light-reflecting properties which are perceived as permanent coloration. However, it must not be assumed that the production of color is the primary function of this unique phenomenon of cytocrine transfer of melanosomes.

Comparative cytological studies seeking the basis for different "skin colors" in *Homo sapiens* have revealed that, for any given region of the body, the same number of epidermal melanocytes are present per unit of area in all "races." Thus, skin color, in common with nearly every other so-called racial characteristic, has a quantitative rather than a qualitative basis, in this case, rate of melanosome synthesis and transfer per unit of time.

Different regions of the human body show differential melanocyte distribution which can be correlated with the probability of their being exposed to incident radiation from the sun at "high noon." These anatomical data suggest that epidermal pigmentation relates to protection from injurious radiation, and this can be substantiated by experience. A light-skinned human being who sits for several hours in the sun in very early summer will almost certainly suffer from sunburn, a pathological condition of the skin tissues reflected in destruction or damage of all constituent cell types. However, gradual exposure to sunlight (or an ultraviolet source) will produce a tan which, once it is acquired, will provide considerable, if not total, protection against future prolonged radiation exposure. However, this protection rarely lasts through the winter season, since the epidermal cells are lost, and the following year a new melanized corneous barrier must be acquired. In mammals with a full hairy pelage, or in birds, the melanization is restricted to hairs and feathers for the most part, whereas the interfollicular epidermis is extremely light, and may be experimentally "burned." Newly captive cetaceans are prone to sunburn since they are disturbed by the restriction of their new environment and may break the water surface with their dorsal body. Although epidermal melanization is seen in reptiles and amphibians, in which the especially prominent "supranuclear capping" emphasizes the need to protect the cells' genetic apparatus, experiments have shown that deeper body tissues are also protected by melanin in dermal and peritoneal tissues.

Other dendritic cells, lacking desmosomal membrane modifications and filaments, may be

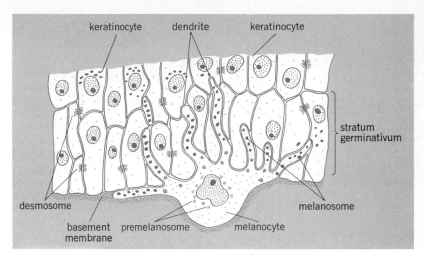

Fig. 4. The concept of the "epidermal-melanin unit" emphasizes the dynamic interaction of melanocytes and keratinocytes. Melanosomes are transferred to the keratinocytes (cytocrine transfer), and may thereafter reside within their cytoplasm in various patterns. Supranuclear capping is represented here by X.

found in the basal layer or in the upper epidermal levels. These Langerhans cells, which have a lobulated nucleus and a characteristic "tennis racket-shaped" organelle, are not related to melanocytes, but their function is unknown.

Regeneration and wound healing. The vertebrate integument demonstrates a variety of restorative properties which have made it an important model system for studies of growth and development in the adult body, and thus it contributes to the knowledge of neoplastic disease.

Physiological regeneration. All tissues in the vertebrate body, except for muscle and neural tissues, are steadily replaced throughout the life of the animal. In the epidermis, continuous basal proliferation, and subsequent keratinization of daughter cells, exactly compensates for the loss of mature cells at the body surface. The epidermis of the sole of a rat's foot, for instance, is completely replaced every 3 weeks. The regional differences in epidermal thickness across the body surface are precisely maintained, thus providing an elegant example of "tissue homeostasis." Epidermal replacement is often considered to be due to surface wear and tear, but regions experimentally protected from physical abrasion continue to show specific rates of proliferation and differentiation with surface accumulation.

The exact functional significance of physiological regeneration in the epidermis, or even in other organ systems, is not really understood. Some vertebrates, such as amphibians, and snakes and lizards, "shed their skins" in a dramatic fashion. In fact, however, only the mature outer epidermal layers are lost, exposing new materials beneath. In birds and mammals, periodic regrowth of feathers and hairs following shedding (molting) may have sexual or ecological significance, but this is certainly a specialized secondary function of an inherently epidermal process.

The periodic shedding of the epidermis and its derivatives is controlled by a variety of factors which are incompletely understood, although various hormones are certainly implicated. Disturbances of epidermal homeostasis may be symptomatic of many somatic disorders of the human body, or the result of viral infection, as, for example, warts. Hyperplastic epidermal disorders such as psoriasis (a common disease, apparently unique to *H. sapiens*, which may be mild or completely debilitating) are of unknown origin. Extreme disturbances of tissue homeostasis in the epidermis, as in other organ systems, may result not only in tissue hyperplasia but in invasion of other tissues, conditions which are recognized as malignant transformations, or cancer. *See* REGENERATION (BIOLOGY).

Cutaneous wound healing. By virtue of its location at the body surface, the integument is constantly exposed to physical insult. Such insults range from minor abrasions to deep wounds which may penetrate to the subcutaneous tissues. The way in which the integument responds to this range of potential trauma varies, and reflects the diverse functions of the system.

Minor abrasions or small cuts leave enough of the system intact for localized increased rates of epidermal proliferation and migration to effect perfect restoration of structure and function in a relatively short space of time, usually about 3 weeks if no complications arise from secondary infections. Such increased proliferative rates, and migration of epidermal cells, represent controlled departures from the normal homeostatic regime discussed above, and thus emphasize the unique harmony of the adult body.

More extensive insult damages the dermis, and typically results in a scar. Although the fundamental epidermal covering of a scarred area resembles that of nontraumatized tissue, and therefore restores the primary barrier function, secondary functions which reside in morphological specializations may not be restored. This is because, beginning in the embryo, there is a complex sequence of dynamic interactions between epidermis and dermis, with the latter having considerable control over patterns of differentiation and morphogenesis in the former. In mammals or birds, hair or feather maintenance is effected by "dermal papillae," and if these are destroyed, the epidermal appendage does not regrow. A similar interaction between constituent tissues is apparently responsible for fingerprints, and although fingerprints may be restored after minor injury, severe burns or tissue ablation will permanently destroy these unique patterns. *See* FINGERPRINT; INTEGUMENTARY PATTERNS.

It is not known why severe damage to the dermis destroys or impairs its function in maintaining epidermal specializations, but scar dermis does have some structural features which distinguish it from the normal tissue. It is thicker and more fibrous, its fibers are not oriented in a typical pattern, and it lacks resilience and mobility. Scarred areas may contract during the restorative process, bringing the margins of the original wound close together. Scarring may be alleviated by skin grafting since the proliferative capacities of untraumatized tissue make it an easy tissue to graft. However, the graft "takes" permanently only if the skin donor is the patient or the identical twin of the patient; otherwise, it will be destroyed ("rejected") by the body's immune system. Temporary xenoplastic skin grafts from cadavers or other mammals have been employed to facilitate initial posttrauma therapy, for even though such grafts are eventually rejected, they function adequately as primary barriers until they are destroyed.

Sometimes hair follicles will be found in scars, but it seems likely that they result from regeneration of portions of the original units, especially the dermal papillae. However, in some mammalian examples, notably the skin of the rabbit ear, and especially annual antler regrowth in deer (covered by "velvet"), full-thickness skin, with hair follicles, may regenerate.   [P. F. A. MADERSON]

Bibliography: W. M. Montagna et al. (eds.), *Advances in Biology of Skin*, 15 vols., 1960–1975; W. Montagna and P. F. Parrakkal, *The Structure and Function of Skin*, 3d ed., 1975; P. F. Parrakkal and N. J. Alexander, *Keratinization: A Survey of Vertebrate Epithelia*, 1972; The Vertebrate Integument (Symposium), *Amer. Zool.*, 12:13–171, 1972; A. S. Zelickson, *Ultrastructure of Normal and Abnormal Skin*, 1967.

Integumentary patterns

These comprise all the features of the skin and its appendages that are arranged in designs, both in man and other animals. Examples are scales, hairs, and feathers; coloration; and epidermal ridges of the fingers, palms, and feet. In its common usage, the term applies to the configurations of epidermal ridges, collectively named dermatoglyphics. Dermatoglyphics are characteristic of primates.

The superficial ridges are associated with a specific inner organization of skin. Skin is composed of two chief layers, the epidermis on the outside and the dermis underlying it (Fig. 1). These two layers are mortised by pegs of dermis, a double row of pegs corresponding to each ridge; these pegs accordingly form a patterning like that of the ridges.

Ridge patterns. The patterning of ridges, including that of the epidermal-dermal mortising, is determined during the third and fourth fetal months. All characteristics of single ridges and of their alignments are then determined definitively. Ridge

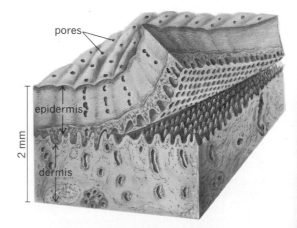

Fig. 1. Structure of ridged skin showing the chief two layers. (*From H. Cummins and C. Midlo, Finger Prints, Palms and Soles, McGraw-Hill, 1943*)

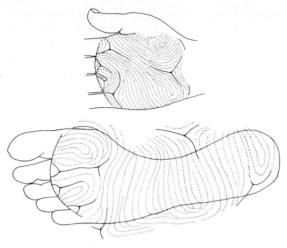

Fig. 2. Dermatoglyphics of palm and sole. (*From H. Cummins and C. Midlo, Finger Prints, Palms and Soles, McGraw-Hill, 1943*)

alignments reflect directions of stress in growth of the hand and foot at the critical period of ridge differentiation. An important element in the production of localized patterns, for example, on the terminal segment of each digit, is the development in the fetus of a series of elevations, the volar pads.

Volar pads. The pads are homologs of the prominent pads on the paws of some mammals, but in primates they attain little elevation and soon tend to subside. The volar pads are disposed in a consistent topographic plan. Localized patterns have the same placement because growth of the pad is the determiner of the specific local pattern. When a pad has subsided before ridges are formed, its area does not present a localized pattern, and the ridges follow essentially straight, parallel courses. Variations in contours of the pads are accompanied by wide variations in the designs formed by the ridges overlying them.

Pattern variability. Variability of patterning is a major feature of dermatoglyphics and the basis for various applications (Fig. 2). In personal identification, prints (Fig. 3), customarily of fingers, are classified for filing in accordance with variables of pattern type and counts of ridges. Systematic filing makes it possible to locate readily and compare the sets of prints corresponding in classification to a set for which an identification is sought. In anthropological and medical investigations, groups of individuals are compared statistically in reference to the occurrence of these varia-

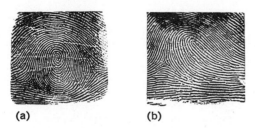

(a)   (b)

Fig. 3. Fingerprints. (a) Whorl pattern. (b) Loop pattern. (*From H. Cummins and C. Midlo, Finger Prints, Palms and Soles, McGraw-Hill, 1943*)

bles. Deductions may be drawn in accordance with likeness or unlikeness in the directions of variation. A few examples are cited. Trends of inheritance have been demonstrated in family groups and in comparisons of the two types of twins, fraternal (two-egg) and identical (one-egg). Dermatoglyphics thus are useful in diagnosing the types of twins and in analyzing cases of questioned paternity. Among different racial groups, the similar or discrepant trends of variation have been used to analyze racial affinities. Trends of variation are unlike in right and left hands, and the fact that they differ in accordance with functional handedness indicates an inborn predisposition of handedness. Departures from normal trends occur in many conditions associated with chromosomal aberrations, such as mongolism. *See* EPIDERMAL RIDGES; FINGERPRINT; HUMAN GENETICS; SKIN (VERTEBRATE).

[HAROLD CUMMINS]

Bibliography: H. Cummins and C. Midlo, *Finger Prints, Palms and Soles*, 1943; S. B. Holt, *The Genetics of Dermal Ridges*, 1968.

Intelligence

The ability to learn, regardless of what is being learned; adaptability to new problems and conditions; ability at abstract thinking.

Nature of intelligence. Rooted in a Latin word meaning "understanding," intelligence is a general capacity underlying human and animal behavior, but investigations of intelligence have tended to focus on differences in intelligence between individuals. However, the question of whether there is only one kind of intelligence as opposed to a variety of kinds has been the subject of extensive theoretical and empirical investigation.

Several theories have been put forward to account for the complexity of intelligence as a trait. One of the first theories was by C. E. Spearman, who viewed it as a single common trait entering into any intellectual task. Later this view was modified by L. L. Thurstone, in his multifactor theory of intelligence. The constituent components of his complex notion of intelligence are: verbal comprehension, verbal fluency, reasoning, ability to handle spatial and numerical relationships, perceptual speed, and rote memory. Thurstone arrived at these "primary mental abilities," as he called them, by analyzing the results of many tests by means of a special statistical technique called factor analysis. Through this technique, factors are assessed in the individual by means of tasks such as the understanding of increasingly difficult words, the number of words an individual can enumerate within a minute's time, the completion of analogies and series, puzzles, and short-term recall of lists of digits. The total performance of an individual on such a battery of tests provides a powerful predictor of scholastic achievement and vocational success. Another model is the hierarchical model suggested by C. Burt and P. E. Vernon, who postulate a general common factor of intelligence, plus a number of more or less related factors. If Thurstone's model can be likened to a bundle of reeds, the hierarchical model can be likened to twigs branching from one stem.

J. P. Guilford has applied both factor analysis and systematic logic to the study of the nature of

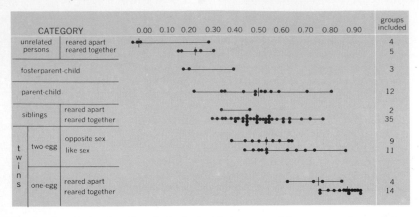

CATEGORY		0.00 0.10 0.20 0.30 0.40 0.50 0.60 0.70 0.80 0.90	groups included
unrelated persons	reared apart		4
	reared together		5
fosterparent-child			3
parent-child			12
siblings	reared apart		2
	reared together		35
twins two-egg	opposite sex		9
	like sex		11
twins one-egg	reared apart		4
	reared together		14

Fig. 1. Graph showing the similarity in IQ (expressed by means of correlation coefficients) between individuals as a function of the closeness of their biological relationship. Correlation coefficients range between 0 (no relationship) and 1 (perfect similarity). The dots on the scales refer to the similarity measures obtained in 52 heredity studies; the vertical lines on the scales indicate the median coefficient of similarity in each row. (*From L. Erlenmeyer-Kimling and L. F. Jarvik, Genetics and intelligence: A review, Science, 142:1477–1479, 1963*)

intelligence. In 1960 he described a three-dimensional model (structure of intellect model) that aims at an exhaustive analysis of intellectual functions. In this model there are five kinds of intellectual operations (cognition, memory, divergent thinking, convergent thinking, and evaluation), which apply to four types of problem contents (figural, symbolic, semantic, and behavioral), and which lead to six types of intellectual products (units, classes, relations, systems, transformations, and implications). Tests have been constructed for more than half of these 120 (that is, $5 \times 4 \times 6$) tasks. A most interesting feature of Guilford's model is his attempt to integrate creativity

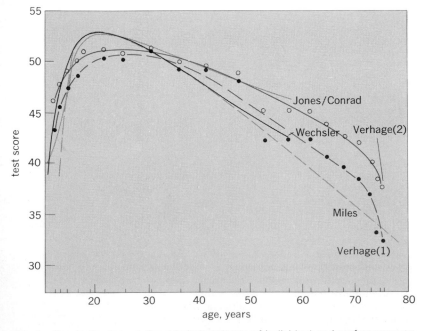

Fig. 2. Graph showing relationship between age of individual and performance on comprehensive intelligence tests in five different studies. (*From F. Verhage, Intelligence and Age, Assen (Holland): van Gorcum, 1964*)

into the study of intelligence. Creativity, or the ability to bring something new into existence, mainly depends upon fluency and variety of ideas, originality, and adaptive flexibility, which fit into the divergent-thinking facet of intellectual operations.

Measurement: intelligence quotient. Since the French psychologists A. Binet and T. Simon published the first comprehensive intelligence test in 1904, it has been customary to express the test performance of the child in terms of the age level for which his performance is characteristic (mental age). Dividing the mental age by the chronological age of the individual and multiplying by 100 yields the intelligence quotient (IQ). This practice has been changed, and the individual's performance is now usually expressed in terms of his deviation from the average performance of his own age group. To facilitate comparisons, a scale with a mean of 100 and a standard deviation of 15 or 16 is used. IQs between 90 and 110 are considered average, those below 65 defective, and those over 120 superior. The distribution of IQs in the population follows the normal curve very closely. *See* STATISTICS.

Heredity and environment. As is the case with many psychological traits, intelligence is not exclusively determined by inheritance or by environmental influences, but is clearly dependent on both. As can be seen from Fig. 1, similarity in intelligence is a function of similarity in biological constitution. However, ample evidence indicates that depressed IQs in culturally disadvantaged children can be raised by means of appropriate education and environmental stimulation. Differences in IQs between the sexes and different races and cultures have been observed, but they can usually be attributed to distortions in measurement. Culture-free or culture-fair tests, which rely less on language and specific environmental background, usually indicate that there are no basic differences in mental ability between the sexes or between various races and cultures. *See* BEHAVIOR AND HEREDITY.

Intelligence and age. The qualitative development of intellectual growth during childhood has been investigated by J. Piaget. On the basis of his observations of intellectual maturation, three major stages are distinguished: sensory-motor intelligence (0–2 years); concrete operations (2–11 years); and formal operations developing into formal logic (11–15 years). Each of these stages is further subdivided into various periods. Additional knowledge concerning the development of intellectual abilities has been gathered by psychometric methods.

By virtue of the metric properties of the IQ scale, there are no mean IQ differences between different age groups, since a correction for age is included in the calculation of the IQ. There are, however, salient differences in absolute test performance. Individual differences in the rate of intellectual growth in children and other factors make predictions of future IQ often fallible in individual cases. Intellectual abilities, on the average, grow at a gradually slowing pace until approximately 15–25 years, then remain at a plateau for approximately 10–15 years, and subsequently begin to decline. At the age of 65 the level of ability is roughly returned to where it was at age 12. Figure 2 shows this relationship between age and performance on composite intelligence test batteries. Some abilities, like verbal comprehension, show little or no decline, but others, such as perceptual speed, decrease faster than the slope of the curve illustrated here. In general, abilities that depend on speed clearly decline with age whereas other intellectual abilities, measured by non-timed tests, show less, or no, decline with age. For optimal mental efficiency older people often rely more on their store of knowledge and experience. *See* LEARNING THEORIES; MENTAL DEFICIENCY; PSYCHOLOGY, PHYSIOLOGICAL AND EXPERIMENTAL.   [G. J. S. WILDE]

Bibliography: A. Anastasi, *Psychological Testing*, 3d ed., 1968; H. J. Butcher, *Human Intelligence: Its Nature and Assessment*, 1968; J. P. Guilford, *The Nature of Human Intelligence*, 1967; D. N. Jackson and S. Messick, *Problems in Human Assessment*, pt. 5, 1967; J. Piaget, *Origins of Intelligence*, 1964; P. E. Vernon, *Intelligence and Attainment Tests*, 1960; P. E. Vernon, *Structure of Human Abilities*, 1960.

Interactions, fundamental

Fundamental forces that act between the elementary particles of which all matter is composed.

Four fundamental interactions. At present, four fundamental interactions are distinguished.

Gravitational interaction. This interaction manifests itself as a long-range force of attraction between all elementary particles. The force law between two particles of masses m_1 and m_2 separated by a distance r is well approximated by the Newtonian expression $G_N(m_1 m_2/r^2)$, where G_N is the Newtonian constant ($G_N = 6.6720 \pm 0.0041 \times 10^{-8}$ cm^3 g^{-1} s^{-2}). The dimensionless constant $(G_N m_e m_p)/\hbar c$ is usually taken as the constant characterizing the gravitational interaction (m_e, m_p are electron and proton masses; $2\pi\hbar$ = Planck's constant, c = velocity of light).

Electromagnetic interaction. This interaction is responsible for the long-range force of repulsion of like and attraction of unlike electric charges. The dimensionless constant characterizing the strength of electromagnetic interaction is the fine-structure constant $\alpha = e^2/\hbar c = 1/(137.03604 \pm 0.00011)$, where e is the (unrationalized electrostatic) electron charge. At comparable distances, the ratio of gravitational to electromagnetic interactions (as determined by the strength of respective forces between an electron and a proton) is given by the ratio of the constants $G_N m_e m_p/e^2 \approx 4 \times 10^{-37}$.

In modern quantum field theory, the electromagnetic interaction and the forces of attraction or repulsion between charged particles are pictured as arising secondarily as a consequence of the primary process of emission of one or more photons (light quanta) by an accelerating electric charge (in accordance with Maxwell's equations) and the subsequent reabsorption of these quanta by a second charged particle. The space-time diagram (first introduced by R. F. Feynman) for this exchange is shown in the illustration. The same picture may also be valid for the gravitational interaction (in accordance with the quantum version of A. Einstein's gravitational equations), however, with the exchange of a zero-rest-mass

particle (the graviton g) rather than the zero-rest-mass photon. (The physical existence of the graviton has, however, not been conclusively demonstrated.)

In accordance with this picture, the electromagnetic interaction is usually represented by Eq. (1), where γ is the photon, emitted by the elec-

$$e + P \rightarrow (e + \gamma) + P \rightarrow e + (P + \gamma) \rightarrow e + P \qquad (1)$$

tron and reabsorbed by the proton. For this interaction, and also for the gravitational interaction $e + P \rightarrow (e + g) + P \rightarrow e + (P + g) \rightarrow e + P$, the nature of the participating particles (electron e and proton P) is the same, before and after the interaction. (Here g is the graviton.) *See* LIGHT; MAXWELL'S EQUATIONS; PHOTON; QUANTUM ELECTRODYNAMICS; QUANTUM FIELD THEORY; QUANTUM MECHANICS.

Weak nuclear interactions. The third fundamental interaction is the weak nuclear interaction, whose characteristic strength for low-energy phenomena, measured by the so-called Fermi constant, G_F (approximately equal to 1.0×10^{-5} $m_p^{-2}\hbar^3/c$), is a thousand times weaker than electromagnetic. Unlike electromagnetism and gravitation, weak interactions are short-range interactions, with a force law of the type $(1/r^2)e^{-Kr}$, the range of the force K^{-1} being much smaller than 10^{-15} cm. Also, unlike electromagnetism and gravitation, weak interactions, until 1973, appeared always to change the nature of the interacting particles, as in reactions (2), where P = proton, N = neutron, μ^- = muon, ν_e and ν_μ are the electron and muon neutrinos.

$$P + e^- \xrightarrow{\text{Weak}} N + \nu_e \quad \text{(this reaction is equivalent to}$$
$$\beta\text{-decay of the neutron } N \rightarrow P + e^- + \bar{\nu}_e) \qquad (2a)$$

$$P + \mu^- \xrightarrow{\text{Weak}} N + \nu_\mu \quad \text{(muon capture by a proton}$$
$$\text{with the emission of a neutrino)} \qquad (2b)$$

$$\mu^- + \nu_e \xrightarrow{\text{Weak}} e^- + \nu_\mu \quad \text{(this reaction is equivalent}$$
$$\text{to muon decay } \mu^- \rightarrow e^- + \bar{\nu}_e + \nu_\mu) \qquad (2c)$$

In reaction (2a), for example, the weak interaction transforms a proton into a neutron and at the same time an electron into a neutrino.

One of the most crucial discoveries in particle physics was the discovery in 1973 of the so-called neutral currents, which manifest themselves through weak interactions of the type where the nature of the interacting particles is not changed during the interaction, as in reactions (3).

$$\nu_\mu + e^- \xrightarrow{\text{Weak}} \nu_\mu + e^-$$
$$\nu_\mu + P \xrightarrow{\text{Weak}} \nu_\mu + P \qquad (3)$$
$$\nu_e + N \xrightarrow{\text{Weak}} \nu_e + N$$

In contrast to gravitation, electromagnetism, and strong nuclear interactions, weak interactions violate conservation of left-right and particle-antiparticle symmetries. *See* INTERACTIONS, WEAK NUCLEAR; PARITY (QUANTUM MECHANICS); SYMMETRY LAWS (PHYSICS).

Strong nuclear interaction. The fourth fundamental interaction is the strong nuclear interaction, which resembles the weak nuclear interaction in being short-range, though the range is approximately 10^{-13} cm rather than $\ll 10^{-15}$ cm. As

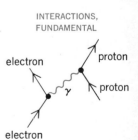

INTERACTIONS, FUNDAMENTAL

Feynman diagram of electromagnetic interaction between an electron and a proton.

the name implies, within this range of distances the strong force overshadows all other forces with a characteristic strength parameter of approximately 1 compared with the electromagnetic strength parameter $\alpha \approx 1/137$.

Analogous to electromagnetism, the strong nuclear interaction was pictured by H. Yukawa as arising through the exchange of mesons in reactions such as (4), where ρ^+, ρ^0, ρ^- are the positively

$$P + N \rightarrow (P + \rho^0) + N \rightarrow P + (\rho^0 + N) \rightarrow P + N$$

$$P + N \rightarrow (N + \rho^+) + N \rightarrow N + (\rho^+ + N) \rightarrow N + P \quad (4)$$

$$P + N \rightarrow P + (\rho^- + P) \rightarrow (P + \rho^-) + P \rightarrow N + P$$

charged, neutral, and negatively charged ρ-mesons (with masses of approximately 750 MeV).

The strong nuclear interaction differs sharply from the other three interactions in the following manner. Gravitational, electromagnetic, and weak interactions act universally between light elementary particles known as leptons (the four particles ν_e, ν_μ, e^-, μ^-) as well as between heavy particles known as hadrons (such as P, N, mesons, and hypothesized quarks, the elusive elementary particles of which all observed hadrons, such as protons, neutrons, and mesons, are assumed to be composed). In contrast, the strong nuclear force appears to act only between hadrons, with leptons exhibiting no strong interactions.

Properties of the four fundamental interactions are summarized in the table.

Unification of interactions. Ever since the discovery and clear classification of these four interactions, particle physicists have attempted to unify these interactions as aspects of one basic interaction between all matter. The work of M. Faraday and J. C. Maxwell in the 19th century, which united the distinct forces of electricity and magnetism as aspects of one single interaction, is a model for the unification idea.

Gravitation and electromagnetism. The first attempt in this direction was that of Einstein, who, having succeeded in understanding gravitation as a manifestation of the curvature of space-time, tried to comprehend electromagnetism as another geometrical manifestation of the properties of space-time. In this attempt, to which he devoted all his later years, he is considered to have failed. *See* UNIFIED FIELD THEORY.

Weak and electromagnetic interactions. A totally different type of unification of weak and electromagnetic interactions was suggested (employing the gauge principle of H. Weyl, C. N. Yang, R. Mills, and R. Shaw) by A. Salam and J. C. Ward

and S. L. Glashow in 1959. This followed a parallel between these two interactions, pointed out by J. S. Schwinger, providing one assumed that the then known classes of weak interactions were mediated by intermediate bosons W^+ and W^-, just as electromagnetism is mediated by the exchange of photons. The empirical properties of weak interaction phenomena, extensively experimented on during 1957, had revealed that if $W^\pm$ existed, these particles must carry an intrinsic spin of magnitude $\hbar$, just like the photon, which also carries spin $\hbar$. If the assumption is made that the intrinsic coupling strength of weak interactions is the same as for electromagnetism, that is, that the characteristic coupling parameter for both these interactions is the fine structure constant $\alpha \approx 1/137$, one may then deduce that the effective constant G_F for weak interactions is approximately $10^{-5} m_p^{-2} \hbar^3/c$, provided the masses of the $W^\pm$ particles are in excess of $\sqrt{\alpha \hbar^3 / G_F c}$, that is, 37 GeV or greater.

One important consequence of this unification is the prediction of the existence of neutral weak interactions represented by reactions (3), or, equivalently, the existence of a new intermediate boson (called Z°) which, like the photon, would be neutral in electric charge.

This unified theory was further elaborated on decisively by S. Weinberg and Salam, who gave estimates of the strength of the predicted neutral weak interactions.

In summary, the unified theory of weak and electromagnetic interactions predicts that, first, the $W^\pm$ bosons must exist with masses in excess of 37 GeV; second, that neutral weak current phenomena of reactions shown in (3) must be observed with strengths comparable to other weak phenomena of reactions shown in (2). The theory is based on a deep fundamental principle: the gauge principle.

Since 1973, the verification of this prediction about neutral-current phenomena of reactions shown in (3) at various accelerator laboratories has given decisive support to the notion of a basic unity between weak and electromagnetic interactions. One of the major predictions for experiments to be carried out some decades in the future, when the necessary energetic particles are available, is that, for energies in excess of $W^\pm$ boson masses, weak interactions will exhibit their primitive strength of the order of α. The weakness of weak interactions is apparently a transitory phenomenon, related to the fact that all experimentation that has been carried out at accelerator laboratories is basically exploring a relatively low-energy

Properties of interactions

Interaction	Characteristic strength	Range	Exchanged quanta
Gravitational	10^{-39}	Long-range	Gravitons*
Electromagnetic	10^{-2}	Long-range	Photons
Weak nuclear	10^{-5}	Short-range, 10^{-15} cm	W^+, Z^0, W^- "intermediate bosons"*†
Strong nuclear	1	Short-range, 10^{-13} cm	Mesons

*No direct experimental evidence.
†Z^0 would be responsible for weak neutral current effects.

regime, below the masses of $W^\pm$ mesons.

Inclusion of strong interactions. The gauge unification of weak and electromagnetic interactions, which started with the observation that the mediating quanta for these interactions possess intrinsic spin $\hbar$, can be carried further to include strong nuclear interactions as well, since these interactions are also known to be mediated by means of particles carrying spin $\hbar$. But such a unification necessarily means that the distinction between leptons and hadrons must disappear and that all interactions (weak, electromagnetic, and strong) must be equally universal and of the same (primitive) strength for experiments carried out at sufficiently high energies. Another consequence of this disappearance of distinction between leptons and hadrons appears to be the possibility, predicted by J. C. Pati and Salam, of proton decays (with a lifetime of approximately 10^{31} years) into leptons. The discovery of proton instability against decay into leptons would be an epic discovery and a direct confirmation of the unification ideas.

Prospects. Future work in particle physics is likely to be concerned with the detailed elaboration of these unification ideas, between strong, weak, and electromagnetic interactions. It is possible that the gravitational interaction will be unified with the other three interactions by other methods. A clue to such methods may be provided by the empirical and theoretical expression (5)

$$\alpha^{-1} \approx \log(G_N m_e^2 / c\hbar) \qquad (5)$$

which relates the fine-structure constant α to the Newtonian constant G_N.

Other interactions. Other interactions besides the four listed above may exist. There is a suspicion that there may be a fifth: an extremely weak interaction, a thousand times weaker than the conventional weak interaction, responsible for time asymmetry observed in the decays $K_L^0 \rightarrow \pi^+\pi^-$. There is no known principle that would limit the existence of interactions to the four that are definitely known at present. *See* ELEMENTARY PARTICLE.   [ABDUS SALAM]

Bibliography: F. J. Hasert et al., Search for elastic muon-neutrino electron scattering, *Phys. Lett.*, 46B:121–124, 1973; J. C. Pati and A. Salam, Is baryon number conserved?, *Phys. Rev. Lett.*, 31: 661–664, 1973; A. Salam, Weak and electromagnetic interactions, Nobel Symposium, Gothenberg, pp. 367–377, 1968; A. Salam and J. Strathdee, Quantum gravity and infinities in quantum electrodynamics, *Lett. Nuovo Cimento*, 4:101–108, 1970; A. Salam and J. C. Ward, Weak and electromagnetic interactions, *Nuovo Cimento*, 11: 568–577, 1959; S. Weinberg, A model of leptons, *Phys. Rev. Lett.*, 19:1264–1266, 1967.

Interactions, weak nuclear

One of the four basic physical interactions (gravitational, strong, electromagnetic, weak). Only the last three are significant for elementary particle and nuclear physics. Among these the weak forces are extremely feeble compared to the other two but manifest themselves because of their special character. Weak interactions include: nuclear beta decay and electron capture; muon capture on nuclei; the slow decays (mean life τ greater than 10^{-10} sec) of unstable elementary particles;

and neutrino-nucleon scattering processes which are necessarily the main tool for investigating weak forces at high energies and large momentum transfers. The weak interaction is also responsible for small parity-violating effects in nuclear forces, and strangeness oscillations in the neutral K-meson system. Some weak processes predicted to exist but not yet observed directly (for example, $e^+ e^- \rightarrow \nu_e \bar{\nu}_e$) may play an important role in astrophysics and cosmology. *See* BETA RAYS; ELEMENTARY PARTICLE; INTERACTIONS, FUNDAMENTAL; MESON; NEUTRINO; NUCLEAR PHYSICS; PARITY (QUANTUM MECHANICS); RADIOACTIVITY.

Weak interactions may be classified as purely leptonic (such as $\mu^- \rightarrow \nu_\mu\, e^-\, \bar{\nu}_e$), semileptonic (such as $n \rightarrow pe^-\, \bar{\nu}_e$, $\Sigma^- \rightarrow n\mu^-\, \bar{\nu}_\mu$), or nonleptonic (such as $\Lambda^0 \rightarrow p\pi^-$). In the first category are reactions involving only the leptons ($\mu^\pm$, $e^\pm$, ν_μ, $\bar{\nu}_\mu$, ν_e, $\bar{\nu}_e$). In the second and third categories, hadrons (strongly interacting particles) appear. Weak transitions involving hadrons may be further classified into those which conserve strangeness (for example, $n \rightarrow pe^-\bar{\nu}_e$, $\pi^+ \rightarrow \mu^+\nu_\mu$) and those in which the total strangeness changes (for example, $K^+ \rightarrow \mu^+\nu_\mu$, $\Sigma^- \rightarrow n\mu^-\bar{\nu}_\mu$). *See* HADRON; LEPTON.

The weak interaction is conveniently described in terms of the coupling of a weak current J to itself, or more precisely to its Hermitian conjugate J^+, at a point or at least an extremely small region in space-time. The current J consists of several parts, as indicated in Eq. (1), where j_e, j_μ, J_0, J_1 are

$$J = j_e + j_\mu + J_0 + J_1 \qquad (1)$$

the electronic-, muonic-, strangeness-conserving hadronic-, and strangeness-changing hadronic weak currents, respectively. Each weak process is generated from a coupling of a particular term in J to another term in J^+. A few of these processes are illustrated by Feynman diagrams in Fig. 1. In each diagram, time advances vertically upward. (The downward-pointed arrow for $\bar{\nu}_e$ in Fig. 1a and c indicates by convention that $\bar{\nu}_e$ is an antiparticle.) Muon decay $\mu^- \rightarrow \nu_\mu e^-\bar{\nu}_e$ (Fig. 1a), occurs through coupling of j_μ^+ (which generates $\mu^- \rightarrow \nu_\mu$) to j_e (vacuum $\rightarrow e^-\bar{\nu}_e$). Neutron decay, $n \rightarrow pe^-\bar{\nu}_e$ (Fig. 1c), is described by a coupling of j_e to $J_0^+ (n \rightarrow p)$. The decay $\Lambda^0 \rightarrow p\pi^-$ (Fig. 1d) is described by a coupling of J_0 (vacuum $\rightarrow \pi^-$) to $J_1^+ (\Lambda^0 \rightarrow p)$. The

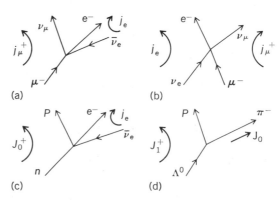

Fig. 1. Schematic (Feynman) diagrams illustrating (a) $\mu^- \rightarrow \nu_\mu e^- \bar{\nu}_e$; (b) $\nu_e + \mu^- \rightarrow \nu_\mu + e^-$; (c) $n \rightarrow pe^-\bar{\nu}_e$; (d) $\Lambda^0 \rightarrow p\pi^-$.

Classification of weak processes by their currents*

		Term of J			
		j_e	j_μ	J_0	J_1
Term of J^+	j_e^+	$(e^- + \nu_e \rightarrow \nu_e + e^-)$ $(e^- + \bar{\nu}_e \rightarrow \bar{\nu}_e + e^-)$ $(e^+ + e^- \rightarrow \nu_e + \bar{\nu}_e)$	$\mu^+ \rightarrow e^+ + \nu_e + \bar{\nu}_\mu$ $(\nu_\mu + e^- \rightarrow \mu^- + \nu_e)$	$(pp \rightarrow de^+ \nu_e)$ $\nu_e p \rightarrow ne$ $\pi^+ \rightarrow e^+ \nu_e$ $\pi^+ \rightarrow \pi^0 e^+ \nu_e$	$K^+ \rightarrow e^+ \nu_e$ $K^+ \rightarrow \pi^0 e^+ \nu_e$ $K_L^0 \rightarrow \pi^- e^+ \nu_e$
	j_μ^+	$\mu^- \rightarrow e^- + \bar{\nu}_e + \nu_\mu$	$(\mu^- + \nu_\mu \rightarrow \nu_\mu + \mu^-)$ $(\mu^- + \bar{\nu}_\mu \rightarrow \bar{\nu}_\mu + \mu^-)$ $(\mu^+ + \mu^- \rightarrow \nu_\mu + \bar{\nu}_\mu)$	$\pi^+ \rightarrow \mu^+ \nu_\mu$ $\mu^- p \rightarrow n \nu_\mu$ $\nu_\mu p \rightarrow n \mu^+$	$K^+ \rightarrow \mu^+ \nu_\mu$ $K^+ \rightarrow \pi^0 \mu^+ \nu_\mu$ $K_L^0 \rightarrow \pi \mu^+ \nu_\mu$ etc.
	J_0^+	$n \rightarrow p\, e^- \bar{\nu}_e$ $\nu_e + n \rightarrow pe^-$ $\pi^- \rightarrow e^- \bar{\nu}_e$	$\pi^- \rightarrow \mu^- \nu_\mu$ $\nu_\mu n \rightarrow p\mu^-$	$n + p \rightarrow p + n$†	$\Sigma^+ \rightarrow n\pi^+$ $\Sigma^+ \rightarrow p\pi^0$ $K \rightarrow 2\pi, 3\pi$
	J_1^+	$K^- \rightarrow e^- \bar{\nu}_e$ $\Lambda^0 \rightarrow pe^- \bar{\nu}_e$ $\Sigma^- \rightarrow ne^- \bar{\nu}_e$ $\Xi^- \rightarrow \Lambda^0 e^- \nu_e$	$K^- \rightarrow \mu^- \bar{\nu}_\mu$ $K_L^0 \rightarrow \pi \mu^- \bar{\nu}_\mu$ etc.	$\Lambda^0 \rightarrow p\pi^-$ $\Sigma^- \rightarrow n\pi^-$ $K \rightarrow 2\pi, 2\pi$	$n + p \rightarrow p + n$†

*Processes in parentheses are conjectured to exist but have not been observed directly. "Neutral weak current" reactions are not included.

†Parity-violating nuclear force.

table classifies weak processes according to the terms in J and J^+ which couple to generate them. *See* NEUTRON; QUANTUM FIELD THEORY.

SELECTION RULES, SYMMETRY LAWS, AND EMPIRICAL REGULARITIES

Study of weak interactions is facilitated by consideration of certain symmetry and conservation laws and empirical regularities. In common with all other interactions, weak interactions conserve energy and linear momentum, angular momentum, electric charge, baryon number B, and lepton numbers L_e (electron number) and L_μ (muon number). Here B is defined as +1 for the spin 1/2 hadrons (p, n, Σ^+, Σ^-, Σ^0, Λ^0, and so on), -1 for their corresponding antiparticles, and zero for mesons and leptons. Also, $L_e(e^-) = L_e(\nu_e) = +1$, $L_e(e^+) = L_e(\bar{\nu}_e) = -1$, $L_e = 0$ for all other particles, $L_\mu(\mu^-) = L_\mu(\nu_\mu) = +1$, $L_\mu(\mu^+) = L_\mu(\bar{\nu}_\mu) = -1$, $L_\mu = 0$ for all other particles. *See* BARYON.

Strangeness. For strangeness-violating weak interactions, the rule $|\Delta S| = 1$ holds (no $|\Delta S| > 1$ transitions are known). Also the change in strangeness of the transforming hadron in a semileptonic $|\Delta S| = 1$ decay is equal to its change in charge ($\Delta S = \Delta Q$). For example, $\Sigma^- \rightarrow ne^- \bar{\nu}_e$, $K^+ \rightarrow \pi^+\pi^- e^+ \nu_e$ conform to the rule and are observed, while $\Sigma^+ \rightarrow ne^+\nu_e$, $K^+ \rightarrow \pi^+\pi^+ e^- \bar{\nu}_e$ violate $\Delta S = \Delta Q$ and are not observed. Finally, the $|\Delta I| = 1/2$ rule, which states that the isospin of the transforming hadron in a semileptonic or nonleptonic $|\Delta S| = 1$ decay is 1/2 unit, is generally valid, but minor exceptions (for example, the unusually slow decay $K^+ \rightarrow \pi^+\pi^0$ and also some other nonleptonic transitions) are known. The $\Delta S = \Delta Q$ and $|\Delta I| = 1/2$ rules are very useful for systematizing the transitions $K \rightarrow 2\pi$, $K \rightarrow 3\pi$, $K \rightarrow \pi l\nu$ (where l is a charged lepton, that is, an electron or a muon), and the hyperon nonleptonic decays. *See* ISOTOPIC SPIN.

Parity. Parity is conserved if and only if a reaction and its space-inverted image (generated by the coordinate transformation $x \rightarrow -x$, $y \rightarrow -y$, $z \rightarrow -z$) have identical transition probabilities. The strong and electromagnetic interactions conserve parity, but the weak interactions violate it maximally. This important discovery is due to T. D. Lee and C. N. Yang in 1956. An example is the decay $\pi^+ \rightarrow \mu^+ \nu_\mu$ as seen in the pion rest-frame (Fig. 2). In observed $\pi^+ \rightarrow \mu^+\nu_\mu$ decay (Fig. 2a), the μ^+ spin is found experimentally to be opposite to its motion (helicity $h = -1$). Since the π^+ spin is zero and angular momentum is conserved, this implies $h(\nu_\mu) = -1$. A parity (P) transformation results in rever-

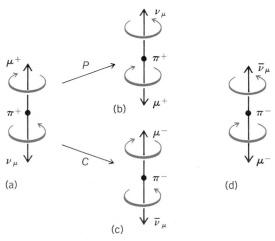

Fig. 2. Schematic diagram of $\pi_{\mu 2}$ decay illustrating P and C violation. (*a*) Decay of π^+. (*b*) Result of P transformation; not observed. (*c*) Result of C transformation; not observed. (*d*) Result of CP transformation: observed π^- decay.

sal of μ^+, ν_μ momenta but leaves spins invariant. Thus under P, $h(\mu^+)$ is reversed (Fig. 2b); but this is never observed, implying maximal P violation. *See* HELICITY (QUANTUM MECHANICS).

Formally, a parity-conserving interaction is described by a scalar Lagrangian density L, that is, a Lagrangian density which is invariant under space inversion as well as proper Lorentz transformations. For example, in quantum electrodynamics (QED) the interaction between an electron and the electromagnetic field is described by Eq. (2),

$$L_{EM}(\mathbf{x},t) = -ej_\lambda(\mathbf{x},t)A^\lambda(\mathbf{x},t) \qquad (2)$$

where $A(\mathbf{x},t)$ is the vector potential of the (quantized) electromagnetic field, and $ej_\lambda(\mathbf{x},t)$ is the electromagnetic four-vector current density of the electron. The operator $j_\lambda(\mathbf{x},t)$ is itself expressed in terms of the quantized electron field $\psi(\mathbf{x},t)$ and its conjugate $\bar{\psi}(\mathbf{x},t) = \psi^+(\mathbf{x},t)\gamma^0$ by Eq. (3), where the

$$j_\lambda = \bar{\psi}_e\gamma_\lambda\psi_e \qquad (3)$$

γ_λ are 4×4 matrices appearing in Dirac's relativistic quantum theory, and in Eq. (2) summation over repeated Greek indices (in this case, λ) is assumed. Since A^λ and j_λ are polar vectors, L_{EM} is a scalar. A parity-violating interaction must be described by a Lagrangian with scalar and pseudoscalar parts (the latter is invariant under proper Lorentz transformations but changes sign under spatial inversion). *See* POTENTIALS; QUANTUM ELECTRODYNAMICS; QUANTUM THEORY, RELATIVISTIC; RELATIVITY.

Charge conjugation. The strong and electromagnetic interactions are charge conjugation (C) invariant, but the weak interactions are not. For example, a C transformation on $\pi^+ \rightarrow \mu^+\nu_\mu$ decay (Fig. 2a) changes the sign of all charges but leaves spins, momenta unchanged, yielding Fig. 2c (a μ^- with negative helicity). This is never observed, that is, weak interactions violate C maximally.

CP transformation. All experimental evidence is consistent with CPT (combined charge conjugation, parity, and time reversal) invariance of strong, electromagnetic, and weak interactions. Violation of CP invariance manifests itself in the appearance of certain "forbidden" weak decays of the neutral K mesons. The K mesons belong to the $J^P = 0^-$ (spin$=0$, negative parity) $SU(3)$ meson octet: (K^+, K^0) and ($\bar{K}^0$, K^-) form two isodoublets in this octet with strangeness $S = +1$ and $S = -1$, respectively. K^+ and K^- are in fact charge conjugates, as are K^0 and $\bar{K}^0$. The states $|K^0\rangle$, $|\bar{K}^0\rangle$ are by definition eigenstates of the strong and electromagnetic Hamiltonians and simultaneously eigenstates of strangeness with eigenvalues ± 1, respectively.

Since the strong and electromagnetic interactions conserve strangeness, the relative phase of $|K^0\rangle$, $|\bar{K}^0\rangle$ is arbitrary and one may define $|\bar{K}^0\rangle \equiv CP|K^0\rangle$. Neither K^0 nor $\bar{K}^0$ has a definite lifetime for weak decay, but one can form two independent linear superpositions of $|K^0\rangle$ and $|\bar{K}^0\rangle$, namely, the states $|K_L^0\rangle$ and $|K_S^0\rangle$ (L and S for "long" and "short," respectively); the particles K_L^0 and K_S^0 do have definite lifetimes but no definite strangeness or isospin. The short-lived K_S^0 decays in only two significant modes: $\pi^+\pi^-$ and $\pi^0\pi^0$ with total in-

verse lifetime $\gamma_S = 1.134 \times 10^{10}$ sec^{-1}. The long-lived component K_L^0 decays in many modes, semileptonic and nonleptonic, and has the total inverse lifetime $\gamma_L = 1.93 \times 10^7$ sec^{-1}. Before the discovery of CP violation in 1964, it was thought that the states $|K_L^0\rangle$ and $|K_S^0\rangle$ could be expressed by Eq. (4). Since the final states $\pi^+\pi^-$ and $\pi^0\pi^0$ each have

$$|K_S^0\rangle = \frac{1}{\sqrt{2}}\left[|K^0\rangle - |\bar{K}^0\rangle\right]$$

$$|K_L^0\rangle = \frac{1}{\sqrt{2}}\left[|K^0\rangle + |\bar{K}^0\rangle\right] \qquad (4)$$

CP eigenvalue $+1$, this "explained" why $K_S^0 \rightarrow \pi^+\pi^-$ and $K_S^0 \rightarrow \pi^0\pi^0$ are allowed ($CP|K_S^0\rangle = +|K_S^0\rangle$) and why the decays $K_L^0 \rightarrow \pi^+\pi^-$, $\pi^0\pi^0$ are forbidden ($CP|K_L^0\rangle = -|K_L^0\rangle$) within the framework of CP invariance. However, it has been shown experimentally that the decays $K_L^0 \rightarrow \pi^+\pi^-$, $\pi^0\pi^0$ actually occur (albeit with low probability). Thus Eq. (4) must be replaced by Eq. (5),

$$|K_S^0\rangle = \frac{1}{\sqrt{2(1+|\epsilon|^2)}}\left[(1+\epsilon)|K^0\rangle - (1-\epsilon)|\bar{K}^0\rangle\right]$$

$$|K_L^0\rangle = \frac{1}{\sqrt{2(1+|\epsilon|^2)}}\left[(1+\epsilon)|K^0\rangle + (1-\epsilon)|\bar{K}^0\rangle\right] \qquad (5)$$

where Re(ϵ) $= 1.42 \times 10^{-3}$, Arg $\epsilon = 42.5^0$. It is not known whether CP violation (as manifested by nonzero ϵ) originates in the strong, electromagnetic, or weak interactions, or possibly even a new "superweak" interaction. *See* EIGENFUNCTION; EIGENVALUE (QUANTUM MECHANICS); UNITARY SYMMETRY.

Time reversal. CPT invariance together with CP violation imply violation of time-reversal (T) invariance at some level. However, except for the foregoing case of $K_L^0 \rightarrow \pi\pi$ decay and a related "charge asymmetry" (that is, differences in rates) for the decays $K_L \rightarrow \pi^+l^-\bar{\nu}_l$ versus $\pi^-l^+\nu_l$, there is no evidence for T violation anywhere in the weak interactions.

Regeneration. Equation (5), or in simpler approximation, Eq. (4), contains the basis for the phenomena of strangeness oscillations and regeneration in the neutral kaon system. For example, consider a beam of K^0 mesons formed at $t = 0$. From Eq. (4). $\psi(0) = (1/\sqrt{2})(K_L^0(0) = K_S^0(0))$. At a later time t, $\psi(t) = (1/\sqrt{2})[K_L(t) + K_S(t)] = 1/\sqrt{2}\,[K_L(0)\,\exp\,(-\lambda_Lt) + K_S(0)\,\exp\,(-\lambda_St)]$ where $\lambda_L = \gamma_L + im_L$ and $\lambda_S = \gamma_S + im_S$, and m_L, m_S are the masses of K_L, K_S, respectively, in units $\hbar = c = 1$. For $(1/\gamma_S) \gg t \gg (1/\gamma_L)$, $\psi \approx (1/\sqrt{2})K_L(0)\,\exp\,(-\lambda_Lt)$, that is, the beam is converted into pure K_L^0. However, if the $K_L^0 = (1/\sqrt{2})(K^0 + \bar{K}^0)$ beam is allowed to pass through matter, the scattering amplitudes for K^0 and $\bar{K}^0$ on matter nuclei are very different. Thus after passage through matter, $\psi \rightarrow (aK^0 + b\bar{K}^0)$ with $a \neq b$; thus $\psi \rightarrow [(a+b)K_L^0 + (a-b)K_S^0]$, namely, the K_S^0 is "regenerated." *See* SELECTION RULES (PHYSICS); SYMMETRY LAWS (PHYSICS).

PHENOMENOLOGICAL THEORY

Attempts to develop a theory of weak interactions have been guided by analogy to QED since Enrico Fermi's original theory of nuclear beta decay in 1934. In QED the simplest imaginable process would be emission of an electromagnetic quantum (photon) by a free electron: $e^- \rightarrow e^- + \gamma$ (although such a process cannot, in fact, occur by conservation of energy-momentum). The simplest nuclear beta decay is neutron decay: $n \rightarrow pe^- \bar{\nu}_e$; to account for it by analogy to Eqs. (2) and (3), Fermi suggested the replacement indicated in Eqs. (6).

$$-e \rightarrow \frac{G}{2}$$

$$\bar{\psi}_e \gamma_\lambda \psi_e \rightarrow \psi_p \gamma_\lambda \psi_n \qquad (6)$$

$$A^\lambda \rightarrow \bar{\psi}_e \gamma^\lambda \psi_{\nu e}$$

Here, ψ_p and ψ_n are assumed to be four-component Dirac fields analogous to ψ_e, ψ_ν; that is, complications arising from strong interactions are ignored. Also G, the so-called weak-interaction, or Fermi, coupling constant is retained in later developments and is found to be given by Eq. (7) or in units $\hbar = c = 1$ (to be used from now on except where noted) by Eq. (8), where m_p is the proton mass.

$$G = 1.43506 \pm 0.00026 \times 10^{-49} \text{ erg cm}^3$$
$$= 1.43506 \pm 0.00026 \times 10^{-62} \text{ Jm}^3 \qquad (7)$$
$$G = 1.03 \times 10^{-5} \, m_p^{-2} \qquad (8)$$

Thus the Fermi beta-decay Lagrangian density of Eq. (9) was obtained where the Hermitian conjugate term (h.c.) was intended to account for nuclear β^+ decay and electron capture.

$$L_\beta = \frac{G}{\sqrt{2}} \bar{\psi}_p \gamma_\lambda \psi_n \cdot \bar{\psi}_e \gamma^\lambda \psi_{\omega e} + \text{h.c.} \qquad (9)$$

In subsequent theoretical developments Fermi's original idea has been modified and generalized, but its basic character has been retained. Thus, to account for weak processes in addition to beta decay, R. Feynman and M. Gell-Mann replaced Eq. (9) with Eq. (10), with J^λ as in Eq. (1). Each

$$L_w = \frac{G}{2} J_\lambda^+ J^\lambda \qquad (10)$$

portion of $J^\lambda (j_e, j_\mu, J_0, J_1)$ corresponds to a transformation in which electric charge Q obeys the rule $\Delta Q = -1$ (for example, j_e corresponds to vacuum $\rightarrow e^- \bar{\nu}_e$ in neutron beta decay); J^λ is thus called a "charged" current. This conforms to the empirical rule that there are "no neutral weak currents," presumed valid until 1973 when events suggesting the existence of neutral weak currents were observed in high-energy neutrino-nucleon scattering experiments. (It is not clear how this development fits into the conventional scheme presented in this discussion.)

Parity violation. In the Fermi theory, parity conservation was assumed, but its actual violation is expressed by the fact that J^λ has been found by experiment to consist of axial vector and polar vector portions. (Axial and polar vectors transform in the same manner under proper Lorentz transformations, but an axial vector undergoes an additional sign change under spatial inversion.) Thus L_w contains pseudoscalar (vector × axial vector) and scalar (vector × vector, axial × axial) parts. Specifically, the electronic and muonic currents are expressed by Eqs. (11) and (12), where $\bar{\psi}\gamma^\lambda \psi$ is a polar

$$j_e^\lambda = \bar{\psi}_e \gamma^\lambda (1 + \gamma^5) \psi_{\nu e} \qquad (11)$$

$$j_\mu^\lambda = \bar{\psi}_\mu \gamma^\lambda (1 + \gamma^5) \psi_{\nu\mu} \qquad (12)$$

vector, $\bar{\psi}\gamma^\lambda \gamma^5 \psi$ is an axial vector. This is called the V-A law, and it has the following significance: the helicity h of a fermion represented by a four spinor u is defined as the projection of its spin direction on its direction of motion; shown in Eq. (13). Defining $a = (1/2)(1 + \gamma^5)$, $\bar{a} = (1/2)(1 - \gamma^5)$, $au = u_L$, and $\bar{a}u = u_R$ (L for left-handed, and R for right-handed), it can be shown that Eqs. (14) and (15) are realized

$$h = \frac{\boldsymbol{\sigma} \cdot \mathbf{p}}{|\mathbf{p}|} \qquad (13)$$

$$hu_L = -u_L \qquad (14)$$

$$hu_R = +u_R \qquad (15)$$

in the limit of zero mass. Thus Eqs. (11) and (12) contain the statement that all (zero-mass) neutrinos ν_e, ν_μ have negative helicity, $h(\nu) = -1$ (are left-handed); and all antineutrinos $\bar{\nu}_e$, $\bar{\nu}_\mu$ have positive helicity, $h(\bar{\nu}) = +1$ (are right-handed); this is called the two-component neutrino theory.

Muon decay. Muon decay, $\mu^+ \rightarrow e^+ \nu_e \bar{\nu}_\mu$ and $\mu^- \rightarrow e^- \bar{\nu}_e \nu_\mu$, is the only purely leptonic weak process which has been observed with certainty. The results of precise experimental observations of the electron energy spectrum, electron helicity, and electron momentum-muon spin correlation in muon decay are available and may be used as a critical test of the theoretical forms of Eqs. (11) and (12). The resulting matrix element for, say, μ^- decay, Eq. (16), where u_e, u_μ, $v_{\bar{\nu}e}$ and $u_{\nu\mu}$ are one-

$$M = \frac{G}{\sqrt{2}} \bar{u}_e \gamma_\lambda (1 + \gamma^5) v_{\nu e} \cdot \bar{u}_{\nu\mu} \gamma^\lambda (1 + \gamma^5) u_\mu \qquad (16)$$

particle Dirac spinors for e^-, μ^-, $\bar{\nu}_e$, and ν_μ, respectively, leads to a theoretical differential transition probability in excellent agreement with the experimental results (Fig. 3). The theoretical total

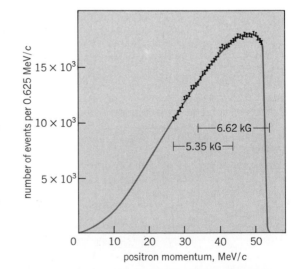

Fig. 3. Results of experiment to determine electron energy spectrum in muon decay. Experimental points are plotted together with V-A theoretical curve corrected for radiative effects and ionization loss. (*From E. D. Commins, Weak Interactions, McGraw-Hill, 1973*)

decay rate, Eq. (17) (where the bracketed factor is

$$\tau^{-1} = W_{\text{Total}} = \frac{G^2 m_\mu^5}{192\pi^3}\left[1 - \frac{\alpha}{2\pi}\left(\pi^2 - \frac{25}{4}\right)\right] \quad (17)$$

a radiative correction; α is the fine structure constant), may be compared to the experimental lifetime; together with a value of muon mass m_μ this leads to the quoted value of G in Eqs. (7) and (8).

Unfortunately, it remains possible to account for the observations of muon decay with more general forms for the leptonic weak currents; this is because the most general possible muon decay amplitude, which is linear in the lepton fields, contains no lepton field derivatives, is invariant under proper Lorentz transformations and obeys lepton conservation, contains 19 real parameters. On the other hand, all possible experiments on muon decay in which the two neutrinos are not observed can determine only six relations between the parameters.

Cabibbo's hypothesis. Unlike j_e and j_μ, the hadronic weak currents J_0 and J_1 cannot be given an explicit form in terms of field operators, because of the complications of strong interactions. However, significant understanding is achieved with the hypothesis formulated by N. Cabibbo in 1963 which states that the hadronic weak current $J_0 + J_1$ can be expressed by Eqs. (18)–(20). Equation (18) states that J_0 and J_1 may each be expressed as the sum of a vector (V) and an axial vector (A) portion. In Eq. (19), $V_0 = \cos\theta\,(j_1 - ij_2)$ and $V_1 = \sin\theta\,(j_4 - ij_5)$ are expressed in terms of an $SU(3)$ octet of vector current operators j_i. In Eq. (20), $A_0 = \cos\theta\,(g_1 - ig_2)$ and $A_1 = \sin\theta\,(g_4 - ig_5)$ are expressed in terms of an $SU(3)$ octet of axial vector current operators g_i. The angle θ (Cabibbo's angle) is a physical constant whose origin is not understood; empirically $\theta \approx 15°$. The vector octet contains not only V_0 and V_1 but also the hadronic electromagnetic current, given by Eq. (21), which is a vector in isospin space (isovector).

$$j_{EM}^\lambda(x) = j_3^\lambda(x) + \frac{1}{\sqrt{3}}j_8^\lambda(x) \quad (21)$$

Cabibbo's hypothesis provided a unified explanation for the following diverse phenomena in hadronic weak interactions.

π_{l2} and K_{l2} decays. The principal mode of charged pion decay is given by Eq. (22). It can be shown that the V-A matrix element for $\pi^- \to l^- \bar{\nu}_l$ must take the form in Eq. (23), where f_π, the pion

$$\pi_{\mu 2}: \quad \pi^+ \to \mu^+\nu_\mu, \pi^- \to \mu^-\bar{\nu}_\mu \quad (22)$$

$$M = \frac{G}{\sqrt{2}}if_\pi \cdot q_\alpha \cdot \bar{u}_l\gamma^\alpha(1+\gamma^5)v_{\nu l} \quad (23)$$

decay constant, cannot be deduced (because of strong interactions) but must be obtained from the observed decay rate, and $q_\alpha = p_{l\alpha} + p_{\bar{\nu}\alpha}$ is the four-momentum transfer to the leptons. This yields a transition probability per unit time given by Eq.

$$J_0 + J_1 = (V_0 + A_0) + (V_1 + A_1) \quad (18)$$

$$V_0 + V_1 = \cos\theta\,(j_1 - ij_2) + \sin\theta\,(j_4 - ij_5) \quad (19)$$

$$A_0 + A_1 = \cos\theta\,(g_1 - ig_2) + \sin\theta\,(g_4 - ig_5) \quad (20)$$

(24). Ratio (25), when corrected for radiative effects, is in excellent agreement with the experimental $\pi_{e2}/\pi_{\mu2}$ branching ratio, in support of "$e-\mu$" universality of the weak interaction.

$$W_{\text{theo}}(\pi^- \to l\bar{\nu}) = \frac{G^2 f_\pi^2}{8\pi}m_l^2 m_\pi\left[1 - \left(\frac{m_l^2}{m_\pi^2}\right)\right]^2 \quad (24)$$

$$\frac{W_{\text{theo}}(\pi^- \to e^-\bar{\nu}_e)}{W_{\text{theo}}(\pi^- \to \mu^-\bar{\nu}_\mu)} \quad (25)$$

The decays $K^+ \to \mu^+\nu_\mu$, $K^- \to \mu^-\bar{\nu}_\mu$ and $K^+ \to e^+\nu_e$, $K^- \to e^-\bar{\nu}_e$ are similar to $\pi_{\mu2}$, π_{e2} decays, respectively, except that the relevant hadronic current is $J_{1\alpha}$ instead of $J_{0\alpha}$; thus one finds Eq. (26),

$$W(K^\pm \to l^\pm\nu(\bar{\nu})) = \frac{G^2}{8\pi}f_K^2 m_l^2 m_K\left(1 - \frac{m_l^2}{m_K^2}\right)^2 \quad (26)$$

where it is expected from Cabibbo's hypothesis that $(f_K^2/f_\pi^2) = (\sin^2\theta/\cos^2\theta)$. Indeed, inserting known masses and transition rates in Eqs. (24) and (26) and comparing for $\pi_{\mu2}$, $K_{\mu2}$, one obtains Eq. (27).

$$\frac{f_K}{f_\pi} = 0.276 = \tan\Theta \quad (27)$$

Nuclear beta decay. The most general matrix element for neutron beta decay is, according to Cabibbo's hypothesis, given by Eq. (28) (and similar expressions can be written to describe the

$$M = \frac{G}{\sqrt{2}}\cos\Theta\,[\bar{u}_p(f_1(q^2)\gamma^\lambda + if_2(q^2)\sigma^{\lambda\nu}q_\nu$$
$$+ f_3(q^2)q^\lambda)u_n + \bar{u}_p(g_1(q^2)\gamma^\lambda\gamma^5 - ig_2(q^2)\sigma^{\lambda\nu}\gamma^5 q_\nu$$
$$+ g_3(q^2)\gamma^5 q^\lambda)u_n] \cdot \bar{u}_e\gamma^\lambda(1+\gamma^5)v_{\overline{\nu e}} \quad (28)$$

other baryon semileptonic decays). Here, u_n, u_p are single-particle Dirac spinors for neutron and proton, respectively, q is the four-momentum transfer to the leptons, $\sigma^{\lambda\nu} = (i/2)(\gamma^\lambda\gamma^\nu - \gamma^\nu\gamma^\lambda)$, and $f_{1,2,3}$, $g_{1,2,3}$ are invariant functions of q^2 ("form factors"). If it were not for strong interactions, these functions would satisfy $f_1 = 1$, f_2, $f_3 = 0$, $g_1 = 1$, $g_2 = g_3 = 0$, and presumably $\cos\Theta \to 1$.

Cabibbo's hypothesis requires $j_1 - ij_2$ and $j_{EM} = j_3 + (1/\sqrt{3})j_8$ to belong to the same vector octet. Thus since $\partial^\alpha j_{EM}/\partial x^\alpha = 0$ (the electromagnetic current is conserved), the same is true for V_0 and V_0^+ (this is the conserved vector current hypothesis, or CVC). Thus it can be shown that $f_3 = 0$. Moreover, the form factors f_1 and f_2 must be identical to their isovector electromagnetic counterparts, satisfying Eqs. (29) and (30), where μ_p, μ_n, and μ_0 are the pro-

$$f_1(0) = 1, \cos\Theta f_1(0) = .975 \quad (29)$$

$$f_2(0) = \frac{1}{2}(\mu_p - \mu_n - \mu_0) \quad (30)$$

ton and neutron nuclear magnetic moments and the nuclear Bohr magneton, respectively. Relation (29) is verified by observation of the ft values of $0^+ \to 0^+$ (initial and final states with spin $= 0$, positive parity) beta decays $0^{14} \to N^{14}$, and so on. The remarkable "weak magnetism" effect, Eq. (30), is verified by detailed observation of the spectral shapes of B^{12}, N^{12} beta decays. Further verification for the CVC hypothesis comes from the branching ratio for the decays $\pi_{e3}(\pi^+ \to \pi^0 e^+ \nu_e)$ and $\pi_{\mu2}$.

An additional restriction on f_1, f_2, f_3, g_1, g_2, g_3 is given by T invariance (which requires all these form factors to be relatively real). Observations of $\text{Arg}(-g_1(0)/f_1(0)) = \text{Arg}(C_A/C_v) = 180° \pm 0.3°$ in neutron and Ne^{19} beta decays confirm T invariance. The term in Eq. (28) in $g_2(q^2)$ is called "second-class" because under a G-parity transformation it has a different character than the other nonvanishing terms. Experimental evidence for nonzero g_2-like terms in $A = 12$, $A = 19$ ($A =$ atomic mass number) beta decays has been uncovered. The induced pseudoscalar term in g_3 is specified by the Goldberger-Treiman relation, based on the so-called partially conserved axial current hypothesis (PCAC), which also establishes a relationship between f_π, $g_1(0) = 1.23$, and the renormalized πNN coupling constant $g_{\pi NN}$ of strong interactions. Terms like that in g_3 in Eq. (28) cannot be uncovered in beta decay (where q^2 is too small), but results of muon capture experiments are in the main consistent with the Goldberger-Treiman relation and CVC. Finally, the Adler-Weisberger relation of current algebra gives a theoretical prediction of $g_1(0)/f_1(0)$ in agreement with experiment. *See* CURRENT ALGEBRA.

Baryon semileptonic decays. Cabibbo's hypothesis provides an accurate prediction of the relative vector and axial vector amplitudes for the various $\Delta S = 0$ and $|\Delta S| = 1$ baryon semileptonic decays.

Selection rules. The $\Delta S = Q$ rule is implicitly contained in the form given by Cabibbo for the hadronic weak current. Likewise the $|\Delta I| = 1/2$ rule is automatically obeyed for leptonic and semileptonic decays since, according to Cabibbo, J_1 transforms as a member of a doublet under isospin transformations (whereas J_0 transforms as a member of an isotriplet). It is not obvious why the $|\Delta I| = 1/2$ rule is approximately valid in nonleptonic decays, and this remains an important unsolved problem.

Parity violation in nuclear forces. Cabibbo's hypothesis provides a starting point for estimating parity violation in nuclear forces (although there are great complications due to nuclear structure effects). Terms of the form $J_0J_0{}^+$ and $J_1J_1{}^+$ in the current-current Lagrangian yield nonzero matrix elements for the two-nucleon interaction $n + p \to p + n$. This generates a weak nuclear force described in the nonrelativistic limit by a two-nucleon potential with a parity-violating (pseudoscalar) component, and leads to the prediction of parity-forbidden nuclear transitions [for example, $O^{16}(2^-) \to {}_6C^{12} + {}_2He^4$] and small degrees of circular polarization in γ-emission from unpolarized nuclei. Such effects are indeed observed and are in order-of-magnitude agreement with calculations.

Intermediate vector boson. The current-current Lagrangian of Eq. (10) describes a contact or point interaction, but quite possibly the weak interaction is nonlocal, and is in reality a second-order process mediated by an intermediate vector boson (IVB). This is suggested from analogy with QED, where the interaction between two electrons is represented in lowest order by the Feynman

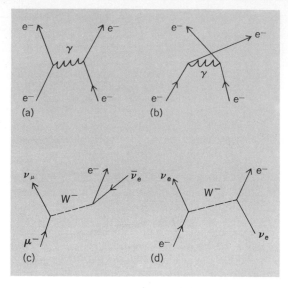

Fig. 4. Feynman diagrams for a: (b) electron-electron scattering by single-photon exchange; (c) muon decay; (d) $e^+\nu_e$ scattering, by W^- exchange.

diagrams of Fig. 4a and b. Each diagram contains two vertices coupled by an intermediate (virtual) photon. Since the latter has zero rest mass, the Coulomb force between the electrons has infinite range ($1/r$ potential); furthermore, the photon has spin-1 and is a vector particle. Similarly, the weak interactions $\mu^- \to e^-\nu_e\nu_\mu$, $e^- \nu_e \to \nu_e e^-$, for example, may be represented in lowest order by Fig. 4c and d where $W^\pm$ is a hypothetical, so far unobserved IVB (charged) which must have unit spin but no definite parity, and must be very massive since the weak interactions certainly have very short range.

The experimental lower limit on m_w is approximately 25 GeV. Because m_w must be so large, even if the IVB exists it can have essentially no practical effect in the decays listed in the table (in all of which the momentum transfer is very small). However, $W^\pm$ (and an analogous neutral IVB associated with neutral weak currents) could manifest themselves in high-energy neutrino-nucleon scattering experiments.

UNSOLVED PROBLEMS

The cross section for the purely leptonic process $e^- + \nu_e \to \nu_e + e^-$ can be calculated using the current-current Lagrangian or the IVB theory and the V-A currents j_e and j_μ given by Eqs. (11) and (12). In the former case, one finds that the cross section in the center-of-mass frame is given by Eq. (31), where w is the neutrino energy. However, unitarity (conservation of probability) requires inequality (32), and results (31) and (32) are contradictory for

$$\sigma(e^-\nu_e) \times \frac{4G^2}{\pi} w^2 \qquad (31)$$

$$\sigma(e^-\nu_e) < \frac{\pi}{w^2} \qquad (32)$$

$w > 300$ GeV. Corrections due to higher-order weak interactions diverge and thus fail to remedy the difficulty; nor can these divergences be removed with a finite number of renormalization constants, as in QED. In the IVB theory the cross section remains bounded as neutrino energy w increases for any finite boson mass, but this cross section also violates unitarity for sufficiently large w. This profound theoretical difficulty is not confined to the example given, but is symptomatic and has not been resolved. By ingenious and sophisticated arguments, an attempt has been made to create a renormalizable theory of weak interactions by establishing basic links to electromagnetic (and strong) interactions through gauge theories. However, these efforts are speculative. One must look to high energy neutrino-nucleon scattering experiments for clues to the solution of unsolved problems in weak interactions. Neutrinos like electrons have been used as probes to demonstrate that the nucleon behaves in many respects as if it contains pointlike scattering centers called partons.   [EUGENE D. COMMINS]

Bibliography: E. D. Commins, *Weak Interactions*, 1973; R. E. Marshak, Riazuddin and C. P. Ryan, *Theory of Weak Interactions in Particle Physics*, 1969; L. B. Okun, *Weak Interactions of Elementary Particles*, 1965; C. S. Wu and S. A. Moszkowski, *Beta Decay*, 1966.

Intercommunicating system

A telephone system providing direct communication between telephones on the same premises. An intercommunicating system, often referred to as an intercom, may be of two general types: (1) those utilizing regular telephones both for the intercom and for association with the nationwide telephone network, and (2) local systems not associated with the nationwide telephone network. Both types range from simple to complex in scope.

Key telephone systems. Intercommunicating systems using instruments associated with the nationwide telephone network usually employ telephones equipped with a number of keys or buttons so that a telephone can be connected either to any one of several central office lines, to Centrex lines or PBX extensions, or to an intercom line. Systems employing such key-equipped telephones, together with any relays, lamps, or other related apparatus or wiring, are called key telephone systems. They are furnished in a wide variety of setups. The keys may be an integral part of the telephone set or separately mounted; signal lamps, which flash to indicate incoming calls, light steadily during conversation, and wink while a line is held, may assume various forms. In most key telephone instruments the key buttons are made of transparent plastic and are illuminated from beneath, thus serving as visual signals. *See* TELEPHONE; TELEPHONE PRIVATE BRANCH EXCHANGE (PBX).

Larger key telephone systems utilize consoles for directing calls and are designed for desk-top usage (Fig. 1). The sets are arranged on a modular basis so that additional strips of key-illuminated pickup buttons can be plugged in if needed for additional line appearances. Other key equipment sets are designed for flush mounting in openings in table or desk tops.

Key telephone systems offer a wide variety of optional features, the more important of which are described as follows.

Hold. One key button, when operated momen-

Fig. 1. Call-director 30-button key telephone. (*American Telephone and Telegraph*)

tarily, places a relay winding across the telephone line so that the user may temporarily transfer his telephone set to any other line that appears at his instrument without disconnecting the held line. He can return to the original call by reoperating the original line key. The held line may also be answered by any other key telephone set within the system that has an appearance of that line.

Intercom signaling. A specific telephone can be called by either dialing a one- or two-digit number or by operating a manual push button.

Preset conference calling. Several different prearranged groups of telephones can be called simultaneously by dialing the specific one- or two-digit code associated with the particular group wanted. In some systems this is accomplished by manually operating several push buttons simultaneously.

Camp-on. When a phone within the same key system equipment is called and found busy, the calling station can wait, without hanging up, and ring that station automatically as soon as it hangs up from its existing call. This camp-on feature enables a calling party to be connected ahead of any others who also may have been trying to reach that particular party.

Add-on. If a call is received from a central office line, a Centrex line, or a PBX extension and the person on the called station wishes to include other telephones within the same key equipment system, he can place the incoming line on hold, call another station within his system, and determine if that station wishes to participate in his call. If so, he then operates a key connecting the incoming call, himself, and the add-on station together. He may elect to drop off the connection, leaving the other parties connected.

Hands-free talking. A small microphone, enclosed in a housing and equipped with an ON-OFF button and a volume control, and a separately enclosed loudspeaker are shown in Fig. 2. The telephone set can be used in the regular way if desired. However, the user can operate on a completely hands-free basis by merely operating the ON button to either originate or answer any telephone

Fig. 2. Key telephone for hands-free talking. (*American Telephone and Telegraph*)

call. To originate a call while in the hands-free mode, the dial of the telephone is used even though the handset is not removed from the cradle.

Privacy. Many arrangements enable a phone, or groups of phones, to have privacy and exclude some or all of the other stations within a key equipment system from a conversation on either a manual key-operated or automatic basis. These arrangements include excluding, or cutting off, all stations in a descending priority order either at all times (including during a conversation) or only when a low-priority station has not answered a call. Most of these features are generally automatically canceled when the main, higher-priority, telephone hangs up.

Executive access. In some key equipment installations, one or more telephones may be arranged so that an executive can obtain access to his subordinates' telephones regardless of whether they are busy.

Other features. There are many additional optional features, too numerous to mention, that can be provided. In general, the relay equipment is designed on a building-block basis so as to accommodate almost any requirement.

Particular applications. Intercommunicating systems are available to meet the particular needs of specific applications for homes, farms, businesses, and hospitals.

Residence (home) intercommunicating systems include such features as local signaling and talking, hands-free talking, and door answering. The latter feature allows anyone in the house to use any telephone to converse with whoever rings the doorbell.

Farm intercommunicating systems are designed to provide communications between the farm residence, barns, and outdoor locations. Such intercoms also include the ability to answer or originate calls by means of the central office line. In general, most of the stations are equipped for hands-free operation. Loud-ringing bells are usually provided for outdoor coverage.

Business intercoms generally allow for completely hands-free operation so as to enable calls to be made and automatically answered at the called location. Additional features include the transference of central office calls throughout the intercommunicating system.

Hospital interphone systems allow a nurse to voice call, monitor, and be called by up to 40 patients. This system provides for complete hands-free operation by all stations with optional handset usage for privacy. Additional features include nurse-call pendant cords, corridor lights, and emergency switches in patients' lavatories.

Local systems. Local intercoms are not connected to, or associated with, the nationwide telephone network. These vary in size, complexity, and method of operation. In various forms they use most of the features of telephone-utilizing systems, although they may accomplish them by different technology.

Many of these systems contain high-gain amplifiers and permit voice paging at all subordinate stations under control of a push-to-talk key at a master station. Usually these loudspeakers also serve as dynamic microphones for speech in the opposite direction. Some systems use carrier principles to superimpose voice communication circuits on ordinary electrical wiring in a building so as to obviate the need for interstation wiring. The intercom may also share facilities with in-plant background music and public address channels. *See* SWITCHING SYSTEMS (COMMUNICATIONS); TELEPHONE SERVICE.   [WILLIAM SCHIAVONI]

Interface of phases

The boundary between any two phases. Among the three phases, gas, liquid, and solid, five types of interfaces are possible: gas-liquid, gas-solid, liquid-liquid, liquid-solid, and solid-solid. The abrupt transition from one phase to another at these boundaries, even though subject to the kinetic effects of molecular motion, is statistically a surface only one or two molecules thick.

A unique property of the surfaces of the phases that adjoin at an interface is the surface energy which is the result of unbalanced molecular fields existing at the surfaces of the two phases. Within the bulk of a given phase, the intermolecular forces are uniform because each molecule enjoys a statistically homogeneous field produced by neighboring molecules of the same substance. Molecules in the surface of a phase, however, are bounded on one side by an entirely different environment, with the result that there are intermolecular forces that then tend to pull these surface molecules toward the bulk of the phase. A drop of water, as a result, tends to assume a spherical shape in order to reduce the surface area of the droplet to a minimum.

Surface energy. At an interface, there will be a difference in the tendencies for each phase to attract its own molecules. Consequently, there is always a minimum in the free energy of the surfaces at an interface, the net amount of which is called the interfacial energy. At the water-air interface, for example, the difference in molecular fields in the water and air surfaces accounts for the interfacial energy of 72 ergs/cm² of interfacial surface. The interfacial energy between the two liquids, benzene and water, is 35 ergs/cm², and between ethyl ether and mercury is 379 ergs/cm². These interfacial energies are also expressed as surface tension in units of dynes per centimeter.

The surface energy at an interface may be altered by the addition of solutes that migrate to the

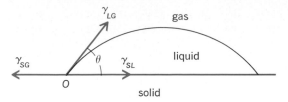

Contact angle θ at interface of three phases.

surface and modify the molecular forces there, or the surface energy may be changed by converting the planar interfacial boundary to a curved surface. Both the theoretical and practical implications of this change in surface energy are embodied in the Kelvin equation, Eq. (1), where P/P_0 is

$$\ln \frac{P}{P_0} = \frac{2M\gamma}{RT\rho r} \qquad (1)$$

the ratio of the vapor pressure of a liquid droplet with diameter r to the vapor pressure of the pure liquid in bulk, ρ the density, γ the surface energy, and M the molecular weight. Thus, the smaller the droplet the greater the relative vapor pressure, and as a consequence, small droplets of liquid evaporate more rapidly than larger ones. The surface energy of solids is also a function of their size, and the Kelvin equation can be modified to describe the greater solubility of small particles compared to that of larger particles of the same solid. *See* Adsorption; Surface-active agent.

Contact angle. At liquid-solid interfaces, where the confluence of the two phases is usually termed wetting, a critical factor called the contact angle is involved. A drop of water placed on a paraffin surface, for example, retains a globular shape, whereas the same drop of water placed on a clean glass surface spreads out into a thin layer. In the first instance, the contact angle is practically 180°, and in the second instance, it is practically 0°. The study of contact angles reveals the interplay of interfacial energies at three boundaries. The illustration is a schematic representation of the cross section of a drop of liquid on a solid. There are solid-liquid, solid-gas, and liquid-gas interfaces that meet in a linear zone at O. The forces about O that determine the equilibrium contact angle are related to each other according to Eq. (2), where the γ

$$\gamma_{SG} = \gamma_{SL} + \gamma_{LG} \cos \theta \qquad (2)$$

terms represent free energies at the interfaces and θ is the contact angle. Since only γ_{LG} and θ can be measured readily, the term adhesion tension is defined by Eq. (3). Adhesion tension, which is the

$$\gamma_{LG} \cos \theta = \gamma_{SG} - \gamma_{SL} = \text{adhesion tension} \qquad (3)$$

free energy of setting, is of critical importance in detergency, dispersion of powders and pigments, lubrication, adhesion, and spreading processes.

The measurement of interfacial energies is made directly only upon liquid-gas and liquid-liquid interfaces. In measuring the liquid-gas interfacial energy (surface tension), the methods of capillary rise, drop weight on pendant drop, bubble pressure, sessile drops, Du Nuoy ring, vibrating jets, and ultrasonic action are among those used. There is a small but appreciable temperature

effect upon surface tension, and this property is used to determine small differences in the surface tension of a liquid by placing the two ends of a liquid column in a capillary tube whose two ends are at different temperatures. The determination of interfacial energies at other types of interfaces can be inferred only by indirect methods. *See* Equilibrium, phase; Flotation; Foam; Free energy; Surface tension.   [WENDELL H. SLABAUGH]

Bibliography: American Chemical Society, *Contact Angle, Wettability, and Adhesion*, Advan. Chem. Ser. no. 43., 1964; J. T. Davies and E. Rideal, *Interfacial Phenomena*, 2d ed., 1963.

Interference, electrical

Any undesired electrical energy that tends to interfere with the reception of desired signals. It may be a man-made signal generated by nearby electric motors, automotive ignition systems, radio transmitters, improperly operating television receivers, or a wide variety of other improperly operating electric devices. The interference signals either may be radiated through space as unguided electromagnetic waves or may travel along power lines as guided radiation. Interference may also be caused by atmospheric phenomena such as lightning. Such natural disturbances are sometimes termed sferics. Interfering signals are sometimes generated intentionally by powerful transmitters for the purpose of jamming communication and radar systems. *See* Noise, electrical.

Man-made interference is best suppressed at its source, by such means as minimizing sparking, connecting capacitors across sparking contacts, shielding the equipment with metal enclosures, and using choke coils or filters that prevent interfering signals from entering power lines. *See* Filter, electric; Grounding, electrical; Grounding, electronic-equipment; Shielding, electrical.   [JOHN MARKUS]

Interference filter, optical

An optical filter in which the wavelengths that are not transmitted are removed by interference phenomena rather than by absorption or scattering. In addition to being able to duplicate most of the spectral characteristics of absorption color filters, these devices can be made to transmit a very narrow band of wavelengths. They can thus be used as monochromators to examine a radiation source at the wavelength of a single spectrum line. For example, the solar disk can be observed in light of the hydrogen line Hα and thus the distribution of excited hydrogen over the disk can be determined.

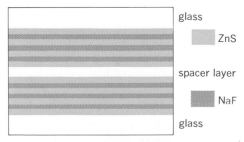

Fig. 1.   Schematic diagram of seven-layer solid Fabry-Perot filter. (*From D. E. Gray, ed., American Institute of Physics Handbook, McGraw-Hill, 1957*)

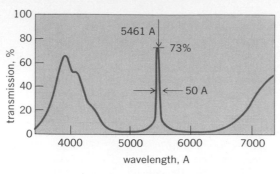

Fig. 2. Transmission of filter shown in Fig. 1 as a function of wavelength. (*From D. E. Gray, ed., American Institute of Physics Handbook, McGraw-Hill, 1957*)

See INTERFERENCE OF WAVES.

Most narrow-band interference filters are based on the Fabry-Perot interferometer. The Fabry-Perot interference filter differs from the interferometer only in the thickness of the space between the partially reflecting layers. In the interferometer, this space can be several centimeters. In the filter, it is normally a few thousand angstroms. In the simplest filter, a glass plate is coated with a layer of silver which is covered by a layer of dielectric followed in turn by another evaporated layer of semitransparent silver. *See* INTERFEROMETRY.

Basic properties. At all wavelengths at which the dielectric layer has an optical thickness of an integral number of half waves, the filter will have a passband. The number of half waves corresponding to a given passband is called the order of the passband. The transmission T of the filter can be represented by Eq. (1), where r is the reflectivity of

$$T = \frac{t^2}{(1-r)^2 + 4r \sin^2 (\delta/2)} \qquad (1)$$

the silver film and t the transmission of the film; δ is defined in Eq. (2), where d is the thickness of the

$$\delta = \frac{4\pi d}{\lambda} (n^2 - \sin^2 \theta)^{1/2} + 2y \qquad (2)$$

dielectric layer, n its refractive index, λ the wavelength, y the phase shift experienced by the light at the metal-dielectric boundary, and θ the angle of incidence.

By inspection of Eq. (1), it is apparent that maxima occur when $\delta/2 = m\pi$, where m is an integer.

Some of the quantities which are of interest to the user of these filters are (1) the peak transmission, (2) the transmission between peaks, (3) the bandwidth, and (4) the angular field of view, that is, the angle through which the filter must be tilted to shift the wavelength of peak transmission a distance equal to the bandwidth.

Each of these quantities can be determined theoretically from Eq. (1). A typical filter has a peak transmission of 40% at its peak wavelength of 5461 A, a transmission between peaks of 0.2%, a bandwidth of 100 A, and an angular field of view of 20°. These numbers represent nearly the best that can be done with the simple metal-dielectric filter.

Multilayer types. An increase in reflectivity can result in narrower bandwidths. There are techniques by which high reflectivities can be achieved which are lossless, that is, which have no absorption. This results in higher peak transmission and lower off-peak transmission. The first is the multi-

layer filter. In this device, the metal layers are replaced by a series of dielectric layers. The boundary between two dielectric layers of refractive indices n_1 and n_2 has a reflecting power of perhaps 4% in the case of glass and air, or less for two dielectrics whose indices are close together. The value of the reflectivity r is given by the standard Fresnel reflection law, Eq. (3). *See* REFLECTION OF ELECTROMAGNETIC RADIATION.

$$r = \left(\frac{n_1 - n_2}{n_1 + n_2}\right)^2 \qquad (3)$$

By making several layers of alternate high- and low-index dielectric, it is possible to reinforce the reflectivity of a single boundary and build it up by multiple reflection to any desired value. It is necessary only that the layers be of such thickness that the reflections from successive boundaries are in phase. When each layer is optically a quarter wavelength in thickness, this reinforcement takes place. A complete filter is sketched in Fig. 1. It might consist of seven alternate layers of high- and low-index dielectric of a thickness of a quarter wavelength apiece, followed by the dielectric spacer which is an integral number of half waves and which is followed by seven more quarter-wavelength layers. The characteristics of such a filter are shown in Fig. 2. For a seven-layer reflection filter, the reflectivity can be built up to 95%. One would expect improvement over the metal filter, and in fact, the peak transmission of such a filter is as high as 80% and the bandwidth as low as 5 A.

Frustrated reflection. A second technique is the use of frustrated total internal reflection for the partially reflecting layers of the filter. Light is totally internally reflected when incident on the hypotenuse of a right angle prism, as in Fig. 3. If another prism is brought up to the first, as in Fig. 4, a part of the light is transmitted through the combination. The fraction transmitted depends on the separation between the prisms. The reflectivity can be made as high as desired. In a filter, the prism hypotenuse is coated with a low-index layer of a thickness chosen to give the proper value of the reflectivity. This is covered with a high-index layer whose thickness determines the wavelength of the passband, as in the normal Fabry-Perot filter. This is followed by a second low-index layer and a second prism. The first frustrated reflection is at the boundary between the first prism and the low-index layer. The second reflection is at the boundary between the first prism and the low-index layer. The second reflection is at the boundary of the high-index layer and the second low-index layer. The thickness of the low-index layer can be adjusted to give any value of reflectivity. The filter is lossless unless there is absorption within the layers. The resulting filter can be made to have a bandwidth of less than 6 A. [BRUCE H. BILLINGS]

Bibliography: D. E. Gray (ed.), *American Institute of Physics Handbook*, 1957; R. Kingslake, *Applied Optics and Optical Engineering*, vol. 1: *Light: Its Generation and Modification*, 1965.

Interference of waves

The process whereby two or more waves of the same frequency or wavelength combine to form a wave whose amplitude is the sum of the amplitudes of the interfering waves. The interfering

INTERFERENCE
FILTER, OPTICAL

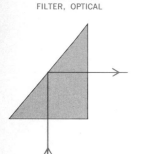

Fig. 3. Total internal reflection in a right-angle prism.

INTERFERENCE
FILTER, OPTICAL

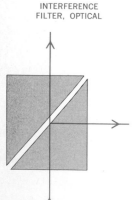

Fig. 4. Frustrated total internal reflection.

waves can be electromagnetic, acoustic, or water waves, or in fact any periodic disturbance.

The most striking feature of interference is the effect of adding two waves in which the trough of one wave coincides with the peak of another. If the two waves are of equal amplitude, they can cancel each other out so that the resulting amplitude is zero. This is perhaps most dramatic in sound waves; it is possible to generate acoustic waves to arrive at a person's ear so as to cancel out noise that is disturbing him. In optics, this cancellation can occur for particular wavelengths in a situation where white light is a source. The resulting light will appear colored. This gives rise to the iridescent colors of beetles' wings and mother-of-pearl, where the substances involved are actually colorless or transparent.

Two-beam interference. The quantitative features of the phenomenon can be demonstrated most easily by considering two interfering waves. The amplitude of the first wave at a particular point in space can be written as Eq. (1), where A_0 is

$$A = A_0 \sin (\omega t + \varphi_1) \qquad (1)$$

the peak amplitude, and ω is 2π times the frequency. For the second wave Eq. (2) holds, where

$$B = B_0 \sin (\omega t + \varphi_2) \qquad (2)$$

$\varphi_1 - \varphi_2$ is the phase difference between the two waves. In interference, the two waves are superimposed, and the resulting wave can be written as Eq. (3).

$$A + B = A_0 \sin (\omega t + \varphi_1) + B_0 \sin (\omega t + \varphi_2) \qquad (3)$$

Equation (3) can be expanded to give Eq. (4).

$$A + B = (A_0 \sin \varphi_1 + B_0 \sin \varphi_2) \cos \omega t$$
$$+ (A_0 \cos \varphi_1 + B_0 \cos \varphi_2) \sin \omega t \qquad (4)$$

By writing Eqs. (5) and (6), Eq. (4) becomes Eq.

$$A_0 \sin \varphi_1 + B_0 \sin \varphi_2 = C \sin \varphi_3 \qquad (5)$$
$$A_0 \cos \varphi_1 + B_0 \cos \varphi_2 = C \cos \varphi_3 \qquad (6)$$

(7), where C^2 is defined in Eq. (8). When C is less

$$A + B = C \sin (\omega t + \varphi_3) \qquad (7)$$
$$C^2 = A_0{}^2 + B_0{}^2 + 2A_0B_0 \cos (\varphi_2 - \varphi_1) \qquad (8)$$

than A or B, the interference is called destructive. When it is greater, it is called constructive. For electromagnetic radiation, such as light, the amplitude in Eq. (7) represents an electric field strength. This field is a vector quantity and is associated with a particular direction in space, the direction being generally at right angles to the direction in which the wave is moving. These electric vectors can be added even when they are not parallel. For a discussion of the resulting interference phenomena *see* POLARIZED LIGHT. *See also* SUPERPOSITION, PRINCIPLE OF.

In the case of radio waves or microwaves which are generated with vacuum tube or solid-state oscillators, the frequency requirement for interference is easily met. In the case of light waves, it is more difficult. Here the sources are generally radiating atoms. The smallest frequency spread from such a light source will still have a bandwidth of the order of 10^7 Hz. Such a bandwidth occurs in a single spectrum line, and can be considered a result of the existence of wave trains no longer than

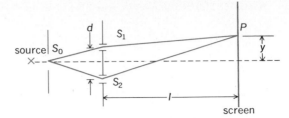

Fig. 1. Young's two-slit interference.

10^{-8} sec. The frequency spread associated with such a pulse can be written as notation (9), where

$$\Delta f \simeq \frac{1}{2\pi t} \qquad (9)$$

t is the pulse length. This means that the amplitude and phase of the wave which is the sum of the waves from two such sources will shift at random in times shorter than 10^{-8} sec. In addition, the direction of the electric vector will shift in these same time intervals. Light which has such a random direction for the electric vector is termed unpolarized. When the phase shifts and direction changes of the light vectors from two sources are identical, the sources are termed coherent.

Splitting of light sources. To observe interference with waves generated by atomic or molecular transitions, it is necessary to use a single source and to split the light from the source into parts which can then be recombined. In this case, the amplitude and phase changes occur simultaneously in each of the parts at the same time.

Young's two-slit experiment. The simplest technique for producing a splitting from a single source was done by T. Young in 1801 and was one of the first demonstrations of the wave nature of light. In this experiment, a narrow slit is illuminated by a source, and the light from this slit is caused to illuminate two adjacent slits. The light from these two parallel slits can interfere, and the interference can be seen by letting the light from the two slits fall on a white screen. The screen will be covered with a series of parallel fringes. The location of these fringes can be derived approximately as follows: In Fig. 1, S_1 and S_2 are the two slits separated by a distance d. Their plane is a distance l from the screen. Since the slit S_0 is equidistant from S_1 and S_2, the intensity and phase of the light at each slit will be the same. The light falling on p from slit S_1 can be represented by Eq. (10) and

$$A = A_0 \sin 2\pi f \left(t - \frac{x_1}{c}\right) \qquad (10)$$

from S_2 by Eq. (11), where f is the frequency, t the

$$B = A_0 \sin 2\pi f \left(t - \frac{x_2}{c}\right) \qquad (11)$$

time, c the velocity of light; x_1 and x_2 are the distances of P from S_1 and S_2, and A_0 is the amplitude. This amplitude is assumed to be the same for each wave since the slits are close together, and x_1 and x_2 are thus nearly the same. These equations are the same as Eqs. (1) and (2), with $\varphi_1 = x_1/c$ and $\varphi_2 = x_2/c$. Accordingly, the square of the amplitude or the intensity at P can be written as Eq. (12).

In general, l is very much larger than y so that Eq. (12) can be simplified to Eq. (13).

Equation (13) is a maximum when Eq. (14) holds

$$I = 4A_0^2 \cos^2 \frac{2\pi f}{c}(x_1 - x_2) \qquad (12)$$

$$I = 4A_0^2 \cos^2 \pi \left(\frac{yd}{l\lambda}\right) \qquad (13)$$

$$y = n\lambda \frac{l}{d} \qquad (14)$$

and a minimum when Eq. (15) holds, where n is an integer.

$$y = (n + \tfrac{1}{2})\lambda \frac{l}{d} \qquad (15)$$

Accordingly, the screen is covered with a series of light and dark bands called interference fringes.

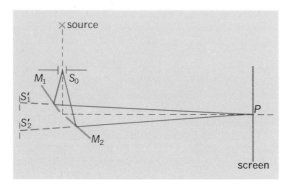

Fig. 2. Fresnel's double-mirror interference.

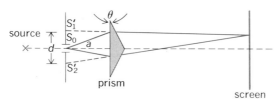

Fig. 3. Fresnel biprism interference.

Fig. 4. Equipment for demonstrating Fresnel biprism interference.

If the source behind slit S_0 is white light and thus has wavelengths varying perhaps from 4000 to 7000 A, the fringes are visible only where $x_1 - x_2$ is a few wavelengths, that is, where n is small. At large values of n, the position of the nth fringe for red light will be very different from the position for blue light, and the fringes will blend together and be washed out. With monochromatic light, the fringes will be visible out of values of n which are determined by the diffraction pattern of the slits. For an explanation of this *see* DIFFRACTION.

The energy carried by a wave is measured by the intensity, which is equal to the square of the amplitude. In the preceding example of the superposition of two waves, the intensity of the individual waves in Eqs. (1) and (2) is A^2 and B^2, respectively. When the phase shift between them is zero, the intensity of the resulting wave is given by Eq. (16).

$$(A + B)^2 = A^2 + 2AB + B^2 \qquad (16)$$

This would seem to be a violation of the first law of thermodynamics, since this is greater than the sum of the individual intensities. In any specific experiment, however, it turns out that the energy from the source is merely redistributed in space. The excess energy which appears where the interference is constructive will disappear in those places where the energy is destructive. This is illustrated by the fringe pattern in the Young two-slit experiment. The energy on the screen from each slit alone is given by Eq. (17), where A_0^2 is the

$$E_1 = \int_0^\infty A_0^2 \, dy \qquad (17)$$

intensity of the light from each slit as given by Eq. (10). The intensity from the two slits without interference would be twice this value. The intensity with interference is given by Eq. (18). The com-

$$E_3 = \int_0^\infty 4A^2 \cos^2\left[2\pi\left(\frac{yd}{l\lambda}\right)\right] dy \qquad (18)$$

parison between $2E_1$ and E_3 need be made only over a range corresponding to one full cycle of fringes. This means that the argument of the cosine in Eq. (18) need be taken only from zero to π. This corresponds to a section of screen going from the center to a distance $y = l\lambda/2d$. From the two slits individually, the energy in this section of screen can be written as Eq. (19).

$$2E_1 = 2\int_0^{l\lambda/2d} A_0^2 \, dy = \frac{A_0^2 l\lambda}{d} \qquad (19)$$

With interference, the energy is given by Eq. (20). Equation (20) can be written as Eq. (21).

$$E_3 = \int_0^{l\lambda/2d} 4A_0^2 \cos^2\left[2\pi\left(\frac{yd}{l\lambda}\right)\right] dy \qquad (20)$$

$$E_3 = \frac{l\lambda}{2\pi d}\int_0^\pi 4A_0^2 \cos^2\varphi \, d\varphi = \frac{A_0^2 l\lambda}{d} \qquad (21)$$

Thus, the total energy falling on the screen is not changed by the presence of interference. The energy density at a particular point is, however, drastically changed. This fact is most important for those waves of the electromagnetic spectrum which can be generated by vacuum-tube oscillators. The sources of radiation or antennas can be made to emit coherent waves which will undergo interference. This makes possible a redistribution of the radiated energy. Quite narrow beams of radiation can be produced by the proper phasing of a linear antenna array. *See* ANTENNA (ELECTROMAGNETISM).

The double-slit experiment also provides a good illustration of Niels Bohr's principle of complementarity. For detailed information on this *see* QUANTUM MECHANICS.

Fresnel double mirror. Another way of splitting the light from the source is the Fresnel double mirror (Fig. 2). Light from the slit S_0 falls on two mirrors M_1 and M_2 which are inclined to each other at an angle of the order of a degree. On a screen where the illumination from the two mirrors overlaps, there will appear a set of interference fringes. These are the same as the fringes produced in the two-slit experiment, since the light on the screen comes from the images of the slits S'_1 and S'_2 formed by the two mirrors, and these two images are the equivalent of two slits.

Fresnel biprism. A third way of splitting the source is the Fresnel biprism. A sketch of a cross section of this device is shown in Fig. 3. The light from the slit at S_0 is transmitted through the two halves of the prism to the screen. The beam from each half will strike the screen at a different angle and will appear to come from a source which is slightly displaced from the original slit. These two virtual slits are shown in the sketch at S'_1 and S'_2. Their separation will depend on the distance of the prism from the slit S_0 and on the angle θ and index of refraction of the prism material. In Fig. 3, a is the distance of the slit from the biprism, and l the distance of the biprism from the screen. The distance of the two virtual slits from the screen is thus $a + l$. The separation of the two virtual slits is given by Eq. (22), where μ is the refractive index of

$$d = 2a(\mu - 1)\theta \qquad (22)$$

the prism material. This can be put in Eq. (14) for the two-slit interference pattern to give Eq. (23) for the position of a bright fringe.

$$y = n\lambda \frac{a + l}{2a(\mu - 1)\theta} \qquad (23)$$

A photograph of the experimental equipment for demonstrating interference with the Fresnel biprism is shown in Fig. 4. A typical fringe pattern is shown in Fig. 5. This pattern was obtained with a mercury-arc source, which has several strong spectrum lines, accounting in part for the intensity variation in the pattern. The pattern is also modified by diffraction at the apex of the prism.

Billet split lens. The source can also be split with the Billet split lens (Fig. 6). Here a simple lens is sawed into two parts which are slightly separated.

Lloyd's mirror. An important technique of splitting the source is with Lloyd's mirror (Fig. 7). The slit S_1 and its virtual image S'_2 constitute the double source. Part of the light falls directly on the screen, and part is reflected at grazing incidence from a plane mirror. This experiment differs from the previously discussed experiments in that the two beams are no longer identical. If the screen is moved to a point where it is nearly in contact with the mirror, the fringe of zero path difference will lie on the intersection of the mirror plane with the screen. This fringe turns out to be dark rather than light, as in the case of the previous interference

experiments. The only explanation for this result is that light experiences a 180° phase shift on reflection from a material of higher refractive index than its surrounding medium. The equation for maximum and minimum light intensity at the screen must thus be interchanged for Lloyd's mirror fringes.

Amplitude splitting. The interference experiments discussed have all been done by splitting the wavefront of the light coming from the source. The energy from the source can also be split in amplitude. With such amplitude-splitting techniques, the light from the source falls on a surface which is partially reflecting. Part of the light is transmitted, part is reflected, and after further manipulation these parts are recombined to give the interference. In one type of experiment, the light transmitted through the surface is reflected from a second surface back through the partially reflecting surface, where it combines with the wave reflected from the first surface (Fig. 8). Here the arrows represent the normal to the wavefront of the light passing through surface S_1 to surface S_2. The wave is incident at A and C. The section at A is partially transmitted to B, where it is again partially reflected to C. The wave leaving C now consists of two parts, one of which has traveled a longer distance than the other. These two waves will interfere. Let AD be the perpendicular from

Fig. 5. Interference fringes formed with Fresnel biprism and mercury-arc light source.

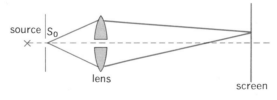

Fig. 6. Billet split-lens interference.

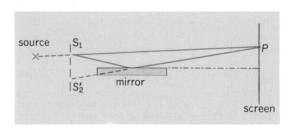

Fig. 7. Lloyd's mirror interference.

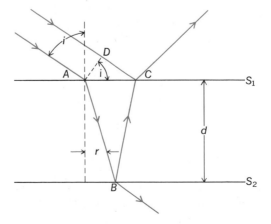

Fig. 8. Dielectric-plate-reflection interference.

the ray at A to the ray going to C. The path difference will be given by Eq. (24), where μ is the

$$\Delta = 2\mu(AB) - (CD) \qquad (24)$$

refractive index of the medium between the surfaces S_1 and S_2, and AB and CD are defined in Eqs. (25) and (26).

$$(AB) = d/\cos r \qquad (25)$$

$$(CD) = 2(AB)\sin r \cos i \qquad (26)$$

From Snell's law, Eq. (27) is obtained and thus Eq. (28) holds.

$$\sin i = \mu \sin r \qquad (27)$$

$$\Delta = \frac{2\mu d}{\cos r} - \frac{2\mu d}{\cos r}\sin^2 r = 2\mu d \cos r \qquad (28)$$

The difference in terms of wavelength and the phase difference are, respectively, given by Eqs. (29) and (30).

$$\Delta' = \frac{2\mu d \cos r}{\lambda} \qquad (29)$$

$$\Delta\varphi = \frac{4\pi\mu d \cos r}{\lambda} + \pi \qquad (30)$$

The phase difference of π radians is added because of the phase shift experienced by the light reflected at S_1. The experimental proof of this 180° phase shift was shown in the description of interference with Lloyd's mirror. If the plate of material has a lower index than the medium in which it is immersed, the π radians must still be added in Eq. (30), since now the beam reflected at S_2 will experience this extra phase shift. The purely pragmatic necessity of such an additional phase shift can be seen by considering the intensity of the reflected light when the surfaces S_1 and S_2 almost coincide. Without the extra phase shift, the two reflected beams would be in phase and the reflection would be strong. This is certainly not proper for a film of vanishing thickness. Constructive interference will take place at wavelengths for which $\Delta\varphi = 2m\pi$, where m is an integer. If the surfaces S_1 and S_2 are parallel, the fringes will be located optically at infinity. If they are not parallel, d will be a function of position along the surfaces and the fringes will be located near the surface. The intensity of the fringes will depend on the value of the partial reflectivity of the surfaces.

Testing of optical surfaces. Observation of fringes of this type can be used to determine the contour of a surface. The surface to be tested is put close to an optically flat plate. Monochromatic light is reflected from the two surfaces and examined as in Fig. 8. One of the first experiments with fringes of this type was performed by Isaac Newton. A convex lens is pressed against a glass plate and illuminated with monochromatic light. A series of circular interference fringes known as Newton's rings appear around the point of contact. From the separation between the fringes, it is possible to determine the radius of curvature of the lens.

Thin films. Interference fringes of this two-surface type are responsible for the colors which appear in oil films floating on water. Here the two surfaces are the oil-air interface and the oil-water

interface. The films are close to a visible light wavelength in thickness. If the thickness is such that, in a particular direction, destructive interference occurs for green light, red and blue will still be reflected and the film will have a strong purple appearance. This same general phenomenon is responsible for the colors of beetles' wings.

Channeled spectrum. Amplitude splitting shows clearly another condition that must be satisfied for interference to take place. The beams from the source must not only come from identical points, but they must also originate from these points at nearly the same time. The light which is reflected from C in Fig. 8 originates from the source later than the light which makes a double traversal between S_1 and S_2. If the surfaces are too far apart, the spectral regions of constructive and destructive interference become so close together that they cannot be resolved. In the case of interference by wavefront splitting, the light from different parts of a source could only be considered coherent if examined over a sufficiently short time interval. In the case of amplitude splitting, the interference when surfaces are widely separated can only be seen if examined over a sufficiently narrow frequency interval. If the two surfaces are illuminated with white light and the eye is used as the analyzer, interference cannot be seen when the separation is more than a few wavelengths. The interval between successive wavelengths of constructive interference becomes so small that each spectral region to which the eye is sensitive is illuminated, and no color is seen. In this case, the interference can again be seen by examining the reflected light with a spectroscope. The spectrum will be crossed with a set of dark fringes at those wavelengths for which there is destructive interference. This is called a channeled spectrum. For large separations of the surfaces, the separation between the wavelengths of destructive interference becomes smaller than the resolution of the spectrometer, and the fringes are no longer visible.

Fresnel coefficient. The amplitude of the light reflected at normal incidence from a dielectric surface is given by the Fresnel coefficient, Eq. (31),

$$A = A_0 \frac{n_1 - n_2}{n_1 + n_2} \qquad (31)$$

where A_0 is the amplitude of the incident wave and n_1 and n_2 are the refractive indices of the materials in the order in which they are encountered by the light. In the simple case of a dielectric sheet, the intensity of the light reflected normally will be given by Eq. (32), where B is the amplitude of the

$$C^2 = A^2 + B^2 + 2AB \cos \varphi \qquad (32)$$

wave which has passed through the sheet and is reflected from the second surface and back through the sheet to join A. The value of B is given by Eq. (33), where the approximation is made that

$$B = \frac{n_2 - n_3}{n_2 + n_3} \qquad (33)$$

the intensity of the light is unchanged by passing through the first surface and where n_3 is the index of the material at the boundary of the far side of the sheet.

Nonreflecting film. An interesting application of Eq. (32) is the nonreflecting film. A single dielec-

White light reflected by a diffraction grating and separated into its visible components of zero, first, and second order. (*Bausch & Lomb Inc.*)

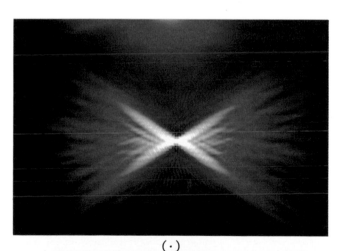

(·)

Patterns which result when light contributions from each point in an opening are added together using the principle of interference. The shapes of the openings used are indicated. (*Courtesy of F.S. Harris, Jr.*)

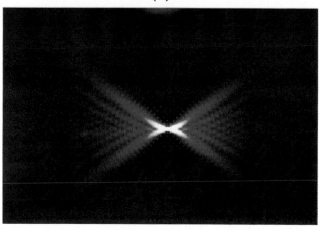

tric layer is evaporated onto a glass surface to reduce the reflectivity of the surface to the smallest possible value. From Eq. (32) it is clear that this takes place when $\cos \varphi = -1$. If the surface is used in an instrument with a broad spectral range, such as a visual device, the film thickness should be adjusted to put the interference minimum in the first order and in the middle of the desired spectral range. For the eye, this wavelength is approximately in the yellow so that such films reflect in the red and blue and appear purple. The index of the film should be chosen to make $C^2 = 0$. At this point Eqs. (34)–(36) hold.

$$(A - B)^2 = 0 \qquad (34)$$

$$\frac{n_1 - n_2}{n_1 + n_2} = \frac{n_2 - n_3}{n_2 + n_3} \qquad (35)$$

$$n_1 n_2 - n_2{}^2 + n_1 n_3 - n_2 n_3$$
$$- n_1 n_2 - n_1 n_3 + n_2{}^2 - n_2 n_3 \qquad (36)$$

Equation (36) can be reduced to Eq. (37). In the case of a glass surface in air, $n_1 = 1$ and $n_3 \cong 1.5$. Magnesium fluoride is a substance which is frequently used as a nonreflective coating, since it is hard and approximately satisfies the relationship of Eq. (37). The purpose of reducing the reflection

$$n_2 = \sqrt{n_1 n_3} \qquad (37)$$

from an optical element is to increase its transmission, since the energy which is not reflected is transmitted. In the case of a single element, this increase is not particularly important. Some optical instruments may have 15–20 air-glass surfaces, however, and the coating of these surfaces gives a tremendous increase in transmission.

Haidinger fringes. When the second surface in two-surface interference is partially reflecting, interference can also be observed in the wave transmitted through both surfaces. The interference fringes will be complementary to those appearing in reflection. Their location will depend on the parallelism of the surfaces. For plane parallel surfaces, the fringes will appear at infinity and will be concentric rings. These were first observed by W. K. Haidinger and are called Haidinger fringes.

Multiple-beam interference. If the surfaces S_1 and S_2 are strongly reflecting, it is necessary to consider multiple reflections between them. For air-glass surfaces, this does not apply since the reflectivity is of the order of 4%, and the twice-reflected beam is much reduced in intensity.

In Fig. 9 the situation in which the surfaces S_1 and S_2 have reflectivities r_1 and r_2 is shown. The space between the surfaces has an index n_2 and thickness d. An incident light beam of amplitude A is partially reflected at the first surface. The transmitted component is reflected at S_2 and is reflected back to S_1 where a second splitting takes place. This is repeated. Each successive component of the waves leaving S_1 is retarded with respect to the next. The amount of each retardation is given by Eq. (38).

$$\varphi = \frac{4\pi n d}{\lambda} \cos \theta \qquad (38)$$

Equation (7) was derived for the superposition of two waves. It is possible to derive a similar expres-

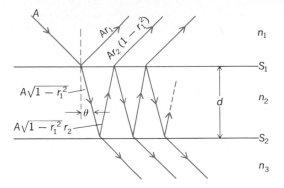

Fig. 9. Multiple reflection of wave between two surfaces.

sion for the superposition of many waves. From Fig. 9, the different waves at a plane somewhere above S_1 can be represented by the following expressions:

Incoming wave $= A \sin \omega t$
First reflected wave $= A r_1 \sin \omega t$
Second reflected wave $= A(1 - r_1{}^2) r_2 \sin(\omega t + \varphi)$
Third reflected wave $= -A(1 - r_1{}^2) r_1 r_2{}^2 \sin(\omega t + 2\varphi)$

By inspection of these terms, one can write down the complete series. As in Eq. (3), the sine terms can be broken down and coefficients collected. A simpler method is to multiply each term by $i = \sqrt{-1}$ and add a cosine term with the same coefficient and argument. The individual terms then are all of the form of expression (39), where m is an integer.

$$B e^{-i\omega t} e^{-im\varphi} \qquad (39)$$

The individual terms of expression (39) can be easily summed. For the reflected wave one obtains Eq. (40). Again, as in the two-beam case, the

$$R = \frac{r_1 + r_2 e^{-i\varphi}}{1 + r_1 r_2 e^{-i\varphi}} \qquad (40)$$

minimum in the reflectivity R is obtained when $\varphi = N\pi$, where N is an odd integer and $r_1 = r_2$. The fringe shape, however, can be quite different from

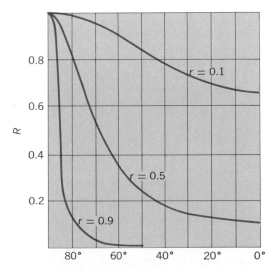

Fig. 10. The shape of multiple-beam fringes for different values of surface reflectivity.

the earlier case, depending on the values of the reflectivities r_1 and r_2. The greater these values, the sharper become the fringes.

It was shown earlier how two-beam interference could be used to measure the contour of a surface. In this technique, a flat glass test plate was placed over the surface to be examined and monochromatic interference fringes were formed between the test surface and the surface of the plate. These two-beam fringes have intensities which vary as the cosine squared of the path difference. It is very difficult with such fringes to detect variations in fringe straightness or, in other terms, variations of surface planarity that are smaller than 1/20 wavelength. If the surface to be examined is coated with silver and the test surface is also coated with a partially transmitting metallic coat, the reflectivity increases to a point where many beams are involved in the formation of the interference fringes. The shape of the fringes is given by Eq. (40). The shape of fringes for different values of r is shown in Fig. 10. With high-reflectivity fringes, the sensitivity to a departure from planarity is increased far beyond 1/20 wavelength.

It is thus possible with partially silvered surfaces to get a much better picture of small irregularities than with uncoated surfaces. The increase in sensitivity is such that steps in cleaved mica as small as 10 A in height can be seen by examining the monochromatic interference fringes produced between a silvered mica surface and a partially silvered glass flat. *See* INTERFERENCE FILTER, OPTICAL; INTERFEROMETRY.   [BRUCE H. BILLINGS]

Bibliography: M. Born and E. Wolf, *Principles of Optics*, 1959; O. S. Heavens, *Optical Properties of Thin Solid Films*, 1955; F. A. Jenkins and H. E. White, *Fundamentals of Physical Optics*, 3d ed., 1957; J. Strong, *Concepts of Classical Optics*, 1958; R. A. Waldron, *Waves and Oscillations*, 1964.

Interferometer, acoustic

A device for measuring the velocity and attenuation of sound waves in a gas or liquid. It operates by sending an ultrasonic beam of sound waves through the medium to be measured in order to obtain a standing-wave system between the driving crystal and a reflector whose distance from the crystal can be varied by a screw system. The illustration shows a typical system for studying the velocity and attenuation of a gas at various temperatures. Illustration *a* shows a system for purifying the gas, which is then admitted to the interferometric chamber shown in illustration *b*. The whole unit placed in a thermostated heating coil which makes it possible to regulate the temperature within close limits.

An oscillator is attached to the crystal and tuned to the crystal's resonant frequency. The variation in the amplitude of current from the oscillator is measured as the distance between the crystal and the reflector is varied. At half-wavelength intervals, sharp dips occur in the received current, indicating that a high mechanical impedance is impressed on the end of the vibrating crystal by the standing-wave system between the crystal surface and the reflecting plate. By counting the number of half wavelengths n occurring in a given displacement l of the reflecting plate and knowing accurately the frequency of vibration f, it is possible to

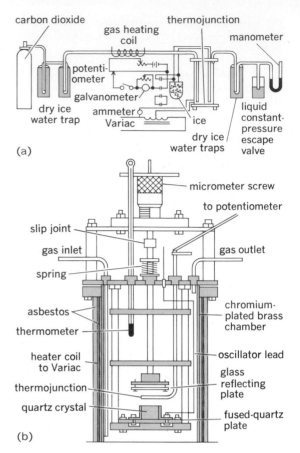

(a)

(b)

Acoustic interferometer for measuring the velocity and attentuation of sound waves in gases. (*a*) System for purifying gas. (*b*) Interferometric chamber.

determine the velocity of propagation v, by the formula given below. The attenuation can be de-

$$v = \frac{2lf}{n}$$

termined by the rate at which the maximum and minimum current values change with distance. Such instruments can measure velocities to within 1 part in 10,000 and attenuation values within a few percent. *See* ULTRASONICS.

[WARREN P. MASON]

Interferometry

The design and use of optical interferometers, devices in which interference of light is used as a tool in metrology and spectroscopy. These uses include precise measurements of wavelength, the measurement of very small distances and thicknesses by using known wavelengths, the detailed study of the hyperfine structure of spectrum lines, the precise determination of refractive indices, and, in astronomy, the measurement of binary-star separations and the diameters of stars. Optical interferometers are based on both two-beam interference and multiple-beam interference. They are also based on wavefront splitting and amplitude splitting.

Rayleigh interferometer. Perhaps the simplest type of interferometer is the Rayleigh interferometer, which is essentially a modification of the instrument used in Young's double-slit interference experiment. In Young's experiment, the two slits

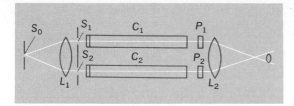

Fig. 1. Rayleigh interferometer.

are on the order of 1 mm apart. In the Rayleigh instrument (Fig. 1),the two slits S_1 and S_2 are of the order of 1 cm apart and 1/2 cm wide. They are illuminated with parallel light from a lens L_1 at whose focus is a single narrow slit S_0. A second lens L_2 spaced several centimeters from the first brings the two beams to a focus, where interference fringes become visible. The instrument is used to measure the refractive indices of liquids and gases, and in this fashion can serve also as a means of chemical analysis. *See* INTERFERENCE OF WAVES.

Cells (C_1 and C_2 in Fig. 1) are placed in front of each slit and matched so that when empty the central bright fringe in the interference pattern is undisplaced. The material of interest is then placed in one cell and the comparison material in the other. If the refractive indices of these two materials are different, the fringes will shift, and this shift is a measure of the index difference.

To make the fringe shift easy to detect, the cells are built to cover only half of the length of each slit. At the focus of the second lens, there will accordingly appear two sets of fringes separated vertically from each other. One set involves a part of the beams which does not pass through the cells. It remains motionless when a gas or liquid is put into either cell and acts thus as a series of reference marks for the displacement. In front of the cells are two plates of glass, P_1 and P_2. These add a retardarion $(\mu - 1)t$ to each beam, where μ is the index of refraction of the glass and t is its thickness. One of the plates can be tilted, thus adding a phase shift to one of the beams. When the phase shift is equal and opposite to the phase difference in the beams caused by the refractive index difference between the contents of the two cells, the fringes will be displaced back to the zero position. They will then match the fringes produced by the beams which do not pass through the cells. In this way, the instrument can be operated in a null fashion and becomes much more precise than if an attempt were made to measure the fringe shifts with a scale.

The Rayleigh interferometer is an extremely sensitive tool for the detection of impurities in gases and solutions. With 100-cm cells, helium in air can be detected at a concentration of 60 parts in 10^6. This represents an index difference of approximately 1×10^{-8}.

Michelson stellar interferometer. This device solves most dramatically the problem of measuring the diameters of stars which are as small as 0.01 second of arc. This task is impossible with an optical telescope, since the resolution obtainable even with the largest is not much better than 1 second of arc.

The Michelson stellar interferometer is a simple adaptation of Young's two-slit experiment. In its

first form, two slits were placed over the aperture of a telescope. If the object being observed were a true point source, the image would be crossed with a set of interference bands. A second point source separated by a small angle from the first would produce a second set of fringes. At certain values of this angle, the bright fringes in one set will coincide with the dark fringes in the second set. The smallest angle α at which the coincidence occurs will be that angle subtended at the slits by the separation of the peak of the central bright fringe from the nearest dark fringe. This angle is given by Eq. (1), where d is the separation of the slits, λ the

$$\frac{\lambda}{2d} = \alpha \qquad (1)$$

dominant wavelength of the two sources, and α their angular separation. The measurement of the separation of the sources is performed by adjusting the separation d between the slits until the fringes vanish.

Consider now a single source in the shape of a slit of finite width. If the slit subtends an angle at the telescope aperture which is larger than α, the interference fringes will be reduced in contrast. For various line elements at one side of the slit, there will be elements of angle α away which will cancel the fringes from the first element. By induction, it is clear that for a separation d' such that the slit source subtends an angle as given by Eq. (2)

$$\alpha' = \frac{\lambda}{d'} \qquad (2)$$

the fringes from a single slit will vanish completely. For additional information on the Michelson stellar interferometer *see* DIFFRACTION.

Michelson interferometer. The Michelson interferometer (Fig. 2) is based on amplitude splitting. Light from a narrow-angle source S is incident at 45° on a 50% partially reflecting plate P_1. Half the light is transmitted to mirror M_1 which reflects it back to the 50% reflecting plate. The light which is reflected proceeds to M_2 which reflects it back to P_1. At P_1, the two waves are again partially reflected and partially transmitted, and a portion of each wave proceeds to the receiver R which may be a screen, a photocell, or a human eye. Depending on the difference between the distances from the beam splitter to the mirrors M_1

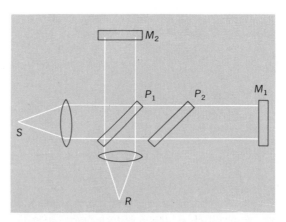

Fig. 2. Michelson interferometer. (*From A. C. Hardy and F. H. Perrin, The Principles of Optics, McGraw-Hill, 1932*)

and M_2, the two beams will interfere constructively or destructively. The plate P_2 compensates for the thickness of P_1.

When the mirrors' images are completely parallel, the interference fringes are circles. The reflectivity of the mirrors in the Michelson interferometer can be made as high as desired, and the interference will still be two-beam interference. The intensity of the fringes can accordingly be made very great. If the mirrors are slightly inclined about a vertical axis, vertical fringes are formed across the field of view. These fringes can be formed in white light if the path difference in part of the field of view is made zero. Just as in other interference experiments, only a few fringes will appear in white light, because the difference in path will be different for wavelengths of different colors. Accordingly, the fringes will appear colored close to zero path difference, and will disappear at larger path differences where the spectral separation of the successive regions of constructive interference is too close for the eye to see colors. If there is one-half cycle relative phase shift at the beam splitter, the fringe of zero path difference is black, and can be easily distinguished from the neighboring fringes. This makes use of the instrument relatively easy. The sensitivity to weak lines and resolution of the interferometer is thus potentially very much greater than that of an optical spectrometer.

Etalons. As a measuring device, the mirror M_1 is usually mounted on a screw so that the path difference can be altered. As the screw is turned, the fringes appear to move and may be counted. For light of the required degree of monochromaticity, the path difference between M_1 and M_2 can be made several centimeters before the fringes become too washed out to count.

The red line of cadmium was discovered by A. A. Michelson to be narrower than any other source available at the time. One of the early uses Michelson made of his device was in the measurement of the standard meter in terms of this red cadmium line. The mirror M_2 was replaced for the purpose by a pair of mirrors mounted as in Fig. 3. This arrangement is termed an etalon. The etalon mirrors were of such size that two etalons could be placed side by side in the field of view.

Several of these etalons were built. Each was twice the length of the next smaller. The largest was 10 cm and the smallest 0.0391 cm. First, the smallest was set up alone and the mirror M_1 adjusted with white light until the black fringe was

centered. At this point, the mirror M_1 and one mirror of the etalon were equidistant from the beam splitter. With red cadmium light, the mirror M_1 was then moved and fringes counted until the distance of mirror M_1 from the beam splitter was equal to the distance from the beam splitter of the mirror at the other end of the etalon. This operation gave the length of the smallest etalon in terms of the red cadmium line. The second etalon was then placed beside the first and their end mirrors aligned with the interferometer until both showed the black white-light fringe. This signified that the two mirrors were exactly the same distance from the beam splitter. The movable mirror M_1 was then shifted until the white-light fringe appeared at the other end of the smallest etalon. This etalon was then moved until the white-light fringe appeared in the mirror at its other end. It had then been shifted a distance exactly equal to the separation between its two mirrors. Again, mirror M_1 was moved until the black fringe appeared in the mirror at the other end of the etalon. This mirror was then separated from the scan mirror of the second largest etalon by a distance equal to twice the length of the smallest etalon. If the larger etalon were exactly equal to twice the smallest, a white-light fringe would appear in its mirror also. With monochromatic light, it was easy to count the fringes which represented the differential between the two mirrors. A repetition of this procedure with larger etalons gave an exact measure of the length of the 10-cm etalon. The meter was then measured by repeating the mirror alignment procedure with 10 successive shifts of this longest etalon. By a recent repetition of these measurements the length of the meter was determined to be 1,553,164.1 wavelengths of the cadmium red line in standard air.

Interference spectroscope. The Michelson interferometer is also interesting as a spectroscope. Consider first the case of two close spectrum lines as a light source for the instrument. As the mirror M_1 is shifted, fringes from each line will cross the field. At certain path differences between M_1 and M_2 the fringes will be out of phase and will essentially disappear; at other points they will be in phase and will be reinforced. By measuring the distance between successive maxima in fringe contrast, it is possible to determine the wavelength difference between the lines.

This is a simple illustration of a very broad use for any two-beam interferometer. As the path length t is changed, the variation in intensity of the light coming from an interferometer gives information on the basis of which the spectrum of the input light can be derived. The equation for the intensity of the emergent energy can be written as Eq. (3), where β is a constant, and $I(\lambda)$ is the inten-

$$I(t) = \int_0^\infty I(\lambda) \cos^2 \frac{\beta t}{\lambda} d\lambda \qquad (3)$$

sity of the incident light at different wavelengths λ. This equation applies when the mirror M_1 is moved linearly with time from the position where the path difference with M_2 is zero, to a position which depends on the longest wavelength in the spectrum to be examined. From Eq. (3), it is possible mathematically to recover the spectrum $I(\lambda)$. In certain situations, such as in the infrared beyond 1.5 μ, this technique has several advantages over con-

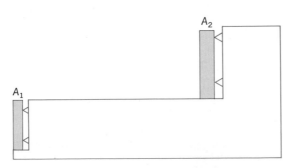

Fig. 3.   An etalon with two adjustable mirrors A_1 and A_2 which are to be used with the Michelson interferometer in distance measurement.

ventional spectroscopy. The detector looks at all wavelengths 50% of the time. The integration time is equal to one-half the length of time taken by the whole scan. In a conventional infrared spectrometer, the integration time is the time taken by the monochromator slit to traverse one resolvable unit of wavelength. For a discussion of the use of the Michelson interferometer in the celebrated Michelson-Morley experiment *see* LIGHT. *See also* INFRARED SPECTROSCOPY.

Fabry-Perot interferometer. A simpler device than the Michelson interferometer is the Fabry-Perot interferometer, sketched in Fig. 4. The two glass plates are partially silvered on the inner surfaces, and the incoming wave is multiply reflected between the two surfaces. Hence, the device is called a multiple-beam interferometer.

The Fabry-Perot interferometer is used in transmission. The expression for transmission, T, is Eq. (4), where d is the separation of the surfaces, θ the

$$T = \frac{t^2}{(1-r)^2 + 4r \sin^2\left[(2\pi nd/\lambda)\cos\theta + y\right]} \quad (4)$$

angle of incidence, y the phase shift for a reflection at the silver surface, t the transmission of the silver surface, r its reflectivity, and n the refractive index of the air between the plates. When a monochromatic light source is viewed through the interferometer, it will appear as a series of rings. These occur at those angles which satisfy Eq. (5),

$$\cos\theta = \frac{(m\pi - y)\lambda_0}{2\pi nd} \quad (5)$$

where λ_0 is the wavelength of the light, and m is an integer.

If a series of spectrum lines is transmitted through the instrument, the angular position of a fringe with order number m may overlap a fringe of order $m+1$ associated with a second line. Two lines which would overlap exactly are given by Eqs. (6). The difference can be written as Eq. (7).

$$\lambda_1 = \frac{2nd\cos\theta}{m} \qquad \lambda_2 = \frac{2nd\cos\theta}{m+1} \quad (6)$$

$$\Delta\lambda = \frac{\lambda_1\lambda_2}{2nd\cos\theta} \quad (7)$$

In terms of wave number and at small angles of incidence, this equation can be simplified to Eq. (8).

$$\Delta\nu = \frac{1}{2nd} \quad (8)$$

An interferometer of separation d can be used as a spectrograph where, for a frequency interval $\Delta\nu$, there is no overlapping. This quantity is called the free spectral range. The resolution is given by the separation of two wavelengths whose fringes overlap at the 50% relative transmission point. This separation can be written as Eq. (9). An interesting

$$d\lambda = \frac{2\lambda \sin^{-1}\left[(1-r)/(2\sqrt{r})\right]}{m\pi - y} \quad (9)$$

expression is the resolution in terms of free spectral range. When y is neglected, this is given by formula (10).

$$\frac{d\nu}{\Delta\nu} = \frac{2}{\pi}\sin^{-1}\left(\frac{1-r}{2\sqrt{r}}\right) \quad (10)$$

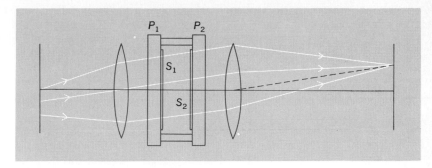

Fig. 4.   Fabry-Perot interferometer. P_1 and P_2 are glass plates; S_1 and S_2 are partially transmitting silver layers.

Reflective coatings. The reciprocal of this number, the resolving power in terms of free spectral range, can be considered a measure of the performance of a set of Fabry-Perot plates used spectrographically. With a routine silver coating, the resolving power in terms of free spectral range is of the order of 20. Higher reflectivities can improve this number, but the transmission of the coating drops to such levels that the exposure time becomes too long. Higher resolving power also begins to put very stringent requirements on the flatness and smoothness of the plates. Temperature and pressure variations during the exposure time will also affect the resolution. One way of improving the interferometer is through the use of multiple dielectric layers in place of the evaporated silver reflecting coat. A typical silver layer has transmission of 4% and a reflectivity of 93%. With a dielectric coating, the reflectivity over a range of several hundred angstroms can be as high as 99% and the transmission 98% at the center of a fringe. *See* REFLECTION OF ELECTROMAGNETIC RADIATION; RESOLVING POWER (OPTICS).

Spectral range. It is difficult to achieve with a spectrograph a resolving power that is higher than a few hundreds of thousands. The spectral range, however, can be several thousand angstroms. A Fabry-Perot interferometer is frequently used in series with a spectrograph so that a single spectral line from the spectrograph represents the smallest free spectral range required of the interferometer. In this fashion, resolving powers of several million can be achieved and the hyperfine structure and isotope shifts of atomic spectra have been discovered and studied.

Mach-Zehnder interferometer. The Mach-Zehnder interferometer, sketched in Fig. 5, is a variation of the Michelson interferometer and, like the Michelson instrument, depends on amplitude splitting of the wavefront. Light enters the instrument and is reflected and transmitted by the semitransparent mirror M_1. The reflected portion proceeds to M_3 where it is reflected through the cell C_2 to the semitransparent mirror M_4. Here it combines with the light transmitted by M_1 to produce interference. The light transmitted by M_1 passes through a cell C_1 which is similar to C_2 and is used to compensate for the windows of C_1.

The major application of this instrument is in studying airflow around models of aircraft, missiles, or projectiles. The object and associated airstream are placed in one arm of the interferometer. Because the air pressure varies as it flows

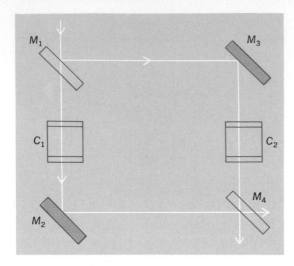

Fig. 5. Mach-Zehnder interferometer.

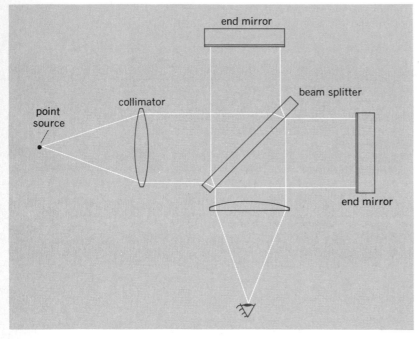

Fig. 6. Conventional Twyman-Green interferometer.

over the model, the index of refraction varies and thus the effective path length of the light in this beam is a function of position. When the variation is an odd number of halfwaves, the light will interfere destructively and a dark fringe will appear in the field of view. From a photograph of the fringes the flow pattern can be mathematically derived.

This application shows the two main features which distinguish this instrument from the Michelson interferometer. The light goes through each path in the instrument only once, whereas in Michelson's device the light traverses each path twice. This double traversal makes the Michelson interferometer extremely difficult to use in applications where spatial location of index variations is desired. The incoming and outgoing beams tend to travel over slightly different paths, and this lowers the resolution because of the index gradient across the field. In the Michelson interferometer the fringes are located at the virtual intersection of the two reflecting mirrors. In the Mach-Zehnder device the mirrors can be adjusted so that the fringes are located at the plane of the model being studied. They thus can be photographed simultaneously with the model so that their geometric position with respect to the model can be precisely determined. See SHOCK-WAVE DISPLAY.

[BRUCE H. BILLINGS]

Laser interferometer. Because of the extremely long coherence length and high intrinsic brilliance of the continuous-wave gas lasers, an important extension to the field of classical interferometers has developed. Classical interferometers have been limited to path differences in the interfering beams of some 20 cm, but with a laser as light source they can operate at hundreds of meters path difference with negligible loss in contrast of the fringes. See LASER.

With the Twyman-Green interferometer (similar to the Michelson interferometer, except that it is illuminated with a point source of light instead of an extended source, Fig. 6) an extremely versatile, compact instrument results from replacing one of the end mirrors with a well-corrected lens (Fig. 7). This converts the plane wave leaving the beam splitter into a spherical wavefront producing a point image at the focus of the lens. As this wave diverges, it becomes capable of providing a test

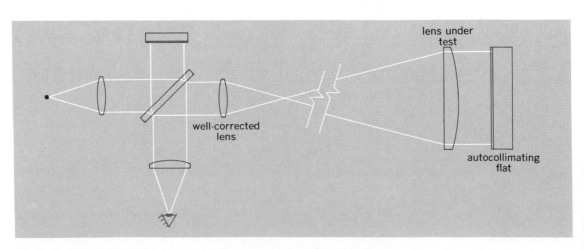

Fig. 7. Twyman-Green interferometer modified for testing spherical wavefronts.

wavefront for interferometrically evaluating any image-forming optical system or surface. Because the path difference between the two arms of the interferometer can be so great, all the optical components of the interferometer can be quite small and closely spaced, while the distance from the focus of the lens in the test arm to the optical system being tested can be arbitrarily large.

The Fizeau multiple-beam interferometer (Fig. 8) markedly increases in versatility when illuminated with laser light. Conventionally, this interferometer requires an extremely small air gap and is capable of resolution comparable to that of the Fabry-Perot interferometer. With laser illumination, the gap width may be increased to several meters, enabling one to test afocal or zero-power optical systems, such as telescopes or optical windows, in the gap.

One of the penalties paid for increasing the gap width in the Fizeau interferometer is in the increased asymmetry in the profile of the fringe. Because of the unequal optical path differences of successively reflected beams, all Fizeau fringes have some asymmetry. With a narrow gap, this results in a negligible loss in resolution. However, since the asymmetry increases with the gap width, the position of a fringe becomes less well defined and the resolution suffers. A very simple concept can be used to eliminate this loss of symmetry in the test of a telescope (Fig. 9). If the objective of the telescope (its entrance pupil) is placed near one mirror, this mirror will be imaged by the telescope in the vicinity of its exit pupil. If the other mirror is placed in coincidence with this image, the asymmetry in the fringes disappears and symmetrical Fabry-Perot type fringes result. By imaging one mirror on the other, the property of the Fabry-Perot interferometer, of having the optical path of each successively reflected beam increase by a constant value, is imparted to the wide-gap Fizeau interferometer (Fig. 10).

A further application of this principle can be made to a spherical wavefront multiple-beam interferometer (Fig. 11). This is similar to the modified Twyman-Green interferometer in that it utilizes a spherical wave for testing, and also to the Fizeau interferometer in that the reference surface (spherical in this case) is highly reflecting to produce multiple beams. If an infinity-focused optical system is tested, its focal point is made coincident with the center of curvature of the reference sphere and a plane mirror placed in the collimated beam. The distance of this mirror from the lens

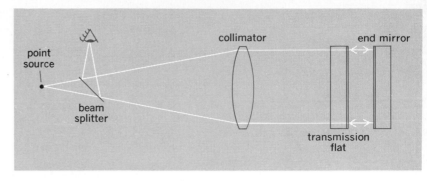

Fig. 8. Fizeau multiple-beam interferometer in reflection.

can be arbitrarily assigned, but if it is placed so that the optical system images it on the spherical reference surface, symmetrical multiple-beam fringes result. An important characteristic of such a multiple-beam interferometer is that aberrations in the source optics (all optical elements between the source and spherical reference surface) produce negligible errors in the fringe pattern. The effect of these aberrations is to produce a wavefront incident on the reference sphere whose normals are slightly inclined to the radii of the sphere. Such angle-of-incidence errors produce distortions in the fringe pattern that increase with the width of the interferometer gap. This is because the optical path between mirrors for successively reflected beams is altered more for a large gap than for a small gap in the presence of a given changed angle of incidence. But so long as the mirrors are imaged on each other, the optical path of any ray between them is independent of its angle of incidence on the reference surface, making the interferometer immune to source optics aberrations.

Another application of the gas laser takes advantage of the several wavelengths simultaneously emitted (if the laser cavity is sufficiently long) to increase the sensitivity of an interferometer. These wavelengths are equally spaced in wave number $\left(\nu_m = \dfrac{m}{2d}\right)$, their separation being the reciprocal of twice the cavity length. Since the number of different wavelengths emitted is determined by the spectral width of the radiating atoms, the longer the cavity the greater the number of wavelengths. If such a laser illuminates a multiple-beam Fizeau interferometer set up for viewing fringes in transmission (Fig. 12), and the ratio of the laser gap to that of the interferometer is equal to

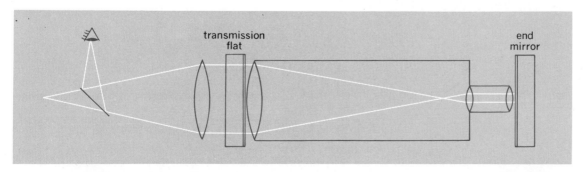

Fig. 9. Wide-gap Fizeau interferometer for testing telescope.

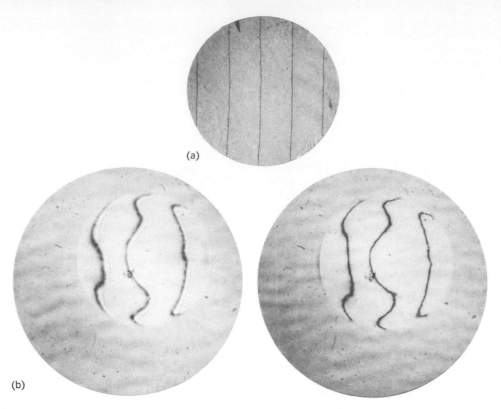

Fig. 10. Fizeau multiple-beam fringe patterns. (*a*) Narrow gap plane mirrors. (*b*) Wide gap lens test without (on the left) and with (on the right) end mirrors imaged on each other.

the number of wavelengths emitted, the spacing between fringes will correspond to path differences reduced by a factor equal to the number of wavelengths. Normally, the optical path difference between two fringes is $\lambda/2$ when the interferometer is illuminated by a single wavelength; but if it is illuminated by a laser emitting p wavelengths, the separation of the fringes in such an interferometer corresponds to a path difference $\lambda/2p$.

Hologram interferometer. A very important type of interferometer made possible by the advent of the laser is the hologram interferometer. A hologram is, in general, a high-resolution photographic plate which has been exposed simultaneously to

the direct radiation of a laser beam and the light of a part of the beam reflected by an object. These two beams produce an interference fringe pattern on the plate which, when processed and returned to its initial position, acts as a diffraction grating on the original two beams. It diffracts the light in such a way that the first-order diffracted beam from either of the two beams reproduces the zero-order beam of the other. That is, if the directly incident beam is blocked (without blocking the light that illuminates the object or is reflected by the object to the plate) the first-order diffraction of this beam from the object will reproduce a wavefront identical to that of the directly transmitted direct beam. If one looks through the plate in the direction of

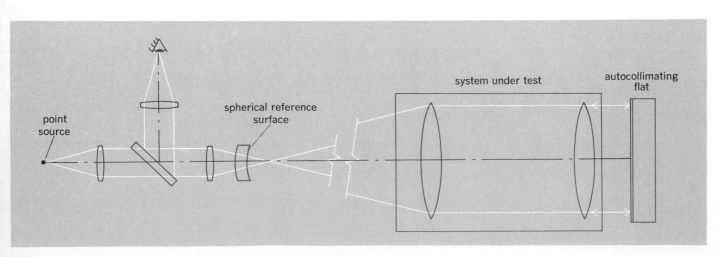

Fig. 11. Spherical wavefront multiple-beam interferometer.

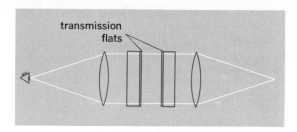

Fig. 12. Fizeau multiple-beam interferometer set up for viewing fringes in transmission.

the direct beam, he will be unaware of the fact that the beam is blocked. Likewise, if the reflected light from the object were blocked, an observer looking through the plate in its direction would see it as if it were unblocked.

This ability of a hologram to reproduce through the diffraction of one beam a wavefront identical to that of the other beam is referred to as wavefront reconstruction. It provides a basis for an extremely versatile interferometer. For in looking through a hologram in the direction of the object, without blocking either beam, two beams will appear to emanate from the object, one directly transmitted through the hologram, the other diffracted by the hologram. These two beams will be identical if the object is unchanged from the time the hologram was exposed; but if any change has occurred subsequent to the exposure, the directly transmitted beam will be altered. The interference between the two beams will produce a fringe pattern that may be interpreted identically to that from a classical interferometer to measure the change in the object.

This is the basis of the hologram interferometer. It differs from the conventional interferometer in that it requires no special reference surface. The object itself provides the reference wavefront. Therefore, one is no longer limited in the shapes of wavefronts that can be evaluated interferometrically. One is also freed from the restriction of evaluating only specularly reflecting surfaces, for hologram interferometry can apply to diffusely reflecting objects as well as specularly reflecting ones.

Hologram interferometry gives information only as to the changes in a surface, not its absolute shape, as other interferometers do. But there are many applications for which it provides heretofore unattainable data. These include mechanical or thermal distortions imparted to any surface. A hologram is made of the undistorted surface. After it is processed and returned to its initial position, the object is subject to the mechanical or thermal loads of interest. The resulting interferogram gives quantitative information on the distortion of the entire surface. In the same way living things can be followed in their growth through hologram interferometry. *See* HOLOGRAPHY.

[HARRY D. POLSTER]

Bibliography: J. M. Burch et al., Dual- and multiple-beam interferometry by wavefront reconstruction, *Nature*, 209:1015–16, 1966; J. B. De-Velis and G. O. Reynolds, *Holography*, 1967; H. A. Elson, *Laser Systems and Applications*, 1967; L. R. Heintze et al., A multiple-beam interferometer for use with spherical wavefronts, *Appl. Opt.*, 6: 1924–9, 1967; D. J. Herriott, Long-path multiple-wavelength fringes, *J. Opt. Soc. Amer.*, 56:719–23, 1966; W. V. Smith and P. P. Sorokin, *The Laser*, 1966; G. W. Stroke, *Coherent Optics and Holography*, 1966; J. Strong, *Concepts of Classical Optics*, 1958; S. Tolansky, *An Introduction to Interferometry*, 1955; W. E. Williams, *Applications of Interferometry*, 1930.

Interferon

A virus inhibitor produced by intact animals or cultured cells when infected with viruses. Within 12–48 hours after maximum virus production in the animal host is reached, interferon is produced in large quantities, and virus production decreases rapidly. Antibody does not appear in the blood of the animal until several days later. This temporal relationship of virus production to the appearance of interferon and then of antiviral antibody strongly suggests that interferon plays a major role in the defense of the animal against virus infections.

Specificity. Interferon is a protein (possibly a glycoprotein) and is acid-stable (pH 2.0), trypsin-sensitive, nondialyzable, and nonsedimentable by ultracentrifuge forces sufficient to pellet viruses. With important exceptions, it is effective as an antiviral substance only on cells from the species of animal in which it was produced but not on cells from other species. Thus, interferon produced by the intact mouse or by mouse cells in tissue culture will protect mouse cells from virus infection but has no protective effect for chicken cells. Although interferon is specific for the species of cells on which it is effective, it is not specific for the virus inhibited. Interferon production stimulated by one virus will effectively inhibit a wide variety of viruses.

Production and action. Interferon is produced (1) by cells in tissue culture when stimulated with viruses or synthetic double-stranded polynucleotides, and (2) by the intact animal when stimulated with viruses, rickettsiae, protozoa, bacterial endotoxins, or synthetic anionic polymers or polynucleotides (for example, polyinosinic polycytidilic copolymer, or poly I:C). The antiviral activity of mold products such as statolon and helenine is related to the presence of an RNA virus in the mold. Many agents that are potent interferon stimulators have in common the presence of double-stranded RNA in the material inoculated, or the production of double-stranded RNA during the replication of the agent. The double-stranded RNA of reovirus extracted from the virion has excellent ability to stimulate interferon. When stimulated, the host cell genome directs the synthesis of interferon, which is released from the infected cell. *See* RIBONUCLEIC ACID (RNA).

Endotoxin induces interferon in animals and in cultured macrophages but not in other tissue culture cells. Other substances exhibit similar characteristics, including synthetic polyanions such as pyran copolymer and polyacrylic acid. The latter are toxic plastics which are not metabolized. They provoke low and short-lived interferon titers but provide prolonged antiviral protection.

Interferon and its inducers are also active against infections caused by intracellular parasites other than viruses (chlamydia, toxoplasma, malaria). Poly I:C is active against certain experimental tumors.

When interferon is added to cells prior to infec-

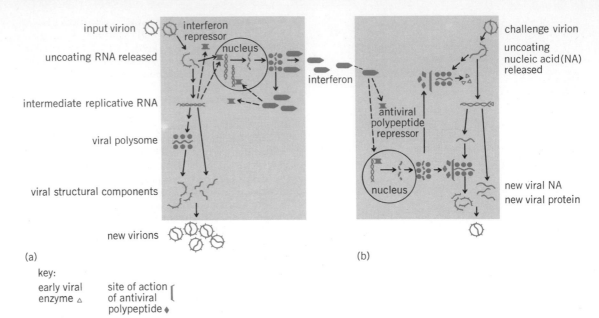

key:
early viral enzyme △   site of action of antiviral polypeptide ◆

Schematic representation of concepts in the mechanisms of interferon activity showing (a) production and (b) action. (From S. E. Grossberg, The interferons and their inducers: Molecular and therapeutic considerations, New Engl. J. Med., 287:79–85. 1972)

tion, there is marked inhibition of virus replication. The presence of interferon stimulates the cell to produce another protein, translational inhibitory protein (TIP). It is thought that TIP binds to cellular ribosomes, altering them in such a way that viral RNA is translated. Cellular messenger RNA is translated normally. This permits normal cell functions to continue but prevents the synthesis of virus-directed protein. Without the production of necessary enzymes and coat protein for progeny virus, new virus is not formed. See RIBOSOMES.

A schematic diagram of the mechanism of interferon action is presented in the illustration. In production (illustration a), an infectious single-stranded RNA virus exemplifies an active inducer of interferon. Its single RNA strand may act as the inducing principle, which, however, is more likely to be the intermediate replicative double-stranded RNA form which is produced after infection. Subsequent steps in viral replication indicate that cellular products may be new virions as well as interferon. The interferon inducer binds to the repressor of the interferon gene which is located in the host DNA. The binding of repressor permits derepression of the host genome for the production of the specific messenger RNA, which is then translated into interferon proteins by normal ribosomal machinery. Then interferons are rapidly released from cells to affect neighboring cells, or they may induce antiviral resistance in the same cell. In action (illustration b), interferon proteins either act at the cell surface or may enter cells to bind the repressor of the host cistron which codes for the antiviral polypeptide. After derepression by the interferon protein, the antiviral polypeptide is produced by the usual cell machinery to inhibit viral replication, probably at the ribosomal level. The production of early viral enzymes, viral nucleic acids, and viral structural proteins is thereby inhibited, and reduced viral yield results.

All of the cells of the intact animal are probably capable of producing interferon; however, the elements of the reticuloendothelial system seem to provide the bulk of interferon during most virus infections.

Classes. Stimulation of animals with various substances has revealed at least two classes of interferons. One class appears to be preformed and is released into the bloodstream within 2 hr after injection of endotoxins. This interferon has a higher molecular weight than the interferon that appears in the serum later (18 hr or longer after stimulation). Interferons of still other molecular weights have been found under various experimental conditions, and it is evident that interferon is a class of proteins that shares the biologic property of inhibiting virus replication by altering cell metabolism.

Disease control. The low antigenicity and potent antiviral effect of interferon has created much interest in its possible application in controlling viral diseases of man. In experimental animals it has been possible to demonstrate the efficacy of exogenous interferon in preventing or decreasing the severity of virus infections. However, the difficulties in producing sufficient quantities of the material for human use have not been overcome. Interferon does not alter the course of viral diseases once the infection has become firmly established.

Present hope for application of interferon to the control of human disease lies in the development of interferon inducers. There have been some clinical trials of attenuated live virus vaccines as inducers of interferon, and a number of synthetic inducers of interferon activity are being investigated. Among the most promising is the analog poly I:C. It, other synthetic polynucleotides, and other inducers have been extensively tested in animals. Poly I:C has had clinical trials with indefinite results. No inducer is yet available for general use. *See* ANIMAL VIRUS; VIRUS INTERFERENCE.   [JOSEPH L. MELNICK]

Bibliography: S. E. Grossberg, The interferons and their inducers: Molecular and therapeutic considerations, *New Engl. J. Med.*, 287:79–85, 1972.

Interhalogen compounds

The elements of the halogen family (fluorine, chlorine, bromine, and iodine) possess an ability to react with each other to form a series of binary interhalogen compounds (or halogen halides) of general composition given by XY_n, where n can have the values 1, 3, 5, and 7, and where X is the heavier (less electronegative) of the two elements. All possible diatomic compounds of the first four halogens have been prepared. In other groups a varying number of possible combinations is absent. Although attempts have been made to prepare ternary interhalogens, they have been unsuccessful; there is considerable doubt that such compounds can exist. *See* HALOGEN ELEMENTS.

Formation. In general, interhalogen compounds are formed when the free halogens are mixed as gases, or, as in the case of iodine chlorides and bromides, by reacting solid iodine with liquid chlorine or bromine. Most of the nonfluorinated interhalogens also readily form when solutions of the halogens in inert solvent (for example, carbon tetrachloride) are mixed. It is also possible to form them by the reaction of a halogen with a salt of a more electropositive halogen, such as $KI + Cl_2 \rightarrow KCl + ICl$. Higher polyhalides can also be prepared by reacting more electronegative halogen with a corresponding halogen halide, for example, $ICl + Cl_2 \rightarrow ICl_3$ or $ClF_3 + F_2 \rightarrow ClF_5$. Chlorine pentafluoride can also be prepared by reacting a $MClF_4$ salt with fluorine (M = alkali metal), $MClF_4 + F_2 \rightarrow MF + ClF_5$. A list of known interhalogen compounds and some of their physical properties is given in Table 1. Interhalogen compounds containing astatine have not been isolated as yet, although the existence of AtI and AtBr has been demonstrated by indirect measurements.

Stability. Thermodynamic stability of the interhalogen compounds varies within rather large limits. In general, for a given group the stability increases with increasing difference in electronegativity between the two halogens. Thus for XY group the free energy of formation of the interhalogens, relative to the elements in their standard conditions, falls in the following order: IF > BrF > ClF > ICl > IBr > BrCl. It should be noted, however, that the fluorides of this series can be obtained only in minute quantities since they readily undergo disproportionation reaction, for example, $5IF \rightarrow 2I_2 + IF_5$. The least-stable compound, bromine chloride, has only recently been isolated in the pure state. Decrease of stability with decreasing difference of electronegativity is readily apparent in the higher interhalogens since compounds such as $BrCl_3$, IBr_3, or ICl_5 are unknown. The only unambiguously prepared interhalogen containing eight halogen atoms is IF_7. *See* ELECTRONEGATIVITY.

Reactivity. The reactivity of the polyhalides reflects the reactivity of the halogens they contain. In general, they behave as strong oxidizing and halogenating agents. Most halogen halides (especially halogen fluorides) readily attack metals, yielding the corresponding halide of the more electronegative halogen. In the case of halogen fluorides the reaction results in the formation of the fluoride, in which the metal is often found in its highest oxidation state, for example, AgF_2, CoF_3, and so on. Noble metals, such as platinum, are resistant to the attack of the interhalogens at room temperature. Halogen fluorides are often handled in nickel vessels, but in this case the resistance to attack is due to the formation of a protective layer of nickel (II) fluoride.

All halogen halides readily react with water. Such reactions can be quite violent and, with halogen fluorides, they may be explosive. Reaction products vary depending on the nature of the interhalogen compound. For example, in the case of chlorine trifluoride, reaction with an excess of water yields HF, Cl_2, and O_2 as reaction products.

The same reactivity is observed with organic compounds. The nonfluorinated interhalogens do not react with completely halogenated hydrocarbons, and solutions of ICl, IBr, and BrCl are quite stable in dry carbon tetrachloride or hexachloroethane, as well as in fluorocarbons, as long as the solvents are very dry. They readily react with aliphatic and aromatic hydrocarbons and with oxygen- or nitrogen-containing compounds. The reaction rates, however, can be rather slow and dilute solutions of ICl can be stable for several hours in solvents such as nitrobenzene. Halogen fluorides usually react vigorously with chlorinated hydrocarbons, although IF_5 can be dissolved in carbon tetrachloride and BrF_3 can be dissolved in Freon 113 without decomposition. All halogen fluorides react explosively with easily oxidizable organic compounds. *See* FLUORINE.

Halogen halides, like halogens, act as Lewis acids and under proper experimental conditions may form a series of stable complexes with various organic electron donors. For example, mixing of

Table 1. Known interhalogen compounds

	XY	XY$_3$	XY$_5$	XY$_7$
mp	ClF −154°C	ClF$_3$ −76°C	ClF$_5$ −103°C	IF$_7$
bp	−101°C	12°C	−14°C	4.77 (sublimes)
mp	BrF ≈−33°C	BrF$_3$ 8.77°C	BrF$_5$ −62.5°C	
bp	≈20°C	125°C	40.3°C	
mp	IF −	IF$_3$ −28°C	IF$_5$ 10°C	
bp	−	−	101°C	
mp	BrCl ≈−54°C	ICl$_3$* 101°C		
bp	−	−		
mp	ICl† 27.2°C(α)			
bp	≈100°C			
mp	IBr 40°C			
bp	119°C			

*In the solid state the compound forms a dimer.
†Unstable β-modification exists, mp 14°C.

carbon tetrachloride solutions of pyridine and of iodine monochloride leads to the formation of a solid complex, $C_5H_5N \cdot ICl$. The same reaction can occur with other heterocyclic amines and ICl, IBr, or ICl_3. Addition compounds of organic electron donors with IF, IF_3, and IF_5 have been reported. In all cases it is the iodine atom which is directly attached to the donor atom.

A number of interhalogen compounds conduct electrical current in the liquid state. Among these are ICl and BrF_3. For example, electrical conductance of molten iodine monochloride is comparable to a concentrated aqueous solution of a strong electrolyte (4.52×10^{-3} ohm^{-1} cm^{-1} at 30.6°C). The conductances, however, are much smaller than those of fused salts and, therefore, it can be concluded that the bonding in these compounds is largely covalent. Electrical conductance is due to self-ionization reactions, as shown in Eqs. (1) and (2).

$$3ICl \rightleftharpoons I_2Cl^+ + ICl_2^- \qquad (1)$$

$$2BrF_3 \rightleftharpoons BrF_2^+ + BrF_4^- \qquad (2)$$

The above behavior leads to the possibility of studying acid-base reactions. In these systems an acid is any compound which generates the solvo-cation, while a base would generate solvo-anions. Thus SbF_5 would be an acid in liquid bromine trifluoride, Eq. (3), while an electrovalent fluoride would be a base, Eq. (4). *See* SUPERACIDS.

$$SbF_5 + BrF_3 \rightleftharpoons BrF_2^+ + SbF_6^- \qquad (3)$$

$$KF + BrF_3 \rightleftharpoons K^+ + BrF_4^- \qquad (4)$$

Analogy with acid-base reactions in water is obvious, as shown by Eqs. (5) and (6). Such relations

$$K^+OH^- + H_3O^+Cl^- \rightarrow K^+Cl^- + 2H_2O \qquad (5)$$

$$K^+BrF_4^- + BrF_2^+SbF_6^- \rightarrow KSbF_6 + 2BrF_3 \qquad (6)$$

have been studied in BrF_3, ClF_3, and IF_5. Numerous salts of the interhalogen acid-base systems have been isolated and studied.

Thus, numerous compounds have been formed containing either interhalogen anions (solvo-anions) or interhalogen cations (solvo-cations) simply by adding the appropriate acid or base to a liquid halogen halide or a halogen halide in an appropriate nonaqueous solvent. In addition, cations derived from previously unknown compounds can be prepared by using powerful oxidizing agents, such as KrF$^+$ salts. For example, even though BrF_7 has not been unambiguously prepared to date, a compound containing BrF_6^+ has been prepared by R. J. Gillespie and coworkers according to Eq. (7).

$$BrF_5 + KrF^+AsF_6^- \rightarrow BrF_6^+AsF_6^- + Kr \qquad (7)$$

Table 2. Known interhalogen anions

Three-membered	Five-membered	Seven-membered	Nine-membered
ClF_2^-	ClF_4^-	ClF_6^-	IF_8^-
BrF_2^-	BrF_4^-	BrF_6^-	
ICl_2^-	IF_4^-	IF_6^-	
IBr_2^-	ICl_4^-		
$IBrCl^-$	$I_2Cl_3^-$		
$BrCl_2^-$	$I_2Cl_2Br^-$		
I_2Cl^-	$I_2ClBr_2^-$		
Br_2Cl^-	I_4Cl^-		
I_2Br^-			

Table 3. Known interhalogen cations

Three-membered	Five-membered	Seven-membered
Cl_2F^+	ClF_4^+	ClF_6^+
ClF_2^+	BrF_4^+	BrF_6^+
BrF_2^+	IF_4^+	
ICl_2^+		
I_2Cl^+		
IBr_2^+		
I_2Br^+		
$BrCl_2^+$		
Br_2Cl^+		
$IBrCl^+$		

Pentahalides can also be formed by the addition of an interhalogen compound to a trihalide ion as shown in Eq. (8).

$$ICl + ICl_2^- \rightarrow I_2Cl_3^- \qquad (8)$$

A compilation of interhalogen cations and anions which have been previously prepared is given in Tables 2 and 3.   [TERRY SURLES]

Bibliography: R. J. Gillespie and G. J. Schrobilgen, The hexafluorobromine (VII) cation, BrF_6^+, *Inorg. Chem.*, 13:1230, 1974, and references therein; A. I. Popov, Interhalogen compounds and polyhalide anions, in V. Gutmann (ed.), *MTP International Review of Science: Inorganic Chemistry*, ser. 1, vol. 3, chap. 2, 1972; A. I. Popov, Polyhalogen complex salts, in V. Gutmann (ed.), *Halogen Chemistry*, vol. 1, 1967; A. I. Popov and T. Surles, Interhalogen compounds and polyhalide anions, in V. Gutmann (ed.), *MTP International Review of Science: Inorganic Chemistry*, ser. 1, vol. 3, chap. 6, 1975, and references therein; L. Stein, Physical and chemical properties of halogen fluorides, in V. Gutmann (ed.), *Halogen Chemistry*, vol. 1, 1967; W. W. Wilson, B. Lands, and F. Aubke, The new interhalogen cations $BrCl_2^+$ and Br_2Cl^+, *Inorg. Nucl. Chem. Lett.*, 11:529, 1975, and references therein.

Interhemispheric integration

The process that is responsible for the exchange of information between the two half brains. The human brain is a bilaterally symmetrical structure which is for the most part richly interconnected by two main bridges of neurons called the corpus callosum and anterior commissure. These structures can be surgically sectioned in humans in an effort to control the spread of epileptic seizures. Although there is no apparent change in everyday behavior of these patients, dramatic differences in cognitive function can be demonstrated under specialized testing conditions. Because of these studies it now can be said that in normal humans these cerebral commissures are largely responsible for behavioral unity; the neural mechanism keeps the left side of the body up to date with the activities of the right, and vice versa.

Results of tests involving speech. Changes in behavioral responses of persons whose cerebral commissures have been sectioned are almost undetectable. The person walks, talks, and behaves in a normal fashion. Dramatic effects are observed only under testing conditions which utilize stimuli that are lateralized exclusively to one hemisphere or the other. When visual stimuli are flashed to the left of fixation of the eye, all information is exclusively projected to the right hemisphere, and vice versa. Likewise, when the left hand manipulates objects that are held out of sight, all relevant sensory information projects to the opposite, right hemisphere. Similarly, information from objects held in the right hand is relayed exclusively to the left hemisphere. This is the normal state and permits the separate input-output testing of each half-brain.

Using these sensory avenues, remarkable "deconnection" effects are seen in split-brain conditions. For example, if a picture of an apple is flashed in the right visual field, the person describes the object normally. However, if the same picture is flashed in the left visual field, in the early days of postoperative testing the person denies that the stimulus was presented at all. Gradually, after many test sessions the person may have the impression that something was flashed, but is unable to say what. In brief, this disparity of recognition in the two sides of the visual field occurs because the information is projected to the right hemisphere, which is incapable of speech. Because the right hemisphere is now disconnected from the left, information arriving in the right hemisphere cannot be communicated by means of speech. *See* VERBAL LEARNING.

Results of tests not involving speech. When tests are used which do not require a spoken response, numerous mental abilities are observed to be present in the "disconnected" right hemisphere. If, for example, instead of being asked for a verbal response, the person is required with the left hand to retrieve from a series of objects the one most closely related to the flashed visual stimulus, good performance is seen. Even though the person is unable to describe a picture of an orange flashed to the left field, when the left hand searches through a field of objects placed out of view, it correctly retrieves the orange. If asked what the object is, the person would say he does not know. Here again the left hemisphere controls speech but cannot solve the problem. The right hemisphere solves the problem but cannot elicit speech. *See* SENSORY LEARNING.

Right hemisphere. Through the use of this kind of test procedure, a variety of other mental abilities have been observed to go on in the separated right hemisphere. It can read simple nouns but not verbs. It can recognize some adjectives. It can recognize negative constructions but cannot differentiate constructions in the active or passive voice. It cannot pluralize nouns. In brief, it can manage some aspects of language only.

Left hemisphere. Despite its linguistic superiority, the left hemisphere does not excel the right in all tasks. Tests have demonstrated that in some specialized functions, such as arranging blocks to match a pictured design or drawing a cube in three dimensions, the right hemisphere is decidedly superior to the left.

Emotional reactions. In the area of emotional reactions there appears to be equal reactivity in the two hemispheres. Although a brain-bisected person cannot describe an emotional stimulus presented to the right hemisphere, an emotional response such as laughter is nonetheless elicited. In one instance, when a picture of a nude was flashed to the right hemisphere, the person said she did not see anything and then chuckled. When asked what the laughing was about, she said that it was "a funny machine." Here the right hemisphere clearly had enough ability to elicit laughter but not speech. The left hemisphere simply observed the changed emotional state and said that something must be funny, but did not know what.

Localization of the speech processor. Experimental psychological research has developed a procedure commonly called dichotic listening. It is a process where two auditory messages are played at the same time—one going to the left ear and one going to the right ear. Previous research, primarily by Doreen Kimura, has shown that under these test conditions normal subjects tend to repeat only the message presented to the right ear and to suppress the message presented to the left. It turns out this phenomenon is exceedingly dramatic in the split-brain person. In a study by S. Springer and M. S. Gazzaniga, when two phonemes (one to each ear) were presented to the split-brain persons, they were totally unable to name phonemes presented to the left ear. This was true despite the fact the left ear sends projections to the left hemisphere that are not affected by split-brain surgery. Indeed, when a phoneme or any auditory signal is presented to the left ear alone, the split-brain person experiences no difficulty whatsoever in describing the auditory event in detail. The suppression comes about only when both ears are simultaneously stimulated, and reflects some kind of dynamic process involving the two cerebral hemispheres that is not well understood.

A more surprising aspect of this study, however, was that the right hemisphere using its own manual response system could not respond at all to phonemes, the basic unit of speech perception. This finding is unexpected in light of the above described language capacities of the right hemisphere. Can the right hemisphere understand simple spoken and written words without being able to understand one of the basic building-block units that make it up? Perhaps the manner in which the right hemisphere comprehends spoken language is drastically different than the way the left processes its information.

Specificity of cortical circuitry. Tests have been conducted in which not all of the subjects have had their entire cerebral commissure sectioned. This is because it is now believed total commissure section is not necessary to stop the interhemispheric spread of some kinds of seizure activity. In neuropsychological testing, these partially sectioned persons showed dramatic breakdown in interhemispheric transfer. When the posterior part of the callosum is sectioned, visual aspects of the syndrome appear. When it is spared and more interior regions are cut, however, tactile and auditory communications are blocked, but not visual. It also appears that no fundamental reorganization of the interhemispheric transfer system takes place, since years after surgery these same deficits are present and are not compensated for in any way.

Individual variation and age of surgery. Another emerging fact is that there would appear to be a large variation in the lateralized talents of each half-brain. Thus, while the right hemisphere frequently appears to have some language talent, it is clear that not all split-brain persons have language skills in the right hemisphere. Similarly, visual spatial skills, which are usually present exclusively in the right hemisphere, are frequently bilaterally represented and sometimes represented only in the left speech hemisphere. There is even some evidence that the commissure system itself varies in what is transferred where. In some cases an intact anterior commissure in those persons with a full callosal section will allow for the complete interhemispheric transfer of visual information. Yet, in other cases visual transfer is blocked with a section of only the posterior callosal area. It is still too early to determine the mechanism of this behavior, but one possible important factor is the age of the surgical section and the extent of early brain damage experienced by the person. *See* BRAIN (VERTEBRATE); PSYCHOLOGY, PHYSIOLOGICAL AND EXPERIMENTAL.   [MICHAEL S. GAZZANIGA]

Bibliography: M. S. Gazzaniga, *The Bisected Brain*, 1970; M. S. Gazzaniga et al., Psychologic and neurologic consequences of partial and complete cerebral commissurotomy, *Neurology*, 25: 10–15, 1975; M. S. Gazzaniga, J. E. Bogen, and R. W. Sperry, Dyspraxia following division of the cerebral commissures in man, *Arch. Neurol.*, 16: 606–612, 1967; M. S. Gazzaniga and R. W. Sperry, Language after section of the cerebral commissures, *Brain*, 90:131–148, 1967; S. Springer and M. S. Gazzaniga, Dichotic testing of partial and complete commissurotomized patients, *Neuropsychologia*, 1976.

Intermediate-frequency amplifier

A tuned amplifier employed in the amplification of the signals produced by the mixer in a radio receiver. Because the carrier frequency of the modulated signal from the mixer is essentially constant, the resonant frequency of the amplifier is fixed.

The proper design of the intermediate-frequency (i-f) amplifier is essential for good selectivity and reproduction of the original transmitted signal. If the amplifier is tuned too sharply, the high-frequency components of the modulating signal will be lost. To avoid this, stagger-tuning of the individual stages may be used. In a stagger-tuned amplifier the resonant frequency of each stage is slightly different from the carrier frequency, with the result that the gain is essentially constant over the bandwidth of the modulated signal. The gain decreases rapidly at frequencies outside this band.

The standard i-f frequency for broadcast radio receivers is 455 kHz; other frequencies used depend upon the application, such as television receivers or radar receivers. *See* AMPLIFIER.

[HAROLD F. KLOCK]

Intermetallic compounds

Materials composed of two or more types of metal atoms, which exist as homogeneous, composite substances and differ discontinuously in structure from that of the constituent metals. They are also called, preferably, intermetallic phases. Their properties cannot be transformed continuously into those of their constituents by changes of composition alone, and they form distinct crystalline species separated by phase boundaries from their metallic components and mixed crystals of these components; it is generally not possible to establish formulas for intermetallic compounds on the sole basis of analytical data, so formulas are determined in conjunction with crystallographic structural information.

The term "alloy" is generally applied to any homogeneous molten mixture of two or more metals, as well as to the solid material that crystallizes from such a homogeneous liquid phase. Alloys may also be formed from solid-state reactions. In the liquid phase, alloys are essentially solutions of metals in one another, although liquid compounds may also be present. Alloys containing mercury are usually referred to as amalgams. Solid alloys may vary greatly in range of composition, structure, properties, and behavior. *See* ALLOY; ALLOY STRUCTURES.

Phase transformations. Much of the accumulated experimental information about the nature of the interaction and the phase transformations in systems composed of two or more metals is contained in phase or equilibrium diagrams, such as Fig. 1, which depicts the phase relationships in the copper-zinc (brass) system. Such phase diagrams, even for binary metal systems, may be of all degrees of complexity, ranging from systems showing the formation of simple solid solutions to dozens of intermetallic phases exhibiting structural (polymorphic), order-disorder, magnetic, bond-type, or deformation-type transformations as a function of composition or temperature, or both. Intermetallic compounds are composed of two or more metals. They may be stable over only a very narrow or over a relatively wide range of composition, which may be stoichiometric, as for the compounds GaAs, $PdCu_{13}$, $Zr_{57}Al_{43}$, and $Nb_{48}Ni_{39}Al_{13}$, or nonstoichiometric, as for the compounds $Co_{1-x}Te_x$ (where x extends continuously from 0.5 to 0.67) and $Al_{\sim 0.8}Ge_{\sim 0.2}Nb_3$ (an important superconducting alloy). *See* EQUILIBRIUM, PHASE.

Crystal structure. The crystal structures found for intermetallic compounds may likewise range from the simple rock-salt structure displayed by BaTe to the extremely complex arrangement found for $NaCd_2$, $Mg_{32}(Al,Zn)_{49}$, and Cu_4Cd_3. Thus, a continuous range of solid solutions is observed in the K-Rb system; a simple eutectic (mixture with the minimum melting point) in the Sn-Pb system; a single intermetallic compound, $PbMg_2$, in the Pb-

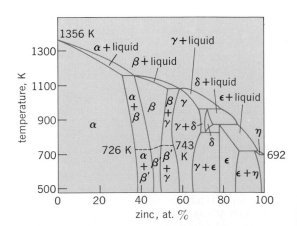

Fig. 1. Equilibrium-phase diagram for the copper-zinc (brass) system, showing the succession of phases which result with increasing concentration of zinc. The ordered phases β' and γ' exist only below 726 K and 743 K, respectively. (*J. L. T. Waugh, The Constitution of Inorganic Compounds, pt. M: Metals and Intermetallic Compounds, Wiley-Interscience, 1972*)

Mg system; and three different intermetallic compounds, $CuZn$, Cu_5Zn_8, and $CuZn_3$ (Hume-Rothery phases), in the brass system. Examples of disordered and ordered structures (superlattices or superstructures), stable above and below a certain temperature, respectively, are shown in Fig. 2. *See* CRYSTAL STRUCTURE; EUTECTICS.

All of the metallic elements, with the exception of Mn and Sn, crystallize with at least one polymorphic modification having a face-centered cubic (fcc), body-centered cubic (bcc), or close-packed hexagonal (cph) structure. Substitutional intermetallic compounds are preferentially formed between metal pairs that do not differ in metallic radii by more than about 15%, that adopt the same crystal structure, and that are of similar electronic structure and electronegativity. More stable compounds are formed between metal pairs with substantially different electrochemical characteristics. Many intermetallic compounds have compositions and structures that are determined largely by the relative sizes of the atoms involved. For example, $MgCu_2$, $NaAu_2$, $CaAl_2$, KBi_2, ZrW_2, $AgBe_2$, $MgZnNi$, and $BiAu_2$ all crystallize with the same structure; bismuth behaves as the smaller atom in KBi_2, and as the larger one in $BiAu_2$.

Another very large class of intermetallic compounds results from one kind of metal atom occupying the interstitial cavities in the close-packed structure of the other, usually in some preferential set or subset of lattice sites. If all the octahedral cavities in an fcc array of metal atoms are occupied by a dissimilar type of metal atom, the rocksalt structure results (CaSe, SrTe, UC); the nickel arsenide structure results when the corresponding set of octahedral holes in the cph structure is similarly occupied (CrSb, FeTe, PtSn); if exactly half of the tetrahedral cavities are occupied by one kind of metal atom, the sphalerite structure is derived from the fcc (GaAs, InSb), and the wurtzite structure from the cph (CdSe, ZnTe), arrangements. Among the more complex intermetallic structures, a large number involve icosahedral coordination of a larger metal atom about a smaller one ($MoAl_{12}$, $MgCu_2$, $Mg_{32}\{Al,Zn\}_{49}$). Coordination numbers greater than 12 (14, 15, and 16) are common among the more complex structures. The Hume-Rothery electron compounds with closely related structures but apparently unrelated stoichiometry are determined by electron-to-atom ratios. Thus, metal pairs for which the electron-atom ratio is 3:2 crystallize with either bcc (Ag_3Al, NiAl), β-Mn complex cubic (Cu_5Si, $CoZn_3$) or cph (Cu_3Ga, Ag_7Sb) structures; if the ratio is 21:13, the complex γ-brass structure results ($Na_{31}Pb_8$, Cu_5Zn_8); and if the ratio is 7:4, cph structures (Au_5Al_3, Ag_3Sn) are obtained. Several groups of intermetallic compounds have been classified on the basis of the results of x-ray crystallographic determinations, such as the three types of Laves phases represented by $MgZn_2$, $MgCu_2$, and $MgNi_2$, which are derived from fcc, cph, and a combination of these structures.

Other characteristics. In addition to the electronic structure of the constituent metal atoms (which determines electron-atom ratios), electronegativity and metallic radii differences, packing and structural considerations (which determine the symmetry and extent of the so-called Fermi surfaces), and thermodynamic factors (such as the

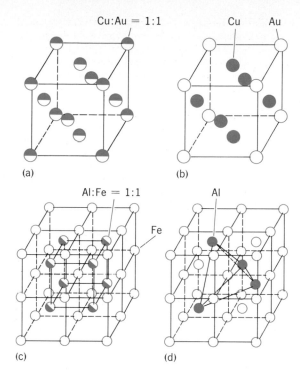

Fig. 2. Superstructure formation in the Cu-Au and the Al-Fe systems. (a) Disordered CuAu and (b) ordered Cu_3Au structures; (c) disordered Fe_3Al and (d) ordered Fe_3Al structures. The ordering of the atoms in *b* compared with *a* involves their redistribution over the lattice sites of an fcc unit cell; ordering of atoms in *d* relative to *c* involves their rearrangement over the lattice sites of a bcc structural unit. (*J. L. T. Waugh, The Constitution of Inorganic Compounds, pt. M: Metals and Intermetallic Compounds, Wiley-Interscience, 1972*)

entropy and enthalpy changes on interaction between different metals) must be taken into account in order to understand why some intermetallic compounds exist as ordered or disordered phases, exhibit stoichiometric or nonstoichiometric compositions, or exist over a range of compositions. *See* NONSTOICHIOMETRIC COMPOUNDS; SEMICONDUCTOR; SOLID-STATE CHEMISTRY.

[JOHN L. T. WAUGH]

Bibliography: A. P. Cracknell and K. C. Wong, *The Fermi Surface*, 1973; S. Samson, D. P. Shoemaker, and C. B. Shoemaker, in *Structural Chemistry and Molecular Biology*, pp. 638–730, 1968; J. L. T. Waugh, *The Constitution of Inorganic Compounds*, pt. M: *Metals and Intermetallic Compounds*, 1972.

Intermolecular forces

Attractive or repulsive interactions that occur between all atoms and molecules. These forces, which become significant at molecular separations of about 1 nm or less, are much weaker than forces associated with chemical bonds or electrostatic interactions of charged particles. They are important, however, since they are responsible for many of the physical properties of solids, liquids, and pressurized gases. Intermolecular forces also determine to an important extent the three-dimensional arrangement of biological molecules, polymers, and even smaller molecules.

Description. A simple description of intermolecular forces can begin with the example of two interacting argon atoms. The atoms are electrically neutral and do not undergo chemical bond formation.

Figure 1 shows the potential energy of two argon atoms as a function of their separation. At distances of about 1 nm or greater this energy is essentially zero and the atoms exert no forces on each other. (The force is the negative gradient, or slope, of the potential energy.) Between 0.4 and 0.8 nm the potential energy decreases and the atoms experience forces of attraction. For distances less than 0.3 nm the potential energy rises sharply as the atoms repel each other. At a distance of 0.38 nm the forces of attraction and repulsion balance each other. The potential energy (and corresponding intermolecular forces) between other pairs of atoms exhibits the same general shape as shown in Fig. 1, although the quantitative values of energy and separation are somewhat different. For intermolecular forces between molecules the relative orientations as well as distances are important and the description is more complex. In general, for either atoms or molecules at separations of 0.3 nm or less, the intermolecular forces are repulsive. At longer range, usually greater than 0.3 nm, the intermolecular forces are attractive. And at some intermediate distance, usually 0.3–0.4 nm (which depends on orientation in the case of molecules), the intermolecular forces of attraction and repulsion just balance.

Origin. The origin of intermolecular forces is again most simply discussed by considering two interacting atoms. Quantum mechanics indicates that the rapid motion of the electrons causes instantaneous fluctuations in the charge density around the nucleus. For atoms far apart the electrons in one atom move independently of electrons

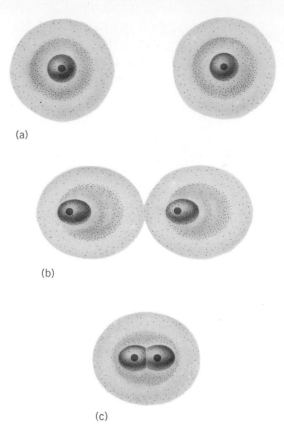

(a)

(b)

(c)

Fig. 2. Schematic illustration of intermolecular interactions. (a) There is no interaction between the atoms that are 1 nm or more apart. (b) For atoms separated by about 0.8 nm or less, dispersion forces which are attractive result from correlated fluctuations of the electron charge distribution in the atoms. (Distribution shown is greatly exaggerated.) (c) For the atoms closer together, 0.3 nm or less, exchange forces which are repulsive cause a permanent distortion of the electron charge distribution. (Distribution shown is greatly exaggerated.)

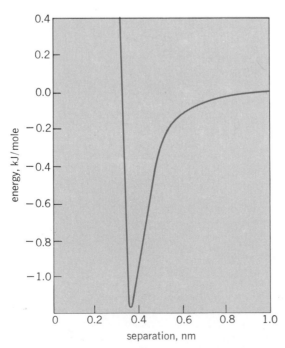

Fig. 1. The intermolecular potential energy of two argon atoms. (*From J. M. Parson, P. E. Siska, and Y. T. Lee, Intermolecular potentials from cross-beam differential elastic scattering measurements, IV:Ar + Ar, J. Chem. Phys., 56: 1511, 1972.*)

in the other atom, and on the average the charge distribution is symmetric as shown in Fig. 2a. At distances where attractive forces become important, the *average* charge distribution is still symmetric. However, an *instantaneous* fluctuation in the electron distribution in one atom can now affect its neighbor nearby. A charge separation in one atom occurs when the electron cloud shifts toward one side of the atom, barring its nucleus to a slight extent. In the other atom the electrons have moved in concert toward this barred nucleus, and an electrostatic attraction is set up. This is illustrated schematically in an exaggerated fashion in Fig. 2b. At another instant the electron clouds may shift in the opposite direction, and the other atom has its nuclear charge partially exposed to the electrons of its neighbor. The electron motions in both atoms are correlated so that an attractive electrostatic force is maintained while the averaged motions assure a symmetric distribution about each atom. These attractive forces are often called London or dispersion forces.

At small separations the electron clouds can overlap, and repulsive forces are set up. These are called Pauli or exchange forces and are also explained by quantum mechanics. They are essentially a consequence of the reluctance of electrons to be confined into the same small region of space. Atoms or molecules brought close together will respond to exchange forces by a permanent distortion of their electron distribution as shown in Fig. 2c.

All atoms and molecules experience dispersion and exchange forces, which thus are a common component of intermolecular forces. Neutral molecules, in addition, may interact with each other because they possess permanent electrical polarity expressed as a dipole, quadrupole, or higher multipole moments. The electrostatic forces associated with these interacting multipole moments depend on the orientation of the molecules and may be either attractive or repulsive. The corresponding energies are usually somewhat less than dispersion or exchange energies. The dispersion, exchange, and permanent multipole electrostatic forces taken together are usually called van der Waals forces. Energies associated with the formation of hydrogen bonds (that is, between two HF or H_2O molecules) are somewhat larger than van der Waals energies.

Interactions considerably stronger than those just discussed sometimes occur between atoms or molecules. The energies of chemical bond formation are hundreds of times greater than that shown by the intermolecular potential well of Fig. 1. Electrostatic interactions between charged particles are likewise relatively strong. These interactions are usually not classified as intermolecular forces. *See* CHEMICAL BINDING.

Occurrence. Intermolecular forces are responsible for many of the bulk properties of matter in all its phases. A realistic description of the relationship among pressure, volume, and temperature of a gas must include the effects of attractive and repulsive forces between molecules. At increased pressures and sufficiently low temperatures the attractive forces between molecules in the gas will cause it to liquefy. The viscosity, surface tension, and diffusion of liquids are examples of physical properties which are a consequence of intermolecular forces. Repulsive forces prevent the molecules from approaching one another too closely and account for the high compressibility of liquids. Intermolecular forces between near and distant neighbors dictate the ordered molecular arrangements in crystalline solids. These forces also account for the elasticity of solids. A detailed accounting of the intermolecular forces in the condensed phase is complex since it must include the interactions of each molecule with many of its neighbors. Nevertheless, the energy of each pair of atom interactions is approximately described by an intermolecular potential of the sort shown in Fig. 1.

Intermolecular forces are also important between atoms within a molecule. In ethane, for example, strong chemical bonds connect C—C and C—H atoms within the molecule as in Fig. 3. The chemical bonds direct the C—C—H and H—C—H angles to be nearly 109°. The van der Waals interactions (indicated by the dots) produce the final stable three-dimensional arrangement, with the H atoms chemically bonded to one C atom staggered with respect to the H atoms on the other C atom. In an analogous way for larger molecules,

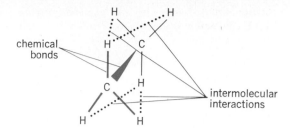

Fig. 3. Chemical bonds and intermolecular interactions in ethane.

proteins, and other biological molecules, the complex spatial arrangements assumed are determined in part by the balance of attractive and repulsive intermolecular forces between atoms that are chemically bonded within the molecule. *See* PROTEINS.

Study methods. The importance of intermolecular forces has been responsible for their extensive study for many decades. In the early 1970s most of the information on intermolecular forces was inferred from the study of matter in bulk. Measurements of the viscosity of gases, or crystal structure of solids, for example, were used to estimate the quantitative nature of the intermolecular interactions that must produce these physical properties. However, it has since been found that studies of individual molecular interactions yield the information more directly.

In molecular beam experiments, low-density streams of atoms or molecules are directed so that individual particles collide. The way in which the molecules rebound as a result of their collision is determined by their initial velocities which can be controlled, and intermolecular forces which can be extracted from the experimental data. The intermolecular potential energy curve shown in Fig. 1 was obtained from studies of the collision dynamics of argon atoms. Mappings of the potential energy surfaces of other atoms and molecules are being determined by this technique.

Another approach is to study van der Waals molecules. In these experiments clusters of atoms or molecules are formed at low temperatures in the gas phase because of their intermolecular attractions for each other. Clusters of two or three atoms or molecules are called van der Waals molecules. For example, gaseous argon at the temperature of the boiling liquid (−186°C) contains about 98% Ar atoms, and the remaining 2% are Ar_2 van der Waals molecules. The ultraviolet spectrum of the gas at low temperatures reveals features due to Ar_2 which can be used to characterize the bond strength and the intermolecular forces between the argon atoms in the van der Waals molecule.

Spectroscopy of van der Waals molecules formed by clusters of chemically bonded molecules has also revealed much about intermolecular forces which depend on the orientation of the molecules within the cluster. Gaseous H_2, O_2, or HF contains small concentrations of $(H_2)_2$, $(O_2)_2$, or $(HF)_2$. The structures of these van der Waals molecules are shown in Fig. 4. The chemical bonds in H_2, O_2, or HF are about 0.1 nm long and not affected by the formation of the 0.3–0.4 nm intermolecular bond of the van der Waals molecule. In $(H_2)_2$ the intermolecular forces do not depend much on

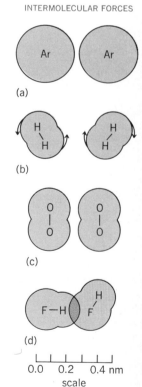

Fig. 4. The structures of some van de Waals molecules. (a) Argon. (b) Hydrogen. (c) Oxygen. (d) Hydrogen fluoride. (From G. Ewing, Structure and properties of van der Waals molecules, Accounts Chem. Res., 8:185, 1975)

the orientation of either H_2, and as a consequence each H_2 molecule, while weakly bound to its neighbor, rotates freely within the cluster. The arrows shown in Fig. 4 are meant to represent this freedom of internal rotation. The $(O_2)_2$ van der Waals molecule appears to reside in a rectangular configuration, while $(HF)_2$ exhibits a bent structure characteristic of hydrogen bond formation. While chemical bonds produce rigid molecules with well-defined geometries, intermolecular forces maintain rather floppy structures of the van der Waals molecules. Internal motions in $(O_2)_2$ or $(HF)_2$ produce considerable distortions of the static structure representations in Fig. 4. The structures of several dozen van der Waals molecules are now known. The determination of properties of this new class of compounds promises to provide a deeper insight into the nature of intermolecular forces. *See* MOLECULAR STRUCTURE AND SPECTRA.

Recent theoretical approaches to intermolecular interactions have taken two directions. Detailed quantum-mechanical calculations have been performed on the interactions of very simple systems, for example, two He atoms. These calculations seek to determine the wave functions, importance of the correlated motions of the electrons, and the precise nature of the energy of the interaction. This theoretical approach then seeks a deeper understanding of the quantum-mechanical origin of intermolecular forces. A more pragmatic approach uses the electron distributions of the isolated molecules from previous calculations. These distributions are treated as an "electron gas" which interacts as the molecules are brought together. Calculations of the electron gas model appear to produce reliable intermolecular energies for both interacting atoms and molecules with a modest amount of computational effort. *See* QUANTUM THEORY, NONRELATIVISTIC; SOLUTION; STATISTICAL MECHANICS; VALENCE.

[GEORGE E. EWING]

Bibliography: A. D. Buckingham and B. D. Utting, Intermolecular Forces, *Annu. Rev. Phys. Chem.* 21:287, 1970; P. R. Certain and L. W. Bruch, Intermolecular forces, in W. B. Brown (ed.), *MTP International Review of Science: Theoretical Chemistry*, ser. 1, 1:113, 1972; J. O. Hirschfelder, C. F. Curtiss, and R. B. Bird, *Molecular Theory of Gases and Liquids*, 1954; Y. S. Kim and R. G. Gordon, Unified theory for intermolecular forces between closed shell atoms and ions, *J. Chem. Phys.* 61:1, 1974; H. Margenau and N. Kestner, *Theory of Intermolecular Forces*, 1971.

Internal combustion engine

A prime mover, the fuel for which is burned within the engine, as contrasted to a steam engine, for example, in which fuel is burned in a separate furnace. *See* ENGINE.

The most numerous of internal combustion engines are the gasoline piston engines used in passenger automobiles, outboard engines for motor boats, small units for lawn mowers, and other such equipment, as well as diesel engines used in trucks, tractors, earth-moving, and similar equipment. This article describes these types of engines. For other types of internal combustion engines *see* GAS TURBINE; ROCKET ENGINE; ROTARY ENGINE; TURBINE PROPULSION.

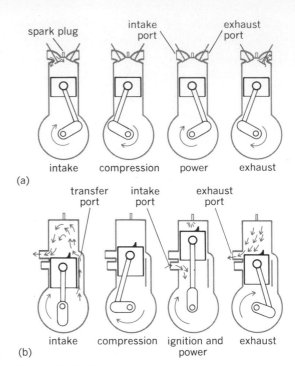

(a)

intake   compression   power   exhaust

(b)

intake   compression   ignition and power   exhaust

Fig. 1. (*a*) The four strokes of a modern four-stroke engine cycle. For intake stroke the intake valve (left) has opened and the piston is moving downward, drawing air and gasoline vapor into the cylinder. In compression stroke the intake valve has closed and the piston is moving upward, compressing the mixture. On power stroke the ignition system produces a spark that ignites the mixture. As it burns, high pressure is created, which pushes the piston downward. For exhaust stroke the exhaust valve (right) has opened and the piston is moving upward, forcing the burned gases from the cylinder. (*b*) The same action is accomplished without separate valves and in a single rotation of the crankshaft by a three-port two-cycle engine. (*From M. L. Smith and K. W. Stinson, Fuels and Combustion, McGraw-Hill, 1952*)

The aircraft piston engine is fundamentally the same as that used in automobiles but is engineered for light weight and is usually air cooled. *See* AIRCRAFT ENGINE, RECIPROCATING.

ENGINE TYPES

Characteristic features common to all commercially successful internal combustion engines include (1) the compression of air, (2) the raising of air temperature by the combustion of fuel in this air at its elevated pressure, (3) the extraction of work from the heated air by expansion to the initial pressure, and (4) exhaust. William Barnett first drew attention to the theoretical advantages of combustion under compression in 1838. In 1862 Beau de Rochas published a treatise that emphasized the value of combustion under pressure and a high ratio of expansion for fuel economy; he proposed the four-stroke engine cycle as a means of accomplishing these conditions in a piston engine (Fig. 1). The engine requires two revolutions of the crankshaft to complete one combustion cycle. The first engine to use this cycle successfully was built in 1876 by N. A. Otto. *See* OTTO CYCLE.

Two years later Sir Dougald Clerk developed the two-stroke engine cycle by which a similar combustion cycle required only one revolution of the crankshaft. In this cycle, exhaust ports in the cylinder were uncovered by the piston as it approached the end of its power stroke. A second cylinder then pumped a charge of air to the working cylinder through a check valve when the pump pressure exceeded that in the working cylinder.

In 1891 Joseph Day simplified the two-stroke engine cycle by using the crankcase to pump the required air. The compression stroke of the working piston draws the fresh combustible charge through a check valve into the crankcase, and the next power stroke of the piston compresses this charge. The piston uncovers the exhaust ports near the end of the power stroke and slightly later uncovers intake ports opposite them to admit the compressed charge from the crankcase. A baffle is usually provided on the piston head of small engines to deflect the charge up one side of the cylinder to scavenge the remaining burned gases down the other side and out the exhaust ports with as little mixing as possible.

Engines using this two-stroke cycle today have been further simplified by use of a third cylinder port which dispenses with the crankcase check valve used by Day. Such engines are in wide use for small units where fuel economy is not as important as mechanical simplicity and light weight. They do not need mechanically operated valves and develop one combustion cycle per crankshaft revolution. Nevertheless they do not develop twice the power of four-stroke cycle engines with the same size working cylinders at the same number of revolutions per minute (rpm). The principal reasons for this are (1) the reduction in effective cylinder volume due to the piston movement required to cover the exhaust ports, (2) the appreciable mixing of burned (exhaust) gases with the combustible mixture, and (3) the loss of some combustible mixture through the exhaust ports with the exhaust gases.

Otto's engine, like almost all internal combustion engines developed at that period, burned coal gas mixed in combustible proportions with air prior to being drawn into the cylinder. The engine load was generally controlled by throttling the quantity of charge taken into the cylinder. Ignition was accomplished by a device such as an external flame or an electric spark so that the timing was controllable. These are essential features of what has become known as the Otto or spark-ignition combustion cycle.

Ideal and actual combustion. In the classical presentation of the four-stroke cycle, combustion is idealized as instantaneous and at constant volume. This simplifies thermodynamic analysis, but fortunately combustion takes time, for it is doubtful that an engine could run if the whole charge burned or detonated instantly.

Detonation of a small part of the charge in the cylinder, after most of the charge has burned progressively, causes the knock which limits the compression ratio of an engine with a given fuel. See COMBUSTION KNOCK.

The gas pressure of an Otto combustion cycle using the four-stroke engine cycle varies with the piston position as shown by the typical indicator card in Fig. 2a. This is a conventional pressure-volume (PV) card for an 8.7:1 compression ratio. For simplicity in calculations of engine power, the

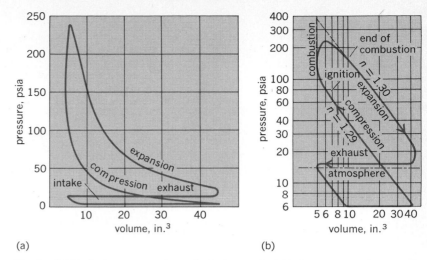

(a)   (b)

Fig. 2. (a) Typical pressure-volume indicator card plotted on rectangular coordinates. (b) The same data plotted on logarithmic coordinates.

average net pressure during the working stroke, called the mean effective pressure (mep), is frequently used. It may be obtained from the average net height of the card, which is found by measurement of the area with a planimeter and by division of this area by its length. Similar pressure-volume data may be plotted on logarithmic coordinates as in Fig. 2b, which develops expansion and compression relations as approximately straight lines, the slopes of which show the values of exponent n to use in equations for PV relationships.

The rounding of the plots at peak pressure, with the peak developing after the piston has started its power stroke, even with the spark occurring before the piston reaches the end of the compression stroke, is due to the time required for combustion. The actual time required is more or less under control of the engine designer, because he can alter the design to vary the violence of the turbulence of the charge in the compression space prior to and during combustion. The greater the turbulence the faster the combustion and the lower the antiknock or octane value required of the fuel, or the higher the compression ratio that may be used with a given fuel without knocking. On the other hand, a designer is limited as to the amount he can raise the turbulence by the increased rate of pressure rise, which increases engine roughness. Roughness must not exceed a level acceptable for automobile or other service. See COMBUSTION CHAMBER; COMPRESSION RATIO.

Compression ratio. According to classical thermodynamic theory, thermal efficiency η of the Otto combustion cycle is given by Eq. (1), where

$$\eta = 1 - \frac{1}{r^{n-1}} \qquad (1)$$

the compression ratio r_c and expansion ratio r_e are the same ($r_c = r_e = r$). When theory assumes atmospheric air in the cylinder for extreme simplicity, exponent n is 1.4. Efficiencies calculated on this basis are almost twice as high as measured efficiencies. Logarithmic diagrams from experimental data show that n is about 1.3. Even with this value, efficiencies achieved in practice are

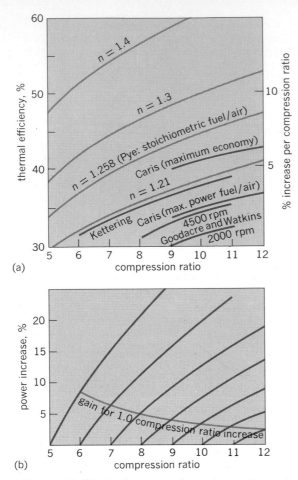

(a)

(b)

Fig. 3.   (a) Effect of compression ratio on thermal efficiency as calculated with different values of *n* and compared with published experimental data. (b) Increase in power from raising compression ratio as calculated with *n*=1.3. Percentage values are but little altered by calculating with different values of *n*.

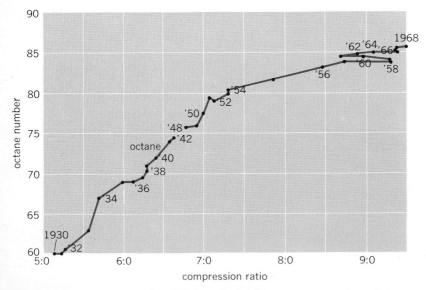

Fig. 4.   Year-to-year relation between average compression ratio of cars and average octane number (ASTM research method) of regular-grade summer gasolines sold in the United States. (*Automotive Industries, March, 1954, 1969*)

less than given by Eq. (1). This is not surprising considering the differences found in practice and assumed in theory, such as instantaneous combustion and 100% volumetric efficiency.

Attempts to adjust classical theory to practice by use of variable specific heats and consideration of dissociation of the burning gases at high temperatures have shown that this exponent should vary with the fuel-air mixture ratio, and to some extent with the compression ratio. G. A. Goodenough and J. B. Baker have shown that, for an 8:1 compression ratio, the exponent should vary from about 1.28 for a stoichiometric (chemically correct) mixture to about 1.31 for a lean mixture. Similar calculations by D. R. Pye showed that, at a compression ratio of 5:1, *n* should be 1.258 for the stoichiometric mixture, increasing with excess air (lean mixture) to about 1.3 for a 20% lean mixture and to 1.4 if extrapolated to 100% air. Actual practice gives thermal efficiencies still lower than these, which might well be expected because of the assumed instantaneous changes in cyclic pressure (during combustion and exhaust) and the disregard of heat losses to the cylinder walls. These theoretical relations between compression ratio and thermal efficiency, as well as some experimental results, are shown in Fig. 3a. The data published by C. F. Kettering and D. F. Caris are about 85% and 82%, respectively, of the theoretical relations for the corresponding fuel-air mixtures.

Figure 3b gives the theoretical percentage gain in indicated thermal efficiency or power from raising the compression of an engine from a given value. They were plotted from Eq. (1) with *n* = 1.3, but would differ only slightly if obtained from any of the curves shown in Fig. 3a. The dotted line crossing these curves shows the diminishing gain obtainable by raising the compression ratio one unit at the higher compression ratios.

Experimental data indicate that a change in compression ratio does not appreciably change the mechanical efficiency or the volumetric efficiency of the engine. Therefore, any increase in thermal efficiency resulting from an increase in compression ratio will be revealed by a corresponding increase in torque or mep; this is frequently of more practical importance to the engine designer than the actual efficiency increase, which becomes an added bonus.

Compression ratio and octane rating. For years compression ratios of automobile engines have been as high as designers considered possible without danger of too much customer annoyance from detonation or knock with the gasoline on the market at the time (Fig. 4). Engine designers continue to raise the compression ratios of their engines as suitable gasolines come on the market.

Little theoretical study has been given to the effect of engine load on indicated thermal efficiency. Experimental evidence reveals that it varies little, if at all, with load, provided that the fuel-air-ratio remains constant and that the ignition time is suitably advanced at reduced loads to compensate for the slower rate of burning which results from dilution of the combustible charge with the larger percentages of burned gases remaining in the combustion space and the reduced turbulence at lower speeds.

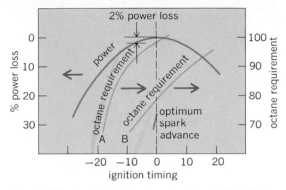

Fig. 5. Effects of advancing or retarding ignition timing from optimum on engine power and resulting octane requirement of fuel in an experimental engine with a combustion chamber having typical turbulence (A) and a highly turbulent design (B) with the same compression ratio. Retarding the spark 7° for a 2% power loss reduced octane requirement from 98 to 93 for design A.

Ignition timing. Designers obtain high thermal efficiency from high compression ratios at part loads, where engines normally run at automobile cruising speeds, with optimum spark advance, but avoid knock on available gasolines at wide-open throttle by use of a reduced or compromise spark advance. The tendency of an engine to knock at wide-open throttle is reduced appreciably when the spark timing is reduced 5–10° from optimum, as is shown in Fig. 5. Advancing or retarding the spark timing from optimum results in an increasing loss in mep for any normal engine as shown by the solid curve. The octane requirement falls rapidly as the spark timing is retarded, the actual rate depending on the nature of the gasoline as well as on the design of the combustion chamber. The broken-line curves A and B show the effects on a given gasoline of the use of moderate- and high-turbulence combustion chambers, respectively, with the same compression ratio. Because the mep curve is relatively flat near optimum spark advance, retarding the spark for a 1–2% loss is considered normally acceptable in practice because of the appreciable reduction in octane requirement.

In Fig. 6 similar data are shown by curve A for another engine with changes in mep plotted on a percentage basis against octane requirement as the spark timing was charged. Point a indicates optimum spark timing, where 85 octane was required of the gasoline to avoid knock. By raising the compression ratio, the power and octane requirement were also raised as shown by the broken-line curve B. Although optimum spark required 95 octane (point b), retarding the spark timing and thus reducing the octane requirement to 86 (point c) developed slightly more power than with the original compression ratio at its optimum spark advance. The gain may be negligible at wide-open throttle, but at lower loads where knock does not develop the spark timing may be advanced to optimum (point b), where appreciably more power may be developed by the same amount of fuel.

In addition to the advantages of the higher compression ratio at cruising loads with optimum spark advance, the compromise spark at full load

may be advanced toward optimum as higher-octane fuels become available, and a corresponding increase in full-throttle mep enjoyed. Such compromise spark timings have had much to do with the adoption of compression ratios of 10:1 to 13:1.

Fuel-air ratio. A similar line of reasoning shows that a fuel-air mixture richer than that which develops maximum knock-free mep will permit use of higher compression ratios. However, the benefits derived from compromise or superrich mixtures vary so much with mixture temperature and the sensitivity of the octane value of the particular fuel to temperature that it is not generally practical to make much general use of this method. Nevertheless it has been the practice with piston-type aircraft engines to use fuel-air mixture ratios of 0.11 or even higher during takeoff, instead of about 0.08, which normally develops maximum mep in the absence of knock.

Compression-ignition engines. About 20 years after Otto first ran his engine, Rudolf Diesel successfully demonstrated an entirely different method of igniting fuel. Air is compressed to a pressure high enough for the adiabatic temperature to reach or exceed the ignition temperature of the fuel. Because this temperature is in the order of 1000°F, compression ratios of 12:1 to 20:1 are used commercially with compression pressures generally over 600 psi. This engine cycle requires the fuel to be injected after compression at a time and rate suitable to control the rate of combustion. *See* FUEL INJECTION.

Conditions for high efficiency. The classical presentation of the diesel engine cycle assumes combustion at constant pressure. Like the Otto cycle, thermal efficiency increases with compression ratio, but in addition it varies with the amount of heat added (at the constant pressure) up to the cutoff point where the pressure begins to drop

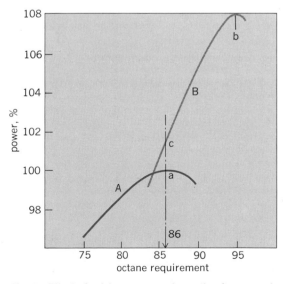

Fig. 6. Effect of raising compression ratio of an experimental engine on the power output and octane requirement at wide-open throttle. While an 86-octane fuel was required for optimum spark advance (maximum power) with the original compression ratio, the same gasoline would be knock-free at the higher compression ratio by suitably retarding the ignition timing.

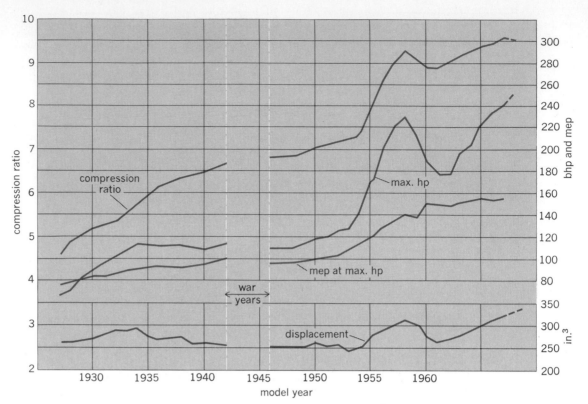

Fig. 7. Trend toward considerable increases in compression ratio, mean effective pressure (mep), and power of average United States automobile engines weighted for production volume. (*Ethyl Corp. data*)

from adiabatic expansion. *See* DIESEL CYCLE; DIESEL ENGINE.

Practical attainments. Diesel engines were highly developed in Germany prior to World War I, and made an impressive performance in submarines. Large experimental single-cylinder engines were built in several European countries with cylinder diameters up to 1 m. As an example, the two-stroke Sulzer S100 single-acting engine with a bore of 1 m and a stroke of 1.1 m developed 2050 gross horsepower at 150 rpm. Multiple-cylinder engines developing 15,000 hp are in marine service. Small diesel engines are in wide use also.

Fuel injection. In early diesel engines, air injection of the fuel was used to develop extremely fine atomization and good distribution of the spray. But the need for injection air at pressures in the order of 1500 psi required expensive and bulky multistage air compressors and intercoolers.

A simpler fuel-injection method was introduced by James McKechnie in 1910. He atomized the fuel as it entered the cylinder by use of high fuel pressure and suitable spray nozzles. After considerable development it became possible to atomize the fuel sufficiently to minimize the smoky exhaust which had been characteristic of the early solid-injection engines. By 1930 solid or airless injection had become the generally accepted method of injecting fuel in diesel engines.

Contrast between diesel and Otto engines. There are many characteristics of the diesel engine which are in direct contrast to those of the Otto engine. The higher the compression ratio of a diesel engine, the less the difficulties with ignition time lag. Too great an ignition lag results in a sudden and undesired pressure rise which causes an audible knock. In contrast to an Otto engine, knock in a diesel engine can be reduced by use of a fuel of higher cetane number, which is equivalent to a lower octane number. *See* CETANE NUMBER; COMBUSTION WAVE MEASUREMENT; OCTANE NUMBER.

The larger the cylinder diameter of a diesel engine, the simpler the development of good combustion. In contrast, the smaller the cylinder diameter of the Otto engine, the less the limitation from detonation of the fuel.

High intake-air temperature and density materially aid combustion in a diesel engine, especially of fuels having low volatility and high viscosity. Some engines have not performed properly on heavy fuel until provided with a super charger. The added compression of the supercharger raised the temperature and, what is more important, the density of the combustion air. For an Otto engine, an increase in either the air temperature or density increases the tendency of the engine to knock and therefore reduces the allowable compression ratio.

Diesel engines develop increasingly higher indicated thermal efficiency at reduced loads because of leaner fuel-air ratios and earlier cutoff. Such mixture ratios may be leaner than will ignite in an Otto engine. Furthermore, the reduction of load in an Otto engine requires throttling, which develops increasing pumping losses in the intake system.

TRENDS IN AUTOMOBILE ENGINES

Cylinder diameters of average American automobile engines prior to 1910 were over $4\frac{1}{4}$ in. By 1917 they had been reduced to only a little over $3\frac{1}{4}$ in., where they stabilized until after 1945. Since then the increased demand for more power,

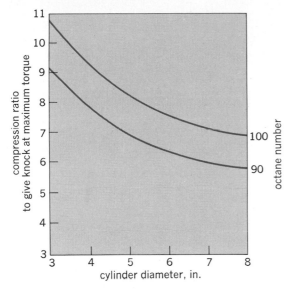

Fig. 8. Relation between cylinder diameter and limiting compression ratio for engines of similar design, one using 90- and the other 100-octane gasoline. (*From L. L. Brower, unpublished SAE paper, 243, 1950*)

with the number of cylinders limited to eight for practical mechanical reasons, the diameters have been increased from year to year until they averaged 3.98 in. in 1969, with a maximum of 4.36 in.

Stroke-bore ratio. Experimental engines differing only in stroke-bore ratio show that this ratio has no appreciable effect on fuel economy and friction at corresponding piston speeds. Practical advantages which result from the short stroke include (1) the greater rigidity of crankshaft from the shorter crank cheeks, with crankpins sometimes overlapping main bearings, and (2) the narrower as well as lighter cylinder block which is possible. On the other hand, the higher rates of crankshaft rotation for an equivalent piston speed necessitate greater valve forces and require stronger valve springs. Also the smaller depth of the compression space for a given compression ratio increases the surface-to-volume ratio and the proportion of heat lost by radiation during combustion. Nevertheless, stroke-bore ratios have been decreasing for more than 25 years and in 1969 reached 0.9 for the average automobile in the United States.

Cylinder number and arrangement. Engine power may be raised by increasing the number of cylinders as well as the power per cylinder. The minimum number of cylinders has generally been four for four-cycle automobile engines, because this is the smallest number that provides a reasonable balance for the reciprocating pistons. Many early cars had four-cylinder engines. After 1912 six-cylinder in-line engines became popular. They have superior balance of reciprocating forces and more even torque impulses. By 1940 the eight-cylinder 90° V engine had risen in popularity until it about equaled the six-in-line. After 1954, the V-8 dominated the field for American automobile engines. There are several important reasons for this besides the increased power. For example, the V-8 offers appreciably more rigid construction with less bearing deflection at high speeds, provides more uniform distribution of fuel to all cylinders

from centrally located downdraft carburetors, and has a short, low engine that fits within the hood demanded by style trends. With the introduction of the smaller "compact" cars in 1959, where the power and cost of eight-cylinder V-type engines were not required, six-cylinder designs increased. By 1969 about 38% of all engine designs were of the six-cylinder in-line type. The evolution of cylinder arrangements included for a short period the V-12 and even a V-16 cylinder design, but experience showed that in their day there was too much practical difficulty in providing good manifold distribution of fuel, especially when starting cold, and too much difficulty in keeping all spark plugs firing. *See* AUTOMOBILE.

Compression ratio. The considerable increase in power of the average automobile engine over the years is shown in Fig. 7 together with the compression ratios which have had much to do with the increased mep. Such ratios approach practical limits imposed by phenomena other than detonation, such as preignition, rumble, and other evidences of undesirable combustion.

The modern trend toward high compression ratios, with their small compression volumes, has dictated the universal use of overhead valves in all American engine designs. High compression ratios also tend to restrict cylinder diameters because the longer flame travel increases the tendency to knock (Fig. 8).

Improved breathing and exhaust. Added power output has been brought about by reducing the pressure drop in the intake system at high speeds and by reducing the back pressure of the exhaust systems (Fig. 9). These results were accomplished by larger valve areas and valve ports, by larger venturi areas, and by more streamlined manifolds. Larger valve areas were achieved by higher lift of the valves and by larger valves. Larger venturi areas in the carburetors were achieved by use of one or more two-stage carburetors; in these, sufficient air velocity was developed to meter the

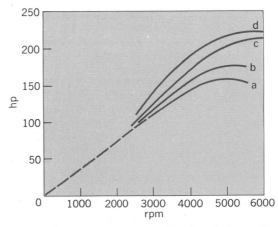

Fig. 9. Increases in peak power of a six-cylinder engine (210 in.³ displacement) by successive reductions in resistance to airflow through it. Curve a, power developed by original engine with two 1¾ in. carburetors; curve b, higher valve-lift valves and dual exhaust systems; curve c, larger valves, smoother valve ports, and two 2-in. carburetors; curve d, three double-barrel carburetors. (*Adapted from W. M. Heynes, The Jaguar engine, Inst. Mech. Eng., Automot. Div. Trans., 1952–1953*)

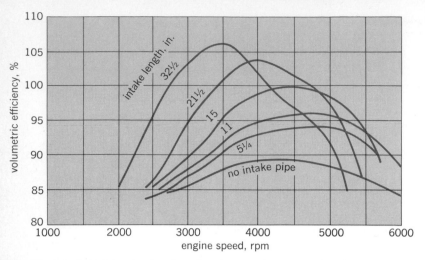

Fig. 10. Effect of intake-pipe length and engine speed on volumetric efficiency of one cylinder of a six-cylinder engine. (*From E. W. Downing, Proc. Automot. Div., Inst. Mech. Eng., no. 6, p. 170, 1957–1958*)

fuel on one venturi at low power; the second venturi was opened for high power. Better streamlining of the manifold passages between carburetors and valves, especially at the cylinder ports, and the use of dual exhausts and mufflers with reduced back pressure have also improved engine breathing and exhaust.

Valve timing. The times of opening and closing of the valves of an engine in relation to the piston position are usually selected to develop maximum power over a desired speed range at wide-open throttle. For convenience the timing of these events is expressed as the number of degrees of crankshaft rotation before or after the piston reaches the end of one of its strokes. Because of the time required for the flow of the burned gas through the exhaust valve at the end of the power or expansion stroke of a piston, it is customary to start opening the valve considerably before the end of the stroke. If the valve should be opened when the piston is nearer the lower end of its stroke,

power would be lost at high engine speeds because the piston on its return (exhaust) stroke would have to move against gas pressure remaining in the cylinder. On the other hand, if the valve were opened before necessary, the burned gas would be released while it is still at sufficient pressure to increase the work done on the piston. Thus for any engine there is a time for opening the exhaust valve which will develop the maximum power at some particular speed. Moreover, the power loss at other speeds does not increase rapidly. It is obvious that, when an engine is throttled at part load, there will be less gas to discharge through the exhaust valve and there will be less need for it to be opened as early as at wide-open throttle.

The timing of intake valve events is normally selected to trap the largest possible quantity of air or combustible mixture in the cylinder when the valve closes at some desired engine speed and at wide-open throttle. The intermittent nature of the flow through the valve subjects it to alternate accelerations and retardation which require time. During the suction stroke, the mass of air moving through the pipe leading to the intake valve is given velocity energy which may be converted to a little pressure at the valve when the air mass still in the pipe is stopped by its closure. Advantage of this phenomenon may be obtained at some engine speed to increase the air mass which enters the cylinder. The engine speed at which the maximum volumetric efficiency is developed varies with the relative valve area, closure time, and other factors, including the diameter and particularly the length of this pipe (Fig. 10). These curves reveal the characteristic falling off at high speeds from the inevitable throttling action as air flows at increased velocities through any restriction such as a valve or intake pipe or particularly the venturi of a carburetor.

The curves shown in Fig. 10 are smoothed averages drawn through data obtained from experiments, but if they had been drawn through data taken at many more speeds, they would have revealed a wavy nature (Fig. 11) due to resonant oscillations in the air column entering the cylinder; these oscillations are somewhat similar to those which develop in an organ pipe.

Volumetric efficiency has a direct effect on the mep developed in a cylinder, on the torque, and on the power that may be realized at a given speed. Since power is a product of speed and torque, the peak power of an engine occurs at a higher speed than for maximum torque, where the rate of torque loss with any further increase in speed will exceed the rate of speed increase. This may be seen in Fig. 12, where the torque and power curves for a typical six-cylinder engine have been plotted. This engine developed its maximum power at a speed about twice that for maximum torque. The average 1969 V-8 engine developed its maximum power at a speed about 60% higher than its maximum torque.

Table 1 shows that the engine speeds at which maximum torque and power are desired in practice require closure of the intake valve to be delayed until the piston has traveled almost half the length of the compression stroke. At engine speeds below those where maximum torque is developed by this valve timing, some of the combustible charge

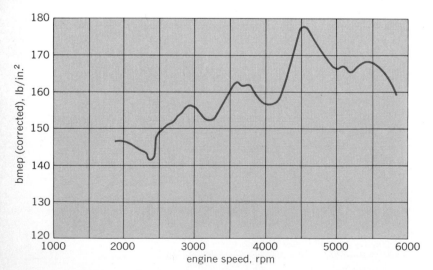

Fig. 11. Evidence of resonant oscillations in the intake pipe to a single cylinder shown by readings taken at small speed increments. (*From E. W. Downing, Proc. Automot. Div., Inst. Mech. Eng., no. 6, p. 170, 1957–1958*)

Table 1. Valve timing of 1969 American automobile engines in degrees crankshaft rotation*

Engine	Intake valve		Exhaust valve	
	Opens before top center	Closes after bottom center	Opens before bottom center	Closes after top center
Six-cylinder				
Average	13	58.5	50.3	16.6
Range	9–26	48–62	42–60	6–28
Eight-cylinder V				
Average	20	71	72	28
Range	10–31	50–114	44–90	10–58

*Adapted from *Automotive Industries*, March 15, 1969.

which has been drawn into the cylinder on the suction stroke will be driven back through the valve before it closes. This reduces the effective compression ratio at wide-open throttle—thus the increasing tendency of an engine to develop a fuel knock as the speed and the resulting gas turbulence are reduced.

Another result of this blowback into the intake pipe is the possible reversal of the flow of some of the fuel-air mixture through the carburetor, which will draw fuel each time it passes through the metering system, thereby producing a much richer mixture than if it had passed through the same carburetor only once. The increased fuel supplied to the air reaching the intake valve because of this reversed air flow may be over 100% greater for a single cylinder engine at wide-open throttle than if there were no reversal of flow. Throttling the engine reduces and almost eliminates the blowback through the metering system and the ratio of fuel to air approaches that which would be expected from air flow in one direction only. Manifolding more than one cylinder to the metering system of one carburetor also reduces the blowback through a carburetor because it averages the blowback from individual cylinders. When six cylinders are supplied by a single carburetor, there may be practically no blowback through the carburetor or enrichment of the combustible mixture by this phenomenon. However, when three carburetors are installed on the same six-cylinder engine, with two cylinders supplied by each, the enrichment may be 80–90%. When four cylinders are supplied by a single metering system, as with most V-type eight-cylinder engines, the enrichment may be 10–50% and varies with many factors besides the closing time of the intake valve, such as the exhaust valve events and exhaust back pressure, so that carburetor settings and compensation must usually be made on the engine which it is to supply, and preferably as installed in the car, if optimum power and economy are to be realized. Unfortunately, such settings can not be predicted by simple calculations of fuel flow through an orifice into an airstream flowing at constant velocity.

Intake manifolds. Intake manifolds for multicylinder engines should meet several requirements for the satisfactory performance of spark-ignition engines. They should (1) distribute fuel equally to all cylinders at temperatures where unvaporized fuel is present, as when starting a cold engine or during the warm-up period; (2) supply sufficient heat to vaporize the liquid fuel from the carburetor as soon after starting as possible; (3) distribute the vaporized fuel-air mixture evenly to all cylinders during normal operation and at low speeds; (4) offer minimum restriction to the mixture flow at high power; and (5) provide equal ram or dynamic boost to volumetric efficiency of all cylinders at some desired part of the engine speed range. This requires that each branch from the carburetor to the valve port should be equal in length, as may be inferred from Fig. 10. Accordingly, no cylinder port should be siamesed with another at the end of a leg of the manifold.

For the warming-up period with liquid fuel present, rectangular sections are desirable to impede spiraling of liquid fuel along the walls, and right-angle bends should be sharp, at least at their inner corner, so as to throw the liquid flowing along the inner wall back into the air stream, and there should be an equal number in each branch.

Manifold heat. Intake manifolds of most American automobile engines are heated to the temperature required to vaporize the fuel from the carburetor (120–140°F) by exhaust gas passing through a suitable passage in the manifold casting, particularly at the first T beyond the carburetor where

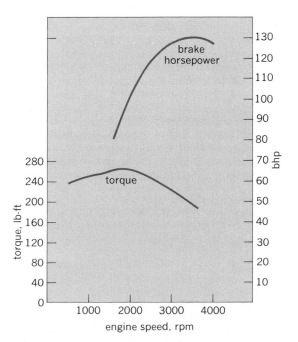

Fig. 12. Typical relation between engine speeds for maximum torque and maximum horsepower.

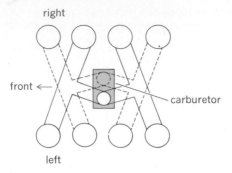

Fig. 13. Schematic of typical intake manifold with dual carburetor on a V-8 engine.

the liquid fuel impinges before turning to side branches.

To speed the warm-up process, thermostatically operated valves are generally placed in the engine exhaust system so as to force most of the exhaust gases through the intake manifold heater passages when the engine is cold. After the intake manifold has reached the desired temperature, such valves are intended to open and permit only the necessary small portion of exhaust gases to continue passing through the heater. This is an important feature, for too much heat causes a loss of engine power and aggravates the tendency for the engine toward knock and vapor lock.

On some engines, the intake manifolds are heated by water jackets taking hot water from the engine cooling system. This gives uniform heating over a wide range of operating conditions without danger of the overheating that might result from exhaust gas heat if the thermostatic exhaust valve should fail to open. It has the disadvantage, however, of requiring more time to reach normal manifold temperature, even though the water supply from the cylinder heads is short-circuited through the manifold jacket by a suitable water thermostat during warm-up.

One of the advantages of the V-8 engine is the excellent intake manifold design permitted by the centrally located carburetor with but small differences in the lengths of the passage between the carburetor and each cylinder, and an equal number of right-angle bends in each, as in a typical

intake manifold using a dual carburetor (Fig. 13). With the usual firing order shown in Fig. 20, the firing intervals for each of the lower branches (shown dotted) are evenly spaced 360 crankshaft degrees apart, but for each of the upper branches two cylinders fire 180° and then 540° apart.

Icing. Because gasoline has considerable latent heat of vaporization, it lowers the air temperature as it evaporates. This is true even at the low temperatures, where only a small part of it is vaporized. It is therefore possible for moisture which may be carried by the air to freeze under certain conditions. Ice is most likely to form when the atmosphere is almost saturated with moisture at temperatures slightly above freezing and up to about 40°F. When ice forms around or near the throttle, it can seriously interfere with the operation of an engine. For this reason small passages have been provided on some engines for jacket water, or exhaust gas from the heating supply for the intake manifold, to warm at least the flange of the carburetor. Here, again, too much heat would produce vapor lock and this would interfere with normal fuel metering. This is one of the reasons for designing some carburetors with separate casting for the throttle bodies which are heated by the manifold through only a thin gasket, while a thick gasket acting as a heat barrier is inserted between it and the float chamber containing the fuel metering systems.

FUEL CONSUMPTION AND SUPERCHARGING

Fuel consumption at loads throughout the operating range of an engine provide insight into such characteristics as friction loss within the engine. Volumetric efficiency of an engine can be increased by use of supercharging.

Part-load fuel economy. When the fuel consumption of a spark-ignition engine is plotted against brake horsepower, straight lines may generally be drawn through the test points at given speeds, as shown in Fig. 14, provided that the tests are run with optimum spark advances and at constant fuel-air ratios. Such lines are similar to the Willans lines long used for the steam consumption of steam engines.

For practical purposes the lines at various speeds may be considered parallel over a wide range of speeds. The assumption that the negative power indicated by extrapolating these lines to zero fuel consumption reveals the power absorbed by internal friction of the engine would be justified only when the thermal efficiency remains constant over the load range. On these coordinates, lines radiating from the origin represent constant ratios of fuel consumed to power developed and therefore constant specific fuel consumption (sfc). Several such lines are indicated in Fig. 14, from which the sfc at various loads may be read directly where they cross the performance lines at the various speeds.

Similar plots of even greater utility may be drawn on an indicated horsepower basis, as has been done in Fig. 15 for the same data. For many engines, a single performance line may be drawn through all test points at a given fuel-air ratio over a considerable range of speeds. When extrapolated, the performance line passes through the origin as it does for the engine shown in Fig. 15; the indi-

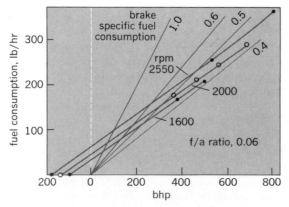

Fig. 14. Fuel consumption of an engine at part loads, plotted as typical Willans lines against brake horsepower. Data were taken with optimum spark advance and with fuel-air ratio adjusted at each test point.

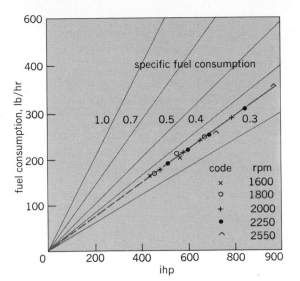

Fig. 15. The fuel consumption data of Fig. 14 plotted against the indicated horsepower, by which speed differences are neutralized.

cated sfc and thermal efficiency of the engine remain constant over the load range covered. Frequently a performance line for an engine passes a little to the left of the origin because of conditions causing a decrease in thermal efficiency as the load is reduced, such as insufficient turbulence or too low a manifold velocity. For a more complete picture of the fuel consumption performance of an engine, similar plots may be made on an mep basis. When this is done and both fuel consumption and horsepower are divided by the engine factor which converts horsepower to mep, the slope and nature of the fuel performance line remain unchanged. The fuel consumption scale then becomes equivalent to the product of mep and sfc. Such plots may be on the basis of either indicated mep (imep) or brake mep (bmep). Figure 16 shows the same data as Figs. 14 and 15 plotted on an imep basis for two different fuel-air ratios.

The fuel consumption performance of diesel engines at part loads may be shown on similar bases, but the plots should not be expected to be straight because the effective fuel-air ratio varies with load. This is illustrated in Fig. 17. It is characteristic of most diesel engines that the curvature of the plot generally flattens out at low loads so that it becomes tempting to extrapolate it to zero fuel consumption, and to consider the negative power intercept as friction. Such an intercept would represent friction only if the efficiency did not change with load, as for the engine characteristics shown in Fig. 14. If the thermal efficiency of a diesel engine improves as the load is reduced, as it would in theory for the classical diesel cycle, the zero fuel intercept for a curve such as shown in Fig. 17 would be to the right of the negative power representing engine friction. Although these fuel consumption performance plots are of considerable utility for recording such data for an engine, the fact that they are curved requires at least three or even more points to fix their location on the plot.

Supercharging spark-ignition engines. Volumetric efficiency and thus the mep of a four-stroke

spark-ignition engine may be increased over a part of or the whole speed range by supplying air to the engine intake at higher than atmospheric pressure. This is usually accomplished by a centrifugal or rotary pump. The indicated power of an engine increases directly with the absolute pressure in the intake manifold. Because fuel consumption increases at the same rate, the indicated sfc is generally not altered appreciably by supercharging.

The three principal reasons for supercharging four-cycle spark-ignition engines are (1) to lessen the tapering off of mep at higher engine speed; (2) to prevent loss of power due to diminished atmospheric density, as when an airplane (with piston engines) climbs to high altitudes; and (3) to develop more torque at all speeds.

In a normal engine characteristic, torque rises as speed increases but falls off at higher speeds because of the throttling effects of such parts of the fuel intake system as valves and carburetors. If a supercharger is installed so as to maintain the volumetric efficiency at the higher speeds without increasing it in the middle-speed range, peak horsepower can be increased.

The rapid fall of atmospheric pressure at in-

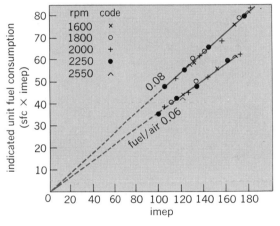

Fig. 16. The fuel consumption data of Fig. 15 plotted on the basis of imep by dividing the fuel scale and the power scale by the same factor ($k=$ ihp/imep). Line for 0.08 fuel-air ratio has been added to show effect of fuel-air ratio on slope of such plots.

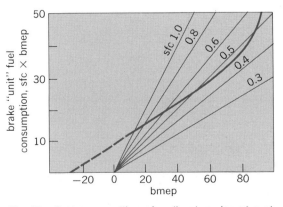

Fig. 17. Fuel consumption of a diesel engine at part loads showing the curvature, typical of such engines on these coordinates, caused by changing effective fuel-air ratios as the loads are increased.

creased altitudes causes a corresponding decrease in the power of unsupercharged piston-type aircraft engines. For example, at 20,000 ft the air density, and thus the absolute manifold pressure and indicated torque of an aircraft engine, would be only about half as great as at sea level. The useful power developed would be still less because of the friction and other mechanical power losses which are not affected appreciably by volumetric efficiency. By the use of superchargers, which are usually of the centrifugal type, sea-level air density may be maintained in the intake manifold up to considerable altitudes. Some aircraft engines drive these superchargers through gearing which may be changed in flight, from about 6.5 to 8.5 times engine speed. The speed change avoids oversupercharging at medium altitudes with corresponding power loss. Supercharged aircraft engines must be throttled at sea level to avoid damage from detonation or excessive overheating caused by the high mep which would otherwise be developed. *See* SUPERCHARGER.

Normally an engine is designed with the highest compression ratio allowable without knock from the fuel expected to be used. This is desirable for the highest attainable mep and fuel economy from an atmospheric air supply. Any increase in the volumetric efficiency of such an engine would cause it to knock unless a fuel of higher octane number were used or the compression ratio were lowered. When the compression ratio is lowered, the knock-limited mep may be raised appreciably by supercharging but at the expense of lowered thermal efficiency. There are engine uses where power is more important than fuel economy, and supercharging becomes a solution. The principle involved is illustrated in Fig. 18 for a given engine. With no supercharge this engine, when using 93-octane fuel, developed an imep of 180 psi at the border line of knock at 8:1 compression ratio. If the compression ratio were lowered to 7:1, the mep could be raised by supercharging along the 7:1 curve to 275 imep before it would be knock-limited by the same fuel. With a 5:1 compression ratio it could be raised to 435 imep. Thus the imep

could be raised until the cylinder became thermally limited by the temperatures of critical parts, particularly of the piston head.

Supercharged diesel engines. Combustion in a four-stroke diesel engine is materially improved by supercharging. In fact, fuels which would smoke badly and misfire at low loads will burn otherwise satisfactorily with supercharging. The imep rises directly with the supercharge pressure, until it is limited by the rate of heat flow from the metal parts surrounding the combustion chamber, and the resulting temperatures. A practical application of this limitation was made on a locomotive built by British Railways where the powers, and thus the heats developed, were held reasonably constant over a considerable speed range by driving the supercharger at constant speed by its own engine. In this way the supercharge pressure varied inversely with the speed of the main engine. The corresponding torque rise at reduced speed dispensed with much gear-shifting which would have been required during acceleration with a conventional engine.

When superchargers of either the centrifugal or positive-displacement type are driven mechanically by the engine, the power required becomes an additional loss to the engine output. Experience shows that there is a degree of supercharge for any engine which develops maximum efficiency; too high a supercharge absorbs more power in the supercharger than is gained by the engine, especially at low loads. Another means of driving the supercharger which is becoming quite general is by an exhaust turbine, which recovers some of the energy that would otherwise be wasted in the engine exhaust. This may be accomplished with so small an increase of back pressure that little power is lost by the engine. This type of drive results in an appreciable increase in efficiency at loads high enough to develop the necessary exhaust pressure.

Supercharging a two-cycle diesel engine requires some means of restricting or throttling the exhaust in order to build up cylinder pressure at the start of the compression stroke, and is used on a few large engines. Most medium and large two-cycle diesel engines are usually equipped with blowers to scavenge the cylinders after the working stroke and to supply the air required for the subsequent cycles. These blowers, in contrast to superchargers, do not build up appreciable pressure in the cylinder at the start of compression. If the capacity of such a blower is greater than the engine displacement, it will scavenge the cylinder of practically all exhaust products, even to the extent of blowing some air out through the exhaust ports. Such blowers, like superchargers, may be driven by the engine or by exhaust turbines.

Engine balance. Rotating masses such as crank pins and the lower half of a connecting rod may be counterbalanced by weights attached to the crankshaft. The vibration which would result from the reciprocating forces of the pistons and their associated masses is usually minimized or eliminated by the arrangement of cylinders in a multicylinder engine so that the reciprocating forces in one cylinder are neutralized by those in another. Where these forces are in different planes, a corresponding pair of cylinders is required to counteract the resulting rocking couple.

If piston motion were truly harmonic, which

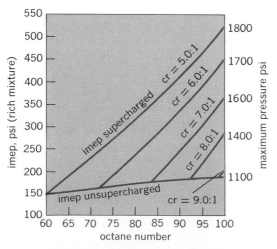

Fig. 18. Graph showing the relationship between compression ratio and knock-limited imep for given octane numbers, obtained by supercharging a laboratory engine. (*From H. R. Ricardo, The High-Speed Internal Combustion Engine, 4th ed., Blackie, 1953*)

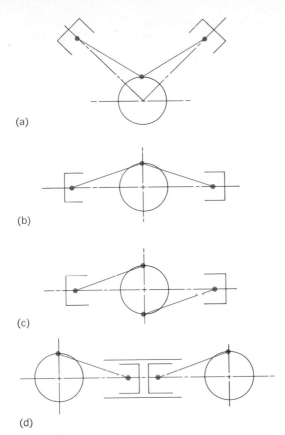

Fig. 19. Arrangements of two cylinders. (*a*) A 90° V formation with connecting rods operating on the same crankpin. (*b*) Opposed cylinders with connecting rods operating on the same crankpin. (*c*) Opposed cylinders with pistons operating on crankpins 180° apart. (*d*) Double-opposed pistons in the same cylinder but with pistons operating on separate crankshafts.

they are so small that they may generally be neglected. Thus, for a connecting rod with the average angularity of the 1969 automobile engines, where $a = 0.291$, the inertia force caused by a piston at firing dead center is about 1.29 times the pure harmonic force, and at inner dead center it is about 0.71 times as large.

Where two pistons act on one crankpin, with the cylinders in 90° V arrangements, as in Fig. 19*a*, the resultant primary force is radial and of constant magnitude and rotates around the crankshaft with the crankpin. Therefore, it may be compensated for by an addition to the weight required to counterbalance the centrifugal force of the revolving crankpin and its associated masses. The resultant of the secondary force of the two pistons is 1.41 times as large as for one cylinder, and reciprocates in a horizontal plane through the crankshaft at twice crankshaft speed.

In engines with opposed cylinders, if the two connecting rods operate on the same crankpin, as in Fig. 19*b*, the primary forces are added and are twice as great as for one piston, but the secondary forces cancel. If the pistons operate on two crankpins 180° apart, as in Fig. 19*c*, all reciprocating forces are balanced. However, as they will be in different planes, a rocking couple will develop unless compensated by an opposing couple from another pair of cylinders. Double-opposed piston pairs, operating in a single cylinder on two crankshafts as in Fig. 19*d* (with a cross shaft to maintain synchronism), are in perfect balance for primary and secondary reciprocating forces as well as for rotating masses and torque reactions.

In the conventional four-cylinder in-line engines with crankpins in the same plane, the primary reciprocating forces of the two inner pistons (2 and 3) cancel those of the two outer pistons (1 and 4), but the secondary forces from all pistons are added. They are thus equivalent to the force resulting from a weight about $4a$ times the weight of one piston and its share of the connecting rod, oscillating

would require a connecting rod of infinite length, the reciprocating inertia force at each end of the stroke would be as in Eq. (2), where W is the total

$$F = 0.000456 W N^2 s \qquad (2)$$

weight of the reciprocating parts in one cylinder, N is the rpm, and s is the stroke in inches. Both F and W are in pounds. But the piston motion is not simple harmonic because the connecting rod is not infinite in length, and the piston travels more than half its stroke when the crankpin turns 90° from firing dead center. This distortion of the true harmonic motion is due to the so-called angularity a of the connecting rod, shown by Eq. (3), where r is the

$$a = \frac{r}{l} = \frac{s}{2l} \qquad (3)$$

crank radius, s the stroke, and l the connecting rod length, all in inches.

Reciprocating inertia forces act in line with the cylinder axis and may be considered as combinations of a primary force—the true harmonic force from Eq. (2)—oscillating at the same frequency as the crankshaft rpm and a secondary force oscillating at twice this frequency having a value of Fa, which is added to the primary at firing dead center and subtracted from it at inner dead center. In reality there is an infinite but rapidly diminishing series of even harmonics at 4, 6, 8, . . . , times crankshaft speed, but above the second harmonic

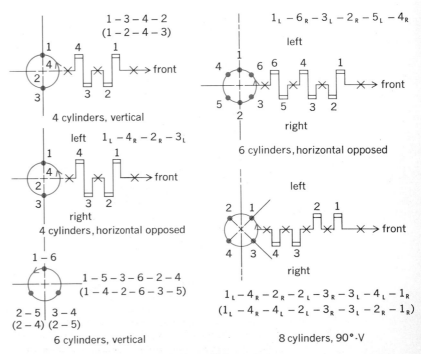

Fig. 20. Typical cylinder arrangements and firing orders.

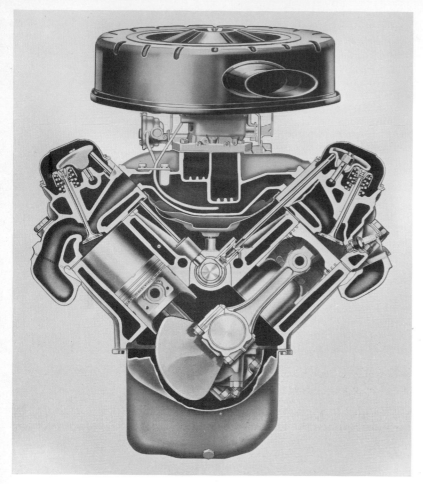

Fig. 21.   Cutaway view of a typical V-8 overhead valve automobile engine with air filter. (*Pontiac Div., General Motors Corp.*)

parallel to the piston movement, having the same stroke, but moving at twice the frequency. A large *a* for this type of engine is advantageous. Where the four cylinders are arranged alternately on each side of a similar crankshaft, and in the same plane, both primary and secondary forces are in balance. Six cylinders in line also balance both primary and secondary forces.

Early eight-cylinder 90° engines with four crankpins in the same plane, like those of the early four-cylinder engine, had unbalanced horizontal secondary forces acting through the crankshafts, which were four times as large as those from one pair of cylinders.

In 1927 Cadillac introduced a crank arrangement for its V-8 engines, with the crankpins in two planes 90° apart. Staggering the 1 and 2 crankpins 90° from each other equalizes secondary forces, but the forces are in different planes. The couple thus introduced is cancelled by an opposite couple from the pistons operating on crankpins 3 and 4. This arrangement of crankpins is now universally used on V-8 engines.

Torsion dampers. In addition to vibrational forces from rotating and reciprocating masses, vibration may develop in an engine from torsional resonance of the crankshaft at various critical speeds. The longer the shaft for given bearing diameters, the lower the speeds at which these

vibrations may develop. Such vibrations are dampened on most six- and eight-cylinder engines by a vibration damper which is similar to a small flywheel on the crankshaft at the end opposite to the main flywheel but coupled to the shaft only through rubber so arranged as to reduce the torsional resonances. Such vibration dampers are usually combined with the pulley driving the cooling fan and generator. *See* MECHANICAL VIBRATION.

Even though the majority of American automobile engines are now dynamically balanced, it has been general practice for several years to mount them in the chassis frame on rubber blocks. This reduces the transmission of the small-amplitude high-frequency vibrations in torque reaction as well as small unbalance of reciprocating parts in individual cylinders, so that a low noise level is developed in the car from the operation of the engine.

Firing order. Cylinder arrangements are generally selected for even firing intervals and torque impulses, as well as for balance. As a result, the cylinder arrangements and firing orders shown in Fig. 20 may be found in automobile use. It is generally customary to identify cylinder banks as left or right as seen from the driver's seat, and to number the crankpins from front to rear. Manufacturers do not agree on methods of numbering individual cylinders of V-type engines. However, the arrangements and firing orders shown are in general use with the addition in parentheses of alternate arrangements only occasionally used.

Typical American automobile engine. American six- and eight-cylinder engines manufactured in 1969 had many features in common. All had overhead valves located in a removable cylinder head and cylinder blocks cast integral with the upper crankcase. The valves were operated from a chain-driven camshaft in the crankcase through push rods and rockers having arm lengths giving a valve lift about 50–75% greater than the cam lift. With the exception of one six-cylinder design, all silenced the valve action by use of hydraulic valve lifters. *See* HYDRAULIC VALVE LIFTER.

Many designs provided means for rotating at least the exhaust valve to improve valve life. All cylinder barrels were completely surrounded by the jacket cooling water, and most designs extended it the full length of the bore. Main bearings supported the crankshaft between each crankpin. Almost all designs locked the wrist pin to the connecting rod; the others permitted it to "float" in both the piston and connecting rod. The use of three piston rings above the wrist pin had become general practice, two being narrow compression rings and the lowest an oil scraper. Compression rings were generally of cast iron about 0.078 in. wide and provided with a coating such as chromium or tin to prevent scuffing the surface during the wearing-in period of a new engine and when it is started cold with but little oil on the cylinder wall. Oil-scraper rings were about 3/16 in. wide, were provided with a nonscuffing surface, and had drain holes through the piston for the return of the excess oil scraped from the cylinder wall.

The largest six-cylinder engine developed a maximum of 155 hp, with a cylinder displacement of 250 in.[3] Where higher power was required, the

Table 2. Selected average dimensions (in inches) of 1969 engines*

Category	Six-cylinder	Eight-cylinder V
Cylinder bore	3.68	4.07
Cylinder stroke	3.32	3.68
Stroke-bore ratio	0.90	0.91
Displacement	212	385
Maximum hp/rpm	131/4100	294/4640
bmep at maximum hp	120	131
Maximum torque (lb/ft)/rpm	194/2000	409/2870
bmep at maximum torque	139	160
Connecting-rod length	5.69	6.32
Crank radius/rod length ratio	0.292	0.292
Crankpin diameter	2.08	2.27
Wrist-pin diameter	0.91	0.99
Main-bearing diameter	2.41	2.79
Intake-valve-head diameter (max)	1.67	1.96
Exhaust-valve-head diameter (max)	1.42	1.60
Piston-head area/intake-valve-head area ratio	4.85	4.31

*Adapted from *Automotive Industries*, March 15, 1969.

eight-cylinder V-type was installed. A typical engine of this type is shown in cross section in Fig. 21. The right and left banks of cylinders were staggered to enable connecting rods of opposing cylinders to be located side by side on the same crankpin. The V arrangement provides a short and very ridged structure, which is important for high engine speeds because of the minimized deflection of the main bearings. It also makes possible efficient intake-manifold designs and almost symmetrical and equal-length branches to each cylinder port from a centrally located downdraft carburetor. The short length and low height of these engines are also important features for car styling. Some of the principal dimensions and other statistics of these engines have been averaged in Table 2.

[NEIL MAC COULL]

Bibliography: T. Baumeister (ed.), *Standard Handbook for Mechanical Engineers*, 7th ed., 1967; W. H. Crouse, *Automotive Engines*, 2d ed., 1959; W. H. Crouse, *Automotive Mechanics*, 4th ed., 1960; L. C. Lichty, *Combustion Engine Processes*, 1967; E. F. Obert, *Internal Combustion Engines*, 3d ed., 1968; H. R. Ricardo, *The High-Speed Internal Combustion Engine*, 1931, 4th ed., 1953; C. F. Taylor and E. S. Taylor, *The Internal Combustion Engine*, rev. ed., 1948.

Internal energy

A characteristic property of the state of a thermodynamic system, introduced in the first law of thermodynamics. For a static, closed system (no bulk motion, no transfer of matter across its boundaries), the change ΔE in internal energy for a process is equal to the heat Q absorbed by the system from its surroundings minus the work w done by the system on its surroundings. Only a change in internal energy can be measured, not its value for any single state. For a given process, the change in internal energy is fixed by the initial and final states and is independent of the path by which the change in state is accomplished.

The internal energy includes the intrinsic energies of the individual molecules of which the system is composed and contributions from the interactions among them. It does not include contributions from the potential energy or kinetic energy of the system as a whole; these changes

must be accounted for explicitly in the treatment of flow systems. Because it is more convenient to use an independent variable (the pressure P for the system instead of its volume V), the working equations of practical thermodynamics are usually written in terms of such functions as the enthalpy $H = E + PV$, instead of the internal energy itself. *See* ENTHALPY; THERMODYNAMICS, CHEMICAL.

[PAUL J. BENDER]

International Date Line

The 180° meridian, where each day officially begins and ends. As a person travels eastward, against the apparent movement of the Sun, he gains 1 hr for every 15° of longitude; as he travels westward, he loses time at the same rate. Two people starting from any meridian and traveling around the world in opposite directions at the same speed would have the same time when they meet, but would be 1 day apart in date. If there were no international agreement as to where each day should begin and end, there could be any number of places so designated. To eliminate such confusion, the International Meridian Conference, in 1884, designated the 180° meridian as the location for the beginning of each day. Thus, when a traveler goes west across the line, he loses a day; if it is Monday to the east, it will be Tuesday immediately as he crosses the International Date Line. In traveling eastward he gains a day; if it is Monday to the west of the line, it will be Sunday immediately after he crosses the line.

An interesting example can be taken from conditions now nearly attainable with jet aircraft. If one could board such a plane, say at noon in Washington, D.C., and fly westward at that latitude around the world to return in 24 hr, the rate would match the rotation of the Earth. Although constantly under the noontime Sun, this traveler would need to adjust his calendar 1 day ahead upon crossing the International Date Line, because he would arrive in Washington at noon, 24 hr after embarking. Thus his calendar day would agree with that of the Washingtonians.

The 180° meridian is ideal for serving as the International Date Line (see illustration). It is exactly halfway around the world from the zero, or Greenwich, meridian, from which all longitude is reckoned. It also falls almost in the center of the largest ocean; consequently there is the least amount of inconvenience as regards population centers. A few deviations in the alignment have been made, such as swinging the line east around Siberia to keep that area all in the same day, and westward around the Aleutian Islands so that they will be within the same day as the rest of Alaska. Other minor variations for the same purpose have been made near the Fiji Islands, in the South Pacific. *See* GEOGRAPHY, MATHEMATICAL.

[VAN H. ENGLISH]

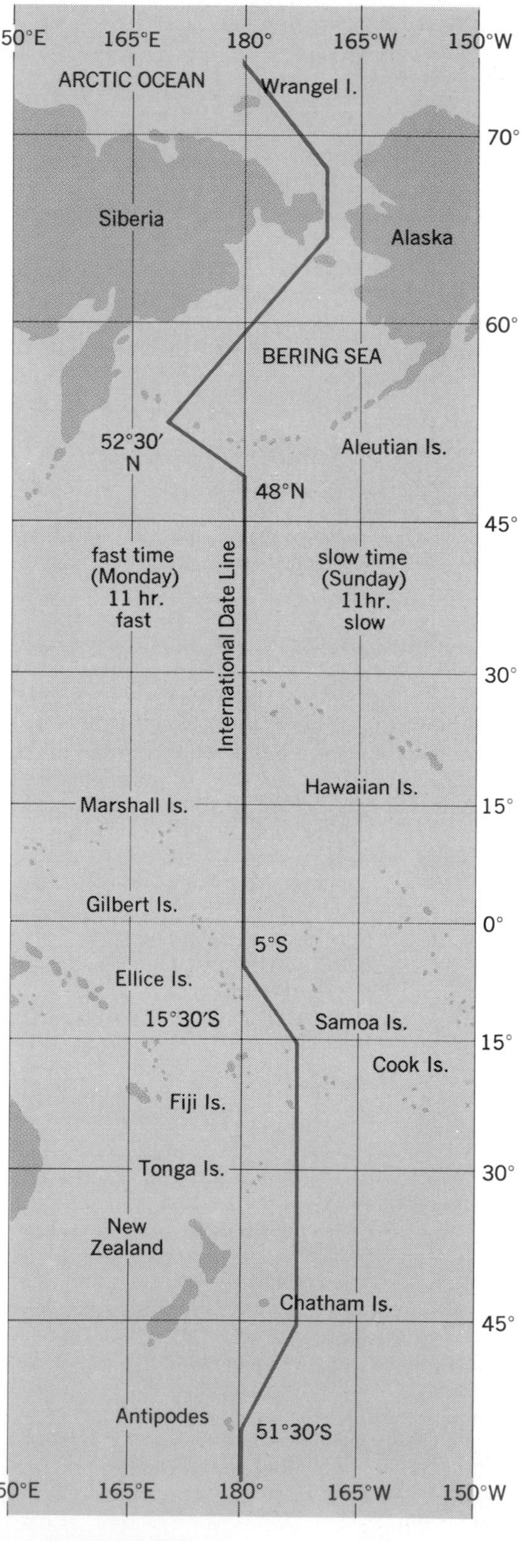

The International Date Line.

Interplanetary propulsion

Means of providing propulsive power for flight to the Moon or to a planet. A variety of different propulsion systems can be used. The space vehicles for these missions consist of a series of separate stages, each with its own set of propulsion systems. When the propellants of a given stage have been expended, the stage is jettisoned to prevent its mass from needlessly adding to the inertia of the vehicle. *See* ROCKET ENGINE; ROCKET STAGING.

By expelling mass at high velocities, the propulsion systems provide impulse which permits the vehicle to execute its flight trajectory and the necessary maneuvers. Interplanetary or lunar trajectories, as well as the associated maneuvers, are complex, and different kinds of propulsion systems are needed to meet the requirements of the various phases of flight. Several basically different propulsion systems have been developed or are under development, each with specific advantages and applications. For example, for take-off from Earth, high thrust and acceleration are needed for a relatively short time, and the chemical combustion–type rocket engine performs this function best. By contrast, the interplanetary elliptical transfer orbit from Earth to, say, Jupiter may be achieved by the application of very low thrust and accelerations for a duration of several years, and an electrical propulsion system currently appears to be most suitable for this purpose.

Engine types. Of the basic propulsion techniques considered suitable for interplanetary propulsion, several are rockets in the strict sense that they expel and carry their own expulsive material or "working" fluid, which is heated and expanded to a high velocity in a nozzle. Only the ion and magnetoplasma devices do not use thermodynamic expansion of gas to obtain a thrust or force.

Table 1 presents data for the principal types of engines. Figure 1 shows the relation of acceleration (which is also the thrust-weight ratio) to specific impulse (or exhaust velocity) and to

Table 1. Ranges of specific impulse, thrust-to-weight ratio, and thrust duration

System	Specific impulse, sec*	Thrust-to-weight ratio	Typical thrust duration
Chemical (liquid bipropellant)	200–450	10^{-2} to 10^2	Minutes (high thrust) Hours (low thrust)
Chemical (solid)	200–310	10^{-2} to 10^2	Minutes
Chemical (monopropellant)	120–220	10^{-3} to 10^2	Minutes
Fission	500–1,100	10^2 to 10	Same as chemical system
Isotope decay	400–800	10^{-5} to 10^{-3}	Days
Electrothermal	120–2,000	10^{-4} to 10^{-2}	Days
Electromagnetic	5,000–25,000	10^{-5} to 10^{-3}	Weeks
Electrostatic (ion)	5,000–60,000	10^{-5} to 10^{-3}	Months
Solar heating	400–700	10^{-3} to 10^{-2}	Days

*Vacuum expansion.

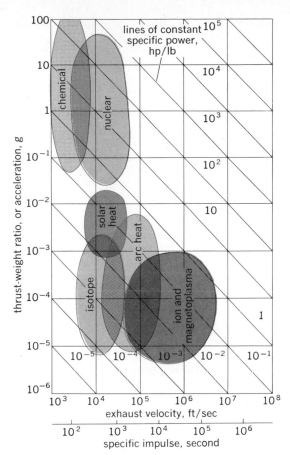

Fig. 1. Regions of accelerations and specific powers for various rockets. (*J. Aerosp. Sci.*)

specific power (which is here the ratio of kinetic energy in the jet per unit time to the overall weight of the vehicle, including its source of energy).

Chemical and nuclear rockets provide high thrust for short durations. The other engines operate at low thrust for long durations. For short trips, rapid acceleration is desirable; for long trips, small, continuously applied acceleration will result in an overall flight time which may be acceptable. For such service as interplanetary freight, the high specific impulse of the electrical units would allow a vehicle so powered to carry a much heavier payload than if driven by a chemical rocket; however, its time of flight may be considerably longer than that desired for the exposure of a crew to space conditions. *See* ROCKET.

Chemical rocket. Chemical rockets are divided broadly into solid- and liquid-propellant types. Both types convert chemical energy into thermal energy of a working fluid by combustion, then convert the thermal energy of the hot gas into kinetic energy in a nozzle, and thereby expel the fluid at high velocity to provide forward thrust. *See* PROPELLANT.

The solid-propellant rocket is the simplest type. Once ignited, it usually burns until all its fuel is consumed. A liquid-propellant rocket using a liquid fuel with a liquid oxidizer (the bipropellant rocket) permits higher performance, is restartable, and is the type most used for current space vehicles. A low-pressure liquid monopropellant rocket is simpler but develops a specific impulse of only

200–245 sec, compared with a theoretical limit of 450 sec for a high-energy bipropellant. Monopropellant units have been used for attitude control rockets in space vehicles. *See* SPECIFIC IMPULSE.

Of the various types of space propulsion, the liquid-propellant and the solid-propellant chemical rockets are the most highly developed and are available in many types and sizes. The chemical rocket presently offers the most practical method of propulsion for space flight. Except for a few experimental devices, operational space vehicles in use and those under development through 1980 will utilize this method. Other types of propulsion systems offer advantages, however, particularly in performance, and if proved practical some of them will find operational application in interplanetary and lunar missions, probably after 1980.

Nuclear propulsion. Two types of nuclear rockets are under development. In the fission type, a uranium-fueled, solid-core reactor heats a working fluid, usually hydrogen. The working fluid then accelerates through a nozzle as in other types of rockets. The use of a working fluid of low molecular weight (namely, hydrogen) enables this rocket to produce a higher specific impulse than a chemical rocket, in which the working fluid is a mixture of the products of combustion. However, the temperature limit of 4000–5000°F imposed by existing reactor and nozzle materials prevents the full potentialities of the nuclear rocket from being realized at present.

The second type of nuclear rocket being investigated uses energy from the decay of radioactive materials. Such an engine would be suitable primarily for low thrust. However, with hydrogen as the working fluid and at sufficiently high temperatures, a system of this type could be expected to develop a specific impulse of 600–700 sec. Disadvantages are the high expense and limited life of isotope fuel and the special cooling provisions required because heat is evolved by the fuel at all times.

The safety provisions needed for the protection of crew, service personnel, and equipment from being exposed to excessive neutron or gamma rays have caused considerable and expensive delays in the programs. For example, the potential emergency of inadvertently landing radioactive material from a nuclear reactor or an isotope decay source requires the development of a rugged container capsule which will not crack on impact and thus can prevent radioactive material from leaking out.

In flight, sensitive equipment and a human crew can be shielded against radiation effects by heavy shadow shields; complete enclosure is unnecessary in interplanetary space because there is no air to scatter the radiation. Alternatively, reactor and sensitive payload can be widely separated, providing considerable saving in weight but introducing additional design problems. Hardware adjacent to the nuclear source must be cooled.

Fusion reactions may someday form the basis for development of another type of nuclear rocket. These reactions would avoid the production of undesirable by-products and could release their energy directly to the working fluid. However, techniques for obtaining sustained fusion have yet to be discovered.

While a reactor and several isotope energy

sources have been flown in experimental space vehicles, these were designed to provide power and not propulsion. Successful ground experiments have been performed on graphite-core nuclear rockets (projects Rover and Nerva) with up to 1500 Mw of power, also on small, radioactive, isotope decay propulsion devices. The application of nuclear propulsion to space flight is awaiting favorable results of current development effort and a clearer definition of the flight mission. *See* NUCLEAR ROCKET.

Electrical propulsion. The expelled material can be given a high velocity by electrical means. Three basic types of electrical propulsion are electrothermal, electrostatic, and electromagnetic. In electrothermal propulsion a working fluid such as nitrogen, ammonia, or hydrogen is passed over hot metal surfaces (often called resistojet) or through an electric arc (often called arc jet), and then expanded in a supersonic nozzle.

In ion propulsion or electrostatic propulsion, as well as in electromagnetic propulsion, the working fluid is accelerated electrically instead of thermodynamically. Specific impulses of more than 60,-000 sec are possible, but values of 2000–20,000 sec appear more practical. The propellant for ion propulsion should be, preferably, a material of high atomic weight, such as cesium, which is easily ionized. *See* ION PROPULSION.

In an electromagnetic rocket a fluid of ions and electrons (electrically neutral as a whole and called a plasma) is accelerated by reaction with a changing magnetic field or by the interaction of an electrical current and a steady magnetic field. With a fluid of low atomic weight, such as hydrogen, accelerated by a changing magnetic field, a specific impulse in excess of 15,000 sec should be possible. In its present form the electrical equipment necessary for electromagnetic propulsion is heavy and complex, and this method does not appear as promising as the other two types of electrical propulsion. *See* ELECTROMAGNETIC PROPULSION.

By nature these electrical propulsion units are low-thrust devices (producing 0.001–10 lb of thrust) which operate at high performance levels (specific impulse values of 1000–20,000 sec) for long periods of time (several months). In some cases they also could be used for attitude control. Experimental resistojet and ion propulsion thrustors have been successfully flown for short durations in small test space vehicles and have been used for attitude control in application where low thrust and relatively long maneuver duration are

Table 2. Typical interplanetary maneuvers

	Type of maneuver	Thrust level, lb	Duration of propulsion	Duration of flight phase	Vehicle stage which is operating	Rocket engine best suited	Remarks
A	Takeoff and ascent from launch planet (Earth)	2,000,000 to 40,000,000	1–2 min	Minutes	Booster	Large liquid or propellant chemical	Several rocket engines operate simultaneously in a cluster
B	Ascent into orbit and/or escape into space	100,000 to several million	Minutes	1 hr.	Sustainer second and third stage	High-energy liquid propellant chemical; in future possibly nuclear fission	Restart 2–6 times; could be clustered engines
C	Interplanetary transfer (overcome Sun's gravity, accelerate away from launch planet, match spacecraft velocity with that of target planet, or change orbital plane)	0.1 to 100 or 1000 to 100,000	Hours to weeks Less than a minute	Months to years	Third stage	Electrical (low thrust) Liquid or fission (high thrust)	Apply corrections to trajectory; many restarts
D	Approach target planet, and/or attain orbit about target planet	1000 to 100,000	Minutes	Minutes to hours	Third or fourth stage	Liquid or fission	Several restarts
E	Descend to surface of target planet	10,000 to 100,000	Minutes	Less than 30 min	Third, fourth, or fifth stage	Chemical	May have variable thrust feature
F	Attitude control (turn spacecraft)	0.1 to 100	0.02 to several seconds	Variable	All stages except booster or first sustainer	Multiple chambers; liquid propellant, heated or cold gas; in future may use electrical, solar heating, or solar sail for some maneuvers	Many hundreds or thousands of reuses
G	Rendezvous or limited trajectory correction such as station keeping	0.001 to 100	0.02 to several seconds	Variable	Only stages equipped for rendezvous contact	Multiple chambers; liquid propellant, heated or cold gas; in future may use electrical, solar heating, or solar sail for some maneuvers	Many restarts
H	Stage separation	100 to 10,000	Seconds	Seconds	All stages except last	Chemical	Usually 2, 4, 6, or 8 units operating simultaneously
I	Ullage settling	100 to 10,000	Seconds	Seconds	Stage with large restartable sustainer	Chemical	Usually 2, 4, 6, or 8 units operating simultaneously

useful. Propulsion by such devices is like pushing an elephant with a feather while he stands on a frictionless plate of ice; if pushed long enough, the elephant can be accelerated to a very high velocity.

The amount of electrical power needed for accelerating the working fluid is relatively great (200 kw/lb of thrust for a typical ion rocket), and the power source required is therefore heavy. For very small thrusts (a few thousandths of a pound) a solar cell array has been demonstrated to be satisfactory. For larger thrusts and therefore larger amounts of power, the only practical energy source appears to be a compact, flight-weight nuclear reactor with a suitable power conversion system, including a mechanism for rejecting waste heat to space. The development of such power sources that will function reliably for long durations under difficult flight conditions unfortunately is not as far advanced as that of the electrical propulsion devices themselves. This factor will delay the practical application of this promising mode of interplanetary propulsion.

Solar heating. The working fluid of a rocket can be heated by solar radiation to temperatures of 1000–2000°F. At Earth's distance from the Sun, a reflector area of 8 ft² is needed to give 1 kw of power if 100% of the captured radiation is converted. With all losses taken into account—optical, radiation, and those resulting from such factors as the efficiency of the solar boiler and the accuracy with which the reflector is pointed at the Sun—the reflector area may need to be 80 ft²/kw. At Mars's distance from the Sun, the reflector would need to be 200 ft²/kw. Just how such a reflector for a high-powered system would be transported into space remains to be determined. *See* SOLAR CONSTANT.

Photon propulsion. The radiation pressure of the photons from the Sun is about 10^{-7} lb/ft² at Earth's distance. The total pressure due to action and reaction on a perfect perpendicular reflecting surface would be twice this value. Solar photons travel in a direction away from the Sun; inclined reflectors create thrusts in various directions. Solar "sailing" (radiation pressure effects on large, thin, metallized, stretched diaphragms) can provide low thrusts and attitude control, turning a spaceship completely around in a few hours. Because no working fluid is carried in the vehicle, this type of propulsion has the advantage of being economical in weight.

A photon rocket appears to be beyond current capabilities. In essence it would be a device incorporating a strong light source with the reflector on which the bouncing of photons would create a small reaction force. The development of such a potent light source and the cooling of the reflector, for example, would present problems for which solutions do not seem to be within the present state of the art.

Trajectory and maneuver requirements. The types of maneuvers or flight phases involved in interplanetary or lunar flight are basically those described in Table 2. A schematic diagram of the Apollo-Saturn Moon vehicle with its several types of propulsion is shown in Fig. 2. Because it is necessary to provide attitude control about three axes of rotation for each maneuvering stage and because the main rocket engines are usually clus-

tered, the number of rocket propulsion devices is large. A typical interplanetary multistage vehicle may have 50–150 separate thrust-producing devices, varying from the million-pound-thrust booster engine to attitude control units producing a thrust of a few ounces.

It is convenient to distinguish two basic types of maneuvers in Table 2. First, there are those (A – E) in which the propulsion systems are used primarily for adding impulse in the intended flight direction (and thus contribute to changing the effective flight velocity). Second, there are those (F – G) in which the propulsion systems are used primarily for control of the vehicle's attitude or rotational position. They usually operate in pairs or multiples and for very short durations, but perhaps for several hundred or thousand times. The number and types of maneuvers and stages listed are for a typical single interplanetary transfer with a parking orbit around both the departure and target planets, as shown in Fig. 3. A round trip would require additional propulsion and stages. The omission of parking orbits would change neither the energy requirements nor the propulsion requirements appreciably but would require different navigation and guidance. It is possible to perform several different maneuvers or flight phases with the same stage or the same propulsion systems.

For the takeoff and the ascent maneuvers, high-thrust chemical rockets are most suitable. The booster rockets usually are capable of the highest thrust. The climb-maneuver rocket engines may also be used for an apogee push to attain orbit and thus usually have restart capability.

The basic interplanetary maneuvers are presented in Fig. 4, together with the engines most suitable for each. Here it is assumed that the interplanetary flight can actually start from a satellite orbit beyond the greater part of Earth's atmosphere. To minimize the energy necessary to overcome the relatively strong gravitational field at this distance

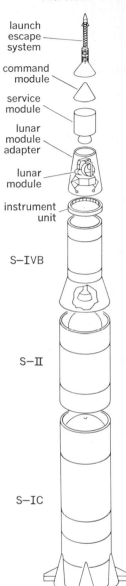

INTERPLANETARY PROPULSION

launch escape system
command module
service module
lunar module adapter
lunar module
instrument unit
S–IVB
S–II
S–IC

Fig. 2. Apollo Saturn 5.

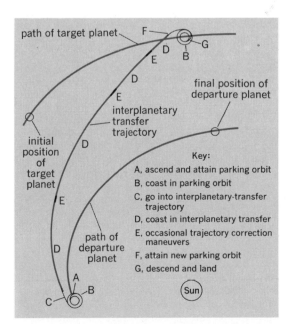

path of target planet

F
D
G
E
B
D
E
D
interplanetary transfer trajectory
initial position of target planet
E
D
final position of departure planet

Key:
A, ascend and attain parking orbit
B, coast in parking orbit
C, go into interplanetary-transfer trajectory
D, coast in interplanetary transfer
E, occasional trajectory correction maneuvers
F, attain new parking orbit
G, descend and land

path of departure planet

A
B
C

Sun

Fig. 3. Simplified trajectory, a schematic diagram of an interplanetary mission (the heavy lines indicate powered-flight portions).

maneuver	chemical	fission	isotope decay	solar heating	solar sail
escape from orbit	○	○	◉	◉	◉
interplanetary transfer	◉	◉	○	○	◉*
change of plane	○	◉	○	○	◉*
orientation	○	◉	○	○	○
trajectory corrections	○	◉	○	○	●*
target approach	○	◉	○	○	●*
capture	○	○	◉	◉	●

Key:
suitable ○
acceptable ◉
undesirable or impossible ●
*forces in general direction of radiation

Fig. 4. Suitability of various propulsion systems to different interplanetary maneuvers. (*J. Aerosp. Sci.*)

from Earth, the thrust should be applied during a brief period. Chemical and nuclear rockets, therefore, are the most suitable engines for the escape-from-orbit maneuvers.

Interplanetary orbit transfer is the next maneuver. Basically it involves overcoming the gravitational field of the Sun and correcting for differences in the tangential velocity of the target and launching planets. The idealized minimum energy is attained when the trajectory is a Hohmann ellipse; it is tangent at its perihelion to the orbit of the inner planet and has the Sun at its prime focus. *See* ELLIPSE.

Vehicular orientation must be controlled for such operations as pointing telescopes for astronomical or navigational observation, accurately orienting directional radio antennas, or orienting solar cells toward the Sun. Accuracy of thrust level and precise pulsing duration is of great importance. The exact duty cycle required of an attitude control rocket (that is, the firing frequency, the sequence of starts and restarts, and the respective durations) is usually not readily predictable. It is thus difficult to preset a given engine endurance or propellant mass for a specific flight, since the flight may call for very many or very few attitude control maneuvers. Therefore, attitude control rocket engines must be designed to withstand the most severe duty cycles that may be necessary (many restarts and long duration). These duty cycles can easily be 10 times as severe as those required on an average flight.

Some of the trajectory corrections may be made slowly, and these can use electric propulsion in future application. The energy required for such corrections is usually small and depends upon vehicular factors, especially errors in navigation devices and in guidance and control, as well as upon the pilot and the accuracy of the map of the solar system.

It is often necessary to make small, repeated, station-keeping trajectory corrections in order to keep a satellite at a predetermined position. For example, a stationary satellite has to have a very precise velocity in order to appear truly stationary to an observer on Earth (the satellite period of revolution around the Earth has to be equal exactly to the Earth's own rotational period, namely, 24 hr), and repeated corrective maneuvers have to be made to prevent the satellite from drifting out of its orbit and thus to keep it on its intended station. Similarly, in a system of several vehicles on a joint Mars expedition, it would be necessary to keep the relative positions or distances of the various vehicles in the squadron. Again, such station-keeping maneuvers may require many thousands of pulsed operations or small thrust control rockets.

As the vehicle approaches the target planet, thrust is applied for the change in velocity that will be needed to match the orbital velocity of the planet about the Sun. The amount of thrust necessary depends on the timing of the maneuver.

Propulsion requirements D and E in Table 2 are similar to A and B but are applied in reverse. The vehicle is decelerated from its solar transfer orbit to a satellite orbit around the target planet. This maneuver requires sufficient acceleration to overcome a high gravitational field, although less than that required for a landing maneuver since the landing vehicle may use the atmosphere for braking. The speed with which the maneuver is to be carried out determines, to some extent, the required thrust.

For the landing maneuver on the Moon (where there is no atmosphere to permit drag or lift forces to be used), a variable thrust or throttlable rocket engine is required. During the landing the balance between thrust and gravitational pull must be carefully controlled.

Emergency maneuvers. In addition to the two classes of maneuvers described in Table 2, there are also emergency maneuvers. These are particularly important in manned space flight if vehicle malfunctions should make it necessary to abandon the intended mission objectives and to return the crew to safety quickly. Because of the excessive weight, it is essentially impossible to carry a spare set of propulsion systems in the various stages for use if the original propulsion systems fail.

First, for those portions of the flight within the vicinity of the launch planet, one can add an emergency-escape rocket system. In the case of a major accident on the launch tower or during the first portion of the ascent, it would quickly pull the manned stage away from the other stages, still loaded with dangerous propellants, and thus would permit the manned stage to be safely landed at some distance from the hazardous launch area. In a normal flight this emergency-escape rocket system would be discarded from the vehicle once it rose above the atmosphere, and it therefore would not have to be accelerated to orbital velocities. In the Apollo vehicle this escape system consists of three different sets of solid rockets: (1) a large rocket for quickly pulling the manned capsule vertically off the vehicle, (2) a pair of small rockets which start later to pull the capsule from the vertical flight path off to the side toward a safe landing area, and (3) a rocket for safely discarding the emergency-escape system during a normal flight.

The second solution for safe emergency escape applies to other phases of the flight, such as the parking orbit or the interplanetary transfer phase. Here the available propulsion systems are used in a manner and sequence other than that intended for normal flight. For example, an emergency occurring during the parking orbit around the launch planet can be solved by abandoning a stage or two in this orbit and using the final stage and propulsion system (intended for landing on the target planet) for an emergency return to the departure planet. The thrust variation and the duration of the acceleration necessary for this emergency return would most likely be somewhat different from those intended for landing on the target planet. All the propulsion systems, particularly those of the upper stages, are therefore designed to fulfill not only the intended mission but also some emergency or abort mission with rather different performance, restart, or response characteristics.

The third solution applies only to multivehicle interplanetary expeditions, in which a squadron of several spacecraft fly in proximity to each other. Here it is possible to abandon a vehicle which has failed during any phase of the flight (except, perhaps, the ascent, descent, or landing phases) and to transfer the crew to one or more of the remaining vehicles by rendezvous technique.

Mission requirements. Interplanetary missions can be grouped in three categories, each with a different range of required energy for the interplanetary transfer orbits: (1) missions in the immediate vicinity of Earth and Moon, (2) missions within the inner solar system or out to the orbit of Mars, and (3) missions to the outer solar system or beyond Mars.

For the first category of missions, accelerations below 0.01 g are too small to be useful. Chemical systems are usually used in this category, but future nuclear engines might also be usable for various transfer maneuvers.

For missions within the inner solar system, accelerations of 0.0001 g are satisfactory for the interplanetary transfer maneuvers. The same propulsion systems mentioned above are also suitable for these flights, and very efficient electrical engines may be added in the future.

For missions to the outer solar system, accelerations greater than 0.00001 g would produce satisfactory interplanetary transfer trajectories which applied over the periods of several months (or years) necessary for such flights. Because of their high specific impulses, electrical propulsion systems will be uniquely qualified for these missions.

Development status. Figure 5 shows several milestones in the development history of interplanetary rocket propulsion engines. Only the chemical rocket types (liquid- and solid-propellant) and the cold gas attitude control jets have repeatedly been flown and are considered to be sufficiently reliable for use in interplanetary applications. The electrical propulsion systems will find some specialized applications in a few missions at very low thrust levels but must await the development of a reliable, large electric power source before it will find wider use.

The engine types that deserve the greatest efforts are those showing the greatest promise, both in efficient use of their energy sources and in efficient conversion of energy to high thrust, as summarized in Fig. 6. The chemical and nuclear engines are the only engines with high-accelera-

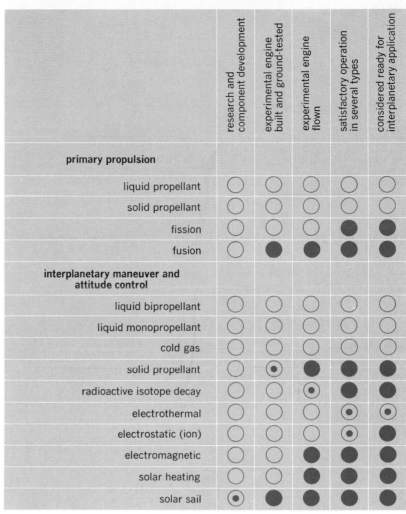

	research and component development	experimental engine built and ground-tested	experimental engine flown	satisfactory operation in several types	considered ready for interplanetary application
primary propulsion					
liquid propellant	○	○	○	○	○
solid propellant	○	○	○	○	○
fission	○	○	○	●	●
fusion	○	●	●	●	●
interplanetary maneuver and attitude control					
liquid bipropellant	○	○	○	○	○
liquid monopropellant	○	○	○	○	○
cold gas	○	○	○	○	○
solid propellant	○	◉	●	●	●
radioactive isotope decay	○	○	◉	●	●
electrothermal	○	○	○	◉	◉
electrostatic (ion)	○	○	○	◉	●
electromagnetic	○	○	●	●	●
solar heating	○	○	●	●	●
solar sail	◉	●	●	●	●

Key:

○ completed, available

◉ under consideration, or program underway

● not accomplished or available

Fig. 5. Development status of various rocket engines.

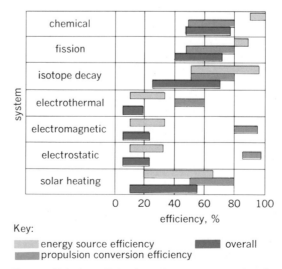

Key:
energy source efficiency   overall
propulsion conversion efficiency

Fig. 6. Relative efficiencies of power conversion for propulsion systems. (*J. Aerosp. Sci.*)

tion capabilities. If an efficient, lightweight electrical generator could be found, permitting accelerations higher than 0.00001 g to be attained, electrical systems, with their versatility, would be highly attractive. *See* MANNED SPACE FLIGHT; SATELLITE, ARTIFICIAL; SPACE FLIGHT; SPACE TECHNOLOGY; SPACECRAFT PROPULSION; SPACECRAFT STRUCTURE.   [GEORGE P. SUTTON]

Bibliography: W. R. Corliss, *Propulsion Systems for Space Flight*, 1960; S. Glasstone, *Source Book of the Space Sciences*, 1965; H. H. Koelle, *Handbook of Astronautical Engineering*, 1961; G. P. Sutton, *Rocket Propulsion Elements*, 3d ed., 1963.

Interpolation

A process in mathematics used to estimate an intermediate value of one (dependent) variable which is a function of a second (independent) variable when values of the dependent variable corresponding to several discrete values of the independent variable are known.

Suppose, as is often the case, that it is desired to describe graphically the results of an experiment in which some quantity Q is measured, for example, the electrical resistance of a wire, for each of a set of N values of a second variable v representing, perhaps, the temperature of the wire. Let the numbers Q_i, $i = 1, 2, \ldots, N$, be the measurements made of Q and the numbers v_i be those of the variable v. These numbers representing the raw data from the experiment are usually given in the form of a table with each Q_i listed opposite the corresponding v_i. The problem of interpolation is to use the above discrete data to predict the value of Q corresponding to any value of v lying between the above v_i. If the value of v is permitted to lie outside these v_i, the somewhat more risky process of extrapolation is used. *See* EXTRAPOLATION.

Graphical interpolation. The above experimental data may be expressed in graphical forms by plotting a point on a sheet of paper for each pair of values (v_i, Q_i) of the variables. One establishes suitable scales by letting 1 in. represent a given number of units of v and of Q. If v is considered the independent variable, the horizontal displacement of the ith point usually represents v_i and its vertical displacement represents Q_i.

If, for simplicity, it is assumed that the experimental errors in the data can be neglected, then the problem of interpolation becomes that of drawing a curve through the N data points P_i having coordinates (x_i, y_i) that are proportional to the numbers v_i and Q_i, respectively, so as to give an accurate prediction of the value Q for all intermediate values of v. Since it is at once clear that the N measurements made would be consistent with any curve passing through the points, some additional assumptions are necessary in order to justify drawing any particular curve through the points. Usually one assumes that the v_i are close enough together that a smooth curve with as simple a variation as possible should be drawn through the points.

In practice the numbers v_i and Q_i will contain some experimental error, and, therefore, one should not require that the curve pass exactly through the points. The greater the experimental uncertainty the farther one can expect the true curve to deviate from the individual points. In some cases one uses the points only to suggest the type of simple curve to be drawn and then adjusts this type of curve to pass as near the individual points as possible. This may be done by choosing a function that contains a few arbitrary parameters that may be so adjusted as to make the plot of the function miss the points by as small a margin as possible. For a more complete discussion of this topic *see* CURVE FITTING.

For many purposes, however, one uses a French curve and orients it so that one of its edges passes very near a group of the points. Having drawn in this portion of the curve, one moves the edge of the French curve so as to approximate the next group of points. An attempt is made to join these portions of the curve so that there is no discontinuity of slope or curvature at any point on the curve.

Tabular interpolation. This includes methods for finding from a table the values of the dependent variable for intermediate values of the independent variable. Its purpose is the same as graphical interpolation, but one seeks a formula for calculating the value of the dependent variable rather than relying on a measurement of the ordinate of a curve.

In this discussion it will be assumed that x_i and y_i $(i = 1, 2, \ldots, N)$, which represent tabulated values of the independent and dependent variables, respectively, are accurate to the full number of figures given. Interpolation then involves finding an interpolating function $P(x)$ satisfying the requirement that, to the number of figures given, the plot of Eq. (1) pass through a selected number of

$$y = P(x) \qquad (1)$$

points of the set having coordinates (x_i, y_i). The interpolating function $P(x)$ should be of such a form that it is made to pass readily through the selected points and is easily calculated for any intermediate value of x. Since many schemes are known for determining quickly the unique nth degree polynomial that satisfies Eq. (1) at any $n + 1$ of the tabulated values and since the value of such a polynomial may be computed using only n multiplications and n additions, polynomials are the most common form of interpolating function.

If the subscripts on x and y are reassigned so that the points through which Eq. (1) passes are now (x_0, y_0), (x_1, y_1), $\ldots$, (x_n, y_n), the polynomial needed in Eq. (1) may be written down by inspection, and one has Eq. (2), where Eq. (3) applies.

$$y = \sum_{k=0}^{n} \frac{L_k(x)}{L_k(x_k)} y_k \qquad (2)$$

$$L_k(x) = \frac{(x - x_0)(x - x_1) \cdots (x - x_n)}{(x - x_k)} \qquad (3)$$

Equation (3) is Lagrange's interpolation formula for unequally spaced ordinates. Since $L_k(x)$ vanishes for all x_s in the set $x_0, x_1, \ldots, x_n$ except x_k, substituting $x = x_s$ in the right-hand side of Eq. (1) gives rise to only one nonzero term. This term has the value y_s, as required.

For $n = 1$ Eqs. (2) and (3) give rise to Eq. (4),

$$y = \frac{x - x_1}{x_0 - x_1} y_0 + \frac{x - x_0}{x_1 - x_1} y_1 \qquad (4)$$

whose plot is a straight line connecting the points (x_0, y_0) and (x_1, y_1). Such an interpolation is referred to as linear interpolation and is used in all elementary discussions of interpolation; however, another equivalent form of this equation, given below in Eq. (12), is more often used in these cases.

Suppose the table were obtained from the equation $y = f(x)$, in which $f(x)$ is some mathematical function having continuous derivatives of all order up to and including the $(n+1)$th. It is then possible to obtain an accurate expression for intermediate values of $f(x)$ by adding to the right-hand side of Eq. (2) a so-called remainder term, expression (5),

$$\frac{(x-x_0)(x-x_1) \cdots (x-x_n)}{(n+1)!} f^{(n+1)}(\xi) \quad (5)$$

where $f^{(n+1)}(\xi)$ is the $(n+1)$th derivative of $f(x)$ at some point $x = \xi$ lying between the smallest and largest of the values $x_0, x_1, \ldots, x_n$. Since the value of ξ is not known, the remainder term is used merely to set an upper limit on the truncation error introduced by using Lagrange's interpolation formula.

If the ordinates are equally spaced, that is, $x_s - x_0 + sh$ where h is the interval of tabulation, Lagrange's interpolation formula simplifies considerably and may be written as Eq. (6), where n is

$$y = \sum_{k=-p}^{n-p} A_k(u) y_k = A_{-p}(u) y_{-p} + \cdots$$
$$+ A_0(u) y_0 + \cdots + A_{n-p}(u) y_{n-p} \quad (6)$$

the degree of interpolating polynomial that now passes through the $n+1$ points $(x_{-p}, y_{-p}), \ldots, (x_{n-p}, y_{n-p})$. Here p is the largest integer less than or equal to $n/2$, and the $A_{-p}(u), \ldots, A_{n-p}(u)$ are polynomials in the variable shown as Eq. (7). The

$$u = \frac{x - x_0}{h} \quad (7)$$

polynomials of the variable in Eq. (7) have been tabulated as functions of u.

Inverse interpolation. If the value of y is known and the value of the corresponding independent variable x is desired, one has the problem of inverse interpolation. Since the polynomials $A_k(u)$ in Eq. (6) are known functions of u, and the values of y and y_k are also known, the only unknown in this equation is u. Thus the problem reduces to that of finding a real root u of an nth-degree polynomial in the range $0 < u < 1$. Having found u, one may find x from Eq. (7). For a discussion of numerical methods for solving for such a root and for more information on interpolation see NUMERICAL ANALYSIS.

One may also perform an inverse interpolation by treating x as a function of y. Since, however, the intervals between the y_i are not equal, it is necessary to employ the general interpolation formula of Eq. (2) with the xs and ys interchanged.

Round-off errors. In the tabulated values, round-off errors ε_i resulting from the need to express the entries y_i of the table as finite decimals will cause an additional error in the interpolated value of y that must be added to the truncation error discussed before. The effect of these errors on the application of Lagrange's interpolation formula is seen by Eq. (6) to be a total error ε_T in y given by Eq. (8). Letting e be the smallest positive

$$\varepsilon_T = \sum_k A_k(u) \varepsilon_k \quad (8)$$

number satisfying the condition $e > |\varepsilon_k|$ for all k,

one knows from Eq. (8) that relation (9) holds.

$$|\varepsilon_T| \le \sum_k |A_k(u)||\varepsilon_k| \le e \sum_k |A_k(u)| \quad (9)$$

Since the sum of the $A_k(u)$ is equal to 1, the factor

$$\sum_k |A_k(u)|$$

in Eq. (9) is usually not much larger than 2 or 3 and thus the interpolated value of y has about the same round-off error as the individual entries.

Use of finite differences. For some purposes it is more convenient to use an interpolating formula based not so much on the entries y_i of a central difference table as upon their differences.

Each difference $\delta^k y_s$ is obtained by subtracting the quantity immediately above and to the left of it from the quantity immediately below and to the left; thus, Eq. (10) can be written, where k and $2s$

$$\delta^k y_s = \delta^{k-1} y_{s+1/2} - \delta^{k-1} y_{s-1/2} \quad (10)$$

are required to be integers. For example,

$$\delta y_{1/2} = y_1 - y_0 \quad \text{and} \quad \delta^2 y_0 = \delta y_{1/2} - \delta y_{-1/2}$$

An interesting property of a difference table is that, if y, the dependent variable tabulated, is a polynomial of the nth degree in x, its kth difference column will represent a polynomial of degree $n-k$. In particular, its nth differences will all be equal and all higher differences will be zero. For example, consider a table of cubes and the difference table formed from it by the rule given above.

x	$y = x^3$			
0	0			
		1		
1	1		6	
		7		6
2	8		12	0
		19		6
3	27		18	0
		37		6
4	64		24	0
		61		6
5	125		30	
		91		
6	216			

Most functions $f(x)$, when tabulated at a small enough interval $\Delta x = h$, behave approximately as polynomials and therefore give rise to a difference table in which some order of difference is nearly constant. Consider, for example, the difference table of log x, in which the third differences fluctuate between 7 and 9 times 10^{-7}.

x	$y = \log x$	δy	$\delta^2 y$	$\delta^3 y$
1.00	0.0000 000			
		43 214		
1.01	0.0043 214		−426	
		42 788		8
1.02	0.0086 002		−418	
		42 370		9
1.03	0.0128 372		−409	
		41 961		8
1.04	0.0170 333		−401	
		41 560		7
1.05	0.0211 893		−394	
		41 166		
1.06	0.0253 059			

Experimental data, if taken at small enough interval of the independent variable, would be expected to exhibit much the same behavior as a mathematical function except for the presence of experimental error. The presence of the latter will cause the differences to have a more or less random fluctuation. The size of the fluctuation may, in fact, be used to indicate the number of significant figures in the data.

The constancy of the third differences for log x indicates that for the accuracy and the interval used, a third-degree polynomial may be employed as an interpolating function. Since such a polynomial is determined by the choice of four coefficients, one would expect the interpolation formula to involve four numbers derivable from the difference table. Thus the forward-interpolation formula of Gauss, Eq. (11), can be written. If

$$y = y_0 + u\,\delta y_{1/2} + \frac{1}{2}\,u(u-1)\,\delta^2 y_0$$

$$+ \frac{1}{3!}\,u(u^2-1)\,\delta^3 y_{1/2} + \frac{1}{4!}\,u(u^2-1)(u-2)\,\delta^4 y_0$$

$$+ \frac{1}{5!}\,u(u^2-1)(u^2-4)\,\delta^5 y_{1/2} + \cdots \quad (11)$$

terminated after the fourth term, it represents a third-degree polynomial in $u = (x-x_0)/h$, and hence in x. It involves the four constants y_0, $\delta y_{1/2}$, $\delta^2 y_0$, and $\delta^3 y_{1/2}$. Since any one of the entries in the y column may be chosen as y_0, the differences required are picked from a central difference table, for example, in relationship to this entry. The interpolating polynomial obtained passes through the four points (x_{-1}, y_{-1}), (x_0, y_0), (x_1, y_1), and (x_2, y_2). In general, the interpolating polynomial will pass through only those points whose y coordinate is needed to form the differences used in the formula.

If one terminates the series in Eq. (11) after the second term, one obtains Eq. (12). This is the linear

$$y = y_0 + u\delta y_{1/2} = y_0 + u(y_1 - y_0) \quad (12)$$

ear interpolation formula most often used when making a simple interpolation in a table.

There are a great variety of interpolation formulas, such as Gregory-Newton's, Stirling's, and Bessel's, that differ mainly in the choice of differences used to specify the interpolating polynomial.

Difference equations. Repeated application of Eq. (10) may be used to express any difference in terms of the tabulated values, for example, Eqs. (13). Expressed in a more general form Eqs. (13)

$$\delta y_{1/2} = y_1 - y_0$$
$$\delta^2 y_0 = y_1 - 2y_0 + y_{-1} \quad (13)$$

become Eqs. (14). If one sets the second difference

$$\delta f(x) = f\left(x + \frac{h}{2}\right) - f\left(x - \frac{h}{2}\right)$$
$$\delta^2 f(x) = f(x+h) - 2f(x) + f(x-h) \quad (14)$$

equal to zero, one obtains a so-called difference equation for $f(x)$. In general, a difference equation is any equation relating the values of $f(x)$ at discrete values of x.

Difference equations play much the same role in analytical work as differential equations. Because they can be interpreted in terms of only those values of $f(x)$ tabulated at some interval $\Delta x = h$, they are admirably adapted to numerical computations. For this reason most numerical solutions of differential equations involve approximating the equation by a suitable difference equation.

For ordinary differential equations the transformation to a difference equation can be made by replacing each derivative in the equation by the appropriate difference expression according to formula (15), where $f^{(n)}(x)$ designates the nth deriv-

$$f^{(2k)}(x) \to \frac{1}{h^{2k}}\,\delta^{2k} f(x)$$

$$f^{(2k+1)}(x) \to \quad (15)$$

$$\frac{1}{2h^{2k+1}}\left[\delta^{2k+1} f\left(x + \frac{h}{2}\right) + \delta^{2k+1} f\left(x - \frac{h}{2}\right)\right]$$

ative of $f(x)$. The difference equation resulting can then, as mentioned before, be used to express the relationship between the values of $f(x)$ at the discrete points $x_s = x_0 + sh, s = 0, 1, 2, \ldots, n$.

Partial difference equations. Suppose one chooses to specify a function of two variables $f(x,y)$ by giving its value at some regular array of points in the xy plane having coordinates (x_m, y_n). Then, in place of a linear partial differential equation for $f(x,y)$ one has a linear partial difference equation, Eq. (16), where $g(x,y)$ is a known function and i

$$\sum_{s,t} A_{st} f(x_{i+s}, y_{j+t}) = g(x_i, y_j) \quad (16)$$

and j any of the set of integers for which the difference equation has significance. A difference equation in which some of the $f(x_{i+s}, y_{j+t})$ occur to a power other than the first is termed a nonlinear difference equation.

If one employs a square lattice makeup of the points, Eqs. (17), then Laplace's differential equa-

$$x_m = x_0 + mh$$
$$y_n = y_0 + nh \quad (17)$$

tion is approximated by difference equation (18),

$$f_{i+1,j} + f_{i,j+1} + f_{i-1,j} + f_{i,j-1} - 4f_{ij} = 0 \quad (18)$$

where, for simplicity, Eq. (19) holds.

$$f_{mn} = f(x_m, y_n) \quad (19)$$

See GRAPHIC METHODS; LATTICE (MATHEMATICS).

[KAISER S. KUNZ]

Bibliography: K. S. Kunz, *Numerical Analysis*, 1957; A. Ralston, *A First Course in Numerical Analysis*, 1965.

Matter exists both as gas and as dust grains in the space between stars. This interstellar matter is observed in a variety of ways. For example, the North American Nebula in Cygnus (*to left*), which is a cloud of gas having a density possibly a thousand times greater than the average density of interstellar gas, fluoresces by ultraviolet light from adjacent stars; it is called an emission nebula. Dust in the space between the nebula and Earth scatters all but the red light. Density of this interstellar dust is about 30 dust grains per cubic kilometer. A much denser cloud of dust between the North American Nebula and Earth obscures portions of the emission nebula to create the appearances of the "Gulf of Mexico" and the "Atlantic Ocean." The Veil Nebula (*above*), also in Cygnus, consists of filaments of high-velocity gas ejected from an exploding star more than 50,000 years ago. As this gas collides with atoms in interstellar space, it ionizes and glows. High-velocity collisions along the leading edge create blue light; less-energetic collisions along the trailing edge create red light. Collisions also decelerate the gas filaments so that the nebula will cease to glow in another 25,000 years. Both photographs were made with the 48-in. Schmidt telescope, California Institute of Technology.

Interstellar matter

The material between the stars. Being the reservoir from which new stars are born in the Galaxy, interstellar matter is of fundamental importance in understanding both the processes leading to the formation of stars, including the solar system, and ultimately the origin of life in the universe. Among the many ways in which interstellar matter is detected, perhaps the most familiar are attractive photographs of bright patches of emission-line or reflection nebulosity. However, these nebulae furnish an incomplete view of the large-scale distribution of material, because they depend on the proximity of one or more bright stars for their illumination. Radio observations of atomic hydrogen, the dominant form of interstellar matter, reveal a widespread distribution throughout the thin disk of the Galaxy, concentrating in the spiral arms. Mixed in with the gas are small solid particles, called dust grains, of characteristic radius 0.1 μm. Although by mass the grains constitute less than 1% of the material, they have a pronounced effect through the extinction of starlight. Striking examples of this obscuration are the dark rifts seen in the Milky Way. On average, the density of matter is only one hydrogen atom per cubic centimeter (equivalently 10^{-24} gm cm^{-3}), but because of the long path lengths over which the material is sampled, this tenuous medium is detectable. Radio and optical observations of other spiral galaxies show a similar distribution of interstellar matter in spiral arms in the galactic plane.

A hierarchy of interstellar clouds, concentrations of gas and dust, exists within the spiral arms. Many such clouds or cloud complexes are recorded photographically. However, the most dense, which contain interstellar molecules, are often totally obscured by the dust grains and so are detectable only through their infrared and radio emission. These molecular clouds contain the birthplaces of stars. *See* GALAXY; GALAXY, EXTERNAL.

Gas. Except in the vicinity of hot stars, the interstellar gas is cold, neutral, and virtually invisible. However, collisions between atoms lead to the production of the 21-cm radio emission line of hydrogen. Because the Galaxy is quite transparent at 21 cm, surveys with large radio telescopes have produced a hydrogen map of the entire Galaxy. Different emission regions in the Galaxy along the same line of sight are moving with different velocities relative to Earth and so are distinguishable by their different Doppler shifts. Supplemental information is obtained from 21-cm absorption-line measurements when hydrogen is located in front of a strong source of radio emission. These radio studies show that the gas is concentrated in clouds within the spiral arms, with densities typically 30 times the average of one atom per cubic centimeter. The cold clouds (70 K) appear to be in near pressure equilibrium with a more tenuous, hotter phase (7000 K) of neutral hydrogen in which they are embedded. *See* RADIO ASTRONOMY.

Other species in the gas are detected by the absorption lines they produce in the spectra of stars, and so are observable only in more local regions of the Galaxy. Interstellar lines are distinguished from stellar atmospheric lines by their extreme narrowness and different Doppler shifts. High dispersion spectra show the lines are composed of several components possessing unique velocities which correspond to the individual clouds detected with the 21-cm line. Elements such as Ca, Na, and Fe are detected in optical spectra, but the more abundant species in the cold gas, such as H, C, N, and O, which produce lines only in the ultraviolet require observations from satellites outside the Earth's atmosphere. The satellite measurements also reveal the molecules H$_2$ and CO in relatively dense clouds.

In the earliest stages of the Galaxy only H and He were present, but by nuclear burning processes in stellar interiors and supernova explosions the abundances of heavier elements have been built up through many cycles of star formation and mass loss from dying stars. The present relative abundances in interstellar matter are expected to be the same as seen in the atmospheres of the later-generation stars like the Sun. However, examination of the abundances obtained from interstellar lines reveals a considerable depletion of heavy atoms relative to hydrogen. The atoms missing in the interstellar gas can be accounted for by the interstellar matter seen in solid particle form.

Particles. The light of stars near the galactic disk is dimmed by dust grains which both absorb the radiation and scatter it away from the line of sight. The amount of extinction at optical wavelengths varies approximately as the reciprocal of the wavelength, resulting in a reddening of the color of a star, much as molecular scattering in the Earth's atmosphere reddens the Sun, especially near sunrise and sunset. The dependence of extinction on wavelength is much less steep than it is for molecular scattering, indicating the solid particles have radii about 0.1 μm. Satellite observations show a continued rise in the extinction at ultraviolet wavelengths which seems to require a component of smaller interstellar grains.

By comparison of the observed color of a star with that predicted from its spectral features, the degree of reddening or selective extinction can be fairly accurately determined, but the total extinction at any given wavelength is more difficult to measure. The best estimates suggest that extinction by dust over a pathlength of 1000 parsecs (1 parsec = 3.26 light-years) in the galactic plane will, on average, reduce a star's brightness by 80%. This requires the mass density of grains to be about 1% of the gas density. Since pure H or He grains cannot exist in the interstellar environment, a major fraction of the heavier elements must be in the solid particles. The number density of 0.1-μm grains would be about 2000 km^{-3}. Studies of reddening in conjunction with measurements of the 21-cm and ultraviolet Lyman-α lines of hydrogen show that dust and gas concentrations are well correlated.

The light of reddened stars is partially linearly polarized, typically by 1% but reaching 10% for the most obscured stars. The broad peak in polarization at yellow light, together with the correlation of the degree of polarization with reddening, suggests that the polarization and extinction are caused by the same dust grains. The grains must be both nonspherical and spinning about a preferred direction in space to produce polarization. The agent for the large-scale ordering required to explain the strong tendency of the planes of polari-

Fig. 1. Quadrant of the shell-shaped Rosette Nebula, showing dense globules of obscuring dust and gas silhouetted on the bright emission-line background of an H II region. Central hole may have been swept clear of gas by radiation pressure from central star (lower left) acting on the dust grains. Photographed in red light with the 48-in. (122 cm) Schmidt telescope of the Hale Observatories.

zation of stars in some directions in the Milky Way to lie parallel to the galactic plane is believed to be the galactic magnetic field. Minute amounts (0.01%) of circular polarization have also been used to study the topology of the magnetic field.

The possible types of grain material can be restricted through considerations of relative cosmic abundances, but detailed identification is difficult because of the paucity of spectral features in the extinction curve. Although ice has been detected by its 3.1-μm absorption band, it is not nearly as abundant as expected. Silicates are suggested by 10-μm absorption in front of strong infrared

Fig. 2. NGC 6611 (M16), a complex H II region in which the exciting stars are members of a cluster. Note dark globules and elephant-trunk structures, and the bright rims where ionizing radiation is advancing into more dense neutral gas. Photographed in Hα + [N II] with the 200-in. (508 cm) telescope of the Hale Observatories. (*Hale Observatories*)

sources, and in the ultraviolet an absorption peak at 2200 A (220 nm) could be explained by a component of small graphite particles. A popular theory of grain formation begins with the production of small silicate particles in the extended atmospheres of red supergiant stars. Radiation pressure ejects these particles into the interstellar gas where accretion of the most abundant elements, H, C, N, and O, builds up an icy mantle. Such core-mantle particles, while not unique, do provide an explanation for many of the observations.

Because of the high concentration of interstellar material toward the galactic plane, it is extremely difficult to detect radiation with a wavelength longer than 1 μm coming a large distance through the plane. Conversely, a line of sight to a distant object viewed out of the plane is much less obscured because the disk is so thin. The zone of avoidance, corresponding roughly to the area occupied by the Milky Way, is that region of the sky in which essentially no extragalactic object can be seen because of intervening dust. The dark rifts in the Milky Way result from the same obscuration. The component of starlight scattered rather than absorbed by the grains can be detected as a diffuse glow in the night sky near the Milky Way. However, it must be carefully separated from other contributions to the night sky brightness: the integrated effect of faint stars, zodiacal light from dust scattering within the solar system, and airglow (permanent aurora). *See* AIRGLOW; AURORA; INTERSTELLAR SPACE.

Dark nebulae. A cloud of interstellar gas and dust can be photographed in silhouette if it appears against a rich star field. The largest and most dense clouds are most easily detected because of the large contrast produced in the apparent star density. A distant cloud is difficult to find because of many foreground stars. Many large dark nebulae or groups of nebulae can be seen in the Milky Way where the material is concentrated. The distance to a dark nebula can be estimated using the assumption that statistically all stars are of the same intrinsic brightness. When counts of stars within a small brightness range are made in the nebula and an adjacent clear region, the dimming effect of the cloud will appear as a sudden relative decrease in the density of stars fainter than a certain apparent brightness, which corresponds statistically to a certain distance. Alternatively, a lower limit to the distance is provided by the distance to the most distant unreddened stars in the same direction. One of the best-known and nearest dark nebulae is the Coal Sack, situated at a distance of 175 parsecs. Another example is the "Gulf of Mexico" area in the North America Nebula. *See* COAL SACK.

Obscuring clouds of all sizes can be seen against the bright H II regions described below. In many cases the H II regions and dark nebulae are part of the same cloud. The bay in the Orion Nebula is one such region, but perhaps even more familiar is the spectacular Horsehead Nebula. Even smaller condensations, called globules, are seen in the Rosette Nebula (Fig. 1) and NGC 6611 (M16; Fig. 2). The globules, which are almost completely opaque, have masses and sizes which suggest they might be the last fragments preceding the birth of stars. *See* NEBULA; ORION NEBULA.

Bright nebulae. An interstellar cloud can also become visible as a bright nebula if illuminated by a nearby bright star. Whether or not an H II region or a reflection nebula results depends on the quantity of ionizing radiation available from the star. To be distinguished from H II regions, but often also called bright gaseous nebulae, are shells of gas that have been ejected from stars. Included in this latter category are planetary nebulae, nova shells and supernova remnants which have a bright emission-line spectrum similar to that of an H II region. *See* CRAB NEBULA; NEBULA, GASEOUS.

H II regions. A star whose temperature exceeds about 25,000 K emits sufficient ultraviolet radiation to completely ionize a large volume of the surrounding hydrogen. The ionized regions (Figs. 1 and 2), called H II regions, have a characteristic red hue resulting from fluorescence in which hydrogen, ionized by the ultraviolet radiation, recombines and emits the Hα line at 6563 A (656.3 nm). Optical emission lines from many other elements have been detected, including the well-known "nebulium" line of oxygen at 5007 A (500.7 nm).

An H II region can be extended with a relatively low surface brightness if the local density is low, as in the North America Nebula. However the best-known regions, such as the Orion Nebula, are in clouds that are quite dense (10^3 to 10^4 cm^{-3}) compared to the average; dense clouds use up the ionizing radiation in a region closer to the star and consequently are smaller with a higher surface brightness. Since the brightest stars are also the youngest, it is not surprising to find them still embedded in the dense regions from which they formed. Later in its evolution an H II region can develop a central hole if radiation pressure on the dust grains is sufficient to blow the dust and gas away from the star, as in the Rosette Nebula (Fig. 1). H II regions are also conspicuous sources of radio emission characteristic of close electron-proton encounters in the 10,000 K gas. Some H II regions are seen only as radio sources because their optical emission is obscured by dust grains. Radio recombination lines of H, He, and C, which result when the respective ions recombine to highly excited atoms, are also important.

Reflection nebulae. In the absence of sufficient ionizing flux, the cloud may still be seen by the light reflected from the dust particles in the cloud. The scattering is more efficient at short wavelengths, so that if the illuminating star is white the nebula appears blue. The absorption or emission lines of the illuminating star appear in the nebular spectrum as well. The Orion Nebula has an underlying reflection nebula arising from the dust in the gas cloud which produced the H II region. However, in this and other H II regions the emission-line radiation rather than the reflected light dominates the nebulosity.

Reflection nebulae are strongly polarized, by as much as 40%. Both the color and the polarization can be explained by dust grains similar to those which cause interstellar reddening. Some reflection nebulae appear to be intimately related to stars in early or late stages of stellar evolution. Other reflection nebulae, such as those in the Pleiades, result from a chance close passage of a cloud and an unrelated bright star, providing a unique look at interstellar cloud structure.

Molecular clouds. Inside interstellar clouds, molecules form on the surfaces of dust grains or, in some cases, by gas-phase reactions. In dense clouds where molecules are effectively shielded from the ultraviolet radiation that would dissociate them, the abundances become appreciable. Hydrogen is converted almost completely to its molecular form, H_2, which is detected in ultraviolet spectra along with the next most abundant molecule, CO. The high abundance of CO may be attributed to the high cosmic abundances of C and O and the great stability of CO.

Most molecules, with the notable exception of H_2, are discovered by their millimeter-line absorption or emission. Dark nebulae are found to contain H_2, CO, OH, and H_2CO. The Orion molecular cloud, which contains the Orion Nebula as a fragment, has a mass of at least 200 solar masses and is one of the chief regions for detection of a large number of more complex polyatomic organic molecules. The other region, the Sagittarius B2 cloud near the galactic center, has a mass exceeding 200,000 solar masses. Altogether, over three dozen species have been detected, many of which could be classed as prebiotic. It is interesting that the first stages of organic evolution can occur in such low-density environments.

Star formation. Superluminous stars such as those exciting H II regions cannot be very old because of the tremendous rate at which they are exhausting their supply of hydrogen for nuclear burning; the most luminous are under 100,000 years old. With this clear proof of recent star formation, observations have been directed toward discovering stars even closer to their time of formation. Compact H II regions, such as the Orion Nebula, appear to be the first fragments of much larger clouds to have formed stars. Ultracompact H II regions, seen at radio wavelengths in molecular clouds but totally obscured optically, appear to be an earlier stage in which the protostellar core has just become highly luminous. These are often called cocoon stars. Even earlier still are the compact infrared sources in which hot dust grains in a protostellar cloud are being detected at wavelengths of 5–100 μm. These earliest phases are often associated with intense H_2O and OH molecular maser emission. Examples of all of these stages are often found in the same region of space. In addition to the Orion molecular cloud, many regions such as the W3 radio and infrared sources associated with the visible H II region IC 1795 have been studied extensively. *See* ASTRONOMY, INFRARED; STELLAR EVOLUTION.   [P. G. MARTIN]

Bibliography: P. A. Aannestad and E. M. Purcell, Interstellar dust, *Annu. Rev. Astron. Astrophys.*, vol. 11, 1973; F. J. Kerr and S. C. Simonson III (eds), *Galactic Radio Astronomy*, 1974; B. M. Middlehurst and L. H. Aller (eds.), *Nebulae and Interstellar Matter*, 1968; D. E. Osterbrock, *Astrophysics of Gaseous Nebulae*, 1975; L. Spitzer, Jr., *Diffuse Matter in Space*, 1968; B. E. Turner, Interstellar Molecules, *Sci. Amer.*, 228, 1973.

Interstellar space

The space between the stars and other celestial bodies. For convenience of reference the space in the immediate vicinity of Earth is termed near space. The intervening region from Earth to the

Moon is cislunar space; farther from Earth is interplanetary space. Interstellar space is strictly the region beyond the planets. This article deals with both interplanetary and interstellar space. *See* SPACE.

Nowhere in the universe is space absolutely empty; it is occupied by extremely tenuous matter, generally gas and solid particles from dust to meteors of 0.1–1 cm diameter, and large meteorite debris.

In the solar system, at least 99% of the solid mass filling space occurs in the form of tiny particles of dust, having diameters of 0.001–0.1 mm. The sunlight reflected from this dust cloud is visible as zodiacal light, showing a strong concentration near the plane of Earth's orbit. The dust is dark, its reflecting power being about the same as that of soot. Its total density is about 1 g/500,000 km³. There appears to be a smaller amount of interplanetary gas, mostly completely ionized hydrogen, at least partly consisting of a continuous stream from the solar corona. *See* ZODIACAL LIGHT.

The meteoric and meteoritic component of matter in interplanetary space represents a much smaller mass, about 1 g/10⁸ km³. It is more conspicuous because individual meteors and meteorite falls are observable from Earth.

Comets with nuclei diameters of 0.1–10 km and asteroids (0.1–500 km) form the next step. During millions of years, meteor craters on Earth and on the Moon have been produced by collisions with these bodies. Unlike meteorites, they can be observed individually in space from a distance, with the aid of telescopes.

Interstellar space near the plane of the Milky Way is filled with gas, mainly hydrogen, of an average density of 1 g/10⁹ km³. Everywhere it is accompanied by dust grains having diameters of 10⁻⁵–10⁻⁴ cm. The total mass of dust is about 0.8% of the gas, and the gas accounts for about 2% of the mass of the Galaxy. The dust obstructs light, but, unlike interplanetary dust, it is white and scatters light without much absorption. Gas and dust appear to be gathered into cosmic clouds 10–50 light years in diameter and with densities 10–50 times the average density in interstellar space. Interstellar gas is transparent; its temperature depends greatly on the radiation of nearby stars and varies from 10,000°C near hot stars (H II regions) to −170°C (H I regions). The temperature of the dust is everywhere very low, −250 to −170°C. *See* GALAXY.

In the space between galaxies the density of matter is perhaps a millionth the density in interstellar space.

Interplanetary and interstellar space also contains primary cosmic-ray particles, which are atomic nuclei traveling with nearly the velocity of light and contributing a weight of about 5 g/10¹⁸ km³.

Electromagnetic radiation, including x-rays, ultraviolet, visible, and infrared light, and radio waves, travels everywhere throughout space. Visible light is emitted by stars, infrared radiation by the dust grains, and radio waves chiefly by ionized gas. *See* RADIO ASTRONOMY.   [ERNST J. OPIK]

Intestine (vertebrate)

The tubular portion of the digestive tract, usually between the stomach and the cloaca or anus. The detailed functions vary with the region, but are

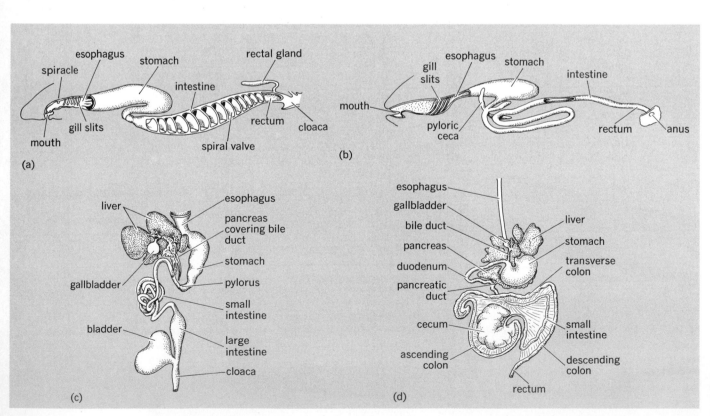

Vertebrate digestive tracts, showing structure of fish, amphibian, and mammalian intestines. (*a*) Shark. (*b*) Perch. (*c*) Frog. (*d*) Guinea pig. (*From A. S. Romer, The Vertebrate Body, 3d ed., Saunders, 1962*)

primarily digestion and absorption of food.

The structure of the intestine varies greatly in different vertebrates (see illustration), but there are several common modifications, mainly associated with increasing the internal surface area. One, seen in many fishes, is the development of a spiral valve; this turns the intestine into a structure resembling a spiral staircase. Another, seen in some fish and most tetrapods, is simply elongating and then coiling the intestine. This can reach extremes in large herbivores: Oxen have intestinal lengths of over 150 ft. In numerous forms there are blind pouches, or ceca, off part of the intestine. In fish these are commonly at the anterior end; in tetrapods they generally lie at the junction between the large and small intestines. In all vertebrates the inner surface of the intestine is irregular, with ridges and projections of various sorts; these reach their maximum development in the extremely fine and numerous finger-shaped villi found in mammals.

In man the intestine consists of the small and large intestines. The small intestine is further divided into three major parts: the duodenum, the jejunum, and the ileum. The duodenum, 10–12 in. long, begins at the pyloric sphincter of the stomach and curves around the head of the pancreas on the right side of the anterior part of the abdomen. It receives the ducts of the biliary system and the pancreas. The jejunum and ilium are about 19 ft long and form a much-coiled tube that empties at right angles into the large intestine through the ileocolic valve. The large intestine, or colon, consists of five parts: the ascending, transverse, descending, and sigmoid regions, and the terminal rectum which empties into the anal canal.

The microscopic structure of the intestine comprises an inner glandular mucosa, a muscular coat, and an outer serosa of connective tissues which is covered in most areas by peritoneum.

The intestine is supported by dorsal mesenteries of varying extent, which contain an extensive system of arteries, veins, lymphatics, and nerves to the various regions. *See* DIGESTIVE SYSTEM (VERTEBRATE).   [THOMAS S. PARSONS]

Intestine disorders

Disorders of the intestine include congenital defects and vascular, mechanical, degenerative, inflammatory, and neoplastic disorders.

The most common congenital defect is Meckel's diverticulum, a pouch or tract which persists as the remnant of the fetal umbilical duct. Other malformations or failures in development are uncommon. *See* CONGENITAL ANOMALIES.

Vascular lesions include mesenteric thrombosis, a blocking of the arterial supply or venous drainage of the intestine by a thrombus or embolus. It is frequently associated with trauma, heart failure, or arteriosclerosis. *See* EMBOLUS; THROMBOSIS.

Common mechanical disorders are hernias, the protrusion of a loop of bowel through a defect or weakness of the abdominal wall, and peritoneal adhesions, resulting from recent or old peritonitis, which often lead to intestinal obstruction. Intussusception is an infrequent disorder, seen most often in infants, in which one part of the intestine telescopes into the next segment. The blood supply may be reduced and obstruction may result. Similar results may follow volvulus, the twisting of

a loop of intestine on its own mesenteric stalk, the supporting fold of peritoneum. *See* HERNIA.

The three degenerative lesions with the highest incidence are sprue, mucoviscidosis, and Whipple's disease, but all are relatively uncommon. The first is marked by fatty, foul-smelling, bulky stools (steatorrhea) and a multiplicity of other clinical symptoms. The second is an inborn deficit of pancreatic secretion that causes steatorrhea; other symptoms include salivary and sweat gland abnormality and asthma. Whipple's disease is a rare disease of middle age in which large fatty deposits accumulate in the intestinal walls and in the mesenteries.

Inflammations include nonspecific and specific diseases. Tuberculosis, typhoid fever, amebiasis, cholera, and bacillary dysentery, specific infectious diseases produced by invasion of microorganisms, may produce intestinal lesions. Various chemicals and poisons may also produce inflammations of the small intestine (enteritis) or of the large bowel (colitis). Ulcerative colitis refers to a disease of unknown cause which produces ulcerations of the intestine, particularly of the colon. *See* AMEBIASIS; BACILLARY DYSENTERY; CHOLERA VIBRIO; ILEITIS; PEPTIC ULCER; TUBERCULOSIS.

Appendicitis is the most frequent provocation of surgery of the abdomen. It probably arises from a combination of causes, including infection and obstruction of the lumen, and also from parasitic infestations. The prompt surgical removal reduces the number and severity of complications encountered in untreated cases. Appendicitis may be an acute fulminating disease or may persist as a chronic, recurrent inflammation with much scarring and final obliteration of the lumen.

Malignant tumors of the small intestine are not common, but the large intestine has a high incidence of carcinomas which account for 15% of all cancer deaths each year in the United States. Other malignancies are lymphomas, Hodgkin's disease, malignant melanomas, sarcomas, and leukemic infiltrations. *See* ONCOLOGY.

Benign tumors are represented by polyps, lipomas, myomas, and fibromas and usually occur in the colon, the small intestine rarely being involved.

Disorders of the peritoneum and other abdominal organs may directly or indirectly involve the intestine. Other rare conditions exist that may affect the intestine; any systemic disorder may also produce intestinal changes or dysfunction.

[EDWARD G. STUART/N. KARLE MOTTET]

Inulin

A reserve polysaccharide in some plants, especially those in the family Compositae, where it is found in the roots and tubers of the dahlia, Jerusalem artichoke, and dandelion. Inulin is only slightly soluble in cold water but readily dissolves in hot water. It does not give a color reaction with iodine, has little reducing power, and is levorotatory, $[\alpha]_D^{20}$ −40° (in water). Its molecular weight is approximately 5000, which corresponds to a chain length of about 30 fructofuranose residues. It is hydrolyzed with acid or the enzyme inulase from *Aspergillus niger* to D-fructose. The inulin molecule (see illustration) is made up principally of fructofuranose units linked β-glycosidically from the hydroxyl group on carbon 2 to the primary alco-

Structural formula for inulin.

holic group on carbon 1. *See* FRUCTOSE; OPTICAL ACTIVITY; POLYSACCHARIDE.

Hydrolysis of inulin with inulase or baker's yeast invertase produces about 1.5% D-glucose. It is not certain whether this glucose is an integral part of the inulin molecule or whether it is a constituent of an associated substance which is hydrolyzed at approximately the same rate as inulin.

[WILLIAM Z. HASSID]

Inventory management

The work of minimizing annual inventory expenses, which usually maximizes the return on the inventory investment. Annual inventory expenses are the annual total of the expenses and losses for carrying inventory, ordering material, and running out of stock. They are controlled by the quantity and location of the materials used in carrying out business.

One phase of inventory management is to minimize the quantity of material needed to conduct business. This is done by standardization to reduce the number of different items that must be kept in stock, by choice of plant locations, by coordinating the number of products and models in the line, and by extending the use of interchangeable parts and materials.

Another phase is to minimize the period that each item is kept in stock by scheduling and by choice of delivery lot quantity based on lengths of production cycles and forecasts of demand.

Effective inventory management requires records of expenses, costs, and losses that come from carrying inventory, from ordering stock, and from running out of stock. From such records one estimates the effects of changing delivery lot quantities, lengths of reserve cycles, and quantities of reserve stock. Reserve cycle is the interval between the scheduled or expected dates of delivery and use. Reserve stock is material that is on hand because a reserve cycle is allowed in scheduling the delivery of material. Delivery ratio is the percentage of deliveries made on or before the day the material is required.

Changes are planned by finding what the expenses were for one set of delivery lot quantities and lengths of reserve cycles, and estimating what the

expenses would be for other conditions. Care must be taken to see that the data used in the calculations represent expenses that actually will change. For example, space costs are a part of the cost of carrying inventory, but a change in the amount of material kept in stock may not change the space costs.

The delivery lot quantities and quantities of reserve stock can be calculated by examining the effects of changes in these quantities on costs and profits. The delivery lot quantities usually are economical when the related costs per year for carrying inventory equal the ordering costs per year plus the average costs of stock-outs per delivery. The amount of reserve stock usually is economical when the cost per year of saving one stock-out per 100 deliveries equals the value that will be paid for this benefit. The practicality of the figures obtained from these computations should be reviewed by someone familiar with purchasing and manufacturing problems. The feasibility of planning deliveries analytically depends on the accuracy of the available data, including plant needs and vendor performance. *See* LINEAR PROGRAMMING.

Inventory management also includes installing and operating a material control system which ensures that material is usually where it is required when it is required. Material control includes developing a signal that material should be ordered to meet known or expected requirements, ordering material, notifying expediters of the need to expedite delivery, and performing related tasks. These tasks include setting up and posting perpetual inventory records and compiling records of quantities used, where used, prices, sources, kind and quantities of material used to make the item, associated items, and similar information. Material-control personnel also gather, store, process, and distribute inventory-management data. Effective inventory management helps a company improve its competitive position. *See* INDUSTRIAL COST CONTROL.

[LAUREN F. SARGENT]

Bibliography: J. W. Prichard, *Modern Inventory Management*, 1965; *Programmed Instruction in Inventory Management*, Entelek, Inc., 1964.

Inverse-square law

Any law in which a physical quantity varies with distance from a source inversely as the square of that distance. When energy is being radiated by a point source (see illustration), such a law holds, provided the space between source and receiver is filled with a nondissipative, homogeneous, isotropic, unbounded medium. All unbounded waves become spherical at distances r, which are large compared with source dimensions so that the angular intensity distribution on the expanding wave surface, whose area is proportional to r^2, is fixed. Hence emerges the inverse-square law. *See* POINT SOURCE.

Similar reasoning shows that the same law applies to mechanical shear waves in elastic media and to compressional sound waves. It holds statistically for particle sources in a vacuum, such as radioactive atoms, provided there are no electromagnetic fields and no mutual interactions. The term is also used for static field laws such as the law of gravitation and Coulomb's law in electrostatics.

[WILLIAM R. SMYTHE]

INVERSE-SQUARE LAW

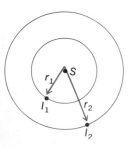

Point source S emitting energy of intensity I. The inverse-square law states that $I_1/I_2 = r_2^2/r_1^2$.

Invertebrata

A division of the animal kingdom which has no taxonomic status. Included under the heading Invertebrata are all those animals which lack a spinal column or backbone, in contrast to Vertebrata. With the exception of the phylum Chordata, all phyla of the subkingdom Metazoa as well as those of the subkingdoms Protozoa, Mesozoa, and Parazoa are invertebrates. *See* ANIMAL KINGDOM; MESOZOA; METAZOA; PARAZOA; VERTEBRATA.

[CHARLES B. CURTIN]

Invertebrate embryology

The study of the development or morphogenesis and growth of the invertebrates. The same general principles of development apply to the invertebrates as do to the vertebrates. Actually, much of the basic knowledge of embryology has been the result of studies on the invertebrates. A common phenomenon in the invertebrates is the release of a free and independent form, the larva, before development is completed. The larvae vary considerably and are characteristic of the different animal groups.

The lives of the vast majority of invertebrate animals begin with the union of an egg and a spermatozoon. While these reproductive cells, known collectively as gametes, could hardly be more unlike each other in most respects, they have two important features in common: the possession of a single set of chromosomes, and the characteristic of being triggered cells. The spermatozoon trigger causes a form change at the egg surface which enables it to penetrate and stimulate the egg. The latter is triggered to respond to this stimulation by beginning the series of complicated processes which draw the spermatozoon into the egg, bring together the two single sets of chromosomes to make up the double set characteristic of the species, and eventually enable this fertilized egg cell to translate the hereditary message into a faithful replica of the ancestral organ systems.

Aside from these two common properties, however, eggs and spermatozoa are highly dissimilar. While the nuclei of both undergo the process of meiosis, which prepares them for eventual union by reducing their characteristic chromosome number to one-half, the courses of preparation followed by their cytoplasms lead in exactly opposite directions.

Spermatogenesis. Number, compactness, and mobility are important for spermatozoa. Toward this end, the process of spermatogenesis consists of a stage of cell proliferation, followed by a period of progressive concentration and streamlining. The essential, heredity-determining material of the chromosomes is packed tightly into a tiny nucleus. The cytoplasm forms the locomotor apparatus, usually a single long flagellum with a centriole at its base and a mitochondrion nearby, as well as an organelle (acrosome) for penetrating the egg coverings. Excess cytoplasm is finally discarded, and the mature spermatozoon (Fig. 1a), ready to take part in fertilization, is a self-contained, stripped-down unit, carrying the hereditary message in code, and provided with enough energy source to propel it in a sustained burst of activity on its one-way trip to join an egg of its species. Millions upon millions of such cells are produced in the testis,

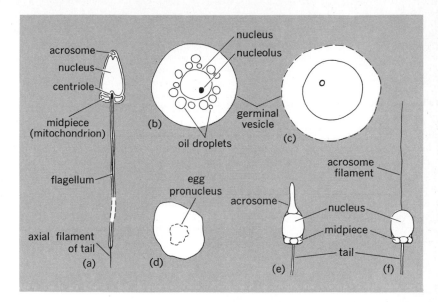

Fig. 1. Spermatozoa and fertilizable eggs. (*a*) Sea urchin spermatozoon. (*b*) Marine annelid (*Nereis*) egg, with intact germinal vesicle containing a nucleolus. Spheres surrounding germinal vesicle are oil droplets. (*c*) Marine mussel (*Mytilus*) egg, germinal vesicle broken down but polar bodies not formed. (*d*) Mature egg of sea urchin (*Arbacia*), containing egg pronucleus. (*e*) Mussel (*Mytilus*) spermatozoa, acrosome intact. (*f*) *Mytilus* spermatozoa, acrosomal reaction.

where they remain quiescent until they are spawned. *See* SPERM CELL; SPERMATOGENESIS.

Oogenesis. The egg is designed for a very different role. It must contain enough substance to provide structural material for the formation of a set of simple organs of locomotion and digestion so that the young animal can secure food to carry on its further growth. It must also contain enough energy-yielding material to perform the work of dividing the single egg cell into body cells from which such organs can be formed and to synthesize the complex chemical substances needed to provide each of these new cells with a new nucleus. *See* OOGENESIS; OVUM.

The egg, therefore, is specialized for large size and protection of its contents, with less concern for numbers and none at all for motility. In addition, its cytoplasm possesses intrinsic capacities for differentiation and building in exact accordance with the specifications contained in its chromosomes, so that a spider egg, for example, always produces a spider and never a fly. The fact that the physical bases of these capacities have so far eluded most of the efforts directed toward their detection in no way casts doubt on their existence.

The reserve building and energy-yielding materials are stored in the egg cytoplasm as minute spheres or platelets of yolk, a stable lipoprotein substance. Eggs are large cells even without this inert material. At the end of their growth period, when they have accumulated the full amount of yolk, they are huge in comparison to the body cells of the parent animal. No invertebrate eggs, however, achieve the spectacular dimensions of bird eggs. The largest are found among the arthropods (crayfish eggs are 2.5 mm in diameter), while some marine animals have very small eggs (oyster eggs are about 65 μ).

During the growth period, while the egg cell is actively synthesizing yolk and increasing the amount of cytoplasm, it has a very large nucleus, the germinal vesicle (Fig. 1b). When it reaches full size, however, and this synthetic activity subsides, the nuclear membrane breaks down, releasing its contents into the cytoplasm (Fig. 1d). The two successive nuclear divisions of meiosis follow, but the cytoplasm, instead of dividing equally, pushes out one of the daughter nuclei each time as a polar body. These two minute bodies have no further function in development. The chromosome material left in the egg forms the egg pronucleus (Fig. 1d), which is ready to unite with the sperm pronucleus. The zygote nucleus, formed by their union, is comparable in size to those of the body cells.

Egg polarity. Many types of eggs show structural departures from radial symmetry which indicate that the unfertilized egg is organized around a bipolar axis, one end of which is called the animal pole and the other the vegetal pole. The polar bodies are given off from the animal pole, and the egg pronucleus remains in this region. When an egg contains conspicuous amounts of yolk, it is usually concentrated in the vegetal half of the egg.

Egg membranes. Since the eggs of invertebrates are often shed directly into the water of oceans and streams, or laid to develop in places where they are exposed to the drying action of air and sunlight, they are always surrounded by a protective covering. In some forms the eggs are laid in batches which may be enclosed in a leathery sac or embedded in a mass of jelly. In other cases each egg has its own separate membranous case, a layer of jelly, or a more complex system of protective structures.

If the young animal is to begin its development under exposed conditions, as do many insects, the egg is provided with a tough covering (chorion) which is impenetrable even to its spermatozoa. In such cases there is a minute hole (micropyle) in the chorion near the animal pole through which the fertilizing spermatozoon can enter. Among echinoderms, the delicate vitelline membrane of the unfertilized egg is lifted away from the surface after a spermatozoon has entered and is strengthened by the addition of materials from the egg.

Preliminaries to fertilization. Sperm cells must complete all the nuclear and cytoplasmic changes and be fully mature before they can take part in fertilization. On the other hand, while this is true of the cytoplasm of egg cells, in most species the nuclear preparation for fertilization is incomplete when sperm entrance takes place. Moreover, the degree of incompleteness varies widely and apparently at random.

The marine annelid *Nereis* sheds its eggs with the germinal vesicle intact (Fig. 1b). The entrance of a spermatozoon stimulates the egg to begin the reduction divisions, and the sperm pronucleus waits within the egg cytoplasm while the egg nucleus carries out its preparation for union.

Sea urchin eggs are at the other extreme in this respect. Their reduction divisions are completed in the ovary (Fig. 1d), and the union of egg and sperm pronuclei follows immediately upon sperm entry. In many other species meiosis begins shortly before the eggs are spawned and stops at one stage or another of the process until the entrance of a spermatozoon sets it going again.

Reproduction among the invertebrates takes place in a variety of ways which differ widely from phylum to phylum. For example, following copulation in *Drosophila* the spermatozoa are stored in a special part of the female reproductive tract and released a few at a time as each ovum passes down the oviduct to be laid. The spermatozoa of squids are packaged in small bundles (spermatophores). These are placed by the male, using one of its arms which is modified for the purpose, in spermatophore receptacles on the body of the female. The eggs are fertilized as they leave the oviduct and are laid singly or in finger-shaped egg masses fixed to underwater objects.

Most of the echinoderms and many mollusks, ascidians, annelids, and coelenterates shed their eggs in tremendous numbers into the sea water where fertilization takes place. The young larvae must usually fend for themselves. In these same groups, however, some species shelter the young in sea water – containing chambers and pockets which are not actually within the body of the parent animal. This is also the case with arthropods such as crabs, which carry the larvae in masses fixed to the abdomen of the female, and some bivalves in which the young develop for a time within the mantle chamber. But whether the fertilized eggs are thus protected by the parent animal, laid in jelly masses or leathery cases or carefully constructed brood cells (bees, hunting wasps), or simply thrown into the water, each egg is an independent unit capable of developing into an adult without any further contribution from the parents.

Fertilization. Because fertilization in many invertebrates occurs within the reproductive tract of the female, or after spawning has taken place in response to stimuli which are not well understood or experimentally reproducible, it is hardly surprising that the actual process has been observed with the microscope in only a relatively few species. There are, however, a number of invertebrates among the forms which shed their gametes into the sea in which it is possible to add spermatozoa to eggs under the microscope and observe the process of sperm entry very clearly. In the most favorable cases the course of events within the egg as the two pronuclei unite and the zygote proceeds to undergo successive division and differentiation can be followed. Ready microscopical observation requires small transparent eggs which are normally fertilized outside the body of the female, are available in large numbers, and remain fertilizable in the laboratory. The eggs of some sea urchins best fulfill these conditions and have provided much of the available information relating to the fertilization process.

The sea urchin egg offers the further advantage of having completed its reduction divisions within the ovary so that it is in a relatively stable state when it is spawned. It also responds to sperm entry in certain easily recognizable ways which facilitate study of the fertilization process. As a result it has come to be generally thought of as a model for all fertilization, although it is actually a rather special case.

The unfertilized sea urchin egg (Fig. 1d) is a spherical cell, about 90 μ in diameter, surrounded by an invisible layer of jelly about 30 μ thick. Its small pronucleus lies eccentrically in the cytoplasm, which is crowded with small particles. About half of these are yolk spheres; the rest are

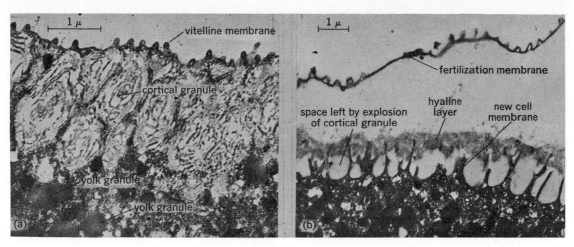

Fig. 2.   Electron micrographs, sections through *Hemicentrotus* egg surfaces. (*a*) Unfertilized. (*b*) Fertilized.

mitochondria, pigment granules, and various other minute bodies whose function is not yet well understood. Directly under the surface membrane is a compact layer of special granules 1 μ in diameter, called cortical granules (Fig. 2*a*). Electron microscopy shows these granules to be composed of at least two kinds of material, and they remain firmly in place under centrifugal forces which move other particles to different parts of the cell.

The surface of this cell is covered by a very delicate vitelline membrane which lies directly over the layer of cortical granules. When the fertilizing spermatozoon makes contact with this surface, an impulse is transmitted around the egg from the point of contact. The first detectable result of this impulse is a change in the cortical granules, which again appears at the point of sperm attachment and sweeps around the egg, reaching the opposite side in about 20 sec. This change consists of a swelling of one of the components of the granules, causing them to burst and separate the vitelline membrane from the rest of the cytoplasm. A colloidal substance released into the space below the raised membrane takes up water and swells, so that by about a minute after sperm attachment the membrane is free from the cytoplasmic surface all around the egg (Fig. 3*a* and *b*).

While this occurs, the spermatozoon is being drawn into the egg. By the time the sperm head has entered, a small bulge of hyaline cytoplasm, the fertilization cone, is formed, and the sperm tail gradually disappears into the cytoplasm through this cone (Fig. 3*c*).

The separated membrane continues to expand for a minute or two. During this period the substance released from the cortical granules and dissolved in the fluid immediately surrounding the egg unites with calcium from the sea water and forms a compound which accumulates on the inner surface of the lifted membrane. This is now called the fertilization membrane, and the added material strengthens it so remarkably that it is able to resist dissolving actions of all kinds except that exerted by a particular enzyme which the sea urchin larva synthesizes when it is ready to hatch.

The new egg surface replacing the elevated vitelline membrane follows the highly irregular contour of the pockets which formerly held the cortical granules (Fig. 2*b*). The second component of

the granules, which was left behind in these pockets, gradually moves outward within a few minutes after fertilization. Taking up calcium from the medium, its substance forms a hyaline layer close to the egg surface which is without a bounding membrane but is supported by delicate cytoplasmic projections. This layer holds the blastomeres together after cleavage and acts as a protective covering after the larva has left the fertilization membrane.

While these changes are occurring at the surface, the sperm head is carried a short way into the egg cytoplasm. For some three minutes, it is possible to distinguish it among the granular inclusions in very transparent eggs. After this, and while part of the tail is still outside the egg, the sperm nucleus begins to swell and can no longer be recognized in living eggs. Within two more minutes, a halo of hyaline rays begins to appear in this region. This is the sperm aster (Fig. 3*c*), formed by the egg cyto-

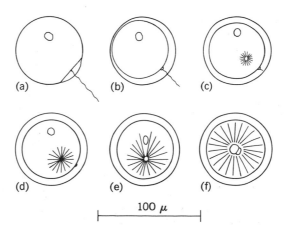

Fig. 3.   Fertilization of the sea urchin egg. (*a*) Forty seconds after insemination. Sperm head beginning to enter cytoplasm; fertilization membrane lifted around sperm entrance point. (*b*) Three minutes. Fertilization membrane free around egg surface; fertilization cone formed. (*c*) Five minutes. Sperm aster appearing; end of sperm tail about to enter egg. (*d*) Six minutes. Sperm aster enlarging and moving toward center of egg; fertilization cone disappearing. (*e*) Eight minutes. Egg pronucleus moving toward center of sperm aster; sperm pronucleus visible. (*f*) Ten minutes. Syngamy complete; sperm astral rays filling cytoplasm.

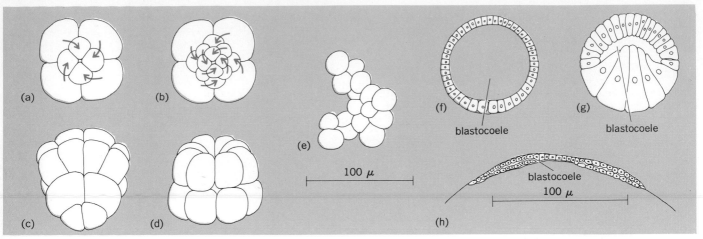

Fig. 4.   Symmetry of cleavage patterns and invertebrate blastulae. (*a, b*) Spiral cleavage. (*c*) Bilateral cleavage. (*d*) Radial cleavage. (*e*) Irregular cleavage. (*f*) Sea urchin blastula. Relatively uniformly sized blastomeres and large blastocoele. (*g*) Annelid blastula. Blastomeres at vegetal side large, yolky; small blastocoele. (*h*) Squid blastula. Blastomeres in animal pole region only; slitlike or no blastocoele.

plasm under the organizing influence of the sperm centriole.

Extending in all directions around this point, the rays soon reach the cell surface (Fig. 3*d*). Probably as the result of the push exerted here by their continuing elongation, the center of the aster containing the sperm pronucleus moves slowly toward the center of the egg. At the same time the egg pronucleus begins to move (Fig. 3*e*) toward the center of the sperm aster, that is, toward the sperm pronucleus, which is now visible as a clear sphere, still somewhat smaller than the egg pronucleus. Ten minutes at most after sperm entrance the two pronuclei lie in contact (Fig. 3*f*), surrounded by the rays of the sperm aster. These continue to elongate, carrying the apposed pronuclei into the center of the egg, and extend to the membrane on all sides.

The union of the two pronuclei (syngamy) marks the completion of the fertilization process. The fusion forms the zygote nucleus, with the full complement of chromosomes, and the dormant egg cell has been aroused to set in motion the series of changes which will produce a new sea urchin.

A number of the events which have just been described are peculiar to sea urchins. With different time schedules and allowance for the individual characteristics of each species, however, these basic processes of sperm entry, aster formation, and syngamy make up the complex phenomenon of the fertilization reaction as it occurs in all animals. *See* FERTILIZATION.

Such a descriptive presentation suggests many questions about what is actually going on in terms of cellular actions and reactions. One such question, for example, concerns the nature of the contact between spermatozoon and egg surface. It is known that in many invertebrates the acrosome at the anterior tip of the sperm cell undergoes an explosive reaction at the egg surface which transforms it into a very slender filament (Fig. 1*e* and *f*). In at least some cases an enzymelike substance is also released from the acrosome which has a dissolving effect on the egg membrane. There is evidence that the penetration of the egg surface by this filament activates the egg mechanically. Whether it is also useful in drawing the spermato-

zoon inside the egg cytoplasm has yet to be proved.

The nature of the specificity which ensures that eggs will be entered only by spermatozoa of their own species has also been the object of a great deal of research and has not received a thoroughly satisfactory explanation. The mechanism by which an egg that has received one spermatozoon rejects all later arrivals is another problem that resists solution, but perhaps more difficult to discover than the answers to any of these questions are those concerning the cytoplasmic differences between unfertilized and fertilized eggs.

Cleavage. The fertilized egg, or zygote, sets about at once to divide the huge mass of the egg into many small cells in order to restore the usual ratio between the amounts of nuclear and cytoplasmic substances. The energy for these repeated mitoses comes from the yolk, which also furnishes at least part of the materials required for synthesis of new nuclear structures. During this cleavage period, which commonly occurs during the first 12 hr after fertilization, the blastomeres, as the cleavage stage cells are called, divide more or less synchronously. Generally, cleavage follows one of several patterns characteristic for large groups of animals and often correlated with the amount and mode of distribution of the yolk.

Whatever cleavage pattern is followed, the plane of the first cleavage passes through the animal pole. When the vegetal region contains a large proportion of yolk, cleavage is retarded in this area and the blastomeres tend to be larger than in the animal pole region.

Small eggs, which contain little yolk, divide completely and usually very regularly, forming a mass of cells that shows spiral (mollusks, Fig. 4*a* and *b*), bilateral (ascidians, Fig. 4*c*), or radial symmetry (echinoderms, Fig. 4*d*). Some coelenterates, however, cleave into what appear to be random masses of cells (Fig. 4*e*).

The very large eggs of squid contain a great deal of yolk concentrated at the vegetal pole. The cleavage furrows do not cut all the way through this part but restrict their activity to the living cytoplasm at the animal pole.

Insect eggs also contain a large store of yolk, which occupies the center of the elongate cells and is surrounded by a thin layer of living cytoplasm containing the egg pronucleus. Following fertilization, the nuclei alone divide and move apart in the layer of cytoplasm after each division so that they distribute themselves all around the egg. After nine such nuclear divisions have taken place (producing 512 nuclei), the cytoplasm also cleaves at the next division, forming a single layer composed of about 1000 cells surrounding the central yolk mass.

Blastula stage. Among all the invertebrate forms except the insects, the result of 6–10 successive cleavage cycles is the formation of a sphere (blastula) composed of small cells which lie in a single compact layer around a central cavity (blastocoele). If the egg has contained relatively little yolk, the blastocoele is rather large (Fig. 4f), while it may be very small if the egg includes much yolk (Fig. 4g) and is little more than a slit in the squid blastula (Fig. 4h).

Gastrula stage. The end of the brief blastula stage occurs when the process of gastrulation begins. In its simplest form, this consists in an indenting (invagination) of the blastula wall in the vegetal region (Fig. 5a). Meanwhile cell division is going on steadily, and since the larva has as yet no way of taking in solid food from the outside, all the form changes which occur during this period are accomplished with the material originally present in the fertilized egg. The only addition is water (blastocoele fluid) and such dissolved substances, mostly salts, from the environment as can enter through the cell membranes. As the blastomeres become smaller and the blastular wall becomes correspondingly thinner, cells are provided to extend the vegetal indentation into a pocket (Fig. 5b). With the appearance of this structure (primitive digestive tract) the larva becomes two-layered, possessing an outer layer, the ectoderm, which will later produce the nervous system as well as the outermost body covering, and an inner layer, the endoderm, from which will be formed the lining of the functional digestive tract and its associated organs and glands. As the primitive digestive tract extends into the blastocoele, its opening to the outside becomes smaller and is known as the blastopore.

A modification of this process of endoderm formation occurs among some species having large, yolk-filled vegetal blastomeres (Fig. 5c). The small, actively dividing cells of the animal pole region spread down to cover these more inert blastomeres (Fig. 5d), which become the endoderm and later form the digestive organs, while the overlying ectoderm leaves a small opening in the vegetal region which corresponds to the blastopore (Fig. 5e).

Mesoderm formation. At this time the first few cells belonging to a third body layer, the mesoderm, make their appearance by slipping from the ectoderm layer into the blastocoele. These early mesoderm cells are of a primitive sort (mesenchyme), possessing pseudopodia and often moving about freely between the ectoderm and endoderm. In sponges and coelenterates, no more highly organized middle layer is formed even in adult animals, but in the other phyla the so-called "true" mesoderm is endodermal in origin, either being formed by successive divisions of a cell which originally belonged to the endoderm (Fig. 5f and g), as in annelids and mollusks, or separating

off from the primitive digestive tract, as in *Branchiostoma* (*Amphioxus*) (Fig. 5h and i).

In either case this mesodermal tissue spreads out between the ectoderm and endoderm, and in all phyla more advanced than the flatworms it splits through its center into an inner and an outer layer. The cavity thus formed within the mesoderm is the true body cavity in which the various internal organs lie. The outer layer of mesoderm becomes closely applied to the inner side of the ectoderm, forming body-wall muscles and other supporting layers, while the inner layer of mesoderm surrounds the endoderm with layers of muscle. The organs of circulation, excretion, and reproduction, as well as all muscles and connective tissue, are eventually formed from this mesoder-

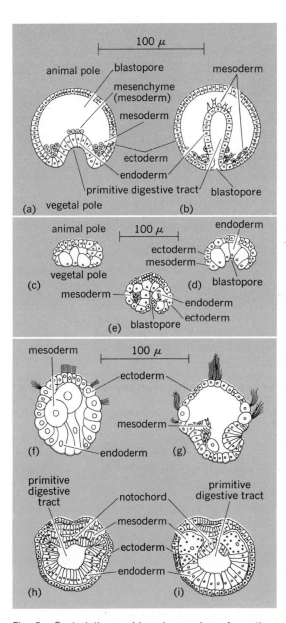

Fig. 5. Gastrulation and larval mesoderm formation. (a) Early and (b) later stage in nonyolky eggs. (c) Late blastula, (d) early gastrula, and (e) late gastrula in yolky egg of the snail. (f) Mesoderm formation in the limpet (*Patella*) shown in sections through center of blastula and (g) trochophore larva. (h) Cross section of *Branchiostoma* embryo immediately following gastrulation and (i) somewhat later.

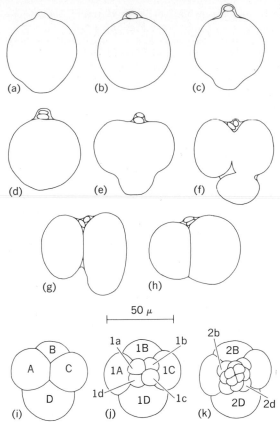

50 μ

Fig. 6. Maturation and early cleavage in the mussel (*Mytilus*). (*a*) First polar body being formed at animal pole; polar lobe at vegetal pole. (*b*) First polar body completely extruded; polar lobe withdrawn. (*c, d*) Second polar body formation. (*e–g*) First cleavage. (*h*) Two-cell stage. (*i*) Four-cell stage. (*j*) Eight-cell stage; first quartet of micromeres (1a, 1b, 1c, and 1d). (*k*) Sixteen-cell stage with three quartets of micromeres.

mal layer which surrounds the endoderm.

Later development. So far it is possible to summarize the development of invertebrate animals as a group but beyond this point each subgroup follows its own course, and these are so widely divergent that every one must be considered separately. Meaningful generalizations are not even possible within a single class in some cases, as attested to by the various modes of development occurring among the Insecta, some of which proceed directly from egg to adult form, while others go through an elaborate series of changes. *See* INSECT PHYSIOLOGY; INSECTA.

In very many species there is a sharp break in the life history when the larva, after passing through a number of morphological phases which lead from one to the next with a steady increase in size and complexity, abruptly forms a whole new set of rudimentary adult organs which take over the vital functions. This metamorphosis represents the end of the larval period. The tiny animal which it produces is for the first time recognizable as the offspring of its parents.

For more or less arbitrary reasons, the developmental processes of certain invertebrate forms have been studied very carefully so that their life histories are fully known. A few of these will be outlined in the following sections.

MOLLUSCAN DEVELOPMENT

The eggs of *Mytilus*, the common mussel, are fertilizable just after the germinal vesicle breaks down (Fig. 1c). As the first polar body is given off from the animal pole, the vegetal surface of the egg bulges out to form the so-called polar lobe (Fig. 6*a–d*). The bulge disappears shortly, to reappear at the time of second polar body formation.

Cleavage. When the egg cleaves, the vegetal cytoplasm is segregated into a more extreme polar lobe (Fig. 6*e* and *f*) and the cleavage furrow divides the remaining material equally between two blastomeres. The constriction forming the polar lobe disappears, returning the polar lobe material to one of the blastomeres (Fig. 6*g* and *h*). The vegetal material is again segregated at the second cleavage and again mixed with one of the four blastomeres.

It is characteristic of this type of cleavage that the mitotic spindles lie aslant in the blastomeres and, moreover, regularly change the direction of their slant by 90° at each division so that a spiral pattern of blastomeres results. Such spiral cleavage is found in the mollusks and in the flat, round, and segmented worms.

Since the animal-vegetal axis is easy to recognize in such eggs, it has been possible to record the course of cleavage very accurately and to determine the role of particular blastomeres in normally developing embryos. The four-cell stage blastomere containing the polar lobe material is designated as D, and proceeding in a clockwise direction, the others become A, B, and C (Fig. 6i). At the third cleavage, these divide very unequally (Fig. 6*j* and *k*) into four large vegetal macromeres (1A, 1B, 1C, and 1D) and four micromeres at the animal side (1a, 1b, 1c, and 1d). *See* CELL LINEAGE.

Blastula stage. After two more such unequal divisions the resulting 28 micromeres have formed a hollow blastula with the four macromeres, 3A, 3B, 3C, and 3D, at its vegetal side. These then extend into the blastular cavity where their descendants will form the digestive tract, except for one of the D daughter cells produced at the next cleavage, 4d, which is set aside as the mesoderm mother cell (Fig. 5*f*).

Trochophore stage. During the succeeding cleavages some of the cells develop cilia, the blastular symmetry becomes bilateral instead of radial, and the micromeres extend down almost to the vegetal pole, thus covering the macromeres except at the small opening of the blastopore. After 24 hr of development the cilia are organized into an encircling girdle and an apical tuft at the animal pole, and the larva, now called a trochophore, begins its free-swimming stage (Fig. 7*a*). The blastopore is shifted forward by the faster proliferation of the ectodermal cells of the other side (Fig. 7*b*) and then closed, but the larval mouth is later formed at this place. Behind it the endoderm forms a stomach, and a narrow tube gradually extends from this to make the intestine. The anus forms later at the place where the intestine reaches the ectoderm. *See* PROTEROSTOMIA.

At this stage a group of ectodermal cells is forming the shell gland which will secrete the shell (Fig. 7*c*). Two small protuberances will unite and develop into the foot, and a pair of elongated pits be-

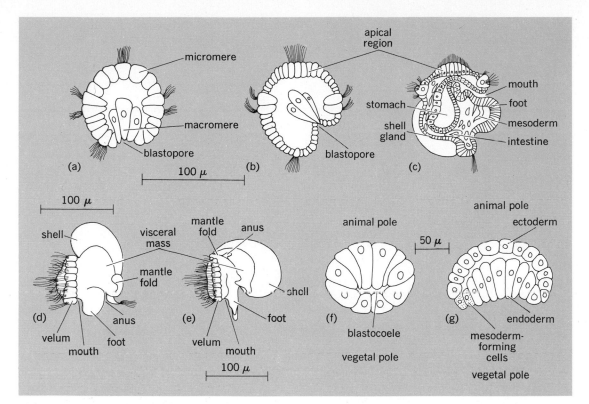

Fig. 7. Stages in the development of *Patella* and gastrulation in a tunicate. (*a*) Late blastula of *Patella*; macromeres being covered by micromeres, and blastopore at vegetal pole. (*b*) Early trochophore of *Patella*; blastopore shifted forward. (*c*) Matured trochophore larva of *Patella*. (*d*) Veliger larva of *Patella*. (*e*) Late veliger stage of *Patella*. Twisting of body has carried mantle fold and position of anus forward and turned shell about. (*f*) Late blastula of tunicate embryo. (*g*) Beginning of gastrulation in tunicate embryo.

side the mouth will form the balancing organs. The 4d blastomere has cleaved into two cells located on either side of the mouth which are giving rise, at this stage, to two rows of mesoderm cells called the mesodermal bands.

Veliger stage. Within a week the shell gland has grown and begun to secrete the shell, and the foot is projecting prominently. The stomach increases in size and bulges into the shell cavity, and cells from the ends of the mesodermal bands form muscular attachments for the stomach and esophagus. The girdle of ciliated cells (velum) enlarges, and the rudiments of a nervous system, including eye cups, appear near the apical tuft. The larva is now called a veliger (Fig. 7*d*).

Metamorphosis. Following further development (especially of the alimentary tract, which becomes U-shaped with the mouth and anus separated from each other only by the foot), there is a sudden period of unequal growth in the two sides of the larva so that the anus is moved around to open on its neck (Fig. 7*e*). Eyes and tentacles have already been formed, and finally the young animal discards its cilium-bearing velum and takes up the adult habit of creeping about on its foot.

SEA URCHIN DEVELOPMENT

The first and second cleavage planes divide the fertilized egg (Fig. 3) into four equal blastomeres, intersecting each other at right angles through the animal-vegetal axis. The third cleavage cuts through these four blastomeres horizontally. The fourth cleavage plane divides the upper four cells into eight equal-sized mesomeres, and the lower group into four very small micromeres at the vegetal pole and four large macromeres (Fig. 8). These 16 blastomeres are each divided about equally at the fifth cleavage, forming 32 cells, but the eight micromeres fail to divide at the sixth cleavage, so that there are 56 instead of 64 blastomeres at this stage. By removing certain of these blastomeres (such as the micromeres) and following the later development, it has been found that the eight micromeres of the 56-cell stage give rise to the first group of mesenchyme cells, the ring of eight cells just above them produces mesenchyme and endoderm, and all the other blastomeres form ectoderm.

Blastulation. The blastomeres continue to divide successively, forming the hollow sphere of the blastula stage. By the tenth cleavage (Fig. 4*f*), each of the thousand or so blastomeres has developed a cilium, and the blastula has also secreted enough hatching enzyme to dissolve the fertilization membrane. After about 12 hr of development at 20°C, the larva begins its free-swimming period.

Gastrulation. Shortly afterward the larva elongates somewhat toward the animal pole, where a tuft of long, immobile cilia appears, and flattens on the vegetal side. Just before invagination begins, the cells descended from the eight micromeres slip out of the vegetal blastular wall into the blastocoele (Fig. 8*g*), forming the primary mesenchyme cells.

The gastrula stage begins about 20 hr after fertilization when the center of the vegetal wall bulges

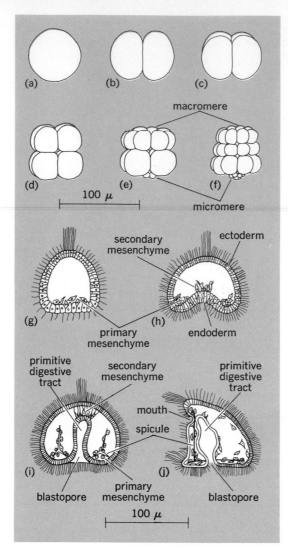

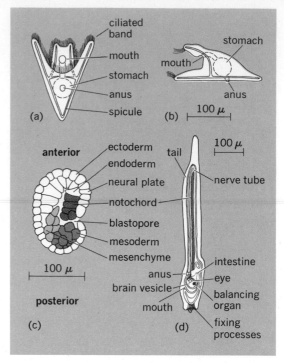

Fig. 9. Larval stages. (a) Ventral and (b) side view of pluteus larva of sea urchin. (c) Section through tunicate gastrula. (d) Tadpole stage of tunicate larva.

Fig. 8. Cleavage and gastrulation in the sea urchin. (a) Zygote. (b) Two-cell stage. (c) Four-cell stage. (d) Eight-cell stage. (e) Sixteen-cell stage. (f) Thirty-two–cell stage. (g) Vegetal region of blastula flattened; primary mesenchymal cells in blastocoele. (h) Beginning of invagination; secondary mesenchymal cells appearing in blastocoele. (i) Primitive digestive tract deepening; beginning of spicules. (j) Late gastrula, from side.

inward (Fig. 8h). Most of the cells which form this pocket will become endoderm, but there are also some mesenchymal cells among them (secondary mesenchyme) which will develop pseudopodia that stretch across the blastocoele and, making contact with the opposite body wall, direct the inward expansion of the primitive digestive tract (Fig. 8i). As this deepens, the primary mesenchyme cells are building calcareous spicules with calcium taken from the sea water. These skeletal rods extend in three directions from two points beside the blastopore; as they lengthen they determine the characteristic shapes of the larval stages (Fig. 8j).

Pluteus stage. In the pluteus stage (Fig. 9a and b), a mouth opening is formed where the tip of the primitive digestive tract joins the body wall. The tract begins to function, and the blastopore is changed into an anus as the larva is first able to take in food from outside. *See* DEUTEROSTOMIA.

Metamorphosis. During the month which the larva spends as a pluteus, the body increases markedly in size and two more pairs of arms are added. In preparation for metamorphosis, a structure called the echinus rudiment forms the beginnings of the adult organ systems. Within a relatively short time most of the larval body is incorporated into these organs, which are recognizable as those of a young sea urchin. Its metamorphosis is complete when it casts off the unusable parts of its larval skeleton (Fig. 10).

TUNICATE DEVELOPMENT

The fact that certain structures, characteristic of vertebrates and found in no other invertebrates, appear during the larval life of the tunicates forms the basis for giving these otherwise unprepossessing animals their high status at the top of the invertebrate subkingdom and makes them especially interesting from the evolutionary aspect.

Fertilization. The eggs of the tunicate *Styela* begin meiosis as they are laid, going as far as the metaphase of the first reduction division where they stop until they are fertilized. The spermatozoon penetrates the thick chorion, enters the egg at the vegetal pole, and stimulates it to proceed with meiosis. While the polar bodies are being given off, cytoplasmic streaming segregates the cell components into a yellow-pigmented region, a clear yolk-free region, and a gray yolky mass. It is possible to recognize these differently colored materials later in development and determine the role of each in body layer formation.

Cleavage. The first cleavage divides the egg into similar blastomeres. Because of the arrangement of the colored cytoplasm, its bilateral symmetry is already visible. The 16-cell stage consists of two layers of eight cells each (Fig. 4h), with the yellow cytoplasm contained in four of the vegetal cells. At the stage with about 40 cells a blastula is formed. The prospective ectoderm making up its animal side consists of thick columnar cells, while the future endoderm cells at the vegetal side are relatively flat (Fig. 7f). This difference is reversed before gastrulation begins (Fig. 7g).

Gastrulation. The gastrula is formed by the movement into the blastocoele of the vegetal-side cells, followed by an overlapping growth of the prospective ectoderm. Within this enveloping layer the yellow cells produce mesoderm; the other vegetal cells form endoderm. As the gastrula develops, the surface layer anterior to the blastopore (Fig. 9c) forms neural tissue which is organized into a brain and spinal cord, while the mesoderm beneath it forms a notochord, a precursor of the vertebral column characteristic of vertebrate animals. This notochord elongates as the axis of a tail, and the larva hatches from its chorion and begins a free-swimming stage.

Tadpole stage. During this stage (Fig. 9d), the tadpole acquires an extensive but nonfunctional digestive tract, two pairs of gill slits (also characteristic of vertebrates), a "cerebral eye," and a balancing organ. At its anterior end it has three papillae with which it will fix itself to a substratum when its short tadpole stage ends.

Metamorphosis. When metamorphosis begins (Fig. 11), the tail ectoderm contracts strongly, bending and breaking up the notochord, nerve cord, and tail muscles which are consumed by

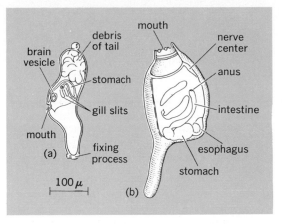

Fig. 11.   Metamorphosis in tunicate. (a) Beginning of metamorphosis. (b) Young adult.

phagocytes. The "chin" region next to the organ of fixation elongates greatly, carrying the mouth upward. A new nervous system replaces the larval one. The intestine elongates, fuses with the ectoderm to open as an anus, and forms a stomach and a liverlike gland.

A circulatory system which was started during the tadpole stage develops a muscular heart. Four new pairs of gill slits open into the pharynx. These later divide and give rise to further rows of smaller slits. The reproductive organs are formed from two masses of mesoderm cells lying near the pharynx. These develop into an ovary and a testis.

[JEAN C. DAN]

Invertebrate pathology

All studies having to do with the principles of pathology as applied to invertebrates. It includes studies of the disease process in its broadest sense, that is, investigations of any departures from the normal state of health. Disease in an invertebrate is initiated by the same factors causing disease in vertebrates: infection with microbes; parasites; and noninfectious factors. Various approaches are used in invertebrate pathology. These include (1) studies of the cause of disease, the symptomatology, the resulting pathology on a gross or microscopic level, and pathophysiology; (2) defense mechanisms used by the host to prevent infection, to overcome microbes and parasites, or to repair damaged body parts by wound healing or regeneration; and (3) the effects and progress of infectious disease in a population (epizootiology).

Pathology in arthropods other than insects and in all the remaining invertebrate groups has been poorly documented in the past except for consideration of a few economically or medically important species. However, the field is growing rapidly; invertebrates are used increasingly as models for study of the basic biochemistry and physiology of living things. In order to study the normal animal, researchers must establish what is normal and what procedures or conditions cause deranged physiology, genetic abnormalities, and other disease conditions in their experimental animals; also, they must determine how disease affects the general physiology of the organism. *See* INSECT PATHOLOGY.

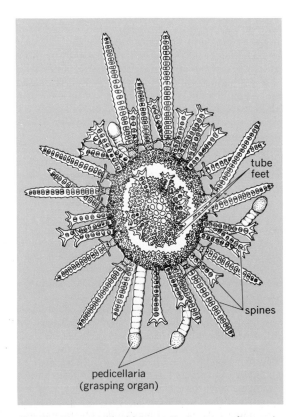

Fig. 10.   Young echinoid *Peronella japonica*, after metamorphosis. (*From K. Okazaki and J. C. Dan, Metamorphosis of partial larvae of Peronella japonica Morteusen, a sand dollar, Biol. Bull., 106:83–99, 1954*)

Approach and applications. One-celled animals, phylum Protozoa, are used for studies on the causes and processes involved in growth of neoplastic tumors. They also serve as some of the simplest models for demonstrating basic defense and immune reactions against microorganisms. Though the fruit fly (*Drosophila*) has long been the invertebrate most extensively used in studies of genetics, some protozoans are excellent substitutes for examining the transmission of lethal or damaging genes from parent to offspring. Protozoans have been used even in studies of aging and senescence.

Species from all the other major invertebrate phyla are being used in comparative pathology for the investigation of mechanisms of natural and acquired immunity, and also for studies of the processes of repair of tissues and organs injured chemically, by invading microorganisms, and through wounding, since much of this research is applicable to the situation in vertebrates as well. Attention is being given to growth-inhibiting substances extracted from the tissues and body fluids of mollusks. These substances not only inhibit growth of certain bacteria and viruses but also may delay growth of vertebrate tumor cells in tissue culture.

Outside of experimental biology, invertebrate pathology finds application to medicine, commercial fisheries, and agriculture. Biological control through microorganisms and parasites has been used successfully to reduce damaging insect populations. Its use to control other invertebrates adversely affecting man, domestic animals, food crops, and commercially important invertebrates is being actively investigated. As the world population grows and man turns increasingly to the sea for food, the need for understanding and control of the diseases and parasites of shellfish becomes apparent. Many studies in invertebrate pathology are directed to these ends. *See* INSECT CONTROL, BIOLOGICAL.

History. Insect pathology has a history extending from the time of Aristotle; however, the pathology of invertebrates other than insects did not come in for serious attention prior to the 1800s. Early papers on invertebrate pathology dealt largely with description of teratologies (abnormal body form). Discussion of diseased animals usually was included only incidentally in works on ecology, taxonomy, anatomy, and parasitology.

Two of the earliest reports of microorganisms which may have been related to disease were those of P. Duchassaing de Fonbressin and O. Schmidt, who both reported, in 1864, the occurrence in sponges of filaments which resembled fungi. E. Haeckel briefly described blood cells and blood clotting in crustaceans in 1857, and later observed the process of phagocytosis in the blood of the European crayfish; this was the first report of phagocytosis in any animal. In 1884 the pathologist E. Metchnikoff used the crustacean *Daphnia* and its fungus parasite *Monospora* in his classic experiments on phagocytosis. Also in the 1880s P. Geddes investigated the clotting mechanism of invertebrate coelomic and blood cells. During the 1890s A. Giard studied the phenomenon of parasitic castration in mollusks and arthropods, and A. Kowalevsky investigated the occurrence and function of phagocytic organs in scorpions, millipeds, spiders, and annelids. In the first quarter of the 20th century, S. S. Metalnikov, L. Cuénot, and J. Cantacuzène studied immune processes in invertebrates; R.-Ph. Dollfus, M. Caullery, F. Mesnil, and P. Grassé concerned themselves with the pathogenic effects of parasites; and the hematologist L. Loeb did his classic research on arthropod blood-clotting mechanisms.

Infectious disease in invertebrates other than insects was not a subject of serious investigation until the middle of the 20th century. Now biologists in the laboratories of the U.S. Bureau of Commercial Fisheries and in various universities and governmental agencies in the United States and abroad are making a particular study of invertebrate pathology, rather than incidentally studying diseased invertebrates in connection with other fields of interest. Together with the insect pathologists, those biologists who study epizootics, parasites, disease organisms, and environmental damage as they relate to shellfish furnish the first large group of invertebrate pathologists.

INFECTIOUS DISEASE

With the exception of some protozoan diseases of mollusks and common metazoan parasites of mollusks, annelids, and arthropods, the scattered and incomplete information on infectious disease in invertebrates other than insects makes generalizations about them premature.

Viral diseases. Diseases caused by viruses are well known in insects but have been reported only rarely in other invertebrate groups. However, invertebrate-infecting viruses may be more common than it appears, for the methods of study used with viruses seldom have been applied to diseased invertebrates other than insects.

Viral diseases have been reported from three acarine arthropods: the European red mite (*Panonychus ulmi*), the citrus red mite (*P. citri*), and a spider mite (*Tetranychus* sp.). The red mites suffer from disease caused by a noninclusion virus which apparently affects the nervous system, since diseased mites become paralyzed. Disorders in metabolism also occur; the body becomes filled with birefringent crystals which are not viral in origin. Although a virus has not been isolated from the spider mite, breeding experiments suggest that a factor like a virus, causing abnormal morphology, may be passed from one generation to another through the egg.

One other virus infection has been found in arthropods. Crabs taken from the French coast of the Mediterranean are sometimes infected with a noninclusion virus. As in the red mite, a major symptom is the development of paralysis.

Members of the Aschelminthes and Annelida probably also suffer from viral diseases. A filterable virus has been reported to cause fatal disease in larvae of a root-knot nematode (*Meloidogyne incognita*). Infected animals became sluggish and were filled with prominent oil-like droplets. Viruslike particles have been isolated from earthworms (*Enchytraeus fragmentosus*) which had a disease causing them to lyse, or disintegrate (Fig. 1). *See* VIRUS.

Bacterial diseases. Few bacteria are known to cause disease in invertebrates other than insects.

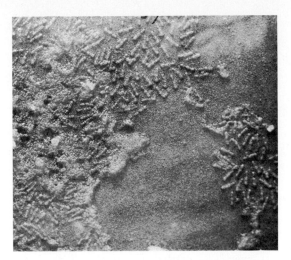

Fig. 1. Electron micrograph of rod-shaped particles, possibly a virus, found in material taken from diseased earthworms (*Enchytraeus fragmentosus*). (From E. A. Steinhaus and G. A. Marsh, *Report of diagnoses of diseased insects 1951–1961, Hilgardia, 33:349–490, 1962*)

The intracellular Rickettsiae are extremely common in insects but never have been identified positively from invertebrates other than arthropods.

Protozoans. Possibly some of the intracytoplasmic particles which occur in the protozoan *Paramecium* (kappa, lambda, nu, and sigma) are rickettsial in origin; some of these particles cause death in nonimmune paramecia. Bacteria, some of which were very small and may have been rickettsias, have been reported to cause fatal infections in soil amebas.

Certain bacteria, after they have gained entrance to an ameba, are thought to multiply in the food vacuoles. Others may infect the nucleus and cause it to enlarge greatly; or, by their presence in the nucleus and cytoplasm, they may inhibit cell division, resulting in giant forms with multiple nuclei. *See* RICKETTSIALES.

Mollusks. Bacteria have been suggested as causes of mortality in some oyster beds, but information is lacking as to their role in oyster disease. Freshwater snails occasionally are found infected with an organism related to *Mycobacterium* which causes abcesses and fibrous masses in various parts of the snail's body. The infection is chronic and seldom kills the snails. Artificial cultures of bivalve mollusk larvae are sometimes killed by bacteria present in the culture water.

Annelids. Earthworms being reared on worm farms in the southern United States sometimes suffer serious mortality due to a disease probably caused by *Bacillus thuringiensis* (which is used for microbial control of insects). Infected worms develop yellow blisters containing the bacteria. These blisters rupture, causing ulcers, and eventually the worm breaks into pieces and dies.

Crustaceans. Luminescent bacteria have been reported to cause death in ocean shrimp, and unclassified bacteria may kill small freshwater crustaceans. The importance of these bacteria as pathogens of crustaceans has not been studied.

There are two species of bacteria which cause disease and mortality in lobsters being held in pounds before sale. The first: a chitinivorous bacterium which extracts chitin from the exoskeleton, causes an infection called soft-shell disease; this bacterium does not attack soft tissues. Among the parts affected are the gill supports, which may be so damaged that infected lobsters suffer serious or total impairment of respiration and thus die from suffocation.

Gaffkemia, another lobster disease, is caused by the second organism, a very small gram-positive micrococcus called *Gaffkya homari*. In the lobster pounds of New England and the Maritime provinces of Canada, epizootics of gaffkemia may cause almost complete mortality of the lobsters. *Gaffkya homari* has been found in about 50% of the wild lobsters taken from various points along the North Atlantic coast. The lack of disease symptoms in wild lobsters suggests that environmental stress is necessary for the disease to manifest itself.

Fungus diseases. Parasitic or disease-causing fungi have been reported to occur in all the major invertebrate phyla except the Coelenterata. Almost all fungi parasitizing invertebrates other than insects belong to the group Phycomycetes and mainly to the order Chytridiales. Nematode-trapping fungi, which are predacious rather than being parasitic or disease-causing, include one basidiomycete and many deuteromycetes as well as members of the Phycomycetes. Three of the most devastating diseases of invertebrates economically important to man are caused by fungi. *See* PHYCOMYCETES.

Fungi commonly attack small animals, such as rotifers, the eggs and immature stages of parasitic worms, and various arthropods and mollusks, especially those being reared in laboratory cultures. The result of this type of fungus infection is usually complete replacement of egg or body contents with hyphae of the growing fungus.

Protozoans. Many protozoans are hosts of fungi which attack the nucleus or the cytoplasm, often causing great nuclear enlargement and abnormalities in cell division. Fungi of the genera *Sphaerita* and *Nucleophaga* are not uncommon in amebas which are endoparasitic (parasitic within the body) in mammals.

Poriferans. Since 1864 it has been known that certain sponges contain filaments that are not a normal part of the animal. Present knowledge indicates that the filaments are hyphae of fungi, and several epizootics in sponges have been associated in retrospect with the fungus *Spongiophaga communis*. In 1939 Caribbean sponges, many of which were fished commercially, underwent a serious epizootic of wasting disease caused by *Spongiophaga*, which destroyed the sponge skeleton as well as killing the animal. Within 5 years, because of the disease and overfishing of the remaining healthy sponges, the Caribbean sponge industry had collapsed. Years after the height of the epizootic dead sponges still littered the ocean floor. Only in the 1950s did sponges begin to reappear in quantity. The spread of wasting disease in the sponge beds of the Gulf of Mexico closely followed the pattern of the water currents which transported the causative fungus.

Nematodes. The predacious species known as nematode-trapping fungi are a most interesting group. Their trapping activities were first de-

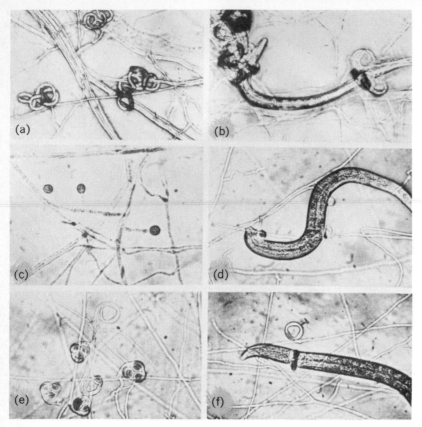

Fig. 2.   Nematode-trapping fungi. (a) *Arthrobotrys conoides* forms networks of hyphal loops coated with a sticky substance. (b) Loops of *A. conoides* capture nematodes by adhesion. (c) *Dactylella drechsleri* produces spherical knobs on short stalks. (d) Knob surfaces of *D. drechsleri* are adhesive and trap nematodes on contact. (e) Constricting rings formed by *A. dactyloides* are the most highly developed organelles of capture; each is composed of three cells. (f) Nematode is trapped by *A. dactyloides*. Nematode entered the ring and the cells swelled rapidly, gripping the worm so that it could not escape. (*From D. Pramer, Nematode-trapping fungi, Science, 144:382–388, 1964*)

scribed in 1888, and now over 50 species representing three major fungal groups are known to be predacious. The method of trapping small nematodes varies according to the species of fungus, but all are dependent either on the principle of adhesion or that of occlusion (Fig. 2). The hyphal loops of *Arthrobotrys conoides* capture nematodes with their sticky coating much as flies are caught on fly paper. The adhesive traps of *Dactylella drechsleri* are sticky knobs borne on short extensions of the hyphae. They are sometimes referred to as "lethal lollipops." *Arthrobotrys dactylloides* has specialized hyphal loops which are stimulated to swell when a nematode enters them and thus hold the worm so tightly it cannot escape. All the fungi which trap nematodes can live on other food. When kept in pure culture without contact with nematodes, many of the species do not form organelles of capture; if nematodes are added to the culture, the organelles are differentiated. The protein nemin, which was isolated from tissues of nematodes, is the active principle causing trap formation. Some workers believe nematode-trapping fungi may be useful in the biological control of plant-feeding nematodes which damage crops.

Mollusks. A fungus disease of oysters caused by *Labyrinthomyxa marina* (formerly *Dermocystidium marinum*) is of great importance in the oyster-growing areas of the South Atlantic and Gulf states. This disease often occurs in epizootic form during the summer months. The major symptom of *L. marina* infection is a drastic weight loss, averaging 33% of the total body weight. In the period of a single summer, in combination with other factors, *L. marina* can destroy virtually all the seed oysters planted along the Louisiana coast. Since this fungus requires high salinity, oyster growers can reduce losses by establishing their oyster beds farther up estuaries in areas of low salinity. *L. marina* may be a specific parasite of the oyster unable to live in other animals or free in the water. No other fungi are known to cause major disease in mollusks, though fungus disease sometimes causes freshwater snails to lose their shells.

Annelids. There are two reports of possible fungus disease in annelids. On occasion individuals of two species of earthworms are found dead and filled with fungus hyphae. Whether the fungi or another factor kills the worm is not known.

Arthropods. Arthropods other than insects are infected by many types of fungi. Mites and ticks are killed by the species of *Beauveria* which also cause disease in insects. Infections of unidentified fungi in spiders and millipedes lead to the production of intersexes or prevent the host from attaining all its normal adult characters. Spiders are killed by fungi of the genus *Cordyceps*, which fill the body with hyphae and, after the death of the spider, produce a stroma, or stem, bearing the fruiting body. This stroma grows out of the spider's body and is often the only visible sign of infection. Pelagic copepods, which are small crustaceans drifting in the ocean currents, are subject to fungus disease. In the summer of 1950 *Eurytemora hirundoides*, the most common summer copepod of the inner Baltic Sea, was attacked by a saprolegnid fungus. It died in such numbers that the nets of herring fishermen were clogged with dead copepods.

The fungus-caused epizootic of most far-reaching economic impact was that called Krebspest or crayfish disease. During the 1860s a disease appeared which destroyed almost all the crayfish in the streams and lakes of northern Italy. During the next 40 years Krebspest spread throughout continental Europe and to Sweden and England, and even today prevents large-scale culture of European crayfish. Before the epizootic, Germany exported 100,000 crayfish per year; by 1881 crayfish exports dropped to practically nothing. This long-lasting epizootic was transported from lake to lake by infected shipping containers used in crayfish export. Not until the 1930s was the fungus *Aphanomyces mystaci* isolated, studied, and proved to be the cause of Krebspest. *A. mystaci* enters the crayfish through the exoskeleton and grows along nervous tissue. One of the symptoms is paralysis, and it is thought that the fungus produces a toxin that acts on the nervous system.

Algal diseases. Parasitic algae have been reported in two groups of echinoderms, the sea urchins and starfish. Some of the parasitic algae live in the tissues and may kill the infected animal or cause abnormal spine development. Others are confined to the epidermis, where they cause green patches. Destructive algae have not been reported in other invertebrates.

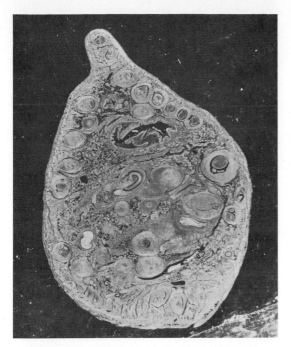

Fig. 3. Cross section of littleneck clam (*Venerupis staminea*) showing numerous encysted tapeworms (*Echeneibothrium*) larvae throughout tissues. (*After A. K. Sparks and K. K. Chew, Gross infestation of the littleneck clam, Venerupis staminea, with a larval cestode (Echeneibothrium sp.), J. Invertebr. Pathol., 8:413–416, 1966*)

Protozoan diseases. Parasitic protozoans are known from members of almost all the invertebrate phyla. While some cause overt disease in animal populations, most are of scattered occurrence; although effects on their hosts may be severe, they do not cause death. A common result of infection with protozoans is the phenomenon known as parasitic castration. Although the gonads themselves may not be infected, they fail to develop either because nutriment usually used for gonad maturation is taken by the parasites or because somehow the parasite disrupts the hormone system of the host. *See* PARASITIC CASTRATION.

Protozoans. Protozoans themselves have protozoan parasites. If the host protozoan is parasitic, the infecting species is known as a hyperparasite. A majority of hyperparasites are microsporidians of the class Sporozoa. Practically nothing is known of their effect on the host. *See* SPOROZOA.

Coelenterates. Hydramoeba hydroxena is a parasitic ameba which causes epizootics in the freshwater coelenterate *Hydra. Hydramoeba hydroxena* is the only protozoan known to be pathogenic to coelenterates.

Worms and related organisms. Parasitic flatworms and nematodes have a number of microsporidian hyperparasites. Those which parasitize the immature stages of worms often prevent infected larvae from reaching the adult stage. In the adults, species of *Nosema* and related genera may cause parasitic castration. Rotifers, which are related to nematodes, also are parasitized by microsporidians. The main symptoms are usually paralysis and enlargement of certain body parts. Other sporozoans, called gregarines, are common parasites of annelids, both earthworms and marine polychaetes. Generally these parasites do not cause severe damage to their hosts.

Mollusks. Despite the number of protozoans known to occur in mollusks, except for oysters their effects on the host have not been studied. Most protozoan oyster parasites, such as the flagellate *Hexamita* and the sporozoan *Nematopsis*, cause relatively little damage. However, two sporozoans, *Minchinia costalis* and *M. nelsoni* (the latter causes MSX disease), are believed to be important causes of death in oysters along the mid-Atlantic coast of the United States. The resulting pathology in their hosts has yet to be studied in detail.

Arthropods. Among the arthropods other than insects, protozoans are reported only rarely to have fatal effects on their hosts. Some species of *Nosema* may kill or cause serious injury to their crab and fresh-water crustacean hosts; *N. steinhausi* kills the tyroglyphid mite *Tyrophagus noxius*. The ellobiopsid *Thalassomyces* attacks and injures the nervous system of its shrimp host; ellobiopsids are a group considered to be fungi by some and protozoans by others.

Echinoderms. Of the remaining invertebrate phyla, only the echinoderms are known to have consistently damaging protozoan parasites. The common starfish *Asterias* is infected with a ciliate, *Orchitophrya stellarum*, which causes parasitic castration and is thought responsible for population declines in starfish.

METAZOAN PARASITES

Multicelled parasites seldom kill their invertebrate hosts though they may cause severe injury. As with protozoan parasites, a common result of infection with metazoan parasites is parasitic castration of the host.

Although not actually parasitic, sponges may bore into the shells of mollusks, making pathways for entrance of bacteria and fungi which can kill the animal; other sponges which grow on oysters actually smother them. *See* BORING SPONGES.

Species of trematodes are noted for causing parasitic castration in snails and bivalve mollusks. Cestodes may infect mollusks so heavily that they cause serious interference mechanically with the life processes of the host (Fig. 3). Sometimes nematodes disrupt host body function severely by completely filling the body cavity.

Knowledge of fossil parasites of any variety is limited to descriptions of scars on fossil crinoids (sea lilies, members of the Echinodermata) which are assumed to have been caused by very small annelid parasites called myzostomes. Myzostomes attack modern crinoids; another annelid group, the leeches, causes injury in mollusks by sucking body fluids.

Many parasitic barnacles and isopods which attack crustaceans profoundly change the metabolism and development of their hosts. Parasitic castration or the production of intersexes is a common result of infection. Some isopods disrupt the host hormone system so that a male host becomes feminized in structure. The crab-parasitizing barnacle *Sacculina* is the best studied of the group. The larva penetrates the host exoskeleton and lives in the blood sinuses, entirely within the body. When ready to molt to the adult, it again penetrates the exoskeleton to live as an ectoparasite (exterior

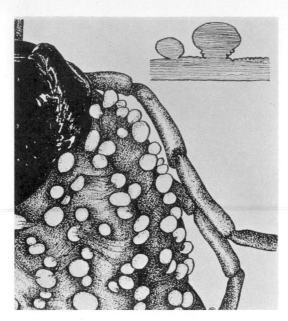

Fig. 4. Portion of a tick, *Hyalomma* sp., dorsal aspect, showing milky-white elevations, or papules. Inset is section of a papule from the ventral integument showing that it consists of chitin. (*From E. A. Steinhaus, Principles of Insect Pathology, Academic, 1949*)

parasite) with an extensive system of roots within host tissues. *Sacculina* interferes with molting, causes parasitic castration, and changes the basic proportions of fats and carbohydrates present in the crab's blood.

Sciomyzid flies, which are members of the insect order Diptera, "parasitize" freshwater snails. Actually the larvae of these flies slowly eat their hosts, and thus are predators not parasites. There are indications that the Sciomyzidae might be used as agents for biological control of snails which serve as intermediate hosts for human blood flukes.

NONINFECTIOUS DISEASES

Invertebrates are subject to pathological conditions caused by mechanical injury or wounding; deranged physiology or metabolism; physical and chemical injury caused by factors in the environment such as temperature changes, dessication, and noxious chemicals; developmental (congenital) and genetic abnormalities; accidents in regeneration; deficiency disease caused by imbalanced or insufficient nutriment; aging; and

tumors. Tumors in invertebrates are the subject of intensive research with the aim of discovering their similarities to, and differences from, vertebrate neoplasms. Many invertebrate tumors are caused by a reaction of host tissue to the presence of a parasite or other microorganism. Papules, such as those illustrated in Fig. 4, are of unknown origin and perhaps qualify as tumors. However, they may have been due to an infectious process, although microorganisms were not associated with their presence. *See* ONCOLOGY; TUMOR.

MASS MORTALITY AND EPIZOOTICS

Mass mortality is a term applied here to the sudden demise of an entire animal population in the sea. Common causes are physical or chemical, such as increased water temperature due to volcanic action, lack of oxygen in the water, devastating storms, and "red water," or "red tide," which causes death through the toxic metabolites released by rapidly multiplying populations of dinoflagellates. Epizootics, on the other hand, are due to the presence in an animal population of infectious disease caused by microorganisms. Information on epizootics in invertebrates other than insects is limited to those affecting animals utilized by man; some of these epizootics were discussed above. Sometimes it is difficult to distinguish the sudden death of a population due to epizootic disease from a mass mortality due to environmental factors. Figure 5 illustrates the fact that certain populations of marine invertebrates may suddenly disappear in a period of only a few days or weeks for unknown causes.

DEFENSE MECHANISMS AND IMMUNITY

Cells which occur in the blood and coelomic fluid and wander in the tissues are responsible for destroying or limiting the spread of intruding parasites and other microorganisms. These cells also close off wounds by clotting and sometimes enter into tissue repair. Phagocytosis by cells, variously called leukocytes or phagocytic amebocytes, is one of the important mechanisms used by invertebrates for removal of small foreign bodies or microorganisms present in the tissues. The phagocytic cells also join together to encapsulate parasites, microorganisms, and injured tissue, and they may congregate in infected areas in a process sometimes referred to as leukocytosis.

Phagocytosis, encapsulation, and leukocytosis are the most easily demonstrated reactions to microorganisms. However, as well as performing these functions, leukocytes and related cells release substances which lyse or otherwise destroy microorganisms. So far as is known, antibodies such as those found in vertebrates are not produced. Under some circumstances, however, an invertebrate infected with a microorganism can develop a heightened response against the invader, as indicated by increased powers of phagocytosis or lysis. Substances which kill microorganisms are normally present in the body fluid of many invertebrates, so that defense against infecting organisms includes both cellular and humoral factors. *See* ANNELIDA; ANTIBODY; ARTHROPODA; ECHINODERMATA; IMMUNITY; MOLLUSCA; NEMATODA; PLATYHELMINTHES; PORIFERA.

[PHYLLIS T. JOHNSON]
Bibliography: F. B. Bang (ed.), Defense reac-

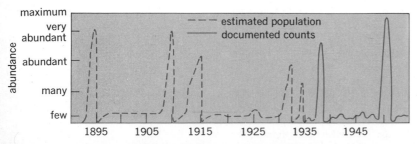

Fig. 5. Mortality in populations of the bean clam (*Donax gouldi*) at La Jolla, Calif., 1895–1950. Abundance defined as follows: "few," less than 5 per square meter; "many," 5–20; "abundant," more than 100; "very abundant," more than 1000; "maximum," more than 10,000. (*From W. R. Coe, Ecology of the bean clam Donax gouldi on the coast of Southern California, Ecology, 36:512–514, 1955*)

tions in invertebrates: A symposium, *Fed. Proc.*, 26:1664–1715, 1967; T. C. Cheng, *Marine Molluscs as Hosts for Symbiosis*, vol. 5 of F. S. Russell (ed.), *Advances in Marine Biology*, 1967; C. G. Huff, Immunity in invertebrates, *Physiol. Rev.*, 20:68–88, 1940; L. H. Hyman, *The Invertebrates*, vol. 1–6, 1940–1967; P. T. Johnson, *Annotated Bibliography of Pathology in Invertebrates Other than Insects*, 1968; B. Scharrer and M. S. Lockhead, Tumors in the invertebrates: A review, *Cancer Res.*, 10:403–419, 1950.

Involute

A term applied to a curve C' that cuts at right angles all tangents of a curve C (see illustration). Each curve C has infinitely many involutes and the distance between corresponding points of any two involutes is constant. If $x_i = x_i(s)$, with $i = 1, 2,$

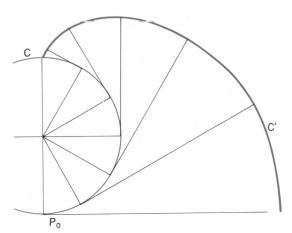

An involute C' of curve C.

3, are parametric equations of C, with parameter s arc length on C, all involutes C' of C have parametric equations $X_i = x_i(s) + (k-s)x'_i(s)$, with $i = 1, 2, 3$, where $x'_i = dx_i(s)/ds$, with $i = 1, 2, 3$, and k denotes an arbitrary constant. Let a length of string be coincident with a curve C, with one end fastened at a point P_0 of C. If the string is unwound, remaining taut, the other end of the string traces an involute C' of C. By varying the length of the string, all involutes of C are obtained. *See* ANALYTIC GEOMETRY.

[LEONARD M. BLUMENTHAL]

Involutional melancholia

A severe depressive reaction, occurring at the time of the female climacteric and during the corresponding age period in men. It is characterized by symptoms of sadness, anxiety, guilt, and low self-esteem, similar to the depressed states of manic-depressive psychosis. Patients often have hypochondriacal and nihilistic delusions; a combination with paranoid reactions is not infrequent. Usually such patients do not exhibit any intellectual deterioration. *See* MANIC-DEPRESSIVE PSYCHOSIS.

The etiology of the disorder is unclear. Endocrine processes seem to be of importance, but detailed evidence for such an assumption does not exist. Organic mechanisms are probably similar to the depressive states in manic-depressive psychoses. Psychological observations have shown that most involutional depressions react to real and imaginary deprivation of love and esteem with severe depression. Often a deprivation of a similar type can be detected in an analysis of the patient's relationship with his mother during his infancy and early childhood.

In the differential diagnosis, organic reactions, presenile disorders, paranoid states, and schizophrenia need to be considered. The differential diagnosis of neurotic and reactive depression may be very difficult. *See* PARANOID STATE; SCHIZOPHRENIA.

Electric convulsive treatment, tranquilizing drugs, and psychic energizers, in combination with psychotherapy, are the usual methods of treatment. In mild cases, electric convulsive treatment can and should be avoided. Hospitalization is usually indicated, and the prevention of suicide and self-destructive behavior is particularly important. *See* PSYCHIC ENERGIZER; PSYCHOTHERAPY; TRANQUILIZER.   [FREDRICK C. REDLICH]

Bibliography: F. C. Redlich and D. X. Freedman, *The Theory and Practice of Psychiatry*, 1966.

Iodate

A negative ion having the formula IO_3^-, and derived from iodic acid, HIO_3. Some salts such as $KH_2I_3O_9$ and $KH(IO_3)_2$ indicate that the acid may exist as polymers of the ion indicated by the empirical formula. Sodium and potassium iodates are the most important salts and are used in medicine.

Iodates occur along with $NaNO_3$ in Chile saltpeter. They are prepared in an electrolytic reaction similar to that for the preparation of chlorates or by oxidation of iodides with chlorine.

The iodates are more stable and are weaker oxidizing agents than bromates and chlorates. *See* BROMATE; CHLORATE; IODINE.

[E. EUGENE WEAVER]

Iodide

A compound which contains the iodine atom in 1− oxidation state and which is derived from hydriodic acid, HI.

The chemistry of iodine and its ability to form covalent and ionic iodides is very similar to the properties described for chloride. *See* CHLORIDE.

In comparing the iodide ion with the other halide ions, it should be pointed out that the iodides are more covalent, are the best reducing agents of the group, and form the least stable complexes. The aqueous solubilities of the metal iodides are much the same as the chlorides but in general a little lower. In organic solvents the order of solubility is frequently reversed. Bismuth and mercuric iodides are only slightly soluble. Iodide ion combines with free iodine to form the triodide ion, I_3^-.

Sodium or potassium iodide is added to table salt to prevent malfunction of the thyroid gland. Silver iodide is used in photographic films and papers.

Iodide can be detected in solution by oxidizing it to the free element with chlorine. It imparts a violet color to the solvent when extracted into carbon tetrachloride. *See* COMPLEX COMPOUNDS; HALIDE; HALOGENATED HYDROCARBON; IODINE; PHOTOGRAPHIC MATERIALS; THYROID GLAND DISORDERS.   [E. EUGENE WEAVER]

Iodine

A nonmetallic element, symbol I, atomic number 53, relative atomic mass 126.9045, the heaviest of the naturally occurring halogens, which form group VIIa of the periodic table. Under normal conditions iodine is a black, lustrous, volatile solid; it is named after its violet vapor. *See* HALOGEN ELEMENTS.

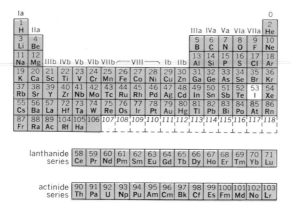

The chemistry of iodine, like that of the other halogens, is dominated by the facility with which the atom acquires an electron to form either the iodide ion I^- or a single covalent bond —I, and by the formation, with more electronegative elements, of compounds in which the formal oxidation state of iodine is +1, +3, +5, or +7. Iodine is more electropositive than the other halogens, and its properties are modulated by: the relative weakness of covalent bonds between iodine and more electropositive elements; the large sizes of the iodine atom and iodide ion, which reduce lattice and solvation enthalpies for iodides while increasing the importance of van der Waals forces in iodine compounds; and the relative ease with which iodine is oxidized. Some properties of iodine are listed in the table. *See* ASTATINE; BROMINE; CHLORINE; FLUORINE.

Occurrence, isolation, and production. Iodine occurs widely, although rarely in high concentration and never in elemental form. The average concentration in the Earth's crust is estimated to be 0.3 parts per million (ppm). Despite the low concentration of iodine in sea water (about 0.05 ppm), certain species of seaweed (for example *Laminaria*) can extract and accumulate the element.

In the form of calcium iodate, iodine is found in

Some important properties of iodine

Property	Value
Electronic configuration	$[Kr]4d^{10}5s^25p^5$
Relative atomic mass	126.9045
Electronegativity (Pauling scale)	2.66
Electron affinity, eV	3.13
Ionization potential, eV	10.451
Covalent radius, —I, Å	1.33
Ionic radius, I^-, Å	2.12
Boiling point, °C	184.35
Melting point, °C	113.5
Specific gravity (20/4)	4.940

the caliche beds in Chile, which at one time were the major commercial source of the element; the iodine content is 0.02−1%. In the isolation process, sodium iodate, extracted from the caliche into solution, is treated with the exact amount of sodium bisulfite needed to reduce all of the iodate to iodide, as shown in Eq. (1). The resulting acid

$$2IO_3^- + 6HSO_3^- \rightarrow 2I^- + 6SO_4^{2-} + 6H^+ \qquad (1)$$

solution is then mixed with just enough fresh mother liquor to effect precipitation of iodine, according to Eq. (2). The product is purified by sublimation.

$$5I^- + IO_3^- + 6H^+ \rightarrow 3I_2 + 3H_2O \qquad (2)$$

Oxidation of iodides is the other way in which iodine is prepared. In 1812 the element was discovered during the treatment of calcined ash of kelp (containing iodides) with concentrated sulfuric acid (an oxidizing agent). In France, the Soviet Union, and Japan, where alginic products are isolated from seaweed, quantities of iodine are similarly obtained as a valuable by-product. Iodine also occurs as iodide ion in commercially exploitable concentrations (typically 40 ppm) in some oil well brines in California, Michigan, and Japan; the extraction processes generally utilize chlorine as the oxidant, and the volatile iodine is blown out of solution in a stream of air.

Nuclear properties. The sole stable isotope of iodine is ^{127}I (53 protons, 74 neutrons), whose properties allow the investigation of iodine derivatives by nuclear quadrupole resonance spectroscopy and Mössbauer spectroscopy. Of the 22 artificial isotopes (masses between 117 and 139), the most important is ^{131}I, with a half-life of 8 days. It finds wide application in radioactive tracer work and in certain radiotherapy procedures. *See* TRACER, RADIOACTIVE.

Physical and chemical properties. Iodine exists as diatomic I_2 molecules in solid, liquid, and vapor phases, although at elevated temperatures (> 200°C) dissociation into atoms is appreciable. Short intermolecular I . . . I distances in the crystalline solid indicate strong intermolecular van der Waals forces; consistent with this picture, iodine has a band gap of 1.3 eV (comparable with some forms of phosphorus and selenium, which are semiconductors) and under high pressure exhibits the electrical characteristics of a metal.

Iodine is moderately soluble in nonpolar liquids [for example, CCl_4, aliphatic hydrocarbons], and the violet color of the solutions suggests that I_2 molecules are present, as in iodine vapor; I_4 aggregates have also been detected in some concentrated solutions. Other liquids (aromatic hydrocarbons, alcohols, ethers, amines, organic sulfides) dissolve larger amounts of iodine to give deep red or brown solutions. The increased solubility and change of color result from charge transfer interactions, in which a solvent molecule donates electron density to a vacant orbital of the I_2 molecule. Although the bonding involved is rather feeble, many such crystalline adducts of definite composition have been isolated [for example, $N(CH_3)_3, I_2$], and some complexes even persist in the vapor phase [such as C_6H_6, I_2]. Charge transfer complexes are often important intermediates in halogenation reactions.

The reactions of iodine in water are usually rapid and are governed by the reduction potentials given in Fig. 1. Iodine is only slightly soluble in water, and the brown solution is acid because of a disproportionation reaction, shown in Eq. (3). In

$$I_2 + H_2O = HOI + H^+ + I^- \qquad (3)$$

$$K_{eq} = \frac{[H^+][I^-][HOI]}{[I_2(aq)]} = 2 \times 10^{-13}$$

alkaline solution, extensive conversion of iodine to iodide and hypoiodite ions may be postulated, as in Eq. (4), but the IO^- ion is so unstable as to have

$$I_2 + 2OH^- = I^- + IO^- + H_2O \qquad (4)$$

$$K_{eq} = \frac{[I^-][IO^-]}{[I_2(aq)][OH^-]^2} = 30$$

eluded detection by conventional methods; disproportionation to iodate and iodide, as in Eq. (5), is

$$3IO^- = 2I^- + IO_3^- \qquad (5)$$

$$K_{eq} = \frac{[I^-]^2[IO_3^-]}{[IO^-]^3} = 10^{20}$$

kinetically and thermodynamically very favorable, and the overall alkaline hydrolysis is shown in Eq. (6). That the reverse reaction, namely the com-

$$3I_2 + 6OH^- \rightarrow IO_3^- + 5I^- + 6H_2O \qquad (6)$$

bination of iodate and iodide to form iodine, as in Eq. (2), occurs in acid solution demonstrates the importance of pH in aqueous iodine chemistry.

Treatment of iodine in alkaline solution with strong oxidizing agents (for example, hypochlorite) generates iodate, which by an excess of oxidant is converted to the +7 oxidation state as periodate. By contrast, oxidation of iodine in anhydrous, very acidic media (for instance, oleum, fluorsulfuric acid) produces variously the cations IO_2^+, IO^+, I_3^+, and blue paramagnetic I_2^+, all of which have been isolated as salts. While the discrete I^+ cation does not exist, complexes with pyridine and thiourea have been prepared respectively as $(py)_2I^+I^-$ and $[(H_2N)_2CS]_2I^+I^-$; the $N-I-N$ and $S-I-S$ units are linear and symmetrical.

Reduction of iodine (to I^-) in aqueous solution is more difficult than reduction of the other halogens, and strong reducing agents are necessary. Hydrogen sulfide, H_2S, effects the transformation, as does the thiosulfate ion, $S_2O_3^{2-}$, which is oxidized by iodine to tetrathionite $S_4O_6^{2-}$, as in Eq. (7); this

$$2S_2O_3^{2-} + I_2 \rightarrow S_4O_6^{2-} + 2I^- \qquad (7)$$

reaction is used in the analytical determination of iodine. Iodide ion is readily oxidized to iodine by moderate oxidants such as bromine; in acid solution it is slowly oxidized by atmospheric oxygen.

Although it is usually less vigorous in its reactions than the other halogens, iodine combines directly with most elements. Important exceptions are the noble gases, carbon, nitrogen, and some noble metals. The readiness with which reaction occurs depends critically on the conditions, especially temperature, phase, and the presence of impurities. Typical organic reactions of iodine include: electrophilic substitution of aromatic compounds to form aryl iodides; iodination of the carbon atom adjacent to a carbonyl function; and addition of I_2 across the multiple bonds of unsaturated hydrocarbons. Hydrogen iodide, a frequent by-product, may inhibit these reactions unless it is continuously removed.

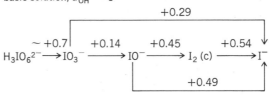

Fig. 1. Standard reduction potentials for iodine species in aqueous solution, E^0 in volts.

Inorganic compounds. The inorganic derivatives of iodine may be grouped into three classes of compounds: those with more electropositive elements, that is, iodides; those with other halogens; and those with oxygen. The last two classes, especially, show close analogies with the chemistry of neighboring elements tellurium and xenon.

Iodides. The properties of iodides depend on the identity of the more electropositive element. Iodine rarely brings out the highest oxidation state in its partner: thus PI_3 and ReI_4 are known, but not PI_5 or ReI_7. Conversely, low oxidation states may be stabilized as iodides, for example, the +2 oxidation state of the lanthanide elements.

Iodides of the more electropositive metals, groups Ia and IIa, are typical ionic solids with three-dimensional or layer lattices; they dissolve in water with dispersion of iodide ion. Lithium iodide is soluble in ether, a phenomenon often attributed to covalency but actually due to a favorable balance of solvation and lattice energies.

Nonmetals give iodides which exist as covalent molecules in solid, liquid, and vapor, for example, BI_3; similar derivatives are found for metals in high oxidation states, for instance, TiI_4. Typically, they are low-melting solids and are rapidly hydrolyzed by water with evolution of hydrogen iodide, as in Eq. (8). The colorless gas HI (bp

$$PI_3 + 3H_2O \rightarrow H_3PO_3 + 3HI \qquad (8)$$

$-35.3°C$) can also be prepared by direct combination of the elements over a platinum catalyst; its solution in water (called hydriodic acid) may contain up to 70% by weight of HI.

Iodides of the less electropositive metals, such as HgI_2, or metalloids, such as SbI_3, have layer or chain structures in the solid state, using partially covalent bonds between the atoms. They are usually insoluble in water, but sublime readily on heating and dissolve in organic solvents as molecular species. The decomposition of some iodides into the elements at high temperature has been utilized in the van Arkel process for the production of metals and metalloids of very high purity (for instance, titanium, hafnium, and silicon).

Iodide ion forms complex anions in solution with many metal and metalloid cations, and with other

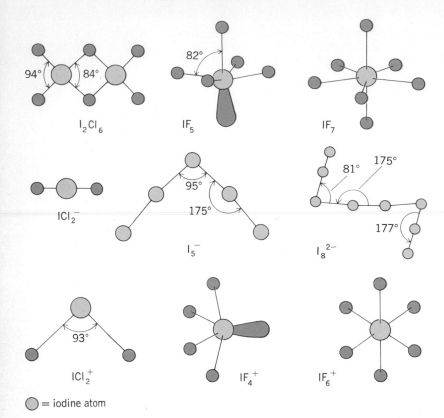

○ = iodine atom

Fig. 2. The shapes of some interhalogen molecules and polyhalide ions.

at the center of these molecules. Detailed knowledge of their structures, summarized in Fig. 2, has been crucial to the development and refinement of theories of molecular structure and bonding. The prevalence of angles close to 90 and 180° shows the clear relation to noble-gas compounds. *See* MOLECULAR STRUCTURE AND SPECTRA; XENON.

Polyiodide ions such as I_5^- or I_8^{2-}, isolable as black crystalline salts of large cations, are best treated as complexes of I_2 and I^-. A charge transfer complex involving polyiodide anions and starch has an intense blue-black color, and starch-iodide mixtures are used as indicators in titrations involving iodine.

Oxygen compounds. The +5 oxidation state is the most stable. Solid iodic acid contains $(HO)IO_2$ molecules joined by hydrogen bonds. Dehydration of the acid is reversible and occurs in two stages, giving at 100°C a hemihydrate, HI_3O_8, and at 200°C iodine pentoxide, I_2O_5. Strong intermolecular $I \cdot \cdot O$ bonding is noted in all these compounds. Acid solutions of iodates are strong oxidizing agents. *See* IODATE.

White crystalline periodic acid is a genuine ortho acid, $(HO)_5IO$. It is a weak acid, and the anions present in its solutions include $H_4IO_6^-$, $H_3IO_6^{2-}$, $H_2IO_6^{3-}$, $H_2I_2O_{10}^{4-}$, and IO_4^-; the latter is found only in strong alkali. Salts of all these anions have been isolated, as well as derivatives such as Ag_3IO_5, Na_5IO_6, and $K_4I_2O_9$. The adoption of six-coordination about iodine in periodates is attributable to the large size of iodine, which allows high coordination numbers.

The periodates are powerful oxidizers in acid solution, converting manganese(II) to permanganate; the cleavage of 1,2-diols by periodic acid (Fig. 3) is stereospecific to cis units, and is a valuable tool in the chemistry of carbohydrates and nucleic acids. Aqueous solutions of periodic acid slowly evolve oxygen and ozone, while the salts lose oxygen on heating to form iodates or, occasionally, novel paramagnetic derivatives of iodine(VI). *See* PERIODATE.

The intermediate acids HIO and HIO_2 are very unstable, but two lower oxides are known: I_2O_4 is characterized as iodosyl iodate $IO^+IO_3^-$, and I_4O_9 as $I(IO_3)_3$. Several unstable covalent compounds have been prepared, as in Eqs. (13) and (14), which

$$I_2 + S_2O_6F_2 \rightarrow 2IOSO_2F \qquad (13)$$

$$I_2 + 6ClOClO_3 \rightarrow 2I(OClO_3)_3 + 3Cl_2 \qquad (14)$$

are formally iodine(I) or iodine(III) esters of strong oxyacids, and include a nitrate, $IONO_2$, a perchlorate, $I(OClO_3)_3$, and fluorosulfates $IOSO_2F$ and $I(OSO_2F)_3$. The relation to interhalogen compounds is clearly shown by the square planar IO_4 unit of the $[I(OClO_3)_4]^-$ anion (compare ICl_4^-). Iodine triacetate, $I(OCOCH_3)_3$, made from iodine, acetic acid, and fuming nitric acid as oxidant, undergoes electrolysis with the generation of iodine (trapped as silver iodide) at a silvered platinum cathode.

Organic compounds. Organoiodine compounds fall into two categories: the iodides; and the derivatives in which iodine is in a formal positive oxidation state by virtue of bonding to another, more electronegative element. The chemistry of all these compounds is summarized in Fig. 4.

Iodides. The simple organoiodides resemble the

acceptor molecules; the complexes are often deeply colored, and are readily isolated as salts, such as K_2HgI_4, $KAgI_2$, $[(C_6H_5)_4P]_2TeI_6$, and $[(CH_3)_4N]BCl_3I$.

Compounds with other halogens. The simple binary compounds ICl and IBr are low-melting solids with halogenlike properties; at ordinary temperatures they are somewhat dissociated into the parent halogens in the vapor phase. Iodine trichloride crystallizes as a dimer I_2Cl_6, but is apparently completely dissociated (into ICl and Cl_2) in the vapor. Of the fluorides, IF_5 and IF_7 are reactive, moisture-sensitive, thermally stable, and respectively liquid and gas under normal conditions; intermediate fluorides IF and IF_3 disproportionate to iodine and IF_5 well below room temperature, as in Eq. (9).

$$5IF \rightarrow 2I_2 + IF_5 \qquad (9)$$

Most interhalogen compounds react under suitable conditions to gain or lose halide ions, giving a series of anions and cations, shown in Eqs. (10)–(12). Other species include $ClIBr^-$, ICl_4^-, IF_6^-,

$$I_2 + I^- \rightarrow I_3^- \qquad (10)$$

$$IF_7 + NOF \rightarrow [NO]^+[IF_8]^- \qquad (11)$$

$$IF_5 + SbF_5 \rightarrow [IF_4]^+[SbF_6]^- \qquad (12)$$

ICl_2^+, and IF_6^+. The heaviest halogen always lies

Fig. 3. Mechanism of oxidation of an α-glycol by periodate.

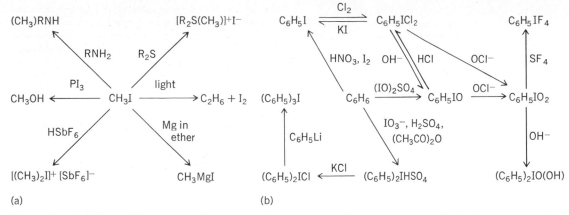

Fig. 4. The chemistry of the organoiodine compounds. (a) Reactions of iodomethane. (b) Reactions of those organoiodine compounds in which iodine is in a positive oxidation state.

other halides in their properties. Because the C-I bond is the weakest of the carbon-halogen bonds, organoiodides are the least stable and most reactive of the organohalides: they are thus useful intermediates. The large mass of the iodine atom makes iodides relatively dense and involatile.

Iodoalkanes are made from the corresponding alcohol with either HI or PI$_3$. Iodomethane, so prepared, is a colorless liquid [bp 42.5°C, $d(20/4)$ 2.28] which decomposes in daylight to iodine and ethane; it reacts (Fig. 4a) with, for example, primary amines, dialkyl sulfides, and magnesium (yielding a Grignard reagent). Iodoarenes may be made from the hydrocarbon, iodine, and nitric acid. *See* GRIGNARD REACTION; HALOGENATED HYDROCARBON; HALOGENATION.

Positive oxidation states. An extensive chemistry of such compounds exists for iodine (Fig. 4b) but not for chlorine or bromine. Even for iodine, alkyl derivatives are unstable; an electronegative organic group (CF$_3$, C$_6$H$_5$) is necessary for a stable compound.

A simple alkyl iodine(III) compound, dimethyliodonium(III) hexafluoroantimonate, has been prepared from iodomethane and hexafluoroantimonic acid. It is a potent methylating agent and unstable except in very electrophilic media. The related diphenyliodonium cation, (C$_6$H$_5$)$_2$I$^+$, is rather stable, and many of its salts may be made by double decomposition reactions; the hydroxide is a strong base and ionic, but the chloride, (C$_6$H$_5$)$_2$ICl, crystallizes as a chlorine-bridged dimer, resembling I$_2$Cl$_6$.

The iodoso compounds RIO and iodoxy compounds RIO$_2$ are white solids, sparingly soluble in water, and powerful oxidizing agents which explode on heating. Iododichlorides RICl$_2$ have found some application as chlorinating agents. The CIF$_4$ skeleton of C$_6$H$_5$IF$_4$ is shaped like IF$_5$, but with a carbon atom in the unique axial position.

Biological importance. Iodine appears to be a trace element essential to animal and vegetable life. Iodide and iodate in sea water enter into the metabolic cycle of most marine flora and fauna, while in the higher mammals iodine is concentrated in the thyroid gland, being converted there to iodinated amino acids (chiefly thyroxine and iodotyrosines). They are stored in the thyroid as thyroglobulin, and thyroxine is apparently secreted by the gland. Iodine deficiency in mammals leads to goiter, a condition in which the thyroid gland becomes enlarged. *See* THYROID GLAND (VERTEBRATE).

Uses. The bactericidal properties of iodine and its compounds bolster their major uses, whether for treatment of wounds or sterilization of drinking water. Also, iodine compounds are used to treat certain thyroid and heart conditions, as a dietary supplement (in the form of iodized salt), and for x-ray contrast media. *See* ANTIMICROBIAL AGENTS; ANTISEPTIC; NUTRITION; SALT (FOOD).

Major industrial uses are in photography, where silver iodide is a constituent of fast photographic film emulsions, and in the dye industry, where iodine-containing dyes are produced for food processing and for color photography. Two important catalysts are NiI$_2$ (for the addition of CO to organic compounds) and TiI$_4$ (for the production of stereospecific polymers). *See* DYE; PHOTOGRAPHIC MATERIALS.   [CHRIS ADAMS]

Bibliography: F. A. Cotton and G. Wilkinson, *Advanced Inorganic Chemistry*, 3d ed., 1972; A. J. Downs and C. J. Adams, *The Chemistry of Chlorine, Bromine, Iodine, and Astatine*, 1974.

Iodoform

A yellow, hexagonal solid with a penetrating odor also called triiodomethane, CHI$_3$. Its specific gravity is 4.08 and melting point 119°C. It is prepared by the action of iodine in a basic solution, NaOI, on ethanol or acetone or by the electrolysis of an alkaline I$_2$-KI solution in the presence of ethanol or acetone. It serves as a qualitative test for the groups

$$CH_3-\underset{\underset{OH}{|}}{\overset{\overset{H}{|}}{C}}- \quad \text{and} \quad CH_3-\underset{\underset{O}{\|}}{C}-$$

It is soluble in organic solvents and insoluble in water. It has weak bactericidal properties and exerts antiseptic action when applied to raw wounds, because of the liberation of free iodine. Iodoform also acts as an inhibitor for wound secretion, and this inhibits bacterial growth. Its chief use is in ointments for minor skin diseases. It is toxic when taken internally. *See* ANTIMICROBIAL AGENTS; HALOGENATED HYDROCARBON.

[ELBERT H. HADLEY]

Bibliography: G. L. Jenkins et al., *Chemistry of Organic Medicinal Products*, 4th ed., 1957; C. R. Noller, *Textbook of Organic Chemistry*, 3d ed., 1966.

Ion

An atom, or group of atoms, which by loss or gain of one or more electrons has acquired an electric charge. If the ion is formed from an atom of hydrogen or an atom of a metal, it is usually positively charged; if the ion is formed from an atom of a nonmetal or from a group of atoms, it is usually negatively charged. The number of electronic charges carried by an ion is called its electrovalence. The charges are denoted by superscripts which give their sign and number; for example, a sodium ion, which carries one positive charge, is denoted by Na^+; a sulfate ion, which carries two negative charges, by SO_4^{--}. *See* ATOMIC STRUCTURE AND SPECTRA; CHEMICAL BINDING.

Salts are usually composed of orderly arrangements of ions which are not free to move easily in the solid. However, when the salt is fused or dissolved in water, the ions become free, and when an electric field is applied to the salt in solution, the positively charged cations move to the cathode and the negatively charged anions move to the anode. At the electrodes the ions lose their electric charge. This process is called electrolysis. *See* ELECTROLYSIS; IONIC CRYSTALS; SALT (CHEMISTRY); VALENCE.

[THOMAS C. WADDINGTON]

Ion exchange

A chemical reaction in which mobile hydrated ions of a solid are exchanged, equivalent for equivalent, for ions of like charge in solution. The solid has an open fishnetlike structure, and the mobile ions electrically neutralize charged, or potentially charged, groups attached to the solid matrix, called the ion exchanger. Cation exchange occurs when the fixed charged groups (functional groups) of the exchanger are negative; anion exchange occurs when the immobilized functional groups are positive. Ion exchange is used for water softening, water demineralizing, separation of substances, and purification of chemicals.

Equilibria for ion-exchange reactions are usually expressed by the generalized equations (1*a*) for cation exchange and (1*b*) for anion exchange, with

$$a B_{(s)}^{+b} + b A_{(r)}^{+a} \rightleftharpoons a B_{(r)}^{+b} + b A_{(a)}^{+a} \qquad (1a)$$

$$a B_{(s)}^{-b} + b A_{(r)}^{-a} \rightleftharpoons a B_{(r)}^{-b} + b A_{(s)}^{-a} \qquad (1b)$$

the corresponding mass action as in expression Eq. (2). Brackets indicate the concentrations of the

$$K_{AB} = \frac{[B^{+b}]_{(r)}^{a}[A^{+a}]_{(s)}^{b}}{[A^{+a}]_{(r)}^{b}[B^{+b}]_{(s)}^{a}} = \frac{[B^{-b}]_{(r)}^{a}[A^{-a}]_{(s)}^{b}}{[A^{-a}]_{(r)}^{b}[B^{-b}]_{(s)}^{a}} \qquad (2)$$

exchanging ionic species A and B (neglecting activity coefficient effects for simplicity), a and b represent their charges, r and s designate the resin (or exchanger) and solution phases, and K_{AB} corresponds to the ion-exchange equilibrium constant. *See* EQUILIBRIUM, CHEMICAL.

Differences often exist between the distribution of various ions between the solid and liquid phases (that is, K values differ for selected pairs of ions). Usually these differences can be enlarged by selectively complexing ions in solution and by repeating the ion-exchange reaction many times. This technique, ion-exchange chromatography, is the basis for many ion-exchange separation procedures that are employed by the analytical chemist. The elements to be separated are sorbed from solution in a narrow band at the top portion of an ion-exchange resin column and are then washed down the column (eluted) by passing a selective complexing agent through the column until separation is achieved. The ion-exchange reaction for each ion is repeated many times by this operation and in combination with the effect of the complexing agent, results in the separation of the individual elements into bands in the column. *See* CHROMATOGRAPHY.

Other important applications of ion exchange are water softening and the demineralization of water. In the first application, calcium, magnesium, and other undesirable cationic impurities are exchanged for sodium ion during percolation of the water through an ion-exchange resin column in the sodium form. In the second, all ionic impurities contained in water are removed by exchange for hydrogen and hydroxyl ions during passage of the water successively through cation- and anion-exchange resins in the acid and base form. *See* WATER SOFTENING.

Zeolites. These are natural clay minerals which exhibit ion-exchange properties. These materials have an open three-dimensional network structure (the alumino-silicate skeleton) containing regular channels to imbibe water. Cations which neutralize the excess negative charges on their skeleton diffuse in and out of the water contained in the uniform interstices of the zeolite crystals during an ion-exchange reaction.

The exchange capacities of these materials are usually expressed as the number of milliequivalents of cations per 100 g of dry material which are exchangeable in neutral solution. Their exchange capacities, as well as their capabilities for exchange, differ considerably. In the more compact structures, exchange is limited to the smaller univalent hydrated ions, and in some instances no exchange can occur. In the case of orthoclase (feldspar), $K_2O \cdot Al_2O_3 \cdot 6SiO_2$, for example, the potassium ions are imprisoned and unavailable for exchange.

Other complicating features in zeolites are their instability in acid or alkaline systems, the presence of different sites with varying binding strengths and rates of exchange, and their uncertain composition.

Artificial zeolites of well-defined composition are now manufactured with perfectly uniform interstices of controlled dimensions and with regularly distributed charged sites which demonstrate equivalent ion-exchange behavior. For a discussion of these materials *see* MOLECULAR SIEVE.

Other interesting inorganic ion-exchange materials with both cation- and anion-exchange properties have been prepared. Examples of these compounds are zirconium phosphate and zirconium oxide, which, when prepared by precipitation from solution and dried at elevated temperatures, behave as cation and anion exchangers, respectively. Presumably, the precipitation of polymeric constituents in solution provides a network with a fixed residual charge. Mobile ions within this network neutralize this charge.

Ion-exchange resins. These are made up of three-dimensional organic networks, including charged or potentially charged groups which are neutralized by mobile ions of opposite charge. Freedom of these mobile, or counter, ions to move in and out of the resin is provided by water imbibed by the resin on immersion in an aqueous solution. The water opens the resin structure, permitting diffusion of ions into and out of the resin-water (gel) phase during ion exchange.

Many natural products exhibit cation-exchange properties. For example, the cell contents of plant tissue are characterized by ion-exchange structure because of the presence of carboxylic acid groups. Other acidic units found are ethereal sulfate groups ($—C—OSO_3Na$) in certain seaweed carbohydrates, phosphoric acid groups in nucleic acid, weakly acidic phenolic $—OH$ groups, and thiol, $—SH$, groups in proteins. Proteins, being amphoteric, behave as anion exchangers in acid systems by virtue of amine, $—NH_2$, side groups.

Since 1935 organic ion exchangers have been synthesized by incorporating functional groups in natural products such as coal, lignin, and peat. An example of this type of organic exchanger is the product from sulfonation of soft coal. The sulfonic acid, $—SO_3H$, functional groups are attached to the aromatic matrix of the coal. Also, cationic and anionic groups, respectively, have been incorporated in cellulose fabrics by phosphorylation with urea and phosphoric acid and by reaction with 2-aminoethylsulfuric acid to provide ion-exchange properties.

The synthetic ion-exchange resins were first prepared by the reaction of polyhydric phenols with formaldehyde, the weakly acidic phenolic groups providing cation-exchange properties to the product. Cation-exchange resins containing strongly acidic sulfonic acid groups were prepared later by the condensation of phenols and formaldehydes in the presence of sodium sulfite. Anion-exchange resins were prepared by the polymerization of amines with formaldehyde.

The most recent technique for the production of ion-exchange resins consists of first forming the polymer unit, followed by incorporation of the functional ionic group. For example, the polymerization of styrene produces linear polystyrene chains. These are held together (cross-linked) by divinyl benzene to produce a network structure. The sulfuric acid groups are then attached to this network by sulfonating with concentrated sulfuric acid. Quaternary amines may be attached to the same matrix by an analogous treatment (chloromethylation of the copolymer followed by reaction with a tertiary amine).

These resins, by far the most popular, are offered commercially with various bead sizes (mesh) and with different porosity or cross-linking (percentage of divinylbenzene). The degree of cross-linking controls their swelling properties. The low cross-linked resins swell to many times their dry volume in aqueous solutions, while the highly cross-linked resins show little volume change. In the highly cross-linked resin the more rigid network structure resists the sizable osmotic pressure due to the concentrated salt solution contained in the resin-gel phase. As a result of such constriction, ion exchange occurs more slowly because of increased resistance to the movement of hydrated counter ions in the resin-gel phase. Differences in resin affinities between pairs of ions are also affected—increasing with increased cross-linking and thereby compensating for slowing the exchange process.

High porosity in an ion-exchange resin results in easier accessibility in ions of high molecular weights, thereby providing a basis for their effective separation from smaller ionic species.

The control of resin porosity possible in this method of manufacture permits the research worker to select the most suitable tool for his specific needs. In addition, these resins are chemically and physically stable and possess high capacity.

The functional units most often employed in synthetic cation-exchange resins are sulfonic, phosphonic, carboxylic, and phenolic groups. The sulfonic resins are strongly acidic as one would expect from their normal behavior in aqueous systems. The different exchange behavior of sodium ion with a carboxylic and sulfonic acid resin illustrates this acidity function. Sodium ion in a neutral solution of sodium chloride will not exchange appreciably with the hydrogen ion of a carboxylic resin in the acid form since the $—CO_3H$ groups are not appreciably dissociated. Only in an alkaline system can exchange occur readily between sodium ion and hydrogen ion. The continuous removal of hydrogen ion by hydroxyl ion to form water necessitates continuous replenishment of the small amounts of hydrogen ion normally replaced by sodium ion. In the case of the sulfonic acid cation exchanger, the $—SO_3H$ groups are essentially completely dissociated, and exchange between sodium ion and hydrogen ion can occur normally in neutral systems.

The functional groups for the synthetic anion-exchange resins are usually amines. When primary, secondary, and tertiary amines are incorporated, the exchanger is weakly basic since these amines are strongly associated in aqueous systems. If the functional group is a quaternary amine, the exchanger behaves like a strong base.

The incorporation of other functional groups to produce unusually selective ion-exchange resins has been attempted with some success. For example, thiol, $—SH$, groups have been used to provide a resin which exhibits a special affinity for metals that form mercaptans; incorporation of a chelating group, *m*-phenylene diglycine, produces a resin with a high selectivity for the transition elements.

The presence of immobilized functional groups and the neutralization of their charge by mobile ions of opposite polarity provide the resin with a preponderance of mobile ions of one charge. Even in the presence of moderately concentrated electrolyte solution, electrolyte diffusion (invasion) into the resin-gel phase does not alter this situation. As a result, there is selective transfer of the mobile counter ions under the influence of an electrical potential gradient. Ion-exchange membranes, coherent sheets of ion-exchange resin, have been prepared to take advantage of this characteristic property. *See* ION-PERMEABLE MEMBRANE.

A principal application of ion-exchange membranes has been the development and improvement of electrodialysis techniques of separation. For example, by arranging cation- and anion-exchange membranes in an alternating pattern between two electrodes, the imposition of a potential gradient will result in the concentration and dilution of electrolyte contained in the neighboring compartments. Cationic species are selectively transferred through the cation-exchange membranes, and anionic species are selectively transferred through the anion-exchange membranes to achieve this result. For a discussion of this separation technique *see* DIALYSIS.

Insoluble liquid cation and anion exchangers have been prepared and studied. The ion-exchange behavior of these analogs of the resins described above is remarkably similar to that of their cross-linked relatives. Successful application of their selective exchange properties for affecting separations by solvent extraction techniques is a certainty.

Ion exclusion. By this process, the penetration of soluble electrolyte into the resin-gel phase is prevented. The distribution of electrolyte between two phases, in this case the resin and external solution, is given by the Donnan equilibrium, Eq. (3),

$$[A^{+a}]^b_{(r)}[B^{-b}]^a_{(r)} = [A^{+a}]^b_{(s)}[B^{-b}]^a_{(s)} \qquad (3)$$

where the symbols are the same as employed in Eq. (2). This behavior of ion-exchange resins is derived from the preponderance of mobile ions in the gel phase that electrically neutralize the immobilized charged functional groups attached to the resin.

At moderate external electrolyte concentrations, mobile ions of one charge already exist in large quantity within the resin, so that only minor quantities of electrolyte can diffuse from the external solution to satisfy Eq. (3). At high electrolyte concentrations, significant electrolyte invasion should and does occur.

This behavior of resins is the basis of a special separations process in which electrolyte is excluded from the ion exchanger while nonpolar materials are not. For example, salt can be separated from glycerin by the selective sorption of the nonpolar glycerin.   [JACOB A. MARINSKY]

Bibliography: C. B. Amphlett, *Inorganic Ion Exchangers,* 1964; F. Helfferich, *Ion Exchange,* 1962; J. A. Marinsky (ed.), *Ion-Exchange,* vol. 1, 1966; F. C. Nachod and J. Schubert (eds.), *Ion Exchange Technology,* 1956; O. Samuelson, *Ion Exchange Separations in Analytical Chemistry,* 1963.

Ion implantation

A process of introducing impurities into the near-surface region of solids by directing a beam of energetic ions at the solid (Fig. 1). When ions of sufficient energy are directed toward a crystal surface, they will penetrate the surface and slow to rest within the solid. Ion implantation differs from the normal thermal equilibrium means of introducing atoms into a solid, which is usually carried out either by adding the elements during growth of a solid from a liquid or vapor phase, or by diffusing the atoms into the solid at elevated temperatures. The advantages of the implantation process are: precise control of the type of impurity to be introduced, the amount of impurity introduced, and the impurity distribution in depth. In addition, since any atoms can be added to any solid, mixtures can be formed which would not normally be found in nature, that is, systems which are not in thermal equilibrium. Most of the work in developing and understanding the ion implantation processes has taken place since the middle 1960s. A major area of application of ion implantation has been to semiconductor device fabrication, where the process is a standard technique. Additional applications have been developed in metals and insulators, and research in ion implantation has been undertaken for the purpose of developing scientific understanding and technological applications. *See* CRYSTAL DEFECTS; CRYSTAL GROWTH; DIFFUSION IN SOLIDS; ION.

Producing an ion beam. Any atom can be implanted into a solid. An ion implantation system is shown schematically in Fig. 2. First the atom must be ionized, which is the process of changing the number of electrons associated with an atom and thus leaving it with a net positive or negative charge. This is done in an ion source, usually by a plasma discharge. As the ions are formed, they are continuously extracted from the plasma and accelerated through a voltage difference V. This gives each ion an energy $E = qV$, where q is the charge of the ion. Typical acceleration voltages are 10,000 to 500,000 V. The ion beam is then passed through a transverse magnetic field so that the different mass ions of energy E will be deflected by different angles, and in this way ions of particular energy and mass are selected. Varying electric fields are often used to sweep the selected ion beam laterally so that any given area can be uniformly implanted with the ions. The region between the ion source and the solid target is kept under vacuum since the ions would travel only very short distances in air. The total number of ions incident on the target is determined by measuring the current to the sample during ion implantation. By integrating this current to obtain the total ion charge and using the known charge per ion, the number of ions implanted in the target can be precisely controlled. *See* ION SOURCES; MASS SPECTROSCOPE.

Implantation process. As the energetic ions penetrate a solid, they encounter electrons and atoms with which they have collisions. In this manner they transfer their energy to the electrons and atoms in the solid until they finally come to rest. In the process of transferring energy to electrons and atoms, radiation damage is produced in the target. For example, in a single crystal target some atoms will be given enough energy to be

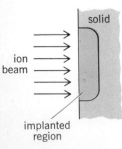

Fig. 1. Ion implantation into a solid.

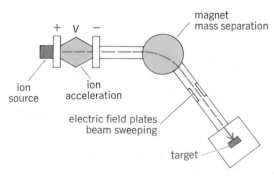

Fig. 2. Ion implantation system.

knocked off normal lattice sites, and the ion thus leaves a track of damage as it comes to rest. Each of the collision events usually results in a loss of only a small fraction of the ion energy, and in many cases the collisions can be treated as statistically independent events. In this case, the final profile of the ions after they come to rest will be very close to a Gaussian distribution in depth. The concentration of implanted atoms n as a function of depth x can then be described by the equation shown,

$$n = \frac{N}{\sqrt{2\pi}\,\Delta R_p} \exp\left[-\left(\frac{x - R_p}{\sqrt{2}\,\Delta R_p}\right)^2\right]$$

where N is the total number of atoms implanted per unit surface area (Fig. 3). In this description there are two important parameters which describe the final distribution of the ions which have been implanted into the solid; R_p is called the projected range of the ion, and ΔR_p is referred to as the spread in this projected range. For a given ion energy, ion mass and target material R_p and ΔR_p have unique values, which can be predicted theoretically. For example, boron ions implanted into silicon with an energy of 100,000 eV would have a projected range of $R_p = 290$ nm and a range spread of $\Delta R_p = 71$ nm. This description of the composition in the implanted region is important in understanding the changes which result in an ion-implanted solid. *See* RADIATION DAMAGE TO MATERIALS.

Applications. The ion implantation process changes the chemical composition of the near-surface region of a solid and introduces radiation damage into this region. These changes provide the possibility of modifying an extremely wide range of near-surface physical and chemical properties of solids. Ion implantation is most widely used in the controlled doping of semiconductors for the microelectronics industry. Here ion implantation is used instead of diffusion to dope semiconductors chemically, for the fabrication of such things as resistors, diodes, and MOS transistors. The advantage of implantation over diffusion is its ability to control very precisely the number of impurities that are introduced, and to form depth distributions of the impurities which may not be easily obtainable by diffusion. A complication introduced by implantation is the introduction of radiation damage, but this has been overcome for most applications by heating the semiconductor material to sufficiently high temperatures. The compatibility with the planar technology used in forming integrated circuits has allowed rapid acceptance of ion implantation in semiconductor device manufacturing. Examples of use include the formation of more accurate and higher-value resistors within smaller areas, and more precise control of the gate voltages of MOS transistors. The selected use of ion implantation to form better devices in terms of their type of operation, speed, reliability, and precision has been an important contribution to the steady advance of microelectronics technology. *See* DIODE, SEMICONDUCTOR; INTEGRATED CIRCUITS; SEMICONDUCTOR; TRANSISTOR.

Another area of application is ion implantation of metals. One important use of ion beams has been to simulate neutron damage anticipated in

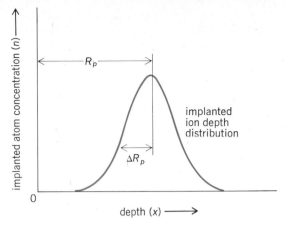

Fig. 3. Implanted ion depth profile.

reactors. In short times, one can approximately simulate many of the radiation effects anticipated in reactors which have not yet been built or for which many years of reactor irradiation time would be required. This has been particularly important in the search for more swelling-resistant alloys for use in breeder power reactors. In other uses, implantation of certain ions has been shown to bring about desirable changes such as increased oxidation and corrosion resistance, increased superconducting transition temperatures, and improved catalytic activity of surfaces. Also ion implantation has been useful for determining basic metallurgical parameters such as the solubility and diffusivity of impurities in alloys formed by ion implantation. Most applications to metals require the introduction of many more ions than for semiconductors, so that the final impurity-atom concentrations may be as high as 1 to 10% of the target, in contrast to semiconductor applications where impurity concentrations in the parts per million are usually desired. *See* CATALYSIS; CORROSION; REACTOR, NUCLEAR; SUPERCONDUCTIVITY.

Changes in properties of insulators can also be accomplished by ion implantation. One potential application involves a class of materials known as magnetic garnets. Films of these materials are used for storing and moving reversed magnetic domains called magnetic bubbles and can be used for memories in much the same way that certain forms of large silicon integrated circuits are used. The implantation process allows better control of the nature and paths of motion of magnetic bubbles. Implantation in optical materials involving both insulators and semiconductors may find important applications as the technology of integrated optics is developed. Here the processing and handling of light signals is envisioned in much the same way that electrical signals are stored and processed in integrated circuits. *See* COMPUTER MEMORY; FERRIMAGNETIC GARNETS.

[SAMUEL T. PICRAUX]

Bibliography: G. Dearnaley et al., *Ion Implantation*, 1973; J. W. Mayer, L. Eriksson, and J. A. Davies, *Ion Implantation in Semiconductors*, 1970; F. F. Morehead and B. L. Crowder, Ion implantation, *Sci. Amer.* 228(4):64–71, 1973; S. T. Picraux, E. P. EerNisse, and F. L. Vook, *Applications of Ion Beams to Metals*, 1974.

Ion propulsion

Vehicular motion caused by reaction from the high-speed discharge of a beam of electrically equally charged minute particles. The charged particles, or ions, are accelerated in an electrostatic field produced within the vehicle that is propelled. Because a given polarity of charge cannot be ejected from the vehicle over an extended period without producing a substantial opposite polarity of charge on the vehicle, the vehicular charge must be removed as part of the propulsive process. Thus separate beams of positive and negative electric charges are ejected and recombined behind the vehicle to eliminate electric fields beyond the outlets. A general representation of a nuclear-powered electrostatic propulsion engine is shown in Fig. 1. The three principal portions of the engine are power generation, propellant feed, and thrust device. Power generation can be broken down into power source, power-conversion equipment, and power-conditioning equipment.

Power source. The power source can be a nuclear reactor or a solar-radiation energy collector. In the first case, protection of the electrical equipment (and of personnel in the case of manned vehicles) is required against neutron and γ-radiation; protection of the main radiator against radioactivity induced in the working fluid by the reactor is also necessary. Effective neutron-shielding materials are rich in hydrogen or carbon or both. In the solid material class these are metal hydrides, particularly lithium hydride and boron carbide. For systems with large electrical-power outputs (megawatt range), boron carbide results in lower neutron-shield weights. For smaller power outputs (tens to hundreds of kilowatts), the smaller shield thickness attainable with metal hydrides contributes to greater compactness of the power package. Typical solid materials for γ-radiation shields are titanium, stainless steel, tungsten, and aluminum; tungsten is advantageous where high radiation intensities and high resulting operational temperatures are involved. *See* NUCLEAR SPECTRA.

To simplify the power-conversion equipment, it is desirable to use the reactor coolant as working fluid driving a turbine. However, this is only possible if the working fluid is relatively inert, such as a noble gas (argon). If metal vapor is used, radioactivity is induced by the reactor, leading to a contamination of turbine, pipelines, and radiation cooler. This spreading of radioactivity would make shielding difficult and would be weight-consuming. Therefore, two metal cycles are preferable, as indicated in Fig. 1. The first cycle carries reactor heat to a heat exchanger in which the metal of a second cycle is evaporated to drive the turbine. For this case, the γ-shield must be located on the far side of the heat exchanger to provide protection from the γ-radiation of the first-cycle metal as well as of the reactor. If only a single cycle is used, a heat exchanger is not required, and the γ-shield can be located directly behind the neutron shield.

If solar radiation is the power source, no radiation shield is required. Solar radiation can be used to provide electric power directly, through photovoltaic solar cells, or indirectly, through a solar heater in which the nuclear reactor is replaced by a solar-radiation collector and a heat exchanger. Only one fluid cycle is required. Solar cells are a simple and useful power source for propelling small probes in the inner solar system and, perhaps, as far out as Jupiter (with some reservations). Their principal disadvantage is that they deliver electric power at low voltage. Power-conditioning equipment, to achieve the high voltage required for ion propulsion, is a major obstacle in attaining low power-specific weights (unit weight per unit watt) even though considerable advances were made during the late 1960s in reducing the weight of solar cells proper. *See* SOLAR BATTERY.

Greater thrusts can be obtained with solar collectors. However, except for intra-Venus space, solar energy is dispersed so much that its use as a power source requires large collectors to concentrate an adequate amount of solar energy (for power levels of hundreds or thousands of kilowatts) on the heat exchanger. Disadvantages of solar heating include requirement of collector alignment with the Sun in addition to thrust alignment in a

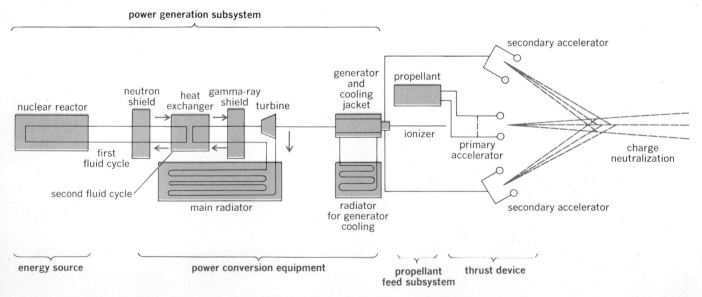

Fig. 1. Schematic presentation of nuclear-powered electrostatic propulsion system.

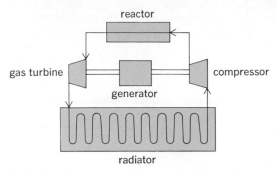

Fig. 2. Brayton gas cycle power-generation system.

different direction, the loss of solar power in the shadow of a celestial body, and the difficulty of collector maintenance in the presence of inclement space environment, such as micrometeorites. The latter are also a problem with solar cells. Because of the reduction of solar power with the square of solar distance and because space in the outer solar system appears to be considerably more filled with dust and micrometeorites, the principal field of mission application of solar-powered ion propulsion systems is the inner solar system and the asteroid belt.

Power conversion. The best developed power converter so far is the turbine generator. Gas cycle and vapor cycle may be distinguished on the basis of the working fluid. The two most interesting cycles, distinguished on the basis of the conversion mechanism, are the Brayton gas-turbine cycle and the Rankine cycle.

The Brayton cycle appears most advantageous for small power plants (up to a few hundred thousand electron volts) using a gas cycle; the preferred fluid is argon, a heavy inert gas. A typical Brayton gas cycle is shown in Fig. 2. Heat is transferred from the reactor to high-pressure gas; the gas is then expanded in a turbine which drives a generator and compressor. Upon leaving the turbine, the gas is cooled in a radiator, then recompressed and fed back into the reactor. *See* BRAYTON CYCLE.

In the direct Rankine cycle a liquid is used for heat transfer from the reactor, leading to evaporation with subsequent expansion of the vapor in the turbine. Because this would spread radioactivity, the indirect Rankine cycle provides for a heat exchanger between a primary all-liquid cycle and a secondary liquid-vapor cycle. However, the uncertainties in the boiling heat-transfer rate under weightless conditions in space have made the intermediate-link Rankine cycle (Fig. 3) appear most attractive; evaporation does not occur in the heat exchanger but in a boiler separator. The saturated liquid from the heat exchanger is expanded to high velocity in the boiler separator. A portion of the liquid evaporates. In subsequent spiral ducting, liquid and vapor are separated by centrifugal effects. The vapor is ducted to the turbine and the liquid is returned to the heat exchanger, together with liquid from the radiator in which the vapor is condensed and cooled following expansion in the turbine. Suitable working fluids are mercury for smaller power plants and sodium or rubidium for larger ones. *See* RANKINE CYCLE.

The radiator is the largest and heaviest compo-

nent of the power-generation package. Its purpose is to radiate excess heat into space. This heat is the result of limited efficiency of energy conversion. The efficiency of conversion, defined as ratio of electrical power output in kilowatts (ekw) to thermal power output in kilowatts (kw) of the reactor is 6–7% for a 3-ekw system using mercury in the Rankine cycle, and 20–23% for a 1000- to 20,000-ekw system using rubidium or sodium.

Figure 4 shows a survey of the achieved and expected conversion efficiencies of various cycles and methods. The thermal power output of the reactor must, therefore, be 4–16 times larger than the ekw output. The energy difference must be radiated into space. For the gas cycle the radiator area required is at least 9–10 ft²/ekw at 1300–1400°F radiator-inlet temperature, whereas for mercury and sodium the values are 3–4 ft²/ekw at 1300–1400°F, 1 ft²/ekw at 2000°F, and 0.8 ft²/ekw at 2300°F. Depending on the requirements for radiator design, temperature (steel above 1000°F, titanium 450–1000°F, aluminum below 450°F), and micrometeor protection, the radiator weight will vary greatly. An aluminum design with tubes ducting the working fluid will weigh around 1.5 lb/ft², the tubes being protected by a bumper (outer shell) against micrometeorite penetration. Aluminum radiators for gas cycles weigh 0.5–0.7 lb/ft² with meteor bumper. For steel the corresponding values are 1.1 and 1.5 lb/ft² for tube-fluid arrangements and gas-cycle radiators, respectively. For large radiators of 10,000–20,000 ft², a weight of 4–5 lb/ft², including long-term micrometeor protection by bumpers, must be expected. Thus radiators represent the heaviest and least certain component of the propulsion package.

Alternate methods of power conversion are semiconductors and thermionic diodes. Semiconductors (thermocouples) provide direct conversion of heat to electricity. Present materials provide an efficiency of 5–6%, imposing a great burden on the heat-rejecting radiator. It is hoped with future high-temperature thermocouples to reach an efficiency of 10–13%. In the thermionic diode, electrons are boiled off a hot plate and condensed on a cold plate. Cesium vapor is stored between the plates to neutralize space charge and to adjust the work functions of the electrodes. Thermionic converters, if materials for high-temperature operation can be developed, can attain acceptable efficiencies at high radiator temperature (small

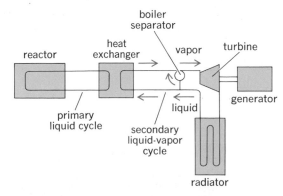

Fig. 3. Power-generation system using the intermediate-link Rankine vapor cycle.

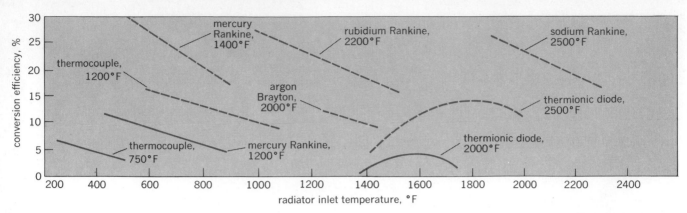

Fig. 4. Curves showing the efficiency of electrical conversion techniques as a function of operating and radiator-inlet temperatures; the broken lines represent projected values.

radiator area), as shown in Fig. 4. Because of these characteristics and because of the absence of large rotating equipment, the thermionic converter is the most attractive technique for the future (Fig. 5). The radiator is the most critical part of the ion propulsion system so far as sensitivity to inclement space environment is concerned. Sensitivity to environmental conditions renders long-term operation questionable, at least in regions of nonbenign environment, such as the asteroid belt, near major planets, and in outer-solar-system space. However, these are precisely the mission regions in which ion propulsion would be most useful in terms of mission performance. More environmental information is needed from space probes to resolve this question. *See* THERMOCOUPLE; THERMOELECTRICITY.

Thrust device. Ion or electrostatic thrust devices consist of three components: an ionizer, which furnishes the ions or other charged particles; an accelerator, which provides the properly shaped electric or electromagnetic field for accelerating and properly shaping the charged-particle beam; and a neutralizer, usually an electron emitter, which neutralizes the exhaust beam after ejection. Electrostatic thrust devices can be classified by the type of charged particle employed; such devices can be further classified according to the method which is utilized to produce these particles.

The ionizer or ion source converts the propellant from its original stored form to charged corpuscles. These can be either atoms, each deprived of

an electron (positive ions), or small liquid droplets or solid particles to which a negative charge (electron) is attached. The charged heavy particles that are collectively referred to as colloids require higher voltage for the same acceleration or, at the same accelerating voltage, yield a lower exhaust velocity. It is important to keep the energy spent for generating and expelling an ion as low as possible. Thus, a readily ionized propellant is preferred; but low ionization is not the only criterion. Cesium, the heaviest member of the alkali metals, requires lowest ionization energy, 3.9 electron volts (ev), and therefore is an attractive propellant. But cesium is a highly reactive metal with high heat of evaporation. It must therefore be handled in a high-temperature environment from vaporizer to ionizer without being contaminated. As a result, it may take as much as 100 times the nominal ionization energy to produce a cesium ion. Ionization of mercury, another attractive, dense propellant requires 10 ev per atom; however, a lower evaporation temperature results in a total requirement of 40–50 times that amount. The energy per discharged ion is, therefore, comparable for these two metals. Promising methods of ionization are surface ionization, high-pressure arcs, and low-pressure arcs. Colloids can be generated by various methods explained below.

The exhaust velocity attainable with any charged particle is equal to $2Vq/m$, where V is the accelerating potential (volts) and q/m is the charge-mass ratio. Typical charge-mass ratios of colloids range from 1 to 100 coulombs/g; those for mercury and cesium are about 500 and 710 coulombs/g, respectively. For a specific impulse of 1000 sec, colloids require an accelerating voltage of 50–0.5 kilovolts (kv) for the range 1–100 coulombs/g, whereas a specific impulse of 10,000 sec requires about 10 and 8 kv for mercury and cesium, respectively. Practically foreseeable limits of accelerating voltage are about 300 kv.

In the surface-ionization system, cesium vapor is impinged on a hot platinum or tungsten grid. The ionization potential of cesium (3.87 volts) is less than the voltage required to remove an electron from the grid surface (work function of the platinum or tungsten). The impinged cesium atoms evaporate; that is, cesium vapor impinges on the hot grid and reevaporates in singly ionized form, having lost one electron for each ion. The boiling

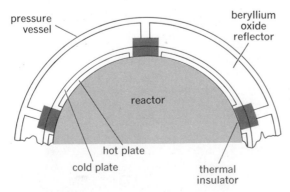

Fig. 5. Basic structure of thermionic converter.

temperature of cesium is 1130°F. The grid temperature must be properly adjusted. If it is too low, not enough cesium evaporates, with the resultant formation of a coat that reduces the ionization efficiency of the grid. If the temperature is too high, electrons escape from the surface together with the ions, reducing the number of ions entering the accelerating electrostatic field. In view of the fairly high boiling temperature of cesium, a high grid temperature, on the order of 2200°F, is required.

To avoid undue power losses by radiation over long propulsion periods, a proper geometric arrangement of the grid elements is necessary. In the acceleration chamber, the (positive) ions are accelerated toward the cathode (Fig. 6).

An example of the high-pressure arc method is the duoplasmatron, a magnetically confined arc operating at high pressure (also named the von Ardenne plasmotron ion source after the physicist M. von Ardenne, who developed the basic concept). The duoplasmatron ion source uses electric discharge to produce a neutral plasma. A strong magnetic field is employed to generate high plasma density, thereby increasing the degree of ionization (Fig. 7). The crucial difficulty of this approach is the charge (electron) exchange between ionized and neutral atoms. This process leads to waste of energy, hence loss in efficiency, and to damage of the accelerating electrode by high-speed impact of particles which cause impact erosion or sputtering. Thus, while the duoplasmatron approach is efficient, sputtering is the main cause of inadequate durability for the long powered flight period needed to make ion propulsion worthwhile. *See* PLASMA PHYSICS.

Low-pressure arcs generate ions by bombarding neutral atoms with high-speed electrons (Fig. 8). Bombardment ionization differs from contact ionization by producing ions in a magnetically constrained arc operating at low gas pressure (about 0.01 atm). The arc consists of a thermionic cathode at the center of a cylindrical anode. An axial magnetic field is imposed by a solenoid around the outside of the anode. The magnetic field causes electrons, emitted by the central cathode and moving radially toward the anode, to follow curved paths (along magnetic field lines) leading back to the cathode. The electrons encounter neutral atoms, which they ionize by collision. In this process the electrons tend to drift to the anode. The magnetic field causes electrons to follow more extended paths than would otherwise be the case, thereby causing them to ionize more atoms before impinging on the anode. The ions are drawn to the aft end of the arc chamber. As in the contact-ion thruster, they are accelerated by a voltage potential and ejected. Cesium and mercury have been tested in this bombardment thruster. Mercury is more convenient to handle, but erosion of the arc cathode by sputtering is a more serious problem than with cesium. The low ionization potential of cesium can be used to replenish the cathode material; deposits of impinging cesium are formed which subsequently evaporate in ionized form and are ejected. *See* MAGNETOHYDRODYNAMICS.

In colloid thrusters particles with atomic masses of 5–50 times that of mercury are a reasonable upper limit; these represent charge-mass ratios of

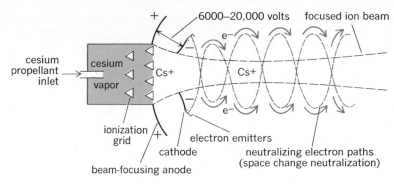

Fig. 6. Acceleration geometry for a surface-ionization thrust device.

100–10 coulombs/g and require potentials between 100 and 10 kv to attain specific impulses around 5000 sec. Colloids can be generated by droplet condensation on ions in supersonic nozzles, by contact charging of microspheres, and by electrostatic spraying.

The droplet-condensation method utilizes the fact that gases expanding adiabatically in a supersonic nozzle can be cooled below the condensation point. When ions are introduced into this stream, nucleation is induced on the ions, which are natural condensation nuclei, and charged droplets are formed. However, this method is not promising because a significant portion of the gas stream is not condensed. This not only represents a loss in propellant efficiency but also interferes seriously with the application of the required strong acceleration potential. Contact charging of microspheres (fine dust particles) was also found not promising, because of fundamental difficulties in raising the rate of solid-particle ejection to a practical level for colloid thrusters.

Electrostatic spraying avoids the basic problems associated with the previously mentioned methods and represents the most promising way of generating charged colloids found so far. This method is based on the exposure of an electrically conducting fluid, such as glycerol, to an electric field. The surface charge in the fluid can, in a sufficiently strong field, overcome the surface tension of the fluid, which breaks into droplets whose size generates a total fluid surface sufficiently large to assure an excess of surface tension over electrostatic pressure. The diameter of these droplets is in the order of a hundred molecules. Colloid thrusters promise useful efficiencies (55–50%) for specific impulses between 600 and 1000 sec. At higher specific impulses overall efficiency (product of mass utilization and power utilization efficiency) of

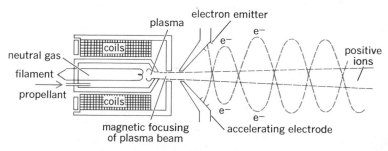

Fig. 7. Basic configuration of high-pressure arc duoplasmatron.

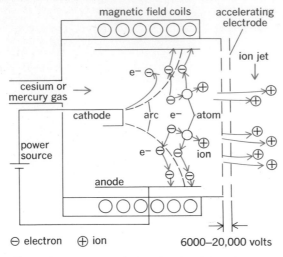

⊖ electron   ⊕ ion   6000–20,000 volts

Fig. 8. Low-pressure electron-bombardment thruster.

the technique drops below 50%. Specific impulses of 2000 sec may eventually be attained.

The space charge represents one of the serious obstacles of any ion-propulsion system. Because the beam produced is singly charged, repulsive Coulomb forces tend to diverge it. Therefore, high current density calls for fast beam acceleration by high voltage in a short accelerator. Although space charge limits ion beams to a density in the order of $10^9–10^{11}$ ions/mm^3, space-charge effects require rapid neutralization of the beam after it has left the propulsion structure. Continuous ejection of one charge would gradually charge the vehicle in the opposite direction, making high-speed ejection of the ions increasingly difficult. Fortunately, the field of an ion beam extracts electrons from nearby hot filaments; this process keeps the vehicle neutral and provides the electrons needed for space-charge neutralization. Figure 6 shows the general arrangement of the electron sources and their oscillating path through the low-density ion beam under the combined effect of their own thermal velocity (which is comparable to the velocity of the ions, 100,000–300,000 ft/sec) and the attracting electric field of the ion beam.

Low ion masses, high voltage, and short acceleration chambers are required for high current densities. On the other hand, for a given exhaust velocity (or specific impulse), high thrust density requires large ion mass. Especially for lower specific impulses (under 2000 sec), emphasis must be on colloids if low thrust density and poor overall efficiency are to be avoided.

The specific impulse cannot simply be selected to be a maximum; it must be optimized as a function of propulsion time, propulsion-system weight, and efficiency of converting electrical power into power of the exhaust jet. For round-trip missions to the Moon, specific impulses of 1000–2000 sec are desirable, depending on the mass required of the propulsion system per kilowatt of power generated and on the powered flight time; for round-trip missions to Venus or Mars, specific impulses range from 3500 to 7000 sec; for missions into the outer solar system, representative specific impulses for maximum payload fraction range from 6000 to 10,000 sec. *See* ELECTROMAGNETIC PROPULSION;

INTERPLANETARY PROPULSION; MAGNETOGAS DYNAMICS; NUCLEAR AIRCRAFT PROPULSION; SPACECRAFT PROPULSION.

[KRAFFT A. EHRICKE]

Bibliography: M. P. Ernstene, *Progress and Prospects of Electrical Propulsion*, AIAA 3d Annual Meeting, Boston, Mass., November–December, 1966; H. R. Kaufman, *An Ion Rocket with an Electron-Bombardment Source*, NASA Lewis Research Center, NASA TND-2172, May, 1964; H. S. Seifert (ed.), *Space Technology*, 1959; E. Stuhlinger, *Ion Propulsion for Space Flight*, 1964.

Ion sources

Devices which produce electrically charged atoms or molecules.

PROTON SOURCES

The principal use for protons in the past has been in particle accelerators of various kinds. A gas discharge in hydrogen in a vessel with some sort of hole in it is a proton source, and nearly every proton ion source in existence begins with this as a minimum requirement. Usually, much more is required. If the protons are simply allowed to stream out of the hole, they will be accompanied by equal numbers of electrons so that the discharge maintains its charge neutrality. In accelerators the protons must be formed into a beam by accelerating them. In cyclotron accelerators the beams desired are flat or ribbonlike, and so the hole in the discharge vessel is usually in some form of a slot; in Van de Graaff, Cockcroft-Walton, and other types of noncyclic or dc accelerators, the beam is cylindrical and drawn from a circular hole. *See* COCKCROFT-WALTON ACCELERATOR; ELECTRICAL CONDUCTION IN GASES; PARTICLE ACCELERATOR; PROTON; VAN DE GRAAFF GENERATOR.

A proton source should have good gas economy: the accelerator must be maintained at low gas pressures to avoid beam loss due to gas scattering and possible internal high-voltage breakdown. Other desirable qualities for a proton source are: good power economy; beam currents which match the current-carrying capacity of the accelerator; high beam brightness (or current per unit area); a low percentage of unwanted ions; compactness; stability and long life; and low energy spread.

The discharge is usually kept in some sort of magnetic-field configuration which conserves the electrons and prevents their loss to the walls until they have a chance to ionize some of the gas atoms in the discharge. The electron economy thus achieved makes it possible to run the discharges at low gas pressure without loss of efficiency, and this contributes to better gas economy for the source.

Early sources. Before 1945, proton accelerators were severely limited by inadequate beam currents (typically less than 10 microamperes, μA, for cylindrical beams) and a short source-life. The discharges were usually supported by electrons fed by a filament-anode arrangement, and filament life was usually of the order 10 hr. The ions were mostly H_2^+ or molecular ions rather than H_1^+; the H_1^+ and H_2^+ parts of the beams were separated in a magnetic analyzer and H_2^+ ions were discarded.

Radio-frequency sources. In the late 1940s the radio-frequency (rf) source was discovered. At first these sources delivered typical beams of 100 μA, but further developments culminated in 1951 with rf sources delivering over 1000 μA. Recombination, the cause of high percentages of H_2^+ in the beams, was avoided by the use of an electrodeless discharge in a Pyrex glass enclosure, at the same time eliminating the filament problem. Radio-frequency power from a small rf oscillator was coupled into the discharge through the glass walls either by using capacitive electrodes on the outside of the enclosure or by placing the glass vessel inside the coil of the rf oscillator. Small, carefully shaped electrodes were used to form and concentrate a beam through a narrow canal which separated the discharge from the accelerator. Pyrex does not catalyze the recombination of H atoms into H_2 molecules, and, by minimizing the exposure of the discharge to materials other than Pyrex, the ratio of H^+ to H_2^+ ions could be kept very high. Sources of this type (Fig. 1) are in use in Van de Graaff accelerators. These combine gas and power efficiency, reliability, and low energy spread, and are relatively simple and inexpensive. A small magnetic field serves the purpose of intensifying the discharge by impeding escape of electrons from the discharge to the walls. Typical operating parameters are: rf power, 60 W; magnet power, 100 W; gas consumption, 6 atmospheric cm³/hr; output current, 1.25 ma (90% H_1^+, 5% H_2^+, and 5% H_3^+); energy spread, approximately 65 eV; beam diameter in extraction canal, approximately 1 mm. *See* GLASS AND GLASS PRODUCTS; OSCILLATOR.

Duoplasmatron. The development of the duoplasmatron ion source in 1956 by Von H. Fröhlich and M. Von Ardenne marks the beginning of the high-current era for proton sources. Although these sources (Fig. 2) make use of an electron-emitting cathode filament, the life is extended through the use of barium-oxide dispenser cathodes, which dispense barium through deep layers toward the surface as they wear away; these were developed for high-current gas-discharge applications. The duoplasmatron makes use of an arc discharge which is constricted as it passes into a very strong magnetic field shaped by iron or mild steel inserts in an intermediate electrode and anode. The beam is extracted at the point where the arc has reached a very small diameter and a very high brilliance. Sources of this type have been developed for accelerators.

Large beams. Since 1960, development has centered upon how to form, shape, and control the large beams which have become available. Very large beams have become a necessity in various fields, including thermonuclear research where beams of more than 1 A are required. Development work is being carried out in the United States, Soviet Union, and Europe on sources for thermonuclear work in the ampere range, as well as development of very-high-current proton sources for other configurations. *See* FUSION, NUCLEAR; ION.   [C. D. MOAK]

HEAVY-ION SOURCES

The term "heavy ion" is used to designate atoms or molecules of elements heavier than helium which have been ionized, that is, those which carry an electrical charge. Thus heavy ions range from lithium to uranium. A heavy ion can be singly ionized (one electron removed or added), can be

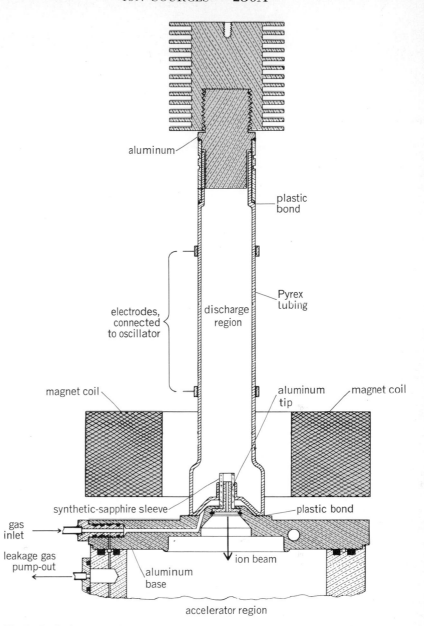

Fig. 1. Radio-frequency ion source.

fully stripped, as in argon 18+, or can have any intermediate charge state.

Applications. Heavy-ion sources are used principally in particle accelerators where they furnish the ions to be accelerated. Accelerated (high-energy) heavy ions are used in scientific research (nuclear, atomic, and materials science studies), in commercial applications (ion implantation), and in scientific applications (radiation effects research). Heavy-ion sources are also used with electromagnetic isotope separators and with mass spectrometers. *See* CHARGED PARTICLE BEAMS; ION IMPLANTATION; ISOTOPE (STABLE) SEPARATION; MASS SPECTROSCOPE.

Positive and negative sources. All ion sources can be divided into two broad classes characterized by the sign of electrical charge of the ions produced. In positive sources the ions carry a net positive charge, electrons having been removed from the neutral atom. In negative sources an electron has been attached to the neutral atom. Posi-

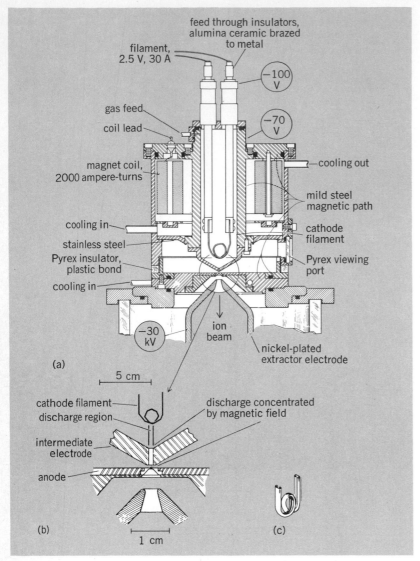

filament,
2.5 V, 30 A

feed through insulators,
alumina ceramic brazed
to metal

−100 V

−70 V

gas feed
coil lead

cooling out

magnet coil,
2000 ampere-turns

mild steel
magnetic path

cooling in
stainless steel
Pyrex insulator,
plastic bond
cooling in

cathode
filament

Pyrex viewing
port

−30 kV

ion
beam

nickel-plated
extractor electrode

(a)

5 cm

cathode filament
discharge region

discharge concentrated
by magnetic field

intermediate
electrode

anode

(b)

1 cm

(c)

Fig. 2. Duoplasmatron ion source for accelerators. (a) Diagram of entire source. (b) Detail of discharge and extraction regions. (c) Drawing of cathode filament.

can be accounted for by the primary electrons from the cathodes. The average energy of plasma electrons may range from a few volts to a few tens of volts. Electrons travel parallel to the magnetic field, are reflected from the opposite cathode, and make many traversals of the length of the hollow chamber. The electrons confined by the magnetic field and the cathode potential have thereby a high probability of making ionizing collisions with any gas present in the chamber. The operating arc potential may vary from less than a hundred volts to several thousand volts, a few hundred usually being satisfactory. The arc current varies from a fraction of an ampere to 10 or 20 A.

The process taking place inside the arc is very complex and consists of electron-atom, atom-atom, electron-ion, and atom-ion collisions in all possible combinations. Collisions result in both ionization and in recombination. It is believed that both single-impact electron-atom collisions resulting in multiple ionization, and step-by-step sequential single ionization collisions take place. Ionization potentials are well known both experimentally and from computer calculation for nearly all atomic configurations. However, the cross sections (effective size of the atom) for the various collision processes for partially ionized atoms are not well known and are difficult to measure. Nevertheless, a good deal of experimental and theoretical work

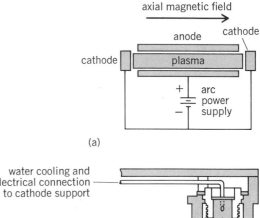

axial magnetic field

anode   cathode

cathode   plasma

+ arc
power
− supply

(a)

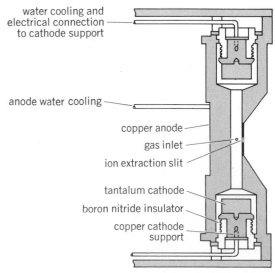

water cooling and
electrical connection
to cathode support

anode water cooling

copper anode
gas inlet
ion extraction slit

tantalum cathode
boron nitride insulator
copper cathode
support

(b)

Fig. 3. Penning ion source. (a) Schematic diagram illustrating basic principles. (b) Section showing geometry. (*From J. R. J. Bennett, A review of PIG sources for multiply charged heavy ions, IEEE Trans. Nucl. Sci., NS-19(2):48–68, 1972*)

tive sources are used ordinarily in several broad classes of accelerators: cyclotrons, linear accelerators, synchrotrons, and single-stage electrostatic accelerators. However, tandem electrostatic accelerators, often called tandem Van de Graaffs, require negative ion sources.

Penning ion sources. Sources of positively charged heavy ions used in accelerators are preponderantly Penning ion sources or variations or modifications of them. This source is based on a high-current gaseous discharge in a magnetic field with gas at a relatively low pressure (10^{-3} torr or 0.1 pascal). The source (Fig. 3) consists of a hollow anode chamber, cathodes at each end, a means for introducing the desired element (usually as a gas), and electrodes for accelerating the ions (not shown). The cathode may be heated to emit electrons, which then help to initiate the arc discharge current, creating the plasma in which the atoms are ionized. The discharge column between the cathodes (the plasma) consists of approximately equal numbers of low-energy electrons and positive ions. The electron density is much larger than

has been undertaken to unravel the complexities of ion sources. *See* IONIZATION POTENTIAL; PLASMA PHYSICS; SCATTERING EXPERIMENTS, ATOMIC AND MOLECULAR.

The net result of all the processes in the arc plasma is that some partially stripped atoms diffuse perpendicular to the magnetic field out of the arc, experience the field of the accelerating electrode, and are moved into the accelerator. The arc potential may be constant with time, or it may be pulsed in the millisecond range so that ions are produced as needed by the accelerator. *See* ARC DISCHARGE.

The intense arc requires water-cooling of both the anode and cathode structures. Tantalum is a good cathode material, but source-life is limited by sputtering of cathode surfaces. *See* SPUTTERING.

Yields of charge states. Good yields of charge states 1+ through 8+ (and for heavier elements, perhaps up to 12+) have been obtained for many elements of the periodic table. Table 1 shows relative yields for selected elements. Each of the currents I^{Q+} for the charge state $Q+$ has been normalized to the value for the 1+ state for each respective element. Thus for argon 5+ the observed current is three times as much as for argon 1+. For argon 8+ there is about 1% as much current as for the reference state. These results represent remarkable progress; as late as 1960 ionized currents of any charge states higher than 1+ or 2+ were virtually unknown. (High-charge states in optical spectroscopy have been known for many years but the technologies used are not applicable to ion sources for accelerators.)

Extracted beam currents. The results of acceleration of heavy ions from a Penning source in a cyclotron are summarized in Table 2 for the same elements as in Table 1. While these results are from similar sources, they are not to be compared directly because of numerous differences, including the characteristics of the cyclotron. Shown in Table 2 are extracted beam currents for a number of charge states. Each beam will have its own characteristic energy, the higher charge states having considerably higher energies. These energies are not shown. While some of the currents shown are quite small (for example, xenon), they are large enough to be very valuable in nuclear studies.

Negative sources. Under proper conditions it is possible to attach an extra electron to a neutral atom, thus making a negative ion. Sources of negatively charged heavy ions fall into two main types: In one type, a primary beam of positive ions is produced and then converted to negative ions by charge exchange. This process occurs in an electron donor vapor target (typically, hydrocarbons, mercury, or an alkali vapor). In a second type, negative ions are produced directly from the source. Two versions of this second type are: (1) a low-powered radial-extraction Penning source to which cesium vapor is added using a cesium boiler, and (2) a sputter source in which a beam of cesium ions bombards a solid surface of the desired element. Both sources have generated large intensities of ions of many elements. Some results are shown in Table 3. It is seen that beam currents of the elements shown vary from 1 to 80 microamperes. This is a most important development for tandem heavy-ion accelerators.

Table 1. Penning heavy-ion source performance

Element	Charge state distributions (I^{Q+}/I^{1+})								
	1+	3+	4+	5+	6+	7+	8+	9+	12+
Argon	1	8.5	8	3	0.8	0.09	0.01		
Calcium	1	23	22	15	3	0.3	.035		
Krypton	1	5	5.5	6	4.5	4	2.2	0.6	
Xenon	1	7	9	9	7.5	5.5	4	1.5	0.025

Table 2. Extracted cyclotron currents from Penning source

Element	Extracted current for charge state, μA							
	3+	4+	5+	6+	7+	8+	9+	12+
Argon	3	34	1	6	1.6	1.2	0.015	
Calcium				0.3	0.7	0.012		
Krypton	0.032	1	2.2	0.15			0.020	
Xenon			0.003		0.09	0.013	0.004	0.0013

Table 3. Negative-ion sources performance

Element	Current for 1− state, μA	Element	Current for 1− state, μA
Lithium	3	Sulfur	40
Boron	3	Copper	10
Carbon	50	Nickel	10
Oxygen	80	Gold	10
Silicon	40	Lead	1

Classification of positive sources. The major types of positive heavy-ion sources can be classified under four broad headings: classical, plasma electron heating, pulsed plasma, and ion confinement.

Classical sources. The principal sources in this category are the Penning source already described and the duoplasmatron (Fig. 2), a variant of the gaseous discharge source. The latter is very successful and is widely used but, for heavy-ion production, is not equal to the Penning source.

Plasma electron heating sources. This class was suggested from studies of controlled thermonuclear fusion. Plasmas are generated by microwave heating using magnetic mirrors to confine the plasma. Microwave power varies from 2 to 20 kW. The electron temperature varies from 10 to 500 keV. Ionization up to argon 10+ and xenon 18+ has been observed. These sources show considerable promise but are not used in actual accelerators.

Pulsed plasma sources. Large amounts of energy concentrated in a small volume can produce a high-temperature high-density plasma and thus highly charged ions.

For many years vacuum spark sources have produced ions for spectroscopic studies. Copper 14+, for example, has been obtained, but intensities are low and ions cannot be extracted. *See* SPECTROSCOPY.

In principle, a beam of coherent light (laser) can deliver a large amount of energy to a small piece of an element in solid form. The resulting rapid heat-

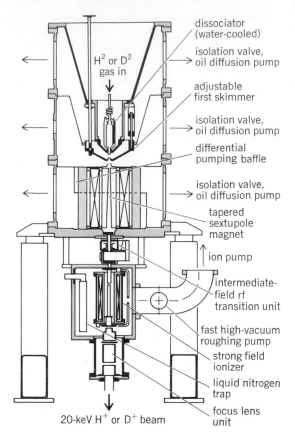

dissociator (water-cooled)

H² or D² gas in

isolation valve, oil diffusion pump

adjustable first skimmer

isolation valve, oil diffusion pump

differential pumping baffle

isolation valve, oil diffusion pump

tapered sextupole magnet

ion pump

intermediate-field rf transition unit

fast high-vacuum roughing pump

strong field ionizer

liquid nitrogen trap

focus lens unit

20-keV H⁺ or D⁺ beam

Fig. 4. Conventional or ground-state polarized ion source. (*From H. F. Glavish, Polarized ion sources, in Proceedings of the 2d Symposium on Ion Sources and Formation of Ion Beams, Berkeley, pages IV-1-1 through IV-1-7, 1974*)

ing will, according to calculations, produce significant numbers of very highly ionized atoms. Using a powerful laser (30 joules in 3.5 nanoseconds), aluminum 13+ and iron 24+ have been detected. Problems of slow pulsing rate and an extraction method remain to be solved. *See* LASER.

A large current from a capacitor bank produces a "pinch" discharge which has been studied as a part of fusion research. Argon 17+ and iron 25+ have been attained, but repetition rates are low and an ion extraction system has not been developed. *See* PINCH EFFECT.

Ion confinement sources. Devices of this class use the potential well of an electron beam to trap positive ions. High ionization is achieved by a step-by-step process which requires extended interaction times. Theory shows that if neutral or slightly ionized atoms can be held in close proximity to energetic electrons, the probability of highly ionized states being reached is a function of the length of time the particles can be held.

In one such scheme, a beam of electrons in a magnetic field is run through a series of cylinders on which different potentials are maintained. The potential of the electron beam keeps the ions from drifting at right angles to the axis of the apparatus. The potentials of the cylinders tend to prevent their drift longitudinally. Some success with this scheme has been achieved, but the ion currents are very small and difficult to inject into an accelerator. This is, however, a promising avenue. Ar-

gon 14+ and xenon 20+ have been produced.

Future sources. It appears that heavy-ion charge states up to 10+ or 12+ can be supplied with good intensities using classical Penning sources. For higher charge states the development of new source types will be required.

[ROBERT S. LIVINGSTON]

POLARIZED ION SOURCES

A polarized ion source is a device that generates ion beams in such a manner that the spins of the ions are aligned in some direction. The usual application is to inject polarized ions into a particle accelerator; however, it also has applications in atomic physics. The possible types of polarized sources are numerous, for in theory the nuclei of all kinds of atoms can be polarized, provided their spin is not zero. *See* SPIN (QUANTUM MECHANICS).

The first polarized ion source was reported in 1956 by G. Clausnitzer, R. Fleischmann, and H. Schopper. The original type of source generated only positive ions but was effectively used on many types of accelerators. However, the development of the tandem Van de Graaff accelerator and its general acceptance for nuclear research created a demand for a polarized source capable of generating negative ions. This stimulated the development of the metastable-state or Lamb-shift type of polarized ion source that produces directly a high-quality negative ion beam with a high degree of polarization. The older type of source is referred to as the conventional or ground-state type of polarized ion source. Its output current of positive ions is an order of magnitude larger than the negative-ion output from the Lamb-shift source. With these two types of sources and their variants, polarized ions have been obtained from hydrogen, deuterium, tritium, helium having mass three, both isotopes of lithium, and others. The extra complication involved in producing polarized ions is such that the output is a factor of a thousand or more below the output of a moderately sized unpolarized ion source.

Conventional or ground-state source. In this type of source (Fig. 4) the first step consists in forming a beam of atoms in the ground state by a technique similar to that used in molecular beams. In the case of hydrogen or deuterium, this is done by dissociating the molecular gas in a glass discharge tube and allowing the atoms to escape through a nozzle. The atoms escaping at thermal energies are collimated into a well-directed beam by plates with holes or an adjustable skimmer. High-capacity diffusion pumps sweep away the large quantity of excess hydrogen or deuterium. *See* MOLECULAR BEAMS; VACUUM PUMP.

The beam is then passed along the axis of an inhomogeneous magnetic field, which is most commonly generated by a sextupole magnet. This type of magnet consists of six magnets arranged in a circular pattern with alternating polarities.

In a sextupole magnet the absolute magnitude of the field increases as the distance squared from the axis. The atoms are subjected to a radial force that is proportional to their magnetic moment times the gradient of the absolute magnitude of the field strength. In the case of a sextupole magnet the force is proportional to the first power of the distance from the axis. The sign of the force does not depend upon the direction of the magnetic lines, but only on the projection of magnetic moment along these lines of force or m_j where j is the spin value of the electron. (The atomic magnetic moment results almost entirely from the electron, since its magnetic moment is 662 times larger than that of the proton and 2156 times larger than that of the deuteron.) The result is that atoms with a positive value of m_j are subjected to a force that is directed radially inward and pass through the sextupole magnet, while atoms with a negative m_j experience a force that is directed outward and are rapidly lost from the beam. Out of the sextupole comes a beam of atomic hydrogen that is polarized with respect to the orientation of its electrons but is, as yet, unpolarized in its nuclear spin. *See* ELECTRON SPIN; MAGNETIC MOMENT; MAGNETON.

Since aligned nuclei rather than aligned electrons are desired, it is necessary to subject the atomic beam to other fields. Each hydrogen atom is in one of two pure states. It is possible to apply an oscillating magnetic field in combination with a dc magnetic field that will flip the sign of m_I (the projection of the nuclear spin) of one of the pure states and not the other. That aligns the spins of the nuclei but may depolarize the electrons. That does not matter, however, since they will be removed.

The final stage is to send the atomic beam into a strong solenoidal magnetic field. As the atoms from the sextupole field—having all orientations in each cross-sectional plane—enter the solenoid, they adiabatically come into alignment with the parallel lines of force within the solenoid since their m_j components of spin are conserved. In the solenoid the atoms are ionized by energetic electrons as in an arc discharge. The ionizer is actually the most difficult part of this type of polarized source to make function efficiently, even though it is conceptionally simple. The ionizer is followed by electric fields that accelerate and focus the ions to get a beam that can be accepted by the accelerator.

Lamb-shift or metastable-atom source. The polarization process in the Lamb-shift type of source (Fig. 5) is also performed upon atoms, in this case, metastable ones. The process is most efficient if the atoms have a velocity of approximately 10^{-3} of light rather than thermal velocity as in the case of the ground-state type of source. To get the beam, hydrogen, deuterium, or tritium can be used, but only hydrogen is discussed in this article. The hydrogen is ionized in a conventional ion source such as a duoplasmatron. The H^+ ions are then accelerated and focused into a beam at about 500 eV. The beam is passed through cesium vapor where cesium atoms donate electrons which are resonantly captured in an $n = 2$ state by the hydrogen ions.

Atoms are formed in both the $2p$ and the $2s$ states in the cesium vapor. However, those in the $2p$ state decay almost immediately to the ground state by the emission of a Lyman alpha photon, energy 10.15 eV. The small energy difference between the $2p$ and the $2s$ states is the Lamb shift. The lifetime of the $2p$ atoms is 1.6×10^{-9} s, while the lifetime of the $2s$ atoms is 0.15 s because two photons must be emitted simultaneously in their decay to the ground state. Actually few $2s$ atoms

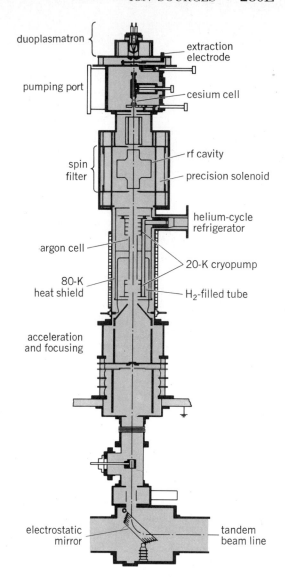

Fig. 5. Lamb-shift polarized ion source. (*From R. A. Hardekopf, Operation of the LASL polarized triton source, in Proceedings of the 4th International Symposium on Polarization Phenomena in Nuclear Reactions, Birkhäuser Verlag, Basel, 1976*)

decay by emission of two photons for they are necessarily subjected to small electric fields which mix into the $2s$ the $2p$ wave function and its tendency to decay to the ground state. To take advantage of this tendency to decay to the ground state, apparatus can be built so that those atoms having the undesired value of m_I are stimulated to decay, while those with the desired value of m_I are allowed to pass on without decay. *See* ATOMIC STRUCTURE AND SPECTRA; FINE STRUCTURE (SPECTRAL LINES).

The polarized H^- ions are formed in argon because its atoms are capable of donating electrons to metastable atoms but have a very weak capability of forming H^- ions out of ground-state atoms. This fact was discovered by B. L. Donnally and coworkers along with the cesium reaction. The ground-state charge-changing cross section appears to be lower by a factor of about 400; however, ground-state atoms outnumber the metastable atoms at this region by a factor of 40 so that the net polarization is 90%. The remainder of the apparatus consists of electric fields that accelerate and focus the beam so it can be accepted by an accelerator.

The electron spins are polarized by applying a transverse field of about 100 V/mm while the atomic beam of metastables is passing along the axis of a solenoid at a field strength of about 57.5 millitesla. The transverse electric field couples the 2s and 2p levels through the Stark effect, and the magnetic field is just sufficient to bring the levels with $m_j = -1/2$ very close together in energy, while those with $m_j = +1/2$ have their energy separation doubled, so that 2s atoms with $m_j = +1/2$ are transmitted without loss. See STARK EFFECT; ZEEMAN EFFECT.

There are several methods of going on to polarize the nuclei, but a device known as the spin filter, developed by G. G. Ohlsen, J. L. McKibben, R. R. Stevens, and G. D. Lawrence, seems to be gaining favor. To produce the spin filter, a longitudinal electric field of about the same strength as the transverse field is added to the apparatus that polarizes the electrons, with the longitudinal field oscillating at about 1.60 GHz. The complete explanation is complicated. However, results are that if the magnetic field is adjusted so the Larmor frequency of the electron in the metastable atom is made equal to the oscillating electric field, then the lifetime of the atom for decay becomes very long exactly at resonance, yet short not far off resonance. The magnetic field that determines the Larmor frequency of the electron in the metastable atom is the sum of that due to the solenoid and that due to the proton aligned in the solenoidal field. These two fields have opposite signs in the case of $m_I = -1/2$ and it is found that the two resonances for transmission are at 54.0 mT for $m_I = +1/2$ and at 60.5 mT for $m_I = -1/2$. In the case of deuterons, there are three resonances, and they are well resolved even though $m_I = +1$ is at 56.5 mT, 0 is at 57.5 mT, and −1 is at 58.5 mT.

[JOSEPH L. MC KIBBEN]

Bibliography: A. J. Bayly and A. G. Ward, Positive ion source, Can. J. Research, A26(2):69–78, 1948; D. Fick (ed.) Polarization Nuclear Physics: Proceedings of a Meeting Held at Ebermannstadt, 1973, Lecture Notes in Physics, 1974; C. D. Moak, H. Rees, Jr., and W. M. Good, Design and operation of radio-frequency ion source for particle accelerators, Nucleonics 9(3):18–23, 1951; O. B. Morgan, G. G. Kelley, and R. C. Davis, Technology of intense dc ion beams, Rev. Sci. Instr., 38:467–480, 1967; Proceedings of the International Conference on Multiply Charged Heavy Ion Sources and Accelerating Systems, IEEE Trans. Nucl. Sci., NS-19(2), 1972; Proceedings of the 1975 Particle Accelerator Conference, IEEE Trans. Nucl. Sci., NS-22(3), 1975; Proceedings of the 2d International Conference on Ion Sources, SGAE, Vienna, 1972; Proceedings of the 2d Symposium on Ion Sources and Formation of Ion Beams, Berkeley, 1974; Proceedings of the Symposium on Ion Sources and Formation of Ion Beams, Brookhaven National Laboratory, 1971; Proceedings of the 3d International Symposium on Polarization Phenomena in Nuclear Reactions, 1970; J. G. Rutherglen and J. F. I. Cole, A radio-frequency ion source with high percentage yield of protons, Nature, 160: 545–546, 1947; P. C. Thonemann et al., The performance of a new radio-frequency ion source, Proc. Phys. Soc. (London), 61:483–485, 1948.

Ionic crystals

A class of crystals in which the lattice-site occupants are charged ions held together primarily by their electrostatic interaction. Such binding is called ionic binding. Empirically, ionic crystals are distinguished by strong absorption of infrared radiation, good ionic conductivity at high temperatures, and the existence of planes along which the crystals cleave easily. See CRYSTAL STRUCTURE.

Compounds of strongly electropositive and strongly electronegative elements form solids which are ionic crystals, for example, the alkali halides, other monovalent metal halides, and the alkaline-earth halides, oxides, and sulfides. Crystals in which some of the ions are complex, such as metal carbonates, metal nitrates, and ammonium salts, may also be classed as ionic crystals.

As a crystal type, ionic crystals are to be distinguished from other types such as molecular crystals, valence crystals, or metals. The ideal ionic crystal as defined is approached most closely by the alkali halides. Other crystals often classed as ionic have binding which is not exclusively ionic but includes a certain admixture of covalent binding. Thus the term ionic crystal refers to an idealization to which real crystals correspond to a greater or lesser degree, and crystals exist having characteristics of more than one crystal type. See CHEMICAL BINDING.

Ionic crystals, especially alkali halides, have played a very prominent role in the development of solid-state physics. They are relatively easy to produce as large, quite pure, single crystals suitable for accurate and reproducible experimental investigations. In addition, they are relatively easy to subject to theoretical treatment since they have simple structures and are bound by the well-understood Coulomb force between the ions. This is in contrast to metals and covalent crystals, which are bound by more complicated forces, and to molecular crystals, which either have complicated structures or are difficult to produce as single crystals. Being readily available and among the simplest known solids, they have thus been a frequent and profitable meeting place between theory and experiment. These same features of ionic crystals have made them attractive as host crystals for the study of crystal defects: deliberately introduced impurities, vacancies, interstitials, and color centers. See COLOR CENTERS; CRYSTAL DEFECTS.

Crystal structure. The simplest ionic crystal structures are those of the alkali halides. At standard temperature and pressure the 16 salts of Li, Na, K, and Rb with F, Cl, Br, and I, have the sodium chloride structure of interpenetrating face-centered cubic lattices (Fig. 1). CsF also has this structure but otherwise the cesium halides have the cesium chloride structure of interpenetrating simple cubic lattices (Fig. 2). The sodium chloride structure is also assumed by the alkaline-earth oxides, sulfides, and selenides other than those of Be and by AgF, AgCl, and AgBr. Other crystal structures, such as the wurtzite structure (Fig. 3) assumed by BeO, β-ZnS, and ZnO and the zinc-blende structure (Fig. 4) assumed by CuCl, CuBr,

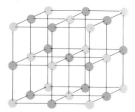

Fig. 1. Sodium chloride lattice. The darker circles represent positive ions and the lighter circles negative ions. (After F. Seitz, The Modern Theory of Solids, McGraw-Hill, 1940)

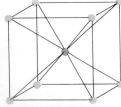

Fig. 2. Cesium chloride lattice. (After F. Seitz, The Modern Theory of Solids, McGraw-Hill, 1940)

CuI, BeS, and α-ZnS are also typical of the ionic crystals of salts in which the atoms have equal positive and negative valence. Ionic compounds consisting of monovalent with divalent elements crystallize typically in the fluorite structure (Fig. 5) assumed by CaF_2, BaF_2, CdF_2, Li_2O, Li_2S, Na_2S, Cu_2S, and Cu_2Se or the rutile structure (Fig. 6) assumed by TiO_2, ZnF_2, and MgF_2.

Cohesive energy. It is possible to understand many of the properties of ionic crystals on the basis of a simple model originally proposed by M. Born and E. Madelung. In the simplest form of this model, the lattice sites are occupied by spherically symmetric ions having charges corresponding to their normal chemical valence. These ions overlap only slightly the ions at neighboring sites and interact with one another through central forces. In NaCl, for example, the spherically symmetric closed shell configurations which the free Na$^+$ and Cl$^-$ ions possess are considered to be negligibly altered by the crystalline environment and to have charges $+e$ and $-e$ respectively, where $-e$ is the charge on the electron. Using this model, together with certain assumptions about the forces between the ions, Born and M. Göppert-Mayer calculated the cohesive energy of a number of ionic crystals. This cohesive energy is defined as the energy necessary to take an ionic crystal from an initial state, in which the crystal is at 0°K and zero pressure, to a final state which is an infinitely dilute gas of its constituent ions at 0°K and zero pressure. While it cannot be measured directly, this cohesive energy can be deduced from experimental quantities by the use of the Born-Haber cycle. Thus the validity of the simple model of Born and Madelung can be tested by comparing the calculated cohesive energy of Born and Mayer with values which have been experimentally determined. *See* COHESION (PHYSICS).

Born-Haber cycle. This is a sequence of processes leading from the initial to the final state specified in the definition of the cohesive energy. Because in most of the processes in this cycle heat changes at constant pressure are measured, it is convenient to consider the change in heat content or enthalpy $H = U + PV$, where P is the pressure, V is the volume, and U is the cohesive energy, rather than the change in E in each step. Since the total change in H is independent of the intermediate steps taken to accomplish it, the ΔH for the change in state specified in the definition of cohesive energy will be given by the sum of the ΔH in the steps of the cycle. Furthermore, because when $P = 0$, $\Delta H = \Delta U$, the ΔH thus calculated will be the cohesive energy. In the following enumeration of the steps in the Born-Haber cycle (A) indicates element A in a monatomic gaseous state and [A] indicates A in a solid state, and so on. The B without brackets in step 3 refers to the natural form of B at the given temperature and pressure. The ionic compound is taken to be AB, where A is the electropositive and B the electronegative element.

The steps of the Born-Haber cycle (all temperatures in °K and pressures in atmospheres) are:

1. $[AB]\,^{0°}_{P=0} \rightarrow [AB]\,^{0°}_{P=1}$. The value of ΔH_1 in this isothermal compression is very small and can be neglected in comparison with other heat content changes in the cycle.

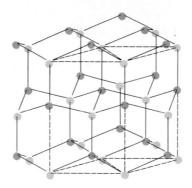

Fig. 3.   Wurtzite lattice. (*After F. Seitz, The Modern Theory of Solids, McGraw-Hill, 1940*)

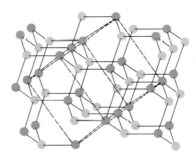

Fig. 4.   Zincblende lattice. (*After F. Seitz, The Modern Theory of Solids, McGraw-Hill, 1940*)

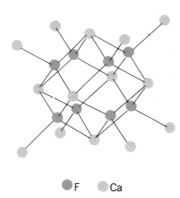

● F   ● Ca

Fig. 5.   Calcium fluoride lattice. (*After F. Seitz, The Modern Theory of Solids, McGraw-Hill, 1940*)

IONIC CRYSTALS

2. $[AB]\,^{0°}_{P=1} \rightarrow [AB]\,^{298°}_{P=1}$. In this step the crystal is warmed to room temperature. The value of ΔH_2 can be calculated from the specific heat at constant pressure for the crystal.

3. $[AB]\,^{0°}_{P=1} \rightarrow [A]\,^{298°}_{P=1} + B\,^{298°}_{P=1}$. The value of ΔH_3 is given by the heat of formation of the compound AB which is referred to substances in their natural forms at standard temperature and pressure.

4. $B\,^{298°}_{P=1} \rightarrow (B)\,^{298°}_{P=1}$. The value of ΔH_4 is the dissociation energy necessary to form a monatomic gas from B in its natural state at standard temperature and pressure. For chlorides, for example, this is the dissociation energy of a Cl_2 molecule into Cl atoms.

5. $[A]\,^{298°}_{P=1} \rightarrow (A)\,^{298°}_{P=1}$. In this step, ΔH_5 is the

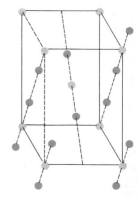

Fig. 6.   Rutile lattice. (*After F. Seitz, The Modern Theory of Solids, McGraw-Hill, 1940*)

heat of sublimation of the metal A. It can be deduced from the heat of fusion, the specific heats of the solid, liquid, and gaseous phases, and the vapor pressure data for the metal. *See* SUBLIMATION.

6. $(A) \underset{P=1}{298°} \rightarrow (A) \underset{P=0}{0°}$; $(B) \underset{P=1}{298°} \rightarrow (B) \underset{P=0}{0°}$. An adiabatic expansion of the gases, considered as ideal, to a very large volume results in a state in which $P=0$, $T=0$, and $\Delta H_6 = -\frac{5}{2}RT$/mole, where R is the gas constant.

7. $(A) \underset{P=0}{0°} \rightarrow (A^+) \underset{P=0}{0°} + e^-$. The ionization of the A atoms gives a ΔH_7 per atom equal to their first ionization energy.

8. $(B) \underset{P=0}{0°} + e^- \rightarrow (B^-) \underset{P=0}{0°}$. The electrons from step (7) are placed on the B atoms. The value of ΔH_8 per atom is given by the electron affinity of the B atom. *See* ELECTRONEGATIVITY.

As an example, for sodium chloride $\Delta H_1 \cong 10^{-4}$ (kilocalories/mole), $\Delta H_2 = 2.4$, $\Delta H_3 = 98.3$, $\Delta H_4 = 26.0$, $\Delta H_5 = 28.8$, $\Delta H_6 = -2.9$, $\Delta H_7 = 11.9$, and $\Delta H_8 = -80.5$. Experimental cohesive energies for a number of the ionic crystals are given in the table. *See* CHEMICAL STRUCTURES.

Born-Mayer equation. By use of the Born-Madelung model, the cohesive energy of an ionic crystal can be related to its measured compressibility and lattice spacing. Because of the opposite signs of electric charges which they carry, the unlike ions in such a crystal model attract one another according to Coulomb's law. However, such a charge distribution cannot be in equilibrium if only Coulomb forces act. In addition to their Coulomb interaction, the ions exhibit a repulsion which, compared with the Coulomb interaction, varies rapidly with interionic separation. The repulsion becomes strong for small separations and diminishes rapidly for increasing separation. The static equilibrium configuration of the crystal is determined by a balance of these forces of attraction and repulsion. *See* COULOMB'S LAW.

The short range repulsion between ions must be described by quantum mechanics. When the electron orbits of two ions overlap, the electron charge density in the region of overlap is diminished as a consequence of the Pauli exclusion principle. This charge redistribution results in a repulsion between the ions in addition to the Coulomb interaction which they have at all interionic distances. In early work, the energy V_{rep} due to repulsion of two ions at a distance r was assumed to have the form of Eq. (1), where B and n are constants to be determined.

$$V_{rep} = B/r^n \qquad (1)$$

mined. Quantum-mechanical calculations of the interaction of atoms with closed shells of electrons indicate that the interaction of repulsion is better approximated by an exponential dependence on interionic distance given in Eq. (2), where A and ρ

$$V_{rep} = Ae^{-r/\rho} \qquad (2)$$

are constants. Both forms for V_{rep} give almost the same calculated cohesive energy; the exponential form gives slightly better agreement with experiment. *See* EXCLUSION PRINCIPLE.

Using the exponential form for the repulsive interaction energy, the potential energy $\varphi(r_{ij})$ of a

pair of ions i and j can be written as Eq. (3), where

$$\varphi(r_{ij}) = \frac{Z_i Z_j e^2}{r_{ij}} + Ae^{-r_{ij}/\rho} \qquad (3)$$

$Z_i e$ and $Z_j e$ are the net charges of the ions i and j, and r_{ij} is the distance of separation of their centers. The assumption that the ions are spherically symmetric has been used here in writing the Coulomb interaction as that of point charges.

The cohesive energy U of an ionic crystal due to the Coulomb and repulsive interactions of its ions is the sum taken over all pairs of ions in the crystal as in Eq. (4), where in the summation the lattice

$$U = \frac{1}{2}\sum_{i,j}{}' \varphi(r_{ij}) \qquad (4)$$

site indices, i and j, range over all sites of the crystal. The prime on the summation sign indicates the exclusion from the sum of terms for which $i=j$ and the factor of 1/2 avoids counting pairs of ions twice.

For crystals in which there are only two types of ion, the Coulomb or electrostatic part of U, U_e, can be written in a simple form given in Eq. (5), where

$$U_e = \frac{1}{2}\sum_{i,j}{}' \frac{Z_i Z_j e^2}{r_{ij}} = -\frac{N\alpha_M (Z_+ Z_-)e^2}{r} \qquad (5)$$

$+Z_+ e$ and $-Z_- e$ are the charges of the positive and negative ions, N is the number of ion pairs in the crystal, r is the nearest anion-cation separation, and α_M is the Madelung constant. *See* MADELUNG CONSTANT.

By anticipating that ρ will be small compared to the nearest neighbor separation, the interactions of repulsion may be neglected for pairs of ions other than nearest neighbors. The energy of the crystal model for arbitrary nearest neighbor separation r is then given by Eq. (6), where M is the

$$U(r) = N[-\alpha_M Z_+ Z_- e^2/r + MAe^{-r/\rho}] \qquad (6)$$

number of nearest neighbors which each ion has in the crystal.

The parameter ρ may be evaluated for a given crystal by requiring that (1) U be a minimum for the observed value of r and (2) that the compressibility of the model equal the measured compressibility of the crystal. It follows from these requirements that Eq. (7) holds. This is the Born-Mayer

$$U(r_0) = N\alpha_M Z_+ Z_- (1-\rho/r_0)\, e^2/r_0 \qquad (7)$$

equation for the cohesive energy, where r_0 refers to the nearest neighbor distance at static equilibrium. Further, in this equation, ρ is given in terms of experimental quantities by Eq. (8), where K is

$$\frac{r_0}{\rho} = \frac{18r_0^4}{\alpha_M e^2 K} + 2 \qquad (8)$$

the measured compressibility of the crystal.

Cohesive energies for some alkali halides and crystals of other structures calculated in this way are shown in the accompanying table, where they can be compared with the experimental values for the cohesive energy. The agreement is considered to be support for the essential validity of the Born-Madelung model. The model has been applied with some success even to the ammonium halides, assuming spherically symmetric ions.

The Born-Mayer theory has been refined, with resulting improvement in the agreement between the calculated and experimental cohesive energies for alkali halides. The refinements have considered the small (a few kilocalories per mole or less) corrections to the cohesive energy arising from van der Waals interactions and zero-point vibrational energy. The van der Waals forces are weak attractive forces between ions due to mutually induced fluctuating dipoles. Similar forces, even weaker, due to dipole-quadrupole interactions have also been considered. Both these interactions make small positive contributions to the cohesive energy. At 0°K the lattice is not in static equilibrium but as a consequence of quantum mechanics is in a state of zero-point vibration with non-zero energy. The energy of these vibrational modes cannot be further reduced. The zero-point vibration energy gives a small negative contribution to the cohesive energy. The results of these refinements of the Born-Mayer theory are also shown in the table. *See* LATTICE VIBRATIONS.

While it has had success in calculating the cohesive energy, the shortcomings of this simple model become evident in its failure to predict correctly the elastic shear constants of ionic crystals. This requires interionic forces of a noncentral character, which are absent from the model. There are also other instances in which the simple model is found to be inadequate.

More elaborate models which take into account features absent from the Born-Mayer model and which may be regarded as extensions of it have had considerable success in accounting for the elastic and dielectric properties of alkali halides. More important has been the ability of these models to account for the lattice phonons in these crystals. This has given the alkali halides a prominent place in the study of phonons in insulators, where they are among the few reasonably well-understood solids.

Ionic conductivity. Just below their melting points, ionic crystals have conductivities of the order of 10^{-4} (ohm-cm)$^{-1}$. Below the melting point the conductivity falls rapidly with decreasing temperature. In a temperature range sufficiently near the melting point, the temperature dependence of the conductivity σ is exponential and of the form $\sigma = ce^{-\phi/T}$, where c and ϕ are constants. In this so-called intrinsic range, the conductivity is characteristic of the material and relatively unaffected by small concentrations of impurities and the previous thermal history of the crystal. At lower temperatures, σ departs from the exponential behavior of the intrinsic region in a manner which does depend on impurities and history. For positive divalent impurities in alkali halides, these departures have been extensively studied.

Ionic conductivity involves the transport of ions, as shown in experiments in which the deposit of metal from the ionic compound on the cathode is measured after the passing of current. Faraday's law of electrolysis relating the charge transport and the matter transport is shown to be valid. If the current were carried by free electrons, this would not be the case.

In a number of ionic crystals, it has been established which of the ions are responsible for the conduction and how they move. In a crystal of

Cohesive energies*

Crystal	Structure	U_{exp}, kcal/mole	U_{calc}, kcal/mole	U_{calc} refined, kcal/mole
LiCl	NaCl	201.5	196.3	200.2
LiBr	NaCl	191.5	184.4	189.5
LiI	NaCl	180.0	169.1	176.1
NaCl	NaCl	184.7	182.0	183.5
NaBr	NaCl	175.9	172.7	175.5
NaI	NaCl	166.3	159.3	164.3
KCl	NaCl	167.8	165.7	167.9
KBr	NaCl	161.2	158.3	161.3
KI	NaCl	152.8	148.2	152.4
RbCl	NaCl	163.6	159.1	162.0
RbBr	NaCl	159.0	151.9	156.1
RbI	NaCl	149.7	143.1	148.0
CaF$_2$	Fluorite	618.0	617.7	
CuCl	Zincblende	226.3	206.1	
ZnS	Wurtzite	851	816	
PbO$_2$	Rutile	2831	2620	
AgCl	NaCl	207.3	187.3	

*The cohesive energies in the last two columns are calculated using the Born-Mayer equation and the refined Born-Mayer theory, respectively. The refined calculations for the last five crystals have not been made.

the type A^+B^-, an experiment to determine the relative contributions of A^+ and B^- ions to the ionic conductivity is done in the following manner (Fig. 7).

Two slabs, 1 and 2, of the crystal A^+B^- are pressed together between electrodes of metal A, and a current is passed through them. If the positive ions alone carried the current, the cathode would increase in weight and the anode would decrease, but the slabs would remain unchanged in size. If, on the other hand, negative ions alone carried the current, these ions would be combined with the metal of the anode to increase the thickness of slab 1 while removal of negative ions from the region of the cathode would cause it to increase in weight, the position of the boundary between 1 and 2 not moving. In general, the case will be intermediate between these extremes. By weighing the electrodes and the slabs before and after passing current, one can in principle determine the fraction of the current carried by the positive and negative ions. These fractions, which are called the transport numbers of the ions, differ for different crystals.

Ionic conductivity requires the motion of ions in the lattice and thus is concerned with departures from the ideal crystal described thus far. In thermal equilibrium at temperatures above 0°K there are always positive and negative ion vacancies and interstitial ions present. A vacancy formed by

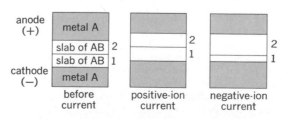

Fig. 7. Experimental arrangement for measuring ionic conductivity.

removing a lattice-site occupant to the surface of the crystal, leaving behind a vacancy, is called a Schottky defect. If the vacancy is formed by removing an ion to an interstitial position far from the vacancy, the vacancy-interstitial pair is called a Frenkel defect. The number of each of these defects present depends on the temperature, the formation energy of the defect, the number of interstitial sites, and considerations of electrical neutrality. Rather than direct interchange of ions in the ideal crystal, it is the motion of the vacancies and the interstitial ions, or both, that is responsible for ionic conductivity. The mechanisms responsible for ionic conductivity are also responsible for self-diffusion of ions in the crystal, and studies of these two phenomena have helped to clarify the mechanisms of ion migration. Experiments have shown that in alkali halides Schottky defects predominate and that ionic conductivity occurs chiefly through positive-ion vacancy migration. In the silver halides, on the other hand, Frenkel defects predominate and positive-ion interstitial migration by the interstitialcy mechanism is the most significant process involved. *See* CONDUCTION, HEAT; DIFFUSION IN SOLIDS.

Ionic conductivity is only one of many properties of ionic crystals which depend critically on the defects in the crystal.

Dielectric properties. The polarization of an ionic crystal in an applied time-varying electric field depends on the frequency ν of the field. If $\nu = 0$ the field is static, and a static equilibrium will be achieved in which the ions are displaced from their lattice sites a distance determined by a balance between the force of repulsion due to their nearest neighbors and the electrical force on the ions. Oppositely charged ions move in opposite directions with a resulting ionic polarization. In addition, the electron clouds of the ions are deformed in the local electric field at the ions and, to a much lesser extent, by the repulsive forces, to give an electronic polarization. As the frequency ν increases, inertial effects must be considered. When ν is sufficiently high, that is, when $\nu > \nu_0$ (ν_0 is the natural frequency of vibration of ions about lattice sites), the heavy ions can no longer follow the rapid variations of the applied field. The ionic polarization ceases to contribute to the total polarization, although the less massive electron clouds continue to give an electronic polarization. Eventually, with increasing frequency, ν exceeds the frequency at which the electron cloud deformations can follow the field, and the electronic polarization becomes altered. The frequency ν_0, called the restrahlung or residual-ray frequency, lies in the infrared region, and the associated vibrations of the charged ions are responsible for the strong absorption of infrared radiation in ionic crystals. The term residual-ray frequency comes from the fact that the reflectivity of visible and infrared radiation is highest for radiation of a frequency near to that of maximum absorption, ν_0. Thus multiple reflection will yield radiation composed predominantly of residual rays of frequency near to ν_0. For additional information on residual rays *see* REFLECTION OF ELECTROMAGNETIC RADIATION.

At frequencies below ν_0 and above it up to the ultraviolet, an ideal alkali halide is transparent to electromagnetic radiation. Impurities or defects often introduce absorption bands into these extensive and otherwise transparent regions and can thus be studied optically. This fact accounts for much of the wealth of experimental detail available on defects and impurities in alkali halides and for the interest which they have attracted.

Ionic radii. It has been found that ionic radii can be chosen such that the lattice spacings of ionic crystals are given approximately by the sum of the radii of the constituent ions. Thus, for instance, eight radii suffice to give the approximate lattice spacings of the 16 alkali halides having sodium chloride structure. This is true because of the rapid variation of the repulsive interaction with distance, so that the ions interact somewhat like hard spheres. There is the implication that a given ion, having the same ionic radius in a variety of crystals, has an electron cloud which is only slightly altered by the differing crystalline environments of these crystals. Several tables of ionic radii appear in the references given in the bibliography.

[B. GALE DICK]

Bibliography: M. Born and K. Huang, *Dynamic Theory of Crystal Lattices*, 1954; C. Kittel, *Introduction to Solid State Physics*, 3d ed., 1966; L. C. Pauling, *The Nature of the Chemical Bond*, 3d ed., 1960; F. Seitz, *The Modern Theory of Solids*, 1940; F. Seitz and D. Turnbull (eds.), *Solid State Physics*, 19 vols. and 8 suppls., 1955–1966.

Ionization

The process by which an electron is removed from an atom, molecule, or ion. This process is of basic importance to electrical conduction in gases and liquids. In the simplest case, ionization may be thought of as a transition between an initial state consisting of a neutral atom and a final state consisting of a positive ion and a free electron. In more complicated cases, a molecule may be converted to a heavy positive ion and a heavy negative ion which are separated.

Ionization may be accomplished by various means. For example, a free electron may collide with a bound atomic electron. If sufficient energy can be exchanged, the atomic electron may be liberated and both electrons separated from the residual positive ion. The incident particle could as well be a positive ion. In this case the reaction may be considerably more complicated, but may again result in a free electron. Another case of considerable importance is the photoelectric effect. Here a photon interacts with a bound electron. If the photon has sufficient energy, the electron may be removed from the atom. The photon is annihilated in the process. Other methods of ionization include thermal processes, chemical reactions, collisions of the second kind, and collisions with neutral molecules or atoms. *See* ELECTRICAL CONDUCTION IN GASES; ELECTRODE POTENTIAL.

[GLENN H. MILLER]

Bibliography: A. Albert and E. P. Serjeant, *Ionization Constants of Acids and Bases*, 1962; G. Francis, *Ionization Phenomena in Gases*, 1960; R. Gomer, *Field Emission and Field Ionization*, 1961; A. H. Von Engel, *Ionized Gases*, 2d ed., 1965.

Ionization chamber

An instrument for determining the amount of electrical charge liberated by the interaction of ionizing radiation with suitable gases, liquids, or solids. The simplest ionization chamber is the gold-leaf

electroscope. *See* ELECTROSCOPE; MONITORING
OF IONIZING RADIATION.

Such devices can be used as radiation detectors
and have found widespread application in nuclear
physics research and radiation monitoring. *See*
PARTICLE DETECTOR.

Signal formation. The simplest form of ionization chamber consists in essence of two electrically conducting electrodes in a container filled
with a gas, as indicated schematically in Fig. 1. A
battery, or other power supply, maintains an electric field between the two electrodes, the positive
one being the anode and the negative one the cathode. When ionizing radiation falls on the chamber,
entering for example through a thin gas-tight window, the incident radiation liberates electrons
from the gas atoms, leaving positively charged
gaseous ions. On the average, approximately 30 eV
of energy are required to generate each electron-
ion pair for most types of filling gas.

The basic signal information consists of the electrical charge liberated in the chamber by the incident radiation. A 6-MeV α-particle, for example,
would produce approximately $(6 \times 10^6)/30$ or
2×10^5 electron-ion pairs if it deposited all its energy in the gas filling the chamber. Since the
charge on the electron is $\sim 1.6 \times 10^{-19}$ coulombs, it
follows that the total negative charge produced
would be $(2 \times 10^5) \times (1.6 \times 10^{-19}) = 3.2 \times 10^{-14}$ coulombs. The quantity of positive charge produced is
always exactly equal in magnitude to the quantity
of negative charge in order to maintain net charge
neutrality.

Excluding the relatively minor probability of
electrons and ions recombining in the gas to form
neutral atoms, the signal charge is rapidly swept
out of the gas by the electric field existing in the
chamber. The electrons move to the anode and the
positive ions to the cathode.

The time taken for the signal charge to reach the
electrodes is a complicated function of the chamber geometry, operating voltage, and the nature
and purity of the filling gas. Using the most widely
employed filling gas, namely argon with a small
admixture of methane, and assuming electrode
spacings of a few centimeters with fields of a few
hundred volts per centimeter, the electron collection times would be $\sim 1~\mu s$. The positive ions travel
much more slowly since they are thousands of
times more massive, exhibiting corresponding
collection times measured in milliseconds.

During the time that the signal charge is moving
in the chamber, a corresponding current flows in
the external circuit. For the simplified case shown
in Fig. 1, for example, and assuming 6-MeV α-particles as before, the signal current waveform would
be substantially as indicated in Fig. 2. The total
negative and positive charges are necessarily of
equal magnitude, and the area of the electron current curve equals that of the positive ion current
curve.

Energy spectrum. While the ionization chamber
generates only small quantities of signal charge,
for incident particles or photons of $\sim$MeV energies,
nevertheless the resulting signals are well above
the noise level of modern low-noise electronic
amplifiers. Consequently the signals generated by
individual incident particles or photons can be individually measured. If this is done, and a histo-
gram is plotted representing the magnitude of a

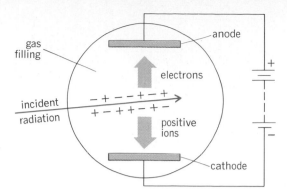

Fig. 1. Schematic form of a gas ionization chamber consisting of two metal electrodes in a sealed gas-filled enclosure.

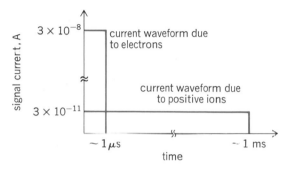

Fig. 2. Idealized current waveforms resulting from the
motion of electrons and positively charged gas ions in an
ionization chamber.

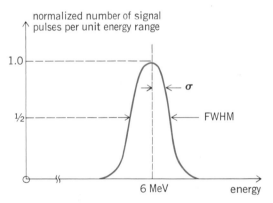

Fig. 3. An example of an idealized energy spectrum
produced by a monoenergetic α-particle source in an
ionization chamber. The spectrum consists of a single
Gaussian peak with standard deviation σ.

signal pulse versus the total number of pulses with
that magnitude, then an energy spectrum results.
Such a spectrum, smoothed out, is sketched in Fig.
3. Again considering the case of a 6-MeV α-particle, the spectrum would consist of an essentially
Gaussian distribution with standard deviation σ.
Assuming a negligible contribution from amplifier
noise, it might at first sight appear that σ should
correspond to the square root of the average number of electron-ion pairs produced per incident
particle. In fact, σ is usually found to be less than
this by a substantial amount, usually designated
F, where F is the Fano factor. *See* PROBABILITY;
STATISTICS.

It is usual to express the width of an energy dis-

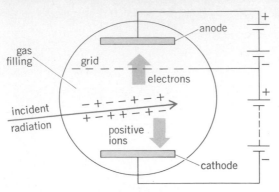

Fig. 4. Schematic form of a gridded ionization chamber in which the anode is electrostatically shielded by a metallic grid.

tribution such as that of Fig. 3 not in terms of σ but in terms of the "full width at half maximum," usually designated FWHM, or Δ. It can be shown that, for situations in which the width of the energy spectrum is governed by statistics alone, the FWHM is given by the equation $\Delta = 2.36\sqrt{F\epsilon E}$, where ϵ is the average energy required to create an electron-ion pair and E is the energy deposited in the chamber by each incident particle or photon. Values of the Fano factor F as low as 0.1 have been observed for certain gases.

Gaseous chamber types. In the previously cited example the incident radiation was envisaged as entering the chamber through a thin window, and in certain applications the presence of such a window can create problems. In such situations it is advantageous to introduce the radiation source—a radioactive material, for example—directly into the inside of the chamber. This leads to the so-called windowless ionization chamber.

In other applications it can be advantageous to employ a grid to electrostatically shield the anode from the cathode. This leads to the gridded ion chamber, whose advantage is that the anode signal is due only to the electron contribution since the incident ionizing radiation is intentionally restricted to the grid-cathode region, as indicated in Fig. 4. Chambers of this type were used extensively for high-resolution α-particle spectroscopy before the advent of semiconductor radiation detectors. *See* JUNCTION DETECTOR.

Gas ionization chambers for nuclear research applications reached a pinnacle of developmend about 1960, at which time they were widely employed for high-resolution α-spectroscopy. Since that time they have been largely superseded for such applications by semiconductor junction detectors, although they remain of substantial importance in other areas. Foremost among these is the use of gas ionization chambers for radiation monitoring. Portable instruments of this type usually employ a detector containing approximately 1 liter of gas, and operate by integrating the current produced by the ambient radiation. They are calibrated to read out in convenient units such as milliroentgens per hour.

A growing application of ionization chambers is the use of air-filled chambers as domestic fire alarms. These employ a small ionization chamber containing a low-level radioactive source, such as ^{241}Am, which generates ionization at a constant

IONIZATON CHAMBER

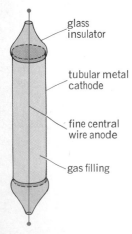

glass insulator

tubular metal cathode

fine central wire anode

gas filling

Fig. 5. Basic form of a simple gas proportional counter.

rate, the resulting current being monitored by a small solid-state electronic amplifier. On the introduction of smoke into the chamber (which is open to the ambient air), the drifting charged ions tend to attach themselves to the smoke particles. This reduces the ionization chamber current, since the moving charge carriers are now much more massive than the initial ions and therefore exhibit correspondingly reduced mobilities. The observed reduction in ion current is used to trigger the alarm.

Another development in ion chamber usage is that of two-dimensional imaging in x-ray medical applications to replace the use of photographic plates. This depends on the fact that if a large flat parallel-plate gas ionization chamber is illuminated with x-rays (perpendicular to its plane), the resulting charges will drift to the plates and thereby form an "image" in electrical charge of the point-by-point intensity of the incident x-rays. This image can be recorded xerographically by arranging for one plate to be a suitably charged insulator. This insulator is preferentially discharged by the collected ions. The resulting charge pattern is recorded by dusting the insulator with a fine powder and transferring this image to paper as in the usual xerographic technique. Alternatively, the xerographic insulator may be a photoconductor, such as selenium, which is preferentially discharged by the ionization produced in the solid material. This is then an example of a solid ionization chamber, and its action closely parallels the operation of the optical xerographic copying machines. Such x-ray imaging detectors provide exceedingly high-quality images at a dosage to the patient substantially less than when photographic plates are used.

Gaseous ionization chambers have also found application as total-energy monitors for high-energy accelerators. Such applications involve the use of a very large number of interleaved thin parallel metal plates immersed in a gas inside a large container. An incident pulse of radiation, due for example to the beam from a large accelerator, will produce a shower of radiation and ionization inside the detector. If the detector is large enough, essentially all of the incident energy will be dissipated inside the detector (mostly in the metal plates) and will produce a corresponding proportional quantity of charge in the gas. By arranging that the plates are alternately biased at a positive and negative potential, the entire device operates like a large interleaved gas ion chamber. The total collected charge is then a measure of the total energy in the initial incident pulse of radiation.

A further development in the application of ion chambers to high-energy physics lies in the use of large interleaved metal plate chambers, conceptually somewhat similar to those described above but employing a filling of purified liquid argon. Such chambers are capable of determining the energy deposited by the passage of a single high-energy particle. By contrast, the large interleaved gas-filled chambers are employed to measure the total integrated charge from a shower due to a large number of particles.

Solid ionization chambers. It is possible, in principle, to replace the gas filling of an ionization chamber by a large single crystal of suitably chosen solid material. In this case the incident radiation creates electron-hole pairs in the crystal, and this constitutes the signal charge. In practice it has been found that only very few materials can be produced with a sufficiently high degree of crystalline perfection to allow this signal charge to be swept out of the crystal and collected. Although many attempts were made in this direction in the 1940s in crystal counters, using such materials as AgCl, CdS, and diamond, these were all failures due to the crystals not having adequate carrier transport properties. In the late 1950s, however, new attempts were made in this direction using single crystals of the semiconductors silicon and germanium. These were highly successful and have led to detectors that have revolutionized low-energy nuclear spectroscopy. *See* CRYSTAL COUNTER; SEMICONDUCTOR.

Liquid ionization chambers. In a manner similar to that described for gases and solids, it has also been found possible to use certain exceedingly pure liquids as the medium of ionization chambers. Although some highly purified organic liquids have exhibited adequate charge transport properties, by far the most successful detectors of this type have employed liquefied noble gases.

Proportional counters. If the electric field is increased beyond a certain point in a gas ionization chamber, a situation is reached in which the free electrons are able to create additional electron-ion pairs by collisions with neutral gas atoms. For this to occur, the electric field must be sufficiently high so that an electron in one mean free path can pick up an energy that exceeds the ionization potential of the neutral gas atoms. Under these circumstances gas multiplication, or avalanche gain, occurs, thereby providing additional signal charge from the detector.

A variety of electrode structures have been employed to provide proportional gas gain of this type. The most widely used is shown in Fig. 5. Here a fine central wire acts as the anode, and the avalanche gain takes place in the high field region immediately surrounding this wire. In practice, under suitable circumstances, it is possible to operate at gas gains of up to approximately 10^6.

The gas gain is a function of the bias voltage applied to the proportional counter and takes the general form shown in Fig. 6.

Similar avalanche multiplication effects can occur in semiconductor junction detectors, although there the situation is less favorable, and such devices have not found very widespread use except as optical detectors.

The large gas gains realizable with proportional counters have made them extremely useful for research applications involving very-low-energy radiation. In addition, their flexibility in terms of geometry has made it possible to construct large area detectors, of the order of 1 m², suitable for use as x-ray detectors in space. Essentially all that has been learned to date regarding x-ray astronomy has involved the use of such detectors aboard space vehicles.

Further exceedingly useful applications of gas proportional counters involve their use as position-sensitive detectors. In Fig. 5, for example, if the anode wire is grounded at both ends, then the signal charge generated at a point will split and flow to ground in the ratio of the impedances seen be-

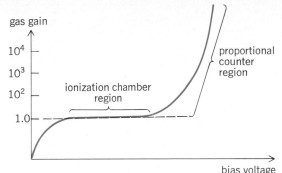

Fig. 6. Plot showing the dependents of the collected charge on the applied voltage in a gas-filled radiation detector.

tween the point of origin and the two ends of the wire. This device therefore comprises a one-dimensional position-sensitive detector. Similar position-sensitive operation can be obtained by taking account of the rise time of the signals seen at each end of the wire. Further extension of such methods allows two-dimensional detectors to be produced, a wide variety of which are under investigation for medical and other imaging uses.

The relatively large signals obtainable from gas proportional counters simplifies the requirements of the subsequent amplifiers and signal handling systems. This has made it economically feasible to employ very large arrays, of the order of thousands, of such devices in multidimensional arrays in high-energy physics experiments. By exploiting refinements of technique, it has proved possible to locate the tracks of charged particles to within a fraction of a millimeter in distances measured in meters. Such proportional counter arrays can operate at megahertz counting rates since they do not exhibit the long dead-time effects associated with spark chambers. *See* SPARK CHAMBERS.

Geiger counters. If the bias voltage across a proportional counter is increased sufficiently, the device enters a new mode of operation in which the gas gain is no longer proportional to the initial signal charge but saturates at a very large, and constant, value. This provides a very economical method of generating signals so large that they need no subsequent amplification. *See* GEIGER-MÜLLER COUNTER.

The most widespread use of Geiger counters continues to be in radiation monitoring, where their large output signals simplify the readout problem. They have also found extensive use in cosmic-ray research, where again their large signals have made it feasible to use arrays of substantial numbers of detectors without excessive expenditures on signal-processing electronics.

[G. L. MILLER]

Bibliography: P. W. Nicholson, *Nuclear Electronics*, 1974; W. J. Price, *Nuclear Radiation Detection*, 2d ed., 1964; J. Sharpe, *Nuclear Radiation Detectors*, 1964; A. H. Snell, *Nuclear Instruments and Their Uses*, 1962.

Ionization gage

An instrument for measuring vacuum by ionizing a gas and measuring the ion current. Ionization gages may be classified by the means used to ion-

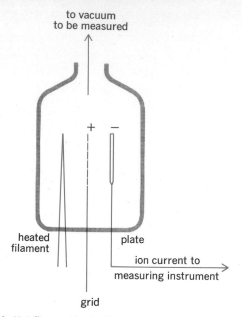

Fig. 1. Hot-filament ionization gage.

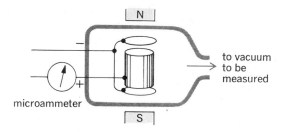

Fig. 2. Cold-cathode (Philips) ionization gage.

ize the gas. See VACUUM MEASUREMENT.

In the hot-filament ionization gage, electrons emitted by an incandescent filament are attracted toward a positively charged grid electrode. Collisions of electrons with gas molecules produce ions. The ions are attracted to a negatively charged electrode. The ion current is a measure of the number of molecules of gas, hence of the vacuum (pressure).

The hot-filament gage (Fig. 1) is most useful at pressures less than 1 μ. Use is limited because the filament will burn out if the pressure rises much above 10 μ. Unwanted chemical reactions or adsorption may take place on the heated surface.

In the cold-cathode (Philips) ionization gage (Fig. 2), a high voltage is applied between two electrodes. Fewer electrons are emitted but a strong magnetic field deflects the electron stream, increasing the length of the electron path and thus increasing the chance for ionizing collisions of electrons with gas molecules. This type is widely used for pressures of 0.01 – 10 μ.

In another type of ionization gage, the gas is ionized by high-energy alpha particles emitted by a radioactive source such as radium.

All ionization gages must be calibrated for the gas to be measured because different gases have different ionizing properties. Accuracy of 10% of full-scale value is attainable.

[BRUCE D. HAINSWORTH; HAROLD G. PAYNE]

Ionization potential

The potential difference through which a bound electron must be raised to free it from the atom or molecule to which it is attached. In particular, the ionization potential is the difference in potential between the initial state, in which the electron is bound, and the final state, in which it is at rest at infinity.

The concept of ionization potential is closely associated with the Bohr theory of the atom. Although the simple theory is applicable only to hydrogenlike atoms, the picture furnished by it conveys the idea quite well. In this theory, the allowed energy levels for the electron are given by the equation below, where E_n is the energy of the state

$$E_n = -k/n^2 \quad n = 1, 2, 3, \ldots$$

described by n. The constant k is about 13.6 ev for atomic hydrogen. The energy approaches zero as n becomes infinite. Thus zero energy is associated with the free electron. On the other hand, the most tightly bound case is given by setting n equal to unity. By the definition given above, the ionization potential for the most tightly bound, or ground, state is then 13.6 ev. The ionization potential for any excited state is obtained by evaluating E_n for the particular value of n associated with that state. For a further discussion of the energy levels of an atom see ATOMIC STRUCTURE AND SPECTRA; ELECTRON VOLT.

The ionization potential for the removal of an electron from a neutral atom other than hydrogen is more correctly designated as the first ionization potential. The potential associated with the removal of a second electron from a singly ionized atom or molecule is then the second ionization potential, and so on.

Ionization potentials may be measured in a number of ways. The most accurate measurement is obtained from spectroscopic methods. The transitions between energy states are accompanied by the emission or absorption of radiation. The wavelength of this radiation is a measure of the energy difference. The particular transitions that have a common final energy state are called a series. The series limit represents the transition from the free electron state to the particular state common to the series. The energy associated with the series limit transition is the ionization energy.

Another method of measuring ionization potentials is by electron impact. Here the minimum energy needed for a free electron to ionize in a collision is determined. The accuracy of this type of measurement cannot approach that of the spectroscopic method. See CHEMICAL STRUCTURES; ELECTRON CONFIGURATION; ELECTRONEGATIVITY.

[GLENN H. MILLER]

Bibliography: R. M. Hochstrasser, Behavior of Electrons in Atoms: Structure, Spectra, and Photochemistry of Atoms, 1964; J. Millman and S. Seely, Electronics, 2d ed., 1951; F. K. Richtmyer and E. H. Kennard, Introduction to Modern Physics, 5th ed., 1955.

Ionosphere

That part of the upper atmosphere which is sufficiently ionized by solar ultraviolet radiation so that the concentration of free electrons affects the propagation of radio waves. Existence of the ionosphere was suggested simultaneously in 1902 by O. Heaviside in England and A. E. Kennelly in the United States to explain the transatlantic radio communication that was demonstrated the previous year by G. Marconi, and for many years it was commonly referred to as the Kennelly-Heaviside layer. The existence of the ionosphere as an electrically conducting region had been postulated earlier by Balfour Stewart to explain the daily variations in the geomagnetic field. The first direct observation was accomplished in 1924, when E. W. Appleton and M. A. F. Barnett in England and G. Breit and M. A. Tuve in the United States independently observed the direct reflection of radio waves by the ionosphere. *See* IONIZATION; RADIO BROADCASTING; RADIO-WAVE PROPAGATION.

The ionosphere has been extensively explored. The earliest technique involved the ionosonde, which utilizes a pulsed transmitter to send radio signals vertically upward while slowly sweeping the radio frequency. For normal incidence upon the ionosphere, a pulse is reflected at that level in the ionosphere where the plasma frequency equals the radio frequency. Pulses reflected by the ionosphere are received at the transmitter and recorded; the elapsed time between pulse transmission and reception can be interpreted as an apparent distance to the point of reflection. The resulting presentation is a curve of apparent height versus plasma frequency. This technique is still used extensively. Many rockets have flown through the ionosphere to record data, and satellites have orbited in and above the ionosphere for similar purposes. Ionosondes in satellites have been utilized to explore the topside of the ionosphere (that portion above the region of maximum electron concentration) and have provided extensive data. *See* PLASMA PHYSICS.

Data from ground-based ionosondes showed that the ionosphere was structured in the vertical direction. It was first thought that discrete layers were involved, referred to as the D, E, F_1, and F_2 layers; however, rocket measurements have shown that the "layers" merge with one another to such an extent that they are now normally referred to as regions rather than layers. Since a vertically incident radio wave is reflected from the ionosphere at that level where the natural frequency of the plasma equals the radio frequency, it is possible to identify the electron concentration at the point of reflection in terms of the radio frequency according to the relation in Eq. (1), where N_e is the elec-

$$N_e = 1.24 \times 10^{10} f^2 \qquad (1)$$

tron concentration per cubic meter and f is the radio frequency in megahertz. The maximum electron concentrations in the various regions are designated N_mD, N_mE, N_mF_1, and N_mF_2. The critical frequency is the frequency that will just penetrate a given region; for the F_2 peak this is designated f_oF_2. From a careful analysis of the indices of refraction of the ionosphere at various levels, the apparent heights h'_mD, h'_mE, h'_mF_1, and h'_mF_2 of the maximum electron concentration in each region can be converted to true heights h_mD, h_mE, h_mF_1, and h_mF_2.

The ionosphere shows important geographic and temporal variations; the latter include regular diur-

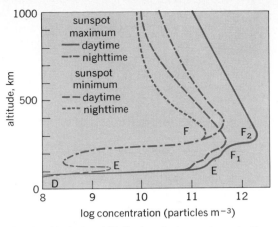

Fig. 1. Typical mid-latitude electron concentration profiles for daytime and nighttime conditions near maximum and minimum of sunspot cycle.

nal, seasonal, and sunspot-cycle components and irregular day-to-day components associated mainly with variations in solar activity and atmospheric motion. Typical electron concentration profiles through the atmosphere are shown in Fig. 1. Curves are shown for daytime and nighttime conditions near the maximum and the minimum of the sunspot cycle. These curves are typical of temperate latitude conditions.

D region. The D region is the lowest ionospheric region, extending approximately from 60 to 85 km. The upper portion is normally caused mainly by the ionization of nitric oxide by hydrogen Lyman-alpha radiation in sunlight, and the lower portion is mainly due to ionization by cosmic radiation. The daytime electron concentrations are about $10^8 - 10^9$ per cubic meter. The region virtually disappears at night even though the cosmic-radiation source of ionization continues, because attachment of electrons to molecules (that is, formation of negative ions) quickly removes free electrons; this effect is suppressed during the daytime by photodetachment. This is the only ionospheric region in which negative ions are thought to be of significance.

The collision frequency for electrons with heavier particles in the D and lower E regions is relatively high, and this condition causes absorption of energy from radio signals traveling through the regions. This severely limits radio propagation and is responsible for the very limited daytime range for stations in the normal broadcast band, where daytime propagation by ionospheric reflection is rendered ineffective by the strong D-region absorption.

The D region is susceptible to disturbance during certain solar events. Sudden ionospheric disturbances (SIDs) occur over the daylight hemisphere with some solar flares; these disturbances last about 1/2 hr and are apparently caused by bursts of solar x-rays with wavelengths near 5 A. These penetrate into the lower D region and cause enhanced electron concentrations there. Polar-cap absorption events (PCAs) are caused by solar cosmic rays that are occasionally emitted by solar flares; these events last for a few days and occur only at high latitudes because of the guiding properties of the geomagnetic field on charged particles. A third type of D-region disturbance occurs regularly near active auroras. All these disturbances are characterized by abnormally intense absorption of radio waves. *See* COSMIC RAYS; SOLAR WIND; SUN.

E region. Soft x-rays and the more penetrating portions of the extreme ultraviolet radiation from the Sun are absorbed in the altitude region from 85 to 140 km, where they cause daytime electron concentrations of the order of 10^{11} per cubic meter in what is known as the E region. This is the region from which ionospheric reflections were first identified. The principal ions have been observed to be O_2^+, O^+, and NO^+, the last presumably being formed in chemical reactions involving the primary ions. The soft x-rays that are principally responsible for the formation of the E region must also produce N_2^+ ions; however, these are not observed because they are removed very rapidly by the reactions in Eq. (2a) and (2b). Of these two reactions,

$$N_2^+ + O \rightarrow NO^+ + N \qquad (2a)$$
$$N_2^+ + O_2 \rightarrow O_2^+ + N_2 \qquad (2b)$$

the former is the more important, but both dominate over direct recombination of N_2^+ ions with electrons.

F region. The most strongly absorbed of the solar extreme ultraviolet radiations (200 A $< \lambda <$ 900 A) produce the F region, which is the region above 140 km. These radiations are most strongly absorbed near 160 km, where they produce a daytime peak in ionization referred to as the F_1 peak or ledge; the F_1 region extends approximately from 140 to 200 km. Above the region of maximum photoionization, the loss rate for electrons and ions decreases with altitude more rapidly than does the source of ionization, so the equilibrium electron concentrations increase with altitude. This decrease in loss rate occurs because O^+ ions recombine directly with electrons only very slowly in a two-body radiative-recombination process, and another loss process predominates instead, an ion-atom interchange reaction, shown in Eq. (3), followed by a dissociative recombination, shown in Eq. (4). The loss rate is controlled by the ion-atom

$$O^+ + N_2 \rightarrow NO^+ + N \qquad (3)$$
$$NO^+ + e \rightarrow N + O \qquad (4)$$

interchange reaction, and it is the decreasing N_2 concentration with altitude that is responsible for lowering the loss rate so rapidly as to cause equilibrium concentrations of O^+ to increase with altitude above the region of maximum production.

The increase in O^+ and electron concentration with altitude finally stops because the atmospheric density becomes so low that there is a rapid downward diffusive flow of ions and electrons by ambipolar diffusion in the gravitational field. The region of maximum ionization concentration normally occurs near 300 km, and it is known as the F_2 peak. Most of the photoionization that occurs above the peak involves atomic oxygen and it is lost by downward diffusion through the peak into the denser atmosphere below, where the ionization can be lost by the ion-atom interchange reaction followed by dissociative recombination, as described above.

The peak electron concentration in the F_2 region is in the vicinity of 10^{12} ions per cubic meter. Above the peak, the distribution of ionization is in diffusion equilibrium in the gravitational field. When a single ion species is dominant, its distribution with altitude is a negative exponential, but with a lesser rate of fall with altitude than for neutral particles of the same mass. The gradient of the logarithm of the concentration is given by Eq. (5),

$$\frac{d}{dz} \ln n = -\frac{mg}{k(T_i + T_e)} \qquad (5)$$

where m is the ion mass, g the acceleration of gravity, k Boltzmann's constant, T_i the temperature of the ions, and T_e the temperature of the electrons. The corresponding expression for neutral particles is Eq. (6), where T is the neutral particle tempera-

$$\frac{d}{dz} \ln n = -\frac{mg}{kT} \qquad (6)$$

ture and the neutral particles are assumed to have the same mass as the ions. If $T = T_i = T_e$, the rate of decrease in the logarithm of the ion concentration with altitude is only half as rapid as for neutrals. However, due to the energy with which photoelectrons are released, it is frequently the case that $T_e > T_i > T$, in which case the rate of fall of ion concentration is still less rapid.

There is an anomaly in F-region behavior, known as the winter anomaly, in that the electron concentrations in winter are higher than in summer, especially if compared under comparable conditions of solar illumination. This is due to a change in atmospheric composition, in particular an increase in atomic oxygen concentration relative to molecular nitrogen. The effect of this change is to reduce the rate of recombination, which is controlled by reaction (3), thus allowing ion and electron concentrations to increase. The change in atmospheric composition results from complications associated with large-scale atmospheric circulation at ionospheric altitudes.

Heliosphere and protonosphere. Helium and atomic hydrogen are important constituents of the upper atmosphere, and their ions are also important at levels above 500 km or so. These gases become important, and finally predominant, constituents of the upper atmosphere because of their low masses. In diffusion equilibrium, each neutral gas is distributed in the gravitational field just as if the other gases were not present, and the lighter ones therefore finally come to predominate over the heavier ones above some sufficiently high altitude. Diffusion equilibrium for ions is more complicated than for neutrals. If the ions and electrons were free to act as independent gases, the electrons would fall off hardly at all with altitude, because of their very small mass. However, even a very small tendency in this direction would set up an electric field sufficient to stop further charge separation, thus requiring that the ions and electrons adopt almost identical distributions. The electric field that is established is sufficient to support half the ion mass and hold the electrons down in the same degree as if they had masses equal to half the ion mass. Therefore, ions and electrons are both distributed as if they had half the ion mass (assuming that $T_e = T_i$; if $T_e > T_i$, the effective mass is even less). Now, if there is in addition, a minor constituent lighter ion, it will experience the same electric field that is established by the major constituent heavier ion and that is sufficient to support half the mass of the heavier ion. This field may be more than sufficient to support the entire mass of the lighter ion.

For example, when the predominant ion is O^+, the field is sufficient to support a mass of 8 atomic mass units with a single charge on it. A helium ion will therefore experience a net upward force equivalent to 4 m_p, where m_p is the mass of a hydrogen atom; a hydrogen ion will experience a net upward force of 7 m_p; and these ions will therefore actually increase in concentration with altitude in a diffusion equilibrium situation. This increase with altitude will continue until the lighter ions finally come to predominate, at which point the electric field decreases and the lighter ions adopt a distribution in which their concentration falls with altitude. This is illustrated in Fig. 2. *See* UPPER-ATMOSPHERE DYNAMICS.

The terms heliosphere and protonosphere are sometimes used to designate the regions in which helium and hydrogen ions respectively are predominant. Sometimes, there may be no region in which helium ions dominate; the transition being from oxygen ions to hydrogen ions.

Magnetosphere and plasmasphere. The magnetosphere is that region in which the movement of ionized plasma is dominated by the geomagnetic field. It extends roughly from the F_1 region upward, as the movement of ions below about 150 km is dominated by collisions with neutral particles. Electrons are not dominated by collisions with neutral particles except below about 90 km.

The outer limit of the magnetosphere is set by interaction of the geomagnetic field with the solar wind, which compresses the magnetic field on the daytime side and limits its outer extension to about 10 earth radii. At that point, the compressed geomagnetic field is sufficiently strong to turn aside the solar wind, and this causes a shock front to develop in the solar wind, as indicated in Fig. 3. *See* GEOMAGNETISM.

On the nighttime side of the Earth, the magnetosphere extends far out into space as a long tail. A notable feature is the separation of the magnetic tail into two segments separated by a current sheet, as indicated in Fig. 3; such a current system must exist between two regions of oppositely directed magnetic field. The bundle of field lines in each section of the tail reaches the Earth's surface in a somewhat smaller area than that encompassed by the auroral zone. *See* AURORA; MAGNETOSPHERE; SOLAR MAGNETIC FIELD.

There is a sudden decrease in hydrogen ion concentration that occurs at a distance of about 4 Earth radii from the Earth's center in the equatorial plane. Ion concentrations fall by a factor of about two orders of magnitude, from values near 10^8 to about 10^6 per cubic meter. Since ions can move more or less freely along magnetic-field lines but not across them, this boundary is projected downward along magnetic-field lines and reaches the region of the F_2 peak near 60° geomagnetic latitude. The region of relatively high ion concentration within this boundary is known as the plasmasphere, and the boundary itself is called the plasmapause. The plasmapause is not entirely regular, as it moves in and out diurnally and with geomagnetic activity.

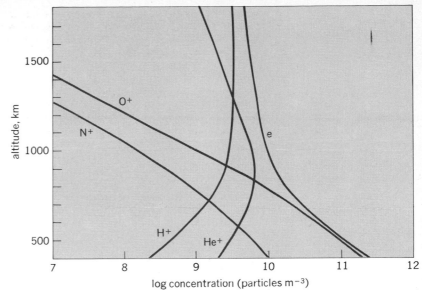

Fig. 2. Diffusion equilibrium distributions of N^+, O^+, He^+, and H^+ ions. Curve e is the total ion concentration, which equals the electron concentration.

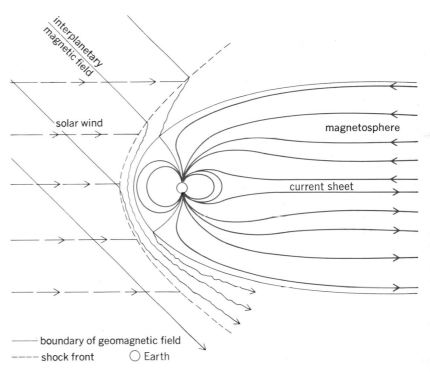

Fig. 3. The Earth's magnetosphere.

Spread F and sporadic E. Since the ionosphere is produced by solar radiation, one might expect it to be very uniform in its properties, but this is not so. There are many irregularities in the ionosphere with scale sizes varying at least from a few meters to a few kilometers. The cause of these is not known. In the F_2 region and above, the irregularities must have a structure aligned along the magnetic field. The phenomenon of spread F is apparently due to such irregularities, and the phenomenon is so named because it gives rise to a widened echo trace on the ionosonde record.

Sporadic E is a condition in which there is a thin region of greatly enhanced electron concentration—a region a kilometer or two thick in which the concentration increases by a factor of two or more. This is apparently caused by a strong wind shear that causes ionization to converge on the shear region by transport from above and below. If the ionization transported into the sporadic E layer includes several constituents, the species with higher recombination coefficients will be preferentially removed by recombination, and the species with the smallest recombination coefficients will accumulate. Sporadic E layers have been observed at times to consist of metallic ions, and it is thought that these are of meteoric origin. Such ions are probably always present as minor constituents in the E region, and they have lesser recombination coefficients than the normal atmospheric ions.

Electrodynamic drift. An important physical process in the atmosphere is electrodynamic drift and dynamo effects. Winds and the resultant movement of ionization in the geomagnetic field give rise to induced currents, which in turn perturb the geomagnetic field. These dynamo effects are recognizable in geomagnetic records made at the Earth's surface.

The magnetosphere is an active magnetoelectrodynamic system driven in part by electric fields generated by atmospheric motions in the dynamo region and in part by the solar wind. Although the processes of energy and momentum transfer from the solar wind are not satisfactorily understood in detail, there is no doubt about the facts that a general pattern of magnetospheric motion or convection exists, that it is due mainly to the solar wind, and that it can be described in terms of electric fields. Auroras and some ionospheric currents constitute a load on this system. Magnetospheric convection is probably responsible for the reduction in ion concentration outside the plasmasphere. The plasmasphere corotates with the Earth and is not involved in the pattern of magnetospheric convection.

Two important ionospheric effects have been recognized as resulting from the transport of ionization under influence of electric fields produced by winds and dynamo action at lower altitudes. One effect is the production of the equatorial anomaly in the F_2 region. The ionization maximum in the afternoon does not occur at the Equator or at the subsolar latitude; instead it is divided into two maxima lying about 10° north and south of the geomagnetic equator. The cause of this has been identified as an eastward-directed field that causes an upward drift of ionization at near equatorial latitudes. This is followed by diffusion downward along the magnetic-field lines under action of gravity. The net effect is a transport of ionization from near the geomagnetic equator to the region of the maxima about 10° north and south of it.

The other important effect is the maintenance of the nighttime F region. Recombination should cause a much more rapid decay than is actually observed; the lesser decay is due to upward electrodynamic drift into a region in which the loss processes are less rapid, due to the lesser concentrations of molecular species there. *See* ATMOSPHERE.   [FRANCIS S. JOHNSON]

Bibliography: P. M. Banks and G. Kocharts, *Aeronomy*, 1973; K. Davies, *Ionospheric Radio Propagation*, 1967; E. R. Dyer, *Solar-Terrestrial Physics/1970*, 1972; J. A. Ratcliffe, *An Introduction to the Ionosphere and Magnetosphere*, 1972; H. Rishbeth and O. K. Garriott, *Introduction to Ionospheric Physics*, 1969; R. C. Whitten and I. G. Poppoff, *Fundamentals of Aeronomy*, 1971.

Ion-permeable membrane

A film or sheet of a substance which is preferentially permeable to some species or types of ions. This selectivity is usually exhibited during electrical transport of ions across the membrane under an electrical potential, but the term is also applied to membranes showing selective mass transport under pressure or concentration gradients. Techniques based on these membranes may be used to deionize water or other solvents, to separate various ion species, or to carry out electrochemical reactions while isolating the products of reaction.

Principles. Membranes showing ion selectivity in electrical transport are said to be permselective and are usually composed of substances having ion-exchange properties. Such substances have a chemical structure incorporating fixed charge sites of one charge type, positive or negative, each one of which is electrically balanced by a counter ion of opposite charge type. The counter ion is readily exchanged for another mobile ion of similar charge type existing in the solution surrounding and permeating the ion-exchange material. The number of mobile ions from the solution which may exist in the membrane phase, in addition to the counter ions, is governed by a Donnan-type membrane equilibrium. *See* DONNAN EQUILIBRIUM; ION EXCHANGE.

After establishment of a Donnan equilibrium, the concentration of one charge type of ion within the membrane will exceed the other by the concentration of the counter ions. Therefore, the proportion of any electrical current crossing the membrane carried by the ions of the most numerous charge type may be very high, and the electrical mass transfer of these ions in the direction of the oppositely charged electrode correspondingly high. Membranes whose fixed charges are negative and counter ions positive are, under an electrical potential, preferentially permeable to cations; those having positive fixed charges and negative counter ions are preferentially permeable to anions.

The electrical transport number of one charge type of ion in a membrane depends on the electrolyte concentration of the surrounding solution in equilibrium with the membrane. This is a necessary consequence of the Donnan-membrane theory and leads to the important practical consequence that the selective ionic permeability of the membrane decreases as the external electrolyte concentration increases. At sufficiently high concentrations, the membrane will have no electrolytic selectivity and will serve simply as a diffusion barrier. Membranes with the highest fixed-charge concentration will be selective at the highest external concentration.

In addition to the electrical transport of ions across ion-permeable membranes, electroosmotic transport of water or solvent also takes place, with the net transport being in the direction of the counter-ion movement. The amount of water trans-

For the study of electric and magnetic fields in the ionosphere, clouds of barium atoms are released by rockets at high altitudes and become ionized by solar radiation, forming artificial plasma clouds. As shown in these sequences, the cloud changes color and shape as its atoms undergo ionization and as the charged particles respond to the electric and magnetic fields of the ionosphere. (*R. Lüst, Max-Planck-Institut für Physik und Astrophysik*)

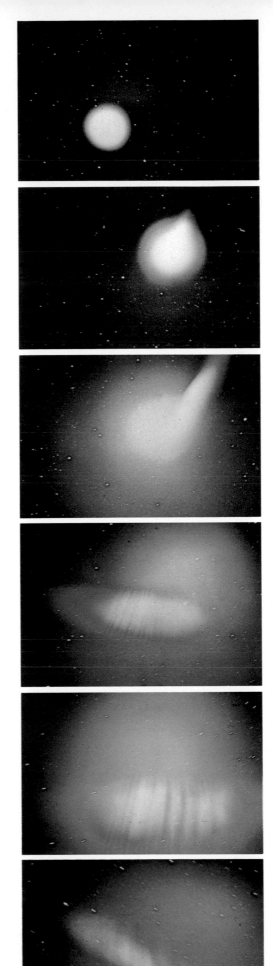

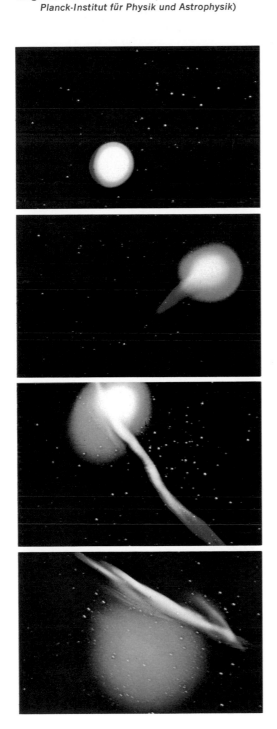

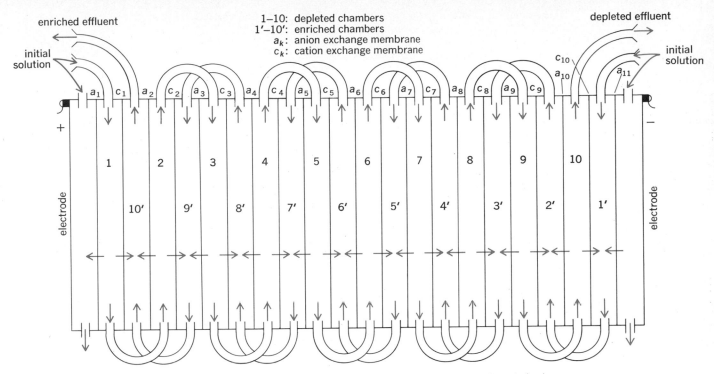

Schematic diagram of a multiple ion-exchange membrane electrodialysis cell. Movement of the ions through the membranes separating the compartments is shown by the small horizontal arrows.

1–10: depleted chambers
1'–10': enriched chambers
a_k: anion exchange membrane
c_k: cation exchange membrane

ported per faraday will depend on the properties of the ions simultaneously transported and on their transport numbers in the membrane. Thus, the net water transport is a maximum at low external solution concentrations.

Measurement of transport numbers within the membranes can be made by direct measurement of ionic transport in electrolytic cells or by measurement of membrane potentials across a membrane separating solutions of known and different electrolyte concentration. The departure of the measured potential from the ideal Nernst potential is a measure of the transport number of the counter ions in the membrane.

Membranes. The physical properties of ion-permeable membranes vary considerably. Ideally, they should be tough and flexible, readily hydrated, yet good diffusion barriers. Practical membranes for electrodialysis should have low resistivity and are usually less than 0.025 in. thick. For analytical measurement of ionic activities, low resistivity is not essential. The membranes may be of heterogeneous or homogeneous physical structure and may be supported internally by inert material.

The principal application of ion-permeable membranes is to electrodialytic deionization using electrolytic cells having multiple compartments separated by the membranes, as shown in the illustration.

Membranes labeled a_i are preferentially permeable to anions, and those labeled c_i are preferentially permeable to cations. Under an applied potential, the ions in compartments 1, 2, . . . , 10 will migrate through the ion-permeable membranes toward the electrodes into compartments 1', 2', . . . , 10'. Hence, alternate compartments will be depleted and the others enriched as electrical current flows. The alternate compartments may be joined as shown, the current traversing the continuous streams n times, n being the number of unit cells. A unit cell is defined as consisting of a depleted chamber and an adjacent concentrated chamber with the corresponding anion-permeable and cation-permeable membranes. The electrode chambers are fed independently to keep the electrode reaction products separate from the concentrated or depleted electrolyte streams. *See* DIALYSIS; ELECTROLYTIC CONDUCTANCE; ELECTROPHORESIS; SALINE WATER RECLAMATION.

[ALVIN G. WINGER; ROBERT KUNIN]

Bibliography: R. Kunin, *Ion Exchange Resins*, 2d ed., 1958; F. C. Nachod and J. Schubert (eds.), *Ion Exchange Technology*, 1956; S. B. Tuwiner, *Diffusion and Membrane Technology*, 1962.

Ipecac

A low, perennial shrub or half-shrub of the tropical forest in Brazil and Colombia. Several species are used, but the dried roots and rhizomes of *Cephaelis ipecacuanha* (Rubiaceae) constitute the material recognized as the official drug ipecac (see illustration). Medically the principal use is as an emetic and an expectorant. *See* RUBIALES.

[P. D. STRAUSBAUGH/EARL L. CORE]

IR drop

That component of the potential drop across a passive element (one which is not a seat of electromotive force) in an electric circuit caused by resistance of the element. This potential drop, by definition, is the product of the resistance R of the element and the current I flowing through it. The IR drop across a resistor is the difference of potential between the two ends of the resistor.

In a simple direct-current circuit containing a battery and a number of resistors, the sum of all the IR drops around the circuit (including that of the internal resistance of the battery itself) is equal

IPECAC

(a)

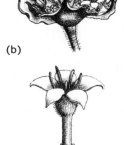
(b)

(c)

Ipecac (*Cephaelis ipecacuanha*). (a) Entire plant. (b) Inflorescence. (c) Single flower.

to the electromotive force of the battery. This is an important circuit theorem useful in the analytic solution of electrical networks. *See* KIRCHHOFF'S LAWS OF ELECTRIC CIRCUITS; NETWORK THEORY, ELECTRICAL; POTENTIALS; RESISTANCE, ELECTRICAL.   [JOHN W. STEWART]

Iridium

A chemical element, Ir, atomic number 77, and atomic weight 192.2. Iridium in the free state is a hard, white metallic substance.

Physical properties of iridium

Property	Value
Atomic weight, C^{12} = 12.00000	192.2
Naturally occurring isotopes and percent abundance	191 (37.3), 193 (62.7)
Crystal structure	Face-centered cubic
Lattice constant a at 25°C, angstrom units	3.8394
Thermal neutron capture cross section, barns	440
Common chemical valence	3, 4
Density at 25°C, g/cm³	22.55
Melting point, °C	2447
Boiling point, °C	4500
Specific heat at 0°C, cal/g	0.0307
Thermal conductivity 0–100°C, cal cm/cm² sec °C	0.35
Linear coefficient of thermal expansion 20–100°C, micro-in./in./°C	6.8
Electrical resistivity at 0°C, microhm-cm	4.71
Temperature coefficient of electrical resistance 0–100°C/°C	0.00427
Tensile strength (1000 psi) Soft	160–180
Hard	300–360
Young's modulus at 20°C psi, static	75.0×10^6
psi, dynamic	76.5×10^6
Hardness, DPN Soft	200–240
Hard	600–700

Uses. At ordinary temperatures iridium is the most corrosion-resistant element known. Iridium exhibits catalytic activity for many reactions—hydrogenation, dehydrogenation, oxidation, and others. However, it has not been widely applied since usually another precious metal is found superior. Difficulties in fabrication limit use of iridium in the pure form at ordinary temperatures. For such use it is usually alloyed with platinum as the base, since a platinum–30% iridium alloy, for instance, is almost as corrosion-resistant as iridium and is a great deal easier to fabricate. *See* PLATINUM.

Pure iridium is used in special aircraft spark plugs, where its resistance to lead corrosion is outstanding. Iridium crucibles are used in the high-temperature growth of laser crystals and high-temperature glass melting. This may be done in reducing or inert atmosphere or air. In the latter case, the iridium is often protected from oxidation by a ceramic coating. Iridium with rhodium is used in very-high-temperature thermocouples, the only such thermocouples which can be used in air above 1800°C. Iridium alloyed with 5% tungsten may be used as a spring material up to 800°C. Iridium is also used at elevated temperatures as wires in vacuum gages which may be exposed to air at times. In this and other applications it may be reinforced by a thorium oxide dispersion through the metal. Iridium may be applied over other materials by cladding, plating from aqueous or fused salt baths, and deposition from organometallic media. *See* RHODIUM.

Chemical and physical properties. When hot, iridium is reasonably ductile, and is worked hot for this reason. Iridium retains its ductility in the cold state after hot working as long as it is not annealed. When annealed, it loses its cold ductility and is then rather brittle. It may be fabricated into fine wire and thin sheet if care is taken to avoid annealing.

Iridium has considerably less oxidation resistance than platinum or rhodium, but more so than ruthenium or osmium. A thin, adherent oxide film, IrO_2, forms above about 600°C. This oxide dissociates around 1100°C, and above this temperature a volatile oxide, possibly IrO_3, is responsible for iridium weight loss. Although such weight loss is greater than that of platinum or rhodium, iridium is the only metal which can be used unprotected in air up to 2300°C with any degree of life expectancy. Iridium is resistant to many molten metals such as sodium, potassium, mercury, bismuth, and lithium. It is only slowly attacked by molten lead, cadmium, tin, silver, and gold. It is rapidly attacked by molten copper, aluminum, zinc, and magnesium. Iridium is not attacked by any acid, including aqua regia. A 30% iridium–platinum alloy is practically insoluble in aqua regia. Iridium is slightly attacked by fused sodium and potassium hydroxides and by fused sodium bicarbonate. Fusion with alkaline oxidizing fluxes is necessary to convert iridium to a soluble form. Dissolution of iridium may also be accomplished by fusing it with sodium chloride while treating the melt with chlorine. The extreme inertness of iridium-rich alloys creates problems in their analysis and refining. Iridium has a strong tendency to form coordination compounds. The table gives values for important physical properties of iridium.

Metallurgical extraction. Osmiridium, obtained in the extraction of platinum, is fused with zinc and subsequently digested with hydrochloric acid to convert the material to a fine powder. This powder is then fused with an alkaline oxidizing flux which converts the iridium to an acid-soluble form. The procedure is rarely quantitative, so that the insoluble residues must be recycled. Various hydrolytic separations are available to separate the iridium from other metals. Often the insolubility of iridium in lead is used to separate it from other precious metals. The relative insolubility of ammonium chloroiridate in water may also be used to effect separation. When heated, this compound yields metallic iridium. Separations are not quantitative; thus, lengthy recycling of tailings is required. *See* OSMIUM.

Principal compounds. Iridium trichloride, $IrCl_3$, is a green, water-insoluble compound, made by treating iridium powder with chlorine at 500°C. Sodium iridium(IV) chloride, $Na_2IrCl_6·6H_2O$, is a black, water-soluble crystalline solid made by heating a fused mixture of iridium and sodium chloride in chlorine, and then dissolving the resulting melt in water. Sodium iridium(III) chloride, $Na_3IrCl_6·12H_2O$, is an olive-green, water-soluble crystalline solid made by reducing a solution of sodium iridium(IV) chloride. Ammonium iridium-(IV) chloride, $(NH_4)_2IrCl_6$, is a red-black, relatively insoluble crystalline solid made by adding ammonium chloride to a solution of sodium iridium-(IV) chloride. Iridium trihydroxide, $Ir(OH)_3·xH_2O$, is a green-black, insoluble solid made by hydrolyzing a solution of iridium(III) chloride.

[HENRY J. ALBERT]

Iron

A chemical element, Fe, atomic number 26, and atomic weight 55.847. Iron is the fourth most abundant element in the crust of the Earth (5%). It is a malleable, tough, silver-gray, magnetic metal. It melts at 1540°C, boils at 2800°C, and has a density of 7.86 g/cm³. Each atom has 26 electrons, and the four stable, naturally occurring isotopes have masses of 54, 56, 57, and 58. The two main ores are hematite, Fe_2O_3, and limonite, $Fe_2O_3·3H_2O$; other ores are magnetite, Fe_3O_4, taconite (an iron silicate), and siderite, $FeCO_3$. Pyrites, FeS_2, and chromite, $Fe(CrO_2)_2$, are mined as ores for sulfur and chromium, respectively. Iron is found in many other minerals, and it occurs in ground waters and in the red hemoglobin of blood.

The greatest use of iron is for structural steels; cast iron and wrought iron are made in quantity, also. Magnets, dyes (inks, blueprint paper, rouge pigments), and abrasives (rouge) are among the other uses of iron and iron compounds.

The free metal is obtained in bulk by reduction of the ore by coke. Pure iron is difficult to obtain because other elements are held tenaciously; a pure material, may, however, be obtained by reduction of the oxide with hydrogen or by electrolysis. The chemistry of the zero oxidation state is chiefly that of the alloys; addition of trace impurities (carbon, phosphorus, silicon, nickel, manganese, chromium, and cobalt) has a marked effect on the properties of the metal. *See* CAST IRON; HEAT TREATMENT (METALLURGY); IRON ALLOYS; IRON EXTRACTION FROM ORE; STAINLESS STEEL; STEEL MANUFACTURE; WROUGHT IRON.

Properties of the metal. There are several allotropic forms of iron. Ferrite or α-iron is stable up to 760°C. The change to β-iron involves primarily a loss of magnetic permeability because the lattice structure (body-centered cubic) is unchanged. The allotrope called γ-iron has the cubic close-packed arrangements of atoms and is stable from 910 to 1400°C. Little is known about δ-iron except that it is stable above 1400°C and has a lattice similar to that of α-iron.

The metal is a good reducing agent and, depending on conditions, can be oxidized to the 2+, 3+, or 6+ state. In most iron compounds, the ferrous ion, iron(II), or ferric ion, iron(III), is present as a distinct unit. Iron(II) is found in simple compounds such as FeO, FeS, $FeBr_2$, and $FeCO_3$, and in hydrated compounds such as $FeSO_4·7H_2O$, $FeF_2·8H_2O$, $FeCl_2·4H_2O$, and ferrous acetate tetrahydrate. Ferrous sulfate and ferrous ammonium sulfate (Mohr's salt), $(NH_4)_2SO_4·FeSO_4·6H_2O$, have been employed as standards for oxidimetric titrations. The oxide, FeO, and hydroxide, $Fe(OH)_2$, are quite basic; they react with strong and weak acids to form salts. The ferrous compounds are usually light yellow to dark green-brown in color; the hydrated ion, $Fe(H_2O)_6^{2+}$, which is found in many compounds and in solution, is light green. This ion has little tendency to form coordination complexes except with strong reagents such as cyanide ion, polyamines, and porphyrins.

Oxidation of ferrous ion to ferric ion is moderately difficult in acid solution, but occurs readily in basic solution because of the insolubility of ferric hydroxide. The electrode potential for the ferrous-ferric reaction is dependent on the complexing species present in the solution. Anions such as CN^-, F^-, and PO_4^{3-} stabilize the 3+ oxidation state, whereas amines such as phenanthroline stabilize the 2+ state.

Coordination compounds. The ferric ion, because of its high charge (3+) and its small size (0.53 A as compared with 0.75 A for ferrous ion), has a strong tendency to hold anions. Ferric hydroxide is only weakly basic. Salts are formed with strong acids, but they are normally hydrated, as in $Fe(NO_3)_3·9H_2O$ and ferric ammonium alum, $NH_4Fe(SO_4)_2·12H_2O$. Because of similarities in size and charge, aluminum ion and ferric ion have quite analogous chemistries, although ferric hydroxide is not amphoteric. The anhydrous chlorides and bromides have dimeric vapors, for example, Fe_2Cl_6, and catalyze certain organic reactions. The hydrated ion, $Fe(H_2O)_6^{3+}$, which is found in solution, combines with OH^-, F^-, Cl^-, CN^-, SCN^-, N_3^-, $C_2O_4^{2-}$, and other anions to form coordination complexes. As the concentration of anion increases, more groups combine with each cation until combinations such as $FeCl_4^-$ and $Fe(CN)_6^{3-}$ are reached.

The ferrate ion FeO_4^{2-}, which contains iron in the 6+ oxidation state, is obtained by reaction with strong oxidants such as hypochlorite in alkaline solution. Salts such as Na_2FeO_4 and $BaFeO_4$ are readily prepared and are fairly stable. Ferrate solutions are reddish purple. They release oxygen slowly in alkaline media, rapidly in acid.

An interesting aspect of iron chemistry is the array of compounds with bonds to carbon. Cementite, Fe_3C, is a component of steel. The cyanide complexes of both ferrous and ferric iron are very stable and are not strongly magnetic in contradistinction to most iron coordination complexes. The cyanide complexes form colored salts, including the famous prussian blue, $KFe_2(CN)_6$, made from ferric ion and potassium ferrocyanide, $K_4Fe(CN)_6$. The compound Turnbull's blue, made from ferrous ion and potassium ferricyanide, $K_3Fe(CN)_6$, is believed to be identical to prussian blue. There are many compounds containing five cyanide groups and one other group (such as NO, CO, SO_3^{2-}, NO_2^-, NH_3, and H_2O) about the iron. The compound $Na_2Fe(NO)(\check{C}N)_5 \cdot 2H_2O$ is one such compound; it can be used to test for sulfide ion.

The three carbonyls $Fe(CO)_5$, $Fe_2(CO)_9$, and $Fe_3(CO)_{12}$ also have iron-to-carbon bonds. Iron pentacarbonyl is formed directly by the reaction of iron with carbon monoxide under pressure; the other carbonyls are synthesized by indirect means. The pentacarbonyl compound is used as a source of pure iron. See METAL CARBONYL.

About 1950, a new type of iron compound was discovered. Cyclopentadiene can react with iron oxide to give dicyclopentadienyl iron, Fe$(C_5H_5)_2$. Many alternative syntheses of this compound, called ferrocene, are available. The structure is that of an iron atom between two symmetrical pentagonal rings, thus bringing to mind the name sandwich compound. Ferrocene has remarkable thermal stability. See ORGANOMETALLIC COMPOUND.

Many derivatives of ferrocene with substituent groups attached to the ring carbons are known; their chemistries are more in the line of organic reactions than reactions of iron. Ferrocene can be oxidized to the ferrocinium ion, $Fe(C_5H_5)_2^+$, and salts containing this ion have been prepared. See COBALT; FERROCENE; NICKEL; TRANSITION ELEMENTS.   [JOHN O. EDWARDS]

Iron alloys

Solid solutions of metals, one metal being iron. A great number of commercial alloys have iron as an intentional constituent. Iron is the major constituent of wrought and cast iron and wrought and cast steel. Alloyed with usually large amounts of silicon, manganese, chromium, vanadium, molybdenum, niobium (columbium), selenium, titanium, phosphorus, or other elements, singly or sometimes in combination, iron forms the large group of materials known as ferroalloys that are important as addition agents in steelmaking. Iron is also a major constituent of many special-purpose alloys developed to have exceptional characteristics with respect to magnetic properties, electrical resistance, heat resistance, corrosion resistance, and thermal expansion. Table 1 lists some of these alloys. See ALLOY; FERROALLOY; STEEL.

Because of the enormous number of commercially available materials, this article is limited to some of the better-known types of alloys. Emphasis is on special-purpose alloys; practically all of these contain relatively large amounts of an alloying element or elements referred to in the classification. Alloys containing less than 50% iron are excluded, with a few exceptions.

Iron-aluminum alloys. Although pure iron has ideal magnetic properties in many ways, its low electrical resistivity makes it unsuitable for use in alternating-current (ac) magnetic circuits. Addition of aluminum in fairly large amounts increases the electrical resistivity of iron, making the resulting alloys useful in such circuits.

Three commercial iron-aluminum alloys having moderately high permeability at low field strength and high electrical resistance nominally contain 12% aluminum, 16% aluminum, and 16% aluminum with 3.5% molybdenum, respectively. These three alloys are classified as magnetically soft materials; that is, they become magnetized in a magnetic field but are easily demagnetized when the field is removed.

The addition of more than 8% aluminum to iron results in alloys that are too brittle for many uses because of difficulties in fabrication. However, addition of aluminum to iron markedly increases its resistance to oxidation. One steel containing 6% aluminum possesses good oxidation resistance up to 2300°F.

Iron-carbon alloys. The principal iron-carbon alloys are wrought iron, cast iron, and steel.

Wrought iron of good quality is nearly pure iron; its carbon content seldom exceeds 0.035%. In addition, it contains 0.075–0.15% silicon, 0.10 to less than 0.25% phosphorus, less than 0.02% sulfur, and 0.06–0.10% manganese. Not all of these elements are alloyed with the iron; part of them may be associated with the intermingled slag that is a characteristic of this product. Because of its low carbon content, the properties of wrought iron cannot be altered in any useful way by heat treatment. See WROUGHT IRON.

Cast iron may contain 2–4% carbon and varying amounts of silicon, manganese, phosphorus, and sulfur to obtain a wide range of physical and mechanical properties. Alloying elements (silicon, nickel, chromium, molybdenum, copper, titanium, and so on) may be added in amounts varying from a few tenths to 30% or more. Many of the alloyed cast irons have proprietary compositions. See CAST IRON; CORROSION.

Steel is a generic name for a large group of iron alloys that include the plain carbon and alloy steels. The plain carbon steels represent the most important group of engineering materials known. Although any iron-carbon alloy containing less than about 2% carbon can be considered a steel, the American Iron and Steel Institute (AISI) standard carbon steels embrace a range of carbon contents from 0.06% maximum to about 1%. In the early days of the American steel industry, hundreds of steels with different chemical compositions were produced to meet individual demands of purchasers. Many of these steels differed only slightly from each other in chemical composition.

Table 1. Some typical composition percent ranges of iron alloys classified by important uses*

Type	Fe	C	Mn	Si	Cr	Ni	Co	W	Mo	Al	Cu	Ti
Heat-resistant alloy castings	Bal.	0.30–0.50		1–2	8–30	0–7						
	Bal	0.20–0.75		2–2.5	10–30	8–41						
Heat-resistant cast irons	Bal.	1.8–3.0	0.3–1.5	0.5–2.5	15–35	5 max						
	Bal.	1.8–3.0	0.4–1.5	1.0–2.75	1.75–5.5	14–30			1		7	
Corrosion-resistant alloy castings	Bal.	0.15–0.50	1 max	1	11.5–30	0–4	0.5 max					
	Bal.	0.03–0.20	1.5 max	1.5–2.0	18–27	8–31						
Corrosion-resistant cast irons	Bal.	1.2–4.0	0.3–1.5	0.5–3.0	12–35	5 max			4 max		3 max	
	Bal.	1.8–3.0	0.4–1.5	1.0–2.75	1.75–5.5	14–32			1 max		7 max	
Magnetically soft materials	Bal.			0.5–4.5								
	Bal.								3.5	16		
	Bal.									16		
	Bal.									12		
Permanent-magnet materials	Bal.						12		17			
	Bal.						12		20			
	Bal.					20	5			12		
	Bal.					17	12.5			10	6	
	Bal.					25				12		
	Bal.					28	5			12		
	Bal.					14	24			8	3	
	Bal.					15	24			8	3	1.25
Low-expansion alloys	Bal.		0.15	0.33		36						
	Bal.		0.24	0.03		42						
	61–53	0.5–2.0	0.5–2.0	0.5–2.0	4–5	33–35		1–3				

*This table does not include any AISI standard carbon steels, alloy steels, or stainless and heat-resistant steels or plain or alloy cast iron for ordinary engineering uses; it includes only alloys containing at least 50% iron, with a few exceptions. Abbreviation bal. indicates balance percent of composition.

Studies were undertaken to provide a simplified list of fewer steels that would still serve the varied needs of fabricators and users of steel products. The Society of Automotive Engineers (SAE) and the AISI both were prominent in this effort, and both periodically publish lists of steels, called standard steels, classified by chemical composition. These lists are published in the *SAE Handbook* and the AISI's *Steel Products Manuals*. The lists are altered periodically to accommodate new steels and to provide for changes in consumer requirements. There are minor differences between some of the steels listed by the AISI and SAE. Only the AISI lists will be considered here. The standard steels represent a large percentage of all steel produced and, although considerably fewer in number, have successfully replaced the large number of specialized compositions formerly used.

A numerical system is used to indicate grades of standard steels. Provision also is made to use certain letters of the alphabet to indicate the steel-making process, certain special additions, and steels that are tentatively standard, but these are not pertinent to this discussion. Table 2 gives the basic numerals for the AISI classification and the corresponding types of steels. In this system the first digit of the series designation indicates the type to which a steel belongs; thus 1 indicates a carbon steel, 2 indicates a nickel steel, and 3 indicates a nickel-chromium steel. In the case of simple alloy steels, the second numeral usually indicates the percentage of the predominating alloying element. Usually, the last two (or three) digits indicate the average carbon content in points, or hundredths of a percent. Thus, 2340 indicates a nickel steel containing about 3% nickel and 0.40% carbon.

All carbon steels contain minor amounts of manganese, silicon, sulfur, phosphorus, and sometimes other elements. At all carbon levels the mechanical properties of carbon steel can be varied to a useful degree by heat treatments that alter its microstructure. Above about 0.25% carbon steel can be hardened by heat treatment. However, most of the carbon steel produced is used without a final heat treatment. *See* HEAT TREATMENT (METALLURGY).

Alloy steels are steels with enhanced properties attributable to the presence of one or more special elements or of larger proportions of manganese or silicon than are present ordinarily in carbon steel. The major classifications of alloy steels are:

High-strength, low-alloy
AISI alloy
Alloy tool
Stainless
Heat-resisting
Electrical
Austenitic manganese

Some of these iron alloys are discussed briefly in this article; for more detailed attention *see* STAINLESS STEEL; STEEL.

Iron-chromium alloys. An important class of iron-chromium alloys is exemplified by the wrought stainless and heat-resisting steels of the type 400 series of the AISI standard steels, all of which contain at least 12% chromium, which is about the minimum chromium content that will confer stainlessness. However, considerably less than 12% chromium will improve the oxidation resistance of steel for service up to 1200°F, as is true of AISI types 501 and 502 steels that nominally contain about 5% chromium and 0.5% molybdenum. A comparable group of heat- and corrosion-resistant alloys, generally similar to the 400 series of the AISI steels, is covered by the Alloy Casting Institute specifications for cast steels.

Corrosion-resistant cast irons alloyed with chromium contain 12–35% of that element and up to 5% nickel. Cast irons classified as heat-resistant contain 15–35% chromium and up to 5% nickel.

During World War I a high-carbon steel used for

Table 2. AISI standard steel designations

Type	Series designation*	Composition
Carbon steels	10xx	Nonresulfurized carbon steel
	11xx	Resulfurized carbon steel
	12xx	Rephosphorized and resulfurized carbon steel
Constructional alloy steels	13xx	Manganese 1.75%
	23xx	Nickel 3.50%
	25xx	Nickel 5.00%
	31xx	Nickel 1.25%, chromium 0.65%
	33xx	Nickel 3.50%, chromium 1.55%
	40xx	Molybdenum 0.25%
	41xx	Chromium 0.50 or 0.95%, molybdenum 0.12 or 0.20%
	43xx	Nickel 1.80%, chromium 0.50 or 0.80%, molybdenum 0.25%
	46xx	Nickel 1.55 or 1.80%, molybdenum 0.20 or 0.25%
	47xx	Nickel 1.05%, chromium 0.45%, molybdenum 0.20%
	48xx	Nickel 3.50%, molybdenum 0.25%
	50xx	Chromium 0.28 or 0.40%
	51xx	Chromium 0.80, 0.90, 0.95, 1.00, or 1.05%
	5xxxx	Carbon 1.00%, chromium 0.50, 1.00, or 1.45%
	61xx	Chromium 0.80 or 0.95%, vanadium 0.10 or 0.15% min
	86xx	Nickel 0.55%, chromium 0.50 or 0.65%, molybdenum 0.20%
	87xx	Nickel 0.55%, chromium 0.50%, molybdenum 0.25%
	92xx	Manganese 0.85%, silicon 2.00%
	93xx	Nickel 3.25%, chromium 1.20%, molybdenum 0.12%
	98xx	Nickel 1.00%, chromium 0.80%, molybdenum 0.25%
Stainless and heat-resisting steels	2xx	Chromium-nickel-manganese steels; nonhardenable, austenitic, and nonmagnetic
	3xx	Chromium-nickel steels; nonhardenable, austenitic, and nonmagnetic
	4xx	Chromium steels of two classes: one class hardenable, martensitic, and magnetic; the other nonhardenable, ferritic, and magnetic
	5xx	Chromium steels; low chromium heat resisting

*The "x's" are replaced by actual numerals in defining a steel grade, as explained in the text.

making permanent magnets contained 1–6% chromium (usually around 3.5%); it was developed to replace the magnet steels containing tungsten that had been formerly but could not then be made because of a shortage of tungsten. See MAGNET.

Iron-chromium-nickel alloys. The wrought stainless and heat-resisting steels represented by the type 200 and the type 300 series of the AISI standard steels are an important class of iron-chromium-nickel alloys. A comparable series of heat- and corrosion-resistant alloys is covered by specifications of the Alloy Casting Institute. Heat- and corrosion-resistant cast irons contain 15–35% chromium and up to 5% nickel.

Iron-chromium-aluminum alloys. Electrical-resistance heating elements are made of several iron alloys of this type. Nominal compositions are 72% iron, 23% chromium, 5% aluminum; and 55% iron, 37.5% chromium, 7.5% aluminum. The iron-chromium-aluminum alloys (with or without 0.5–2% cobalt) have higher electrical resistivity and lower density than nickel-chromium alloys used for the same purpose. When used as heating elements in furnaces, the iron-chromium-aluminum alloys can be operated at temperatures of 2350°C maximum. These alloys are somewhat brittle after elevated temperature use and have a tendency to grow or increase in length while at temperature, so that heating elements made from them should have additional mechanical support. Addition of niobium (columbium) reduces the tendency to grow.

Because of its high electrical resistance, the 72% iron, 23% chromium, 5% aluminum alloy (with 0.5% cobalt) can be used for semiprecision resistors in, for example, potentiometers and rheostats. See RESISTANCE, ELECTRICAL; RESISTANCE HEATING.

Iron-cobalt alloys. Magnetically soft iron alloys containing up to 65% cobalt have higher saturation values than pure iron. The cost of cobalt limits the use of these alloys to some extent. The alloys also are characterized by low electrical resistivity and high hysteresis loss. Alloys containing more than 30% cobalt are brittle unless modified by additional alloying and special processing. Two commercial alloys with high permeability at high field strengths (in the annealed condition) contain 49% cobalt with 2% vanadium, and 35% cobalt with 1% chromium. The latter alloy can be cold-rolled to a strip that is sufficiently ductile to permit punching and shearing. In the annealed state, these alloys can be used in either ac or dc applications. The alloy of 49% cobalt with 2% vanadium has been used in pole tips, magnet yokes, telephone diaphragms, special transformers, and ultrasonic equipment. The alloy of 35% cobalt with 1% chromium has been used in high-flux-density motors and transformers as well as in some of the applications listed for the higher-cobalt alloy.

Although seldom used now, two high-carbon alloys called cobalt steel were used formerly for making permanent magnets. These were both high-carbon steels. One contained 17% cobalt, 2.5% chromium, 8.25% tungsten; the other contained 36% cobalt, 5.75% chromium, 3.75% tungsten. These are considered magnetically hard materials as compared to the magnetically soft materials discussed in this article.

Iron-manganese alloys. The important commercial alloy in this class is an austenitic manganese steel (sometimes called Hadfield manganese steel after its inventor) that nominally contains 1.2% carbon and 12–13% manganese. This steel is highly resistant to abrasion, impact, and shock.

Iron-nickel alloys. The iron-nickel alloys discussed here exhibit a wide range of properties related to their nickel contents.

Nickel content of a group of magnetically soft materials ranges from 40 to 60%; however, the highest saturation value is obtained at about 50%. Alloys with nickel content of 45–50% are characterized by high permeability and low magnetic losses. They are used in such applications as audio transformers, magnetic amplifiers, magnetic shields, coils, relays, contact rectifiers, and choke coils. The properties of the alloys can be altered to meet specific requirements by special processing techniques involving annealing in hydrogen to minimize the effects of impurities, grain-orientation treatments, and so on.

Another group of iron-nickel alloys, those containing about 30% nickel, is used for compensating changes that occur in magnetic circuits due to temperature changes. The permeability of the alloys decreases predictably with increasing temperature.

Low-expansion alloys are so called because they have low thermal coefficients of linear expansion. Consequently, they are valuable for use as standards of length, surveyors' rods and tapes, compensating pendulums, balance wheels in timepieces, glass-to-metal seals, thermostats, jet-engine parts, electronic devices, and similar applications.

The first alloy of this type contained 36% nickel with small amounts of carbon, silicon, and manganese (totaling less than 1%). Subsequently, a 39% nickel alloy with a coefficient of expansion equal to that of low-expansion glasses and a 46% nickel alloy with a coefficient equal to that of platinum were developed. Another important alloy is one containing 42% nickel that can be used to replace platinum as lead-in wire in light bulbs and vacuum tubes by first coating the alloy with copper. An alloy containing 36% nickel and 12% chromium has a constant modulus of elasticity and low expansivity over a broad range of temperatures. Substitution of 5% cobalt for 5% nickel in the 36% nickel alloy decreases its expansivity. Small amounts of other elements affect the coefficient of linear expansion, as do variations in heat treatment, cold-working, and other processing procedures.

A 9% nickel steel is useful in cryogenic and similar applications because of good mechanical properties at low temperatures. Two steels (one containing 10–12% nickel, 3–5% chromium, about 3% molybdenum, and lesser amounts of titanium and aluminum and another with 17–19% nickel, 8–9% cobalt, 3–3.5% molybdenum, and small amounts of titanium and aluminum) have exceptional strength in the heat-treated (aged) condition. These are known as maraging steels.

Cast irons containing 14–30% nickel and 1.75–5.5% chromium possess good resistance to heat and corrosion. *See* THERMAL EXPANSION.

Iron-silicon alloys. There are two types of iron-silicon alloys that are commercially important: the magnetically soft materials designated silicon or electrical steel, and the corrosion-resistant, high-silicon cast irons.

Most silicon steels used in magnetic circuits contain 0.5–5% silicon. Alloys with these amounts of silicon have high permeability, high electrical resistance, and low hysteresis loss compared with relatively pure iron. Most silicon steel is produced in flat-rolled (sheet) form and is used in transformer cores, stators and rotors of motors, and so on that are built up in laminated-sheet form to reduce eddy-current losses. Silicon-steel electrical sheets, as they are called commercially, are made in two general classifications: grain-oriented and nonoriented.

The grain-oriented steels are rolled and heat-treated in special ways to cause the edges of most of the unit cubes of the metal lattice to align themselves in the preferred direction of optimum magnetic properties. Magnetic cores are designed with the main flux path in the preferred direction, thereby taking advantage of the directional properties. The grain-oriented steels contain about 3.25% silicon and are used in the highest efficiency distribution and power transformers and in large turbine generators.

The nonoriented steels may be subdivided into low-, intermediate-, and high-silicon classes. Low-silicon steels contain about 0.5–1.5% silicon and are used principally in rotors and stators of motors and generators; steels containing about 1% silicon are also used for reactors, relays, and small intermittent-duty transformers. Intermediate-silicon steels contain about 2.5–3.5% silicon and are used in motors and generators of average to high efficiency and in small- to medium-size intermittent-duty transformers, reactors, and motors. High-silicon steels contain about 3.75–5% silicon and are used in power transformers and communications equipment and in highest efficiency motors, generators, and transformers.

High-silicon cast irons containing 14–17% silicon and sometimes up to 3.5% molybdenum possess corrosion resistance that makes them useful for acid-handling equipment and for laboratory drain pipes.

Iron-tungsten alloys. Although tungsten is used in several types of relatively complex alloys (including high-speed steels not discussed here), the only commercial alloy made up principally of iron and tungsten was a tungsten steel containing 0.5% chromium in addition to 6% tungsten that was used up to the time of World War I for making permanent magnets.

Hard-facing alloys. Hard-facing consists of welding a layer of metal of special composition on a metal surface to impart some special property not possessed by the original surface. The deposited metal may be more resistant to abrasion, corrosion, heat, or erosion than the metal to which it is applied. A considerable number of hard-facing alloys are available commercially. Many of these would not be considered iron alloys by the 50% iron content criterion adopted for the iron alloys in this article, and they will not be discussed here. Among the iron alloys are low-alloy facing materials containing chromium as the chief alloying element, with smaller amounts of manganese, silicon, molybdenum, vanadium, tungsten, and in some cases nickel to make a total alloy content of up to 12%, with the balance iron. High-alloy ferrous materials containing a total of 12–25% alloying elements form another group of hard-facing alloys;

a third group contains 26–50% alloying elements. Chromium, molybdenum, and manganese are the principal alloying elements in the 12–25% group; smaller amounts of molybdenum, vanadium, nickel, and in some cases titanium are present in various proportions. In the 26–50% alloys, chromium (and in some cases tungsten) is the principal alloying element, with manganese, silicon, nickel, molybdenum, vanadium, niobium (columbium), and boron as the elements from which a selection is made to bring the total alloy content within the 26–50% range.

Permanent-magnet alloys. These are magnetically hard ferrous alloys, many of which are too complex to fit the simple compositional classification used above for other iron alloys. As already mentioned in discussing iron-cobalt and iron-tungsten alloys, the high-carbon steels (with or without alloying elements) are now little used for permanent magnets. These have been supplanted by a group of sometimes complex alloys with much higher retentivities. The ones considered here are all proprietary compositions. Two of the alloys contain 12% cobalt and 17% molybdenum and 12% cobalt and 20% molybdenum.

Members of a group of six related alloys contain iron, nickel, aluminum, and with one exception cobalt; in addition, three of the cobalt-containing alloys contain copper and one has copper and titanium. Unlike magnet steels, these alloys resist demagnetization by shock, vibration, or temperature variations. They are used in magnets for radio speakers, watt-hour meters, magnetrons, torque motors, panel and switchboard instruments, and so on, where constancy and degree of magnet strength are important. *See* ALLOY STRUCTURES; MAGNETIC MATERIALS; MAGNETISM.

[HAROLD E. MC GANNON]

Bibliography: American Chemical Society, *Corrosion Resistance of Metals and Alloys*, 2d ed., 1963; American Iron and Steel Institute, *Alloy Steel*, 1964; American Iron and Steel Institute, *Flat-Rolled Electrical Steel*, 1964; American Iron and Steel Institute, *Stainless and Heat Resisting Steels*, 1963; American Iron and Steel Institute, *Steel Products Manual: Carbon Steel*, 1957; E. C. Bain and W. H. Paxton, *Functions of the Alloying Elements in Steel*, 2d ed., 1961; Engineering Alloys Digest, *Alloy Digest*, annual; H. E. McGannon (ed.), *The Making, Shaping and Treatment of Steel*, 1964; Society of Automotive Engineers, *SAE Handbook*, 1970; N. E. Woldman (ed.), *Engineering Alloys*, 4th ed., 1962.

Iron extraction from ore

Iron is extracted from iron ore and concentrates by pyrometallurgical processes based upon the application of heat and reducing gases. At temperatures of 700–1100°C, iron oxides such as hematite (Fe_2O_3) and magnetite (Fe_3O_4), the principal iron-bearing minerals, are readily reduced to metallic iron. Metallic iron (Fe) is formed from hematite by reactions (1) and (2) and from magnetite by similar reactions.

$$Fe_2O_3 + 3CO \rightarrow 2Fe + 3CO_2 \quad (1)$$

$$Fe_2O_3 + 3H_2 \rightarrow 2Fe + 3H_2O \quad (2)$$

Reduction is carried out in blast furnaces and in so-called direct-reduction processes. The blast-furnace process produces molten pig iron containing about 4.0% carbon and a total of approximately 3.0% of manganese, silicon, and phosphorus, which must be largely removed in the conversion of pig iron into steel, the principal outlet for iron.

Direct-reduction processes (in rotary kilns, fluidized beds, or traveling grates) operate below the melting point of iron. They produce solid reduced iron that may be converted to steel with little further refining. The iron is relatively free of the carbon, silicon, and manganese that form alloys with iron in the hotter blast-furnace operation. The reducing agent in direct-reduction processes is solid carbon, hydrogen, or carbon monoxide; for blast furnaces it is coke, supplemented with natural gas or oil.

Blast-furnace process. The blast-furnace process is the most widely used method for production of molten pig iron which can be cast into useful shapes or refined into steel. Short primitive furnaces, which were operated prior to about 1350, produced low-carbon and, therefore, malleable iron, called wrought iron, which could be fashioned into useful articles by reheating and hot working. The operation of the short furnaces was interrupted periodically to remove the solid metal, which contained a small percentage of occluded slag. Higher temperatures, stronger reducing conditions, and the longer time required for smelting reactions to occur in larger furnaces produced carburized and liquefied iron and a liquid slag from the ore gangue and the flux. The production of liquid pig iron and slag in a continuous operation marked the beginning of the blast-furnace process. This transition from the production of wrought iron

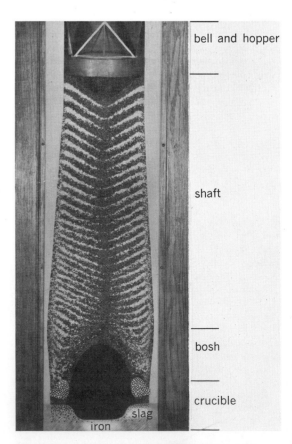

Fig. 1. Arrangement of materials in furnace model.

to pig iron was the result of the development of equipment for forcing a strong blast of air into a deeper bed of fuel and driving the products of combustion upward through an overlying column or bed of raw materials consisting of ore, coke, and flux. The blast furnace consists, therefore, of a heat-transfer reduction and melting chamber superimposed on a gas producer.

The general shape and arrangement of the charges in a blast furnace are shown in Fig. 1, a half section of a model furnace. In order to avoid mixing of sizes, hence loss of void space and permeability, solids are charged in layers. The dark layers represent the relatively coarse coke, and the white layers the charges of relatively fine ore. This layered arrangement is inherent in finer size of the ore, its steeper angle of repose, and the flow of material off the conical-shaped bell when it is lowered to discharge the raw materials from the hopper shown at the top of the model. Liquid iron and slag collect in a bottom cylindrical section called the hearth. The lighter, pear-shaped areas near the top of the crucible correspond to combustion zones. Preheated air is forced into these combustion zones through water-cooled copper castings called tuyeres.

An inverted frustrum of a cone, located immediately above the hearth, is known as the bosh. The formation of molten pig iron and slag in the bosh permits a gradual reduction in cross-sectional area at lower levels. The upper part of the furnace, called the shaft, is conical except for short cylindrical sections. Widening of the shaft at lower levels permits the charge to move downward without mechanical bridging as it expands upon heating. Additional cross-sectional area at lower levels of the shaft also permits a slower descent of the charge and a longer residence time at temperatures where ore reduction and other metallurgical reactions occur. Similarly, more area in the wider section of the furnace reduces the velocity of the ascending gases, prolonging gas-solid contact.

Structurally, the furnace consists of a steel shell lined with fire brick to conserve heat. The bosh and hearth are water-cooled to maintain the refractory lining. The trend is to use carbon linings in the hearth and bosh and to insert bronze cooling plates in the lining of the lower part of the shaft. The hearth diameter of modern furnaces ranges from 28 to 37 ft and the working volume from about 50,000 to 100,000 ft^3. Raw materials of more uniform size have permitted some lowering of overall height to about 95 ft.

Ore of high iron content reduces the amount of coke and stone (fluxing materials) required per ton of pig. Replacement of coke and stone with ore increases the weight of iron per cubic foot of charge. An increase from 50 to 60% in the iron content of the ore will, for example, introduce about 30% more iron per cubic foot of overall charge.

Permeable charges permit larger volumes of gases to move through the stock column more uniformly without building up pressures which support the charge and prevent it from settling regularly. Under such conditions, a larger volume of blast can be applied, more coke can be burned, the charge will settle more rapidly, and more pig iron will be produced. In general, beds composed of particles of uniform size will offer less resistance to the flow of gases.

The value of uniformity in the size of the coke has been fully appreciated for many years because coke occupies about 60% of the volume of the charge. Pieces smaller than 3/4 in. are normally discarded for furnace use by screening the coke immediately before charging. Care is taken to blend the coals available to obtain a strong coke which will not fracture into small sizes upon handling prior to charging or in the furnace. The steps involved in the extraction of iron in the blast furnace are shown in Fig. (2).

Size preparation of raw materials. Since 1956, the large capital expenditures required to build new blast furnaces and the pressure, due to inflation, to increase the productivity of existing furnaces have stimulated a pronounced interest in the size preparation of iron ore and concentrates. The depletion of high-grade deposits has necessitated the beneficiation of low-grade deposits and the production of concentrates that must be agglomerated before they can be used in the blast furnace. During agglomeration, small sizes are converted to larger lumps by the application of heat, which forms porous masses by grain growth and causes slag bonding without complete fusion. Although the use of agglomerated concentrates is increasing, large tonnages of direct-shipping ore containing 52–60% Fe will be available for at least several decades. This ore can be used effectively in the blast furnace providing coarse sizes are crushed to about 1 in. and particles smaller than 3/8 in. are agglomerated by sintering. This procedure eliminates particles in a size range which offers high resistance to gas flow and also greatly reduces the amount of fine ore carried off as flue dust.

A recent development is to incorporate fluxes such as calcium oxide or magnesium oxide into agglomerated material in the form of irregularly shaped sinter or spherical pellets. Pellets are produced by heating wet balls made from finely divided concentrates. The production of self-fluxing agglomerates permits two-component charges consisting of coke and self-fluxing agglomerates or to three-component charges of coke, screened ore, and self-fluxing agglomerates. In comparison with the unsized ore charges used in the past, these more ideal charges of sized materials will make it possible to burn about 30% more coke per day per furnace and to increase production accordingly.

From 1950 to 1970, daily output of single, leading furnaces increased more than 4-fold. Larger furnaces, more permeable charges, and additions of oxygen and steam to the blast led to a 2.33-fold increase in the amount of coke consumed per furnace. Concurrently, higher blast temperatures, richer ores or agglomerates, partially fluxed burdens, and tuyere injections of hydrocarbon fuels increased the number of pounds of hot metal made per pound of coke 1.85 times. Thus, the burning of more coke per day along with the more efficient use of coke permitted a 4.33-fold increase in daily production ($2.33 \times 1.85 = 4.31$; $1500 \times 4.31 = 6465$ tons per day).

Reactions in the furnace. Heat and reducing gases are released near the base of the furnace by the combustion of coke. Because air contains about 79% nitrogen, N_2, and 21% oxygen, O_2, combustion in a deep, full bed can be expressed as shown

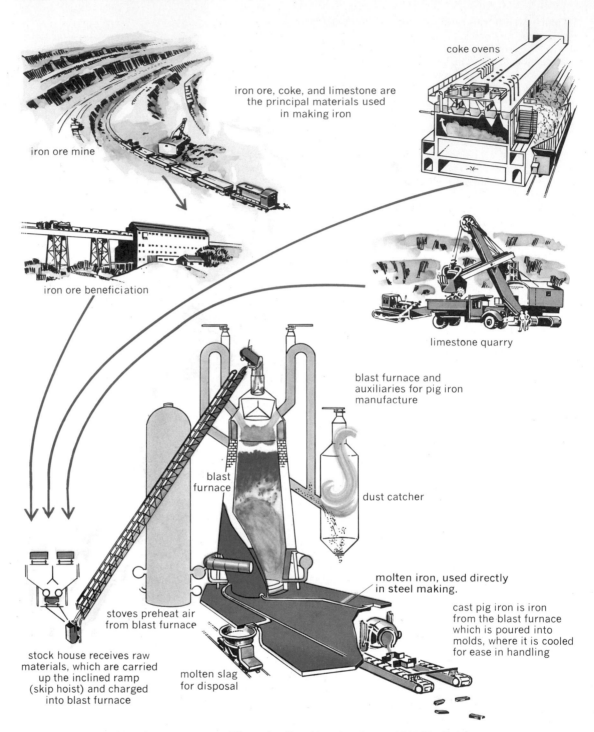

iron ore, coke, and limestone are
the principal materials used
in making iron

coke ovens

iron ore mine

iron ore beneficiation

limestone quarry

blast furnace and
auxiliaries for pig iron
manufacture

blast
furnace

dust catcher

molten iron, used directly
in steel making.

cast pig iron is iron
from the blast furnace
which is poured into
molds, where it is cooled
for ease in handling

stoves preheat air
from blast furnace

stock house receives raw
materials, which are carried
up the inclined ramp
(skip hoist) and charged
into blast furnace

molten slag
for disposal

Fig. 2. Steps in the blast-furnace process of iron extraction. (*American Iron and Steel Institute*)

by reaction (3). The products of combustion when

$$2C + O_2 + 3.76N_2 \rightarrow 2CO + 3.76N_2 + Heat \quad (3)$$
$$(^{79}/_{21} = 3.76)$$

using dry air consist of 34.7% carbon monoxide, CO, and 65.3% $N_2[(5.76/2.0) \times 100 = 34.7\%]$. Two moles of coke carbon require 4.76 moles of air. If the air is enriched from 21 to 30% O_2 by the addition of manufactured oxygen, the combustion reaction will be as in Eq. (4). With this degree of oxygen

$$2C + O_2 + 2.33N_2 \rightarrow 2CO + 2.33N_2 + Heat \quad (4)$$
$$(^{70}/_{30} = 2.33)$$

enrichment, only 3.33 moles of blast will be re-

quired to gasify 2 moles of carbon, C, and the products of combustion will consist of 46.2% CO and 53.8% nitrogen, as shown by Eq. (5). By the elimi-

$$(2.0/4.33) \times 100 = 46.2 \quad (5)$$

nation of inert N_2, oxygen-enriched air provides a means for burning C at faster rates, for increasing the reducing power of the gas, and for increasing flame temperatures.

Air normally contains some moisture which reacts with carbon as shown by Eq. (6). Experience

$$C + H_2O \rightarrow CO + H_2 - Heat \quad (6)$$

has proved that, when moisture is added and

higher blast temperatures are used to compensate for the heat absorbed by reaction (6), smoother furnace operation and higher productivity result. The hydrogen, H_2, and CO produced by reaction (6) dilute the N_2 and increase the reducing power of the gas, permitting the ore to be properly reduced even though it descends more rapidly. Oxygen-enriched air has a similar but more pronounced effect.

Substantially all of the heat for the process is either introduced in the blast, which is heated to temperatures of 550–950°C, or is released by the combustion of coke in localized zones before the tuyeres. Nitrogen, carbon monoxide, and hydrogen, which leave the combustion zones at temperatures of 1700–1900°C, carry heat to the overlying column of stock. Gases leave the top of the furnace at 150–200°C. More permeable areas receive more gas and are thus heated to a higher temperature. Efficiency of heat transfer as well as efficiency of reduction and other smelting reactions depends upon the degree of uniformity of gas flow through the stock column. The recent trend toward the use of higher percentages of sinter, pellets, and sized ore with the gradual elimination of all unscreened ore will increase efficiency and productivity, and permit closer control over the composition of the metal.

Control of pig-iron composition. The carbon content of the iron is fairly well fixed at about 4.0% because the iron is saturated with carbon. However, higher percentages of silicon tend to lower the solubility of carbon in iron. Because all the phosphorus and about 75% of the manganese in the charge are recovered in the pig iron, these elements are largely controlled through the selection of the ore. Silicon and sulfur are controlled by adjustments in the temperature in the crucible and by regulating the composition of the slag through the addition of varying amounts of limestone. An increase in the temperature of the hearth will raise the silicon and lower the sulfur content of the iron. Hearth temperatures are regulated by the amount of fuel used, by the temperature of the blast, and by the addition of steam to the blast, Eq. (6). Variations in gas flow and in the transfer of heat from gases to solids result in uneven heating of the charge, making it difficult to control the temperature of the hearth.

Direct processes. Although direct processes have been subjected periodically to extensive study, research, and investigation on a pilot-plant scale since 1920, they have thus far failed to compete with the blast furnace in supplying an economical source of iron for steelmaking. However, the present outlook for such processes to succeed in more favorable localities is brighter than in the past. Imported ores and concentrates of low gangue content are now more readily available for producing a solid reduced iron that will not produce excessive amounts of slag in steelmaking furnaces. Techniques for reforming nonreducing hydrocarbon gases such as methane into CO and H_2 have been improved. Substantial progress has also been made in the development of fluidized-bed techniques which permit the flow of reducing gases through beds of finely divided ore or concentrates in counterflow processes which are more efficient thermally than older processes. The generally high prices of steel scrap after World War II

stimulated interest in direct processes as a means for producing a cheaper source of metallic iron for steelmaking. Current developments which show great promise in increasing the productivity of blast furnaces will have an important bearing upon the price of steel scrap and the future competitive positions of steel scrap, direct-process iron, and molten pig iron or hot metal. Further replacement of open-hearth furnaces with oxygen converters will tend to keep scrap prices at a relatively low level. *See* CAST IRON; COKE; IRON; IRON ALLOYS; MINING, OPEN-PIT; ORE DRESSING; PYROMETALLURGY; STEEL MANUFACTURE; WROUGHT IRON.   [THOMAS L. JOSEPH]

Bibliography: American Institute of Mining, Metallurgical, and Petroleum Engineers, *Ironmaking Proceedings*, annual; G. D. Elliott and J. A. Bond, *Practical Ironmaking*, 1959; J. W. Franklin, Industry looks at direct reduction, *Eng. Mining J.*, 158(12):84–93, 1957; Metallurgical Society of the American Institute of Mining, Metallurgical and Petroleum Engineers, *History of Iron and Steelmaking in the United States*, 1961; N. J. Pounds, *The Geography of Iron and Steel*, 1959; C. W. Spencer and F. E. Werner (eds.), *Iron and Its Dilute Solid Solutions*, 1963; U.S. Steel Corp., *The Making, Shaping and Treating of Steel*, 7th ed., 1957.

Iron metabolism

The metabolism of iron is discussed with respect to its role in the erythrocyte or red blood cell and its absorption from the intestinal tract.

Erythrocytes. The red cells of blood constitute about 40–50% of the total blood. The mammalian erythrocyte contains approximately 32% hemoglobin, 60–65% water, and about 8% stroma, the meshwork within which the hemoglobin is intimately held. The erythrocytes are formed in the red bone marrow and pass through several phases before entrance into the blood. The physiological stimulus for the production of erythrocytes is a lowering of the oxygen content of the blood in the bone marrow. This response of the bone marrow constitutes one of the most important aspects of acclimatization to life at high altitudes. The erythrocytes are undergoing continual disintegration, mostly in the spleen, but to some extent in the bone marrow and other reticuloendothelial tissues capable of phagocytic action. The breakdown of hemoglobin in this process leads to the release of iron and the formation of bile pigments. The average life-span of erythrocytes is about 120 days for man, 100 days for the dog, and 32 days for the chicken. *See* BLOOD; HEMATOPOIESIS.

Hemoglobin, the respiratory protein of vertebrate erythrocytes, is a conjugation of an iron-porphyrin compound (heme) with a basic protein (globin). Four atoms of iron and four heme groups are present in the hemoglobin molecule.

The porphyrins are all derivatives of porphin, a ring structure composed of four pyrrole groups linked by —CH= (methene) bridges. The ferroprotoporphyrin of hemoglobin contains methyl groups (M) in positions 1, 3, 5, and 8 (see figure), vinyl groups (V) in 2 and 4, and propionic acid groups (P) in 6 and 7. In each of the ferroprotoporphyrin groups of hemoglobin, an iron atom is centrally located, bound to the four nitrogens and to imidazole groups of globin. Combination of hemoglobin

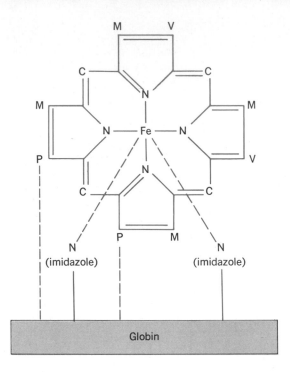

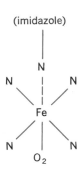

oxyhemoglobin

Hemoglobin fragments; M = methyl, V = vinyl, and P = propionic acid groups. (*A. White, P. Handler, and E. L. Smith, Principles of Biochemistry, 4th ed., McGraw-Hill, 1968*)

with oxygen probably involves a displacement of one imidazole group in each ferroprotoporphyrin moiety. Different hemoglobins among different vertebrates differ in the amino acid structure of the respective globins.

Synthesis of the protoporphyrin of hemoglobin requires only two precursors, glycine and acetate. The glycine supplies all of the nitrogen and eight of the 36 carbons. The utilization of acetate involves the intermediate formation of a succinyl derivative arising from the decarboxylation of α-ketoglutaric acid, a component of the citric acid cycle, with which the synthesis is closely linked. The mechanism for the incorporation of iron is probably an enzyme reaction. The essentiality of copper to iron utilization is well established.

Metabolic breakdown of hemoglobin released on disintegration of the erythrocytes involves oxidative cleavage of the porphin ring at a methene bridge to form linear pyrroles, the bile pigments. Very little of the iron of heme is excreted, most of it being used again in hemoglobin formation or

being stored as ferritin. Biliverdin (green) is first formed and is the main pigment in the bile of some animals, but in the human, bilirubin (red) is the main pigment. The bile pigments are removed from the blood by the liver, and are secreted into the intestinal tract in the bile where they undergo further chemical changes. Some of the intermediates in these reactions are reabsorbed and either returned to the liver or excreted in the urine to form its pigment, urochrome. *See* BILIRUBIN; LIVER (VERTEBRATE).

Absorption and transport. Iron metabolism relates to iron liberated from functional combinations (for example, the heme proteins) in the tissues, and that absorbed from the intestinal tract. The efficiency of iron absorption is not rated highly; however, it is impossible to measure it accurately because of the bilateral exchange of iron between blood and the intestinal lumen. Whatever its origin, iron is transported in combination with a plasma globulin (transferrin) in the ferric state to the liver, spleen, or bone marrow, where it is released as ferrous iron to unite with a protein, apoferritin, in the presence of oxygen to form ferritin. Ferritin contains tightly bound micelles of a ferric hydroxide having the approximate composition $[Fe(OOH)_8 \cdot (FeOPO_3H_2)]$; its iron content is variable and may reach values as high as 23%. The iron released from transferrin may be incorporated into protoporphyrin to form heme, and into the prosthetic groups of various enzymes as an iron-porphyrin complex. Hemosiderin, another storage form of iron in combination with protein, is found in the tissues in increasing amounts as the tissue iron increases beyond the normal physiological levels.

The excretion of iron in the urine is normally inconsiderable. Its excretion from the skin is continuous, and may be important in the daily iron economy of the body. Iron released from transferrin in the blood of the subcutaneous capillary bed combines with the proteins in the dermis, and is carried slowly to the surface of the skin as the cells of the epidermis desquamate. The replacement of the epidermal layers in man is a slow process. *See* HEMOGLOBIN.

[HAROLD H. MITCHELL/RALPH ENGLE]

Bibliography: J. S. Fruton and S. Simmonds, *General Biochemistry*, 2d ed., 1958; National Research Council, *Conference on Hemoglobin*, NASNRC Publ. no. 557, 1958; A. White, P. Handler, and E. L. Smith, *Principles of Biochemistry*, 4th ed., 1968.

Iron transport compounds, microbial

Microbial sources have yielded a number of compounds that appear to be involved in iron transport and metabolism. These substances can be classified in two groups: the hydroxamic acids (see table) and the 2,3-dihydroxybenzoic acid conjugates of amino acids.

One, two, or three hydroxamic acid groupings, $R' \cdot CO \cdot N(OH)R$, may be present in one molecule, and in nearly all cases R is a substituent other than hydrogen (H); that is, the hydroxamic acid is said to be of the secondary variety. When a single molecule contains three hydroxamate groups capable of clustering around a central ferric ion (see illustration), very stable complexes result.

The importance of the hydroxamic acid compounds in microbial physiology is apparent from the observation that some function as growth factors, others as antibiotics. The biological potency is extremely high. For example, the growth re-

Microbial hydroxamic acids

Compound	Remarks
Albomycin	Antibiotic probably identical to grisein
Aspergillic acids	Diketopiperazine-like antibiotics derived from amino acids such as leucine and valine and containing a hydroxamic acid grouping within the heterocyclic ring
Coprogen	Growth factor for coprophylic fungi and certain other microbial species; probably closely related to ferrirhodin
Ferrichromes	Ferrichrome is a cyclic hexapeptide containing three residues glycine, three residues L-2-amino-5-hydroxy-aminovaleric acid (L-N^δ-hydroxyornithine), three residues acetic acid, one ferric ion; ferrichrome A is similar to ferrichrome except that two serines are exchanged for two glycines and the acyl part is three residues of trans-β-methyl glutaconic acid
Ferrimycins	Antibiotics closely related to the ferrioxamines
Ferrioxamines	Compounds containing repeating units of succinic acid, acetic acid, and 1-amino-ω-hydroxyamino pentane (or butane) clustered around a central ferric ion
Fusarinines	Compounds containing one or more of N^δ-cis-5-hydroxy-3-methylpent-2-enoyl-N^δ-hydroxy-L-ornithine
Fusigen	Three residues of fusarinine condensed head to tail in ester linkages to form a cyclic structure containing 36 atoms and three hydroxamic acid groups
Grisein	An antibiotic containing iron, a cyclic hexapeptide structure similar to that of the ferrichromes, and additional chemical substituents
Hadacidin	N-formyl-N-hydroxyglycine, a tumor inhibitor
Mycelianamide	Antibiotic containing a diketopiperazine ring in which both amide linkages are oxidized to the hydroxamic acid stage
Mycobactins	Compounds produced by some, and required for growth by other, strains of *Mycobacteria*; two residues of N^δ-hydroxylysine are present in hydroxamate linkage
Nocardamin	Identical with iron-free ferrioxamine E
Pulcherriminic acid	Structurally related to aspergillic acid, with one hydroxamic acid linkage further oxidized; the ferric complex is known as pulcherrimin
Rhodotorulic acid	A diketopiperazine composed of two molecules of N^δ-acetyl-N^δ-hydroxyornithine; a potent growth factor for *Arthrobacter* species, and other organisms auxotrophic for hemin or iron transport compounds
Schizokinen	A monohydroxamic acid which promotes cell division
Terregens factor	Unidentified hydroxamic acid with growth factor properties

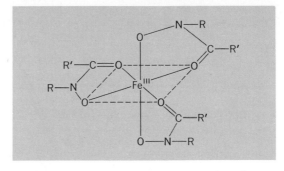

Iron-binding center of the ferrichrome compounds.

quirements of *Arthrobacter* strain JG-9 can be satisfied with about 10 molecules of ferrichrome per cell. The nutritional requirement for iron transport compounds can usually be replaced by hemin supplied at much higher levels. Overproduction of most microbial hydroxamic acids is induced by culturing the source organisms, most of which are fungi, under conditions of iron starvation.

The 2,3-dihydroxybenzoyl conjugates of glycine and serine have been identified as products from *Bacillus subtilis* and *Escherichia coli*, respectively, when the organisms are grown at low levels of iron. These substances form complexes with ferric ion, and they may play a role in iron transport. *See* BACTERIAL METABOLISM.

[J. B. NEILANDS]

Bibliography: J. B. Neilands, Hydroxamic acids in nature, *Science*, 156:1443, 1967; J. B. Neilands, Naturally occurring non-porphyrin iron compounds, *Structure and Bonding*, vol. 1, 1966.

Iron-silicon alloy

A soft (easily magnetized) magnetic material. Iron-silicon alloys are used principally as the magnetic core materials in power transformers. The yearly product is valued in the hundreds of millions of dollars.

Although silicon contents up to 6% have been employed, the material most used contains about 3% silicon. The highest quality is obtained by a combination of (1) purification, especially with respect to nonmetallic impurities such as carbon, oxygen, and sulfur, and (2) control of the processes of rolling and annealing so that the crystallites composing the sheet are aligned with a favorable crystal axis parallel to the length of the sheet. A newer development in processing produces material in which a favorable axis is perpendicular as well as parallel to the length of the sheet.

For a list of properties of iron-silicon alloys *see* MAGNETIC MATERIALS.

[RICHARD M. BOZORTH]

Ironwood

The name given to any of at least 10 kinds of tree in the United States. Because of uncertainty as to just which tree is indicated, the name ironwood has been abandoned in the checklist of the native and naturalized trees of the United States. Probably the best known of the 10 is the American hornbeam (*Carpinus caroliniana*). Some of the others are *Ostrya virginiana*, eastern hophornbeam; *Bumelia lycioides*, buckthorn bumelia; *B. tenax*, tough bumelia; *Cliftonia monophylla*, buckwheat tree; and *Cyrilla racemiflora*, swamp cyrilla or swamp ironwood. All of these species except *Ostrya* are restricted to the southeastern United States. Others commonly called ironwood are *Eugenia confusa*, redberry eugenia of southern Florida and the Florida Keys; *Exothea paniculata*, butterbough of southern Florida; and *Ostrya knowltonii*, Knowlton hophornbeam of southwestern United States. Leadwood (*Krugiodendron ferreum*) a native of southern Florida, has the highest specific gravity of all woods native to the United States and is also known as black ironwood. *See* FOREST AND FORESTRY; HOPHORNBEAM; HORNBEAM; TREE.

[ARTHUR H. GRAVES/KENNETH P. DAVIS]

Irradiation, isotopic

The subjection of a material to radiation from radioactive isotopes (radioisotopes) for therapeutic and other purposes. Although only irradiation by radioactive isotopes is treated in this article, many other forms of radiation are used for irradiation. *See* RADIOISOTOPE; ULTRAVIOLET RADIATION (BIOLOGY).

In the first type of isotopic irradiation, the source of radiation is placed in a capsule. In the second, the radioisotope is dispersed in the material to be irradiated. Encapsulated radiation sources are used to sterilize foodstuffs and biologicals, to produce changes in plastics and other materials, and to aid in medical therapy. Irradiation with dispersed isotopes is largely confined to certain procedures of medical therapy in which a short-lived isotope is administered to a patient.

The radiation from radioactive isotopes produces essentially the same effect as the radiation from high-voltage particle accelerators; thus the choice of a radioisotope or a high-voltage machine depends primarily on convenience and cost. A radioisotope radiation source does not require the extensive and complex machinery necessary for a high-voltage radiation source. Radioisotope radiation, however, cannot be turned on and off, and consequently requires elaborate shielding for health protection.

Cobalt-60 is by far the most commonly used radioisotope for encapsulated radiation sources, although cesium-137 offers considerable promise, primarily because it is an abundant fission product. Cesium-137 is available in large quantities in the reprocessing of fuel elements from nuclear reactors. Indeed, the application of cesium-137 to irradiation may solve some of the problems of disposing of radioactive wastes from the operation of nuclear reactors. *See* NUCLEAR FUELS REPROCESSING.

Iodine-131 and gold-198 are widely used in the dispersal technique. A short-lived radioisotope is administered in highly purified form, orally or intravenously in the case of iodine-131, or injected in colloidal form directly into a tumor in the case of gold-198. The specific advantage of internal therapy with iodine-131 is that the thyroid gland concentrates the element iodine. Consequently, certain forms of thyroid disorders and cancers may be treated by the administration of large amounts of this isotope. *See* RADIATION BIOLOGY; RADIOISOTOPE (BIOLOGY).

[GORDON L. BROWNELL]

Irregularia

The name given by G. Cuvier in 1817 to an assemblage of echinoids in which the anus and periproct lie outside the apical system, the ambulacral plates remain simple, the primary radioles are hollow, and the rigid test shows more or less bilateral symmetry. J. Durham and R. Melville (1957) have shown that the pygasteroid Irregularia stem from Late Triassic pedinids, whereas the other Irregularia evidently arose from Early Jurassic Echinacea. The Irregularia are therefore an artificial assemblage of similar but unrelated forms, and the name can have no taxonomic validity. *See* DIADEMATACEA; ECHINOIDEA; PYGASTEROIDA; REGULARIA.

[HOWARD B. FELL]

Irrigation of crops

The artificial application of water to the soil to produce plant growth. Irrigation also cools the soil and atmosphere, making the environment favorable for plant growth. The use of some form of irrigation is well documented throughout the history of humankind. Over 50,000,000 acres (202,300 km²) are irrigated in the United States.

Use of water by plants. Growing plants use water almost continuously. Growth of crops under irrigation is stimulated by optimum moisture, but retarded by excessive or deficient amounts. Factors influencing the rate of water use by plants include the type of plant and stage of growth, temperature, wind velocity, humidity, sunlight duration and intensity, and available water supply. Plants use the least amount of water upon emergence from the soil and near the end of the growing period. Irrigation and other management practices should be coordinated with the various stages of growth. A vast amount of research has been done on the use of water by plants, and results are available for crops under varying conditions. *See* PLANT, WATER RELATIONS OF; PLANT GROWTH.

Consumptive use. In planning new or rehabilitating old irrigation projects, consumptive use is the most important factor in determining the amount of water required. It is also used to determine water rights.

Consumptive use, or evapotranspiration, is defined as water entering plant roots to build plant tissues, water retained by the plant, water transpired by leaves into the atmosphere, and water evaporated from plant leaves and from adjacent soil surfaces. Consumptive use of water by various crops under varying conditions has been determined by soil-moisture studies and computed by other well-established methods for many regions of the United States and other countries. Factors which have been shown to influence consumptive use are precipitation, air temperature, humidity, wind movement, the growing season, and latitude, which influences hours of daylight. Table 1 shows how consumptive use varies at one location during the growing season. When consumptive-use data are computed for an extensive area, such as an irrigation project, the results will be given in acre-feet per acre for each month of the growing season and the entire irrigation period. Peak-use months determine system capacity needs. An acre-foot is the amount of water required to cover 1 acre 1 ft deep (approx. 1214 m² of water).

Soil, plant, and water relationships. Soil of root-zone depth is the storage reservoir from which plants obtain moisture to sustain growth. Plants take from the soil not only water, but dissolved minerals necessary to build plant cells. How often this reservoir must be filled by irrigation is determined by the storage capacity of the soil, depth of the root zone, water use by the crop, and the amount of depletion allowed before a reduction in yield or quality occurs. Table 2 shows the approximate amounts of water held by soils of various textures.

Water enters coarse, sandy soils quite readily, but in heavy-textured soils the entry rate is slower. Compaction and surface conditions also affect the rate of entry.

Soil conditions, position of the water table, length of growing season, irrigation frequency, and other factors exert strong influence on root-zone depth. Table 3 shows typical root-zone depths in well-drained, uniform soils under irrigation. The depth of rooting of annual crops increases during the entire growing period, given a favorable, unrestricted root zone. Plants in deep, uniform soils usually consume water more slowly from the lower root-zone area than from the upper. Thus, the upper portion is the first to be exhausted of moisture. For most crops, the entire root zone should be supplied with moisture when needed.

Maximum production can usually be obtained with most irrigated crops if not more than 50% of the available water in the root zone is exhausted during the critical stages of growth. Many factors influence this safe-removal percentage, including the type of crop grown and the rate at which water is being removed. Application of irrigation water should not be delayed until plants signal a need for moisture; wilting in the hot parts of the day may reduce crop yields considerably. Determination of the amount of water in the root zone can be done by laboratory methods, which are slow and costly. However, in modern irrigation practice, soil-moisture-sensing devices are used to make rapid determinations directly with enough accuracy for practical use. These devices, placed in selected field locations, permit an operator to schedule periods of water application for best results. Evaporation pans and weather records can be used to estimate plant-water use. Computerizing these data also helps farmers schedule their irrigations. The irrigation system should be designed to supply sufficient water to care for periods of most rapid evapotranspiration. The rate of evapotranspiration may vary from 0 to 0.4 in. per day (10 mm per day) or more.

Water quality. All natural irrigation waters contain salts, but only occasionally are waters too saline for crop production when used properly. When more salt is applied through water and fertilizer than is removed by leaching, a salt buildup can occur. If the salts are mainly calcium and magnesium, the soils become saline, but if salts predominantly are sodium, a sodic condition is possible. These soils are usually found in arid areas, especially in those areas where drainage is poor. Rainfall in humid areas usually carries salts downward to the groundwater and eventually to the sea.

Saline soils may reduce yields and can be especially harmful during germination. Some salts are toxic to certain crops, especially when applied by sprinkling and allowed to accumulate on the plants. Salt levels in the soil can be controlled by drainage, by overirrigation, or by maintaining a high moisture level which keeps the salts diluted.

Sodic soils make tillage and water penetration difficult. Drainage, addition of gypsum or sulfur, and overirrigation usually increase productivity.

Ponding or sprinkling can be used to leach salts. Intermittent application is usually better and, when careful soil-moisture management is practiced, only small amounts of excess irrigation are needed to maintain healthy salt levels.

Diagnoses of both water and soil are necessary for making management decisions. Commercial laboratories and many state universities test both water and soil, and make recommendations.

Methods of application. Water is applied to crops by surface, subsurface, sprinkler, and drip irrigation. Surface irrigation includes furrow and flood methods.

Furrow method. This method is used for row crops (Fig. 1). Corrugations or rills are small furrows used on close-growing crops. The flow, carried in furrows, percolates into the soil. Flow to the furrow is usually supplied by siphon tubes, spiles, gated pipe, or valves from buried pipe. Length of furrows and size of stream depend on slope, soil type, and crop; infiltration and erosion must be considered.

Flood method. Controlled flooding is done with border strips, contour or bench borders, and ba-

Table 1. Example of consumption of water by various crops, in inches (1 in. = 25.4 mm)

Crop	April	May	June	July	Aug.	Sept.	Oct.	Seasonal total
Alfalfa	3.3	6.7	5.4	7.8	4.2	5.6	4.4	37.4
Beets		1.9	3.3	5.3	6.9	5.8	1.1	24.3
Cotton	1.1	2.0	4.1	5.8	8.6	6.7	2.7	31.0
Peaches	1.0	3.4	6.7	8.4	6.4	3.1	1.1	30.0
Potatoes			0.7	3.4	5.8	4.4		14.0

Table 2. Approximate amounts of water in soils available to plants

Soil texture	Water capacity, in inches for each foot of depth
Coarse sandy soil	0.5-0.75
Sandy loam	1.25-1.75
Silt loam	1.75-2.50
Heavy clay	1.75-2.0 or more

Table 3. Approximate effective root-zone depths for various crops

Crop	Root-zone depth, ft (1 ft = 0.3 m)
Alfalfa	6
Corn	3
Cotton	4
Potatoes	2
Grasses	2

Fig. 1. Furrow method of irrigation. Water is supplied by pipes with individual outlets, or by ditches and siphon tubes.

sins. Border strip irrigation is accomplished by advancing a sheet of water down a long, narrow area between low ridges called borders. Moisture enters the soil as the sheet advances. Strips vary from about 20 to 100 ft (6 to 30 m) in width, depending mainly on slope (both down and across), and amount of water available. The border must be well leveled and the grade uniform; best results are obtained on slopes of 0.5% or less. The flood method is sometimes used on steeper slopes, but maldistribution and erosion make it less effective.

Bench-border irrigation is sometimes used on moderately gentle, uniform slopes. The border strips, instead of running down the slope, are constructed across it. Since each strip must be level in width, considerable earth moving may be necessary.

Basin irrigation is well adapted to flatlands. It is done by flooding a diked area to a predetermined depth and allowing the water to enter the soil throughout the root zone. Basin irrigation may be utilized for all types of crops, including orchards where soil and topographic conditions permit.

Subirrigation. This type of irrigation is accomplished by raising the water table to the root zone of the crop or by carrying moisture to the root zone by perforated undergound pipe. Either method requires special soil conditions for successful operation.

Sprinkler systems. A sprinkler system consists of pipelines which carry water under pressure from a pump or elevated source to lateral lines along which sprinkler heads are spaced at appropriate intervals. Laterals are moved from one location to another by hand or tractor, or they are moved automatically. The side-roll wheel system, which utilizes the lateral as an axle (Fig. 2), is very popular as a labor-saving method. The center-pivot sprinkler system (Fig. 3) consists of a lateral carrying the sprinkler heads, and is moved by electrical or hydraulic power in a circular course irrigating an area containing up to 135–145 acres (546,200–586,700 m²).

Extra equipment can be attached in order to irrigate the corners, or solid sets can be used. Solid-set systems are systems with sufficient laterals and sprinklers to irrigate the entire field without being moved. These systems are quite popular for irrigating vegetable crops or other crops requiring light, frequent irrigations and, in orchards, where it is difficult to move the laterals.

Sprinkler irrigation has the advantage of being adaptable to soils too porous for other systems. It can be used on land where soil or topographic conditions are unsuitable for surface methods. It can be used on steep slopes and operates efficiently with a small water supply.

Drip irrigation. This is a method of providing water to plants almost continuously through small-diameter tubes and emitters. It has the advantage of maintaining high moisture levels at relatively low capital costs. It can be used on very steep, sandy, and rocky areas and can utilize saline waters better than most other systems. Clean water, usually filtered, is necessary to prevent blockage of tubes and emitters. The system has been most popular in orchards and vineyards, but is also used for vegetables, ornamentals, and for landscape plantings.

Automated systems. Automation is being used with solid-set and continuous-move types of systems, such as the center-pivot and lateral-move. Surface-irrigated systems are automated with check dams, operated by time clocks or volume meters, which open or close to divert water to other areas. Sprinkler systems, pumps, and check dams can all be activated by radio signals or low-voltage wired systems, which, in turn, can be triggered by soil-moisture-sensing devices or water levels in evaporation pans.

Automatically operated pumpback systems, consisting of a collecting pond and pump, are being used on surface-irrigated farms to better utilize water and prevent silt-laden waters from returning to natural streams.

Fig. 2. A side-roll sprinkler system which uses the main supply line (often more than 1000 ft, or 300 m, long) to carry the sprinkler heads and as the axle for wheels.

Multiple uses. With well-designed and -managed irrigation systems, it is possible to apply chemicals and, for short periods of time, to moderate climate. Chemicals which are being used include fertilizers, herbicides, and some fungicides. Effectiveness depends on uniformity of mixing and distribution and on application at the proper times. Chemicals must be registered to be used in this manner.

Solid-set systems are frequently used to prevent frost damage to plants and trees, since, as water freezes, it releases some heat. A continuous supply of water is needed during the protecting period. However, large volumes of water are required, and ice loads may cause limb breakage. Sequencing of sprinklers for cooling is being practiced for bloom delay in the spring and for reduction of heat damage in the summer.

Humid and arid regions. The percentage of increase in irrigated land is greater in humid areas than in arid and semiarid areas, although irrigation programs are often more satisfactory where the farmer does not depend on rainfall for crop growth. Good yields are obtained by well-timed irrigation, maintenance of high fertility, keeping the land well cultivated, and using superior crop varieties.

There is little difference in the principles of crop production under irrigation in humid and arid regions. The programming of water application is more difficult in humid areas because natural precipitation cannot be accurately predicted. Most humid areas utilize the sprinkler method.

To be successful, any irrigation system in any location must have careful planning with regard to soil conditions, topography, climate, cropping practices, water quality and supply, as well as engineering requirements.

Outlook. As mentioned, there are over 50,000,000 acres of land irrigated in the United States. Studies of land that can be developed for irrigation are becoming obsolete with the improvements in irrigation systems. Limitations of water supplies and economics will prevent future developments—not land resources. The limit of simple diversions of natural rivers and streams has been reached, with few exceptions. Future development of large acreages of irrigated land must come through extensive storage; high-lift pumping projects, characteristic of Federal programs and large corporations; transportation of supply water sources to water-poor areas; and better conservation of water supplies.

Since the most economical irrigation projects have already been developed, future development will be more costly, depending upon many factors. Major diversions of stream flow to regions outside the watershed will involve many complicated interstate problems and compacts. As the population increases, there will be greater competition for water by industry, municipalities, power generators, recreational facilities, and wildlife reserves. Some underground water supplies are being depleted and must, at some time, be replenished by transported water if the irrigated area is to remain under cultivation.

Better water conservation could assist in expanding the irrigated acreage. It is estimated that phreatophites (water-loving plants) along streams

Fig. 3.   Center-pivot systems are very popular in new irrigation developments.

and irrigation canals transpire 25,000,000 acre-feet of water to the atmosphere. Evaporation from reservoirs accounts for the loss of millions of additional acre-feet. Other losses include seepage from canals and water lost through the soil when more water is applied than the plants can use. *See* AGRICULTURE, SOIL AND CROP PRACTICES IN; LAND DRAINAGE (AGRICULTURE); TERRACING (AGRICULTURE); WATER CONSERVATION.

[MEL A. HAGOOD]
Bibliography: R. M. Hagan, H. R. Haise, and T. W. Edminster, *Irrigation of Agricultural Lands*, 1967; C. H. Pair, *Sprinkler Irrigation*, 4th ed., 1975.

Ischnacanthiformes

One of three orders of the class Acanthodii. Members of this order are slender, lightly armoured predators with sharp teeth anchylosed to the upper and lower jawbones, deeply inserted fin spines, and two dorsal fins. They are found from the Upper Silurian to the Upper Carboniferous.

Little is known about this group, which is mainly represented in the fossil record by detached spines and jawbones. The Lower Devonian *Ischnacanthus* has been described from articulated specimens. Intermediate spines, ancillary gill covers, and dermal plates in the shoulder region are all wanting, but the spines of the median and paired fins are long and slender, and the pectoral girdle is moderately developed. The jaw teeth are multicuspid, there is a large tooth whorl at the symphysis of the mandibles, and numerous small teeth line the mouth cavity. In later forms like the Carboniferous *Acanthodopsis*, the marginal jaw teeth are much developed at the expense of other teeth in the mouth cavity. The mouth has a wide gape, and probably the upper jaw articulates directly with the braincase and is braced by the hyomandibula. Peculiar scales with an extensive pore-canal system are associated with some of the laterosensory lines of the head in *Ischnacanthus* and Silurian species. Generally the skeleton lacks bone cells, and the body scales are of acellular bone and dentine like those of Acanthodiformes. The acellular condition is secondary, however, as cell spaces are found in Silurian tooth whorls and jawbones.

The Ischnacanthiformes probably evolved from primitive Climatiiformes. They have yet to be subdivided satisfactorily into families because of the fragmentary nature of most of the fossils. The Lower Devonian *Xylacanthus* appears to have reached a length of some 2½ m and is the largest known acanthodian. *See* ACANTHODIFORMES; ACANTHODII; CLIMATIIFORMES.

[ROGER S. MILES]

Isentropic flow

The flow of a fluid is isentropic when its entropy is identical at all points in the flow. Isentropic flow can be approached for fluids flowing either in a duct or over the outside of a body. Because the entropy of the fluid is a thermodynamic property, similar to the enthalpy or energy of a fluid, the value of the entropy is fixed by the state of the fluid. For a pure substance, in the absence of external forces, entropy is a function of two independent properties. For example, in the absence of gravity, capillarity, electricity, and magnetism, the entropy of a single-phase fluid is a function of the pressure and temperature. *See* ENTHALPY; ENTROPY; ISENTROPIC PROCESS; THERMODYNAMIC PRINCIPLES.

One of the simplest examples of isentropic flow is the flow of a fluid through a nozzle wherein the fluid is accelerated by means of a pressure gradient. This flow can be easily computed for the situation shown in the illustration from the conservation of energy (first law of thermodynamics). Per unit mass of fluid, Eq. (1) holds, where v_1 and

$$\frac{1}{2}v_1{}^2 + h_1 = \frac{1}{2}v_2{}^2 + h_2. \qquad (1)$$

v_2 are the entering and exiting velocities of the fluid and h_1 and h_2 are the corresponding fluid enthalpies. Here the cross-sectional areas and pressures are assumed constant in each half of the nozzle. Thus Eq. (2) holds, where P_1 is the

$$h_1 = P_1 V_1 + U_1 \qquad (2)$$

fluid pressure, V_1 is the volume per unit mass, and U_1 is the internal energy per unit mass.

In an actual nozzle, the fluid flow is not completely isentropic because (1) the fluid shear stress at the walls is not zero, thereby introducing some friction; (2) a significant rate of heat transfer can occur between fluid and walls, as in a rocket nozzle; (3) a significant rate of mass transfer or diffusion may occur normal to the streamlines, thus producing local changes of entropy in the real flow; and (4) chemical reactions can occur in the flow, thus causing local changes of entropy.

Isentropic flow is often used as a basis of comparison of the real flow with the ideal flow. The figure of merit for flow in a nozzle is defined by Eq. (3), where v_a is the actual measured velocity issu-

$$\text{Nozzle efficiency} \equiv v_a/v_s \qquad (3)$$

ing from the nozzle, and v_s is a hypothetical velocity for isentropic flow of the same fluid from the same initial state to the same exit pressure as the real flow. The concept of isentropic flow is useful for fluid flow inside ducts and outside of variously shaped bodies. Isentropic flow is also used for

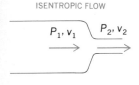

ISENTROPIC FLOW

P_1, v_1 P_2, v_2

Flow through a nozzle.

predicting such flows as those of perfect gases; real gases; dissociating and chemically reacting systems; liquids; two-phase, single-, and multicomponent systems; and plasmas.

Isentropic fluid flow can be obtained in irreversible processes by selecting a process in which the local entropy could increase and then providing sufficient heat transfer to maintain the entropy constant at all points. *See* FLUID-FLOW PRINCIPLES; GAS DYNAMICS.   [PHILIP E. BLOOMFIELD]

Bibliography: F. W. Sears, *An Introduction to Thermodynamics*, 1953; V. L. Streeter (ed.), *Handbook of Fluid Dynamics*, 1961.

Isentropic process

In thermodynamics a change that is accomplished without any increase or decrease of entropy is referred to as isentropic. Since the entropy always increases in a spontaneous process, one must consider reversible or quasistatic processes. During a reversible process the quantity of heat transferred, dQ, is directly proportional to the system's entropy change, dS, as in Eq. (1), where T is the absolute

$$dQ = TdS \qquad (1)$$

temperature of the system. Systems which are thermally insulated from their surroundings undergo processes without any heat transfer; such processes are called adiabatic. Thus during an isentropic process there are no dissipative effects and, from Eq. (1), the system neither absorbs nor gives off heat. For this reason the isentropic process is sometimes called the reversible adiabatic process. *See* ADIABATIC PROCESS; ENTROPY; THERMODYNAMIC PROCESSES.

Work done during an isentropic process is produced at the expense of the amount of internal energy stored in the nonflow or closed system. Thus, the useful expansion of a gas is accompanied by a marked decrease in temperature, tangibly demonstrating the decrease of internal energy stored in the system. For ideal gases the isentropic process can be expressed by Eq. (2), where P is the

$$P_1 V_1{}^k = P_2 V_2{}^k = \text{constant} \qquad (2)$$

pressure in pounds per square foot, V is the volume in cubic feet, and k is the ratio between the specific heat at constant pressure and the specific heat at constant volume for the given gas. It can be closely approximated by the values of 1.67 and 1.40 for dilute monatomic and diatomic gases, respectively. For a comparison of various processes involving a gas *see* POLYTROPIC PROCESS.

[PHILIP E. BLOOMFIELD]

Isentropic surfaces

Surfaces along which the entropy and potential temperature of air are constant. Potential temperature, in meteorological usage, is defined by the relationship given below,

$$\theta = T\left(\frac{1000}{P}\right)^{\frac{c_p - c_v}{c_p}}$$

in which T is the air temperature, P is atmospheric pressure expressed in millibars, c_p is the specific heat of air at constant pressure and c_v is the specific heat of air at constant volume. Since the potential temperature of an air parcel does not change if the processes acting on it are adiabatic (no exchange of heat between the parcel and its environment), a surface of constant potential temperature is also a surface of constant entropy. The slope of isentropic surfaces in the atmosphere is of the order of 1/100 to 1/1000. An advantage of representing meteorological conditions on isentropic surfaces is that there is usually little air motion through such surfaces, since thermodynamic processes in the atmosphere are approximately adiabatic. *See* ADIABATIC PROCESS; ATMOSPHERIC GENERAL CIRCULATION. [FREDERICK SANDERS]

Ising model

A model which consists of a lattice of "spin" variables with two characteristic properties: (1) each of the spin variables independently takes on either the value +1 or the value −1; and (2) only pairs of nearest neighboring spins can interact. The study of this model (introduced by Ernst Ising in 1925) in two dimensions, where many exact calculations have been carried out explicitly, forms the basis of the modern theory of phase transitions and, more generally, of cooperative phenomena.

The two-dimensional Ising model was shown to have a phase transition by R. E. Peierls in 1936, and the critical temperature or Curie temperature, that is, the temperature at which this phase transition takes place, was calculated by H. A. Kramers and G. H. Wannier and by E. W. Montroll in 1941. Major breakthroughs were accomplished by Lars Onsager in 1944, by Bruria Kaufman and Onsager in 1949, and by Chen Ning Yang in 1952. Onsager first obtained the free energy and showed that the specific heat diverges as $-\ln|1 - T/T_c|$ when the temperature T is near the critical temperature T_c; Kaufman and Onsager computed the short-range order; and Yang calculated the spontaneous magnetization. Since then several other properties have been obtained, and since 1974 connections with relativistic quantum field theory have been made. *See* QUANTUM FIELD THEORY.

Cooperative phenomena. A macroscopic piece of material consists of a large number of atoms, the number being of the order of Avogadro's number (approximately 6×10^{23}). Thermodynamic phenomena all depend on the participation of such a large number of atoms. Even though the fundamental interaction between atoms is short-ranged, the presence of this large number of atoms can, under suitable conditions, lead to an effective interaction between widely separated atoms. Phenomena due to such effective long-range interactions are referred to as cooperative phenomena. The simplest examples of cooperative phenomena are phase transitions. The most familiar phase transition is either the condensation of steam into water or the freezing of water into ice. Only slightly less familiar is the ferromagnetic phase transition that takes place at the Curie temperature, which, for example, is roughly 1043 K for iron.

Of the several models which exhibit a phase transition, the Ising model is the best known. In three dimensions the model is so complicated that no exact computation has ever been made, while in one dimension the Ising model does not undergo a phase transition. However, in two dimensions the Ising model not only has a ferromagnetic phase transition but also has very many physical properties which may be exactly computed. Indeed, despite the restriction on dimensionality, the two-dimensional Ising model exhibits all of the phenomena peculiar to magnetic systems near the Curie temperature. *See* CURIE TEMPERATURE, MAGNETIC; FERROMAGNETISM; MAGNETISM; SECOND-ORDER TRANSITION.

Definition of model. The mutual interaction energy of the pair of spins σ_α and $\sigma_{\alpha'}$ when α and α' are nearest neighbors may be written as $-E(\alpha,\alpha') \cdot \sigma_\alpha \sigma_{\alpha'}$. The meaning of this is that the interaction energy is $-E(\alpha,\alpha')$ when σ_α and $\sigma_{\alpha'}$ are both +1 or −1, and is $+E(\alpha,\alpha')$ when $\sigma_\alpha = +1$, $\sigma_{\alpha'} = -1$, or $\sigma_\alpha = -1$, $\sigma_{\alpha'} = 1$. In addition, a spin may interact with an external magnetic field H with energy $-H\sigma_\alpha$. From these two basic interactions the total interaction energy for the square lattice may be written as Eq. (1), where j specifies the row and k

$$E = -\sum_j \sum_k \left[E_1(j,k)\sigma_{j,k}\sigma_{j,k+1} + E_2(j,k)\sigma_{j,k}\sigma_{j+1,k} + H\sigma_{j,k} \right] \quad (1)$$

specifies the column of the lattice. In this form the interaction energies $E_1(j,k)$ and $E_2(j,k)$ are allowed to vary arbitrarily throughout the lattice. A special case of great importance is the translationality invariant case (E_1 and E_2 independent of j and k) which was studied by Onsager in 1944. This is the model needed to study a pure ferromagnet without impurities. *See* MAGNETIC FIELD.

Several generalizations of Ising's original model have been considered. For example, σ can be allowed to take on more values than just ±1, and interactions other than nearest neighbor can be used. For these generalizations no exact calculations have been performed in two or three dimensions. However, various approximate calculations indicate that the phase transition properties of these models are the same as those of the Onsager lattice.

The extension to the nontranslationally invariant case where $E_1(j,k)$ and $E_2(j,k)$ are treated as independent random variables is important for studying the effects of impurities in ferromagnetics. Some extensions in this direction were made in 1968 by Barry McCoy and Tai Tsun Wu.

Thermodynamic properties. The basic simplification in framing the definition of the Ising model of the preceding section is the choosing of the fundamental variables to be the numbers $\sigma_{j,k}$ which can be only +1 or −1. Because of this choice there can be no terms in the interaction energy which refer to kinetic energy or to angular momentum. Consequently, the $\sigma_{j,k}$ do not change with time, and study of the system is, by necessity, confined to those physical properties which depend only on the distribution of energy levels of the system. When the number of energy levels is large, this study requires the use of statistical mechanics.

Statistical mechanics allows the calculation of average macroscopic properties from the microscopic interaction E. If A is some property of the spins σ of the system, then the thermodynamic average of A is given by Eq. (2), where T is the tem-

$$\langle A \rangle = \lim_{N \to \infty} \frac{1}{Z} \sum_{\{\sigma\}} A e^{-E(\sigma)/kT} \quad (2)$$

perature, k is Boltzmann's constant, Z is given by

Eq. (3), the sums are over all values of $\sigma_{j,k} = \pm 1$,

$$Z = \sum e^{-E(\sigma)/kT} \qquad (3)$$

and N is the number of rows and the number of columns. It is mandatory that the thermodynamic limit $N \to \infty$ be taken for these thermodynamic averages to have a precise meaning. *See* BOLTZMANN CONSTANT.

The most important thermodynamic properties of a ferromagnet are the internal energy per site $u = \langle E/N^2 \rangle$, the specific heat $c = \partial u/\partial T$, the magnetization per site $M = \langle \sigma \rangle$, and the magnetic susceptibility $\chi = \partial M/\partial H$. These quantities have all been computed for the two-dimensional Ising model $E_1 \neq E_2$, but for convenience this discussion is restricted to $E_1 = E_2 = E$.

Onsager studied the two-dimensional square lattice at $H = 0$ and computed the specific heat exactly. From that calculation he found that the specific heat was infinite at the critical temperature of Kramers and Wannier given as the solution of Eq. (4). When T is close to the critical temperature T_c, the specific heat is approximated by Eq. (5).

$$\sinh 2E/kT = 1 \qquad (4)$$

$$c \sim -\frac{8E^2}{kT_c^2\pi} \ln|1 - T/T_c| \qquad (5)$$

The behavior of the specific heat for any temperature is plotted in Fig. 1.

The spontaneous magnetization is defined as $M(0) = \lim_{H \to 0^+} M(H)$. For $T > T_c$, $M(0) = 0$. For $T < T_c$, Yang found that $M(0)$ is given by Eq. (6). When T is near T_c, $M(0)$ is approximated by Eq. (7). The be-

$$M(0) = [1 - (\sinh 2E/kT)^{-4}]^{1/8} \qquad (6)$$

near T_c, $M(0)$ is approximated by Eq. (7). The be-

$$M(0) \sim \left[\frac{8\sqrt{2}}{kT_c}(1 - T/T_c)\right]^{1/8} \qquad (7)$$

havior $M(0)$ as a function of T is plotted in Fig. 2.

The magnetic susceptibility χ at $H = 0$ is much more difficult to compute than either the specific heat or the spontaneous magnetization. Indeed, no closed form expression is known for χ over the entire range of temperature. However, near T_c it is known that as $T \to T_c^+$, χ is approximated by Eq. (8), and, as $T \to T_c^-$, is approximated by Eq. (9):

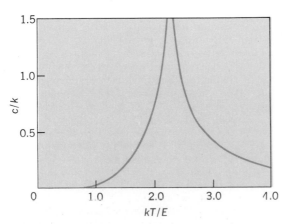

Fig. 1. Specific heat of Onsager's lattice for $E_2/E_1 = 1$.

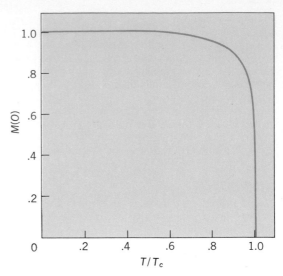

Fig. 2. Spontaneous magnetization $M(0)$ of Onsager's lattice for $E_2/E_1 = 1$.

$$\chi(T) \sim C_0^+ |1 - T_c/T|^{-7/4} + C_1^+ |1 - T_c/T|^{-3/4} + C_2 \qquad (8)$$

$$\chi(T) \sim C_0^- |1 - T_c/T|^{-7/4} + C_1^- |1 - T_c/T|^{-3/4} + C_2 \qquad (9)$$

where $C_0^+ = 0.96258\,17322 \cdots$
$\quad C_0^- = 0.02553\,69719 \cdots$
$\quad C_1^+ = 0.07498\,81538 \cdots$
$\quad C_1^- = -0.0198\,94107 \cdots$

See INTERNAL ENERGY; MAGNETIZATION; SPECIFIC HEAT OF SOLIDS; STATISTICAL MECHANICS; SUSCEPTIBILITY, MAGNETIC; THERODYNAMIC PRINCIPLES.

Random impurities. A question which can be very usefully studied in the Ising model is the generalization of statistical mechanics to deal with the experimental situation in which the interaction energy of the system is not completely known because of the presence of impurities. The term "impurity" refers not only to the presence of foreign material in a sample but to any physical property such as defects or isotopic composition which makes lattice sites different from one another. The distribution of these impurities is governed by spin-independent forces. At least two different situations can be distinguished.

1. As the temperature changes, the distribution of impurities may change; such a situation will occur, for example, near the melting point of a lattice.

2. The distribution of impurities may be independent of temperature, at least on the time scale of laboratory measurements; such a distribution will obtain when the temperature of a lattice is well below the melting temperature. Impurities of this sort are said to be frozen in. *See* CRYSTAL DEFECTS.

For a study of frozen-in impurities to be realistic, the impurities must be distributed at random throughout the lattice. The translational invariance of the system is now totally destroyed, and it is not at all clear that the phase transition behavior of the pure and random system should be related to each other at all.

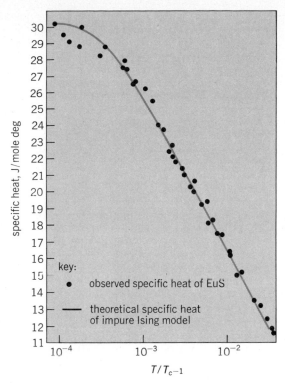

Fig. 3. Comparison of the impure Ising model specific heat with the observed specific heat of EuS for $T > T_c$.

These problems were studied for a special case of the Ising model by McCoy and Wu in 1968. They let $E_2(j,k)$ depend on j but not k and kept $E_1(j,k)$ independent of both j and k. Then the variables $E_2(j)$ were chosen with a probability distribution $P(E_2)$. When $P(E_2)$ was of narrow width, they showed that logarithmic divergence of Onsager's specific heat is smoothed out into an infinitely differentiable essential singularity. Such a smoothing out of sharp phase transition behavior may in fact have been observed. The results of one such experiment, carried out by B. J. C. van der Hoeven and colleagues, are compared with the result of the Ising model random specific heat calculation in Fig. 3.   [BARRY M. MC COY; TAI TSUN WU]

Bibliography: C. Domb and M. Green, *Phase Transitions and Critical Phenomena*, 1972; B. McCoy and T. T. Wu, *The Two Dimensional Ising Model*, 1973; L. Onsager, Crystal statistics 1: A two-dimensional model with order-disorder transition, *Phys. Rev.*, 65:117–149, 1944; T. T. Wu et al., Spin-spin correlation functions for the two-dimensional Ising model: Exact theory in the scaling region, *Phys. Rev.*, B, 13:316–374, 1976; C. N. Yang, The spontaneous magnetization of two-dimensional Ising model, *Phys. Rev.*, 85:808–816, 1952.

Island faunas and floras

Islands generally have fewer species of animals and plants than do comparable continental areas, at least if the islands lie some distance from continents. Often there are rather few genera, but some genera have many local species that sometimes exhibit strong adaptive radiation. The proportion of endemic species that are present generally increases with the degree of isolation and with the complexity of the habitat. Gigantism, flightless-

ness, and other unusual characteristics are relatively frequent.

Island floras and faunas are of far greater interest than either the number of species or their economic importance might seem to justify. How the plants and animals came to be on the islands, why so many of them are found nowhere else, why so many have special or strange forms, and where their ancestors came from, are fascinating questions which present problems of great scientific importance to the biologist. The theory of organic evolution emerged from studies of island faunas by Charles Darwin and Alfred Russell Wallace, and islands are still among the most advantageous places to study evolutionary processes. They also offer uniquely suitable sites for the investigation of the nature and functioning of ecosystems.

Nature and classification of islands. Islands are themselves very diverse, and their biotas (floras and faunas) are correspondingly varied. Islands are commonly separated into high and low islands; into volcanic, limestone, granitic, metamorphic, and mixed islands; and, perhaps most important from a biogeographic standpoint, into continental and oceanic islands. Naturally, most of these types are not sharply separated. On the contrary, as with wet and dry islands, there is a continuous series of intermediates between any pair or set of extremes. Usually significantly different biotas are associated with these categories. Low islands have small, impoverished biotas; high islands, richer ones. Oceanic islands are those considered never to have been part of, or connected with, any continental land mass. Their biotas are commonly poor in genera and unbalanced or disharmonic, that is, with the families—or larger groups—very unevenly represented, compared with those of continents or continental islands. The continental islands are believed to be the remnants of former continental land masses or at least to have had land connections with continents.

Oceanic islands or groups with outstanding or much-studied biotas are Rapa, the Hawaiian Islands, the Galapagos, Juan Fernandez, the Fiji Islands, the Marianas, the Carolines, the Marshalls, the Society Islands, the Azores, the Canary Islands, Tristan d'Acunha, Kerguelen, the Seychelles, and Aldabra.

Continental islands or groups are Madagascar, Japan, Formosa, the Philippines, New Guinea, the Malay Archipelago, Ceylon, New Zealand, the Antilles, the California Islands, and the British Isles.

Endemism. On almost all islands, except those immediately contiguous to other land and most low coral islands, are species and varieties that are endemic to the particular island or group, that is, found nowhere else. Even coastal islands, such as those off southern California, have a few endemics. On high islands the percentage of the total biota that is endemic usually increases with the degree of isolation from larger land masses. Thus, the endemics form over 95% of the indigenous vascular flora in the Hawaiian Islands, which are very remote from other land. On all islands there are some widespread species. They are usually seacoast species, aquatics, or marsh dwellers; and, of course, there are many introduced weeds.

The first basic type of endemics includes species that have differentiated on the islands where they are now found, usually the products of relatively recent evolution. In certain genera there may be several to many species of a given genus adapted to particular habitats — a phenomenon called adaptive radiation. Examples of this are the famous Darwin finches (Geospizidae) of the Galapagos; *Hedyotis* (Rubiaceae) in the Hawaiian Islands; and many insect genera with species adapted to different plant hosts. Other genera have differentiated into several or many species which occupy very similar habitats, apparently an evolutionary result of geographic isolation only, for example, *Cyrtandra* (Gesneriaceae) in the Hawaiian group.

The other basic type of endemics includes relicts, isolated remnants of populations that were much more widespread in the past. A good example is *Lyonothamnus floribundus*. Though it is now confined to the California Islands, fossil evidence shows that it was widespread in California during the Tertiary Period. Islands provide refuges for these species where they may not be exposed to the competition or other unfavorable circumstances that eliminated them elsewhere. For example, islands may provide a more equable climate when severe conditions develop in continental areas. Disjunct species, those found on widely separated islands, may likewise be relicts, or they may be the result of successful long-distance transport. A case in point is possibly *Charpentiera obovata* (Amaranthaceae), a shrub found only in the Hawaiian and Austral islands, separated by almost 3000 mi.

Geography of floras and faunas. The geographical distribution of insular biotas shows very definite patterns, making it possible to group species into "elements" with similar distributions and, in some cases, apparently common geographic origins. Relationships of island genera and species may often be guessed at on the basis of their taxonomic relationships with groups in other areas. Thus a preponderance of the Hawaiian vascular flora and terrestrial invertebrate fauna have their affinities to the southwest in the Indo-Malaysian region. Another element has its connections, and likely its derivation, in the Australia-New Zealand area; still another, but smaller, element finds its relationships in America; another in eastern Polynesia. The numbers of species in the predominant Indo-Pacific element are very high in the western islands of the Pacific, but fall off very markedly from island to island, eastward, except where there has been strong local evolution of species, as in Hawaii. *See* BIOGEOGRAPHY.

Dispersal. One of the central problems in the study of island biotas is that of dispersal. How did the species, or their ancestors, reach the islands where they are now found? This problem is especially fascinating on oceanic islands. To solve it, those with little faith in the effectiveness of transoceanic dispersal would assume that very few islands are truly oceanic, that at different times in the past land bridges have existed in all directions, or that ancient continents foundered or drifted away, leaving the islands and their biotas as scattered remnants. A more plausible theory is that given time enough, rare events, such as accidental transport of individuals or propagules (seeds, eggs, or spores) by wind, water, or birds, will account for the ancestors of all present indigenous biotas of oceanic islands. It is also reasonable to assume that there were many former islands, now marked only by shoals and underwater mountains, that could have served as stepping stones along which some species may have traveled to their present isolated homes. *See* POPULATION DISPERSAL.

Morphological peculiarities. A notable and little-understood feature of island biotas is the frequency of unusual morphologic features — flightlessness in birds and insects, gigantism and woody habit in usually small, herbaceous groups of plants, as well as gigantism in animals, for example, the giant tortoises of the Galapagos and Aldabra and the moas of New Zealand, and the "rosette-tree" habit in many island plant genera. Some of these features are also found in the biotas of tropical mountains. Another feature of the plants of oceanic islands is that, having evolved in the absence of large herbivores, they have weak defenses against grazing and trampling, if any. Spines and thorns are infrequent on indigenous species.

Evolution on islands. The origin of these morphological peculiarities, as well as of the more ordinary diversity found in island biotas, poses evolutionary problems that are extremely interesting and that seem amenable to solution. Islands not only provided early information on which the theory of organic evolution was founded; much study has since been devoted to insular evolution to test various concepts.

In addition to generic and specific diversity, island organisms frequently show an extraordinary degree of polymorphism within species. For example, the tree *Metrosideros collina* of Polynesia, especially Hawaii, shows a complexity of forms bewildering for taxonomic arrangement.

At the other extreme are species with almost no genetic plasticity, usually existing in small populations. The extinct *Clermontia haleakalae* and *Hedyotis cookiana* of Hawaii were presumably of this nature, as are the Hawaiian hawk, crow, and goose.

The adaptive radiation and geographic speciation mentioned above are important evolutionary patterns notable on islands. Isolation, as well as the existence of many unoccupied niches and reduced competition during the early stages in evolution of insular biotas, is probably responsible for the evolutionary persistence of certain features that would be speedily eliminated in a more complex continental biotic situation.

The evolution of major ecosystems themselves may also be elucidated by careful and long-continued study of islands, as island ecosystems are simpler and better defined than those of larger land masses, and the development of their biotas can be better correlated with that of their physical features. Also, a great range of size and complexity is exhibited. *See* ECOSYSTEM; SPECIATION.

Biotas of volcanic islands. New volcanic substrata are bare of all plant and animal life. But from the time the lava or ash surface is cool, it is subject to colonization by plants and animals. The biotas of volcanic islands are the products of a sequence: sporadic colonization; repeated partial destruction by new eruptions; isolation of local populations, with evolution of local races; breaking down of such isolation by mixing of populations, development of genetic diversity, and renewed isolation by dissection of volcanic domes which leads to further evolution of local forms, species, and even genera with time; new arrivals may occupy any open habitats available or may displace current tenants. New open habitats become available as long as volcanic activity continues and erosion makes still other habitats.

As volcanic islands grow older some subside, or get partly flooded by changing sea level. Parts or all may be eroded down to sea level by wave action. In the tropics coral reefs grow up around their shores and provide habitats for still other species, those characteristic of limestone shores and strands, mostly species of very wide distribution. As the high volcanic mountains gradually subside and wear away, the species dependent on high elevations, orographic rainfall, or rain shadows gradually disappear. New ones evolve, or colonize from elsewhere, that are adapted to the old worn-down topography and deeply weathered volcanic substrata. Even most of these finally disappear as subsidence gradually changes the island to an atoll. *See* SUCCESSION, ECOLOGICAL; VOLCANO.

Biotas of coral atolls and limestone islands. Scattered through most tropical seas are flat, often ring-shaped, islands made up of the limestone skeletons of marine plants and animals and resulting from the processes described above. Most rise just above sea level. In this relatively uniform environment, varying principally in rainfall, biotas are very impoverished and are largely made up of strand species. Numbers of species are lowest on the drier atolls. As an example of the variation, the native vascular flora of the Marshall Islands ranges from 9 species per atoll in the driest northern ones, to 75 or 100 in the very wet southern atolls. Endemics are very few. With even a slight elevation, as on Aldabra Atoll, the number of species increases very sharply. Strand species are still prominent, but species requiring moderate habitats increase in numbers. There are some endemics. High limestone islands have rich, strongly endemic biotas. *See* ATOLL; LIMESTONE.

Interesting and important as insular floras and faunas are, they are disappearing with distressing rapidity. The introduction of large herbivorous animals, rats, and aggressive exotic plants capable of quickly invading disturbed situations, as well as the complete destruction of whole habitats by human activities, has resulted in the reduction or total disappearance of many species. The growth of human populations and development, disturbance, and destruction even in remote islands are accelerating. More and more species are becoming rare and threatened with extinction as man achieves greater capacity to change his environment. Many fascinating features of island floras and faunas will not long be available for study and enjoyment unless adequate measures for protecting substantial areas of natural habitats on islands are taken at once. Conservation is an immediate necessity if the study of island biotas is to have a future. *See* EVOLUTION, ORGANIC; PLANT GEOGRAPHY.

[F. R. FOSBERG]

Bibliography: R. I. Bowman, *The Galapagos*, 1966; S. Carlquist, *Island Life*, 1965; F. R. Fosberg, *Man's Place in the Island Ecosystem*, 1963; J. L. Gressitt, *Insects of Micronesia: Introduction*, 1954; J. L. Gressitt (ed.), *Pacific Basin Biogeography*, 1963; D. Lack, *Darwin's Finches*, 1947; R. N. Philbrick, *Biology of the California Islands*, 1967; M. H. Sachet and F. R. Fosberg, *Island Bibliographies*, 1955; E. C. Zimmerman, *Insects of Hawaii*, 1948.

Isoantigen

A serologically active protein or polysaccharide present in some but not all individuals in a particular species. These compounds initiate the formation of antibodies when introduced into other individuals of the species that lack the isoantigen. Like all antigens, they are also active in stimulating antibody production in heterologous species. The ABO, MN, and Rh blood factors in man constitute important examples. Consequently, elaborate precautions for typing are required in blood transfusion.

Analogous situations exist for the bloods of most other animal species. Isoantigens are also believed responsible for the ultimate failure of tissue grafts between individuals of the same species, except those of the same genetic constitution or those that have been rendered tolerant, a situation of consequence for the surgical procedure of skin grafting. *See* IMMUNOLOGICAL TOLERANCE, ACQUIRED; TRANSPLANTATION BIOLOGY.

Isoantigens are to be distinguished from autoantigens, which are antigens active even in the species from which they are derived and in individuals who already possess the antigen. Brain and lens tissue, as well as sperm, constitute examples. These exceptions to the usual rule of nonantigenicity for self-constituents may be more apparent than real, however, since the substances cited are all protected to some degree from contact with the blood, and thus normally do not reach the sites of antibody formation except after experimental manipulation.

Autoantibodies may also be produced in various disease states, perhaps as a result of modification of normal tissue by the infecting microorganism or by altered host metabolism. Examples are the paroxysmal hemoglobinuria observed in syphilis, acquired hemolytic anemia, or some of the manifestations in rheumatic fever. *See* ANTIBODY; ANTIGEN; BLOOD GROUPS; POLYSACCHARIDE; PROTEIN.   [HENRY P. TREFFERS]

Isobar (atomic physics)

One of two or more atoms which have a common mass number A but which differ in atomic number Z. Thus, although isobars possess approximately equal masses, they differ in chemical properties; they are atoms of different elements. Isobars whose atomic numbers differ by unity cannot both be stable; one will inevitably decay into the other by β^--emission $(Z \rightarrow Z+1)$, β^+-emission $(Z \rightarrow Z-1)$, or electron capture $(Z \rightarrow Z-1)$. There are many examples of stable isobaric pairs, for instance, Ti^{50} $(Z=24)$ and Cr^{50} $(Z=26)$, and four examples of stable isobaric triplets. At most values of A the number of known radioactive isobars exceeds the number of stable ones. *See* ELECTRON CAPTURE; RADIOACTIVITY.

[HENRY E. DUCKWORTH]

Isobar (meteorology)

A line passing through points at which a constant value of the air pressure exists, within a specified surface of reference. Central regions of closed iso-

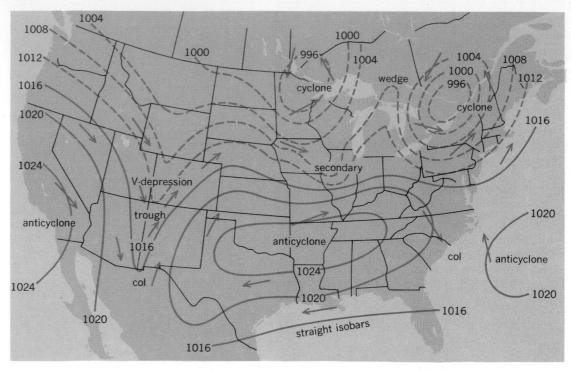

1008
1004
1012
1016
1020
1024
1024
1020
1016

1000
996
cyclone
1004
1000
wedge
1004
1008
996
cyclone
1000
1012
1016
secondary
V-depression
trough
1016
anticyclone
col
anticyclone
1024
col
anticyclone
1020
1020
1020
1016
straight isobars
1016

Hypothetical weather chart designed to show the fundamental configurations of isobars. Arrows fly with surface wind. (*After R. Abercromby, in L. P. Harrison, Meteorology, National Aeronautics Council, Inc., 1940*)

bars on the globe reveal systems of relative high and low pressure as shown on synoptic weather charts based upon simultaneous barometric observations at many stations. For such charts the surface of reference is usually the geoid (mean sea level). In this case the data represent pressures reduced to sea level, which yield unreal isobars over land. Systems of relatively high and low pressure are called anticyclones and cyclones, respectively.

Horizontal pressure gradients determined from real isobars correlate well with the wind velocity about 300–800 m above surface. The illustration presents a fictitious weather chart designed to portray the configurations of isobars generally observed. See AIR PRESSURE.

[LOUIS P. HARRISON]

Isobaric process

A frictionless thermodynamic process of a gas in which the heat transfer to or from the gaseous system causes a volume change at constant pressure. This process can be illustrated by the expansion of gas when it is heated to lift a weight or do other work on its surroundings. The work W done by the system on its surroundings is given by Eq. (1), where P is absolute pressure and V is volume.

$$W = \int_1^2 P\,dV = P \int_1^2 dV = P(V_2 - V_1) \qquad (1)$$

For isobaric processes it is useful to introduce the enthalpy H, which is defined as the total heat content of the system and is given by the sum of the internal energy U and PV. Then Eq. (2) can be

$$Q_p = H_2 - H_1 = m \int_1^2 C_p\,dT \qquad (2)$$

formulated to represent Q_p, the transferred heat at

constant pressure, where C_p is the specific heat at constant pressure and m is mass. By the first law of thermodynamics the change of the internal energy in any process is equal to the difference between the heat gained and the work done by the system; thus Eq. (3) holds.

$$Q_p = U_2 - U_1 + W \qquad (3)$$

If the isobaric process is also carried out quasistatically, Eq. (4) is obtained, where S is the en-

$$Q_p = \int_1^2 T\,dS \qquad (4)$$

tropy. See ISOMETRIC PROCESS; THERMODYNAMIC PROCESSES. For a comparison of the isobaric process with other processes involving a gas see POLYTROPIC PROCESS.

[PHILIP E. BLOOMFIELD]

Isobryales

An order of mosses in which the plants are slender to robust and, in *Fontinalis*, up to 90 cm in length. In some families, they are strongly flattened, mostly with creeping primary stem, and irregularly to pinnately divided into erect, ascending, dendroid, or pendulous branches. In Leucodontineae, the foliated branches are often julaceous.

The leaves are in two, three, or many rows, generally symmetrical. The leaves in Rhacopilaceae are often dimorphous with smaller dorsal leaves; in Neckeraceae they are frequently undulate; in Phyllogoniaceae they are equitant; and in Ptychomniaceae they are often plicate. The costa is absent (illustration *a*), single or double, short to excurrent. Cells vary from quadrate or rounded to linear, smooth to papillose, with walls thin to incrassate (illustration *b*, and *d*). The calyptra is cucullate or mitriform and commonly smooth. In

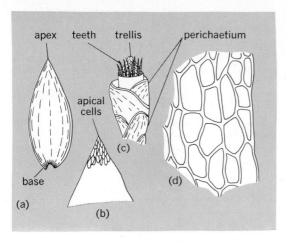

Fontinalis duriaei, an example of the order Isobryales. (a) Leaf. (b) Apex of leaf. (c) Trellis and teeth. (d) Alar cells. (*From W. H. Welch, Mosses of Indiana, Indiana Department of Conservation, 1957*)

Cryphaeaceae it is more or less rough, while in Rhacopilaceae and Meteoriaceae it is frequently hairy.

The sporophytes are pleurocarpous or cladocarpous. The capsule varies from immersed to exserted and is commonly smooth; however, it is eight-ribbed in Ptychomniaceae and usually plicate when dry in Rhacopilaceae. The peristome, frequently absent in Hedwigiaceae, may be single with the endostome lacking, or double with the endostome imperfectly to well developed. In Fontinalaceae the cilia are free or united into a conical trellis (illustration *c*). *See* BRYOPSIDA.

[WINONA H. WELCH]

Isocyanate

One of a group of neutral derivatives of primary amines, formula $R—N{=}C{=}O$, that are obtained by reaction with excess phosgene and loss of hydrogen chloride, Eq. (1). Two double bonds on carbon presage a very reactive system, useful in the identification of alcohols and phenols as urethanes (RNHCOOR'), Eq. (2); amines as ureas, Eq. (3); alkyl halides as anilides (C_6H_5 NHCOR), Eq. (4); and carboxylic acids as amides (RNHCOR'), Eq. (5). *See* AMINE.

$$RNH_2 + COCl_2 \rightarrow RNHCOCl \rightarrow$$
$$R—N{=}C{=}O + HCl \quad (1)$$

$$RNCO + R'OH \rightarrow RNHCOOR' \quad (2)$$

$$RNCO + R'NH_2 \text{ (or } R'R''NH) \rightarrow$$
$$RNHCONHR' \text{ (or } RNHCONR'R'') \quad (3)$$

$$RX \rightarrow RMgX + C_6H_5NCO \rightarrow$$
$$C_6H_5N{=}CR(OMgX) \xrightarrow{H_2O} C_6H_5NHCOR \quad (4)$$

$$RNCO + R'COOH \rightarrow RNHCOOCOR' \rightarrow$$
$$RNHCOR' + CO_2 \quad (5)$$

Isocyanates hydrolyze in air to the unstable carbamic acids (RNHCOOH) which revert to the original amines by loss of carbon dioxide, Eq. (6). Unhydrolyzed isocyanate then reacts with the amine to give a symmetrical urea, RNHCONHR. Nor-

$$RNCO + H_2O \rightarrow RNHCOOH \rightarrow RNH_2 + CO_2 \quad (6)$$

mally an important reaction only because the nuisance must be avoided, controlled hydrolysis of free isocyanate groups to liberate carbon dioxide is the basis of one type of foamed plastic. *See* POLYURETHANE RESINS.

To avoid hydrolysis by atmospheric moisture, isocyanates are best kept in sealed glass containers.

Replacement of oxygen by sulfur, as occurs in isothiocyanates, reduces the activity considerably.

[LEALLYN B. CLAPP]

Isoelectric point

The pH value of the dispersion medium of a colloidal suspension at which the colloidal particles do not move in an electric field; that is, the particles are electrophoretically inert (see illustration). The isoelectric point is often employed to characterize

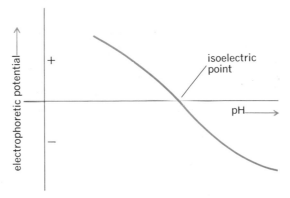

Graph showing the isoelectric point.

colloidal material such as the proteins. However, a range of values is usually necessary, since the isoelectric point varies detectably with (1) the size of the particles, (2) the purity, and (3) the concentration of other than hydrogen ions.

If the pH value of an extrinsic sol (stability attributed primarily to electrical charge) is adjusted toward the isoelectric point, coagulation will occur at or near the isoelectric point. Intrinsic sols (stability attributed primarily to solvation) may be carried to the isoelectric point without coagulation, but such sols will be in a region of minimum stability, so that a minimum concentration of desolvating agent will cause coagulation. Likewise, viscosity changes often reach a minimum at or near the isoelectric point. *See* COLLOID; ELECTROPHORESIS; ION-PERMEABLE MEMBRANE; PROTEIN.

[W. O. MILLIGAN]

Isoelectronic sequence

A term used in spectroscopy to designate the set of spectra produced by different chemical elements ionized in such a way that their atoms or ions contain the same number of electrons. The sequence in the table is an example. Since the neutral atoms of these elements each contain Z electrons, removal of one electron from scandium, two from titanium, and so forth, yields a series of ions all of which have 20 electrons. Their spectra are therefore qualitatively similar, but the spectral terms (energy levels) increase approximately in proportion to the square of the core charge, just as they depend on Z^2 in the one-electron sequence H, He+,

Example of isoelectronic sequence

Designation of spectrum	Emitting atom or ion	Atomic number, Z
CaI	Ca	20
ScII	Sc$^+$	21
TiIII	Ti^{2+}	22
VIV	V^{3+}	23
CrV	Cr^{4+}	24
MnVI	Mn^{5+}	25

Li^{++}, and so forth. As a result, the successive spectra shift progressively toward shorter wavelengths, soon reaching the vacuum ultraviolet region. Isoelectronic sequences are useful in predicting unknown spectra of ions belonging to a sequence in which other spectra are known. See ATOMIC STRUCTURE AND SPECTRA.

[F. A. JENKINS/W. W. WATSON]

Isoetales

A monotypic order of the plant division Lycopodiophyta in the class Isoetopsida, containing only one genus, *Isoetes*. These plants are called quillworts because in all 64 species the leaves are long and narrow with a spoonlike base, spirally arranged upon an underground cormlike structure (see illustration). Most species are semiaquatic, although a few are terrestrial. They are confined to the cooler regions of the world. See STEM (BOTANY).

Morphologists differ in their opinions regarding the relationship of the group. They are like the Selaginellales in having ligules, two kinds of spores (heterosporous), and in producing two kinds of gametophytes. They are different from the other Lycopodiophyta in having multiciliate sperms, in lacking a suspensor (a chain of cells which serves

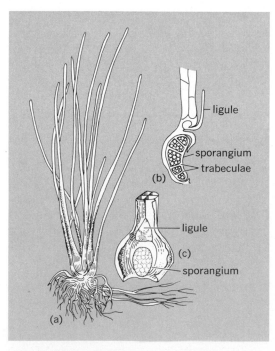

Isoetes. (*a*) Entire plant. (*b*) Longitudinal section of leaf base. (*c*) Face view of ventral surface of leaf base. (*From A. J. Eames, in E. W. Sinnott and K. S. Wilson, Botany: Principles and Problems, 6th ed., McGraw-Hill, 1963*)

to put the embryo in a favorable position in relation to its food supply), and in possessing a secondary meristem which develops some secondary tissue. The presence of a special root-producing region which develops new dichotomous (forked) roots each year, together with their anatomy, suggests a close relationship with the fossil Lycopodiophyta, especially the genus *Pleuromeia*. The phylogenetic connection between *Pleuromeia* and *Isoetes* is accepted by many botanists, and for this reason the families Pleuromeiaceae and Isoetaceae are often included in the Isoetales. See LYCOPODIOPHYTA; MERISTEM, LATERAL; PALEOBOTANY; PLEUROMEIALES; TRACHEOPHYTA.

[PAUL A. VESTAL]

Isoetopsida

One of the two classes of the division Lycopodiophyta, the other being the Lycopodiopsida. Like all Lycopodiophyta, the Isoetopsida are vascular cryptogams with narrow, single-veined leaves and with the sporangia more or less distinctly axillary to the sporophylls. They differ from the Lycopodiopsida in being heterosporous (as opposed to homosporous), and in having a small appendage, the ligule, on the upper side of the leaf near the base. There are two living orders, the Isoetales and Selaginellales, and one fossil order, the Lepidodendrales. The Lepidodendrales were woody plants, commonly of tree size, that formed an important part of the vegetation of the Paleozoic Era from the Upper Devonian to the Permian. The Isoetales and Selaginellales are smaller, herbaceous plants which were never as important as the Lepidodendrales but which had a long fossil history as separate groups. The Isoetales go back at least to the early part of the Cretaceous Period, and the Selaginellales at least to the Upper Carboniferous. There are probably not more than about 600 living species of Isoetopsida. See ISOETALES; LEPIDODENDRALES; LYCOPODIALES; LYCOPODIOPHYTA; LYCOPODIOPSIDA; SELAGINELLALES.

[ARTHUR CRONQUIST]

Isoleucine

An amino acid considered essential for normal growth of animals. The amino acids are characterized physically by the following: (1) the pK_1, or the dissociation constant of the various titratable groups; (2) the isoelectric point, or pH at which a

Physical constants of the L isomer at 25°C:
pK_1 (COOH): 2.36; pK_2 (NH$_3^+$): 9.68
Isoelectric point: 6.02
Optical rotation: [α]$_D$(H$_2$O): +12.4; [α]$_D$(5 N HCl): +39.5
Solubility (g/100 ml H$_2$O): 4.12

Isoleucine

dipolar ion does not migrate in an electric field; (3) the optical rotation, or the rotation imparted to a beam of plane-polarized light (frequently the D line of the sodium spectrum) passing through 1 dm of a solution of 100 g in 100 ml; and (4) solubility, See EQUILIBRIUM, IONIC; ISOELECTRIC POINT; OPTICAL ACTIVITY; SPECTROPHOTOMETRIC ANALYSIS.

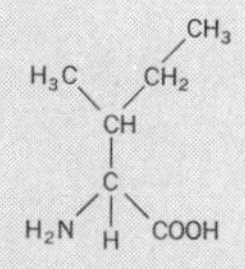

The biosynthesis of isoleucine occurs when pyruvate and α-ketobutyrate react to form α-aceto-α-hydroxybutyrate, which undergoes rearrangement

CH₃—CH₂—CH—C—S—CoA

$$CH_3-CH_2-\underset{\underset{CH_3}{|}}{CH}-\underset{\underset{O}{\|}}{C}-S-CoA$$

α-Methylbutyryl-CoA

| −2H

$$CH_3-CH=\underset{\underset{CH_3}{|}}{C}-\underset{\underset{O}{\|}}{C}-S-CoA$$

Tiglyl-CoA

| H₂O

$$CH_3-\underset{\underset{OH}{|}}{CH}-\underset{\underset{CH_3}{|}}{CH}-\underset{\underset{O}{\|}}{C}-S-CoA$$

α-Methyl-β-hydroxybutyryl-CoA

| DPN

$$CH_3-\underset{\underset{O}{\|}}{C}-\underset{\underset{CH_3}{|}}{CH}-\underset{\underset{O}{\|}}{C}-S-CoA$$

α-Methylacetoacetyl-CoA

| CoA

$$CH_3-\underset{\underset{O}{\|}}{C}-S-CoA+CH_3-CH_2-\underset{\underset{O}{\|}}{C}-S-CoA$$

Acetyl-CoA   Propionyl-CoA

Pathway for the metabolic degradation of isoleucine after its conversion to α-methylbutyryl-coenzyme A.

and reduction to α,β-dihydroxy-β-methylvalerate. Dehydration to the α-keto acid and transamination complete the biosynthesis. Most or all of the enzymes concerned also catalyze the analogous reactions in valine biosynthesis. See AMINO ACIDS; VALINE.

During metabolic degradation the first steps are deamination and oxidative decarboxylation, forming α-methylbutyryl-coenzyme A (see illustration). [EDWARD A. ADELBERG]

Isomerism, molecular

The term isomerism was introduced by J. J. Berzelius for different chemical compounds having the same elementary percentage composition, that is, the same relative proportions of constituent elements. The possibility that the same elementary composition can denote two or more substances, isomers, derives from organic structure theory, which requires only that the valence number of each element (carbon, 4; oxygen, 2; hydrogen, 1) be fully satisfied.

Thus the identity of any given compound is a function of the arrangement, or order of linkage, of the component elements.

Chain isomerism. Among the alkanes, C_nH_{2n+2}, isomerism results from the possibility of linking the same number of carbon atoms to produce either a straight chain or branched chains; for example, C_4H_{10} represents n-butane,

$CH_3CH_2CH_2CH_3$, and isobutane, $CH_3CH(CH_3)_2$. As the number of carbons increases, the number of isomers increases: C_5H_{12} represents CH_3CH_2-$CH_2CH_2CH_3$, $CH_3CH_2CH(CH_3)_2$, and $C(CH_3)_4$. Chain isomers, then, differ from each other by virtue of possessing different carbon skeletons.

Position isomerism. The position of an element (or elements) other than carbon and hydrogen which takes the place of one or more hydrogens may give rise to isomeric structures. The number of such isomers is determined by the number of different types of hydrogen in a given hydrocarbon. Thus in molecules of methane, ethane, and neopentane, all the hydrogen atoms are equivalent (no monosubstituted isomers are possible, and all derivatives of formula $C_nH_{2n+1}Z$ are identical), but propane, butane, and isobutane each have two different types of hydrogen and are capable of providing two different monosubstituted derivatives, $C_nH_{2n+1}Z$ (for example, $CH_3CH_2CH_2Z$ and CH_3CHZCH_3).

If more than one hydrogen atom is replaced by atoms of one substituent element, more position isomers are possible, and even ethane may provide these. The principle, however, remains the same: The number of different types of hydrogen in a given monosubstituted alkane determines how many isomeric disubstituted isomers are possible. Thus, C_2H_4ZX (where Z and X may be the same or different elements) represents CH_3CHZX and CH_2ZCH_2X; and C_3H_6ZX represents CH_3CH_2-$CHZX$, CH_3CHXCH_2Z, CH_3CHZCH_2X, CH_2-XCH_2CH_2Z, and CH_3CZXCH_3, the second and third being identical if Z and X represent the same element.

In unsaturated compounds (simple alkenes and alkynes), the number of unique locations for the double or triple bond determines the number of position isomers. Thus a four-carbon alkene (or alkyne) is the simplest system for which positional isomerism is possible: $CH_3CH_2CH=CH_2$ and $CH_3CH_2C\equiv CH$, or $CH_3CH=CHCH_3$ and $CH_3C\equiv CCH_3$.

In symmetrical cyclic systems (cycloalkanes and benzene), all hydrogens are equivalent, and so, as with ethane, at least two substituents must be present for the existence of positional isomers. The number of such isomers depends on the size of the ring when the number of substituents (for example, Z and CH_3) is limited to two. Thus, referring to Fig. 1, there are two disubstituted cyclopropanes (1,1- and 1,2-), three disubstituted cyclobutanes and cyclopentanes (1,1-, 1,2-, and 1,3-), and four disubstituted cyclohexanes (1,1-, 1,2-, 1,3-, and 1,4-). The nature of the benzene ring limits disubstituted benzenes to three: 1,2- (ortho), 1,3- (meta), and 1,4- (para). Where there are three identical substituents, $C_6H_3Z_3$, three positional isomers are possible (1,2,3-, 1,2,4-, and 1,3,5-); but for $C_6H_3Z_2X$ there are five (1-Z-, 2-Z-, 3-X-; 1-Z-, 2-Z-, 4-X-; 1-Z-, 3-Z-, 2-X-; 1-Z-, 3-Z-, 5-X-; 1-Z-, 4-Z-, 2-X-). A method for determining the absolute position of such substituents in benzene depends upon the experimental production of positional isomers from a given disubstituted benzene (Körner's method).

When a cyclic system is not symmetrical (cycloalkene, heterocyclic compounds, bi- and polycyclic systems such as decalin and naphthalene) the parent hydrocarbon possesses more than

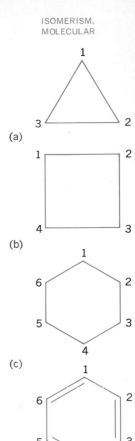

Fig. 1. Symmetrical cyclic systems.
(a) Cyclopropane.
(b) Cyclobutane.
(c) Cyclohexane.
(d) Benzene.

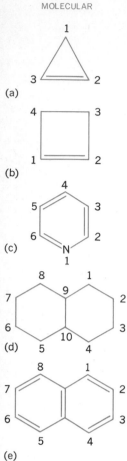

Fig. 2. Unsymmetrical
cyclic systems.
(a) Cyclopropene.
(b) Cyclobutene.
(c) Pyridine. (d) Decalin.
(e) Naphthalene.

one type of hydrogen atom, and hence positional isomerism is possible with but one substituent, whereas polysubstitution leads to a greater multiplicity of positional isomers than in symmetrical cyclic systems. Thus, referring to Fig. 2, there are two monosubstituted cyclopropenes and cyclobutenes (1-Z-, and 3-Z-), three monosubstituted pyridines (2-Z-, 3-Z-, and 4-Z-), three monosubstituted decalins (1-Z-, 2-Z-, and 9-Z-), and two monosubstituted naphthalenes (1-Z-, and 2-Z-). Naphthalene illustrates the effect of polysubstitution on positional isomerism in unsymmetrical cyclic systems. In the parent hydrocarbon, positions 1, 4, 5, and 8 are equivalent and are designated α; positions 2, 3, 6, and 7 are designated β. The introduction of a substituent, Z, in either an α or β position destroys the equivalence of all other positions; thus there are no fewer than 10 positional isomers for a naphthalene bearing two identical substituents Z (1,2-, 1,3-, 1,4-, 1,5-, 1,6-, 1,7-, 1,8-, 2,3-, 2,6-, and 2,7-). When the two substituents Z and X are not identical, four of the listed isomers each provide a pair, since an interchange of substituents in the 1,2, 1,3, 1,6, and 1,7 structures each provides a new isomer, making 14 in all. Higher order substitution enormously increases the number of positional isomers, but in naphthalene, as well as in the other types of unsymmetrical cyclic systems, the principle governing the total number of positional isomers remains the same. It is determined by the number of positionally different hydrogens which may be replaced by other atoms or groups of atoms.

Functional group isomerism. The presence of multiple bonds or of atoms other than carbon and hydrogen (for example, oxygen, nitrogen) in an organic compound may give rise to isomers whose functional groups or reactive centers exhibit chemically distinguishable properties. Thus the formula C_4H_6 may represent 1-butyne, 2-butyne (positional isomers), methylallene ($CH_3CH{=}C{=}CH_2$), 1,3-butadiene ($CH_2{=}CHCH{=}CH_2$), 1- or 3-methylcyclopropene (positional isomers), or methylenecyclopropane. In the same manner, the monoalkenes are isomeric with cycloalkanes; and dialkenes with alkynes, bicycloalkanes, and spiranes.

The introduction of oxygen to give compounds of the molecular formula C_3H_8O may produce alcohols, $CH_3CH_2CH_2OH$ and $CH_3CHOHCH_3$ (positional isomers), or an ether, $CH_3CH_2OCH_3$. When the carbon content is greater, positionally isomeric ethers are possible as well. Likewise, the introduction of nitrogen may lead to functional group isomers; thus C_3H_9N represents n- and isopropylamine, $CH_3CH_2CH_2NH_2$ or $(CH_3)_2CHNH_2$, chain isomers; ethylmethylamine, $CH_3CH_2NHCH_3$; or trimethylamine, $(CH_3)_3N$. These three amines are functionally different and are classed as primary, secondary, and tertiary, respectively.

Where the molecular formula is $C_nH_{2n}O$, further possibilities for functional group isomers obtain. Thus C_3H_6O represents an aldehyde, CH_3CH_2CHO; a ketone, CH_3COCH_3; two cyclic ethers, $CH_3\underset{O}{CHCH_2}$ and $CH_2\underset{O}{CH_2CH_2}$; and an unsaturated alcohol, $CH_2{=}CHCH_2OH$. In the same manner C_3H_7N represents three aldimines, $CH_3CH_2CH{=}NH$, $CH_3CH{=}NCH_3$, and $CH_2{=}NCH_2CH_3$; a ketimine, $CH_3C{=}NHCH_3$; azetidine, $(CH_2)_3NH$; two ethyleneimines, $(CH_2)_2NCH_3$ and

$[CH_2CH(CH_3)]NH$; and two unsaturated amines, $CH_2{=}CHCH_2NH_2$ and $CH_2{=}CHNHCH_3$.

When two or more heteroatoms (atoms other than carbon and hydrogen) are present, the possibilities for functional group isomerism may be illustrated by structures (I), (II), and (III), where Z represents OH, O-alkyl, halogen, or NH_2. Where Z

$$CH_3CH_2COZ \qquad CH_3CHZCHO \qquad CH_3COCH_2Z$$
$$(I) \qquad\qquad (II) \qquad\qquad (III)$$

is OH, (I) represents an acid, (II) a hydroxyaldehyde, and (III) a hydroxyketone; where Z is O-alkyl, (I) is an ester, (II) an alkoxyaldehyde (etheraldehyde), and (III) an alkoxyketone; where Z is halogen, (I) is an acid halide, (II) a haloaldehyde, and (III) a haloketone; and where Z is NH_2, (I) is an amide, (II) an aminoaldehyde, and (III) an aminoketone. Furthermore, within each series, regroupings of the constituent atoms are possible; for example, where Z is OH, the atoms of (I) may be rearranged to represent various ethylene and propylene oxides (three- and four-membered cyclic ethers) with a hydroxyl substituent, or the OH group may be broken up to form methoxyacetaldehyde, CH_3OCH_2CHO, or the esters, methyl acetate, CH_3OCOCH_3, and ethyl formate, CH_3CH_2OCHO.

Geometrical isomerism. In the molecules of these isomers, the atoms are attached to each other in the same order but with different spatial, or geometrical, orientation. The explicit geometry imposed upon a molecule by the presence of a double bond between carbon atoms or between carbon and nitrogen or by the presence of a ring system (which for convenience may be considered planar) makes possible the existence of these isomers. For a discussion of ring systems see CONFORMATIONAL ANALYSIS.

Thus, two atoms or groups of atoms, attached to each of two different carbons, may be relatively closer to or farther from each other, depending on the respective directions of the bonds from the carbons to which they are attached. The rigid double bond, or planar ring, serves as a reference point in the molecule, and such isomers are called geometrical isomers.

Cis-trans isomerism. In any alkenic system in which the doubly bonded carbons each carry two different atoms or groups, cis-trans isomers are possible; that is, two similar atoms or groups may lie on the same side (*cis*) or opposite sides (*trans*) of a plane bisecting the alkenic carbons and perpendicular to the plane of the alkenic system, for example, *cis*-1,2-dichloroethylene (IV) and *trans*-1,2-

dichloroethylene (V). The rigidity of the double bond prevents free rotation, and thus the two geometrically different isomers constitute distinct entities which cannot be interconverted without the expenditure of sufficient energy to destroy the essential character of the double bond.

Such isomers will differ both in chemical properties and in energy content; the *trans* isomer is usually the more stable thermodynamically. Evidence in support of the latter statement is found in the composition of the equilibrium mixture, which contains more of the trans isomer whenever either form is subjected to conditions permitting interconversion (high temperatures, or catalysis involving photochemical, free-radical, or ionic mechanisms).

In designating a particular isomer as *cis* or *trans*, the proper reference points must be used; preferably each of the alkenic carbons should carry an identical atom or group or should carry atoms or groups so closely related that no ambiguity is possible (Br and Cl, CO_2H and CO_2R, or CO_2H and CN). Thus the α-phenylcinnamic acid (VI) derived

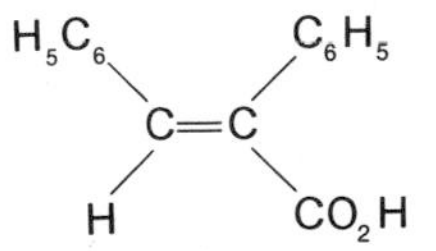

$$H_5C_6 \quad C_6H_5$$
$$C=C \qquad (VI)$$
$$H \quad CO_2H$$

from ordinary (*trans*) cinnamic acid (VII) has erro-

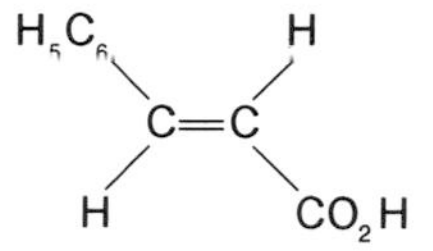

$$H_5C_6 \quad H$$
$$C=C \qquad (VII)$$
$$H \quad CO_2H$$

neously been called *trans* because of its derivation. Actually the *cis* designation is correct since the only group common to both alkenic carbons is the phenyl group.

Syn-anti isomerism. Geometrical isomers that do not carry identical or closely related atoms or groups on the alkenic carbons cannot be named unambiguously using *cis-trans* nomenclature. For example, the α-methylcinnamic acids (VIII) and (IX) and the oximes of acetophenone (X) and (XI)

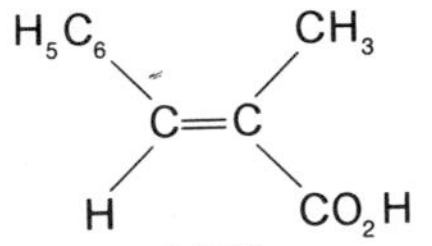

$$H_5C_6 \quad CH_3$$
$$C=C$$
$$H \quad CO_2H$$
$$(VIII)$$

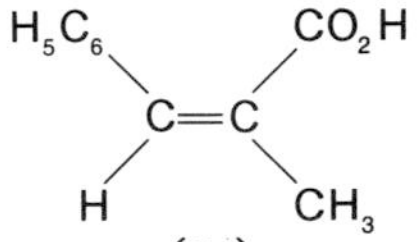

$$H_5C_6 \quad CO_2H$$
$$C=C$$
$$H \quad CH_3$$
$$(IX)$$

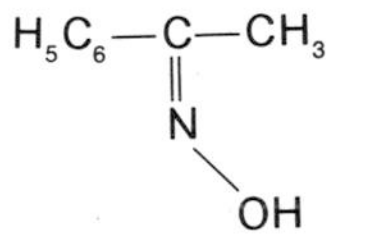

$$H_5C_6-C-CH_3$$
$$\|$$
$$N$$
$$OH$$
$$(X)$$

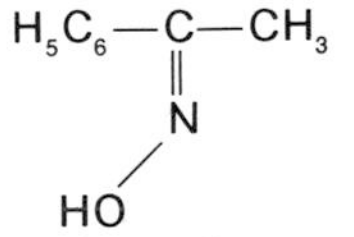

$$H_5C_6-C-CH_3$$
$$\|$$
$$N$$
$$HO$$
$$(XI)$$

cannot properly be called *trans* or *cis*. Therefore, the terms *anti* (on opposite sides) and *syn* (on the same side) are used. The larger (or chemically more reactive) group is taken as the reference point (C_6H_5) and the functional group, CO_2H in (VIII) and (IX) and OH in (X) and (XI), is said to be *anti* (VIII) and (X) or *syn* (IX) and (XI) to it. Thus, (VIII) is *anti*-α-methylcinnamic acid, (X) is *anti*-acetophenoneoxime, and (IX) and (XI) are the corresponding *syn* isomers. *Syn-anti* isomerism will be encountered throughout imine chemistry insofar as the doubly bonded carbon carries two different substituents, but it may be necessary to specify to which substituent the nitrogen substituent is *syn* or *anti*; thus in (XII) the phenylamino

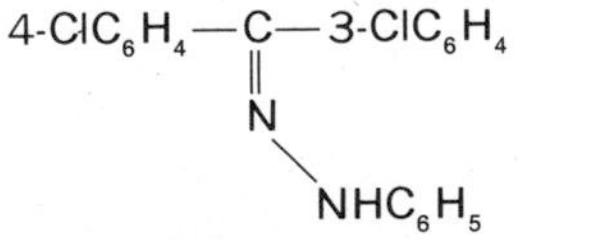

$$4\text{-}ClC_6H_4-C-3\text{-}ClC_6H_4$$
$$\|$$
$$N \qquad (XII)$$
$$NHC_6H_5$$

group is *anti* to the 4-chlorophenyl and *syn* to the 3-chlorophenyl. Similar situations may obtain in syn-

anti isomerism among alkenic compounds.

Ring isomerism. If substituents are attached to alkenic carbons which are part of a cyclic system the ring of which contains fewer than eight members, for example, 1,2-dichlorocyclohexene, bond lengths and bond angles prevent the existence of the *trans* structure. In naming such substances, it is not necessary to be explicit, and the *cis* designation is not employed.

However, cyclic structures admit of geometrical isomerism in another sense: Substituents may be placed on the same (*cis*) or opposite (*trans*) sides of the plane (arbitrarily) defined by the atoms of the ring. Thus, there are the geometrical isomers, *cis*-4-methylcyclohexanol (XIII) and *trans*-4-methylcyclohexanol (XIV). The geometrical designations are referred to the principal (lowest-numbered) substituent, in the present instance, the hydroxyl.

Bicyclo[x,y.0]alkanes, for example, bicyclo[4.4.0]decane, decalin, may have either a *cis* or *trans* ring junction if the rings are sufficiently large. Structural formulas are shown for *cis*-decalin (XV) and *trans*-decalin (XVI).

However, in bicyclo[x.y.z]alkanes, notably Diels-Alder adducts of cyclopentadiene with substituted alkenes, only *cis* ring junction is possible, and the designations *endo* and *exo* are used respectively to locate substituents *trans* (*anti*) or *cis* (*syn*) to a bridge. Thus in (XVII) the carboxyl is

(XIII)

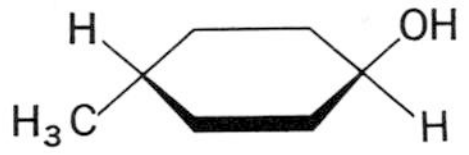

(XIV)

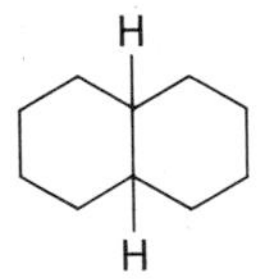

cis-Decalin
(XV)

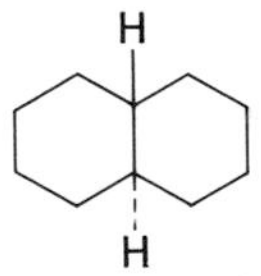

trans-Decalin
(XVI)

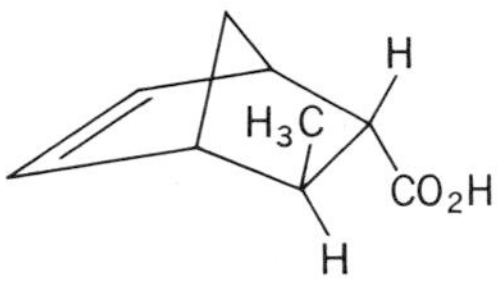

(XVII)

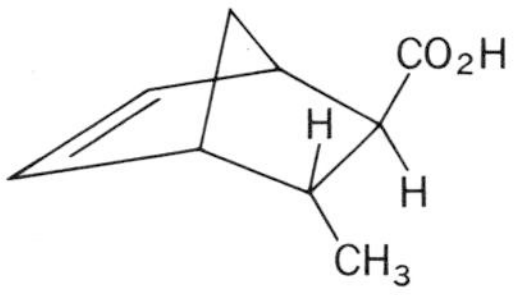

(XVIII)

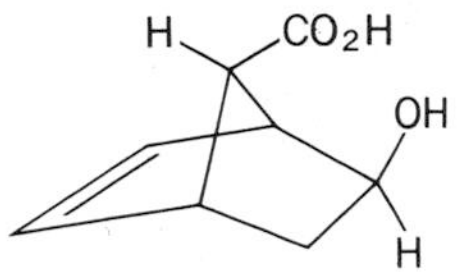

(XIX)

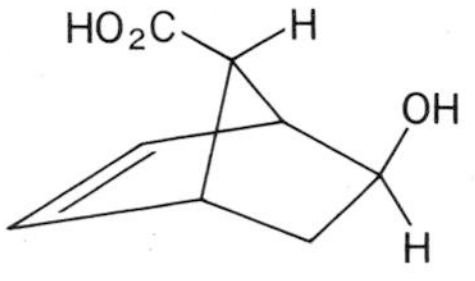

(XX)

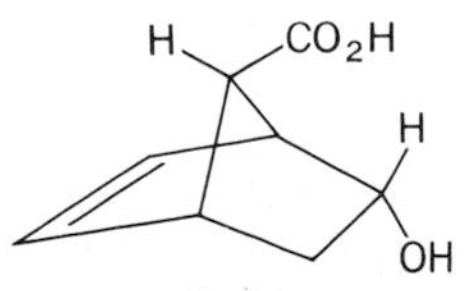

(XXI)

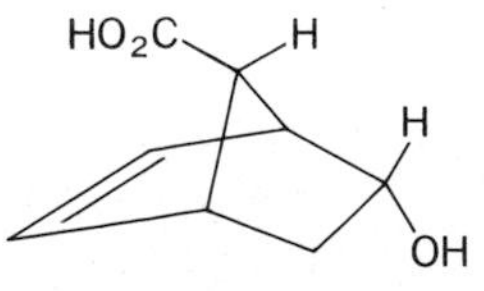

(XXII)

endo and the methyl *exo*, whereas in (XVIII) they are reversed. Two further geometrical isomers are possible in this system: one in which both substituents are *endo* and one in which both are *exo*.

The terms *cis* and *trans*, and *syn* and *anti* also find application in the bicyclic systems. Thus (XIX) and (XX) and (XIX) and (XXI) may be considered *syn* (XIX) and *anti* (XX, XXI) isomeric pairs, and the prefix *anti* may be affixed to both carboxyl and hydroxyl in naming (XXII). All four are geometrical isomers. Compound (XIX) is termed *exo*-2-hydroxybicyclo[2.2.1]-5-heptene-*syn*-7-carboxylic acid or *syn*-2-hydroxybicyclo[2.2.1]-5-heptene-*syn*-7-carboxylic acid. Compound (XX) is termed *exo*-2-hydroxybicyclo[2.2.1]-5-heptene-*anti*-7-carboxylic acid or *syn*-2-hydroxybicyclo[2.2.1]-5-heptene-*anti*-7-carboxylic acid. Compound (XXI) is termed *endo*-2-hydroxybicyclo[2.2.1]-5-heptene-*syn*-7-carboxylic acid or *anti*-2-hydroxybicyclo-[2.2.1]-5-heptene-*syn*-7-carboxylic acid. Lastly, compound (XXII) is termed *endo*-2-hydroxy-bicyclo[2.2.1]-5-heptene-*anti*-7-carboxylic acid or *anti*-2-hydroxybicyclo[2.2.1]-5-heptene-*anti*-7-carboxylic acid. *See* OPTICAL ACTIVITY; STEREOCHEMISTRY; TAUTOMERISM.

[WYMAN R. VAUGHAN]

Bibliography: H. Gilman, *Organic Chemistry*, 2d ed., vol. 1, 1943; J. D. Roberts and M. C. Caserio, *Basic Principles of Organic Chemistry*, 1964.

Isomerism, nuclear

The existence of metastability of excited states of atomic nuclei. The existence of well-defined excited states in atomic nuclei indicates that a nucleus, like a molecule, can exist in any one of several physical configurations of the constituent protons and neutrons. Therefore, metastable states of the same nucleus may exist, differing from the lowest energy state or ground state by energies ranging from a few thousand to a few million electron volts. This excitation energy corresponds to that energy required to change the configuration of nucleons from that giving the lowest potential energy to a new configuration existing at some higher potential. The implication is that a nucleon or nucleons in the excited states occupy energy levels or quantum states not occupied in the ground-state or lowest energy configurations. *See* ISOMERISM, MOLECULAR.

Lifetimes. By definition, the existence of isomeric configurations and their classification as "isomeric states" depends solely on the ability to observe experimentally a measurable lifetime for the individual excited state or states. Lifetimes as short as 10^{-11} sec and as long as several hundred years have been measured for isomeric states. The lifetime of the excited state is the result of a combination of effects related to the energy and angular momentum difference between the isomeric state and the next-lowest energy state of the nucleus. In general, the larger the difference in angular momentum between states (equal to the multiplicity of the radiative transition), and the lower the energy difference, the longer the lifetime. Hence, the existence of isomerism is fundamentally related to the properties associated with the radiation of electromagnetic energy by the circulating charge within the nucleus. *See* ELECTROMAGNETIC RADIATION; MULTIPOLE RADIATION; NUCLEAR SPECTRA; SELECTION RULES (PHYSICS).

The existence of a metastable excited state or isomer is not dependent on the lifetime of the ground state. That is, the lowest energy configuration of nucleons may in itself be unstable toward radioactive decay much like the isomer. The existence, by definition, of the isomeric pair (or even triplet) depends only on the observable lifetimes of the states in question.

Decay. The characteristic decay rate of the isomeric states is a pure first-order kinetic process with a well-defined rate constant and half-life. Decay to the lower energy states of the same nucleus (isomeric transition) via gamma-ray emission or interaction with the orbital electrons of the atom may be in competition with other modes of radioactive decay, should they be energetically possible. Isomeric states may decay via beta decay (either negatron or positron emission), electron capture, emission of alpha particles, or even fission, with one or more of these decay modes in competition with the isomeric transition. Each of the possible decay modes of the isomeric state results in a lowering of the total potential energy of the system. The total decay rate of the isomeric state is then dependent on both the multipole order of the isomeric transition and the energy associated with this and other possible decay modes. *See* KINETICS, CHEMICAL; NUCLEAR RADIATION; RADIOACTIVITY.

The lifetime of the isomeric state τ is related to the total decay rate λ and half-life $T_{1/2}$ by Eq. (1).

$$\tau = \frac{1}{T_{1/2}} = \frac{\lambda}{.693} \tag{1}$$

The observable lifetime of the isomeric state (τ_{obs}) is related to the lifetimes associated with each of the possible individual modes of decay by Eq. (2), where τ_{IT} is the lifetime of the isomeric

$$\frac{1}{\tau_{obs}} = \frac{1}{\tau_{IT}} + \frac{1}{\tau_{\beta^+}} + \frac{1}{\tau_{\alpha}} + \cdots \tag{2}$$

transition, τ_{β^+} that of decay via positron emission, τ_{α} associated with alpha decay, and so forth.

Formation. Isomeric states can be formed through any of several nuclear processes. Just as the probability of the isomeric transition is dependent on angular momentum change and energy difference, the population of the metastable states is governed, for the most part, by the same selection rules and probabilities. The isomeric state (or states) may be formed via beta or alpha decay of a parent nucleus, where energy and angular momentum conservation favor the population of a particular excited state in the daughter. Simply, since the existence of isomerism implies a large angular momentum difference between states, this same angular momentum difference enhances the probability of populating states in radioactive decay processes that closely resemble the properties of the parent. Isomers may also be populated preferentially in nuclear reactions where again consideration of angular momentum and energy conservation favor the formation of the excited metastable state over that of other states in the same nucleus. Typical of such processes are those induced in reactions where the intrinsic high angular momentum of the system of target nucleus and projectile favors population of high angular momentum states in the products, where the lowest energy state may have relatively low angular momentum, or vice versa. *See* NUCLEAR REACTION; NUCLEAR STRUCTURE.

Classical interpretation. Classically, the existence of isomerism has been related to a change in the ordering of the discrete energy levels occupied by individual or groups of nucleons within the nucleus in a manner analogous to isomerism in chemical systems. Evidence points to the possibility that isomerism may also result from a departure from spherical nuclear shapes to some aspherical shape. The transition from one elipsoidal configuration to another can induce isomerism since each shape may have a unique and measurable lifetime, dependent on the exact position of the specific angular momentum states within the deformed potential.   [I. L. PREISS]

Bibliography: B. L. Cohen, *Concepts of Nuclear Physics*, 1971; M. Goldhaber and R. D. Hill, *Rev. Mod. Phys.*, 24:179, 1952; I. L. Preiss et al., *Nucl. Phys.*, A205:619, 1973; I. L. Priess, P. M. Strudler, and D. A. Bromley, *Phys. Rev.*, 141(3).1097, 1966, K. Siegbahn (ed.), *Alpha, Beta and Gamma-Ray Spectroscopy*, vols. 1 and 2, 1965.

Isomerization

Rearrangement of the atoms within hydrocarbon molecules. Isomerization processes of practical significance in petroleum chemistry are (1) migration of alkyl groups, (2) shift of a single-carbon bond in naphthenes, and (3) double-bond shift in olefins.

Migration of alkyl groups. An example of alkyl group migration (skeletal isomerization) is Eq. (1).

$$
\begin{array}{c}
\text{C—C—C—C—C—C} \rightleftharpoons \underset{\substack{|\\ \text{C}}}{\text{C—C—C—C}} \quad \text{C} \rightleftharpoons \\
\text{\textit{n}-Hexane} \qquad\qquad \text{2-Methylpentane}
\end{array}
$$

$$
\underset{\substack{|\\ \text{C}}}{\overset{\substack{\text{C}\\ |}}{\text{C—C—C—C}}} \tag{1}
$$

2,2-Dimethylbutane

Isomerization to more highly branched configurations has commercial importance since it results in improvement in combustion quality in the automobile engine as measured by octane number and increased chemical reactivity because tertiary carbon atoms result. The unleaded, motor-method octane numbers of the hexane isomers shown in Eq. (1) are 26.0, 73.5, and 93.4, respectively. Normal butane is converted to isobutane (which has a tertiary carbon atom) to attain chemical reactivity with olefins in alkylation reactions where *n*-butane is inert.

Isomerization of paraffins is a reversible first-order reaction limited by thermodynamic equilibrium which favors increased branching at lower temperatures. Undesirable cracking reactions leading to catalyst deactivation occur at higher temperatures. They are controlled by adding a cracking suppressor such as hydrogen.

Conversion of normal butane to isobutane is the major commercial use of isomerization. Usually, it is carried out in either liquid- or vapor-phase over aluminum chloride catalyst promoted with hydrogen chloride. In the vapor-phase process (250–300°F), the aluminum chloride is often supported on bauxite. In the liquid-phase processes (180°F), it is dissolved in molten antimony trichloride or used in the form of a liquid complex with hydrocarbon. A second type of catalyst for vapor-phase isomerization (300–850°F) is a noble metal, usually platinum, supported on a carrier. This may be alumina with halide added to provide an acidic surface. All the processes are selective (95–98% to isobutane). Approximately 60% of the *n*-butane feed is converted per pass to isobutane in the liquid-phase process.

Isopentane, a high-octane component used in aviation gasoline, is made commercially by isomerization of *n*-pentane. Petroleum naphthas containing five- and six-carbon hydrocarbons also are isomerized commercially for improvement in motor-fuel octane numbers. Noble-metal catalyst is normally used with higher-molecular-weight feeds. Isomerization of paraffins above six carbon atoms is of less importance, since octane improvement is limited by predominance of monomethyl branching at equilibrium. Skeletal isomerization is an important secondary reaction in catalytic cracking and catalytic reforming. Aromatics and olefins undergo skeletal isomerization as do paraffins.

Single-carbon bond shift. This process, in the case of naphthenes, is illustrated by Eq. (2). Cy-

$$
\square\text{—C} \rightleftharpoons \hexagon \tag{2}
$$

Methyl-   Cyclo-
cyclopentane hexane

clohexane and methylcyclohexane have been produced commercially by liquid-phase isomerization of the five-carbon ring isomers over aluminum chloride–hydrocarbon-complex catalyst promoted by hydrogen chloride. Conversion per pass is high, selectivity excellent, and reaction conditions mild (190°F). Cyclohexane is a raw material for making nylon, and it may be dehydrogenated to benzene. Methylcyclohexane has been used to make synthetic nitration-grade toluene.

Shift of a double-bond. This process is usefully applied when a specific olefin is needed for chemical synthesis, for example, Eq. (3). Double-bond

$$
\text{C—C—C}{=}\text{C} \rightleftharpoons \text{C—C}{=}\text{C—C} \tag{3}
$$
1-Butene   2-Butene

shift occurs selectively over acidic catalysts at temperatures below 450°F. However, the proportion undergoing skeletal isomerization increases as temperature is increased until, at temperatures in the range of 600–950°F, equilibrium is approached at fairly high space velocities. Equilibrium favors movement of double bonds to the more stable internal positions (85.6% 2-butene at 400°F), and octane improvement accompanies this shift; however, the increase of octane number normally is insufficient to justify the cost of processing thermally cracked gasolines solely for this purpose. This type of isomerization occurs as a secondary reaction in the catalytic cracking and catalytic polymerization processes, in part accounting for the high octane numbers of the gasolines. *See* ALKYLATION; AROMATIZATION; CRACKING; ISOMERISM, MOLECULAR; PETROLEUM PROCESSING.

[GEORGE E. LIEDHOLM]

Isometric process

A constant-volume, frictionless thermodynamic process in which the system is confined by mechanically rigid boundaries. No direct mechanical work can be done on the surroundings by a system with rigid boundaries; therefore the heat transferred into or out of the system equals the change of internal energy stored in the system. This change in the internal energy, in turn, is a function of the specific heat and the temperature change in the system as in Eq. (1), where Q_V is the heat trans-

$$Q_V = U_2 - U_1 = m \int_1^2 C_V \, dT \qquad (1)$$

ferred at constant volume, U is the internal energy, m is the mass, C_V is the specific heat at constant volume, and T is the absolute temperature. If the process occurs quasistatically (the system going through a continuous sequence of equilibrium states), Eq. (2) holds, where S is the entropy. There

$$Q_V = \int_1^2 T \, dS \qquad (2)$$

is an increase in both the temperature and the pressure of a constant volume of gas as heat is transferred into the system. For a comparison of the isometric process with other processes involving a gas see POLYTROPIC PROCESS.

[PHILIP E. BLOOMFIELD]

Isomorphism (crystallography)

A similarity of crystalline form between substances of similar composition. Two substances which are isomorphous have a similar chemical formula, an equal or nearly equal ratio of cation to anion radius, and comparable polarizabilities of their ions. Isomorphism is morphotropism in a narrower, more precise sense. Similarity in the macroscopic characteristics of isomorphous crystals becomes so close that extreme precision is needed to distinguish between them. See MORPHOTROPISM.

Examples of isomorphous substances are $NaNO_3$ and $CaCO_3$, $CaAl_2Si_2O_8$ and $NaAlSi_3O_8$, and $BaSO_4$, $SrSO_4$, and $PbSO_4$. Substances such as ThO_2 and LiO_2 are anti-isomorphous; they both have the calcium fluoride structure, but the positions of the anions and cations are interchanged in the two structures because of the relative sizes of the ions. Isomorphous substances form mixed crystals, while anti-isomorphous substances do not. For a discussion of the chemical composition and crystal structure of isomorphous minerals see MINERALOGY. See also CRYSTAL STRUCTURE.

[WILLY C. DEKEYSER]

Bibliography: M. J. Buerger, *Crystal Structure Analysis*, 1960; R. C. Evans, *Introduction to Crystal Chemistry*, 1939.

Isonicotinic acid hydrazide

A chemical compound used as a chemotherapeutic agent in the treatment of tuberculosis. The observation in 1945 that the vitamin nicotinamide inhibits the growth of the bacillus (*Mycobacterium tuberculosis*) which causes tuberculosis led to a search for similar activity among compounds chemically related to nicotinamide; at a later date, two drugs with outstanding potency against the

Structural formulas for three antitubercular drugs.

tubercle bacillus were discovered: isonicotinic acid hydrazide (isoniazid) and its isopropyl derivative (iproniazid) (see illustration).

Isoniazid and iproniazid are highly specific in their action, exerting an effect in the animal body only against the tubercle bacillus. They have proved inactive against a large number of other bacteria, protozoa, and viruses against which they have been tested. Isoniazid is now the mainstay in the treatment of tuberculosis since it is highly effective and does not cause serious toxic reactions. However, since the tubercle bacilli can develop resistance to isoniazid, a second antitubercular substance, usually *para*-aminosalicylic acid, is given concomitantly in order to prevent or delay the development of resistance. Frequently, in the initial treatment of tuberculosis, the potent antibiotic streptomycin is added to the regimen. See PARA-AMINOSALICYLIC ACID; DRUG RESISTANCE; STREPTOMYCIN; TUBERCULOSIS.

Isoniazid is also used for long-term prophylactic treatment of individuals who have had close contact with a tuberculous patient or are considered to be highly susceptible to the disease. Iproniazid is no longer used in the treatment of human tuberculosis because of toxic effects produced during its chronic administration. This toxicity may be due to the inhibition of monamine oxidase, an enzyme with which isoniazid does not interfere. However, iproniazid has been used as a psychic energizer to treat both mild and severe depressions and psychoses associated with severe repressions or regressions. See CHEMOTHERAPY; PSYCHIC ENERGIZER; PSYCHOSIS.

[NICHOLAS J. GIARMAN/EMANUEL GRUNBERG]

Isopoda

An order of malacostracan crustaceans characterized by a cephalon, or head, bearing one pair of maxillipeds in addition to the antennae, mandibles, and maxillae. The maxillipeds are derived from the legs of the first postcephalic somite which is almost completely fused with the cephalon. A carapace is lacking. The peraeon, or thorax, consists of seven distinct somites each bearing a pair of peraeopods, the legs. The first pair of peraeopods is never chelate, and typically all pairs of peraeopods are similar in structure. The pleon, or abdomen, has six somites. The first five pairs of pleonal appendages are foliaceous. The last pair is modified into hardened appendages called uropods and the last somite is fused with the telson into a pleotelson. Attached to the inner base of the peraeopods of the female are platelike appendages called oostegites. Prior to egg deposition, the oostegites enlarge and interleave to form a brood pouch into which the eggs are deposited (see illustration). Larval development occurs in the egg while it is in the brood pouch, or marsupium. The young hatch from the eggs into the brood pouch from which they emerge as miniature adults, lacking only the seventh pair of peraeopods. These are added during a few postembryonic molts. There is no parental care of the hatched young. The old theory that the young receive nourishment from the parent has been abandoned. The sexes are separate, fertilization is internal, and the number of eggs produced is usually less than 50. The female is the sexually heterozygous individual, but little is known of the genetics of isopods.

Ecology. The most familiar isopods are the terrestrial sow bugs or pill bugs. Many of these animals roll up into a compact ball when disturbed. Land isopods are usually found in moist environments, under decaying leaves and wood, and under rocks. Isopods range in size from 1 mm (*Munna*) to around 4 cm (*Bathynomus*). In addition to being found on land, isopods have been found in hot springs (*Exosphaeroma thermophilum*) and in subterranean fresh waters (*Caecidotea*) and caves. The majority are marine where they range from the intertidal areas to the greatest depths of the sea. Their food varies from wood (*Limnoria*) and seaweeds (*Limnoria* and *Idothea*) to animal flesh (*Cirolana* and *Cymothoa*). *Cirolana* has been known to inflict small but painful wounds on swimmers. *Cymothoa* and its allies are predaceous, commensal, and parasitic on fish. The destructive marine wood-boring isopod *Limnoria*, the gribble, causes annual damage estimated at $5,000,000 to wharf piling in the United States, and one species is reported to attack treated timbers.

Geologically, isopods are an ancient group, ranging from the Devonian to the Recent. The genus *Cyclosphaeroma* from the Jurassic of England is the best-known isopod fossil and belongs to the contemporary family Sphaeromidae.

Classification. The classification of the Isopoda is not stabilized; for example, the gnathiids and anthurids are included in the suborder Flabellifera by some investigators and not by others. There is a need for critical research on the generalities of their classification. Most students divide the isopods into seven equivalent suborders, the Oniscoidea, Valvifera, Flabellifera, Bopyroidea, Gnathiodea, Asellota, and Phreatoicidea.

The Oniscoidea are the pill bugs, sow bugs, and wood lice which are terrestrial. Some have pseudotrachea on pleopods, and the uropods are terminal. *Oniscus*, *Porcellio*, and *Deto* are common examples.

Valvifera are marine and fresh-water species with valvelike uropods inflexed under the pleon and covering the pleopods. The mandibles lack a palp. These animals, such as *Idothea* and *Arcturus*, are free-living.

The Flabellifera are marine and fresh-water species with the uropods expanded laterally with the pleon to form a tail fan. These animals are free-living, predaceous, commensal, and parasitic. *Cirolana*, *Cymothoa*, *Anthura*, and *Limnoria* are common genera.

Bopyroidea are marine parasites on crustaceans. The young are much like flabelliferans, but the adult female is metamorphosed into a leaflike sac, and the male is minute with many parts reduced. Examples are *Bopyrus* and *Ione*.

Gnathioidea are 10-legged marine parasites of fish with lateral uropods. The female mouthparts

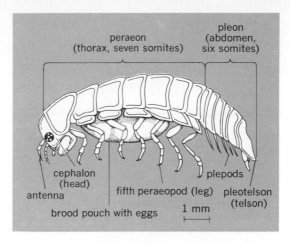

Limnoria, female. (*From R. J. Menzies, The comparative biology of the wood-boring isopod crustacean Limnoria, Museum of Comparative Zoology, Harvard Coll. Bull., 112(5):363–388, 1954*)

are suctorial, while the male mandibles are pincerlike and project beyond the frontal margin of the cephalon as in *Gnathia*.

Asellota are small marine and fresh-water species. The uropods are styliform and terminal. External pleopods are operculate and cover the other pleopods. *Asellus*, *Ianira*, and *Jaera* are genera of this suborder.

Phreatoicidea are laterally compressed isopods living in the fresh-water streams of New Zealand and South Africa. In general aspect they look much like amphipods.

About 3000 species of isopods are known today, but it may be estimated that only one-half of the existing species have been described. *See* CRUSTACEA.   [ROBERT J. MENZIES]

Isoprene

A five-carbon, conjugated diolefin, or diene, having the structure shown. It does not occur naturally

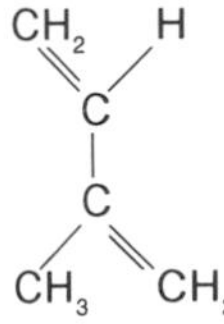

$$
\begin{array}{c}
CH_2 \quad H \\
\backslash \; / \\
C \\
| \\
C \\
/ \; \backslash \\
CH_3 \quad CH_2
\end{array}
$$

but is obtained by the destructive distillation of gas oil, naphthas, and rubber. It is also prepared by the catalytic decomposition of dipentene. It is commercially available in about 96% purity and is used in the production of butyl rubber.

The terpenes may be regarded as multiples of the isoprene unit, C_5H_8. Indeed, isoprene may also be the foundation for other important plant products such as phytol, the sterols, and the carotenoids.

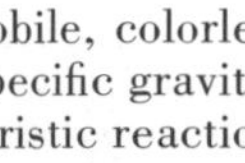

Isoprene is a mobile, colorless liquid boiling at 34.1°C, having a specific gravity of 0.862. It exhibits all the characteristic reactions of dienes of this type. Isoprene polymerizes readily to form dimers and high-molecular-weight resins. The principal dimer formed is isopropenyl methyl cyclohexene as shown below. Polymerization of isoprene during storage can be controlled by avoiding contact with oxygen, by using inhibitors such as *tert*-butyl catechol, and by keeping it at low temperatures. See DIENE; TERPENE.

[WILLIAM MOSHER]

Isopropylphenol

One of three isomeric compounds, having structures shown in the figure. The nomenclature of the derivatives of these compounds is directly parallel to that of the cresol analogs.

Production. These compounds can be prepared by a number of syntheses. Starting from cumene, the same processes may be used as those used for the production of phenol from benzene. Of little interest or use prior to the introduction of the cumene hydroperoxide process of phenol manufacture, supplies of these compounds are very largely derived as by-products of this process.

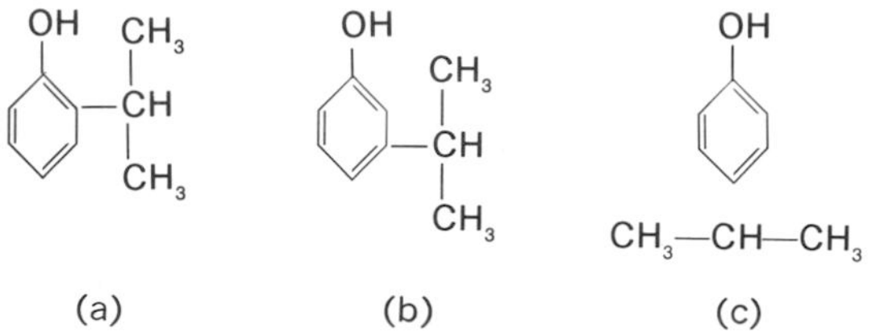

Structural formulas for isopropylphenol. (*a*) Ortho, (*b*) meta, and (*c*) para configurations.

Uses. In many of the applications of which cresols have historically been used, one or another of the isopropylphenols can be used. As a by-product of an extremely large industrial manufacturing operation, that is, phenol from cumene, isopropylphenols are potentially plentiful and are free of many troublesome impurities commonly found in the cresols or cresylic acids of commerce, which are largely derived from coal tars or waste products of petroleum-refining operation. These advantages have actually led at least one large European supplier of triarylphosphates to undertake a wholly synthetic route to these popular plasticizers, using isopropylphenol instead of the traditional cresol. The uniformity and improved properties of the end-use derivatives are relied upon to justify the somewhat higher costs. There are as yet no reliable published statistics on prices or production and sales of isopropylphenol, but substantial quantities are known to move in commerce. See CRESOL; CUMENE; PHENOL.

[ROBERT I. STIRTON]

Isoptera

An order of Insecta into which are grouped certain morphologically primitive forms commonly referred to as termites. They have a gradual metamorphosis, lacking true larval and pupal stages; the antennae are moniliform; the mouthparts are biting and prognathous; there are two pairs of subequal wings; and the abdomen is broadly joined to the thorax. All members of the Isoptera live in colonies. Termites perform a very important role in the breakdown of all types of cellulose debris. Certain species are very destructive when they extend their areas of activity into man's structures. Ter-

mites are most abundant in the tropics, often in very moist situations, although many species are typical of semiarid or even of desert regions. A few species extend into North Temperate areas in Europe, Asia, and North America; the genus *Reticulitermes* is restricted to these regions.

Approximately 1900 species of Isoptera are recognized. Termites are often grouped into higher and lower forms. The higher termites include the family Termitidae, which contains about 80% of the species. The lower termites include all other families, which are generally considered to be relatively primitive. Included in the lower termites are the Mastotermitidae (with a single living species in Australia), Hodotermitidae, Kalotermitidae, and Rhinotermitidae. Some investigators consider the Hodotermitidae to include two distinct groups distinguished as the family Hodotermitidae (harvester termites of Africa) and the Termopsidae (damp wood–dwelling forms). A. E. Emerson has elevated the monotypic genus *Serritermes* to family rank (Serritermitidae).

General characteristics and castes. The mature termite (alate or imago) has membranous wings which extend beyond the end of the abdomen. There is a pair of compound eyes, and a pair of ocelli is present in most groups. The wings are superimposed flat on the abdomen when the insect is not in flight. The venation is relatively simple and varies from family to family (Fig. 1). Flight is weak and fluttering and is usually short. When the alate alights, the wings are shed along a basal suture with the base of each wing (the wing scale) being retained. The alates vary in color from yellow, through brown, to coal black. Some species (usually lightly pigmented) fly during the night; others (usually heavily pigmented) fly during the day. The time of flight varies from species to species both with respect to the season of the year and the time of the day or night.

Soldier caste. In almost all termite species a second type of individual is produced in the colony. This is the soldier which lacks wings, is nonreproductive, and is variously modified for defense. There are four rather distinct types of soldiers: mandibulate, phragmotic, nasutoid, and nasute. In mandibulate soldiers the head and mandibles are greatly enlarged and heavily sclerotized (Fig. 2*a*). The mandibles may be biting, snapping, or pincherlike and more or less symmetrical or strongly asymmetrical. In phragmotic soldiers the mandibles are not as conspicuously enlarged as in mandibulate forms. The head is high and truncate in front and is used to plug openings in the workings.

In the families Rhinotermitidae and Termitidae there is a cephalic gland which opens via a small pore on the dorsal head surface. In some groups this gland and its opening have been variously modified for defense in the soldiers. In some rhinotermitid soldiers the opening of the gland lies at the anterior margin of the head, and the fluid is discharged into an open troughlike structure which extends forward from the head capsule. These have been termed nasutoid soldiers. Certain species have both mandibulate and nasutoid forms. Finally, in the termitid subfamily Nasutitermitinae the cephalic gland opens at the tip of an elongated tube or nasus which projects anteriorly, giving the head the appearance of a pear-shaped

ISOPTERA

2 mm

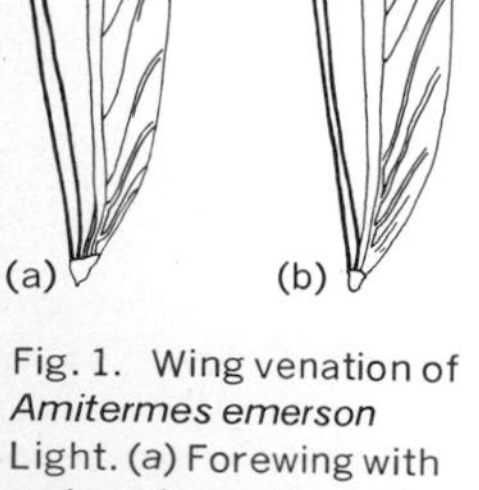

Fig. 1. Wing venation of *Amitermes emerson* Light. (*a*) Forewing with reduced venation in fore area, typical of Rhinotermitidae and Termitidae. (*b*) Hindwing.

syringe (Fig. 2b). These are the nasute soldiers.

All types of soldiers are preceded during development by an intermediate stage (the soldier-nymph or white soldier) which is soldierlike in form but is unsclerotized. In general, mandibulate soldiers are rather large and occur in relatively small numbers in the colonies, whereas nasute soldiers are relatively minute and constitute as much as 30% of the population.

Worker caste. In the more advanced termites there is a third caste, the worker. True workers usually have some pigmentation as opposed to the immature termites, which are generally white. Workers lack wings, are nonreproductive, and have mandibles which resemble those of the imagoes; they are usually blind. In many lower termites there is no distinct worker caste, and the work of the colony is performed by young individuals which have undergone at least two stages of development. These are still capable of becoming either alates or soldiers, but may continue to molt without becoming differentiated. Eventually they may undergo stationary molts with no appreciable change in size or structure. These "stabilized" individuals, which function as workers but are still capable of differentiation into other castes, have been termed pseudergates by P.-P. Grassé and Ch. Noirot.

Replacement reproductives. In addition to the definitive castes (alate, soldier, and worker) another type of individual may occur in the colony under certain circumstances. These individuals are the supplementary or replacement reproductives. Although the original pair (king and queen) may live for two or three decades, the life of the colony itself is not limited by their survival. If one or both are lost, other individuals in the colony become reproductive. If these new reproductives have wing buds, they are termed nymphoid (or brachypterous or second-form) reproductives. If they lack wing buds, they are termed ergatoid (or apterous or third-form) reproductives. Occasionally such reproductives may be produced even in the presence of the original pair, but this appears to be the

exceptional rather than the usual situation. Replacement reproductives may be produced in large numbers. In the kalotermitids the number of supplementary reproductives may be reduced to a single pair through cannibalism, but in other groups multiple reproductives are retained.

Caste development and determination. The actual mechanisms of caste development vary in the different families of termites, particularly between the lower and higher groups. In general the lower termites have a very gradual development through an indefinite number of stages or instars before reaching a terminal form or caste. Apparently all individuals in the colony, except terminal types and soldier-nymphs, are capable of developing into any type of individual normally produced by the colony. They are also capable of becoming replacement reproductives. The actual development of individuals of a particular caste appears to be due to environmental factors related to the makeup of the colony and its general vitality. This has been termed the extrinsic theory of caste determination. The presence of individuals of a certain caste, in certain proportions, appears to limit the production of additional individuals of the same caste. This observation led to the development of the ectohormonal inhibition theory of caste determination. M. Lüscher has proposed that the castes are regulated by substances termed pheromones, which are produced by individuals in the colony in response to contact with other individuals of a particular caste. When a certain caste is lacking or falls below a certain proportion, the pheromones are lacking, and the most plastic individuals develop toward the required type until the inhibition is reestablished.

In the higher termites the caste lines are much more rigid and generally are fixed after the first stage of development, at least with respect to the individuals giving rise to alates in one case, and the workers and soldiers in the other. Noirot has found that in some groups of higher termites, workers and soldiers are of both sexes, as in the lower termites. In other groups, however, males are soldiers and females are workers, while in still other groups the situation is reversed. In some instances there is a sexual dimorphism, with one sex producing small workers or soldiers and the opposite sex producing large workers or soldiers.

The colony. Basically the termite colony is a family assemblage, with the original group being the offspring of a pair of alates. The development of the termite colony may be summarized as follows.

Periodically (usually once each year) a mature termite colony produces alates which leave the colony in which they developed, fly for a short distance, alight, and shed their wings. The male usually seeks out the female, who may assume a "calling" position with the tip of the abdomen held at a right angle to the rest of the body. When a male contacts a female, she moves away with the male following closely behind. The pair seek out a nesting site in wood or soil, depending upon the species. They make a chamber large enough for both of them, seal the entrance, and then mate. The male continues to live with the female, and insemination is repeated at intervals throughout the life of the pair.

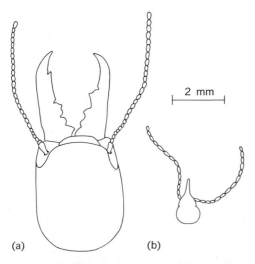

Fig. 2. Head capsule and antennae of (a) mandibulate soldier of *Zootermopsis nevadensis* (Hagen) showing enlarged mandibles, and (b) nasute soldier of *Lacessititermes palawanensis* (Light).

The initial group of eggs is usually small, with rarely more than 20 being deposited during the first 6 months after pairing. The pair care for the eggs and the young termites as they hatch from the eggs. When the young termites have passed through two or more stages of development, they become active in the colony and begin caring for other young and the eggs, as well as extending the workings, and grooming the king and the queen. Over a period of years the colony gradually increases in size as the queen matures and lays more eggs. The immature colony consists of the king and queen, the young, a few soldiers, and the workers or functional workers.

After several years the colony can be termed mature since it produces alates which will depart under suitable conditions to start new colonies. As the colony grows, more and more alates are produced each year in addition to the other castes. The ultimate size of the colony varies. Many of the Kalotermitidae deposit a small number of eggs, even when the queen is mature. Their colonies may consist of only a few hundred, or at most a few thousand, individuals. Many of the Termitidae have a much greater egg-laying capacity. A single queen may lay thousands of eggs per day and the colony may contain a million or more individuals.

Foods and feeding. Termites vary greatly in their food habits. Many consume wood which may be sound or in varying stages of decay. Many, however, are humivores, that is, they feed on humus; and others forage above ground, collecting grasses, leaves or wood fragments, or lichens and algae. Such materials are carried back to the nest and stored. In one termitid subfamily (Macrotermitinae) elaborate fungus gardens are developed within the nests. The ability of termites to utilize cellulose is associated in the lower termites with the presence of complex flagellate protozoons in the hindgut. Similar flagellates occur in the woodroach *Cryptocercus*, and the roaches and termites are thought to have evolved from a common ancestral stock. In the higher termites such flagellates are few or absent. There is, however, a rich flora of bacteria in these species, and the bacteria may play a role in cellulose digestion in this group.

Nests. The nests of some termites consist of simple excavations in the wood upon which they are feeding. In others the nests are simple excavations in the soil. Most of the conspicuous aboveground termitaria are built by the higher termites. These nests may be massive domes or chimneylike shapes or may be wedge-shaped; or they may be relatively small and of precise form, such as the mushroom shaped nests of *Cubitermes*. Many species construct arboreal nests which may be compact and often rounded; or they may be sheetlike, extending over tree trunks and having overhanging structures which are thought to be rain-shedding devices. Many nests are constructed of a variable mixture of saliva, fecal material, and either soil or wood fragments or both, making up a tough material referred to as "carton." Among the most remarkable termite nests are the subterranean structures of *Apicotermes*. These have a series of level floors interconnected by straight or spiraled ramps and surrounded by a discrete wall which contains circular galleries. These in-wall galleries open to the interior and to the exterior by precisely arranged and shaped pores, which are too small to allow passage of the termites and are believed to aid in ventilation of the nest. *See* INSECTA; SOCIAL INSECTS; TERMITE.

[FRANCES M. WEESNER]

Bibliography: K. Krishna and F. M. Weesner (eds.), *Biology of Termites*, vol. 1, 1969, vol. 2, in press; T. E. Snyder, *Annotated, Subject-heading Bibliography of Termites 1350* B.C. *to* A.D. *1954*, Smithsonian Miscellaneous Collections, vol. 130, 1956; T. E. Snyder, *Catalog of the Termites (Isoptera) of the World*, Smithsonian Miscellaneous Collections, vol. 112, 1945; T. E. Snyder, *Supplement to the Annotated, Subject-heading Bibliography of Termites, 1955–1960*, Smithsonian Miscellaneous Collections, vol. 143, 1961; F. M. Weesner, *The Termites of the United States: A Handbook*, 1965.

Isopycnic

The line of intersection of an atmospheric isopycnic surface with some other surface, for instance, a surface of constant elevation or of constant atmospheric pressure. An isopycnic surface is a surface in which the density of the air is constant. Such surfaces are also designated isosteric surfaces because these surfaces, in which the specific volume is constant, coincide with the reciprocal conditions of isopycnic surfaces since density is the reciprocal of specific volume. On a surface of constant pressure, isopycnics coincide with isotherms, because on such a surface, density is a function solely of temperature. On a constant-pressure surface, isopycnics lie close together when the field is strongly baroclinic and are absent when the field is barotropic. *See* BAROCLINIC FIELD; BAROTROPIC FIELD; SOLENOID, METEOROLOGICAL.

[FREDERICK SANDERS]

Isoquinoline

One of a group of organic compounds containing a benzene ring fused to the 3,4 positions of pyridine. Isoquinoline (I) is a representative member of the

(I)

group. Quinoline produced from coal tar contains approximately 1–4% of isoquinoline, and it is an important source of the latter material. Separation is effected by selective extraction of the more basic isoquinoline with acid, or by selective precipitation and fractional crystallization of salts. Repeated fractional freezing and distillation have furnished pure isoquinoline. Many plant alkaloids, especially those in the cactus, opium, and curare groups, are isoquinoline derivatives. *See* HETEROCYCLIC COMPOUNDS; QUINOLINE.

Properties and preparation. Isoquinoline is a colorless, odorous, liquid with bp 243.3°C, mp 26.5°C, and n_D^{30} 1.62078. Its stability to acid, base, or heat is high. Isoquinoline, which is somewhat more basic than quinoline (the pK_a's are, respectively, 5.14 and 4.51), can be protonated to form simple salts and alkylated to form quaternary salts. Substitution reactions such as nitration, sulfonation, bromination, and mercuration are observed.

Phenylethylamines (II) are starting points in two general syntheses. The Bischler-Napieralski method cyclizes an *N*-acylated derivative (III) with phosphorus oxychloride to a 3,4-dihydroisoquinoline (IV). The Pictet-Spengler sequence converts phenylethylamine (II) by reaction with an aldehyde to an imine (V), which cyclizes in the presence of mineral acid to a 1,2,3,4-tetrahydroisoquinoline (VI). Dehydrogenation of either 3,4-dihydro or 1,2,3,4-tetrahydroisoquinoline by chemical or catalytic methods is generally uncomplicated, so that fully unsaturated isoquinolines (VII) are

readily obtained. In another kind of synthesis (Pomeranz-Fritsch), sulfuric acid is used to cyclize anils (IX) produced by the reaction of aminoacetal (VIII) and an aromatic aldehyde.

When the aromatic ring in phenylethylamine (II) is activated by a phenolic hydroxyl group para to the point of cyclization, the Pictet-Spengler reaction (V → VI) occurs under exceptionally mild conditions. Biosynthesis of isoquinoline plant products proceeds in this way.

Important derivatives. 1,2,3,4-Tetrahydroisoquinolines are obtained in Pictet-Spengler syntheses or by reduction of 3,4-dihydroisoquinolines, 1,2-dihydroisoquinolines, or fully aromatic isoquinolines. Further hydrogenation over platinum catalyst can lead to completely saturated decahydroisoquinolines. 1,2,3,4-Tetrahydroisoquinolines are orthosubstituted benzylamines, the basicity of which is comparable to that of aliphatic amines.

Sulfonation of isoquinoline with oleum gives mainly isoquinoline-5-sulfonic acid. Higher temperatures give increasing amounts of isoquinoline-8-sulfonic acid and isoquinoline-7-sulfonic acid. The reaction of isoquinoline with nitric-sulfuric acid gives mainly 5-nitroisoquinoline together with some 8-nitroisoquinoline. Bromination and mercuration favor the 4 position. Sodamide reacts to yield 1-aminoisoquinoline. Nucleophilic activity at the isoquinoline 1 position is shown by the sodamide process as well as by combination of isoquinoline with organometallic compounds to give

1-substituted derivatives, and by the enhanced reactivity of 1-haloisoquinolines to displacement processes. 1-Chloroisoquinolines have been obtained by diazotization of 1-aminoisoquinolines in concentrated hydrochloric acid. 1-Hydroxyisoquinoline, or isocarbostyril, exists in aqueous solution almost entirely in the 1-oxo form (X). Other hydroxyisoquinolines deviate little from the normal behavior of phenols. 1-Methylisoquinoline, and to a much lesser degree, 3-methylisoquinoline, possess reactive methyl groups. Quaternization enhances this reactivity.

Isoquinoline-1-carboxylic acid (XII) has been obtained directly or by way of isoquinoline-1-carboxaldehyde by oxidation of 1-methylisoquinoline (XI). The same acid is accessible from isoquinoline through the Reissert compound (XIII). 3-Methylisoquinoline can be oxidized with selenium dioxide first to isoquinoline-3-carboxaldehyde and thence with hydrogen peroxide to isoquinoline-3-carboxylic acid. Other isoquinoline carboxylic acids are available as hydrolysis products from the corresponding cyano derivatives, all of which can be obtained by interaction of the bromoisoquinolines with cuprous cyanide. Isoquinoline-1-carboxylic acids are readily decarboxylated.

With some exceptions, the reactions of isoquinoline have analogous counterparts in the reactions of quinoline.   [WALTER J. GENSLER]

Bibliography: R. Adams (ed.), *Organic Reactions*, vol. 6, 1951; G. M. Badger, *The Chemistry of Heterocyclic Compounds*, 1961; R. C. Elderfield (ed.), *Heterocyclic Compounds*, vol. 4, 1952.

Isostasy

A theory of hydrostatic equilibrium in the various parts of the Earth's crust. Many scientists judge this theory so well confirmed as to be considered a basic law of geology. B. F. Howell, Jr., in 1959, stated it in such explicit terms as: "All large land masses on the Earth's surface tend to sink or rise so that, given time for readjustment to occur, their masses are hydrostatically supported from below, except where local stresses are acting to upset equilibrium." *See* EARTH; TERRESTRIAL GRAVITATION.   [CHARLES V. CRITTENDEN]

Bibliography: W. A. Heiskanen and F. A. Vening Meinesz, *The Earth and Its Gravity Field*, 1958; B. F. Howell, Jr., *Introduction to Geophysics*, 1959; E. N. Lyustikh, *Isostasy and Isostatic Hypotheses*, 1960.

Isotach

A line along which the speed of the wind is constant. Isotachs are customarily represented on surfaces of constant elevation or atmospheric pressure, or in vertical cross sections. The closeness of spacing of the isotachs is indicative of the intensity of the wind shear on such surfaces. In the region of a jet stream the isotachs are approximately parallel to the streamlines of wind direction and are closely spaced on either side of the core of maximum speed. The term isovel is used synonymously with the term isotach. *See* JET STREAM; WIND.

[FREDERICK SANDERS]

Isothermal chart

A map showing the distribution of air temperature (or sometimes sea surface or soil temperature) over a portion of the Earth's surface, or at some level in the atmosphere. On it, isotherms are lines connecting places of equal temperature. The temperatures thus displayed may all refer to the same instant, may be averages for a day, month, season, or year, or may be the hottest or coldest temperatures reported during some interval.

Maps of mean monthly or mean annual temperature for continents, hemispheres, or the world sometimes show values reduced to sea level to eliminate the effect of elevation in decreasing average temperature by about 3.3°F/1000 ft (see illustration). Such adjusted or sea-level maps represent the effects of latitude, continents, and oceans in modifying temperature; but they conceal the effect of mountains and highlands on temperature distributions. The first isothermal chart, prepared by Alexander von Humboldt in 1817 for low and middle latitudes of the Northern Hemisphere, was the first use of isopleth methods to show the geographic distribution of a quantity other than elevation.

These maps are now varied in type and use. Isothermal charts are drawn daily in major weather forecasting centers; 5-day, 2-week, and monthly charts are used regularly in long-range forecasting; mean monthly and mean annual charts are compiled and published by most national weather services, and are presented in standard books, on, for example, climate, geography, and agriculture. *See* AIR TEMPERATURE; TEMPERATURE INVERSION.

[ARNOLD COURT]

Isothermal process

Any thermodynamic process, such as the frictionless expansion or contraction of a gas, accompanied by heat addition or removal from the system at a rate just adequate to maintain a constant temperature while the process is going on. For an ideal gas the internal energy U is a function of the temperature alone; therefore in an isothermal process, the internal energy stored in an ideal gas also remains constant.

In an isothermal expansion process the heat added to the system is precisely equal to the work W done by the isothermal expansion, for there is no change in the internal energy of the system, as shown in Eq. (1). Here Q_T is the heat transferred at

$$Q_T = W = \int_1^2 P\,dV \qquad (1)$$

constant temperature, P is the pressure, and V is the volume. From the perfect gas law in Eq. (2),

$$PV = nRT \qquad (2)$$

where n is the number of moles of gas, T the absolute temperature, and R the gas constant, Eq. (3) is

$$Q_T = nRT \int_1^2 \frac{dV}{V} = nRT \log_e \left(\frac{V_2}{V_1}\right) \qquad (3)$$

obtained. Here $\log_e$ indicates the natural logarithm (to the base $e = 2.71828 \ldots$). For an isothermal reversible compression or expansion process (quasistatic process), Eq. (4) holds, where S is the

$$Q_T = T(S_2 - S_1) \qquad (4)$$

entropy. *See* ISOMETRIC PROCESS; THERMODYNAMIC PROCESSES. For a comparison of the isothermal process with other processes involving a gas *see* POLYTROPIC PROCESS.

[PHILIP E. BLOOMFIELD]

Isotone

One of two or more atoms which display a constant difference $A - Z$ between their mass number A and their atomic number Z. Thus, despite differences in the total number of nuclear constituents, the numbers of neutrons in the nuclei of isotones are the same. The numbers of naturally

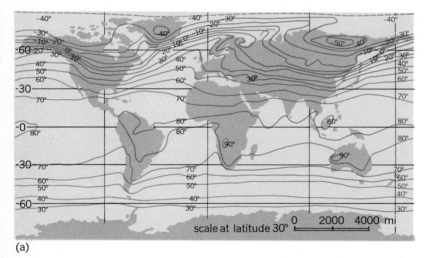

(a)

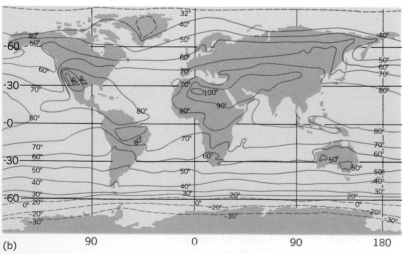

(b)

Isothermal charts of mean air temperature reduced to sea level (°F). (a) January. (b) July. (*From G. T. Trewartha, Introduction to Climate, 4th ed, McGraw-Hill, 1968*)

occurring isotones provide useful evidence concerning the stability of particular neutron configurations. For example, the relatively large number (six and seven, respectively) of naturally occurring 50- and 82-neutron isotones suggests that these nuclear configurations are especially stable. On the other hand, from the fact that most atoms with odd numbers of neutrons are anisotonic, one may conclude that odd-neutron configurations are relatively unstable. *See* NUCLEAR STRUCTURE.

[HENRY E. DUCKWORTH]

Isotope

One of two or more nuclidic species of an element having identical number of protons (Z) in the nucleus but different number of neutrons (N). Isotopes differ in mass but chemically are the same element. All naturally occurring elements have radioactive isotopes, and the majority have at least one stable nuclide. Some elements which occur in nature, such as uranium, are radioactive but have isotopes with long half-lives. For a discussion of artificially produced radioisotopes *see* RADIOISOTOPE.

The isotopic composition of an element is generally determined by mass spectrometry. Of the 83 elements present on Earth in significant amounts, 20 possess only a single stable nuclide and are referred to as mononuclidic or anisotopic. The others have 2 to 10 stable isotopes. *See* MASS SPECTROSCOPE.

Nuclear stability. Of the 287 nuclidic species listed in the table, 168 have even-even structure (even number of protons and even number of neutrons in the nucleus), 57 have even-odd, 53 have odd-even, and only 9 have odd-odd. This indicates the pairing tendency of the nuclear constituents. The extra stability of a 50-proton configuration is indicated by the existence of 10 isotopes of tin. *See* NUCLEAR STRUCTURE.

Isotopic abundance. This refers, unless otherwise specified, to the isotopic composition of the naturally occurring terrestrial element (see table). Some elements are observed to vary in isotopic composition. The variability ranges from a few per mil to a percent or two, although variations greater than this are observed in some samples. This variation occurs for several reasons. In the lighter elements—hydrogen, lithium, and boron, for instance—the isotopes differ enough in mass and to some extent in chemical reactivity that processes of distillation or chemical exchange between different chemical compounds of the element can produce significant differences in isotopic composition. Indeed, exchange reactions are used in the case of hydrogen, lithium, boron, carbon, nitrogen, and oxygen to separate isotopes of these elements on a relatively large scale.

Elements which take part in the life cycle of living organisms will vary somewhat in isotopic composition because of exchange reactions and diffusion through membranes. Slight differences in reaction rates are also important.

The composition of some elements may be variable because one or more of the isotopes are stable products of radioactive decay. Thus three of the four lead isotopes, ^{208}Pb, ^{207}Pb, and ^{206}Pb, are end products of the decay of thorium and of ^{235}U and ^{238}U respectively. The fourth lead isotope, ^{204}Pb,

does not come from any known decay chain. The rare potassium isotope, ^{40}K, which is present in only 0.012%, has a half-life of 1.28×10^9 years. It decays by beta-particle emission to stable ^{40}Ar and by electron capture to ^{40}Ca. The argon in the atmosphere, approximately 1.1%, is 99.6% in ^{40}Ar. Argon in potassium-bearing minerals will differ in composition from atmospheric argon just as the composition of lead will depend upon its past association with thorium and uranium. These decays and the decay of rubidium, ^{87}Rb, to ^{87}Sr are the basis of methods used in the determination of geological age. *See* ROCK, AGE DETERMINATION OF.

The three nuclides ^{40}K, ^{40}Ar, and ^{40}Ca are examples of "isobars" in that they have the same mass number, $N + Z$, but differ in the number of protons in the nucleus. The radioactive potassium isotope is an example of an odd-odd nucleus, while the stable ^{40}Ar and ^{40}Ca are both even-even. *See* ISOBAR (ATOMIC PHYSICS).

Anomalous isotopic compositions in some elements occur because of nuclear processes in nature. The discovery, in 1972, of a "fossil reactor" in the Oklo uranium deposit in Gabon (West Africa) is such a case. As a result of a chain reaction, perhaps 1.7×10^9 years ago, much of the uranium in this formation is depleted in the fissionable isotope, ^{235}U. Fission products are present in the composition found in reactor waste, and some elements in the surrounding rocks have been modified in isotopic composition by neutron absorption and subsequent radioactive decay.

The only nonterrestrial materials which are available for comparison of isotopic composition are meteorites and the lunar materials returned to Earth by the several Apollo manned missions and the Soviet unmanned lunar probes. Isotopic compositions in these are generally identical with terrestrial samples within the precision of the measurements. Differences, when they are found, can be identified as caused by radioactive decay, cosmic-ray bombardment and, in the case of lunar surface materials, bombardment by solar "wind" particles. In iron meteorites the spallation of iron nuclei by very energetic particles from cosmic rays produces helium, neon, and some other light elements with anomalous isotopic composition. The helium ^{3}He/^{4}He ratio is used as an indicator of the cosmic ray "bombardment" age. *See* GEOCHEMISTRY; GEOPHYSICS; RADIOACTIVITY.

Use of separated isotopes. Isotopes of certain elements possess unique or peculiar properties, and their separation or enrichment is desirable. Deuterium, ^{2}H, which occurs in an abundance of about 0.16% in terrestrial hydrogen, is useful for moderating neutrons in heavy-water reactors and is expected to form the fuel of fusion reactors in the future. Very large quantities of deuterated water have been produced by a variety of distillative, exchange, and electrolysis processes. The desirable fissionable isotope of uranium, ^{235}U, is enriched by gaseous diffusion in uranium hexafluoride gas in very large plants. *See* ISOTOPE (STABLE) SEPARATION.

Of the biologically important elements, only carbon and hydrogen have radioisotopes of sufficiently long half-lives to be used in tracer studies in living organisms. The use of large amounts of ^{3}H and ^{14}C is not desirable and, in any case, the mass

Natural isotopic abundances of the elements

#	Element	Mass no.	Atom %	#	Element	Mass no.	Atom %	#	Element	Mass no.	Atom %	#	Element	Mass no.	Atom %
1	H*	1	99.985	30	Zn	64	48.9	50	Sn	112	1.0	66	Dy	156	0.06
		2	0.015			66	27.8			114	0.7			158	0.1
		3	0.00013			67	4.1			115	0.4			160	2.34
2	He*	4	≈100.			68	18.6			116	14.7			161	18.9
3	Li*	6	7.5			70	0.6			117	7.7			162	25.5
		7	92.5	31	Ga	69	60.0			118	24.3			163	24.9
4	Be	9	100.			71	40.0			119	8.6			164	28.2
5	B*	10	19.8	32	Ge	70	20.7			120	32.4	67	Ho	165	100.
		11	80.2			72	27.5			122	4.6	68	Er	162	0.1
6	C*	12	98.89			73	7.7			124	5.6			164	1.6
		13	1.11			74	36.4	51	Sb	121	57.3			166	33.4
7	N*	14	99.64			76	7.7			123	42.7			167	22.9
		15	0.36	33	As	75	100.	52	Te	120	0.1			168	27.0
8	O*	16	99.756	34	Se	74	0.9			122	2.5			170	15.0
		17	0.039			76	9.0			123	0.9	69	Tm	169	100.
		18	0.205			77	7.6			124	4.6	70	Yb	168	0.1
9	F	19	100.			78	23.5			125	7.0			170	3.1
10	Ne*†	20	90.51			80	49.8			126	18.7			171	14.3
		21	0.27			82	9.2			128	31.7			172	21.9
		22	9.22	35	Br	79	50.69			130	34.5			173	16.2
11	Na	23	100.			81	49.31	53	I	127	100.			174	31.7
12	Mg	24	78.99	36	Kr†	78	0.35	54	Xe†	124	0.1			176	12.7
		25	10.00			80	2.25			126	0.1	71	Lu	175	97.4
		26	11.01			82	11.6			128	1.9			176	2.6
13	Al	27	100.			83	11.5			129	26.4	72	Hf	174	0.2
14	Si	28	92.2			84	57.0			130	4.1			176	5.2
		29	4.7			86	17.3			131	21.2			177	18.5
		30	3.1	37	Rb	85	72.17			132	26.9			178	27.1
15	P	31	100.			87	27.83			134	10.4			179	13.8
16	S*	32	95.00	38	Sr†	84	0.56			136	8.9			180	35.2
		33	0.76			86	9.84	55	Cs	133	100.	73	Ta	180	0.012
		34	4.22			87	7.0	56	Ba	130	0.1			181	99.988
		36	0.02			88	82.6			132	0.1	74	W	180	0.1
17	Cl	35	75.77	39	Y	89	100.			134	2.4			182	26.3
		37	24.23	40	Zr	90	51.4			135	6.6			183	14.3
18	Ar†	36	0.34			91	11.2			136	7.9			184	30.7
		38	0.07			92	17.1			137	11.2			186	28.6
		40	99.59			94	17.5			138	71.7	75	Re	185	37.40
19	K	39	93.26			96	2.8	57	La	138	0.09			187	62.60
		40	0.01	41	Nb	93	100.			139	99.91	76	Os†	184	0.02
		41	6.73	42	Mo	92	14.8	58	Ce	136	0.2			186	1.58
20	Ca	40	96.937			94	9.1			138	0.3			187	1.6
		42	0.65			95	15.9			140	88.4			188	13.3
		43	0.14			96	16.7			142	11.1			189	16.1
		44	2.08			97	9.5	59	Pr	141	100.			190	26.4
		46	0.003			98	24.4	60	Nd	142	27.1			192	41.0
		48	0.19			100	9.6			143	12.2	77	Ir	191	37.4
21	Sc	45	100.	44	Ru	96	5.5			144	23.9			193	62.6
22	Ti	46	8.0			98	1.9			145	8.3	78	Pt	190	0.01
		47	7.5			99	12.7			146	17.2			192	0.79
		48	73.7			100	12.6			148	5.7			194	32.9
		49	5.5			101	17.1			150	5.6			195	33.8
		50	5.3			102	31.6	62	Sm	144	3.1			196	25.3
23	V	50	0.25			104	18.6			147	15.0			198	7.2
		51	99.75	45	Rh	103	100.			148	11.2	79	Au	197	100.
24	Cr	50	4.35	46	Pd	102	1.0			149	13.8	80	Hg	196	0.2
		52	83.79			104	11.0			150	7.4			198	10.1
		53	9.50			105	22.2			152	26.7			199	16.9
		54	2.36			106	27.3			154	22.8			200	23.1
25	Mn	55	100.			108	26.7	63	Eu	151	47.8			201	13.2
26	Fe	54	5.85			110	11.8			153	52.2			202	29.7
		56	91.7	47	Ag	107	51.83	64	Gd	152	0.2			204	6.8
		57	2.14			109	48.17			154	2.2	81	Tl	203	29.5
		58	0.31	48	Cd	106	1.2			155	14.9			205	70.5
27	Co	59	100.			108	0.9			156	20.6	82	Pb†	204	1.4
28	Ni	58	68.3			110	12.4			157	15.7			206	24.1
		60	26.1			111	12.8			158	24.7			207	22.1
		61	1.1			112	24.0			160	21.7			208	52.4
		62	3.6			113	12.3	65	Tb	159	100.	83	Bi	209	100.
		64	0.9			114	28.8					90	Th	232	100.
29	Cu	63	69.2			116	7.6						92	U*	234	0.0054
		65	30.8	49	In	113	4.3							235	0.7200
						115	95.7							238	99.2746

*Isotopic composition of the element may be somewhat variable with specific geological or biological origin of the sample. Commercial chemicals may have, in some cases, quite anomalous composition as the result of processes of isotope separation.

†The element may vary in isotopic composition in some samples because one or more of the isotopes result from radioactive decay, or from nuclear processes in nature, such as spontaneous fission of uranium or α,n reactions on light elements.

differences are so great that they do not act precisely like the isotopes ^1H and ^{12}C. Studies of metabolism, drug utilization, and other reactions in living organisms are best done with stable isotopes like ^{13}C, ^{15}N, ^{18}O, and ^2H. Compounds are tagged by introducing a high concentration of the isotope into the molecular structure, and the metabolized products are studied using the mass spectrometer to measure the altered isotopic ratios. *See* TRACER, RADIOACTIVE.

The process of isotope dilution consists of adding a known amount of material containing the tracer isotope, allowing the system to reach chemical equilibrium and then recovering a small sample sufficient in size for a mass spectrometric measurement of the new isotopic composition. This is a method of very wide applicability. *See* ISOTOPE DILUTION TECHNIQUES.

Atomic weight. Atomic weight is the ratio of the average mass per atom of the natural isotopic composition of an element to 1/12 of the mass of an atom of the nuclide ^{12}C. It is a dimensionless number. The atomic weight of a mononuclidic element is simply the atomic mass of that nuclide relative to 1/12 of the mass of $^{12}C = 12$ exactly. These masses can be measured with very high precision. Thus the atomic weight of the mononuclidic element beryllium is 9.01218. Atomic weights of the elements having two or more isotopic species are increasingly based upon calculations from isotopic composition and atomic masses. The mass spectrometer is not an absolute instrument because of certain inherent discriminations. Comparisons must be made with isotopic standards carefully prepared by gravimetric procedures from separated isotopes of high chemical and isotopic purity. This is hardly less exacting than the chemical determination of atomic weights used early in this century by T. W. Richards and coworkers at Harvard University. The "Harvard method" involved the careful precipitation and weighing of silver chloride formed by the stoichiometric reaction of silver nitrate with the chloride of the element being investigated. Thus, one would determine the germanium chloride/silver chloride ratio ($GeCl_4/4AgCl$) and thus the ratio of the atomic weight to that of silver. Even in the 1975 Table of Atomic Weights some of the polynuclidic elements, such as germanium, tin, and mercury, are still based upon chemical ratios measured by the Harvard method. *See* ATOMIC WEIGHT; ELEMENTS (CHEMISTRY).

[A. E. CAMERON]

Bibliography: J. F. Duncan and G. B. Cook, *Isotopes in Chemistry*, 1968; S. Glasstone, *Sourcebook on Atomic Energy*, 3d ed., 1967; N. E. Holden and F. W. Walker, (eds.), *Chart of the Nuclides*, Educational Relations, General Electric Co., 1972; J. Robos, *Introduction to Mass Spectrometry*, 1968; A. Romer (ed.), *Radiochemistry and the Discovery of Isotopes*, 1970.

Isotope dilution techniques

The introduction of radioisotopes into stable isotopes of an element in order to make volume, mass, and age measurements of the element. The technique may best be described by giving a simple example. If knowledge of the total amount of water present in a container is desired, a small amount of labeled water, D_2O or T_2O, is added to the H_2O in the container. For the purposes of the measurement, this amount can be considered to be 100%. After the labeled deuterium or tritium is completely mixed with the water in the container, a sample of the mixture of unit volume, perhaps 1 cm³, is removed, and the amount of labeled water in it determined. If this amount is found to be $a\%$, then the total amount of water in the container will be $100/a$.

In other applications the mass of an element may be determined. In this case the radioisotope is introduced and mixed, a sample is withdrawn, and both the amount of radioisotope and the amount of stable chemical present are measured. The ratio of these two amounts, called the specific activity, can be expressed as % dose/mg. For example, sodium-24 is administered to a patient and the specific activity of sodium in the blood (the amount of labeled sodium divided by the amount of stable sodium) determined. This permits measurement of the total amount of exchangeable sodium within the body.

All isotope dilution techniques may be considered to be an extension of these simple measurements. As more compartments or containers are added to the system, however, the analysis becomes increasingly difficult. Furthermore, the assumption of immediate and complete mixing will no longer be valid. Despite these difficulties, the underlying principles are identical with the simple one-volume case, and isotope dilution techniques have found a wide range of application in medicine, biology, and industry.

Isotope dilution is often used in analytical chemistry in cases where a quantitative extraction of the substance to be assayed is difficult. A small quantity of labeled substance is added to the sample before separation and analysis. The extraction is then carried out without regard to completeness, and an arbitrary sample is removed for analysis of the chemical substance and of the labeled substance. From the ratio of the initial and final activity, the chemical assay may be corrected to give the correct amount of chemical in the sample before extraction.

Isotope dilution techniques have also found increasing applications in geology. Here they are used to determine the abundance and distribution of trace constituents, especially in meteorites, and to make geological age determinations; for example, the determination of the age of a mica by the potassium-argon method uses Ar^{38} as a "spike" to determine the radiogenic Ar^{40} in the sample. The Ar^{40} is a decay product of K^{40} retained within the mica. In these applications, mass spectrometers are used to make the sample analysis. *See* ROCK, AGE DETERMINATION OF.

[GORDON L. BROWNELL]

Isotope (stable) separation

The physical separation of different stable isotopes of an element from one another. Many chemical elements always occur in nature as a mixture of several isotopes. The isotopes of any given element have identical chemical properties, but there are slight differences in their physical properties because of the differences in mass of the individual isotopes. Thus it is possible to separate physically the isotopes of an element to produce material of isotopic composition different from that which occurs in nature. Although these separation processes are all quite difficult and expensive to carry out, they are not inherently different from the usual operations employed in the chemical process industries. *See* ISOTOPE.

The separation of isotopes is particularly important in the nuclear energy field because individual isotopes may have completely different nuclear properties. For example, uranium-235 is used as a fuel for nuclear chain reactors, heavy water (deuterium oxide) is used as a neutron moderator in nuclear chain reactors, and deuterium gas is a possible fuel for thermonuclear reactors. Separated isotopes are also used widely for research on the structure and properties of the nucleus.

The process which is best suited for separating the isotopes of a given element depends upon the mass of the element and the desired quantity of separated material. Research quantities of separated isotopes are best prepared by electromagnetic separation in a mass spectrometer. For example, gram quantities of many separated isotopes have been prepared at Oak Ridge National Laboratory using the large electromagnetic separators which were built during World War II. The electromagnetic process has the advantage that a fairly complete separation of two isotopes can be obtained in one operation.

When moderate quantities of a separated isotope are desired, thermal diffusion may be used. Although thermal diffusion requires a large energy input, this is more than offset by the simplicity of the equipment, absence of moving parts, and high separation obtained in a small volume.

In the large-scale separation of stable isotopes, the best processes are those which have the highest thermodynamic efficiencies. Reversible processes involving distillation and chemical exchange are best for separating the light isotopes such as deuterium. For heavy isotopes such as those of uranium, however, no appreciable separation is obtained by the reversible processes and some type of irreversible process such as gaseous diffusion must be used. Although reversible processes have in general higher efficiencies than irreversible ones, the absolute efficiency of any isotope separation process is very small and the cost is very high in comparison to the usual operations employed in the chemical process industries.

Gaseous diffusion. This process has turned out to be the most economical for the separation of the isotopes of uranium. It is based on the fact that in a mixture of two gases of different molecular weights, molecules of the lighter gas will on the average be traveling at higher velocities than those of the heavier gas. If there is a porous barrier with holes just large enough to permit passage of the individual molecules but without permitting bulk flow of the gas as a whole, the probability of a gas molecule passing through the barrier will be directly proportional to its velocity. From kinetic theory it can be shown that the velocity of a gas molecule is inversely proportional to the square root of its molecular weight, so that the efficiency of gaseous diffusion will depend on the ratio of the square roots of the molecular weights of the two gases present. *See* DIFFUSION IN GASES AND LIQUIDS; KINETIC THEORY OF MATTER.

The only uranium compound which is a gas at a hexafluoride, UF_6. The two isotopes to be separated are $U^{235}F_6$ and $U^{238}F_6$, and the efficiency of separation depends on the quantity in the equation below. Since this number is close to unity, the

$$\sqrt{U^{238}F_6/U^{235}F_6} = 1.0043$$

separation is very small in any one step of the process.

The separation of the isotopes of uranium in the United States is carried out in the three plants operated for the Atomic Energy Commission which are located at Oak Ridge, Tenn.; Paducah, Ky.; and Portsmouth, Ohio. In each of these installations natural uranium containing 0.71% U^{235} and the balance U^{238} in the form of UF_6 gas is separated into an enriched uranium product containing more than 90% U^{235}, and a waste containing about 0.3% U^{235}. It is also possible to use "depleted" uranium (uranium recovered from plutonium production reactors has lower U^{235} content than the natural) as feed to a gaseous diffusion plant. Britain has a gaseous diffusion plant at Capenhurst, and the Soviet Union has facilities at an undisclosed location.

The success of the gaseous diffusion process is dependent on the performance of the single diffusion stage. In each stage, UF_6 gas is compressed, passed through a cooler to remove the heat of compression, and then admitted to the vessel containing the porous barrier (Fig. 1). About half the gas entering the vessel diffuses through the barrier and passes to the next higher stage. This diffused gas contains slightly more of the U^{235} isotope. The undiffused gas is slightly depleted in the U^{235} isotope, and passes to the next lower stage.

Several thousand individual stages are required to bring about the necessary overall change in composition. The combination of stages is known as a cascade, and the cascade which brings about the separation with the least work is known as an ideal cascade. The size of the stages varies tremendously; those feeding the natural uranium into the cascade are the largest and the final product stages are the smallest.

Nickel-clad piping and process equipment are used to handle the UF_6 gas. Thousands of pumps, coolers, and control instruments are required. The electric power requirements of the gaseous diffusion process are large; for many years approximately 10% of the total electric power output of the United States was required to operate the three diffusion plants.

The gaseous diffusion process was originally developed as a means of producing highly en-

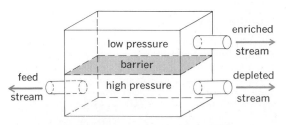

Fig. 1.   Gaseous diffusion stage. (*From H. Etherington, ed, Nuclear Engineering Handbook, McGraw-Hill, 1958*)

riched uranium for atomic bombs. At present, the gaseous diffusion plants are being modified to produce partially enriched uranium to fuel nuclear power reactors. There has also been a substantial decrease in the output from the three plants.

Chemical exchange. The chemical exchange process has proved to be the most efficient for separating isotopes of the lighter elements. This process is based on the fact that if equilibrium is established between, for example, a gas and a liquid phase, the composition of the isotopes will be different in the two phases. Thus, if hydrogen gas is brought into equilibrium with water, it is found that the ratio of heavy hydrogen (deuterium) to light hydrogen is several times greater in the water than in the hydrogen gas. By repeating the process in a suitable cascade, it is possible to effect a substantial separation of the isotopes with a relatively small number of stages.

The chief use of chemical exchange is in the large-scale production of heavy water. A dual-temperature exchange reaction between water, which contains HDO and D_2O molecules as well as H_2O molecules, and hydrogen sulfide (H_2S) gas for the primary separation of heavy water is used at the Savannah River plant. The separation is carried out in a series of hot towers operating at the boiling point and cold towers operating at room temperature (Fig. 2). In each tower liquid water passes downward countercurrent to the rising H_2S gas. The relative distribution of heavy and light water is affected by temperature, and the success of the process is determined by the difference between the concentrations in the hot and the cold towers.

Chemical exchange has also been used for the large-scale separation of other isotopes. For example, the isotopes of boron have been separated by fractional distillation of the boron trifluoride–dimethyl ether complex.

Distillation. The separation of isotopes by distillation is much less efficient than separation by other methods. Distillation was used during World War II to produce heavy water, but the cost was high and the plants are no longer in existence. Fractional distillation has been used at Savannah River to concentrate the product from the dual-temperature process (12–16% D_2O) up to 95–98% D_2O. *See* DISTILLATION.

Electrolysis. Electrolysis of water is the oldest large-scale method of producing heavy water. Under favorable conditions, the ratio of hydrogen to deuterium in the gas leaving a cell in which water is electrolyzed is eight times the ratio of these isotopes in the liquid. In spite of this high degree of separation, electrolysis can be used only where electricity is very cheap, as in Norway, because of the large power consumption per pound of D_2O produced. Electrolysis is used in the United States only as a finishing step to concentrate to final-product specifications.

Electromagnetic process. Electromagnetic separation was the method which was first used to prove the existence of isotopes. The mass spectrometer and mass spectrograph are still widely used by physicists as a research tool. In the electromagnetic process, vapors of the material to be analyzed are ionized, accelerated in an electric field, and enter a magnetic field which causes the ions to be bent in a circular path. Since the light

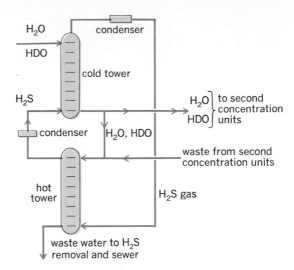

Fig. 2. Dual-temperature H_2S–HDO exchange. The end product is a mixture of H_2O, HDO, and D_2O, with the relative abundance of D_2O increasing in the later stages. (*From R. Stephenson, Introduction to Nuclear Engineering, 2d ed., McGraw-Hill, 1958*)

ions have less momentum than the heavy ions, they will be bent through a circle of smaller radius, and the two isotopes can be separated by placing collectors at the proper location. *See* MASS SPECTROSCOPE.

During World War II, a large electromagnetic separation plant was built on Oak Ridge to separate the isotopes of uranium. The large mass spectrometers used there were referred to by the code name Calutron, a contraction of California University cyclotron (Fig. 3). The first kilogram quantities of U^{235} were produced in 1944. With the completion of the gaseous-diffusion plant at Oak Ridge, the electromagnetic process was found to be uneconomical and was abandoned in 1946. However, some of the equipment is still being used to produce gram quantities of separated isotopes for research purposes.

Thermal diffusion. The separation of isotopes by thermal diffusion is based on the fact that when a temperature gradient is established in a mixture of uniform composition, one component will concentrate near the hot region and the other near the cold region. Thermal diffusion is carried out in the

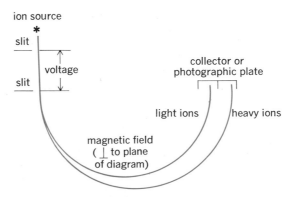

Fig. 3. Diagrammatic representation of Calutron mass spectrometer. (*From R. Stephenson, Introduction to Nuclear Engineering, 2d ed., McGraw-Hill, 1958*)

annular space between two vertical concentric pipes, the inner one heated and the outer one cooled. Because of thermal convection of the fluid, there is a countercurrent flow which greatly increases the separation obtained in simple thermal diffusion and makes possible substantial separations in a reasonable column height.

Thermal diffusion has been used to separate small quantities of isotopes for research purposes. In 1944 a plant was built at Oak Ridge to separate the isotopes of uranium by thermal diffusion. However, the steam consumption was very large, and the plant was dismantled when the gaseous-diffusion facilities were completed.

Centrifugation. The use of a centrifuge to separate isotopes has one major advantage, namely that the separation depends only on the difference in masses of the two isotopes, and not on the ratio of their masses. Thus it is no more difficult to separate the isotopes of uranium than those of the light elements. A disadvantage of the centrifuge method is that a very high speed of rotation is required to obtain any substantial separation in a single unit.

A centrifuge pilot plant was built during World War II, but further work on the process was discontinued because of engineering problems involved in the operation of high-speed rotors, the low capacity of the individual machines, and the large power input required to overcome friction.

There has been a renewal of interest in centrifugation, particularly in Europe where several new approaches are being studied. There is some evidence that this will eventually lead to a simple and cheap way to produce enriched U^{235}. Politically, this has been a very sensitive subject since, if successful, it would enable even the smallest nations to manufacture nuclear weapons for possible military use. *See* CENTRIFUGATION.

Nozzle process. Isotopes can be separated by allowing a gaseous compound to exhaust through a properly shaped nozzle. Preliminary calculations indicate that this relatively new method may be competitive with gaseous diffusion for separating the isotopes of uranium. *See* DEUTERIUM; HEAVY WATER; NUCLEAR FUELS; RADIOISOTOPE PRODUCTION.

Laser methods. Security-classified for many years, laser methods for separating isotopes are under intensive study throughout the world. Most of the effort is directed toward the separation of U^{235}. For example, by the use of a tunable dye laser, U^{235} atoms (or possibly molecules) can be raised to an excited state while the accompanying U^{238} atoms remain unaffected. The U^{235} can then be separated out by conventional electromagnetic means. Laser methods have the advantage of a high rate of separation in one step, but they also have the potential problems of other beam processes such as the obsolete electromagnetic process. *See* LASER.

[RICHARD M. STEPHENSON]

Bibliography: M. Benedict and T. H. Pigford, *Nuclear Chemical Engineering*, 1957; K. P. Cohen, *The Theory of Isotope Separation as Applied to the Large-scale Production of U^{235}*, 1951; H. Etherington, *Nuclear Engineering Handbook*, 1958; R. Stephenson, *Introduction to Nuclear Engineering*, 2d ed., 1958.

Isotope shift

A displacement of spectral lines which come from the different isotopes of an element. This shift in wavelength is caused mainly by differences in nuclear masses and differences in nuclear volumes. The first effect predominates in the lightest elements but becomes negligible for atomic numbers greater than about 40. The mass effect in one-electron atoms results from the variation of the reduced mass, $mA/(m+A)$, where m is the electron mass and A is the nuclear mass, and was important in establishing the Bohr theory. The shift of the

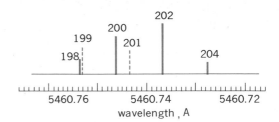

Isotope shifts in the green line of mercury. Dashed lines indicate the hyperfine structure components of the isotopes having odd mass numbers.

first line of the Balmer series between hydrogen and deuterium is 1.325 A, and this led to the discovery of deuterium. *See* ATOMIC STRUCTURE AND SPECTRA; RYDBERG CONSTANT.

For one of the heavy elements the volume of the atomic nucleus should increase as neutrons are added in the progression from the lightest to the heaviest of its isotopes. The strength of the electrostatic interaction between an s electron, whose charge distribution is mostly close to the nucleus, and the nucleus of charge Ze therefore varies slightly, and produces a small shift of this energy level in the order of the nuclear masses. The resulting displacement of the P-S or S-P spectral lines is called the volume effect. It is important for heavy elements and gives information about nuclear structure. Deviations from the theoretical volume effect allow an evaluation of nuclear quadrupole moments and nuclear shapes. A typical volume effect is shown in the illustration. Though the lines from the isotopes with even mass numbers are seen to be almost equally spaced, the lines for the odd ones do not lie even approximately halfway between.

[F. A. JENKINS/W. W. WATSON]

Isotopic spin

A quantum-mechanical variable introduced to describe charge independence of nuclear forces, with a more general utility in describing the properties of nuclei and of elementary particles.

The properties of neutrons and protons—especially those which determine nuclear characteristics—show greater similarities than differences. Mass, spin, and particularly the specifically nuclear interactions between pairs of nucleons are essentially the same: the nuclear force is observed to be charge-independent. This suggests that the neutron and proton are simply different states of the same particle, the nucleon. The fact that the states are two in number suggests that the state variable which distinguishes neutron and proton is dichotomic, like the mechanical spin of the electron, and hence this new variable is designated isotopic spin. The most obvious difference between the two isotopic spin states is the charge, and so the formalism which is developed to treat the isotopic spin must exhibit a connection with the charge.

The formalism to be used is suggested by the designation: all of the apparatus for treating mechanical spin can be adapted without change except for relabeling the meaning of variables. Hence if t_3 represents the third component (in the "z" direction of this abstract isotopic spin space) of the isotopic spin, then $t_3 = 1/2$ for a proton and $t_3 = -1/2$ for a neutron. (The value choice as well as the sign is arbitrary at this stage. Sometimes the opposite-sign convention is used, that is $t_3 = 1/2$ for a neutron and $t_3 = -1/2$ for a proton.) Extending the analogy to the mechanical spin, the existence of a total isotopic spin projection for a collection of N neutrons and Z protons given by Eq. (1) may be

$$T_3 = 1/2(Z - N) \qquad (1)$$

assumed. For a collection of $A = Z + N$ nucleons, the charge is clearly given by Eq. (2), where e is the

$$Q = (T_3 + A/2)e \qquad (2)$$

magnitude of the electron charge and is positive. *See* CHARGE, ELECTRIC; NEUTRON; NUCLEON; NUCLEUS, ATOMIC; PROTON; SPIN (QUANTUM MECHANICS).

The symmetry of the wave function of A nucleons under exchange of space and mechanical spin variables is governed by the Pauli principle: the wave function must be antisymmetric for such exchanges between two protons or two neutrons. If the isotopic spin is added to these variables, and the Pauli principle is extended to include it, then the symmetry with respect to exchange of neutron and proton may also be treated within the formalism. *See* EXCLUSION PRINCIPLE.

Nuclear reactions. Charge conservation is guaranteed by any interaction which preserves T_3. However, if it is assumed that the total isotopic spin T is conserved, there are new consequences of the theory. Nuclear reactions between states of different total isotopic spin, for example, are then prohibited. Since the Coulomb force between two protons violates the predicted symmetry, strict prohibition does not occur, but an inhibition of such T-changing reactions is observed. For example, in dueteron-alpha scattering (a $T = 0$ state), no resonance corresponding to the $T = 1$ state of Li^6 is seen, but those corresponding to $T = 0$ for Li^6 occur. *See* NUCLEAR REACTION; SELECTION RULES (PHYSICS).

Description of isobaric nuclei. It is a consequence of the charge independence of nuclear forces (and hence of the isotopic spin theory) that mirror nuclei (those with the numbers of neutrons and protons interchanged) will have the same energy levels. For light nuclei, where the Coulomb force is not too large, this is observed. But one can go somewhat further. In the simplest case of two

nucleons, there are three possible pairs: two neutrons, two protons, and one neutron, one proton. The first and second possibilities correspond to a state of $T = 1$, since for isotopic spin as for mechanical spin, T_3 takes on $2T + 1$ values from $-T$ to $+T$. The np system may be either in the $T_3 = 0$ projection of the $T = 1$ state, or in the only $T = 0$ state one can form. Total symmetry of the nucleon-nucleon state requires that the $T = 1$ state have total mechanical spin $S = 0$. The Pauli principle *without* isospin requires that nn and pp be in a state with mechanical spin zero. But the isospin theory now requires that the $T = 1$, $T_3 = 0$ state of np also be in a state with $S = 0$, and that all the $S = 0$ states of two nucleons have the same energy. Since the dineutron and diproton are not bound, neither should the spin singlet np state be bound. This is indeed the case: the bound state of the deuteron is the spin triplet ($S = 1$) state.

Thus isotopic spin symmetry is a useful organizing principle for describing the states of isobaric (same mass) nuclei, but is limited in its application by the symmetry-breaking Coulomb force (although the latter obviously does not change T_3). Conservation of isotopic spin occurs only for strong interactions: weak interactions leading to beta decay, for example, obviously do not conserve isospin. *See* ANALOG STATES, NUCLEAR; NUCLEAR STRUCTURE.

Extension to pi mesons. The extension of the isotopic spin concept to other elementary particles is even more interesting. The pi meson, for example, since it exists in three charge states, may be treated as a single particle with $T = 1$ which can exist in three states with $T_3 = -1$, 0, 1 (corresponding, respectively, to π^-, π^0, π^+). The observed mass difference between π^0 and the charged pions is expected to be due to electromagnetic interactions. If one assumes that $A = 0$ for pions, formula (2) for Q is still valid. *See* ELEMENTARY PARTICLE; MESON.

A meson-nucleon interaction can be proposed which leads to charge independence of the nucleon-nucleon force (it is also charge-independent in pion-nucleon interactions) and exhibits charge conservation. The interaction term can be a scalar in isotopic spin space, and hence is T-conserving as well. This has consequences for statistical weight factors of pion-nucleon interactions including, for example, the annihilation of nucleon-antinucleon states into multipion states. (The antinucleons have reversed isotopic spin projections compared to the nucleons.) These consequences are verified experimentally. *See* QUANTUM FIELD THEORY; SCATTERING EXPERIMENTS, NUCLEAR.

Other elementary particles. The other mesons and hyperons are also capable of description consistently in the isotopic spin formalism. For example, the K mesons form an isospin doublet like the nucleons, Λ is a singlet (and hence cannot be an excited state of a nucleon, which is an isospin doublet), the Σ particles form an isospin triplet, and the cascade particles (Ξ) a doublet. The properties of these multiplets exhibit T conservation: the masses and spins of the components are the same, and their observed interactions and decay schemes are consistent with its assumption.

The relation bewteen T_3 and the charge requires

development, however. This is readily seen in the K meson case. Since the K's are mesons, like the π meson, A should be zero. Formula (2) for Q then gives half-integral charge for K, since $T_3 = 1/2$, $-1/2$. This is clearly incorrect. Equation (2) must be replaced by the extended relationship given by Eq. (3), where S is the strangeness quantum num-

$$Q = T_3 + \tfrac{1}{2}A + \tfrac{1}{2}S \qquad (3)$$

ber ($S = 0$ for nucleons and pions, 1 for K mesons, Λ and Σ hyperons, and 2 for the Ξ hyperon). Nucleons and hyperons have $A = 1$ (antinucleons and antihyperons have $A = -1$). Other meson states, nucleon resonances, and new particles (predicted as components of isospin multiplets) can be described in the isotopic spin formalism with the treatment of nucleon number A and strangeness S as new conserved variables. *See* BARYON; HYPERON.

Unitary spin. Since the mid-1960s, the concept of isotopic spin has been generalized with great benefit to understanding of the elementary particles. One introduces the unitary spin, with eight components (three of which are the components of isotopic spin, one is the hypercharge $Y = A + S$, and the other four are new). Symmetry with respect to unitary spin implies the existence of super-multiplets of mesons or baryons (nucleons and hyperons), the components of which have the same spin and parity. The components of isotopic spin multiplets have masses close together, but the unitary spin multiplets exhibit mass splitting of considerable magnitude. This is thought of as being the result of unitary spin symmetry-breaking terms in the total fundamental interaction, just as the Coulomb force breaks isotopic spin symmetry. Even so, certain relationships among the masses of the members of a multiplet are observed. One triumph of the unitary spin theory was the prediction of a new particle, the Ω^-, with $S = -3$, which was later discovered. *See* SYMMETRY LAWS (PHYSICS); UNITARY SYMMETRY.

[MC ALLISTER H. HULL, JR.]

Bibliography: S. de Benedetti, *Nuclear Interactions*, 1964; J. M. Blatt and V. Weisskopf, *Theoretical Nuclear Physics*, 1952; H.A. Enge, *Introduction to Nuclear Physics*, 1966; M. Gell-Mann and Y. Neéman, *The Eightfold Way*, 1964; D. Lurie, *Particles and Fields*, 1968; P. Roman, *Theory of Elementary Particles*, 1960.

Isotropy (physics)

A body is said to be isotropic if its physical properties are not dependent upon the direction in the body along which they are measured. A body displaying isotropy has only one refractive index, one dielectric constant, and so on. Most but not all liquids and aggregates made up of many small crystals randomly oriented in space are isotropic in all their properties. Depending on their symmetry, single crystals may or may not be isotropic with respect to a given property. For example, single crystals with cubic structure are isotropic with respect to electrical resistivity but not with respect to elastic deformability. *See* ANISOTROPY (PHYSICS); ELASTICITY.

[DAVID TURNBULL]

Itaconic acid

An organic acid, methylene succinic acid, which decomposes at 165°C, having the structural formula shown below.

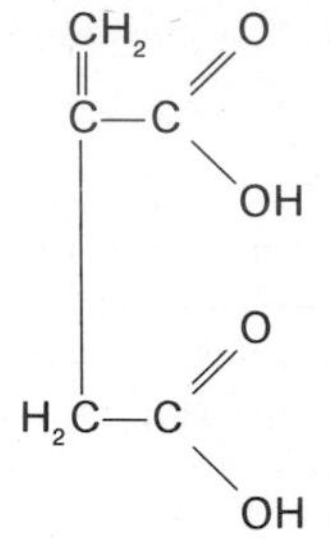

Fermentation is a potential microbiological, large-scale method for production of this unsaturated, branched-chain dicarboxylic acid as a raw material for certain plastics with special properties. A medium, consisting of commercial grade glucose, 15–20%, and inorganic salts, is adjusted with hydrochloric acid to pH 1.8–2.0. The pH of the medium is critical. The medium inoculated with spores or mycelium of a selected strain of the filamentous ascomycetous fungus *Aspergillus terreus* is aerated or agitated in a stainless steel fermentor at 28–35°C for 2–5 days. Itaconic acid can also be produced by the surface, or tray, process. The yield is approximately 40% of the sugar consumed. The free acid is crystallized from the filtered fermentation liquor after vacuum concentration and cooling. *See* FERMENTATION.

[JACKSON W. FOSTER/R. E. KALLIO]

Itatartaric acid

An organic acid, dihydroxyitaconic acid, which has been produced experimentally by fermentation. Its structural formula is shown below.

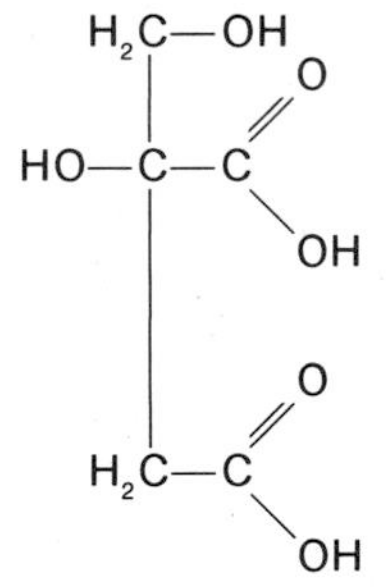

The compound is formed as a minor product, 5.8% of the total acidity produced, of an itaconic acid–producing strain of *Aspergillus niger*. The strain is an ultraviolet-induced mutant. Itatartaric acid occurs as a salt and a lactone in approximately equal quantities. *See* FERMENTATION; INDUSTRIAL MICROBIOLOGY; ITACONIC ACID; MUTATION.

[JACKSON W. FOSTER/R. E. KALLIO]

Itch

A pattern of cutaneous sensation allied to pain. A variety of different stimuli—mechanical, electrical, chemical, and thermal—may produce it. The most vivid and persistent itching is associated with pathological skin conditions (prurigo) and with the irritant action of itch powder, the spicules of the tropical woody plant cowage (cowitch).

The most convincing evidence that itch derives from the operation of the same nervous mechanism as that responsible for pain comes from experiments on human skin. In these experiments a brief train of high-voltage sparks imparted to the skin at a rate of five or more per second, each too weak to elicit any sensation singly, may evoke the itching sensation. The same stimuli at higher intensity yield pain. The presumption is that itch is associated with a slow but enduring succession of low-frequency impulses in the cutaneous nerve fibers leading to the central nervous system from that portion of the skin. Pain of a stronger and more disagreeable quality results from nervous discharges of higher frequency.

In skin rendered irritable by disease or inflammation, itching may sometimes be evoked by simply touching or stroking it mechanically. Temperature changes may produce itch under some conditions. However aroused, the itch sensation is a powerful stimulator of action, leading promptly to scratching or some other device to break up the persistent, steady pattern of nervous impulses underlying it. *See* PAIN, CUTANEOUS.

[FRANK A. GELDARD]

Bibliography: C. A. Keele and D. Armstrong, *Substances Producing Pain and Itch*, 1964; S. Rothman, *Physiology and Biochemistry of the Skin*, 1954; G. E. W. Wolstenholme and M. O'-Conner, *Pain and Itch*, 1959.

Ixodides

A suborder of the Acarina, class Arachnida, comprising the ticks. Ticks differ from mites, their nearest relatives, in their larger size and in having a pair of breathing pores, or spiracles, behind the third or fourth pair of legs. They have a gnathosoma (or so-called head or capitulum), which consists of a base (basis capituli), a pair of palps, and a rigid, elongated, ventrally toothed hypostome which anchors the parasite to its host. They also have a pair of protrusible cutting organs, or chelicerae, which permit the insertion of the hypostome. The stages in the life cycle are egg, larva, nymph, and adult. The larvae have three pairs of legs; nymphs and adults, four. The 600 or so known species are all bloodsucking, external parasites of vertebrates including amphibians, reptiles, birds, and mammals.

Ticks are divided into three families, Argasidae, Ixodidae, and Nuttalliellidae. The last contains but one exceedingly rare African species, *Nuttalliella namaqua*, morphologically intermediate between the Argasidae and the Ixodidae. It is of no known importance, either medically or economically.

Argasidae. Argasids, or the soft ticks, differ greatly from ixodids in that the sexes are similar; the integument of adults and nymphs is leathery and wrinkled; there is no dorsal plate, or scutum; the gnathosoma is ventral in adults and nymphs but anterior in larvae; and the spiracles are small and anterior to the hindlegs (Fig. 1). These ticks frequent nests, dens, and resting places of their hosts. Adults feed intermittently and eggs are laid a few at a time in niches where the females seek shelter. Larvae feed for a few minutes to several days, then detach and transform to nymphs which

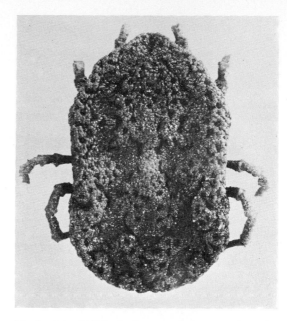

Fig. 1. Argasid tick, *Ornithodoros coriaceus,* enlarged to about 8 times natural size.

feed and molt several times before transforming to adults. Nymphs and adults are notably resistant to starvation; some are known to live 10 years or longer without feeding.

The family contains about 85 species, with 20 in the genus *Argas* and 60 in *Ornithodoros.* Several are of medical or veterinary importance. *Argus persicus* (Oken), *A. miniatus* Koch, *A. radiatus* Railliet, *A. sanchezi* Dugès, and *A. neghmei* Kohls and Hoogstraal, are serious pests of poultry. Some of these species carry fowl spirochetosis, a disease with a high mortality rate. Larvae and nymphs of *Otobius megnini*, the spinose ear tick, feed deep in the ears of domesticated animals in many semiarid regions. Heavy infestations cause intense irritation which leads to unthriftiness and sometimes to death. The life history is unusual in that the adults do not feed. *Ornithodoros moubata* (Murray) transmits relapsing fever in East, Central, and South Africa. It is a highly domestic parasite and man is probably the chief host. *O. turicata* (Dugès), *O. hermsi* (Wheeler, Herms, and Meyer), *O. talaje* (Guérin-Méneville), and *O. rudis* (Karsch) are important vectors of relapsing fever in the Western Hemisphere, as is *O. tholozani* (Laboulbène and Mégnin) in Asia. The bites of most species that attack man produce local and systemic reactions; the bites of some species, especially *O. coriaceus* (Koch) of California and Mexico, are extremely venomous.

Ixodidae. In contrast to argasids, Ixodidae have a scutum covering most of the dorsal surface of the male but only the anterior portion of females, nymphs, and larvae (Fig. 2). They are known as the hard ticks. The sexes are thus markedly dissimilar. The gnathosoma extends anteriorly, and the large spiracles are posterior to the hindlegs. Instead of frequenting nesting places, these ticks are usually more or less randomly distributed throughout their hosts' environment. Larvae, nymphs, and adults feed but once, and several days are required

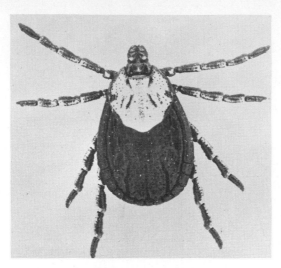

Fig. 2.   Ixodid tick, *Dermacentor andersoni*, female, enlarged to about 15 times natural size.

for engorgement. The immature stages of most species drop to the ground for molting but those of the genus *Boophilus* and a few others molt on the host. The female lays a mass containing up to 10,-000 or more eggs. The life cycle is usually completed in 1–2 years.

The family consists of about 11 well-defined genera with 500 species. Many species transmit disease agents to man and animals; included are viruses, rickettsiae, bacteria, protozoa, and toxins. Transmission is by bite or by contact with crushed tick tissues or excrement. Virus diseases of man include Colorado tick fever of western United States, transmitted by *Dermacentor andersoni*

Stiles, and Russian spring-summer encephalitis and related diseases in Europe and Asia, transmitted by *Ixodes persulcatus* Schulze. Some important rickettsial diseases of man are Rocky Mountain spotted fever, widely distributed in the Western Hemisphere, of which *D. andersoni, D. variabilis* (Say), *Rhipicephalus sanguineus* (Latreille), and *Amblyomma cajennense* (Fabricius) are important vectors; boutonneuse fever and related diseases of the Mediterranean region and Africa transmitted by *R. sanguineus* and some other species; and Q fever, the agent of which, although occurring in many ticks throughout the world, is not commonly transmitted directly by them. Tularemia (rabbit fever) is a bacterial disease transmitted by several species of ticks in the Northern Hemisphere. Tick paralysis, believed to be toxin-caused, occurs in many parts of the world and is produced by several species of ticks during the feeding process.

The numerous tick-borne diseases of animals cause vast economic losses, especially in tropical and subtropical regions. Examples are babesiasis, a protozoan disease caused by species of *Babesia* including *B. bigemina*, the agent of the widely distributed Texas cattle fever, which is transmitted principally by species of *Boophilus*; East Coast fever of Africa, another protozoan disease, caused by *Theileria parva*, is carried by several species of *Rhipicephalus*. Aside from carrying disease, several species are extremely important pests of man and animals. Heavy infestations of ticks produce severe anemia in livestock, and even death, from loss of blood alone. *See* ACARINA; FIEVRE BOUTONNEUSE; Q FEVER; SPOTTED FEVER, ROCKY MOUNTAIN; TICK FEVER; TICK PARALYSIS; TICK TYPHUS, QUEENSLAND; TICK TYPHUS, SIBERIAN; TICKBITE FEVER, SOUTH AFRICAN.   [GLEN M. KOHLS]

J particle

An elementary particle with an unusually long lifetime and large mass which does not fit into any of the schemes for classifying the large number of previously known particles. The discovery of J particles in proton-proton (p,p) and electron-positron (e^-,e^+) collisions created excitement in elementary particle physics in the mid-1970s.

Discovery of J particle. There has been much theoretical speculation on the existence of long-lived neutral (no electrical charge) particles with superheavy masses larger than $10 \text{ GeV}/c^2$ (ten billion electron volts divided by the speed of light squared, or about ten times the mass of the hydrogen atom). These are thought to play the role in weak interactions that photons play in electromagnetic interactions. There is, however, no theoretical justification, and no predictions exist, for long-lived particles in the mass region $1-10 \text{ GeV}/c^2$.

The J particles are rarely produced in p-p collisions. Statistically, they occur once after many millions of subnuclear reactions, in which most of the particles are "ordinary" elementary particles, such as kaon (K), pion (π), or proton (p). One searches for the J particle by detecting its e^+e^- decays. A two-particle spectrometer was used by a group from Massachusetts Institute of Technology (MIT) at the Brookhaven National Laboratory (BNL) to discover this particle. A successful experiment must have: (1) A very-high-intensity incident proton beam to produce a sufficient amount of J particles for detection; the Alternating Gradient Synchrotron (AGS) accelerator at BNL provides a beam of 10^{12} 30-GeV protons per second for this experiment. (2) The ability, in a billionth of a second, to pick out the $J{\to}e^-e^+$ pairs amidst billions of other particles through the detection apparatus.

The detector is called a magnetic spectrometer. A positive particle and a negative particle each traversed one of two 70-ft-long (21 m) arms of the spectrometer. The e^+ and e^- were identified by the fact that a special counter, called a Cerenkov counter, measured their speed as being slightly greater than that of all other charged particles. Precisely measured magnetic fields bent them and measured their energy. Finally, as a redundant check, the particles plowed into high-intensity lead-glass and the e^+ and e^- immediately transformed their energy into light. When collected, this light "tagged" these particles as e^+ and e^-, and not heavier particles such as π, K, or p. The simultaneous arrival of an e^- and e^+ in the two arms

indicated the creation of high-energy light quanta from nuclear interactions. The sudden increase in the number of e^+e^- pairs at a given energy (or mass) indicated the existence of a new particle.

The trajectory of electrons was measured by precision devices called multiwire proportional chambers. They consisted of 10,000 very fine gold-plated wires of 2-mm spacing, each with its own amplifier and recording system. The signals from the Cerenkov counters were collected by thin spherical and elliptical mirrors measuring about $3\frac{1}{2}$ ft (1.1 m) in diameter. The counters were filled with gaseous hydrogen so that only energetic e^- (and e^+) would produce the light which is due to the Cerenkov effect. To measure the arrival time of e^+e^- pairs to one-billionth of a second, there were 100 elements of thin plastic scintillation counters, each less than 2 mm thick. *See* CERENKOV RADIATION; SCINTILLATION COUNTER.

By August 1974 the MIT group began to observe abundant numbers of e^+e^- pairs with a total combined mass of 3.112 GeV (see illustration), and line width Γ much smaller than the resolution of the detector: $\Gamma < 5$ MeV (five million electron volts). From August through October, many experimental checks on the detector were made. The most important of these checks consisted of changing the magnetic field of the detecting magnets. This moved the particle trajectories to different regions of the detector. Still the abundance of e^+e^- pairs did not change, indicating a real particle had been discovered. This new particle was called the J particle since J is the symbol used to denote electromagnetic current and spin in elementary particle physics. A joint announcement of the discovery was made together with a team from Stanford Linear Accelerator Center (where it was called the psi particle) in November 1974.

Properties of J particle. The $J \to e^-e^+$ production rate from proton-proton reactions at an incident energy of 30 GeV is 10^{-34} cm^2, or one part in 10 million of "ordinary" particle yields. The yield increases by a factor of about 50 with a 300 GeV/c incident neutron beam. This increase of yield with energy is very similar to that of the K meson and antiproton productions. In the Brookhaven experiment, the yield of J decreased by almost a factor of 10 when the incident beam energy was reduced to 20 GeV/c. J particles have also been produced by bombardment of complex nuclear targets with high-energy photons. Production rates seem to be consistent with diffractive production like photoproduction of ordinary vector particles (the ρ, ω,

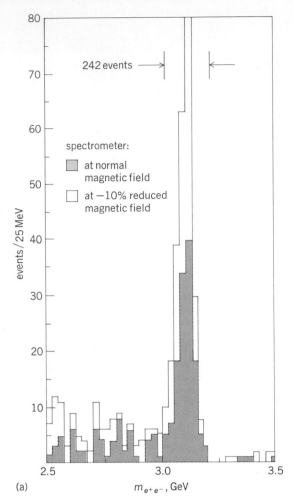

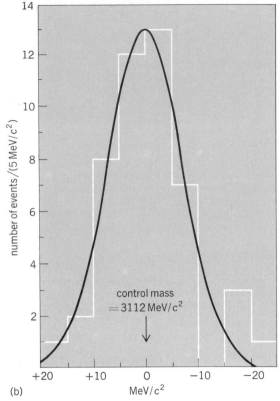

Observation of the J particle. (a) On-line data from August and October 1974, showing the existence of the J particle. (b) Measurement of the width of the J particle, showing it has a width less than 5 MeV. (*From J. J. Aubert et al., Discovery of the new particle, J. Nuc. Phys., B89(1):1–18, 1975*)

and ϕ). Analyses of photoproduction data indicate the J is not an ordinary intermediate vector boson. Rather, it belongs to a strong-interaction family.

Most of the properties of J particles have been measured at various electron-positron storage rings. The data show that the measured line width of J is less than 2 MeV. By measuring $e^+e^- \rightarrow J \rightarrow e^-e^+$ decay rate, one obtains a total width of less than 100 keV. The observed mass, m_J, varies from 3090 MeV/c^2 to 3112 MeV/c^2 from one laboratory to another. The spin (intrinsic angular momentum) is the same as that of the photon. Modes of decay into $\pi^+\pi^-$ and K^+K^- were not found. From this it has been concluded that the J particle does not belong to families of particles of about equal mass (like the π-mesons, π^+, π^0, π^-, or the two nucleons p, n), but that these particles are "single." Decays of $J \rightarrow \bar{p}p$, $\Lambda\bar{\Lambda}$, $n\pi^0$ have been found. *See* ISOTOPIC SPIN.

A sister state of the J particle was found by the Stanford group, and later by the DESY (Deutsches Elektronen-Synchrotron) groups at Hamburg with an observed mass varying between 3695 MeV/c^2 and 3680 MeV/c^2. This high-mass particle has many properties similar to those of the J particle and decays into J particles via the modes shown below, where 3.1 stands for the J particle and 3.7 for its sister state (according to their masses).

$$3.7 \rightarrow 3.1 + \pi^+\pi^-$$
$$\pi^0\pi^0$$
$$\eta \quad \text{(the eta particle)}$$

Origin of J particle. The most striking feature of the J particle is its very long lifetime. The J particle lives from a hundred to a thousand times longer than all other known mesons of heavy mass. Whenever objects in nature are found much more stable than expected, there is reason to be curious. There must be some hidden cause, some yet unknown effects of new principles that change the anticipated course of events.

Among the hundreds of theoretical models, two stand out as most attractive. The first model assumes that the J particles, because of their unusually long lifetime, carry a new conserved quantum number; thus they are produced either in pairs like $p + p \rightarrow J\bar{J} + \ldots$ or via associated productions like $p + p \rightarrow J\bar{\theta} + \ldots$ in the same way as the $K\Lambda$ system. The θ-particle decays to $\pi^- + p$, and there should be a long-lived $\pi^- p$ state around the mass of 2 GeV. The second model assumes that instead of three quarks as the building block of nucleons, there is an additional charmed quark. The J particle is a bound $C\overline{C}$ state (the ortho state), and there should be a para state near 3.1 GeV which decays into $p\bar{p}$ states. *See* CHARM; QUANTUM NUMBERS; QUARKS; SYMMETRY LAWS (PHYSICS).

To search for these new states, a large-scale experiment was performed by the same MIT group at BNL, in which all nine final states, π^-p, $\pi^+\pi^-$, p^-p, K^-p, $K^+\pi^-$, $K^-\pi^+$, K^+K^-, $K^+\bar{p}$, $\pi^+\bar{p}$, were measured simultaneously from pp reactions. Based on 20,000,000 pairs analyzed as of June 1975, no additional long-lived particles have been found to the level of $J \to e^-e^+$ yields, in contradiction to theoretical model predictions.

Measurements at DESY and SLAC have found a state, known as P_c, which came from $3.7 \to P_c + \gamma \to J + \gamma$. The state was detected by setting the colliding beam at a total energy of 3.7 BeV and observing two monoenergetic gamma rays plus J. The state P_c is consistent with charm model predictions. The model predicted there could be quite a few such states. However, until a charmed particle is found directly, there could be other explanations for such a state. *See* ELEMENTARY PARTICLE.   [SAMUEL C. C. TING]

Bibliography: J. J. Aubert et al., *Phys. Rev. Lett.*, 33:1404 and 1624, 1974; E. J. Augustin et al., *Phys. Rev. Lett.*, 33:1406, 1974; W. Braunschweig et al., *Phys. Lett.*, 53B:393, 1974.

Jade

A name that may be applied correctly to two distinct minerals. The two true jades are jadeite and nephrite. In addition, a variety of other minerals are incorrectly called jade. Idocrase is called California jade, dyed calcite is called Mexican jade, and green grossularite garnet is called Transvaal or South African jade. The most commonly encountered jade substitute is the mineral serpentine. It is often called "new jade" or "Korean jade." The most widely distributed and earliest known true type is nephrite, the less valuable of the two. Jadeite, the most precious of gemstones to the Chinese, is much rarer and more expensive.

Nephrite. Nephrite is one of the amphibole group of rock-forming minerals, and it occurs as a variety of a combination of the minerals tremolite and actinolite. Tremolite is a calcium-magnesium-aluminum silicate, whereas iron replaces the magnesium in actinolite. Although single crystals of the amphiboles are fragile because of two directions of easy cleavage, the minutely fibrous structure of nephrite makes it exceedingly durable. It occurs in a variety of colors, mostly of low intensity, including medium and dark green, yellow, white, black, and blue-gray. Nephrite has a hardness of 6 to $6\frac{1}{2}$ on Mohs scale, a specific gravity near 2.95, and refractive indices of 1.61 to 1.64. On the refractometer, nephrite gem stones show a single index near 1.61. Nephrite occurs throughout the world; important sources include the Soviet Union, New Zealand, Alaska, several provinces of China, and a number of states in the western United States. *See* AMPHIBOLE; GEM; TREMOLITE.

Jadeite. Jadeite is the more cherished of the two jade minerals, because of the more intense colors it displays. It is best known in the lovely intense green color resembling that of emerald (caused by a small amount of chromic oxide). In the quality known as imperial jade, the material is at least semitransparent. White, green and white, light reddish violet, bluish violet, brown, and orange colors are also found. Jadeite also has two directions of easy cleavage, but a comparable random fibrous structure creates an exceptional toughness. Although jadeite has been found in California and Guatemala, the only important source of gem-quality material ever discovered is the Mogaung region of upper Burma. The hardness of jadeite is $6\frac{1}{2}$ to 7, its specific gravity is approximately 3.34, and its refractive indices are 1.66 to 1.68; on a refractometer, only the value 1.66 is usually seen. *See* JADEITE.   [RICHARD T. LIDDICOAT, JR.]

Jadeite

The name given to the monoclinic sodium aluminum pyroxene, $NaAl(SiO_3)_2$. Jadeite forms green, fibrous crystals that are colorless in thin sections and exhibit the 87° pyroxene (110) cleavages. Jadeite is a rare metamorphic mineral found in some

Specimen of massive jadeite in its dark-green variety chloromelonite from Susa in Sardinia. (*Specimen from Department of Geology, Bryn Mawr College*)

serpentine masses associated with other dense minerals such as lawsonite, $CaAl_2Si_2O_7(OH)_2$; glaucophane; garnet; and other minerals such as diopside, tremolite serpentine, natrolite (a zeolite), and chlorite. The presence of albite and quartz or albite and nepheline mineral pairs in some jadeite specimens, and the rough similarity of the rock to the eclogites, implies a high-pressure, low-temperature modification of albite and nepheline: albite + nepheline = 2 jadeite. This reaction proceeds to the right at 100°C and 5 kilobars (1 kbar = 10^8 N/m²) pressure, or 500°C and 11 kbar pressure. The reaction albite = jadeite + quartz occurs at 100°C and 10 kbar pressure, or 200°C at 12.5 kbar pressure. The mineral associations in the natural occurrences of jadeite deny the presence of high temperatures. Jadeite could be the sodium mineral in the lower regions of the Earth's crust. Jadeite is rare; the few places where it occurs include

Burma, Celebes, Central America, Japan, Sardinia (see illustration), and San Benito County, CA. In plate tectonic theory, jadeite-bearing rocks are interpreted as high-pressure, low-temperature transformations in graywacke and altered basaltic materials in the oceanic plate portion of the subduction zones. The jadeite-bearing rocks may be characteristic of zones of interaction between certain continents and ocean basins. The absence of jadeite in rocks of Precambrian age suggests that crustal temperatures were higher in the past at the depths that jadeite forms in more recent rocks.

The albite + nepheline reactions to form jadeite at 1000°C requires 20 kbar pressure or a depth of 60 km. Accordingly, jadeite may be a mineral in the upper mantle of the Earth if albite or nepheline compositions are present. Jadeite is valued as a precious stone for carvings. *See* GLAUCOPHANE; HIGH-PRESSURE PHENOMENA; JADE; PYROXENE.

[GEORGE W. DE VORE]

Jahn-Teller effect

A small distortion in the lattice structure of certain crystal defects and, in some cases, of entire crystals, and in the structure of certain molecules, which reduces their symmetry and removes electronic degeneracy. The Jahn-Teller effect was predicted theoretically in 1937 and was first observed experimentally in 1952, but only since 1965 have its most commonly observable consequences come to be understood. Whereas in 1965 it was regarded as a "mystical effect" to be invoked when all other explanations of anomalous data failed, now it is as well understood as most other phenomena in solid-state physics.

The Jahn-Teller effect is most commonly observed in the optical and microwave spectra of point defects and localized impurity centers in solids. Such a center has a certain point symmetry, that is, there is a group of coordinate transformations which leave the center and its surroundings unaltered. If this symmetry is high enough, some of the electronic states of the center will be orbitally degenerate. The Jahn-Teller theorem states that any such degenerate system is unstable against small distortions of the lattice framework which remove the degeneracy. *See* DEGENERACY (QUANTUM MECHANICS).

Because of the great complexity of the interatomic forces in a real solid, it is customary to think of the center as a molecule consisting of the central impurity or defect and its nearest neighbor atoms. The rest of the crystal is regarded as a featureless heat bath. This "cluster" model preserves the symmetry of the center, which is its most essential attribute for purposes of this discussion, and is appropriate when the electronic wave function is sufficiently localized not to be greatly influenced by motion of atoms outside the nearest neighbor shell. In practice, this is also a condition for electron-lattice coupling to be strong enough for the Jahn-Teller effect to be important.

Simple model for Jahn-Teller distortion. As an illustrative example, consider the center shown in Fig. 1, which is a greatly simplified model of the first excited level of the F-center in an alkali halide. (The F-center is an electron trapped at a nega-

tive ion vacancy.) The electron is in a p state and is surrounded by a regular octahedron of positive ions. Because of the cubic symmetry, the three possible p states, p_x, p_y, and p_z (of which only p_z is shown), are obviously degenerate. Now suppose that the positive ions move a small distance in the directions indicated by the arrows. If their mean distance from the center is fixed, the mean energy of the p level is unaltered in first order. The p_z state, in which the electron approaches the positive ions more closely, is lowered in energy by the distortion, while the p_x and p_y states are raised, as shown in Fig. 2. The splitting is initially linear in the displacement Q. The surrounding crystal resists the distortion, and by Hooke's law the additional energy due to this resistance is initially quadratic in Q. Thus the total energy of the center in the p_z state goes through a minimum, and equilibrium is reached at some finite value of Q. It is easy to see that this equilibrium value of Q is A/k, where A is the initial (downward) slope and k the opposing force constant, and that the stabilization energy is $E_{JT} = A^2/2k$. *See* COLOR CENTERS; HOOKE'S LAW.

Strong electron-lattice coupling or weak interatomic forces favor a strong Jahn-Teller effect. E_{JT} can range from several electron volts (eV) for deep states in diamond and silicon to less than 10^{-3} eV for rare-earth ions in ionic crystals. In transition-metal ions, for which most of the data have been obtained, E_{JT} ranges from 0.01 to 1 eV.

The distortion shown in Fig. 1 could equally well have been along the x or y axes, and the lowest state would have been p_x or p_y, respectively. Thus each electronic state is associated with its own distortion, and if one considers the lowest vibrational state only, the threefold electronic degeneracy is replaced by threefold "vibronic" degeneracy. If no transitions were possible between the different directions of distortion, any one center would remain indefinitely in one state and its corresponding distorted configuration. This distortion might be observable, for instance, in a spin resonance experiment and in a concentrated crystal, by x-rays. Such an observable reduction in symmetry is called the static Jahn-Teller effect. However, the more common case in isolated centers is the dynamic Jahn-Teller effect, in which transitions between different distorted configurations are rapid compared with the characteristic measurement time. Such transitions can occur through thermal activation, or through quantum mechanical tunneling due to the zero point vibrational motion about the different equilibrium configurations. The observed spectrum now has the same symmetry as if there were no Jahn-Teller effect, and it might be thought that the effect has "disappeared." *See* ELECTRON PARAMAGNETIC RESONANCE (EPR) SPECTROSCOPY; X-RAY CRYSTALLOGRAPHY.

Observable consequences. In 1965, F. S. Ham pointed out that this is not the case; there are in fact pronounced observable consequences of the dynamic Jahn-Teller effect. These show up in the effects of off-diagonal electronic operators, that is, operators which connect electronic states associated with different distortions. For instance, consider the angular momentum operator $\hbar L$. The p level in the undistorted octahedron of Fig. 1 has

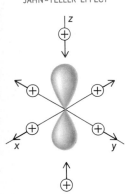

Fig. 1. A p-electron in an octahedral site. The arrows show one possible tetragonal mode of distortion. Only a p_z wave function is shown.

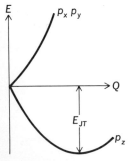

Fig. 2. Energy of the center shown in Fig. 1 as a function of the tetragonal distortion.

$L = 1$. When a magnetic field H is applied parallel to the z axis, the three-fold degenerate level is split into three states with wave functions $\frac{1}{2}\sqrt{2}(p_x + ip_y)$, $\frac{1}{2}\sqrt{2}(p_x - ip_y)$, p_z, corresponding respectively to the $+1$, -1, and 0 eigenvalues of L_z. If one ignores electron spin, the energies are given by $-e\hbar H L_z/2mc$ (in cgs gaussian units). The $(p_x + ip_y)$ state corresponds to clockwise rotation about the field, and the $(p_x - ip_y)$ state to counterclockwise rotation. In the Jahn-Teller distorted case the electron has to "drag" its distortion with it and can no longer rotate freely. Its mass is thus effectively increased and the orbital contribution to the magnetic splitting correspondingly decreased. At 0K, where rotation is only possible at all because of zero point motion, the reduction factor can be shown to be approximately $e^{-3E_{JT}/2\hbar\omega}$, where ω is an effective frequency of vibration for the cluster. Note that the spin operator S is not "quenched" as L is, since the spin direction has no distortion associated with it (except insofar as spin-orbit coupling causes L to follow S).

This reduction in the orbital contribution to magnetic splittings has been observed in the spin resonance and Zeeman spectra of many transition-metal impurity ions in crystals. Other off-diagonal operators, such as the spin-orbit coupling (which can often be written in the form $-\lambda L \cdot S$), are quenched in the same way. A detailed comparison of the Ham theory with experiment has been possible in the optical spectra of $3d$ ions in some cubic crystals. *See* ZEEMAN EFFECT.

If the electronic state is doubly degenerate, it has no orbital momentum or spin-orbit coupling in first order. If the Jahn-Teller interaction is strong, the small local strains which are inevitable in a real crystal are sufficient to stabilize the center in one or other distorted configuration at low temperature. In a crystal containing many centers, three uniaxial spin resonance spectra are seen, corresponding to the three possible directions of distortion. As the temperature is raised, transitions from one distorted configuration to another become possible, and the three anisotropic resonances collapse into one isotropic motionally averaged resonance. This process is exactly analogous to motional narrowing due to diffusion in nuclear magnetic resonance. If the Jahn-Teller effect is sufficiently weak, tunneling will average out the spectrum, even at 0K. The spectrum in this case still retains a cubic anisotropy, qualitatively similar to that expected in the absence of Jahn-Teller interaction. Even if tunneling is too slow to produce this averaging, it can still profoundly affect relaxation processes. For instance, it causes the rapid dephasing of magnetic free induction, and it produces strong damping of acoustic waves whose period is of the order of the tunneling time. *See* NUCLEAR MAGNETIC RESONANCE (NMR).

Cooperative Jahn-Teller effect. In a crystal containing a large concentration of ions with degenerate ground states, a Jahn-Teller distortion of the whole crystal can occur through the strain-mediated interaction between ions. Many rare-earth compounds have been found to undergo second-order phase transitions, in which the crystal symmetry is reduced and electronic degeneracy removed in the manner expected for the Jahn-Teller

effect. The electron-lattice interaction is so weak in these ions that, even though their ground states are degenerate, Jahn-Teller effects are not normally detected in the ion as an isolated impurity. However, the ion-ion interaction greatly enhances the effect by involving the cooperation of many ions, and transition temperatures of a few kelvin are typically observed.

Four adjacent lattice cells of a highly simplified two-dimensional model of such a crystal are shown in Fig. 3. Each positive rare-earth ion is surrounded by four negative ions which can distort as shown by the arrows, producing a splitting of its ground state analogous to that in Fig. 2. Because of the shared negative ions, distortions on adjacent cells must be related in phase. It follows that there is an effective coupling between ions analogous to magnetic exchange but of much greater range, since it is mediated by the strain field of the lattice. As an initially symmetric crystal is cooled, a spontaneous distortion begins to appear at a critical temperature T_c, just as spontaneous magnetization appears in a ferromagnet at the Curie temperature. This distortion can be measured directly by x-rays, and by the energy-level splitting that it produces; it also manifests itself in a specific-heat anomaly (see Fig. 4). As T_c is approached from above, the lattice becomes "soft" with respect to the preferred mode of distortion. If this distortion involves a macroscopic strain, the associated elastic (compliance) constant, which is analogous to the magnetic susceptibility, becomes very large at T_c, as illustrated in Fig. 5. If the distortion is not macroscopic, it will still show up as a corresponding "softness" in a phonon mode, detectable by Raman effect or neutron diffraction. *See* FERROMAGNETISM; LATTICE VIBRATIONS; NEUTRON DIFFRACTION; PHONON; RAMAN EFFECT; STRESS AND STRAIN.

When the electron-lattice interaction is strong, as in nonmetallic compounds containing Cu^{2+} or Mn^{3+} ions, values of T_c up to 1000 K can be obtained. In metallic compounds, if E_{JT} is less than

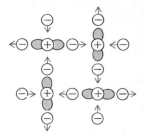

Fig. 3. Part of a two-dimensional model of a cooperative Jahn-Teller system. Arrows indicate the motion in the soft mode.

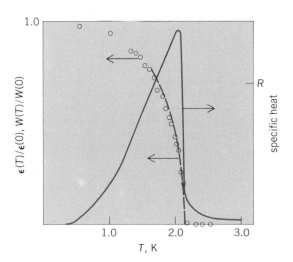

Fig. 4. Normalized lattice distortion (ratio $\epsilon(T)/\epsilon(0)$ of lattice distortion at temperature T to distortion at absolute zero, dashed line); normalized splitting of the lowest energy levels $W(T)/W(0)$ (points); and specific heat (full line) of $TmVO_4$ below T_c.

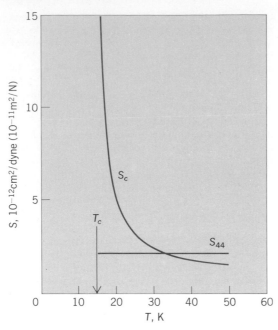

Fig. 5. Elastic constants $S_c = 2(S_{11} - S_{12})$ and S_{44} of DyVO₄ just above T_c. S_c couples to the soft mode.

the bandwidth due to overlap of wave functions, the Jahn-Teller effect is suppressed. However, in some narrow-band metals, such as Nb₃Sn and V₃Si, E_{JT} is so large that Jahn-Teller distortion does occur. The same strong electron-lattice coupling contributes to the high superconducting temperatures of these compounds. *See* SECOND-ORDER TRANSITION; SUPERCONDUCTIVITY.

Vibrational spectra of molecules. The intimate coupling between electronic and nuclear motion which is a consequence of the Jahn-Teller interaction should have pronounced effects on the vibrational spectra of molecules. However, it has proved quite difficult to pin down these effects, chiefly because of the difficulty of establishing what the vibrational spectrum would be in the absence of the Jahn-Teller interaction. *See* ATOMIC STRUCTURE AND SPECTRA; CRYSTAL DEFECTS; CRYSTAL STRUCTURE; CRYSTALLOGRAPHY; MOLECULAR STRUCTURE AND SPECTRA; QUANTUM THEORY, NONRELATIVISTIC.   [M. D. STURGE]

Bibliography: R. Englman, The Jahn-Teller effect in molecules and crystals, 1972; G. A. Gehring and K. A. Gehring, Cooperative Jahn-Teller effects, *Rep. Progr. Phys.* 38:1–150, 1975; F. S. Ham, The Jahn-Teller effect in EPR spectra, in S. Geschwind (ed.), *Electron Paramagnetic Resonance*, 1971; G. Herzberg, *Electronic Spectra of Polyatomic Molecules*, 1966; M. D. Sturge, The Jahn-Teller effect in solids, *Advan. Solid State Phys.*, 20:91–211, 1967.

Jamming

Intentional generation of interfering signals by powerful transmitters as a countermeasure intended to block a communication or radar system or to impair its effectiveness appreciably. Radio broadcasts or radio messages can be jammed by beaming a more powerful signal on the same frequency at the area in which reception is to be impaired, using carefully selected noise modulation to give maximum impairment of intelligibility of reception. When stations on many different frequencies are to be jammed or when an enemy is changing frequencies to avoid jamming, the jamming transmitter is correspondingly changed in frequency or swept through a range of frequencies over and over again. Similar techniques are used at radar frequencies to jam early-warning and gunfire-control radar systems. *See* ELECTRONIC COUNTERMEASURES.

[JOHN MARKUS]

Japanning

In the industrial sense, the finishing of metal objects with a black baking varnish that consists of a hard asphalt such as gilsonite. Modern technology has made the practice obsolete by production of enamels and other finishes based on synthetic resins that provide the same properties as gilsonite, with the added advantage of being heat- and solvent-resistant and being available in a variety of colors. *See* ENAMEL, NONVITREOUS.

The term survives as an art form. Lacquers and high-gloss varnishes are hand-applied in layers of varying thickness for unusual effect to wood, ceramic, and metal creations in a variety of hues, often with filigree tracing of white or highlight color. *See* LACQUER; VARNISH.

[C. MARTINSON; C. W. SISLER]

Jasper

An opaque, impure type of massive fine-grained quartz that typically has a tile-red, dark-brownish-red, brown, or brownish-yellow color. The color of the reddish varieties of jasper is caused by admixed, finely divided hematite, and that of the brownish types by finely divided goethite. Jasper has been used since ancient times as an ornamental stone, chiefly of inlay work, and as a semiprecious gem material. Under the microscope, jasper generally has a fine, granular structure, but fairly large amounts of fibrous or spherulitic silica also may be present. *See* GEM; QUARTZ.

Jasper has a smooth conchoidal fracture with a dull luster. The specific gravity and hardness are variable, depending upon particle size and the nature and amount of the impurities present; both values approach those of quartz. The color of jasper often is variegated in banded, spotted, or orbicular types. Heliotrope is a translucent greenish chalcedony containing spots or streaks of opaque red jasper, and jaspagate contains bands of chalcedonic silica alternating with jasper. Jaspilite is a metamorphic rock composed of alternating layers of jasper with black or red bands of hematite. *See* CHALCEDONY.

[CLIFFORD FRONDEL]

Jaundice

The yellow staining of the skin and mucous membranes associated with the accumulation of bile pigments in the blood plasma. Bile pigments are the normal result of the metabolism of blood pigments, and are normally excreted from the blood into the bile by the liver. An increase in circulating bile pigments can, therefore, come about through increased breakdown of blood (hemolytic jaundice), through lack of patency of the bile ducts (obstructive jaundice), through inability or failure of the liver to clear the plasma (parenchymal jaundice), or through combinations of these. *See* GALLBLADDER (VERTEBRATE); LIVER (VERTEBRATE).

Metabolic pathway. Metabolism of the hemoglobin from destroyed red blood cells is carried on in organs of the reticuloendothelial system, such as the spleen, and the resulting bilirubin is liberated to the plasma. The plasma then circulates through the liver, where the bilirubin is conjugated enzymatically with glucuronic acid, and is excreted in the bile. Bile travels through the bile ducts to the small intestine, whence a small amount of altered bilirubin, termed urobilinogen, may be reabsorbed into the plasma. Excessive destruction of red blood cells causes accelerated production of bilirubin, overloads the ability of the liver to remove the pigment from the circulation, and produces jaundice. Blockage of the bile ducts causes elevation of plasma bilirubin glucuronide level because of the inability to dispose of this material in the usual channel. Damage to liver cells may cause elevation of the plasma bilirubin or bilirubin glucuronide or both, depending on the type and severity of liver cell damage.

Although the major portion of circulating bilirubin is derived from the breakdown of red cells, some is also contributed through inefficient or incomplete utilization of precursors of hemoglobin, which spill into the plasma bilirubin pool without having been used for hemoglobin synthesis. In addition, metabolism of cytochromes and similar pigmented compounds found in all cells of the body yields small quantities of bilirubin. *See* HEMOGLOBIN.

Jaundice occurs when the level of these circulating pigments becomes so high that they are visible in the skin and mucous membranes, where they are bound by a reaction which has not been identified. In the normal adult, total bilirubin, that is, the total bilirubin and bilirubin glucuronide, levels rarely exceed 0.8–1.0 mg/100 ml of plasma, while jaundice usually becomes visible when total bilirubin approaches 1.5 mg. *See* BILIRUBIN.

Hemolytic jaundice. Destruction of red blood cells in the normal human adult proceeds at a rate at which about 0.8% of the circulating hemoglobin is broken down each day. This can be increased in states of excessive hemolysis up to 10- to 15-fold without overtaxing the remarkable ability of the liver to clear bilirubin from the plasma. Even this rate of clearing can be exceeded, however, in certain morbid states which result in hemolytic jaundice; such states include various hemolytic anemias, hemolysis resulting from incompatible blood transfusion, severe thermal or electric injuries, or introduction of hemolytic agents into the bloodstream. Similar jaundice occurs in pulmonary infarction.

In infants, and especially in premature infants, the ability of the liver to conjugate bilirubin with glucuronide is much less than in adults, apparently because of the lack of suitable enzymes. Jaundice appears in many infants shortly after birth, then disappears within a few days, with development of the appropriate enzyme structure. In the uncommon constitutional hepatic dysfunction, this enzyme defect apparently persists into adult life. The infantile jaundice accompanying erythroblastosis

fetalis (Rh babies) is due to the inability of the infantile liver to metabolize the bilirubin resulting from markedly accelerated hemolysis. *See* BLOOD GROUPS; ERYTHROBLASTOSIS FETALIS.

A related form of jaundice occurs when an abnormality exists in the hemoglobin-forming cells of the bone marrow in which inadequate utilization of hemoglobin precursors occurs. Moderate degrees of jaundice can result from the accumulation of these substances in the bloodstream.

Obstructive jaundice. The highest levels of total bilirubin are seen in chronic obstructive jaundice, in which plasma levels may reach 50–60 mg/100 ml, and the skin may take on a remarkable deep-yellow hue. This condition may be brought about through a variety of means. In the infant there may be a severe maldevelopment of the bile ducts such that no channel for the flow of bile exists, while in the adult obstructive jaundice is most commonly caused by impaction of a gallstone in the ducts. Benign and malignant tumors of the gallbladder, bile ducts, pancreas, lymph nodes, and other organs may also cause compression of the bile ducts with loss of patency, and bile duct stricture may follow surgery or inflammation in the region. A similar, less severe, reversible picture is seen as a hypersensitivity response to the administration of some drugs, the most common of which are chlorpromazine and related drugs, and methyl testosterone. In the uncommon benign disorder known as idiopathic familial jaundice, there appears to be decreased ability to excrete conjugated bilirubin into the bile ducts, giving rise to a constant, but usually quite slight, elevation in the plasma bilirubin glucuronide.

Parenchymal jaundice. A wide variety of diseases exists in which part of the jaundice can be accounted for by actual damage to liver cells, with consequent decrease in their ability to conjugate bilirubin and excrete the glucuronide, causing an elevation of both fractions in the plasma. This group comprises such conditions as inflammations of the liver, including viral hepatitis, Weil's disease, syphilis, parasitic infestations, and bacterial infections; toxic conditions, including poisoning from a wide variety of organic and inorganic compounds and in a broader sense the toxemias associated with severe systemic diseases; tumorous conditions, including primary hepatic tumors and those metastatic from other organs; and other miscellaneous conditions, the most common of which is congestive heart failure. Some of these conditions have an added component of obstructive or hemolytic jaundice, which confuses the picture for the clinician.

Symptomatology. The appearance and symptomatology of subjects suffering from jaundice vary from case to case and depend on the underlying disease process producing the jaundice. Elevation of bilirubin by itself has very limited deleterious effects on overall physiology, with two exceptions: The brain of very young infants is subject to damage by high levels of circulating bilirubin (a condition termed kernicterus); and, under certain conditions in the adult, high levels of circulating bilirubin appear to contribute to kidney damage.

Aside from the effects of the jaundice itself, however, patients with hemolytic jaundice usually have an accompanying anemia. Those with ob-

---

structive jaundice commonly note that their stools are not brown, a symptom caused by lack of bile pigments (acholic stools), while bilirubin glucuronide appears in the urine and causes it to turn dark. Patients with malignancies demonstrate the usual signs of weight loss and anemia, while those with inflammatory conditions commonly have fever chills and prostration. *See* CLINICAL PATHOLOGY; FECES; PIGMENTATION.   [ROLLA B. HILL]

Java man

The Indonesian human fossils belonging to the species *Homo erectus*. The first remains of Java man were discovered in 1890–1891 by Eugene Dubois in central Java and consist of a skull cap, a fragment of a lower jaw, a few teeth, and a femur. Originally called "Pithecanthropus erectus" (meaning erect ape-man), these and related fossils were later recognized as sufficiently similar to modern humans to be called *Homo*, hence the taxonomic name *H. erectus*. Through the work of Dubois and later of G. H. R. von Koenigswald, S. Sartono, and T. Jacob, Java man is now represented by the remains of more than 31 individuals, including skull fragments, jaws, teeth, and femora.

The Java *H. erectus* fossils come from deposits containing the Djetis and Trinil faunal assemblages, which are considered to be Lower Pleistocene and early Middle Pleistocene, respectively; hence these fossils are roughly contemporaneous with other *H. erectus* fossils. Morphologically, Java man (see illustration) exhibits a long, low skull with massive brows; large, strongly built jaws with relatively large teeth; somewhat small braincases; and limb bones indistinguishable from those of modern humans. The Java *H. erectus* sample shows considerable variation in individual size and robustness, which has led scholars to refer some of the specimens to other taxonomic groups. All of the fossils, however, probably belong to *H. erectus* and are ancestral to the more recent Solo fossils and modern *H. sapiens*. *See* FOSSIL MAN; HEIDELBERG MAN; PEKING MAN; SOLO MAN; TERNIFINE MAN.   [ERIK TRINKAUS]

Bibliography: W. W. Howells, *Evolution of the Genus Homo*, 1973.

Jerboa

The common name for 25 species of rodents which make up the family Dipodidae. All species occur in desert or semiarid regions of Asia; three species extend into North Africa. All of these animals are adapted for jumping, and the hindlegs and feet are extremely long. They can cover 6 ft in a single leap. There are usually only three functional toes. The long tail terminates in a tuft of hair and serves as a balancing organ when the animal squats (see illustration). The short forelimbs are used for grasping and holding food. The animals search for food at night because of the high temperatures of their semiarid habitat. During the day they retire to their burrows, plugging the tunnel entrance to maintain a lower temperature in the burrow. The number of teeth varies from 16 to 18, dependent

Jerboa, with body 3–6 in. long and tail up to 8 in. long.

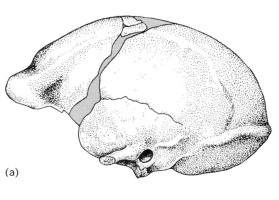

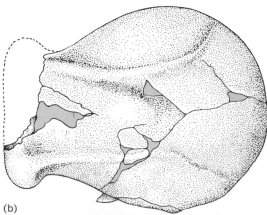

(a)

(b)

Cranium of *Pithecanthropus erectus II*, after von Koenigswald, shown in (*a*) lateral view and (*b*) dorsal view. (*Carnegie Institution of Washington, as used in M. F. Ashley Montagu, An Introduction to Physical Anthropology, 2d ed., Charles C. Thomas, 1951*)

upon the species. The dental formula is I 1/1 C 0/0 Pm 0–1/0 M 3/3. *Jaculus orientalis*, the greater Egyptian jerboa, was referred to as the desert rat during World War II. The most common species is *J. jaculus*, the lesser Egyptian jerboa. *See* RODENTIA.

[CHARLES B. CURTIN]

Jet (gemology)

A black, opaque material that takes a high polish. Jet has been used for many centuries for ornamental purposes. It is a compact variety of lignite coal. It has a refractive index of 1.66 (which accounts for its relatively high luster), a hardness of 3–4 on Mohs scale, and a specific gravity of 1.30–1.35. Jet is compact and durable, and can be carved or even turned on a lathe. The principal source is Whitby, England, where it occurs in hard shales. Although popular from Pliny's day until the 19th century, the use of jet for jewelry purposes declined markedly for centuries until a resurgence of interest occurred in the 1960s. *See* GEM; LIGNITE.

[RICHARD T. LIDDICOAT, JR.]

Jet flow

A local high-velocity stream in a relatively stationary surrounding fluid. Fluid may be caused to flow in a jet for a variety of reasons. Propulsion of a body through a fluid may be by jet propulsion, by a propeller, or by a pump or compressor. The impulse wheel used in hydroelectric power plants takes energy from a jet of water and converts it into torque applied to a rotating shaft. In fighting fires, a smooth jet of water is produced by a nozzle to carry water to the fire without separating it into droplets. Water jets are also used to move earth in gold mining, and as a means of dissipating fluid energy. *See* JET PROPULSION.

A jet may be formed by flow out of a closed conduit, or downstream from a propeller. Flow through an orifice or nozzle causes a jet to form. In the case of power development, a needle nozzle is used to convert the pressure and kinetic energy in the penstock (pipe leading from the reservoir to the turbine) into a smooth jet of variable diameter and discharge but practically constant velocity (Fig. 1). The change in size of jet is accomplished by movement of the needle forward or backward. With flow through a nozzle of fixed diameter a change in discharge causes a corresponding change in velocity of the jet.

The formation of a jet requires that a force be exerted on the fluid in the direction of the jet. An equal and opposite force is exerted on the machine causing the jet, and this is the propulsive force in the case of an airplane or a ship.

When a jet of liquid issues into a gas, such as a water jet entering into air, turbulence within the liquid and friction between gas and liquid cause

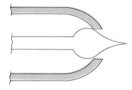

Fig. 1. Needle nozzle.

the liquid jet eventually to pull gas along with it and to allow penetration of the gas into the stream. At length the jet breaks into a spray.

When a jet issues into fluid of the same density, fluid is brought or inducted into the jet (Fig. 2). The jet spreads at a linear rate with $b = x/8$. Turbulent shear forces reduce the jet velocity within the central cone, and equal turbulent shear forces act to increase velocity in the outer portions of the jet. The momentum of the jet must remain substantially constant in the axial direction, because no forces act external to the jet.

[VICTOR L. STREETER]

Jet fuel

Fuel blended from the light distillates fractionated from crude petroleum. There are two general types, a wide-cut heavy naphtha-kerosine blend used by the U.S. Air Force as JP-4 (or commercially as Jet B) and a kerosine used by the world's airlines as Jet A (or Jet A-1) or by the U.S. Navy as JP-5. Since 1970, commercial kerosine of 38°C flash point has grown from a small-volume household-heating fuel into a major product of commerce rivaling gasoline in importance as a source of transportation energy. During the 1960–1975 period, JP-4 diminished in relative importance, and commercial use of Jet B also declined to a small percent. *See* KEROSINE.

All jet fuels must meet the stringent performance requirements of aircraft turbine engines and fuel systems, which demand extreme cleanliness and freedom from oxidation deposits in high-temperature zones. Combustors require fuels that atomize and ignite at low temperatures, burn with adequate heat release and controlled radiation, and produce neither smoke nor attack of hot turbine parts. The operation of the aircraft in long-duration flights at high altitude imposes a special requirement of good low-temperature flow behavior; this need establishes Jet A-1 which has a freezing point of −50°C (wax) as an international flight fuel; Jet A which has a freezing point of −40°C (wax) can serve shorter domestic routes. *See* AIRCRAFT ENGINE.

Fuels pumped through long multiproduct pipelines or delivered by tanker are usually clay-filtered to ensure freedom from surfactants. Many stages of filters operate to ensure clean, dry product as the fuel moves into airport tanks, hydrant

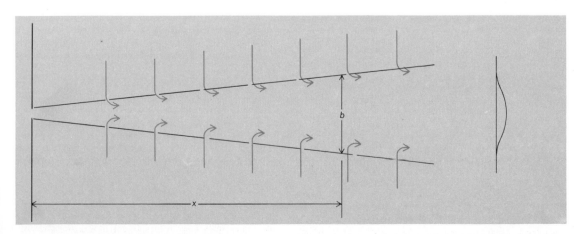

Fig. 2. Fluid jet issuing into some fluid medium.

systems, and finally aircraft. Because high-speed filtration can generate static charges, fuels may contain an electrical conductivity additive to ensure rapid dissipation of charge.

[W. G. DUKEK]

Jet propulsion

Propulsion of a body by means of force resulting from discharge of a fluid jet. This fluid jet issues from a nozzle and produces a reaction (Newton's third law) to the force exerted against the working fluid in giving it momentum in the jet stream. Turbojets, ramjets, and rockets are the most widely used jet-propulsion engines.

Jet nozzles. In each of these propulsion engines a jet nozzle converts potential energy of the working fluid into kinetic energy. Hot high-pressure gas escapes through the nozzle, expanding in volume as it drops in pressure and temperature, thus gaining rearward velocity and momentum. This process is governed by the laws of conservation of mass, energy, and momentum and by the pressure-volume-temperature relationships of the gas-state equation. *See* FLUID DYNAMICS; JET FLOW; NOZZLE.

For propulsion systems in which the pressure of the working fluid is not more than approximately twice the absolute ambient pressure, a converging nozzle is used (Fig. 1a). The mass flow from this nozzle in terms of conditions at sections 1 and 2 is

given in Eq. (1), and the velocity of the jet leaving the nozzle is given in Eq. (2).

$$m = A_2 \rho_1 \sqrt{2gJC_pT_{1t}} \sqrt{\left(\frac{p_2}{p_1}\right)^{2/\gamma} - \left(\frac{p_2}{p_1}\right)^{(\gamma+1)/\gamma}} \quad (1)$$

$$v_2 = \sqrt{2gJC_pT_{1t}} \sqrt{1 - \left(\frac{p_2}{p_1}\right)^{(\gamma-1)/\gamma}} \quad (2)$$

Here m = mass flow, slug/sec
 A = cross-sectional area, ft²
 ρ = density, slug/ft³
 J = work equivalent of heat, ft-lb/Btu
 T_t = total temperature, °R
 C_p = specific heat at constant pressure, Btu/(°F)(lb)
 γ = ratio of specific heats
 p = static pressure, lb/in.²
 v = velocity, ft/sec

Maximum flow P occurs when Eq. (3a) holds. This is the critical pressure at which flow velocity in the nozzle throat is equal to local sound velocity. For air and most combustion gas mixtures, Eq. (3b) holds.

$$p_2 = p_1 \left(\frac{2}{\gamma+1}\right)^{\gamma/(\gamma-1)} = p_c \quad (3a)$$

$$p_c \cong 0.5p_1 \quad (3b)$$

For propulsion systems in which the working fluid pressure is high compared to ambient pressure, a converging-diverging nozzle is used (Fig. 1b). In this nozzle the working fluid continues to expand from the critical throat pressure to ambient pressure at section 3 with a further increase in velocity beyond the sonic throat velocity, as in Eq. (4).

$$v_3 = \sqrt{2gJC_pT_{1t}} \sqrt{1 - \left(\frac{p_3}{p_1}\right)^{(\gamma-1)/\gamma}} \quad (4)$$

Turbojet. The turbojet is an air-breathing propulsion engine used in most military fighters, bombers, and transports and in modern commercial airline transports. Thrust ratings range from a few hundred pounds to more than 50,000 lb for engines for the experimental stage. The engine operates best at high subsonic or supersonic flight speeds, where the high-velocity jet achieves good propulsion efficiency. Specialized versions of turbojets are used for subsonic flight and others for supersonic flight. The turbofan, or bypass engine, operates with lower exhaust-gas velocity and provides improved efficiency for subsonic flight. In some turbojets more heat is added to the exhaust stream in an afterburner to increase propulsion power output for efficient supersonic flight. *See* AFTERBURNER; PROPULSION; TURBOFAN.

The turbojet is a heat engine. Air enters the inlet diffuser and is compressed adiabatically there and in the rotating compressor (Figs. 2 and 3). Heat is added by burning fuel at constant pressure in the combustor. The hot gas expands in the turbine, where energy is extracted to drive the rotating compressor. Further expansion through the jet nozzle converts the remaining available energy of the gas stream into high velocity, producing thrust for propulsion power. Additional propulsion power can be realized by heat added in an afterburner. *See* BRAYTON CYCLE; TURBOJET.

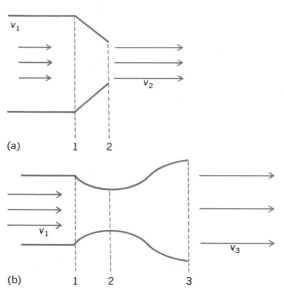

(a)   1   2

(b)   1   2   3

Fig. 1. Propulsion nozzles. (a) Low-pressure converging nozzle. (b) High-pressure converging-diverging nozzle.

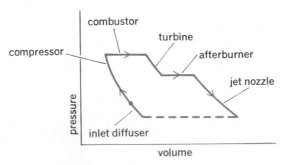

Fig. 2. Turbojet pressure-volume diagram.

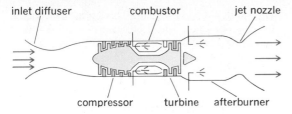

inlet diffuser   combustor   jet nozzle

compressor   turbine   afterburner

Fig. 3. Simplified schematic diagram of turbojet.

During high-speed flight the inlet diffuser decreases the relative velocity of the entering air, increasing its pressure in a process that is the reverse of nozzle expansion (Fig. 3). For supersonic speeds, a converging-diverging passage is required. The diffuser inlet and throat areas may be varied mechanically to match the airflow requirements of varying flight speed. *See* SUPERSONIC DIFFUSER.

The axial-flow compressor has alternate rows of rotating and stationary blades which compress the air further. The individual action of the blades is like that of an airplane wing in deflecting the air passing over it. The rotating blades add kinetic energy to the airstream, and the stationary blades convert some of this kinetic energy into a pressure rise. Normally, each rotating and stationary blade row produces a pressure rise as the air moves through the tapered annular passage between the rotor drum and stator casing.

Most of the total air compression is provided by the rotating compressor during subsonic flight. As the flight speed increases, the inlet diffuser performs an increasing amount of the total compression. At high supersonic speeds, the inlet diffuser does most of the compression work and the rotating compressor becomes less important in producing efficient engine performance.

In the combustor, jet fuel (a kerosinelike petroleum fraction) is sprayed, vaporized, and burned. Air from the compressor discharge is fed into the combustion space through various shaped openings in sheet-metal combustion liners. These openings serve to mix the air and fuel vapor for efficient combustion and to cool the liner which protects the outer combustion casing from the heat of the flame.

The stationary turbine nozzle vales expand the hot gas to a high-velocity stream which is directed toward the rotating blades in a near-tangential direction. The gas is turned in the blades and imparts rotational energy to the turbine wheel by this momentum change. Energy absorbed by the turbine wheel is used to provide power to drive the compressor through a connecting shaft (Fig. 4). The number of turbine stages is determined by the amount of power required to drive the compressor. The trend in modern jet engines is toward higher pressure ratio, higher work compressors, and turbines for improved economy in fuel consumption. The implied increase in number of stages has been offset by increasing the work potential of individual stages.

In the afterburner, following the turbine, fuel is injected into the hot gas stream through spray bars. V-section flame-stabilizing channels are mounted downstream from the fuel spray bars. These channels produce eddies in the gas stream to promote stable burning and prevent flame blowout at high altitude. A louvered sheet-metal liner protects the afterburner casing from overheating. In some turbojets the extra propulsion power from the afterburner is not required; the hot gas from the turbine flows directly into the exhaust-jet nozzle. *See* TURBORAMJET.

For subsonic flight applications, a converging jet nozzle of fixed dimensions is used. For supersonic flight the converging-diverging type jet nozzle is required for good performance. The dimensions of the throat and exit may be varied to match the flow and expansion requirements of varying flight speeds and altitudes.

Accessories to the engine are driven by an arrangement of shafts and gears taking power from the compressor-turbine shaft. These accessories include fuel and lube-oil pumps, tachometer, and in some installations an electric starter-generator.

An engine control system senses air pressure and temperature from the inlet diffuser, rotor speed, and throttle setting. It computes from these, by mechanical or electric analog means, the required fuel rates and meters fuel flow to the combustor and afterburner. The system senses gas temperature (entering the turbine and in the afterburner), from which it operates to limit fuel rates

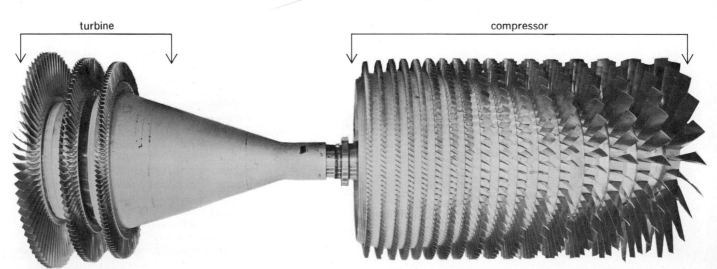

turbine   compressor

Fig. 4. Compressor and turbine rotors assembled.

Table 1. Characteristics of medium-sized turbojet

Characteristics	With afterburner	Without afterburner
Weight, lb	3,600	2,900
Length, in.	200	110
Diameter, in.	38	37
Compressor stages	17	17
Turbine stages	3	3
Takeoff thrust, lb	15,500	11,000
Takeoff SFC*	2.0	0.8
Best cruise thrust, lb	2,600	2,500
Best cruise SFC (Mach 0.9 at 35,000 ft altitude)	1.0	0.9
Military thrust, lb	6,000	
Military SFC (Mach 2.0 at 55,000 ft altitude)	2.3	

*Specific fuel consumption, pounds of fuel per pound of thrust per hour.

so as to prevent overheating or overspeeding. Engine characteristics representative of a modern turbojet are given in Table 1.

Ramjet. The ramjet is the simplest air-breathing propulsion system and is used principally in guided missiles (Fig. 5). It travels at high velocity, compressing air in its inlet diffuser, burning fuel in the air, and discharging it through a jet nozzle. There are no rotating compressor and turbine elements in the ramjet. The ram air-pressure ratios achieved by its forward velocity are high enough for efficient operation on the Brayton cycle; therefore, no additional air compression is required from a rotating compressor. However, the ramjet must be accelerated up to operating speeds by other means. In most missile applications this is accomplished by a rocket-powered booster stage. In other applications the acceleration of the ramjet can be accomplished by turbojet power.

Characteristics of a modern ramjet are a weight of 1000 lb, a length of 150 in., and a diameter of 36 in.; traveling at a speed of Mach 4 at 75,000 ft, it would have a thrust of 2000 lb and a specific fuel consumption of 2.0.

The ramjet is usually designed to cruise within a fairly narrow speed range. The combustor is similar to a turbojet afterburner with fuel injectors, V-channel flame holders, and a louvered liner for heat insulation. Controls and accessories, housed in the bullet nose of the inlet, are powered by a very small high-speed turbine drive by ram air bled from the inlet. *See* Pulse jet; Ramjet.

Liquid-propellant rocket engine. The liquid-propellant rocket engine is a propulsion system used to power missiles and space vehicles. It is used primarily to accelerate a load to high velocity, and to do this it delivers a large thrust for a relatively short time.

Liquid fuel (usually a hydrocarbon resembling kerosine) and oxidizer (usually liquid oxygen) are pumped from tanks to a fuel injector, which sprays them into a combustion chamber, in which they burn. The hot combustion gases escape through a converging-diverging jet nozzle, leaving at very high velocity (Fig. 6). The reaction forces developed on nozzle wall and injector head are transmitted to the body of the vehicle. Rocket vehicles are sometimes steered by mounting the entire engine on gimbals.

Fuel and oxidizer pumps have high-speed centrifugal impellers driven through speed-reduction gears by a small, single-stage gas turbine. High-pressure gas to drive the turbine usually is obtained by burning fuel and oxidizer bled from the pump discharge to a small combustion chamber feeding the turbine. Starting is accomplished by burning these propellants from gas-pressurized tanks. In some rocket engines the turbine working fluid is decomposed hydrogen peroxide. *See* Propellant; Rocket engine.

The effectiveness of a rocket propellant combination is indicated by its theoretical specific im-

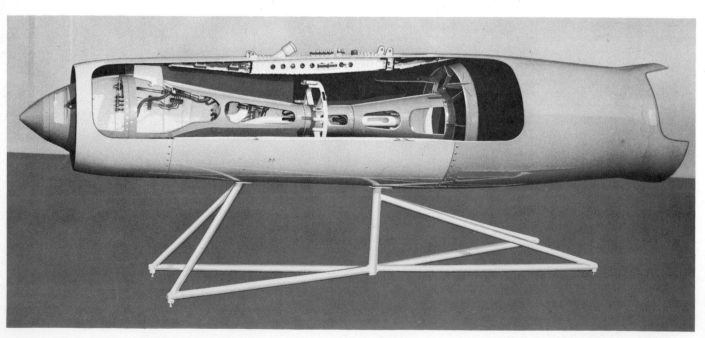

Fig. 5. Cutaway model of ramjet. (*Marquardt Aircraft Co.*)

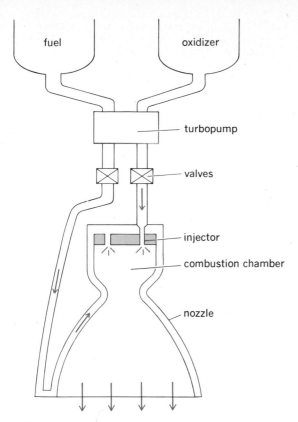

Fig. 6. Liquid-propellant rocket engine. Arrows represent direction of flows. (*General Electric Co.*)

pulse. Fuels and oxidizers of low molecular weight are preferred because, for a given energy release and mass flow, the lightweight molecules achieve greatest velocity and consequently highest specific impulse (I_{sp}) (Table 2).

Typical characteristics of a large, liquid-propellant rocket engine follow.

Propellant	Liquid oxygen – JP1
Chamber pressure, psia	900
Thrust, lb	200,000
Height, in.	120
Nozzle exit diameter, in.	60
Overall I_{sp}, sec (at 100,000 ft altitude)	284
Burn time, sec	200
Weight (not including fuel or fuel tanks), lb	2000

Solid-propellant rocket engine. The solid-propellant rocket engine is a propulsion system used

Table 2. Theoretical values for several fuel-oxidizer combinations (1000 psi combustion chamber pressure)

Combinations	I_{sp}, sec
Liquid oxygen – JP1*	286
Liquid oxygen – liquid hydrogen	388
Hydrogen peroxide – JP1	266
N_2O_4 – UDMH†	274

*JP1 is a kerosinelike hydrocarbon.
†Nitrogen tetroxide – unsymmetrical dimethylhydrazine.

in missiles and space vehicles. It can deliver large thrust and is instantly ready for operation.

This engine consists of a casing filled with a mixture of propellant chemicals in solid form. These burn, generating hot high-pressure gases which escape through nozzles as high-velocity jets, creating a forward-thrust reaction. To steer the vehicle, some nozzles are constructed so that they can be tilted or swiveled during firing.

A large engine, such as the first stage of a 1500-mi ballistic missile, might have the following general characteristics.

Propellant	Potassium perchlorate – polyurethene and additives
Length, in.	220
Diameter, in.	50
Gross weight, lb	20,000
Average thrust, lb	80,000
Burn time, sec	60
Overall I_{sp}, sec	230

Propellant combinations are selected to contain as much fuel and oxidizer and as little inert substance as possible. As with liquid propellants, combinations with low molecular weights are preferred because they result in greater specific impulse for a given energy release. The propellant mixture must burn at a relatively slow, uniform rate; certain inhibitors are added to control this.

There are now in use a variety of propellants of two general classes. Double-base types contain nitrocellulose and nitroglycerin plus additives for stability and for control of combustion rates. Composite types contain ammonium or potassium perchlorate granules embedded in a rubberlike hydrocarbon compound. *See* ELECTROMAGNETIC PROPULSION.   [J. W. BLANTON]

Bibliography: N. E. Borden, Jr., *Jet Engine Fundamentals*, 1967; J. V. Casmassa and R. D. Bent, *Jet Aircraft Power Systems*, 3d ed., 1965; J. W. Hesse and N. V. S. Mumford, Jr., *Jet Propulsion for Aerospace Applications*, 2d ed., 1964; C. W. Smith, *Aircraft Gas Turbines*, 1956; G. P. Sutton, *Rocket Propulsion Elements*, 2d ed., 1956; P. H. Wilkinson, *Aircraft Engines of the World*, rev. ed., 1967; M. J. Zucrow, *Aircraft and Missile Propulsion*, vol. 2, 1958.

Jet stream

A relatively narrow, fast-moving wind current flanked by more slowly moving currents. Jet streams are observed principally in the zone of prevailing westerlies above the lower troposphere and in most cases reach maximum intensity, with regard both to speed and to concentration, near the tropopause. At a given time, the position and intensity of the jet stream may significantly influence aircraft operations because of the great speed of the wind at the jet core and the rapid spatial variation of wind speed in its vicinity. Lying in the zone of maximum temperature contrast between cold air masses to the north and warm air masses to the south, the position of the jet stream on a given day usually coincides in part with the regions of greatest storminess in the lower troposphere, though portions of the jet stream occur over regions which are entirely devoid of cloud. *See* CLEAR-AIR TURBULENCE (CAT).

Characteristics. The specific characteristics of the jet stream depend upon whether the reference is to a single instantaneous flow pattern or to an averaged circulation pattern, such as one averaged with respect to time, or averaged with respect both to time and to longitude.

If the winter circulation pattern on the Northern Hemisphere is averaged with respect to both time and longitude, a westerly jet stream is found at an elevation of about 13 km near latitude (lat) 25°. The speed of the averaged wind at the jet core is about 148 km/hr (80 knots). In summer this jet is displaced poleward to a position near lat 42°. It is found at an elevation of about 12 km with a maximum speed of about 56 km/hr (30 knots). In both seasons a speed equal to one-half the peak value is found approximately 15° of latitude south, 20° of latitude north, and 5–10 km above and below the location of the jet core itself.

If the winter circulation is averaged only with respect to time, it is found that both the intensity and the latitude of the westerly jet stream vary from one sector of the Northern Hemisphere to another. The most intense portion, with a maximum speed of about 185 km/hr (100 knots), lies over the extreme western portion of the North Pacific Ocean at about lat 22°. Lesser maxima of about 157 km/hr (85 knots) are found at lat 35° over the east coast of North America, and at lat 21° over the eastern Sahara and over the Arabian Sea. In summer, maxima are found at lat 46° over the Great Lakes region, at lat 40° over the western Mediterranean Sea, and at lat 35° over the central North Pacific Ocean. Peak speeds in these regions range between 74 and 83 km/hr (40–45 knots). The degree of concentration of these jet streams, as measured by the distance from the core to the position at which the speed is one-half the core speed, is only slightly greater than the degree of concentration of the jet stream averaged with respect to time and longitude. At both seasons and at all longitudes the elevation of these jet streams varies between 11 and 14 km.

Variations. On individual days there is a considerable latitudinal variability of the jet stream, particularly in the western North American and western European sectors. It is principally for this reason that the time-averaged jet stream is not well defined in these regions. There is also a great day-to-day variability in the intensity of the jet stream throughout the hemisphere. On a given winter day, speeds in the jet core may exceed 370 km/hr (200 knots) for a distance of several hundred miles along the direction of the wind. Lateral wind shears in the direction normal to the jet stream frequently attain values as high as 100 knots/300 nautical miles (185 km/hr/556 km) to the right of the direction of the jet stream current and as high as 100 knots/100 nautical miles (185 km/hr/185 km) to the left. Vertical shears below and above the jet core are often as large as 20 knots/1000 ft (37 km/305 m). Daily jet streams are predominantly westerly, but northerly, southerly, and even easterly jet streams may occur in middle or high latitudes when ridges and troughs in the normal westerly current are particularly pronounced or when unusually intense cyclones and anticyclones occur at upper levels.

Insufficiency of data on the Southern Hemisphere precludes a detailed description of the jet stream, but it appears that the major characteristics resemble quite closely those of the jet stream on the Northern Hemisphere. The day-to-day variability of the jet stream, however, appears to be less on the Southern Hemisphere.

It appears that an intense jet stream occurs at high latitudes on both hemispheres in the winter stratosphere at elevations above 20 km. The data available, however, are insufficient to permit the precise location or detailed description of this phenomenon. *See* AIR MASS; AIR WAVES, UPPER SYNOPTIC; ATMOSPHERE; GEOSTROPHIC WIND; STORM; VORTEX.   [FREDERICK SANDERS]

Jet velocity

The velocity of the engine exhaust gases relative to the exhaust nozzle. In ideal air-breathing cycles, it is assumed that the exhaust gas mass rate equals the inlet air mass rate, the mass of fuel burned being neglected, and that the exhaust gases are expanded to ambient pressure in the nozzle. Under these conditions the thrust of an engine is directly proportional to the airflow rate and to the difference between the jet and vehicle velocities; thrust is greatest at zero flight speed and becomes zero when the flight speed is the same as the jet velocity.

Both engine power and efficiency are proportional to the product of thrust times flight velocity. Therefore, both power and efficiency are zero at zero flight speed and also zero when flight speed equals jet velocity. Between these limits there is a ratio of jet velocity to flight speed that will give maximum power and efficiency, and jet velocity becomes a factor in the design of engine exhaust nozzles. In subsonic turbojets, jet velocity can be controlled to some extent by varying the open area of the exhaust nozzle by any of several arrangements. The clamshell is an early form of variable nozzle. Iris or multiple-leaf designs are more recent. Internal plugs are also used to vary nozzle area. *See* TURBOJET.

In practice, on a turbojet with afterburner, a simple convergent variable-area exhaust nozzle is used more to control the turbine back pressure in the tail pipe than to optimize exhaust jet efficiency. In a turbojet without afterburner, a variable-area nozzle may be used to regulate internal engine conditions in the presence of such variations of flight conditions as extreme changes in altitude. On supersonic engines, fully variable convergent-divergent nozzles are used to optimize both internal engine conditions and jet efficiency. *See* AFTERBURNER.

In the non-air-breathing rocket cycle, the above assumptions do not apply; thrust of a rocket is independent of flight speed. Rather, at constant mass flow, thrust is directly proportional to nozzle exhaust velocity plus a pressure times area term. *See* PROPULSION; ROCKET ENGINE.

[ROBERT R. HIBBARD]

Jetty

An artificial barrier at river mouths and harbor entrances to deflect and regulate river flow and tidal currents, particularly in areas where littoral drift causes formation of bars across channel entrances. Jetties are built singly or in pairs, curved or straight, to prevent littoral drift across a channel and to increase velocity of tidal flow, which keeps the channel open. Jetties are constructed to suit natural conditions and often resemble breakwaters, although solid sheet-pile walls of timber, concrete, and steel have been used extensively. A fine example is the $4\frac{1}{2}$-mi jetty at the mouth of the Columbia River, Oregon. *See* COASTAL ENGINEERING; RIVER ENGINEERING.   [EDWARD J. QUIRIN]

Jewel bearing

A bearing used in quality timekeeping devices, gyros, and instruments, usually made of synthetic corundum (crystallized Al_2O_3) which is more commonly known as ruby or sapphire. The extensive use of such bearings in the design of precision devices is mainly due to the outstanding qualities of the material. Sapphire's extreme hardness $(1520-2200 \text{ kg/mm}^2)$ imparts to the bearing excellent wear resistance, as well as the ability to withstand heavy loads without deformation of shape or structure. The crystalline nature of sapphire lends itself to very fine polishing and this, combined with the excellent oil- and lubricant-retention ability of the surface, adds to the natural low-friction characteristics of the material. Sapphire is also nonmagnetic and oxidization-resistant, and has a very high melting point (3685°F). Ruby has the same properties as sapphire; the red coloration is due to the introduction of a small amount of chromium oxide. *See* BEARING, ANTIFRICTION; GEM; GYROSCOPE; WATCH.

Types. Jewel bearings, classified as either instrument or watch jewels, are also categorized according to their configuration or function. The ring jewel is the most common type. It is basically a journal bearing which supports a cylindrical pivot. The wall of the hole can be either left straight (bar hole) or can be imparted a slight curvature from end to end (olive hole). This last configuration is designed to reduce friction, compensate for slight misalignment, and help lubrication. A large variety of instrument and timing devices are fitted with such bearings, including missile and aircraft guidance systems. *See* GUIDANCE SYSTEMS.

Vee, or V, jewels are used in conjunction with a conical pivot, the bearing surface being a small radius located at the apex of a conical recess. This type of bearing is found primarily in electric measuring instruments.

Cup jewels have a highly polished concave recess mated to a rounded pivot or a steel ball. Typical are compass and electric-meter bearings.

End stone and cap jewels, combined with ring jewels, control the end play of the pivot and support axial thrust. They consist of a disk with highly polished flat or convex ends. Other relatively common jewel bearings are pallet stones and roller pins; both are part of the timekeeping device's escapement.

Dimensions. Minute dimensions are a characteristic of jewel bearings. A typical watch jewel may be .040 in. in diameter with a .004-in. hole, but these dimensions may go down to .015 and .002 in., respectively. Jewels with a diameter of more than 1/16 in. are considered large. It is usual for critical dimensions, such as hole diameter and roundness, to have a tolerance of .0001 in. or less. In some instances these tolerances may be as low as .0000020 in.

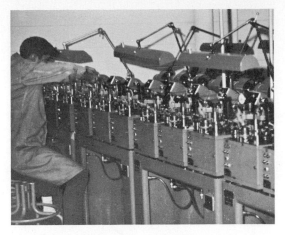

Fig. 1. Automatic cupping machines for the manufacture of jewel bearings. (*Bulova Watch Co.*)

Manufacturing. Because of its hardness sapphire can only be worked by diamond, which is consequently the main tool for the production of jewel bearings. Both natural and synthetic diamond are used, mostly under the form of slurry, broaches, and grinding wheels.

The machining of the blanks, small disks, or cylinders of varied diameter and thickness is the first step in the manufacturing process, and is common to most types of jewel bearings. The boules (pear-shaped crystals of synthetic ruby or sapphire) are first oriented according to the optical axis to ensure maximum hardness of the bearing working surface. They are then sliced, diced, and ground flat, and rounded by centerless grinding to the required blank dimensions. From this point on, the process varies considerably, according to the type of bearing.

Ring jewels are drilled with a steel or tungsten wire and coarse diamond slurry, or bored with a small grinding tool. The hole is then enlarged and sized by reciprocating wires of increasingly larger sizes through a string of jewels, until the required hole size is achieved. Fine diamond powder and a very slow rate of material removal permit the respect of strict tolerances and high-finish quality requirements. The jewels, supported by a wire

Fig. 2. Individual head of an automatic cupping machine. (*Bulova Watch Co.*)

Fig. 3. Technician shown operating an automatic cupping machine. (*Bulova Watch Co.*)

strung through the hole, are then ground in a special centerless-type grinding machine to the desired outside diameter dimension. After the cutting of a concave recess, which functions as an oil reservoir, the top and bottom of the bearing are polished and beveled by lapping and brushing. Finally, the "olive" configuration is obtained by oscillating a diamond charged wire through the hole of the rotating jewel. Between each operation the boiling of the jewels in a bath of sulfuric and nitric acid disposes of remaining slurries and other contaminating products.

The conical recess in vee jewels is first roughly shaped by a pyramidal diamond tool. The wall of the vee and the radius are then polished and blended with an agglomerated diamond broach, and a succession of brushing operations. Lapping of the top of the jewel and brushing a bevel around the upper outside edge conclude the process.

Most other types of jewel bearings, such as end stones, pallet stones, and roller pins, are shaped by a succession of grinding, lapping, and brushing operations.

A full line of automatic and semiautomatic high-precision equipment has been developed to handle and machine jewel bearings efficiently, permitting mass production and relatively low cost (Figs. 1–3). Traditionally, a large proportion of the labor involved is devoted to in-process and final quality control (Fig. 4).

Consumption and history. In 1964 the world production of jewel bearings was estimated at 2,-

Fig. 4. Quality-control section of William Langer Jewel Bearing Plant. (*Bulova Watch Co.*)

100,000,000 units each year, 50% of which were turned out in Switzerland and 26% in the Soviet Union. The remaining quantity was produced in other western European countries and Japan, and a lesser proportion in the United States. American consumption that year exceeded 89,100,000 units, an increase of about 40% over 1959.

It was around 1700 that the first handmade jewel bearings were incorporated in watch movements. With the advent of the mass production of watches in the middle of the 19th century, a considerable jewel bearing industry developed in Switzerland. The jewels were then made of natural gemstone, but the development of the Verneuil process for synthesis of corundum in the early 1900s gave further impetus to the industry, and jewel bearings started to be used in electric meters and other instruments. At that time and until the end of World War II, most of the world production was concentrated in Switzerland.

United States production. In the United States some efforts were made during World War I to establish a domestic industry, but these efforts were abandoned after the war. When Swiss imports were cut off in 1940, the United States faced a very serious and dangerous shortage of jewel bearings. The government then turned to the watch industry to develop and build specialized manufacturing equipment. By 1943 the watch companies had succeeded in producing the required quantities of instrument jewels necessary for defense, and they turned their activity to the production of watch jewels, still in short supply. However, at the end of the war jewel bearing production was halted.

New problems developed during the Korean War, and steps were taken to establish a domestic industry to take care of defense requirements. In 1952 a pilot plant was established in Rolla, N. Dak., to train and maintain a nucleus of skilled specialists. This plant, the William Langer Jewel Bearing Plant, operated for the Federal government by the Bulova Watch Co. under a contract with General Services Administration, is the only manufacturing facility in the United States with a high-precision jewel bearing manufacturing capability (Fig. 4). *See* MACHINING OPERATIONS.

[ROBERT M. SCHULTZ]

Jig, fixture, and die design

The planning of tools for use in making production parts. Jigs position parts for machining operations such as drilling, milling, reaming, and boring, and they physically guide the cutting tools. Fixtures position parts for machining, welding, or assembly operations, but permit cutting tools to find their own path. Dies are tools which, when mounted in a press, produce parts by punching and forming.

Jig and fixture construction. Fabricated steel, strain-relieved after welding and with hardened steel inserts at the wear points, has generally superseded castings in jig and fixture construction. Aluminum and magnesium are used for many applications because they are lighter to handle and less costly to fabricate and machine. When large parts are involved, such as those in automobile or aircraft industries, the epoxy resins, reinforced with fiberglass or other suitable materials, can be used in conjunction with hardened steel inserts and bushings. Kirksite and other low-melting-point alloys are used where difficult contours must be duplicated.

Locating methods. The accuracies obtained by jigs or fixtures are dependent upon the accuracy with which parts can be consistently positioned in the first and subsequent jigs or fixtures. The dimensional consistency of the part itself is a factor. One desirable practice is to have the first jig or fixture in a series of operations create locating points on the part in the form of milled surfaces or holes which can be used in subsequent operations. The same locating points should be used for as many operations as possible.

Clamping may deform the part if the locating points are not selected advantageously. Locating points should be spaced as far apart as the part design permits and be as small as possible considering the need for adequate support and anticipated wear. Three buttons or small pads determine a plane, and stops or pins determine locations within the plane (Fig. 1).

The use of two or more straight pins is not advisable for locating from holes because a slight dimensional variation between holes will make it difficult to load or unload the part. The most common method is to use one round pin and one diamond or elliptical pin (Fig. 2).

Self-centering devices must be used when it is necessary to locate centrally from two surfaces that can vary. When locating round parts, it is preferable to work from hardened V-shaped locators. The angle of the V should be 90°. Sight holes should be provided to permit the operator to observe visually whether or not the part is down on stops and buttons. Where practical, multiple station jigs and fixtures should be made in such a way that parts can be loaded while others are being machined. The cost of loading or unloading parts is often greater than the machining cost.

Clamping. The clamping arrangement should be simple. Clamping pressure is applied over or between the pads or buttons, and standard vises are adapted as machining fixtures where practical.

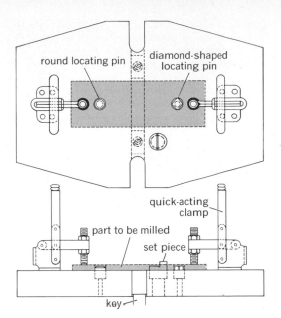

Fig. 2.   Round pin paired with diamond pin locates part and allows it to be loaded readily in milling fixture.

Machining is done toward the fixed jaw. Standard air cylinders are used. If more than three supports in one plane are necessary, all in excess of three should be adjustable. Where a wrench must be used, all nuts should be standardized. Large hand knobs ensure ease of operation, and cam-type clamps are usually faster and more positive than other methods.

Tool guides. In jigs, tool guides consist of bushings that guide drills, reamers, and boring bars. American Standards Association sizes should be used when possible. Bushings normally extend as close to the part as practical to assure accuracy of hole locations and to eliminate chip tangles within the jig. Use of slip-removable bushings in hardened liners permits secondary operations such as tapping, reaming, and counterboring. Where accuracy demands, reamers may also be guided in slip bushings either by pilots or by their cutting edges.

Tool guides in fixtures are limited to set pieces for establishing a relationship between the cutting tools and the fixture. These are permanently located on the fixture and clearance is allowed between the set pieces and the cutting tools. A feeler gage, the thickness of the clearance, is then used to locate and set the cutting tools.

Chip control. During drilling, chips are either brought out through the bushing by allowing zero clearance between the bushing and the part, or permitted to stay in the jig by allowing sufficient clearance for that purpose. The former method is the more desirable, but if tolerances are not critical and the chips are discontinuous, wear on the bushings can be reduced by the latter method. To compromise between zero and ample clearance creates the disadvantages of both with none of their advantages.

Jigs and fixtures are designed with sufficient openings for chips to be removed by gravity, coolant flow, brush, or air blast. Dust collectors are used to collect dry grinding chips or chips of nonmetallic or toxic materials.

Safety. Practice dictates the use of no. 1/2 to 13 or larger tapped holes for eyebolts in all jigs or

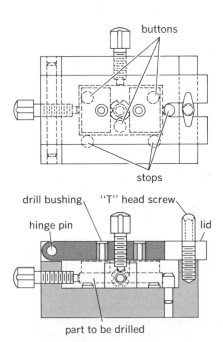

Fig. 1.   Buttons and stops locate part in drill jig.

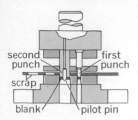

second punch   first punch
scrap
blank   pilot pin

Fig. 3. Progressive die performs several operations sequentially; material advances from first to second punch between strokes.

fixtures weighing 50 lb or more and the removal of all sharp corners that could injure operators.

Stamping dies. Dies produce parts either by stressing the part material in shear until fracture occurs, or by stressing the part material beyond the yield point where it takes a permanent set but not beyond the rupture point where it breaks. Dies using the former method include blanking, piercing, cutoff, and lancing dies, and those employing the latter method include drawing, bending, embossing, and forming dies. Any practical combination of the above or other operations can be included in one die and can be performed either simultaneously or in steps using a progressive-type die (Fig. 3).

Die construction. Conventional stamping dies are composed of the following main parts: the die set, consisting of top and bottom shoes, guide pins, and bushings; the punch, which is the male cutting

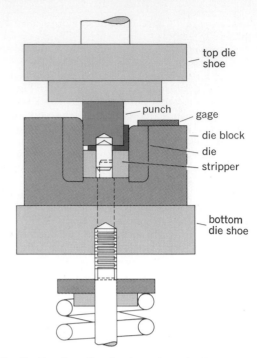

Fig. 5. Bending die, showing adequate clearance between hole and bend.

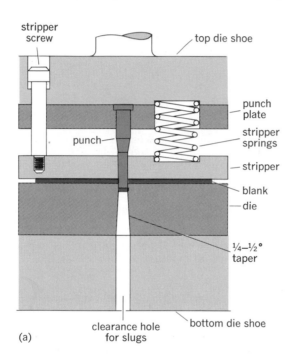

(a)

or forming member; the die, which is the female cutting or forming member; strippers that remove the part from the punch or die; gages to locate the part or part material; and blank holders that remove wrinkles and cause an even flow of material in drawing (Fig. 4).

Die selection. The die type selected for a job is determined by part design and economic considerations, such as the number of pieces to be produced, cost of alternate methods of production, and maintenance of the tool itself.

Very low activities or liberal part tolerances often justify special types of dies, such as steel rule dies or plow steel dies for cutting operations and plastic or rubber dies for forming operations. Most applications involving moderate to high activities require dies employing hardened tool steels.

Large activities can sometimes justify the high cost of dies made from tungsten carbide. Well-designed and well-built tungsten carbide dies can turn out a large volume of extremely accurate parts at high speeds with little sharpening or repair.

Materials for stampings. Any material softer than the punch and die and which will not shatter with impact may be cut, pierced, or blanked in a die. In drawing, bending, and forming operations, ductility is the prime requisite. Springback and deformations, including changes in material thickness occurring during the forming operation, are taken into consideration. Material is used either in individually sheared pieces or in strip or coil form. Strip or coiled material makes possible the use of various feeding mechanisms with resultant labor economies not usually practical with sheared material unless a fully automated setup is provided.

Stamping tolerances. Obtainable accuracies are broad in range, depending on the part material, thickness of stock being punched, sizes and part

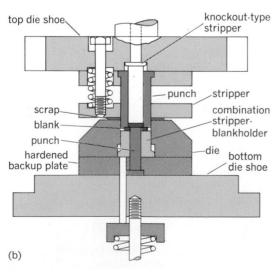

(b)

Fig. 4. Main parts of conventional stamping dies. (a) For piercing. (b) For blanking and drawing.

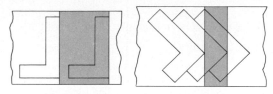

Fig. 6. Blank layout compares the amount of material which is used to produce one blank (shaded portions) by two different die positions.

configuration, and the proximity of bends or holes. Accuracies of ±0.0005 in. are obtainable on thin materials using compound-type dies; however, under normal operating conditions, practical tolerances on light material can be expected to vary from ±0.002 to ±0.010 in. and on heavy material from 0.010 to 0.035 in. To avoid distortion, holes are placed a distance no less than the material thickness from the edge of another hole and no less than three times the material thickness from a bend radius. Hole locations established previous to bending operations are subject to material thickness tolerances and the stretch of the material which occurs during the bending operations (Fig. 5).

General design. A tool lineup is made to determine the sequence of operations and the type and number of dies required to complete the part or blank. A blank layout is made to obtain optimum material usage and to reduce scrap (Fig. 6), and the material is checked to make sure the burr is on an acceptable side of the blank. Material is normally fed into the die from right to left or front to back. When tolerances permit, the finished edge of the stock can be used as a side of the blank.

Tonnage to shear or form is computed and the die is designed for a press that will handle this capacity. It is generally possible to improve die life by using a press rated for twice the required capacity. Work can be performed on presses of less than the computed capacity by staggering the lengths of punches or by adding shear to the punch or die. Safety requires that tapped holes be provided by eyebolts in dies 50 lb or more in weight and that safety bolts be provided to hold the die together when not in use. Guards and undercuts are applied to dies to protect the hands of the operator. Standard punches and die inserts for round, oblong, square, or rectangular holes can be purchased in small increments of size and are readily available for replacement when breakage occurs. *See* MACHINING OPERATIONS; METAL FORMING; SHEET-METAL FORMING.

[R. L. CAMMACK]

Bibliography: H. E. Grant, *Jigs and Fixtures*, 1967; F. W. Wilson (ed.), *ASTME Die Design Handbook*, American Society of Tool and Manufacturing Engineers, 1965; F. W. Wilson (ed.), *ASTME Handbook of Fixture Design*, 1962; F. W. Wilson (ed.), *ASTME Tool Engineer's Handbook*, 2d ed., 1959.

Johne's disease

A specific enteritis, or inflammation of the gastrointestinal tract, of cattle, sheep, and deer, caused by the bacterium *Mycobacterium paratuberculosis*. The disease is spread by animals having eaten the droppings of sick animals, and is diagnosed by the clinical symptoms, the presence of *M. paratuberculosis* in the droppings, and by a positive skin reaction to johnin, a preparation similar to tuberculin. *M. paratuberculosis* is a short acid-fast rod, which can be cultured only in the presence of an iron chelator, mycobactin, found in hot alcohol extracts of mycobacteria cultivated on media containing minimal iron. The disease has an incubation period of a year or more. Sick animals have intermittent diarrhea without fever and gradually become emaciated. The mucosa of the small intestine shows gross thickening and the mesenteric lymph nodes enlarge without ulceration. Caseation (death of tissue with change to a cheeselike consistency) and calcification do not occur. The drugs streptomycin, viomycin, 4,4′-diaminosulfone, and isonicotinic acid hydrazide, although used successfully in the treatment of tuberculosis, have been used without success in the treatment of Johne's disease. Animals that have Johne's disease must be destroyed because they rarely recover and present a focus of infection for other animals in the herd. *See* TUBERCULOSIS.

[GARDNER MIDDLEBROOK]

Joint (anatomy)

The contact surface between two individual bones, also known as an articulation, whose detailed structure depends largely upon the movement between the bones. Articulations can be grouped according to the intervening tissues: diarthrosis, when cartilaginous articular pads are present and an articular cavity with a capsule separates the two bones; and synarthrosis, when the connecting material is continuous. The synarthrosis can be divided into three types: synchondrosis (cartilage between the bones), syndesmosis (collagen fibers between the bones), and synostosis (bone is continuous). Articulations may be freely movable, of limited movement (amphiarthrosis), or rigid (some sutures). Bones that are completely fused to one another are ankylosed, in which the articulation is obliterated.

The type of motion permitted by various joints may be restricted to one kind of movement or may be a combination of movements depending largely upon the configuration of the joint surfaces, intervening tissues, and ligaments associated with the articulation; in many joints, the ligaments are the major or sole limitations to the movements of bones. Some individuals possess a wider range of motion, particularly if the joints are malformed or diseased or if the ligaments vary in length or position. Moreover the ligaments provide most of the strength against tensile forces acting on the joint. *See* LIGAMENT (VERTEBRATE); MUSCULAR SYSTEM (VERTEBRATE); SKELETAL SYSTEM (VERTEBRATE).

Synarthrosis. Although most synarthroses are of limited motion (amphiarthrosis), some permit a wide range of movement, such as in the joint between the avian mandible and base of the skull. The bony surfaces in synarthroses are separated by small amounts of fibrous tissue or cartilage, but in some cases a larger gap exists between bones. When the gap is very narrow, the bones may be immovable, such as in sutures of the adult skull. Synchondroses may serve as a special type of developmental articulation, for example, those seen in children before the cartilage is converted into

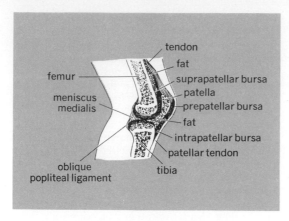

Lateral view of knee articulation, a diarthrosis. (*From N. L. Hoerr and A. Osol, eds., Blakiston's New Gould Medical Dictionary, 2d ed., Blakiston, 1956*)

bony tissue of the adult structures. Many other special forms of synarthroses occur, such as gomphosis, the insertion of teeth into the bony sockets (alveoli of the jaws). The existence of continuous bone between two structural bones does not imply a rigid joint. In some cases, as in the avian nasal-frontal hinge between the upper mandible and brain case, a thin continuous sheet of bone connects the two elements; the thin bone is flexible, just as is a thin band of steel.

Amphiarthrosis. In these slightly movable joints, the opposing bony surfaces may be connected by a continuous mass of tissue or may be separated by an articular cavity. Intervertebral articulations are diarthroses with a flattened fibrocartilaginous disk present within the articular cavity, and are amphiarthroses of very limited movement. The ligamentous attachment between the tibia and the fibula at the ankle or between the two mandibular rami in some mammals are examples of amphiarthroses that are synarthroses.

Diarthrosis. These articulations are characterized by a synovial cavity between the bones; the cavity is formed by a fibrous synovial capsule connecting the bones. Moreover, the ends of each bone are covered by cartilage, usually hyaline, but fibrocartilage is found in those articulations subject to large compression forces. The fluid within the synovial cavity acts as a lubricant. An articular disk, fatty pads, and other special structures may be present between the two principal surfaces to facilitate movement.

Diarthroses are named according to the type of movement permitted by them. The hinge joint, or ginglymus, permits motion only in one plane, such as in the finger, knee, and elbow articulations (see illustration). The pivot joint, or trochoid, allows a rotation of one bone around another. The rotation of the atlas about the axis, or second cervical vertebra, is typical. The ball-and-socket joint, or enarthrosis, is characteristic of the hip and shoulder articulations, in which a wide range of motion on all planes is possible. Gliding joints, or arthrodia, permit only restricted motion between a concave and convex surface, as seen in some of the wrist and ankle articulations. *See* ANKLE (HUMAN); WRIST (HUMAN).

Condyloid. Condyloid articulations are formed by an ovoid surface that fits into an elliptical cavity

so that all movement except rotation is permitted. The wrist joint is an example. [WALTER BOCK]

Joint (geology)

A fracture that traverses a rock and does not show any discernible displacement of one side of the fracture relative to the other. The term joint refers primarily to the actual fracture as represented by a fine line or trace marking the intersection of the fracture and rock surfaces. Commonly, however, the joint is represented superficially by a cleft or fissure resulting from weathering, mechanical separation, or by one face of the fracture on an outcrop (Fig. 1).

Varieties of joints. Joints fall into two major classes based on their spatial relations, geometry, and surface structures: systematic and nonsystematic joints.

Systematic joints. Systematic joints are arranged in groups of regularly spaced parallel or subparallel planar fractures called joint sets. Two or more joint sets comprise a joint system. Two joint sets believed related genetically constitute a conjugate joint set or, if more than two, a conjugate joint system.

Specific terms applied to individual systematic joints are both descriptive and genetic. Descriptive terms are based on the geometric relations between joints and other structures. In sedimentary rocks, joints that parallel the strike or dip of strata are called strike and dip joints, respectively. Joints having an intermediate angular relation are called diagonal or oblique joints (Fig. 2). Where one set of joints appears dominant, the joints may be called primary joints, and those of less prominent sets then are termed secondary. Particularly large or prominent joints of any set are called major or master joints.

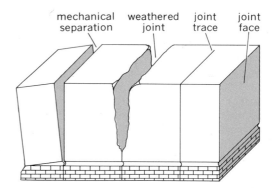

Fig. 1. Aspects of a systematic joint.

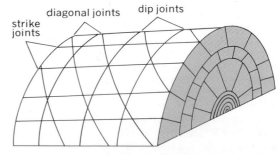

Fig. 2. Diagram showing geometric relations between systematic joints and folded or tilted strata.

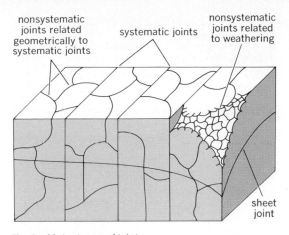

Fig. 3. Major types of joints.

In igneous and metamorphic rocks, systematic joints are described on the basis of their geometric relations with linear or planar structures within the rock. Joints lying perpendicular to lineation are called cross or Q joints and, where parallel with lineation, longitudinal or S joints. Those with intermediate angular relations are diagonal joints.

Genetic terms denote the geometric relations between joints and forces that have deformed the host rock. Where the acute angle of intersection between two sets of joints is bisected by the direction of the maximum principal stress, as inferred from other structural features such as folds or faults, the joints are called shear joints and, where parallel or normal to this stress direction, tension joints. Joints produced directly by bending of strata into folds are termed release joints where parallel to the fold axis, and extension joints where normal to the axis. Joints that originate as a direct result of deformation may be called first-order joints. Joints that result from a readjustment of stresses following first-order jointing, faulting, or folding may be called second-order joints.

Nonsystematic joints. Nonsystematic joints fall into two classes: those related geometrically to systematic joints, and those related to weathering of rock surfaces.

Joints of the first class are highly curved or roughly planar fractures that extend across the interval between systematic joints. They probably result from adjustment of the rock between systematic joints to local tectonic stresses. Joints of the second class commonly are of shallow depth, occur normal to the weathered rock surface, and show random curvilinear patterns (Fig. 3).

Sheeting is a variety of nonsystematic joint that lies roughly parallel to the weathered rock surface. Spacing between these joints is irregular and increases with depth. They are generally believed to result from near-surface rock expansion due to removal of overburden or to chemical weathering. They may possibly result also from stresses related to residual strain energy remaining in the rock from previous deformations. Similar flat-lying joints occur in igneous intrusions and are believed to result from a decrease in magmatic pressure as the magma solidifies, or from removal of overburden.

Size and spacing. Joints of all types have a very great range in size. The surface area of individual fractures ranges from a few square inches to several thousand square feet. Despite this range in size, all classes of joints are entirely similar in their other morphologic aspects. The spacing of joints tends to be more close in thin-layered rocks than in thick-layered ones. Spacing between systematic joints ranges from a few inches to as much as 50 ft.

Columnar or prismatic joints. Columnar joints occur in sheetlike or pluglike intrusive and extrusive igneous rock bodies. They also may be found on occasion in sedimentary rocks that have been heated adjacent to igneous intrusions. Individual columns ordinarily are five- or six-sided but may have from three to eight sides in some examples. The columns range from a few inches to several feet in diameter and from a few feet to many tens of feet in length. Dish-shaped cross joints (cup-and-ball joints) divide the columns into segments.

Columnar joints invariably are oriented normal to cooling surfaces and result from the contraction of the igneous rock on cooling (Fig. 4). The columnar structure is what may be expected if the strain energy generated by contraction of the rock is to be dissipated by the least amount of work.

Plumose structure. Joint surfaces commonly show a patterned structure composed of ridges of low relief. These plumose ridges converge at a point or straight axis near the center of the joint face on systematic joints. Curvilinear ridges of

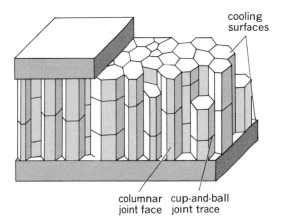

Fig. 4. Columnar jointing.

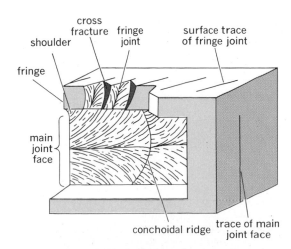

Fig. 5. Schematic block diagram showing primary surface structures of a systematic joint.

Various theories of jointing

Theory	Origin
Systematic joints	
Cleavage	Genetically similar to cleavage in minerals (theory no longer maintained)
Magnetic forces	Genetically related to and controlled by Earth's magnetic field (theory no longer maintained)
Torsion	Result of local and regional warping of Earth's crust
Earthquake	Seismic shock produces and controls direction of jointing
Torsion-earthquake	Rock warped to breaking point; fracturing triggered by earthquakes
Tension	Result of contraction of sediments due to compaction or loss of water or both
Tension	Result of local or regional (or both) vertical forces involved in uplift or folding of Earth's crust
Tension and shear	Result of local or regional (or both) tangential forces involved in uplift or folding of Earth's crust
Tidal	Result of cyclic tidal forces acting tangential to Earth's crust; forces control direction of jointing
Tidal fatigue	Result of rock fatigue engendered by cyclic semidiurnal tidal forces; direction of jointing inherited from preexisting fracture pattern
Residual stresses	Residual rock stresses modified during uplift in such a manner as to produce shear and tension joints
Nonsystematic joints	
Tension	Resulting from contraction of sediments due to compaction or loss of water or both
Tension	Resulting from contraction of igneous rock upon cooling
Compression-tension	Resulting directly or indirectly from removal of overburden or from surface weathering of rock

proportionally greater relief and disposed concentrically about the central axis or point occur on some systematic joint faces (Fig. 5). These are referred to as conchoidal ridges. Irregular, curvilinear plumes characterize the faces of nonsystematic joints. Plumose structures suggest that joints are tensional brittle fractures that originate at some small structural inhomogeneity within the rock and are propagated outward from that point, until the causal forces are dissipated.

Occurrence. Joints are the most abundant structures in the Earth's crust and are present everywhere. They constitute structural inhomogeneities in rock bodies and as such are influential in determining details of form in topographic relief. Locally, systematic joints determine the azimuth of topographic features such as hills, valleys, stream courses, and cliff faces and underground features

such as cave passages. Joints also serve as loci for limonite, quartz, calcite, or other vein and ore minerals deposited from solution.

Origin. At present no one theory for the origin of systematic joints is accepted universally. The majority of theories are based on analyses of systematic joint patterns and their relations to other structures, such as folds and faults. The interpretations of these relations is based primarily on theoretical and experimental work on rock fracture and deformation.

Most geologists believe systematic joints are related genetically to folding, faulting, and uplift and attribute them to tangential compressional or tensional forces acting regionally or locally in the Earth's crust. A minority holds that joints are produced by some universal long-acting force, such as earth tides, and are not directly the result of local folding or faulting. The several major theories of jointing are summarized in the table. *See* FAULT AND FAULT STRUCTURES; FOLD AND FOLD SYSTEMS; STRUCTURAL PETROLOGY; TECTONOPHYSICS.   [ROBERT A. HODGSON]

Bibliography: M. P. Billings, *Structural Geology*, 1954; L. U. De Sitter, *Structural Geology*, 2d ed., 1964; E. S. Hills, *Elements of Structural Geology*, 1963; F. H. Lahee, *Field Geology*, 6th ed., 1961; C. R. Longwell and R. F. Flint, *Introduction to Physical Geology*, 2d ed., 1962; N. J. Price, *Fault and Joint Development in Brittle and Semi-Brittle Rock*, 1966.

Joint (structures)

The surface at which two or more mechanical or structural components are united. Whenever parts of a machine or structure are brought together and fastened into position, a joint is formed. *See* STRUCTURAL CONNECTIONS.

Mechanical joints can be fabricated by a great variety of methods, but all can be classified into two general types, temporary (screw, snap, or clamp, for example), and permanent (brazed, welded, or riveted, for example). The following list includes many of the more common methods of forming joints.

1. Screw threads. Bolt and nut, machine screw and nut, machine screw into tapped hole, threaded parts (rod, pipe), self-tapping screw, lockscrew, studs with nuts, threaded inserts, coiled-wire inserts, drive screws. *See* BOLT; BOLTED JOINT; NUT, MECHANICAL; SCREW; SCREW FASTENER; WASHER.

2. Rivets. Solid, hollow, explosive and other blind side types. *See* RIVET; RIVETED JOINT.

3. Welding. *See* WELDED JOINT; WELDING AND CUTTING OF METALS.

4. Soldering. *See* SOLDERING.

5. Brazing. *See* BRAZING.

6. Adhesive. *See* ADHESIVE; ASPHALT AND ASPHALTITE; EPOXY RESIN; GLUE; MUCILAGE.

7. Friction-held. Nails, dowels, pins, clamps, clips, keys, shrink and press fits.

8. Interlocking. Twisted tabs, snap ring, twisted wire, crimp.

9. Other. Peening, staking, wiring, stapling, retaining rings, magnetic. Also, pipe joints are made with screw threads, couplings, caulking, and by welding or brazing; masonry joints are made with cement mortar.   [WILLIAM H. CROUSE]

Joint (anatomy) disorders

Local or generalized conditions that affect the joints and related tissues. About one person in 25 has some form of joint disturbance. The most common are forms of arthritis, or inflammation of joint structures. These may follow specific infections, injury, generalized disease states, or degenerative changes which largely parallel the aging processes in the body.

The arthritis of rheumatic fever is fairly characteristic in that joint involvement is usually migratory and coincides with the clinical state. Although the exact cause is unknown, a form of hypersensitivity to a prior streptococcal infection is apparently contributory. *See* RHEUMATIC FEVER.

Rheumatoid arthritis, one of the most common diseases which involve joints, is also of unknown etiology, but again hypersensitivity is related, as well as other factors. It occurs most often in young or middle-aged adults and has a marked tendency to undergo spontaneous remission. Multiple joints are involved, especially those of the small bones in the hands, feet, wrists, or jaw. In many affected persons there is a close correlation between symptoms and emotional state.

Other so-called collagen diseases such as lupus erythematosus and periarteritis nodosa frequently are accompanied by arthritis much like the rheumatoid variety. *See* LUPUS ERYTHEMATOSUS.

Degenerative joint disease is most common in older people. The major form, osteoarthritis, is marked by deterioration of the articular surfaces of the weight-bearing joints of the knees, hips, and spine. The ends of the fingers are often involved.

Arthritis may be produced by trauma sustained from a single tissue-damaging injury or from repeated physical insults to a particular joint. *See* ARTHRITIS.

Infectious arthritis may be caused by gonorrhea, or invasion by pneumococci, streptococci, and other pathogenic organisms. Special forms, more chronic in nature, are seen in tuberculosis and syphilis. *See* GONORRHEA; PNEUMOCOCCUS; STREPTOCOCCUS; SYPHILIS.

Gout is an example of arthritis caused by a metabolic disorder. In this instance there is an abnormality of purine breakdown that produces accumulations of uric acid and urates in joint tissues. The physical findings and clinical course are usually distinctive. *See* GOUT.

Joint disorders may be found in certain nervous system diseases, particularly those which affect the spinal cord. Syringomyelia, syphilis (tabes dorsalis), and leprosy are examples in which destructive joint lesions may be secondary to nerve damage. *See* LEPROSY.

Many systemic diseases of unknown or quite different etiology may produce some form of arthritis or joint degeneration. Hemophilia, a hereditary blood disorder; acromegaly, an endocrine dysfunction; and Raynaud's disease, a disease of blood vessels, are examples which indicate the diversity.

Local disturbances, either specific or nonspecific, may also lead to joint involvement. These include bursitis, tenosynovitis, formation of cysts in the synovium, and other conditions.

Tumors of joint tissues are usually benign chondromas, fibromas, and lipomas, but the malignant synoviosarcoma is not rare.

Rheumatism is a nonspecific, predominantly lay term which includes local pain and tenderness. Most causes are chronic inflammations or mildly progressive degenerations, and many, when investigated, fall into one of the previously mentioned categories. *See* RHEUMATISM.

Other relatively rare forms of arthropathy are occasionally encountered. *See* HYPERSENSITIVITY; IMMUNOPATHOLOGY; TUBERCULOSIS; URIC ACID.

[EDWARD G. STUART/N. KARLE MOTTET]

Josephson effect

Tunnel passage of electron pairs through a thin insulating barrier (on the order of 15 A, or 1.5 nm, thick) between two superconducting metals (Fig. 1). The tunnel process itself is a well-known wave-penetration phenomenon which is basic to cold electron emission of metals, radioactivity, and so on. In the ac Josephson effect, alternating currents may occur within the barrier when an external steady voltage is applied to the system. In the dc Josephson effect, a steady current may flow across the barrier when a steady magnetic field is applied. Useful applications and knowledge have resulted from study of the effect, such as: (1) devices for generation and detection of extremely short-wavelength electromagnetic radiation; (2) detectors of very small magnetic fields; (3) a new accuracy of measurement of the ratio of constants e/h, where e is the electron charge and h is Planck's constant; (4) understanding of superconductivity in terms of long-range quantum-wave phase coherence. It is extremely difficult to give a simple explanation of B. Josephson's prediction; some scientists believe that the effect defies intuitive physical interpretation. Josephson was given the Nobel prize for its discovery. *See* FIELD EMISSION; QUANTUM THEORY, NONRELATIVISTIC; RADIOACTIVITY.

Electrons in superconductors. When the group of metals known as superconductors (lead, tin, mercury, and so on) are studied from the point of view of understanding what happens as the temperature is raised (or lowered) through the transition temperature T_c, it is found that there is no appreciable change of the volume of the metal and no change of the atomic ordering on the lattice sites. Nevertheless, the specific heat of the metal specimen undergoes a considerable enhancement as the temperature is raised through T_c; this means that some form of transition has occurred in which the degree of order among the electrons has been changed to disorder (just as the melting of a snow flake increases the disorder among its molecules). Below T_c, the electron order gives the metal new electric and magnetic properties. There is an entropy increase in passing up through T_c; and M. Tinkham has shown that electrons require extra

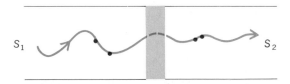

Fig. 1. Josephson junction showing two superconductors S_1 and S_2 separated by a thin insulator.

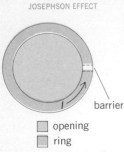

☐ opening
☐ ring

Fig. 2. Superconducting ring with circulating current I.

energy (about 10^{-8} eV/atom) to jump an energy gap ϵ in order to reach the normal conducting state. *See* ENTROPY.

Fritz London sought to give superconducting electrons a long-range order with respect to their local mean value of momentum. This long-range order would be due to the wide extension in the crystal of the wave representing the electron. This special wave would be somewhat rigid and not be influenced (bent about) by small magnetic fields. London could thus make plausible the enormous diamagnetism of superconductors by showing that a steady external magnetic field causes supercurrents (identical to undamped eddy currents) to flow such that the magnetic induction **B** inside the bulk metal is equal to zero (a phenomenon known as the Meissner effect). The current is made up of electrons moving in pairs but separated by as much as 10^{-5} cm. *See* FREE-ELECTRON THEORY OF METALS; MEISSNER EFFECT; SUPERCONDUCTIVITY.

Theory of the effect. The open-ended geometry of Fig. 1 permits the use of a battery as a steady voltage source. Josephson predicted that such a voltage would produce an alternating current across the thin insulator barrier as in Eq. (1).

$$\nu = 2eV/h \qquad (1)$$

Here ν is current frequency, e the electron charge, V the voltage, and h Planck's constant. The ratio $2e/h$ is numerically equal to 483.6 MHz/μV. This is a unique ac generator. F. Bloch showed that the Josephson effect is the direct result of fundamental facts of electrodynamics and of quantum mechanics applied to superconductors. Only for details does one need to use the approximations inherent to the superconductivity theories of J. Bardeen, L. N. Cooper, and J. R. Schreiffer (or V. L. Ginzburg and L. D. Landau) or to consider the restriction to current flow caused by the barrier. Bloch considered a superconduct-

ing ring (Fig. 2) with a very thin barrier and a circulating current I. The barrier may restrict the magnitude of the current, but those electrons which tunnel through may retain their special wave character of the superconducting state—that is, a phase coherence.

By use of a time-dependent magnetic flux ϕ through the opening O of the ring, a voltage may be induced, and the time dependence of the supercurrent I with periodic variation having a frequency ν can be deduced from fundamental physical principles. An outline of Bloch's proof follows. The closed geometry of Fig. 2 requires the superconducting wave function ψ for the electron within, R, to be simply multiplied by a phase factor exp$[-2\pi i e\phi/h]$ when the particle is brought around the ring full-circle. Here all quantities are expressed in SI (mksa) units, and $i = \sqrt{-1}$ for the periodic exponential function. The phase factor repeats itself whenever the external magnetic flux increases by the amount $\Delta\phi = h/e$. The same repetition occurs in the set of energy levels E_n obtained from the solution of the Schrodinger wave equation, $\mathscr{H}\psi = E\psi$, where the Hamiltonian operator $\mathscr{H}$ includes all electron momenta, effects of the magnetic vector potential, and the perturbations of the lattice ion vibrations on the electrons. These electron energy levels E_n determine the partition function of statistical mechanics, Q, from which the free energy, $F = -kT \ln Q$, may be written. Thus the free energy also must be a periodic function of the external flux ϕ with period h/e. The free energy may therefore be written as a Fourier series, Eq. (2). The Fourier coefficients F_n have a magnitude

$$F = \sum_{n=0}^{\infty} F_n \cos 2\pi n \left[\frac{\phi}{h/e}\right] \qquad (2)$$

determined by details inherent in the microscopic approach to the theory of superconductivity, and indeed only even integer values, $n = 2$, 4, etc., may be allowed. *See* FOURIER SERIES AND INTEGRALS; FREE ENERGY; STATISTICAL MECHANICS.

The current is obtained from the free energy by the relation $I = dF/d\phi$, which gives Eq. (3), with

$$I = \sum_{n=1}^{\infty} I_n \sin 2\pi n \left[\frac{\phi}{h/e}\right] \qquad (3)$$

$I_n = 2\pi n e F_n/h$. So long as ϕ is finite and constant, there is a set of finite coefficients I_n and a circulating direct current exists. The application of a constant voltage V corresponds to the linear time dependence $\phi = -Vt$ because the voltage is the line integral of the electric field, and from Maxwell's equations of electrodynamics $V = d\phi/dt$. Thus, Eq. (4) gives the current in this case. The allowed current has a periodic variation with a spectrum of

$$I = -\sum_{n=1}^{\infty} I_n \sin(2\pi n e V/h) t \qquad (4)$$

rent has a periodic variation with a spectrum of frequencies $\nu_n = neV/h$, which is the Josephson result, Eq. (1), in general form. *See* FARADAY'S LAW OF INDUCTION; MAGNETIC FLUX; MAXWELL'S EQUATIONS.

Experimental results. Measurements by W. H. Parker and coworkers of the frequency and voltage (together with accepted values of e/h) show that the integer $n = 2$. The particular choice of $n =$

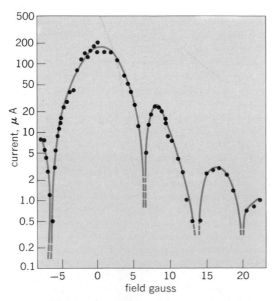

Fig. 3. The field dependence of the Josephson current in a Pb-*I*-Pb junction at a temperature of 1.3°K. The insulator *I* is lead oxide. (*From J. M. Rowell, Magnetic field dependence of Josephson tunnel currents, Phys. Rev. Lett., 11:200–206, 1963*)

2 for the Josephson relation requires further considerations arising from details of the interaction of superconducting electron pairs with the lattice vibrations. *See* LATTICE VIBRATIONS.

Further experimental results explained by the theory are shown in Fig. 3. J. M. Rowell applied a very small external current source to a junction like that in Fig. 1 and measured the variation of an observed direct current as a function of the steady magnetic field. In this case there is no voltage drop between the superconductors, and this is the dc Josephson effect modulated by the external magnetic flux as demanded by Eq. (3) with the integer $n = 2$. The current is reduced to a minimal value whenever the junction contains integral numbers of flux units, $h/2e = 2.068 \times 10^{-15}$ weber. (In cgs gaussian units, the flux unit is $hc/e = 2.068 \times 10^{-7}$ gauss cm².) *See* QUANTIZED VORTICES AND MAGNETIC FLUX.

As the dc Josephson current is increased beyond a critical current I_c, a finite voltage appears across the junction, and the supercurrent oscillates as described above. H. H. Zappe has indicated that the current I_c varies by one order of magnitude for a variation in the thickness of the insulating oxide barrier of only 3.5 Angstrom units (0.35 nm). I. Giaever received the Nobel prize, together with Josephson, for early experiments (see Fig. 4) on current-voltage characteristics of electron tunneling through a thin insulating film between two metals with (*a*) both metals in the normal state, (*b*) one metal in the normal state, the other in the superconducting state, and (*c*) both metals in the superconducting state; ϵ_1 and ϵ_2 are one-half the respective energy gaps of the two metals.

Applications. All scientific evidence indicates that Eq. (1) holds *exactly* and therefore, since frequency can be easily measured with an accuracy of 1 part per million, the Josephson effect provides a very useful technique for maintaining and comparing standards of emf (electromotive force). The Josephson effect has played a role in adjustment of the value of fundamental constants with important consequences for the fine structure constant. Josephson tunnel junctions are considered potential high-speed computer elements. Finally, the alternating-current frequency and electron wavelength in the junction can be adjusted by voltage and by an applied magnetic field, respectively, so that frequency and wavelength are matched to the junction mode considered as an electromagnetic resonant cavity. Radiation may then be emitted from the cavity into a waveguide, and it has been found to be coherent (all in phase) and highly monochromatic, according to Parker, D. N. Langenberg, A. Denenstein, and B. N. Taylor. The region of the electromagnetic spectrum between microwaves and infrared waves has thus been made available. Power transfer techniques may need to be improved for the demands of future applications. *See* ATOMIC CONSTANTS; ELECTRICAL UNITS AND STANDARDS; SUPERCONDUCTING DEVICES.

[CHARLES F. SQUIRE]

Bibliography: F. Bloch, Josephson effect in a superconducting ring, *Phys. Rev.*, B, 2:109–121, 1970; B. D. Josephson, *Superconductivity*, edited by R. D. Parks, 1969; D. N. Langenberg, D. J. Scalapino, and B. N. Taylor, The Josephson effects, *Sci. Amer.*, 214(5):30–39, 1966; F. London, *Superfluids*, 1950; W. H. Parker et al., Radiation from Josephson junctions, *Phys. Rev.*, 177:639–648, 1969; J. M. Rowell, Magnetic field dependence of Josephson tunnel currents, *Phys. Rev. Lett.*, 11: 200–206, 1963; M. Tinkham, *Introduction to Superconductivity*, 1975; H. H. Zappe, Josephson tunnel currents in computer circuits, *Phys. Rev.*, B, 11:2535–2546, 1975.

Joule

A unit of energy or work in the meter-kilogram-second (mks) system of units, being equal to the work done by a force of magnitude 1 newton when the point at which the force is applied is displaced 1 meter in the direction of the force. Joule is a short name for "newton-meter of energy or work" and hence is equivalent to 10^7 ergs and also to 1 watt-sec. The term joule should never be used as a synonym for "newton-meter of torque." *See* CALORIE; UNITS, SYSTEMS OF; WORK.

[DUANE E. ROLLER/LEO NEDELSKY]

Joule's law

A quantitative relationship between the quantity of heat produced in a conductor and an electric current flowing through it. As experimentally determined and announced by J. P. Joule, the law states that when a current of voltaic electricity is propagated along a metallic conductor, the heat evolved in a given time is proportional to the resistance of the conductor multiplied by the square of the electric intensity. Today the law would be stated as $H = RI^2$, where H is rate of evolution of heat in watts, the unit of heat being the joule; R is resistance in ohms; and I is current in amperes. This statement is more general than the one sometimes given that specifies that R be independent of I. Also, it is now known that the application of the law is not limited to metallic conductors.

Although Joule's discovery of the law was based on experimental work, it can be deduced rather easily for the special case of steady conditions of current and temperature. As a current flows through a conductor, one would expect the observed heat output to be accompanied by a loss in potential energy of the moving charges that constitute the current. This loss would result in a descending potential gradient along the conductor in the direction of the current flow, as usually defined. If E is the total potential drop, this loss, by definition, is equal to E in joules for every coulomb of charge that traverses the conductor. The loss conceivably might appear as heat, as a change in the internal energy of the conductor, as work done on the environment, or as some combination of these. The second is ruled out, however, because the temperature is constant and no physical or chemical change in a conductor as a result of current flow has ever been detected. The third is ruled out by hypothesis, leaving only the generation of heat. Therefore, $H = EI$ in joules per second, or watts. By definition, $R = E/I$, a ratio which has positive varying values. Elimination of E between these two equations gives the equation below,

$$H = RI^2$$

which is Joule's law as stated above. If I changes to a new steady value I', R to R', and H and H', then $H' = R'I'^2$ as before. The simplest case oc-

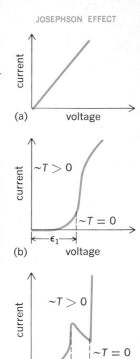

(a)   voltage

(b)   voltage

(c)   voltage

Fig. 4. Current-voltage characteristics of electron tunneling through a thin insulating film between two metals. (*After I. Giaever and K. Megerle, Electron tunneling in superconductors, Phys. Rev., 120: 1101–1111, 1961*)

curs where R is independent of I. If the current is varying, the resulting variations in temperature and internal energy undoubtedly exist and, strictly speaking, should be allowed for in the theory. Yet, in all but the most exceptional cases, any correction would be negligible.

This phenomenon is irreversible in the sense that a reversal of the current will not reverse the outflow of heat, a feature of paramount importance in many problems in physics and engineering. Thus the heat evolved by an alternating current is found by taking the time average of both sides of the equation. Incidentally, the changes in internal energy, if they were included in the theory, would average out. Hence the equation continues to have a similar form, $\overline{H} = \overline{RI^2}$, for ac applications. *See* HEATING, ELECTRIC; OHM'S LAW.

[LLEWELLYN G. HOXTON/JOHN W. STEWART]

Juglandales

An order of flowering plants, division Magnoliophyta (Angiospermae), in the subclass Hamamelidae of the class Magnoliopsida (dicotyledons). The order consists of three families: the Juglandaceae with a little over 50 species, the Picrodendraceae with 3 species, and the Rhoipteleaceae with only 1 species. Within its subclass the order is sharply set off by its compound leaves. *Juglans* (walnut and butternut) and *Carya* (hickory, including the pecan, *C. illinoensis*) are familiar genera of the Juglandaceae. The West Indian genus *Picrodendron* has walnutlike fruits and is called the Jamaica walnut. *See* HAMAMELIDAE; HICKORY; MAGNOLIOPHYTA; MAGNOLIOPSIDA; PLANT KINGDOM.

[ARTHUR CRONQUIST]

Juncales

An order of flowering plants, division Magnoliophyta (Angiospermae), in the subclass Commelinidae of the class Liliopsida (monocotyledons). The order consists of the family Juncaceae, with about 300 species, and the family Thurniaceae, with only 3. Within its subclass the order is marked by its reduced, mostly wind-pollinated flowers and capsular fruits with one to many anatropous ovules per carpel. The flowers have six sepals arranged in two more or less similar whorls, both sets chaffy and usually brown or green. The ovary is tricarpellate, with axile or parietal placentation. The pollen grains are borne in tetrads, and the embryo is surrounded by endosperm. The order is most unusual among higher plants in that, together with at least some members of the Cyperaceae in the related order Cyperales, it has chromosomes with diffuse centromeres. *See* COMMELINIDAE; CYPERALES; FLOWER; LILIOPSIDA; MAGNOLIOPHYTA; PLANT KINGDOM.

[ARTHUR CRONQUIST]

Junction detector

A reverse-biased semiconductor junction functioning as a solid ionization chamber to produce an electrical output pulse whose amplitude is linearly proportional to the energy deposited in the junction depletion layer by incident ionizing radiation. *See* IONIZATION CHAMBER.

Introduced into nuclear studies in 1958, the junction detector or, more generally, the nuclear semiconductor detector has revolutionized the field. Typically these devices have improved experimentally attainable energy resolutions in the

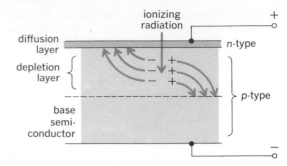

Fig. 1.   Silicon junction detector.

detection of both charged particles and of γ-radiation by a factor of at least 100 over any previously attainable. To this they have added unprecedented flexibility of utilization, speed of response, miniaturization, freedom from deleterious effects of extraneous electromagnetic (and often nuclear) radiation fields, low-voltage requirements, and effectively perfect linearity of output response.

Fabrication of diodes. Figure 1 illustrates an early junction device fabricated by diffusing phosphorus into p-type silicon to produce a pn junction. The applied reverse bias sweeps all free carriers from the depletion layer; collection of electrons or holes produced in this depletion layer by incident ionizing radiation results in the output signal. Equivalent devices have been fabricated by diffusion of zinc or gallium into n-type silicon.

The first practical detectors were prepared by evaporating a very thin gold electrode onto a polished wafer of n-type germanium, Ge, or silicon, Si, to produce a surface barrier junction. Reflecting the fact that the band gaps in Ge and Si are 0.67 and 1.106 ev at 300°K, respectively, it has been necessary to operate Ge devices at liquid-nitrogen temperatures, or below, to minimize impurity-generated carriers, while the Si devices operate satisfactorily at room temperature except under requirements of maximum energy resolution where these too are cooled. For this reason the Si-based devices have supplanted all earlier devices in the study of charged particle radiations and of x-radiation. Typical full width at half maxima for the pulse output spectra are 700 ev for x-radiation and 5 kev for charged particle radiation. The latter almost certainly reflects surface effects since much superior results are attained with incident β- and γ-radiation, which is more sensitive to the bulk characteristic of the detectors.

These resolutions reflect the fact that the energy required to produce an electron-hole pair in Ge and in Si is 2.96 and 3.66 ev, respectively, as compared to ~30 ev in a perfect gas-ionization detector. In consequence the attainable resolution is exceedingly rarely determined by statistical considerations, and almost always by the associated electronic circuitry which characteristically uses field-effect transistors to minimize extraneous electronic noise contributions. Reflecting the much higher carrier mobilities in a semiconductor as compared to an ionization chamber, as well as the smaller physical size, output-pulse rise times are characteristically $< 3 \times 10^{-9}$ sec for Ge and $< 10^{-8}$ sec for Si junction devices.

The effective thickness of the depletion layer is given by $0.32\,(\rho V)^{1/2}\,\mu$, where ρ is the resistivity in ohm-centimeters of the base material and V is the applied bias in volts. The capacitance of the device is given by $3.3 \times 10^4 (\rho V)^{-1/2}$ picofarad/cm². In the case of the Si-based junctions in common use, limitations imposed by available base-material resistivity and bias voltages before breakdown limit the effective thickness of the depletion layer to ≤ 2500 μ corresponding to the range of a proton of ~25 Mev, of an α-particle of ~90 Mev, and of an electron of ~1.5 Mev. Much greater effective depletion depths are possible through stacking of a number of fully depleted units, or through irradiation of the detector through the edge of the depletion layer rather than the top (Fig. 1).

For use with much more penetrating γ-radiation and higher-energy electrons, it has become customary to prepare much larger sensitive volumes by drifting lithium ions through the base crystal material—usually Ge since its higher atomic number and density increase its effective stopping power for nuclear radiation. This effectively deactivates all impurity-based carrier generation centers. Typically active volumes ~ 20 cm³ are in common use, while single units ~ 100 cm³ have been prepared under special laboratory conditions. Even the former have detection efficiencies between 5 and 10% of that of a 3 in. diameter × 3 in.-deep NaI spectrometer crystal. They have enormous advantages over the NaI crystals in all respects save overall efficiency in terms of resolution, size, response time, power supply requirements, freedom from magnetic field effects, and relative freedom from neutron activation effects. Their major disadvantage is that they must be both utilized and stored at liquid-nitrogen temperatures, or below, to prevent redrifting or precipitation of the lithium. There is high promise, on the basis of current (1968) work directed toward production of superpure Ge (<10¹⁰ impurity atoms/cm³), that the requirement for lithium drift as well as for consequent refrigeration may be relaxed.

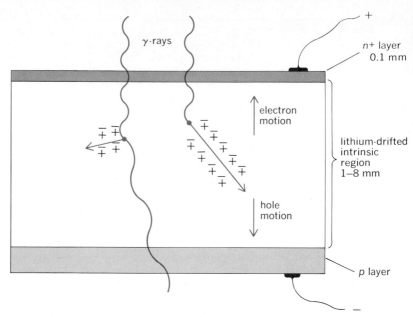

Fig. 2.   Typical lithium-germanium α-ray spectrometer.

Figure 2 shows a typical lithium-germanium device in section. Figure 3 compares the response of a 3 in. diameter × 3 in. deep NaI and a 21 cm³ LiGe spectrometer to γ-radiation from the osmium isotope of mass 188, illustrating the fact that essentially every spectrum earlier obtained with NaI spectrometers has revealed important additional transitions when examined with the Li-Ge devices. Figure 4 illustrates the response of a smaller device to x- and γ-radiation from a radioactive source of the gadolinium isotope of mass 153. An electronic pulser has been superimposed in the spectrum to illustrate that the dominant resolution limitations are electronic in origin.

Among their many other advantages is the ease with which special semiconductor detector con-

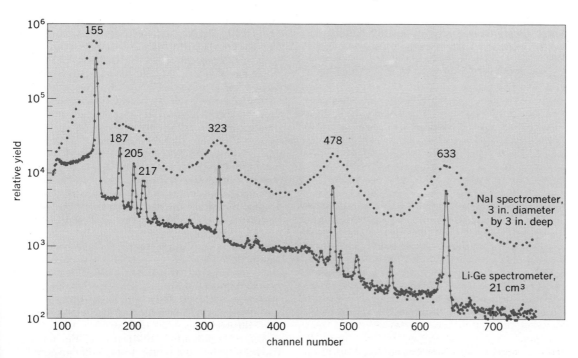

Fig. 3.   Gamma-radiation spectra from Os¹⁸⁸ as detected in NaI and Li-Ge spectrometers.

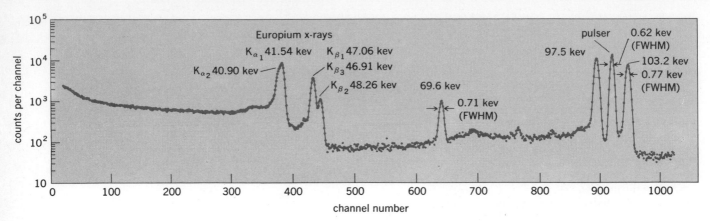

Fig. 4.   Spectrum of x- and γ-radiation from radioactive decay of Gd¹⁵³. Pulser shows dominant resolution limitations are electronic in nature; FWHM refers to energy width of peak at half the maximum height.

figurations may be fabricated. Figure 5 illustrates one of the simple yet very important examples of this in the form of an annular detector, which is characteristically used to detect nuclear reaction products from a bombarded target in a tight cone around the incident beam. By examining the decay γ-radiation in coincidence with such products, studies may be carried out only on residual nuclei which have had their spins very highly aligned in the nuclear reaction; this has been shown to provide an extremely powerful nuclear spectroscopic probe and is extensively used in laboratories throughout the world.

Composite detector systems are very readily assembled with the semiconductor devices. For example, it is standard in charged particle detection to use a very thin detector and a very thick detector (or even two thin and one thick) in series. Multiplication of the resultant signals readily provides a characteristic identification signature for each nuclear particle species in addition to its energy. Three-crystal γ-ray spectrometers are readily assembled, wherein only the output of the central detector is examined whenever it occurs in time coincidence with two correlated annihilation

quanta escaping from the central detector. These systems essentially eliminate background from Compton scattering of other more complex electromagnetic interactions and yield sharp single peaks for each incident photon energy. This is illustrated in Fig. 6.

Similarly neutrons may be indirectly detected through examination of recoil protons from a hydrogenous radiator in the case of high-energy neutrons, or through examination of fission fragments resulting from slow neutrons incident on a fissile converter foil mounted with the semiconductor detectors. (It should be noted that the response of the detectors is essentially perfectly linear all the way from electrons and photons to fission fragments.) Neutrons also may be detected and their energy spectra studied through examination of the charged products of the (nα) reaction (where α-particles are emitted from incident neutrons) induced in the Si or Ge base material of the detector itself.

Fabrication of triodes. Whereas the detectors thus far discussed are electrically nothing more than diodes, it has been possible to construct equivalent triodes which have extremely important uses in that they provide not only an output which is linearly proportional to the energy deposited in them, but also a second output which in combination with the first establishes the precise location on the detector itself where the ionizing radiation was incident. This has very obvious advantages in the construction of simple systems for the measurement of angular distributions, where such positron-sensitive detectors are located about a bombarded target. Their most important impact however has been in terms of their on-line use in the focal planes of large nuclear magnetic spectrographs. Simultaneous determination of the energy and location of a particle in the focal plane, together with the momentum determination by the magnet itself, establishes unambiguously both the mass and energy of the particle, and does so instantaneously so that additional logical constraints may be imposed through a connected on-line computer—something totally impossible with the earlier photographic plate focal-plane detectors. Figure 7 illustrates such an application.

A further important utilization of the nuclear triodes has followed their fabrication in an annular

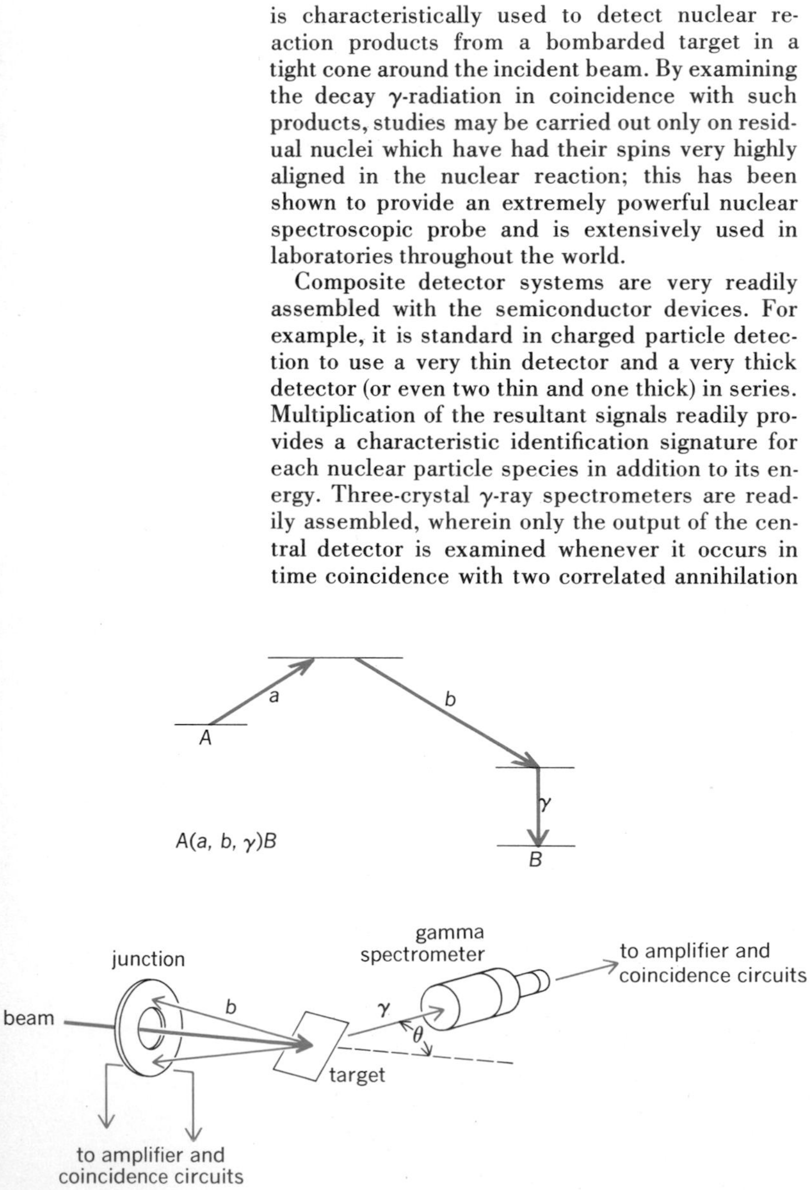

$A(a, b, \gamma)B$

Fig. 5.   Schematic of use of annular detector in nuclear reaction studies.

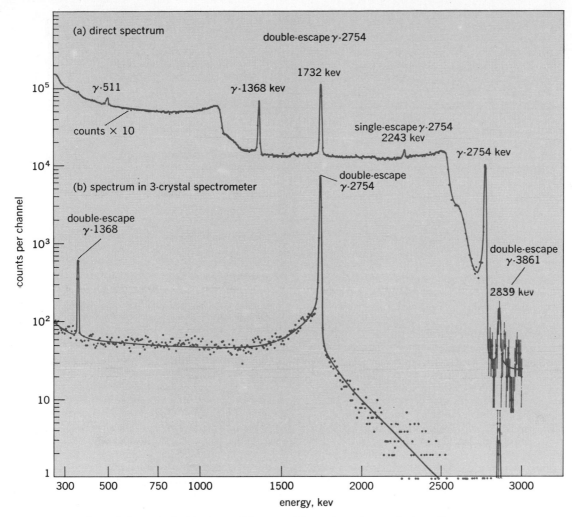

Fig. 6. Comparison of direct single detector and three-crystal spectrometer spectra from Na²⁴ source.

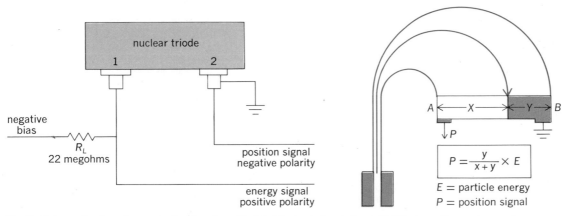

Fig. 7. Schematic of positron-sensitive (nuclear triode) detector in focal plane of 180° magnetic spectrograph.

geometry similar to that shown in Fig. 5. With radial position sensitivity it becomes possible to correct on-line, and event by event, for the kinematic variation of particle energy with angle over the aperture of the detector. Without this correction possibility all particle group structures in the detector spectrum are smeared beyond recognition.

Role in space program. The junction and semiconductor detectors have played—and continue to

play—a fundamental role in the space program, ranging from studies of the radiation fields in the solar system to the composition of the lunar surface. Here their low mass, ruggedness, minimal power-supply requirements, and insensitivity to ambient conditions are of paramount importance.

Active development programs are in progress throughout the world directed toward production of larger, hence more efficient room-temperature devices of ever increasing resolution. *See* PARTI-CLE DETECTOR; SEMICONDUCTOR.

[D. ALLAN BROMLEY]

Bibliography: G. Dearnaley and D. C. Northrop, *Semiconductor Counters for Nuclear Radiations*, 1963; G. T. Ewan, in F. J. M. Farley (ed.), *Progress in Nuclear Techniques and Instrumentation*, vol. 3, 1968; W. J. Price, *Nuclear Radiation Detection*, 2d ed., 1964; J. M. Taylor, *Semiconductor Particle Detectors*, 1963.

Junction diode

A semiconductor rectifying device in which the barrier between two regions of opposite conductivity type produces the rectification (Fig. 1). Junction diodes are used in computers, radio and television, brushless generators, battery chargers, and electrochemical processes requiring high dc current and low voltage. Lower-power units are usually called semiconductor diodes, and the higher-power units are usually called semiconductor rectifiers. For a discussion of conductivity types, carriers, and impurities *see* SEMICONDUCTOR.

Junction diodes are classified by the method of preparation of the junction, the semiconductor material, and the general category of use of the finished device. By far the great majority of modern junction diodes use silicon as the basic semiconductor material. Germanium is used where a low forward voltage is necessary, and a III–IV compound crystal such as gallium arsenide is used where its relatively large band gap energy is needed. A partial list of silicon types include diffused silicon switching diode, alloyed silicon voltage reference diode, epitaxial planar silicon photodiode, and diffused silicon rectifier. Other types are the alloy-diffused germanium rectifier and planar gallium arsenide light-emitting diode.

In silicon units nearly all categories of diodes are made by self-masked diffusion, as shown in Fig. 2a. Exceptions are varactor diodes where both alloying and epitaxy are used in control of the doping profile to give the desired voltage dependence of capacitance. Large-area junctions are seldom made today. If high current-carrying capacity is required, two or more smaller units are usually paralleled in a single package. The mesa structure shown in Fig. 2b is used for some varactor and switching diodes if close control of capacitance and voltage breakdown is required. *See* DETECTOR; DIODE, SEMICONDUCTOR; RECTIFIER; SWITCH, ELECTRONIC.

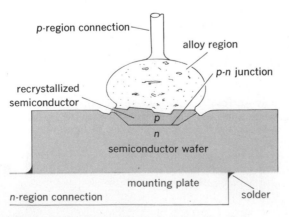

Fig. 1. Section of a bonded or fused junction diode.

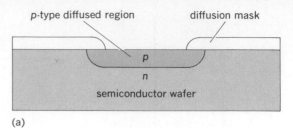

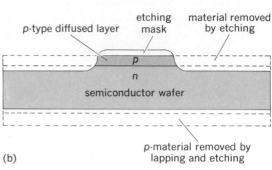

Fig. 2. High-speed diffused silicon diodes. (a) Mesaless structure. (b) Mesa structure.

Fabrication methods. The alloy and mesa techniques are largely historical but were important in the development of junction diodes. The alloy junction section (Fig. 1) is produced by placing a pill of doped alloying material on the clean flat surface of a properly oriented semiconductor wafer and heating it until the molten alloy dissolves a portion of the semiconductor immediately beneath it. Upon cooling, the dissolved semiconductor, now containing the doping impurity, recrystallizes upon the surface of the undissolved semiconductor, reproducing its crystal structure and creating a pn junction at the position marking the limit of the solution of the original wafer. A junction formed by this method is usually abrupt, the doping type changing precisely at the limit of solution of the alloy. It is also heavily doped on the recrystallized side, where the alloy usually contains the limit of solid solubility of the doping impurity. If such a junction is held at the peak temperature of its alloying cycle for sufficient time to allow diffusion of the alloy impurity beyond the limit of the dissolved semiconductor into the solid semiconductor, the junction produced is called alloy-diffused and is not nearly so abrupt. It will now have a significantly higher breakdown voltage with more scatter. The simple alloy junction has been used for voltage reference or voltage regulator applications.

The planar diffused junction section (Fig. 2a) is produced in silicon by first polishing the top surface of a large silicon wafer and then oxidizing the surface by heating the wafer at about 1000°C in the presence of wet oxygen. After about 5×10^{-4} cm of oxide has grown on the surface, the wafer is cooled and an array of holes is opened in the oxide by high-precision etching geometrically controlled by a photoresist technique. The wafer is then subjected at high temperature to a vapor containing the doping impurity. The impurity diffuses into the solid semiconductor through the holes in the oxide, thereby forming diffused pn junctions beneath each hole. Subsequently the individual junctions are separated out of the large wafer by scribing and breaking and are encapsulated as individual

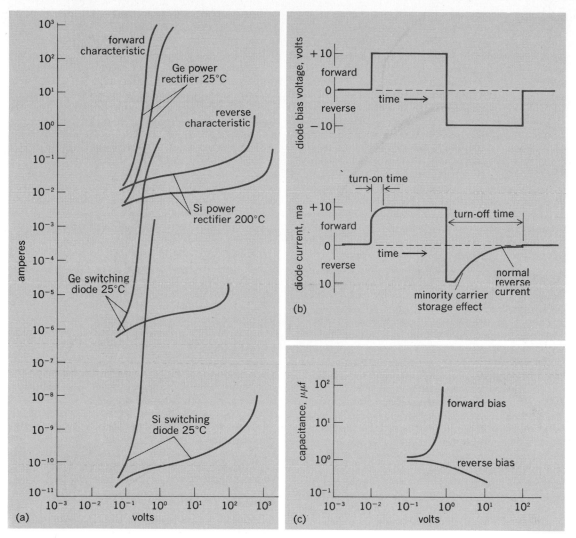

Fig. 3. Junction diode characteristics. (a) Rectification. (b) Switching. (c) Silicon switching diode capacitance.

diodes. Such planar diffused diodes have relatively high breakdown voltages and low leakage currents. The ends of the junction are automatically protected by the oxide mask so that such diodes show long-term stability. This protection by the oxide is often referred to as passivation.

The mesa structure (Fig. 2b) is produced by diffusing the entire surface of the large wafer and then delineating the individual diode areas by a photoresist-controlled etch that removes the entire diffused area except the island or mesa at each diode site.

Still another method of doping control used in modern diodes is through epitaxially deposited material. In this process the polished wafer is subjected at an elevated temperature to a vapor containing a compound of the semiconductor together with a compound containing the appropriate doping element. These compounds decompose upon contact with the surface of the wafer and cause the semiconductor to grow a layer of doped material on its surface. Under proper conditions of cleanliness and growth rate, the underlying crystal structure is propagated into the growing layer, which is then said to be epitaxial in character. In this way either mesas or entire surfaces of either conductivity type may be produced. In diode fabrication it is

typical to use the epitaxially grown material as a lightly doped layer over the original substrate material of the same conductivity type. The junction is then formed in the lightly doped layer by masked diffusion of the opposite-conductivity-type material. By this means the thickness of the web of lightly doped material immediately beneath the diffusion can be controlled to give both a desired reverse breakdown voltage and a relatively constant capacitance. Forward-bias recovery time can be controlled in a trade-off with reverse breakdown voltage in such a structure.

Junction rectification. Rectification occurs in a semiconductor wherever there is a relatively abrupt change of conductivity type. In any semiconductor the product of the concentrations of the majority and minority current carriers is a temperature-dependent equilibrium constant. The conductivity is proportional to the majority carrier concentration, and inversely proportional to the minority carrier concentration. When a pn junction is reverse-biased (p-region negative with respect to n-region), the majority carriers are blocked completely by the barrier and only the minority carriers can flow under the barrier. This minority carrier current is the sum of the individual currents from the n- and p-regions and each component is inversely proportional to the conductivity of its region. In addition, there is a thermal regeneration current of minority carriers generated in the depletion region of the reverse-biased junction. In germanium of about 10 ohm-cm resistivity, a pn junction reverse leakage current is about 10^{-3} amp/cm^2 with regeneration current negligible. In silicon of the same resistivity it is about 10^{-7} amp/cm^2 with regeneration current dominant, because of the much lower equilibrium constant of silicon.

When a pn junction is forward-biased (p-region positive with respect to the n-region), the majority hole and electron distributions can flow into the opposite region because the bias has markedly lowered the barrier. Since electrons flowing into a p-region or holes flowing into an n-region represent a great increase in minority carrier concentration, the thermodynamic equilibrium of the holes and electrons is disturbed and the product of their concentrations increases as the junction is approached. The resistivity of both the n- and p-type regions is considerably lowered by these excess minority carriers, and the forward current is greater than the current through a geometrically equivalent bar of material containing no pn junction.

The electrons in an n-type semiconductor are given up to the conduction process by donor impurity atoms which remain as fixed, positively charged centers. Similarly, the holes of a p-region are created by the capture of electrons by acceptor impurity atoms which remain as fixed, negatively charged centers. In both cases the space charge of the ionized impurity centers is neutralized by the space charge of the majority carriers.

At a pn junction the barrier that keeps majority carriers away consists of a dipole layer of charged impurity centers, positive on the n-type side and negative on the p-type side. When a reverse bias is applied, the barrier height increases and requires more charge in the dipole layer to produce the required step in voltage. To add to the charge, the layer must widen, because ionized impurities are in fixed positions in the crystal. As the layer widens, the capacitance of the junction decreases since the plates of the capacitor are farther apart. Therefore, a pn junction acts as a variable capacitance as well as variable resistance. *See* VARACTOR.

Optical properties. When light of sufficient energy is absorbed by a semiconductor, excess minority carriers are created. In a pn-junction device these excess carriers will increase the reverse-bias leakage current by a large factor if they are within diffusion distance of the junction. If the junction is open-circuited, a forward voltage will develop to oppose the diffusion of the excess carriers generated by the light absorption. This photovoltaic response is the basis of the operation of most photographic exposure meters. *See* EXPOSURE METER; PHOTOVOLTAIC CELL; PHOTOVOLTAIC EFFECT.

The inverse of the above effect also exists. When a pn junction in a so-called direct-gap semiconductor is forward-biased, the electrically injected excess minority carriers recombine to generate light. This is the basis of light-emitting diodes and injection lasers. Typical direct-gap semiconductors are compounds between III and V

group elements of the periodic table such as gallium arsenide. *See* LASER.

Characteristics. Figure 3 shows typical rectification, switching, and capacitance characteristics of a junction diode. Rectification characteristics (Fig. 3a) show that silicon units provide much lower reverse leakage currents and higher voltage breakdowns. Germanium units present lower forward-bias voltage drops at a given current density. Silicon units can be operated up to 200°C, while germanium units must be operated below 100°C.

For switching purposes, turn-on and turn-off times are most important (Fig. 3b). The turn-on time of a diode is governed by its junction capacitance and is usually short. The turn-off time, usually the critical characteristic, is governed by the time required to remove all of the excess minority carriers injected into the n- and p-regions while the diode was in the forward-bias state. This is called the minority carrier storage effect, and it is of the order of a few microseconds for good switching diodes. Silicon diodes are usually somewhat superior to germanium units in this respect. The limits of operation of present junction diodes are about 2500 volts reverse-standoff voltage and 1500 amperes forward current in power rectifiers; about 1.0 nanosecond reverse recovery time and 100 picoseconds rise time for fast switching diodes; a minimum reverse leakage current in a small signal diode is about 0.01 nanoampere.

For further discussion of the properties of pn junctions *see* JUNCTION TRANSISTOR; TRANSISTOR. *See also* POINT-CONTACT DIODE.

[LLOYD P. HUNTER]

Bibliography: J. Lindmeyer and C. Wrigley, *Fundamentals of Semiconductor Devices*, 1965.

Junction transistor

A transistor in which emitter and collector barriers are formed between semiconductor regions of opposite conductivity type. The junction transistor is by far the most widely used. Junction transistors range in power rating from a few milliwatts (mw) to about 200 watts, in characteristic frequency from 0.5 to 2000 MHz, and in gain from 10 to 50 dB. Silicon is the most widely used semiconductor material, although germanium is still used for some applications. Junction transistors are applicable to any electronic amplification, detection, or switching problem not requiring operation above 200°C, 300 volts, or 2000 MHz. Not all these limits can be achieved in one device, however. Junction transistors are classified by the number and order of their regions of different conductivity type, by the method of fabricating and structure, and sometimes by the principle of operation. Most modern transistors are silicon-fabricated by the self-masked planar double-diffusion technique. The alloy technique and the grown-junction technique are no longer widely used but are of historical importance.

Alloy-junction transistors. Also called fused-junction transistors, these are made in the pnp and npn forms. The emitter and collector regions are formed by recrystallization of semiconductor material from an alloy of semiconductor material dissolved in some suitable metal mixture. For the germanium pnp alloy-junction transistor the metal is usually indium with a small percentage of gallium. For the germanium npn alloy-junction transistor the metal may be a mixture, such as lead with antimony or bismuth with arsenic. The major metal of each of the alloys (indium, lead, or bismuth) serves as the solvent for germanium. The minor element serves as a source of the impurity needed to render the recrystallized germanium opposite in conductivity type to the original wafer.

Alloy junctions are abrupt and allow for bidirectional operation. They usually show a low series resistance and are therefore used in high-power transistors. Today germanium is almost exclusively used for alloy-junction transistors because the technology is well developed and because there are many alloys compatible with germanium with regard to thermal expansion. For a general description and definition of terms used here *see* TRANSISTOR.

Figure 1 compares several transistor profiles which show how the impurity content varies through the structure. In these profiles C_p is the concentration of the p-type impurity; C_n is the concentration of the n-type impurity. The net impurity content determines the conductivity type and magnitude. The profile of the alloy transistor shows that there are abrupt changes of impurity concentration at emitter and collector junctions and that the conductivities of emitter and collector regions are therefore high compared to those of the base region. Such a structure shows good emitter-injection efficiency but only moderate collector-voltage rating and relatively high collector capacitance. *See* SEMICONDUCTOR.

Grown-junction transistors. These are made in the *pnp* and *npn* forms, as well as in more complicated forms. There are several variations of the grown-junction technique. The simplest consists of successively adding different types of impurities to the melt from which the semiconductor crystal is being grown.

A semiconductor crystal is usually grown by dipping the end of a seed crystal into molten semiconductor and by arranging the thermal gradients so that new semiconductor solidifies on the end of the seed as it is slowly withdrawn. The solid-liquid interface is roughly a plane perpendicular to the axis of withdrawal. A *pnp* structure can be grown by starting with a *p*-type melt; by adding, at one point in the crystal growth, enough *n*-type impurity to give a slight excess over the *p*-type impurity originally present; and, after growth has continued for a few microns, by adding an excess of *p*-type impurity to the melt. The last grown region will be the emitter region, and the original *p*-type crystal will be the collector region. The impurity profile of such a structure is shown in Fig. 1*b*.

The high-conductivity emitter region gives a good injection efficiency, and the junction between the base and collector regions is gradual enough so that the unit will show a relatively low collector capacitance and a high breakdown voltage. The one disadvantage of this method is that both the collector and base regions show relatively high series resistances.

Planar diffused epitaxial transistors. The structure of this transistor is shown in section in Fig. 2, and the doping profile through the emitter, base, and collector is shown in Fig. 1*c*. In this structure both collector and emitter junctions are formed by diffusion of impurities from the top surface, as

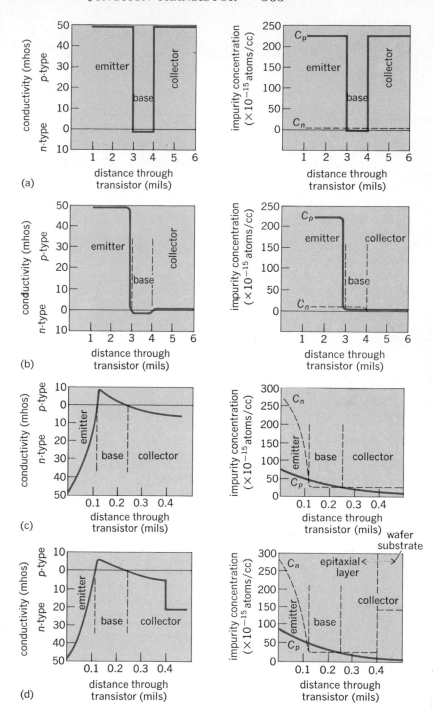

Fig. 1. Conductivity and impurity profiles of typical junction transistors. (*a*) *pnp* alloy-junction type. (*b*) *pnp* grown-junction type. (*c*) *npn* double-diffused-junction type. (*d*) *npn* epitaxial double-diffused-junction type.

shown in Fig. 2. Using silicon, the structure is formed by growing a diffusion mask of native oxide (silicon dioxide) on the previously polished wafer. A hole is opened in the oxide by a photoresist etch technique (Fig. 2*a*) to define the area of the collector buried layer. For a *p*-type substrate a heavy concentration (n^+) of *n*-type impurity such as phosphorus is diffused into the substrate through the opening in the masking oxide. The oxide is etched away, and an epitaxial layer of lightly doped (n^-) silicon is grown over the entire wafer by vapor decomposition at a temperature low enough

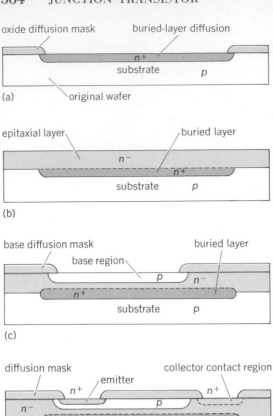

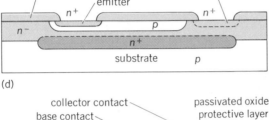

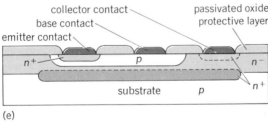

Fig. 2. Double-diffused planar epitaxial transistor structure and method of fabrication. (*a*) Buried layer. (*b*) Epitaxial layer. (*c*) Collector junction formation. (*d*) Emitter junction. (*e*) Contact stripe placement.

to prevent significant diffusion of the n^+ material in the buried layer (Fig. 2*b*). A new oxide layer is grown on the surface of the epitaxial layer, and an opening is etched in it to define the limits of the p-type base diffusion (Fig. 2*c*). (This automatically controls the collector junction geometry and capacitance.) The masking oxide is again stripped and regrown for the masking of the n^+ diffusion used to form the emitter and collector contact region (Fig. 2*d*). Next the emitter mask is removed, and an impervious layer of oxide is formed over the surface of the crystal. A layer of glass is bonded to the crystal by means of the oxide layer. The glass must match the expansion coefficient of silicon fairly well, and the oxide must be sufficiently impervious to the glass at the bonding temperature to prevent the diffusion of impurities from the glass into the silicon transistor structure. Finally, holes are etched in the glass-oxide struc-

ture so that electrical contact can be made to the various regions of the transistor (Fig. 2*e*).

In this transistor formation of the base region by diffusion from the emitter side produces a steep doping gradient and thereby a strong electric field in the base region. In the typical alloy-junction transistor (uniform base doping) the minority-carrier transport across the base is achieved by a relatively slow diffusion process. In this diffused base type (sometimes called a drift transistor) the base region shows a high conductivity gradient, decreasing from the emitter to the collector (Fig. 1*d*). This conductivity gradient means that the majority-carrier concentration is much greater near the emitter than near the collector. In order to cancel the natural diffusion of majority carriers from the high- to the low-concentration region, an electric field must exist of such a polarity as to tend to drive majority carriers back toward the emitter. This polarity of field then tends to drive minority carriers from the emitter to the collector; when normal bias is applied to the device, excess injected minority carriers will be accelerated across the base by this field.

The buried layer of n^+ doped material has very low resistance and acts as a shorting bar between the area immediately beneath the emitter to the area immediately beneath the collector contact stripe, thus maintaining a low collector series resistance even if the n^- material of the epitaxial region is of quite high resistivity. The collector breakdown voltage may be maintained at a reasonably high value and the collector capacitance at a low value by controlling the thickness of the n^- material between the base and the buried layer and by keeping the doping level of the n^- material quite low.

Mesa transistors. These transistors minimize the collector capacitance by limiting the collector junction area. This area limitation is achieved by etching away the surrounding material so that the entire transistor structure stands up from the body of the wafer like a small mesa. This structure and its associated planar variation have yielded the highest frequency cutoff and the fastest switching times of any of the junction transistors.

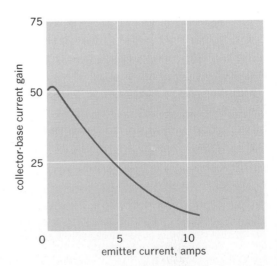

Fig. 3. Variation with emitter current of common emitter current gain for an alloy power transistor.

Power transistors. These are used in the output stage of an electronic circuit both as switches and as amplifiers. Depending on the load, a high voltage rating, a high current rating, or a high power rating may be required. With any of these, heat dissipation within the device is a serious limitation. Figure 3 shows the variation of grounded-emitter current gain with emitter current for a typical *pnp* junction power transistor with a circular emitter and collector. This curve shows that high current rating is obtained at a serious sacrifice in gain. This drop in gain primarily results from the crowding of the emitter current density toward the rim of the emitter because of the *IR* drop of the base current. Modern silicon power transistors minimize this effect by using emitter-base geometries that maximize their perimeter-to-area ratios. In addition, high-current capacity is achieved by paralleling several individual units in a single encapsulation. *See* CONTROLLED RECTIFIER.

Unijunction transistor. This device is really a diode in which the resistance of a portion of the base region is modulated by minority carriers injected by forward-biasing its single junction. Its structure typically consists of a lightly doped base region with ohmic contacts at opposite ends. The single junction is formed over a narrow range near the center of the base region by a shallow diffusion of heavily doped material of the opposite conductivity type. If a bias current is set up from end to end in the base, the potential at the junction can be set at a desired reverse bias relative to ground. If a signal is applied to the junction electrode, the device will turn on when the signal exceeds the original reverse-bias potential of the base at that point. Once forward-biased, the junction injects sufficient minority carriers into the base to short the region beneath the junction to the ground end of the base, and the device remains conducting until reset, either by the base bias or by the emitter signal. These devices show a typical negative resistance characteristic and are used for timing, control, and sensing circuits.

Summary. Silicon planar passivated transistors show a wide range of performance with characteristic frequencies up to 1000 MHz, voltage ratings of 12–300 volts, and power dissipation ratings of 100 mw–50 watts. The highest-frequency devices range up to 4000 MHz. Silicon planar technology is used in fabricating integrated circuit chips. The general form of the transistor structure displayed in Fig. 2 is used in integrated circuits. Such a structure is used for diodes as well as transistors since, for example, it is necessary only to connect the base and collector contacts to use the collector junction as a diode. *See* INTEGRATED CIRCUITS.

[LLOYD P. HUNTER]

Bibliography: L. P. Hunter, *Handbook of Semiconductor Electronics*, 3d ed., 1970; R. H. Mattson, *Basic Junction Devices and Circuits*, 1963.

Jupiter

The largest planet in the solar system, and the fifth in the order of distance from the Sun. It is visible to the naked eye, except for short periods when in near conjunction with the Sun. Usually it is the second brightest planet in the sky; only Mars at its maximum luminosity and Venus appear brighter. Jupiter is brighter than Sirius, the brightest star.

Planet and its orbit. The main orbital elements are a semimajor axis, or mean distance to Sun, of 485×10^6 mi; an eccentricity of 0.048, causing the distance to the Sun to vary about 47×10^6 mi between perihelion and aphelion; sidereal revolution period of 11.86 years; mean orbital velocity of 8.2 mi/sec; and inclination of orbital plane to ecliptic of 1°.3. *See* PLANET.

The apparent equatorial diameter of its disk varies from about 47″ at mean opposition (50″ at perihelic opposition, 44″ at aphelic opposition) to 32″ at conjunction. The polar flattening due to its rapid rotation is considerable and is easily detected by visual inspection; the ellipticity is $(r_e - r_p)/r_e = 0.065$, where r_e is the equatorial radius and r_p is the polar radius. The equatorial diameter is

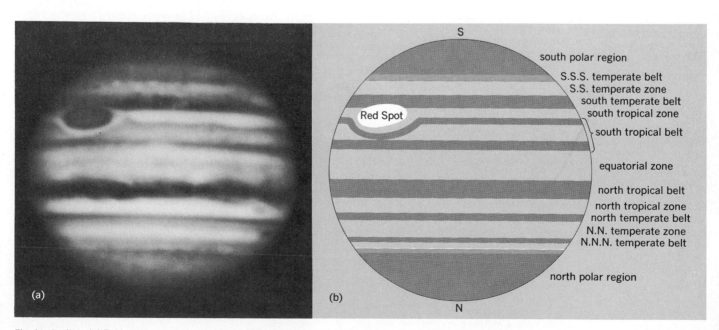

Fig. 1.   Jupiter. (*a*) Telescopic appearance (*Mount Wilson and Palomar Observatories*). (*b*) Principal bands.

Table 1. Mean latitudes and rotation periods of Jupiter's bands (after B. M. Peek)

Band	Region	Latitude	Period, 9h+	Remarks
North polar region		>+48°	55m42s	
N.N.N. temperate belt	Center	+43°	55m20s	Temporary
N.N. temperate belt	Center	+38°	55m42s	Temporary
	South edge	+35°	53m55s	
North temperate belt	North edge	+31°	56m05s	
	Center	+27°	53m17s	
	South edge	+23°	49m07s	
North tropical zone	Center	+18°	55m29s	
North tropical belt	Center	+13°	54m09s	Temporary
	South edge	+6°	50m24s	
Equatorial zone	Middle	0°	50m24s	
South tropical belt	North edge	−6°	50m26s	
	Center	−10°	51m21s	
	South edge	−19°	55m39s	
South tropical zone	Center	−23°	55m36s	
Great Red Spot	Center	−22°	55m38s	
South temperate belt	North edge	−27°	55m02s	Temporary
	Center	−29°	55m20s	
	South edge	−31°	55m07s	
S.S. temperate zone	Center	−38°	55m07s	
S.S.S. temperate zone	Center	−45°	55m30s	Temporary
South polar region		<−45°	55m30s	

JUPITER

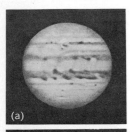

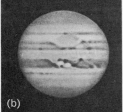

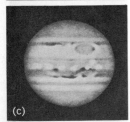

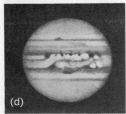

Fig. 2. Variable visibility of the Red Spot. (a) 1929. (b) 1933. (c) 1936. (d) 1938. (After L. Rudaux)

about 88,700 mi, and the polar diameter is 82,800 mi. The volume is about 1317 (Earth = 1) with an uncertainty of a few percent. The mass is about 318.4 (Earth = 1), and is accurately determined from the motion of the four major satellites. The mean density is 1.34 g/cm³, a low value characteristic of the four major planets; the corresponding value of the mean acceleration of gravity at the visible surface is about 26 m/sec²; however, because of the large radius and rapid rotation, the centrifugal force at the equator amounts to 2.25 m/sec², reducing the effective acceleration of gravity to about 24 m/sec².

Phases. As an exterior planet, Jupiter shows only gibbous phases. Because of the large size of Jupiter's orbit compared to that of the Earth, the maximum phase angle is only 12° at quadratures and the phase effect shows up only as a slightly increased darkening of the edge at the terminator. The apparent visual magnitude at mean opposition is −2.4, and the corresponding value of the reflectivity (geometrical albedo) is about 0.4; the physical albedo is 0.45, with some uncertainty due to the small range of phase angle observable. The high value of the albedo, characteristic of the four major planets, indicates the presence of a dense, cloud-laden atmosphere. *See* ALBEDO.

Telescopic appearance. Through an optical telescope Jupiter appears as an elliptical disk, strongly darkened near the limb, and crossed by a series of bands parallel to the equator (Fig. 1). Even fairly small telescopes show a great deal of complex structure in the bands and disclose the rapid rotation of the planet. The period of rotation, determined from long series of observations of transits of spots at the central meridian, is very short, about 9h55m, the shortest of the main planets. The details observed, however, do not correspond to the solid body of the planet but to clouds in its atmosphere, and the rotation period varies markedly with latitude. The nomenclature and mean rotation periods of the main belts of clouds are given in Table 1. The rotation period

of any given zone is not exactly constant but suffers continual fluctuations about a mean value. Occasionally, short-lived atmospheric phenomena may depart more strongly from the mean rotation period of the zone in which they appear and thus drift rapidly with respect to other details in the zone. The rotation axis is inclined only 3° to the perpendicular to the orbital plane, so that seasonal effects are practically negligible.

Red Spot. Apart from the constantly changing details of the belts, some permanent or semipermanent markings have been observed to last for decades or even centuries, with some fluctuations in visibility. The most conspicuous and permanent marking is the great Red Spot, intermittently recorded since the middle of the 17th century and observed continually since 1878, when its striking reddish color attracted general attention. It was conspicuous and strongly colored again in 1879–1882, 1893–1894, 1903–1907, 1911, 1914, 1919–1920, 1926–1927, and especially in 1936–1937, 1961–1968, and in 1973–1974; at other times it has been faint and only slightly colored, and occasionally only its outline or that of the bright "bay" or "hollow" of the south temperate zone which surrounds it has remained visible (Fig. 2).

The mean rotation period of the Red Spot between 1831 and 1955 was 9h55m37.5s, with a range

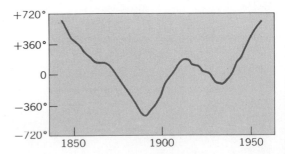

Fig. 3. Drift in longitude of the Red Spot between 1840 and 1955. (After B. M. Peek)

of variation of about ±6 sec. The mean dimensions of the Red Spot are about 30,000 mi in longitude and 10,000 mi in latitude.

Other markings. Another remarkable, semipermanent marking, the South Tropical Disturbance, was intermittently observed on Jupiter between 1901 and 1935 and possibly also in 1940. This marking circulated in the same zone as the Red Spot, but with a shorter mean rotation period, about 9h55m16s, and periodically came into conjunction with it. A large number of temporary features of shorter duration have been observed, but their mechanisms of formation and laws of motions are not understood.

In 1972 a smaller spot was seen to appear in the northern hemisphere, and 18 months later *Pioneer 10* observed it to be similar in size and shape to the Great Red Spot. *Pioneer 11*, passing the planet a year later, showed that the spot had disappeared, thereby suggesting that its lifetime was about 2 years.

The locations of spots on the visible surface of Jupiter are recorded in two longitude systems: System I, based on a rotation period of 9h50m30.003s, is used for spots in the equatorial zone; System II, based on a rotation period of 9h55m40.632s, is used for spots outside this zone. The period of rotation of any given spot being variable and not necessarily equal to the mean period of either longitude system, there is a general drift which after sufficient time can exceed one or more complete turns, that is, can exceed 360°. The drift in longitude of the Red Spot with respect to System II for 1831–1955 is shown in Fig. 3.

Atmosphere. Limits on atmospheric pressure at the top of the clouds can be set by observing pressure broadening of spectral lines; the pressure is probably a bit less than 3 atm. The atmosphere must be largely composed of hydrogen and helium. Detection of weak spectral lines arising from molecular hydrogen indicates that hydrogen constitutes most of this atmosphere.

The radiometrically determined temperature of the visible disk of Jupiter, about 130° ± 30°K, is in fairly good agreement with the value theoretically estimated from the assumption that the visible

cloud layer is mainly composed of ice crystals of solidified ammonia (about 160°K). *Pioneer 10* and *11* flybys showed that Jupiter's zones and belts consist of gases at different altitudes and temperatures. The coloring agents of the bands are unknown; the only substances positively identified spectroscopically in Jupiter's atmosphere have been hydrogen, helium, ammonia, methane, and water, while the presence of hydrogen sulfide is inferred. These are all colorless, and so other molecules have been proposed as coloring agents; among these are ammonium sulfide, ammonium hydrosulfide, free radicals, various organic compounds, and complex inorganic polymers.

The cloud tops of the near-white zones have been found to be the highest and coldest features, consisting probably of ammonia crystals. At lower altitudes, the melting point of ammonia is reached and the colored compounds are found. At the next lower level, it is likely that water is present, first as ice crystals, and then, at still lower levels, as droplets.

Interior composition and structure. Although the low density of Jupiter has in the past led to speculation that it might be a very small star with a hot gaseous interior, the lack of extremely pronounced thermal emission shows it to be a "cold" planet in the sense that thermally induced pressure gradients contribute little to its support against gravitational collapse. Model calculations made on this basis prove that hydrogen must constitute the bulk of the planet; any other element would give too high a mean density. Specifically, Jupiter probably consists of about 80% hydrogen, 15% helium, and a few percent of all the remaining elements (these values are similar to estimates of the composition of the pre-solar nebula, from which the solar system was formed). From Pioneer observations, a model of Jupiter's interior has been made which is consistent with present knowledge of the planet's magnetic and gravitational fields, and with extrapolations of laboratory studies of the behavior of hydrogen at very high temperatures and pressures. The model projects a small rocky core of iron and silicates at the center of the planet, where the temperature is

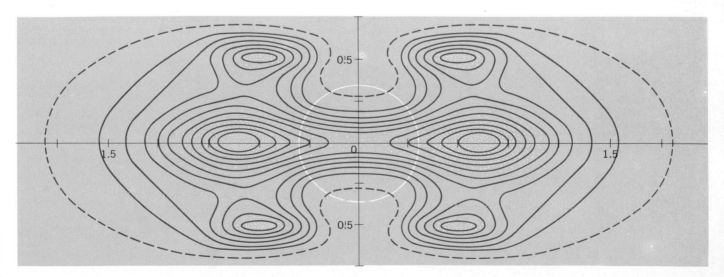

Fig. 4.   Contours of radio emission from Jupiter originating from Van Allen belts, and extending far beyond the planet's disk.

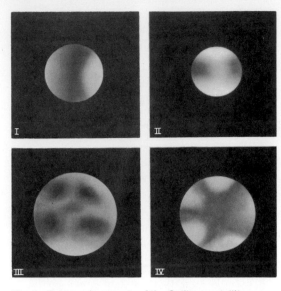

Fig. 5. Telescopic aspects of the Galilean satellites.

Radio astronomy. Jupiter is known to produce three distinct types of radio emission. Thermal radiation from the high stratosphere is detectable at wavelengths below about 10 cm and indicates temperatures in the upper emitting layers of 100–130°K. Microwave nonthermal emission is observed in the band from about 3 to 70 cm, and is known to arise from synchrotron radiation from relativistic electrons in extended Jovian Van Allen belts (Fig. 4). Sporadic decametric radio noise storms are heard on wave-lengths longer than 8 m; these seem consistent with gyrofrequency radiation produced by spiraling electrons, perhaps near the magnetic poles of Jupiter. The existence of both kinds of these nonthermal emissions suggests a Jovian polar surface magnetic field. Except near Jupiter, the major component of this magnetic field is dipolar, though opposite in direction to the Earth's magnetic field. At a distance of three Jupiter radii, the field has a strength of 0.16 gauss (G). Closer to the planet, *Pioneer 11* has shown that the field is quadrupolar and octopolar and ranges in field strength from 3 to approximately 14 G. This complex field structure may be attributed to the complicated circulation that is thought to exist in the planetary interior. *See* RADIO ASTRONOMY; VAN ALLEN RADIATION.

Periodicities in the radio-noise storms and rocking of the polarization plane of the microwave nonthermal emission lead to the well-determined radio rotation period 9h55m29.7s. The difference between the radio and the various other observed Jovian rotation periods suggest that the core of Jupiter is rotating about 13s faster than the mantle, that the atmosphere has vast wind currents with relative velocities up to 500 km/hr, and that angular momentum may be significantly exchanged among these regions over periods of years.

Satellites. Jupiter has 13 known satellites and a suspected fourteenth, of which the first four, I Io, II Europa, III Ganymede, and IV Callisto, discovered by Galileo in 1610, are by far the most important (Table 2).

The four Galilean satellites are of fifth and sixth stellar magnitude and would be visible to the naked eye if they were not so close to the much brighter parent planet. All the others are faint telescopic objects. The fifth satellite, discovered visually by E. E. Barnard in 1892, moves inside the orbit of Io; the others move in much larger orbits outside that of Callisto and have been discovered

likely to be about 30,000°K. Above this core is the thick layer in which hydrogen is the predominant element and which makes up virtually all the mass and volume of the planet. This layer is segregated into two strata. In the first, where the pressure is estimated as 3,000,000 Earth atmospheres and the temperature as 11,000°K, hydrogen is in its liquid metallic state wherein the molecules are dissociated into separate atoms, forming an electrically conductive fluid. This stratum extends from the core to approximately 46,000 km from the center. The second layer extends to 70,000 km and is thought to consist of liquid molecular hydrogen. The atmosphere then extends another 1000 km above this to the cloud tops.

Jupiter emits, in the infrared, several times as much energy as it absorbs from the Sun. This thermal leakage from the interior presumably taps the heat source of gravitational potential energy liberated by the original formation of the planet which is then conveyed to the surface by large-scale convection currents in the liquid hydrogen.

Jupiter's diameter is about as large as possible for a nonstellar object; central pressures would induce increasing degeneracy with higher density, leading to a decrease in radius with the further addition of mass.

JUPITER

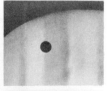

Fig. 6. Satellite phenomena. (*After L. Rudaux and G. de Vaucouleurs, Astronomie, Larousse*)

Table 2. Satellites of Jupiter

Satellite	Mean distance from Jupiter, 10^3 mi	Sidereal period, days	Diameter, mi	Magnitude at mean opposition	Albedo (visual)	Mass (Moon = 1)
I Io	262	1.769	2300	5.5 (visual)	0.37	1.05
II Europa	417	3.551	1950	5.7 (visual)	0.39	0.65
III Ganymede	666	7.155	3200	5.1 (visual)	0.20	2.1
IV Callisto	1,170	16.689	3200	6.3 (visual)	0.03	1.2
V Amalthea	113	0.498	150	13.0 (visual)		
VI Himalia	7,120	250.6	35	17.0 (photo)		
VII Elara	7,290	259.6	20	18.0 (photo)		
VIII Pasiphae	14,600	739	35	17.0 (photo)		
IX Sinope	14,700	758	17	18.6 (photo)		
X Lysithea	7,300	260.5	15	18.8 (photo)		
XI Carme	14,000	700	19	18.4 (photo)		
XII Ananke	13,000	625	14	18.9 (photo)		
XIII Leda	7,000	260	—	20 (photo)		

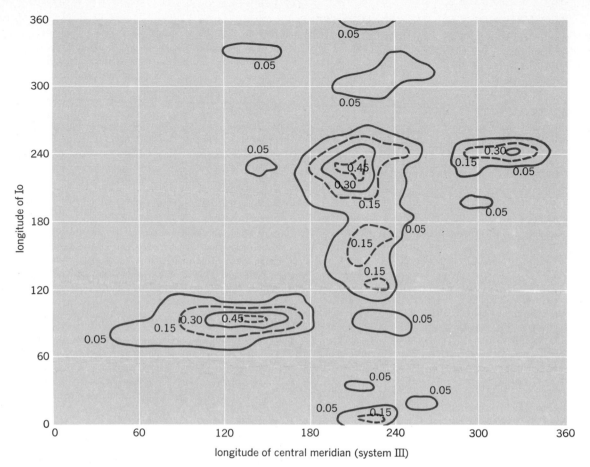

Fig. 7.   The intensity of radio bursts from Jupiter as a function of longitude of central meridian and orbital longitude of Io, the first satellite, frequency 22.2 MHz. Its nearness to surface of Jupiter results in modification of the radio propagation properties of the planet's ionosphere and radiation belts. (*Yale data, 1956–1961*)

by photography since 1904, the twelfth in 1951, the thirteenth in 1974, and the suspected fourteenth in 1975.

The masses of the major satellites can be roughly estimated from their mutual perturbations; in terms of the mass of the Moon as a unit, the mass of III is about 2, of I about 1, of IV about 4/3, and of II about 2/3. The four Galilean satellites show measurable disks easily seen with telescopes of 6- to 8-in. aperture; larger telescopes show distinct markings on the disks (Fig. 5). The apparent diameters are of the order of $1-2''$, and the corresponding linear diameters are listed in Table 2. Ganymede and Callisto are larger than Mercury but smaller than Mars; their densities, about $2.5-3$ g/cm^3, are approximately the same as the density of the Moon and of the Earth's crust. Observations of their surface markings and of the regular variations of brightness as they move along their orbits indicate that the four major satellites (and probably also V) always present the same face to Jupiter, much as the Moon does to the Earth; that is, their periods of rotation are equal to their periods of revolution. The albedos for I, II, and III are relatively high; IV is very dark and its albedo is smaller than that of the Moon.

The planes of the orbits of the major satellites are inclined less than $0°5$ to the equatorial plane of Jupiter, so that with the occasional exception of IV, they are eclipsed in Jupiter's shadow at each revolution and also project their shadows on Jupi-

ter and transit in front of its disk near conjunction. The eclipses, transits, and occultations of Jupiter's satellites (Fig. 6) led to the discovery of the finite velocity of propagation of light by O. Roemer in 1675. The first satellite, Io, is so close to the surface of Jupiter that it actually moves within the ionosphere and radiation belts of the planet and modifies the radio propagation properties of these regions in a most remarkable way, affecting the reception of the radio noise bursts (Fig. 7). Satellites VIII, IX, XI, and XII have a retrograde motion. The very small outer satellites of Jupiter may be captured asteroids; their orbits are subject to large perturbations by the Sun. Thus at least the more distant satellites probably form part of a fluctuating population gained and lost over very long time spans. *See* ECLIPSE, ASTRONOMICAL; OCCULTATION; TRANSIT (ASTRONOMY).

[ELAINE M. HENDRY]

Bibliography: G. P. Kuiper and B. M. Middlehurst (eds.), *Planet and Satellites*, 1961; B. M. Peek, *The Planet Jupiter*, 1958; J. H. Wolfe, Jupiter, *Sci. Amer.*, 233(3):118–126, 1975.

Jurassic

The system of rocks deposited during the middle part of the Mesozoic Era. The Jurassic System is normally underlain by the Triassic and overlain by the Cretaceous System. It was named after the Jura Mountains in Switzerland and consists mainly of sedimentary rocks. Volcanic rocks are re-

PRECAMBRIAN	PALEOZOIC								MESOZOIC		CENOZOIC	
					CARBON-IFEROUS							
	CAMBRIAN	ORDOVICIAN	SILURIAN	DEVONIAN	Mississippian	Pennsylvanian	PERMIAN	TRIASSIC	JURASSIC	CRETACEOUS	TERTIARY	QUATERNARY

stricted to certain regions, particularly the North American Cordillera. The duration of the Jurassic Period, which according to M. Howarth ended about 135,000,000 years ago, is estimated to be about 55,- to 60,000,000 years.

Subdivision. The Jurassic System consists of three main divisions: the Lower, Middle, and the Upper Jurassic. Either Lias, Dogger, and Malm or Black, Brown, and White Jurassic are the terms used in some countries, particularly Germany. Their boundaries coincide largely with those of the Lower, Middle, and Upper Jurassic. Finer subdivisions into stages (chronostratigraphic units) and zones (biostratigraphic units) are based mainly on the northwest European standard section. The Hettangian, Sinemurian, Pliensbachian, and Toarcian stages form the Lower Jurassic; the Bajocian, Bathonian and Callovian form the Middle Jurassic; and the Oxfordian, Kimmeridgian, Portlandian, and Purbeckian form the Upper Jurassic. In some countries, particularly France and Germany, a special stage name, the Aalenian, is used instead of the Lower Bajocian, most commonly applied in the United States, Canada, and Great Britain. Up to the Kimmeridgian these names are in worldwide use, but because of regional differentiation of the faunas in the younger Jurassic, other stage names have been introduced for these beds, for example, Volgian in Russia and the Arctic and Tithonian in Mediterranean regions. The stages are groupings of zones which are based on certain species of ammonites (order Ammonoidea of Cephalopoda) characteristic of certain beds and, in many cases,

of worldwide distribution. See CEPHALOPODA.

The northwest European standard section is now subdivided into about 60 ammonite zones. Local faunal differentiations have resulted in separate zone sequences for certain regions, and some zones may be based on fossils other than ammonites, for example, the pelecypod genus *Buchia* (*Aucella*) in the Upper Jurassic. See BIVALVIA.

Other subdivisions are formations mainly defined by lithology, which form the basis of geological mapping. In North America formation names are extensively used. In Great Britain formation names such as Lias, Inferior Oolite, Cornbrash, and Oxford Clay are used, and as some of these rock units are traceable to other countries, the same names have been applied. The use of formation names for mapping has great advantages; however, worldwide correlations are based on stages and zones.

Paleogeography. Different opinions have been expressed on the configuration and position of continents and oceans during the Jurassic Period. The existence of large continents or land bridges in the region of present-day oceans is now denied by many authorities. W. Arkell considered as possible the existence of an African-Brazilian continent that formed the western half of Gondwanaland. He also supposed that the present-day Indian Ocean was land in the Jurassic. In this interpretation the Australo-Indo-Madagascan continent, or Lemuria or eastern part of Gondwanaland, was separated from the western part of Gondwanaland by a Jurassic sea on the east side of Africa. The distribution of both land and neritic faunas makes land connections between continents—as between Africa and South America—necessary. But as there is no indication of any former continents or land bridges in the Atlantic and Indian oceans, and as most if not all of the ocean floor seems to be younger than Jurassic, the hypothesis of continental drift is now widely accepted. See CONTINENT FORMATION.

In recent years continental drift has been seen as a result of movement of lithospheric plates away from oceanic spreading centers. A. Hallam's tentative reconstruction of Gondwanaland for the Callovian by reassembly of Gondwanaland illustrates this hypothesis (Fig. 1). Active drifting apart of various parts of Gondwanaland did not take place before the Middle or Late Cretaceous. The distribution of faunas in the Arctic and Scandic suggest that the Arctic area was occupied by an oceanic basin in Jurassic times. The basin extended southward into the North Atlantic, where marine Jurassic sediments are now also known to be present in the offshore areas of eastern Canada. Figure 2 illustrates the paleogeography in these northern areas for the time of the Oxfordian. All known marine Jurassic deposits occur in parts of the continents, where they were deposited in comparatively shallow water. There is hardly a single epeiric (shallow inland) sea which remained in one and the same region throughout all of the estimated 55,- to 60,000,000 years of the Jurassic Period. The duration of the Jurassic marine transgressions is dependent on the character of the part of the continent which was flooded. Stable shelves, such as those in the European part of Russia, generally have fairly thin and incomplete covers of Jurassic sediments, whereas less stable shelves, such as

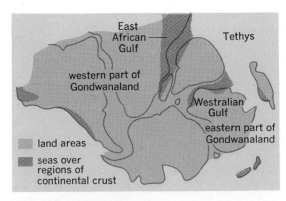

Fig. 1. Tentative reconstruction of Gondwanaland for the Callovian. (*After A. Hallam, The bearing of certain palaeozoogeographic data on continental drift, Palaeogeography, Palaeoclimatology, Palaeoecology, vol. 3, 1967*)

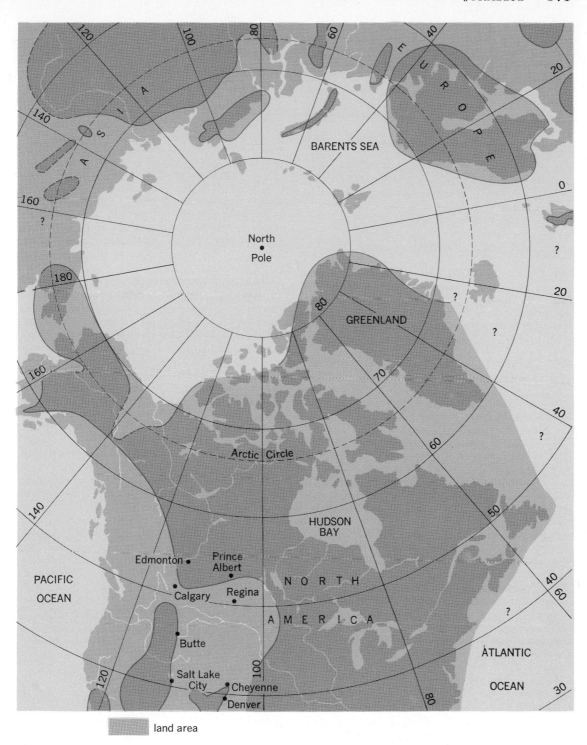

Fig. 2. Paleogeography of the boreal regions for the time of the Oxfordian of the Jurassic Period. (*Modified* *from H. Frebold, The Jurassic faunas of the Canadian Arctic, Geol. Surv. Can. Bull., no. 74, 1961*)

those in northwest Europe, have considerably thicker deposits.

Great thicknesses of sediments were accumulated in the mobile belts—for example, the American Cordillera, which formed part of the circum-Pacific mobile belt. Another mobile belt, the Tethys, extended from the Mediterranean to Persia, the Himalaya, and Indonesia. Such mobile belts included were rapidly and extensively sinking troughs. The erosion of the equally fast-rising adjacent lands made vast quantities of sediments available for filling them. In spite of their rapid sinking, the depths of these geosynclinal seas were not (with some exceptions) considerably greater than those of the more stable areas, as the rapid sinking was compensated by rapid filling. Typical shallow-water deposits were common. *See* GEO-SYNCLINE.

The Precambrian shields, which form the oldest and most stable parts of the Earth's crust, were land areas subject to erosion during the Jurassic Period. Thus in North America the Canadian

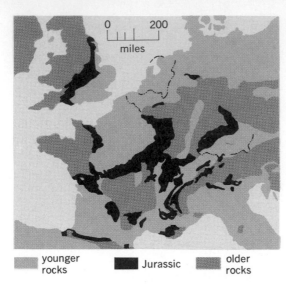

younger rocks   Jurassic   older rocks

Fig. 3.   Distribution of Jurassic outcrops in west central Europe. (From R. C. Moore, Introduction to Historical Geology, 2d ed., McGraw-Hill, 1958)

Shield was land bordered by seas both to the north and west, and the Scandinavian Shield was surrounded by seas in Late Jurassic times (Fig. 2). Smaller areas underlain by older rocks, such as the Bohemian Mass, the French Central Plateau, the Belgian Ardennes, and the German Hartz Mountains, were lands, the coastlines of which are indicated in many places by conglomerates and sandstones (Fig 3). See PRECAMBRIAN.

The faunas of the Early Jurassic, particularly the ammonites, seem to have been universal, and distinct faunal realms can scarcely be recognized for this time. During the early part of the Middle Jurassic (middle Bajocian) a Pacific province seems to be indicated and, at about the same time and particularly during the Bathonian and Callovian, a boreal realm is developing in the Arctic and parts of Russia. Its ammonite faunas are clearly distinguished from those of the realm of the Tethys. These faunistic differences do not necessarily indicate any differentiation of climate, which apparently was very uniform as indicated by the fossil vegetation. Jurassic glaciations are unknown.

Tectonic history. The Jurassic Period was for most parts of the world a time of comparative tectonic passivity: diastrophism of a major order, connected with the intrusion of batholiths, occurred in the North American Cordillera. Mountain building connected with volcanic activity took place in the Crimea and Caucasus, and minor folding and faulting (the germano-type folding) is known from northwest Germany and other areas. In the North American Cordillera, as the Pacific plate was subducted along the western margin of the American plate, two episodes of volcanism, plutonism, and deformation occurred — the Nassian in the Middle Jurassic and the Nevadan or the Columbian in the Late Jurassic or Early Cretaceous. While the Nassian seems to have been restricted to parts of southern British Columbia, the Nevadan resulted in great uplifts throughout the mobile belt. This new land was consequently subjected to erosion,

and the eroded material was deposited in the East, where it contributed to the sediments of the Lower Cretaceous..

Germanotype folding affected many parts of the world within belts of earlier folding which had not yet become entirely stable. Orogenic movements of this type occurred during the Late Triassic and perhaps earliest Jurassic (early Cimmerian phase). The late Cimmerian phase of many regions is about equivalent to the Nevadan orogeny F. See BATHOLITH.

Of utmost importance in the history of the Jurassic Period were the epirogenic movements. They were the cause of some transgressions and regressions of the sea, of the openings and closings of seaways, and of the migrations of faunas. They also influenced the rate and character of the sedimentation. See DIASTROPHISM; OROGENY; TECTONIC PATTERNS.

Life of the Jurassic. The Jurassic flora is characterized by ferns, which occur abundantly in certain regions, and by numerous gymnosperms. Among these the cycadeoids (Bennettitales), such as *Taeniopteris*, *Nilssonia*, and *Pterophyllum*, extinct palmlike plants related to the modern cycads, were dominant. Ginkgos were widely distributed; pines, cedars, and other conifers were abundant. The presence of angiosperms is indicated by pollen grains. Calcareous algae built reefs in some marine areas. The comparatively great uniformity of the Jurassic land plants does not suggest the existence of climatic zones.

Reptiles were represented by a great number of different forms. On the land the dinosaurs were the dominating forms, represented by the orders Saurischia, with hipbones comparable to those of other reptiles, and the Ornithischia, with more birdlike hipbone structure. Among the Saurischia both small and giant forms were present. *Compsognathus* was about the size of a chicken, while the bipedal carnivore *Allosaurus* attained gigantic size. *Brontosaurus*, an amphibious, four-footed, plant-eating dinosaur, reached a length of more than 60 ft. The plant-eating Ornithischia were represented by bipedal Ornithopoda and the heavily plated quadruped *Stegosaurus*. In the sea *Plesiosaurus*, the fishlike *Ichthyosaurus*, and crocodiles, such as *Stenosaurus* (up to 18 ft long), were locally abundant; in the air the pterosaurs were represented by *Pterodactylus* and others.

Skeletons of the first bird, *Archaeopteryx*, which still had reptilian jaws and teeth and strong hip and breast girdles like the pterosaurs, were found in the Late Jurassic lithographic stones at Solenhofen, Bavaria. Practically all groups of modern fishes were present in the Jurassic. Of the ganoids, homocercal forms such as *Gyrodus*, *Lepidotus*, and *Dapedius* were abundant; the teleosts were represented by *Leptolepis* and the sharks by *Hybodus* and *Acrodus*. Of the amphibians, frogs and toads were present. In both England and North America fragments, mostly jaws and teeth of small mammals belonging to the Marsupialia, occur.

The most important marine invertebrates of the Jurassic are the ammonites, which developed many new families and genera unknown in the preceding Triassic Period. Their short vertical and wide horizontal range and their appearance in almost every facies have made them the best guide fossils of the Jurassic. Other Jurassic cephalopods are nautiloids and the very abundant belemnoids. Both gastropods and pelecypods are well represented, the latter with a number of genera (such as *Buchia, Gryphaea, Inoceramus*, and *Trigonia*) of importance for age determinations and correlations. The echinoids are abundant in certain regions, but the crinoids are represented only by a few genera, such as *Pentacrinus*. The brachiopods are restricted to a few groups. Siliceous sponges formed reefs locally, and the lime-secreting hydrozoan colonies were abundant in the Mediterranean region. Corals lived in certain regions at various times during the Jurassic. Land invertebrates are represented by such insects as flies, butterflies, and moths. Among the crustaceans, ostracods are known from both marine and fresh-water beds and are used for age determinations. *See* PALEOBOTANY; PALEONTOLOGY.

Economic products. Iron ores of Jurassic origin are present in western Europe. Those of Lorraine belong to the Middle Jurassic, whereas those in Britain occur in both the Lower and Upper Jurassic and the underlying beds of the Rhaetic in many parts of the world, including Arctic regions. Upper Jurassic coal is known from Scotland, the Lofoten Islands, and northeast Greenland. Not all of the Jurassic coal occurrences are workable under present conditions. Highly bituminous rocks and oil shales occur in western Europe. Lime and cement are made from calcareous beds in England and Germany and bricks from the clays. Lithographic stone comes from the Upper Jurassic of Solenhofen, Bavaria; many good building stones from various beds are quarried in the Jurassic limestones of western Europe. Petroleum is obtained from various Jurassic horizons in the United States, Canada, Mexico, Brazil, France, Germany, Morocco, and Saudi Arabia.   [HANS FREBOLD]

Bibliography: W. J. Arkell, *Jurassic Geology of the World*, 1956; R. H. Dott, Jr., and R. L. Batten, *Evolution of the Earth*, 1971; H. Frebold, The Jurassic faunas of the Canadian Arctic, *Geol. Surv. Can. Bull.*, no. 74, 1961; A. Hallam, The bearing of certain palaeozoogeographic data on continental drift, *Palaeogeography, Palaeoclimatology, Palaeoecology*, vol. 3, 1967; H. Hoelder, Jura, in F. Lotze (ed.), *Stratigraphische Geologie*, vol. 4, 1964; M. K. Howarth, The Jurassic period, *Quart. J. Geol. Soc. London*, vol. 122S, 1964; R. W. Imlay, Paleoecology of Jurassic seas in the western interior of the United States, in H. S. Ladd (ed.), Treatise on marine ecology and paleoecology, *Geol. Soc. Amer. Mem.*, no. 67, vol. 2, 1957.

Jute

A natural fiber obtained from two Asiatic species, *Corchorus capsularis* and *C. olitorius*, of the plant family Malvaceae (Fig. 1). These are tall, slender, half-shrubby annuals, 8–12 ft tall.

When harvested, the stems are retted in tanks or pools to rot out the softer tissues. The strands of jute fiber are then loosened by beating the stems on the surface of the water. The fibers are not very strong and deteriorate quickly in the presence of moisture, especially salt water. Despite these weaknesses, jute is much used. It is inexpensive

Fig. 1. Morphological features of *Corchorus capsularis*.

and easily spun and converted into coarse fabrics. It is made into gunny, burlap bags, sacks for wool, potato sacks, covers for cotton bales, twine, carpets, rug cushions, curtains, and a linoleum base. It is also used in making coarse, cheap fabrics, such as novelty dress goods. Most of the commercial supply comes from plants grown in the Ganges and Brahmaputra valleys in Bangladesh and India. *See* FIBER, NATURAL; FIBER CROPS; MALVALES.

[ELTON G. NELSON]

Fig. 2. Disease lesions on jute stems.

Diseases. A number of diseases of jute cause losses in yield and reduce fiber quality. "Runner" and "specky" fiber are primarily due to disease-producing organisms.

The fungus *Macrophominia phaseoli* is believed to cause the most serious disease of the two species of jute. It is seed-borne and soil-borne, and pycnidiospores from susceptible plants besides jute also serve as sources of infection. The stem, leaves, and roots of both young and older plants are subject to attack. Stem infection usually takes place through a leaf petiole or at a node (Fig. 2). Root rot is complicated in that severity is increased when *M. phaseoli* is in combination with other fungi, bacteria, or nematodes, such as *Fusarium solani*, *Pseudomonas* sp., and *Meloidogyne incognita*, respectively. *See* LEAF (BOTANY); MELANCONIALES; MONILIALES; MYCELIA STERILIA; NEMATODA; PSEUDOMONADACEAE; ROOT (BOTANY); SEED (BOTANY); SPHAEROPSIDALES; STEM (BOTANY).

In contrast to *Macrophominia phaseoli*, *Colletotrichum capsici* causes lesions on the stem internodes and may also attack seedlings and capsules of *C. capsularis*. *Macrophoma corchori* and *Diplodia corchori* cause stem diseases. Two species of bacteria, *Xanthomonas makatae* and *X. makatae* var. *olitorii*, attack the stem and leaves of both *C. capsularis* and *C. olitorius*. *Pythium splendens* causes a root rot and subsequent wilt of *C. capsularis*, and indications are that other species of *Pythium* also are root pathogens of jute. *See* FRUIT (BOTANY).

Other fungi which attack jute are *Sclerotium rolfsii*, *Curvularia subulata*, *Cercospora corchori*, *Rhizoctonia solani*, *Helminthosporium* sp., and *Alternaria* sp.

Seed treatment with an organomercuric material such as Ceresan, Granosan, or Agrosan GN should be practiced for control of seed-borne pathogens and seedling diseases. Stem rot may be prevented by spraying with bordeaux or a colloidal copper compound. The excessive use of nitrogenous fertilizers increases the incidence of stem diseases. Root-rot control requires the use of crop rotation and, in some areas, the use of recently developed varieties of *C. capsularis* which are more tolerant of certain root-rot pathogens than *C. olitorius*. *See* AGRICULTURAL SCIENCE (PLANT); PLANT DISEASE.   [THOMAS E. SUMMERS]

Kale to Kyanite

Kale

Either of two cool-season biennial crucifers, *Brassica oleracea* var. *acephala* and *B. fimbriata*, of Mediterranean origin and belonging to the plant order Capparales. Kale is grown for its nutritious green curled leaves which are cooked as a vegetable (see illustration). Distinct varieties (cultivars) are pro-

Kale (*Brassica oleracea* var. *acephala*), cultivar Vates. (*Joseph Harris Co., Rochester, N.Y.*)

duced in Europe for stock feed. Kale and collards differ only in the form of their leaves; both are minor vegetables in the United States. Cultural practices are similar to those used for cabbage, but kale is more sensitive to high temperatures. Strains of the Scotch and Siberian varieties are most popular. Kale is moderately tolerant of acid soils. Monthly mean temperatures below 70°F favor best growth. Harvesting is generally 2–3 months after planting. Virginia is an important producing state. The total annual farm value in the United States from approximately 1200 acres is $600,000. *See* CABBAGE; CAPPARALES; COLLARD; KOHLRABI; VEGETABLE GROWING.

[H. JOHN CAREW]

Kaliophilite

A rare mineral tectosilicate found in volcanic rocks high in potassium and low in silica. Kaliophilite is one of three polymorphic forms of $KAlSiO_4$; the others are the rare mineral kalsilite and an orthorhombic phase formed artificially at about 500°C. It crystallizes in the hexagonal system in prismatic crystals with poor prismatic and basal cleavage. The hardness is 6 on Mohs scale, and the specific gravity is 2.61. At high temperatures a complete solid-solution series exists between $KAlSiO_4$ and $NaAlSiO_4$, but at low temperatures the series is incomplete. The principal occurrence of kaliophilite is at Monte Somma, Italy. *See* SILICATE MINERALS. [CORNELIUS S. HURLBUT, JR.]

Kalsilite

A rare mineral described in 1942 from volcanic rocks at Mafuru, in southwest Uganda. Kalsilite has since been synthesized. It is one of the three polymorphic forms of $KAlSiO_4$; the others are kaliophilite and an orthorhombic phase formed artificially at about 500°C. The mineral as shown by x-ray photographs is hexagonal. The specific gravity is 2.59. In index of refraction and general appearance in thin section it resembles nepheline and is difficult to distinguish from it. Structurally the two minerals are similar but belong to different crystal classes. The rock in which kalsilite was found has a dark-colored, fine-grained matrix with large olivine crystals and greenish-yellow patches. These patches are intimate mixtures of diopside, calcite, and kalsilite. *See* FELDSPATHOID.

[CORNELIUS S. HURLBUT, JR.]

Kame

A moundlike hill of stratified glacial drift, commonly sand and gravel (see illustration). Kames range from a few meters to more than 100 m in height and are built commonly during the process of deglaciation. Some consist of sediment deposited by superglacial streams in crevasses and other openings in nearly stagnant ice, which then melted, leaving the sediment as a mound. Others were deposited as small fans or deltas built out-

Kames in southeastern Wisconsin. (*USGS*)

ward from ice or inward against ice which, in melting, isolated the body of sediment as an irregular mound. Some kames grade into eskers; many are closely associated with kettles. Large fields of kames are common in western New York and in many parts of New England. *See* GLACIATED TERRAIN.

[RICHARD F. FLINT]

Kanamycin

An antibiotic produced by an actinomycete which has found utility in medicine. Kanamycin was discovered in Japan in 1957 by Hamao Umezawa and coworkers, who isolated the producing organism, *Streptomyces kanamyceticus*. Collaborative research between scientists at Bristol Laboratories in the United States and scientists in Japan indicated therapeutic usefulness against certain organisms which cause severe medical problems.

Chemistry. Chemical studies of kanamycin have shown it to be a water-soluble, stable, basic antibiotic composed of three moieties: 2-deoxystreptamine (a component of neomycin), 3-D-glucosamine, and 6-D-glucosamine. The latter two components are attached to the deoxystreptamine in an alpha linkage, and represent compounds not previously found in natural products. The structure of kanamycin A is shown in the illustration. Commercial kanamycin sulfate USP is almost exclusively kanamycin A but may contain up to 5% kanamycin B, a related metabolite.

Microbiological properties. The antibiotic has bactericidal activity against a wide variety of gram-positive, gram-negative, and acid-fast bacteria. Gram-negative organisms sensitive to kanamycin include *Escherichia coli*, *Aerobacter*, *Klebsiella*, *Shigella*, and *Salmonella* species. Kanamycin is also effective against mycobacteria and some strains of *Proteus* and *Neisseria*. Most strains of *Pseudomonas* are resistant. Among the gram-positive bacteria, this agent is active against *Staphylococcus aureus*, but other gram-positive pathogens, such as streptococci and pneumococci, are usually resistant.

In general, its antimicrobial spectrum is similar to that of neomycin. Organisms resistant to neomycin are also resistant to kanamycin, and vice versa. Resistance to kanamycin develops slowly, and in stepwise fashion, except for the mycobacteria, which develop resistance rapidly. *See* DRUG RESISTANCE; NEOMYCIN.

Pharmacological properties. The drug is absorbed rapidly after intramuscular injection, peak serum concentrations occurring after 1 hr. It is very poorly absorbed through the gastrointestinal tract and, consequently, is only used orally for preoperative bowel sterilization or treatment of gastrointestinal infections. High urine concentrations are obtained after parenteral (injected) administration, making the drug useful in the treatment of urinary tract infections. Kanamycin is less toxic than neomycin, both acutely and chronically, when given by the parenteral route. High dosages given for prolonged times to experimental animals can cause kidney, vestibular, and auditory damage.

Clinical results. Parenteral therapy with kanamycin has been successful in the case of gram-negative infections of the respiratory tract, soft tissues, and the urinary tract, as well as with osteomyelitis and bacteremia (bacteria growing in the blood). Kanamycin may be given orally for infections of the gastrointestinal tract due to sensitive organisms, such as salmonella and shigella. It may also be used for preoperative preparation of the large intestine by this route. Staphylococcal infections accompanied by bacteremia have responded. Studies on chronic tuberculosis indicate usefulness when the organism is resistant to streptomycin and isonicotinic acid hydrazide. *See* ISONICOTINIC ACID HYDRAZIDE; STREPTOMYCIN; TUBERCULOSIS.

Adverse reactions. The major toxic effect of parenterally administered kanamycin is its action on the auditory portion of the eighth nerve. Decreased auditory acuity (hearing) in the high-frequency ranges has occurred occasionally and usually is demonstrable only by audiometric measurements. A small proportion of patients had subjective loss and, in rare situations, complete deafness occurred. The effect on the eighth nerve is apparently related to both the concentration and duration of this agent in the blood, and toxic reactions occur infrequently in well-hydrated patients with normal kidney function who receive less than 20 g of kanamycin in 10–14 days. Evidence of renal irritation sometimes occurs in parenteral therapy but appears to be reversible on cessation of therapy. *See* ANTIBIOTIC; AUDIOMETRY.

[JOSEPH LEIN]

Kangaroo

The name for a number of Australian marsupials that are members of the family Macropodidae. This family also includes the wallabies. The kangaroos and their relatives occur principally in Australia, but are found in Tasmania and New Guinea as well.

Kangaroos have a long, thick tail that is used as a balancing organ, and enlarged hindlegs that are adapted for jumping in many species. The forelimbs are quite short, except in arboreal species such as the blacktree kangaroo (*Dendrolagus ursinus*) and its relatives, in which all four limbs are about the same length. The two largest species are

Structural formula of kanamycin A.

The red kangaroo at rest, with its weight carried by a tripod consisting of hindlegs and tail.

the red kangaroo (*Macropus rufus*) (see illustration), and the great gray kangaroo (*M. giganteus*). The males of these species may weigh as much as 220 lb and grow to a length of 5 ft, excluding the 2-ft tail. These animals have a total of 32 teeth, with the two lower medium incisors being enlarged and projected forward. The dental formula is I 3/1 C 0/0 Pm 2/2 M 4/4. The six grinding teeth on each half jaw are flat and are well suited for the vegetarian diet. The red kangaroo prefers the highlands of southern and eastern Australia, while the gray kangaroo inhabits the prairies. *See* DENTITION (VERTEBRATE).

Kangaroos usually have one offspring each year. After the uterine gestation period of about 6 weeks, the very immature young is born and crawls into the marsupium. At this stage the forelimbs are well developed and are used for crawling, while the hindlimbs are still embryonic buds. After an uninterrupted period of 2 months, it ventures out to find food and then returns to the safety of the marsupium. It may seek the protection of the pouch for up to 9 months. *See* MARSUPIALIA; OPOSSUM.   [CHARLES B. CURTIN]

Kaolinite

The principal mineral of the kaolinite group of clay minerals. Kaolinite is composed of a single silica tetrahedral sheet and a single alumina octahedral sheet. These two units are combined so that a common layer is formed by the tips of the silica tetrahedrons and one layer of the octahedral sheet (Fig. 1). *See* CLAY MINERALS.

Kaolinite is important in the ceramics industry because of its excellent firing properties and refractoriness. It is also used extensively as a filler in rubber products and for coating and filling paper products. *See* CERAMIC TECHNOLOGY.

Structure. In the common structural layer, two-thirds of the atoms become O instead of OH because they are shared by the silicon and alumi-

num. The aluminum atoms which are present occupy only two-thirds of the possible positions in the octahedral sheet and are hexagonally distributed in a single plane in the center of the sheet. All charges within the structural unit are balanced, and the formula for kaolinite is $(OH)_8Si_4Al_4O_{10}$. Any replacements within the lattice are of very small magnitude.

The sheet units of the kaolinite minerals are continuous in both the a and b crystallographic directions. They are stacked one above the other in the c direction with superposition of O and OH planes in adjacent units. These units are held firmly together by hydrogen bonding.

All kaolinites do not have the same order of crystallinity. Poorly crystallized kaolinites are disordered along the b axis with the unit layers randomly displaced by multiples of $b_0/3$. These poorly crystalline varieties also have a slightly higher first-order spacing than the well-crystallized mineral. Kaolinite is usually referred to as a 7-A clay mineral because its first-order basal spacing is of this approximate magnitude. Dickite and nacrite, although rarely found in clay materials, have structures similar to kaolinite. They are composed of the same structural units and differ only in the stacking of the layers. The unit cell of dickite is made up of two kaolinite layers. The nacrite unit cell is composed of six unit layers.

Morphology. Electron micrographs of well-crystallized kaolinite show well-formed, six-sided flakes, frequently with an elongation in one direction (Fig. 2). Particles with less distinct six-sided flakes have been observed in poorly crystallized kaolinite. The latter have ragged and irregular edges, and a very crude hexagonal outline. In general, poorly crystallized kaolinite occurs in smaller particles than the well-crystallized mineral.

Dickite particles are very well formed and have a distinct hexagonal outline. They differ from kaolinite in that they are much larger and can sometimes be studied with a light microscope.

Nacrite particles are somewhat irregular, rounded, flake-shaped units. Some of them show a crude hexagonal outline.

Properties. The mineral kaolinite has a low cation-exchange capacity (5–15 meq/100 g). Broken bonds around the edges of the silica-alumina

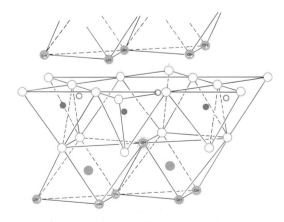

○ oxygens ⦿ hydroxyls ⬤ aluminums ○ ● silicons

Fig. 1. Diagram of the structure of kaolinite layer. (*From R. E. Grim, Clay Mineralogy, McGraw-Hill, 1968*)

Fig. 2. Electron micrograph of kaolinite, Macon, Ga. (*From R. E. Grim, Clay Mineralogy, McGraw-Hill, 1968*)

units are the major cause of this exchange capacity. Anion-exchange capacity is also low, except when structural hydroxyls are replaced or the anion is adsorbed because of its epitaxial fit. The rate of the exchange reaction is very rapid and the exchange capacity increases as the particle size decreases.

Kaolinite contains no interlayer water between the unit-cell layers. It does, however, have the ability to adsorb water and develop plasticity. This water has a definite configuration and is referred to as nonliquid water.

When kaolinites are heated they begin to lose their OH structural water at about 400°C with the dehydration being complete at 550–600°C. The precise temperature for the loss of this OH water varies from kaolinite to kaolinite, and may be explained by variations in particle size and crystallinity. The loss of OH water in poorly crystallized kaolinite is accompanied by a fairly complete loss of structure, but in well-crystallized kaolinite some structural remnants persist until subjected to a higher temperature.

A sharp exothermic reaction occurs at 950°C in poorly crystallized kaolinite. The explanation for the exothermic reaction has been attributed to the formation of γ-Al_2O_3 by some workers, and to the nucleation of mullite by others. Recent work strongly favors the latter explanation.

Formation. The kaolinite type of mineral forms under acid conditions at low temperatures and pressures. It can form under hydrothermal conditions or through the alteration of other clay minerals. Kaolinite may form from any constituents if the alkalies and alkaline earths are removed as fast as they are liberated from the parent rock or if the environment is acid and the temperature is moderate. It is a principal component of lateritic-type soils. When calcium is present in the environment, the formation of kaolinite is retarded.

The present-day marine environment does not favor the formation of kaolinite. The presence of a kaolinitic marine sediment is evidence of a kaolinitic source area, since this mineral does not form in the sea. It is also indicative of relatively rapid accumulation of material. *See* CLAY, COMMERCIAL.

[FLOYD M. WAHL; RALPH E. GRIM]

Kapok tree

Also called the silk-cotton tree (*Ceiba pentandra*), a member of the bombax family (Bombacaceae). The tree has a bizarre growth habit and produces pods containing seeds covered with silky hairs called silk cotton (see illustration). It occurs in the

Pods and leaves of kapok tree (*Ceiba pentandra*).

American tropics, and has been introduced into Java, Philippine Islands, and Ceylon. The silk cotton is the commercial kapok used for stuffing cushions, mattresses, and pillows. Kapok has a low specific gravity and is impervious to water, making it excellent for filling life preservers. *See* MALVALES.   [PERRY D. STRAUSBAUGH/EARL L. CORE]

Karman vortex street

A double row of line vortices in a fluid. Under certain conditions a Karman vortex street is shed in the wake of bluff cylindrical bodies when the relative fluid velocity is perpendicular to the generators of the cylinder, as illustrated. This periodic shedding of eddies occurs first from one side of the body and then from the other, an unusual phenomenon because the oncoming flow may be perfectly steady. Vortex streets can often be seen, for example, in rivers downstream of the columns supporting a bridge. The streets have been studied most completely for circular cylinders at low subsonic flow speeds. Regular, perfectly periodic, eddy shedding occurs in the range of Reynolds number (Re) of 50–300, based on cylinder diameter. Above a Re of 300, a degree of randomness begins to occur in the shedding and becomes progressively greater as Re increases, until finally the wake is completely turbulent. The highest Re at which some slight periodicity is still present in the turbulent wake is about 10^6.

Vortex streets can be created by steady winds blowing past smokestacks, transmission lines, bridges, missiles about to be launched vertically, and pipelines aboveground in the desert. The streets give rise to oscillating lateral forces on the shedding body. If the vortex shedding frequency is near a natural vibration frequency of the body, the resonant response may cause structural damage.

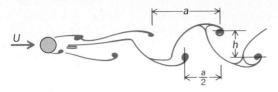

Karman vortex street.

The aeolian tones, or singing of wires in a wind, is an example of forced oscillation due to a vortex street. T. von Kármán showed that an idealized infinitely long vortex street is stable to small disturbances if the spacing of the vortices is such that $h/a = 0.281$; actual spacings are close to this value. A complete and satisfying explanation of the formation of vortex streets has, however, not yet been given. For $10^3 < \text{Re} < 10^5$ the shedding frequency f for a circular cylinder in low subsonic speed flow is given closely by $fd/U = .21$, where d is the cylinder diameter and U is stream speed; h/a is approximately 1.7. A. Roshko discovered a spanwise periodicity of vortex shedding on a circular cylinder at $\text{Re} = 80$ of about 18 diameters; thus, it appears that the line vortices are not quite parallel to the cylinder axis. *See* FLUID-FLOW PRINCIPLES; VORTEX.

[ARTHUR E. BRYSON, JR.]

Bibliography: L. Prandtl, *Essentials of Fluid Dynamics*, 1953; A. Roshko, On the wake and drag of bluff bodies, *J. Aeronaut. Sci.*, 22(2):124–132, 1955; V. L. Streeter, *Handbook of Fluid Dynamics*, 1961.

Karst topography

Characteristic associations of minor, third-order, destructively developed land features resulting from subaerial and underground solution of limestone under conditions of humid climate. These pattern features are progressively carved into second-order structural forms, such as plains, plateaus, or even hilly and mountainous uplands, containing limestone layers at or near the surface. Virtually all limestone formations are products of biological or chemical precipitation of calcium carbonate, $CaCO_3$, from aqueous solution. They are therefore susceptible in varying degrees to resolution in surface waters and groundwaters, particularly when such waters are charged with carbon dioxide, CO_2, from vegetative sources or when the waters are under pressure. *See* LIMESTONE.

Favoring combinations of the factors involved produce a landscape unlike those resulting from other agents of land sculpture. Where a karst topography has developed, most surface water disappears quickly by entering sinkholes (see illustration) and other entrance ways into underground tubes, passages, and caverns that are enlarging because of the presence and passage of these waters. Most streams—generally all but a few trunk streams in their deeper valleys—are short and scarce; ill-defined valleys end abruptly in sinks or swallow holes. The lack of a surface stream system is reflected in hills without the linear divides and intervalley character of normally stream-eroded regions. The name karst is taken from the regional designation of a coastal zone of hills and low mountains along the northeastern shore of the Adriatic Sea. In this region of limestone rocks the landscape is nearly every-where characterized by a predominance of solution-made topography.

Sinkhole types. Two contrasted types of sinkholes, the doline and the ponor, may be present.

Doline type. A doline is a saucer-shaped, solution-made depression, commonly called sinkhole, perhaps hundreds of feet in diameter with soil-covered slopes. It generally has no open entrance to the underground and usually does not discharge into an enterable cave. The bottom may be sufficiently clogged with waste to maintain a pond for years, then may abruptly lose all its water in one abrupt down-gulp.

Ponor type. These generally result from local failure of a solution-chamber roof, whose collapse produces the surface ponor sinkhole. This type tends to be steeply walled, commonly with bare rock, and leads into a cave large enough to have developed instability in roof rock. Coalescence of such collapse openings may eventually expose most of the former cave at the Earth's surface, with remnant portions of the roof in some places as natural bridges or tunnels. A surface valley which, by coincidence, crossed the course of an existing cave may have its stream swallowed down by formation of a ponor and thus become a sinking creek or lost river. Return of this stream water to surface drainage is likely along some major river valley. It there appears as a resurgence which, turbid and contaminated, is unlike a spring of filtered groundwater.

Soil, lapiés, and shore conditions. In a karst topography much limestone is dissolved by rainwater and soil water before runoff reaches sinkholes or streamways. Insoluble constituents like clay, sand grains, and chert nodules may remain on gentler slopes as a continually thickening cover of residual soil, named terra rossa from its red color.

Contact of such soil with the underlying limestone is generally sharp and may be very irregular because of varying structures in the rock, variations in supply of water, and varied solubility of the rock. If erosion strips off this soil mantle, a miniature karst may come to light, a solution-made surface with relief of a few inches to a few feet that is known as lapiés. Even without a soil cover, the attack of rainwater and snow water can produce a kind of lapiés.

Seawater, even when saturated with $CaCO_3$, may produce a ragged lapiés surface on limestone shorelines that are subject to constant wave wash. This may be a result of slight changes in tempera-

Sinkholes in Meade County, Kans. Depth to the underground water is about 50 ft. (*U.S. Geological Survey*)

ture and pressure in breaking waves. Marine algae growing on such wave-swept surfaces may also be a favoring factor.

Related features. A sinkhole-riddled surface in noncalcareous rock may result from extensive groundwater solution in a subjacent limestone or dolomite. Consequent subsidence of the overlying insoluble rock into depressions occurs with solutional removal. This is commonly termed solution subsidence, the filled cavities never having been open caverns. Even the partial removal of buried bodies of gypsum and rock salt may similarly produce a solutional subsidence topography.

Cyclical landform development. The sequence of changes in landforms of karst regions, from initiation of sinkholes through stages of their increase in number, enlargement, and coalescence, with collapse of caverns, and eventual wasting away of intervening hills, leaves broad-bottomed lowlands (uvalas) and scattered surviving elevations. Although this karst cycle is still subject to theoretical debate, the concept seems valid because different stages of landform development are identifiable in various karsts of differing ages.

Noteworthy karst regions are relatively uncommon in spite of extensive limestone terrains on the lands of the Earth. The best showing in Europe are the Karst area of Yugoslavia, the Causse region of France, and areas in Greece and Andalusia. North American examples are found in northern Yucatan, in central Florida, the Appalachian Great Valley in Tennessee and Virginia, in southern Indiana, and in west-central Kentucky. Jamaica, Puerto Rico, and Cuba also have well-developed karsts. *See* CAVE; GEOMORPHOLOGY; GROUNDWATER; WEATHERING PROCESSES.   [J. HARLEN BRETZ]

Bibliography: A. H. Doerr and D. R. Hoy, Karst landscapes of Cuba, Puerto Rico, and Jamaica, *Sci. Mon.*, vol. 85, no. 4, 1957; O. D. von Engeln, *Geomorphology*, 1942; B. W. Sparks, *Geomorphology*, 1960; W. D. Thornbury, *Principles of Geomorphology*, 1954.

Kelvin bridge

A specialized version of the Wheatstone bridge network designed to eliminate, or greatly reduce, the effect of lead and contact resistance and thus permit accurate measurement of low resistance. The circuit shown in the figure accomplishes this by effectively placing relatively high-resistance-ratio arms in series with the potential leads and contacts of the low-resistance standards and the unknown resistance. In this circuit R_A and R_B are the main ratio resistors, R_a and R_b the auxiliary ratio, R_x the unknown, R_s the standard, and R_y a heavy copper yoke of low resistance connected between the unknown and standard resistors.

By applying a delta-wye transformation to the network consisting of R_a, R_b, and R_y, the equivalent Wheatstone bridge network shown in the illustration is obtained, where Eqs. (1) hold. By an analysis similar to that for the Wheatstone bridge, it can be shown that for a balanced bridge Eq. (2)

$$R_x = \frac{R_B}{R_A} R_s + R_y \left(\frac{R_b}{R_a + R_b + R_y} \right)\left(\frac{R_B}{R_A} - \frac{R_b}{R_a} \right) \quad (2)$$

holds. If Eq. (3) is valid, the second term of Eq. (2)

$$\frac{R_B}{R_A} = \frac{R_b}{R_a} \quad (3)$$

is zero, the measurement is independent of R_y, and Eq. (4) is obtained.

$$R_x = \frac{R_B}{R_A} R_s \quad (4)$$

As with the Wheatstone bridge, the Kelvin bridge for routine engineering measurements is constructed using both adjustable ratio arms and adjustable standards. However, the ratio is usually continuously adjustable over a short span, and the standard is adjustable in appropriate steps to cover the required range. *See* WHEATSTONE BRIDGE.

Sensitivity. The Kelvin bridge sensitivity can be calculated similarly to the Wheatstone bridge. The open-circuit, unbalance voltage appearing at the detector terminals may be expressed, to a close degree of approximation, as in Eq. (5).

$$e = E \frac{r}{(r+1)^2} \left[\frac{\Delta R_x}{R_x + R_y \left(\frac{r}{r+1} \right)} \right] \quad (5)$$

The unbalance detector current for a closed detector circuit may be expressed approximately as in Eq. (6).

$$I_G = \frac{E \left(\frac{\Delta R_x}{R_x} \right)}{\frac{R_G}{r/(r+1)^2} + R_A + R_B + R_a + R_b} \quad (6)$$

The Kelvin bridge requires a power supply capable of delivering relatively large currents during the time a measurement is being made. The total voltage applied to the bridge is usually limited by the power dissipation capabilities of the standard and unknown resistors.

Errors. Kelvin bridge resistance-measurement errors are caused by the same factors as for the Wheatstone bridge. However, additional sources of error, as implied by the second term of Eq. (2), must be evaluated since these factors will seldom be reduced to zero. For minimum error the yoke resistance should be made as low as possible by physically placing the commonly connected current terminals of the unknown and standard as close together as possible and connecting with a low-resistance lead.

The ratio resistors each include not only the resistance of the resistors but also that of the interconnecting wiring and external leads and the contact resistance of the potential circuit contacts. The external leads are most likely to cause errors, and they should therefore be of the same resistance so that Eq. (3) will be fulfilled as nearly as possible. In addition, they should be relatively short, since the addition of a large resistance (long leads) will introduce an error in the calibrated ratio R_B/R_A. For precise measurements, trimmer adjust-

KELVIN BRIDGE

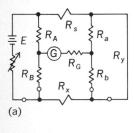

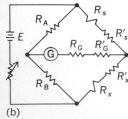

Kelvin bridge.
(*a*) Actual circuit.
(*b*) Equivalent Wheatstone bridge circuit.

$$R'_s = \frac{R_y R_a}{R_y + R_a + R_b}$$

$$R'_x = \frac{R_y R_b}{R_y + R_a + R_b} \quad (1)$$

$$R'_G = \frac{R_a R_b}{R_y + R_a + R_b}$$

ments are required in the ratio-arm circuits and provision is made to connect the bridge resistors into two different Wheatstone bridge configurations. By successively balancing first the Kelvin network and then each of the Wheatstone networks, these additive errors are virtually eliminated. *See* BRIDGE CIRCUIT; RESISTANCE MEASUREMENT.   [CHARLES E. APPLEGATE]

Kelvin's circulation theorem

A theorem in fluid dynamics that pertains to an incompressible, inviscid fluid. A direct consequence of this theorem is a great simplification in understanding and analyzing a large class of fluid flows called irrotational flows. The circulation Γ about a closed curve in a fluid is defined as the line integral of the component of velocity along the contour, as in Eq. (1), where $\mathbf{v}$ is the fluid velocity and

$$\Gamma = \oint \mathbf{v} \cdot ds \qquad (1)$$

ds is a length element along the curve. (Circulation bears a strong analogy to the work done in moving a particle around a closed curve in a force field.) Kelvin's theorem states that in an incompressible, inviscid fluid the circulation along a closed curve, always consisting of the same fluid particles, does not change with time. An important consequence of this theorem relates to fluid motions starting from rest or uniform motion; in such flows the circulation is initially zero for every possible closed curve and hence remains equal to zero thereafter according to Kelvin's theorem. This implies that the line integral of the velocity taken from fixed point A to fixed point B is independent of the path from A to B. Consider, for example, two different paths, C_1 and C_2, from A and B; then $C_1 - C_2$ forms a closed contour, and the line integral around it vanishes, showing that the line integrals along C_1 and C_2 are equal. If the line integral is independent of path, the integrand must be a perfect differential; that is, the velocity must equal the gradient of some scalar function $\mathbf{v} = \nabla\phi$ so that Eq. (2) holds, where ϕ is called the velocity potential

$$\int_A^B \mathbf{v} \cdot ds = \int_A^B \nabla\phi \cdot ds = \phi_B - \phi_A \qquad (2)$$

and is a function of position in the fluid and time, $\phi = \phi(\mathbf{r}, t)$. Because Eq. (3) is valid, the fluid motion

$$\mathrm{curl}\ \mathbf{v} = \mathrm{curl}\ (\mathrm{grad}\ \phi) = 0 \qquad (3)$$

is irrotational. The fluid motion is also divergence-free (solenoidal) from consideration of continuity. Thus it follows that div $\mathbf{v}$ = div (grad ϕ) = 0, which is Laplace's equation. *See* BERNOUILLI'S THEOREM; FLUID-FLOW PRINCIPLES; LAPLACE'S IRROTATIONAL MOTION.   [ARTHUR E. BRYSON, JR.]

Kelvin's minimum-energy theorem

A principle of fluid mechanics which states that the irrotational motion of an incompressible, inviscid fluid occupying a simply connected region has less kinetic energy than any other fluid motion consistent with the boundary condition of zero relative velocity normal to the boundaries of the region. This remarkable theorem is easily proved as follows. Let T be the kinetic energy of the irrotational motion with velocity potential ϕ. Let T_1 be the kinetic energy of another motion with velocity field $\mathbf{v} = \nabla\phi + \mathbf{v}_0$. From continuity it follows that $\nabla \cdot \mathbf{v}_0 = 0$, and from the boundary condition,

$\mathbf{n} \cdot \mathbf{v}_0 = 0$ on the boundary, where $\mathbf{n}$ is unit vector normal to the boundary surface. It follows that Eq. (1) is valid. However $\mathbf{v}_0 \cdot \nabla\phi = \nabla \cdot \phi\mathbf{v}_0$ because

$$T_1 = T + \frac{\rho}{2} \int\int\int \mathbf{v}_0 \cdot \mathbf{v}_0\, d\tau + \rho \int\int\int \mathbf{v}_0 \cdot \nabla\phi\, d\tau \qquad (1)$$

$\phi\nabla \cdot \mathbf{v}_0 = 0$, and applying Gauss' divergence theorem to the last integral above, it becomes integral (2), which vanishes, since $\mathbf{v}_0 \cdot \mathbf{n} = 0$ on the bound-

$$\rho\int\int \phi\mathbf{v}_0 \cdot \mathbf{n}\, dA \qquad (2)$$

ary. The second integral is a positive quantity, since its integrand is everywhere positive, thus proving that $T_1 > T$. *See* D'ALEMBERT'S PARADOX; FLUID-FLOW PRINCIPLES; LAPLACE'S IRROTATIONAL MOTION.   [ARTHUR E. BRYSON, JR.]

Kentucky coffee tree

A large, strikingly distinct tree, *Gymnocladus dioica*, which usually grows 80–90 high ft, but sometimes attains a height of 110 ft and a diameter of 5 ft. The species name, *dioica*, means that the tree is dioecious; that is, male and female flowers are on

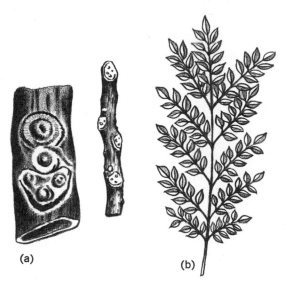

(a)   (b)

Kentucky coffee tree (*Gymnocladus dioica*). (*a*) Parts of branch showing leaf scars and winter buds. (*b*) Branch.

different individuals. It grows from eastern Nebraska, Kansas, and Oklahoma to southern Ontario, western New York, and Pennsylvania, and southwestward to Louisiana. It can readily be recognized when in fruit by its leguminous pods containing hard, heavy, red-brown seeds, which were used by early settlers as a substitute for coffee, hence the name coffeetree. The branches are stout and thick, and the bark has thin, twisted ridges. The leaves are twice pinnate, and the winter buds, sunken in the bark, are superposed, two or three together (see illustration). Never a common tree, it is sometimes cultivated in parks and gardens of the eastern United States and northern and central Europe. It is sometimes used as a street tree. The wood is hard and reddish, and is used for construction. It is durable in contact with the soil and is also used for railroad ties and fenceposts. *See* FOREST AND FORESTRY; ROSALES; TREE.

[ARTHUR H. GRAVES/KENNETH P. DAVIS]

Kenyapithecus

The genus name given by L. S. B. Leakey to several fossil Hominoidea from the Miocene deposits of Kenya. The older specimens, referred to as "Kenyapithecus africanus," come from Lower Miocene deposits on Rusinga Island and at Songhor. These specimens are now included in the genus *Dryopithecus*, which includes most of the Miocene and Pliocene Pongidae, or apes. The younger specimens, initially called "Kenyapithecus wickeri," derive from Upper Miocene deposits at Fort Ternan. Moderate-sized with short faces, these primates are the East African representatives of the genus *Ramapithecus* and are considered by many to be the earliest known hominids. *See* FOSSIL MAN; RAMAPITHECUS.   [ERIK TRINKAUS]

Bibliography: E. L. Simons, *Primate Evolution*, 1972.

Kepler's equation

The mathematical relationship between two different systems of angular measurement of the position of a body in an ellipse; specifically the relation between the mean anomaly M and eccentric anomaly θ', $M = \theta' - e \sin \theta'$, where e is the eccentricity of the ellipse. *See* ELLIPSE; PLANET.

The true position of a planet in an elliptical orbit can be represented by the angle θ (true anomaly) measured at the focus between the line directed to the planet and the line directed to the perihelion. The radius vector r from the focus to the planet can be expressed by $r = a(1 - e^2)/(1 + e \cos \theta)$, where a is the semimajor axis. The radius vector may also be expressed in terms of the eccentric anomaly θ' by $r = a(1 - e \cos \theta')$, where θ' is the angle at the center of the ellipse measured from perihelion along the circumscribed circle of radius a to the point whose projection perpendicular to the major axis passes through the planet. The true anomaly θ may be expressed in terms of the eccentric anomaly θ' by $\tan (\theta/2) = [(1 + e)/(1 - e)]^{1/2} \tan (\theta'/2)$. In actual practice, however, it is more convenient to describe the angular position of a planet in an elliptical orbit at anytime t by means of its average angular velocity n (called mean motion) and the time T of last perihelion passage. This angle M (mean anomaly) is expressed by $M = n(t - T)$. Therefore, given the orbital elements a, e, n, and T, it is possible by means of Kepler's equation and the intermediary angle θ' to evaluate the true anomaly θ and the actual position of the planet in the orbit for any instant t.

Several practical methods for the iterative solution of this transcendental equation exist: (1) Starting with the value of M and denoting approximate values of θ' and $\sin \theta'$ by θ'_0 and $\sin M$, solve $\theta'_0 = M + e \sin M$. Denoting a second approximation to θ' by θ'_1, solve $\theta'_1 = M + e \sin \theta'_0$. Indicating the third approximation to θ' by θ'_2 solve $\theta'_2 = M + e \sin \theta'_1$, continuing the iteration until the required convergence of θ'_i is obtained. (2) Starting with M and an approximate value of θ' solve $M_0 = \theta'_0 - e \sin \theta'_0$, where M_0 corresponds to θ'_0. Set $M - M_0 = \Delta M_0$ and let $\Delta \theta'_0$ be a first correction to θ'_0, then $\Delta \theta'_0 = \Delta M_0 / 1 - e \cos \theta'_0$. Set $\theta'_1 = \Delta \theta'_0 + \Delta \theta'_0$, evaluate M_1 corresponding to θ'_1, form $M - M_1 = \Delta M_1$, and solve $\Delta \theta'_1 = \Delta M_1 / 1 - e \cos \theta'_1$.

Then $\theta'_2 = \theta'_1 + \Delta \theta'_1$, and the process is repeated until the required convergence of $\Delta \theta'_i$ is reached.

[RAYNOR L. DUNCOMBE]

Keratitis

Any inflammation involving the cornea of the eye. It may involve the cornea alone (keratitis), or more commonly, both cornea and conjunctiva (keratoconjunctivitis). The various forms of keratitis and keratoconjunctivitis together constitute a major portion of all eye disease. Fortunately many forms are not severe and do not result in visual impairment; nevertheless, keratitis is a disease of major importance, being second only to cataract and equal to glaucoma as a cause of blindness.

Corneal structure and injuries. The outer surface of the cornea is covered by a delicate membrane (ephithelium) composed of flattened epithelial cells and closely resembling the mucous membrane lining the nasal passages and mouth. The substance (stroma) of the cornea is composed of interlacing layers of collagen fibrils similar to those seen in tendons, ligaments, and other connective tissues of the body but more regularly arranged. It is this regularity of structure that gives to the cornea its property of optical transparency. Any alteration of structural regularity due to epithelial or stromal destruction, the presence of pus, edema fluid, or scar formation results in an opaque region in the affected cornea.

Such an opacity will usually be temporary if the injury is limited to the epithelium but permanent if the stroma is also damaged. Although the response of the cornea to injury is similar in many respects to that of other body tissues, it is modified by the fact that the normal cornea does not contain blood vessels. Blood vessels from the surrounding conjunctiva are stimulated to grow into the cornea only if injury is severe enough to cause destruction and swelling of the corneal stroma. Since scar formation requires the presence of blood vessels, this means that injuries of only the epithelial membrane ordinarily heal without scarring. In such cases healing occurs by growth and spread of epithelial cells from adjacent uninjured zones over the injured area, leaving little or no residual opacity. These epithelial cells have a remarkable capacity for such growth and can completely recover the entire cornea within a few days. Deeper injuries which cause ingrowth of blood vessels leave a permanent mark in the form of a scar, visible as a gray-white opaque region in the cornea. The degree of visual impairment is determined by the position as well as the size of the scar.

Etiology. Keratitis may be caused by mechanical injury, infectious agents (bacteria, viruses, fungi, and protozoa), extremes of heat or cold, drying, radiation, corrosive or irritating chemicals, or allergy, or it may occur in association with diseases affecting the entire body, such as a vitamin deficiency. Keratitis often accompanies other eye diseases, such as glaucoma. *See* GLAUCOMA.

Since the cornea reacts to injury in only two basic ways, either by regrowth of epithelium in superficial injuries or by varying degrees of scar formation in more severe injuries, the degree of permanent impairment of vision is determined primarily by the extent of the injury rather than by the nature of the injurious agent. Of the various agents, bacteria are the most important, since a cornea injured by any method is particularly susceptible to superimposed bacterial infection, frequently resulting in damage far greater than the initial injury. *See* EYE (VERTEBRATE); EYE DISORDERS.   [W. ROBERT ADAMS]

Keratoconjunctivitis

A viral disease characterized by an acute conjunctivitis, followed by keratitis, which in some cases leaves round, superficial opacities in the cornea for up to 2 years. It is also known as shipyard eye.

The etiological agent is adenovirus type 8. Diagnosis is made by isolation of virus from ocular secretions inoculated into tissue culture, or by the demonstration of antibody responses in the blood. *See* ADENOVIRUS; ANTIBODY; CULTURE, TISSUE.

Sporadic cases and epidemics have occurred in many countries, particularly in shipyards and industrial plants. The mode of spread is not positively known. Rigid asepsis in medical divisions of industrial institutions is recommended to prevent possible spread by medical attendants or by instruments. *See* ANIMAL VIRUS; KERATITIS.

[JOSEPH L. MELNICK]

Kernite

A hydrated borate mineral with chemical composition $Na_2B_4O_7 \cdot 4H_2O$. It occurs in large quantity at the Kramer borate deposit near Boron, Calif., chiefly associated with borax, $Na_2B_4O_7 \cdot 10H_2O$. Clear, single crystals of kernite 2–3 ft thick and somewhat longer than this have been found at the deposit. The mineral kernite, earlier called rasorite, is a major source of boron and boron compounds. *See* BORATE MINERALS; BORON.

Kernite is a monoclinic, colorless mineral with vitreous luster; it gives a white streak, its hardness is 2.5 on Mohs scale, and its specific gravity is 1.91; twinning occurs on {110}. The crystals are characterized by perfect cleavages parallel to the planes (001) and (100). Hence, when handled, they tend to break down into a mass of parallel fine fibers resembling tremolite asbestos (see illustration). Because of this behavior, and from considerations of the physical chemistry of kernite, it was

1 in.

Kernite crystal, Kern county, Calif. (*Specimen from Department of Geology, Bryn Mawr College*)

considered that the crystals contain infinite chains of composition $[B_4O_6(OH)_2]_n^{-2n}$ and that the structural formula of kernite is $Na_2[B_4O_6(OH)_2] \cdot 3H_2O$. These postulates have been verified by a detailed x-ray crystal structure analysis of kernite.

The minerals kernite, $Na_2[B_4O_6(OH)_2] \cdot 3H_2O$, tincalconite, $Na_2[B_4O_5(OH)_4] \cdot 3H_2O$, and borax, $Na_2[B_4O_5(OH)_4] \cdot 8H_2O$, are closely related, as might be expected from their structural formulas. Borax and tincalconite both contain polyions of composition $[B_4O_5(OH)_4]^{-2}$ and differ only in molecular H_2O content; thus they convert readily to one another with changes in the humidity. On the other hand, kernite, containing infinite chains, is formed from borax or tincalconite by the more difficult process of polymerization. In nature, kernite is apparently formed from a primary deposit of borax under temperature-pressure conditions of the order of 58°C and 150 atm. *See* POLYMERIZATION.   [CHARLES L. CHRIST]

Bibliography: R. F. Geise, Jr., Crystal structure of kernite, $Na_2B_4O_6(OH)_2 \cdot 3H_2O$, *Science*, 154: 1453, 1966.

Kerogen

A name given to the complex organic matter present in carbonaceous shales and oil shales. It is insoluble in all common solvents but on destructive distillation yields oil, gas, and acidic and basic compounds. Kerogen is formed by the biochemical and dynamochemical conversion of plant and animal remains; both may be present in variable proportions. It is the most common form of organic carbon on Earth, and it has been estimated that there is 1000 times as much kerogen as coal. Because of its diversified origin, kerogen varies in composition, consisting of approximately 77–83% carbon, 5–10% hydrogen, 10–15% oxygen, and some nitrogen. *See* OIL SHALE.

[IRVING A. BREGER]

Kerosine

A refined petroleum fraction used as a fuel for heating and cooking, jet engines, lamps, and weed burning, and as a base for insecticides. Kerosine, known also as lamp oil, is recovered from crude oil by distillation. It boils in the approximate range of 350–550°F. Most marketed grades, however, have narrower boiling ranges. The specific gravity is about 0.8. Determined by the Abel tester, the flash point is not below 73°F, but usually a higher flash point is specified. Down to a temperature of −25°F, kerosine remains in the liquid phase. Components are mainly paraffinic and naphthenic hydrocarbons in the C_{10}–C_{14} range. A low content of aromatics is desirable except when kerosine is used as tractor fuel.

Specifications are established for specific grades of kerosine by government agencies and by refiners. Since these specifications are developed from performance observations, they are adhered to rigidly to assure satisfactory operation. For use in lamps, for example, a highly paraffinic oil is desired because aromatics and naphthenes give a smoky flame; and for satisfactory wick feeding, a viscosity no greater than 2 centipoises is required in this application. Furthermore, the nonvolatile components must be kept low. In order to avoid atmospheric pollution, sulfur content must be low;

a minimum flash point of 100°F is desirable to reduce explosion hazards.

Today, kerosine represents only a little over 4% of the total petroleum-products production in the United States, whereas it was the major product in the 1800s. Kerosine production and kerosine sales as range oil have remained fairly constant over the past decade. Tractor fuel now represents an insignificant percentage of the total production. Use of kerosine as a jet fuel continues to increase as more jet planes are put into operation. Kerosine also is the principal hydrocarbon fuel for rockets.

The price of kerosine has followed the price of crude oil, and this cost continues to be the dominant price factor; however, changing use patterns and specifications may exert an additional upward force on the price of kerosine. *See* DISTILLATE FUEL; JET FUEL; PETROLEUM PRODUCTS.

[HAROLD C. RIES]

Kerr effect

Electrically induced birefringence that is proportional to the square of the electric field. When a substance (especially a liquid or a gas) is placed in an electric field, its molecules may become partly oriented. This renders the substance anisotropic and gives it birefringence, that is, the ability to refract light differently in two directions. This effect, discovered in 1875 by John Kerr, is called the electrooptical Kerr effect, or simply Kerr effect.

When a liquid is placed in an electric field, it behaves optically like a uniaxial crystal with the optical axis parallel to the electric lines of force. The Kerr effect is usually observed by passing light between two capacitor plates inserted in a glass cell containing the liquid. Such a device is known as a Kerr cell. There are two principal indices of refraction, n_o and n_e (known as the ordinary and extraordinary indices), and the substance is called a positively or negatively birefringent substance, depending on whether $n_e - n_o$ is positive or negative.

Light passing through the medium normal to the electric lines of force (that is, parallel to the capacitor plates) is split into two linearly polarized waves traveling with the velocities c/n_o and c/n_e, respectively, where c is the velocity of light, and with the electric vector vibrating perpendicular and parallel to the lines of force.

The difference in propagation velocity causes a phase difference δ between the two waves, which,

for monochromatic light of wavelength λ_0, is $\delta = (n_e - n_o)x/\lambda_0$, where x is the length of the light path in the medium.

Kerr constant. Kerr found empirically that $(n_e - n_o) = \lambda_0 B E^2$, where E is the electric field strength and B a constant characteristic of the material, called the Kerr constant. Havelock's law states that $B\lambda n/(n-1)^2 = k$, where n is the refractive index of the substance in the absence of the field and k is a constant characteristic of the substance but independent of the wavelength λ. Roughly speaking, the Kerr constant is inversely proportional to the absolute temperature. The phase difference δ is determined experimentally by standard optical techniques. If the wavelength λ_0 is expressed in centimeters, and the field strength E in statvolts/cm (1 statvolt = 300 volts), the Kerr constant for carbon disulfide, which has been determined most accurately, is $B = 3.226 \times 10^{-7}$. Values of B range from -23.00×10^{-7} for paraldehyde to $+346.0 \times 10^{-7}$ for nitrobenzol.

The theory of the Kerr effect is based on the fact that individual molecules are not electrically isotropic but have permanent or induced electric dipoles. The electric field tends to orient these dipoles, while the normal agitation tends to destroy the orientation. The balance that is stuck depends on the size of the dipole moment, the magnitude of the electric field, and the temperature. This theory accounts well for the observed properties of the Kerr effect.

In certain crystals there may be an electrically induced birefringence that is proportional to the first power of the electric field. This is called the Pockels effect. In these crystals the Pockels effect usually overshadows the Kerr effect, which is nonetheless present. In crystals of cubic symmetry and in isotropic solids (such as glass) only the Kerr effect is present. In these substances the electrically induced birefringence (Kerr effect) must be carefully distinguished from that due to mechanical strains induced by the same field. *See* ELECTROOPTICS.

Kerr shutter. An optical Kerr shutter or Kerr cell consists of a cell containing a liquid (for example, nitrobenzene) placed between crossed polarizers. As such, its construction very much resembles that of a Pockels cell. With a Kerr cell (see illustration), an electric field is applied by means of an electronic driver v and capacitorlike electrodes in contact with the liquid; the field is perpendicular to the axis of light propagation and at 45° to the axis of either polarizer. In the absence of a field, the optical path through the crossed polarizers is opaque. When a field is applied, the liquid becomes birefringent, opening a path through the crossed polarizers. In commercial Kerr cell shutters, the electric field (a typical value is 10 kV/cm) is turned on and off in a matter of several nanoseconds (1 ns = 10^{-9} s). For laser-beam modulation the Pockels cell is preferred because it requires smaller voltage pulses. Kerr cell shutters have the advantage over the Pockels cell of a wider acceptance angle for the incoming light. *See* OPTICAL MODULATORS.

A so-called ac Kerr effect or optical Kerr effect has also been observed and put to use in connection with lasers. When a powerful plane-polarized

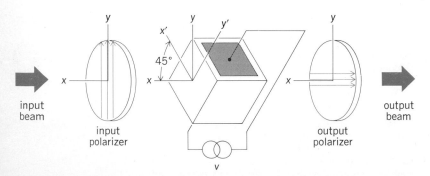

Kerr shutter. The arrows designate a light beam propagating through the cell. Polarizer axes are parallel to *x* and *y* axes; electric field of cell is parallel to *x'* or *y'* axis. (*Adapted from Amnon Yariv, Quantum Electronics, John Wiley & Sons, p. 315, 1967*)

laser beam propagates through a liquid, it induces a birefringence through a mechanism that is very similar to that of the ordinary, or dc, Kerr effect. In this case, it is the ac electric field of the laser beam (oscillating at a frequency of several hundred terahertz) which lines up the molecules. By using laser pulses with durations of only a few picoseconds (1 ps = 10^{-12} s), extremely fast optical Kerr shutters have been built in the laboratory. *See* LASER.   [MICHEL A. DUGUAY]

Bibliography: J. M. Stone, *Radiation and Optics: An Introduction to the Classical Theory*, 1963.

Ketene

A colorless, highly reactive gas, bp −56°C, mp −151°C. It is soluble in ether and acetone, and decomposes in water and alcohol. The ketene molecule shown below contains a highly reactive

carbonyl group $>C=O$.

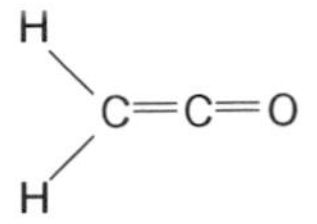

$$\underset{H}{\overset{H}{>}}C=C=O$$

Ketene is produced by the pyrolysis of acetone or acetic acid vapor, as in Eqs. (1) and (2).

$$CH_3-\overset{O}{\overset{\|}{C}}-CH_3 \xrightarrow{700°C} CH_2=C=O + CH_4 \quad (1)$$

$$CH_3\overset{O}{\overset{\|}{C}}-OH \xrightarrow[\substack{\text{Phosphate} \\ \text{catalyst}}]{700°C} CH_2=C=O + H_2O \quad (2)$$

Carboxylic acids add to ketene to give acid anhydrides, as in Eq. (3), and this is the basis of an

$$CH_2=C=O + RCOOH \rightarrow$$

$$\left[\underset{CH_2=C}{\overset{OH}{|}}-O-\overset{O}{\overset{\|}{C}}-R\right] \rightarrow CH_3\overset{O}{\overset{\|}{C}}-O-\overset{O}{\overset{\|}{C}}-R \quad (3)$$

important commercial process for the manufacture of acetic anhydride. Ketene reacts with the enol form of ketones to produce enol acetates, as in Eq. (4). Molecules containing O—H or N—H

$$CH_3\overset{O}{\overset{\|}{C}}-R \rightleftharpoons \underset{CH_2=C}{\overset{OH}{|}}-R \xrightarrow{CH_2=C=O}$$

$$\underset{CH_2=C-R}{\overset{O}{\overset{\|}{O-C-CH_3}}} \quad (4)$$

bonds add to the carbonyl group, for example, Eq. (5). Dimerization of ketene to diketene occurs on

$$CH_2=C=O + R_2NH \rightarrow \left[\underset{CH_2=C}{\overset{OH}{|}}-NR_2\right] \rightarrow$$

$$CH_3-\overset{O}{\overset{\|}{C}}-NR_2 \quad (5)$$

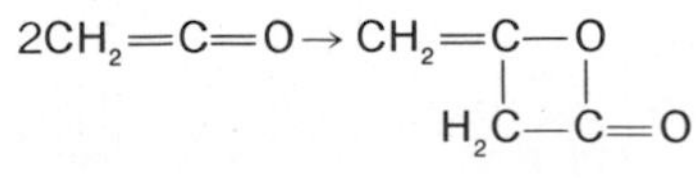

$$2CH_2=C=O \rightarrow \underset{H_2C-C=O}{\overset{CH_2=C-O}{|}} \quad (6)$$

storage of ketene even at low temperature, as in Eq. (6). *See* KETONE.

[DAVID A. SHIRLEY]

Bibliography: J. D. Roberts and M. C. Caserio, *Basic Principles of Organic Chemistry*, 1964.

Ketone

One of a class of chemical compounds of the general formula given below. In the formula,

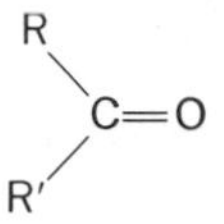

$$\underset{R'}{\overset{R}{>}}C=O$$

R and R′ are alkyl, aryl, or heterocyclic radicals. The groups R and R′ may be the same or different or incorporated into a ring as in $CH_2CH_2CH_2CH_2C=O$ (cyclopentanone). The ketones, acetone and methyl ethyl ketone, are used as solvents. Ketones are important intermediates in the syntheses of organic compounds.

By common nomenclature rules, the R and R′ groups are named followed by the word ketone, for example, $CH_3CH_2COCH_2CH_3$ (diethyl ketone), $CH_3COCH(CH_3)_2$ (methyl isopropyl ketone), and $C_6H_5COC_6H_5$ (diphenyl ketone). IUPAC nomenclature uses the hydrocarbon name corresponding to the maximum number of carbon atoms in a continuous chain in the ketone molecule, followed by "one" and preceded by a number designating the position of the carbonyl group in the carbon chain. The first two ketones above are named 3-pentanone and 3-methyl-2-butanone.

Reactions of ketones. Addition to the carbonyl group is the most important type of ketone reaction. Ketones are generally less reactive than aldehydes in addition reactions. Methyl ketones are more reactive than the higher ketones because of steric group effects.

Hydrogen adds catalytically to the carbonyl group, and lithium aluminum hydride gives the same type of product—the secondary alcohol, as shown in Eq. (1). The Grignard reagent adds to the

$$RCOR' \xrightarrow[\text{or LiAlH}_4]{H_2(\text{Ni or Pt})} RCHOHR' \quad (1)$$

carbonyl group, and tertiary alcohols are formed by hydrolysis, Eq. (2). The Reformatsky reagent,

$$RCOR' \xrightarrow{R''MgX} \underset{R''}{\overset{OMgX}{\underset{|}{RCR'}}} \xrightarrow[H^+]{H_2O} \underset{R''}{\overset{OH}{\underset{|}{R-C-R'}}} \quad (2)$$

$XZnCH_2COOR$, reacts similarly. Hydrogen cyanide and sodium bisulfite will add to methyl ketones, as in Eq. (3). Alcohols do not add readily to

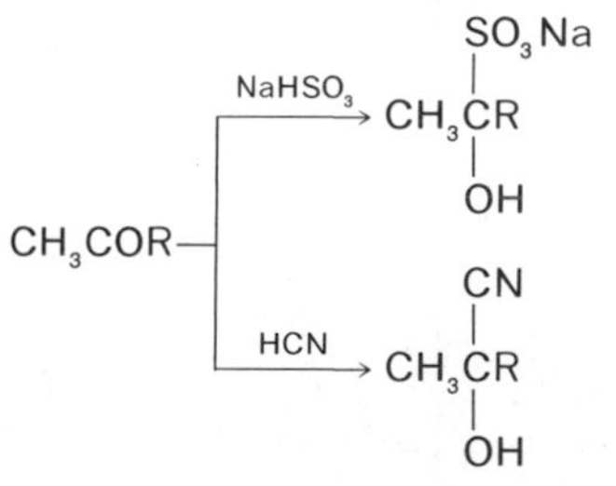

$$CH_3COR \begin{cases} \xrightarrow{NaHSO_3} \underset{OH}{\overset{SO_3Na}{\underset{|}{CH_3CR}}} \\ \xrightarrow{HCN} \underset{OH}{\overset{CN}{\underset{|}{CH_3CR}}} \end{cases} \quad (3)$$

the ketone carbonyl as they do to the aldehyde carbonyl, but ketals may be formed by the action of orthoformates, as in Eq. (4). Amine deriva-

$$RCOR' + HC(OC_2H_5)_3 \xrightarrow{H^+}$$
$$R-C(OC_2H_5)_2R' + HCOOC_2H_5 \quad (4)$$

tives such as hydroxylamine (NH_2OH), phenyl-hydrazine ($C_6H_5NHNH_2$), and semicarbazide ($NH_2CONHNH_2$) add to the carbonyl by breaking an N—H bond and with subsequent loss of water, as in Eq. (5). The resulting oximes, phenylhydra-

$$RCOR' + C_6H_5NHNH_2 \xrightarrow{H^+} R\overset{\overset{NNHC_6H_5}{\|}}{C}R' + H_2O \quad (5)$$

zones, and semicarbazones are useful derivatives for the identification and characterization of ketones.

Ketones supply alpha hydrogen in aldol type condensation reactions, as in Eq. (6), but can sup-

$$H_2CO + CH_3COR \xrightarrow{OH^-} RCOCH_2CH_2OH \xrightarrow{H^+}$$
$$RCOCH=CH_2 + H_2O \quad (6)$$

ply a carbonyl group only to a limited extent because of its lower reactivity. An exception is the self-condensation of acetone to diacetone alcohol represented by Eq. (7).

$$2CH_3COCH_3 \xrightarrow{Ba(OH)_2} CH_3\overset{\overset{OH}{|}}{\underset{\underset{CH_3}{|}}{C}}CH_2COCH_3 \quad (7)$$

The lower cyclic ketones generally have more reactive carbonyl groups for all reactions because of the exposed position of the group and the rigidity of the ring.

Ketones are oxidized less readily and less selectively then aldehydes to give oxidation products such as carboxylic acids, by cleavage of the bonds from the carbonyl carbon atom to the adjacent atom. They fail to give the Tollen's and Fehling's tests, although α-hydroxy ketones will give these tests.

Methyl ketones give the haloform reaction with solutions of iodine in aqueous potassium hydroxide, Eq. (8).

$$CH_3COR \xrightarrow[KOH]{I_2} CHI_3 + RCOOK \quad (8)$$

The oxygen atom of the carbonyl group of ketones may be replaced by two chlorine atoms by treatment with phosphorus pentachloride, PCl_5, although, in general, the yields are poor. Equation (9) represents the reaction. Chlorine or bromine

$$RCOR' + PCl_5 \rightarrow RCCl_2R' + POCl_3 \quad (9)$$

will substitute the alpha hydrogen atoms of ketones, as in Eq. (10). The resulting α-haloketones

$$RCH_2COR' + Cl_2 \rightarrow R\overset{\overset{Cl}{|}}{C}HCOR' + HCl \quad (10)$$

are quite reactive in displacement reactions of the halogen atom.

Preparation. The dehydrogenation or oxidation of secondary alcohols at elevated temperature

yields ketones, as in Eq. (11), and this is a valuable

$$RCH(OH)R' \xrightarrow{Cu} RCOR' + H_2 \quad (11)$$

commercial method in which the secondary alcohol is of low cost, for example, those obtained by olefin hydration.

Aromatic ketones may be prepared by the Friedel-Crafts reaction by acylation with acid halides or anhydrides, Eqs. (12) and (13). The Gri-

$$(12)$$

$$(13)$$

gnard reagent adds to nitriles to give ketimines which form ketones on hydrolysis, Eqs. (14) and (15). Organozinc and organocadmium reagents

$$RC\equiv N \xrightarrow{R'MgX} R-\overset{\overset{R'}{|}}{C}=N-MgX \xrightarrow{H_2O}$$
$$R-\overset{\overset{R'}{|}}{C}=NH \quad (14)$$

$$R-\overset{\overset{R'}{|}}{C}=NH + H_2O \rightarrow RCOR' + NH_3 \quad (15)$$

give ketones upon reaction with acid halides, Eq. (16). These organometallic compounds are best

$$RCOCl + R'CdX \rightarrow RCOR' + CdXCl \quad (16)$$

prepared from the Grignard reagent and the metal halide as in Eq. (17).

$$RMgCl + ZnCl_2 \rightarrow RZnCl + MgCl_2 \quad (17)$$

Beta-ketoesters formed in the Claisen condensation are cleaved by aqueous sodium hydroxide solution to form ketones, Eqs. (18) and (19). The

$$2RCH_2COOEt \xrightarrow{EtONa}$$
$$RCH_2COCHRCOOEt + EtOH \quad (18)$$

$$RCH_2COCHRCOOEt + H_2O \xrightarrow{NaOH}$$
$$RCH_2COCH_2R + CO_2 + EtOH \quad (19)$$

Dieckmann condensation of esters of dibasic acids leads in similar fashion to cyclic ketones, Eqs. (20) and (21).

$$EtOOC(CH_2)_4COOEt \xrightarrow{NaOEt}$$
$$\overline{CH_2CH_2CH_2COCHCOOEt} \quad (20)$$

$$\overline{CH_2CH_2CH_2COCHCOOEt} + H_2O \xrightarrow{NaOH}$$
$$\overline{CH_2CH_2CH_2CH_2CO} + CO_2 + EtOH \quad (21)$$

Both cyclic and open chain ketones are formed by pyrolysis of calcium or thorium salts of dibasic acids or monobasic acids, Eqs. (22) and (23).

$$\tfrac{1}{2}CaOOC(CH_2)_4COO\tfrac{1}{2}Ca \xrightarrow{\text{Heat}}$$

$$CH_2CH_2CH_2CH_2CO + CaCO_3 \quad (22)$$

$$2RCOO\tfrac{1}{2}Ca \xrightarrow{\text{Heat}} RCOR + CaCO_3 \quad (23)$$

Esters of phenols rearrange when heated with aluminum chloride to form phenyl ketones (Fries rearrangement) in which the carbonyl group is ortho or para to a phenolic hydroxyl, Eq. (24). *See*

$$\text{(structure)} \quad (24)$$

Aldehyde; Condensation reaction; Friedel-Crafts reaction; Grignard reaction; Haloform reaction; Reformatsky reaction; Steric effect (chemistry)   [DAVID A. SHIRLEY]
   Bibliography: R. Adams (ed.), *Organic Reactions*, vols. 5 and 8, 1949 and 1954; J. D. Roberts and M. C. Caserio, *Basic Principles of Organic Chemistry*, 1964.

Ketosis

An excessive accumulation of ketone bodies in the tissues, blood, and urine. Ketone bodies are acetoacetic acid, β-hydroxybutyric acid, and acetone, all of which are formed by the oxidation of fatty acids by the liver. *See* KETONE.

When a disturbance in carbohydrate metabolism is present, ketone bodies are produced in excess. This results when the tissue requirements for energy are not met by a readily available or utilizable supply of glucose. Fat and protein must then be mobilized to the liver, which transforms these substances into glucose, a process called gluconeogenesis. In the series of biochemical shifts involved, fatty acids are oxidized in large amounts so that their normal breakdown products, the ketone bodies, accumulate. *See* CARBOHYDRATE; CARBOHYDRATE METABOLISM; GLUCOSE.

The ketone bodies, by virtue of their acid properties, hold an equivalent amount of base, thus producing a form of acidosis. This is seen clinically in certain cases of diabetes, in starvation, occasionally in hyperthyroidism, and in a variety of other disease states in which there is an absolute or relative decrease in carbohydrate availability. *See* DIABETES; THYROID GLAND DISORDERS.

In ketosis, the breath has a fruity odor, and there is usually evidence of dehydration; weakness, malaise, headache, nausea, and vomiting are common. *See* DISEASE; PATHOLOGY.

[EDWARD G. STUART/N. KARLE MOTTET]

Kidney (vertebrate)

An organ involved with the elimination of water and waste products from the body. In vertebrates the kidneys are paired organs located close to the spine dorsal to the body cavity and are covered ventrally by the coelomic epithelium. They consist of a number of smaller functional units which are called urinary tubules or nephrons. The nephrons open to larger ducts, the collecting ducts, which open into a ureter. The two ureters run backward to open into the cloaca or into a urinary bladder.

The shape and the location of the kidneys varies in different animals. In fish, they are extremely elongated and may reach forward almost to the pericardium. In reptiles, the kidneys are smaller and are located in the pelvic region. In mammals, they are bean-shaped and found between the thorax and the pelvis. The number, structure, and function of the nephrons vary with evolution and, in certain significant ways, with the adaptation of the animals to their various habitats.

Nephron. In its most primitive form, found only in invertebrates, the nephron has a funnel opening into the coelomic cavity followed by a urinary tubule leading to an excretory pore. In amphibians, some of the tubules have this funnel, but most of the tubules have a Bowman capsule (Fig. 1). In all higher vertebrates, the nephron has the Bowman capsule, which surrounds a tuft of capillary loops, called the glomerulus, constituting the closed end of the nephron. The inner epithelial wall of the Bowman capsule is in intimate contact with the endothelial wall of the capillaries. Electronmicroscopy reveals that the wall of the capillaries, together with the inner wall of the Bowman capsule, forms a membrane ideally suited for filtration of the blood.

The nephron is differentiated into several parts which vary in different groups of vertebrates. Following the glomerulus comes a short neck segment which may be ciliated (this segment is absent in mammals). This is followed by the proximal tubule. Following the proximal tubule in lower vertebrates, there is a short intermediary segment; in mammals, there is the so-called thin segment comprising part of the loop of Henle. This is followed by the distal tubule, which is followed by the collecting duct. In lower vertebrates, all of the nephrons are convoluted and arranged perpendicular to the collecting ducts. In birds, a few of the nephrons form a straight loop, the loop of Henle, which runs parallel to the collecting ducts. In mammals, all of the nephrons have loops of Henle (Fig. 2). The glomerulus and convoluted proximal tubule are found in the cortex of the kidney. The loops of Henle run through the outer and inner medulla; some are short, others are long. The thin limb runs into a straight part of the distal tubule, which continues into the distal convoluted tubule found in the cortex. The capillaries are arranged in bundles which form true retia mirabilia, ideally suited for counter-current exchange. The collecting ducts run from the cortex through the medulla to the tip of the papilla, where they open into the renal pelvis. Through their entire course the collecting ducts are parallel to the loops of Henle and to the capillary loops.

The blood pressure in the capillaries of the glomerulus causes filtering of blood by forcing fluid, small molecules, and ions through the membrane into the lumen of Bowman's capsule. This filtrate contains some of the proteins and all of the smaller molecules in the blood. As the filtrate passes down through the tubule, the walls of the tubule extract those substances not destined for excretion and return them to the blood in adjacent capillaries. Many substances which are toxic to the organism are moved in the opposite direction from the blood into the tubules. The urine thus produced by each nephron is conveyed by the collecting duct

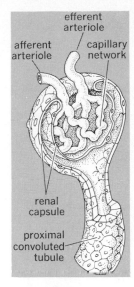

Fig. 1. Nephron, from frog kidney dissected to show glomerulus within Bowman capsule.

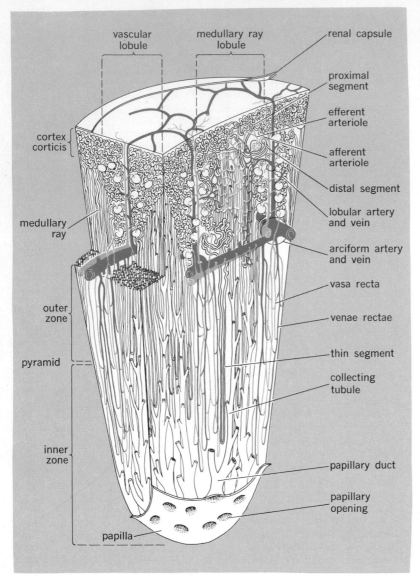

vascular lobule

medullary ray lobule

renal capsule

cortex corticis

proximal segment

efferent arteriole

afferent arteriole

distal segment

lobular artery and vein

medullary ray

arciform artery and vein

vasa recta

outer zone

venae rectae

pyramid

thin segment

collecting tubule

inner zone

papillary duct

papillary opening

papilla

Fig. 2. Diagram of a small section of a human kidney, showing the loops of Henle with their thick and thin segments, and the vasa recta and the collecting ducts as they are arranged parallel to one another in the outer and inner medulla. (*From Homer Smith, Principles of Renal Physiology, Oxford University Press, 1956*)

and ureter to the cloaca or bladder from which it can be eliminated.

Blood supply. In all classes of vertebrates the renal arteries deliver blood to the glomeruli and through a second capillary net to the tubules. The major blood supply to the kidney tubules comes, however, from the renal portal vein, which is found in all vertebrates except mammals and cyclostomes (Fig. 1). Venous blood from the tail and hindlimbs of the animal drains into the renal portal vein. It then splits up into a second capillary net which surrounds the renal tubules. Waste products from the venous blood can thus be secreted directly into the urinary tubules.

Evolutionary and adaptive aspects. It is assumed that the ancestors of vertebrates were marine forms, with body fluids at osmotic equilibrium with sea water. Migration into fresh water posed the threat of rapid dilution of body fluids by osmosis. The kidneys in fresh-water forms arose as the

major organ for the removal of excess water. A large glomerulus is present in amphibians and fishes found in fresh water today. Their nephrons have a long distal tubule which reabsorbs ions from the filtrate and returns them to the blood. Through this process the urine becomes dilute, and very little salt is lost hereby. However, when fishes reinvaded the ocean, or evolved into land forms, their kidneys had to be modified to their new habitats.

The salt-water teleost fishes maintain an osmotic blood concentration lower than that of the sea water. They are thus in danger of becoming dehydrated, and water conservation becomes a prime objective. Some marine teleosts have lost their glomeruli and the distal tubules entirely, and produce urine by tubular secretion alone. In other marine fishes the glomeruli have been reduced in size.

To maintain osmotic pressure of the blood lower than that of the sea water, the gills secrete salt into the sea water.

In land animals such as reptiles, birds, and mammals, water conservation is also a major objective. The primary function of the kidneys is to excrete salt and nitrogenous waste products. Reptiles and birds save water by excreting their nitrogenous wastes in the form of a nearly dry paste of uric acid crystal. The kidney of reptiles can only make urine that is isoosmotic or hypoosmotic to the blood, but not hyperosmotic. Reptiles faced with a large salt intake, such as marine turtles and lizards, excrete salt by specialized salt glands located in the head. Birds can produce a urine hyperosmotic to the blood because of the countercurrent system formed by the few loops of Henle. They excrete uric acid. Salt-water birds excrete excess salt through salt glands similar to those of the reptiles.

Mammals are peculiar in that, regardless of habitat, they always excrete almost all of their nitrogenous wastes in the form of urea. It may be for this reason that the mammal has developed its unique kidney, which can produce urine with several times the osmotic concentration of the blood. The ability of the mammalian kidney to concentrate urine is inherent in the spatial arrangement of renal tubules, blood vessels, and collecting ducts which permits the tubules to function together as a countercurrent multiplier system (Fig. 2). *See* EXCRETION; OSMOREGULATORY MECHANISMS.

Embryology. Although nephrons are assembled into kidneys in a variety of ways, the tubules always arise in the intermediate mesoderm along the upper border of the body cavity (Fig. 3). The first tubules develop at the anterior end of this mesoderm, followed by others in sequence behind. Variations in this plan permit the following kidney types to be distinguished: holonephros, pronephros, mesonephros, opisthonephros, and metanephros.

Holonephros. There is one nephron beside each somite along the entire length of the coelom. Supposedly, this is the ancestral condition and is thus referred to as an archinephros. It is seen today only in the larvae of a few primitive fishes, the myxinoid cyclostomes.

Pronephros. Nephrons are formed only in the anterior part of the intermediate mesoderm. The

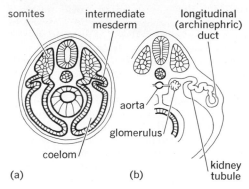

somites   intermediate   longitudinal
           mesderm   (archinephric)
                           duct

aorta

glomerulus

coelom

(a)   (b)   kidney
                        tubule

Fig. 3.   Diagrammatic cross section of the embryos to show origin of nephron. (a) Early stage with mesoderm differentiation. (b) Later stage.

pronephros is rarely found as an adult excretory organ, but is important because in all embryos it initiates the formation of the archinephric duct used by subsequent tubules for drainage. The pronephros is of interest also because it shows features presumed to be primitive, suggesting a series of stages in evolution of the nephron.

Mesonephros. The nephrons form behind the pronephros along the middle zone of the coelom. The tubules usually become involved in sperm transport. Drainage is through the archinephric duct. The mesonephros never forms an entire adult kidney. It is functional in the chick and other amniote embryos.

Opisthonephros. Nephrons develop from the middle and posterior portions of intermediate mesoderm. The middle, or mesonephric, tubules sometimes form only a minor part. The archinephric duct may be supplemented or even replaced by ureterlike accessory drainage ducts. This is the functional adult kidney in amphibians and fishes.

Metanephros. Nephrons form from only the extreme posterior part of the intermediate mesoderm, and are drained by a distinctive new canal, the ureter, which grows forward from the base of the archinephric duct before the tubules have begun to develop. This is the kidney of adult reptiles, birds, and mammals. Their embryos show a distinct sequence of pro-, meso-, and metanephric kidneys. *See* URINARY SYSTEM (VERTEBRATE).

[BODIL SCHMIDT-NIELSEN]

Kidney disorders

Probably because of their complex embryological derivation, the genitourinary organs are frequently the site of developmental abnormalities. Congenital absence of both kidneys permits only a brief postnatal life. Absence of one kidney is common and produces no symptoms, a single kidney being perfectly adequate for normal life. Horseshoe kidney, a descriptive term for fusion of the two kidneys at one pole, is not rare. Also familiar are double ureters, double pelves, and aberrant arteries. These conditions are significant only if they produce obstruction to the flow of urine. Obstruction due to any cause leads to dilatation of the proximal structures, known as hydroureter if the ureter is involved or hydronephrosis if the kidney pelvis is affected. One characteristic form of maldevelop-

ment is the polycystic kidney in which both kidneys become greatly enlarged and converted into many thin-walled cysts, apparently because of a diffuse defect in development of the tubules. Persons so afflicted usually develop renal failure and die in middle adult life.

Calculi. Stones in the kidney pelvis occur frequently. There is a deposition of chemicals present in urine, apparently starting in the tips of the papillae, later protruding into the renal pelvis, and ultimately breaking free. There are several different chemical types of calculi occurring under different circumstances. Uric acid and urate calculi reflect excessive metabolic production of these substances and appear particularly when the person suffers from gout. Calcium oxalate calculi commonly occur if there are excessive oxalates in the diet, derived particularly from fruits and vegetables. Crystalline calcium carbonate and phosphate stones may occur when an overactive parathyroid gland increases the excretion of calcium, when too much milk is ingested, or when alkalosis decreases the solubility of calcium salts. Amorphous carbonates and triple phosphates are characteristically found in calculi as a result of infection and stasis, probably being deposited around a core of pus and bacteria. Small calculi may be composed of a single component; large calculi usually are of mixed composition.

Stones may pass into the ureters when they are small or may remain in the pelvis and enlarge until, in some instances, they assume a stag-horn shape. Calculi are sometimes harmless and "silent," but more often cause some difficulty. They may lead to infection by eroding through the protective epithelium and providing a foothold for bacteria, or by causing obstruction to urine flow. When a stone passes into the ureter the spasm of the ureteral muscle in its efforts to expel the stone causes a violent pain known as renal colic. This is often accompanied by bloody urine as the stone tears small blood vessels. Most calculi are passed out of the body via the ureter, bladder, and urethra. Those which are not passed may require surgical treatment lest the obstruction and infection permanently damage the kidney.

Infections. Most infections of the kidney are bacterial and related to the pelvis. Classification according to degree of severity is as follows: pyelitis when only the pelvic lining is involved; pyelonephritis when the kidney substance is also affected; and pyonephrosis when the pelvis is greatly dilated and filled with pus. These infections are characteristically associated with some form of obstruction which has caused stasis of urine and a favorable milieu for bacterial growth. Common causes of obstruction are congenital abnormalities of the ureters in women; prostatic overgrowth in elderly men; and calculi, tumors, or loss of proper nervous control of bladder-emptying following spinal cord disease in either sex. The organisms most often involved are the colon bacillus and related gram-negative rods. It is thought that the organisms usually ascend the ureters to reach the kidneys; less frequently bacteria enter the bloodstream and infect the kidneys while being excreted. Acute pyelonephritis is accompanied by chills, fever, and pus in the urine. Recovery from single attacks is the rule, with healing by scarring. Repeated attacks or chronic infections may destroy so much kidney tissue that renal failure ensues.

Two other forms of infection are important, although less common. Cortical abscesses occur in patients with staphylococcus septicemia as single or more often multiple lesions. If such an abscess ruptures to the outside through the capsule of the kidney, a perinephric abscess is formed. Tuberculosis organisms may enter the kidney by the blood or lymph from a primary site in the lung or other infected organ. In untreated cases the tuberculous process may destroy the kidney; it may also spread to the bladder and other kidney.

Glomerulonephritis. This is a specific form of renal inflammation particularly involving the glomeruli. Apparently it is a result of a reaction between circulating streptococcal toxins and antibodies present in the cells of the glomeruli. The bacteria, which themselves do not enter the kidneys, are found chiefly in the throat. There are two main variants of the disease. Type I cases have an abrupt onset, usually 1 or 2 weeks after an acute streptococcal infection. The urine is smoky or bloody, the blood pressure high, and nitrogenous waste products are markedly increased in the blood. Most patients recover completely, although a few die in the early stages, and a few others develop chronic nephritis. Type II cases have an insidious onset with no obvious preceding infection. They manifest edema of the legs and eyelids, anemia, proteinuria, and later hypertension. Complete recovery is rare and there is usually gradual progression to complete kidney failure over a period of several years. In the early stages, the kidneys are enlarged, later becoming increasingly scarred and shrunken. The glomerular tufts are first enlarged and cellular with later scarring and final disappearance. The urine contains red blood cells and dissolved protein which leak through the damaged glomeruli, as well as precipitated protein in the form of casts of the tubules.

Nephrosis (lipoid nephrosis). This is a disease of small children. At the onset there is a gradual accumulation of tissue fluid, noticed first as swelling of the eyelids and legs. There is a marked loss of protein in the urine. The resultant very low level of serum albumin is inadequate to perform the normal osmotic functions of pulling water from the tissue spaces into the bloodstream, hence the edema. The blood cholesterol rises to a very high level, possibly as the result of the body's attempt to compensate for the lowered protein levels. The renal lesions are subtle, consisting chiefly of swelling of the walls of the glomerular capillaries. These swollen walls are excessively porous and the proteins leak through them. The cause of this ailment is unknown, although most cases are probably a form of glomerulonephritis and related to streptococcal toxins. Some of the children make complete recoveries. Others die from bacterial infections to which they are extremely susceptible, or from interference with body functions caused by excessive accumulations of fluid. Others develop chronic glomerulonephritis with hypertension, hematuria, and renal failure.

Nephrosclerosis. This term is frequently applied to kidneys damaged by arteriosclerosis or arteriolosclerosis. Under these conditions, as the arterial vessels, either large or small, become progressively narrower, the portions of renal substance which they supply are gradually damaged by anoxia and eventually die, to be replaced by scar tissue. The kidneys slowly shrink and the outer surfaces become roughened by depressed scars which alternate with the hypertrophied, still healthy intervening tissue. Ultimately so much parenchyma may be destroyed that renal failure occurs. Nephrosclerosis is only part of a generalized disorder of arteries, although the renal involvement may be the dominant feature. In the malignant form of nephrosclerosis, the arterioles become necrotic and rupture, giving a flea-bitten appearance to the kidney surface. In this form of the disease there is very high blood pressure and death occurs early.

If diabetes complicates nephrosclerosis, there are two distinctive features. One is an extreme narrowing of the arterioles in the region of the glomeruli. The other is the presence of peculiar round deposits in the glomeruli. This is called intracapillary glomerulosclerosis and is associated with marked albuminuria.

Hypertension. The kidneys are almost invariably diseased when high blood pressure is present, but there is not a simple cause and effect relationship. In ordinary essential hypertension, kidney tissue is often entirely normal after hypertension is well established, strongly suggesting that the lesions of nephrosclerosis are the result rather than the cause of the elevated blood pressure. Apparently the prolonged arteriolar spasm necessary to maintain the high blood pressure eventually leads to organic damage to arterioles followed by scarring of the kidneys. On the other hand, it is well established that damaged renal tissue may release pressor substances capable of causing hypertension. Pressors may be produced in a kidney whose major arteries are markedly narrowed, or in which pyelonephritis is present. The hypertension so produced can be eliminated by removal of the diseased kidney.

Lower-nephron nephrosis. This is the most widely used appellation for an extremely important acute reversible form of severe kidney shutdown. It occurs following prolonged shock and hypotension or after absorption of certain hemoglobin compounds following incompatible blood transfusion, or after destructive injuries of muscle. The sequence of events following the original injury is suppression of urinary secretion, usually complete, followed by rising blood levels of nitrogenous wastes. After about 10 days, urine secretion usually commences again but the urine first produced consists chiefly of water with relatively little nitrogen. In due time, which may be as long as 2 months, the renal function returns to normal. Death is not common unless the original injury is itself fatal or the patient receives too much fluid which he cannot excrete. Because of its reversibility this particular renal lesion is one of the best suited for treatment by the artificial kidney.

The renal lesions are relatively subtle and inconstant, consisting of edema of the tissues between the tubules, heme casts in the tubules, and focal injury to tubular epithelium, chiefly of the distal nephron.

Renal failure. When the kidneys are unable to carry on their normal functions of excreting wastes and balancing the internal chemical environment of the body, renal failure is present. This may occur abruptly, as in lower-nephron nephrosis or obstruction of the urinary tract, but most often develops gradually because of progressive destruction of renal tissue by disease. The sequence of events is about the same whether parenchymal destruction is the result of pyelonephritis, glomerulonephritis, nephrosclerosis, or other cause.

In the early or compensated stage, the kidneys lose their functional reserve, that is, the ability to meet stress. The power to excrete concentrated urine when little fluid is taken in, or to excrete a dilute urine when much fluid is ingested, gradually disappears and the urine specific gravity ultimately becomes fixed at 1.010, the specific gravity of an ultrafiltrate of blood serum. Similarly, nitrogenous wastes cannot be selectively concentrated but are merely washed through. Also, ability to secrete bases or acids to compensate for shifts in intake and to maintain the normal pH (7.4) of blood gradually vanishes.

In the later or decompensated stage, complete excretion of certain substances is no longer possible and their blood concentrations rise. Most important of these substances are potassium, phosphates, and nitrogenous wastes such as urea and creatinine. Sodium and water accumulate in the tissues in edema fluid. Edema is also aggravated by fall in serum albumin, caused both by loss in the urine and poor intake in the diet. In children with chronic renal failure the phosphate retention produces severe disturbances of calcium metabolism which sometimes results in a bone disorder called renal rickets. Anemia is usually present for several reasons, of which probably the most important is destruction of red blood cells by circulating poisons.

The end stage of renal failure is uremia. Symptoms include convulsions, vomiting, diarrhea, and coma. Lesions are found in many organs. They include fatty droplet accumulations in heart muscle, acute fibrinous pericarditis, ulcers of the intestinal tract, a white powdery "frost" on the skin, and edema of the brain. These changes are reversible if the renal failure can be relieved. Extensive studies have not found the particular poison or poisons responsible for the individual lesions.

Cysts. Single or multiple cysts with thin walls containing clear fluid are extremely common. They often result from obstruction of tubules by scars and are not important except in polycystic kidneys.

Tumors. There are three important malignant tumors of the kidney, all having a high mortality rate. Wilms tumor, found in infants, is made up of a mixture of small glands and of sarcoma cells particularly of muscle type; it is often termed an adenosarcoma. The tumor usually occurs as a large abdominal mass and is rapidly fatal if untreated. Some children are cured by radiation and surgery or chemotherapy with cancericidal agents. Renal-cell carcinoma or hypernephroma grows in the renal substance of adults as a hemorrhagic, yellow, usually encapsulated tumor. The first symptoms may be irregular fever, hematuria, or those of distant metastasis. Because it invades the veins early, it tends to metastasize through the bloodstream, particularly to lungs, brain, and bone. Metastases may appear many years after the kidney tumor has been removed. Sometimes there may be solitary metastases curable by surgery. Transitional-cell carcinoma arises in the epithelium of the pelvis, produces hematuria early, and tends to spread down the epithelium of the ureters. See KIDNEY (VERTEBRATE); METABOLIC DISORDERS; ONCOLOGY; URINARY SYSTEM (VERTEBRATE).

[ROY N. BARNETT]

Bibliography: A. C. Allen, *The Kidney: Medical and Surgical Diseases*, 2d ed., 1962; R. H. Heptinstall, *Pathology of the Kidney*, 1966; M. Strauss and L. G. Welt (eds.), *Diseases of the Kidney*, 1963.

Killing equations

A set of special, linear, first-order partial differential equations involving the coordinate variables x of an N-dimensional Riemannian space, the metric tensor g_{ab} of this space, and a vector whose components V_a, V^a $(V^a = g^{ab}V_b)$ are functions of x. Specifically, it is given by either Eq. (1) or Eq. (2).

$$V^c g_{ab;c} + g_{ac}V^c_{;b} + g_{bc}V^c_{;a} = 0, \ (;c \equiv \partial/\partial x^c) \quad (1)$$

$$V_{a,b} + V_{b,a} = 0 \quad (2)$$

The subscript semicolon indicates partial differentiation with respect to a coordinate variable x, while the comma specifies covariant differentiation. The equivalence of Eqs. (1) and (2) may be established as follows. From the definitions of covariant differentiation and the structure of the Christoffel symbols {}, Eq. (3) follows, and one

$$V_{a,b} = V_{a;b} - V_d\{^d_{ab}\}$$
$$= (V^c g_{ac})_{;b} - V_d g^{cd}\tfrac{1}{2}[-g_{ab;c} + g_{cb;a} + g_{ac;b}] \quad (3)$$

obtains Eq. (4) by virtue of the symmetry of g_{ab}.

$$V_{a,b} + V_{b,a} = (V^c g_{ac})_{;b} + (V^c g_{bc})_{;a} - \tfrac{1}{2}V^c \cdot$$
$$[-g_{ab;c} - g_{ba;c} + g_{cb;a} + g_{ca;b} + g_{ac;b} + g_{bc;a}]$$
$$= g_{ac}V^c_{;b} + g_{bc}V^c_{;a} + V^c g_{ab;c} \quad (4)$$

The tensor components $V_a(x)$ are supposedly of the type (0,1,0). Specifically, transformation equation (5) for V_a is induced by the change of coordinates $x^a = x^a(\bar{x})$, $\bar{x}^r = \bar{x}^r(x)$. The quantities $V_{a,b}$

$$\overline{V}_r = V_a \partial x^a / \partial \bar{x}^r \quad (5)$$

obtained by covariant differentiation are tensor components of the type (0,2,0). Specifically, they obey Eq. (6), and therefore Eq. (7) is valid. Hence

$$\overline{V}_{r,s} = V_{a,b} \partial x^a/\partial \bar{x}^r \partial x^b/\partial \bar{x}^s \quad (6)$$

$$\overline{V}_{r,s} + \overline{V}_{s,r} = V_{a,b} \partial x^a/\partial \bar{x}^r \partial x^b/\partial \bar{x}^s + V_{b,a}\partial x^b/\partial \bar{x}^s \partial x^a/\partial \bar{x}^r$$
$$= (V_{a,b} + V_{b,a})\partial x^a/\partial \bar{x}^r \partial x^b/\partial \bar{x}^s \quad (7)$$

if for a given tensor V_a [type (0,1,0)] and a certain coordinate system (x) of the infinite collection of tensor analysis the Killing equations are valid, then these equations are valid for each coordinate system of the infinite collection. Briefly, the Killing equations are invariant for class C' coordinate transformations.

The Killing equations are fundamental with regard to the question of the existence of certain mappings (or, synonymously, transformations) of a space onto itself. These particular transformations are frequently called infinitesimal groups of motions and are related to the ordinary groups of motions, that is, the motions which have the characteristic group properties.

Groups of motions. The one-parameter motions determined by mapping equations of the type in

Eq. (8) will be said to constitute a one-parameter

$$x^{*a} = f^a(x,t), \quad (x \equiv \text{set } x^1, x^2, \ldots, x^N) \quad (8)$$

group of motions if the following group properties hold: (1) there exists an admissible value of t, t_0, such that for each admissible x, $x^{*a}(t_0) = f^a(x,t_0) = x^a$ (the identity transformation); (2) for each admissible set of x-values x^{*a} and each admissible value for t_1, the equations $x^{*a} = f^a(x,t_1)$ have a unique solution expressible in the form $x^a = f^a(x^*,t_2)$ (existence of an inverse); (3) for each admissible set of x-values and each admissible pair of t-values t_1, t_2, there exists an admissible parameter value t_3, such that $x^{**a} = f^a(x^*,t_2) = f^a[f(x,t_1),t_2] = f^a(x,t_3)$ (the closure or principal group property). The term "groups of motions" has been used at times under lighter conditions than these; for example, condition 1 or 2, or both, has been omitted. *See* GROUP THEORY.

A very simple example of a group of motions for an ordinary euclidean plane E_2 and a rectangular cartesian coordinate system is given by Eq. (9),

$$x^{*a} = f^a(x,t) = x^a + t \quad (9)$$

with t restricted to real numbers. Group property 1 is obviously met by $t = 0$. Property 2 is satisfied since the inverse of $x^{*a} = x^a + t_1$ for each admissible value of t_1 is $x^a = x^{*a} + (-t_1) = f^a(x^*,-t_1)$. Property 3 and Eq. (9) require that for each admissible choice of t_1 and t_2 there exists an admissible t-value t_3 such that $x^{**a} = f^a(x,t_3) = x^a + t_3$ with $x^{**a} = f^a(x^*,t_2) = x^{*a} + t_2 = (x^a + t_1) + t_2$. Obviously the requirement of property 3 is fulfilled by $t_3 = t_1 + t_2$. Thus the group requirements are satisfied. The motion criterion is also met since $g_{ab} = \delta_{ab}$ ($=1$ for $a = b$, $=0$ for $a \neq b$) throughout the plane. Thus a change in the evaluation point does not cause a change in the value of g_{ab}, and the metric magnitudes (such as arc lengths and angles between intersecting curves) are preserved. A slight generalization of the motion of Eq. (9) is given by Eq. (10), with k^1 and k^2 given constant

$$f^a(x,t) = x^a + k^a t \quad (10)$$

real numbers, not both zero.

As another simple example of a motion, consider a plane E^2 which has a point O deleted and contains a rectangular cartesian coordinate system (x) with center at O; and impose the point transformation in Eqs. (11). The identity transformation is

$$x^{*1} = f^1(x,\theta) = \cos\theta\, x^1 - \sin\theta\, x^2 \quad (11a)$$

$$x^{*2} = \sin\theta\, x^1 + \cos\theta\, x^2 \qquad 0 \leq \theta < 2\pi \quad (11b)$$

given by $\theta = 0$; and for $0 < \theta_1 < 2\pi$, the inverse is $x^a = f^a(x^*,\theta_2)$, $\theta_2 = 2\pi - \theta_1$. The group requirement is satisfied provided there is a functional relation $\theta_3 = \theta_3(\theta_1,\theta_2)$ for $0 \leq \theta_1 < 2\pi$, $0 \leq \theta_2 < 2\pi$, the relation having the following properties (1): $0 \leq \theta_3 < 2\pi$; (2) from $x^{**a} = f^a(x^*,\theta_2)$, $x^{*a} = f^a(x,\theta_1)$, it follows that $x^{**a} = f^a(x,\theta_3)$. From the rotational interpretation of mapping equations (11), it is evident that $\theta_3(\theta_1,\theta_2)$ may be defined by $\theta_3(\theta_1,\theta_2) = \theta_1 + \theta_2$ for $0 \leq \theta_1 + \theta_2 < 2\pi$; and $\theta_3(\theta_1,\theta_2) = \theta_1 + \theta_2 - 2\pi$ for $\theta_1 + \theta_2 \geq 2\pi$. Also, $g^*_{ab}(dx^{*a}/du)(dx^{*b}/dv) = g_{ab}(dx^a/du)(dx^b/dv)$, and it is evident that the metric entities are preserved in value by the mapping.

Formulation of motions. The definitive require-

ment for a mapping to be a motion, namely, the preservation of the magnitudes of the metric entities, is of course very stringent. Given a Riemannian space R, with a coordinate system (x), one method of attack for searching for a motion is to seek a suitable coordinate transformation which preserves the structure of the metric tensor. If such a coordinate transformation $\bar{x}^r = \bar{x}^r(x)$ exists, then the tensor transformation in Eq. (12) can be

$$g_{ab}(x) = \bar{g}_{rs}[\bar{x}(x)]\partial\bar{x}^r/\partial x^a \partial\bar{x}^s/\partial x^b \quad (12)$$

altered in structure by replacing $\bar{g}_{rs}[\bar{x}(x)]$ with $g_{rs}[\bar{x}(x)]$ since the functional structure of $\bar{g}_{rs}(\bar{x})$ is the same as that of $g_{rs}(x)$ and therefore $\bar{g}_{rs}(\bar{x}) = g_{rs}(\bar{x})$. If this change is made, Eq. (13) results.

$$g_{ab}(x) = g_{rs}[\bar{x}(x)]\partial\bar{x}^r/\partial x^a \partial\bar{x}^s/\partial x^b \quad (13)$$

Conversely, if the coordinate transformation is not known, then this last equation may be regarded tentatively as a set of $N(N+1)/2$ first-order partial differential equations in N unknown functions $\bar{x}^r(x)$. Thus, while motions can be formulated in certain very special cases, the desirability of replacing the motion requirement with another which is less demanding is obvious.

Modified motion requirement. If C_1 and C_2 are two class C' arcs given by $x^a = x_1{}^a(u)$, $x^a = x_2{}^a(v)$ which intersect at a point P for $u = u_0$, $v = v_0$ and if, in addition, the motion requirement is met, then $g_{ab}(x^*) \cdot dx_1{}^{*a}/du\, dx_2{}^{*b}/dv|_{u_0}, v_0$ is a constant function of t and therefore its derivative vanishes identically in t. A natural reduction in the severity of this requirement is obtained by imposing the alternate condition that the derivative vanish at t_0, the identity value for t, which is now assumed to be $t_0 = 0$. It follows that $f^a(x,0) = x^a$ and hence, by the McLaurin expansion for $f^a(x,t)$ that $x^{*a} = x^a + t \cdot [(\partial f^a/\partial t)|t=0] + , \ldots , .$ These facts suggest consideration of the simpler point transformation, Eq. (14). For each fixed point $P(x)$, Eq. (14)

$$x^{*a} = x^a + tV^a(x) \quad (14)$$

represents a parameterized arc through P, and the tangent vector at P has the components $\partial x^{*a}/\partial t|P$ which are equal to $V^a(x)$ at P. In the literature Eq. (14) is frequently written in the modified form of Eqs. (15), and referred to as an infinitesimal

$$x^{*a} = x^a + \delta t V^a(x) \quad (15)$$

transformation.

With regard to the motion requirement: (1) for a parameterized arc C given by $x^a = x^a(u)$ and of class C', $dx^{*a}/du = dx^a/du + V^a{}_{;b}(dx^b/du)t$ and therefore $\partial/\partial t(dx^{*a}/du) = V^a{}_{;b}dx^b/du$; and (2) $g_{ab}|MP = g_{ab}(x^*) = g_{ab}(x^c + V^c t)$ and hence $\partial g_{ab}/\partial t = \partial g_{ab}/\partial x^{*c}V^c$, and $\partial g_{ab}/\partial t_{t=0} = g_{ab;c}(x)V^c$. Hence for the case of the two arcs C_1 and C_2, previously introduced, Eq. (16) follows. Thus the Killing equa-

$$d[g_{ab}(x^*)(dx_1{}^{*a}/du)(dx_2{}^{*b}/dv)]/dt|_{t=0}$$
$$= g_{ab;c}(x)V^c(dx_1{}^a/du)(dx_2{}^b/dv)$$
$$\quad + g_{ab}(x)V_{;c}{}^a(dx_1{}^c/du)(dx_2{}^b/dv)$$
$$\quad + g_{ab}(x)(dx_1{}^a/du)V_{;c}{}^b dx_2{}^c/dv$$
$$= (g_{ab;c}V^c + g_{cb}V_{;a}{}^c + g_{ac}V_{;b}{}^c)(dx_1{}^a/du)(dx_2{}^b/dv)$$
$$(16)$$

tions constitute a necessary and sufficient condition for a point transformation, Eq. (14), to be an infinitesimal motion. *See* CALCULUS OF TENSORS; DIFFERENTIAL EQUATION; GEOMETRY, RIEMANNIAN.

[HOMER V. CRAIG]

Bibliography: L. P. Eisenhart, *Continuous Groups of Transformations*, 1933; L. P. Eisenhart, *Riemannian Geometry*, 1926; W. Killing, *Ueber die Grundlagen der Geometrie, Journal für die reine und angew. Math. (Crelle)*, 109:121–186, 1892.

Kiln

A device or enclosure to provide thermal processing of an article or substance in a controlled temperature environment or atmosphere, often by direct firing, but occasionally by convection or radiation heat transfer. Kilns are used in many different industries, and the type of device called a kiln varies with the industry.

"Kiln" usually refers to an oven or furnace which operates at sufficiently high temperature to require that its walls be constructed of refractory materials. The distinction between a kiln and a furnace is often based more on the industry than on the design of the device. For instance, an electrically heated refractory tunnel oven equipped with a stainless mesh conveyor belt to carry the work through is referred to as a tunnel kiln if it is used for sintering small ceramic electronic parts such as ferrite transformer cores. The same device used to sinter small metal parts from powdered aluminum alloys is called a sintering furnace.

Generally the word "kiln" is used when referring to high-temperature treatment of nonmetallic materials such as in the ceramic, the cement, and the lime industries. When melting is involved as in steel manufacture, the term "furnace" is used, as in blast furnace and basic oxygen furnace. In glass manufacture, the melting furnace is often called a glass tank when the process is continuous. *See* FURNACE CONSTRUCTION.

Rotary kiln. This is the largest type of kiln, being used for heating of loose bulk materials such as cement and lime. Rotary kilns are used exclusively for the production of cement and for 85% of the commercial quicklime produced in the United States. Over the years, rotary cement kilns have increased in size to gain greater capacity and productivity from a single unit: kilns having a shell as large as 21 ft (6.3 m) diameter by 700 ft (210 m) long have been constructed. This is probably approaching the practical size limit. *See* CEMENT; LIME (INDUSTRY).

A rotary kiln (Fig. 1) is a long, refractory-lined cylinder supported on steel hoops, or "tires," riding on rollers, or "trunnions." The kiln is inclined at a slight slope, often 1/2–5/8 in. (12.7–15.9 mm) fall per foot, and rotated at a slow speed (about 0.8–1.0 rpm for cement and lime kilns). For calcining ores, there is a tendency to use flatter slopes and higher rotational speeds because of the improved heat transfer obtained. A burner (usually using pulverized coal, natural gas, fuel oil, or a combination of fuels) is placed at the lower end to supply the necessary heat. Raw feed enters the upper end and slowly tumbles through the kiln, being heated by the countercurrent flow of hot combustion gases. Physical and chemical changes take place in the charge, such as drying and dehydration, decomposition of carbonates, and fusing

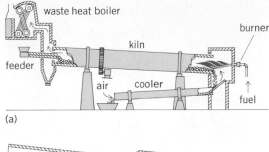

(a)

(b)

Fig. 1. Rotary kilns. (*a*) Single-diameter; (*b*) two-diameter. (*From R. H. Perry and C. H. Chilton, Chemical Engineers' Handbook, 5th ed., McGraw-Hill, 1973*)

of materials into a clinker.

A rotary kiln tends to have poor thermal efficiency. To conserve fuel, auxiliary devices are usually added. The hot product drops into a cooler which preheats the combustion air. Chain curtains at the feed end and feed-preheating devices are frequently used to extract heat from the flue gas and preheat the kiln feed.

Rotary kilns are also used for calcining or agglomerating phosphate rock; alumina and bauxite; magnesia; clays; carbon and petroleum coke; lightweight aggregates; iron, chromium, lithium, and uranium ores; burning of elemental sulfur; and incineration or pyrolysis of combustible wastes.

Shaft kiln. A shaft kiln (Fig. 2) consists of a vertical, stationary, refractory-lined steel tube, frequently tapered to grow slightly larger toward the bottom to permit free downward movement of the

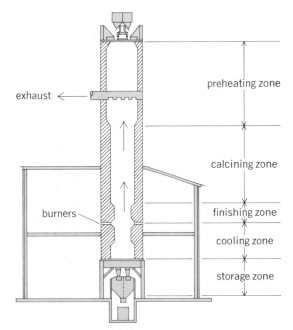

Fig. 2. Vertical shaft kiln. (*From C. J. Lewis and B. B. Crocker, The lime industry's problem of airborne dust, J. APCA, 19:31–39, 1969*)

Fig. 3. Interior of periodic kiln, with electrical porcelain insulators for firing. (*Swindell-Dressler, Inc.*)

charge. These kilns have been used for centuries for calcining limestone and cement and are still used worldwide (except in the United States) for cement production; 15% of the commercial quicklime made in the United States is made in shaft kilns. Shaft kilns are now also used for calcining taconite pellets and as an oil shale retort. The raw charge is added at the top and is discharged through grates at the bottom. Originally, coal or coke was mixed with the charge to supply fuel and was burned out by hot air introduced through the walls above the grates.

Most shaft kilns are now equipped with external burners to supply the heat in order to eliminate product contamination caused by in-place firing of solid fuels. Shaft kilns have much better heat economy than rotary kilns since the gases flowing up through the kiln have more intimate contact with the solid charge. However, shaft kilns generally have lower production rates than rotary kilns, and accurate control of the charge time-temperature relationship is less precise because of gas channeling.

Newer lime kilns. Two developments for lime calcining are the fluo-solids lime kiln, in which the

limestone is calcined in a bed fluidized with hot combustion products, and the Calcimatic kiln, which is a horizontal, rotating, circular refractory hearth kiln. The charge is deposited continuously on a pie-shaped element of the hearth as the hearth rotates about its center. The charge advances below radiant burners on the underside of a refractory dome roof. Rotational speed is set so that the charge is calcined when the hearth has rotated about 350°, at which point the charge is scraped from the hearth with a conveyor and transferred to a shaft cooler which preheats combustion air. One advantage of the Calcimatic is the low dust emission in the flue gas.

Kilns for formed ceramic products. Several types of kilns, either batch or continuous, are used for firing formed ceramic products. An early batch kiln is the scove kiln, in which unfired bricks are stacked to provide fireboxes and flues. Thus the basic material to be fired forms the kiln structure, which is covered on the outside with fired brick and a clay wash. Such kilns holding 30,000–200,000 brick were fired with wood to produce all early American brick. Later, coal or oil fuel was used. *See* CERAMIC TECHNOLOGY.

Periodic kiln. This is the general name given to a batch-type kiln in which the ware passes through a cycle consisting of heat-up, soaking (holding at peak temperature for some time), cooling, and removing or "drawing" the ware at the end of the cycle (Fig. 3). Such kilns vary in size from small laboratory or art-pottery kilns to large domed structures 40–50 ft (12–15 m) in size. Periodic kilns have poor thermal efficiency since all the heat used to heat the ware is lost during cooling. Such units are also wasteful of manpower in loading and unloading.

Another early kiln, the beehive kiln, is a type of periodic kiln. It consists of a permanent, hemispherical chamber surrounded by a number of fuel beds, often burning coal. The hot gases pass up inside the walls and down through the material to be fired to a central flue in the floor.

Chamber kiln. This kiln, widely used in Europe, but not generally adopted in the United States, is one of the earliest attempts to improve heat efficiency and approach continuous firing. The kiln is built in the form of a ring or long oval, divided into separate chambers connected with flues. All chambers are loaded with ware. Firing takes several days, the fire being started in one chamber and then moved progressively around the circuit. The combustion gas passes through chambers yet to be fired, where the ware is preheated. After completion of firing in a chamber, combustion air is passed through to cool the ware and preheat itself.

Tunnel kilns. These are long structures in which the ware is moved through the kiln (Fig. 4). Combustion air is introduced countercurrent to the unfired ware. Thus the air is preheated by cooling the fired ware. Fuel is burned in the center section, and the hot combustion gases preheat the incoming ware and remove its moisture. Tunnel kilns may be as small as a few inches wide and high by 15–30 ft (4.5–9 m) long for small ceramic parts, to 6 by 8 ft (1.8 by 2.4 m) in cross section and several hundred feet long for continuous brick processing.

A tunnel kiln is often named after the method of

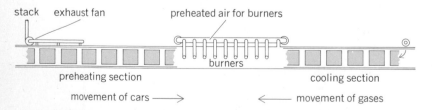

stack exhaust fan   preheated air for burners

burners

preheating section   cooling section

movement of cars ⟶   ⟵ movement of gases

Fig. 4. Section of direct-fired tunnel kiln showing principle of operation. (*From T. J. W. van Thoor , ed., Chemical Technology: An Encyclopedic Treatment, vol. 2, Barnes and Noble, 1971*)

moving the work through the kiln. Brick is loaded on a refractory-topped small car which rolls through the tunnel on a track. The refractory hearth of the car forms the bottom of the kiln. Such kilns are called car tunnel kilns. The car is frequently pushed into the tunnel with a hydraulic ram. Movement of the cars may be either continuous or intermittent at spaced time intervals. Small tunnel kilns may have a traveling mesh belt (mesh belt tunnels), or the ware may be placed in metal or refractory boxes called saggers. The boxes may be pushed through on a stationary hearth (pusher tunnel) or may roll on top of ceramic rollers (roller hearth tunnel).

Muffle kiln. In this type of kiln, which may be either periodic or continuous, the ware is protected from direct contact with the fire by a radiant refractory wall if the kiln is fuel-fired. In many special small muffle tunnels, electric heat is used. The tunnel refractory lining may be wound with a chrome-nickel resistance element, or silicon carbide glowbar units may be suspended from the roof and buried in the hearth. A muffle kiln may be used for firing high-priced dinnerware. Special electrically heated muffle tunnels have been developed for the electronics industry to fire ceramic capacitors, ceramic magnets, and printed ceramic circuits, often with carefully controlled atmospheres such as dry hydrogen.

Lumber and tobacco kilns. Kilns used in these industries are basically low-temperature dryers and do not require refractories for their construction. Lumber-drying kilns are of two basic types: a box or compartment type in which the lumber is loaded in batches for drying, and a progressive kiln in which the lumber travels through on cars much as in a tunnel kiln. The lumber is stacked horizontally with "stickers" (uniform wood spacers) between successive layers. Air of controlled temperature and humidity is passed over and through the lumber, often with circulating fans. Older kilns may have natural circulation produced by rising currents of hot air from steam heating coils. Fan kilns have better thermal efficiency because the air is recirculated more times before venting. Air temperatures vary from 150 to 200° F (66 to 93°C), depending on lumber size and species. *See* LUMBER MANUFACTURE.

A tobacco-drying kiln is an airtight barn used for flue-curing of tobacco. Leaf strings of tobacco are hung from racks. Wood, coal, oil, or natural gas is burned in a firebox, or "kiln." The hot flue gases circulate through sheet-iron flues near the floor of the barns. The leaf is subjected to a carefully controlled temperature-time cycle (usually 3–5 days long) to produce the desired characteristics in the leaf.

Heating. Gas and oil have been the preferred fuels for all types of kilns and will probably remain so for moderate-priced ceramics. Pulverized coal has been used in the past for firing rotary kilns and will probably become the preferred fuel of the future because of the scarcity of hydrocarbon fuels. Coal has been burned on an external grate to fire shaft kilns when oil or gas is not available. Coal gas and producer gas have also been generated for this purpose. Electric heat is generally limited to smaller kilns processing valuable wares where precise temperature control is needed or where convenience is important, as in laboratory and pilot plant kilns.

Pollution control. Kilns can be sources of air pollution arising either from the combustion of the fuel or from entrainment of dust particles from the charge. The latter source applies only to kilns where loose bulk materials are processed, such as rotary and shaft kilns. The exhaust gases from these kilns are relatively hot, and cyclonic collectors have often been used to remove dust particles down to 5–10 μm in size. This collection does not meet today's requirements, and such kilns are being equipped with electrostatic precipitators or bag filters. Wet scrubbing can be used efficiently, but this results in a condensing steam plume in cool weather, which may be esthetically objectionable.

Combustion pollution can take the form of incompletely burned fuel (hydrocarbons, carbon monoxide, soot, and smoke) or oxides of nitrogen. The former is prevented by good combustion practices—adequate excess air, proper mixing of fuel and air, and adequate combustion temperatures. Nitrogen oxides are produced only when combustion is carried out at very high temperatures (usually 2000–2800°F; 1093–1538°C) to produce high peak temperatures in the kiln. NO_x is controlled by reducing flame temperature, using less excess air, two-stage combustion, and recirculation of exhaust gases. *See* ATMOSPHERIC POLLUTION; DUST AND MIST COLLECTION.

[BURTON B. CROCKER]

Bibliography: C. Carmichael, *Kent's Mechanical Engineers Handbook*, 12th ed., vol. 1, Chap. 25, 1950; R. H. Perry and C. H. Chilton, *Chemical Engineers' Handbook*, 5th ed., chap. 20, 1973; T. J. W. van Thoor (ed.), *Chemical Technology: An Encyclopedic Treatment*, vol. 2: *Non-metallic Minerals and Rocks*, 1971, vol. 4: *Wood*, 1973; W. Trinks and W. H. Mawhinney, *Industrial Furnaces*, vol. 1, 5th ed. 1961.

Kilogram

A unit of mass. The standard of mass in both the mks and cgs systems of units is a platinum-iridium cylinder called the standard kilogram (kg). The unit of mass in the mks system is the mass of the standard kilogram. The unit of mass in the cgs system is one-thousandth of this mass and is called a gram. A kilogram weighs 2.20462 lb, and a metric ton contains 1000 kg.

A kilogram may also be defined as a unit of force—the weight of a 1-kg mass at sea level at 45° latitude, also called kilogram-force. *See* MASS; UNITS, SYSTEMS OF.

[LEO NEDELSKY]

Kinematics

That branch of mechanics which deals with the motion of a system of material particles without reference to the forces which act on the system. Kinematics differs from dynamics in that the latter takes these forces into account. *See* DYNAMICS.

For a single particle moving in a straight line (rectilinear motion), the motion is prescribed when the position x of the particle is known as a function of the time. For uniformly accelerated motion,

Eq. (1) holds, where a is the constant acceleration, and hence x is defined as in Eq. (2), where x_0 and

$$\frac{d^2x}{dt^2} = a \qquad (1)$$

$$x = \tfrac{1}{2}at^2 + v_0 t + x_0 \qquad (2)$$

v_0 are the initial position and velocity of the particle, respectively. *See* ACCELERATION; MOTION, RECTILINEAR; VELOCITY.

Plane kinematics of a particle is concerned with the specification of the position of a particle moving in a plane by means of two independent variables, usually the rectangular cartesian coordinates x and y, but often the polar coordinates $r = (x^2 + y^2)^{1/2}$, $\theta = \tan^{-1}(y/x)$, or even other coordinates that may be especially convenient for a particular problem. Here the x and y components of velocity are dx/dt, dy/dt, and of acceleration d^2x/dt^2, d^2y/dt^2, respectively, but the components of velocity in the directions of increasing r and increasing θ are dr/dt and $r(d\theta/dt)$, and are therefore not so simply given in terms of the time derivatives of the coordinates.

The kinematics of a particle in space is concerned with the ways in which three independent coordinates may be chosen to specify the position of the particle at a given time, and with the relations between the first and second time derivatives of these coordinates and the components of velocity and acceleration of the particle. The motion is specified if three such coordinates are given as functions of the time, and the path of the particle in space is then obtained by eliminating the time between these equations.

For describing the simultaneous position in space of N particles, $3N$ coordinates are necessary, and the configuration of the system may be represented by a point in a $3N$-dimensional space called the configuration space of the system. As the particles move, the corresponding representative point in configuration space traces out a curve called the trajectory of the system. If the particles do not move independently but are subject to m constraints, that is, if m relations between the $3N$ coordinates continue to hold throughout the motion, then it is possible to represent the trajectory in a space of $f = (3N - m)$ dimensions, f being the number of degrees of freedom of the system. For a rigid body, even though N and m are very large, $f = 6$, and the configuration at any time may be represented by six independent variables (usually three to represent the position of the center of mass and three angles to specify the orientation of the body). Kinematics describes the relations between the time derivatives of such angles and the components of the angular velocity of the body. *See* CAYLEY-KLEIN PARAMETERS; CONSTRAINT; DEGREE OF FREEDOM (MECHANICS); EULER ANGLES.

Among the coordinate systems studied in kinematics are those used by observers who are in relative motion. In nonrelativistic kinematics the time coordinate for each such observer is assumed to be the same, but in relativistic kinematics proper account must be taken of the fact that lengths and time intervals appear different to observers moving relative to each other. *See* RELATIVITY.

[H. C. CORBEN/BERNARD GOODMAN]

Kinesthetic sensation

The kind of feeling originating in bodily movement and arising from stimulation of the sensory nerves that terminate in muscles, tendons, and joints. It is by means of kinesthesis (literally, feelings of motion) that it becomes possible to walk without looking at the feet, eat without watching the hand convey food to the mouth, or estimate the weight of objects merely by hefting them.

For well over a century, since the time of Sir Charles Bell (around 1826), the muscle sense, as it was then called, has been recognized as a sixth sense to be added to the classical five of Aristotle: sight, sound, touch, smell, and taste. At first sensory returns from muscles were thought to be exclusively responsible for supplying information about postural changes. However the evidence gradually came to favor the nerves ending in tendinous and articular (joint) tissues as providing the bulk of such sense data. Now it is known that appreciation of limb movement, hence of the majority of postural adjustments, is conveyed by way of receptors at or near bony articulations. The most crucial evidence comes from abnormalities. Joint sensibility may be retained in the presence of muscular and cutaneous anesthesia of the same region. The converse also occurs; for example, it is possible to retain sensibility of skin and muscle but fail to discriminate passive movement of the limb because articular sensitivity has been impaired.

Kinesthetic receptors. Four types of sensory nerve ending are involved in the appreciation of bodily movement. They are (1) free nerve endings, (2) Pacinian corpuscles, (3) Golgi tendon organs, and (4) muscle spindles.

Free nerve endings. These are the most commonly occurring sensory nerve terminations in the body. They are found in deep tissues—subcutaneous layers, fascia (musclesheath), periosteum (bone-covering)—as well as the cutaneous level, where their distribution is almost ubiquitous and endings of neighboring nerve fibers overlap and interdigitate with one another in some profusion. The presumption is that it is the free endings of the periosteum and the fascia that contribute so heavily to kinesthetic sensation, though this is by no means firmly established. Free nerve endings are also suspect as initiators of pain, which can be most exquisite in the periosteal and fascial regions.

Pacinian corpuscles. Pacinian corpuscles are the largest of the specialized sensory nerve endings in the body, being situated suitably to serve as kinesthetic receptors. They are found in the sheaths covering muscles and tendons and are scattered through deep fatty tissue. Relatively few in number, Pacinian corpuscles are encapsulated, consisting of many concentric layers of fibrous tissue containing nerve endings. They are attached to large diameter sensory fibers and are known to respond to any nearby distorting force, whether it be muscular contraction or mechanical deformation transmitted from the body surface.

Golgi tendon organs. These organs (see illustration), situated near the junction of muscle fibers and tendon, would seem to be most advantageously placed to report on muscle movement. Recordings of electrical changes in their attached fibers

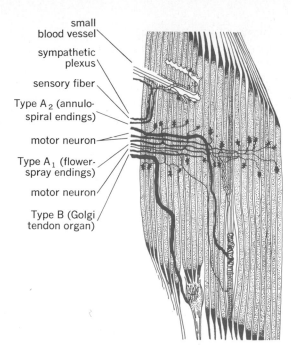

small
blood vessel
sympathetic
plexus
sensory fiber
Type A$_2$ (annulo-
spiral endings)
motor neuron
Type A$_1$ (flower-
spray endings)
motor neuron
Type B (Golgi
tendon organ)

The neural control of movement: spinal reflex mechanisms. (*R. F. Thompson, Foundations of Physiological Psychology, Harper and Row, 1967*)

show them to signal both muscle stretch and active contraction. Impulse frequency has been found to be roughly proportional to the common logarithm of the tension imposed on the tendon. Golgi organs may therefore be thought of as muscle-tension recorders.

Muscle spindles. The most complexly organized of the kinesthetic receptors are the muscle spindles, which have two distinctly different kinds of terminations (see illustration). These have been designated type A$_1$ (flowerspray endings) and type A$_2$ (annulospiral endings) by the British physiologist B. H. C. Matthews, who originally described them. Typically, A$_1$ fibers are of smaller diameter, and are therefore slower-conducting than A$_2$ fibers. Both types initiate impulses when their muscle bundle is stretched and generally cease firing impulses when the muscle contracts. The fibers leading from muscle spindles, therefore, have been termed stretch afferents. Type A$_2$, additionally, generates kinesthetic impulses when its muscle is very strongly contracted.

Kinesthetic discriminations. Since it is the receptors situated in the neighborhood of bony articulations that supply the most exact information about bodily movement, the finest kinesthetic discriminations are associated with joint displacements. Measurement of the ability to discriminate such movements is accomplished by arranging for a limb or appendage to be moved mechanically and passively at a uniform rate; the least detectable amount of movement, based solely on feel, is then determined. In such experiments, it is typically found that the greatest sensitivity is associated with the larger joints (hip and shoulder), whereas the least is found in the ankle and toes. At a speed of displacement of 10°/min, thresholds vary roughly between 0.2° (hip) and 0.7° (main joint of big toe).

Since appreciation of postural change seems to depend mainly on articular sensitivity, the roles of muscles and tendons must be secondary in kinesthesis. Muscular and tendinous sensitivities presumably add feelings of strain when resistance to limb movement is encountered, as in lifting weights or pushing against stationary objects. *See* SOMESTHESIS.   [FRANK A. GELDARD]

Bibliography: T. C. Ruch and J. C. Fulton (eds.), *Medical Physiology and Biophysics*, 1960.

Kinetic methods of analysis

The measurement of reaction rates for the analytical determination of the initial concentrations of the species of interest taking part in chemical reactions. This technique can be used since, in most cases, the rates or velocities of chemical reactions are directly proportional to the concentrations of the species taking part in the reactions.

The rate of a chemical reaction is measured by experimentally following the concentration of some reactant or product involved in the reactions as a function of time as the reaction mixture proceeds from a nonequilibrium to an equilibrium or static state (steady state). Thus, kinetic techniques of analysis have the inherent problem of the increased experimental difficulty of making measurements on a dynamic system, as time is now a variable which is not present in making measurements on an equilibrium system. However, kinetic methods often have advantages over equilibrium techniques in spite of the increased experimental difficulty. For example, the equilibrium differentiations or distinctions attainable for the reactions of very closely related compounds are often very small and not sufficiently separated to resolve the individual concentrations of a mixture without prior separation. However, the kinetic differentiations or distinctions obtained when such compounds are reacted with a common reagent are often quite large and permit simultaneous analysis.

A further advantage of kinetic methods is that they permit a larger number of chemical reactions to be used analytically. Many reactions, both inorganic and organic, are not sufficiently well-behaved to be employed analytically by equilibrium or thermodynamic techniques. Many reactions attain equilibrium too slowly; side reactions occur as the reactions proceed to completion, or the reactions are not sufficiently quantitative (do not go to completion) to be applicable. However, a kinetic-based technique can often be employed in these cases simply by measuring the reaction rate of these reactions during the early or initial portion of the reaction period. Also, the measurement of the rates of catalyzed reactions generally is a considerably more sensitive analytical method for the determination of trace amounts of a large number of species than equilibrium methods. *See* CATALYSIS; KINETICS, CHEMICAL.

Kinetic methods of analysis can be divided into three basic categories: (1) methods employing uncatalyzed reactions, (2) methods employing catalyzed reactions, and (3) methods for the simultaneous in-place determination of mixtures.

Uncatalyzed reactions. In order to determine the concentration of a single species in solution, the rate is measured of an irreversible uncatalyzed

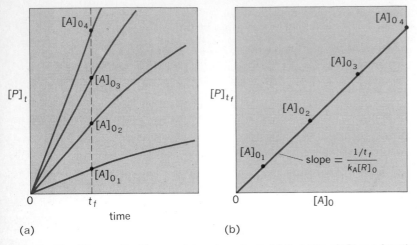

Fig. 1. Fixed-time method for uncatalyzed reactions. (a) Variation of [P] as a function of time for different initial concentrations of $[A]_0$. (b) Variation of amount of P formed at fixed time t_f with initial concentration of A.

reaction of the type shown by Eq. (1), where the

$$A + R \xrightarrow{K_A} P \qquad (1)$$

single species A reacts with the reagent R to form a product or products P. In general, the types of reaction employed for kinetic analysis are second-order irreversible reactions. The rate of the reaction is most conveniently measured by following the rate of formation of the product P as a function of time (although the change in concentration of either A or R could also be followed). Any method of measuring concentrations, such as titrimetry, spectrophotometry, and polarography, can be employed. The rate of formation of the product, $d[P]_t/dt$, is given by differential equation (2). In this rela-

$$\frac{d[P]_t}{dt} = k_A([R]_0 - [P]_t)([A]_0 - [P]_t) \qquad (2)$$

tion $[A]_0$ is the initial or original concentration of the species to be determined, $[R]_0$ is the initial concentration of the reagent (added in known concentration), and $[P]_t$ is the concentration of product formed at any time t; thus, the terms $([R]_0 - [P]_t)$ and $([A]_0 - [P]_t)$ equal the concentration of A and R, respectively, remaining at any time t, and k_A is the second-order rate constant for the reaction of A with R.

If the rate of the reaction given by Eq. (1) is measured only during the initial portion of the reaction (during a time period chosen so that the reaction is only 2–3% complete), the concentration of product formed $[P]_t$ is small compared to $[A]_0$ and $[R]_0$. Thus, on rearrangement, Eq. (2) becomes Eq. (3).

$$[A]_0 = \frac{d[P]_t/dt \,(\text{initial})}{k_A[R]_0} \qquad (3)$$

Integration of Eq. (3) between the time interval $t = 0$ to t and for an initial concentration of $P = 0$ at $t = 0$ gives Eq. (4). Both fixed- and variable-

$$[A]_0 = \frac{[P]_t/t}{k_A[R]_0} \qquad (4)$$

time methods may be used to calculate $[A]_0$ when employing Eq. (4).

When using the fixed-time method, the concentration of P formed at a chosen fixed time t_f is measured experimentally. As t_f and $[R]_0$ are thus kept constant for all measurements and k_A is a constant for all conditions, the term $\frac{1/t_f}{k_A[R]_0}$ is the proportionately constant relating $[A]_0$ to $[P]_{t_f}$, as shown in Fig. 1. For the variable-time method, with $[R]_0$ constant for all solutions, the time $t_{[P]_f}$ necessary for the reaction to form a fixed amount of product $[P]_f$ is measured. As $[P]_f$ is now a constant, $[A]_0$ is related to $1/t_{[P]_f}$ by the proportionately constant, $P_f/k_A[R]_0$, as shown in Fig. 2.

Catalyzed reactions. A catalyst may be broadly defined as an agent which alters the rate of a chemical reaction without shifting the equilibrium of the reaction. Although the catalyst undoubtedly enters into the reaction mechanism at a critical stage, it does so in a cyclic manner and hence does not undergo a permanent change. Therefore a catalyst speeds up the rate of attainment of the equilibrium of a system, but it does not change the position of the equilibrium and it is not consumed during the reaction.

In many reactions involving a substance which acts as a catalyst, the concentration of the catalyst is directly proportional to the rate of the reaction. Thus the rates of these reactions can be employed for the analytical methods and are extremely sensitive since the catalytic agents are not consumed but participate in the mechanism in a cyclic manner. The amount of the catalyst that can be determined employing catalyzed reactions is several orders of magnitude smaller than can be found by most direct equilibrium methods. Also, in many cases—enzyme-catalyzed (biological catalyst) reactions in particular—the catalyzed reaction of a reactant is extremely specific with respect to the chemical nature of the reactant. Such reactions, therefore, can be employed for the in-place analysis of a particular reactant in the presence of either a large excess or a large number of other species which would interfere with conventional equilibrium techniques, unless separation was performed prior to the analysis reaction.

The general mechanism for catalyzed reactions involves the combination of the catalyst C and the reactant, called the substrate S, to form a "complex" X, which then decomposes to form the product. Thus the catalyst is regenerated and combines

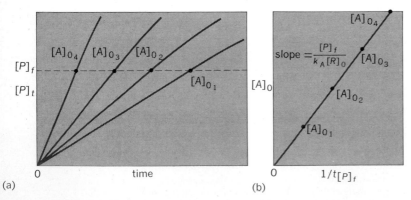

Fig. 2. Variable-time method for uncatalyzed reactions. (a) Variation of [P] as a function of time for various initial concentrations of $[A]_0$. (b) Variation of $1/t_{[P]_f}$ required to reach a fixed value of product formed $[P]_f$ with initial concentration of A.

once more with a substrate molecule. The general mechanism is given by Eqs. (5) and (6), where R is

$$C + S \underset{k_{-1}}{\overset{k_1}{\rightleftharpoons}} X \qquad (5)$$

$$X + R \overset{k_2}{\to} P + C \qquad (6)$$

a reagent molecule (which in many catalyzed reactions is not needed) that reacts with the complex to give the product P, and k_1 and k_{-1} are the rate constants for the formation of the complex and for the conversion of the complex back to C and S, respectively; the equilibrium constant K for the reaction given by Eq. (5) is equal to k_1/k_{-1}, and k_2 is the rate constant for the decomposition of X to form the product and C, as in Eq. (6). In many catalyzed reactions the rate of the decomposition reaction, Eq. (6), is slow compared to the rate of the conversion back to C and S ($k_2 \ll k_{-1}$). Thus, in this case, Eq. (6) is the rate-determining step. Under these conditions, the rate of reaction of S in a catalyzed process is given by Eq. (7).

$$-\frac{d[S]_t}{dt} = \frac{k_2[C]_0[S]_0[R]_t}{[S]_0 + 1/K} \qquad (7)$$

When the equilibrium constant K is relatively small, so that $[S]_0$ is small compared to $1/K$, and when R is in a nonlimiting excess (or nonexistent) such that its concentration does not change during the initial 2–3% of the reaction, Eq. (7) becomes Eq. (8), where the constant K' equals $k_2 K[R]_0$. Thus

$$-\frac{d[S]_t}{dt} = (K')[C]_0[S]_0 \qquad (8)$$

the initial rate, $-(d[S]_t)/(dt)$, is directly proportional to the initial concentration of either the catalyst or the substrate, and the initial concentration of either can be determined. If a catalyst is to be determined, the substrate is added to each sample in nonlimiting excess (so that the concentration of S does not change during the initial 2–3% of the reaction). If a substrate is to be measured, the catalyst concentration added is not critical in any way because the catalyst is not consumed during the reaction. The choice of the catalyst concentration is dictated only by the range of initial rates that are conveniently measured. Either the fixed- or variable-time method of analysis can be employed using catalyzed reactions.

Simultaneous determinations. Often the thermodynamic properties of closely related species are such that conventional equilibrium analytical techniques are unable to resolve the analytical concentrations of the components of the mixture without prior separation. However, in general the rates of reaction of such species in a mixture are sufficiently different to enable one to determine the initial concentration of each species without resorting to separation, which is a considerable saving of time and labor. Such techniques are called differential kinetic analysis methods. There are a large number of differential kinetic methods applicable to both first- and second-order reactions. Each of these methods has special conditions under which it is most applicable. The principles of these methods are complex and an explanation of each is beyond the scope of this section. However, the detailed explanation of the logarith-

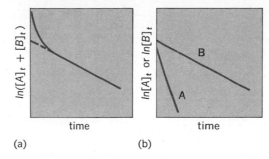

Fig. 3. Logarithmic extrapolation method for a mixture of species A and B reacting by first-order kinetics. (a) Rate data obtained for the mixture. (b) Rate data obtained for each component separated.

mic extrapolation method given below illustrates the general concepts of differential kinetic methods.

Consider two competing second-order irreversible reactions of the type shown by Eqs. (9) and (10),

$$A + R \overset{k_A}{\longrightarrow} P \qquad (9)$$

$$B + R \overset{k_B}{\longrightarrow} P \qquad (10)$$

where k_A and k_B are the second-order rate constant for the reaction of A and B with R, respectively. If the concentration of R is 50 to 100 times greater than the sum of the concentration of A plus B (the components of interest in the analysis), the reaction is pseudo first order because $[R]_t$ is constant and is equal to the initial concentration $[R]_0$. The differential rate expressions for the reactions given by Eqs. (9) and (10) then become Eqs. (11) and (12).

$$-\frac{d[A]_t}{dt} = k_A[R]_0[A]_t = k'_A[A]_t \qquad (11)$$

$$-\frac{d[B]_t}{dt} = k_B[R]_0[B]_t = k'_B[B]_t \qquad (12)$$

The sum of the concentrations of A and B which react competitively to form a common product P is given at any time t by the rate expression, as shown by Eq. (13).

$$\frac{d[P]_t}{dt} = -\left[\frac{(d[A]_t}{dt} + \frac{d[B]_t)}{dt}\right] = k'_A[A]_t + k'_B[B]_t \quad (13)$$

On integration of Eq. (13) between the limits $t = 0$ and $t = \infty$, Eq. (14) is formed. In the case where the rate of reaction of component A is larger than that of B, the term $[A]_0 e^{-k'_A t}$ eventually becomes very small compared to $[B]_0 e^{-k'_B t}$ at some time t after A has reacted essentially to completion ($[A]_t \cong 0$) and can be considered negligible. Thus, by taking the logarithm of both sides of Eq. (14),

$$[P]_\infty - [P]_t = [A]_t + [B]_t$$
$$= [A]_0 e^{-k'_A t} + [B]_0 e^{-k'_B t} \quad (14)$$

Eq. (15) is obtained, which predicts that a plot

$$\ln([A]_t + [B]_t) = \ln([P]_\infty - [P]_t)$$
$$= -k'_B t + \ln[B]_0 \quad (15)$$

$\ln([A]_t + [B]_t)$ or $\ln([P]_\infty - [P]_t)$ versus time t will yield a straight line with a slope of $-k'_B$ and an intercept (at $t = 0$) equal to $\ln[B]_0$. The value of $[A]_0$ is then obtained by subtracting $[B]_0$ from the total initial concentration of the mixture, $[A]_0 + [B]_0$, which must be determined either by independent methods or from $[P]_\infty$. A typical reaction rate curve of this type is illustrated in Fig. 3. *See* CHAIN REACTION, CHEMICAL; ELECTROCHEMICAL TECHNIQUES; INHIBITOR (CHEMISTRY); KINETIC THEORY OF MATTER.

[HARRY B. MARK, JR.]

Bibliography: A. A. Frost and R. G. Pearson, *Kinetics and Mechanisms*, 2d ed., 1961; O. A. Hougen et al., *Chemical Process Principles*, pt. 3, 1947; K. J. Laidler, *Chemical Kinetics*, 1950; H. B. Mark, Jr., L. J. Papa, and C. N. Reilley, *Advances in Analytical Chemistry*, vol. 2, 1963; H. B. Mark, Jr., G. A. Rechnitz, and R. A. Greinke, *Kinetics in Analytical Chemistry*, 1968; K. B. Yatsimerskii, *Kinetic Methods of Analysis*, 1966.

Kinetic theory of matter

A theory which states that the particles of matter in all states of aggregation are in vigorous motion. In computations involving kinetic theory, the methods of statistical mechanics are applied to specific physical systems. The atomistic or molecular structure of the system involved is assumed, and the system is then described in terms of appropriate distribution functions. The main purpose of kinetic theory is to deduce, from the statistical description, results valid for the whole system. The distinction between kinetic theory and statistical mechanics is thus of necessity arbitrary and vague. Historically, kinetic theory is the oldest statistical discipline. Today a kinetic calculation refers to any calculation in which probability methods, models, or distribution functions are involved.

For information which is related to and supplements the present article *see* STATISTICAL MECHANICS. *See also* BOLTZMANN STATISTICS; BOLTZMANN TRANSPORT EQUATION; QUANTUM STATISTICS.

Classes of problems. Kinetic calculations are not restricted to gases, but occur in chemical problems, solid-state problems, and problems in radiation theory. Even though the general procedures in these different areas are similar, there are a sufficient number of important differences to make a general classification useful.

Classical ideal equilibrium problems. In these, there are no interactions between the constituents of the system. The system is in equilibrium, and the mechanical laws governing the system are classical. The basic information is contained in the Boltzmann distribution f (also called Maxwell or Maxwell-Boltzmann distribution) which gives the number of particles in a given momentum and positional range $(d^3x = dxdydz, \quad d^3v = dv_xdv_ydv_z,$ where x, y, and z are coordinates of position, and v_x, v_y, and v_z are coordinates of velocity). In Eq. (1)

$$f(xyz, v_xv_yv_z) = A \, e^{-\beta\epsilon} \tag{1}$$

ϵ is the energy, $\beta = 1/kT$ (where k is the Boltzmann constant and T is the absolute temperature), and A is a constant determined from Eq. (2). The calcula-

$$\int\int d^3x \, d^3v \, f = N \qquad \text{total number of particles} \tag{2}$$

tions of gas pressure, specific heat, and the classical equipartition theorem are all based on these relations.

Classical ideal nonequilibrium problems. Many important physical properties refer not to equilibrium but to nonequilibrium states. Phenomena such as thermal conductivity, viscosity, and electrical conductivity all require a discussion starting from the Boltzmann transport equation, Eq. (3a), where $\mathbf{X}$ is the force per unit mass, ∇ is the gradient on the space-dependence, ∇_v is the gradient on the velocity-dependence, and $(\partial f/\partial t)_{\text{coll}}$ is the change due to collisions. If one deals with states that are near equilibrium, the exact Boltzmann equation need not be solved; then it is sufficient to describe the nonstationary situation as a small perturbation superimposed on an equilibrium state. Even though the rigorous discussion of the nonequilibrium processes is difficult, appeal to simple physical pictures frequently leads to quite tractable expressions in terms of the equilibrium distribution function. Of special importance is the example of the so-called electron gas in a metal. The kinetic treatment of this system forms the basis for the classical (Lorentz) conductivity theory. For states far from equilibrium, no general simple theory exists.

Classical nonideal equilibrium theory. The basic classical procedure for arbitrary systems (systems with interactions taken into account) that allows the calculation of macroscopic entities is that using the partition function, as shown in Eq. (3b), where $\lambda = h/\sqrt{2\pi mkT}$. Here Ψ is the thermo-

$$\frac{\partial f}{\partial t} + \mathbf{v} \cdot \nabla \, f + \mathbf{X} \cdot \nabla_v f = \left(\frac{\partial f}{\partial t}\right)_{\text{coll}} \tag{3a}$$

$$Z_{cl} = e^{-\psi/kT} = \frac{1}{N! \, \lambda^{3N}} \int \cdots \int d^3x_1$$
$$\cdots \, d^3x_N e^{-(1/kT)U(x_1, \ldots, z_N)} \tag{3b}$$

dynamic free energy, and h is Planck's constant. Although Eq. (3b) is written so that it may be applied to gases, the partition function may also be written for classical spin systems, such as ferromagnetic and paramagnetic solids. The mathematical problems of evaluating the integrals or the sums are difficult. Equation (3b), with appropriate modifications, is the starting point for all these considerations.

Classical nonideal nonequilibrium theory. This is the most general situation that classical statistics can describe. In general, very little is known about such systems. The Liouville equation applies, and has been used as a starting point for these studies, but no spectacular results have yet been obtained. For studies of the liquid state, however, the results have been quite promising.

Quantum problems. There are quantum counterparts to the classifications just described. In a quantum treatment a distribution function is also used for an ideal system in equilibrium to describe its general properties. For systems of particles which must be described by symmetrical wave functions, such as helium atoms and photons, one has the Bose distribution, Eq. (4), where $\beta = 1/kT$,

$$f(v_xv_yv_z) = \frac{1}{(1/A)e^{\beta\epsilon} - 1} \tag{4}$$

and A is determined by Eq. (2). *See* BOSE-EINSTEIN STATISTICS.

For systems of particles which must be described by antisymmetrical wave functions, such as electrons, protons, and neutrons, one has the Fermi distribution, Eq. (5). Use of these functions

$$f(v_x v_y v_z) = \frac{1}{(1/A)e^{\beta\epsilon}+1} \quad (5)$$

in calculations is actually not very different from the use of the classical distribution function, but the results are quite different, as are the analytical details. As in the classical case, the treatment of the nonequilibrium state can be reduced to a treatment involving the equilibrium distribution only. The application to electrons as an (ideal) Fermi-Dirac gas in a metal is the basis of the Sommerfeld theory of metals. *See* FERMI-DIRAC STATISTICS; FREE-ELECTRON THEORY OF METALS.

In quantum theory, nonideal systems in equilibrium are described in terms of the quantum partition function, Eq. (6). Here E_n indicates the energy

$$Z_g = \sum_n g_n e^{-E_n/kT} = e^{-\psi/kT} \quad (6)$$

levels of the system, and g_n indicates the weights of these levels. If one defines a Slater sum by Eq. (7), where $U_n(x_1, \ldots, x_N)$ is the wave function of

$$S(x_1, \ldots, x_N) = \sum_n e^{-E_n/kT}|U_n(x_1, \ldots, x_n)|^2 \quad (7)$$

the state n, Z_g may be written as an integral similar to Z, as in Eq. (8). For the applications, the energy

$$Z_g = \int \cdots \int d^3x_1 \cdots d^3x_N S(x_1, \ldots, x_N) \quad (8)$$

levels and the wave functions must be known. In the evaluation of S, given by Eq. (7), the symmetry character of the wave functions must be explicitly introduced. It is sometimes easier to use the grand partition function, Eq. (9). Here μ is the chemical

$$Z_{q \cdot m \cdot gr} = \sum_N \sum_n e^{(\mu N - E_{N,n})/kT} \quad (9)$$

potential, and $E_{N,n}$ is the nth level of a system having N particles. The current theories of quantum statistics, as for example the hard-sphere Bose gas, are for the most part concerned with questions in this area.

The only technique now available is that of the density matrix. It is possible to express certain entities which characterize transport properties, such as the conductivity tensor, in terms of the unperturbed stationary density matrix. A complete discussion of the validity of the approximations is still lacking. In addition, once the conductivity tensor is obtained in terms of the density matrix of the stationary (but still interacting) system, a problem of the same order of difficulty as the evaluation of the quantum mechanical partition function remains, if explicit expressions for these quantities in terms of the forces between atoms are desired.

Classical examples. Kinetic theory gave the first insight into many of the phenomena that take place in gases as well as in metals, where the free (conduction) electrons can be considered as an ideal gas of electrons. The following examples illustrate some of the more fundamental calculations that have been made.

Ideal-gas pressure. A classical ideal gas is described by the Boltzmann distribution of Eq. (1). The constant A is given by Eq. (10), where m is the

$$A = \frac{N}{V}\left(\frac{m\beta}{2\pi}\right)^{3/2} \quad (10)$$

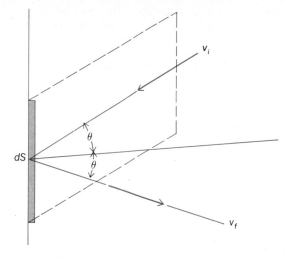

Fig. 1. Change of momentum of gas molecule as it strikes a wall; v_i, initial velocity; v_f, final velocity.

mass of an individual molecule and V is the total volume of the gas. A gas exerts a force on the wall by virtue of the fact that the molecules are reflected by it. The component of momentum normal to the wall changes its sign as a consequence of this collision. Hence if the normal velocity of the molecule is v_n, the momentum given off to the wall is $2mv_n$. To calculate the total force on a wall, one needs the total momentum transferred to the wall per unit time. Let dS be a small section of the wall, and call θ the angle made by the molecule's velocity vector with the normal to dS (Fig. 1). From Eq. (1), the number of molecules per unit volume that have a speed $c = (v_x^2 + v_y^2 + v_z^2)^{1/2}$ and whose velocity vector makes an angle θ with a given axis (call these c,θ molecules) is given by Eq. (11a). The number of such molecules that in time dt will collide with dS (assuming spatial homogeneity) is given by notation (11b). From this information an in-

$$f(c,\theta) = 2\pi A e^{(-1/2)\beta mc^2} c^2 \sin\theta \, d\theta \, dc \quad (11a)$$

$$dS \, c \cos\theta \, dt \, f(c,\theta) \quad (11b)$$

teresting side result may be calculated, namely, that the number of all collisions with a unit area of the wall per second may be written as expression (12a). If one introduces the average speed $\bar{c}$, defined in Eq. (12b), the total number of collisions

$$2\pi A \int_0^{\pi/2} d\theta \sin\theta \cos\theta \int_0^\infty dc \, c^3 e^{-(1/2)\beta mc^2} \quad (12a)$$

$$\bar{c} = \frac{1}{N} \int\int d^3x \, d^3v \, c f \quad (12b)$$

may be written as $1/4 \, n\bar{c}$. Here $n = N/V$, the number density. This relation is of importance in calculating the efflux of gases through orifices. Each (c,θ) molecule contributes a momentum of $2mc \cos\theta$ to the wall per collision. Since the pressure p is given by the force per unit area, or the total momentum transferred per unit time per unit area, one obtains for p Eqs. (13) and (14). Since it is

$$p = \int_0^\infty \int_0^{\pi/2} 2mc \cos\theta$$

$$\cdot 2\pi A c \cos\theta \, e^{-(1/2)\beta mc^2} c^2 \sin\theta \, d\theta \, dc \quad (13)$$

$$p = \frac{1}{3} N m \overline{c^2} = (N/V)\beta^{-1} \quad (14)$$

known experimentally that $pV = NkT$, where k is the Boltzmann constant, $\beta = 1/kT$ is now identified. This pressure calculation is a typical example of a kinetic theory calculation. *See* GAS.

Equipartition theorem and specific heat. In a classical ideal gas, the average kinetic energy associated with any translational degree of freedom is expressed by Eq. (15). It is quite straightforward

$$\tfrac{1}{2}m\overline{v_x^2} = \tfrac{1}{2}m\overline{v_y^2} = \tfrac{1}{2}m\overline{v_z^2} = \frac{1}{2\beta} = \tfrac{1}{2}kT \quad (15)$$

to show that the average energy associated with any degree of freedom, which occurs as a quadratic term in the expression for the mechanical energy, is given by $1/2\ kT$. Formally, the result one proves is that, for a Boltzmann distribution written in terms of momenta and generalized coordinates p_i and q_i, one has Eq. (16), which is called the equi-

$$\overline{q_i \frac{\partial \epsilon}{\partial q_i}} = \overline{p_i \frac{\partial \epsilon}{\partial p_i}} = kT \quad (16)$$

partition theorem. From the equipartition theorem the specific heat C, defined by Eq. (17), may be immediately obtained.

$$C = \frac{\partial \overline{E}}{\partial T} \quad (17)$$

If one has just the three translational degrees of freedom, the average energy per degree of freedom per molecule is $1/2\ kT$, hence the average energy $\overline{E} = 3/2\ NkT$ and $C = 3/2\ R$. The specific heat is thus constant and independent of T. In the case of a diatomic molecule, one usually has three translational and two rotational degrees of freedom. Therefore $\overline{E} = 5/2\ NkT$, and $C = 5/2\ Nk = 5/2\ R$. Suppose there are N atoms in a solid, each bound by elastic forces to a center. In that case, the mechanical energy will be expressed by Eq. (18), where k_F is the eleastic force instant. Accord-

$$\epsilon = \frac{1}{2m}\ (p_x^2 + p_y^2 + p_z^2) + \frac{1}{2}\ k_F^2\ (x^2 + y^2 + z^2) \quad (18)$$

ing to the equipartition theorem, these six terms in ϵ will give a specific heat $C = 3\ Nk$, the Dulong and Petit value for a monatomic solid. *See* SPECIFIC HEAT OF SOLIDS.

The equipartition theorem is especially useful when a gas is at a temperature so high that it becomes necessary to use relativistic mechanics to describe the system properly. *See* RELATIVISTIC MECHANICS.

This happens at temperatures for which the thermal energy kT is of the same order as mc^2, where c denotes the speed of light. The appropriate relation between energy and momentum is then given by Eq. (19a).

The equipartition relation (16) may then be written as Eq. (19b). From this relation one can obtain the relativistic correction to the average energy of a gas, as shown in Eq. (19c).

$$\epsilon = c[\ (p_x^2 + p_y^2 + p_z^2) + m^2c^2]^{1/2} \quad (19a)$$

$$\frac{\overline{c^2 p_x^2}}{\epsilon} = kT \quad (19b)$$

$$\overline{E} = \tfrac{3}{2}NkT\left(1 + \frac{5}{4}\frac{kT}{mc^2} + \cdots\right) \quad (19c)$$

Electrical conductivity of metals. Experimentally, one observes a proportionality between the applied electric field and the current produced in a metallic conductor. Kinetic procedures give an explanation of this general connection. Consider the electrons in a metal as an ideal gas of electrons of mass m and charge e. In equilibrium, the electrons are described by the Boltzmann distribution, Eq. (1), or the Fermi-Dirac distribution, Eq. (5). Call this distribution f. Because f depends on $v_x^2 + v_y^2 + v_z^2$ only, it is clear that $\bar{v}_x = 0$; no net current can flow in an equilibrium state. The application of an electric field therefore results in the destruction of the spherical symmetry in the velocities. This effect is described by the Boltzmann transport equation, Eq. (20).

$$\frac{\partial f}{\partial t} + \mathbf{v} \cdot \nabla f + \mathbf{X} \cdot \nabla_v f = \left(\frac{\partial f}{\partial t}\right)_{coll} \quad (20)$$

The collision term causes considerable difficulty. In many problems one is justified in introducing a relaxation time τ which may depend on x and v and which is defined by Eq. (21). Here f_0

$$\left(\frac{\partial f}{\partial t}\right)_{coll} = -\frac{f - f_0}{\tau} \quad (21)$$

is the distribution function at thermal equilibrium. The introduction of such a relaxation time presupposes that a nonequilibrium state will decay exponentially into an equilibrium state as a consequence of the action of the collisions. This is undoubtedly a good approximation near equilibrium.

Assume the existence of an electric field in the x direction which distorts the initial distribution a small amount. By using Eq. (21), Eq. (20) may be written as Eq. (22). Here E_e is the external electric

$$\frac{eE_e}{m}\frac{\partial f}{\partial v_x} + v_x\frac{\partial f}{\partial x} = -\frac{f - f_0}{\tau} \quad (22)$$

field. If it is now assumed that $(f - f_0)/f_0 \ll 1$, with f very near an equilibrium state, so that quadratic terms may be neglected, and that the parameters A and β in f_0 are independent of x, then an explicit expression for f can be obtained, as shown in Eq. (23).

$$f = f_0 - \tau e E_e v_x \frac{\partial f_0}{\partial \epsilon} \quad (23)$$

The electric current in the x-direction is always given by Eq. (24).

$$j_x = \int e v_x f\ d^3v = -\tau e^2 E_e \int v_x^2 \frac{\partial F_0}{\partial \epsilon} d^3v \quad (24)$$

The first term in Eq. (23) does not contribute to the current, as has already been pointed out. This is an example of a formal result in transport theory. The result for j_x is in a form in which only a knowledge of the equilibrium distribution is required to obtain an explicit answer.

In the case of Boltzmann statistics, use of Eqs. (1), (2), (10), and (24) gives Eq. (25a). The conductivity $\sigma = Ne^2\tau/m$ cannot be compared with experiment unless the relaxation time τ is known. In the Fermi case the evaluation of Eq. (24) is facilitated by observing that $\partial f_0/\partial \epsilon$ has a δ-function character. Call $A = e^{\beta\mu}$, so that Eq. (5) reads as Eq. (25b).

$$j_x = \frac{Ne^2\tau}{m}E_e \qquad (25a)$$

$$f_0 = \frac{1}{e^{\beta(\epsilon-\mu)}+1} \qquad (25b)$$

It is easy to show from Eq. (25b) that expression (26) holds for sufficiently low temperatures. This

$$\frac{\partial}{\partial\epsilon}\frac{1}{e^{\beta(\epsilon-\mu)}+1} \cong -\delta(\epsilon-\mu) \qquad (26)$$

allows the immediate calculation of Eq. (24) also for the case in which the relaxation time depends on the velocity.

Viscosity and mean free path. One of the early successes of kinetic theory was the explanation of the viscosity of a gas. Strictly speaking, this is again a transport property, and as such it should be obtained from the Boltzmann transport equation, Eq. (20). It is possible, however, to give an elementary discussion. Consider a gas that is contained between two walls or plates, the lower one $(y-0)$ at rest and the upper one constrained to move with a given velocity in the x direction (Fig. 2). A force is necessary to maintain the constant velocity of the plate. This force is given by Eq. (27). Here dv_x/dy

$$\mathbf{X}_x = \eta A\frac{dv_x}{dy} \qquad (27)$$

is the velocity gradient, and $\mathbf{X}$ is the viscous force on the area A, which is perpendicular to the y-axis ($\mathbf{X}_x/A$ is sometimes called the shear stress). Equation (27) defines η, the viscosity coefficient. The physical reason for this force stems from the fact that molecules above the surface S have a greater flow velocity than those below this surface. (This will certainly be true on the average.) Thus molecules crossing from above to below will carry a larger amount of momentum in the positive x direction than those crossing from below S upwards. Hence the net effect is a transport of momentum in the x direction across the surface. By Newton's second law, this will yield a force. The computation can be carried out in this manner. Consider an area in the xz plane. The number of molecules passing through per second is $fv_y dS$, where f is the distribution function.

The amount of momentum transported in the x direction is (per collision) $fv_y dS \cdot mv_x$.

The force per unit area is the sum of these terms, given by expression (28). For the evaluation

$$fv_y v_x m \qquad (28)$$

of this sum, the notion of mean free path is useful. The mean free path is the average distance traveled by a molecule between collisions, and is usually designated by λ. To investigate this entity, imagine each molecule to be a hard sphere with radius a. If a molecule moves with average speed $\bar{c}$, it sweeps out a volume $4\pi a^2 \bar{c}t$ in time t. If there are $n = N/V$ molecules per unit volume, the number of collisions per second is given by the collision frequency z in Eq. (29).

$$z = n4\pi a^2\bar{c} \qquad (29)$$

For a typical gas (oxygen) under standard conditions $n = 3\times10^{25}$, $\bar{c} = 4.5\times10^4$ cm/sec, and $a \cong 1.8\times10^{-12}$ cm. Hence, numerically, $z = 5.5\times10^9$ collisions/sec. The average distance between collisions, that is, the mean free path, is given by Eq. (30a). Numerically, $\lambda \cong 8\times10^{-6}$ cm. This discussion is, of course, exceedingly crude. Making the calculation on the basis of a Boltzmann distribution gives Eq. (30b).

$$\lambda = \frac{\bar{c}t}{n(4\pi a^2)\bar{c}t} = \frac{1}{n(4\pi a^2)} \qquad (30a)$$

$$\lambda = \frac{1}{\sqrt{2}n(4\pi a^2)} \qquad (30b)$$

Using similar methods, it may be shown that the distribution of the mean free paths (that is, the number of molecules whose mean free path lies between x and $x + dx$) is given by Eq. (31).

$$dN = \frac{N_0}{\lambda}e^{-x/\lambda}\,dx \qquad (31)$$

There is an interesting connection between the mean free path and the relaxation time introduced previously. It should be stressed, however, that this connection follows more from a qualitative discussion than from a rigorous calculation. One would guess that Eq. (32) is valid. This means that

$$\tau = \lambda/\bar{c} \qquad (32)$$

a relaxation time describes the decay from a state so near an equilibrium state that, when the molecules have traveled (on the average) one mean free path, the equilibrium is reestablished. Stated differently, the nonequilibrium state is, on the average, one collision per molecule removed from the equilibrium state.

Expression (28) may now be evaluated in terms of the mean free path as shown in Eq. (33a). From this, the viscosity coefficient follows directly, as shown in Eq. (33b). Introducing the mean free path, as given by Eq. (30a), one sees that Eq. (33c) holds. The remarkable result is that the viscosity

$$\text{Force per unit area} = \tfrac{1}{3}nm\lambda\bar{c}\frac{dv_x}{dy} \qquad (33a)$$

$$\eta = \tfrac{1}{3}nm\bar{c}\lambda \qquad (33b)$$

$$\eta = \frac{1}{3}\frac{m\bar{c}}{4\pi a^2} \qquad (33c)$$

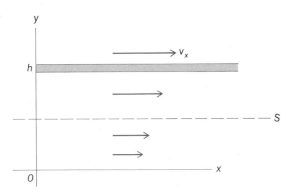

Fig. 2. Explanation of viscosity. The plate at $y = h$ is moving with velocity v_x, and the plate at $y = 0$ is stationary. The horizontal component of velocity of the gas molecules varies from 0 at $y = 0$ to v_x at $y = h$. Momentum therefore must be transferred across surface S by the vertical component of velocity.

is indeed independent of the pressure. Since Eq. (34) holds, the viscosity depends on the tempera-

$$\bar{c} = \sqrt{8kT/\pi m} \qquad (34)$$

ture but not on the pressure. This somewhat un-intuitive result was one of the first triumphs of kinetic theory. Both aspects of Eq. (33c) are in good agreement with experiment; for not too low densities η is indeed independent of the pressure; η is also proportional to the square root of the temperature.   [MAX DRESDEN]

Bibliography: R. H. Fowler, *Statistical Mechanics*, 2d ed., 1936; C. Kittel, *Elementary Statistical Physics*, 1958; R. D. Present, *Kinetic Theory of Gases*, 1958; D. Ter Haar, *Elements of Statistical Mechanics*, 1954.

Kinetics (classical mechanics)

That part of classical mechanics which deals with the relation between the motions of material bodies and the forces acting upon them. It is synonymous with dynamics of material bodies. *See* DYNAMICS.

Basic concepts. Kinetics proceeds by adopting certain intuitively acceptable concepts which are associated with measurable quantities. These essential concepts and the measurable quantities used for their specification are as follows:

1. Space configuration refers to the positions and orientations of bodies in a reference frame adopted by the observer. It is expressed quantitatively by an arbitrarily chosen set of space coordinates, of which cartesian and polar coordinates are examples. All space coordinates rest on the motion of distance measurement.

2. Duration is expressed quantitatively by time measured by a clock or comparable mechanism.

3. Motion refers to change of configuration with time and is expressed by time rates of coordinate change called velocities and time rates of velocity change called accelerations. The classical assumption that coordinates behave as analytic functions of time permits representation of velocities and accelerations as first and second derivatives, respectively, of the space coordinates with respect to time.

4. Inertia is an attribute of bodies implying their capacity to resist changes of motion. A body's inertia with respect to linear motion is denoted by its mass.

5. Momentum is an attribute proportional to both the mass and velocity of a body. Momentum of linear motion is expressed as the product of mass and linear velocity.

6. Force serves to designate the influence exercised upon the motion of a particular body by other bodies, not necessarily specified. A quantitative connection between the motion of a body and the force applied to it is expressed by Newton's second law of motion, which is discussed later.

Distance, time, and mass are commonly regarded as fundamental, all other dynamical quantities being definable in terms of them.

Newton's second law. A primary objective of classical kinetics is the prediction of the behavior of bodies which are subject to known forces when only initial values of the coordinates and momenta are available. This is accomplished by use of a principle first recognized by Isaac Newton. New-

ton's statement of the principle was restricted to the linear motion of an idealized body called a mass particle, having negligible extension in space.

The basic dynamical law set forth by Newton and known as his second law states that the time rate of change of a particle's linear momentum is proportional to and in the direction of the force applied to the particle. This statement, although special in form, serves as a basis for more comprehensive statements of the principle which have since appeared.

Stated analytically, Newton's second law becomes the differential equation, Eq. (1), in which m

$$\frac{d(mv)}{dt} = F \qquad (1)$$

represents the particle's mass, v its velocity, F the applied force, and t the time. Equation (1) provides a definition of force and of its units if units of mass, distance, and time have previously been adopted. The classical assumption of constancy of mass permits Eq. (1) to be expressed as Eq. (2), where a

$$ma = F \qquad (2)$$

represents the linear acceleration. A particle in physical space requires three cartesian coordinates, x, y, and z, to specify its position. Its linear acceleration is a vector with three cartesian components, the second time derivatives of x, y, and z. Equation (2) therefore equates two vectors, requiring equality of their components expressed in detail by Eqs. (3a)–(3c). These are the Newtonian

$$m\frac{d^2x}{dt^2} = F_x \qquad (3a)$$

$$m\frac{d^2y}{dt^2} = F_y \qquad (3b)$$

$$m\frac{d^2z}{dt^2} = F_z \qquad (3c)$$

equations of motion of an unconstrained particle in space. If the three force components F_x, F_y, and F_z are expressed functions of the coordinates and time, the dependence of each coordinate upon the time is implied and can in favorable cases be found as solutions of the equations of motion in the form of Eqs. (4). The primary objective of kinetics is achieved in the discovery of such functions.

$$x = x(t) \qquad y = y(t) \qquad z = z(t) \qquad (4)$$

One-dimensional particle motion. The motion of a particle which remains on the x axis, either because of constraints or initial conditions, is determined by Eq. (3a) alone, whose solution is simplified by the absence of y and z. Such one-dimensional dynamical problems provide an attractively simple introduction to the subject. Examples are the motion of a body falling vertically, subject to gravitational force, and linear harmonic motion.

Two-dimensional particle motion. The motion of a particle remaining in the plane of the x and y axes is determined by Eqs. (3a) and (3b), from which z is absent. Two-dimensional problems are reasonably tractible and include many of physical interest such as the motion of a projectile (exterior ballistics), and of a body attracted toward a central point, as in planetary motion. Solution of a two-dimensional problem is frequently simplified by change of variables which reduces it to a pair of one-dimensional problems.

Three-dimensional particle motion. All three equations of motion, Eqs. (3a)–(3c), apply to an unconstrained particle in space. Complete solutions are possible only when the functions expressing force components are relatively simple in character. Fortunately, many of the solvable cases correspond to important physical examples in which simplicity of the forces allows separation into one- and two-dimensional motions. Three-dimensional projectile motion without friction is an example.

Newton's third law. The behavior of systems composed of two or more interacting particles is treated by Newtonian dynamics augmented by Newton's third law of motion which states that when two bodies interact, the forces they exert on one another are equal and oppositely directed. The important laws of momentum and energy conservation are derivable for such systems (the latter only for forces of special type) and useful in solution of problems. The equations of motion for systems of more than two interacting particles in space are mathematically intractible in the absence of geometrical constraints or special initial conditions, but assumptions approximating the physical situation permit solution of many problems of physical interest. The principles of particle dynamics are transferred to extended bodies by regarding them as systems of particles subject to specified mutual constraints and mutual forces. *See* ACCELERATION; BALLISTICS, EXTERIOR; FORCE; GRAVITATION; HARMONIC MOTION; MASS; MOMENTUM; ORBITAL MOTION; RIGID-BODY DYNAMICS; VELOCITY.   [RUSSELL A. FISHER]

Bibliography: H. Goldstein, *Classical Mechanics*, 1950; C. Kittel, W. D. Knight, and M. A. Ruderman, *Mechanics*, 1965; J. C. Slater and N. H. Frank, *Mechanics*, 1947; E. T. Whittaker, *A Treatise on the Analytical Dynamics of Particles and Rigid Bodies*, 4th ed., 1944.

Kinetics, chemical

The branch of physical chemistry concerned with the mechanisms and rates of chemical reactions. It includes the study of activation energies of reactions.

Reaction rate. The rate at which chemical reactants are used up, or at which chemical products are formed, is an important topic, particularly in organic chemistry and biochemistry, where many different products can be obtained, all of which are possible according to thermodynamic calculations. The product actually obtained is determined by the relative rates of the competing reactions. The more abundant products are the ones produced by the faster reactions. Therefore, a knowledge of reaction rates is important in controlling chemical reactions.

Rate constant. The specific rate constant k is useful in calculating reaction rates at various concentrations. If in a simple chemical reaction, $A \rightarrow B$, the rate of reaction is directly proportional to the concentration of A, the proportionality constant k is called the specific reaction rate constant. Thus Eq. (1) is formed, where $-dc_A/dt$ is the rate of decrease in concentration of A with time t. The

rate of change will decrease as A is used up and its concentration decreases, but the rate will always be proportional to the concentration. Thus, if k has

$$\frac{-dc_A}{dt} = kc_A \qquad (1)$$

a value of 0.01 per minute, 1% of A which is present at any time t will react per minute. Over long periods of time, the concentration will be changing during the time interval, and it is necessary to integrate the equation. This gives Eq. (2), where c_{A_1} is

$$k = \frac{2.303}{t_2 - t_1} \log \frac{c_{A_1}}{c_{A_2}} \qquad (2)$$

the concentration at time t_1, and c_{A_2} is the concentration at time t_2. Thus it is possible to determine the rate constant k from the measurements of the concentration of A at two different times. When the rate constant has been evaluated by experimental measurements or by theoretical calculations, it is possible to calculate accurately the concentration c_{A_2} at any later time when it is known at one time. Frequently the known concentration c_{A_1} is taken as the initial concentration starting at zero time.

Order of reaction. The order of reaction depends on the exponent of the concentration which determines the rate of the reaction. Thus in Eq. (1) the reaction is first order. In another type of reaction which depends on the collision of two molecules, $A + A \rightarrow A_2$, the rate is given by Eq. (3). Such a

$$\frac{-dc_A}{dt} = kc_A^2 \qquad (3)$$

reaction is called a second-order reaction. If the reaction is $A + B \rightarrow AB$, the rate of the reaction may be given by Eq. (4). The reaction rate is first

$$\frac{-dc_A}{dt} = \frac{-dc_B}{dt} = \frac{dc_{AB}}{dt} = kc_A c_B \qquad (4)$$

order with respect to A and first order with respect to B, but the overall reaction rate is said to be second order. The reaction order is equal to the sum of the exponents of the concentrations of all the reacting materials. In a first-order reaction, k is merely a ratio and is independent of the units of concentration used. In a second-order reaction, which depends on the product of the concentrations, the constant k includes a term for the concentration. Its numerical value depends on the concentration units used, such as moles per liter.

When the time interval is long, it is necessary to use integrated formulas. Thus Eq. (3) becomes Eq. (5) and Eq. (4) becomes Eq. (6). Where a is the

$$k = \frac{1}{t} \frac{x}{a(a - x)} \qquad (5)$$

$$k = \frac{2.303}{t(a - b)} \log \frac{b(a - x)}{a(b - x)} \qquad (6)$$

concentration of A, and b is the concentration of B at the start of the reaction when $t = 0$, and x is the change in concentration during time t.

Third-order reactions involving three substances A, B, and C are known, but they are rather rare. The rate of the reaction is proportional to the product of the concentrations of A, B, and C.

Zero-order reactions, given by Eq. (7), are known. In these the rate is independent of the concentration and is constant over a long period of time. Such zero-order reactions are found in photo-

$$\frac{-dc_A}{dt} = k \qquad (7)$$

chemistry, where the intensity of light, rather than the concentration, is the rate-determining factor. They are found also in saturated solutions where the concentrations of the reacting materials are kept constant by the solution of more of the solid.

If the order of a reaction is known, it is a simple matter to calculate, with the formulas just listed, the amount of reactant remaining or the amount of product formed at any time. If the order is not known, it must be evaluated from experimental data. For example, in the rate expression for the reaction $A + B$, Eq. (8), it is necessary to evaluate

$$\frac{-dc_A}{dt} = \frac{-dc_B}{dt} = kc_A{}^m c_B{}^n \qquad (8)$$

the exponents m and n. Usually they will not be the whole numbers 1, 2, or 3, but rather numbers such as 1.3, which then do not give first-, second-, or third-order reactions. A reaction with fractional exponents involves several reactions which take place simultaneously.

In evaluating these exponents, the data may be substituted into formulas to determine if a first-, second-, or third-order formula will agree with the experimental data. Also, the order may be determined by graphing. The reaction is first order if $\log c_A$ gives a straight line when plotted against time, second order if $1/c_A$ gives a straight line, and third order if $1/c_A{}^2$ gives a straight line, provided all reactants start with the same concentration. The order of each reactant can be determined also by increasing its concentration and measuring the effect on the reaction rate while the concentrations of all other reactants are kept constant, or are present in large excess.

Experimental measurements. Measurements of reaction rates are made by determining the concentrations of reactants or products as a function of time. In a closed system, samples can be withdrawn and analyzed immediately (or chilled quickly) so that the concentration does not change after the time of sampling. The analysis may be carried out by chemical means, such as titration, or by physical measurements of such properties as light absorption, refractive index, or volume.

If there is a change in pressure of a gas, electrical conductance, volume of a solution, electrical conductance, or other physical property, the course of the reaction can be followed in the whole reaction vessel without withdrawing samples. If only one of the reactants or products absorbs light of a given wavelength, the changing concentration may be followed by measuring the absorption of light.

In some reactions the rate is determined conveniently by passing the reacting material in a stream through the reaction chamber and determining the concentrations before and after entering the chamber. The time is estimated from the rate of flow and the volume of the chamber.

Activation energy. Energy must be supplied to prepare the molecules for reaction. Otherwise the

reactions would take place almost instantly. In ordinary thermal reactions, this extra energy is provided by collisions with molecules that are moving at very high velocities. Energies of about $10-100$ kcal/mole are usually required to activate the molecules in reactions which proceed with measurable rates.

The activation process usually involves the weakening or breaking of a chemical bond. Thus in a typical reaction, Eq. (9), energy is required to ac-

$$\begin{array}{ccc} \text{A—B} & \text{A—B} & \text{A} \quad \text{B} \\ + & \rightarrow \; |\text{X}| \; \rightarrow & | \; + \; | \\ \text{C—D} & \text{C—D} & \text{C} \quad \text{D} \end{array} \qquad (9)$$

tivate the molecules AB and CD and bring them so close to each other that there is a force of attraction exhibited between all the atoms. The activated complex which results may break up in different ways, for example, into the products AC and BD.

Complex molecules with many atoms may be regarded as composed of many different atom pairs held together either by electrical charges or by electron pairs, and the energies required to loosen or break these pairs and to move the atoms around varies greatly in different chemical reactions.

If the activation energy is low, there are many effective molecular collisions, and the reaction is fast. If the activation energy is high, only a small fraction of the molecules will be moving fast enough to give sufficient energy, on collision, to activate the molecule and bring about the reaction. Then the reaction is slow. The distribution of molecular velocities at a given temperature is given by the Maxwell-Boltzmann distribution.

When the temperature is raised, the average kinetic energy of the molecules increases at a rate proportional to the absolute temperature, but the number of molecules with very high velocities increases very greatly. Accordingly, a chemical reaction proceeds very much more rapidly when the temperature is raised, because there are more collisions of high energy. The effect is so large that most chemical reactions in the neighborhood of room temperature will double or treble in rate for a 10-degree Celsius rise in temperature.

The Arrhenius equation, Eq. (10), gives the rela-

$$k = se^{-\Delta H_{act}/RT} \qquad (10)$$

tion between reaction rate and temperature. Here k is the specific reaction rate constant, e is the base of natural logarithms (2.71828), R is the gas constant, and T is the absolute temperature. The constant ΔH_{act} is known as the heat of activation, and the constant s is called the frequency factor.

According to this equation, when the logarithm of the specific rate constant is plotted against the reciprocal of the absolute temperature, a straight line is produced. The slope of this line is multiplied by the gas constant R and the conversion factor 2.303 to convert $\log_{10}$ into $\log_e$ to give the value of the heat of activation. This relation holds remarkably well for most chemical reactions, even for complex reactions. An example is shown in Fig. 1 for the decomposition of nitrogen pentoxide. When the constants are evaluated by measuring k at two or more different temperatures, it is possible to calculate the specific reaction rate constant at any

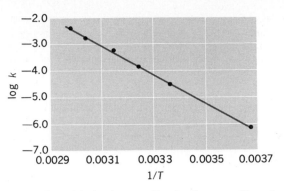

Fig. 1. Plot of Arrhenius equation for decomposition of nitrogen pentoxide. Logarithm of specific reaction rate plotted against reciprocal of the absolute temperature.

temperature over which the equation holds.

The Eyring equation is similar to the empirical Arrhenius equation, but it is based on statistical mechanics and is more exact. It takes the form shown in Eq. (11), where ΔS_{act} is the entropy of activation;

$$k = \frac{RT}{Nh} e^{\Delta S_{act}/R} e^{\Delta H_{act}/RT} \qquad (11)$$

tivation; N is the Avogadro number, 6.02×10^{23}; and h is Planck's constant, 6.62×10^{-27} erg-sec. The entropy of activation can be determined in simple reactions from statistical mechanics using data of spectroscopy and molecular structure. The term RT/Nh has a value of about 10^{13} sec^{-1} near room temperature.

Prediction of reaction rates. Rate predictions are difficult, but a few general rules are available for the simpler reactions. The chief difficulty lies in the fact that most chemical reactions are complex, and involve several different reactions taking place simultaneously. Fortunately, most of the reactions that make up the complex reaction are themselves either simple unimolecular reactions which involve only one molecule and follow the first-order rate equation, or bimolecular reactions which involve two molecules and give a second-order reaction.

In simple unimolecular reactions the frequency factor s has a value of about 10^{13} sec^{-1}. If it is much larger, a chain reaction is probably present. If it is smaller, a change in entropy is probably involved in activation. In a simple bimolecular reaction which gives a second-order reaction rate, the frequency factor is close to the frequency of molecular collision, a quantity which can be calculated by the kinetic theory from a knowledge of the concentration of molecules and the molecular diameters. This is not the case in bimolecular reactions involving complex molecules, and it is necessary to calculate the entropy of activation as indicated in the Eyring equation. *See* CHAIN REACTION, CHEMICAL.

The heats of activation can be determined experimentally from the rates at two temperatures, and attempts have been made to estimate them on the basis of molecular constants.

In the estimation of activation energy, potential energy curves are sometimes drawn, in which the energy holding a pair of atoms together is plotted against the interatomic distance. There is a minimum in the curve at the interatomic distance in the normal molecule, and the energy requirements rise sharply as the atoms are forced closer together. As the atoms are pulled farther apart, the energy requirements increase also, but finally an interatomic distance is reached when the atoms break apart giving a horizontal line and indicating that no further energy is required for further separation. The shape of this curve can be determined from spectroscopic data and quantum numbers. If these potential energy curves can be plotted in three dimensions with energy along the vertical axis and interatomic distance between the two different pairs of atoms along the horizontal axes, a potential energy surface is obtained. The height of the lowest energy pass, separating two valleys corresponding to stable reactants and products, is a measure of the activation energy.

This type of calculation for activation energy has been attempted in a semiempirical method proposed by Henry Eyring. It involves oversimplified calculations based on the energy required to break the atom-pair bonds, as determined from calorimetry or from spectroscopy, on the vibration frequency as determined from infrared spectroscopy, and on the interatomic distances in the stable molecule. A knowledge of the type of binding is necessary, that is, the fraction of the binding energy which is electrostatic in nature and the fraction which is homopolar or electron-pair binding. The results have not been very satisfactory.

In bimolecular reactions, according to a crude empirical rule proposed by J. O. Hirschfelder, the activation energy is roughly equal to 28% of the energy required to break the bonds of the two reacting pairs of atoms.

In endothermic bond-breaking reactions, if the rupture of the bond produces atoms or free radicals which recombine without the requirement of activation energy, then the activation energy is equal to the energy required to break the bond.

In exothermic reactions there is no relation between the heat evolved and the activation energy. It is thus necessary to rely on experimental data or rough estimates based on molecular structure, or on the Eyring or Hirschfelder rules.

Many chemical reaction rates are greatly affected by catalysts and by light and high-energy radiation.

Complex reactions. These are chemical reactions in which more than one reaction is going on at the same time. Simple first-order reactions, in which the rate depends directly on the concentration, or second-order reactions, in which the rate depends on the square of the concentration, can be described in simple mathematical terms. However, when several reactions occur simultaneously, even if they are all first- or second-order reactions, the mathematical description often becomes very complicated. The most common types of complex reactions involve reverse reactions, side reactions, and consecutive reactions.

Reverse reactions. In this type of reaction, products reunite to give the original reactants, Eq. (12).

$$A + B \rightleftharpoons AB \qquad (12)$$

All reactions are reversible, but often the reverse reaction is so slight that it can be neglected. Even if the reverse reaction is appreciable, it can usually be ignored in the early stages of the reaction, before a sufficient amount of the products have accumulated.

An example of a reversible reaction is the reaction between acetic acid and ethanol to give ethyl acetate and water, shown by Eq. (13).

$$CH_3COOH + C_2H_5OH \underset{k_2}{\overset{k_1}{\rightleftharpoons}} CH_3COC_2H_5 + H_2O \quad (13)$$

At first the rate of formation of ethyl acetate is given by Eq. (14), but in later stages by Eq. (15),

$$-\frac{dc_{C_2H_5OH}}{dt} = k_1 c_{CH_3COOH} c_{C_2H_5OH} \quad (14)$$

$$-\frac{dc_{C_2H_5OH}}{dt} = k_1 c_{CH_3COOH} c_{C_2H_5OH} - k_2 c_{CH_3COC_2H_5} c_{H_2O} \quad (15)$$

where k_1 is the forward reaction and k_2 is the reverse reaction. It is possible to express k_2 in terms of k_1 and the equilibrium constant K and to obtain by integration a complete formula for the complex reaction.

Side reactions. When these occur, there are several ways in which the starting materials can react to give different products. All are possible according to thermodynamics, and the faster reactions will predominate to give the larger yields of products. In the reaction of A with both B and C, Eqs. (16) and (17), the ratio of AB to AC will depend on

$$A + B \overset{k_1}{\rightarrow} AB \quad (16)$$

$$A + C \overset{k_2}{\rightarrow} AC \quad (17)$$

the relative rates. The rate expressions are Eqs. (18) to (20). An example of competing or side reac-

$$\frac{-dc_A}{dt} = k_1 c_A c_B + k_2 c_A c_C \quad (18)$$

$$\frac{dc_{AB}}{dt} = k_1 c_A c_B \quad (19)$$

$$\frac{dc_{AC}}{dt} = k_2 c_A c_C \quad (20)$$

tions is the nitration of chlorobenzene. Three different reactions take place simultaneously to give products in which the NO_2 and Cl groups are adjacent in the benzene ring, where they are separated by one carbon atom, or where they are separated by two carbon atoms. Most of the materials react to give the third product.

Consecutive reactions. In this type of reaction the product of the first reaction reacts to give a second product, and this in turn reacts to give a third product, as shown in Eq. (21). The rate at which B changes is given by Eq. (22), where k_1 is the

$$A \overset{k_1}{\rightarrow} B \overset{k_2}{\rightarrow} C \quad (21)$$

$$\frac{dc_B}{dt} = k_1 c_A - k_2 c_B \quad (22)$$

specific reaction-rate constant for the decomposition of A, and k_2 is the constant for the decomposition of B.

Figure 2 shows a graph in which the quantities A, B, and C are each given as a function of time, when $k_1 = 0.1$, $k_2 = 0.05$, and 1 mole of A is taken at the beginning of the reaction. The concentration of

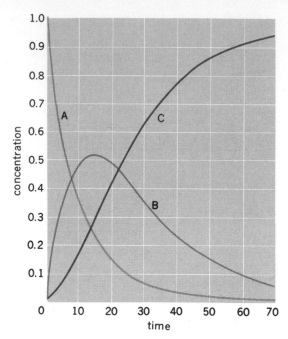

Fig. 2. Graph of a complex reaction, illustrated by consecutive first-order reactions, $A \rightarrow B \rightarrow C$.

A decreases rapidly at first and then more slowly, but always at a rate proportional to the concentration of A. It approaches zero as the time increases. The concentration of B builds up rapidly at first, goes through a maximum, then decreases and eventually approaches zero. The product C is slow in getting started, but after B has accumulated, C increases rapidly and finally approaches 1 mole, which was the original amount of A. It is clear that the concentration of B is a complicated function of time.

There are many types of still more complex reactions, for example, Eqs. (23) and (24). All are

$$\begin{array}{c} A \overset{k_1}{\rightarrow} B \\ k_2 \updownarrow k_3 \\ C \end{array} \quad (23)$$

$$A + B \underset{k_2}{\overset{k_1}{\rightleftharpoons}} AB \overset{k_3}{\rightarrow} C \quad (24)$$

made up of simple unimolecular and bimolecular reactions in various combinations, and the differential equations can be written down easily. In only a few cases, however, can these differential equations be solved by integration. The rate expressions of many of the complex reactions cannot be integrated, but can be solved graphically or by modern mathematical computing machines.

It is clear that in complex reactions the order of the reaction may be given by fractions rather than by whole number integers, and these may change during the course of the reaction.

Often in bimolecular reactions involving the reaction between two molecules, the rate appears to be that of a first-order reaction. For example, in the hydrolysis of ethyl acetate in water solution, the rate of the reaction appears to depend only on the concentration of the ethyl acetate because the solvent water is in such large excess that it does not appear to change appreciably in amount.

The intermediate steps in a complex reaction can be studied with the help of kinetic measurements made by changing the concentrations over wide limits, by changing the temperature, or by using specific catalysts and inhibitors. Sometimes the concentration of an intermediate substance can be followed by measuring its absorption of light at a particular wavelength which is not absorbed by the other materials present.

Many chemical reactions are complex. For example, a simple reaction such as the oxidation of organic materials by air at elevated temperatures may go through a series of reactions in which peroxides, alcohols, ketones, acids, and finally carbon dioxide and water are formed.

Important progress is being made in the study of very rapid chemical reactions with half-lives of milli- and microseconds and less. Such reactions can be initiated by powerful electrical currents, by absorption of intense beams of light, and by shock waves. The course of the reaction is usually followed by measuring the absorption of light by one of the reactants or products using photocells and electronic equipment.

Shock waves are produced either by flames or explosions, or by the rupture of a diaphragm which separates a low pressure gas in a tube from a high-pressure gas. The sudden release of the carrier gas at a very high pressure can produce temperatures of many thousands of degrees, and the thermodynamic and kinetic properties of the mixing gases can be calculated.

Fundamental measurements of collisions and energy transfer between molecules and within molecules are being made by colliding two beams of unidirectional molecules in a vacuum. *See* CATALYSIS; FREE RADICAL; HALF-LIFE; INHIBITOR (CHEMISTRY); KINETIC THEORY OF MATTER; PHOTOCHEMISTRY; POTENTIAL BARRIER.

[FARRINGTON DANIELS]

Bibliography: P. G. Ashmore, F. S. Dainton, and T. M. Sugden, *Photochemistry and Reaction Kinetics*, 1962; S. W. Benson, *Foundation of Chemical Kinetics*, 1960; H. Eyring and E. M. Eyring, *Modern Chemical Kinetics*, 1963; A. A. Frost and R. G. Pearson, *Kinetics and Mechanism*, 1961; E. E. Green and J. P. Toennies, *Chemical Reactions in Shock Waves*, 1964; O. A. Hougen et al., *Chemical Process Principles*, 1947; E. L. King, *How Chemical Reactions Occur: An Introduction to Chemical Kinetics and Reaction Mechanisms*, 1963; K. J. Laidler, *Chemical Kinetics*, 1965; E. E. Petersen, *Chemical Reaction Analysis*, 1965.

Kinetoplastida

An order of the class Zoomastigophorea in the phylum Protozoa, also known as Protomastigida, containing a heterogeneous group of colorless flagellates possessing one or two flagella in some stage of their life cycle. These small organisms (5–89 μ in length) typically have pliable bodies. Some species are holozoic and ingest solid particles, while others are saprozoic and obtain their nutrition by absorption. Their life cycles are usually simple but some species have two or more recognizably distinct stages. The species may be either free-living or parasites of vertebrates, invertebrates, and plants. Reproduction is by longitudinal fission, although multiple fission occurs in some

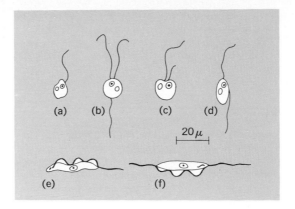

Representative genera of families of order Kinetoplastida. (*a*) *Oikomonas* (family Oikomonadidae), one anterior flagellum. (*b*) *Amphimonas* (family Amphimonadidae), two equally long anterior flagella. (*c*) *Monas* (family Monadidae), two unequally long anterior flagella. (*d*) *Bodo* (family Bodonidae), two unequally long flagella, one of them trailing. (*e*) *Trypanosoma* (family Trypanosomatidae), one flagellum with undulating membrane. (*f*) *Cryptobia* (family Cryptobiidae), two flagella, one free and one with undulating membrane.

species. Several are important disease-producing parasites of man and of domestic and wild animals.

Taxonomy. The order Kinetoplastida includes all of the Protozoa with only one or two flagella that are not markedly ameboid, do not contain chloroplasts, and are not closely related to chloroplast-bearing flagellates. Organisms that possess one or two flagella and are decidedly ameboid belong to the family Mastigamoebidae in the order Rhizomastigida. Those which are colorless but considered to be close relatives of protozoans possessing chloroplasts belong to the family Chlamydomonadidae of the order Volvocida. *See* RHIZOMASTIGIDA; VOLVOCIDA.

There is disagreement on the division of the order into families. However, the five or more families can be divided into two general groups. The first group contains simple organisms with no distinctive features save one or two flagella of equal or unequal length. This includes the families Oikomonadidae, Amphimonadidae, Monadidae, and Bodonidae (illustration *a-d*). The second group contains organisms which have an undulating membrane in addition to one or two flagella. The families included in this group are Trypanosomatidae, and Cryptobiidae (illustration *e* and *f*). A third group, including organisms possessing a peculiar collar surrounding a single flagellum, once assigned to this order, now makes up the order Choanoflagellida. *See* CHOANOFLAGELLIDA.

Trypanosomatidae. The most important family of the Kinetoplastida is the Trypanosomatidae, since it includes several species that infect man and his domestic animals with serious diseases, such as African sleeping sickness. The organisms in this family are polymorphic, changing their form in various stages of their development. Their life cycles may involve one or two hosts (invertebrate, vertebrate, or plant). The trypanosome form possesses a single flagellum and an undulating membrane extending the full length of the body. The other related forms are simpler in that they lack one or both of these structures. *See* CILIA AND FLA-

GELLA; MASTIGOPHORA; PORIFERA; PROTOZOA;
TRYPANOSOMATIDAE.   [M. M. BROOKE]

Bibliography: R. P. Hall, *Protozoology*, 1953;
S. H. Hutner and A. Lwoff (eds.), *Biochemistry
and Physiology of Protozoa*, 3 vols, 1961–1964;
T. L. Jahn and F. F. Jahn, *How to Know Protozoa*,
1949.

Kingfisher

Bird which is a member of the family Alcedini-
dae, of the order Coraciformes. Most are tropical
Old World species characterized by short legs,
long bills, bright plumage, and short wings.

The only members of the family found in the
New World are the fishing kingfishers, of which
the most common species is the belted kingfisher
(*Magaceryle alcyon*) (see illustration). It is distributed over a wide range, breeding from the southern Pacific coast area north to the Arctic regions,
and wintering as far south as northern South
America. The green kingfisher (*Chloroceryle
americana*) is a small species found from the
southwestern United States to southern South
America. One of the smallest species, about 5 in.
long, is the least kingfisher (*C. aenea*), which
occurs in Central America and South America.
The crested Amazon kingfisher (*C. amazona*) is
the largest species of the New World group.

The best-known European representative is the
common kingfisher (*Alcedo atthis*), a small solitary
bird found along quiet streams in Europe and Asia.
It remains motionless on a tree branch and then
suddenly dives for a fish, which it takes back to its
perch and swallows head first. Like the European
bee-eater, it usually nests in a burrow, made in a
river bank, that has a long entrance passage and a
terminal enlarged chamber. When the hen incu-
bates her eggs, the cock provides her with food
and removes all excrement from the nest. There
are two clutches of 2–7 eggs each year and the
incubation period is about 20 days. After the eggs
have hatched, both parents care for the brood.

An interesting species found in Africa and
southwestern Asia is the pied kingfisher (*Ceryle
rudis*), which is especially fond of crayfish, and will
pound them to pieces prior to eating them. The
pygmy kingfisher (*Ispidina picta*), an arboreal species of Africa, is insectivorous, although other arboreal species eat reptiles, small birds, and mam-

The belted kingfisher with white underparts, a blue-gray
back, and a band across the chest.

mals. One of the best known of these species is the
laughing jackass (*Dacelo novaeguineae*), which is
also called the kookaburra. It is the largest Old
World species, reaching the size of a crow, and
makes its nest on termite mounds. It is often seen
scavenging in towns or eating snails and slugs. *See*
AVES; CORACIIFORMES.   [CHARLES B. CURTIN]

Kinorhyncha

A class of the phylum Aschelminthes, consisting of
superficially segmented microscopic marine animals lacking external ciliation. All members of the
class are benthonic, so called because they generally dwell on mud bottoms in shallow water in the
littoral zone. Three suborders are generally recognized, Cyclorhagae, Conchorhagae, and Homalorhagae.

The body is completely enclosed in a transparent cuticle secreted by an underlying epidermis.
As is the case among the Nematoda, the cuticle is
periodically molted during growth. Three major
regions of the body are recognizable, head, neck,
and a jointed trunk. The apparent segments of the
body are termed zonites; the head and neck each
consist of one zonite (illustration *a*). Except for one
genus in which the trunk has 12 zonites, the usual
number is 11. *See* NEMATODA.

The head is completely retractable. When protruded, it bears 5–7 circles of spines called scalids
(illustration *b*).

The neck is covered by a varying number of
large plates called placids. Only among the Cyclorhagae do the placids close over the end of the body
when the head is withdrawn. Among the Conchorhagae, the closing apparatus consists of a pair of
lateral plates on the third zonite, while in the
Homalorhagae, it consists of a single dorsal plate
and three ventral plates on the third zonite.

A pair of ventral adhesive tubes occurs on either
the third or fourth zonite. Each of the remaining
trunk zonites usually bears a pair of lateral spines
and a single dorsal spine. The terminal zonite may
bear, in addition, a pair of large, movable lateral
spines.

The musculature is segmentally arranged. There
are two pairs of longitudinal muscle bands, a
dorsolateral and a ventrolateral. Anteriorly, these
form the retractors of the head. Protrusion of
the head is accomplished by contraction of the ring
muscles of the first two zonites and the paired dor-
soventral muscle bands in the trunk zonites.

Locomotion is accomplished in the following
manner: The head is protruded and the surface
gripped by the scalids. The trunk is then advanced
by contraction of the longitudinal muscles and the
head is retracted. Repetition of this sequence results in creeping locomotion.

The nervous system consists of a brain, which
encircles the anterior end of the pharynx, and a
ventral ganglionated cord; in addition, ganglion
cells in each zonite are located in the lateral and
dorsal epidermal chords.

The mouth is terminally located on the mouth
cone, which encloses a short buccal cavity. Posterior to the buccal cavity is a muscular pharynx. The
pharynx is similar to that of the Nematoda and
Gastrotricha but differs in being lined by a syncy-
tial epithelium. The pharynx is followed by a short,
slender esophagus whose epithelial lining is con-

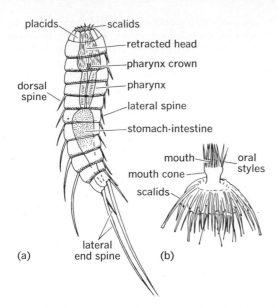

Echinoderella sp., a cyclorhagous kinorhynch. (*a*) Side view. (*b*) Head. (*After L. H. Hyman, The Invertebrates, vol. 3, McGraw-Hill, 1951*)

tinuous with that of the pharynx. The stomach-intestine is a straight, simple tube covered externally by a loose network of muscle fibers. The short endgut is separated from the stomach-intestine by a sphincter. A second, more posterior sphincter is also present. The anus is terminal. *See* GAS-TROTRICHA.

The fluid-filled body cavity between the digestive tract and the epidermis is unlined and is presumably a pseudocoele, or false coelom.

A single pair of protonephridial excretory organs is present in the tenth zonite. The flame bulbs are multinucleate and generally each contains a single long flagellum. Among the Homalorhagae, a second short flagellum is present. The protonephridial canals, which lead to nephridiopores on the eleventh zonite, contain driving flagella. The pores consist of sieve plates.

The Kinorhyncha are dioecious. In each sex there is one pair of gonads. The genital pores are located on the thirteenth zonite. The ovary contains both ova and nutritive cells. A short oviduct extends from the posterior end of each ovary to the genital pore. The male genital pore is armed with two or three penial spicules which may serve as a copulatory apparatus.

The early embryology of the Kinorhyncha is unknown. In the Cyclorhagae, the eggs hatch into a minute larva which lacks external evidence of zonites, head, placids, pharynx, or anus. This larva passes through several successive stages, separated by molts, during which the number of zonites increases and adult morphology is attained. Among the Homalorhagae, the first larva has 6−7 zonites and is, in general, a more advanced larva than that of the Cyclorhagae. The development of the Conchorhagae is unknown. *See* ASCHEL-MINTHES.   [MARTIN SACKS]

Kirchhoff's laws of electric circuits

Fundamental natural laws dealing with the relation of currents at a junction and the voltages around a loop. These laws are commonly used in the analy-

sis and solution of networks. They may be used directly to solve circuit problems, and they form the basis for network theorems used with more complex networks.

In the solution of circuit problems, it is necessary to identify the specific physical principles involved in the problem and, on the basis of them, to write equations expressing the relations among the unknowns. Physically, the analysis of networks is based on Ohm's law giving the branch equations, Kirchhoff's voltage law giving the loop voltage equations, and Kirchhoff's current law giving the node current equations. Mathematically, a network may be solved when it is possible to set up a number of independent equations equal to the number of unknowns. *See* CIRCUIT (ELECTRICITY); NETWORK THEORY, ELECTRICAL.

When writing the independent equations, current directions and voltage polarities may be chosen arbitrarily. If the equations are written with due regard for these arbitrary choices, the algebraic signs of current and voltage will take care of themselves.

Kirchhoff's voltage law. One way of stating Kirchhoff's voltage law is: "At each instant of time, the algebraic sum of the voltage rise is equal to the algebraic sum of the voltage drops, both being taken in the same direction around the closed loop."

The application of this law may be illustrated with the circuit in Fig. 1. First, it is desirable to consider the significance of a voltage rise and a voltage drop, in relation to the current arrow. The following definitions are illustrated by Fig. 1.

A voltage rise is encountered if, in going from 1 to 2 in the direction of the current arrow, the polarity is from minus to plus. Thus, E is a voltage rise

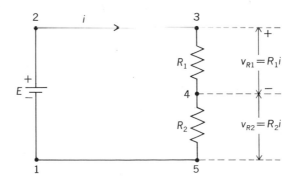

Fig. 1. Simple loop to show Kirchhoff's voltage law.

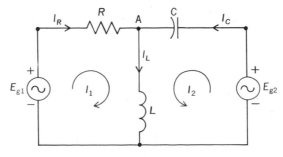

Fig. 2. Two-loop network demonstrating the application of Kirchhoff's voltage law.

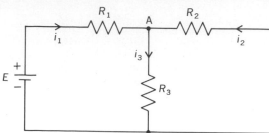

Fig. 3. Circuit demonstrating Kirchhoff's current law.

from 1 to 2.

A voltage drop is encountered if, in going from 3 to 4 in the direction of the current arrow, the polarity is from plus to minus. Thus, $v_{R1} = R_1 i$ is a voltage drop from 3 to 4. The application of Kirchhoff's voltage law gives the loop voltage, Eq. (1).

$$E = v_{R1} + v_{R2} = R_1 i + R_2 i \qquad (1)$$

In the network of Fig. 2 the voltage sources have the same frequency. The positive senses for the branch currents I_R, I_L, and I_C are chosen arbitrarily, as are the loop currents I_1 and I_2. The voltage equations for loops 1 and 2 can be written using instantaneous branch currents, instantaneous loop currents, phasor branch currents, or phasor loop currents.

The loop voltage equations are obtained by applying Kirchhoff's voltage law to each loop as follows.

By using instantaneous branch currents, Eqs. (2)

$$e_{g1} = R i_R + L \frac{d i_L}{dt} \qquad (2)$$

and (3) may be obtained. By using instantaneous

$$e_{g2} = \frac{1}{C} \int i_C \, dt + L \frac{d i_L}{dt} \qquad (3)$$

loop currents, Eqs. (4) and (5) are obtained. Equa-

$$e_{g1} = R i_1 + L \frac{d(i_1 + i_2)}{dt} \qquad (4)$$

$$e_{g2} = \frac{1}{C} \int i_2 \, dt + L \frac{d(i_2 + i_1)}{dt} \qquad (5)$$

tions (6) and (7) are obtained by using phasor

$$\mathbf{E}_{g1} = R \mathbf{I}_R + j \omega L \mathbf{I}_L \qquad (6)$$

$$\mathbf{E}_{g2} = -j \frac{1}{\omega C} \mathbf{I}_C + j \omega L \mathbf{I}_L \qquad (7)$$

branch currents. By using phasor loop currents, Eqs. (8) and (9) may be obtained.

$$\mathbf{E}_{g1} = R \mathbf{I}_1 + j \omega L (\mathbf{I}_1 + \mathbf{I}_2) \qquad (8)$$

$$\mathbf{E}_{g2} = -j \frac{1}{\omega C} \mathbf{I}_2 + j \omega L (\mathbf{I}_2 + \mathbf{I}_1) \qquad (9)$$

Kirchhoff's current law. Kirchhoff's current law may be expressed as follows: "At any given instant, the sum of the instantaneous values of all the currents flowing toward a point is equal to the sum of the instantaneous values of all the currents flowing away from the point."

The application of this law may be illustrated with the circuit in Fig. 3. At node A in the circuit in Fig. 3, the current is given by Eq. (10).

$$i_1 + i_2 = i_3 \qquad (10)$$

The current equations at node A in Fig. 2 can be written by using instantaneous branch currents or phasor branch currents.

By using instantaneous branch currents, Eq. (11) is obtained.

$$i_R + i_C = i_L \qquad (11)$$

By using phasor branch currents, Eq. (12) is obtained.

$$\mathbf{I}_R + \mathbf{I}_C = \mathbf{I}_L \qquad (12)$$

See DIRECT-CURRENT CIRCUIT THEORY.

[K. Y. TANG/ROBERT T. WEIL]

Kite

A tethered flying device that supports itself and the cable that connects it to the ground by means of the aerodynamic forces created by the relative motion of the wind. This relative wind may arise merely from the natural motions of the air or may be caused by towing the kite through the agency of its connecting cable.

Kites take many forms (see illustration); the bow and box kites are common in the United States. In many countries, particularly in Asia, kites are frequently used in rituals and festivals; their bizarre forms and shapes are traditional, some having been developed centuries ago.

The lifting force of all kites is produced by deflecting the air downward, the resulting change in momentum producing an upward force. To be successful, a kite must have an extremely low wing loading (weight/area) so that it can fly even on days when the wind velocity is not high. It must be completely stable, since the only controls available to the operator are the length of cable and the rate at which it is taken in or let out. Efficient design requires that its lift-to-drag ratio be as high as possible. *See* AERODYNAMIC FORCE; AERODYNAMICS.

Experiments on the possible application of efficient aircraft-type lifting surfaces, in which most of the lift arises from the low pressures created by the air flowing over the upper surfaces, have shown them to be too sensitive to changes in wind force and direction. Under normal atmospheric conditions the use of this type of lifting surface results in a kite that behaves in a violent and unpredictable manner. For this reason, the higher drag associated with a surface from which the flow has separated is tolerated, and most of the lifting force is obtained from pressure on the lower surface, because stalled surfaces are much less sensitive to wind changes.

Both the lift-to-drag ratio and the stability of the kite are functions of the length of cable. The more cable released, the more drag created. The increased drag, combined with the increase in weight being supported, causes the kite to sag off downwind, reducing the flight angle, which is the angle formed between the horizontal and a line passing through the kite and the operator.

Most kites with a properly located cable pivot point, generally slightly ahead of the center of gravity, demonstrate longitudinal stability. Lateral and directional instabilities generally couple to produce violent motions. The longer the cable, the more these motions are damped. Lateral and directional stability are improved by the use of effective dihedral (the bow of the bow kite) and a flexible tail, which provides both directional stabil-

KITE

(a)

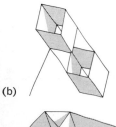

(b)

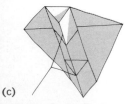

(c)

Common forms of kites.
(a) Bow kite. (b) Box kite.
(c) Modified bow kite.

ity and yaw damping.   [DAVID C. HAZEN]

Bibliography: H. W. Fowler, Jr., *Kites*, 1963; C. L. Strong, The lore and aerodynamics of making and flying kites, *Sci. Amer.*, 220(4):130–136, 1969.

Klebsiella

A genus of the bacterial family Enterobacteriaceae. Its members do not decarboxylate ornithine, are nonmotile, and almost invariably produce mucoid colonies. Three species are recognized at present. *K. pneumoniae* is Voges-Proskauer positive, ferments lactose, produces gas from glucose, utilizes citrate, frequently splits urea, and rarely produces indole and never hydrogen sulfide. *K. ozaenae* is Voges-Proskauer negative and gives variable results in the above tests. *K. rhinoscleromatis* is biochemically quite inert. Seventy-two serotypes exist, based on the capsular (K) antigen. *K. ozaenae* form part of serotypes 3, 4, 5, and 6; *K. rhinoscleromatis* are of type 3. *See* ENTEROBACTERIACEAE.

K. pneumoniae may be found in soil, in water, and on plants. Some strains are able to fix atmospheric nitrogen. The organism occasionally occurs in human and animal feces. In men over 40 years (often alcoholics), it may cause a pneumonia characterized by rusty sputum and involvement of several lobes. More frequently, it is found as a hospital-acquired bacterium which may colonize the throat or intestine or cause urinary tract infection, pneumonia, septicemia (mostly in debilitated patients), and meningitis (in newborns). *K. pneumoniae* is the second most frequent gram-negative rod isolated in clinical laboratories. Originally resistant to ampicillin and carbenicillin, some strains have now become resistant to more antimicrobials.

K. ozaenae occurs in a certain type of chronic rhinitis called ozaena. *K. rhinoscleromatis* causes rhinoscleroma, a granulomatous disease of the upper respiratory tract occurring in Eastern Europe. *See* ENTEROBACTER; SERRATIA.

[ALEXANDER VON GRAEVENITZ]

Bibliography: E. B. Edmondson and J. P. Sanford, The *Klebsiella-Enterobacter (Aerobacter)-Serratia* group, *Medicine*, 46:323, 1967; P. R. Edwards and W. H. Ewing, *Identification of Enterobacteriaceae*, 3d ed., 1972.

Klystron

An evacuated electron-beam tube in which an initial velocity modulation imparted to electrons in the beam results subsequently in density modulation of the beam. A klystron is used either as an amplifier in the microwave region or as an oscillator. For use as an amplifier, a klystron receives microwave energy at an input cavity through which the electron beam passes. The microwave energy modulates the velocities of electrons in the beam, which then enters a drift space. Here the faster electrons overtake the slower to form bunches. In this manner, the uniform current density of the initial beam is converted to an alternating current. The bunched beam with its significant component of alternating current then passes through an output cavity to which the beam transfers its ac energy. *See* MICROWAVE.

Klystron amplifier. In a typical klystron (Fig. 1), a stream of electrons from a concave thermionic cathode is focused into a smaller cylindrical

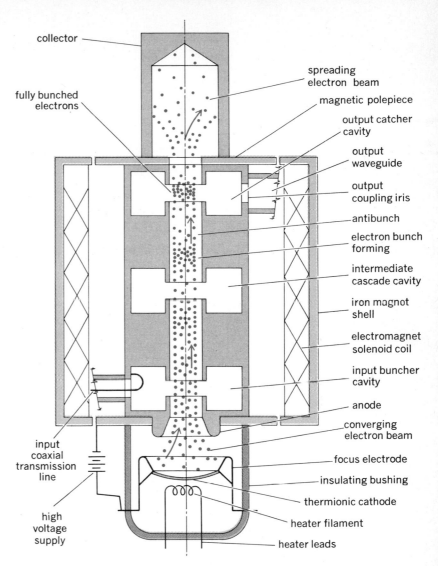

Fig. 1. Cross section of cascade klystron amplifier. (*Varian Associates*)

beam by the converging electrostatic fields between the anode, cathode, and focusing electrode. The beam passes through a hole in the anode and enters a magnetic field parallel to the beam axis. The magnetic field holds the beam together, overcoming the electrostatic repulsion between electrons which would otherwise make the beam spread out rapidly. The electron beam goes through the cavities of the klystron in sequence, emerges from the magnetic field, spreads out, and is stopped in a hollow collector where the remaining kinetic energy of the electrons is dissipated as heat. *See* ELECTRON MOTION IN VACUUM.

The signal wave to be amplified is introduced into the first, or buncher, cavity through a coaxial transmission line. This hollow metal cavity is a resonant circuit, analogous to the familiar inductance-capacitance combination, with the electric field largely concentrated in the reentrant noses so that the highest voltage occurs between them. The inductance may be considered as a single-turn conductor formed by the outer metal walls. In Fig. 1 the current in the center conductor of the input transmission line flows through a loop inside the cavity and back to the outer conductor. The magnetic flux generated in the loop links

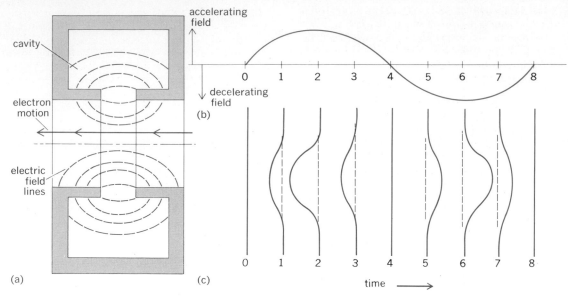

Fig. 2. Cavity concentrates electric field between the reentrant noses. As field varies sinusoidally with time, an electron crossing in the gap between the noses experiences an electric field whose strength and direction depend on the instantaneous phase of the field. (a) Map of instantaneous field. (b) Cycle variation of field. (c) Profile of field on the axis at various times in the cycle. (*Varian Associates*)

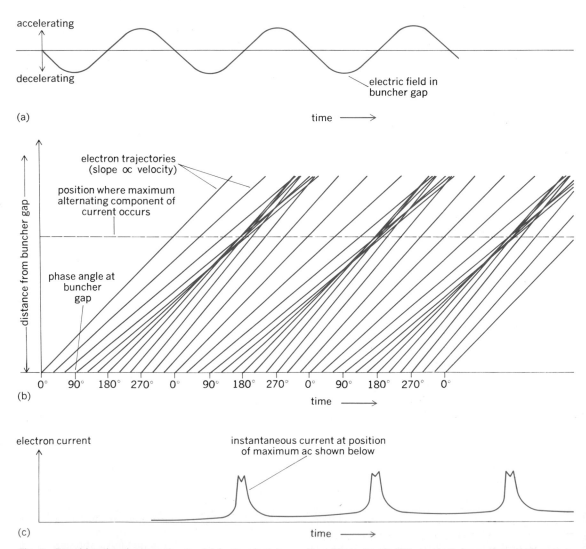

Fig. 3. Bunching the electron stream. (a) As the electric field varies periodically, electrons traversing the cavity are speeded up or slowed down. (b) Distance-versus-time lines graph the gradual formation of bunches. (c) Current passing a fixed point becomes periodic, that is, it becomes alternating. (*Varian Associates*)

through the cavity inductance, as in a transformer. At the resonant frequency of the cavity, the voltage across the reentrant section through which the electron beam passes is built up by the cavity configuration to 10–100 times the voltage in the input line. *See* CAVITY RESONATOR.

Figure 2 shows the pattern of the electric field in the cavity and how it varies cyclically with time. As electrons pass through the gap in which the cavity field is concentrated (Fig. 2a), they are accelerated or decelerated (Fig. 2b), depending on the instantaneous direction of the field.

Figure 3 illustrates graphically the effect of these velocity changes. Each slanted line represents the flight of an electron as a function of time measured in electrical degrees as the electron travels from the buncher gap. The slope of a line is thus the velocity of that electron. The velocities leaving the buncher vary sinusoidally with time, as determined by the instantaneous field. The horizontal broken line represents a fixed point beyond the buncher. The flow of electrons past this point is given by the time sequence in which the electron paths cross the broken line. Figure 3 shows how the electrons have gathered into bunches. The rate of current flow is now periodic with time so that the current has an alternating component.

When the bunched beam passes through a second cavity, its space charge induces in the walls of this cavity an alternating current of opposite sign to the electron current. The cavity is tuned to the input frequency so that it has a high resonant impedance; the induced current flowing through this impedance generates voltage in the cavity.

In Fig. 1 the second cavity is not coupled to any outside circuits. Voltage built up here by the beam current produces further velocity modulation in the beam. The resulting alternating current component is about 10 times greater than the initial current. More of these uncoupled cascade cavities can be added for increased amplification.

The final output cavity is coupled into a transmission line (a waveguide in Fig. 1) which carries off the generated power to its useful destination. Because the cavity is tuned to resonance, its reactance is canceled. The induced current flowing through its pure resistive impedance generates in-phase voltage in the direction opposing the current flow. Thus the field in the gap is at its maximum decelerating value at the time a bunch of electrons passes. Most of the electrons therefore are slowed down, and there is a net transfer of kinetic energy of the electrons into electromagnetic energy in the cavity. Klystron amplifiers are used in transmitters for radar and one-way radio communication, for driving particle accelerators, and for dielectric heating. The useful range of frequencies is from 400 MHz to 40 GHz. Power levels range from a few watts up to 400 kw of continuous power or 20 Mw for short pulses. Amplification is about 10 dB for a two-cavity tube. With more cavities, gains up to 60 dB are practical.

Figure 4 shows the construction of a four-cavity amplifier rated at 2 Mw pulsed output at 2.8 GHz. It operates in a solenoid magnet, as in Fig. 1. The cavities are tuned to the operating frequency by moving one flexible inner wall of the box-shaped cavity, changing its volume and its effective inductance. In Fig. 5 the details of the input cavity are enlarged.

Reflex oscillator. Klystrons may be operated as oscillators by feeding some of the output back into the input circuit. More widely used is the reflex oscillator in which the electron beam itself provides the feedback. Figure 6 illustrates the operation. The beam is focused through a cavity, as in the amplifier. No magnetic field is needed to keep the beam focused because the total travel distance is short and the amount of natural spreading is tolerable. The cavity usually has grids with open mesh through which the electrons can penetrate. The purpose of the grids is to concentrate the electric field in a short space so that the field can interact with a slow, low-voltage electron beam.

In the cavity the beam is velocity-modulated as in the amplifier. Leaving the cavity, the beam en-

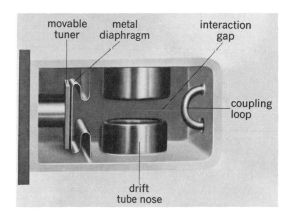

Fig. 5. Detail of input cavity for four-cavity klystron amplifier of Fig. 4. (*Varian Associates*)

Fig. 4. Cutaway view of four-cavity amplifier. (*Varian Associates*)

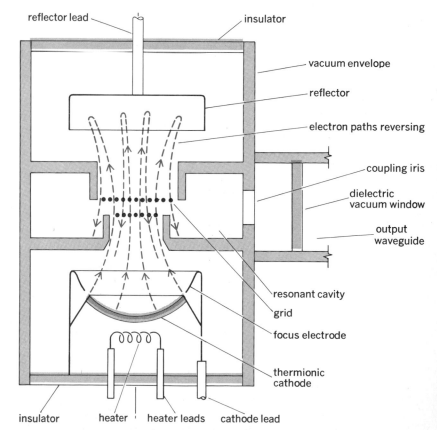

Fig. 6. Schematic cross section of reflex oscillator. (*Varian Associates*)

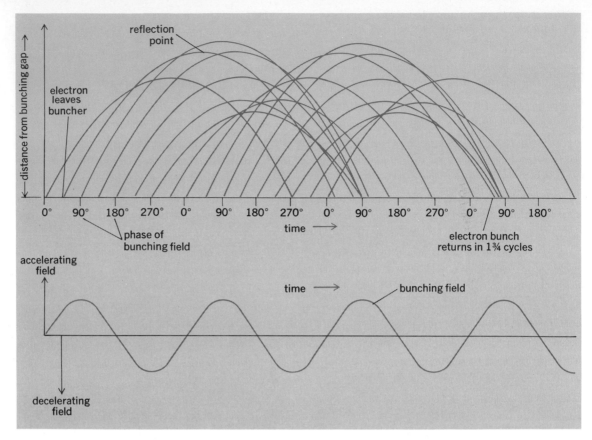

Fig. 7. Trajectories show bunching in a reflex oscillator. Electrons are turned back by a retarding field, faster ones going farther and taking longer. The bunched beam returns through the cavity. (*Varian Associates*)

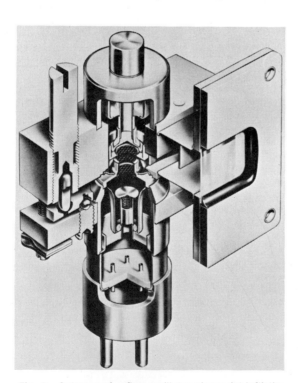

Fig. 8. Cutaway of reflex oscillator shows (at left) the screw that deforms the flexible bottom wall of the cavity to change its resonant frequency. The beam passes through honeycomb grids in the cavity. The cathode leads come out at the bottom socket, and the reflector lead comes out at the top. The output waveguide has a mica vacuum window. (*Varian Associates*)

ters a region of dc electric field opposing its motion, produced by a reflector electrode operating at a potential negative with respect to the cathode. The electrons do not have enough energy to reach the electrode, but are reflected in space and return to pass through the cavity again. The points of reflection are determined by electron velocities, the faster electrons going farther against the field and hence taking longer to get back than the slower ones.

A trajectory plot for the reflex oscillator is shown in Fig. 7. In a uniform retarding field the space-versus-time curves are parabolas. As in Fig. 3, velocity modulation produces bunches of electrons. If the voltages are adjusted so that the average time to return is $n + 3/4$ cycles ($n =$ integer), the bunches cross the cavity when the alternating field is maximum in the decelerating direction. This transfers beam energy to the cavity.

Because the reflex klystron has only one cavity, it is easy to tune its frequency with a single adjustment. Power output is from 10 mw to a few watts. Reflex oscillators are used as signal sources from 3 to 200 GHz. They are also used as the transmitter tubes in line-of-sight radio relay systems and in low-power radars.

A cutaway of a typical reflex klystron is shown in Fig. 8. This tube is tuned by deforming the upper cavity wall, which varies the spacing between the grids and hence the effective capacitance of the cavity resonator.   [RICHARD B. NELSON]

Bibliography: M. A. Atwater, *Introduction to Microwave Theory*, 1962; A. H. W. Beck, *Velocity*

Modulated Thermionic Tubes, 1948; L. Ginzton, The klystron, *Sci. Amer.*, March, 1954; J. R. Pierce and W. G. Shepherd, Reflex oscillators, *Bell Syst. Tech. J.*, 26:663, July, 1947.

Knudsen number

In fluid mechanics, the ratio l/L of the mean free path length l of the molecules of the fluid to a characteristic length L of the structure in the fluid stream. When the mean free path of the fluid particles is short relative to the size of the object being considered, the fluid can be treated as a continuum. If the path length between molecular encounters is comparable to or larger than a significant dimension of the flow region, the gas must be treated as consisting of discrete particles. The usual classifications of flow according to Knudsen number are as follows: For $l/L \leq 0.01$ the flow can be dealt with by the methods of gas dynamics; for $l/L \approx 1$ the behavior is termed slip flow; for $l/L \geq 10$ the behavior is termed free-molecular flow or rarefied gas dynamics. *See* FIELD THEORY, CLASSICAL; GAS DYNAMICS; STATISTICAL MECHANICS; SUPERAERODYNAMICS.

[FRANK H. ROCKETT]

Koala

A single species, *Phascolarctos cinereus*, which is a member of the family Phalangeridae in the mammalian order Marsupialia (pouch-bearing animals). It is a small, clean, affectionate animal that weighs from 11 to 17 lb when mature. They are restricted to eastern Australia, where the eucalyptus grows. Not only do they have a specialized diet of eucalyptus leaves, but the leaves must be of a certain age from a specific species of tree, and the tree must grow upon a certain type of soil. Each adult will eat an average of 2–3 lb of leaves each day. While the diet is a definite limiting factor to an increase in the population, millions died from

The koala, a slow-moving arboreal animal with opposable digits and a rudimentary tail.

epidemic diseases at the turn of the 20th century, and later thousands were killed for their fur. These animals are now strictly protected by the Australian government because their numbers are small. Exportation is prohibited.

The koala is well-adapted to its arboreal habitat, since both the fore- and hindlimbs have two of the clawed digits opposing the other three, thus constituting an efficient grasping organ (see illustration). The koala breeds once each season, and the usual number of offspring is one. It remains in the pouch for 6 months; then it clings to the back of its mother and is carried around in this manner until 1 year old. *See* MAMMALIA; MARSUPIALIA.

[CHARLES B. CURTIN]

Kohlrabi

A cool-season biennial crucifer, *Brassica caulorapa* and *B. oleracea* var. *caulo-rapa*, of northern European origin belonging to the plant order Capparales. Kohlrabi is grown for its turniplike enlarged stem, which is usually eaten as a cooked vegetable (see illustration). Kohlrabi is a German

Kohlrabi (*Brassica caulorapa*), cultivar Early White Vienna. (*Joseph Harris Co., Rochester, N. Y.*)

word meaning cabbage-turnip and reflects a similarity in taste and appearance to both vegetables. Cultural practices for kohlrabi are similar to those used for turnips. White Vienna and Purple Vienna are popular varieties (cultivars). Harvesting, when the enlarged stems are 2–3 in. in diameter, is usually 2 months after planting. A common cooked vegetable in Europe, especially Germany, kohlrabi is of minor importance in the United States. *See* CABBAGE; CAPPARALES; KALE; TURNIP; VEGETABLE GROWING.

[H. JOHN CAREW]

Kojic acid

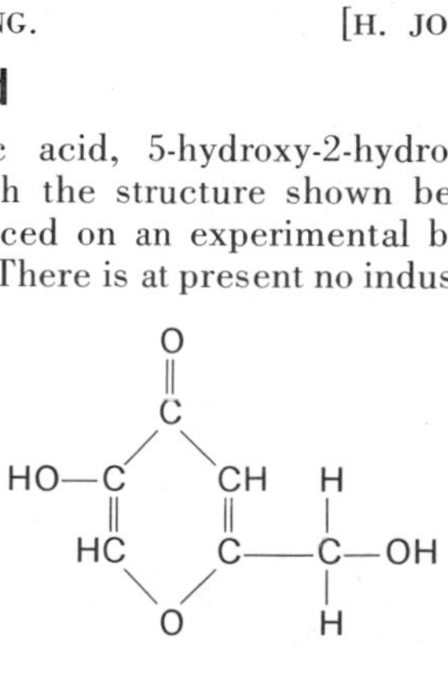

An organic acid, 5-hydroxy-2-hydroxymethyl-γ-pyrone, with the structure shown below. It has been produced on an experimental basis by fermentation. There is at present no industrial use for kojic acid. Several species of filamentous fungi of the genus *Aspergillus* can produce this noncarboxylic, ring-type acid from various carbohydrates. Selected strains belonging to the *Aspergillus flavus-oryzae* group give the highest weight yields, producing 50–65% of glucose utilized. A medium containing mineral salts and 10–25% commercial glucose is adjusted to pH 1.8–2.0. The pH of the medium is critical. The medium is inoculated with spores or mycelium of the mold and continuously aerated and agitated under submerged conditions for 2–5 days. The organism also produces good yields in the slower surface or tray process. Production of kojic acid is followed, quantitatively, by the red color it gives with ferric chloride. *See* INDUSTRIAL MICROBIOLOGY.

[JACKSON W. FOSTER/R. E. KALLIO]

Kolbe hydrocarbon synthesis

The production of an alkane by the electrolysis of a water-soluble salt of a carboxylic acid. The carboxylate ions are discharged at the anode, yielding carbon dioxide and alkyl radicals which couple to form the saturated hydrocarbon. Hydrogen is liberated at the cathode, as in the equation below.

$$2RCOONa + 2H_2O \rightarrow RR + 2CO_2 + 2NaOH + H_2$$

Good yields of alkanes are obtained with the straight-chain acids containing 5–18 carbon atoms. Alkenes rather than alkanes are formed if the acid contains an alkyl branch at the carbon atom adjacent to the carboxyl group, that is, at the α position of the acid. *See* ALKANE.

[LOUIS SCHMERLING]

Kondo effect

The large anomalous increase in the resistance of certain dilute alloys of magnetic materials in nonmagnetic hosts as the temperature is lowered. This behavior is contrary to the decrease in electrical resistance with temperature observed in almost all other systems. It is believed that the interaction of the conduction electrons with the spin of the magnetic impurity creates rather long-range correlations in the electrons in the vicinity of the impurity and that this strong interaction tends to inhibit conductivity. This interaction is an exchange interaction similar to that in ferromagnets. *See* FERROMAGNETISM.

For temperatures above a few degrees absolute, however, thermal agitation tends to wash out these fairly delicate correlations, and on heating the sample further the contribution to the resistance diminishes. The temperature below which this effect predominates is referred to as the Kondo temperature, and it depends on the magnetic impurity and the host material.

It has been suggested that Kondo temperatures may range from millidegrees to hundreds of degrees, but for materials studied so far they are below about 30°. *See* CURIE TEMPERATURE, MAGNETIC; RESISTIVITY, ELECTRICAL.

Importance. The importance of studying the Kondo effect lies in the fact that it is a property of what is, presumably, the simplest possible magnetic system—a single magnetic atom in a nonmagnetic environment. (The alloys used are sufficiently dilute that interaction between different magnetic impurities can be safely ignored. For example, this is true for iron impurities in copper if the iron is less than one part in 10,000 of the alloy.) It is hoped that an understanding of this system will provide a key to the far more complex and important problem of the electronic structure of magnetic materials themselves, one of the greatest challenges in physics.

Theory. Although the Kondo effect had been experimentally observed many years earlier, the first satisfactory theoretical explanation was not given until J. Kondo's work in 1964. Newer theories predict quite accurately the temperature variation of the resistance, but associated phenomena such as the specific heat and magnetic susceptibility of Kondo-type alloys are not so well understood. Experimentally, these quantities are very hard to measure accurately, and furthermore seem to vary a great deal from one impurity-host system to another. Theoretically, the problem has a deceptive air of simplicity, and all solutions given so far are based on rather broad assumptions and approximations whose importance is hard to assess. Although some understanding of the system has certainly been gained, the picture is far from complete.

Kondo's explanation. Bearing in mind that no similar effect is observed in scattering from nonmagnetic impurities and that the magnetic field associated with the impurity was far too weak to cause the observed effect, Kondo assumed the crucial new feature to be the spin associated with the magnetic impurity. For simplicity he assumed further that the impurity spin had the same magnitude as the electron spin. The impurity can interact in a spin-dependent way with a conduction electron: When the impurity spin flips over, the electron spin flips at the same time, so that the total spin in any direction remains constant. This possible scattering mode causes the total scattering by an impurity, and hence the electrical resistance of the alloy, to depend on the temperature. To understand why this is so, one first considers ordinary (no spins involved) scattering theory and gives an argument showing why this is not temperature-dependent. It can then be demonstrated that the argument breaks down if the concept of "spin-flip" scattering is introduced. *See* ELECTRON SPIN; EXCLUSION PRINCIPLE; FERMI-DIRAC STATISTICS; UNCERTAINTY PRINCIPLE.

Scattering from impurity. In scattering theory the electron is considered to undergo a succession of interactions with the scatterer causing the electron to be in a sequence of so-called intermediate states before finally emerging from the scattering process. In these intermediate states the energy of the electron does not have to be identical with its initial energy. This is a consequence of the Heisenberg uncertainty principle: Since the intermediate states are short-lived (lifetime Δt, say) there is a corresponding arbitrariness in their energy which is given by $\Delta E \, \Delta t \sim h$, Planck's constant. However, there is one important constraint on possible intermediate-state energies—there are many other electrons present, and (from Pauli's exclusion principle) the electron cannot go to an intermediate state already occupied by another electron. Hence the presence of the other electrons modifies the scattering of the electron under consideration. Furthermore, when the temperature changes, the distribution of the other electrons over available energy levels alters, and thus the scattering becomes temperature-dependent. For ordinary (non-spin-dependent) scattering, this temperature dependence is cancelled by another contribution to the scattering referred to as hole intermediate states. This corresponds to a slightly different sequence of events. For normal scattering (particle intermediate states) an electron with momentum p is scattered to an intermediate state m, then scattered again by the impurity to a state p' (in general, of course, there is a succession of intermediate states). It is also possible for an electron in a state m to be scattered into the state p'; the original electron p subsequently falls into the hole, or vacancy, in state m. Provided the events follow each other closely, that is, the hole in state m is short-lived, the possible energy of the hole is spread by the uncertainty principle; this scattering process will also be temperature-dependent. However, for a given state m the probability of a hole appearing in m and an electron appearing in state p' is just proportional to the probability of the state m having an electron in it in the first place; whereas the probability of the same m being an intermediate state in the normal scattering sequence is proportional to the probability of its being initially vacant (and hence available). Thus when the two types of scattering are taken together, the probability of the state m being initially occupied cancels out and the full scattering becomes temperature-independent. *See* BAND THEORY OF SOLIDS; HOLES IN SOLIDS.

Spin-flip scattering. Kondo demonstrated that if the impurity can flip the spin of the electron, these arguments are no longer valid. The hole intermediate states no longer cancel out the temperature dependence of the particle intermediate states. In fact, he proved that for his model the imbalance was such that the scattering increased as $-\log T$ for T (the temperature) going to zero. It has been shown since that, with further refinement, the scattering levels off below a certain temperature, but at a considerably higher value than the minimum reached at higher temperatures.

In the illustration $\Delta\rho$ is the extra resistance of a piece of copper resulting from the addition of iron impurities. The continuous line is the theoretical prediction. The different experimental points are for different concentrations. $\Delta\rho$ is measured in ohm-centimeters per part per million of added iron. Since all the points fall on the same curve, the increase in resistance is directly proportional to the amount of added iron. Thus, adding the thousandth iron atom causes exactly the same increase as adding the first iron atom—the iron atoms are not influenced detectably by each other's presence; this suggests that Kondo's idea of considering "one atom" scattering is reasonable.

One example is given here to demonstrate how spin flipping restricts intermediate states differently for particles and holes. Consider an electron in an initial state $p\uparrow$ ($\uparrow$ means spin up) scattered by an impurity initially $I\downarrow$ into a state $p'\uparrow$. Possible particle intermediate states are of the type $m\uparrow I\downarrow$ and $m\downarrow I\uparrow$. (The total spin remains constant. The interaction in general may or may not flip the spin. A full, rigorous treatment is given by Kondo.)

Now consider the hole intermediate states. Examining the sequence of events in this case, both $p\uparrow$ and $p'\uparrow$ are already present at the same time as the hole m. Hence, since the total initial

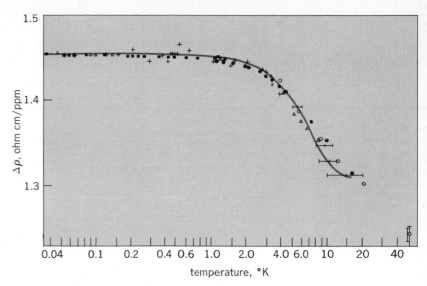

Relationship between impurity contribution to the electrical resistivity ($\Delta\rho$) and the absolute temperature. (*After M. Daybell and W. Steyert, Phys. Rev. Lett., 18:398, 1967*)

spin of the system was zero, both the hole m and the impurity I must be in states $m\downarrow$, $I\downarrow$, because at intermediate times also the total spin of the two electrons, one hole, and the impurity must equal zero. Thus the only allowed hole intermediate states are those in which the impurity has its spin pointing down. In contrast, for the particle intermediate states the impurity spin could be up or down. Since this means that there are twice as many possible particle intermediate states as hole intermediate states, the two sets cannot cancel out as in ordinary scattering, and this results in the scattering being temperature-dependent.

[MICHAEL FOWLER]

Bibliography: J. R. Schrieffer, Kondo effect, for links between magnetic and non-magnetic impurities in metals, *J. Appl. Phys.*, 38:1143, 1967; H. Suhl, in W. Marshall (ed.), *Theory of Magnetism in Transition Metals*, 1967.

Krebs cycle

A sequence of enzymatic reactions that is involved in the oxidative metabolism of many organisms, especially those that carry out respiration. The cycle is also known as the citric acid cycle and the tricarboxylic acid cycle. Through this sequence, or pathway, the two-carbon acetyl unit (Fig. 1), an important metabolic intermediate derived from a variety of substrates, can be oxidized to two molecules of CO_2. In aerobic metabolism, two molecules of oxygen are used for this oxidation. Organic acids containing four, five, and six carbons are formed in the course of the oxidation. These are consumed and regenerated continually through the operation of the entire sequence. Hence, this oxidative pathway is a catalytic mechanism which operates in a cyclic manner and is known as the citric acid cycle, because this tricarboxylic acid is one of the metabolic intermediates. The cycle is also named after H. A. Krebs, who recognized its functional role. *See* METABOLISM.

There are four oxidative steps in the cycle. In each of these, two atoms of hydrogen or two protons and two electrons are removed from the substrate. In aerobic respiration, the electrons are transferred to molecular oxygen through the cyto-

KREBS CYCLE

$$(CH_3\overset{\overset{\displaystyle O}{\|}}{C}—)$$

Fig.1. The two-carbon acetyl unit, an important metabolic intermediate.

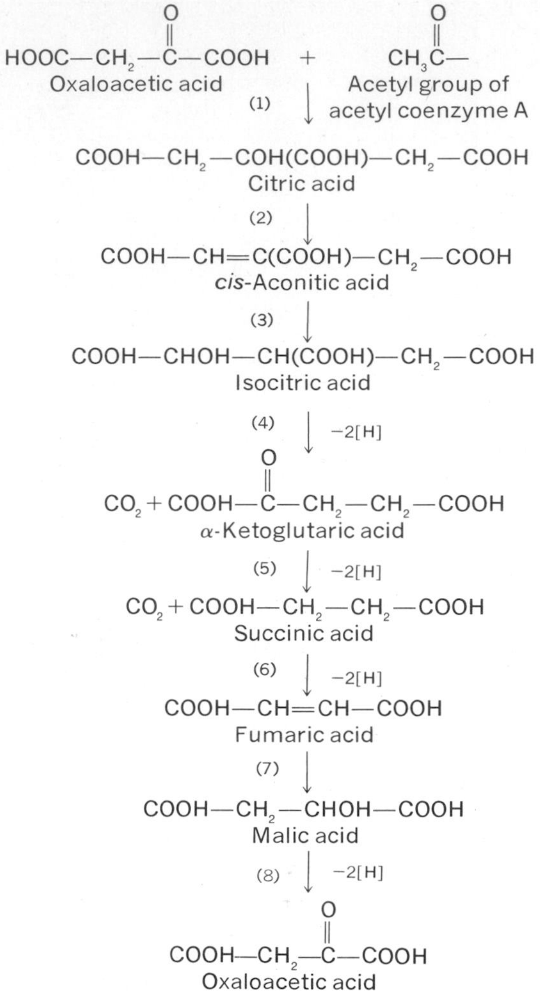

$$\underset{\text{Oxaloacetic acid}}{\text{HOOC—CH}_2\text{—}\overset{\displaystyle O}{\overset{\|}{\text{C}}}\text{—COOH}} \quad + \quad \underset{\substack{\text{Acetyl group of}\\\text{acetyl coenzyme A}}}{\text{CH}_3\overset{\displaystyle O}{\overset{\|}{\text{C}}\text{—}}}$$

(1)

$$\underset{\text{Citric acid}}{\text{COOH—CH}_2\text{—COH(COOH)—CH}_2\text{—COOH}}$$

(2)

$$\underset{\textit{cis}\text{-Aconitic acid}}{\text{COOH—CH}=\text{C(COOH)—CH}_2\text{—COOH}}$$

(3)

$$\underset{\text{Isocitric acid}}{\text{COOH—CHOH—CH(COOH)—CH}_2\text{—COOH}}$$

(4) $\downarrow$ −2[H]

$$\underset{\alpha\text{-Ketoglutaric acid}}{\text{CO}_2 + \text{COOH—}\overset{\displaystyle O}{\overset{\|}{\text{C}}}\text{—CH}_2\text{—CH}_2\text{—COOH}}$$

(5) $\downarrow$ −2[H]

$$\underset{\text{Succinic acid}}{\text{CO}_2 + \text{COOH—CH}_2\text{—CH}_2\text{—COOH}}$$

(6) $\downarrow$ −2[H]

$$\underset{\text{Fumaric acid}}{\text{COOH—CH}=\text{CH—COOH}}$$

(7) $\downarrow$

$$\underset{\text{Malic acid}}{\text{COOH—CH}_2\text{—CHOH—COOH}}$$

(8) $\downarrow$ −2[H]

$$\underset{\text{Oxaloacetic acid}}{\text{COOH—CH}_2\text{—}\overset{\displaystyle O}{\overset{\|}{\text{C}}}\text{—COOH}}$$

Fig. 2. Principal reactions of the Krebs cycle. Oxidative steps are indicated by the symbol −2[H].

chrome system of iron porphyrin enzymes. Energy-rich phosphate bonds in the form of adenosinetriphosphate (ATP) are generated in the course of electron transport which therefore serves as the major source of energy for cellular metabolism. Besides serving as a means for the complete, or terminal, oxidation of organic substrates which can be converted to acetyl fragments, the reactions of the Krebs cycle are used by the organism for the biosynthesis of important cell constituents which can be derived from the metabolic intermediates participating in the cycle. The principal reactions of the Krebs cycle, which will be discussed below, are outlined in Fig. 2.

(1) The acetyl group, which may be derived from pyruvic acid or from other metabolic intermediates or substrates, enters into the cycle as acetyl coenzyme A (acetyl CoA). The so-called condensing enzyme catalyzes the transfer of the acetyl group to the four-carbon compound, oxaloacetic acid, with the formation of the six-carbon tricarboxylic acid, citric acid, and the liberation of CoA.

(2) and (3) Citric acid is dehydrated to cis-aconitic acid, which is then rehydrated to isocitric acid. Both these reactions are catalyzed by the enzyme aconitase.

(4) Isocitric acid is next oxidatively decarboxylated to yield CO_2 and the five-carbon compound,

α-ketoglutaric acid by enzyme systems known as isocitric dehydrogenases. Depending on the biological source of the enzyme, either diphosphopyridine nucleotide (DPN), also known as nicotinamide adenine dinucleotide (NAD), or triphosphopyridine nucleotide (TPN), also known as nicotinamide adenine dinucleotide phosphate (NADP), may act as the coenzyme and immediate hydrogen acceptor for this reaction. There is evidence that an unstable intermediate, oxalosuccinic acid, is formed in the course of the reaction.

(5) α-Ketoglutaric acid is then oxidized and decarboxylated to yield the four-carbon compound succinic acid. This involves a sequence of reactions in which DPN serves as the hydrogen acceptor and inorganic phosphate is taken up with the formation of ATP from adenosinediphosphate (ADP). The first step, catalyzed by the α-ketoglutaric dehydrogenase system, requires DPN and CoA. The products are CO_2, reduced DPN (DPNH) and succinyl CoA. In the next step, the P enzyme decomposes the succinyl CoA with the simultaneous formation of ATP from inorganic phosphate and ADP. Succinic acid is produced and CoA is regenerated.

(6) and (7) Succinic acid is next oxidized to fumaric acid by the enzyme succinic dehydrogenase, and fumaric acid is hydrated to malic acid by fumarase.

(8) Malic acid is oxidized to oxaloacetic acid by the enzyme malic dehydrogenase with the concomitant reduction of DPN. Thus, oxaloacetic acid is regenerated through the entire sequence of reactions of the cycle and becomes available for reacting with acetyl CoA.

The enzymatic reactions of the Krebs cycle are important, not only as a source of energy for living cells, but also as a source of essential metabolic intermediates which serve as starting materials for biosynthetic processes. Thus, α-ketoglutaric acid becomes aminated to yield the amino acid, glutamic acid. This compound, in turn, takes part in the synthesis of other amino acids. Oxaloacetic acid is converted to aspartic acid, which is also a precursor of a number of amino acids as well as of pyrimidines. Some of the intermediates of the Krebs cycle are excreted by the cells. Citric acid, for example, appears in the body fluids of animals and, under some conditions, is a major product of the metabolism of molds. Similarly, α-ketoglutaric, fumaric, and succinic acids are excreted by some microorganisms. See AMINO ACIDS.

The removal of the intermediates in the Krebs cycle through biosynthesis or excretion destroys the integrity of the cycle as a catalytic mechanism for the oxidation of acetyl units. Therefore, living systems must possess auxiliary mechanisms for replenishing their supply of such intermediates so that oxaloacetic acid can be constantly available for the reactions. One such mechanism is the addition of carbon dioxide to pyruvic acid with the oxidation of reduced triphosphopyridine nucleotide (TPNH). This reaction, which yields malic acid, is catalyzed by the malic enzyme. Oxaloacetic acid can by synthesized from phosphoenolpyruvic acid and CO_2 in the presence of inosinediphosphate by a different enzyme.

Another important means by which four-carbon dicarboxylic acids may be synthesized is a sequence of reactions in which isocitric acid is split to succinic and glyoxylic acids and the latter compound reacts with acetyl CoA to yield malic acid. As a consequence of these reactions, two molecules of four-carbon acids can be produced with the consumption of one molecule of oxaloacetic acid and two molecules of acetyl CoA. *See* ADENOSINEDIPHOSPHATE (ADP); ADENOSINETRIPHOSPHATE (ATP); BIOLOGICAL OXIDATION; CELL (BIOLOGY); COENZYME; CYTOCHROME; DIPHOSPHOPYRIDINE NUCLEOTIDE (DPN); ENZYME; PORPHYRIN; TRIPHOSPHOPYRIDINE NUCLEOTIDE (TPN).

[MICHAEL DOUDOROFF]

Kronig-Penney model

An idealized, one-dimensional model of a crystal which exhibits many of the basic features of the electronic structure of real crystals. Consider the potential energy $V(x)$ of an electron illustrated in Fig. 1 with an infinite sequence of potential wells of depth $-V_o$ and width a, arranged with a spacing b. The Schrödinger wave equation can be readily solved for such an arrangement to give electron energy as a function of wave number.

The energy bands thus obtained with the choice of constants given in Eqs. (1) and (2) are shown in

$$(2ma^2V_o/h^2)^{1/2} = 12.0 \tag{1}$$

$$b/a = 0.1 \tag{2}$$

Fig. 2. Notice that the width and the curvatures of the allowed bands increase with energy. The

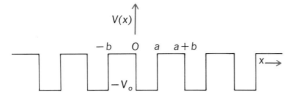

Fig. 1. Potential energy which is assumed for the one-dimensional Kronig-Penney model.

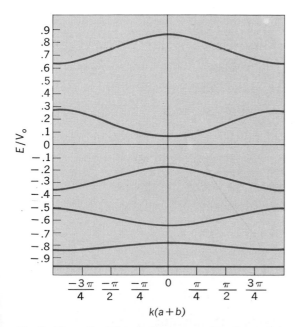

Fig. 2. The ratio of the electron energy E to the magnitude of the well depth V_o shown as a function of electron wave number k times the "lattice constant" $(a+b)$ for the Kronig-Penney model.

Kronig-Penney model has been extended to include the effects of impurity atoms. A solution can also be derived from the Dirac equation. *See* BAND THEORY OF SOLIDS; QUANTUM THEORY, NONRELATIVISTIC. [JOSEPH CALLAWAY]

Bibliography: C. Kittel, *Elementary Solid State Physics*, 1962; C. Kittel, *Introduction to Solid State Physics*, 2d ed., 1956; R. de L. Kronig and W. G. Penney, Quantum mechanics of electrons in crystal lattices, *Proc. Roy. Soc. (London) ser. A*, 130: 499–513, 1931; G. Wannier, *Elements of Solid State Theory*, 1959.

Krypton

A gaseous chemical element, Kr, atomic number 36, and atomic weight 83.80. Krypton is one of the noble gases in group 0 of the periodic table. *See* INERT GASES.

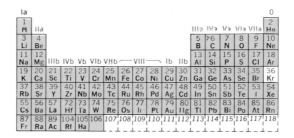

Uses. The principal use for krypton is in filling electric lamps and electronic devices of various types. Krypton-argon mixtures are widely used to fill fluorescent lamps.

The higher the molecular weight of the inert gas used to fill tungsten-filament lamp bulbs, the less the tendency of the tungsten to evaporate when heated; thus, krypton is a better filling gas, although more expensive, than argon. Two different uses are made of this advantage: Krypton-filled lamps can be operated at the same brightness as those filled with argon, in which case the life of the lamp is a good deal longer; or they can be operated at higher filament temperatures to give a brighter light at greater efficiency (more lumens per watt), but the latter alternative reduces the life of the lamp somewhat. Brighter lamps are used in slide projectors and projectors for home movies.

Krypton is also used to fill electric-arc lamps. An example is the lamp which will pierce fog for 1000 ft or more. Lamps of this type are arranged in rows to mark airplane runways at night. They flash 40 times a minute; each flash lasts only 17 μsec so as not to blind the pilots.

Radioactive krypton-85 is relatively inexpensive and is finding a number of uses. One is in leak-testing of sealed containers. Another is in the continuous measurements of the thicknesses of materials, such as sheets of metal and plastics. Another use is in lamps which give off light for several years with no source of energy other than the radioactivity of the krypton; in these, the invisible radiation from the krypton activates a phosphor coated on the inner glass walls of the lamp, and the phosphor gives off light continuously until the krypton has decayed to a low level of radioactivity. Krypton-85 can be used to detect abnormal heat openings; the introduction of a small quantity of this gas into the human body is practical because the krypton is not retained by the body; it remains there for only a short time, working its way out by way of the bloodstream and lungs.

Occurrence. The only commercial source of stable krypton is the air, although traces of krypton are found in minerals and meteorites. Krypton constitutes 1.14 ppm by volume of the Earth's atmosphere, and this krypton is almost entirely a mixture of the following isotopes, none of which is radioactive: 78 (0.35%), 80 (2.27%), 82 (11.56%), 83 (11.55%), 84 (56.90%), and 86 (17.37%). The relative abundance of the particular isotope is given in parentheses after each mass number.

A mixture of stable and radioactive isotopes of krypton is produced in nuclear reactors by the slow-neutron fission of uranium.

It is estimated that about $2 \times 10^{-8}\%$ of the weight of the Earth is krypton. Krypton also occurs outside the Earth; the best estimate is that there are about 51 atoms of krypton for each 1,000,000 atoms of silicon in the visible universe, silicon being used as a standard of abundance.

Discovery. Krypton was discovered in England in 1898 by Sir William Ramsay and M. W. Travers; they found it in the less volatile part of the inert-gas mixture left after the oxygen and nitrogen had been removed chemically from a sample of air. The fact that a new element was present was ascertained by the discovery of new lines in the emission spectrum of the residual gas.

Radioactive isotopes. The following radioactive isotopes of krypton are known: Kr^{76}, Kr^{77}, Kr^{79}, Kr^{81}, Kr^{83m} (metastable form of Kr^{83}), Kr^{85}, Kr^{87}, Kr^{88}, Kr^{89}, Kr^{90}, Kr^{91}, Kr^{92}, Kr^{93}, Kr^{94}, Kr^{95}, and Kr^{97}. As mentioned above, they are produced as by-products of the nuclear fission of uranium in nuclear reactors. They can also be formed in particle accelerators, such as the cyclotron, or by the neutron bombardment of the appropriate atomic species. The only radioactive isotope which is produced in more than traces and also has a reasonably long lifetime (half-life about 10 years) is krypton-85.

Properties. Krypton is a colorless, odorless, and tasteless gas. Table 1 gives some physical properties of krypton. The outer shell of the krypton atom is filled with electrons in a stable structure. For this reason there is only one atom in a molecule, and no one has yet been able to prepare a chemical compound of krypton which is stable at room temperature. However, the compound KrF_4 has been prepared; it is stable only at $-80°C$

Table 1. Physical properties of krypton

Property	Value
Atomic number	36
Atomic weight (atmospheric krypton only)	83.80
Melting point, triple point °C	−157.20
Boiling point at 1 atm pressure, °C	−153.35
Gas density at 0°C and 1 atm pressure, g/liter	3.749
Liquid density at its boiling point, g/ml	2.413
Solubility in water at 20°C , ml krypton (STP) per 1000 g water at 1 atm partial pressure krypton	59.4

Table 2. Isotopes in reactor-produced krypton

Mass number	Percentage by volume	
	Stable	Radioactive
78, 80, 82	1–3	
83	13–14	
84	29–33	
85		5–6.5
86	46–50	

and below. *See* KRYPTON COMPOUNDS.

Krypton can be trapped in crystals of hydroquinone, phenol, and other host compounds to form clathrates. The clathrate of hydroquinone is stable at room temperature and has found a use as a convenient method of supplying radioactive Kr^{85}. *See* CLATHRATE COMPOUNDS.

Production. Stable krypton is produced in air-separation plants. Air is liquefied and distilled; krypton and xenon remain with the oxygen. The liquid oxygen is redistilled to concentrate the krypton and xenon from a few parts per million to a few percent. The rare gases are then adsorbed from the liquid oxygen onto silica gel, desorbed, separated, and purified. Final purification is carried out by passing the krypton over hot titanium metal, on which all except inert gas impurities are removed.

Of the radioactive krypton isotopes produced in nuclear reactors, the only one that has a half-life of over about 3 hr is krypton-85; this isotope has a half-life of about 10 years. Thus, when krypton from a nuclear reactor has been stored for several days, the only radioactive isotope left is krypton-85. Table 2 gives the approximate isotopic composition of the krypton thus produced. The composition of the krypton mixture varies somewhat with conditions in the reactor. As nuclear reactors come into wide use for power production, a large amount of radioactive krypton will become available. Since no cheap method of separating the radioactive from the stable isotopes of krypton is known, the nuclear reactor is not at present a source of stable krypton, even though about 95% produced in it consists of stable isotopes.

Analytical methods. The principal modern methods of detecting and quantitatively determining the krypton content in gases are mass spectrometry and gas chromatography. Until these methods were developed, it was necessary to separate krypton from other gases by selective low-temperature adsorption on activated carbon in order to determine how much krypton was present in a mixture. The older method of detecting krypton makes use of its characteristic emission spectrum, obtained by passing a gas sample through an electric discharge tube at low pressures and analyzing the light with a spectrometer. *See* ATMOSPHERIC GASES, PRODUCTION OF.

[ARTHUR W. FRANCIS]

Bibliography: I. Asimov, *The Noble Gases*, 1966; G. A. Cook, *Argon, Helium and the Rare Gases*, 2 vols., 1961; F. P. Gross, Jr., Rare gases in everyday use, *J. Chem. Educ.*, 18:533–539, 1941.

Krypton compounds

Until 1962, krypton, like the other noble gases, was thought to be chemically inert. Even now the only well-characterized krypton compounds are the diflouride, KrF_2, and its complexes. An early report of a tetrafluoride, KrF_4, has not been confirmed, and there is no convincing evidence for the existence of krypton compounds in aqueous solution.

Krypton difluoride can be prepared by passing an electric discharge through a 1:1 mixture of gaseous fluorine and krypton at $-188°C$. After the reaction is complete, any excess krypton and flurrine are pumped away, the reactor is slowly warmed, and the KrF_2 is sublimed into a glass vessel. KrF_2 can also be prepared by the action of ionizing radiation on a gaseous mixture of Kr and F_2 at -60 to $-150°C$ and by the action of near-ultraviolet light on a solution of krypton in liquid fluorine. See FLUORINE.

KrF_2 is a volatile solid at room temperature and is easily sublimed to give clear, colorless crystals. It decomposes slowly at room temperature to krypton and fluorine, but it can be stored at $-78°C$ without decomposition. Reaction of KrF_2 with water is very rapid, giving krypton, oxygen, and aqueous hydrogen fluoride. Reaction with chlorine compounds yields ClF_5. Krypton difluoride reacts with antimony pentafluoride to give the addition compound $KrF_2 \cdot 2SbF_5$, which is somewhat more stable than the parent molecule. Spectral studies show that the KrF_2 molecule is linear and symmetrical.

[EVAN H. APPELMAN; JOHN G. MALM]

Bibliography: N. Bartlett and F. O. Sladky, The chemistry of krypton, xenon, and radon, *Comprehensive Inorganic Chemistry*, vol. 1, 1973; H. H. Claassen, *The Noble Gases*, 1966; J. G. Malm and E. H. Appelman, The chemical compounds of xenon and other noble gases, *At. Energ. Rev.*, 7(3), 1969.

Kudzu

A perennial vine legume, capable of rapid growth in a warm temperate, humid subtropical climate. The name Kudzu has a Japanese origin. Kudzu (*Pueraria thunbergiana*) was introduced into the United States in 1876 and used as a shade plant until 1906, when a few enthusiastic growers in the southeastern United States began to use it as a forage crop, a practice that continued for 30 years. It was then promoted as a soil-conserving plant. However, much prejudice has developed against its use because of its spread into forest borders, drainage ditches, and other areas.

Kudzu is not adapted to tropical or arid climates or to alkaline soils, and probably requires winter cold for growth rejuvenation. The technique for growing kudzu successfully is to set out a few well-developed plants, use enough commercial fertilizer or equivalent for good growth, and protect the plants from weeds, insects, and grazing animals. The vines will spread over the area to be covered, producing roots at the nodes, which, unless killed by severe winter freezing, become independent plants. Thus, a few plants produce many in 2 or 3 years if adequately protected. See FERTILIZING.

Kudzu produces moderate yields of forage, but it must be grazed with care to prevent loss of stand. Since the viney stems are difficult to harvest for hay, adapted grasses and clovers are preferred as forage crops.

A simple, economical method is needed for keeping kudzu from spreading into areas where it is not wanted. It may be stopped by a permanent pasture at the border of a kudzu field, or by repeated cultivation or harrowing the borders of cropland. No acceptable method of spread prevention has been developed for uncultivated or non-pastured areas.

Tropical kudzu (*P. phaseolides*) is one of the most important and widely planted cover and green manure crops of the tropics. It makes rapid vigorous growth, providing quick ground cover and suppressing most other vegetative growth. It is increasingly used as a forage crop although careful management is required to prevent complete domination of mixtures with grasses and other species. Its habit of growth is similar to kudzu of subtropical areas. See LEGUME FORAGES; ROSALES.   [PAUL TABOR]

Kumquat

Shrubs or small trees that are members of the genus *Fortunella*, which is one of the six genera in the group of true citrus fruits. Kumquats are believed to have originated in China and the Malay Peninsula, but are now widely grown in all citrus areas of the world. Of the several species the most common are *F. margarita*, which has oval-shaped fruit (see illustration), and *F. japonica*, which has round fruit.

The kumquat's stems, leaves, flowers, and fruits resemble those of *Citrus*. Kumquats bear numerous flame- to orange-colored fruits of small size, often less than 1 in. in diameter, having three to five locules filled with an acid pulp and a sweet, edible pulpy rind. The trees are among the most cold-hardy of the citrus fruits; they stay dormant even during protracted warm spells in the winter months, which enables them to withstand low temperatures.

Kumquats, with their brilliant orange-colored fruits and dense green foliage, are highly ornamental and are most frequently grown for this reason. Sprays of foliage and fruit are commonly used for decoration, particularly in gift packages of ordinary citrus fruits. Kumquat fruits can be eaten whole without peeling; they are also used in marmalades and preserves and as candied fruits. See FRUIT, TREE; SAPINDALES.   [FRANK E. GARDNER]

Fruit cluster and foliage of Nagami kumquat (*Fortunella margarita*). (*J. Horace McFarland Co.*)

Kutnahorite

A rare carbonate of calcium and manganese, $CaMn(CO_3)_2$. Kutnahorite is never found in nature without some magnesium and iron substituting for manganese. Investigations of mineral synthesis show complete solid solution between calcite and kutnahorite at somewhat elevated temperature.

Kutnahorite has hexagonal (rhombohedral) symmetry and the same structure as dolomite. It is pink and has a specific gravity of 3. It is found in Czechoslovakia, Sweden, and at Franklin, N.J. See CARBONATE MINERALS.   [ROBERT I. HARKER]

Kwashiorkor disease

A nutritional deficiency disease in infants and young children, caused primarily by diet in which protein is of poor quality or inadequate quantity, or both, and carbohydrate intake is normal or high. The disease has frequently been referred to as protein-calorie malnutrition or deficiency. Kwashiorkor occurs throughout the world, but especially among the underdeveloped areas. According to the World Health Organization, it is the most widespread and important dietary disease in the world today. It is estimated that between 100- and 270,000,000 children, commonly in the 6-month to 3-year age group, are afflicted, many of whom die. Clinical features consist of poor growth and development, edema, dyspigmentation of skin and hair, and psychological changes characterized by apathy and irritability. Pathological changes consist of enlarged fatty liver; atrophy of pancreas, salivary glands, and intestinal glands; and wasting of muscles. Because of its worldwide importance, investigators have studied experimental models of kwashiorkor in animals to understand better the mechanisms involved in the metabolism of dietary constituents, especially protein and its amino acids. This article concerns itself with research with experimental kwashiorkor-like syndromes induced in laboratory animals.

Experimental conditions. It has been frequently demonstrated that the feeding of an amino acid–deficient or imbalanced protein diet to an experimental animal invariably leads to reduced food consumption when the animal is allowed to eat as much as it desires. Under these conditions it has not been possible to demonstrate pathologic changes simulating those found in humans with kwashiorkor. Consequently, investigators have adopted the force-feeding technique on rats and monkeys to assure an adequate intake of an amino acid– or protein-deficient diet. With this technique a kwashiorkor-like syndrome has been induced in animals within days or weeks. The chief morphologic alterations consist of an enlarged fatty liver, increased liver glycogen, and atrophy of the pancreas, salivary glands, gastrointestinal glands, thymus, and spleen.

Conclusions. Based on a number of these experimental studies, the conclusion is that a variety of factors are important in the induction of pathologic changes resembling kwashiorkor in animals: Animals must ingest a sufficient amount of food in order to develop the condition; progressive emaciation or marasmus, which is quite different from kwashiorkor, develops if the food intake is decreased; the caloric intake derived mainly from carbohydrates must be quantitatively adequate; adrenal-cortical hormone stimulation is not important, since adrenalectomized animals maintained on low doses of corticosteroids still develop the condition; and finally, single deficiencies of all essential amino acids except arginine and leucine, as well as poor-quality proteins, can produce most of the features of the kwashiorkor-like experimental model.

Among biochemical changes in the livers of these experimental animals were increases in lipid, glycogen, and ribonucleic acid. The results of experiments in which young rats were force-fed an essential amino acid–devoid diet for 3–7 days indicated that protein synthesis is enhanced in the liver but diminished in skeletal muscle of the experimental animals. This has been demonstrated by measuring the incorporation of leucine-C^{14} into proteins of these organs, or of cell-free preparations from these organs. The increased protein synthesis in the liver is related predominantly to enhanced activity of ribosomes, and the aggregates of liver ribosomes (polyribosomes) as measured in a sucrose gradient indicated a shift from lighter toward heavier polyribosomes with a decrease in monomers. Ultrastructural changes in the hepatic cells of experimental animals revealed nucleolar enlargement; increased amounts of lipid and glycogen; increased number of lysosomes, many of which contained glycogen; and an increased amount of free (unbound) polyribosomes. *See* METABOLIC DISORDERS.   [HERSCHEL SIDRANSKY]

Bibliography: H. N. Munro and J. B. Allison, *Mammalian Protein Metabolism*, vol. 2, 1964.

Kyanite

A nesosilicate mineral, composition Al_2SiO_5, crystallizing in the triclinic system. Crystals are usually long and tabular and commonly occur in bladed aggregates (see illustration). There is one perfect cleavage; the specific gravity is 3.56–3.66. The hardness of kyanite, one of its interesting features, is 5 (Mohs scale) parallel to the length of the crystals but 7 across the length.

├─ 29 mm ─┤

Kyanite on paragonite schist, St. Gotthard, Switzerland. (*American Museum of Natural History Specimens*)

The luster is vitreous to pearly, and the color is usually a shade of blue but may be white, gray, or green. *See* SILICATE MINERALS.

Kyanite is one of three polymorphic forms of Al_2SiO_5, the other two being andalusite and sillimanite. All three forms are commonly found in pelitic (aluminous) schists and hornfelses. Experimental work on Al_2SiO_5 has made it possible to use the presence of a specific polymorph as an indicator of the temperature and pressure of formation of the rock in which the polymorph occurs. Kyanite is formed at low temperature and high pressure relative to andalusite and sillimanite. The transitions from one mineral to another are so sluggish that they may coexist in the same rock. *See* ANDALUSITE; SILLIMANITE.

Kyanite is characteristically found in mica schists in association with garnet, staurolite, quartz, muscovite, and biotite. Fine crystals, some of gem quality, are found at St. Gothard, Switzerland. Kyanite has been mined in India and in the United States in Georgia and North Carolina for the manufacture of highly refractory porcelain.

[CORNELIUS S. HURLBUT, JR.]

L

Labradorite to Lytic reaction

Labradorite

A plagioclase feldspar with a composition ranging from $Ab_{50}An_{50}$ to $Ab_{30}An_{70}$, where $Ab = NaAlSi_3O_8$ and $An = CaAl_2Si_2O_8$ (see illustration). In the high-

Labradorite from Labrador, Canada. (*Specimen from Department of Geology, Bryn Mawr College*)

temperature state, labradorite has albite-type structure. In the course of cooling, natural material develops a peculiar structural state which, when investigated by x-rays, shows reflections that indicate the beginning of an exsolution process sometimes accompanied by a beautiful variously colored luster (labradorizing). *See* FELDSPAR; IGNEOUS ROCKS. [FRITZ H. LAVES]

Labyrinthodontia

An important subclass of fossil amphibians descended from crossopterygian fishes, ancestral to reptiles, and probably antecedent to part, if not all, the other amphibian types. The labyrinthodonts ranged in time from Late Devonian to Late Triassic and included all the larger ancient amphibians as well as a variety of smaller types. The labyrinthodonts were characterized by vertebral centra which include distinct intercentra and pleurocentra of variable construction. Earliest of labyrinthodonts were the Ichthyostegalia of the Upper Devonian; they were succeeded by the Rhachitomi, dominant in the Carboniferous and Permian, which were succeeded in turn by the Trematosauria and Stereospondyli in the Triassic. In these four groups, classed as Temnospondyli, the pleurocentra were small and sometimes absent. In a derived group, the Anthracosauria, found in the Carboniferous and Permian, the pleurocentra had expanded to complete rings; included are the Embolomeri and forms leading to the Sey-

mouriamorpha and to the reptiles. *See* AMPHIBIA; ANIMAL EVOLUTION; ANTHRACOSAURIA; TEMNOSPONDYLI. [ALFRED S. ROMER]

Labyrinthulia

Protozoa forming a subclass of Rhizopodea with obscure relationships to rest of the class. There is one order, Labyrinthulida (see illustration). The mostly marine, ovoid to spindle-shaped, uninucleate organisms secrete a network of filaments (slime tubes) along which they glide, usually singly, at rates of $4-150\,\mu$/min. This network inspired the name "net slime molds" sometimes applied to them. The mechanism of locomotion is unknown. The unique *Labyrinthula minuta* may move for about $20\,\mu$, then stop, and later move again, either forward or backward. The slime filaments, which show an amorphous matrix, may or may not be tubular within a single culture. Individual organisms may move within or along the outside of a slime tube and occasionally may leave

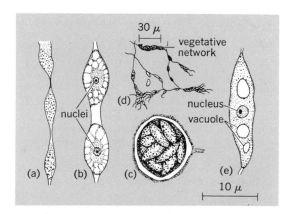

Labyrinthulia. *Labyrinthula zopfi:* (*a*) portion of living network (*after Valkanov*), (*b*) two organisms stained (*after Valkanov*), (*c*) encysted stage (*after Valkanov*). *L. macrocystis:* (*d*) vegetative network (*after Cienkowski*), (*e*) single organism stained (*after Cienkowski*). (*From R. P. Hall, Protozoology, Prentice-Hall, 1953*)

the track. A thin pellicle permits changes in diameter, but pseudopodia and phagotrophy have not been seen. Fission has been described and, in certain species, formation of a sorus by aggregation of individuals. Some produce a pseudosorus by enlargement of one organism; inside, uninucleate pseudospores are produced. The organisms occur on algae and eelgrass, and may be associated with a disease of the latter. *See* PROTOZOA; RHIZOPO-

DEA; SARCODINA; SARCOMASTIGOPHORA.

[RICHARD P. HALL]

Bibliography: K. L. Pokorny, *J. Protozool.*, 14: 697–708, 1967.

Laccolith

A geologic name for a body of igneous rock intruded into sedimentary rocks in such a way that the overlying strata have been notably lifted by the force of intrusion. In the most simple example the laccolith is circular or oval in plan, the subjacent sedimentary rocks and the floor of the laccolith are nearly horizontal, and the roof is dome-shaped, with the sedimentary rocks arching over the dome. Type examples in the Henry Mountains, Utah (described by G. K. Gilbert in 1877), and many supposed examples subsequently described in the western United States have been shown by more detailed work to be much more complex than originally described. *See* PLUTON.

[JAMES A. NOBLE]

Lacquer

A fast-drying, hard, high-gloss surface coating. Lacquers are made by dissolving a cellulose derivative and other modifying materials in a solvent and adding pigment if desired.

The cellulose derivative most commonly used is nitrocellulose, but a cellulose ester such as cellulose acetate or cellulose butyrate, or a cellulose ether such as ethyl cellulose is often used in formulation. Nitrocellulose is not soluble in conventional paint thinners, and so a mixture of solvents with high volatility and extremely fast drying times is used. The mixture usually contains esters (ethylacetate), aromatic hydrocarbons (toluene), and petroleum thinners. Solvent mixtures are formulated carefully in order to produce the combination of solvency and evaporation rate which is required.

Nitrocellulose is extremely hard. Its flexibiligy is enhanced by addition in the formulation of plasticizers such as vegetable oils (castor or linseed) modified for compatibility with nitrocellulose, or chemical compounds such as dibutyl phthalate or tricresyl phosphate.

Adhesion of lacquers can be improved by the addition of other resins, and these also may reduce the cost by permitting the use of less expensive solvents or, alternatively, by allowing the application of heavier films.

Lacquers dry by evaporation of the solvent. They usually are applied by spray because of their rapid drying properties. The drying time may be extended by altering the formulation to allow application by brushing, but brushing lacquers seldom are used.

Lacquers have been used extensively as fast-drying, weather-resistant finishes for automobiles and as coatings for furniture and other factory-finished items. Their use is diminishing, however, as less expensive coating materials with improved properties appear on the market. *See* CELLULOSE DERIVATIVES; SURFACE COATINGS.

[C. R. MARTINSON; C. W. SISLER]

Bibliography: R. Kirk and D. Othmer (eds.), *Encyclopedia of Chemical Technology*, vol. 12, 1967.

Lacrimal gland (vertebrate)

A tubuloalveolar or acinous skin gland, also known as the tear gland. The lacrimal glands develop from the skin epithelium which folds inward over the developing eye. Two types occur among the vertebrates, the lacrimal proper and the Harderian. These eye glands are first found in the amphibians, associated with the inside of the lower eyelid. In urodele amphibians the eye gland extends along the inner aspect of the lower eyelid. In *Salamandra* it becomes divisible into an anterior Harderian gland associated with lower eyelid structures and a posterior lacrimal gland below the upper eyelid. In frogs and toads only the Harderian gland is present and is associated with the nictitating membrane or third eyelid which develops in relation to the lower lid.

The eye glands drain into the nasal cavity by means of the lacrimal duct. In higher vertebrates, with the exception of snakes, certain lizards, and geckos, Harderian and lacrimal glands are present. In the higher vertebrates the Harderian gland functions in relation to the third eyelid or nictitating membrane, and the lacrimal gland becomes located dorsally above the eyeball near the outer angle of the palpebral fissure. In aquatic mammals, such as seals, whales, and sea cows, the Harderian gland is the more prominent of the two sets of eye glands and it secretes a sebaceous or oily substance. In land mammals the lacrimal gland proper is highly developed as a complex tubuloalveolar structure with several ducts which pour their copious fluid into the outer, upper part of the conjunctival sac or cavity. The tear substance washes across the eyeball to the inner palpebral fissure or commissure. Eventually it passes through two small openings, one on the margin of each lid, into the lacrimal ducts. The latter converge to form the lacrimal sac, from which the nasolacrimal duct leads into the nasal passageway. Tears contain a considerable quantity of the common salt, sodium chloride. *See* EPITHELIUM; EYE (VERTEBRATE); EYE DISORDERS; GLAND.

[OLIN E. NELSEN]

Lactam and lactim

Lactams are internal (cyclic) amides, formed by heating γ- and δ-amino acids; structurally, they are nitrogen analogs of the corresponding γ- and δ-lactones. Thus, γ-aminobutyric acid readily forms γ-butyrolactam, known as pyrrolidone, as in Eq. (1).

$$H_2NCH_2CH_2CH_2C\overset{O}{\overset{\|}{O}}H \rightarrow CH_2 \begin{matrix} CH_2-CH_2 \\ | \quad | \\ NH-C=O \end{matrix} \rightleftharpoons$$

γ-Butyrolactam
(keto form)

$$CH_2 \begin{matrix} CH_2-CH_2 \\ | \quad | \\ N=C-OH \end{matrix} \qquad (1)$$

γ-Butyrolactim
(enol form)

The formulas show lactims to be tautomeric, enol forms of lactams with which they form an equilibrium whenever the lactam nitrogen carries a free hydrogen.

Pyrrolidone has weakly acidic and basic properties, forming a hydrochloride and a sodium salt. Alkylation of the latter gives *N*-alkyl pyrrolidones. With acetylene, *N*-vinylpyrrolidone forms, polymerizing to polyvinylpyrrolidone; the latter is used in aerosol sprays for hair setting.

Commercial production of pyrrolidone utilizes either treatment of γ-butyrolactone with ammonia or the partial reduction of succinimide. Piperidone (δ-valerolactam) is made from δ-valerolactone and ammonia. Smaller rings (β-lactams) and larger (ε- and higher) must be made by indirect methods, such as addition of ketenes to imines for β-lactams and ring expansion via Beckmann rearrangement of cyclic ketoximes for ε- and higher lactams.

The tautomeric equilibrium existing between lactam and lactim structures is well exemplified in isatin, an oxidation product of indigo. Thus, as shown in Eq. (2), isatin, as the lactam, reacts with

Isatin (lactam)   Isatin (lactim)

(2)

N-Acetylisatin   Isatin chloride

acetic anhydride to form *N*-acetylisatin; with phosphorus pentachloride, it reacts as the lactim, giving isatin chloride.

Although γ- and δ-amino acids are nontoxic themselves, the corresponding lactams show pronounced strychnine-type toxicity. *See* AMINO ACIDS; LACTIDE AND LACTONE.

[EVANS B. REID]

Lactase

An enzyme found in mammals, honeybee larvae, and some plants. It is a β-galactosidase which hydrolyzes lactose to galactose and glucose. In mammals, lactase appears in the intestinal secretion from the intestinal villi, and exerts its effect on lactose in chyme. *See* ENZYME; GLUCOSE.

[DANIEL N. LAPEDES]

Lactate

A salt or ester of lactic acid (2-hydroxypropanoic acid, α-hydroxypropionic acid) in which the acidic hydrogen of the carboxyl group has been replaced by a metal or an organic radical (symbol X). The structural formulas of the two optical isomers are

Optical isomers of lactic acid. (*a*) D(−) Form. (*b*) L(+) Form.

given in the illustration. Alkali metal salts have practical applications, for example, as blood coagulants in calcium therapy (calcium), as antiperspirants (aluminum), in treatment of anemia (iron), and as plasticizers (sodium). Lower-molecular-weight esters are water-soluble liquids. Esters are used as plasticizers and as solvents for lacquers, cellulose acetate, and cellulose nitrate. *See* CARBOXYLIC ACID; ESTER; LACTIC ACID; SOLVENT.

[ELBERT H. HADLEY]

Bibliography: W. L. Faith et al., *Industrial Chemicals*, 1965.

Lactation

The functional activity of the mammary gland, which is the secretion of milk. Soon after birth, the young begin periodic nursing, or milk removal. A very important mechanism involving both the nervous and endocrine systems comes into play. The act of nursing sends nerve impulses to the brain and posterior pituitary, which cause the discharge of oxytocin into the blood. This hormone flows to the mammary glands, causing the contraction of myoepithelial cells which surround each alveolus, forcing the milk from the lumen of the alveoli into the ducts for removal by the nurslings. If the mother is disturbed at the time of nursing, a hormone, adrenalin, is discharged from the adrenal gland. If adrenalin reaches the mammary gland before oxytocin is discharged, it prevents the contraction of the myoepithelium, and milk removal does not occur. A nursing mother must be content and free from stress or excitement.

Milk production. The first milk, or colostrum, secreted before parturition, is especially rich in gamma globulin, which contains immune bodies. The digestive tract of the newborn absorbs this protein intact into the blood; this confers a temporary immunity to common diseases. With the regular removal of milk, the amount of milk secreted increases for a period of time, then gradually declines until the animal is said to dry up. In dairy animals such as the cow, pregnancy may be induced about the ninetieth day of the lactation period. Lactation and pregnancy with additional growth of the mammary glands occurs so that a second lactation period is initiated in a single year. The total yield of milk increases each year until about the seventh or eighth year, then declines with old age. *See* MILK.

Control of secretion. As mentioned, the secretion of milk is under hormonal control. At the time milk is removed by the mediation of oxytocin, the nursing stimulus also causes a discharge of lactogenic hormone into the blood, which stimulates milk manufacture during the interval between nursings. Without this hormone the cells cease to secrete milk. However, there are a number of other hormones which influence the intensity of secretion. Of these, thyroxine from the thyroid gland and growth hormone of the pituitary gland have been shown to be effective in stimulating increased milk secretion in dairy cattle. A synthetic thyroid hormone, thyroprotein, produced by the iodination of casein, is now fed to dairy cattle to increase milk production. For the increase in milk production to be sustained, it is necessary to feed a ration high in energy, not only to supply nutrients for the extra milk, but also to produce an increase in percentage of fat.

In extensive studies in normal lactating rats, milk yield has been increased by the injection of lactogenic, thyroxine, growth, insulin, parathyroid, and corticosterone hormones, separately and in combination. These studies suggest that the inheritance of intense milk secretion (high production) is dependent upon the optimal secretion of the hormones from these various endocrine glands. *See* ADRENAL GLAND (VERTEBRATE); MAMMARY GLAND; PITUITARY GLAND (VERTEBRATE); THYROID GLAND (VERTEBRATE).

[CHARLES W. TURNER]

Lactic acid

A slightly hygroscopic syrup with a specific gravity of 1.294 at 25°C. The acid, also known as α-hydroxypropionic acid, contains one asymmetric carbon atom and exists as two optically active isomers that can be converted by an enzyme, racemase, into the racemic mixture. L(+)-Lactic acid, also called sarcolactic and paralactic acid, is metabolized by animals, whereas D(−)-lactic acid is not. The lactic acid of commerce is the racemic mixture, commonly known as *i*-lactic acid (optically inactive). Annually, 8,000,000 lb is produced for use in pharmaceutical preparations, in the textile and leather industry, in the production of inks, solvents, lacquers, and plastics, as the bread additive, calcium stearyl-2 lactylate, and as a food preservative. *See* OPTICAL ACTIVITY; RACEMIZATION.

Lactic acid can be prepared chemically, although industrial production is by fermentation (see illustration).

Fermentation. Industrial production of lactic acid is accomplished by fermentation of refined glucose, hydrolyzed starch, whey, and molasses. The more refined substrates are used for production of the higher grades to reduce purification costs. Sulfite waste liquor, juice of culled citrus fruits, enzymatically hydrolyzed potatoes and acid-hydrolyzed wood, mill sawdust, straw, corncobs, extracted beet slices, and Jerusalem artichokes have also been proposed as substrates.

Bacteria suitable for commercial production of lactic acid (Table 1) may be isolated from souring foods, grains, malt sprouts, or soil. These organisms have complex nutritional requirements. For example, *Lactobacillus delbrueckii* requires 14 amino acids and 4 vitamins and is stimulated by a number of other substances. These growth factors can be provided by adding small amounts of malt sprouts, corn steep liquor, distillers' grains or solubles, koji, rice bran, peanut oil cake, soybean cake, undenatured milk, or extracts of liver and yeast. Maintenance of the culture requires frequent transfers in media of low sugar content, such as 5% corn mash.

In one commercial process, *L. delbrueckii* is transferred serially from test tube to flasks, to seed tanks, and finally to the wooden or stainless steel fermentor, an inoculum level of 10% being maintained. Each transfer is made after 16–20 hr of growth at 49°C, a temperature that is assiduously controlled. The medium, consisting of 15% glucose, 0.4% malt sprouts, 0.25% diammonium phosphate, and 10% calcium carbonate, is not sterilized. The industry relies on cleanliness, the high temperature, and low pH to restrict contaminants. Butyric bacteria are the most troublesome, particularly because they produce volatile acids in addition to a racemase. Although the batch process is usually employed, continuous processes have been devised. Corrosion is a major problem.

The fermentation requires 4–6 days, when the concentration of sugar falls to 0.1% or less, and the yield reaches 90–95%. Small changes in pH adversely affect the fermentation, and automatic control is advantageous. The optimum pH of 5.6–5.8 is maintained by adding sufficient calcium carbonate initially, or intermittently, as required. Although not used industrially, the submerged cultivation of *Rhizopus oryzae* in media containing 13% glucose gives a 75% yield of lactic acid in 35 hr. This fermentation has the disadvantages of lower yield and requiring aseptic conditions but the advantage of producing L-lactic acid in a clear broth.

Recovery of lactic acid from fermented liquors is rather difficult because of the low vapor pressure of lactic acid and its tendency to form anhydrides and undergo self-esterification when heated and concentrated above 20%, and also because its solubility is similar to that of water. The important contaminants are proteins, unfermented sugars, inorganic salts, and colored materials. The usual practice is to coagulate the proteins and completely neutralize the lactic acid by heating to 90°C with excess lime. The insoluble matter is removed by filtration; colored materials and soluble proteins, by treatment with charcoal. The calcium

Table 1. Characteristics of lactic acid organisms

Organism	Morphology	Substrates	Optimum temperature, °C	Acid produced
Lactobacillus bulgaricus	Rod	Lactose, whey	45–50	Racemic
L. delbrueckii	Rod	Glucose, molasses	45–50	L(+)
L. brevis	Rod	Pentoses, hydrolyzed wood	30	Racemic
L. plantarum	Rod	Pentoses, sulfite liquor	30	Racemic
L. leichmannii	Rod	Sucrose, glucose	30	D(−)
Streptococcus lactis	Coccus	Lactose, whey	35	L(+)
Bacillus coagulans	Rod, forms spores	Glucose, lactose	45–50	L(+)
Rhizopus oryzae	Mold	Glucose, starch	30	L(+)

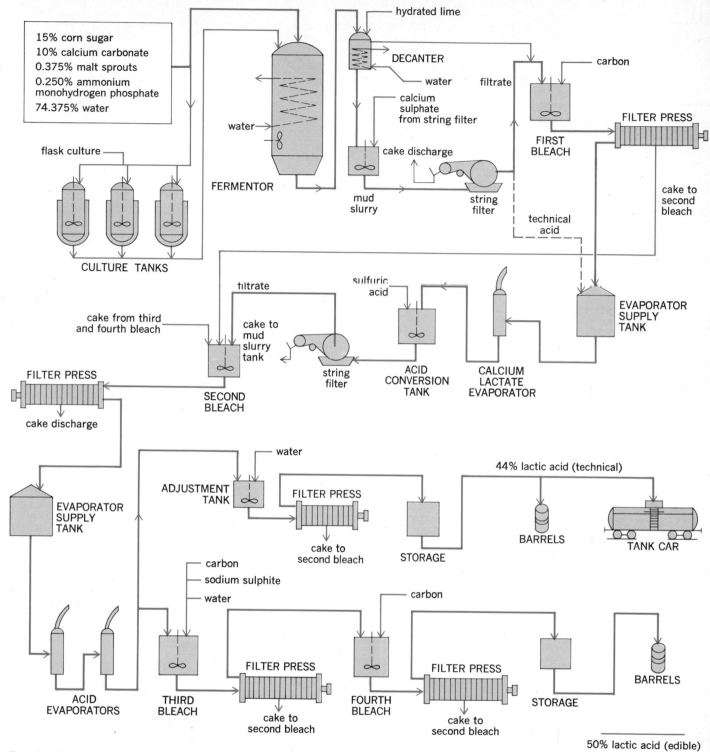

Flow sheet for the production of lactic acid.

lactate is then either concentrated and spray-dried, or cooled and allowed to crystallize. Lactic acid is prepared by treating a solution of calcium lactate with sulfuric acid and filtering off the insoluble calcium sulfate. The acid is then repeatedly bleached until the desired purity is attained. Some organic impurities in lactic acid can be destroyed by treatment with oxidizing agents, such as hypochlorites or nitric acid. Table 2 shows the chemi-

Table 2. Grades of lactic acid

Category	Technical	Edible	Plastic	USP
Total acidity	44%	50%	50%	85%
Volatile acids	1–2%	1–2%	1–2%	2–3%
Ash	0.6–0.7%	0.4–0.5%	0.005–0.01%	0.05–0.1%
Carbonizable matter	Present	Present	None	None
Color	Brown	Pale yellow	Colorless	Colorless

cal characteristics of the various grades of lactic acid, including USP (U.S. Pharmacopeia).

Alternative methods proposed for purifying lactic acid include (1) steam distillation under vacuum, (2) crystallization as the zinc or aniline salts, (3) extraction of the lactic acid with nitroparaffins, isopropyl ether, or isoamyl alcohol, and (4) conversion to an ester. The last-mentioned process has been successfully applied to the recovery of methyl lactate from fermented potato mashes.

Chemical production. Lactic acid may be prepared chemically by (1) reacting silver hydroxide with bromopropionic acid, (2) oxidizing 1,2-propane diol with permanganate, (3) deaminating alanine with nitrous acid, (4) reducing pyruvic acid, (5) combining acetaldehyde with carbon monoxide at 200°C and 900 atm pressure, (6) hydrolyzing acetaldehyde cyanohydrin, and (7) degrading sugars with alkali. These methods produce racemic lactic acid and, being more expensive than the fermentation process, are not widely employed. D(−)-Lactic acid may be prepared from methyl α-D-6-deoxymannopyranoside and L(−)-lactic acid from methyl α-L-5-deoxyarabinofuranoside by oxidation with periodic acid and bromine in aqueous solution.

Production of lactic acid by alkaline degradation of sugars according to the equation below is the subject of numerous patents.

$$C_{12}H_{22}O_{11} + 2CaO \rightarrow$$

$$2Ca(-O-CO-CHOH-CH_3)_2 + H_2O$$

In practice, however, the sugars undergo many other kinds of rearrangements, fragmentation, and oxidation. The actual yield of lactate is therefore of the order of 50% of the theoretical. In alkaline solutions sugars undergo rearrangement, fragmentation, and oxidation to various acids. The yield of lactic acid depends on such factors as structure of the sugar, concentration and type of alkali, and temperature. Sucrose is most efficiently degraded, and aldonic acids are superior to aldoses. When molasses is diluted to 18% sugar and treated with 13% calcium oxide for 2 hr at 230−235°C, the yield of lactic acid is about 40−50%. *See* CARBOXYLIC ACID. [F. J. SIMPSON]

Bibliography: L. E. Casida, Jr., *Industrial Microbiology*, 1968; Chemistry and metabolism of L- and D-Lactic acids, *Ann. N. Y. Acad. Sci.*, 119: 851−1165, 1965; R. Montgomery, *The Chemical Production of Lactic Acid from Sugars*, Sugar Res. Found. Sci. Rept. Ser. no. 11, 1949; S. C. Prescott and C. G. Dunn, *Industrial Microbiology*, 3d ed., 1959; L. A. Underkofler and R. J. Hickey (eds.), *Industrial Fermentations*, 1954.

Lactide and lactone

Lactides are cyclic, intermolecular double esters formed from α-hydroxy acids. Thus, lactic acid, $CH_3CHOHCOOH$, on heating forms the lactide shown in (1). Glycolic acid, $HOCH_2COOH$, behaves analogously. In each case bimolecular interaction occurs, forming the strain-free 6-membered ring. Most lactides are relatively low-melting solids and are easily hydrolyzed by base to form salts of the parent acids, for example, sodium lactate.

Lactones are internal cyclic mono esters formed by γ- or δ-hydroxy acids spontaneously; thus, γ-hydroxybutyric acid, $HOCH_2CH_2CH_2COOH$,

forms γ-butyrolactone, with structural formula (2).

$$
\begin{array}{c}
\text{H} \quad \text{CH}_3 \\
\text{O} - \text{C} \\
\text{O} = \text{C} \qquad \text{C} = \text{O} \\
\text{H}_3\text{C} - \text{C} - \text{O} \\
\text{H}
\end{array}
\qquad (1)
$$

$$
\begin{array}{c}
\text{H}_2\text{C} - \text{CH}_2 \\
\text{H}_2\text{C} \qquad \text{C} = \text{O} \\
\text{O}
\end{array}
\qquad (2)
$$

Other lactones of smaller or greater ring size are prepared specially.

The γ- and δ-lactones are commonly prepared by either hydrolysis or distillation of γ- or δ-halo acids, by treatment of unsaturated acids with aqueous hydrobromic or sulfuric acids, or by partial reduction of cyclic acid anhydrides. β-Lactones result from the reaction of ketene with aldehydes or ketones. Reaction of ketene with formaldehyde is shown in Eq. (3). Large-ring lactones can

$$
CH_2{=}C{=}O + \begin{array}{c} \text{H} \\ \text{C}{=}\text{O} \\ \text{H} \end{array} \rightarrow \begin{array}{c} \text{H}_2\text{C} - \text{C}{=}\text{O} \\ \text{H}_2\text{C} - \text{O} \end{array} \qquad (3)
$$
$$\text{β-Propiolactone}$$

be made by oxidation of cyclic ketones with Caro's acid; thus, cyclohexanone yields ε-caprolactone.

The lower lactones are neutral liquids that react with bases (alkalies, ammonia, and amines) to give open-chain derivatives of the parent hydroxy acids. Very-large-ring (macrocyclic) lactones, for example 15 or 16 carbons, have pronounced musk odors (perfumes).

Unsaturated lactones (butenolides) are widely distributed: for example, the angelica lactones; penicillic acid and protoanemonine (mold metabolites); coumarin (tonka bean); and dicoumarol (spoiled sweet clover), used medicinally as a hemorrhaging or anticlotting agent in coronary thrombosis. *See* ESTER. [EVANS B. REID]

Lactobacillaceae

A family of bacteria of the order Eubacteriales. Although they are primarily saprophytic, a few species are pathogenic. The bacteria are generally known as sugar fermenters, producing lactic acid as a major product. They are found in fermenting food products, in the mouth, in the intestinal tract, and in body lesions. The saprophytes usually are nonmotile and show neither reduction of nitrates, gelatin liquefaction, catalase production, nor surface growth in liquid media. The species are microaerophilic or anaerobic, inactive toward protein sources but fastidious in their requirements. Some strains have become valuable in biological assay because of their exacting nutritional requirements. *See* BIOASSAY; VITAMIN.

The family is divided morphologically into a tribe of spherical or slightly elongated coccus forms, Streptococceae, and a tribe of rod forms, Lactobacilleae (Fig. 1).

Morphology remains the reliable guide to classification, but suitable data are scanty. Physiological characters are therefore indispensable, even though some taxonomists prefer not to use such characters for differentiation beyond the species level.

Streptococceae. The tribe Streptococceae is divided into five genera based upon morphological and physiological characters. The species of four of these, *Diplococcus, Streptococcus, Leuconostoc,* and *Peptostreptococcus,* include typical spherical coccus forms which divide in one plane and usually occur in pairs or short chains (Fig. 2). The fifth genus, *Pediococcus,* includes species which often occur in tetrad grouping, necessitating division in two planes (Fig. 3).

Four of the genera obtain their energy by fermentation of carbohydrates or related compounds. These include the species of the genera *Diplococceus* and *Streptococcus* which produce dextrorotatory lactic acid, species of the genus *Pediococcus* which produce a racemic mixture of lactic acids, and species of *Leuconostoc* which produce considerable amounts of carbon dioxide, ethyl alcohol and acetic acid, in addition to levorotatory lactic acid. Species of the fifth genus, *Peptostreptococcus,* are anaerobic, utilize protein-decomposition products as well as carbohydrates, and produce acids other than lactic and acetic.

The genus *Diplococcus* includes the parasitic species *D. pneumoniae,* a causative agent of lobar pneumonia.

The genus *Streptococcus* includes numerous parasites and saprophytes. The genus is divided into four physiological groups: pyogenic, viridans, enterococcus, and lactic. In general, the first two consist of low acid-producing species present in the alimentary tract and in infections. The second two groups are of the higher acid-producing type associated with fermentation of food products. Separations of species by serological methods correlate well with temperature-growth ranges, salt tolerance, and growth in milk. *See* SEROLOGY; STREPTOCOCCUS.

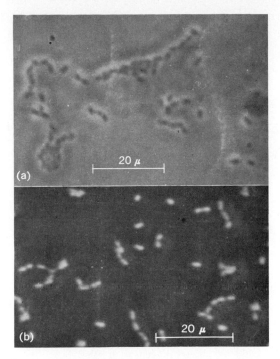

Fig. 2. Photomicrographs of *Leuconostoc mesenteroides,* tribe Streptococceae. (*a*) Encapsulated cells. (*b*) Chained, paired, and single cells.

The genus *Leuconostoc* includes the heterofermentative species associated with food. They were first observed in the slime masses of sugar factories. They convert glucose to levo-lactic acid, alcohol, carbon dioxide, and acetic acid and, in addition, partially convert fructose to mannitol and sucrose to dextran.

The species of the genus *Pediococcus,* first isolated as contaminants in beer, have been observed in other fermenting foods. They include the highest acid-producing species in the tribe.

The species of the genus *Peptostreptococcus* are usually anaerobic, associated with septic and gangrenous conditions or found in the respiratory tract, and may be pathogenic. The species differ physiologically from those in the other genera in their growth on protein-decomposition products, as well as on organic acids and carbohydrates, with production of carbon dioxide, hydrogen, hydrogen sulfide, and various acids in addition to lactic acid.

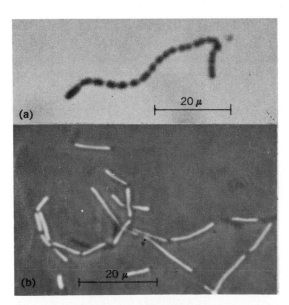

Fig. 1. Photomicrographs showing the morphology of the two tribes of the Lactobacillaceae. (*a*) *Streptococcus faecalis,* tribe Streptococceae. (*b*) *Lactobacillus brevis,* tribe Lactobacilleae.

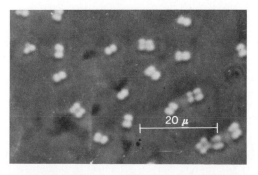

Fig. 3. Photomicrograph of tetrads and paired cells of *Pediococcus cerevisiae.*

Lactobacilleae. The tribe Lactobacilleae is divided into five genera, based upon morphological and physiological characters.

The species of the genus *Lactobacillus* include the high acid-producing bacteria associated with fermenting foods and the intestinal tract. They include two general groups. The homofermentative species, like those of the genera *Streptococcus* and *Pediococcus* among the cocci, produce lactic acid as a major fermentation end product. The heterofermentative species are comparable to those of the genus of cocci, *Leuconostoc*, in their growth end products. Species of the genera *Eubacterium*, *Catenobacterium*, and *Cillobacterium* are low acid-producing bacteria from the intestinal tract and lesions and are comparable with species of *Peptostreptococcus* among the cocci. In addition to lactic acid, formic, acetic, propionic, butyric, and other acids are formed during fermentation. Some of these may be protein-degradation products. They are divided into the four genera on the basis of morphological characters, motility, and cell branching. The physiology of these strict anaerobes has not been completely elucidated, and further work may show that some of these species should be placed in genera of other families. *See* EUBACTERIALES. [CARL S. PEDERSON]

Lactose

Milk sugar or 4-O-β-D-galactopyranosyl-D-glucose. This reducing disaccharide is obtained as the α-D anomer (see formula below, where the asterisk indicates a reducing group); the melting point (mp) is 202°C and the optical activity is $[\alpha]_D^{20} + 85.0 \rightarrow +52.6°$. Crystallization at higher temperatures

(above 93.5°C) gives the β-D anomer; mp 252°C, and $[\alpha]_D^{20} + 35 \rightarrow +55.4°$. Lactose is found in the milk of mammals to the extent of approximately 2–8%. It is usually prepared from whey, which is obtained as a by-product in the manufacture of cheese. Upon concentration of the whey, crystalline lactose is deposited. Lactose is not fermentable by ordinary baker's yeast. In the souring of milk, *Lactobacillus acidophilus* and certain other microorganisms bring about lactic acid fermentation by transforming the lactose of the milk into lactic acid, $CH_3CHOHCOOH$. *See* CHEESE; LACTIC ACID; MILK; OPTICAL ACTIVITY.

Chemical evidence shows that the glycosidic linkage involves the carbon atom 1 of D-galactose and carbon 4 of D-glucose. Enzymatic studies indicate that the galactosidic linkage has the β-configuration. *See* OLIGOSACCHARIDE.

The mammary glands of lactating animals, and their milk, contain an enzyme, lactose synthetase, capable of transferring the D-galactose unit from uridine diphosphate D-galactose to D-glucose, forming lactose according to the scheme: Uridine diphosphate D-galactose + D-Glucose → Lactose + Uridine diphosphate.

The lactose synthetase can be resolved into two protein components, A and B, which individually do not exhibit any catalytic activity. Recombination of these fractions, however, restores full lactose synthetase activity. The B fraction was identified as α-lactalbumin. *See* LACTATION; METABOLISM IN RUMINANTS; URIDINE DIPHOSPHOGLUCOSE (UDPG).

[WILLIAM Z. HASSID]

Bibliography: D. M. Greenberg (ed.), *Metabolic Pathways*, vol. 1, 3d ed., 1967; W. W. Pigman (ed.), *The Carbohydrates*, 2d ed., 1957.

Lagomorpha

The order of mammals including rabbits, hares, and pikas. They were formerly considered relatives of the rodents because both have ever-growing incisors, but it is now universally agreed that there is no relationship between the two orders. *See* RODENTIA.

Morphology. There are many differences between rodents and lagomorphs. Rodents have one pair of upper and one pair of lower incisors, the enamel limited to the anterior side; lagomorphs have two pairs of upper incisors (the second pair minute), and enamel surrounds the tooth, which does not form a sharp chisel. In rodents the lower jaw is capable of backward-and-forward motion, the grinding teeth being disengaged when the incisors are in use, and vice versa. In lagomorphs motion is vertical or transverse. Lagomorphs have three upper and two lower premolars (Fig. 1); even the earliest fossil rodents have one less of each. The detailed tooth structure of lagomorphs is much more like that of various extinct ungulates than like that of any rodent. The tibia and fibula are fused, the fibula articulating with the calcaneum as in artiodactyls (Fig. 2). There is a spiral valve in the cecum, and the scrotum is prepenial. *See* ARTIODACTYLA; DENTITION (VERTEBRATE); TOOTH (VERTEBRATE).

Classification and fossil record. The order includes three families: Leporidae (rabbits and hares); Ochotonidae (pikas, whistling hares, or American coneys); and Eurymylidae, an extinct family from the Paleocene of Mongolia.

Leporidae. These are the most familiar members of the order. There are, in general, two kinds:

LAGOMORPHA

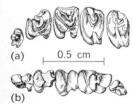

(a)

0.5 cm

(b)

Fig. 1. Cheek teeth of Oligocene rabbit, *Palaeolagus haydeni*. (a) Lower (Pm 2– M 2). (b) Upper (Pm 3– M 3). (*From A. E. Wood, The mammalian fauna of the White River Oligocene, pt. 3: Lagomorpha, Trans. Amer. Phil. Soc., n.s., 28(3):271–362, 1940*)

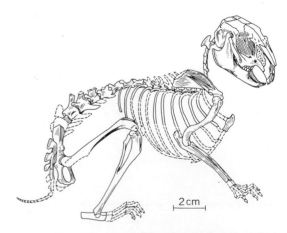

2 cm

Fig. 2. Oligocene rabbit, *Palaeolagus haydeni*, earliest known rabbit skeleton. (*From A. E. Wood, The mammalian fauna of the White River Oligocene, pt. 3: Lagomorpha, Trans. Amer. Phil. Soc., n.s., 28(3):271–362, 1940*)

rabbits (such as the American cottontail), which are relatively small, with shorter hindlegs, shorter ears, and short tails; and hares, larger forms with longer legs, ears, and tails. Rapid locomotion is by leaps, using the hindlegs, combined (especially in rabbits) with abrupt changes of direction. Both types occur in the same region, with rabbits inhabiting brush, scrub, or woods and hares living in open grassland. In North America, hares are usually called jackrabbits.

Leporids occur naturally in all continents but Antarctica and Australia; introduced in Australia by Europeans, they became an outstanding pest, partly controlled by the introduction of a viral infection, myxomatosis.

Although fossils that may be either leporids or ochotonids are present in the late Eocene of North America and Mongolia, unquestioned leporids first appear in the early Oligocene. Since then, they have remained very abundant as individuals, but with no great differentiation of one type from another, even though the family has continued to evolve as a whole. *See* RABBIT.

Ochotonidae. Members of this family are much smaller animals, and often are not seen even in areas where they are relatively abundant. All four legs are about equally long, and the animals run rather than leap. They live at high altitudes in western North America; in central Asia, some forms are plains dwellers. They extend across Asia and just into Europe, in the Soviet Union. Pikas make burrows under or between rocks in the mountains and underground in the plains. The animals rarely go far from their burrows, using them as shelters from enemies. Pikas collect and store grass and other plants in haystacks for winter food.

Fossil ochotonids occur throughout the Northern Hemisphere, being quite abundant in selected regions of North America and Europe, but their history is poorly known. *See* PIKA.

Eurymylidae. The known eurymylids, found as fossils only in the Paleocene of Mongolia, are too specialized to be ancestral to either the leporids or ochotonids, but it seems probable that unknown members of the family were the ancestral stock of the order. *See* EUTHERIA; MAMMALIA.

[ALBERT E. WOOD]

Bibliography: S. Anderson and J. K. Jones, Jr., *Recent Mammals of the World*, 1967; E. R. Hall and K. R. Nelson, *The Mammals of North America*, 1959; A. E. Wood, What, if anything, is a rabbit?, *Evolution*, 11:4, 1957.

Lagrange's equations

Equations of motion of a mechanical system for which a classical (non-quantum-mechanical) description is suitable, and which relate the kinetic energy of the system to the generalized coordinates, the generalized forces, and the time. If the configuration of the system is specified by giving the values of f independent quantities $q_1, \ldots, q_f$, there are f such equations of motion.

In their usual form, these equations are equivalent to Newton's second law of motion and are differential equations of the second order for the q's as functions of the time t.

Derivation. Let the system consist of N particles. The masses of the particles are m_ρ and their cartesian coordinates are $x_{\rho i}(\rho = 1, 2, \ldots, N; i =$

1, 2, 3). These cartesian coordinates are expressible as functions of $f(\leq 3N)$ generalized coordinates $q_1, \ldots, q_f$ between which there are no constraints. The time does not appear here if the constraints are fixed (as will be assumed), as in Eq. (1).

$$x_{\rho i} = f_{\rho i}(q_1, \ldots, q_f) \tag{1}$$

Then, denoting time differentiation by a dot, Eq. (2) holds. The cartesian velocity components are linear functions of the generalized velocities $\dot{q}_j$.

$$\dot{x}_{\rho i} = \sum_{j=1}^{f} \frac{\partial x_{\rho i}}{\partial q_j} \dot{q}_j \tag{2}$$

Let δq_j represent a small displacement of the system. It is automatically consistent with the constraints. In this displacement, the forces of constraint do no work; their only action is to prevent motion contrary to the constraint, so they have no components in the direction of a displacement consistent with the constraints. If W is the work done during the displacement, it is done entirely by the externally applied forces, as in Eq. (3).

$$W = \sum_{\rho,i} F_{\rho i}\, \delta x_{\rho i} = \sum_{\rho,i} m_\rho \ddot{x}_{\rho i}\, \delta x_{\rho i} \tag{3}$$

Each term in the sum of Eq. (3) may receive contributions from forces of constraint, but they cancel when the summation is made. From Eq. (1), Eqs. (4) and (5) are obtained.

$$\delta x_{\rho i} = \sum_j \frac{\partial x_{\rho i}}{\partial q_j} \delta q_j \tag{4}$$

$$\frac{\partial \dot{x}_{\rho i}}{\partial \dot{q}_j} = \frac{\partial x_{\rho i}}{\partial q_j} \tag{5}$$

It is readily verified that Eq. (6) is valid. The quantity in the brackets is thus, for each j, the noncartesian analog of the cartesian $m_\rho \ddot{x}_{\rho i}$.

$$W = \sum_j \left[\frac{d}{dt} \frac{\partial}{\partial \dot{q}_j} \left(\frac{1}{2} m_\rho \dot{x}_{\rho i}^2 \right) - \frac{\partial}{\partial q_j} \left(\frac{1}{2} m_\rho \dot{x}_{\rho i}^2 \right) \right] \delta q_j \tag{6}$$

The only quantity which appears differentiated in Eq. (6) is the total kinetic energy T of the system. This is easily calculated in generalized coordinates because the connection between the cartesian velocities and the generalized velocities is linear and homogeneous. Usually, the kinetic energy can be written by inspection without using Eq. (2) explicitly. Thus Eq. (7) holds.

$$W = \sum_j \left(\frac{d}{dt} \frac{\partial T}{\partial \dot{q}_j} - \frac{\partial T}{\partial q_j} \right) \delta q_j \tag{7}$$

Transforming the right-hand side of Newton's equation is simpler, as shown in Eqs. (8) and (9),

$$W = \sum_{\rho,i} F_{\rho i}\, \delta x_{\rho i} = \sum_j Q_i\, \delta q_j \tag{8}$$

$$Q_j = \sum_{\rho,i} F_{\rho i} \frac{\partial x_{\rho i}}{\partial q_j} \tag{9}$$

where Eq. (9) is the jth component of the generalized force. By the preceding argument, Q_j depends only on the externally applied forces, the forces of constraint necessarily cancelling in the summation.

The displacement δq_j was entirely arbitrary.

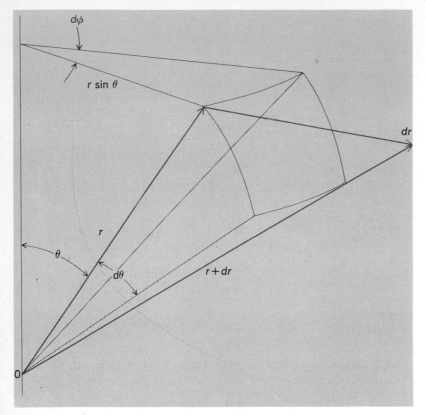

Fig. 1. The vector *dr* is decomposed into three orthogonal vectors of magnitudes *r* sin θ *d*φ, *rd* θ, and *dr*, respectively, *r*, θ, and φ being spherical coordinates of the terminus of the vector *r*.

Thus, it follows from equating expressions (7) and (8) for W that Eq. (10) is valid. Equation 10 sums up

$$\frac{d}{dt}\frac{\partial T}{\partial \dot{q}_j} - \frac{\partial T}{\partial q_j} = Q_j \qquad (10)$$

Lagrange's equations of motion. They are valid also when moving constraints are present. *See* CONSTRAINT.

Examples. Use of this form of Lagrange's equations is shown in the two following examples.

Particle in central force field. Here, the force acting on a particle acts always through a fixed point. Choose this point as origin of a spherical coordinate system with coordinates r, θ, ϕ (Fig. 1). Only radial displacements involve work, so that only Q_r differs from zero, or Eq. (11) holds.

$$Q_\theta = Q_\phi = 0 \qquad (11)$$

The kinetic energy is given by inspection of Fig. 1 and is expressed by Eq. (12).

$$T = \frac{m}{2}(\dot{r}^2 + r^2\dot{\theta}^2 + r^2\sin^2\theta\,\dot{\phi}^2) \qquad (12)$$

Lagrange's equations are Eqs. (13). These may

$$\frac{d}{dt}\frac{\partial T}{\partial \dot{r}} - \frac{\partial T}{\partial r} = m(\ddot{r} - r\dot{\theta}^2 - r\sin^2\theta\,\dot{\phi}^2) = Q_r$$

$$\frac{d}{dt}\frac{\partial T}{\partial \dot{\theta}} - \frac{\partial T}{\partial \theta} = m\left[\frac{d}{dt}(r^2\dot{\theta}) - r^2\sin\theta\cos\theta\,\dot{\phi}^2\right] = 0$$

$$\frac{d}{dt}\frac{\partial T}{\partial \dot{\phi}} - \frac{\partial T}{\partial \phi} = m\frac{d}{dt}(r^2\sin^2\theta\,\dot{\phi}) = 0 \qquad (13)$$

be compared with the cartesian equations, Eqs.

(14), in which the force function F appears in all

$$m\ddot{x} = \frac{x}{r}F(x,y,z)$$
$$m\ddot{y} = \frac{y}{r}F(x,y,z) \qquad (14)$$
$$m\ddot{z} = \frac{z}{r}F(x,y,z)$$

three of the equations of motion, while two of the three Lagrange equations are independent of the detailed nature of the force.

Two particles, fixed separation. In this system, there is one constraint, the particles being a constant distance d apart, and so there are five degrees of freedom instead of six. Choose as generalized coordinates the cartesian coordinates X,Y,Z of the center of mass and the polar angles θ, ϕ of the line joining the two particles, as in Fig. 2. Then, because the kinetic energy is the sum of the kinetic energy of the center of mass and the kinetic energy relative to the center of mass, Eq. (15) holds.

$$T = \frac{m_1+m_2}{2}(\dot{X}^2 + \dot{Y}^2 + \dot{Z}^2)$$
$$+ \frac{1}{2}\left(\frac{m_1 m_2}{m_1+m_2}\right)d^2(\dot{\theta}^2 + \sin^2\theta\,\dot{\phi}^2) \qquad (15)$$

The equations of motion may be written down immediately, the generalized forces being evaluated from Eq. (9). There is one equation for each degree of freedom, and the constraint is automatically satisfied.

Conservative systems. In many problems, the forces Q_j are derivable from a potential. Then Eq. (16), and thus Eq. (17), is valid. When this is so, it is

$$W = \sum_{j=1}^{f} Q_j\,dq_j = -dV \qquad (16)$$

$$Q_j = -\frac{\partial V}{\partial q_j} \qquad \frac{\partial V}{\partial \dot{q}_j} = 0 \qquad (17)$$

convenient to define a function, called the Lagrangian, by Eq. (18). Then the equations of motion

$$L(q,\dot{q},t) = T(q,\dot{q},t) - V(q,t) \qquad (18)$$

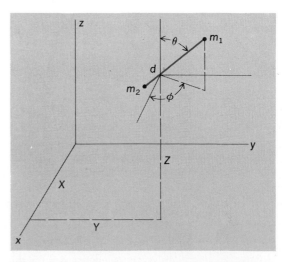

Fig. 2. *X, Y,* and *Z* are cartesian coordinates of the center of mass of dumbbell $m_1 m_2$; θ and φ are the polar angles of the dumbbell axis in a set of axes parallel to fixed set of axes and with origin at center of mass.

become simply Eq. (19), which is the most commonly encountered form of Lagrange's equations.

$$\frac{d}{dt}\frac{\partial L}{\partial \dot{q}_j} - \frac{\partial L}{\partial q_j} = 0 \qquad (19)$$

Example of use of L. A simple vibration problem illustrates the use of generalized coordinates and shows the ease of application of Lagrange's equations when T and V are relatively easy to obtain. Consider the one-dimensional system of two equal masses m connected by a spring of restoring constant k. Let the displacements of the masses from their equilibrium positions be x_1 and x_2, respectively.

Now introduce new coordinates, $q_1 = x_2 - x_1$ and $q_2 = x_2 + x_1$. Then Eqs. (20) hold.

$$V = kq_1^2/2$$
$$T = m\,(\dot{q}_1^2 + \dot{q}_2^2)/4 \qquad (20)$$

Using $L = T - V$, Lagrange's equations for the two variables are written as Eqs. (21).

$$\tfrac{1}{2}m\dot{q}_1 - kq_1 = 0$$
$$\tfrac{1}{2}m\dot{q}_2 = 0 \quad \text{or} \quad \dot{q}_2 = 0 \qquad (21)$$

The solution to the first of these differential equations is $q_1 = A \sin \sqrt{2k/m}\ t$ (setting $q_1 = 0$ at $t = 0$), which is a simple harmonic vibration of the two masses. The second equation merely states that the acceleration of the center of mass of the system is zero.

Nonconservative systems. Form (19) is also correct even if the system is not conservative, provided one can write Eq. (22), as in the case of a

$$Q_j = \frac{d}{dt}\frac{\partial V}{\partial \dot{q}_j} - \frac{\partial V}{\partial q_j} \qquad (22)$$

charged particle in an electromagnetic field. Here V is defined in Eq. (23), where ϕ and $\mathbf{A}$ are the scalar and vector potentials of the field, respectively, e

$$V = e\left(\phi - \frac{\mathbf{v}\cdot\mathbf{A}}{c}\right) \qquad (23)$$

is the charge of the particle, $\mathbf{v}$ is the particle's velocity, and c is the velocity of light.

Cyclic coordinate. If L does not depend explicitly on a particular coordinate, say q_k, then Eqs. (24)

$$\frac{d}{dt}\frac{\partial L}{\partial \dot{q}_k} = 0 \qquad \frac{\partial L}{\partial \dot{q}_k} = \text{constant} \qquad (24)$$

hold. Here q_k is called an ignorable or cyclic coordinate, and $\partial L/\partial \dot{q}_k$ is an integral of motion. In the example of the central force field given previously, ϕ may be ignored.

Total energy. By use of Eq. (19), Eq. (25) is obtained. Hence, if L does not depend explicitly on

$$\frac{d}{dt}\left(\sum_{j=1}^{f} \frac{\partial L}{\partial \dot{q}_j}\dot{q}_j - L\right) = -\frac{\partial L}{\partial t} \qquad (25)$$

the time, expression (26) is an integral of motion.

$$\sum_{j=1}^{f}\left(\frac{\partial L}{\partial q_j}q_j - L\right) \qquad (26)$$

Usually, it is the total energy of the system. *See* HAMILTON'S EQUATIONS OF MOTION.

Conjugate momentum. The quantity in Eq. (27) is defined to be the momentum conjugate to the coordinate q_k. (If a Lagrangian does not exist because of the presence of dissipative forces, the conjugate momentum is sometimes defined as $\partial T/\partial \dot{q}_k$.) This

$$p_k = \frac{\partial L}{\partial \dot{q}_k} \qquad (27)$$

momentum is not necessarily a linear momentum as defined in Newtonian mechanics, since its character depends both on the system and on the nature of the coordinate q_k. If q_k is an ignorable coordinate, its conjugate momentum is a constant of motion.

Kinetic momentum. For a charged particle in an electromagnetic field, the Lagrangian may be written as Eq. (28), and the momentum conjugate to

$$L = \tfrac{1}{2}m\mathbf{v}^2 - e\phi + \frac{e}{c}\,\mathbf{v}\cdot\mathbf{A} \qquad (28)$$

the cartesian coordinate x is given by Eq. (29). The

$$p_x = \frac{\partial L}{\partial \dot{x}} = m\dot{x} + \frac{e}{c}A_x \qquad (29)$$

quantity $m\dot{x}$ is called the kinetic momentum. The kinetic momentum is related to p_x much as the kinetic energy is related to the total energy. If ϕ and $\mathbf{A}$ are independent of x, it is the momentum p_x and not the kinetic momentum $m\dot{x}$ which is a constant of motion.

Relativistic systems. The equations of motion of a relativistic particle may be written in Lagrangian form. The simplest way to do this is to replace the kinetic energy T by another function τ of the mass and velocity so as to get the desired form, namely, Eq. (30). Here m_0 is the rest

$$\frac{d}{dt}\frac{\partial \tau}{\partial \dot{x}_k} - \frac{\partial V}{\partial x_k} = \frac{d}{dt}\left(\frac{m_0\dot{x}_k}{\sqrt{1 - v^2/c^2}}\right) - \frac{\partial V}{\partial x_k} = 0 \qquad (30)$$

mass of the particle. This is accomplished by setting τ as in Eq. (31), which reduces to T in the

$$\tau = (1 - \sqrt{1 - v^2/c^2})m_0 c^2 \qquad (31)$$

limit $\mathbf{v}/c \to 0$. The equations of motion in this form are still valid in only one reference frame because the time and the coordinates are treated on different bases. *See* RELATIVISTIC MECHANICS; RELATIVITY.   [PHILIP M. STEHLE]

Bibliography: H. C. Corben and P. Stehle, *Classical Mechanics*, 1950; H. Goldstein, *Classical Mechanics*, 1950; W. Hauser, *Introduction to the Principles of Mechanics*, 1965; R. B. Lindsay, *Concepts and Methods of Theoretical Physics*, 1951.

Lagrangian function

A function of the generalized coordinates and velocities of a dynamical system from which the equations of motion in Lagrange's form can be derived. The Lagrangian function is denoted by $L(q_1, \ldots, q_f;\ \dot{q}_1, \ldots, \dot{q}_f;\ t)$. *See* LAGRANGE'S EQUATIONS.

For systems in which the forces are derivable from a potential energy V, if the kinetic energy is T, Eq. (1) holds.

$$L = T - V \qquad (1)$$

If the system is continuous rather than discrete, the Lagrangian function L is the integral of a Lagrangian density $\mathfrak{L}$, as in Eq. (2), where $\eta(x_1,x_2,x_3)$

$$L = \int \mathfrak{L}\,(\eta, \text{grad}\,\eta, x_1, x_2, x_3, t)\,dx_1\,dx_2\,dx_3 \qquad (2)$$

describes the displacement of the medium at the point (x_1, x_2, x_3). The equations of motion are in this case written as Eq. (3).

$$\frac{\partial}{\partial t}\frac{\partial \mathfrak{L}}{\partial (\partial \eta / \partial t)} + \sum_{j=1}^{3}\frac{\partial}{\partial x_j}\frac{\partial \mathfrak{L}}{\partial (\partial \eta / \partial x_j)} - \frac{\partial \mathfrak{L}}{\partial \eta} = 0 \quad (3)$$

This formulation of Lagrange's equations applies to the motion of a gas containing sound waves, to a vibrating jelly, or to any medium where discrete masses are replaced by a continuum.

[PHILIP M. STEHLE]

Bibliography: *See* LAGRANGE'S EQUATIONS.

Lake

An inland body of water, small to moderately large in size, with its surface exposed to the atmosphere. Most lakes fill depressions below the zone of saturation in the surrounding soil and rock materials. Generically speaking, all bodies of water of this type are lakes, although small lakes usually are called ponds, tarns (in mountains), and less frequently pools or meres. The great majority of lakes have a surface area of less than 100 square miles (mi^2). More than 30 well-known lakes, however, exceed 1500 mi^2 in extent, and the largest freshwater body, Lake Superior, North America, covers 31,180 mi^2 (see table).

Most lakes are relatively shallow features of the Earth's surface. Even some of the largest lakes have maximum depths of less than 100 ft (Winnipeg, Canada; Balkash, Soviet Union; Albert, Uganda). A few, however, have maximum depths which approach those of some seas. Lake Baikal in the Soviet Union is about a mile deep at its deepest point, and Lake Tanganyika, Africa, is about 0.9 mi.

Because of their shallowness, lakes in general may be considered evanescent features of the Earth's surface, with a relatively short life in geological time. Every lake basin forms a bed onto which the sediment carried by inflowing streams is deposited. As the sediment accumulates, the

Dimensions of some major lakes

Lake	Area, mi²	Volume (approx), 1000 acre-ft	Shore-line, mi	Depth Av, ft	Depth Max, ft
Caspian Sea	169,300	71,300	3,730	675	3,080
Superior	31,180	9,700	1,860	475	1,000
Victoria	26,200	2,180	2,130		
Aral Sea	26,233*	775			
Huron	23,010	3,720	1,680		
Michigan	22,400	4,660			870
Baikal	13,300*	18,700		2,300	5,000
Tanganyika	12,700	8,100			4,700
Great Bear	11,490*		1,300		
Great Slave	11,170*		1,365		
Nyasa	11,000	6,800		900	2,310
Erie	9,940	436			
Winnipeg	9,390*		1,180		
Ontario	7,540	1,390			
Balkash	7,115				
Ladoga	7,000	745			
Chad	6,500*				
Maracaibo	4,000*				
Eyre	3,700*				
Onega	3,764	264			
Rudolf	3,475*				
Nicaragua	3,089	87			
Athabaska	3,085				
Titicaca	3,200	575			
Reindeer	2,445				

*Area fluctuates.

storage capacity of the basin is reduced, vegetation encroaches upon the shallow margins, and eventually the lake may disappear. Most lakes also have surface outlets. Except at elevations very near sea level, a stream which flows from such an outlet gradually cuts through the barrier forming the lake basin. As the level of the outlet is lowered, the capacity of the basin is also reduced and the disappearance of the lake assured.

Variations in water character. Lakes differ as to the salt content of the water and as to whether they are intermittent or permanent. Most lakes are composed of fresh water, but some are more salty than the oceans. Generally speaking, a number of water bodies which are called seas are actually salt lakes; examples are the Dead, Caspian, and Aral seas. All salt lakes are found under desert or semiarid climates, where the rate of evaporation is high enough to prevent an outflow and therefore a discharge of salts into the sea. Many lesser arid-region lakes are intermittent, sometimes existing only for a short period after heavy rains and disappearing under intense evaporation. These lakes are called playas in North America, shotts in North Africa, and other names elsewhere. In such regions the surface area and volume of permanent lakes may differ enormously from wet to dry season. *See* EVAPORITE, SALINE; PLAYA.

The water of the more permanent salt lakes differs greatly in the degree of salinity and the type of salts dissolved. Compared to typical ocean water (approximately 35 parts per thousand, ‰) some salt lakes are very salty. Great Salt Lake water has a dissolved solids content about four times that of sea water (150‰) and the Dead Sea about seven times (246‰). Some of the larger salt lakes have a much lower dissolved solids content, as the Aral Sea (11‰) and the Caspian Sea (6‰). The composition of salts depends in part on the geological character of the drainage area discharging into the lake, in part on the age of the lake, and in part on the excess of evaporation over inflow. As saturation is approached, the salts common in surface waters are precipitated in such an order that magnesium chloride and calcium chloride remain in solution after other salts have precipitated.

Lakes with fresh waters also differ greatly in the composition of their waters. Because of the balance between inflow and outflow, fresh lake water composition tends to assume the composite dissolved solids characteristics of the waters of the inflowing streams—with the lake's age having very little influence. Lakes with a sluggish inflow, particularly where inflowing waters have much contact with marginal vegetation, tend to have waters with high organic content. This may be observed in small lakes or ponds in a region where drainage moves through a topography of glacial moraines. Lakes formed within drainage areas having a crystalline, metamorphic, or volcanic country rock tend to have low dissolved solids content. Thus Lake Superior, with its major drainage from the Laurentian Shield, has a dissolved solids content of 0.05‰. The water of Grimsel Lake in the high Alps, Switzerland, has a dissolved solids content of only 0.0085‰. Lakes within limestone or dolomitic drainage areas have a pronounced calcium carbonate and magnesium carbonate content. As in all surface water, dissolved gases, notably oxygen, also are present in lake waters. Under a few special situations, as crater lakes in volcanic areas, sulfur or other gases may be present in lake water, influencing color, taste, and chemical reaction of the water. *See* FRESH-WATER ECOSYSTEM; HYDROSPHERE, GEOCHEMISTRY OF; LAKE, MEROMICTIC; SURFACE WATER.

Basin and regional factors. Most lakes are natural, and a large proportion of them lie in depressions of glacial origin. Thus alpine locations and regions with ground moraine or glacially eroded exposures are the sites of many of the world's lakes. The lakes of Switzerland, Minnesota, and Finland are illustrations of these types.

Lakes may be formed in depressions of differing glacial origin: (1) terrain eroded by continental glaciers, with the surface differentially deepened by ice abrasion of rocks of varying hardness and resistance; (2) valleys eroded differentially by valley glaciers; (3) cirques (glacially eroded valley heads in mountains); (4) lateral moraine barriers; (5) frontal moraine barriers; (6) valley glacier barriers; (7) irregularities in the deposition of glacial drift or ground moraine. *See* GLACIATED TERRAIN.

Lakes are particularly important surface features in the peneplaned ancient rocks of the Laurentian Shield and on the Fennoscandian Shield of northern Europe. The lake region of northern North America, which centers on the Laurentian Shield, probably has one-fifth to one-quarter of its surface in lake. Many streams in these areas are interrupted over more than half their total length by lakes. Lakes on the shields as a rule are island-studded and have extremely irregular outlines. The most permanent lakes lie on the shields themselves. Many such lakes on recently ice-scoured shields have fresh hard-rock rims with high resistance to erosion at their outlets. Because of generally low stream gradients and little sediment carried, the abrading and depositing stream actions are slow. As a result, the life of all but the small lakes in these areas probably will be measured in terms of a whole geologic period.

Some lakes in glacially formed depressions, as well as in other basins, may be considered barrier lakes. These glacial lakes are formed on glacial drift behind lateral or frontal moraine barriers. In addition, depressions of sufficient depth to contain a lake may be formed by (1) sediment deposited by streams (alluvium), and also stream-borne vegetative debris, such as tree dams on the distributaries and braided river courses of the lower Mississippi Valley; (2) landslides in mountainous areas; (3) sand dunes; (4) storm beaches and current-borne sediments along the shores of large bodies of water; (5) lava flows; (6) artificial or man-made barriers.

A large percentage of lakes is found in either the glacially formed or barrier-formed depressions. However, a few other types of depressions contain lakes: (1) craters of inactive volcanoes, or calderas (Crater Lake in Oregon is a famous example); (2) depressions of tectonic or structural origin (Great Rift Valley of Africa includes Lakes Albert, Tanganyika, and Nyasa); (3) solution cavities in limestone country rock; (4) shallow depressions cause a dotting of lakes in many parts of the tundra of high latitudes.

Conservation and economic aspects. Lakes created behind manmade barriers are becoming common features and serve multiple purposes. Examples are Lake Mead behind Hoover Dam on the Colorado River, Lake Roosevelt behind Grand Coulee Dam on the Columbia, Kentucky Lake and other lakes of the Tennessee Valley, and Lake Tsimlyanskaya on the Don.

Both natural and manmade lakes are economically significant for their storage of water, regulation of stream flow, adaptability to navigation, and recreational attractiveness. A few salt lakes are significant sources of minerals. Recreational utility, long important in the alpine region of Europe and in Japan, is now a major economic attribute of many American lakes. Economic value is generally increased by location near substantial human settlement. Most of the world's lakes, however, are located in regions where they have only minor economic significance at present. *See* EUTROPHICATION, CULTURAL.   [EDWARD A. ACKERMAN]

Bibliography: R. Gresswell (ed.), *Standard Encyclopedia of the World's Rivers and Lakes*, 1966; International Association for Great Lakes Research, *Proceedings of the 10th Conference on Great Lakes Research*, 1967; P. H. Kuenen, *Realms of Water*, 1955; R. H. Lowe-McConnell (ed.), *Man Made Lakes*, 1965.

Lake, meromictic

A lake whose water is permanently stratified and therefore does not circulate completely throughout the basin at any time during the year. In the temperate zone this permanent stratification occurs because of a vertical, chemically produced density gradient. There are no periods of overturn, or complete mixing, since seasonal fluctuations in the thermal gradient are overridden by the stability of the chemical gradient.

The upper stratum of water in a meromictic lake is mixed by the wind and is called the mixolimnion. The bottom, denser stratum, which does not mix with the water above, is referred to as the monimolimnion. The transition layer between these chemically stratified strata is called the chemocline (Fig. 1).

In general, meromictic lakes in North America are restricted to (1) sheltered basins that are proportionally very small in relation to depth and that often contain colored water, (2) basins in arid regions, and (3) isolated basins in fiords. *See* LAKE.

Prior to 1960 only 11 meromictic lakes had been reported for North America. However, during 1960–1968 that number was more than doubled. There are seven known meromictic lakes in the

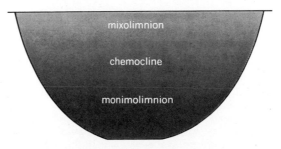

Fig. 1. Cross-sectional diagram of a meromictic lake.

state of Washington, six in Wisconsin, four in New York, three in Alaska, two in Michigan, and one each in Florida, Nevada, British Columbia, Labrador, and on Baffin Island. Research activity on meromictic lakes has been focused on studies of biogeochemistry, deepwater circulation, and heat flow through the bottom sediments.

Chemical studies. Very few detailed chemical studies have been done on meromictic lakes, particularly on a seasonal basis. The most typical chemical characteristic of meromictic lakes is the absence of dissolved oxygen in the monimolimnion. Large quantities of hydrogen sulfide and ammonia may be associated with this anaerobic condition in deep water. J. Kjensmo proposed that the accumulation of ferrous bicarbonate in the deepest layers of some lakes may have initiated meromixis under certain conditions.

Meromictic lakes are exciting model systems for many important biogeochemical studies. The isolation of the monimolimnetic water makes these studies quite interesting and important. E. S. Deevey, N. Nakai, and M. Stuiver studied the biological fractionation of sulfur and carbon isotopes in the monimolimnion of Fayetteville Green Lake, N.Y. They found the fractionation factor for sulfur to be the highest ever observed. T. Takahashi, W. S. Broecher, Y. Li, and D. L. Thurber have undertaken detailed, comprehensive geochemical and hydrological studies of meromictic lakes in New York.

Sediments from meromictic lakes are among the best for studies of lake history, since there is little decomposition of biogenic materials. G. J. Brunskill and S. D. Ludlam have obtained information on seasonal carbonate chemistry, sedimentation rates, and sediment composition and structure in Fayetteville Green Lake. Also, deuterium has been used by several workers in an attempt to unravel the history of meromictic lake water.

Physical studies. Radioactive tracers have been used by G. E. Likens and A. D. Hasler to show that the monimolimnetic water in a small meromictic lake (Stewart's Dark Lake) in Wisconsin is not stagnant but undergoes significant horizontal movement (Fig. 2). While the maximum radial spread was about 16–18 m/day, vertical movements were restricted to negligible amounts because of the strong density gradient. W. T. Edmondson reported a similar pattern of movement in the deep water of a larger meromictic lake (Soap Lake) in Washington. V. W. Driggers determined from these studies that the average horizontal eddy diffusion coefficient is 3.2 cm²/sec in the monimolimnion of Soap Lake and 17 cm²/sec in Stewart's Dark Lake.

Because vertical mixing is restricted in a meromictic lake, heat from solar radiation may be trapped in the monimolimnion and can thereby produce an anomalous temperature profile. G. C. Anderson reported monimolimnitic water temperatures as high as 50.5°C at a depth of 2 m in shallow Hot Lake, Wash.

N. M. Johnson and Likens pointed out that some meromictic lakes are very convenient for studies of geothermal heat flow because deepwater temperatures may be nearly constant. Studies of terrestrial heat flow have been made in the sediments of Stewart's Dark Lake, Wis., by Johnson and Li-

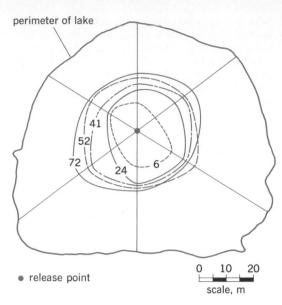

Fig. 2.   The outlines show the maximum horizontal displacement of a radiotracer (sodium-24) following its release at the 8-m depth in Stewart's Dark Lake, Wis. The numbers indicate the hours elapsed after release. (*After A. D. Hasler and G. E. Likens*)

kens and in Fayetteville Green Lake by W. H. Diment. Steady-state thermal conditions were found in the sediments of Stewart's Dark Lake, and the total heat flow was calculated to be 2.1×10^{-6} cal/cm² sec. However, about one-half of this flux was attributed to the temperature contrasts between the rim and the central portion of the lake's basin. R. McGaw found that the thermal conductivity of the surface sediments in the center of Stewart's Dark Lake was 1.10×10^{-3} cal cm/cm² sec °C, a value substantially lower than that for pure water at the same temperature but consistent with measurements on colloidal gels.

Biological studies. Relatively few kinds of organisms can survive in the rigorous, chemically reduced environment of the monimolimnion. However, anaerobic bacteria and larvae of the phantom midge (*Chaoborus* sp.) are common members of this specialized community. Using an echo sounder, K. Malueg has been able to observe vertical migrations of *Chaoborus* sp. larvae in meromictic lakes. The sound waves are reflected by small gas bladders on the dorsal surface of the larvae. The migration pattern is similar to that shown by the deep-scattering layers in the sea. The larvae come into the surface waters at night when the light intensity is low and sink into deeper waters during the daylight hours. Hasler and Likens found that biologically significant quantities of a radioisotope (iodine-131) could be transported from the deep and relatively inaccessible part of a meromictic lake to the surface and thence to the adjacent terrestrial environment by these organisms. The radiotracer appeared in flying adult *Chaoborus* sp. along the shoreline of the lake within 20 days after it had been released within the region of the chemocline. *See* FRESH-WATER ECOSYSTEM; LIMNOLOGY.

[GENE E. LIKENS]

Bibliography: E. S. Deevey, N. Nakai, and M. Stuiver, Fractionation of sulfur and carbon isotopes in a meromictic lake, *Science*, 139:407–408, 1963; W. T. Edmonson and G. C. Anderson, Some features of saline lakes in central Washington, *Limnol. Oceanogr.*, suppl. to vol. 10, pp. R87–R96, 1965; D. G. Frey (ed.), *Limnology in North America*, 1963; G. E. Hutchinson, *A Treatise on Limnology*, vol. 1, 1957; D. Jackson (ed.), *Some Aspects of Meromixis*, 1967; N. M. Johnson and G. E. Likens, Steady-state thermal gradient in the sediments of a meromictic lake, *J. Geophys. Res.*, 72:3049–3052, 1967; J. Kjensmo, Iron as the primary factor rendering lakes meromictic, and related problems, *Mitt. Int. Ver. Limnol.*, 14:83–93, 1968; G. E. Likens and P. L. Johnson, A chemically stratified lake in Alaska, *Science*, 153:875–877, 1966; F. Ruttner, *Fundamentals of Limnology*, 3d ed., 1963; K. Stewart, K. W. Malueg, and P. E. Sager, Comparative winter studies on dimictic and meromictic lakes, *Verh. Int. Ver. Limnol.*, 16:47–57, 1966.

Lambert

A unit of luminance (photometric brightness) that is equal to $1/\pi$ candela per square centimeter. It is also defined as the uniform luminance of a perfectly diffusing surface emitting or reflecting light at the rate of 1 lumen per square centimeter. The lambert, a large unit, is satisfactory for expressing high values of luminance from bright light sources. For more moderate values of luminance, as from fluorescent lamps or reflecting surfaces, the millilambert (0.001 lambert) or the footlambert is generally used.

The luminance B' in lamberts is given by the equation below, where I is the luminous intensity

$$B' = \pi \, dI/(dA \cos \theta)$$

in candelas, A is the surface area, and θ is the angle between the normal to the surface and the line of sight. *See* LUMINANCE; PHOTOMETRY.

[RUSSELL C. PUTNAM]

Lamiales

An order of flowering plants, division Magnoliophyta (Angiospermae), in the subclass Asteridae of the class Magnoliopsida (dicotyledons). The order consists of 5 families and more than 7800 species: Labiatae (about 3200 species), Verbenaceae (about 2600 species), and Boraginaceae (about 2000 species) make up the bulk of the order. Within its subclass the order is marked by its characteristic gynoecium, consisting of usually two biovulate carpels, each carpel divided between the ovules by a "false" partition, or with the two halves of the carpel seemingly wholly separate. Except in some of the more primitive species, the fruit usually consists of four separate or separating nutlets. In the Labiatae and most of the Boraginaceae, the four segments of the ovary are nearly or quite distinct from each other and are united chiefly by the gynobasic style. *See* FLOWER.

Boraginaceae. Members of this family, making up about 100 genera, are widely distributed throughout the world but are especially abundant in the Mediterranean region and in western North America. Most species are herbs, but some of the tropical ones are trees. Within the order the family is characterized by its mostly alternate and entire leaves and mostly regular flowers with five stamens. Forget-me-not (*Myosotis*), heliotrope

Monarda fistulosa, an eastern North American species of wild bergamot, showing the characteristic square stem, opposite leaves, and irregular corolla of the family Labiatae, the largest family of the order Lamiales. (*Courtesy of A. W. Ambler, National Audubon Society*)

(*Heliotropium*), and hound's-tongue are familiar members of the Boraginaceae.

Verbenaceae. This family, also with about 100 genera, is most abundant in tropical and subtropical regions, and is also well developed in temperate regions of the Southern Hemisphere; relatively few species occur in temperate regions. The Verbenaceae have opposite or whorled leaves and regular or irregular flowers, usually with four or two functional stamens. The ovary is only slightly, if at all, lobed at the summit, so that the style appears to be terminal. The plants are variously woody or herbaceous and seldom notably aromatic. The garden *Lantana* is a familiar member of the Verbenaceae. Teakwood is obtained from *Tectona grandis*, another member of the family.

Labiatae. This family, with approximately 200 genera, is widely distributed throughout the world, but is especially well represented in the Mediterranean region and adjacent Asia. The Labiatae, also known as Lamiaceae or Menthaceae, are clearly derived from the Verbenaceae, and the distinction between the two families is not sharp, but the flowers of the Labiatae are more consistently irregular (see illustration) and the ovary is more or less deeply four-cleft, so that the style is gynobasic or nearly so. The plants are typically aromatic and usually herbaceous or merely shrubby. Catnip (*Nepeta cataria*), horehound (*Marrubium vulgare*), and especially the various species of mint (*Mentha*) are familiar members of the Labiatae. *See* ASTERIDAE; MAGNOLIOPHYTA; MAGNOLIOPSIDA; PLANT KINGDOM.   [ARTHUR CRONQUIST]

Laminar flow

Streamline flow of a viscous fluid which satisfies the Navier-Stokes equations of motion. In laminar flow, the fluid moves in layers without large irregular fluctuations. Laminar flow occurs at a low Reynolds number. This corresponds to the conditions of small velocities and dimensions of bodies, to very large viscosity, or to small density of the fluid. Laminar flow plays an important role in several practical problems.

The flow of oil in the bearings for lubrication is laminar. The theory of laminar flow shows that under great normal pressure, the oil in the bearing has only slight frictional resistance. *See* TURBULENT FLOW.

The motion of a minute particle in a viscous fluid produces laminar flow. The drag coefficient of such a body is inversely proportional to its Reynolds number.

Flow on the surface of modern aircraft and missiles flying at extremely high altitude may be laminar. The Reynolds number for this case is usually moderate, and the viscous effect is confined to the boundary layer region of the body. Laminar boundary-layer flow determines the skin friction and the aerodynamic heating of these bodies. *See* NAVIER-STOKES EQUATIONS; REYNOLDS NUMBER.

[SHIH I. PAI]

Bibliography: F. K. Moore, *Theory of Laminar Flows*, 1964; S. I. Pai, *Viscous Flow Theory*, vol. 1, 1956; V. L. Streeter (ed.), *Handbook of Fluid Dynamics*, 1961.

Lamp, electric

A device for converting electrical energy into illumination. In the common incandescent lamp a resistance wire, such as tungsten, is heated to incandescence by an electric current. In a vapor lamp the passage of electricity through mercury vapor or sodium vapor serves to ionize the gas and produce a brilliant visible glow discharge. Inert gases may be used in place of vapors to give other colors of light, as in neon lamps and in luminous tubing for advertising signs. A fluorescent lamp is a type of vapor lamp in which the radiation from ionized mercury vapor is converted into a more suitable white light by fluorescent coating on the inside of the glass tubing. In an arc lamp the light is produced by an electric arc passing through the space between two electrodes. *See* ARC LAMP; FLUORESCENT LAMP; INCANDESCENT LAMP; MERCURY-VAPOR LAMP; VAPOR LAMP.   [JOHN MARKUS]

Lampridiformes

An order of teleost fishes including the ribbonfishes, oarfishes, opah, and their allies, also known as the Allotriognathi. Although these fishes are diverse in form, they share characters that define this order of actinopterygian fishes, which appears to be related to the Beryciformes. In most, the body is notably compressed, often ribbonlike (see illustration). The fins are composed of soft rays, or the dorsal has one or two anterior spines; an orbitosphenoid bone is present; the pelvic fin, if present, is thoracic, with 1–17 spineless rays; the pelvic girdle is inserted between the coracoids, or attached to them, or is lost; scales are small and cycloid or, commonly, absent; the swim bladder is without a duct; mesocoracoid, intercalar, and subocular shelf are absent; and the maxillae are protractile, a distinctive feature. Most species are large and colorful; some furnish the source material of sea serpent stories. *See* BERYCIFORMES.

The lampridiform fishes date from the Cretaceous. They comprise 4 suborders, 7 families, 12 Recent genera, and 21 or more species. These rare fishes are all oceanic, some living at shallow

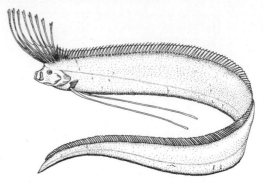

Oarfish (*Regalecus glesne*), length to over 20 ft. (*After D. S. Jordan and B. W. Evermann, The Fishes of North and Middle America, U.S. Nat. Mus. Bull. no. 47, 1900*)

depths but most inhabiting the mesopelagic zone. *See* ACTINOPTERYGII.

[REEVE M. BAILEY]

Bibliography: V. Walters and J. E. Fitch, *The Families and Genera of the Lampridiform (Allotriognath) Suborder Trachipteroidei*, in *California Fish and Game*, vol. 46, no. 4, 1960.

Lamprophyre

Any of a group of igneous rocks characterized by a porphyritic texture in which abundant large crystals (phenocrysts) of dark-colored (mafic) minerals (biotite, amphibole, pyroxene, or olivine) appear set in an aphanitic (not visibly crystalline) matrix. As a group, these dark rocks are characterized by (1) an abundance of mafic minerals in association with alkali-rich feldspar, (2) the presence of mafics both as phenocrysts and in the matrix, and (3) an abundance of mafic phenocrysts in the absence of feldspar phenocrysts. Lamprophyres are chemically unusual. They have a low silica content and a high iron, magnesium, and alkali content. *See* PHENOCRYST.

Varieties. Many varieties of lamprophyre are known, but only the more common are shown in the table.

Minette, vogesite, kersantite, and spessartite. The four most common types are minette, vogesite, kersantite, and spessartite. The first two are commonly referred to as syenitic lamprophyres, the last two as dioritic lamprophyres. Under the microscope large hexagonal plates of biotite show zonal structure with pale yellow, magnesium-rich centers and red-brown iron-rich borders. In some rocks these crystals are corroded. Slender prisms or needles of green or brown hornblende are common. Pyroxene is generally diopsidic augite, but in some rocks a titanium-rich augite is abundant. Olivine is rare in most types but is common in spessartite.

A large proportion of the rock matrix is com-

Common lamprophyres

Principal or diagnostic mafic	Principal feldspar		Without feldspar	
	Alkali feldspar	Plagioclase	With olivine	Without olivine
Biotite	Minette	Kersantite	Alnoite	Ouachitite
Hornblende	Vogesite	Spessartite		
Barkevikite		Camptonite	Monchiquite	Fourchite

posed of feldspar. Plagioclase (albite-andesine) forms irregular grains or poorly developed laths and may show zonal structure with calcium-rich cores and sodium-rich borders. Potash feldspar, usually orthoclase or sanidine, occurs as irregular to rectangular grains and commonly encloses other minerals as abundant tiny grains. Quartz may be a minor constituent and, when present, is interstitial. Accessory minerals include apatite, sphene, and magnetite.

Camptonite. The uncommon lamprophyre camptonite is characterized by the presence of barkevikite and labradorite or andesine feldspar. Pyroxene, biotite, and olivine may or may not be present.

Alnoite and ouachitite. These varieties are rare biotite lamprophyres. They are feldspar-free but carry melilite or feldspathoid. Alnoite is characterized by lepidomelane phenocrysts and by the presence of melilite, perovskite, olivine, and carbonate in the matrix. Ouachitite is devoid of olivine and may carry more or less glass and considerable augite as phenocrysts.

Monchiquite and fourchite. The rare types monchiquite and fourchite lack feldspar but carry more or less barkevikite in addition to augite, biotite, analcite, and glass. Monchiquite is distinguished from fourchite largely by the presence of olivine.

Alteration products. Lamprophyres are highly susceptible to weathering, and many are so completely decomposed that it is impossible to do more than approximate their original mineral composition. Common products of alteration include carbonate, chlorite, serpentine, and limonite.

Occurrence. Lamprophyres occur most commonly in small or shallow intrusives (dikes, sills, and plugs) and are frequently associated with large bodies of granite and diorite. Lamprophyre dikes may form parallel swarms or, as at Spanish Peaks, Colo., may form groups which radiate from a common center.

Formation. Lamprophyres form in a variety of ways. Some are products of direct crystallization of lamprophyric magma (rock melt); others represent older rock which has been converted by metamorphic or metasomatic action. Normal basaltic magma may be made lamprophyric by assimilation of foreign material. That such a process has operated is suggested by the presence, in some lamprophyres, of abundant foreign rock and mineral fragments. The mafic phenocrysts commonly show strong resorption, indicating they were not in equilibrium with the adjacent liquid. Early formed mafic crystals may settle out of a deep, slowly crystallizing magma. Clusters of these may be reincorporated in late, alkali-rich fraction of the melt, just before it is erupted, to form lamprophyres at higher levels.

Some normal basaltic dikes appear to have been transformed to lamprophyre after solidification. This metamorphic or metasomatic change could have been accomplished by vapors or fluids which were driven out from the deeper crystallizing portions to permeate and alter the solidified portions above. Similar emanations from deeply buried granitic masses may be channeled along dikes of basalt or diabase in the overlying rocks and convert them to lamprophyres.

Some bodies of lamprophyre which resemble dikes may not actually be intrusive. They may have formed when solution or fluids from depth moved up along fractures and reacted with the adjacent rock and converted it to lamprophyre. *See* DIABASE; IGNEOUS ROCKS; METASOMATISM.

[CARLETON A. CHAPMAN]

Lancefield differentiation scheme

An accurate means of identifying most streptococci. The procedure was determined by Rebecca C. Lancefield, and it applies to those streptococci, the β-hemolytic types, which have major significance. Strains from many sources can be classified in terms of their natural host with ability to produce disease in this host as well as in other hosts. However, the immunochemical (serologic) differentiation system of Lancefield depends on the presence within the bacterial cell of a specific carbohydrate, the so-called C substance, which determines the group. The definition of the group depends on the precipitation of the C substance in a clear bacterial extract with appropriate rabbit antisera. Designations of the groups, now 13 in number, extend from A through O, with two alphabetic omissions.

Those streptococci of group A are almost always responsible only for human illness. Most strains of group B are from animals and are especially important in producing mastitis, or udder infection, in cows. Group C is unusual in that it includes both animal and human pathogens and distinction cannot be made by this method alone. Group D streptococci are found in dairy products and in the intestinal tracts of man and animals. They are a cause of urinary tract infection and of subacute bacterial endocarditis, which is an infection of the inner lining of the heart, especially the heart valves. Subacute bacterial endocarditis is a complication of congenital or rheumatic heart disease. Both human diseases represent the ability of ordinarily benign organisms to be pathogenic when there is some basic deformity which interferes with normal function. Streptococci of groups E and N are found in milk, cream, and cheese but have no relation to infection. Groups F and G are occasionally human pathogens, whereas H and K strains do not provoke illness. People may harbor streptococci of groups H, K, and O, which are almost uniformly innocuous. Dogs likewise merely carry group M streptococci, but canine sickness can be caused by strains of groups L and G.

F. Griffith, at the same time the Lancefield system was developed, utilized agglutination to define types within each group. Agglutination is a serologic method in which intact organisms are clumped by appropriate typed rabbit antisera. There was a time when both methods were used, precipitin test for the determination of group, the Lancefield and Griffith's agglutination test for type selection. It was later noted that the type found by both methods need not be in complete agreement. This led to the study by Dr. Lancefield and her associates of the bacterial antigen or chemical substance concerned. It was found that two proteins, M and T, were involved in differentiating types within groups. Only the M substance of the Lancefield method was specific for each type of microorganism while more than one T substance could be present in a single type.

Serologic classification is not commonly employed since it is time-consuming and not always necessary. In a clinical hospital laboratory, for example, streptococci can be satisfactorily and rapidly recognized by their typical appearance on suitable growth mediums. Appropriate signs and symptoms in the patient are supportive evidence. Specific treatment is quite as effective as with the finer method of identification. There are times when streptococci must be labeled. These occasions include epidemics and studies on the reaction pattern of the host, especially such an undesirable one as rheumatic fever, where prevention is sought. For such research studies the Lancefield serologic techniques are of great value. *See* MASTITIS; RHEUMATIC FEVER; SCARLET FEVER; STREPTOCOCCUS. [PAUL L. BOISVERT]

Land drainage (agriculture)

The removal of water from land to improve the soil as a medium for plant growth and a surface for crop management operations. Water in excess of that needed by the plants may inhibit growth or the production of the economically important portion of the plant. High water content also lubricates the soil particles and frequently leads to unstable conditions unsuitable to machine and other crop operations. Drainage needs, or the amount of excess water, therefore, varies depending upon the soil, the demands of the crop, and the stability needs of the management practices. If the crops are water-tolerant and only light equipment is needed to manage the crop, the water excess may be small, but for an identical location where either the crop is not tolerant or the management practices place heavy loads on the soil, the water excess may be great.

Excess water creates problems in agricultural production over vast areas. Estimates of the acreage in need of drainage in the United States vary widely. G. D. Schwab stated that 22% of the total cropland, or 94,000,000 acres, has a dominant drainage problem, and Q. C. Ayres indicated that reclamation by drainage would be a benefit on about 216,000,000, or about 24% of all potential agricultural land in the United States. Ayres also

indicated that about 33,000,000 acres have already been drained in the humid regions, and about 17,000,000 acres at least partly drained in the irrigated lands of the arid western states. Similar drainage problems exist in other countries.

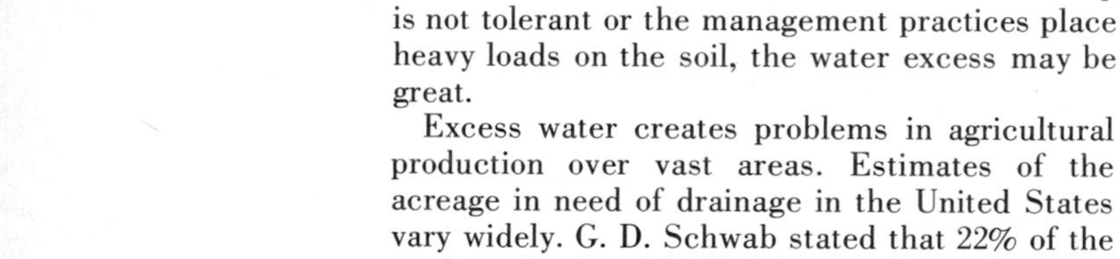

Water source and disposal. The excess water may be due to rainfall overflow from streams, swamps, or other bodies of water; seepage or runoff from higher areas; or irrigation. The source of water should be identified before a solution is proposed. In general, the solution must fit the soil, the topography, the source of water, the crops being grown, and the management scheme used, including machinery, and must be economically feasible. Obviously, no one ideal solution exists, but a range of solutions may be proposed which vary in advantage with the individual situation.

Before discussing drainage systems, water disposal must be considered. All drainage systems must have an outlet for disposal of the water collected (Fig. 1); the outlet places restrictions on the type of system that may be used. A good outlet is low enough to permit water removal from the lowest area needing drainage; is stable, neither eroding nor filling rapidly; and is capable of accepting all design flows. Such an outlet is not always easily found; frequently deficiencies in the outlet must be corrected before a drainage system may be designed. In general, drainage outlets may be either natural or artificial, and the water may flow naturally by gravity or be moved by pumps; again, many combinations are possible.

Methods. There are two basic methods of draining land, and these may be combined to form a third. The first method, surface drainage, attempts to remove excess water before it enters the soil; the second, subsurface drainage, attempts to remove it after it is within the soil. When both are used, the system is called combined drainage. In practice, surface drainage is difficult to isolate from subsurface flow or vice versa, but separation is useful for a discussion of principles.

Surface drainage. This is usually accomplished by using shallow (less than 2 ft deep) open ditches to collect the surface water; the land surface, either between or along the ditches, is either graded or smoothed, or both, to promote movement of water into the ditches (Fig. 2). The ditches are constructed so that they slope toward a collector ditch or the outlet, and water flows naturally down the slope. Surface systems are usually less costly than subsurface or combined systems, and because the ditches are shallow, an adequate outlet is easier to find. Heavy soils that are slow to absorb rainfall, soils that are shallow over impermeable layers, or drainage problems due to surface flow are ideally suited to surface drainage.

Subsurface drainage. This is usually accomplished by burying conduits within the soil. They are buried so that they slope toward a collector or the outlet, and flow is the result of gravity. Outlets for subsurface systems must be lower in elevation than outlets for surface systems and thus are more difficult to locate. Conduits must be buried at least 2 ft below the soil surface to prevent damage by machines traveling over the surface. They must also be buried deep enough to promote water movement toward the drains at a rapid enough rate

Fig. 1. Parts of surface-drainage system. (*From Engineering Handbook for Work Unit Staffs, USDA Soil Conservation Service, 1964*)

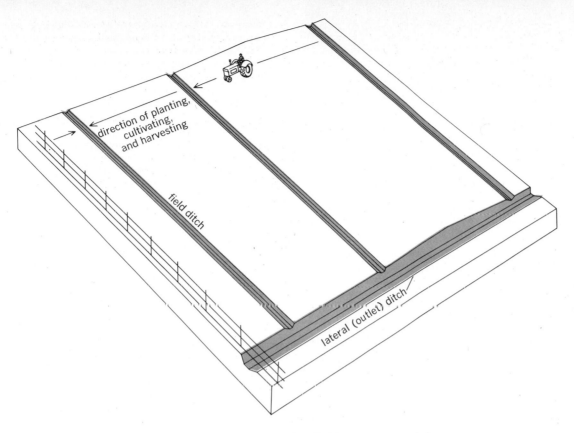

typical cross section of ground surface that has some general slope
in one direction and is covered with many small depressions and pockets

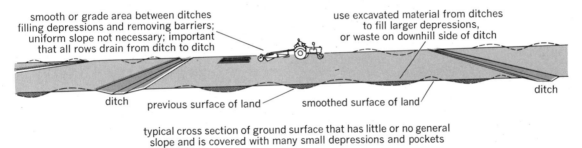

smooth or grade area between ditches
filling depressions and removing barriers;
uniform slope not necessary; important
that all rows drain from ditch to ditch

use excavated material from ditches
to fill larger depressions,
or waste on downhill side of ditch

ditch   previous surface of land   smoothed surface of land   ditch

typical cross section of ground surface that has little or no general
slope and is covered with many small depressions and pockets

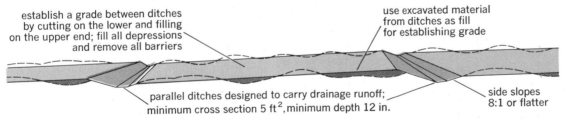

establish a grade between ditches
by cutting on the lower and filling
on the upper end; fill all depressions
and remove all barriers

use excavated material
from ditches as fill
for establishing grade

parallel ditches designed to carry drainage runoff;
minimum cross section 5 ft^2, minimum depth 12 in.

side slopes
8:1 or flatter

Fig. 2. Shallow-ditch system for surface drainage. Field ditches should be about parallel but not necessarily equidistant; and the outlet should be about 1 ft deeper than field ditches. It is necessary to clean ditches after each farming operation.

to prevent damage to crops and stabilize the soil for machinery operation. The water movement within the soil toward a drainage conduit is primarily by gravity; thus, unless the ability of the soil to conduct water is restricted as depth increases, a deeper drain will provide more rapid drainage over a wider area than will a shallower drain. Soils which are deep and permit rapid movement of water into and through the soil can be said to be ideally suited to subsurface drainage.

Practical applications. Ideal conditions suited exclusively to either surface or subsurface drainage are rare. Most systems operate, by either design or nature, as combined drainage systems. Ditches, even shallow ones, promote some subsurface drainage, and water in excess of that which may infiltrate the soil frequently flows over the surface to some outlet.

Drainage-system patterns. Patterns for drainage systems are of two general types. Where the

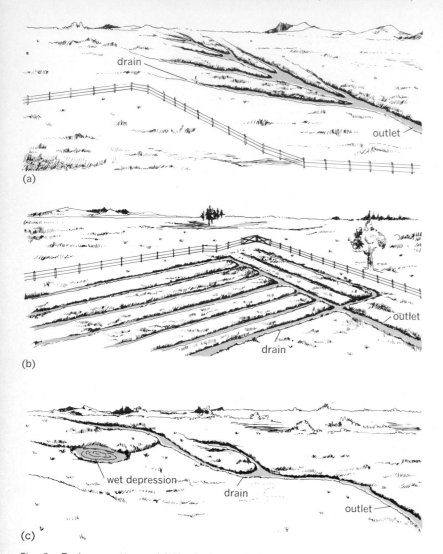

(a)

(b)

(c)

Fig. 3.   Drainage patterns. (a) Herringbone. (b) Parallel. (c) Random. (*From Engineering Handbook for Work Unit Staffs, USDA Soil Conservation Service, 1964*)

drainage problem is general over an area, drainage channels may be provided at regular intervals (Fig. 3a and b); where the problem exists in isolated areas within a larger area, channels may be provided at random to include only those areas which need drainage (Fig. 3c). Random systems are usually less expensive to install, and a high proportion of drainage problems first appear as isolatable areas within a larger block. As time passes, however, the second-wettest areas become the limiting factor, and random systems are extended until they look like regularly spaced systems. If expansion at a future date is considered in the initial design, the first cost is increased, but the final system is adequate. Conversely, if expansion is not considered, the initial cost may be much lower, but much of the system may have to be replaced or avoided in order to improve it in the future.

The quantity of water which must be removed by a drainage system is variable and based primarily upon experience. Most drainage system designs permit temporary flooding and require that the system dispose of a quantity of water expressed in inches of depth over the area to be drained in a period of 24 hr (called drainage coefficient). Increased protection is provided in design by increasing the drainage coefficient.

Systems construction. Most drainage systems are constructed with power equipment. Scrapers, graders, bulldozers, draglines, backhoes, plows, and special trenching machines are used to dig open ditches, and the land surface may be shaped and smoothed with some of the same equipment. There are also special machines, land levelers, for smoothing the land. Surface drainage channels are designed with gentle side slopes to permit machines and equipment to cross easily. The earth removed from the channels is deposited where it will not interfere with drainage. Soil from the channels and high spots in the field is used to fill holes or depressions, or it may be used to raise the level of the ground surface to create increased slope into the channel (Fig. 2).

Subsurface drains are of two types. The most common type consists of buried pipes. Special pipes, made of ceramic materials, concrete, plastic, bituminous impregnated paper, or zinc-coated steel, are constructed for subsurface drainage. Openings are provided into the pipes by holes and slots cut through the pipe walls or by space left between pipe sections and, in rare cases, by permeable wall materials. A machine digs a trench to the required depth with the bottom sloping toward the outlet. The pipes are then placed into the trench and the earth returned over the pipe. In particular soils, special materials may be placed immediately over or around the pipes to prevent soil particles from entering the opening or to promote more rapid drainage. Gravel, sand, fiber glass sheets, and organic materials such as corncobs, hay, grass, or sawdust are used for this purpose. These materials are more open than the surrounding soil but present smaller openings than those present in the pipes. They also contact a larger area of soil than the openings in the pipe and thus promote better drainage.

In some soils, subsurface channels may be provided by pulling a solid object through the soil at the proper slope and depth. This type of drain is called a mole drain and the object used is called a mole. The mole is usually a steel cylinder formed to a wedge-shaped point on one end and attached to a chain or metal plate. The mole is placed in the outlet and then pulled through the soil. The mole is shaped so that it will stay in the ground at a fixed depth; therefore, the land surface must have the desired slope. If the soil has sufficient clay for a binder and enough silt, sand or stones for stability, the channel created will stay open for several years and forms an inexpensive method of subdrainage.

Systems maintenance. All drainage systems require care in design, construction, and maintenance. Erosion or silting may occur in open ditches. Vegetation such as grass may be used along or within the ditch to stabilize the soil against either erosion or siltation, but the grass must be mowed, fertilized, and occasionally replaced. Outlets from subsurface drains may erode, removing support from around the outlets and permitting the pipe to break away, causing further erosion. Animals must be prevented from entering the pipes, and occasionally pipe sections break or collapse and require replacement. Surface entrances into buried conduits may become plugged and need cleaning; roots of perennial plants, for instance, may enter and clog a pipe.

A surface drainage system should last at least 10 years without major improvement if yearly maintenance is provided. Subsurface drainage systems should last 50 years or more if the pipe material is durable. Adequate agricultural drainage, however, is not static. As crops, management, and machines change, different demands are placed upon drainage systems and changes must be made. The soils, crops, water, and technology involved in drainage are dynamic, and the assistance of specialists is needed in devising well-designed drainage systems. The investment in drainage systems is frequently as great or greater than the original price of agricultural land. Such designs should, therefore, receive careful attention. *See* AGRICULTURE, SOIL AND CROP PRACTICES IN; IRRIGATION OF CROPS; TERRACING (AGRICULTURE).

[RICHARD D. BLACK]

Bibliography: J. N. Luthin, *Drainage Engineering*, 1966; J. N. Luthin, *Drainage of Agricultural Lands*, 1957; G. O. Schwab et al., *Soil and Water Conservation Engineering*, 1966.

Landing gear

Those parts of an aircraft structure which serve to support the aircraft when it is not in the air. Tires, wheels, brakes, shock struts, drag struts, and miscellaneous equipment, such as retracting mechanisms, steering mechanisms, shimmy dampers, and doors, are the components of an aircraft landing gear. For special purposes the wheels and tires may be replaced with skids or skis. A landing gear is required on all piloted aircraft. It provides a means for supplying a force normal to the ground of sufficient magnitude and duration to divert the flight path from one intersecting the ground to one parallel to the ground during landing, a means of retarding forward motion by use of brakes, and a means for supporting, positioning, and moving the airplane on the ground. Requirements for landing gear impose a weight penalty of 10–20% of the structural weight of the airplane from which the vehicle receives no benefit once it is airborne. The loss in performance and payload because of this weight requires a designer to give a great deal of attention to the landing gear when a new airplane

Fig. 1.   Typical tricycle landing gear on F8U-2 Crusader. (*Chance Vought Aircraft, Inc.*)

Fig. 2.   Typical tail-wheel landing gear on F4U-4 Corsair. (*United Aircraft Corp.*)

is at the stage of being designed.

Arrangement. Currently the most widely accepted arrangement is the tricycle landing gear (Fig. 1). This arrangement places a nose gear well forward of the center of gravity on the fuselage and two main gears slightly aft of the center of gravity, with a sufficient distance between them to provide stability against rolling over during a yawed landing in a cross wind, or during ground maneuvers. The nose gear prevents nosing over when brakes are applied. The nose gear, which is always castored, is often equipped with power steering to allow ground maneuvering.

Additional benefits from a tricycle configuration are that (1) the center of gravity is forward of the main gear, resulting in a configuration which is inherently directionally stable during landing; (2) the pilot's visibility is good; (3) the floor of the aircraft is nearly level when the airplane is on the ground, contributing to ease in loading and to passenger comfort; and (4) at touchdown the position of the main gear aft of the center of gravity results in nosing the airplane over onto the nose gear, reducing the angle of attack, and reducing the possibility of the airplane's rebounding into the air.

Another arrangement still used extensively is the tail-wheel landing gear (Fig. 2). Two main-gear struts are located slightly ahead of the center of gravity, well-spaced laterally to provide lateral stability; a third wheel is mounted on the fuselage near the aft end of the airplane. This arrangement is usually lighter than the tricycle landing gear. It suffers in comparison with a tricycle gear because of the increased problems associated with vision, ground maneuvering, and braking.

Other arrangements are often used because of special problems associated with a particular design. The bicycle landing gear used on the Boeing B-47 is an example; this gear requires outriggers for lateral stability. A gear of this type will usually be heavier than the more conventional type but often is attractive for overall performance. Research airplanes such as the X-2 and X-15 often use skids instead of wheels, because ground maneuvering is less important on airplanes of this type; space and temperature considerations control the choice. Helicopters and VTOL aircraft often use unconventional arrangements because many of the considerations for ground maneuvering are less important.

Design. The primary working part of any landing gear is the shock strut, which supplies the force as

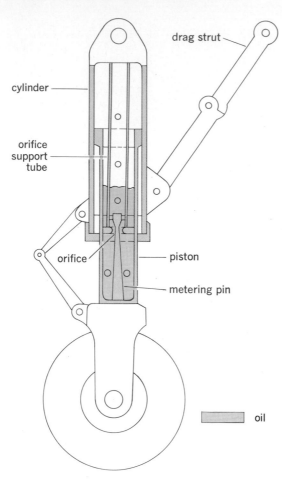

cylinder

orifice
support
tube

orifice

drag strut

piston

metering pin

oil

Fig. 3. Diagram of oleopneumatic shock strut.

the airplane sinks toward the ground, turning the flight path from one intersecting the ground to one parallel to the ground. The most efficient shock strut is the oleopneumatic strut (Fig. 3). It operates by generating a force as oil is pushed through an orifice. When the wheel of the landing gear first contacts the ground, the tire deflects to stop the unsprung mass of the landing gear (tire, wheel, brake, and piston of the oleo). As the airplane continues to sink, the piston in the shock strut forces oil through an orifice, causing a force which changes the flight path of the airplane.

To obtain a better force-time relationship, a metering pin is often used to change the orifice size as the shock strut strokes. Air pressure in the strut extends the gear after it is compressed and also acts as an air-spring suspension system for the airplane during taxiing.

As alternates to the oleo, leaf springs, rubber springs, and oil springs have been successfully used.

In order to provide for the drag load during braking and the impulsive drag load caused by spin-up of the wheel after initial contact, strength must be provided, usually by means of a drag strut. If the wheel can be castored, as a nose wheel or tail wheel is likely to be, provision must be made to prevent shimmy. This is usually accomplished by the use of a shimmy damper. *See* AIRPLANE; WING STRUCTURE.

[JOHN E. STEVENS]

Bibliography: H. G. Conway, *Landing Gear Design*, 1958; W. T. Gunston (ed.), *Flight Handbook*, 6th ed., 1962; B. Milwitsky and F. E. Cook, *Analysis of Landing-Gear Behavior*, NACA Rep. no. 1154, 1953.

Landing ships and craft

Combat vessels employed in amphibious warfare to transport mobile equipment, amphibious vehicles, tanks, general cargo, and personnel, and to discharge such directly onto the beach. These vessels are generally designated by the letters LC for landing craft and LS for landing ship. Landing craft, ranging from about 36 to 135 ft in length, are differentiated from landing ships in that they are not designed for long transoceanic voyages. Landing craft are carried to the unloading area aboard ships (usually LPDs, LSTs, LSDs, LPAs, and LKAs). Landing ships and landing craft were first used during World War II, when a wide variety of types were developed. Of the many types, the following have been in operation since World War II.

Fig. 1. The LST-1173, designed to discharge vehicles directly onto a beach. (*Official U. S. Navy photograph*)

Fig. 2. The LST-1179 landing vehicle on a causeway. (*Official U. S. Navy photograph*)

Fig. 3. The LSM-201. (*Official U. S. Navy photograph*)

Landing ships. The Landing Ship Tank (LST) is primarily designed for transporting and landing tanks or other vehicles on a beach. Because of its versatility and size it may be used to transport almost any type of cargo. With modification it has been used as a refueling vessel, a general cargo ship, and a mine squadron flagship. A large tank deck extends aft from the bow doors and permits cargo to be discharged directly onto the beach from a built-in ramp. It is capable of carrying pontoons that are lashed to each side and can be launched and utilized as causeways for vehicles when beaching is required in relatively deep water. The LST-1173 shown in Fig. 1 is 446 ft long and displaces 7100 tons. The newest class of LSTs (LST-1179) features a revolutionary unloading method. Instead of the bow doors which slowed the speed of the older LSTs, this class has an over-the-bow ramp for disembarking vehicles and a full-length well with greater parking space. Elimination of bow doors gives this new type of LST a speed comparable to that of other ships in an amphibious assault. The LST-1179 shown in Fig. 2 is 518 ft long and displaces 8400 tons.

The Landing Ship Medium (LSM) is designed to transport and land amphibious vehicles on a beach. Its design was devised from a combination of the LST and the LCU; it has an open well and a bow ramp. The LSM-201 shown in Fig. 3 is 204 ft long and displaces 1040 tons.

The Landing Ship Medium – Rocket (LSMR) is a conversion of the LSM and provides close-in fire support with a barrage of rocket bombardment for an assault landing operation. *See* ARMAMENT, NAVAL.

The Landing Ship Dock (LSD) has an LS designation, but it is not a beaching ship. *See* SHIP, NAVAL.

Landing craft. The Landing Craft Utility (LCU), which has the largest capacity of all the landing craft, discharges tanks, mobile equipment, general cargo, and personnel directly onto the beach. Most of the LCUs have the means of marrying with LSTs, which permits their use as causeways for unloading LSTs when beaching conditions prevent the LST from "hitting the beach." The LCU-1613 shown in Fig. 4 is 135 ft long and has a combat landing displacement of 370 tons.

The Landing Craft Mechanized (LCM) is made in three different sizes: the LCM-3, which is 50 ft long; the LCM-6, which is 56 ft long; and the LCM-8, which is 74 ft long. Their full load displacements are 112,000, 124,000, and 254,000 lb, respectively. Their primary function is to land heavy mechanized equipment. An LCM-6 is shown in Fig. 5.

The Landing Craft Vehicle Personnel (LCVP) is used to land and retrieve personnel or equipment during amphibious operations. The LCVP is 36 ft long and capable of carrying 8100 lb of equipment or 36 troops with personal arms and equipment. An LCVP-Mark 7 is shown in Fig. 6.

The Landing Craft Personnel, Large (LCPL) is a dual-purpose boat for guiding other landing craft in an assault wave in a ship-to-shore operation and for use as a personnel boat. The LCPL shown in Fig. 7 is 36 ft long and as a personnel boat carries 17 men and a crew of 3.

Amphibious vehicles. The Landing Vehicle, Tracked (LVT) is an armored amphibian. It can

Fig. 4. The LCU-1613, used to discharge vehicles, mobile equipment, general cargo, and personnel directly onto a beach. (*Official U. S. Navy photograph*)

Fig. 5. The LCM-6. (*Official U. S. Navy photograph*)

Fig. 6. The LCVP-Mark 7. (*Official U. S. Navy photograph*)

Fig. 7. The LCPL-Mark 4. (*Official U. S. Navy photograph*)

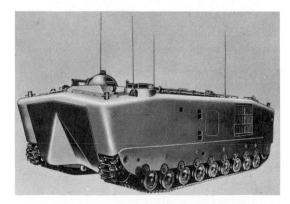

Fig. 8. The LVTP-5. (*Official U. S. Navy photograph*)

Fig. 9. The DUKW, developed for landing on sandy beaches. (*Official U. S. Signal Corps photograph*)

negotiate a surf up to 15 ft high and discharge 34 men inland. The LVTP-5 (Fig. 8) is 30 ft long and has a cargo capacity of 6 tons in water or 9 tons on land.

The Barge, Amphibious, Resupply, Cargo (BARC) is a wheeled amphibious vehicle for over-the-beach handling of cargo. Used by the U.S. Army, it is 63 ft long and has a cargo capacity of 60 tons.

The Truck, Amphibian, $2\frac{1}{2}$ Ton (DUKW), commonly called duck, is used by the U.S. Army to transport cargo or personnel on land or water (Fig. 9). It is 36 ft long and has a cargo capacity of 5175 lb.   [THOMAS J. DI MASI]

Landscape architecture

Landscape architecture was defined by Charles Eliot, one of America's first landscape architects, as the art of arranging and fitting land for human use and enjoyment. It is an applied art founded on the premise that use and beauty are compatible and that neither is complete without the other.

The services of the landscape architect, similar to those of a building architect, involve consultation; preparation of reports, plans, specifications, and estimates; assistance in letting contracts; and supervision of the work done by the contractors.

Projects undertaken by landscape architects include large parks and recreation areas, real estate subdivisions, highways and parkways, and landscaping for industries, institutions, cemeteries, and large private estates.

Training required. Most landscape architects are college-trained specialists. Their training includes design drawing, building architecture (usually a 4-year course), engineering, surveying, building and construction materials, cost estimates, and problems in landscape design. Courses in soil and plant sciences are sometimes included.

Landscape architecture is primarily an American profession. In England most projects involving institutional and public developments are handled by engineers and architects; landscaping of private residences and the design of decorative plantings is ordinarily done by the landscape gardener. The term landscape gardening originally applied to the designing of naturalistic settings. It was a school that arose in England during the 19th century in opposition to the form, or geometric, design that dominated land development before then.

Landscape-nursery service. This is a specialization of the field of landscape architecture. The service, limited to small civic and public problems, concentrates on the development of home proper-

LANDSLIDE

slump with earthflow toe

debris-fall

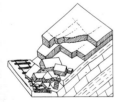

rockslide

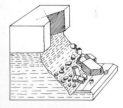

rockfall

Fig. 1. Some principal types of landslides.

ties and small estates. Operators, usually associated with a nursery, handle all phases of a project from design plans through construction and planting and supply all necessary materials. Training for landscape-nursery service usually includes courses in the basic sciences plus advanced work in soils and plant sciences, nursery management, surveying and earthwork, and minor construction of such structures as terraces, pools, and walls. *See* CIVIL ENGINEERING.   [JOSEPH P. PORTER]

Landslide

The perceptible downward sliding or falling of a relatively dry mass of earth, rock, or combination of the two. The term is sometimes extended to cover related flowage movements, including earthflow, mudflow, and debris-avalanche. All belong to the family of mass movement processes which range from slow soil creep to abrupt rockfall.

Landslides have been given greatly increased attention because of a better understanding of the place of mass movement in the shaping of the landscape, and because of increase in magnitude and frequency of landslide problems as ever larger cuts and fills are made for dams, superhighways, foundations, and other engineering projects. Landslides cost the highways and railroads of the United States and Canada over $10,000,000 a year in construction and maintenance and an unknown but large additional amount in indirect losses. In the world at large many lives are lost each year directly or indirectly because of landslides, and major catastrophes sometimes wipe out whole villages.

Landslides, also called landslips, range from low-angle, rather slow slides to vertical falls. Some flowage may accompany sliding, for example, following initial slippage of a wet soil mass or in the very rapid outward spreading of the tongue of a large rockslide or rockfall which may consist almost entirely of dry rock. *See* EROSION; MASS WASTING.

Types. Based on type of movement, relative rate, and kind of material involved, landslides can be separated into five main types: slump, debris-slide, debris-fall, rockslide, and rockfall (Fig. 1).

Slump. The downward slipping of a mass of rock or unconsolidated debris, moving as a unit or several subsidiary units, characteristically with backward rotation on a horizontal axis parallel to the slope, is known as slump. Movement is usually rather slow and may be intermittent; displacement is small relative to the size of the mass. The major slip surface is typically spoon-shaped and concave toward the slip block in both vertical and horizontal section. The moved area at the head of a slump may be broken into many steplike, irregularly tilted blocks. A marsh, pond, or lake may form between a backward tilted block and the cliff from which it descended. Slumping usually results from removal of support lower on the slope. This may occur by natural or artificial undercutting or by flowage, either subaqueously or as an upward-bulging earthflow. Slumps are common on natural cliffs and banks and on the sides of artificial cuts and fills.

Debris-slide. This type of landslide is a rapid downward sliding and forward rolling of unconsolidated earth and rocky debris, usually with the

Fig. 2. Madison Canyon landslide, showing slide scar on south wall and slide debris damming the Madison River, forming new lake, in foreground. (*J. R. Stacy, USGS*)

formation of an irregular hummocky deposit. At the time of movement the material must be fairly dry or the mass would take on the characteristics of an earthflow. Debris-slides are usually rather small and often result from natural or artificial undercutting of a slope.

Debris-fall. A relatively free downward or forward falling of unconsolidated or poorly consolidated earth or rocky debris constitutes this type. Debris-falls are common along undercut banks of rivers, from walls of rapidly eroding gullies, and in steep excavations.

Rockslide. This type applies to any downward and usually rapid movement of newly detached segments of the bedrock, sliding on bedding or any other plane of separation. Rockslides may form wherever dipping strata or jointed rocks are interrupted downslope by any kind of cut. They include some of the greatest of recorded landslides. A rockslide in the valley of the Madison River in Montana accompanying the Hebgen Lake earthquake on Aug. 17, 1959 (Fig. 2), displaced more than 35,000,000 yd³ of slide material. The slide moved a maximum of about 1300 ft vertically and extended to a distance of 3000 ft, killing 25 persons and damming the Madison River to a depth of over 150 ft.

Rockfall. The relatively free falling of a newly detached segment of bedrock of any size from a cliff or steep slope is called rockfall. Rockfalls are common along headwalls of glacial cirques and on wave-cut cliffs. They are a constant hazard on vertical rock cuts along transportation routes, where the fall of a block weighing even a few pounds may disable a vehicle or kill its occupants.

Many of the world's largest landslides have been combinations of rockslide and rockfall. Landslides into fiords, rivers, lakes, or reservoirs sometimes produce enormous waves capable of demolishing waterfront villages or doing extensive damage even at a considerable distance. On the night of Oct. 9, 1963, what was probably the largest landslide in Europe in historical time fell from Monte Toc into the reservoir behind the Vaiont Dam, near Belluno in northern Italy. The mass of about one-third billion cubic yards of rock and earth almost filled the reservoir and dashed a 300-ft-high wall of water over the crest of the 858-ft-high concrete dam. Although the dam held, the water rushing from the mountain canyon into the valley of the

Piave surged hundreds of feet up the opposite wall, wiping out almost all of the town of Longarone and a number of smaller villages down the valley, with a loss of about 2200 lives.

Prevention and control. This depends primarily on avoidance of unsuitable construction in areas of old slides or recognizable mass movement hazard. Other basic measures for prevention and control include excavation to remove fallen or unstable material; drainage of unstable or potentially unstable material to reduce weight and increase shear resistance, and to prevent additional water from gaining access to dangerously placed masses; placement of restraining structures, such as piling, buttresses, retaining walls, cribbing, and wire fences or netting, to keep fallen rocks off communication routes. Warning devices are sometimes used to close railroad blocks when fallen rocks or debris-slides enter a right of way.

[C. F. STEWART SHARPE]

Bibliography: Landslides and Engineering Practice, NAS-NRC Publ. 544, Highway Research Board Spec. Rept. 29, 1958; C. F. S. Sharpe, *Landslides and Related Phenomena*, 1938; reprint, 1960.

Land-use classes

Categories into which land areas can be grouped according to present use of suitability or potential suitability for specified use or according to limitations which restrict their use. The term is most commonly applied to land uses for productive purposes, such as agriculture or forestry, but may be applied for any use including engineering, architecture, urban development, wildlife, and recreation.

A complete description and assessment of land involves climate, land form, surface details, rock type, soil, vegetation, subsurface characteristics, hydrological features, and geographically associated factors, such as availability of water for irrigation purposes, location, and accessibility. Components of the land complex vary considerably from place to place, individually and in combination. The many attributes of land interact and are not equally important for all use purposes in all situations. This makes both the assessment of land potential and the specification of land requirements for a particular purpose difficult and often highly subjective. There is a trend toward, and a need for, more quantitative precision in methods. Approaches to land-use classification are based on present land use, land component surveys, landscape analysis, mathematical procedures, or a combination of these.

Present land use. The World Land Use Survey sponsored by the International Geographical Union aimed to encourage countries to map their own lands in terms of a uniform series of nine broad categories of use: settlements and associated nonagricultural land, horticulture, tree and other perennial crops, crop lands, improved permanent pasture, unimproved grazing land, woodland, swamps and marshes, and unproductive lands. Many countries which have reported on land-use surveys have introduced numerous subdivisions appropriate to the local situation.

More detailed classifications in terms of present land use required for statistical, administrative, or management purposes may be based on a variety of criteria, such as areas of individual crops or forest types, yields, disease and pest occurrence, climatic hazards, pasture types, animal grazing capacity, land-management systems, or purely economic factors (for example, input-output ratios and land values). If used in conjunction with potential land-use classifications, these can be used for comparative purposes or, in the area studied, to indicate where further productivity can be achieved or present land use modified.

Land-component surveys. The common basis for most methods is subdividing the land surface into unit areas which, at the scale of working, are essentially homogeneous in relation to use possibilities and follow with an assessment of use potential by comparison with known situations or responses. The characteristics used for subdivision are the inherent features of land, such as geology, land form, soil, climate, and vegetation. Although classifications on the basis of individual factors such as vegetation or land form have their use, single components of land are inadequate to determine land usefulness precisely. Most attention has been given to classification based on soil surveys, but associated features such as climate and topography are usually taken into account as well.

A widely used system for grouping soil taxonomic or mapping units into land-use classes is that of the USDA handbook *Land-Capability Classification.* The USDA system aims to assess suitability of soils for adapted or native plants and the kind of management the soils require to maintain continued productivity. Assessments are made largely on the kind and degree of hazards or limitations to productivity. This system provides for eight capability classes with a number of subclasses and units identified by these limitations. *See* SOIL, SUBORDERS OF.

The first four classes include groups of soils suitable for cultivation, but from class I to class IV the choice of suitable plants becomes more restricted or the need for more careful management or conservation practices increases. Classes V to VII soils are generally restricted to pasture, range, woodland, and wildlife use. Class VIII soils are restricted to nonagricultural purposes.

As moderately high levels of land management and of inputs and a favorable ratio of inputs to outputs are assumed, the method as a whole can be applied with precision only in areas where knowledge of land use is well advanced. In less developed areas, classification must be limited to the broader categories. In undeveloped countries with low economic ceilings, judgments need to be modified according to local standards.

The same general principles are applied, with appropriately selected criteria, for grouping soils into land classes with different degrees of suitability for a variety of purposes, such as woodland establishment, recreation, wildlife, and engineering. *See* SOIL, ZONALITY OF.

A method adopted by the Canada Land Inventory for forestry purposes illustrates a variant of this approach. It adopts a division of the land surface into homogeneous units determined by physical characteristics, followed by a rating of these units into seven capability classes according to environmental factors which influence their inherent ability to grow commercial timber. These factors are the subsoil, soil, surface, local and regional climate, and the tree species. A feature of this sytem is that a productivity rating is set in quantitative terms for each class expressed as volume of merchantable timber per acre per year. Regional inventories include a reference to the indicator tree species present for each class.

A system of land classification for the specific purpose of establishing the extent and degree of suitability of lands for sustained profitable production under irrigation has been developed by the U.S. Bureau of Reclamation. Suitability of land is measured in terms of payment capacity and involves consideration of potential productive capacity, costs of production, and costs of land development. These are assessed from soil characteristics and topographic and drainage factors. Six basic classes of land are recognized. Classes I to III are all arable lands, but they decrease in payment capacity and become more restricted in usefulness in that order. Class IV includes lands with special uses. Class V lands are at least temporarily nonirrigable, and Class VI lands are unsuitable for irrigation use.

Landscape analysis. An alternative to the subdivision of land according to single components is the subdivision of the landscape itself into natural units, each characterized by a combination of geologic, land-form, soil, and vegetation features. This approach, referred to as the land-system approach, has been developed in a number of countries but especially in Australia. It is particularly well suited to the use of aerial photographs and has special value in the reconnaissance survey of little-known lands. The approach is based on the concept that there are discernible natural patterns of landscape covering areas with common and distinctive histories of landscape genesis. The boundaries of land systems coincide with major changes in geology or geomorphology. The pattern of a land system is formed by a number of associated and recurring land units. Each occurrence of the same land unit within a specific land system represents a similar end product of land-surface evolution from common parent material by common processes through the same series of past and present climates over the same period of time. Thus, in addition to being described similarly in terms of observable slopes and surfaces, vegetation cover, and soils, they are assumed to have a similar array of natural and potential habitats for land use. This record of the inherent features of the landscape can be interpreted in terms of land-use classes in the light of the technical knowledge available at any subsequent time. Surveys are made by concurrent and integrated studies of all the observable land features by a team of specialists in the fields of geomorphology, soils, and plant ecology. Conclusions to be drawn about immediate land use must be derived by analogy with known areas or from basic principles. The method has particular value in less developed regions. It is also being applied to special purposes such as forestry and engineering. *See* AERIAL PHOTOGRAPH; FOREST MANAGEMENT AND ORGANIZATION; GEOMORPHOLOGY.

Mathematical approaches. Land-use classification dependent upon descriptive data of land characteristics and analog processes of assessment suffer from inherent inadequacies of descriptive processes and involve a good deal of intuition and subjectivity. For this reason, effort has been made to introduce more quantitative approaches to the assessment of potentials. Foresters have developed methods of site evaluation based on the measurement of environmental factors and productivity at different sites. Key parameters are identified from multiple regressions. This information can then be applied to classification of areas in terms of potential production. *See* FOREST MEASUREMENT.

With the advent of modern computers capable of handling and storing masses of data, there is a rapidly growing trend toward quantifying land characteristics and land-use responses and using mathematical models which relate the numerous land parameters to a variety of use responses in agricultural, forestry, and engineering fields.

The automatic scanning equipment for aerial photography and the remote sensing and automatic recording devices used with Earth satellites are opening up completely new approaches to land description and subdivision and hence to land classification. *See* AGRICULTURE; LAND-USE PLANNING; LIFE ZONES; SATELLITES, APPLICATIONS.

[C. S. CHRISTIAN]

Bibliography: C. S. Christian and G. A. Stewart, Methodology of integrated surveys, *Aerial Surveys and Integrated Studies: Proceedings of the Toulouse Conference 1964*, UNESCO, 1968; *Field Manuals for Land Capability Classifications*, Canada Land Inventory (ARDA), Department of Forestry and Rural Development, various years; G. A. Hills, *The Ecological Basis for Land-Use Planning*, Ont. Dep. Lands Forests Res. Rep. no. 46, December, 1961; *Irrigated Land Use, Bureau of Reclamation Manual*, U.S. Department of the Interior, vol. 5, pt. 2, 1951; A. A. Klingebiel and P. H. Montgomery, *Land-Capability Classification*, Soil Conservation Service, USDA Handb. no. 210, 1961; D. S. Lacate, *Forest Land Classification for the University of British Columbia Research Forest*, Can. Dep. Forest. Publ. no. 1107, 1965; R. J. McCormack, *Land Capability for Forestry: Outline and Guidelines for Mapping*, Canada Land Inventory (ARDA), Department of Forestry and Rural Development, 1967; D. L. Stamp, *Land Use Statistics of the Countries of Europe*, World Land Use Surv. Occas. Pap. no. 3, 1965; D. L. Stamp, *Our Developing World*, 1960; G. A. Stewart (ed.), *Land Evaluation*, in press.

Land-use planning

Humanity has gained unprecedented physical and technical resources to regulate use and misuse of lands and resources. Planning for the distant as well as immediate consequences of man's actions is essential for long-run economy, and such investigations can warn of irreversible and irreparable deterioration in the quality of environment and life. See ECOLOGY; ECOLOGY, APPLIED; ECOSYSTEM; VEGETATION MANAGEMENT.

The objective of land-use planning on an individual ownership may exceed maximizing the net income of the owner. The objective of land-use planning by a public agency is even more complex and is intended to maximize long-range community benefits. Three characteristics of social and economic history nurtured contemporary land-use problems.

First, strong competitive forces, promoted by a philosophy of free private enterprise, urged rapid uncoordinated exploitation of land and its resources. This exploitation was characterized by extensive use instead of intensive methods known to modern technology.

Second, early land exploitation demonstrated real but primitive understanding of natural land capability. The plow followed the ax in many areas where agriculture cannot do as well as forests; the plow broke the plains in many places where crops are not as compatible with the climate as more drought-resistant forage; and the plow has turned the sod of fields whose soil and slope cannot support cultivation without rapid erosion and soil depletion.

Third, early city development, in the absence of a land-use policy and plan, resulted in an indiscriminate mixture of residential, industrial, and commercial land uses. Residential areas often suffered deflated values, and commercial and industrial enterprises failed to achieve possible economies in transportation, power, and waste disposal. As deterioration of the central city set in, residences, stores, service establishments, and industries favored the suburbs for new location. This situation suggested needs for public decisions and controls on major aspects of land use in the central city and in the suburbs of every metropolitan center.

Rural land planning emphasizes the development of land-use patterns which reflect the physical and biological limits beyond which long-run depletion of the land resource will result. Increasingly, private land operators are learning that production according to land capability is good business. Since the great depression of the 1930s, both Federal and state governments have cooperatively developed plans for improved productivity of land and related resources of various large natural regions of the nation. The Great Lakes Cut-Over Region and the Northern Great Plains were areas early subject to such regional studies. The Tennessee Valley Authority (1933) marked a further emphasis upon regional resource planning. TVA and Appalachia development programs have had as their primary planning objective the discovery of economic opportunities and social amenities.

The importance of regional and major drainage basin resource development is indicated by rapid expansion since World War II. In addition, hundreds of local organizations such as watershed associations, various special districts, and intergovernmental planning committees have sprung up in response to land-use problems. These smaller regions may be dominated by a vigorous urban center. Here rural-type resource development problems intermingle with the problems of space allocation and design of the spreading urban area. Land-use planning during the later years of the 20th century is challenged to meet the needs of a planning area, composed of admixtures of rural and urban land-use problems, and recreation fringes.

Planning and the future. Trends in land-use planning suggest four developments in concept and practice. First, integrating the space-use considerations of urban-exurban planning and the resource-use considerations of wild-land, rural planning will demand increasing attention. More interchange and adaptation should take place between the space design orientation of the city planner, architect, and landscape designer and the resource capability orientation of the resource planner. The overlapping metropolitan areas must be planned to meet the basic regional needs for water, waste disposal, transportation, and open spaces for light, air, and recreation. More regard to the natural capability of each environment is needed to sustain an optimum level of regional economic opportunity and social values. Community requirements and the environmental capability should find reconciliation in a new type of integrated regional design.

Second, standards and criteria to guide land-use allocations must be developed by the students and practitioners of land-use planning and become incorporated in public land policies. *See* FOREST AND FORESTRY; RANGELAND CONSERVATION; SOIL CONSERVATION; WATER CONSERVATION; WILDLIFE CONSERVATION.

Third, if land planning is to be made an instrument of democracy, improvements are required in the methods by which social choices of the majority can find expression in the planning process, without ignoring individualism and diverse goals of a pluralistic society. Approving of general criteria for land-use allocation or of specific plans, should be made by people of the community through the operation of the political process. New communication channels can be more articulate in serving the political process and in responding to it. Experimentation should be applied to assure representation of the community and provide protection against misuse from self-seeking interests.

Fourth, to assure that plans will be carried out, closer relationships must be established between the planning authority and the agencies which exercise powers of implementation. A city planning commission may be established somewhat apart from the government authority available to implement the plan. Regional planning bodies, whether oriented primarily to large metropolitan regions or to large resource development regions, have a planning area and a scope of functions for which no single governmental body can serve as the implementing authority. Both of these situations make it difficult to mesh the "planning gears" with the "administrative gears." Political and administrative sciences are challenged to discover institutional forms and procedures through which the early and final planning process and the public power of implementation can interact responsibly.

By the 21st century, science and technology can wrest from the earth practically anything that is left over from prodigal use in the 20th. Critical problems in land-use planning of the future will arise from the difficulty of determining what is sought. These problems are not scientific and technological but they are human ones. They involve reconciling different human desires and organizing for cooperative decisions and programs of action. The challenge is to mold the natural sciences and technologies into an environmental framework with the applied social sciences to produce a land-use scheme to serve man best.

Zoning of land. Industrial societies often have zoned their lands (if at all) according to current economic values, forgetting long-range costs until struck by catastrophe. Residential areas and fac-

tories are built on floodplains, maybe unwittingly, because development is easier and cheaper on level lands than on hills. If costs of flood damage or of flood control are considered, major investments might well be allocated to uplands.

Farms, especially on rich soil, in large blocks suitable for mechanized treatment are frequently removed from best natural uses by urban sprawl. Many need legal protection from encroachment and fragmentation, and taxation policies which unfairly penalize open land or continued farming. See AGRICULTURE; SOIL.

Developed parks and protected reserves, demonstrating natural plant and animal communities typical of diverse sites and regions, can include some areas too wet, dry, or shallow-soiled for other purposes. For educational and scientific as well as recreational purposes they also deserve high priority in the face of competing uses, according to each area's quality and location. National parks and monuments preserve unique scientific and scenic treasures, but even these are becoming degraded through lack of internal park zoning or control. Trampling of plants, compaction of soil, other abuse by pack animals or vehicles, wind or water erosion, noise and other disturbance, and unwise attraction of animals by feeding increasingly upset Nature's balances in the very places that attract most public use.

American legislation protecting wilderness areas in national forests and other areas as well as parks is a landmark of policy recognizing different levels and kinds of use, and sometimes calls for hard choices of clear priority. Since the first hearings under this legislation (for the Great Smoky Mountains National Park) the record of widespread public preference has commonly been for more inclusion of areas, and more exclusion of highways and resorts than were favored by local developers, and by officials who feel committed to the latter. To deflect commercialization within parks has demanded long-range plans in a larger regional context, commonly favoring wider private service to visitors while firmly limiting massive encroachment and destruction in the heartland.

At village, city, county, and state levels as well, there is a need for planning and implementing a balance of the most suitable areas for (1) direct urban and suburban uses, (2) use and, where pertinent, renewal of natural resources, and (3) acquisition or at least options allowing future control of landscapes and water areas. Nature conservancies (private in the United States, part of the Natural Resources Research Council in the United Kingdom) have been alert in finding and acting on opportunities.

Awareness and promptness. Urban blight and depressing obsolescence of huge, monotonous suburban developments are more symptoms — very expensive ones — of complex problems of a technologically oriented society. Great cities, stimulated to haphazard growth by industry in the 19th century, now may put on a front of greenery along modern roadways and developments; but this often masks failure to adapt the mix of land uses to the possibilities and real limitations of resources. *See* LAND-USE CLASSES.

Population explosions may be slower in industrial countries or regions than elsewhere; still their pressures and mobility commit space for industry, highways, airports, shopping centers, asphalt parking "deserts," and housing developments too fast for collecting and weighing relevant factors.

Democracies assert the right of each person, from the open country to the concrete city, to learn the choices that affect his life, to use expert counsel on the repercussions of the choices, and to exert his responsibility for influencing decisions. Yet even where mechanisms exist to implement these rights, irrevocable choices are often committed legally, or turned into reality, before their full consequences are either explored or identified. Even where conscientious citizens warn legislators or administrators of dangers or preferred alternatives, lobbying often speaks fastest and loudest from the sources having financial interest in a particular scheme. *See* CIVIL ENGINEERING.

Pollution of air, water, and landscapes is a problem which recently gained a public spotlight, without yet having wise enforcement against obvious abuses, or recognition of subtle ones. Disposal of residues in ways that will not upset the environment is a growing social problem. Effective management requires knowing the path and potential effects of each waste product until it becomes neutralized. So much public attention was rightly drawn to increasing radioactive waste in a nuclear-powered economy and to the analysis of "maximum credible" accidents (and of some incredible contingencies as well) that the far-reaching effects have been subjected to study as never before. *See* RADIOECOLOGY.

Modern man cannot escape costs of containment of by-products, of monitoring the inevitable releases to maintain low and tolerable levels, and of providing for emergency action in case of accidents. Such costs have been charged to governments and to increasing numbers of other producers of nuclear energy and materials. To control sulfur dioxide (from coal), exhaust effluents (from internal combustion engines), petroleum, and other chemical wastes, it is necessary to decide on passing along similar costs to consumers or taxpayers as a payment for maintaining or improving environmental quality. Land-use planning in site selection of major facilities, such as electric generators (which all produce waste heat, regardless of radioactive or combustion effluents), is essential for limiting these costs and maintaining compatibility with social values.   [JERRY S. OLSON]

Bibliography: S. Chase, *Rich Land, Poor Land*, 1936; Environmental Pollution Panel, President's Science Advisory Committee, *Restoring the Quality of Our Environment*, 1965; R. Lord, *The Care of the Earth: A History of Husbandry*, 1962; V. Obenhaus, L. Walford, and J. Olson, Technology and man's relation to his natural environment, in C. P. Hall (compiler), *Human Values and Advancing Technology*, 1967.

Langevin function

A mathematical function which is important in the theory of paramagnetism and in the theory of the dielectric properties of insulators. The analytical expression for the Langevin function (see illustration) is shown in the equation below. If $x \ll 1$, $L(x) \cong x/3$. The paramagnetic susceptibility of a classical (non-quantum-mechanical) collec-

$$L(x) = \coth x - 1/x$$

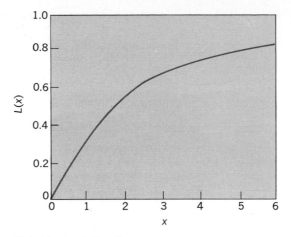

Plot of Langevin function.

tion of magnetic dipoles is given by the Langevin function, as is the polarizability of molecules having a permanent electric dipole moment. In the quantum-mechanical treatment of paramagnetism, the Langevin function is replaced by the Brillouin function. For further discussion *see* DIELECTRIC CONSTANT; PARAMAGNETISM.

[ELIHU ABRAHAMS; F. KEFFER]

Langmuir-Child law

A law governing space-charge-limited flow of electron current between two plane parallel electrodes in vacuum when the emission velocities of the electrons can be neglected. It is often called the three-halves power law, and is expressed by the formula shown below.

$$j(\text{amp/cm}^2) = \frac{\epsilon}{9\pi}\left(\frac{2e}{m}\right)^{1/2}\frac{V^{3/2}}{d^2}$$
$$= 2.33 \times 10^{-6}\,\frac{V(\text{volts})^{3/2}}{d(\text{cm})^2}$$

Here ϵ is the dielectric constant of vacuum, $-e$ the charge of the electron, m its mass, V the potential difference between the two electrodes, d their separation, and j the current density at the collector electrode, or anode. The potential difference V is the applied voltage reduced by the difference in work function of the collector and emitter. The Langmuir-Child law applies, to a close approximation, to other electrode geometries as well. Thus for coaxial cylinders with the inner cylinder the cathode, it leads to a deviation from the true value of the current density of 13% at most. *See* SPACE CHARGE.   [EDWARD G. RAMBERG]

Lanolin

The hydrous sheep's-wool wax (primarily cholesterol esters of higher fatty acids) derived as a by-product from the preparation of raw wool for the spinner. The crude wool wax is purified, melted, mixed with about 30% water, and allowed to harden into a soft, pale-yellow, unguentlike substance.

Lanolin is widely used as a base for emollients in cosmetics and shampoos. Perhaps because of its slight antiseptic effect and high resistance to rancidity it has continued wide use as a base for skin ointments. Actually, lanolin hinders absorption of the medication through the skin, as compared to other substances such as olive oil or lard. Lanolin is soluble in alcohol, ether, chloroform, and benzene. *See* FAT AND OIL, NONEDIBLE.

[FRANK H. ROCKETT]

Bibliography: D. T. C. Gillespie, Wool wax: A review of its properties, recovery, and utilization, *J. Text. Inst.*, 39(2):45–85, 1948; D. Swern (ed.), *Bailey's Industrial Oil and Fat Products*, 3d ed., 1964.

Lantern slides

Positive transparent pictures on glass or film which can be viewed by projection onto reflecting screens or in hand viewers or devices in which the picture is projected onto the rear of a diffusing screen. Standard sizes are United States, $3\frac{1}{4}$ by 4 in.; British, $3\frac{1}{4}$ by $3\frac{1}{4}$ in.; Continental, 85 by 100 mm; intermediate Continental, 70 by 70 mm; $2\frac{1}{4}$ by $2\frac{1}{4}$ in.; miniature (the most popular), 2 by 2 in. The last two sizes named are used everywhere. The aperture sizes of the masked pictures have been standardized. Although black-and-white transparencies had been made in 2-by-2 in. size from miniature camera negatives since the early days of that type of camera, the great increase in popularity of the 2-by-2-in. slide and the appropriate projectors and viewers (and, in fact, the miniature camera itself) stems from the introduction of color film and its return from processing in the cardboard mounts as positive transparencies.

Slides are projected in conventional projectors (optical lanterns) consisting of a concentrated source of light, a condenser system, a holder (or changer) for the slide, a projection lens, and (usually) a blower for cooling the slide. Tungsten filament lamps, generally with a prefocused base, are used in wattages from 150 to 1000, depending upon the projected picture size; medium wattages are for home use. The focal length of the lens is governed by the projection distance and screen size. Many variations of slide changer are made, including hand-feeding slides one at a time, a changer which carries two slides (one is changed while the other is projected), a cassette which carries a group of slides and feeds slides in and out semiautomatically, and a magazine of slides which are associated with a timer and are fed and changed fully automatically. Some holders carry long rolls of film and film strips. *See* PROJECTION SYSTEMS, OPTICAL.

Slides, particularly in the miniature sizes, can be viewed in hand viewers or table viewers. The former have a means for illuminating the slide, pointing it to the sky or to a lamp, or using a lamp included in the viewer for enclosed battery or wall-outlet operation, and a simple magnifying lens. They are made for single or stereo slides. Table viewers are really small projectors giving an image of moderate magnification for viewing on a rear-projection screen, and have manual or automatic slide changers. Large display transparencies use banks of lights giving uniform illumination over a diffusing screen immediately behind the picture.

[WALTER CLARK]

Lanthanide contraction

The name given to an unusual phenomenon encountered in the rare-earth series of elements. The radii of the atoms of the members of this series decrease slightly as the atomic number increases. Starting with element 58 in the periodic table, the balancing electron fills in an inner incomplete $4f$ shell as the charge on the nucleus increases. According to the theory of atomic structure, this shell can hold 14 electrons; so starting with element 58, cerium, there are 14 true rare earths. Lanthanum has no electrons in the $4f$ shell, cerium has 1, and lutetium, 14. The $4f$ electrons play almost no role in chemical valence; therefore, all rare earths can have three electrons in their valence shell and they all exist as trivalent ions in solution. As the charge on the nucleus increases across the rare-earth series, all electrons are pulled in closer to the nucleus so that the radii of the rare-earth ions decrease slightly as the compounds go across the rare-earth series. Any given compound of the rare earths is very likely to crystallize with the same structure as any other rare earth. However, the lattice parameters become smaller and the crystal denser as the compounds proceed across the series. This contraction of the lattice parameters is known as the lanthanide contraction. For many compounds the lattice parameters decrease only partway across the series, and when the contraction has progressed to that point, a new crystalline form develops. Frequently, both crystalline forms can be observed for a number of the elements. For this reason, the rare-earth series is of particular interest to scientists because many of the parameters determining the properties of a substance can be kept constant while the lattice spacings can be varied in small increments across the series.

The atomic and ionic radii of atoms are not clearly defined. The atoms can be polarized by the neighboring atoms and there is no clear-cut boundary between the electrons associated with one atom and another. Therefore, the atomic radii will vary somewhat from compound to compound, and the absolute values depend on the method of cal-

Atomic and ionic radii of rare-earth metals

Element	Radius, A $3+$ ion[a]	Metal crystal structure[b]	Metallic radii, A c	Metallic radii, A d
Sc		hcp	1.6545	1.6280
Y		hcp	1.8237	1.7780
La	1.061	hcp	1.8852	1.8694
Ce	1.034	fcc	1.8248	
Pr	1.013	hcp	1.8363	1.8201
Nd	0.995	hcp	1.8290	1.8139
Pm	0.979			
Sm	0.964	rhom-hcp	1.8105	1.7943
Eu	0.950	bcc	1.994	
Gd	0.938	hcp	1.8180	1.7865
Tb	0.923	hcp	1.8005	1.7626
Dy	0.908	hcp	1.7952	1.7515
Ho	0.894	hcp	1.7887	1.7428
Er	0.881	hcp	1.7794	1.7340
Tm	0.869	hcp	1.7688	1.7237
Yb	0.858	fcc	1.9397	
Lu	0.848	hcp	1.7516	1.7171

[a]Data from D. H. Templeton and Carol H. Dauben, *J. Amer. Chem. Soc.*, 76:5237–5239, 1954.

[b]Data from F. H. Spedding, A. H. Daane, and K. W. Herrmann, *Acta Cryst.*, 9(7):559–563, 1956; hcp, hexagonal close-packed; fcc, face-centered cubic; rhom, rhombic; bcc, body-centered cubic.

[c]Data from K. W. Herrmann, Doctoral thesis; radii calculated from atoms in basal plane.

[d]Data from K. W. Herrmann, Doctoral thesis; radii between layers.

culation. However, if most of the parameters are assumed constant, and the difference in lattice parameters in the rare-earth crystalline series is attributed to the rare-earth ion or atom, then the lanthanide contraction becomes clearly evident. Although scandium and yttrium are not members of this series, the information is usually wanted at the same time and is given for completeness. The atomic radii of the trivalent ion and the metal atoms are given in the table. *See* PERIODIC TABLE; RARE-EARTH ELEMENTS.   [FRANK H. SPEDDING]

Lanthanum

A chemical element, La, atomic number 57, atomic weight 138.91. Lanthanum, the second most abundant element in the rare-earth group, is a metal. The naturally occurring element is made up of the isotopes La^{138}, 0.089%, and La^{139}, 99.91%. La^{138} is a radioactive positron emitter with a half-life of 1.1×10^{11} years. The element was discovered in 1839 by C. G. Mosander and occurs associated with other rare earths in monazite, bastnasite, and other minerals. It is one of the radioactive products of the fission of uranium, thorium, or plutonium. Lanthanum is the most basic of the rare earths and can be separated rapidly from other members of the rare-earth series by fractional crystallization. Considerable quantities of it are separated commercially, since it is an important ingredient in glass manufacture. Lanthanum imparts a high refractive index to the glass and is used in the manufacture of expensive lenses. The metal is readily attacked in air and is rapidly converted to a white powder. For other properties of the metal *see* RARE-EARTH ELEMENTS.

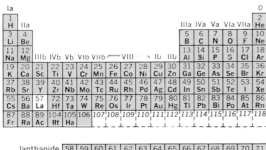

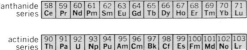

Lanthanum becomes a superconductor below about 6°K in both the hexagonal and face-centered crystal forms.   [FRANK H. SPEDDING]

Laplace transform

An integral transform extensively used by P. S. Laplace in the theory of probability. In simplest form it is expressed as Eq. (1). It is thought of as

$$f(s) = \int_0^\infty e^{-st}\phi(t)\,dt \qquad (1)$$

transforming the determining function $\phi(t)$ into the generating function $f(s)$. The variable t is real, the variable s may be real or complex, $s = \sigma + i\tau$. As an example, if $\phi(t) = 1$ the integral converges for $\sigma > 0$, and $f(s) = 1/s$.

The Laplace transform is used for the solution of differential and difference equations, for the evaluation of definite integrals, and in many branches of abstract mathematics (functional analysis, operational calculus, and analytic number theory).

Method. Extensive tables of Laplace transforms exist, and these are used as any table of integrals. To illustrate how a differential equation may be solved, two excerpts (A and B) from such a table can be used.

$$A. \quad f(s) = 1/(s-a) \qquad \phi(t) = e^{at}$$
$$B. \quad f(s) = 1/(s^2+1) \qquad \phi(t) = \sin t$$

Suppose it is required to find a solution $y(t)$ of Eq. (2) such that $y(0) = 1$, $y'(0) = 2$. Denote the

$$y''(t) + y(t) = 2e^t \qquad \left(y'' = \frac{d^2y}{dt^2}, y' = \frac{dy}{dt}\right) \quad (2)$$

Laplace transform of the unknown function $y(t)$ by $Y(s)$. Integration by parts gives Eq. (3) on the as-

$$\int_0^\infty e^{-st}y''(t)\,dt = -y'(0) - y(0)s + s^2\int_0^\infty e^{-st}y(t)\,dt$$
$$= -2 - s + s^2Y(s) \qquad (3)$$

sumption that the integrated part is zero at $t = \infty$. Applying the Laplace transform to Eq. (2) and using A for the right-hand side, one obtains Eq. (4).

$$-2 - s + s^2Y(s) + Y(s) = \frac{2}{s-1} \qquad (4)$$

The differential equation has become an algebraic one, whose solution is Eq. (5). However, a further

$$Y(s) = \frac{1}{s-1} + \frac{1}{s^2+1} \qquad (5)$$

use of the table shows that the Laplace transform of $y(t) = e^t + \sin t$ is precisely the right-hand side of Eq. (5). Assuming uniqueness, one has thus obtained the required solution. Because its properties can be checked directly, the unproved assumptions need not be verified.

This example illustrates the general method. The unknown function is taken as the determining function and the Laplace transform is applied to the differential (or difference) equation. There results an equation with the generating function as unknown, and this must be solved. Finally the determining function must be determined from the generating function, either from tables or by use of an inversion formula. In general, if the original differential equation is partial in any number of independent variables, one application of the Laplace transform reduces the number of these variables by one. If the equation was ordinary (one independent variable), the transformed equation is algebraic, as in the above example.

Properties. Here are the fundamental properties of the Laplace transform:

I. There exists a number σ_c (perhaps $+\infty$ or $-\infty$) called the abscissa of convergence such that the integral in Eq. (1) converges for $\sigma > \sigma_c$, diverges for $\sigma < \sigma_c$. That is, the region of convergence is a half-plane (a half-line if s is real).

II. The generating function is holomorphic for $\sigma > \sigma_c$.

III. The determining function is uniquely determined by the generating function. (Ambiguity is possible only on sets of measure zero.)

IV. The product of two generating functions is in general a generating function. Thus, if Eq. (1) holds for two pairs of functions $f_1(s), \phi_1(t)$ and $f_2(s), \phi_2(t)$, then the product $f_1(s) f_2(s)$ is the transform of the convolution

$$\phi_1(t) * \phi_2(t) = \int_0^t \phi_1(u)\phi_2(t-u)\,du$$

As was evident in the above example, it is very important to be able to derive the determining function $\phi(t)$ from the generating function $f(s)$. This is especially true when tables are unavailable or inadequate. Any expression of $\phi(t)$ in terms of $f(s)$ is called an inversion formula. Many are known. The classical one is Eq. (6). Here the integration is along any line $\sigma = c$ of the complex s-plane on which the integral in Eq. (1) converges

$$\phi(t) = \frac{1}{2\pi i}\int_{c-i\infty}^{c+i\infty} f(s)e^{st}\,ds \qquad 0 < t < \infty \qquad (6)$$

gration is along any line $\sigma = c$ of the complex s-plane on which the integral in Eq. (1) converges absolutely. Another inversion which employs the real variable only is Eq. (7). Here $f^{(k)}(x)$ means the

$$\phi(t) = \lim_{k\to\infty} \frac{(-1)^k}{k!} f^{(k)}\left(\frac{k}{t}\right)\left(\frac{k}{t}\right)^{k+1} \qquad 0 < t < \infty \qquad (7)$$

kth derivative of $f(x)$. Equation (7) can be illustrated by the example A above. For that pair $f^{(k)}(x)$ is easily computed and Eq. (7) becomes Eq. (8), a familiar result of calculus. *See* INTEGRAL TRANSFORM.

$$e^{at} = \lim_{k\to\infty}\left(1 - \frac{at}{k}\right)^{-k-1} \qquad (8)$$

Generalizations. Certain generalizations of Eq. (1) are in frequent use. The transform shown as Eq. (9) is called the bilateral Laplace transform.

$$f(s) = \int_{-\infty}^{\infty} e^{-st}\phi(t)\,dt \qquad (9)$$

An inversion is still provided by Eq. (6), which now holds for $-\infty < t < \infty$. If one sets $s = iy$ in Eq. (9), the result is Eq. (10). This equation defines $g(y)$ as

$$g(y) = f(iy) = \int_{-\infty}^{\infty} e^{-iyt}\phi(t)\,dt \qquad (10)$$

the Fourier transform of $\phi(t)$. That is, the Laplace transform (9), if considered along a single line, becomes a Fourier transform. By setting $c = 0$ and $s = iy$ in formula (6) one obtains Eq. (11), the classical inversion of the Fourier transform.

$$\phi(t) = \frac{1}{2\pi i}\int_{-i\infty}^{i\infty} f(s)e^{st}\,ds$$
$$= \frac{1}{2\pi}\int_{-\infty}^{\infty} g(y)e^{iyt}\,dy \qquad (11)$$

Another generalization of Eq. (1) is the Laplace-Stieltjes integral, Eq. (12), where now the integral is a Stieltjes integral with respect to the "integrator" function $\alpha(t)$. If $\alpha(t)$ has a derivative $\phi(t)$ the integral (12) becomes the integral (1). On the other hand, if $\alpha(t)$ is a step-function, Eq. (12) reduces to a

$$f(s) = \int_0^{\infty} e^{-st}\,d\alpha(t) \qquad (12)$$

Dirichlet series, Eq. (13), a type of series of great importance in analytic number theory.

$$f(s) = \sum_{k=1}^{\infty} a_k e^{-\lambda_k s} \qquad (13)$$

It must not be supposed that one may choose $\phi(t)$ or $f(s)$ arbitrarily in Eq. (1) and expect its

mate to exist. For example if $\phi(t) = e^{t^2}$, the integral (1) diverges for all s and $\sigma_c = +\infty$. Again if $f(s) = s$, no corresponding determining function $\phi(t)$ exists, since it is easily seen that every generating function must approach a limit as $s \to +\infty$ along the real axis. Hence it is clearly important to know what functions $\phi(t)$ and $f(s)$ may be used in Eq. (1). So far as $\phi(t)$ is concerned the problem is completely solved by the formula, with $\sigma_c > 0$, shown as Eq. (14). The other problem has been partially

$$\sigma_c = \overline{\lim_{t\to\infty}} \frac{\log|\alpha(t)|}{t} \qquad (14)$$

solved by representation theorems for integral transforms, one striking example of which is presented below.

A function $f(s)$ of the real variable s is said to be completely monotonic on $a < s < \infty$ if and only if relations (15) hold. Examples are $f(s) = 1$, $f(s) = 1/(s-a)$, and $f(s) = e^{-s}$. A theorem of S. Bernstein

$$f(s) \geq 0, f'(s) \leq 0, f''(s) \geq 0, f'''(s) \leq 0, \ldots$$
$$(a < s < \infty) \qquad (15)$$

states that $f(s)$ has a representation (12) converging for $s > a$ and with integrator function $\alpha(t)$ nondecreasing if and only if $f(s)$ is completely monotonic for $a < s < \infty$. For example, if $f(s) = 1$, then $\alpha(t) = 1$, $t > 0$, $\alpha(0) = 0$; if $f(s) = 1/(s-a)$, then $\phi(t) = e^{at}$ and $\alpha(t) = (e^{at}-1)/a$; if $f(s) = e^{-s}$, $\alpha(t) = 0$ for $0 < t < 1$ and $\alpha(t) = 1$ for $1 < t < \infty$. In each case $\alpha(t)$ is nondecreasing, as predicted by Bernstein's theorem. This result is particularly remarkable because the mere signs of the successive derivatives of a function on the real axis determine not only its holomorphic character (property II above) in a half-plane but also its representation in the form of Eq. (12).

[DAVID V. WIDDER]

Bibliography: A. Erdélyi (ed.), *Tables of Integral Transforms*, vol. 1, 1954; E. D. Rainville, *The Laplace Transform: An Introduction*, 1963; C. J. Savant, *Fundamentals of the Laplace Transformation*, 1962; D. V. Widder, *The Laplace Transform*, 1941.

Laplace's differential equation

Laplace's equation in two independent variables x and y is given as Eq. (1) and is of central impor-

$$\frac{\partial^2 u(x,y)}{\partial x^2} + \frac{\partial^2 u(x,y)}{\partial y^2} = 0 \qquad (1)$$

tance in both pure mathematics and mathematical physics. A function $u(x,y)$ having continuous first and second partial derivatives and satisfying Laplace's equation in a neighborhood of a point is called harmonic at that point. If a plane piece of tinfoil has its edges kept at a temperature which varies from point to point but does not change with time, and if the flow of heat in the tinfoil is steady (that is, independent of the time), the temperature $u(x,y)$ at interior points of the foil is harmonic. Likewise Laplace's equation dominates the flow of electricity (the potential is similarly harmonic) and the flow of any incompressible fluid.

Two-dimensional relations. If $f(z) \equiv u(x,y) + iv(x,y)$ is an analytic function, $u(x,y)$ and $v(x,y)$ are conjugate functions and are harmonic; conversely, if $u(x,y)$ is harmonic in a simply connected region D, one may write Eq. (2), where (x_0,y_0) is fixed in D

$$v(x,y) \equiv \int_{(x_0,y_0)}^{(x,y)} \left(-\frac{\partial u}{\partial y} dx + \frac{\partial u}{\partial x} dy \right) \qquad (2)$$

and (x,y) arbitrary in D. It follows from Green's theorem that the integral over a path in D is independent of the path, so $v(x,y)$ is uniquely defined throughout D; the functions $u(x,y)$ and $v(x,y)$ are conjugate in D, and $f(z) \equiv u + iv$ is analytic there. Under these conditions, let C now be a regular Jordan curve in D; if n denotes the interior normal of C, the equation $\partial u/\partial n = -\partial v/\partial s$ follows from the Cauchy-Riemann equations, whence obtains Eq. (3). The first and last members of this equation

$$\int_C \frac{\partial u}{\partial n} ds = -\int_C \frac{\partial v}{\partial s} ds = -v(x,y) \Big|_C = 0 \qquad (3)$$

form the flux theorem, namely that the total flux (of heat if u is temperature) over C is zero. *See* COMPLEX NUMBERS AND COMPLEX VARIABLES.

If $u(x,y)$ is harmonic in the closed disk bounded by the circumference γ, and $f(z)$ the corresponding analytic function, one can take the real parts of both members of the equations expressing Cauchy's integral formula, as in Eqs. (4).

$$z - z_0 = \rho(\cos\theta + i\sin\theta) \qquad (4a)$$

$$dz = i(z - z_0) d\theta \qquad (4b)$$

$$f(z_0) = u(x_0,y_0) + iv(x_0,y_0) = \frac{1}{2\pi i} \int_\gamma \frac{f(z) dz}{z - z_0} \qquad (4c)$$

$$= \frac{1}{2\pi} \int_\gamma f(z) d\theta \qquad (4d)$$

$$u(x_0,y_0) = \frac{1}{2\pi} \int_\gamma u(x,y) d\theta \qquad (4e)$$

Equation (4e) expresses Gauss's mean value theorem, that the average of $u(x,y)$ over γ is the value at the center of γ. From this theorem it follows that a function harmonic at a point (x_0,y_0) cannot have a strong local maximum (or minimum) there, and can have a weak local maximum (or minimum) only if identically constant throughout a neighborhood of (x_0,y_0). If $u(x,y)$ is harmonic in a bounded region D, continuous in the corresponding closed region $\bar{D}$, the maximum and minimum of $u(x,y)$ occur on the boundary of D; if a maximum

or minimum occurs interior to D, then $u(x,y)$ is identically constant throughout D.

If D is a bounded region with boundary B, and if continuous values $U(x,y)$ are assigned on B, the Dirichlet problem is the problem of determining a function $u(x,y)$ harmonic in D, continuous on $D + B$, equal to $U(x,y)$ on B. If D is a circular region, a Jordan region, or any nonpathological region, the Dirichlet problem has a solution, necessarily (by the absence of nontrivial maxima and minima interior to D) unique. If D is a circular disk of radius a (see illustration), the Dirichlet problem for D is solved by Poisson's integral, Eq. (5), using polar

$$u(r,\theta) = \frac{1}{2\pi} \int_0^{2\pi} \frac{(a^2 - r^2) U(\psi) d\psi}{a^2 - 2ar\cos(\theta - \psi) + r^2} \qquad (5)$$

coordinates (r,θ) with pole the center of D. If D is a less elementary region but with smooth boundary B, the Dirichlet problem is solved by Green's formula, Eq. (6), where n indicates the interior normal in this formula.

$$u(x,y) = \frac{1}{2\pi} \int_B U(\xi,\eta) \frac{\partial g}{\partial n} ds(\xi,\eta) \qquad (6)$$

Green's function $g(x,y;\xi,\eta)$ is harmonic in D except at (x,y), continuous and equal to zero on B, and in the neighborhood of (x,y) has the form $\frac{1}{2} \log [(\xi - x)^2 + (\eta - y)^2] + g_1(\xi,\eta)$, where $g_1(\xi,\eta)$ is harmonic at (x,y). If the boundary B is not smooth, this formula can be expressed in terms of harmonic measure instead of $(\partial g/\partial n) ds$.

Numerous series expansions (for example, Fourier's series) can be used for the solution of the Dirichlet problem for various regions.

n-Dimensional relations. The foregoing remarks apply to Laplace's equation with two independent variables; the facts (but not the methods of proof using analytic functions) apply also in three or more dimensions. Thus, in three dimensions, a point distribution of matter of masses m_k at points (x_k,y_k,z_k) has a potential defined by Eq. (7),

$$u(x,y,z) \equiv \Sigma m_k [(x - x_k)^2 + (y - y_k)^2 + (z - z_k)^2]^{-1/2} \qquad (7)$$

which is harmonic except in the points (x_k,y_k,z_k). Except at such points, the force (Newtonian law of gravitation) exerted by the distribution on a unit exploratory particle at (x,y,z) has the components $(\partial u/\partial x, \partial u/\partial y, \partial u/\partial z)$, and the component of the force in any direction is the directional derivative of $u(x,y)$ in that direction. *See* POTENTIALS; SPHERICAL HARMONICS.   [JOSEPH L. WALSH]

Bibliography: G. H. D. Duff and D. Naylor, *Differential Equations of Applied Mathematics*, 1966; O. D. Kellogg, *Foundations of Potential Theory*, 1929.

Laplace's irrotational motion

Laplace's equation for irrotational motion of an inviscid, incompressible fluid is partial differential equation (1), where x_1, x_2, x_3 are rectangular carte-

$$\frac{\partial^2 \phi}{\partial x_1{}^2} + \frac{\partial^2 \phi}{\partial x_2{}^2} + \frac{\partial^2 \phi}{\partial x_3{}^2} = 0 \qquad (1)$$

sian coordinates in an inertial reference frame, and Eq. (2) gives the velocity potential. The fluid

$$\phi = \phi(x_1,x_2,x_3,t) \qquad (2)$$

velocity components u_1, u_2, u_3 in the three respec-

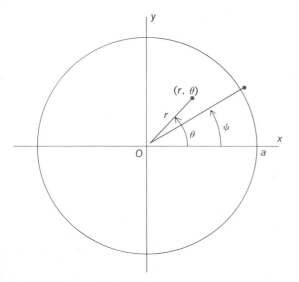

Circular disk D of radius a.

LARCH

(a)

(b)

Characteristic structures of two species of larch. Cone and needles of (a) tamarack (*Larix laricina*) and (b) western larch (*L. occidentalis*).

tive rectangular coordinate directions are given by $u_i = \partial\phi/\partial x_i$, $i = 1,2,3$. More generally, in any inertial coordinate system, the equation is div (grad ϕ) = 0 and the velocity vector is $\mathbf{v} = $ grad ϕ.

Irrotational motion implies that the fluid particles translate without rotation (like the cars on a ferris wheel) and is stated mathematically by saying curl $\mathbf{v} = 0$ where $\mathbf{v} = \mathbf{v}(\mathbf{r},t)$ is the velocity vector, $\mathbf{r}$ is the position vector of a particular point in the fluid flow, and t is the time. If the fluid motion is at any time irrotational it will stay irrotational. Thus any motion starting from rest will be irrotational. If curl $\mathbf{v} = 0$ then $\mathbf{v}$ may be written as grad ϕ because curl (grad ϕ) is identically zero. For an incompressible fluid, the continuity equation is div $\mathbf{v} = 0$; hence, combining this relation with irrotationality gives Laplace's equation, div (grad ϕ) = 0. *See* FLUID-FLOW PRINCIPLES; KELVIN'S CIRCULATION THEOREM.

The velocity field $\mathbf{v}(\mathbf{r},t)$ in a certain region is determined by Laplace's equation with a boundary condition given on the entire surface surrounding the region. The two most common boundary conditions are those at a solid surface and at a free surface. At a solid surface the fluid velocity normal to the surface must match the velocity of the surface normal to itself, $\mathbf{v} \cdot \mathbf{n} = \mathbf{v}_s \cdot \mathbf{n}$; that is, $\partial\phi/\partial n = \mathbf{v} \cdot \mathbf{n}$ is given on the boundary. At a free surface, such as one occurring between two fluids of different density, the pressure must be continuous; this boundary condition, in general, involves the use of the nonstationary Bernoulli equation and usually leads to wave motion. *See* BERNOULLI'S THEOREM; WAVE MOTION IN FLUIDS.

[ARTHUR E. BRYSON]

Laplacian

The differential operator $\partial^2/\partial x^2 + \partial^2/\partial y^2 + \partial^2/\partial z^2$, in which the symbols x,y,z denote the variables of a rectangular cartesian coordinate system. The Laplacian is frequently denoted by the symbol ∇^2 (read del square) in accordance with the fact that the Laplacian of a scalar function $S(x,y,z)$ is the divergence of the gradient of S, that is, the equation below applies.

$$\partial^2 S/\partial x^2 + \partial^2 S/\partial y^2 + \partial^2 S/\partial z^2 = \nabla \cdot (\nabla S)$$

The Laplacian operator is involved in some of the most fundamental equations of mathematical physics, namely, Laplace's equation ($\nabla^2 u = 0$), Poisson's equation, various wave equations (as those of electricity and magnetism, sound, vibrations, the Schrödinger equation of quantum mechanics), and the heat flow and diffusivity equations. *See* CALCULUS OF VECTORS; GAUSS' THEOREM; GRADIENT OF A SCALAR; GREEN'S THEOREM.

[HOMER V. CRAIG]

Lapping

A precision abrading process used to bring a surface to a desired state of refinement or dimensional tolerance by removal of an extremely small amount of material. Lapping is accomplished by abrading a surface with a fine abrasive grit rubbed about it in a random manner. Usually less than 0.0005 in. of stock is removed.

A loose unbonded grit is used. It is traversed about with a mating piece or lap of a somewhat softer material than the workpiece. The unbonded

grit is mixed with a vehicle such as oil, grease, or soap and water compound. When a bonded grit is used, it may be in the form of a bonded abrasive lap or a charged lap such as cast iron or copper with the lapping compound embedded in it. In some cases, abrasive-covered paper laps are used.

Although some lapping is done by hand, most production work is done on a lapping machine. Various types are designed for work on flat, cylindrical, and spherical surfaces. *See* GRINDING; MACHINING OPERATIONS.   [ALLEN H. TUTTLE]

Larch

A genus, *Larix*, of the pine family, with deciduous needles and short spurlike branches, which annually bear a crown of needles. The cones are small and persistent, varying by species in size, number, and form of the cone scales. The tamarack (*L. laricina*), also called hackmatack, is a native species, has erect, narrowly pyramidal habit, and grows in wet and moist soils in the northeastern United States, west to the Lake states, and across Canada to Alaska. The cones are $\frac{1}{2}$ to $\frac{3}{4}$ in. long (illustration a). The tough resinous wood is durable in contact with the soil and is used for railroad ties, posts, sills, and boats. Other uses include the manufacture of excelsior, cabinet work, interior finish, and utility poles. *See* PINALES.

The western larch (*L. occidentalis*), the most important and largest of all the species, grows in the northwestern United States and southeastern British Columbia. The cones of this species are larger, 1 to $1\frac{1}{2}$ in. long, with bracts growing out beyond the cone scales (illustration b). The trunk is tall and erect, sometimes attaining a height of 200 ft and a diameter of 6–7 ft. The western larch has an estimated stand of 20,000,000,000–30,000,000,000 board ft, mostly in the national forests. The annual production of the lumber usually ranges from 200,000,000 to 300,000,000 board ft. More than one-half of the production comes from Montana, the remainder from Idaho, Washington, and Oregon.

The European larch (*L. decidua*) has cones about twice the size of those of the tamarack and 40–50 scales to a cone, whereas tamarack has only 12–15. The European larch does better in drier soil and is the species usually planted in parks and private grounds.

Golden larch (*Pseudolarix amabilis*), from China, is occasionally cultivated. Its leaves, also deciduous, are golden yellow in the fall. The cone scales fall off one by one, leaving the central axis of the cone on the tree. *See* FOREST AND FORESTRY; TREE.   [ARTHUR H. GRAVES/KENNETH P. DAVIS]

Larmor precession

A precession in a magnetic field of the motion of charged particles or of particles possessing magnetic moments.

Charged particles. The Larmor theorem (J. Larmor, 1897) states that, for electrons moving in a single central field of force, the motion in a uniform magnetic field H is, to first order in H, the same as a possible motion in the absence of H except for the superposition of a common precession of angular frequency given by Eq. (1). Here e/c is

$$\omega_L = eH/2mc \qquad (1)$$

the magnitude of the electronic charge in electromagnetic units, and m is the electronic mass. The frequency ω_L is called the Larmor frequency and is numerically equal to 2π times 1.40 MHz per oersted or 2π times 111 MHz per SI unit of magnetic field strength (ampere-turn per meter). *See* PRECESSION.

The Larmor theorem is derived in numerous texts. For the special case of an electron moving in a circular orbit of radius r about a fixed nucleus, with H applied normal to the plane of the orbit, the derivation is as follows: The centripetal force holding the electron in orbit must equal $m\omega^2 r$ and is the sum of the Coulomb force Ze^2/r^2 and the Lorentz force, $(e/c)\omega r H$. Therefore Eq. (2) is valid, where

$$\omega = \pm[\,(eH/2mc)^2 + (Ze^2/mr^3)\,]^{1/2} + (eH/2mc)$$
$$= \pm(\omega_L{}^2 + \omega_0{}^2)^{1/2} + \omega_L \qquad (2)$$

ω_0 is the angular frequency in the absence of H. If $\omega_0 \gg \omega_L$ (bound electron, and first order in H) the approximate angular frequency is given by Eq. (3),

$$\omega = \pm\omega_0 + \omega_L \qquad (3)$$

which is the Larmor theorem. For a free or unbound electron (no Coulomb force), the approximation breaks down, but direct solution of the equation involving $m\omega^2 r$ and the Lorentz force yields $\omega = eH/mc$. This is twice the Larmor frequency and is called the cyclotron frequency. *See* PARTICLE ACCELERATOR.

In stating the Larmor theorem, use was made of the phrase "a possible motion." If H is applied sufficiently slowly, it can be proved that the motion is the same as in the absence of H, except for the superposition of the Larmor precession. However, a sudden application of H may change, for example, a circular orbit into an elliptical one. For an important application of the Larmor theorem *see* DIAMAGNETISM. *See also* ELECTRON MOTION IN VACUUM.

Magnetic moments. According to elementary electromagnetic theory, a current loop of area A and of current I possesses a magnetic moment μ of magnitude IA and of direction normal to the loop. Thus an electron moving with a velocity v in a circular orbit of radius r, and hence with current $(-e/c)(v/2\pi r)$ in emu, has an orbital magnetic moment of magnitude as given by Eq. (4). *See* MAGNETIC MOMENT.

$$\mu = (-e/c)\,(v/2\pi r)\,(\pi r^2) = -(evr/2c) \qquad (4)$$

The electron also has orbital angular momentum mvr, which by quantum theory must equal $\hbar J$, where J is an integer and $\hbar$ is Planck's constant h divided by 2π. The ratio of magnetic moment to angular momentum (Eq. 5) is called the magneto-

$$\gamma_J \equiv \frac{\mu}{\hbar J} = \frac{-e}{2mc} \qquad (5)$$

gyric (and often the gyromagnetic) factor γ_J. *See* ANGULAR MOMENTUM; GYROMAGNETIC RATIO; QUANTUM THEORY, NONRELATIVISTIC.

In terms of the equivalent magnetic moment, Eq. (1) may be written in the form of Eq. (6). In this

$$\omega_L = -\gamma_J H = -(\mu/\hbar J)\,H \qquad (6)$$

form the Larmor precession is exhibited by any magnetic moment μ, including magnetic moments associated with spin angular momentum as well as those associated with orbital angular momentum. Equation (6) may also be derived from equating the time rate of change of angular momentum (d/dt) $(\hbar J)$ to the magnetic torque $\mu \times H$, as in Eq. (7). In

$$d(\hbar J)/dt = \mu \times H = \gamma_J \hbar J \times H \qquad (7)$$

this form the Larmor precession applies to experiments in molecular beams, electron paramagnetic resonance (EPR), and nuclear magnetic resonance (NMR). *See* ELECTRON PARAMAGNETIC RESONANCE (EPR) SPECTROSCOPY; ELECTRON SPIN; MAGNETIC RESONANCE.

Rotating coordinate system. Let $(\partial/\partial t)$ represent differentiation with respect to a coordinate system rotating with angular velocity ω. Then differentiation with respect to a stationary observer (d/dt) is given by Eq. (8). Here J is mea-

$$(dJ/dt) = (\partial J/\partial t) + (\omega \times J) \qquad (8)$$

sured by the stationary observer. This equation may be combined with Eq. (7) in the form of Eq. (9). Here H_r is the effective field in the rotating coordinate system as given by Eq. (10). Therefore,

$$\frac{\partial(\hbar J)}{\partial t} = \gamma_J \hbar J \times \left(H + \frac{\omega}{\gamma_J}\right) = \gamma_J \hbar J \times H_r \qquad (9)$$

$$H_r = H + (\omega/\gamma_J) \qquad (10)$$

in a frame which is rotating at the Larmor frequency, the effect of a constant field H is reduced to zero.

This result, which is an extension of Larmor's original theorem, also holds in quantum mechanics. It is the basis of simplified analyses of the effects of oscillating magnetic fields on particles with charges and magnetic moments.

[ELIHU ABRAHAMS; FREDERIC KEFFER]

Bibliography: R. P. Feynman, *The Feynman Lectures on Physics*, 1964; N. F. Ramsey, *Molecular Beams*, 1956.

Larnite

A rare nesosilicate mineral with composition Ca_2SiO_4, originally described in 1929 from Scawt Hill, County Antrim, Ireland. Well-formed crystals have not been observed, but the presence of two mutually perpendicular cleavages with polysynthetic twinning parallel to one of them indicates monoclinic symmetry. Three artificial calcium silicates of this same composition are designated as α, β, and γ. Larnite probably corresponds to α-Ca_2SiO_4. Transformation to the γ-phase may be produced by heating or shock. At Scawt Hill, larnite occurs intimately associated with spurrite, melilite, merwinite, and spinel in a limestone contact zone. It has also been described from Crestmore, near Riverside, Calif. *See* SILICATE MINERALS.

[CORNELIUS S. HURLBUT, JR.]

Larvacea

A class of the subphylum Tunicata consisting of minute planktonic animals in which the tail, with dorsal nerve cord and notochord, persists throughout life. Two gill slits are commonly present, but an atrium is lacking. The epidermis of the anterior

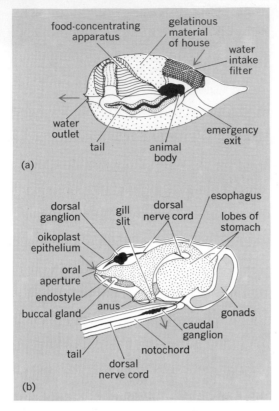

Oikopleura dioica, a larvacean. (*a*) In its house, a gelatinous tunic with strainers to filter food. (*b*) Body and anterior tail region. Arrows indicate the direction of water flow. (*Modified after Körner*)

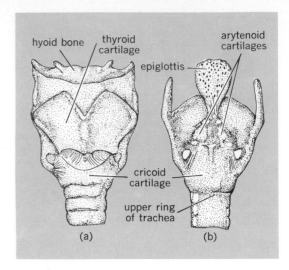

Human laryngeal cartilages and ligaments. (*a*) Front view. (*b*) Back view. Epiglottis acts as lid to larynx. (*After Sappey, from J. Symington, ed., Quain's Elements of Anatomy, vol. 2, pt. 2, Longmans, Green, 1914*)

region secretes a gelatinous tunic, which is composed of polysaccharides bearing amino groups, but contains no cellulose. In some species this tunic is highly complex and encases the whole body in a "house" equipped with chambers and strainers which enable the inhabitant to filter and concentrate very small plankton from the sea. About 60 species are known; *Oikopleura* (see illustration) and *Fritillaria* are the largest genera. *See* TUNICATA.

[DONALD P. ABBOTT]

Larynx (vertebrate)

The complex of cartilages and related structures at the opening of the trachea, or windpipe, into the pharynx, or throat.

The cartilages of the larynx (see illustration) are derivatives of the primitive gill-bar system. Their original function is to act as a protective sphincter at the entrance of the air passage into the throat, and, when this is the only function, the structure is quite simple, with only one or two pairs of cartilages.

In higher vertebrates the additional function of phonation has been acquired; it is almost limited to mammals since in birds the sound is produced in the syrinx lower in the trachea, but some frogs and reptiles can produce sounds with their larynges.

In man and most other mammals, the signet-shaped cricoid cartilage forms the base of the larynx and rests upon the trachea. The thyroid carti-

lage, which forms the prominent Adam's apple ventrally, lies anterior to the cricoid. Dorsally there are paired pivoting cartilages, the arytenoids. Each is pyramid-shaped and acts as the movable posterior attachment for the vocal cords and the laryngeal muscles that regulate the cords. Two other small paired cartilages, the cuneiform and the corniculate, also lie dorsal to the thyroid cartilage. The epiglottis, a leaf-shaped elastic cartilage with its stem inserted into the thyroid notch, forms a lid to the larynx. *See* LARYNX DISORDERS; SPEECH; THYROID GLAND (VERTEBRATE).

[THOMAS S. PARSONS]

Larynx disorders

Diseases of the larynx manifest themselves by hoarseness and by stridor, a form of noisy breathing caused by localized narrowing in the larynx or trachea.

Laryngitis is an inflammation of the mucous membrane of the larynx always associated with hoarseness. It frequently occurs with common colds and as a complication of other inflammatory diseases of the upper respiratory system. In diphtheria, the formation of a membrane of fibrin, leukocytes, destroyed tissue, and bacteria can cause severe respiratory difficulties, which might demand tracheotomy. The development of chronic laryngitis is favored by a chronic irritation such as that caused by smoking.

Hoarseness is also a manifestation of paralysis of the recurrent laryngeal nerve of the vagus, which is easily damaged upon surgical removal of a goiter. *See* NERVOUS SYSTEM (VERTEBRATE).

Benign tumors of the larynx occur in the younger age group. A tumorlike formation of the vocal cord, known as singer's node, often accounts for the hoarseness of people who abuse the voice. Cancer of the larynx is not uncommon in man, but is frequently of a slowly growing type and the outlook is often good. The highest incidence is in males over 60.

[EWALD R. WEIBEL]

Laser

A device that uses the maser principle of amplification of electromagnetic waves by stimulated emission of radiation and operates in the infrared, optical, or ultraviolet region. For these short wavelengths, it is not practical to construct amplifiers or oscillators by scaling down conventional electronic devices. However, atoms, ions, molecules, and crystals have resonances through this region. These substances ordinarily absorb light of some characteristic wavelength, but when suitably excited they can emit light spontaneously or be stimulated to emit by a light wave. If the excitation is vigorous enough, amplification of light by stimulated emission of radiation can predominate. Thus these molecular resonances can be used in laser amplifiers and as generators of coherent radiation throughout the entire wavelength region from submillimeter through the infrared, visible, and ultraviolet, down to wavelengths as short as 110 nm in the vacuum ultraviolet. With a suitable structure, a laser can produce a highly directional, intense light beam. The term "laser" is derived from light amplification by stimulated emission of radiation. *See* MASER.

Structure. Although the basic resonance in a maser is that of the atom, the microwave maser also uses a cavity resonator. The dimensions of the cavity resonator are usually the same order as the wavelength (about a few centimeters). The resonator stores the wave long enough so that it has time to interact strongly with the excited atoms. If such a box were scaled down to match the dimensions of a light wave, it would be impossibly small. Therefore, resonators of more convenient dimensions, many times the wavelength of the light, are used. In one such resonator (Fig. 1), two small end walls or plates face each other at a distance which is large compared with their diameters. A wave starting out near one plate can grow as it passes through the excited atoms. If the wave moves straight along the axis, it can bounce back and forth between the plates and stay in the resonator long enough to build up a strong oscillation. A wave oriented in any other direction soon passes off the edge of the plates and is lost before it has been much amplified.

One of the end plates is partially transparent, so that part of the wave can emerge through it. The output wave, like those being amplified between the plates, travels almost exactly along the axis and is thus very nearly a plane wave.

Comparison with other light sources. In contrast to lasers, all conventional light sources are basically hot bodies. The electrons in the tungsten filament of an incandescent lamp are agitated by, and acquire excitation from, the high temperature of the filament. Once excited, they emit light and revert to a lower-energy state. Similarly, in a gas lamp the electron current excites the atoms to high quantum states, and they give up this excitation energy by radiating it as light quanta. This spontaneous radiation takes place quite rapidly at optical wavelengths. However, the rate of spontaneous emission is proportional to $1/\lambda^3$, where λ is the wavelength, so that spontaneous emission occurs more and more slowly at longer wavelengths. Thus, conventional light sources, which depend on

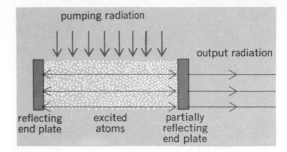

Fig. 1. Structure of a parallel-plate laser.

spontaneous radiation, are weak in the infrared region.

At any wavelength, the most that a hot body can radiate is given by the blackbody radiation curve of M. Planck. The surface of the Sun radiates like a blackbody at a temperature of 6000 K, and emits a total of 7 kW/cm². This is spread out over all wavelengths, however, and only a very tiny fraction is emitted at any particular wavelength. *See* HEAT RADIATION.

Moreover, spontaneous emission from each excited electron or atom takes place independently of emission from the others. The overall wave produced by a conventional light source is a jumble of waves from the numerous individual atoms. The phase of the wave from one atom has no relation to the phase from any other atom. Thus, the overall phase of the light emitted fluctuates randomly from moment to moment and from point to point on the wavefront; that is, the light from conventional sources is incoherent.

In contrast to this, an ideal plane wave would exhibit spatial coherence in that the phase would be the same all across any one of its plane wavefronts. Similarly on a spherically converging or diverging wave, the spherical wavefront surface would have an instantaneous phase that is everywhere the same.

The output of the parallel-plate laser is extremely directional and thus very nearly a plane wave; consequently, it is spatially coherent. This coherence arises because the atoms are stimulated to emit in phase with the wave rather than independently. The output is powerful because the atoms can be stimulated to emit much faster than they would spontaneously. It is also very nearly monochromatic because stimulated emission is a resonance process, which occurs most rapidly at the center of the range of wavelengths that would be emitted spontaneously. The atoms are stimulated to emit in phase with the existing wave, and so the phase is preserved over many cycles. Thus, the laser radiation has a high degree of time coherence.

Optically pumped lasers. To amplify by stimulated emission, a laser needs a supply of atoms in the proper excited state. Since these atoms decay spontaneously, the supply must be provided in a short time by some rapid pumping mechanism. Lasers are usually classified by their type of excitation.

Many lasers are three-level lasers. That is, atoms are excited by the absorption of light to a broad upper level, from which they very quickly relax to

Fig. 2. Tunable dye laser, pumped by ultraviolet pulse from a nitrogen laser. Light from nitrogen laser (not shown) enters from right through quartz lens. Diffraction grating acts as a tunable mirror. (*Photo by Frans Alkemade, Stanford University*)

the emitting state. Laser action occurs as they are stimulated to emit radiation and so return to the original ground level. Because of the short wavelengths involved, the pumping radiation comes from a powerful lamp. Solid, glass, liquid, and gaseous systems have been found suitable, but many possible materials remain to be explored.

Solid three-level lasers usually make use of ions of a rare-earth element or of a transition metal, such as chromium, dispersed in a transparent crystal or glass. For example, ruby, which is crystalline aluminum oxide containing a fraction of a percent of trivalent chromium ions in place of aluminum ions, has been used for optical masers producing red light with wavelengths of 693–705 nm. The chromium ions, like several others suitable for optical masers, have broad absorption bands in which pumping radiations can be supplied. Thus broad-band or white light can be used to excite the atoms.

Many rare-earth ions, such as neodymium, use a fourth level above the ground level. This level serves as the terminal level for the laser transition. It is kept empty by rapid nonradiative relaxation to the ground state. Such materials require relatively low pumping light intensity for laser action. Neodymium ions can provide laser action in many host materials, producing outputs around 1 μm in the near infrared. In glass, which can be made in large sizes, neodymium ions can generate high-energy pulses or very high peak powers. Lasers using neodymium ions in crystals such as yttrium aluminum garnet can provide continuous output powers up to a kilowatt. The output of either type (in glass or in crystals) can be converted to visible light near 500 nm by a harmonic generator crystal such as barium sodium niobate.

The structure of a solid-state, three-level optical maser can be especially simple. Essentially, it is a rod of the material with parallel ends polished and coated to reflect light. Pumping radiation enters through the transparent sides. *See* OPTICAL PUMPING.

Other structures can be used. The mirror ends can be spherical rather than plane, with the common focal point of the two mirrors lying halfway between them. Still other structures make use of

internal reflection of light rays that strike the surface of a crystal at a high angle.

Liquid lasers have structures generally like those of optically pumped solid-state lasers, except that the liquid must be contained in a transparent cylindrical cell. Some liquid lasers use rare-earth ions in suitable dissolved molecules, while others make use of organic dye solutions. The dyes can provide a range of output wavelengths, which are adjustable by altering the composition or concentration of the dye or solvent. Thus some degree of tunability is obtained. Fine adjustment of the output wavelength can be provided by using a diffraction grating in place of one of the laser end mirrors. The grating acts as a good mirror for only one wavelength, which depends on the angle at which it is set. With further refinements, liquid dye lasers can be made extremely monochromatic as well as broadly tunable. They may be pumped by various gas lasers to generate either short, intense pulses or continuous-wave output (Fig. 2). Dyes may also be incorporated in solid media, such as plastics or gelatin, to provide tunable laser action. Then the tuning may be controlled by a regular corrugation in the refractive index of the host medium, which acts as a distributed Bragg reflector. The reflection from any one layer is small, but when the successive alternations of refractive index provide reflections in phase, the effect is that of a strong, sharply tuned reflection.

In several infrared regions, tunable laser action can be obtained by using an infrared gas laser to pump a semiconductor crystal in a magnetic field, giving amplification by stimulating spin-flip Raman scattering from the electrons in the semiconductor. Tuning is achieved by varying the magnetic field.

High-power, short-pulse lasers. Optically pumped solid-state lasers provide relatively high peak-output powers. Tens of kilowatts can easily be obtained in a pulse lasting a few hundred microseconds. Much higher peak powers and correspondingly shorter pulse lengths can be obtained by the Q-switch technique. In this method, the optical path between one mirror and the amplifying rod is blocked by a shutter. The rod is then excited much beyond the degree ordinarily needed, but the shutter prevents laser action. At this time the shutter is abruptly opened and the stored energy is released in a giant pulse of 1–100 MW peak power, lasting 1–30 nanoseconds. Still higher peak powers can be obtained by passing this output through a traveling-wave laser amplifier without mirrors. Peak powers in excess of 1000 MW have been obtained in this way.

Still shorter pulses can be generated by the mode-locking techniques. Without mode locking, a typical laser often oscillates simultaneously and independently at several very closely spaced wavelengths. These modes of oscillation can be synchronized so that the peaks of their waves occur simultaneously at some instant. The result is a very short, sharp pulse, which quickly ends as the waves of different frequencies get out of step. Mode-locked lasers have generated pulses as short as 1 picosecond. Since such brief pulses tend to produce somewhat less damage to materials, they can be amplified to very high peak intensities. Power outputs of picosecond pulses as high as 10^{12} W have been obtained.

For the highest peak powers, the output is fur-

Gratings and filters being tested by laser beam for use in the manufacture of advanced state-of-the-art electrooptical systems and components. (*Conductron Corp.*)

Research worker adjusts a dichromatic gelatin grating mechanism of an argon laser. As the laser beam passes through the grating, the light is dispersed into its color components. (*Conductron Corp.*)

ther intensified by one or more stages of traveling-wave laser amplification. The beam diameter is increased by some optical arrangement such as a telescope so as to prevent damage to the laser material. Sometimes the amplifying medium is divided into flat slabs separated by cooling liquid (Fig. 3). The open faces of the slabs present a large area to receive pumping light from flash lamps.

Development of very large multistage lasers has been undertaken for research on thermonuclear fusion. In a particularly large one, at the Lawrence Livermore Laboratory, a single neodymium glass oscillator is designed to drive 20 amplifier chains of glass rods and disks, each delivering a pulse of more than 10^{12} W. Focusing of all these pulses onto the surface of a small pellet of heavy hydrogen is designed to heat and compress the pellet by ablation, until it is so hot that hydrogen nuclei fuse to produce helium and release large amounts of energy. Ultimately, controlled laser fusion may become an important source of thermal, electrical, and chemical energy. *See* FUSION, NUCLEAR.

Gas-discharge lasers. Another large class of lasers makes use of nonequilibrium processes in a gas discharge. At moderately low pressures (of the order of 1 torr) and fairly high currents, the population of energy levels is far from an equilibrium distribution. Some levels are populated especially rapidly by the fast electrons in the discharge. Other levels empty particularly slowly and so accumulate large numbers of excited atoms. Thus, laser action can occur at many wavelengths in any of a large number of gases under suitable discharge conditions. For some gases, a continuous discharge, using either direct or radio-frequency current, gives continuous laser action. Output powers of continuous gas lasers range from less than 1 μW up to about 100 W in the visible region. Wavelengths generated span the ultraviolet and visible regions and extend up to more than 700 μm in the infrared. They thus provide the first intense sources of radiation in much of the far-infrared region of the spectrum.

The earliest, and still most widely used, gas-discharge laser utilizes a mixture of helium and neon. Various infrared and visible wavelengths can be generated, but most commonly they produce red light at 632.8-nm wavelength, with power outputs of a few milliwatts or less, although they can be as high as about a watt. Helium-neon lasers can be quite small and inexpensive. Argon- and krypton-ion gas-discharge lasers provide a number of visible and near-ultraviolet wavelengths with continuous powers typically about 1 to 10 W, but ranging up to more than 100 W.

Many molecular gases, such as hydrogen cyanide, carbon monoxide, and carbon dioxide, can provide infrared laser action. Carbon dioxide lasers can be operated at a number of wavelengths near 10.7 μm on various vibration-rotation spectral lines of the molecule. They can be relatively efficient, up to about 30%, and have been made large enough to give continuous power outputs up to tens of thousands of watts.

Many gas discharges produce only very small optical gain; thus losses must be kept low. Consequently, mirrors with very high reflectivity are used, and diffraction losses can be kept low by using spherical mirrors. One common arrangement, which combines relatively low diffraction

Fig. 3. High-power laser amplifier stage, 30-cm aperture, using liquid-cooled slabs of neodymium glass. (*From Lawrence Livermore Laboratory*)

losses and good mode selection, uses a flat mirror at one end and a spherical mirror at the other. The spacing between mirrors is made equal to the radius of a curvature of the spherical mirror.

On the other hand, in some of the higher-power lasers even the plane-parallel mirror structure does not given sufficient discrimination against those undesired modes of oscillation which cause the beam to be excessively divergent. It is then helpful to use "unstable" resonators with at least one of the mirrors convex toward the other.

A smooth, small-bore dielectric tube can guide a light wave in its interior with little loss. Thus a light wave can be amplified by a long, narrow medium without spreading by diffraction. A gas discharge in a hollow optical waveguide can be run at higher pressure and benefit from cooling by the nearby walls, so that relatively high-power outputs can be obtained from a small volume. Waveguide structures are also useful when a medium is pumped optically by another laser, whose narrow beam can be confined within the bore of the tube. For example, pumping of various molecules such as methyl fluoride by a carbon dioxide laser has been used to generate coherent light in the very-far-infrared (submillimeter wavelength) region.

Gas discharges can sometimes give laser action at several wavelengths simultaneously. Individual wavelengths can be selected from among these by using a dispersing prism between the discharge tube and one mirror.

Gas lasers have been constructed with nearly ideal monochromaticity, wavelength stability, and narrowness of beam.

Pulsed-gas lasers. Pulsed-gas discharges permit a further departure from equilibrium. Thus pulsed-laser action can be obtained in some additional gases. In some of them, the length of the laser pulse is limited when the lower state is filled by stimulated transitions from the upper level, and so introduces absorption at the laser wavelength. Nitrogen lasers give 5–10-ns pulses with peak powers from tens of kilowatts up to a few megawatts at a wavelength of 337 nm in the ultraviolet. They are much used for pumping tunable dye lasers throughout the wavelength range from about 350 to 1000 nm. Very powerful laser radiation in the vacuum ultraviolet region, between 100 and

200 nm, can be obtained from short-pulse discharges in hydrogen, and in rare gases such as xenon at high pressures. When the gas pressure is too high to permit an electric discharge, the excitation may be provided by an intense burst of fast electrons from a small accelerator.

Some gases, most notably carbon dioxide, which can provide continuous laser action, also can be used to generate intense pulses of length about 1 μs. For this purpose, gas pressures about 1 atm (1 atm = 101,325 Pa) are used, and the electrical discharge takes place across the diameter of the laser column, hence the name TEA laser, from transverse-electrical-atmospheric.

Chemical lasers. It is also possible to obtain laser action from the energy released in some fast chemical reactions. Atoms or molecules produced during the reaction are often in excited states. Under special circumstances there may be enough atoms or molecules excited to some particular state for amplification to occur by stimulated emission. Usually the reacting gases are mixed and then ignited by ultraviolet light or fast electrons. Both continuous infrared output and pulses up to several thousand joules have been obtained in reactions which produce excited hydrogen fluoride molecules.

Photodissociation lasers. Intense pulses of ultraviolet light can also dissociate molecules in such a way as to leave one constituent in an excited state capable of sustaining laser action. The most notable examples are iodine compounds, which have given peak infrared pulse powers above 10^9 W from the excited iodine atoms.

Nuclear lasers. Laser action in several gases has also been excited by the fast-moving ions produced in nuclear fusion. These fusion products excite and ionize the gas atoms, and make it possible to convert directly from nuclear to optical energy.

Gas dynamic lasers. When a hot molecular gas is allowed to expand suddenly through a nozzle, it cools quickly, but different excited states lose energy at different rates. It can happen that, just after cooling, some particular upper state has more molecules than some lower one, so that amplification by stimulated emission can occur. Very high continuous power outputs have been generated from carbon dioxide in large gas dynamic lasers.

Semiconductor lasers. Another method of providing excitation for lasers can be employed with certain semiconducting materials. Laser action takes place when free electrons in the conduction band are stimulated to recombine with holes in the valence band. In recombining, the electrons give up energy corresponding nearly to the band gap. This energy is radiated as a light quantum. Suitable materials are the direct-gap semiconductors, such as gallium arsenide. In them, recombination occurs directly without the emission or absorption of a quantum of lattice vibrations. A flat junction between p-type and n-type material may be used. When a current is passed through this junction in the forward direction, a large number of holes and electrons are brought together. A light wave passing along the plane of the junction can be amplified by stimulating recombination of these electrons and holes. The ends of the crystal provide the mirrors to complete the laser structure. In indirect-gap semiconductors such as germanium and silicon, because of their requirement for interaction with the lattice vibrations, only a small amplification by stimulated emission is usually possible.

Semiconductor lasers can be very small, less than 1 mm in any dimension. They can also have efficiencies as high as several tens of percent. Power densities are high, but the thinness of the active layer tends to limit the total output power. Even so, maximum continuous powers are comparable with those of other moderate-size lasers. Since semiconductor lasers are so small, they can be assembled into compact arrays of many units, so as to generate higher peak powers. An alternative excitation method, using bombardment of the semiconductor by a high-voltage beam of electrons, may provide laser action in larger crystal volumes, but it is likely to cause damage to the crystal.

Performance. Lasers are powerful, directional, coherent, and monochromatic. Pulsed-power outputs of some kilowatts have been obtained with wavelength spreads of less than 1 ppm. The very short wavelengths produced by lasers in the optical region permit very narrow beam angles. Lasers have produced beams parallel to within a small fraction of a degree. Such sharply directed beams will be useful for communication in space over very long distances. With the aid of an auxiliary telescope used backward, a beam could be sent to a spot on the Moon only a mile or so in diameter. An optical echo was observed when light from a pulsed ruby laser was scattered on the Moon's surface. Stronger reflections have been obtained from arrays of reflecting prisms placed at several locations on the Moon, so that the round-trip time of a short light pulse can be measured very accurately. Laser beams can be made so very powerful and directional that it is even possible that they could be used to communicate with other intelligent beings, if there are any, on planets of other stars. Over shorter ranges, the high carrier frequency of light waves would permit very-broad-band modulation, so that a single channel could carry enormous amounts of information. Modulated signals from small lasers can be transmitted over low-loss glass fibers, conveying broad-band telephone and picture signals. *See* OPTICAL COMMUNICATIONS.

A parallel beam can be focused by a lens or mirror to a point, or at least to a small area comparable in dimensions with the wavelength. Since this is a very small area, a laser beam can produce an enormous power density in such a focal spot. Pulsed-power densities of the order of 10^{15} watts/cm² have been attained.

The laser principle of amplification by stimulated emission can also be used to intensify a weak image of the proper wavelength (Fig. 4). Thousand-fold intensification with good resolution has been attained in a dye cell 1.3 mm long, pumped by a nitrogen laser. The bandwidth was about 0.15 nm, but this was sufficient to demonstrate that recognizable color differences were preserved during amplification. Laser image amplification is necessarily accompanied by considerable spontaneous emission, so that it is more suited to projecting high-intensity displays than for aiding vision under low illumination.

At the high power densities attainable by focusing laser beams, the electric field of the light can be quite large. Thus when the light intensity is 10^{10} W/cm², the corresponding electric field is 10^6 V/cm. To such large fields, many transparent materials have a nonlinear dielectric response. This nonlinearity can be large enough to permit the generation of optical harmonics. It is possible, with careful design and good nonlinear materials, to obtain substantially complete conversion of a laser's output to the second harmonic, at twice the frequency, even for continuous lasers near 1-watt output power. Nonlinear dielectrics can be used as mixers to give the sum or difference of two laser frequencies. They also permit the construction of optical parametric oscillators which, when pumped by a laser, can generate coherent light tunable over a wide range of wavelengths. *See* OPTICS, NONLINEAR. [ARTHUR L. SCHAWLOW]

Applications. Since lasers are stable sources of nearly monochromatic optical frequency waves — that is, they are highly coherent sources — their output radiation can be collimated to form a sharply directed beam or may be focused to a very small spot. Important related dimensions of an ideal single-mode laser beam are (1) w_0, the radius of the beam waist or focus; (2) $\pi w_0^2/\lambda$, the "near-field distance" — at this distance from its waist the beam has spread hyperbolically by diffraction to twice its minimum area; and (3) $\theta = \lambda/\pi w_0$, the half-apex angle of divergence of the beam in the far field.

By (3), a laser beam can be focused by a sharply convergent lens to a spot only a few wavelengths across. The near-field distance is the distance over which the focus is maintained, and a compromise must often be made between these two. Lenses and mirrors can change w_0 and θ; in the absence of aberrations, however, the product $\theta w_0 = \lambda/\pi$ remains constant. Many lasers show multimode operations; for these, the product θw_0 is larger than λ/π. When focused to a small spot, pulsed laser beams can melt or vaporize small volumes of any known material, including diamond. Since nothing but the light beam need touch the spot to be melted, small welds or holes can be made, for example, inside a sealed transparent container in a controlled atmosphere or in vacuum. Since the evaporated material is relatively free from contamination, clean thin films can be made this way. Minimum damage to surrounding areas is also thereby achieved. In a similar manner, laser beams have been used for welding detached retinas in eyes, a potentially important medical application.

When the beam waist w_0 is relatively large, say 1 cm, the near-field distance is quite long (0.3 km for $\lambda = 1\ \mu$), and the far-field divergence angle quite small (3×10^{-5} radian for the same example). In addition, the laser beam can be periodically refocused by positive lenses or mirrors. Such a beam can be used for communications. With available moderate continuous powers (of the order of 1 watt), communications with gigahertz (10^9 Hz) bandwidths and comparable information rates will be possible over distances of many miles. In addition, point-to-point communications through distances considerably greater than the dimensions of the solar system appear possible at more moderate information rates. Several systems for transmission of voice and television signals on laser beams have already been demonstrated.

Laser beams can also be transmitted along tiny

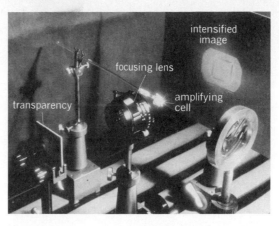

Fig. 4. Dye laser image amplifier. Light from a first dye cell (not shown) illuminates a transparency and then is focused through the amplifying cell to produce the intensified image. (*From T. W. Hansch, F. Varsanyi, and A. L. Schawlow, Image amplification by dye lasers, Appl. Phys. Lett., 18:108–110, 1971; photo by Frans Alkemade*)

(diameters of several microns) transparent fibers embedded in transparent material whose refractive index (ability to deflect light as a prism does) is slightly smaller than that of the fiber. The diffraction of light which spreads the beam in space is canceled by the fiber, and it can guide the light beam around corners. Use of such fibers may allow the development of a variety of optical frequency electronic circuits.

Many lasers can be "mode-locked" to produce a continuous sequence of extremely short optical pulses, each lasting less than 1 nanosecond (10^{-9} sec). Such short (traveling at the velocity of light, a pulse lasting 10^{-10} sec is only 3 cm long!) pulses can be used for communication, for example, by canceling out appropriate ones according to a binary coded message. Also of possible importance is the discovery that short intense optical pulses can under certain circumstances penetrate absorbing material by a process called self-induced transparency.

Interference effects. Because of the monochromaticity of laser light, interference effects can be observed over large distances (theoretically, over many miles). Thus, distances of many feet can be measured without too much difficulty to an accuracy of a fraction of an optical wavelength with an interferometer, and this ability is being put to use in the control of machine tools, for example. The National Bureau of Standards has remeasured the length of its standard meter bar using a laser interferometer as a secondary standard. To accomplish this operation, the laser wavelength must first be calibrated against a primary standard, in this case the wavelength of the mercury 198 line at $436\ \mu$. Interference effects can be obtained with the light from the primary standard only over about a decimeter, because of its spectral width.

The famous Michelson-Morley ether drift experiment has been repeated using laser light. The result, although more accurate, was the same as Michelson's. The experiment again demonstrated that the ether drift, if it exists at all, is less than a small fraction of the Earth's orbital velocity. *See* LIGHT; RELATIVITY.

Interference effects are also being applied to the problem of information storage and retrieval. If two laser beams of the same wavelength cross in photographic material, such as film, an interference pattern is recorded. When the developed film is illuminated again by one beam, the other beam is recreated. When one of the two original beams is replaced by coherent laser light reflected from an object, a more complicated interference pattern called a hologram is created, and a truly three-dimensional view of the object is seen when the recorded hologram is again illuminated by the second beam. Many different holograms which can be independently viewed can be recorded in a single thick piece of photographic material by multiple exposures using laser beams going in different directions. In an allied application, a laser beam can be deflected by sound waves in a transparent solid, the sound waves playing the role of the interference pattern in a hologram. In addition to information storage applications, these methods may pave the way for two or three dimensional television displays using lasers. *See* HOLOGRAPHY; INTERFERENCE OF WAVES; INTERFEROMETRY.

Spectroscopic analysis. Lasers are being used as monochromatic sources for the spectroscopic analysis of materials. In the visible and near-infrared region, white light sources in combination with suitable frequency filtering (with a monochromator) already provide adequate sources for most such work, but in the far-infrared ($\lambda > 10 \mu$) lasers or parametric oscillators driven by lasers may ultimately prove advantageous. Gas lasers now produce a multitude of wavelengths in the far-infrared. Continuous wavelength coverage is, however, not yet available. Of considerable importance for spectroscopic analysis is Raman spectroscopy. A portion of the light scattered from a laser beam passing through a transparent sample has its frequency lowered (Stokes lines) or raised (anti-Stokes lines) by the molecular vibration frequencies of the material of the sample. Light scattered from the needlelike laser beam can be conveniently focused on the entrance slit of a spectrometer, which can then measure the displacements of these Raman lines. Because of the high light intensity that can be achieved in a monochromatic laser beam, a substantial increase in sensitivity (the ability to detect weak light scattering) has been achieved over previous apparatus. This field of research has blossomed, with important results in investigations of the propagation of sound and other vibrational waves (phonons) as well as other elementary excitations of solids and liquids, and in investigations of the various phase transitions (liquid-gas and so on) that materials undergo as their temperature is varied. *See* RAMAN EFFECT; SPECTROSCOPY.

Nonlinear effects. When light passes through material, the charged particles (electrons, ions) that make up the material move in response to the oscillating electric field of the light wave. When the light field is strong enough that the amplitude of the particle motion is no longer proportional to the field strength, various important nonlinear effects make their appearance. Among these effects are the generation of power at twice the frequency of the laser beam (harmonic generation), the "mixing" of two laser beams to produce the sum and difference frequencies, and parametric amplification, wherein an optical or infrared signal at a lower frequency than that of the laser beam can draw power from the beam and be amplified. The importance of parametric amplification is that, unlike the parent laser, it can be varied over a broad range of wavelengths. Parametric amplifiers can be used also as optical generators, and while they work uncertainly at present, they most likely will soon provide us with broadly tunable generators of continuous coherent light.

[JAMES P. GORDON]

Bibliography: G. Birnbaum, Optical masers, *Advan. Electron. Electron Phys.*, suppl. no. 2, 1964; A. L. Bloom, *Gas Lasers*, 1968; B. Bova, *The Amazing Laser*, 1971; M. Brotherton, *Masers and Lasers: How They Work, What They Do*, 1964; R. Brown, *Lasers: Tools of Modern Technology*, 1968; J. P. Cedarholm et al., New experimental test of special relativity, *Phys. Rev. Lett.*, 1:342, 1958; S. S. Charschan (ed.), *Lasers in Industry*, 1972; C. C. Eaglesfield, *Laser Light*, 1967; D. Fishlock (ed.), *A Guide to the Laser*, 1967; J. A. Giordmaine, The interaction of light with light, *Sci. Amer.*, 210(4):38, 1964; A. F. Harvey, *Coherent Light*, 1970; O. S. Heavens, *Optical Masers*, 1964; H. A. Klein, *Masers and Lasers*, 1963; Lasers and light, *Readings from Scientific American*, introductions by A. L. Schawlow, 1969; B. A. Lengyel, *Introduction to Laser Physics*, 1966; B. A. Lengyel, *Lasers*, 1969; A. Maitland and M. H. Dunn, *Laser Physics*, 1969; J. F. Ready, *Effects of High Power Laser Radiation*, 1971; D. Röss, *Lasers: Light Amplifiers and Oscillators*, 1969; M. Sargent, III, M. O. Scully, and W. E. Lamb, Jr., *Laser Physics*, 1974; F. P. Schäfer (ed.), *Dye Lasers*, 1973; A. E. Siegman, *Lasers and Masers*, 1969; W. V. Smith and P. P. Sorokin, *The Laser*, 1966; C. S. Willett, *Introduction to Gas Lasers*, 1974; A. Yariv, *Introduction to Optical Electronics*, 1971; A. Yariv, *Quantum Electronics*, 1975.

Laser photobiology

The interaction of laser light with biological molecules, and the applications to biology and medicine. The spatial coherence of laser emissions makes it possible to focus beams in small volumes. Consequently, high-power-density irradiations can be obtained, especially with Q-switched lasers whose energy is delivered in a few nanoseconds or even picoseconds. These two characteristics combined with the great variety of wavelengths now available and the readiness of use have opened up many possibilities of applications to biology and medicine.

Physical aspects. In the visible range, absorption of laser light by biological material is caused only by pigments such as chlorophyll, hemoglobin, cytochromes, or melanin, while ultraviolet laser light is absorbed by nucleic acids and proteins. At low intensity, the absorption of photons produces photochemical effects similar to those obtained with ordinary light sources. By increasing the absorbed energy, heating becomes important, and the induced thermochemical effects (that is, thermal denaturations) are the most useful in biological applications. When the energy density on the target becomes very high, some nonlinear phenomena may occur, such as optical saturation,

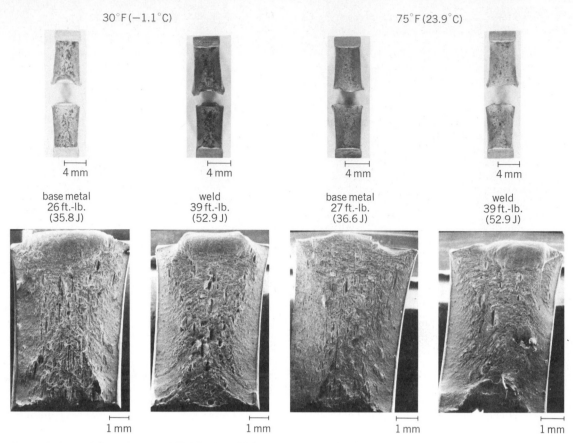

30°F (−1.1°C)   75°F (23.9°C)

| 4 mm | 4 mm | 4 mm | 4 mm |

| base metal 26 ft.-lb. (35.8 J) | weld 39 ft.-lb. (52.9 J) | base metal 27 ft.-lb. (36.6 J) | weld 39 ft.-lb. (52.9 J) |

| 1 mm | 1 mm | 1 mm | 1 mm |

Fig. 1. A change in the shape and distribution of the inclusions on the fracture surfaces of HY-130 weld and base metal impact specimens is shown. The top row comprises low-magnification macrophotographs of both halves of the fractured specimens. The bottom row comprises higher-magnification scanning electron micrographs of one-half of the fracture. Welds were made at 5 kW and a speed of 45 ipm. (*From E. M. Breinan and C. M. Banas, Fusion zone purification during welding with high power CO$_2$ lasers, Proceedings of the 2d International Symposium of the Japan Welding Society, Osaka, Aug. 24–28, 1975.*)

either oscillator/amplifier (Gaussian output beam) or unstable resonator (hollows output beam) optics. These lasers, available in output powers ranging from approximately 1000 to 15,000 W, have been used to demonstrate specific welding accomplishments in a variety of metals and alloys. In addition, several high-power laser systems have been designed, produced, and delivered for actual commercial applications. Substantial advances in laser technology made possible the production of fully automated multikilowatt industrial laser systems which can be operated on a continuous production basis. These systems can be used for a variety of development programs and on-line production applications.

Welding of HY-130 steel. A significant metallurgical phenomenon was reported following the laser welding of HY-130 alloy steel. After welding with a 7-kW continuous-wave coaxial-flow CO$_2$ laser with oscillator-amplifier optics (Gaussian mode beam), concurrent increases in the hardness, tensile strength, and impact energy of the welds were noted when compared with base metal properties. Welding was accomplished in the open atmosphere under gas shielding with no preheating, postheating, or filler metal additions. A decrease in the visible inclusion content was noted both on the impact fracture surface (Fig. 1) and in metallographic sections. Chemical analysis revealed significant decreases in the oxygen and nitrogen contents of the weld metal as compared to the base metal. It was concluded that during laser welding of this alloy a purification of the fusion zone with respect to inclusions, and perhaps a change to a more favorable inclusion size distribution, occurs.

Welding of pipeline steel. Single- and dual-pass laser welds have been made in an alloy of the 80,000-psi (5.5×10^8 N/m²) yield strength class. Figure 2 shows cross sections of these welds. The welds were formed with a 12-kW beam from a continuous-wave CO$_2$ laser operating in the TEM$_{00}$ mode. Welding speeds of 25 and 30 ipm (1.06 and 1.27 cm/sec) and 60 and 65 ipm (2.54 and 2.75 cm/sec) were used for the single- and dual-pass welds respectively. The welds were evaluated by visual inspection, x-ray, metallography, hardness tests, cross-weld tensile tests, transverse root, face, and side bend tests, longitudinal face bend tests, and Charpy impact tests at temperatures ranging from 70 to −100°F (21.1 to −73.3°C). The laser welds exhibited excellent overall mechanical properties and a Charpy shelf energy greater than 264 ft-lb (358 J), which is substantially greater than that of the base material. Dual-pass welds exhibited a ductile-to-brittle transition temperature below −60°F (−51.1°C). The increased shelf energy was attributed to a reduction in the visible inclu-

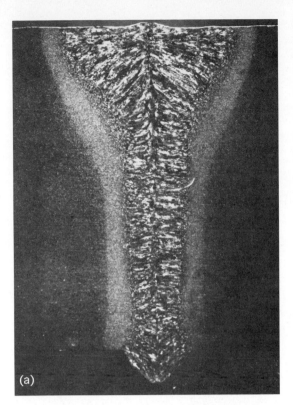

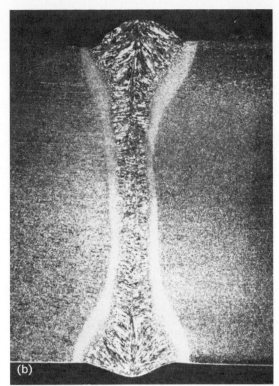

Fig. 2. Cross sections (0.52 in., or 1.32 cm, thick plate) of laser welds in X-80 arctic pipeline steel. (a) Single-pass weld. (b) Dual-pass weld. (*From E. M. Breinan and C.*

M. Banas, Weld. Res. Counc. Bull. No. 201, pp. 47–57, December 1974)

sion content of the fusion zone, while transition temperature was shown to be strongly dependent on grain size. These results indicated that laser welding should be a strong candidate for future pipeline welding in certain alloys.

Titanium alloys. Two in-depth studies involving laser welding of Ti-6Al-4V alloy were conducted. In one study, laser welding was compared with electron-beam and plasma-arc welding. Sound welds were made by all processes, and the tensile properties of these welds exceeded base metal properties. The laser and electron-beam welds, because of their lower specific energies for fabrication, were substantially finer-grained and possessed finer substructures than the plasma-arc welds. Due to the finer microstructure of the beam welds (fine acicular alpha), they exhibited lower K_{IC} (fracture toughness in plane strain) values than the plasma-arc welds, which had a coarser, Widmanstatten-type structure. The finer structure of the laser and electron-beam welds showed potential for good fatigue crack initiation resistance.

A comparison study of the fatigue behavior of laser and plasma-arc welds was also conducted. It was found that the distributions of fatigue lives at the 60,000–80,000 psi ($4.1-5.5 \times 10^8$ N/m²) stress level with $R = 0.1$ were quite similar for the laser welds and the state-of-the-art plasma-arc welds with which they were compared. Both techniques were capable of making weld specimens where fatigue failures initiated in the base metal; however, there were a number of specimens welded by each process in which fractures initiated at pores in the weld which were below the radiographic detectability limit of ~0.006 in. (0.015 cm).

While both the laser and plasma-arc processes appear to be capable of making useful welds in Ti-6Al-4V, the process which exhibits the best reproducibility and shows the least tendency to produce fine porosity will ultimately be favored for critical applications.

Aluminum alloys. While successful weld penetrations have been made in some aluminum alloys, and reasonable-looking beads have been produced, no extensive weld-quality studies or mechanical-property studies have been reported. Most weld penetrations in aluminum alloys made to date show some porosity, but it appears that the potential of welding aluminum exists if proper effort is devoted to solving the problems associated with aluminum laser welding. It was expected that considerable effort would be devoted toward laser welding of aluminum alloys during 1975.

Industrial applications. Five multikilowatt laser welding systems are in use for industrial welding and heat-treating applications. Ford Motor Company purchased an automated underbody welding system powered by a 6-kW laser from Hamilton Standard Division of United Technologies Corporation. Hamilton Standard also sold two 3-kW battery welding systems to the Western Electric Company. Caterpillar Tractor and General Motors Corporation both purchased 10-kW lasers (Model HPL-10) from AVCO Everett Research Laboratories, primarily for use in laboratory development of heat-treating applications.

The system which displays the advantages of laser processing most clearly is the underbody welding system at Ford (Fig. 3). As illustrated, a total of five degrees of freedom are achieved in

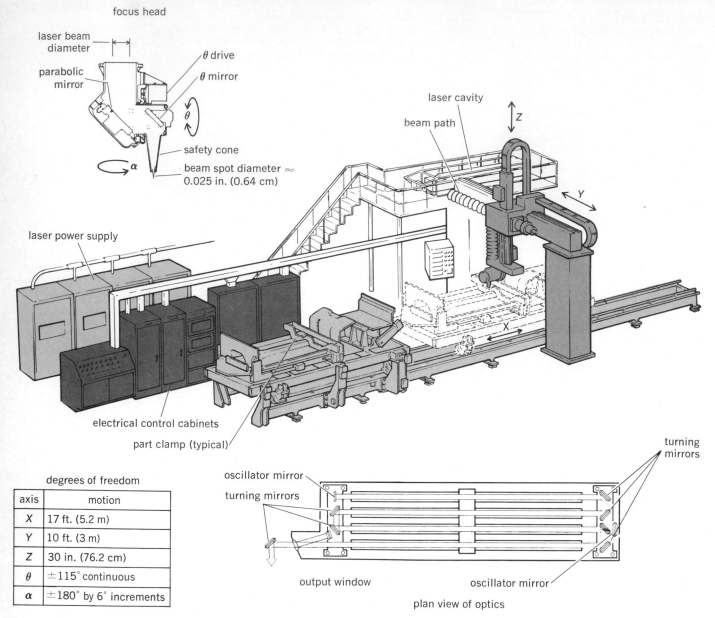

focus head

laser beam diameter

parabolic mirror

θ drive

θ mirror

θ

α

safety cone

beam spot diameter $\approx$ 0.025 in. (0.64 cm)

laser cavity

beam path

laser power supply

Z

Y

X

electrical control cabinets

part clamp (typical)

turning mirrors

oscillator mirror

turning mirrors

output window

oscillator mirror

plan view of optics

degrees of freedom	
axis	motion
X	17 ft. (5.2 m)
Y	10 ft. (3 m)
Z	30 in. (76.2 cm)
θ	$\pm 115°$ continuous
α	$\pm 180°$ by 6° increments

Fig. 3. Schematic diagram of 6-kW multiaxis laser welding system at Ford Motor Company; arrows indicate direction of motion along various axes. This is the first complete high-power laser welding system designed to perform a high-production-rate industrial fabrication operation. (*Hamilton Standard Division, United Technologies Corporation*)

movement of the focused beam relative to the intricately curved underbody panels. Only one of these, the X direction, is achieved by mechanical actuation of the parts; all the rest are achieved by rotation or translation of water-cooled copper mirrors. A unique 90° off-axis parabolic focusing mirror turns the beam 90° as it focuses the beam, thus allowing easy implementation of the α and θ axes. The focused beam creates a continuous lap weld in the thin steel sheet to a depth of 0.060 in. (0.15 cm) and at a speed of 500 in./min (21 cm/sec). The continuous weld achieves much greater strength than spot welding techniques, and assures leak-tightness of the joint. At that welding speed, four large panels are joined together into an underbody in a total cycle time of 1 min.

Future applications. As the power available in commercial lasers increases, the scope of applica-

tions is also increasing. In addition to assembly-line operations, several opportunities are seen in the area of welding of heavy sections for industrial purposes. Autogenous welds have been made at depths up to 1 in., or 2.54 cm (welding from both sides), and further research and development work should increase laser welding capabilities into weld depths of several inches. These capabilities will be of great utility in a number of industries, including shipbuilding, pipeline welding, and general heavy steel fabrication. The use of added filler metal, which has already proved feasible with laser, promises to extend the thickness capability of the laser welding process.

Laser welding has several potential advantages for deep welding tasks. For one thing, high welding speeds are achievable. For example, single-pass butt welds have been made in 0.5-in. (1.27 cm) steel at a speed of 50 ipm (2.1 cm/sec) with a 12-kW beam. In addition, the lack of necessity for filler metal or substantial reduction in the amount of filler required, promises economic advantages for the laser process, particularly when welding arctic pipeline steels and in other applications normally requiring large quantities of expensive filler metal. Reductions in edge preparation cost may also be realized due to the elimination of the edge chamfer required, although some machining of the butt edges is usually necessary to achieve a joint fitup which is suitable for laser welding. Another important factor is the ability of a single laser installation to supply a beam to several different welding stations, by appropriate indexing of turning mirrors. This capability, along with the high welding speed, allows high productivity on a laser welding installation.

A final advantage to laser welding is in the quality of the joint, apart from the low cost of welding. For quality-critical applications, such as welding of nuclear reactor components or arctic pipeline steel, the joint quality alone may prove to be an important factor in the selection of laser welding. *See* WELDING AND CUTTING OF METALS.

[EDWARD M. BREINAN]
Bibliography: E. M. Breinan and C. M. Banas, *Weld. Res. Counc. Bull. No. 201*, pp. 47–57, December 1974; F. F. Gagliano, *SME Pap. MR74-954*, 1974; E. V. Locke, *SME Pap. MR74-952*, 1974; F. D. Seaman, *SME Pap. MR74-957*, 1974; M. Yessick and D. J. Schmatz, *SME Pap. MR74-962*, 1974.

Latent image

An invisible image produced by a physical or chemical effect of light on the individual crystals (usually silver halide) of photographic emulsions. This image can be rendered visible by the process known as development. For details of latent image formation and development *see* PHOTOGRAPHY; PHOTOLYSIS.

[R. H. NOBLE]

Laterite

The name given by F. Buchanan in 1807 to the iron-rich weathering product of basalt in southern India. The term is now used in a compositional sense for weathering products composed principally of the oxides and hydrous oxides of iron, aluminum, titanium, and manganese. Iron-rich or ferruginous laterite is largely hematite, Fe_2O_3, and goethite, $HFeO_2$, and may be an ore of iron and nickel (Cuba, New Caledonia). Aluminous laterite is composed of gibbsite and boehmite, and is the principal ore of aluminum. Clay minerals of the kaolin group are typically associated with, and are genetically related to, laterite. Laterites range from soft, earthy, porous material to hard, dense rock.

Concretionary forms of varying size and shape commonly are developed. The color depends on the content of iron oxides and ranges from white to dark red or brown, commonly variegated. *See* BAUXITE; CLAY MINERALS; KAOLINITE; WEATHERING PROCESSES.

Origin. Laterite is formed by weathering under conditions that lead to the removal of silica, alkalies, and alkaline earths. The resulting concentrations of iron and aluminum oxides sharply differentiate lateritization from temperate-climate weathering in which the end product is largely clay minerals (hydrous aluminum silicates). Early workers from temperate regions considered lateritization as profound leaching (desilication) in a stage beyond ordinary kaolinization. Studies in tropical regions, however, show that weathering of alkaline silicates may yield gibbsite directly without passing through an intermediate clay stage.

Investigations in many parts of the world stress certain genetic factors. A tropical to subtropical climate with high temperature and abundant rainfall, seasonal or at least with periods of marked dryness, is fundamental. Relief sufficient to ensure good drainage is requisite: lateritic soils are permeable and do not erode by sheet wash like clay or shale. Aluminous laterite forms above the water table and may grade to clay in depth. It may be found on hills and on well-drained slopes, but the residual deposits in adjacent valleys usually are kaolin.

Parent materials. The parent material controls or greatly influences the composition of laterites, which may be developed from a variety of igneous, metamorphic, and sedimentary rocks. Iron-rich rocks (peridotite) yield iron ore; aluminous rocks (syenite) produce bauxite; whereas andesite or basalt give intermediate products. Commonly textural and structural features of the parent material are preserved and the more resistant insoluble minerals remain.

Mature lateritic soils lack fertility for most systems of agriculture. Savannas or parklike grasslands are typical on laterite. Clay, not laterite, is found beneath rainforests and jungle vegetation. *See* VEGETATION ZONES, WORLD.

[SAMUEL S. GOLDICH]
Bibliography: *See* BAUXITE.

Lathe

A machine for the removal of metal from a workpiece by gripping it securely in a holding device and rotating it under power against a suitable cutting tool. The tool may be moved radially or longitudinally in respect to the turning axis of the workpiece either manually or by attached power. Forms such as cones, spheres, and related concentric-shaped workpieces as well as true cylinders can be turned on a lathe. Machining operations such as facing, boring, and threading, which are variations of the turning process, can also be performed on a lathe.

Turning equipment may be classed as being either of the horizontal or vertical type referring to the turning axis of the workpiece in the machine. The basic engine lathe is primarily a manually operated machine. Filling the gap between the engine lathe and fully automatic turning equipment is the turret lathe. Fastest and best suited to high-quantity production is the automatic screw machine.

Engine lathe. The versatile engine lathe ranges in size and design from small bench and speed lathes to large floor types. The workpiece may be held between tapered centers and rotated with the power spindle by means of a clamping device, or it may be held in a chuck, or even fixed to a rotating plate. When the work is swung between centers, the tailstock, which holds the stationary tapered center, is clamped firmly in place. Long work-

Fig. 1.   Saddle-type turret lathe. (*Jones and Lamson Machine Co.*)

pieces may be supported in the middle by either a steady rest or a mechanically driven follow rest, which moves with the cutting tool.

The single-point tool is held in a tool post or block, which is supported on a cross slide and carriage. The tool may be moved radially or longitudinally in relation to the workpiece.

Angular cuts are possible, and longitudinal tapers may be cut by offsetting the tailstock, or by adjustment of a taper attachment on the rear of the lathe which actuates the cross slide.

Fig. 2.   Vertical turret lathe. (*Bullard Co.*)

Turret lathe. The tailstock of the engine lathe can be replaced with a multisided, indexing tool holder or turret designed to hold several tools, the machine becoming a turret lathe. The single-tool post and compound is usually replaced by a four-position indexing tool post. Power cuts may be taken from both of these tool mountings either individually or simultaneously. Turret lathes are constructed in vertical models and also in horizontal models.

Horizontal turret lathes are classed as either bar or chucking machines referring to the manner in which the workpiece is held. The headstock of the bar machine is constructed so that bar stock may be slid through it and the collet on line with the turning axis of the lathe. The collet closes, holding the piece firmly. Chucking-type machines grasp a unit size workpiece in a chuck or jawed device. Chucking devices permit work of relatively large diameters to be machined.

Horizontal machines may be further classed as being either of the ram or saddle type, a designation referring to the manner in which the turret is mounted on the machine. On the ram machine, the turret is mounted on a ram or slide which rides on a saddle. When the turret is indexed for successive operations, the saddle acts as a guide for the ram in its strokes to and from the work. On the saddle-type machine, designed without a ram, the turret mounts directly on the saddle, which slides on the bedways of the lathe (Fig. 1). Ram-type lathes with their short turret travel are generally of lighter construction than the saddle type. Excessive overhang of the ram reduces tool rigidity. Fast in operation, they perform best on small-diameter work and light chucking jobs. The rigid construction and the longer stroke of the saddle-type lathe enable it to handle both longer and heavier bar and chuck work than the ram-type machine.

The vertical turret lathe, similar in principle to the horizontal machine, is capable of handling heavier and bulkier workpieces. The vertical machine is constructed with a rotary, horizontal worktable whose diameter normally designates the capacity of the machine. Machine tables range from 30 to 74 in. in diameter. A crossrail mounted above the table carries a turret, which indexes in a vertical plane with tools that may be fed either across or downward (Fig. 2). The crossrail may also carry a vertical swiveling ram with a nonindexing tool holder which feeds in a manner similar to the turret. Below the crossrail, a side head with an indexing tool holder is sometimes provided. This tool may be fed in horizontally or moved vertically. Tools may be operated simultaneously either manually or by power.

Automatic screw machines. When a high production rate of relatively small turned parts is required, automatic screw machines are used. The screw machine was originally developed from the lathe for the purpose of more economical manufacture of screws and bolts. The name has persisted even though numerous partially or completely finished products are produced on them. Generally these machines are classified as single-spindle or multiple-spindle automatics. The usual machine is of the horizontal type.

The single-spindle machine is constructed to feed bar stock through the hollow machine spindle and collet similar to a turret lathe. When the bar meets a stop, the collet closes on the piece. The manner in which the cutting tools are held may vary. Usually a small five- or six-position turret or drum indexes and feeds tools longitudinally against the end of the rotating workpiece. Turret tools may include drills, reamers, hollow mills, and counterboring tools; at times a single-threading die mounted on line with the spindle is used.

Usually two independent cam-actuated cross slides, front and rear, are provided to hold forming, grooving, or cutoff tools. Turret indexing, actuation of the collet, feeding of stock, and spindle clutch operations are automatic. Machining operations commonly performed include facing, drilling, reaming, forming, and knurling. Special attachments permit such operations as milling, index drilling, or thread chasing to be performed.

Multiple-spindle automatics employ several machine spindles arranged in a circular pattern. Each spindle is equipped to hold stock in a manner similar to a single-spindle machine. In some instances automatically operating chucks capable of holding irregularly shaped workpieces replace the collet-type chucks. Standard machines may have as many as eight rotating spindles. These in turn index in a carrier about a nonrotating turret, which holds a variety of cutting tools. As each spindle indexes and progresses around the carrier, successive machining operations are performed by the turret tools. Simultaneously, tools located in successive positions are performing their respective operations on the various workpieces. Cross slides mounted at right angles to the spindles carry the forming, grooving, or cutoff tools.

The time for the longest cut plus the allowance for such programming actions as indexing and tool traversing sets the time required to produce one finished piece. By dividing long cuts over two or more operations, the cycling time may be reduced. Machining operations performed on these machines are generally the same as those done on single-spindle automatics.

Vertical multiple-spindle machines of the semiautomatic chucking type are constructed with as many as 16 spindles. A horizontal, circular table holding the rotating chucks indexes under the vertical spindles with a different operation being performed at each station. This machine is constructed to handle much larger workpieces than horizontal automatics; on many models the speed, feed, and direction of rotation for each spindle combination may be varied from one position to the next. This permits selection of the correct feed and speed for the operation being performed at each station. *See* MACHINING OPERATIONS.

[ALAN H. TUTTLE]

Bibliography: J. Anderson and E. E. Tatro, *Shop Theory*, 5th ed., 1968; H. D. Burghardt and A. Axelrod, *Machine Tool Operation*, part 1: *Safety, Measuring Instruments, Bench Work, Drill Press, Lathe, and Forge Work*, 5th ed., 1959; P. S. Houghton, *Lathes*, vol. 2, 1963.

Latite

An aphanitic (not visibly crystalline) rock of volcanic origin, composed chiefly of sodic plagioclase (oligoclase or andesine) and alkali feldspar (sanidine or orthoclase) with subordinate quantities of dark-colored (mafic) minerals (biotite, amphibole, or pyroxene). Latite is intermediate between trachyte and andesite. Plagioclase is dominant over alkali feldspar in latite but is subordinate to alkali feldspar in trachyte. Andesite carries little or no alkali feldspar. *See* ANDESITE; TRACHYTE.

[CARLETON A. CHAPMAN]

Latitude (astronomy)

Latitude determined by astronomical observations, as distinguished from geographic latitude, which is that shown on a map. Astronomical latitude is determined by the definition that the angular altitude from the horizon of the celestial pole is identical with the latitude of the observer. The altitude of the pole is ascertained by observing the meridian altitude of any celestial object whose declination is known. This is done by measuring the angle between the line of sight and a horizontal surface, which is established either by the visible horizon, a liquid surface, or a spirit level, and ideally is perpendicular to the direction of gravity. The direction of gravity varies locally, however, these deviations being called deflections of the vertical. A map constructed exclusively from astronomical observations would be distorted by a few hundred feet to more than a mile. *See* GEODESY; LONGITUDE (ASTRONOMY).

[GERALD M. CLEMENCE]

Lattice (mathematics)

Lattice theory deals with properties of order and inclusion, much as group theory treats symmetry. As a generalization of Boolean algebra, lattice theory was first applied around 1900 by R. Dedekind to algebraic number theory; however, its recognition as a major branch of mathematics, unifying various aspects of algebra, geometry, and functional analysis, as well as of set theory, logic, and probability (to which Boolean algebra had already been applied), dates from the years 1933–1938. *See* BOOLEAN ALGEBRA; GROUP THEORY; SET THEORY.

The most basic concept of lattice theory is that of a partial ordering of a set S of elements x, y, z, ... By this is meant a binary relation, usually denoted $\leqq$ (or $\geqq$), with the following properties:

(P1) $x \leqq x$ for all $x \in S$
(P2)   If $x \leqq y$ and $y \leqq x$, then $x = y$
(P3)   If $x \leqq y$ and $y \leqq z$, then $x \leqq z$

If $\leqq$ is any partial ordering of S, then its converse or dual $\geqq$, defined by statement (1), is also a par-

$$x \geqq y \text{ if and only if } y \leqq x \qquad (1)$$

tial ordering of S. This easily verified fact provides a fundamental Duality Principle, which is useful in many connections.

Suppose, for example, the join $x \cup y$ of two elements x and y of a partially ordered set S is defined by conditions (2a) and (2b). (It is easily shown that

$$x \leqq x \cup y \qquad y \leqq x \cup y \qquad (2a)$$

$$\text{If } x \leqq z \text{ and } y \leqq z, \text{ then } x \cup y \leqq z \qquad (2b)$$

there is at most one such $x \cup y$.) Then the duality principle suggests defining the meet $x \cap y$ by Eqs. (3a) and (3b), and shows that there is at most one such $x \cap y$.

$$x \geqq x \cap y \qquad y \geqq x \cap y \qquad (3a)$$

Fig. 1. Ordinal number 4.

If $x \geqq z$ and $y \geqq z$, then $x \cap y \geqq z$ (3b)

A lattice is defined as a partially ordered set in which any two elements x and y have a meet $x \cap y$ and a join $x \cup y$. These binary operations satisfy the four basic identities:

(L1) $x \cap x = x \cup x = x$
(L2) $x \cap y = y \cap x$ and $x \cup y = y \cup x$
(L3) $x \cap (y \cap z) = (x \cap y) \cap z$ and $x \cup (y \cup z)$
$= (x \cup y) \cup z$
(L4) $x \cap (x \cup y) = x \cup (x \cap y) = x$

The operations $\cap$ and $\cup$ are connected with the relation $\leqq$ by the condition that $x \leqq y$, $x \cap y = x$, and $x \cup y = y$ are three equivalent statements. Conversely, if L is an algebraic system with operations $\cap$ and $\cup$ satisfying (L1) to (L4) for all x, y, z, then the preceding condition defines $\leqq$ as a partial ordering of L, with respect to which $\cap$ and $\cup$ have the meanings defined above. This principle was discovered in 1880 by C. S. Peirce.

Kinds of lattices. There are many different kinds of lattices. Thus the real numbers form a lattice if $x \leqq y$ is given its usual meaning. This lattice is simply ordered, in the sense that:

(P4) Given x and y, either $x \leqq y$ or $y \leqq x$

Any such simply ordered set (or chain) is a lattice, in which $x \cup y$ is simply the larger of x and y, and dually.

Again, the set J of positive integers forms a lattice, if one lets $m \leqq n$ mean "m divides n" (usually denoted $m|n$). In this case, $m \cap n = \gcd (m,n)$ and $m \cup n = \mathrm{lcm}\ (m,n)$. Still again, one can let Σ consist of all subsets $S, T, \ldots$ of a fixed ensemble I, and let $S \leqq T$ mean that every point in S is in T. Then Σ is a lattice, in which $S \cap T$ is the intersection of S and T, whereas $S \cup T$ is their union. Actually, Σ is a Boolean algebra.

In all the preceding lattices, the distributive laws hold:

Fig. 2. The projective line over the field J_2 of integers mod 2.

(L5) $x \cap (y \cup z) = (x \cap y) \cup (x \cap z)$ and
$x \cup (y \cap z) = (x \cup y) \cap (x \cup z)$, for all x, y, z

Such lattices are called distributive lattices. Any chain is a distributive lattice; so is any Boolean algebra. More generally, a ring of sets is defined as a family Φ of subsets of a fixed set I which contains, with any S and T, also their intersection $S \cap T$ and their union $S \cup T$. Then any ring of sets is a distributive lattice.

It is obvious that each of the two identities of (L5) is dual to the other. It is a curious fact that, in a lattice, each also implies the other.

If G is any group, then its subgroups form a lattice, and its normal subgroups also form a lattice, in both cases under inclusion. The normal subgroups satisfy the (self-dual) modular law:

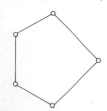

Fig. 3. The simplest nonmodular lattice.

(L6) If $x \leqq z$, then $x \cup (y \cap z) = (x \cup y) \cap z$

In general, lattices satisfying (L6) are called modular, and every distributive lattice is modular.

The lattice of all linear subspaces of the n-dimensional vector space $\mathbf{V}_n(F)$ over any field (or division ring) F is also a modular lattice, usually called the $(n-1)$-dimensional projective geometry $P_{n-1}(F)$ over F. This lattice contains special elements $\mathbf{0}$ (the zero vector) and $I = V_n (F)$ (the whole space), such that:

(P5) $\mathbf{0} \leqq x \leqq I$ for all x

Such special elements always exist in any lattice whose chains all have finite length, but they need not exist in general; for example, they do not in the simply ordered set of real numbers.

The lattice $P_{n-1}(F)$ is complemented, in the sense that each subspace x has at least one complement x', with the property:

(L7) $x \cap x' = \mathbf{0}$ and $x \cup x' = I$

Thus $P_{n-1}(F)$ is a complemented modular lattice. Similarly, it may be verified that the class of Boolean algebras is precisely the class of complemented distributive lattices. This principle enables one to consider Boolean algebra as a branch of lattice theory.

Lattices L containing few elements can be conveniently visualized by diagrams. In these diagrams, small circles represent elements of L, a being higher than b whenever $a > b$. A segment is then drawn from a to b whenever $a > b$, but no x exists such that $a > x > b$. Any such diagram defines L up to isomorphism: $a > b$ if and only if one can travel from a to b along a descending broken line. Figures 1–6 are typical of such diagrams.

Such graphs often give useful information very simply. For example, let a finite lattice be called semimodular if any two elements a and b immediately above (covering) a given element c are also immediately under (covered by) another element $d = a \cup b$. This condition can easily be tested by inspection. Dedekind showed that a finite lattice L was modular if and only if it and its dual were both semimodular. For L to be distributive, the extra condition of containing no subgraph such as that of Fig. 2 is necessary and sufficient.

Applications to algebra and geometry. Lattices, like groups and rings, can be defined as abstract algebras, that is, as systems of elements combined by universally defined operations. These operations may be unary, binary, or ternary. In any such abstract algebra A, define a subset S to be a subalgebra of A when the result of performing any operation of A on elements in S is again in S. Call an equivalence relation $a \equiv b \pmod{\theta}$ on A a congruence relation when, for any n-ary operation f of A, $a_i \equiv b_i \pmod{\theta}$ for $i \equiv 1, \ldots, n$ implies that $f(a_1, \ldots, a_n) \equiv f(b_1, \ldots, b_n) \pmod{\theta}$. Then the subalgebras of A form one (complete) lattice, and the congruence relations form another.

Results like the preceding, which are true of abstract algebras in general, are called theorems of "universal algebra." Another such result is the theorem that any algebra can be decomposed into subdirectly irreducible algebras. From this theorem, it follows easily that any distributive lattice is isomorphic with a ring of sets.

In groups, rings, and many other algebras, all congruence relations are permutable; it follows that the congruence relations form a modular lattice. This fact, combined with the existence of one-element subalgebras, permits the development of an extensive structure theory that includes a unique factorization theorem under appropriate finiteness conditions. Curiously, lattices themselves satisfy a unique factorization theorem, but for another reason.

It is important to note that the lattice products LM, involved in the statement of the preceding unique factorization theorem, are one of the six basic operations of a general arithmetic of partially ordered sets, which contains the usual cardinal and ordinal arithmetic as a special case. Thus, the most general finite Boolean algebra is just 2^n, where 2 is the ordinal number two and n is a finite cardinal number.

In any lattice with O, a "point" is an element that covers O. In most lattices arising in geometry, every element is a join of a finite number of points; semimodular lattices with this property are called geometric lattices. The lattice of all partitions of a finite set into nonoverlapping subsets is such a geometric lattice. Figure 6 depicts the geometric lattice of all partitions of four objects. Any geometric lattice is complemented; conversely, any complemented modular lattice in which all chains are finite is geometric.

It was mentioned above that the $(n-1)$-dimensional projective geometry $P_{n-1}(F)$ over a division ring F was always a complemented modular lattice. Many interesting applications of lattice theory to geometry are extensions of this observation. For instance, let P be any abstract projective geometry, defined by incidence relations. Then P is also a complemented modular lattice. Conversely, any finite-dimensional complemented modular lattice is a product of projective geometries and a Boolean algebra.

The preceding abstract combinatorial approach to projective geometries led to the construction in 1936, by J. von Neumann, of his continuous-dimensional projective geometries (so-called continuous geometries). O. Frink, in 1946, developed a parallel theory of projective geometries of discretely infinite dimension. By analogy, affine geometries can also be regarded as lattices, and any affine geometry in which all chains are finite is a geometric lattice.

Although important applications of geometric lattices to combinatorial problems have been made by G.-C. Rota and others since 1960, the theory of semimodular lattices in general is quite limited. A negative result is Dilworth's theorem, which states that every finite lattice is isomorphic with a sublattice of a finite semimodular lattice. Similarly, P. M. Whitman has shown that every lattice is isomorphic with a sublattice of the lattice of all partitions of some (infinite) class. *See* ALGEBRA, ABSTRACT; COMBINATORIAL THEORY.

Relation to set theory and logic. In set theory, one is frequently concerned with various special families of sets such as closed sets, open sets, measurable sets, and Borel sets. Some of these families form Boolean algebras. But the closed sets (and, dually, the open sets) of a topological space X usually form just an uncomplemented distributive lattice $L(X)$. In $L(X)$, the complemented elements (which necessarily form a Boolean algebra) are those which disconnect X.

Both the theory of finite sets and the theory of finite distributive lattices are fairly trivial. Each such lattice L has a unique representation as the cardinal power $L = 2^P$ of the two-element Boolean algebra 2, with a general finite partially ordered set P as exponent. Moreover any finite-dimensional distributive lattice is finite.

If L is an infinite-dimensional lattice, distributive or not, then one can define several topologies on L, which may or may not be equivalent. Most of these are suggested by a consideration of the special case that L is the rational number system.

Adapting Dedekind cuts, as first shown in 1935 by H. M. MacNeille, one can extend any lattice (or partially ordered set!) to a complete lattice, in which any subset A of elements x_α has a least upper bound (join) $\vee x_\alpha$ and greatest lower bound $\wedge x_\alpha$. In such a complete lattice, one can define lim sup $\{x_n\}$ and lim inf $\{x_n\}$ for any sequence x_1, x_2, x_3, . . . of elements. One can define a convergence topology in L by letting $x_n \to x$ mean that lim sup $\{x_n\} = $ lim inf $\{x_n\} = x$. Alternatively, one can define an interval topology by taking closed intervals $[a,b]$, each consisting of all $x \in L$ with $a \leq x \leq b$, as a subbasis of closed sets.

The concept of convergence, in lattices and other spaces, can be extended by considering directed sets of indices α, this concept itself being lattice-theoretic. To describe convergence in topological spaces not satisfying countability axioms, such directed sets are essential.

Using the concepts of lattice completeness and lattice topology, one can give interesting characterizations of various lattices associated with topological spaces. Thus, this was done for the lattice of regular open sets by S. Ulam and G. Birkhoff. It was done for the (distributive) lattice of continuous functions by Irving Kaplansky. Finally, it was done for closed (and open) sets by J. C. C. McKinsey and Alfred Tarski, all since 1935.

An interesting analogy between set theory and logic is provided by the equivalence between the algebra of open sets and Brouwerian logic, which arose entirely from a consideration of formal logic. Abstractly, both deal with relatively pseudo-complemented distributive lattices.

Another type of logic, directly suggested by quantum theory, was developed in 1936 by G. Birkhoff and von Neumann. In this logic, properties form an orthocomplemented modular lattice. The Aristotelian logic of classical mechanics appears as the special, distributive (hence Boolean) case of permutable observations. The formal properties of the probability $p[x]$ of an event x are shown to be similar to those of dimension in projective geometry.

Lattice theory also gives perspective on other kinds of formal logic, including modal logic and "strict implication." However, since the deepest problems of formal logic concern the foundations of set theory, these applications will not be discussed here.

Lattice-ordered groups. The first study of lattice-ordered groups was made in 1897 by Dede-

LATTICE (MATHEMATICS)

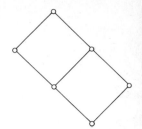

Fig. 4. The lattice of divisors of 12, under divisibility.

LATTICE (MATHEMATICS)

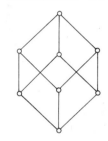

Fig. 5. The Boolean algebra of order 8.

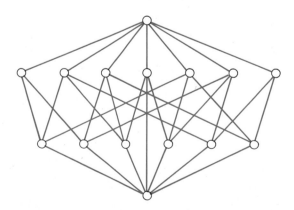

Fig. 6. Lattice of partitions of [a, b, c, d].

kind, in analyzing the properties of the rational number system R. Letting $a \leqq b$ signify that $ma = b$ for some integer m (that is, that b is a multiple of a), the nonzero rational numbers form a lattice in which $a \cap b$ and $a \cup b$ again mean the lcm and gcd of a and b. Also, with respect to ordinary multiplication, the nonzero rationals form a (commutative) group. Finally, in R, implication (4a)

$$a \leqq b \text{ implies } ac \leqq bc \qquad (4a)$$

holds; that is, every group translation is a lattice automorphism. Any lattice L which is also a group, and which shares this property, is called a lattice-ordered group, or l-group.

Though many noncommutative l-groups exist, it is a striking fact that every complete l-group is necessarily commutative.

Lattice-ordered groups arise in function theory, as well as in number theory. If $f \leqq g$ means that $f(x) \leqq g(x)$ for all (or almost all) x, then most function spaces of real functions form lattices. They are also commutative groups under addition. Moreover, implication (4b) holds for any $c(x)$; this is simply (4a) in additive notation. Hence most (real)

$$f(x) \leqq g(x) \text{ implies}$$
$$f(x) + c(x) \leqq g(x) + c(x) \qquad (4b)$$

function spaces are l-groups. In addition, they are vector spaces, in which implication (4c) holds. The

$$f \geqq g \text{ and } \lambda \geqq 0 \text{ imply } \lambda f \geqq \lambda g \qquad (4c)$$

l-groups with these additional properties are vector lattices. Although E. H. Moore and F. Riesz had discussed related ideas earlier, the first systematic analysis of vector lattices as such was made in 1937 by L. V. Kantorovitch.

The application of vector lattice concepts to function theory is still not very far advanced. Using the intrinsic lattice topologies defined earlier, and others related to them, one can avoid the necessity of introducing a distance function, in many function spaces. Thus, the notion of metric boundedness is equivalent to order boundedness for functionals on any Banach lattice, and metric convergence is equivalent to relative uniform star-convergence, which is a purely lattice-theoretic concept.

An additive l-group which is also a ring is called a lattice-ordered ring or l-ring when its multiplication satisfies the partial analog of implication (4b) shown in formula (5). Such rings have been

$$\text{If } f \geqq 0 \text{ and } g \geqq 0, \text{ then } fg \geqq 0 \qquad (5)$$

studied systematically only since 1955; a typical theorem about them is the following: An l-ring is a product of simply ordered rings if and only if it satisfies implication (6).

$$a \cap b = 0 \text{ and } c \geqq 0 \text{ imply}$$
$$ca \cap b = ac \cap b = 0 \qquad (6)$$

Further applications. It is clear that the concepts of vector lattice and of l-ring are essential in various physical applications. This was first apparent in connection with the ergodic theorem, as proved in 1931 by Birkhoff and von Neumann, for the deterministic processes of classical mechanics. A generalization of this theorem to stochastic processes, whose natural formulation is based on the concept of a vector lattice, was proved in 1939–

1941 by Shizuo Kakutani and Kosaku Yosida.

A second application is to the theory of Reynolds operators, or averaging operators, arising in turbulent fluid motions. The essential connection with the order relation is simply the obvious principle that any average of nonnegative quantities is nonnegative. Using this principle and the theory of l-rings, one can decompose (subdirectly) any vector-averaging operator into scalar components.

A third application is to the concept of criticality in nuclear reactor theory. Neutron chain reactions involve the birth (through fission), migration, and death (through absorption) of neutrons. The laws governing the evolution of the statistical distributions of neutrons (as functions of position x, velocity v, and time t) evidently carry nonnegative distributions into nonnegative distributions. To deduce the mathematical principle that the neutron distribution must satisfy the asymptotic relation $N(x,v;t) \sim e^{\lambda t} N_0(x,v)$, λ^{-1} being called the reactor "period," it is again most convenient to reformulate the problem in lattice-theoretic language. One can then apply results of Oskar Perron, G. Frobenius, and R. Jentzsch on positive linear operators to prove the desired result. *See* LOGIC; RING THEORY.

[GARRETT BIRKHOFF]

Bibliography: G. Birkhoff, *Lattice Theory*, 3d ed., 1967; G. Birkhoff and T. C. Bartee, *Modern Applied Algebra*, 1970; P. Crawley and R. P. Dilworth, *Algebraic Theory of Lattices*, 1973; R. P. Dilworth (ed.), *Lattice Theory*, Symposia in Pure Mathematics, American Mathematics Society, vol. 2, 1961; G. Grätzer, *Universal Algebra*, 1968; H. H. Schaefer, *Banach Lattices and Positive Operators*, 1974; G. Szasz, *Introduction to Lattice Theory*, 1963.

Lattice constant

A parameter defining the unit cell of a crystal lattice, that is, the length of the edges of the cell and the angle between edges. If the unit cell is a cube, the side of it is the lattice constant, and it is usually in this sense that the term is used. The lattice constants of simple compounds of the type A^+X^-, $A^+X_2^-$ having the sodium chloride, cesium chloride, zinc sulfide, or calcium fluoride structures have values between 5 and 10 A. The lattice constants of the elements and of a variety of compounds are listed in the Wyckoff reference given in the bibliography. *See* CRYSTALLOGRAPHY.

[WILLY C. DEKEYSER]

Bibliography: F. A. Wade and R. B. Mattox, *Elements of Crystallography and Mineralogy*, 1960; R. W. G. Wyckoff, *Crystal Structures*, 3 vols., 1948–1957.

Lattice vibrations

Periodic oscillation of the atoms in a crystal lattice about their equilibrium positions. As the crystal is heated, the amplitude of the vibrations increases. If the heating is continued, the temperature of the crystal eventually reaches a value at which the vibrations are so violent that the atoms break away from their lattice sites, and the solid melts. On the other hand, if the crystal is cooled to absolute zero, the amplitude of the vibrations does not subside entirely. A residual vibration of the atoms, which is quantum-mechanical in origin, remains. It is called the zero-point vibration.

Fig. 1. Schematic diagram of a simple one-dimensional lattice of equidistant identical particles.

Lattice vibrations are involved in many of the temperature-dependent phenomena of a solid. For example, the electrical resistance of a metal at room temperature arises primarily from the scattering of the conduction electrons by the vibrating atoms. The higher the temperature, and hence the more violent the vibrations of the atoms, the more the electrons are scattered, and the higher the electrical resistance becomes. A ferromagnetic solid becomes paramagnetic at temperatures above the Curie temperature because the lattice vibrations are then sufficiently energetic to overcome the interatomic magnetic forces. The Bardeen-Cooper-Schrieffer theory of superconductivity postulates that subtle interactions between lattice vibrations and conduction electrons are mainly responsible for this phenomenon. *See* SUPERCONDUCTIVITY.

Lattice vibrations conduct heat through a crystal; thus a knowledge of the vibration mechanism is essential to an understanding of the heat conductivity of crystalline solids.

The theory of lattice vibrations began as an effort to explain specific heat. Theory shows that the classical treatment of vibrations of atoms in solids leads to the Dulong-Petit law of specific heats, a law which is not obeyed at all temperatures. To account for this failure, Albert Einstein formulated a theory which assumes that each atom vibrates with the same frequency and that the energy of vibration is quantized. Then followed the improvements on the theory made by P. Debye and by M. Born and T. von Kármán. In both of these theories, the single Einstein frequency is replaced by an acoustical spectrum. The theory of Born and von Kármán requires that this spectrum be obtained from the study of the vibrations of a system of point particles distributed in the same manner as the atoms in a crystal lattice. The various models representing a given solid differ from one another essentially in the assumptions made concerning the nature of the forces between the constituent particles. It is usually in the sense of the Born–von Kármán theory and its extensions that one speaks of lattice vibrations. *See* SPECIFIC HEAT OF SOLIDS.

One-dimensional lattice. Some of the main concepts of the Born–von Kármán theory are illustrated by the vibrations of a simple one-dimensional lattice. Figure 1 shows such a lattice composed of equidistant identical particles of mass M. The equilibrium separation of neighboring particles is a. They are located along an x axis, the location of the nth particle being $x = na$. Assume that the forces between the particles are effective only between nearest neighbors and that these forces obey Hooke's law. In Fig. 1, the forces are represented schematically by springs with Hooke's constant α.

Let u_n denote the displacement along the x axis of the nth particle from its equilibrium position at $x = na$. The force exerted on the nth particle depends upon the distance which the two springs connected to it are stretched. The extensions are $(u_n - u_{n-1})$ for the left-hand spring and $(u_{n+1} - u_n)$ for the right-hand spring. Thus the equation of motion of the nth particle is Eq. (1).

$$M(d^2 u_n/dt^2) = -\alpha(u_n - u_{n-1}) - \alpha(u_n - u_{n+1}) \quad (1)$$

Let the length of the lattice be $L = Na$, so that there are $(N + 1)$ particles in all. The number N is taken to be large, comparable with Avogadro's number, so that whether the end particles are free to move or are fixed in position is of no practical importance here. To solve Eq. (1), write tentatively the equation of a wave of frequency ν and amplitude A, namely, Eq. (2), where $k = 1/\lambda$, the reciprocal of the wavelength. However, the lattice is discrete and the wave has no real existence between the atoms. This leads one to put $x = na$. The trial solution is now given by Eq. (3). Substitution shows

$$u_n = A \exp\left[2\pi i(\nu t - kx)\right] \quad (2)$$

$$u_n = A \exp\left[2\pi i(\nu t - kna)\right] \quad (3)$$

that this expression satisfies Eq. (1) provided that Eq. (4) holds, where $\nu_0 = \sqrt{a/\pi^2 M}$. It is evident

$$\nu = \nu_0 \sin \pi ak \quad (4)$$

that ν_0 is the highest permissible frequency. If the end particles are assumed to be immovable, then standing-wave solutions are appropriate and are analogous to those of the continuous string with fixed ends. The restrictions on the wavelengths are $j\lambda/2 = L$, where j is an integer indicating the number of loops in the particular standing wave. Thus Eq. (4) becomes Eq. (5), the range of j being

$$\nu = \nu_0 \sin (\pi j/2N) \quad (5)$$

$1 < j < N$. Each kind of vibration is called a mode of vibration, and there is a mode of vibration for each j. Thus there are N modes. The total number of modes is also equal to the number of particles, because N and $(N + 1)$ are about equal.

The phase velocity U_p and the group velocity U_g can be calculated from Eq. (3) by means of the definitions $U_p = \nu/k$ and $U_g = d\nu/dk$. The results

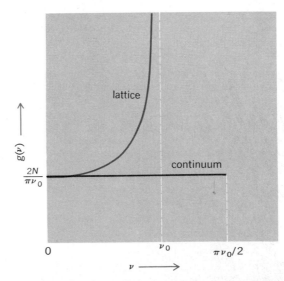

Fig. 2. The function $g(\nu)$ for the one-dimensional lattice and the one-dimensional continuum.

are given by Eqs. (6) and (7). Here U_∞ is the veloc-

$$U_p = U_\infty \frac{\sin \pi ak}{\pi ak} \qquad (6)$$

$$U_g = U_\infty \cos \pi ak \qquad (7)$$

ity of sound at long wavelengths and has the value $\pi a \nu_0$. By long wavelengths is meant wavelengths which are great compared with the equilibrium distance a.

An important characteristic revealed by this treatment is the dependence of the velocity on wavelength. Such a dependence is called dispersion. It becomes increasingly important as the wavelength approaches the value $2a$. At the cutoff wavelength $\lambda = 2a$, the group velocity is zero. A continuum theory, such as Debye's, gives simply $U_p = U_g = U_\infty = \pi a \nu_0 = $ constant, and the medium is dispersionless.

The number of modes $g(\nu)d\nu$ with frequencies between ν and $\nu + d\nu$ can be calculated from Eq. (5) by noticing that this number is also given by dj. From this, the distribution function $g(\nu)$ is given by Eq. (8). Figure 2 shows a plot of $g(\nu)$ for the lattice

$$g(\nu) = 2N/(\pi\sqrt{\nu_0^2 - \nu^2}) \qquad (8)$$

as given by Eq. (8) and for the Debye continuum theory. The latter is simply $g(\nu) = 2N/\pi\nu_0$, with the range $0 < \nu \leq \pi\nu_0/2$.

The energy E of the lattice is given by Eq. (9),

$$E = \int_0^\infty \frac{g(\nu)h\nu\,d\nu}{e^{(h\nu/kT)} - 1} \qquad (9)$$

where h is Planck's constant, k is Boltzmann's constant, and T is the absolute temperature.

Three-dimensional lattices. In three-dimensional models, some attempt is usually made to approximate the conditions in actual crystals. At the same time, the considerable mathematical complexity of the problem requires that the model be kept as simple as possible. A decision has to be made concerning both the number of neighbors to be included in the calculation and the types of forces acting between the particles before the equations of motion can be formulated.

The forces are of several kinds. A force which arises when the distance between a pair of neighbors changes is called a central or radial force. This is the type of force assumed for the one-dimensional lattice discussed earlier. If a change in the angle between the pair of lines joining a given particle to two neighbors (bond lines) gives rise to a force, the force is called an angular force. A more general force compounded of radial and angular forces is often referred to as a noncentral force. The most general combination of noncentral forces compatible with symmetry requirements about the particles is commonly called a tensor force.

The types of forces just named are defined for a pair of particles or a pair of bond lines. The motions of the other particles or bond lines in the vicinity of the pair do not affect the force between the pair. In metals, however, the conduction electrons give rise to a type of force which cannot be defined in this manner. To visualize this, imagine that a group of particles (positive ions) move toward each other, thus increasing the local density. The corresponding increase in positive charge density is immediately compensated by a flow of

conduction electrons into this region. However, the compressibility of this assembly of electrons, called an electron gas, is small, and the gas resists this increase in its density. The electrons would thus screen the ions less completely, the ions would repel one another more strongly, and consequently a "stiffening" would occur in the elastic constants involved. To summarize, additional forces arise to oppose any change in particle density. This is clearly a cooperative or collective effect not covered by the types of forces just listed. The forces produced by this collective effect are sometimes called volume forces.

A common objective of the theory of lattice vibrations is the calculation of the distribution function $g(\nu)$. Analytic methods generally cannot be used and the calculation must be done numerically. R. B. Leighton (1948) calculated $g(\nu)$ for a model of a face-centered cubic lattice with central forces between nearest neighbors and between next-nearest neighbors. Figure 3 shows his result, obtained after much labor, for a selection of force constants approximating copper. The availability of high-speed digital computers has reduced the labor to the extent that calculations of distribution functions are becoming increasingly common.

Figure 3 shows three branches of the distribution function, labeled I, II, and III. In an elastically isotropic solid, branches I and II would correspond to transverse waves and branch III to longitudinal waves, for $\lambda \gg 2a$. In an elastically anisotropic crystal, very few waves are purely transverse or longitudinal, even for $\lambda \gg 2a$. Thus in actual crystals, the waves constituting branches I and II are only predominantly transverse in character and those in branch III only predominantly longitudinal. Lattices in which each atom has exactly the same environment as any other atom are Bravais lattices, and these always have three branches in the distribution function. The distribution functions of non-Bravais lattices such as NaCl and diamond have additional high-frequency branches, called optical branches. The lower-frequency branches, corresponding to the three branches

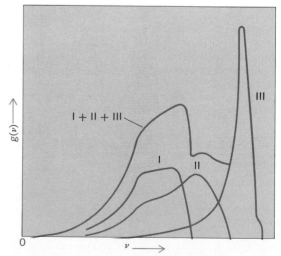

Fig. 3. The function $g(\nu)$ obtained by Leighton for a face-centered cubic lattice. (*After R. B. Leighton, Rev. Mod. Phys., 20:165–174, 1948*)

of the Bravais lattices, are called acoustical branches.

Experimental studies. The results of the calculations of $g(\nu)$ for a given model are often tested by computing the specific heat and then comparing the result with the measured specific heat. The comparisons which have been made leave little doubt that the Born–von Kármán theory can account satisfactorily for the discrepancies between the Debye theory of specific heats and the experimental measurements. A particular disadvantage of the specific-heat approach is the insensitivity of the specific heat to the detailed shape of the distribution function for frequencies beyond the first peak.

Another experimental approach is the study of the scattering of x-rays by lattice vibrations. A static lattice should produce sharp diffraction patterns conforming to the regularity of the lattice. If a wave now passes through such a lattice, a periodic error is produced in the lattice and a pair of spots (ghosts) may occur on either side of the principal diffraction maximum. If the combined effect of all possible waves is considered, the result is a broad diffused spot centered at the diffraction maximum. The theory of this effect was first given in 1918 by H. Faxén and experimentally observed in 1938 by J. Laval. The theory and experimental methods are developed to the extent that dispersion curves for lattice vibrations can be determined from the observations. These are determined for special directions in the crystal, using monochromatic x-rays. The dispersion curves so obtained are then compared with those given by various models. Primary difficulties are the correction for Compton scattering and the determination of the dispersion curve near its ends. Generally, the method seems to be poorest where the specific-heat method is best, so that the two methods can be used advantageously to supplement one another.

A third experimental approach is the study of the inelastic scattering of cold (very slow) neutrons by lattice vibrations. It is possible to use an almost monochromatic beam of neutrons with energies considerably less than those acquired by thermal scattering on passage through the lattice. This is contrary to what happens to x-rays, for which the photon energy is so great as to be scarcely changed by thermal scattering. The theory of neutron scattering becomes simplified if the scattering is also incoherent. *See* NEUTRON DIFFRACTION.

[JULES DE LAUNAY]

Bibliography: W. H. Bragg, *The Crystalline State*, 1933–1953; L. Brillouin, *Wave Propagation in Periodic Structures*, 1946; S. Fluegge (ed.), *Handbuch der Physik*, vol. 7, pt. 1, 1955; C. Kittel, *Introduction to Solid State Physics*, 1966; F. Seitz and D. Turnbull (eds.), *Solid State Physics*, vol. 2, 1956.

Launch complex

The composite of facilities and support equipment needed to assemble, check out, and launch a rocket-propelled vehicle. The term usually is applied to the facilities and equipment required to launch larger vehicles for which a substantial amount of prelaunch preparation is needed. Small operational rockets may require similar but highly simplified resources on a much smaller scale. For these, the term launcher is usually used. *See* SPACE FLIGHT.

Prelaunch processing. A rocket vehicle consists of one or more stages and a payload. These elements usually are manufactured at different locations and shipped separately to the launch site. The assembly process consists of properly mating these elements in the launch configuration, assuring that all mechanical and electrical interconnections are properly made. Components that have been shipped separately from the main vehicle elements also are installed. These include items requiring special handling or safety precautions, such as batteries and ordnance devices. *See* GUIDED MISSILE; MANNED SPACE FLIGHT; MISSILE; SPACECRAFT STRUCTURE.

The check-out process consists of detailed testing of all elements of the vehicle to assure they are functioning properly and ready for launch. Checkout of individual components or subsystems may begin prior to or during assembly when easy access for repair or replacement is available. After completion of assembly, the vehicle is given a detailed overall test, usually called combined systems test or integrated systems test, to verify launch readiness.

The launch process includes the final countdown and lift-off. During countdown, propellants are loaded, all vehicle systems are activated and given a final readiness check, ordnance devices are armed, and first-stage motors are ignited.

The term facilities applies to the larger permanent and usually fixed structures on the complex. The support equipment, often called ground support equipment (GSE), includes all ground equipment necessary to transport, service, test, control, and launch the vehicle, providing that this equipment is furnished as a part of the vehicle system. In this grouping are handling equipment, transporter trailer, the service tower and erecting mechanisms, assembly and launch test equipment and consoles, weighing mechanisms, umbilical accessories, control and instrumentation cables, and other external equipment tailored to the missile and required for its preparation and operation.

Launch-complex elements. A typical launch complex (Fig. 1) for a vehicle of the Thor-Delta class has two launch pads serviced by a single blockhouse and by common support equipment, providing some economy. The spacing between pads, blockhouse, and other elements of the complex is based on safety considerations. The overall size of a complex therefore depends primarily on the quantities and types of propellants used.

Pad. The launch pad itself is usually a massive concrete structure designed to withstand the heat and pressure of the rocket exhaust. The pad provides physical support for the vehicle prior to launch, but it also has a number of other functions. A propellant-transfer room contains equipment for remote control of propellant loading. An environmental control system supplies conditioned air to the vehicle to maintain normal operating temperatures. A terminal-connection room provides for data and remote-control links between the vehicle and the launch control center or blockhouse.

A flame deflector, located underneath the vehicle, diverts the rocket exhaust in a direction that does not cause harm. A high-pressure, high-vol-

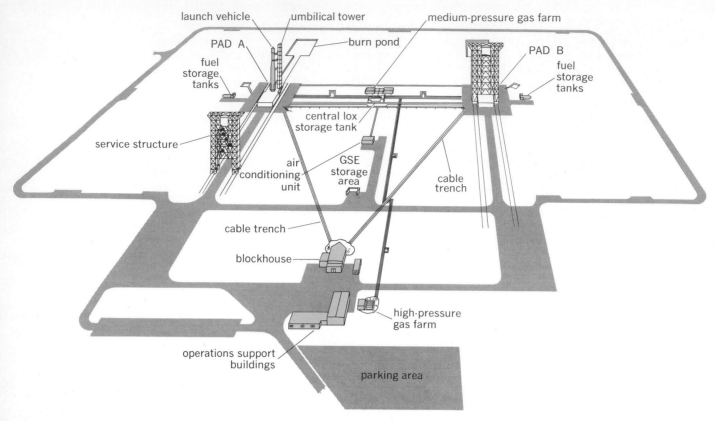

Fig. 1.   Overall perspective of a typical launch complex in which one blockhouse serves two launch pads.

Fig. 2.   Launch pad with Soyuz spacecraft and rocket. Gantry and hold-down mechanisms fold back at proper moments in launch sequence. (*Tass, from Sovfoto*)

ume water supply may be used to cool the flame deflector and otherwise minimize damage to the pad. The water supply also provides fire protection.

Many pads have a hold-down-and-release mechanism that holds the vehicle in place during ignition and releases it when full thrust is built up (Fig. 2). The hold-down feature also can be used for test firings of the rocket motors.

The weighing system is a particularly important part of on-stand equipment; it can be used for weighing the missile, determining the amount of propellants loaded, measuring the thrust in static tests, and establishing acceptable thrust performance prior to launching of a missile. Weighing mechanisms are generally load-cell strain gages.

Umbilical connections. During the check-out and countdown phases until just prior to lift-off, the vehicle is serviced from the ground by electrical and mechanical connections. The electrical connections provide electrical power, control signals for remote operation, and data links. The mechanical connections provide for propellant loading, high-pressure gas transfer, and air conditioning. These connections usually are supported by an umbilical tower located on the pad adjacent to the vehicle. Attachment to the vehicle is made by quick-disconnect devices that are pulled away just after rocket motor ignition but prior to lift-off.

Service structure. During assembly and check-out, technicians require access to all parts of the vehicle. This is provided by a service structure, or gantry, which can be moved up to the vehicle when needed and pulled away prior to launch. The service structure has movable platforms to provide access to all levels of the vehicle (Fig. 3). It may also have means for partially or totally enclosing the vehicle to provide a clean, air-conditioned environment. Figure 4 shows an environmental enclosure near the top of a *Titan 3* service structure intended for sheltering only the payload. Figure 5 shows the inside of this structure with the hole through which the vehicle will project covered with a safety net.

Fig. 3. A Minuteman ICBM on the launch pad with its service structure in place. (*Boeing Aircraft*)

When the vehicle is assembled on the launch pad, the service structure also provides such necessary mechanical equipment as cranes and hoists, as well as electrical power, gas supplies, and other utilities needed for the process.

Fueling systems. The launch complex for liquid-fueled vehicles has fuel and oxidizer storage and loading systems. Cryogenic (low-temperature) propellants such as liquid oxygen and liquid hydrogen contribute greatly to the complexity of a launch complex. The cryogens must be stored close enough to the point of use so they will not vaporize in the transfer lines but far enough back from the launch point so that the takeoff blast will not endanger the storage tanks. *See* CRYOGENICS; PROPELLANT; ROCKET ENGINE.

Extreme cleanliness is necessary in handling liquid oxygen, the most common of oxidizers. Tanks and lines must be thoroughly clean because, under proper conditions, hydrocarbons may combine with an oxidizer to cause a fire explosion. Filters are employed in the system to trap particles which may have been accidentally introduced. Liquid oxygen has a normal temperature of $-297°F$; to minimize losses, the cryogenic tanks are double-jacketed, with a vacuum in the annular space. The entire system is usually made of stainless steel for cleanliness and strength at low temperatures.

Liquid hydrogen, used as a fuel in some vehicles, is even more difficult to handle. It not only has a much lower temperature ($-422.9°F$), but hydrogen systems are highly susceptible to leaks. When mixed with air, hydrogen is extremely explosive.

Noncryogenic propellants, such as hydrazine and fuming nitric acid, are used in some vehicles. These are sometimes called storable propellants because they do not boil off at normal ambient temperatures as do cryogens. However, they pose other handling problems because they are extremely toxic, and fuming nitric acid is highly corrosive.

Fixed fuel storage and pumping areas are ordinarily located on the opposite side of the launch stand from the oxidizer area for safety. For many fueling facilities, the more dangerous and corrosive propellants are handled and transferred by means of specially designed trailers. Even liquid oxygen may be transported in field trailers with their own pumping and pressurization equipment and introduced into the launch vehicle through a loading manifold. Both fuel and oxidizer systems are usually operated remotely from the blockhouse.

A method of dumping propellants is usually provided. Liquid oxygen may be dumped into a concrete evaporation pond, sometimes called a burn pond, on the launch complex. Fuel is rarely dumped on the ground; it is either fed back into its storage tank or piped to a fuel-holding pond where it can be disposed of later. More elaborate closed storage is required for highly toxic or otherwise dangerous propellants.

Control center. The launch operation is controlled from a launch control center, often contained in a blockhouse on the launch complex, as in Fig. 1. The blockhouse itself is a massive concrete structure (Fig. 6) designed to protect personnel from a possible inadvertent impact of the vehicle on the blockhouse with consequent explosion or fire. Thickness of the walls, sometimes $5-7$ ft of concrete, is designed to withstand the force of the highest-order explosion that the vehicle propellants could produce. Other precautions, such as baffled underground escape hatches and tunnels and remote air intakes for air conditioning, may be provided.

Fig. 4. Upper part of a *Titan 3* service structure showing the enclosure which provides shelter for the payload. (*TRW Systems*)

Fig. 5. Inside view of the structure of the enclosure shown in Fig. 4. Movable platforms at upper left are temporarily folded away. (*TRW Systems*)

Fig. 6. External view of typical blockhouse. Massive concrete construction protects workers during launch.

The launch control center contains both control and monitoring systems. The control systems usually are consoles or panels that enable the operator to remotely operate and control specific functions on the launch pad and in the vehicle. For example, the propellant-transfer console enables the operator to load propellants from the storage tanks to the vehicle. Within the propellant system and the vehicle are sensors that measure flow rates, temperatures, pressures, and other parameters critical to the propellant-loading process. These measurements are transmitted via cable back to the control console, where they are displayed. The operator controls the process and also monitors it to assure satisfactory performance.

Figure 7 shows the control center inside a typical blockhouse. The many controls and instruments on the consoles and panels connect to a terminal room on the floor below. There connections are made to the cables which lead to a similar terminal room at the pad, where in turn connections are made to the vehicle. Closed-circuit TV allows direct viewing of many critical operations.

Because many people participate in a launch operation, the communications system that links them together is important to operational success. An operations intercom system may have a dozen or more channels, each channel having a specific assignment. For example, channel 1 may be assigned to stage 1 propulsion, channel 2 to guidance, channel 3 to instrumentation, channel 4 to pad safety, and so on. This scheme of assignment allows each subsystem crew to communicate with each other. The subsystem crew chief also has a link with the system test conductor, for example, with the stage 1 test conductor. The stage 1 test conductor has an additional channel to the chief test conductor, who has overall launch responsibility. The chief test conductor also has direct links to important functions outside the launch complex, such as range control, range safety, and weather forecasting.

Check-out computers. The use of digital computers has become increasingly important in check-out and launch operations. A typical comprehensive test or launch countdown may require many hundreds of test sequences, each switched on in proper sequence. In addition, many hundreds or even thousands of data measurements must be made and evaluated to verify the test results. If this were done manually by human operators, the total time required would be unreasonably long. Computers can control test sequences and data readouts many times faster and with less chance for error. Computers can also promptly evaluate whether a measurement is within a specified tolerance and, if an out-of-tolerance measurement occurs, notify the test conductor. Without the increased speed and accuracy that computers provide, the check-out and launch of a complex space vehicle, such as a manned Apollo-Saturn lunar vehicle, would be virtually impossible. Many rocket vehicle components are deliberately designed for a short life but high performance to achieve weight savings. Such components could wear out faster then they could be checked out in a manual operation. *See* COMPUTER; DIGITAL COMPUTER.

Mobile launcher. Many launch complexes provide for assembly of the rocket vehicle on the launch pad, using the service structure as the assembly facility. Large, complex vehicles may require an assembly time of many weeks or months.

Fig. 7. Typical launch control center inside a blockhouse. Closed-circuit TV system provides personnel with close-up viewing of distant hazardous operations.

Fig. 8.   Aerial view of the *Titan 3* complex at Cape Kennedy.

For them, the launch site is inadequate because the service structure does not usually provide adequate shelter from weather, salt-air corrosion, and other environmental factors; for safety, service shops, stockrooms, engineering offices, and other needed services cannot be located near the launch pad; and the relatively long vehicle preparation time makes the pad unavailable for launchings, thus limiting launch schedules.

To overcome these problems, the mobile-launch concept has been used as the basis for launch-complex design on several programs (*Titan 3* and *Apollo–Saturn 5* are notable examples). In this procedure, the vehicle is assembled and checked out in a specially designed facility that provides shelter as well as other facilities and support equipment needed for the process. Because propellants are not loaded in the assembly building, several vehicles can be safely processed at once, allowing much support equipment to be shared. Usually the vehicle is assembled directly on a transporter which, after the vehicle is assembled and checked out, carries the vehicle to the launch pad and then serves as a launch platform.

Figure 8 is an aerial view of the *Titan 3* launch complex at the Eastern Test Range, Cape Kennedy. The large building in the center is the vehicle integration building; it can accommodate up to four *Titan 3* core vehicles. (*Titan 3* is made up of a three-stage core vehicle using liquid propellants and two solid-propellant strap-ons.) After assembly and initial check-out, the core vehicle is transported to the solid motor assembly building (SMAB), to the left in Fig. 8, where the solid-propellant rockets are attached. The two launch pads, one shown in use, are separated from the SMAB and from each other by nearly 2 mi. The large, low building at the bottom of Fig. 8 is used for receiving inspection and processing of subassemblies prior to vehicle assembly. The structures to the left of the vehicle integration building are used for processing and temporary storage of the solid-propellant rocket segments.

Fig. 9.   *Titan 3* on its mobile platform leaving the solid motor assembly building. The tall structure on platform next to the vehicle is the umbilical tower.

Figure 9 shows the *Titan 3* leaving the SMAB on its mobile platform. The platform runs on rails and is propelled by two locomotives.

Figure 10 is a view of the *Saturn 5* vehicle assembly building on Complex 39 at the Kennedy Space Center. This structure is 525 ft high, 716 ft long, and 518 ft wide, making it the largest (in volume)

Fig. 10.   *Saturn 5* vehicle assembly building on Complex 39 at the Kennedy Space Center. The launch control center is at lower right. (*NASA*)

Fig. 11.   Fully assembled *Apollo – Saturn 5* on the mobile launcher leaving the vehicle assembly building. (*NASA*)

building in the world. The low-bay area in front is used for preassembly check-out of the second stage, third stage, and instrument unit. The high-bay area has four bays, of which three are activated. Each bay can accommodate the total height of the *Saturn 5* launch vehicle and Apollo spacecraft (365 ft). Retractable platforms in each bay provide

direct access to the vehicle at all work levels. The *Saturn 5* vehicle is seen on its mobile launch platform on the right side of the building. Diagonal to the vehicle assembly building on the lower right side is the launch control center. This facility contains four firing rooms (three active). Each firing room has some 450 consoles which contain controls and displays required for check-out and launch. Two launch pads, one active and one planned, are located about 5 mi distant.

Figure 11 shows *Saturn 5* on its mobile launcher. The platform which supports the vehicle is 160 ft long, 135 ft wide, and 25 ft deep. One end of the platform supports the 380-ft umbilical tower which carries propellant, electrical, and pneumatic lines to the vehicle and provides access at various levels. The other end of the platform supports the vehicle over a 45-ft square opening. Two levels within the mobile launcher contain check-out equipment, including a check-out computer, propellant-loading accessories, electrical equipment, hydraulic servicing units, and other equipment required for launch. The mobile launcher serves as the basic support structure for the vehicle during the entire assembly check-out and launch process. The structure, with vehicle aboard, is moved the 5 mi from vehicle assembly building to launch pad by means of a huge tractor-like device which moves under the mobile launcher, lifts it, and carries it along a specially built roadway.

Figure 12 shows the mobile launcher in place on the launch pad. The mobile service structure on the right is moved up to the vehicle to provide additional access at all work levels. The 402-ft-high mobile service structure is the world's largest portable structure.

Solid-propellant complex. Solid-propellant rocket vehicles require, in general, a launch complex similar to those just described for liquid-propellant vehicles. The most important difference is the absence of facilities and equipment needed for propellant handling and transfer. This greatly simplifies design of the complex. However, solid-propellant stages are a severe fire or explosion hazard at all times during prelaunch operations. Extreme precautions are taken to eliminate the possibility of accidentally igniting the solid propellants by static electricity, sparks from electrical equipment or machinery, or lightning. Work areas and storage areas are usually separated by revetments.

Operational systems. The launch complex for an operational ballistic missile weapon has essentially the same features as the complexes described above. An operational system is one which is deployed and ready for use by the armed forces. The major differences in the launch complexes are therefore concerned with the operational requirement to maintain a large number of missiles in a launch-ready condition at all times, to have a high probability of successful launches in a short time, and to have a high probability of striking designated targets.

Facilities and equipment for operational systems are designed to have extremely high reliability. Much of the instrumentation used in research-and-development launches is eliminated, and only that which is essential to monitoring launch readi-

Fig. 12. *Apollo–Saturn 5* vehicle in place on its launch pad. Mobile service structure on right is moved up to vehicle to provide access at all work levels. (*NASA*)

ness is retained. A single launch control center may be used for operational control of a large number of missiles, with launch sites for the individual missiles well dispersed around the complex. For weapon systems, solid propellants and storable propellants are much favored over cryogens, which are difficult to store and handle and which cannot be loaded into the vehicle until just prior to launch, greatly increasing the reaction time of the system.

Fig. 13. Minuteman ICBM emplaced in silo.

Large weapon-system launch sites may be hardened; that is, they are designed to withstand all but a direct hit from a nuclear weapon in the event of an enemy attack. Hardening usually consists of locating the entire complex underground. The launch pad is a deep hole, called a silo, which contains the vehicle. Figure 13 shows a Minuteman missile in place in its silo. Tunnels provide limited access for umbilicals and at key work levels. As the Fig. 13 shows, a technician can be lowered from the surface for additional access, although, on an operational system, access requirements would be minimal.

Perhaps the most unique type of operational launch complex is the Polaris-type submarine. These nuclear-powered vessels can launch four Polaris intermediate-range ballistic missiles with nuclear warheads while submerged. The relatively short, squat solid-propellant missiles are mounted vertically in injection tubes in a compartment of the submarine. At launch, compressed air ejects the missile to the surface, where the rocket motor is ignited. Complete check-out gear and targeting computers are contained in the submarine. Successful use of the Polaris depends on the ability to obtain an accurate geodetic location of the submarine relative to its target. This can be accomplished to a high degree of accuracy anywhere in the world by means of navigation satellites. *See* SATELLITES, NAVIGATION BY.

[M. G. WENTZEL]

Bibliography: A. H. Bagnulo, *AIAA Pap.*, no. 67–247, AIAA Flight Test, Simulation and Support Conference, Cocoa Beach, Fla., Feb. 6–8, 1967; K. Brown and P. Weiser (eds.), *Ground Support Systems for Missiles and Space Vehicles*, 1961; D. H. Driscoll, Jr., *AIAA Pap.*, no. 67–234, AIAA Flight Test, Simulation and Support Conference, Cocoa Beach, Fla., Feb. 6–8, 1967; W. V. Foley, *AIAA Pap.*, no. 67–267, AIAA Flight Test, Simulation and Support Conference, Cocoa Beach, Fla., Feb. 6–8, 1967; B. A. Hehman and J. E. Wambolt, *AIAA Pap.*, no. 67–887, AIAA Annual Meeting and Technical Display, Anaheim, Calif., Oct. 23–27, 1967; M. G. Wentzel and J. H. McCurley, *AIAA Pap.*, no. 66–865, AIAA 3d Annual Meeting, Nov. 29–Dec. 2, 1966, Boston, Mass., 1966; R. L. Wilkinson, *ISA Pap.*, no. A67–11117, Instrument Society of America, 12th National Aerospace Instrumentation Symposium, May 2–4, 1966, Philadelphia, Pa., 1967.

Lava

Molten rock material that reaches the Earth's surface through volcanic vents and fissures; also, the igneous rock formed by consolidation of such molten material. Relatively rapid cooling at the Earth's surface may transform fluid lava into a dense-textured volcanic rock composed of tiny crystals or glass or both. Molten rock material below the Earth's surface, however, is usually known as magma and, upon cooling, gives rise to coarse-grained igneous rock, such as granite or gabbro.

Magma and lava are mutual solutions of silicate minerals with more or less dissolved gases. When magma is brought to the surface from regions of high pressure within the Earth, the gases expand and form bubbles in the fluid lava. If this lava quickly congeals, many of these bubbles may be trapped to form a highly porous rock.

The temperature of liquid lava ranges widely but generally does not exceed 1200°C. Basaltic lavas are usually hotter than rhyolitic ones. The viscosity of lava depends largely upon the temperature, composition, and gas content. Lavas poor in silica (basaltic) are the most fluid and may flow down very gentle slopes for many miles. The Hawaiian flows advance commonly at a rate of about 2 mph. When descending the courses of steep valleys, local velocities of up to 40 mph may be attained. Silica-rich lavas (rhyolite) are highly viscous. They move slowly and for relatively short distances. As lava cools, it becomes more viscous and the rate of flow decreases. Rapid cooling, as at the surface of a flow, promotes the formation of glass. Slower cooling, as near the center of a flow, favors the growth of crystals.

During many volcanic eruptions the lava is so rapidly ejected that it is blown to bits by the explosive force of expanding gases. The small masses rapidly congeal and settle to the earth to form thick blankets of volcanic tuff and related pyroclastic rock. Lava flows and volcanic tuffs cover large areas of the Earth's surface and may form more or less alternating layers totaling many thousands of feet in thickness. *See* IGNEOUS ROCKS; MAGMA; PYROCLASTIC ROCKS; TUFF; VOLCANIC GLASS; VOLCANO.   [CARLETON A. CHAPMAN]

Lawrencium

A chemical element, symbol Lr, atomic number 103. Lawrencium, named after E. O. Lawrence, is the eleventh transuranium element; it completes the actinide series of elements. The symbol Lw was originally proposed by the discoverers: it was changed to Lr in 1963. *See* ACTINIDE ELEMENTS; TRANSURANIUM ELEMENTS.

The element was discovered in February 1961 at the HILAC (Heavy-Ion Linear Accelerator) in the Lawrence Radiation Laboratory in Berkeley, CA, by A. Ghiorso, T. Sikkeland, A. E. Larsh, and R. M. Latimer, when a 2-μg target including the californium isotopes 249, 250, 251, and 252 was bombarded with boron-11 ions. The recoiling atoms of the new element were attracted electrically to a metallized mylar tape which was pulled successively in front of four semiconductor alpha detectors. In many runs, a total of about 100 events of an alpha activity with energy 8.6 MeV and half-life about 8 sec was detected. By means of excitation functions and cross-bombardment techniques

(that is, the use of other heavy ions and other targets) as well as by the use of α-systematics, it was concluded that an isotope of element 103 with a probable mass number of 257 or 258, or both together, was the source of the α-particle activity. (Later Berkeley work showed that the activity was due to the isotope 258, and a better half-life of 4.2 sec was determined). *See* CALIFORNIUM.

In 1965 at the Dubna laboratories near Moscow, teams under the general direction of G. N. Flerov discovered another isotope, ^{256}Lr, produced by the reaction of oxygen-18 ions with americium-243. By recoil-milking a known daughter product from alpha decay, they measured a half-life of about 45 sec. Later work in their laboratory measured a complex α-spectrum (8.35–8.60 MeV) and a 25–35-sec half-life for this nuclide.

In 1968 at Berkeley, R. J. Silva and coworkers used this isotope in a tour de force of chemistry to make a crucial observation of the behavior of element 103 in aqueous solution. Whereas nobelium, element 102, had been found to have a stable +2 state and was indeed very difficult to oxidize to a +3 valency in aqueous solution, lawrencium was shown to have a stable +3 state just as the other heavy actinide elements do. This finding is what was expected for the last element in the actinide transition series.

By 1971 work in the Berkeley laboratory had established the nuclear properties of all the isotopes of lawrencium from mass 255 to mass 260. ^{260}Lr is an alpha emitter with a half-life of 3 min and consequently is the longest-lived isotope known.   [ALBERT GHIORSO]

Bibliography: E. D. Donets et al., *Sov. J. At. Energy*, 19:995–999, 1965; G. N. Flerov et al., *Study of Alpha Decay of ^{256}103 and ^{257}103*, Dubna Preprint no. E7-3257, 1967; A. Ghiorso et al., New element: Lawrencium, atomic number 103, *Phys. Rev. Lett.*, 6(9):473–475, 1961.

Lawson criterion

A necessary but not sufficient condition for the achievement of a net release of energy from nuclear fusion reactions in a fusion reactor. As originally formulated by J. D. Lawson, this condition simply stated that a minimum requirement for net energy release is that the fusion fuel charge must combust for at least enough time for the recovered fusion energy release to equal the sum of energy invested in heating that charge to fusion temperatures, plus other energy losses occurring during combustion.

The result is usually stated in the form of a minimum value of $n\tau$ that must be achieved for energy break-even, where n is the fusion fuel particle density and τ is the confinement time. Lawson considered bremsstrahlung (x-ray) energy losses in his original definition. For many fusion reactor cases, this loss is small enough to be neglected compared to the heating energy. With this simplifying assumption, the basic equation from which the Lawson criterion is derived is obtained by balancing fusion energy release against heat input to the fuel plasma. Assuming hydrogenic isotopes, deuterium and tritium at densities n_D and n_T respectively, with accompanying electrons at density n_e, all at a maxwellian temperature T, one obtains Eq. (1), where the recovered fusion energy release is set equal to or greater than the energy input to heat the

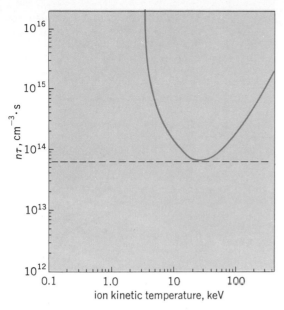

Typical plot of minimum value of $n\tau$ necessary for net release of energy versus ion kinetic temperature in a mixture containing equal amounts of deuterium and tritium. *(From R. F. Post, Nuclear fusion, in J. M. Hollander, ed., Ann. Rev. Energy, 1:213–255, 1976)*

$$n_D n_T \langle \sigma v \rangle Q \tau \eta_r \geq \left[\frac{3}{2} kT\left(n_D + n_T + n_e\right)\right]\frac{1}{\eta_h} \qquad (1)$$

fuel. Here $\langle \sigma v \rangle$ is the product of reaction cross section and relative ion velocity, as averaged over the velocity distribution of the ions, Q is the fusion energy release, η_r is the efficiency of recovery of the fusion energy, η_h is the heating efficiency, and k is the Boltzmann constant.

For a fixed mixture of deuterium and tritium ions, Eq. (1) can be rearranged in the general form of Eq. (2). For a 50-50 mixture of deuterium and tri-

$$n\tau \geq F(\eta_r, \eta_h, Q)\left[\frac{T}{\langle \sigma v \rangle}\right] \qquad (2)$$

tium (see illustration), the minimum value of $T/\langle \sigma v \rangle$ occurs at about 25 keV ion kinetic temperature (mean ion energies of about 38 keV). Depending on the assumed efficiencies of the heating and recovery processes, the lower limit values of $n\tau$ range typically between about 10^{14} and 10^{15} cm^{-3}s. These values serve as a handy index of progress toward fusion, although their achievement does not alone guarantee success. Under special circumstances (unequal ion and electron temperatures, unequal deuterium and tritium densities, and nonmaxwellian ion distributions), lower $n\tau$ values may be adequate for nominal break-even.

The discussion up to this point has been oriented mainly to situations in which the fusion reactor may be thought of as a driven system, that is, one in which a continuous input of energy from outside the reaction chamber is required to maintain the reaction. Provided the efficiencies of the external heating and energy recovery systems are high, a driven reactor generally would require the lowest $n\tau$ values to produce net power. An important alternative operating made for a reactor would be an ignition mode, that is, one in which, once the initial heating of the fuel charge is accomplished, energy

directly deposited in the plasma by charged reaction products will thereafter sustain the reaction. For example, in the D-T reaction, approximately 20% of the total energy release is imparted to the alpha particle; in a magnetic confinement system, much of the kinetic energy carried by this charged nucleus may be directly deposited in the plasma, thereby heating it. Thus if the confinement time is adequate, the reaction may become self-sustaining without a further input of energy from external sources. Ignition, however, would generally require $n\tau$ products with a higher range of values, and is thus expected to be more difficult to achieve than the driven type of reaction. However, in all cases the Lawson criterion is to be thought of as only a rule of thumb for measuring fusion progress; detailed evaluation of all energy dissipative and energy recovery processes is required in order properly to evaluate any specific system. *See* FUSION, NUCLEAR; PLASMA PHYSICS.

[RICHARD F. POST]

Layout drawing

A design drawing or graphical statement of the overall form of a component or device, which is usually prepared during the innovative stages of a design. Since it lacks detail and completeness, a layout drawing provides a faithful explanation of the device and its construction only to individuals such as designer and draftsmen who have been intimately involved in the conceptual stage. In a sense, the layout drawing is a running record of ideas and problems posed as the design evolves. In the layout drawing, for instance, considerations of kinematic design of a mechanical component are explored graphically in incomplete detail, showing only those aspects of the elements and their interrelationships to be considered in the design. In most cases the layout drawing ultimately becomes the primary source of information from which detail drawings and assembly drawings are prepared by other draftsmen under the guidance of the designer. *See* DRAFTING; ENGINEERING DRAWING.

[ROBERT W. MANN]

Lazurite

The chief mineral constituent in the ornamental stone lapis lazuli. It crystallizes in the isometric system, but well-formed crystals, usually dodecahedral, are rare (see illustration). Most commonly, it is granular or in compact masses. There is imperfect dodecahedral cleavage. The hardness is 5–5.5 on Mohs scale, and the specific gravity is 2.4–2.5. There is vitreous luster and the color is a deep azure, more rarely a greenish-blue. Lazurite is a tectosilicate, the composition of which is expressed by the formula $Na_4Al_3Si_3O_{12}S$, but some S may be replaced by SO_4 or Cl. Lazurite is soluble in HCl with the evolution of hydrogen sulfide.

Lazurite is a feldspathoid but, unlike the other members of that group, is not found in igneous rocks. It occurs exclusively in crystalline limestones as a contact metamorphic mineral. Lapis lazuli is a mixture of lazurite with other silicates and calcite and usually contains disseminated pyrite. It has long been valued as an ornamental material. Lazurite was formerly used as blue pigment, ultramarine, in oil painting. Localities of occurrence are in Afghanistan; Lake Baikal, Siberia;

Lazurite from Afghanistan. (*Specimen from Department of Geology, Bryn Mawr College*)

Chile; and San Bernardino County, Calif. *See* FELDSPATHOID; SILICATE MINERALS.

[CORNELIUS S. HURLBUT, JR.]

Leaching

The removal of a soluble fraction, in the form of a solution, from an insoluble, permeable solid with which it is associated. The separation usually involves selective dissolving, with or without diffusion, but in the extreme case of simple washing it consists merely of the displacement (with some mixing) of one interstitial liquid by another with which it is miscible. The soluble constituent may be solid (as the metal leached from ore) or liquid (as the oil leached from soybeans).

Leaching is closely related to solvent extraction, in which a soluble substance is dissolved from one liquid by a second liquid immiscible with the first. Both leaching and solvent extraction are often

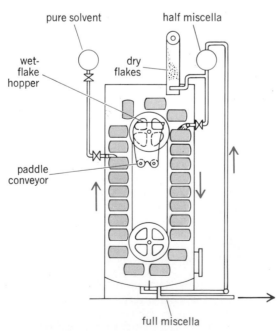

Fig. 1. Bollman extractor. (*From W. L. McCabe and J. C. Smith, Unit Operations of Chemical Engineering, 2d ed., McGraw-Hill, 1967*)

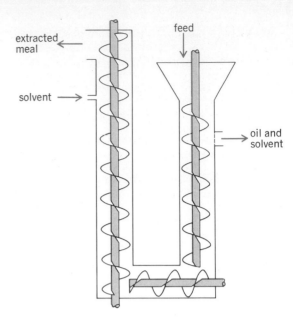

Fig. 2. Hildebrandt screw-conveyor extractor. (*From W. L. McCabe and J. C. Smith, Unit Operations of Chemical Engineering, 2d ed., McGraw-Hill, 1967*)

called extraction. Because of its variety of applications and its importance to several ancient industries, leaching is known by a number of other names: solid-liquid extraction, lixiviation, percolation, infusion, washing, and decantation-settling. The liquid used to leach away the soluble material (the solute) is termed the solvent. The resulting solution is called the extract or sometimes the miscella.

The mechanism of leaching may involve simple physical solution, or dissolution made possible by chemical reaction. The rate of transportation of solvent into the mass to be leached, or of soluble fraction into the solvent, or of extract solution out of the insoluble material, or some combination of these rates may be significant. A membranous resistance may be involved. A chemical reaction rate may also affect the rate of leaching. The general complication of this simple-appearing process results in design by chiefly empirical methods. Whatever the mechanism, however, it is clear that the leaching process is favored by increased surface per unit volume of solids to be leached and by decreased radial distances that must be traversed within the solids, both of which are favored by decreased particle size. Fine solids, on the other hand, cause mechanical operating problems during leaching, slow filtration and drying rates, and possible poor quality of solid product. The basis for an optimum particle size is established by these characteristics.

Leaching processes fall into two principal classes: those in which the leaching is accomplished by percolation (seeping of solvent through a bed of solids), and those in which particulate solids are dispersed into the extracting liquid and subsequently separated from it. In either case, the operation may be a batch process or continuous. *See* EXTRACTION; MASS-TRANSFER OPERATION; SOLVENT EXTRACTION.

Percolation. In addition to being applied to ores and rock in place and by the simple technique of heap leaching, percolation is carried out in batch tanks and in several designs of continuous extractors.

The batch percolator is a large circular or rectangular tank with a false bottom. The solids to be leached are dumped into the tank to a uniform depth. The are sprayed with solvent until their solute content is reduced to an economic minimum and are then excavated. A simple example is the brewing of coffee in a percolator (repeated extraction) or a drip pot (once-through). Countercurrent flow of the solvent through a series of tanks is common, with fresh solvent entering the tank containing the most nearly exhausted material. Some leach tanks operate under pressure, to contain volatile solvents or increase the percolation rate.

Continuous percolators employ the moving-bed principle, implemented by moving baskets that carry the solids past solvent sprays, belt or screw conveyors that move them through streams or showers of solvent, or rakes that transport them along a solvent-filled trough. In a revolving-basket type like the Rotocel extractor, bottomless compartments move in a circular path over a stationary perforated annular disk. They are successively filled with solids, passed under solvent sprays connected by pumps so as to provide countercurrent flow of the extracting liquid, and emptied through a large opening in the disk. Alternating perforated-bottom extraction baskets may be arranged in a bucket-elevator configuration, as in the Bollman extractor (Fig. 1). On the up cycle, the partially extracted solids are percolated by fresh solvent sprayed at the top; on the down cycle, fresh solids are sprayed with the extract from the up cycle. A screw-conveyor extractor like the Hildebrandt extractor (Fig. 2) moves solids through a V-shaped line in a direction opposite to the flow of solution. The solids may be conveyed instead by rakes or paddles along a horizontal or inclined trough counter to the direction of solvent flow, as in the Kennedy extractor (Fig. 3). In the last two types, the action is predominantly percolation but involves some solids dispersal because of agitation by the conveyors. Horizontal continuous vacuum filters of the belt, tray, or table type sometimes are used as leaching equipment. See FILTRATION.

Dispersed-solids leaching. Equipment for leaching fine solids by dispersion and separation, a particularly useful technique for solids which disintegrate during leaching, includes batch tanks and continuous extractors.

Inasmuch as the purpose of the dispersion is usually only to permit exposure of the particles to unsaturated solvent, the agitation in a batch-stirred extractor need not be intense. Air agitation is often used. Examples are Pachuca tanks, large cylinders with conical bottoms and an axial air nozzle or air-lift tube. If mechanical agitation is employed, a slow paddle is sufficient. The Dorr agitator (Fig. 4) combines a rake with an air lift. In all cases, the mixture of solids and liquid is stirred until the maximum economical degree of leaching has occurred. The solids are then allowed to settle. The extract is decanted, and the solids, sometimes after successive treatments with fresh solvent, are removed by shoveling or flushing.

Continuous dispersed-solids leaching is accomplished in gravity sedimentation tanks or in vertical plate-extractors. An example of the latter is the Bonotto extractor shown in Fig. 5. Staggered open-

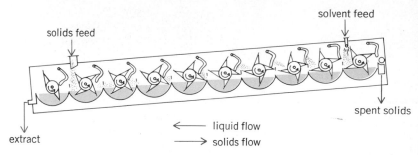

Fig. 3. Kennedy extractor. (*From R. H. Perry and C. H. Chilton, eds., Chemical Engineer's Handbook, 5th ed., McGraw-Hill, 1973*)

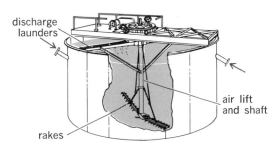

Fig. 4. Dorr agitator for batch washing of precipitates. (*From W. L. Badger and J. T. Banchero, Introduction to Chemical Engineering, McGraw-Hill, 1955*)

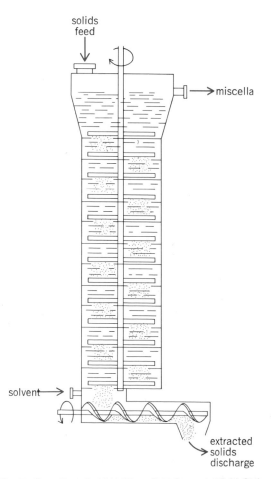

Fig. 5. Bonotto extractor. (*From R. H. Perry and C. H. Chilton, eds., Chemical Engineer's Handbook, 5th ed., McGraw-Hill, 1973*)

ings in the plates allow the solids, moved around each plate by a wiping radial blade, to cascade downward from plate to plate through upward-flowing solvent.

Gravity sedimentation thickeners can serve as effective continuous contacting and separating devices for leaching fine solids. A series of such units properly connected provides true continuous countercurrent washing (known as CCD for continuous countercurrent decantation) of the solids. *See* COUNTERCURRENT MASS-TRANSFER OPERATION; THICKENING. [SHELBY A. MILLER]

Bibliography: W. L. Badger and J. T. Banchero, *Introduction to Chemical Engineering*, 1955; *Kirk-Othmer Encyclopedia of Chemical Technology*, 2d ed., vol. 8, 1965; W. L. McCabe and J. C. Smith, *Unit Operations of Chemical Engineering*, 3rd ed., 1976; R. H. Perry and C. H. Chilton (eds.), *Chemical Engineers' Handbook*, 5th ed., 1973; R. E. Treybal, *Mass-Transfer Operations*, 2d ed., 1968.

Lead

A chemical element, Pb, atomic number 82 and atomic weight 207.19. Lead is a heavy metal (specific gravity 11.34 at 16°C), of bluish color, which tarnishes to dull gray. It is pliable, inelastic, easily fusible, melts at 327.4°C, and boils at 1740°C. The normal chemical valences are 2 and 4. It is relatively resistant to attack by sulfuric and hydrochloric acids but dissolves slowly in nitric acid. Lead is amphoteric, forming lead salts of acids as well as metal salts of plumbic acid. Lead forms many salts, oxides, and organometallic compounds.

Industrially, the most important lead compounds are the lead oxides and tetraethyllead. Lead forms alloys with many metals and is generally employed in the form of alloys in most applications. Alloys formed with tin, copper, arsenic, antimony, bismuth, cadmium, and sodium are all of industrial importance. *See* LEAD ALLOYS; TETRAETHYLLEAD.

Lead compounds are toxic and have resulted in poisoning of workers from misuse and overexposure. However, lead poisoning is presently rare because of the industrial application of modern hygienic and engineering controls. The greatest hazard arises from the inhalation of vapor or dust. In the case of organolead compounds, absorption through the skin may become significant. Some of the symptoms of lead poisoning are headaches, dizziness, and insomnia. In acute cases there is usually stupor, which progresses to coma and terminates in death. The medical control of employees engaged in lead usage involves precise clinical tests of lead levels in blood and urine. With such control and the proper application of engineering control, industrial lead poisoning may be entirely prevented. *See* TOXICOLOGY.

Lead is one of the oldest known metals, and the earliest archeological specimens date from about 3000 B.C. It is mentioned several times in the Old Testament, and in ancient Egypt it was used to glaze pottery and make ornamental objects. Lead was used extensively by the Romans for water pipes, even to the extent of being standardized by size and length. Some ancient Roman pipe is still intact and in serviceable condition, attesting to lead's outstanding corrosion resistance.

Natural occurrence. Yearly world production of lead is about 2,500,000 tons. The leading countries in lead mining and production are Australia, the United States, Canada, and the Soviet Union. However, the United States is unique in that it consumes much more lead than it produces (about 50% of the world's consumption). Other important sources of lead are Mexico, Peru, Yugoslavia, Germany, Morocco, South-West Africa, and Spain. Most of the lead produced in the United States is mined in Missouri, with additional lead from Idaho, Utah, Arizona, Colorado, and Montana.

Lead rarely occurs in its elemental state. The most common ore is the sulfide, galena. The other minerals of commercial importance are the carbonate, cerussite, and the sulfate, anglesite, which are much more rare.

Lead also occurs in various uranium and thorium minerals, arising directly from radioactive decay. Because certain isotopes are concentrated in lead derivatives from such sources, both the atomic weight and density of such samples vary significantly from normal lead.

Lead ores generally occur in nature associated with silver and zinc. Other metals commonly occurring with lead ores are copper, arsenic, antimony, and bismuth. Most of the world production of arsenic, antimony, and bismuth arises from their separation from lead ores.

Commercial lead ores may contain as little as 3% lead, but a lead content of about 10% is most common. The ores are concentrated to 40% or greater lead content before smelting. A variety of mechanical separation processes may be employed for the concentration of lead ores, but the sulfide ores are generally concentrated by flotation processes, for which they are particularly suitable. *See* LEAD METALLURGY.

More than one-third of the lead produced in the United States is derived from reclaimed lead or lead alloy. The chief source of this secondary lead is scrapped automobile storage batteries. The secondary lead is obtained by melting of scrap and is generally not purified but rather reused as lead alloy. Secondary lead from storage batteries contains antimony and is reused in battery manufacture. The lead scrap which contains tin is reused in the manufacture of solder.

Uses. As implied in the discussion of scrap lead above, the largest single use of lead is for the manufacture of storage batteries. Other important applications are for the manufacture of tetraethyllead, cable covering, construction, pigments, solder, and ammunition.

The lead storage battery consists of a negative plate of porous lead, a positive plate of lead peroxide, and an electrolyte of sulfuric acid solution. The plates are made by casting perforated grids from an alloy of lead containing 7–12% antimony and small amounts of tin. The antimony is added to give the grids better hardness and corrosion resistance, whereas the tin is added to give better casting properties. The negative grid is coated with a paste of litharge (PbO) or a mixture of litharge and finely divided metallic lead. The positive grid is coated with a mixture of litharge and finely divided lead and in some cases red lead (Pb_3O_4). After the batteries are assembled and charged with sulfuric acid, the plates are "formed" by passing an electric current through them to convert the lead oxide at the negative plate into finely divided lead sponge and convert the lead oxide at the positive plate to lead peroxide. The chemical reactions which take place on charging or discharging a battery are very complex but may be indicated in simple form by Eq. (1). Because sulfuric acid is

$$PbO_2 + 2H_2SO_4 + Pb \xrightleftharpoons[\text{Charge}]{\text{Discharge}}$$
$$2PbSO_4 + 2H_2O \quad (1)$$

consumed on discharge of the battery, the sulfuric acid concentration is used as a measure of the state of charge of lead batteries.

In 1957 an important addition to the family of lead batteries was the introduction of a chargeable dry cell. The battery is composed of an alkali–lead oxide–silver system and is capable of repeated discharge and recharge without affecting its capacity. Also noteworthy are its operability at both low and high temperatures and its long-term shelf life. *See* STORAGE BATTERY.

The manufacture of tetraethyllead is the second largest use of lead. When 2–3 ml of this antiknock agent is added to 1 gal of gasoline, the octane rating of the fuel is raised about 10 octane numbers. This increase in octane number permits the use of higher compression ratios in automobile engines, thereby improving engine efficiency. For example, it has been shown that an average decrease in gasoline consumption of about 45% can be obtained in going from 5.25:1 to 10:1 compression ratio under conditions of constant performance. *See* ANTIKNOCK AGENTS; INTERNAL COMBUSTION ENGINE.

Tetraethyllead is manufactured by the reaction of ethyl chloride with an alloy of sodium and lead. The chemical reaction is given by Eq. (2).

$$4PbNa + 4C_2H_5Cl \rightarrow$$
$$(C_2H_5)_4Pb + 4NaCl + 3Pb \quad (2)$$

Lead is melted and mixed with sodium to form an alloy containing about 10% sodium. The solid alloy is treated with ethyl chloride in autoclaves at moderate temperature and pressure in the absence of solvent. The crude product is separated by steam distillation and is purified by air-blowing to remove organic bismuth compounds arising from bismuth impurities in the original lead metal. The tetraethyllead is then blended with ethylene dichloride and ethylene dibromide, which serve as scavengers to aid in the removal of lead deposits from engines. It is sold in this form to petroleum refiners as antiknock fluid for addition to their automotive gasoline stocks. In aviation gasoline, ethylene dibromide alone is used as the scavenger.

Other commercial organolead antiknock agents which have been introduced include tetramethyllead, mixtures of tetramethyllead and tetraethyllead, and redistributed mixtures containing all the possible methylethyllead compounds as well as the original tetraethyllead and tetramethyllead. Tetramethyllead is manufactured both by a process based on methyl chloride and sodium-lead alloy, analogous to the manufacture of tetraethyllead, and by a new electrochemical process using a lead anode and an ether–methylmagnesium chloride electrolyte.

A much smaller commercial use for tetraethyllead arises from its ability to alkylate mercury compounds. The resulting organomercury compounds are used as fungicides. *See* FUNGISTAT AND FUNGICIDE.

In addition to these traditional uses of organolead compounds, some high-value applications are being developed. These applications are in such diverse fields as catalysts for polyurethane foams, marine antifouling paint toxicants, biocidal agents against gram-positive bacteria, protection of wood against marine borers and fungal attack, preservatives for cotton against rot and mildew, molluskicidal agents to kill snails which act as intermediate hosts for the serious tropical disease "snail fever" (bilharzia, or schistosomiasis), anthelmintic agents against tapeworms in domestic animals, wear-reducing agents in lubricants, and corrosion inhibitors for steel. *See* SCHISTOSOMIASIS.

Because of its excellent resistance to corrosion, lead finds extensive use in construction, particularly in the chemical industry. It is resistant to attack by many acids because it forms its own protective oxide coating. Because of this advantageous characteristic, lead is used widely in the manufacture and handling of sulfuric acid. The corrosion resistance of lead is also utilized in the cathodic protection of metal structures. This is accomplished by impressing an electric charge on the structure to be protected; the structure serves as the cathode and lead as the anode. Such protection is useful for large structures such as ships, pipelines, and bridges. *See* CORROSION.

Because of the poor structural strength of lead, it is generally used in the form of its alloys, particularly in combination with antimony, or is used as coatings or plates on stronger structural metals. Tin alloys are frequently used in protective plates to impart mechanical strength and better corrosion resistance. To some extent, the lead that has been used in traditional architectural and plumbing applications in the past, such as roofing and flashings, for example is now being displaced by other materials. However, lead is being used increasingly in construction because of its excellent sound attenuation properties. Changes in building codes have established maximum permissible noise levels for multistoried structures; these have accelerated the trend toward the use of lead sheet as either an integral part of the wall partition or as a barrier in the air space above suspended ceilings. Lead has also been used in construction for many years because of its vibration-damping properties. This use also is growing. Heavy machinery and even large buildings are isolated from vibration by placing them on lead and asbestos pads. A splendid example in antivibration building techniques is the Pan American Building built over Grand Central Terminal in New York. In inertial guidance research, the experimental chambers are isolated from adjacent ground and structures by the use of massive lead pads.

Lead has long been used as protective shielding for x-ray machines. Because of the expanded applications of atomic energy, radiation-shielding applications of lead have become increasingly important. Basically, shielding effectiveness against radiation depends on density, and lead has the highest density of the commonly available materials. Wherever glass windows are required in radiation equipment, a type of glass containing large amounts of lead is used. *See* RADIATION SHIELDING.

Lead sheathing for telephone and television cables continues to be a sizable outlet for lead. The unique ductility of lead makes it particularly suitable for this application because it can be extruded in a continuous sheath around the internal conductors. The lead used in this application is generally alloyed with small amounts of arsenic and bismuth.

The use of lead in pigments has been a major outlet for lead but is decreasing in volume. White lead, $2PbCO_3 \cdot Pb(OH)_2$, is the most extensively used lead pigment. It is prepared from metallic lead by treatment with acetic acid, air, and carbon dioxide. It is an excellent pigment because of its outstanding chemical affinity for paint vehicles and its great hiding power. Pigments such as red lead (Pb_3O_4) and blue lead (a combination of basic lead sulfate, $PbSO_4 \cdot PbO$, zinc oxide, and carbon) are used as metal protective pigments in paints. Other lead pigments of importance are basic lead sulfate and lead chromates. The lead chromates are frequently used as dyes in formulating yellow, orange, red, and green paints. *See* PIGMENT.

The high density of lead, which permits a maximum of striking power with a minimum of air resistance, has made lead the ideal metal for bullets and shot. Lead shot is manufactured by a unique process which involves dropping molten lead into water from heights up to 125 ft, thus freezing lead in the spherical form assumed by its droplets.

Principal compounds. From a commercial standpoint, the most important lead compounds have been mentioned above. In addition, a considerable variety of lead compounds, such as silicates, carbonates, and salts of organic acids, are used as heat and light stabilizers for polyvinyl chloride plastics. These compounds are both inexpensive and effective. They function as hydrogen chloride acid acceptors, thereby preventing the autocatalytic breakdown of the plastic by the acid.

Certain inorganic lead compounds find specialized use. Lead silicates are used for the manufacture of glass and ceramic frits, which are useful in introducing lead into glass and ceramic finishes. Lead azide, $Pb(N_3)_2$, is the standard detonator for explosives. Lead arsenates are used in large quantities as insecticides for crop protection. Organic insecticides have displaced lead arsenate to some extent, but not completely because they have not been found as effective in certain applications.

Among the newer uses for inorganic lead compounds, litharge (lead oxide) is widely employed at a level of approximately 2% lead to improve the magnetic properties of barium ferrite ceramic magnets. High lead ferrites, containing approximately 20% lead, have been developed with superior magnetic properties. Also, a calcined mixture of lead zirconate and lead titanate, known as PZT in the trade, is finding increasing markets as a piezoelectric material. The most important application is in ultrasonic cleaning equipment. Finally, lead telluride is finding some use as the active component of thermoelectric generators.

[HYMIN SHAPIRO; JAMES D. JOHNSTON]
Bibliography: D. S. Carr, Some market potentials for organolead compounds, *Chem. Ind.*, pp. 1854–1857, Nov. 4, 1967; Lead Industries Association, *Lead in Modern Industry*, 1952; E. J. Mullarkey, Lead and its alloys, *Ind. Eng. Chem.*, 51: 1185–1191, 1959; E. J. Mullarkey, Lead: Its alloys and compounds, *Ind. Eng. Chem.*, 53:82–86, 1961; H. Shapiro and F. W. Frey, The chemistry of organolead compounds, in R. E. Kirk and D. F. Othmer (eds), *Encyclopedia of Chemical Technology*, vol. 12, 2d ed., 1967; N. V. Sidgwick, *Chemical Elements and Their Compounds*, 1950; S. W. Turner and B. A. Fader, New uses for lead, *Ind. Eng. Chem.*, 54:52–55, 1962; L. C. Willemsens, *Organolead Chemistry*, 1964; R. I. Ziegfeld, Lead and its applications, *Metals Handbook*, vol. 1, 1961.

Lead alloys

Substances formed by the addition of one or more elements, usually metals, to lead. Lead alloys may exhibit greatly improved mechanical or chemical properties as compared to pure lead. The major alloying additions to lead are antimony and tin. The solubilities of most other elements in lead are small, but even fractional weight percent additions of some of these elements, notably copper and arsenic, can alter properties appreciably.

Cable-sheathing alloys. Lead is used as a sheath over the electrical components to protect power and telephone cable from moisture. Alloys containing 1% antimony are used for telephone cable, and lead-arsenical alloys, containing 0.15% arsenic, 0.1% tin, and 0.1% bismuth, for example, are used for power cable. Aluminum and plastic cable sheathing are replacing lead alloy sheathing in many applications, but improvements in methods of applying a lead sheathing (continuous extrusion) may offset this trend somewhat.

Battery-grid alloys. Lead alloy grids are used in the lead-acid storage battery (the type used in automobiles) to support the active material composing the plates. Lead grid alloys contain 6–12% antimony for strength, small amounts of tin to improve castability, and one or more other minor additions to retard dimensional change in service. No lead alloys capable of replacing the lead-antimony alloys in automobile batteries have been developed, although research in this area has been extensive. An alloy containing 0.03% calcium for use in large stationary batteries has had success.

Chemical-resistant alloys. Lead alloys are used extensively in many applications requiring resistance to water, atmosphere, or chemical corrosion. They are noted for their resistance to attack by sulfuric acid. Alloys most commonly used contain 0.06% copper, or 1–12% antimony, where greater strength is needed. The presence of antimony lowers corrosion-resistance to some degree.

Type metals. Type metals contain $2\frac{1}{2}$–12% tin and $2\frac{1}{2}$–25% antimony. Antimony increases hardness and reduces shrinkage during solidification. Tin improves fluidity and reproduction of detail. Both elements lower the melting temperature of the alloy. Common type metals melt at 460–475°F.

Bearing metals. Lead bearing metals (babbitt metals) contain 10–15% antimony, 5–10% tin, and for some applications, small amounts of arsenic or copper. Tin and antimony combine to form a compound which provides wear-resistance. These alloys find frequent application in cast sleeve bearings, and are used extensively in freight-car journal bearings. In some cast bearing bronzes, the lead content may exceed 25%.

Solders. A large number of lead-base solder compositions have been developed. Most contain large amounts of tin with selected minor additions to provide specific benefits, such as improved wetting characteristics.

Free-machining brasses, bronzes, and steels. Lead is added in amounts from 1 to 25% to brasses and bronzes to improve machining characteristics. Lead remains as discrete particles in these alloys. It is also added to some construction steel products to increase machinability. Only about 0.1% is needed, but the tonnage involved is so large that this forms an important use for lead. *See* ALLOY; LEAD; LEAD METALLURGY; SOLDERING; TIN ALLOYS.   [DUDLEY WILLIAMS]

Bibliography: E. J. Minarcik, *The Melting and Casting of Lead and Lead Alloys*, in American Society for Metals, *Metals Handbook*, 1948; N. E. Woldman, *Engineering Alloys*, 4th ed., 1962.

Lead isotopes, geochemistry of

The study of the isotopic composition of lead in minerals and rocks in order to relate it to past associations of the lead with uranium and thorium. Lead has four isotopes, of relative mass 204, 206, 207, and 208. Pb^{206}, Pb^{207}, and Pb^{208} are produced by the radioactive decay of uranium and thorium. The decay relations are shown in Eqs. (1), where α

U^{238}(half-life 4.5×10^9 years)→

$$Pb^{206} + 8\alpha + 6\beta$$

U^{235}(half-life 0.71×10^9 years)→   (1)

$$Pb^{207} + 7\alpha + 4\beta$$

Th^{232}(half-life 13.9×10^9 years)→

$$Pb^{208} + 6\alpha + 4\beta$$

denotes an alpha particle (doubly charged helium nucleus) and β denotes a beta particle (electron). For a given system such as a mineral or rock, primary lead is defined as the lead present in the system at the time it was formed. Radiogenic lead is lead produced in the system by the decay of uranium and thorium since it was formed. Since Pb 204 is not produced by the decay of any naturally occurring radioactive species, it can be used as an index to determine the amount of primary lead in a system. Primary lead will include all of the Pb^{204} and variable amounts of Pb^{206}, Pb^{207}, and Pb^{208}, depending on its origin. *See* RADIOACTIVITY.

Common rocks, for example, granites and basalts, or a system such as the crust of the Earth have sufficiently high ratios of thorium and uranium to lead to cause measurable changes in the isotopic composition of their lead over time intervals as short as some tens of millions of years. Some minerals have practically all radiogenic lead, while others have such low ratios of uranium and thorium to lead that the isotopic composition of their lead will not have changed appreciably over thousands of millions of years. Since the isotopic composition of lead observed in rocks and minerals is related directly to the amounts of uranium and thorium the lead was associated with in the past and the length of time of the association, it is possible to gain information from coordinated studies of the geochemistry of uranium, thorium, and lead which could not be obtained from elements not involved in radioactive-decay chains. Some examples will illustrate the unique value of this type of investigation.

Variation with time. If the uranium, thorium, and lead contents and lead isotopic composition are known for a particular mineral, it is possible to calculate the isotopic composition of the lead at any time in the past if the mineral represents a closed system. A closed system is one in which the uranium-lead and the thorium-lead ratios have not been changed in the past by processes other than radioactive decay; that is, these ratios were never changed by chemical processes. It is possible in principle to treat the outer portion of the Earth as a geologic system and to study the isotopic composition of its lead as a function of time. Meaningful results can be obtained if the uranium-lead and thorium-lead ratios are reasonably uniform throughout the system or if sampling is extensive enough to average out the local variations which may exist. In either case it is possible then to determine the uranium-lead and thorium-lead ratios of the system producing the lead.

This time variation has been studied in two different sources, potassium feldspar and galena. Potassium feldspar ($KAlSi_3O_8 + NaAlSi_3O_8$) is a common mineral in igneous rocks, such as granites and pegmatites. Galena, PbS, commonly occurs in veins of hydrothermal origin. Both minerals have such low uranium-lead and thorium-lead ratios that the isotopic composition of their lead will not have changed appreciably since the time of crystallization, providing there has been no contamination from external sources. Although potassium feldspar contains relatively small amounts of lead (5–100 ppm), it has the great advantage that the time of formation of the mineral can be accurately determined by the rubidium-strontium method or by measuring the age of cogenetic minerals, such as mica, uraninite, or zircon. On the other hand, the veins from which galena is obtained do not as a rule contain minerals which are suitable for age determination, so that the time of formation of a galena must be inferred from general geological relations. Since lead is easily obtained from galena, many more data exist for it than for igneous minerals, although the data given by the latter can be interpreted with greater certainty. *See* ROCK, AGE DETERMINATION OF.

The isotopic composition of lead has been determined in about 50 potassium feldspars of known age, covering a time span of 3×10^8 to 27×10^8 years. Representative data are illustrated in Fig. 1, in which the ratios of Pb^{206}, Pb^{207}, and Pb^{208} to Pb^{204} are plotted as a function of age. The experimental uncertainties in the ratios are about 1% or less; thus the variations are many times the errors. It is seen that the more recently a feldspar was formed, the more Pb^{206}, Pb^{207}, and Pb^{208} it contains

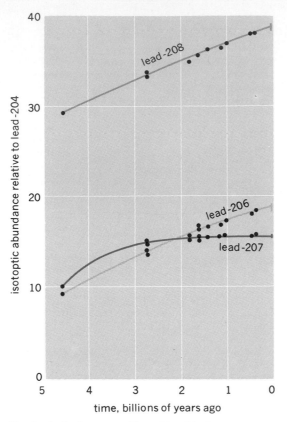

Fig. 1. Isotopic composition of lead in igneous rocks plotted as a function of age.

al lead 4.5×10^9 years ago is represented by the troilite lead. That is, starting 4.5×10^9 years ago with lead of the same isotopic composition as troilite lead, it is possible to select values for the uranium-lead and thorium-lead ratios which will give growth curves for the lead isotopes which fit the observed data quite closely. The calculated growth curves in Fig. 1 correspond to present-day ratios in the source environment shown in Eqs. (2).

$$U^{238}/Pb^{204} = 9.3 \text{ (atom ratio)}$$
$$U/Pb = 0.18 \text{ (weight ratio)} \qquad (2)$$
$$Th^{232}/U^{238} = 4.0 \text{ (atom ratio)}$$

This simple model assumes a closed-system source environment in the sense defined above. *See* EARTH, AGE OF; METEORITE.

It is not entirely clear just what source material the feldspar lead characterizes. The feldspars occur in pegmatites which were associated with processes of mountain building in the past. The lead may have been derived either from lead brought toward the surface from the mantle, from sediments which were buried and melted or partially melted during the process, or from both of these sources. However, it has been shown above that lead in rocks of deep-seated origin, such as basalts, does not differ greatly in isotopic composition from lead in oceanic materials which are of surface origin and have undergone erosion. Thus the curves in Fig. 1 approximate the past variation of isotopic composition of lead in both the outer mantle and average crustal material.

Model age. The isotopic composition of lead in several hundred galenas has been published. Some difficulty arises in determining the time of formation of galena, as already mentioned; accordingly, a different method of plotting these leads is used. If Pb^{206}/Pb^{204} is plotted against Pb^{207}/Pb^{204} from the curves in Fig. 1, a new curve will result on which each point corresponds to a model age for a particular lead. The model age is the time at which the lead appears to have been separated from uranium in the source environment. This treatment may be expanded to allow for varying uranium-lead ratios in the source environment. Assume that a group of galenas originated at a time $t \times 10^6$ years ago from source environments which all had the isotopic composition of meteoritic troilite lead 4.5×10^9 years ago and which were closed systems with differing uranium-lead ratios. Since the ratio of radiogenic Pb^{207} to radiogenic Pb^{206} will be the same in all the leads, regardless of the uranium-lead ratios in the sources, a plot of Pb^{207}/Pb^{204} against Pb^{206}/Pb^{204} for these leads will yield a straight line. Such a line of constant model age and different uranium-lead ratios is called an isochron. Figure 2 shows isochrons for modern lead and for leads which were separated from uranium 1×10^9, 2×10^9, and 3×10^9 years ago. Knowledge of the Pb^{207}/Pb^{204} and Pb^{206}/Pb^{204} ratios for any lead will determine its model age.

The model age is based on a number of assumptions, the most stringent being the requirement of a closed system in the source environment up to the time of separation of the lead and absence of contamination of the lead with lead of a different age after separation. It is not surprising that these conditions are not always met (see following section in this article on anomalous leads). In favorable cases, model ages agree with the ages of sup-

with respect to Pb^{204}, indicating an accumulation of radiogenic lead in the source environment.

All points in Fig. 1 represent feldspars except those at 0 and 4.5×10^9 years. The points at the present (modern primary lead) represent the range in isotopic composition of lead from three basalts, a gabbro, red clay from the Pacific Ocean, and 26 manganese nodules from the Atlantic and Pacific oceans. The basalts and the gabbro probably originated from partial fusion of silicate minerals at great depths in the Earth. Defining the crust as the region above the Mohorivičić discontinuity, which has a depth of about 35 km under the continents and 10 km under the oceans, and the mantle as the region between the Mohorovičić discontinuity and the boundary of the core at 2900 km, basalts and gabbros originate from the deep crust or outer mantle. The red clay and manganese nodules, on the other hand, represent a sampling of lead from a large number and variety of sources on the surface of the Earth. Their lead may thus represent a reasonable average for the isotopic composition of lead at the Earth's surface today. Note that the oceanic and basaltic leads have quite similar isotopic compositions, indicating that average surface lead is not greatly different in isotopic composition from lead of deep-seated origin.

The points at 4.5×10^9 years represent lead in troilite, FeS, from two iron meteorites. The troilite contains so little uranium with respect to lead that the isotopic composition of its lead would not have changed measurably in the last 4.5×10^9 years, the presently accepted age of the Earth. The data from igneous rocks are indeed compatible with the assumption that the isotopic composition of terrestri-

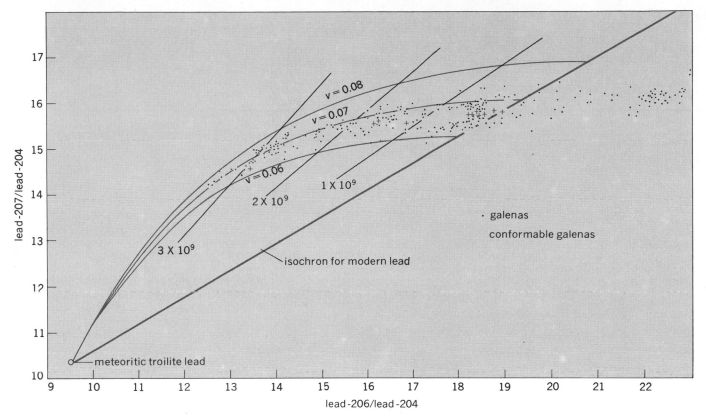

Fig. 2. Pb^{206}/Pb^{204} as a function of Pb^{207}/Pb^{204} for lead from galenas. Conformable galenas are defined in the text. The age of the Earth is taken as 4.55×10^9 years. Diagonal lines are isochrons giving model ages in years; $v = U^{235}/Pb^{204}$ (atom ratio) at the present time in hypothetical sources of ores. The growth curves are calculated for various values of v, assuming closed systems prior to the time of ore formation.

posedly contemporaneous igneous rocks within $\pm 100 \times 10^6$ years, about the limit of accuracy of the method. The method is better suited for distinguishing between two leads of greatly differing age than for determining precise ages.

The Pb^{206}/Pb^{204} and Pb^{207}/Pb^{204} ratios for approximately 200 galena specimens taken from all continents except Antarctica are shown in Fig. 2. To a first approximation the galenas plot on a closed-system growth curve, as do the feldspar leads in Fig. 1. The U^{238}/Pb^{204} ratio of 9.3 used in Fig. 1 corresponds to a value of 0.067 for v in Fig. 2. In general there is a close relationship between the isotopic composition of lead in igneous rocks and in lead ores of the same age. In fact, detailed studies in several localities have shown that the isotopic composition of lead in ores and in neighboring, approximately contemporaneous igneous rocks are nearly identical. It is striking that the galena leads form such a closely defined curve, since they represent random sampling on a worldwide scale.

The conformable galenas, delineated in Fig. 2, show much less scatter about a single closed-system growth curve than do the remaining samples, and deserve special mention. Conformable galenas are recognized from geological criteria and are believed to be derived from volcanic activity in island-arc environments off the margins of continents. The Aleutian Islands of Alaska are an example of such an environment at the present time. If the theory is correct, lead in conformable galenas is the same as that in the igneous rocks formed by

the volcanism. Whether the theory is correct or not, the conformable galenas closely fit a single closed-system growth curve. Their model ages agree with the estimated times of emplacement within 100×10^6 years in all cases—substantially better agreement than is obtained from ores in general. It is widely believed that the igneous rocks in island arcs originate in the outer mantle of the Earth, because in these areas it is difficult to generate enough heat within the crust to melt rocks. For this reason the conformable galenas probably record the evolution of the isotopic composition of lead in some portion of the Earth's outer mantle.

Anomalous leads. These leads give model ages which are clearly incorrect. For instance, if the model used to determine ages is correct, no leads should be found to the right of the modern lead isochron in Fig. 2 (negative model age). Figure 2 does not cover the full range of anomalous leads; Pb^{206}/Pb^{204} ratios as high as 93 and Pb^{207}/Pb^{204} ratios as high as 24 have been observed. These extreme cases are rare, however. Anomalous leads are generally characterized by high Pb^{206}/Pb^{204} and sometimes by high Pb^{208}/Pb^{204} ratios as well. Pb^{207}/Pb^{204} may not be high since the abundance of U^{235} is presently so low that only small amounts of Pb^{207} with respect to Pb^{206} have been made over the past 2×10^9 years. Note that the Pb^{207}/Pb^{204} curve in Fig. 1 is nearly flat over that time interval.

A significant feature of anomalous leads is that the isotopic composition varies considerably from one sample to another. The Mississippi Valley

Table 1. Isotopic composition of lead from the Lake Athabaska district, Saskatchewan

Type	Atom ratios			
	$\dfrac{Pb^{206}}{Pb^{204}}$	$\dfrac{Pb^{207}}{Pb^{204}}$	$\dfrac{Pb^{208}}{Pb^{204}}$	$\dfrac{\text{Radiogenic } Pb^{207}}{\text{Radiogenic } Pb^{206}}$
Ordinary	14.36	14.96	34.49	
Anomalous	40.01	19.36	37.10	0.17
Anomalous	43.5	18.7	35.7	0.12

leads of Arkansas, Missouri, Illinois, Iowa, and Wisconsin have variable Pb^{206}/Pb^{204} ratios; they range from 18.5 to 23. Several examples are known of variable isotopic composition in different parts of the same mine. Nonanomalous lead, on the other hand, does not display this variation, giving almost the same isotopic composition throughout a mine or even a geologic province. Thus, some sort of localized mixing is indicated to be the cause of anomalous lead.

An occurrence of anomalous lead at Lake Athabaska, Saskatchewan, has been well studied; in this article it will be used to illustrate a solution to the problem. Uraninite in a pegmatite in the district is reliably dated by concordant lead ages at 1.9×10^9 years. Pitchblende of hydrothermal origin occurs widely, and isotopic ages, all discordant, have been determined for 28 of these. A systematic study of the discordant pitchblende ages indicates that they could have been formed from a 1.9×10^9-year-old uraninite by dissolving parts of it 1.1×10^9 to 1.2×10^9 years ago and again about 0.2×10^9 years ago, transporting the uranium hydrothermally to new sites in veins and losing variable fractions of the accumulated radiogenic lead in the process. Three kinds of lead are found in lead ores, an ordinary lead and two kinds of anomalous lead. Examples of their isotopic compositions are given in Table 1.

The ordinary galena may be used as a basis to calculate the ratios of excess or radiogenic Pb^{207} and Pb^{206}, the values of which appear in Table 1. Also, it has been calculated that a uranium mineral formed 1.9×10^9 years ago would contain radiogenic lead with a Pb^{207}/Pb^{206} ratio of 0.17, 1.2×10^9 years ago and 0.12, 0.2×10^9 years ago. Thus, the uraninite, pitchblendes, and lead ores tell a consistent story: Pitchblendes were formed 1.2×10^9 and 0.2×10^9 years ago from 1.9×10^9-year-old uranium minerals, probably uraninite. Varying proportions of highly radiogenic lead separated from uranium during the formation of the pitch-

blendes, the freed radiogenic lead finding its way into lead ores which now contain anomalous lead. The mixing of radiogenic lead with ordinary lead might be expected to have varied considerably on a local scale, which fits observation.

Other occurrences of pitchblendes together with lead ores having anomalous lead are the Colorado Plateaus and the Blind River district in Ontario, north of Lake Huron. These districts have not been studied as completely as that at Lake Athabaska, but a similar mechanism undoubtedly applies. The source of radiogenic lead for the anomalous Mississippi Valley leads has not yet been identified. No pitchblendes are observed, and the Pb^{208}/Pb^{204} ratios are high, along with the Pb^{206}/Pb^{204} ratios. The radiogenic lead in the galena is probably derived from the Precambrian $(1.4 \times 10^9$ years old) granitic rocks in the area. Such a source is plausible according to data given in the following section.

Lead isotopes in a granite. Another application of isotopic-lead data to geochemical problems is illustrated by a study of the distribution of uranium, thorium, and lead isotopes in minerals of a Precambrian granite from the Canadian Shield, collected near Tory Hill, Ontario. (Precambrian rocks are those having ages greater than 6×10^8 years). Zircon contained lead which was entirely radiogenic, enabling the mineral to be dated accurately at 1.05×10^9 years. Perthite, plagioclase, and quartz contained lead which appeared to be entirely primary. Determination of the uranium-lead and thorium-lead ratios in perthite indicated that the ratios were so low that the isotopic composition of the lead would not have changed appreciably in the last 10^9 years if the mineral was a closed system. The other minerals studied, sphene, apatite, and magnetite, all had mixtures of primary and radiogenic lead. The composite rock was analyzed for uranium, thorium, and lead concentration and lead isotopic composition.

From the age of the rock, it is possible to make material balance calculations to study possible migrations of lead, uranium, and thorium within the rock. If each mineral has been a closed system, then the isotopic composition of lead calculated for the rock 1.05×10^9 years ago from the present-day uranium, thorium, and lead data should be the same as that found in the feldspars. This comparison is shown in Table 2. It is obvious that some type of migration has occurred. The leads in the feldspars have model ages of about 10^8 years, which suggests that they are contaminated with radiogenic lead and are anomalous. A large crystal of perthite from a neighboring pegmatite was found to contain ordinary lead with a model age of about 10^9 years. Several galenas with similar lead are also known from the district.

The uranium-lead ratio found for the granite gives calculated Pb^{206}/Pb^{204} and Pb^{207}/Pb^{204} ratios in the rock 1.05×10^9 years ago, which agree rather closely with the isotopic composition of lead in the feldspar crystal in a pegmatite. This could indicate that the rock as a whole closely approximated a closed system since it was formed, with respect to uranium, Pb^{206}, and Pb^{207}. However, a similar test of the Pb^{208}/Pb^{204} ratio shows that the granite cannot have been a closed system for thorium and Pb^{208}. The close balance calculated for uranium and lead may thus be accidental. The fact that lead

Table 2. Isotopic composition of lead in a granite and an associated pegmatite

Source of lead	Atom ratios		
	$\dfrac{Pb^{206}}{Pb^{204}}$	$\dfrac{Pb^{207}}{Pb^{204}}$	$\dfrac{Pb^{208}}{Pb^{204}}$
Granite, today	20.3	15.7	48.7
Granite, 1.05×10^9 years ago	16.4	15.4	30.0
Granite, perthite	18.6	15.7	39.5
Granite, plagioclase	18.2	15.5	40.0
Pegmatite, perthite crystal	16.8	15.3	36.0

from both feldspars in the granite has an isotopic composition close to that of an ordinary modern lead is further suggestive of external contamination. At any rate, the lead in the feldspars is anomalous, although it is not clear whether the source of contamination was from within the granite or outside it. Only two granites have been studied in this manner by use of lead isotopes. However, many analogous studies based on the radioactive decay of rubidium—87 to strontium—87 have been made. Many times these studies can show whether or not the rubidium and strontium in granites have been subject to contamination from external sources.

Note that the present-day lead in the granite is anomalous; that is, it has a negative model age when plotted in an isochron diagram such as Fig. 1. All Precambrian granites for which the isotopic composition of lead is known contain anomalous lead. This indicates that granites have higher uranium-lead ratios than that which exists in the source environment which produced the lead in the galenas and potassium feldspars plotted in Figs. 1 and 2. Granites are possible sources of anomalous lead if a mechanism exists for concentrating their lead into ore bodies or pegmatites.

Lead isotopes and magmatic differentiation. Isotopic lead studies have been used to demonstrate a genetic relation between several different rock types in the southern California batholith. Gabbro, tonalite, granodiorite, and granite were analyzed. The Pb^{206}/Pb^{204} ratios varied from 18.72 in the gabbro to 19.44 in the granite. The age of the batholith is accurately known to be 10^8 years. Treating each of the above rocks as closed systems and calculating the isotopic composition of the lead 10^8 years ago from the uranium-lead ratios, it was found that all of the Pb^{206}/Pb^{204} ratios agree within limits of error at 18.6 to 18.7. Thorium was not determined, so that similar calculation for Pb^{208}/Pb^{204} ratios could not be made. The data suggest that this sequence of rocks originated from a common reservoir in which the isotopes of lead were homogeneously distributed. The rubidium-87 − strontium-87 system can also be applied to this kind of problem.

[GEORGE R. TILTON]

Bibliography: H. Brown, The age of the solar system, *Sci. Amer.*, April, 1957; H. Faul, *Ages of Rocks, Planets and Stars*, 1966; R. D. Russell and R. M. Farquhar, *Lead Isotopes in Geology*, 1960; G. R. Tilton et al., Isotopic composition and distribution of lead, uranium and thorium in a Precambrian granite, *Geol. Soc. Amer. Bull.*, 66(9):1131–1148, 1955.

Lead metallurgy

The extraction of lead from ore and its subsequent purification and processing. As evidenced by lead ornamental figures which date back to about 4000 B.C., this is probably the oldest metal known to man. Lead is mentioned in the Bible, and lead pipes, installed by the early Romans for transporting water, have been uncovered. From these early beginnings, the world consumption of lead reached about 3,340,000 short tons in 1967.

By far the most important ore of lead is galena, PbS. Cerussite, $PbCO_3$, and anglesite, $PbSO_4$, are also commercial sources. Zinc is usually present in significant amounts in these ores; silver and gold also occur with lead and the recovery of these metals contributes in a large way to its profitable processing.

The metallurgy of lead comprises three distinct areas of technical knowledge and experience. These are concentrating, smelting, and refining; for further details see the flowsheet and the table of typical analyses (the figures in parentheses beside certain materials in the flowsheet refer to those items whose typical compositions are given in the table).

Concentrating. The beneficiation of ores to raise the lead content, and to separate the lead from the zinc and iron minerals, is an important step in lead metallurgy. The usual process consists of crushing; wet grinding, with the product being returned to the mill by a classifier, to a particle size of about 75% minus 325 mesh; conditioning the resulting slurry with certain reagents to establish a proper alkalinity; further mixing with flotation chemicals which collect the lead minerals in a froth; and thickening and filtrating this concentrate. While each individual type of ore is unique and requires some adjustment in the quantities and properties of the reagents added, those most suitable are sodium carbonate, lime, copper sulfate, pine oil, cresylic acid, xanthate, and sodium cyanide; generally all of these are used in amounts ranging from 0.05 to 5.0 lb per ton of ore. Reagents other than those named are also employed.

Zinc minerals, iron compounds, and the earthy components of the original ore are depressed instead of floated by this treatment. After further separations, the zinc (and sometimes the iron) is usually recovered. Copper, silver, and gold, if present, normally remain with the lead and are removed and recovered in the refinery; should the copper be a major constituent, it may be desirable and economical to separate it from the mixed lead-copper concentrates by special flotation procedures. Concentrating plants of this nature treating up to 12,000 tons of ore per day are in operation.

Minor amounts of lead ores are concentrated by utilizing their high specific gravity. This may be done on jigs and tables, in Humphreys spirals (an inclined chute in the form of a helix, in which centrifugal force throws the lighter ore particles toward the outside edge), or by heavy media sink-float techniques. More recently developed, but not yet widely practiced, is the water leaching of lead ores under air or oxygen pressure; this treatment converts the sulfur in the ores to sulfuric acid and the lead into lead sulfate which is suitable for feed to a smelter.

Smelting. Lead concentrates plus occasionally high-grade raw ores and returned intermediates such as flue dusts, limerock, and other fluxes suitable for slag formation are blended on mixing beds or drawn from proportioning bins to form a smelter feed. In present practice this mixture is formed into pellets (pelletized) so that a homogeneous and carefully sized material is provided.

This is then sintered in order to eliminate most of the sulfur and to agglomerate the particles into relatively large and hard lumps that will not be blown out of a blast furnace.

Sinter machines consist of a chain of moving pallets on which the porous feed is ignited. Pellets are subjected to a downdraft of air which burns the sulfur to sulfur dioxide and at the same time oxi-

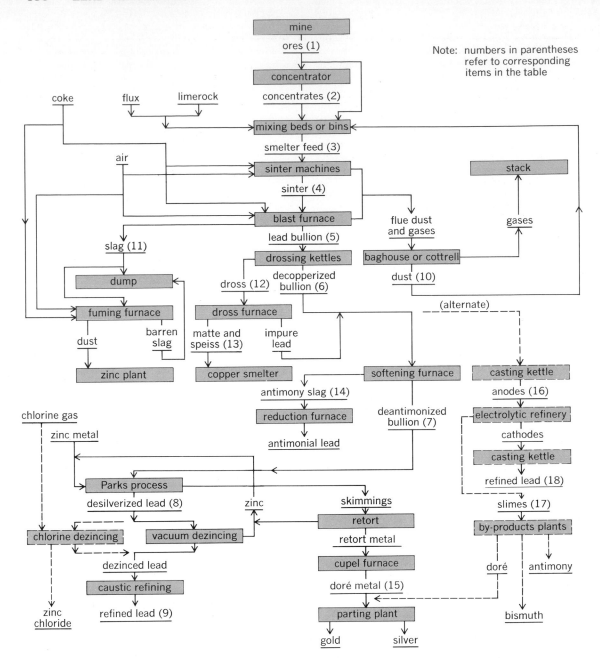

An abridged flowsheet of lead metallurgy.

dizes the metallic elements. Modern design has increased the dimensions of sinter machines to 100×10 ft (from a former $22 \times 3\frac{1}{2}$ ft).

The sinter, together with about 9% of its weight of carbon in the form of coke, is charged into the top of blast furnaces. These are simply vertical shafts, some 18 ft tall, with a typical cross section of 20×5 ft, lined with refractory brick and equipped near the base with tuyeres through which low-pressure air is blown. The coke supplies both the fuel for melting the charge and also the reducing gas which reacts with the lead oxide to form metallic lead. As the charge descends in the furnace, the molten metal flows to the bottom from where it is withdrawn for further treatment. The remainder of the charge forms slag that floats on the lead lead is removed from the furnace at a higher level; a modern development, the continu-

ous tapping of slag, is an improvement over the usual intermittent operation with its high labor demand. The lead bullion collects the silver, gold, copper, bismuth, antimony, arsenic, tin, and other minor metals; the slag carries the zinc and iron as well as the silica, lime, and other gangue; the dust contains a variety of elements including cadmium and indium, plus considerable lead and zinc. A typical blast furnace will take about 750 tons of sinter per day and produce 250 tons of lead bullion.

The impure bullion is cooled in kettles to about 350°C, whereupon a dross forms which carries almost all of the copper along with major amounts of lead. This material is skimmed and treated with soda ash in a small reverberatory furnace which produces a matte and speiss, high in copper and low in lead, which is sent to a copper smelter. Additions of sulfur remove the final traces of copper,

Typical analyses

Item no.	Type of material	Percent									Troy ounces per short ton	
		Pb	Cu	Zn	Bi	As	Sb	Fe	SiO$_2$	S	Ag	Au
1	Ores	12.7	0.7	7.0	0.03	1.60	0.65	16.7	30.0	13.1	18.0	0.02
2	Concentrates	59.8	3.6	4.4	0.12	0.60	0.75	8.2	1.5	19.5	85.0	0.10
3	Smelter feed	37.2	2.6	5.6	0.05	0.90	0.70	12.4	9.2	7.1	40.8	0.05
4	Sinter	38.8	2.7	6.1	0.05	0.82	0.71	13.5	10.6	1.7	42.0	0.05
5	Lead bullion	91.5	6.0	—	0.15	0.20	1.50	—	—	—	96.6	0.12
6	Decopperized	97.5	0.01	—	0.15	0.25	1.70	—	—	—	99.0	0.12
7	Deantimonized	99.5	—	—	—	—	0.01	—	—	—	99.0	—
8	Desilverized	98.8	—	1.0	—	—	—	—	—	—	0.3	—
9	Refined lead	*	4	4	45	2	1	6	—	—	12	—
10	Dust	45.1	0.4	5.4	0.50	4.0	1.5	1.5	1.8	9.8	4.9	—
11	Slag	2.2	0.2	9.0	0.05	0.10	0.10	24.0	23.7	0.30	0.2	—
12	Dross	57.0	24.6	0.6	0.50	6.3	2.0	1.5	1.0	2.8	75.5	0.05
13	Matte/speiss	11.0	48.0	—	0.01	11.0	3.5	6.0	3.0	6.0	65.0	0.05
14	Sb slag	55.0	0.1	—	—	4.0	25.0	—	—	—	0.5	—
15	Doré	—	1.5	—	—	—	—	—	—	—	97.0%	1.0%
16	Anodes	96.5	0.06	—	1.04	0.25	1.6	—	—	—	116.0	0.11
17	Slimes	15.5	1.5	—	21.8	5.2	33.6	—	—	—	2500.0	2.5
18	Refined lead	*	2	3	12	1	1	2	—	—	2	—

*Items 9 and 18, refined lead, are both 99.99+% Pb; impurities shown are parts per million.

after which the decopperized lead bullion proceeds to the refinery. Should significant quantities of tin be present, the lead bullion may be reheated to 600°C with the introduction of air, and a second dross containing the tin is removed prior to refining proper.

Slag from the blast furnace is frequently treated by fuming out the zinc and remaining lead; the dust from this operation is sent to a zinc plant while the now barren slag is discarded. The dusts from the sinter plant and blast furnace are collected in a baghouse or Cottrell precipitator and returned to the mixing beds or bins.

Mention should be made of the Imperial smelting process which, though still minor in magnitude with respect to lead, is of increasing importance. This unique process was developed primarily for the smelting of zinc ores and concentrates in a special type of blast furnace which can treat a mixed zinc-lead feed. The zinc is driven off as a vapor and later condensed; the lead in the charge is reduced to metal and tapped as in a standard blast furnace.

Refining. Decopperized lead bullion still contains significant amounts of arsenic, antimony, tin, bismuth, silver, and gold, and must be further refined by one of two principal processes. About 88% of the world's output is produced by means of pyrometallurgical techniques, while the remainder is electrolytically refined. The latter process is used only when the bismuth content of the lead bullion is relatively high. See ELECTROLYSIS; PYROMETALLURGY, NONFERROUS.

The first step in the pyro operations is to "soften" the metal by removing the arsenic, antimony, and tin. This may be accomplished by air oxidation of these elements in a small reverberatory furnace;

the slag, which is normally high in antimony and lead content, is reduced with coke and the resulting antimonial lead alloy is marketed. Another widely used method of removing these impurities is the Harris process. In this operation liquid lead bullion is sprayed through molten caustic soda and molten sodium nitrate; the latter reagent oxidizes the arsenic, antimony, and tin, which are converted into sodium salts and skimmed from the bath. The three elements can be recovered from these compounds by wet chemical reactions.

Silver and gold are next separated by the Parkes process. This technique is based on the principle that, when molten lead is saturated with an excess of zinc, the precious metals become insoluble in lead. Intermetallic compounds with zinc are formed, and these float to the surface as a solid crust which can be skimmed off. After two or three such treatments the lead is free of silver and gold and proceeds to the subsequent operation. The zinc crusts are heated in a retort and the zinc is vaporized and then condensed for reuse, while the residue is essentially an impure alloy of silver and gold. After an oxidation treatment in a cupel furnace the now partially refined alloy, termed doré metal, is electrolytically parted into refined silver and gold.

Excess zinc from the Parkes process is most commonly removed from the lead by distilling it under vacuum at 600°C. The zinc is condensed on the cool cover of the vacuum chamber and recycled. Some plants dezinc the lead by passing chlorine gas through the metal, whereupon zinc chloride is formed and separated; in this case the zinc cannot be reused.

Occasionally lead refined by pyrometallurgical means will carry a minor quantity of bismuth. This may be eliminated by the addition of small amounts of calcium and magnesium which will combine with the bismuth and result in a dross which is readily skimmed. This reaction, which is similar to the Parkes desilverizing, is known as the Betterton-Kroll process.

Finally, the lead is given a finishing treatment with caustic soda, which cleans the metal of minute amounts of impurities that may remain. It is then cast into the usual 100-lb bars or into the more modern 1-ton blocks.

As previously mentioned, some lead is refined electrolytically. In this case the decopperized and detinned bullion is cast into anodes which are electrolyzed in a solution containing 75 g/liter of lead as fluosilicate and 70 g/liter of free hydrofluosilicic acid. The cells, each made up of several anodes and corresponding cathodes, are electrically connected in series and operate at a current density of approximately 16 amp/ft^2 of cathode surface and about 0.5 volt. Starting sheets, on which the lead is deposited initially, are of cast lead. After normal anodic dissolution for 4 days, the anode scrap is remelted with new bullion while the cathodes are washed, melted, and cast into bars or blocks as usual. The anode residue, or slimes, is treated by complex processes for the recovery of silver, gold, bismuth, and antimony.   [I. L. BARKER]

Bibliography: American Bureau of Metal Statistics, *1967 Yearbook*, 1968; American Institute of Metallurgical and Mining Engineers, *Metallurgy of Lead and Zinc*, Proc. AIME no. 121, 1936; Australian Institute of Mining and Metallurgy, *Sintering Symposium*, 1958; Broken Hill Associated Smelters, *The Production of Nonferrous Metals at B.H.A.S.*, 1963; J. W. Hanley, Lead electrorefining, *Encyclopedia of Electrochemistry*, 1964; J. E. McKay, Lead, *Encyclopedia of Chemical Technology*, 1967.

Lead poisoning

The toxicity of lead has been known since antiquity. Lead and lead-containing products are extremely important industrially and are of great health significance. Exposure of humans may result from inhalation of fumes and dust in the smelting of lead, from the manufacture of insecticides, pottery, and storage batteries, or from contact with gasoline containing lead additives. Intoxication by the inhalation of lead fumes is a most serious mode of exposure. The ingestion of soluble lead compounds accidentally, or with suicidal intent, is another portal of entry of the body. Only tetraethyllead in gasolines can be absorbed through the intact skin. Traces of lead occur in the diet; small amounts are absorbed into the body when it is present in the food as a soluble salt. Lead is absorbed mainly through the small and large intestines.

Irrespective of the route of entry, lead is absorbed very slowly into the human body. Even this slow and constant chronic absorption is sufficient to produce lead poisoning (plumbism) because the rate of elimination of lead is even slower and a slight excess in intake may result in its accumulation in the body. Much of the lead is taken up by red blood cells and circulated throughout the body. Organic lead compounds such as tetraethyllead become distributed throughout the soft tissues, with especially high concentrations in the liver and kidneys. Over a period of time, the lead may be redistributed, becoming deposited in bones, teeth, or brain. In bones, lead is immobilized and does not contribute to the general toxic symptoms of the patient. Organic lead compounds have an affinity for the central nervous system and produce lesions there.

Acute form. Acute plumbism is ordinarily seen as a result of the ingestion of inorganic soluble lead salts for suicidal, accidental, or abortion-inducing reasons. They produce a metallic taste, a dry burning sensation in the throat, cramps, retching, and persistent vomiting. The gastrointestinal tract is encrusted with the coagulated proteins of the necrotic mucosa, thereby hindering further absorption of the lead. Muscular spasms, numbness, and local palsy may appear.

Chronic form. Chronic lead poisoning is much more common. Two general patterns of symptoms relate to the gastrointestinal and nervous systems. One or the other may predominate in any particular patient; in general, the central nervous system changes, which predominate in children, are of greater significance. The abdominal symptoms in chronic lead poisoning are similar to those for acute cases, but are less severe: loss of appetite, a feeling of weakness and listlessness, headache, and muscular discomfort. Nausea and vomiting may result. Chronic excruciating abdominal pain, sometimes referred to as lead colic, may be the most distressing feature of plumbism.

Microscopic examination reveals inclusion bodies within cells of the kidney, brain, and liver. Focal necrosis of the liver is found in some cases. Lesions in the central nervous system are primarily vascular and consist of scattered hemorrhages, often in the perivascular tissue; degenerative and necrotic changes in the small vessels, surrounded by a zone of edema; and sometimes a fibrinous exudate. *See* LEAD.

[N. KARLE MOTTET]

Leaf (botany)

A modified aerial appendage which develops from a plant stem at a node (stem "joint") and usually has a bud in its axil. In most plants leaves are flattened in form, although they may be needlelike, as in pine; scalelike, as in arborvitae, or nearly cylindrical, as in onion. Leaves usually contain chlorophyll and are the principal organs in which the important processes of photosynthesis and transpiration occur. *See* BUD (BOTANY); CHLOROPHYLL; PHOTOSYNTHESIS; PLANT, WATER RELATIONS OF; STEM (BOTANY).

GENERAL CHARACTERISTICS

A complete dicotyledon leaf consists of three parts: the expanded portion, or blade; the petiole which supports the blade; and a pair of stipules, small appendages attached at the base of the petiole (Fig. 1). Stipules may be green and bladelike, as in pea; coarse, rigid spines, as in black locust; sheaths, as in smartweed; tendrillike, as in greenbrier; or mere temporary hairs or bristles, as in Lespedeza. Leaves that have a blade and petiole but no stipules are said to be exstipulate. Some leaves have no apparent petioles and are described as sessile. The leaves of grasses have neither petioles nor stipules; the blades are attached to the stem by an encircling sheath. At the junction of the sheath and the blade is a collarlike structure called the ligule. In pines, needlelike leaves are borne in fascicles (clusters of 2–5, rarely 1) at the ends of short dwarf branches. *See* GRASS CROPS; MAGNO-LIOPSIDA; PINE.

Margins. The margin, or edge, of a leaf may be entire (without indentations, or teeth); serrate (with sharp teeth pointing forward), or serrulate (finely serrate); dentate, with coarse teeth pointing outward, or denticulate (finely dentate); crenate or scalloped, with broad rounded teeth; undulate, with a wavy margin; incised, cut into irregular or jagged teeth or segments (if segments are narrow and pointed, laciniate; if directed backward, runcinate); pinnatifid, deeply pinnately parted (featherlike); dissected, cut into numerous slender, irregularly branching divisions (Fig. 2).

When the blade is deeply cut into fairly large portions, these are called lobes. The degree of such lobing may be designated by the following terms: lobed, with sinuses usually not more than halfway from margin to midrib (midvein) or base, and with lobes and sinuses more or less rounded; cleft, when incisions extend halfway or more from margin to midrib, and especially when they are sharp; parted, cut so deeply that the sinuses extend almost to the midrib or base; divided, cut entirely to the midrib, making a leaf compound.

Tips and bases. The tip of a leaf may be acuminate, gradually tapering to a sharp point; acute, tapering more abruptly to a sharp point; obtuse, with a blunt or rounded tip; truncate, seeming to be cut off square or nearly so; emarginate, decidedly notched at tip but not lobed; mucronate, abruptly tipped with a small short point; cuspidate, ending in a sharp rigid point (Fig. 3).

The base of the blade may be cuneate, wedge-shaped; oblique, the two sides of the base unequal; cordate, heart-shaped, base with a conspicuous sinus; auriculate, with a small earlike lobe on either side of petiole; sagittate, arrow-shaped, with a pair of basal lobes turned inward; hastate, halberd-shaped, with basal lobes turned outward; clasping, when sessile and partly investing the stem.

Shapes. A leaf may be linear, long and narrow, with the sides parallel or nearly so (Fig. 4); lanceolate, narrow but tapering from base toward apex; oblanceolate, broader at apex, tapering toward base; spatulate, broad and obtuse at apex tapering to narrow base; ovate, egg-shaped, broadest toward base; obovate, reverse of ovate, broadest toward apex; elliptic, broadest at middle and tapering slightly to a broadly rounded base and apex; oblong, somewhat rectangular, with nearly straight sides and rounded base and apex; deltoid, triangular; reniform, kidney-shaped, broader than long; orbicular, circular or nearly so; peltate, shield-shaped, usually a circular leaf with petiole attached at or near the center of the lower surface; perfoliate, having the stem apparently passing through it; connate, when the bases of two opposite leaves seem to have fused around the stem.

Types. A leaf is simple when it has but one blade; compound, when the blade is divided into two or more separate parts called leaflets; trifoliolate, if it has three leaflets (Fig. 5); palmately compound, if leaflets originate from a common point at end of petiole; pinnately compound, if leaflets are borne on the rachis (continuation of petiole); odd-

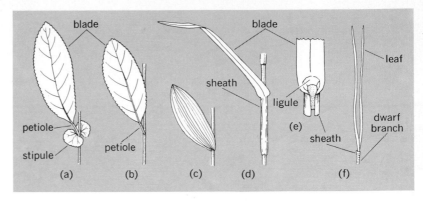

Fig. 1. Leaf parts in different structural patterns. (*a*) Complete. (*b*) Exstipulate. (*c*) Sessile. (*d*) Grass leaf. (*e*) Detail of *d*. (*f*) Pine leaves.

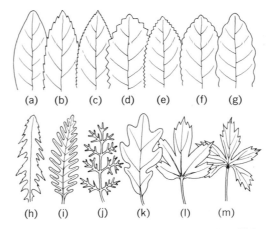

Fig. 2. Leaf margins of various types. (*a*) Entire. (*b*) Serrate. (*c*) Serrulate. (*d*) Dentate. (*e*) Denticulate. (*f*) Crenate. (*g*) Undulate. (*h*) Incised. (*i*) Pinnatifid. (*j*) Dissected. (*k*) Lobed. (*l*) Cleft. (*m*) Parted.

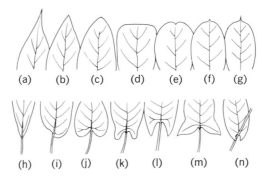

Fig. 3. Leaf tips and bases. (*a*) Acuminate. (*b*) Acute. (*c*) Obtuse. (*d*) Truncate. (*e*) Emarginate. (*f*) Mucronate. (*g*) Cuspidate. (*h*) Cuneate. (*i*) Oblique. (*j*) Cordate. (*k*) Auriculate. (*l*) Sagittate. (*m*) Hastate. (*n*) Clasping.

pinnate, if such a compound leaf is terminated by a leaflet; even-pinnate, if without terminal leaflet; decompound (bipinnate), if twice compound.

Venation. The arrangement of the veins, or vascular bundles, of a leaf is called venation (Fig. 6). There are three basic types: parallel, the main veins extending through most of the blade in a parallel manner, as in grasses, lilies, and most monocotyledons; net, the main veins branching, forming an irregular network (pinnately if featherlike, palmately if fanlike), as in most dicotyledons; and dichotomous, forked, each vein dividing at inter-

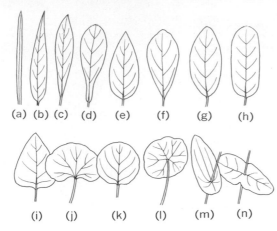

Fig. 4. Leaf shapes. (*a*) Linear. (*b*) Lanceolate. (*c*) Oblanceolate. (*d*) Spatulate. (*e*) Ovate. (*f*) Obovate. (*g*) Elliptic. (*h*) Oblong. (*i*) Deltoid. (*j*) Reniform. (*k*) Orbicular. (*l*) Peltate. (*m*) Perfoliate. (*n*) Connate.

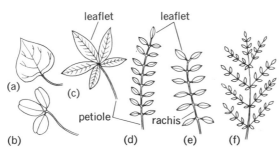

Fig. 5. Leaf types. (*a*) Simple. (*b*) Trifoliate. (*c*) Palmately compound. (*d*) Odd-pinnately compound. (*e*) Even-pinnately compound. (*f*) Decompound.

vals into smaller veins of approximately equal size, as in most ferns and in ginkgo. *See* GINKGOALES; LILIOPSIDA; POLYPODIALES.

Arrangement (phyllotaxy). Leaves occur on a stem in a definite, fixed order according to species (Fig. 7). They usually appear in one of three ways: alternate, if only one leaf occurs at a node; opposite, if two leaves appear at a node on opposite sides; and whorled (verticillate), if more than two appear. Alternate leaves are usually arranged spirally on the stem; if they occur in but two rows, they are two-ranked, if in five rows, five-ranked.

In a cluster of flowers, each flower arises normally in the axil (armpit) of a leaf; this leaf, however, is usually much reduced in size and is called a bract. The bract is said to subtend the flower. *See* FLOWER.

Surface. Surfaces of leaves provide many characters used in identification. A surface is glabrous if it is smooth or free from hairs; glaucous, if covered with a whitish, waxy material, or "bloom"; scabrous, if rough or harsh to the touch; pubescent, a general term for hairiness of any kind, as opposed to glabrous; puberulent, with very fine downlike hairs; tomentose, with matted woolly hairs; villous, with long, soft, shaggy hairs; hirsute, covered with short, erect, stiff hairs; and hispid, dense with bristly, harshly stiff hairs.

Texture. The texture may be described as succulent when fleshy and juicy; hyaline, if thin and almost wholly transparent; chartaceous, papery, opaque but thin; scarious, thin and dry, appearing

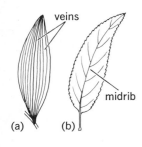

LEAF (BOTANY)

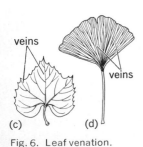

Fig. 6. Leaf venation.
(*a*) Parallel.
(*b*) Net, pinnate.
(*c*) Net, palmate.
(*d*) Dichotomous.

shriveled; and coriaceous, tough, thickish, and leathery.

Duration. Leaves may be fugacious, falling nearly as soon as formed; deciduous, falling at the end of the growing season; marcescent, withering at the end of the growing season but not falling until toward spring; or persistent, remaining on the stem for more than one season, the plant thus being evergreen. *See* DECIDUOUS PLANTS; EVERGREEN PLANTS; PLANT ORGANS; PLANT PHYSIOLOGY; PLANT TAXONOMY.

[NELLE AMMONS]

ANATOMY

In a true leaf, the veins are continuous with one or more vascular strands of the stele, the primary vascular cylinder of the stem. The strands that connect the leaf with the vascular system of the stem are the leaf traces. A part of such a leaf trace passes through the outer stem tissue, or cortex. Above the point of divergence of the leaf trace from the stele, an interruption of the vascular tissue occurs, and this parenchyma-filled area is the leaf gap. Although there are flattened leaflike structures among the lower plants, such as algae, liverworts, and mosses, these are not true leaves because they lack a well-defined vascular system. Some of the primitive vascular plants have microphylls (small leaflike organs), but these have a single unbranched leaf trace, and there is no leaf gap in the stele. *See* ALGAE; BRYOPSIDA; CORTEX (PLANT); MARCHANTIATAE; PARENCHYMA; STELE; VASCULAR BUNDLES.

Internal structure in relation to function. The foliage leaf represents the chief photosynthetic component of most vascular land plants (Fig. 8). Although leaves vary greatly in size, shape, and complexity, they consist of the same three tissues as does the stem: the dermal or surface layer; the vascular, including the xylem and phloem, or conductive tissues; and the ground tissues, parenchyma and sclerenchyma. In the leaf, however, the arrangement and modifications of these tissues have obviously evolved along lines favorable for photosynthesis within various habitats and climates. The ground tissue, or mesophyll of the leaf, is primarily chlorenchyma, that is, tissue with cells containing chloroplasts. With a maze of intercellular spaces in the chlorenchyma, each cell is exposed to an internal moist atmosphere. The sheetlike distribution of the chlorenchyma in the blade exposes each individual cell to light. The vascular tissue is an intricately branched network of veins embedded in a midplane within the chlorenchyma. Every cell of the mesophyll therefore is relatively close to mass movement of water and solutes through the xylem and phloem of the veins—water and mineral ions

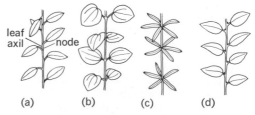

Fig. 7. Leaf arrangement: (*a*) Alternate. (*b*) Opposite. (*c*) Whorled. (*d*) Alternate, two-ranked.

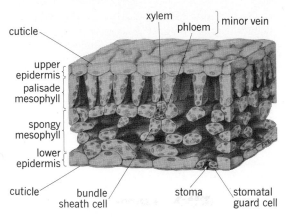

Fig. 8.   Three-dimensional diagram of internal structure of a bifacial mesomorphic leaf.

move chiefly through the xylem; dissolved foods, chiefly through the phloem. The nongreen compact epidermis is similar to the epidermis of stems and other exposed plant organs (Fig. 9). The stomatal guard cells alone among the epidermal cells contain chloroplasts, and the stomata constitute openings in the otherwise continuous epidermis, the cells of which are covered by a waxy or fatty cuticle (Fig. 10). Thus, the epidermis is a translucent, waterproofed, but partially perforated covering through which light penetrates, but through which the diffusion of water vapor and gases is limited. These gases escape principally through the stomata. The opening and closing of stomata within the epidermis of many leaves result essentially from light and temperature sensitivity. Thus, during periods of rapid photosynthesis, when diffusion pressure gradients of gases (between the internal spaces and the external atmosphere) are steep, the stomata are open. Although water loss is high during these periods, water is conserved at other times when the stomata are closed. The intercellular spaces of the mesophyll are lined with a thin fatty film similar to the cuticle in composition. This film no doubt inhibits evaporation to some extent from the mesophyll cells to the internal spaces. *See* CELL PLASTIDS; EPIDERMIS (PLANT); EVOLUTION, ORGANIC; PHLOEM; PLANT, MINERAL NUTRITION OF; PLANT TRANSLOCATION OF ORGANIC SOLUTES; SCLERENCHYMA; XYLEM.

In well-differentiated broad leaves, the chlorenchyma consists of two contrasting tissue systems, palisade and spongy mesophyll. Palisade cells are typically columnar in shape, closely spaced, and oriented parallel with one another but with their longest diameters at right angles to the surface of the blade. Spongy mesophyll cells, on the contrary, are variously shaped, and are oriented with their longest diameters chiefly in the paradermal plane (the plane of the blade parallel with the epidermal layer or blade surface). The greater part of photosynthesis occurs within the palisade cells where the chloroplasts are most numerous. Transfer of water and solutes from living cell to living cell and between veins occurs chiefly through the spongy mesophyll and dermal tissues, because these are the tissues having consistent lateral contiguity. There is only limited lateral contiguity among the palisade cells.

Among leaves in which guttation (secretion of water from the tips and margins of leaves) is known to take place, specialized tissues called hydathodes occur near the vein endings in the blade tips and margins. Most leaf movements are caused by irregular growth rates, but certain reversible movements are caused by changes in water content of cells within specialized structures such as the pulvinus (a swelling at the base of the petiole). Ergastic substances (waste products and crystals of mineral elements) accumulate in various living cells of foliage leaves. With leaf abscission (fall), these waste materials are eliminated from the living plant. *See* PLANT GROWTH.

Development. A foliage leaf, whether simple or compound, arises as a leaf primordium (protrusion composed of relatively superficial cells; Fig. 11) at the apical meristem or region of cell division at the shoot tip. As the leaf primordium elongates and thickens by numerous cell divisions, a primordial axis forms which later differentiates, in part, as midrib and petiole. On either side of the axis, marginal meristems containing marginal and submarginal initials give rise to cell layers from which the blade tissues differentiate. *See* MERISTEM, APICAL; MITOSIS.

In woody plants, tiny folded leaves with recognizable shapes can be distinguished in winter buds. These young undifferentiated leaves develop in buds during the growing season preceding the winter. Thus, in the spring, as the buds open and the bud axis elongates forming the new twig, the tiny leaves unfold and expand to their mature size within a few days. Although during this surge of spring growth a limited amount of cell division occurs, the major activity is concerned with cell enlargement and differentiation. In some woody

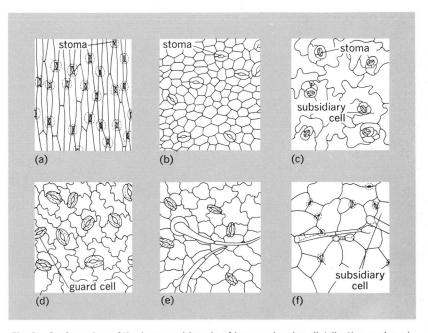

Fig. 9.   Surface view of the lower epidermis of leaves showing distribution and position of stomata and relative size and shape of subsidiary cells. Dotted lines show the outline of guard cells or subsidiary cells that are obscured from surface view by overlapping cells. (a) *Iris*. (b) *Vitis* (grape). (c) *Sedum*. (d) *Capsicum* (red pepper). (e) *Lycopersicon* (tomato). (f) *Oxalis*. Broken lines indicate that the stomata are sunken in *Iris* and *Oxalis*; raised in *Sedum, Capsicum*, and *Lycopersicon*; and flush in *Vitis*. (*From K. Esau, Plant Anatomy, 2d ed., Wiley, 1965*)

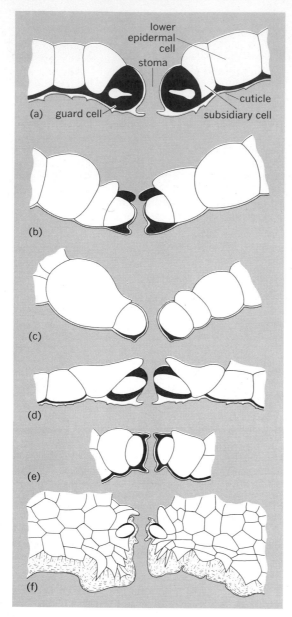

Fig. 10. Transverse sections of stomata in the lower epidermis of leaves showing various degrees of raised and sunken positions of the guard cells relative to subsidiary epidermal cells. The cuticle forms protrusions or ledges at the stomatal opening. (a) *Prunus* (peach). (b) *Pastinaca* (parsnip). (c) *Solanum* (potato). (d) *Hedera* (English ivy). (e) *Musa* (banana). (f) *Euonymus*. (*Adapted from K. Esau, Plant Anatomy, 2d ed., Wiley, 1965*)

through several growing seasons. Leaf abscission occurs within a definite zone near the base of the petiole. After, or sometimes during abscission, a film of suberin and a gumlike substance is secreted by the living cells, forming a leaf scar or cicatrice tissue. A corky tissue, periderm, then forms by cell divisions under the cicatrice. Within the leaf scar are bundle scars which show where the vascular bundles abscised. Abscission of the stipules and bracts usually occurs very early in the growing season. Leaf scars, bundle scars, and stipule scars are distinguishing features of taxonomic importance. *See* PERIDERM.

Factors that influence structure. Both heredity and environment are responsible for the great variation in leaf structure. Although heredity accounts for a great array of forms, the plants usually have some of the structural features associated with the habitat (environment). Striking similarities can be observed among unrelated but ecologically convergent plant forms (species that have evolved similarities of form and behavior within a given habitat).

Therefore, not all structural characteristics reflect phylogenetic relationships, that is, relationships that indicate a similar evolutionary history.

Although ecological equivalence among unrelated species is not fully understood, some general leaf forms are commonly associated with certain well-defined environments or habitats. For example, moist boreal forests are dominated by trees with needlelike, evergreen leaves; temperate rainy regions, by trees with broad deciduous leaves.

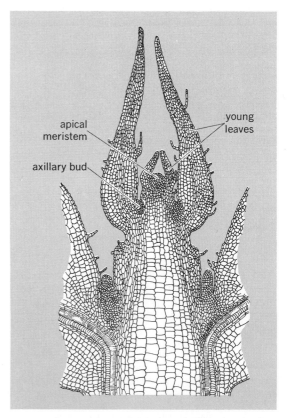

Fig. 11. Longitudinal section through a shoot tip (terminal bud) showing stages in the development of foliage leaves from the apical meristem. (*From E. N. Transeau, H. C. Sampson, and L. H. Tiffany, Textbook of Botany, rev. ed., Harper, 1953*)

plants, additional leaves arise and mature during the same growing season. Foliage leaves consist mainly of primary tissues (tissues that arise from the apical meristem or the embryonic stem tip) and, except for some ferns and a few seed plants, they have determinate growth (as contrasted with the continued growth of a stem). Leaves of some monocotyledons complete terminal growth and expansion in width, but they continue to grow for a considerable time from meristematic cells in the leaf base.

Leaf abscission occurs at the end of a single growing season in deciduous woody plants. Among evergreen species, however, the leaves persist

Fig. 12. Transverse sections of the xeromorphic leaf blades of eleven species of evergreen chaparral shrubs from the Santa Monica coastal mountain range of Cali-fornia. Although the mesophyll of most is predominantly palisade, the structure is bifacial for ten and centric for one. Note the relatively thin epidermal layers.

Mediterranean climates are dominated by shrubs and small trees with hard, broad, or needlelike leaves. Perhaps the ecological equivalents most difficult to understand are the desert shrubs. Some of these shrubs have small scurfy, or hairy, early deciduous leaves; a few have small varnished evergreen leaves; and still others are succulent and practically leafless.

Xerophytes and hydrophytes are the terms used for plants that normally grow in dry (xeric) and moist habitats, respectively. Most plants that grow in temperate climates are mesophytes; for example, the common crop plants are considered to be mesophytes which are intermediate between the xerophytes and hydrophytes in their moisture requirements. Through many anatomical studies of plants growing in extreme and temperate habitats,

certain structural features have become known as xeromorphic, mesomorphic, or hydromorphic. Since some of the structural features are quantitative, such as vein spacing and thickness of cuticle, various degrees of xeromorphy or hydromorphy can be observed in otherwise mesomorphic plants. For example, among leaves on typically mesomorphic plants, those at the higher levels of insertion on the plant usually show some xeromorphic tendencies as compared with leaves at lower levels, and the leaves that have developed in full sunlight are more xeromorphic than the ones that develop in the shade. Among leaves of mesomorphic plants growing in dry and moist habitats, the leaves of plants from the drier habitat show a tendency toward xeromorphy. Xeromorphic features are not always easily interpreted. Some aspects of xeromorphy are associated with higher rather than lower rates of transpiration, and xeromorphy is characteristic of plants growing in wet peaty soil, as well as in dry habitats (Fig. 12).

Dermal tissues. Epidermal cells are typically tabular in shape, but the surface and inner walls may be flat, rounded, or papillose (with conical protuberances), and the anticlinal walls (walls at right angles to the surface) are flat or undulate (wavy). The surface shapes of epidermal cells are polygonal, ameboid, or sinuous. Epidermal cells on the lower surface are usually smaller and more sinuous than are the upper epidermal cells. Cells of shade leaves and hydromorphic leaves are more sinuous than those of sun leaves and xeromorphic leaves.

The upper or lower epidermis may consist of multiple layers. It is termed a multiple epidermis if all of its cells arise from the protoderm (the surface meristem from which the epidermis is derived). If the subdermal cells are derived from an interior meristem, the subdermal tissue is termed hypodermis. Subdermal cells are usually fewer in number and larger in volume than are the accompanying epidermal cells, and they tend to be less variable in shape (usually isodiametric or elongated polyhedrons). The multiple epidermis and hypodermis are characteristic of tropical evergreen leaves. Dermal tissues compose 20–30% of the total blade volume in mesomorphic leaves, but some xeromorphic leaves have very shallow epidermal layers.

The cuticle that covers the outer surface of the epidermis is composed of varying amounts of cutin and pectin. The outer walls are not only cuticularized (covered by a layer of cutin) but, in addition, they are cutinized (impregnated with cutin). Radial epidermal walls are also frequently cutinized.

Heavily cutinized epidermal cells tend to be rather uniform and isodiametric in shape. *See* PECTIN.

The stomata may be flush with adjacent epidermal cells, or they may be raised or sunken. Stomata are classified chiefly upon the basis of shape and orientation of subsidiary cells (cells adjacent to the guard cells). In dorsiventral blades of mesomorphic leaves and tropical evergreen leaves, the stomata are confined principally to the lower epidermis. In floating leaves of water plants, well-developed stomata are found only within the upper epidermis. In xeromorphic isolateral leaves and cylindrical leaves, the stomata are distributed over the entire surface. Stomatal frequencies have been reported for a number of species, and they range from 14 to 1540 per square millimeter.

An epidermal appendage (trichome or hairline structure) may be a simple outgrowth of an epidermal cell or a unicellular or multicellular structure. The separate cells may be relatively undifferentiated or secretory (glandular) in nature. Idioblasts (cells markedly different in shape or function from other cells within the same tissue) and cells with crystals and other inclusions, as well as trichomes, are constant features of certain species. *See* SECRETORY STRUCTURES (PLANT).

Mesophyll. In bifacial leaves (those with dissimilar tissues above and below), the chlorenchyma consists of one or more layers of palisade cells near the upper (adaxial) surface of the blade, and one or more layers of spongy mesophyll cells in the lower (abaxial) portion. There is relatively little lateral contiguity among the palisade cells, although some of the cells are tangent along portions of their length. Palisade cells subtend epidermal or hypodermal cells and are contiguous with the spongy mesophyll cells below. In most mesomorphic leaves the spongy mesophyll includes large intercellular spaces, and its cells are oriented chiefly within the paradermal (horizontal) plane of the blade. If the cells are branched, one or more arms are extended within the paradermal plane. In the midplane of the blade, which includes the intricate system of minor veins, the spongy mesophyll is in the form of continuous nets of living cells, closely integrated with the small veins. The intercellular space system in the entire mesophyll tissue of measured leaves accounts for 3.5–71% of the entire blade volume. Although the greater volumes of space are within the spongy mesophyll, larger internal surfaces (cell wall surfaces lining the intercellular spaces) are exposed within the palisade mesophyll. Among measured leaves, the internal exposed surface is 4.6–31 times as great as the external blade surface of the same leaves. In a few species, the palisade cells are H- or U-shaped; in other species, the columnar cells are laterally contiguous by adjoining protuberances. The number of layers of palisade may vary in leaves on the same plant. In isolateral blades (blades with similar tissues on both sides) the mesophyll consists entirely of palisadelike cells; or palisade cells occur in both adaxial and abaxial portions, on either side of a midregion of spongy mesophyll. In the all-palisade leaf, the cells in the midregion of the blade have some lateral contiguity. In some foliage leaves, the palisade and spongy mesophyll are not highly contrasting tissues. The degree of contrast between the closely spaced,

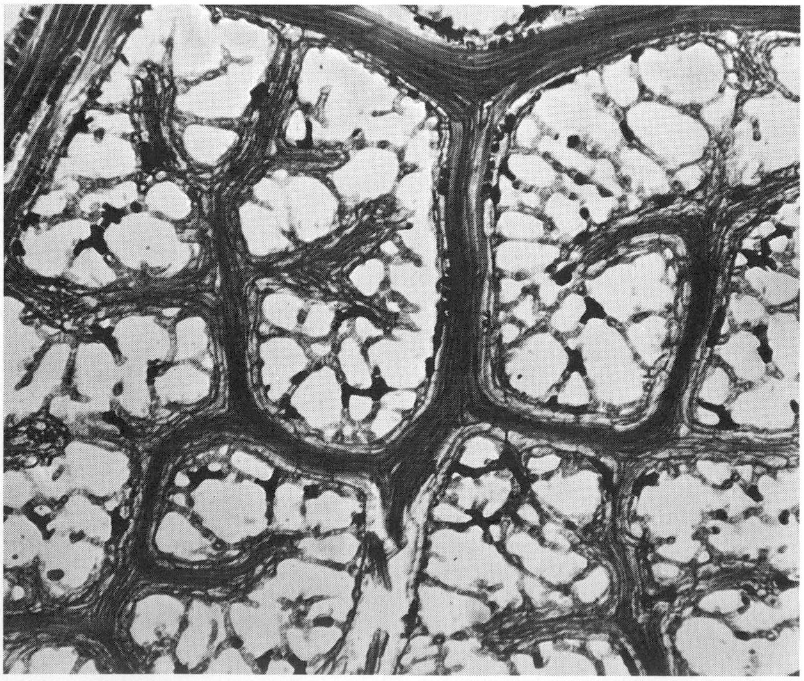

Fig. 13.   Photomicrograph of the intricate network of minor veins and associated spongy mesophyll of *Ficus elastica*. The tissue section was cut from a midplane of the blade, parallel with the surface of the leaf.

vertically oriented palisade cells and the more diffusely spaced, laterally oriented spongy mesophyll cells varies with the species as well as the environment.

Vascular system. The two main patterns of venation (arrangement of the prominent veins) in broad foliage leaves are netted and parallel. Most dicotyledons have netted veins, and the most prominent veins are arranged pinnately (one midvein with secondary veins diverging from it) or palmately (several large veins spreading out from the base of the blade). Among the larger veins, an intricate network of minor veins is well distributed within a median plane of the mesophyll (Fig. 13). Most monocotyledons have parallel veins (main veins arranged longitudinally within a linear blade) interconnected by smaller veins.

The primary and secondary veins, along with associated mesophyll tissues, form prominent ridges (vein ribs) on the abaxial side of many foliage leaves. The midrib region of such leaves is similar in structure to the petiole; it consists primarily of one or more vascular bundles of xylem and phloem (the xylem on the adaxial side) and supportive tissue. Xylem or phloem fibers or both may be present, and some of the ground tissue on either side of the vein proper may be in the form of collenchyma, a tissue consisting of elongate cells with unevenly thickened walls. The parenchyma associated with the major veins has relatively few chloroplasts. The epidermal cells ensheathing the major veins are frequently elongate with the vein axis. *See* COLLENCHYMA.

The minor veins that are embedded in the mesophyll consist of xylem and phloem elements closely surrounded by a bundle sheath, a sheath of thin-walled parenchyma cells (Fig. 14). There is no supportive tissue about the minor veins. Some vein endings consist of a single tracheid (a xylem element), no phloem, and a bundle sheath. Bundle sheaths vary in complexity, and may have only few or no chloroplasts. However, in a few xeromorphs the bundle sheath is the principal chlorophyll-bearing tissue. In many dicotyledons, bundle sheath extensions (panels of parenchyma cells) extend from the bundle sheath to one or both epidermal layers. Since the bundle sheath extensions usually have little chlorophyll, they appear like partitions between blocks of photosynthetic mesophyll within the blade. Bundle sheath extensions are of frequent occurrence in deciduous foliage

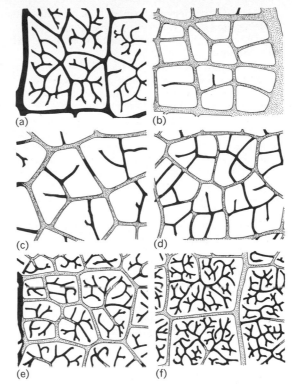

Fig. 15. Diagrams of minor vein patterns of six species of *Ficus*. Stippled veins are those with bundle sheath extensions. Veins represented by solid broad lines are embedded in the mesophyll and lack bundle sheath extensions. (*a*) *F. asperrima.* (*b*) *F. indica.* (*c*) *F. macrophylla.* (*d*) *F. pavapa.* (*e*) *F. lacor.* (*f*) *F. religiosa.*

leaves. The minor veins ramify through the midplane of the mesophyll in various complex patterns (Fig. 15). In the paradermal plane, interconnecting minor veins may form closed nets, or they may be dendriform (arranged in intricate branching systems) with numerous vein endings.

[JANE PHILPOTT]

Bibliography: *See* PLANT ANATOMY.

Learning, neural mechanisms of

An important research area in physiological psychology (neuropsychology) that, broadly speaking, may include any of the contributions made to all types of learning by any part or activity of the nervous (neural) system. Common usage of the term, however, as well as the research itself, is restricted almost entirely to considerations of the central nervous system, that is, the brain and spinal cord, and especially of the brain. The research may be directed toward an understanding of the neural mechanisms important in gathering and processing information, or it may be concerned with determining the location and nature of changes in the brain which presumably correspond to the storage and retrieval (memory and recall) of this information. This article focuses on the second category, since most investigators are involved in studies of memory and recall.

Historically, research in this area usually has required the close cooperation of workers from psychology, neurophysiology, and neuroanatomy. However, as appreciation develops for the immense variety and complexity of the most relevant

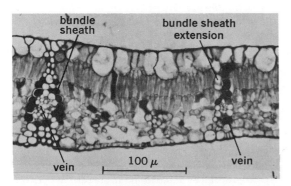

Fig. 14. Transverse section through the bifacial foliage leaf of *Ficus sycamorus* showing veins with bundle sheath extensions of thin-walled cells.

problems, scientists are being drawn into the field from a growing number of specialties, including biochemistry, biophysics, molecular biology, and mathematics.

Most experiments are performed on subhuman organisms, because investigations into the neural mechanisms of learning commonly involve brain surgery or the administration of powerful drugs. Representatives of every animal class may be employed, but most research involves a small number of mammalian species, typically rats, mice, cats, dogs, monkeys, and chimpanzees. An important exception to this rule arises in the case of humans who, as a result of injury, senility, or disease, manifest disturbances in language, problem solving, or memory. Valuable data are often obtained when these persons are available as research subjects. *See* APHASIA; PROBLEM SOLVING (PSYCHOLOGY).

RESEARCH METHODS

Technological advances and improved training have resulted in the use of more sophisticated instruments by increasingly skillful workers in neuropsychological investigations. However, two century-old procedures continue to be employed, although they have been augmented substantially by modern techniques, especially since about 1930. These "classical" procedures for investigating the contributions of various brain structures to behavior are the method of ablation, or production of experimental lesions in which an attempt is made to assess the functions of a structure by destroying (ablating) it; and the method of stimulation, in which behavior is elicited by the activation (stimulation) of some neural tissue, usually by an electric current but often by the introduction of some chemical or drug.

Ablation method. Many ablation studies have involved attempts to localize learning or memory somewhere in the brain; an assumption implicit in those studies is that memory is confined to a discrete population of nerve cells (neurons). (A term frequently used to denote the functional or structural unit of the cerebral memory store is "engram.") Characteristically, an investigation of the neural apparatus which sustains the process of learning (acquisition) will begin by selectively destroying a specific portion of an animal's brain. Following an appropriate recovery period, the animal is required to learn some task, and its learning ability is compared with that of an animal (control) whose brain has not been damaged in this way. (The comparison, or control, animal may have sustained a different lesion, however.) An essentially similar logic is employed in studies of the learning ability of brain-damaged humans, except, of course, that the investigator usually has not produced the lesion and is barred from "sacrificing" the subject for post mortem verification of the locus and extent of the damage.

In investigations concerned with memory, the ablation is carried out after the task has been learned. The animal is then tested for retention, and if it appears to have "forgotten" the task, retraining may be carried out to determine whether some residual effect of the original training is present; that is, if relearning is accomplished in fewer trials or with fewer errors than original learning,

the animal is said to exhibit "savings," and it can be concluded that the "engram" is at least partly intact. It must be emphasized that if the animal is unable to relearn the task, it cannot be automatically concluded that the storage location for that memory has irretrievably been lost or even damaged. The lesion may have produced deficits in any of several processes, such as sensation, attention, or motivation, which are essential to the acquisition or performance, or both aspects, of many learned behaviors.

Ablation methods include knife cuts, direct and high-frequency electric currents, aspiration, ionizing radiation, ultrasound, and certain chemicals as the lesion producing agents; the brain damage they produce is generally permanent. However, it has been found that some chemicals, for example, potassium chloride, will depress temporarily the normal electrical activity of the cerebral cortex, producing a "reversible functional lesion" in the affected tissue, and thereby permitting the investigator to compare acquisition and retention effects in the same animal. This technique also has been used extensively in research examining the transfer of information between the two cerebral hemispheres.

Stimulation method. In some neuropsychological investigations, the effects of brain stimulation precisely complement the results of ablation studies. For example, the fact that an animal will persistently refuse to ingest any liquids following destruction of a particular region of the hypothalamus, but will drink avidly—even with a stomachful of water—when the same region is stimulated, leads to a relatively straightforward interpretation concerning the role of that region in the regulation of the animal's fluid economy. However, such convenient reciprocal relations seldom are in evidence where research is directed toward the description of brain learning mechanisms. In fact, about the only safe conclusion that can be made about the effect of electrical stimulation on learning is that it either is ineffective or disruptive. Many studies have employed very weak electrical stimulation of the brain as substitutes for the cue (conditioned stimulus, or CS) or the reinforcement (unconditioned stimulus, or UCS), or both components of a classical conditioning procedure. Experiments of this kind assume that learning consists of an association between the CS and UCS and that this association must have some corresponding neural representation. It is hoped that by stimulating some portions (and bypassing others) of the pathways belonging to the neural circuits of CS and UCS, the essential elements of this representation may be identified. *See* REFLEX, CONDITIONED; REFLEX, UNCONDITIONED.

Psychopharmacology. Increased knowledge about the chemistry and molecular biology of the central nervous system has resulted in the expansion of research aimed at understanding and manipulating relations between essential chemical processes and learned behavior. Much of this research involves the oral or parenteral administration of various drugs to research animals, followed by appropriate behavioral observations. Other experiments, to be described in greater detail below, are characterized by the chemical analysis and comparison of neural tissues extracted from

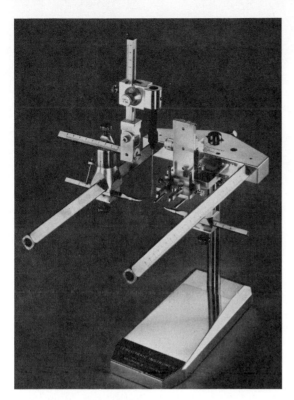

Fig. 1. Universal stereotaxic apparatus. (*Courtesy of H. Neuman and Co., Skokie, Ill.*)

derive from a view of the brain which regards the engram as being organized primarily into anatomical units. These are studies, in other words, which first ask "Where?" and only later, if at all, "How?"

Beginning with the pioneering research of K. S. Lashley on the cerebral neocortex and continuing through the current attack on the limbic system, it is possible to summarize at least one outcome of all these studies: No single structure or group of structures in the brain appears to be responsible or even essential for all kinds of learning. Accordingly, modern ablation studies have been concerned primarily with the identification of brain structures which appear to be specifically involved either in particular types of learning or in particular stages of the learning process.

Sensory learning. This statement is especially true regarding investigations of the neural mechanisms of sensory learning. Most of these studies have involved restricted ablations of the cerebral neocortex (Fig. 2), the sheet of nerve cells and fibers which covers most of the mammalian forebrain. By destroying portions of the neocortical areas associated with vision or hearing, for example, it has been possible to demonstrate the selective impairment of memory for visual or auditory discriminations, respectively. After some lesions, and with some tasks, relearning may be possible.

trained and untrained animals. In still other studies, attempts are made to stimulate, and sometimes to block, normal metabolic processes by the introduction of appropriate chemical agents directly into the brain; chemical analyses of the brain may also be performed, subsequent to this treatment. In a typical experiment, a cannula, or tube, which acts as a guide for the injection needle, is chronically implanted in the anesthetized animal's brain. The cannula may be aimed, with the aid of a stereotaxic device (Fig. 1), at a specific structure deep within the brain for purposes of restricting chemical effects to the immediate locus of the injection. If relatively homogeneous effects are to be produced throughout the entire brain, then the cannula may be directed toward one of the cerebral ventricles, vessels conducting the cerebrospinal fluid which bathes every neuron in the central nervous system.

The injection, usually performed after the animal has recovered from the implant surgery, appears to cause little discomfort. The material injected may take the form of a liquid (where fairly rapid diffusion is desired) or minute amounts of a fine powder. The latter is allowed to dissolve in the tissue adjacent to the tip of the cannula, and it tends to produce more localized metabolic effects. *See* PSYCHOPHARMACOLOGIC DRUGS.

ANATOMICAL LOCALIZATION

An important distinction may be made between the kinds of speculation about learning mechanisms which lead to procedures like those described in the preceding paragraph, and notions concerning the neural basis of learning which usually are evaluated through electrical stimulation and ablation experiments. These studies, as a rule,

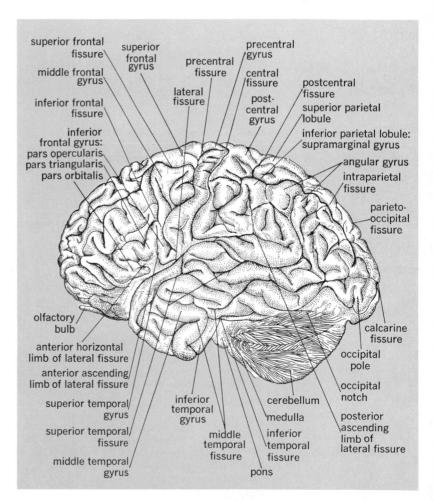

Fig. 2. Fissures and convolutions (gyri) of the human cerebral cortex, left hemisphere. (*From E. Crosby, T. Humphrey, and J. Lauer, Correlative Anatomy of the Nervous System, Wiley, 1962*)

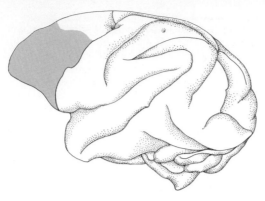

Fig. 3. Frontal granular cortex of the rhesus monkey. (*From J. M. Warren and K. Akert, The Frontal Granular Cortex and Behavior, McGraw-Hill, 1964*)

Other ablations of sensory cortex result in total loss of memory and a complete inability to relearn the discrimination.

Unfortunately, it often is equally likely that the animal fails postoperatively to perform the discrimination, not because it has "forgotten," but because it now lacks the necessary sensory apparatus to do so. Since it is almost impossible to test an animal's capacity for discriminating between two stimuli without also making demands upon its ability to memorize the difference between them, a rigorous test of the above proposition is extraordinarily difficult to arrange. Similarly, cortical lesions which produce learning deficits may also involve damage to motor mechanisms; the animal is rendered incapable of performing the appropriate response, even though the neural apparatus for acquiring and storing the relevant information is undamaged.

Lesions in other parts of the brain, however, reliably impair certain kinds of learning, or memory, or both, without noticeably affecting sensory or motor capacities of the organism. At least one such region exists within the cerebral cortex, and there are others in structures underlying the cortical mantle.

Monkeys that have learned a delayed-response discrimination are unable to perform or relearn this task following ablation of the "granular" cortex of the frontal lobes (Fig. 3), a region which has no direct role in sensory or motor functions. In a delayed-response experiment, the animal watches a reward being placed under one of two stimulus objects. An opaque door is lowered between the subject and the stimuli for a short duration and is then removed (Fig. 4). The subject is then allowed to choose between the stimulus objects. This was at one time regarded as a test for short-term memory.

It has been suggested, however, that delayed-response deficit following frontal lesions may be caused by an inability on the part of the experimental animal to attend to the critical discriminative stimuli long enough for the experience to register in the animal's memory. It is significant that similar behavioral results are obtained following ablation of certain subcortical structures and that important connections have been shown to exist between these structures and the frontal areas.

Limbic system. The forebrain structures referred to above are among several which lie along the inner edge of the cerebral cortex in the medial and temporal regions of the cerebral hemispheres. Collectively, they are known as the limbic system (Fig. 5), after the suggestion by the 19th-century French surgeon Paul Broca that they form a border (limbus) around the junction between the cerebral hemispheres and the upper end of the brain stem. The various parts of the limbic system are interconnected by a network of fiber bundles. In addition, they have rich connections with many subcortical parts of the brain.

The most suggestive experimental findings have come from clinical observations indicating that damage to the human hippocampus, a structure within the temporal lobe, resulted in severe deficits in short-term or recent memory. The patient would experience no difficulty in recalling such well-learned information as his address, his birthdate, or the names of old friends; but he might be unable to remember, for more than a few minutes at a time, the name of his neurologist, despite having seen the doctor and having been told his name scores of times. It has been suggested, as a result of these findings, that the hippocampus might function to "hold" new learning temporarily until a more permanent storage location has been prepared for it elsewhere in the brain.

Ablation studies on the brains of lower animals, however, have not always supported this conclusion. Instead, it appears that only certain kinds of learning are affected by hippocampal lesions, while other kinds are not. As in the case of frontal lobe damage, deficits in attention have been invoked to explain some of the experimental findings. Other experiments suggest that hippocampal lesions produce a perseverative tendency; that is, the animal appears unable to keep from making previously adaptive responses that have become maladaptive in a changed test situation. Finally, as is almost always the case with limbic system ablations, the effects on learning may be entangled (confounded) with effects on motivational or emotional mechanisms.

With respect to lesions in other parts of the limbic system, roughly comparable findings have

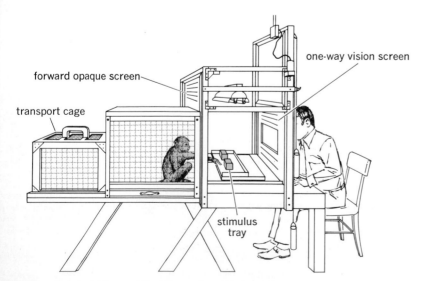

forward opaque screen

one-way vision screen

transport cage

stimulus tray

Fig. 4. Wisconsin General Test Apparatus for delayed-response experiment. (*From Harlow; copyrighted by Regents of the University of Wisconsin in S. Grossman, Textbook of Physiological Psychology, Wiley, 1967*)

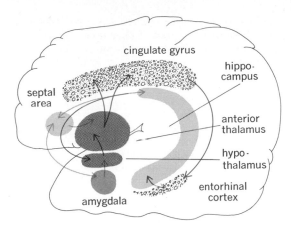

cingulate gyrus

hippo-
campus

septal
area

anterior
thalamus

hypo-
thalamus

entorhinal
cortex

amygdala

Fig. 5. The limbic system, projected on an outline of the
brain. Dark arrows represent the neural connections
referred to by Papez as the limbic pathways. Light arrows represent connections discovered since that time.
*(From F. Leukel, Introduction to Physiological Psychology,
C. V. Mosby, 1968)*

been obtained. Different lesions affect different
kinds of learning in different ways. In almost every
case the effect on learning per se cannot experimentally be distinguished from the effect on some
fundamental response system; and although the
limbic system undoubtedly plays an important
accessory role in many kinds of learning, there
seems to be little reason for believing that mechanisms of memory or learning of any kind are
confined there.

MECHANISM OF MEMORY

Ablation experiments seem to have established
the validity of the assertion that "there are no special cells for special memories"; it is very likely
that learning and memory require the same neurons to be used over and over again. However, the
fact that separate memories are behaviorally demonstrable implies that these neurons must operate in different ways or in different combinations,
as circumstances may require.

The relationship between the individual or cooperative behavior of brain cells, and learning and
memory is perhaps the central problem of neuropsychology.

Consolidation. There is fairly widespread agreement among psychologists that the neural consolidation of memory, that is, the translation of a
learning experience into a relatively enduring
neural trace, does not take place all at once. Rather, there seems to be a transitional period during
which a new memory is highly labile, or susceptible of disruption. The duration of this short-term
memory phase probably varies somewhat from
species to species and with the type of learning,
but a number of estimates put it at about 30 min.
The available evidence suggests that the integrity
of short-term memory is highly dependent upon
the maintenance of normal electrical activity in
the brain. Disruption of this activity leads to disruption of recent but not of well-established
memory. The phenomenon of retrograde amnesia
in humans is often considered an example of disturbance of short-term memory. Hence, it is widely
held that the coding and storage of information in
the brain must require the participation of at least

two separate (but not necessarily autonomous)
mechanisms. *See* AMNESIA; ELECTROENCEPHA-
LOGRAPHY.

Theories of memory. The relation between
short-term memory and the integrity of the normal
electroencephalogram (EEG) has suggested to
many scientists that this early phase of learning
probably resides in the electrical activity itself.
But older, well-established memories seem to be
impervious to agents, such as electroconvulsive
shock (ECS), which produce violent electrical
"storms" in the brain. Therefore, the "electrical
memory" must be translated (consolidated) into a
"structural memory" by appropriate alteration of
the brain substance. The crucial questions pertain
to what is altered, and how.

Deterministic. Many experts have long insisted
that a given memory trace must be distributed
among many neurons, each of which encodes a
fragment of the total message. Thus, for a given
memory trace to endure, the connections
(synapses) among the participating neurons (and
only among these) must be so modified that an
appropriate event, for example, exposure to the
learning situation, will activate the entire complex
circuit, with each component responding precisely
as and when its role in the overall mechanism
demands.

As yet, very little evidence exists that the synapse can be modified in ways consistent with the
schema just described.

Theories of the neural engram (such as that outlined above) which make use of fixed components
are called deterministic. The great preponderance
of ablation experiments discredit any deterministic theory. E. R. John, an American neurophysiologist, has cited a number of phenomena which
seem to be irreconcilable with the deterministic
viewpoint, including the following: (1) Most brain
cells, far from being responsive only to specific
inputs, are incessantly active, discharging impulses while the organism is active, resting, or even
sleeping. It is difficult to see, therefore, how a neuron might distinguish between inputs possessing
informational properties and those arising from the
spontaneous activity of connecting cells. (2) A
deterministic theory must eventually explain how
some cells are selected for memory storage, while
others are not. This is a serious problem, since the
initial occurrence of an event-to-be-remembered
may change the activity patterns of extremely
large numbers of neurons, of which only a relatively small proportion will undergo lasting modification in the coding and storage processes.

Probabilistic. E. R. John has proposed the following hypothesis of memory mechanism: The
firing pattern of an individual "memory" neuron is
never coherent; that is, a single brain cell is incapable of receiving, storing, or transmitting a recognizable unit of information.

John postulates that experience (an incoming
stimulus) sets up an oscillating pattern or wave of
electrical excitation in a group of cells. Each neuron contributes to the oscillation in a manner analogous to the role of a single molecule of water in a
ripple. Information resides only in the pattern, and
it is impossible to discern the features of the wave
from the behavior of any one of its component
units. Each learning experience generates its own
pattern of excitation. A given neuron may partici-

pate in thousands of separate memories, but its removal will not appreciably diminish any of them. Thus, John offers a plausible explanation for the persistence of memory in the face of discrete brain damage.

Biochemical. Even though they may incorporate deterministic or probabilistic elements, or both, biochemical theories are considered separately for two reasons: They have generated a good deal of research activity; and they present unique problems of their own.

Physiologists have always maintained that any structural changes (such as those implied by learning) ultimately must be understood as chemical changes. However, it was not until the middle 1950s, with the discovery of deoxyribonucleic acid (DNA), that a mechanism was discovered which linked chemical processes to the directed storage and transfer of information within a single cell. DNA, through the mediating activity of a related molecule, ribonucleic acid (RNA), specifies the nature of all of the proteins in a cell, including those which form its structural elements and the all-important enzymes which regulate virtually all of its metabolic processes. *See* GENETIC CODE; MOLECULAR BIOLOGY; NUCLEIC ACID.

By the early 1960s, there were a number of proposals concerning various analogies and linkages between the genetic code and memory. Perhaps the best known of these was presented by the Swedish neurophysiologist Holger Hydén. Hydén's idea was that a novel experience might be capable of effecting a subtle alteration in the DNA of a neuron such that a new, or stimulus-specific, protein (an enzyme, perhaps) would be manufactured by the cell. The specified protein, in turn, could alter the discharge characteristics of the neuron and thus initiate the formation of a memory trace. One of the distinguishing characteristics of this theory is the idea that a single brain cell might be capable of encoding substantial quantities of information and that the same information would be carried by other neurons. Thus, the theory incorporates both diffusion of the engram and resistance to the effects of brain damage.

Hydén's theory is basically deterministic, and susceptible, therefore, to the difficulties cited earlier. The theory is seriously weakened by the inclusion of several unlikely hypothetical processes, including the specification of DNA by a particular sensory event. This and all other biochemical theories of memory remain unverified. *See* LEARNING THEORIES; MEMORY.   [LEONARD J. GOLDSMITH]

Bibliography: S. P. Grossman, *A Textbook of Physiological Psychology*, 1967; E. R. John, *Mechanisms of Memory*, 1967; C. T. Morgan, *Physiological Psychology*, 3d ed., 1965; R. F. Thompson, *Foundations of Physiological Psychology*, 1967.

Learning theories

A field of psychology embracing habits, skills, memories, and problem solving, including their acquisition, retention, and utilization. Theories of learning have been evolved by psychologists working on the stimulus response, or conditioned reflex, concept and the cognitive, or awareness, concept. *See* MEMORY; PROBLEM SOLVING (PSYCHOLOGY); SENSORY LEARNING; VERBAL LEARNING.

Stimulus response concept. Conditioning experiments have yielded certain lawful relationships, such as favorable time intervals between stimuli, effects of rate of presentation upon acquisition, and spontaneous recovery of response strength following reduction through extinction procedures. It is assumed that these relationships can serve as the basis for predicting more complex habit phenomena, such as those involved in learning a language or operating a typewriter. Application of conditioning principles is not solely by analogy, but the effort is made to predict effects that, superficially, are unlike the relationships used in prediction. For example, the fact that items in the middle of a list are more difficult to memorize than items at the beginning or the end is not reflected in simple conditioned reflex experiments. Yet it is possible to use relationships from these experiments to predict this fact. C. L. Hull was a leading theorist in this field, and his proposals have been extended and modified by N. E. Miller and K. W. Spence.

Hull recommended the hypothetico-deductive method. That is, he preferred to set up a small set of "postulates," which are relationships from simple conditioning experiments, usually empirically established. From these he proceeded to "deduce" more complex phenomena as "theorems." These theorems, if confirmed by experiments, tend to confirm the postulates. In order to simplify the total number of postulates, he departed somewhat from purely empirical relationships in order to infer "intervening variables" as simplifying constructs. These intervening variables, given such names as "drive" or "habit strength," lie between the observed stimulus, or input, variables and the measured response, or output, variables. In Hull's most basic formula, shown below, each of the

$$_sE_R = {_sH_R} \times D$$

terms is an intervening variable. Thus $_sE_R$ is the inferred reaction potential which, if above threshold, leads to the measurable overt response R to the stimulus S. This reaction potential is some multiplicative function of habit strength ($_sH_R$), acquired as a result of repeatedly rewarded experiences, and the motivational conditions or drive (D) active at the time. Thus habit is activated into behavior only when the drive conditions are adequate. Habit strength depends upon rewarding circumstances known as reinforcements. The complete formula for response evocation includes a number of factors in addition to habit and drive, for example, stimulus intensity and the amount of reward used in prior reinforcement.

B. F. Skinner also bases his interpretations on the data from conditioning experiments and accepts reinforcement as the basic operation for response strengthening. He takes a more positivistic approach than Hull but disclaims the notion of intervening variables and refuses to use the hypothetico-deductive method. He has been successful in demonstrating a great deal of lawfulness in learned behavior by using a variety of schedules of intermittent reinforcement, that is, reward on only a predetermined fraction of the trials. Thus the reinforcement can come after every tenth response (fixed-ratio schedule) or every 10 min (fixed-interval schedule). Other complications can be introduced, as in the "clock" experiments. In one of these, an experiment in which the spot pecking of a pigeon is reinforced with food, the food is deliv-

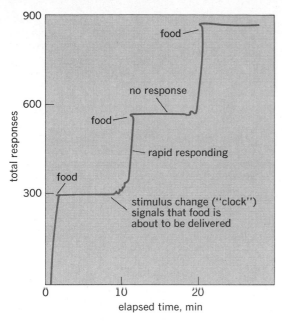

Fig. 1. Control of response by reinforcement. (*Adapted from C. B. Ferster and B. F. Skinner, Schedules of Reinforcement, Appleton-Century Crofts, 1958*)

ered only once every 10 min. The target at which the pigeon pecks changes slowly so that it serves as a kind of clock. The pigeon begins to respond 7–8 min following prior reinforcement and is responding more than 10 times per second by the time food reinforcement is delivered (Fig. 1). While his basic work has been on lever-pressing by rats and spot pecking by pigeons, Skinner has had some success in applying his analysis also to verbal behavior, schoolroom learning, and the behavior of psychotic patients.

There is also a nonreinforcement variety of learning theory, according to which the basic relationship for habit formation is merely the contiguous presence of a stimulus with a response (Fig. 2). A sophisticated theory, derived from the contiguity principle, has been proposed by E. R. Guthrie and given mathematical expression by W. K. Estes and C. J. Burke. The mathematical model is based on the assumption that the learner is sensitive to some fraction of the total stimulus complex

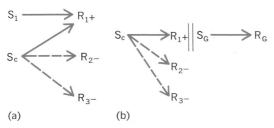

(a)   (b)

Fig. 2. Nonreinforcement and reinforcement S-R theories. (*a*) Nonreinforcement S-R theory. Conditioned stimulus (S_c) evokes R_1 rather than R_2 or R_3 because another stimulus (S_1) causes R_1 to occur in the presence of S_c. This is also called contiguity theory because the conjunction of S_c and R_1 is all that is needed to produce an association between them. (*b*) Reinforcement S-R theory. S_c evokes R_1 at a higher probability than R_2 or R_3 (other responses in its repertory) because R_1 has been followed (rewarded or reinforced) by S_G leading to R_G, a goal stimulus and response.

available to him and that this fraction becomes attached, or conditioned, to the response that occurs. Hence, as a response is repeated, an increasing fraction of the possible stimulus components will have been conditioned to it, so that the response becomes increasingly probable on later trials. The model describing this increasing probability makes use of the mathematics of set theory and is known as a stochastic model. The model and its parameters are logically specified, and the parameters are then empirically derived by curve fitting. Considerable success has been achieved, especially with relatively simple probability learning, as in guessing which of two lights will come on when the experimenter uses a random order but with one light relatively more frequent than the other. *See* BIOMETRICS; MATHEMATICAL BIOLOGY; STOCHASTIC PROCESS.

Mathematical models are not limited to the contiguity interpretation; in fact, by providing transformation equations they may reveal that some of the distinctions between reinforcement and contiguity interpretations are scientifically unimportant. Other varieties of model have been provided by R. R. Bush and F. Mosteller, F. Restle, and others.

All of the viewpoints thus far mentioned classify as habit or association theories and are commonly grouped together as stimulus-response, or S-R, theories. Some S-R psychologists have devoted more attention to complex human learning, as in rote serial memorization or in perceptual-motor skills. These experiments reveal lawful properties at their own level of complexity, such as serial position effects, short-circuiting or place-skipping tendencies, and intralist and interlist effects. Those whose theoretical writings include statements deriving from this level of complexity are commonly known as functionalists, of whom A. L. Irion, A. W. Melton, and B. J. Underwood are contemporary representatives.

Cognitive concept. Set in opposition to the S-R theorists are a number of psychologists who emphasize the importance of cognitive factors in learning, that is, perceptual or ideational processes, as distinct from sensorimotor habits. They see learning not merely as a chain of conditioned reflexes but as a sign-following process in which the learner somehow knows "what leads to what" and can modify his behavior accordingly. What is learned is more nearly like following a map than like the unfolding of a movement pattern fixed by habit. Although the pigeon's responses in Fig. 1 are used to illustrate reinforcement theory, they also represent an adaptation to the patterned nature of the stimulation in time; the use of the clock by the pigeon is coherent with cognitive theory. Cognitive theorists tend to use such concepts as "insight" or "expectation" when they describe the control of behavior. The names associated with the cognitive position include the Gestalt psychologists, especially Wolfgang Köhler and Kurt Lewin, and some neobehaviorists, for instance, Edward C. Tolman and Karl S. Lashley.

While for a time the interest in behavioral laws overshadowed interest in neurophysiological correlates of learning, since 1950 there has been renewed interest in brain processes underlying learning. Thus far the specific neural changes related to learning are unknown. *See* LEARNING,

NEURAL MECHANISMS OF; PSYCHOLOGY, PHYSIO-
LOGICAL AND EXPERIMENTAL.

[ERNEST R. HILGARD]

Bibliography: J. E. Deese and S. Hulse, *Psychology of Learning*, 3d ed., 1967; E. R. Hilgard and G. H. Bower, *Theories of Learning*, 3d ed., 1966.

Least action, principle of

Like Hamilton's principle, the principle of least action is a variational statement that forms a basis from which the equations of motion of a classical dynamical system may be deduced. Consider a mechanical system described by coordinates $q_1, \ldots, q_f$ and their canonically conjugate momenta $p_1, \ldots, p_f$. The action S associated with a segment of the trajectory of the system is defined by Eq. (1), where the integral is evaluated along the

$$S = \int_c \sum_j p_j \, dq_j \qquad (1)$$

given segment c of the trajectory. The action is of interest only when the total energy E is conserved. The principle of least action states that the trajectory of the system is that path which makes the value of S stationary relative to nearby paths between the same configurations and for which the energy has the same constant value. The principle is misnamed, as only the stationary property is required. It is a minimum principle for sufficiently short but finite segments of the trajectory. *See* HAMILTON'S EQUATIONS OF MOTION; HAMILTON'S PRINCIPLE; MINIMAL PRINCIPLES.

Assume that Eq. (2) holds, where $p_j + \delta p_j$ is ca-

$$S + \Delta S = \int_c \sum_j (p_j + \delta p_j) \, d(q_j + \delta q_j) \qquad (2)$$

nonically conjugate to $q_j + \delta q_j$. Neglecting second order term in Eq. (2), one obtains Eq. (3), where an

$$\Delta S = \int_c \sum_j (p_j \, d\,\delta q_j + \delta p_j \, dq_j)$$

$$= \int_c \sum_j (\delta p_j \, dq_j - \delta q_j \, dp_j) \qquad (3)$$

integration by parts has been made, the integrated parts vanishing.

The vanishing of ΔS requires the integrand to be a perfect differential of a quantity whose end variations vanish. The coefficients of the variations $\delta q_j, \delta p_j$ need not vanish separately because the variations are not independent, the varied qs and ps necessarily being canonically conjugate, as in Eq. (4), with terms defined by Eqs. (5).

$$\sum_j (\delta p_j \, dq_j - \delta q_j \, dp_j) = dU(q,p) \qquad (4)$$

$$\delta p_j = \frac{\partial U}{\partial q_j} \qquad \delta q_j = -\frac{\partial U}{\partial p_j} \qquad (5)$$

Writing $U = -H \, \delta t$ leads to Hamilton's equations of motion, Eqs. (6). The quantity $H(q,p)$, known as

$$\dot{p}_j = -\frac{\partial H}{\partial q_j} \qquad \dot{q}_j = \frac{\partial H}{\partial p_j} \qquad (6)$$

the Hamiltonian function, does not contain the time explicitly because $U(q,p)$ cannot be a function of the time since the end times are not fixed and in general will vary as the path is varied. Thus, the principle is useful only for conservative systems, where H is constant.

If $H(q,p)$ consists of a part H_2 quadratic in the momenta and a part H_0 independent of the momenta, then Eq. (7) holds by Euler's theorem on

$$S = \int_{t_1}^{t_2} \sum_j p_j \dot{q}_j \, dt$$

$$= \int_{t_1}^{t_2} \sum_j p_j \frac{\partial H}{\partial p_j} \, dt$$

$$= 2 \int_{t_1}^{t_2} H_2 \, dt \qquad (7)$$

homogeneous functions. Usually H_2 is the kinetic energy of the system so that the principle of least action may be written as Eq. (8), where v is the potential energy.

$$\Delta \int_{t_1}^{t_2} 2T \, dt = \Delta \int_{t_1}^{t_2} 2(E-V) \, dt = 0 \qquad (8)$$

The principle of least action derives much importance from the fact that it is the action which is quantized in the quantum form of the theory. Planck's constant is the quantum of action. *See* QUANTUM THEORY, NONRELATIVISTIC.

[PHILIP M. STEHLE]

Bibliography: See LAGRANGE'S EQUATIONS.

Least squares, method of

A method, developed originally by A. M. Legendre, of obtaining the best values (the ones with least error) of unknown quantities supposed to satisfy a system of linear equations of the form shown as notation (1), where $n > m$. Since there are more

$$M_{11}a_1 + M_{12}a_2 + \cdots + M_{1m}a_m = b_1$$
$$M_{21}a_1 + M_{22}a_2 + \cdots + M_{2m}a_m = b_2 \qquad (1)$$
$$\cdots \cdots \cdots \cdots \cdots \cdots$$
$$M_{n1}a_1 + M_{n2}a_2 + \cdots + M_{nm}a_m = bn$$

equations than unknowns, the system is said to be overdetermined. Furthermore, the values obtained for the unknowns by solving a given selection, m in number, of the equations will differ from the values obtained by solving another selection of equations. In the physical situation, the b_i are measured quantities, the M_{ij} are known (or assumed) quantities, and the a_i are to be adjusted to their best values.

Consider a simple example. A quantity y of interest is supposed (perhaps for theoretical reasons) to be a linear function of an independent variable x. For a series of selected values $x_1, x_2, \ldots$ of x the values $y_1, y_2, \ldots$ of y are measured. The expected relation is shown as notation (2), and the

$$x_1\alpha + \beta = y_1$$
$$x_2\alpha + \beta = y_2 \qquad (2)$$
$$x_3\alpha + \beta = y_3$$
$$\cdots \cdots \cdots$$

problem is to find the best values of α and β, that is, respectively, the slope and intercept of the line which graphically represents the function. The best values of α and β, in the least squares sense, are obtained by writing Eq. (3) and asserting that

$$\eta_i = y_i - (x_i\alpha + \beta) \qquad (3)$$

term (4) shall be minimized with respect to α and

$$\sum_{i=1}^{n} \eta_i^2 \qquad (4)$$

β; that is, that Eqs. (5) hold. This leads to the two

equations, Eqs. (6) and (7), which may be solved

$$\frac{\partial}{\partial \alpha} \sum_{i=1}^{n} \eta_i{}^2 = 0$$

$$\frac{\partial}{\partial \beta} \sum_{i=1}^{n} \eta_i{}^2 = 0 \tag{5}$$

$$\alpha \sum_{i=1}^{n} x_i + n\beta - \sum_{i=1}^{n} y_i = 0 \tag{6}$$

$$\alpha \sum_{i=1}^{n} x_i{}^2 + \beta \sum_{i=1}^{n} x_i - \sum_{i=1}^{n} x_i y_i = 0 \tag{7}$$

for α and β. For m, rather than two, unknowns the generalization is obvious in principle, although the labor of solution may be great if m is large unless a high-speed electronic computer is available.

It should be noted that the measurements y_i in the example have all been assumed to be equally good. If it is known that the measurements are of variable quality, a weight may be attached to each value of y_i. The least squares equations are readily modified to take this into account, as in Eq. (8), where w_i is the weight of measurement y_i.

$$\alpha \sum_{i=1}^{n} w_i x_i + \beta \sum_{i=1}^{n} w_i - \sum_{i=1}^{n} w_i y_i = 0$$

$$\alpha \sum_{i=1}^{n} w_i x_i{}^2 + \beta \sum_{i=1}^{n} w_i x_i - \sum_{i=1}^{n} w_i x_i y_i = 0 \tag{8}$$

The least squares equations can be shown to lead to the most probable (in the statistical sense) values of the unknowns under a variety of assumptions about the measurements and their weights. In applications in the physical sciences, however, it is rarely possible to show that one's observations satisfy all, or even any, of the assumptions. However, the conditions may be approximately satisfied in many instances, and the method is widely used because of its convenience. The empirical result is that the unknowns so determined lead to excellent representations of the data in the usual case. *See* CURVE FITTING.

[MC ALLISTER H. HULL, JR.]

Bibliography: Y. V. Linnik, *Method of Least Squares and Principles of the Theory of Observations*, 1961.

Leather and fur processing

The technology of processing hides and skins. The leather industry is probably the oldest in continuous production, since leather and furs have been made for thousands of years; it is still among the most complex. Leather and fur processing was originally a home industry, but the huge increase in meat-packing operations around the turn of the century supplied the market with large quantities of cattlehides and sheepskins. Since these by-products are perishable, large tanneries were built

to handle them. Development of chemical and physical controls in the past few years ensures uniformity lacking in earlier leathers. Processing time has been reduced from months to days by improved chemical processes and complete and ingenious mechanization.

Besides cattle- and horsehides and sheep-, kid- and goatskins, specialties such as snakeskins and lizard skins and crocodile and camel hides are tanned. Trappers and breeders supply fur pelts on demand. Table 1 lists statistics on animals used for hides and skins.

About 85% of all leather produced is used for shoes, and lesser amounts for clothing, handbags, luggage, transmission belting, mechanical goods, picture frames, billfolds, and keycases. Furs are used for luxury garments. In a sense, leather is also a luxury material, because if all the world's population could afford shoes, the leather supply would be grossly inadequate.

Sales of leather and furs depend primarily on their beauty and "feel" and secondarily on chemical and physical properties. Table 2 summarizes properties of principal leathers. Their excellent abrasion resistance, flexing life, dimensional stability, and vapor permeability are notable.

Special leathers, such as chamois (oil-tanned), possess high water absorption, and buffalo and rawhide possess great abrasion resistance.

Economics. The leather and fur industries face sharp competition from synthetic substitutes—all of which are advertised as "like leather." A large part of the shoe-sole business has been lost to rubber-type synthetics, and synthetic materials for the upper shoe pose another threat. Tanners have used new tanning agents, finishes, and impregnants to improve leathers and furs, and have organized industry-wide quality controls. They have emphasized the luxury properties of natural products and, in the case of shoes, their comfort and foot protection. The widespread popularity of suede and grain leather garments is largely due to these steps. Hides and skins represent approximately two-thirds of leather factory costs, but competition could drive their cost down to levels where leather could still compete with synthetic substitutes.

Raw stock preservation. Since hides and skins are subject to putrefaction, steps to preserve them are essential. Thus, as soon as animal heat is dissipated, cattlehides are washed to remove blood, and packed carefully in dry salt or immersed in brine to partially dehydrate them. After salt cure and draining, they are shipped all over the world, and may be kept for up to 2 years if slightly refrigerated.

As such hides have large amounts of flesh adhering to them, many new plants at or near packing houses have been built to remove flesh before salt curing. This yields more uniform results and reduces shipping weights. However, the method requires large investment in special machines, skilled crews, and large-volume operation.

Sun or shade drying of hides and skins is used in remote areas as an alternative to salt cure, but with less satisfactory results. In the case of sheepskins, wool, the valuable primary by-product, is removed or "pulled" at the packing house and the skins are pickled prior to worldwide shipment.

Table 1. Animals used for leather

Animal	World population	Annual kill	
		World	United States
Cattle; calf	813,000,000	170,000,000	20,000,000
Water buffalo	76,700,000	8,900,000	None
Sheep; lamb	855,000,000	160,000,000	15,000,000
Goat; kid	306,700,000	115,000,000	433,000

Table 2. Leather properties and uses*

Properties and uses	Cowhide		Calfskin	Sheepskin (lambskin)	Goatskin (kidskin)	Horsehide
	Light (upper)	Heavy (sole)				
Unit of sale	Side (1/2 hide)	Bends Shoulders Bellies Heads	Whole skin	Whole skin	Whole skin	Butts Fronts Sides
Area or weight	10–25 ft²	30–60 lb per hide	3–20 ft²	3–10 ft²	2–8 ft²	3–7 ft² per cut
Thickness: 1 oz = 1/64 in. 1 iron = 1/48 in.	2–6 oz	5–12 iron	2–4 oz	1–4 oz	1–3 oz	3–6 oz
Main uses (indicated by x)						
Shoe uppers	x		x		x	x (Cordovan)
Shoe linings			x	x	x	
Shoe soles		x				
Garments	x		x	x		x
Gloves	x (work)			x	x	x (work)
Handbags	x		x	x		
Billfolds	x		x	x	x	
Luggage	x	x				
Transmission belting		x				
Mechanical	x	x				
Special characteristics†						
Abrasion resistance	E	E	E	E	G	E (Cordovan)
Flexing life	E	E	E	G	E	E
Appearance	E		S	G	E	S (Cordovan)
Finish (polish)	E		S	G	E	S (Cordovan)
Comfort	E	E	E	E	E	G (Cordovan)
Drape	G		G	E	G	G
Protection	E	E	E	G	G	E
Dimensional stability	E	E	E	E	E	E
Vapor permeability	E	M	G	E	G	L (Cordovan)
Relative cost	M	M	H	M	H	H (Cordovan)
Tannages (indicated by x)						
Chrome	x		x	x	x	x
Vegetable	x	x	x	x		
Alum				x		
Formaldehyde				x		
Tensile strength 1-in. width sample; thickness, as used	250 lb	700 lb	200 lb	90 lb	125 lb	70 lb
Bursting strength	300 lb		300 lb	60 lb	250 lb	300 lb

*Hide connotes larger and heavier leathers; skin, smaller and thinner leathers.

†E, excellent; G, good; M, medium; L, low; S, superior; H, high.

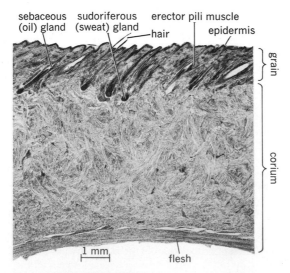

Fig. 1. The histology of cattlehide.

With the advent of close controls on water and atmospheric pollution in densely populated areas, and because of difficulty of obtaining workers for heavy, wet operations, there is an increasing trend toward processing cattlehides and goat skins through hair removal and pickling, or even chrome tanning, with equipment installed at or near packing plants. This is done extensively in South America, Africa, and India, where labor costs are advantageous, as well as in the United States. The resulting products are shipped wet or damp throughout the world. Tanneries utilizing them require much less floor space; processing time is greatly reduced; money is saved; and messy operations are eliminated.

Structure of hides. Figure 1 illustrates the histology of cattlehide. The dark area from epidermis to base of hair pockets is known as grain. In its arrangement of large or small hair pockets, and smooth or rough surface, the grain gives each leather its characteristic appearance.

The lighter, deeper area, characterized by the pronounced three-dimensional mixture of fibers in fiber bundles, is the corium, composed mainly of the protein collagen. Besides proteins, hides and skins contain lipids, carbohydrates, inorganic salts, and water. *See* INTEGUMENT.

Processing. The tanner's skill consists in processing hides and skins of varying size, thickness, and texture to produce leathers of reasonably uniform quality, while perpetuating or enhancing their appearance, long flexing life, abrasion resistance, and vapor permeability.

Wet operations (beamhouse). If salted hides are used, they are first soaked in clean, cool water to dissolve the salt, loosen dirt, and rehydrate protein. Excessive flesh is removed mechanically after soaking, as necessary.

Hides are next placed in pits containing solutions of alkali—hydrated lime, $Ca(OH)_2$, with or without sodium sulfide, Na_2S. Because of the loosening or dissolving action of the alkali at the hair roots, hair can be removed mechanically after this treatment. Furs and sheepskins (shearlings), which are tanned with the wool on, are not given the alkali treatment. After application of a sodium sulfide paste on the flesh, sheepskins are hung overnight, and the wool is pulled off the next day.

Sole-leather hides and many others are relimed after dehairing to prepare fibers for tanning. Cattlehides for upper leather are split after liming (alternatively after chrome tanning) to reduce thickness to that acceptable to modern wearers. The stock is next delimed, that for chrome tanning in a bate of pancreatic enzyme activated with a solution of an ammonium salt, which also selectively removes certain proteins. Such stock is immediately pickled in a solution of salt and sulfuric acid. Hides for heavy, vegetable-tanned leathers are immersed in solutions of weak organic acids, such as lactic or hydroxyacetic, for surface lime removal without loss of hide substance. The stock now leaves the beamhouse for further treatment.

Tanning. Tanning is the conversion of hides and skins into insoluble, nonputrescible leather without destruction of the original structure. Furs are partially tanned or dressed without loosening or damaging the filaments, or fur.

Pickled stock to be chrome tanned is saturated with a solution of basic chrome sulfate, $Cr(OH)SO_4$, or an equivalent, in a revolving drum. After penetration is complete, the unstable chrome compound is precipitated by addition of a small quantity of mild alkali, such as sodium bicarbonate, $NaHCO_3$. Stock is allowed to stand 24 hr to complete chemical combination of chrome with hide substance. Many modern tanneries conduct all beamhouse operations and tanning without removal of stock from large revolving drums.

In vegetable tanning, heavy hides are given a series of baths of gradually increasing strengths of vegetable extracts. These extracts are mainly imported wattle or quebracho. This countercurrent method assures complete penetration of the fibers, without distortion of the hide. Vegetable extracts are adsorbed but do not form strong chemical bonds with collagen.

In a process known as tawing, alum is used as a partial tannage, supplementing or replacing chrome. Formaldehyde is sometimes used to produce white, fluffy leathers. Zirconium salts make pure white leather of great strength. Glutaraldehyde is a tanning agent widely employed to increase perspiration resistance in combat-boot leathers and to impart washability in wool skins (shearlings).

Dyeing and lubrication. Upper leathers are dyed with acid or basic aniline dyes in drums. Close control of pH is essential to ensure uniform and correct shade. While in the drum, lubricants in the form of sulfated animal-, vegetable-, or mineral-oil emulsions are added. These are taken up selectively by the leather.

Heavy leathers are rarely dyed, but are given an oil load. The function of the lubricant in both upper and heavy leathers is to protect the fibers during drying and subsequent processing and to lubricate the finished leather. Lubrication is necessary because the natural oils have been saponified and removed by alkaline treatment during wet processing.

Development work since 1945 on tanning and wet operations in acetone, rather than water, has shown that the time required for these operations can be cut to minutes.

Drying. Both heavy and upper leathers are next passed through revolving cylinders with radial blades to remove water and smooth the grain. Most upper leathers are then "pasted" on plate glass or enameled iron sheets and dried there to improve smoothness and uniformity. Conveyor driers or dry rooms are used to remove a controlled amount of moisture, with exacting schedules of temperature and humidity control to avoid hardening or shrinkage. Upper leathers require hours, and heavy leathers days to dry. Rapid vacuum drying is also employed to a limited extent in special equipment.

Finishing. Upper leathers are partially rehydrated and softened (staked) by machines which separate individual fibers without rupturing fiber bundles. Impregnation is practiced extensively at this point to increase scuff and abrasion resistance of

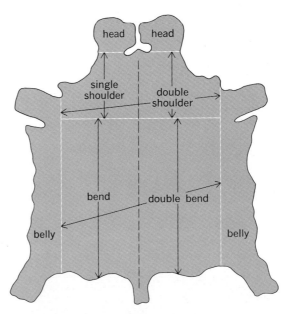

Fig. 2. Sole-leather cuts from cattlehide.

Fig. 3. Cattlehides being cured with salt in a pack containing about 800 hides.

shoe leathers. Several coats of finishing materials are next applied by hand or spray guns, on automatic conveyors. Pyroxylin lacquers, polyacrylate emulsions, acrylonitriles, or pigment finishes of casein-shellac-pigment are employed as indicated, but they must not obscure the grain. Boiled linseed oil is the usual coating for patent leather, but in some cases synthetic resins are used.

Glazing or polishing may be used to enhance luster. A good finish is lustrous, water-repellent, and vapor-permeable. The flesh side of upper leather is made attractive by buffing on machines with oscillating cylinders covered with fine sandpaper. Suede leathers are finished similarly on the flesh side, to produce a soft, fine nap. Nubuck is the trademark for cowhide with a sueded grain surface. Recently developed semiautomatic machinery is now replacing many hand-finishing operations, with improvement in quality and uniformity. Sole and heavy leathers are rolled under high pressure (after dampening) to compact fibers for maximum wear; then they are waxed. Figure 2 shows the locations on cattlehide from which sole leather cuts are taken.

Furs. Dressed animal pelts used for garments or adornment are known as furs. Sometimes the filaments or hairs are known as fur. Rabbit fur is

universally used for felting hats. *See* TEXTILE.

Originally fur clothing was worn for protection; today its style and luxury features predominate. The pelts of mink, beaver, fur seal, leopard, silver and other foxes, marten, and ermine come from wild or ranch animals obtained from trappers or breeders. In addition, wool skins, skunk, rabbit, and muskrat are processed to imitate rarer or more costly furs in demand. As with leather, furs are sold on the basis of beauty and feel, not because of technical considerations.

Fur pelts are collected all over the world by organizations such as the Hudson Bay Co. and sold at auction in London, Leipzig, New York, Leningrad, Montreal, or St. Louis (Alaskan seal). The finest pelts come from cold climates.

Fur filaments are barbed lengthwise, but are soft and downy, and tend to curl. Guard hairs keep the filaments apart and prevent matting, and are removed in processing.

As in the case of leather, the quality of the original pelt and the care observed in its preservation determine the final value of the product.

The processor or dresser rehydrates the pelt by soaking it in water and removes adhering flesh with a knife or machine. In contrast to leather production, every precaution is taken to avoid the loosening of fur filaments; for this reason, alkalies are not employed. Furs are often tacked or toggled in stretched position before leathering, or tawing. Formaldehyde can be used as either the complete or the partial tannage; also, natural aldehydes from the decomposition of oxidizing oils or alum are used. Glutaraldehyde, mentioned earlier, is used extensively in tanning wool skins (shearlings). Such shearlings have found a wide market as comfortable bed pads for bedridden patients, as material for paint rollers, and so on. The dressed pelt must be soft and drapey.

Subsequent operations are closely guarded secrets, but dyeing is often employed to alter color or shade, or to make them even. Imitation furs are usually dyed to simulate the natural coloring. Dye must not transfer to the person or clothing of the wearer, a process called crocking.

When necessary, guard hairs are removed by special machines. Furs are repeatedly tumbled in

Table 3. Short-time preservation of fresh calfskin with BAC*

No.	Treating solution†	3-Day incubation			7-Day incubation		
		Soluble N % on skin wt‡	Bacteria per gram of skin‡	Condition	Soluble N % on skin wt	Bacteria per gram of skin	Condition
1	Control, water only	0.21	6×10^9	Slimy growth, putrid odor	0.39	2×10^{10}	Very putrid odor, all hair loose
2	0.1% BAC	0.13	2.7×10^9	Small amounts of yellowish growth, slight odor, not objectionable	0.14	7.2×10^8	Ammoniacal odor, hair very loose
3	0.2% BAC	0.09	2.4×10^9	As above	0.19	3.6×10^9	Ammoniacal odor, hair fairly loose
4	0.3% BAC	0.05	1.8×10^9	As above	0.14	2.8×10^9	As above
5	0.4% BAC	0.07	2.4×10^8	As above	0.16	3.8×10^8	As above

*From T. C. Cordon, H. W. Jones, and J. Naghski, *J. Amer. Leather Chem. Ass.*, 59:317–326, 1964.
†Skin pieces tumbled 1 hr in treating solution; 1–6 float.
‡Determinations made immediately after treatment on skin pieces treated with 0.1% BAC showed 0.01% soluble nitrogen and 1×10^4 bacteria per gram.

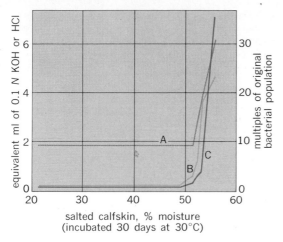

Fig. 4. Bacterial population and decomposition of stored salted calfskins. A= carboxyl groups as equivalent milliliters of 0.1 *N* KOH. B= ammonia nitrogen as equivalent milliliters of 0.1 *N* HCl. C= bacterial counts in multiples of original population. (*From L. S. Stuart and R. W. Frey, J. Amer. Leather Chem. Ass., 33:198–203, 1938*)

Fig. 5. A piece of leather which has been damaged by bacteria during the curing process.

closed or screen-covered drums to soften them, polish the filaments, and remove excess dye. Sawdust or pulverized walnut shells are often added to assist in the polishing. Luster is enhanced by "electrifying" machines, in which rapidly revolving cylinders, with alternating quadrants of clothing card and hot iron, comb and iron the filaments.

Skilled sorters select matching skins, and from these the furrier pieces together a garment. Dropping machines cut and stitch the pieces so that no joint can be detected along the surface of the fur.

Mouton, closely resembling beaver, is made from sheepskins with the wool attached. Only 5% of the pelts available have fine enough wool and adequate fiber density for the purpose. Raw pelts are washed with detergent solution, tanned with alum, dyed, combed, clipped, and ironed repeatedly to straighten the filaments. Formaldehyde is used to set filaments and to enhance luster.

Persian lamb is made from the skins of unborn lambs from Afghanistan. It possesses fine filaments and tight curls. Rabbit, skunk, opossum, and muskrat are given appropriate variations of the treatment outlined above to produce the desired imitations, such as imitation ermine from rabbit. Labeling restrictions protect the customer from misrepresentation. [KENNETH E. BELL]

Microbial aspects of tanning. Animal hides and skins, the raw material from which leather is made, are readily attacked by a variety of microorganisms; therefore, care must be taken to protect them while they are being processed into leather. Under moist, humid conditions finished leather is also susceptible to microbial attack, especially by molds.

Curing. After removal from the animal (flaying), hides are allowed to cool and are then layered in packs with salt (Fig. 3) or immersed in saturated salt brine for about 18 hr. The salt reduces the moisture content of the hides and removes unwanted blood and other minor constituents. Low moisture content helps to maintain the keeping quality of calfskins, as shown in Fig. 4. The formation of ammonia and carboxyl groups indicates

degradation of hide substance. The values given are not absolute but are indicative of what happens. Red-pigmented bacteria sometimes cause a reddening of salted hides and skins known as red heat. Bacteria can also cause damage to the grain such as that shown on the piece of leather in Fig. 5. In severe infections the hide may be ruined completely. Certain chemicals added to curing salts will prevent bacterial growth, but none of these have gained commercial acceptance because of increased cost.

It has become desirable to collect small lots of hides from small-scale butchers and to process them into leather without curing. This often necessitates holding the hides several days, during which spoilage can occur. It has been found that calfskins treated in a solution containing 0.1–0.2% benzalkonium chloride (BAC), a quaternary ammonium chloride, can be safely held for a few days without being damaged by bacteria. Table 3 shows the results of laboratory tests with this chemical. BAC also decreases bacterial growth and activity in the brining and salt-curing procedures, as confirmed by pilot-scale tanning of whole calfskins.

Soaking. Cured hides and skins must be soaked in water before being processed into leather. Preservatives, usually chlorinated phenols, added to the soak water prevent microbial attack on hide substance.

Bating. This treatment with proteolytic enzymes opens the fiber structure so that soft mellow leather can be produced. Some of the enzymes used are produced by bacteria or fungi.

Pickling. Dehaired hides are often treated in salt and acid (pH 1.5–2.0) for storage and for shipping long distances. To prevent permanent staining from mold growth, the skins can be treated with penta- or trichlorophenol, but the cost of the mold inhibitors has prevented their widespread use.

Tanning. During vegetable tannage, molds will attack and destroy the tannins unless a preservative is added. Chrome-tanned leather may also be attacked by molds if held in a damp condition, especially in the summer months.

Finished leather. Under humid conditions molds may grow on leather. Although the actual damage may be negligible, the leather becomes unsightly

Fig. 6. Leather carrying cases, right and left ones treated for mold resistance; center one untreated. (*From T. C. Cordon et al., J. Amer. Leather Chem. Ass., 44:473–503, 1949*)

and esthetically undesirable. If leather is kept dry or treated with a preservative, molds are unable to grow on it. Figure 6 shows mold growth on an untreated leather binocular carrying case and lack of growth on a treated case. The United States military services specify that leather bought for use in the tropics must contain *para*-nitrophenol to prevent mold growth.

[THEONE C. CORDON]

Bibliography: T. C. Cordon, H. W. Jones, and J. Naghski, *J. Amer. Leather Chem. Ass.*, 59:317–326, 1964; J. Naghski and I. D. Clarke, *USDA Misc. Publ.*, no. 857, August, 1961; F. O'Flaherty, W. T. Roddy, and R. M. Lollar, *The Chemistry and Technology of Leather*, ACS Monogr. no. 134, 4 vols., 1956–1965; J. Tancous, W. T. Reddy, and F. O'Flaherty, *Skin, Hide, and Leather Defects*, Tanners Council Laboratory, Cincinnati, 1956.

Lecanicephaloidea

An order of tapeworms of the subclass Cestoda. All species are intestinal parasites of elasmobranch fishes. These tapeworms are distinguished by having a peculiar scolex divided into two portions. The lower portion is collarlike and bears four small suckers; the upper portion may be discoid or tentacle-bearing and is provided with glandular structures (see illustration). The scolex is usually

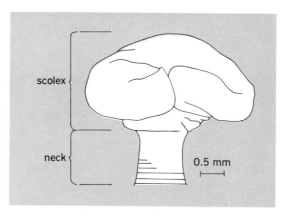

Anterior end of a lecanicephaloid tapeworm.

buried in the intestinal wall of the host and may produce local pathology. The anatomy of the segments is very similar to that of the Proteocephaloidea. Some authorities place the lecanicephaloids in the order Tetraphyllidea, to which they are obviously closely related. Essentially nothing is known of the life history of lecanicephaloids. *See* CESTODA; TETRAPHYLLIDEA.

[CLARK P. READ]

Lecanorales

An order of the Ascolichenes, also known as the Discolichenes. Lecanorales is the largest and most typical order of lichens and parallels closely the fungal order Helotiales. The apothecia are open and discoid, with a typical hymenium and hypothecium. There are four growth forms—crustose, squamulose, foliose, and fruticose—all showing greater variability than any other order of lichens.

Reproduction. Details of sexual reproduction are at best poorly known. Meiotic and mitotic stages leading to spore formation seem normal, but earlier stages of fertilization have not been studied. It is not known if fertilization is somatogamous, or if the microconidia act as spermatia and fertilize the egg in the ascogonium. Mature spores burst the ascal wall and are ejaculated from the apothecial disk to heights of several centimeters. When the spores are germinated in pure culture, a distinct though slow-growing mycelium is often formed. The mycelium has no resemblance to the growth form of the original parent lichen, nor have apothecia or pycnidia ever been seen in culture. Furthermore, it is doubtful whether anyone has ever synthesized a lichen by combining a fungal culture with an alga. In nature vegetative reproduction is highly developed. Isidia, minute coralloid outgrowths of the cortex, and soredia, powdery excrescences from the medulla, are present in many species. They are apparently capable of growing into new lichens when dislodged from the thallus.

Taxonomy. The Lecanorales is divided into 25 families, about 160 genera, and 8000–10,000 species. Family divisions are based on growth form of the thallus, structure of the apothecia, the species of symbiotic algae present, and spore characters. Species are separated by such characters as isidia, soredia, rhizines, and pores, and by chemistry. The larger families are described below.

Cladoniaceae. This family includes the reindeer mosses and cup lichens. The main thallus is a hollow structure in the shape of a stalk, cup, or is richly branched. There are about 200 species of *Cladonia*; the most highly developed are found in the boreal and arctic regions.

Lecanoraceae. The thallus is typically crustose, and the apothecia have a distinct thalloid rim. This family is primarily temperate and boreal. The largest genus, *Lecanora*, has over 500 species.

Lecideaceae. This family differs from the Lecanoraceae in lacking a thalloid rim around the apothecia. It is also very common in temperate and boreal zones and includes one of the largest and most difficult lichen genera to diagnose, *Lecidea*, with over 1000 species.

Parmeliaceae. This family, known as the foliose shield lichens, is common on trees throughout the world. *Parmelia* has almost 1000 species; *Cetraria*, a common genus on pine trees, includes the well-known Iceland moss (*Cetraria islandica*).

Umbilicariaceae. The thallus is circular in outline, quite large, and umbilicate. The family occurs on acidic rocks in temperate, boreal, and arctic regions. The members are known as the rock tripes.

Usneaceae. The beard lichens comprise perhaps the outstanding lichen family because of their conspicuous fruticose growth form. The largest genus is *Usnea*, some species of which grow to 5 ft in length. *See* ASCOLICHENES.

[MASON E. HALE]

Lecythidales

An order of flowering plants, division Magnoliophyta (Angiospermae), in the subclass Dilleniidae of the class Magnoliopsida (dicotyledons). The order consists of the single family Lecythidaceae, with about 450 species. They are tropical, woody plants with alternate, entire leaves, valvate sepals, separate petals, numerous centrifugal stamens, and a syncarpous, inferior ovary with axile placentation. Brazil nuts are the seeds of *Bertholletia excelsa*, a member of the Lecythidaceae. *See* BRAZIL NUT; DILLENIIDAE; FLOWER; MAGNOLIO-PHYTA; PLANT KINGDOM.   [ARTHUR CRONQUIST]

Leg (human)

The portion of the lower, or inferior, limb between the thigh and the foot. The leg bones are the tibia (a large, weight-supporting shinbone) and the slender, rodlike fibula. Compartmentalized muscle groups include functional sets which act on the foot, ankle, and thigh. Such muscles may either move the point to which they are attached or immobilize it to prevent movement, as when standing. Blood is supplied to the leg by branches of the femoral artery; two sets of veins, superficial and deep, drain blood to the thigh. Innervation is by branches of the femoral and sciatic nerves of the lumbosacral plexus.   [THOMAS S. PARSONS]

Legendre functions

Solutions to the differential equation $(1-x^2)y'' - 2xy' + v(v+1)y = 0$.

Legendre polynomials. The most elementary of the Legendre functions, the Legendre polynomial $P_n(x)$ can be defined by the generating function in Eq. (1). More explicit representations

$$(1-2xr+r^2)^{-\frac{1}{2}} = \sum_{n=0}^{\infty} P_n(x) r^n \qquad (1)$$

are Eq. (2), and the hypergeometric function,

$$P_n(x) = \frac{(-1)^n}{2^n n!} \frac{d^n}{dx^n} (1-x^2)^n \qquad (2)$$

Eq. (3). *See* HYPERGEOMETRIC FUNCTIONS.

$$P_n(x) = {}_2F_1[-n, n+1; 1; (1-x)/2] \qquad (3)$$

Generating function (1) implies Eq. (4). The

$$(a^2 - 2ar\cos\theta + r^2)^{-\frac{1}{2}} = \frac{1}{a}\sum_{n=0}^{\infty} P_n(\cos\theta)(r/a)^n \quad (4)$$

$$0 < r < a$$

function $(a^2 - 2ar\cos\theta + r^2)^{-\frac{1}{2}}$ represents the potential in an inverse square field at a point P of a source at A, where r and a are the distances from P and A to a fixed point O, and θ is the angle between the segments PO and OA. These functions were extensively studied by A. M. Legendre and P. S. Laplace because they could be used in the study of the celestial mechanics. They had arisen slightly earlier in probabilistic work of J. Lagrange. *See* POTENTIALS.

Since Legendre polynomials are the zonal spherical harmonics on the surface of the unit sphere in three-space, they arise when studying physical phenomena associated with spherical geometry. They arise in many other applications. One application is to the estimation of the smallest eigenvalue of the truncated Hilbert matrix $(a_{i,j})_0^N$, $a_{i,j} = (i+j+1)^{-1}$. This matrix is very hard to invert numerically because the condition number, or the ratio of the largest to the smallest eigenvalue, is

very large. One of the essential parts of the proof is an asymptotic formula for Legendre polynomials off the interval $[-1,1]$. *See* MATRIX THEORY; SPHERICAL HARMONICS.

Relation to trigonometric functions. When $x = \cos\theta$, Legendre polynomials have a relation with trigonometric functions, given by Eq. (5), where

$$(\sin\theta)^{\frac{1}{2}}\left(n+\frac{1}{2}\right)^{\frac{1}{2}} P_n(\cos\theta)$$

$$= \sqrt{\frac{2}{\pi}}\cos\left[\left(n+\frac{1}{2}\right)\theta - \frac{\pi}{4}\right] + R(n,\theta) \quad (5)$$

$|R(n,\theta)| \leq A/(n\sin\theta)$, A being a fixed constant, when $c/n \leq \theta \leq \pi - c/n$ for $c > 0$. *See* TRIGONOMETRY, PLANE.

Properties. A graph of $P_n(x)$ (Fig. 1) illustrates a number of properties. $P_n(x)$ is even or odd as n is even or odd, that is, $P_n(-x) = (-1)^n P_n(x)$. All the zeros of $P_n(x)$ are real and lie between -1 and 1. The zeros of $P_{n+1}(x)$ separate the zeros of $P_n(x)$. $P_n(x)$ satisfies the inequality $|P_n(x)| \leq 1, -1 \leq x \leq 1$. The successive relative maxima of $|P_n(x)|$ increase in size as x increases over the interval $0 \leq x \leq 1$. The closest minimum to $x = 1$ of $P_n(x)$, called $\mu_{1,n}$, satisfies $\mu_{1,n} < \mu_{1,n+1}$. Similar inequalities hold for the first maxima of $P_n(x)$ to the left of $x = 1$, and for the kth relative maxima or minima. All of these results have been used in applications. The last two results about the relative maxima were used to obtain bounds on the phase of scattering amplitudes. From an important limiting relation of F. G. Mehler, Eq. (6), it is easy to show that $\mu_{1,n}$ ap-

$$\lim_{n \to \infty} P_n\left(\cos\frac{\theta}{n}\right) = J_0(\theta) \qquad (6)$$

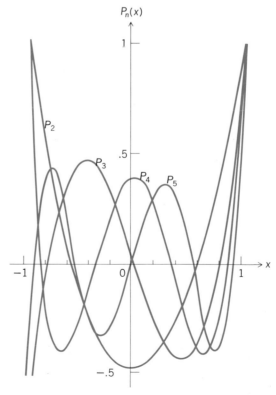

Fig. 1. Legendre polynomials $P_n(x)$, $n = 2, 3, 4, 5$.

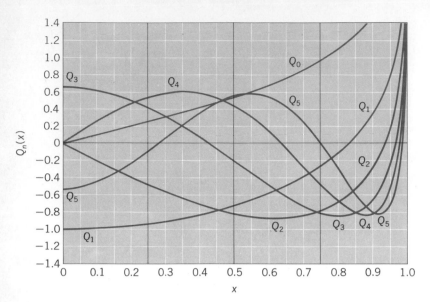

Fig. 2. Legendre functions of the second kind, $Q_n(x)$. (*From E. Jahnke and F. Emde, Tables of Functions with Formulae and Curves, 4th ed., Dover Publications, 1945*)

proaches the minimum value of the Bessel function J_0. See BESSEL FUNCTIONS.

Differential equation. One of the main reasons for the occurrence of Legendre polynomials is a differential equation they satisfy, Eq. (7). This

$$(1-x^2)y'' - 2xy' + n(n+1)y = 0 \qquad y = P_n(x) \tag{7}$$

equation arises from the solution of Laplace's equation by separation of variables in spherical coordinates. There is a second solution $Q_n(x)$ given by Eq. (8), when x is not in the interval

$$Q_n(x) = \frac{1}{2} \int_{-1}^{1} \frac{P_n(t)}{x - t} \, dt \tag{8}$$

$-1 < x < 1$. For $-1 < x < 1$, $Q_n(x)$ is defined by Eq. (9). $Q_n(x)$ has many of the same properties as

$$Q_n(x) = \lim_{\epsilon \to 0^+} \frac{Q_n(x + i\epsilon) + Q_n(x - i\epsilon)}{2} \tag{9}$$

$P_n(x)$, as illustrated by Fig. (2). See DIFFERENTIAL EQUATION; LAPLACE'S DIFFERENTIAL EQUATION.

Addition formula. One of the most important formulas satisfied by Legendre polynomials is the addition formula, Eq. (10). The functions $P_n{}^m(x)$ are

$$P_n(\cos\theta\cos\varphi + \sin\theta\sin\varphi\cos\chi)$$

$$= P_n(\cos\theta)P_n(\cos\varphi)$$

$$+ 2\sum_{m=1}^{n} \frac{(n-m)!}{(n+m)!} P_n{}^m(\cos\theta)P_n{}^m(\cos\varphi)\cos m\psi \tag{10}$$

defined by Eq. (11). These functions, often called

$$P_n{}^m(x) = (-1)^m(1-x^2)^{m/2}\frac{d^m}{dx^m}P_n(x) \tag{11}$$

$$-1 < x < 1 \qquad m = 0, 1, \ldots, n$$

associated Legendre functions, satisfy differential equation (12).

$$(1-x^2)\frac{d^2y}{dx^2} - 2x\frac{dy}{dx} + \left\{n(n+1) - \frac{m^2}{1-x^2}\right\}y = 0 \tag{12}$$

General Legendre functions. In work of F. G. Mehler on electrical distribution with conical symmetry, solutions arose for Eq. (12) with $m=0$ and $n = -\frac{1}{2} + it$, where t is a parameter that can assume any real value. When Laplace's equation is solved by separation of variables in toroidal coordinates, this equation occurs with $n = \nu - \frac{1}{2}$ and $m = \mu$, where μ and ν are separation parameters. Thus there are reasons for considering solutions to Eq. (12) when n and m are arbitrary complex numbers. In the general context, n and m are usually called ν and μ. $P_\nu{}^\mu(z)$ is defined by Eq. (13) for

$$P_\nu{}^\mu(z) = \frac{1}{\Gamma(1-\mu)}\left(\frac{z+1}{z-1}\right)^{\mu/2} \cdot$$

$$_2F_1[-\nu, \nu+1; 1-\mu; (1-z)/2] \tag{13}$$

$|z-1| < 2$, with the cut $z < 1$ removed. By analytic continuation, this function is extended to the plane cut from 1 to $-\infty$ and then defined for $-1 < z < 1$ by Eq. (14). There is a second solution to dif-

$$P_\nu{}^\mu(x) = \lim_{\epsilon \to 0^+} \frac{P_\nu{}^\mu(x + i\epsilon) + P_\nu{}^\mu(x - i\epsilon)}{2} \tag{14}$$

ferential equation (12) which is called $Q_\nu{}^\mu(z)$. The easiest way to carry out the analytic extension of Eq. (13) and to define $Q_\nu{}^\mu(z)$ is to use the general theory of hypergeometric functions. Legendre functions are just appropriate algebraic functions times a hypergeometric function of the type that has a quadratic transformation. The theory of Legendre functions up to but not including the addition formula can be developed very easily in this fashion.

Addition formulas are best derived by interpreting the functions as spherical harmonics and using the rotation groups which operate on spheres and then analytically continuing the resulting formulas in the appropriate parameters. Addition formula (10) can be extended not only to the general Legendre function of the first kind, $P_\nu{}^\mu(z)$, but to the general function of the second kind $Q_\nu{}^\mu(z)$. A number of results can be obtained from this, such as Eq. (15). This is an extension of the formula $\cos^2 x + \sin^2 x = 1$ to Legendre functions.

$$[P_n(x)]^2 + \frac{4}{\pi}[Q_n(x)]^2$$

$$= \frac{4}{\pi^2}\int_1^\infty Q_n[x^2 + (1-x^2)z](z^2-1)^{-\frac{1}{2}}dz \tag{15}$$

$$-1 < x < 1$$

Gegenbauer functions. There is a related set of functions, called Gegenbauer and associated Gegenbauer functions, which arise as spherical harmonics on higher-dimensional spheres. They are really equivalent to general Legendre functions, since they are algebraic functions times hypergeometric functions which have a quadratic transformation. The only difference is that different algebraic functions are used as multipliers since the functions satisfy different second-order differential equations. A completely analogous theory has been developed for them. There are two extreme cases of Gegenbauer functions which are very important, and in many respects their theory is easier to develop and is used as a model to suggest further developments. One is connected with the unit sphere in two-space (or the circle $x^2 + y^2 = 1$), and the resulting functions are $\cos\lambda\theta$ and $\sin\lambda\theta$.

The other comes from an "infinite dimensional sphere" (actually an appropriate limit of finite dimensional spheres), and the spherical functions are Hermite polynomials and their extensions to Hermite functions. *See* ORTHOGONAL POLYNOMIALS.   [RICHARD ASKEY]

Bibliography: A. Erdélyi et al., *Higher Transcendental Functions*, vols. 1–3, 1953–1955; E. W. Hobson, *The Theory of Spherical and Ellipsoidal Harmonics*, 1931; C. Müller, *Spherical Harmonics*, 1966; L. Robin, *Fonctions spheriques de Legendre et fonctions spheroidales*, vols. 1–3, 1957–1959; N. Ja. Vilenkin, *Special Functions and the Theory of Group Representations*, 1968.

Legume

A dry, dehiscent fruit derived from a single, simple pistil. When mature, it splits along both dorsal and ventral sutures into two valves. The term also designates any plant of the order Rosales that bears this type of fruit.

The family Leguminosae characteristically contains a single row of seeds attached along the lower or ventral suture of the fruit. The seeds are highly nutritious and several species of legumes furnish a large amount of food for both man and animals. Nitrogen-fixing bacteria living symbiotically in the roots of legumes accumulate nitrogenous materials and are beneficial for the soil.

Some more common and important legumes are alfalfa, beans, clovers, kudzu, lespedeza, locust, peas, peanuts, soybeans, and vetch. See articles on these individual legumes. *See* FRUIT (BOTANY); NITROBACTERIACEAE; ROSALES.

[PERRY D. STRAUSBAUGH/EARL L. CORE]

Legume forages

Plants of the legume family used for livestock feed, grazing, hay, or silage. Legume forages are usually richer in protein, calcium, and phosphorus than other kinds of forages, such as grass. The production, preservation, and use of forage legumes require special skills on most soils. One important requirement is a supply of the needed symbiotic nitrogen-fixing bacteria if these are not already in the soil; commercial cultures for various strains of these bacteria can be purchased and applied to the legume seed just before planting. Additional lime and commercial fertilizers may be needed on all except fertile soils. Protection from weeds, injurious insects, diseases, and other harmful influences is often required. *See* LEGUME; NITROGEN FIXATION.

Legume forages may be preserved for future use as dry hay or silage. To obtain high quality, most legume hay crops must be harvested before the mature stage and thoroughly cured without the loss of leaves; high-moisture hay spoils in warm weather. Legumes are more difficult to preserve as silage than grasses due to their high protein content. However, the addition of a high-carbohydrate material, such as molasses, or, as is usually preferred, the mixing of grass and legume are helpful measures. Legume silage often has an objectionable odor if fermentation is not sufficiently rapid to prevent breakdown of the proteins.

Much of the legume forage is grazed. Some crops can be grazed continuously without injury while others require intermittent grazing with rest periods to permit recovery or frequent very light use. A lush growth of a palatable legume crop, such as white clover or alfalfa, often causes bloat in grazing animals, but this may be prevented by restricting intake. The use of drenches to prevent or break up foaming in the stomach act as temporary cures.

Alfalfa is the most important legume forage crop in the United States; it is used mainly for hay but is often grazed. White clover and the annual lespedezas are the most extensively grown legumes for grazing particularly in the southeastern United States. Red clover was an important crop prior to 1930 but is minor now. About a dozen other species of legumes are used for cultivated forage in the United States and a large number for range grazing. *See* ALFALFA; CLOVER; COVER CROPS; COWPEA; KUDZU; LESPEDEZA; LUPINE.

[PAUL TABOR]

Leishmaniasis

An infection caused by any one of three pathogenic hemoflagellates of the genus *Leishmania*, transmitted to man by blood-sucking sand flies (*Phlebotomus*).

Etiology. Three diseases are distinguished, representing different clinicopathological manifestations of *Leishmania* infections: visceral leishmaniasis, known also as kala azar, caused by *L. donovani*; cutaneous leishmaniasis (Oriental sore, Baghdad boil), caused by *L. tropica*; and mucocutaneous leishmaniasis (espundia), caused by *L. braziliensis*.

The three causative organisms are morphologically identical but may be distinguished by their invasiveness and tissue tropism. The flagellates (leptomonads), introduced into the body by the bite of an infected *Phlebotomus*, are engulfed by phagocytic cells. The parasites develop and multiply within the host cells, destroying them and causing proliferation of new phagocytic cells.

Pathology. In cutaneous leishmaniasis the parasites invade only the skin and subcutaneous tissue, causing deep ulcers which upon healing confer a lasting immunity. The disease is widespread in the Near and Middle East, stretching to Asiatic Russia. Mucocutaneous leishmania affects the mucous membranes of the nose and mouth causing severe ulceration and disfigurement. It is prevalent in some areas of Central and South America. Visceral leishmaniasis is a severe, generalized, and often fatal infection affecting the organs rich in reticuloendothelial cells. It is accompanied by fever, spleen and liver enlargement, anemia and leukopenia, skin pigmentation, and changes in plasma proteins. There are endemic areas in India and China, in Central and North Africa, and along the Mediterranean littoral, where it affects mainly the young (infantile kala azar). Sporadic cases are also found in South America. Dogs are the natural reservoir of the infection.

Diagnosis. The incubation period is usually prolonged (several months). Diagnosis can be made by serological tests, by the study of the typical fever curve, and, above all, by the demonstration of the parasites in stained smears from the bone marrow, spleen, or liver. The organisms are cultivated in appropriate media at room temperature, and by animal inoculation; the hamster is a most susceptible laboratory animal for research in leishmaniasis. *See* CLINICAL PATHOLOGY; PARASITOLOGY, MEDICAL; SEROLOGY.

Treatment. Antimony compounds (Neostibosan) and the aromatic diamidines (Stilbamidine) produce a cure. Preventive measures are mainly against sand flies. Insecticides and sand-fly nets are essential precautions in endemic areas. *See* DIPTERA; TRYPANOSOMATIDAE.

[MEIR YOELI]

Leitneriales

An order of flowering plants, division Magnoliophyta (Angiospermae) in the subclass Hamamelidae of the class Magnoliopsida (dicotyledons). The order consists of a single family, genus, and species (*Leitneria floridana*) of the southeastern United States. The plants are simple-leaved, dioecious shrubs with the flowers in catkins. The ovary is superior and pseudomonomerous, with a single ovule ripening into a small drupe. *See* FLOWER; HAMAMELIDAE; MAGNOLIOPHYTA; PLANT KINGDOM.

[ARTHUR CRONQUIST]

Lemming

The name applied to 11 species of rodents in the subfamily Microtinae, family Muridae. These animals (see illustration) have a northern circumpolar distribution. Lemmings have been extensively studied from the viewpoint of population dynamics, since their population increases cyclically, resulting in mass migration to the sea.

The Norway lemming (*Lemmus lemmus*) is found in the mountainous wastelands of northern Norway and Lapland. It is usually nocturnal and timid

The lemming, a small burrowing rodent with short legs and stout claws adapted for digging.

in its habits, except when a population explosion occurs with its resultant migration. Despite many reasons advanced to account for these migrations, the causative factors are not known. Cyclic variations in fertility may be a factor. Usually there are two litters of five offspring each year, but often four litters of two to eight offspring occur. The collard lemmings (*Dicrostonyx*) of Arctic Asia and America and the bog lemmings (*Synaptomys*) of North America have similar migrations.

L. trimucronatus is found in the Arctic regions of North America. It is a heavy-bodied, short-tailed rodent with short legs, large forefeet with a rudimentary thumb, and short hindfeet. The animal burrows extensively. This species has 16 teeth, with the dental formula I 1/1 C 0/0 Pm 0/0 M 3/3, and eats vegetation. The sexes are identical in

appearance. The female may have from two to four litters a year with four to six offspring in each. *See* MUSKRAT; RODENTIA.

[CHARLES B. CURTIN]

Lemniscate of Bernoulli

A curve shaped like the figure eight (see illustration), referred to by Jacques Bernoulli in 1694. Let F_1, F_2 be points of a plane π, with $F_1F_2 = 2a$, $a > 0$. The locus of a point P of π which moves so that $PF_1 \cdot PF_2 = b^2$, where b is a positive constant, is

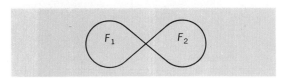

Curve known as a lemniscate.

called an oval of Cassini. The lemniscate is obtained when $b = a$. Its equation in rectangular coordinates is $(x^2 + y^2)^2 = a^2(x^2 - y^2)$ and in polar coordinates $\rho^2 = a^2 \cos 2\theta$. It is the locus of the point of intersection of a variable tangent to a rectangular hyperbola with the line through the center perpendicular to the tangent. The area enclosed by the lemniscate $\rho^2 = a^2 \cos 2\theta$ is a^2. *See* ANALYTIC GEOMETRY.

[LEONARD M. BLUMENTHAL]

Lemon

The fruit *Citrus limon*, commercially the most important of the acid citrus fruits. Its origin is somewhat obscure, but evidence points to southern China and upper Burma as its native home. There are many varieties with wide diversity in tree vigor, size, shape, and character of fruit. Many of these diverse lemon types are no doubt hybrids with other citrus. The rough lemon is probably such a hybrid and, though scarcely edible, it is quite important because it is frequently used as a rootstock on which other citrus is budded. *See* BUDDING.

The true lemons (see illustration) are small evergreen trees, the leaves of which have petioles with very narrow wings and give off a lemon odor when crushed. The fruits are medium-sized and elongated, with 9 or 10 segments and few seeds. *See* FRUIT (BOTANY).

No other citrus fruit has such a wide variety of

Fruit and branch of lemon tree (*Citrus limon*).

uses. Lemon juice, very high in vitamin C, is used in beverages and to garnish meats and fish. It has many culinary uses, especially in pies, cakes, ices, candies, jellies, and marmalades. Citric acid, pectin, and lemon oil are by-products of the fruit. The average yearly value of the lemon crop in the United States for 1965–1967 was approximately $57,-000,000. *See* CITRIC ACID; FAT AND OIL, EDIBLE; FRUIT, TREE; PECTIN; SAPINDALES.

[FRANK E. GARDNER]

Lemur

A primate of the family Lemuridae; all 16 species are indigenous to Madagascar. All lemurs have long tails, foxlike faces, and scent glands on the shoulder region and wrists. By constantly running its tail through its arms, the lemur creates a scent flag.

Lemurs are largely dependent upon fruit as their diet; they also eat insects and eggs. They have 36 teeth and the dental formula is I 2/2 C 1/1 Pm 3/3 M 3/3. A litter containing one or two young is born in the early spring after a gestation period of 2 months. The young cling to the underside of the mother for about the first 3 weeks and are then carried on her back for a period of time. An interesting behavioral feature of the lemur is that the young are raised in communal nurseries and cared for by females other than the mother.

Of all the species, the ring-tailed lemur (*Lemur catta*) is probably the best known (see illustration). Unlike most lemurs, which inhabit forests, it is found among rocks in the southern part of Madagascar. It is vocal and gregarious, participates in such social activities as grooming, and is active diurnally.

The ruffed lemur (*L. variegatus*), the largest species, is on the verge of extinction. The brown lemur (*L. fulvus*) is a dull-colored animal, difficult to distinguish from other species such as the mongoose lemur (*L. mongos*) and the red-bellied lemur (*L. rubriventer*). The dwarf lemurs make up a subfamily of the Lemuridae. Included are species such as *Microcebus murinus*, the smallest living primate. It is a little larger than a mouse, squirrel-like in its activity, and nocturnal and insectivorous in habit. *See* DENTITION (VERTEBRATE); LORIS AND ALLIES; PRIMATES.

[CHARLES B. CURTIN]

Length

Extension in space. Length is one of the three fundamental physical quantities (the other two being mass and time), and therefore cannot be defined in terms of simpler quantities. It is measured by comparison with an arbitrary standard called the international meter, defined as 1,650,763.73 times the wavelength of the orange light emitted when a gas of pure krypton-86 is excited in an electrical discharge. Sticks or tapes are calibrated by direct comparison with krypton light by means of an interferometer; multiples and submultiples are also indicated by calibration marks. Lengths are calculated by direct comparison with such sticks or tapes. Decimal multiples and submultiples of the meter are frequently used in specifying length; in English-speaking countries the foot (0.3048 m) is a length unit. *See* METER (UNIT).

[DUDLEY WILLIAMS]

Lens, optical

A curved piece of ground and polished or molded material, usually glass, used for the refraction of light. Its two surfaces have the same axis. Usually this is an axis of rotation symmetry for both surfaces; however, one or both of the surfaces can be toric, cylindrical, or a general surface with double symmetry. The intersection points of the symmetry axis with the two surfaces are called the front and back vertices and their separation is called the thickness of the lens.

LENS TYPES

There are three lens types, namely, compound, single, and cemented. These are described in the following sections.

Compound lenses. A compound lens is a combination of two or more lenses in which the second surface of one lens has the same radius as the first surface of the following lens and the two lenses are cemented together. Compound lenses are used instead of single lenses for color correction, or to introduce a surface which has no effect on the aperture rays but large effects on the principal rays, or vice versa. Sometimes the term compound lens is applied to any optical system consisting of more than one element, even when they are not in contact. A group of lenses used together is a lens system. A symmetrical lens is a lens system consisting of two parts, each of which is the mirror image of the other. If one part is a mirror image of the other magnified m times, the system is called hemisymmetric. When $m = 1$, the system is often said to be holosymmetric.

Single lenses. The lens diameter is called the linear aperture, and the ratio of this aperture to the focal length is called the relative aperture. This latter quantity is more often specified by its reciprocal, called the f-number. Thus, if the focal length is 50 mm and the linear aperture 25 mm, the relative aperture is 0.5 and the f-number is $f/2$. *See* FOCAL LENGTH.

In precalculation formulas, the lens thicknesses (but not the separations of the lenses) can frequently be neglected. This leads to the convenient fiction of a thin lens.

If ρ_1 and ρ_2 are the front and back curvatures of a lens of refractive index n and thickness d, its power is given by Eq. (1). The curvature of the surface is the reciprocal of its radius.

$$\phi = (n-1)(\rho_1 - \rho_2) + \frac{d(n-1)^2}{n}\rho_1\rho_2 \qquad (1)$$

The distances from the back vertex to the back nodal point and to the back focal point, respectively, are given by Eq. (2). The last distance is often called the back focus, especially in photographic optics.

$$S'_N\phi = -(d/n)(n-1)\rho_1$$
$$S'_F\phi = 1 + S'_N\phi \qquad (2)$$

The bending of a lens is a change in the curvature of the two surfaces by the same amount. It does not change the power of a thin lens, which is $(n-1)(\rho_1 - \rho_2)$. Bending is an important tool of the designer, for it permits the replacing of one lens by another without changing the data of Gaussian optics.

LEMUR

Ring-tailed lemur, measuring 4 ft, including tail.

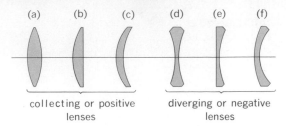

Fig. 1. Common lenses. (*a*) Biconvex. (*b*) Plano-convex. (*c*) Positive meniscus. (*d*) Biconcave. (*e*) Plano-concave. (*f*) Negative meniscus. (*From F. A. Jenkins and H. E. White, Fundamentals of Optics, 3d ed., McGraw-Hill, 1957*)

LENS, OPTICAL

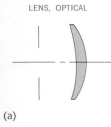

(a)

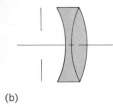

(b)

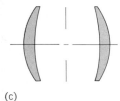

(c)

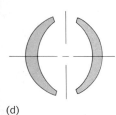

(d)

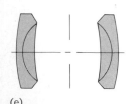

(e)

Fig. 2. Older camera lenses. (*a*) Meniscus. (*b*) Simple achromat. (*c*) Periskop. (*d*) Hypergon wide-angle. (*e*) Symmetrical achromat.

When thick lenses are involved, Gaussian optics remains constant only if both the powers of the thick lenses and the distance between the back nodal point of the first lens and the front nodal point of the second remain unchanged. Thus a bending of a thick lens should be accompanied by such an adjustment.

The optical center of a thick lens is the image of the nodal point produced inside the lens. All finite rays through the optical center emerge parallel to their respective directions at their entrance.

An optical center exists also in a hemisymmetric system. It is the point of symmetry which divides the separation of the two parts in the ratio $1/m$. If negative values of m are permitted, any single lens is a hemisymmetric system and the point dividing the thickness of the lens (the separation of the two vertices) in a ratio equal to the ratio of the two radii is the optical center of the lens.

A lens is said to be a collecting lens if $\phi > 0$ and a diverging lens if $\phi < 0$. When $\phi = 0$, the lens is afocal. Several types of collecting and diverging lenses are shown in Fig. 1.

The surfaces of most lenses are either spherical or planar, but nonspherical surfaces are used on occasion to improve the corrections without changing the power of the lens. *See* SPHERICAL AND ASPHERICAL SURFACES, OPTICAL.

A concentric lens is a lens whose two surfaces have the same center. If the object to be imaged is also at the center, its axis point is sharply imaged upon itself, and since the sine condition is fulfilled, the image is free from asymmetry. Such a lens can be used as an additional system to correct meridional errors.

Another type of lens consists of an aplanatic surface followed by a concentric surface, or vice versa. Such a lens divides the focal length of the original lens to which it is attached by n^2, thus increasing the f-number by a factor of n^2 without destroying the axial correction of the preceding system. It does introduce curvature of field which makes a rebalancing of the whole system desirable. *See* ABERRATION, OPTICAL.

Cemented lenses. Consider a compound lens made of two or more simple thin lenses cemented together. Let the power of the κth simple lens be ϕ_κ and its Abbe value ν_κ. The difference between the powers of the combination for wavelengths corresponding to C and F is given by Eq. (3),

$$\Phi_F - \Phi_C = \Phi/N = \Sigma\phi_\kappa/\nu_\kappa \quad (3)$$

where N may be considered to be the effective ν-value of the combination. The ν-values of optical

glasses vary between 25 and 70, with the ν-value of fluorite being slightly larger ($\nu = 95.1$). By using compound lenses, effective values of N can be obtained outside this range. Color correction is achieved as N becomes infinite, so that $\Phi_F - \Phi_C = 0$. A lens so corrected is called an achromat. In optical design, it is sometimes desirable to have negative values of N to balance the positive values of the rest of the system containing collecting lenses. Such a lens is said to be hyperchromatic. A cemented lens corrected for more than two colors is said to be apochromatic. A lens corrected for all colors of a sizable wavelength range is called a superachromatic lens. *See* CHROMATIC ABERRATION; OPTICAL MATERIALS.

LENS SYSTEMS

Optical systems may be divided into four classes: telescopes, oculars (eyepieces), photographic objectives, and enlarging lenses. *See* EYEPIECE; MICROSCOPE, OPTICAL.

Telescope systems. A lens system consisting of two systems combined so that the back focal point of the first (the objective) coincides with the front focal point of the second (the ocular) is called a telescope. Parallel entering rays leave the system as parallel rays. The magnification is equal to the ratio of the focal length of the first system to that of the second.

If the second lens has a positive power, the telescope is called a terrestrial or Keplerian telescope and the separation of the two parts is equal to the sum of the focal lengths.

If the second lens is negative, the system is called a Galilean telescope and the separation of the two parts is the difference of the absolute focal lengths. The Galilean telescope has the advantage of shortness (a shorter system enables a larger field to be corrected); the Keplerian telescope has a real intermediate image which can be used for introducing a reticle or a scale into the intermediate plane.

Both objective and ocular are in general corrected for certain specific aberrations, while the other aberrations are balanced between the two systems. *See* TELESCOPE.

Photographic objectives. A photographic objective images a distant object onto a photographic plate or film. *See* PHOTOGRAPHY.

The amount of light reaching the light-sensitive layer depends on the aperture of the optical system, which is equivalent to the ratio of the lens diameter to the focal length. Its reciprocal is called the f-number. The smaller the f-number, the more light strikes the film. In a well-corrected lens (corrected for aperture and asymmetry errors), the f-number cannot be smaller than 0.5.

The larger the aperture (the smaller the f-number), the less adequate may be the scene luminance required to expose the film. Therefore, if pictures of objects in dim light are desired, the f-number must be small. On the other hand, for a lens of given focal length, the depth of field is inversely proportional to the aperture.

Since the exposure time is the same for the center as for the edge of the field, it is desirable for the same amount of light to get to the edge as gets to the center, that is, the photographic lens should have little vignetting.

The camera lens can be considered as an eye looking at an object (or its image), with the diaphragm corresponding to the eye pupil. The Gaussian image of the diaphragm in the object (image) space is called the entrance (exit) pupil. The angle under which the object (image) is seen from the entrance (exit) pupil is called the object (image) field angle. For most photographic lenses, the entrance and exit pupils are close to the respective nodal points; for such lenses, the object and the image field angles are equal.

In general, photographic objectives with large fields have small apertures; those with large apertures have small fields. The construction of the two types of systems is quite different. One can say in general that the larger the aperture, the more complex the lens system must be.

There exist cameras (so-called pinhole cameras) that do not contain any lenses. The image is then produced by optical projection. The aperture in this case should be limited to $f/22$.

Other types of lenses. A single meniscus lens, with its concave side toward the object and with its stop in front at its optical center, gives good definition at $f/16$ over a total field of 50° (Fig. 2*a*). The lens can be a cemented doublet for correcting chromatic errors (Fig. 2*b*). For practical reasons, a reversed meniscus with the stop toward the film is often used.

Combining two meniscus lenses to form a symmetrical lens with central stop makes it possible to correct astigmatic and distortion errors for small apertures as well as large field angles (Fig. 2*c*).

The basic type of wide-angle objective is the Hypergon, consisting of two meniscus lenses concentric with the regard to stop (Fig. 2*d*). This type of system can be corrected for astigmatism and field curvature over a total field angle of 180° but it can only be used for a small aperture ($f/12$), since it cannot be corrected for aperture errors. The aperture can be increased to $f/4$ at the expense of field angle by thickening and achromatizing the meniscus lenses and adding symmetrical elements in the center or at the outside of the basic elements.

Two positive achromatic menisci symmetrically arranged around the stop lead to the aplanatic type of lens (Fig. 2*e*). This type was spherically and chromatically corrected. Since the field could not be corrected, a compromise was achieved by balancing out sagittal and meridional field curvature so that one image surface lies in front and the other in back of the film.

Anastigmatic lenses. The discovery of the Petzval condition for field correction led to the construction of anastigmatic lenses, for which astigmatism and curvature of field are corrected. Such lenses must contain negative components.

The Celor (Gauss) type consists of two airspaced achromatic doublets, one on each side of the stop (Fig. 3*a*). The Cooke triplet combines a negative lens at the aperture stop with two positive lenses, one in front and the other in back. It is called a Tessar (Fig. 3*b*) if the last positive lens is a cemented doublet, or a Heliar if both positive lenses are cemented. The Dagor type consists of two lens systems that are nearly symmetrical with respect to the stop, each system containing three or more lenses (Fig. 3*c*).

Modern lenses. To increase the aperture, the field, or both, it is frequently advantageous to replace one lens by two separated lenses, since the same power is then achieved with larger radii and this means that the single lenses are used with smaller relative apertures. The replacing of a single lens by a cemented lens changes the color balance, and thus the designer may achieve more favorable conditions. Moreover, the introduction of new types of glass (first the glasses containing barium, later the glasses containing rare earths) led to lens elements which for the same power have weaker surfaces and are of great help to the lens designer, since the errors are reduced.

Of modern designs the most successful are the Sonnar, a modified triplet, one form of which is shown in Fig. 4*a*; the Biotar (Fig. 4*b*), a modified Gauss objective with a large aperture and a field of about 24°; and the Topogon (Fig. 4*c*), a periscopic lens with supplementary thick menisci to permit the correction of aperture aberrations for a moderate aperture and a large field. One or two plane-parallel plates are sometimes added to correct distortion.

Special objectives. It is frequently desirable to change the focal length of an objective without changing the focus. This can be done by combining a fixed near component behind the stop with an exchangeable set of components in front of the stop. The designer has to be sure that the errors of the two parts are balanced out regardless of which front component is in use. For modern ways to change the magnification *see* ZOOM LENS.

The telephoto objective is a specially constructed objective with the rear nodal point in front of the lens, to combine a long focal length with a short back focus. *See* TELEPHOTO LENS.

The Petzval objective is one of the oldest designs (1840) but one of the most ingenious. It consists in general of four lenses ordered in two pairs widely separated from each other. The first pair is cemented and the second usually has a small air space. For a relatively large aperture, it is excellently corrected for aperture and asymmetry errors, as well as for chromatic errors and distortion. It is frequently used as a portrait lens and as a projection lens because of its sharp central definition. Astigmatism can be balanced but not corrected.

Enlarger lenses and magnifiers. The basic type of enlarger lens is a holosymmetric system consisting of two systems of which one is symmetrical with the first system except that all the data are multiplied by the enlarging factor m. When the object is in the focus of the first system, the combination is free from all lateral errors even before correction. A magnifier in optics is a lens that enables an object to be viewed so that it appears larger than its natural size.

The magnifying power is usually given as equal to one-quarter of the power of the lens expressed in diopters. *See* DIOPTER; MAGNIFICATION.

Magnifying lenses of low power are called reading glasses. A simple planoconvex lens in which the principal rays are corrected for astigmatism for a position of the eye at a distance of 10 in. is well suited for this purpose, although low-power magnifiers are often made commercially with biconvex lenses. A system called a verant consists of two

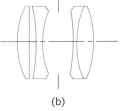
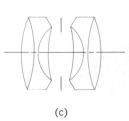

Fig. 3. Types of anastigmats. (*a*) Celor. (*b*) Tessar. (*c*) Dagor.

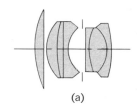
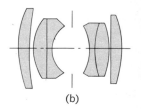
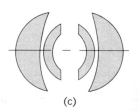

Fig. 4. Modern camera lenses. (*a*) Sonnar. (*b*) Biotar. (*c*) Topogon.

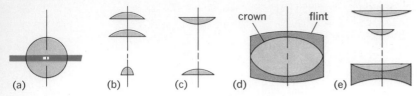

Fig. 5. Typical magnifiers. (a) Sphere with equatorial diaphragm. (b, c) Planoconvex lens combinations. (d) Steinheil triple aplanat. (e) Chevalier type.

lenses corrected for color, astigmatism, and distortion. It is designed for stereoscopic vision at low magnification.

For higher magnifications, many forms of magnifiers exist. One of the basic designs has the form of a full sphere with a diaphragm at the center, as shown in Fig. 5a. The sphere may be solid or it may be filled with a refracting liquid. When it is solid, the diaphragm may be formed by a deep groove around the equator. Combinations of thin planoconvex lenses as shown in Fig. 5b and c are much used for moderate powers. Better correction can be attained in the aplanatic magnifier of C. A. Steinheil, in which a biconvex crown lens is cemented between a pair of flint lenses (Fig. 5d).

A design by C. Chevalier (Fig. 5e) aims for a large object distance. It consists of an achromatic negative lens combined with a distant collecting front lens. A magnifying power of up to $10 \times$ with an object distance up to 3 in. can be attained.

[MAX HERZBERGER]

Bibliography: K. Brandt, *Das Photo Objectiv*, 1956; B. Carroll and A. E. Murray, *Photographic Lenses*, 1965; A. Cox, *A System of Optical Design*, 1964; M. Herzberger, *Modern Geometrical Optics*, 1958; R. Kingslake, *Applied Optics and Optical Engineering*, vol. 1, 1965; R. Kingslake, *Lenses in Photography*, rev. ed., 1963.

Lentil

A semiviny annual legume with slender tufted and branched stems 18–22 in. long (Fig. 1). The lentil plant (*Lens esculenta*) was one of the first plants brought under cultivation. They have been found in the Bronze Age ruins of the ancient lake dwellings of St. Peter's Island, Lake of Bienne, Switzerland. Lentils have been discovered in Poland dating back to the Iron Age. In the Bible the "red pottage" for which Esau gave up his birthright to his brother, Jacob, was probably lentil soup. Large-seeded lentils originated in the Mediterranean region; medium-sized lentils originated in the inner mountains of Asia Minor; and Afghanistan was the original home of the smallest-seeded lentils. *See* LEGUME.

Production. The world's lentil production is centered in Asia, with nearly two-thirds of the production from India, Pakistan, Turkey, and Syria. Whitman and Spokane counties in Washington, and Latah, Benewah, and Nez Perce counties in Idaho grow about 95% of the lentils produced in the United States. Total production for the United States averaged about 60,000,000 lb for 1962–1967.

Description. Lentil leaves are pinnately compound and generally resemble vetch. The plant has tendrils similar to those of pea plants.

The seeds grow in short broad pods, each pod producing two or three thin lens-shaped seeds (Fig. 2). Seed color varies from yellow to brown and may be mottled, although mottled seed are not desirable for marketing. The lentil flowers, ranging in color from white to light purple, are small and delicate and occur at many different locations on the stems.

Lentil seed is used primarily for soups but also in salads and casseroles. Lentils are more digestible than meat and are used as a meat substitute in many countries.

Culture. Lentils require a cool growing season; they are injured by severe heat. Therefore, they are planted in April, when soil moisture is adequate and temperatures are cool. A fine firm seedbed is required; the land is usually plowed in the fall and firmed by cultivation in the spring before seeding. Lentils are usually planted in rotation with winter wheat. Seeds are planted in 7–12 in. rows at depths of $\frac{1}{2}$ to 1 in., on an average of 60–75 lb/acre. Applications of sulfur, molybdenum, and phosphorus are used to increase yields.

Wild oats often infest lentil fields, but chemicals are available to aid weed control. Cowpea and black-bean aphids are the two most important insect pests on lentils. Predators such as ladybird beetles and syrphid-fly larvae usually keep these insects under control. However, when populations of predators are too low to control the aphids, chemical insecticides are used.

Harvesting. Lentils are mowed or swathed when the vines are green and the pods have a golden

Fig. 1. Lentil plant showing growth habit.

Fig. 2. Lens-shaped lentil seeds.

color. About 10 days later, lentils are ready to combine harvest using the same combines that are used for wheat, oats, and barley. A pick-up attachment picks up the material from the windrows. The combines must be operated at a maximum speed of 1½ mph to prevent loss of, and damage to, the lentil seed. Average yields are about 900 lb/acre. *See* AGRICULTURE, MACHINERY IN; AGRICULTURE, SOIL AND CROP PRACTICES IN.

[KENNETH J. MORRISON]
Bibliography: F. M. Entenmann, K. J. Morrison, and V. E. Youngman, *Growing Lentils in Washington*, Wash. State Univ. Ext. Bull. no. 590, 1968; E. T. Field and G. E. Marousek, *Lentil Production and Marketing in the U.S.A.: A Description and Trends*, Idaho Agr. Res. Progr. Rep. no. 128, 1968; V. E. Youngman, Lentils: A pulse of the Palouse, *Econ. Bot.*, 22:135–139, 1968.

Lenz's law

A law of electromagnetism which states that, whenever there is an induced electromotive force (emf) in a conductor, it is always in such a direction that the current it would produce would oppose the change which causes the induced emf. If the change is the motion of a conductor through a magnetic field, as in the illustration, the induced current must be in such a direction as to produce a force opposing the motion. If the change causing the emf is a change of flux threading a coil, the induced current must produce a flux in such a

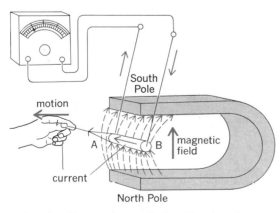

Induced emf in a moving conductor. Direction of current induced in wire AB is indicated by the arrows. (*From M. W. White, K. V. Manning and R. L. Weber, Practical Physics, 2d ed., McGraw-Hill, 1955*)

direction as to oppose the change. That is, if the change is an increase of flux, the flux due to the induced current must be opposite in direction to the increasing flux. If the change is a decrease in flux, the induced current must produce flux in the same direction as the decreasing flux.

Lenz's law is a form of the law of conservation of energy, since it states that a change cannot propagate itself. *See* CONSERVATION OF ENERGY; INDUCTION, ELECTROMAGNETIC.

[KENNETH V. MANNING]

Leo

The Lion, in astronomy, is a magnificent zodiacal constellation appearing during spring and early summer. It is the fifth sign of the zodiac. Leo is well defined and bears a close resemblance to the creature it represents. The head is outlined by a group of six stars, $\alpha,\eta,\gamma,\zeta,\mu$, and ϵ, called the Sickle. The first-magnitude star, Regulus (Little Ruler), forms the handle of the sickle. Three stars to the east, β,δ, and θ, forming a small triangle, constitute the Lion's haunches, with the bright star, Denebola (tail of the lion), a navigational star (see

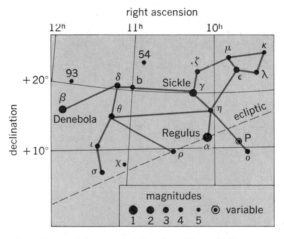

Line pattern of the constellation Leo. The grid lines represent the coordinates of the sky. Right ascension (E-W) in hours, and declination (N-S) in degrees, corresponding to the longitude and latitude of the Earth. The apparent brightness, or magnitudes, of the stars is shown by the sizes of the dots, which are graded by appropriate numbers as indicated.

illustration). Associated with this constellation are the famous Leonids shower of meteors, which can be seen radiating from Leo in November of each year and appearing especially brilliant at intervals of about 33 years. *See* CONSTELLATION; METEOR.

[CHING-SUNG YU]

Lepadomorpha

A suborder of the Thoracica. These barnacles have a peduncle and a capitulum which is usually protected by calcareous plates (Fig. 1). Caudal furca and filamentary appendages often are present. These crustaceans are either hermaphroditic or the sexes may be separate.

Eight families are differentiated, based in part on the number and structure of the plates. The most primitive family, Scalpellidae, has more than five plates. The other families have five or less plates, which vary both in number, from five to none, and in size.

Males occur in some species of Scalpellidae and Iblidae. These vary from small replicas of the larger form to mere sacs. Usually, the least modified species are associated with hermaphrodites and the most modified, with females.

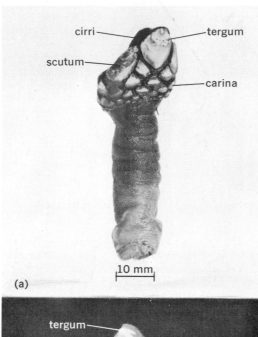

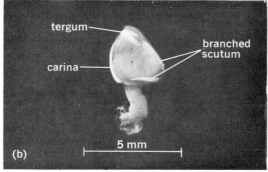

Fig. 1.   Morphology of Lepadomorpha. (a) *Mitella polymerus* Sowerby. A primitive Lepadomorpha, with numerous lateral plates and calcareous scales on peduncle (*from D. P. Henry, The Cirripedia of Puget Sound with a key to the species, Univ. Wash. Publ. Oceanogr., 4(10):1–48, 1940*). (b) *Octolasmis lowei* (Darwin) from crab gills.

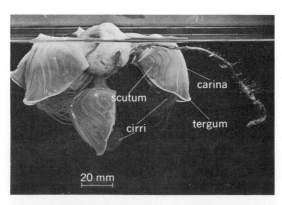

Fig. 2.   *Lepas fascicularis* Ellis and Solander, on float formed by barnacles after object (*Velella*, feather, and the like) on which cyprids settled becomes unable to supply adequate buoyancy. (*Photograph by D. P. Wilson*)

Many species attach to floating objects (Fig. 2) or to motile animals; others attach to fixed objects or animals, often in deep water. The species of one genus excavates burrows in calcareous rocks and corals. One species approaches parasitism, in that the peduncle is deeply embedded in the skin of a shark. *See* THORACICA.   [DORA P. HENRY]

Lepidocentroida

This name was proposed by T. Mortensen (1934) for an assemblage of echinoids now regarded as polyphyletic. J. Durham and R. Melville (1957) refer the included families to Echinocystitoida and Echinothurioida.   [HOWARD B. FELL]

Lepidodendrales

Giant club mosses, an order of extinct lycopods consisting primarily of arborescent forms, which constituted an important element of the Carboniferous flora. The group is characterized by dichotomous branching, by spirally arranged microphyllous leaves, by the production of small amounts of secondary vascular tissues relative to the total bulk of the plant (several species are not known to have produced secondary tissues), by the possession of a ligule, and by heterospory (bearing spores of two types). In at least two genera heterospory attained a level of specialization approaching the seed habit. Several genera are characterized by stems that bear raised, diamond-shaped mounds of tissue called leaf cushions, to which the leaves were attached. The order had a world-wide distribution and existed from Late Devonian through Permian times. Well-known genera are *Lepidodendron*, *Lepidophloios*, *Sigillaria*, *Bothrodendron*, *Stigmaria*, *Lepidostrobus*, *Lepidocarpon*, *Miadesmia*, and *Mazocarpon*. *See* LYCOPODIOPSIDA.

Vegetative morphology. The arborescent genera were large trees (Fig. 1), some attaining heights in excess of 100 ft. The basal parts consisted of dichotomous systems of rhizomelike organs called *Stigmaria*, originating as nearly horizontal branches at the base of the trunk. These organs, which bore numerous slender roots (Fig. 2), had a stemlike structure. Stigmarian axes were borne by *Lepidodendron*, *Lepidophloios*, *Bothrodendron*, *Sigillaria*, and possibly others.

Fig. 1.   Reconstructed *Lepidodendron*. (*From M. Hirmer, Handbuch der Paläobotanik, R. Oldenbourg, Verlag, 1927*)

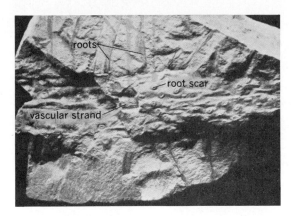

Fig. 2. Fossil of stemlike sigmarian axis with attached roots. (*From J. Walton, An Introduction to the Study of Fossil Plants, A. and C. Black, 1940*)

The tall, straight trunks usually branched by unequal dichotomy into a crown of successively smaller branches (Fig. 1). *Sigillaria*, however, branched only slightly, or not at all. Anatomically, the stem (Fig. 3) contained a solid, or tubular, central strand of primary xylem surrounded by primary phloem. Leaf traces branched from the surface of the central strand and frequently were accompanied by two parenchymatous parichnos strands of unknown function. In many species secondary xylem and phloem were produced. Frequently a wide peripheral zone of elongated, thick-walled cells was produced by meristematic activity within the cortex. This secondary cortex, or periderm, was the major supporting tissue in the arborescent species since the secondary wood comprised only about one-tenth to one-fifth of the total bulk of the stem. The inner cortical tissues are frequently lacking in fossils. *See* MERISTEM, LATERAL; STELE.

Leaf cushions are a prominent feature of some genera (Figs. 4 and 5a). Those from which leaves have fallen are usually characterized by a leaf scar, a ligule pit scar, and in *Lepidodendron* a vertical ridge beneath the leaf scar called the keel (Fig. 5a). Abscission of the leaf in *Lepidophloios* and *Sublepidophloios* did not occur flush with the surface of the cushion but from an outward extension called the leaf base (Fig. 4). The long axis of the cushion in *Lepidophloios* is horizontally oriented in contrast to the vertical orientation of *Lepidodendron*. Some genera, such as *Bothrodendron* and *Sigillaria*, either lacked leaf cushions (Fig. 5b) or bore slightly raised ones. Stem compressions of *Sigillaria* (Fig. 5b) are easily distinguished from *Lepidodendron* (Fig. 5a) by the arrangement in the former of leaf scars in vertical rows.

The linear to awl-shaped leaves varied from about 1 cm to 1 m in length and usually contained a single vein. However, leaves of *Sigillaria*, which also attained the greatest length, may have always contained two veins, at least in their bases. On vegetative parts of these plants, the ligule, a small unvascularized appendage, was located above the leaf in a depression called the ligule pit (Fig. 4). In strobili, the ligule was located on the sporophyll just distal to the sporangium (Fig. 6a).

Reproductive structures. The Lepidodendrales were heterosporous. Strobili named *Lepidostrobus* were produced by *Lepidodendron* and *Lepidoph-*

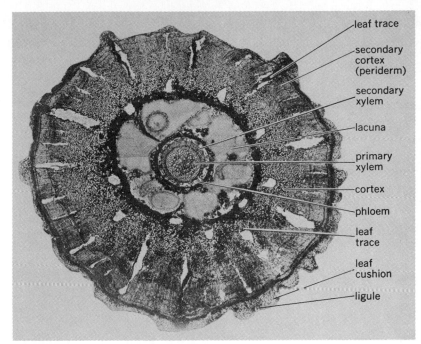

Fig. 3. *Lepidodendron vasculare*. Transverse section of the stem showing lacunae, formed by breakdown of cortical tissue. (*From M. Hovelacque, Recherches sur lepidodendron selaginaides, Stern. Mem. Linn. Soc. Normandie, 2d ser., vol. 17, 1892*)

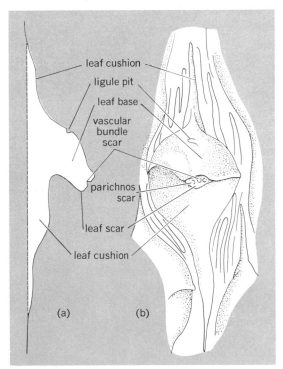

Fig. 4. *Sublepidophloios ventricosus*. Leaf cushions shown in (a) sectional view, and (b) frontal view. (*From C. A. Hopping, A note on the leaf cushions of a species of Palaeozoic arborescent lycopod, Proc. Roy. Soc. Edinburgh, sec. B, vol. 66(1), pt. 1, 1956*)

LEPIDODENDRALES

Fig. 5. Stem compressions, showing leaf scar arrangement. (a) Lepidodendron aculeatum. (b) Sigillaria boblayi. (From A. Renier et al., Flore et Fauna Houillères de la Belgique, Mus. Roy. Hist. Natur. Belg., 1938)

loios, *Lepidocarpon* by *Lepidodendron*, and *Bothrodendrostrobus* by *Bothrodendron*. *Sigillariostrobus* and *Mazocarpon* were strobili of *Sigillaria*. The plant which bore *Miadesmia* is unknown. *Lepidostrobus* was bisexual, containing both megaspores

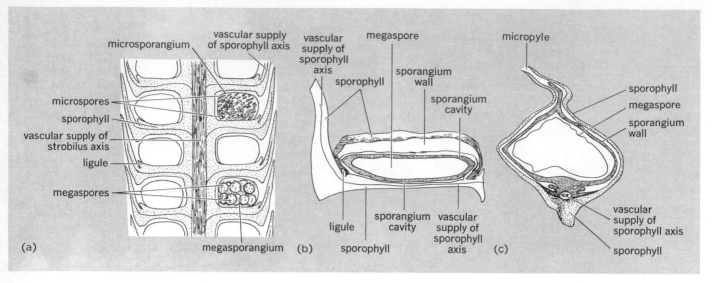

Fig. 6. Reproductive structures of Lepidodendrales. (a) *Lepidostrobus*, idealized drawing of a longitudinal section of part of a strobilus (*from C. A. Arnold, An Introduction to Paleobotany, McGraw-Hill, 1947*). (b) *Lepidocarpon ioense*, reconstruction of a sporophyll and spor-angium, cut away along a median longitudinal plane and (c) cut at a right angle to the long axis (*from J. H. Hoskins and A. T. Cross, A consideration of the structure of Lepidocarpon Scott, based on a new strobilus from Iowa, Amer. Midland Natur., vol. 25, 1941*)

and microspores (Fig. 6a). *Lepidocarpon* produced either all microspores or all megaspores. The horizontally elongated megasporangium (Fig. 6b and c) was enclosed by the sporophyll, sometimes called an integument. The cavity formed by this integument had a slit (micropyle) opening along the upper surface. One of the four megaspores developed into the gametophyte; the remaining three atrophied. Archegonia have been observed in the gametophytes, and microspores were discovered within the sporophyll. Presumably fertilization occurred while the gametophyte and sporangium were contained within the enclosing sporophyll, but no embryos have been found. These structures are frequently called seeds, although there is little evidence that the so-called integument is more than a highly specialized sporophyll. *Miadesmia* produced even more highly specialized seedlike structures. *Mazocarpon* differed from *Lepidocarpon* in producing eight megaspores, several or all of which were functional, embedded in a mass of parenchymatous tissue. The megasporangium and sporophyll were shed as a unit. *Mazocarpon* and *Sigillariostrobus* (a compression fossil), which are probably synonymous, were borne laterally on the trunk or large branches of *Sigillaria*; *Lepidodendron* strobili were probably always terminal.

The lepidodendralean genera were common inhabitants of the extensive Pennsylvanian coal swamps and are found commonly as fossils in coal balls and in shales associated with coal seams. *See* COAL BALLS; PALEOBOTANY. [CHARLES B. BECK]

Bibliography: C. A. Arnold, *An Introduction to Paleobotany*, 1947; T. Delevoryas, *Morphology and Evolution of Fossil Plants*, 1962; H. Scott, *Studies in Fossil Botany*, 3d ed., 1962; J. Walton, *An Introduction to the Study of Fossil Plants*, 2d ed., 1953.

Lepidolite

A mineral of variable composition that is also called lithium mica and lithionite, $K_2(Li,Al)_{5-6}(Si_{6-7},Al_{2-1})O_{20-21}(F,OH)_{3-4}$. Rubidium, Rb, and cesium, Cs, may replace potassium, K; small amounts of Mn, Mg, Fe(II), and Fe(III) normally are present; and the OH/F ratio varies considerably. Polithionite is a silicon- and lithium-rich, and thus aluminum-poor, variety of lepidolite.

Lepidolite is uncommon, occurring almost exclusively in structurally complex granitic pegmatites, commonly in replacement units. Common associates are quartz, cleavelandite, alkali beryl, and alkali tourmaline. Lepidolite is a commercial source of lithium, commonly used directly in lithium glasses and other ceramic products. Important deposits occur in the Karibib district of South-West Africa and at Bikita, Rhodesia.

The structural modifications show some correlation with lithium content: the six-layer monoclinic form contains 4.0–5.1% Li_2O; the one-layer monoclinic, 5.1–7.26% Li_2O. A three-layer hexagonal form is also found. There is a compositional gradation to muscovite, intermediate types being called lithian muscovite, containing 3–4% Li_2O, and having a modified two-layer monoclinic muscovite structure.

Lepidolite usually forms small scales or fine-grained aggregates (see illustration). Its colors,

Group of lepidolite crystals found in Zinnwald, Czechoslovakia. (*Specimen from Department of Geology, Bryn Mawr College*)

pink, lilac, and gray, are a function of the Mn/Fe ratio. It is fusible at 2, yielding the crimson (lithium) flame. It has a perfect basal cleavage. Hardness is 2.5–4.0 on Mohs scale; specific gravity is 2.8–3.0. *See* MICA; SILICATE MINERALS.

[E. WILLIAM HEINRICH]

Lepidoptera

The order of scaly-winged insects, including the butterflies, skippers, and moths. This is one of the largest orders in the class Insecta. It has more than 100,000 species, about 10,000 in North America, and between 125 and 175 families. The adults have a covering of hairs and flattened setae, or scales, on the wings, legs, and body, and are often beautifully colored. With minor exceptions, the adults are also characterized by two pairs of mem-

branous wings and sucking mouthparts, featuring a prominent, coiled proboscis, formed by the grooving together of the maxillary galeae. Butterflies and skippers usually fly in the daytime, and most moths are nocturnal. The adults usually take liquid food, such as nectar and juices of fruits. The caterpillars are almost always herbivorous. This article discusses the morphology, development, classification, and biological aspects of this insect order.

MORPHOLOGY

The most unusual feature of the head of the adult animal is the form of the mouthparts (Fig. 1b and c). The proboscis is extended by blood pressure created by the retraction of the stipites of the maxillae. Diagonal muscles within each proboscis

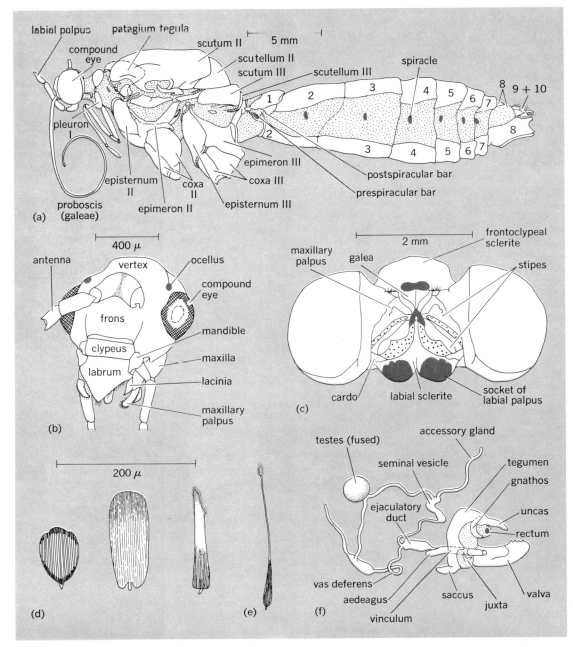

Fig. 1. Adult Lepidoptera anatomy. (a) Lateral view of *Danaus plexippus* L. (Nymphalidae) with vestiture, wings, and pterothoracic legs removed. (b) Front view of head of *Epimartyria* (Micropterygidae). (c) Ventral view of head of *D. plexippus* L. (d) Unspecialized scales. (e) Androconia. (f) Male genital system (diagrammatic).

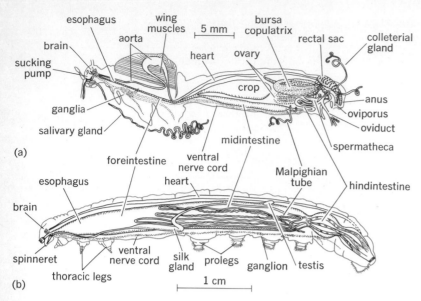

Fig. 2. Internal anatomy of *Danaus plexippus* L. (Nymphalidae), tracheal system and most of musculature omitted. (*a*) Adult female. (*b*) Larval male.

unit cause the proboscis to coil. The liquid food is sucked up by means of a muscular pump, formed from the pharynx, buccal cavity, and cibarium, a food pocket of the mouth cavity (Fig. 2). Ocelli are absent in many groups, such as the Hesperioidea and Papilionoidea. Antennae are quite variable in the Lepidoptera (Fig. 3).

Thorax. The prothorax is well developed in some lower groups, such as the Hepialoidea, but it is considerably smaller than the pterothoracic, or wing-bearing, segments, and is largely membranous in most Lepidoptera. The most prominent feature of the dorsum of the prothorax, in most groups, is a pair of protuberant sclerotized lobes, or patagia. In some butterflies, these structures are absent, and the prothorax consists of a sclerotized ring, connected dorsally to the mesothorax by only a small plate of the pronotum and ventrally by the thin sclerotized ribbon of the spinasternum.

The coxae of the prothorax are cylindrical and functional. Those of the pterothorax are fused with the thoracic capsule and are immobile. In each segment of the thorax the midventral suture, the discrimen, is represented internally by a thin, transparent lamella. The conformation of this lamella and its relationship to the furca, especially in the mesothorax, have proved to be characters of considerable systematic value in the butterflies. It is known to vary in the moths, being lower in the more generalized groups. The mesothorax is the largest of the three segments and may completely overlap the metathorax.

There is considerable variation throughout the order in the details of the sutures and sclerites of the integumental anatomy. In the butterflies, interesting variation has been found in the structure of the gut, but this has not been studied in the rest of the order.

Scales. The scales are very variable in form (Fig. 1*d*). Generally, they are flat, thin, sclerotized sacks, with striated outer surfaces. They have a basal pedicel which fits a socket in the wing membrane. In the males, some scales, the androconia,

have feathered tips which serve for the dissemination of scents (Fig. 1*e*). The vast spectrum of colors seen in the Lepidoptera can be grouped into two categories, pigmentary and structural colors. Pigmentary colors result from pigments which are present in the scales. Structural colors are the result of either fine surface ridges on the scales or layers within the cuticle, which interfere with or diffract the light. The structural colors are generally metallic or iridescent.

Wings. In most moths, the wings are coupled by a single spine formed by a number of fused setae, usually occurring in males, or a group of setae, usually found in females, which project forward from the base of the hindwing and are held by a clasp on the forewing. The spine is known as the frenulum and the clasp is the retinaculum (Fig. 4*a*). In the Homoneura, there is a lobe, the jugum, at the base of the forewing, which engages the hindwing, or the frenulum, when it is present (Fig. 4*b*). In some of the higher moths, skippers, and butterflies, the humeral angle of the hindwing is expanded and strengthened by one or more humeral veins (Fig. 4*c*). In these groups, the frenulum is usually lost and the wings are coordinated by the overlapping lobe.

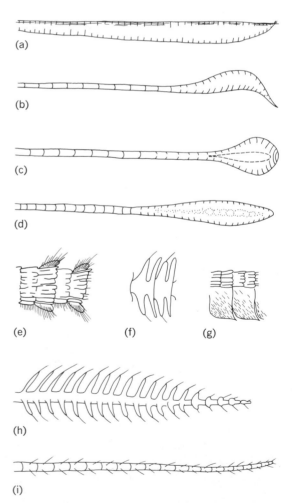

Fig. 3. Types of Lepidoptera antennae. (*a*) Fusiform. (*b*) Clubbed and hooked. (*c, d*) Clubbed. (*e*) Serrate and fasciculate, dorsal view. (*f*) Doubly bipectinate (quadripectinate). (*g*) Laminate, lateral view. (*h*) Bipectinate. (*i*) Simple and ciliate.

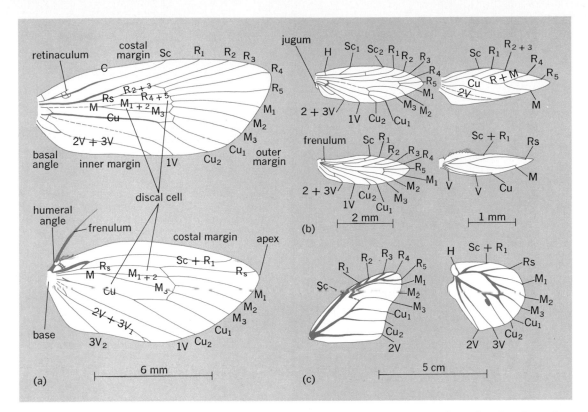

Fig. 4. Wing venation patterns, with the veins labeled. (a) Male *Acrolophus popeanellus* Clemens (Tineidae). (b) *Epimartyria* (Micropterygidae) and *Nepticula nyssae-foliella* Chambers (Nepticulidae). (c) *Danaus plexippus* L.

(Nymphalidae). C= costa; Sc= subcosta; R= radius; R$_s$= radial sector; M= media; Cu= cubitus; 1V, 2V, 3V= vannal veins; H= humeral vein; subscripts refer to branches (R$_2$ is second branch of radius).

Genitalia. The form of the external genitalia, especially that of the male, has been widely used in the separation of species (Fig. 1*f*).

DEVELOPMENTAL STAGES

Metamorphosis is complete. The larvae, commonly called caterpillars, are mandibulate and cylindrical, with short thoracic legs and a variable number of abdominal prolegs. They have one pair of thoracic and eight pairs of abdominal spiracles. Pupae are variable in form and have appendages that are usually firmly cemented down (obtect), though they are sometimes partly or completely free (exarate). Pupae often are enclosed in a silken cocoon.

Egg. There are two general types of eggs, flattened and upright. In both types, the surface is usually sculptured, and the sculpturing of the upright type is generally more complex (Fig. 5). There is a microscopic entrance for the sperm, the micropyle. The eggs are usually laid on the food plant either singly or in clusters. They are normally attached by a cement but may be merely scattered in the vicinity of the food plant.

Larva. These are quite variable in superficial characters, such as color, size, and shape; presence of warts, hairs, and setae; arrangement of hooks, or crochets, on the prolegs; and other minor details. They are rather constant, however, in basic morphology (Fig. 6). There are usually six pairs of ocelli, and compound eyes are absent. At most, eight pairs of prolegs are present, but almost every degree of reduction is found. The usual number is five.

Silk is produced by the labial glands. They are long, coiled, simple tubes which unite anteriorly in a cylindrical spinneret, which opens at the front of the labium.

Pupa. There are three types, but the vast majority are obtect, with appendages fastened down and immobile and with only the terminal abdominal segments movable. Those of the Hepialoidea, Cossoidea, and some other lower moths, are incomplete, in that the abdominal segments, cephalad

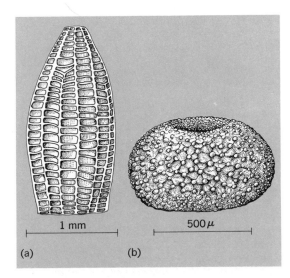

Fig. 5. Butterfly eggs. (a) *Strymon acadica* Edwards (Lycaenidae). (b) *Anthocaris midea* Hübner (Pieridae).

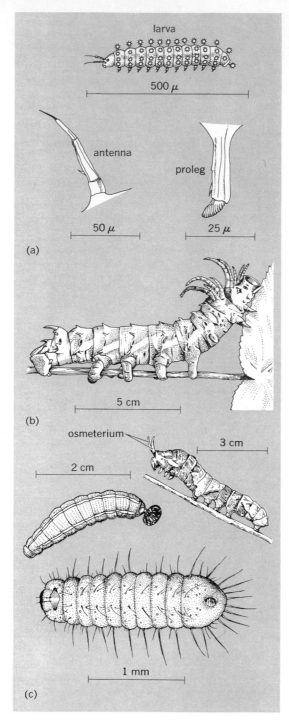

Fig. 6. (a) Newly hatched larva of *Micropteryx calthella* L. (Micropterygidae). (b) Hickory horned devil, larva of *Citheronia regalis* Fabricius (Saturniidae). (c) Skipper and butterfly larvae. Mature *Thorybes pylades* Scudder (Hesperiidae). Mature *Papilio cresphontes* Cramer (Papilionidae), osmeterium extruded. First instar, *Strymon liparops* Boisduval and Leconte (Lycaenidae).

of the fourth, are movable (Fig. 7). The appendages are partly free but sheathed. The Micropterygoidea and Eriocranioidea have "free" pupae in which the mandibles are functional; the appendages of head, thorax, and segments of the abdomen are movable; and the integument is soft.

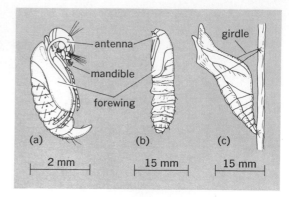

Fig. 7. Pupae. (a) *Mnemonica auricyanea* Walshingham (Eriocraniidae). (b) *Sthenopis thule* Strecker (Hepialidae). (c) *Papilio troilus* L. (Papilionidae).

CLASSIFICATION

The Lepidoptera were long divided into two suborders, the Rhopalocera, with clubbed antennae, such as butterflies and skippers; and the Heterocera, with antennae of other forms, as in the moths and millers. The butterflies and skippers form a major natural unit, with about 10,000 species, and are often referred to as the series Rhopalocera. The basic dichotomy in the order, however, appears to be between the Homoneura and Heteroneura, and those suborders are recognized here.

The terms Macrolepidoptera and Microlepidoptera often appear in zoological literature as names of former divisions. The microlepidoptera are the Homoneura and the superfamilies Incurvarioidea through Pyraloidea of the Heteroneura, which are generally small moths. The macrolepidoptera are the larger moths and butterflies. Some microlepidoptera, such as many cossids, hepialids, and others, are much larger than many macrolepidoptera.

Unfortunately, the classification of the Lepidoptera is the subject of considerable controversy; much will doubtless be resolved when more is known of the anatomy and life history of many groups. Because of these difficulties, the relationships expressed in Fig. 8 are at best provisional.

A rather conservative classification has been used here. It is similar to that of C. L. Remington (in Brues, Melander, and Carpenter), except that the classification of the Papilionoidea follows P. R. Ehrlich (1958). In many superfamilies only the more important families are mentioned. Table 1 lists the important families.

Homoneura (Jugatae). Fore- and hindwings are similar in shape and venation. They are connected by a jugum and, sometimes, also by a frenulum. Mouthparts are mandibulate, with mandibles vestigial or absent, and the galeae forms a rudimentary proboscis (Fig. 9). The female has a single genital opening. The pupae are free or partially free.

This small suborder, including less than 1% of the species in the order, contains a diverse group of primitive forms showing certain features in common with the Trichoptera, or caddis flies.

Superfamily Micropterygoidea. One small family, the Micropterygidae, are minute moths possessing toothed, functional mandibles and lacking even

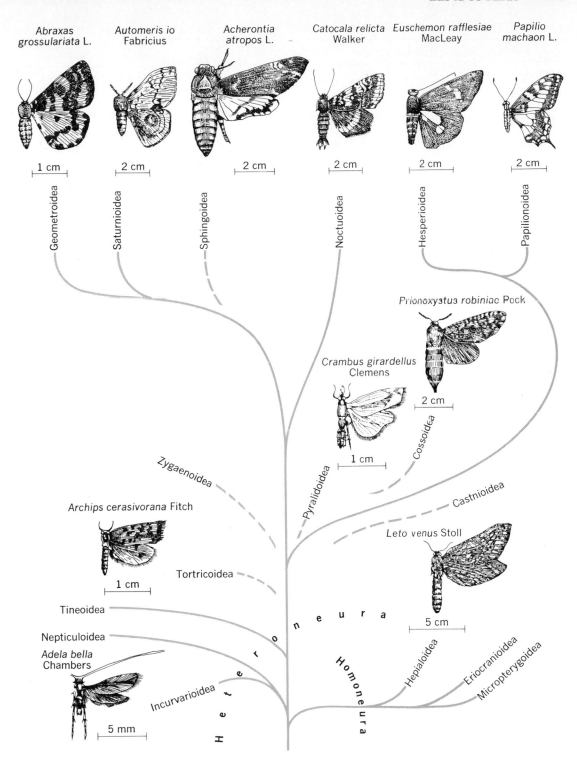

Fig. 8. Relationships of superfamilies of Lepidoptera.

the most rudimentary proboscis. The galea of the maxilla is short and the adults feed on pollen. The larvae, which feed on mosses, are unusual in having eight pairs of abdominal prolegs.

It has been suggested by various authorities that these moths are actually terrestrial Trichoptera, or that they should be placed in a separate order, Zeugloptera. Characters of the wing venation, tracheation, the presence of broad, well-developed scales with numerous ridges, and other features,

indicate that these insects are best included in the Lepidoptera.

Superfamily Eriocranioidea. In this group of tiny moths the mandibles are greatly reduced and untoothed, and the galeae of the maxillae form an abbreviated proboscis. Three families, the Eriocraniidae, Neopseustidae, and Mnesarchaeidea, have been recognized within the superfamily. The leaf-mining larva is essentially apodous (lacks legs). The adults reportedly do not feed. The fe-

Table 1. Size, distribution, and common names of some families of Lepidoptera

Classification	Common name	Distribution	No. of species*
Suborder Homoneura			
Micropterygidae	Micropterygids	Holarctic and Australia	3 (35)
Eriocraniidae	Eriocraniids	Holarctic	5 (20)
Mnesarchaeidae	Mnesarchaeids	New Zealand	
Hepialidae	Swift or ghost moths	Cosmopolitan	18 (200)
Suborder Heteroneura			
Incurvariidae	Yucca moths and relatives	Cosmopolitan	60
Nepticulidae	Serpentine leaf miners	Cosmopolitan	75
Cossidae	Goat or carpenter moths	Cosmopolitan	45
Aegeriidae	Clearwing moths	Cosmopolitan	120
Coleophoridae	Case bearers	Cosmopolitan	110 (900)
Gelechiidae	Gelechiids	Cosmopolitan	590 (3800)
Gracilariidae	Gracilariids	Cosmopolitan	235
Heliodinidae	Heliodinids	Cosmopolitan	21
Oecophoridae	Oecophorids	Cosmopolitan; largely Australian	225 (3000)
Orneodidae	Many-plume moths	Cosmopolitan	1
Psychidae	Bagworms	Cosmopolitan	25
Tineidae	Clothes moths and relatives	Cosmopolitan	135 (2500)
Yponomeutidae	Ermine moths	Cosmopolitan	65 (800)
Olethreutidae	Olethreutids	Cosmopolitan	715 (2500)
Tortricidae	Tortricids	Cosmopolitan	210 (1500)
Thyrididae	Window-winged moths	Tropical	10
Pyralididae	Pyralids; snout moths	Cosmopolitan	1135 (12,000)
Pterophoridae	Plume moths	Cosmopolitan	130
Eucleidae	Slug moths	Cosmopolitan	50 (900)
Megalopygidae	Flannel moths	Mostly American; a few African	11
Zygaenidae	Foresters and burnets	Palearctic, African, and Indo-Australian	—
Castniidae	Castniids	Neotropical and Indo-Australian	
Drepanidae	Hooktips	Holarctic	6
Geometridae	Measuring worms, loopers, cankerworms, carpets, waves, and pugs	Cosmopolitan	1200 (4000)
Uraniidae	Uraniids	Tropical	—
Sphingidae	Sphinx, hawk, or hummingbird moths	Cosmopolitan	106 (1000)
Lasiocampidae	Tent caterpillars, lappet moths	Cosmopolitan except New Zealand; mainly tropical	30 (1400)
Saturniidae	Giant silkworms	Cosmopolitan	60
Bombycidae	Silkworm and allies	Tropical	1 (introduced)
Arctiidae	Tiger moths	Cosmopolitan	200 (3600)
Lymantriidae	Tussock moths	Largely African and Indo-Malayan, but with important Holarctic species	27
Notodontidae	Prominents, puss moths	Cosmopolitan except New Zealand	120
Noctuidae	Noctuids, owlets, underwings, millers	Cosmopolitan	2700 (20,000)
Hesperiidae	Skippers, agave worms	Cosmopolitan	240 (3000)
Papilionidae	Swallowtails, bird-wings, parnassians	Cosmopolitan	27 (600)
Pieridae	Whites, sulfurs, orangetips	Cosmopolitan	61 (1000)
Nymphalidae	Four-footed butterflies	Cosmopolitan	211 (5000)
Libytheidae	Snout butterflies	Cosmopolitan	1 (17)
Lycaenidae	Blues, coppers, hairstreaks, metal marks	Cosmopolitan	138 (3500)

*The first figure is the number of described species in North America north of Mexico. This figure is reasonably accurate. The second figure, in parenthesis, is a rough estimate of the number of described species in the world. It is difficult to postulate the actual total of Lepidoptera species from these figures since in some groups, such as the Papilionidae, probably more than 90% of the existing species have been described, while in others, such as many families of microlepidoptera, the figure may be well under 25%.

males have a piercing ovipositor.

Superfamily Hepialoidea. These are medium- to large-sized moths which possess rudimentary mouthparts. The larvae are borers. The rapid flying adults are mostly crepuscular, thus the common name swift, or ghost, moths. The only family of importance is the Hepialidae, which has about 200 species. The Nearctic species belong to the genera *Hepialus* and *Sthenopis*. See ZOOGEOGRAPHY.

Heteroneura (Frenatae). Fore- and hindwings are markedly different in shape and venation.

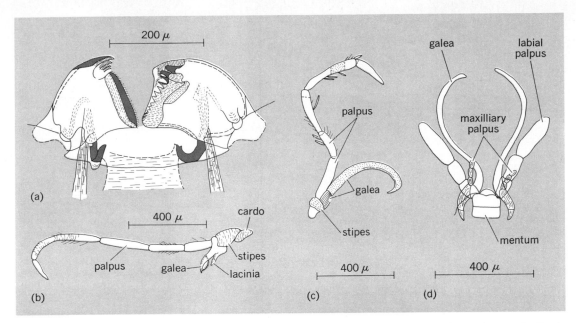

Fig. 9. Mouthparts of adult Homoneura. (*a*) Mandibles of *Sabatinca incongruella* Walker (Micropterygidae). (*b*) Maxilla of *Micropteryx aruncella* Scop. (Microptery-gidae). (*c*) Maxilla of *Eriocrania semipurpurella* Steph. (Eriocraniidae). (*d*) Maxillae and labium of *Mnesarchaea paracosma* Meyrick (Mnesarchaeidae).

Usually they are connected by a frenulum and retinaculum. Mouthparts are haustellate (formed for sucking) or, rarely, are vestigial. Adults with functional mouthparts feed on nectar of flowers, juices of rotten fruits, and other liquids. The female usually has two genital openings. Pupae are usually obtect.

Superfamily Incurvarioidea. One family, the Incurvariidae, comprises the superfamily. The wings are covered with microscopic spines, or aculeae, and the females have a single genital opening, as in the Homoneura. The venation is almost complete. The basal segment of the antenna is not enlarged to form an eye cap. The larvae are seed, leaf, stem, or needle miners. In the subfamily Incurvariinae, the larva is first a miner and then a case bearer. The pupa is not completely obtect.

This superfamily includes the yucca moth, *Tegeticula* (= *Pronuba*) *yuccasella* Riley. This small, white moth has an obligatory mutualistic relationship with the yucca plant. The female gathers pollen with its specially adapted mouthparts and fertilizes the yucca flower. The eggs are laid in the plant ovary by means of a piercing ovipositor; the larvae eat some of the developing seeds.

Superfamily Nepticuloidea. One family is included, the Nepticulidae. These tiny moths have wing spines and the females have a single genital opening, but they differ from the Incurvarioidea in having a reduced venation and a large eye cap at the base of the antenna. The larvae, with the exception of some gall-making species of the genus *Ectoedemia*, are miners in leaves, bark, and rarely, in fruits. Many species of the genus *Nepticula* have a wing expanse in the 3- to 5 mm range, being the smallest insects in the order.

Superfamily Cossoidea. One family, the Cossidae, commonly called the carpenter or goat moths, are heavy-bodied moths, with the abdomen extending well beyond the hindwings. Mouthparts are rudimentary except for labial palpi. The median

vein (M, in Fig. 5) stem extends to the base of the wing and is forked within the discal cell. The larvae are borers, often tunneling in the hardwood of tree trunks. *Prionoxystus robiniae* Peck is the best-known American species. It is very destructive to a large variety of deciduous trees.

Superfamily Tineoidea. There are 16 – 39 families; the number varies with the author. This is a very large group, of uncertain composition. These moths are of small size, usually with well-developed maxillary palpi. The labial palpi have a slender, pointed third segment. In the hindwing the subcosta and radius (Sc + R_1) are free or are joined to the cell by a bar. Venation may be reduced and the wings may be divided into plumes.

Aegeriidae is the family commonly called the clearwing moths because of the large, transparent, scaleless areas on the wings. The Sc + R_1 vein in the hindwing is apparently absent, but actually it is concealed in a costal fold. Many species are excellent mimics of wasps. They are diurnal and often brightly colored. Many of the boring larvae are economic pests. Among these are the currant borer (*Ramosia tipuliformis* Clerck), the peach-tree borer (*Sanninoidea exitiosa* Say), and the squash-vine borer (*Melittia satyriniformis* Hübner).

Coleophoridae are small, narrow-winged moths whose larvae are case bearers, carrying "shells" made out of silk and bits of leaves. The adults lack maxillary palpi.

Gelechiidae is a large family of minute to small moths, usually with rounded or, rarely, pointed forewings and trapezoidal, often pointed, hindwings. Venation is variable and sometimes reduced. The R_s, radial sector, and media, M_1, of the hindwing are stalked or close together at the base. The larvae include seed feeders, miners, borers, gall makers and foliage feeders. The family includes a number of economically important insects. The Angoumois grain moth (*Sitotroga cer-*

ealella Olivier) infests grain both in the field and in storage. The pink bollworm (*Pectinophora gossypiella* Saunders) is an extremely important worldwide pest of cotton.

Gracilariidae are small moths with leaf-mining larvae. Both pairs of wings are lanceolate and widely fringed. The young larvae are flattened and have bladelike mandibles with which they slash the cells of the leaf, sucking up the exuding juices. The full-grown larvae are quite different, being "normal" in appearance and feeding on parenchyma, either in a leaf mine or externally. The best-known genera are *Gracilaria* and *Lithocolletis*.

Heliodinidae is a small family of tiny moths, which are often brilliantly colored. A number of genera such as *Stahtmopoda* and *Euclemensia* contain species whose larvae attack coccids.

Oecophoridae are small to moderately small moths with a comb of bristles, the pecten, on the scape of the antenna. The larvae are varied in habits; some feed in webs or rolled leaves, others are scavengers or predators on coccids. The genera *Blastobasis*, *Holcocera*, and others are often placed in a separate family, the Blastobasidae. *Holcocera pulverea* Meyrick is an important predator on the lac insect in India.

Orneodidae is a small family whose adults have each wing divided into six featherlike plumes. Some authors place them in the Pyralidoidea with the Pterophoridae, but the majority believe that the resemblance between the two families is a case of convergence.

Psychidae are the bagworms. The males are hairy, strong-bodied, swift-flying moths with reduced mouthparts. They are large, with a wing expanse of about 25 mm for the tineoids. The females are degenerate, wingless, legless, and often sluglike. They live concealed in bag-shaped cases made by the caterpillars. The best-known North American representative is *Thyridopteryx ephemeraeformis* Haworth (Fig. 10). In this species, the larva fastens the bag to a twig and pupates within it. The vermiform female emerges from the pupa and moves to the bottom of the bag, where she is fertilized. This is accomplished from the outside by the highly specialized, extrusible genitalia of the male. The female deposits her eggs in the bag, then drops to the ground and dies.

Tineidae is a family of small moths which usually has an erect, bristling vestiture on the head. The proboscis is reduced, or vestigial, and the maxillary palpi are usually well developed. The venation is generally primitive, with most veins being present and unstalked. Most larvae are saprophagous and many are case bearers. The best-known species are the clothes moths, which include the casemaking clothes moth (*Tinea pellionella* L.), the webbing clothes moth (*Tineola bisselliella* Hummel), and the carpet moth (*Trichophaga tapetzella* L.) These are important pests, whose larvae devour wool and other animal products. The three species all have wing expanses of less than 1 in.

Yponomeutidae is a heterogeneous assortment of small, often brightly colored moths which usually have smooth heads and reduced or absent ocelli.

Superfamily Tortricoidea. These small, wide-winged moths belong to two families, the Olethreutidae and the Tortricidae. The maxillary palpi are vestigial or absent, and the third segment of the labial palpus is short and usually obtuse. In the hindwing, $Sc + R_1$ is contiguous with the cell and then diverges from R_s. The hair fringes of the wings are always shorter than the width of the wing.

Olethreutidae is a family of moths whose hindwings usually have a fringe of long hairs on the upper side along the basal part of the cubitus. The larvae are generally hidden feeders and live in rolled leaves, in foliage webbed together, or inside fruits. The family contains a number of agriculturally undesirable species. Paramount among these is the codling moth (*Carpocapsa pomonella* L.), which is a very serious pest of apples and other fruits. The large genus *Laspeyresia* contains the interesting Mexican jumping bean moth (*L. saltitans* Westwood). The violent movements of the larvae of this moth are responsible for the action of the beans which they inhabit.

Tortricidae is a family which generally lacks the fringe of long hairs along the cubitus, characteristic of the Olethreutidae. The spruce budworm (*Choristoneura fumiferana* Clemens) is probably the most important injurious tortricid. In many places, especially in eastern Canada, it has defoliated vast areas of coniferous forest.

Superfamily Pyralidoidea. These moths are moderately small to medium-sized, long-legged, and slender-bodied. $Sc + R_1$ of the hindwings is almost always united for a considerable distance with R_s. The maxillary palpi are usually well developed.

Pyralididae is the second largest family of moths. They are small and medium-sized, and a wing expanse of 20–35 mm is not uncommon. The labial palpi are well developed, and the broad vannal regions in the hindwings often have three vannal veins. The legs are usually long and slender.

The subfamily Crambiinae, the snout moths, contains small forms which are common in marshes and grasslands. The labial palpi are quite long and porrect, giving the adults a beaked appearance.

The small subfamily Galleriinae contains the bee moth, or wax worm (*Galleria mellonella* L.), which lives in beehives. The larvae feed on the wax at night and destroy the combs. The species occurs throughout the range of the honeybee.

The subfamily Nymphulinae is notable because some of the included species are aquatic. Some larvae develop tracheal gills; others do not. In some species the pupa is enclosed in a cocoon below the surface of the water, and the adult emerges from the water.

The subfamily Phycitinae is a large group of moths in which the frenulum of the female is a simple spine rather than a bundle of bristles. The larvae have very diverse habits, being leaf rollers, case bearers, borers, stored-products pests, and so on. One species is predacious on coccids. The Indian meal moth (*Plodia interpunctella* Hübner) is described by A. D. Imms as "one of the most important economic insects known." The species is cosmopolitan, feeding on a wide variety of stored products, especially cereals. Another extremely important pest in this group is the Mediterranean flour moth (*Ephestia kuehniella* Zeller), which in-

LEPIDOPTERA

1 cm

(a)

2 cm

(b)

Fig. 10. *Thyridopteryx ephemeraeformis* Haworth (Psychidae). (a) Male. (b) Male case.

fests cereals throughout the world.

In contrast to these harmful species, the Phycitinae also contains *Cactoblastis cactorum* Berg, which was imported into Australia from South America to help control *Opuntia* cactus, which had spread rapidly and caused the ruin of millions of acres of pasture. The success of this program is an outstanding example of biological control.

The subfamily Pyralidinae reaches its richest development in the tropics. The meal moth *(Pyralis farinalis* L.) is a cosmopolitan pest of stored products.

The subfamily Pyraustinae is another very large group. It contains relatively large moths, many extending 30 mm or more, in which vein R_s of the forewing arises unstalked from the cell. To this subfamily belong the infamous European corn borer (*Pyrausta nubilalis* Hübner), the grape leaf folder (*Desmia funeralis* Hübner), and other economically important species.

The subfamily Schoenbiinae includes the genus *Acentropus*, the most completely aquatic Lepidoptera. One form of female adult never emerges from the water; it uses its reduced wings for swimming.

Pterophoridae is the family known as the plume moths. The wings are divided into featherlike plumes, of which there are usually two in the forewing and three in the hindwing, as in the tineoid Orneodidae (Fig. 11). The moths lack maxillary palpi and have slender bodies and long legs. The larvae feed exposed or are borers.

Superfamily Zygaenoidea. These moderately small- to medium-sized moths have complete venation, rudimentary palpi and, usually, a rudimentary proboscis. The wings are broad with short fringes. In the hindwing $Sc + R_1$ is separate from R_s beyond the cell. The larvae are short and more or less sluglike and are exposed feeders.

Eucleidae are the slug caterpillars, a small family of heavy-bodied hairy moths (Fig. 12); the larva is short and sluglike, with a large head concealed by the thorax. The prolegs have been replaced by midventral suckers. The best-known Nearctic form is the saddleback caterpillar (*Sabine stimulea* Clemens), which has urticating hairs.

Zygaenidae is a diverse family of small, often brightly colored moths in the subfamily Himantopterinae, which is primarily African. The hindwings are very narrow with long ribbonlike tails, and the body and wings are covered by long hairs. The larvae of this subfamily live within termite colonies, and the newly emerged moths escape from the nests under the attacks of the termites. The hairs pull free readily and the tails are expendable.

Superfamily Castnioidea. One family, the Castniidae, is included in this group. They are large, diurnal, butterflylike moths with clubbed antennae, upright eggs, and boring larvae. A proboscis may be either present or absent. These moths are considered by some to be distantly related to the butterflies, but the resemblances may very well be due to convergence. They are found in the Neotropical and Indo-Australian regions.

Superfamily Geometroidea. These are small to large moths with reduced maxillary palpi and tympanal organs at the base of the abdomen. In the forewing, the base of the media M_2 is usually about the same distance from the bases of M_3 and M_1, and the vannal veins 2V and 3V anasto-

mose to form a basal fork. The frenulum may be present or absent.

Geometridae includes the measuring worms, loopers, and cankerworms, which make up a very large family of small- and medium-sized moths with slender bodies and relatively broad wings. The females are occasionally apterous. The larvae have the anterior prolegs reduced or absent; usually only those on segments 6 and 10 are well developed. They proceed with a characteristic looping motion, which is the basis for the scientific name. The larvae ordinarily are exposed feeders, but they sometimes live in folded leaves, and often bear a striking resemblance to twigs. This resemblance is enhanced by their habit of resting with the body held rigid at an angle to the branch while holding on with the terminal prolegs. Pupation takes place in the ground or in a weak, silken cocoon.

Among the economically important Nearctic geometrids are the spring and fall cankerworms, *Paleacrita vernata* Peck and *Alsophila pometaria* Harris, which attack a large variety of trees. Also of interest is *Sterrha bonifata* Hulst (= *Eois ptelearia*), the herbarium moth, whose larvae are pests in collections of dried plants.

Uraniidae is a tropical family that includes some slender-bodied, brilliantly colored diurnal insects which lack a frenulum and are often mistaken for butterflies. The multiple-tailed metallic *Chrysiridia madagascariensis* Lesson from Madagascar is considered by many to be the most beautiful lepidopteran. The larvae are diverse in form but have a complete set of prolegs.

Superfamily Sphingoidea. Sphinx, hawk, or hummingbird moths constitute the one family Sphingidae. These medium-sized to very large, heavy-bodied moths have extremely rapid flight. The adults are mostly crepuscular or nocturnal, but a few genera are diurnal. The antennae are thickened and have a pointed apex. The proboscis is well developed and often extremely long. The wings are narrow, with the hindwing much shorter than the forewing. The frenulum is present. In the hindwing, $Sc + R_1$ is connected to R_s at the middle of the cell by a strong vein, and remains parallel to it, at least to the end of the cell. The larvae are external feeders and usually have a characteristic caudal horn. The pupa is in a cell in the ground or in a loose cocoon at the surface, and its long proboscis is often in a projecting case resembling a pitcher's handle.

One of the most widely distributed species is the white-lined sphinx (*Celerio lineata* Fabricius). Well known also is the Old World death's-head sphinx (*Acherontia atropos* L.). Of economic importance in the Western Hemisphere are the tomato hornworm, *Phlegethontius* (=*Protoparce*) *quinquemaculatus* Haworth, and the tobacco hornworm (P. *sextus* Johansson), both of which are pests of solanaceous plants.

Superfamily Saturnioidea. These are medium-sized to very large moths with the frenulum almost always reduced or absent. There is no tympanum and the mouthparts are usually reduced. The antennae are ordinarily pectinate, especially in the males.

Lasiocampidae are moths which may be separated from the Saturniidae by the presence of a

1 cm

Fig. 11. *Platypilia carduidactyla* Riley (Pterophoridae).

1 cm

Fig. 12. *Prolimacodes badia* Hübner (Eucleidae).

pair of strong humeral veins in the hindwings. This widespread, largely tropical family contains the familiar tent caterpillars of the genus *Malacosoma*. In this genus, the larvae live together in large silk-web nests. Those of *M. americana* Fabricius are a common sight on trees and shrubs in the eastern United States.

Saturniidae are the giant silkworms which are medium-sized to extremely large moths having a single, often weak humeral vein. The mouthparts are reduced or vestigial, and the antennae are strongly bipectinate. The rami of the male antennae are conspicuously longer than those of the female. The frenulum is absent. The larvae usually bear spiny processes known as scoli. Pupation takes place in large silken cocoons, which are suspended from twigs or formed in leaf litter on the ground.

Attacus edwardsi White of Australia, *A. atlas* L. of the Himalayan foothills, and *Coscinoscera hercules* Misken of Australia reach an expanse in the female of more than 10 in. They have the largest wing areas in the Insecta.

A number of saturniids produce usable silk, which has long been gathered in the Orient. Among these are the Japanese oak silkworm (*Antheraea yamamai* Guérin), and the Muga silkworm (*A. assamensis* Helfer). In North America the best-known species of the subfamily Saturniinae are the cecropia (*Samia cecropia* L.), the polyphemus (*Antheraea polyphemus* Cramer), the luna (*Actias luna* L.), the promethea (*Hylaphora promethea* Drury), and the cynthia (*Samia cynthia* Drury).

The regal moth, or hickory horned-devil (*Citheronia regalis* Fabricius) and the imperial moth (*Eacles imperialis* Drury) are the most familiar members of the subfamily Citheroniinae.

In the subfamily Hemileucinae, consisting of the buck moths and relatives, the io moth (*Automeris io* Fabricius) is noteworthy because its common larvae, like those of some other members of the subfamily, are equipped with urticating (poisonous) hairs which can cause considerable irritation to the unwary who handle the caterpillars.

Bombycidae is a small family of silkworms in which the frenulum is either present, as in *Eupterote*, or vestigial, as in *Bombyx*. The family contains the commercial silkworm (*Bombyx mori* L.). The larvae of this thoroughly domesticated species feed on mulberry leaves and resemble the larvae of sphingids in that they possess a small caudal horn. The silk from the white or yellow cocoons has been used since prehistoric times. The caterpillars are subject to the ravages of a disease known as pébrine, caused by the protozoan *Nosema bombycis* Nägeli. They are also attacked by two viral diseases known as grasserie and flacherie. *See* MICROSPORIDEA.

The genus *Eupterote* and its relatives are often placed in a separate family.

Superfamily Noctuoidea. This is a large, rather uniform superfamily of more than 20,000 species. Most of them are moderately large moths with reduced maxillary palpi. Tympanal organs are present in the metathorax. In the forewing the base of M_2 is closer to the base of M_3 than to that of M_1 (except in the Notodontidae), and 2V and 3V do not form a basal fork.

Arctiidae is a relatively large family of strikingly colored, heavy-bodied moths, the tiger moths. The larvae are generally very hairy and feed either exposed or in webs. Larval hairs are incorporated into the silken cocoon. The best-known arctiid is probably the banded woolly bear caterpillar (*Isia isabella* J. E. Smith). There is a widespread misconception that the banding pattern of this larva predicts the severity of the coming winter.

Lymantriidae (= Liparidae) are the tussock moths, which are of medium size. The antennae of the males are broadly pectinate. The end of the female abdomen has a tuft of hairs, which is deposited on top of the eggs. The hairy larvae usually have prominent "toothbrush" tufts. The most familiar species is the infamous European gypsy moth (*Porthetria dispar* L.), which, along with the destructive brown-tail moth (*Nygmia phaeorrhoea* Donovan), was imported into New England in the last half of the 19th century.

Notodontidae are commonly called prominents, or puss moths. They are distinguished from the rest of the Noctuoidea by the apparently three-branched cubitus. The larvae are external feeders and the pupa is formed in a cell in the ground or in a loose cocoon on the surface. The Nearctic *Datana ministra* Drury, the yellow-necked caterpillar, and *Schizura concinna* J. E. Smith, the red-humped caterpillar, are pests on apples and other trees.

Noctuidae (= Phalaenidae) are the owlet moths, an extremely large family of mostly dull-colored, medium-sized moths. The vast majority of moths which are attracted to lights belongs to this family. As exemplified by the genus *Catocala*, the forewings are almost always dully and cryptically colored. They cover the hindwings, which may or may not be strikingly colored, when the insect is at rest. The larvae are mostly exposed foliage feeders, but a few, such as *Papaipema*, are borers. Some of those of the genus *Eublemma* prey on scale insects, one being an important enemy of the commercially valuable lac insect. Pupation is usually in the ground.

The family includes many agricultural pests. The cutworms, *Euxoa* and *Peridroma*, attack a large variety of plants, as does the army worm (*Leucania unipunctata* Haworth). The former derive their name from the habit of cutting off shoots at the surface of the soil without consuming them, the latter from the fact that they often appear in vast numbers. The exceedingly important pest *Heliothis armigera* Hübner is variously known as the corn earworm, cotton bollworm, and tomato fruitworm. Its diet may be surmised from its names.

Superfamily Hesperioidea. There is one rather large family, the Hesperiidae. The skippers are small to moderately large, heavy-bodied, mostly diurnal insects with a clubbed antenna, which is bent, curved, or reflexed at the tip. Forelegs are fully developed and bear an epiphysis. The forewings have all veins arising separately from the cell; rarely there is some slight stalking. The frenulum is absent except in the male of *Euschemon*. The larvae have a prominent constriction, or neck, behind the head, and often live in leaves drawn together by silk. Those of the giant skippers, Megathymiinae, are borers in yucca and agave. The pupa is usually enclosed in a slight cocoon.

The name skipper refers to the rapid, erratic

flight of most species. One familiar American species is the silver-spotted skipper (*Epargyreus clarus* Cramer); another is the sachem (*Atalopedes campestris* Boisduval). The Australian *Euschemon rafflesiae* MacLeay is placed in a separate subfamily, largely because of the presence of a frenulum in the male. In Mexico the caterpillars of the Megathymiinae are fried and canned for human consumption.

Superfamily Papilionoidea. These butterflies are small to large diurnal insects with clubbed antennae, which are rounded at the tip and not bent or reflexed. The forewings always have two or more veins, which are stalked. The frenulum is absent. The larvae have no constriction behind the head and are usually exposed feeders. The pupa is naked, with the exception of *Parnassius* and relatives, and is often suspended by caudal hooks, the cremaster. Pupae are either inverted or held in an upright position by a silken girdle.

Papilionidae is a family of which swallowtails and parnassians are typical common forms. These are the only butterflies with fully developed forelegs bearing an epiphysis. They are also unique in having the cervical sclerites joined beneath the neck. The hindwings have only one well-developed vannal vein, except in the anomalous Mexican *Baronia brevicornis* Salvin. The larvae have an eversible forked organ on the prothorax, the osmeterium. This organ disperses a disagreeable odor and presumably functions as a defensive mechanism. The pupa is girdled. In the boreal genus *Parnassius* and some other members of the family a horny pouch, or sphragis, is found. This is secreted by the male during copulation and covers the genital opening of the impregnated female. The commonest genera, *Papilio*, *Graphium*, *Battus*, and *Parides* (=*Atrophaneura*), all contain large and attractive species, many of which possess the characteristic tails which give the family its name. The bird-wing butterflies, *Ornithoptera* and *Troides*, are among the largest and most beautiful species. In North America the eastern tiger swallowtail (*Papilio glaucus* L.) has dichromatic females, one form being black-and-yellow-striped like the male, and the other being entirely dark brown or black, the latter presumably mimicking the protected Aristolochia swallowtail (*Battus philenor* L.). The larva of the orange dog (*Papilio cresphontes* Cramer) is sometimes injurious to citrus.

Pieridae is a family of which common members are whites, sulfurs, and orange tips. These butterflies are unique in lacking the prespiracular bar at the base of the abdomen. The forelegs are completely developed in both sexes, but lack the epiphysis. The tarsal claws are bifid. A great many species exhibit striking sexual dimorphism. Most are basically white, yellow, or orange. The larvae are partial to plants of the families Cruciferae (mustards), Leguminosae (beans, peas, and so on), and Capparidaceae (capers). The pupae are suspended by a girdle, as in the Papilionidae.

The cabbage butterfly (*Pieris rapae* L.) is one of the most economically important butterflies, attacking cultivated crucifers in Europe and North America. The alfalfa butterfly (*Colias eurytheme* Boisduval) is a major pest of alfalfa in America.

Butterflies of this family are often extremely abundant. The giant sulfurs (*Phoebis*) and others often migrate in huge swarms. The larvae of the Mexican *Eucheira socialis* Westwood are gregarious, living together in a nest.

Nymphalidae is a family of four-footed butterflies. The prothoracic legs are greatly atrophied in both sexes, and the patagia are always well developed and heavily sclerotized.

In the subfamily Danainae the adults are tough-bodied and have an acrid taste to humans. They are supposedly distasteful to most predators and are the models in a great many mimicry complexes. The widely distributed and migratory monarch, or milkweed, butterfly (*Danaus plexippus* L.) is the best-known representative of this subfamily. In North America it is mimicked by the viceroy (*Limenitus archippus* Cramer), a member of the subfamily Nymphalinae.

The weak-flying Ithomiinae, or glassy-wings, are structurally very similar to the danaids. Many of them have broad transparent areas in the wings, in which the scales are reduced to short hairs.

The large, cosmopolitan subfamily Satyrinae, containing the woodnymphs, meadow browns, graylings, and arctics, is characterized by the weak, bouncing flight of the imagines and the bladderlike swellings of the bases of the veins on the forewings of most species. The larvae feed primarily on grasses. The genera *Erebia* and *Oeneis* penetrate extremely inhospitable arctic and alpine situations.

The subfamily Morphinae is a tropical group of large butterflies allied to the satyrids. The brilliant, metallic species of the genus *Morpho* are among the most beautiful insects. The well-known owl butterflies, certain *Caligo* species whose undersides bear conspicuous eyespots, belong to this group.

The subfamily Nymphalinae is extremely large and superficially varied, although structurally quite homogeneous. Well-known genera of the temperate regions include *Argynnis* and *Boloria* (fritillaries), *Melitaea* and *Euphydryas* (checkerspots), *Polygonia* (angle-wings), *Nymphalis* (tortoise-shells), *Vanessa* (thistle butterflies), *Limenitis* (admirals), and *Asterocampa* (hackberry butterflies). The subfamily is represented in the tropics by numerous genera and species. The long-winged genus *Heliconius*, often placed in a separate subfamily, contains complexes of Müllerian mimics, and the genus *Kallima*, the leaf butterflies, contains species whose undersides bear a very close resemblance to dead leaves. *See* PROTECTIVE COLORATION.

Libytheidae is a small family of less than 20 species, divided into two genera, *Libythea* and *Libytheana*, called snout butterflies. The forelegs of the males are reduced and nonfunctional; those of the female are smaller than the pterothoracic legs, but are used in walking. The palpi are long and porrect. These butterflies are separated from the nymphalids by the structure of the prothoracic legs and by features of the thorax, but their affinities are clearly with the Nymphalidae.

Lycaenidae is a family which includes the blues, gossamers, hairstreaks, coppers, and metal marks. These butterflies are unique in that the lamella of the mesodiscrimen is not complete to the furca. In the male, the prothoracic legs, which always lack tarsal claws, are functional in one subfamily, the

Table 2. Important injurious species of Lepidoptera

Name	Damage by larvae	Name	Damage by larvae
Aegeriidae		**Geometridae**	
Peach-tree borer (*Sanninoidea exitiosa*)	Feeds under bark; the most important insect injurious to peach	Spring cankerworm (*Paleacrita vernata*), fall cankerworm (*Alsophila pometaria*)	Defoliate fruit and shade trees in outbreak years; spring cankerworm over-winters as a pupa, fall cankerworm over-winters as an egg
Squash-vine borer (*Melittia satyriniformis*)	Infests various cucurbits; often very destructive		
Gelechiidae		**Sphingidae**	
Angoumois grain moth (*Sitotroga cerealella*)	Feeds on grain in storage and in the field; imported from Europe	Tomato hornworm (*Phlegethontius quinquemaculatus*), tobacco hornworm *P. sextus*)	Large larvae are most conspicuous pests of tomato and tobacco; also feed on other solanaceous plants
Pink bollworm (*Pectinophora gossypiella*)	Feeds on bolls, causes failure of blossoms to open; most important insect injurious to cotton		
Tineidae		**Lasiocampidae**	
Case-making clothes moth (*Tinea pellionella*)	Clothes moth larvae infest woolen products, upholstery, furs, and other dried animal products	Tent caterpillars (*Malacosoma* sp.)	Serious defoliators of forest, shade, and orchard trees
Webbing clothes moth (*Tineola bisselliella*)		**Arctiidae**	
Carpet moth (*Trichophaga tapetzella*)	Damage similar to that by clothes moths but less common	Fall webworm (*Hyphantria cunea*)	Feeds on wide variety of forest, shade, and fruit trees
Olethreutidae		**Lymantriidae**	
Codling moth (*Carpocapsa pomonella*)	Larva bores in apples and some other fruits; most important pest of apples	European gypsy moth (*Porthetria dispar*)	Important defoliator of forest, shade, and fruit trees in New England
Oriental peach moth (*Laspeyresia molesta*)	Feeds on fruits and twigs of peaches, plums, and so on	Brown-tail moth (*Nygmia phaeorrhoea*)	Similar to gypsy moth
Tortricidae		**Noctuidae**	
Spruce budworm (*Choristoneura fumiferana*)	Defoliator of vast areas of coniferous forest	Cutworms (*Euxoa* sp., *Peridroma* sp., etc.)	Numerous species damage many crop plants
Ugly-nest tortricids, leaf rollers, etc. (*Cacoecia* sp., *Tortrix* sp.)	Feed on leaves and fruits of apples and other fruits, strawberry, shade trees, ornamental plants, and so on	Army worm (*Leucania unipunctata*)	Varies tremendously in abundance; in epidemic years it does great damage to corn and other grasses
Pyralidae		Corn earworm, cotton bollworm, tomato fruitworm, tobacco budworm, *Heliothis armigera*	Exceedingly important cosmopolitan pest attacking many cultivated plants
Oriental rice borer (*Chilo simplex*)	Very destructive to rice in Asia		
Bee moth (*Galleria mellonella*)	Destroys combs in neglected beehives		
Indian meal moth (*Plodia interpunctella*)	Widespread and very important pest of dried fruit and animal products; imported from Europe	**Pieridae**	
		Cabbage butterflies, (*Pieris* sp., especially *P. rapae* in United States)	Pests of various cultivated crucifers
Mediterranean flour moth (*Ephestia kuehniella*)	Infests flour, stored grain, cereals		
Grape leaf folder (*Desmia funeralis*)	Sometimes an important defoliator of grapevines	Alfalfa butterfly (*Colias eurytheme*)	Important pest of alfalfa in southwestern United States

Lycaeninae, and nonfunctional in the other two subfamilies, Riodininae and Styginae. Those of the female are smaller than the pterothoracic legs, but are functional. The eyes are emarginate at the antennal bases, or at least contiguous with them. The patagia are never sclerotized. The larvae are mostly flattened, and the pupa is usually girdled. The larvae generally have a mutualistic relationship with ants, which tend and protect them in return for honeydew which they secrete; a few are predacious.

The subfamily Riodininae, or metal marks, is primarily Neotropical. They are small butterflies, often brilliantly colored.

The subfamily Lycaeninae includes the blues, coppers, hairstreaks, and harvesters, minute to medium-sized insects which are often a metallic blue or green on the upper surface. The largest species is the Australian *Liphyra brassolis* Westwood, with an expanse of about 75 mm. The caterpillars live in ant nests and prey on ant larvae. The harvesters (*Gerydus*, *Spalgis*, and *Fenesica*) have predacious larvae that feed on homopterans, such as aphids, membracids, and coccids, which are attended by ants. In the United States the bean lycaenid (*Strymon melinus* Hübner) has at times been an important pest of hops.

BIOLOGICAL ASPECTS

The Lepidoptera are a group of insects on which much biological research remains to be done. A great deal is still unknown about the genetics, physiology, and ecology of this group. Moreover, butterflies and moths have proved useful as experimental animals in genetical research. The larvae of many species are injurious to certain crops, causing severe economic losses (Table 2).

Ecology and distribution. The variation in larval habits is discussed under the different families. Lepidoptera of all stages are subject to the attacks of a large number of predators, including birds, mammals, lizards, frogs, and spiders. They also must be wary of rapacious insects, such as dragonflies, mantids, phymatids, pentatomids, asilids, and vespoids. Some wasps (Sphecidae) paralyze caterpillars, oviposit in them, and place them in specially constructed cells. Others (brachonids and ichneumonids) place their eggs in the caterpillars without paralyzing them. In either case the caterpillars serve as food for the growing wasp larvae inside them. Some chalcids oviposit in the eggs of Lepidoptera.

One mite, *Myrmonyssus phalaenodectes* Treat, infests the tympanum of a variety of moths, and another, *Otopheidomenis zalestes* Treat, is restricted, as far as is known, to the genus *Zale*. Pseudoscorpions of the genera *Atemnus*, *Stenowithius*, and *Apocheiridium* have been found on adults of various species. Lepidoptera are also subject to viral, bacterial, protozoan, and fungal infections. *See* INSECT PATHOLOGY.

The Lepidoptera penetrate almost every section of the globe, with the major exception of Antarctica. Butterflies of the genus *Boloria* have been taken at Alert on northern Ellesmere Island, about 400 mi from the North Pole. Arctic and alpine tundra areas normally support a lepidopteran fauna which, although relatively poor in species, is rich in numbers. After rains, deserts are often alive with butterflies and moths. Tropical areas are by far the richest in species. One of the strangest habitats occupied by a lepidopteran is the hair of the neotropical three-toed sloth, where the sloth moth (*Bradypodicola hahneli* Spüler), a pyralid, passes its entire life cycle presumably feeding on algae which grow in the hair. The following outline indicates some of the habitats of larval Lepidoptera.

1. Plant associations, terrestrial
   a. Feeding exposed on foliage, flowers, or plant lice
   b. Feeding on foliage in web or case
   c. Leaf and needle mining
   d. Living under bark
   e. Boring in stem, root, fruit, or seeds
   f. In soil, feeding on roots
   g. Boring in or feeding on fungi, mosses, liverworts, club mosses, lichens, and ferns
   h. Living in dried plant products, such as cereals, flour, dried fruit, refuse, and herbarium specimens
2. Associations with social insects
   a. In beehives
   b. In ant nests
   c. In termite nests
3. Associations with other animals
   a. In stored animal products, such as woolens and feathers
   b. In hair of three-toed sloths, feeding on algae
   c. In plant lice as parasitoids
4. Aquatic
   a. Feeding on, boring in, or mining in aquatic plants
   b. On rocks, feeding on microorganisms

In temperate zones, species are known which hibernate in every stage of the life cycle.

Individual, seasonal, and geographic variations are common and striking in the Lepidoptera. Color and size are characters most frequently affected, but wing shape, venation, genitalia, and other structures are often involved. The number of broods per season often differs geographically.

Behavior and physiology. Migration is the most spectacular behavior occurring in the order. It is most frequent in the butterflies *Phoebis*, *Danaus*, *Libytheana*, and others, but is also known in moths such as *Chrysiridia*. Huge migratory swarms are frequently reported in many parts of the world, but this phenomenon, as well as the related communal roosting of adults, daily use of flyways, hilltopping, and so forth, is poorly understood.

Aggregations of butterflies at a mud puddle are a frequent sight, and moths have been observed "pumping," an act which consists of sipping water steadily from a puddle and ejecting it as a stream of drops or fine spray from the anus.

In the field of lepidopteran physiology, the work of C. M. Williams and his associates on growth and differentiation hormones in large saturniids and that of D. E. Beck on larval nutrition and behavior in *Pyrausta nubilalis* should be mentioned. Also noteworthy are the investigations of A. Treat and others of the function of the tympanum, which serves primarily as a warning mechanism by detecting ultrasonic pulses of approaching bats.

Evolution and genetics. There is no known extinct family in this order. N. B. Tindale's family Eosetidae, based on *Eoses triassica* from the Triassic of Australia, was originally placed in the Lepidoptera, but its position is very doubtful. The earliest fossil Lepidoptera are microlepidoptera from the early Tertiary, but it seems likely that the order existed well back in the Mesozoic, perhaps first differentiating in the Jurassic.

The phenomena of mimicry and protective resemblance are widespread in the Lepidoptera.

A. Blest, in 1957, demonstrated that eyespot patterns in certain species elicit escape responses in passerine birds. Thus, they are an effective protective device.

Lepidoptera have advantages for many kinds of genetic investigation, especially for the study of population genetics. J. Gerould first demonstrated a sex-limited character in his work on *Colias*. E. B. Ford, C. L. Remington, and others contributed to genetics with studies of various butterflies and moths.

The appearance of dark forms of various moths in heavily industrialized, and thus heavily sooted, areas is a widespread phenomenon. This "industrial melanism" is doubtless due to shifting selection pressures and is one of the best-known examples of evolution in action. *See* ARTHROPODA; BUTTERFLY; CATERPILLAR; EVOLUTION, ORGANIC; GENETICS; INSECT PHYSIOLOGY; INSECTA; MOTH; PROTECTIVE COLORATION; SEXUAL DIMORPHISM.

[PAUL R. EHRLICH]

Bibliography: J. V. Brower, Experimental studies of mimicry in some North American butterflies, *Evolution*, 12:32–47, 123–136, 273–285, 1958; C. T. Brues, A. L. Melander, and F. M. Carpenter, *Classification of Insects*, 2d ed., Mus. Comp. Zool. Bull. no. 108, 1954; P. R. Ehrlich, The comparative morphology, phylogeny, and higher classification of the butterflies, *Univ. Kans. Sci. Bull.*, 39:305–370, 1958; P. R. Ehrlich, The integumental anatomy of the monarch butterfly, *Danaus plexippus* L., *Univ. Kans. Sci. Bull.*, 38(pt. 2): 1315–1349, 1958; E. B. Ford, *Butterflies*, 1945; E. B. Ford, *Moths*, 1955; A. B. Klots, *The World of Butterflies and Moths*, 1958; R. T. Mitchell and H. S. Zim, *Butterflies and Moths*, 1962–1964.

Lepidosauria

A subclass of reptiles, both living and extinct, in which the structure of the skull is characterized by two temporal openings (diapsid condition) on each side (Fig. 1). Lepidosauria differ from Archosauria, which are also diapsid, in that the bony arcades bordering their temporal openings may suffer reduction or loss, causing the apparent disappearance of one or both openings. They differ also in that no lepidosaur skull has any antorbital opening in front of the orbit, and in that their teeth are typically fused to the jaw (acrodont or pleurodont) rather than implanted in sockets (thecodont). *See* ARCHOSAURIA.

In the classification adopted here the Lepidosauria include three orders. The Eosuchia were mainly small lizardlike reptiles which ranged from

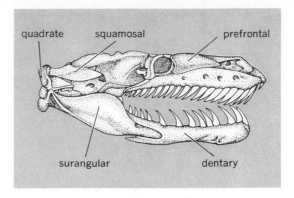

Fig. 3. Skull of a primitive living snake, the python. *(From A. S. Romer, Vertebrate Paleontology, 3d ed., University of Chicago Press, 1966)*

the Middle Permian to the lower Eocene and which doubtless gave rise to the later orders. The Rhynchocephalia, with their beaky skulls, appeared in the Early Triassic, reached the apex of their development in the Middle and Upper Triassic with the herbivorous and often very large rhynchosaurs, and then declined; they survive, however, as the single species *Sphenodon punctatum*, the tuatara of New Zealand. The Squamata, which also appeared in the Triassic, include the lizards, in which the lower temporal arcade is lost (Fig. 2), and the amphisbaenians and snakes, in which both temporal arcades are lost (Fig. 3); thus the Squamata include the great majority of living reptiles, whether counted by numbers of species or of individuals. The gliding lizards (Kuehneosauridae) of the Upper Triassic are the earliest known aerial vertebrates. *See* EOSUCHIA; REPTILIA; RHYNCHOCEPHALIA; SQUAMATA.   [ALAN J. CHARIG]

Bibliography: E. H. Colbert, *The Age of Reptiles*, 1965; A. S. Romer, *Osteology of the Reptiles*, 1956.

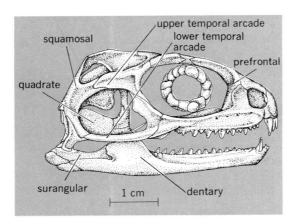

Fig. 1. Skull of a living rhynchocephalian, *Sphenodon*, a typical diapsid. *(From A. S. Romer, Vertebrate Paleontology, 3d ed., University of Chicago Press, 1966)*

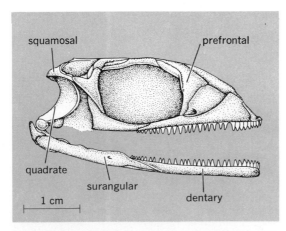

Fig. 2. Skull of an early, but typical, lizard, *Kuehneosaurus*. *(From A. S. Romer, Vertebrate Paleontology, 3d ed., University of Chicago Press, 1966)*

Lepospondyli

A term used to characterize Paleozoic amphibian groups in which the vertebral centra, basically spool-shaped, are formed by ossification directly around the notochord. This is in contrast to the structure seen in the Labyrinthodontia, where the centra are formed in blocks of cartilage found well outside the notochord. The term Lepospondyli has been used in the past to include only a restricted number of older amphibians, but is currently used as a subclass name to include all older amphibians whose vertebrae are formed in this general fashion. *See* LABYRINTHODONTIA; VERTEBRA.

In the Paleozoic, the larger amphibians were labyrinthodonts, but small lepospondyls appear to have been very abundant, particularly as water dwellers in the coal swamps. They are commonly considered in three orders: Nectridia, Aistopoda, and Microsauria. All three are quite distinct from one another, and there is no guarantee that they form a truly natural group descended from a common lepospondylous ancestor. Except for the Devonian ichthyostegids, the oldest known fossil amphibians (Carboniferous) are lepospondylous, and it is thought by some that the group may have derived from fish ancestors independently of the labyrinthodonts. The vertebral structure of the living amphibians is basically lepospondylous, and it is possible that part or all of the modern orders (unknown in the Paleozoic) have descended from lepospondyls, particularly the microsaurs. *See* AISTOPODA; AMPHIBIA; MICROSAURIA; NECTRIDEA.

[ALFRED S. ROMER]

Leprosy

A chronic infectious disease of humans, also known as Hansen's disease, of worldwide occurrence but most commonly found in tropical areas where the incidence may be as high as 5 per 1000 population. Infection appears most often to result from intimate contact with a leprous individual. The incubation period is long, ranging from a few months to several years. The presence of bacilli in peripheral nerves is a hallmark of leprosy.

The most debilitating form of the disease, lepromatous leprosy, is characterized by multiple nodular lesions (lepromata) of the skin (see illustration). Microscopically these lesions contain masses of rodlike microorganisms, leprosy bacilli, which in sections of fixed tissue have the capacity to retain certain basic dyes in the presence of a decolorizing agent consisting of 3% mineral acid in 95% ethanol; thus they are called acid-fast organisms. *See* ACID-FAST STAIN; MYCOBACTERIACEAE.

Bacilli extracted from ground nodules, washed, packed, resuspended in saline, autoclaved, and preserved with 0.5% phenol make up a reagent called lepromin, important in the clinical characterization of the various types of leprosy. Skin may be tested for response to lepromin in the same manner as for tuberculin, brucella, or diphtherial toxin-toxoid. Two classes of positive reaction to

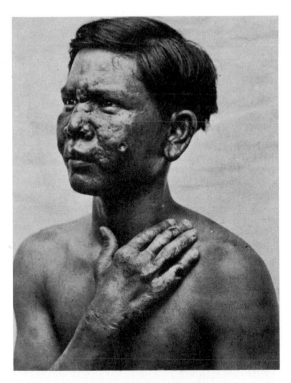

Filipino male with lepromatous leprosy showing the classical ear, cheek, and nose nodular lesions (lepromata) and burn scars on the anesthetic right hand, which also shows muscular atrophy and typical eversion of the thumb. (*Photograph by H. L. Arnold, Jr.*)

lepromin are known: a delayed type of hypersensitivity manifesting itself in 24–72 hr, the Fernandez reaction; and a nodular reaction, a granulomatous response, occurring in 2–4 weeks, the Mitsuda reaction. A positive Mitsuda reaction is given by individuals with tuberculoid leprosy (TL) and by a large percentage of persons living in areas where leprosy is not endemic. A negative Mitsuda reaction is given by persons suffering with the more severe lepromatous leprosy (LL). In fact, a negative reaction is required for establishing a diagnosis of LL. Leprologists arrange the forms of leprosy in a gradient between two extremes: TL (relatively resistant) and LL (showing little or no resistance). Circumstantial evidence suggests that persons most likely to experience LL possess an immunological peculiarity found in only a small number of individuals. *See* BRUCELLOSIS; DIPHTHERIA; SKIN TEST; TUBERCULOSIS.

Bacilli isolated from cases of leprosy do not appear as the bacilli seen in lepromata. Therefore leprologists believe that the leprosy bacillus is noncultivable. By agreement, these experts have given the leprosy bacillus the name *Mycobacterium leprae*. It is a *species forma* outstanding for being acid-fast in place and noncultivable. Research has revealed that the acid-fastness of leprosy bacilli can be taken away with pyridine, whereas the acid-fastness of other mycobacteria cannot be taken away with this solvent. It has also been found that leprosy bacilli from human leprosy multiply and cause a lepromatous disease (complete with nerve involvement) in the nine-banded armadillo, *Dasypus novemcinctus*. The acid-fastness of *M. leprae* from the leprous armadillo, like that of leprosy bacilli from humans, can be taken away with pyridine. As yet, no *M. leprae* has been cultivated from infected armadillos.

The use of sulfones, particularly 4,4′-sulfonyldianiline (dapsone), in the treatment of leprosy leads to apparent improvement. Relapses associated with sulfone-resistant strains have been encountered. While Rifampin is known to be a satisfactory antileprosy drug, its cost has limited the extent of its use.

BCG (Bacillus Calmette-Guérin) vaccine has been used on an experimental basis in several parts of the world for the vaccination of persons exposed to leprosy. Final evaluation of these trial studies had not yet been made by 1974.

[LANE BARKSDALE]

Bibliography: J. Convit and M. E. Pinardi, Leprosy: Confirmation in the armadillo, *Science*, 184: 1191–1192, 1974; C. A. Fisher and L. Barksdale, Cytochemical reactions of human leprosy bacilli and mycobacteria: Ultrastructural implications, *J. Bacteriol.*, 113:1389–1399, 1973; W. H. Jopling, *Handbook of Leprosy*, 1971; E. E. Storrs et al., Leprosy in the armadillo: New model for biomedical research, *Science*, 183:851–852, 1974.

Leptolepiformes

An order of small ray-finned fishes of the Mesozoic that are important to the understanding of the early evolution of the Teleostei. The principal family included in this order, the Leptolepidae, has been reported from the Middle Triassic of Europe and ranges into the Cretaceous. Leptolepids represent the first teleosts as defined on the structure of the

Leptolepis dubia (Blainville) restoration, scales omitted. (*From A. S. Woodward, Catalogue of the Fossil Fishes in the British Museum (Natural History), pt. 3, 1895*)

caudal skeleton. They were either derived from the earliest Pholidophoridae (known from the European Middle Triassic sediments) or shared a common ancestry with them. In size, body shape, fin position, structure of head and jaws, and general habitat, leptolepids are similar to the pholidophorids. In many ways these two families appear to parallel one another.

The structural differences separating the two groups are related to locomotion since the leptolepids have the better ossified axial skeleton, including the advanced type of tail support characterized by a series of paired uroneural bones bracing the caudal upturn of the vertebral column and the first two hypurals articulated to one centrum. Correlated with the stronger axial skeleton in leptolepids is the scalation composed of thin cycloid scales instead of the ganoid type of the early weak-backboned pholidophorids. Another structural difference may be present in the cheek region. Leptolepids have a preopercle with an elongated dorsal portion that is intimately associated with the underlying hyomandibula and provides a base for jaw muscle attachment. The relationships of the bones of the cheek, jaw suspension, and palate reflect the beginning of improvements in feeding and gill ventilation which is so greatly elaborated in advanced teleosts and is made possible by more efficient muscle placement and muscle differentiation. *See* PHOLIDOPHORIFORMES.

Certain leptolepid characters such as enamel-covered dermal bone, independent supraorbital sensory canal, and one pair of hypohyals are archaic features (mostly holostean). For this reason the group is sometimes placed in the Holostei or in a separate subclass (Halecostomi) with the Pholidophoriformes.

Within the order it may be possible to find evidence of an early radiation into the major teleostean lines. The oldest member, *Leptolepis*, is quite generalized (see illustration). Its mouth and dentition suggest it was a plankton feeder or lived by predation on small organisms. There is some fossil evidence to indicate it was a schooling fish in shallow marine waters. *Clupavus*, Jurassic-Cretaceous of Africa, is similar to *Leptolepis* on many points, especially jaw structure and dentition. *Allothrissops*, an early member of the line that gave rise to the Cretaceous Ichthyodectidae, also shares a resemblance with *Leptolepis*. Lycopterids, small, fresh-water forms of the Jurassic and Cretaceous of Asia, are usually in-

cluded with uncertainty in the Leptolepiformes. Many authors have derived the tarpons (Elopidae) and herrings (Clupeidae) from the extinct leptolepids on the basis of similar osteological structures.

It appears likely that the first teleosts were able to take advantage of a ready (perhaps largely unexploited) food source through a pelagic existence. This was possible because of improvements in their swimming and feeding mechanisms and also probably through larval adaptations. Some of the improvements were already underway at the pholidophoriform level. *See* TELEOSTEI.

[TED M. CAVENDER]

Bibliography: D. V. Obruchev, *Fundamentals of Paleontology*, vol. 11: *Agnatha, Pisces*, 1967; J. Piveteau, *Traité de Paléontologie*, vol. 4, 1966; A. S. Romer, *Vertebrate Paleontology*, 1966.

Lepton

A collective name for fermions having a mass smaller than the proton mass. (A fermion is a particle which obeys Fermi-Dirac statistics.) There are eight known leptons: the electron (e^-), the negatively charged μ meson, or muon (μ^-), and the neutrinos (ν_e, ν_μ), plus their antiparticles — the positron (e^+), the positively charged muon (μ^+), and the antineutrinos $(\bar{\nu}_e, \bar{\nu}_\mu)$. Being fermions, these particles have half-odd-integral spin; in fact, they have spin $\frac{1}{2}\hbar$.

The leptons interact with electromagnetic and gravitational fields by virtue of their charge and energy, respectively, but beyond this they are only known to interact through the β-interactions in the broad sense (weak interactions). These interactions appear to conserve leptons of each kind (e or μ); that is, just as many e-leptons (electrons and e-neutrinos) come out of an interaction as went in (antileptons being counted negatively), and similarly, the number of μ-leptons (muons and μ-neutrinos) is conserved. Thus the e^- and the ν_e can be considered to be a single particle in two guises, as can the μ^- and the ν_μ. Except for the difference of the electron and muon masses, the e-leptons and the μ-leptons have identical properties, so far as has been determined; no reason for this "coincidence" is known. *See* ELEMENTARY PARTICLE; MESON; NEUTRINO.   [CHARLES GOEBEL]

Leptospira

A genus of spirochetes of the family Spirochaetales. Many species produce an acute disease in man, dogs, horses, and certain other animals. Rodents are frequently carriers. In morphology, biologic characteristics, and antigenic structure, *Leptospira* have little in common with either the *Treponema* or the *Borrelia* group of spirochetes. The organism is $4-20$ μ in length, $0.1-0.2$ μ in width, and has closely wound spirals. *Leptospira* grow well aerobically on artificial media and in the chick embryo. Classification of pathogenic species is primarily on an antigenic basis. *See* SPIROCHAETALES; WEIL'S DISEASE.

[THOMAS B. TURNER]

Leptostraca

A small and unimportant group of Crustacea, formerly ranked with the Branchiopoda but now regarded as the most primitive of the Malacostraca.

Only a few genera, such as *Nebalia*, have survived to the present time, but numerous apparently related Paleozoic fossils are known.

The Leptostraca differ from the rest of the higher Crustacea, the Eumalacostraca, chiefly in having an additional abdominal somite which never bears appendages; a telson bearing two movably articulated prongs, the furcal rami; and an adductor muscle connecting the two halves of the shell or carapace. These are doubtless primitive features and indicate that the Leptostraca diverged from the main Malacostracan stock before the emergence of the typical "caridoid" form. The lamellar thoracic limbs are highly specialized, not primitive. The alliance of the group with the other Malacostraca is amply justified by the agreement in number of the appendages, by the sharp distinction between the thoracic and abdominal series, and by the position of the genital openings. They are most closely related to the Mysidacea, especially in their mode of development. Mysids, like *Hemimysis*, are now known to have an embryonic caudal furca which is shed at the first molt, and a seventh abdominal somite which later fuses more or less completely with the sixth one.

The fossil Leptostraca are the oldest known Malacostraca, ranging from the Cambrian to the Triassic epochs (see illustration). Since they are

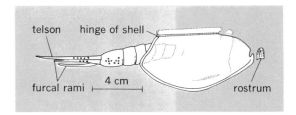

Ceratiocaris stygia Salter, a fossil leptostracan. (*After T. R. Jones and H. Woodward, A Monograph of British Palaeozoic Phyllopoda (Phyllocarida Packard), Palaeontographical Society, 1888–1889*)

already well differentiated in the earliest fossiliferous rocks, they shed no light on the possible origin of the Malacostraca. See MALACOSTRACA; NEBALIACEA; PHYLLOCARIDA. [ISABELLA GORDON]

Bibliography: T. R. Jones and H. Woodward, *A Monograph of the British Palaeozoic Phyllopoda (Phyllocarida Packard)*, 1888–1889; V. Van Straelen and G. Schmitz, *Crustacea Phyllocarida (= Archaeostraca)*, Fossilium Catalogus, pt. 64, 1934.

Leptostromataceae

A family of fungi of the order Sphaeropsidales. Although most species are saprophytes, some are fruit-tree pathogens. Pycnidia, the fruit bodies containing conidia, are black, shield-shaped, circular or oblong, and slightly asymmetrical (see illustration); the opening of the pycnidium is long and slitlike. There are 80 genera and 375 species recognized. Many of these are conidial stages of Hemisphaeriales. See SPHAEROPSIDALES.

The important genera of the Hyalosporae, whose spores are one-celled and bright (hyaline), follow:

Leptothyrium, with approximately 100 species.

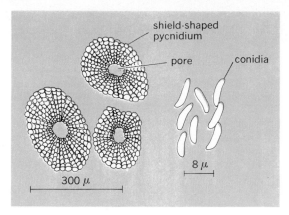

Pycnidia and conidia of *Leptothyrium vulgare*.

The pycnidium is typically shieldlike. *Leptothyrium pomi* causes apple flyspeck.

Leptostroma, with oblong and thin pycnidia. *Leptostroma pinastri* is a stage of *Lophoderium pinastri* (Ascomycetes).

Melasmia, with pycnidia lying in a flattened stroma. *Melasmia* species are imperfect stages of *Rhytisma* (Ascomycetes).

A representative genus of the Hyalodidymae (spores two-celled and hyaline) is *Gloeodes*. It is similar to *Leptothyrium*, but the pycnidia of *Gloeodes* are gelatinous. *G. pomigena* causes sooty blotch of apple. See FUNGI IMPERFECTI; PLANT DISEASE. [NIELS F. BUCHWALD]

Lernaeopodoida

A group of Crustacea known as the fish maggots, which are ectoparasites partially buried in the flesh of marine and fresh-water fish. They are characterized by (1) a modified postembryonic development, reduced to two or three recognizable stages, (2) a unique free-swimming larva, and (3) failure when sexually mature to show signs of external physical maturity typical of other copepods. The females, lacking distinct segmentation and functional thoracic appendages, appear in fish as comparatively large, wormlike, sometimes tufted outgrowths. The male (illustration *a*) is dwarfed and, though more distinctly segmented, does not exceed female morphological development. Using modified head appendages, the male clings to the female (illustration *b*) but retains some freedom of movement.

As a consequence of R. Gurney's discussion of Lernaeopodoida in 1933 and C. B. Wilson's account of the free-swimming stage of Sphyriidae in 1932, the Lernaeopodoida are limited in this account to the families Lernaeopodidae and Sphyriidae. Characteristics distinguishing these families are associated with their larval development and specializations for attachment to the host.

Life cycle. The unusual lernaeopodid life cycle begins with the eggs undergoing both embryonic and nauplial phases of development while retained in the elongated external ovisacs of the female. The free-swimming larva, equivalent to the first copepodite stage, emerges from the ovisac. Externally, it resembles the chalimus larva of Caligoida, but internally it is unique in having a preformed

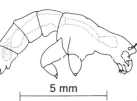
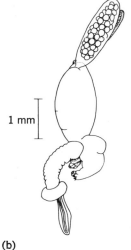
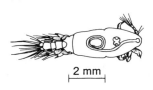

Lernaeopodoida. (*a*) Male and (*b*) female *Salmincola thymalli* (Kessler) (*based on R. Gurney, British Freshwater Copepoda, vol. 3, Ray Society Monogr., 1933*). (*c*) Free-swimming larva of *Paeon elongatus* (Wilson) (*after C. B. Wilson, The copepods of the Woods Hole region, Massachusetts, Bull. U.S. Nat. Mus., 158:509–531, 1932*).

frontal tube originating in the forehead and terminating in a coil above the esophagus. The larva, viable for only a short period, must find a host rapidly. A suitable host may be any one of perhaps several species of fish; the degree of host-parasite specificity is usually not pronounced. Adhering to an appropriate site with hooklike appendages, the larva makes a wound in the host tissue; the coiled tube is everted and becomes embedded in the wounded area as the tissue regenerates. The tube, a temporary method for attachment, degenerates when the larva metamorphoses into the vermiform stage (illustration *c*).

Attachment and nutrition. Permanent attachment is accomplished in two ways. Lernaeopodidae are attached to the walls of the fishes' gill chamber by strongly modified second maxillae. These appendages assume an elongated sausage-like appearance obscuring normal segmentation. The ends fuse and the apex, a buttonlike swelling or bulla, lies buried in host tissue. In Sphyriidae, the head and part of the thorax become embedded either within the gill chamber or externally, often near the dorsal fin. *Salmincola salmonea*, parasitizing salmon, withstands the anadromous habits of its host and has a life expectancy in excess of 1 year. In both families typical hosts include sharks, skates, rays, and most bony fishes.

Lernaeopodoida appear to feed on body fluids of the host. They are not considered a threat to fish populations except possibly under conditions of excessive crowding. *See* COPEPODA.

[ABRAHAM FLEMINGER]

Bibliography: R. Gurney, *British Fresh-water Copepoda*, vol. 3, Ray Soc. Monogr., 1933; C. B. Wilson, The copepods of the Woods Hole region, Massachusetts, *Bull. U.S. Nat. Mus.*, 158:509–531, 1932.

Lespedeza

A warm-season legume with trifoliate leaves, small purple pea-shaped blossoms, and one seed per pod. There are 15 American and more than 100 Asiatic species; two annual species, *Lespedeza striata* and *L. stipulacea*, and a perennial, *L. cuneata*, from Asia are grown as field crops in the United States. The American species are small shrubby perennials (see illustration) found in open woods and on idle land, rarely in dense stands; they are harmless weeds. Common lespedeza, once known as Japanese clover, is a small variety of *L. striata*. This variety, unintentionally introduced in the 1840s and used widely until the late 1920s, has been replaced by a larger variety, Kobe. Korean lespedeza (*L. stipulacea*) has been widely grown since the mid-1920s and is preferred in the northern part of the lespedeza belt, from Missouri

Lespedeza. (*Soil Conservation Service*)

eastward, while Kobe is preferred in the lower part of the belt (south to the Gulf of Mexico and across northern Florida).

The annual lespedezas can grow on poorer wetter land than most other forage crops. They give a moderate yield of good quality. The plants reseed well and the young seedlings at the two-leaf stage are very tolerant of cold. Most crop damage is from parasitic nematodes and severe drought, especially in the spring.

The perennial lespedeza (*L. cuneata*), commonly known as sericea, is adapted to the same growing conditions as the annuals but persists better on steep slopes and for this reason has been used widely in soil conservation. The young growth from well-established sericea is a moderate-quality forage, but the older growth is poor. The perennial is difficult to establish because the seed does not germinate readily unless scarified or exposed to cold and moisture for several months.

Several large species of lespedeza, such as the widely used *L. bicolor*, are used in plantings for wild life to provide seed and cover. *See* LEGUME FORAGES; NEMATODA. [PAUL TABOR]

Lethal dose 50

One special form of the effective dose 50, or median effective dose. The lethal dose 50, or median lethal dose, is also written as LD_{50}. It is used when the response that is being investigated is the death of the experimental animal. The median lethal dose is therefore the dose which is fatal to 50% of the test animals.

The LD_{50} for strychnine given subcutaneously to rabbits is 0.5 mg/kg of body weight. This description shows that in specifying the LD_{50} attention has been given to the species of animal, the route of injection, and the weight of the animals used. For other toxic substances a different set of specifications might be important.

The use of the concept of LD_{50} is not restricted to toxicology, because there are many therapeutic substances, such as digitalis, which can be assayed more conveniently by their lethal effect than by their therapeutic effect. The same active principle is involved in both effects. *See* BIOASSAY; EFFECTIVE DOSE 50; TOXICOLOGY. [COLIN WHITE]

Lethal gene

A gene which brings about the death of the organism carrying it. Lethal genes constitute the most common class of gene changes (that is, mutations) and are reflections of the fact that the fundamental function of genes is the control of processes essential to the growth and development of organisms. In higher diploid forms lethals are usually recessive and expressed only in homozygotes. Dominant lethals, expressed in heterozygotes, are rapidly eliminated and thus rarely detected. Recessive zygotic lethals are retained with considerable frequency in natural populations of cross-fertilizing organisms, while gametic lethals (those affecting normal functioning of eggs and sperm among animals, or the pollen and ovules of plants) are subject to stringent selection and are accordingly rare.

Evaluation and analysis. Some lethal mutations have proved to be losses of small or large sections of chromosomal material, rather than gene changes in the strictest sense. The presence of lethal mutants or losses is expressed as a failure of growth, leading to the premature death of the organism carrying them. As most detectable lethal genes are recessive, their effects are observed only in homozygotes. If the processes controlled by a particular gene come early in the developmental sequence, the disturbance produced by its loss or mutation is often severe and extensive; if the processes occur later in the sequence, the disturbance usually has fewer ramifications. An analysis of the effects of lethal genes provides a valuable means of investigating the complex processes of embryonic development and cellular differentiation and, in the case of man has practical medical implications as well. In microorganisms the study of lethal biochemical mutants opened up a whole new era in the unraveling of biosynthetic processes and led to the establishment of modern biochemical genetics. *See* CHROMOSOME; MOLECULAR BIOLOGY; MUTATION.

Conditional mutants. Although typically lethal genes express themselves independently of internal and external environments, there are cases in which expression is dependent on some particular condition. The temperature-sensitive (*ts*) and amber (*am*) mutants in bacteriophage T4D provide instances where the mutant behaves as a lethal under one set of conditions (restrictive) while under another set of conditions the phenotype of the mutant resembles wild type. Such conditional mutants have been used to analyze gene function in the synthesis and assembly of the macromolecules making up bacteriophage particles. Over 40 conditional lethal mutants have been shown to affect the processes leading to the complete morphogenesis of T4D phage particles. Such conditional lethals are thus invaluable tools for full analysis of gene action in the organization and functioning of biological systems.

Obtaining test materials. One of the difficulties in establishing the mechanism of action of lethal genes in higher organisms is that only one-quarter of the progeny in crosses of heterozygous lethals are homozygous. Analysis at the biochemical level can be greatly simplified if progenies consisting wholly of homozygous lethal individuals are readily obtained. Conditional lethal mutants of the temperature-sensitive type, readily produced in the fly *Drosophila* with the aid of chemical mutagens, viable and fertile when raised at 18°C and completely lethal at 28°C, have been the subject of active investigation in several laboratories. By maintaining flies at 18°C and transferring eggs to 28°C, any quantity of lethal tissue becomes available for biochemical study.

Occurrence in natural populations. Lethal genes in natural populations of plants and animals have received considerable attention from geneticists. In certain instances these may form the bases of polymorphisms in such populations. In some organisms they have led to what are known as balanced lethal systems: When two different lethal genes (or losses) are located at different points in different members of a pair of homologous chromosomes in such positions, or circumstances, that recombination by crossing over is prevented, only those individuals heterozygous for both of the two different lethals survive. In this way strains, or even species, which are permanent heterozygotes

have arisen as in the evening primroses of the genus *Oenothera*. In human populations the accumulation of lethal mutations has been a matter of some concern in the face of increases in mutagenic agents in the environment and the lowering of selective pressures associated with modern life. The human wastage and misery which accompany high frequencies of lethal genes are considerable. *See* POLYMORPHISM (GENETICS).

Examples. Lethal genes include the following: biochemical lethals in microorganisms such as *Neurospora*, *Escherichia*, and *Salmonella*; lethal chlorophyll mutants of higher plants; notch and many others in the fruit fly *Drosophila*; creeper in fowl; yellow, brachyury, W-anemia in the mouse; gray-lethal in the rat; Dexter in cattle; thalassemia, sickle-cell anemia, and many others in man. *See* GENE ACTION; HUMAN GENETICS.

[DONALD F. POULSON]

Bibliography: D. M. Bonner and S. E. Mills, *Heredity*, 2d ed., 1964; T. Dobzhansky, *Mankind Evolving*, 1962; E. Hadorn, *Developmental Genetics and Lethal Factors*, 1961; W. Hayes, *The Genetics of Bacteria and Their Viruses*, 2d ed., 1968; M. W. Strickberger, *Genetics*, 1968; R. P. Wagner and H. K. Mitchell, *Genetics and Metabolism*, 2d ed., 1964.

Lettuce

A cool-season annual, *Lactuca sativa*, of Asian origin and belonging to the plant order Asterales. Lettuce is grown for its succulent leaves, which are eaten raw as a salad. Four subspecies of this leading salad crop are head lettuce (*L. sativa* var. *capitata*) (see illustration), leaf or curled lettuce (*L. sativa* var. *crispa*), cos or romaine lettuce (*L. sativa* var. *longifolia*), and stem or asparagus lettuce (*L. sativa* var. *asparagina*). There are two types of head lettuce, butterhead, and crisphead or iceberg.

Propagation. The outdoor crop is propagated by seed usually planted directly in the field, but also occasionally planted in greenhouses for later transplanting. Field spacing varies, but usually plants are grown 10–16 in. apart in 14–20-in. rows. Greenhouse lettuce, predominantly leaf and butterhead varieties, is transplanted to ground beds with plants placed 7–12 in. apart.

Uniformly cool weather promotes maximum

Head lettuce (*Lactuca sativa*), cultivar Fulton, of the plant order Asterales. (*J. Harris Co., Rochester, N.Y.*)

yields of high-quality lettuce; 55–65°F is optimum. Heading varieties (cultivars) are particularly sensitive to adverse environment. High temperatures prevent heading, encourage seed-stalk development, and result in bitter flavor and tip-burned leaves. However, varieties vary considerably in their resistance to these high-temperature effects and to diseases. Commercial production of lettuce is extensive in several California and Arizona valleys where mild winter and cool summer climates prevail.

Crisphead or iceberg lettuce is the most widely grown type; strains of the Great Lakes and Imperial varieties account for most of the acreage. Popular varieties of other types of head lettuce are butterhead, Bibb, and White Boston. Grand Rapids, cos or romaine, and White Paris are popular types of leaf lettuce.

Harvesting. The harvesting of heading varieties begins when the heads have become firm enough to satisfy market demands, usually 60–80 days after planting. Most western-grown lettuce is field-packed in paperboard cartons and chilled by vacuum cooling.

Leaf lettuce and cos lettuce are harvested when full-sized, but before the development of seed stalks or a bitter taste. This varies from 40 to 70 days after planting.

California raises more lettuce than any other state; Arizona and Texas are next in importance. Total annual farm value in the United States from approximately 220,000 acres is about $200,000,000. *See* ASTERALES; VEGETABLE GROWING.

[H. JOHN CAREW]

Diseases. Lettuce diseases of greatest importance are caused by fungi and bacteria which disrupt the structure and function of the invaded parts causing malformation, discoloration, and breakdown of tissues. The injuries produced are usually classified as spots, rots, scabs, wilts, or blights.

The most serious fungus diseases and their causal organisms are sclerotinia rot, *Sclerotinia sclerotiorum*; gray mold rot, *Botrytis cinerea*; downy mildew, *Bremia lactucae*; anthracnose, *Marssonina panattoniana*; and bottom rot, *Pelli-*

cularia filamentosa. Several fungi of minor importance cause leaf spots.

The most serious of the bacteria parasitic on lettuce are *Erwinia carotovora* and *Pseudomonas marginalis*, both of which cause soft rots. Soft rots are especially damaging when they follow injuries and diseases, such as spotted wilt and tip burn.

Virus diseases of lettuce are usually transmitted by insects that feed on the plants. Big vein disease, however, is believed to be caused by a soil-inhabiting virus that enters the plant through its root system. Other virus diseases of importance are mosaic, yellows, and spotted wilt.

Tip burn, a physiological disease, is especially damaging to head lettuce. The exact cause of this trouble is not known. *See* PLANT DISEASE; PLANT PHYSIOLOGY; PLANT VIRUS.   [GLEN B. RAMSEY]

Leucettida

An order of the subclass Calcinea in the class Calcarea. These sponges have either a radiate arrangement of the flagellated chambers or a leuconoid structure. A distinct dermal membrane or cortex is present. The spongocoel is not lined with choanocytes; these cells are restricted to the flagellated chambers. Two families are recognized, the Leucascidae and Leucaltidae; and the genera *Leucascus*, *Leucetta*, and *Leucaltis* are examples. *See* CALCAREA; CALCINEA.

[WILLARD D. HARTMAN]

Leucine

An amino acid considered essential for normal growth of animals. The amino acids are characterized physically by the following: (1) the pK_1, or the dissociation constant of the various titratable groups; (2) the isoelectric point, or pH at which a dipolar ion does not migrate in an electric field; (3) the optical rotation, or the rotation imparted to a beam of plane-polarized light (frequently the D line of the sodium spectrum) passing through 1 dm of a solution of 100 g in 100 ml; and (4) solubility. *See* EQUILIBRIUM, IONIC; ISOELECTRIC POINT; OPTICAL ACTIVITY; SPECTROPHOTOMETRIC ANALYSIS.

In leucine biosynthesis, carboxyl and some carbon atoms of leucine are derived from acetate; the

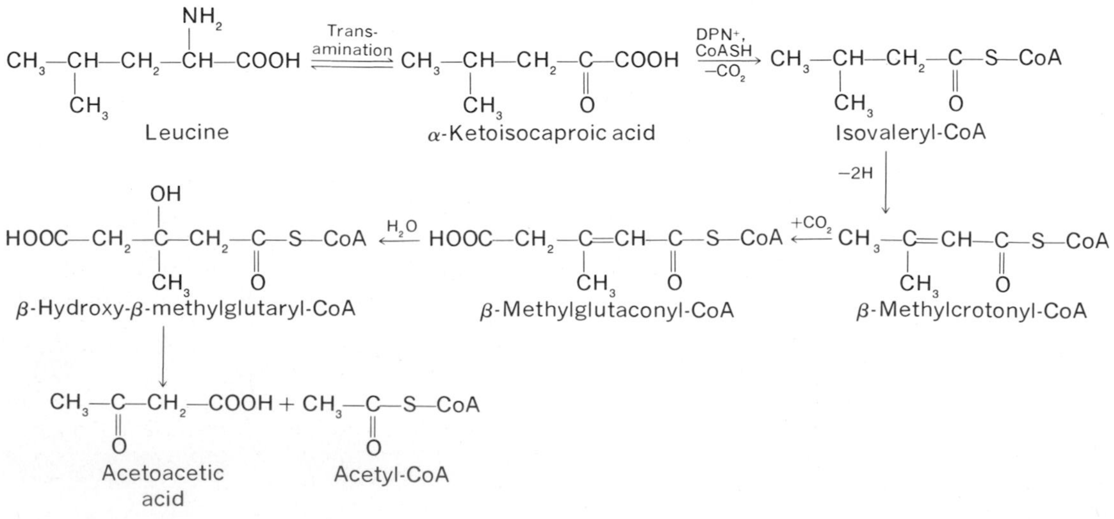

Metabolic degradation pathway of leucine.

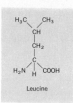

Physical constants of the L isomer at 25°C:
pK₁ (COOH): 2.36; pK₂ (NH₃⁺): 9.60
Isoelectric point: 5.98
Optical rotation: [α]_D(H₂O): −11.0; [α]_D(5 N HCl): +16.0
Solubility (g/100 ml H₂O): 2.19

rest of the molecule is furnished by α-ketoisovalerate, which arises in the biosynthesis of valine. A condensation of acetylcoenzyme A (acetyl-CoA) with α-ketoisovalerate, followed by a series of reactions analogous to the tricarboxylic acid cycle, including isomerization, dehydrogenation, and decarboxylation, leads to the formation of α-ketoisocaproic acid. This compound is known to yield leucine by transamination. *See* AMINO ACIDS; TRANSAMINATION.

During metabolic degradation the first steps are deamination and oxidative decarboxylation, forming isovalerylcoenzyme A (isovaleryl-CoA). This compound is ultimately converted to acetoacetic acid and acetyl-CoA (see illustration).

[EDWARD A. ADELBERG]

Leucite

A framework structure silicate of the feldspathoidal mineral group. It is commonly found as euhedral phenocrysts in K₂O-rich and SiO₂-poor volcanic rocks, consistent with its chemical composition, (K,Na)AlSi₂O₆, where K is more abundant than Na. The crystals are typically white, and the luster varies from dull to vitreous. Mohs hardness lies between 5.5 and 6. Density ranges from 2.45 to 2.5. Crystals, exhibiting the characteristic isometric trapezohedral form (see illustration), as large as 7 cm are common in the lavas of Nyiragongo volcano, Congo. Other well-known areas include: West Kimberley, Australia; Mount Vesuvius, Italy; Highwood Mountains and Bearpaw Mountains, Montana; and the Leucite Hills, Wyoming.

At low temperatures leucite is tetragonal; heating to 625°C results in a gradual change to the isometric symmetry consistent with the external morphology. This is presumably due to structural collapse of the (Al,Si)-O framework around the potassium ions which are not large enough to fill the cavities in the structure under low-temperature conditions.

Leucite found in the natural environment is usually nearly pure KAlSi₂O₆. Experimental phase equilibrium studies have demonstrated that large amounts of Na can substitute for K at high temperatures. If cooled rapidly, these crystals can persist metastably and finally react to form pseudoleucite, a mixture of nepheline (nearly [Na.₇₅ K.₂₅]AlSiO₄) and potassium feldspar (KAlSi₃O₈). Pseudoleucite may also form through reaction of magmatic leucite crystals with residual silicate liquid, suggested by the existence of crystals that have cores of leucite and rims of pseudoleucite. Leucite, as pure KAlSi₂O₆, is not thermodynamically stable at low temperature and high pressure, a mixture of potassium feldspar and kalsilite (KAlSiO₄) being favored. The boundary between the fields of leucite and kalsilite plus potassium feldspar is defined by the equation $P = 3.230T - 1583$, where pressure, P, is in kilobars, and the temperature, T, is in de-

grees celsius, with leucite existing only on the high-temperature and low-pressure side of the boundary. This observation, based on experimental data, is consistent with the absence of leucite in plutonic rocks that have crystallized at relatively high pressures. [WILLIAM LUTH]

Leucite rock

Igneous rocks rich in leucite but lacking or poor in alkali feldspar. Those types with essential alkali feldspar are classed as phonolites, feldspathoidal syenite, and feldspathoidal monzonite. The group includes an extremely wide assortment both chemically and mineralogically.

The rocks are generally dark-colored and aphanitic (not visibly crystalline) types of volcanic origin. They consist principally of pyroxene and leucite and may or may not contain calcic plagioclase or olivine. Types with plagioclase in excess of 10% are called leucite basanite (if olivine is present) and leucite tephrite (if olivine is absent). Types with 10% or less plagioclase are called leucitite (if olivine is absent) and olivine leucitite or leucite basalt (if olivine is present).

The texture is usually porphyritic with large crystals (phenocrysts) of augite and leucite in a very fine-grained or partly glassy matrix. If plagioclase occurs as phenocrysts, it is generally labradorite or bytownite and is slightly more calcic than that of the rock matrix. It may be zoned with more calcic cores surrounded by more sodic margins. Leucite appears in two generations. As large phenocrysts it forms slightly rounded to octagonal grains with abundant tiny inclusions of glass or other minerals zonally arranged. Small, round grains of leucite with tiny glass inclusions also occur in the rock matrix. Augite or diopside (sometimes rimmed with aegirine-augite) and aegirine-augite form the mafic phenocrysts. Pyroxene of the matrix is commonly soda-rich. Olivine may occur as well-formed phenocrysts. Other minerals present may include nepheline, sodalite, biotite, hornblende, and melilite. Accessories include sphene, magnetite, ilmenite, apatite, and perovskite.

Leucite rocks are rare. They occur principally as lava flows and small intrusives (dikes and volcanic plugs). Well known are the leucite rocks of the Roman province, in Italy, and the east-central African province. In the Italian area the feldspathoidal lavas are essentially leucite basalts and may have developed by differentiation of basalt magma (rock melt). In the African province the leucitic rocks are associated with ultramafic (peridotite) rocks and may have been derived from peridotitic material. This may have been accomplished by the abstraction of early formed crystals from a peridotitic magma or by the mobilization of peridotitic rocks by emanations from depth. Assimilation of limestone by basaltic magma may help to decrease the silica content and promote the formation of leucite instead of potash feldspar. The crystallization of leucite, however, is in large part a function of temperature and water content of the magma. Conditions of formation, therefore, may strongly influence the formation of leucitic rocks. *See* IGNEOUS ROCKS; NEPHELINE SYENITE; PETROGRAPHIC PROVINCE; PHONOLITE.

[CARLETON A. CHAPMAN]

LEUCITE

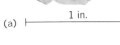
(a) 1 in.

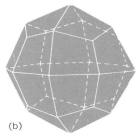
(b)

Leucite. (a) Crystals from Wiesenthal, Saxony (*specimen from Department of Geology, Bryn Mawr College*). (b) Crystal habit (*from C. S. Hurlbut, Jr., Dana's Manual of Mineralogy, 17th ed., Wiley, 1959*).

Leucosoleniida

An order of the subclass Calcaronea in the class Calcarea. These sponges have an asconoid structure. A true dermal membrane or cortex does not develop in this order, and the spongocoel is lined with choanocytes. Leucosoleniidae is the one family recognized, and the genera *Leucosolenia* and *Ascyssa* are examples of this order. *See* CALCAREA; CALCARONEA.

[WILLARD D. HARTMAN]

Leukemia

A disease of humans and animal characterized by generalized neoplastic proliferation of leukocytes, which are usually present in increased numbers in the bloodstream. As a result of uncontrolled proliferation, leukocytes or their precursors progressively replace the blood-forming tissues of the body, crowding out normal cellular elements so that anemia and thrombocytopenia ensue. Leukemic cells also tend to infiltrate certain other organs and tissues of the body, producing symptoms referable to the specific structures involved.

Classification. Leukemia includes a spectrum of clinical and cytological entities which may be subdivided according to three main criteria: the type of leukocyte affected; the degree of cytological differentiation; and the number of leukemic cells in the peripheral blood.

Type of leukocyte affected. Since only one cell type undergoes neoplastic transformation in a given individual, identification of the cell line involved provides a major basis for classification. Three major types of leukemia occur, corresponding to the three main varieties of leukocyte in the blood: granulocytes or polymorphonuclears, lymphocytes, and monocytes (Fig. 1). Types of leukemia are termed granulocytic (or myelogenous), lymphocytic, and monocytic (Fig. 2). Since eosinophilic and basophilic leukemias are rare and most cases of granulocytic leukemia affect the neutrophilic series of granulocytes, the unmodified term granulocytic leukemia implies that it is the neutrophilic series which is neoplastic. A transitional form of leukemia with cytological features suggestive of both myelogenous and monocytic differentiation is occasionally observed and is termed myelomonocytic leukemia. One explanation for this type of leukemia is that the granulocytic and monocytic series of leukocytes may be derived from a common precursor (stem) cell. On rare occasions red blood cell precursors in the bone marrow undergo a neoplastic proliferation; such a condition is termed erythroleukemia. *See* HEMATOPOIESIS.

Degree of cytological differentiation. In many cases of leukemia the majority of abnormal cells are incapable of differentiation beyond the stem cell, or blast stage. Such cases, usually seen in children, commonly pursue a more malignant course than other leukemias. They ordinarily cause death within a few months, unless responsive to therapy, and are referred to as acute leukemias. When the leukocytes are so undifferentiated that it is impossible to determine the type of leukocyte affected, the term stem-cell leukemia is customarily used. If the abnormal cells are identifiable as blasts of one of the three major cell types listed above, the disease is termed acute lymphocytic (or lymphoblastic), acute granulocytic (or myeloblastic), or acute monocytic (or monoblastic) leukemia. Sometimes the neoplastic cells may differentiate and approach full maturation, being almost indistinguishable from mature leukocytes. In this instance the disease is usually more slowly progressive, often lasting for several years, and the term chronic leukemia is applied.

Between these two extremes are intermediate forms for which the designation subacute leukemia is sometimes used. It is important to recognize that, although in general the degree of differentiation closely parallels duration of clinical course, many exceptions occur. This is particularly true since therapy of acute leukemia has proven capable of prolonging life for several years in many instances.

Number of circulating leukemic cells. Most often, leukemia is characterized by a markedly increased white blood cell count, which often attains levels of 5–50 times normal values. In fact, the term leukemia literally means white blood, a characteristic appearance noted by early observers of the disease. However, there may be a normal or low white blood cell count with no abnormal cells in peripheral blood. These uncommonly encoun-

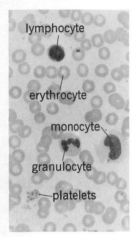

LEUKEMIA

lymphocyte

erythrocyte

monocyte

granulocyte

platelets

Fig. 1. A normal stained blood smear showing a granulocyte, lymphocyte, monocyte, some erythrocytes, and a clump of platelets.

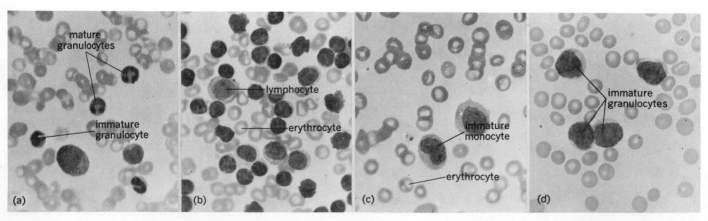

Fig. 2. Stained blood smears in different types of leukemias. (*a*) Chronic granulocytic leukemia showing mature and immature granulocytes. (*b*) Chronic lymphocytic leukemia showing many mature and some immature lymphocytes. (*c*) Monocytic leukemia showing two immature monocytes. (*d*) Acute granulocytic leukemia showing four very immature granulocytes.

tered cases are given the illogical but widely accepted name of aleukemic leukemia. In other cases, also with a low or normal total count, leukemia cells may be readily recognizable in smears of peripheral blood; the term subleukemic leukemia is often applied. *See* BLOOD.

Lymphoma. A group of neoplastic diseases known as the lymphomas are closely related to lymphocytic leukemia. In lymphomas there is malignant proliferation of immature lymphoid cells in a localized, nodular fashion, usually in lymph nodes or other lymphoid organs of the body. Transitional forms between lymphomas and leukemia occur, and in children approximately one-third of the cases of lymphoma eventuate as lymphocytic leukemia. Like lymphomas, the lymphocytic leukemias can be classified as B-lymphocyte (bursal equivalent–derived) or T-lymphocyte (thymus-derived) leukemias. Nearly all chronic lymphocytic leukemias are of B-lymphocyte origin, whereas acute lymphocytic leukemias usually display the cellular characteristics of T lymphocytes.

Incidence and epidemiology. Leukemia affects most mammalian species; it is found in domestic and laboratory animals, providing a convenient source for experimental studies of the disease. Mouse leukemia has proven an especially useful investigative tool, since hereditary and transmissible forms have been discovered which provide a means for accumulating large numbers of affected animals. A disease closely resembling, if not identical to leukemia, occurs in birds. This disorder, termed avian leukosis, occasionally occurs in epidemic form, especially among poultry. Much evidence suggests that avian leukosis is due to a transmissible virus. *See* AVIAN LEUKOSIS.

In humans leukemia is a major cause of death; approximately 0.6% of newborn infants are destined to develop leukemia at some time during life. In the United States the annual death rate from leukemia is about 7 for every 100,000 population, and the incidence has been increasing steadily since the 1920s.

Leukemia occurs at all ages and has even been reported in the newborn infant; the type varies with age. Acute leukemias are characteristically disorders of children and young adults, being the most frequent type of malignancy encountered during the childhood years. Chronic granulocytic leukemia becomes increasingly common after age 25, whereas chronic lymphocytic leukemia, the commonest form in elderly Caucasians, is rare before age 45. Monocytic leukemia, in either the acute or chronic form, is usually a disease of the middle adult years. Significant racial and geographic differences are also observed. Chronic lymphocytic leukemia, for example, is rare among Orientals of any age. Hence, the overall death rate from leukemia in Japan is less than half that observed in the United States. Mortality rates for leukemia are lower among Negroes than among Caucasians. Sex differences are inconsequential except for chronic lymphocytic leukemia, which is about three times more frequent among men than women.

Leukemia is not a hereditary disease in the usual sense, and even a pregnant woman ill with leukemia almost never bears an infant with the disease. However, there is a slight tendency toward familial aggregation of cases. A patient with leukemia is more likely to have relatives with leukemia than is the general population. Identical twins of children with leukemia have been reported to have a 1 in 5 risk of developing the disease.

A few "epidemics" of leukemia have been observed over the years. Perhaps the best known of these occurred in Niles, Ill., a suburb of Chicago, between 1957 and 1960. During this period eight cases of childhood leukemia occurred, representing an increase of 15–20 times over the expected rate for a population of that size. It must, however, be emphasized that despite intensive search few such instances have been encountered and their statistical significance remains uncertain.

Etiology. A great many factors have been suspected of causing leukemia, but few are supported by a substantial body of evidence.

Physical and chemical agents. Probably the most firmly established causative factor for human leukemia is ionizing radiation. This was first suspected when it became apparent that radiologists, who were exposed in their daily work to radiation in the form of x-rays, had an incidence of leukemia approximately 20 times that of the general population. Additional protective measures have been instituted which seem to have abolished this excessive risk for radiologists. Further evidence of the leukemogenic effect of radiation comes from the increased incidence of the disease in patients who have undergone radiotherapy of the spinal column for a chronic rheumatic condition known as ankylosing spondylitis. Finally, a decided increase in leukemia has been observed in survivors of the atomic-bomb attacks on Hiroshima and Nagasaki during World War II. The rise in leukemia incidence was found to be directly related to the distance from the center of the explosions. Hiroshima survivors who were less than 1000 m from the center of the blast suffered an incidence of leukemia approximately 50 times that for survivors between 2000 and 10,000 m away. *See* RADIATION INJURY (BIOLOGY); X-RAYS, PHYSICAL NATURE OF.

A number of chemical agents have been suspected of being leukemogenic, but proof is lacking, except for benzol, which seems capable of causing leukemia but only after years of heavy exposure.

Viral agents. Studies on the etiology of leukemia have been particularly concentrated on possible viral agents. It has been known for many years that fowl leukosis can be produced by viruses, and cell-free infectious material obtained from leukemic mice has also been repeatedly demonstrated to cause leukemia when inoculated into newborn suckling mice. In other mammals viruses have also been shown to be associated with leukemia, but for humans the evidence is unconvincing. In the late 1960s research centered upon (1) attempts to isolate viruses from leukemic patients; (2) attempted transfer of the disease from affected cases to animals; (3) immunological studies, aimed at identification of recent virus infection in cases of leukemia; (4) search for viral particles in specimens from cases, usually with the help of the electron microscope; and (5) search for "epidemics" of leukemia.

During the 1960s considerable interest arose in the demonstration of viruslike particles in blood and marrow specimens of leukemic patients; the particles are visible under the high resolution pos-

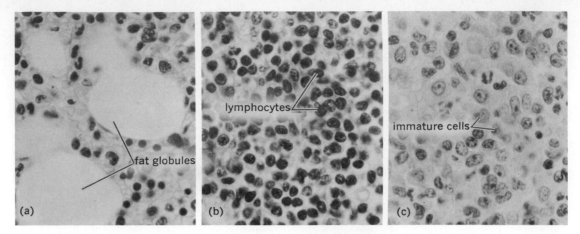

Fig. 3. Bone-marrow sections from normal and leuke-mic individuals. (*a*) Normal, showing variety of cells and globules of fat. (*b*) In chronic lymphocytic leukemia, showing almost complete replacement with lympho-cytes. (*c*) In acute granulocytic leukemia, showing near-ly complete replacement by very immature cells.

sible with the electron microscope. However, it was found that similar particles are equally fre-quent in specimens from nonleukemic controls. Despite the lack of positive proof of a viral etiology for human leukemia, the Epstein-Barr virus is known to cause infectious mononucleosis and has consistently been identified in and isolated from Burkitt lymphoma cells.

In 1973 R. Gallo and associates identified an enzyme, reverse transcriptase, in human acute leukemic cells. This enzyme was not found in nor-mal cells and has been postulated to be derived from a tumor virus. An antibody against a viral reverse transcriptase derived from subhuman primate sarcoma cells has been found to react with reverse transcriptase from human leukemia cells. These data strongly support a viral etiology for some cases of human leukemia. *See* VIRUS.

Chromosomal abnormalities. In 1956 individuals with Down's syndrome (Mongolism), were discov-ered to have an incidence of leukemia up to 30 times that which would be expected on a random basis. Three years later the cause of Down's syn-drome was traced to the presence of an extra chro-mosome in the cells of the body, giving a total of 47 instead of the normal human chromosome count of 46. The extra chromosome was found to corre-spond to the number 21 (smallest) pair of autosom-al chromosomes; instead of the normal comple-ment of two mongoloid individuals have three number 21 chromosomes.

The number 21 chromosome pair was subse-quently found also to be abnormal in many cases of chronic myelogenous leukemia: One of the num-ber 21 chromosomes is normal, but the other is incomplete, being represented only by a partial chromosome. This rudimentary chromosome has been termed the Philadelphia, or Ph-1, chromo-some after the city in which it was discovered. Hence two types of abnormality of the same pair of chromosomes have been found to be associated with leukemia.

Two diseases known to be related to excessive chromosomal breakage, Bloom's syndrome of he-reditary telangiectatic dwarfism and Fanconi's anemia, are also associated with an exaggerated risk of developing leukemia. Thus there is a sub-stantial body of evidence to suggest that chromo-somal abnormalities of certain types may predis-pose an individual toward leukemia.

Genetic imbalance. It is possible that ionizing radiation, viral infection, chemical damage, and chromosomal lesions may all produce a particular type of imbalance of genetic material that predis-poses to leukemia. For this reason, particular emphasis is being given to the search for other potential factors which disturb the distribution of genetic material in the cell. One of the most perti-nent of these is the drug lysergic acid diethylamide (LSD), which seems capable of causing a number of abnormal chromosomal ties resembling those of Bloom's syndrome and Fanconi's anemia. A search for leukemia, not only among adults who have taken this agent but also among fetuses who have been exposed before birth, is being under-taken. *See* ONCOLOGY.

Pathology and pathogenesis. Leukemia is char-acterized by abnormal proliferation of immature or incompletely differentiated leukocytes. Nor-mally the rates of cell production and cell de-struction are balanced, so that the number of leu-kocytes in the body remain reasonably constant. In leukemia this balance is disturbed, and the pro-duction rate exceeds the destruction rate; the ex-cess cells must be accommodated by the body. The earliest effects of this overproduction are usually seen in the blood-forming tissues, particularly the bone marrow (Fig. 3). Fatty tissue, which normal-ly occupies about one-half the volume of the blood-forming regions of the marrow, is progressively replaced by the excessive leukocytes. When this replacement is complete, the proliferating cells then crowd out normal cellular elements of the marrow, resulting in anemia, due to decreased red cell production, and thrombocytopenia, due to re-placement of the platelet-forming cells, or mega-karyocytes. The production of normal leukocytes capable of effectively resisting infection is also hindered. Usually abnormal leukemic cells reach the circulating blood, where they may produce striking elevation in the white blood cell count.

In addition to the peripheral blood and bone marrow, certain other organs of the body, such as the lymph nodes, spleen, and liver (Fig. 4*a*), are almost always infiltrated by and filled with large numbers of leukemic cells; this produces enlargement of these organs and disturbances of their normal functions. Almost any organ of the body may be infiltrated: the meninges surrounding the brain and spinal cord may be involved, as may the kidneys (Fig. 4b), adrenal glands, and even the heart. Symptoms referable to interference with normal function of infiltrated organs and tissues are seen in most cases.

Clinical signs and symptoms. The chief symptoms of leukemia are weakness and listlessness, severe anemia, hemorrhages into the skin or mucosal surfaces, and enlargement of spleen, liver, or peripheral lymph nodes. In a patient with symptoms suspiciously like those of leukemia, the usual first step in diagnosis is examination of the peripheral blood. Usually, however, additional diagnostic measures are utilized, the most important of which is examination of the bone marrow. For this procedure a specimen is withdrawn through a needle introduced into the pelvis, sternum, or other readily accessible, superficially placed bone. In many cases biopsies of lymph nodes or liver are also obtained, and in some institutions puncture of the spleen in a manner similar to that used for the bone marrow has proven useful. *See* CLINICAL PATHOLOGY.

Anemia, usually prominent at the outset, is a severe complication in most cases of leukemia. This is due not only to crowding out of red-cell precursors in the marrow but often also to increased destruction of circulating red cells by the enlarged spleen. Hemorrhage usually contributes to the anemia and is one of the major causes of death in leukemia patients. This results chiefly from thrombocytopenia (a decrease in blood platelets), but in some cases severe infiltration of the liver by leukemic cells contributes to the problem by interfering with the synthesis of proteins used in forming blood clots.

Susceptibility to infection is a particular problem for the leukemic patient, primarily due to the reduced numbers of normally functional leukocytes, the body's first line of defense against infection. The patient's ability to withstand infection is also reduced by other contributory factors, including anemia, general debility, and side effects of drugs used in treating leukemia. Many organisms which are relatively innocuous for normal persons are capable of causing severe problems and even death in leukemic patients. Fungi, in particular, are in this category, but bacteria, viruses, and protozoans all present major hazards to such patients.

Other symptoms referable to infiltration of specific organs and tissues are usually present at some time during the course of the disease; these symptoms are extremely variable from patient to patient. For example, the overcrowding of bone marrow by leukemic cells may produce bone destruction, leading to fractures or collapsed vertebrae; infiltrates of the membranes surrounding the brain and spinal cord may cause neurological symptoms; enlarged lymph nodes may compress vital structures, such as the intestinal tract, bile ducts, or urinary tract; and infiltration of the liver may produce jaundice or hepatic failure.

It should be emphasized that during periods of clinical remission, such as may be obtained with the use of modern therapeutic measures, the patient is usually entirely well and free of symptoms and often is capable of more or less normal activity.

Course and treatment. Leukemia is usually a fatal disease, but remarkable variations in duration of course and severity of symptoms are seen. These variations are to a large extent determined by the type of leukemia; however, within any given category of leukemia there are individual variations. Thus, a patient with chronic leukemia may succumb to a fatal hemorrhage within a few days after the diagnosis is established, yet another individual with the same disease may survive with few symptoms for over a decade. Precise survival figures are thus misleading when applied to a specific individual, and are constantly being revised upward with each new advance in therapy. In general, the median survival for patients with acute leukemia, in the absence of specific chemotherapy, is 2.4 months for children and 1.7 months for adults, regardless of cell type affected. For chronic leukemia the median survival is about 2.7 years; however, very long remissions of chronic leukemia are frequently observed. About 20% of leukemia patients survive at least 5 years from onset of symptoms to death, and about 6% can be expected to survive for longer than 10 years.

Advances in the therapy of leukemia have brought about a constant upward revision of the survival figures given above. A number of agents have been developed which are capable of producing disease-free intervals, or remissions, some of which may last for several years. During complete remissions patients with leukemia are able to lead an almost normal life. The effect of modern therapy is perhaps most dramatic in acute leukemia of children. Complete remissions of over a year are frequently produced and some have lasted for over 10 years. A median survival of 22 months in acute lymphocytic or stem-cell leukemia of children is now being reported. Not infrequently, following relapse from an initial remission, additional remissions may be obtained by changing therapeutic agents. However, the initial remission is usually the longest. Technological advances such as antibiotics and advances in blood transfusion have increased the benefits derived from specific antileukemic therapy and often are capable of sustaining life until a remission can be obtained. Such measures, and the use of x-ray therapy for local control of disease, are capable not only of prolonging life but also of making the patient infinitely more comfortable. Many of the serious complications of leukemia can be prevented entirely, and others can be rendered less serious.

Chemotherapy. Several categories of drugs are in common use in the chemotherapy of leukemia, and new agents are being discovered at an impressive rate. Drugs in all the categories are capable of inducing remissions in individual cases. Each agent has a specific spectrum of effectiveness, and no one drug is useful in the treatment of all types of leukemia. These agents may be employed singly or in combination.

Antimetabolites. These are compounds which are capable of blocking specific metabolic processes essential for the survival and reproduction of

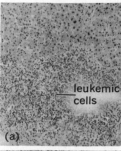

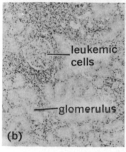

Fig. 4. Tissue sections from (a) liver and (b) kidney of individual with chronic granulocytic leukemia, showing infiltration by leukemic cells.

leukemic cells. Many antimetabolites act by interfering with the metabolism of folic acid, purines, and pyramidines, all of which are necessary in the synthesis of deoxyribonucleic acid (DNA); notable among these are 6-mercaptopurine and amethopterin, or methotrexate. These antimetabolite compounds were synthesized to substitute as normal metabolic precursors of DNA. Structurally they are almost identical to the normal metabolite and are therefore accepted by the cell. However, once incorporated into the cell, they block its activity because of the presence of specific chemical groups differing from the normal precursor. *See* DEOXYRIBONUCLEIC ACID (DNA); ENZYME INHIBITION.

Alkylating agents. These are drugs which are unstable in solution; they break down to liberate active alkyl groups capable of combining with, and inactivating, certain important cellular components. Over 25 alkylating agents are of proven value in the treatment of leukemia, including the nitrogen mustards, busulfan, and cyclophosphamide.

Antimitotic drugs and hormones. A third major group are often termed the antimitotic drugs; these interfere with mitotic cellular division. Two derivatives of the periwinkle plant, vincristine and vinblastine, are in this category, and both are highly effective in certain cases of leukemia. Finally, adrenocorticotropic hormone (ACTH) and adrenal steroids form a major line of defense in the specific chemotherapy of leukemia.

Bone marrow transplantation. Bone marrow transplantation has become an accepted form of therapy for some cases of leukemia. Patients who receive bone marrow transplants are first treated with total-body irradiation and high-dose chemotherapy to destroy the leukemic cells (and also normal bone marrow cells). The bone marrow to be transplanted is usually obtained from a close relative with the same cellular antigenic characteristics as the patient. The bone marrow obtained from the donor is filtered and injected intravenously into the patient. The injected bone marrow cells circulate in the blood and eventually lodge in the marrow cavities where they proliferate to form new circulating blood cells.

Complications and limitations. Unfortunately, every drug in current use has dangerous side effects. Most are toxic to normal marrow cells and to other rapidly proliferating normal cells of the body, such as those lining the intestinal tract. Therefore great skill and experience is required in the proper administration of these powerful drugs, and cancer chemotherapy is rapidly becoming a field for medical specialization. Not infrequently, leukemic patients who are receiving chemotherapy develop infections with organisms that usually do not cause the infection in persons who do not have the disease. Such infections require very aggressive antibiotic therapy and may be difficult to eradicate. Radiant energy in the form of conventional x-rays has a role in the therapy of leukemia; however, this therapy is limited by the fact that radiation of the entire body is highly toxic, and leukemia is a generalized disease. More useful, however, are certain radioactive isotopes which are selectively taken up by certain cell types, including leukemia cells. Unfortunately, remis-

sions cannot be induced in every case, and no agent has yet been discovered which is capable of inducing permanent remission of leukemia. Eventually, even after a remission lasting many years, the disease recurs. Many workers suspect that leukemic cells bypass the inhibitory effects of a given agent by an adaptive evolutionary process comparable to the appearance of antibiotic-resistant strains of bacteria. *See* BACTERIAL GENETICS; CIRCULATION DISORDERS; DRUG RESISTANCE.

[J. BRUCE BECKWITH]

Bibliography: G. D. Amromin, *Pathology of Leukemia*, 1968; W. Dameshek and R. M. Dutch, *Perspectives in Leukemia*, 1968; W. Dameshek and F. Gunz, *Leukemia*, 2d ed., 1964; J. B. Miale, *Laboratory Medicine: Hematology*, 4th ed., 1972; R. W. Miller, Persons with exceptionally high risk of leukemia, *Cancer Res.*, 27:2420–2423, 1967; H. Temin, Malignant transformation of cells by viruses, *Perspect. Biol. Med.*, 14:11, 1970; M. W. Wintrobe, *Clinical Hematology*, 7th ed., 1974.

Leukocidin

A substance that destroys leukocytes, the white cells of blood. Leukocidins are extracellular substances which appear in the environment of certain species of growing bacteria. They are released into culture media, and their physical and chemical properties suggest that they are proteins.

Leukocidins can be detected by a variety of methods, for example, by impairment of capacity of leukocytes to reduce methylene blue or other hydrogen acceptors, reduction or loss of locomotion of leukocytes, alterations in cellular morphology as revealed by examination of living cells or stained preparations, release of lysozyme or other enzymes by damaged leukocytes, and failure of leukocytes to agglutinate in the presence of kaolin and alkali. The most thoroughly studied leukocidins are those produced by staphylococci and streptococci. In most instances their role in the genesis of disease has not been clearly established. *See* STAPHYLOCOCCUS; STREPTOCOCCUS.

[ALAN W. BERNHEIMER]

Bibliography: American Physiological Society, *Handbook of Physiology*, 2 vols., 1962, 1963; G. P. Gladstone and W. E. van Heyningen, Staphylococcal leucocidins, *Brit. J. Exp. Pathol.*, 38:123–137, 1957; E. W. Todd, The leucocidin of group A haemolytic streptococci, *Brit. J. Exp. Pathol.*, 23: 136–145, 1942.

Leukopenia

The condition of the blood characterized by a decrease in the number of circulating white cells, or leukocytes. It is not a disease in itself, but rather a symptom that may occur in the course of a great variety of diseases. Its importance in any given case depends on its severity and on the nature of the underlying pathologic process. In some conditions, leukopenia may be an incidental concomitant; in others, it may be the outstanding feature, and in itself a serious threat to life.

The leukocytes play an essential role in the body's mechanism of defense against infection by many kinds of microorganisms. A marked decrease of available white cells weakens this defense, and this threat of uncontrollable infection constitutes the menace of leukopenia.

The normal white cell count in man ranges between 5000 and 10,000 cells per cubic millimeter of blood. Counts between 2000 and 5000 are common in a number of conditions and may not be of great significance; however, counts below 2000 are of grave import. Of the three kinds of white cells found in normal blood, the granulocytes, the lymphocytes, and the monocytes, it is the granulocytes that are largely affected in most forms of leukopenia, and these cells are the most important in resistance to infection. *See* BLOOD; INFECTION.

Etiology and pathogenesis. Leukopenia occurs as an incidental finding in a vast and heterogeneous group of pathologic conditions. Among these are infections, including those of bacterial, viral, and protozoal origin, for example, typhoid fever, influenza, and malaria. However, in most of the common bacterial infections the white count is elevated, reflecting the increased production of white cells by the marrow and their transport to the involved tissues. But even these infections, if overwhelming, may be associated with leukopenia if the demand for white cells is so great that the resources of the marrow are exhausted.

Leukopenia is regularly found in a group of diseases whose only common factor is enlargement of the spleen, for example, cirrhosis of the liver. The mechanism of this form of leukopenia is not entirely understood. Some investigators postulate a hormonelike substance, produced in increased amounts by an enlarged spleen, that specifically inhibits development of leukocytes in the marrow. Attempts to isolate and identify this hormone have not been successful.

Advanced and widespread cancer may produce leukopenia when it invades the bone marrow extensively and effectively crowds out the white-cell-producing tissues. In this condition, known as myelophthisis, there is an associated anemia due to the same cause.

Leukopenia, sometimes severe and life-threatening, is regularly produced by certain physical and chemical agents when they are given in sufficient dosage. Among these are x-rays, benzene, arsenic, and a group of synthetic compounds used in the treatment of leukemia. These agents all have the power of selectively injuring rapidly dividing and proliferating cells. They thus impair multiplication of the white cell precursors in the marrow, which are among the most actively dividing cells in the body. Accidental exposure to sufficient dosage of radiation—as from the products of nuclear weapons, or of benzene, which may occur in industrial exposure—may result in a severe, progressive, and fatal leukopenia. Smaller doses produce a milder leukopenia or none at all. The drugs used in the treatment of leukemia and other neoplastic diseases are given in dosages carefully controlled to keep the bone marrow depression within tolerable limits. However, the leukopenia induced by therapy occasionally becomes a clinical problem.

Another chemically and pharmacologically diverse group of drugs, many of them in common use, rarely or occasionally produce a severe leukopenia in especially sensitive patients, although they have no effect on the white cells of the great majority of the patients who receive them. Amidopyrine is a well-known offender, and several sulfonamides, antihistamines, anticonvulsant agents, and a variety of others are occasionally implicated. The mechanism whereby these drugs can produce such devastating effects in an occasional patient, while having no effect on the majority, even when given in large doses, is not known. There is some evidence that it is a hypersensitivity phenomenon, related to allergy. The leukopenia usually appears after the drug has been given for some considerable period of time, or often during a second course of treatment with the same drug, suggesting that sensitivity is acquired, in the manner of an allergy.

A considerable number of cases of severe leukopenia occur spontaneously, without any discoverable cause, and are designated as being idiopathic. *See* ALLERGY, ATOPIC; LYMPHATIC SYSTEM (VERTEBRATE); X-RAYS, PHYSICAL NATURE OF.

Course and manifestations. The milder, incidental forms of leukopenia usually produce no clinical manifestations and will not be discussed further. The more severe forms, as produced by radiation or chemical agents, often present the clinical pictures of pancytopenia or agranulocytosis.

In pancytopenia, all circulating blood cells, red and white as well as platelets, are diminished, reflecting depressed production of all of these elements by the marrow. This condition can be produced both by the regular marrow-depressing agents and those producing hypersensitivity, as well as occurring spontaneously. The clinical findings are those of an aplastic anemia, leukopenia, and purpura, a bleeding tendency due to diminished platelets. Often anemia is the most serious problem, requiring repeated transfusion, but in other cases repeated and poorly resisted infections due to the leukopenia predominate. The cases induced by drugs frequently recover after a variable period of time, if the patient does not succumb to infection early in the course, but many of those occurring without known cause follow a chronic course for many years. Microscopic examination of the marrow in pancytopenia may show either hypoplasia, with generally reduced numbers of blood cell percursors, or a normal degree of cellularity with failure of young cells to mature.

Agranulocytosis is the name applied to the acute illness resulting from drug hypersensitivity, or, occasionally, from an undiscoverable cause. It is manifested by a rather sudden onset of severe leukopenia, often with complete disappearance of the granulocytes. There is usually no associated anemia or platelet deficiency. Most of the symptoms and the high mortality are due to infection by bacteria. The clinical onset is often sudden, with chills, fever, marked prostration, soon followed by evidence of infection which often takes the form of ulcerative lesions in the mouth. The patient may die in a few days of pneumonia or dissemination of the infectious process throughout the body. With modern antimicrobial treatment, infection may be controlled for a time, but even so the condition has a high mortality. Bone marrow findings are variable and nonspecific, but there is always a lack of the more mature white cell precursors.

The only treatment that is undoubtedly useful is the administration of antibiotic drugs to control infection. These drugs have been life-saving in some cases, but their effectiveness is hampered by the almost complete lack of the patient's natural mechanisms of defense. *See* CIRCULATION DISORDERS.   [PETER B. HUKILL]

Levan

A polysaccharide consisting of D-fructose (fructan) produced by a range of microorganisms such as *Bacillus mesentericus*, *B. vulgatus*, *B. subtilis*, *B. megatherium*, and *Aerobacter levanicum*. Although the molecular size of the polysaccharides formed by different microorganisms varies, it is always high, being of the order of 10^7 or more. The levans are polyfructose chains in which 9–10 D-fructose units are joined through carbon atom 2 (glycosidal hydroxyl) and carbon atom 6 positions. These chains are combined with similar polyfructose short chains, which are mutually joined through carbon atoms 2 and 1, thus forming highly branched molecules. *See* FRUCTOSE.

The microorganisms use sucrose as a substrate but polymerize only the fructose molecule to form the levan, as given in the equation below.

$$(n)C_{12}H_{22}O_{11} \rightarrow (C_6H_{10}O_5)_n + (n)C_6H_{12}O_6$$
Sucrose   Levan   D-Glucose

Various tissues of a number of plants, particularly those of the Compositae and Gramineae families, contain D-fructose polymers. These polymers are similar inasmuch as they are composed of D-fructose units having the furanose configuration, but differ from each other in molecular structure. They are classified into two general groups: the inulin type, which has its D-fructose units joined through carbon atom 2 (glycosidal hydroxyl) and carbon atom 1 positions; and the plean group, united through carbon atom 2 and carbon atom 6. The polysaccharides of both groups have a number of properties in common, such as solubility in warm water, susceptibility to acid hydrolysis, and a negative optical rotation. *See* OPTICAL ACTIVITY.

The polyfructoses occurring in plants are usually designated as fructans, while those elaborated by microorganisms are called levans. The bacterial levans in which the D-fructose units are joined by 2,6 glycosidic linkages are structurally similar to the plant polyfructoses of the plean group. However, while the microbial levans are giant molecules comprising many thousands of D-fructose units, the plant fructans have low molecular weights consisting of 8 to 30 fructose units. *See* POLYSACCHARIDE.   [WILLIAM Z. HASSID]

Bibliography: D. M. Greenberg (ed.), *Metabolic Pathways*, vol. 1, 3d ed., 1967; R. L. Whistler and C. L. Smart, *Polysaccharide Chemistry*, 1953.

Level (communication and acoustics)

Logarithm of the ratio of a given quantity to a reference quantity of the same kind. The base of the logarithm, the reference quantity, and the kind of level must be indicated. Frequently used levels, usually expressed in decibels, are power level, voltage level, and sound pressure level. The decibel is a unit of level when the base of the logarithm is the tenth root of 10, and the quantities concerned, such as voltage squared, are proportional to power. Some other units of level are the neper (Np) and bel (B); 1 Np = 8.686 dB and 1 B = 10 dB. *See* DECIBEL; NEPER.   [R. W. YOUNG]

Level, surveying

An instrument for establishing a horizontal line of sight. The level usually consists of a telescope and attached spirit level, mounted on a tripod for rotation about a vertical axis. The line of sight is fixed by cross hairs. Stadia hairs may be provided for

Engineer's dumpy level. (*W. and L. E. Gurley Co.*)

convenience in keeping foresights and backsights equal. The tubular spirit level vial is long, and its upper interior surface is ground to a circular arc. A tangent to the center of the ground arc is parallel with the line of sight; thus the line of sight is horizontal when the spirit level bubble is centered. The instrument is leveled by manipulation of footscrews in the mounting.

Engineer's levels include the dumpy (see illustration), with telescope rigidly attached to the supporting yoke, and the wye, with the telescope removable from the yoke. A precise level has a circular level for approximate leveling and a micrometer screw for final centering of the tubular level bubble. The bubble can be viewed through a series of prisms while the rod is read.

A different principle is applied in self-leveling instruments. The magnifying optical system for establishing the line of sight is suspended within a housing by fine wires. After preliminary leveling with footscrews and a circular level, the line of sight remains horizontal for all pointings. For the use of levels in surveying *see* SURVEYING.

[ROBERT H. DODDS]

Level measurement

The determination of the linear vertical distance between a reference point or datum plane and the surface of a liquid or the top of a pile of divided solids. Since there is little similarity between liquid and solid level measurements they are treated separately in this article.

LIQUID LEVEL MEASUREMENT

Satisfactory measurements are possible only when the liquid is undisturbed by turbulence or wave action. When a liquid is too turbulent for the average level to be read, a baffle or stilling chamber is inserted in the tank or vessel to provide a satisfactory surface.

Stick, hook, and tape gages. These are used in open vessels where the surface of the liquid can readily be observed. The stick gage is a suitably

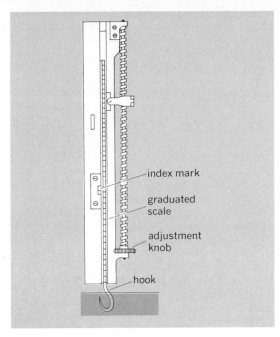

Fig. 1. Hook gage. (*From W. Unwin, A Treatise on Hydraulics, Macmillan, 1912*)

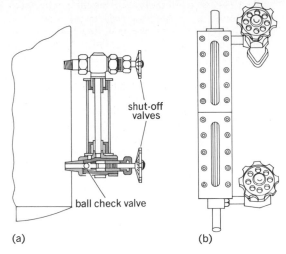

Fig. 3. Gage glasses. (*a*) Low-pressure type. (*b*) High-pressure type. (*From D. M. Considine, ed., Process Instruments and Controls Handbook, McGraw-Hill, 1957*)

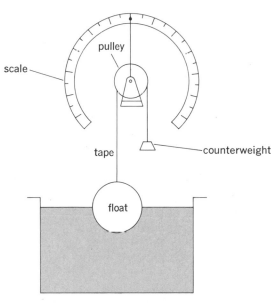

Fig. 4. Float, tape, and pulley gage for measuring large changes in level. (*D. M. Considine, ed., Process Instruments and Controls Handbook, McGraw-Hill, 1957*)

divided vertical rod, or stick, anchored in the vessel so that the magnitude of the rise and fall of the liquid level may be observed directly. The hook gage (Fig. 1) provides a needle point, which is adjusted to produce a very tiny pimple in the liquid surface at the level reading, thereby minimizing the meniscus error. The tape gage reads the correct elevation when the point of the bob just touches the liquid surface, as in Fig. 2. In one important variation of the tape gage, an electric circuit is completed when the bob point touches the liquid surface and thus actuates some type of signal. This requires that the liquid be a conductor of electricity, which eliminates its use with oils.

Gage glasses. Many forms of gage glass are available for the measurement of liquid level. Liquid in a tank or vessel is connected to the gage glass by a suitable fitting, and when the tank is under pressure the upper end of the glass must be connected to the tank vapor space. Thus the liquid rises to substantially the same height in the glass as in the tank, and this height is measured by suitable scale. To avoid errors due to capillarity, gage glasses with an internal diameter smaller than

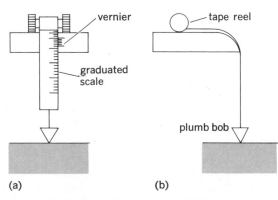

Fig. 2. Tape gage. (*a*) Front view. (*b*) Side view.

1/2 in. are not recommended unless a correction is applied. Most gage glasses are equipped with shut off valves to permit cleaning and replacement of the gage glass without emptying the tank, and with check valves to prevent loss of liquid in the event of glass breakage. Low-pressure gages are cylindrical glass or plastic tubes; high-pressure gages are metal tubes with thick glass windows (Fig. 3).

Float mechanisms. Various types of float mechanism are also used for liquid level measurement. The float, tape, and pulley gage provides an excellent method of measuring large changes in level with accuracy (Fig. 4). It has the advantage that the scale can be placed for convenient reading at any point within a reasonable distance of the tank or vessel. It can be applied to the measurement of liquid in tanks under pressure, but frictional errors are introduced in bringing the tape through the wall of the tank. One solution is to maintain the

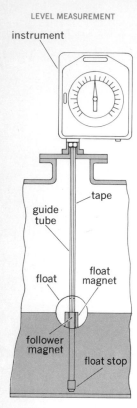

instrument

Fig. 5. Magnetic-type float gage. (*Fischer and Porter Co.*)

tank pressure over the entire length of the tape and read the tape through a glass. Another solution is to introduce a magnetic coupling between the float, which is under pressure, and the tape actuator, which is at atmospheric pressure (Fig. 5).

Float and lever mechanisms. These may be used advantageously in closed tanks under pressure when the liquid measured is viscous, or when it would foul a gage glass. The internal type is located directly within the tank and is connected to an external indicator through a stuffing box (Fig. 6). The external type incorporates an auxiliary cage or housing with connections to the liquid and vapor space in the main tank (Fig. 7).

Float-operated hydraulic level gage. This utilizes the float-and-lever principle to actuate a remote gage through a hydraulic coupling (Fig. 8). Float movements transfer liquid between a transmitting bellows and a remote bellows in the receiving gage. A dual, opposed, hydraulic system permits the mechanism to operate under pressure and compensates for temperature changes in the connecting hydraulic lines.

Buoyant displacer elements. The change in buoyancy of a solid as its immersion in a liquid is varied is used to measure liquid level. This principle is used for liquid level measurement only when the densities of the liquid and vapor are substantially constant. Temperature changes will produce errors of greater magnitude than with the float mechanisms. However such errors may be largely eliminated by using a displacer with a somewhat flexible wall and filled with a gas so that it will expand and contract with changes of temperature. Thus the buoyant force will be nearly independent of variations of the density of the tank liquid with temperature. Buoyant cylinders or displacers are of prime importance for use in closed tanks under pressure. As the level changes on the displacer, a force is developed without an appreciable move-

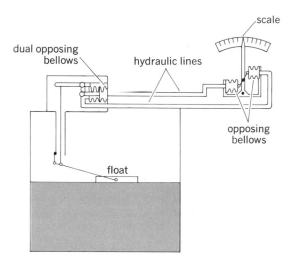

Fig. 8. Float-operated hydraulic level gage, with dual, opposed, hydraulic system. (*D. M. Considine, ed., Process Instruments and Controls Handbook, McGraw-Hill, 1957*)

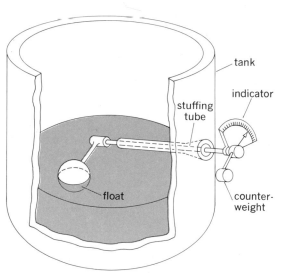

Fig. 6. Internal float-and-lever mechanism.

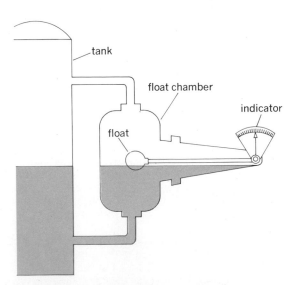

Fig. 7. External float-and-lever mechanism.

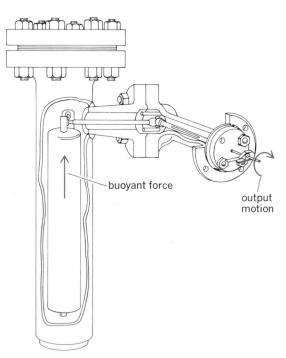

Fig. 9. Torque-tube level unit. (*Fisher Governor Co.*)

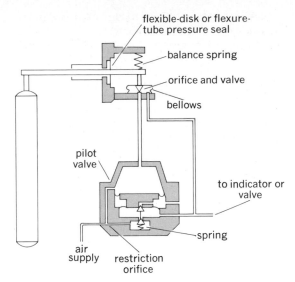

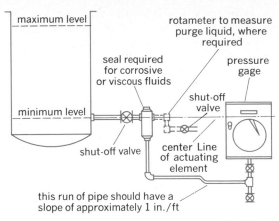

Fig. 11. Pressure gage measuring tank level. (*D. M. Considine, ed., Process Instruments and Controls Handbook, McGraw-Hill, 1957*)

Fig. 10. Force-balance level unit for external tank installation. (*D. M. Considine, ed., Process Instruments and Controls Handbook, McGraw-Hill, 1957*)

ment of the displacer. This force is transmitted through this vessel wall by a shaft and a torque tube, flexible tube, or a flexible diaphragm. Because mechanical friction is minimized, displacer units are sensitive to smaller changes in level than float units, which require stuffing boxes. Another advantage is that long displacers can be supported on short lever arms, because of the negligible travel with level change; thus good designs can be provided for larger level changes than can be handled advantageously by the float-and-lever system.

Displacer-type level elements are available in many designs for both internal and external tank installations. In Fig. 9 the small motion of the displacer is brought out of the tank, for measuring or transmission purposes, by a small shaft within a torque tube which supports the displacer. In Fig. 10 the change in weight of the displacer is detected externally by a nozzle and baffle and is balanced by a pneumatic servo.

Pressure gages. Hydrostatic head may also be used to measure liquid level. The pressure exerted by a column of liquid varies directly with its density as well as with its height, and thus this method of measurement requires that density be substantially constant. Densities of liquids vary with temperature; errors are therefore introduced with temperature changes, or the measuring element must be temperature compensated. Errors also result from appreciable flow in the pipe line to which the pressure-measuring element is connected.

Most pressure-measuring elements may be used in liquid level service. The pressure gage is limited to relatively large level changes in open tanks. Mercury and aneroid manometers can be used for intermediate ranges in closed as well as open tanks, because the instruments are basically differential pressure–measuring devices. The bell manometer and slack diaphragm gage can be adapted to the measurement of small liquid level changes. For details on pressure-measuring devices *see* PRESSURE MEASUREMENT.

Open tanks. In open tanks the pressure-measuring element is connected directly to the tank at or

near its bottom (the reference or datum point). The pressure element may be located at any reasonable distance from the tank and at any desired distance below the reference point, but the element responds to the additional head due to the liquid in the lead line. Gas or vapor must be vented from the lead lines. Pressure-measuring elements, when used in this way, normally have suppressed ranges, so that the zero reading occurs at the pressure corresponding to the bottom of the tank level. With careful installation, the pressure-measuring element may be located above the reference point. This installation requires a gas-tight lead line, venting only when the tank level is high and the maximum safe elevation of the pressure-measuring element is the head corresponding to one-half of atmospheric pressure less the vapor pressure of the liquid (Fig. 11). Temperature changes on sloping lead lines cause density change, and errors are introduced unless the pressure element has special compensation for this effect.

To minimize lead-line errors and venting, trapped-gas and gas-purge systems have been developed. These systems make it possible to locate the pressure-measuring device at any reasonable distance above the reference point. The diaphragm box may be installed directly in the tank or connected to it at the reference level. A slack diaphragm separates the liquid from the trapped air

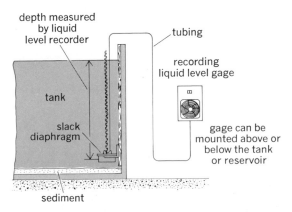

Fig. 12. Trapped-gas system using diaphragm box, to minimize lead-line errors and venting. (*Bristol Co.*)

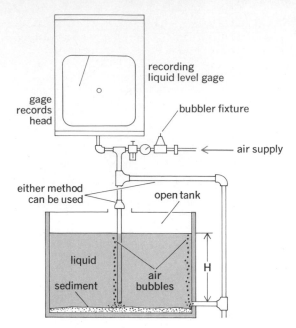

Fig. 13.   Gas-purge level system. (*Bristol Co.*)

or gas leading to the pressure gage. The diaphragm has no spring constant; hence the gas pressure equals that at the reference point and the remote gage indicates the pressure of the liquid column with respect to the reference level (Fig. 12). The gas volume in the pressure-measuring system is kept small to minimize the temperature and compressibility effects with liquid level changes.

On some installations a trapped-gas system eliminating the slack diaphragm can be used. On these installations the gas must not be absorbed in the liquid, and the liquid must not vaporize and condense in the lead line to the pressure-measuring device. Provision is generally made for periodic gas checking and replacement. In the gas-purge system (Fig. 13), a small quantity of gas is bled into the system continuously. This gas is permitted to bubble from the box or a tube with an opening at the reference level. Purge rates of about 1 standard ft³/hr are common. Precautions are taken to

avoid measurement of the pressure drop due to the purge gas flow. The bottom edge of the dip tube or box is designed to minimize the size of the bubbles released.

When the liquid being measured is highly corrosive or contains solid particles and the gas purge is not satisfactory, a liquid purge, a combination gas and liquid purge, or a combination air and steam purge is possible.

The pressure of a liquid column above a reference line may also be measured by a force-balance diaphragm system. As the level of liquid varies, the diaphragm operates a sensitive detector, such as a nozzle and baffle, and a simple servo to produce a force equal to that developed by the liquid head (Fig. 14). The diaphragm is maintained in a substantially constant position. The simplest systems are pneumatically operated, and the pressure opposing the liquid pressure is equal to it. Other systems oppose the hydrostatic pressure on the diaphragm with pressure in a bellows of reduced area so that the pressure change for a given level change is magnified. The force-balance diaphragm system eliminates the purge system and its maintenance problem.

Closed tanks. When manometers or other differential-pressure elements are applied to level measurement in closed tanks, a second connection must be made to the tank at a higher level (Fig. 15). Normally, the distance between the two tank connections is equal to the calibrated range of the manometer divided by the difference between the density of the liquid and the vapor in the tank. The external leg is filled with liquid to a seal chamber at the upper connection in the tank to avoid uncertainty due to condensation and possible overflow into an unfilled leg. Thus zero pressure occurs when the liquid in the tank is at the upper level, and the maximum pressure is read when the liquid in the tank is at the lower connection or the reference level. To avoid confusion in this service, the

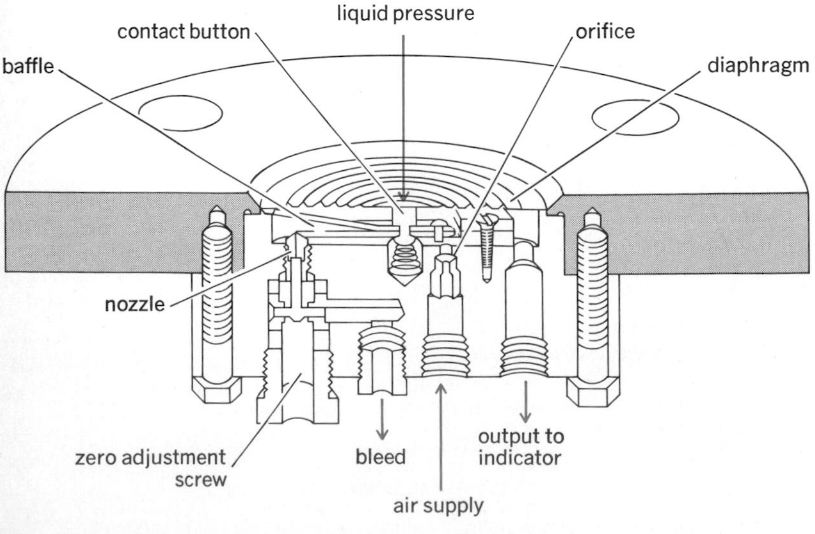

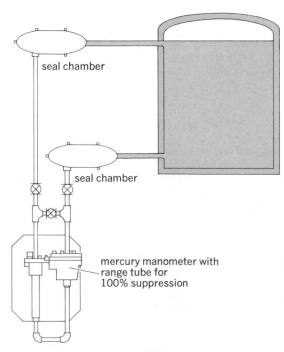

Fig. 14.   Force-balance diaphragm-type transmitter for level measurement. (*Taylor Instrument Co.*)

Fig. 15.   Mercury manometer with two liquid seals at different levels on tank. (*Taylor Instrument Co.*)

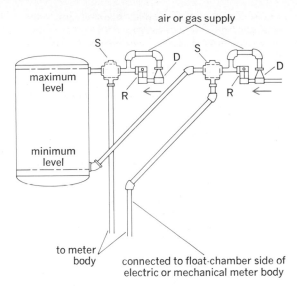

Fig. 16. Gas-purge system for level measurement in closed tanks. D is a differential pressure regulator, R is a rotameter with needle valve, and S is a sediment chamber, if required. (*From D. M. Considine, ed., Process Instruments and Controls Handbook, McGraw-Hill, 1957*)

manometer is calibrated in reverse so that it reads zero when the tank liquid is at the reference level. Temperature changes on the external leg cause reading errors, and internal legs have been used to avoid the trouble. The differential pressure–measuring element may be remotely located from the tank so that lead-line errors are avoided.

Gas-purge systems (Fig. 16) are also used for level measurement in closed tanks. Both tank connections are supplied with gas at a pressure slightly higher than the maximum which can occur in the tank. The gas flows are limited by restrictions or needle valves and are often measured by rotameters. The gas flow in each line must be large enough to prevent liquid from entering the lead lines and to ensure accurate readings. It must be small enough to keep the pressure drop between the manometer taps and the tank at a minimum. The manometer connections (taps) are located as close to the tank connections as possible. The lower tap connection is piped up to the upper tap connection to prevent liquid from entering the meter body in the event of failure of the purge system.

Liquid seals are occasionally used on closed-tank liquid level measurements. Seal chambers

must be installed at suitable locations, and consideration must be given to the differential in the specific gravity between the sealing liquid and the liquid in the tank.

Thermal elements may be used for level control when the temperatures of a liquid and of the vapor space above it differ appreciably. Although standard thermal elements are used in some services, a number of special designs, offered for boiler service, are particularly designed to handle the problem of boiler swell on increased steam demand. The level devices in this group are not precision devices and are not used for measurement purposes, but as operating guides.

Electrode or probe systems. These are used in various forms for level indication and control (Fig. 17). The number of electrodes and their design depend upon the characteristics of the liquid and the application. Fundamentally, a circuit through a relay coil is closed (or opened) when the liquid contacts a probe. Electronic circuits are used when

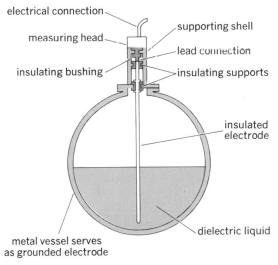

Fig. 18. Capacitance probe-type level system for dielectric liquid. (*D. M. Considine, ed., Process Instruments and Controls Handbook, McGraw-Hill, 1957*)

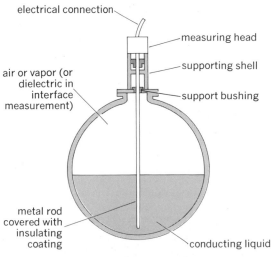

Fig. 19. Capacitance probe-type level system for conducting liquid. (*D. M. Considine, ed., Process Instruments and Controls Handbook, McGraw-Hill, 1957*)

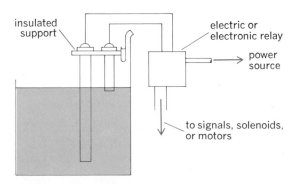

Fig. 17. Conductivity probe-type system, used for level indication and control. (*D. M. Considine, ed., Process Instruments and Controls Handbook, McGraw-Hill, 1957*)

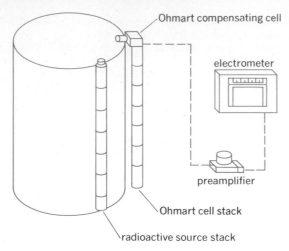

Fig. 20. Gamma-ray level gage, used for difficult applications. (*D. M. Considine, ed., Process Instruments and Controls Handbook, McGraw-Hill, 1957*)

the liquid resistance is high.

Capacitance-measuring devices. If the liquid being measured is a dielectric, one or two probes or rods, extending nearly to the bottom of the tank, are supported in an insulating mounting. The probes may be bare or covered with insulation. Also if only one probe is used, the metal tank may serve as the grounded electrode. The capacitance of the assembly will vary as the rise and fall of the liquid changes the depth of the probe immersion. This capacitance can be measured and expressed in terms of liquid level (Fig. 18). The dielectric constant of the liquid must be different from that of the vapor, and errors will occur if either of these changes appreciably. If the liquid is a conductor of electricity, only one probe is necessary but it must be covered with an insulating coating (Fig. 19). Although capacitance methods of measuring liquid have certain advantages (no moving parts and corrosion resistance), they have certain disadvantages such as high cost, susceptibility to dielectric changes, and susceptibility to electrical interference. Therefore this method of level measurement is limited to those applications which cannot be readily handled by other means.

Nuclear level gages. These are used for difficult applications. Basically, all of the units involve a source of gamma (γ) radiation and a detector separated by the vessel or a portion of the vessel in which a liquid level varies. As the level rises, the detector receives less γ-radiation and thus the level is measured. Numerous designs are available, one of which is illustrated (Fig. 20).

Sonic level detector. This is based on the time increment between the emission of a sound wave pulse and its reflection from liquid surface. The sound wave pulse is generated electronically, and its time in transit is measured very accurately by electronic means. If the speed of sound in the liquid or vapor is known accurately, the liquid level is known (Fig. 21).

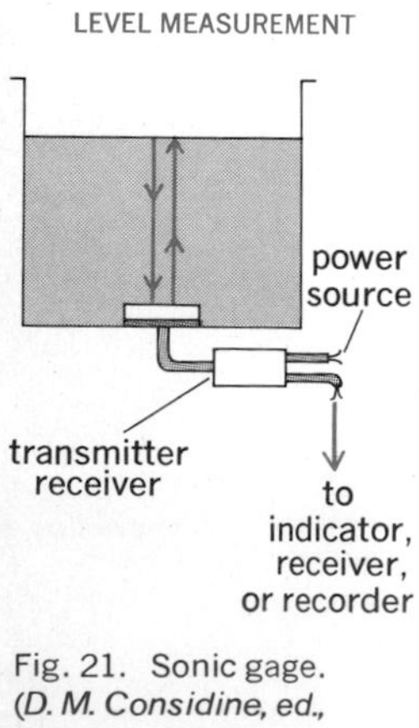

SOLIDS LEVEL MEASUREMENT

Solids level detectors are used to locate the top of a pile of divided solids in large vessels or processing equipment. The instruments are de-

LEVEL MEASUREMENT

Fig. 21. Sonic gage. (*D. M. Considine, ed., Process Instruments and Controls Handbook, McGraw-Hill, 1957*)

signed for the different solids handled, and the installation must be carefully made to ensure proper measurements. Because solids funnel, cone, and vary in average density with the particle size, shape, distribution, moisture content, and other factors, these detectors provide only an approximate indication of the volume present or the top of the pile. Solids level detectors are classified as continuous or fixed point.

Continuous detectors. These provide a continuous measurement of the level over the range for which they were designed. Their output is an analog representation of the level of the solids.

The grid-type element (Fig. 22) is used in a bed of finely divided moving particles and provides a continuous indication of the solid surface. The grid is supported and weighed by a conventional pneumatic system, the weight increasing as the immersion increases. This construction may be used in vessels under pressure.

The γ-ray emitter and detector, discussed under liquid level measurement, is particularly useful in this service, because all the equipment may be located outside the vessel.

The beam scale and other weighing devices are used particularly on small vessels and processing equipment. *See* WEIGHT MEASUREMENT.

The capacitance-type level detector, also discussed under liquid level measurement, is satisfactory for use on many finely divided solids.

A system that can be used with liquids, slurries, and bulk granular materials has a resistance sensing unit, which may be a resistance element in a helix or corrugated form, enclosed within a long flexible jacket. This jacket acts as a pressure-receiving diaphragm pressing the resistance element against a return line (within the jacket), thus changing the resistance of the system as the level of the material rises or falls. The signal from this system can be transmitted a considerable distance from the tank or bin.

Fixed-point detectors. These indicate when a specific level has been reached and are used mainly for actuating alarm signals. By installing a number of these, however, at different points, the combined response can be made to approach that of a continuous detector. Four fixed-point detectors are shown in Fig. 23.

Diaphragm-operated level detectors are widely

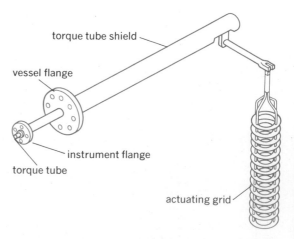

Fig. 22. Grid-type element. (*Union Oil Co. of California*)

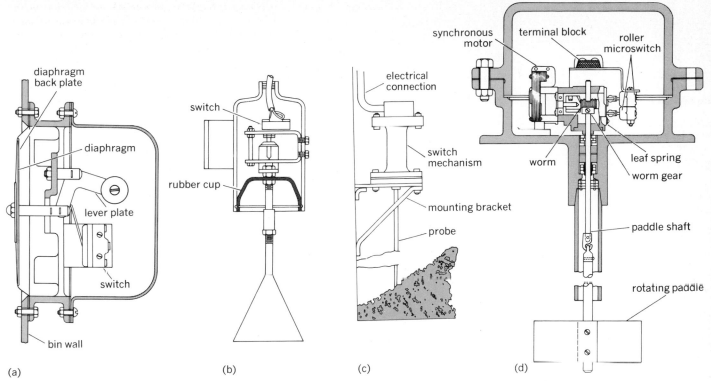

(a) (b) (c) (d)

Fig. 23. Fixed point solids level detectors, used mainly for actuating alarm signals. (a) Diaphragm-operated level detector (*Bin-Dicator Co.*). (b) Pendant cone-type level detector (*Stephens-Adamson Manufacturing Co.*). (c) Probe deflection-type detector (*G. B. Linderman Co.*) (d) Rotating paddlewheel-type detector (*Fuller Co.*).

used for finely divided solids, such as wheat, in open vessels.

The pendant-cone and movable-probe units, used in large open vessels, are actuated by the surface of the cone resulting from the piling of the material.

The rotating paddlewheel level detector is driven continuously by a small motor until the paddle is stopped by the presence of solid material. This device is widely used on finely divided solids in open vessels. [HOWARD S. BEAN]

Bibliography: T. G. Beckwith and N. L. Buck, *Mechanical Measurements*, 1961; D. M. Considine (ed.), *Process Instruments and Controls Handbook*, 1957.

Lever

A pivoted rigid bar used to multiply force or motion, sometimes called the lever and fulcrum (Fig. 1). The lever uses one of the two conditions for static equilibrium, which is that the summation of moments about any point equals zero. The other

condition is that the summation of forces acting in any direction through a point equals zero. *See* INCLINED PLANE.

If moments acting counterclockwise around the fulcrum of a lever are positive, then, for a frictionless lever, $F_B b - F_A a = 0$, which may be rearranged to give Eq. (1).

$$F_B = \frac{a}{b} F_A \qquad (1)$$

If F_B represents the output and F_A represents the input, the mechanical advantage, MA, is then given by Eq. (2).

$$MA = \frac{F_B}{F_A} = \frac{a}{b} \qquad (2)$$

Applications of the lever range from the simple nutcracker and paper punch (Fig. 2) to complex multiple-lever systems found in scales and in testing machines used in the study of properties of materials. *See* SIMPLE MACHINE.

[RICHARD M. PHELAN]

Libra

The Balance, in astronomy, appearing during the spring. It is the seventh sign of the zodiac. The constellation consists of faint stars and is not conspicuous. It lies just west of the claws of Scorpius. The principal stars, α, β, γ, and σ, outline a four-sided figure resembling a balance with beam and pans (see illustration). The balance might have been held originally in the hand of Virgo, a zodiacal constellation nearby, who was identified with the Goddess of Justice. Another possible reason for identifying the constellation with the balance is

LEVER

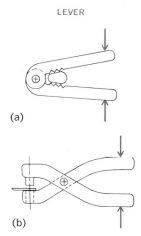

(a)

(b)

Fig. 2. (a) Nutcracker. (b) Paper punch.

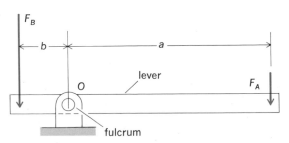

Fig. 1. The lever pivots at the fulcrum.

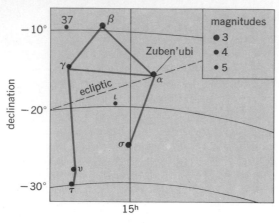

Line pattern of the constellation Libra. The grid lines represent the coordinates of the sky. The apparent brightness, or magnitude, of the stars is shown by the sizes of the dots, graded by appropriate numbers.

that 2000 years ago the Sun was in Libra at the then autumnal equinox, at which time days and nights are of equal length. *See* CONSTELLATION.

[CHING-SUNG YU]

Lichenes

A group of organisms consisting of fungi and algae growing together symbiotically. The fungal component is usually an ascomycete or, less commonly, a basidiomycete. The algal component is green or blue-green. When the two components grow together, they form a distinct, long-lived plant body (see illustration) that may be foliose (leaflike), crustose (crustlike), or fruticose (branched, shrubby, or bearded). The fungus makes up the vast bulk of the thallus and is usually differentiated into an upper cortex, a thick spongy medulla, and in foliose lichens a lower cortex. The algae occur in a thin layer between the medulla and the upper cortex, tightly clasped by the fungal hyphae. Most lichens are mineral gray, but a few are a brightly pigmented orange or red, and some are black. *See* ASCOLICHENES; BASIDIOLICHENES; CHLOROPHYTA; CYANOPHYCEAE.

Reproduction. Lichens reproduce vegetatively by fragmentation of the thallus or by dispersal of unique soredia, isidia, or other diaspores. The fungal component also produces sexual spores that can lichenize symbiotic algae and form a new lichen, although this apparently happens only rarely.

Classification. Because their dual nature was not discovered until 1868, lichens were long classified as a separate division of the plant kingdom between the fungi and bryophytes. They are presently considered to be fungi, remarkable in their ability to parasitize yet live symbiotically with algae. Eventually the lichens will be assimilated among the orders and families of the true fungi, but since their relationships are so poorly known, they are still maintained out of necessity as a separate group. Three classes and five orders with nearly 18,000 species are recognized:

Class Ascolichenes
    Order: Pseudosphaeriales
                Hysteriales (Lichenized)

Lichen formations. (*a*) Foliose (*Parmelia*). (*b*) Crustose, growing upon a rock. (*c*) Fruticose (*Usnea barbata*), growing upon a tree branch. (*From H. J. Fuller and O. Tippo, College Botany, rev. ed., Holt, 1954*)

Pyrenulales
Lecanorales
Caliciales
Class Basidiolichenes
Class Lichenes Imperfecti

See separate articles on the classes and orders.

Economic importance. Lichens are widely distributed in the tropical rainforests, the temperate zones, and even on the nunataks of Antarctica and the Arctic. They are often pioneers in dry exposed habitats where other plants cannot survive. Hence they are basically important as soil builders and stabilizers. Fruticose lichens on soil are utilized as fodder for reindeer and sheep, and even man has eaten them in emergencies. Lichens have limited commercial importance as sources of dyestuffs (especially litmus), essential oils in perfumes, and the antibiotic usnic acid.

[MASON E. HALE]

Bibliography: H. des Abbayes, *Traité de lichenologie*, 1951; B. Fink, *Lichen Flora of the United States*, 1935; M. E. Hale, *The Biology of Lichens*, 1967; M. E. Hale, *Lichen Handbook*, 1961; A. L. Smith, *Lichens*, 1921.

Lichenes Imperfecti

A class of the Lichenes containing species with no known method of sexual reproduction. This class parallels the Fungi Imperfecti, but differs significantly in absence of any asexual conidiospores. The plant body is usually an undifferentiated crustose mat of hyphae clasping separate colonies of unicellular symbiotic algae. Reproduction is entirely by dispersal of fragments of the crusts by wind, water, or animals. These crusts are worldwide in distribution and often occur in shady, rather moist habitats on rocks and tree bases. The largest genus, with about six species, is *Crocynia* (*Amphiloma*), a mineral-gray crust. Another genus, *Lepraria*, frequently has a bright sulfur-yellow thallus. *See* FUNGI IMPERFECTI; LICHENES.

[MASON E. HALE]

Licorice

A product obtained from the licorice plant (*Glycyrrhiza glabra*) of the legume family (Leguminosae). It is a perennial herb which grows wild and is cultivated in southern Europe and in western and central Asia. The roots are dried for several months and then packaged for shipment. Spain leads in the production of cultivated licorice roots. Licorice is used in medicine to mask objectionable taste and as a laxative; as a flavoring material in the brewing, tobacco, and candy industries; and in the manufacture of shoe polish. *See* ROSALES; SPICE AND FLAVORING.

[PERRY D. STRAUSBAUGH/EARL L. CORE]

Lie detector

The lie detector is a useful application of the physiological responses of emotion. The machine is known as the polygraph lie detector and is widely used in crime detection and in security-sensitive areas of employment. It operates on the principles or the assumptions that lying involves an emotional reaction and that emotional reactions are characterized by physiological changes. The physiological variables measured by lie detectors are changes in relative blood pressure, heart rate, respiration movements, and frequently palmar sweating or galvanic skin response. The apparatus is attached to a suspect who is then asked a series of questions to which he can simply answer yes or no. Some of the questions are neutral and serve to establish the subject's characteristic responses to known truthful answers. The significant questions, or those which pertain to the crime, are phrased so the suspect must either admit knowledge of the crime, or lie.

Lying is believed to involve an emotional response because of the conflict between the strong cultured conditioning for truthfulness and the desire to escape detection. Persons lacking a "conscience" may still exhibit an emotional response to lying if they believe in the infallibility of the machine.

According to a review made by Dael Wolfle in 1946, the accuracy of the polygraph in practice is 80% correct, 3% wrong, and 17% not interpretable. The accuracy depends to a great extent on the skill of the examiner, who must be thoroughly trained and experienced.

Most courts do not as yet accept lie detector reports as evidence, although confessions elicited by means of a lie detector examination may usually be accepted. *See* EMOTION.   [ALBERT F. AX]

Bibliography: The polygraphic truth test: A symposium, *Tenn. Law Rev.* vol. 22, February, 1953; J. E. Reid and F. E. Inbau, *Truth and Deception: The Polygraph ("Lie Detector") Technique*, 1966; D. Wolfle, Summary of report of subcommittee on methods of lie-detection, *Amer. J. Psychol.*, 59:544, 1946.

Lie group

A topological group with only countably many connected components whose identity component is open and is an analytic group. An analytic group or connected Lie group is a topological group with the additional structure of a smooth manifold such that multiplication and inversion are smooth. Many groups that arise naturally as groups of symmetries of physical or mathematical systems are Lie groups. The study of Lie groups has applications to analytic function theory, differential equations, differential geometry, Fourier analysis, algebraic number theory, algebraic geometry, quantum mechanics, relativity, and elementary particle theory. *See* GROUP THEORY; MANIFOLD, SMOOTH; TOPOLOGY.

Examples of Lie groups. Euclidean n-dimensional space with the standard differentiable structure and with vector addition as the group operation is an analytic group $\mathbf{R}^n$ of dimension n.

The nonsingular n-by-n real matrices form a Lie group $GL(n,\mathbf{R})$ with the differentiable structure obtained as an open set of n^2-dimensional space and with the group operation given by matrix multiplication. Similarly $GL(n,\mathbf{C})$, the group of nonsingular n-by-n complex matrices under matrix multiplication, forms a Lie group. The latter group is connected.

Each closed subgroup of a Lie group, in its relative topology, admits one differentiable structure on its identity component that makes it a Lie group. With $GL(n,\mathbf{R})$ and $GL(n,\mathbf{C})$, the subgroups of upper triangular matrices, matrices of determinant

1, rotation matrices, orthogonal matrices, and unitary matrices are all examples. *See* MATRIX THEORY.

Lie algebra. Let G be a Lie group. The identity component of G is an open subset of G and is a smooth manifold. Thus it makes sense to speak of tangent spaces. Let $\mathfrak{g}$ be the tangent space to G at the identity element. A multiplication $[X,Y]$ within $\mathfrak{g}$ is defined as follows.

By means of the differential of left translation by p, one can associate a tangent vector X_p to the point p of G. If f is a smooth real-valued function on G, let $\widetilde{X}f$ be the function on G with $\widetilde{X}f(p)=X_pf$. Then $\widetilde{X}f$ is smooth. The tangent vector $[X,Y]$ is defined on the function f by Eq. (1).

$$[X,Y]f=X(\widetilde{Y}f)-Y(\widetilde{X}f) \tag{1}$$

The resulting multiplication satisfies the properties:

(i) $[X,Y]$ is linear in X and in Y.
(ii) $[X,X]=0$.
(iii) $[X,[Y,Z]]+[Y,[Z,X]]+[Z,[X,Y]]=0$.

The vector space $\mathfrak{g}$ equipped with this multiplication is called the Lie algebra of the Lie group G.

The Lie algebra $\mathfrak{g}$ determines the multiplication in G near the group identity by a result known as the Campbell-Hausdorff formula. Consequently, many properties of the identity component of G can be deduced from the simpler object $\mathfrak{g}$.

Homomorphisms. If φ is a smooth homomorphism from an analytic group G to an analytic group G', the differential $d\varphi$ of φ at the identity is a Lie algebra homomorphism in the sense that it is linear and satisfies $d\varphi([X,Y])=[d\varphi(X),d\varphi(Y)]$. If φ is one-to-one or onto, so is $d\varphi$. Conversely, if G is simply-connected, each homomorphism of Lie algebras arises as the differential of some unique smooth group homomorphism.

Abstract Lie algebras. An abstract real Lie algebra is a finite-dimensional real vector space with a multiplication satisfying (i), (ii), and (iii). Each abstract real Lie algebra can be identified with the Lie algebra of some analytic group.

Examples. The Lie algebra of $\mathbf{R}^n$ is an n-dimensional vector space with all $[X,Y]$ equal to 0.

For $GL(n,\mathbf{R})$, the Lie algebra can be identified with the vector space of all n-by-n real matrices with multiplication given in terms of matrix multiplication as $[X,Y]=XY-YX$. The identification of a member L of the Lie algebra with a matrix X is done entry by entry. Namely, the i-jth entry of X is the result of applying L to the i-jth entry function, which is a smooth real-valued function on the group.

For $GL(n,\mathbf{C})$, the Lie algebra can be identified with all n-by-n complex matrices with multiplication $[X,Y]=XY-YX$.

For a closed subgroup G of $GL(n,\mathbf{R})$ or $GL(n,\mathbf{C})$, inclusion is a smooth one-to-one homomorphism and allows one to identify the Lie algebra of G with a Lie subalgebra of matrices. The matrices of the Lie subalgebra can be computed conveniently as all entry-by-entry derivatives $c'(0)$ of smooth curves $c(t)$ that lie completely within the subgroup and have $c(0)$ equal to the identity matrix. For unitary matrices, for example, the result is skew-Hermitian matrices. (A skew-Hermitian matrix is equal to the negative of the transpose of its complex conjugate.)

Analytic subgroups. Let G be a Lie group. An analytic subgroup H of G is a subgroup of G that admits the structure of an analytic group in such a way that the inclusion mapping of H into G is smooth. The topology of H as a manifold need not be the relative topology inherited from G.

If H is an analytic subgroup of G, the differential of the inclusion mapping identifies the Lie algebra of H with a Lie subalgebra of the Lie algebra of G. Distinct analytic subgroups give rise to distinct Lie subalgebras. Every Lie subalgebra arises from some analytic subgroup by this construction.

Exponential map. Let G be a Lie group with Lie algebra $\mathfrak{g}$. Let $c(t)$ be a curve in G, that is, a smooth mapping from the real line into G, and suppose that $c(0)$ is the group identity. The image of d/dt under the differential dc of $c(t)$ at 0 is a member of $\mathfrak{g}$. For each X in $\mathfrak{g}$ there is one and only one such curve c_X such that $dc_X(d/dt)=X$ and $c_X(s+t)=c_X(s)c_X(t)$ for all s and t. The exponential mapping is the function from $\mathfrak{g}$ to G defined by $\exp X=c_X(1)$. It is smooth and, in a small open neighborhood of the group identity, has a smooth inverse function that provides local coordinates for G near the identity.

For $GL(n,\mathbf{R})$ or $GL(n,\mathbf{C})$ the exponential mapping is given by the series expansion for the ordinary exponential function e^x if Lie algebra is identified with Lie algebra of all matrices. *See* SERIES.

Representations. In many applications of Lie groups the notion of group representation is a basic tool. A unitary representation of the Lie group G on a Hilbert space H is a function R from G to the set of unitary operators on H such that $R(g_1g_2)=R(g_1)R(g_2)$ for all g_1 and g_2 in G and such that the mapping $g\rightarrow R(g)x$ is continuous from G into H for each x in H. [A unitary operator U on H is a linear operator from H onto H that preserves inner products; that is, $(Ux, Uy)=(x,y)$ for each x and y in H.] The representation R is irreducible if there is no proper nonzero closed subspace S such that $R(g)$ maps S into itself for all g in G.

Because of the applications, it is an important mathematical problem to classify the irreducible unitary representations of a given Lie group G and to investigate properties of decompositions of general unitary representations into irreducible ones.

In the analysis of the unitary representations, use of the Lie algebra plays an important role. If X is in the Lie algebra, it is possible to define $R(X)$ as a skew-Hermitian operator on a dense subspace by differentiating $R(\exp tX)$ appropriately. Then $iR(X)$ extends to a self-adjoint operator that is easier to deal with than $R(\exp tX)$.

In many applications, H is a space of functions, and some linear operator or set of linear operators is acting in H. Typically, the representation of G is given by a group of transformations that commute with these linear operators. When the representation of G on H is decomposed suitably, one can expect the linear operators to act separately in each of the constituent spaces of H and to be easier to understand on each constituent space than on H. *See* OPERATOR THEORY.

Sample applications. Three examples will illustrate this principle.

Rotations and the Fourier transform. Let H be the Hilbert space of square-integrable functions on $\mathbf{R}^n$, let G be the rotation group, and let the representation of G on H be given by rotation of the coordinates. The Fourier transform is a linear operator on H that commutes with each member of G. The space H decomposes under the representation of G into a sequence of simpler spaces, the kth space being all sums of products of a radial function and a harmonic polynomial homogeneous of degree k. The Fourier transform has a simple form on the kth space, given as a one-dimensional integral involving a Bessel function. *See* BESSEL FUNCTIONS; FOURIER SERIES AND INTEGRALS.

Solutions of differential equations. Some information about solutions in $\mathbf{R}^n$ to elliptic partial differential equations with constant coefficients can be obtained by using the Fourier transform. Here the differential operator commutes with the representation of $G = \mathbf{R}^n$ given by translation in the space of square-integrable functions on $\mathbf{R}^n$, and the Fourier transform is the device for decomposing the representation. For an operator with nonconstant coefficients, one can apply a freezing principle, replacing a given equation by one with constant coefficients and obtaining an approximate solution near a point.

For many equations of subelliptic type, which are elliptic except for certain degeneracies, a similar analysis is possible by means of representations of a more complicated Lie group. In the simplest new case, the Lie group is the (nonabelian) group of 3-by-3 real upper-triangular matrices with ones on the diagonal, and the differential equation is Eq. (2) in $\mathbf{R}^3$. Other equations can be analyzed by

$$\left(\frac{\partial^2 f}{\partial x^2} + \frac{\partial^2 f}{\partial y^2}\right) + \left(x\frac{\partial^2 f}{\partial y\partial z} - y\frac{\partial^2 f}{\partial x\partial z}\right)$$
$$+ \frac{1}{4}\left(x^2 + y^2\right)\frac{\partial^2 f}{\partial z^2} = 0 \quad (2)$$

a combination of this method and a freezing principle. *See* DIFFERENTIAL EQUATION.

Quantum mechanics. In quantum theory the things one observes from experiments are eigenvalues (or the spectrum, if there are not discrete eigenvalues) of certain self-adjoint operators on Hilbert spaces of wave functions. Conservation laws come from self-adjoint operators whose associated one-parameter groups of unitary operators commute with the Hamiltonian. In this way a system of conservation laws leads to a group of symmetries, namely, a group of unitary operators commuting with the Hamiltonian. Frequently this group is a Lie group. Analysis of the representation of the Lie group given by the unitary operators gives information about the physical system. *See* ELEMENTARY PARTICLE; LORENTZ TRANSFORMATIONS; QUANTUM THEORY, NONRELATIVISTIC; QUANTUM THEORY, RELATIVISTIC; SYMMETRY LAWS (PHYSICS).   [ANTHONY W. KNAPP]

Bibliography: S. Helgason, *Differential Geometry and Symmetric Spaces*, 1962; L. Pukanszky, *Leçons sur les représentations des groupes*, 1967; V. S. Varadarajan, *Lie Groups, Lie Algebras, and Their Representations*, 1974; G. Warner, *Harmonic Analysis on Semi-Simple Lie groups*, vols. 1 and 2, 1972; H. Weyl, *The Classical Groups, Their Invariants and Representations*, 1939.

Liesegang rings

Periodic precipitation zones obtained by counterdiffusion of soluble substances which form a slightly soluble precipitate. In the case of one-dimensional diffusion these zones are called Liesegang bands.

When placing a drop of a concentrated silver nitrate solution upon a glass plate covered with a gelatin layer containing potassium dichromate, R. E. Liesegang observed in 1896 that the resulting precipitation of silver chromate did not spread continuously or quasi-continuously, but that individual precipitation zones separated by zones free of precipitate evolved in the form of concentric circles. Similar phenomena have been observed with many other reactions.

Periodic precipitation is due to the interplay of diffusion and nucleation, which requires a certain degree of supersaturation. Assume that solute A diffuses from a fairly concentrated solution into a dilute solution of B. The precipitate AB starts to evolve, the initially high degree of supersaturation drops, and accordingly, nucleation ceases. Upon diffusion of A into farther regions not yet depleted with respect to B, a sufficient degree of supersaturation is reached once more, a second precipitation zone evolves, and so forth. *See* PRECIPITATION (CHEMISTRY).   [CARL WAGNER]

Bibliography: E. S. Hedges, *Liesegang Rings and Other Periodic Structures*, 1932; A. E. Nielsen, *Kinetics of Precipitation*, 1964; K. H. Stern, Liesegang phenomenon, *Chem. Rev.*, 57(1):79–99, 1954.

Life, origin of

The origin of life refers to the means by which living organisms arose on the Earth, which contained no life when it was formed. While there is little difficulty in telling whether a higher organism is alive, there is no agreement as to what characteristics would be required for the most primitive organisms in order to call them living. The most outstanding characteristic of living organisms is their ability to reproduce. Therefore, the definition used here for a living organism will be an entity which is capable of making a reasonably accurate reproduction of itself, with the copy being able to accomplish the same task. Furthermore, this organism must be subject to a low rate of changes (mutations) that are transmitted to its progeny.

Spontaneous generation. Until the 19th century it was almost universally held that life could arise spontaneously, as well as by sexual or nonsexual reproduction. The appearance of various insects and animals from decaying organic materials in the presence of warmth or sunlight was a common observation, but this observation was misinterpreted to mean that the insects and animals arose spontaneously.

The Tuscan physician Francesco Redi performed the first effective experiments (1668) to disprove the hypothesis of spontaneous generation. He showed that the development of maggots in meat did not occur when the flask containing the meat was covered with muslin so that flies could not lay their eggs on the meat. Lazzaro Spallanzani performed similar experiments (1765) showing that microorganisms did not appear in various nutrient broths if the vessels were sealed and boiled. Objections were raised that the heating had destroyed in the broth and air the "vital force" which was postulated as necessary for life to develop. By readmitting air it was possible to show that the broth could still support the growth of microorganisms. But Spallanzani could not demonstrate that the air in the sealed flask had not been altered, and the doctrine of spontaneous generation persisted widely.

The problem was finally solved by Louis Pasteur (1862) who used a flask with broth, but instead of sealing the flask, drew out a long S-shaped tube with its end open to the air. The air was free to pass in and out of the flask, but the particles of dust, bacteria, and molds in the air were caught on the sides of the S-shaped tube. When the broth in the flask was boiled and allowed to cool, no microorganisms developed, but when the S-shaped tube was broken off at the neck of the flask, microorganisms developed. These experiments were extended by Pasteur and by J. Tyndall to answer all objections that were raised, and the doctrine of spontaneous generation was finally disproved.

Theories. Shortly before 1858, Charles Darwin and A. R. Wallace had published, simultaneously and independently, the theory of evolution by natural selection. This theory could account for the evolution from the simplest single-celled organism to the most complex plants and animals, including man. Therefore, the problem of the origin of life was no longer how each species developed, but only how the first living organism arose. To answer this question a number of proposals have been advanced.

It has been proposed that life was created by a supernatural event. This has been a common belief of many people based on a literal interpretation of the first chapter of Genesis, which describes the creation of all living organisms by a direct act of God. This type of proposal is not considered by most scientists since it is not subject to scientific investigation.

In 1903 S. Arrhenius offered a second theory, that life developed on the Earth as a result of a spore or other stable form of life coming to the Earth in a meteorite from outer space or by the pressure of sunlight driving the spores to the Earth. One form of this theory assumes that life had no origin, but like matter, has always existed. The presence of long-lived radioactive elements shows that the elements were formed anywhere from 5×10^9 to 12×10^9 years ago. If the elements have not always existed, it is difficult to understand how life could have always existed. Another form of this theory assumes that life was formed on another planet and traveled to the Earth. This hypothesis does not answer the question of how life arose on the other planet. In addition, most scientists doubt that any known form of life could survive for very long in outer space and fall through the Earth's atmosphere without being destroyed. Therefore, while this theory has not been disproved, it is held to be highly improbable.

A third hypothesis held that the first living organism arose from inorganic matter by a very improbable event. This organism, in order to grow in an inorganic environment, would have to synthesize all of its cellular components from carbon dioxide, water, and other inorganic nutrients. Present knowledge of biochemistry shows that even the simplest bacteria are extremely complex and that the probability of the spontaneous generation of a cell from inorganic compounds in a single event is much too small to have occurred in the approximately 5×10^9 years since the Earth's formation.

A more plausible proposal is that life arose spontaneously in the oceans of the primitive Earth, which contained large quantities of organic compounds similar to those which occur in living organisms. This theory was outlined mainly by A. I. Oparin in 1938 and forms the basis of most of the present ideas on the origin of life. Oparin argued that, if large amounts of organic compounds were in the oceans of the primitive Earth, these organic compounds would react to form structures of greater and greater complexity, until a structure would form which would be called living. In other words, the synthesis of the first living organism would involve many nonbiological steps, and none of these steps would be highly improbable.

Oparin also proposed that organic compounds might have been formed on the primitive Earth if there were a reducing atmosphere of methane, ammonia, water, and hydrogen, instead of the present oxidizing atmosphere of carbon dioxide, nitrogen, oxygen, and water. In 1952 H. C. Urey placed the reducing-atmosphere theory on a firm foundation by showing that methane, ammonia, and water are the stable forms of carbon, nitrogen, and oxygen if an excess of hydrogen is present. Cosmic dust clouds, from which the Earth is believed to have been formed, contain a great excess of hydrogen. The planets Jupiter, Saturn, and Uranus are known to have atmospheres of methane and ammonia. Oxidizing conditions have developed on Mercury, Venus, Earth, and Mars due to the escape of hydrogen followed by the production of oxygen by the photochemical splitting of water. There has not been sufficient time for the hydrogen to have escaped from Jupiter, Saturn, Uranus, Neptune, and Pluto owing to their lower temperatures and higher gravitational attraction.

Origin of organic compounds. Many attempts have been made to synthesize organic compounds under oxidizing conditions using carbon dioxide and water with various sources of energy, such as ultraviolet light, electric discharges, and high-energy radiation (see illustration). All of these experi-

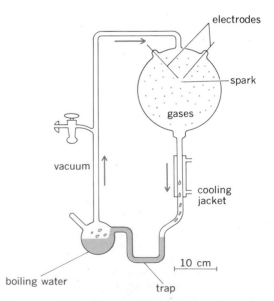

Apparatus used for production of amino acids by electric discharges. Water is boiled in small flask and mixes with gases in large flask where spark is located. Products of the discharge are condensed, pass through U-tube, and accumulate in small flask.

ments failed to give organic compounds except in extremely small yield.

Support for the ideas of Oparin and Urey came from experiments by S. L. Miller in 1953. He showed that a mixture of methane, ammonia, water, and hydrogen, when subjected to an electric discharge, gave significant yields of simple amino acids, hydroxy acids, aliphatic acids, urea, and possibly sugars. Ultraviolet light gives similar results.

On the basis of these experiments, it is thought that various organic compounds (such as amino acids, hydroxy acids, aliphatic acids, and sugars) were present in the oceans of the primitive Earth. It is still necessary to show mechanisms for the formation of polypeptides, purines and pyrimidines (the bases which occur in nucleic acids), nucleotides and polynucleotides, a source of phosphate bond energy (as in adenosinetriphosphate), a synthesis of polypeptides with catalytic activity (enzymes), and the development of polynucleotides capable of self-duplication.

Nature of first organisms. In present living organisms the duplication proceeds by duplicating the genes, followed by synthesis of more enzymes and other cell constituents, and division of this cell into two fragments. The genes, which are located on the chromosomes, are composed of polynucleotides called deoxyribonucleic acid. *See* CHROMOSOME; GENE; NUCLEIC ACID.

Since the essential characteristic of living organisms is the ability to duplicate, it has been proposed that the first living organisms were simply strips of deoxyribonucleic acid (or perhaps strips of ribonucleic acid) which, together with the necessary enzymes, could duplicate. This organism would be similar to a virus except that it would have the necessary enzymes for its reproduction.

The virus consists of several strips of nucleic acid surrounded by a coat of protein. The virus is capable of duplication, but only within another living cell where the virus makes use of the cell's enzymes and metabolism. Since the virus can reproduce itself, but only within a living cell, there is considerable dispute as to whether a virus should be called living or not. In any case many virologists believe that the virus was not the first type of living organism, but that it is a regression from a more advanced type of organism. *See* VIRUS.

It has been proposed by Oparin that the first organisms were coacervate particles instead of strips of polynucleotides. A coacervate is a type of colloid which forms two phases, one of the solution and one of the coacervate, instead of a uniform dispersion as with most colloids. It is assumed that there would be some coacervate particles which could absorb proteins and other substances from the environment, grow in size, and then split into two or more fragments which would repeat this process. In time the duplication would become more accurate, and the genetic apparatus of nucleic acids would then develop. *See* COACERVATION.

These ideas about the nature of the first organisms are based on the assumption that the first forms of life were similar to present organisms in their basic chemical composition. All present organisms contain proteins, nucleic acids, carbohydrates, and lipids. While the assumption may be incorrect, it is the best working hypothesis until it can be shown to be inadequate. Other possibilities for the development of the first organisms can be enumerated, but so little is known of the composition of the primitive oceans that such speculation is not profitable.

Heterotrophic organisms. While little can be said about the development of the first living organisms, reasonable hypotheses can be made for the development of the simple bacteria, algae, and protozoa from the most primitive organisms. The theory that the primitive oceans contained large quantities of organic compounds implies that the first organisms were heterotrophic. Heterotrophic organisms do not synthesize their basic constituents, such as amino acids, nucleotides, carbohydrates, and vitamins, but obtain them from the environment. Autotrophic organisms synthesize all their cell constituents from CO_2, H_2O, and other inorganic materials. Heterotrophic organisms are simpler than autotrophic organisms in that they contain fewer enzymes and specialized structures to carry out their metabolism. Hence heterotrophic organisms would have been formed first.

A mechanism by which heterotrophic organisms could acquire various synthetic abilities was proposed by N. H. Horowitz in 1945. It has been found that the presence of an enzyme in an organism is often dependent on a single gene. This is known as the one gene–one enzyme hypothesis. Suppose that the synthesis of A involves the steps

$$D \xrightarrow{c} C \xrightarrow{b} B \xrightarrow{a} A$$

where a, b, and c are the enzymes and A, B, and C are compounds which the organism cannot synthesize. If the necessary compound A becomes exhausted from the environment, then the organism must synthesize A in order to survive. But it is extremely unlikely that there would be three simultaneous mutations to give the enzymes a, b, and c. However, a single mutation to give enzyme a would not be unlikely. If compounds D, C, and B were in the environment when A was exhausted, the organism with enzyme a could survive while the others would die out. Similarly, when compound B was exhausted, enzyme b would arise by a single mutation, and organisms without this enzyme would die out. By continuing this process the various steps of a biosynthetic process could be developed, the last enzyme in the sequence being developed first, and the first enzyme last.

Energy and biosynthetic processes. It is necessary for all living organisms to have a source of energy to drive the biochemical reactions that synthesize the various structures of the organism. The quantity that measures the available energy for a chemical reaction at constant temperature and pressure is termed the free energy. Animals obtain their free energy from the oxidation of organic compounds by molecular oxygen. Plants and other photosynthetic organisms obtain their free energy from the energy of light. There are also many microorganisms that obtain their free energy from fermentation reactions. For example, the lactic acid bacteria obtain their free energy from glucose as shown in reaction (1). Yeasts also ferment glu-

$$C_6H_{12}O_6 \rightarrow 2CH_3CH(OH)COOH \qquad (1)$$
Glucose   Lactic acid

cose, but they produce ethyl alcohol and CO_2 instead of lactic acid, and there are bacteria which carry out many other types of fermentation reactions. It is likely that organisms obtained their free energy from fermentation reactions until the supply of fermentable compound was exhausted. At that point the development of photosynthetic organisms which obtain their free energy from light would become necessary. The porphyrin substance chlorophyll seems to be necessary for all types of photosynthesis and would have had to be present in the environment or its biosynthesis developed in some way. The first type of photosynthesis was probably similar to that of the sulfur bacteria and blue-green algae which carry out reactions ($2a$) or ($2b$), where (H_2CO) means carbon on

$$2H_2S + CO_2 \xrightarrow{\text{Light}} 2S + (H_2CO) + H_2O \quad (2a)$$

$$H_2S + 2CO_2 + 2H_2O \xrightarrow{\text{Light}} H_2SO_4 + 2(H_2CO) \quad (2b)$$

the oxidation level of formaldehyde (carbohydrate). It is much easier to split H_2S than to split H_2O, so that it would seem likely that organisms would develop photosynthesis with sulfur first. When the H_2S and S were exhausted, it would become necessary to split water and evolve O_2, the hydrogen being used for the reduction of CO_2. *See* CHLOROPHYLL; FREE ENERGY; PHOTOSYNTHESIS.

When the methane and ammonia of the primitive atmosphere had been converted to carbon dioxide and nitrogen by photochemical decomposition in the upper atmosphere, water would be decomposed to oxygen and hydrogen. The hydrogen would escape, leaving the oxygen in the atmosphere and thereby resulting in oxidizing conditions on the Earth. It is likely, however, that most of the oxygen in the atmosphere was produced by the photosynthesis of plants rather than by the photochemical splitting of water.

Multicellular organisms. The evolution of multicellular organisms probably occurred after the development of photosynthesis. The evolution of primitive multicellular organisms to more complex types and the development of sexual reproduction can be understood on the basis of the theory of evolution. *See* EVOLUTION, ORGANIC.

[STANLEY L. MILLER]

Bibliography: E. S. Barghoorn, *Origin of Life*, Geol. Soc. Amer. Mem. no. 67, 2:75–86, 1957; J. D. Bernal, *The Physical Basis of Life*, 1951; G. Ehrensvärd, *Life: Origin and Development*, 1962; J. Keosian, *The Origin of Life*, 1964; S. L. Miller, A production of amino acids under possible primitive earth conditions, *Science*, 117:528–529, 1953; S. L. Miller, Production of some organic compounds under possible primitive earth conditions, *J. Amer. Chem. Soc.*, 77:2351–2361, 1955; S. L. Miller and H. C. Urey, Organic compound formation on the primitive earth, *Science*, 130:245–251, 1959; A. I. Oparin, *The Chemical Origin of Life*, 1964; M. Rutten, *Geological Aspects of the Origin of Life on Earth*, 1962; G. Wald, The origin of life, *Sci. Amer.*, 191(2):44–53, 1954; Symposium: Modern ideas on spontaneous generation, *Ann. N.Y. Acad. Sci.*, 69(art. 2):257–376, 1957.

Life zones

Large portions of the Earth's land area which have generally uniform climate and soil and, consequently, a biota showing a high degree of uniformity in species composition and adaptations to environment. Related terms are vegetational formation and biome.

Merriam's zones. Life zones were proposed by A. Humboldt, A. P. DeCandolle and others who emphasized plants. Around 1900 C. Hart Merriam, then chief of the U.S. Biological Survey, related life zones, as observed in the field, with broad climatic belts across the North American continent designed mainly to order the habitats of America's important animal groups. The first-order differences between the zones, as reflected by their characteristic plants and animals, were related to temperature; moisture and other variables were considered secondary.

Each life zone correlated reasonably well with major crop regions and to some extent with general vegetation types (see table). Although later studies led to the development of other, more realistic or detailed systems, Merriam's work provided an important initial stimulus to bioclimatologic work in North America. *See* DENDROLOGY; VEGETATION AND ECOSYSTEM MAPPING.

Work on San Francisco Mountain in Arizona impressed Merriam with the importance of temperature as a cause of biotic zonation in mountains. Isotherms based on sums of effective temperatures correlated with observed distributions of certain animals and plants led to Merriam's first law, that animals and plants are restricted in

Characteristics of Merriam's life zones

Zone name	Example	Vegetation	Typical and important plants	Typical and important animals	Typical and important crops
Arctic-alpine zone	Northern Alaska, Baffin Island	Tundra	Dwarf willow, lichens, heathers	Arctic fox, muskox, ptarmigan	None
Hudsonian zone	Labrador, southern Alaska	Taiga, coniferous forest	Spruce, lichens	Moose, woodland caribou, mountain goat	None
Canadian zone	Northern Maine, northern Michigan	Coniferous forest	Spruce, fir, aspen, red and jack pine	Lynx, porcupine, Canada jay	Blueberries
Western division					
Humid transition zone	Northern California coast	Mixed coniferous forest	Redwood, sugar pine, maples	Blacktail deer, Townsend chipmunk, Oregon ruffed grouse	Wheat, oats, apples, pears, Irish potatoes

Characteristics of Merriam's life zones (cont.)

Zone name	Example	Vegetation	Typical and important plants	Typical and important animals	Typical and important crops
Western division (cont.)					
Arid transition zone	North Dakota	Conifer, woodland sagebrush	Douglas fir, lodgepole, yellow pine, sage	Mule deer, whitetail, jackrabbit, Columbia ground squirrel	Wheat, oats, corn
Upper Sonoran zone	Nebraska, southern Idaho	Piñon, savanna, prairie	Junipers, piñons, grama grass, bluestem	Prairie dog, blacktail jackrabbit, sage hen	Wheat, corn, alfalfa, sweet potatoes
Lower Sonoran zone	Southern Arizona	Desert	Cactus, agave, creosote bush, mesquite	Desert fox, four-toed kangaroo rats, roadrunner	Dates, figs, almonds
Eastern division					
Alleghenian zone	New England	Mixed conifer and hardwoods	Hemlock, white pine, paper birch	New England cottontail rabbit, wood thrush, bobwhite	Wheat, oats, corn, apples, Irish potatoes
Carolinian zone	Delaware, Indiana	Deciduous forest	Oaks, hickory, tulip tree, redbud	Opossum, fox, squirrel, cardinal	Corn, grapes, cherries, tobacco, sweet potatoes
Austroriparian zone	Carolina piedmont, Mississippi	Long-needle conifer forest	Loblolly, slash pine, live oak	Rice rat, woodrat, mocking bird	Tobacco, cotton, peaches, corn
Tropical zone	Southern Florida	Broadleaf evergreen forest	Palms, mangrove	Armadillo, alligator, roseate spoonbill	Citrus fruit, avocado, banana

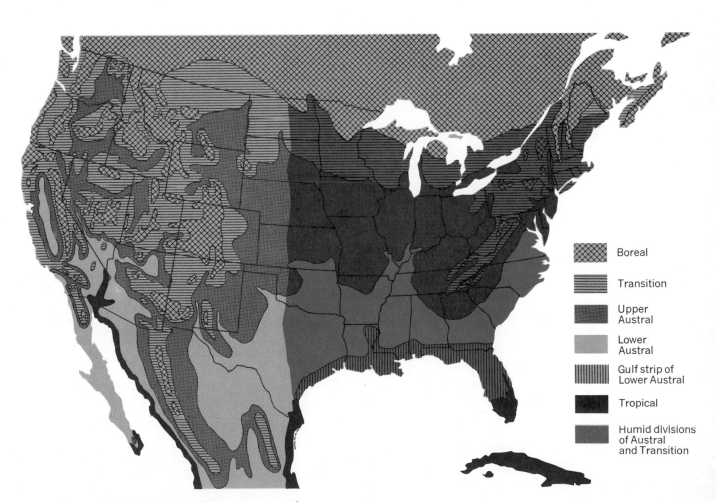

Fig. 1. Life zones of the United States. (*C. H. Merriam, 1898*)

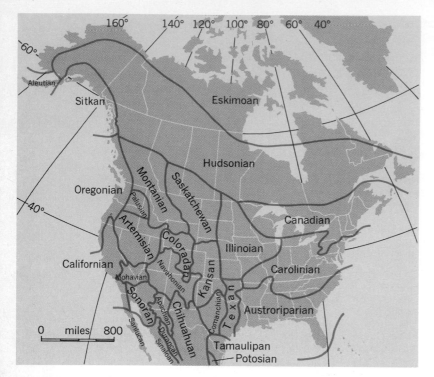

Fig. 2. Biotic provinces of part of North America (Veracruzian not shown). (*From L. R. Dice, The Biotic Provinces of North America, University of Michigan Press, 1943*)

northward distribution by the sum of the positive temperatures (above 43°C) during the season of growth and reproduction. The mean temperature for the six hottest weeks of the summer formed the basis for the second law, that plants and animals are restricted in southward distribution by the mean temperature of a brief period covering the hottest part of the year. Merriam's system emphasizes the similarity in biota between arctic and alpine areas and between boreal and montane regions. It was already recognized that latitudinal climatic zones have parallels in altitudinal belts on mountain slopes and that there is some biotic similarity between such areas of similar temperature regime.

In northern North America Merriam's life zones, the Arctic-Alpine, the Hudsonian, and the Canadian, are entirely transcontinental (Fig. 1 and table). Because of climatic and faunistic differences, the eastern and western parts of most life zones in the United States (the Transition, upper Austral, and lower Austral) had to be recognized separately. The western zones had to be further subdivided into humid coastal subzones and arid inland ones. The Tropical life zone includes the extreme southern edge of the United States, the Mexican lowlands, and Central America.

Although once widely accepted, Merriam's life zones are little used today because they include too much biotic variation and oversimplify the

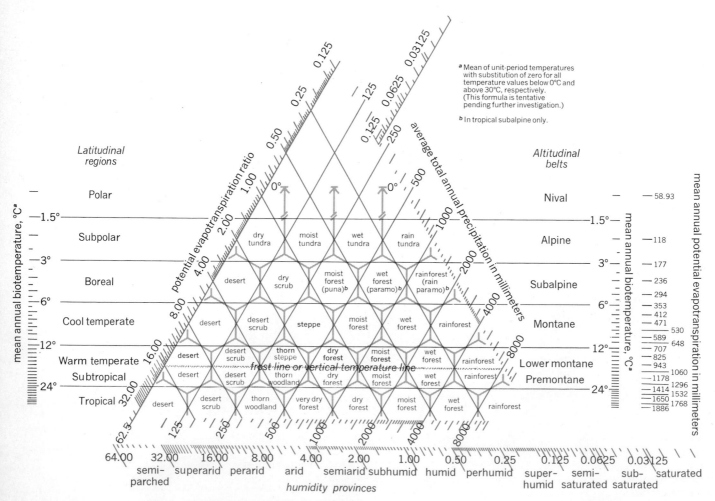

Fig. 3. World life zones. (*From L. R. Holdridge, Life Zone Ecology, rev. ed., Tropical Science Center, 1967*)

situation. However, much of the terminology he proposed persists, especially in North American zoogeographic literature.

Dice's zones. Another approach to life zones in North America is the biotic province concept of L. R. Dice. Each biotic province covers a large and continuous geographic area and is characterized by the occurrence of at least one important ecological association which is distinct from those of adjacent provinces. Each biotic province is subdivided into biotic districts which are also continuous, but smaller, areas distinguished by ecological differences of lesser extent or importance than those delimiting provinces. Life belts, or vertical subdivisions, also occur within biotic provinces. These are not necessarily continuous but often recur on widely separated mountains within a province where ecological conditions are appropriate.

Boundaries between biotic provinces were largely subjective, supposedly drawn where the dominant associations of the provinces covered approximately equal areas. In practice, however, too few association data including both plants and animals were available, and vegetation generally offered the most satisfactory guide to boundaries. The Dice system recognizes 29 biotic provinces in North America (Fig. 2). The two northernmost (Eskimoan and Hudsonian) are transcontinental, reflecting broad, high-latitude climatic belts. Those in the eastern United States (Canadian, Carolinian, and Austroriparian) reflect both latitudinal temperature change and physiography, whereas province boundaries in the remainder of North America are strongly influenced by physiography.

Holdridge's zones. A life-zone system with boundaries more explicitly defined is that of L. R. Holdridge. These life zones are defined through the effects of the three weighted climatic indexes: mean annual heat, precipitation, and atmospheric moisture. Each axis of the triangle (Fig. 3) represents one climatic component, and the three sets of lines parallel to the axes define the life-zone framework. Climatic values which the lines represent progress geometrically. Mean annual biotemperature is computed by summing the monthly mean temperatures from 0° (setting negative winter months equal to 0° when they occur) to about 30°C and dividing by 12. Values increase from top to bottom in Fig. 3 and, by extension to either margin, establish equivalent latitudinal regions and altitudinal belts. Precipitation is computed as mean annual precipitation in millimeters and increases from the apex toward the lower right margin of the diagram. The humidity factor is calculated by dividing the value of the mean annual potential evapotranspiration by the mean annual precipitation in millimeters. Values of this ratio increase toward lower left side of the diagram, signifying a decrease in effective humidity.

Associations of vegetation, animals, climate, physiography, geology, and soils are interrelated in unique combinations, mostly having distinct physiognomy, in Holdridge's climatic grid. Local investigation should show how well the framework represents the most common life zones of Earth, both north and south of the Equator. The system has been applied with considerable success to the mapping of life zones, primarily in the American tropics. Names of Holdridge's life zones are based on the dominant vegetation of each climatic type; animal life of the zone is supposedly distinctive. *See* BIOME; PLANT FORMATIONS, CLIMAX; VEGETATION ZONES, ALTITUDINAL; VEGETATION ZONES, WORLD; ZOOGEOGRAPHY.   [ARTHUR W. COOPER]

Bibliography: J. R. Carpenter, *An Ecological Glossary*, 1956; C. H. Merriam, *Life Zones and Crop Zones of the United States*, USDA Bull. no. 10, 1898; J. E. Weaver and F. E. Clements, *Plant Ecology*, 2d ed., 1938.

Ligament (vertebrate)

A strong, flexible connective tissue band usually found between two bony prominences. Most ligaments are composed of dense fibrous tissue formed by parallel bundles of collagen fibers. They have a shining white appearance and are pliable, strong, and noncompliant. These ligaments are closely associated with all articulations and are largely responsible for the exact movement of the bones. Other ligaments may span larger gaps between bones and may serve as important controlling elements in complex bone-linkage systems.

A second kind of ligament, composed either partly or almost entirely of yellow elastic fibers, is extensible or compliant, thereby allowing the connected bones to move apart. The large ligamenta nuchae between the neural spines of cervical vertebrae and the skull in large ungulates are of this type, and permit lowering and raising the head with less muscular effort. *See* BONE (BIOPHYSICS); CONNECTIVE TISSUE; JOINT (ANATOMY).

[WALTER BOCK]

Light

The term light, as commonly used, refers to the kind of radiant electromagnetic energy that is associated with vision. In a broader sense, light includes the entire range of radiation known as the electromagnetic spectrum. The branch of science dealing with light, its origin and propagation, its effects, and other phenomena associated with it is called optics. Spectroscopy is the branch of optics that pertains to the production and investigation of spectra.

Any acceptable theory of the nature of light must stem from observations regarding its behavior under various circumstances. Therefore, this article begins with a brief account of the principal facts known about visible light, including the relation of visible light to the electromagnetic spectrum as a whole, and then describes the apparent dual nature of light. The remainder of the article discusses various experimental and theoretical considerations pertinent to the study of this problem.

PRINCIPAL EFFECTS

The electromagnetic spectrum is a broad band of radiant energy which extends over a range of wavelengths running from trillionths of inches to hundreds of miles; wavelengths of visible light are measured in hundreds of thousandths of an inch. Arranged in order of increasing wavelength, the radiation making up the electromagnetic spectrum is termed gamma rays, x-rays, ultraviolet rays, visible light, infrared waves, microwaves, radio and TV waves, and very long electromagnetic waves. Detailed descriptions of these radiations are given separately; *see* ELECTROMAGNETIC RADIATION and the articles listed therein. *See also* OPTICS; PHO-TOCHEMISTRY; PHOTOSYNTHESIS; POLARIZED LIGHT; RAMAN EFFECT; SPECTROSCOPY.

Finite velocity. The fact that light travels at a finite speed or velocity is well established, both theoretically and experimentally. This speed is so high, however, and the experimental complexities of measuring it are so great, that this is not a phenomenon the average man can check for himself. The first successful measurement of the speed of light was made in 1676 by the Danish astronomer Ole Roemer, and since that time, numerous increasingly precise experiments have been conducted, leading to the currently accepted value of 299,792.8 km/sec in a vacuum. In round numbers, the speed of light in vacuum or air may be said to be 186,000 mi/sec or 300,000 km/sec.

Diffraction and reflection. One of the most easily observed facts about light is its tendency to travel in straight lines. Careful observation shows, however, that a light ray spreads slightly when passing the edges of an obstacle. This phenomenon is called diffraction. The reflection of light is also well known. The Moon, as well as all other satellites and planets in the solar system, are visible only by reflected light; they are not self-luminous like the Sun. Reflection of light from smooth optical surfaces occurs so that the angle of reflection equals the angle of incidence, a fact that is most readily observed with a plane mirror. When light is reflected irregularly and diffusely, the phenomenon is termed scattering. The scattering of light by gas particles in the atmosphere causes the blue color of the sky. When a change in frequency (or wavelength) of the light occurs during scattering, the scattering is referred to as the Raman effect.

Refraction. The type of bending of light rays called refraction is caused by the fact that light travels at different speeds in different media—faster, for example, in air than in either glass or water. Refraction occurs when light passes from one medium to another in which it moves at a different speed. Familiar examples include the change in direction of light rays in going through a prism, and the bent appearance of a stick partially immersed in water.

Interference and polarization. In the phenomenon called interference, rays of light emerging from two parallel slits combine on a screen to produce alternating light and dark bands. This effect can be obtained quite easily in the laboratory, and is observed in the colors produced by a thin film of oil on the surface of a pool of water. Polarization of light is usually shown with Polaroid disks. Such disks are quite transparent individually. When two of them are placed together, however, the degree of transparency of the combination depends upon the relative orientation of the disks. It can be varied from ready transmission of light to almost total opacity, simply by rotating one disk with respect to the other.

Chemical effects. When light is absorbed by certain substances, chemical changes take place. This fact forms the basis for the science of photochemistry. Rapid progress has been made toward an understanding of photosynthesis, the process by which plants produce relatively complex substances such as sugars in the presence of sunlight. This is but one example of the all-important response of plant and animal life to light.

HISTORICAL APPROACH

Early in the 18th century, light was generally believed to consist of tiny particles. Of the phenomena mentioned in the preceding section, reflection, refraction, and the sharp shadows caused by the straight path of light were well known, and the characteristic of finite velocity was suspected. All of these phenomena except refraction clearly could be expected of streams of particles, and Isaac Newton showed that refraction would occur if the velocity of light increased with the density of the medium through which it traveled.

This theory of the nature of light seemed to be completely upset, however, in the first half of the 19th century. During that time, Thomas Young studied the phenomena of interference, and could see no way to account for them unless light were a wave motion. Diffraction and polarization had also been investigated by that time. Both were easily understandable on the basis of a wave theory of light, and diffraction eliminated the "sharp-shadow" argument for particles. Reflection and finite velocity were consistent with either picture. The final blow to the particle theory seemed to have been struck in 1849, when the speed of light was measured in different media and found to vary in just the opposite manner to that assumed by Newton. Therefore, in 1850, it seemed finally to be settled that light consisted of waves.

Even then, however, there was the problem of the medium in which light waves traveled. All other kinds of waves required a physical medium, but light traveled through a vacuum—faster, in fact, than through air or water. The term ether was proposed by James Clerk Maxwell and his contemporaries as a name for the unknown medium, but this scarcely solved the problem because no ether was ever actually found. Then, near the beginning of the 20th century, came certain work on the emission and absorption of energy that seemed to be understandable only if one assumed light to have a particle or corpuscular nature. The external photoelectric effect, that is, the emission of electrons from the surfaces of solids when light is incident on the surfaces, was one of these. At that time, then, science found itself in the uncomfortable position of knowing a considerable number of experimental facts about light, of which some were understandable regardless of whether light consisted of waves or particles, others appeared to make sense only if light were wavelike, and still others seemed to require it to have a particle nature. *See* ETHER HYPOTHESIS.

THEORY

The study of light deals with some of the most fundamental properties of the physical world and is intimately linked with the study of the properties of submicroscopic particles on the one hand and with the properties of the entire universe on the other. The creation of electromagnetic radiation from matter and the creation of matter from radiation, both of which have been achieved, provide a fascinating insight into the unity of physics. The same is true of the deflection of light beams by strong gravitational fields, such as the bending of starlight passing near the Sun.

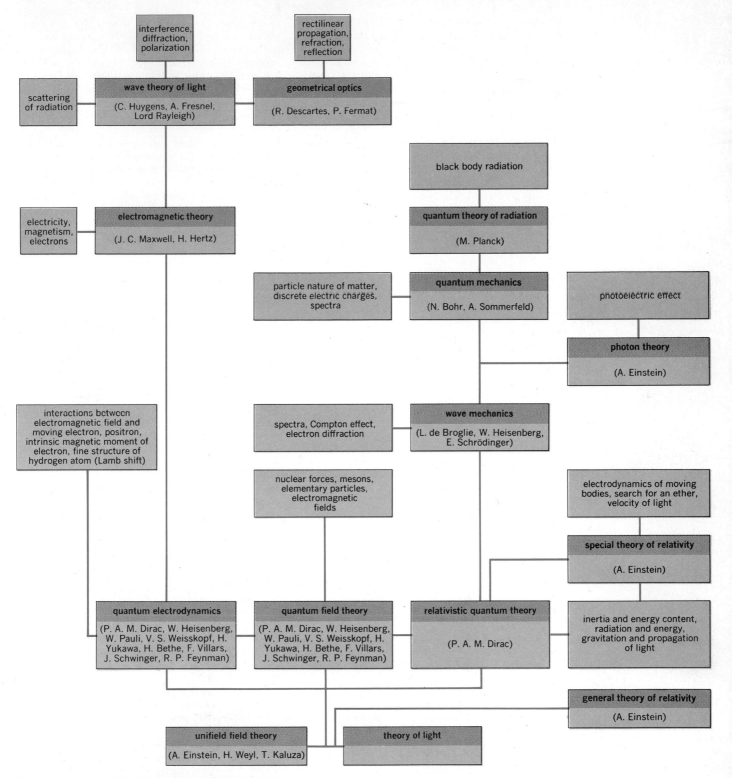

Fig. 1. Developments in the theory of light. (*G. W. Stroke*)

A classification of phenomena involving light according to their theoretical interpretation provides the clearest insight into the nature of light. When a detailed accounting of experimental facts is required, two groups of theories appear which, in the majority of cases, account separately for the wave and the corpuscular character of light. The quantum theories seem to obviate questions concerning this dual character of light, and make the classical wave theory and the simple corpuscular theory appear as two very useful limiting theories. It happens that the wave theories of light can cope with a considerable part of the phenomena involving electromagnetic radiation. Geometrical optics, based on the wave theory of light, can solve many of the more common problems of the propagation of light, such as refraction, provided that the limitations of the underlying theory are not disregarded. *See* OPTICS, GEOMETRICAL.

Phenomena involving light may be classed into

three groups: electromagnetic wave phenomena, corpuscular or quantum phenomena, and relativistic effects. The relativistic effects appear to influence similarly the observation of both corpuscular and wave phenomena. The major developments in the theory of light closely parallel the rise of modern physics. These developments are charted in Fig. 1, and are discussed in the remainder of this article.

Wave phenomena. Interference and diffraction, mentioned earlier, are the most striking manifestations of the wave character of light. Their fundamental similarity can be demonstrated in a number of experiments. The wave aspect of the entire spectrum of electromagnetic radiation is most convincingly shown by the similarity of diffraction pictures produced on a photographic plate, placed at some distance behind a diffraction grating, by radiations of different frequencies, such as x-rays and visible light. The interference phenomena of light are, moreover, very similar to interference of electronically produced microwaves and radio waves.

Polarization demonstrates the transverse character of light waves. Further proof of the electromagnetic character of light is found in the possibility of inducing, in a transparent body that is being traversed by a beam of plane-polarized light, the property of rotating the plane of polarization of the beam when the body is placed in a magnetic field. *See* FARADAY EFFECT.

The fact that the velocity of light had been calculated from electric and magnetic parameters (permittivity and permeability) was at the root of Maxwell's conclusion in 1865 that "light, including heat and other radiations if any, is a disturbance in the form of waves propagated . . . according to electromagnetic laws." Finally, the observation that electrons and neutrons can give rise to diffraction patterns quite similar to those produced by visible light has made it necessary to ascribe a wave character to particles. *See* ELECTRON DIFFRACTION; NEUTRON DIFFRACTION.

Velocity determination. The finite velocity of transmission of signals by means of electromagnetic waves is a fact well established for all wavelengths of radiation in a variety of terrestrial and astronomical experiments. Astronomical observation of so-called eclipsing binaries, both visual and spectroscopic, is taken to indicate that the velocity of light, denoted by c, does not depend on the velocity of the source. *See* BINARY STARS.

Measurements of c fall basically into two groups, (1) group velocity, or signal velocity, determinations, using the relation $c = \text{distance/time}$; and (2) phase velocity, or wave velocity, determinations, using $c = \text{frequency/wavelength}$.

The group velocity is the average time for a light signal, that is, a modulated electromagnetic wave train, to traverse a given distance. Roemer's original determination of the velocity of light was based on careful observations of the time of eclipse of the moons of Jupiter, from various points in the Earth's orbit. Subsequent terrestrial determinations, making use of revolving mirrors (J. L. Foucault and A. A. Michelson), a toothed wheel (H. L. Fizeau), or an electronically modulated light beam (E. Bergstrand), also measured the time required for light to traverse the distance between a source

and a reflecting mirror. These are all group velocity measurements, and furnish a value of c only if the experiments are carried out in a nondispersive medium, in which the velocity is the same for all wavelengths. Otherwise, the group velocity is smaller than the phase velocity by 0.007% (2.2 km/sec) in air, 1.5% in water, and 2.4% in ordinary crown glass. *See* DISPERSION (RADIATION); GROUP VELOCITY.

Determinations of the phase velocity, which is simply the speed of the wavefront, are indirect and make the assumption that $c = \lambda f$, where f is the frequency, and λ the free-space wavelength of the electromagnetic radiation. Most of the measurements of this type involve microwave interference in various forms: E. F. Florman with two sources; K. D. Froome in an apparatus similar to a Michelson interferometer; L. Essen, J. P. Gordon, and H. M. Smith in a microwave resonance cavity. D. H. Rank and also E. K. Plyer determined c by calculation from microwave and infrared spectroscopic measurements of the ratio of hertz to waves per centimeter for a certain molecular rotation frequency. The phase velocity can also be calculated from the ratio of electromagnetic to electrostatic units. A detailed description of these experiments is found in the bibliographical references. *See* PHASE VELOCITY.

The measured magnitudes of the velocity of transmission and the phase velocity seem to be in reasonably good agreement among themselves, possibly to a few parts in 10^5. However, the experimental values determined by the two methods are at an admitted and disturbing variance with one another. The velocity of light is not as constant as it is sometimes thought to be. While according to the theory of special relativity, the velocity of light will be the same in any frame of reference, independent of its state of motion, this is true only for frames of the same gravitational potential. Conceivably, local variations of the gravitational potential could lead to variations of the measured velocity of light, although experiments for detecting these variations remain to be devised. The effect of the gravitational potential at the Sun's surface or at other points in the universe is quite appreciable compared with the precision of many measurements of c. *See* POTENTIALS.

Electromagnetic-wave propagation. Electromagnetic waves can be propagated through free space, devoid of matter and fields, and with a constant gravitational potential; through space with a varying gravitational potential; and through more or less absorbing material media which may be solids, liquids, or gases. Radiation can be transmitted through waveguides with cylindrical, rectangular, or other boundaries, the insides of which can be either evacuated or filled with a dielectric medium. *See* WAVEGUIDE.

From electromagnetic theory, and especially from the well-known equations formulated by Maxwell, a plane wave disturbance of a single frequency f is propagated in the x direction with a phase velocity $v = \lambda f = \lambda\omega/2\pi$, where $\omega = 2\pi f$. The wave can be described by the equation $y = A \cos \omega(t - x/v)$. Two disturbances of same amplitude A, of respective angular frequencies ω_1 and ω_2 and of velocities v_1 and v_2, propagated in the same direction, yield the resulting disturbance y', defined in

Eq. (1). Here $\Delta\omega=\omega_1-\omega_2$ and $\omega=1/2(\omega_1+\omega_2)$.

$$y'=y_1+y_2=2A\cos\tfrac{1}{2}[(\Delta\omega)t$$
$$-x\Delta(\omega/v)]\cos(\omega t-\omega x/v) \quad (1)$$

The ratio $u=\Delta\omega/\Delta(\omega/v)$ is defined as the group velocity, just as the ratio $\omega/(\omega/v)$ is identical with the phase velocity. In the limit for small $\Delta\omega$, $u=d\omega/d(\omega/v)$. Noting that $\omega=2\pi v/\lambda$, $d\omega=2\pi(\lambda\,dv-v\,d\lambda)/\lambda^2$, and $d(\omega/v)=-2\pi\,d\lambda/\lambda^2$, an important relation, Eq. (2), between group and

$$u=v-\lambda\,dv/d\lambda \quad (2)$$

phase velocity is obtained. This shows that the group velocity u is different from the phase velocity v in a medium with dispersion $dv/d\lambda$. In a vacuum, $u=v=c$. With the help of Fourier theorems, the preceding expression for u can be shown to apply to the propagation of a wave group of infinite length, but with frequencies extending over a finite small domain. Furthermore, even if the wave train were emitted with an infinite length, modulation or "chopping" would result in a degrading of the monochromacity by introduction of new frequencies, and hence in the appearance of a group velocity. Considerations of this nature are not trivial in measurements of the velocity of light, but are quite fundamental to the conversion of instrumental readings to a value of c. Similar considerations apply to the incorporation of the effects of the medium and the boundaries involved in the experiments. Complications arise in the regions of anomalous dispersion (absorption regions), where the phase velocity can exceed c and $dv/d\lambda$ is positive. *See* MAXWELL'S EQUATIONS; WAVE EQUATION; WAVE MOTION.

Refractive index. A plane wavefront, in going from a medium in which its phase velocity is v into a second medium where the velocity is v', changes direction at the interface. By geometry, it can be shown that $\sin i/\sin i'=v/v'$, where i and i' are the angles which the light path forms with a normal to the interface in the two media. It can also be shown that the path between any two points in this system is that which minimizes the time for the light to travel between the points (Fermat's principle). This path would not be a straight line unless $i=90°$ or $v=v'$. Snell's law states that $n\sin i=n'\sin i'$, where n is the index of refraction of the medium. It follows that the refractive index of a medium is $n=c/v'$, because the refractive index of a vacuum, where $v=c$, has the value 1.

Dispersion. The dispersion $dv/d\lambda$ of a medium can easily be obtained from measurements of the refractive index for different wavelengths of light. The weight of experimental evidence, based on astronomical observations, is against a dispersion in a vacuum. Observation of the light reaching the Earth from the eclipsing binary star Algol, 120 light-years distant (1 light-year $\cong 6\times10^{12}$ mi is the distance traversed in vacuum by a beam of light in 1 year), shows that the light for all colors is received simultaneously. The eclipsing occurs every 68 hr 49 min. Were there a difference of velocity for red light and for blue light in interstellar space as great as 1 part in 10^6, this star would show a measurable time difference in the occurrence of the eclipses in these two colors.

Corpuscular phenomena. In its interactions with matter, light exchanges energy only in discrete amounts, called quanta. This fact is difficult to reconcile with the idea that light energy is spread out in a wave, but is easily visualized in terms of corpuscles, or photons, of light.

Blackbody (heat) radiation. The radiation from theoretically perfect heat radiators, called blackbodies, involves the exchange of energy between radiation and matter in an enclosed cavity. The observed frequency distribution of the radiation emitted by the enclosure at a given temperature of the cavity can be correctly described by theory only if one assumes that light of frequency ν is absorbed in integral multiples of a quantum of energy equal to $h\nu$, where h is a fundamental physical constant called Planck's constant. This startling departure from classical physics was made by Max Planck early in the 20th century.

Photoelectric effect. When a monochromatic beam of electromagnetic radiation illuminates the surface of a solid (or less commonly, a liquid), electrons are ejected from the surface in the phenomenon known as photoemission, or the external photoelectric effect. The kinetic energies of the electrons can be measured electronically by means of a collector which is negatively charged with respect to the emitting surface. It is found that the emission of these photoelectrons, as they are called, is immediate, and independent of the intensity of the light beam, even at very low light intensities. This fact excludes the possibility of accumulation of energy from the light beam until an amount corresponding to the kinetic energy of the ejected electron has been reached. The number of electrons is proportional to the intensity of the incident beam. The velocities of the electrons ejected by light at varying frequencies agree with Eq. (3),

$$\tfrac{1}{2}mv^2_{max}=h(\nu-\nu_0) \quad (3)$$

where m is the mass of the electron, v_{max} the maximum observed velocity, ν the frequency of the illuminating light beam, and ν_0 a threshold frequency characteristic of the emitting solid.

In 1905 Albert Einstein showed that the photoelectric effect could be accounted for by assuming that, if the incident light is composed of photons of energy $h\nu$, part of this energy, $h\nu_0$, is used to overcome the forces binding the electron to the surface. The rest of the energy appears as kinetic energy of the ejected electron.

Compton effect. The scattering of x-rays of frequency ν_0 by the lighter elements is caused by the collision of x-ray photons with electrons. Under such circumstances, both a scattered x-ray photon and a scattered electron are observed, and the scattered x-ray has a lower frequency than the impinging x-ray. The kinetic energy of the impinging x-ray, the scattered x-ray, and the scattered electron, as well as their relative directions, are in agreement with calculations involving the conservation of energy and momentum. *See* COMPTON EFFECT; HEAT RADIATION; PHOTOELECTRICITY; PHOTON.

Quantum theories. The need for reconciling Maxwell's theory of the electromagnetic field, which describes the electromagnetic wave character of light, with the particle nature of photons, which demonstrates the equally important corpuscular character of light, has resulted in the formulation of several theories which go a long way to-

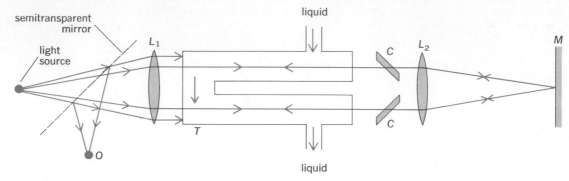

Fig. 2.   Fizeau's experiment. *C*, compensator plates; *M*, mirror; *L₁*, *L₂*, lenses; *T*, tube; *O*, interference fringes.

ward giving a satisfactory unified treatment of the wave and the corpuscular picture. These theories incorporate, on one hand, the theory of quantum electrodynamics, first set forth in articles by P. A. M. Dirac, P. Jordan, W. Heisenberg, and W. Pauli, and on the other, the earlier quantum mechanics of L. de Broglie, Heisenberg, and E. Schrödinger. Unresolved theoretical difficulties persist, however, in the higher-than-first approximations of the interactions between light and elementary particles. The incorporation of a theory of the nucleus into a theory of light is bound to call for additional formulation.

Dirac's synthesis of the wave and corpuscular theories of light is based on rewriting Maxwell's equations in a Hamiltonian form resembling the Hamiltonian equations of classical mechanics. Using the same formalism involved in the transformation of classical into wave-mechanical equations by the introduction of the quantum of action $h\nu$, Dirac obtained a new equation of the electromagnetic field. The solutions of this equation require quantized waves, corresponding to photons. The superposition of these solutions represents the electromagnetic field. The quantized waves are subject to Heisenberg's uncertainty principle. The quantized description of radiation cannot be taken literally in terms of either photons or waves, but rather is a description of the probability of occurrence in a given region of a given interaction or observation. *See* HAMILTON'S EQUATIONS OF MOTION; QUANTUM ELECTRODYNAMICS; QUANTUM FIELD THEORY; QUANTUM MECHANICS; QUANTUM THEORY, NONRELATIVISTIC; QUANTUM THEORY, RELATIVISTIC; UNCERTAINTY PRINCIPLE.

Relativistic effects. The measured magnitudes of such characteristics as wavelength and frequency, velocity, and the direction of the radiation in a light beam are affected by a relative motion of the source with respect to the observer, occurring during the emission of the signal-carrying electromagnetic wave trains. *See* DOPPLER EFFECT.

A difference in gravitational potential also affects these quantities. Several important observations of this nature are listed in this section, followed by a discussion of several important results of general relativity theory involving light. For extended discussions of both the special theory of relativity and the general theory of relativity *see* RELATIVITY.

Velocity in moving media. In 1818 A. Fresnel suggested that it should be possible to determine the velocity of light in a moving medium, for example, to determine the velocity of a beam of light traversing a column of liquid of length d and of refractive index n, flowing with a velocity v relative to the observer, by measuring the optical thickness nd. The experiment was carried out by Fizeau in a modified Rayleigh interferometer, shown in Fig. 2, by measuring the fringe displacement in O corresponding to the reversing of the direction of flow. If v' is the phase velocity of light in the medium (deduced from the refractive index by the relation $v' = c/n$), it is found that the measured velocity v_m in the moving medium can be expressed as $v_m = v' + v(1 - 1/n^2)$ rather than $v_m = v' + v$, as would be the case with a Newtonian velocity addition.

Aberration. J. Bradley discovered in 1725 a yearly variation in the angular position of stars, the total variation being 41 seconds of arc. This effect is in addition to the well-known parallax effect, and was properly ascribed to the combination of the velocity of the Earth in its orbit and the speed of light. Bradley used the amplitude of the variation to arrive at a value of the velocity of light. George Airy compared the angle of aberration in a telescope before and after filling it with water, and he discovered, contrary to his expectation, that there was no difference in angle. *See* ABERRATION OF LIGHT.

Michelson-Morley experiment. The famous Michelson-Morley experiment, one of the most significant experiments of all time, was performed in 1887 to measure the relative velocity of the Earth through inertial space. Inertial space is space in which Newton's laws of motion hold. Dynamically, an inertial frame of reference is one in which the observed accelerations are zero if no forces act. A point in an orbit is the center of such a frame. *See* FRAME OF REFERENCE.

The rotation of the Earth about its axis, with tangential velocities never exceeding 0.5 km/sec, is easily demonstrated mechanically (Foucault's pendulum, precession of gyroscopes) and optically (Michelson's rectangular interferometer). The surface of the Earth is not an inertial frame. In its orbit around the Sun, on the other hand, the Earth has translational velocities of the order of 30 km/sec, but this motion cannot be detected by mechanical experiments because of its orbital nature. The hope existed, however, that optical experiments would permit the detection (and measurement) of the relative motion of the Earth through inertial space by comparing the times of travel of two light beams, one traveling in the direction of the translation through inertial space, and the other at right angles to it. The hope was based on the now disproven notion that the velocity of a light beam would be equal to the constant c only when measured with respect to the inertial space, but would be measured as smaller $(c-v)$ or greater $(c+v)$ with respect to a reference frame, such as the earth, moving with a velocity v in inertial space if a light beam were projected respectively in the direction and in the opposite direction of translation of this frame. According to classical velocity addition theorems (which, as is now known, do not apply to light), a velocity difference of $2v$ would be detected under such circumstances. *See* EARTH, ORBITAL MOTION OF.

Not only does the Earth move in an orbit around the Sun, but it is carried with the Sun in the galactic rotation toward Cygnus with a velocity of several hundreds of kilometers per second, and the Galaxy itself is moving with a high speed in its local spiral group. Speeds of hundreds and possibly thousands of kilometers per second should be detectable by measurements on Earth in two orthogonal directions, assuming of course that the Earth motion is itself with respect to inertial space, or indeed that such a space has the physical meaning ascribed to it. The unexpected result of the experiment was that no such velocity difference could be detected, that is, no relative motion could be detected by optical means.

The Michelson-Morley apparatus (Fig. 3) consists of a horizontal Michelson interferometer with its two arms at right angles. The mirrors are adjusted so that the central white-light fringe falls on the cross hair of the observing telescope. This indicates equality of optical phase, and therefore an equality of the times taken by the light beams to travel from the beam-splitting surface to each of the two mirrors and back. Rotation of the entire system by 90°, or indeed by any angle, as well as repetition of the experiment at various times of the year all are found to leave the central white-light fringe and associated fringe system undisplaced, indicating no change in the time required by the light to traverse the two arms of the interferometer when their directions relative to the direction of the Earth's motion are varied. Had there been a difference in the velocity of light in the two directions OM and ON, the two arms would be of unequal length in the initial adjustment. For example, if the light traveled faster in the direction OM (on the average, going back and forth), then the corresponding arm would have to be longer so as to make the time of travel equal in both arms. If the apparatus were turned through 90°, the shorter arm would take the place of the longer arm, and the "faster" light would now travel in the shorter arm, and the "slower" light in the longer arm; a noticeable fringe displacement would, but actually does not, take place. *See* INTERFEROMETRY.

Einstein's theory accounts for the null result by the simple explanation that no relative motion between the apparatus and the observer exists in the experiment. No change in measured length has occurred in either direction, and because the propagation of light is isotropic, no velocity difference or detection of relative motion should be expected.

Gravitational and nebular red shifts. Two different kinds of shifts, or displacements of spectral lines toward the red end of the spectrum, are

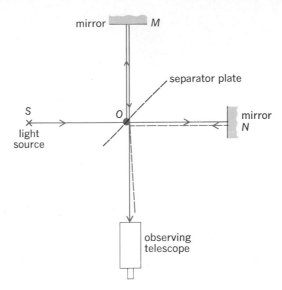

Fig. 3. Michelson-Morley experiment.

observed in spectrograms taken with starlight or light from nebulae. One is the rare, but extremely significant, gravitational red shift or Einstein shift, which has been measured in some spectra from white dwarf stars. The other, much more widely encountered, is the red shift in spectra from nebulae, usually described as being caused by a radial Doppler effect characterizing an "expansion of the universe."

The most famous example of the gravitational red shift is that observed in the spectrum of the so-called dark companion of Sirius. The existence of this companion was predicted by F. W. Bessel in 1844 on the basis of a perturbation in the motion of Sirius, but because of its weak luminosity (1/360 that of the Sun, 1/11,000 that of Sirius), it was not observed until 1861. This companion is a white dwarf of a mass comparable to that of the Sun (0.95), but having a relatively small radius of 18,000 km and the fantastically high density of 61,000 times that of water. This companion shows a shift in its spectral lines relative to the ones emitted by Sirius itself; the shift of 0.3 A for the Balmer β-line of hydrogen was reliably determined by W. S. Adams.

The nebular red shift is a systematic shift observed in the spectra of all nebulae, best measured with the calcium H and K absorption lines. Distances of nebulae are determined photometrically by measurements of their intensities, and it is found that the wavelength shift towards the red increases with the distance of the nebulae from the Earth. (The Earth does not have a privileged position if expansion of the universe is indeed involved, rather its position is somewhat like that of a man in a crowd which is dispersing—every individual in the crowd will find himself at an ever-increasing distance from every one of his neighbors.) The change of wavelength with distance has been given by E. G. Hubble as shown in Eq. (4), where d

$$\Delta\lambda/\lambda = 1.7 \times 10^{-9}d \qquad (4)$$

is in parsecs (1 parsec $= 3 \times 10^{18}$ cm). Red shifts in nebulae up to 1.1×10^9 light-years distant have been measured in a comprehensive program carried out at the Mount Wilson Observatory by Hubble and M. L. Humason. *See* EINSTEIN SHIFT; RED SHIFT.

Results of general relativity. The propagation of light is influenced by gravitation. This is one of the fundamental results of Einstein's general theory of relativity which has been subjected to experimental tests and found to be verified. Three important results involving light need to be singled out.

1. The velocity of light, measured by the same magnitude c independently of the state of motion of the frame in which the measurement is being carried out, depends on the gravitational potential Φ of the field in which it is being measured according to Eq. (5). Here $\Phi = -GM/R$, where G is the

$$c = c_0 \left(1 + \frac{\Phi}{c^2}\right) \qquad (5)$$

universal constant of gravitation (6.670×10^{-8} cgs units), M the mass of the celestial body in grams, R the radius of the body in centimeters, and c_0 the velocity of light in a vacuum devoid of fields.

For example, the absolute value of the term Φ/c^2 is about 3000 times greater on the Sun than on Earth, making the measurements of c smaller by two parts in 10^6 on the Sun as compared to measurements on Earth.

2. The frequency ν of light emitted from a source in a gravitational field with the gravitational potential Φ is different from the frequency ν_0 emitted by an identical source (atomic, nuclear, molecular) in a field-free region, according to Eq. (6).

$$\nu = \nu_0 \left(1 + \frac{\Phi}{c^2}\right) \qquad (6)$$

Spectral lines in sunlight should be displaced toward the red by two parts in 10^6 when compared to light from terrestrial sources.

3. Light rays are deflected when passing near a heavenly body according to Eq. (7), where α is the

$$\alpha = \frac{4GM}{c^2 R} \qquad (7)$$

angular deflection in radians, and R the distance of the beam from the center of the heavenly body of mass M. The deflection is directed so as to increase the apparent angular distance of a star from the center of the Sun when starlight is passing near the edge of the Sun. The deflection according to this equation should be 1.75 seconds of arc, a value which compares favorably with eclipse measurements of the star field around the Sun in 1931. These measurements indicated values up to 2.2 seconds of arc when compared with photographs of the same field 6 months earlier. This most sensational prediction of Einstein's theory might seem less surprising today when the corpuscular-photon character of light is widely known, and when a Newtonian M/R^2 attraction might be considered to be involved in the motion of a corpuscle with the velocity c past the Sun. However, application of Newton's law predicts a deviation only half as great as the reasonably well verified relativistic prediction.

Matter and radiation. The possibility of creating a pair of electrons—a positively charged one (positron) and a negatively charged one (negatron) —by a rapidly varying electromagnetic field (γ-

rays of high frequency) was predicted as a consequence of Dirac's wave equation for a free electron and has been experimentally verified. I. Curie and F. Joliot, as well as J. Chadwick, P. M. S. Blackett, G. P. Occhialini, and others have compared the number of positrons and negatrons ejected by γ-rays passing through a thin sheet of lead (and other materials) and have found them to be the same, after accounting for two other groups of electrons also appearing in the experiment (photoelectrons and recoil electrons). Other examples of negatron-positron pair production include the collision of two heavy particles, a fast electron passing through the field of a nucleus, the direct collision of two electrons, the collision of two light quanta in a vacuum, and the action of a nuclear field on a γ-ray emitted by the nucleus involved in the action.

Evidence of the creation of matter from radiation, as well as that of radiation from matter, substantiates Einstein's equation, Eq. (8), which

$$E = mc^2 \qquad (8)$$

was first expressed in the following words: "If a body [of mass m] gives off the energy E in the form of radiation, its mass diminishes by E/c^2."

In regard to exchanges of energy and momentum, electromagnetic waves behave like a group of particles with energy as in Eq. (9) and momentum as in Eq. (10).

$$E = mc^2 = h\nu \qquad (9)$$

$$p = h\nu/c = h/\lambda \qquad (10)$$

Finally, many experiments with photons show that they also possess an intrinsic angular momentum, as do particles. Circularly polarized light, for example, carries an experimentally observable angular momentum, and it can be shown that, under certain circumstances, an angular momentum can be imparted to unpolarized or plane-polarized light (plane wave passing through a finite circular aperture). In any case, the angular momentum will be quantized in units of $h/2\pi$.

The inverse process to the creation of electron pairs is the annihilation of a positron and a negatron, resulting in the production of two γ-ray quanta (two-quantum annihilation). Nuclear chain reactions are known to involve similar processes. *See* CHAIN REACTION, NUCLEAR; ELEMENTARY PARTICLE; INERTIA OF ENERGY; PAIR PRODUCTION, ELECTRON-POSITRON.

Unified field theories. In conclusion, one might mention the so-called unified field theories. These are classical theories that attempt to represent electromagnetism and gravitation as aspects of space-time geometry. Although there is some experimental evidence for a connection between electromagnetism and gravitation, the subject remains speculative. For a discussion of the better-known unification proposals *see* UNIFIED FIELD THEORY. *See also* SPACE-TIME.

[GEORGE W. STROKE]

Bibliography: P. G. Bergmann, *Introduction to the Theory of Relativity*, 1950; M. Born and E. Wolf, *Principles of Optics*, 1959; R. W. Ditchburn, *Light*, 2d ed., 1963; S. Fluegge (ed.), *Handbuch der Physik*, vol. 24, 1956; W. Heitler, *The Quantum Theory of Radiation*, 3d ed., 1954; F. A. Jenkins and H. E. White, *Fundamentals of Optics*,

3d ed., 1957; H. H. Skilling, *Fundamentals of Electric Waves*, 2d ed., 1948; A. J. W. Sommerfeld, *Lectures on Theoretical Physics*, vol. 4, 1954; J. Strong, *Concepts of Classical Optics*, 1958.

Light amplifier

In the broadest sense, a light amplifier is a device which produces an enhanced light output when actuated by incident light. A simple photocell relay—light source combination would satisfy this definition. To make the term more meaningful, common usage has introduced two restrictions: (1) A light amplifier must be a device which, when actuated by a light image, reproduces a similar image of enhanced brightness, and (2) the device must be capable of operating at very low light levels without introducing spurious brightness variations (noise) into the reproduced image. The term is used synonymously for image intensifier. The light amplifier will increase the brightness of an image which is below visual threshold to a level where it can be readily seen with the unaided eye.

The light amplifier, or image intensifier, is the fundamental component in a scotoscope, which is a telescope for seeing in the dark. It is, of course, impossible to see under conditions of complete darkness. Indeed, there is a fundamental lower limit of illumination under which an image of a given quality can be recognized. This limitation arises because of the corpuscular nature of light. *See* PHOTON.

Photons arriving through a lens, or other optical system, onto an image area are random in time. If the number of photons per unit area arriving during the time allotted for image formation (for example, the period of the persistence of vision) is too low, the statistical fluctuation will be greater than the variation in number due to true differences in image brightness. Under these circumstances, image recognition is impossible. The quantitative expression given below relates the minimum image illumination L (lumens) for ideal image recognition with contrast C, resolution N (number of picture elements), integration time t, and quantum efficiency κ.

$$L = \text{const}\, \frac{N}{\kappa C^2 t}$$

With this relationship and a knowledge of the optical systems employed in a night-viewing telescope or scotoscope, the limiting performance can be estimated, assuming an ideal image intensifier or light amplifier. Modern intensifiers come within a factor of two or three of reaching this limit.

Image-intensifier tubes. Intensifier tubes used in scotoscopes consist of a semitransparent photocathode which emits electrons with a density distribution proportional to the distribution of light intensity incident on it. Thus, when a light image falls on one side of the cathode, an electron-current image is emitted from the other side. An electron optical system, which acts on electrons in much the same way as does a glass lens system on light, focuses the electron image onto an intensifier element. The intensified electron image from the other side of the intensifier element is again focused by a second electron lens onto a second intensifier element (if additional amplification is required) or onto a fluorescent viewing screen

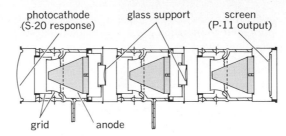

Three-stage electrostatically focused image tube.

where the electron energy is converted to visible light. The spectral response of the image-intensifier tube depends upon the type of photocathode employed, and the tube may serve as a wavelength converter as well as an intensifier.

Two types of intensifier elements are ordinarily employed. One consists of a transparent support (either a thin film or a fiber optics plate) with a phosphor screen on the side on which the electron image is incident and a photocathode on the other. An electron striking the phosphor produces a flash of light which causes a release of 50 or more electrons from the photoemitter directly opposite the point of impact. Thus, the intensifier element increases the electron-image current density by a factor of 50. Two such intensifier elements in cascade give a gain of 2500. The illustration shows diagrammatically a three-stage electrostatic image-intensifier tube. The second type of intensifier element is a transmission secondary-emission film which releases 4 to 6 electrons from one side when bombarded by an energetic (~5000-volt) electron from the other. Five such screens in cascade are required to give an image-current gain of the order of 2500.

The intensifier image device can be combined with television pickup tubes to make a television camera whose sensitivity is very close to the threshold determined by the photon "shot noise" from the scene being televised. This arrangement permits making television pictures with only one-hundredth the light required with a conventional image orthicon camera tube.

Solid-state image intensifiers. A great deal of exploratory work has been done on solid-state light amplifiers. The form which has been most extensively investigated consists of a photoconductive film in contact with an electroluminescent screen. Current flows through the photoconductor at illuminated areas, causing light to be generated in the electroluminescent layer. This type of intensifier gives considerable image enhancement at intermediate light levels, but fails at the very low light levels because of extreme time lag. Almost certainly in the future solid-state intensifiers will be developed that perform as well or better than image-tube intensifiers. *See* ELECTROLUMINESCENCE.

Applications. In addition to their application for night vision, these devices have been useful in many fields of science. Image intensifiers have given valuable results in the x-ray field, both for fluoroscopy and cinematography. They are of considerable interest in astronomy and permit a reduction of exposure time by a factor of nearly 50. They have also been used in nuclear physics, for observing the tracks of high-energy particles, and in microbiology, for observing organisms that are vulnerable to radiation.   [WARREN B. BOAST]

Bibliography: G. A. Morton, Image intensifiers and the scotoscope, *Appl. Opt.*, 3:615, 1964; A. Rose, The sensitivity performance of the human eye on an absolute scale, *J. Opt. Soc. Amer.*, 38: 196, 1948; Symposium on Photoelectronic Image Devices, *Advances in Electronics and Electron Physics*, 1960; S. Weber, *Optoelectronic Devices and Circuits*, 1964; V. K. Zworykin and G. A. Morton, *Television*, 1954.

Light curves

Graphical descriptions of the light changes of variable stars. If the light changes occur in a regular manner, that is, if they are periodic or nearly so, a plot can be made of the brightness versus the phase. The phase is the time interval between an observation and the instant when the star was at some easily determined portion of the light curve. For eclipsing variables, this "epoch" will be at the time of greatest light loss when the star is faintest; for intrinsic variables the phase is usually reckoned from the time of maximum light when the star is brightest. *See* ECLIPSING VARIABLE STARS.

Interpretation. Figures 1 and 2 are examples of light curves. In these the phases are expressed in fractions of the period, and the brightness changes are in magnitudes—the standard method in astronomy for describing the brightness of a star. As is customary, the brightness is plotted as difference of magnitude, Δm_y or $V - C_y$, between the variable and a nearby comparison star; ΔC is the difference in color. For other examples of light curves *see* VARIABLE STAR.

The magnitude scale is a logarithmic one. This means that large changes in brightness can be expressed with relatively few numbers. A change of brightness of 5 magnitudes is equivalent to a factor of 100 in the total luminosity output. That is to say that if one star is described as being 5 magnitudes brighter than another, it actually emits 100 times as much light. A star 10 magnitudes brighter than another would thus emit 100×100 or 10,000 times as much radiation. Such great changes of light occur in certain red variable stars. Even greater changes occur in novae and far greater yet

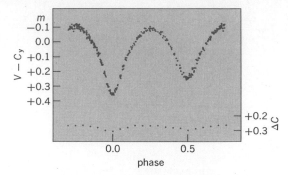

Fig. 2. Variable star XY Leonis. The top observations represent the light curve; the lower ones represent the color curve, showing the changes of color during the light variation. (*Based on photoelectric observations made by R. H. Koch at the Steward Observatory*)

in supernovae. *See* NOVA; SUPERNOVA.

Applications. Studies of light curves are helpful in many ways. Careful analysis of those of eclipsing stars combined with spectrographic data enables one to determine the sizes, masses, and densities of the stars themselves. When a star has been identified as a certain type of pulsating variable (Cepheid or RR Lyrae star) with the help of the shape of the light curve, the period tells the intrinsic brightness of the star, that is, its real output of light. Measures of its apparent brightness combined with this enables one to compute its distance and, hence, the distance to any association within which it lies. The distances to the nearer external galaxies are based on this kind of analysis of the variable stars found in them.

The precise appearance of a light curve will frequently depend on the color in which the observations are made. In eclipsing stars when the hotter star passes behind the cooler, a much greater fraction of the light is lost when observed in the shorter wavelengths (for example, ultraviolet or blue light) than when observed in the longer wavelengths (for example, yellow or red). The reverse is true when the cooler star is eclipsed by the hotter. In certain pulsating stars (Cepheid variables), the exact time when the light is a maximum depends on the color in which the observations are made. *See* CEPHEIDS.

The measures from which light curves are formed are nearly always expressed as the differences of magnitude between the variable star and a nearby star assumed to be of constant light. This is because some light is always lost in passage through the Earth's atmosphere. This extinction varies not only with the angular distance of the star above the horizon but also with atmospheric conditions. By observing the variable and comparison stars in rapid succession—or even simultaneously in some extremely sophisticated light-measuring instruments—the errors of measurement caused by extinction can be made a minimum.

When the light changes do not repeat in a regular manner, no phase can be computed and the brightness must simply be plotted against the date and time of observation. This is necessary in the case of almost all red variable stars, as well as explosive variables, and certain stars which at irregular intervals lose a great deal of light only to regain it after an unpredictable interval.

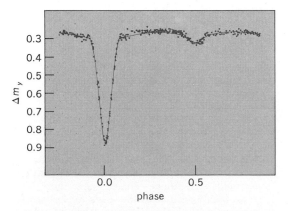

Fig. 1. Light curve of the variable star R Canis Majoris. Each dot represents one measure of brightness. The phase is expressed in fractions of the period of light variation. (*Based on photoelectric observations made by R. H. Koch at the Steward Observatory*)

Light curves can be obtained by visual, photographic, or photoelectric methods. Visual observations are now confined chiefly to amateurs, but they are useful in studies of long-period variables whose light changes are large; there are not enough professional astronomers to keep these stars under constant scrutiny. Photographic means are chiefly used for discovery and classification of variable stars, and for preliminary studies to determine which merit more precise observation. The most accurate studies of light curves are carried on by photoelectric techniques, in which the precision considerably exceeds that possible by visual or photographic methods. It is customary to use color filters to isolate certain regions of the spectrum; most studies are now made in two, three, or even more such color bands.

[FRANK B. WOOD]

Bibliography: C. Payne-Gaposchkin, *Variable Stars and Galactic Structure*, 1954; H. Weaver, Photographic photometry, in S. Flugge (ed.), *Encyclopedia of Physics*, vol. 54, 1962; A. E. Whitford, Photoelectric techniques, in S. Flugge (ed.), *Encyclopedia of Physics*, vol. 54, 1962; F. B. Wood, *Photoelectric Astronomy for Amateurs*, 1963.

Light panel

A surface-area light source that employs the principle of electroluminescence to produce light. Light panels are composed of two sheets of electrically conductive material, one a thin conducting backing and the other a transparent conductive film, placed on opposite sides of a plastic or ceramic sheet impregnated with a phosphor, such as zinc sulfide, and small amounts of compounds of copper or manganese. When an alternating voltage is applied to the conductive sheets, an electric field is applied to the phosphor. Each time the electric field changes, it dislodges electrons from the edges of the phosphor crystals. As these electrons fall back to their normal atomic state, they affect the atoms of the slight "impurities" of copper or manganese, and radiation of the wavelength of light is emitted. *See* ELECTROLUMINESCENCE.

In contrast to incandescent, vapor-discharge, and fluorescent lamps, which are essentially point or line sources of light, the electroluminescent light panel is essentially a surface source of light. Complete freedom of size and shape is a fascinating aspect of luminescent cells (see illustration).

Brightness of the panel depends upon the voltage applied to the phosphor layer and upon the electrical frequency. In general, higher voltage and higher frequency both result in a brighter panel. Blue, green, red, or yellow light can be produced by the choice of phosphors, and the proper blend of these colors produces white light. Color can be varied for a particular phosphor by changing the frequency of the applied voltage. Increasing the frequency shifts the color toward the blue end of the spectrum.

The efficiency of these light panels is only a fraction of that of the most efficient fluorescent lamps. Theoretical limits indicate, however, that the efficiency can be further improved, probably to exceed that for fluorescent lamps. Because panel lights employ no filaments and no evacuated or gas-filled bulbs, replacement of units is virtually eliminated. Glareless uniform distribution of light from large-area sources is possible without shades, louvers, or other control devices. *See* ILLUMINATION.   [WARREN B. BOAST]

Lighthouse

A distinctive structure, built on or near a shore, which exhibits a light of distinctive characteristics to serve as an aid to navigation (see illustration). Lesser lights may be displayed from fixed structures called beacons or from floating buoys or lightships.

The characteristics of the lights displayed by lighthouses are given in light lists available to mariners and, in abbreviated form, on charts. Some lights have one or more sectors in which the light appears red, usually to warn of some danger in this sector. In other sectors most lights are white.

Lighthouses have been diverse in structure and type of light. Towers up to 200 ft were constructed

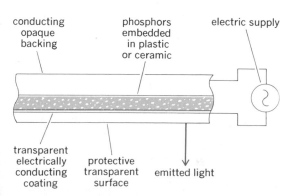

Simplified diagram of an electroluminescent cell; the sketch is not drawn to scale.

Lighthouse overlooking sea from a coastal headland. (*U.S. Coast Guard*)

along the Mediterranean coast of Egypt many centuries before Christ, with beacon fires maintained by priests. Logs, coal, oil, gas, and finally electricity have been used to provide lights. An attendant is continuously on duty at many lighthouses, but some are unattended, and some recent installations are controlled remotely from a convenient location. *See* PILOTING.   [ALTON B MOODY]

Bibliography: N. Bowditch, *American Practical Navigator*, U.S. Naval Oceanographic Office, H.O.9, 1962; U.S. Coast Guard, *Aids to Navigation Manual*, CG-222, 1964.

Lightning

The large spark produced by an abrupt discontinuous discharge of electricity through the air generally under turbulent conditions of the atmosphere.

Cumulonimbus clouds, which produce lightning, contain strongly upward-moving air and frequently extend to altitudes higher than 40,000 ft. Within these clouds, various factors act to cause the creation and separation of electric charge. Conflicting theories have been proposed to explain the electrification. For the most part, they involve cloud and precipitation particles and lead to essentially the same charge distribution. The upper

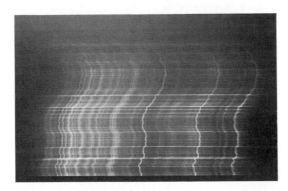

Fig. 1. Slitless spectrum of a ground flash, obtained by placing a diffraction grating in front of a camera lens with focal length 61 cm. Range of wavelengths is 3900 A (left) to 6165 A (right). (*Photograph by L. E. Salanave, Institute of Atmospheric Physics, University of Arizona*)

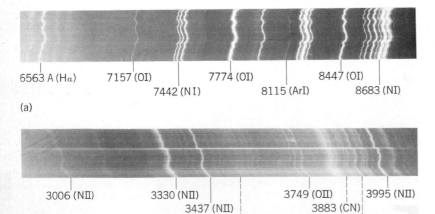

6563 A (Hα)   7157 (OI)   7774 (OI)   8447 (OI)
7442 (N I)   8115 (ArI)   8683 (NI)

(a)

3006 (NII)   3330 (NII)   3749 (OII)   3995 (NII)
3437 (NII)   3883 (CN)
3530 (N_2, N_2^+)   3914 (N_2^+)

(b)

Fig. 2. Lightning spectra. (*a*) Vertical section of a red and infrared slitless spectrum. Very strongly exposed spectra show absorption features due to molecular oxygen and water vapor; the A band of oxygen near 7600 angstrom units is especially prominent. (*b*) The ultraviolet spectrum. Atomic lines are due to ionized nitrogen and oxygen, emitted when the temperature (and the current) is near its peak.

parts of thunderstorms are positively charged, whereas the central and lower parts are predominantly negative. In some clouds, after rain has formed, a small positive charge center is found in the lower part of the cloud.

When cloud electrification has proceeded to the point where the potential gradient between adjacent charge centers, or between the cloud and ground, reaches the breakdown value, the large spark called lightning occurs. A breakdown potential of about 10,000 volts/cm has been suggested as an appropriate value for air containing water droplets.

Leaders and return strokes. It has been established that a lightning stroke to ground is initiated by a step leader. A surge of electrons moves downward about 50 m in about 1 μsec and then stops. After a pause of perhaps 50 μsec another step takes place. This sequence is repeated until the leader reaches the ground. When this occurs, a surge of charge moves rapidly up the path taken by the step leader. Whereas the average speed of the step leader is of the order of 10^6 m/sec, the main stroke moves up the preionized path at a speed of the order of 10^7 m/sec. The luminosity of this so-called return stroke is quite intense.

After an interval of the order of 0.01 sec, another stroke usually occurs in the same channel. These secondary return strokes are preceded by downward-moving dart leaders which do not show step characteristics. As many as 30–40 strokes have been observed to occur in the same channel.

Cloud-to-cloud strokes also involve a leader and main return stroke.

The current in the return strokes increases in an average of about 6 μsec to a maximum value frequently exceeding 30,000 amp, but sometimes as high as 200,000 amp. It decreases to about one-half the maximum value in an average of about 24 μsec. Average currents during the strokes are of the order of 10,000 amp, and about 500 amp may flow through the ionized channel in the interval between strokes. A reasonable estimate of the diameter of the channel appears to be about 10 cm. The high currents lead to thunder and intense heating. For a brief instant the temperature is probably near 30,000°C. The high currents and temperatures are mainly responsible for the damage caused by lightning. *See* CLOUD PHYSICS; LIGHTNING AND SURGE PROTECTION; TERRESTRIAL ELECTRICITY; THUNDERSTORM.

[LOUIS J. BATTAN]

Optical spectrum. Lightning is a thermal source of radiation emitting a very broad spectrum of electromagnetic frequencies. If boundaries are set by practical considerations of atmospheric transmission and the spectral sensitivity of photographic film, optical observations extend from approximately 3000 A (ozone absorption) to nearly 10,000 A (limit of ordinary infrared photographic film). This is the "spectrum" that will be discussed here.

Ground flashes in the clear air below a cloud provide the best opportunity to study the optical properties of lightning. In such cases the source is a narrow line at a considerable distance from the observer, and the technique of slitless spectroscopy can be used to advantage. In this method of observation, the dispersive element (a prism or diffraction grating or both) is placed in front of a camera focused at infinity.

Photographs of lightning flashes obtained with a transmission diffraction grating in front of the camera lens. Most of the light passes straight through the grating and is imaged by the lens on the film. The remaining light is deviated by the grating and split into characteristic spectral colors, which compose the "signature" of the atoms and ions within the hot lightning channel. (*R. E. Orville, State University of New York at Albany*)

Figure 1 is the reproduction of a slitless spectrum of lightning from 3900 to 6200 A obtained on panchromatic film. With a few minor exceptions, the spectrum in this region is composed of emissions from singly ionized nitrogen. When the spectrum is recorded on film with extended red sensitivity, one finds a very strong line due to hydrogen, termed the Hα line, with a wavelength of 6563 A. The atomic hydrogen is derived from water vapor completely dissociated by the electrical discharge. Still further extensions with film sensitive to infrared reveal a spectrum dominated by strong lines of neutral nitrogen and oxygen and a few faint lines due to neutral argon (Fig. 2a).

The photographic ultraviolet spectrum of lightning extends from the visible limit (approximately 4000 A) to slightly below 3000 A, where absorption bands due to ozone become significant and eventually cut down the transmitted radiation to very low levels. (Ozone is generated around the electrical discharge channel and in the intervening air path during thunderstorms.) Emission lines due to singly ionized nitrogen and oxygen predominate, as in the visible spectrum, but in addition there may occur emission bands due to molecules of cyanogen (CN) and nitrogen (N_2 and N_2^+). These molecular features are notably variable in their relative intensities from flash to flash; their excitation is presumably dependent upon the amount of electrical current and its duration (Fig. 2b).

Bright horizontal streaks are produced by the comparatively weak continuous spectrum of heated air—enhanced where short sections of the discharge channel lie along the direction of dispersion or in a line toward the spectrograph. Some of the faintest streaks may be caused by regions (beads) of intrinsic brightening along the channel, but this is not established with certainty.

The spectrum of each stroke in a flash can be obtained by placing a mask with a narrow, horizontal slot close to the photographic film which is moving vertically (usually by means of a rotating drum). Since strokes typically occur 50 msec apart or more, a film transport of only 20 cm/sec suffices with a masking slot 1 cm high.

Figure 3 shows the time-resolved spectrum of a three-stroke flash. The similarity among the component spectra shows that the discharge attained substantially the same temperature at each stroke, though the overall luminosity of individual strokes clearly differ.

In some cases a stroke continues for 50 msec or more at a very low discharge rate, called a continuing current. This produces a continuing luminosity whose spectrum is dominated by Hα and N_2^+ emissions, with lesser contributions from neutral oxygen and nitrogen.

An application of a high-speed drum camera and extremely sensitive film produces time-resolved spectra of individual strokes, on which changes occurring in microseconds can be measured.

A comparison of the relative intensities of certain emissions of ionized nitrogen (N II) shows that temperatures of the order of 25,000–30,000°C are attained in a few microseconds, dropping to half that in about 30 μsec. Measurements of the broadened profile of Hα indicate that electron densities are somewhere between 10^{17} and 10^{18} cm^{-3} (almost complete ionization) at peak temperature, at which

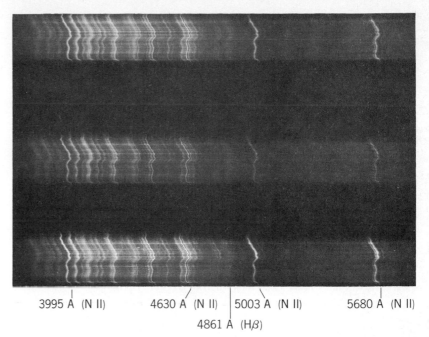

3995 A (N II)   4630 A (N II)   5003 A (N II)   5680 A (N II)
4861 A (Hβ)

Fig. 3. Spectra of three consecutive strokes of a lightning flash. Time sequence upward; intervals 45 and 55 msec, respectively. (*Photograph by L. E. Salanave, Institute of Atmospheric Physics, University of Arizona*)

time the pressure inside the lightning channel is of the order of 10 atm. The physical basis for these rather spectacular values can be intuitively grasped in terms of the very high, rapidly rising currents described in the previous section.

[LEON E. SALANAVE]

Bibliography: E. P. Krider, Lightning spectroscopy, *Nucl. Instrum. Methods*, vol. 110, July 1973; R. E. Orville and L. E. Salanave, Lightning spectroscopy: Photographic techniques, *Appl. Opt.*, vol. 9, August 1970; M. A. Uman, *Understanding Lightning*, chap. 11, 1971.

Lightning and surge protection

Means of protecting electrical systems, buildings, and other property from lightning and other high-voltage surges.

The destructive effects of natural lightning are well known. Studies of lightning and means of either preventing its striking an object or passing the stroke harmlessly to ground have been going on since the days when Franklin first established that lightning is electrical in nature. From these studies, two conclusions emerge: (1) Lightning will not strike an object if it is placed in a grounded metal cage. (2) Lightning tends to strike, in general, the highest objects on the horizon. *See* ATMOSPHERIC ELECTRICITY; LIGHTNING.

One practical approximation of the grounded metal cage is the well-known lightning rod or mast (Fig. 1). The effectiveness of this device is evaluated on the cone-of-protection principle. The protected area is the space enclosed by a cone having the mast top as the apex of the cone and tapering out to the base. Laboratory tests and field experience have shown that if the radius of the base of the cone is equal to the height of the mast, equipment inside this cone will rarely be struck. A radius equal to twice the height of the mast gives a cone of shielding within which an object will be

Fig. 1.   Lightning rods on a house.

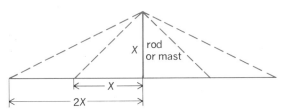

Fig. 2.   Cone of protection of a lightning rod or mast.

struck occasionally. The cone-of-protection principle is illustrated in Fig. 2. *See* LIGHTNING ROD.

A building which stands alone, like the Empire State in New York City, is struck many times by lightning during a season. It is protected with a mast and the strokes are passed harmlessly to ground. It is interesting to note, however, that lightning has been observed to strike part way down the side of this building (Fig. 3). This shows that lightning does not always strike the highest object but rather chooses the path having the lowest electrical breakdown.

The probability that an object will be struck by

Fig. 3.   Multiple lightning stroke A to Empire State and nearby buildings, Aug. 23, 1936. Single, continuing stroke B to Empire State Building, Aug. 24, 1936.

lightning is considerably less if it is located in a valley. Therefore, electric transmission lines which must cross mountain ranges often will be routed through the gaps to avoid the direct exposure of the ridges. *See* ELECTRIC DISTRIBUTION SYSTEMS; ELECTRIC POWER SYSTEMS.

Overhead lines of electric power companies are vulnerable to lightning. Lightning appears on these lines as a transient voltage, which, if of sufficient magnitude, will either flash over or puncture the weakest point in the system insulation.

Many of the troubles that cause service interruptions on electrical systems are the results of flashovers of insulation wherein no permanent damage is done at the point of fault, and service can be restored as soon as the cause of the trouble has disappeared. A puncture or failure of the insulation, on the other hand, requires repair work, and damaged apparatus must be removed from service.

There are a number of protective devices to limit or prevent lightning damage to electric power systems and equipment. The word protective is used to connote either one or two functions: the prevention of trouble, or its elimination after it occurs. Various protective means have been devised either to prevent lightning from entering the system or to dissipate it harmlessly if it does.

Overhead ground wires and lightning rods. These devices are used to prevent lightning from striking the electrical system.

The grounded-metal-cage principle is approached by overhead ground wires, preferably two, installed over the transmission phase conductors and grounded at each tower. The ground wires must be properly located with respect to the phase conductors to provide a cone of protection and have adequate clearance from them, both at the towers and throughout the span. If the resistance to ground of the tower is high, the passing of high lightning currents through it may sufficiently elevate the tower in potential from the transmission line conductors so that electrical flashover can occur. Since the magnitude of the lightning current may be defined in terms of probability, the expected frequency of line flashover may be predicted from expected storm and lightning stroke frequency, the current magnitudes versus probability, and the tower footing resistance. Where the expected flashover rate is too high, means to lower the tower footing resistance are employed such as driven ground rods or buried wires connected to the base of the tower (Fig. 4).

The ground wires are often brought in over the terminal substations. For additional shielding of the substations, lightning rods or masts are installed. Several rods are usually used to obtain the desired protection. *See* ELECTRIC POWER SUBSTATION.

Lightning arresters. These are protective devices for reducing the transient system overvoltages to levels compatible with the terminal-apparatus insulation. They are connected in parallel with the apparatus to be protected. One end of the arrester is grounded and also connected to the case of the equipment being protected; the other end is connected to the electric conductor (Fig. 5).

Lightning arresters provide a relatively low discharge path to ground for the transient overvoltages, and a relatively high resistance to the power system follow current, so that their operation does not cause a system short circuit.

In selecting an arrester to protect a transformer, for example, the voltage levels that can be maintained by the arrester both on lightning surges and surges resulting from system switching must be coordinated with the withstand strength of the transformers to these surges. Arrester gap spark-over voltages, discharge voltages on various magnitudes of discharge currents, the inductance of connections to the arrester, voltage wave shape, and other factors must all be evaluated. *See* SURGE ARRESTER.

Rod gaps. These also are devices for limiting the magnitude of the transient overvoltages. They usually are formed of two 1/2-in.^{2} rods, one of which is grounded and the other connected to the line conductor but may also have the shape of rings or horns. They have no inherent arc-quenching ability, and once conducting, they continue to arc until the system voltage is removed, resulting in a system outage.

These devices are applied on the principle that, if an occasional flashover is to occur in a station, it is best to predetermine the point of flashover so that it will be away from any apparatus that might otherwise be damaged by the short-circuit current and the associated heat.

The flashover characteristics of rod gaps are such that they turn up (increase of breakdown voltages with decreasing time of wavefront) much faster on steep-wavefront surges than the withstand-voltage characteristics of apparatus, with the result that if a gap is set to give a reasona-

Fig. 5. Installation of three single-pole station lightning arresters rated 276 kv for lightning and surge protection of large 345-kv power transformer. (*General Electric Co.*)

ble margin of protection on slow-wavefront surges, there may be little or no protection for steep-wavefront surges. In addition, the gap characteristics may be adversely affected by weather conditions and may result in undesired flashovers.

Immediate reclosure. This is a practice for restoring service after the trouble occurs by immediately reclosing automatically the line power circuit breakers after they have been tripped by a short circuit. The protective devices involved are the power circuit breaker and the fault-detecting and reclosing relays.

This practice is successful because the majority of the short circuits on overhead lines are the result of flashovers of insulators and there is no permanent damage at the point of fault. The fault may be either line-to-ground or between phases. Reclosing relays are available to reenergize the line several times with adjustable time intervals between reclosures.

If the relays go through the full sequence of reclosing and the fault has not cleared, they lock out. If the fault has cleared after a reclosure, the relays return to normal.

Permanent faults must always be removed from a system and the accepted electric protective devices are power circuit breakers and suitable protective relays. *See* CIRCUIT TESTING (ELECTRICITY); ELECTRIC PROTECTIVE DEVICES.

[GLENN D. BREUER]

Bibliography: E. Beck, *Lightning Protection for Electric Systems*, 1954; Institute of Electrical and Electronics Engineers, *Revised Report on Standards for Value and Expulsion Type Lightning Arresters*, IEEE Stand. no. 28B, 1956; W. W. Lewis, *The Protection of Transmission Systems Against Lightning*, rev. ed., 1965.

Lightning rod

A metallic rod set up on an exposed elevation of a structure and connected to a low ohmic ground to intercept and to provide a direct conducting path

Fig. 4. Double-circuit suspension tower on the Olive–Gooding Grove 345-kv transmission line, with two circuits strung and with two ground wires. (*Indiana and Michigan Electric Co., American Electric Power Co.*)

Outdoor substation at Hudson River water-pumping station in Chelsea, N.Y. Lightning rods are located on the four raised corners of the steel structure.

to ground for lightning discharges that might otherwise damage the structure being protected. Lightning rods are also called lightning masts (see illustration).

The effectiveness of the protection obtained depends upon (1) sufficient height of the air terminal end of the rod for the lightning stroke to terminate on it, rather than on the protected structure, (2) sufficient conductor (rod) cross section to carry the lightning stroke current to ground, and (3) a low-resistance ground connection. *See* LIGHTNING AND SURGE PROTECTION.   [GLENN D. BREUER]

Light-year

A unit of measurement of astronomical distance. A light-year is the distance light travels in 1 sidereal year. One light-year is equivalent to 9.461×10^{12} km, or 5.879×10^{12} mi. Distances to some of the nearer celestial objects, measured in units of light time, are shown in the table.

Distances from the Earth to some celestial objects

Object	Distance from Earth (in light time)
Moon (mean)	1.3 sec
Sun (mean)	8.3 min
Mars (closest)	3.1 min
Jupiter (closest)	33 min
Pluto (closest)	5.3 hr
Nearest star (Proxima Centauri)	4.3 years
Andromeda Nebula (M31)	2,300,000 years

This unit, while useful for its graphic presentation of the enormous scale of stellar distances, is seldom used technically. *See* ASTRONOMICAL UNIT; PARALLAX (ASTRONOMY); PARSEC.

[JESSE L. GREENSTEIN]

Lignite

Soft and porous carbonaceous material intermediate between peat and subbituminous coal, with a heat value less than 8300 Btu on a mineral-matter-free (mmf) basis. In the United States, lignitic coals are classified as lignite A or lignite B, depending upon whether they have calorific value of more or less than 6300 Btu. Elsewhere than in North America, lignites are called brown coals, two major classes being recognized, namely, hard and soft brown coals. The soft brown coals are described as either earthy, resembling peat, or as fragmentary; the hard brown coals are regarded as dull (matte) or bright (glance). The soft brown coals correspond in a general way with lignites of class B, and the hard brown coals with lignites of class A, but the two classifications should not be regarded as strictly interchangeable. Brown coals generally have a brown color, compared with the usual black color of bituminous coal. At the present time there is a vigorous effort being made by the International Committee of Coal Petrology that will make possible a satisfactory differentiation on a petrologic basis of bituminous and lignitic coals and provide a satisfactory classification of varieties of lignite on a basis other than calorific.

Reserves. The lignite of North America is mainly of lignite class A. The known reserves of lignite in 15 states have been estimated as 224,000,000,000 short tons, as compared with an estimated total coal reserve of 828,000,000,000 tons. About 98% of the lignite reserves in the United States are in the northern Great Plains area, which includes the western half of North Dakota, eastern Montana, northeastern Wyoming, and northeastern South Dakota. A single lignite mine in California operates for the production of montan wax. The reserves in the northern Great Plains are believed to represent a total of 99% of the total reserves of lignite in the United States. The production of lignite in the United States in 1963 was 2,700,000 tons, excluding the production from two mines in Texas, amounting to less than 1% of the world production. Of these 2,700,000 tons, all but about 7,000 tons was recovered by strip mining.

Uses. The main use of lignite in the United States is in residential and industrial heating and in the generation of power. Because of the competition with bituminous coal, the market area of lignite in the United States is relatively small. When used at any great distance from its source, lignite requires dehydration and even briquetting.

Some investigation is under way that may lead to the production of pipeline gas from lignite. The U.S. Bureau of Mines regards the total gasification of lignite a potential for a large tonnage outlet for this variety of coal. *See* COAL; MINERAL FUEL AREAS.

[GILBERT H. CADY]

Bibliography: *Mineral Facts and Problems*, U.S. Bur. Mines Bull. no. 360, 1965.

Lignumvitae

A tree, *Guaiacum sanctum*, also known as holywood lignumvitae, which is cultivated to some extent in southern California and tropical Florida. Lignumvitae is native in the Florida Keys, Bahamas, West Indies, and Central and South America. It is an evergreen tree of medium size with abruptly pinnate leaves. The tree yields a resin or gum known as gum guaiac or resin of guaiac which is used in medicine. The very heavy black heartwood is used in bowling balls, blocks and pulleys, and parts of instruments. *See* FOREST AND FORESTRY; SAPINDALES; TREE.

[ARTHUR H. GRAVES/KENNETH P. DAVIS]

Liliales

An order of flowering plants, division Magnoliophyta (Angiospermae), from whose name the names of the subclass Liliidae and class Liliopsida (monocotyledons) are derived. The order consists of some 13 families and nearly 7700 species, more than half of which belong to the family Liliaceae (4200 species). The Iridaceae (about 1500 species), Dioscoreaceae (about 650 species), Agavaceae (about 550 species), and Smilacaceae (about 300 species) are other good-sized families of the order. The order includes the families of its subclass that do not have the specialized, mycotrophic habit of the related order Orchidales. The ovary is either superior or inferior, the seeds are of ordinary number and structure, and most families and genera have septal nectaries in the flowers. *See* FLOWER; ORCHIDALES.

The Liliaceae (including Amaryllidaceae) and Iridaceae are mostly herbs from a bulb or rhizome, usually with narrow, parallel-veined leaves that do not persist from one season to the next. The stomates lack subsidiary cells, and the shoot lacks vessels. The Liliaceae usually have six stamens; the ovary is either superior or inferior. The Iridaceae have only three stamens and an inferior ovary. These two families include many showy ornamentals. Tulip, hyacinth, narcissus, trillium (unusual in its broad leaves and green sepals), and the various kinds of true lilies (see illustration), as well as onion, garlic, and asparagus, belong to the Liliaceae. Iris, crocus, and gladiolus are familiar Iridaceae. *See* ASPARAGUS; COLCHICINE; GARLIC; LILIIDAE; LILIOPSIDA; MAGNOLIOPHYTA; ONION; ORNAMENTAL PLANTS; PLANT KINGDOM; SISAL.

[ARTHUR CRONQUIST]

Liliidae

A subclass of the class Liliopsida (monocotyledons) of the division Magnoliophyta (Angiospermae), the flowering plants, consisting of 2 orders (Liliales and Orchidales), 17 families, and nearly 28,000 species. The Liliidae are syncarpous monocotyledons with both the sepals and the petals usually petaloid. The seeds are either nonendospermous or have an endosperm with various sorts of reserve foods such as cellulose, fats or protein, and only seldom starch. The stomates are without subsidiary cells or, in a few small families, have two subsidiary cells. The pollen is binucleate. The vast majority of the species have the vessels chiefly or wholly confined to the roots. The flowers generally have well-developed nectaries, and pollination is usually by insects or other animals. *See* FLOWER; LILIALES; LILIOPSIDA; MAGNOLIOPHYTA; ORCHIDALES; PLANT KINGDOM. [ARTHUR CRONQUIST]

Liliopsida

One of the two classes which collectively make up the division Magnoliophyta (Angiospermae), the flowering plants. The Liliopsida, often known as Monocotyledoneae, or monocotyledons, embrace 4

subclasses (Alismatidae, Commelinidae, Arecidae, and Liliidae), 18 orders, 61 families, and about 55,000 species. *See* ALISMATIDAE; ARECIDAE; COMMELINIDAE; LILIIDAE.

All of the characters which collectively distinguish the Liliopsida from the Magnoliopsida (dicotyledons) are subject to exception, but most of the Liliopsida have parallel-veined leaves, and when the embryo is differentiated into recognizable parts, there is only a single cotyledon. The vascular bundles are generally scattered or borne in two or more rings, so that the stems and roots do not have a well-defined pith and cortex. Monocotyledons never have an intrafascicular cambium, and most of them have no secondary growth at all. Some few of them have an unusual type of cambium which produces vascular bundles toward the inside and usually only a small amount of parenchyma toward the outside. Woody monocots such as palms become woody by lignification of the ground tissues in which the vascular bundles of the stem are embedded. The mature root system of monocots is wholly adventitious. The floral parts of monocots, when of definite number, are most often borne in sets of 3, seldom 4, never 5. The pollen is uniaperturate or of uniaperturate-derived type. *See* CORTEX (PLANT); FLOWER; LEAF (BOTANY); PITH; STEM (BOTANY); VASCULAR BUNDLES.

It is generally agreed that the monocotyledons are derived from primitive dicotyledons. The solitary cotyledon, the parallel-veined leaves, the absence of a cambium, the dissected stele, and the adventitious root system of monocotyledons are all regarded as secondary rather than primitive characters in the angiosperms as a whole, and any plant which was more primitive than the monocotyledons in these several respects would certainly be a dicotyledon. The monocotyledons are more primitive than the bulk of the dicotyledons in having uniaperturate pollen, but several of the primitive families of dicotyledons also have uniaperturate pollen. *See* MAGNOLIOPHYTA; MAGNOLIOPSIDA; PLANT KINGDOM. [ARTHUR CRONQUIST]

Limburgite

A dark, glass-rich igneous rock with abundant large crystals (phenocrysts) of olivine and pyroxene and with little or no feldspar.

The glass is brown and usually alkali-rich. Phenocrysts are well formed, and the pyroxene shows zonal structure (diopsidic cores and titanium-rich margins). Small quantities of feldspathoid, hornblende, and biotite may be present. Granules of titanium-iron oxide are abundant and widespread.

Limburgite is a rare rock and forms lava flows and small intrusive bodies (dikes, sills, and plugs). It is associated with basanite and alkali-basaltic rocks. It probably originates from basaltic magmas and lavas by accumulation of olivine and pyroxene crystals under the influence of gravity. Thus, limburgite may grade through oceanite to basalt. *See* BASALT; IGNEOUS ROCKS.

[CARLETON A. CHAPMAN]

Lime (botany)

An acid citrus fruit, *Citrus aurantifolia*, usually grown in tropical regions because of its low cold resistance. The two principal groups of limes are the West Indian or Mexican, and the Tahiti (see

LILIALES

Carolina lily (*Lilium michauxii*), a characteristic member of the Liliaceae. The six-membered perianth is typical of the family, order, and subclass. (*Courtesy of John H. Gerard, from National Audubon Society*)

Tahiti (Persian) lime
(*Citrus aurantifolia*).

illustration). The West Indian lime is a small tree with irregular branches having short, stiff, sharp spines. The fruit is small (walnut-sized) and strongly acid. It is more sensitive to cold than the Tahiti lime, which is a more vigorous tree, bearing fruits of lemon size. *See* SAPINDALES.

In the United States limes are grown mainly in southern Florida, but also in the warmer areas of southern California. Lime acreage in Florida (chiefly Tahitian type) has remained fairly constant for the period 1958–1967 at approximately 3500 acres, although the production of fruit increased during this period because of greatly improved cultural practices. Prior to 1958 a considerably larger acreage existed but produced less fruit. The average yearly production during 1958–1967 was 377,000 boxes, roughly two-thirds of which was utilized fresh, the remaining one-third being diverted to processing. Lime fruits are generally harvested just before the rind turns from green to yellow, because at this stage they have their maximum aromatic bouquet.

The West Indian lime is grown in most tropical countries in small, widely scattered plantings. Thus statistics on the number of trees or yields are not available, although they undoubtedly constitute a very sizable aggregate.

Limes are used in frozen sherbets and in beverages such as ades and in alcoholic drinks. The juice has many culinary uses such as flavoring in jellies, jams, and marmalades, and as garnishing for fish and meats. *See* FRUIT (BOTANY); FRUIT, TREE.   [FRANK E. GARDNER]

Lime (industry)

A general term for burned (or calcined) limestone, also known as quicklime, hydrated lime, and unslaked or slaked lime. Use of lime as a building material in mortar and plaster coincides with earliest recorded history. However, since development of the chemical process industries, its predominant usage is as a basic industrial chemical, where it ranks second to sulfuric acid in tonnage. It still enjoys its traditional building uses, but 92% is used as a chemical. Uses in order of decreasing size are: steel fluxing (45%), water treatment, nonferrous metals (alumina, magnesium, copper, and others), pulp and paper, refractories, soil stabilization, sewage and trade waste treatment, chemicals, and glass manufacture.

Lime is not a mineral; it is manufactured from a mineral—limestone, coral, oystershell, all being sources of calcium carbonate. Dolomite, a calcium-magnesium carbonate, is used to produce dolomitic (magnesium) lime. Only the purest types of stone or shell are used for lime. *See* LIMESTONE.

The quarried or mined limestone raw material is crushed and screened to produce either pebble size (1/2–2 in.; 1.3–5 cm) for charging into rotary kilns or lump size (3–8 in.; 7.5–20 cm) for vertical kilns. At high temperatures of 2000–2500°F (1100–1400°C), the limestone is decomposed, evolving carbon dioxide gas that constitutes 44% of the weight of the limestone, with calcium oxide remaining. *See* KILN.

The resulting lime, depending upon degree of burn, is chemically reactive with water and acids. Soft-burned limes literally explode on contact with water. If the lime is hard-burned, its reactivity is retarded. One lime product is dead-burned dolo-

mite that has no reactivity with water, making it useful as a refractory material. Unlike conventional limes, hydraulic lime is made from impure limestone that contains considerable silica, alumina, and iron. Calcination causes the lime to combine chemically with these impurities, forming silicates and aluminates. Such limes when hydrated will harden under water, somewhat like cement. *See* PORTLAND CEMENT.

When lime hydrates in water, considerable heat is generated, enough for 1 lb of quicklime to heat 2 lb of water from 0°F (−18°C) to boiling. The lime manufacturer produces hydrated lime (calcium hydroxide), an ultrafine, fluffy, dry powder by controlled addition of water to ground quicklime. Hydrated lime is 25% chemically combined water by weight. Lime users also hydrate or slake quicklime with excess water into slurries or "milk of lime" of various concentrations. Dolomitic lime does not slake as readily as high-calcium types. Under normal conditions, only the calcium oxide component hydrates, yielding $Ca(OH)_2 \cdot MgO$. If slaking occurs under pressure, a highly hydrated dolomitic lime is produced.

Hydrated lime, as a base, is a strong alkali and neutralizes the strongest acids, forming neutral salts, such as calcium sulfate, magnesium sulfate, and chloride.

[ROBERT S. BOYNTON]

Bibliography: R. Boynton, *Chemistry and Technology of Lime and Limestone*, 1967; R. Boynton and K. Gutschick, Lime, in *Industrial Minerals and Rocks*, AIME, 1975; National Lime Association, *Chemical Lime Facts*, 1973; U.S. Bureau of Mines, Lime, in *Minerals Yearbook*, 1973.

Limestone

A sedimentary rock composed dominantly of carbonate minerals, principally carbonates of calcium and magnesium. Limestones are the most abundant of the nonclastic rocks. They are overwhelmingly the largest reservoir of the element carbon at or near the surface of the Earth. Much knowledge of invertebrate paleontology and consequently of the evolution of life and earth history comes from the fossils contained in them. Although the word limestone is used in the general sense above, specifically it refers to carbonate rocks dominated by the mineral calcite, $CaCO_3$, as opposed to dolomite, a term for carbonate rocks dominated by the mineral dolomite, $CaMg(CO_3)_2$. Although the mineralogical composition of most limestones is similar, their textures are not because they are formed under a great variety of conditions. For a discussion of commercial uses of limestone *see* LIME (INDUSTRY); STONE AND STONE PRODUCTS. *See also* DOLOMITE ROCK.

Chemical composition. The chemical composition of limestones is largely calcium oxide, CaO, and carbon dioxide, CO_2. Magnesium oxide, MgO, is a common constituent; if it exceeds 1 or 2%, the rock may be termed magnesian limestone. Small amounts of silica and alumina may also be present as a result of the presence of clastic materials, quartz, and clay. Iron oxide may be present, either as carbonate (siderite, ferroan-dolomite) or in other minerals, such as clays. Strontium may be present as an important trace element, probably derived from original fossil materials in which it was incorporated into aragonite, a form of $CaCO_3$.

Mineralogy. The chief minerals of limestones are calcite and aragonite, and in the dolomitic limestones, dolomite. Calcite and aragonite have the same composition but different crystal structures. Aragonite is unstable with reference to calcite in surface environments and is transformed into the calcite with time. Even though aragonite is unstable, it forms as a precipitate from sea water and some fresh waters inorganically, for example, oolites, and biogenically through the intermediate mechanism of biological secretion by shell-forming organisms. The conversion of aragonite to calcite is accomplished in a relatively short time—just a few years for some fossils. Large crystals may take much longer. Ancient rocks always contain calcite rather than aragonite. *See* ARAGONITE; CALCITE.

Although dolomite is stable in surface environments, it is formed as a primary precipitate only under special conditions, such as high salinity or high alkalinity. Petrographic and O^{18}/O^{16} isotopic evidence suggests that most dolomite is formed postdepositionally by the interaction of magnesium ions in interstitial waters with calcite. Studies of modern environments in which dolomitization is taking place imply that much of the alteration of dolomite must have taken place relatively soon after its original deposition.

Siderite, the iron carbonate, is found in some limestones, but iron occurs in carbonate form chiefly as ferroan-dolomite. Silica may be present, either finely disseminated throughout the rock or segregated into nodules of chert. Silica also occurs as small crystals of quartz that have grown in place during diagenesis. Feldspar occurs in the same way but is a little less common than silica. Other minerals found in limestones are glauconite, collophane, and pyrite. A host of other minerals may be found as small amounts of clastic material brought in by currents, including almost always a small amount of fine-grained clay. Organic matter may also be found. *See* DIAGENESIS.

Textures. The textures of limestones reveal much about their origin. The chief textural elements in a great many limestones are fossils, whole or fragmental. The species represented and their state of preservation are a guide to the ecology and environment of the site of deposition. In addition, the nature of the fossil debris, whether complete articulated shells or fragments in all stages of disarticulation, fragmentation, and comminution, is a guide to the presence of bottom currents. Shells may be broken or destroyed by bottom-dwelling organisms that ingest sediment in search of organic matter for food or by parasitic boring algae. Scavenging organisms turn the sediment over and destroy much of its original structure while producing tubes or burrows and leaving behind spherical or ovoid fecal pellets of fine-grained lime mud. All of these textures can be recognized in ancient limestones.

Some mechanically deposited limestones show well sorted particles of calcite cemented by clear intergranular calcite. If the particles are sand-size, the rock is termed calcarenite. In general, the mechanically deposited limestones show the same kinds of sedimentary structures as do clastic rocks. Cross-bedding, stratification, current lineation, and even graded bedding may be displayed. Oolitic and pisolitic textures are most abundant in limestones. *See* OOLITE AND PISOLITE.

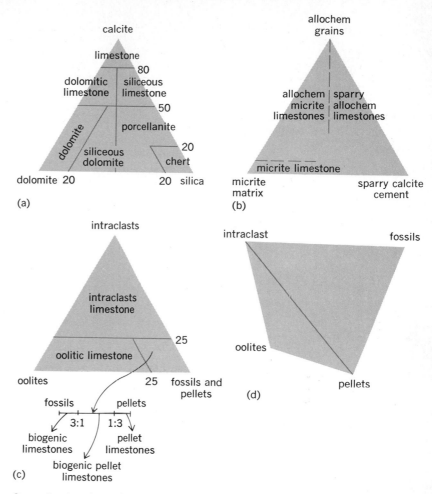

Generalized end-member groupings of carbonate rocks. (*a*) Limestone-dolomite-chert grouping (*adapted from an unpublished chart by F. J. Pettijohn, 1944*). (*b*) Grouping based on coarse and fine components; (*c*) Folk's basic classification of limestones, (*b* and *c* from *R. L. Folk, 1959*). (*d*) Tetrahedral grouping of Folk's end members. (*From W. C. Krumbein and L. L. Sloss, Stratigraphy and Sedimentation, 2d ed., Freeman, 1963*)

A prominent textural constituent of many limestones is a featureless, finely crystalline mosaic of calcite. R. L. Folk termed this material micrite and emphasized its importance as the lithified equivalent of the fine-grained lime mud found in many modern carbonate depositional environments.

Secondary textural features that are found in limestones include stylolites, cross-cutting veinlets filled with calcite, and replacement effects (typically that of rhombohedrons of dolomite replacing calcite). Because the carbonate minerals are relatively soluble in aqueous solutions and because of the transformation of aragonite to calcite and calcite to dolomite, recrystallization and other diagenetic effects are particularly common in limestones. *See* STYLOLITES.

Classification. Many schemes of limestone classification have been proposed and used, but since 1959, when R. L. Folk showed the utility of textural elements as a primary basis for classification, most workers have used systems similar to his (see illustration). Folk used the general concepts developed in 1949 by F. J. Pettijohn, who saw limestones as mixtures of three pure end-member components, calcite, dolomite, and silica. The textural classification recognized limestones as made up of varying proportions of allochem grains (oolites, intraclasts, fossils, and pellets), micrite (fine-grained mosaic), and sparry calcite cement (clear, coarsely crystalline, postdepositional pore fillings). Recognition of these textures in ancient limestones and comparison with their analogs in modern carbonate sediments have lead to vigorous activity in mapping of ancient limestone terrains in order to reconstruct paleogeographies and deduce environments of deposition.

Origin. Limestones may be referred to two types of origin, autochthonous and allochthonous. Autochthonous limestones have been formed in place by biogenic precipitation from the water of the environment, usually sea water; allochthonous limestones have been transported from the site of original precipitation by current action. The total distance of transport of allochthonous particles may not be great; the transportation is mainly a process of moving a chemical precipitate from one part of a sedimentary basin to another. Primary agents in the formation of autochthonous limestones are the lime-secreting organisms. Some of the most important of these are the calcareous coralline algae, such as *Lithothamnion*, which, together with corals, make up the major part of modern reef limestones. Foraminiferans also contribute large quantities of carbonate. Other groups of carbonate-rock-forming invertebrates are mollusks and echinoderms (in particular the crinoids). *See* ALGAE.

The commonest autochthonous limestone is the normal marine limestone. Although whole fossils or parts of them are plentiful, the bulk of the rock may not be recognizable as fossil debris. Much of the now unrecognizable material probably came from calcareous algae or other biogenic carbonate broken up into very small particles by the action of bottom-dwelling scavengers. Inorganic precipitation may account for some of the fine-grained micrite. The rock is light-colored and normally has moderately well-developed bedding. Dolomitization is common, as are chert nodules.

Biohermal limestones are reefs or reeflike mounds of carbonate that accumulated in much the same fashion as the modern reefs and atolls of the Pacific Ocean. The mounds may range from a few feet up to several thousand feet in diameter and hundreds of feet thick. Some of the best described fossil reefs are those of the Silurian system in Illinois, Indiana, and Wisconsin. The central core of the reef is fossiliferous, dolomitized, and massive. It grades radially outward into sparsely fossiliferous, well-bedded reef flank strata that commonly dip away from the core. Farther away from the core, the reef flank beds grade into fine-grained, well-bedded, relatively unfossiliferous, normal carbonate interreef sediment. *See* BIOHERM; REEF.

Biostromal limestones are biogenic carbonate accumulations that are laterally uniform in thickness, in contrast to the moundlike nature of bioherms. The fossils may be of many different kinds or they may be dominated by a single group. Particularly common are crinoidal and algal biostromes. The algal biostromes may show very few recognizable fossils, but stromatolites and algal laminations are common. Many of the biostromes are of mixed autochthonous and allochthonous origin; that is, some of the fossil debris of many biostromes shows evidence of transport. *See* BIOSTROME; STROMATOLITE.

Pelagic limestones are formed from the accumulation of the limy parts of pelagic, or floating, organisms such as foraminiferans. The resulting limestones are fine-grained and contain very few fossils of bottom-dwelling faunas. Since the foraminiferans are chiefly responsible for pelagic limestones and the lime-secreting pelagic foraminiferans did not evolve until the Cretaceous, pelagic limestones are restricted to Cretaceous and later systems. *See* FORAMINIFERIDA.

The allochthonous, or transported, limestones show clastic textures typical of detrital rocks. The clastic particles may be of fossils, as in coquinas or coquinoid limestones; of inorganically precipitated carbonate particles, as in oolites; or of earlier deposited limestones. *See* CALCARENITE; CHALK; COQUINA; MARL; SEDIMENTARY ROCKS; SEDIMENTATION (GEOLOGY); TRAVERTINE; TUFA.

[RAYMOND SIEVER]

Bibliography: W. E. Ham (ed.), *Classification of Carbonate Rocks*, 1962; W. C. Krumbein and L. L. Sloss, *Stratigraphy and Sedimentation*, 2d ed., 1963; F. J. Pettijohn, *Sedimentary Rocks*, 2d ed., 1957.

Limiter circuit

An electronic circuit used to prevent the amplitude of an electrical waveform from exceeding a specified level while preserving the shape of the waveform at amplitudes less than the specified level. The limiting action takes place by effectively shunting a normal load resistance with a much lower resistance at and above the specified limiting level.

Diode limiters. Limiting action is usually accomplished by use of the highly nonlinear or "switching" voltage-current characteristic of an electronic device. A semiconductor or vacuum diode used as a clamp is often the essential element in a limiter circuit. The action of such a diode may be understood with reference to the voltage-current characteristic of a typical semiconductor diode shown in Fig. 1. In the first quadrant, the diode is forward-biased, conduction depends on forward resistance r_F, and the characteristic can be approximated by a linear resistance of a very low value, ranging from a few ohms for some junction diodes to several hundred ohms for some vacuum diodes. In the third quadrant, the diode is reverse-biased. Until the breakdown point is reached, this portion of the characteristic can be approximated by a very large resistance r_R, ranging up to several megohms. At the breakdown point a sharp change occurs. For the semiconductor diode, this point is known as the Zener or avalanche breakdown region and can range from a few volts upward. After the breakdown, current becomes essentially independent of voltage. Either the sharp transition between reverse- and forward-bias regions or the avalanche transition may be used as a limit. *See* CLAMPING CIRCUIT.

Use of both limits. Use of both limits simultaneously is illustrated in Fig. 2, where input voltage v_s and output voltage v_2 are shown as functions of time t. The positive amplitude of the output waveform is limited at reference voltage V_L; negative amplitude is limited at a level $V_L - V_B$. Sharp limiting in the forward-bias direction is achieved when

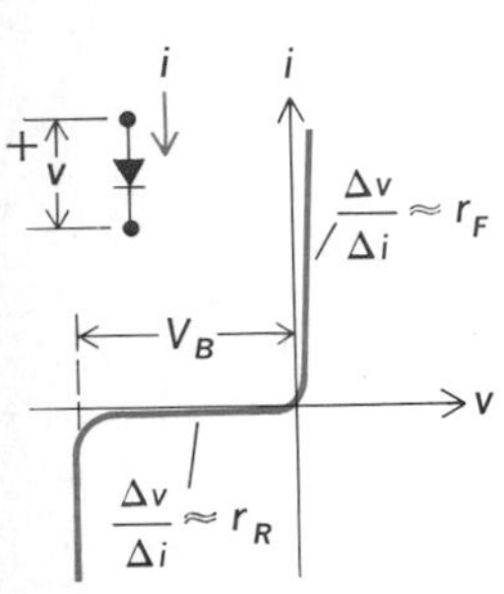

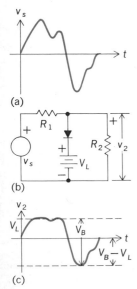

Fig. 1. Voltage-current characteristic of junction diode shows instantaneous diode current i as a function of instantaneous voltage v.

Fig. 2. In combined forward-bias and reverse-breakdown limiting, (a) input voltage applied to (b) limiting circuit produces (c) output voltage.

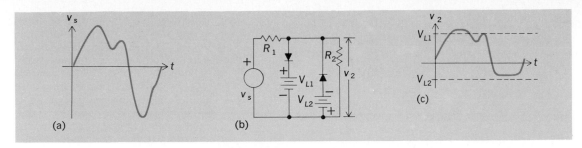

Fig. 3. For bidirectional limiting (a) input voltage can be limited in both directions by (b) circuit. (c) Bias voltage V_{L1} sets positive limit level and independent voltage V_{L2} sets negative level to produce output voltage.

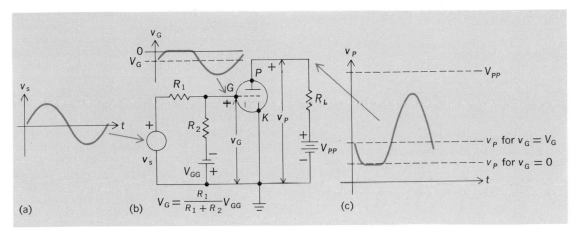

$$V_G = \frac{R_1}{R_1 + R_2} V_{GG}$$

Fig. 4. In grid-current limiter using a triode vacuum tube, (a) input voltage v_s is limited in (b) the grid circuit to produce grid voltage v_G. This voltage is then amplified and inverted in the plate circuit to produce output (c) voltage v_p. Grid bias voltage V_{GG} and divider composed of the resistances R_1 and R_2 determine the quiescent level V_G which corresponds to $v_s = 0$.

forward resistance r_F is small compared to reverse-bias resistance r_R and when r_F is small compared to series limit resistance R_1. Any shunting load resistance R_2 is assumed large compared to R_1. A vacuum diode may be used as the limiter at its transition between reverse- and forward-bias regions, but it has no well-defined breakdown level comparable to the avalanche region of a junction diode. Either positive or negative limiting at the reverse–forward-bias transition can be obtained with semiconductor or vacuum diodes by choice of polarities, voltage, and diode. If bidirectional limiting is desired, the circuit of Fig. 3 can be used with two diodes and two separate bias voltages.

Transistor and tube limiters. Transistors or vacuum grid-controlled tubes also can be used as limiters. Grid current limiting in a vacuum tube is shown in Fig. 4. Normally, when used as an amplifier, the grid is negative with respect to the cathode, and input resistance is always high. If the grid becomes positive or forward-biased, grid current flows and input resistance is low. In this respect, the grid-cathode circuit behaves as a vacuum diode to limit the waveform as shown. Limiting level with respect to the input waveform depends upon the particular grid-bias circuit used. In the plate circuit, the resultant waveform is inverted and amplified before appearing at the output. Similar input circuit limiting can be obtained by use of the junction field-effect transistor if the gate is forward-biased enough respective to the source.

Limiting by saturation. Saturation characteristics of the output circuit of a field-effect transistor, pentode vacuum tube, or bipolar transistor can also produce limiting action. Figure 5 illustrates this action for an n-channel junction field-effect transistor. The characteristics shown are those of drain current i_D as a function of drain voltage v_D for successive values of input, or gate voltage V_G through the normal range. The input voltage bias level and the load resistance are so chosen that limiting takes place in the output circuit when drain current saturation level is reached. The corresponding output voltage waveform is plotted.

General shapes of characteristic curves for vacuum tube pentodes are similar to those for the field-effect transistor shown. Consequently, in a pentode limiter the action is similar, with the grid replacing gate G, the plate replacing drain D, and the cathode replacing the source S, while screen and suppressor grids are held at their normal fixed voltages.

The bipolar transistor functions in a similar manner, with the output characteristics that are similar to those of Fig. 5 being collector current as a function of collector voltage, for a range of values of input or base current (rather than input voltage) because the bipolar transistor is a low-input-resistance device. Such a circuit is shown in Fig. 6 for an npn transistor with the base-emitter circuit forward-biased.

All polarities can be reversed by using p-channel field-effect transistors or pnp transistors in place of the transistors of Figs. 5 and 6, respectively. This flexibility is not possible with vacuum tubes.

The term limiter is often defined in sufficiently

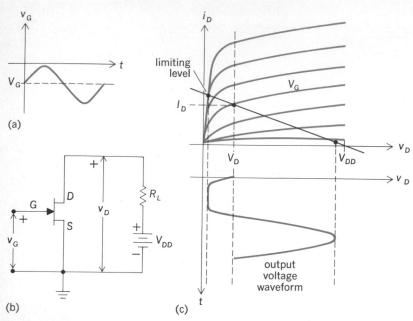

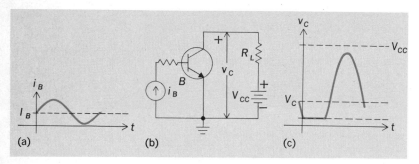

Fig. 5.   Output saturation limiting using *n*-channel junction field-effect transistor limits (*a*) input voltage in (*b*) circuit because of saturation of drain current, as shown on (*c*) the $v_D - i_D$ characteristic curves, to produce output voltage.

Fig. 6.   When (*a*) the input voltage is applied to (*b*) the circuit with an *npn* transistor, the collector current limiting produces (*c*) the output voltage.

broad terms to include the related operation of clipping, which also results in deletion of a portion of a waveform. Some authors even use the terms interchangeably. Limiters and clippers are often used together, rather than bidirectional limiters alone, where both negative and positive peaks of a waveform are to be removed. *See* CLIPPING CIRCUIT.   [GLENN M. GLASFORD]

Bibliography: G. M. Glasford, *Electronic Circuits Engineering*, 1970; G. M. Glasford, *Fundamentals of Television Engineering*, 1955; J. Millman and H. Taub, *Pulse, Digital, and Switching Waveforms*, 1965; S. Seely, *Electronic Circuits*, 1968.

Limnology

The science of the life and conditions for life in lakes, ponds, and streams. It involves the study of the physical and chemical conditions of bodies of water and the production of all organisms subject to these conditions. The ultimate result of these studies is the determination of the productivity in terms of the amounts and kinds of aquatic organisms.

Limnological studies are of importance to any activity involving the use of bodies of fresh water, such as the management of fisheries, water supplies, pollution, sewage, reservoirs, and impoundments. Bodies of water are classified according to their conditions for productivity. Lakes with poor physical conditions for the production of organisms and with low nutrient levels are known as oligotrophic lakes. Lakes with excellent physical conditions for the production of organisms and with high nutrient levels are known as eutrophic lakes. Lakes, in general, tend to become more productive through the activities of their organisms, and in the process pass from oligotrophic to eutrophic types.

Factors influencing productivity. The primary factors necessary for the production of aquatic organisms are sunlight and certain inorganic and organic substances dissolved in the water or contained in the associated bottom materials. These substances include the dissolved gases, most important of which are oxygen and carbon dioxide, and various dissolved solids, such as compounds of nitrogen, phosphorus, and other elements, all of which have interrelated chemical reactions. The secondary factors of a body of water controlling the velocity and limiting the production of aquatic life are the shape, area, various dimensions, bottom types, temperatures, currents, turbidity, amount and depth of light penetration, and other physical conditions.

Food cycles. The aquatic organisms develop from the same chemical nutrients as those that support terrestrial organisms. The basic organisms are the plants which use the raw nutrient substances and support the animals by complicated food chains. Definite cycles are maintained between the organisms and the nutrients, the chemical elements being converted into living organisms which die and decompose back into the organic and inorganic nutrients (see illustration). The whole process is based on the photosynthetic activity of all aquatic plants utilizing solar energy received by radiation, and is controlled by the vari-

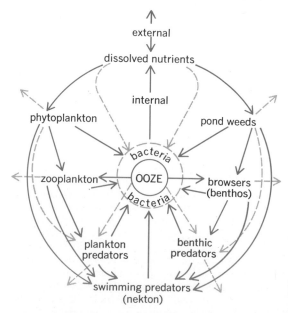

Diagram of food chains in a lake, showing the interdependence of organisms. Broken lines depict the energy pathways and solid lines the food source.

ous physical factors. The transfer of energy can be traced through various levels of food chains.

General habitats of organisms. The aquatic organisms are grouped into various categories according to where they live. The most outstanding groups are the microscopic plants and animals living suspended in the open water, the plankton; organisms living in or on the bottom, the benthos; larger free-swimming animals, the nekton; and rooted plants or pond weeds living in shallow water.

Zonation in aquatic habitats. Standing waters are divided into zones according to the conditions for life. The littoral zone is the shallow area limited by the depth of sufficient light penetration for growth of rooted plants. The limnetic zone is the open water beyond the littoral zone and is inhabited by plankton and some nekton. The profundal zone exists below the limnetic zone in waters deep enough to develop temperature stratification which, for part of the year, restricts the lower waters from circulating with the upper waters. Running waters are divided mainly into zones created by conditions of current and by related bottom types: swift-water (rapids) zones and sluggish-water (pool) zones, each with characteristic organisms. *See* ECOLOGY; FRESH-WATER ECOSYSTEM; LAKE; MARINE ECOSYSTEM; OCEANOGRAPHY; PHOTOSYNTHESIS.   [SAMUEL EDDY]

Bibliography: B. Arnov, Jr., *Secrets of Inland Waters*, 1965; S. A. Forbes, The Lake as a microcosm, *Bull. Ill. Natur. Hist. Surv.*, 15:557–560, 1925; D. G. Frey (ed.), *Limnology in North America*, 1963; G. E. Hutchinson, *A Treatise on Limnology*, vol. 1, 1957, and vol. 2, 1967; R. L. Lindeman, Seasonal food-cycle dynamics in a senescent lake, *Amer. Midland Natur.*, 26(3):636–673, 1941; F. Ruttner, *Fundamentals of Limnology*, 3d ed., 1963; P. S. Welch, *Limnology*, 2d ed., 1952.

Limonite

A natural amorphous-appearing material constituting in part the material known as brown iron ore. Limonite occurs in mammillary, stalactitic, and earthy masses (see illustration). The hardness is 5–5.5 (Mohs scale) and the specific gravity is 3.6–4. The luster is vitreous and the color dark brown to black. Limonite is essentially $FeO(OH) \cdot nH_2O$, but with admixed hematite, clay, and manganese oxides. X-ray analysis has shown that much of the material formerly thought to be limonite is goethite. Limonite forms by the oxidation of preexisting iron minerals and occurs with goethite. It is the pigmenting material in yellow soils and mixed with fine clays is known as yellow ocher. *See* GOETHITE; IRON EXTRACTION FROM ORE; ORE AND MINERAL DEPOSITS.   [CORNELIUS S. HURLBUT, JR.]

Limpet

Any of several species of marine mollusks which make up two families in the class Gastropoda: the Patellidae, or true limpets, and the Acmaeidae. The shell is conical and tentlike, with ridges extending from the apex to the border of the shell (see illustration). Most limpets are small, 2 in. or less in length. They feed on algae, organic particulate matter, and other vegetation. Most species have become adapted to life on rocky shores and have a mode of life similar to that of the chitons.

The common limpet (*Patella vulgata*) has a circular shell and adheres closely to the substratum, on which it moves slowly by means of the sucker-like foot. Between the muscular foot and the edge of the shell is the pallial groove, which contains numerous secondary gills, since the original gills have disappeared. *P. vulgata* is the commonest intertidal gastropod of the eastern Atlantic Coast.

The flat, cone-shaped shell resists wave action on rocky shores, particularly in the blue peacock limpet (*Patina pellucida*), which is attached to the seaweed *Laminaria*. Some of the small limpets of the family Acmaeidae live in quiet pools of water. Species of *Patella* and *Acmaea* shed their eggs into the water singly, and the larval life of *Patella* is about 10 days as a planktonic organism. Most limpets are monoecious; however, an interesting condition exists in *A. rubella*, which is an ambisexual hermaphrodite; that is, all individuals have well-developed male and female gonads, and both ova and sperm are produced by two individuals side by side, each displaying identical sexual behavior. Limpets have a relatively long life, and some species live as long as 17 years. A few species are used as food; otherwise, they are not economically important.

The limpets have a homing behavior that has been studied in *P. granularis*, which may travel and return a distance of 5 ft after a feeding sortie at night or during high tide. This European species was found to have less accuracy of performance with an increase in distance. *See* CHITON; GASTROPODA.   [CHARLES B. CURTIN]

Linales

An order of flowering plants, division Magnoliophyta (Angiospermae), in the subclass Rosidae of the class Magnoliopsida (dicotyledons). The order consists of three families: the Linaceae, with about 450 species; the Erythroxylaceae, with about 200 species; and the Humiriaceae, with about 50 species. They are simple-leaved herbs or woody plants, with hypogynous, regular, syncarpous flowers that have five to many stamens which are connate at the base. The pollen is trinucleate. *Linum usitatissimum* (the source of flax fibers and linseed oil) and *Erythroxylon coca* (the source of cocaine) are well-known members of the Linales. *See* COCA; FLAX; FLOWER; MAGNOLIOPHYTA; MAGNOLIOPSIDA; PLANT KINGDOM; ROSIDAE.

[ARTHUR CRONQUIST]

Line

In axiomatic geometry, a line frequently is taken as a primitive (undefined) concept: an element in line geometry; a class of points, in a geometry with point as an element, whose essential properties are stated as axioms. In analytic geometry, a line is identified with (or is the graph of) a linear equation $Ax + By + C = 0$, with A, $B \neq 0$, 0, in cartesian coordinates (x,y). As a derived concept, line is defined in terms of primitive notions such as distance or betweenness. For example, a line L is a set of points in one-to-one correspondence with the set of real numbers, such that if points p, q of L correspond to numbers x, y, respectively, then distance $(pq) = |x - y|$, or if p, q are points, the line $L(pq)$ is the set of points consisting of p, q and all points x, such that one of the points p, q, x is between the other two. *See* ANALYTIC GEOMETRY; GEOMETRY, EUCLIDEAN; PLANE; POINT.

[LEONARD M. BLUMENTHAL]

1 in.

Limonite in stalactitic masses, which are found in the Tintic District of Utah. (*Specimen from Department of Geology, Bryn Mawr College*).

LIMPET

Rough keyhole limpet (*Diodora aspera*) of the Pacific Coast, maximum length 2 in.

Line integral

The line integral of a vector function **F** of position over a path C is represented by Eq. (1), where F_x,

$$\int_C \mathbf{F} \cdot d\mathbf{r} = \int_C F_x(x,y,z)\,dx$$
$$+ \int_C F_y(x,y,z)\,dy + \int_C F_z(x,y,z)\,dz \quad (1)$$

F_y, F_z are the scalar components of **F** along the coordinate axes. The path C is supposed to be a curve, smooth at least in part, defined parametrically by equations of form (2) for each smooth por-

$$x = x(p) \qquad y = y(p) \qquad z = z(p) \quad (2)$$

tion. The functions $F_x(x,y,z)$, etc., must be defined at all points of C. When this is so, the line integral can be evaluated by writing Eq. (3), where the

$$\int_C \mathbf{F} \cdot d\mathbf{r} = \int_{p_1}^{p_2} F_x(p)x'(p)\,dp$$
$$+ \int_{p_1}^{p_2} F_y(p)y'(p)\,dp + \int_{p_1}^{p_2} F_z(p)z'(p)\,dp \quad (3)$$

prime means differentiation with respect to the parameter, and p_1, p_2 are values of the parameter at the end points of the path C or of a smooth piece. The integral has been converted into an ordinary definite integral.

When C is a closed curve, the line integral is called a circuit integral, and is written as notation (4).

$$\oint \mathbf{F} \cdot d\mathbf{r} \quad (4)$$

Some physical applications of the line integral follow. If **F** is a force, the line integral is the work done in moving a mass along the curve C. If **F** is the velocity of flow of a fluid, the line integral is the circulation of the fluid along the curve. If **F** is the electrostatic field strength, the integral is the electric potential difference between the end points of the curve. If **F** is the electric field strength of an electromagnetic field, the circuit integral is the electromotive force of the circuit. In each example $d\mathbf{r}$ is physically a length. *See* INTEGRATION.

[MC ALLISTER H. HULL, JR.]

Line spectrum

A discontinuous spectrum characteristic of excited atoms, ions, and certain molecules in the gaseous phase at low pressures, to be distinguished from band spectra, emitted by most free mole-

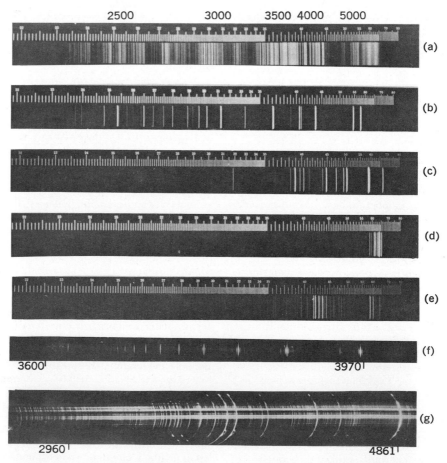

Photographs of common line spectra, wavelengths given in angstroms; emission spectra (a–e) all taken with the same quartz spectrograph. (a) Spectrum of iron arc. (b) Mercury spectrum from an arc enclosed in quartz. (c) Helium in a glass discharge tube. (d) Neon in a glass discharge tube. (e) Argon in a glass discharge tube. (f) Balmer series of hydrogen in the ultraviolet, photographed with a grating spectrograph. (g) Emission spectrum from gaseous chromosphere of the Sun, a grating spectrum taken without a slit at the instant immediately preceding a total eclipse, when the rest of the Sun is covered by the Moon's disk. Two strongest images, H and K lines of calcium, show marked prominences, or clouds, of calcium vapor. Other strong lines are caused by hydrogen and helium. (*From F. A. Jenkins and H. E. White, Fundamentals of Optics, 3d ed., McGraw-Hill, 1957*)

cules, and continuous spectra, emitted by matter in the solid, liquid, and sometimes gaseous phase. If an electric arc or spark between metallic electrodes, or an electric discharge through a low-pressure gas, is viewed through a spectroscope, images of the spectroscope slit are seen in the characteristic colors emitted by the atoms or ions present.

To avoid the overlapping of close spectral images, the slit illuminated by the light source is made very narrow. The spectrum then appears as an array of bright line slit images on a dark background (see illustration). Under certain conditions spectra show dark absorption lines against a bright background. *See* ATOMIC STRUCTURE AND SPECTRA; SPECTROSCOPY.   [GEORGE R. HARRISON]

Linear accelerator

A particle accelerator which accelerates electrons, protons, or heavy ions in a straight line by the action of alternating voltages. It has been successful in accelerating electrons to velocities very near that of light. For an extended discussion *see* PARTICLE ACCELERATOR. [WOLFGANG K. H. PANOFSKY]

Linear energy transfer (biology)

A physical factor which is significant in biology because, in many biological actions of high-energy radiations, linear energy transfer (LET) strongly and characteristically modifies the quantitative relations between dose and effect. *See* RADIATION BIOLOGY.

When a moving charged particle (for example, an electron or proton) traverses matter, it progressively loses kinetic energy which is gained, in diverse amounts, by individual molecules along its path. The amount of energy thus transferred, per unit length of path, is the LET. It is roughly proportional, at least in gases, to ionization per unit path (specific ionization; ion density). LET increases as the velocity of the particle decreases; accordingly it varies greatly along any individual path and is maximal near the end. Also, LET increases as the square of the charge on the particle; for example, the LET of an α-particle is four times that of an electron or proton of the same instantaneous velocity.

For experimental purposes, different values of LET can be obtained in two general ways. In the track-segment method, a beam is used made up of particles (protons) that have uniform initial velocity and produce substantially straight tracks. The experiment is so arranged that they all traverse a sample of biological objects with the same linear segment of their paths. Different LET values are obtained by using suitably different path segments. In the track-average method, either entire paths are spent within the biological object or random linear segments of different paths traverse it. Different average values of LET are obtained by using radiations that produce suitably different spectra of LET. Although the average usually used is the arithmetic mean, other averages will probably be found to be more meaningful as greater understanding of radiobiological mechanisms is achieved.

In some radiobiological actions the shape of the dose-effect curve is changed as LET is varied. However, these cases are rare in comparison with those in which only the slope is changed. In the latter case the effect of LET can be expressed by a single parameter (relative biological effectiveness, RBE) which varies as the slope (or as the reciprocal of the dose, in rads, required to produce a given degree of effect).

For some radiobiological actions, RBE decreases with increase in LET; this is exemplified by inactivation of macromolecules, viruses, and certain bacteria. For other actions, RBE increases with LET; in some of these cases, RBE has been shown to pass through a maximum; there are theoretical reasons to believe that maxima would be observed for all of them if larger values of LET were available. *See* RADIATION MICROBIOLOGY.

The distribution of individual energy transfers along a path corresponds to the distribution of ionized and excited molecules, production of which is believed to initiate the radiobiological mechanism. Because of diffusion, this distribution in space lasts only 1 μsec or less. Thus the observed variation of RBE with LET demonstrates that some essential part of the radiobiological mechanism is completed in a very short time.

Among other interesting facts are the following: (1) In at least some radiobiological actions, the influence of LET is modified by other factors, such as concentration of molecular oxygen. (2) Influence of LET is also encountered in radiation chemistry. For instance, the decomposition of liquid water depends strongly on LET, and this dependence is modified by oxygen. Facts such as these make LET of prime interest in attempts to understand radiobiological mechanisms. *See* RADIATION; RADIATION BIOCHEMISTRY; RADIATION INJURY (BIOLOGY).   [RAYMOND E. ZIRKLE]

Bibliography: M. Errera and A. Forssberg (eds.), *Mechanisms in Radiobiology*, 2 vols., 1960, 1961; A. Hollaender (ed.), *Radiation Biology*, vol. 1, 1954.

Linear programming

A mathematical subject whose central theme is finding the point where a linear function defined on a convex polyhedron assumes its maximum or minimum value. Although contributions to this question existed earlier, the subject was essentially created in 1947, when G. B. Dantzig defined its scope and proposed the first, and still most widely used, method for the practical solution of linear programming problems. The largest class of applications of linear programming occurs in business planning and industrial engineering. Its basic concepts are so fundamental, however, that linear programming has been used in almost all parts of science and social science where mathematics has made any penetration. Furthermore, even in cases where the domain is not a polyhedron, or the function to be maximized or minimized is not linear, the methods used are frequently adaptions of linear programming. *See* NONLINEAR PROGRAMMING; OPERATIONS RESEARCH.

General theory. A typical linear programming problem is to maximize expression (1), where

$$c_1x_1 + \cdots + c_nx_n \qquad (1)$$

$x_1, \ldots, x_n$ satisfy the conditions shown by notation (2). The a_{ij}, c_j, and b_i are constants; $x_1, \ldots, x_n$ are variables.

$$x_j \geqq 0$$
$$a_{11}x_1 + \cdots + a_{1n}x_n \leqq b_1$$
$$\cdots\cdots\cdots\cdots\cdots\cdots \quad (2)$$
$$a_{m1}x_1 + \cdots + \alpha_{mn}x_n \leqq b_m$$

Sometimes the problem may be to minimize rather than to maximize; sometimes some of the variables may not be required to be nonnegative; sometimes some of the inequalities $a_{i1}x_1 + \cdots + a_{in}x_n \leqq b_i$ may be reversed, or be equalities.

The linear inequalities, which the variables satisfy, correspond algebraically to the fact that the variable point $x = (x_1, \ldots, x_n)$ lies in a convex polyhedron. Hence, the principal mathematical bases for linear programming are the theory of linear equalities, a part of algebra, and the theory of convex polyhedra, a part of geometry. The most important theoretical foundations are as follows.

If the function $c_1x_1 + \cdots + c_nx_n$ does not get arbitrarily large for points $x = (x_1, \ldots, x_n)$ on the convex polyhedron, then a maximum is attained at a vertex of the polyhedron.

If there is a maximum in the stated problem, then there is a minimum in the dual problem of minimizing $b_1y_1 + \cdots + b_my_m$, where notation (3)

$$y_i \geqq 0$$
$$a_{11}y_1 + \cdots + a_{m1}y_m \geqq c_1$$
$$\cdots\cdots\cdots\cdots\cdots\cdots \quad (3)$$
$$a_{1n}y_1 + \cdots + a_{mn}y_m \geqq c_n$$

applies. Further, the maximum and minimum are the same number. This duality theorem is essentially equivalent to the minimax theorem. Both the duality theorem and the minimax theorem are algebraic paraphrases of the geometric fact that if a point is not contained in a convex polyhedron then a hyperplane can be found which separates the point from the polyhedron. The duality theorem is also closely related to the concept of Lagrange multipliers. *See* CALCULUS OF VARIATIONS; GAME THEORY.

In an important class of linear programming problems, there is imposed an additional requirement that some or all of the variables must be whole numbers. If the numbers $b_1, \ldots, b_n$ are whole numbers, and if every square submatrix of (a_{ij}) has determinant 0, 1, or −1, then the integrality requirement will be automatically satisfied. In any case, the faces of the convex hull of the integral points inside a polyhedron can be found by a finite process in which, at each stage, a new inequality is determined by taking a nonnegative linear combination of previous inequalities and replacing each coefficient by the largest integer not exceeding it.

Methods of calculation. The most popular method for solving linear programs, that is, the finding of the $x = (x_1, \ldots, x_n)$ which maximizes $c_1x_1 + \cdots + c_nx_n$, is the simplex method, developed by Dantzig in 1947. The method is geometrically a process of moving from vertex to neighboring vertex on the convex polyhedron, each move attaining a higher value of $c_1x_1 + \cdots + c_nx_n$, until the vertex yielding the greatest value is reached. Algebraically, the calculations are similar to elimination processes for solving systems of algebraic equations. The computer programs used try to take advantage of the fact that almost all the matrices (a_{ij}) arising in practice have very few nonzero entries.

The popularity of the simplex method rests on the empirical fact that the number of moves from vertex to vertex is a small multiple (about 2–4) of the number of inequalities in most of the thousands of problems handled, although it has never been proved mathematically that this behavior is characteristic of "average" problems.

There have also been developed methods of calculation specifically tailored for special classes of problems. One such class (network flows) is exemplified by the transportation problem described below. Another important class is where most of the columns of (a_{ij}) are not known explicitly in advance, but consist of all columns which satisfy some particular set of rules. The idea used here (known in different contexts as the column-generation or the decomposition principle) is that the new vertex prescribed by the simplex method can be found if a solution is achieved for the problem of maximizing a linear function defined on the set of columns satisfying the particular rules.

The technology of solving integer linear programming problems is not stabilized. Methods vary from testing integral points near the optimum fractional point to systematic methods for generating relevant faces of the convex hull of the integral points satisfying the given inequalities.

Applications. Let a_{ij} denote the number of units of nutrient j present in one unit of food i. Let c_j be the minimum amount of nutrient j needed for satisfactory health, b_i be the unit price of food i. Let the variables y_i denote respectively the number of units of food i to be bought. Then the dual problem is the so-called diet problem, to maintain adequate nutrition at least cost. It has been used in planning feeding programs for several varieties of livestock and poultry. With a different interpretation of the symbols, the same format describes the problem of combining raw materials in a chemical process to produce required end products as cheaply as possible.

In another situation, let c_{ij} be the cost of shipping one unit of a given product from warehouse i to customer j for all m warehouses i and all n customers j pertaining to a business. Let a_i be the amount available at warehouse i, and b_j the amount required by customer j. The transportation problem is to ship the required amounts to the customers at least cost. Thus, if x_{ij} is the amount to be shipped from i to j, then minimize notation (4),

$$c_{11}x_{11} + \cdots + c_{mn}x_{mn} \quad (4)$$

where $x_{i1} + \cdots + x_{in} \leqq a_i$ $(i = 1, \ldots, m)$ and $x_{1j} + \cdots + x_{mj} = b_j$ $(j = 1, \ldots, n)$.

The foregoing are representative of the business applications of linear programming, several hundred of which have now been reported. The best-known applications outside business have been to economic theory and to combinatorial analysis. In both instances, the duality theorem has been used to illuminate and generalize previous results.

[ALAN J. HOFFMAN]

Bibliography: S. I. Gass, *Linear Programming*, 2d ed., 1964.

Linear system analysis

The study of a system by means of a model consisting of a linear mapping between the system inputs (causes or excitations), applied at the input terminals, and the system outputs (effects or responses), measured or observed at the output terminals. A system is a set or an arrangement of things or devices so related or connected as to form a unity or organic whole. For example, the inputs or controlling factors of the national economy (the system) are government spending, interest rates, tax levels, and monetary policy. The outputs are the gross national product, unemployment rate, inflation rate, and the consumer price index. Similar correspondences can be established for most systems, ranging from the more common variety of physical systems to the less-well-defined socioeconomic variety whose structure and control represent some of the major problem areas of society. *See* SYSTEMS ANALYSIS; SYSTEMS ENGINEERING.

State of a system. Many dynamic systems exhibit the Markovian property; that is, the total effect of all the past inputs, applied prior to time t_0 can be completely characterized by a set of numbers called the state of the system at t_0. More formally if $u_{(t_0, \infty)}$ denotes the input applied from time t_0 on, the state at t_0 is defined as the set of information which, together with $u_{(t_0, \infty)}$, will uniquely determine the behavior of the system for all $t > t_0$.

As an example, consider the network shown in Fig. 1. The state of the system at any time t consists of the capacitor voltages $x_1(t)$ and $x_2(t)$, and the inductor current $x_3(t)$. A knowledge of the state x_1, x_2, x_3 at time t, and the systems inputs $u^1_{(t, \infty)}$ and $u^2_{(t, \infty)}$ uniquely determine the future behavior of the system. If any of the values of x_1, x_2, or x_3 is not known at time t, the future behavior cannot be determined even if the inputs $u^1_{(t, \infty)}$ and $u^2_{(t, \infty)}$ are known. In this example, the state vector x is represented by Eq. (1). Each component of the

$$x(t) = \begin{bmatrix} x_1(t) \\ x_2(t) \\ x_3(t) \end{bmatrix} \qquad (1)$$

vector x is called a state variable of the system. *See* ALGEBRA, LINEAR.

As another example, the state of the mechanical system shown in Fig. 2 consists of the position $x_1(t)$ and the velocity $x_2(t)$ of the mass m, denoted by Eq. (2), where $\dot{y}(t) \equiv dy(t)/dt$.

$$x(t) = \begin{bmatrix} x_1(t) \\ x_2(t) \end{bmatrix} = \begin{bmatrix} y(t) \\ \dot{y}(t) \end{bmatrix} \qquad (2)$$

Systems which have a finite number of state variables, such as the electrical and mechanical system shown in Fig. 2 consists of the position eter systems, whereas systems which are characterized by an infinite number of state variables, such as transmission lines and flexible structures, are called distributed parameter systems. Attention in this article will be limited to lumped parameter systems.

State variable description. Every system can be dichotomously classified as being linear or nonlinear, time-invariant or time-varying. Roughly speaking, if the sum of responses of any two excitations of a system is equal to the response of the sum of the two excitations, the system is said to be linear, otherwise it is nonlinear. If the characteristics of a system do not change with time, the system is said to be time-invariant, otherwise it is time-varying.

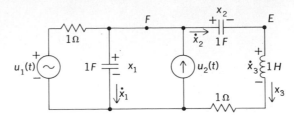

Fig. 1. Network example depicting the state variables $x_1, x_2,$ and x_3.

It can be established that every linear, time-varying system can be described by a set of equations of the form (3), where the system state x

$$\dot{x}(t) = A(t)\,x(t) + B(t)\,u(t) \qquad (3a)$$

$$y(t) = C(t)\,x(t) + D(t)\,u(t) \qquad (3b)$$

is an n-vector, the system input u is a p-vector, and the system output y is an m-vector. Equation (3a) governs the behavior of the state and is called the state equation. Since the state x has n elements or state variables, the state equation consists of a set of n first-order differential equations driven by the input u. Output equation (3b) relates the instantaneous value of the output to the instantaneous values of the system state and input and consists of a set of m linear algebraic equations. If the system is linear and time-invariant, the system matrices A, B, C, D are constant and do not vary with time. *See* DIFFERENTIAL EQUATION; LINEAR SYSTEMS OF EQUATIONS; MATRIX THEORY.

Consider the development of the state equations which characterize the network of Fig. 1. If the voltage across a capacitor is $x(t)$, the current through it is $C\,\dot{x}(t)$, and similarly, if the current through an inductor is $x(t)$, the voltage across it is $L\,\dot{x}(t)$. Using these basic properties and applying Kirchhoff's current law to nodes E and F, one obtains Eqs. (4). The application of Kirchhoff's voltage

$$\dot{x}_2 = x_3$$
$$(u_1 - x_1) - \dot{x}_1 + u_2 - x_3 = 0 \qquad (4)$$

law yields Eq. (5). Arranging these three equations

$$\dot{x}_3 + x_3 - x_1 + x_2 = 0 \qquad (5)$$

in matrix form yields Eq. (6) as the state variable

$$\begin{bmatrix} \dot{x}_1 \\ \dot{x}_2 \\ \dot{x}_3 \end{bmatrix} = \begin{bmatrix} -1 & 0 & -1 \\ 0 & 0 & 1 \\ 1 & -1 & -1 \end{bmatrix} \begin{bmatrix} x_1 \\ x_2 \\ x_3 \end{bmatrix} + \begin{bmatrix} 1 & 1 \\ 0 & 0 \\ 0 & 0 \end{bmatrix} \begin{bmatrix} u_1 \\ u_2 \end{bmatrix} \qquad (6)$$

equation of the network. The output equation is given by Eq. (7). *See* KIRCHHOFF'S LAWS OF ELEC-

$$y = \dot{x}_3 = [1 \, -1 \, -1] \begin{bmatrix} x_1 \\ x_2 \\ x_3 \end{bmatrix} + [0 \ 0] \begin{bmatrix} u_1 \\ u_2 \end{bmatrix} \qquad (7)$$

TRIC CIRCUITS; NETWORK THEORY, ELECTRICAL; SUPERPOSITION THEOREM (ELECTRIC NETWORKS).

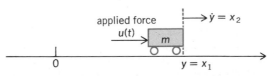

Fig. 2. Mechanical example with position and velocity as state variables.

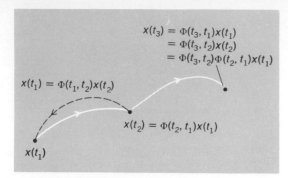

Fig. 3. State transitions of an unforced system.

The relationship between the state variable description of a linear, time-invariant system and the transfer function description can be readily established by taking the Laplace transform of the state and output equations, for zero initial state, and solving for the transformed output $y(s)$. This yields Eq. (8). Hence the system transfer function

$$y(s) = [C(sI - A)^{-1}B + D]u(s) \qquad (8)$$

is given by the $m \times p$ matrix of Eq. (9). Although

$$G(s) = [C(sI - A)^{-1}B + D] \qquad (9)$$

the transition from a state variable description to a transfer function description is a straightforward one, the opposite is not true, especially for multi-input multi-output systems. *See* FILTER, ELECTRIC; LAPLACE TRANSFORM.

Solution of the state equation. One of the attractive features of the state variable description of linear dynamical systems is that the solution of the state equation can be written in the closed form of Eq. (10), where the $n \times n$ transition matrix

$$x(t) = \Phi(t,t_0)x(t_0) + \int_{t_0}^{t} \Phi(t,\tau)B(\tau)u(\tau)d\tau \qquad (10)$$

$\Phi(t,t_0)$ satisfies matrix differential equation (11).

$$\frac{d}{dt}\Phi(t,t_0) = A(t)\Phi(t,t_0) \quad \Phi(t_0,t_0) = I \qquad (11)$$

The transition matrix can be shown to possess the useful properties (see Fig. 3) given by Eqs. (12). *See* INTEGRAL EQUATIONS.

$$\Phi(t_1,t_2)\Phi(t_2,t_3) = \Phi(t_1,t_3)$$

$$\Phi^{-1}(t_1,t_2) = \Phi(t_2,t_1) \qquad (12)$$

Except for some special cases, it is usually impossible to solve for $\Phi(t,t_0)$ in closed form. In the time-invariant case, however, the state transition matrix can be explicitly expressed in terms of the matrix exponential given by Eq. (13). The matrix

$$\Phi(t,t_0) = e^{A(t-t_0)} \qquad (13)$$

e^{At} can be numerically determined from the power series expansion of Eq. (14) or expressed in closed

$$e^{At} = I + At + \frac{A^2t^2}{2!} + \frac{A^3t^3}{3!} + \cdots \qquad (14)$$

form through use of a similarity transformation. In particular, if the eigenvalues λ_i of A are distinct, A can be written in terms of the diagonal matrix $\Lambda = \text{diag}(\lambda_1, \ldots, \lambda_n)$ [where diagonal elements are $\lambda_1, \ldots, \lambda_n$, and whose off-diagonal elements are all 0] and the transformation Q, as in Eq. (15).

$$A = Q \Lambda Q^{-1} \qquad (15)$$

The columns of Q are the corresponding eigenvectors of A. Equation (16) follows immediately from

$$e^{At} = Qe^{\Lambda t}Q^{-1} = Q \text{ diag}[e^{\lambda_1 t}, \ldots, e^{\lambda_n t}]Q^{-1} \qquad (16)$$

Eqs. (14) and (15). Thus, each entry of e^{At} is composed of a linear combination of terms $e^{\lambda_1 t}, \ldots, e^{\lambda_n t}$. Each of these terms is called a mode of the system and exhibits a time behavior which is characteristic of the eigenvalue associated with it (see Fig. 4). If the eigenvalue has a nonnegative real part, the magnitude of the corresponding time response increases with time or remains oscillatory. Conversely, if an eigenvalue has a negative real part, then the magnitude of the corresponding time response will approach zero as time tends to infinity. Thus if all of the eigenvalues of an unforced system have negative real parts, the natural response, from any initial state, will decay to zero. In this case, the system is said to be asymptotically stable. If a system is asymptotically stable, it can be shown that every applied bounded input will produce a bounded output. *See* COMPLEX NUMBERS AND COMPLEX VARIABLES; EIGENFUNCTION.

Application to feedback control. One approach, then, to the design of a linear feedback controller is to employ the state feedback $u = Kx$, in which case the closed loop dynamics are described by Eq. (17) or Eq. (18). If K is chosen so as to cause the

$$\dot{x} = Ax + B(Kx) \qquad (17)$$

$$\dot{x} = (A + BK)x \qquad (18)$$

eigenvalues of $(A + BK)$ to have negative real parts, the feedback law $u = Kx$ will achieve stabilization of the open-loop system.

Discrete-time systems. The closed-form solution of $x(t)$ given by Eq. (10) can also be used to develop the state propagation equations when the system is to be controlled in a discrete-time fashion. If the control input $u(t)$ is constant over intervals of T seconds, Eq. (10), when evaluated at $t_0 = nT$, and $t = (n+1)T$, yields Eq. (19), or, as it is

$$x[(n+1)T] = \Phi[(n+1)T, nT]x(nT)$$

$$+ \int_{nT}^{(n+1)T} \Phi[n+1)T, \tau]B(\tau)d\tau \cdot u(nT) \qquad (19)$$

usually written, Eq. (20), where $x_n \equiv x(nT)$ and

$$x_{n+1} = \Phi_n x_n + \Gamma_n u_n \qquad (20)$$

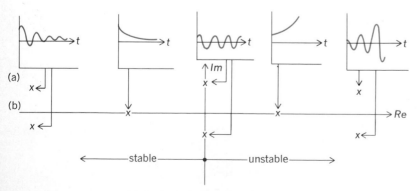

Fig. 4. Mode behavior as a function of eigenvalue position. (a) Graphs showing time behavior of various modes. (b) Diagram of complex number of plane showing locations of corresponding eigenvalues.

$u_n = u(nT) = u(t)$ for $nT \leq t < (n+1)T$. Since no approximations have been made, Eq. (20) exactly describes the evolution of the state at discrete instants of time, in response to any piecewise constant input function $u(t)$. If the system is time-invariant, the matrices Φ_n and Γ_n are constant, that is, independent of the discrete-time index n.

The discrete-time form of the state equations are particularly useful for designing digital feedback algorithms for on-line digital computer control. In a typical application, data are collected over the time interval T and used to determine an estimate of x_n at the end of the interval. The control u_n, to be applied over the next interval, is then determined as a function of x_n. Again, one particularly simple control law is the linear state feedback $u_n = K x_n$, which for time-invariant systems, yields the closed-loop behavior of Eq. (21). The

$$x_{n+1} = \Phi x_n + \Gamma(K x_n) = (\Phi + \Gamma K) x_n \quad (21)$$

closed-loop state response is thus given by Eq. (22).

$$x_n = (\Phi + \Gamma K)^n x_0, n = 0, 1, 2, \cdots \quad (22)$$

In order for x_n to converge to zero, it is necessary that the feedback gain K produce eigenvalues of $(\Phi + \Gamma K)$ which have magnitude less than one.

Another type of discrete-time control which is frequently used is the "deadbeat" control, in which a sequence of control values is determined so as to cause the state to transfer from a given initial state to a final specified state in a fixed number of sample intervals. The required sequence can be easily calculated using discrete-time equation (20).

Controllability and observability. The relationship between a system transfer function and the state variable description has been established in Eq. (9). In order to discuss this in more detail, it is necessary to introduce the concepts of controllability and observability. A dynamical system is said to be controllable if any initial state can be transferred to any final state in a finite amount of time by proper choice of the system input $u(t)$. It can be shown that for linear, time-invariant systems characterized by the state and control matrices A and B, the necessary and sufficient condition for controllability is that the composite matrix U given by Eq. (23) have rank n.

$$U = \left[B \mid AB \mid A^2B \mid \cdots \mid A^{n-1}B \right] \quad (23)$$

A dynamical system is said to be observable if knowledge of the input and output over a finite interval is sufficient to uniquely determine the system's initial state. Again it can be shown that for a linear, time-invariant system characterized by the system matrix A and the measurement matrix C, a necessary and sufficient condition for observability is that the composite matrix given by Eq. (24) have rank n.

$$V = \left[C^T \mid A^T C^T \mid (A^2)^T C^T \mid \cdots \mid (A^{n-1})^T C^T \right] \quad (24)$$

If a system is not controllable and observable, the state equation can be decomposed into four parts, as shown in Fig. 5, separating its uncontrollable and unobservable substates. Since a transfer function describes only the relationship between the input and output of a system, from Fig. 5 it is clear that the transfer function represents only the controllable and observable part of a dynamical system.

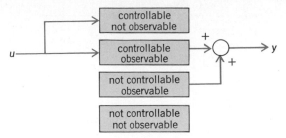

Fig. 5. Decomposition of a dynamical system.

Realization. Consider a system described by the transfer function matrix $G(s)$. If it is possible to find a dynamical, and output, equation set which has $G(s)$ as its transfer function, then $G(s)$ is said to be realizable, and the state and output equation set is called a realization of $G(s)$. A study of the realizability problem is important for three reasons. First, a dynamical equation can be readily simulated on an analog or a digital computer; hence if a realization is found, the transfer function can then be simulated by using the realization. Second, the realization provides a method of synthesizing a transfer function by use of operational amplifier circuits. Finally, there are many design techniques and computational algorithms developed exclusively for dynamical equations. In order to apply these techniques and algorithms, the transfer function must be realized in state variable form. *See* AMPLIFIER; SIMULATION.

Since a transfer function describes only the controllable and observable part of a dynamical equation, a realization is generally required to be controllable and observable. This kind of realization is called a minimal-dimensional realization. The minimal-dimensional realization of general transfer-function matrices is a difficult problem, hence only the realization of scalar transfer functions ($m = p = 1$) will be discussed in this article.

Consider the transfer function of Eq. (25), with

$$g(s) = \frac{\delta_0 s^n + \delta_1 s^{n-1} + \cdots + \delta_n}{\gamma_0 s^n + \gamma_1 s^{n-1} + \cdots + \gamma_n} \quad (25)$$

$\gamma_0 \neq 0$. The other coefficients are not necessarily nonzero. It can be shown that the transfer function $g(s)$ can be realized in the standard-state variable form if and only if the degree of the numerator of $g(s)$ is equal to or less than that of the denominator. In the realization, Eq. (25) is first reduced to the form of Eq. (26). Then state and output equations (27) are called the controllable-form realization of

$$g(s) = d + \frac{\beta_1 s^{n-1} + \beta_2 s^{n-2} + \cdots + \beta_n}{s^n + \alpha_1 s^{n-1} + \cdots + \alpha_n} \quad (26)$$

$$\dot{x} = \begin{bmatrix} 0 & 1 & 0 & \cdots & 0 \\ 0 & 0 & 1 & \cdots & 0 \\ \cdot & \cdot & & \cdot & \cdot \\ \cdot & \cdot & & \cdot & \cdot \\ 0 & 0 & 0 & & 1 \\ -\alpha_n & -\alpha_{n-1} & -\alpha_{n-2} & & -\alpha_1 \end{bmatrix} x + \begin{bmatrix} 0 \\ 0 \\ \cdot \\ \cdot \\ 0 \\ 1 \end{bmatrix} u$$

$$y = \begin{bmatrix} \beta_n & \beta_{n-1} & \beta_{n-2} & & \beta_1 \end{bmatrix} x + du$$

(27)

$g(s)$. This equation is controllable for any α_i and β_i. If the denominator and numerator of $g(s)$ have no common factor, the realization is also observable.

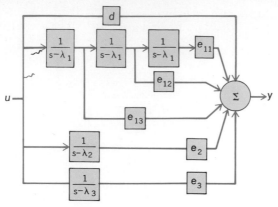

Fig. 6. Block diagram of $g(s)$ given by Eq. (29).

The realization can be obtained directly from the coefficients of $g(s)$.

The realization of Eqs. (28) is called the observ-

$$\dot{x} = \begin{bmatrix} 0 & 0 & \cdots & 0 & -\alpha_n \\ 1 & 0 & & 0 & -\alpha_{n-1} \\ 0 & 1 & & 0 & -\alpha_{n-2} \\ \cdot & & & & \cdot \\ \cdot & & & & \cdot \\ \cdot & & & & \cdot \\ 0 & 0 & \cdots & 1 & -\alpha_1 \end{bmatrix} \begin{bmatrix} \beta_n \\ \nu_{n-1} \\ \nu_{n-2} \\ \cdot \\ \cdot \\ \cdot \\ \beta_1 \end{bmatrix} u$$

$$y = [\, 0 \; 0 \cdots 0 \quad 1 \;] x + du \qquad (28)$$

able-form realization of $g(s)$. It is always observable, and is controllable if there is no common factor between the denominator and numerator of $g(s)$. The state vectors in Eqs. (27) and (28) represent entirely different variables.

In addition to the above two realizations, there are many other forms of realization which are useful in the analysis of linear time-invariant systems. The Jordan-form realization is particularly useful and is most easily illustrated by means of an example. Let $g(s)$ be a transfer function with three distinct poles λ_1, λ_2, and λ_3 with respective multiplicities 3, 1, and 1. It is assumed that the $g(s)$ can be represented by the partial fraction expansion of Eq. (29). The realization of Eqs. (30) is called the

$$g(s) = \frac{e_{11}}{(s-\lambda_1)^3} + \frac{e_{12}}{(s-\lambda_1)^2} + \frac{e_{13}}{(s-\lambda_1)}$$

$$+ \frac{e_2}{(s-\lambda_2)} + \frac{e_3}{(s-\lambda_3)} + d \qquad (29)$$

$$\begin{bmatrix} \dot{x}_{11} \\ \dot{x}_{12} \\ \dot{x}_{13} \\ \dot{x}_2 \\ \dot{x}_3 \end{bmatrix} = \begin{bmatrix} \lambda_1 & 1 & 0 & 0 & 0 \\ 0 & \lambda_1 & 1 & 0 & 0 \\ 0 & 0 & \lambda_1 & 0 & 0 \\ 0 & 0 & 0 & \lambda_2 & 0 \\ 0 & 0 & 0 & 0 & \lambda_3 \end{bmatrix} x + \begin{bmatrix} 0 \\ 0 \\ 1 \\ 1 \\ 1 \end{bmatrix} u$$

$$y = [\, e_{11} \; e_{12} \; e_{13} \; e_2 \; e_3 \,] x + du$$

Jordan-form realization of $g(s)$. It is directly obtained from the block diagram shown in Fig. 6 by associating a state variable with the output of each dynamic block.

The usefulness of these forms is not limited to transfer-function realizations, and in many situations, systems which are already in state variable form are transformed into an appropriate canonical form depending on the type of analysis or design task being undertaken.

Linear feedback control. The simplest type of control problem is the so-called regulator problem, in which it is required to drive the system from an arbitrary initial condition to the zero state. For a linear, time-invariant system, if all of the states are available in the form of measurements, this can be easily achieved by employing state feedback of the form $u = Kx$ resulting in closed-loop equation (31).

$$\dot{x} = (A + BK)x \qquad (31)$$

The feedback matrix K must result in stable eigenvalues for the matrix $(A + BK)$ if $x(t)$ is to decay to zero. It can be shown that if the system is controllable, there exists a K such that the eigenvalues of $(A + BK)$ take on any set of specified values. Thus the behavior of the transient response can be made arbitrarily fast by assigning closed-loop eigenvalues with sufficiently large negative real parts. In practice, care must be exercised in use of this procedure, as the large feedback gains required will amplify measurement noise as well as the signal, and may in addition cause unacceptably large peaks in the transient response.

Output feedback. If the entire state is not available for feedback purposes but only the lower-dimensional output vector $y = Cx$, it may or may not be possible to stabilize the system by simple output feedback of the form $u = My = MCx$. For example, the single-output single-input double integrator plant (a plant which performs integration twice in succession on the control signal) cannot be stabilized by direct output feedback. It can be shown that in general if there are m independent outputs, output feedback will usually allow the placement of only m poles in the transfer function of the closed-loop system. If, however, the sum of the number of inputs and outputs is greater than the dimension of the system, in almost all cases, all of the closed loop poles can be made arbitrarily close to any set of preassigned values using only output feedback.

Compensator techniques. In those cases in which

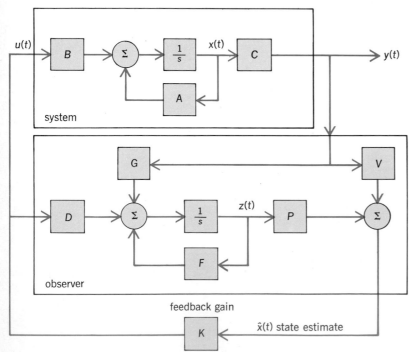

Fig. 7. Controller structure for a multi-input, multi-output system using an observer with linear feedback on the state estimate.

acceptable closed-loop response cannot be achieved by direct output feedback, a compensator must be employed. In its most general form, a compensator is another linear system which is driven by both the system input u and the system output y and whose state z can be used for feedback stabilization purposes. The state equation of an r-dimensional compensator is represented by Eq. (32), where the matrices F, G and J are of

$$\dot{z} = Fz + Gy + Ju \qquad (32)$$

appropriate dimension. If the matrices F, G and J are chosen properly, the system represented by Eq. (32) can be used to estimate the state of the original system in the sense that the state z will converge to a linear function Tx of the system state. Such systems are called observers or state estimators.

The required observer constraint equations are Eqs. (33), where it is assumed that $y = Cx$. If these

$$\begin{aligned} FT - TA + GC &= 0 \\ J - TB &= 0 \end{aligned} \qquad (33)$$

are satisfied, and F has stable eigenvalues, the error $e \equiv (z - Tx)$ will converge to zero according to Eq. (34).

$$\dot{e} = Fe \qquad (34)$$

If the original system is observable, Eqs. (33) can always be solved with F having arbitrary eigenvalues, such that a linear combination of z and y will converge to the system state x, if the observer dimension is equal to or greater than $n - m$ ($m =$ dimension of y). This estimate of x can then be used as if it were a measurement of x and thus allow the use of the feedback law $u = Kx$. The overall closed-loop system, whose structure is shown in Fig. 7, can be shown to have closed-loop eigenvalues equal to those of $(A + BK)$ plus those of F, all of which can be arbitrarily assigned.

More sophisticated use of these techniques reveals that, for a controllable and observable system, the minimum order of the dynamic compensator required for almost arbitrary assignment of the closed-loop eigenvalues is never greater than $n - m - p + 1$, where n, m, and p are the respective values of the state dimension, output dimension, and input dimension.

There are other applications in which it is required to generate not the entire state x but some specific set of linear functions of the state, say Ex. In these cases, function observers can be constructed of lower dimension than that required for full state estimation.

Although attention in this article has been primarily directed at continuous linear systems, all of the topics and techniques discussed apply to discrete-time systems as well. *See* CONTROL SYSTEMS.   [C. T. CHEN; PATRICK E. BARRY]

Bibliography: C. T. Chen, *Analysis and Synthesis of Linear Control Systems*, 1975; C. T. Chen, *Introduction to Linear System Theory*, 1970; B. C. Kuo, *Automatic Control Systems*, 1975; K. Ogata, *State Space Analysis of Control Systems*, 1967; L. A. Zadeh and C. A. Desoer, *Linear System Theory*, 1963.

Linear systems of equations

Systems of mathematical equations of the form of system (1), where the a_{ij}, $i = 1, 2, \ldots, m$, $j = 1, 2, \ldots, n$, and the b_i, $i = 1, 2, \ldots, m$, are constants, or fixed numbers, and the x_i, $i = 1, 2, \ldots, n$ are called unknowns. System (1) is referred to as m linear equations in n unknowns. A solution of system (1) is a set of numbers $c_1, c_2, \ldots, c_n$ such that when x_1 is replaced by c_1, x_2 by $c_2, \ldots,$ x_n by c_n, every equation of system (1) becomes a

$$\begin{aligned} a_{11}x_1 + a_{12}x_2 + \cdots + a_{1n}x_n &= b_1 \\ a_{21}x_1 + a_{22}x_2 + \cdots + a_{2n}x_n &= b_2 \\ \cdots\cdots\cdots\cdots\cdots\cdots \\ a_{m1}x_1 + a_{m2}x_2 + \cdots + a_{mn}x_n &= b_m \end{aligned} \qquad (1)$$

true equality. The problem posed by such a system of equations is to find criteria for the existence of a solution and, when solutions exist, to obtain systematic methods for finding the solutions.

An example of such a linear system of equations is system (2). System (2) has infinitely many solutions which can be given in the form $x_1 = -c_5 - \tfrac{4}{7}c_3 + \tfrac{6}{7}$, $x_2 = \tfrac{1}{3}c_5 - \tfrac{8}{7}c_3 + \tfrac{8}{21}$, $x_3 = c_3$, $x_4 = -\tfrac{1}{3}c_5 - \tfrac{2}{3}$, $x_5 = c_5$, where c_3 and c_5 are arbitrarily chosen numbers.

$$\begin{aligned} 2x_1 - x_2 \qquad - x_4 + 2x_5 &= 2 \\ x_1 + 3x_2 + 4x_3 - 3x_4 - x_5 &= 4 \\ 3x_1 + 2x_2 + 4x_3 + 2x_4 + 3x_5 &= 2 \end{aligned} \qquad (2)$$

Two linear systems of equations are equivalent if every solution of one of the systems is a solution of the other and vice versa. System (1) is replaced by an equivalent system, for which the solutions, if they exist, are easily found by performing certain elementary operations on the equations. These operations are interchanging two equations, adding to one equation another equation multiplied by a constant, or multiplying an equation by a nonzero constant. By a sequence of such operations, system (1) can be replaced by equivalent system (3). If $e_{r+1} = e_{r+2} = \cdots = e_m = 0$, then the solutions of the system can be obtained

$$\begin{aligned} x_{n_1} + d_{1,n_1+1}x_{n_1+a} + d_{1,n_1+2}x_{n_1+2} + \cdots \\ + d_{1,n}x_n &= e_1 \\ x_{n_2} + d_{2,n_2+1}x_{n_2+1} + \cdots + d_{2,n}x_n &= e_2 \\ \cdots\cdots\cdots\cdots\cdots\cdots\cdots \\ x_{n_r} + d_{r,n_r+1}x_{n_r+1} + \cdots + d_{r,n}x_n &= e_r \\ 0 &= e_{r+1} \\ \cdots \\ 0 &= e_m \end{aligned} \qquad (3)$$

by choosing arbitrary values for the unknowns other than

$$x_{n_1}, x_{n_2}, \ldots, x_{n_r}$$

and solving the equations in turn, beginning with the last equation, for

$$x_{n_r}, x_{n_r-1}, \ldots, x_{n_2}, x_{n_1}$$

When the system has solutions it is said to be consistent. If any one of $e_{r+1}, e_{r+2}, \ldots, e_m$ is not zero, then the equations do not have a solution and are inconsistent. In example (2), the system is equivalent to the system

$$\begin{aligned} x_1 + 3x_2 + 4x_3 - 3x_4 - x_5 &= 4 \\ x_2 + \tfrac{8}{7}x_3 - \tfrac{5}{7}x_4 - \tfrac{4}{7}x_5 &= \tfrac{6}{7} \\ x_4 + \tfrac{1}{3}x_5 &= -\tfrac{2}{3} \end{aligned}$$

Arbitrary values c_5 and c_3 can be chosen for x_5

and x_3, respectively, and the equations can be solved successively for x_4, x_2, and x_1.

The operations used in reducing system (1) to system (3) can be most effectively performed on the rows of the matrix of system (4), and they are

$$A^* = \begin{pmatrix} a_{11} & a_{12} & \cdots & a_{1n} & b_1 \\ a_{21} & a_{22} & \cdots & a_{2n} & b_2 \\ \cdots & \cdots & \cdots & \cdots & \cdots \\ a_{m1} & a_{m2} & \cdots & a_{mn} & b_m \end{pmatrix} \qquad (4)$$

precisely the elementary row transformations of a matrix. *See* MATRIX THEORY.

Matrix (4) is called the augmented matrix of the system, and matrix (5) is called the matrix of the coefficients. The rank of a rectangular matrix is defined, and the method for determining the rank is given in the article on matrix theory. A necessary and sufficient condition for the existence of a

$$A = \begin{pmatrix} a_{11} & a_{12} & \cdots & a_{1n} \\ a_{21} & a_{22} & \cdots & a_{2n} \\ \cdots & \cdots & \cdots & \cdots \\ a_{m1} & a_{m2} & \cdots & a_{mn} \end{pmatrix} \qquad (5)$$

solution of system (1) is that the rank of the augmented matrix A^* be the same as the rank of the matrix of coefficients A. In example (2), the rank of the augmented matrix

$$A^* = \begin{pmatrix} 2 & -1 & 0 & -1 & 2 & 2 \\ 1 & 3 & 4 & -3 & -1 & 4 \\ 3 & 2 & 4 & 2 & 3 & 2 \end{pmatrix}$$

as well as the rank of

$$A = \begin{pmatrix} 2 & -1 & 0 & -1 & 2 \\ 1 & 3 & 4 & -3 & -1 \\ 3 & 2 & 4 & 2 & 3 \end{pmatrix}$$

is equal to 3. Therefore the system is consistent and has the solutions given earlier.

If $m = n$ in system (1) and if the determinant $|A|$ of the matrix of the coefficients A is not zero, then system (1) has a unique solution which can be obtained by Cramer's rule, which gives the values of the unknowns x_1, x_2, . . . , x_n as the ratio of two determinants. *See* DETERMINANT.

Specifically, Cramer's rule gives

$$x_1 = \begin{vmatrix} b_1 & a_{12} & \cdots & a_{1n} \\ b_2 & a_{22} & \cdots & a_{2n} \\ \cdots & \cdots & \cdots & \cdots \\ b_n & a_{n2} & \cdots & a_{nn} \end{vmatrix} \Big/ |A|$$

$$x_2 = \begin{vmatrix} a_{11} & b_1 & a_{13} & \cdots & a_{1n} \\ a_{21} & b_2 & a_{23} & \cdots & a_{2n} \\ \cdots & \cdots & \cdots & \cdots & \cdots \\ a_{n1} & b_n & a_{n3} & \cdots & a_{nn} \end{vmatrix} \Big/ |A|$$

$$\cdots \cdots \cdots \cdots \cdots \cdots \cdots \cdots \cdots$$

$$x_n = \begin{vmatrix} a_{11} & \cdots & a_{1n-1} & b_1 \\ a_{21} & \cdots & a_{2n-1} & b_2 \\ \cdots & \cdots & \cdots & \cdots \\ a_{n1} & \cdots & a_{nn-1} & b_n \end{vmatrix} \Big/ |A|$$

In every case the determinant in the denominator is $|A|$, and for the value of the unknown x_i, the determinant in the numerator is the determinant of the matrix obtained from A by replacing the ith column of A by b_1, b_2, . . . , b_n. If $|A| = 0$ in this case of n equations in n unknowns, the system may either be inconsistent or it may have infinitely many solutions. In the example

$$x_1 - 4x_2 + 5x_3 = 6$$

$$2x_1 - 11x_2 + 10x_3 = -1$$

$$4x_1 - x_2 + 5x_3 = 0$$

$$|A| = \begin{vmatrix} 1 & -4 & 5 \\ 2 & -11 & 10 \\ 4 & -1 & 5 \end{vmatrix} = 45$$

and the unique solution is given by

$$x_1 = \begin{vmatrix} 6 & -4 & 5 \\ -1 & -11 & 10 \\ 0 & -1 & 5 \end{vmatrix} \Big/ 45 = -285/45 = -19/3$$

$$x_2 = \begin{vmatrix} 1 & 6 & 5 \\ 2 & -1 & 10 \\ 4 & 0 & 5 \end{vmatrix} \Big/ 45 = 194/45 = 13/3$$

$$x_3 = \begin{vmatrix} 1 & -4 & 6 \\ 2 & -11 & -1 \\ 4 & -1 & 0 \end{vmatrix} \Big/ 45 = 267/45 = 89/15$$

If in system (1), $b_1 = b_2 = \cdots = b_m = 0$, system (6) is called homogeneous. Since the augmented

$$a_{11}x_1 + a_{12}x_2 + \cdots + a_{1n}x_n = 0$$

$$a_{21}x_1 + a_{22}x_2 + \cdots + a_{2n}x_n = 0 \qquad (6)$$

$$\cdots \cdots \cdots \cdots \cdots \cdots \cdots \cdots$$

$$a_{m1}x_1 + a_{m2}x_2 + \cdots + a_{mn}x_n = 0$$

matrix of system (6) differs from the matrix of the coefficients by an additional column of zeros, these matrices always have the same rank, so that a homogeneous system is always consistent. It has the obvious solution $x_1 = 0$, $x_2 = 0$, . . . , $x_n = 0$, which is called the trivial solution. Any solution with at least one $x_i \neq 0$ is called nontrivial. A homogeneous system of m equations in n unknowns has a nontrivial solution if and only if the rank of the matrix of coefficients A is less than n. In particular, if m is less than n, the rank of A is at most m, and the system has a nontrivial solution. Further, if $m = n$, the system has a nontrivial solution if and only if the determinant $|A| = 0$. In system (7)

$$3x_1 + 4x_2 - 2x_3 = 0$$

$$-2x_1 + 3x_2 - 4x_3 = 0 \qquad (7)$$

$$5x_1 + x_2 + 2x_3 = 0$$

$$-9x_1 + 5x_2 - 10x_3 = 0$$

the rank of A is 2, so that system (7) has a non-trivial solution.

$$A = \begin{pmatrix} 3 & 4 & -2 \\ -2 & 3 & -4 \\ 5 & 1 & 2 \\ -9 & 5 & -10 \end{pmatrix}$$

The rank of A is seen to be 2 by performing elementary row transformations on A to reduce it to the form

$$\begin{pmatrix} 1 & 4/3 & -2/3 \\ 0 & 1 & -16/17 \\ 0 & 0 & 0 \\ 0 & 0 & 0 \end{pmatrix}$$

System (7) is equivalent to the system

$$x_1 + 4/3 x_2 - 2/3 x_3 = 0$$

$$x_2 - 16/17 x_3 = 0$$

The solutions are given by $x_1 = -10/17 c_3$, $x_2 = 16/17 c_3$, $x_3 = c_3$, where c_3 is an arbitrarily chosen number.

In the event that a linear system (1) has infinitely many solutions, they can be given as in example (2) or, alternatively, a finite number of solutions can be listed in terms of which all solutions can be conveniently obtained. Any one fixed solution of system (1) is called a particular solution. In example (2), $x_1 = 6/7$, $x_2 = 8/21$, $x_3 = 0$, $x_4 = -2/3$, $x_5 = 0$ is a particular solution. If system (1) is consistent, every solution is of the form $x_1 = c_1 + c_1^{(0)}, x_2 = c_2 + c_2^{(0)}, \ldots, x_n = c_n + c_n^{(0)}$, where $c_1, c_2, \ldots, c_n$ is a fixed particular solution of system (1) and $c_1^{(0)}, c_2^{(0)}, \ldots, c_n^{(0)}$ is any solution of the corresponding homogeneous system (6). By virtue of this result, it is sufficient to investigate the solutions of the homogeneous system (6).

Let $C^{(1)} = (c_1^{(1)}, c_2^{(1)}, \ldots, c_n^{(1)})$ and $C^{(2)} = (c_1^{(2)}, c_2^{(2)}, \ldots, c_n^{(2)})$ be solutions of the homogeneous system (6), written as n-tuples. Then $C^{(1)} + C^{(2)} = (c_1^{(1)} + c_1^{(2)}, c_2^{(1)} + c_2^{(2)}, \ldots, c_n^{(1)} + c_n^{(2)})$ is the sum of the solutions, and $eC^{(1)} = (ec_1^{(1)}, ec_2^{(1)}, \ldots, ec_n^{(1)})$, where e is a fixed number and is a scalar multiple of the solution $C^{(1)}$. It is evident that any linear combination $e_1 C^{(1)} + e_2 C^{(2)} + \cdots + e_s C^{(s)}$ of solutions $C^{(1)}, C^{(2)}, \ldots, C^{(s)}$, of system (6) is again a solution of system (6). A set of solutions $C^{(1)}, C^{(2)}, \ldots, C^{(s)}$ of the homogeneous system (6) is called a linearly independent set if $e_1 C^{(1)} + e_2 C^{(2)} + \cdots + e_s C^{(s)} = (0, 0, \ldots, 0)$ implies that $e_1 = e_2 = \cdots = e_s = 0$. Otherwise the solutions form a linearly dependent set, that is, there exist numbers $e_1, e_2, \ldots, e_s$, not all zero such that $e_1 C^{(1)} + e_2 C^{(2)} + \cdots + e_s C^{(s)} = (0, 0, \ldots, 0)$. A set of solutions is a maximal linearly independent set if the set is linearly independent and if a set formed by adding any other solution of system (6) is linearly dependent. If the rank of the matrix A of system (6) is r, then the number of solutions in a maximal independent set is $n - r$. Every solution of system (6) is a linear combination $e_1 C^{(1)} + e_2 C^{(2)} + \cdots + e_{n-r} C^{(n-r)}$ of the solutions in a maximal independent set. In example (2), the corresponding homogeneous system is system (8).

$$2x_1 - x_2 \quad\quad - x_4 + 2x_5 = 0$$
$$x_1 + 3x_2 + 4x_3 - 3x_4 - x_5 = 0 \quad\quad (8)$$
$$3x_1 + 2x_2 + 4x_3 + 2x_4 + 3x_5 = 0$$

The rank of the matrix A of system (8) is 3. Thus there are $5 - 3 = 2$ solutions in a maximal linearly independent set. They can be given as $C^{(1)} = (-4, -8, 7, 0, 0)$ and $C^{(2)} = (-3, 1, 0, -1, 3)$. Every solution of system (8) is given by $e_1 C^{(1)} + e_2 C^{(2)} = (-4e_1 - 3e_2, -8e_1 + e_2, 7e_1, -e_2, 3e_2)$ for arbitrary e_1 and e_2. Every solution of the original system is given by $(6/7 - 4e_1 - 3e_2, \; 8/21 - 8e_1 + e_2, \; 7e_1, \; -2/3 - e_2, 3e_2)$.

In example (7), the rank of the matrix is 2, and there is $3 - 2 = 1$ solution in a maximal linearly independent set. It can be given as $C^{(1)} = (-10, 16, 17)$, and every solution of example (7) is a multiple of $C^{(1)}$.

If the system of equations (1) has a solution, then it has a rational solution. This means that there is a solution which is obtained by employing only the rational operations of addition, subtraction, multiplication, and division on the coefficients of the system. The particular solution as well as the lin-

early independent solutions of the corresponding homogeneous system can always be selected rationally. This is evident since the operations performed on the system to obtain a solution are rational operations. *See* EQUATIONS, THEORY OF; POLYNOMIAL SYSTEMS OF EQUATIONS.

[ROSS A. BEAUMONT]

Bibliography: R. A. Beaumont, *Linear Algebra*, 1965.

Linearity

The relationship that exists between two quantities when a change in one of them produces a directly proportional change in the other. Thus, an amplifier has good linearity when doubling of its input signal strength always doubles the output signal strength. A transistor displays good linearity when doubling of the instantaneous base voltage serves to double the instantaneous value of collector current; a vacuum tube displays good linearity when doubling the instantaneous control-grid voltage doubles the instantaneous value of the plate current.

In television, uniform linearity of scanning is essential for uniform spacing of picture elements in each horizontal line and uniform vertical spacings between scanning lines. Poor linearity causes the picture to be compressed or stretched in some area, so that circles of test patterns become egg-shaped or even more greatly distorted. *See* SAWTOOTH-WAVE GENERATOR.   [JOHN MARKUS]

Linen

A widely used cloth made from flax fibers. Linen is noted for its evenness of thread, fineness, and density. Its uses include garments, tablecloths, sheeting, towels, and thread.

The flax plant which supplies the fiber needs a short, cool growing season with plenty of rainfall. The Soviet Union leads the world in growing flax for fiber. The manufacture of the fiber into fabric requires unusual care during each process to retain the strength and beauty of the fiber. *See* FLAX.

Processing the fiber. Seeds and leaves are removed from the stems of the flax plant by passing the stalks through coarse combs, a process called rippling. Bundles of stems weighted down with heavy stones to ensure complete immersion are then steeped in water. The wetting allows the bark surrounding the hairlike flax fibers to decompose, thus loosening the gums that hold the fibers. This decomposing, or fermentation, is called retting.

After removal from water the stems may be spread out on the ground for a week or more. After the woody tissue has become dry, it is crushed by fluted iron rollers. After this breaking operation there remain only small pieces of wood called shives. The shives are removed by means of rotating beaters, thus finally releasing the flax fibers. This operation can be done either by hand or with machinery. The simple combing process, known as hackling, straightens the flax fibers, separates the short from the long staple, and leaves the longer fibers in parallel formation. For coarse linen, hackling is usually done by hand. Fine linen is also hackled by machine, a finer comb being used with each hackling treatment.

Properties of flax fiber. Under a microscope the flax fiber shows hairlike cylindrical filaments with fine pointed ends. The filaments are cemented together by a gummy substance called pectin. The long flax filaments contain a lumen, or central canal. The fiber resembles a straight, smooth bamboo stick, with its joints producing a slight natural unevenness that cannot be eliminated. Chemically the flax fiber is composed of about 70% cellulose, 25% pectin, and the remainder of woody tissue and ash. From such fibers linen yarns are produced that have a smooth, straight, compact, and lustrous appearance. *See* CELLULOSE; FIBER, NATURAL; PECTIN; TEXTILE.   [ELTON G. NELSON]

Bibliography: *See* AGRICULTURAL SCIENCE (PLANT).

Lines of force

Imaginary lines in fields of force whose tangents at any point give the direction of the field at that point and whose number through unit area perpendicular to the field represents the intensity of the field. The concept of lines of force is perhaps most common when dealing with electric or magnetic fields.

Electric lines of force. These are lines drawn to represent, or map, an electric field graphically in the space around a charged body. They are of great help in visualizing an electric field and in quantitative thinking about such a field. The direction of an electric line of force at any point is drawn parallel to the direction of the electric field intensity $\mathbf{E}$ at that point. The quantitative convention is that the number of electric lines of force drawn through an imaginary unit area of surface perpendicular to the field shall be numerically equal to $\mathbf{E}$ at that area. From this quantitative convention, and using the rationalized mks system of units, q/ϵ_0 lines of force originate on a positive charge q (in coulombs) and a like number terminate on a negative charge q of the same magnitude. Here $\epsilon_0 = 8.85 \times 10^{-12}$ coulomb²/newton-m² is the permittivity of empty space. *See* ELECTRIC FIELD.

Magnetic lines of force. A magnetic field of strength $\mathbf{H}$ may also be represented by lines called lines of force. The number of lines of force per unit area of a surface perpendicular to $\mathbf{H}$ is numerically equal to the value of $\mathbf{H}$. The lines of magnetic force due to currents are closed curves, as are the lines of magnetic induction. Magnetic lines of force due to magnets originate on north poles and terminate on south poles, both inside and outside the magnet. *See* MAGNETIC FIELD.

[RALPH P. WINCH]

Bibliography: A. E. Kip, *Fundamentals of Electricity and Magnetism*, 1967; R. P. Winch, *Electricity and Magnetism*, 1955.

Lingulida

An order of inarticulate brachiopods that is first known in Lower Cambrian rocks and attains maximum diversity in the Ordovician, and that is represented in present-day oceans by species of *Lingula* and *Glottidia*. These two genera belong to the Lingulacea, the principal superfamily of the order. *See* INARTICULATA.

Nearly all lingulaceans have shells of calcium phosphate and are biconvex and commonly elongate in outline. The majority have a pedicle, which emerges from between the valves. In Recent species the pedicle is extensile and is used to anchor the animal to the base of its burrow, which is exca-

vated in soft sediments in shallow water by a combination of relative rotary and longitudinal sliding movement of the valves. The musculature needed to perform these movements is complex and, in addition to oblique muscles comparable with those found in other inarticulate stocks, there is a pair of transmedian muscles which cross the body cavity laterally (Fig. 1).

In the interior of the pedicle valve, a pedicle groove divides the pseudointerarea medially. The pseudointerarea of both valves is typically much

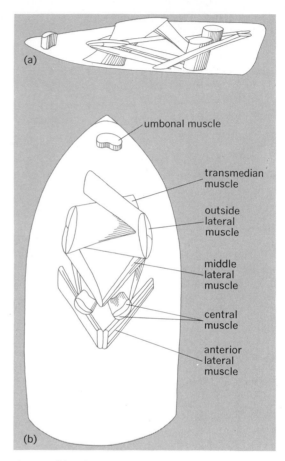

Fig. 1. *Lingula* muscle system viewed (*a*) laterally, (*b*) dorsally. (*R. C. Moore, ed., Treatise on Invertebrate Paleontology, pt. H, Geological Society of America and University of Kansas Press, 1965*)

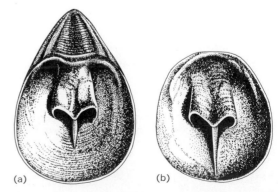

Fig. 2. Internal molds of Silurian trimerellacean *Trimerella*: (*a*) pedicle valve; (*b*) brachial valve. (*C. D. Walcott, Cambrian Brachiopoda, USGS Monogr. no. 51, 1912*)

more strongly developed in Cambrian and Ordovician forms than in post-Ordovician representatives, and commonly is approximately in the plane of commissure of the valves.

The Trimerellacea are the second group of lingulide taxa believed to be descended from lingulacean ancestors in Upper Ordovician times. As a group they had a brief existence and were extinct by the end of the Silurian. Later species are amongst the biggest known inarticulates with heavy valves and well-developed muscle platforms. The shell is of calcium carbonate, and they are also unusual in having a unique form of articulation consisting of a simple transverse ridge or tooth in the brachial valve which is associated with a corresponding median groove in the pedicle valve. A pedicle seemingly was not developed and the animals probably lay free on the sea floor (Fig. 2). See BRACHIOPODA. [A. J. ROWELL]

Link

An element of a mechanical linkage. A link may be a straight bar or a disk, or it may have any other shape, simple or complex. It is assumed, for simple analysis, to be made of unyielding material; that is, its shape does not change. The frame, or fixed member, of a linkage is one of the links, whether the frame is fixed relative to the Earth or relative to a movable body such as the chassis of an automobile. See VELOCITY ANALYSIS.

Two links of a kinematic chain meet in a joint, or pair, by which these links are held together. Just as each joint is a pair having two elements, one from each link, so a link in a kinematic chain is the rigid connector of two or more elements belonging to different pairs. See SLIDING PAIR.

Each free link has three degrees of freedom in the plane (two translational and one rotational). If the tail of one link is pin-joined to the head of its predecessor two degrees of freedom are lost. If several links are joined so as to form a closed loop, $2g$ degrees of freedom are lost, where g is the number of lower pair joints. If one member of an n-member chain is fixed, all three degrees of freedom of the fixed member are eliminated, the remaining degrees of freedom F being given by Eq. (1). Hence for a triangle $F=0$ and for a chain of

$$F=3(n-1)-2g \qquad (1)$$

four links called simply a four bar linkage, $F=1$.

Assuming that $F=1$ is stipulated for a mechanism, and investigating n in Eq. (1), it is found that Eq. (2) results. Because 4 is even and $2g$ is even, n

$$3n=4+2g \qquad (2)$$

must always be even. There can be no pin-joint mechanism with five links; the next highest must have six links. See FOUR-BAR LINKAGE; MECHANISM. [DOUGLAS P. ADAMS]

Bibliography: R. S. Hartenberg and J. Denavit, *Kinematic Synthesis of Linkages*, 1964; J. Hirschhorn, *Kinematics and Dynamics of Plane Mechanisms*, 1962.

Linkage, genetic

Failure of two or more genes to recombine at random as a result of their location on the same chromosome pair. Among the haploid products of a cell which has gone through meiosis, two genes located in the same chromosome pair remain in their two original combinations of alleles ("parental") unless an odd number of exchanges of homologous segments occurred within the interval bounded by their loci. The incidence of exchanges of homologous segments at meiosis is roughly proportional to the length of the chromosome segment between two loci. The percentage of recombinants thus provides an estimate of this length and a basis for constructing gene maps on which linked loci are arranged in linear order and spaced out in proportion to the recombination percentages between them. The analogy is with a linear map of the stations along a railway line spaced out according to the time of travel between any two or more of them. By extension of the meaning of linkage to bacterial transduction or transformation, two bacterial genes are said to be linked when both are incorporated together in the chromosome of recipient cells more often than would be expected from mere coincidence of two separate incorporations. This is the consequence of their being frequently carried on the same transducing or transforming DNA fragment. See DEOXYRIBONUCLEIC ACID (DNA); MEIOSIS; TRANSDUCTION, BACTERIAL; TRANSFORMATION, BACTERIAL.

If at meiosis in an individual heterozygous at two loci on the same chromosome pair—for example, AB—less than one exchange occurs on the average between them, the proportions of haploid "recombinant" meiotic products—aB and Ab—will be less than 50%, and that of the "parental" meiotic products—AB and ab—correspondingly more than 50%. This defines the two loci as linked. If, however, two loci on the same chromosome pair are so far apart that on the average one or more exchanges occur between them at meiosis, the proportions of "parental" and "recombinant" meiotic products will be equal, that is, 50%. This is the same as for two loci on different chromosome pairs. Thus for two loci between which recombination is 50% there are no means of deciding, on the basis of the recombination percentage alone, whether they are on one and the same chromosome pair but sufficiently far apart, or on different pairs. (For an explanation of why 50% is the upper limit of recombination see RECOMBINATION, GENETIC.)

Thus while evidence of linkage—less than 50% recombination—implies that two loci are on the same chromosome pair, absence of linkage—50% recombination—does not imply that they are not. Further evidence may reveal that two loci showing no linkage to each other are both linked to a third locus, and so on. This additional evidence permits the construction of linkage maps of all the loci belonging to one and the same "linkage group." Linkage groups can then be homologized to chromosome pairs.

In human genetics, since the introduction of the techniques of somatic cell genetics, very often loci can be assigned to one and the same chromosome pair well before knowing anything about the recombination percentages between any two of them at gametogenesis, that is, before knowing whether or not they are linked in the precise arithmetical sense of the word. To overcome this ambiguity, the term "syntenic," proposed by J. Renwick, is now in general use to define two loci carried on the same chromosome pair, irrespective of what is the recombination percentage between them at meiosis. *See* GENETICS, SOMATIC CELL.

Genetic linkage is a structural relationship between two genes, that is, a restriction on their recombination at meiosis. It does not necessarily imply a functional relationship. However, in bacteria some of the structural and regulatory genes controlling enzymes sequentially involved in a metabolic pathway often turn out to be closely linked. With very few exceptions, this is not so in eukaryotes.

The biochemical and evolutionary significance of linkage is poorly understood. In the cases just mentioned in bacteria, linkage permits coordinated regulation of enzyme synthesis. In eukaryotes several linkages have been preserved over long periods of evolution, but it cannot be excluded that for some at least this is simply a matter of historical accident. The fact that more and more cases of linkage disequilibrium are coming to light, on the other hand, reflects selective advantages, possibly related to the preservation of genetic variation in natural populations.   [G. PONTECORVO]

Linkage, mechanical

A set of rigid bodies, called links, joined together at pivots by means of pins or equivalent devices. A body is considered to be rigid if, for practical purposes, the distances between points on the body do not change. Linkages are used to transmit power and information. They may be employed to make a point on the linkage follow a prescribed curve, regardless of the input motions to the linkage. They are used to produce an angular or linear displacement $f(x)$, where $f(x)$ is a given function of a displacement x. *See* LINK.

If the links are bars the linkage is termed a bar linkage. In first approximations a bar may be treated as a straight line segment or a portion of a curve, and a pivot may be treated as a common point on the bars connected at the pivot. A common form of bar linkage is then one for which the

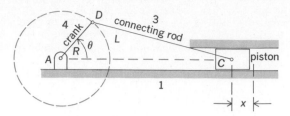

Fig. 3. Slider crank mechanism.

bars are restricted to a given plane, such as a four-bar linkage (Fig. 1). A body pivoted to a fixed base and to one or more other links is a crank. As crank 4 in Fig. 1 rotates, link 2 (also a crank) oscillates back and forth. This four-bar linkage thus transforms a rotary motion into an oscillatory one, or vice versa. Link AB, marked 1, is fixed. Link 2 may be replaced by a slider 5 (Fig. 2). As end D of crank 4 rotates, pivot C describes the same curve as before. The slider moves in a fixed groove. *See* FOUR-BAR LINKAGE.

A commonly occurring variation of the four-bar linkage is the linkage used in reciprocating engines (Fig. 3). Slider C is the piston in a cylinder, link 3 is the connecting rod, and link 4 is the crank. This mechanism transforms a linear into a circular motion, or vice versa. The straight slider in line with the crank center is equivalent to a pivot at the end of an infinitely long link. Let R be the length of the crank, L the length of the connecting rod, while θ denotes the angle of the crank as shown in Fig. 3. Also, let x denote the coordinate of the pivot C measured so that $x=0$ when $\theta=0$. Length L normally dominates radius R, whence the approximate relation in Eq. (1) holds.

$$x = R(1 - \cos\theta) + \frac{R^2}{2L}\sin^2\theta \qquad (1)$$

A Scotch yoke is employed to convert a steady rotation into a simple harmonic motion (Fig. 4). The angle of the crank 2 with respect to the horizontal is denoted by θ, and the coordinate of the slider 4 by x. If R again denotes the length of the crank, Eq. (2) holds. In computer mechanisms, sli-

$$x = R\cos\theta \qquad (2)$$

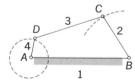

Fig. 1. Four-bar linkage

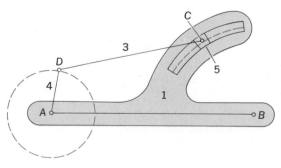

Fig. 2. Equivalent of four-bar linkage.

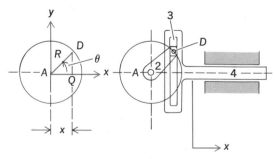

Fig. 4. Scotch yoke.

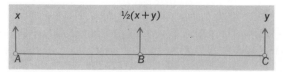

Fig. 5. Addition with a lever.

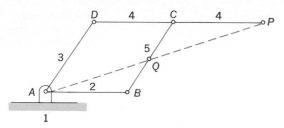

Fig. 6. Pantograph.

der 3 can be moved along the crank so that the component of a vector can be obtained mechanically. *See* SLIDER CRANK MECHANISM.

A lever is normally understood to be a bar connected to three other links. A lever is often used for addition (Fig. 5). Point *B* is at the center of lever *AC*. It is assumed that *x* and *y* are numerically small compared to the length of the lever. Levers with a fixed pivot (fulcrum) are particularly useful for amplifying or attenuating linear displacements and forces. *See* LEVER.

The pantograph shown in Fig. 6 is a five-link mechanism employed in drafting to magnify or reduce diagrams. As point *Q* on link 5 describes a curve, point *P* on link 4 will describe a similar but enlarged curve. Here *ABCD* is a parallelogram. Pivot *A* is fixed. *See* PANTOGRAPH.

A mechanism composed of six links, introduced by Joseph Whitworth (1803–1887), makes the return stroke of a body in oscillatory linear motion faster than the forward stroke. Universal joints for transmitting motion between intersecting axes and escapements permitting rotation about an axis in one direction only are examples of the many kinds of linkages used in machines. However, complicated linkages are becoming obsolete because of the ease with which information and power can be transmitted by electrical, hydraulic, and penumatic means. *See* MECHANISM.

[RUFUS OLDENBURGER]

Bibliography: A. S. Hall, *Kinematics and Linkage Design*, 1961; R. S. Hartenburg and J. Denavit, *Kinematic Synthesis of Linkages*, 1964; H. H. Mabie and F. W. Ocvirk, *Mechanisms and Dynamics of Machinery*, 1957; A. Svoboda, *Computing Mechanisms and Linkages*, 1948; D. C. Tao, *Applied Linkage Synthesis*, 1964.

Linoleum

A floor covering made by applying a mixture of gelled linseed oil, pigments, fillers, and other materials to a burlap backing and curing to produce a hard, resilient sheet.

Linoleum has been supplanted to a large extent by adhesive-backed vinyl or other resinous tiles that can be readily arranged in a variety of patterns; and by improved floor paints with excellent color retention and wear-resistant properties. *See* DRYING OIL; PAINT.

[C. R. MARTINSON; C. W. SISLER]

Linseed oil

A product made from the seeds of the flax plant (*Linum usitatissimum*) by crushing and pressing, either with or without heat. After the linseed oil is extracted, it is formulated either raw or boiled, in various grades and with various drying agents, such as litharge, minium, and zinc sulfate. Its most common uses are as the vehicle in oil paints and as a component of oil varnishes; after a longer period of heating (called boiling but actually carried out below the boiling point) and further processing, it is used as a vehicle in printing and lithographic ink, and in the manufacture of oilcloth, linoleum, and some soaps. Oil cake, the residue of the crushed seeds, is a protein- and mineral-rich material used in cattle feed. Most linseed raised for oil comes from plants cultivated for their many seeds rather than for their long fibers in Argentina, the Soviet Union, and India, with lesser quantities from the United States and Canada. It was first used as a drying oil in 12th-century paintings. *See* DRYING OIL; FLAX; SURFACE COATING.

[FRANK H. ROCKETT]

Lip

A fleshy fold above and below the entrance to the mouth of mammals. Even the lower vertebrates with movable jaws, such as fishes, amphibians, and most reptiles, have folds of skin external to the jaws and teeth. These folds, covered superficially with epithelium, are usually soft and supported internally by connective tissue. By contrast, in turtles, birds, and the lowest mammals (monotremes) the epithelium is cornified and hard, so as to form a beak or bill. Nevertheless, only the typical mammals (marsupials and placental mammals) possess a highly developed type of true lips. These are separated by deep clefts from the jaws and are made mobile by containing muscles belonging to the facial group. The separating cleft or space constitutes the vestibule of the mouth. The increased size and mobility of the lips of mammals are adapted to such functions as sucking, drinking, grasping food, and holding it against the teeth. The opening into the mouth is expanded or restricted at will. The lips of man are put to important use in producing articulate speech, and they aid in facial expression.

Cheeks. Since the mouth opening of mammals lies well in front of the joints of the jaws, a region is created in each side known as the cheek. Each is a continuation backward of the lips and is essentially similar in structure. Some rodents and Old World monkeys have expanded these spaces into cheek pouches which are used for the temporary storage and carriage of food. The pocket gophers

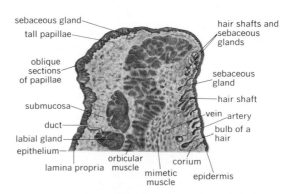

Section through the lower lip of man. (*From R. O. Greep, ed., Histology, 2d ed., McGraw-Hill, 1966*)

have expansive pouches in the cheek region, but in this instance they are hair-lined sacs opening onto the external surface and not into the mouth.

Morphology. Structurally a lip is a plate of voluntary muscle, the outer side surfaced with skin and the inner side with a mucous membrane (see illustration). At the red margin between these surfaces there is a zone that is transitional between these two coverings. The skin of the outer side of the lip is quite typical, with a thin epidermis which is, however, both opaque and cornified. Beneath the epidermal cover lie hair follicles, sweat glands, and sebaceous glands. Centrally within the lip is a plate of skeletal muscle, which is formed mostly by the orbicular muscle. Slips of muscle insert among the connective-tissue bundles of the skin, and this relation renders the skin mobile. Hence the muscle here belongs to the facial muscles of expression, the mimetic muscles. The inner side of the lip is a typical mucous membrane. Its stratified squamous epithelium is markedly thicker than the epidermis, but is not cornified. This type of soft epithelium is characteristic of wet surfaces, subject to friction, and not absorptive. Groups of rounded labial glands occur deep in what may be called a submucosal layer. Their secretion is a slimy mucus and their ducts open into the vestibule. The red margin is intermediate in structure between the outer and inner surfaces. The epithelium is translucent. Tall, vascular papillae of connective tissue indent the epithelium far toward the surface. Tiny blood vessels extend into these papillae and their contained blood shows through, thus giving the lip a red color. Similarly, nerve endings are brought close to the surface, so that the marginal region is highly sensitive. Glands are absent in this zone, and the soft epithelium has to be moistened frequently by wiping with the tongue. Accordingly, the labial margin is subject to chapping or cracking.

Development. The lower lip and jaw develop from the fusion of two components, the mandibular processes, while the upper lip and jaw develop from four, the two maxillary and two median nasal processes. The upper lip is more prone to mishaps in development, the commonest of which is "hare lip." At times, this cleftlike defect involves the upper jawbone, and then may be combined with a cleft palate. Lips become separate entities when the epithelium of each primitive jaw region grows inward as a plate and then cleaves in two, thereby producing the vestibule and separating the lip from the gum. *See* ORAL GLANDS.

[LESLIE B. AREY]

Lipid

One of a class of compounds which contains long-chain aliphatic hydrocarbons (cyclic or acyclic) and their derivatives, such as acids (fatty acids), alcohols, amines, amino alcohols, and aldehydes. The presence of the long aliphatic chain as the characteristic component of lipids confers distinct solubility properties on the simpler members of this class of naturally occurring compounds. This led to the traditional definition of lipids as substances which are insoluble in water but soluble in fat solvents such as ether, chloroform, and benzene. However, those lipids (particularly glycolipids and phospholipids) which contain polar components in the molecule may be insoluble in these solvents, and some are even soluble in water. Most of the phosphatides are good emulsifying agents. *See* EMULSION; SOLVENT.

Classification. The lipids are generally classified into the following groups:

A. Simple lipids
1. Triglycerides or fats and oils are fatty acid esters of glycerol. Examples are lard, corn oil, cottonseed oil, and butter.
2. Waxes are fatty acid esters of long-chain alcohols. Examples are beeswax, spermaceti, and carnauba wax.
3. Steroids are lipids derived from partially or completely hydrogenated phenanthrene. Examples are cholesterol and ergosterol.

B. Complex lipids
1. Phosphatides or phospholipids are lipids which contain phosphorus and, in many instances, nitrogen. Examples are lecithin, cephalin, and phosphatidyl inositol.
2. Glycolipids are lipids which contain carbohydrate residues. Examples are sterol glycosides, cerebrosides, and plant phytoglycolipids.
3. Sphingolipids are lipids containing the long-chain amino alcohol sphingosine and its derivatives. Examples are sphingomyelins, ceramides, and cerebrosides.

The classification scheme is not rigid since sphingolipids are found which contain phosphorus and carbohydrate, and glycolipids which contain phosphorus. Those lipids not included in the above groupings are of the chemically simpler type and include the fatty acids, alcohols, ethers such as batyl and chimyl alcohols, and hydrocarbons such as the terpenes and carotenes.

The fatty acids found in lipids may be saturated or unsaturated, cyclic or acyclic, and contain substituents such as hydroxyl or keto groups. They are combined in ester or amide linkage.

The alcohols, likewise, may be saturated or unsaturated and are combined in ester or ether linkages; in the case of α-β unsaturated ethers, these are best considered as aldehyde derivatives.

The amino alcohols found in lipids are linked as amides, glycosides, or phosphate esters. Sphingosine is the predominant amino alcohol found in animal tissue lipids, and phytosphingosine is its equivalent in the plant kingdom.

Occurrence. Lipids are present in all living cells, but the proportion varies from tissue to tissue. The triglycerides accumulate in certain areas, such as adipose tissue in the human being and in the seeds of plants, where they represent a form of energy storage. The more complex lipids occur closely linked with protein in the membranes of cells and of subcellular particles. More active tissues generally have a higher complex lipid content; for example, the brain, liver, kidney, lung, and blood contain the highest concentration of phosphatides in the mammal. Fish oils are important sources of vitamins A and D, and seed-germ oils contain quantities of vitamin E. The vitamins K occur chiefly in plants and microorganisms, and carotenoids are widely distributed in both the plant and animal kingdoms. Waxes are found in insects and as protective agents on plant leaves and the cuticles of fruits and vegetables (for instance, the bloom on grapes and plums). Soybeans are used for the commercial preparation of phosphatides. *See* CELL MEMBRANES.

Extraction. Many of the lipids, particularly glycolipids, phosphatides, and steroid esters, are present in the tissue in association with other cell components, such as protein. The nature of the linkages in these complexes is not fully understood although they are most certainly not covalent. Combinations of lipid and protein, like lipoproteins and proteolipids, have been isolated from sources such as blood serum and egg yolk. For the efficient extraction of lipids from tissues, these complexes must first be disrupted, and this is usually accomplished by dehydration procedures such as freeze-drying or acetone extraction, or by denaturation with alcohol before extracting the tissue with suitable lipid solvents. The use of mixed solvents, such as ethanol and ether, ethanol and benzene, or chloroform and methanol, allows disruption, dehydration, and extraction in one operation. Partial fractionation of lipids may be effected by successive extractions with different solvents.

Separation. Prior to 1948, solvent fractionations or the use of specific complexing agents, such as cadmium chloride for the separation of lecithin, were the only methods available for the separation of lipids, and progress in the chemical identification of individual lipids was slow. Since then, the use of column chromatography (particularly on alumina, silicic acid, and diethylaminoethyl cellulose), thin-layer chromatography, counter-current distribution, and gas-phase chromatography has allowed rapid progress in separation, and many new classes of lipid have been discovered as a result of the application of these techniques. Other methods which allow selective degradation of certain lipids either chemically (for example, the use of dilute alkali to cause preferential hydrolysis of esters) or enzymatically are also being used successfully for the separation of certain lipids. *See* CHROMATOGRAPHY.

Identification. Most of the conventional analytical procedures are used in the identification of lipid samples. Physical methods, such as infrared and ultraviolet spectroscopy, optical rotation, proton magnetic resonance, and mass spectrometry, have been used. The advent of gas-phase chromatography allowed a major advance in lipid chemistry by providing a complete analysis of the fatty acid content of lipid mixtures. The high resolving power of thin-layer chromatography has accelerated the routine analysis of the mixture of lipids from many natural sources; qualitative and quantitative methods for the determination of esters, aldehydes, carbohydrates, long-chain bases, and the other important components of lipids such as glycerol, glycerophosphate, ethanolamine, choline, serine, amino sugars, and inositol; and elementary analyses for nitrogen and phosphorus. The nitrogen-to-phosphorus ratio of a phosphatide sample was once used as a criterion of purity because carbon and hydrogen analyses on a single phosphatide species are not highly revealing, owing to the spectrum of fatty acids that may be present. The degree of unsaturation is determined by hydrogenation or iodine number determination,

and free hydroxyl groups are expressed by the acetyl value. Lipid samples also may be hydrolyzed under a variety of acid or alkaline conditions and the products investigated by paper chromatography. *See* CARBOXYLIC ACID; FAT AND OIL, EDIBLE; FAT AND OIL, NONEDIBLE; GLYCOLIPID; LIPID METABOLISM; PHOSPHATIDE; SPHINGOLIPID; STEROID; TERPENE; TRIGLYCERIDE; VITAMIN; WAX, ANIMAL AND VEGETABLE.   [ROY H. GIGG]

Bibliography: G. B. Ansell and J. N. Hawthorne, *Phospholipids*, 1964; D. Chapman, *Introduction to Lipids*, 1969; H. J. Deuel, *Lipids*, 3 vols., 1951–1957; D. J. Hanahan, *Lipide Chemistry*, 1960; R. T. Holman, W. O. Lundberg, and T. Malkin (eds.), *Progress in the Chemistry of Fats and Other Lipids*, 10 vols., 1952–1969; G. V. Marinetti (ed.), *Lipid Chromatographic Analysis*, 1967; E. J. Masoro, *Physiological Chemistry of Lipids in Mammals*, 1968; R. Paoletti and D. Kritchevsky (eds.), *Advances in Lipid Research*, 5 vols., 1963–1969.

Lipid metabolism

Lipids form a class of compounds composed of neutral fats (triglycerides), phospholipids, glycolipids, sterols, and fat-soluble vitamins (A, D, E, and K). As a group they share the common physical property of insolubility in water. This article describes the assimilation of dietary lipids and the synthesis and degradation of lipids by the mammalian organism.

Fat digestion and absorption. The principal dietary fat is triglyceride, an ester of three fatty acids and glycerol. This substance is not digested in the stomach and passes into the duodenum, where it causes the release of enterogastrone, a hormone which inhibits stomach motility. The amount of fat in the diet, therefore, regulates the rate at which enterogastrone is released into the intestinal tract. Fat, together with other partially digested foodstuffs, causes the release of hormones, secretin, pancreozymin, and cholecystokinin from the wall of the duodenum into the bloodstream.

Secretin causes the secretion of an alkaline pancreatic juice rich in bicarbonate ions, while pancreozymin causes secretion of pancreatic enzymes. One of these enzymes, important in the digestion of fat, is lipase. Cholecystokinin, which is a protein substance chemically inseparable from pancreozymin, stimulates the gallbladder to release bile into the duodenum. Bile is secreted by the liver and concentrated in the gallbladder and contains two bile salts, both derived from cholesterol: taurocholic and glycocholic acids. These act as detergents by emulsifying the triglycerides in the intestinal tract, thus making the fats more susceptible to attack by pancreatic lipase. In this reaction, which works best in the alkaline medium provided by the pancreatic juice, each triglyceride is split into three fatty acid chains, forming monoglycerides. The fatty acids pass across the membranes of the intestinal mucosal (lining) cells. Enzymes in the membranes split monoglyceride to glycerol and fatty acid, but triglycerides are reformed within the mucosal cells from glycerol and those fatty acids with a chain length greater than eight carbons: Short- and medium-chain fatty acids are absorbed directly into the bloodstream once they pass through the intestinal mucosa.

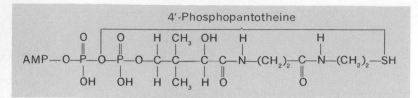

Fig. 1. Structural formula for coenzyme A.

The triglycerides formed in the mucosal cells are associated with proteins and phospholipids and pass into the intestinal lymphatics as the lipoproteins called chylomicrons. The thoracic duct, the main lymphatic channel, carries the chylomicrons to the great veins, where they enter the circulatory system and can be taken up and metabolized by most tissues. *See* BILE ACID; CHOLESTEROL; DIGESTIVE SYSTEM (VERTEBRATE); GALLBLADDER (VERTEBRATE); LIVER (VERTEBRATE); PANCREAS (VERTEBRATE).

The bile salts are absorbed from the ileum, together with dietary cholesterol. They enter the portal circulation, which carries them to the liver, where they can be reexcreted into the duodenum. The bile also contains pigments which are derived from hemoglobin metabolism. *See* BILIRUBIN; HEMOGLOBIN.

Abnormalities in the pancreatic secretions, bile secretion, or intestinal wall can lead to either partial or complete blockage of fat absorption and an abnormal excretion of fats in the stool.

Adipose tissue. Accumulations of fat cells in the body are called adipose tissue. This is found beneath the skin, between muscle fibers, around abdominal organs and their supporting structures called mesenteries, and around joints. The fat cell contains a central fat vacuole so large that it fills the cell, pushing the nucleus and cytoplasm to the periphery. The adipose tissue of the body is of considerable magnitude and is quite active metabolically, functioning as a buffer in energy metabolism. *See* ADIPOSE TISSUE.

During meals carbohydrates, amino acids, and fats which are absorbed in excess of immediate energy requirements are converted and stored in the fat-cell vacuoles as triglycerides. These fat depots provide an economical storage form for energy requirements and serve as the source of energy between meals and during periods of fasting.

Triglycerides cannot traverse the cell membrane until they are broken down into glycerol and fatty acids. Lipoprotein lipase, an enzyme which catalyzes the hydrolysis of lipoprotein triglycerides, controls entry of triglycerides into the cell; it is induced by feeding and disappears during periods of fasting. The lipase which breaks down triglycerides before they leave the cell is controlled by hormones. Norepinephrine, a hormone released from sympathetic nerve endings, is probably the most important factor controlling fat mobilization. Epinephrine from the adrenal medulla also activates lipase, as do adrenocorticotropic hormone (ACTH) and glucagon. Growth hormone and a separate fat-mobilizing factor released from the pituitary gland of some species also stimulate lipolysis and release of fatty acids from adipose tissue. While the fat-mobilizing factor may activate the

same enzyme as the hormones cited above, growth hormone appears to function through a different mechanism, probably requiring protein synthesis. *See* ADRENAL MEDULLA; ENDOCRINE MECHANISMS; GLUCAGON; HORMONE, ADENOHYPOPHYSIS.

Obesity is a condition in which excessive fat accumulates in the adipose tissue. One factor responsible for this condition is excessive caloric intake. The metabolic and psychological factors are under investigation. In starvation, uncontrolled diabetes, and many generalized illnesses the opposite occurs and the adipose tissue becomes markedly depleted of lipid. *See* DIABETES; METABOLIC DISORDERS.

Blood lipids. Lipids are present in the blood in the form of lipoproteins and as free fatty acids (FFAs), which are bound to albumin. The FFAs originate in adipose tissue and are mobilized for oxidation in other tissues of the body. The lipoproteins are small droplets of lipids complexed with proteins dispersed in the blood. They originate in the liver, with the exception of the small droplets, called chylomicrons, derived from the intestinal tract. The lipoproteins have been divided into arbitrary classes based upon their physical properties. The two methods of classification separate them according to density by ultracentrifugation and according to electrical charge by electrophoresis. The classes of lipoproteins defined by either of these methods differ in their content of triglycerides, cholesterol, and phospholipids. Some disorders of lipid metabolism are manifested in the blood by an abnormality in the pattern of lipoprotein classes. *See* CENTRIFUGATION (BIOLOGY); CLINICAL PATHOLOGY; ELECTROPHORESIS.

Fatty acid oxidation. Fatty acids are oxidized to meet the energy requirements of the body by a β-oxidation pathway, that is, the second or β-carbon from the carboxyl group is oxidized. Before oxidation can proceed, the fatty acid must be coupled with coenzyme A (CoA), a compound containing adenosinemonophosphate (AMP), two molecules of phosphate, and the complex nitrogen derivative pantotheine (Fig. 1).

In Eq. (1), showing activation, the fatty acid is as shown in Fig. 2, in which only the terminal 4-carbon atoms are given. The initial activation proceeds in two steps. First, as shown in Eq. (1a), the fatty acid reacts with adenosinetriphosphate (ATP) to form fatty acid adenylate (fatty acid—AMP). This in turn reacts with CoA to form acyl CoA, as shown in Eq. (1b). The enzyme that cata-

$$(CH_3)_3-N^+-CH_2-CHOH-CH_2-COO^-$$

Fig. 3. Structural formula for carnitine.

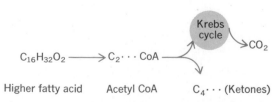

Fig. 4. Fates of acetyl CoA from fatty acid oxidation.

LIPID METABOLISM

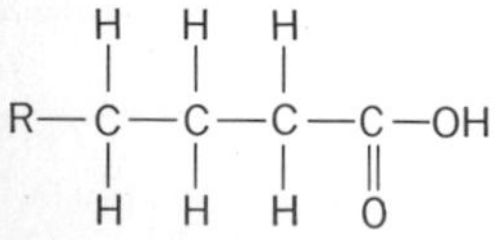

Fig. 2. Structural formula for a fatty acid. R represents chain of up to 14 carbon atoms.

lyzes reactions (1*a*) and (1*b*) is called the activating enzyme or thiokinase. The fatty acid—CoA cannot penetrate the mitochondrial membrane to gain access to the oxidative enzymes. To traverse the membrane, fatty acid must be esterified with the hydroxyl group of carnitine (Fig. 3), a reaction catalyzed by carnitine acyl transferases. The carnitine—fatty acid ester traverses the membrane, and the fatty acid is again coupled to CoA by a transferase on the other side.

In reaction (2), the first of two oxidation reactions, one hydrogen is removed from the α-carbon and one is removed from the β-carbon, producing an unsaturated fatty acid analogous to crotonic acid. The enzyme is called acyl CoA dehydrogenase, and the cofactor which acts as the oxidizing agent and accepts the hydrogen is flavin adenine dinucleotide (FAD).

In reaction (3), hydration, hydrogen and the hydroxyl group (from water) are introduced at the α- and β-carbons, respectively; a β-hydroxy acid is formed; the enzyme is crotonase.

In (4), the second oxidation, two hydrogens are removed from the α- and β-carbons of the β-hydroxy acid to form a keto acid. The oxidizing agent is diphosphopyridine nucleotide (DPN), also called nicitinamide adenine dinucleotide (NAD).

In reaction (5), showing cleavage, acetyl CoA is split off. Thus a fatty acid two carbons shorter than the original one is formed. Simultaneously, the shortened fatty acid combines with another mole-

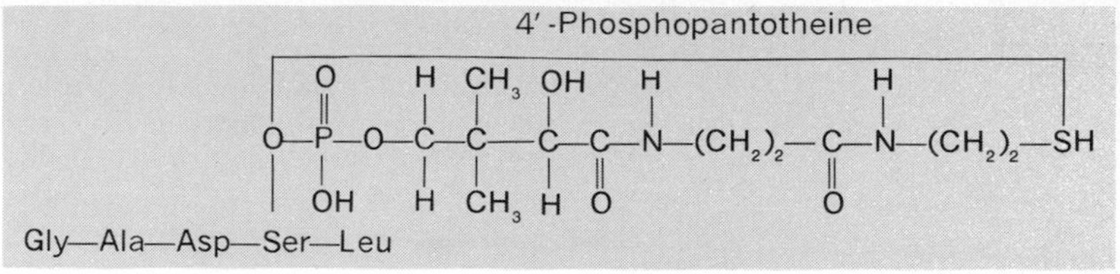

Fig. 5. Active site of acyl carrier protein (ACP). Gly, glycine; Ala, alanine; Asp, aspartic acid; Ser, serine; and Leu, leucine.

cule of CoA at the newly formed carboxyl group to yield a new activated fatty acid. The enzyme is β-ketothidase. The new activated fatty acid undergoes the same series of reactions, and this process is repeated until the entire fatty acid molecule has been reduced completely to acetyl CoA. Oxidation of the higher fatty acids, such as palmitic acid with 16 carbon atoms, yields eight molecules of acetyl CoA. Two fates of acetyl CoA formed in the liver are of immediate interest (Fig. 4): (1) oxidation through the Krebs cycle to carbon dioxide and (2) condensation of two molecules of acetyl CoA with the splitting off of the CoA forming so-called ketone bodies. *See* BIOLOGICAL OXIDATION; DIPHOSPHOPYRIDINE NUCLEOTIDE (DPN); KREBS CYCLE.

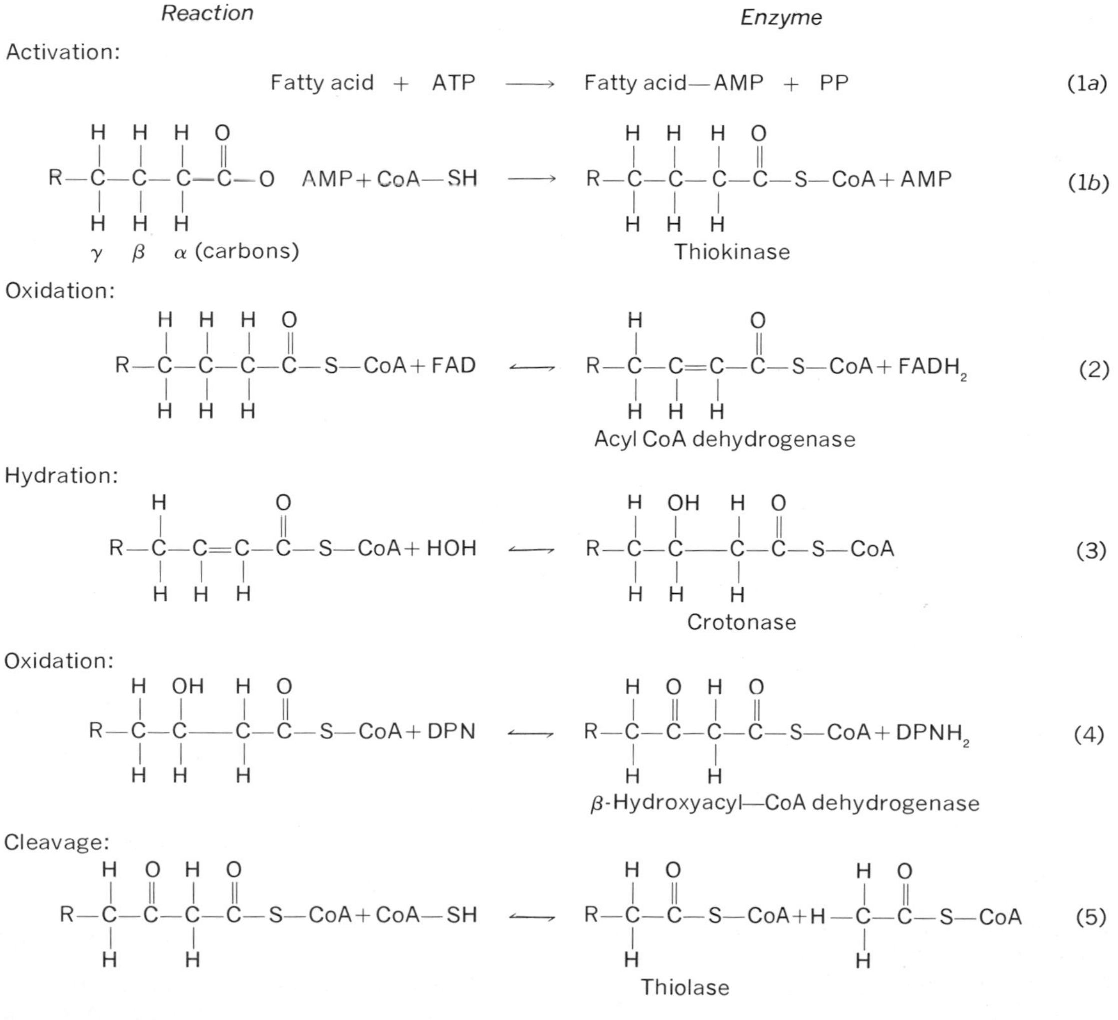

Ketone body formation. If excessive amounts of fatty acids are mobilized from the depots and metabolized by the liver, as occurs in starvation or diabetes, all the acetyl CoA formed cannot be oxidized. Instead, two molecules of acetyl CoA may form acetoacetic acid, which cannot be further oxidized by the liver but is transformed into β-hydroxybutyric acid or acetone according to reactions (6) and (7). These three compounds, aceto-

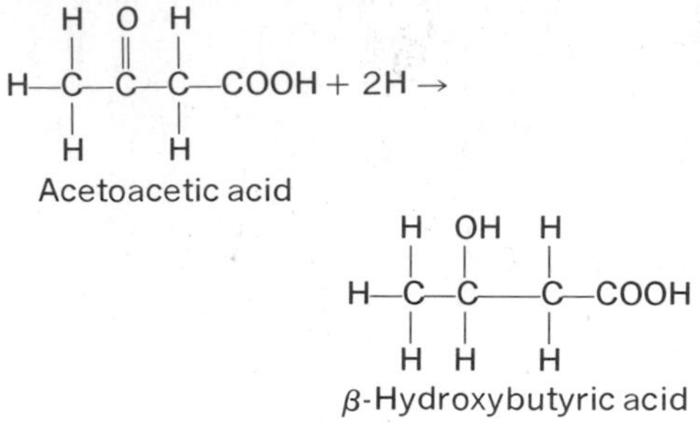

$$\underset{\text{Acetoacetic acid}}{H-\underset{\underset{H}{|}}{\overset{\overset{H}{|}}{C}}-\overset{\overset{O}{\|}}{C}-\underset{\underset{H}{|}}{\overset{\overset{H}{|}}{C}}-COOH} + 2H \rightarrow$$

$$\underset{\beta\text{-Hydroxybutyric acid}}{H-\underset{\underset{H}{|}}{\overset{\overset{H}{|}}{C}}-\underset{\underset{H}{|}}{\overset{\overset{OH}{|}}{C}}-\underset{\underset{H}{|}}{\overset{\overset{H}{|}}{C}}-COOH} \quad (6)$$

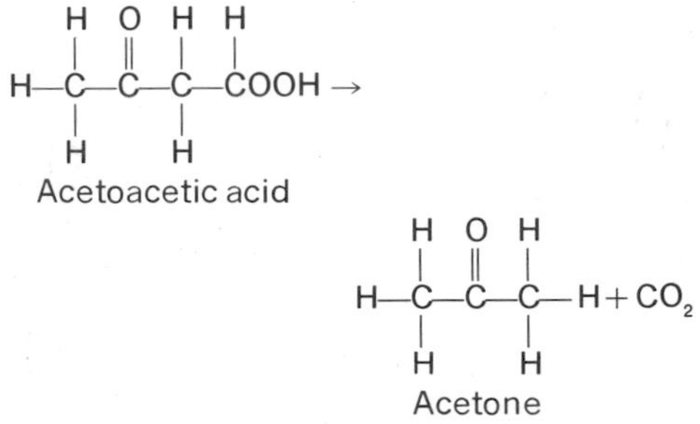

$$\underset{\text{Acetoacetic acid}}{H-\underset{\underset{H}{|}}{\overset{\overset{H}{|}}{C}}-\overset{\overset{O}{\|}}{C}-\underset{\underset{H}{|}}{\overset{\overset{H}{|}}{C}}-COOH} \rightarrow$$

$$\underset{\text{Acetone}}{H-\underset{\underset{H}{|}}{\overset{\overset{H}{|}}{C}}-\overset{\overset{O}{\|}}{C}-\underset{\underset{H}{|}}{\overset{\overset{H}{|}}{C}}-H} + CO_2 \quad (7)$$

acetic acid, acetone, and β-hydroxybutyric acid, called the ketone bodies, are formed in excessive amounts in the liver not only because of the excess acetyl CoA present but also because the metabolic disturbances caused by such conditions reduce the capacity of the Krebs cycle for oxidation by diverting oxaloacetate, a key intermediate, to gluconeogenesis (synthesis of glucose from noncarbohydrate precursors).

The excessive formation of ketone bodies leads to a high concentration in the blood—a condition called ketonemia. This causes ketone body excretion in the urine, together with water, sodium, and potassium. These losses, along with the excessive acidity of the blood caused by the high concentration of ketone bodies (ketoacidosis), can lead to coma and death if uncorrected. Such fatal acidosis was quite common in diabetics before the advent of insulin. During starvation or diabetic ketosis, muscles are capable of deriving the major portion, if not all, of their energy from oxidizing the ketone bodies. However, in the uncontrolled diabetic subject, the formation of ketone bodies may greatly exceed the utilization of them by the muscle or other tissues, so that ketoacidosis results.

Lipotropic factors. A diet high in fat results in fat accumulation in the liver. This can be prevented by feeding choline, a phosphatide base; the effect has been termed a lipotropic effect. If methionine is present in the diet, it can substitute for choline, since enzymes are present in the liver which can transfer methyl groups from methionine to ethanolamine to form choline, as shown in Eq. (8).

$$\underset{\substack{\text{Methyl (from} \\ \text{methionine)}}}{3CH_3{}^-} + \underset{\beta\text{-Ethanolamine}}{H_2NCH_2\cdot CH_2OH} \rightarrow$$

$$\underset{\text{Choline}}{(CH_3)_3\cdot\overset{+}{N}CH_2\cdot CH_2OH} \quad (8)$$

Fatty livers can also occur when large amounts

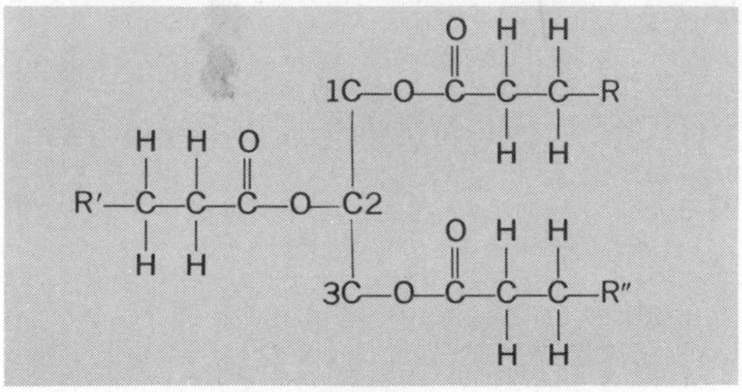

Fig. 6. General structural formula for a triglyceride.

of depot fat are mobilized in starvation and diabetes. This type of fatty liver is not affected by choline. Since choline is a constituent of the phospholipid lecithin, it has been suggested that the lipotropic effect of choline is exerted through lecithin. Evidence to support this suggestion has not been forthcoming.

Fatty acid synthesis. Certain carbohydrates, such as glucose and fructose, yield simpler substances, such as lactic acid, pyruvic acid, and acetic acid, as do proteins, which are potential precursors of acetyl CoA. The acetyl CoA molecules thus formed are used to synthesize higher fatty acids.

The initial reaction in fatty acid synthesis is the carboxylation of acetyl CoA to form malonyl CoA. This reaction is catalyzed by acetyl CoA carboxylase, a biotin-containing enzyme (enz-biotin) and probably proceeds in two steps. In the first, reaction (9a), biotin is carboxylated in the 1′-nitrogen position in an ATP-dependent reaction to form 1′-N-carboxamide biotin. In the second step, reaction (9b), the carboxyl group is transferred to acetyl CoA to form malonyl CoA.

For the synthesis to proceed, the malonyl group must be transferred from CoA to a specific carrier protein called the acyl carrier protein (ACP). The compound with which the malonyl group is to be condensed is similarly transferred from CoA to ACP to form an acyl ACP. The acyl group can be acetate or any fatty acid less than 16 carbons in length, depending upon the species. The enzyme catalyzing the transfer from CoA to ACP is specific for each compound and is called a CoA-ACP transacylase. The active group of both CoA and ACP is 4′-phosphopantetheine (Fig. 5). This group is linked to the protein through the hydroxyl group of serine by a phosphodiester bond. The sulfhydryl group is the functional group of both CoA and ACP linkage occurring through sulfur (thiol ester).

The joining of malonyl ACP with an acyl ACP,

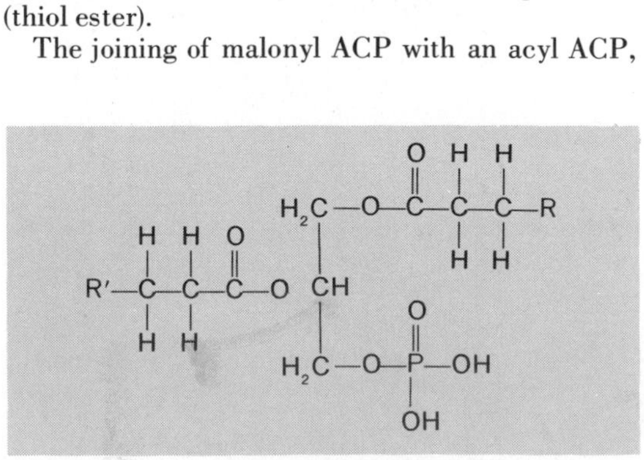

Fig. 7. Structural formula for a phosphatidic acid.

as shown in reaction (10), is catalyzed by the acyl-malonyl ACP condensing enzyme. Since carbon dioxide is released in this reaction, the acyl carbon chain is lengthened by two, resulting in a β-keto acid linked to ACP. This compound is converted to a fatty acid by the reverse of reactions (4), (3), and (2). These reactions are catalyzed by a series of enzymes which are associated in a complex protein called fatty acid syntase in yeast and pigeon liver. The individual enzymes have been isolated from bacterial systems and are specific for ACP compounds, in contrast to the specificity of oxidative enzymes for CoA compounds. The reverse reaction (4) is catalyzed by β-keto ACP reductase and the cofactor is triphosphopyridine nucleotide (TPN), also known as nictinamide adenine dinucleotide phosphate (NADP), rather than NAD. Reverse reaction (3) is catalyzed by enoyl ACP dehydrase and reverse reaction (2) by enoyl ACP reductase. *See* TRIPHOSPHOPYRIDINE NUCLEOTIDE (TPN).

Repetition of this series of reactions leads to a building up of the fatty acid molecule to a chain containing 16 or 18 carbons. When the fatty acid reaches this length, it is transferred from ACP to CoA. The CoA is split off by an enzyme, deacyclase, yielding the free fatty acid, which in turn combines with glycerol to form phospholipids and the triglyceride that is stored in depots and circulates in lipoproteins.

Triglyceride synthesis. Triglycerides, which consist of three fatty acids joined to glycerol by ester bonds (Fig. 6), are synthesized in the liver from glycerol phosphate and CoA derivatives of fatty acids. A series of enzymes called transferases catalyze the joining of the first two fatty acids with glycerol phosphate to form a phosphatidic acid (Fig. 7). The phosphate group must be removed to form a 1,2-diglyceride before the third fatty acid can be transferred to form a triglyceride. The removal of phosphate is catalyzed by a specific phosphatase. In the intestinal mucosa, triglycerides can be synthesized from monoglycerides without passing through phosphatidic acid as an intermediate.

Modification of dietary lipids. The fatty acids in the depots of each mammalian species have a characteristic composition. The chain length of ingested fat is either lengthened or shortened by the same pathways used for the synthesis and degradation of fatty acids so that the composition of the depots remains constant. However, when large amounts of a fat of markedly different composition from the animal's depot fat is ingested, the depots slowly change to reflect the characteristics of the ingested fat. The ability to introduce double bonds or reduce them to form an unsaturated or a saturated fatty acid is limited; highly unsaturated fatty acids can only be derived from dietary sources.

Essential fatty acids. Most mammals, including man, probably require arachadonic acid for normal growth and for maintenance of the normal condition of the skin, although the mechanism of these effects has not been elucidated. Linoleic acid, a precursor of arachadonic acid in mammals, is the usual dietary source of unsaturated fatty acid. Highly unsaturated fatty acids cannot be synthesized by mammals from either saturated fatty acids or their precursors. Hence, the so-called essential fatty acids must be present in the diet.

Role of phosphatides. The phosphatides are compounds of glycerol, fatty acids, phosphate, and certain bases containing nitrogen, namely, ethanolamine, serine, and choline, the so-called phosphatide bases. When combined with fatty acids, they tend to make the insoluble fatty acids soluble in water. They are important constituents of both outer cell membranes (plasma membranes) and the network of cytoplasmic membranes (endoplasmic reticulum). *See* CELL MEMBRANES.

Phospholipid synthesis. In the liver, phospholipid synthesis follows the same pathway as triglyceride synthesis to the point of diglyceride formation. The nitrogenous base to be coupled at the 3 position of glycerol must first be phosphorylated by ATP and then reacted with cytidine triphos-

$$\text{ATP} + \text{CO}_2 + \text{Enz-biotin} \longrightarrow \text{Enz-biotin}-\text{CO}_2 + [\text{1'-}N\text{-Carboxamide biotin}] + \text{ADP} + \text{P}_i \quad (9a)$$

(1'-*N*-Carboxamide biotin)

$$\text{Enz-biotin}-\text{CO}_2 + \text{CH}_3-\overset{O}{\underset{}{C}}-\text{SCoA} \longrightarrow \text{HO}-\overset{O}{\underset{O}{C}}-\text{CH}_2-\overset{O}{\underset{}{C}}-\text{SCoA} + \text{Enz-biotin} \quad (9b)$$

Acetyl CoA   Malonyl CoA

$$\text{R}-\text{C}-\text{C}-\text{C}-\text{S}-\text{ACP} + \text{HO}-\text{C}-\text{C}-\text{C}-\text{S}-\text{ACP} \rightleftharpoons \text{R}-\text{C}-\text{C}-\text{C}-\text{C}-\text{C}-\text{S}-\text{ACP} + \text{CO}_2 + \text{ACPSH} \quad (10)$$

Acyl ACP   Malonyl ACP β-Keto acid—ACP

phate to form a cytidine diphosphoryl derivative. This derivative can then be coupled with the diglyceride in a reaction catalyzed by a specific transferase to form a phospholipid. *See* LIPID.

[MARTIN A. RIZACK]

Bibliography: D. S. Fredrickson, R. I. Levy, and R. S. Lees, Fat transport in lipoproteins: Integrated approach to mechanisms and disorders, *New Engl. J. Med.*, 276:34–44, 1967; D. M. Greenberg (ed.), *Metabolic Pathways*, 1968; H. A. Krebs, The regulation of the release of ketone bodies by the liver, *Advan. Enzyme Regul.*, 4:339–353, 1966; A. E. Renold and G. F. Cahill, Jr. (eds.), *Handbook of Physiology*, 1965.

Lipostraca

An order of the subclass Branchiopoda erected by D. J. Scourfield for the Middle Devonian fossil *Lepidocaris rhyniensis*. This organism was small and blind, and was an inhabitant of fresh water. It resembled present-day anostracans in some details but differed in others; for example, it had biramous antennae, the trunk appendages lacked gills, and the first three pairs of appendages were modified for feeding, whereas the remaining ones were adapted for swimming. *See* BRANCHIOPODA.

[CHARLES B. CURTIN]

Bibliography: D. J. Scourfield, On a new type of crustacean from the Old Red Sandstone (Rhynie chert bed, Aberdeenshire), *Phil. Trans. Roy. Soc. London*, Ser. B, 214:153–187, 1926.

Liquefaction of gases

Liquefied gases are used in industry, in various military operations, and in scientific research. Although the steel, chemical, and refrigeration industries are the largest users of such liquids, the space development programs account for an increasingly significant share of the annual production. Relatively small amounts of the various liquefied gases are used in scientific investigations, primarily for establishing low-temperature environments in which studies at temperatures below ambient may be carried out. *See* CRYOGENIC ENGINEERING; CRYOGENICS; GAS.

Several processes have been developed for liquefying gases. Although significant differences in efficiency exist among them, difficulties encountered in reducing the more efficient to practice have made them all roughly competitive. Consequently, the choice of the refrigerating cycle to be used for liquefaction of a given gas is made largely on the basis of convenience. In large industrial liquefaction plants, however, a significant reduction in production costs may be realized by use of the more efficient cycles. This has provided an incentive for improving the actual efficiencies of these cycles, with the result that high-capacity plant designs are usually based upon the most efficient thermodynamic process.

Although any thermodynamic process which results in a lowering of the temperature could in principle be utilized in the liquefaction of some gas, only four have found practical employment. These are, in order of increasing complexity, vapor compression; use of refrigerators to cool the gas at constant pressure until liquefaction occurs; utilization of the adiabatic performance of work by the gas in a cyclic process with regenerative cooling;

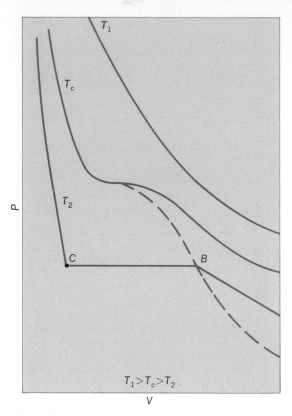

Fig. 1.   Pressure-volume (*P-V*) diagram for typical gas.

and use of the Joule-Thomson process in a cycle with regenerative cooling. Combinations of these four processes have also often been found effective.

Vapor compression. In order to liquefy any gas by compression, its temperature must be below its critical temperature, a requirement first recognized by Thomas Andrews (1863). The process of vapor compression can be most easily elucidated with the help of the pressure-volume (*P-V*) diagram shown in Fig. 1, in which several typical isotherms are plotted. The upper one for a temperature in excess of the critical temperature $T > T_c$ shows that as the gas is compressed, the pressure rises monotonically with no change of phase no matter how much the gas is compressed. The $T = T_c$ isotherm divides the *P-V* diagram such that at constant temperatures below T_c, and at certain definite pressures and volumes for each isotherm (the locus of these points lies on the dotted curve, for example, point *B*), no further increase in pressure with decreasing volume is observed. Instead, a liquid phase progressively forms with decreasing volume until, at point *C*, the system is all liquid. Further decrease of the volume then leads to a sharply rising value of the pressure due to the much smaller compressibility of the liquid. Assuming the critical temperature of a gas to lie above ambient, liquefaction of a gas by vapor compression consists simply of compressing it isothermally until it liquefies.

A procedure whereby a liquefied gas at a temperature below ambient can be produced using a vapor-compression process is best described with the help of a temperature-entropy diagram for a gas. In Fig. 2, the decrease in entropy *S* with tem-

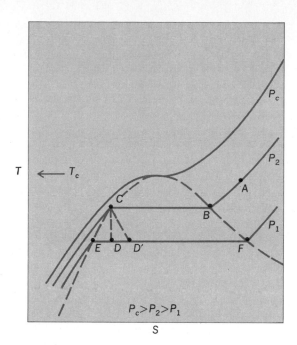

Fig. 2.   Temperature-entropy (*T-S*) diagram for a gas.

perature T for a gas held at constant pressure is shown by the various curves labeled P_1, P_2, and P_c. At some temperature less than T_c, for every pressure less than P_c (for example, point B), liquid will begin to form at constant temperature with further decrease in volume and continued withdrawal of heat from the gas. Let it now be assumed that the vapor has been entirely compressed to a liquid (path $A \to B \to C$ in Fig. 2). If at point C the confining pressure, that is, the vapor pressure, on the compressed liquid is in excess of atmospheric pressure, the liquid can be isolated thermally from its surroundings and the confining pressure slowly reduced. The liquid will then boil, but due to its thermal isolation, the heat of vaporization can be supplied only by the sensible heat of the liquid itself. The whole liquid is therefore cooled at the expense of vaporizing a part of it. On the T-S diagram, Fig. 2, this result is illustrated by the path $C \to D$, assuming the pressure reduction is carried out slowly and hence quasi-reversibly. For such an isentropic expansion, this diagram provides a direct determination of the remaining amount of liquid. The fraction of gas produced is given by the length ratio ED/EF; that for the liquid is DF/EF. In practice, the expansion of the compressed liquid is carried out irreversibly through a valve. Under these conditions, a larger fraction of liquid is evaporated as shown by the path $C \to D'$. *See* ENTROPY.

When refrigeration is the objective of the process outlined above (which is frequently the case), the gas formed in the throttling stage as well as that originating from evaporation of the cold liquid is returned to the compressor intake, and the cycle just described is repeated. By balancing liquid evaporation due to heat leak with liquid production by expansion, steady conditions can be achieved in the low-temperature reservoir. Refrigerators of this type are useful in the liquefaction of still other gases. If the cold liquid produced is withdrawn as product and used elsewhere, a liquefaction rather than a refrigeration cycle is the re-

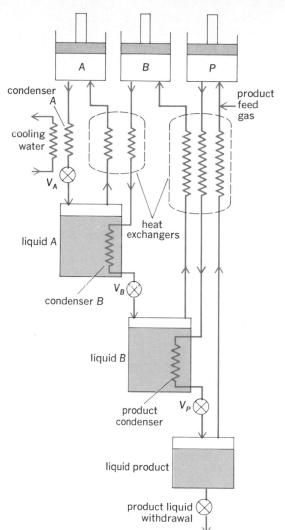

Fig. 3.   Schematic of gas liquefaction by a cascade of vapor-compression refrigerators.

sult. Make-up gas in an amount equivalent to that liquefied must then be continuously introduced to the cycle at the compressor intake.

Refrigeration. When the critical point of a gas lies below ambient temperature, a refrigerator can always be used to cool the gas sufficiently to liquefy it by the vapor-compression process described above. Refrigerators are also employed to precool the feed gas in other liquefaction cycles, producing thereby a more nearly reversible and hence more efficient process. Because not only different uses but also many types of refrigerators exist, the particular one chosen for any given application requires a careful matching of process requirements to refrigerator characteristics. In this section, therefore, it will be possible only to outline the basic aspects of gas liquefaction using refrigerators.

In any refrigerating machine with gas streams flowing from high- to low-temperature regions and conversely, it is necessary to prevent these circulating gases from transporting heat energy, thus introducing large and undesirable heat leaks into the cycle. By using a device called a heat exchanger, heat energy can be transferred across the

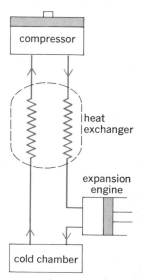

Fig. 4.   Schematic flow diagram of a gas refrigerator.

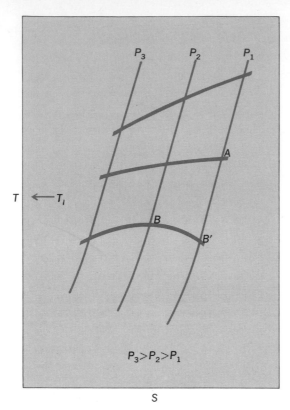

Fig. 5. Temperature-entropy (*T-S*) diagram for a typical gas (high-temperature region).

medium separating the gas streams in such a way that the warm stream is cooled and the cold stream warmed over essentially the same temperature interval while, ideally, no heat energy flows along the temperature gradient of the exchanger. A simple heat exchanger consists of two concentric copper tubes through which countercurrent gas streams flow, with heat being transferred across the copper surface adjacent to both streams.

A particularly simple refrigerator is a continuously replenished bath of a liquefied gas boiling under reduced pressure, and hence at a lower temperature than the normal boiling point, as described in the previous section. By using several of these baths, each boiling at a lower temperature and each acting in turn as a condenser for the one with the next lowest boiling point, a cascade-type liquefier may be assembled, as illustrated in Fig. 3. *See* REFRIGERATION.

Performance of external work. The maximum possible cooling effect resultant from any spontaneous change in the state of a substance is obtained when that change is caused to occur isentropically, that is, adiabatically and reversibly. This means that the change in question must be carried out in such a way that no heat exchange takes place between the substance and its surroundings, and the internal forces tending to produce the change are at all times balanced by corresponding external forces; that is, throughout the change, the system remains in equilibrium. In practice, such processes can be approximated but never actually realized. The isentropic process most often utilized in refrigerators and liquefiers

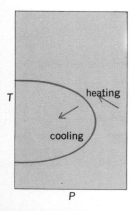

Fig. 6. Inversion temperature versus pressure for a gas.

as a basis for cycle design is the reversible adiabatic expansion of a gas. The adiabatic requirement may be satisfied relatively easily by standard methods of thermal insulation as well as by causing the process to occur so rapidly that there is insufficient time available for transfer of heat from surroundings to the substance. A reversible expansion of a gas from a high to a low pressure can be approximated with either a piston- or turbine-type expander. The former has been the more widely used; the latter, although the more efficient, handles such large volumes of gas in proportion to its size that its development and actual use has depended upon the need for large production facilities for liquefied gases. Widespread use of both types of expanders had at one point been retarded by lack of suitable methods of lubrication for moving parts and bearings at very low temperatures. In order to maintain the high theoretical efficiency inherent in this type of cycle, heat generation by friction in these expanders must be minimized, because all temperature increases produced by a given amount of heat are usually appreciably larger at low temperatures than at high.

The simplest type of refrigerating machine employing the above principles is the gas refrigerator shown in Fig. 4. In an actual machine of the type shown, expansion of air from 40 to 1 atm pressure when the gas temperature at the expander intake is 200 K results in an exit temperature of about 100 K. In principle, this refrigerating machine could also serve as a liquefier for the working fluid. If the expander were adequately insulated, its temperature would fall progressively with continued operation, and eventually the gas would liquefy in the expander itself. While this is usually avoided to prevent loss of efficiency and erosion, some expanders are operated in this way. The production of the actual liquid is usually carried out either by vapor compression or the Joule-Thomson process. Discussion of the latter is reserved until after the next section, but an application of the principles developed is illustrated by the A. D. Little hydrogen liquefier. In this compact device, the working substance, helium gas, circulates through two reciprocating expansion engines operating at two different temperature levels. The exit temperature from the lower-temperature engine is maintained at approximately the boiling point of hydrogen. The hydrogen gas to be liquefied is cooled by passing it through the heat exchanger counter to the expanded cold helium until its temperature is reduced sufficiently to permit liquefaction by vapor compression.

Joule-Thomson expansion. Under certain conditions, an irreversible adiabatic expansion of a gas (or a liquid) through a throttling device such as a valve can result in a decrease in temperature of the effluent gas. A thermodynamic analysis shows that in this process, called a Joule-Thomson expansion, the heat content or enthalpy of the gas is unchanged. If the temperature at which the expansion is carried out is above a particular temperature for each gas, the gas effluent from the throttle valve will be warmer than that at the valve inlet; below this temperature, called the inversion temperature, expansion will result in a cooling of the gas. In order to use this cooling effect in gas

Inversion temperatures of common gases

Gas	T_i,K
Helium	51
Hydrogen	205
Neon	242
Nitrogen	621
Argon	723
Oxygen	893

liquefaction, the expansion temperature must always be below the upper inversion temperature. The table shows the inversion temperatures of a number of common gases, from which it will be noted that neon, hydrogen, and the helium isotopes are the only gases not below their inversion points at ordinary temperatures. Precooling before expansion is necessary only with these three gases in order to obtain refrigeration from the Joule-Thomson process. In Fig. 5, a simplified TS plot of a typical gas is presented. On it is shown the variation of S with T for three different pressures. The heavy lines represent constant enthalpy lines, and hence, display the temperature history (as well as the entropy history) of a gas in an isenthalpic expansion. Above the point marked T_i, it is seen that a Joule-Thomson expansion always results in a temperature rise, no matter what the initial pressure. At T_i, and here only for very low pressures, for example, below point A, there is no change in T upon expansion. Below T_i, there is a drop in the temperature

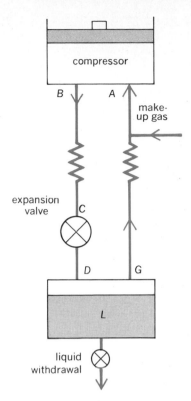

Fig. 8. The Linde gas liquefaction cycle.

upon expansion first at very low pressures, and then with further decrease of the temperature at increasing pressures (path $B \rightarrow B'$). At low temperatures, the pressure at which cooling occurs upon expansion again decreases. If these temperatures at which the inversion occurs are plotted against the pressure, a roughly parabolic curve such as is shown in Fig. 6 is obtained. Expansion from any point within the curve always produces cooling. From a point outside, a heating occurs initially until the pressure and temperature of the gas are reduced sufficiently to bring it within the curve.

In considering this process for gas liquefaction, it must be mentioned that, at least until the expansion is made to occur at very low temperatures, the Joule-Thomson process is a far less effective cooling process than isentropic expansion. By cycling the gas successively through a compressor, a heat exchanger, and an expansion valve, however, the cooling obtained from the passage of each mass of gas through the valve can be made cumulative. In this way, the temperature of the expansion valve is continuously reduced until finally enough refrigeration is produced to liquefy a fraction of the effluent gas. By separating this liquid and returning the remainder to the compressor through the heat exchanger, the rate of cooling of the expansion valve begins to decrease, and eventually, a steady state is reached at which the refrigeration removed as liquid just balances that produced in the expansion. In Fig. 7, this steady state is represented on a T-S diagram. The ratio DG/LG is the fraction of liquid produced per mole of gas passing through the Joule-Thomson valve; the fraction DL/LG represents moles of gas returning to the

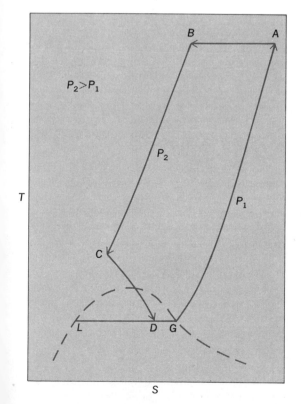

Fig. 7. Joule-Thomson liquefaction cycle represented on a temperature-entropy (T-S) diagram (high-temperature and two-phase region).

compressor, at which point make-up gas is added to the system in an amount necessary to maintain steady state conditions throughout the cycle. The great attraction of the Joule-Thomson process is its lack of moving parts at low temperatures. Furthermore, it provides a satisfactory method of producing the liquid and separating it from the gas stream.

The Joule-Thomson process is used in many different ways in the practice of gas liquefaction. The simplest is that first developed by Carl von Linde, shown in Fig. 8. The letters attached to the drawing correspond to the state of the gas as shown in Fig. 7. In practice, gas purifiers must be introduced downstream of the compressor in order to prevent condensible impurities from clogging the heat exchanger. This basic cycle has been adapted in many ways, of which a few are (1) use of an expansion engine, a vapor compression, or a gas refrigerator to precool the gas feed to the Joule-Thomson valve, (2) use of independent high- and low-pressure circuits with or without additional precooling, and (3) use of closed-circuit refrigeration with liquefaction by heat exchange with the working substance. The irreversible expansion of the liquid through a throttle valve mentioned in the section on vapor compression is an isenthalpic or Joule-Thomson expansion.

In the discussion of the four basic methods of gas liquefaction given above, no association of a particular gas with one or more liquefaction schemes has been made. This omission has been deliberate because, with few exceptions, all schemes will work with all gases, and therefore, as mentioned earlier, the choice of method is governed by other considerations.

Storage of liquefied gases. Liquefied gases at temperatures below ambient could be stored indefinitely if it were possible to prevent heat influx to the liquid from the warmer surroundings. No perfect heat insulators exist, however; hence, heat constantly flows, at a rate dependent upon the quality of the insulation, into all such liquefied gas reservoirs. If the container is vented to the atmosphere, this heat leak causes continuous evaporation of the reservoir contents. If the container is sealed, the heat leak will cause the temperature of the liquid to rise, with concomitant increase in the pressure, until either the temperatures of the liquid and surroundings have equalized or until the increased pressure has ruptured the container. Because sufficiently effective insulation is available to reduce boil-off losses to economically tolerable levels, atmospheric venting is common practice in most large storage facilities. Occasionally, however, because of hazard, value of the product, or liquefaction costs, no-loss storage of liquefied gases is required. This may be accomplished for short periods of time by simply plugging the vent line and allowing the temperature of (and pressure in) the reservoir to rise. For longer periods, refrigeration at the temperature level of the liquefied gas must be supplied to balance the heat leak to the container.

Liquefied gas containers range in size from tiny glass laboratory dewars for specialized studies with small amounts of rare liquids to reservoirs containing hundreds of thousands of gallons for use in industry. The insulation employed depends entirely upon the proposed usage. In a missile tank, for example, the storage time is short; insulation can therefore be almost eliminated, especially because minimization of propellant tank weight is a prime consideration. For storage of liquid helium, however, where reliquefaction is costly and the gas supply itself is not unlimited, use of the best insulation available is usually justified. *See* ATMOSPHERIC GASES, PRODUCTION OF; HELIUM, LIQUID; LOW-TEMPERATURE PHYSICS.

[EDWARD F. HAMMEL]

Bibliography: M. M. Davies, *The Physical Principles of Gas Liquefaction and Low Temperature Rectification*, 1949; S. Fluegge (ed.), *Handbuch der Physik*, vol. 14, 1956; K. D. Timmerhaus, *Advances in Cryogenic Engineering*, vols. 1–20, 1954, 1956–1975.

Liquefied natural gas (LNG)

A product of natural gas which consists primarily of methane. Its properties are those of liquid methane, slightly modified by minor constituents. One property which differentiates liquefied natural gas (LNG) from liquefied petroleum gas (LPG), which is principally propane and butane, is the low critical temperature, about −100°F. This means that natural gas cannot be liquefied at ordinary temperatures simply by increasing the pressure, as is the case with LPG; instead, natural gas must be cooled to cryogenic temperatures to be liquefied and must be well insulated to be held in the liquid state.

Peak shaving system. The earliest commercial use for LNG was for storage to meet winter home-heating peak demands. In the construction of long-distance gas pipelines normal practice is to size the line somewhat larger than the average yearly demand rate. Capacity can be increased by 15–20% by operating all compression equipment at the maximum possible flow rate. However, during the coldest days of winter the demand for heating fuel is extremely high and some temporary additional supply of gas is required. Several different techniques are available to supply this temporary peak demand. One reasonable approach is to collect gas during the summer months when demand is low and to store it until winter when it can be withdrawn to meet demands. Since 600 ft³ of natural gas condenses to less than 1 ft³ of liquid, this form of storage is relatively convenient.

Such a peak shaving system was constructed in Cleveland, Ohio, in the early 1940s and operated successfully for several years until one of the storage tanks developed a leak and the escaping liquid caught fire, producing a very destructive conflagration. It was almost a generation before this approach to peak shaving was used again. In the many LNG peak shaving systems that have been built since then, careful attention has been given to separation of storage tanks and other equipment so that even complete destruction by fire of one storage tank would not be likely to cause damage in any other area of the plant.

The large insulated storage tanks constitute the main cost element in LNG peak shaving. Conventional double-walled metal or prestressed concrete tanks with perlite powder insulation between the walls have proved to be satisfactory for units of all sizes. In the development of LNG technology, a number of novel storage schemes have been attempted, but all have weaknesses that result in unsafe or uneconomical operations. The most radical of these attempts utilized frozen earth to form the tank lining for an excavated reservoir. Thermal insulation would theoretically be provided by the earth itself. With large enough tanks, the ratio of surface area to volume declines sufficiently so that the heat-conduction rate of natural soil results in a tolerably low evaporation rate of the liquefied gas. Several attempts to exploit this scheme have failed because the intrusion of an underground stream into the frozen earth area created an unacceptable increase in heat input and liquid evaporation rate. A second weakness of this scheme is that the soil around the tank becomes saturated with hydrocarbon vapors, greatly increasing the potential for an uncontrollable conflagration should a fire start in the tank area.

Systems relying on reservoirs lined either with special concrete or with plastic films or foams have all failed. In these experiments the unequal contraction of the liner during cool-down to the cryogenic temperature of the LNG has caused cracks to open in the liner. Subsequent loss of LNG into the insulation space or into the soil around the tank has made the projects both unsafe and uneconomical.

A tragic accident occurred at an installation on Staten Island, NY, in 1973. Workers were attempting to repair the plastic film liner in the tank. The tank had been emptied and held out of service for a year before the repair work was started. Somehow, a fire originated in the equipment being used inside the tank. The fire spread, and eventually the roof collapsed; all of the men working in the tank were killed. The circumstances were unlike those of the earlier accident in Cleveland, for no LNG was in the tank or anywhere in the plant site at the time. No LNG facility has had a fire or other serious problem since the Cleveland disaster. With more than 100 LNG storage facilities operating around the world, this is an exemplary safety record.

Liquefaction equipment represents a relatively small portion of total project cost and has not produced much variation in design. One novel approach has been used successfully by the San Diego Gas and Electric Co. The peak shaving plant is located adjacent to a large electrical generating station, which is fueled by natural gas supplied at transmission-line pressure of about 600–800 psi. The gas is first processed by the liquefier through heat exchangers and expansion turbines. A fraction of the gas is liquefied, and the remainder returns through the heat exchangers to pass on to the electric generating station at low pressure. In this way the liquefaction plant uses the otherwise wasted pressure of the pipeline. The maximum amount of LNG which can be produced in any given location is limited to at most 20% of the quantity of gas which drops from pipeline pressure to low pressure.

Ocean transport. In Algeria, Libya, Alaska, and other oil-producing areas there is not much demand for the natural gas that is produced along with the oil. One economical way of transporting the surplus gas to markets in industrial centers is to liquefy the gas and ship it in specially designed insulated tankers.

The original transoceanic LNG system transported liquefied gas from Algeria to England and northern France. Since 1965 about 150,000,000 ft³ of gas has been liquefied daily for shipment aboard three specially insulated tank ships. Larger systems have been built to supply Italy and Spain from Libya, Japan from Alaska and Borneo, and southern France from Algeria. Smaller quantities have been supplied to cities in the eastern United States from Libya and Algeria. Liquefaction capacity is expanding in these source areas and in other areas rich in gas resources.

Other uses. As peak shaving and LNG tanker systems have developed, attention has been directed to the potential use of LNG as a fuel in different types of vehicles. LNG has a number of attractive characteristics as an engine fuel. It has an antiknock value well over 100, without any additives. It is extremely clean-burning, resulting in low maintenance costs and a minimum of air pollution from engine exhaust. In gas turbines it provides a large heat sink and burns with a relatively nonluminous flame, both of which assist in engine cooling. Its specific energy per pound of fuel is 15% higher than for gasoline or kerosine. In rocket engines it provides the highest specific impulse of any hydrocarbon fuel.

The primary limitation on the development of LNG as an engine fuel is the general shortage of natural gas in the United States, Europe, and Japan. LNG imported into these areas is intended to supplement the available supplies of natural gas for home heating and cooking. Studies on the use of LNG engine fuel for a variety of transport vehicles have shown its advantages. However, manufacturers are reluctant to undertake engine and vehicle redesign efforts since adequate supplies of LNG cannot be assured. *See* CRYOGENIC ENGINEERING; LIQUEFACTION OF GASES; LIQUEFIED PETROLEUM GAS (LPG).   [ARTHUR W. FRANCIS]

Bibliography: R. Blakely, *Oil Gas J.*, 66(1):60–62, 1968; W. B. Emery, E. L. Sterrett, and C. F. Moore, *Oil Gas J.*, 66(1):55–59, 1968; C. H. Gatton, *Liquefied Natural Gas Technology and Economics*, 1967; M. Sittig, *Cryogenics: Research and Applications*, 1963; C. M. Sliepcevich and H. T. Hashemi, *Chem. Eng. Progr.*, 63(6):68–72, 1967; J. W. White and A. E. S. Neumann (eds.), *Proceedings of 1st International Conference on Liquefied Natural Gas*, Chicago, 1968.

Liquefied petroleum gas (LPG)

A product of petroleum gases, principally propane and butane, which must be stored under pressure to keep it in a liquid state. At atmospheric pressure and above freezing temperature, these substances would be gases. Large quantities of propane and butane are now available from the gas and petroleum industries. These are often employed as fuel for tractors, trucks, and buses and mainly as a domestic fuel in remote areas. Because of the low boiling point (−44 to 0°C) and high vapor pressure of these gases, their handling as liquids in pressure cylinders is necessary. Owing to demand from industry for butane derivations, LPG sold as fuel is made up largely of propane. On a gallonage basis, production of LPG in the United States exceeds that of kerosine and approaches that of diesel fuel.

Operating figures for gasoline, diesel, and LPG fuels show that LPG compares favorably in cost per mile. LPG has a high octane rating, making it useful in engines having compression ratios above 10:1.

Another factor of importance in internal combustion engines is that LPG leaves little or no engine deposit in the cylinders when it burns. Also, since it enters the engine as a vapor, it cannot wash down the cylinder walls, remove lubricant, and increase cylinder-wall, piston, and piston-ring wear. Nor does it cause crankcase dilution. All these factors reduce engine wear, increase engine life, and keep maintenance costs low. However, allowances must be made for the extra cost of LPG-handling equipment, including relatively heavy pressurized storage tanks, and special equipment to fill fuel tanks on the vehicles. *See* GAS TURBINE; INTERNAL COMBUSTION ENGINE; PETROLEUM PRODUCTS.   [MOTT SOUDERS]

Liquid

A state of matter intermediate between that of crystalline solids and gases. Macroscopically, liquids are distinguished from crystalline solids in their capacity to flow under the action of extremely small shear stresses and to conform to the shape of a confining vessel. Liquids differ from gases in possessing a free surface and in lacking the capacity to expand without limit. On the scale of molecular dimensions liquids lack the long-range order that characterizes the crystalline state, but nevertheless they possess a degree of structural regularity that extends over distances of a few molecular diameters. In this respect, liquids are wholly unlike gases, whose molecular organization is completely random.

Thermodynamic relations. The thermodynamic conditions under which a substance may exist indefinitely in the liquid state are described by its phase diagram, shown schematically in the figure. The area designated by L depicts those pressures and temperatures for which the liquid state is energetically the lowest and therefore the stable state. The areas denoted by S and V similarly indicate those pressures and temperatures for which only the solid or vapor phase may exist. The connecting lines OC, OB, and OA define pressures and temperatures for which the liquid and its vapor, the solid and its liquid, and the solid and its vapor, respectively, may coexist in equilibrium. They are usually termed phase boundary or phase coexistence lines. The intersection of the three lines at O defines a triple point which, for the three states of matter under discussion, is the unique pressure and temperature at which they may coexist at equilibrium. Other triple points exist in the phase diagram of a substance that possesses two or more crystalline modifications, but the one depicted in the figure is the only triple point for the coexistence of the vapor, liquid, and solid. Line OA has its origin at the absolute zero of temperature and OB, the melting line, has no upper limit. The liquid-vapor pressure line OC is different from OB, however, in that it terminates at a precisely

reproducible point C, called the critical point. Above the critical temperature no pressure, however large, will liquefy a gas. *See* TRIPLE POINT (THERMODYNAMICS).

Along any of the coexistence curves the relationship between pressure and temperature is given by the Clausius-Clapeyron equation: $dP/dT = \Delta H/T\Delta V$, where ΔV is the difference in molar volume of the corresponding phases (gas-liquid, gas-solid, or liquid-solid) and ΔH is the molar heat of transition at the temperature in question. By means of this equation the change in the melting point of the solid or the boiling point of the liquid as a function of pressure may be calculated. When a liquid in equilibrium with its vapor is heated in a closed vessel, its vapor pressure and temperature increase along the line OC. ΔH and ΔV both decrease and become zero at the critical point, where all distinction between the two phases vanishes. *See* EQUILIBRIUM, PHASE.

Transport properties. Liquids possess important transport properties, notably their capacity to transmit heat (thermal conductivity), to transfer momentum under shear stresses (viscosity), and to attain a state of homogeneous composition when mixed with other miscible liquids (diffusion). These nonequilibrium properties of liquids are well understood in macroscopic terms and are exploited in large-scale engineering and chemical-process operations. Thus, the rate of flow of heat across a layer of liquid is given by $\dot{Q} = \kappa \, dT/dx$, where $\dot{Q}$ is the heat flux, dT/dx is the thermal gradient, and κ is the coefficient of thermal conductivity. Similarly, the shearing of one liquid layer against another is resisted by a force equal to the momentum transfer: $F = \dot{p} = \eta \, dv/dx$, where dv/dx is the velocity gradient and η is the coefficient of viscosity. Likewise, the rate of transport of matter under nonconvective conditions is governed by the gradient of concentration of the diffusing species: $J = -D \, dC/dx$, where J is the matter flux and D is the coefficient of diffusion. Each of these transport coefficients depends upon temperature, pressure, and composition and may be determined experimentally. An a priori calculation of κ, η, and D is a very difficult problem, however, and only approximate theories exist.

Theoretical explanations. In fact, although a great deal of effort has been expended, there still exists no satisfactory theory of the liquid state. Even so commonplace a phenomenon as the melting of ice has no adequate theoretical explanation. The reason for this state of affairs lies in the tremendous structural and dynamical complexity of the liquid state. To understand this, it is useful to compare the structural and kinetic properties of liquids with those of crystalline solids on the one hand and with those of gases on the other.

In crystals, atoms or molecules occupy well-defined positions on a three-dimensional lattice, oscillating about them with small amplitudes; their kinetic energy is entirely distributed among these quantized vibrational states up to the melting point. This nearly perfect spatial order is revealed by diffraction techniques, which utilize the coherent scattering of x-rays or particles having wavelengths comparable with interatomic spacings. The structural and dynamical properties are sufficiently tractable mathematically so that the

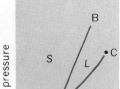

LIQUID

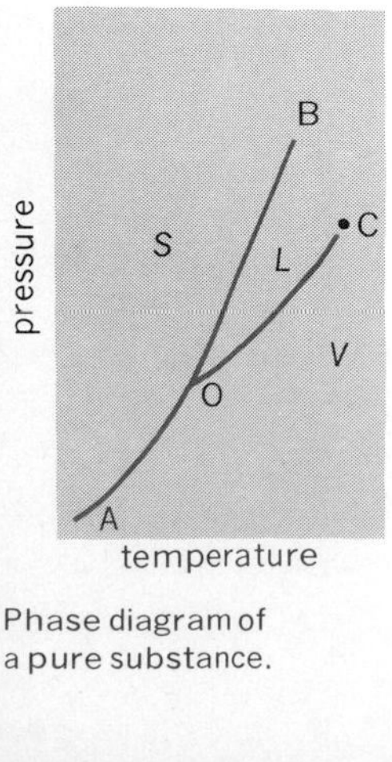

Phase diagram of a pure substance.

theory of solids is quite well understood.

The theory of gases is also simple, but for quite a different reason. No vestige of positional regularity of atoms remains in gases, and their energy resides entirely in high-speed translational motion. Except for collisions, which deflect their motions into new straight-line trajectories, atoms in gases do not interact with one another; vibrational modes in monatomic gases are absent.

Liquids, by contrast, lie intermediate between gases and crystals from both a structural and dynamic point of view. Kinetic energy is partitioned among translational and vibrational modes, and diffraction studies reveal a degree of short-range order that extends over several molecular diameters. Moreover, this "structure" is continually changing under the influence of vibrational and translational displacements of atoms. Physical reality may be attributed to this short-range structure, nevertheless, in the sense that a time average over the huge number of possible configurations of atoms may show that a fairly definite number of neighboring atoms lie close to any arbitrary atom. At a somewhat greater distance from this reference atom, the density of neighbors oscillates above and below the average density of atoms in the liquid as a whole.

Information about the degree of local order is contained in the radial distribution function, a mathematical property which may be deduced from diffraction measurements. This is the starting point for a theory of the liquid state, and although efforts by J. G. Kirkwood and his students, by H. Eyring and his collaborators, and by J. E. Lennard-Jones and A. F. Devonshire have yielded partial successes, prodigious mathematical difficulties lie in the path of an entirely satisfactory solution. *See* BOILING POINT; KINETIC THEORY OF MATTER; LIQUEFACTION OF GASES; MELTING POINT; VISCOSITY OF LIQUIDS; X-RAY DIFFRACTION.   [NORMAN H. NACHTRIEB]

Bibliography: D. Dreisbach, *Liquids and Solutions*, 1966; P. A. Egelstaff, *An Introduction to the Liquid State*, 1967; S. A. Rice and P. Gray, *The Statistical Mechanics of Simple Liquids*, 1965.

Liquid chromatography

A separation technique involving passage of a liquid moving phase through a solid or a liquid stationary phase. Interest in modern liquid chromatography derives from the fact that with this approach separation problems can be attacked that are difficult or even impossible to solve by other techniques. The major separation method up to the present time has been gas chromatography; yet this method is limited to volatile components (mol. wt < 300) that are thermally stable. Most biochemical mixtures are therefore chromatographed only with great difficulty (in most cases after the preparation of a volatile derivative). Liquid chromatography, however, is compatible with the samples commonly encountered in biochemical and organic mixtures (that is, mol. wt > 300, ionic, heat-sensitive, and so on), and this is its greatest attribute. *See* GAS CHROMATOGRAPHY.

Historically, column liquid chromatography was the first chromatographic method. For the most part, however, it has remained as a preparative

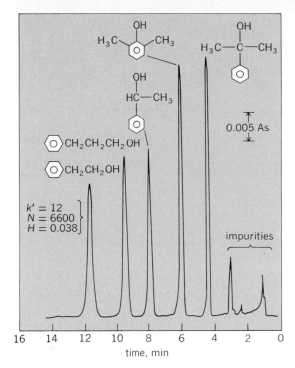

Fig. 1. Liquid-liquid chromatography with porous microspheres. Capacity factor is represented by *k'*, *N* is the number of theoretical plates, *H* the height equivalent to a theoretical plate, and As the absorbance unit. (*From J. J. Kirkland, J. Chromatogr. Sci., 10:583, 1972*)

tool to obtain milligram-to-gram quantities of purified material. It is only since about 1970 that major advances in speed and efficiency have occurred such that performance characteristics similar to those of gas chromatography have been achieved. Another major development has been the introduction of new instrumentation, especially high-sensitivity detectors and high-pressure pumps. Advances in selectivity and flexibility of liquid chromatographic systems also have taken place. Finally, many new applications have broadened the scope of the method.

High performance. From theoretical considerations developed first with reference to gas chromatography, it became clear in the late 1960s that liquid chromatographic conditions were not optimized for speed and efficiency. Typically at that time, columns were packed with support particles of 100–200-μm diameter, and mobile phase flow was controlled by gravity feed of the eluent (velocity about 0.01 cm/sec). A 50-cm column might generate 100 theoretical plates and 0.01 effective plate per second.

To improve performance, much smaller particle diameters (about 5 μm), velocities a hundredfold greater (about 1 cm/sec), and columns with narrower diameters (about 2–3 mm) are now employed. Performance characteristics of 10,000 plates per meter and 35 effective plates per second are possible. These modifications have resulted in the need for high-pressure pumps to drive the mobile phase through the column. Pressures of 1000 to 6000 psi (where 1 psi is about 6890 N/m^2) are now typically employed. Of course, the low compressibility of liquids minimizes the safety

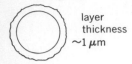

layer
thickness
~1 μm

Fig. 2. Diagrammatic illustration of porous layer beads. Particle diameter about 30 μm, porous shell about 1 μm.

hazards with this equipment. Figure 1 shows the separation of a series of alcohols by liquid-liquid chromatography on 5–6-μm porous silica.

Ion-exchange resin particles in the diameter range of 10–15 μm have been used since about 1963 for the separation of amino acid and peptide mixtures, as well as complex body fluid mixtures (for example, urine). Indeed, major advances have been made in the application of this method to clinical diagnosis. However, workers found difficulty in efficiently and reproducibly packing porous silica or alumina particles less than 20 μm in diameter. In 1972 slurry packing methods were introduced which produce excellent columns with these materials down to 5 μm. Because of the high pressures involved, column lengths are typically no greater than 30 cm; however, more than 10,000 theoretical plates can be generated with these short lengths. These columns have been used for the separation and analysis of a variety of mixtures, with both liquid-liquid and liquid-solid chromatography employed. *See* ION EXCHANGE.

Although support columns packed with small-diameter particles will be used with increasing frequency in the future, at present porous layer bead supports provide the major load in liquid chromatography. As shown in Fig. 2, these materials consist of a solid spherical core (about 30 μm diameter) coated with a thin porous layer of adsorbent (about 1 μm thick). The larger particle diameter and high density allow easy and reproducible packing characteristics. The specially designed supports result from the need to reduce pore depths and thus enhance mass transfer into and out of the particles. The quite slow diffusion of components in liquids would severely limit the velocity of the mobile phase by the particle, if fully porous materials of > 30 μm were to be used. The drawback to the exclusive application of porous layer beads is the low sample capacity (and peak capacity), a result of the small amount of stationary phase. Yet, since their introduction in 1969–1970, porous layer beads have been used quite extensively, and, as noted above, they represent the most popular support materials at the present

time for liquid-liquid, liquid-solid, and ion-exchange chromatography. Small-diameter support (about 5 μm) columns provide greater speed characteristics by a factor of 10 and somewhat greater sample capacity; however, porous layer bead columns make fewer demands on equipment and are easier to manufacture.

Instrumentation. Detectors are available which fulfill the demands of high sensitivity, low dead volume, rapid response, and selectivity. The ultraviolet detector is most frequently selected because of its high sensitivity. Previously, commercial devices used only the 254-nm line of a mercury discharge tube. In 1972 it became possible to use other Hg lines (for example, 280 nm), and indeed several instruments now allow the simultaneous detection of the chromatogram at two wavelengths. This is valuable when complex mixtures are employed. The differential refractive index is the second most widely used detector. Although its sensitivity is generally less than that of the ultraviolet detector, it is a more universal sensor. More interest is being shown in the moving wire detector, in which the effluent from the column is deposited onto a moving platinum wire. The mobile phase is evaporated, leaving a residue of the solute which is then pyrolyzed into a flame ionization detector. This device finds increasing use with sophisticated multisolvent gradients, as indicated by Fig. 3. Other detectors are valuable in specialized areas, for example, polarographic and radiometric applications.

The development of pumps specifically designed for high-performance liquid chromatography has also played an important role in the advancement of the field. It is possible to obtain relatively pulse-free pumps operating a constant flow rate up to 6000 psi. Moreover, these devices permit simple digitization of the flow rate and with simultaneous operation of two pumps allow the convenient dialing of a variety of solvent gradient programs.

Modes of liquid chromatography. Much activity occurred during the early 1970s in the use of chemically bonded phases in liquid chromatography. Here, organic substances are bonded in either a monomeric or a polymeric fashion to a solid support such as silica or alumina. One advantage of these phases is that selectivity can be "tailor-made" for a particular application. Moreover, certain of these phases act as bonded liquids, and as such they require no precolumn to maintain saturation of the mobile phase, and they can be use in gradient elution.

The development of techniques to enhance the peak capacity (that is, number of solutes that can be resolved) of the liquid chromatographic system has also been important. Coupled columns are finding increasing use in this regard. Here, two columns with different capacities or selectivities are placed in series. By proper use of switching valves the sample components are resolved on the column best suited to their separation.

Finally, R. P. W. Scott developed a 15-solvent series for gradient elution which overcomes the problem of displacement effects with silica or alumina as support. A particularly striking example of the separation of a complex mixture is shown in Fig. 3.

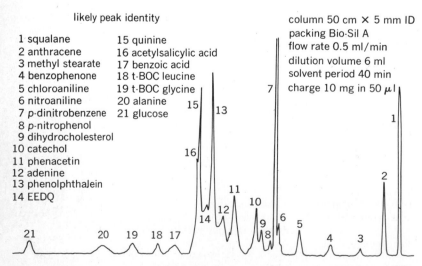

likely peak identity

1 squalane
2 anthracene
3 methyl stearate
4 benzophenone
5 chloroaniline
6 nitroaniline
7 p-dinitrobenzene
8 p-nitrophenol
9 dihydrocholesterol
10 catechol
11 phenacetin
12 adenine
13 phenolphthalein
14 EEDQ

15 quinine
16 acetylsalicylic acid
17 benzoic acid
18 t-BOC leucine
19 t-BOC glycine
20 alanine
21 glucose

column 50 cm × 5 mm ID
packing Bio-Sil A
flow rate 0.5 ml/min
dilution volume 6 ml
solvent period 40 min
charge 10 mg in 50 μl

Fig. 3. Chromatogram of a mixture containing solutes of widely differing polarities using incremental gradient elution. (*From R. P. W. Scott and P. Kucera, Anal. Chem., 45:749, 1973*)

Applications. The application of modern liquid chromatography to various separation problems is increasing at a rapid pace. Pesticides, drugs of abuse, antioxidants, and nonionic surfactants are only a few of the substances to which the method can be applied. Liquid chromatography, moreover, played a central role in the successful synthesis of vitamin B_{12}. Finally, it is interesting to note that 20 or more phenylthiohydantoin derivatives of amino acids can now be separated in less than 1 hr. This development will prove of great value to workers in protein sequencing. *See* ADSORPTION; CHROMATOGRAPHY; SEPARATION, CHEMICAL AND PHYSICAL.   [BARRY L. KARGER]

Bibliography: N. Hadden et al., *Basic Liquid Chromatography*, 1972; J. F. K. Huber, *5th International Symposium on Separation Methods: Column Chromatography, Lausanne, 1969*, Supplement to *Chimia*, ed. by E. Kovats, p. 24, 1970; J. J. Kirkland (ed.), *Modern Practice of Liquid Chromatography*, 1971; R. E. Majors, *Anal. Chem.*, 45:755, 1973.

Lissajous figures

Plane curves traced by a point which executes two independent harmonic motions in perpendicular directions, the frequencies of the motion being in the ratio of two integers. Such figures, produced on the screen of a cathode-ray tube, are widely used in frequency and phase measurements. *See* HARMONIC MOTION.

Methods of constructing. Mechanical, optical, and electronic methods may be used to construct Lissajous figures. Blackburn's pendulum consists of one ordinary pendulum hung from another with the two axes of rotation at right angles. The Lissajous figure is traced by the bob. An example of an optical method is afforded by a beam of light reflected successively from two small mirrors, each mounted on the prong of a separate tuning fork; the forks vibrate in perpendicular directions, and the image of the light source, focused on a screen, traces out the figure.

The cathode-ray oscilloscope furnishes the most important and practical means for the generation of the figures. The *x*-deflection plates of the tube are supplied with one alternating voltage, and the *y*-deflection plates with another. If the frequencies are incommensurable, the figure is not a closed curve and, except for very low frequencies, will appear as a patch of light because of the persistence of the screen. On the other hand, if the frequencies are commensurable, the figure is closed and strictly periodic; it is a true Lissajous figure, stationary on the screen and, if the persistence is sufficient, visible continuously as a complete pattern. *See* OSCILLOSCOPE, CATHODE-RAY.

Frequency measurements. Measurement of frequency in sinusoidal wave motion (for example, in alternating electric current) may be accomplished with an oscilloscope if one of the alternating voltages is derived from a variable-frequency calibrated oscillator. The standard is adjusted until a stationary figure results; if f_v is the frequency of the signal on the vertical deflection plates, and f_h that on the horizontal plates, then f_v/f_h equals the number of times a side of the figure is tangent to a vertical line divided by the number of times the top or bottom is tangent to a horizontal line.

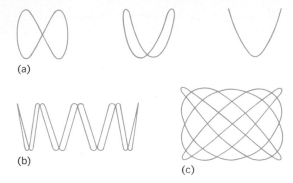

(a)

(b)   (c)

Fig. 1. Typical Lissajous figures for ratios of vertical frequency to horizontal frequency. (*a*) 2:1, with various phase relations. (*b*) 8:1. (*c*) 5:4. (*From F. E. Terman, Radio Engineers' Handbook, McGraw-Hill, 1943*)

The shape of the figure depends also on the relative phase of the signals (Fig. 1), and in the case where one-half of the figure coincides with the other, the rule just stated does not hold. This situation is not troublesome in practice because unavoidable slow-frequency drifts of the signals prevent the figure from being exactly stationary; instead, the shape of the figure changes as though the relative phase of the signals were slowly changing.

In some cases, the standard is of fixed frequency and the unknown is nearly equal to the standard (or to some multiple or submultiple thereof). The figure will drift slowly through the various aspects it can assume for the nominal ratio involved. The actual ratio can be obtained from the nominal ratio, and a correction determined by the rate and direction of drift of the figure. This method is

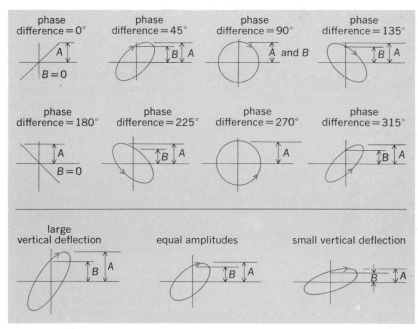

equal amplitudes and varying phase differences

constant phase difference of 45° but varying amplitude on vertical deflection

Fig. 2. Typical Lissajous figures of 1:1 frequency ratio. (*From F. E. Terman, Radio Engineers' Handbook, McGraw-Hill, 1943*)

particularly valuable for the intercomparison of standards.

Phase measurements. These usually involve two signals of the same frequency. In this case, the Lissajous figure is an ellipse of which the shape and orientation depend on the relative phase and relative amplitudes of the signals, as shown in Fig. 2. The relative phase θ is given by the equation below, in which A is the maximum half-height

$$\sin \theta = \pm\, B/A$$

and B is the intercept on the y axis. Figure 2 shows how the quadrant in which θ lies is determined by the orientation of the major axis of the ellipse and the direction of motion of the spot. If the latter is unknown, the resulting ambiguity amounts to that in the sign of $\sin \theta$, once it is decided which signal is the reference. The ambiguity can be resolved by shifting the phase of either signal in a known direction and noting whether the phase shift indicated by the figure increases or decreases.

The most precise comparison can be made when the ellipse degenerates into a straight line. This condition is therefore used as a null in measurements made with a calibrated phase shifter. The unknown phase shift disturbs the null, and the calibrated phase shifter restores it.

More complicated figures are employed in the calibration of phase shifters. A signal at the desired frequency is connected through the phase shifter to the y plates of an oscilloscope and also through a frequency multiplier to the x plates. If the frequency multiplication is 8, for example, the Lissajous figure will appear as in Fig. 1b. The phase shifter is then successively set to the various positions for which one-half of the figure coincides with the other; these positions will correspond to phase differences of, in this case, $\pi/8$. *See* PHASE-ANGLE MEASUREMENT.   [MARTIN GREENSPAN]

Bibliography: Lord Rayleigh, *Theory of Sound*, vol. 1, reprint, 1945; F. E. Terman and J. M. Pettit, *Electronic Measurements*, 2d ed., 1952; R. P. Turner, *Practical Oscilloscope Handbook*, 1964.

Lissamphibia

The subclass of Amphibia including all living amphibians (frogs, toads, salamanders, and apodans). The other two subclasses are the Labyrinthodontia and the Lepospondyli, predominantly Paleozoic fossil groups lacking recent representatives. *See* LABYRINTHODONTIA; LEPOSPONDYLI.

Morphology. Living amphibians are grouped together by possession of a unique series of characters, the most important of which are (1) pedicellate teeth; (2) an operculum-plectrum complex of the middle ear; (3) the papilla amphibiorum, a special sensory area of the inner ear; (4) green rods in the retina of the eye; (5) similar skin glands; and (6) a highly vascular skin used in respiration (cutaneous respiration). Only the first two characters can be used to separate the Lissamphibia from the other two subclasses, because the other characters are soft anatomy features not ordinarily preservable as fossils. *See* EAR (VERTEBRATE); EYE (VERTEBRATE); RESPIRATION.

Dentition. The pedicellate condition of the lissamphibian tooth is formed by a separation of the tooth into two segments, a crown and a pedicel (Fig. 1). The pedicel is usually fused to the jaw,

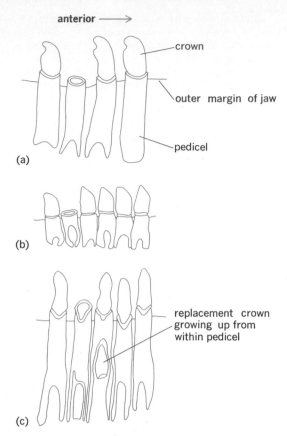

anterior ⟶

Fig. 1.   Pedicellate teeth in Lissamphibia. (*a*) Salamander. (*b*) Apodan. (*c*) Frog. Teeth are from the lower jaw and are viewed from inside the mouth. (*From T. Parsons and E. Williams, J. Morphol., 110:375–390, 1962*)

but crown and pedicel are separated by uncalcified or fibrous material, which in effect functions as a hinge in the tooth. In at least some lissamphibians the tooth bends inward more easily than outward. This indicates that it may be an adaptation for allowing prey to enter the mouth easily, but tends to prevent its escape. Some teleost fishes also have such a hinged tooth that bends inward toward the throat and is relatively rigid when bent back; however, there the hinge is between tooth and jaw, without an intervening pedicel, and is not comparable with the lissamphibian condition except perhaps in function. *See* TOOTH (VERTEBRATE).

Sound perception. The operculum-plectrum complex of the middle ear is an unusual feature unknown in other tetrapods. In the characteristic tetrapod condition a single sound-transmitting bone, the stapes, fills the opening (fenestra ovalis) of the middle ear and connects to the inner ear. In lissamphibians two structures occur in the fenestra ovalis: plectrum (stapes) and operculum (Fig. 2). The plectrum has the usual tetrapod connections to tympanic membrane or to quadrate. The operculum is connected to the shoulder girdle by a specialized shoulder muscle that may provide passage to the ear for vibrations traveling through the front limb from the substrate. Thus the animal may receive both airborne and ground-borne sounds.

Phylogeny and fossil record. Apodans lack a fossil record; frogs and salamanders go back to the Jurassic. There is a Triassic fossil from Madagascar that may be near the ancestry of frogs, but

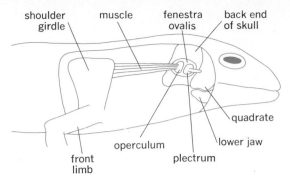

Fig. 2. Operculum-plectrum complex of a primitive salamander (skin removed to show internal structures).

it does not indicate the group of origin. Except for this Triassic fossil, the ancient amphibians and the lissamphibians do not overlap in their geological ranges. On the basis of vertebral similarities (solid, spool-shaped centrum), lissamphibians have often been thought to be lepospondyl descendants, but in the skull (especially the palate) there are similarities to temnospondyl labyrinthodonts.

Pedicellate teeth and the operculum-plectrum complex are characters that form the best evidence for close relationship of modern amphibians to each other. Pedicellate teeth and the possible presence of an operculum have been found in one family of primitive rhachitome labyrinthodonts. These two occurrences may indicate that lissamphibians are derived from the latter group, although new fossil finds are necessary to clarify lissamphibian ancestry. *See* AMPHIBIA; ANURA; APODA; URODELA.   [RICHARD ESTES]

Bibliography: T. Monath, The opercular apparatus of salamanders, *J. Morphol.*, 116:149–170, 1965; T. Parsons and E. Williams, The relationships of the modern Amphibia, *Quart. Rev. Biol.*, 38:26–53, 1963; A. Romer, *Vertebrate Paleontology*, 3d. ed., 1966.

Listeriosis

A disease of man and animals caused by a small gram-positive bacillus, *Listeria monocytogenes*. The microorganism, a member of the family Corynebacteriaceae, has been isolated from man and 27 species of animals and birds in 30 countries from the Arctic to the tropics. The epidemiology is not well understood.

The organism measures 0.5 by $1-3\,\mu$. Small

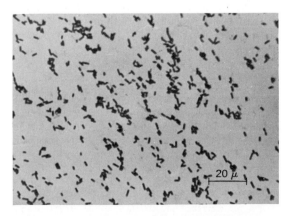

Listeria monocytogenes, family Corynebacteriaceae.

rods with rounded ends predominate (see illustration), showing palisade formation and short chains. Young broth cultures display characteristic "tumbling" motility. Growth on simple media is enhanced by enrichment with blood or glucose. There are four serotypes. *See* IMMUNOLOGY.

Listeriosis in animals takes many forms, such as meningoencephalitis in swine, goats, and sheep; distemperlike disease in dogs and foxes; and generalized infection in rabbits and guinea pigs, with focal liver necrosis and marked monocytosis. Abortion with extensive fetal necrosis may occur in ruminants such as cattle.

Listeriosis in man occurs primarily as meningitis or granulomatosis infantiseptica with high fetal or newborn mortality and necrosis. Septicemia is less common. *See* SEPTICEMIA.

Diagnosis depends on isolation of *Listeria* from appropriate clinical specimens. Identification is certain if a pure culture instilled into the rabbit conjunctiva produces severe keratoconjunctivitis. An agglutination titer of over 1/200 on patients' serum is diagnostically significant. *See* KERATOCONJUNCTIVITIS.

Treatment should be based on antibiotic sensitivity tests on laboratory cultures of the strain isolated. *See* BACTERIOLOGY, MEDICAL; CORYNEBACTERIACEAE; EPIDEMIOLOGY.   [ROGER W. REED]

Bibliography: R. J. Dubos and J. G. Hirsch (eds.), *Bacterial and Mycotic Infections of Man*, 4th ed., 1966; H. P. Seelinger, *Listeriosis*, 1961.

Literature of science and technology

The accumulated body of scientific-technical writings published to serve the informational needs of, primarily, scientists, engineers, and research workers, and, to the extent that it can be understood, the general public. This literature is worldwide in origin, international in language, diverse in subject content, complex in form, uneven in quality, and tremendous in amount. It is also expensive.

Science-technology and its literature are essentially inseparable. This immutable interrelationship between the two puts the nonscientist at a singular disadvantage in fully understanding either one without considerable preliminary study or experience. Because of the discrete and specialized symbols, terminologies, and word meanings that are peculiar to most branches of science and engineering, even scientists and engineers themselves have difficulty in fully understanding everything that goes on in specialties other than their own without some further study.

The primary sources of scientific-technical information are the first (and often the only) published records of original research and development and accounts of new applications of science and engineering to technology and industry. These unorganized and usually unrelated contributions appear almost exclusively in periodicals, separate research reports, patents, dissertations, or manufacturers' technical bulletins.

Organized works and compilations that derive from or refer to the primary-source literature make up the secondary sources of scientific-technical information: the handbooks, encyclopedias, treatises, bibliographies, reviews, abstracting and indexing serials, and other reference works, and, of course, all translations.

Guides to the literature, directories (of persons, organizations, or products, for example), and textbooks are often looked upon as tertiary sources. Popularizations, biographies, histories, and similar relatively nontechnical publications are thought of as being about science rather than of science.

PRIMARY SOURCES

These are the records of original research, patents, and manufacturers' literature.

Periodicals. These publications traditionally make up the bulk of the primary-source literature of science and technology. Periodicals include journals, bulletins, transactions, proceedings, or other serial publications that appear regularly and continuously in numbered sequence, but not newspapers or annuals. Unfortunately, there are no complete international (or even national) lists of all the scientific periodicals that are being published in the world today, nor for that matter is there complete agreement on what exactly is a scientific, technical, or research periodical. However, it has been reliably estimated that in 1975 the world total of scientific and technical periodicals was about 40,000 separate titles and that this number was growing slowly every year. It has also been estimated that each one of these periodicals publishes an average of 85 articles each year. Literature experts believe that, while births and deaths of periodicals just about offset each other, the numbers and the lengths of articles published are both increasing at a steady annual rate, so that the total amount of this literature is doubling every 10–15 years. Other studies have shown that almost 95% of all the cited literature of the basic sciences is published in serials.

At least 100 countries regularly publish research journals in science-technology, and these may appear in any one of 50 different languages irrespective of the country of origin. Over 50% of the material appears in English. Another 20% appears in Russian, 7% in German, 5% in French, 4% in Japanese, and 3% each in Italian and Spanish, but the importance of language varies from subject to subject and the importance of subject varies from country to country.

The contents of periodicals vary considerably in the kind of material included as well as in the technical level. Professional, scientific, and technical societies, for example, tend to emphasize basic research and the more technical aspects of a subject, while industrial and trade associations and private publishers lean toward the practical, personal, and popular side. Societies, too, almost uniformly make use of highly qualified subject experts, called referees, to evaluate critically all original contributions before they are published, thus imparting reliability, authoritativeness, and prestige to their publications. Other organizations, such as research institutes, university experiment stations, and government agencies, publish periodicals comparable with those published by the societies but such organizations are more likely to emphasize their own work. There are, of course, noteworthy exceptions.

Conference papers. Papers presented at national and international scientific conferences, symposia, colloquia, seminars, and other technical meetings often are important contributions to the primary-source literature of science and technol-

ogy but are also often uneven and elusive. At widely varying intervals following their presentation they may be published as complete (or partial) collections by the convening bodies themselves, by the parent scientific societies, or by enterprising private publishers; they may appear as articles in, or supplements to, regular journals; or they may only be noted or at most abstracted. Sometimes they are preprinted and not later published at all, thus being made available only to those who attend the meetings or who write for them in time. Announcements of forthcoming scientific meetings as well as of resulting technical publications are now being published regularly.

Research monographs. These are separately published reports on original research that are too long, too specialized, or otherwise unsuitable for publication in one of the standard journals. Each monograph is self-contained, frequently summarizes existing theory or practice before presenting the author's original and previously unpublished work, and is likely to be one of a series of such research monographs in the same field.

Research reports. Reports from one or another of the United States government's research and development projects make up another important part of the primary-source literature of science and technology. Reports are a more primitive form of literature because they are produced earlier in the research program, often as progress reports. While a progress report is only a temporary reference, it is important because it may include negative results and incidental data which may be missing from a final report.

Research reports originate in all research organizations doing work under contract to the United States government and are in turn distributed to all installations (including designated depository libraries) that have an interest in them. Classified reports, those containing data having some bearing on national security, are distributed only to properly cleared individuals or organizations. Since distribution of even the unclassified reports is somewhat limited, few of them are indexed, abstracted, or even listed by the standard services which cover other parts of the primary literature.

The National Technical Information Service (NTIS) of the U.S. Department of Commerce (Spring, VA 22161) is the agency responsible for the announcement, sales, and distribution of unclassified research and development reports, translations, or other analyses prepared by Federal agencies, by their contractors or grantees, or by Special Technology Groups, including many Information Analysis Centers. Established in 1970, it supersedes the Clearinghouse for Federal Scientific and Technical Information, which itself superseded (in 1965) the Office of Technical Services.

Summaries of unclassified government research reports and of available translations of foreign research reports are published in the comprehensive biweekly *Government Reports Announcements & Index* (*GRA&I*) and in the 26 *Weekly Government Abstract* (*WGA*) newsletters covering 26 selected subject areas. Copies of most of the reports are available from NTIS as paper documents or in micro form, but patents and journal article reprints, for example, are not, although both *GRA&I* and *WGA* list a number of patents and reprints. Tapes containing the complete NTIS Bibliographic Data

File back to 1964 are available for custom on-line searches and may be acquired or leased.

While *GRA&I* and the 26 *WGA*s cover most government research reports, a number of government agencies also publish their own more specialized announcement and abstracting services. Examples include *Scientific and Technical Aerospace Reports* (*STAR*), published by the National Aeronautics and Space Administration (NASA); *ERDA Energy Research Abstracts,* published by the Energy Research and Development Administration; and *Technical Abstracts Bulletin* (*TAB*), covering classified Department of Defense reports, available on a restricted basis from the Defense Documentation Center (DDC).

Preprints. Preprints are research-in-progress reports to the small but often international group of fellow scientists working in the same relatively narrow research specialty, somewhat after the fashion of their 17th-century predecessors. Like government research reports, the preprint is a primitive form of technical literature, unrefereed and unannounced but historical. Preprints are far more elusive than conference papers since only those on particular "preprint circuits" know of their existence, even though they may be cited as references (for example, "in preparation"). They are rapid and timely but extra literature or preprimary publications that are not picked up by indexing or abstracting services unless formally published at some later date. To scientists on the research frontiers preprints are vital (and quite conventional), but to others they may not be so important. The clue to their existence is in knowing who is doing what right now in relevant fields of research. One source of such information is the Science Information Exchange (SIE) of the Smithsonian Institution (Washington, D.C. 20560).

Patents. Patents are considered part of the primary-source literature, since an invention must be new in order to be patented and more often than not there will be no other published description either of the idea or of its application. Patents are of particular interest to chemists and to engineers, but the chemical patents are the only ones as a group that are well indexed, abstracted, or listed by standard services. Announcement publications, abstracts journals, and the individual patents are available from the various national patent offices and in selected depository libraries, but extensive patent searches require knowledge of both the subject field and patent law, as well as much time.

During fiscal year 1975 the U.S. Patent and Trademark Office issued 74,315 new patents, including 3631 for designs and 155 for plants (flowers, vegetables). A total of 3,943,570 (including 239,144 for designs and 3840 for plants) has been issued through Mar. 9, 1976, plus some 10,000 from 1790 to 1836. Other important patent-producing countries include Great Britain, France, West Germany, Canada, and Belgium.

Dissertations. These frequently contain data on important original work of such special interest as to make it unlikely that a report will ever be published in the normal literature channels; or if the essential portion of a thesis is published, it may be that certain minor but significant details must still be obtained from the original source.

Manufacturers' literature. These publications are often the only source of specific information about particular products or their development; if the information is not published elsewhere, this source becomes a primary source. Technical bulletins, data sheets, price lists, and other specialized publications describing a manufacturer's equipment or processes are typical examples.

SECONDARY SOURCES

Reference works, which make up the bulk of the secondary literature, are of three types:

1. Those which in effect index selected portions of the primary literature and which thus aid in finding what has been published on a given subject (and where) generally or specifically, recently or retrospectively; for example, indexes themselves, bibliographies, indexing serials, and under certain conditions, abstracting serials.

2. Those which survey selected portions of the primary literature and which thus aid in acquiring state-of-the-art, recent background, or comprehensive and definitive information on a given subject; for example, reviews, treatises, monographs, and again under certain conditions, abstracting serials.

3. Those which themselves contain the desired information (but rather concisely presented), such as established facts, formulas, procedures, equations, meanings, theories, history, and biography. All this information is compiled selectively from the primary literature and arranged in some definite manner for convenience in finding; for example, dictionaries, encyclopedias, handbooks, and compilations of tables.

Technical translations are another important (and expensive) part of the secondary literature of science and technology. For all practical purposes these take on the characteristics of the particular primary-, secondary-, or tertiary-source literature that was translated, and they are often treated as if they were in fact the original publications. Translations, of course, are not new to science and technology, but since 1957 a crash program to disseminate scientific-technical information has encouraged many more government, society, academic, and private organizations throughout the world to produce English translations of works published originally in Russian, Chinese, Japanese, and other languages; translation from English into these languages is also occurring.

Some 350 periodicals translated into English are now available on subscription in cover-to-cover, selected-articles, or abstracts form, and additional thousands of separate translated articles, monographs, and other works are available from commercial sources, from the National Translations Center (John Crerar Library, Chicago, IL 60616), from NTIS (including reports of JPRS, the Joint Publications Research Service), and from similar agencies in Canada and elsewhere. Specialized journals regularly index or announce new additions to this growing mass of hybrid (primary-secondary) literature-in-translation which is being assimilated only slowly with the main body of scientific-technical literature. Many authors of research papers, however, still prefer to cite original sources, so translations tend to serve largely as a kind of monitoring system on non-Western science and technology.

Index types. The following subsections discuss the various indexing services.

Indexes. An index is a detailed alphabetical list

of the names, terms, topics, places, formulas, numbers, or other significant items in a completed work (such as a book, set, or bound journal) with exact page references to material discussed in that work. It may be a part of the work or it may be published separately. Rather than alphabetically, it may be arranged chronologically, geographically, numerically, or in some other fashion. No reference work except, of course, a dictionary and possibly an encyclopedia is any better than its index; a properly compiled index is the information hunter's most useful tool.

Bibliographies. A bibliography is a list of references to primary or other sources relating to some one given subject or person; it is usually arranged chronologically or aphabetically by authors. It may be comprehensive or selective and annotated or otherwise evaluative. It may be published separately or as a part of a larger work to acknowledge information sources, recommend additional reading, or direct attention to more detailed treatment elsewhere. Almost always the time span and the subject coverage are well defined, so that a good bibliography represents a definitive coverage of the literature within the defined limits and makes retrospective literature searching that much easier.

Indexing serials. An indexing serial, sometimes called a current bibliography, is a regularly issued compilation of titles of articles (or of subsequent references to them) that appear in current primary-source journals; titles of new books and of other separately published material are often also included. In each issue the titles usually are listed alphabetically by subject; but they may be listed by both subject and author in one alphabet, or the authors may be indexed separately. Issues are published with a specific frequency, weekly or monthly, for example, and they may cumulate. Some indexing serials aim to cover only selected journals in their fields but to do it comprehensively, listing all articles of any possible interest; others aim to cover all journals in their fields but to do it selectively, listing only articles of real substantive interest. Current issues of typical indexing serials and of express services which reproduce (and often translate) the tables of contents of selected primary-source journals aid in finding recent articles. A number of these are now being compiled and printed with the aid of computers, and the tapes may be available for custom current literature searches or by regular subscription. Bound sets of indexing serials aid in retrospective literature searching.

Abstracting serials. An abstracting serial, also at times called a current bibliography, is a regularly issued compilation of concise summaries of (1) significant articles (often in a very limited subject field) that appear in current primary-source journals and (2) important new research monographs, reports, patents, and other primary-source publications in that field. In each issue the summaries usually are arranged by rather broad subjects followed by an author index, but some abstracting serials also include report number, corporate author, and more exact subject indexes as well. Most abstracting serials aim at selective substantive coverage of all known and available journals in their fields. Issues may be published with almost any specific frequency, but usually they do not cumulate. Instead, exhaustive annual or multiple-year indexes according to author, subject, patent, formula, or report number are published.

Besides being printed in journal form grouped by subject, abstracts also are published on cards, on lined or perforated card-sheets, and as sections of a regularly issued primary-source journal. The abstracts published in any of these ways may be combined annually into bound, well-indexed volumes or may be indexed annually by the journals in which they appeared. A number of abstracting serials are now also being compiled and printed with the aid of computers and the tapes may be available for custom searches or by subscription. Long runs of certain abstracting serials are available on microfilm.

Abstracts, however published, are likely to be farther behind the current primary literature than title compilations are, but they also cover much more primary literature and give much more subject information per reference. Serious research workers therefore find lengthy runs of abstracts, properly indexed, particularly suitable for retrospective subject literature searching in depth, and the current issues particularly suitable for keeping abreast of worldwide scientific progress in their chosen fields. Abstracts thus serve both as indexes to aid in retrieving specific subject information and as surveys to aid in acquiring current state-of-the-art knowledge about a subject field.

Survey types. The following subsections discuss the various survey types of reference works.

Reviews. A review is a subject survey of the primary literature usually covering the material that was published since the most recent review on the same subject. A good review accumulates, digests, and correlates the current literature and indicates the direction future research might take. A critical or evaluative review by a competent author is an important aid in keeping up with the work being done in a particular field and in spotting the outstanding developments. A review may appear as one of a collection of more or less related papers in an annual, a quarterly, or a monthly series, or it may appear as an article in a regularly issued primary-source subject journal.

Review reference lists are often extensive and, in a way, form a definitive bibliography of the subject for the period covered. One shortcoming of a review, though, is the inevitable delay before comprehensive and critical coverage of the primary literature can be obtained; one hindrance in getting a good review is the difficulty in finding a qualified specialist with both the time and the literature-searching competence to do the job. Nevertheless, information experts believe that reviews are and will continue to be an important part of the literature of science and technology, and that if the time lag and manpower problems can be overcome, reviews may well become the publications of choice for the entire scientific community.

Treatises. A treatise is a comprehensive, authoritative, systematic, well-documented compilation or summary of known information on a subject, covering it so completely that the work becomes definitive at the time of writing. Thus a treatise is an aid in acquiring a foundation in a subject adequate to enable a trained man to carry on advanced research; at the same time, it is a source of facts, procedures, theories, and other important data, presented in such a way as to show their development, correlation, and probable reliability. A treatise combines characteristics of an advanced textbook and a handbook; in fact, certain multivolume German *Handbücher* and certain multivolume British textbooks are good examples of treatises. One drawback to a treatise is that it is soon out of date. By the time the last part of a multipart treatise is published, the first parts are already quite old. Here, too, the time lag is unavoidable because of the scholarly inventory nature of the work itself. Periodic supplements and occasional new editions of the whole or of consecutive parts help overcome this lack of recency of the basic work, but so long as the dates of publication and the probable value of the original sources are clearly brought out, the usefulness of a treatise will not be impaired.

Monographs. A monograph, in effect, is a short treatise on a particular subject, a single division or part of a larger branch of knowledge. Hence it can be more up to date when it is published and can be brought up to date more easily for a revised edition; it frequently is published as a part of a series. Almost any book on a special topic may be called a monograph unless obviously it is something else, for example, a research monograph, part of the primary literature already discussed, or a textbook, discussed later.

Reference books. Four major types of reference books are encyclopedias, dictionaries, handbooks, and critical tables.

Encyclopedias. An encyclopedia is an idea book. It deals with concepts and is usually arranged alphabetically by separate subjects, those parts of all knowledge or of a branch of knowledge which are considered to be significant by the compilers. Treatment is as thorough as space will permit and may be descriptive, explanatory, statistical, historical, or evaluative, but it should be impartial and should summarize the information thought to be of the most general value at the time of writing. Illustrations, tables, drawings, and photographs are often used, lavishly on occasion. Source references may or may not be given, so authoritativeness depends largely on the reputations of the publisher and of the authors of the individual articles. The large-subject approach usual in a multivolume set requires a competent index, while the small-subject approach usual in a one-volume work does not; however, if the subject arrangement is not alphabetic, an index is indispensable.

An encyclopedia is designed to give a summary of background knowledge and a discussion of concepts in a particular field, useful for orientation in a strange subject area and for determining key words for more specific searching. One-volume encyclopedias give something less than this and tend to be considered and used as expanded dictionaries.

Dictionaries. A dictionary is a word book. In science and technology its purpose is to define commonly used terms as simply as possible, preferably without much recourse to other technical terms and, in addition, with some indication of the specific subject field in which a given definition is the accepted one. Etymology and pronunciation of terms may also be included. Terms may be illustrated to show shape or use, and the treatment of terms may be expanded beyond simple definitions to include information bordering on encyclopedic or handbook coverage. A good technical dictionary always gives the generally accepted correct spellings of the words it defines and it often notes variants. Bilingual and polylingual dictionaries normally give just the direct equivalent word in the other language with no definition other than identifying the subject field in which it is used, but a growing number of otherwise standard technical dictionaries are overcoming this deficiency by appending polylingual indexes to their main vocabularies.

Handbooks. A handbook (or manual) in science and engineering usage is a compact, fairly up-to-date, relatively complete, authoritative compilation of specific data, procedures, and professional principles of a subject field. Much of the information is given in tables, graphs, and diagrams, and illustrations are freely used. Symbols, equations, formulas, abbreviations, and concise technical language all help condense much practical information into a handbook, but they also require that the reader already have rather broad knowledge of the field in order to use the work effectively. Hallmarks of a good handbook are an exhaustive index, up-to-date references, expert editorial staff, easy-to-read printing, and convenient format. The term handbook implies ready reference in a technical subject field. The term manual in common usage implies instruction in how to do something with the aid of very explicit step-by-step directions.

Critical tables. A collection of standard reference data (for example, physical and chemical properties of materials) that has been compiled, documented, evaluated, published, and kept up-to-date by reputable and responsible specialists is an essential reference tool to the working scientist or engineer. The compilation of critical (that is, critically evaluated) tables of standard reference data is now being undertaken by an International Committee on Data for Science and Technology (CODATA) set up in 1966 by the International Council of Scientific Unions (ICSU) to identify, encourage, coordinate, and publicize existing (and possible future) national programs around the world for compiling, evaluating, and publishing numerical data. The National Standard Reference Data Program of the U.S. National Bureau of Standards is the United States affiliate.

TERTIARY SOURCES

These sources include textbooks, directories, and literature guides.

Textbooks. A textbook is a standard work used for instruction so arranged as to develop an understanding of a branch of knowledge rather than to impart information for its own sake. Textbooks characteristically are graded in difficulty and are, therefore, suitable for either introductory or advanced study at almost any level, depending on the background of the reader. The most elementary textbooks assume no prior subject knowledge or experience at all; they work with and build on long-established (and sometimes already superseded) theories and concepts and proceed logically to newer and more advanced ideas as the student learns to master what he is reading. The most advanced textbooks, on the other hand, take on the aspects of monographs or treatises in the thorough coverage, systematic development, and rigorous treatment of their subjects.

Directories. A directory is an alphabetical listing of at least the names and addresses of persons, organizations, manufacturers, or periodicals belonging to a particular class or group. The directory may include indexes or supplementary lists by subject field, geographical location, product, or some other desirable classification. It may be published separately in one or more volumes, or it may appear as a part in, an issue of, or a supplement to another work such as a journal or the proceedings of a society; it may be revised from time to time or not at all.

Biographical directories vary in coverage from annual membership lists with only the briefest information about the persons listed to the five- or ten-yearly specialized compendiums of the *Who's Who* type. Trade and product directories, often appearing as annual buyers' guide issues of certain industrial and trade periodicals, frequently contain an unsuspected variety of information, such as sources, producers, dealers, properties, hazards, shipping instructions, trade names, and trademarks. Industrial directories — city, state, national, and international — often contain considerable data on manufacturers and their officials, subsidiaries, plants, dealers, capitalization, trade names, and products.

Organization directories of learned societies, professional associations, research establishments, trade groups, educational institutions, and government agencies, for example, usually give at least the officers, the address, the functions, and the date of establishment of each organization listed; coverage varies from community-size to worldwide. Periodicals directories include full bibliographical information, frequency, publisher, and price of each title listed; some also include index coverage, advertising arrangements, and availability in libraries. The more complete periodicals directories and lists are often considered to be national or subject bibliographies rather than directories.

Literature guides. A literature guide is a reference manual designed to aid research workers or other prospective users in finding their way into and through the literature of a specific subject field. Its purpose is to acquaint its readers with the important types of information sources available to satisfy their needs and to help them make fruitful current or retrospective literature searches. Library resources and services are discussed, and all important specific and related titles of works in the field are cited and many are described quite fully. The essential aim of every guide to the literature of science or engineering is to enable the alert scientist or engineer to make the most of the technical literature of his subject field.

INFORMATION SCIENCE

Assembling, systematizing, and providing easy access to literature is generally performed by libraries and science itself.

Role of libraries. The library has long been the institution most concerned with the methods, skills, and systems for the acquisition, storage, preservation, retrieval, and use of literature. The depth of a library's concern has always been determined primarily by the needs of those it serves and by the fields of literature it handles.

A subject-specialized research library, for example, collects subject-specialized research literature; it acquires, classifies, catalogs, stores, and supplies both whole works and bits of information as precisely as possible, combining library practice with the techniques of documentation or information science.

But science-technology literature is vast, polylingual, expensive, and only imperfectly distributed or indexed (although better than most other subject literatures). No large research library can have everything published in even a relatively narrow subject field, to say nothing of the broader subject disciplines. Often an inquirer must settle for something less than he requests.

To the serious research worker, to industry, and to science itself, this present inability of libraries (however large, rich, or dedicated) to cope with the literature of science and technology is a very critical matter. One promising attempt at a solution to the problem is the development of a network of specialized information centers, each one of which concentrates its entire resources in a very few quite restricted subject areas (and which therefore may have to depend on the large research libraries for more general or for even related subject materials). Users with special information needs may be referred to appropriate information centers through various referral services.

Other aspects of the problem are being worked on in a number of countries. Progress has been good in several areas: microform publication, reprography, computer-aided indexing and printing, and complete mechanized information systems using film, tape, or electronic equipment.

Documenting progress. Science and technology themselves are being called on more and more to help manage their own literature and to aid in its use. How well they are doing is reported regularly in the National Science Foundation's annual *Progress Report: Federal Scientific and Technical Communication Activities*; in the *Annual Review of Information Science and Technology*, sponsored by the American Society for Information Science (ASIS); in *Information Science Abstracts*, sponsored jointly by the ASIS, the Special Libraries Association, and the Division of Chemical Information of the American Chemical Society; in the *Journal of Library Automation*, sponsored by the Information Science and Automation Division of the American Library Association; in *The LARC Reports* of the LARC (Library Automation Research & Consulting) Association; in other journals published by these organizations; and in publications of similar organizations in other countries.

Searching aids. The following list gives examples of specialized literature-searching aids.

A. Literature guides
  R. T. Bottle and H. V. Wyatt, *The Use of Biological Literature*, 2d ed., 1971
  H. Coblans, *Use of Physics Literature*, 1975
  D. Grogan, *Science and Technology: an Introduction to the Literature*, 2d ed., 1973

M. Hyslop, *A Brief Guide to Sources of Metals Information*, 1973

J. W. Mackay, *Sources of Information for the Literature of Geology*, 1973

L. T. Morton, ed., *Use of Medical Literature*, 1974

A. J. Walford, ed., *Guide to Reference Material*, 3d ed., vol. 1: *Science and Technology*, 1973

H. M. Woodburn, *Using the Chemical Literature*, 1974

B. Indexing serials

Applied Science and Technology Index, H. W. Wilson, 1958– (supersedes in part *Industrial Arts Index*, 1913–1957)

Bibliography of Agriculture (compiled at the National Agricultural Library), Oryx Press, 1970– (Government Printing Office, 1942–1970)

Current Programs: Life Sciences, Chemistry, Physical Sciences, Geosciences, Engineering, World Meetings Information Center, 1973–

Index Medicus (compiled at the National Library of Medicine), Government Printing Office, 1880–

Science Citation Index, Institute for Scientific Information, 1963–

Translations Register-Index, National Translations Center, 1971– (SLA Translations Center, 1967–1971; supersedes OTS's *Technical Translations*, published 1959–1967, and SLA's *Translation Monthly*, published 1955–1958)

C. Abstracting serials

Atomindex [title varies], International Atomic Energy Agency, 1959– (cf. *Nuclear Science Abstracts*, Government Printing Office, 1947–1976)

Biological Abstracts, Biosciences Information Service of Biological Abstracts, 1926–

Chemical Abstracts, Chemical Abstracts Service, 1907–

Engineering Index, Engineering Index, Inc., 1884–

Government Reports Announcements & Index [title varies], NTIS, 1971– (supersedes *U.S. Government Research Reports*, 1954–1964; and *Bibliography of Technical Reports* [title varies], 1946–1954)

Official Gazette—Patent Abstracts Section, Government Printing Office, 1968–

D. Directories

Scientific Meetings, Special Libraries Association, 1957– (regional, national, international)

U.S. Library of Congress, *A Directory of Information Resources in the United States: Biological Sciences*, 1972; *Federal Government*, 1974; *Physical Sciences and Engineering*, 1971

World Guide to Technical Information and Documentation Services, UNESCO, 1975 (lists 476 in 93 countries)

World List of Scientific Periodicals Published in the Years 1900–1960, 4th ed., 1963 (almost 60,000 entries); supplemented by *New Periodical Titles*, 1960–

[GEORGE S. BONN]

Bibliography: J. Gray and B. Perry, *Scientific Information*, 1975; *IEEE Trans. Prof. Comm.*, PC-16(3):47–194, PC-16(4):204, 1973; *Information*, pt. 2: *Reports, Bibliographies*, 3(3):1–52, 1974; *Information Hotline*, pp. 17–21, March 1976; *J. Doc.*, pp. 110–130, June 1967, and pp. 234–245, December 1975; *Libr. Trends*, pp. 793–908, April 1967; National Technical Information Service, *PR-174*, January 1976.

Lithium

A chemical element, Li, atomic number 3, and atomic weight 6.939. Lithium heads the alkali metal family in group Ia in the periodic table. In nature it is a mixture of the isotopes Li[6] and Li[7]. Much of the lithium processed in the United States has

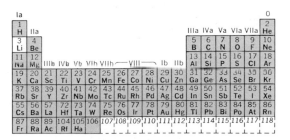

been subjected to isotope separation to obtain the pure lighter isotope, Li[6], important in thermonuclear processes; lithium available in the market may thus have a different atomic weight. Lithium, the lightest solid element, is a soft, low-melting, reactive metal. In many physical and chemical properties it resembles the alkaline-earth metals in group IIa as much as, or more than, it does the alkali metals. First discovered by J. A. Arfvedson in Sweden during an analysis of petalite ore, it was first isolated by Sir Humphry Davy in 1818 by electrolysis.

Uses. The major industrial use of lithium is in the form of lithium stearate as a thickener for lubricating greases. Lithium-base greases combine high water resistance with high-temperature resistance and good low-temperature properties. These all-purpose greases replace a multitude of specialized greases, and they have captured about one-third of the total automotive grease market.

Another important use of lithium compounds is in ceramics, specifically in porcelain enamel formulation. The major action of the lithium is as a flux. In ceramic mixtures, lithium carbonate readily undergoes the reaction in the equation below,

$$Li_2CO_3 \rightarrow Li_2O + CO_2$$

yielding the Li_2O as one constituent of the oxide mixtures of which most ceramics are composed. Alternatively, lithium aluminum silicate ores can be used directly.

There are a number of lesser uses for lithium and its compounds. Lithium hydroxide is used as an additive to give longer life and higher output in the alkaline storage batteries known as Edison cells. Lithium chloride and fluoride are used in

Table 1. Commercially important lithium minerals

Mineral	Simplified formula	% Li$_2$O, theoretical
Spodumene	LiAlSi$_2$O$_6$	8.03
Lepidolite	LiKAl$_2$F$_2$Si$_3$O$_9$	4.09
Petalite	LiAlSi$_4$O$_{10}$	4.89
Amblygonite	LiAlFPO$_4$	10.10

welding and brazing fluxes; lithium-copper and lithium-silver alloys are used as self-fluxing brazing alloys. Lithium perchlorate has been suggested as an oxidizer for solid-propellant rocket mixtures; it offers a higher percentage of available oxygen than any other perchlorate due to the low atomic weight of the lithium.

Occurrence. Lithium is a moderately abundant element and is present in the Earth's crust to the extent of 65 parts per million (ppm). This places lithium a little below nickel, copper, and tungsten, and a little above cerium and tin in abundance. In sea water there is about 0.1 ppm of lithium. It is found in human and animal organisms, in soil, in mineral waters, and in plants such as cocoa, tobacco, and seaweed. Its presence in the Sun's atmosphere has been established.

Although lithium occurs widely in nature, the concentrations in ores are generally quite low, unlike sodium and potassium whose relatively pure halides may be mined or pumped from brine wells. Lithium resembles rubidium and cesium in its manner of occurrence as part of complex silicate minerals. Lithium ore assays are usually expressed in terms of percent of lithium oxide, Li$_2$O, and 1.0% of Li$_2$O affords a commercially processable ore. Most commercial ores run from 1 to 3% Li$_2$O; virgin ores rarely contain as much as 6%. In North America the most common lithium ore is spodumene which generally occurs in pegmatite formations in association with feldspar. The largest known reserves are in the Kings Mountain area of North Carolina, in Quebec, and in Manitoba. Substantial quantities of lepidolite and petalite occur in Africa, particularly in Rhodesia. Amblygonite, the other commercially important lithium mineral, is found in Europe, Africa, and South America. A nonmineral source of lithium is the brine in Searles Lake, Calif. Table 1 lists the four commercially important lithium ores.

Metallurgical extraction. Lithium ores are concentrated from 1–3% Li$_2$O to 4–6% Li$_2$O by heavy media separation using dense, nonaqueous liquids, and by froth flotation. The silicate ores, which are those processed most widely, are then chemically cleaved by acid or alkaline processes.

In the acid process, spodumene ore is first heated in a kiln to about 1000–1100°C to change the naturally occurring alpha spodumene to beta spodumene, a more friable, less dense crystal form which is readily attacked by the acid. The ore is then roasted in a second kiln with an excess of sulfuric acid. Water leaching of the kiln product yields lithium sulfate which is treated with sodium carbonate to give lithium carbonate. The carbonate is converted to the chloride by reaction with hydrochloric acid.

In the alkaline process, spodumene or lepidolite ores are ground and calcined with about 3 parts limestone to 1 part lithium ore at 900–1000°C. Water leaching of the kiln product yields lithium hydroxide which is converted to the chloride by reaction with hydrochloric acid.

Dry lithium chloride is the feed material for the manufacture of lithium metal by electrolysis. The cell bath is actually a molten mixture of lithium chloride and potassium chloride. The cell is operated at 400–420°C with a voltage across the cell of 8–9 volts. The current consumption in electrolysis is about 18 kwhr per pound of lithium produced. The illustration shows a typical lithium cell.

Thermochemical processes for the manufacture of lithium have been studied but have not been applied commercially.

Physical properties. The physical properties of lithium metal which make it attractive for many of its present and potential uses are summarized in Table 2. (The values cited are the best available; there are many values in the literature which disagree with them.)

Noteworthy among the physical properties are the high specific heat (heat capacity), large temperature range of the liquid phase, high thermal conductivity, low viscosity, and very low density.

Lithium metal is soluble in liquid ammonia and is slightly soluble in the lower aliphatic amines, such as ethylamine. It is insoluble in hydrocarbons. There are no published data on the solubility of lithium metal in molten inorganic salts.

Chemical properties. Lithium undergoes a large number of reactions with both organic, and inorganic, reagents.

Inorganic reactions. Lithium reacts with oxygen to form the monoxide, Li$_2$O, and the peroxide, Li$_2$O$_2$. The superoxide, LiO$_2$, is not known and may not exist because of insufficient room for two oxygen atoms around the small lithium atom.

Lithium is the only alkali metal that reacts with nitrogen at room temperature to form a nitride, Li$_3$N, which is black.

Lithium reacts readily with hydrogen at about 500°C to form lithium hydride, LiH. Of all the alkali metals, lithium reacts most readily with hydrogen, and is the only alkali metal forming a hydride stable enough to be melted without decomposition. An important reaction of LiH is that with boron fluoride diethyletherate, BF$_3$·(C$_2$H$_5$)$_2$O, to yield diborane, B$_2$H$_6$, or lithium borohydride, LiBH$_4$.

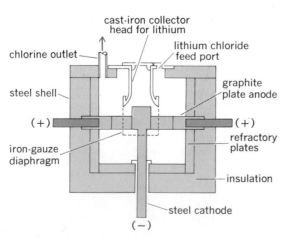

Cross section of lithium cell.

[Labels on figure:]
cast-iron collector head for lithium
lithium chloride feed port
chlorine outlet
steel shell
graphite plate anode
(+)
(+)
iron-gauze diaphragm
refractory plates
insulation
steel cathode
(−)

Table 2. Physical properties of lithium metal

Property	Temperature °C	Temperature °F	Metric (scientific) units	British (engineering) units
Density	20	68	0.534 g/cm³	33.7 lb/ft³
	400	752	0.490 g/cm³	30.6 lb/ft³
	800	1472	0.457 g/cm³	28.6 lb/ft³
Melting point	179	354		
Boiling point	1317	2403		
Heat of fusion	179	354	103.2 cal/g	186 Btu/lb
Heat of vaporization	1317	2403	4680 cal/g	8420 Btu/lb
Viscosity	200	392	5.62 millipoises	11.1 kinetic units
	400	752	4.02 millipoises	8.2 kinetic units
	600	1112	3.17 millipoises	6.7 kinetic units
Vapor pressure	727		0.78 mm	0.015 lb/in.²
	1077		91.0 mm	1.76 lb/in.²
Thermal conductivity	216	420	0.109 cal/(sec)(cm²)(cm)(°C)	26.5 Btu/(hr)(ft²)(°F)
	539	1002	0.073 cal/(sec)(cm²)(cm)(°C)	17.6 Btu/(hr((ft²)(°F)
Heat capacity	100	212	0.90 cal/(g)(°C)	0.90 Btu/(lb)(°F)
	300	573	1.02 cal/(g)(°C)	1.02 Btu/(lb)(°F)
	800	1472	0.99 cal/(g)(°C)	0.99 Btu/(lb)(°F)
Electrical resistivity	0	32	1.34 microhm·cm	
	100	212	12.7 microhm·cm	
Surface tension	200–500	392–932	About 400 dynes/cm	

The reaction of lithium metal with water is exceedingly vigorous. Lithium reacts directly with carbon to form the carbide, Li_2C_2. Lithium combines readily with the halogens at elevated temperatures, forming halides with the emission of light. Lithium does not burn in chlorine unless heated, and it reacts only superficially with liquid bromine, presumably because of the formation of protective halide coatings.

With ammonia, lithium reacts to form the amide, $LiNH_2$. The reaction may be carried out by heating lithium in a stream of ammonia at 400°C, or by reaction with liquid ammonia in the presence of a catalyst such as iron. On heating, the amide loses ammonia to form the imide, Li_2NH; lithium is the only alkali metal which forms an imide.

Lithium and carbon monoxide react to give lithium carbonyl, LiCO, a rather unstable product.

Organic reactions. While lithium does not react with paraffin hydrocarbons, it does undergo addition reactions with arylated alkenes and with dienes. The addition of lithium to diene systems is the basis for the catalytic polymerization of dienes to synthetic elastomers having natural rubber properties. One recently developed synthetic rubber employs lithium metal dispersed in petroleum jelly as the catalyst to polymerize isoprene to *cis*-polyisoprene, the exact structural counterpart of natural hevea rubber from *Hevea brasiliensis* (rubber trees). Lithium also reacts with acetylenic compounds, replacing the acetylenic hydrogen atom, and forming lithium acetylides, which are important in the synthesis of vitamin A.

Lithium metal reacts with alcohols forming lithium alkoxides, LiOR, and liberating hydrogen. Alkoxides form less readily with lithium than they do with sodium, and their commercial utility is much less than that of the sodium alkoxides.

Lithium adds to carbonyl compounds such as aldehydes and ketones giving a lithium addition compound as an intermediate which can be hydrolyzed to an alcohol. Similarly, organolithium compounds add to aldehydes and ketones giving addition products which yield alcohols upon hydrolysis. Organolithium compounds react with organic acids to give addition products which can be hydrolyzed to ketones; this reaction is used in one step of the vitamin A synthesis.

Lithium can also be used in conjunction with ammonia or amines to effect unique reductions of various organic compounds, as for example the reduction of aromatic hydrocarbons to monoolefins.

Not only lithium metal but also compounds derived directly from lithium metal are important in organic chemistry. Thus, lithium amide, $LiNH_2$, made from lithium and ammonia, is used to introduce amino groups, as a dehalogenating agent, and as a catalyst. Lithium aluminum hydride, $LiAlH_4$, made by the reaction of lithium hydride and aluminum chloride, is a powerful reducing agent for specific linkages in complex molecules. This reagent has become important in many organic syntheses.

Availability. Lithium is available in two grades, a regular grade containing 99.8% lithium and a low-sodium grade containing 0.005% sodium or less.

The metal is available commercially in the form of shot, wire, ribbon, and in cylindrical bricks. The shot are shipped under hydrocarbon and the wire and ribbon are coated with petrolatum. The bricks (usually 1 lb in weight) are shipped in hermetically sealed metal cans.

Lithium salts are many times more expensive than the corresponding sodium or potassium compounds.

Handling. Handling of lithium metal is much the same as handling sodium metal with a few specific exceptions:

1. Because lithium reacts with nitrogen, argon or helium must be used instead of nitrogen to blanket molten lithium.

2. Because molten lithium reacts with glass and ceramics, metal equipment must be used.

3. Lithium fires are most easily extinguished by the use of graphite powder, in contrast to the use of the metal halide or carbonate powders for the extinguishing of other alkali metal fires.

4. Lithium behaves differently in the presence of other materials. *See* SODIUM.

Principal compounds. The most important lithium compound in terms of major uses and pounds produced is lithium hydroxide. As discussed earlier, it is the end product resulting from water leaching of the clinker produced by limestone roasting of lithium ores. It is also produced from the lithium sulfate from the sulfuric acid roasting process. It is a white powder, and the material of commerce is actually lithium hydroxide monohydrate, $LiOH \cdot H_2O$. Tallow, or another natural fat, is cooked with the lithium hydroxide, producing lithium stearate, which is used as a thickener or gelling agent to transform oil into lithium-base lubricating grease. The U. S. Atomic Energy Commission also makes substantial purchases of the hydroxide.

Lithium carbonate, $LiCO_3$, finds application in the ceramic industries, particularly in the manufacture of frits (powdered glass) for porcelain enamel formulation.

Both lithium halides, lithium chloride and lithium bromide, form concentrated brines having the ability to absorb moisture over a wide temperature range; these brines are used in commercial air conditioning systems. The chloride and fluoride have low melting points, high boiling points, and high solvent power for metal oxides; these properties lead to commercial applications in welding and brazing fluxes for aluminum, magnesium, and titanium.

Lithium hydride, discussed earlier under inorganic reactions of the metal, is also of major importance. It was used in large amounts during World War II as a compact source of hydrogen for military balloon inflation. The deuteride formed by lithium-6 can be converted in a thermonuclear reaction to two atoms of helium with the evolution of large amounts of energy; this energy may provide the key to thermonuclear power production.

Analytical methods. In most quantitative analytical separations, all other elements are removed leaving the alkali metal ions in solution. Lithium chloride is separated from the other alkali chlorides by virtue of its greater solubility in organic solvents, and then converted to lithium sulfate, dried, and weighed.

For qualitative identification of lithium, the characteristic crimson color produced when lithium compounds are introduced into a hot flame is very satisfactory. A flame spectrophotometer may be used to measure the intensity of the characteristic lithium color at 670.8 millimicrons wavelength. This technique is the most suitable for routine determination of lithium because it is fast and accurate. *See* ALKALI METALS.

[MARSHALL SITTIG]

Bibliography: F. A. Cotton and G. Wilkinson, *Advanced Inorganic Chemistry*, 2d ed., 1966; R. N. Lyon, *Liquid Metals Handbook*, 2d ed., Navexos P-733, 1954.

Lithosphere

A term initially used in a general way to distinguish the rock phase of the Earth from the hydrosphere, biosphere, and atmosphere at the surface of the Earth. The term lithosphere has been used variously in recent geological literature to designate the crust ("upper lithosphere"), the crust plus the mantle (that is, the total silicate phase of the Earth), or the entire solid part of the Earth (crust + mantle + core). The core is thought to consist mainly of a mixture of iron and nickel, the mantle mainly of iron and magnesium silicates, and the crust of siliceous rocks with an average composition somewhere near that of granodiorite. *See* ATMOSPHERE; BIOSPHERE; EARTH; ELEMENTS, GEOCHEMICAL DISTRIBUTION OF; HYDROSPHERE; LITHOSPHERE, GEOCHEMISTRY OF.

[HEINRICH D. HOLLAND]

Lithosphere, geochemistry of

The study of the distribution of the elements and of their isotopes in the lithosphere and of the processes, both past and present, which affect this distribution. The lithosphere contains more than 99% of the mass of the Earth; it is therefore by far the most important reservoir of almost all of the chemical elements, and the processes in the lithosphere exert a dominant influence on many surface phenomena.

Knowledge of the distribution of the elements and of their isotopes in the lithosphere is very uneven. Almost all of the samples that have actually been analyzed have come from the present surface of the Earth or from a depth of less than 8 km. Some materials from depths of several kilometers have been brought to the surface in kimberlite pipes and are yielding important clues regarding the nature of the upper mantle. The average radius of the Earth is, however, 6371 km; analyzed samples therefore represent only scratchings from the outermost portions of the globe. The evidence for the chemical composition of the Earth as a whole is largely indirect and is based on the data of geophysics and inferences from the composition of meteorites. Since the mass and volume of the Earth are known, its average density can be calculated. On the basis of travel-time curves for sound waves generated by earthquakes and man-made explosions, convincing models of the changes with depth in the density and in the velocity of compressional and shear waves can be constructed. These data, together with information on the magnetic field of the Earth and the mechanical and thermal properties of the Earth, contribute to the fund of information used in constructing the presently accepted model for the chemical structure of those portions of the Earth which cannot be observed directly. *See* SEISMOLOGY.

Data regarding the chemical composition of meteorites complement the geophysical data. Models based on geophysical data and those based on meteorite data are surprisingly similar, and present concepts of the major features of the chemistry of the Earth's interior are almost certainly correct for the major elements.

Such a degree of certainty cannot be claimed for most of the minor elements. Fewer than 10% of the chemical elements account for over 98% of the mass of the lithosphere. The abundance of the majority of the elements in the lithosphere is therefore well below 0.1%. The advances in trace element analysis since 1920 have opened the study of the distribution even of elements present in rocks in amounts less than 1 part per million (ppm), or $10^{-4}\%$, so that the geochemistry of these elements in the upper few kilometers of the lithosphere is reasonably well known. However, geophysical

Average chemical composition of sediments and rocks*

Sediments and rocks	Index	SiO$_2$	TiO$_2$	Al$_2$O$_3$	Fe$_2$O$_3$	FeO	MnO	MgO	CaO	Na$_2$O	K$_2$O	P$_2$O$_5$	CO$_2$	H$_2$O§
Sediments and sedimentary rocks														
Calcareous sands, oozes†	1	18.8	0.3	5.1	3.8		0.4	1.4	39.0	0.5	0.7	0.2	29.8	
Red clay†	2	52.8	0.9	16.4	9.0		1.0	2.9	7.7	1.7	2.7	0.3	4.6	
Siliceous oozes†	3	55.5	0.5	15.1	5.8		0.7	2.3	9.7	1.0	2.2	0.3	6.9	
Average pelagic sediment†	4	28.5	0.4	8.1	5.0		0.6	1.8	30.5	0.8	1.2	0.2	22.9	
Composition of suboceanic sediments†	5	43.2	0.7	11.6	4.6	0.6	0.3	2.4	21.1	1.1	1.8	0.1	12.5	
Average sediment†	6	44.5	0.6	10.9	4.0	0.9	0.3	2.6	19.7	1.1	1.9	0.1	13.4	
Average shale‡	7	58.10	0.65	15.40	4.02	2.45		2.44	3.11	1.30	3.24	0.17	2.63	5.00
Average sandstone‡	8	78.33	0.25	4.77	1.07	0.30		1.16	5.50	0.45	1.31	0.08	5.03	1.63
Average limestone‡	9	5.19	0.06	0.81	0.54			7.89	42.57	0.05	0.33	0.04	41.54	0.77
Average sediment‡	10	57.95	0.57	13.39	3.47	2.08		2.65	5.89	1.13	2.86	0.13	5.38	3.23
Igneous rocks														
546 granites† 4	11	70.8	0.4	14.6	1.6	1.8	0.1	0.9	2.0	3.5	4.1	0.2		
137 granodiorites† 6	12	67.2	0.6	15.8	1.3	2.6	0.1	1.6	3.6	3.9	3.1	0.2		
635 intermediate igneous rocks† 13	13	54.9	1.5	16.7	3.3	5.2	0.2	3.8	6.6	4.2	3.2	0.4		
198 basalts† 19	14	49.9	1.4	16.0	5.4	6.5	0.3	6.3	9.1	3.2	1.5	0.4		
182 ultramafic rocks† 25	15	44.0	1.7	6.1	4.5	8.8	0.2	22.7	10.2	0.8	0.7	0.3		
Average igneous rocks† 11	16	60.1	1.1	15.6	3.1	3.9	0.1	3.5	5.2	3.9	3.2	0.3		
Metamorphic rocks														
250 quartzofeldspathic gneisses† 50	17	70.7	0.5	14.5	1.6	2.0	0.1	1.2	2.2	3.2	3.8	0.2		
103 mica schists† 61	18	64.3	1.0	17.5	2.1	4.6	0.1	2.7	1.9	1.9	3.7	0.2		
61 slates† 65	19	61.8	0.7	19.1	3.3	5.4	0.2	2.9	1.0	1.7	3.8	0.1		
200 amphibolites† 68	20	50.3	1.6	15.7	3.6	7.8	0.2	7.0	9.5	2.9	1.1	0.3		
Crust of the Earth														
Average composition†	21	55.2	1.6	15.3	2.8	5.8	0.2	5.2	8.8	2.9	1.9	0.3		
Average composition‡	22	60.18	1.06	15.61	3.14	3.88		3.56	5.17	3.91	3.19	0.30		

*Based on summaries by A. Poldervaart, *Crust of the Earth*, Geol. Soc. Amer. Spec. Pap. no. 62, 1955, and by B. Mason, *Principles of Geochemistry*, 3d ed., Wiley, 1966.

†From Poldervaart; figure, if given, refers to index number as used in Poldervaart.

‡From Mason.

§Values omitted where samples were calculated water-free.

techniques are not suitable for such studies in the deeper parts of the Earth, and the use of meteorite data is always beset by uncertainties regarding the assumption that there is necessarily a close correspondence between the abundance of elements in the analyzed meteorites and in the Earth's interior.

Chemical composition of crust. Analyses numbering in the tens of thousands have been made on geologic materials. All or nearly all of the analyzed samples have come from the crust of the Earth, that is, from above the Mohorovičić discontinuity, which lies at a depth of about 30–50 km under the continents and at a depth of about 5–10 km under the oceans. Although many of these chemical analyses are of somewhat questionable accuracy, the large mass of available chemical data is amply sufficient to give an accurate description of the average composition and of the compositional variation of the various types of geologic materials. This information can be combined with data on the relative abundance of the various rock types in the Earth's upper crust to give a fairly complete description of the average chemical composition of directly accessible parts of the lithosphere.

Some data on the chemical composition of various sediment and rock types are given in the table. A comprehensive statement of all the elements in each geochemical phase is given elsewhere. *See* ELEMENTS, GEOCHEMICAL DISTRIBUTION OF.

One of the most striking aspects of these data is the relatively small range of chemical composition in the upper part of the lithosphere. The same dozen or so elements predominate in all of the major sediment and rock types. All of these, except the carbonate rocks, contain roughly between 40 and 75% silica, SiO$_2$. The carbonate rocks are almost always calcium or calcium-magnesium carbonates, often containing an admixture of siliceous material. The crustal abundances of the major elements, and also of the minor elements, therefore do not depend in a very critical way on the relative proportions assigned to the various sediment and rock types in the Earth's crust. Nevertheless, it should be remembered that the crustal abundances listed in the table and in Figs. 1 and 2 must be regarded as averages subject to revision.

The crustal abundance data have been plotted

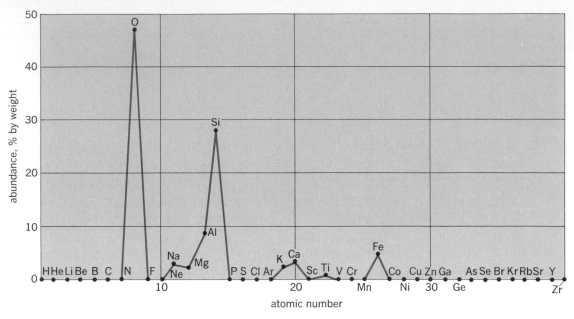

Fig. 1. Crustal abundance of elements of atomic number 1–40.

on a linear scale in Fig. 1 and on a logarithmic scale in Fig. 2. The first plot emphasizes the fact of the preponderance of certain of the elements of low atomic number. No element with atomic number greater than 26 is present in the Earth's crust with an abundance greater than 0.1%. The logarithmic plot of Fig. 2 permits the representation of the abundance of all of the naturally occurring elements and emphasizes the frequency of crustal abundance values between 10 ppm (10^{-3}%) and 0.1 ppm (10^{-5}%), particularly among the elements of higher atomic number. The broad features of these

figures are now reasonably well understood. They are the product of the processes of element formation, of the processes operating during the formation of the Earth, and of the processes within the earth, which have operated both in the direction of homogenization and diversification of the chemical composition of the various parts of the Earth. *See* ELEMENTS, COSMIC ABUNDANCE OF; ELEMENTS AND NUCLIDES, ORIGIN OF.

Chemical composition of mantle. The Earth's mantle consists of material between the Mohorovičić discontinuity, marking the base of the crust,

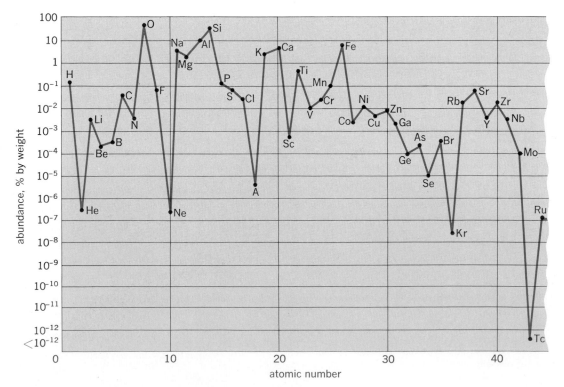

Fig. 2. Crustal abundance of elements of atomic number 1–92.

and the discontinuity at a depth of 2898 km, marking the outer limit of the Earth's core. Present thinking regarding the chemical composition of this portion of the Earth is based largely on geophysical evidence and on the circumstantial meteorite evidence. The latter line of evidence suggests that the composition of the mantle is not very different from that of ultramafic igneous rocks (see index 15 in the table); the geophysical evidence is in substantial agreement with this hypothesis. However, the evaluation of the density and elastic wave propagation data is at present somewhat tenuous, since observations on the state of silicate minerals at temperatures and pressures characteristic of the middle and lower portions of the mantle are still fragmentary. It is very likely that silicon, oxygen, magnesium, iron, and calcium are the major elements in the Earth's mantle and that its bulk composition corresponds roughly to a 1:3 mixture of basalt and periodotite. It is also likely that the mantle, at least in its upper portion, is more heterogeneous than has usually been assumed, and that its structure may be almost as complicated as that of the crust.

Knowledge of the abundance of the minor elements in the mantle is even more fragmentary than that regarding the major elements. Only in the case of the radioactive elements can something be said independent of the meteorite model. The abundance of these elements must be low enough to prevent melting of the mantle, and this must mean that the abundance of potassium, uranium, and thorium must be very much less in the mantle than in the crust. In theory, heat flow measurements can give more definite indications about maximum concentrations of these elements in the mantle. However, the complexities introduced by radiative and possibly convective heat flow in the mantle introduce serious complications into the interpretation of surface heat flow measurements. *See* EARTH, HEAT FLOW IN.

Chemical composition of core. The core of the Earth consists of the material between a depth of 2898 km and the center of the Earth. Problems similar to those besetting the study of the composition of the mantle also limit study of the core. Again, all data are circumstantial. The meteorite data suggest a core of metallic iron with an admixture of nickel. The geophysical data indicate that this is distinctly probable, but that the average atomic number of core material is less than that of iron. Silicon has been suggested as an important core constituent, but several other alternatives, including sulfur, are at least possible, and the problem remains open. A small amount of potassium may be present and may supply the energy for convection in the core. The concentration of other minor constituents can be estimated roughly from the chemistry of meteorites. *See* METEORITE.

Important crystalline phases. The major portion of the lithosphere is solid. In the crust the solids are predominantly well crystallized, and this is almost certainly true for the mantle as well. Oxygen and silicon are the most abundant elements in the crust and in the mantle, and the solid phases are almost all either oxides or silicates. Since the processes described below depend to a considerable extent on the properties of these compounds, the crystal chemistry of the major mineral phases is briefly reviewed at this point. *See* MINERALOGY.

Oxide structures. The structure of most of the rock-forming oxides is simple. In all of them O^{--} is the largest ion and determines the structural framework. In periclase, MgO, the O^{--} ions are cubic close-packed and the Mg^{++} ions occupy the octahedrally coordinated holes between neighbor-

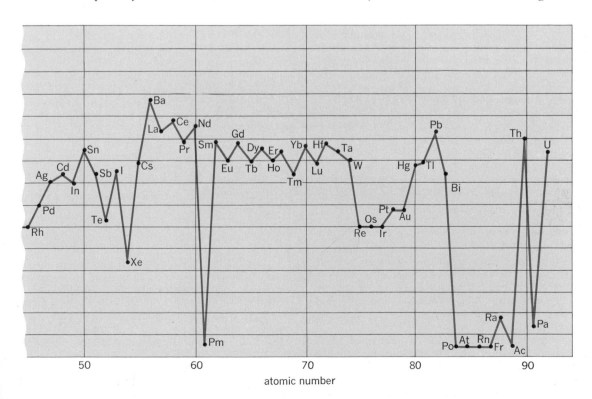

ing O^{--} ions. In the spinel group, $R''R'''_2O_4$, the divalent R'' ions and the trivalent R''' ions again occupy either tetrahedrally or octahedrally coordinated holes between cubic close-packed O^{--} ions. Hexagonal close packing of O^{--} ions is found in corundum, Al_2O_3, and in related minerals such as ilmenite, $FeTiO_3$, and geikielite, $MgTiO_3$. In these structures the cations again occupy 4- or 6-coordinated holes in the O^{--} framework.

Silicate structures. The rather improbable formulas suggested for many silicate minerals by chemical analyses before the advent of x-ray crystallography were shown to be quite natural results of their often complicated silicon-oxygen framework structures, and the development of the crystal chemistry of the silicates has been one of the most significant contributions of mineralogy to geology since 1920. The structure of silicate minerals is normally more complicated than that of the oxide. The basic unit of structure is the silicon-oxygen tetrahedron, SiO_4^{4-}, consisting of a Si^{4+} ion at the center of a tetrahedron whose corners are occupied by the center of O^{--} ions. These tetrahedrons may be joined by cations, as in the olivines $(Mg, Fe)_2SiO_4$, zircon, $ZrSiO_4$, or the garnets, $R''_3R'''_2(SiO_4)_3$. In these structures the $[SiO_4^{4-}]$ group serves a function similar to that of the $[SO_4^{2-}]$ group in the sulfates. The structure of the olivines is similar to that of the spinels; in fact, olivines invert to a spinel form at high pressures.

In the majority of silicate compounds the $[SiO_4^{4-}]$ groups are polymerized. Such polymerization may lead to groups of two, three, or six tetrahedrons, to effectively infinite chains or double chains, to sheet structures, or to three-dimensional networks of $[SiO_4^{4-}]$ groups. The physical properties of silicate minerals are often related in a simple manner to the type of polymerization of the $[SiO_4^{4-}]$ groups. The cleavage planes of the pyroxenes, the amphiboles, and the feldspars are parallel to the effectively infinite chains or double chains characteristic of these mineral groups. Sheets of $[SiO_4^{4-}]$ tetrahedrons are parallel to the pronounced cleavage of the micas, clays, and chlorite group minerals. The uniform, three-dimensional network of quartz is reflected in the lack of cleavage in this mineral.

At pressures in excess of roughly 10,000 atm the structure of many silicates transforms to higher-density modifications. Quartz collapses to coesite and then to stishovite; it is very likely that the mineralogy of the mantle is strongly influenced by such phase changes. *See* SILICATE MINERALS.

Processes involving lithosphere. The Earth is not at rest. Earthquakes, volcanism, and mountain building, the three major manifestations of disequilibrium at depth, are evidence of widespread physical and chemical change below the surface of the Earth. Erosion and chemical weathering are evidence of the pronounced disequilibrium between the atmosphere and the lithosphere. Chemical changes thus are taking place at various levels in the lithosphere and at the boundary between the lithosphere and the atmosphere, biosphere, and hydrosphere. The processes taking place at the Earth's surface can be observed directly and therefore are more completely understood than the processes taking place within and below the Earth's crust. Yet it is just the processes at depth that are of primary interest as a key to the history of the crustal evolution of the Earth.

For most of the elements, the Earth can be considered a closed or an almost closed system. The two major exceptions to this rule are hydrogen and helium, which escape at significant rates from the upper atmosphere. Hydrogen escape has probably played an important role in the evolution of the atmosphere. A number of elements are entering the atmosphere as constituents of meteorites and as cosmic-ray particles, or as interplanetary material which is swept up by the Earth. It seems unlikely that these additions have affected the chemistry of the Earth materially.

The Earth therefore can be regarded as an essentially closed system consisting of a number of reservoirs. The exchange between these reservoirs may be appreciable and is often a complex function of time. During the course of history, the mass and composition of various reservoirs have changed radically. The basic problem in the study of the dynamics of geochemistry consists of defining suitable reservoirs and of determining the nature and the rates of the processes that are operating on them today and have done so in the past.

The primary division of the Earth into the lithosphere, hydrosphere, and atmosphere is convenient since the limits of these reservoirs are reasonably well defined and their contents are appreciably different. The biosphere has a somewhat different status since in space it overlaps the other spheres completely. The primary division into four spheres is too coarse for many problems in earth science. The subdivision of the lithosphere into crust, mantle, and core is evidence of this, and the further subdivision of the crust into sediments and sedimentary rocks, metamorphic rocks, and igneous rocks was implied in the discussion of the average chemical composition of materials in these categories. Further subdivisions of these large units into regional reservoirs, or local reservoirs, is demanded by many problems. For instance, in studies of the age of an igneous rock unit the systems of interest are individual crystals of the mineral or minerals used for dating purposes. The problem, in general terms, thus consists of determining the events that have taken place within the system of interest and the extent of exchange that has occurred between the system and its environment. *See* ROCK, AGE DETERMINATION OF.

Changes in systems are of various kinds. Certain systems probably have changed rather little in size and composition during much of earth history. The core of the Earth may well be one such system. Other systems probably have grown steadily throughout earth history. The hydrosphere is almost certainly one of these, and the Earth's crust itself may well have increased appreciably in volume at the expense of the Earth's mantle. Still other systems, the biosphere for one, have probably fluctuated appreciably in volume during geologic time. All these systems are at least partially interdependent, so that the dynamics of the Earth are exceedingly complicated and are generally difficult to reduce to manageable proportions.

The major cycle. Many of these processes participate in what has come to be known as the major geochemical cycle, whose most important aspects are shown in Fig. 3. The weathering of igneous, sedimentary, and metamorphic rocks at the lithosphere-atmosphere boundary produces sediments. These are buried and compacted at depth to produce sedimentary rocks. At more elevated temperatures or pressures or both, sedimentary rocks recrystallize to metamorphic rocks. At still higher temperatures partial or complete melting may produce rock melts, which cool to form igneous rocks. Exposure of these rocks to weathering processes then renews the cycle.

Weathering processes. The chemical disequilibrium between the atmosphere and many of the rock types exposed to it is one of the important facts of geochemistry. The reactions of hydration, solution, oxidation, and carbonation, characteristic of the zone of weathering, are part of the continuing process of the approach to equilibrium at the lithosphere-atmosphere boundary.

The effect of the atmosphere, rain, ice, and groundwater on various minerals depends on their stability in this type of environment and on the rate at which unstable minerals disappear in favor of stable ones. The feldspars, pyroxenes, and olivine are among the most readily destroyed rock-forming minerals. In this type of environment, the removal of the larger cations (Na^+, K^+, Ca^{++}, Mg^{++}) is usually rapid, and the remaining minerals are usually hydrated aluminum silicates containing none of these cations or containing them in appreciably smaller quantities than the original minerals. Some unstable rock-forming minerals react much more slowly with the atmosphere. Magnetite, Fe_3O_4, is one of these. The stable oxide of iron in equilibrium with atmospheric oxygen is hematite, Fe_2O_3, yet magnetite is a common constituent of beach sands, accumulates of the residue of weathering processes. Finally, minerals such as quartz, SiO_2, and calcite, $CaCO_3$, are stable in the zone of weathering. These minerals will only dissolve in rainwater and groundwater until these waters are saturated with respect to SiO_2 and $CaCO_3$. Weathering and transport processes thus alter exposed units of the lithosphere in the direction of equilibrium with the atmosphere but often do not reach this state completely. *See* WEATHERING PROCESSES.

Sedimentation and diagenesis. The processes that take place as sediments enter the oceans are complicated but normally alter their overall chemistry and mineralogy only slightly. However, major ore deposits of iron, potentially major deposits of copper and nickel in manganese nodules, and very large, low-grade uranium deposits have formed in the oceans, and evaporation in closed or partially closed basins has led to the formation of evaporite beds of composition very different from that of the normal detrital sediments. Sequences of sulfates, carbonates, and halides formed in such environments therefore are interspersed with sediments consisting largely of silicates and oxides. The overall composition of sedimentary rocks is appreciably richer in carbon dioxide, water, chlorine, nitrogen, sulfur, and boron than average igneous rock. This is not unexpected since CO_2 and H_2O, as well as geologically important compounds of the

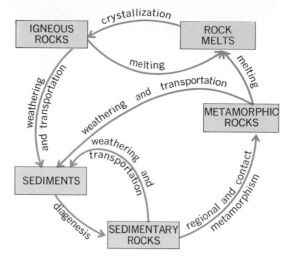

Fig. 3. Important features of major geochemical cycle.

other elements, are in a gaseous state at the temperature of rock melts; they are undoubtedly lost preferentially to the Earth's surface and finally incorporated in sediments. *See* EVAPORITE, SALINE; FUSED-SALT PHASE EQUILIBRIA.

After deposition most sediments become buried. They therefore are subjected to progressively higher temperatures and pressures. At temperatures up to 200°C and pressures up to 2000 atm cementation, a loss of porosity, and a certain amount of recrystallization take place, and sandstones, limestones, and shales (the major types of sedimentary rocks) are formed. Such changes are perhaps most pronounced in the case of carbonaceous sediments. Peat becomes dehydrated and passes through the various grades of coal, approaching the composition of graphite with increasing depth of burial. Organic compounds in marine sediments can break down to form hydrocarbons; in suitable settings, these accumulate to form gas and oil fields. *See* DIAGENESIS.

Metamorphism. At higher temperatures and pressures important changes take place in the mineralogy and texture of most sedimentary rocks. These changes are termed metamorphism and lead to the formation of metamorphic rocks. There is no sharply defined boundary between diagenetic and metamorphic processes. Metamorphism may be either local in nature, as in the vicinity of a body of intrusive igneous rock, or may be regional, occasioned by a rise in temperature, pressure, or both over a considerable area. Reactions in both contact and regional metamorphism are determined by the prevailing pressure and temperature, by the composition of the rock units, and by the kinetics of the possible reactions. Understanding of these reactions has been increasing, especially since the introduction of water carbon dioxide, and salts as components in systems studied in the laboratory. Many metamorphic reactions are nearly isochemical, but volatile compounds such as H_2O, CO_2, CH_4, and H_2S are frequently lost, and cation-exchange reactions involving the chlorides of Na^+, K^+, Ca^{++}, and Mg^{++} may be important. *See* METAMORPHISM; SILICATE PHASE EQUILIBRIA.

Melting and crystallization. At temperatures somewhat above 600°C, melting of rock units begins. The composition of rock melts depends on the composition of the parent material, on the ratio of liquid volume to that of the solid residuum, on the temperature and pressure during melting, and on the degree of equilibrium attained between the liquid and solid phases. The liquid fraction may move into higher levels of the Earth's crust and may break through to the surface as lava or other volcanic products. Crystallization normally takes place at the lower temperatures higher in the crust; the nature of the resulting solids is determined largely by the initial composition of the liquid and by the conditions prevailing during cooling. With the aid of pertinent laboratory studies the relationships between many different igneous rock types from the highly silicic granites through the intermediate gabbros to the ultramafic peridotites and dunites have been satisfactorily established, although the major question of the origin of the bulk of the siliceous rocks of the upper crust is still largely unanswered. *See* MAGMA.

This question leads directly to the problem of the nature and extent of exchange between mantle and crustal material. The validity of the proposal of ocean-floor spreading by H. H. Hess has been supported dramatically by studies of the intensity and direction of magnetization of oceanic rocks. F. J. Vine was the first to demonstrate the marked symmetry of the magnetization of rocks on opposite sides of mid-oceanic ridges and to suggest that this symmetry is related to ocean-floor spreading due to upwelling of mantle material under mid-oceanic ridges. W. J. Morgan has shown that this process is consistent with the present evidence of worldwide faulting and crustal movement.

The rapidity of this ocean-floor spreading implies that exchange of material between the mantle and the crust may be quantitatively important. The data suggest that sediments and volcanics on the ocean floor are returned to the mantle, that the reaction of sea water with oceanic basalt is quantitatively important, and that mantle-crust exchange during geologic time must be considered a two-way process.   [HEINRICH D. HOLLAND]

Bibliography: L. H. Ahrens (ed.), *Origin and Distribution of the Elements*, 1969; W. Eitel, *Structural Conversions in Crystalline Systems and Their Importance for Geological Problems*, Geol. Soc. Amer. Spec. Pap. no. 66, 1958; V. M. Goldschmidt and A. Muir (eds.), *Geochemistry*, pt. 1, 1954; K. Krauskopf, *Introduction to Geochemistry*, 1967; B. Mason, *Principles of Geochemistry*, 3d ed., 1966; A. Poldervaart, *Crust of the Earth*, Geol. Soc. Amer. Spec. Pap. no. 62, 1955; A. E. Ringwood, in P. M. Hurley (ed.)., *Advances in Earth Science*, 1966; H. J. Rösler and H. Lange, *Geochemical Tables*, 1972; O. F. Tuttle and N. L. Bowen, *Origin of Granite in the Light of Experimental Studies in the System* $NaAlSi_3O_8$-$KAlSi_3O_8$-H_2O, Geol. Soc. Amer. Mem. no. 74, 1958; K. H. Wedepohl (exec. ed.), *Handbook of Geochemistry*, 1969; H. G. F. Winkler, *Petrogenesis of Metamorphic Rocks*, 2d ed., 1968.

Litopterna

Hoofed herbivores confined to the Cenozoic of South America. The order was well represented from the Paleocene to the Pleistocene, and apparently arose on that continent from a condylarth

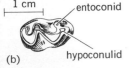

LITOPTERNA

Fig. 1. Molars, (*a*) upper and (*b*) lower, of *Proterotherium*, an early Miocene proterotheriid litoptern from the Santa Cruz Formation of Patagonia, Argentina.

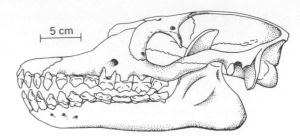

Fig. 2.   Skull and jaw of *Theosodon garrettorum*, an early Miocene macraucheniid litoptern, Santa Cruz Formation, Patagonia, Argentina. (*After W. Scott, 1910*)

ancestry. By later Paleocene time two main lines of descent were clearly demarcated. The Proterotheriidae (Fig. 1) displayed a remarkable evolutionary convergence with the horses in their dentition and in reduction of the lateral digits of their feet. In one group the foot was reduced to a single median toe by early Miocene time. The members of the Macraucheniidae (Fig. 2) were proportioned much as in the camels and by late Tertiary time had similarly lost the vertebral arterial canal of the cervical vertebrae. They early developed a tendency toward retraction of the nasal bones; by late Tertiary time, the nasal opening was situated above the eyes on top of the skull. This specialization has been thought to indicate the presence of a trunk or perhaps dorsally situated nostrils adapted for a submerged amphibious existence. Three functional digits remain in the feet of even the most advanced macraucheniids. *See* CONDYLARTHRA.

Both groups share the following characters held typical for the order. The skull was without expansion of the temporal or squamosal sinuses, as in the notoungulates; a postorbital bar was present. Dentition was primitive, with full eutherian formula and a tendency to reduce incisors and upper canine teeth; cheek teeth were unreduced, low- to high-crowned, the posterior premolars becoming molariform. The upper molars were quadrate, the paracone and metacone were crescentic, the protoloph was present and the metaloph was absent. The lower molars were bicrescentic; the entoconid was not isolated but joined to the posterior end of the tooth. The feet were functionally three-toed or reduced to a single digit, the weight borne by digit three. *See* DENTITION (VERTEBRATE).

The earliest known mammalian faunas in South America (later Paleocene) contained representatives of both groups of litopterns. By Oligocene and Miocene time the litopterns reached their greatest recorded abundance. The proterotheres did not survive the Pliocene, and of the macraucheniids only the large *Macrauchenia* survived until Pleistocene time. *See* MAMMALIA.

[RICHARD H. TEDFORD]

Liver (vertebrate)

A large gland found in all vertebrates. It consists of a continuous parenchymal mass arranged to form a system of walls (muralium) through which venous blood emanating from the gut must pass. This strategic localization between nutrient-laden capillary beds and the general circulation is associated with hepatic regulation of metabolite levels in the blood through storage and mobilization mechanisms controlled by liver enzymes.

The liver is the sole source of such necessary constituents of the blood as fibrinogen, serum albumin, and cholinesterase (an enzyme which may prevent harmful buildup of the neuronal transmitter agent acetylcholine). In the embryonic stage of most vertebrates the liver serves as the major manufacturing site of erythrocytes, a process known as erythropoiesis. The liver also removes toxins from the systemic circulation and degrades them, as well as excess hormones, by oxidation, hydroxylation, or conjugation. Particulate material may be removed through a phagocytic action of specialized cells (Kupffer cells) lining the lumen of the hepatic "capillary spaces," or sinusoids. In addition to the products which the liver delivers directly to the general circulation (endocrine function), it secretes bile through a duct system which, involving the gallbladder as a storage chamber, eventually passes into the duodenum (exocrine function). Bile functions as an emulsifier of fats to facilitate their digestion by fat-splitting lipases, and may also activate the lipase directly.

The liver does not exist in invertebrates except as the smaller analogous hepatopancreas of the crustaceans. In the vertebrate embryo the liver is a rapidly growing organ, but it rarely undergoes cellular division in the adult unless a portion is removed. Excision leads to such rapid restoration that regenerating mammalian liver has been used as a model in growth studies. As the largest discrete internal organ, it may make up more than 5% of total body weight in rodents and up to 3% in man. Though highly variable in shape in different vertebrate groups, the basic features of circulation are probably similar in all livers.

The major afferent blood supply (70%) is from the hepatic portal vein, carrying blood from the spleen and digestive organs, with a smaller contribution from the hepatic artery. The smaller branches of these vessels join with the bile ductules and, as portal canals, reach the sinusoids. The major efferent vessel is the hepatic vein, whose smaller processes are known as central veins. Drainage is into the inferior vena cava.

Within the liver there are small branches of both the hepatic artery and portal vein supplying an elaborate network of capillaries apposed to the bile canaliculi (minute channels coursing in a continuous network between the parenchymal cells of the liver). These capillaries drain into the portal vein to form a minute portal system. This subsidiary system, supplying the liver cells directly with nutrient and oxygen, may represent a significant site for hepatic artery function in some species. In a number of mammalian species, for example, the pig, a discrete connective tissue boundary demarcates the liver mass into hexagonal bundles or lobules with portal canals arranged peripherally and central veins occupying the center. Even those forms lacking patent lobular arrangements demonstrate lobular distribution of key metabolites.

COMPARATIVE ANATOMY

The liver is found only in vertebrates and is ubiquitous in this group. Despite the variations in embryonic hepatic development encountered in the various vertebrate prototypes, all adult livers

are essentially similar. This phenomenon is at sharp variance with Haeckel's biogenetic law, which stresses the similarity of embryonic structures and the divergence of adult structures. It also attests to the evolutionary success of the basic pattern of hepatic architecture: a continuous convoluted wall of functional parenchymal cells tunneled by passageways which contain a network of blood vessels and biliary tracts in homogeneous fashion within the organ.

In *Branchiostoma* (*Amphioxus*) a hollow outpocketing of the foregut, called a hepatic cecum, receives veins from the intestine. This may presage the portal system and liver of higher forms, and many consider this structure the first true homolog of the vertebrate liver.

Humans. Knowledge of the precise connections between liver and intestine and of liver and systemic circulation provides the mechanical basis for understanding hepatic function (Fig. 1). Although complex biochemically, the human liver is extremely simple structurally; it will be used here as a model for the vertebrates in terms of gross anatomic features, blood supply, and intrahepatic morphology. The liver is a remarkably plastic structure, and its general contour varies greatly from one class to another; however, the underlying

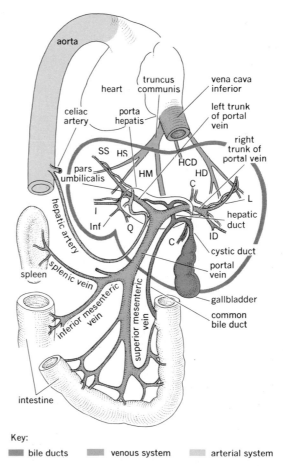

Key:

▆ bile ducts   ▨ venous system   ░ arterial system

Fig. 1. Schematic diagram showing the human liver and its major dorsal connections. Rami venae portae: C, centralis; I, intermedius; ID, inferior dexter; Inf, inferior sinister; L, lateralis; Q, quadratus; and SS, superior sinister. Venae hepaticae: HCD, dorsocaudalis; HD, dextra; HM, media; and HS, sinister.

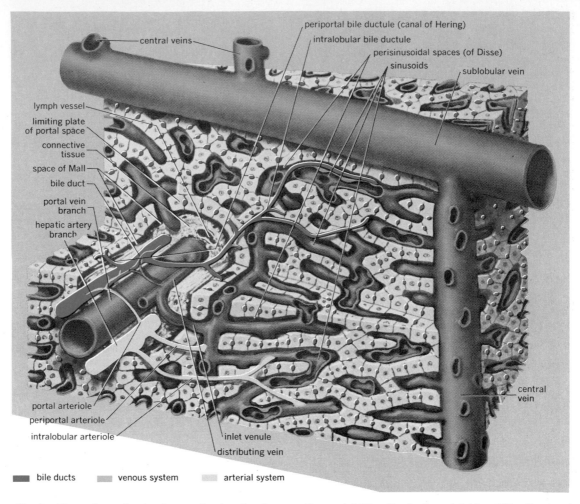

Fig. 2. Three-dimensional schema showing the liver structure (*after H. Elias*). (*From E. Oppenheimer, ed., prepared by F. Netter, The Ciba Collection of Medical Illustra-* *tions, vol. 3: Digestive System, pt. 3: Liver, Biliary Tract and Pancreas, Ciba, 1957*)

features that determine function are sufficiently similar to justify selection of a single prototype.

General features. The human liver is a massive wedge-shaped organ divided into a large right lobe and a smaller left lobe. Its anterior surface underlies the diaphragm. The upper portion of the liver is partially covered ventrally by the lungs, whereas the lower portion overhangs the stomach and intestine. Both liver and gallbladder are innervated by the right phrenic nerve and by branches of the sympathetic fibers arising from the celiac ganglion. The vagus constitutes the parasympathetic source. The entire liver is covered by Glisson's capsule, an adherent membranous sheet of collagenous and elastic fibers.

Blood supply. Venous blood from the intestine, and to a lesser extent from spleen and stomach, converges upon a short broad vessel, called the hepatic portal vein, which enters the liver through a depression in the dorsocaudal surface termed the porta hepatis. There the hepatic portal vein divides (Fig. 1) into a short right branch (truncus dexter) and a longer left branch (truncus sinister). These vessels then ramify into the small branches which actually penetrate the functional parenchymal mass as the inner tubes of the portal canals (Fig. 2).

The hepatic artery, which is a branch of the larger celiac artery, also enters at the porta hepatis and ramifies into smaller branches, which flank the portal venules within the portal canals (Figs. 1 and 2). Anastomoses of these arterial pairs result in the formation of a loose arterial plexus around each portal-vein branch. Arterial blood is also brought to the bile ducts and the lining cells of the portal canals by other sets of arterioles branching from tributaries of the hepatic artery.

The branches of the portal vein and hepatic artery then empty into the sinusoids (Fig. 2), the specialized capillaries lying within the spaces (lacunae hepatis) of the hepatic muralium. The sinusoids, which are major regions of hepatovascular exchange, communicate with small branches of the hepatic veins and, through the hepatic vein, the blood is returned to the heart by way of the vena cava. *See* CARDIOVASCULAR SYSTEM (VERTEBRATE).

The hepatic canals house the hepatic venous branches in a manner similar to the containment of portal-vein and hepatic-artery branches within the portal canals. These respective canal systems are independent and are separated by the parenchymal cells (Fig. 3). Sinusoidal elements provide a functional continuum between afferent and efferent vascular systems. The absence of anastomoses within the portal canal system, as well as within the hepatic canal system, leads to a highly discrete pattern of vascular supply and drainage. This is of particular importance in surgical manipulations of the liver, since collateral circulation is of little significance in providing connections to those lobes whose vessels have been ligated.

Bile ducts. The tiny bile canaliculi, which lie between grooves in adjacent parenchymal cells, communicate with tiny intralobular bile ducts (cholangioles or ductules), termed the canals of Hering (Figs. 2 and 4). These intralobular bile ducts empty into increasingly larger interlobular bile ducts which lie within the portal canals and make up the third element of the so-called portal triad (Fig. 5).

Although the biliary system is intimately associated with the portal system anatomically, even in terms of egress through the porta hepatis, it should be stressed that bile flow within this biliary system is in the opposite direction from that of blood flow. The hepatic duct arises in the region of the porta from the confluence of the largest bile ducts.

Gallbladder. A large outpouching of the hepatic duct forms the gallbladder (vesica fellea) and its cystic duct. Beyond the bifurcation the common bile duct enters the duodenum, thus serving as a conduit for bile from both the liver and its bile storage organ. Several mammals (for example, the horse and rat) do not possess gallbladders, and large numbers of humans have had their gallbladders removed surgically, thus indicating that the gallbladder is not essential for life. Electron microscopy studies indicate that the tall columnar cells which line the lumen of the gallbladder contain many microvilli. These may function in the concentration of bile, a significant function of gallbladder epithelium involving the active transport of sodium and potassium.

In addition to acting as a reservoir for the bile synthesized by the liver, the gallbladder plays a significant role in controlling the flow of bile to the intestine, when needed, because of its exquisite sensitivity to ingested fat or meat. Following a high-fat meal, up to three-fourths of gallbladder bile is expelled in less than 1 hr as a result of contraction of the musculature of that organ and the concomitant relaxation of the sphincter choledochus (a ring of muscle) at the terminus of the common bile duct. Spasms of the sphincter may lead to abdominal pain and respiratory distress, symptoms which may be confused with those of heart seizures.

Vertebrata. The primitive vertebrate group Agnatha is represented by the extant cyclostomes: hagfishes and lampreys. The cyclostome liver is extremely small and consists of only a single lobe in the adult lamprey. The bilobed structure of the hagfish is more typical of the vertebrate archetype. The livers of higher fish are relatively large and lobated.

In some forms the gallbladder and even the bile duct may degenerate during metamorphosis and thus may be absent in the adult. A gallbladder is generally present, although a few species of cartilaginous fish (shark) may lack this structure. In the teleosts (bony fish), the gallbladder is often quite large.

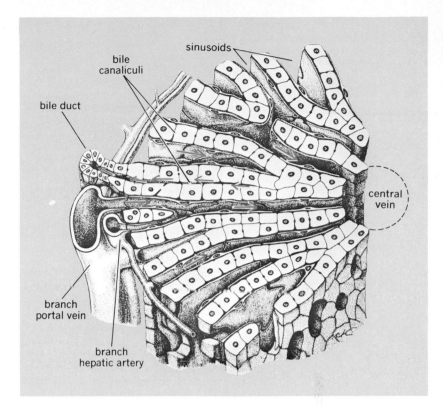

Fig. 3. Diagram of radial disposition of the liver-cell plates and sinusoids around the terminal hepatic venule or central vein, showing centripetal flow of blood from branches of the hepatic artery and portal vein and centrifugal flow of bile (small arrows) to the small bile duct in the portal space. (*From W. Bloom and D. W. Fawcett, A Textbook of Histology, 9th ed., Saunders, 1968*)

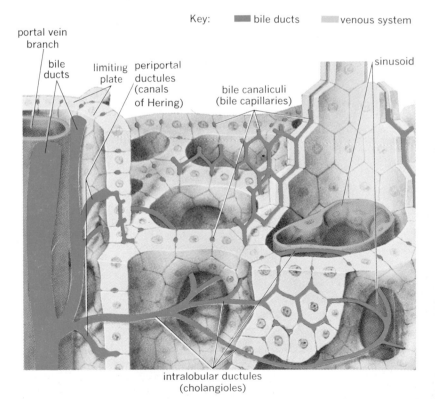

Fig. 4. Schema of intrahepatic biliary system showing the arrangement of the bile ducts and the venous system (*after H. Elias*). (*From E. Oppenheimer, ed., prepared by F. Netter, The Ciba Collection of Medical Illustrations, vol. 3: Digestive System, pt. 3: Liver, Biliary Tract and Pancreas, Ciba, 1957*)

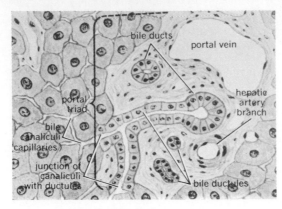

Fig. 5. Liver section, with components of portal triad. (*From E. Oppenheimer, ed., prepared by F. Netter, The Ciba Collection of Medical Illustrations, vol. 3: Digestive System, pt. 3: Liver, Biliary Tract and Pancreas, Ciba, 1957*)

The economic importance of fish liver (oils and vitamins) stems in part from the capacity of this organ to synthesize and store lipids. George Fried and coworkers hypothesized that in lower vertebrates the liver carries out synthetic functions, particularly of a lipid nature, which are delegated to the increasingly important adipose-tissue mass in birds and mammals. Shape and pancreatic relationship are quite variable for teleost liver.

Amphibians. The large liver of the amphibians, which is lobed and possesses a gallbladder, receives venous blood from the abdominal vein as well as the hepatic portal vein. Some variation in lobation patterns are encountered in the members of this class. *See* AMPHIBIA.

Reptiles. Reptiles demonstrate typical patterns of hepatic topography. In snakes (Squamata) one elongated lobe is found, whereas turtles (Chelonia) and crocodiles (Crocodilia) have two lateral masses connected by a thin isthmus. The gallbladder is invariably present. *See* REPTILIA.

Birds. The liver of birds is invariably lobed, but quite compact. The gallbladder is often absent but may exist as a primordium in the embryo. In the pigeon two separate bile ducts lead from each lobe of the liver to the duodenum. *See* AVES.

Mammals. A great deal of variation characterizes the lobation of mammals (Fig. 6). Two major lobes may be subdivided in a variety of ways. If the lobations are superficial, the liver is classed as unified, as in the case of man, whales, and many of the ruminants. In these livers the portal and hepatic canals tend to run at right angles to each other, whereas markedly lobated livers show a parallel arrangement of the large branches of these canals. Removal of a portion of a unified liver is a more difficult procedure than similar extirpation from a lobated liver.

Many mammals lack a gallbladder (rat, whale, coney, some of the artiodactyla, and all hoofed animals). In humans, those individuals who carry typhoid fever usually harbor the typhoid bacilli within the gallbladder and discharge the organisms in their feces. *See* ARTIODACTYLA; MAMMALIA.

HISTOLOGY

In all vertebrates, with the exception of the adult lamprey, the liver is essentially a three-dimensional lattice of interconnected walls or plates made up

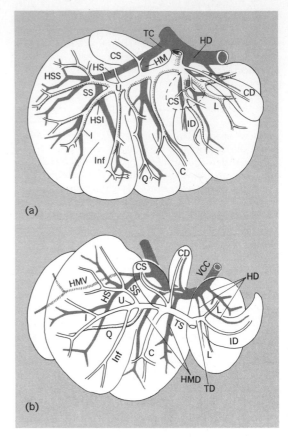

Fig. 6. Dorsal aspect of (*a*) liver of dog, (*b*) liver of rat. Rami venae portae: C, centralis; I, intermedius; ID, inferior dexter; Inf, inferior sinister; L, lateralis; Q, quadratus; SS, superior sinister. Venae hepaticae: HD, dextra; HM, media; HMV, ventral division of media; HMD, dorsal division of media; HS, sinister; HSI, sinistrae radix inferior; HSS, sinistrae radix superior. Ramuli: CD, caudatus dexter; CS, caudatus sinister; TC, truncus communis. Trunci venae portae: TD, dexter; TS, sinister; U, pars umbilicalis; VCC, vena cava inferior.

of epithelial cells (hepatic parenchyma) (Fig. 7). The cylindrical spaces (lacunae hepatis) which tunnel within this parenchymal mass interconnect to form a continuous maze of corridors known as the labyrinthus hepatus. Suspended in the lacunae hepatis are the sinusoids, or liver capillaries. These are relatively large, irregular, anastomosing vessels lined discontinuously with a specialized endothelial layer. Each functional unit of the liver is in contact with terminal branches of the two pervasive tunnel systems traversing the liver: the portal and hepatic canals. The branches of these two systems never communicate directly, but each is dovetailed into the sinusoid, the site of functional linkage between the two arborescent systems.

Limiting plate. The external surface of the liver is covered by a single layer of thin parenchymal cells lying immediately beneath the fibrous Glisson's capsule. This limiting plate is continuous with the hepatic muralium. It is reflected inward at the porta hepatis to form a fenestrated wall around the portal canals. In a similar manner the hepatic canals are also lined with a limiting plate (Figs. 2 and 4). Thus the external and internal limiting plates make up a continuum which is in intimate association with the internal muralium. The fibrous capsule is also reflected into the interior of both canal systems and provides a matrix for bringing blood vessels, lymph channels, nerves, and ducts into the interior of the liver.

Sinusoids. The lining of these specialized capillaries has been the subject of considerable debate. Older theories postulated a syncytium (absence of individual cell boundaries), but electron micrographs show separate cells present. These lining cells were once hypothesized to be of only one type, the Kupffer cell, with a great capacity for engulfing foreign material (phagocytosis). It is now thought that there may be as many as three distinct types making up the lining, although transformation of one type to another is possibile. The large macrophages (Kupffer cells) engulf degenerative erythrocytes as well as foreign or unassimilated material, thus "straining" the blood which moves through the liver sinusoids from the vessels of a portal canal to the central veins. Once again electron microscopy has clearly demonstrated that, although some of these lining cells tend to overlap one another, others have margins which are separated by as much as 0.5 μ. Some animals, for example, the calf, do not demonstrate this discontinuity; this raises the interesting problem of possible species differences. The regulation of blood flow within the hepatic lobule is achieved in part through a swelling of the macrophages lining the sinusoid.

A patent space, called the space of Disse, stands between sinusoid cells and the liver parenchymal cells (Fig. 2). Irregularly arranged projections, or microvilli, extend from the parenchymal cell and lie within the space of Disse. No limiting basement membrane blocks the free interchange between sinusoid gaps and the parenchymal cell. Furthermore, the villi of the parenchymal cell lying within the perisinusoidal space increase parenchymal cell surface across which exchanges may occur with the lymphlike fluid emanating abundantly from the sinusoid. This increase in the area of exchange is particularly important in terms of the rapid transfer of metabolites that must occur between the liver and portal blood.

Biliary system. The bile canaliculi are found between parenchymal cells throughout the liver. They are composed of a continuous network of polyhedral links associated with the convoluted muralium. If a single parenchymal cell is isolated, a hexagonal mesh of bile canaliculi appears to surround it. Although the specialized membranes of adjacent liver cells appear to extend into the lumen of the bile canaliculus in the form of microvilli, it is quite clear that no intracellular channels extend from the canaliculus into the parenchymal cytoplasm. Older studies suggested that the canalicular apparatus contained discrete walls which reinforced the histoarchitecture of the parenchymal mass. However, it is now believed that the lumen of the canaliculus is merely an expansion of the parenchymal intercellular space, with fibrous reinforcement of the cytoplasm immediately surrounding the lumen. In the normal liver no communication exists between the canaliculi and the Disse space; under pathological conditions, however, the microvilli may swell and obstructive pressure may then lead to passage of bile into the Disse spaces. The high levels of the enzyme adenosinetriphosphatase (ATPase) found around the lumen of the bile canaliculi serve as a histochemical marker for these tiny structures. *See* HISTOCHEMISTRY.

Comparative histology. In early radiations of the vertebrate group, a parenchymal muralium two cells in thickness developed. In mammals, songbirds, and several primitive bird groups, a single-cell-thick muralium (simplex) is found. This mammalian pattern is more efficient in increasing the blood–parenchymal cell interface area and affords a more rapid exchange between blood and liver, which is appropriate to the metabolically more active homeotherms. Exchange is also facilitated by the microvilli which extend from parenchymal cell surfaces apposed to sinusoidal spaces.

Cyclostomes. This group is sufficiently different from other vertebrate groups to justify a separate categorization of their cellular arrangement: large bile canaliculi lined by parenchymal cells demonstrating polarity (structural and functional speciali-

(a)

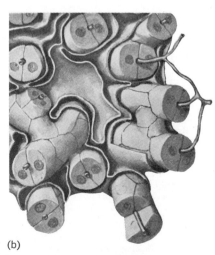

(b)

Fig. 7. Three-dimensional schemas showing the liver structure. (*a*) Stereogram of liver-cell plates after removal of ducts, vessels, and connective tissue (*after H. Elias*). (*b*) A former concept of liver structure which was thought to be composed of cell cords. (*From E. Oppenheimer, ed., prepared by F. Netter, Ciba Collection of Medical Illustrations, vol. 3: Digestive System, pt. 3: Liver, Biliary Tract and Pancreas, Ciba, 1957*)

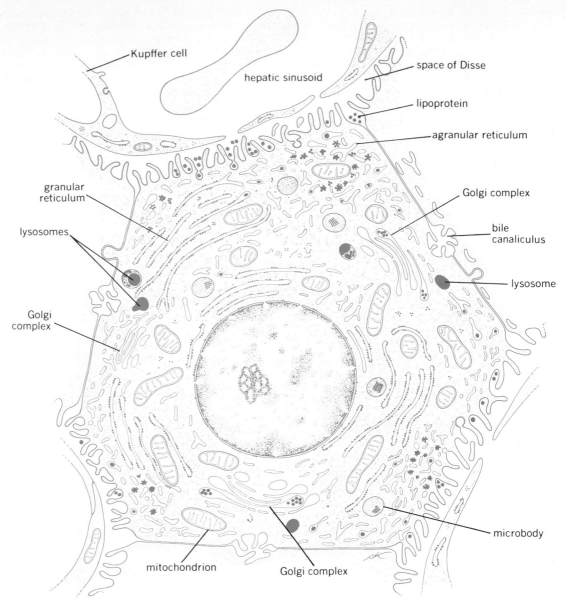

Fig. 8. Relationship of the liver cells to each other and to the sinusoids, with principal components of the hepatic cell as seen in electron micrographs. (*Drawing by S. C. Keene, from W. Bloom and D. W. Fawcett, A Textbook of Histology, 9th ed., Saunders, 1968*)

zation in relation to the cell axis).

Fishes. The elasmobranchs (sharks) show bile canaliculi tubules of smaller bore lined by large, fat-laden cells. Some degree of polarity persists in the hepatocytes of teleosts. The lobular arrangement of higher forms, involving centripetal flow of blood through the sinusoid from peripheral vessels of portal triad to central veins, is absent. During vitellogenesis (egg development) or pregnancy in many teleosts, a marked reduction of hepatic lipid and hepatocyte volume is found, which suggests a calorie storage function for the liver in lower vertebrates.

Great variations in liver cell nuclei size have been reported in several teleosts, with some indications of a rhythmic alternation. This alternation of size may be related to the mitotic periodicities of higher vertebrate forms, such as the circadian (24-hr) rhythms of the cortical liver cells in certain

amphibians, namely, salamanders and newts.

Amphibians and reptiles. The capillaries between portal and hepatic channels are not arranged in typical radial fashion in the frog, so that a lobular pattern is not discernible. In addition, there are marked seasonal variations in the appearance of the liver cells and in their content of glycogen and fat. Large lymphatic channels are a characteristic of many amphibian livers, and cells within these ducts, as well as the hepatocytes themselves, are often pigmented. Similar pigmentation is encountered in the reptiles, whose livers also lack the clear-cut lobular pattern of birds and mammals.

Birds and mammals. In birds and mammals the classical lobular pattern is usually a distinguishing feature, although the absence of connective tissue boundaries may obscure this basic recurrent form of peripheral portal triads, radial sinusoids, and central veins. Some investigators maintain that the classical lobule is merely an artifact of blood flow patterns within the liver, and attempts to construct a functional unit in which the portal triad is the center of a smaller "acinar" lobule have received considerable support. In a number of mammalian species vinyl-injection techniques have provided histological evidence for such a pattern; this would also accord with the more usual glandular arrangement of secretory drainage ducts standing in the center of the lobule.

CYTOLOGY

The significant functional cell of the liver is the hepatic parenchymal cell (Fig. 8). The sinusoids and a small amount of connective tissue also contribute distinctive cell types.

Parenchymal cells. These are polyhedral units of at least six sides, varying in volume from less than 10,000 to 35,000 μ^3. The surfaces may be in contact with the perisinusoidal space, the lumen of a bile canaliculus, or an adjacent parenchymal cell.

The round nucleus may be quite large and variable, a consequence of the polyploidy (containing more than two sets of chromosomes) which characterizes many liver cells. Occasionally two nuclei are found. Nucleoli are prominent, and scattered chromatin clumps may be seen within the smooth nucleus. The cytoplasm varies with the nutritional state and general health of the animal, particularly in terms of glycogen space and fat vacuoles. Numerous mitochondria (filamentous) are present. A compact series of vesicles, the Golgi complex, is discernible near the bile canaliculus, where it probably acts to concentrate and "package" the bile. See CELL (BIOLOGY).

Electron microscopy. Electron micrographs have extended the understanding of hepatic ultrastructure. Thus both smooth- and rough-surfaced types of endoplasmic reticulum (tubular cytoskeleton) are present, with the rough-surfaced type being a major site of protein synthesis. The endoplasmic reticulum is continuous with the external membrane of the double nuclear envelope and possibly with the cytoplasmic membrane, so that a continuous system of channels may completely traverse the cell. Glycogen appears as single granules 400 A thick, or in conglomerates up to 0.2 μ across, in close association with the smooth endoplasmic reticulum. See MICROSCOPE, ELECTRON.

Distributed in the region of the bile canaliculus are dense rounded bodies, called lysosomes, containing several hydrolytic enzymes. The peroxisomes are similar in structure but contain crystalline inclusions rich in the enzyme uricase, which splits uric acid. They are also rich in enzymes which split hydrogen peroxide or use it to effect oxidations. See CELL-SURFACE DIFFERENTIATIONS.

Adjacent parenchymal cells are reciprocally attached through tiny projections in one that fit snugly into recesses of the other in a manner analogous to snap fasteners.

Fractionation techniques. The region of cytoplasm immediately surrounding the bile canaliculus is fibrillar and probably gelated and may persist as tubular fragments even when the liver is macerated or homogenized. The homogenizing process involves the breakdown and dispersal of the cellular elements of a tissue to yield a homogeneous mixture for assay of enzymes, metabolites, and so forth. Differential centrifugation may then yield fractions which are particularly rich in mitochondria, other cellular organelles, or the nonparticulate cell sap. These fractionation techniques were originally developed with mammalian liver as the test material and sharply increased understanding of the biochemical patterns of liver organelles. See CENTRIFUGATION (BIOLOGY).

Sinusoidal lining cells. A great deal of variation is characteristic of these "littoral" cells, and even the shape and texture of the nucleus differs from one preparation to the next. Mitochondria have been noted, but not in all cells. Lipid has been demonstrated in these cells, but the existence of glycogen is in doubt. Lysosome activity has been indirectly demonstrated. The phagocytosing cells are believed to contain lysosomes and are characterized by long cytoplasmic extensions which prompted earlier investigators to call them the stellate (starlike) cells of Kupffer. Some of the living endothelial cells are vacuolated and are believed to serve as fat-storing cells, whereas others are highly granular. About 15% of the human liver cell population is made up of littoral cells.

Connective tissue. Although the liver is the largest internal organ in the body, it contains only slight amounts of supporting connective tissue, averaging less than 1%. Connective tissue is found in the capsular material, the sheath which surrounds the vessels and ducts of the portal canal system, and in lesser amounts within the hepatic canal system. See CONNECTIVE TISSUE.

Collagenous fibers are found within the canal systems, whereas delicate reticular fibers extend between the parenchymal cells of the muralium and the endothelial cells within the space of Disse. These fibers, appearing early in embryonic development, probably arise from the sinusoidal living cells. The connective tissue in the perisinusoidal space, which is continuous with the dense connective material of the portal sheath, is believed to function in channeling the large amount of lymph which arises as tissue fluid in the liver.

EMBRYOLOGY

The liver arises initially as a thickening and then an outpocketing of the ventral surface of that portion of the gut destined to become the duodenum. Two branches are quickly formed: an anterior portion from which the liver parenchyma develops; and a posterior portion giving rise to the gallbladder, its cystic duct, and the common bile duct.

Morphogenesis. The liver parenchyma develops as a series of cords or strands which are essentially tubular. At first these proliferating cords lie alongside the vitelline vessels of the yolk sac; then they are found ramified within these vessels to form an intermingled mass of epithelial cords and tiny channels, or sinusoids, representing subdivisions of the original larger venules. See FETAL MEMBRANE; YOLK SAC.

Alternative explanations for sinusoid development have been offered. Branching of the cords takes place in a manner characteristic of each vertebrate group and results in the varied hepatic architectures found in the adult. Work by Hans Elias indicates that there may be as many as 12 basic patterns of hepatic development among vertebrates and that the proliferation of hollow cords may be characteristic of only a few species.

The parenchymal cells, which are arranged in a continuous muralium (lamina) either two cells thick (lower vertebrates) or one cell thick (mammals and songbirds), are mainly derived from the endoderm. In early stages the muralium may be several cells thick. The sinusoidal elements are chiefly derived from mesenchyme, but the origin of the blood-forming cells that characterize the embryonic liver is not fully known.

The bile canaliculi, which also form a continuous network in the liver, arise as a result of parenchymal cell differentiation. They are probably formed even before the liver is capable of producing bile and ramify into ductules and ducts which join the hollow posterior pouch, or future gallbladder. One contribution of electron microscopy has been the finding that the walls of the bile canaliculi do not exist as separate structures but simply represent differentiated surfaces of apposed hepatic cells.

Metabolic aspects. Although the liver plays a key role in assuring carbohydrate homeostasis (dynamic steady-state conditions) through the storage of sugar as glycogen, the embryonic livers of most species are incapable of synthesizing glycogen until a precise point is reached, rather late in development, which varies with different species. In most mammals a maximum reserve is achieved at birth, with a rapid falling off soon after. In amphibians glycogen build up is associated with metamorphosis, whereas in birds several cycles are apparent. Control is probably achieved through hormonal action of the pituitary-adrenal axis, although the role of enzyme activation has also been explored.

Much less is known about the handling of lipids, although it is clear that hepatic fat appears quite early in most mammalian embryos and is extremely high in the newborn. High lipid reserves exist in chick livers at hatching but diminish sharply thereafter. Protein incorporation, as measured by the uptake of radioactive-labeled amino acids, is extremely high in the livers of rat embryos despite the generally lower protein levels in comparison to adults. These livers also demonstrate an early competence for the synthesis of histamine.

Many of the enzymes which play a key role in the metabolic transformations of the adult mammalian liver are either low or absent at birth. The synthesis of these enzymes at term may explain the sharp changes at birth in the levels of such metabolites as glycogen. Iron reserves also tend to increase late in embryonic life, and the storage in the liver of histologically identifiable iron is associated with the gradual diminution of the erythropoietic function. Electron microscopy studies have revived an interest in the nature of the hemopoietic (blood-forming) function in mammalian embryos, particularly in terms of the nature and origin of the blood-forming island cells which proliferate within the parenchymal elements early in embryonic life.

PHYSIOLOGY

The large size of the liver is matched by its functional complexity and involvement in a diverse array of regulatory mechanisms. Although said to perform over 100 specific functions, the liver's key physiological role may be delineated under seven major headings.

Control of carbohydrate metabolism. Following the ingestion of food, blood entering the liver from the hepatic portal vein is rich in glucose and other monosaccharides (simple sugars). The parenchymal cells, which are freely permeable to these sugars, remove them from the general circulation and store them as a complex insoluble polymer, glycogen. Since glycogen is a polysaccharide made up of glucose subunits, the liver must first convert other dietary monosaccharides (fructose, galactose) to glucose prior to glycogen synthesis (glycogenesis). As a result of this storage mechanism, the blood leaving the liver through the hepatic veins contains normal levels of glucose. In the intervals between ingestion of food, liver glycogen is broken down. This process, known as glycogenolysis, tends to maintain blood sugar levels between 80 and 100 mg per 100 ml of blood. *See* CARBOHYDRATE METABOLISM; GLUCOSE; GLYCOGEN; MONOSACCHARIDE.

Blood sugar level. Since the brain and other vital central nervous system structures are almost entirely dependent on the glucose fuel of the blood, hepatic regulation of circulating carbohydrate is necessary to life. Under conditions of prolonged fast, where glycogen stores are exhausted, the liver is capable of converting noncarbohydrate metabolites into glucose to maintain blood sugar levels. This process, by which amino acids, fats, glycerol, and other metabolites are converted enzymatically to glucose, is called gluconeogenesis. It represents a significant metabolic pathway for the regulation of carbohydrate metabolism by the liver, although the kidney is also capable of carrying out gluconeogenesis in mammals.

All of the processes involved in the manipulation of carbohydrate by the liver are active constantly, but the relative rates of individual pathways vary in response to food intake, tissue uptake, hormone levels, and so on. The maintenance of constancy through the dynamic balance of opposing metabolic forces is called homeostasis, a concept which was first developed a century ago by Claude Bernard in his studies of the biochemical competency of the liver. *See* HOMEOSTASIS.

Endocrine regulation. It has long been recognized that the pancreatic hormone insulin is necessary for the utilization and storage of glucose, while hormones of the adrenal cortex are involved in the gluconeogenic process. The actions of epinephrine and glucagon in enhancing hepatic glycogenolysis have been elucidated. Thus the complex steps involved in maintaining carbohydrate homeostasis are subject to endocrine control, with the liver serving as a particularly sensitive target organ for the hormone regulators. *See* HORMONE, ADRENAL CORTEX; INSULIN.

Enzyme systems. The precise nature of these hepatic responses are being explored, and emphasis is placed upon the role of specific enzyme systems which are rate-limiting in some overall pathway or which occupy a branch position serving to divert metabolites from one major pathway to another. *See* ENZYME.

1. Gene activation. George Weber showed that

insulin, which is known to enhance uptake of glucose by muscle and adipose tissue, tends to increase the formation of three "bottleneck" enzymes in the liver which act to degrade glucose (glycolysis) and prepare its carbon skeletons for eventual oxidation by the mitochondria. On the other hand, insulin inhibits formation of a quartet of hepatic enzymes which act to bring the degradation products of glucose back to free monosaccharide. Hormones of the adrenal cortex have a completely opposite effect on these same target enzymes, thus explaining the opposing actions of cortisone (which increases blood sugar) and insulin (which decreases blood sugar) on carbohydrate balance. Presumably these hormones act on a specific gene site which is responsible for the synthesis of a particular group of enzymes influencing a single effect. See GENE ACTION.

2. Enzyme targets. Direct action on an enzyme already present in the liver cell is also utilized. Thus epinephrine and glucagon (a pancreatic hormone antagonistic to insulin) each act to produce increased quantities of cyclic adenosinemonophosphate (AMP), which activates the hepatic enzyme involved in the cleavage of glycogen bonds to yield glucose. The rise in blood sugar associated with increased circulating levels of epinephrine is effected through this mechanism. Thyroxine also influences blood sugar levels, but its mode of action is unclear. See ADENYLIC ACID, CYCLIC.

Negative feedback mechanisms. These provide a fine-control system for the hormones themselves in carbohydrate regulation. Thus insulin release is enhanced by high blood sugar levels (hyperglycemia), but its release brings into play mechanisms which tend to lower blood sugar levels and hence dampen the evoking stimulus. Long-term heavy ingestion of carbohydrate may exhaust the insulin-producing tissue of the pancreas and lead to the production of diabetes — a complex disease usually involving insulin deficiency with concomitant disturbances of carbohydrate and fat metabolism. See DIABETES; ENDOCRINE MECHANISMS; LIPID METABOLISM.

In genetic obesity in mice, characterized by a seemingly paradoxical situation of high insulin levels and hyperglycemia, elevations in hepatic glycolytic enzymes are found. This is consistent with Weber's findings of a similar hepatic glycolytic enzyme increase in animals injected with insulin. The relatively high fasting levels of liver glycogen also associated with this syndrome may be attributed to the induced synthesis of glycogen synthetase (glycogen-forming enzyme). This synthesis has been shown to occur with insulin and is related to the action of insulin in promoting carbohydrate storage in the liver. The liver possesses an enzyme, called insulinase, which degrades insulin. Its importance under physiological conditions is not clear; however, with overproduction of insulin, insulinase may prevent excessive buildup of the hormone in the blood stream.

Interconversion of metabolites. The liver is a major site of the production of both fatty acids and neutral fats (triglycerides — three fatty acids joined to glycerol). Under conditions of adequate feeding and high carbohydrate intake, a significant proportion of the food entering the liver through the portal vein, or coming in from the systemic circulation through the hepatic artery, will be converted to fatty acids and neutral fat. The liver possesses enzyme systems which activate both fatty acids and glycerol to facilitate their conversion to the less soluble triglycerides. In addition, the manipulation of carbohydrate in the liver leads to the formation of substrates and coenzymes (accessory molecules for enzyme action) necessary for fat synthesis, although an increasingly important role in fat metabolism is being assigned to the adipose tissue mass in mammals.

The link between the pathways of carbohydrate and fat metabolism is the probable basis for the disruption of fat synthesis and storage which occurs when carbohydrate metabolism is impaired. Synthesis of the phospholipids, which are readily transportable lipids, also occurs in the liver. The production of fatty infiltrations in the livers of animals deficient in choline or methionine may be associated with an impairment in phospholipid metabolism, since these substances are necessary for their formation. Inability to form lipoproteins has also been suggested as a factor. Agents which increase fatty acid transport to the liver or decrease hepatic fatty acid oxidation may also bring about fatty degeneration. Chronic accumulation of lipid in the liver leads to fibrotic changes and eventual cirrhosis, a condition in which the liver loses its resiliency and gradually ceases to function as the parenchymal mass is replaced by connective tissue. A variety of hepatotoxins (alcohol, chloroform, carbon tetrachloride, lead, and others) also produce fatty livers. See COENZYME; LIPID; LIVER DISORDERS.

Ketone bodies. Representing partial oxidation products of fatty acid degradation, these are produced by the liver and utilized as a rapid short-term energy source by extrahepatic tissues. In starvation and advanced cases of diabetes, the excess production of acidic ketone bodies poses a serious threat to life. It has been shown that hepatic glutamine production is markedly augmented in these circumstances, presumably to meet the increased needs of extrahepatic tissues. See KETONE.

Sterols and steroids. Cholesterol is also synthesized by the liver, although small amounts are ingested. High levels of cholesterol are believed to be associated with atherosclerosis (hardening of the arteries) by some investigators, but the relationship is a complex one. Cholesterol is excreted in the feces following its conversion to bile salts, but considerable recycling of these sterol compounds may occur between intestine and liver. Steroid hormones, which resemble cholesterol in their chemical structure, are degraded in the liver and may then be excreted in the bile. See CHOLESTEROL; STEROID; STEROL.

Amino acids. From the gut amino acids reach the liver through the portal circulation, and a small portion may be incorporated into hepatic structural protein to serve as a labile reserve. Protein is not generally regarded as being stored in animal tissues. During gluconeogenesis, which is enhanced by glucagon as well as the adrenal steroids, amino acids are stripped of their nitrogen and the remainder of the carbon skeleton is utilized for the production of glucose. The nitrogen arising from this deamination process may be secreted into the blood as ammonia, or the ammonia may be converted by the liver to urea or uric acid. The latter two nitrogenous end products are associated with vertebrates inhabiting marine or terrestrial environments. Nitrogen, or more specifically amino groups, may also be shifted from one organic molecule to another through the process of transamination. High levels of the enzymes mediating these transfers (transaminases) are found in the liver, and they may pass into the bloodstream during hepatic degeneration, where they serve as a clinical sign of liver damage. *See* AMINO ACIDS; CLINICAL PATHOLOGY; EXCRETION; PROTEIN METABOLISM; TRANSAMINATION.

Blood protein manufacture. The liver is the only source of plasma cholinesterase, albumin, prothrombin, and fibrinogen. Most of the globulins also arise in the liver, but gamma globulin is produced in plasma cells found principally in the spleen with only a minor contribution coming from the lymphoid tissue of the liver. On the basis of the hepatic contribution to the pool of blood proteins, a significant role may be averred for the liver in maintaining the osmotic pressure of the blood, preventing hemorrhage, providing a reserve of body protein (7 g per 100 ml of blood), promoting the transport of lipids, and preventing excess build up of acetylcholine. *See* ACETYLCHOLINE; BLOOD; GLOBULIN; HEMORRHAGE.

Erythropoietic function. In almost all vertebrates the liver is the predominant embryonic source of red blood cells. Thrombocytes (platelets) are also produced in this organ, and in embryonic forms other than man the liver may be an important region for lymphocyte and granulocyte production as well. In urodeles the cortical region of the liver continues its erythropoietic function throughout the life cycle. *See* HEMATOPOIESIS.

Detoxification. Many poisons are rendered harmless by degradation reactions in the liver. These reactions involve conjugation, methylation, oxidation, reduction, and complex syntheses. The microsomes (endoplasmic reticulum and ribosomes) are particularly active in neutralizing potentially harmful drugs, for example, barbiturates, tranquilizers, and hallucinogens such as lysergic acid diethylamide (LSD). Detoxification mechanisms are also involved in the formation of bile pigments, the esterification of cholesterol, and the conjugation or oxidation of steroid hormones. A diseased liver cannot carry out these functions, which lowers the body threshold to most toxic materials.

Bile formation. Bile is formed at the rate of about a quart per day. It consists of bile pigments, bile salts, protein, cholesterol, and inorganic salts. The pigments, bilirubin and biliverdin, arise from the breakdown of heme in those tissues (spleen, bone marrow) where erythrocytes are usually broken down, and then are conjugated with glucuronic acid by the liver. Bile salts and cholesterol are formed in the liver. The significance of bile release into the intestine lies in the ability of the bile salts to emulsify fats, activate lipase, and promote the uptake of the fat-soluble vitamins by the intestine. *See* BILE ACID; BILIRUBIN.

Vitamin storage. The liver stores vitamins A and D and is apparently a reservoir for iron as well. Some of the B vitamins are also at high levels in liver tissue, suggesting a possible storage mechanism. Carotene in the liver is converted to vitamin A. *See* DIGESTIVE SYSTEM (VERTEBRATE); VITAMIN A.   [GEORGE H. FRIED]

Bibliography: W. Bloom and D. W. Fawcett, *A Textbook of Histology*, 9th ed., 1968; H. Harper, *Review of Physiological Chemistry*, 11th ed., 1967; C. Rouiller (ed.), *The Liver: Morphology, Biochemistry, Physiology*, vol. 1, 1963; J. Tepperman, *Metabolic and Endocrine Physiology*, 2d ed., 1968; G. Weber (ed.), *Advances in Enzyme Regulation*, vol. 4, 1966.

Liver disorders

Disorders of the liver are of particular importance because of the many essential physiological functions of this large organ. The liver is intimately concerned with the metabolism of proteins, fats, and carbohydrates and with the removal or detoxification of many hormones and many injurious metabolic waste products. The liver receives about 25% of all of the blood pumped out from the heart, including all of the blood draining from the intestine. *See* LIVER (VERTEBRATE).

Cirrhosis. Cirrhosis is the term applied to the end stage of scarring of the liver. It may follow repeated minor insults or a single sudden massive injury. The more common causes are nutritional deficiency (Laennec's, or portal, cirrhosis), shown in the illustration; viral infections and injury by chemical agents (postnecrotic cirrhosis); obstruction of the drainage of bile from the liver (biliary cirrhosis); and prolonged and severe heart failure (congestive or cardiac cirrhosis). Since the use of antibiotics, syphilis has become relatively uncommon as a cause of cirrhosis, at least in those regions where antisyphilitic therapy is available. Because the liver has a large functional reserve, minor degrees of cirrhosis may have little physiological effect. More severe instances of the disease, however, lead to early death.

Malnutrition. Nutritional disease of the liver is seen particularly after prolonged and excessive use of alcoholic beverages, but may occur as a result of malnutrition from any cause and may even develop spontaneously in animals. The dietary deficiency is not a lack of sufficient calories, but an inadequacy of a few essential foodstuffs, particularly the amino acid choline. Under these conditions, mas-

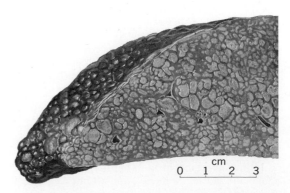

Portal (Laennec's) cirrhosis of the liver. The normal uniform architectural pattern has been destroyed and replaced by large and small nodules of regenerated liver tissue separated by bands and septa of connective tissue. This is the edge of a large slice from the liver.

sive amounts of fat first accumulate in the liver cells, enlarging the cells and the liver itself. Such large, fatty livers may have impaired function and are unduly susceptible to further injury from many toxic agents to which the liver may be exposed. Sudden, unexplained death of many liver cells may occur at this fatty stage, lowering the function of the organ below the level essential for life of the individual. More often, the disease progresses insidiously and persistently with the development of fibrous scarring and with disorganized regeneration of liver tissue.

After a period of many months or years, the entire architectural pattern of the liver is distorted and replaced by small and large nodules of disorganized liver cells separated by bands or septa of connective tissue. With this progressive fibrosis and distortion, there also occur significant changes in the small blood vessels in the liver, seriously altering its blood supply. The small veins draining blood from the liver become narrowed, and the flow of blood through the liver is impeded. Behind this obstruction, pressure rises in the portal veins which carry the blood from the gastrointestinal tract to the liver, and new channels develop to carry the blood around rather than through the liver. Particularly important among these latter channels are veins beneath the mucosa of the esophagus (varices), which may become eroded and be the source of massive hemorrhage. In late stages of the disease the amount of blood reaching liver cells may be insufficient to maintain the barest essential function. The patient then dies in hepatic failure, with alterations in the blood proteins, accumulation of fluid in the body cavities, and the accumulation in the blood of toxic metabolic products such as ammonia. See MALNUTRITION; METABOLIC DISORDERS.

Viral infections. Viral hepatitis, the most important infectious disease of the liver, may be caused by either of two closely related viruses. These two agents produce identical changes in the liver but differ in their mode of spread and in the latent period between the invasion of the host and the manifestation of disease. In one case the organisms are acquired by the oral route (infected food or water). In the other case they are acquired by the injection of serum from a patient with the disease or carrying the virus. The former disease is called infectious hepatitis; the latter, homologous serum hepatitis. Jaundice is the striking clinical feature of viral hepatitis, although other symptoms such as malaise, abdominal pain, and loss of appetite may also be distressing. The effects of the virus on the liver are characterized by injury and death (necrosis) of isolated liver cells or groups of liver cells and by an accumulation throughout the liver of large numbers of mononuclear white blood cells. Drainage of bile through the tiniest bile channels, the canaliculi, is impaired, and bile accumulates there and in the liver cells themselves. In the more severe cases, necrosis of liver cells may be so widespread that death of the patient occurs early in the illness. Usually, however, complete healing occurs after a few weeks, with no residual alteration in the structure or function of the organ. In a few instances when the disease is protracted in its course or is associated with severe but nonfatal necrosis, postnecrotic cirrhosis may result. The architecture is then distorted by wide bands of connective tissue and large bizarre nodules of regenerated liver tissue. This postnecrotic cirrhosis is also associated with vascular obstruction and its complications, just as is the portal cirrhosis of nutritional liver disease. See JAUNDICE.

Yellow fever is another virus disease affecting the liver. It is spread by certain mosquitoes and is geographically restricted to Africa and to South and Central America, where in recognized and unrecognized form it is relatively common. Jaundice resulting from the necrosis of liver cells is responsible for its name. Hepatitis of viral etiology occurs in dogs and in pigs, and it is possible that the latter disease, porcine hepatitis, is caused by the same virus that is responsible for infectious hepatitis in humans. In South Africa an obscure fatal disease of horses, called staggers or malziekte, is associated with severe necrosis of the liver and may also be of viral origin. See YELLOW FEVER.

Bacterial infections. Bacterial infections of the liver are often secondary to infections primary in other sites. Infection may spread from the gallbladder up into the small bile ducts within the liver, producing inflammation there (cholangiolitis). Infections in the intestinal tract, particularly appendicitis, may spread to the liver through the portal vein. Abscesses are the usual consequence of this so-called pylephlebitis. Almost any generalized infection of the body may involve the liver secondarily, and in some, such as typhoid fever, brucellosis, and tularemia, this is the rule. In leptospirosis, the spirochetes responsible for the disease produce liver injury and jaundice, as well as damage to other organs. Dogs and other animals are frequently infected by these same and similar bacterial organisms. See BRUCELLOSIS; TULAREMIA.

Parasites. Parasitic infections involve the liver of man and animals. Amebic abscesses of the liver are the most common and most serious complications of the intestinal infection by this protozoan. Amebic liver abscesses are usually single and may become very large. They may become secondarily infected with bacteria or may rupture into the pleural, pericardial, or peritoneal cavities and even into the lungs or bowel. The tiny blood flukes known as schistosomes invade the skin of individuals bathing in infected tropical waters. They eventually find their way to the small blood vessels in the liver, where they mature and cause chronic inflammation and eventually vascular obstruction. Other parasitic diseases of the liver are more common in animals, but man is often infected incidentally. The larval stage of the dog tapeworm (*Echinococcus granulosus*) has a predilection for the liver in the intermediate hosts, which include man, sheep, cattle, and many other animals. In the liver these parasites produce large cysts, usually with thick, calcified walls. Other important parasites infecting the liver of man and animals are the liver flukes which inhabit and obstruct the small bile ducts. These specific bacterial, viral, and parasitic diseases are discussed in detail elsewhere under the appropriate headings. See AMEBIASIS.

Chemical agents. Toxic hepatitis is the type of liver injury caused by chemical agents ingested or inhaled into the body. There are many compounds, called hepatotoxins, which are capable of damaging the liver. The most important are chloroform, carbon tetrachloride, phosphorus, and the poison of the mushroom *Amanita phalloides*. They produce fatty accumulation and necrosis in liver cells, often with a distinctive pattern. Because the liver has great powers of regeneration, complete healing without residual effects may follow moderately severe toxic injury. If too severe, however, such insults may be rapidly fatal or may be followed by the development of postnecrotic cirrhosis. Many drugs may be injurious to the liver in susceptible or sensitive individuals. Fortunately, drug injuries are not often severe enough to be fatal and are rarely followed by cirrhosis.

Obstruction. Biliary cirrhosis is the end result of prolonged obstruction to the drainage of bile from the liver. Behind the obstruction, the bile ducts become greatly dilated. Bile accumulates there, in the tiny canaliculi, and within the liver cells themselves. Bacterial infection of the bile ducts is a frequent complication and is often recurrent. Eventually extensive connective tissue proliferation results, and liver function progressively fails. Impaction of a gallstone in the common bile duct may precipitate such a process, and tumors of the common bile duct or pancreas are also common causes. Biliary atresia, or failure of the bile ducts to develop in the embryo, has the same consequence; unless surgically correctable it leads to biliary cirrhosis and death in infancy or early childhood.

Other diseases. There are several obscure diseases which may involve the liver along with other organs. In amyloidosis the peculiar chemical deposits characteristic of this disease are laid down in the liver, as well as in certain other tissues. The obscure chronic inflammatory disease called sarcoidosis commonly involves the liver. Wilson's disease, or hepatolenticular degeneration, is a rare disorder of copper metabolism in which there occurs cirrhosis of the liver along with severe damage to the central nervous system. Hemochromatosis, a disorder of iron metabolism, is characterized by excessive accumulation of iron in the liver and other tissues and by the development of severe and often fatal cirrhosis. In prolonged or recurrent heart failure, sufficient liver damage may be produced so that connective tissue scarring occurs (congestive or cardiac cirrhosis).

Tumors. Tumors of the nonmalignant type are rare in the liver except for the neoplasm of blood vessels called hemangiomas. These, as well as less common benign tumors of the liver, are usually small and of little significance. Malignant tumors of the liver appear to be increasing in frequency. They are not amenable to surgical treatment and are rapidly fatal. Such tumors usually arise from the liver cells themselves (hepatomas) or from the bile ducts within the liver (cholangiomas). They are particularly common in livers with portal or postnecrotic cirrhosis and hemochromatosis. They have little tendency to spread outside the liver, but cause death by the massive replacement of functioning hepatic tissue by nonfunctioning tumor tissue. The liver is probably the most common location for the secondary spread of neoplasms. It is the growth of these metastatic tumor nodules in the liver which is the ultimate cause of death in many individuals with cancer of the bowel, breast,

lung, or other tissues. See NEOPLASIA; ONCOLOGY.

[MILTON R. HALES]

Bibliography: W. A. D. Anderson (ed.), *Pathology*, 5th ed., 1966; L. Schiff (ed.), *Diseases of the Liver*, 2d ed., 1963; S. Sherlock, *Diseases of the Liver and Biliary System*, 3d ed., 1963; H. A. Smith and T. C. Jones, *Veterinary Pathology*, 1957.

Living fossils

Living species very closely resembling fossil relatives in most anatomical details. The term is a relative one and, applied loosely, could embrace nearly all extant animals and plants. In its more restricted usage, the term applies to living species with four additional characteristics: (1) truly close anatomical similarity to (2) an ancient fossil species—generally at least 100,000,000 years old; (3) living members of the group are represented by only a single or at best a few species, which are (4) often found in a very limited geographic area. Examples are horseshoe crabs (found on the eastern shores of the two Northern Hemisphere continental landmasses, with closely similar relatives over 200,000,000 years old); ginkgo trees, dating from the Mesozoic era, which, until their recent introduction over the world, were restricted to Asia; and coelacanth fish, which have not changed much since the Devonian and which today are found only between eastern Africa and Madagascar and the Comoro Islands.

Living fossils are important because they retain many primitive traits, hence their study helps elucidate the anatomy of their fossil relatives. Living fossils also raise the issue of the lack of evolutionary change. Survival of primitive organisms in an unchanged state over many millions of years implies a low rate of evolution. Evidence suggests that slowly evolving groups speciate at a lower rate than faster-evolving groups. Such slowly evolving groups are frequently ecologically generalized, that is, able to exploit a wide variety of food resources and withstand large-scale fluctuations in the physical and chemical characteristics of their habitats. Organisms so adapted are unable to withstand competition from closely related species with similar habits. This factor reduces the number of available habitats and hence reduces the chance of new species evolving. Without speciation there can be little evolution, and as long as the habitat persists, so will the organisms, surviving virtually unchanged for millions of years. See SPECIATION.   [NILES ELDREDGE]

Bibliography: N. Eldredge, Survivors from the good old, old, old days, *Natur. Hist.*, 74(2):60–69, 1975.

Lizard

Any reptile in the suborder Sauria, order Squamata; there are about 3000 living species in 20 families. Included in the group are a variety of diverse forms such as the iguanas, geckos, monitors, and chameleons. They have a cosmopolitan distribution but are most abundant in tropical regions and are absent in Antarctica. Lizards are found in all types of habitats, to which they are well adapted; although a few species are herbivorous, the majority are insectivorous or carnivorous.

The body ranges from 2 in. to 12 ft in length and, although most species have paired limbs, some lizards have elongated bodies with reduced limbs and thus resemble snakes. Typical of the lizards are movable eyelids and a fused lower jaw. The tongue may be extensile, even prehensile, or thick and attached to the floor of the mouth with only the anterior edge extensile. Dentition is homodont, and the teeth are inserted on the jaw ridge (acrodont condition) or on the inner surface of the jaws (pleurodont). Lizards have epidermal scales which may be modified to form tubercles or reinforced with bony plate. *See* DENTITION (VERTEBRATE).

Lacertidae. The typical lizards, including about 150 species, belong to the family Lacertidae. They are numerous in Europe, Africa, and Asia but are absent in Australia and the New World. Characteristic members of this family are the wall lizard (*Lacerta muralis*), the European green lizard (*L. viridis*), the sand lizard (*L. agilis*), and the European common lizard (*L. vivipara*), an unusual species that is midway between the oviparous and ovoviviparous forms (sometimes the young are born alive after hatching in the mother).

Gekkonidae. The Gekkonidae are represented by more than 300 species in all warm regions of the world. Common geckos are the tokay (*Gecko gecko*), wall gecko (*Tarentola mauritanica*), Mediterranean gecko (*Hemidactylus turcicus*), and the leaf-tailed gecko (*Uroplatus fimbriatus*). All species have a flattened body, and most are pentadactyl. *See* GECKO.

Agamidae. Agamid lizards are included in the family Agamidae. The more than 200 species in 34 genera peculiar to warm areas of Africa, Asia, southern Europe, and Australia resemble the South American iguanas. This is a diverse family: *Agamus* species are terrestrial and polygamous, *Phrynocephalus* and *Leiolepis* are monogamous, and the Indian and Malayan *Calotes* are arboreal.

Iguanidae. About 700 species are commonly called iguanas. They are cosmopolitan in distribution and variously modified for terrestrial, arboreal, burrowing, semiaquatic, and even semimarine adaptations. The common iguana (*Iguana iguana*) is an arboreal species in South American forests; green anoles (*Anolis carolinensis*) are treedwellers in the Carolinas; the fence lizard (*Sceloparus undulatus*) has a terrestrial-arboreal habitat in the eastern United States (see illustration); and the marine iguana (*Amblyrhynchus cristatus*), peculiar to the Galapagos Islands, is the only marine lizard. *See* IGUANID.

Teiidae. Tegus lizards are the American equivalents of lacertids. This diverse family is especially abundant and widespread in South America and includes the race runners (*Cnemidophorus*), also found in the United States, Mexico, and Central America, and the common tegu (*Tupinambis teguixin*), largest representative of the family.

Helodermatidae. This family includes only two species: the gila monster (*Heloderma suspectum*) and the beaded lizard (*H. horridum*), both of Mexico and the United States. *See* GILA MONSTER.

Varanidae. Monitors make up the family Varanidae. There are four species, all with long, forked tongues: the Komodo dragon (*Varanus komodoensis*), Nile monitor (*V. niloticus*), desert monitor (*V. griseus*), and Malayan monitor (*V. salvator*). These lizards, some terrestrial and some aquatic, are found in the hot regions of Africa, Asia, Australia, and Malaya. *See* MONITOR.

Chamaeleontidae. The 80 species of chameleons occur mainly in Africa and Madagascar. They are excellent tree-climbers because of prehensile tails and opposing digits. The common chameleon (*Chamaeleon chamaeleon*) is typical of the family. *See* CHAMELEON.

Scincidae. Skinks have reduced limbs and snakelike bodies. There are more than 600 species; their distribution is cosmopolitan. Common United States species are the five-lined skinks (*Eumeces fasciatus*), the broad-headed skink (*E. laticeps*), the coal skink (*E. anthracinus*), and the ground skink (*Lygosoma laterale*). *See* SKINK.

Cordylidae. There are approximately 20 species of girdle-tailed lizards, the most common being the zonure (*Cordylus giganteus*), a large species that grows to 2 ft; it inhabits South Africa.

Anguidae. Anguid lizards, commonly known as slowworms or glass snakes because of their elongated bodies and reduced limbs, include 60 species which have a worldwide distribution. The slowworm (*Anguis fragilis*) is the only legless form, but there are vestiges of shoulder and pelvic girdles. Other species are the alligator lizard (*Gerrhonotus liocephalus*) and the glass snake (*Ophisaurus apodus*).

Amphisbaenidae. The taxonomic status of the worm lizards is questionable since they have greatly reduced structural features. The hindlimbs are absent and so are the forelimbs in many species. They occur mainly in South America and Africa but are found in Mexico, southern Europe, the southwestern United States, and Florida. This family includes the two-handed worm lizard (*Bipes canaliculatus*), a species with small forelimbs; the worm lizard (*Rhineura floridana*); and the white amphisbaena (*Amphisbaena alba*).

Other families. The remaining families are small and geographically limited: Pygopodidae are flap-footed lizards, with 20 snakelike species restricted to New Guinea, Australia, and Tasmania. Xantusiidae are night lizards, with 11 species resembling iguanas found in Central America, Mexico, Cuba, and the southwestern United States. Dibamidae are the flap-legged skinks, with three species (*Dibamus*) in southeastern Asia. Feyliniidae are the limbless skinks, with four rare species (*Feylinia*) in tropical Africa. Anelytropsidae are known by a few specimens of the single species *Anelytropsis papillosus*, found in Mexico. Gerrhosauridae are the gerrhosaurid lizards, with 25 species, related to the skinks and lacertids and found in Africa and Madagascar. Anniellidae are burrowing, snakelike (limbless) relatives of the anguids, including two species (*Anniella*) in California. Xenosauridae are a family of four rare species, composed of the Chinese lizard (*Shinisaurus crocodilurus*) and three Central American species (*Xenosaurus*). The earless monitor, *Lanthanotus borneensis*, is the only representative of the family Lanthanotidae. It is a terrestrial species found in Borneo.

[CHARLES B. CURTIN]

Llama

A member of the camel family (Camelidae) found only in South America. The llama is an artiodactyl, or even-toed ungulate, with two toes on each foot. The upper lip is cleft and prehensile. The animal has a long neck, and attains a maximum length of less than 8 ft and a maximum weight of nearly 300

LIZARD

Fence lizard (*Sceloporus undulatus*).

The llama of South America.

Hybrids produced by crossing different llamas

Male	Female	Hybrid
Llama	Alpaca	Huarizo
Alpaca	Llama	Misti; machurga
Vicuña	Llama	Llamavicuña
Guanaco	Llama	Llamahuanaco
Vicuña	Alpaca	Vacovicuña

lb (see illustration). A single young is born after a gestation period of about 11 months. The maximum life-span is about 20 years. Like other members of the family, the llama is herbivorous. It has 36 teeth with the dental formula I 1/3 C 1/1 Pm 3/3 M 3/3. These animals lack a gallbladder. Many interesting crosses have occurred among the different breeds in South America (see table).

Lama glama has been domesticated since the Inca civilization by the Peruvian Indians, who still keep large herds. While the fur provides material for rugs, rope, and cloth, the main use of this animal is as a beast of burden, especially in mining areas. The males are used as pack animals and the females are kept for breeding and wool production. Another race is the alpaca, which is more restricted in its distribution and is specialized in wool production.

There are two species of wild llama, the guanaco or huanaco (*L. guanicoe*) and the vicuña (*L. vicugna*). The most widespread is the guanaco, which stands about 4 ft at the shoulder and just over 5 ft at the top of the head. The guanacos live in small herds composed of a male, several females, and the young. It was formerly thought that this stock represented the forerunner of the domestic breeds. The vicuña is becoming rare because of overshooting for its pelt of soft wool. *See* ALPACA; ARTIODACTYLA.   [CHARLES B. CURTIN]

Load line

A line drawn on the output characteristic curves of a vacuum tube or transistor, used to determine the operating range and the quiescent point (the operating point for zero signal input voltage or current). In the case of vacuum tubes, this graphical con-

struction is necessary because the mathematical function relating plate current to plate voltage and grid voltage is not known analytically. A set of plate current curves with a load line drawn is illustrated. *See* VACUUM TUBE.

If the tube has fixed bias, as shown in the circuit diagram in the illustration, the procedure for drawing the load line is simple. The load line is drawn to intersect the plate-voltage axis at a value equal to V_{PP}, the plate supply voltage, and the vertical axis at a value of plate current which is equal to V_{PP}/R_L, where R_L is the load resistance. The quiescent plate current I_P and plate voltage V_P are read at the point Q, where the curve for the value of grid bias intersects the load line.

If the tube is self-biased with a cathode resistor, R_K, then the load line is drawn to intersect the plate-voltage axis at V_{PP} and the vertical axis at $V_P/(R_L + R_k)$. A bias line is constructed corresponding to the equation $v_c = -i_p R_k$ (assuming that $V_{GG} = 0$ and a grid-leak resistor is used). The point of intersection of the bias line and load line gives the quiescent point on the tube characteristics.

Load lines similarly constructed may also be used with transistor circuits and other nonlinear devices to determine the quiescent point. *See* TRANSISTOR.   [CHRISTOS C. HALKIAS]

Loading, electrical

The addition of inductance to a transmission line in order to improve its transmission characteristics over the required frequency band. Loading coils are often inserted in telephone lines, at spacings as close as 1 mi, to counteract the capacitance of the line and thus make the line impedance more closely equivalent to a pure resistance. Similar coils are used between sections of lines used for carrier transmission. *See* TRANSMISSION LINES.

With coaxial cables and waveguides, loading is placed at the end of the line to absorb all power reaching the end, thereby achieving a nonreflecting termination. The lossy sections used for this purpose may be tapered metal vanes, wedges of lossy dielectric material, or tapered sections of iron-dust core material that partly or completely fill the end of the line. *See* WAVEGUIDE.

When an antenna is too short to give resonance at the desired frequency, a coil can be inserted in series with the antenna to give the required amount of loading needed for resonance. As an example, practically all auto radios have a loading coil in the antenna circuit, because the usual whip antenna is much too short for broadcast-band frequencies. *See* ANTENNA (ELECTROMAGNETISM).   [JOHN MARKUS]

Loads, dynamic

A force exerted by a moving body on a resisting member is called a dynamic or energy load. Loads suddenly applied with appreciable striking velocity, as when a falling weight strikes another body, are called impact loads. Such loads are produced by moving machine parts, cars on a bridge, airplane landing shock, falling weights, and other nonstationary loading conditions where forces are applied in relatively short time intervals. Dynamic load is expressed in terms of the amount of energy transferred or by an equivalent static load which has the same stress-producing effect on the member.

LOAD LINE

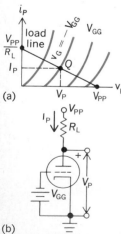

Graphical construction for load line. (*a*) Load line drawn on plate current characteristics. (*b*) Tube with load resistor.

Impact or load factor is the measure of the increased static load required to produce the same stress as the dynamically applied load. It is thus the ratio of the equivalent static load to the weight of the moving body. Impact affects both the magnitude of the stress and the properties of the material. Increased strain rates associated with impact increase the elastic and ultimate strength.

Elastic behavior. If a material or structure has sufficient elastic energy capacity as a resisting member to absorb completely the energy load through elastic action and recover completely from the deformation, it behaves elastically. Elastic stresses associated with stored strain energy depend on the kind of deformation produced. Part of the kinetic energy is dissipated by deformation of connections, friction, and inertia effects. *See* SHOCK ABSORBER.

Dynamic stresses and deformations can be found by equating externally applied energy to the internal strain energy expressed in terms of stresses and dimensions. Dynamic force produced by weight W falling through height h onto a beam or spring is approximately indicated in the equation shown below, where Δ_{st} is elastic deflection due to

$$P_{dyn} = W + W \sqrt{1 + \frac{2h}{\Delta_{st}}} \ (c)$$

gradual application, and c is an inertia-correcting factor. Stresses and deflection are found by expressions of the same form.

Inelastic behavior. An energy load exceeding the elastic energy capacity of the member produces inelastic deformation. For a known amount of energy, the dynamic load and deformation can be found approximately from the boundaries of the area under a statically determined load deformation curve, which represents absorbed energy. Similarly, the inelastic stress produced by impact or the dimensions necessary to limit the stress can be found from areas in the static stress-strain curve.

Toughness is the energy absorbed per unit volume when the material is stressed to fracture, and is represented by the total area under the stress-strain diagram. Ability to store strain energy is increased by uniformity of stress and ductility. Capacity to dissipate energy overloads by plastic deformations is desirable in members subject to shock or impact. *See* STRESS AND STRAIN.

[W. J. KREFELD/W. G. BOWMAN]

Loads, repeated

Forces reapplied many times and causing varying stresses as the load changes. Repeated loads exist in crankshafts, axles, springs, piston rods, rails, bridge members, and many other machine and structural elements. Repeated loads considerably smaller than similarly applied static loads cause failure by progressive fracture. Designs for repeated loads depend on the maximum value of repeated stress which the material can resist, called the fatigue strength, and the magnitude of localized stress concentrations. *See* STRESS CONCENTRATION.

Cycle of stress. The variation of stress S during one typical fluctuation when the load is repeatedly applied and removed or when the load fluctuates continuously from tension to compression is a stress cycle (as illustrated). A fluctuating or pulsating stress varies between maximum and minimum values of the same sign; this condition is called repeated if the stress reduces to zero in each cycle. Reversed stress varies between values of opposite signs and is partly reversed when the opposite stresses are unequal, and completely reversed when the opposite stresses are equal.

Stress variation is described either by the maximum stress, its kind (tension or compression), and the range of fluctuations, or by the mean or steady stress (S_m in illustration) together with the magnitude of the superimposed alternating stress producing the cyclic variation. Repeated loads superimposed on a constant load, such as dead weight of the structure, may produce fluctuation without reversal. Engine crankshafts and parts, rails, and axles are subjected to the reversed-type cyclic loading, which may involve billions of repetitions in the life of the member.

Stress raisers. Dimensional changes, internal discontinuities, or other conditions causing localized disturbance of the normal stress distribution, with increase of stress over that normally produced, are termed stress raisers. The maximum stress is called a stress concentration. High local stresses exist at bolt threads, abrupt change of shaft diameter, notches, holes, keyways, or fillet wells. Internal discontinuities, such as blowholes, inclusions, seams or cracks, ducts and knots in wood, voids in concrete, and variable stiffness of component constituents, cause stress concentrations. At points of contact of ball or roller bearings, gear teeth, or other local load applications, the stress may be greatly increased. Maximum stress is obtained by multiplying the nominal stress, computed without regard to the modifying effect of the stress raiser, by a stress concentration factor. This factor, found experimentally, depends on the type of discontinuity and the properties of the material under static load. Ductile material relieves most of the stress concentration by plastic yielding. Stress raisers contribute to failure under high stress rates and low temperatures, which tend to inhibit plastic flow, and are of great importance under repeated loads, which produce progressive fracture, called fatigue failure.

[W. J. KREFELD/W. G. BOWMAN]

Loads, transverse

Forces applied perpendicularly to the longitudinal axis of a member. Transverse loading causes the member to bend and deflect from its original position, with internal tensile and compressive strains accompanying change in curvature.

Concentrated loads are applied over areas or lengths which are relatively small compared to the dimensions of the member. A single resultant force is used to analyze effects on the member. Examples are a heavy machine occupying limited floor area, wheel loads on a rail, or a tie rod attached locally. Loads may be stationary or they may be moving, as with the carriage of a crane hoist or with truck wheels.

Distributed loads are forces applied continuously over large areas with uniform or nonuniform intensity. Closely stacked contents on warehouse floors, snow, or wind pressures are uniform loads. An equivalent uniform load may represent multiple closely spaced, concentrated loads. Variably distributed load intensities include foundation

LOADS, REPEATED

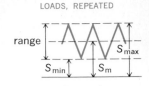
fluctuating

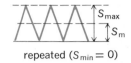
repeated ($S_{min} = 0$)

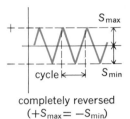
completely reversed
($+S_{max} = -S_{min}$)

random reversed
(range and stress varies)

Types of repeated stress.

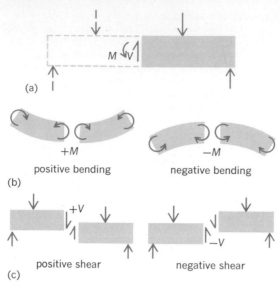

Fig. 1. Beam loading analysis. (a) By free-body analysis of beam. (b) Conventions of sign for bending moment. (c) Conventions of sign for shear.

soil pressures and hydrostatic pressure.

Bending and shear. Transverse forces produce bending moments and transverse shearing forces at every section which must be balanced by an internal couple or resisting moment M and an internal tangential shear force V (Fig. 1a). These component forces are evaluated by free-body analysis. The bending moment and resultant shear force can be expressed analytically as functions of the loads and the distance locating the section from a reference origin. The magnitude and distribution of internal stresses associated with the external force system are found by the theory of flexure.

Bending moment diagram. A graphical representation shows the bending moment produced at all sections of a member by a specified loading. The bending moment is the sum of the moments of external forces acting to the left of the section, taken about the section. The sign of the curvature is related to the sign of the bending moment (Fig. 1b).

Typical bending moment and shear diagrams are shown in Fig. 2 for concentrated and uniformly

distributed loads on a simply supported beam. The reactions and moments are found for equilibrium. See STATICS.

Shear diagram. A graphic representation of the transverse shearing force at all sections of a beam produced by specified loading is called a shear diagram. The shear at any section is equal to the sum of the forces on a segment to the left of the section considered (Fig. 1c). Positive shears are produced by upward forces (Fig. 2).

Relationships between shear and bending moment that assist in construction of diagrams are: (1) Maximum moment occurs where shear is zero. (2) Area in the shear diagram equals change of bending moment between sections. (3) Ordinates in the shear diagram equal slope of moment diagram. (4) Shear is constant between concentrated loads and has constant slope for uniformly distributed loads. Combined loading can be represented by conventional composite moment diagrams or by separate diagrams, called diagrams by parts.

Beams. Members subjected to bending by transverse loads are classed as beams. The span is the unsupported length. Beams may have single or multiple spans and are classified according to kind of support, which may permit freedom of rotation or furnish restraint (Fig. 3).

The degree of restraint at supports modifies the stresses, curvature, and deflection. A beam is statically determinate when all reaction components can be evaluated by the equations of statics. Two equations are available for transverse loads and only two reaction components can be found. Fixed-end and continuous beams are statically indeterminate, and additional load deformation relationships are required to supplement statics.

Bending stresses are the internal tensile and compressive longitudinal stresses developed in response to curvature induced by external loads. Their magnitudes depend on the bending moment and the properties of the section. The theory of flexure assumes elastic behavior. No stresses act along a longitudinal plane surface within the beam called the neutral surface. A segment subjected to constant bending moment, as when bent by end couples, is in pure bending, and stresses vary linearly across the section which remains plane. Simultaneous shear causes warping and nonlinear stress distribution. The stress is maximum at boundary surfaces of sections where bending moment M is greatest. Maximum stress S at distance c from the neutral axis to the extreme element of a section having a moment of inertia I is $S = Mc/I = M/Z$, where $Z = I/c$ is called the section modulus, a measure of the section depending upon shape and dimensions. Greatest economy results when the section provides the required section modulus with least area. The common theory of flexure applies only to elastic stresses. Stresses exceeding the elastic limit involve plastic strains producing permanent deflections upon load removal.

Inelastic stresses are first produced at surface elements of sections resisting maximum moment. A distinct yield point in materials, such as structural grade steel, causes stresses near the surface to remain constant, while interior stresses increase with increasing load. Redistribution of stress con-

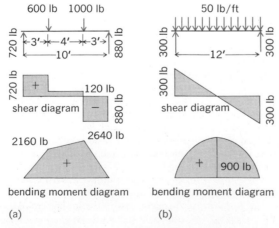

Fig. 2. Typical shear and bending moment diagrams for (a) concentrated and (b) uniformly distributed loads.

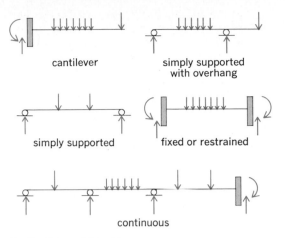

Fig. 3.   Types of beams.

tinues until the entire section behaves plastically. The fully plastic resisting moment of a standard I section is about 15% greater than the moment just producing yield point at the surface. Where small increases in deflection can be tolerated, this plastic-hinge moment is used in design.

Shearing stresses are intensities of the tangential shear force distributed over a beam section. At any point in the beam, the vertical and longitudinal shear stresses are equal. They are zero at the boundary surfaces and maximum at the neutral axis. The shear stress according to the common theory of flexure is $S_s = VQ/Ib$, where V is total shear force, Q is statical moment about the neutral axis of that area of the beam section between the level of investigation and the top or bottom of the section, I is moment of inertia, and b is width. In rectangular sections the distribution is parabolic, and maximum S_s is one and one-half times the average stress.

Shear flow is the shear force per unit length acting along component areas of the section. The concept of flow is derived from the distribution of unit shear forces represented by vectors pointing in the same direction along the median line of a thin element, and the similarity of the expression for unit shear to the equation for quantity of liquid flowing in a pipe or channel. The direction of unit shear forces acting in typical structural shapes is shown in Fig. 4. See TORSION.

Deflection. A beam is said to undergo deflection when the displacement is normal to the original unloaded position and is due to curvature produced by loads. Deflections may be limited to less than structurally safe values in buildings to avoid cracking the ceiling or in machines because of necessary clearances.

Elastic curve is the curved shape of the longitudinal centroidal surface when loads produce wholly elastic stresses. Deflection at any point is found from the equation of this curve.

Radius of curvature depends on the moment and stiffness of the beam and is constant, with bending to a circular arc, when the moment is constant. This is simple bending. Curvature varies under variable moment, and shear forces contribute additional deflection.

Statically indeterminate beam. In the presence of more reaction components than can be deter-

mined by the equations of statics alone, a beam is statically indeterminate. The first step in design is to find the reactions. Equations of statics are supplemented by additional equations relating external forces to slope and deflection.

The curvature and accompanying stresses depend on the loads and the restraint imposed by the supports. In a simply supported beam only vertical displacement is prevented at the supports, whereas when the beam is fixed, rotation is also prevented at the ends. Similarly, beams with overhanging ends and beams continuous over multiple supports are partially restrained. Introduction of the boundary conditions describing the conditions imposed by the supports in the solution of the basic differential equation for the elastic curve furnishes relations between the loads, reaction components, and properties of the beam. During elastic behavior these relations, together with the equations of statics, evaluate the indeterminate reactions. Inelastic behavior is caused by plastic deformations at critical sections of high flexural stress when the yield point of the material is exceeded. When a fully plastic condition is reached, the section acts as a plastic hinge, with constant resisting moment during unrestrained rotation. These known moments at supports or intermediate points permit determination of the reactions by the equations of statics. Advantage is taken of this inelastic action in plastic design.

Unsymmetrical bending. Loads that produce bending moments in planes oblique to the principal axes of a section cause unsymmetrical bending. Components of the applied moment produce curvature and deflection in planes perpendicular to these axes; the stresses produced by these components acting separately can be superimposed. To avoid torsional twist, the oblique forces must pass through either the geometric center of symmetrical sections or a particular point called the shear center for unsymmetrical sections.

Lateral buckling of beams is caused by instability of the compression flange, which results in lateral deflection and twist of the section. The action corresponds to buckling of long columns. The critical compressive stress producing buckling depends on the sectional dimensions and type of loading, which in turn determine the flange stresses and torsional rigidity of the section. Beams having small moments of inertia about the weak axis with low torsional resistance are susceptible to buckling. For rolled beams, buckling resistance is related approximately to the Ld/bt ratio, where L is span, d is beam depth, and b and t are width and thickness of the flange, respectively.

Buckling of beam webs is caused by excessive diagonal compression induced by the complementary shear stresses in the web or by the combined shear and bending stress in the web acting as an edge-supported plate. Because the web is relatively thin, it tends to buckle laterally in a series of waves. For deep beams the shear stress must be reduced by increased web thickness or reinforcement in the form of stiffener angles or plates added to increase resistance to lateral buckling.

Variable cross sections make more efficient use of material in beams by providing resisting moment at all sections more nearly equal to the applied moment. If the section modulus varies

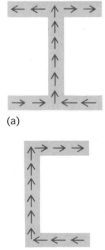

(a)

(b)

Fig. 4.   Shear flow in (a) an I section and (b) a channel section.

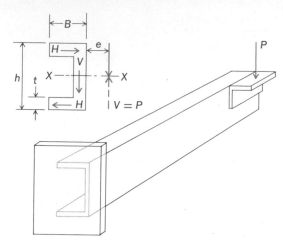

Fig. 5. Shear center under P at distance e from symmetrical axis of member.

directly as the bending moment, stress M/Z will be constant at all sections, and the beam is said to have constant strength. Dimensions of the section can be varied by tapering the width and depth or by adding cover plates to a built-up section to vary the section modulus. Common examples of beams with variable section are tapered cantilevers, plate girders with multiple cover plates, forgings, and other machine elements. A leaf spring is a tapered beam in which uniform stress increases energy absorption.

Shear center. A point on a line parallel to the axis of a beam through which any transverse force must be applied to avoid twisting of the section is called the center of twist or shear center. A beam section twists when the resultant of the internal shearing forces is not in the same plane as the externally applied shear. The shear center can be found by locating the resultant of the internal shear forces. A channel section has one axis of symmetry and is subject to twist unless the external shear passes through a point located e from the center of the web so that $e = B^2 h^2 t / 4 I_x$, approximately, when the flange and web thickness are small, where B is flange width, h is depth of section, t is thickness, and I_x is moment of inertia about the symmetrical axis (Fig. 5).

Beams on elastic foundation are members subjected to transverse loads while resting either on a continuous supporting foundation or on closely spaced supports which behave elastically. The curvature, deflection, and bending moment depend upon the relative stiffness of the beam and supporting foundation. Beams of this type include railroad rails on ties, a timber resting on level ground, a long pipe supported by closely spaced hanger rods or springs, and a concrete footing on a soil foundation. The unknown reactive force per unit length of beam is assumed to be proportional to the deflection, and the solution of this statically indeterminate problem requires derivation of the equation of the elastic curve whose differential equation is $EI(d^4y/dx^4) = -ky$, where k is the spring constant per unit length of support and ky is the elastic force exerted per unit length of beam. Solution of this equation determines the distribution of reactive forces from which the shear and moments can be found. *See* STRENGTH OF MATERIALS.

[W. J. KREFELD/W. G. BOWMAN]

Lobata

An order of the Ctenophora in which the body is helmet-shaped. A pair of large lobe-shaped processes extend from the oral end in the pharyngeal plane. A pair of "auricles" are attached to the base of each process (Fig. 1). The body is com-

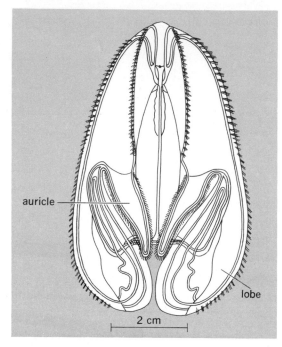

auricle

lobe

2 cm

Fig. 1. Late cydippid stage of *Mnemiopsis leidyi.*

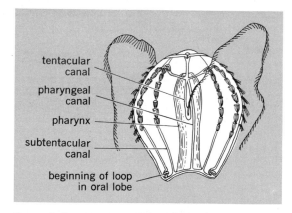

tentacular canal

pharyngeal canal

pharynx

subtentacular canal

beginning of loop in oral lobe

Fig. 2. *Bolinopsis mikado,* of the Lobata.

pressed, with the tentacular axis being shorter than the pharyngeal axis. The ribs are of unequal length, with the longest being the subpharyngeal pairs. The primary tentacles are degenerate and are replaced by secondary tentacles. The canal system is complex (Fig. 2). Examples of this order include *Bolinopsis, Ocyropsis, Mnemiopsis,* and *Leucothea. See* CTENOPHORA; TENTACULATA.

[TAKU KOMAI]

Lobosia

A subclass of Rhizopodea characterized by lobopodia predominantly, although certain of these protozoan species also may form slender pseudopodia, or even develop several different kinds. Lobosia are divided into two orders which differ in presence or absence of a test. The Amoebida include the amebas, Lobosia which do not produce a test. The Arcellinida possess a one-chambered test ranging from a predominantly chitinous to a mainly siliceous covering. Pseudopodia emerge through the aperture of the test. *See* AMOEBIDA; ARCELLINIDA; PROTOZOA; SARCODINA; SARCOMASTIGOPHORA.

[RICHARD P. HALL]

Lobster

The name for a number of species of decapod crustaceans which are members of the family Homaridae and are commercially important because of their food value. Three species are commonly marketed: the European lobster (*Homarus vulgaris*), the Norwegian lobster (*Nephrops norvegicus*), and the American lobster (*Homarus americanus*). The American lobster is found in greatest abundance along the eastern coast of North America from Maine to New York, but ranges from Labrador to North Carolina. Migrations are chiefly to deeper waters (to 600 ft) in the fall, with a return to shallower waters (to 6 in.) in the spring.

Characteristics. The average length of lobsters taken for market is about 10 in.; however, the record is for one which weighed over 40 lb and measured over 3 ft. At the other extreme, the South African cape lobster (*H. capensis*), measures about 5 in. Like most of the large marine crustaceans, the lobster is a bottom dweller, and seldom moves away from the floor of the ocean. However, it swims well and when alarmed will swim backward rapidly, using its muscular tail (see illustration). It is a nocturnally active animal and waits in rock crevices or mud burrows to capture its prey. Although it is a scavenger, it prefers fresh food, such as fish, mollusks, and small amounts of vegetation.

Molting, a process necessary for growth, occurs during the summer months, and is repeated during the life of the lobster. In the female it occurs once every 2 years when she hatches her eggs. After molting, it takes about 6–8 weeks for the new skeleton to harden.

Life cycle. Mating occurs after the female molts, and spawning usually occurs the following July to September. The male transfers the sperm to the female seminal receptacle, where it remains until spawning, when the female lies on her back and flexes the abdomen to form a pouch. The eggs are extruded from the genital apertures at the base of the third walking legs, and as they pass down to the abdominal region, sperm cells are released from the receptacles and fertilization occurs. These fertile eggs adhere to the pleopods or swimmerets, where they are carried by the female throughout the period of incubation. Following this period, which lasts somewhat over 10 months, hatching occurs and the female moves the swimmerets vigorously so that the newly hatched larvae are dispersed. There are a number of larval stages which last for varying periods of time. The larvae are voracious and may resort to cannabalism. When they assume the adult form, they sink to the bottom.

Conservation. There has been a gradual decline in the lobster industry since 1889, even though a number of measures have been taken to maintain and increase the lobster population. Beginning as early as 1874 unsuccessful attempts were made to

The lobster, showing the stout pincers on the front legs which are used to crush prey.

transplant the American lobster to the Pacific Coast. Conservation laws and measures have been passed which protect the egg-bearing females, and limit the size of lobsters which can be taken. In addition, rearing stations have been established for artificial propagation. *See* CRAYFISH; CRUSTACEA; DECAPODA (CRUSTACEA).

[CHARLES B. CURTIN]

Location fit

Mechanical sizes of mating parts such that, when assembled, they are accurately positioned in relation to each other. Locational fits are intended only to determine the orientation of the parts. For normally stationary parts that require ease of assembly or disassembly, for parts that fit snugly, and for parts that move yet fit closely as in spigots, slight clearance is provided between parts. Where accuracy of location is important, transition fits are used. In these fits the holes and shafts are normally nearly the same diameter. For greater accuracy of location, the shafts are made slightly larger than the holes; such fits are termed location interference fits. *See* ALLOWANCE. [PAUL H. BLACK]

Locomotive

A machine on flanged wheels, capable of propelling itself and other vehicles on railroad tracks by converting heat or electric energy into mechanical power. Most of the 28,000 locomotive units in use on United States railroads are diesel-electric units. There are less than 400 all-electric locomotives in service, and the steam locomotive has almost completely disappeared.

Diesel. Diesel units in production today range in capacity from 1000 to 3600 hp. Yard switching locomotives are generally lower horsepower units, with cabs designed for wide visibility and gearing for low speeds but high traction for acceleration. Passenger units are geared for higher speeds and must have oil-fired steam generators or auxiliary power plants for train heating. Freight diesels must deliver as much horsepower as possible to the rail. The largest units measure almost 90 ft long and 17 ft high and weigh more than 200 tons.

Each locomotive unit is a self-sufficient portable power plant, complete with diesel engine, electric generator, auxiliaries, and operating cab housed within a car body (Fig. 1). The car body is mounted

Fig. 1. Diesel-electric freight unit, with 16-cylinder turbocharged diesel engine, an ac traction generator, and six traction motors. Unit is rated at 3300 hp and weighs 182 tons. (*General Electric Co.*)

on swiveling wheel assemblies called trucks, which also contain electric traction motors wired to the main generator. Most diesel units have four traction motors, but some have six. Two, three, or more of these diesel units can be coupled in multiple to operate as one locomotive under the control of one man in the cab of the lead unit. Locomotives totaling 10,000 or even 15,000 hp are frequently used on coal and ore trains, weighing up to 15,000 tons, or on lighter but faster piggyback and merchandise trains.

In recent years there have been experimental diesel locomotives built with hydraulic transmission in place of electric generator and traction motors. One diesel-hydraulic model operates successfully with a rating of 4300 hp. There are numerous small industrial locomotives with mechanical or hydraulic drives.

Engine and auxiliaries. Most modern diesel locomotives have V-type engines, either two- or four-cycle, with top speed in the neighborhood of 1000 rpm. The engine has between 8 and 20 cylinders, with typical dimensions of 9-in. bore and $10\frac{1}{2}$-in. stroke. Turbochargers are in common use. A few in-line and opposed-piston engines remain in service in the older units. *See* DIESEL ENGINE.

The lubricating oil system on a diesel unit ensures a supply of clean, cool oil to the engine, and provides for extracting carbon deposits left by combustion along with some heat from the engine. The oil is circulated by engine-driven scavenging and pressure pumps, through removable filter elements, strainers, and a water-cooled heat exchanger.

The engine cooling system also uses engine-driven centrifugal pumps to circulate treated water through cylinder heads, liners, and oil cooler to banks of radiator units. The large fans which force air through the radiators are driven mechanically from the engine or by individual electric motors.

The fuel system includes a motor-driven pump to deliver fuel oil from a storage tank, usually mounted under the car body between the trucks, with a capacity of $4000 \pm$ gal. The oil is forced through a strainer and filter to the injectors, where it is sprayed into the cylinders by high-pressure pumps and nozzles. The quantity of fuel delivered for each stroke of the engine piston is regulated by a mechanical linkage connected to the engine governor. This device keeps engine speed constant for each throttle setting, regardless of load.

Electric power units. The main generator on a diesel locomotive unit is driven directly from the diesel engine; that is, the armature is bolted directly to the engine crankshaft flywheel (Fig. 2). Until recently, the generator was always a direct-current (dc) machine, but now alternators are often used, with solid-state rectifiers to supply dc to the traction motors. The dc generator has an exciter, or battery field, which can be strengthened or weakened by control circuitry to vary power output. The generator also has a separate starting field, which can be connected across the locomotive battery to make the generator act as a motor and start the diesel engine.

Traction motors on a diesel locomotive are series-wound dc machines, each with a continuous rating of over 1100 amp and up to 800 hp.

They are usually cooled by motor-driven blowers located in the car body, forcing air to the traction motors through ducts. Each traction motor is mounted on the truck between the wheels with its armature parallel to the axle adjacent to it. A pinion is mounted on the motor's armature shaft which drives a gear on the axle. A typical gear ratio for a freight locomotive is 62 teeth in the bull gear and 15 teeth on the pinion. High-speed passenger units have bull gears with fewer teeth and pinions with more, so that the traction motor armatures will not fly to pieces by centrifugal force at high axle rpm. These locomotives, however, run the risk of armature overheating at low speeds when counter electromotive force (emf) is too low to prevent high currents through the traction motors. *See* ELECTRIC ROTATING MACHINERY; ELECTROMOTIVE FORCE (EMF).

Electric controls. The typical diesel locomotive unit has a 74-volt dc lighting and control circuit, powered by storage batteries in the car body. The battery is kept charged through a reverse current relay by an auxiliary generator geared to the diesel engine. A voltage regulator keeps voltage constant. Connected to this circuit are the main generator or exciter field, governor solenoids, fuel-pump motor, lighting, warning devices, and magnet valves which respond to the cab throttle and reverser control to admit compressed air to operate the main power contactors and reverser drum.

Power contactors are rugged switches in the high-voltage circuits between main generator and traction motors. They open and close automatically in various combinations during acceleration or deceleration of the locomotive, during the process known as transition. For example, when a four-motor unit first starts its train, the traction motors are connected in series-parallel; that is, one pair of motors is connected in series across the main generator, and the other series-connected pair, also connected across the generator, is in parallel with the first pair. Thus, half the current output of the main generator flows through each traction motor, furnishing high torque to start the train moving. As speed picks up, higher voltage must be supplied by the generator to overcome counter emf in the traction motors. When the voltage reaches the allowable maximum, contactors operate to connect

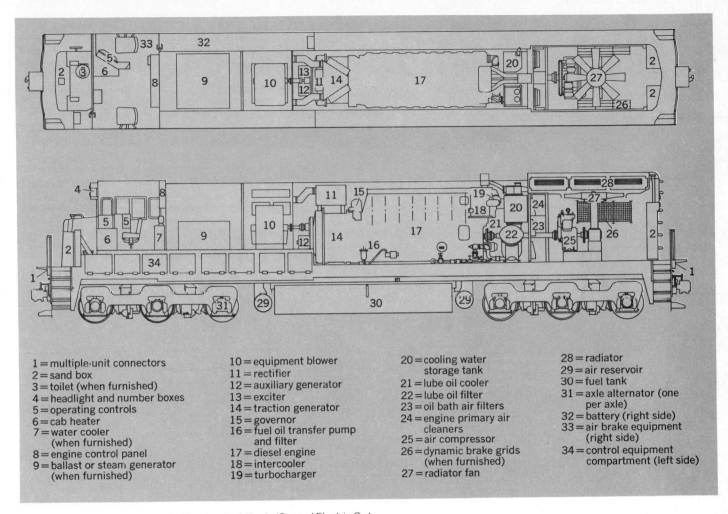

Fig. 2.   Location of equipment in diesel unit of Fig. 1. (*General Electric Co.*)

1 = multiple-unit connectors
2 = sand box
3 = toilet (when furnished)
4 = headlight and number boxes
5 = operating controls
6 = cab heater
7 = water cooler
    (when furnished)
8 = engine control panel
9 = ballast or steam generator
    (when furnished)

10 = equipment blower
11 = rectifier
12 = auxiliary generator
13 = exciter
14 = traction generator
15 = governor
16 = fuel oil transfer pump
     and filter
17 = diesel engine
18 = intercooler
19 = turbocharger

20 = cooling water
     storage tank
21 = lube oil cooler
22 = lube oil filter
23 = oil bath air filters
24 = engine primary air
     cleaners
25 = air compressor
26 = dynamic brake grids
     (when furnished)
27 = radiator fan

28 = radiator
29 = air reservoir
30 = fuel tank
31 = axle alternator (one
     per axle)
32 = battery (right side)
33 = air brake equipment
     (right side)
34 = control equipment
     compartment (left side)

shunting circuits around the traction motor fields, thereby weakening them enough to lower voltage requirements. When the main generator voltage limit is reached again, the shunt field contactors drop out and the power contactors operate to reconnect all four traction motors in parallel across the main generator, with full voltage output across each motor. As speed increases again, the field shunt contactors pick up.

During train deceleration on a grade, the contactors operate in reverse sequence to reconnect motors for high amperage to furnish high torque at low speeds.

Many diesel units are also equipped with dynamic braking, which is used to slow down the locomotive without using the air brakes. The equipment consists of power contactors which close to insert heavy resistors into the traction motor circuits so that the motors can act as generators, dissipating energy in resistor grids, which are cooled by blower fans.

Among the many protective features on most diesel locomotive units are engine high-temperature switch, low-oil-pressure switch, overspeed trip, wheel slip relay, ground relay, no-alternating-current-voltage relay, and emergency fuel cutoff.

Air brakes. Air for the brakes is supplied by an air- or water-cooled compressor driven by the diesel engine. It usually has low- and high-pressure

cylinders, with an intercooler unit between them. Air pressure is maintained in the locomotive main air reservoirs to supply air throughout the train. Other reservoirs on each car are kept charged except when the engineman desires to set the

Fig. 3.   Electric locomotive rated at 5000 hp. Retractable pantograph ahead of cab roof collects high-voltage alternating current from overhead wires. Power is fed through a transformer and smoothing reactor to silicon rectifiers, where it is converted to direct current for use by the traction motors. (*General Electric Co.*)

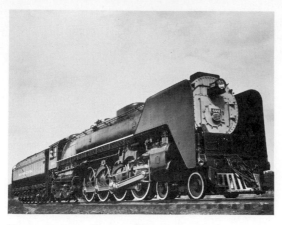

Fig. 4. Steam locomotive. (*Penn Central Co.*)

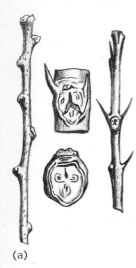

(a)

(b)

The black locust
(*Robinia pseudoacacia*).
(*a*) Branches, with
and without thorns,
and enlarged views
of leaf scars.
(*b*) Odd-pinnate leaves.

brakes. He then exhausts air from the train brake pipe, and a control valve on each car automatically diverts air from its reservoir to the car's brake cylinders. When the engineman desires to release the brakes, he recharges the brake pipe, and the control valve on each car then automatically releases air from the brake cylinders. *See* AIR BRAKE.

Electric. Under certain conditions, such as very dense traffic, operation of trains by electric power becomes economically possible. Only a few segments of United States railroads are operated electrically, but there is increasing electrified mileage in Europe every year.

Electric locomotives are similar in many respects to diesel electrics, except that the electrics pick up their electricity from a system of continuous overhead wires called a catenary (Fig. 3). Usually the current used is 11,000-volt single-phase alternating current (ac), which is relatively cheap to distribute great distances from permanent power plants. However, since this voltage is much too high for use in traction motors, it must be reduced in large transformers carried in the locomotive car body. Some locomotives convert the ac to dc in rectifiers for use in more conventional dc traction motors.

A few electric locomotives use 660-volt dc picked up directly from a third rail mounted alongside the track. Still others pick up 3000-volt dc from overhead catenary.

Gas turbine. One of the western United States railroads operates a small fleet of 8500-hp locomotives powered by gas turbines. Actually they are three-unit locomotives with cab and auxiliaries in the lead unit, the gas turbine and four generators in middle unit, and a tender for fuel as trailing unit. Traction motors are geared to all axles on the first two units. The same railroad has also been experimenting with a gas-turbine locomotive using powdered coal as fuel. *See* GAS TURBINE.

Steam. In the United States the only steam locomotives in use are those on excursion or tourist lines, but overseas thousands remain in service, notably in India. These locomotives have fire-tube boilers through which hot exhaust gases flow from the firebox in the rear to the smokebox and stack at the front. High-pressure steam from the boiler is allowed to expand in the two large cylinders beneath the smokebox to move pistons back and forth. Connecting rods then force the driving

wheels to rotate. Water and coal or fuel oil are carried in a tender coupled behind the cab. Relatively high maintenance and servicing costs associated with the boiler, water supply, and fuel handling forced the steam locomotive out of service in favor of the diesel. Thermal efficiency of the diesel locomotive is approximately 25% at the rail, compared with only 5 or 6% for the steam locomotive (Fig. 4). *See* STEAM ENGINE.

Performance. In addition to horsepower, locomotives are rated on the basis of tractive effort (measured at the rail) or drawbar pull (measured at the rear coupler of locomotive). A diesel or electric locomotive is designed to develop tractive effort equal to about 25% of the weight on its driving wheels when starting a train on clean dry rail. Greater tractive effort than this results in wheel slip. Three types of train resistance must be overcome by this tractive effort: resistance on grades, on curves, and during acceleration. Grade resistance amounts to approximately 20 lb per ton of train weight for each percent of grade (the ratio of altitude rise to track distance, times 100). Curve resistance varies up to 0.8 lb per ton per degree of track curvature. Accelerating at a rate of 1 mph/sec requires a force of approximately 100 lb per ton of train weight. Tonnage ratings for locomotives on a given portion of railroad are calculated from these values. *See* RAILROAD ENGINEERING.

[W. D. EDSON]

Bibliography: N. E. Bateson, Progress in railway mechanical engineering, *Mech. Eng.*, May, 1968; A. W. Bruce, *The Steam Locomotive in America*, 1952; *Car and Locomotive Cyclopedia*, 1966; D. P. Morgan, Locomotive showroom 67/68, *Trains Mag.*, December, 1967; J. A. Pinkepank, *Diesel Spotter's Guide*, 1967; H. Sampson, *Jane's All the World's Railways*, 10th ed., 1968; H. L. Smith, Jr., The diesel locomotive, *Mech. Eng.*, December, 1967.

Locust (forestry)

A name commonly applied to two trees, the black locust (*Robinia pseudoacacia*) and the honey locust (*Gleditsia triacanthos*). Both of these commercially important trees have podlike fruits similar to those of the pea or bean. *See* LEGUME.

The black locust, which may attain a height of 75 ft, is native in the Appalachian and the Ozark regions and is now widely naturalized in the eastern United States, southern Canada, and Europe. It is characterized by odd-pinnate leaves, fragrant white blossoms like those of pea plants in long clusters, flat brown pods 2–4 in. long, and a pair of stipular thorns at the base of each leaf. In winter these paired thorns and the superposed buds hidden under the leaf scar are important distinguishing features (see illustration).

The wood is hard and durable in the soil and is used for fence posts, mine timbers, poles, and railroad ties. It is the principal wood for insulator pins on utility poles. Because of its wide-spreading, fibrous root system which supports nitrogen-fixing nodules, the black locust is well adapted to erosion control and soil reclamation. Many varieties are planted as shade or ornamental trees. *See* SOIL CONSERVATION.

The honey locust, which reaches a height of 135 ft, is native in the Appalachian and the Mississippi

Valley regions, but is also widely naturalized in the eastern United States and southern Canada. The flowers are inconspicuous, but the large brown fruit pods (12–18 in. long) are striking. The tree may be easily recognized at any season of the year by the branched thorns growing on the trunk or branches. The leaves are either pinnate or bipinnate. The reddish wood is hard, strong, and coarse-grained and takes a high polish. Because it is durable in contact with soil, it is used for fence posts and railroad ties; it is also used for construction, furniture, and interior finish. Although it belongs to the legume family, the roots do not possess nitrogen-fixing bacteria. The tree is often used for hedges, as an ornamental, and for shade. *See* FOREST AND FORESTRY; TREE.

[ARTHUR H. GRAVES/KENNETH P. DAVIS]

Loess

An essentially unconsolidated, unstratified, calcareous silt. Most commonly it is homogeneous, permeable, buff to gray in color, and contains calcareous concretions and fossils. In natural and artificial excavations the loess maintains notably stable, vertical faces.

Texture and composition. Mechanical analyses of loess give the following composition: fine sand (grains > 0.074 mm), 0–10%; silt (0.074–0.005 mm), 50–85%; and clay (grains < 0.005 mm), 15–45%. Variations in published analyses may result in part from the different grade-size limits used by the investigators. Although dominantly silt and often referred to as well sorted, loess is actually only moderately well sorted. The silt (and sand) grains are usually angular to subangular. Clay occurs as silt-size aggregates, as coatings on silt grains, and as interstitial filling.

The silt and sand fraction of the loess has the following mineral composition: quartz, 50–70%; feldspars, 15–30%; carbonates (mainly calcite), 0–11%; and heavy minerals, 5–15%. X-ray identification of the minerals in the < 0.005-mm fraction show abundant quartz plus varying proportions of montmorillonoids, illites, and chlorites. In texture and mineral composition, loess is comparable to the average mudstone or shale. It is susceptible to authigenic changes such as the transformation of clay minerals, formation of chlorites, and the movement of soluble materials, especially carbonates. *See* AUTHIGENIC MINERALS; CLAY MINERALS; SEDIMENTARY ROCKS.

Occurrence and origin. Loess is typically developed in areas peripheral to those covered by the last ice sheets in North America and Europe (see illustration). Greatest thicknesses (more than 150

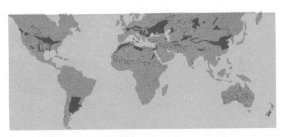

Loess deposits of the world. (*From A. K. Lobeck, Geomorphology, McGraw-Hill, 1939*)

ft) occur in the uplands bordering the valleys of the major streams. There is a general, but irregular, thinning of the loess in one or more directions away from each valley. Other notable areas are northern China and Argentina. Most of the loess seems to have been deposited during the latest (Wisconsin) glacial stage. Little or no loess is found associated with earlier glacial deposits or covering the area actually occupied by the latest ice sheet. No loess deposits older than the Pleistocene are known.

Many theories have been proposed to explain the deposition of loess. The most widely accepted theory is that the materials were transported and deposited by wind. Source areas were nearby floodplains and till plains. Controversy continues, however, for some investigators present evidence and arguments in favor of the alluvial origin of some loess. Others hold that primary deposits, by wind or water, have been greatly modified by colluviation to produce loess. *See* SEDIMENTATION (GEOLOGY).   [CHALMER J. ROY]

Bibliography: W. C. Krumbein and L. L. Sloss, *Stratigraphy and Sedimentation*, 2d ed., 1963; A. K. Lobeck, *Geomorphology*, 1939; F. J. Pettijohn, *Sedimentary Rocks*, 2d ed., 1957.

Logarithm

An exponent of a suitably chosen positive number (base) larger than unity. Logarithms are of value in mathematical computation and in the equations and formulas used in expressing natural phenomena.

The invention of natural logarithms (base $e =$ 2.718 . . .) is usually attributed to John Napier, who published a table of logarithms in Edinburgh in 1614 under the title *Mirifici Logarithmorum Canonis Descriptio*, even though the functions tabulated by Napier are merely related to, and not identical with, natural logarithms.

The invention of common logarithms (base 10) is generally attributed to Henri Briggs; Napier's share is acknowledged, however, by Briggs himself in his *Arithmetica logarithmica* in 1624.

Theory. If $b^l = n$, where b is a positive number larger than unity, then l is called the logarithm of n to the base b and is written $l = \log_b n$. From this definition it follows at once that all positive numbers larger than unity have positive logarithms, all positive numbers smaller than unity have negative logarithms, and the logarithm of unity is equal to 0. Since $b^0 = 1$ irrespective of the value of b, it follows that the logarithm of unity is equal to 0. From the known properties of exponentials expressed by relations (1), it follows immediately that (1) the log-

$$b^{l_1} \cdot b^{l_2} = b^{l_1 + l_2} \qquad b^{l_2} \div b^{l_2} = b^{l_1 - l_2}$$
$$(b^l)^m = b^{lm} \qquad \sqrt[m]{b^l} = b^{l/m} \tag{1}$$

arithm of a product of two or more factors is equal to the sum of the logarithms of the factors; (2) the logarithm of the ratio of two numbers is equal to the difference between the logarithm of the numerator and the logarithm of the denominator; (3) the logarithm of the mth power of a number is the product of m and the logarithm of the number; and (4) the logarithm of the mth root of a number is equal to the logarithm of the number divided by m. These properties are responsible for the great simplification in the task of carrying out numerical computations involving multiplication, division, raising numbers to certain powers, or extracting the roots of a certain order of given numbers. It also follows from these properties that in constructing a table of logarithms to a certain base from the beginning, the computer need only compute the logarithms of prime numbers, because the logarithm of any number which is not prime can be obtained from the logarithms of its prime factors by simple operations of addition and multiplication. *See* EXPONENT.

Although any positive number b larger than unity might have been chosen as the base of a system of logarithms, actually two numbers have been chosen in the construction of tables of logarithms, namely, the number $b = 10$ and the number e defined as the sum of infinite series (2). The system

$$1 + \frac{1}{1} + \frac{1}{1 \cdot 2} + \frac{1}{1 \cdot 2 \cdot 3} + \cdots$$
$$+ \frac{1}{1 \cdot 2 \cdot 3 \cdots n} + \cdots \quad (2)$$

of logarithms to the base 10 is usually referred to as common logarithms; the system of logarithms to the base e is called natural logarithms.

Common logarithms have certain obvious advantages not shared by natural logarithms. All numbers between 1 and 10 have logarithms between 0 and 1, all numbers between 10 and 100 have logarithms between 1 and 2, and so on. Thus the integral part of the common logarithm of a number larger than unity is one unit less than the number of digits before the decimal point. This integral part is called the characteristic; the decimal part is called the mantissa. Similarly, because the common logarithms of 0.1, 0.01, and 0.001 are -1, -2, and -3, it follows that the logarithm of a number smaller than unity having p zeros after the decimal point may be expressed as the sum of a negative characteristic equal to $-(p+1)$ and a positive mantissa.

Natural logarithms. Because the number $e = 2.718 \ldots$, the base of the system of natural logarithms is irrational; that is, it cannot be expressed as the ratio of two integers (and therefore when expressed as a decimal number, it involves an infinite number of decimals with no repeating groups of decimals). It might seem odd, therefore, that it has been chosen as the base of a system of logarithms. The primary motivation for this choice lies in the fact that the solutions of numerous problems in applied mathematics are most naturally expressed in terms of powers of e. Thus, for instance, the solutions of the problems of equilibrium of a flexible cable, the transient flow of electric current in a circuit, and the disintegration of radioactive elements are expressed in terms of e^x, where x is either positive or negative and depends on the physical parameters of the problem in question. Thus, the tabulation of the function e^x for both positive and negative values of x was an indispensable aid in obtaining the solutions of many physical problems. Because the logarithmic function is the inverse of the exponential function, a table of natural logarithms can be constructed with relative ease from a table of e^x by the process of inverse interpolation. Another motivation for constructing tables of natural logarithms is the fact that natural logarithms arise in the integration of

Table of logarithms

Number	Natural logarithm	Common logarithm
0.01	$5.586 - 10$	$-2\ (8 - 10)$
1.0	0	0
$e = 2.718 \ldots$	1	0.4343
10	2.303	1
1000	6.909	3

quotients of polynomials and of functions which may be approximated by quotients of polynomials. Indeed, it is a well-known fact that if the n roots of a polynomial $Pn(x)$ are known so that $Pn(x) = a_0(x - x_1)(x - x_3) \cdots (x - x_n)$ and if $Q(x)$ is a polynomial of degree smaller than n, then the quotient $Q(x)/Pn(x)$ may be expressed in the form of expression (3) when the A_ks may be easily determined.

$$\sum_{k=1}^{n} \frac{A_k}{x - x_k} \quad (3)$$

Accordingly Eq. (4) may be written.

$$\int \{Q(x)/Pn(x)\}\, dx = \sum_{k=1}^{n} A_k \int \frac{dx}{x - x_k}$$
$$= \sum_{k=1}^{n} A_k \log_e (x - x_k) \quad (4)$$

Relation between logarithm bases. Let l_1 be the known natural logarithm of a number N. What is the common logarithm of N? Clearly $N = e^{l_1}$, and therefore $\log_{10} N = l_l \log_{10} e$. Since $\log_{10} e = 0.43429$ (to five decimals) it can be concluded that natural logarithms are converted into common logarithms by multiplying the natural logarithm by the factor 0.43429, which is called the modulus. Similarly the common logarithm is converted into a natural logarithm by division by the modulus 0.43429 or multiplication by 2.303. The table shows natural and common logarithms for some numbers from 0.01 to 1000.

Logarithms of complex numbers. Since b^l is positive when b is positive, negative numbers have no real logarithms. Nevertheless, Euler's famous formula $e^{i\theta} = \cos \theta + i \sin \theta$, where $i = \sqrt{-1}$, makes it possible to define not only the logarithm of a negative number but also the logarithm of a complex number. Indeed, any complex number $a + ib$ may be written in the form $\rho e^{i\theta}$, where $\rho = \sqrt{a^2 + b^2}$ and $\theta = \arctan (b/a)$. It follows that the logarithm of a complex number $a + ib$ is a complex number whose real part is the logarithm of its modulus ρ and whose imaginary part is its phase (or argument) θ. Because $e^{i\theta} = e^{i(\theta + 2n\pi)}$, it follows that a complex number has an infinite number of logarithms given by Eq. (5), where n is a positive or negative integer.

$$\log (a + ib) = \log \sqrt{a^2 + b^2} + i(\theta + 2n\pi) \quad (5)$$

As previously mentioned, the evaluation of natural logarithms may be based on inverse interpolation in a table of e^x. For small values of x, positive or negative, it is possible to use expansion (6). This

$$\log_e (1 + x) = x - \frac{x^2}{2} + \frac{x^3}{3} - \cdots$$
$$+ (-1)^{n+1} \frac{x^n}{n} + \cdots \quad (6)$$

expansion may be derived from Taylor's series, notation (7), by writing $f(x) = \log_e x$. After a

$$f(x+a) = f(a) + \frac{x}{1} f'(a) + \frac{x^2}{1 \cdot 2} f''(a)$$
$$+ \cdots \frac{x^n}{1 \cdot 2 \cdot 3 \cdot m} f^{(n)}(a) + \cdots \quad (7)$$

sufficiently accurate table of $\log_e (1 \pm x)$, for $x = k10^{-n}$, $k=1,2,3 \ldots 9$, and $n=1,2,3 \ldots N$, where N is sufficiently large (perhaps 15) has been computed, this table may be used to compute the logarithm of any number. From the expansion (6) for $\log_e (1+x)$ it is easy to derive expansion (8),

$$\log_e p = \tfrac{1}{2}\{\log_e (p-1) + \log_e (p+1)\} + S(p) \quad (8)$$

where Eq. (9) holds; this has the virtue that $S(p)$

$$S(p) = \frac{1}{2p^2-1} + \frac{1}{3(2p^2-1)^3}$$
$$+ \cdots \frac{1}{(2n+1)(2p^2-1)^{2n+1}} + \cdots \quad (9)$$

converges quite rapidly. Thus for $p > 10,000$ the error resulting from neglecting all but the first term in $S(p)$ is considerably less than 10^{-20}. See e (MATHEMATICS); NUMBER THEORY; NUMERICAL ANALYSIS.

[ARNOLD N. LOWAN/SALOMON BOCHNER]

Logging (forestry)

The cutting and removal of the woody stem portions of forest trees. In effect, logging is the harvesting of timber from forests. Originally, logging produced only sawlog lengths for lumber sawmills, but with increased utilization of wood for other products such as pulp and paper, veneers, handles, crossties, and poles and piles, the term logging includes practically all operations conducted in the forest in connection with felling trees, cutting the trunks to desired lengths, and transporting the merchantable material to processing mills or to shipping points. Most lumber log lengths are 12, 14 or 16 ft, corresponding to the standard lengths of lumber used in general construction; or in multiples of these lengths, in which event final log lengths are cut at the mill. Pulpwood generally is cut in lengths of 4, 5, or 8 ft, although sawlog lengths may be produced for efficient mechanized handling of larger diameters. For other products, various lengths are produced. The steps usually followed in logging are: (1) selection of trees; (2) felling the trees; (3) bucking the trunks (cutting into appropriate lengths) and removing limbs and unmerchantable tops; (4) skidding or forwarding the bucked lengths to roadside or waterside; (5) loading on hauling vehicles, usually trucks; and (6) hauling to destination. See LUMBER MANUFACTURE.

Equipment. Felling is commonly accomplished with gasoline-powered chain-type saws operated by one man. These saws have cutting teeth inserted in a sprocket chain that moves rapidly around the edge of an oval-shaped blade (Fig. 1). Saws of this type with blades 3 ft long can fell and buck trees up to 6 ft in diameter; saws with shorter blades can sever trees that are proportionally smaller in diameter. Several types of tree-felling machines are in use which can cut trees up to a stump diameter of 18 in. with heavy-duty, hydraulic-powered shears. Some only fell; others both fell and buck pulpwood. All tree-felling ma-

chines are mounted on or incorporated in heavy tractors, either wheeled or crawler type.

Skidding is the short-distance movement of tree lengths or segments over unimproved terrain to loading points on transportation routes. The loading points may be along railroads, truck haul roads, public highways, winter sled roads, or stream banks for floating. Skidding requires considerable power, as do all subsequent steps in logging. A longleaf pine log 16 ft long and 18 in. in diameter, for example, weighs 1500 lb. Logs are skidded by mobile power units, such as a horse or a tractor, or by stationary diesel power units with drums and cables. The mobile unit is taken to the logs and it then pulls them to roadside. A stationary unit at a loading point operates by the use of cables fastened to the logs; as the cable is wound on a drum, the logs are pulled to it. A wide variety of types and sizes of skidding equipment has been perfected to meet variations of timber size, terrain, and density of the stand to be cut.

Skidding methods. Attachments generally used to fasten logs to skidding devices are tongs and chain or cable chokers resembling slip nooses placed around the ends of logs. Cable skidding is used when the ground is too swampy to support mobile power units or when the terrain is too steep. The high fixed cost of cable skidding installations requires large daily log output and an operating life of several years, and stands averaging several thousand feet per acre. The simplest form of cable skidding consists of dragging logs along the ground. A much more widely used method is the high lead with rehaul. With it the pulling cable is led from the pulling drum through a pulley block fastened 30–150 ft up a tree or spar and then out to the felling area. A lighter cable is rigged to pull the skidding cable rapidly out to the felling area after a load has been skidded in. When skidding has been completed in one such segment, the rehaul blocks are relocated and the skidding line is moved clockwise to the adjacent segment. In all

Fig. 1. A logger bucking a felled tree trunk with a one-man gasoline-powered chain saw.

Fig. 2. Four-wheel skidder hauling logs with an arch mounted over the rear axle.

cable skidding a similar pattern is followed; that of moving around the circle until the entire area with logs has been covered.

Where terrain permits, mobile power skidding is favored because of its versatility and because it permits skidding without damaging seed trees and young trees; cable skidding is destructive to all uncut trees in the skidding area. Animal logging has disappeared from all but the smallest operations. Tractors without winches simply drag logs on the ground. Tractors with arches, available in a wide range of sizes, are adaptable to large or small

Fig. 3. A self-propelled heavy-duty log loader loading pine and fir logs in Idaho for truck haul. (*Photograph by Washington Iron Works, Seattle, Wash.*)

logs, to steep or muddy terrain, and to snow-covered ground. The arch supports an elevated pulley through which the cable from the tractor winch is extended to the logs. The logs are fastened to the cable, pulled up under the arch, and then dragged with their front ends off the ground to reduce friction (Fig. 2). A crawler-tread arch tractor is generally used for heavy loads and on steep or muddy ground. Wheeled tractors are used on firm ground in the eastern and southern United States for small logs. Tractor skidding distances vary, but practical considerations usually limit skidding to 1/2 mi. For longer skidding, as in getting logs out of wide river bottoms, log wagons with wide wheels or with crawler treads are pulled by tractors. This procedure is often referred to as pre-hauling, a term used when skidding is done by assembling cut products at the stump (or felling location) and loading them on slides, dollies, runners, or more sophisticated conveyances for moving to loading points.

Pneumatic-tired, four-wheel skidders (Fig. 2), suitable for travel over unimproved ground with heavy loads, such as the vehicles used in major earth-moving operations, have proven highly adaptable to logging. They have replaced crawler-type tractors in many areas because of greater maneuverability and travel speed, and more economical operating and maintenance costs. On these skidders the arch and pulley block are mounted over the rear axle so that the arch is not pulled behind as on most crawlers; thus the weight of the logs is on the skidder and not on the arch. Some of these four-wheeled skidders are steered by wheel movement, others by fixed wheels and axles but with a pin-pivoted frame so that the skidder "bends" at its mid-point, allowing for extreme maneuverability.

Tests with helicopters and balloons show that aerial transportation of logs is possible, but it has not yet been shown to be economically practicable. One important advantage of such transportation is that most of the heavy expense of ground transport systems is eliminated.

Loading. After logs have been skidded, they are loaded onto vehicles or put into water for delivery to mills or shipping points. Loading methods vary greatly with the size of the operation, size of trees being harvested, and preference of the operators. Except when products can be rolled onto vehicles by gravity, which is seldom, they must be elevated over the sides or ends of the vehicles upon which they are placed. This is accomplished by cranes, booms, or derricks, all specially designed and ranging from very lightweight machines to massive, heavy-duty, self-propelled equipment (Fig. 3).

Simpler and older loading equipment requires manual placement of tongs or slings on the logs prior to their elevation and, also manual removal of these attachments once the logs are in place on the carrier.

Newer large-scale and more sophisticated machines which do not require hookers or unhookers have been adopted and proven thoroughly applicable. These machines are self-propelled and equipped with movable booms and hydraulic or cable-operated grabs. The machine operator performs all the labor required, spotting the machine, positioning the boom, manipulating the grab mechanism, and finally placing the log on the vehicle. Lift-type loaders that elevate logs by inserting prongs beneath them and then lifting them have come into wide use. The jobs these machines have eliminated are among the most hazardous.

Transportation systems. Large forest areas require special transportation systems for hauling logs in adequate volume to supply mills; daily capacities range from 50,000 to several hundred thousand board feet of lumber per day. Standard-gage logging railroads formerly were widely used to log virgin forests with volumes of 10,000–50,000 ft of timber per acre over tens of thousands of acres. A few are still in use. They consist of main lines with spurs that ultimately cover all the timber property. Spur spacing depends on the topography and the skidding method, but it usually averages about 1/2 mi. Diesel or steam locomotives move empty log cars to the loading area and assemble trains of loaded cars for hauling to the mill. Spur tracks are removed after the logs have been loaded out and relaid on new spur locations.

At present, however, most log hauling is by motor trucks ranging 5–100 tons of gross vehicle weight. The largest types are used only on private haul roads. Logs may be transferred from one transport to another, such as from a large off-highway truck to one acceptable for public highway hauling, or from a truck to rail car, or to water for floating or barging.

Water transportation is desirable where available because less power is required to move floating logs. In the spring, when northern freshets flow from the logging area to rivers that, in turn, flow to mills, stream driving is an effective means of log transport. Logs are floated unconfined but are kept from jamming in narrows or stranding on shores by a crew that follows the drive downstream. Water from melting snows is held upstream by small dams and released when the drive starts so as to flush the logs downstream. Stream driving can be used only in nonnavigable waters. In larger, calm bodies of water, logs can be floated within a boom of floating logs fastened end-to-end that serves as a floating fence. In navigable rivers and inland waterways, logs are fastened together and transported in rafts which are pushed or towed by tugs. Barges also can be loaded with logs and then moved by river craft from loading points to destination. The logging operation is completed when the logs are delivered to the mill or a shipping point.

Prelogging and relogging. In the western forests of large trees, prelogging is sometimes done, in which the smaller trees are cut and removed with light tractors or even teams before the large trees are logged by cable systems or by heavy tractors. This reduces waste caused by breakage of the smaller trees. When the large trees are logged first and the smaller ones salvaged later, often for pulpwood, the second step is known as relogging. Integrated logging means the harvesting of several kinds of products from the same stand at about the same time. They are removed either prior to, subsequent to, or at the time of logging, and may include poles, pulpwood, veneer bolts, crossties, or other products. *See* FOREST RESOURCES; WOOD PRODUCTS.

[A. E. WACKERMAN]

Logic

The subject that investigates, formulates, and establishes principles of valid reasoning.

The first attempt to investigate systematically acceptable modes of reasoning was made by Aristotle, in whose *Organon* reasoning was recognized as the subject of a special science. He formulated three basic "laws of thought": (1) the law of contradiction (no proposition is both true and false), (2) the law of excluded middle (each proposition is either true or false), and (3) the law of identity (each proposition implies itself). Advanced as his views of logic were, it seems doubtful that the idea of axiomatizing it ever occurred to him, despite the success of his contemporary Euclid in organizing geometry. That aspect of logic which enunciates or establishes valid reasoning (rather than merely investigating it) began with the publication in 1854 of George Boole's *An Investigation of the Laws of Thought*. The partly abstract treatment of logic presented in this work initiated the completely abstract developments that were to follow. In it the laws of thought are regarded as mere conventions which, like the postulates of euclidean geometry, might be modified or even rejected to create new logics. The only requirement that a "new" logic must satisfy is the one demanded of every deductive system—consistency. Just as different geometries are useful for different purposes, a logic that is appropriate in one environment might not be so in another. There are a growing number of competent individuals, for example, who consider that any logic containing Aristotle's law of excluded middle is not suitable for mathematics.

This article does not attempt to discuss all the topics that are traditionally grouped under the term logic. Because it is assumed that the reader's principal interest in the subject is to acquaint himself with some of the principles for testing inferences, to acquire a knowledge of the use of formal procedures in clarifying arguments, and to learn how important parts of logic have, in recent years, been developed axiomatically, the classical theory of the syllogism will be examined, and the deductive theory of truth-functions (the calculus of propositions) will be presented in some detail. For the most part, the discussion will be limited to what is called symbolic, or mathematical, logic, for a special set of symbols is employed to exhibit in the clearest way the (at times) complicated logical relations involved. The advantages of a thoroughgoing use of a good symbolization are enormous. It is by such means that modern logic is able to solve problems that were quite beyond the power of logicians of an earlier time, and to render some of the achievements of those logicians almost trivial by comparison.

The syllogism. The formulation of the syllogism is contained in Aristotle's *Organon*. It had a great fascination for medieval logicians, for almost all their work centered about ascertaining its valid moods. The three characteristic properties of a syllogism are as follows:

1. It consists of three statements, the third one (conclusion) being a logical consequence of the first two (the premises).

2. Each of the three sentences has one of the following four forms:

(A) All X is Y (universal affirmative judgment)
(E) No X is Y (universal negative judgment)
(I) Some X is Y (particular affirmative judgment)
(O) Some X is not Y (particular negative judgment)

3. The three sentences contain altogether three terms, S (the subject of the conclusion), P (the predicate of the conclusion), and M (the middle term), that occur in each of the premises. P is called the major term, S the minor term, and a premise is called major or minor depending on whether it contains P or S, respectively. It is by virtue of the middle term, M, which occurs in both premises, that a relation between S and P can be asserted. Thus the three statements

> No professors are rich
> All handsome men are professors
> No handsome men are rich

form a syllogism in which M stands for professors, P for rich, and S for handsome men. This particular syllogism is valid (that is, the conclusion is a consequence of the premises), and hence the validity of a syllogism is seen to be independent of the truth or falsity of its conclusion. Notice that the major premise, minor premise, and conclusion of the above syllogism are statements of type (E), (A), and (E), respectively.

The four figures of the syllogism are represented by

(I)	(II)	(III)	(IV)
MP	PM	MP	PM
SM	SM	MS	MS
SP	SP	SP	SP

Each symbol MP, SM, and so on stands for one of the four statements (A), (E), (I), and (O); for example, MP might symbolize the statement "No M is P." Hence each figure has $4^3 = 64$ possible forms, and consequently there are $4 \cdot 4^3 = 256$ possible forms for a syllogism. The great attention given the syllogism by logicians of the Middle Ages resulted in establishing that only 19 of the possible 256 forms that a syllogism might take are admissible (that is, yield valid conclusions). It should be noted that the result they obtained is correct provided one accepts the Aristotelian principle that the statement "All X is Y" implies that "There are X's." It was also maintained that "No X is Y" implies that "Some X is not Y." Thus, according to early logicians, universals have existential implications. Modern logicians do not take this point of view. This results in eliminating 4 of the 19 admissible forms (moods). Three-syllable mnemonic words were coined in which the vowels indicated, in order, those combinations of the statements (A), (E), (I), and (O) that resulted in valid syllogisms. For figure (I) they are *Barbara, Celarent, Darii, Ferio*; figure (II), *Cesare, Camestres, Festino, Baroco*; figure (III), *Datisi, Feriso, Disamis, Bocardo, Darapti, Felapton*; and figure (IV), *Calemes, Fresison, Dimatis, Bamalip*, and *Fesapo*. The illustrative syllogism given above is Celarent, figure (I). The four moods that depend for their validity on the classical assumption that universals imply existence (and which must be eliminated if that assumption is not made) are *Darapti, Bamalip, Fesapo*, and *Felapton*.

The syllogism is now of little more than historical interest. Its limitations were recognized quite early, for it was clear that few of the inferences needed in mathematics could be put in syllogistic form. As remarked by Hans Reichenbach, "It is only the combination of a logic of functions of several variables with the logic of propositions that leads to the full understanding of conversational language as well as of mathematics."

Propositional calculus. The unit of the propositional calculus is the proposition, which may be defined as the meaning of an indicative or declarative sentence. Each proposition has a truth-value; it is either true or false in the classical logic, but may assume other truth-values (for example, uncertain) in some of the new logics. Thus,

> To err is human

is a proposition, whereas

> Lead us not into temptation

is not a proposition. It is convenient to use letters p, q, r, and so on to denote propositions. Propositions are combined by logical connectives to form other propositions. Chief among these are (1) negation (the negation of a proposition p, symbolized by p', may be formed verbally by writing "It is not the case that" before p); (2) conjunction (a binary operation symbolized by $\cdot$ or &, for example, $p \cdot q$, read "p and q"); (3) disjunction (a binary operation symbolized by $+$ or $\vee$; for example, $p \vee q$, read "p or q"); (4) implication (a binary operation symbolized by $\supset$ or $\rightarrow$, for example, $p \rightarrow q$, read "p implies q"); and (5) equivalence (a binary operation denoted by $\equiv$; for example, $p \equiv q$, read "p is equivalent to q," means that each proposition implies the other). The proposition $p \cdot q$ asserts that both p and q are true. It is important to note that the disjunctive "or" is always used in the inclusive sense; that is, $p \vee q$ asserts that at least one, possibly both, of the propositions p and q is true. In addition, the meaning of "implication" in logic differs markedly from the ordinary sense of that term. The proposition $p \rightarrow q$ asserts merely that it is not the case that p is true and q is false; that is, in classical logic $p \rightarrow q$ means that either p is false or q is true. Thus any false proposition (for example, Napoleon Bonaparte was born in the year 1600) implies any proposition whatever. It implies, New York is a large city, or Cleopatra was queen of France. Similarly, any true proposition is implied by any proposition whatever. In an implication $p \rightarrow q$, p is called the antecedent and q the consequent. The logical connectives introduced above are examples of logical constants. They are not independent notions. Implication may be defined in terms of negation and conjunction, and this is true for all of the connectives. By use of the symbol $\equiv_D$ to mean "is defined," it follows, for example, that

$$p \cdot q \equiv_D (p' \vee q')' \qquad p \supset q \equiv_D p' \vee q$$
$$p \equiv q \equiv_D [(p' \vee q)' \vee (q' \vee p)']'$$

In fact, all the connectives can be defined in terms of one binary operation, the so-called Scheffer stroke function $p|q$, which may be read "p is false or q is false." Therefore, it follows

$$p' \equiv_D p|p \quad p \cdot q \equiv_D (p|q)|(p|q)$$
$$p \vee q \equiv_D (p|p)|(q|q)$$
$$p \supset q \equiv_D [(p|p)|(p|p)]|(q|q)$$

The logical connectives, the letters p, q, r, and so on, that stand for propositions (propositional variables), and the parentheses, brackets, and braces needed for punctuation are formal concepts. A propositional function is a combination of concepts, formal or nonformal, involving at least one variable, which is not a proposition but becomes one when all the variables are given specified values. Examples are (1) p and q, (2) $x^2 + y^2 = 1$, and (3) x is president of the United States. Unlike a proposition, a propositional function has no truth-value. On the other hand, the statement "For every integer x, $x^2 + 1 = 0$" is a proposition (not a propositional function) even though it contains a (bound) variable. A formal function is a propositional function that contains only formal concepts—for example $[(p \vee q) \cdot r] \supset s$. The form of a given proposition is the formal function obtained from it by substituting formal concepts for all nonformal ones. Thus $p \cdot q$ is the form of the proposition "Leonard is happy and the moon is blue." Notice that in formalizing the above proposition, different propositions were replaced by different propositional variables. A truth function is a propositional function in which only propositional variables occur, and such that the truth value of each proposition obtained from the function by substituting specific propositions for the variables depends only on the truth values of those propositions. Examples are (1) $(p \cdot q) \vee r'$, (2) Napoleon was a great general and p (where p is a propositional variable). On the other hand, "Jones stated that p" is not a truth function.

This article is concerned principally with the class of formal truth functions (that is, those containing only formal concepts) and with an important subclass called tautologies [that is, those formal truth functions such that every proposition obtained from them by substituting specific values (propositions) for the variables they contain has the truth value T (true)]. A truth function is contingent provided it assumes both truth values T and F (false)—here only the classical two-valued (Aristotelian logic) truth functions are considered—and self-inconsistent provided it has only the value F. The character of a truth function is determined when it is shown to be self-inconsistent, contingent, or a tautology.

Character of a truth function. A truth function may be analyzed by means of a truth table, or matrix. Such tables are made first for the elementary truth functions p', $p \vee q$, $p \cdot q$, $p \supset q$ as follows:

(I)		(II)			(III)			(IV)		
p	p'	p	q	$p \vee q$	p	q	$p \cdot q$	p	q	$p \supset q$
T	F	T	T	T	T	T	T	T	T	T
F	T	T	F	T	T	F	F	T	F	F
		F	T	T	F	T	F	F	T	T
		F	F	F	F	F	F	F	F	T

In every table each row of the last column gives that truth value of the function which is determined by those truth values (assigned to the variables) that are entered in the preceding columns of that row. For example, the second row of table (IV) states that the truth function $p \supset q$ has value F (that is, is false) when p has value T (that is, is true) and q has value F. Since all logical connectives used in the logic of propositions are expressible in terms of negation and disjunction, only tables (I) and (II) are really essential, but tables (III) and (IV) are convenient when analyzing complicated truth functions. Although the character of any truth function may be determined by constructing its truth table, the method is tedious when the function contains a large number of variables, since the truth table of a truth function of n variables has 2^n rows. As an example, the character of the truth function $(p \supset q) \supset (q' \supset p')$ may be determined. Using tables (I) and (IV), the truth table of the function is found to be

p	q	p'	q'	$p \supset q$	$q' \supset p'$	$(p \supset q) \supset (q' \supset p')$
T	T	F	F	T	T	T
T	F	F	T	F	F	T
F	T	T	F	T	T	T
F	F	T	T	T	T	T

Since every entry in the last column is T, the function $(p \supset q) \supset (q' \supset p')$ is seen to be a tautology. It may also be observed that in each row the entries in columns 5 and 6 are the same; that is, for any given values of p and q, the functions $p \supset q$ and $q' \supset p'$ assume the same truth value. Hence $(q' \supset p') \supset (p \supset q)$ and the two functions are logically equivalent. The principle embodied in this equivalence (the law of contraposition) is often used in mathematics.

Another device for ascertaining the character of a truth function is the method of negation. It sometimes happens that the character of the function obtained by negating a given function can be easily established, and since a function is contingent if and only if its negation is, self-inconsistent if and only if its negation is a tautology, and a tautology provided its negation is self-inconsistent, a knowledge of the character of the negated function solves the problem. The method makes use of the so-called De Morgan formulas $(p \vee q)' = p' \cdot q'$, $(p \cdot q)' = (p' \vee q')$, in which the sign $=$ signifies that an expression appearing on one side of it may be replaced by the expression appearing on the other side, and the Boolean expansion of a truth function. This is an expansion that exhibits all of the conjunctive possibilities for truth value T; it is a disjunction of mutually exclusive (incompatible) conjuncts. For example, $p \vee q$ assumes truth value T if and only if the function $(p \cdot q') \vee (p \cdot q) \vee (p' \cdot q)$ does. The latter function is the Boolean expansion of $p \vee q$. The method of negation can be illustrated first by considering the same truth function $(p \supset q) \supset (q' \supset p')$ analyzed by the matrix method. Writing

$$[(p \supset q) \supset (q' \supset p')]' = (p \supset q) \cdot (q' \supset p')'$$
$$= (p' \vee q) \cdot (q \vee p')'$$
$$= (p' \vee q) \cdot (p' \vee q)'$$

and noting that $(p' \vee q) \cdot (p' \vee q)'$, as the conjunction of a proposition and its negation, is self-inconsistent, it follows that $[(p \supset q) \supset (q' \supset p')]'$ is self-inconsistent, and so $(p \supset q) \supset (q' \supset p')$ is a tautology. A second example, using the Boolean expansion of $p \vee q$, is

$$
\begin{aligned}
{[(p \vee q) \supset q]'} &= (p \vee q) \cdot q' \\
&= [(p \cdot q) \vee (p \cdot q') \vee (p' \cdot q)] \cdot q' \\
&= p \cdot q'
\end{aligned}
$$

Clearly, the truth function $p \cdot q'$ is contingent, and consequently so is the function $(p \vee q) \supset q$.

Tautologies and forms of argumentation. One truth function T_1 necessarily implies another T_2 provided the truth function $T_1 \supset T_2$ is a tautology. For example, $(p \supset q) \supset (q' \supset p')$ is a tautology, and so the truth function $p \supset q$ necessarily implies the truth function $q' \supset p'$. On the other hand, neither of those two truth functions is a tautology, and so the implications they exhibit are not necessary implications. Valid forms of argumentation are based on the notion of necessary implication. Two such forms were known to classical logic as *modus ponendo ponens* and *modus tollendo tollens*. The first states that if an implication $p \supset q$ and its antecedent p are asserted, then the consequent q may be asserted; that is, the truth function $(p \supset q) \cdot p$ necessarily implies q. This is established by proving that the truth function $[(p \supset q) \cdot p] \supset q$ is a tautology. The *modus tollendo tollens* form of argument permits the denial of the antecedent of an asserted implication when its consequent is denied. The validity of the argument rests upon the tautology $[(p \supset q) \cdot q'] \supset p'$. A common fallacy (fallacy of the consequent) is to assert the antecedent of an implication whose consequent has been asserted. This is not permissible since the truth function $[(p \supset q) \cdot q] \supset p$ is not a tautology. Although such an argument is logically unwarranted, it is frequently invoked in testing experimentally a hypothesis p, for if q follows from p, numerous instances of the validity of q are sometimes taken to furnish strong evidence of the validity of p.

A traditional form of argumentation known as the dilemma is an application of *modus ponendo ponens* or *modus tollendo tollens*. Consider the following example:

(1) If I speak the truth, men will hate me; if I lie, the gods will hate me
(2) I must either speak the truth or lie
(3) Therefore, either men will hate me, or the gods will hate me

The form of this dilemma is as follows:

(1) $(p \supset q) \cdot (r \supset s)$
(2) $p \vee r$
(3) $q \vee s$

The argument is valid by *modus ponendo ponens*, for since premise (2) affirms the antecedent of at least one of the implications of premise (1), at least one of the consequents may be asserted, which is what the conclusion (3) does.

Deductive theory. The intent of this section is to sketch the procedure by which all tautologies are deducible from a set of just four of them. This is analogous to what is done in an axiomatic geometry in which all the theorems of the geometry are deduced from a selected set of axioms or postulates that are assumed to be true. The nature of the geometry is fully determined by the set of statements forming the postulates.

Let P denote a nonempty abstract set whose elements, denoted by p, q, r, and so on, are called propositions (for suggestiveness), and suppose P is closed with respect to a unary operation denoted by $(')$, and a binary operation, denoted by $\vee$; that is, if p and q are any elements of P, then p' and $p \vee q$ must be unique elements of P. An additional binary operation, denoted by $\supset$ and read "implies" is defined in P:

Definition. If p, q are elements of P, $p \supset q \equiv_D p' \vee q$.

An expression formed by elements of P and the operations $(')$, $\vee$ is a truth function if and only if it conforms to the following syntactical rules: (1) p, q, r, and so on are truth functions; (2) if A and B are truth functions, then so are A' and $A \vee B$; (3) no sequence of symbols denotes a truth function unless it is constructed according to (1) or (2), or unless it is replaceable by definition by a truth function. A subclass of the class of truth functions constitutes the class of tautologies. It is assumed that the following four truth functions are tautologies (these are the postulates):

P1 $(p \vee p) \supset p$
P2 $q \supset (p \vee q)$
P3 $(p \vee q) \supset (q \vee p)$
P4 $(q \supset r) \supset [(p \vee q) \supset (p \vee r)]$

These simple tautologies were selected by Alfred North Whitehead and Bertrand Russell in their monumental work *Principia Mathematica* as the basic set from which all tautologies can be derived. They had an additional postulate

P5 $[p \vee (q \vee r)] \supset [q \vee (p \vee r)]$

which was later shown by Paul Bernays to be deducible from the first four postulates, and hence may be deleted as a postulate. It is interesting to note that none of the three Aristotelian principles (the laws of identity, contradiction, and excluded middle) are in this basic set. They are all deduced as theorems.

Before any theorems can be deduced there must be available rules that allow one to go from one step in a proof to another. Among such rules are those of formal substitution, according to which any truth function may be substituted for any of the variables occurring in a postulate or theorem, provided the substitution is carried out completely, and any expression may be replaced by one definitionally identical. Among the rules of formal assertion are the principle of inference (that is, if P and $P \supset Q$ are postulates or proved theorems, then Q is a theorem) and the principle of adjunction (that is, if P, Q are postulates or theorems then $P \cdot Q$ is a theorem).

It was proved by E. L. Post that postulates P1–P4 are consistent; that is, it is impossible to deduce from them truth functions X and X'. He also proved that the four postulates form an independent set; that is, no one of the postulates can be deduced from the others.

Non-Aristotelian logics. In these logics the principle of the excluded middle (that is, each proposition is either true or false) is not valid. There are, then, at least three truth values possible for a proposition, and for this reason such logics are known as many-valued. In a many-valued logic the *reductio ad absurdum* procedure that is so frequently used in mathematics is much less effective than it is in the two-valued classical logic, for if it has been shown that a given proposition is not false, one may not conclude (as is permissible in Aristotelian logic) that the proposition is true. The Aristotelian tautology $(p')' \equiv p$ is not a tautology in a logic with at least three truth values. Doubts that the classical logic is an appropriate one to use in mathematics stem from the modern view that the concepts and relationships dealt with in mathematics have no real existence in the external world. It would seem to follow that the a priori nature of "true" and "false" then disappears and with it the law of excluded middle! If "true" be interpreted in mathematics to mean "provable," and a proposition be called false provided its negation is provable, then the law of excluded middle is demonstrably false, since it has been shown that mathematics contains propositions p such that neither p nor its negation p' is provable. The intuitionistic school of the Dutch mathematician L. E. J. Brouwer rejects the excluded middle law, and it forms no part of the logic of that school, formulated by A. Heyting. *See* BOOLEAN ALGEBRA; GEOMETRY, EUCLIDEAN; MATHEMATICAL NOTATION, CONTEMPORARY; MATHEMATICS; SCIENCE; SCIENTIFIC METHODS.

[LEONARD M. BLUMENTHAL]

Bibliography: A. Ambrose and M. Lazerowitz, *Fundamentals of Symbolic Logic*, rev. ed., 1962; A. Church, *Introduction to Mathematical Logic*, 1956; H. B. Curry, *Foundations of Mathematical Logic*, 1963; E. L. Post, *Introduction to a general theory of elementary propositions*, *Amer. J. Math.*, 43:163–185, 1921; A. Tarski, *Introduction to Logic*, 3d ed., 1965; A. N. Whitehead and B. Russell, *Principia Mathematica*, 3 vols., 2d ed., 1925–1927.

Logic circuits

Essentially, simple switching circuits used in large numbers in modern digital computers and other digital electronic equipment. These circuits are designed so that certain output signals are produced upon application of certain input signal conditions. The output signal is dependent upon the combination of the presence or absence of several input signals. Both voltage and current levels are used to identify the input and output signals.

Extensive use of the vacuum tube was made in earlier digital equipment. Theoretically, a variety of electronic components can be used to construct logic circuits. For instance, magnetic elements can and have been used. However, advances in semiconductor technology have put transistors and diodes into the forefront. Since most logic circuits use semiconductors extensively, only transistor logic circuits are described in this article.

Binary logic functions. The familiar decimal number system, based upon ten digits (zero through nine), is not very suitable for electronic computing systems or other digital equipment. Use of the decimal system would necessitate using ten different signal levels to identify the ten digits. This would require extensive electronic hardware. By using a number system which utilizes a smaller number of digits, the electronic hardware required can be substantially reduced, thereby enhancing circuit simplicity and system reliability. *See* BINARY NUMBER SYSTEM; NUMBER SYSTEMS.

The binary number system, which uses only two digits, zero and one, is ideally suited for these applications. Only two circuit-condition states are required to define both the input and output signals. Consequently, circuit configurations of maximum simplicity, using a minimum number of electronic components, are realized.

Three basic functions are performed by logic circuits. They are (1) the AND, (2) the OR, and (3) the NOT, or inverter, functions. An AND circuit is one whose output signal is a logical "1" only if a logical "1" signal is present on each input simultaneously. Otherwise the output is a logical "0." Figure 1a shows the symbolic logical representation of this condition. Theoretically the circuit could have any number of inputs. In reality the number of inputs is limited by practical considerations. Only three inputs are shown for convenience. Figure 1b shows one of the possible input conditions which would result in a "0" output.

An OR circuit is one whose output signal is a logical "1" if a logical "1" signal is present on at least one of its inputs. Otherwise the output is a "0." Figure 2a shows one of the input conditions which will result in a "1" output and Fig. 2b shows the only input signal condition for a "0" output.

The NOT, or inverter, circuit inverts the input signal at its output. Figure 3a shows the condition for a "1" output and Fig. 3b that for a "0" output.

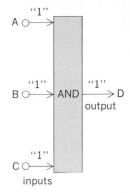

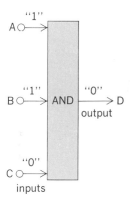

Fig. 1. AND circuit. (a) "1" signal on all three inputs gives a "1" output. (b) Output is "0" if all three inputs are not "1" simultaneously.

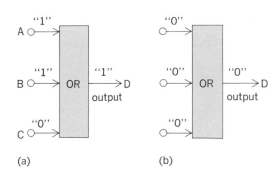

Fig. 2. OR circuit. (a) "1" signal on at least one input gives a "1" output. (b) When all the inputs are simultaneously "0," the output is "0."

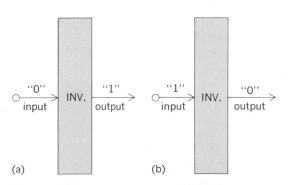

Fig. 3. NOT circuit, also known as inverter circuit. (a) A "0" input produces a "1" output. (b) A "1" input produces a "0" output.

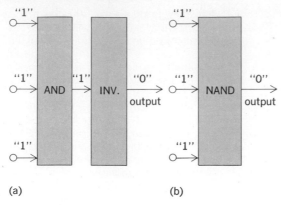

(a)   (b)

Fig. 4. NAND circuit. (a) AND gate followed by an inverter. If all inputs are "1," the output is "0." (b) AND-inverter combination shown as a single symbol.

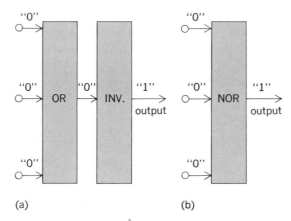

(a)   (b)

Fig. 5. NOR circuit. (a) OR gate followed by an inverter. If all inputs are "O," the output is "1." (b) OR-inverter (NOR) combination shown as a single symbol.

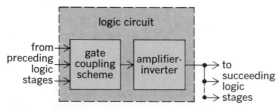

Fig. 6. Fan-ins and fan-outs.

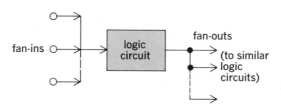

Fig. 7. Logic circuit.

Circuits. In computer parlance, the AND and OR circuits are very often referred to as switches or gates. In applications involving transistors, these gates are usually followed by an inverter. Figure 4a shows an AND gate followed by an inverter, and the logical functions involved. Such a combination is called a NAND circuit (NOT-AND) and is shown symbolically as a single block, as indicated in Fig. 4b.

Figure 5a shows an OR circuit followed by an in-

verter, and the logic functions they perform. This combination is called a NOR circuit (NOT-OR), and is also shown as a single block, as in Fig. 5b.

In the preceding examples only three inputs are shown in each case. The inputs to any logic circuit are commonly referred to as "fan-ins," and the number of such inputs to the circuit as the "fan-in." Similarly, the term "fan-outs" is used in referring to the outputs of the circuit, and the number of similar logic circuits which it is capable of driving at its output is called the "fan-out." Figure 6 shows the symbolic representations. In theory, logic circuits can have any number of fan-ins and fan-outs. In practice, they are limited by electronic considerations.

In this discussion, logic circuits refer to NOR and NAND circuits, since OR and AND gates are very seldom used by themselves. The reason is that most signals, in passing through gates, suffer amplitude deterioration. A transistor amplifier is used to bring the output signal back to the desired level, and a simple single-stage amplifier inverts the logical sense of the signal.

The elements used in the gate couple the output of one logic circuit to the input of the next stage. Since various electronic components can be used for the gate, it is the method of coupling by which logic circuits are usually classified. Figure 7 shows a block diagram of a typical logic circuit.

Obviously, the functions of Fig. 7 can be implemented by several different circuit configurations which would have different properties. Systems requirements, such as speed of signal propagation, type and number of components available, signal levels, and circuit impedances, usually govern the choice of logic circuit configurations for particular applications. *See* DIGITAL COMPUTER; SWITCHING CIRCUIT; SWITCHING THEORY.

Saturating transistor circuits. Logic circuits are broadly classified by the mode of operation of the amplifier-inverter transistor. One type of circuit uses the saturating mode for the transistor; that is, the transistor is either completely turned off or fully turned on (saturated). These circuits, which are the more commonly used, are called saturating types. In the nonsaturating types, the transistors involved are not fully turned on. They operate in what is called the nonsaturating, or linear, mode. Generally, in the saturating-type logic circuits logical signals are established by voltage levels, while in the nonsaturating types they are established essentially by current levels. *See* TRANSISTOR.

Resistor-transistor logic (RTL). Figure 8 shows one of the simplest logic circuits, one which is relatively cheap. Resistors R_1, R_2, and R_3 are the elements coupling this circuit to the preceding similar circuits. Along with resistor R_4 and the positive supply voltage $+V_B$ they form the gate portion of the circuit referred to in Fig. 7. The rest of the components make up the amplifier-inverter portion of the circuit. If a negative-going voltage at the gate input is defined as "1," then the gate is an OR circuit and is operated by voltage levels. The application of a negative signal at any one or more of the inputs will drive the base of the transistor negative, thereby turning the transistor on and providing a near ground (or relatively more positive) voltage at the output. This then is a NOR circuit for a negative "1" signal. Conversely, it is a NAND circuit if "1" is defined by a positive-going signal. RTL circuits are usually used in applications requiring only low or moderate speeds of operation. Because resistors are used as coupling elements, fairly large signal swings can be used. This in turn improves the noise susceptibility of the circuit. Also, larger fan-ins and fan-outs are possible because of the larger signal swings. The speed of operation of this curcuit is limited by the switching speed of the transistor used.

Direct-coupled transistor logic (DCTL). Figure 9 shows another simple form of logic circuit, the direct-coupled transistor logic circuit. The three inputs are to the bases of the three transistors. The output of the circuit, which is the common-collector node of the transistors, is directly coupled to the input transistor bases of the succeeding stages. In this circuit, the gate function and the amplifier-inverter functions are performed by the same component, namely each transistor. Once again, definition of the signal voltage levels defines the NOR or NAND nature of the circuit. For positive-going signals defined as "1" the circuit performs the NOR function, while for the negative-going "1" it is a NAND circuit.

Another version of this configuration is the series DCTL gate, where the transistors are connected in series as shown in Fig. 10. The parallel transistor DCTL is more commonly used.

In DCTL, low signal voltage swings are involved, and the circuit is thus more susceptible to noise. However, because of the low signal swings, it is possible to use transistors with lower voltage breakdown ratings. The transistors are usually driven hard into saturation, and the switching speed is correspondingly low, although the speed of signal propagation through this circuit is reasonably high. The principal drawback of DCTL is a phenomenon called current hogging. Because of the fact that the input characteristics of the transistors of the succeeding stages are not identical, the current supplied by the driving stage will be divided unequally among the bases of the following stages. This condition may result in one transistor hogging all, or most, of the available current, thereby starving the rest. The result may be the failure of some transistors to turn on. In spite of this drawback, however, the use of only one power supply and the absence of interstage coupling elements make DCTL a relatively simple and useful configuration.

Diode-transistor logic (DTL). One of the most widely used configurations is the diode-transistor logic circuit (DTL), shown in Fig. 11.

The noise immunity of DTL is very good, as are

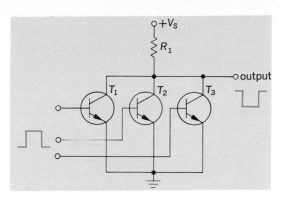

Fig. 9. Direct-coupled transistor logic circuit (DCTL).

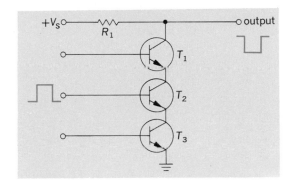

Fig. 10. DCTL circuit series gate.

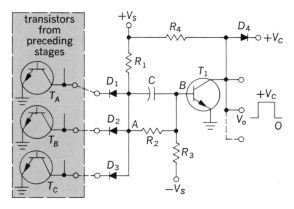

Fig. 11. Diode-transistor logic circuit (DTL).

its signal propagation speed and power dissipation. The fan-out capability is moderately good, but the limitations on its fan-in capability constitute its principal shortcoming. For transistor T_1 to be turned off, at least one of the preceding stage transistors T_A, T_B, or T_C must be saturated. Suppose that only T_A is saturated while T_B and T_C are off. Under this condition diode D_1 is forward-biased and current flows from $+V_S$ through R_1, D_1, and T_A to ground. The voltage at node A is slightly positive and is the sum of the saturation voltage drop of T_A and the forward voltage drop of D_1. The voltage at A is not sufficiently positive to turn T_1 on. Now consider the case when T_B and T_C are also turned on. D_2 and D_3 would then also become forward-biased. Current would now be flowing through all the three input diodes and the three transistors of the preceding stages. This additional current is supplied by $+V_S$ through R_1, and there is a limit to the amount of current that can be supplied. In

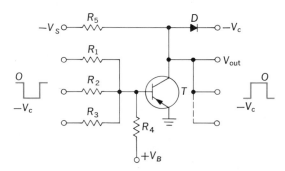

Fig. 8. Resistor-transistor logic circuit (RTL).

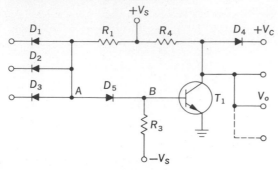

Fig. 12. Low-level logic circuit (LLL).

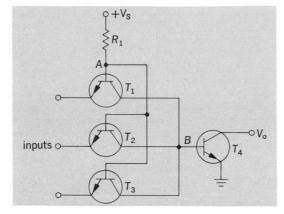

Fig. 13. Transistor-transistor logic circuit (TTL or T²L).

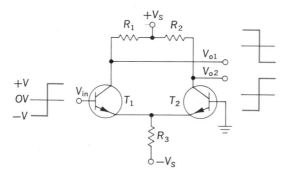

Fig. 14. Current-mode logic circuit (CML), nonsaturating circuit, with complementary outputs.

practice resistor R_1 can be reduced only to a certain value to supply the additional current as the number of inputs are increased. This then limits the fan-in capabilities of the DTL circuit.

Low-level logic (LLL). The previously discussed DTL can be modified to another configuration called the low-level logic circuit, as shown in Fig. 12. The LLL circuit is the same as the DTL circuit of Fig. 11 but with the combination of base resistor R_2 and capacitor C replaced by a diode, D_5. The reason for this change is that only a relatively small voltage swing is required at the base of transistor T_1 to switch it on or off. In the DTL circuit a relatively large voltage swing is required at node A even though the desired voltage swing at node B is small. This is because a fairly large portion of the voltage developed at node A is dropped across the R_2C combination. By replacing this combination of diode D_5 as in Fig. 12, this condition can be greatly

improved. This circuit switches a constant direct current into the base of the transistor T_1, or out of it, and accomplishes this with small voltage swings at node A.

The LLL circuit has all the advantages of the DTL circuit and a few of its own. The circuit can be operated at lower signal levels and therefore can be designed for lower power dissipation. Lower power dissipation is further aided by the use of one less resistor than in DTL. While only one diode, D_5, is shown in Fig. 12, two or more in series can be used. The use of additional diodes, often referred to as offset diodes, improves the noise immunity of the circuit. Also, by proper design it is possible to return resistor R_3 to ground, thereby eliminating the need for an additional power supply.

Transistor-transistor logic (TTL or T²L). This is an extension of the LLL circuit which offers a significant advance in low-power, high-speed operation. In this circuit (Fig. 13) the gating diodes of Fig. 12 are replaced by three transistors whose bases and collectors are respectively tied together. The circuit inputs are to the three emitters of the transistors. Because the collectors are tied to a common node B, the circuit is called a contiguous-collector TTL circuit. A contiguous-emitter circuit is also possible. In such a circuit all the emitters would be tied to node B, while the circuit inputs would be tied to the three collectors.

In TTL, the gating transistors are always on, and this fact increases the switching speed considerably. Although more transistors are used, the number of other components is reduced. The use of only one resistor and one power-supply voltage, in addition to the absence of diodes, is a significant advantage. In TTL circuits, the noise margins are usually low. However, lower operating currents are involved and noise-margin requirements are therefore also lower. The low-level operation of this circuit allows it to be used in conjunction with the direct-coupled transistor logic (DCTL) circuits.

Nonsaturating circuits. In logic circuits in which the transistors are operated in the saturated mode, storage time is a very significant portion of the signal propagation delay. Storage time is the time required for the transistor to come out of saturation. It is the delay which a saturated transistor inherently exhibits before its collector current begins to turn off. By operating the transistor in the nonsaturating or linear mode, this portion of the signal propagation delay can be greatly reduced and the circuit operated at higher speeds.

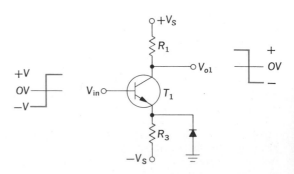

Fig. 15. Current-mode logic circuit (CML), nonsaturating circuit, with single output.

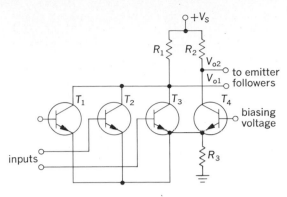

Fig. 16.   Emitter-coupled transistor logic circuit (ECTL).

Current-mode logic (CML). The most commonly used nonsaturating logic circuit is the current-mode logic circuit of Fig. 14. This circuit, which is a differential-type split-load amplifier with an emitter follower output driving a grounded-base stage uses a signal swing about a biased operating point. There is no on-and-off switching. Since the outputs of transistors T_1 and T_2 are complementary, they provide a convenient scheme for logic operations in some systems. If complementary outputs are not required, then T_2 can be removed and the emitter of T_1 can be clamped to ground by a diode as shown in Fig. 15.

Very high operating speeds can be obtained with CML circuits. Since this circuit is basically a differential amplifier, the common-mode voltage gain can be made less than one. Consequently, power-supply line noise can be readily attenuated and prevented from interfering with the signal. The principal disadvantage of this configuration is that the quiescent operating levels of the input and the output signals are different. Thus, when similar stages are cascaded in a logical chain, level-restoring circuits have to be inserted after each stage. Sometimes this problem is also solved by alternating each *npn* transistor stage with a *pnp* transistor stage in the logical chain.

Emitter-coupled transistor logic (ECTL). A modification of the CML circuit is shown in Fig. 16. It is also a current-mode switch operated in the nonsaturated mode. Transistor T_4 is the reference transistor. A fixed bias applied to its base from an external source keeps it conducting when the input transistors are logically off. The problem of signal level restoration, inherent in CML, is also present in ECTL. In ECTL this is usually solved by connecting two emitter-follower circuits to the two outputs V_{01} and V_{02}. The outputs of the emitter-followers are at the desired signal levels and are used for logical inputs to the following stages.

The emitter-follower outputs provide many advantages, besides giving signal level compatibility. The output impedance of the emitter followers is low, giving the circuit a large fan-out capability. It also helps to minimize the noise and cross-talk problems. The circuit has relatively high input impedance, providing good fan-in capabilities. Since transistor T_4 is independently biased, drifts in output signal levels caused by variations in supply voltage or temperature can be easily corrected by providing a self-regulating circuit for the bias voltage of T_4.   [ADI J. KHAMBATA]

Bibliography: T. C. Bartee, *Digital Computer Fundamentals*, 2d ed., 1967; R. L. Fogleson, Transistor-transistor logic, *Electron. Equip. Eng.*, November, 1962; A. J. Khambata, *Introduction to Integrated Semiconductor Circuits*, 1963; E. F. Kvamme and U. S. Davidsohn, ECTL: A new tool for the digital circuit designer, *Electron. Equip. Eng.*, September, 1962; W. D. Roehr, Techniques of current-mode logic switching, *Electron. Des.*, Sept. 13, 1962; H. W. Ruegg, First design details: Transistor-transistor logic circuits, *Electronics*, Mar. 22, 1963; Texas Instruments, Inc., *Transistor Circuit Design*, 1963.

Longitude (astronomy)

Longitude determined by astronomical observations, as distinguished from geographic longitude, which is that shown on a map. Astronomical longitudes are determined by the definition that the longitude from Greenwich is identical with the difference between Greenwich Mean Time (GMT) and local mean time. Greenwich Mean Time is ascertained by reception of radio time signals. Local mean time is determined by observing either meridian transits of celestial objects or their altitudes near the prime vertical. In either case, the angle between the line of sight and the direction of gravity is measured, the direction of gravity being established with the aid of a spirit level, a liquid surface, or observation of the horizon corrected for the elevation of the observer. However, the direction of gravity is not, in general, precisely toward the axis of rotation of the Earth, as it would be if the Earth were concentrically homogeneous. The deviations are called deflections of the vertical; they commonly amount to some seconds of arc, occasionally reach a minute of arc, and vary from place to place. A map constructed exclusively from astronomical observations would in consequence be distorted by amounts varying from a few hundred feet to more than a mile. *See* GEODESY.

[GERALD M. CLEMENCE]

Lophiiformes

The anglerfishes and their relatives. This highly modified order of antinopterygian fishes, also known as the Pediculati, is thought to have been derived from a stock close to the Batrachoidiformes, or toadfishes, at least as early as Eocene time. The first dorsal fin is reduced to a few flexible rays, of which the first is placed on top of the head and bears a terminal bulb or tassel and functions as a fishing lure (see illustration). The epiotics are large and meet behind the supraoccipital. There are no ribs or epipleurals, and the gill opening is small and placed far back. Pelvic fins, if present, are thoracic, and the pectoral girdle is notably modified. *See* BATRACHOIDIFORMES.

There are 3 suborders, 15 families, nearly 60 genera, and about 195 known species. All are marine. Most species inhabit deep seas or tropical shore waters; the shore forms are commonly sedentary in habit. Luminescent organs may be present, and in some deep-sea species the males are dwarfed and attached as ectoparasites on the females. *See* ACTINOPTERYGII. [REEVE M. BAILEY]

Bibliography: E. Bertelsen, *The Ceratioid Fishes: Ontogeny, Taxonomy, Distribution and Biology*, Dana Rept. no. 39, 1951.

LOPHIFORMES

Anglerfish (*Cryptopsaras couesi*). (*After G. B. Goode and T. H. Bean, Oceanic Ichthyology, U.S. Nat. Mus. Spec. Bull. no. 2, 1895*)

Lophophore

A food-gathering organ, found in the Bryozoa, Phoronida, and Brachiopoda. The expression tentacular crown is sometimes used as a synonym. Located anteriorly (occasionally termed dorsally), it consists of a circumoral, fleshy basal ridge or lobe, from which numerous ciliated tentacles arise.

The basal ridge of the lophophore surrounds the mouth symmetrically; however, the ridge usually does not also surround the nearby anus. In the entoproct bryozoans, and in the stenolaemate and gymnolaemate ectoproct bryozoans, this basal ridge as seen in plan view (from in front of the mouth) is essentially circular. In the phylactolaemate ectoproct bryozoans and phoronids, the ridge is horseshoe-shaped. In the brachiopods, the ridge is usually drawn out laterally into a complex set of arms or lobes, which often contain an internal skeletal support consisting of either hard calcareous shell material or semirigid cartilagelike material.

The tentacles of the lophophore are flexible and are provided with muscle fibers and nerves. They are essentially hollow or tubular, and their central cavity is filled either with colorless fluid (as in the eucoelomate ectoproct bryozoans, phoronids, and brachiopods) or with gelatinous material (as in the pseudocoelomate entoproct bryozoans). In bryozoans and phoronids, the tentacles are long and thin, while in brachiopods, they are short and very thin. The total number of tentacles borne on the basal ridge ranges from about 10 (in some bryozoans) to several hundred (in some brachiopods).

During the embryonic period, the lophophore develops from a foldlike outgrowth of the anterior body wall. When the lophophore is fully developed, its principal function is to gather food. The beating of the cilia on the tentacles generates currents of water bearing food particles; the general conformation of the tentacles and basal ridge channels these currents to the animal's mouth. In addition to its primary food-gathering function, the lophophore in some forms may also have some secondary respiratory function. Generally, the lophophore is protected by being located permanently within, or by being retracted into, a cavity within the shell, body, or tube of the animal.

The term lophophore is occasionally applied only to the fleshy basal ridge described above; in contrast, the term is rarely expanded to include the introvertible soft body wall (tentacle sheath) below the ridge and tentacles. Moreover, some workers define lophophore as enclosing the mouth but never enclosing the anus also; according to that usage, the entoproct bryozoans do not possess a true lophophore. Finally, the term lophophore is applied only to the food-gathering organ seen in the so-called lophophorate phyla (that is, the Bryozoa, Phoronida, and Brachiopoda); it is not used to denote the similar-appearing tentacle-bearing structures found in the Pogonophora and Pterobranchia (Hemichordata). *See* BRACHIOPODA; BRYOZOA; PHORONIDA.

[ROGER J. CUFFEY]

Loran

A navigation system from which hyperbolic lines of position are determined by measuring the difference in times of arrival of pulses from widely

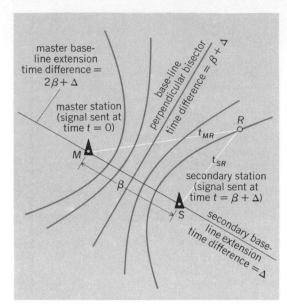

Fig. 1. Principle of position fixing by hyperbolic navigation systems. (*From J. P. VanEtten, Navigation systems: Fundamentals of low and very-low-frequency hyperbolic techniques, Electr. Commun., 45(3):192–212, 1970*)

spaced, synchronized transmitting stations. Since radio waves travel at the speed of light, a line of position represents a constant range difference from two transmitters.

In Fig. 1, it is assumed that the master station M transmits a pulse signal at time $t = 0$, and that the secondary station S transmits a similar pulse signal Δ microseconds after receiving the master pulse signal. The secondary station receives the master signal at time $t = \beta$ microseconds, and therefore transmits its pulse at time $t = \beta + \Delta$ microseconds. A receiver at R measuring the difference in the times of arrival of the signals from the secondary and master stations measures the time difference $TD = (\beta + \Delta) + t_{SR} - t_{MR}$. The locus of all points with this common time difference is a hyperbola through the receiver position R. Similarly, every hyperbolic line of position is uniquely defined by a time difference. The intersection of two such lines of position gives a navigational fix, or location. Loran is used by commercial and military ships and aircraft. The name is derived from "long-range navigation." *See* HYPERBOLIC NAVIGATION SYSTEM.

The loran position-fixing system was originally developed during World War II as an aid to the navigation of Allied aircraft and the North Atlantic convoys. Following the war, its use was extended by the U.S. Coast Guard to aid marine navigation. This system, operating at a frequency of about 2 MHz, is now known as loran A.

Loran A. Transmitters for this service send out trains of pulses having a rise time of 21 μs and a width of 40 μs. Pulses recur at 20, 25, or 33 1/3 pulses per second and at rates very slightly removed from these basic rates. Standard peak power for a transmitter in this service is 160 kW, although the averate power is considerably less. Amplifiers rated at 1 MW are used at about half the stations. The radio frequencies employed are at 1850, 1900, and

1950 kHz. Bandwidth of the transmission is 75 kHz.

One station of a pair is designated a master and normally sends an uninterrupted series of pulses. The second station, located about 200–300 nautical miles (360–540 km) away, is called a secondary station and must maintain its radio frequency and pulse recurrence rate exactly in accord with those of the master. In addition, the secondary station must maintain a fixed time difference (usually 1000 μs) between its reception of the master signal and its transmission of the secondary pulse. If either the master or secondary operator notices a change in these relationships, the operator periodically interrupts (blinks) the signal until proper adjustments have been made. The blinking signal warns navigators of possible error.

Loran C. The loran C system has evolved from loran A to provide greater range and more precise, repeatable, and accurate measurements. Loran C is a hyperbolic system that uses ground wave transmissions at low frequencies, to give an operating range in excess of 1000 n mi (1800 km). Sky-wave contamination is avoided by using pulse techniques. High accuracy is obtained by carrier-phase comparison, and cycle identification is accomplished by measurement of the pulse envelope. Operation is possible in poor signal-to-noise conditions by using correlation techniques and time-domain filtering.

Loran C chains are composed of a master transmitting station and two or more secondary transmitting stations. Master and secondary stations are generally separated by about 600 mi (970 km). Peak power of up to 4 MW is generated by the transmitter and is radiated by a single vertical antenna tower which may be as high as 1350 ft (410 m).

All loran C transmitters operate at a fixed frequency of 100 kHz and confine 99% of their radiated energy within the 90–110-kHz band. Each radiated pulse is designed, therefore, to build up quickly and decay slowly (Fig. 2). The radio frequency within each pulse is coherent with the repetition frequency.

Each loran C transmitting station transmits a group of these pulses at a specified group repetition interval, or GRI (Fig. 3). The master pulse group consists of eight pulses spaced 1000 μs apart, and a ninth pulse 2000 μs after the eighth. Secondary pulse groups contain eight pulses spaced 1000 μs apart.

Multiple pulses are used so that more signal energy is available at the receiver, improving significantly the signal-to-noise ratio without an increase in the peak transmitted power capability of the transmitters. The master's ninth pulse is used for visual identification of the master and for blinking. Blinking, used to warn users that there is an error in the transmissions of a particular station, consists of turning the ninth pulse on and off in a specified code. The secondary station of the unusable pair blinks by turning its first two pulses on and off.

Each pulse within a group may have its radio-frequency cycles in phase or 180° out of phase with an established reference. The phase coding identifies master or secondary transmissions so that automatic signal acquisition can be accomplished

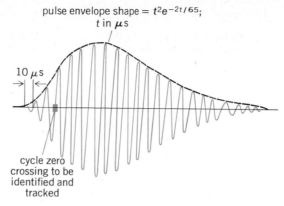

pulse envelope shape = $t^2 e^{-2t/65}$; t in μs

10 μs

cycle zero crossing to be identified and tracked

Fig. 2. Loran C pulse. (*From U.S. Coast Guard, Department of Transportation, Loran-C User Handbook, CG-462, 1974*)

unambiguously; coding is chosen so that the effects of long sky-wave pulse trains (over 1000 μs) cancel, thus permitting ground-wave accuracy to be retained under all conditions.

In 1975 there were nine loran C chains operating in the world comprising a total of 35 transmitting stations, 27 of which were United States–sponsored and 8 of which were in the Soviet Union.

After much evaluation, the United States officially selected loran C in 1974 as its national system for navigation in the coastal confluence region. As a consequence, five transmitting stations were planned for the West Coast of the United States and three for the Gulf of Alaska, and scheduled for operation by Jan. 1, 1977. Three new transmitters were planned for the Gulf of Mexico with operation by July 1, 1978. Modernization and expansion of loran C along the East Coast of the United States was planned to be completed by July 1, 1978, and expansion on the Great Lakes was to be finished by Feb. 1, 1980.

High accuracy is obtained in the receiver by measuring the crossover time of individual radio-frequency cycles on the leading edge of the pulse envelope. Because of the longer propagation path of the sky wave, sky-wave contamination does not start until at least 30 μs after the beginning of the ground-wave pulse. Therefore, the first three cycles of the received signal are always stable ground waves. Tracking is accomplished on the third rf cycle of each pulse in the group to obtain the greatest measurement precision and stability.

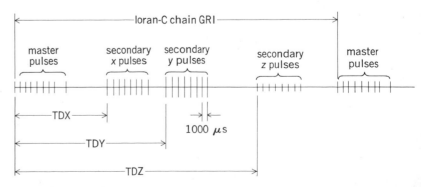

loran-C chain GRI

master pulses   secondary x pulses   secondary y pulses   secondary z pulses   master pulses

TDX

1000 μs

TDY

TDZ

Fig. 3. Loran C signal format. (*From U. S. Coast Guard, Department of Transportation, Loran-C User Handbook, CG-462, 1974*)

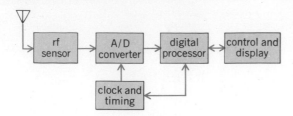

Fig. 4. Loran C navigation receiver.

Cycle identification requires integration or filtering of measurements of the pulse envelope to permit a reliable decision as to whether the correct third cycle or an earlier or later cycle is being tracked. Once this decision is made, the correct cycle can be selected. Continuous monitoring of this cycle selection process is generally implemented, with an alarm if correction is required. *See* RADIO-WAVE PROPAGATION.

There are five major sections of a modern loran C automatic acquisition and tracking, cycle-matching receiver (see Fig. 4): the rf sensor, the analog-to-digital (A/D) converter, a loran clock and timing circuits, a digital processor, and appropriate operator controls and displays. *See* ANALOG-TO-DIGITAL CONVERTER.

The rf sensor receives and processes the signal to provide adequate overall signal-to-interference protection, to "notch out" severe continuous-wave (cw) interferences, and if necessary, to provide amplification to a suitable level for digital sampling. The bandpass filter characteristic is typically a three-pole configuration with a 3-dB bandwidth of 20 kHz.

The A/D converter is used to convert the analog rf signal information from the rf sensor to digital samples associated with each loran pulse or pulse group. Typically, three samples of each loran pulse are required: one for phase tracking and two for cycle identification. The digital samples are taken at sample times determined by the loran clock and timing circuits. The digital samples so obtained are then processed in the digital-processing circuits.

Technology that emerged during 1972–1975 made it possible to incorporate tracking-loop algorithms, search algorithms, cycle identification algorithms, and so on, in digital software using inexpensive read-only memory for control of an inexpensive microprocessor. As a result, it is now possible to produce a fully automatic search-and-track loran C receiver at a user cost comparable to that of the manual, and comparatively crude, predecessor loran A equipment.

Operator controls and displays can take many forms depending on the application; operator control is very simple since signal acquisition and track is accomplished automatically. The navigation display, in its simplest form, is time-difference line-of-position numbers in microseconds for two signal pairs. More useful navigation out-puts in terms of latitude-longitude coordinates, course-to-steer and distance-to-go to a selected destination, along-track and cross-track error, and so on, are generally provided in avionics configurations.

Loran D. Loran D is a tactical version of the loran C system and is designed for short-range service. The advantage of loran D over loran C is that the ground stations are transportable and can, therefore, be quickly deployed. This mobility is gained at the expense of the substantially reduced radiated power (30 kW peak) inherent in the somewhat smaller transmitters and lower antenna towers (400 ft, or 120 m). To partially compensate for this reduction in radiated power, loran D is configured to use 16 phase-coded pulses 500 μs apart, in each group. Loran D phase code is different from, but compatible with, loran C. Alternate pulses in the loran D group (odd-numbered pulses) are phase-coded identically to the loran C system format and are 1000 μs apart as in loran C. Pulse sampling is carried out near the signal peak because sky-wave delays are greater at short range.

[JAMES P. VAN ETTEN]

Bibliography: E. R. Blood, Navigation aid requirements for the U.S. coastal confluence region, *J. Inst. Nav.*, 20(1), Spring 1973; J. A. Pierce, A. A. McKenzie, and R. H. Woodward, *Loran*, 1948; U.S. Coast Guard, Department of Transportation, *Loran-C User Handbook*, CG-462, 1974; U.S. Department of Transportation, *National Plan for Navigation, Annex 1974*, July 1974; J. P. VanEtten, Loran-C system and product development, *Electr. Commun.*, 45(2), 1970; J. P. VanEtten, Navigation systems: Fundamentals of low and very-low-frequency hyperbolic techniques, *Electr. Commun.*, 45(3):192–212, 1970.

Lorentz transformations

The family of mathematical transformations used in the special theory of relativity to relate the space and time variables of uniformly moving (inertial) reference systems. They were discovered in part by H. A. Lorentz during his studies of matter moving in an electromagnetic field and were formulated more completely by H. Poincaré and by Albert Einstein. If S and S' are two inertial reference frames with space-time coordinates (x,y,z,t) and (x',y',z',t'), respectively, the equations connecting these variables have the form ($x^0 = ct$, $x^1 = x$, $x^2 = y$, $x^3 = z$) of Eq. (1), where a^0, a^1, a^2, a^3 are

$$x'^\alpha = \sum_{\beta=0}^{3} L_\beta{}^\alpha (x^\beta - a^\beta) \qquad (\alpha = 0, 1, 2, 3) \quad (1)$$

arbitrary real constants. Forming the two real matrices (2) and (3), the matrix L is subject to the con-

$$L = \begin{bmatrix} L_0{}^0 & L_1{}^0 & L_2{}^0 & L_3{}^0 \\ L_0{}^1 & L_1{}^1 & L_2{}^1 & L_3{}^1 \\ L_0{}^2 & L_1{}^2 & L_2{}^2 & L_3{}^2 \\ L_0{}^3 & L_1{}^3 & L_2{}^3 & L_3{}^3 \end{bmatrix} \quad (2)$$

$$\eta = \begin{bmatrix} 1 & 0 & 0 & 0 \\ 0 & -1 & 0 & 0 \\ 0 & 0 & -1 & 0 \\ 0 & 0 & 0 & -1 \end{bmatrix} \quad (3)$$

dition shown as Eq. (4), where L^{-1} and L_t are the inverse and transposed matrices of L.

$$L^{-1} = \eta L_t \eta \quad (4)$$

The Lorentz transformation group is composed of four classes of transformations, which can be organized as indicated in the table, depending on the algebraic signs of the pivotal element $L_0{}^0$ of the matrix L, and the determinant of L, which is des-

Classes of Lorentz transformations

	$L_0{}^0$	$D(L)$
Proper	+	+
Improper	+	−
	−	+
	−	−

ignated by $D(L)$. Positive signs are indicated by + and negative signs by − in the table.

The improper Lorentz transformations involve changes of sign (reflections) of the space coordinates, the time coordinate, or both. For additional discussion of improper Lorentz transformations *see* SPACE-TIME.

The proper Lorentz transformations form a 10-parameter group which can be generated from translations of the space-time coordinates, rotations of the space reference frame, and transformations to uniformly moving systems. *See* RELA-TIVITY.   [E. L. HILL]

Bibliography: R. Brauer and H. Weyl, *Amer. J. Math.*, 57:425, 1935; R. Katz, *An Introduction to the Special Theory of Relativity*, 1964; H. P. Robertson and T. W. Noonan, *Relativity and Cosmology*, 1968.

Loris and allies

Primate mammals that make up the family Lorisidae, all of which are inhabitants of southeastern Asia and Africa. There are 11 species in 6 genera, including animals such as the angwantibo, the bushbaby, and the potto (see table).

Loris. The best-known species of this family are the slender loris (*Loris tardigradus*) and the larger slow loris (*Nycticebus coucang*), which occur in southern Asia (see illustration). These are slow-moving, arboreal, nocturnal animals which stalk insects and small animals. They rarely leave the arboreal habitat, and sleep during the day rolled

Species of family Lorisidae and their geographic distribution

Scientific and common name	Geographic distribution
Loris tardigradus Slender loris	Southeastern Asia
Arctocebus calabarensis Angwantibo	West Africa
Nycticebus coucang Slow loris	Malay Archipelago
Nycticebus pygmaeus Lesser slow loris	Southern Asia
Perodicticus potto Potto	Central Africa
Euoticus elegantulus Needle-clawed bushbaby	West Africa
Euoticus inustus Eastern needle-clawed bushbaby	Central Africa
Galago crassicandatus Thick-tailed bushbaby	Africa
Galago senegalensis Senegal bushbaby	Africa
Galago alleni Allen's bushbaby	Central West Africa
Galago demidovi Demidoff's bushbaby	Central Africa

The slow loris (*Nycticebus coucang*), with no tail and with vestigial index fingers.

up in tree hollows. The slender loris has two breeding seasons each year, the spring and late winter; the female has one young after a gestation period of about 25 weeks. The term slow lemurs has been used to describe the lorises and their African relatives, the potto and the angwantibo.

Angwantibo. *Arctocebus calabarensis* is found in forested areas of western Africa but is extremely rare. It is a small animal, similar to the potto but without a tail, and there are no defensive spines on the neck vertebrae. The hands show an extreme specialization; the first and second fingers are reduced in size, the first finger being represented by a knoblike, nailless stump, and are used like pincers to provide a powerful grip when climbing. These animals are thought to be omnivorous since they feed on insects, fruit, and small birds when observed in captivity. In their natural habitat they live in high branches of deciduous forests, and sleep clinging to a horizontal branch.

Potto. *Perodicticus potto* is the only species and is quite common in the forest areas of West, Central, and East Africa. It is nocturnal, omnivorous, arboreal, and slow-moving. An unusual characteristic is the row of tubercles covering the spiny prolongations of the last cervical and first two thoracic vertebrae, which serve as protective structures and are used in self-defense. The tail is short and stumpy; there is a rudimentary index finger.

Bushbaby. There are six species of bushbaby in two genera. Also known as night apes or galagos, they are very-fast-moving, arboreal animals, which hop like a kangaroo when on the ground. All have long bushy tails, short muzzles, large eyes, and hindlegs which are much longer than the forelimbs. The smallest species, *Galago demidovi*, fits into the hand of a man. Members of *G. senegalensis*, frequently kept as pets, are gregarious, grooming each other with the comblike lower incisors. On the other hand, the thick-tailed bushbaby or great galago (*G. crassicaudatus*) is more aggressive and solitary, and grows to over 1 ft in length with an equally long tail.

These animals are nocturnal and feed mostly on

insects, although they also eat fruit and eggs. The gestation period for the bushbaby is about 17 weeks; usually, only one or two young are born, though females have four mammary glands, two of which are inguinal. The young remain in the nest and are not carried about by the mother. These animals mature in about 20 months. They have a peculiar habit of constantly licking their hands and feet as they climb in the trees, but the significance of this action is unknown. *See* LEMUR; MAMMALIA; PRIMATES.   [CHARLES B. CURTIN]

Loschmidt's number

The number of molecules in 1 ml of a perfect gas at 1 atm pressure and 0°C. It is found by dividing the Avogadro number by the normal molar volume or normal gram-molecular volume in milliliters. Its value is $2.68719 \pm 0.0001 \times 10^{19}$ cm³/ml. *See* AVOGADRO NUMBER; MOLAR VOLUME.

[THOMAS C. WADDINGTON]

Loudness

The psychological property of sound characterized by strength or weakness. It varies most directly with the physical intensity of sound, increasing as intensity is increased and decreasing as intensity is decreased. Loudness also varies with frequency of sound waves. Tones of very low and very high frequencies require much more intensity than those in the middle range of 1000–5000 hertz (Hz) in order to be perceived as equally loud.

Loudness measurement is significant in all technical and theoretical measurement of hearing, especially in noise investigation and control, analysis of communication instrumentation, and in systematic study of the means by which the auditory mechanism resolves differences in sound intensity.

Decibel. In practical work dealing with sound, the strength of sounds is usually expressed in decibel (dB) units of physical intensity or sound pressure level. Decibels are relative units expressing a ratio between a given sound intensity and a reference sound. A bel is the logarithm to the base 10 of the ratio of a sound of given intensity to the reference intensity; a decibel is one-tenth of a bel. One convenient reference sound which is widely used as a standard for a decibel scale is the sound pressure of a tone of 1000 Hz (0.0002 dynes/cm²) which is just barely heard by the normal ear. This value is sometimes called zero loudness. A sound 10 times as intense in energy has a value of 10 dB; one 100 times as intense, 20 dB. The intensity levels of some ordinary sounds are as follows: a whisper, 10 dB; a quiet office, 30 dB; street noises, 60–70 dB; a plane, 110 dB. Very loud thunder, at 120 dB, would be 1,000,000,000,000 times as intense as a barely perceptible sound of 1000 Hz (Fig. 1).

Phon. Decibels are useful units of physical intensity, but do not specify the psychological loudness of a sound. A scale of loudness level, based on intensity level, uses a unit called the phon. The loudness level of a tone is the intensity level of the 1000-Hz tone to which it sounds equal in loudness. Thus a 1000-Hz tone at 40 dB has a loudness of 40 phons. Other frequencies which are judged equal in loudness also have a loudness level of 40 phons, although their intensity level may be considerably higher than 40 dB. Figure 2 shows the relationship

between decibels and phons in terms of several equal loudness contours. The lowest curve represents the absolute threshold, a loudness level of 0 phon. It should be noted that equal loudness contours at progressively higher loudness levels become more and more flattened out. In other words, the higher the intensity, the more closely does loudness level correspond with intensity level throughout the range of frequencies.

Sones. Phons are based on psychological judgments of equal loudness, but do not constitute a true loudness scale. That is, a sound of 80 phons is not necessarily twice as loud as a sound of 40 phons. To obtain a loudness scale based on equal psychological intervals, observers were asked to adjust sounds to be "twice as loud" or "half as loud" as reference sounds, or to be midway between two reference sounds. Such a scale has been established for units called sones. One sone is defined as the loudness of a 1000-Hz tone at 40 dB (Fig. 2). If both loudness and intensity are represented logarithmically, loudness increases very rapidly at low intensities, and less and less rapidly at higher intensities.

The loudness of complex sounds varies with the distribution of their components along the frequency range. If the components are sufficiently separated, the loudness of the total complex sound is equal to the sum of the loudnesses of the components presented separately. If the components are not widely separated, the total loudness is less than the sum of the component loudnesses. This is apparently due to the fact that regions of cochlear or neural activity set up by the components overlap when they are close together.

Noise pollution. The most significant phenomenon of sound loudness is related to excessive noise in the human environment. Excessive noise has been called noise pollution. Exact standards of noise pollution have not been defined, but working standards based on extensive studies since the 1930s can be established which encompass both noise level and time duration specifications. Generally, standards of noise pollution must be defined in human factor terms or tolerance levels. These levels are (1) deafness-producing or intolerance level; (2) dangerous or deafness-promoting; (3) disturbing or deafness-related; and (4) excessive or

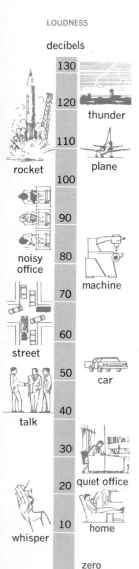

Fig. 1. Decibel scale of common sounds.

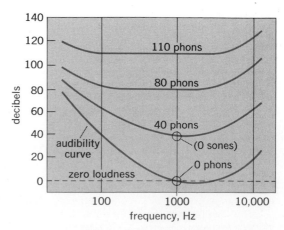

Fig. 2. The definitive relationship between loudness level, in phons, and intensity level, in decibels.

harmful exposure. Sounds of 100 dB, that is, 20 dB above the noise of a busy street, can be defined as deafness-producing under any circumstances. Sounds of 85–100 dB may be classed as deafness-producing if exposure lasts for more than a few minutes each day and as deafness-promoting if exposure is momentary and periodic. Sounds of 65–85 dB may be classed as of moderate noise level; they are disturbing or deafness-related if persistent in daily work, and excessive or constitute harmful exposure if periodic in day-to-day activity. Workers should be protected from all levels of noise pollution. Guides to these levels can be obtained in terms of common sounds indicated in the intensity scale in Fig. 1. *See* NOISE CONTROL.

A certain amount of noise in the environment is essential to the feeling of well-being. The physiological basis of background sound is that the auditory system utilizes such noise to keep the visual system maximally sensitive. Although the ear system does not habituate to sound, noise detection is closely related to maintaining alertness and neural activation. *See* AUDIOMETRY; DEAFNESS; HEARING (HUMAN).   [KARL U. SMITH]

Bibliography: G. A. Briggs and J. Moir, *Audio and Acoustics*, 1964; I. J. Hirsh, *The Measurement of Hearing*, 1952.

Loudspeaker

A device that converts electrical signal energy into acoustical energy, which it radiates into a bounded space, such as a room, or into outdoor space. The essential characteristic of the signal energy must be preserved in the energy-conversion process. The shorter term speaker is also used. The term unit, as in horn or driver unit, is applied to devices comprising a motor and diaphragm but lacking the acoustic element such as a horn to complete the speaker. The term speaker system is normally reserved for a plurality of speaker units and the associated accessories, horns, cabinets, and electrical networks required for an integrated system.

Requirements for speech and music. The two most important variables in a sound wave are pressure and the frequency of the pressure alternations, because the auditory response of the listener is most closely related to these variables. The approximate auditory limits of the ear are shown in Fig. 1. These limits cover an enormous range. Frequencies differ by a factor of 10^3, pressures by a factor of 10^6, and intensities (power) by a factor of 10^{12}. Fortunately, commonly reproduced sounds such as music and speech cover the more limited ranges in Fig. 1. The need for reproducing even these ranges is further modified by psychoacoustic factors and noise. *See* HEARING (HUMAN).

Tests have shown that frequency ranges of 80–11,000 Hz for music and 150–8000 Hz for speech, if uniform throughout the listening area, give high-quality reproduction in an otherwise distortionless system. The pressure range requirements are set primarily by listener preference. The listener is seldom interested in hearing a wanted sound in the presence of a competing unwanted sound (by definition, noise) of the same pressure. Noise, therefore, commonly limits the minimum pressure of interest, particularly in music reproduction.

Alternatively, the requirements of the speaker may be considered from the viewpoint of the

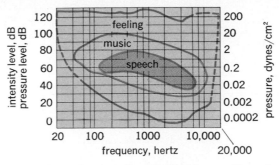

Fig. 1. Approximate auditory limits of the ear. The colored areas shown indicate the intensity and pressure levels that are found in speech and music. The standard reference levels are 10^{-16} watt/cm² intensity (plane wave) and 0.0002 dyne/cm² pressure.

complete-system designer, who normally desires a speaker that does not noticeably narrow the frequency bandwidth, lower the signal-to-noise ratio, or increase the distortion of the system. Probably the most important requirement is that the speaker not introduce noises that are clearly spurious and unrelated to the signal, such as rattles and buzzes. Some physical performance characteristics can be adequately measured and specified; others cannot, however, and a thorough listening test is an essential part of any speaker-testing program. The final judge of the value and adequacy of the speaker (and system) is the ear of the ultimate user. *See* PSYCHOACOUSTICS; SOUND-REPRODUCING SYSTEMS.

Physical performance characteristics. Because of the importance of pressure and frequency in hearing, the loudspeaker output in terms of these variables is also important.

Pressure response. Measured at a specified frequency, the pressure response is the sound pressure in dynes/cm² at a designated location with respect to the speaker, per volt input. The pressure is commonly measured under actual or simulated outdoor free-field conditions in the absence of unintended reflecting or diffracting surfaces.

Pressure response–frequency characteristic. Commonly called simply the frequency response, this is the pressure response obtained as the test-signal frequency is varied slowly over the audible frequency range (20–20,000 Hz). The response characteristic should be free of abrupt changes, which are accompanied by transient and sometimes nonlinear distortion. A smooth response curve also simplifies electrical equalization or compensation.

Directional characteristics. These are determined by measuring the pressure response at a sufficient number of points in the intended listening region and at a sufficient number of frequencies to predict the frequency response at any listener's position with the desired precision. The most common practice is to obtain frequency-response characteristics at a few points equidistant from the speaker but subtending different angles with the principal axis of the speaker. On-axis curves are usually published because the pressure is normally a maximum, particularly at high frequencies when focusing is marked. In typical environments the response characteristic at 30° is much more useful for predicting the average re-

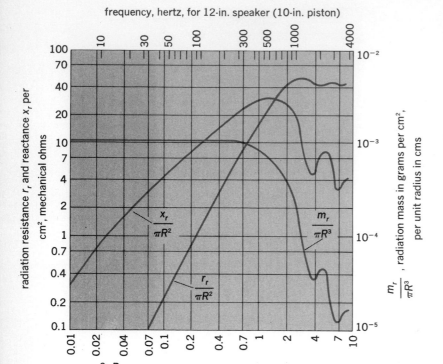

frequency, hertz, for 12-in. speaker (10-in. piston)

$$\frac{2\pi R}{\lambda} = 1.833 \times 10^{-4} Rf, \text{ or frequency in kilohertz}$$

for R of 5.47 cms (5-in. speaker)

Fig. 2. Average radiation resistance, reactance, and mass per square centimeter of a flat, rigid piston vibrating in an infinite, rigid, nonabsorbing baffle. Piston radiates into a solid angle $\Omega = 2\pi$ steradians (hemisphere). λ = length of radiated wave in centimeters. (*From K. Henney, ed., Radio Engineering Handbook, 5th ed., McGraw-Hill, 1959*)

sponse throughout the listening region. In use, the speaker is positioned and oriented to give the most favorable average pressure response–frequency characteristic in the listening region. *See* SOUND-REINFORCEMENT SYSTEM.

Distortion. Although distortion may be more broadly defined, loudspeaker distortion is commonly limited to nonlinear, transient and frequency-modulation distortion.

Nonlinear distortion arises from lack of proportionality between the electrical input and acoustical output signal with a sustained (steady-state) input signal. With a sinusoidal input, integral multiples (harmonics) and rarely submultiples (subharmonics) of the fundamental frequency are generated. If two or more sinusoidal signals are applied, the nonlinearity gives rise to interaction between the two (intermodulation distortion), resulting in the generation of harmonics and sum and difference frequencies of the fundamentals and harmonics.

Transient distortion arises from the inability of the speaker to generate an acoustic wave which exactly follows sudden changes in the wave shape of the electrical signal. Sharp peaks and valleys in the pressure-frequency response curve are to be avoided, because they indicate the presence of inadequately damped resonances which give rise to this distortion. These poorly damped resonances color the reproduction and, when present at high frequencies, can be detected as a ringing sound.

Frequency modulation of a signal is the alteration, or modulation, of its apparent frequency by a second, or modulating, signal. It occurs when the radiating diaphragm is vibrated toward and away from the listener by the modulating signal. The velocity component with respect to the listener produces a Doppler shift. *See* DOPPLER EFFECT.

Electrical speaker impedance. This is the complex ratio of the applied sinusoidal voltage across the signal terminals to the resulting current. The normal or loaded impedance Z_N, the free or unloaded impedance Z_f, and the blocked impedance Z_b correspond respectively to operation with the normal acoustic load, in a vacuum, and with the motor and diaphragm blocked or immobilized.

Force factor. The force factor M of the motor is (1) the complex ratio of the force required to block or immobilize the motor to the signal input current to the motor, and (2) the complex ratio of the resulting open-circuit voltage to the velocity of the mechanical system. In a moving-coil speaker the first quantity equals Bl, where B is the average radial flux density through the conductor and l is the conductor length. The second quantity equals $-Bl$. When consistent units are used, these quantities have the same magnitude but opposite signs indicating an antireciprocal relationship, that is, $M_{21} = -M_{12}$.

The normal impedance of a speaker is given by Eq. (1), where Z_m is the total effective mechanical

$$Z_N = Z_b + \frac{M^2}{Z_m} \tag{1}$$

impedance of the moving system, including diaphragm and air load. In a moving-coil speaker $M^2 = B^2 l^2$. In a two-electrode electrostatic speaker $M^2 = -(E_0/\omega d_0)^2$, where E_0 is the dc bias voltage, d_0 the average biased electrode separation, and $\omega = 2\pi$ times frequency. The second term on the right side of Eq. (1) arises from motion of the moving system and is called the loaded or normal motional impedance.

Efficiency. The speaker efficiency is the ratio of the useful acoustic-energy output to the electrical signal-energy input. The system, or available power, efficiency is the ratio of the useful acoustic-energy output to the energy the signal system will supply to a designated resistance.

The acoustic output may be obtained by determining the intensity (see below) of the sound wave at a number of equidistant points from the speaker, under free-field (outdoor) conditions, and summing or integrating the intensity in all directions. The distance should be several times the maximum dimension of the speaker and a sufficient number of points must be used to permit accurate interpolation and integration. Because sound-intensity meters are rarely available, the pressure is measured. Under the above test condition, but not in a normal listening room, the intensity I in the direction of propagation is given by Eq. (2), where p is

$$I = p^2/\rho c = 2.42 \times 10^{-9} p^2 \text{ watt/cm}^2 \tag{2}$$

the pressure in dynes/cm^2; ρc is the characteristic impedance (here resistance) of air, ρ is the density of dry air (0.0012 g/ml), and c is the velocity of propagation of the sound wave (34,280 cm/sec), both at 20°C and 760 mm Hg pressure. The electrical input power is the rms input current squared times the resistive part of the electrical normal impedance. *See* SOUND INTENSITY; SOUND PRESSURE.

Speaker power rating (input). Unless otherwise stated, this is the maximum output power rating of a speaker's intended associated amplifier when the signal is speech and music and the amplifier is operated in its linear (undistorted) rated region. Under these conditions the average power over 15-sec intervals will be 1–2%, and the maximum power at any one frequency over a 2-sec interval will be 10–15% of the rated value. As an example, a 10-watt speaker is intended to work with a normally operated 10-watt amplifier with speech and music input. It is not expected to operate continuously with 10 watts input at all frequencies in its transmitted frequency band without distortion or failure.

Speaker types. Speakers are commonly classified by terms which describe their three important functional parts: the motor, diaphragm, and acoustic radiation-controlling element. The motor converts electrical energy into mechanical energy and couples the electrical signal source, commonly an amplifier, to the diaphragm. Common motor types are moving-conductor (principally moving-coil or dynamic, rarely ribbon conductor) and electrostatic. Moving-conductor speakers are those in which the mechanical forces result from magnetic reactions between the field of the current in a moving conductor and a steady magnetic field. In a moving-coil speaker the conductor is in the form of a coil conductively connected to a source of electrical energy and mechanically attached to the diaphragm. An electrostatic speaker is one in which the mechanical displacements are produced by the action of electrostatic fields. Other motor types, which include the magnetic-armature, pneumatic, piezoelectric, and ionic types, are discussed in the references listed in the bibliography.

Speakers are also classified by their radiation-controlling elements into direct-radiator (hornless) and horn types. The direct-radiator types commonly employ baffles or enclosures.

The diaphragm is the element which, vibrated by the motor, causes the air to vibrate; hence it couples the air load (radiation impedance) to the motor.

Radiation impedance. The added impedance to motion of a diaphragm or sound-radiating surface, arising from its contact with air, is called the radiation impedance. Radiation resistance is a component arising from energy radiation. Radiation reactance is the reactive component. A major problem to the speaker designer and one of importance to the user is that of obtaining efficient energy transfer or coupling from a relatively high motor-and-diaphragm mechanical impedance to a low air-radiation impedance.

The average radiation impedance per unit area of a flat piston is shown in Fig. 2. When the length of the radiated wave λ exceeds the perimeter of the diaphragm (when $\lambda > 2\pi R$ at low frequencies), the radiation reactance is mass reactance and proportional to frequency, and the radiation resistance per unit area is proportional to the frequency squared. The acoustic power P_a radiated by a piston is given by Eq. (3), where r_r is the total radiation resistance (obtained by multiplying the per

$$P_2 = r_r v^2 \times 10^{-7} \text{ watt} \qquad (3)$$

tion resistance (obtained by multiplying the per

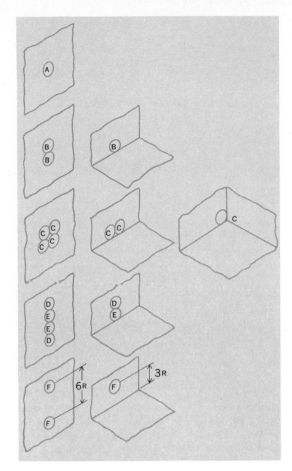

Fig. 3. The effect of adding pistons and reflecting planes on radiation impedance. All pistons that are marked with the same letter see the same radiation impedance. *(From K. Henney, ed., Radio Engineering Handbook, 5th ed., McGraw-Hill, 1959)*

unit area value from Fig. 2 by the area) and v the rms piston velocity, cm/sec.

If two or more identical speakers are mounted close together in a common plane and supplied with signal energy from the same source, the combined diaphragms will have substantially the same radiation impedance at low frequencies as a single diaphragm of the combined area. Differences occur as the wavelength diminishes (as frequency rises) and approaches half the diaphragm perimeter. By mounting a speaker adjacent to one or two large, rigid, nonabsorbing surfaces, such as a floor or floor-wall combination, the radiation resistance may be made that of one speaker of a two- or four-speaker combination, as shown in Fig. 3. Each surface, by complete wave reflection, adds an image (by analogy with an optical mirror) which the diaphragm cannot distinguish from a true speaker or pair of speakers. An alternate way of considering this effect is to note that the solid angle into which the speaker radiates is 2π, π, and $\pi/2$ steradians, or half, quarter, or eighth spheres, respectively. A further advantage of locating a speaker in the corner is the narrower angle over which uniform high-frequency response is required. *See* IMAGE, ACOUSTICAL.

Direct-radiator speakers. These are the most widely used speakers. Of these, substantially all of the general-purpose and special-purpose low-fre-

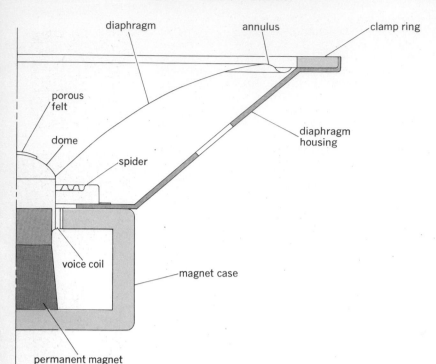

Fig. 4.   Sectional view of small moving-coil permanent magnet speaker.

quency (woofer) speakers are of the moving-coil type. Electrostatic speakers are sometimes used as high-frequency speakers (tweeters) in broadband systems and infrequently as general-purpose speakers.

Moving-coil speakers. Widely used because they can cover a broad frequency range (5–8 octaves), these speakers are compact, reliable, and low in cost (Fig. 4). Their available power or system efficiency is low, typically 0.5–4%.

Their relatively uniform efficiency up to 1000 Hz or more is attained by placing the fundamental mechanical resonant frequency near the lowest frequency of interest. This is the frequency at which the diaphragm and air mass are resonated by the diaphragm suspension and air stiffness, if present. At frequencies that are higher than this by a factor of from 2 (1 octave) to 10 or so, the normal impedance is substantially the blocked impedance, which in a moving-coil speaker is largely

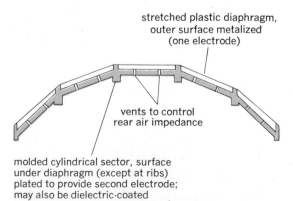

Fig. 5.   Simplified cross-sectional view of a two-electrode electrostatic speaker of the type that is used to obtain an approximately cylindrical wave front at high frequencies. (*From K. Henney, ed., Radio Engineering Handbook, 5th ed., McGraw-Hill, 1959*)

resistive. A constant-voltage and constant-internal-resistance signal source will give a fairly uniform current and hence, force, in this range. The effective moving mass is nearly constant. The impedance is therefore proportional to frequency and the diaphragm velocity varies inversely with frequency. Consequently, the radiated power is independent of frequency.

Electrostatic speakers. In these speakers the mechanical forces are produced by the action of electrostatic fields. Structurally they are capacitors in which one electrode is free to move and serve as a diaphragm. The movable electrode is commonly a thin plastic sheet with a conductive surface, which may be a metallic foil or evaporated film. Two fixed electrodes, one on each side of the movable electrode, are used in push-pull types when the diaphragm amplitude may be relatively large. A single fixed electrode is used in single-ended types (Fig. 5). These are commonly used as high-frequency speakers (tweeters). A high dc bias potential, of the order of kilovolts, is applied between the fixed and movable electrodes to increase sensitivity and reduce distortion. The ac potential is superimposed on the dc, resulting in an alternating force on the diaphragm.

Some advantages of the electrostatic speaker are (1) the diaphragm can be very light, (2) the force is distributed over the diaphragm, (3) the transient and phase distortion may be kept low by proper design, and (4) the cost may be low.

Some of the disadvantages are (1) the permissible diaphragm amplitude is normally low, leading to large-area diaphragms or limiting the lower cutoff frequency, (2) the normal impedance is substantially that of a capacitor and is difficult to match properly, except in a limited frequency range, and (3) because of the high potentials, dielectric problems are involved. *See* ELECTROSTATICS.

Radiation-controlling structures. These are rigid, nonvibrating, normally nonabsorbing structures which control the divergence of the sound wave from one or both sides of the diaphragm and prevent or control interaction between the front and back radiation. Their principal function is to increase the sound pressure in the listening region, both by increasing the energy-conversion efficiency and by directing the emitted wave in the desired direction. Vibration of the diaphragm produces slight alternating increases and decreases in the air pressure above and below atmospheric pressure. Increases occur on one side as decreases occur on the other. At low frequencies the front pressure wave can reach the back of the diaphragm (and the back pressure wave can reach the front of the diaphragm) before the diaphragm has moved appreciably. Hence almost complete pressure neutralization, or destructive interference, occurs unless a rigid surface, or baffle, is interposed as an extension of the diaphragm support.

The term baffle is usually applied to a relatively flat structure of substantial area. Three types are shown in Fig. 6. Either the speaker is mounted eccentrically, or the baffle shape is made irregular so that serious destructive interference does not occur in a narrow frequency range. Destructive interference may be prominent in outdoor environments but is usually obscured indoors by reflected waves.

Enclosures. These are radiation-controlling structures of the hornless type. Enclosures may be classified as total, in which the rear radiation is completely absorbed by the closed rear enclosure, and vented, or reflex, in which the rear radiation is shifted in phase, modified in amplitude, and combined with the front radiation to increase the total radiation in a frequency band of approximately an octave just above the lowest transmitted frequency. At still lower frequencies the total radiation is substantially reduced by pressure neutralization or destructive interference.

Total enclosures are used primarily when the volume is smaller than desirable for a reflex type. Because the air stiffness limits the minimum diaphragm–motor system stiffness, the mass may be raised to reduce the resonant frequency. In the reflex type, higher efficiency is obtained in the lowest octave by using a larger cabinet with less air stiffness and by making use of the reflex, or rear, acoustic by network phase shift and radiation. A simplified sketch of a reflex enclosure is shown in Fig. 7. The cabinet air stiffness couples the diaphragm to the air mass in the vent, or port region. The total enclosure has no port or opening.

Horns. Structurally a horn is a rigid, nonabsorbing, tapered duct or passage. It may be straight, bent, or folded with concentric sound passages. The small end, to which the horn or driver unit (speaker) is attached, is called the throat and the large radiating end is called the mouth.

Functionally it is both a tapered acoustical transmission line used to match the relatively high impedance looking back into the diaphragm to the relatively low radiation impedance seen at the mouth, and a device which by virtue of its mouth size and shape controls directional properties.

V. Salmon has discovered a family of horns with interesting properties to which many common horns belong. In these Salmon or hyperbolic exponential horns, the cross-sectional area A is related to the axial distance by Eq. (4), where A_t is the

$$A = A_t \left[\cosh\left(\frac{x}{x_0}\right) + T \sinh\left(\frac{x}{x_0}\right) \right]^2 \quad (4)$$

throat area, x_0 is a constant fixing the axial scale of length, T is a constant determining a member of the general family, and the cosh and sinh are the hyperbolic exponential cosine and sine functions, respectively. The longitudinal sections of these horns for various values of T are shown in Fig. 8 for straight-axis circular horns.

The performance of a horn depends principally on the throat impedance and its dependence on frequency. Although waves reflected from the mouth introduce variations with frequency into the throat impedance, it has been found that the average impedance is close to that of a horn with no reflected wave. The throat impedance in mechanical ohms of catenoidal horns with rigid, nonabsorbent walls and negligible reflected waves is given by Eq. (5), where ρ, c, and T have been defined

$$Z_t = R_t + jx_t$$
$$= A_t \rho c \frac{[1 - (f_c/f)^2]^{1/2} - j(Tf_c/f)}{1 - (1 - T^2)(f_c/f)^2} \quad (5)$$

previously, f is the frequency, and f_c is the cutoff frequency, given by $f_c = c/2x_0$ and $j = \sqrt{-1}$. Thus

the reference axial length x_0 is of fundamental importance in determining the behavior of the impedance. In Fig. 9 the behavior of R_t for various values of T is shown.

Speaker systems. Two or more identical speakers may be used to improve low-frequency efficiency and increase permissible input and output powers or, by proper relative orientation, improve the directional properties. A source of sound which is very small compared to the length of the radiated wave (point source) is nondirectional. The source becomes increasingly directional as the wave length is diminished; that is, the frequency is increased. Noncircular sources tend to be most directional in the plane of the maximum dimension. Diaphragm and horn mouth shapes are therefore used to control directivity. Acoustic lenses, analogous to optic lenses, are also useful. Several substantially identical speakers may be used in combination to influence the directivity pattern. More commonly, the two or more speakers cover complementary frequency ranges. The more important advantages are (1) improved frequency response, because each type of unit covers a moderate range, (2) higher system efficiency, for the same reason, (3) improved directivity characteristic, because the diaphragm (or horn mouth) for the highest-frequency speaker may be made relatively small, (4) improved transient response, because many of the artifices used to obtain extended fre-

Fig. 6. Irregular baffle shapes used outdoors to broaden frequency band of destructive interference between speaker front and back waves at listener's position.

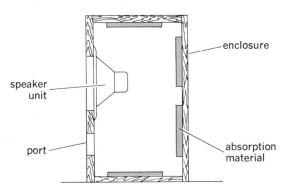

Fig. 7. One type of reflex enclosure in which a large port area is placed near the diaphragm to obtain maximum aid from the port radiation. Phase shift of backside radiation is obtained by choice of enclosure volume and port mass. (*From K. Henney, ed., Radio Engineering Handbook, 5th ed., McGraw-Hill, 1959*)

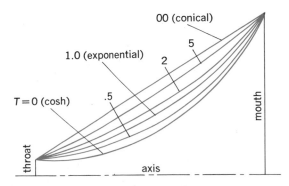

Fig. 8. Longitudinal section of straight-axis circular cross-sectional horns of hyperbolic exponential or catenoidal family. (*From K. Henney, ed., Radio Engineering Handbook, 5th ed., McGraw-Hill, 1959*)

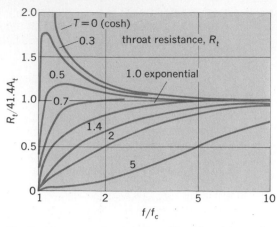

Fig. 9. Frequency dependence of throat resistance for horns of various values of T. Below f_c the resistance is zero. The factor 41.4 is the value of ρc in cgs units at room temperature. (*From K. Henney, ed., Radio Engineering Handbook, 5th ed., McGraw-Hill, 1959*)

quency ranges in single units make the transient response worse, particularly at high frequencies, (5) reduced intermodulation, because large amplitudes are confined to speakers reproducing low frequencies, and (6) reduced frequency modulation.

Coaxial units comprise two, or less commonly three, speaker units mounted on substantially the same axis in an integrated mechanical assembly. Appropriate wave filters or dividing networks control the amplitude and phase of the electrical signal reaching each unit.

The term coaxial speaker should be reserved for a coaxial unit with its required acoustic radiation controlling structure.

Speaker placement. In single-channel, or monophonic, systems speaker placement is largely controlled by the desirability of uniform response in the listening area. A corner location in a room is desirable, improving the low-frequency efficiency and reducing the angle the high-frequency radiation must cover. In stereophonic systems multiple speakers and signal channels are used to reproduce the spatial effects present in the original sound environment. In these, speaker placement is more critical, and the region for satisfactory reproduction is relatively limited. The simpler conventional systems employ two channels and two speakers or speaker systems. These speakers are located symmetrically in the room, preferably along the shorter wall. In typical small listening rooms the speakers are 6–10 ft apart. Properly connected, identical speakers will give identical pressure and phase along the plane of symmetry between the speakers which coincides with a median vertical plane of the room. Optimum listener locations are in the plane of symmetry approximately 1.5–2.5 times farther away from the plane of the speakers than the speakers are from each other. Transverse locations are limited by the need of having the more remote speaker less than 5 ft farther from the listener to prevent the precedence effect, in which the earlier signal takes substantial control of the ear if its arrival precedes the latter by approximately 5–35 msec, corresponding to path-length differences of 5.6–39 ft. For additional

material on speaker placement *see* ARCHITECTURAL ACOUSTICS.

[HUGH S. KNOWLES]
Bibliography: L. L. Beranek, *Acoustics*, 1954; F. V. Hunt, *Electroacoustics: The Analysis of Transduction, and Its Historical Background,* 1954; H. S. Knowles, Loud-speakers and room acoustics, in K. Henney (ed.), *Radio Engineering Handbook*, 5th ed., 1959; H. F. Olson, *Acoustical Engineering*, 3d ed., 1957; H. F. Olson, *Music, Physics and Engineering*, 1967; V. Salmon, Generalized plane wave horn theory, *J. Acoust. Soc. Amer.*, 17(3):199–218, 1946.

Louping ill

A viral disease of sheep, capable of producing central nervous system manifestations. It occurs chiefly in the British Isles. Infections have been reported, although rarely, among persons working with sheep.

In sheep the disease is usually biphasic with a systemic influenzalike phase, followed by encephalitic signs. In man the first phase is usually the extent of illness.

The virus is a member of the Russian tick-borne complex of the group B arboviruses. Characteristics, diagnosis, and epidemiology are similar to those of other viruses of this complex. *See* ANIMAL VIRUS; ARBOVIRAL ENCEPHALITIDES.

[JOSEPH L. MELNICK]
Bibliography: F. L. Horsfall, Jr., and I. Tamm (eds.), *Viral and Rickettsial Infections of Man*, 4th ed., 1965.

Louse

Any insect which is a member of the orders Anoplura (biting and sucking lice) or Mallophaga (biting or chewing lice). Lice are apterous (wingless) ectoparasites of birds and mammals. The leathery body is flattened dorsoventrally, with a head that is more or less distinct, and segmented legs. Development in both groups is referred to as ametabolous; that is, metamorphosis is simple in which the young insects, called nymphs, are similar in appearance to the adults except for size. The eggs, as they are laid, are attached to the hair or feathers of the host.

Mallophaga. The majority of the estimated 2000 species occur on birds and thus are called bird lice. These insects have biting mouth parts and feed on hair, feathers, skin, and exudates from wounds, causing irritation to the host. Several generations may be produced each year even under unfavorable conditions. The body temperature of the host provides a fairly uniform heat for development. Man is not known to be attacked by these lice, and transmission from one host to another is by contact.

Anoplura. There are about 500 species of true or biting lice. They are small permanent ectoparasites on mammals with mouthparts modified for piercing and sucking. The legs are stumpy and terminate in a single claw which is frequently specialized to grasp hair. Two of the three species which infest man and produce lousiness or pediculosis are important vectors of human diseases, such as typhus and relapsing and trench fevers.

There are two varieties of *Pediculus humanus*; the head louse is the variety *capitis* (see illustra-

LOUSE

The head louse of humans, a pale graybrown insect with black margins and a maximum length of 1/10 in.

tion), while the body louse, also known as the cootie or grayback, is the variety *corporis*. The third species is *Phthirus pubis*, called the pubic louse or crab louse, because of its resemblance to a crab. This species is not known to transmit any disease, but it is a cause of skin irritations. The two varieties of *P. humanus* interbreed, and the adult stages cannot be differentiated. From two to six eggs, known as nits, are laid each day by the female. The eggs hatch in about a week and it takes the same period of time for the insects to become adults. The life-span is about 1 month for these species. Other important species are the hog louse (*Haematopinus suis*); sheep foot louse (*Linognathus pedalis*); dog louse (*L. setosus*); and the rabbit louse (*Haemodipsus ventricosus*), which may transmit tularemia to man. *See* ANOPLURA; INSECTA; MALLOPHAGA.   [CHARLES B. CURTIN]

Low-level counting

The measurement of very small amounts of radioactivity. The instruments used are special modifications of standard radioactivity measurement equipment, designed to yield greater detection efficiencies and lower background counting rates than those necessary for ordinary applications. These techniques are applied mainly to the measurement of long-lived natural radioactive isotopes and their daughter products, cosmic-ray-produced isotopes, and isotopes produced artificially during nuclear explosions. *See* RADIOACTIVITY.

Because of the statistical error associated with the measurement of radioactivity, accurate assays of samples with small amounts of radioactivity cannot be made in the presence of large amounts of radiation from other sources. This can be readily seen from the relationship between the sample count rate A_s, its percentage error E_s, the background count rate A_b, and the length of time the sample is counted t. The percentage error, when $A_s \ll A_b$, can be approximated by the expression below. For example, the percentage error for an

$$E_s \cong \sqrt{\frac{2A_b}{tA_s^2}} \times 100$$

overnight measurement of a sample with an activity of 1 count/min in the presence of a background of 90 counts/min is 50%. This error can be decreased by reducing the background or by increasing the sample count rate. Doubling the sample count rate is twice as effective as halving the background. Thus most low-level counters are designed (1) to hold as large a sample as it is practical to obtain and to prepare; (2) to record with as high an efficiency as possible the radiations from the sample; and (3) to have as low a background counting rate as possible. The instruments which have been most widely used are internal gas counters, solid source β-counters, and scintillation counters. For a general discussion of various types of particle counters *see* PARTICLE DETECTOR.

For internal gas counters, the sample itself is used as the filling gas for a discharge tube such as an ionization chamber, proportional counter, or Geiger counter. Because of the high detection efficiency of such chambers, they are ideal for measuring the radiations from α- and weak β-

emitting isotopes. The counters are designed to hold the available sample in a minimum volume. Where large quantities of sample are available, counters as large as 10 liters have been used with gas pressures up to 5 atm.

For α-counting, the background can be largely eliminated by selecting only the most energetic pulses produced in the counter. For weak β-detection, however, it is usually impossible to separate the sample pulses from those caused by β-particles emitted from the materials making up the counter, by γ-rays entering the counter from the outside, and by neutrons and mesons produced by cosmic-ray activity.

Gamma-ray background. The greater part of the γ-ray background can be eliminated by placing the counter in an iron or lead shield. Selected low-level lead and mercury are often used as an inner shield to remove γ-radiation originating from impurities in the iron or lead. Even with an 8-in. iron shield and a 2-in. inner mercury shield, a small γ-ray component remains. Since these residual radiations result mainly from the reaction of cosmic rays which have penetrated the shielding, their contribution to the background may change with time. This is because the cosmic-ray flux at a given point on the Earth's surface varies with the change in shielding introduced by atmospheric pressure changes. The time-variable component of the background that is introduced by cosmic-ray flux changes becomes negligible when the total amount of shielding is increased to 30 in. of iron. Such a heavy shielding is often obtained by placing the whole counting room 10 ft or more underground. This arrangement not only eliminates the time variable component but also reduces the absolute value of the background counting rate. A typical background counting rate of 1 count/min for a 1-liter carbon-14 counter at surface level is reduced to about 0.5 count/min in a room 40 ft underground. Another method is to construct a set of anticoincidence counters within the gas counter itself. This may be done with a series of wires, as shown in Fig. 1. The large dots represent 0.01-in.-diameter wires, used as the cathodes of the individual internal counters, and the small dots the 0.001-in.-diameter wires which form the anodes of these counters. The α- and β-radiations originating at the wall of the counter, as well as electrons ejected by γ-rays striking the wall, are recorded by the internal system, and any corresponding pulses in the central counter are electronically canceled.

Since at moderate gas pressures the majority of the recorded γ-rays are detected due to the fact that electrons are ejected from the counter walls, rejection of pulses occurring simultaneously on one of the wires near the wall and on the center wire eliminates much of the γ-ray component of the background. The loss of sample counts that results from introducing the ring of wires often outweighs the sensitivity gain brought about by the reduction in background. This problem is solved by replacing the inner set of wires with a thin leaktight foil that separates the inner counter from the anticoincidence section. The inner and outer sections are filled, respectively, with sample gas and commercially available counting gas. Only very thin foils can be used, since otherwise the electrons ejected through gamma-ray interaction

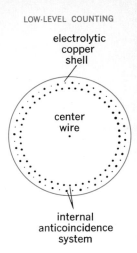

electrolytic copper shell

center wire

internal anticoincidence system

Fig. 1. Cross section of a proportional gas counter with an internal anticoincidence system.

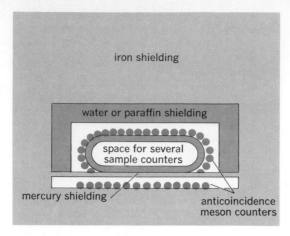

Fig. 2.   Cross section of the shielding used for large-volume ultra-low-level gas counters.

with the foil may contribute again to the background counting rate of the inner counter.

Beta-particle background. This is eliminated by choosing counter materials with low radioactivity. Copper, quartz, and plastics, for instance, have considerably lower radioactivity than brass, aluminum, or glass. The internal anticoincidence system used for γ-rays is effective in removing any residual β-radiation.

Neutrons and mesons. These particles are not effectively removed by iron or lead shielding. A paraffin and boric acid shield will remove the neutrons and the anticoincidence system will remove the mesons. Where no internal system is used, a set of Geiger or proportional counters can be placed around the sample counter in such a manner that any meson entering the sample counter will pass through at least one Geiger counter. Coincident pulses are electronically canceled. A diagram of a typical low-level gas counter and its associated shielding is shown in Fig. 2. *See* GEIGER-MULLER COUNTER.

Gas counters. Internal gas counters are used for the detection of cosmic-ray-produced carbon-14 and tritium. The small quantities of argon-37 and krypton-85 produced by nuclear explosions can also be measured with such devices. Alpha activity

from radon and thoron emanations from uranium and thorium minerals have been measured in unshielded gas counters.

Solid-source counters. For strong β-emitters and for isotopes which cannot be put into a suitable gaseous form, low-level solid-source counters are often used. The counters are usually end window proportional counters or Geiger tubes, but in some cases the sample is mounted on the inside of a cylinder which is placed within the counter. The latter gives a more favorable sample-to-background ratio. Background reduction is accomplished in much the same manner as for the internal gas counter. Since the counters are much smaller, metal shielding and a ring of external anticoincidence tubes produce a sufficient background reduction. A number of the isotopes, such as strontium-90, produced during nuclear tests have been detected with such equipment.

Scintillation counters. Many low-level radioactivity measurements are made with scintillation counters. The energy of the radiations is converted into light which is recorded by a photomultiplier tube. This method has proven especially useful for the detection of small quantities of weak β-radiation from samples in liquid form. The sample is dissolved in a suitable liquid phosphor and placed in a cell designed to reflect all the light produced into one or more adjacent photomultiplier tubes. Such systems have efficiencies as high as 90% for the detection of the weak β-particles from carbon-14 and tritium. *See* PHOTOTUBE, MULTIPLIER; SCINTILLATION COUNTER.

Since this type of counter is more sensitive to γ-radiation than a discharge-type tube, more emphasis must be placed on physical shielding designed to absorb the γ-component of the natural background. In addition to the background originating from nuclear radiations, electrical noise pulses in the photomultiplier tubes must be eliminated. This has been done by facing two phototubes to the same sample. Only pulses seen simultaneously by both tubes are recorded. Fast electronic circuits greatly reduce the possibility of accidental coincidences between the noise pulses from the individual tubes. Cooling the apparatus in a freezer unit also aids in reducing the noise level. A diagram showing the design of a typical low-level liquid scintillation counter is shown in Fig. 3. This system has been successfully used for the measurement of a number of isotopes, including tritium and radiocarbon.

[MINZE STUIVER]

Bibliography: W. B. Mann and S. B. Garfinkel, *Radioactivity and Its Measurement*, 1966; E. H. Quimby and S. Feitelberg, *Radioactive Isotopes in Medicine and Biology: Basic Physics and Instrumentation*, 1963; C. H. Wang and D. L. Willis, *Radioactive Methodology in Biological Science*, 1965.

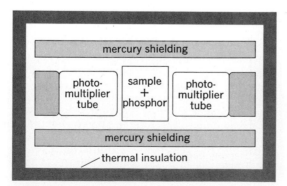

Fig. 3.   Idealized diagram showing a low-level liquid scintillation counting system. The two photomultiplier tubes are connected so that only those pulses occurring simultaneously in both tubes are recorded. The thermal background is further reduced by cooling the whole apparatus in a refrigerating unit.

Low-temperature physics

A study of the properties of gross matter at such low temperature that the quantum character of the substance becomes observable. Most famous among the effects attributed to the quantum mechanical nature of macroscopic matter are (1) superconductivity, (2) superfluid liquid helium, (3) magnetic cooling, and (4) nuclear orientation. The

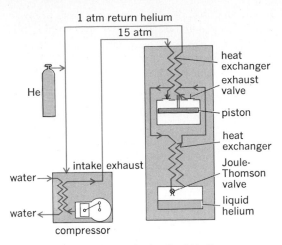

1 atm return helium

15 atm

He

heat exchanger

exhaust valve

piston

heat exchanger

Joule-Thomson valve

liquid helium

water

water

intake exhaust

compressor

Fig. 1. Equipment for producing liquid helium.

topic of cryogenics includes the method of producing low temperatures. In addition to the considerable technology involved in refrigeration, there is a very important section of low-temperature physics concerned with temperature measurement and related instrumentation. *See* CRYOGENICS; HELIUM, LIQUID.

Production of low temperature. Low temperatures such as those of liquid hydrogen (normal boiling point 20.4°K) and liquid helium (normal boiling point 4.2°K) are produced from the gas phase by the use of three mechanical devices: (1) the gas compressor or pump, (2) heat exchangers at several places along the path of flow of the gas, and (3) devices for expanding the gas such as a piston expander and a throttle valve. These are shown schematically in Fig. 1, where the flow of helium gas may be traced from the intake of the compressor, past the water-cooled heat exchanger, and into the liquefier.

A basic concept to grasp in understanding why the temperature drops when the gas is allowed to expand against the piston is that the process has occurred at constant entropy. Entropy S is a thermodynamic property of the system of gaseous molecules, and it is useful to think of its physical meaning as being that of the degree of disorder of the system. The entropy versus temperature plot in Fig. 2 shows, for the gas system, a series of constant-pressure curves. The thermodynamic process at the compressor (with water cooling to keep the temperature constant) is shown as the path from point α to the point β. The entropy decrease $\Delta S = \Delta Q/T$, where ΔQ is the heat removed at temperature T during this reversible process that is assumed to be ideal. The diagram also shows an expansion of the gas against the piston as the path from point β to the point γ. This is done at constant entropy so that one may think of the pressure and temperature variables as having shifted suitably to preserve the degree of disorder. Energy was removed from the gaseous system by the work it did on the piston. Figure 1 shows still another expansion at the throttle valve or Joule-Thomson valve at which a further temperature drop and some liquefaction may occur. The throttle valve expansion is at constant enthalpy and is somewhat more subtle. To sum up briefly, thermodynamic variables may be manipulated in a

straightforward manner to produce a cold temperature, for example, by liquid helium. By vigorous pumping on the liquid helium, the vapor pressure may be greatly reduced, and the temperature of the liquid becomes of the order 1°K. *See* ENTROPY.

To produce temperatures below 1°K, it is necessary to use a system which contains disorder (entropy) and to operate in such a way as to bring about order and so lower the entropy. Such a system is the set of electrons within a paramagnetic salt whose magnetic dipole moments may be oriented in a powerful magnetic field. The entropy versus temperature diagram (Fig. 3) represents a paramagnetic salt such as potassium chrome alum. The process is carried out in two steps represented by the path a to b and then b to c. In the first stage, a large magnetic field is isothermally applied to the salt, and the heat of magnetization is carried off to the surroundings, a liquid helium bath at about 1°K. Then the heat contact of the salt with the bath is removed, and the step b to c, the demagnetization caused by turning off the external magnetic field, is an adiabatic process in which the entropy of the salt remains constant. During this last stage, the magnetic energy density, $BH/2$ joules/m³, contained within the salt has moved out and dissipated as eddy current heating in the external magnet. The temperature has dropped to some 0.04°K, and the basic thermodynamic reasoning is similar to the expansion-engine cooling of a gas in Fig. 2 path β to γ. The order created by the powerful external field at 1°K is maintained by very weak magnetic fields within the salt because of the very substantial drop in temperature. Path b to c is a constant entropy process in which the degree of disorder remains constant among the magnetic dipoles. The lattice vibrations of the salt are already in their zero-point energy state, and there simply is not enough entropy left to the lattice system to alter the order of the electron dipole system. The lattice system cannot go into debt with a negative entropy, because such a thing is absurd. To sum up, by using statistical mechanics, the entropy is expressed by means of the partition function and the Helmholtz potential A as in Eqs. (1) and (2).

$$S = -\left(\frac{\partial A}{\partial T}\right)_B \tag{1}$$

$$A = -kT \ln \sum_i e^{-E_i/kT} \tag{2}$$

The energy levels E_i in the magnetic field B are integer multiples of the quantity βB, where β is the electron dipole moment. In Eq. (2) k is the Boltzmann constant. It can be shown that the factor in the exponential, $E_i/kT \approx \beta B/kT$, must remain constant if the entropy is to remain constant. If the large value of B drops to the small internal dipole-dipole magnetic field within the salt itself because of the demagnetization process, then the temperature T must become small in order to keep the above factor in the exponential a constant. This does not demand that the dipoles be aligned as in ferromagnetism, because they can become anti-parallel aligned as in antiferromagnetism and still maintain the constant entropy requirement. A discussion of the electron spin system and magnetic cooling naturally leads to the matter of nuclear orientation and further reduction in temperature.

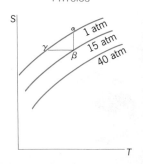

LOW-TEMPERATURE PHYSICS

Fig. 2. Entropy-temperature plot for helium gas.

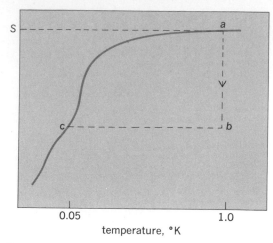

Fig. 3. Entropy-temperature diagram of potassium chrome alum, a paramagnetic salt.

He³-He⁴ dilution refrigerator. A method of producing very low temperatures by dissolving liquid He³ into liquid He⁴ was proposed by H. London, G. R. Clarke, and E. Mendoza in 1962, though it can be traced to earlier ideas by H. London. Experiments in Leiden (1965), England (1966), the Soviet Union (1966), and the United States (1966) led to dilution refrigerators capable of continuously achieving a temperature of 0.010°K. The He³-He⁴ dilution refrigerator has opened up exciting research possibilities, especially for nuclear polarization. It would be very helpful for the reader to become acquainted with the basic properties of the superfluid formed by the isotope He⁴ in which the He³ is dissolved. *See* SUPERFLUIDITY.

Figure 4a shows the dilution refrigerator used in experiments on nuclear polarization at Texas A & M University. The drawing in Fig. 4b indicates the important components of the He³-He⁴ dilution refrigerator. The refrigerator itself is surrounded by liquid helium of the normal and abundant isotope He⁴ at 4.2° K. There is also usually a shield of liquid nitrogen at 78°K. A special chamber (at top of drawing) contains liquid He⁴ at reduced vapor pressure and acts as a thermal dam holding a temperature of about 1°K. The dilution refrigerator suspended in the high-vacuum region below the 1°K thermal dam is to be viewed as a device into which liquid He³ is continuously injected and out of which an equivalent molar volume of He³ gas is extracted. The fact that the circulating He³ is contaminated by some He⁴ isotope need cause no confusion, but attention must be given to the heat load caused by the contamination.

The two chambers marked "still chamber" and "mixing chamber" are connected by a tube which also has the function of a heat exchanger. This tube doubles back on itself for efficient and essential operation of the device. When the refrigerator is put into operation, the still and mixing chambers are filled to the A level in the still with a mixture of 40% He³ and 60% liquid He⁴. The gas phase above the liquid in the still is very rich in isotope He³ because of its high vapor pressure relative to He⁴. When these vapors are pumped vigorously, the still begins to drop in temperature because of He³ evaporation. Because of the very high heat conductivity of superfluid liquid He⁴ along the connecting tube, the mixing chamber also drops in temperature. After a short time this cooling produces a temperature of about 0.6°K in the mixing chamber, and a solution very rich in He³ begins to separate out of the original mixture. Being less dense, the solution rich in He³ floats on top in the mixing chamber. The dotted line shows the separation, with nearly pure He³ on top of the now depleted He³-He⁴ mixture at the bottom of the mixing chamber. Figure 5 shows the phase separation diagram for He³-He⁴ mixtures. Remarkably, as T → 0°K, the lower phase, shown at the left side of Fig. 5,

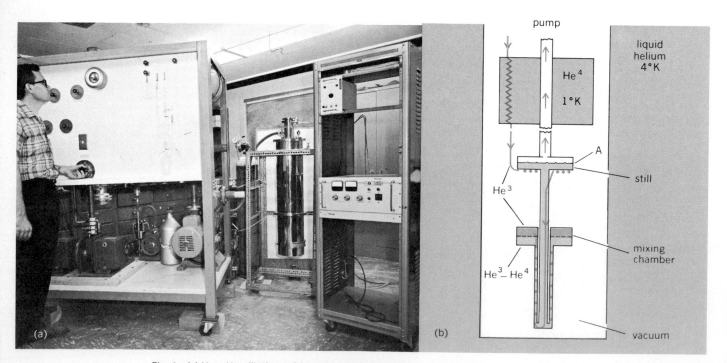

Fig. 4. (a) He³–He⁴ dilution refrigerator. (b) Schematic drawing of a. (*Robert A. Welch Foundation*)

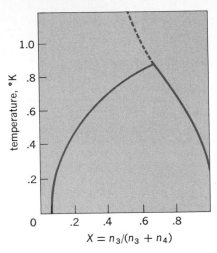

$$X = n_3/(n_3 + n_4)$$

Fig. 5. Phase separation diagram for He³ – He⁴ mixtures given by solid line such that at 0.6°K the heavy liquid will be at the bottom of the mixing chamber with about 3 atoms of He³ to 7 atoms of He⁴. The dotted line represents superfluid properties to the left and normal liquid properties to the right; n_3 and n_4 in the equation refer to the concentrations of He³ and He⁴, respectively.

continues to contain about 6% of He³ in the mixture. Thus even at a very low temperature of, say, 0.015°K, the following may happen in the mixing chamber: If He³ atoms are removed from the lower region via the heat exchanger and the still, then more He³ will dissolve into the He³-He⁴ mixture and the latent heat of dilution will produce refrigeration sufficient to balance heat leaks or other sources of heat and so continuously maintain a temperature of 0.015°K. Thus the refrigeration produced in the mixing chamber by dissolving liquid He³ into liquid He³-He⁴ mixture is the reason that the mixing chamber steadily reduced its temperature from the initial value of about 0.6°K established at the still.

All that remains is to trace the way in which the upper layer of He³ in the mixing chamber can be continuously replenished with nearly pure He³. The schematic diagram shows the supply line of He³ coming through temperature cooling stages until it is so thoroughly precooled that it is delivered to the mixing chamber at very nearly that chamber's final steady low temperature—say about 0.015°K. Of course, the incoming supply of He³ atoms is just the recirculated He³ atoms (contaminated by He⁴) which has been removed instants before from the still. Pumps shown in the photograph do all the external work of sucking on the still and compressing the He³ for the incoming supply line. Baths of liquid nitrogen and liquid helium are heat sinks for removing energy from the incoming He³. The entropy of a small part of a total system has been lowered in order to produce low temperatures. The entropy of the total system, of course, has increased. An excellent review by John C. Wheatley, in the bibliography, will give the advanced reader more scientific details; for example, the 6% dilution of He³ in He⁴ at 0°K is *not* contrary to the third law of thermodynamics. *See* THERMODYNAMIC PRINCIPLES.

Low-temperature phenomena. Consider a solid specimen of matter (for example, a crystal of LiF because the Li⁷ nucleus has a magnetic moment)

having a collection of nuclei with magnetic moment μ, which is taken as parallel to the spin angular momentum characterized by I. A considerable polarization will occur only if the energy μB is large compared to the energy kT because these quantities appear as they did in Eq. (2) as a factor of an exponential. The requirement is most demanding of even the best equipped low-temperature laboratory. Magnetic fields of 10 webers/m² must be combined with temperatures of 0.01°K before significant polarization of the nuclei will be obtained. An alternative approach to nuclear polarization is to obtain a larger magnetic field at the nucleus by using the intrinsic magnetic field of the atomic electrons (of the order 100 webers/m² at such close range). The interaction of these fields with μ gives rise to hyperfine structure in optical spectra. Thus at 0.01°K, a small external field of only 0.01 weber/m² can orient the electron spin system nearly to saturation value, and the field, cast on the nuclei by the spin system, is so large that significant nuclear polarization appears. An example of a paramagnetic salt whose nuclear moments may be polarized in this way are certain cobalt (II) salts. *See* MAGNETOCHEMISTRY.

Nuclear orientation becomes particularly interesting if the oriented nuclei are radioactive. Anisotropy of the intensity of α-, β-, and γ-radiation from oriented nuclei has been observed experimentally. These experiments have led to important conclusions in the field of theoretical physics of nuclear interactions. *See* PARITY (QUANTUM MECHANICS).

The properties of matter at low temperatures can be most surprising. There are a large number of metals which become superconductors at low temperatures such that the electric field **E** within the metal is zero. The magnetic field described by the induction vector **B** is also zero within the metal superconductor. Many of these same metals exhibit at low temperatures (and in the normal state) an unusual diamagnetism with a peculiar periodiclike dependence on the applied magnetic field. The effect was first noted by W. De Haas and P. van Alphen and has been extensively studied by D. Shoenberg. Magnetic nuclear resonance studies on a number of systems have been useful in understanding the properties of matter at low temperature.

N. Osheroff, T. Richardson, and P. Lee discovered the superfluid phases of liquid He³ at temperatures below 2.6 millikelvins (mK) and at pressures just below the approximate 30 bar (3×10^6 Pa) needed to push the liquid into the solid state. Under zero pressure the superfluid transition occurs at 0.93 mK. To obtain these low temperatures below 3 mK, new methods had to be developed, such as adiabatic compression of the liquid-solid mixture of He³ along the melting curve (Pomeranchuk's method) and adiabatic demagnetization of both electronic and nuclear spins, as discussed above. *See* ABSOLUTE ZERO; CRYOGENIC ENGINEERING; HELIUM, LIQUID; LIQUEFACTION OF GASES; LOW-TEMPERATURE THERMOMETRY; REFRIGERATION; STATISTICAL MECHANICS; SUPERCONDUCTIVITY.   [CHARLES F. SQUIRE]

Bibliography: S. Fluegge (ed.), *Handbuch der Physik*, vols. 14 and 15, 1956; C. J. Gorter (ed.) *Progress in Low Temperature Physics*, vols. 1 and 2, 1956, 1957, and vol. 6, 1970; O. V. Lounasmaa,

Solid helium-three, *Contemp. Phys.*, 15:373–422, 1974; O. V. Lounasmaa, *Experimental Principles and Methods below 1 K*, 1974; J. C. Wheatley, Solid helium-three, *Rev. Mod. Phys.*, 47:415–470, 1975.

Low-temperature thermometry

The measurement of temperatures extending approximately from 90 K to 0.001 K and in a very few instances to 0.000001 K. Although the diverse thermometers used for measuring temperatures in this region tend to be somewhat unconventional, they nevertheless meet the three main criteria required of all thermometers: reproducibility, a monotonic variation with temperature (preferably according to some simple law) of the thermometric property being measured, and intimate thermal contact with the body or material for which temperature information is desired. Additional conditions, such as high sensitivity and independence of the thermometer readings from magnetic fields, may also be imposed for special applications. Factors such as these, as well as the degree to which the instrument must satisfy the three basic criteria mentioned above, determine the selection of the thermometer.

For accurate work, low-temperature thermometers must be calibrated, usually before each use, either at a number of well-known fixed and reproducible temperature points or against a suitable standard thermometer. Accurate and relatively simple standard secondary thermometers, which have themselves been calibrated against a primary standard, are readily available. The primary standard serves, as its name implies, chiefly to establish secondary thermometer temperature scales, improve existing ones, and accurately determine useful fixed points of temperature, for example, the triple point of hydrogen.

Hence, low-temperature thermometry is resolved into, first, the basic problem of assigning numbers on the absolute temperature scale to achievable and reproducible low-temperature states, and second, the choice and the calibration of suitable instruments for the practical measurement of low temperatures.

Temperatures on the absolute or thermodynamic scale are defined in terms of the second law of thermodynamics by the equation $T = dQ/dS$. The absolute temperature of any system is thus given by the limiting value of the reversible heat absorbed by the system per unit change of its entropy. To determine temperatures on this basis would in general be a formidable task, however. Fortunately, it can be shown that valid absolute temperatures may also be obtained by other means, for example, from measurements with a gas thermometer. Although a laborious technique, gas thermometry is ordinarily much simpler than a temperature determination based upon reversible heat and entropy measurements. *See* ENTROPY.

Gas thermometry above 1 K. Gas thermometry depends upon the concept of an ideal gas, for which it may be shown that the pressure times the volume equals a constant times the temperature, or $PV = RT$. Hence, if a definite amount of such a gas were enclosed in a fixed and known volume in thermal equilibrium with that object whose temperature is to be determined, measurement of the gas pressure would be sufficient for specification of the temperature, provided the value of the constant in the equation relating P, V, and T were known. The value of this constant is in a sense arbitrary because the number assigned it merely fixes the size of the degree. Use of a single value for this quantity is an obvious desideratum, however, and to assure this it has been decided by international agreement to assign to the triple point of water the absolute temperature 273.16 K, thus defining the Kelvin (K). A measurement of the gas thermometer pressure at this relatively easily reproduced fixed point of temperature then defines the constant for each gas thermometer. The following fundamental difficulty remains. Although certain gases are more nearly perfect than others, an ideal gas as such does not exist. For practical measurements, it is therefore necessary to modify the simple procedures outlined above to correct for the nonideal character of real gases. In addition, many other precautions must be taken to ensure reliable results. The extent to which the measurements have been carefully made and properly corrected determines either the validity of the absolute temperatures which can be assigned low-temperature reference states or the accuracy of appropriate parameter values of standard secondary thermometers. Below 1 K, however, gas thermometers become useless because the maximum pressure at which a gas thermometer can be operated must, in order to minimize nonideality corrections, always be appreciably less than the vapor pressure of the gas being used, and at 1 K the vapor pressure of all gases either is already, or is rapidly becoming, negligible.

Adiabatic demagnetization below 1 K. Since the measurement of temperatures below 1 K is inextricably related to the production of these low temperatures, these two subjects will be jointly discussed. The essential condition for the production of low temperatures is a large entropy resident in the system to be cooled. An isothermal reduction of the system entropy by an appropriate parameter change, followed by an isentropic reversal of the same parameter change to its original value, always results in a lowering of the temperature. At temperatures of 1 K and below, the entropy changes accompanying any parameter change associated with a gas, such as the pressure, become negligible, and no further temperature reduction is possible. To continue down the temperature scale, a different physical system is required, possessing at these temperatures large amounts of entropy. Since the entropy of all systems decreases with temperature, however, there are few such systems available. In 1933, W. F. Giauque and P. Debye independently suggested that paramagnetic salts might satisfy the above requirement, on the premise that the random orientation of the electron magnetic moments of the paramagnetic ions in the crystal would contribute, even at temperatures of the order of 1 K, an appreciable amount of entropy to the system. If such a system is magnetized isothermally at 1 K, the elementary magnets will be aligned parallel to the field direction and the system entropy correspondingly decreased. If the system is then thermally isolated and demagnetized, it will cool isentropically to that temperature which would itself have been sufficient in zero field to have aligned the spins to the same extent as did the isothermal magnetization. This is the technique of low-temperature production by adiabatic demagnetization. *See* CRYOGENICS.

In order to measure the temperatures produced by this procedure, a different kind of thermometer is required. Again, fortunately, the absolute temperature of an ideal paramagnetic salt is related simply to another easily measurable quantity, its susceptibility; and just as in the case of gas thermometry, this parameter may be used to establish a provisional temperature scale for use below 1 K. This relationship, $T = \text{const}/\chi$, where χ is the magnetic susceptibility of the salt and the constant is determined at some known "high" temperature established by gas thermometry, is known as the Curie law. As in the case of a gas thermometer,

no ideal paramagnetic exists any more than does an ideal gas, and therefore, to obtain reliable temperature measurements from 1 to about 0.1 K, small corrections for nonideality must be applied. At lower temperatures, these corrections become less and less certain, and eventually completely unreliable.

To establish correct absolute temperatures in this range, it has proved necessary to utilize the basic thermodynamic definition of temperature given at the beginning of this article. This is done in the following way. Starting at an absolute temperature accurately known by gas thermometry, for example, T_0, the susceptibility of a paramagnetic salt is measured in zero field. From this measurement the Curie law constant is deter-

Low-temperature thermometers

Temperature-measuring instrument	Applicable temperature range, K	Approximate precision at low-temperature limit, K	Remarks
Thermocouples			Simple construction, low cost, wide range, small size, ease of installation, rapid response, no heat introduction at point of measurement; careful selection of wires necessary to avoid spurious thermal electromotive force because of inhomogeneities in lead-in wires; rapid loss of sensitivity occurs with decreasing temperature; calibrations stable except at lowest temperatures
Chromel/constantan	30–1270	0.05	Sensitivity drops off rapidly below 30 K
0.02, 0.03, or 0.07 at. % Au in Fe/Chromel-P	1–300	0.01	Sensitivity of $10\mu V/K$, or better, obtainable over most of range; batch-to-batch reproducibility of alloy about 9%, but calibration data can be scaled from a few points on a given batch; 0.02% and 0.07% alloy available in United States; 0.03% alloy obtainable in Europe
Resistance thermometers			Normally conducting metals, partial superconductors, and semiconductors have been used; all introduce heat at point of measurement
Platinum	4–1100	0.01	Standard interpolation device between fixed points defining the International Practical Temperature Scale of 1968 from 903.15 K to 13.81 K; from 90 to 4 K can be accurately calibrated; calibration stable with respect to thermal cycling if Pt is extremely pure and if wires supported in strain-free configuration; at 20 K and above, sensitivity is of order of 10^{-3} degree; Pt resistance thermometers rapidly lose sensitivity below this temperature.
Leaded phosphor bronze	0.03–7	0.001	Sensitivity has been extended to 10^{-5} degree; measured resistance dependent upon measuring current and upon magnetic field; magnetic field dependence can be reduced by addition of bismuth; difficult to reproduce same temperature coefficient of resistivity in successive wire samples
Carbon	0.1–20	0.001	Highly sensitive, capable of measuring temperature changes of the order of 10^{-5} degree; resistance dependent upon measuring current, but somewhat independent of magnetic field
Germanium	0.03–100	0.0002	Highly sensitive, capable of measuring temperature changes of the order of 10^{-6} degree; highly stable on cycling to room temperature; often used as a secondary standard
Capacitance thermometers	70–400 0.005–60	0.01 0.001	Strontium titanate sensor; capacitance versus T has maximum near 60 K, minimum near 0.070 K, and is linear in region 1–6 K; calibration unaffected by magnetic fields up to 15 tesla or more

Low-temperature thermometers (cont.)

Temperature-measuring instrument	Applicable temperature range, K	Approximate precision at low-temperature limit, K	Remarks
Vapor-pressure thermometers	Triple point to critical point of all liquefied gases	Approx. 0.001	Secondary standard; sensitive, slow reading; following conditions must be fulfilled to ensure reliable results: uniform liquid temperature, uniform and known liquid composition (for example, ortho-para composition in liquid hydrogen), accurate pressure measurement; simple vapor pressure thermometers can be constructed which are rugged, reliable, and sensitive
Gas thermometers Helium	1–800	0.001	Primary standard; requires extreme care in order to obtain accurate results; any gas whose behavior is almost ideal may be used in a gas thermometer; at low temperatures, this requirement restricts permissible gases to hydrogen and helium, the latter being only one suitable for use at lowest temperatures; crude gas thermometers have been constructed, using rough but sturdy pressure indicators to provide reliable low-temperature indicators for engineering uses (gas liquefiers)
Magnetic thermometers	For all temperatures below 1 K (have been used for special purposes at higher temperatures)	Depends upon temperature range; in liquid helium range precision about 0.001°	Secondary standard; different paramagnetic salt samples may require individual calibrations; salts are unstable and must be carefully handled to ensure constancy of chemical composition; useful nevertheless from 0.1 to 1 K using a calibration determined either theoretically or experimentally; below 0.1 K, the susceptibility serves as a thermometric parameter
Electronic paramagnetic materials such as chromium methylamine alum	$10^{-3}-1$		
Nuclear paramagnetic materials such as copper	$10^{-6}-10^{-3}$		

LOW-TEMPERATURE THERMOMETRY

(a)

(b)

(c)

Determination of the thermodynamic temperature below 1°K. (a) Measurement of S as a function of T^*. (b) Measurement of Q as a function of T^*. (c) Slope of Q-S curve gives T.

mined. Then, while the system is maintained at this temperature, the magnetic field is arbitrarily increased to some value, for example, H_1. The entropy of the salt will decrease by an amount accurately calculable (at these temperatures) by a rather complicated but nonetheless tractable theoretical formula. Assume this decrease in entropy to be ΔS_0. If the paramagnetic salt is now thermally isolated and slowly demagnetized, an isentropic process will be carried out as described above, and the temperature will fall. The susceptibility at this new low temperature can be measured to provide a provisional temperature $T_1^* = C/\chi_1$. Next, a measured amount of heat ΔQ_1 is added to the isolated specimen to return its temperature to T_0, as measured by its susceptibility. If this process is repeated, starting now from T_1^* and going to T_2^*, and so on, the variation both of S and Q with T^* can be obtained. T^* can then be eliminated from these relationships to give S as a function of Q. The slope of this curve then yields the correct absolute temperature to be associated with each Q and S, and from this information, the proper corrections to the provisional temperatures as obtained from susceptibility measurements can be made. In the illustration, curves which are typically obtained in such measurements are presented.

Low-temperature thermometers. Low temperatures usually derive from the presence of a liquefied gas or a demagnetized paramagnetic salt, each of which contains a built-in temperature measurement capability in the form of a vapor pressure and a magnetic thermometer, respectively. If, however, the low temperature is produced by a refrigerator, or if use of the thermometers mentioned above is proscribed by some special requirement, other types of thermometers must be utilized. A variety of these are available, the differing properties of which enable an optimum matching of thermometer to application. In the table, some examples of the more successful low-temperature thermometers are listed, together with brief comments on their advantages and limitations. In addition, several specialized techniques are being developed for thermometry at very low temperatures, including acoustic thermometry using He^4 gas, for the range 2 to 30 K; the recoil-free emission and absorption of γ-rays (Mössbauer effect), from about 0.010 to 0.100 K; anisotropic emission of γ-rays or β-rays from oriented radionuclei, from 0.002 to 0.050 K; noise thermometry, from 0.005 to 300 K; the pressure-temperature relation for the melting curve of solid ^{3}He, from 0.002 to 0.35 K. All of these methods can, in principle, serve as absolute or primary thermometers.

See TEMPERATURE MEASUREMENT.

[EDWARD F. HAMMEL]

Bibliography: C. M. Herzfeld (ed.), *Temperature: Its Measurement and Control in Science and Industry*, vol. 3, pts. 1 and 2, 1962, pt. 3, 1963; L. C. Jackson, *Low Temperature Physics*, 4th rev. ed., 1955; N. Kurti, Magnetism at very low temperatures and nuclear orientation, *Nuovo cimento*, suppl. 6(3):1101, 1957; B. Le Neindre and B. Vodar (eds.), *Experimental Thermodynamics*, vol. 2, 1975; H. H. Plumb (ed.), *Temperature: Its Measurement and Control in Science and Industry*, vol. 4, 1972; H. C. Wolfe (ed.), *Temperature, Its Measurement and Control in Science and Industry*, vols. 1 and 2, 1941, 1955.

Lubricant

A gas, liquid, or solid used to prevent contact of parts in relative motion, and thereby reduce friction and wear. In many machines, cooling by the lubricant is equally important. The lubricant may also be called upon to prevent rusting and the deposition of solids on close-fitting parts.

Liquid hydrocarbons are the most commonly used lubricants because they are inexpensive, easily applied, and good coolants. In most cases, petroleum fractions are applicable, but for special conditions such as extremes of temperature, select synthetic liquids may be used. At very high temperatures, or in places where renewal of liquid lubricants is impossible, solid lubricants (graphite or molybdenum disulfide) may be used.

Petroleum lubricants. Crude petroleum is an excellent source of lubricants because a very wide range of suitable liquids, varying in molecular weight from 150 to over 1000 and in viscosity from light machine oils to heavy gear oils, can be produced by various refining processes. Studies carried out by the American Petroleum Institute and others have established that crude petroleum fractions consist of saturates (normal, iso-, and cycloparaffins); monoaromatics which may contain saturated rings as well as saturated side chains; substituted polyaromatics; hetero compounds containing sulfur, nitrogen, and oxygen; and asphaltic material made up of polycondensed aromatics and hetero compounds. Of these, the wax-free saturates and monoaromatics are desired in the finished oil in order to obtain the desired viscometric properties and oxidation and thermal stability. So far as possible, the other types of compounds are generally removed. Classically, only those crudes (so-called paraffinic crudes) relatively rich in the desired type of hydrocarbons were used in lubricating oil manufacture. By atmospheric distillation, neutrals were produced as distillates and the residual fractions were treated with activated clays, which removed asphaltic and hetero material to produce bright stock.

In modern refining, vacuum distillation removes the desired hydrocarbons from the asphaltic constituents (which may be present in amounts up to 40% of the total) and gives oils in the required viscosity and boiling range. Extraction of the distilled fractions with solvents such as liquid sulfur dioxide, furfural, and phenol permits the removal of the polyaromatic and hetero compounds to improve viscosity-temperature characteristics (viscosity index, VI) and stability of the oil. Dewaxing removes the high-melting paraffins.

Viscosity of oils for various applications

Application	Viscosity in centistokes at 25°C (77°F)	Primary function
Engine oils		Lubricate piston rings, cylinders, valve gear, bearings; cool piston; prevent deposition on metal surfaces
SAE 10W	60–90	
SAE 20	90–180	
SAE 30	180–280	
SAE 40	280–450	
SAE 50	450–800	
Gear oils		Prevent metal contact and wear of spur gears, hypoid gears, worm gears; cool gear cases
SAE 80	100–400	
SAE 90	400–1000	
SAE 140	1000–2200	
Aviation engine oils	220–700	Same as engine oils
Torque converter fluid	80–140	Lubricate, transmit power
Hydraulic brake fluid	35	Transmit power
Refrigerator oils	30–260	Lubricate compressor pump
Steam-turbine oil	55–300	Lubricate reduction gearing, cool
Steam cylinder oil	1500–3300	Lubricate in presence of steam at high temperatures

If the viscosity index of the oil is not important for a particular application, the most reactive polyaromatics and hetero compounds may be removed by treatment with concentrated sulfuric acid. In either case, the oil is treated finally with an active clay to remove trace amounts of residual acids and resins (heterocyclics).

Small amounts of heterocyclics, such as substituted benzthiophenes, quinolines, and indoles, may remain in the finished oil. The sulfur-containing compounds serve the useful purpose of acting as natural antioxidants. The nitrogen compounds may be harmful in certain applications because of their propensity to form deposits on hot surfaces and are generally reduced to low concentrations.

The table lists typical specification data for lubricants in several applications. Viscosity is a determining factor in lubricant selection. Machines are generally designed to operate on the lowest practicable viscosity, since the lighter fluids give lower friction and better heat-transfer rates. However, if loads or temperatures are high, more viscous and less volatile lubricants are required. Change of viscosity with temperature is often of considerable practical importance and is customarily expressed in terms of viscosity index, an arbitrarily chosen scale on which an oil from a Pennsylvania crude, high in saturate and monoaromatic content, is assigned a value of 100, and those containing a relatively large amount of cyclohydrocarbons, both paraffinic and aromatic (from naphthenic crudes) are in the 0–50 viscosity-index range. As already described, the refiner can produce oils of high viscosity index from naphthenic stocks, but yields may be low.

Multigrade oils. In order to standardize on nomenclature for oils of differing viscosity, the Society of Automotive Engineers (SAE) has established viscosity ranges for the various SAE designations (see table). By the use of relatively large amounts of additives for improving viscosity index, it is possible to formulate one oil which will fall within the range of more than one SAE viscosity grade, the so-called multigrade oils, illustrated by Fig. 1. Since all viscosity-index improvers also increase viscosity, it is necessary to use a base oil of low

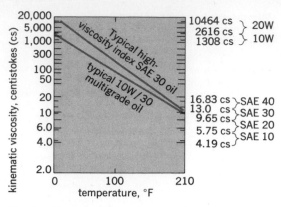

Fig. 1. Viscosity-temperature relationships indicated for additive-thickened multigrade oils.

viscosity to formulate such oils. Since the viscosity-increasing effect of the additive decreases with increase in shear rate, an artificially thickened oil of this type behaves as a light oil in engine parts under high shear, and thereby friction is low, but acts as a heavier oil in low-shear regions or at higher temperatures. However, volatility and ability to protect high shear parts from rubbing set limits to the use of light oils in these formulations.

Additives for lubricating oils. It is often desirable to add various chemicals to lubricating oils to improve their physical properties or to obtain some needed improvement in performance.

Viscosity-index improvers. The fall in viscosity with increase in temperature of oils of a given grade, or viscosity level, can be made less steep by thickening lighter oils with polymeric substances such as polybutenes and copolymers of polymethacrylates. The polymer may increase in solubility as the temperature increases, and correspondingly, the molecules uncoil and thicken the base oil more at high than at low temperatures and thus counteract, to some extent, the natural decrease in viscosity the base oil would undergo (Fig. 1).

Pour-point depressants. The dewaxing process removes the higher melting hydrocarbons, but some components remain which may solidify and gel the oil, and thus reduce its fluidity at low temperatures. Small amounts of chemicals, such as metallic soaps, condensation products of chlorinated wax and alkyl naphthalenes or phenols, polymethacrylates, and a host of others, increase fluidity at low temperatures. The mechanism of their action is still uncertain, but it probably involves adsorption of additive molecules on the surface of the growing wax crystals. The adsorbed layer either reduces intercrystal forces or modifies crystal growth.

Antioxidants. Lubricants are exposed to oxidation by atmospheric oxygen in practically all of their applications. This results in formation of acids and sludges which interfere with the primary function of the lubricant. Substantial increases in service life can be obtained by using small amounts (0.1–1.0%) of antioxidants. For lighter, more highly refined oils such as steam turbine oils, hindered phenols such as dibutyl-*p*-cresol are very effective. In heavy lubricants, as engine oils, amines (phenyl-α-naphthylamine), metal phenates (alkali-earth salts of phenol disulfides), and zinc salts of thiophosphates and carbamates are used.

Antiwear and friction-reducing additives. If sliding surfaces can be completely separated by an oil film, friction and wear will be at a minimum. There are many systems, however, in which the combination of component geometry and operating conditions is such that a continuous oil film cannot be maintained. If no separating film of any kind were interposed, however, complete seizure would result. If the pressures and temperatures between such contacting surfaces are moderate, the provision of a boundary lubricating film will suffice, whereas if, as in some gears, conditions of both temperature and pressure are severe, some form of extreme pressure (EP) lubrication may be necessary.

The form in which wear manifests itself in machine components varies widely with the conditions, from the catastrophic welding together of gear-tooth surfaces to the slow continuous removal of material from an engine cylinder; even within the same mechanical system, it may change profoundly with changes in operating and environmental conditions.

In internal-combustion engines operating at low cylinder temperatures, for instance, condensation of acids from the gaseous combustion products on the cylinder and ring surfaces results in corrosive wear; in such cases, the addition of alkaline-earth phenates to the lubricating oil to neutralize the acids has succeeded in reducing wear rates.

Extreme pressure additives. Certain types of gears, particularly the hypoids used in automotive rear-axle transmissions, operate under such severe conditions of load and sliding speed, with resulting high temperature and pressure, that ordinary lubricants cannot provide complete protection against metal contact; this leads to welding, transfer of surface metal, and ultimate destruction of the gears. Also, in certain machining operations, it is necessary to prevent the chip from welding to the cutting tool. For such applications, lubricants containing sulfur and chlorine compounds are used. At the temperatures developed in the contact, these react chemically with the metal surfaces; the resulting sulfide and chloride films provide penetration-resistant, low-shear-strength films which prevent damage to the surfaces. Care in formulating these lubricants must, however, be exercised to ensure that corrosion of metal at normal temperatures does not occur. Figure 2 shows the damaging results which may occur in the absence of such protective additives.

Dispersants. In internal-combustion engines, some of the products of combustion, which include carbonaceous particles, partially burnt fuel, sulfur acids, and water, enter the oil film on the cylinder walls. These materials may react to form lacquers and sludges which are deposited on the working parts of the engine. The lubricant is called upon to keep engine parts (oil screens, ring grooves) clean, and this is accomplished by the use of so-called detergent oils (Fig. 3).

The most commonly used dispersants are alkaline-earth salts. Of these, the most successful have been the salts of petroleum sulfonic acids, phenates, salicylates, thiophosphates, and oxidized olefinphosphorus pentasulfide reaction products. These materials owe their effectiveness partly to their surface activity, which ensures that foreign particles are kept in suspension in the lubricating

LUBRICANT

new gear

scoring

plowing

Fig. 2. Gear failures.

oil, and partly to their alkalinity, which enables them to neutralize combustion acids that would otherwise catalyze the formation of lacquers.

Polymeric additives have been introduced for dispersancy. They are copolymers of a long-chain methacrylate, for example, lauryl methacrylate, and a nitrogen-containing olefin, such as vinyl pyrrolidone. These nonash dispersants are superior to the metal-containing additives for low-temperature operation and can be used very effectively in motor oils. Because they are less alkaline than metallic additives, the polymeric additives are less effective in diesel engines whose fuels often lead to combustion products which are much more acid than those from gasoline.

Boundary lubricants. Boundary conditions are encountered in many metal-forming processes in which the pressures required to deform the metal are too high to allow an oil film to form. In such applications, fatty oils, such as palm oil, or lubricants containing fatty materials are employed to reduce the friction and wear; the fatty acids react with the metal surface to form a tenacious soap film which provides lubrication up to temperatures near the melting point of the soap, usually about 120°C. *See* MACHINABILITY OF METALS.

Synthetic oils. During World War II, synthetic lubricants were extensively employed by Germany as substitutes for mineral lubricants which were in short supply. Some of these, such as the ester oils, were in fact superior to mineral lubricants in some applications. Since the war, the use of synthetics for special applications, where their performance justifies the higher costs, has steadily increased. Their main advantage is that they have a greater operating range than a mineral oil (Fig. 4). Esters such as 2-ethylhexyl sebacate, containing oxidation inhibitors and sometimes mild extreme pressure additives, are used as lubricants for aircraft jet engines. Silicones, although ideal with regard to thermal stability and viscosity characteristics, are poor lubricants for steel on steel, and this has restricted their use, although they are invaluable in some applications. Another class of widely used synthetic oils are the polyglycols, such as polypropylene and ethylene oxides. These polymers are available in a wide molecular-weight range, and they vary considerably in solubility in water and hydrocarbons. Thus, water-base lubricants may be formed which are used when a fire hazard exists.

Solid lubricants. The most useful solid lubricants are those with a layer structure in which the molecular platelets will readily slide over each other. Graphite, molybdenum disulfide, talc, and boron nitride possess this property.

The principal difficulty encountered with the use of solid lubricants is that of maintaining an adequate lubricant layer between the sliding metal surfaces. If the solid lubricant is applied as a suspension in a fluid, there is a tendency for the particles to settle out, and they may not reach the region where they are required. If they are applied as a thick paste to overcome the tendency to settle out, it is frequently difficult to force the paste through the narrow clearances between the sliding surfaces. A third method is to pretreat the surfaces with a relatively thick film of solid lubricant suspended in a resin and bake it on. Difficulties may then arise through progressively increasing clearances as the film wears, and there is no method for its continuous renewal. Materials such as graphite and molybdenum disulfide oxidize quite rapidly in air at 400°C. Unless air can be excluded, therefore, alternative solid lubricants will be required at temperatures greatly in excess of this.

A unique type of solid lubricant is provided by the plastic polytetrafluoroethylene (PTFE). At low speeds, PTFE slides on itself or on metals with a coefficient of friction in the lowest range observed for boundary lubrication (about 0.05). The friction rises to values common for other plastics after the sliding speed passes a critical value, and this high friction tends to persist when returning to low speeds.

It is likely that this is a thermal effect, the incidence of which is promoted by the extremely low thermal conductivity of PTFE, and it can be combated to a considerable extent by using the PTFE in a copper matrix. The use of such a matrix also reduces the wear which would occur with PTFE alone. This type of structure is used as a self-lubricated bearing material.

Greases. A lubricating grease is a solid or semifluid lubricant comprising a thickening (or gelling) agent in a liquid lubricant. Other ingredients imparting special properties may be included. An important property of a grease is its solid nature; it has a yield value. This enables grease to retain itself in a bearing assembly without the aid of expensive seals, to provide its own seal against the ingress of moisture and dirt, and to remain on vertical surfaces and protect against moisture corrosion, especially during shut-down periods.

The gelling agents used in most greases are fatty acid soaps of lithium, sodium, calcium, and aluminum. Potassium, barium, and lead soaps are used occasionally. The fatty acids employed are oleic, palmitic, stearic, and other carboxylic acids derived from animal, fish, and vegetable oils. The use of soaps in greases is limited by phase changes (or melting points) that occur in the range 50–200°C. Because many applications require operation above this temperature, greases have been introduced which are thickened by high-melting solids such as clays, silica, organic dyes, aromatic amides, and urea derivatives. In these cases, the upper limit of application is set by the volatility or by the oxidation and thermal stability of the base oil. To extend the operating range to extremes of high and low temperatures, synthetic oils may be used as the substrate. Silicones, esters, and fluorocarbons have all been employed for this pur-

(a)

(b)

Fig. 3. Effect of dispersant on cleanliness. (a) Ordinary oil. (b) Oil containing dispersant additive.

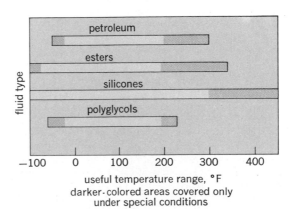

Fig. 4. Useful temperature range of synthetic oils.

pose. As in the case of liquid lubricants, various additives such as oxidation inhibitors, EP agents, and antirust additives are frequently used in greases. Volatility of the base oil is particularly important in greases, because it is constantly exposed as a thin film on the bearing surfaces.

Structure of greases. The gelling agent in most soap-base greases is present as crystalline fibers having lengths of $1-100\ \mu$ and diameters $1/10-1/100$ of their lengths. The fibers thicken the oil by forming a network or brush-heap structure in which the oil is held by capillary forces. For a given concentration of soap, the larger the length-to-diameter ratio of the fibers, the greater is the probability of network junction formation and the harder is the grease. However, not all soaps form such crystalline fibers; aluminum soap is a notable exception. There is evidence that aluminum soap molecules form a network akin to that formed by polymer molecules.

Nonsoap thickeners are present as small isometric particles much smaller than $1\ \mu$ in linear size. Silica particles are round, and some clay particles are plates. Forces of interaction between silica, clay particles, and oil are easily destroyed by water, so it is necessary to waterproof the primary particles by special treatments. Electron micrographs of various types of greases are shown in Fig. 5.

Mechanical properties of greases. When the applied stress exceeds the yield stress, the grease flows and the viscosity falls rapidly with further increase of stress until it reaches a value only a little higher than that of the base oil. This fall in viscosity is largely reversible, since it is caused by the rupture of network junctions which, following the release of stress, can reform. However, continued vigorous shearing usually results in the rupture of fibers and permanent softening of the grease. Here, certain greases such as those derived from lithium salts of hydroxy acids are superior to other types, although the manner of grease preparation exercises considerable influence.

In a major grease application, for example, ball and roller bearings, shearing between the races and rolling elements is extremely severe. The reason that a good bearing grease does not soften and run out lies in the fact that only a very small fraction of the grease put in the bearing is actually subjected to shearing. As soon as a freshly packed bearing is set in motion, most of the grease is redistributed to places where it can remain static in the cover plate recesses and attached to the bore of the cage between the balls or rollers. The small amount of grease remaining on the working surfaces of the bearing is quickly broken down to a soft oily material which lubricates these parts, and because it has only a low viscosity, it develops very little friction.

A number of methods exist for the measurement of yield stress and viscosity at various temperatures. One of the oldest of grease tests, still universally used, is a consistency (hardness) test in which the depth of penetration of a cone of standard weight and dimensions is determined. However, few of the many tests used are helpful in predicting the performance of a grease. Consequently, in development work, recourse is made to actual performance data and to the use of rigs that simulate field conditions. *See* BEARING, ANTIFRICTION; GEL; PETROLEUM PROCESSING.

[ROBERT G. LARSEN]

Bibliography: A. A. Bondi, *Physical Chemistry of Lubricating Oils*, 1951; C. J. Boner, *Gear and Transmission Lubricants*, 1964; E. R. Braithwaite, *Solid Lubricants and Surfaces*, 1963; F. D. Rossini and B. J. Mair, *Composition of Petroleum*, Advances in Chemistry Series, vol. 5, 1951; H. A. Van Westen and K. Van Nes, *Aspects of the Constitution of Mineral Oils*, 1951; H. H. Zuidema, *Performance of Lubricating Oils*, 2d ed., 1959.

Lubrication, engine

Continuous supply of a viscous film between moving surfaces in an engine during its operation. Relative motion between unlubricated surfaces of engine parts would result in excessive wear, overheating, and seizure, with destruction of the contacting surfaces. The energy loss associated with the wear and overheating would be evident in reduced engine power and efficiency. The purpose of lubrication is to minimize these effects by interposing a viscous fluid called the lubricant between the parts. Hydrodynamic forces are produced in the lubricant to keep the surfaces separated, even though heavy loads are tending to bring them into contact. The thickness of the separating film, for a given part geometry, depends upon the relative velocity of the two surfaces, the absolute viscosity of the lubricant, and the load per unit of surface area.

Lubricant. Engines are generally lubricated with petroleum base oils containing chemical additives to improve their natural properties. Refining methods and composition may be varied to suit special engine design operating conditions. Perhaps the most important oil property is the absolute viscosity, which is a measure of the force required to move one layer of the oil film over another. With low viscosity a protecting oil film between the parts is not formed. With high viscosity, too much power is required to shear the oil film; in addition, the flow of oil through the engine is retarded.

The viscosity of oil tends to decrease as its temperature is raised. It is therefore difficult to maintain the proper viscosity as the engine warms up, or as other operating conditions change. Viscosity index is a number which indicates the resistance of an oil to changes in viscosity with temperature. The smaller the change in viscosity with temperature, the higher is the viscosity index of the oil.

Lubrication system. The function of the lubrication system is to supply clean oil cooled to the proper viscosity to critical points in the engine, where the motion of the parts produces hydrodynamic oil films to separate and support the various rubbing surfaces. In all but the most primitive lu-

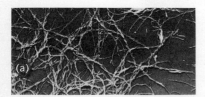

Fig. 5.   Electron micrograph of (*a*) soap-thickened and (*b*) dye-thickened grease.

brication systems, the oil, which has been heated by friction and by heat transfer from the hot engine parts, is collected, cooled, and pumped back into the engine. In automobiles the oil returns by gravity to the crankcase. The air motion over the outside of the crankcase is usually sufficient to cool the oil. It is then pumped from the crankcase back to the oil distributing galleries in the engine. In high-output engines the oil is pumped from the crankcase into a storage reservoir and then through an air- or water-cooled heat exchanger on the way back to the engine.

A positive oil distribution method called force-feed or full-pressure lubrication is used in most engines (see illustration). In this system the pump forces the oil at considerable pressure through passages to each main crankshaft bearing. The crankshaft is drilled to convey the oil from the main bearings to the connecting rod bearings. Sometimes the connecting rods are drilled to carry the oil to the wrist pins. Except in the largest engines, the cylinder walls are lubricated by leakage spray from the ends of the connecting rod bearings. In the force-feed system the pump pressure ensures a steady replacement of the oil which has been heated by friction work, with fresh cool oil from the crankcase or reservoir. In this way the temperature, and therefore the working viscosity of the oil in the bearings, is controlled.

Wear. Since the lubricating oil flows to all parts of the engine, it is important that it does not carry abrasive or corrosive material with it. Such material may come from the combustion of the fuel, from dirt in the inducted air, or from parts of the engine itself. It is common practice to filter all or part of the oil as it flows through the system, to reduce wear. In some engines a small fraction of the oil leaving the pump is continually bypassed through a filter and returned to the crankcase. The rest of the oil flows directly to the bearings. Since considerable time is required for all of the oil to be filtered by this method, most automobile engines pump all of the oil through a full-flow filter placed in the line between the pump and the bearings.

A large part of the cylinder wear occurs during starting and before the engine has warmed up. Cold oil, being quite viscous, is slow to reach the bearings, and the end leakage spray onto the cylinder walls is greatly reduced. At low engine temperatures, there is an increased tendency for corrosive products of combustion to condense on the cylinder walls. Chemical additives are used in modern engine oils to increase the viscosity index, reduce corrosive action, wash away accumulations of sludge or other deposits on engine parts or in oil passages, and improve oil performance in various other ways.

Excessive use of oil may be caused by wear of rings or cylinder bore, oil of improper viscosity, leakage past the shaft seals, or loss of vapor from the crankcase breather. By design and maintenance, oil consumption can be made small. The principal objection to high oil flow into the combustion space is smoky exhaust, fouling of spark plugs, and carbon deposits in the cylinders. *See* INTERNAL COMBUSTION ENGINE; LUBRICANT.

[AUGUSTUS R. ROGOWSKI]

Bibliography: W. H. Crouse, *Automotive Fuel, Lubricating, and Cooling Systems*, 3d ed., 1967;

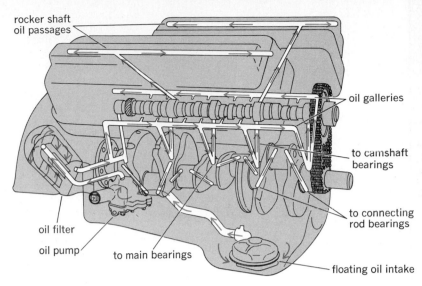

Full-pressure lubrication system in eight-cylinder V-type engine. (*Chrysler Motor Corp.*)

W. G. Forbes, C. L. Pope, and W. T. Everitt, *Lubrication of Industrial and Marine Machinery*, 2d ed., 1954; D. D. Fuller, *Theory and Practice of Lubrication for Engineers*, 1956.

Lumber manufacture

The production of lumber involves sawing logs lengthwise into boards (1 in. thick), dimension lumber (2 in. thick), and timbers of larger dimensions; seasoning or drying the wood; and planing it to exact size.

Softwood lumber is produced from coniferous trees and hardwood lumber from deciduous trees; actually, some "softwood" lumber is harder than some "hardwood" lumber. Most softwood lumber is used in construction for framing, finish, and trim.

Hardwood lumber is used mainly by converting industries, such as furniture, cabinet, and flooring manufacturers; however, some hardwood is used for interior finish when a particular color or grain is desired. Both hardwood and softwood lumber are used for paneling. *See* WOOD, ANATOMY AND IDENTIFICATION OF; XYLEM.

Lumber production in the United States since

Fig. 1. Log deck of a cypress sawmill in Florida showing log conveyor trough at right, carriage with log being sawed at left, and the band saw housing in the rear. (*A. Lescheu and Sons Rope Co.*)

Fig. 2. An air-drying package-stacked pine lumber yard in northern Louisiana. (*Woodard-Walker Lumber Co.*)

1950 has averaged around 38,000,000,000 bd ft per year, of which about 80% is softwood. A board foot is a board 1 ft square and 1 in. thick, or its equivalent in other dimensions. The softwood lumber industry in the United States is concentrated in the West and the South. Hardwood lumber is produced primarily in the Great Lakes states, the Appalachian and Ozark mountains, and the southern swamps and river bottoms, where cypress is also sawed. All forested areas produce some lumber. *See* LOGGING (FORESTRY).

Sawmill operation. At the mills, logs are stored in a log pond or on dry ground, to be moved later to the conveyor chain, hoist, or other system which delivers them into the mill. If not sawed promptly, land-stored logs must be treated with chemicals to avoid deterioration from fungi and insects. In the mill logs are sawed by circular saws, band saws, or log gang saws. A circular saw is a steel disk with cutting teeth around the edge. A band saw is an endless band of steel that runs over two wheels and has cutting teeth on one or both edges. A gang saw consists of a steel frame in which thin saws are individually placed with the desired distances between them. Tension is applied to the saws of the gang saw and the frame is then moved rapidly up and down with a slight forward oscillating ac-

tion on the down, or cutting, stroke. Logs or cants fed through the gang saw are firmly held by pressure rolls to assure correct alignment. The entire log is sawed with a single pass through the gang saw.

Thick circular-head saws with replaceable teeth are used with small portable mills because they are readily kept in good cutting condition. However, they are the most wasteful because their kerf, or cutting thickness, often is as much as 1/4 in., which means that, for every four boards 1 in. thick that are produced, one is lost in the form of sawdust.

Sawing and trimming. In a conventional sawmill logs enter on a conveyor chain and are stored on a deck. Each log is then placed on the carriage, positioned, and moved by the carriage against the saw. The carriage is powered by a cable and drum or by a large steam, air, or hydraulic piston. As slabs and boards are sawed off, the log is repositioned for the next cut. The sawyer, standing by the saw, operates the carriage, the set works, the log-turning mechanism, and the device that moves logs from the deck to the carriage (Fig. 1). When the sawyer is skillful, all of these operations are quickly performed and there is little unproductive time.

Mass-production mills use carriages with setworks that the sawyer operates by remote control. Logs are repositioned as sawing progresses to obtain the maximum amount of the better grades of lumber. Many of the larger mills use resaws to speed production; with resaws, thick cants are conveyed from the head saw to a sash gang or to a band resaw that cuts them to final thicknesses. The boards are then conveyed to the edger to remove uneven or waney edges and then to the trimmer, where they are cut to correct lengths.

Lumber sorting and waste disposal. From the trimmer the lumber goes to the sorting table, where it is separated by grade, size, and species. At some mills the boards are dipped in a solution to prevent fungus stain while drying. Wood waste, if needed for fuel, is hogged (mechanically chipped) and sent to the fuel bin. Wood requirements of the pulp and paper industry have created a demand for sawmill waste in the form of chips, and much of the sawmill waste formerly discarded or used for fuel is now converted into pulp chips and sold.

Since bark is not acceptable for pulping, logs are debarked before being sawed. There are several types of debarking machines. On the West Coast large logs are debarked by revolving them before powerful manipulated jets of water. In the Eastern and Southern mills mechanical bark removers have been perfected for smaller logs.

The widespread use of sawmill leftovers for pulp chips has encouraged the development of machines for converting logs into lumber and chips: of those now in use, some both chip and saw as logs are fed into them lengthwise. Thus, barked logs go in at one end and lumber and chips emerge from the other. Some types have automatic adjustments that vary lumber sizes cut with log diameter; others require manual setting. These developments have proven operable and in the eastern and southern United States have become important components of the lumber industry. They provide for maximum utilization of the wood con-

Fig. 3. A battery of dry kilns at southern pine sawmill in Texas showing lumber stacks on trucks ready for pushing into the kilns. (*Kirby Lumber Corp.*)

tent of logs with a minimum amount of sawdust.

Seasoning. Lumber is not fully processed for use until it has been dried to a moisture content consistent with that of the place where it will be used; if not dried correctly it will shrink or swell, since it is a hygroscopic substance. Lumber can be air-dried in stacks in a yard so that air can circulate through it (Fig. 2). When air-dried, it contains about 18% moisture in terms of its oven-dry weight. For faster drying and to attain lower, specifically desired moisture contents, dry kilns are used. These are chambers in which stacked lumber is placed, air circulation is forced, and temperature and humidity are controlled. For example, 1-in. green (freshly sawn) southern pine lumber can be dried in kilns to 12% moisture content in 72 hr (Fig. 3). Air-drying to only 18% moisture content would require from 2 weeks to 2 months, depending on the weather.

Softwood lumber often is kiln dried not only to give it better utility but to reduce its weight and save on freight costs. Softwood lumber is finished in planing mills, usually connected with sawmills, to standard thicknesses and widths for siding, ceiling, and floor patterns. Most hardwood lumber is only air-dried at sawmills and shipped rough. This is because most hardwoods cannot be kiln-dried from green condition without causing severe damage and also because hardwood lumber users, such as furniture factories, prefer to do the final drying themselves so that all lumber they purchase from various sources will be of uniform moisture content when they use it.

For either air- or kiln-drying, the green lumber must be placed in stacks with separater strips (stickers) between each layer. The lumber is stacked in unit packages for handling with forklift self-propelled equipment that can put the unit packages in stacks up to 20 ft high. The movement of lumber packages to yards or kilns and to planing mills is highly mechanized in mass-production mills; smaller mills still use manual techniques. Much of the stacking is now a mechanical process.

Grading. Lumber is graded into useful classifications according to rules prescribed by manufacturing and marketing trade associations; most rules meet the specifications of American Lumber Standards developed by the U.S. Department of Commerce. Although the rules vary slightly among the lumber-producing regions, all yard and construction lumber is interchangeable in use.

Lumber grades, which are based on the intended use of the lumber, are of three basic classes: yard lumber, which is used in general construction for houses, barns, and similar frame structures; structural lumber, for heavy industrial construction where strength is important; and factory or shop lumber, which is usually cut up by converting mills before ultimate use. Yard and structural lumber is graded for use of the piece as a whole because each is of standard width, thickness, and length. Grades are based on the number and extent of knots, splits, decay, and similar defects. Moisture content and dimension specifications are also specified. When lumber is to be used in structures in which strength is important, grain density and straightness and knot combinations that would impair strength are also evaluated. Nearly all hard-

Fig. 4. Laminated southern pine arches of Jai-Alai Fronton, a 10,000-seat amphitheater in West Palm Beach, Fla. Providing a clear span of 247 ft, only 12 arches were required for the complete structural support of this building. (*Southern Pine Ass.*)

wood lumber, and some softwood, is graded on a factory or shop basis because such lumber is used by furniture and other factories, where it is cut up into clear (nondefective) pieces. Hence, hardwood and softwood shop lumber is graded from the poorer face (surface), whereas yard and structural lumber grades are based on the better face.

Because of the importance of using thoroughly dried and correctly graded lumber, much of it is grade marked by official grading associations and also trademarked by nationally known manufacturers. Lumber of construction quality is stress rated, according to grading rules, so that wood structures can be designed as safely as those of metals or masonry. Since wood is an organic substance, it is subject to insect and fungus attacks, and is thus often treated with preservatives to prevent such attacks both in exposed exterior use and in protected interior use. These developments have greatly increased the usefulness of wood in

Fig. 5. Overall view of Jai-Alai Fronton. Each arch weighs over 15 tons. (*Unit Structures, Inc.*)

permanent structures. Similarly, fire-retardant treatments are used to reduce fire hazards.

Laminated wood. Fabricating wooden panels, beams, and trusses by gluing smaller pieces of lumber together is called laminating. The growing scarcity of large trees from which large timbers can be sawed has stimulated laminating, which was introduced into the United States in the 1930s. Laminating also permits prefabricating large trusses and arches to strength and size specifications (Figs. 4 and 5). Because they both have desirable strength qualities, Douglas fir and southern pine are the most widely used woods for laminating in the western and eastern United States, respectively. Oak is also used, but chiefly for marine laminates. Laminated wood can readily be impregnated with preservatives and fire retardants, thus enhancing its usefulness. *See* ARCH; PLYWOOD; TRUSS; WOOD PRESERVATION.

[ALBERT E. WACKERMAN]

Lumen

The unit of luminous flux. It is defined as the luminous flux emitted within a unit solid angle (1 steradian) by a point source having a uniform intensity of 1 candela. It follows, therefore, that a light source having an intensity of 1 candela in every direction will be emitting a total luminous flux of 4π lumens.

The lumen is also equal to the luminous flux received on a unit surface, all points of which are at a unit distance from a point source having a uniform intensity of 1 candela.

The output of light sources is given in lumens. *See* CANDELA; LUMINOUS FLUX.

[RUSSELL C. PUTNAM]

Lumen-hour

The unit of quantity of light. It is the quantity of light (luminous flux) radiated or received for a period of 1 hr by a flux of 1 lumen. Similarly other units that can be used to measure the quantity of light are the lumen-second or the million-lumen-hour.

The symbol of the lumen-hour is Q. The equation is given below, where t is time in hours and F

$$Q = \int F \, dt$$

is the luminous flux measured in lumens. *See* LUMEN; LUMINOUS ENERGY; PHOTOMETRY.

[RUSSELL C. PUTNAM]

Luminaire

An electric lighting fixture, wall bracket, portable lamp, or other complete lighting unit designed to contain one or more electric lighting sources and associated reflectors, refractors, housing, and such support for these items as is necessary.

Luminaires for general lighting are classified by the International Commission on Illumination, in accordance with the percentages of total luminaire output emitted above and below the horizontal, into direct or indirect types with intermediate classifications of semi-direct, semi-indirect, and general diffuse or direct-indirect.

Luminaires for floodlighting or for street or highway lighting are classified according to beam spread or according to candlepower distribution patterns for the specific applications. *See* ILLUMINATION; INCANDESCENT LAMP; LAMP, ELECTRIC.

[WARREN B. BOAST]

Luminance

The luminous intensity of any surface in a given direction per unit of projected area of the surface viewed from that direction. The International Commission on Illumination defines it as the quotient of the luminous intensity in the given direction of an infinitesimal element of the surface containing the point under consideration, by the orthogonally projected area of the element on a plane perpendicular to the given direction. Simply, it is the luminous intensity per unit area. Luminance is also called photometric brightness.

Since the candela is the unit of luminous intensity, the luminance, or photometric brightness, of a surface may be expressed in candelas/cm², candelas/in.², and so forth.

Mathematically, luminance B may be found from the equation below, where θ is the angle be-

$$B = dI / (dA \cos \theta)$$

tween the line of sight and the normal to the surface area A considered and I is the luminous intensity. *See* LUMEN; LUMINOUS INTENSITY.

The stilb is a unit of luminance (photometric brightness) equal to 1 candela/cm². It is often used in Europe, but the practice in America is to use the term candela/cm² in its place.

The apostilb is another unit of luminance sometimes used in Europe. It is equal to the luminance of a perfectly diffusing surface emitting or diffusing light at the rate of 1 lumen/m². *See* PHOTOMETRY.

[RUSSELL C. PUTNAM]

Luminescence

Light emission that cannot be attributed merely to the temperature of the emitting body. Various types of luminescence are often distinguished according to the source of the exciting energy. When the light energy emitted results from a chemical reaction, such as in the slow oxidation of phosphorus at ordinary temperatures, the emission is called chemiluminescence. When the luminescent chemical reaction occurs in a living system, such as in the glow of the firefly, the emission is called bioluminescence. In the foregoing two examples part of the energy of a chemical reaction is converted into light. There are also types of luminescence that are initiated by the flow of some form of energy into the body from the outside. According to the source of the exciting energy, these luminescences are designated as cathodoluminescence or electronoluminescence if the energy comes from electron bombardment; radioluminescence or roentgenoluminescence if the energy comes from x-rays or from γ-rays; photoluminescence if the energy comes from ultraviolet, visible, or infrared radiation; and electroluminescence if the energy comes from the application of an electric field. By attaching a suitable prefix to the word luminescence, similar designations may be coined to characterize luminescence excited by other agents. Since a given substance can frequently be made to luminesce by a number of different external exciting agents, and since the atomic and electronic phenomena that cause luminescence are basically the same regardless of the mode of excitation, the subdivision of luminescence phenomena into the foregoing categories is essentially only a matter of convenience, not of fundamental distinction.

When a luminescent system provided with a special configuration is excited, or "pumped," with sufficient intensity of excitation to cause an excess of excited atoms over unexcited atoms (a so-called population inversion), it can produce laser action (laser is an acronym for light amplification by stimulated emission of radiation). This laser emission is a coherent stimulated luminescence, in contrast to the incoherent spontaneous emission from most luminescent systems as they are ordinarily excited and used. *See* LASER.

Fluorescence and phosphorescence. A second basis frequently used for characterizing luminescence is its persistence after the source of exciting energy is removed. Many substances continue to luminesce for extended periods after the exciting energy is shut off. The delayed light emission (afterglow) is generally called phosphorescence; the light emitted during the period of excitation is generally called fluorescence. In an exact sense this classification based on persistence of the afterglow is not meaningful because it depends on the properties of the detector used to observe the luminescence. With appropriate instruments one can detect afterglows lasting no more than a few thousandths of a microsecond, which would be imperceptible to the human eye. The characterization of such a luminescence as either fluorescence or phosphorescence therefore depends upon whether the observation is made by eye or by instrumental means. These terms are nevertheless commonly used in the sense defined here, and are convenient for many practical purposes. However, they can be given a more precise meaning. Fluorescence may be defined as a luminescence emission having an afterglow duration which is temperature-independent. Phosphorescence may be defined as a luminescence with an afterglow duration which becomes shorter with increasing temperature. For details on various types of luminescence *see* BIOLUMINESCENCE; CATHODOLUMINESCENCE; CHEMILUMINESCENCE; ELECTROLUMINESCENCE; FLUORESCENCE; PHOSPHORESCENCE; PHOTOLUMINESCENCE; THERMOLUMINESCENCE.

Because of their many practical applications, materials that give a visible luminescence have been studied and developed more intensively than those which emit in other spectral regions. Luminescence, however, may consist of radiation in any region of the electromagnetic spectrum. The production of x-radiation by the bombardment of a metal target by a fast electron beam is an example of luminescence. Certain fluorescent lamps, called black-light lamps, are coated with a luminescent powder chosen for its ability to emit ultraviolet light of approximately 3600 A rather than visible light. A number of luminescent solids have been developed which luminesce in the near-infrared under excitation by visible light or by a cathode-ray beam. *See* ULTRAVIOLET LAMP.

The ability to luminesce is not confined to any particular state of matter. Luminescence is observed in gases, liquids, and solids. The passage of an electrical discharge through a gas will excite the gaseous atoms or molecules to luminesce under certain conditions. An example of such a gaseous luminescence is the mercury-vapor lamp, in which an electrical discharge excites the mercury vapor to emit both visible and ultraviolet light. Many liquids, such as oil or solutions of certain dyestuffs in various solvents, luminesce very strongly under ultraviolet light. A large number of solids, including many natural minerals as well as thousands of synthetic inorganic and organic compounds, luminesce under various types of excitation. The major applications of luminescence involve solid luminescent materials. *See* MERCURY-VAPOR LAMP.

The term phosphor, originally applied to certain solids that exhibit long afterglows, has been extended to include any luminescent solid regardless of its afterglow properties. Other terms sometimes used synonymously with phosphor are luminophor, fluor, or fluorphor. The term luminophor is preferable, since it carries no connotation that afterglow times are long or short. In conformity with current usage, however, the term phosphor is used in the succeeding discussion of solid luminescent materials.

Comparatively few pure solids luminesce efficiently, at least at normal temperatures. In the category of organic solids, the pure aromatic hydrocarbons, such as naphthalene and anthracene, which consist exclusively of condensed phenyl rings, are luminescent, as are many heterocyclic closed-ring compounds. However, the closed phenyl ring structure is not in itself a guarantee of efficient luminescence, since certain substituents for hydrogen in the structure, particularly halogens, tend to reduce luminescence efficiency. Among the pure inorganic solids that luminesce efficiently at room temperature are the tungstates, uranyl salts, platinocyanides, and a number of salts of the rare-earth elements.

Activators. The development of a large number of inorganic phosphors is due to the discovery that certain impurities, called activators, when present in amounts ranging from a few parts per million to several percent, can confer luminescent properties on the compounds (host or matrix compounds) in which they are incorporated. Manganese is a particularly effective activator in a wide variety of matrices when incorporated in amounts ranging from a small trace up to the order of several percent. The emission spectrum of these manganese-activated phosphors generally lies in the green, yellow, or orange spectral regions. Other frequently used activators are copper, silver, thallium, lead, cerium, chromium, titanium, antimony, tin, and the rare-earth elements. The most important host materials are silicates, phosphates, sulfides, selenides, the alkali halides, and oxides of calcium, magnesium, barium, and zinc. Preparation of phosphors and the incorporation of the activators is generally done by high-temperature reaction of well-mixed, finely ground powders of the components.

In order for ultraviolet light to provoke luminescence in a substance, the substance must be able to absorb it. Because of variations in its ability

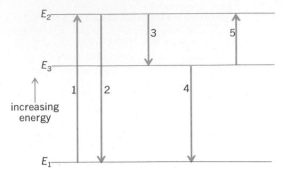

Fig. 1.   Schematic representation of energy levels and electronic transitions in an atomic gas; E_1, ground state; E_2 and E_3, excited states; 1, excitation; 2, emission of resonance luminescence; 3, radiationless transition to lower excited state; 4, luminescence emission, if transition $E_3 \rightarrow E_1$ is allowed (if it is not allowed, 4 does not occur and E_3 is called a metastable state); 5, stimulation of atom back to emitting state.

to absorb ultraviolet light of different wavelengths, a material may be nonluminescent under ultraviolet light of one wavelength and strongly luminescent under a different wavelength in the ultraviolet region. For a similar reason a substance may not luminesce under ultraviolet light at all and yet be strongly luminescent under x-ray irradiation.

LUMINESCENCE IN ATOMIC GASES

The processes that occur in luminescence may be most simply illustrated in the case of an atom in a gas. The atom can exist only in certain specific states of energy, some of which are shown schematically in Fig. 1. The lowest energy level of the atom, E_1, corresponds to the atom in its unexcited, or ground, state, and the higher energy levels, E_2 and E_3, represent electronically excited states of the atom. The excitation of the atom from state E_1 to state E_2 requires the absorption of an amount of energy $\Delta E = E_2 - E_1$. If this excitation is to be produced by the absorption of light, the energy of the acting light photon, E_{absorbed}, must equal ΔE. The frequency of the exciting light must therefore be $v = E_{\mathrm{absorbed}}/h = (E_2 - E_1)/h$, and the wavelength of the exciting light must be $\lambda_{\mathrm{absorbed}} = hc/(E_2 - E_1)$, where h is Planck's constant and c is the velocity of light. In an isolated atom this extra energy cannot be dissipated and is emitted as radiation when the atom eventually returns to its ground state. The emitted light will therefore again correspond in energy to ΔE and it will have a wavelength $\lambda_{\mathrm{emitted}} = \lambda_{\mathrm{absorbed}}$.   When $\lambda_{\mathrm{emitted}} = \lambda_{\mathrm{absorbed}}$,   the emitted light is sometimes referred to as resonance luminescence or resonance radiation. See ATOMIC STRUCTURE AND SPECTRA.

If a large number of atoms are excited and the excitation then removed, the luminescence intensity will decrease with time exponentially according to the equation $I_t = I_0 e^{-t/\tau_l}$, where I_t is the intensity of luminescence at a time t after removal of the excitation, I_0 is the intensity at $t = 0$, and τ_l is the average time required by an atom to make a spontaneous luminescent transition. The quantity τ_l is independent of temperature, and it is this temperature independence that is emphasized in the more precise definition of fluorescence given earlier. If the transition between the energy levels E_1 and E_2 is highly probable (a so-called permitted or allowed transition), τ_l is very small, of the order of 10^{-8} sec for transitions involving visible light.

The excited atom can also lose a certain amount of energy and fall to an energy state E_3, of intermediate energy between E_1 and E_2. This can happen, for example, if the excited atom collides with another atom. If the transition from state E_3 to the ground state E_1 can occur with a reasonably high probability, fluorescence will occur starting from this intermediate · excited state. In this case the fluorescent wavelength $\lambda_{\mathrm{emitted}} = hc/(E_3 - E_1)$. Since $(E_3 - E_1)$ is smaller than $(E_2 - E_1)$, the fluorescence in this case will be of longer wavelength than the resonance radiation.

If, however, a transition between state E_3 and state E_1 is highly improbable (a so-called forbidden transition), state E_3 is known as a metastable state. The atom can remain in this state for long periods of time and cannot return to the ground state with the emission of radiation. Luminescence can occur under these circumstances only if the atom regains the energy $(E_2 - E_3)$ by a collision with another atom or by some other process. Once the atom has been brought back to state E_2, a transition to the ground state is again allowed, and luminescence corresponding to $(E_2 - E_1)$ will be emitted. The existence of metastable states like E_3 explains the delayed emission termed phosphorescence. The atom may spend a considerable amount of time in such a state before some external influence causes it to return to an emitting state such as E_2, in which case the luminescence is correspondingly delayed and appears as an afterglow. In order for the atom to get from E_3 to E_2 it must absorb energy somehow. The rate of return to the emitting state, and hence the duration of the afterglow, will therefore depend to a very large extent on temperature. At high temperatures the atoms will be excited back to the emitting state rapidly, and there will be a bright afterglow of short duration. At lower temperatures the atoms will be raised back to the emitting state very slowly, and the afterglow will be of long duration but of low intensity. This temperature dependence is the basis for the more precise definition of phosphorescence that was given earlier.

The principles illustrated in the foregoing discussion may be extended with little modification to the case where the primary absorption or excitation act completely removes an electron from an atom (that is, ionizes the atom) instead of merely raising the atom to an excited state. Under these conditions the electron can be trapped temporarily by other atoms, and its return to the parent atom can also be delayed by this mechanism.

CONFIGURATION COORDINATE CURVE MODEL

Luminescence in atomic gases is adequately described by the concepts of atomic spectroscopy, but luminescence in molecular gases, in liquids, and in solids introduces two major new effects which need special explanation. One is that the emission band appears on the long wavelength (low-energy) side of the absorption band; the other is that emission and absorption often show as bands hundreds of angstroms wide instead of as the lines found in atomic gases.

Both of these effects may be explained by using

the concept of configuration coordinate curves illustrated in Fig. 2. As in the case of atomic gases, the ground and excited states represent different electronic states of the luminescent center, that is, the region containing the atoms, or electrons, or both, involved in the luminescent transition. On these curves the energy of the ground and excited state is shown to vary parabolically as some configuration coordinate, usually the distance from the luminescent center to its nearest neighbors, is changed. There is a value of the coordinate for which the energy is a minimum, but this value is different for the ground and excited states because of the different interactions of the luminescent center with its neighbors. Absorption of light gives rise to the transition from A to B. This transition occurs so rapidly that the ions around the luminescent center do not have time to rearrange. Once the system is at B it gives up heat energy to its surroundings by means of lattice vibrations and reaches the new equilibrium position at C. Emission occurs when the system makes the transition from C to D, and once again heat energy is given up when the system goes from D back down to A. This loss of energy in the form of heat causes the energy associated with the emission $C \rightarrow D$ to be less than that associated with the absorption $A \rightarrow B$. See LATTICE VIBRATIONS.

When the system is at an equilibrium position, such as C of the excited-state curve, it is not at rest but migrates over a small region around C because of the thermal energy of the system. At higher temperatures these fluctuations cover a wider range of the configuration coordinate. As a result, the emission transition is not just to point D on the ground state curve but covers a region around D. In the vicinity of D the ground state curve shows a rapid change of energy, so that even a small range of values for the configuration coordinate leads to a large range of energies in the optical transition.

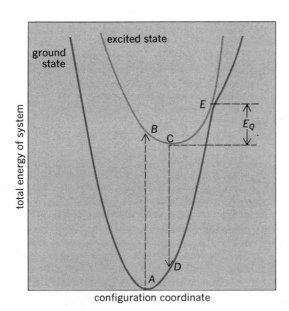

Fig. 2. Configuration coordinate curves for a simple luminescence center. Transition $A \rightarrow B$ shows absorption, and transition $C \rightarrow D$ shows emission. Energy E_Q is that necessary for the excited state to reach point E, from which a transition to the ground state can be made without luminescence.

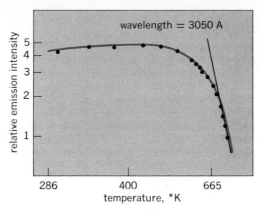

Fig. 3. Variation of the brightness of a thallium-activated potassium chloride phosphor with temperature.

This explains the broad emission and absorption bands that are observed. An analysis of this sort predicts that the widths of the band (usually measured in energy units between the points at which the emission or absorption is half its maximum value) should vary as the square root of the temperature. For many systems this relationship is valid for temperatures near and above room temperature.

Two other phenomena which can be explained on the basis of the model described in Fig. 2 are temperature quenching of luminescence and the variation of the decay time of luminescence with temperature. In Fig. 3 a curve of temperature quenching for the emission in thallium-activated potassium chloride is shown. At low temperatures there is very little change in brightness with temperature, but at elevated temperatures the luminescence decays rapidly. On the scheme of Fig. 2 this is interpreted as meaning that the thermal vibrations become sufficiently intense to raise the system to point E. From point E the system can fall to the ground state by emitting a small amount of heat, or infrared radiation. If point E is at an energy E_Q above the minimum of the excited state curve, it may be shown that the efficiency, η, of luminescence is given by Eq. (1), where C is a con-

$$\eta = 1 + C \exp\left(-E_Q/kT\right)^{-1} \qquad (1)$$

stant, k is Boltzmann's constant, and T is the temperature on the Kelvin scale. By fitting an expression of this form to the data of Fig. 3, a value of 0.60 ev is obtained.

This temperature quenching tends to occur most strongly for centers that would have stayed in the excited state for a relatively long period of time. As a result, the decay time of the emission that occurs in this temperature region is largely characteristic of centers in which transitions to the ground state have been rapid; therefore, the decay time of the luminescence is observed to decrease.

Quantum-mechanical corrections. Although the configuration coordinate curve model of Fig. 2 is successful in describing many aspects of luminescence in solids, it predicts that emission and absorption bands should become narrow lines as the temperature is reduced to absolute zero (0°K). This is not the case, as is shown in Fig. 4, which gives the width of the absorption band of the F-center in potassium chloride as a function of the square root of the temperature. (An *F*-center is the simplest type of color center, a color center being a lattice defect which can absorb light.) At high temperatures the previously quoted results are valid, but at low temperatures the width of the band is constant. *See* COLOR CENTERS.

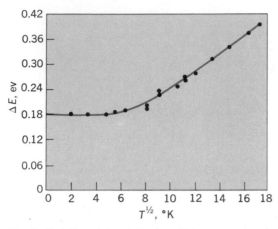

Fig. 4. Variation of the width of the *F*-center absorption band in potassium chloride at its half-maximum points, ΔE, as a function of the square root of the temperature.

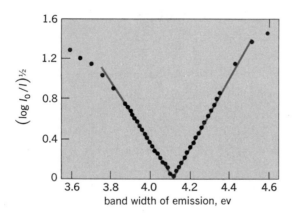

Fig. 5. Emission of thallium-activated potassium chloride at 4°K. In a plot of this particular form, a Gaussian curve would consist of two straight lines which make equal angles with the abscissa.

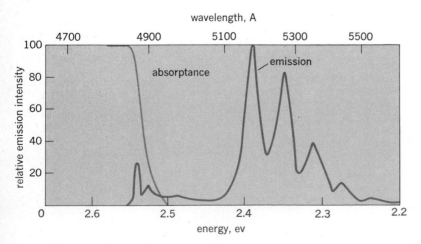

Fig. 6. Graph showing the absorptance (100 minus percent of transmission) and edge emission of cadmium sulfide at 4°K.

This phenomenon may be explained by treating the configuration coordinate curves quantum mechanically. The curves have the energy versus displacement characteristics of the simple harmonic oscillator. For this case the quantum-mechanical analysis shows that there is a series of equally spaced energy levels separated by an energy of $h\nu$, where h is Planck's constant and ν is the frequency of vibration. The lowest of these levels occurs at a value of $1/2\ h\nu$ above the minimum of the classical curve, and this energy is called the zero point energy. Its importance is that even at absolute zero the system is not at rest but varies over a range of configuration coordinates characteristic of this lowest vibrational level. Analysis shows that under these conditions the widths of the bands, ΔE, vary as in Eq. (2), where A is a

$$\Delta E = A[\coth(h\nu/2kT)]^{1/2} \qquad (2)$$

constant. A curve of this form is drawn through the experimentally obtained points of Fig. 4 and shows satisfactory agreement.

One other result of the introduction of quantum mechanics is that this simple model predicts that both absorption and emission bands should be Gaussian in shape at all temperatures; that is, they should be of the form of Eq. (3), where I is the

$$I = I_0 \exp[-A(E-E_0)^2] \qquad (3)$$

emission intensity or absorption strength for light of energy E, and I_0 and E_0 refer to corresponding quantities at the maximum of the curve. Figure 5 shows the emission spectrum of thallium-activated potassium chloride plotted in such a way that the expression of Eq. (3) would give two straight lines making equal angles with the horizontal. Although there is some disagreement in the wings of the emission spectrum, and although the lines are not quite at the same angle, there still is fairly good agreement with the predictions of Eq. (3).

High dielectric constant materials. The use of configuration coordinate curves is justifiable only when the electron taking part in an optical transition is tightly bound to a luminescent center and interacts primarily with its nearest neighbors. This appears to be generally the case in materials with low dielectric constants such as the alkali halides. For high dielectric constant materials the situation is very different. It has been estimated that for boron in silicon the electron is spread over about 500 atoms so that its interaction with the nearest neighbors is small. In these cases the center interacts with the lattice during an optical transition through the creation of many vibrational phonons (sound quanta) at relatively large distances from the center. A case of this sort is illustrated in Fig. 6, which shows edge emission in cadmium sulfide near the absorption edge of the material. The major peaks correspond to an optical transition with simultaneous emission of 0, 1, 2, and 3 phonons, respectively, starting with the highest peak. The peaks are equidistant in energy, and the spacing is just that to be expected for the creation of phonons. Although both short- and long-range interactions of a center with its environment occur in all cases, the short-range interaction dominates for tightly bound electrons in low dielectric constant materials, while the long-range interaction dominates for loosely bound electrons in high dielectric constant materials. *See* PHONON.

(a)

(b)

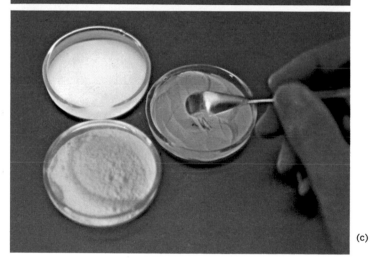

(c)

(d)

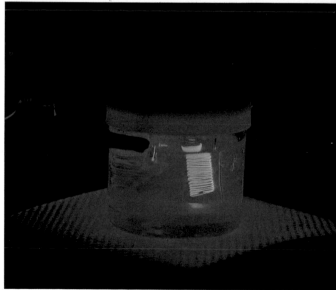

(e)

Various kinds of luminescence are distinguished according to the source of the exciting energy. (a) Cathodoluminescent phosphors under examination in test equipment. (b) Inorganic phosphors excited by ultraviolet light. (c) Multicolored luminescence from several photoluminescent inorganic phosphors. (d) Equipment involving a solvent, a luminescent material, and an electrolyte, with platinum electrodes, for the production of (e) liquid electroluminescence. (*General Telephone & Electronics Laboratories*)

SENSITIZED LUMINESCENCE

A process of considerable interest occurs in systems where one type of activator absorbs the exciting light and transfers its energy to a second type of activator which then emits. The transfer does not involve motion of electrons. This process is often called sensitized luminescence, and the principal phosphors used in fluorescent lamps are of this type. The results on calcium carbonate ($CaCO_3$) shown in Fig. 7 illustrate this kind of system. If divalent manganese (Mn^{++}) alone is incorporated as an activator into $CaCO_3$, the system gives only a very weak Mn^{++} emission for all wavelengths of exciting ultraviolet light. This is because the Mn^{++} ion has only forbidden transitions in this spectral region, as shown by its very low absorption strength and slow decay of luminescence. In other words the Mn^{++} does not absorb very much ultraviolet light. Bright phosphors may be prepared by incorporating divalent lead (Pb^{++}), monovalent thallium (Tl^+), or trivalent cerium (Ce^{+++}) along with Mn^{++} in $CaCO_3$. In each case the Mn^{++} emission is the same, but the wavelengths which excite these phosphors vary as the other activator (called the sensitizer) is changed. All these absorptions occur for energies well below the energy at which free electrons are formed in the solid. The conclusion is that the ultraviolet light is absorbed by the Pb^{++}, Tl^+, or Ce^{+++} sensitizers, which pass the energy on to nearby Mn^{++} activators and excite the Mn^{++} emission. It is important to know over what distances the energy may be transferred from absorber to emitter. Experiments on calcium silicate with Pb^{++} and Mn^{++} as added impurities have shown that if a Mn^{++} ion is in any one of the 28 nearest lattice sites around the Pb^{++} ion, the energy transfer may take place.

Resonant transfer. The transfer of energy has been treated theoretically using quantum mechanics and assuming a model of resonance between coupled systems. A number of cases have been examined; in each case the sensitizer has an allowed transition since only in this case will it absorb the exciting light appreciably.

The first case is one in which the activator undergoes an allowed transition. The probability of energy transfer, P (defined as the reciprocal of the time required for a transition to occur) varies as in expression (4). Here R is the distance between

$$P \propto \frac{1}{R^6 n^4} \int f_s F_A \, dE \qquad (4)$$

sensitizer and activator; n is the index of refraction of the host material, which may be liquid or solid; f_s is a function describing the emission band of the sensitizer as a function of energy, and is normalized so that the area under the curve is unity; and F_A is a function describing the absorption band of the activator, also normalized to unity. The integral of expression (4) measures the overlap of these bands and determines the resonance transfer. In typical cases the sensitizer will transfer energy to an activator if the activator occupies any one of the 1000–10,000 nearest available sites around the sensitizer. One type of forbidden transition is a "quadrupole" transition; in this case the number of sites for transfer would be about 100. If the transition in the activator is even more strongly forbidden, quantum-mechanical "exchange" effects

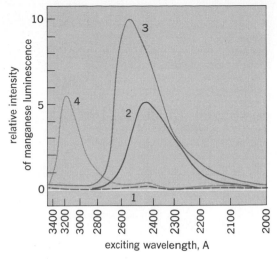

Fig. 7. Excitation spectra for emission of calcium carbonate phosphors, using manganese as an activator, with different impurity sensitizers: 1, no sensitizer; 2, thallium; 3, lead; and 4, cesium.

predominate and the number of available sites for transfer should be reduced to 40 or less. In both of the cases just mentioned the integral measuring the overlap enters in the same way as it does in expression (4). From this theoretical treatment it appears that phosphors with Mn^{++} as the activator probably receive their energy by exchange interactions.

Concentration quenching. Another phenomenon related to the resonant transfer of energy is that of concentration quenching. If phosphors are prepared with increasing concentrations of activators, the brightness will first increase but eventually will be quenched at high concentrations. It is believed that at high concentrations the absorbed energy is able to move from one activator to a nearby one by resonant transfer and thus migrate through the solid or liquid. If there are "poisons," or quenching sites, distributed in the material, the migrating energy may reach one and be dissipated without luminescence. Impurity atoms, vacancies, jogs at dislocations, normal lattice ions near dislocations, and even a small fraction of the activator ions themselves, when associated in pairs or higher aggregates, can act as poisons. As the concentration is increased, the speed of migration is increased, and the quenching process becomes increasingly important.

LUMINESCENCE INVOLVING ELECTRON MOTION

In an important group of luminescent materials the transfer of energy to the luminescent center is brought about by the motion of electrons. Many oxides, sulfides, selenides, and tellurides are of this type, and of these, zinc sulfide has been most widely studied. It is frequently used as the luminescent material in cathode-ray tubes and electroluminescent lamps.

Electronic processes in insulating materials are described by a band model such as that illustrated in Fig. 8. The electron energy increases vertically and the horizontal dimension shows position in a crystal. In the shaded area, called the filled band, or the valence band, all the energy levels are filled with electrons. No electron can be accelerated or

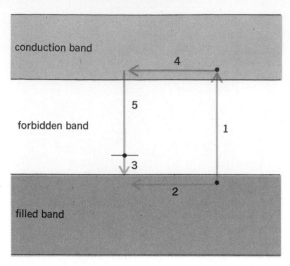

Fig. 8. Energy-band model for luminescence process-es in zinc sulfide. Electron energy increases vertically.

moved to higher energies within this band since the higher levels are already filled. Thus the material is an insulator. Above the valence band is an energy region called the forbidden band, which has no energy levels in it for pure materials. However, imperfections may introduce a local energy level in this region, as illustrated by the short line above 3 in Fig. 8. Above the forbidden band is an energy region called the conduction band. Here there are energy levels but, in an insulator, no electrons. If an electron in the filled band absorbs light of sufficiently high energy, it may jump up into the conduction band, as shown in 1 of Fig. 8. The empty position left behind in the filled band, called a hole, has properties which allow it to be described as an electron except that it has a positive charge. Both the electron in the conduction band and the hole in the filled band are free to move and gain energy. As a result, current can flow when a voltage is applied externally. The electron and hole can also diffuse far from their origin and can thus transport energy to a distant luminescent center. *See* HOLES IN SOLIDS.

A simple luminescent transition is illustrated in Fig. 8. Assume that the impurity level (black dot in the forbidden band) is due to a luminescent center and that there is an electron in the level at the be-

ginning of the process. Transition (1) shows the creation of a free electron and hole due to the absorption of light. The hole migrates to the center (2), and the electron in the impurity level falls into the hole (3), thus destroying it. The free electron now migrates toward the center (4) and falls into it (5), giving off luminescence. The cycle is complete and the center once again has an electron. It is important to note that in this process both electrons and holes must be assumed to move.

In zinc sulfide the normal activators are monovalent metals such as Ag^+ or Cu^+. They replace Zn^{++} ions. To maintain electrical neutrality it is necessary also to incorporate into the lattice ions such as Cl^- for S^{--} or Al^{+++} for Zn^{++}. These additional ions are called coactivators. The situation is now quite complex, and is illustrated for ZnS with Cu and Al as added impurities in Fig. 9. The various arrows in this figure show some of the electronic transitions that give rise to observable effects. Arrows 1 and 4 show two different luminescent transitions. Arrow 2 shows the excitation of an electron into the luminescent center, which may occur by absorption of infrared light or by thermal fluctuations. In either case the center is prevented from luminescing and thus is quenched. Transition 3 is the excitation of an electron into the conduction band. The electron may be excited by thermal energy and its appearance may be detected by the appearance of transition 1. This process is called thermoluminescence.

The wide variety of transitions and effects in zinc sulfide makes it difficult to come to firm conclusions concerning the details of any one system. At the same time this variety makes luminescence involving electron motion one of the most fascinating and challenging fields of study.

[CLIFFORD C. KLICK; JAMES H. SCHULMAN]

Bibliography: D. Curie, *Luminescence in Crystals,* 1960; P. G. Goldberg (ed.), *Luminescence of Inorganic Solids,* 1966; H. W. Leverenz, *An Introduction to the Luminescence of Solids,* 1950; P. Pringsheim, *Fluorescence and Phosphorescence,* 1949; F. Seitz and D. Turnbull (eds.), *Solid State Physics,* vol. 5, 1957.

Luminosity factor

The ratio of luminous flux at a particular wavelength to the corresponding radiant flux at the same wavelength. It is expressed in lumens per watt. *See* LUMEN; PHOTOMETRY.

The luminosity factor is often used in the form of the relative luminosity factor, which is the ratio of the luminosity factor of a certain wavelength to the value at the wavelength of maximum luminosity. The International Commission on Illumination and the International Committee on Weights and Measures have adopted values for relative luminosity factors as a basis for establishment of photometric standards of types differing from the primary standard in spectral distribution of light.

The luminosity factor K is found from the equation below, where F is the luminous flux, P is the

$$K = F_\lambda / P_\lambda$$

radiant flux, and λ designates the wavelength.

The lumen per watt is also the term used to express the efficiency of an electric lamp, which should not be confused with luminosity factor. The efficiency is expressed as the ratio of the output, in

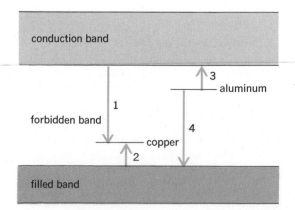

Fig. 9. Energy-band model for luminescence processes in zinc sulfide which uses copper as the activator and aluminum as the coactivator.

lumens, divided by the input, in watts.

[RUSSELL C. PUTNAM]

Luminous efficacy

There are three ways this term can be used: (1) The luminous efficacy of a source of light is the quotient of the total luminous flux emitted divided by the total lamp power input. Light is visually evaluated radiant energy. Luminous flux is the time rate of flow of light. Luminous efficacy is expressed in lumens per watt. (2) The luminous efficacy of radiant power is the quotient of the total luminous flux emitted divided by the total radiant power emitted. This is always somewhat larger for a particular lamp than the previous measure, since not all the input power is transformed into radiant power. (3) The spectral luminous efficacy of radiant power is the quotient of the luminous flux at a given wavelength of light divided by the radiant power at that wavelength. A plot of this quotient versus wavelength displays the spectral response of the human visual system. It is, of course, zero for all wavelengths outside the range from 380 to 760 nanometers. It rises to a maximum near the center of this range. Both the value and the wavelength of this maximum depend on the degree of dark adaptation present. However, an accepted value of 680 lumens per watt maximum at 550 nanometers represents a standard observer in a light-adapted condition. The reciprocal of this maximum spectral luminous efficacy of radiant power is sometimes referred to as the mechanical equivalent of light, with a probable value of 0.00147 watt per lumen.

If the spectral luminous efficacy values at all wavelengths are each divided by the value at the maximum (680 lumens per watt), the spectral luminous efficiency of radiant power is obtained. It is dimensionless.

For purposes of illuminating engineering, light is radiant energy which is evaluated in terms of what is now simply and descriptively the spectral luminous efficiency curve, V_λ. In this case it is the intent that an illuminating engineer have a measure for the capacity or capability for the production of luminous flux from radiant flux, or that radiant flux has efficacy for the production of luminous flux.

Since the human eye is sensitive only to radiations of wavelengths between about 400 and 700 nanometers, it follows that the sensation of sight evoked by a radiator is due only to the radiation within this limited wavelength band. If all the wavelengths within this band were equally effective for the purpose of vision, then the area under the graph of radiant power distribution would be a measure of the light output.

The illustration shows the spectral distribution curve for an electric tungsten filament lamp. The total area under the curve between the wavelength limits $\lambda = 0$ and $\lambda = \infty$ is proportional to the power of the total radiation. In illustration a the shaded area between the limits of $\lambda = 400$ and $\lambda = 700$ nanometers is a measure of the power of visible radiation. However, this is not the same thing as luminous efficacy because of the variable sensitivity of the eye with respect to wavelength. If each ordinate, say at 10-nanometer wavelength intervals, is multiplied by the appropriate value of V_λ, a new distribution is obtained, as shown in the shaded area of illustration b. This area is a meas-

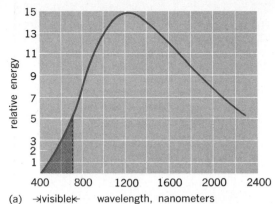

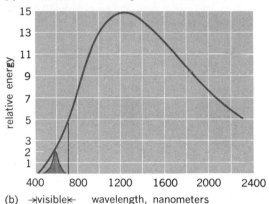

Spectral energy distribution of electric filament lamp. (a) Colored area is measure of power of visible radiation. (b) Colored area is measure of light output.

ure of the light output of the source. The ratio of the shaded area of illustration b to the area under the larger curve is the luminous efficacy in terms of lumens per watt, and this is the method of determination that is generally used. The name luminous efficacy expresses exactly what is meant by the effectiveness of 1 watt of radiant power in producing luminous flux. See CANDELA; LUMEN; LUMINOUS EFFICIENCY; LUMINOUS FLUX; PHOTOMETRY.   [G. A. HORTON]

Luminous efficiency

Visual efficacy of visible radiation, a function of the spectral distribution of the source radiation in accordance with the "spectral luminous efficiency curve," usually for the light-adapted eye or photopic vision, or in some instances for the dark-adapted eye or scotopic vision. Account is taken of this fact in photometry, or the measurement of light. Actually, instead of indicating directly the physical radiant power of light, photometers are designed to indicate the corresponding luminous flux of the visually effective value of the radiation.

The spectral luminous efficiency of radiant flux is the ratio of luminous efficacy for a given wavelength to the value of maximum luminous efficacy. It is a dimensionless ratio. Values of spectral luminous efficiency for photopic vision V_λ at 10-nanometer-wavelength intervals were provisionally adopted by the International Commission on Illumination in 1924 and by the International Committee on Weights and Measures in 1933, as a basis for the establishment of photometric standards of types of sources differing from the primary standard in spectral distribution of radiant flux. These standard values of spectral luminous efficiency were determined by visual photometric observations with a 2° photometric field having a moderately high luminance (photometric brightness); photometric evaluations based upon them, consequently, do not agree exactly to other conditions of observations, such as in physical radiometry. Watts weighted in accordance with these standard values are often referred to as light-watts.

Values of spectral luminous efficiency for scotopic vision V'_λ at 10-nanometer intervals were provisionally adopted by the International Commission on Illumination in 1951. These values of spectral luminous efficiency were determined by visual observations by young, dark-adapted observers, using extra-foveal vision at near-threshold luminance.

The main difference between the terms luminous efficacy and luminous efficiency is that the latter came into use a century or so before the former. It should also be noted that in engineering, general values of efficiency never exceed 1.0, and illuminating engineers were the only ones who had values of efficiency up to 680, but this has now been changed by the usage of the term luminous efficacy. In view of the above discussion, the term luminous efficiency is evidently a misnomer. *See* CANDELA; LUMEN; LUMINOUS EFFICACY; PHOTOMETRY.   [G. A. HORTON]

Bibliography: United States of America Standards Institute, *U.S.A. Standard Nomenclature and Definitions for Illuminating Engineering*, 1967.

Luminous energy

The radiant energy in the visible region or quantity of light. It is in the form of electromagnetic waves, and since the visible region is commonly taken as extending 380–760 nanometers (millimicrons) in wavelength, the luminous energy is contained within that region. It is equal to the time integral of the production of the luminous flux. *See* LUMEN-HOUR; PHOTOMETRY.   [RUSSELL C. PUTNAM]

Luminous flux

The time rate of flow of light. It is radiant flux in the form of electromagnetic waves which affects the eye or, more strictly, the time rate of flow of radiant energy evaluated according to its capacity to produce visual sensation. The visible spectrum is ordinarily considered to extend from 380 to 760 nanometers (millimicrons) in wavelength; therefore, luminous flux is radiant flux in that region of the electromagnetic spectrum. The unit of measure of luminous flux is the lumen. *See* LUMEN; PHOTOMETRY.   [RUSSELL C. PUTNAM]

Luminous intensity

The solid angular luminous flux density in a given direction from a light source. It may be considered as the luminous flux on a small surface normal to the given direction, divided by the solid angle (in steradians) which the surface subtends at the source of light. Since the apex of a solid angle is a point, this concept applies exactly only to a point source. The size of the source, however, is often extremely small when compared with the distance from which it is observed, so in practice the luminous flux coming from such a source may be taken as coming from a point. For accuracy, the ratio of the diameter of the light source to the measuring distance should be about 1:10, although in practice ratios as large as 1:5 have been used without exceeding the permissible error.

Mathematically, luminous intensity I is given by the equation below, where ω is the solid angle

$$I = dF/d\omega$$

through which the flux from the point source is radiated, and F is the luminous flux. The luminous intensity is often expressed as candlepower. *See* CANDLEPOWER; LUMEN; PHOTOMETRY.
[RUSSELL C. PUTNAM]

Luminous paint

A type of paint that glows in the dark. Luminous paints may be either of the self-luminous type (energized by a radioactive salt) or of the type that requires preexcitation by an outside energy source such as light. Both types are made by incorporating luminescent material into the paint formulation.

The self-luminous paints glow continuously for several years, and are used chiefly for painting the dials of watches or compasses. Zinc sulfide–type phosphors are generally used in their preparation, along with traces of some radioactive salt such as radium bromide or mesothorium bromide. Excitation of the phosphor is principally by bombardment of alpha particles (nuclei of helium atoms) from the radioactive salt or from one of its decay products.

The paints that require an outside source for excitation may be classed as either short-afterglow or long-afterglow paints. The luminescent materials in the short-afterglow, or fluorescent, paints are generally organic polynuclear hydrocarbons like anthracene or chrysene, or various luminescent dyes such as rhodamine. Long-afterglow, or phosphorescent, paints generally employ either alkaline-earth sulfide-type phosphors or zinc sulfide–type phosphors. The former type give distinguishable luminescence for periods up to 10 hr, while the useful afterglow of the latter type is generally limited to 1 hr or less. These paints are used, for example, under blackout conditions during wartime. *See* PAINT; RADIUM.

[CLIFFORD C. KLICK; JAMES H. SCHULMAN]

Lung (vertebrate)

Paired, air-filled respiratory sacs, usually in the anterior or anteroventral part of the trunk of most tetrapods. Many elongated tetrapods, such as snakes, possess a simple lung, and one group of salamanders have lost their lungs completely with gas exchange occurring through the skin of the back. Lungs originated, however, early in fish evolution, presumably in the placoderms, as an adaptation to obtain atmospheric oxygen required for life in stagnant, oxygen-depleted waters. Lungs are still found in a few living species of fresh-water fishes (such as lungfish) in which they may fuse to form a single organ and may come to lie well dorsally in the body; in many fishes the lungs have evolved into the swim bladder, a dorsal hydrostatic organ. *See* DIPNOI; SWIM BLADDER.

Lungs develop as outgrowths of the anterior end of the digestive tract, from the pharynx or throat,

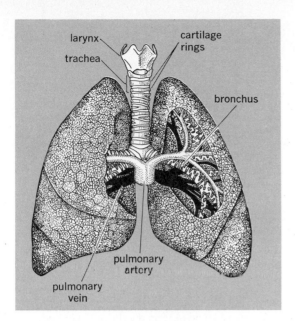

The human lung. (*From T. I. Storer and R. L. Usinger, General Zoology, 4th ed., McGraw-Hill, 1965*)

and remain connected to it by the trachea or windpipe. They lie within the coelom and are covered by peritoneum. In mammals they are within special chambers of the coelom known as pleural cavities and the peritoneum is termed pleura.

Amphibian lungs are often simple sacs, with only small ridges on the internal walls. In higher forms the lungs become more and more subdivided internally, thus increasing greatly the surface areas across which the respiratory exchange takes place. However, even in many reptiles the lungs may be quite simple. Birds have especially complex lungs with a highly differentiated system of tubes leading into and through them to the air sacs which are contained in many parts of the bird's body. Mammalian lungs are simpler, but in them the internal subdivision into tiny sacs or alveoli is extreme; there may be over 350,000,000 of them in one human lung.

In man the two lungs lie within the chest, separated by the heart and mediastinum. The right lung has three lobes and the left lung two. A bronchus, an artery, and a vein enter each lung medially at the hilum; each branches again and again as it enters the lobules and smaller divisions of the lungs (see illustration). The terminal airways or bronchioles expand into small clusters of grapelike air cells, the alveoli. The alveolar walls consist of a single layer of epithelium and collectively present a huge surface. A small network of blood capillaries in the walls of the alveoli affords surfaces for the actual exchange of gases. *See* RESPIRATION; RESPIRATORY SYSTEM (VERTEBRATE).

[THOMAS S. PARSONS]

Lung disorders

Inflammatory processes resulting from infections of bronchi and lungs are daily events in medical practice; scarring and tissue destruction as well as impaired pulmonary circulation are frequently encountered. Since the lungs literally filter the venous blood drainage from the entire body, they often become involved secondarily by blood-borne processes; many malignant tumors eventually metastasize to the lungs. More important still is the steadily rising incidence of primary malignancy of the lung and its controversial causative relationship to cigarette smoking and environmental pollution. *See* ONCOLOGY.

Congestion and edema. Although congestion and edema may occur separately, they are so frequently associated in the lungs that they may be discussed together. Factors predisposing to their development are increased pulmonary venous pressure, increased capillary permeability, decreased lymphatic drainage, and sodium retention.

Acute congestion (passive hyperemia) is the overdistention of the pulmonary capillaries and veins, most frequently subsequent to increased venous pressure because of left heart failure. The lungs are dark blue and heavy. The overdistention of the alveolar capillaries leads to increased pressure and endothelial damage. This, in turn, favors the leak of fluid into the air spaces. Normally the alveoli tend to resorb the escaped fluid, but their resorptive capacity can be overwhelmed by massive amounts of fluid, rapidity of leakage, and cell damage. Then fluid accumulates in the air spaces, a condition termed edema (Fig. 1). The lungs may quadruple their weight. Cut surfaces exude abundant frothy fluid. *See* EDEMA.

Heart failure is the most frequent cause of acute pulmonary congestion and edema. This is often a life-threatening situation that requires prompt medical attention. Shock, lung inflammations, and central nervous system disorders may also lead to acute pulmonary congestion and edema. In mitral valve disease, repeated episodes of congestion and edema cause leakage of blood into alveoli. Blood pigment is taken up by macrophages. This condition is termed brown induration of the lungs (chronic congestion).

Pulmonary embolism. This is the occlusion of one or more pulmonary arteries by a clot (Fig. 2). The clot usually originates in the deep veins of the legs, but femoral, iliac, and pelvic veins may also be the site of origin. Predisposing factors in the

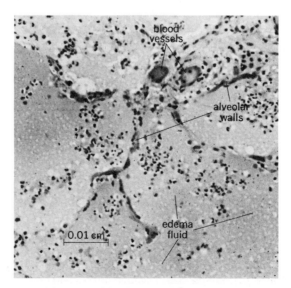

Fig. 1. Section of lung with edema showing dilated alveoli containing abundant edema fluid with desquamated cells and leukocytes; capillaries and venules distended.

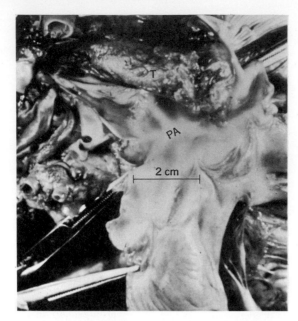

Fig. 2. Gross specimen of lung. Dilated pulmonary artery (PA) is held open with forceps; massive thromboembolus (T) is shown occluding the lumen.

development of vein occlusion (phlebothrombosis) with subsequent embolism are heart failure, postoperative convalescence, and malignant tumors. Massive pulmonary embolism is a life-threatening situation accounting for 2–6% of deaths in a general hospital. *See* CIRCULATION DISORDERS; EMBOLISM.

Clinically, large pulmonary emboli occluding the main pulmonary arteries may kill in a matter of minutes, causing few detectable changes in the lungs. Smaller thrombi will lodge in the smaller, more peripheral pulmonary arteries. In this instance, in patients with adequate circulation, the bronchial arteries will maintain the lung tissue despite occlusion of the pulmonary artery branches. Bleeding into the alveoli may occur, but no infarction (tissue death) will follow. Conversely, in patients with cardiovascular insufficiency, occlusion of pulmonary artery branches will result in areas of hemorrhages with death of the underlying tissue. *See* INFARCTION.

Pulmonary embolism and infarction are frequently, but not invariably, complications of underlying diseases affecting particularly the elderly, debilitated, or immobilized patient. The treatment emphasizes prevention of predisposing factors.

Pulmonary vascular sclerosis. This is a progressive thickening and narrowing of the pulmonary artery branches, associated with increased pressure (pulmonary hypertension). Primary pulmonary hypertension is rare. Most often pulmonary vascular sclerosis is a secondary process resulting from various underlying pulmonary or heart diseases such as left heart failure, multiple pulmonary emboli, pulmonary fibrosis, thoracic deformities, and emphysema. When the vascular narrowing and thickening become diffuse and severe, the reduction of blood circulation and proper oxygenation threatens life. Respiration is difficult, and the low blood oxygen results in dark, bluish skin (black cardiacs). The increased resistance in the thickened pulmonary arteries causes the right

ventricle to become dilated and thickened (cor pulmonale).

Pulmonary fibrosis. This is the widespread thickening of the alveolar septa by fibrous tissue or proliferating cells or both. The resulting alveolocapillary block causes a decrease in pulmonary function manifested with dyspnea (labored breathing), cynosis, and eventually cor pulmonale. Collagen diseases, sarcoidosis, pneumoconiosis, tuberculosis, and irradiation, among many others, may be factors underlying this disease.

Atelectasis. The adult secondary type of atelectasis is the collapse of pulmonary alveoli that were previously expanded. This process can be caused by complete bronchial occlusion with subsequent alveolar air resorption or by increased intrapleural pressure produced by air or fluid in the pleural cavity or by deviation of the diaphragm (Fig. 3). Atelectatic areas appear dark red-blue, rubbery, and airless. Microscopically, the alveolar lumina are reduced to inconspicuous slitlike spaces, and surrounding alveoli may become overdistended (Fig. 4); this change does not constitute true emphysema. Atelectasis is associated with all conditions causing complete bronchial obstruction and all forms of pleural effusions. Massive atelectasis may prove fatal by reducing ventilating function. However, if the causative factor is corrected, the lung tissue will usually reverse to normal.

Atelectasis of the newborn is a failure of the lungs to expand at birth. This often fatal condition is most frequently seen in premature babies and is also known as respiratory distress syndrome of the newborn. Microscopically, peripheral air spaces are often lined by fibrin and debris known

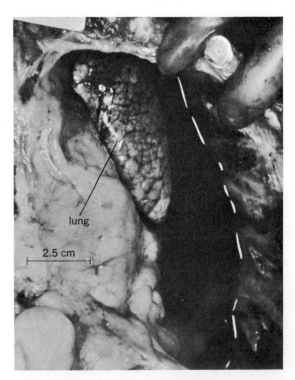

Fig. 3. Chest cavity. Lung has been displaced upward by a massive accumulation of blood in pleural cavity (hemothorax), secondary to traumatic injury; 2600 cm³ of blood were removed. White lines indicate approximate contour the lung would normally occupy.

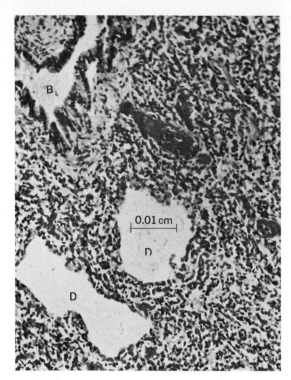

Fig. 4. Photomicrograph of atelectatic lung. The alveolar architecture is obliterated. Some air spaces are markedly dilated (D); B stands for bronchus.

as hyaline membranes.

Pneumonia. This is a general term denoting solidification (consolidation) of the lung tissue caused by inflammation. The disease may be caused by bacteria, viruses, chemical or physical agents, or any combination of these and varies greatly in pattern and severity. Bacterial pneumonias are most frequently caused by pneumococci, streptococci, staphylococci, and *Haemophilus* influenza. *See* HEMOPHILIC BACTERIA; PNEUMOCOCCUS; PNEUMONIA; STAPHYLOCOCCUS; STREPTOCOCCUS.

Lobar. Though now infrequent, lobar pneumonia is still found in alcoholics or unattended, debilitated patients, particularly after exposure to cold weather. It may involve a segment, a lobe, or an entire lung. The disease manifests with chest pain, cough, blood-tinged sputum, and fever. In the initial phase, if untreated, the lung is congested, heavy, and bulky; the capillaries are engorged; and the alveoli contain neutrophils, fibrin, and bacteria. Early consolidation (red hepatization) occurs at 2–4 days: The lung is heavy, dull, red, and friable, and the alveoli contain very abundant neutrophils and fibrin. Late consolidation (gray hepatization) occurs at 4–8 days: The lung may increase several times in weight and the alveoli contain fibrin and disintegrating leukocytes (pus cells) (Fig. 5). Resolution occurs at 8–12 days: The exudate is resorbed by macrophages without destruction of the normal lung architecture. Modern antibiotic therapy alters the course of the disease and may induce resolution in 3–5 days.

Bacterial bronchopneumonia. This denotes the patchy consolidation of the lungs occurring around bronchi from which the infection has extended. This form may be a serious complication of practically any disease. In any bacterial pneumonia, the intralveolar exudate may not be resorbed but rather replaced by granulation tissue (organizing pneumonia), resulting in scar formation. Also, the involved area may be destroyed (necrosis) by the infection, resulting in abscess formation (Fig. 6). Other complications include empyema and sepsis.

Viral. In viral pneumonia, caused by virus of various types of influenza or measles, among others, the inflammatory reaction predominates in the septa. The alveolar lumina contain fibrin, edema fluid, and some inflammatory cells. In severely ill, debilitated patients, superimposed bacterial or fungal infections may occur. These modify the tissue reaction, aggravate the disease, and may have a fatal outcome.

Other causes of pulmonary inflammation. Aspiration of oily substances by infants and children may cause lung consolidation (lipoid pneumonia); fatty droplets tend to be ingested by macrophages. Inhalation of food or gastric contents can produce similar changes. Heavy exposure to radiation also causes pulmonary inflammation. The severity of the pulmonary inflammation varies according to the type and amount of radiation and the duration of exposure.

Tuberculosis. This disease is caused by *Mycobacterium tuberculosis* of various types. The death rate from tuberculosis (TB) in the United States has decreased from 200 per 100,000 population in 1900 to 5.9 per 100,000 in 1960. The number of new cases, however, has not decreased in similar proportion. The identification of acid-fast bacilli (AFB) in tissue sections is not necessarily pathognomonic of tuberculosis. Other AFB may cause pulmonary granulomatous diseases that differ from tuberculosis in course, prognosis, and antibiotic sensitivity. It has been estimated that in some regions 1–10% of patients thought to have pulmonary tuberculosis may actually suffer from nontuberculous mycobacteriosis. *See* MYCOBACTERIAL DISEASES.

Modern therapy is highly effective against tu-

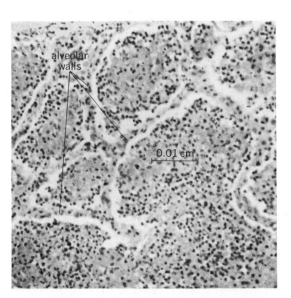

Fig. 5. Focus of acute bacterial pneumonia. The alveoli are distended and contain abundant fibrin and leukocytes. Basic lung architecture is preserved.

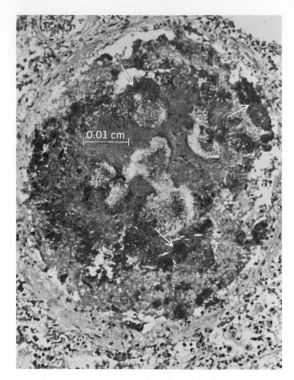

Fig. 6. Small pulmonary abscess. In the irregularly round area the alveolar architecture has been completely destroyed, with necrotic debris remaining. The darkest foci (arrows) represent bacterial colonies.

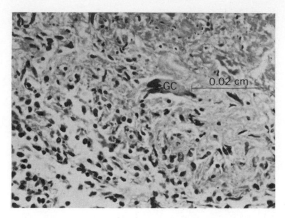

Fig. 7. Active tuberculous granuloma (*Mycobacterium tuberculosis* obtained by culture). The alveolar architecture is obliterated. Caseous necrosis is evident in upper right corner. A typical giant cell (GC), some epithelioid cells, and lymphocytes are evident.

The mechanism whereby particulate matter produces lung disease is not quite clear and, possibly, may not be the same for all materials involved. Most induce alveolar fibrosis, sometimes with granuloma formation.

Anthracosis is coal dust accumulation in lungs. It is almost invariably found in urban dwellers without clinical signs of disease; it may cause emphysema in miners of soft coal. Silicosis is caused by silica particles and may affect mine, stone, and metal workers. It is a slowly progressive but serious disease, leading to pulmonary failure and secondary heart disease. Also, it is frequently

berculosis. However, the disease must be considered serious, and it is not infrequent. Autopsy examinations have revealed active tuberculosis in patients that were not even suspected of having the disease.

Primary pulmonary tuberculosis usually develops in children but may also appear in adults transferred from an area where the disease is rare to another where it prevails. The initial lesion is the tubercle (tuberculous granuloma); it consists of epithelioid cells, occasional giant cells, and lymphocytes and is found in the lung and mediastinal lymph nodes. The center of the lesion may become liquified (caseous necrosis) (Fig. 7). In most cases the lesion heals with scar formation. Occasionally the disease may spread throughout the lungs (tuberculous pneumonia or bronchopneumonia) or invade the bloodstream and disseminate throughout the body (miliary tuberculosis). In other cases the disease progresses slowly, locally destroying large areas of the lung and forming cavities. Many microorganisms may produce granulomatous pneumonias with or without generalized disease. Particularly noteworthy are coccidioidomyocosis, blastomycosis, sarcoidosis, actinomycosis, aspergillosis, tularemia, histoplasmosis, and toxoplasmosis. These diseases and others, not necessarily caused by bacteria or fungi (the cause of sarcoidosis remains obscure), share similar granulomatous lesions. The positive identification of the organism is the only way to establish the diagnosis. *See* ACTINOMYCOSIS; ASPERGILLOSIS; BLASTOMYCOSIS; COCCIDIOIDOMYCOSIS; HISTOPLASMOSIS; TOXOPLASMOSIS; TUBERCULOSIS; TULAREMIA.

Pneumoconioses. These are a group of pulmonary diseases caused by inhalation of specific dusts.

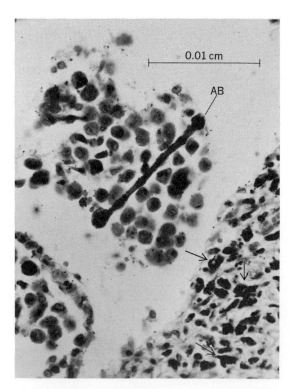

Fig. 8. Asbestosis. The dilated alveolar lumen contains numerous macrophages surrounding a drumstick-shaped asbestos body (AB). The massively thickened alveolar septum (lower right) contains macrophages with carbon particles (arrows).

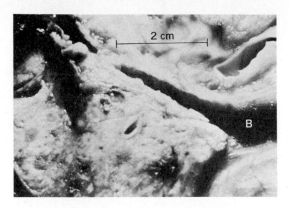

Fig. 9. Gross specimen of lung. Compressed lumen of right main bronchus (B) and branches appears surrounded by infiltrating tumor (white in photograph) that extends into and obliterates the pulmonary tissue.

complicated by bacterial infections, including tuberculosis. Asbestosis is a pneumoconiosis caused by asbestos inhalation. The disease affects asbestos miners and particularly workers processing the material or using it for insulation. The disease evolves in 5–10 years or more to severe pulmonary fibrosis and insufficiency (Fig. 8). Statistically, it seems to predispose to pulmonary and pleural cancer. Other forms of pneumoconiosis include byssinosis (cotton dust), berylliosis (beryllium dust or vapor), and bagassosis (sugarcane fiber).

Tumors. Most primary lung tumors are carcinomas. Evidence suggests that, for several decades prior to the 1960s, the incidence of this malignancy multiplied several times in men and about doubled in women. Lung carcinoma has become the most frequent visceral malignant tumor in males in the Western world. In the period 1955–1961 the number of deaths per year attributed to carcinoma of the lung in the United States increased approximately from 27,000 to 39,000. Some, but not all, of this increase can be attributed to better diagnosis, longevity, and reduction of mortality from other diseases. Statistical evidence pointing to cigarette smoking as a contributory factor appears strong. Other contributing factors may include air pollution from engine and industrial fumes and exposure to specific materials such as asbestos, cadmium, beryllium, and radioactive particles. The Channel Islands, with the world's highest per capita rate of cigarette consumption, has the highest incidence of lung carcinoma. However, in general heavy cigarette smokers living in rural areas show a lower incidence—thus the suggested importance of atmospheric pollution in the heavily industrialized and populated urban areas. *See* AIR-POLLUTION CONTROL; BRONCHIAL DISORDERS; RESPIRATORY SYSTEM DISORDERS.

About 60–75% of lung carcinomas are related to or arise from a tumor of the main bronchus or its bifurcation. Some tumors grow into the bronchial lumen, causing partial or complete occlusion, while others grow outwardly and infiltrate the lung and mediastinal tissues (Fig. 9). Early diagnosis and surgery or radiotherapy offer the best hope for treatment.   [VICTOR E. GOLDENBERG]

Bibliography: W. A. D. Anderson (ed.), *Pathology*, 5th ed., 1966; P. B. Beeson and W. McDermott (eds.), *Textbook of Medicine*, 12th ed., 1967; *The Health Consequences of Smoking*, Public Health Serv. Publ. no. 1696, 1967–1968; S. L. Robbins, *Pathology*, 3d ed., 1967; *Smoking and Health*, Public Health Serv. Publ. no. 1103, 1964–1965; H. Spencer, *Pathology of the Lung*, 1962.

Lupine

A cool-season legume with an upright stem, leaves divided into several digitate leaflets, and terminal racemes of pea-shaped blossoms. Three species, yellow, blue, and white, each named for the color of its blossoms, are cultivated as field crops (see illustration); several hybrids are grown as orna-

Lupine. (a) A field crop near maturity. (b) Uprooted yellow plants grown without (left) and with (right) needed nitrogen-fixing bacteria. Bacteria grow in the nodules seen on the roots at the right.

mentals; and many species occur as wild plants. The yellow crop varieties are usually the earliest and smallest, the whites latest and largest. Field crop lupines have been grown in Europe since early Roman times as a soil-improving crop. The older varieties could not be used as forage since the plants contained a bitter, water-soluble, toxic alkaloid. However, since 1912, plant breeders have developed "sweet" varieties with only traces of alkaloid.

Blue lupine was grown extensively as a winter cover crop in the southeastern United States during the 1940s. Since the beginning of the 1950s, however, lupine culture has declined, partly because cheap nitrogen fertilizer competes with the nitrogen-fixing value of the cover crop, and also because diseases and winter killing of lupines have increased. Some progress has been made in developing disease-resistant varieties.

The ornamental lupines are perennials unsuited to areas with hot summers. The many kinds of wild lupines which occur in the western part of the United States vary greatly in length of life, size, and site requirement. In the East a few small perennial wild lupines are found on deep sands. *See* BREEDING (PLANT); COVER CROPS; LEGUME FORAGES.

[PAUL TABOR]

Lupus erythematosus

One of the collagen diseases in which connective tissues show a similar form of biochemical and structural alteration. Each of the collagen diseases is a distinct clinical and pathologic entity and such a grouping exists more as a matter of convenience than because of a common etiologic factor. *See* CONNECTIVE TISSUE.

Collagen diseases display three characteristic alterations: mucoid degeneration, fibrinoid degeneration, and collagenous fibrosis or hyalinization of the degenerated foci. Each has its own histologic appearance and may occasionally be seen in other diseases. The causes are poorly understood, but hypersensitivity and nucleoprotein derangement appear to be involved. *See* HYPERSENSITIVITY; NUCLEOPROTEIN.

Lupus erythematosus is an acute or subacute febrile disease that may produce widespread damage to the connective tissue components of many organs and tissues. It occurs predominately in young women, but males appear to be increasingly affected, perhaps because of development of better diagnostic procedures. The blood vessels, heart, kidneys, skin, and serous membranes are characteristically involved. Only a minority (about 10%) of the patients show the characteristic erythematous butterfly-shaped rash over the cheeks (see illustration) and even less show the characteristic periungual erythema. A blood serum factor is present in lupus which is responsible for the so-called lupus erythematosus (LE) cell, a polymorphonuclear leukocyte filled with an inclusion body.

There is a significant incidence of involvement of the heart (myocarditis), and also of the lining and valves of the heart (endocarditis). Kidney lesions (nephritis) and lesions of the serous membranes of the heart (pericarditis), lungs (pleuritis), and abdominal peritoneum (peritonitis) are commonly encountered. Small arteries become the

LUPUS ERYTHEMATOSUS

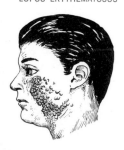

Lupus erythematosus. (*W. A. N. Dorland, American Illustrated Medical Dictionary*, 24th ed., Saunders, 1965)

sites for inflammatory reactions, especially in the kidney.

The clinical picture is quite varied, depending on the organs involved.

Constitutional symptoms and signs (fever, weakness, fatigability, or weight loss) may be the first manifestation of illness and are often insidious. The joints are affected in 90% of the patients. Severe joint pain with little or no objective evidence of arthritis is most common. The heart is involved in over one-half of the cases and there may be heart failure. Lupus nephritis is very frequent and is the most serious feature of lupus. It usually progresses to subacute glomerulonephritis with uremia and usually hypertension. Anemia is also a very common feature of lupus. Lupus often mimics other diseases because of the wide possibility of tissue damage.

The disease is treated with high doses of corticosteroids which often exert a remarkable effect, especially in the early stages of the illness. The antineoplastic drugs are also being used in the treatment of lupus; some favorable results have been reported but, as with steroids, their side effects are severe. The course and prognosis are variable. Even with treatment the disease may be fulminant with a rapid progression of severe symptoms leading to death in a few weeks. More frequently the disease follows an episodic pattern with recurrent involvement of one or more organ systems over a period of many years.

Discoid lupus is a common skin disease that seldom precedes disseminated lupus; many authorities consider the two to be separate diseases entirely. Other collagen diseases include periarteritis nodosa, dermatomyositis, and scleroderma. Rheumatic fever, rheumatoid arthritis, and serum sickness, as well as certain other diseases, may be included in the category. Each of these disorders has its own peculiarities, clinical course, and prognosis. *See* RHEUMATIC FEVER.

[N. KARLE MOTTET]

Lutetium

A chemical element, Lu, atomic number 71, atomic weight 174.97, a very rare metal and the heaviest member of the rare-earth group. The naturally occurring element is made up of the stable isotope Lu^{175}, 97.41%, and the long-life β-emitter Lu^{176} with a half-life of 2.1×10^{10} years. It was discovered in 1906 by G. Urbain, who named it lutecium, after an ancient name for Paris. It was also discovered independently by C. F. Auer von Welsbach,

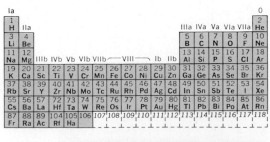

who named the element cassiopeium, but lutetium is the accepted form at present. The salts are colorless and give a trivalent ion in solution.

Lutetium, along with yttrium and lanthanum, is of interest to scientists studying magnetism. All of these elements form trivalent ions with only subshells which have been completed, so they have no unpaired electrons to contribute to the magnetism. Their radii with regard to the other rare-earth ions or metals are very similar so they form at almost all compositions either solid solutions or mixed crystals with the strongly magnetic rare-earth elements. Therefore, the scientist can dilute the magnetically active rare earths in a continuous manner without changing appreciably the crystal environment. *See* MAGNETOCHEMISTRY; RARE-EARTH ELEMENTS.

[FRANK H. SPEDDING]

Lux

The unit of illumination when the square meter is taken as the unit of area. It is defined as the illumination on a surface, 1 m² in area, on which there is a luminous flux of 1 lumen uniformly distributed, or the illumination on a surface all points of which are at a distance of 1 m from a uniform point source of 1 candela. *See* ILLUMINATION; LUMEN.

[RUSSELL C. PUTNAM]

Lychee

The plant *Litchi chinensis*, also called litchi, a member of the soapberry family (Sapindaceae). It is a native of southern China, where it has been cultivated for more than 2000 years. It is now being grown successfully in India, Union of South Africa, Hawaii, Burma, Madagascar, West Indies, Brazil, Honduras, Japan, Australia, and the southern United States.

The fruit is a one-seeded berry. The thin, leathery, rough shell or pericarp of the ripe fruit is bright red in most varieties. Beneath the shell, completely surrounding the seed, is the edible aril or pulp. The fruit may be eaten fresh, canned, or

dried. In China the fresh fruit is considered a great delicacy. *See* SAPINDALES.

[PERRY D. STRAUSBAUGH/EARL L. CORE]

Lychniscosa

An order of the subclass Hexasterophora in the class Hexactinellida. The parenchymal megascleres of these sponges are united to form a rigid framework. The parenchymal megascleres are all or in part lychniscs, hexactins in which the central part of the spicule is surrounded by a system of 12 struts. Examples of this order are *Aulocystis* and *Dactylocalyx*. *See* HEXACTINELLIDA; HEXASTEROPHORA.

[WILLARD D. HARTMAN]

Lycopodiales

An order of the Lycopodiophyta with a fossil history extending back into the Paleozoic. These plants, commonly called club mosses, attained their greatest development during the Carboniferous Period and included several treelike forms comparable in size with the present-day pines (Fig. 1). Today the order contains only two genera, *Lycopodium* and *Phylloglossum*, composed of small, herbaceous, evergreen plants with upright or trailing stems bearing numerous small, simple, spirally arranged leaves. Some of the tropical species are epiphytes (living perched on larger plants). The stems are dichotomously branched (forked). In some species one fork forms a lateral branch while the other becomes the main axis. The sporophylls (sporangium-bearing leaves) may be aggregated together into a cone or strobile, or the sporangia may be borne singly on the upper surface of sporophylls scattered along the stem. The cones are often elevated on upright branches so that they occur above the leafy shoots. Because the cones resemble a club, these plants have been given the erroneous common name, club moss. The similar spores (homospores) are grouped in tetrads, and this is an indication that reduction division occurs in their formation from the spore mother cells. *See* GEOLOGY; PALEOBOTANY.

Fig. 1. Restoration of forest trees, ancestral ferns, club mosses, and equisetum from Carboniferous Period. *(From E. N. Transeau, H. C. Sampson, and L. H. Tiffany, Textbook of Botany, rev. ed., Harper, 1953)*

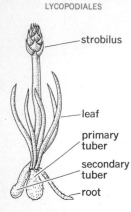

— strobilus

— leaf

— primary tuber

— secondary tuber

— root

Fig. 2. Fertile plant of *Phylloglossum drummondii*, twice natural size. (*From A. W. Haupt, Plant Morphology, McGraw-Hill, 1953*)

Classification. Of the two genera included in this order, *Phylloglossum* is confined to portions of Australia, Tasmania, and New Zealand (Fig. 2). This genus has only one species of simple structure which may be the living descendant of more highly developed ancestors. The other genus, *Lycopodium*, has about 180 species and is worldwide in distribution. The greatest concentration of species is in the tropics and subtropics, but these plants also grow in the arctic region. Many of the species are perennial and inhabit the moist woodland or forest floor, whereas others are epiphytic and grow upon forest trees.

Structure. In most respects the anatomy of *Lycopodium*, especially that of the roots and leaves, is like that of other vascular (higher) plants. The primitive arrangement of the vascular tissues in the stem is called a protostele (a solid cylinder devoid of pith), which may be modified by the relative position and amounts of xylem and phloem in the mature stem. All of the vascular tissue is primary, since it is produced by a cluster of apical meristematic cells with no indication of cambial, or secondary, activity. Development and maturation of the xylem is exarch, or toward the center. A strand of conducting tissue extends from the stele into each leaf and becomes its vein, but no leaf gap is produced where the strand emerges from the stele. Similarly, each branch and root is connected with the primary stele, but in these cases gaps are left in the central cylinder, as is true in all vascular plants. *See* Leaf (botany); Meristem, apical; Phloem; Root (botany); Stem (botany); Xylem.

Alternation of generations. The conspicuous plant with its roots, stem, leaves, and sporangia is the sporophyte or spore-producing generation (Fig. 3). The sporophyte begins with the zygote (fertilized egg) and ends with the formation of spores.

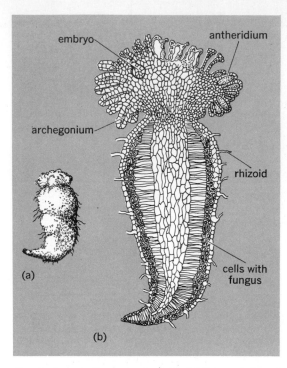

Fig. 4. *Lycopodium* gametophyte. (*a*) External view of lumpy, subterranean gametophyte. (*b*) Longitudinal section of gametophyte, showing cells containing fungus (dark cells). (*After Bruchmann, from H. J. Fuller and O. Tippo, College Botany, rev. ed., Holt, 1954*)

The gametophyte (gamete-producing generation) begins with the spore and ends with the union of the egg and sperm in the formation of the zygote. In *Lycopodium* sporangia are large and somewhat kidney-shaped, and are borne singly in the upper leaf axils. Each sporangium has a short stalk and a thick wall of several cell layers. When mature, it contains a mass of small, yellow, thick-walled spores. The spores are liberated through a transverse slit appearing in the sporangium wall. In this genus there is a series of progressive stages among the numerous species, varying from those in which every leaf on the plant is potentially a sporophyll to those in which the sporophylls are aggregated into compact, well-differentiated strobili (cones). According to species, the liberated spores vary in the speed with which they germinate and develop into gametophytes (gamete-producing generation). Many spores seemingly require the presence of a mycorrhizal fungus for their continued existence and maturation. In some species the gametophytes grow above ground and become green, whereas in other species they develop below ground and are devoid of chlorophyll. The gametophytes may be either tuberous or branching. The subterranean types (Fig. 4*a*) are infected with a fungus and are saprophytic (living on dead organic matter). The sex organs (antheridia and archegonia) are large and are borne on the top surface of the gametophyte (Fig. 4*b*). From a superficial cell, each antheridium develops into an oval mass of sperm mother cells which protrude slightly above the mass of gametophytic tissue. The sperms produced in the antheridium are fusiform (spindle-shaped) and each has two flagella attached at the anterior end. The archegonia (female) are produced on the same thallus, but usually toward the

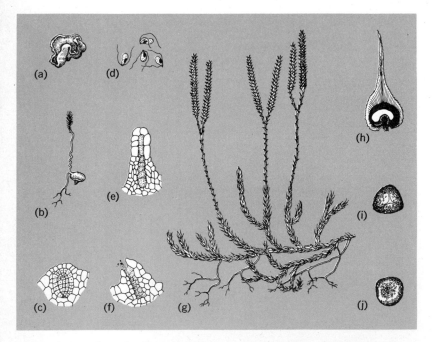

Fig. 3. *Lycopodium clavatum* sporophyte. (*a*) Old prothallus. (*b*) Prothallus shown with young plant attached. (*c*) Antheridium shown in vertical section. (*d*) Spermatozoids. (*e*) Young archegonium, with the neck still closed. (*f*) Open archegonium ready for fertilization. (*g*) Plant bearing cones. (*h*) Sporophyll with an opened sporangium, (*i, j*) Spores shown from two viewpoints. (*After Bruchmann, from E. Strasburger et al., A Textbook of Botany, 3d ed., Macmillan, 1908*)

margins. Many archegonia are formed, but only a few develop after fertilization. In some species several new sporophytes may be produced from the same thallus at intervals over several years. The gametophytic thallus persists and supplies nutrition to the developing sporophyte, which eventually becomes entirely self-supporting. The whole process of sporophyte development is very slow, and it may take several years for the young plant to reach the surface of the soil and take on the features of the mature plant. *See* FUNGI; TRACHEOPHYTA.

[PAUL A. VESTAL]

Lycopodiophyta

A division (formerly the subphylum Lycopsida) of the subkingdom Embryobionta (Embryophyta) having a long fossil history but now restricted to five living genera. The living members, commonly called club mosses, spike mosses, or quillworts, are significant in that the study of their structures and life cycles provides clues to the understanding of the many extinct species. Some of these plants, known only as fossils, were the dominant species of the extensive swampy forests of the Carboniferous and grew to 30 m in height. *See* GEOLOGY; LEPIDODENDRALES; PALEOBOTANY; PLEUROMEIALES.

Classification. The living members of this division are confined to the five genera: *Lycopodium* (Fig. 1), *Selaginella* (Fig. 2), *Phylloglossum* (Fig. 3), *Isoetes* (Fig. 4), and *Stylites*. They are herbaceous plants that grow close to the ground or as small epiphytes. All are widely distributed except *Phylloglossum*, which is confined to portions of Australia, New Zealand, and Tasmania, and *Stylites*, which is found in lakes in the high Andes of Peru. The greatest number of species are found in the tropical zone, although some species of *Lycopodium* may be found in the arctic regions. Nearly all of the genera grow in habitats which are moist and shady.

Structure. The living Lycopodiophyta are characterized by a dominant, independent (self-sustaining) sporophyte, which is differentiated into a root, stem, and leaves. The stems and roots branch dichotomously (fork). The leaves are small, simple, and spirally arranged. Typically they have but one

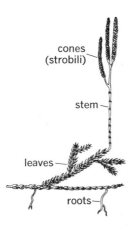

Fig. 1. *Lycopodium*, or club moss, showing prostrate stem with many small leaves, upright stem bearing cones, or strobili, and roots arising from rhizome. (*From H. J. Fuller, The Plant World, rev. ed., Holt, 1951*)

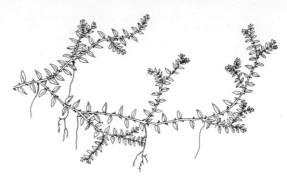

Fig. 2. *Selaginella kraussiana*, part of a sporophyte showing dichotomous branching. (*From G. M. Smith, Cryptogamic Botany, vol. 2, 2d ed., McGraw-Hill, 1955*)

vascular bundle, which produces no leaf gap as it departs from the stele of the stem. The vascular tissue of the stem is basically a protostele consisting of a solid, central core of xylem surrounded by a cylinder of phloem, although some lycopods have the center of the stele composed of parenchyma (pith) rather than tracheids, an organization termed a siphonostele. The phloem is surrounded by a pericycle and a well-defined cortex and epidermis. The tracheids found in the vascular tissue are largely scalariform. Several species have a cambium which forms secondary vascular tissue. The sporangia (spore-producing structures) are solitary at the base of the upper surface of fertile leaves called sporophylls, or are axillary to the sporophylls. The sporophylls may be localized at the tip of a branch and form a structure called a cone, or strobilus. The gametophytes in *Lycopodium* are varied; some are green and photosynthetic, whereas others are subterranean, are infected with a mycorrhizal fungus, and live as saprophytes. In *Selaginella*, *Isoetes*, and *Stylites*, which are heterosporous, the unisexual gametophytes derive their food from a sporophytic origin which has been stored in the spores. The sperm produced in the antheridia of *Lycopodium* and *Selaginella* are biflagellate, while in *Isoetes* and *Stylites* they are multiflagellate. *See* CORTEX (PLANT); EMBRYOBIONTA; EPIDERMIS (PLANT); ISOETALES; LEAF (BOTANY); LYCOPODIOPSIDA; MERISTEM, LATERAL; PARENCHYMA; PERICYCLE; PHLOEM; STELE; STEM (BOTANY); VASCULAR BUNDLES; XYLEM.

[PAUL A. VESTAL]

Bibliography: H. C. Bold, *Morphology of Plants*, 2d ed., 1967; H. J. Fuller and O. Tippo, *College Botany*, rev. ed., 1954; A. W. Haupt, *Plant Morphology*, 1953; R. F. Scagel et al., *An Evolutionary Survey of the Plant Kingdom*, 1965; G. M. Smith, *Cryptogamic Botany*, vol. 2, 2d ed., 1955.

Lycopodiopsida

One of the two classes of the division Lycopodiophyta, the other being the Isoetopsida; another name for this class is Lycopodineae. The plants in this class are commonly referred to as the lycopods. They have a very long history among vascular plants, dating back to the early Devonian and possibly Cambrian. During the Carboniferous the lycopods were among the largest and most numerous of the plant groups. Following the Carboniferous these large, treelike lycopods disappeared

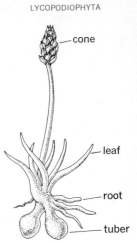

Fig. 3. *Phylloglossum*, showing the fleshy tubers at the base which give rise to roots, quill-like leaves, and a stalk with a terminal cone. (*From H. J. Fuller and O. Tippo, College Botany, rev. ed., Holt, 1954*)

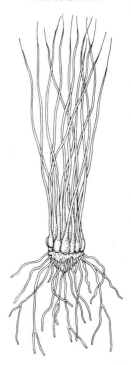

Fig. 4. *Isoetes nuttallii*, natural size. (*From A. W. Haupt, Plant Morphology, McGraw-Hill, 1953*)

and as a group declined in relative importance among vascular plants. The living members of this group are significant in that the study of their structures and life cycles provides clues to the understanding of the now extinct species of previous eras and the numerous changes which have occurred during the evolution of the present land flora. *See* PALEOBOTANY.

The two most common genera and the ones having the widest distribution are *Lycopodium* and *Selaginella*. In *Lycopodium* the spores produced by the sporophyte are all alike (homosporous), whereas in *Selaginella* the spores are unlike (heterosporous). *See* ISOETALES; LEPIDODENDRALES; LYCOPODIALES; PLEUROMEIALES; SELAGINELLALES; TRACHEOPHYTA.   [PAUL A. VESTAL]

Bibliography: *See* LYCOPODIOPHYTA.

Lymph node (vertebrate)

Aggregations of lymphoid tissue found along the course of lymphatic vessels. Lymph nodes vary in length from 1 mm to 2 cm or more, and in shape from spherical to flattened on one or more sides. A fibrous capsule surrounds each gland and sends supporting sheets, or trabeculae, through the lymphoid tissue (see illustration). The lymphoid tissue is composed of germinal centers of lymphocytes together with other cells, particularly those of the reticuloendothelial system. Lymph glands are found in superficial and deep sites, including the neck, axilla, groin, mediastinum, abdomen, pelvis, and intestinal wall (Peyer's patches). At least two functions are apparent: to supply lymphocytes to the circulation, and to remove bacteria and particulate matter carried to the nodes by the

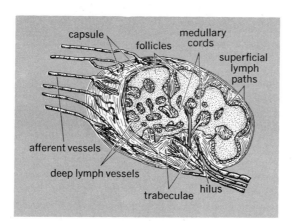

Diagrammatic sketch of a lymph node. (*After C. Toldt, An Atlas of Human Anatomy for Students and Physicians, vol. 1, 2d ed., Macmillan, 1941*)

lymphatic vessels. In addition, it is thought that lymph glands may be the site of antibody formation. *See* ANTIBODY; LYMPHATIC SYSTEM (VERTEBRATE).   [WALTER BOCK]

Lymphatic system (vertebrate)

A system of vessels in the vertebrate body, beginning in a network of exceedingly thin-walled capillaries in almost all the organs and tissues except the brain and bones. This network is drained by larger channels, mostly coursing along the veins and eventually joining to form a large vessel, the thoracic duct, which runs beside the spinal column

to enter the left subclavian vein at the base of the neck. The lymph fluid originates in the tissue spaces by filtration from the blood capillaries. While in the lymphatic capillaries it is clear and watery. However, at intervals along the larger lymphatic vessels, the lymph passes through spongelike lymph nodes, where it receives great numbers of cells, the lymphocytes, and becomes turbid. The lymphatic vessels other than capillaries contain numerous valves preventing backflow of the lymph. The functions of the lymphatic system are to remove particulate materials such as molecular proteins and bacteria from the tissues, to transport fat from the intestine to the blood, and to supply the blood with lymphocytes.

This article contains sections on the comparative anatomy, histology, embryology, and physiology of the lymphatic system.

COMPARATIVE ANATOMY

It is disputable how far down in the animal series lymphoid tissue is found. Some authors speak of lymphocytic cells in annelid worms, in mollusks, and in other invertebrates.

In general, the more primitive and less differentiated blood cells resemble more closely the lymphocytes of vertebrates. This is true also of the lower chordates in which so-called lymphocytes of the tunicates are said to differentiate into any of the other cell types of the body. Lymph nodules are described in some species of tunicates in the body wall and in the gut wall.

In the nonchordates it is difficult or impossible to make a distinction between blood vessels and lymphatic vessels. Probably the two systems had a common type of origin.

In *Branchiostoma* (*Amphioxus*), however, among the lower chordates, lymphatic vessels can be identified surrounding the blood vessels, and are seen also around the central nervous system, in the dorsal fin, and in the metapleural folds.

The lymphatics of vertebrates are distributed in somewhat the same manner as are the veins (Fig. 1). Thus they are divided into a deep and a superficial set. The deep set develops in relation to the cardinal veins and then makes connection with them and with the superficial set. *See* LYMPH NODE (VERTEBRATE).

Fishes. In fishes the lymphatic system varies in degree of development in different species. In some, the veins seem to carry on the function of lymphatic drainage, but in a number of species of ganoids, teleosts, and elasmobranchs a well-developed lymphatic system is seen. The lymphatic vessels surround the veins in the elasmobranchs; in other fishes they are not so closely related to the blood vessels. In the elasmobranchs, also, there may be one or two large trunks which run parallel with the aorta and are comparable to the thoracic duct or ducts of man and other mammals. Pulsating organs called lymph hearts, which assist in driving the lymph along the vessels, occur in some fishes (Fig. 2). The lymph vessels lack valves. Lymph nodes are absent, although there are collections of lymphoid tissue, chiefly close to mucosal surfaces. Nodules are found in these collections but they lack germinal centers.

The organ of Leydig of the selachian fishes consists of two large accumulations of lymphoid tissue which run longitudinally the whole length of the

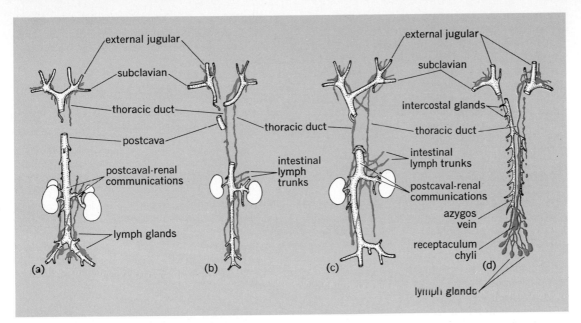

Fig. 1. Chief mammalian lymphatic trunks and their relations to the veins; ventral views. (a) South American monkeys. (b) Mammals (*Lepus*) in which postcaval-renal communications are wanting. (c) Mammals in general. (d) Man (postcava and kidneys not shown). In all mammals the lymph enters the veins at the point of junction between the jugular and the subclavian veins. In most mammals there is also communication between the lymph vessels and the postcaval and renal veins. In man the right thoracic duct degenerates in part and the only communication with veins is at the root of the jugulars. (*From H. V. Neal and H. W. Rand, Chordate Anatomy, Blakiston–McGraw-Hill, 1939*)

esophagus and even extend into the stomach. Although the cells of this organ are primarily lymphocytic, they apparently serve as stem cells for all the types of white blood corpuscles in these fishes, and hence the tissue perhaps should be spoken of as lymphomyeloid, indicating its hematopoietic as well as lymphoid character. *See* GLAND.

In the sturgeon a mass of lymphoid tissue has been described which occurs near the heart and resembles a lymph node, but its homologies are doubtful.

Amphibia. In the Amphibia the lymphatic system often includes very large subcutaneous sinuses, probably to protect against the danger of loss of moisture. Dilatations of the lymphatics, or lymph sacs, occur frequently. Lymph hearts are common in various parts of the body and a thoracic duct is present. Accumulations of lymphoid tissue are often present adjacent to the lymph sacs, but are not comparable in architecture to the lymph nodes of higher vertebrates. They may, however, represent an early stage in the development of such nodes.

Reptiles. Reptiles have a well-developed lymphatic system but lack the special features seen in amphibians. In the crocodiles true lymph nodes have been described in the mesentery.

Birds. The birds present a special condition in relation to the development of a lymphatic system. Here, as in other vertebrates, the more fluid portions of the blood readily leave the capillaries to diffuse among the tissue cells. This exudate is collected into lymphatic capillaries which join to form larger lymphatic vessels. The entire system converges toward the two superior (or anterior) venae cavae into which the main lymphatic trunks (thoracic ducts) empty. These veins bring the blood from the head, neck, and wings of the bird. Thus far the description of the lymphatic system of birds resembles that of mammals, even to the union of lymphatic with venous system.

The lymphatic vessels of birds are similar in structure to those of mammals and are provided with valves to help ensure one-way flow. However, the number of vessels is less than in mammals.

Lymph nodes are absent in vertebrates lower than the birds; that is, they are lacking in cold-blooded animals. According to J. W. Dransfield, lymph nodes are not present in the common fowl (*Gallus domesticus*), but plexuses of lymphatic vessels do occur in the places where nodes are found in some other birds. Nodes are developed only in certain families of birds. This is not to say, however, that lymphatic tissue is lacking in birds without lymph nodes, for such tissue occurs in both diffuse and nodular form in the walls of the alimentary and respiratory systems: "Tonsillar" structures occur in the upper parts of these tracts; a large "esophageal tonsil" is found in the common domestic fowl; the bursa of Fabricius is a lymphoid organ in the wall of the posterior part of the alimentary tract; and the thymus appears to be an important producer of lymphocytes.

Early studies demonstrated that the quantity of lymphatic tissue increases, in general, toward the anal region in all birds studied. The avian cecum, in such mammals as the rabbit, is particularly rich in lymphoid tissue and may be said to constitute an "intestinal tonsil."

According to other findings, true lymph nodes probably occur only in water, shore, and swamp birds. Even when present, they are few, two pairs being found for the cervical-thoracic and two for the lumbar region. The nodes generally are elongated and spindle-shaped and lack a hilum. Capsule and trabeculae are lacking or very poorly developed.

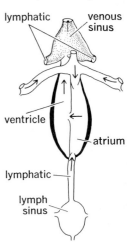

Fig. 2. Lymph heart in the tail of a teleost. (*From G. C. Kent, Comparative Anatomy of the Vertebrates, 2d ed., Blakiston–McGraw-Hill, 1965*)

Lymph nodes arise in birds about midway in embryonic life. The wall of the lymphatic vessel in the region of development becomes surrounded by a thick mass of mesenchyme which sends cordlike growths into the lumen of the vessel and eventually leaves only the central portion free, to serve as a central sinus into which the peripheral spaces open. Infiltrating lymphocytes transform the mesenchyme into truly lymphoid tissue. Nodules are formed, and in the centers of these nodules the germinal, or reaction, center develops, which is not found in fishes, amphibians, or reptiles.

These centers contain many phagocytic cells, or macrophages, as well as lymphocytes, some of which are undergoing mitotic division. These new centers in birds are found not only in the lymph nodes but also in individual nodules in other locations.

The lymph hearts, which are prominent features in cold-blooded vertebrates, were described by A. Stannius in a number of species of birds, including the ostrich, cassowary, goose, swan, stork, and dove. The lymph hearts are paired structures in close relation to the kidneys and are true pulsatile organs provided with striated muscle. Two additional lymph hearts have been described in the goose; these have delicate, bladderlike walls which lie close under the hypogastric veins and are places of convergence for lymphatic vessels coming from several directions. From each of these hearts, a lymphatic vessel passes forward to the lumbar lymph nodes, and in its course receives tributaries from the pelvis and lower extremities.

In the fowl (*Gallus*) genuine lymph hearts with striated muscle are found only in the embryo. They do not degenerate completely, however, until after hatching, and traces may be seen up to the thirty-fifth day after hatching.

According to J. R. Simons, lymph hearts occur in the embryos of all birds and are found on either side of the last sacral vertebrae. After hatching they disappear in some birds, as in the fowl, but are retained to some degree in the majority. They seem to be most prominent in the ostrich.

There does not appear to be a regular pulsation in the lymph hearts of birds but rather a contraction initiated by stretching of the wall due to filling. In the birds as a class the lymph hearts appear to be in a regressive condition when compared with such structures in amphibians and reptiles.

Mammals. The lymph nodes of mammals vary in number, size, form, and structure in different species. There appear to be some rather surprising differences in numbers of lymph nodes in different kinds of mammals. For example, the dog is said to have about 60, the cow 300, man 465, and the horse 8000. In the dog and cow, however, the lymph nodes are relatively large; in man and the horse, they are relatively small. Their shape also varies. In the cow, for instance, the mesenteric nodes are long and narrow, attaining a length of several decimeters. In the rabbit the mesenteric nodes seem to have fused into one huge node, the pancreas aselli, weighing several times as much as all the other nodes combined. There is a similar arrangement for the mesenteric nodes in some other rodents and in some carnivores. These species include the mole, the dolphin, the seal, and the narwhal whale.

The amount of connective tissue of the lymph nodes, that is, the degree of development of the capsule and trabeculae, varies in different mammals. It is weakly developed in the pig, dog, sheep, rabbit, guinea pig, and rat. In man and in the horse the trabeculae of the medullary part are more strongly developed than those of the cortex. Many mammals show a strong development of smooth muscle in the capsule and trabeculae, for instance, the horse, sheep, cow, and dog. The cow has an especially thick capsule and trabeculae.

In some lymph nodes the cortex, with its nodules, is very sharply set off from the medulla, with its more diffuse lymphoid tissue. This is true of the cow. In other animals the nodes seem to vary in this respect in different parts of the body or even in a rather haphazard manner, whereas in still others, such as the pig, there may seem to be almost a reversal of arrangement, with the nodular tissue situated centrally and the diffuse tissue peripherally.

Other lymphoid organs. These include the tonsils, thymus gland, and spleen, and in certain classes and groups of animals, structures which are confined to such groups, for instance, the bursa of Fabricius in the birds, a diverticulum from the lower end of the alimentary canal. *See* SPLEEN (VERTEBRATE); THYMUS GLAND (VERTEBRATE); TONSIL (HUMAN).

Tonsils. The tonsils seem to be represented in the frog and toad by two large fossae palatinae, invaginations of epithelium surrounded by dense masses of lymphoid tissue. Smaller structures of simple construction also are present.

In the posterior part of the buccal cavity of *Proteus* and *Salamandra* there are areas in which the epithelium is thickly infiltrated by lymphocytes.

Among the reptiles, *Lacerta agilis* shows definite tonsils in the pharynx, with elongated lymph nodules occurring along epithelial invaginations and with an apparent lack of basement membrane and a complete intermingling of epithelial and lymphoid tissue near the surface. Tonsils are seen also in the crocodiles.

Birds not only have large aggregations of lymphoid tissue in the pharynx but often also a large esophageal tonsil. The light-staining germinal centers are seen for the first time in birds.

A ring of lymphoid tissue, known as Waldeyer's ring, consisting of palatine, pharyngeal, and lingual tonsils is present in many mammals, for instance, man. Other mammals, however, may lack one or more of its component parts. In the rat and mouse, there are no palatine tonsils. In the hare and rabbit, pharyngeal tonsils are lacking. The rabbit, hare, dog, cat, and sheep apparently lack lingual tonsils.

Tonsils of the horse are peculiar in that they contain structures apparently identical with the Hassall's corpuscles of the thymus gland.

Spleen. The spleen is an organ of dual nature, belonging to both the blood-vascular and the lymphatic system. In the lower vertebrates it is actually within the wall of the alimentary canal. In cyclostomes it lies in the submucosa of stomach and intestine. In the lamprey it is within the spiral valve. Among the Dipnoi it is found within the stomach wall, but in elasmobranchs and ganoids it is a separate structure attached to the mesentery. In fishes and amphibians it is active as an erythrocytopoietic and lymphocytopoietic organ. Its function of forming lymphocytes is important also in higher vertebrates (including man and other mammals), but its blood-cell-forming function is lost and replaced by one of destruction of "old" or worn-out red corpuscles and, to some degree, of blood storage. *See* HEMATOPOIESIS.

The lymphoid tissue of the spleen in higher vertebrates contains many nodules known as Malpighian bodies. These usually contain active germinal centers.

HISTOLOGY

Lymphoid tissue is a tissue in which the predominant cell type, under normal conditions, is the lymphocyte. Lymphocytes are associated with reticular fibroblasts which form a delicate framework of fibers which can be demonstrated by their affinity for silver. There usually are cells of other types present also, including macrophages (fixed and free), plasma cells, and occasional eosinophile granulocytes.

Function. The chief function of the lymphoid tissue is to produce lymphocytes. In efferent lymphatic vessels, those leaving the lymph nodes, lymphocytes are considerably more abundant than in afferent lymphatics. Lymphocytes are added to the lymph as it flows through the lymph nodes. The lymphocyte content of veins coming from the nodes also is greater than that of arteries going to them. Apparently newly formed lymphocytes also migrate into the capillaries and small veins. A second important function of lymphoid tissue is protection against infection, which it carries out by means of its phagocytic and antibody-producing activity.

Cellular composition. Lymphocytes are of different sizes, small, medium, and large. The range in size of these three kinds, however, varies with different classes and even species of animals. The large lymphocyte resembles very closely the stem cell of blood-forming tissue, the hemocytoblast. In fact, lymphoid tissue itself bears a considerable resemblance to hematopoietic tissue, and in lower vertebrates it is not always possible to distinguish one from the other.

The function of the lymphocytes themselves is imperfectly known. The study of comparative histology, however, seems to throw light on their probable role in higher forms. In many invertebrates, the regenerative powers are much greater than in higher vertebrates. It has been shown that primitive mesenchymatous and free lymphoid cells often aggregate to form the various organ primordia and eventually even a completely differentiated animal. In some flatworms it has been found that not only do new parts arise from such elements (cells with conspicuous nucleoli and basophilic cytoplasm comparable to the large lymphocyte) but that the relative power of regeneration in different species may be correlated with the relative numbers of such cells. Thus, where they are numerous, as in *Planaria*, regenerative power is high. When they are not so common, as in *Procotyla*, regenerative power is low. In flatworms these cells will form such varied structures as epithelium of the gut, germ cells, and brain.

Although the lymphocytes of mammals do not have such remarkable potentialities, many workers have shown that they can develop into various other types of cells, including macrophages, which are large phagocytic cells; some types of granular white cells of the blood; red blood cells; and fibroblasts, or fiber-forming cells. The suggestion has been made also that they may metamorphose into epithelial cells in some locations. Thus one of their important roles in the higher vertebrates is still a regenerative or cell-replacing one.

Lymph vessels. The lymph capillaries, the smallest lymph vessels, form a dense network in almost every portion of the body. They appear to end blindly and consist only of an endothelial layer. As they unite to form larger vessels, connective tissue appears in the wall. The blind, rounded ends of lymphatic capillaries are easily seen in the lacteals in the villi, which receive the fatty foods from the intestinal lumen and hence contain chyle, a fluid made milky by the presence of multitudes of minute droplets of fat (Fig. 3).

In larger lymphatic vessels there are more collagenous fibers and also some elastic and smooth muscle fibers. In vessels with a diameter of more than 0.2 mm, the wall generally shows three layers which bear the same names as those in the blood vessels, namely, intima, media, and adventitia. In lymph vessels the adventitia is usually the thickest layer and often contains smooth muscle cells as well as collagenous and elastic connective tissue fibers. In the largest lymphatic vessels, the right lymphatic duct and the thoracic duct, the media is more fully developed and contains considerable smooth muscle.

Lymph nodes. The lymph nodes are masses of lymphoid tissue with a special type of architecture. They are interposed as way stations in the course of the lymphatic stream. The vessels entering the node are called afferent; those leaving it are the efferent vessels.

The cortex usually consists of rather dense lymphoid tissue composed of rounded nodules, made up chiefly of small lymphocytes. In the centers of the nodules are lighter-staining portions,

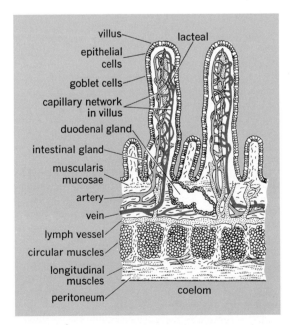

Fig. 3. Cross section of duodenum. (*From T. I. Storer and R. L. Usinger, General Zoology, 4th ed., McGraw-Hill, 1965*)

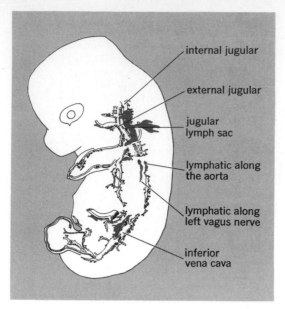

internal jugular

external jugular

jugular
lymph sac

lymphatic along
the aorta

lymphatic along
left vagus nerve

inferior
vena cava

Fig. 4. Lymphatic vessels and veins in a 14.5-mm rabbit embryo. (*From R. O. Greep, ed., Histology, 2d ed., Blakiston–McGraw-Hill, 1966*)

the germinal centers with macrophages and lymphocytes of medium and large size, frequently seen in mitosis.

The medulla of the lymph node consists of anastomosing channels or sinusoids, through which the lymph is coursing toward the efferent vessels and of cords of lymphoid tissue. It is more spongy than the cortex.

The lymph node is surrounded by a capsule consisting of collagenous and elastic connective tissue fibers and often containing some smooth muscle. At the hilum, where the blood vessels of the node enter and leave and from which the efferent lymphatics emerge, the connective tissue of the capsule penetrates well into the node.

[WARREN ANDREW]

EMBRYOLOGY

The mode of development of the lymphatic system, in contradistinction to the arterial and venous systems, has been elucidated only since about 1910. Embryos of representative species from all classes of vertebrates have been studied, and as yet there is no agreement on fundamental questions pertaining to the early embryology of this elusive system. The unsolved status appears to rest more on differences of interpretations than on sufficient observations and facts.

A sharp distinction must be made between the processes of the earliest formation of discrete lymph vessels and those leading merely to an extension of once-established channels. The former entail actual genesis of endothelium (vasculogenesis); the latter merely represent growth. Growth by budding and branching of newly established vessels occurs throughout the vascular system, whether hemal (arterial and venous) or lymphatic. The controversial aspects mainly concern the earliest beginnings of lymph vessels.

Theories. There are two conflicting concepts. One is that all lymph channels develop by centrifugal outgrowth from embryonic veins, specifically from their endothelial lining. In the neck and loin

regions, as well as in the posterior abdominal wall and at the root of the intestinal mesentery, such outgrowths form saclike expansions (Fig. 4; this figure, as well as Figs. 5 and 7, shows general topography and structural plan in schematic form, but in detail these situations are not precisely as shown in the diagrams). These are, respectively, the jugular and iliac (lumbar) lymph sacs, the future cisterna chyli (pool of chyle near the base of the abdominal aorta), and the mesenteric lymph sac. All the body-pervading lymphatics are said to be formed by sprouting from these sacs and thus to be derived indirectly from the lining epithelium of blood vessels. The original points of budding from certain veins (cardinals, for instance) either could be retained or lost, and secondary evacuation taps for the lymph could be established.

The opposing view holds that the lymphatics arise directly from mesenchymal spaces, either adjacent to decadent venules or unassociated with veins. During early embryonic development, some rich and temporary venous plexuses become reorganized. Certain venules atrophy and disappear, particularly in regions of the developing lymph sacs. The drainage function is taken over by a lymphatic network which temporarily may carry blood cells received either from degenerative venules or from blood islands in the surrounding mesenchyme. The ultimate connection with major stem veins is always secondary. The system arises in a discontinuous manner and independently of the endothelium of veins. The overall direction of its development is centripetal. The isolated and discrete spaces converge and fuse to form continuous lymphatic trunks and networks. The main connection of the jugular lymph sacs with the venous system develops early, before the more distal lymphatic primordia have interconnected or joined the sacs. This link between two drainage systems, lymphatic and venous, occurs near the base of the neck (Fig. 4) at the jugulosubclavian junction.

Investigative methods. There are two general methods for the investigation of lymphatic genesis; each has serious limitations. They consist either of the study of serial sections of well-preserved embryos or of injections of the developing lymphatics (Fig. 5). Crucial events in that development take place very early, in human embryos, for instance, when they measure only 10 mm. Obviously it is impossible to detect discontinuous segments of developing lymphatics by the injection method. Serial sections, on the other hand, can be obtained of uninjected as well as of injected embryos. This method in its turn presents difficulties in regard to fixation effects, shrinkage of the delicate embryonic tissues, and interpretive errors in reconstructions. Further progress may depend on newly devised methods.

The latest evidence strongly supports the centripetal view of the origin of the lymphatic system. According to this view, the very first beginnings are identical with those of the arterial and venous components of the vascular system. All vasculogenesis appears to occur through the same fundamental transformation of mesenchymal cells into endothelial cells. The lymphatic precursor stages, succeeding the development of veins, as a rule are less distinct, evanescent, and therefore difficult to recognize. This has caused differing interpretations.

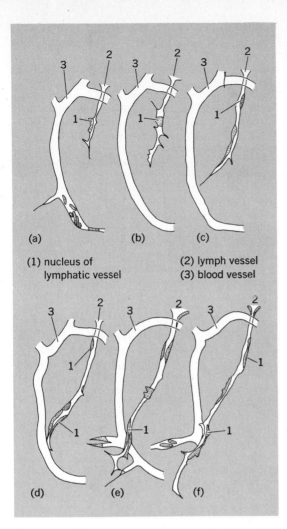

(a)   (b)   (c)

(1) nucleus of
    lymphatic vessel
(2) lymph vessel
(3) blood vessel

(d)   (e)   (f)

Fig. 5. (a–f) Successive stages in growth of lymphatic vessel in tail of tadpole during 72 hr. (*From R. O. Greep, ed., Histology, 2d ed., Blakiston–McGraw-Hill, 1966*)

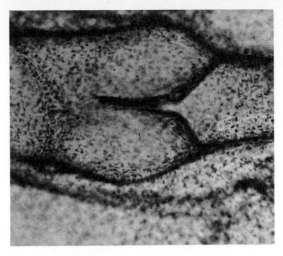

Fig. 6. Valve in lymph vessel to prevent backflow of blood. (*General Biology Supply House, Inc.*)

sels, become provided with valves which arise earlier in lymphatic vessels than in veins. Their structural arrangement enhances centripetal flow of lymph. By the fifth month, valves are present throughout the system (Fig. 6).

By ingrowth of connective tissue strands, the lymph sacs become transformed into a lacy plexus or fenestrated structure. In the meshes between the remaining lymph spaces the mesenchymal cells proliferate rapidly. Blood capillaries pervade the tissue bridges and immature lymphocytes appear in great numbers. The whole represents the beginning of a lymph node (Fig. 7). True lymphocytes further infiltrate the structure and produce compact cellular masses surrounded by lymph vessels. The complex may become subdivided to form groups of smaller lymph nodes. By the middle of fetal life the surrounding lymph vessels form a characteristic peripheral sinus, and an investing capsule is formed by the adjacent connective tissue. The full differentiation of lymph nodes is attained only after birth. [ARNOLD A. ZIMMERMANN]

PHYSIOLOGY

In any given region the fluid which enters a lymphatic capillary may be that which is close to the arteriolar or to the venous end of a blood capillary, or it may be fluid relatively distant from either.

Embryonic phase. In human embryos of about 1 in. (20–30 mm) a connected primitive lymph system is already established. The main lymph channel ascending through the thorax in front of the vertebral column (thoracic duct) has joined the jugular lymph sac. At its distal end, it is continuous with the cisterna chyli, into which converge the lymphatics from the intestinal tract and those of the lower trunk and limbs. The jugular lymph sacs also receive lymphatic drainage from the developing upper limbs through the primitive ulnar lymphatics. Other lymph trunks develop in the head and neck regions, so that the jugular lymph sacs soon receive returning tissue fluids from all body regions. Lymph thereby is returned to the economy of the blood vascular system.

In early stages, the thoracic ducts are paired and symmetrically developed. Some animals retain double thoracic ducts. In many mammals and in man the thoracic duct system becomes asymmetrical through the formation of interconnecting links and atrophy of certain segments. Henceforth, lymph drainage from the lower body regions occurs through a single main trunk, which terminates at the left jugulosubclavian junction.

Fetal phase. During early fetal months, the main lymph trunks, as well as the peripheral lymph ves-

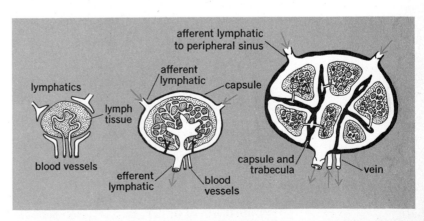

afferent lymphatic
to peripheral sinus

afferent
lymphatic

capsule

lymphatics

lymph
tissue

capsule and
trabecula

vein

blood vessels

efferent
lymphatic

blood
vessels

Fig. 7. Development of a lymphatic gland, with ingrowth of connective tissue strands. (*From L. B. Arey, Developmental Anatomy, 7th ed., Saunders, 1965*)

There exists a gradient of pressure from the tissue fluid to the lumen of the lymphatic capillary. In edema such as can be brought about experimentally by treatment with xylol or by heat, the gradient is increased. This pressure gradient is known to be an important factor in the formation of lymph. *See* EDEMA.

Any fluid in which filtration exceeds absorption through the wall of the blood capillary is normally drained by the lymphatics. Experimentally, this drainage is demonstrated where the filtration rate is increased by intravenous transfusions or by warming of parts.

Lymph collection. The function of the lymphatics is to remove such excess fluid and also to return to the bloodstream the protein which has escaped from the blood capillaries. Lymph collected from lymphatics of the kidney, leg, and cervical region, as well as from the thoracic duct, has been shown to contain this protein which has been "leaked" from the blood-vascular system. These proteins include serum globulin, serum albumin, and fibrinogen. The filtrate from the blood capillaries consists of water and of sugar and salts in concentrations found within the bloodstream, plus these proteins in low concentration. The salts and much of the water are reabsorbed by blood vessels, whereas the protein, together with some water and salt, enters the lymphatics. When the normal drainage of protein from the tissue spaces is prevented, elephantiasis may result, as when the lymphatics are obstructed in filariasis. The result is massive edema of a part, with dilatation and thickening of the lymphatic capillaries and an overgrowth of connective tissue. *See* FILARIASIS.

In addition to water, ions, small molecules, and the large molecules of extravascular protein, particles such as injected graphite or extravasated red cells also enter the lymphatic capillaries. Such relatively huge structures pass between the cells.

Composition and properties of lymph. Basically, the composition of lymph closely resembles that of the plasma. Chemical and electrophoretic studies show that lymph contains all of the types of protein found in plasma, but in lower concentration. The composition of lymph varies to some extent from one part of the body to another. Thus, the lymph from the liver contains more protein than that from the skin.

The ability of lymph to coagulate depends largely on its content of fibrinogen and prothrombin, and this varies rather markedly. In dogs the mean prothrombin level is 7.6 for leg lymph, 51.2 for thoracic duct lymph, and 93.2 for liver lymph. There are no blood platelets in lymph, and it is thought that thromboplastin, generally considered necessary in clotting, is formed at the time of clotting from precursors present in the lymph. Calcium, also needed in clotting, is present in the lymph in slightly lower concentration than in the blood.

The ionic content of lymph does not vary much from that of plasma (Table 1). Chloride and bicarbonate levels are higher in the lymph than in the plasma, whereas total base is slightly lower.

The glucose content usually is essentially the same in the lymph as in the blood. Curiously, the lymph of the kidney has a lower concentration of glucose. When carbohydrate is fed or when an intravenous injection of glucose is made, the concentration of this sugar in the lymph lags behind

Table 1. Average values for the concentrations of electrolytes in lymph and plasma

Substance or property	Source of lymph	Subject[a]	Plasma	Lymph	Anesthetic
Sodium, meq/liter	Thoracic duct[b]	Man (5)	127	127	
	Cervical duct[c]	Dog	163	157	Nembutal
Potassium, meq/liter	Thoracic duct[b]	Man (5)	5.0	4.7	
	Right lymph duct[d]	Dog (5)		5.4	Nembutal
Calcium, meq/liter	Thoracic duct[b]	Man (5)	5.0	4.2	
	Thoracic duct[e]	Dog (1)	5.2	4.6	Amytal
	Cervical duct[f]	Dog (11)	5.8	4.9	Nembutal
Chloride, meq/liter	Thoracic duct[b]	Man (5)	96	98	
	Thoracic duct[e]	Dog (1)	110	116	Amytal
	Cervical duct[f]	Dog (7)	116	122	Nembutal
Inorganic phosphorus, mg%	Thoracic duct[b]	Man (5)	4.5	4.4	
	Thoracic duct[e]	Dog (1)	4.3	3.6	Amytal
	Cervical duct[f]	Dog (3)	5.6	5.9	Nembutal
Carbon dioxide, ml%	Cervical duct[g]	Dog (7)	56.8	58.8	Nembutal
Carbon dioxide tension, mm Hg	Cervical duct[g]	Dog (7)	46.4	40.3	Nembutal
pH	Cervical duct[g]	Dog (7)	7.34	7.41	Nembutal

[a]Numbers in parentheses indicate number of subjects studied.

[b]H. R. Bierman et al., The characteristics of thoracic lymph duct in man, *J. Clin. Invest.*, 32:637–649, 1953.

[c]O. H. Lowry and F. W. Maurer, quoted by C. K. Drinker and J. M. Yoffey, *Lymphatics, Lymph and Lymphoid Tissue*, Harvard University Press, 1941.

[d]F. C. Courtice, unpublished data.

[e]R. M. Arnold and L. B. Mendel, Interrelationships between the chemical composition of the blood and the lymph of the dog, *J. Biol. Chem.*, 72:189–211, 1927.

[f]J. W. Heim, On the chemical composition of lymph from subcutaneous vessels, *Amer. J. Physiol.*, 103:553–558, 1933.

[g]J. W. Heim and O. C. Leigh, The carbon dioxide content and combining power and pH of cervical lymph, *Amer. J. Physiol.*, 112:699–704, 1935.

Table 2. Average flow and protein content of lymph from various lymph ducts in dogs under nembutal anesthesia*

Lymph duct†	Lymph flow, ml/hr	g%	g/day	% of total intravascular plasma protein
Thoracic duct (4)	26.0	4.1	26.1	47.5
Right lymph duct (21)	2.3	3.7	2.0	3.6
Cervical ducts (18)	2.2	2.8	1.3	2.4
Foreleg ducts (5)	3.8	1.3	1.2	2.2
Hindleg ducts (17)	2.2	1.8	1.0	1.8

*From C. K. Drinker and J. M. Yoffey, *Lymphatics, Lymph and Lymphoid Tissue*, 2d ed., Harvard University Press, 1956. Average weight of each animal in each group was approximately 21 kg and average intravascular plasma protein 55 g. Total lymph protein from leg and cervical ducts includes both sides.

†Numbers in parentheses represent the number of animals in each group.

that in the plasma, but eventually catches up when an equilibrium between plasma and lymph is attained. Concentrations of amino acid, creatinine, and urea are approximately equal in lymph and plasma, but in the kidney the urea of the lymph is slightly higher than that of the plasma. It seems probable that some of the renal lymph comes from fluid reabsorbed from the distal and collecting tubules. *See* KIDNEY (VERTEBRATE); URINARY SYSTEM (VERTEBRATE).

The lipids consist of neutral fat forming the chylomicrons, and of cholesterol and phospholipid, the latter two kinds of compounds being associated chiefly with proteins to form lipoproteins. The quantity of neutral fat depends on the fat absorption by the lacteals of the intestine, and the amounts of cholesterol and phospholipid vary with the amounts of protein present.

Lymph flow. Contractility of the walls of lymphatic vessels in man and other mammals appears to be only slight and to play a minor role in lymph flow. This is in marked contrast to the condition in some lower vertebrates, particularly the Amphibia, in which well-developed pulsatile lymph hearts occur along the course of the lymphatic channels (Fig. 2). The lymph flow in mammals appears to be very sluggish compared with that in the frog. The rather slow rate of flow of the lymph in dogs is seen in Table 2. It is calculated that extravascular circulation involves 50–100% of the plasma proteins each day.

The pressure of the lymph flow is variable, even in any given lymphatic vessel. Because lymph flows centripetally, it is clear that the peripheral lymph is under higher pressure than that farther along the vessels. Muscle contractions, such as those of the legs in walking, may greatly increase the pressure of the lymph in an individual region. During contraction of the muscle the lymphatics tend to empty, whereas during relaxation they fill again. In one experiment no pressure was recorded from a lymphatic just below the popliteal lymph node of a dog when the animal was at rest; but during rapid flexion and extension of the leg, the pressure rose to 68 cm of water.

Measurements of lymph pressure usually are made under standard conditions of rest. Pressure in a major division of the thoracic duct of the dog has been recorded as 6.4 cm of water, compared with a pressure of 2.4 cm in the internal jugular vein of the same dog. Such a pressure difference promotes flow from the thoracic duct into the venous system. Pressures have been recorded from a number of different lymphatic vessels. That of the lymphatic capillaries is rather high, having been recorded as 42.6 cm of water (31.3 mm of Hg) in the capillary of the villus.

Immunity. The lymph nodes serve as filtering-out places for foreign particles, including microorganisms, because the lymph comes into intimate contact with the many phagocytic cells of the sinusoids. These macrophages are of both the fixed and free wandering types. *See* PHAGOCYTOSIS.

In addition to the phagocytic function, experimental studies have demonstrated that lymphoid tissue produces antibodies. If antigens such as typhoid bacilli or sheep erythrocytes are injected into the pad of the foot of an experimental animal, antibodies begin to appear in the efferent lymph vessels of the popliteal node a few days later and their titer rapidly increases to a maximum and then declines. Some investigators have found that most of the antibody is within the lymphocytes rather than in the fluid. Germinal centers of the lymph node increase in size, and there is an increased output of lymphocytes. *See* ANTIBODY.

The actual process of antibody formation in lymphoid tissue is not well understood. A considerable body of evidence would assign the chief role to the plasma cell, generally abundant in lymphoid tissue, rather than to the lymphocyte. Antigen injection has been shown to bring about an intense plasma cell response in many instances.

The use of fluorescent antigens and antibodies offers evidence that antigen is fixed by plasma cells in lymph nodes which are forming antibodies, but is not fixed by lymphocytes. The increased numbers of lymphocytes in infection and their passage into the bloodstream may thus possibly have some significance not connected with antibody formation.

Lymphocyte transformations. A large body of evidence obtained over the years supports the view that lymphocytes undergo transformations. Tracer studies combined with tissue culture methods for peripheral blood seem to show conclusively that the large proliferating cells in such cultures originate from small lymphocytes. Phytohemagglutinin acting on cultures of peripheral blood induces a transformation of small lymphocytes into large stem-type cells, which then become active mitotically.

There is a profound modification in the ultrastructure of human lymphocytes cultured with phytohemagglutinin: New organelles increase in number in the cytoplasm of such cells; mitochondria become more numerous; the Golgi apparatus enlarges; and a variety of vesicles is seen. In the living rat a transformation of small lymphocytes (parental strain) into larger cells of pyroninophilic type has been shown in F_1 hybrid rats. Cytochemical studies on lymphocytes and large blast cells in short-term culture show the dehydrogenase system highly active in glycolysis.

Cerebrospinal fluid. The cerebrospinal fluid fills the ventricles of the brain and the space between the arachnoid mater and the pia mater. Cerebrospinal fluid is a clear liquid derived from the capillary vessels of the choroid plexuses which permit the passage of water, electrolytes, and small molecules, but hold back the proteins of the blood. Fluid passing through the walls of the capillaries must also go through a layer of epithelium covering the capillaries. Traces of protein are found in the normal cerebrospinal fluid. *See* NERVOUS SYSTEM (VERTEBRATE).

By far the greater part of the cerebrospinal fluid is absorbed into the veins of the arachnoid villi. The relationship between the cerebrospinal fluid and the lymphatic system is found in certain regions in which the fluid escapes from the meninges and enters lymphatic capillaries. Thus it may pass by way of the arachnoid sheaths of the olfactory nerves into the submucosa of the nasal cavity, and so reach the cervical lymphatic vessels. Arachnoid coverings of the sensory ganglia also are locations where cerebrospinal fluid may reach the lymphatics in the normal animal.

Clinically, such connections may serve as routes by which bacteria or viruses can enter the central nervous system. They may also, in cases of meningitis or of hemorrhage, help to clear the cerebrospinal fluid of cells and of plasma proteins.

Cerebrospinal fluid differs from other lymph in its low protein content, low cell content, and lower specific gravity (1.004–1.008). It contains no fibrinogen.

The cerebrospinal fluid has at least three functions: (1) It acts as a fluid buffer against shocks and jars of the central nervous system. (2) It serves as a reservoir in regulation of the cranial contents. (3) It acts as a mechanism for exchange of gases and nutrients in the nervous system. In relation to these functions, its properties are particular, but it may be considered as a specialized type of lymph.

[WARREN ANDREW]

Bibliography: L. Allen, On the penetrability of the lymphatics of the diaphragm, *Anat. Rec.*, 124(4):639–658, 1956; L. Calhoun, The microscopic anatomy of the digestive tract of *Gallus domesticus*, *Iowa State Coll. J. Sci.*, 7(3):261–381, 1933; J. Clayton, Observations on the leucocytes of the earthworm, *Proc. Penn. Acad. Sci.*, 12:38–40, 1938; E. H. Cooper and D. Inman, The ultrastructure of transformed lymphocytes, *Proceedings of the 9th Congress of the European Society of Hematology*, 1963; J. W. Dransfield, The lymphatic system of the domestic fowl, *Vet. J.*, 101 (8):171–179, 1945; C. K. Drinker and M. E. Field, The protein content of mammalian lymph and the relation of lymph to tissue fluid, *Amer. J. Physiol.*, 97(1):32–39, 1931; W. C. George, A comparative study of the blood of the tunicates, *Quart. J. Microsc. Sci.*, 81(3):391–428, 1938; W. C. George, The function and fate of the lymphocytes, *S. Med. Surg.*, 97(3):122–126, 1935; J. L. Gowans, The fate of parental strain small lymphocytes in F_1 hybrid rats, *Ann. N.Y. Acad. Sci.*, 99:432–455, 1962; J. L. Gowans, D. C. McGregor, and D. M. Cowen, Initiation of immune responses by small lymphocytes, *Nature*, 196:651–655, 1962; H. E. Jordan, The histology of the blood and the blood-forming tissues of the urodele, *Proteus anguineus*, *Amer. J. Anat.*, 51(1):215–251, 1932; A. S. MacKinney, Jr., F. Stohl-man, Jr., and G. Brecher, The kinetics of cell proliferation in cultures of human peripheral blood, *Blood*, 19:349–358, 1962; P. C. Nowell, Phytohemagglutinin: An initiator of mitosis in cultures of normal human leukocytes, *Cancer Res.*, 20:462–466, 1960; D. Quaglino, F. G. Hayhoe, and R. J. Felmans, Cytochemical observations on the effect of phytohemagglutinin in short-term tissue culture, *Nature*, 196:338–340, 1962; Y. Tanaka et al., Transformation of lymphocytes in cultures of human peripheral blood, *Blood*, 22:614–629, 1963.

Lymphocytic choriomeningitis

A viral infection endemic in mice, but occasionally transmitted to man, in whom it may produce a mild influenzalike disease or an aseptic meningitis. Sometimes encephalitis or even a fatal systemic disease may be produced.

Diagnosis of lymphocytic choriomeningitis (LCM) is by isolation of the virus from acute phase spinal fluid or blood, or by increases in serum antibodies.

Mice are the usual reservoir of virus; they may harbor it throughout life and transmit it to their offspring. Dust and food contaminated by urine or feces are probable vehicles of transmission to man; however, *Trichinella spiralis* can be infected experimentally and has been suggested as a possible means of transmission from animal to animal or, through pigs, to man.

In addition to its interest as a cause of human disease, the LCM virus is important for the unusual virus-host relationship demonstrated in mice infected before birth or within a few days after birth. In these animals, a persistent tolerant infection is established which is never manifested by illness; however, virus in large amounts is harbored in the animal throughout life and may be passed on to other mice within a colony or transmitted by infected females to the offspring. This type of virus-host relationship shows promise as a model from which to learn more concerning the immune response, which is of significance for problems of transplantation in human beings, for human cancer, and for certain chronic diseases of man. *See* ANIMAL VIRUS.　　[JOSEPH L. MELNICK]

Bibliography: J. E. Prier (ed.), *Basic Medical Virology*, 1966.

Lymphogranuloma venereum

A venereal disease caused by a microorganism, a member of the PLT-Bedsonia (*Chlamydia*) group. It is also known as venereal bubo. The disease manifests itself as an enlargement of the inguinal lymph node and as such is seldom overlooked in the male; however, mild atypical or asymptomatic infections may be common. In the female the disease is more difficult to diagnose because the lymphadenopathy may be retroperitoneal. Occasionally the disease may be complicated by chronic ulcerative lesions of the vulva or rectal strictures or by systemic diseases such as meningoencephalitis, pneumonitis, and arthritis with or without cutaneous eruptions. Conjunctivitis with oculoglandular fever has been reported.

Fever commonly accompanies the development of the disease. Diagnosis of the disease is based upon the demonstration of a delayed hypersensitivity reaction to Bedsonia antigen—the so-called Frei test—or of significant levels of complement-fixing antibody. Epidemiologic surveys performed in the past, using these tools, indicated that the agent was widespread in the sexually promiscuous population but that the disease was relatively uncommon. The majority of untreated lymphogranuloma venereum cases will be Frei-test and complement-fixation-test positive. Subjective clinical and functional improvement has followed treatment with tetracycline or sulfonamides but responses to either may be irregular. Early diagnosis and treatment, supervision of infected persons, and public education are important in prevention. *See* ANTIBIOTIC; ANTIGEN; PLT-BEDSONIA GROUP; SKIN TEST; VIRUS.

[K. F. MEYER; J. SCHACHTER]

Lymphoma

A group of malignant tumors characterized by neoplastic proliferations of the various cell types endogenous to lymphoid tissue. Hodgkin's disease is considered to be a type of lymphoma. It usually originates in lymph nodes and comprises approximately 50% of such tumors. The remainder of the lymphomas are referred to as non-Hodgkin lymphomas and are classified on a morphological basis according to the cell of origin of the neoplasm, the degree of differentiation of the tumor cells, and the overall architecture of the involved lymphoid tissue. A commonly used classification is as follows: lymphocytic lymphoma, well differentiated; lymphocytic lymphoma, poorly differentiated; histiocytic lymphoma; mixed lymphocytic histiocytic lymphoma; and undifferentiated stem cell lymphoma (includes Burkitt's lymphoma). Many of these tumors have either a nodular or diffuse microscopic appearance.

Etiology. The etiology of lymphomas remains unknown, although current data suggest a viral causation. A number of different viruses have been proven to cause lymphomas in experimental and domestic animals. Burkitt's lymphoma, a common undifferentiated lymphoma of Central African children, is thought to be caused by a viral agent transmitted by mosquitoes. A herpesvirus known as the Epstein-Barr virus has been identified in cultured Burkitt lymphoma cells. Electron-microscopic studies of Burkitt tumor cells have shown intracellular viral particles. Individuals with Burkitt's lymphoma have also been found to have elevated antibody levels in their sera against the Epstein-Barr virus. *See* EPSTEIN-BARR VIRUS.

Pathogenesis. The disease most commonly begins in the lymph nodes of the cervical (neck) and axillary (underarm) regions of the body; occasionally it originates in the lymphoid tissue of the intestine, and in Burkitt's lymphoma starts in the jaw. As the disease progresses, any organ in the body can become involved. Most commonly, the liver, spleen, and other lymph nodes become diseased as evidenced by their increase in size.

Grossly, the involved lymph nodes are usually enlarged, firm and rubbery, grayish-white, and may be focally necrotic. Microscopically, all lymphomas show varying degrees of obliteration of the usual lymph node architecture due to proliferation of the neoplastic cells. In the morphologically diffuse form, sheets of neoplastic lymphocytes and histiocytes, or both, completely obscure the lymphoid follicles and often infiltrate the capsule of the lymph node with extension into the adjacent fibrofatty tissue. In the nodular form, the neoplastic cells are arranged in nodular aggregates which may be associated with distorted residual lymphoid follicles. The classification into diffuse and nodular varieties is of practical importance, since the nodular variety has an overall better prognosis than the diffuse form.

T- and B-lymphocytes. Since 1973, the explosion of knowledge in the discipline of immunology has led to new concepts concerning lymphomas. It is currently accepted that most lymphocytes are produced in the bone marrow from primitive lymphoid cells. These lymphocytes are then acted upon by specific lymphoid organs, namely the thymus gland and the bursal equivalent (equivalent to the bursa of Frabicius in the fowl). Those processed in the thymus gland are referred to as T-lymphocytes, and those acted upon by the bursal equivalent are called B-lymphocytes. T-lymphocytes function primarily in cell-mediated immunity, whereas B-lymphocytes and their derivatives, plasma cells, produce antibodies. These lymphocytes can be differentiated from each other by specific morphological and immunological techniques. For example, with the scanning electron microscope T-lymphocytes have been shown to have relatively smooth surfaces while B-lymphocytes have "hairy" surfaces. With these techniques a typical lymph node can be demonstrated to be composed of regions of B-lymphocytes and T-lymphocytes. The B-lymphocytes are present predominantly in the germinal centers and the T-lymphocytes in the deep cortical regions. *See* ANTIBODY; IMMUNOLOGY.

When the immunological and morphological techniques are applied to malignant lymphoid tissue, many of these tumors can be demonstrated to be predominantly B- or T-cell neoplasms. Most nodular lymphomas and Burkitt's lymphoma are considered to be neoplastic B-lymphocyte tumors. Many lymphomas previously classified as histiocytic have also been shown to be B-lymphocyte tumors. An uncommon mediastinal lymphoma occurring in young females is of neoplastic T-lymphocyte origin. The morphological classification of lymphomas as a whole is undergoing modification to incorporate these immunological data. Careful light-microscopic studies have provided evidence that many lymphomas arise from dedifferentiated germinal center cells and correspond to B-lymphocyte tumors.

Medical care and prognosis. Most persons with lymphomas seek medical advice because of enlarged, nontender lymph nodes. These people usually have no other signs or symptoms, although occasionally they may be anemic and have a fever. As in Hodgkin's disease, lymphomas are staged to determine the extent of involvement of the disease. Treatment is based on the stage of the disease and the tumor's microscopic appearance. The clinical course and prognosis of non-Hodgkin lymphomas are highly variable. When the disease is localized to a single group of lymph nodes in a well-defined region of the body, the therapy of choice is usually radiation. When the disease is disseminated, it is usually treated with a combination of drugs, such as cyclophasphamide, vincristine, methotrexate, and corticosteroids. Some long survivals and apparent cures have resulted from this chemotherapy. *See* HODGKIN'S DISEASE; LYM-

PHATIC SYSTEM (VERTEBRATE).

[SAMUEL P. HAMMAR]

Bibliography: J. A. Hansen and R. A. Good, Malignant disease of the lymphoid system in immunological perspective, *Hum. Path.*, 5:567–599, 1974; R. J. Lukes and R. D. Collins, Immunological characterization of human malignant lymphomas, *Cancer*, 34:1503, 1974.

Lyophilization

Solvent removal from the frozen state by sublimation; commonly referred to as freeze-drying. The dried material, a porous solid, is sealed under vacuum or inert-gas atmosphere and retains its physical and biological characteristics indefinitely. When reconstituted for use, the porous product readily readmits solvent; hence the term lyophile, or solvent-loving.

Lyophilization is accomplished by freezing the material to be dried below its eutectic point and then providing the latent heat of sublimation. Precise control of heat input permits drying from the frozen state without product melt-back. In practical application, the process is accelerated and more precisely controlled under reduced pressure conditions.

Several forms of apparatus are in use, ranging from the small laboratory equipment illustrated

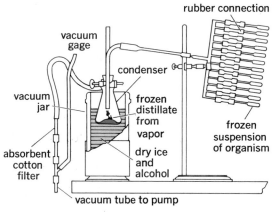

Apparatus for rapid high-vacuum desiccation of frozen suspensions of biologic materials. (*After Stein and Rodgers, in A. J. Salle, Fundamental Principles of Bacteriology, 4th ed., McGraw-Hill, 1954*)

here to large chamber-type installations for preservation of serums, vaccines, blood plasma, soluble beverages, and foodstuffs. *See* MICROBIOLOGICAL METHODS.

[SAMUEL C. TEASE]

Lyra

The Lyre, in astronomy, a summer constellation, small but important. Lyra has a first-magnitude star, Vega, a navigational star and the most brilliant star in this part of the sky. Vega forms, with two faint stars ε and ζ to the east, an almost perfect equilateral triangle (see illustration). The southern one in turn forms, with three brighter stars, β, γ, and δ, to the south, an approximate parallelogram. The resulting overall figure resembles a tortoise more than a stringed musical instru-

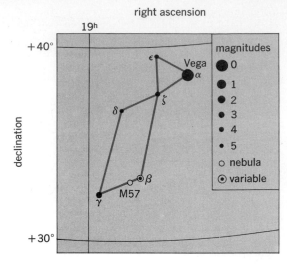

Line pattern of the constellation Lyra. The grid lines represent the coordinates of the sky. The apparent brightness, or magnitudes, of the stars is shown by the sizes of the dots, graded by appropriate numbers.

ment. However, according to legend, Mercury made the first lyre from a turtleshell by placing strings across it. Hence the two different representations are not incompatible. In this constellation lies the famous Ring Nebula. *See* CONSTELLATION; NEBULA, GASEOUS.

[CHING-SUNG YU]

Lysin

A term used in biology to describe substances that will disrupt a cell, with the release of some of its constituents. Unless the damage is minor, this action leads to the death of the cell. Lysins vary in the range of host species whose cells they will attack and in their requirements for accessory factors for lysis; the immune lysins are strictest in their requirements. Erythrocytes are lysed by a wide variety of chemicals, including water and hypertonic salt solutions, which displace the osmotic pressure from that of isotonicity. They are also susceptible to surface-active substances, such as saponin. Many bacteria, such as the staphylococcus and the streptococcus, elaborate one or more hemolysins that will lyse erythrocytes from certain, although not all, species of animals. These patterns may be used in the taxonomic classification of microorganisms. There are a number of other lytic substances, such as lysozyme, an enzyme in the eye aqueous humor and to some extent the bacteriophages, which attack diverse species of bacteria. These actions are not entirely nonspecific since they depend on the presence of common cell constituents or receptors which need not, however, coincide with those involved in immune lysis. *See* BACTERIOPHAGE; STAPHYLOCOCCUS; STREPTOCOCCUS.

Immune lysin is a term used in serology to designate an antibody that, in the presence of complement and cofactors, such as Mg^{++} (magnesium ion) and Ca^{++} (calcium ion), will disrupt a particular type of cell, with the release of some of its constituents. This action is in accord with the principles of serologic specificity and is to be distinguished from the more general actions of other lytic chemicals. Immune lysins are also classified according to the general class of cell attacked: Hemolysins are antibodies lysing erythrocytes, while bacteriolysins cause the lysis of bacterial cells. Hemolysins against various foreign erythrocytes occur normally in many sera; the Forssman and other heterophile antibodies and the blood group isoantibodies constitute special instances. High titers of hemolysins may also be produced by immunization with erythrocytes, and rabbit antibody against sheep erythrocytes is an important laboratory reagent, the so-called amboceptor, in complement-fixation tests. For example, sheep erythrocytes injected into a rabbit produce antibodies in the rabbit against the sheep red blood cells. The lysis of erythrocytes releases hemoglobin, which may be quantitated by eye or by a colorimeter. *See* ANTIBODY; ANTIGEN; COMPLEMENT, SERUM; COMPLEMENT-FIXATION TEST; HEMOGLOBIN; HETEROPHILE ANTIGEN; LYTIC REACTION.

[HENRY P. TREFFERS]

Lysine

An amino acid constituent of proteins that contains a terminal amino group in addition to the usual α-amino group. For discussions of the physical constants shown here *see* EQUILIBRIUM, IONIC; ISOELECTRIC POINT; OPTICAL ACTIVITY; SPECTROPHOTOMETRIC ANALYSIS.

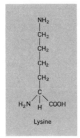

Physical constants of the L isomer at 25°C:
pK$_1$ (COOH): 2.18; pK$_2$ (NH$_3^+$): 8.95; pK$_3$ (ϵ-NH$_3^+$): 10.53
Isoelectric point: 9.74
Optical rotation: [α]$_D$(H$_2$O): +13.5; [α]$_D$(5 N HCl): +26.0
Solubility (g/100 ml H$_2$O): very soluble

Lysine must be supplied in the diet of animals and humans. It is widely distributed in proteins; however, it is absent in some such as zein (a grain protein). It is a precursor of 5-hydroxylysine, an amino acid occurring in only a few proteins such as collagen, and it is found in biocytin (ϵ-*N*-biotinyl-L-lysine), a complexed form of the vitamin biotin, which is found in yeast.

There are two biosynthetic pathways to lysine. The one which occurs in most bacteria and blue-green algae is characterized by intermediates containing seven carbon atoms, including diaminopimelic acid, a constituent of the bacterial cell wall. In most yeast and green algae a pathway of six-carbon intermediates occurs. *See* AMINO ACIDS.

Lysine does not participate in transamination; however, in mammalian tissue and in some microorganisms the major degradative pathway (see illustration) occurs by the irreversible loss of the α-amino group to form α-keto-ϵ-aminocaproic acid. The product of this pathway, glutaric acid, is further metabolized to α-ketoglutaric acid.

In another degradative pathway which occurs in bacteria, lysine is oxidatively decarboxylated and

Major degradative pathway of lysine.

deaminated to form δ-aminovaleric acid, which is metabolized further to glutaric acid.

[DAVID W. E. SMITH]

Lysogeny

Almost all strains of bacteria are lysogenic; that is, they have the capacity on rare occasions to lyse with the liberation of particles of bacteriophage (see illustration). Such particles can be detected by their ability to form plaques (colonies of bacteriophage) on lawns of sensitive (indicator) bacteria. The genetic determinant of the capacity of lysogenic bacteria to produce bacteriophage is a repressed phage genome (provirus) which exists in the bacterium in one of two states: (1) integrated into the bacterial chromosome (most cases), or (2) occupying some extra-chromosomal location (rare cases). Bacteriophages having the potential to exist as provirus are called temperate phages. When the provirus is integrated into the bacterial genome, it is called prophage. When the germinal substance (deoxyribonucleic acid or deoxyribonucleoprotein) of certain temperate phages (for example, wild type of coliphage λ) enters a sensitive bacterium, the outcome may be death (lysis) for the bacterium as a result of phage multiplication, or it may result in the integration of the phage nucleic acid into the host genome (as a prophage), with the formation of a stable lysogenic bacterium. The lysogenic strain is designated by the name of the sensitive strain followed, in parentheses, by the strain of lysogenizing phage, for example, *Escherichia coli* (λ). Such a bacterium differs from its nonlysogenic ancestor in one very special way: It is immune to lysis by phage homologous to its carried prophage. *See* DEOXYRIBONUCLEIC ACID (DNA).

Repression. The conversion of a nonlysogenic cell to an immune lysogenic state by its resident prophage is the essential element of lysogeny, and is related to the regulatory mechanism by which the lysogenic state is maintained stably. When a prophage becomes part of the bacterial genome, most of the prophage genes, especially those having to do with DNA replication and the synthesis of viral protein, are repressed. This repression is attributed to the synthesis, under the direction of the prophage genome, of a repressor substance; thus at least one gene of prophage is expressed. This process of repression is responsible for the immunity of lysogenic bacteria to lytic superinfection.

Immunity. The immunity conferred by prophage explains the fact that in most lysogenic cultures, while the host bacteria multiply normally, mature phage particles of the kind for which the host is lysogenic can be found in the surrounding medium. In a very few cells in any lysogenic population, the repressive mechanism fails to function, and prophage initiates replication and the host cell undergoes productive lysis. Since the population as a whole is immune, free virus merely persists in the environment. Under certain conditions, of which irradiation and the introduction of certain

chemicals into the medium have been the best studied, a lysogenic culture may be induced to shift from the prophage to the lytic cycle; that is, the regulatory mechanism in the majority of cells is caused to break down and mass lysis of the culture occurs, with the production of many infectious phage particles. *See* INFECTION, LYTIC.

Not all genes of the prophage are repressed, and a number of properties of lysogenic bacteria, assumed to be solely under the genetic control of the host cell, have been shown to be the expression of genes of the harbored prophage. Thus, the synthesis of diptherial toxin by the diphtheria bacillus is controlled by the tox$^+$ gene of several corynebacteriophages, which render nontoxinogenic diphtheria bacilli both lysogenic and toxinogenic. Certain strains of *Salmonella* synthesize one kind of antigen when they are nonlysogenic, while when they harbor a specific prophage, antigen synthesis is altered by an expressed prophage enzyme, giving rise to another kind of antigen.

In one well-studied case, the origin of viral genes expressed in the lysogenic host cell is clear. Coliphage λ, liberated from galactose-fermenting *E. coli*(λ) has, sometimes, in the process of prophage integration, exchanged λ genes having to do with capsid production with host cell genes having to do with the enzymes of galactose utilization. When the repression mechanism for prophage fails, copies of λDNA now contain genes for galactose fermentation. The DNA of this mutant, λdg, converts *E. coli* cells unable to utilize galactose into *E. coli* (λdg), the cells of which are lysogenic for λdg and able to utilize galactose. This galactose-positive state results from the expression of bacterial genes that have become an integral part of a viral DNA. *See* BACTERIOPHAGE; COLIPHAGE.

[LANE BARKSDALE]

Bibliography: M. H. Adams, *Bacteriophages*, 1959; B. D. Davis et al., *Microbiology*, 1967; W. Hayes, *The Genetics of Bacteria and Their Viruses*, 1969; G. S. Stent, *Molecular Biology of Bacterial Viruses*, 1963.

Lysosome

A specialized cell part, of variable size and structure, surrounded by a single membrane and containing a mixture of hydrolytic (digestive) enzymes, most of which require an acid medium for optimal activity. More than 40 such enzymes are already known. Each splits a certain type of bond found in natural substances; together they have the ability to break down into small fragments the main constituents of living matter, that is, proteins, nucleic acids, polysaccharides, and lipids. Thus, the lysosomes make up the digestive system of the cell. They are found in most if not all animal cells, and in at least some plant cells.

There are two main types of lysosomes: primary lysosomes, packages of newly made enzymes that have not yet been involved in a digestive event; and secondary lysosomes, vacuoles derived from the primary lysosomes, within which digestion is taking place or has taken place.

Mechanisms of digestive activities. The material undergoing digestion in the lysosomes can be of either extracellular or intracellular origin (Fig. 1). When provided by the environment, the material to be digested is taken into the cell

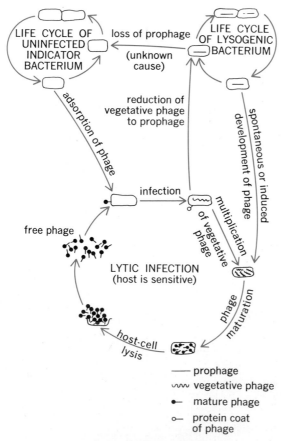

Life cycles of phage and bacterial host. (*E. Jawetz, J. L. Melnick, and E. A. Adelberg, Review of Medical Microbiology, 2d ed., Lange, 1956*)

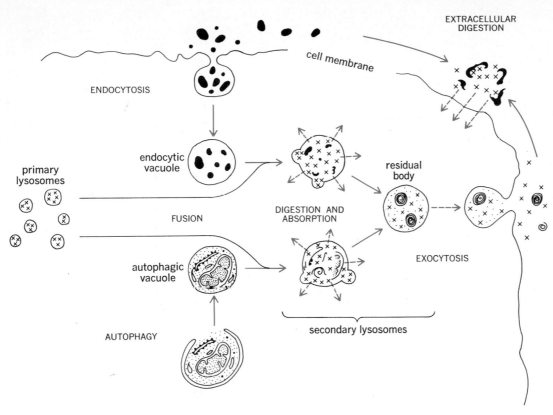

Fig. 1. Schematic representation of lysosome function. Only primary lysosomes are shown to fuse with endocytic or autophagic vacuoles. Secondary lysosomes can do the same, especially in cells lacking the ability to discharge their lysosomes by exocytosis.

by endocytosis, that is, by some form of phagocytosis or pinocytosis. This phenomenon introduces the captured material into the cytoplasm as the content of a closed vesicle, called a phagosome or pinosome, surrounded by a membrane derived from the cell membrane. This vesicle later fuses with a lysosome, either primary or secondary, in such a manner that their respective contents merge, while their membranes coalesce to form a single continuous boundary. When the material is a piece of the substance of the cell itself, it is segregated from the cytoplasm by entrapment within the lumen, either of a lysosome, or of another kind of vesicular structure which subsequently fuses with a lysosome. This process is called cellular autophagy. *See* PHAGOCYTOSIS; PINOCYTOSIS.

In both endocytic and autophagic vacuoles, the lysosomal enzymes and their substrates are brought together without any discontinuity affecting the membranes that surround them. Consequently, the enzymes cannot diffuse into the cytoplasm and attack the cell constituents indiscriminately. Even when lysosomes digest fragments of their own cells, they do so in a piecemeal and orderly fashion, without disruption of the cellular organization, at least under physiological conditions. Enzymes released from disrupted lysosomes can injure cells in pathological situations.

Product and waste discharge. As digestion proceeds within secondary lysosomes, its micromolecular products diffuse, or are transported, into the cytoplasm to be metabolized. If the material undergoing digestion includes substances or structures that are resistant to the action of the lysosomal enzymes, the indigestible constituents remain within the lysosomes. Such residue-laden lysosomes are called residual bodies.

In lower organisms the contents of old lysosomes or digestive vacuoles are eventually discharged outside the cells by a process called cellular defecation or exocytosis, which is essentially the reverse of endocytosis. Many cell types in higher organisms lack the ability to empty their lysosomes in this way. Consequently, they tend to accumulate increasing amounts of residues in their lysosomes as they grow older. This accumulation may have something to do with the aging phenomenon. *See* GERONTOLOGY.

Description and identification. Owing to these diverse functional properties, lysosomes have no unique structure of their own. Primary lysosomes have the unspecific structure of vesicles filled with protein, whereas the structure of secondary lysosomes is determined mainly by the nature and by the extent of digestion of their contents. Most easily recognizable are the residual bodies which are filled with dense amorphous masses, membrane fragments, ferritin particles, or myelin figures (Fig. 2). Also, the size of lysosomes is extremely variable, ranging from that of tiny vesicles that are barely visible under the electron microscope to the enormous dimension of the digestive vacuoles that are seen in many unicellular organisms. Their characteristic enzymes represent the only common feature whereby these diverse structures can all be identified as lysosomes. It is possible in some cases to separate lysosomes in pure form by special centrifugation techniques and to demonstrate their various enzyme activities biochemically. This is how lysosomes were first discovered and characterized. Many workers rely on the cytochemical detection of some of the enzymes of lysosomes, especially acid phosphatase, for the morphological identification of lysosomes in tissue sections. See CYTOCHEMISTRY.

Functions. Phylogenetically, the primary role of lysosomes is related to cellular nutrition. In all lower forms of life, they are the main sites, and their enzymes the main agents, of the digestion of food. When food is unavailable, lysosomes, through the mechanisms of cellular autophagy, play an equally important role by allowing the cells to survive on their own reserves and on fragments of their own substance that can be sacrificed. To the extent that food particles are by chance living organisms, potentially harmful as invaders or predators, lysosomes also serve defensive purposes. They can act as an invasion apparatus and as agents of extracellular digestion when their enzymes are discharged by exocytosis in a stagnant and confined environment.

The same basic functions are subserved by lysosomes in the cells of higher organisms, but with a greater emphasis on the breakdown of endogenous components. Only peripherally located cells (lungs, intestinal tract) can interact directly with objects coming from the outside. Their lysosomes form the first line of defense against invaders of all kinds. The second line is made up by the circulating leukocytes and by the fixed macrophages of the reticuloendothelial system. Most other cells direct their lysosomal digestive activities against their own constituents (autophagy), against macromolecules taken up from their environment (endocytosis), and sometimes, as in bone, against extracellular structures (exocytosis). Innumerable specialized functions depend on these mechanisms. Jointly, the lysosomes of the whole organism accomplish an enormous degradative task. Depending on the manner in which this activity is

compensated by biosynthesis, they allow turnover and renewal, replacement and remodeling, or even regression, involution, and atrophy of biological structures. Pathologically, lysosomes are implicated in a variety of cellular and tissue injuries. See CELL (BIOLOGY). [CHRISTIAN DE DUVE]

Bibliography: J. T. Dingle and H. B. Fell (eds.), *Lysosomes*, vols. 1 and 2, 1969, vol. 3, 1973; T. Hayashi (ed.), *Subcellular Particles*, 1959; A. B. Novikoff, in J. Brachet and A. E. Mirsky (eds.), *The Cell*, 1961; A. V. S. de Reuck and M. P. Cameron (eds.), *Ciba Foundation Symposium on Lysosomes*, 1963.

Lysozyme

An enzyme that was first identified and named by Alexander Fleming, who recognized its bacteriolytic properties. It has been designated muramidase, since it is known to facilitate the hydrolysis of a β-1-4-glycosidic bond between N-acetylglucosamine and N-acetylmuramic acid in bacterial cell walls; it also hydrolyzes similar glycosidic bonds in fragments of chitin. In recognition of this enzyme function, the International Enzyme Commission assigned the official chemical name of N-acetylmuramide glycanohydrolyase, coded as 3.2.1.17.

The most detailed studies have been performed on hen egg-white lysozyme, because this product is readily available. However, enzymes possessing lysozyme activity have been found in bacteria, bacteriophages, and plants and in human leukocytes, nasal secretions, saliva, and tears. People with a certain type of monocytic leukemia excrete large quantities of lysozyme in their urine. This excretion is presumed to be secondary to the synthesis of the enzyme by the large mass of leukemic cells.

The three-dimensional structure of the protein has been defined by x-ray crystallography. Additional data are available for the amino acid sequence of human lysozyme and also for a bacteriophage lysozyme. These results have given rise to speculation concerning the origin of the lysozyme gene during evolution.

Amino acid sequence. The complete amino acid sequence of hen egg lysozyme was determined independently by P. Jollès and by R. E. Canfield with their respective coworkers. The molecule consists of a single polypeptide chain containing 129 amino acid residues. The complete structure is illustrated in Fig. 1. This was determined by a technique which involved splitting the reduced alkylated lysozyme molecule into small peptide fragments with the proteolytic enzymes trypsin, chymotrypsin, and pepsin. The precise sequence of amino acids in each of the isolated fragments was then determined by the Edman degradation technique.

There are eight half-cystine residues in hen egg lysozyme, arranged in the native molecule as four cystine bridges. M. Inouye and A. Tsugita reported the amino acid sequence of a lysozyme synthesized by T4 bacteriophage that facilitates the penetration of phage deoxyribonucleic acid (DNA) through the bacterial cell wall. The protein is a single polypeptide chain containing 160 amino acid residues and no cystine bridges (Fig. 2).

A portion of the structure of human lysozyme, which has been described by Canfield, is illustrated in Fig. 3.

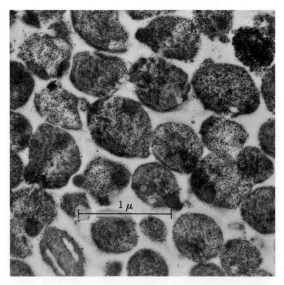

Fig. 2. Isolated rat liver secondary lysosomes, with the typical structure of residual bodies.

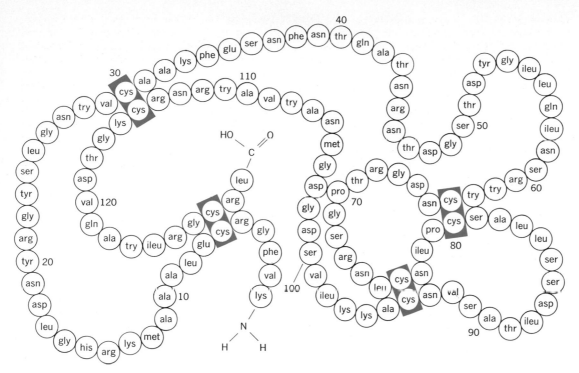

Fig. 1. Amino acid sequence of hen egg lysozyme. Four rectangles indicate cystine bridges. Lys is lysine; Val, valine; Phe, phenylalanine; Gly, glycine; Arg, arginine; Cys, half-cystine; Glu, glutamic acid; Leu, leucine; Ala, alanine; Met, methionine; His, histidine; Asp, aspartic acid; Asn, asparagine; Tyr, tyrosine; Ser, serine; Thr, threonine; Ileu, isoleucine; Try, tryptophan; Pro, proline; and Gln, glulamine. (*From R. E. Canfield and A. K. Liu, J. Biol. Chem., 240:1997, 1965*)

X-ray crystallography. D. C. Phillips and co-workers constructed a model of hen egg lysozyme based on a detailed x-ray crystallographic study of the three-dimensional structure of the enzyme at a resolution that permitted discrimination of structural features that were separated by a distance of 2 A. This model agreed with the primary structure shown in Fig. 1, including the location of the disulfide bonds. The lysozyme molecule appears to be folded in a fashion that accommodates most of the hydrophobic amino acid side chains inside the molecule, where they are removed from contact with the aqueous solvent, while most of the polar groups are distributed at the surface.

A unique aspect of the hen egg lysozyme model is the presence of a long cleft in the surface of the molecule. A trisaccharide inhibitor of lysozyme (tri-*N*-acetylglucosamine) can be bound in the cleft. This binding occurred at the same location when the inhibitor was diffused into crystals of lysozyme and when lysozyme was crystallized in the presence of the inhibitor. This unique binding of the inhibitor in the cleft may serve to localize the region of the active site of lysozyme. Of the various ways in which a model of the enzyme's substrate can be fitted into the region of the active site, the most favorable arrangement places the β-1-4-glycosidic bond that is hydrolyzed by lysozyme directly between the carboxyl group of the aspartic acid in position 52 and the carboxyl group of the glutamic acid in position 35 (Fig. 1). Structural observations indicate that these two amino acids have an important function in the mechanism of action of lysozyme.

Evolution. Certain enzyme functions appear to be widely distributed in nature. The amino acid sequences of proteins possessing these functions reflect changes that have occurred in the course of evolution. The structures of lysozymes from three sources, distant in evolution, have been carefully examined. Hen egg lysozyme (Fig. 1) has no structural elements in common with bacteriophage lysozyme (Fig. 2). Thus it must be concluded that these two enzymes emerged in evolution completely independent of each other. Preliminary studies of the structure of human lysozyme reveal considerable similarity to the structure of hen egg lysozyme. In fact, the resemblance is so great, including the strategically placed aspartic and glutamic acids in the active site, that it can be concluded that these proteins evolved from the same gene and have an essentially identical mechanism of action. Figure 3 illustrates this similarity, or homology, for the first 25 amino acid residues of these two lysozymes, where 18 of the first 25 amino acids are identical.

The amino acid composition of α-lactalbumin, a protein in cow's milk, is quite similar to that of hen egg lysozyme; nearly half of the amino acid positions in these two proteins are identical (Fig. 3). This is surprising since hen egg lysozyme operates to hydrolyze a β-1-4-glycosidic bond between amino sugars in bacterial cell walls, while α-lactalbumin facilitates the synthesis of a β-1-4-glycosidic bond between glucose and galactose to form lactose. One possible explanation is that the lysozyme gene was accidentally duplicated at some point in evolution and that subsequently one of the genes acquired lactose synthetase function. This new function seems to have appeared late in evolution, coincident with the development of breast feeding.

It is postulated from a comparison of the amino acid sequences of hen egg lysozyme, human lysozyme, and α-lactalbumin (Fig. 3) that a "deletion" occurred during evolution in the α-lactalbumin

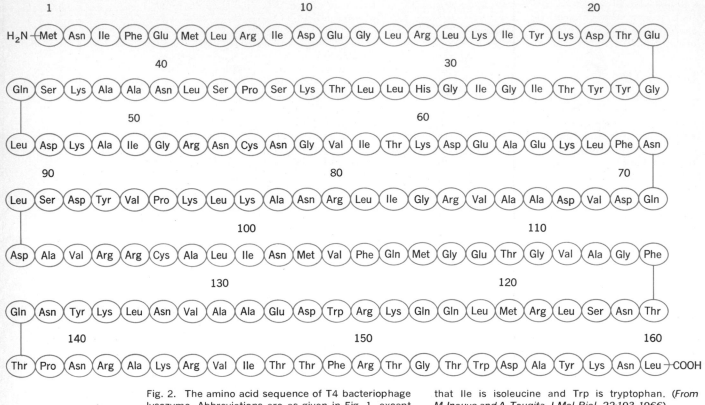

Fig. 2. The amino acid sequence of T4 bacteriophage lysozyme. Abbreviations are as given in Fig. 1, except that Ile is isoleucine and Trp is tryptophan. (*From M. Inouye and A. Tsugita, J. Mol. Biol., 22:193, 1966*)

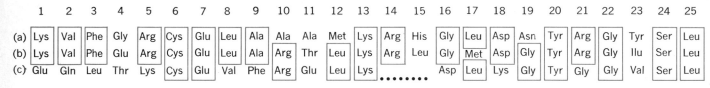

Fig. 3. Amino acid terminal sequences for (*a*) hen egg lysozyme, (*b*) human lysozyme, and (*c*) α-lactalbumin from cow milk. Rectangles identify common structural features. Dotted line for milk α-lactalbumin at positions 14–15 indicates gene deletion. Abbreviations are as in Fig. 1.

gene with a resulting loss of information for two amino acids near position 13. In addition, positions 10, 12, and 19 in human lysozyme and α-lactalbumin are identical, so it is possible to see remnants of a common ancestral gene in all three proteins. These data illustrate the manner in which amino acid sequence information is being used as a molecular reflection of the paths of evolution. *See* PROTEINS (EVOLUTION).

Experimental approach. One approach to the problem of determining the mechanism of action of the enzyme is to determine the structure of other lysozymes that derive their catalytic activity from amino acid sequences different from that of hen egg lysozyme. An examination of the structure of the active sites in these other lysozymes will permit the exploration of the common features that are essential for enzyme activity. If the present hypothesis based on the location of the active site for the mechanism of lysozyme action is correct, then the finding of two carboxyl groups immediately adjacent to the substrate bond that is hydrolyzed is to be expected. The structural data for human lysozyme appear to follow this pattern.

The synthesis of various substrate analogs will also be important in future work, because as these are fitted into the model of the lysozyme cleft, it should be possible to predict their chemical behavior and verify the predictions by experiments. *See* BACTERIA; CRYSTALLOGRAPHY; ENZYME; LYSIN; LYTIC REACTION; POLYSACCHARIDE.

[ROBERT E. CANFIELD]

Bibliography: K. Brew, T. Vanaman, and R. Hill, *J. Biol. Chem.*, 242:3747, 1967; R. E. Canfield and A. K. Liu, *J. Biol. Chem.*, 240:1997, 1965; R. E. Dickerson and I. Geis, *The Structure and Action of Proteins*, 1969; M. Inouye and A. Tsugita, *J. Mol. Biol.*, 22:193, 1966; D. C. Phillips, *Proc. Nat. Acad. Sci.*, 57:484, 1967; D. C. Phillips, *Sci. Amer.*, 215: 78, 1966; J. L. Strominger and J. Ghuysen, *Science*, 156:213, 1967.

Lyssacinosa

An order of the subclass Hexasterophora in the class Hexactinellida. In these sponges the parenchymal megascleres are typically free and unconnected but are sometimes secondarily united. The parenchymal megascleres vary from hexactins to diactins; the latter are always present and sometimes exclusively so. *Asconema, Euplectella, Rossella,* and *Rhabdocalyptus* are examples of this order. *See* HEXACTINELLIDA; HEXASTEROPHORA.

[WILLARD D. HARTMAN]

Lytic reaction

A term used in serology to describe a reaction that leads to the disruption or lysis of a cell. The best example is the lysis of sheep red blood cells by specific antibody and complement in the presence of Ca^{++} (calcium ion) and Mg^{++} (magnesium ion), a reaction that forms the indicator system of the standard Wassermann test for syphilis, as well as other complement-fixation reactions. In this example lysis results in the release of cellular hemoglobin into the medium; the reaction may be followed by visual or instrumental estimation of the decreased cell turbidity or the increased color of the medium due to the free hemoglobin. The initiation of lysis by complement can apparently proceed after the attachment of only one molecule of IgM or two molecules of IgG antibody to the red blood cell. IgM and IgG are both immunoglobulins. *See* ANTIBODY.

Pfeiffer reaction. This is an example of a bacteriolytic reaction, carried out with gram-negative bacteria, such as cholera vibrios, which are especially susceptible to the lytic action of specific antibody, complement, and Mg^{++} and Ca^{++}. This reaction may be carried out in the living subject (Pfeiffer's system) or in the test tube, and is a valuable aid in diagnosis. An analogous reaction is given by the spirochetes of relapsing fever. Although demonstrable instances of bacteriolysis occur, it is not clear whether all the bactericidal actions of antibody and complement are necessarily preceded by lysis. *See* CHOLERA VIBRIO.

Phagocytic reaction. The engulfment of microorganisms by phagocytic cells is often accompanied by their digestion and lysis. The initial stages are aided by specific antibody and by complement. It has been estimated that no more than 8 molecules of IgM or 2200 molecules of IgG are needed for the phagocytosis of a typical enteric bacterium.

Cytotoxic assays. A variety of reactions have come into use because of the necessity of detecting tissue incompatibilities before transplantation. The cell damage, including permeability changes resulting from the actions of antibody and complement, can be estimated from the differential staining by dyes or by assay of the release of tracers. *See* COMPLEMENT, SERUM; COMPLEMENT-FIXATION TEST.   [HENRY P. TREFFERS]

Bibliography: B. Cinader (ed.), *Antibodies to Biologically Active Molecules*, 1967; H. Stolp and M. P. Starr, *Ann. Rev. Microbiol.*, 19:79, 1965; D. M. Weir (ed.), *Handbook of Experimental Immunology*, 1967.